COLLISION REPAIR AND REFINISHING

A Foundation Course for Technicians

Alfred Thomas and Michael Jund

Collision Repair and Refinishing: A Foundation Course for Technicians
Alfred Thomas and Michael Jund

Vice President, Career and Professional Editorial: Dave Garza

Director of Learning Solutions: Sandy Clark

Executive Editor: David Boelio

Managing Editor: Larry Main

Senior Product Manager: Matthew Thouin

Editorial Assistant: Lauren Stone

Vice President, Career and Professional Marketing: Jennifer McAvey

Executive Marketing Manager: Deborah S. Yarnell

Senior Marketing Manager: Jimmy Stephens

Marketing Coordinator: Mark Pierro

Production Director: Wendy Troeger

Production Manager: Mark Bernard

Content Project Manager: Cheri Plasse

Art Director: Benj Gleeksman

Library of Congress Control Number: 2009900602

ISBN-13: 978-1-4018-8994-4

ISBN-10: 1-4018-8994-8

Delmar
5 Maxwell Drive
Clifton Park, NY 12065-2919
USA

Cengage Learning is a leading provider of customized learning solutions with office locations around the globe, including Singapore, the United Kingdom, Australia, Mexico, Brazil, and Japan. Locate your local office at: **international.cengage.com/region**

Cengage Learning products are represented in Canada by Nelson Education, Ltd.

For your lifelong learning solutions, visit **delmar.cengage.com**

Visit our corporate website at **cengage.com**

Notice to the Reader

Publisher does not warrant or guarantee any of the products described herein or perform any independent analysis in connection with any of the product information contained herein. Publisher does not assume, and expressly disclaims, any obligation to obtain and include information other than that provided to it by the manufacturer. The reader is expressly warned to consider and adopt all safety precautions that might be indicated by the activities described herein and to avoid all potential hazards. By following the instructions contained herein, the reader willingly assumes all risks in connection with such instructions. The publisher makes no representations or warranties of any kind, including but not limited to, the warranties of fitness for particular purpose or merchantability, nor are any such representations implied with respect to the material set forth herein, and the publisher takes no responsibility with respect to such material. The publisher shall not be liable for any special, consequential, or exemplary damages resulting, in whole or part, from the readers' use of, or reliance upon, this material.

Printed in Canada

1 2 3 4 5 XX 10 09

Contents

Chapter 14 Straightening and Repairing Structural Damage

Chapter 15 Structural Parts Replacement

Chapter 16 Full Frame Sectioning and Replacement

Preface

Collision Repair and Refinishing is a comprehensive textbook that covers non-structural repair, structural repair, painting, and refinishing for technical students in automotive collision repair programs.

While the text and its supporting materials stand as a complete, turnkey instructional package, the chapter content is also designed to align with and augment leading industry-sponsored programs. While such programs are excellent for working technicians who possess a basic body of knowledge related to collision repair, students in postsecondary programs who do not possess the foundational knowledge cannot fully use and understand a curriculum designed for working professionals. *Collision Repair and Refinishing* provides the missing link—printed foundation-building materials for students who have no background in collision repair and refinishing.

The text has also been designed to cover the entire task lists as outlined by the current NATEF Standards for Structural Analysis & Damage Repair, Non-Structural Analysis & Damage Repair, Plastics & Adhesives, and Painting & Refinishing. This will equip collision repair programs with the appropriate teaching materials for retention of their NATEF-certified status, and enable those programs that wish to pursue first-time NATEF certification.

The text covers all of the basic collision repair techniques of metalworking and welding, as well as using fillers, tools, trim, and hardware. After reading the text, students should fully understand vehicle construction and collision damage analysis procedures. These procedures will enable students to utilize their knowledge to perform measuring as well as unibody structural and body-over-frame realignment. They should also be able to apply their knowledge of new techniques to structural part and exterior part replacement.

Additionally, the text will help students gain proficiency in performing the current corrosion protection procedures. Coverage of current and emerging refinish technology will provide students with adequate theory and understanding to confidently perform the refinishing procedures required by today's vehicles.

Finally, visual learners will benefit from the eye-catching text elements that highlight the chapter content and bring the material to life, including full-color photos, line art, and text boxes that draw attention to safety issues and other information that requires special emphasis. At the end of the chapters, ASE-style multiple choice questions, which will familiarize and focus students on ASE certification, are followed by essay questions, topic related math questions, critical thinking questions, and lab activities.

About the Authors

Al Thomas has been an educator for over 20 years and has served as the collision repair department head at Pennsylvania College of Technology in Williamsport, Pennsylvania, since 2000. In addition to his work experience with Ford Motor Company and independent shops, he is a NATEF evaluation team leader, current I-CAR member, member of the Collision Repair Instructors Network, and former member of the NATEF task list advisory board (2000–2006). He is also a contributing editor for *Auto Body Repair News* (ABRN), for which he has published over 40 collision repair, refinishing, and management articles from 2006 to the present.

Michael Jund has worked in the auto collision repair industry as a technician and educator for many years (over four decades). He has been an instructor, department coordinator, and dean of applied technologies at Scott Community College in Bettendorf, Iowa, for over 30 years. Mike has also been relentless in promoting excellence and professionalism to both his students and colleagues alike, encouraging all instructors to pursue ASE certification and pushing for all training programs to become NATEF certified. In addition to having been an I-CAR instructor and weld qualification administrator, Mike has served on various local, regional, and national advisory councils and boards, as well as the NATEF Standards Review Committee. He currently serves as an evaluation team leader for institutions seeking NATEF certification.

Acknowledgments

From the Authors

Al Thomas

I wish to acknowledge Pennsylvania College of Technology and the many students and faculty who helped during the writing of this book. Without their help this task would have been much more difficult. Special thanks to my former student, Brandon Smith, for his immeasurable help and encouragement. Also, thank you to Lynn M. Thomas for her advice, editorial assistance, and unconditional support.

Michael Jund

I would like to thank my wife, Sharon, for her unwavering indulgence and patience as I completed my lifelong ambition and dream by creating a learning tool such as this textbook. I would also like to acknowledge my work partner and confidante, Roy VenHorst at Scott Community College, who shares my passion for promoting the auto collision repair industry. He contributed countless hours of time proofing the manuscript and helping me clarify some of the content included herein. I would also like to recognize the proofreaders for helping to ensure the content is accurate and current.

The authors and Publisher also acknowledge the contributions of Rodney Kohlhepp, author of *Chapter 3: Hand Tools,* and Peter Pratti, author of *Chapter 1: Introduction to Collision Repair,* and *Chapter 10: Estimating Collision Damage.*

Reviewers

Mark Blohm
Northeast Wisconsin Technical College
Green Bay, WI

David Bradley
Clackamas Community College
Oregon City, OR

Russell Butler
Idaho State University
Pocatello, ID

Bruce Evanstad
Bellingham Technical College
Bellingham, WA

Richard Frey
College of Southern Idaho
Twin Falls, ID

Neal D. Grover
Salt Lake Community College
Salt Lake City, UT

Ross Hardin
Southern High School MCA
Louisville, KY

Jim Hormann
Ivy Tech Community College
Fort Wayne, IN

Michael Janovsky
Hennepin Technical College
Eden Prairie, MN

Tony Jones
Cypress College
Cypress, CA

Loren Larson
Hennepin Technical College
Brooklyn Park, MN

Daniel Norton
Walla Walla Community College
Walla Walla, WA

Archie Watley
Texas State Technical College
Waco, TX

Steve White
Portland Community College
Portland, OR

Features of the Text

The following chapter elements are designed to guide readers through the complex material presented in this text, and help ease both the teaching and learning process:

Objectives Each chapter begins with a list of learning objectives that state the expected outcome that will result from completing a thorough study of the chapter's contents.

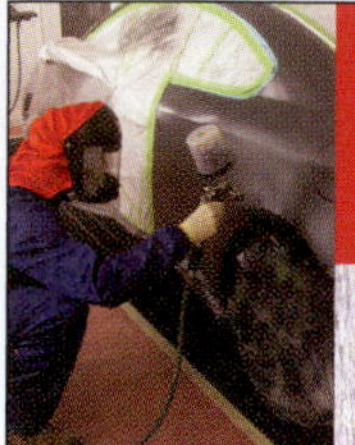

Chapter 20

Paint Spray Guns

OBJECTIVES

Upon completing this chapter, you should be able to:

- Operate a spray gun using proper:
  - Personal safety
  - Environmental safety
  - Fellow worker protection safety
- Describe spray gun types.
- Identify the different gun designs.
- List proper air supply and regulation needs for refinish equipment.
- Describe the function of spray gun components.
- Demonstrate the proper spray gun adjustment.
- Demonstrate correct cleaning and maintenance of a spray gun.

INTRODUCTION

This chapter will be devoted to the study of spray paint application and the spray gun. Although coatings are mostly applied through spraying today, that's not how coatings were applied in the early years of automobile manufacturing. Before the spray gun was invented, automotive coatings were applied by brush. See **Figure 20-1**. This was a very time-consuming and labor intensive operation. Coats of finish, varnish, and India enamels were initially brushed on. Then the vehicle was allowed to dry, after which the brush strokes were sanded smooth. Subsequent coats of finish were applied until the manufacturer reached the desired thickness of finish. After the final finish was polished, the vehicle was assembled and prepared for delivery.

The invention of the spray gun, which occurred in the early 1900s, is generally credited to two different persons, Joseph Binks and Alan DeVilbiss, a physician. See **Figure 20-2**. Dr. DeVilbiss invented a bulb spray device in 1888 that would spray medicine to the back of a patient's throat. See **Figure 20-3**. In 1890, Binks—a painter who painted the basement's storage areas for the Marshall Field & Company department store—figured out how to spray the walls instead of brushing them. He developed his spray finishing equipment, and by 1893 it was in use outside the automotive refinishing industries.

In 1907, Thomas DeVilbiss, son of Alan DeVilbiss and an inventor in his own right, experimented with adapting the original atomizer to create a spray gun to meet the challenges of spray finishing. See **Figure 20-4**. Soon after, spray technology was adapted to the repair

535

Cautions Shop safety is critical, and to emphasize this point, "Caution" statements are placed in the text where issues of personal safety are involved, especially when describing a procedure that requires Personal Protective Equipment, and when discussing topics where potential property/equipment damage is involved.

Chapter 2 Collision Repair Safety **59**

CAUTION

Under NO CONDITION should one initiate the repairs or get under the vehicle until the safety catches have been solidly engaged. They should be checked routinely for proper functioning.

- In the event large assemblies such as the differential, fuel tank, or other heavy parts are to be removed, the technician should use a jack stand tall enough to reach the rear bumper to help stabilize the vehicle.
- Make sure the entire area under the vehicle is clear before lowering it. Check to ensure all tools and equipment that have been used while working under the vehicle have been moved away to a clear area nearby.

Safety for the Vehicle Being Lifted

- Ensure the vehicle transmission is in park (automatic transmission), in reverse gear (standard transmission), or the emergency brake is engaged.
- Check the doors to ensure they are closed and latched.
- Pad the lifting arms with solid blocking items to ensure the vehicle is raised level and remains stable when lifted off the ground. See **Figure 2-37**.
- Raise the vehicle using only lift points that are adequately reinforced to support its weight. One can safely use the vehicle's normal weight-bearing surfaces such as suspension and steering mounting locations, engine cross members, and frame rails. Never use the floor pan or any other "soft" or unreinforced panels. These same locations can be used with nearly all vehicle-lifting equipment.

FIGURE 2-37 The vehicle can safely be lifted off the ground by using the lift points specified by automobile manufacturers as well as any load-bearing surfaces such as the axle housing, suspension mounting locations, and the frame rails.

Floor Jacks

Many times floor jacks are used to raise a vehicle off the ground to gain access to a vehicle's underside. Before raising the vehicle, check to ensure the saddle raises solidly with each pump of the handle. Failure to do so is an indication of a low fluid level and may cause it to collapse when a heavy object is lifted. Before getting under the vehicle raised with a floor jack, one should always place jack stands under the vehicle and let the weight down to fully rest on them. Serious injuries are apt to occur unnecessarily when an impatient technician carelessly gets under a vehicle without properly supporting it.

CAUTION

One should shake the vehicle to ensure it is firmly planted on the jack stands prior to getting under it.

Hydraulic Jacks and Straightening Equipment

Corrective forces as high as 20,000 pounds per square inch (psi) are not uncommon when hydraulic pulling and straightening equipment are used for repairing heavily damaged vehicles. One should stand aside from the line corrective forces travel to avoid serious injury in the event a chain or clamp should break or slip, causing the equipment to lurch or a part to rip loose and become air-borne. Never allow any hydraulic fluid leak to go without being repaired.

PROTECTION FROM VEHICLE-RELATED INJURIES

Air Bags and Hybrid Vehicles

Two areas of concern regarding the potential for serious injuries caused by a vehicle are when working around air bags and their circuitry and when working on hybrid vehicles. In both cases the manufacturer's recommendations should be consulted before proceeding with the repairs.

Hybrid Vehicles. Hybrid vehicles have high voltage circuits that, when not handled properly, could cause serious injury or even death. The manufacturer's precautions should be read and understood BEFORE initiating any work on the vehicle.

Air Bags. Air bags also present numerous opportunities for possible injury while installed in the vehicle and also

186 **Collision Repair and Refinishing**

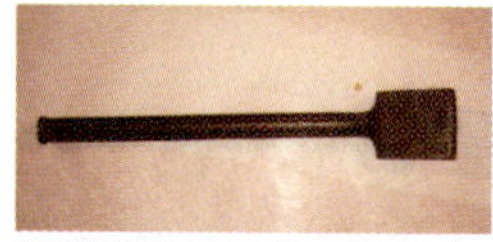

FIGURE 7-41 Dolly spoons are often used to get to places that a technician might not otherwise be able to reach.

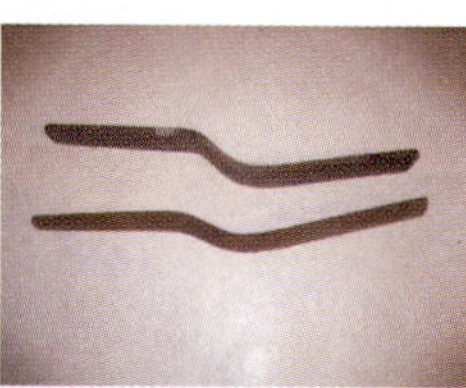

FIGURE 7-42 A bumping file or slapping spoon can eliminate most, if not all, of the hammer marks left from the metal finishing process.

hammer marks left from the metal finishing process. This is the last step before the technician sands and primes the metal for painting.

Picks

Picks are tools used when access to the back side of a vehicle is not possible. They are often inserted through holes in a door, after the taillight has been removed, or under the hood. In the past, and even now sometimes when "paintless dent removal" (PDR) techniques are used, a hole is placed in an area where it is not visible. A plug is then inserted into the hole after the work is completed. This practice remains very controversial among both auto manufacturers and technicians. Some believe that the creation of an access hole is acceptable. Others, however, believe that once the hole is placed in the vehicle's body it causes irreversible damage, and therefore it should never be done.

Picks come in an assortment of shapes and lengths: short and long picks, chisel bit, as well as many tools specifically designed for paintless dent removal. See **Figure 7-43**, **Figure 7-44**, **Figure 7-45**, and **Figure 7-46**.

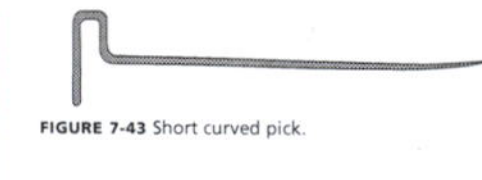

FIGURE 7-43 Short curved pick.

FIGURE 7-44 Long curved pick.

FIGURE 7-45 Chisel bit pick.

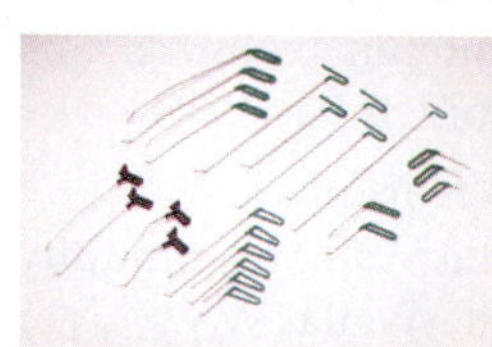

FIGURE 7-46 Paintless dent repair (PDR) tools.

TECH TIP When repairing a vehicle with either a pick hammer or a pick tool, the technician should keep in mind that each time the factory corrosion protection is disturbed, it will create a corrosion hot spot which will cause that damaged area to corrode. Because a technician is not always able to see these damaged areas due to close or limited access, corrosion protection should always be re-applied following the repair. Epoxy paint followed with petroleum or wax base corrosion protection is best.

Tech Tips These text boxes include tips to reduce time on a task, notes on tool operation and storage, and other tips on shop procedures commonly performed by experienced technicians.

Summary These bulleted lists at the end of each chapter highlight key bits of information and are designed to refresh the reader on the main points from the chapter.

Key Terms Each chapter ends with a list of the terms that were introduced in the chapter. These terms are highlighted in the text when they are first used and are also defined in the glossary.

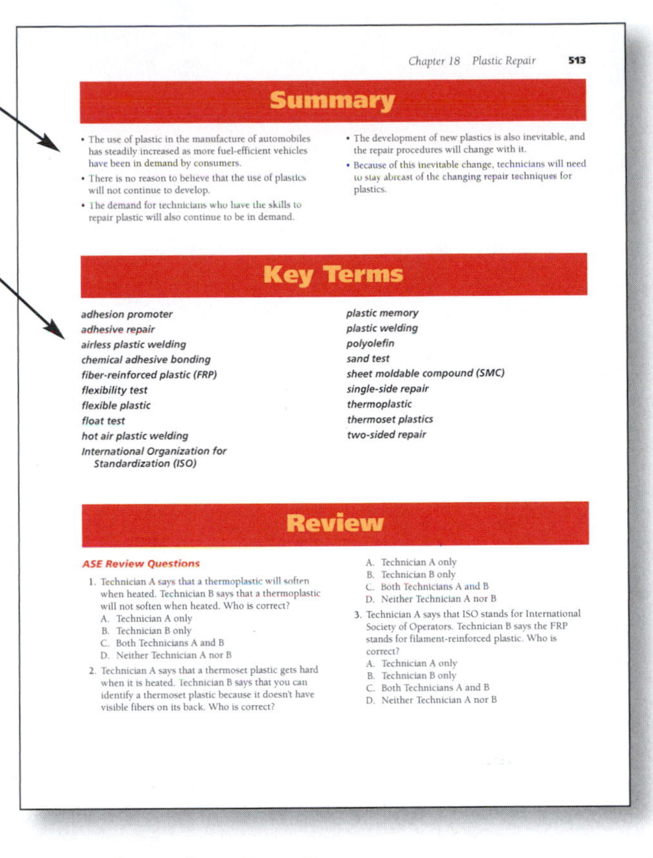

Chapter 18 Plastic Repair 513

Summary

- The use of plastic in the manufacture of automobiles has steadily increased as more fuel-efficient vehicles have been in demand by consumers.
- There is no reason to believe that the use of plastics will not continue to develop.
- The demand for technicians who have the skills to repair plastic will also continue to be in demand.
- The development of new plastics is also inevitable, and the repair procedures will change with it.
- Because of this inevitable change, technicians will need to stay abreast of the changing repair techniques for plastics.

Key Terms

adhesion promoter
adhesive repair
airless plastic welding
chemical adhesive bonding
fiber-reinforced plastic (FRP)
flexibility test
flexible plastic
float test
hot air plastic welding
International Organization for Standardization (ISO)
plastic memory
plastic welding
polyolefin
sand test
sheet moldable compound (SMC)
single-side repair
thermoplastic
thermoset plastics
two-sided repair

Review

ASE Review Questions

1. Technician A says that a thermoplastic will soften when heated. Technician B says that a thermoplastic will not soften when heated. Who is correct?
   A. Technician A only
   B. Technician B only
   C. Both Technicians A and B
   D. Neither Technician A nor B
2. Technician A says that a thermoset plastic gets hard when it is heated. Technician B says that you can identify a thermoset plastic because it doesn't have visible fibers on its back. Who is correct?
   A. Technician A only
   B. Technician B only
   C. Both Technicians A and B
   D. Neither Technician A nor B
3. Technician A says that ISO stands for International Society of Operators. Technician B says the FRP stands for filament-reinforced plastic. Who is correct?
   A. Technician A only
   B. Technician B only
   C. Both Technicians A and B
   D. Neither Technician A nor B

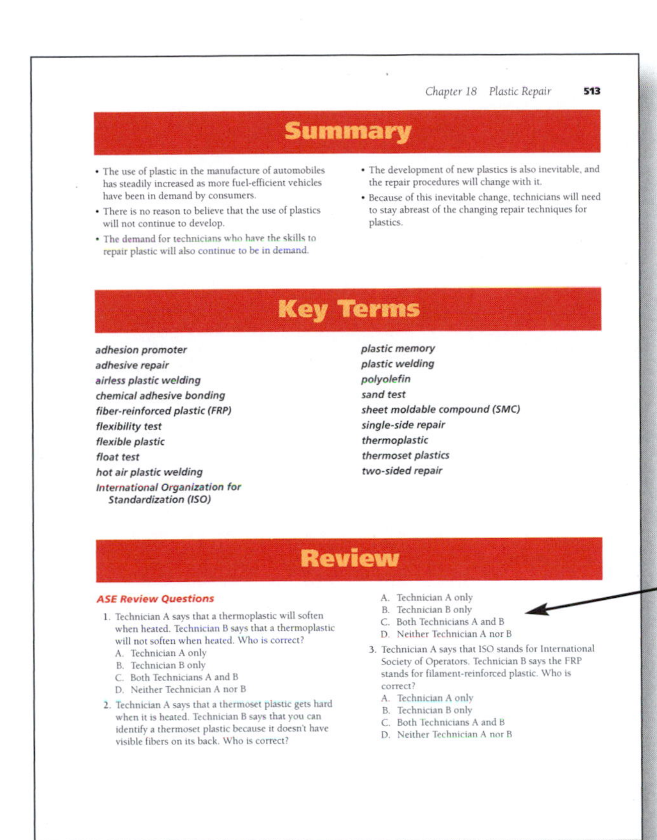

Chapter 18 Plastic Repair 513

Summary

- The use of plastic in the manufacture of automobiles has steadily increased as more fuel-efficient vehicles have been in demand by consumers.
- There is no reason to believe that the use of plastics will not continue to develop.
- The demand for technicians who have the skills to repair plastic will also continue to be in demand.
- The development of new plastics is also inevitable, and the repair procedures will change with it.
- Because of this inevitable change, technicians will need to stay abreast of the changing repair techniques for plastics.

Key Terms

adhesion promoter
adhesive repair
airless plastic welding
chemical adhesive bonding
fiber-reinforced plastic (FRP)
flexibility test
flexible plastic
float test
hot air plastic welding
International Organization for Standardization (ISO)
plastic memory
plastic welding
polyolefin
sand test
sheet moldable compound (SMC)
single-side repair
thermoplastic
thermoset plastics
two-sided repair

Review

ASE Review Questions

1. Technician A says that a thermoplastic will soften when heated. Technician B says that a thermoplastic will not soften when heated. Who is correct?
   A. Technician A only
   B. Technician B only
   C. Both Technicians A and B
   D. Neither Technician A nor B
2. Technician A says that a thermoset plastic gets hard when it is heated. Technician B says that you can identify a thermoset plastic because it doesn't have visible fibers on its back. Who is correct?
   A. Technician A only
   B. Technician B only
   C. Both Technicians A and B
   D. Neither Technician A nor B
3. Technician A says that ISO stands for International Society of Operators. Technician B says the FRP stands for filament-reinforced plastic. Who is correct?
   A. Technician A only
   B. Technician B only
   C. Both Technicians A and B
   D. Neither Technician A nor B

ASE-Style Review Questions In most chapters there is a set of ASE-style review questions to help assess the readers' knowledge of the chapter content, and familiarize future technicians with the question types that they will see on ASE certification exams.

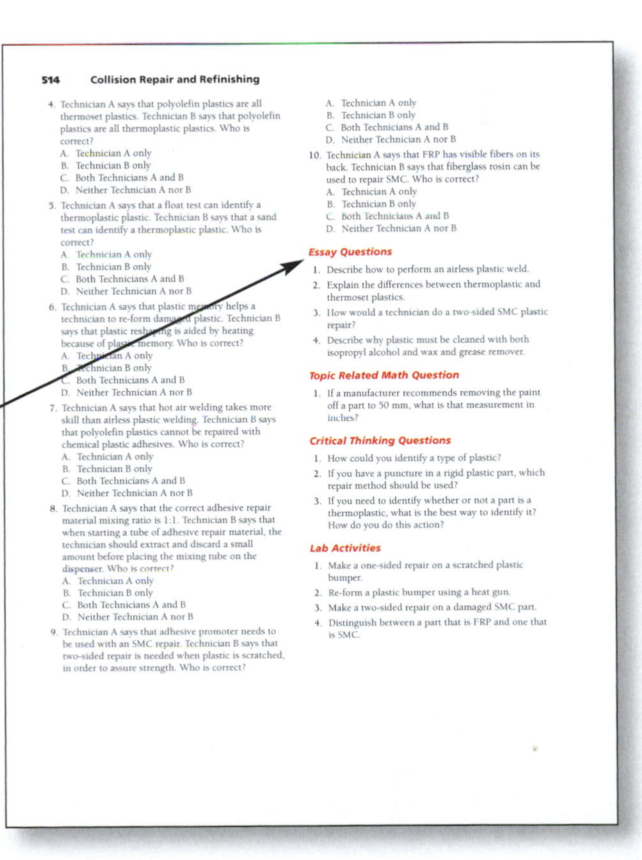

514 Collision Repair and Refinishing

4. Technician A says that polyolefin plastics are all thermoset plastics. Technician B says that polyolefin plastics are all thermoplastic plastics. Who is correct?
   A. Technician A only
   B. Technician B only
   C. Both Technicians A and B
   D. Neither Technician A nor B
5. Technician A says that a float test can identify a thermoplastic plastic. Technician B says that a sand test can identify a thermoplastic plastic. Who is correct?
   A. Technician A only
   B. Technician B only
   C. Both Technicians A and B
   D. Neither Technician A nor B
6. Technician A says that plastic memory helps a technician to re-form damaged plastic. Technician B says that plastic reshaping is aided by heating because of plastic memory. Who is correct?
   A. Technician A only
   B. Technician B only
   C. Both Technicians A and B
   D. Neither Technician A nor B
7. Technician A says that hot air welding takes more skill than airless plastic welding. Technician B says that polyolefin plastics cannot be repaired with chemical plastic adhesives. Who is correct?
   A. Technician A only
   B. Technician B only
   C. Both Technicians A and B
   D. Neither Technician A nor B
8. Technician A says that the correct adhesive repair material mixing ratio is 1:1. Technician B says that when starting a tube of adhesive repair material, the technician should extract and discard a small amount before placing the mixing tube on the dispenser. Who is correct?
   A. Technician A only
   B. Technician B only
   C. Both Technicians A and B
   D. Neither Technician A nor B
9. Technician A says that adhesive promoter needs to be used with an SMC repair. Technician B says that two-sided repair is needed when plastic is scratched, in order to assure strength. Who is correct?
   A. Technician A only
   B. Technician B only
   C. Both Technicians A and B
   D. Neither Technician A nor B
10. Technician A says that FRP has visible fibers on its back. Technician B says that fiberglass rosin can be used to repair SMC. Who is correct?
   A. Technician A only
   B. Technician B only
   C. Both Technicians A and B
   D. Neither Technician A nor B

Essay Questions

1. Describe how to perform an airless plastic weld.
2. Explain the differences between thermoplastic and thermoset plastics.
3. How would a technician do a two-sided SMC plastic repair?
4. Describe why plastic must be cleaned with both isopropyl alcohol and wax and grease remover.

Topic Related Math Question

1. If a manufacturer recommends removing the paint off a part to 50 mm, what is that measurement in inches?

Critical Thinking Questions

1. How could you identify a type of plastic?
2. If you have a puncture in a rigid plastic part, which repair method should be used?
3. If you need to identify whether or not a part is a thermoplastic, what is the best way to identify it? How do you do this action?

Lab Activities

1. Make a one-sided repair on a scratched plastic bumper.
2. Re-form a plastic bumper using a heat gun.
3. Make a two-sided repair on a damaged SMC part.
4. Distinguish between a part that is FRP and one that is SMC.

Review Questions A combination of essay questions, topic related math questions, critical thinking questions, and lab activities make up the end-of-chapter review section. The different question types are provided in order to reinforce comprehension of the chapter material.

Chapter 1

Introduction to Collision Repair

OBJECTIVES

This chapter will outline the types of collision repair service centers, basic facets of the operation, and career opportunities in the field. In addition, it will lay a foundation that students can build upon as the various areas of collision repair are explored, explained, and practiced in greater detail in subsequent chapters. Upon completing this chapter, you should be able to:

- Identify the various types of collision repair centers, such as:
  - Independent
  - Dealership
  - Franchise
- Recognize the basic tasks of collision repair operations, such as:
  - Estimating collision damages
  - Metal work
  - Structural repairs
  - Mechanical/electrical
  - Refinishing
  - Vehicular identification
  - Service information retrieval
  - Post-paint operations
- Distinguish between collision repair career opportunities, such as:
  - Collision repair technician
  - Collision repair shop owner
  - Damage estimator
  - Production manager
  - Parts manager
  - Office manager
  - Receptionist
  - Insurance adjuster or appraiser
  - Related collision industry careers

INTRODUCTION

Each year no less than 8 percent of U.S.-registered motor vehicles are involved in some sort of collision event. As a result, the field of collision repair is vital to the restoration of over 17 million vehicles annually. In other words, nearly 1/12 of the 217 million light-duty vehicles that transport people, products, and cargo inside and across communities each day will require some degree of physical damage repair every 12 months. Conservative estimates have placed the total societal cost of motor vehicle accidents in excess of $200 billion annually, with approximately $10 billion allotted to the physical repairs.

Just how far-reaching are the effects of these collisions on a typical motor vehicle owner? After a house, a passenger vehicle is usually the second most expensive investment that a person will make in his or her lifetime. According to one motor car company's surveys, one out of every four customers that *do not* return to that company for their next new car places the blame on a poor collision repair job performed on their present vehicle—a job that they blame, in part, on the *vehicle's manufacturer* regardless of who repaired it. "Had 'so-and-so' built a good car," they reason, "it could have been repaired properly!"

How important is collision repair? Take a motor vehicle away from anyone—a professional person, a student, a physically challenged or otherwise dependent driver—and his or her quality of life can be seriously affected. Beyond the damaged vehicle, collisions also often leave people physically, if not psychologically, injured. Thus, the collision repair center is in need of an ever more educated and vital workforce revolving around the *collision repair technician*. It services *lives* as well as motor vehicles. In addition to being responsive to the person, a collision repair center has the responsibility of seeing to it that the vehicle is restored safely and completely in *all* aspects, thus returning it to its owner literally "as good as new." See **Figure 1-1**.

TECH TIP I usually begin my I-CAR classes with this question: "What is THE most important occupation needed in society?" With barely an exception, a chorus of voices fills the air with, "*Doctor*." Yes, physicians are extremely important to society. However, when I suggest that doctors can only have one life entrusted to them at any one time, while the collision repair technician can positively (or negatively) affect *many* lives at once with one correct (or incorrect) repair decision, the class sits up a bit taller! A collision repair technician may not *save* lives in the traditional sense, but he or she can certainly have a lasting effect in helping to *preserve* many of them.

FIGURE 1-1 To the average vehicle owner, a traffic accident is an unsettling experience. Although this woman was not injured physically she is hopeful that this collision repair expert will restore her late-model vehicle to its original condition and her emotions to a calm, confident state as well!

This book will serve as a comprehensive introduction to collision repair and refinishing for technical students. At the same time, it will augment the **Inter-Industry Conference on Auto Collision Repair's (I-CAR)** LIVE Curriculum.

WHAT IS COLLISION REPAIR?

Collision repair is the process by which a motor vehicle that has been cosmetically, structurally, or otherwise damaged is restored to the vehicle manufacturer's *original design intent* in all aspects. Utilizing the **original equipment manufacturer (OEM)**, I-CAR, and other approved testing organizations' tested, recommended, and/or essential repair procedures, the collision repair technician has an obligation to restore the vehicle's:

- Safety features
- Structural integrity
- Durability
- Performance
- Proper fit
- **Cosmetic** appearance

The basic physical components of motor vehicle collision repair are:

- **Damage analysis**/creation of an estimate
- Work/repair order and file creation
- Parts/materials ordering
- Disassembly or teardown of the vehicle
- Structural/frame repairs
- Sheet metal and parts repair and/or replacements
- Mechanical/electrical considerations

- Refinishing
- Reassembly of the vehicle
- Cleaning and detailing the vehicle
- Final inspection and safety/road checks
- Delivery of the repaired vehicle to its owner

In addition to eliminating *the physical* damages, a collision repair center strives to restore the owner's *confidence* in the vehicle by reversing the negative *predicament* and *outlook* assigned to it. Among other things, a professional center helps the affected parties *forget the loss entirely* by making the experience overwhelmingly positive in *every* way. That challenge usually starts with the collision repair center meeting, overcoming, and erasing society's age-old mindset of "This (damaged) car will never be the same again!"

Once a customer has been "won over," it can feel like he or she expects the car back shinier, cleaner, quieter—*better*—than the day it was purchased. In addition, the customer wants it back *fast!* Customer service and satisfaction are integral parts of the collision repair equation in our consumer-oriented society. Successful collision repair is by no means an easy task; it is quite challenging but extremely rewarding when accomplished. It is an admirable profession filled with challenges, revolutionary changes, and just rewards. Images, phrases, and gross misconceptions like junkyard dogs, "That's good enough," and uneducated, misfit "grease monkeys" have been—and will continue to be—replaced by spotlessly organized preparation areas, fiber optically connected computers, and educated men and women reading sophisticated technical charts that often define precision to a single *millimeter.* Space-age materials, computerized navigation systems, crash-avoidance equipment, and alternatively fueled electric vehicles—parts of the faraway "future of the industry" merely a few years ago—are now commonplace in the world of passenger vehicles and collision repair. Now more than ever, repairing collision-damaged vehicles requires intelligence, ongoing education, pride, honesty, and commitment. See **Figure 1-2**.

FIGURE 1-2 When handing over the keys of a restored vehicle, the professional repairer also restores the owner's peace of mind, their confidence in the vehicle and the means for the owner to care for their loved ones.

Let us begin to cover the following aspects, operations, and positions that are vital to the ultimate success of safe, efficient, esthetically pleasing, and successful collision repair.

COLLISION REPAIR CENTERS

Collision repair centers exist in a number of different forms, such as the independently owned and operated center, the dealer-owned and operated center, and **franchises**. The average United States collision repair shop is a bit over 8,000 square feet, and has between seven and eight employees. About one-third of them have more than six technicians, and two-thirds of them have at least one direct repair relationship with an insurance company (an agreement whereby the insurer recommends their policyholders to *that* shop when a damage loss is reported).

Approximately half of all registered shops are considered "large" (annual sales between $300,000 and $1 million) and the other half are split about evenly between "small" (less than $300,000 annually) and "super-sized" ($1 million and over in annual sales). With a few minor differences, the final goal of a collision repairer is to restore a damaged motor vehicle to its originally intended build characteristics in every way—commonly referred to as "pre-loss" or "**pre-accident condition**."

Independent Repair Centers

Independently operated collision repairers make up a majority of the shops in the United States. Oftentimes owned by one or two individuals, the typical "independent" is a small-to-mid-sized operation, although they can be larger. See **Figure 1-3**. They are sometimes passed

FIGURE 1-3 A typical independently owned collision repair shop. These shops are usually small-to-mid-sized operations.

down through generations of family members. Although no two independently owned shops are alike, every repair store will have—at a minimum—body repair and refinishing sections. Usually fed by referrals, the typical independent can be as small as two or three bays and as large as 20 or more. They generally favor smaller, faster repair jobs that take up less time and space. The organization and structure of these centers vary depending upon factors like size and space, available equipment and technology, customer pool, and available workforce.

The typical small independently owned repair shop is usually run by a handful of individuals that typically take care of a variety of functions. Oftentimes, the owner and/or a few employees are "**combo repairers**" that work on a vehicle from disassembly to final refinish. The shop may also "farm out" (**sublet**) specialized repairs, and employ few—if any—"**specialists**," although there are independently owned shops that operate with departments that include separate body, frame, refinish, and detailing technicians. The shop may have a secretary or front person handling the office duties and communications.

As the business evolves and economic changes occur, the smaller, independent shop faces challenges similar to the local hardware or drug store that finds itself competing with mega-centers. Across the country, the total number of smaller, independent shops has dropped sharply in recent years. However, the independent shop has advantages that give it hope for the future. With no corporation, partners, or board of directors to answer to, the independent can make changes quickly. This is crucial in an industry that demands the adaptability of sudden change more than ever before. The independent is usually smaller, and located in a rather convenient location. In addition, people tend to like the personalized attention that their long-associated independent shop can give.

Dealership Repair Centers

Approximately one-quarter of the registered repair shops in the United States are dealership owned and operated. These centers have an obvious advantage over most other collision repair centers when relatively new vehicles are in need of collision repairs—a "built-in" customer base. Many new car owners quite naturally return to the dealership where they purchased their vehicle when they have suffered a collision loss since they usually have a relationship with the shop from servicing the vehicle *mechanically*. The dealership—with the brand name recognition that many consumers crave—simply directs the damaged vehicle to its own shop.

There are other advantages as well. Today's certification programs, which promote certain repairs by and at certain centers, are pushing specific models to their dealer-operated centers for collision repairs. In fact, **aluminum-intensive vehicles** like the Audi A8 and the Jaguar XJ have parts that certain non-dealership, non-certified stores cannot even purchase! The dealership-operated collision store also has the advantage of readily obtaining repair methods, OEM-based technical training programs, and any/all parts and materials that fit their repair needs.

Because they usually have more space in which to operate, the funds to support the operation, and sometimes backing from the vehicle's manufacturer, the dealership repair stores generally have a vast array of equipment and employees that can fill the needs of the larger store. In-house mechanical, alignment, electronic, air conditioning, and specialized tools, along with a shop-based parts section, give the dealership store an advantage. Purchasing their parts and some materials "from themselves" also garners the shop a higher profit margin.

The dealership-operated center usually has a staff of office personnel and a "front person" to handle filing, communications, estimates, and advertising. The shop will usually be split into specialized areas—sheet metal and plastic, structural, mechanical, electrical, refinish, and detailing—supported and organized by a shop foreman or production manager.

However, there are some disadvantages to this setup, too. The overhead to run a dealership-sized repair center is considerably higher than that of the independent. With generally higher posted labor rates, the dealership is at a slight disadvantage when it comes to securing **direct repair program (DRP)**-type relationships with insurers—something that many independently owned stores rely on in today's collision repair environment.

Franchise Repair Centers

The franchised collision repair store is a relatively new and growing trend in the industry. Like typical "licensed" businesses, the franchised collision repair center has the advantage of "instant name recognition," a uniformed look, and a set system of conducting business. Having a network of stores that share the same vision gives the franchised operation a proverbial "leg up" in garnering a customer base.

Many franchise organizations offer resources, training programs, and assistance that an independently owned store would have a more difficult time obtaining. Along with specialized training and resources, the franchised center usually strives to develop ties within the collision repair industry such as insurance company relationships, training initiatives, and mass advertising. In some cases, the franchises have initiated or have become involved with work-study programs in order to sponsor entry-level technicians coming into the collision repair industry.

Operated largely within the specialization repair model that promises to reduce repair cycle (down) time for insurers and vehicle owners, you can expect to see more of these centers as time moves on. See **Figure 1-4**.

FIGURE 1-4 This franchise-type repair center boasts a network of collision repair centers that build "name recognition" within their customer base.

COLLISION REPAIR OPERATION

The specific methods and areas the vehicle will travel through during repairs are largely determined by the type and extent of the damages the vehicle has suffered. Typically motor vehicles are damaged in collisions with other vehicles or objects, although occasionally fire-damaged, vandalized, or flooded vehicles are brought in for repairs.

Unless the damaged vehicle can be safely driven, it is towed into the collision repair center. The facility's owner, manager, or estimator will secure the customer's name, address, and contact phone numbers; the insurance carrier; and pertinent details of physical events surrounding the accident or other mishap that damaged the vehicle. How a vehicle is performing since the damages occurred (if drivable) is a crucial part of the initial discussion with a customer. The shop then creates a customer file and stores all related information—estimates, parts orders, sublet invoices, insurer, and personal details. See **Figure 1-5**.

The repair process starts with a pre-production *damage analysis* and a detailed *estimate* of the damages. This includes all relevant numbers from the vehicle, vehicular options, colors, repairs and replacement parts necessary, and any unrelated, old damages that may be present. Most centers are creating estimates utilizing a computerized estimating system. A step known as dismantling or *teardown* of the vehicle may be warranted here, or at the time of the insurance company representative's inspection. A teardown helps the estimator perform a detailed, accurate initial inspection by removing some/all of the damaged parts. This helps the estimator see many damaged parts, panels, and structures that would otherwise be hidden from view. Wiring, harnesses, grounds, and circuits are often routed in and around hidden, hard-to-see areas that a teardown uncovers. In addition to teardowns, many collision repair centers power wash a damaged vehicle before creating an estimate. If not, this is done no later than just before the vehicle enters the repair shop. See **Figure 1-6**.

After—if not during—the estimating process, it is good business practice to explain the repair process to the vehicle's owner in layman's terms. Of particular note are the parts, materials, and methods utilized to repair the vehicle. In some areas, the vehicle's owner must be informed as to what types of parts will be used to repair the vehicle, whether they are supplied by the Original Equipment Manufacturer (OEM), an **aftermarket** company, a salvage yard, or any other parts refurbisher. The common misconceptions associated with collision repair are legendary, and nothing dispels false information, rumors, and future unnecessary conflicts like a thorough education! In addition to educating the customer, the pre-repair communication is an excellent time to hear and set expectations, ascertain what is—and is not—related to the current damages, and "**upsell** the job." Since the average motor vehicle is generally in need of maintenance, a good shop will take the time to sell more repairs while it is in the shop.

If an insurance company is involved, they may want to have their company representative or adjuster create an estimate. If the shop is affiliated with the insurer through a direct repair program (DRP), the shop may serve to function as that company's estimator. Regardless of who writes the estimate, financial considerations are made *before* any repairs begin. Of particular interest is the vehicle's value in relation to the cost of repairs. Most vehicles can be repaired if the owner and/or insurer are willing to pay the price. However, if a vehicle's estimated repair costs are close to, or exceed, the pre-loss value of the vehicle, it may be deemed a **total loss** and go unrepaired.

An estimate of repairs is used to create a ***repair*** or ***work order***. The shop orders replacement parts deemed necessary and the technician(s) assigned to the job follows a basic **repair plan** as the vehicle enters the production phase. That cycle is typically smooth and continuous, much like the original production line where the vehicle was built. *If structural damages* exist, they will be addressed by a *structural technician* using a *measuring system,* which compares specific checkpoints on the vehicle to the original build dimensions. *A frame bench* and/or portable floor equipment is utilized to "pull" a vehicle's structure back to its proper dimensions. See **Figure 1-7**.

Although unrepairable parts will need to be replaced, trained *metal, plastic, mechanical,* and *electrical* technicians can restore some damaged parts to their intended usage and condition. In cases where electronic apparatus was repaired or affected by the repairs, **scan tools** may need to be employed to clear codes and reset electronic functions.

There will be times when additional damages not written during the initial inspection and estimate creation are

10574 Job Number:

PRELIMINARY ESTIMATE
Written By: Supervisor
Adjuster:

Insured: **Claim #**
Owner: **Policy #**
Address: **Deductible:**
Date of Loss:
Day: **Type of Loss:**
Evening: **Point of Impact:** 12. Front

Inspect Location:

Insurance Company: Days to Repair

2000 SATU SL2 4-1.9L-FI 4D SED black Int:tan
VIN: 1G8ZK5271YZ114471 **Lic:** **Prod Date:** **Odometer:** 141326
Condition: Good

Air Conditioning	Rear Defogger	Tilt Wheel
Intermittent Wipers	Dual Mirrors	Clear Coat Paint
Power Steering	Power Brakes	AM Radio
FM Radio	Stereo	Search/Seek
Driver Air Bag	Passenger Air Bag	Cloth Seats
Bucket Seats	Automatic Transmission	Overdrive

NO.	OP.	DESCRIPTION	QTY	EXT.	PRICE	LABOR	PAINT
1		FRONT BUMPER					
2		O/H front bumper				1.8	
3	Repl	Bumper cover black	1	286.03		Incl.	3.0
4		Add for Clear Coat				1.2	
5#	Repl		1				
6	Repl	Absorber	1	63.82		Incl.	
7	Repl	Lower support	1	29.68		Incl.	
8	Repl	License bracket	1	18.33		0.2	
9		RADIATOR SUPPORT					
10	Repl	Radiator support from XZ152252	1	316.22	s	7.5	1.0
11		Aim headlamps				0.5	
12		Evacuate & recharge			m	1.4	
13		Refrigerant recovery			m	0.4	
14		Add for auto trans			m	0.2	
15		Add for AC option			m	0.6	
16	Repl	RT Bumper bracket	1	40.07			
17	Repl	LT Bumper bracket	1	40.07			
18	Repl	Shield w/AC	1	45.72		0.3	
19	Repl	Air deflector	1	16.04		0.3	
20	Repl	Air deflector spring	1	1.31			
21	Repl	Air duct	1	40.15		0.3	

FIGURE 1-5 A repair facility should endeavor to create a detailed estimate of damages prior to beginning repairs to a vehicle.

10574 Job Number:

PRELIMINARY ESTIMATE

2000 SATU SL2 4-1.9L-FI 4D SED black Int:tan

NO.	OP.	DESCRIPTION	QTY	EXT.	PRICE	LABOR	PAINT
22		HOOD					
23	Repl	Hood	1	356.41		0.8	2.8
24		Add for Clear Coat					1.1
25*	Rpr	RT Hinge				0.5	0.3
26*	Rpr	LT Hinge				0.5	0.3
27	Repl	Latch assy	1	29.68		Incl.	
28	Repl	Release cable	1	17.29		0.6	
29	Repl	Support rod	1	14.63		Incl.	
30	Repl	Support rod bushing	1	4.25			
31	Repl	Support rod retainer	1	3.64			
32	Repl	Bumper front	1	1.89		0.1	
33	Repl	Bumper side	1	2.37		0.1	
34	Repl	Insulator	1	53.42		Incl.	
35	Repl	Insulator retainer	1	0.42			
36	Repl	Seal	1	23.47		Incl.	
37		FENDER					
38	Repl	RT Fender	1	247.92		1.3	2.1
39		Overlap Major Adj. Panel					-0.4
40		Add for Clear Coat					0.3
41		Add for Edging					0.5
42		Add for Clear Coat					0.1
43		Deduct for Overlap				-0.3	
44	Repl	LT Fender	1	247.92		1.3	2.1
45		Overlap Major Adj. Panel					-0.4
46		Add for Clear Coat					0.3
47		Add for Edging					0.5
48		Add for Clear Coat					0.1
49		Deduct for Overlap				-0.3	
50	Repl	RT Fender liner	1	50.76		Incl.	
51	Repl	LT Fender liner	1	50.76		Incl.	
52	Repl	RT Emblem	1	8.80		0.2	
53	Repl	LT Emblem	1	8.80		0.2	
54	Repl	LT Rail assy	1	288.43	s	8.0	1.0
55		Overlap Minor Panel					-0.2
56	Repl	LT Upper rail	1	141.58	s	3.5	0.4
57		Overlap Minor Panel					-0.2
58		COOLING					
59	Repl	Cooling module w/AC	1	694.09	m	Incl.	
60		ENGINE / TRANSAXLE					
61	Repl	Air cleaner assy	1	177.26	m	0.4	
62	Repl	Air temp sensor	1	13.09	m	0.4	
63	Repl	Resonator	1	33.31	m	0.3	
64	Repl	Duct assy	1	80.05	m	0.3	
65		FRONT DOOR					
66	Repl	RT Door frame assy	1	523.68		3.5	1.0
67		Overlap Major Non-Adj. Panel					-0.2
68		Add for mirror				0.3	
69	Repl	RT Door w'strip	1	20.42		Incl.	

FIGURE 1-5 Continued

FIGURE 1-6 This estimator is writing an estimate of damages after having removed the parts that were obstructing his view of all the damages. This is called a *teardown*.

discovered during the course of a repair cycle. If so, a supplemental estimate and *parts order* may be required. Ongoing communication with all involved parties continues throughout the repair process.

The vehicle will need to be refinished in the areas that were disturbed by the collision event and subsequent repair processes. Refinish preparation is accomplished by sanding, priming, sealing, and/or corrosion-protecting the areas to be painted. Areas that are not being refinished are taped off or masked, and/or their parts are removed and reinstalled later to facilitate proper refinishing. A technician prepares to match the vehicle's color using a color mixing system, along with his or her considerable training, experience, and "eye" for detail. This may include tinting the paint, making a spray-out panel, and/or blending non-damaged adjacent panels. Spraying is commonly done in a climate-controlled spray booth utilized to keep contaminants off the surfaces being painted.

The vehicle is now ready for reassembly. Parts such as trim, ground effects, nameplates, and parts refinished off the vehicle are now reattached to it. If steering and suspension work was done, or if the vehicle's steering and suspension mounting points were affected, a qualified technician will need to perform a **wheel alignment**.

Many shops now have full-time detailing technicians and areas where the completed vehicle is washed and cleaned, inside and out. Polish, vinyl and leather cleaners, conditioners, and window dressings are applied. If minor imperfections in the paint are discovered, they are removed by a process known as color sanding.

After final inspections for safety and cosmetic finish, the shop often employs a road test. All paperwork is completed and any insurance forms are processed, the file is closed, and the customer takes *delivery* of the now-completed vehicle. This is sometimes referred to as "selling the car back to the owner." The repair process is not over until a *completely satisfied* owner leaves the repair facility with his or her restored vehicle.

WORK ORDER

1995 MITS ECLIPSE GS TURBO 4-2.0L-T   3D H/B Green Int:Gray
VIN: 4A3AK54F2SE092063 **Lic:**   PA **Prod Date:** 11/1994   **Odometer:** 136000
Estimator: Amos Albright   Adjuster:
Ins. Co.

#	ASN ID	ACT ID	OPERATION	DESCRIPTION OF DAMAGE	ASN HRS	ACT HRS	DATE COMPLETE
BODY	LABOR						
1				DOOR			
2	___	___	Repair	RT Outer panel	1.0	___.__	______
3	___	___	R&I	RT R&I trim panel	0.4	___.__	______
4				QUARTER PANEL			
5	___	___	Repair	RT Quarter panel GS, GS-T, GSX	2.0	___.__	______
6	___	___	R&I	RT Qtr trim panel upper gray	1.0	___.__	______
7	___	___	R&I	RT Qtr trim panel lower gray	Incl	___.__	______
00	___	___	______	______	___.__	___.__	______
00	___	___	______	______	___.__	___.__	______
				TOTAL BODY LABOR ==>	4.4		

FIGURE 1-7 A repair or work order serves as a guide for the repair technicians.

Estimating Collision Damages

One of the most important steps in the collision repair process is the *first one*—creating an accurate, detailed **estimate** or **damage report** of the collision damages. This step serves many functions, not the least of which is being the first impression you will make on a prospective customer. (Note that the terms *damage report* and *damage estimate* are often used interchangeably.)

A damage estimator must be certain that his or her estimate will *at least* provide:

- An accurate, detailed overview of damages that need to be repaired that can be communicated to the customer
- A tool by which he or she can "sell" the repair center's services
- A plan or "road map" to a safe, factory-recommended, cosmetically pleasing restoration of the vehicle
- All necessary information to create accurate parts and work orders
- The means to price his or her business product (the repairs) accurately and competitively
- An increase in the shop's productivity and the assurance of the speediest "on-time" delivery possible

The estimate often serves as an agreement between the shop, customer, and the insurer involved. The first, or administration, page includes the name, address, and contact numbers of the vehicle's owner and insurance information such as policy and claim numbers, the loss date, and any **deductibles** that may apply. It also contains information such as the **vehicle identification number (VIN)**, year, make, model and production date, trim and paint codes, mileage, and all of the vehicle's options.

The body of an estimate is an itemized listing of all labor operations necessary to restore the damaged vehicle. This section usually includes parts prices and numbers, labor times, the operations involved, and shop, paint, and/or any other materials needed; any work or operations that will need to be subletted to a specialty repairer; and documentation concerning any parts, operations, or procedures.

The estimate may also contain a listing of any unrelated old damages, various disclaimers such as those pertaining to the use of salvage or aftermarket parts, instructions that advise a customer and/or repairer or insurer of rights, and a power of attorney or permission for the repairer to act as the vehicle owner's designated representative.

The estimate concludes with the total amount necessary to repair all related (and sometimes unrelated) damages, by listing labor rates, operations, parts, additional costs, applicable taxes, and any customer-responsible charges and/or additions such as a deductible, depreciation, and appearance allowances.

Damage reports can be created manually utilizing collision damage estimating guides that list vehicles by model. The estimator lists each labor operation and its labor units, part repair or replacement, all part prices, and refinishing entries on a blank sheet by hand. The guides contain **procedural** or **"P" pages** that assist the estimator in writing the estimate. Here, all procedural explanations are found, including definitions; use of the crash guides; additions to labor times; identities of structural, body, mechanical, and refinishing operations; and which operations are included or not included in various times allowed.

Damage reports can also be created using a computer software system. A computerized system helps an estimator by completing many functions like totaling costs, listing part numbers, and determining what is and is not included in the operations selected automatically. These systems are designed to streamline the business, saving time and guaranteeing numerical accuracy. Computerized systems are now utilized beyond the creation of an estimate; **job costing**, workflow, **cycle time**, productivity and management systems, and even sales trend tracking are among the "automatic" advantages of computerization.

Many shops are now able to correspond electronically with insurers. The compilation of claims-related data, electronic supplements, and digital photography make faster, long distance business relationships the norm. Digital imaging is often utilized by taking electronic digital photos or video pictures of a damaged vehicle and sending them to insurance adjusters and back, via the worldwide Web. This system allows repairers and insurers to create, evaluate, store, and retrieve damage estimates or entire claim files without having to physically leave their respective premises.

Whether one creates a damage report manually or using a computer, experience—as well as technical and estimating knowledge—are essential. A damage estimate is only as good as its writer. Ultimately, repair/replace, judgment time, and even repairability decisions will remain with the professional damage estimator. See **Figure 1-8**.

Metal Work

Working with various metals in collision repair can be considered an artistic endeavor, as well as a technical one. When sheet metal on a motor vehicle is damaged and considered to be repairable, a variety of methods, techniques, power and hand tools, materials, and experience will be employed to restore a misshapen part or panel to its original shape, contour, functionality, and durability. Today's vehicles employ a variety of metals (and non-metal **substrates**) in the building of structural and non-structural parts. Aluminum, magnesium, boron, and a bevy of mild to high tensile strength, martensitic (ultra) high-strength steels are a few of the metal substrates that will commonly be encountered in today's collision repair environment.

To perform high-quality repairs to sheet metal, the damaged panels must be thoroughly cleaned, making

WE CARE INSURANCE
234 Maple Drive, Anywhere, USA 11234
555-4000

Damage Assessed By: Joseph Smith | Claim Rep: Sally Jones

*Product Type: Auto
*Date of Loss: 04/30/2008
*Deductible: 1,000.00
Days to Repair: 4
Insured: PETER PRATTI
Address: 123 Main Street, USA
Telephone: Work Phone: 555-1000 Home Phone: 555-2000

Mitchell Service: 911754

Description: 2005 Toyota Corolla S
Body Style: 4D Sed
VIN: 2T1BR32E75C10000
Mileage: 8,745
OEM/ALT: A
Color: GRAY
Options: ANTI-LOCK BRAKE SYS. (ABS), ALUM/ALLOY WHEELS, AIR CONDITIONING, POWER STEERING POWER BRAKES, POWER WINDOWS, POWER DOOR LOCKS, TILT STEERING WHEEL CRUISE CONTROL, ELECTRIC DEFOGGER, AUTOMATIC TRANSMISSION AM-FM STEREO/CDPLAYER(SINGLE), PASSENGER-FRONT AIR BAG, POWER REMOTE MIRROR 4 WHEEL DISC BRAKES, FRONT WHEEL DRIVE, L-4 ENGINE, 4-DOOR, DRIVER-FRONT AIR BAG

Vehicle Production Date: 10/04
Drive Train: 1.8L Inj 4 Cyl 4A FWD
License: DBU-1357 TN

Line Item	Entry Number	Labor Type	Operation	Line Item Description	Part Type/ Part Number	Dollar Amount	Labor Units
				FRONT BUMPER			
1	100027	BDY	OVERHAUL	FRT COVER ASSY			1.8 #
2	102343	BDY	REPAIR	FRT BUMPER COVER	Existing		1.5* #
3		REF	REFINISH/REPAIR	FRT BUMPER COVER		C	1.3*
4	102367	BDY	REMOVE/REPLACE	R FRT BUMPER SPOILER	76081-02030-B1	350.10	INC
				FRONT LAMPS			
5	102382	BDY	REMOVE/REPLACE	R FRT COMBINATION LAMP ASSEMBLY	81110-02370	214.87	INC #
6		BDY	CHECK/ADJUST	HEADLAMPS			0.4
				FRONT FENDER			
7	100220	BDY	REMOVE/INSTALL	R FENDER ASSY			1.4 #
8	100224	BDY	REPAIR	R FENDER PANEL	Existing		2.0* #
9				refinish in repaired panel			
10		REF	REFINISH/REPAIR	R FENDER PANEL		C	1.8*
				FRONT INNER STRUCTURE			
11	100255	BDY	REPAIR	UPR FRONT BODY TIE BAR -S	Existing		1.0*
12		REF	REFINISH	UPPER TIE BAR			0.5
13	102445	BDY	REPAIR	R FRONT BODY FRONT APRON PANEL -S	Existing		1.0* #
14		REF	REFINISH	R APRON			0.5
				ABS BRAKES			
15	102463	MCH	REMOVE/INSTALL	ABS MODULATOR UNIT -M	Existing		0.5*
16				d & r for apron repair			
				FRONT SUSPENSION			

FIGURE 1-8 A typical computerized estimate.

proper preparation a part of the process. The surrounding undamaged parts and assemblies may have to be removed or protected from harm during these repairs. At times, mangled parts may impede any disassembly of the area and/or the damaged vehicle, necessitating additional time (commonly referred to as **access time [AT]**) and caution on the part of the collision repair metal worker.

After identifying high and low areas of the panel to be restored along with the extent of the distortion, a repair method is chosen based on whether there is access to the backside of the panel, which in turn will determine which tools to employ. Depending upon this and other characteristics of the vehicle and its unique damages, the metal worker will utilize hand and power tools such as body hammers, dollies (or blocks), spoons, pry bars, picks, sanders and grinders, hydraulic rams and pulling devices, welders, and body fillers in the metal repair process. Even heat has its place in some metal repairs. Far too often, a metal repair technician is simply thought of as "banging up the low areas, and hitting down the high spots." If only it were that simple! Analysis and identification of the damages, and specialized techniques such as the "hammer-on-" and "hammer-off-dolly" repair methods, are but a few of the choices that accomplish proper, time-efficient metal repairs. See **Figure 1-9**.

Once the original shape and contour is restored, body filler is used to fill low spots that still exist. This is another specialized talent that requires technical know-how, as well as an artistic approach. Restoring durability is assured by utilizing the proper corrosion protection materials on the restored metal. Finally, any attached new or restored parts that need to be reassembled are tended to at this juncture.

As always, a collision repairer's technical and continual training, hands-on practice, "eye," and "feel" are what will ultimately make the tools and processes successful when working on damaged metal. Cold and heat shrinking of the metal, hammering, pushing, pulling, filing, and filling are a few of the techniques that a successful bodyperson will need to perfect and employ to return sheet metal parts and panels to their pre-loss condition.

FIGURE 1-9 A body repair technician straightening a damaged sheet metal panel.

Structural Repairs

Perhaps one of the most important steps in the collision repair process, structural repairs restore the skeleton or "backbone" of the damaged vehicle to the OEM's original functionality, dimensions, and durability. Without proper structural restoration, a vehicle's parts will not fit correctly, nor will the vehicle run properly, efficiently, or safely.

The technician will be faced with a number of build characteristics and parameters when presented with a structurally damaged vehicle (see **Figure 1-10**):

- Is the structure a unitized (unibody) one, or full-frame? If unibody, is it a space frame?

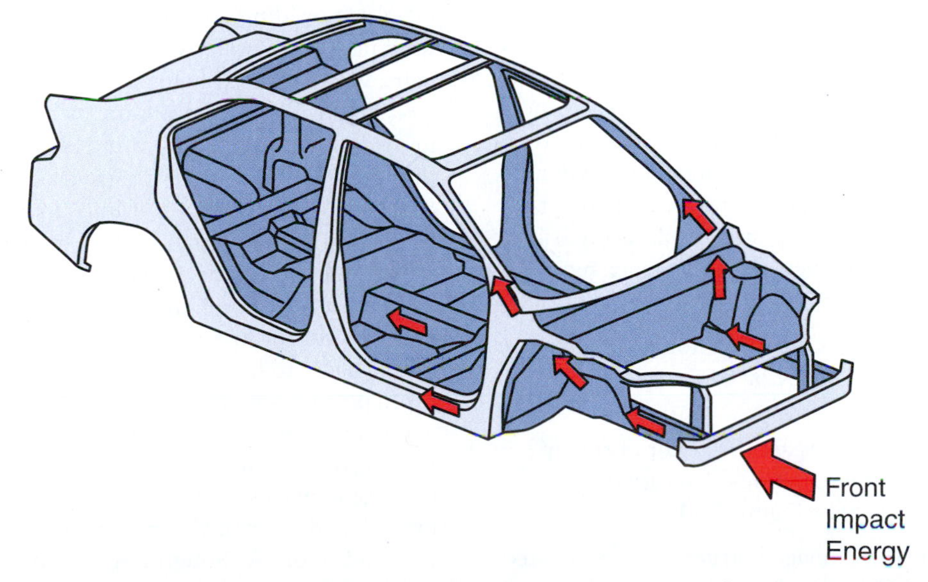

FIGURE 1-10 The way a vehicle absorbs, transmits, and manages energy is no accident. Many vehicles are crash-tested on computers before they are ever manufactured.

- If the structure is a full-frame one, is it a perimeter or ladder style?
- Is it **hydroformed**?
- What is the structure made with—mild steel, high-strength steel, aluminum, or something else?
- Can heat be utilized during the repair process and, if so, how much and for how long?
- Where was the structure designed to crush?

The structurally skewered vehicle must be analyzed to determine what damage conditions are present, the extent of the damages, and exactly how far into the structure damage has traveled, since vehicles are manufactured to absorb and manage the energy of collision forces that rumble through it when it is involved in an accident. That collision energy, or inertia force, leaves a path of evidence for the structural repair technician to find, analyze, and utilize in the development of his repair plan, and then *eliminate*. Examples of collision energy evidence—also referred to as visual indicators—may include:

- Panel gaps that are too wide or too narrow
- Hoods, deck lids, and doors that are misaligned
- Damaged spot welds, split seams and seam sealer, and/or cracked undercoating or paint finish
- Buckles, bows, and ripples in sheet metal parts not directly struck
- Cracked stationary glass that was not hit directly

The amount of distortion must be determined, measured to the millimeter, and a repair plan must be developed. The technician must consider which structural parts, if any, can be restored for reuse and which might necessitate replacement after the structural dimensions have been restored. Note that an absolute determination of repair or replacement of structural parts can only be made after all damaged structural parts have either been successfully restored to proper positioning or pulled as much as possible without successful restoration.

While anchored to a frame repair machine or rack/bench, the damages are methodically pulled out using a variety of hooks, chains, straps, supports, mounting plates, hydraulic pulling posts, and at times pushing rams. Among other aspects, the *type* of damage can factor into the decision to pull, how much, and with what particular method when repairing a vehicle's structure. For example, is the structural damage:

- Direct? Which damages are the result of the *direct impact* with another vehicle or object (also called *primary damages*)?
- Indirect? Which damages are the result of *energy* traveling through the structure (also called *secondary damages*)? See **Figure 1-11**.

As stated earlier, the damaged structure is measured before beginning the pulling and repair process. However, the measuring process proceeds throughout the structural repairs during the pulls, as well as after all pulling has been completed. This requires a three-dimensional measuring system which is manufactured in various forms—including computerized laser or sound wave, as well as dedicated and universal mechanical. In fact, the vehicle's replacement parts—both structural and nonstructural—also serve to help measure the vehicle's structure. Again, this shows the importance of structural repairs—if a door cannot fit its opening or a frame rail cannot be welded into its proper place, the vehicle's repairs cannot proceed to a proper, safe, and pre-loss conclusion.

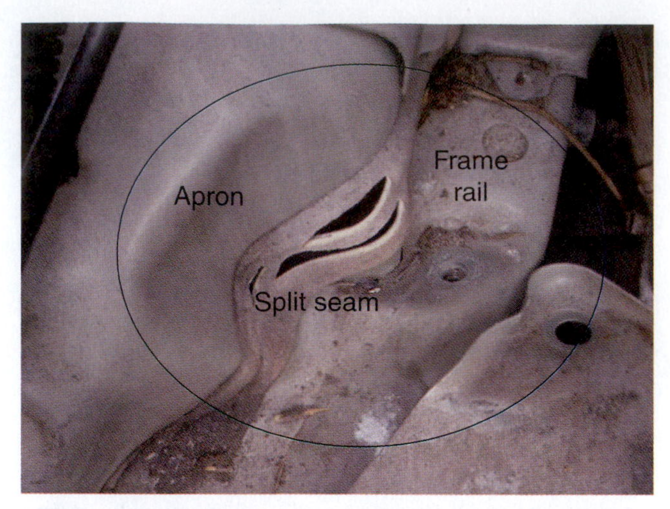

FIGURE 1-11 Although the area pictured was never hit directly, the distortion to the apron, buckles in the frame rail, and splits in the welded seams were all caused by collision energy flowing through the area. This is an example of *indirect damages*.

A variety of welding equipment and techniques may be employed during structural repairs, particularly if the replacement (or repair) of structural parts is necessary. Gas metal arc welding, also known as metal inert gas (MIG) welding, is the preferred choice for today's collision repairs, with thinner gauge metals being utilized.

It may be helpful to picture structural damage repairs in the following context—if the technician had a video of the accident that caused the structural damages, he or she could play the tape *backwards* in super-slow motion, watching the damages "reverse themselves." As the tape ran, the damages that occurred last would "come out" first, gradually making their way to the damages that first occurred during the collision (which would be the very last damages to leave the structure on the film). Now picture a person—or group of persons—strong enough to literally pull the structural damages out of the crippled vehicle; they would each grab hold of a damaged area, and pull the damages back out in the direction they traveled inward, in the very order they went in—from **last-in first-out**, and so on. Although not exactly scientific in nature, the illustration serves a purpose— it is a general blueprint for helping a structural repair technician develop a repair plan, and provides a glimpse of the purpose that structural repairs serve.

Mechanical/Electrical

When one thinks of a collision-damaged motor vehicle, the mind usually visualizes crumpled hoods and fenders, a broken taillight, and a variety of moldings and nameplates hanging off of the body—or missing altogether. However, today's collision-damaged vehicles are loaded with mechanical and electrical equipment that oftentimes wind up in harm's way when an accident occurs.

The drive train—including the oftentimes transverse-mounted engine with transmission and front-wheel-drive axles that are often packed into smaller front compartments than ever before—are subject to physical damages, making mechanical acumen a necessity. The engine's many extremities, like the air induction assembly, manifold, alternator, air conditioning compressor, and power steering unit, are routinely damaged or destroyed from *frontal impact,* which accounts for at least two-thirds of all vehicular damages. Radiator, air conditioning condenser, and attached fans, motors, and lines do not beg the question, "Did they sustain damage?" but rather, "How bad is the damage to them?" Because of space and design factors, even undamaged mechanical and electrical parts and assemblies may have to be removed to make room for structural and sheet metal repairs. Lines, wiring harnesses, low- and high-pressure hoses, mounts, sensors, and brake systems are just a few of the components that may be restricting a portion of a collision repair job—if they are not damaged themselves.

Steering and suspension systems are sometimes tuned to within 1/16 of a degree! Geometric adjustments to caster and camber angles, toe, steering axis inclination (SAI), and computerized systems make steering and suspension diagnosis and repair a "routine challenge" when a collision event occurs. In order to make an exacting diagnosis to unconventional-looking parts like steering knuckles, control arms, a steering rack and a pinion gear, and spring steel torsion bars, a technician must first understand what these—and other steering and suspension components—are supposed to do when they are *not* damaged. Even "after-the-fact" damages to components like tow truck-damaged constant velocity (CV) joints may demand the collision repairer's attention. See **Figure 1-12**.

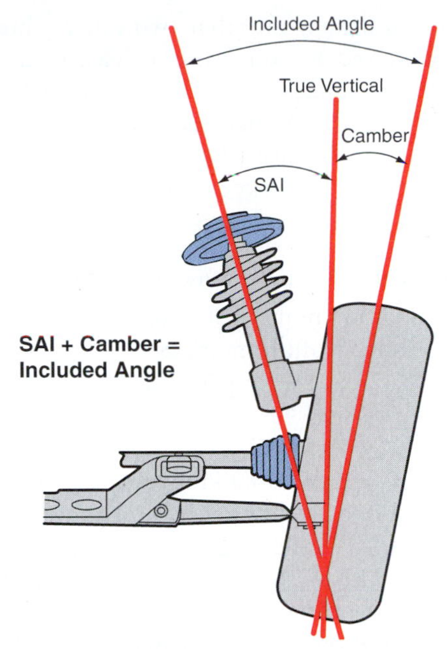

FIGURE 1-12 Proper wheel alignment, as well as diagnosis of steering and suspension parts, are dependent upon checking alignment "geometry" like camber, caster, toe, and included angle (steering axis inclination plus camber) as illustrated here.

The average vehicle contains over 30 miles of wiring, which connects sophisticated and computerized power, safety, and luxury equipment. Think about the last time you manually cranked a window open, or imagine how many air bags the 21st-century automobile will contain inside before the experimental *exterior* (pedestrian) air bag is standard equipment. Soon, a road atlas will likely be found in a museum, as drivers punch in their destination and hear a soothing voice "tell" them exactly where to turn and when! Many luxury vehicles already have "night vision"—infrared technology and camera-like sensors that detect heat-emitting objects (i.e., human beings and animals) up to a mile away that alert the driver on a "head's up" display. Some of these "**collision avoidance systems**" can actually adjust the vehicle's adaptive cruise control system automatically, and/or sound a tone to "wake up" a sleepy driver! Imagine the precise standards that a collision repair technician must adhere to when servicing, repairing, or replacing these and other systems. These are but a few of the collision-damaged electric motors, computers, systems, and navigational apparatus technicians face after the vehicle is presented in a collision repair shop. In fact, one can no longer dismantle a vehicle, much less cut through a rocker panel or pillar, without collision and service manuals to tell the technician what lies ahead!

Air flow meters, relay and fuse boxes, sensors, **antilock brake (ABS) systems** and **traction control units (TCU)**, emission canisters, on-board computerized navigation systems, "smart" air bags, tires, and even glass make electronic and computer training and experience a must for today's collision repair technician.

Computerized wheel alignment equipment, digital volt-ohmmeters (DVOM), electrostatic discharge straps (ESD), electrical wiring diagrams and symptom **flowcharts**, a volt-amp tester (VAT), leak and pressure testers, recycling machines and federally mandated licensing, a box of one-time fasteners, shrink tubing for wire repairs, and vehicle-specific breakout boxes and trouble code scanning tools are becoming as commonplace in the collision shop as a set of jumper cables used to be!

It is interesting to note that two out of three shops surveyed admitted to subletting at least some of their work out to specialized repairers. Out of that subletted amount, two-thirds of them sent out electrical work while three-quarters "farmed out" the mechanical. More technical training and experience in this area would tend to:

- Make technicians more *valuable* to a shop
- Grant the shop more control over the repairs, and receive more profit from them
- Speed up the repair cycle by saving valuable cycle time

Just imagine a harried-looking customer standing beside a collision-damaged vehicle complaining that this little yellow light (the check engine lamp) is now always on, the car pulls to the left (torque steer) when she feeds gas from a stop, and the windshield wipers do not operate since the accident she was just involved in. As the technician reaches for the DVOM, ponders whether the frame or the steering parts (or both) are bent, and wonders if the wipers are connected to a sensor and computer control module that automatically activates them when water, rain, or snow is detected, the repair center can rest assured that the work will be done in-house, will probably be done faster, and will satisfy the customer while garnering more profits!

Collision repair goes far beyond dents and broken windows. In fact, we have not even scratched the surface of mechanical and electrical innovations and repairs on the average collision-damaged vehicle!

Refinishing

Automobiles are now sold largely by *color.* Therefore, is it any wonder when a customer shows up at a collision repair center oblivious to the twisted frame his or her vehicle has sustained, the skewered thrust angle that is now pushing the car out of alignment, or the frozen **seat belt pre-tensioners** that no longer hold their passengers in place? Hardly, since the three-stage *Long Beach Blue Pearl* fender is in need of an expert "eye" and the shop's climate-controlled spray booth!

In all seriousness and with all due respect, the finish of an automobile creates a majority of its all-important cosmetic appeal. If you were to see a Hyundai Sonata, a Mercury Sable, a Ford Taurus, and a Jaguar XJ parked side-by-side minus its lamps, plastic effects, nameplates, and *color,* you would be quite surprised to find that they tend to resemble each other more than any of the OEMs would like to admit. The age of aerodynamic styling and energy efficiency has limited design options. And so, painting a repaired vehicle and creating the illusion that makes the owner unable to detect what, where, and how you repaired and refinished it may not affect safety and handling but *will* affect how the owner continues to view the vehicle—and your repair shop.

As technical as refinishing is, and how "artistic" a refinisher may need to be, "fooling the human eye" can sometimes take precedence over both of them. Many refinishers will say that painting a vehicle often begins by analyzing where good break-off points are, what *undamaged* panels may need to be refinished (blended), and just how fussy the vehicle's owner is and/or what the expectations are. So begins the headaches for the collision repair's refinish technician! As discussed earlier, the variety of materials used in the construction of today's vehicles is growing; that must also be taken into account. Refinishing sheet metal, plastic, aluminum, and various other substrates will help tweak the painter's planning and approach. See **Figure 1-13**.

One of the most important technical steps taken when refinishing a repaired vehicle is the preparation that is necessary before spraying. Painters use a **paint thickness gauge** to decide if any preexisting paint layers will need to be removed, and if any preexisting finish or refinish is defective. The vehicle must be cleaned using soap and water, as well as grease and wax removers. By the time a repaired vehicle reaches the paint prep area, all trim that will need to be removed is already removed; any pieces that can—and will—remain are now taped or masked off. Using sandpaper in a gradual shift from coarse to extremely fine, the surface is made smooth. Metal conditioners and/or conversion coatings, as well as primers, surfacers, and sealers, may be utilized at this juncture. The procedure for refinishing a replacement part is similar to refinishing a repaired one—with the exception of a few steps, most notably a solvent test on the repaired layers of substrate materials. Once primed, parts are ready for topcoats.

Vehicles are finished with a variety of systems. Topcoats can be single-stage, or more likely multiple-stage, finishes that consist of base color coats, and/or an intermediate pearl or mica coating, followed by a clear coating. Collision repair centers have a refinish section, complete

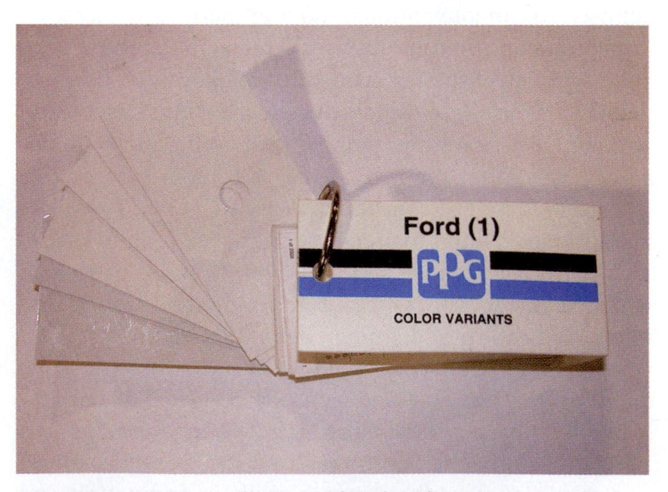

FIGURE 1-13 The collision refinisher has to contend with hundreds of colors. Even white is no longer simple since it often contains as little as a half gram of color. The white pictured here has five variations.

with prep, spray booth, and after-paint care or detailing areas. A compression-driven spray system, complete with a variety of spray guns, a computerized color mixing system, and an array of safety equipment, are essential in today's repair environment.

Before spraying the vehicle, a spray-out or **letdown (test) panel** may need to be prepared. The painter will need to take the paint code off of the vehicle and research the formula for that code, as well as any variants of it. The formula could contain three, four, or more tints or toners that will need to be added to achieve the OEM's original shade. After the test panel is sprayed out, it will need to be clearcoated if the vehicle is clearcoated, and then compared to the vehicle under natural—or simulated natural—light. When satisfied that the color match is as close as possible—sometimes referred to as a "blendable match"—the vehicle can be recleaned, checked for any imperfections, re-taped, and finished inside the spray booth. A spray gun is adjusted based upon the manufacturer's recommendations and the material that will be utilized, before any sealers, adhesion promoters, and topcoats are applied.

After the vehicle has been sprayed to satisfactory coverage utilizing a sufficient number of coats, the vehicle is unmasked and inspected for any refinish defects. A few imperfections are unavoidable regardless of how clean, prepared, and careful the painter is. Many of these rather minor problems can be corrected in the post-paint operations, while the ones that cannot may need to be resprayed.

A professionally painted vehicle is an all-important first step toward helping a vehicle's owner to welcome the car back from an accident. An owner's faith in a vehicle is reestablished—starting with an outstanding refinish job!

Vehicular Identification

Correctly identifying a vehicle is essential to proper collision repairs. There is vehicular information that may need to be researched before, during, and sometimes after a repair job. The Vehicle Identification Number (VIN) most often comes to mind when vehicular identification is discussed, although a number of other identification "plates" and stickers are also important to successful repairs.

The VIN is displayed on passenger vehicles in a variety of places—most notably on the plate that is riveted to the top of the dash pad, viewed through the bottom portion of the windshield at the lower driver's side corner. The information that is ascertained from the VIN includes:

- Nation of origin (i.e., USA)
- Manufacturer and division (i.e., Ford)
- Vehicle type (i.e., passenger car)
- Model and body type (i.e., Taurus GL and four-door sedan)
- Restraint system (i.e., dual front and side air bags; manual front and rear seat belts)
- Engine type (i.e., 3.0L V6 EFI)
- VIN check digit (manufacturer's internal code; no relevance to collision repair)
- Model year (i.e., 2004)
- Assembly plant (i.e., Flatrock, Michigan)
- Serial number (sequential production number, unique to each vehicle)

From the creation of a damage report through the repair cycle, the VIN may be necessary at a number of junctures such as when writing an estimate using a computer, using electronic diagnostic equipment, ordering parts, and occasionally to solve a problem that no one but the vehicle's manufacturer can help with.

Other plates and stickers contain information vital to the collision repair center. (See **Figure 1-14**.) The service parts ID label—which is commonly affixed to the driver's door jam, center pillar, inside of the rear compartment, firewall, or glove box door—can contain information like:

- Production date
- Paint code
- Engine and transmission type/codes
- Interior trim and accent codes
- Vinyl roof color code
- Body side molding code
- Gross weight of the vehicle
- Axle ratios
- Tire inflation information
- Some power options

Other labels that can be of importance to the collision repairer include:

- Emissions information
- Air conditioning capacities

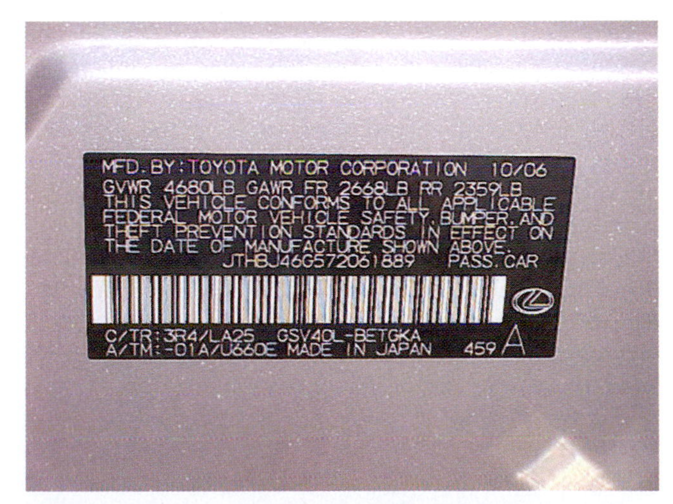

FIGURE 1-14 This identification label contains the production date of the vehicle—a vital piece of information when ordering replacement parts.

- Cautionary/warning information (i.e., air bags, fuel, fan, etc.)
- Radiation

In addition to assisting the technician in the vehicle repairs, these labels are part of the vehicle. They must remain with it for the job to be considered complete and the vehicle returned to "pre-accident condition." In fact, some are federally mandated. In addition to the obvious one (the VIN plate), the emissions label is considered a part of that system—a technician could receive a fine if it is not affixed after a repair.

TECH TIP Take caution when documenting a VIN. Some states have registration stickers that adhere to the windshield near the VIN plate. Oftentimes, an estimator will copy the VIN from a registration sticker rather than take the time to read it off of the VIN *plate*. In cases where the sticker contains an error, it can lead to problems such as errant VIN decoding, wrong parts orders, and other mistakes that could be avoided. Always take the VIN off of the VIN plate!

Service Information Retrieval

Whether or not a vehicle is involved in a collision, computer systems and electronics are extremely important to today's vehicles. The information that is processed and stored in these systems is vital to the safe, pre-loss repairs that a collision repair center is striving to accomplish.

Computers and sensors are continually monitoring a vehicle's operations. Computer control systems have three basic elements—input, process, and output. Scanning tools, diagnostic testers, and flowcharts allow a technician to retrieve vital service information when analyzing, repairing, and maintaining a vehicle.

Sensors are employed to change operating conditions and various other *input* parameters into electrical signals. The main control modules such as the **supplemental restraint system (SRS)**, anti-lock braking system (ABS), and **electronic control module (ECM)** monitor the respective systems and maintain proper operations. These modules take in signals, *process* the information, and supply needed directions to the system it is monitoring. The modules also store diagnostic trouble codes (DTCs) and other information. The *output* signals sent from the various control modules signal various actuators like malfunction lamps, motors, solenoids and relays which take the appropriate actions.

After servicing, damages, deployments, and other events associated with a vehicle occur, a diagnostic tester or scan tool can retrieve performance parameters, DTCs, and other information in order to properly service, repair, or reset the vehicle's systems. Diagnostic connectors are mandatory ports located in every vehicle that a diagnostic tool can be connected to in order to retrieve information and communicate with a specific vehicular system. Utilizing the vehicle's repair manual, electrical wiring diagrams, flowcharts, and technical service bulletins (TSB), a qualified technician can clear trouble codes, reset systems, and ensure that a collision repaired vehicle can function in the manner in which the OEM designed it to.

Post-Paint Operations

A number of operations, procedures, details, and finish areas will be employed between the time a vehicle is refinished and when the customer returns to pick it up. Although some may consider the following steps relatively minor "afterthoughts," a repair job can nonetheless be pushed into the realm of "fantastic!" or lost to a level of mediocrity depending upon the way the following steps are viewed and carried out by the collision repair center.

Reassembly of the Vehicle. After refinish operations are complete and any/all maskings, car covers, tape, and wheel coverings have been removed, the vehicle will need to be reassembled. Parts such as lamps, moldings, grilles, glass, and any parts that may have been refinished off of the vehicle—such as major panels, mirrors, bumper covers and assemblies, and ground effects—will be reinstalled at this time.

Some mechanical adjustments, electronic reconnections and checks described earlier, and drive train maintenance may need to be addressed. In some cases, the repair center will choose to—or need to—send a vehicle out for some of these and other final operations. Procedures like upholstery repairs, dealer-only checks and code clearing, wheel alignment and high speed balancing, and even detailing are sometimes subletted to another specialized center that is equipped and/or specializes in the particular post-paint operation that is necessary.

Detailing the Vehicle. There really is no sole definition of the detailing process. It is best defined as a method of "deep-cleaning" a vehicle that is also meant to protect and restore it, as well as make it look attractive to its owner. Detailing is best accomplished by spreading the process out as the vehicle is repaired, although the majority of what will be done can only be finalized after refinishing is complete. We can define what detailing is NOT—*a cover-up for poor repairs.*

Here are some of the "details of detailing":

- Polishing the exterior surfaces
- Removing surface paint defects
- Checking for, and removing, any overspray
- Conditioning all exterior/interior plastic and rubber surfaces
- Vacuuming the interior and rear compartments; shampooing seats, upholstery, and carpets
- Removing stains and odors
- Cleaning all glass, inside and out

- Checking and filling all fluid levels
- Performing a final wipe-down (removing excess polish from gaps, moldings, and nameplates)
- Resetting all electronic devices, memories, and mirrors
- Checking and adjusting tire pressures and lug nut tightness
- Checking all gaps, alignment, and proper door/lid operation
- Checking and/or replacing burned-out lamps

Additional operations that can fall under the detailing category may include:

- Applying corrosion protectors
- Checking paint film thickness
- Applying decals and pinstriping
- Checking for proper seat belt operation and excess wear
- Replacing missing clips, bolts, and fasteners
- Checking spare tire and jacking tools, and securing them properly
- Cleaning the engine compartment
- Checking and/or replacing sound deadeners
- Checking for water leaks

See **Figure 1-15**.

Although collision repair jobs will differ and customer expectations vary, many shops find it helpful to utilize checklists when detailing vehicles. The process of detailing is tantamount to assuring your shop a smooth "**sell-back**" process of the vehicle to the customer, a repair job that *looks* as good as it *is* and lasts the life of the vehicle. This service should be kept consistent from vehicle-to-vehicle, and from technician-to-technician.

Final Inspection and Safety/Road Checks. Final inspection and safety checks can be considered part of the detailing process. They are, however, a link to guaranteeing safety and therefore go well beyond the appearance and protection of the repaired vehicle and its parts. In many cases, the detailer is the last link between the repair shop and the customer's safety and continued smooth operation of the vehicle. Once again, a checklist can serve to standardize the final inspection, and assure the shop and the vehicle's owner that "no page of the repair job" has been left unturned.

A final inspection would not be complete without a road test. In fact, many of today's motor vehicles *have* to be driven at least a few miles to reset or re-learn the various computer settings that may have been altered or disturbed during—or by—the repairs. "Computer memories" go beyond proper operation of locks and windows; proper and smooth drive train operation is oftentimes dependent upon the "re-learn" process.

Some of the final inspection and safety/road check items and operations may include:

- Dash warning lamps
- Proper engine operation
- Motor mounts
- Power options
- Directional signals, defoggers, and so on
- Seat belt inspection
- Proper wheel alignment, steering, and thrust angle
- Rattles, wind noise, and vibrations
- Proper braking
- Restoration of **memory presets**

Only after an exacting, painstaking review of an entire vehicle is a collision repair center ready to deliver it to a customer.

TECH TIP For any who may be thinking "I can't afford to do all this," think again. *You cannot afford not to!* While discussing detailing in my I-CAR classes, I regularly hear, "It's a great idea, but I can't find the time (to detail a vehicle)!" I love the answer one of my students—a successful collision repair tech—recently volunteered. *"We all know how long it takes, and what it costs to detail a car . . ."* he stated rather matter-of-factly, slumped back in his chair. Then he sat up, his eyes narrowed, and his voice turned emphatic, *". . . but I know how long it took to build a great reputation, and it only takes one car to ruin it!"*

Delivery of the Repaired Vehicle. Can delivery of the vehicle to a customer be included in the repair process? Yes it can! Ours is a society where *how* people handle themselves is as important as *what* it is they handle. Therefore, *how* a vehicle is delivered is very important, to say the least.

A vehicle should be delivered to the customer by a representative of the collision repair center. The old method of "leave the keys under the mat" does not help *explain* the repair, how to *upkeep* the vehicle, or answer any of the customer's *questions*. Make sure the vehicle's finish has been checked in the same light the owner will take charge of it in—preferably daylight. Some of the information that you may want to convey includes:

- How to protect the new finish and care for it
- How to upkeep the detailing work
- Who to contact with any future questions concerning the repair

P&L CONSULTANTS **Final - Inspection Checklist**

R.O. #: ______________________ **Mileage Out:** __________________

___ A/C service caps are installed
___ All panels are lined-up
___ Antenna works properly
___ Anti-corrosion compounds installed
___ Battery is held down and cables are tightened
___ Brake lights work
___ Courtesy lights and door ajar lights function
___ Deck lid opens and closes
___ Defroster, heater, and A/C operate
___ Doors open and close
___ Exhaust system heat shields are in place
___ Exhaust system operates properly
___ Express features reset
___ Fasteners are installed and tight
___ Fasteners are touched up
___ Fluid levels are topped off
___ Glass is cleaned
___ Headlights are aimed—high and low beam
___ Heated seats and power controls operate
___ Handles operate
___ Hood opens and closes
___ Horn honks
___ Keyless entry functions properly
___ Mats and floor coverings are in place
___ Overspray is cleaned off
___ Personal belongings are in place
___ Pinch welds repaired and refinished
___ Power windows, locks, and mirrors work
___ Radio is reset and all speakers work
___ Rear license and back-up lamps work
___ Remote transmitter operates
___ Restraint system lamp operates
___ Seats adjust properly and lock into position
___ Seat belts operate and buckle
___ Security system functions correctly
___ Spare tire, wheel lock, and jack in place and bolted down
___ Splash skirt clips are installed and correct
___ Tailgate/hatch/trunk opens and closes
___ Touch up bottle
___ Turn signals work
___ Tilt mechanisms function properly
___ Undercoating installed
___ Underhood labels are in place
___ Vehicle test drive completed (separate sheet)
___ Vents operate (blow out in all positions)
___ Washer nozzles spray and align; wipers work
___ Water leak test has been performed
___ Wheels are torqued
___ Wires and hoses are secure
___ Wires and hoses are not rubbing anything

☑ Completed **N/A = Not Applicable**

☒ Not Completed

Measuring Printout ______ **Estimate** ______ **Initial** ______ **Final**

Inspected by ______________________________

Inspected by ______________________________

FIGURE 1-15 A typical detailing checklist. Courtesy of P&L Consultants.

In addition to the final itemized bill, a good collision repairer will include the checklist of what was detailed, a warranty card, and any information that will assist the owner to maintain the repaired vehicle, as described previously. The deliverer should be prepared to answer questions and allay any fears the customer may have. It is not surprising to realize that even customers who may have never had any experience with a collision repair center may have misconceptions about the business that can lead to prejudice and distrust.

Finally, a promise to follow-up with a phone call and/or an invitation to stop back for a 30-day check-up instills confidence that the shop will always be there for the customer—as well as the customer's referrals!

CAREER OPPORTUNITIES IN COLLISION REPAIR

Many careers are available in the collision repair field, and the opportunities for having a satisfying, financially rewarding career have never been better. Surveys of the past decade indicate shortages of entry-level technicians. The industry needs qualified, well-educated people. Because the industry recognizes that changes are necessary, new technicians have begun to experience better technical training from OEMs, independent organizations like **I-CAR** and its ***Education Foundation***, technical schools, and other institutions.

Average pay for technicians has increased over the past six years, and incomes are rising faster than the current inflation rate. Income tends to increase with shop sales and frequent training and certifications, with the current average income among the top 10 percent of technicians averaging over $72,000 per year. A collision repair tech's earning potential—on average—is higher than many comparable trades. Benefit packages—largely lacking, if not nonexistent, in years past—have begun to improve, and all signs indicate they will continue to do so. Overall expectations for the average technician have increased, along with the pay scale, framing, and vocational school industry involvement.

Collision repair is becoming increasingly "high-tech." Therefore, quality, entry-level people are needed more than ever before. If you enjoy hands-on building, love cars and other vehicles, possess good reading and mathematical skills, and enjoy helping to make people whole, the time to enter the collision repair industry has never been better.

Collision Repair Technician

Collision repair technicians must create and develop an appropriate repair plan for each collision-damaged vehicle that is as unique as it is routine. Whether repairing or replacing panels or structural parts, painting, diagnosing damages, testing circuits, or writing an estimate, a collision repairer is a combination of technician, artist, problem solver, planner, inventor, and social worker. *All these roles are important!* Overall, the professional repairer is a *customer servant*—a factor oftentimes overlooked by the consumer, the collision repair industry, and even by technicians themselves.

In the previous sections of this chapter, basic overviews of the functions of each area of repair are provided. The following are short synopses of *career opportunities* that exist in the collision repair field, along with a basic overview of some traits, skills, and abilities required.

Metal Tech. Metal technicians straighten and repair damaged sheet metal, aluminum, high-strength steel, and other metal substrate panels. The metal tech is one of three central persons of collision repair, along with structural and refinish technicians. In addition to collision damages, a metal worker is the tech charged with restoring, patching, and/or refabricating a rusted or rotted panel. In any case, bringing a damaged or deteriorated piece back as close to—if not exactly at—its original shape, dimensions, function, and durability is the goal. When within 1/16 of an inch tolerance, the metal tech proceeds to finish the repair with plastic fill.

Ideally, a metal tech should be a hands-on type person who enjoys restoring misshapen panels and structures much the way a sculptor enjoys bringing abstract material to life. He or she also needs to have the dexterity, talent, training, and desire to utilize the variety of power and hand tools discussed in the metal work section. Having a good eye, developing a "feel" for straightening panels, and possessing meticulous attention to detail is essential.

The art of metal working is vital to the collision repair industry for several reasons. The business has seen a growing, steady trend toward the *replacement* of damaged panels and parts. Granted, some of this is due to the use of thinner gauge substrates in building modern vehicles, while part is driven by economic factors. Overstretched HSS panels that prohibit or restrict the use of heat shrinking and the availability of competitively priced replacement parts make *replacement* of damaged parts a viable option. However, the shortage of professionally trained collision repairers that have the talent and training to perform proper, safe metal repairs and restorations on repairable pieces has driven many collision centers to discard what could be repairable parts. This in turn has driven the industry to a point where replacement of parts now constitutes two-thirds of the business—up from approximately one-third of it in the 1970s. In turn, the profit margin of the average repair center has shrunk as we trend toward becoming an industry of "parts replacers." Good metal workers can help to slow—if not reverse—this trend, which could help shops to earn better profits. Better profits mean better wages for all—especially the ever-more-valuable metal technician!

Structural Tech. Although all facets of collision repair are important, perhaps *the* most vital part is the *structural restoration* of a motor vehicle—making the structural technician a very important part of the process. Structural techs need a solid foundation and ongoing education in the area of how vehicles are constructed, how and where they are designed to transmit collision inertia energy, and exactly how a vehicle maker intended the vehicle to react upon absorbing that energy in a collision. A structural tech will need to stay ahead of the ever-changing parameters within structural repairs—an example being heat usage guidelines, as well as what materials are utilized in each structure. As repairers, we are entrusted with repairing and restoring a damaged vehicle with the goal of returning the *exact* specifications and characteristics that the OEM intended for it. Failing to do so would be akin to *redesigning* it!

The structural technician should be a good reader and excel in mathematics, as well as possess many of the traits of a metal tech. Collision repair manuals are important to the structural tech, as mathematical parameters bring one to within a single millimeter in collision repair. They will also need to be efficient, meticulous planners since an exacting repair plan is vital.

The willingness and ability to work with heavy equipment is essential since a frame rack that requires the use of hand and power tools and hydraulic pulling posts averaging 5,000 pounds per square inch (psi) of power is the centerpiece of a structural tech's equipment. Combine this with a sophisticated, oftentimes computerized measuring system, and the structural tech will easily find himself or herself shifting between raw power and precision on a regular basis! Like most technicians in the collision repair business, the structural tech is usually quite *inventive*. This comes, at least in part, from a *creative* nature, as well as the need to adapt to *changing needs*. A list of the various tools, tabs, and devices that are created "on-the-spot" by many techs in order to complete a job—even a partial list—could probably make up a whole chapter of its own!

The possibility of replacing some structural parts requires the structural tech to be a skilled welder. Welding a variety of *substrates,* in a variety of *positions,* with a variety of *methods,* for a variety of *applications* is a skill all its own. As with all parts of the collision repair process, a view to ultimate safety—for oneself, one's coworkers, the shop building, and the customer's vehicle—cannot be overstated when working with welding and structural repairs (as well as all other areas of collision repair).

As alluded to earlier, the entire production line depends upon the structural tech. Leave a strut tower a few millimeters out of specification on a unibody vehicle, and the steering, suspension, and alignment techs cannot do their job. Fail to restore proper height (or any other) dimensions, and the reassemblers will be left scratching their heads over parts that do not fit properly. Even the finest painters and detailers will not be able to close (or open) an abnormal gap in a full-frame, unibody, or any other vehicle. Without absolute accuracy, vehicular systems such as steering, suspension, braking, supplemental restraint (SRS), and drive train cannot function properly or optimally. Incidentally, the close association between unibody vehicle structure and advanced vehicular systems in vehicles makes steering and suspension, mechanical, and electrical systems training and experience an important part of the structural (and metal) tech's expertise. See **Figure 1-16**.

FIGURE 1-16 Whether utilizing a computerized measuring system—such as this one—or a manual one, the structural technician strives to restore dimensions to the millimeter.

Refinish Tech. The finish of a motor vehicle does not affect safety. *No one ever died from a poor refinish job!* (Then again, it might be safe to assume that at least a few customers may have come *close!*) Although refinishing does not affect safety, it does *sell* the collision repair job—much the way the original finish sold the vehicle from a showroom floor, thus making the refinish technician pretty important! Throw in the fact that the finish *protects* the vehicle, as well as completes the *cosmetics* of pre-loss repairs, and we have a vital link in the repair process—one that also leads to the ultimate satisfaction of a customer.

Refinish techs have to be precise, mathematical, and extremely organized. Perhaps the most vital talents the refinisher needs to possess are a great "eye" and love for color, an "always-follow-the-instructions" mentality, and a near-fanatical approach to cleanliness, detail, and preparation. Taping or masking off parts that will not be removed is a refinish preparation art all its own. A good taper tends to preserve OEM build characteristics while saving time and expenses.

Although at present, the collision repair production workforce is made up of only 1 percent women, the qualities and skills necessary for fine refinishing make the job a perfect fit for female techs! Women, for example, tend to see and rate color a bit better than men. They tend to escape variations of color blindness while men do not. In fact, they tend to *care* more about color than men do. Many paint manufacturers employ women—as chemists, artists, and public relations people—for researching, designing, and choosing vehicular colors.

The refinish process in the collision repair center—like the OEM assembly plant—begins, continues, and ends with *cleanliness*. The colors are mixed by the gram—sometimes as little as a half of one. The color is

TECH TIP A woman's place may be in . . . *the refinish department!* Studies indicate that 25 percent of men suffer from some type of color identification deficiency while just 1 out of 1,000 women do. For the mathematicians out there—that is only *1/10 of 1 percent!* It therefore makes sense to consider a career as a refinish technician or consultant if you are a woman. In addition to *seeing* colors better than their male counterparts, women tend to *care* about color a bit more, too. Many OEM and paint companies employ women to decide what their next line of vehicular colors should be.

completed by adding a mixture of tints, reducers, hardeners, and sometimes metallic flakes. In addition to matching the color and/or "fooling the untrained eye" by the process known as blending, even "**flop**"—how the chips lay in the paint suspension **reflecting** and **refracting** light—is a factor a refinish tech must consider. Even if the match is a perfect one to the original paint code (not likely, as at least three to five variations of each color usually leave the OEM assembly plants), a refinish tech is faced with one or more conditions like fading, weathering, poor maintenance, nicks and chips, and poor prior refinish repairs.

The refinisher must be trained in how to assemble, set up, use, and clean up a spray gun. How the paint atomizes at the spray gun tip, the amount of pressure to apply, and the use of high velocity, low pressure guns are as technical in nature as the spraying is artistic. Patience is also needed, as solvents must "flash off" between coats or they could become trapped inside the refinish layers and break out at an inopportune time (destroying the finish).

Refinish technicians are not competing with fellow painters, other repair shops, or even themselves, as much as they are competing with the OEM refinish process—spearheaded by robotics! The meticulous OEM environment and perfect rhythm of the robotic arms sweeping each panel at a 90-degree angle while overlapping each perfect pass make the refinisher's job a challenge. That is why a good refinisher often appears "robotic" in the act of spraying a vehicle, as well as preparing to. However, good things come out of a tough assignment, as there is perhaps no better reward than a satisfied customer standing at a professionally refinished vehicle with a smile from ear-to-ear!

Detailing Tech. Detailing is an important part of the collision repair process for a number of reasons. Among other things, it is often the final check point before the vehicle is delivered back to the customer. Any details, damages, and errant or incomplete repairs are "caught" and/or attended to by the detailing technician. Cleaning, polishing, and final preparation of a vehicle before returning it to its owner helps "sell the vehicle back" to the customer.

Many entry-level techs start off detailing vehicles. Although many full-blown collision repair careers have been launched from there, it is better to get formal training in a technical school, college, or similar program.

As stated earlier in this chapter, detailing could be viewed as *deep-cleaning a vehicle*. One of the intents is to remove as many contaminants and as much dirt, grease, and grime as possible, in order to preserve the repair job—and the vehicle itself—for as long as possible. Along with a knowledge of the products necessary and the tools utilized, an attention to detail (no pun intended!) is a trait that is vital to the detail tech. An overview of the process is found in the section titled Post-Paint Operations.

Detailing is also a business in itself, as a vehicle does not necessarily need to have been involved in an accident or repaired to need a good cleaning, reconditioning, and updating. That makes a professional detailer an asset inside or outside of the repair shop. The more value you add—or can potentially add—to a business, the more valuable *you* are! See **Figure 1-17**.

Helper/Apprentice. Helpers in a collision repair shop assume a number of duties as they shadow repair technicians or painters, assisting them in their daily routines. Additionally, helpers will assist the shop with duties such as disassembling damaged vehicles, cleaning and moving cars, and repairing small dents and damages. Many collision repair techs began their careers as helpers, picking up the skills necessary from the experienced techs they assist, *or* as apprentices in a work-study program sponsored by their school and the shop. Apprenticeship programs are gaining popularity in this business for a number of reasons. The shortage of qualified technicians makes the idea an attractive one if for no other reason but to invite interest in collision repair careers. A shop can also indoctrinate the helper/apprentice into the collision repair business while teaching the specific systems of *their* shop, and the way they wish to conduct business. Nearly 60 percent of

FIGURE 1-17 The detailer does the final preparation of a vehicle before it is returned to its owner.

FIGURE 1-18 With today's technologically advanced vehicles, it is essential to pursue higher education before embarking on a career in automotive collision repair.

shops recently surveyed indicated they would sponsor an apprentice, co-op, or work-study collision repair student.

Students who are interested in a collision repair apprenticeship-type program can seek information on how to enter one, its validity, and its effectiveness in a number of ways, namely by:

- Contacting a technical institution or college and gathering information such as placement and graduation statistics, as well as specific details of their program
- Contacting the National Institute for Automotive Excellence (ASE) to determine which programs are certified by them
- Determining if the program(s) meet the National Automotive Technical Education Foundation's (NATEF) standards
- Seeking information and direction from your state's department of education
- Contacting I-CAR's Education Foundation

Although many techs entered the business out of high school, with little or no formal collision repair training or education, it is a good idea to seek training at a secondary level, following the steps above as a starting point. See **Figure 1-18**.

Other Collision Repair Personnel

Various other positions round out the collision repair process. The following positions are found in many repair centers and/or are an absolute necessity depending upon size, goals, location/community, affiliations with insurance companies, and resources.

Collision Repair Shop Owner. In addition to assuming the responsibility of starting and maintaining a business, a collision repair shop owner sets the vision and goals for his or her establishment. The individual should be concerned with all phases of collision repair. Owners sometimes double as the shop manager, and depending upon size, revenue, and other considerations can be found filling other functions as well. Most importantly, the owner must keep pace with the climate of the industry, the community, and to an extent the national economy. The collision repair center owner will be involved in many of the tactical, day-to-day plans of the business, as well as strategic, long-term planning for it.

Many collision repair shop owners have come from the production lines, and many have been successful. Many former techs—some still working on vehicles—have built fine repair stores, accomplishing much for themselves, their families, and their communities. However, many individuals and groups have begun to realize that this industry is, first and foremost, a *business*—and every business goes beyond mere "nuts and bolts." Shops require consideration in areas such as job costing, the demographics of their potential customer base, advertising strategies, sales analysis, profit margins, and innovative thinking and ideas—much like a *K-Mart* store, *McDonald's* restaurant, or *Microsoft* does. Some have begun to invest in the industry, surrounding themselves with well-informed *business people* as well as collision repair techs. Although these people may seem to be at a disadvantage—having never "hung a fender" themselves—they nonetheless bring a new *outlook* and *innovations* to the industry that many technically trained people would not consider. Although a successful collision repair center owner does not necessarily have to be able to repair a vehicle, he or she needs to know *what needs to be done* to properly serve a collision repair customer.

Estimator. Many repair shops have full-time estimators that create damage reports on every vehicle that enters the shop. As stated earlier, estimating collision damage is one of the most important steps in the collision repair process. Unfortunately, it is also often the most overlooked, underappreciated, and neglected one!

A collision damage estimator should be an extremely detailed, organized, and *investigative* person. In most fields, when one starts off incomplete, disorganized, or worse—*wrong*—the endeavor begins at a disadvantage. Start off with a poor estimate—or none at all—in collision repair and the entire repair process starts in the wrong direction, oftentimes resulting in incomplete parts orders, broken promises (with respect to completion dates), poor repair quality, and loss of profits.

Among other things, the damage estimator must be depended upon to:

- Have technical knowledge, and keep up with industry rules, regulations, and changes
- Function as a "salesperson"
- Be able to negotiate while maintaining open lines of communication with insurers as well as consumers

- Create an estimate that begins the repair process with a "guide" to safe, cost-efficient, pre-loss repairs
- Understand what parts and materials will be necessary
- Help keep the center profitable and liability-free
- Start, continue, and complete speedy, productive, and profitable repair jobs

Production Manager. A production or shop manager runs the productive repair areas of a collision repair center by essentially putting—and keeping—the production cycle together. A production manager creates and delivers a repair or work order from the estimator's damage report. This in turn guides the various technicians throughout the repair process.

The production cycle must be started and fed methodically in a way that keeps the flow of the cycle going *forward* at all times. In addition, all phases of the cycle—including all technicians and departments—must be kept motivated, busy, and working effectively and efficiently; have their orders, parts, and materials delivered in a timely fashion; and assisted to deliver a safe, pre-loss repair in the most timely fashion possible.

A "production shop" is similar to an OEM assembly plant in that it has specific areas, technicians, and duties set up to complete all phases of a repair. A "job shop" is usually smaller, with one or two "combo" techs working on a vehicle from beginning to end. These shops usually do not have a person dedicated to simply filling the production manager position.

In some shops, the production manager also assumes responsibility for shop maintenance scheduling such as spray booth filter changes; cleaning fans, hoses, and burners; checking pulling equipment and hydraulic lifts; and maintaining safety equipment, training, and procedures.

A production manager should be an organized, energetic person with technical skills that will allow the person to be able to work on a vehicle, as well as to guide, train, and assist the technicians in every aspect of the repair cycle. They need to be motivational, serious, focused, and at times a "policeman"—as horseplay, alcohol and drugs, disunity, and contentiousness in the shop can lead to employees getting hurt, as well as less than safe repairs to vehicles.

Parts Manager. Many mid-size to large shops employ a parts manager. As the title suggests, this person is in charge of ordering all parts, checking them in, inspecting them, delivering them to the respective technicians waiting for them, inventorying the parts and materials in-stock, and processing payments and refunds/returns. The position is important since the parts choices that a collision repair facility have available have never been broader—or more diverse—than they are now. In addition to OEM-produced parts and recycled (salvaged) ones, companies have evolved that now offer remanufactured, rebuilt, and otherwise refurbished parts, in addition to aftermarket ones (new parts that are manufactured by companies other than the original equipment manufacturer).

The parts manager needs the VIN, the production date, the assembly plant, and many other details about the vehicle such as the model, the options, even the sequential number that identifies *what place* in the production line the vehicle was made. Part designs have been changed in some models in mid-year. For example, General Motors has changed the design of certain supplemental restraint systems (SRS), necessitating the steering wheel to be changed when replacing the driver's air bag. Further complicating this change was the fact that certain Grand Ams (the affected vehicle) from a particular sequential number were unaffected, while every vehicle after that spot in the production sequence was.

What type of part (OEM, aftermarket, recycled, remanufactured, etc.) will the shop need to utilize? What will the customer request, accept, or reject? Which insurer is the shop negotiating with on which car, and what are their policies toward alternative parts usage? Does the state have a regulation or law? There is at least one state in the Midwest that necessitates the vehicle owner's permission to use aftermarket parts; another northeastern state has mileage and age parameters around their usage.

Materials needed—paint and supplies, polyester and plastic repair fillers, panel-bonding adhesives, hardware, and various other things essential to fine collision repair—are needed, ordered, and stocked regularly. The parts manager—particularly in a large, high volume store, franchise or dealership—is becoming more of a necessity than ever before.

Office Manager. Larger shops generally employ an office manager that oversees the front office, insurance claims handling, record- and bookkeeping, accounts received and payables, scheduling and filing, typing, and all of the functions that accompany a larger, busy office of a production business. Trips to places like the bank, the post office, and the stationary supplier are common as well.

The office will usually maintain scheduling, house the vehicle repair scheduling and delivery charts, and contain the files. As with any business, a repair shop office—particularly with multiple employees like estimators, receptionists, parts managers, and insurance personnel—will need a leader that keeps order, forward movement, and direction.

Typically, the smaller a shop, the smaller the staff. That means some duties become *combined*. For example, the office manager may double as the receptionist and/or the estimator. In some cases, these and other duties extend to the shop's production or parts managers.

Receptionist. Like most other business establishments, the collision store's receptionist is often the very first person to receive and greet a prospective customer. Often, the *voice* of the receptionist is the very first link in the process. Keep in mind that a typical or prospective customer in the collision repair business is *not coming* in to purchase a

FIGURE 1-19 The receptionist is oftentimes the first person the customer encounters—whether in person or on the telephone.

traditional product. The person who calls or stops at a repair shop is usually not very happy. The damaged vehicle's owner is frequently upset, disillusioned, hurt, unfamiliar with the repair process, has preconceived notions, and is downright scared! The collision repair store needs a great "first impression." The first voice/person that receives a potential customer needs to be:

- Kind
- Soothing
- Caring
- Sympathetic
- Reassuring
- Respectful
- Knowledgeable
- Helpful
- Hopeful
- Efficient
- Motivated
- Smart
- Energetic
- Appropriately attired, and
- Well-spoken

If not, the center may lose a potential patient (the vehicle), customer (the vehicle's owner), and future referrals (everyone the dissatisfied lost customer knows). See **Figure 1-19**.

Related Collision Industry Careers

There are a number of additional careers related to the collision repair process. Like many other product manufacturers, there are supporting functions, extraneous assistance, and numerous suppliers. Here are a few.

Insurance Adjuster or Appraiser. Many insurance companies employ damage appraisers and/or adjusters to handle their collision industry-related claims from the insurance claims' end. The insurance damage appraiser creates a damage report concerning the damages to the company's insured vehicle, and/or any **third-party** vehicle that may have been involved in the claim. The insurance adjuster handles the case to conclusion, including payments, rental fees, injury claims, and other aspects.

The insurance appraiser or adjuster negotiates an agreed price with the collision shop on behalf of the company, and will need to have technical knowledge, much like a repair technician. However, the basic difference between the tech and the adjuster/appraiser position is the insurance person needs to know what has to be done to return the vehicle to pre-loss condition, whereas the technician needs to be able to physically perform the work as well. See **Figure 1-20**.

Many technicians have "crossed-over" to the insurance position as they get older, or desire a change of scene. In any case, the insurance person should have many of the same traits as the shop estimator/technician, including good communication, negotiation, and organizational skills. In addition, the insurance representative—traveling from shop-to-shop—better have a valid driver's license and a clean driving record!

Towing Services. Many shops own tow trucks or a fleet of them to tow damaged vehicles from the scene of an accident to the repair shop. Those that do not have trucks hire towing companies to tow for them. Again, a clean license, some technical knowledge, and a hands-on approach are needed to operate a tow truck or flat bed truck in order to transport vehicles.

FIGURE 1-20 The shop owner (left) disassembles a vehicle while the insurance appraiser (right) inputs information into his laptop in order to create a damage estimate that both will negotiate.

Sublet Shops. Upholstery repairs, glass installation, SRS services, plastic dash and vinyl repair, cloth and leather repairs, pinstriping, and locksmithing are a few of the examples of specialty repairers or *sublet shops* that are necessary to complete some collision repair jobs. These and other specialty techs work in their own shops and/or travel to collision repair centers to ply their trade.

Alignment Technician. A mechanically inclined technician that understands steering and suspension, is good with computers and mathematics, is a hands-on individual, and can operate both hand and power tools is needed to restore the geometric angles that comprise the steering and suspension systems of collision repaired vehicles. Occasionally, parts will need to be replaced. Many shops do not have alignment machines or the techs to operate them for lack of space and/or the reluctance to fill the space with alignment facilities. They reason that—financially speaking—they are better off sending their alignment work out.

Porter. Dirt is an enemy of fine collision repair. Dust, dirt, grease, and grime contaminate finishes, clog hand and power tools, disrupt computers, conduct electricity (draining power from drive trains!), and worst of all can harm the employees' health. Therefore, some of the larger centers employ part- and full-time custodians to clean the shop. Retired technicians, school-age helpers, and other individuals looking for part- or full-time work usually fit in nicely.

Remember that most industrialized countries travel, do business, and transport by motor vehicle. Even though innovations like improved handling, better tires, and anti-lock braking (ABS), traction control unit (TCU), and collision avoidance systems have made vehicles safer, as long as people are imperfect they will continue to collide into other objects, as well as each other. *A dedicated workforce will always be needed to repair these damaged vehicles.*

Being a collision repair technician is an honorable, interesting, lucrative career path that requires hands-on proficiency, mathematical ability, and good reading and computer skills. We wish you all the best as you embark on a challenging adventure!

Summary

This chapter covered:

- The definition of collision repair and all of the components that comprise it
- Different types of collision repair centers
- How a collision repair center operates
- How vehicular damages are located in order to create an estimate, and a work order of damages
- The physical aspects of collision repair including metal, structural, mechanical, electrical, refinishing, post-paint, reassembly, and detailing operations and repair work
- How to properly identify a vehicle
- Retrieval of service information
- How and why to conduct post-repair inspections and safety checks
- How a repaired vehicle is delivered back to the customer
- Various collision repair career opportunities including repair, metal, structural, refinish, and detailing technician; helper; estimator; and production, parts, and office management
- Ownership opportunities and apprenticeship programs
- Related industry positions including insurance apprasier, tow operator, sublet repairer, alignment technician, and shop porter

Key Terms

access time (AT)
aftermarket
aluminum-intensive vehicles
anti-lock brake systems (ABS)
collision avoidance systems
combo repairer
cosmetic
cycle time
damage analysis
deductible
direct repair program (DRP)
electronic control module (ECM)

estimate/damage report
flop
flowcharts
franchise
hydroformed
I-CAR (Inter-Industry Conference on Auto Collision Repair)
I-CAR Education Foundation
job costing
last-in first-out
letdown (test) panel
memory presets
original equipment manufacturer (OEM)
paint thickness gauge
pre-accident condition
procedural/"P" pages
reflection
refraction
repair/work order
repair plan
scan tools
seat belt pre-tensioners
sell-back
specialist
sublet repairs
substrate
supplemental restraint system (SRS)
third party
total loss
traction control unit (TCU)
upselling
vehicle identification number (VIN)
wheel alignment

Review

ASE Review Questions

1. All of the following are operations associated with collision repair EXCEPT:
   A. Creating a damage report.
   B. Vacuuming the interior carpets.
   C. Manufacturing a new fender.
   D. Retrieving and clearing DTCs.
2. Technician A says a collision repair technician must be concerned with durability and performance. Technician B says that a collision repair technician is only concerned with structural integrity.
   Who is right?
   A. Technician A only
   B. Technician B only
   C. Both Technicians A and B
   D. Neither Technician A nor B
3. In addition to restoring a collision-damaged vehicle to pre-loss condition, a collision repair shop seeks to restore:
   A. The owner's deductible.
   B. The owner's confidence in the vehicle.
   C. The insurance company's subrogation payments.
   D. The shop's reputation in the industry.
4. Technician A says that a direct repair program rarely if ever involves an insurance company. Technician B says that a combo repairer may work on a collision-damaged vehicle from start to finish.
   Who is right?
   A. Technician A only
   B. Technician B only
   C. Both Technicians A and B
   D. Neither Technician A nor B
5. The collision repair process begins with:
   A. A work input.
   B. A parts order.
   C. A detailing process.
   D. A damage analysis.
6. All of the following are parts of the refinishing process EXCEPT:
   A. Bleeding.
   B. Sealing.
   C. Sanding.
   D. Corrosion protection.
7. Technician A says that a road test is rarely necessary on collision repaired vehicles. Technician B says that most shops do their own wheel alignments.
   Who is right?
   A. Technician A only
   B. Technician B only
   C. Both Technicians A and B
   D. Neither Technician A nor B
8. Technician A says that a crash book usually contains everything needed to create an estimate of damages on a vehicle. Technician B says digital imaging can

be utilized by collision repairers and insurers to correspond with each other.

Who is right?

A. Technician A only
B. Technician B only
C. Both Technicians A and B
D. Neither Technician A nor B

9. Which of the following is a repair method that a typical metal technician would utilize?
   A. Shrink tubing
   B. Hammer-on, hammer-off
   C. Sound deadening
   D. Hammer-up, hammer-down
10. Which of the following is NOT a vehicular frame style?
   A. Unibody
   B. Ladder
   C. Innerbody
   D. Space frame
11. A late-model four-door GM vehicle is brought into a collision repair center. It is found to have a closed gap between the left fender and front door, a misaligned hood, a cracked windshield with no evidence of anything having struck it, and a buckle in the roof. The vehicle most probably has:
   A. Structural misalignment.
   B. Mechanical damage.
   C. Non-OEM parts.
   D. A faulty collision avoidance system.
12. Technician A says that direct damages are the result of energy having traveled through the vehicle's structure. Technician B says that indirect damages are found in the area of the direct impact to the vehicle.

   Who is right?
   A. Technician A only
   B. Technician B only
   C. Both Technicians A and B
   D. Neither Technician A nor B
13. Technician A says that he is a metal repair technician that utilizes hammers, spoons, picks, and pry bars. Technician B says that she is a refinish technician and has painted aluminum, high-strength steel, and plastic parts on vehicles.

   Who is right?
   A. Technician A only
   B. Technician B only
   C. Both Technicians A and B
   D. Neither Technician A nor B
14. TRUE or FALSE: The VIN plate, located on the top of the dash pad, can give a refinish technician the paint code and color.
15. A collision repair tech is attempting to retrieve DTCs. The tech proceeds to hook the diagnostic scanning tool to the vehicle via a:
   A. Diagnostic trouble code.
   B. Hydraulic ram.
   C. Diagnostic connector.
   D. Portable power pack.
16. Detailing can include all of the following actions EXCEPT:
   A. Checking for water leakage.
   B. Filling all fluid levels.
   C. Totaling up the damage estimate.
   D. Removing overspray from the vehicle.
17. Technician A says that one of the benefits of ongoing communication with the customers of a collision repair center is that it helps alleviate preconceived notions and misconceptions about the repair industry. Technician B says that there are more entry-level technicians coming into the collision repair business than are presently necessary.

   Who is right?
   A. Technician A only
   B. Technician B only
   C. Both Technicians A and B
   D. Neither Technician A nor B
18. All of the following are traits that can be of value to a collision repair shop's structural technician EXCEPT:
   A. Good reading skills.
   B. Good mathematics skills.
   C. A tendency to be inventive.
   D. A tendency to be tall.
19. TRUE or FALSE: The collision repair center's refinish technician can have a negative effect on the safety of a vehicle.
20. Technician A says that a collision repair center is, first and foremost, a business. Technician B says that unless someone has worked on collision-damaged vehicles, she can never run a successful collision repair center.

   Who is right?
   A. Technician A only
   B. Technician B only
   C. Both Technicians A and B
   D. Neither Technician A nor B
21. Technician A says that an insurance appraiser has to be able to repair a damaged vehicle. Technician B says any vehicle that still runs after having been involved in a collision will not need to be towed to the collision repair center for repairs.

   Who is right?
   A. Technician A only
   B. Technician B only
   C. Both Technicians A and B
   D. Neither Technician A nor B

22. Gas metal arc welding is commonly referred to as:
    A. MAW welding.
    B. GIA welding.
    C. MIG welding.
    D. AMG welding.
23. After achieving an agreed price with the insurance company, a collision repair center proceeds to start repairs but finds two damaged parts that neither the shop nor the insurance appraiser saw on the initial inspection. The subsequent estimate that is written for the additional damages is called a:
    A. Second estimate.
    B. Additional estimate.
    C. Subsequent estimate.
    D. Supplemental estimate.
24. All of the following are acronyms commonly heard in a collision repair center EXCEPT:
    A. VOT.
    B. DVOM.
    C. SRS.
    D. CV.

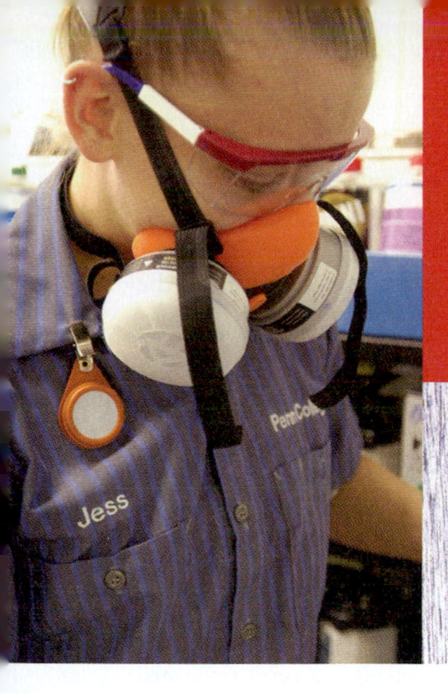

Chapter 2

Collision Repair Safety

OBJECTIVES

Upon completing this chapter, you should be able to:

- Understand hazard communication and employee right-to-know regulations.
- Practice personal safety concerning dress, including proper protective eyewear, clothing, gloves, and respiratory and hearing protection.
- Follow safety practices when using personal hand tools, power tools, and other equipment.
- Apply common sense safety practices when using any force application, lifting, and straightening equipment.
- Know how to properly handle and store all hazardous materials and waste materials generated in the shop and the regulations governing their use.

INTRODUCTION

The underlying purpose of any safety training program is to increase everyone's awareness of the potential dangers that exist in the work environment and the precautions that must be taken to prevent personal injury. Because of the hazardous materials, flammable products, and the potentially dangerous equipment commonly used in the collision repair shop, the chance for accidents and personal injury is always present. Automotive collision repair can be a dangerous occupation because the technician is constantly exposed to hazards such as flying objects and hot sparks created by grinding and welding operations. The work environment is further endangered by the presence of sharp and exposed metal edges on collision damaged vehicles and surfaces exposed during their repair. Safety and accident prevention must be a top priority in all automotive shops. While the collision repair shop environment is thought to be very dangerous, all the hazards can be overcome by employing a safe and common sense attitude toward the job and using the materials and equipment as intended. The training program must also include information concerning the legislated regulations governing the industry.

POSITIVE SAFETY ATTITUDE IS A MUST

When people are exposed to known hazards for any length of time they often become careless about the potential dangers that exist. The existing hazards are often less of a problem than the lack of concern practiced by the technician. When this careless attitude occurs accidents are the inevitable result. Every year thousands of work-related accidents occur unnecessarily, and when they happen it has a widespread effect. The injured party is usually thought of as the only victim in these cases, but many other people are affected as well. Family and loved ones often

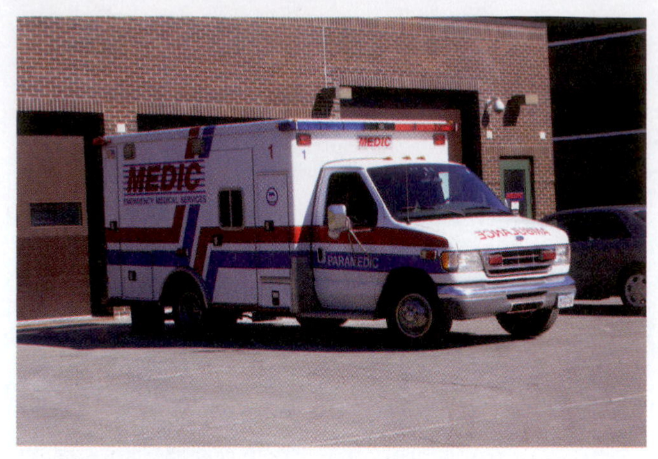

FIGURE 2-1 Every year thousands of work loss injuries occur due to personal negligence.

suffer the consequences as their routine must be changed to accommodate the injured person. Coworkers often must fill the void left because the injured person is not able to fulfill assigned job requirements, and the employer suffers the loss of the worker's production. Many times the individual's careless work habits and attitudes are what make the workplace dangerous. See **Figure 2-1**. The employer and employee must work together to create a safe environment.

OSHA, EPA, AND DOT

In recent years three government agencies have become the key advocates for workplace safety and environmental protection: the **Occupational Safety and Health Administration (OSHA)**, the **Environmental Protection Agency (EPA)**, and the **Department of Transportation (DOT)**. The DOT also plays a vital role in enforcing all the rules and regulations mandated by the federal government. Several legislative acts have been passed granting each agency administrative and enforcement power to ensure compliance for their respective jurisdictions by all industrial entities throughout the United States.

Occupational Safety and Health Administration (OSHA)

OSHA is primarily concerned with creating and ensuring a safe workplace environment for the workforce using the **Code of Federal Regulations (CFR)** for compliance guidelines.

The Environmental Protection Agency (EPA)

The EPA is primarily concerned with protecting our environment: the air, land, and water. The agency is concerned with any hazardous substance that may enter the water stream by being dumped either accidentally or intentionally. Dumping **hazardous materials** can contaminate groundwater and, when in large enough quantities, can render a water supply unfit for consumption. The EPA is also concerned about pollutants emitted into the atmosphere, such as paint overspray during painting operations. These materials are recognized as **volatile organic compounds (VOCs)** and are known to contribute to the formation of smog and deplete the ozone layer.

Department of Transportation (DOT)

The DOT is largely responsible for safely transporting the hazardous materials used—and the hazardous waste generated—by collision repair shops. The manufacturer and the jobber who supply the solvents and other paint materials are compelled to comply with very strict guidelines to ensure the products can be transported safely. After the collision repair shops have used the products until they are no longer usable, the products become hazardous waste. These materials must then be transported to a hazardous waste disposal site for treatment and recycling. The DOT helps to ensure the safety of the workers transporting the materials and the public as well.

Empowered by Legislation

Each of the three agencies discussed is empowered to enforce very specific safety mandates that affect the collision repair industry. Their shared goals are to reduce workplace injuries and illness as well as protect the public. A well-defined safety training program for a collision repair shop must include training on the topics and regulations from each of these agencies that affect them.

TRAINING FOR SAFETY

The first step of any training program is to become aware and knowledgeable of the hazards that exist in the workplace. All employers are required to train their employees in safe practices and make them aware of any existing workplace hazards. The Code of Federal Regulations (CFR) cites and outlines very specific compliance guidelines that must be followed to ensure a business is OSHA compliant. These guidelines include information concerning the specific **personal protective equipment (PPE)** needs for the worker, safety measures that must be used, and specific equipment needs. The guidelines also include many of the precautionary measures that must be exercised to ensure a safe work environment for the employee.

Code of Federal Regulations

The Code of Federal Regulations is a collection of 50 books, each covering a broad segment of industry in the United States that is subject to federal regulation. The section of the CFR used by OSHA, the EPA, and the DOT for enforcement guidelines regarding the collision repair industry is book number 29; hence the reference to 29 CFR 1910. The CFR is made up of numerous subparts, which are regulations taken from various legislative acts. These regulations include very specific guidelines for collision repair shops.

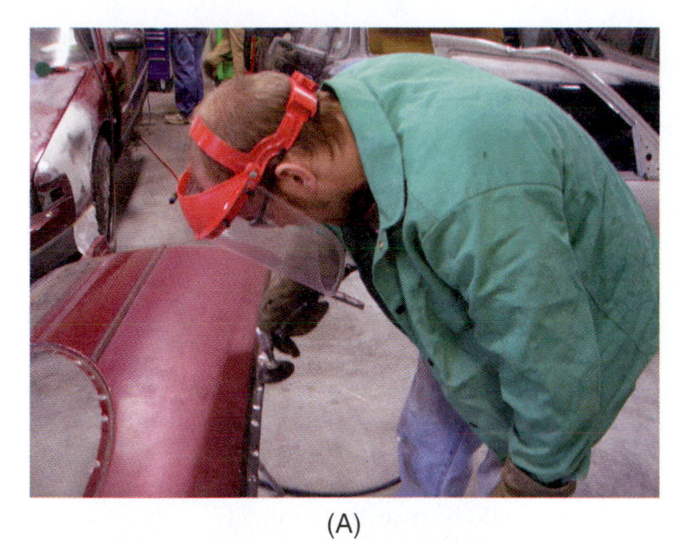

(A)

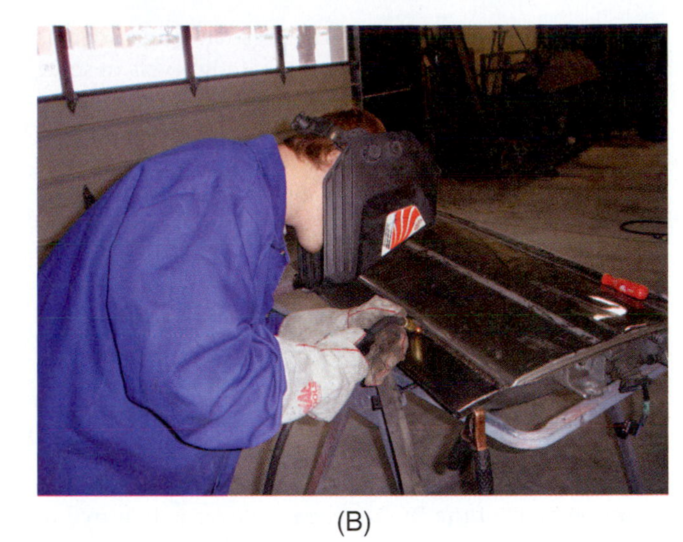

(B)

(C)

FIGURE 2-2 The correct PPE for grinding operations includes wearing safety glasses, a face shield, respiratory protection, and gloves (A). The correct PPE for welding operations includes safety glasses, a welding helmet, respiratory protection when specified, gloves, and a welding jacket (B). The correct PPE for painting operations includes a full paint suit, respiratory protective equipment, and gloves (C).

OSHA General Duty Clause

It would be impossible to anticipate and cite every potential hazard and every possible equipment need that exists in the collision repair shop. Therefore, OSHA uses a set of generic safety guidelines that would apply to all businesses. These guidelines are called the OSHA General Duty Clause. The essence of the OSHA General Duty Clause specifies that both the employer and employee share in the responsibility for creating a safe work environment. The employer must furnish the required PPE for employees and train them in the proper use of it. Employees, in turn, must use and properly maintain the equipment. See **Figure 2-2**. Therefore, the burden of compliance lies with both parties—the employer must create a safe working environment for employees and the employees are required to practice safe work habits and attitudes in carrying out their duties.

ENFORCEMENT THROUGH LEGISLATION

The enforcement guidelines used by the three federal government agencies are taken from several legislative acts: the Clean Air Act (CAA) and its amendments (CAAA) and the Resource Conservation and Recovery Act (RCRA). Other legislative acts that impact the collision repair industry are the Emergency Planning and Community Right-to-Know Act (EPCRA), Clean Water Act (CWA), and the Oil Pollution Act (OPA). The National Emissions Standards for Hazardous Air Pollutants Act of 2008 (NESHAP) is the most recent legislative enactment and specifically targets all paint spray operations in the auto collision repair industry. The law is directed specifically at training personnel using spray painting equipment and implementing the best practices and procedures for reducing the emission of volatile organic compounds (VOCs). Improving spraying techniques used during spray painting operations, cleaning equipment used, and the methods employed to clean the spraying equipment are the principal focus of the law. Its principal objective is to reduce VOC emissions not only by the auto collision repair industry, but any industry that utilizes spray painting equipment for applying surface coatings.

RESOURCE CONSERVATION AND RECOVERY ACT

The Resource Conservation and Recovery Act (RCRA) is the first legislative mandate requiring a proactive attitude about preserving our environment. It came about as a result of the widespread pollution that was occurring nationwide. The water in several large lakes and streams was not fit for human consumption. Smog levels in some of the nation's larger cities and the depletion of the ozone layer became a greater concern than ever before. These factors all contributed to the enactment of the law. It is the first attempt at reducing the amount of pollution generated and takes steps to eliminate the mindless disposal of hazardous materials by dumping them into the ground and waterways. Ultimately the goal is to better preserve our natural resources by making those companies generating any hazardous waste responsible for properly disposing of it as well. For that reason the RCRA is commonly referred to as the Cradle to Grave Act.

The RCRA outlines very specific training requirements for any business using hazardous materials. Under the provisions of the RCRA, the employer is required to develop and maintain a written hazard communication program.

HAZARD COMMUNICATION PROGRAM

The **hazard communication program** is a written document that outlines the business's compliance and training procedures. It must include a number of items pertaining to the specific training needs of each individual job classification within the business. The following items are considered the most critical components of the hazard communication program:

- Written documentation outlining employee training requirements
- Information regarding training employees
- Documentation verifying the completion of training
- Documentation of personal protective equipment training and use
- Hazardous materials and hazardous waste usage, handling, and disposal
- Material safety data sheet use and interpretation
- Methods of placarding and labeling hazardous materials and hazardous waste
- Respiratory safety and training
- Maintaining a manifest
- Lockout/tagout

REQUIRED TRAINING OF CURRENT AND NEW EMPLOYEES

Hazardous Chemicals and Hazardous Waste

The hazard communication must include provisions for training all new and current employees on all the hazardous chemicals and materials they use while performing their jobs. They must also be trained on any new products brought into the shop prior to their usage. This is commonly known as the employee's **right to know**. The training program must include information on the potential dangers associated with the handling, use, storage, and disposal of all hazardous materials found in an employee's work area. It must also include instruction on the PPE requirements for each product.

Employees must also be trained in the proper handling and disposal of hazardous wastes. **Hazardous wastes** are the unusable materials left over after an operation has been completed; for example, the leftover paint materials after painting a vehicle. The cloths used to solvent wipe a vehicle prior to painting are also considered a hazardous waste and should be treated accordingly.

Hazardous Waste Characteristics

Any material that exhibits one or more of the following characteristics is said to be hazardous waste material and must be included in the training.

- Ignitable/flammable. This has a **flash point** below 140°F, catches fire or explodes, and includes paints, solvents, and gasoline.
- Corrosive. This dissolves metal and skin. It includes metal etching materials used for rust removal, as well as battery acid. Solvents are sometimes considered corrosive.
- Reactive. This becomes dangerous when mixed with air, water, heat, or other chemicals. An example is acetylene.
- Toxic. This may create a health risk through exposure. It also contaminates anything it touches such as water, air, your skin, and your lungs.

Personal Protective Equipment (PPE)

PPE training must include information on the proper use of each of the hazardous materials employees will encounter, including the PPE requirements to follow. Proper labeling of all hazardous materials along with the potential health effects resulting from exposure and use of the materials must also be included. Certification verifying the successful completion of the training must also become part of the documentation to be kept on file.

MATERIAL SAFETY DATA SHEETS

The shop's written hazard communication must also include a written inventory of all the hazardous materials kept on hand along with a copy of a corresponding **material**

safety data sheet (MSDS). Chemical manufacturers are required to create an MSDS for each chemical they produce and make it available to the user. The MSDS can be obtained upon request from the manufacturer or by visiting the manufacturer's website online. They must be kept in a file that is logically organized and must be available and accessible to all employees. Shops may choose to access the MSDS on the Internet even though it may not be the fastest means of obtaining one. Locating the correct MSDS may prove to be too time consuming in the event of an emergency. Therefore, it may be advisable to have a binder with printed copies available for each known hazardous material.

Although the MSDS is intended to be a helpful document for planning the shop's training program, and a source of information in the event of an accident or emergency, most people are intimidated by it. The language used is unfamiliar; consequently, people rarely look beyond the first two or three sections. Depending on the manufacturer and the age of the document, the MSDS usually includes information in 12 to 16 individual sections even though only 8 are required. While each section contains valuable information, the collision repair technician is most concerned with five or six of those sections.

An MSDS must be available for every hazardous product.

Interpreting Material Safety Data Sheets

Each of the sections discussed in the following text is taken from an actual MSDS. The information included in each respective section is what the technician and the supervisor should know about the product before using it. The sections not included are equally important, but are intended for use by emergency responders, transporters, and so on.

Section 1. Product and Manufacturing Company Identification. This section includes the following information:

- Manufacturer's name, address, and emergency telephone number
- Product name
- The manufacturer's part or stock number
- The emergency telephone number for the manufacturer
- The emergency telephone number for the product manufacturer and the National Chemtrec Service, which is a clearinghouse capable of giving emergency information in the event of an accident. In an emergency, correctly identifying the product and the manufacturer can be critical to the emergency room staff treating the victim. See Table 2-1.

Section 2. Hazards Identification. This section is usually divided into two parts. The first part gives an emergency overview or quick reference of the hazards involved for emergency responders such as medical emergency attendants and fire fighters. See Table 2-2.

- It is flammable.
- It is an irritant.
- Distinguishing features of the material are color, odor, and form.

The second section gives information on the potential aggravation of pre-existing medical conditions and the principal routes of entry:

- Eyes
- Skin
- Inhalation
- Ingestion
- The potential health effects from exposure to each of the entry routes.

Section 4. First Aid Measures. This section gives information in simple terms on how to respond or what to do when the product accidentally comes into contact with the eyes or skin, or when it is inhaled or ingested. See Table 2-3.

- The instructions may include flushing the affected area such as the eyes or skin, giving artificial respiration, or moving the victim to another area for fresh air.
- This information is helpful for emergency room doctors who must treat the victim. If available, an

1. CHEMICAL PRODUCT AND COMPANY IDENTIFICATION	
Product code	P68551
Product name	Defroster Grid Pen
Recommended Use	Repair Material
Supplier	Kent Automotive 6200 Oak Tree Blvd. Independence, OH 44131 (800) 458-3222
Emergency telephone number	(888) 426-4851

TABLE 2-1 A Material Safety Data Sheet, Section 1.

2. HAZARDS IDENTIFICATION		
	Emergency Overview Flammable. Irritant.	
Color Red	**Odor** Aromatic Hydrocarbon-like	**Form** Liquid

Aggravated Medical Conditions	Pre-existing heart conditions. Pre-existing skin, eye, or respiratory conditions may be aggravated by exposure to this product.
Principal Routes of Exposure	Eyes, Skin, Ingestion, Inhalation.
Potential health effects	
Eyes	Irritation. Tearing. Redness. Swelling. Corneal damage.
Skin	Prolonged skin contact may cause skin irritation and/or dermatitis.
Inhalation	Harmful by inhalation. Drowsiness. Headaches. Nausea. Vomiting. Dizziness. May cause irritation of respiratory tract. May cause irritation to the mucous membranes. Extreme overexposure may cause. Narcosis. Possible unconsciousness.
Ingestion	Harmful if swallowed. Central nervous system effects.

TABLE 2-2 A Material Safety Data Sheet, Section 2.

4. FIRST AID MEASURES	
Eye contact	Immediately flush with plenty of water. After initial flushing, remove any contact lenses and continue flushing for at least 15 minutes. If eye irritation persists, consult a specialist.
Skin contact	Remove and wash contaminated clothing before re-use. Wash off immediately with soap and plenty of water. Seek medical attention if irritation persists.
Ingestion	Drink 1 or 2 glasses of water. Do not induce vomiting. Seek medical attention.
Inhalation	Remove from exposure. Provide oxygen if breathing is difficult. Call a physician. Administer artificial respiration if not breathing. Call a physician immediately.

TABLE 2-3 A Material Safety Data Sheet, Section 4.

MSDS might be sent along with the ambulance attendant to present to the medical staff at the hospital for treatment information. The emergency telephone number in the section can also be used for faxing information to the attending physician.

Section 5. Fire Fighting Measures. This section explains the flammable and explosive characteristics of the product, including the following:

- Flash point. The lowest temperature at which vapors given off become ignitable.
- Flammability limits. Usually described as LEL and UEL or upper and lower.
  - Fire and explosion hazards. Information used by fire fighters.
- Recommended fire extinguishing media. What type of fire extinguisher should be used in the shop or by the fire department: dry chemical, water spray, carbon dioxide, or alcohol foam. See Table 2-4.

Section 7. Handling and Storage. This section identifies the proper methods used to handle and store the material. See Table 2-5. It includes:

- PPE requirements that will prevent personal injury
- Personal hygiene measures to follow after handling material
- How to avoid dangerous reactions with other products and accidental release
- Conditions to avoid when storing material

Section 8. Exposure Controls/Personal Protection. This section is divided into two smaller sections. See Table 2-6. The first section lists the hazardous chemicals and the level or concentration at which one can safely be exposed without the use of PPE during an 8-hour work shift. The following abbreviations are commonly used in this section to express exposure limits:

- PEL—Permissible exposure limit. This is the highest level one can safely be exposed to a hazardous material for an 8-hour work shift.
- TLV—Threshold limit value. This form of exposure limit is similar to that expressed as PEL, although it is usually more restrictive and allows a lower exposure level.
- STEL—Short-term exposure limit. This permissible exposure limit is slightly higher than the PEL or TLV, but it cannot exceed 15 minutes in duration. A chemical has a PEL of 500 PPM but has a STEL of 600 PPM. The technician can be exposed at 600 PPM for 15 minutes without consequence.
- TWA—Time weighted average. This is the average exposure level for an 8-hour shift. A technician

5. FIRE FIGHTING MEASURES

Flash point °C	24
Flash point °F	76
Method	Tag closed cup

Autoignition temperature °C	No data available
Autoignition temperature °F	

Flammability Limits (% in Air)

Upper	10
Lower	1.5

Suitable extinguishing media
Alcohol foam. Water spray. Dry chemical. Foam.

Extinguishing media which must NOT be used for safety reasons
No information available.

Special Fire-Fighting Procedures
Firefighters should wear NIOSH/MSHA approved (or equivalent) self-contained pressure-demand breathing apparatus and full protective clothing.

Fire and Explosion Hazards
Water should be used to cool closed containers to prevent pressure build-up and possible autoignition or explosion when exposed to extreme heat. Vapors may form explosive mixture in air between upper and lower explosive limits which can be ignited by many sources, such as pilot lights, open flames, electrical motors and switches.

Sensitivity to shock
No information available.

Sensitivity to static discharge
No information available.

TABLE 2-4 A Material Safety Data Sheet, Section 5.

7. HANDLING AND STORAGE

Handling
Ensure adequate ventilation. Avoid contact with skin, eyes and clothing. Thoroughly wash hands and exposed skin after handling. Keep container closed when not in use.

Storage
Keep away from direct sunlight. Keep out of the reach of children.

TABLE 2-5 A Material Safety Data Sheet, Section 7.

8. EXPOSURE CONTROLS/PERSONAL PROTECTION

Exposure limits

Chemical Name	OSHA PEL (TWA)	OSHA PEL (Ceiling)	ACGIH OEL (TWA)	ACGIH OEL (STEL)
Silver	0.01 mg/m3	-	0.1 mg/m3	-
Propylene glycol monomethyl ether acetate	-	-	-	-
n-Butyl acetate	150 ppm 710 mg/m3	-	150 ppm	200 ppm
Acrylic Resin	-	-	-	-

Ventilation and Environmental Controls
Adequate ventilation should be provided to keep exposure levels below acceptable exposure limits. Local: as necessary.

Hygiene measures
General industrial hygiene practice.

Personal protective equipment

Respiratory protection
Use NIOSH approved respirator if TLV limit is exceeded.

Hand protection
Chemical resistant gloves. Rubber gloves.

Eye protection
Safety glasses with side-shields.

Skin and body protection
No information available

TABLE 2-6 A Material Safety Data Sheet, Section 8.

works with a chemical with a PEL of 500 ppm and a STEL of 600 ppm. For 7 3/4 hours the exposure level is 400 ppm and for 15 minutes the exposure level is 600 ppm. The TWA is approximately 382 ppm, well below the 500 PEL.

- "C"—Ceiling. This is the highest level of allowable exposure for even a short period of time. It is typically used with materials that are considered more hazardous.
- PPM—Particles per million. This numeric value used to measure the exposure limits or levels is typically used by the EPA.
- MG/M^3—Milligrams per cubic meter. This is another numeric value used to express exposure limits or levels. It is typically used by OSHA.

The second part of section 8 addresses two requirements:

- Engineering controls. This MSDS indicates that adequate ventilation is necessary to keep exposure limits below the recommended levels.
- PPE requirements. This is usually given for each of the routes of entry to prevent the material from getting into the body. This MSDS recommends using safety glasses with side shields, chemical resistant rubber gloves, and a respirator if the TLV is exceeded.

It should be understood that the lower the ppm or MG/M^3 number values shown for the PEL, the more hazardous the product is. A product with a PEL of 50 ppm is more dangerous than one showing a PEL of 100 ppm. A quick scan of the numbers in the permissible exposure column can give a quick glimpse of how hazardous the material is. REMEMBER—SMALL IS BAD!

The six sections discussed include information the technician should be aware of when using the product, how to protect against unnecessary exposure, and how to respond in an emergency. The sections that were not discussed are of equal importance; however, the information included in those sections is used by others concerned with designing the engineering controls, transporting the material, and so on.

HEARING PROTECTION

The single most frequently ignored hazard in the collision repair shop is potential hearing damage. Like many other hazards, hearing loss often occurs as a gradual deterioration and therefore sometimes goes unnoticed. Ideally one should wear some form of hearing protection at all times. It is particularly important when using equipment known to create particularly loud noises (such as the air hammer), or when hammering certain surfaces—such as when straightening damaged metal or stress relieving heavier or thicker metal surfaces. Even though hearing protection is typically used for only a short duration and it may seem like an inconvenience to put on, it should always be used.

LABELING/PLACARDING

Another component of the hazard communication program is the proper labeling of all containers, not just those containing hazardous materials and hazardous waste. The manufacturer's product label or another prescribed label must be attached—and legible—on any container used to store and/or dispense a hazardous material. See **Figure 2-3**. Many times it becomes necessary to transfer bulk products such as cleaning solvents, thinners, and so on, from the larger manufacturers' shipping containers into smaller, more portable containers. When this is done a **secondary container label** must be affixed to the container used to dispense the material in the shop.

The secondary container label must include the following identification information:

- Name of the manufacturer
- Product name and stock number
- The hazard classification (i.e., flammable, corrosive, etc.)
- "MSDS AVAILABLE"

While not required, the manufacturer's emergency 800 telephone number is also useful to include.

Exception to Product Label Requirements

With rare exceptions, all containers must have some form of label attached to them identifying their contents. Failure to do so can result in an accident that may have some serious consequences. Any deviation from the shop's labeling standard must be noted in the hazard communication program.

Alternative labeling or unlabeled containers are permitted only as long as all the employees/students are informed of the labeling criteria and it is documented in the hazard communication document. See **Figure 2-4**. An example is a "community" solvent container used by every-

FIGURE 2-3 Secondary container labels can be in a variety of forms so long as the product is properly identified.

FIGURE 2-4 An approved container should always be used to dispense flammable liquids.

one in the shop that is kept in the same place at all times. Because everyone knows its contents, use it for the same reason, and always find it in the same location, a label is not required. The container should also reflect the type of material inside the container. A solvent-dispensing container with a self-sealing lid indicates that it contains a potentially flammable material. OSHA does not endorse any specific label manufacturer's labeling system so long as the labeling criteria is met.

HAZARDOUS MATERIAL VS. HAZARDOUS WASTE

Once a material has been used for its intended purpose and is no longer usable, it is considered hazardous waste. As an example, after a solvent has been used two or three times to clean a paint gun, it is no longer considered usable. At that point, it becomes hazardous waste. It is then usually poured into a common collection barrel with other waste paints that are also commonly identified as hazardous waste. See **Figure 2-5**.

Hazardous Waste Storage

The containers used for storing these waste materials must be identified with a label stating it is "waste paint" along with the date when the barrel first started being used for collecting the material.

Hazardous Waste Generator Classifications

There are three hazard generator classifications, each determined by the amount of waste generated: large quantity generator (LQG), small quantity generator (SQG), and conditionally exempt small quantity generator (CE-SQG). The amount of waste the shop generates will determine how long the container can be left in the shop after the first hazardous waste material is poured into it.

Automotive service and collision repair shops often offer additional customer service operations such as oil changes, lubrication, and cooling system repairs. See **Figure 2-6**. The by-products—such as waste oil, contaminated coolant, and other materials that are drained from a vehicle while performing the collision repair work—must all be stored in separate containers with the proper markings. The towels used to solvent wipe a vehicle prior to painting and the used floor dry material are also considered hazardous waste in some states. The contaminated towels may need to be stored in an airtight container. In addition, regulations may specify that the towels should not be mixed together with other hazardous waste material such as used floor dry. Technicians should check with their local and state regulations concerning the proper storage and disposal of these materials.

FIGURE 2-5 Hazardous waste storage containers must have the appropriate identification labels attached to them along with the start date of their use as a storage container.

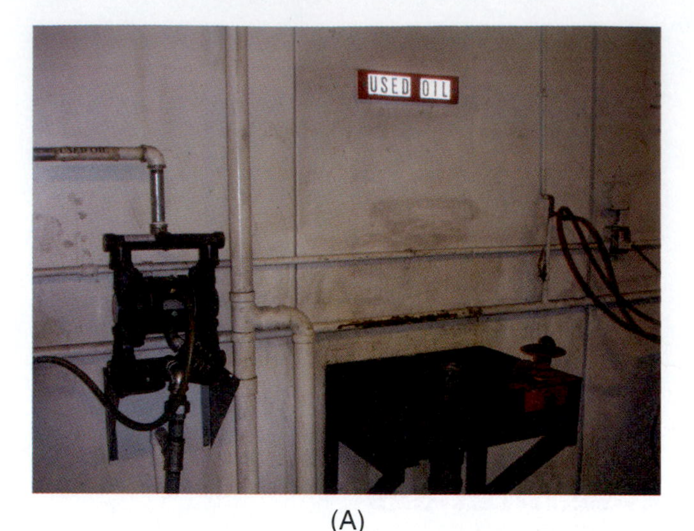

(A)

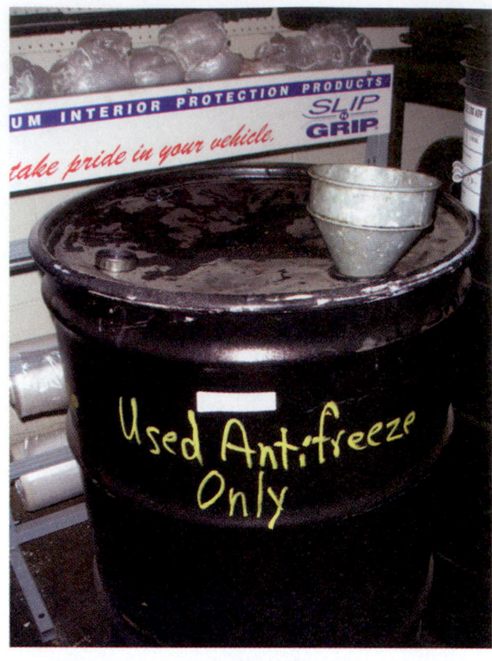

(B)

FIGURE 2-6 Non-hazardous waste materials (A) and hazardous waste materials (B) must be stored in separate containers, each having its own identifying label.

NEVER MIX HAZARDOUS WASTE MATERIALS. Pouring them into a common barrel or "blending" some ingredients with others may create a material with unknown and dangerous properties, making it very unstable and potentially explosive.

HAZARDOUS WASTE DISPOSAL AND MANIFEST

The proper disposal of hazardous waste is as important as protecting the employees when using the materials that create the waste. A collision repair business should be very selective about the waste hauler contracted to remove and dispose of their hazardous waste. The credentials of the waste hauler should be carefully researched to ensure the removal company is properly licensed, bonded, and meets all DOT requirements. The shop should also verify the treatment, storage, and disposal facility (TSDF) where the hazardous wastes are being transported is also properly licensed and meets all the EPA requirements.

RCRA Hazardous Waste Disposal Guidelines

According to the RCRA, the generator of any hazardous waste is responsible for the material forever. In the event the waste transporter is involved in an accident while in transit, causing the waste to be dumped or released into the environment, the generating company—in this case, the collision repair shop—can be held responsible. If the TSDF to which the waste is transported is found to be operating improperly, the generating company can also be held liable for damage that may result from the improper operation. The RCRA is commonly referred to as the Cradle to Grave Act. That means the collision repair shop is responsible for the hazardous waste, starting from the time it is initially poured into the waste storage barrel, during its transportation to the TSDF, and until it is treated and properly disposed of—and even beyond.

The Manifest

One of the most important components of the hazardous waste disposal process is the paper trail that must be generated. See **Figure 2-7**. When the waste hauler removes the hazardous waste from the facility a multiple-copy form identifying the quantity and make-up of material that is being transported is generated and signed by the responsible parties. A copy is left with the generator. When the hazardous waste safely arrives at the TSDF a copy verifying its arrival is sent to the generator. After the material has been properly treated, distilled, and redistributed or properly disposed of, another form is sent to the generator verifying its disposal. These documents are commonly called the shop's manifest and must be kept

TECH TIP The **manifest** is one of the most important documents the shop must be able to provide. Failure to properly maintain it may result in fines and other citations from OSHA and the EPA.

Piano, Texas 75024 www.safety-kleen.com

CUSTOMER NO. 2 2 6 4 4 7 9

CUSTOMER GENERATOR

BILL TO

FOR SERVICE CALL	BRANCH MANAGER	DOC. EXP.	SCHEDULED SERVICE WEEK	SCHEDULED TERRITORY	REFERENCE NUMBER
570-825-8134	PAUL HOLMGREN		08		M005305162

CREDIT CODE | PREVIOUS BALANCE | BAL. OVER 60 DAYS

CUSTOMER SEGMENT | CHAIN | OUTER COUNTY | SVC. P/C | PROD. P/C

LOCATION: 218001 | TAX EXEMPTION NO.

SERVICE DATE	SALES REP NO.	CUSTOMER P.O. NUMBER	CUSTOMER PHONE #	TAX CODE	DATE EQPT/PROD ORDERED	SERVICE TAX	C.O.M.S. TAX	PRODUCT TAX
12/2/08	433562		570-327-4515		PW			

	DEPT	SERVICE/ PRODUCT	SERIAL NUMBER	REMARKS/ UNIT PRICE	QUAN.	CHARGE	SALES TAX	TOTAL CHARGE	WASTE MIN.	SOLVENT/DRUMS CLEAN	SPENT	# OF CONT.	SK DOT	CC	SERVICE TERM	MSDS GIVEN
1		52150	32060	MDL52/150	1	165.00		165.00		15	14	1	557		16	☐
2		52150	25939	MDL52/150	1	110.00		110.00		15	14	1	557		8	☐
3		52150	25938	MDL52/150	1	110.00		110.00		15	14	1	557		8	☐
4		52150	25937	MDL52/150	1	110.00		110.00		15	14	1	557		8	☐
5		52150	25936	MDL52/150	1	110.00		110.00		15	14	1	557		8	☐
6		52150	25935	MDL52/150	1	110.00		110.00		15	14	1	557		8	☐
7		52150	25934	MDL52/150	1	110.00		115.00		15	14	1	557		7	☐
8		10001		Fuel	1	15.46		15.46								☐
9		100411		service	1	10.00		10.00								☐
10																☐
11																☐
12																☐

TOTAL-SERVICE/PRODUCTS

CHECK APPROPRIATE BOXES: MACHINE CONDITION & CLEANLINESS (GOOD ☐ POOR ☐); LAMP ASSEMBLY CONDITION (GOOD ☐ POOR ☐); DECALS IN PLACE AND LEGIBLE (YES ☐ NO ☐); FUSIBLE LINK INSTALLED (YES ☐ NO ☐); EMERGENCY CLOSING OF LID UNOBSTRUCTED (YES ☐ NO ☐); MACHINE PROPERLY GROUNDED (YES ☐ NO ☐); LOCAL PHONE NO. STICKER AFFIXED TO MACHINE (YES ☐ NO ☐); SPENT SOLVENT MEETS ACCEPTANCE CRITERIA (YES ☐ NO ☐)

USEPA TRANSPORTER 1 ID NO.	USEPA TRANSPORTER 2 ID NO.	GENERATOR USEPA ID NO.	GENERATOR STATE ID NO.
TXR000050930		PAD982365793	

11. US DOT DESCRIPTION (INCLUDING PROPER SHIPPING NAME, HAZARD CLASS, AND ID.) | 12. CONTAINERS NO. TYPE | 13. TOTAL QUANTITY | 14. UNIT WT/VOL | SK DOT NUMBER 5163055

A. WASTE COMBUSTIBLE LIQUID, N.O.S. (PETROLEUM NAPHTHA) NA1993,PGIII(D001,D018,D039,D040) (ERG#128) 6.7LBS/GAL

B. WASTE COMBUSTIBLE LIQUID, N.O.S. (PETROLEUM NAPHTHA) NA1993 PG III (D039)(ERG#128) 6.7LBS/GAL

C. WASTE COMBUSTIBLE LIQUID, N.O.S. (PETROLEUM NAPHTHA) NA1993 PG III RQ (D001) (ERG#128)6.7 LBS/GAL (D018,D039,D040)

D. used Cleaning compounds (Petroleum Naphtha) (not US Dot or US RQ regulated) (150 solvent) (No) — 7 DM — G

I CERTIFY THAT MY TOTAL WASTE STREAMS ARE WITHIN ONE OF THE FOLLOWING CATEGORIES. 0 TO 220 LBS./MONTH — INITIALS; 220 LBS. TO 2,200 LBS./MONTH — INITIALS; GREATER THAN 2,200 LBS./MONTH — INITIALS

DESIGNATED FACILITY NAME AND ADDRESS SAFETY-KLEEN SYSTEMS, INC.

I CERTIFY THAT NO MATERIAL CHANGE HAS OCCURRED EITHER IN THE CHARACTERISTICS OF THE WASTE MATERIALS OR IN THE PROCESS GENERATING THE WASTE MATERIALS.

USA EPA ID NO. PAD981737109

STATE ID NO.

PAYMENT RECEIVED SECTION: CASH ☐ | CHECK NUMBER | TOTAL RECEIVED | APPLY PAYMENT TO: ☐ TODAY'S SERVICE/SALE ☐ PREVIOUS BALANCE AS FOLLOWS | INVOICE # | AMOUNT $ | INVOICE # | AMOUNT $

PREVIOUS CREDIT CARD NO. | CREDIT CARD NO. | AMEX VISA MC | EXP. DATE

MANIFEST NO. | LDR MESSAGE | MANIFEST CODE | SEQ # 10 D

IN THE EVENT OF AN EMERGENCY CALL

I AGREE TO PAY THE ABOVE CHARGES AND TO BE BOUND BY THE TERMS AND CONDITIONS SET FORTH ABOVE AND ON THE REVERSE SIDE OF THIS DOCUMENT. PLEASE CHARGE MY ACCOUNT FOR THIS TRANSACTION UNLESS OTHERWISE INDICATED IN THE PAYMENT RECEIVED SECTION. THE INDIVIDUAL SIGNING THIS DOCUMENT IS DULY AUTHORIZED TO SIGN AND BIND CUSTOMER TO ITS TERMS.

"This is to certify that the above-named materials are properly classified, packaged, marked and labeled, and are in proper condition for transportation according to the applicable regulations of the Department of Transportation."

Print Customer Name

By: Customer's Authorized Representative

TOTAL CHARGE (FROM ABOVE) | WASTE MIN. (FROM ABOVE) | TOTAL DUE

DO NOT WRITE IN THE AREA BELOW

M005305162
433562

SERVICE AND SALES ACKNOWLEDGMENT PART 1366 (Rev. 05/07)

FIGURE 2-7 A pickup receipt is provided to the shop each time the hazardous materials hauler picks up hazardous waste from a shop. This must be kept on file as part of the shop's manifest.

in a permanent file readily accessible in the event of an OSHA or EPA inspection. Failure to do so will result in a citation and a substantial fine.

Properly Storing Hazardous Wastes

Two other very important considerations that must be remembered about storing both hazardous materials and hazardous wastes are proper container grounding and ventilation. Most hazardous materials and wastes in the collision repair shop are stored in steel containers and barrels. Something as simple as touching the containers to each other while transferring materials from one container to another can result in a static spark being emitted. This can cause an explosion with devastating results. In order to prevent this, each storage container must be grounded with a grounding strap. See **Figure 2-8**. A removable grounding strap equipped with a pinch clamp should also be attached to the container being filled. The grounding cable should be attached to a grounding rod driven into the ground outside the building.

Storage Area Ventilation

Proper ventilation is equally important to prevent potentially ignitable fumes from accumulating in the storage area. See **Figure 2-9**. All enclosed storage cabinets used for storing paint, solvents, and other flammables must have provisions for proper ventilation. The vapors given off may be either heavier or lighter than air and therefore will accumulate at either the top or bottom of the storage unit. So whether it is an enclosed storage cabinet or a room with open shelves, provisions must be made to evacuate the fumes at all levels. All electrical outlets and light switches and fixtures in the storage areas must also be sealed and explosion proof. See **Figure 2-10**.

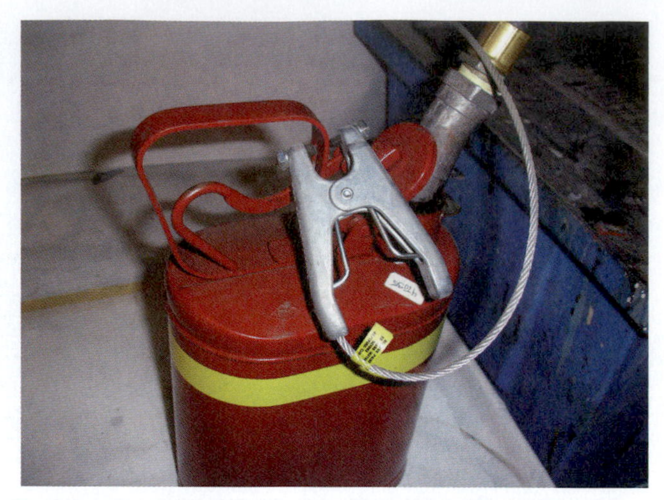

FIGURE 2-8 To avoid a static electricity spark, the container being filled with solvent should be grounded prior to pouring the material into it. A removable grounding clamp is used here.

FIGURE 2-9 All storage cabinets should be properly ventilated to prevent a buildup of fumes. An upper and lower venting outlet should be provided.

NFPA PLACARD

Another placard that is commonly displayed on the doorways of businesses using hazardous materials are the multi-colored **NFPA placards**. See **Figure 2-11**. Each of the four colors represents a different hazard. The red color represents the fire hazard, the blue represents the health hazard, the yellow indicates the reactivity hazard, and the white is used to show a specific reaction hazard. This placard is displayed for the benefit of the first responders in the event of a fire. Each quadrant has a different color representing a different hazard, and the number in each section quickly alerts the emergency personnel of the level of each hazard. See **Figure 2-12**. These placards may also be used inside of a building on

FIGURE 2-10 Explosion-proof light fixtures and outlets should be provided in the areas used to store any flammable and explosive materials.

ACCUFORM SIGNS

FIGURE 2-11 An NFPA placard is commonly posted on the entry door into shops and areas where hazardous materials are commonly used or stored.

HAZARDOUS MATERIAL CODE IDENTIFICATION

HEALTH		FLAMMABLE		REACTIVE	
Recommended Protection		Susceptibility to Burning		Susceptibility to Energy Release	
4	Deadly - Special full protective suit and breathing apparatus must be worn.	4	Below 73°F - Very Flammable.	4	May detonate under normal conditions.
3	Danger - Full protective suit and breathing apparatus should be worn.	3	Below 100°F - Ignites under normal temperature conditions.	3	May detonate with shock or heat.
2	Hazardous - Breathing apparatus with full face mask should be worn.	2	Below 200°F - Ignites with moderate heating.	2	Violent chemical change but does not detonate.
1	Slightly hazardous - Breathing apparatus may be worn.	1	Above 200°F - Ignites when preheated.	1	Not stable if heated use precautions.
0	Normal - No precautions necessary.	0	Will not burn.	0	Normally stable.

FIGURE 2-12 This is an interpretation of the hazardous materials code numbers used on the NFPA placard.

the entrance door to the storage room and on the flammable storage cabinets as well.

Informational Warning Posters

Employers should place informational and warning posters in strategic locations around the shop as reminders to the employees of the potential hazards involved in an area. See **Figure 2-13**. These include dangers and hazards that may occur when operating dangerous equipment or situations where the employee may be working in a hazardous environment such as a spray booth.

STORAGE CONTAINERS

Regardless of the materials that are dispensed from them, all portable or secondary storage containers must meet all local fire codes. Any container used to dispense a flammable liquid should have a spring-loaded self-sealing lid or shutoff when released. This will automatically seal the opening when the handle is released after dispensing the material and will also prevent accidental spills. The container should be properly marked with either the manufacturer's label or a secondary container label to identify its contents and to avoid accidents.

MAINTENANCE SCHEDULE

A routine maintenance schedule should also become part of the shop's hazard communication. The maintenance schedule should include the name of the party responsible for checking each of the areas in question. This schedule may include items that need to be checked daily, weekly, or monthly and should include each of the following:

- Fire extinguishers (tag must be initialed, verifying monthly inspection)
- Grounding straps that are properly attached
- Exhaust fans that are operable and running
- Storage containers that are not leaking
- Warning and informational labels that are properly attached to each container and legible

COMMON PERSONAL SAFETY

The single most effective means to prevent accidents and injuries is for the employee to practice safe work habits. This includes dressing appropriately, using the proper PPE when the task being performed requires it, and simply thinking safety. Many people think personal dress attire is a personal choice and should be left to the discretion of the individual. However, proper dress attire such as correct shoes, shirts, and trousers—and the material they are made of—can be as important to accident prevention as correctly using tools and equipment.

Personal Safety Considerations

- Never attempt to work or allow anyone around you to work while under the influence of alcohol or drugs. Prescription drugs may also have warnings that must be taken into consideration.
- Always advise those around you about any medical condition that may affect your ability to safely perform your job from time to time.

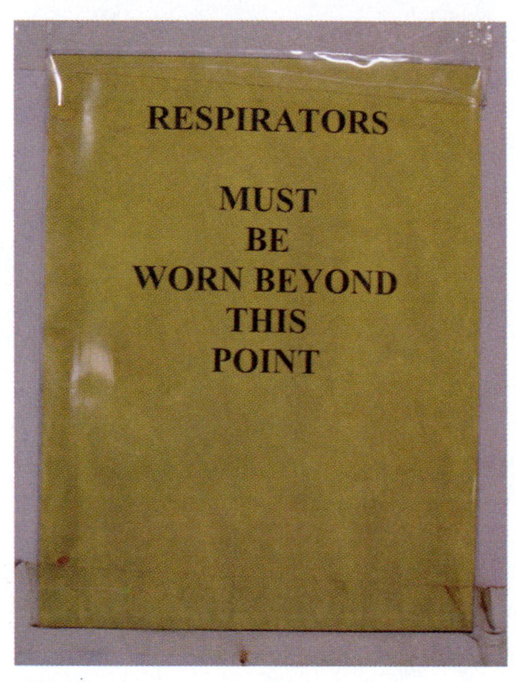

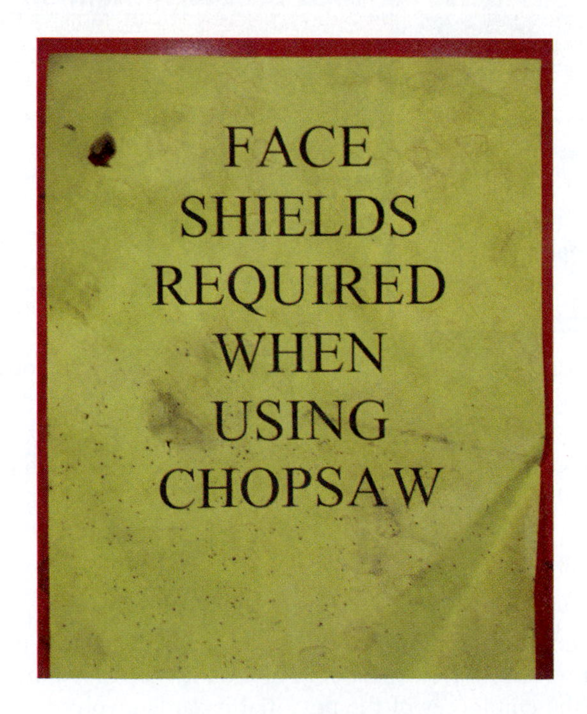

FIGURE 2-13 These warning and informational signs are commonly posted on doorways and on or near equipment to alert and remind workers that they must use PPE when operating the item.

- Never use a piece of equipment you have not been trained to use.
- ALWAYS WEAR EYE PROTECTION! Either prescription lenses or safety glasses should be worn any time one is working in the shop. Eye injuries are the most common—yet preventable—accidents that occur in industry.
- A face shield should be worn whenever the task being performed warrants it. This includes any grinding or sanding operation where material is being removed from the surface. Safety glasses should always be worn along with a safety shield.
- Long hair should be tied or tucked in to prevent it from becoming entangled with moving machinery.

- Never wear loose fitting clothing that can become entangled in equipment or machinery.
- Always keep shoes tied. Long dangling shoelaces are a trip hazard.
- Never wear jewelry in the shop. Jewelry conducts electricity and can cause severe burns if it becomes shorted to ground. It may also get tangled in moving parts of equipment and machinery.
- Always use the appropriate respiratory protection when the task warrants its use. Respiratory protection should never be used before a medical evaluation and the proper fit testing requirements have been fulfilled.
- Always use hearing protection and warn those around you whenever you are using tools or equipment that are known to be noisy.
- Never wear clothing that is soaked in grease or oil or that has been splashed with solvents.
- Always wear a welding helmet along with safety glasses when welding.
- Take the necessary precautions to prevent becoming infected by **blood-borne pathogens**.
- When moving or lifting heavy objects, always position yourself to lift using your legs and not your back.
- Remember, you are not only responsible for your own safety, but your coworkers' safety as well.

HEALTH HAZARDS

Much of the training and information included in the employee right-to-know laws is designed to make the individual aware of the potential hazards that exist in the work environment. Not only is a knowledge of the health hazards a very important segment of the overall training program but it should also include information about how people become exposed, what the effects of these exposures are, and what employees must do to prevent it. The training should include both the short- and long-term effects from being exposed to hazardous chemicals and the precautions that should be taken to avoid health problems of any type.

No one part of the collision repair industry is exempt from exposure to hazardous chemicals because a variety of products with known hazardous ingredients are used in virtually every phase of the industry. Not only must one become familiar with the effect these chemicals have on us, but we must also become familiar with their **route of entry**. Hazardous chemicals knowingly enter our body by four principal routes of entry:

- Inhalation
- Absorption
- Ingestion
- Injection

Inhalation

Inhalation is the most common means for hazardous chemicals to enter the body in the collision repair shop. Harmful chemicals in the collision shop are commonly suspended in the air in the following form:

- Dust. Solid particles that are generated from sanding and grinding operations.
- Mists. Suspended fluid droplets that are produced by atomizing paint.
- Vapors. Gaseous compounds that are given off from the chemical reaction when fillers and paints are mixed.
- Fumes. Solid particles that are given off from welding operations such as zinc oxide when welding zinc clad metals.
- Gases. Carbon monoxide from engines that are running in the shop.

Using the correct respiratory protective equipment is vital to safeguard against unnecessarily inhaling any of the air-borne materials. A respiratory safety training program will identify the correct equipment to use when needed.

Absorption

Absorption is also an inevitable means of entry into the body as many of the chemicals and compounds used for collision repairs are in a liquid or semi-liquid form. Absorption not only occurs through contact with unprotected skin, but also occurs when air-borne materials enter through the mucous membranes such as those near the eyes, nose, and mouth. **Isocyanates**, additives commonly used in paint, are the most significant concern to the technician as they can enter through the skin as a liquid, through the eyes from overspray, and by inhalation from paint overspray. The skin is said to resist or protect the body from absorbing a foreign substance for a maximum of 10 seconds. Any contact, however incidental or short term, is likely to be absorbed. Therefore, appropriate protective gloves should always be worn anytime the possibility of skin exposure is apt to occur. Likewise, respiratory and eye protection should be used when any air-borne mists are present in the environment.

Ingestion

Ingestion is yet another route of entry. Although it is not likely someone will intentionally swallow harmful chemicals, food eaten in the workplace can easily be contaminated; for example, by eating without first washing one's hands and from particles sticking to one's clothing. Another means of ingestion is drinking from a container that does not have a protective covering over the drinking contact surface. A technician drinking from a soda can left open on a toolbox while working is an excellent example of how this can occur. To prevent contamination, any beverage container should have a lid to cover the surface that comes into contact with the mouth.

Injection

Injection is the least likely means of entry into the body. This might occur when a high pressure line such as a hose on a hydraulic system suddenly breaks, allowing the fluids under pressure to escape. This could also occur if one were to break the high pressure line on the air conditioning system, allowing the gas to suddenly escape.

LEVELS OF EXPOSURE

The terms *chronic exposure* and *acute exposure* are commonly used to describe the levels and duration of exposure to a hazardous chemical or material faced by an individual.

Acute Exposure

An **acute exposure** occurs once and develops very quickly at a very high level of exposure. The effects of an acute exposure are normally short term and can be reversed provided the level was not too high to cause serious damage.

Chronic Exposure

A **chronic exposure** occurs at a very low level over an extended period of time, sometimes lasting for a period of many years. The dosage is low enough so the effects are not felt immediately; however, when exposed over an extended period of time, serious medical complications are apt to develop.

Because many of the hazards technicians are exposed to have little or no immediate ill effects, they tend to take them very lightly. For example, over a period of years, a technician will mix and apply plastic filler repeatedly without the benefit of using protective gloves. Each time it is done some of the material may get smeared onto the hands with no noticeable effect, even though each time some of the material is absorbed through the skin. Eventually the body may respond by showing signs of fatigue, extreme lethargy, and in extreme cases permanent damage to vital organs. A classic example of a chronic exposure causing permanent damage to vital organs is isocyanate sensitization. The result is frequently permanent lung damage that can lead to emphysema, a very painful and often debilitating respiratory ailment.

SAFEGUARDS AGAINST EXPOSURE

Manufacturer's Warning Labels

Hazardous material exposure can be reduced significantly by first becoming aware of what the hazardous materials are and then taking the necessary precautions to protect against them. See **Figure 2-14**. The first thing one should do is read the label and observe the precautionary statements included on them. Manufacturers often use one of three words to not only indicate the product is hazardous, but also to note the severity of the hazard.

In the order of their severity, warning labels may indicate:

- CAUTION
- WARNING
- DANGER

Steps to Control Exposure

There are three ways an employee's exposure to hazardous materials can be controlled:

- Administrative controls. This might include scheduling a rotating work assignment to minimize exposure by each employee. For example, the technicians may use a rotating schedule so they may perform tasks on alternate days such as masking and final prepping one day and painting the next. This will reduce the employee's constant exposure to the hazardous environment.
- Engineering controls. This might include purchasing more efficient equipment such as closed cabinet paint gun washers, installing additional ventilation into the mixing room, and so on.
- Personal protective equipment. The use of personal protective equipment is the responsibility of the technician.

PERSONAL PROTECTIVE EQUIPMENT

Precautions for Refinishing Tasks

Refinishing operations require a more extensive use of PPE and more involved training program than any other part of the shop. Fume exposure and inhalation occur while the product is being mixed at the bench as well as when applying the material. The CFR mandates some very specific requirements about respiratory equipment selection, fit testing, and the proper use and maintenance of the equipment. One cannot simply use any respiratory protective device that looks appropriate and think it will adequately offer the necessary protec-

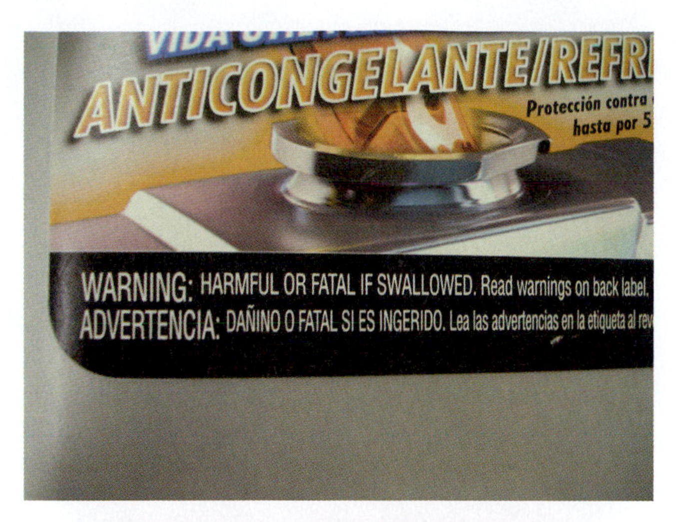

FIGURE 2-14 The words *warning, danger,* and *caution* are commonly used on the label instructions of various products that are potentially dangerous to the user.

tion for the job being done. Wearing the incorrect type or improperly fitting PPE can be as dangerous as not wearing any.

Respiratory Protective Gear

There are essentially three types of respiratory protective gear available. See **Figure 2-15**.

- Dust/mist respirators. These are usually made of a paper-like material and are rarely used for longer than a single shift on the job. They are typically used when sanding plastic filler or putty; when dry sanding paints, primers, and primer surfacers; and so on. In addition to the OSHA or NIOSH approval stamp, they should also have the markings N95, P95, or P100 displayed on them. See **Figure 2-16**. The equipment must be OSHA and NIOSH approved, as indicated by their respective markings printed on the item.
- Air purifying. The face piece can be either a half face mask or a full face mask with permanent or interchangeable organic charcoal cartridges. See **Figure 2-17**. The half mask seals to the face and covers only the mouth and nose. The full face mask covers the eyes and the entire face. After putting on the mask and before entering the spray area, a positive and negative user seal check should be made to ensure a good seal is achieved to prevent contamination inside the mask when breathing. The cartridges must be replaced routinely and a changeout schedule must be documented and become part of the hazard communication program.
- Supplied air system. A supplied air respirator usually has a full face shield covering the entire face and has an air supply either from outside the booth or from a motor-driven pump worn by the painter. See **Figure 2-18**. The supplied air system offers a positive airflow that is pumped into the face mask, creating a pressure higher than the atmospheric pressure. This creates a very safe and desirable system because the possibility of fume penetration is virtually eliminated. This is also the only approved respiratory protection for those with facial hair.

Sampling the Environment

Before any decisions can be made about the type of equipment that is to be selected and used, several preliminary steps must first be taken. See **Figure 2-19**. The first thing about the type of respiratory PPE that must be accomplished is to collect a sample of the environment. To do this, technicians working in the area attach an organic vapor monitoring device to their clothing and wear it for a normal one-day shift. The device is then sealed and sent to an independent lab for analysis. The results will reveal the type of chemicals found in the area and the level of concentration to which the employees are exposed. This

(A)

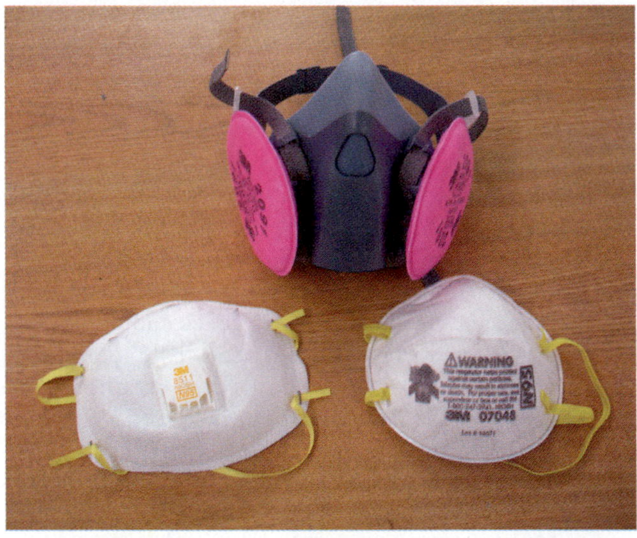

(B)

(C)

FIGURE 2-15 Three types of respiratory protective equipment are the disposable air purifying dust/mist respirators (A), the air purifying respirator which can be used with interchangeable filters to include a HEPA filter (B), and the air supplied system (C).

information is then used to determine the type of respiratory protection equipment that is required to work in that environment. Selecting the PPE to be used without having this information could lead to serious permanent injury.

RESPIRATORY STANDARD OPERATING PROCEDURE

Another component of the hazard communication specifies that every employer must develop a written standard operating procedure for respiratory protection (SOP). This document should include information on each of the following topics:

- Jobs requiring respiratory protection for personnel
- Pre-use medical evaluation
- Respiratory equipment selection for specific applications
- Qualitative fit testing
- Respirator inspection and maintenance
- Respirator cleaning and disinfecting
- Respirator storage
- Respirator use training
- Positive and negative user seal check
- Cartridge changeout schedule

RESPIRATORY PROTECTION NEEDS

After deciding which job classifications require the use of respiratory protective equipment, the next step is to determine the personnel who will be required to use the equipment. An employer is required to furnish the necessary PPE only to those who work in an environment in which they are exposed to hazards. Painters and their helpers are constantly exposed to air-borne contamination resulting from sanding operations and paint overspray. Therefore, respiratory protective equipment is mandatory PPE for everyone involved in performing painting tasks. Repeated exposure to solvents and paint overspray is known to cause industrial asthma and can also affect the central nervous system and other vital organs. The problem is further compounded by isocyanates, an additive used in many paint products that causes isocyanate sensitization. The isocyanates affect the respiratory system, which—in the advanced stages—causes emphysema and other respiratory disorders such as pulmonary fibrosis.

Medical Pre-Evaluation

Before any efforts can be made to select or use any respiratory PPE, the designated personnel are required to submit to a medical pre-evaluation. This medical evaluation is

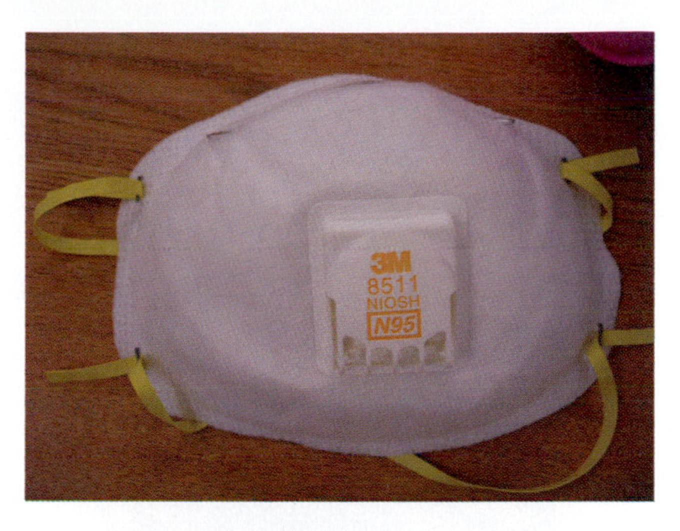

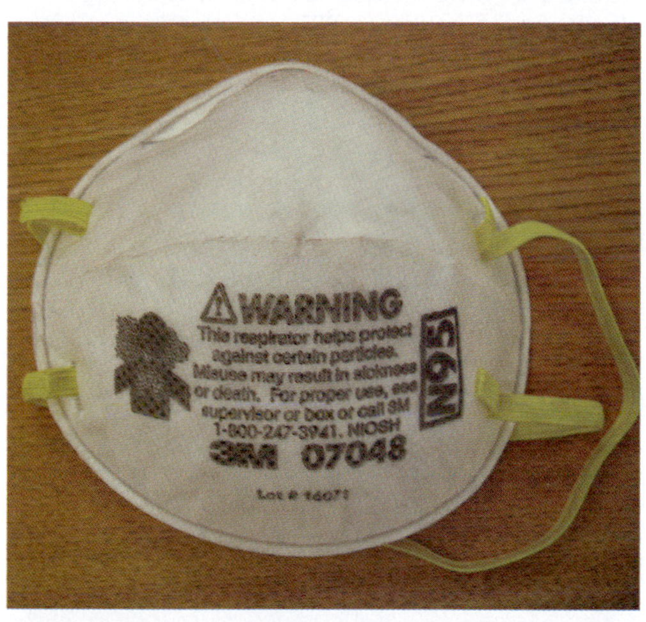

FIGURE 2-16 All respiratory protective equipment used must be OSHA or NIOSH approved and their stamp of approval must be visibly displayed on the PPE being used.

FIGURE 2-17 The air purifying respirator may be either a half mask covering only the nose and mouth or a full face mask which also covers the eyes. Some manufacturers furnish organic cartridges that are interchangeable with both styles of equipment.

FIGURE 2-18 The supplied air respirator is usually available with a full face mask. The air supply may be from an outside supply of at least class D air or it may be from a motor-driven pump worn by the painter.

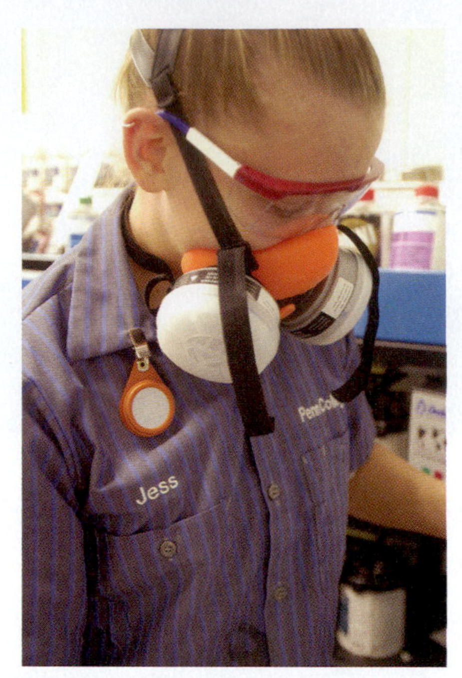

FIGURE 2-19 An environmental sample must be collected using an organic vapor button to determine the type of respiratory protective equipment required in the area.

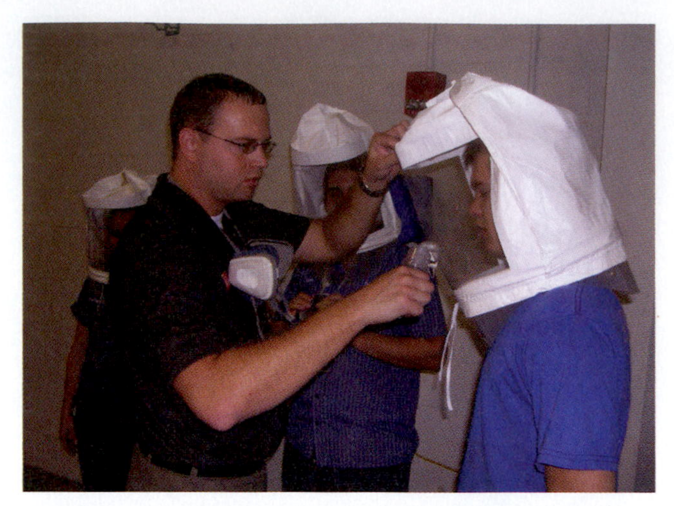

FIGURE 2-20 Once the criteria has been established identifying the respiratory PPE requirements, a qualitative fit test must be performed to determine the size and type of respirator required by each individual working in that area.

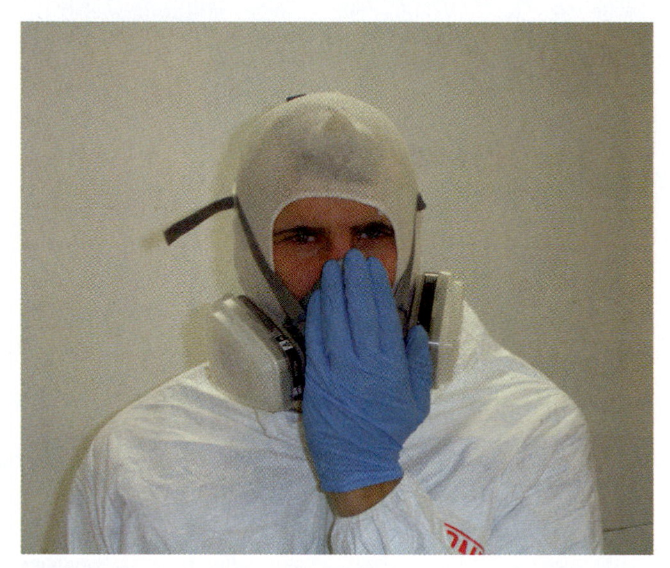

FIGURE 2-21 Each time the individual uses respiratory PPE, a positive seal test must be performed. This is done by covering the exhaust port and breathing out normally, inflating the face piece.

necessary to determine whether an individual is physically capable of wearing and using any type of respiratory equipment. If circumstances warrant it, they may be referred to a medical doctor for further evaluation. People who suffer from various forms of respiratory problems frequently are not able to wear or use the equipment because it is too confining. Past medical conditions may also prevent someone from successfully wearing any of these devices. An individual may be approved to wear the equipment with certain restrictions or environmental limitations.

Qualitative Fit Test

Provided that technicians pass the medical evaluation, they must complete a qualitative fit test using the specific respiratory protective equipment that has been selected. See **Figure 2-20**. The test is used to determine the appropriate size respirator to be issued to ensure their safety and well-being. The criteria for fit testing prohibits the growth of any facial hair in any area where the face piece makes contact with the skin when a half face mask is used. Facial hair will prevent a proper seal, thus compromising the efficiency of the equipment by allowing the inhalation of contaminated air when inhaling. The fit testing must be performed by certified personnel using the criteria prescribed in the CFR.

Positive and Negative Seal Check

The technician must also be trained in the proper use of respiratory PPE, making sure to correctly wear the equipment and check for a proper seal each time before using it. Each time the respiratory PPE is used, a positive and negative user seal check should be performed prior to entering the contaminated environment. See **Figure 2-21**. The positive user seal check is accomplished by placing the palm of one's hand over the exhalation port of the respirator face piece while exhaling normally. While holding one's breath, the respirator face piece should remain inflated and billow out slightly, yet remain in full contact with the face without losing pressure or leaking. See **Figure 2-22**. The negative user seal check is accomplished by placing the palms of one's hands over the inhalation ports of the respirator face piece, being careful not to collapse

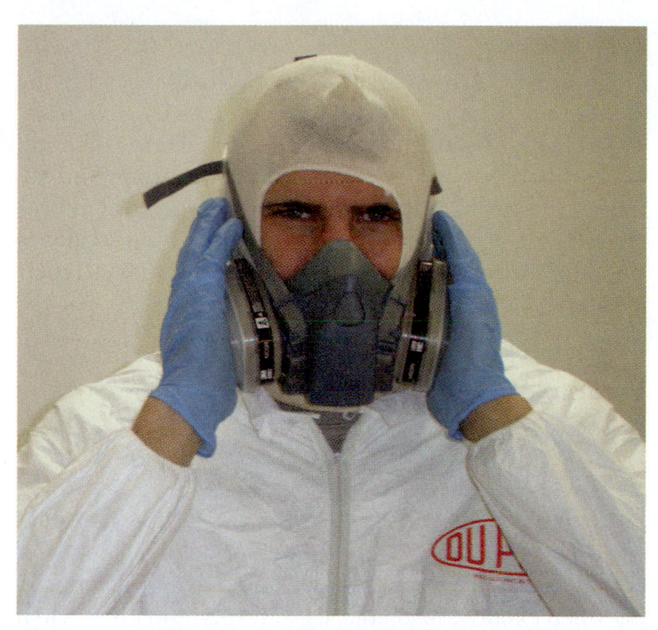

FIGURE 2-22 The user must also perform a negative seal test on the respirator by covering the air intake port and inhaling normally, causing the face piece to deflate.

it, while attempting to inhale in the same manner as when breathing normally. The face piece should be drawn in slightly and remain there while the vacuum is sustained without losing contact with the face or allowing outside air to enter the breathing chamber. If outside air enters the breathing chamber, the straps must be adjusted until a seal is obtained.

Respirator Storage

Immediately after one finishes using the respirator, the charcoal organic cartridge and pre-filters should be removed from the face piece and stored in a clean and airtight container. See **Figure 2-23**. The pre-filter should be separated from the organic charcoal cartridge and the two need to be stored in separate containers to minimize odor penetration into the charcoal. The charcoal organic filter must be stored in an airtight container to preserve and maximize the filter's life expectancy. The activated charcoal in the filter starts to deteriorate immediately upon exposure to the air and continues until it is rendered unsafe to use. The normal life expectancy is approximately 8 hours of use or exposure to the atmosphere. Storing it in an airtight container can extend that time to approximately 2 weeks, at which time the filter must be replaced as it is no longer effective. The face piece should also be stored in a separate container that will protect it from airborne contaminants and from tools and other sharp objects poking holes into it.

Filter Changeout Schedule

Another reason for conducting the environmental analysis is to determine the changeout schedule for the organic cartridge filters when the air purifying respirator

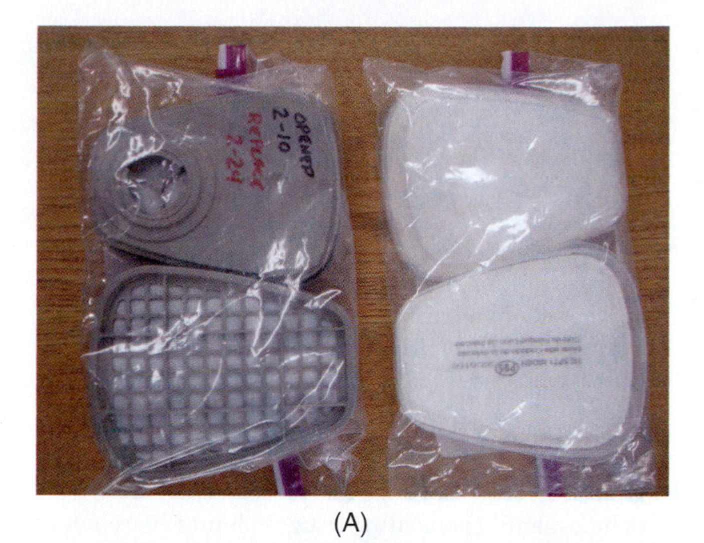

(A)

(B)

FIGURE 2-23 The pre-filter and organic charcoal filter should be stored in separate containers to prevent the infiltration of odors from the material in the pre-filter into the organic charcoal filter (A). The filters can then be stored together in an airtight container (B).

is the PPE selected. The higher the concentration level of contaminants, the more frequently the filters must be replaced. Assuming the ventilation is adequate in the area where the respirators are being used, the normal life span of the charcoal organic cartridge is relatively short and should be changed out after approximately 8 hours of usage or exposure. A changeout schedule must be determined and included in the respiratory safety SOP. It may be advisable to date the cartridges when they are first removed from their sealed packaging and exposed to the air to avoid accidental exposure to unsafe contaminants by the user.

Respirator Inspection and Maintenance

The respirator should be inspected prior to each use to ensure the air intake and exhaust valves are free of any deformations and are functioning properly. The face piece should be inspected to ensure there are no rips or tears to prevent contaminated, unfiltered air from entering the breathing area. The straps are inspected to ensure there are no tears in them that would prevent applying sufficient tension for a proper seal to the operator's face.

Respirator Cleaning and Disinfecting

The organic charcoal cartridges should be removed after each use to ensure a proper cleaning of all the inside face piece surfaces. To clean the face piece, prepackaged respirator cleaning wipes may be used to cleanse the entire inside surface. If the respirator allows removing and interchanging filters, it is advisable to remove them periodically and submerse the entire face piece into an antibacterial soap solution to thoroughly wash the entire face piece. If possible, it may be advisable to remove the retaining straps and face piece housing to avoid soaking them.

PROTECTIVE CLOTHING

Along with the respiratory requirements, the CFR also addresses other protective measures including clothing and other attire such as proper gloves for the operations performed. Because paint overspray becomes air borne, there is an increased possibility of absorption through the skin as the painter is surrounded by the material. See **Figure 2-24**.

Hooded Paint Suit

To protect against exposure, a hooded, airtight or Tyvek® paint suit should be worn to protect the entire body.

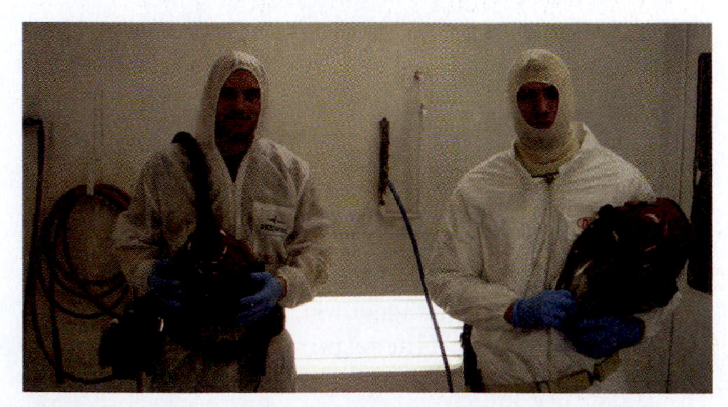

FIGURE 2-24 To prevent paint absorption through the skin from the overspray exposure, the painter should wear a full body paint suit.

Gloves

One must also wear protective gloves because the hands are constantly in contact with paints, chemicals, and solvents while they are being handled. There are many different types of gloves to choose from, and selecting the wrong one can compromise the level of protection once again. See **Figure 2-25**. Four types of gloves are commonly marketed for use in the collision repair industry:

- Latex
- Nitrile
- Neoprene
- Barrier

Holdout to Chemicals. One of the things to consider when selecting gloves is the degree of holdout they offer or how resistant they are to breaking down or deteriorating when exposed to chemicals, solvents, and other harsh substances. Contrary to popular belief, latex gloves do not afford the degree of protection from hazardous materials exposure that is necessary. While they may be a popular choice, their use should be limited to handling things like plastic fillers, oil, and non-solvent base materials. They are relatively effective against acid exposure such as that used for metal etching operations.

Nitrile Gloves for Protection Against Isocyanates. Because many of the paint finishes and other products contain isocyanates, a nitrile glove should be used whenever any material with this additive is used. These products include most paint clearcoats and many of the 2-k urethane primers, primer fillers, and select types of color coats as well. Nitrile gloves should not only be worn when spray painting, but should be used when pouring the ingredients and mixing the paint at the bench. This includes stirring the paint and filling the paint gun with the material. Many of the two-part adhesives dispensed in the side-by-side dispenser also contain isocyanates. A common practice is to smooth these materials by paddling them with the fingers after they are applied. Protective nitrile gloves should also be used to avoid skin contact when these materials are used. The latex and nitrile are often mistaken for each other because they look very similar and their intended purpose is often not understood. One must be certain the correct glove type is selected for the operation being performed.

Neoprene Gloves. Neoprene gloves offer more protection against solvent penetration than nearly any other style available. However, they are often very thick and cumbersome, making them difficult to use for fine or detailed tasks. Sometimes a lighter weight neoprene-covered rubber is available, making them somewhat more flexible. However, they still lack the sensitivity necessary for detailed work. They are an excellent bench glove to be used for long-term exposure to solvents and chemicals such as when cleaning and scrubbing equipment. They are also a

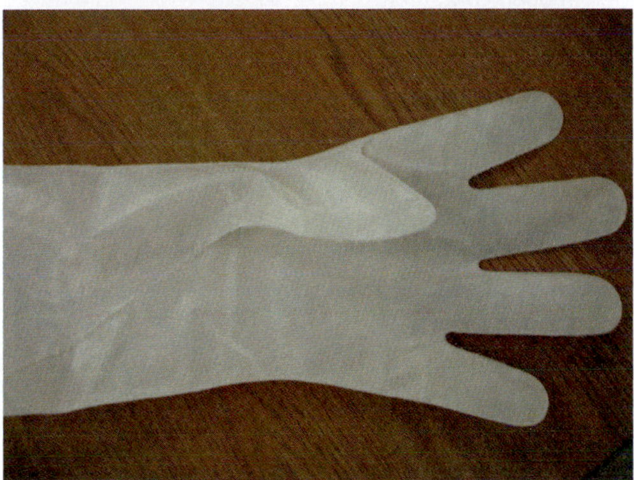

FIGURE 2-25 Because hazardous materials are readily absorbed through the skin, protective gloves should be worn to protect against exposure. The operation being performed will dictate whether a latex, nitrile, neoprene, or barrier glove is selected.

good choice when working with materials such as paint strippers, as they resist breaking down and deteriorating from the extended exposure to the harsh chemicals in the stripping compounds.

Barrier Gloves. Barrier gloves are highly resistant to break through even from the most aggressive solvents and are designed to withstand exposure against nearly any chemical. These are not a fitted glove like most others, making them difficult to use for operations requiring fine dexterity. However, they are an excellent glove to use when manually cleaning a paint spray gun or performing other operations where extended exposure to harsh chemicals is necessary.

BLOOD-BORNE PATHOGENS

Blood-borne pathogens are viruses that are carried in the bloodstream and are transmitted from person to person by contact with an infected person's body fluids. The OSHA Blood Borne Pathogen Standard specifically addresses **hepatitis B (HBV)** and **human immunodeficiency virus (HIV)**. The most serious danger of exposure may occur if a coworker becomes injured on the job, resulting in severe bleeding. Those rendering assistance must take the necessary precautions to protect themselves by wearing protective gloves and avoiding contact with the individual's blood. Damaged skin such as open sores, cuts, abrasions, and even acne must be protected from contact with body fluids as they are potential routes of entry by the infected blood.

Another possibility of exposure exists when dried blood in a collision damaged vehicle must be cleaned. The HBV virus can live in dried blood for seven days. Therefore, precautions must be taken to protect oneself from coming into contact in either case. In the event of a shop accident where bleeding occurred, a mixture of 1 cup of chlorine bleach in a gallon of warm water should be used to scrub and remove fresh blood from the affected area. Any affected tools and surfaces that are contaminated should be sterilized as well. The same solution can be used on a vehicle, but one must check to ensure against discoloring any affected upholstery. *Universal precautions* describe a strategy to treat all blood and body fluids as though they are infected. Any clothing that becomes blood soaked should be removed as soon as possible and the affected area should be thoroughly scrubbed with a disinfectant. The affected clothing must be cautiously handled and placed into a labeled container until cleaned or disposed. Caution must be taken to avoid skin contact; handwashing with a disinfectant is necessary as soon as the protective gloves are removed. A medical specialist or doctor should be consulted if there are any questions or concerns about the exposure.

FIRST AID STATION

Every shop should have a designated area where certain emergency response items are stored. See **Figure 2-26**. These items should include a "community first aid kit," a blood-borne pathogen kit, and a fire blanket. The blood-borne pathogen kit should include all the necessary items to protect the responding individual from exposure to blood and other body fluids. The first aid kit and fire

FIGURE 2-26 The shop's first aid station should include an amply stocked first aid kit, a blood-borne pathogen kit, and a fire blanket.

blanket should be placed in the same area to avoid confusion as to their location in the event of an emergency.

SPILL CONTROL AND CLEANUP

Most collision repair shops purchase and store some of their solvents in large barrels or drums. A spill control plan must be in place in advance so that—in the unlikely event any storage container should spring a leak or a larger, more serious spill occurs—the employees know how to respond. If the quantity of the material spilled is larger than what can easily be treated or eliminated, the local fire department should be called. A little embarrassment is easier to overcome than a building that is destroyed by a fire ignited from the spill. When dealing with large spills one should:

- Alert the supervisor and all coworkers.
- In the event material is flammable, shut down any equipment that could be an ignition source and ventilate the area, if necessary.
- If respiratory protection is required, put it on but only if someone else is available to help in the event of an emergency.
- Contact the local emergency responders.
- Take the necessary precautions to prevent the material from entering the floor drains by using absorbents, damming, or squeegeeing the material away from them.

FIRE SAFETY

During any given day, there are numerous possibilities for accidental fires in a collision repair shop. Nearly all the chemicals and materials used are flammable and many of the materials on the vehicle's interior are very ignitable as well. A clean shop and organized work area can be a very effective fire deterrent. Knowing the correct fire extinguisher to use for the type of material on fire is also very important in minimizing the damage if one occurs. It is necessary to develop an emergency response plan and to make everyone aware of the plan in advance, so the response is more spontaneous in the event of a fire. The following are some common sense steps that can be taken to avoid a fire in the shop:

- Store all flammable liquids and combustible materials in approved fireproof containers.
- Move solvent storage containers out of the shop repair area immediately when finished using them. Never allow any solvent to be brought into the shop in open containers.
- Clean up gas, oil, and grease spills and leaks immediately.
- Remove a leaking gas tank from a vehicle if it or the gas lines are damaged.
- When a gas tank is removed, store it in a well-ventilated area, away from any sparks or heat source—preferably outside in an enclosed area.
- Do not drag the tank across the floor when taking it to a storage or disposal area.
- Never turn on the ignition in a vehicle with a fuel line disconnected.
- Remove all masking materials to an outside waste container as soon as they are removed from the vehicle.
- Remove and dispose of any towels used for cleaning or wiping vehicles in a flameproof container.
- Avoid smoking while working on vehicles.
- Have a water-filled squirt bottle or some other extinguishing media close by when welding on or near the interior of the vehicle.
- Remove or cover any interior upholstery and items that are likely to catch on fire when welding near them. A spark from an arc welder can easily travel across the distance of the vehicle's interior.
- Note that fire blankets are as effective for deterrent as they are for extinguishing a blaze.
- Use spark resistant paper masking to protect the glass on the vehicle.
- Do not allow any sparks near a battery, especially when it is being charged.
- Keep the doors to storage areas and cabinets closed at all times when not being used.
- Keep all solvent containers tightly closed and covered to avoid fumes from accumulating in the storage area.
- Use appropriate removable grounding straps when transferring flammable liquids from bulk containers to secondary containers.

- Ensure all electrical cords are free of cuts and abrasions. Make sure the grounding plug has not been removed to avoid a shocking hazard. Some tools are grounded internally; therefore, the cords require only a two-prong plug.
- Avoid heating the work area with an open flame (such as a wood stove, kerosene heater, or recycled oil burner).

Fire Emergency Plan

A plan should be in place long before the possibility exists of a fire erupting in the shop. It should include an evacuation plan for all personnel; the location of the fire alarm pull boxes, if they exist; and a location for all to congregate outside the shop area to account for all personnel. In the unfortunate event a fire should get started in the shop area, the following steps should be taken immediately.

- Quickly evaluate the size of the blaze and extinguish it if possible using the recommended fire fighting media. If it can't be extinguished immediately, call for help from the fire department. A fire doubles in size every 30 seconds.
- Alert everyone of the fire in the work area and adjacent areas—such as the office—to evacuate the premises. If possible, close the doors as you exit.
- Call the 911 emergency number.
- If you leave the building do not re-enter. Leave the heroics to the fire department.

Fire Extinguishers

Portable fire extinguishers are the first line of defense in the event of a shop fire. What is done in the first 15 to 30 seconds of a fire can often make the difference between exciting conversation tomorrow or a building burning down. REMEMBER THAT A FIRE DOUBLES IN SIZE EVERY 30 SECONDS!

There are essentially four different types or classifications of fire extinguishers. They are marked as either a class A, ABC, BC, or D extinguisher. They are identified as such mostly for the class of fire they can effectively extinguish. The following graph shows the classification and the extinguishing media with which they are charged.

Identifying and Recognizing Extinguishers. In an emergency one should be able to recognize the extinguishers from a distance and not depend on the printed signage on the unit. See **Figure 2-27**. A very quick and relatively

FIGURE 2-27 The classification of each fire extinguisher can quickly be determined by the size and shape of the material outlet. They are class A, class ABC, class BC, and class D.

EXTINGUISHER CLASSIFICATION	CLASS A	CLASS ABC	CLASS BC	CLASS D
USED TO EXTINGUISH TYPE OF FIRE	ORDINARY COMBUSTIBLES—wood, paper, cloth, leaves, etc.	FLAMMABLE LIQUIDS—solvents, gasoline, etc. ELECTRICAL— not recommended where electronic equipment is used AND ORDINARY COMBUSTIBLES	ENERGIZED ELECTRICAL, FLAMMABLE LIQUIDS—ideal for areas where electronic equipment is used	NOT AVAILABLE TO NON-PROFESSIONAL PERSONNEL—used on metal fires such as magnesium, etc.
EXTINGUISHING MEDIA	PRESSURIZED WATER	DRY CHEMICAL	CARBON DIOXIDE	
PRECAUTIONARY MEASURES IN THEIR USE	ELECTRICAL SHOCK HAZARD—do not use on or near electrical fixtures	DIFFICULT TO SEE ONCE DISCHARGE IS INITIATED—only spray directly onto fire as it will fill room with fine powder, making visibility impossible	DEPLETES AREA OF OXYGEN—DANGER OF ASPHYXIATION	
HOW TO USE EXTINGUISHER	BREAK UP WATER STREAM INTO SMALL DROPLETS	SPRAY DRY POWDER ON TOP OF FIRE	SPRAY DISCHARGE AT BASE OF THE FLAME CLOSEST TO YOU	

accurate way to identify them is to simply look at the outlet of the extinguisher. The class A extinguisher has a pointed nozzle at the end of the hose. It should be used by placing a finger over the nozzle outlet to break up the stream of water into small droplets. The water pressure could otherwise spread the fire.

The class ABC extinguisher usually has a small bell outlet at the end of the hose or a short snout opposite the trigger. The extinguisher should be used by spraying the dry powder to settle on top of the blaze, thus suffocating it.

The BC extinguisher has a large cone-shaped bell at the end of the hose. One must hold the hose by the insulated part of the handle as the discharged material is extremely cold and can otherwise freeze the user's hands. The abbreviation "PASS" is commonly used to describe the proper use of the fire extinguisher:

- Pull the pin from the top of the extinguisher
- Aim the extinguisher nozzle at the base of the fire
- Squeeze the trigger to activate the discharge of material
- Sweep the nozzle from side to side at the base of the fire

Most extinguishers found in the shop are small enough to be carried about easily. They have a discharge rate of approximately 15 to 20 seconds, so using it correctly is important to extinguish the blaze. In the hands of an untrained individual, a fire extinguisher can cause more problems than it can correct. An ABC extinguisher emptied by randomly spraying it into the air will do little more than fill the room with a blanket of dust, making it impossible to see the blaze to properly extinguish it. A partially discharged extinguisher should be returned for service once the trigger has been pulled because the propellant will continue to leak out, leaving only the powder in an ABC extinguisher and a completely empty BC extinguisher. The extinguishers should be checked monthly and the tag initialed by the person doing the inspection. They must also be inspected and serviced annually by a professional fire service company.

LOCKOUT/TAGOUT—THE CONTROL OF HAZARDOUS ENERGY

A **lockout/tagout** procedure must also be included in the hazard communication document. A lockout/tagout program makes provisions to render a piece of equipment unusable when it becomes dangerous to operate. This may be the result of a switch not functioning properly, a broken part, or any dysfunction that makes the equipment unsafe to operate. Taking it out of service may require disconnecting the equipment from its power supply, locking the electric breaker in the off position, or any other steps to ensure it cannot be turned on or energized. See **Figure 2-28**. Any of the tags or markings in the previous illustration can be used to identify a piece unsafe to use. The tag should be attached on or near the power source or switch to prevent anyone from turning it on or using it. This applies to ALL equipment, not just electrical items.

FIGURE 2-28 Any piece of portable or stationary equipment that is not safe to be used should be tagged out using a recognizable marking to do so. Any of the tags in this illustration attached to the equipment's power switch or supply cord should identify the equipment as not safe for use.

VEHICLE BATTERY SAFETY

Another frequently forgotten about hazardous waste is used batteries. When they are taken out of service, they must be stored and treated as a hazardous waste. The **electrolyte** contains sulfuric acid, which is extremely corrosive and is also laden with huge amounts of lead. They must be stored in a well-ventilated area on a coated floor which prevents the liquid from leaching into the ground if a leak develops.

When removing and installing a battery, one must disconnect the negative cable first and then the positive cable. When re-installing, the negative cable is connected AFTER the positive cable. When charging the battery or jump-starting a vehicle, the negative cable should always be connected last and disconnected first.

HAND TOOL SAFETY

It is rare that any task in a collision repair shop is performed without the use of either hand tools or power

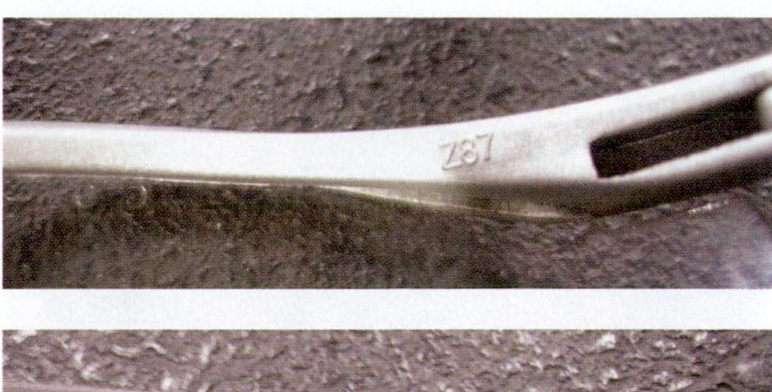

FIGURE 2-29 All safety glasses must have the marking Z87 stamped on their frame to verify they are approved safety glasses.

tools. Due to their constant use, one might have a tendency to become lax about using them with all the safety precautions in mind. While it is not possible to anticipate every possible mishap that may occur, one must always use common sense when using any tools. The following are perhaps the most common sources of injury resulting from improper hand tool use.

- Safety glasses should be worn AT ALL TIMES when in the shop. Even though an individual may not be using any tools, others working in the shop may be and the necessary precautions should be taken to protect everyone from flying objects they may produce. See **Figure 2-29**.
- In addition to safety glasses, a full face shield must be worn when using a high speed grinder, sander, or any other tool that produces flying sparks.
- Leather gloves should always be worn when using tools, especially where sudden directional change may occur. The startup and tightening torque of the air ratchet can unexpectedly slam one's hand against a sharp edge, or an air chisel bit may become disengaged when cutting, resulting in a serious cut.
- Technicians should always keep tools clean and in good working condition. It is important to clean and inspect them after each use to ensure they are in the proper working condition.
- One should never use a hammer or any other tool with a handle that is cracked or loose as the head may fly off, causing personal injury. Many times a loose hammer handle can be stabilized by driving the wedges deeper into the handle inside the hammer. See **Figure 2-30**.

FIGURE 2-30 Hammers with cracked, loose, or damaged handles should be taken out of service due to the danger of a potential injury from the head flying off and striking someone nearby.

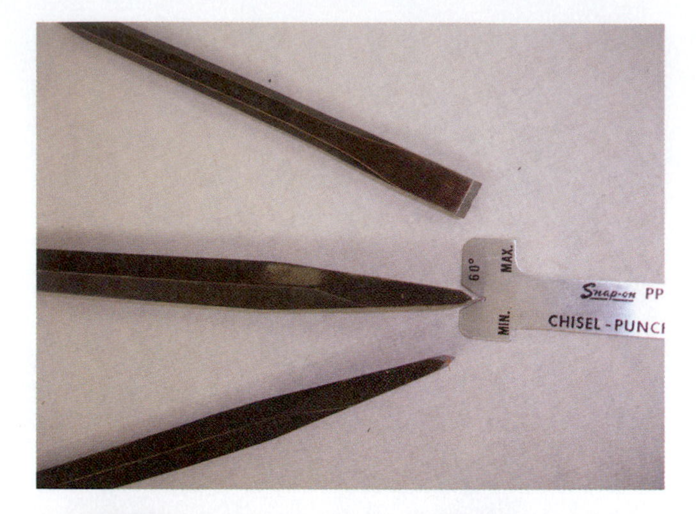

FIGURE 2-31 Cutting tools should always be properly sharpened and any broken ones should be taken out of service.

- Technicians should use professional quality tools as they are intended to be used by the manufacturer. Hobbyist's tools are not designed to be used under the same conditions as professional shop tools.
- Cutting tools such as chisels are usually intended to be sharp; keep the cutting edge honed and maintain the correct shape of the cutting edge. Dull cutting tools are a hazard. One should always move them away from the body when using them. See **Figure 2-31**.
- Technicians should always keep the striking end of the punches and chisels properly dressed and not allow the ends to become mushroomed. A mill file should be used to re-shape the end of the tool at the first sign it is becoming deformed. See **Figure 2-32**.
- One should use tools for their intended use only. A prybar is for prying; a screwdriver is for driving screws, not for chiseling, cutting, or prying. See **Figure 2-33**.

FIGURE 2-32 The striking surface of any chisel or punch should be kept properly dressed and any sign of mushrooming should be removed immediately.

FIGURE 2-33 Tools should never be modified to fit a special need or requirement. Heating them or changing their design will inevitably result in an accident.

- Impact and chrome sockets are not to be used interchangeably with air ratchets and impact tools. See **Figure 2-34**. Contrary to common belief and technician use, chrome-plated sockets are not designed and manufactured for use on air ratchets or impact wrenches. They are hardened differently

FIGURE 2-34 Chrome-plated sockets should never be used with an air ratchet as they may become damaged and cause personal injury.

and may become damaged, crack, or split. The chrome plating can also crack and split, possibly resulting in a serious injury to the user.

SHOP EQUIPMENT SAFETY

The collision repair shop contains a wide variety of equipment commonly used to perform a number of different tasks. The equipment is used to secure and hold vehicles, apply corrective forces, and raise vehicles off the ground, to mention just a few. While each piece of equipment enhances the performance of the repair process, together they also provide numerous opportunities for potential injuries to a careless or untrained user.

Electric Tools and Equipment

Much of the heavy-duty equipment, such as welders, multiple outlet hydraulic pumps, induction heaters, and so on, used in the collision repair shop are electrically powered. Since it is often considered "community property" or shop owned, nobody wants to accept responsibility for its maintenance, so it can become a source of possible accidents. Before using any equipment that is frequently used by others, it should be inspected to protect against personal injury. Check to ensure the ground prong on the plug is still intact and check for cuts or separation where the cord is attached to the machine. Ensure the power switch and all adjustments are functional. This will help protect the user from injury due to electric shock. See **Figure 2-35**. Whenever working in an area with damp floors, an inline ground fault circuit interrupter (GFCI) should be used if it is not hard-wired into the building's electrical system.

Hydraulic Equipment

Before using hydraulic equipment, check to ensure the hydraulic ram extends and retracts correctly. Failure to do so may be caused by a low fluid level or air in the lines.

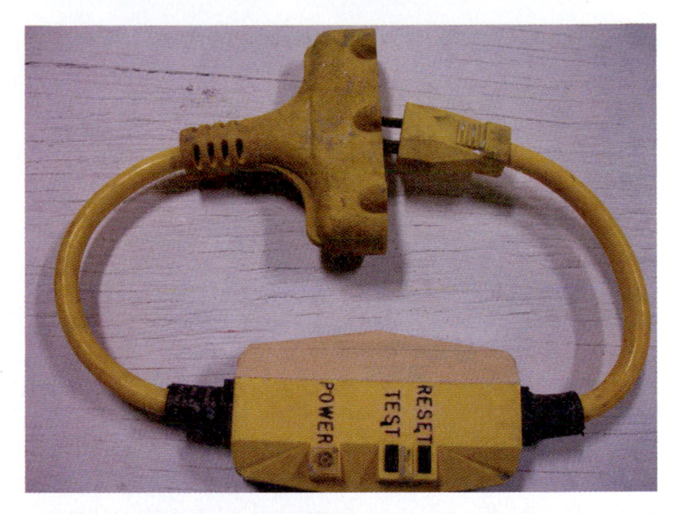

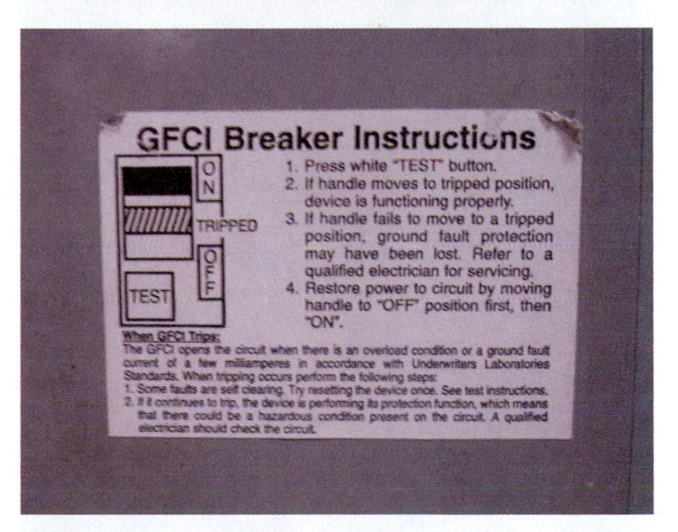

FIGURE 2-35 A ground fault circuit interrupter (GFCI) should be used when using electric tools and working on wet or damp floors. The extension cord should be attached to an inline GFCI when it isn't possible to plug into a protected hard-wired circuit or the breaker panel is not wired with a protected circuit breaker.

This should be adjusted before applying any corrective force to avoid possible injury. Also inspect the hose for cuts or kinks and the connections to ensure they are properly tightened.

Lifting Equipment

Service and repairs on the underside of the vehicle are always performed more easily when it is raised high enough to walk—or at least crouch—underneath it. It is often said as much as 40 percent of the vehicle collision damage goes undetected if the undercarriage is not inspected as part of the damage analysis. Therefore, a lift is considered mandatory equipment for the collision repair shop. See **Figure 2-36**. A number of different types of portable and permanent lifts are available. The most popular unit for body shops is the two post system that has adjustable, rotating swing arms for lifting the vehicle. This makes it possible to use the lift on nearly all vehicle designs and models and allows nearly total access to the undercarriage. A portable lifting device such as the four point swing arm is also a popular choice for smaller shops lacking the space to dedicate a stall for a permanent hoist. Another advantage of these lifts is they can be moved about the shop and used wherever needed. Regardless of the type or style used, some basic common sense safeguards must be practiced to protect the operator, the vehicle, and the lift being used.

FIGURE 2-36 The four-point swing arm lift is the most practical for the collision repair shop as it allows access to nearly the entire undercarriage. Portable lifts are also used to elevate the vehicle, allowing the technician to gain access to the undercarriage and perform other mechanical repair operations.

Safe Use of Vehicle Lifts

- Never attempt to operate a lift unless you have been trained in its proper use.
- Always ensure the lift is properly maintained by lubricating moving parts such as the safety catch and their releases and the swinging arms. The cables used to lift the apparatus should be checked to ensure they are not frayed. In addition, if the system uses it, check the hydraulic fluid. Check the safety catches to ensure they engage and release properly.
- Never attempt to raise a vehicle weighing more than the lift manufacturer's recommended capacity. The lifting capacity must be clearly marked on the equipment.
- Place the vehicle in park or engage the emergency brake prior to lifting it.
- Pad the lift arms as needed to ensure the vehicle is lifted level and the weight is distributed equally on all lifting surfaces.
- Ensure that all gas, fuel vapor recovery, and high pressure hydraulic and brake lines are protected from contacting the lifting pads. These and the high voltage lines on hybrid vehicles are frequently routed in the area of the vehicle where the lift pads are placed. Moving them aside may be necessary to prevent crushing or damaging them.
- When the vehicle is lifted high enough for the tires to clear the ground, the vehicle should be shook vigorously from the front or rear end to ensure it is firmly and safely settled on the lifting arms. If the vehicle moves more than what is normal movement from the lifting arms, it should be lowered back onto the ground and the arms adjusted before lifting once again. At this time one should also check underneath the vehicle to ensure none of the lines mentioned previously—exhaust parts or any other parts—are being crushed or damaged.
- Once the vehicle is raised to the desired height, it should be lowered until the safety catches have engaged before attempting to get under the vehicle or initiating the work.

CAUTION

Under NO CONDITION should one initiate the repairs or get under the vehicle until the safety catches have been solidly engaged. They should be checked routinely for proper functioning.

- In the event large assemblies such as the differential, fuel tank, or other heavy parts are to be removed, the technician should use a jack stand tall enough to reach the rear bumper to help stabilize the vehicle.
- Make sure the entire area under the vehicle is clear before lowering it. Check to ensure all tools and equipment that have been used while working under the vehicle have been moved away to a clear area nearby.

Safety for the Vehicle Being Lifted

- Ensure the vehicle transmission is in park (automatic transmission), in reverse gear (standard transmission), or the emergency brake is engaged.
- Check the doors to ensure they are closed and latched.
- Pad the lifting arms with solid blocking items to ensure the vehicle is raised level and remains stable when lifted off the ground. See **Figure 2-37**.
- Raise the vehicle using only lift points that are adequately reinforced to support its weight. One can safely use the vehicle's normal weight-bearing surfaces such as suspension and steering mounting locations, engine cross members, and frame rails. Never use the floor pan or any other "soft" or unreinforced panels. These same locations can be used with nearly all vehicle-lifting equipment.

FIGURE 2-37 The vehicle can safely be lifted off the ground by using the lift points specified by automobile manufacturers as well as any load-bearing surfaces such as the axle housing, suspension mounting locations, and the frame rails.

Floor Jacks

Many times floor jacks are used to raise a vehicle off the ground to gain access to a vehicle's underside. Before raising the vehicle, check to ensure the saddle raises solidly with each pump of the handle. Failure to do so is an indication of a low fluid level and may cause it to collapse when a heavy object is lifted. Before getting under the vehicle raised with a floor jack, one should always place jack stands under the vehicle and let the weight down to fully rest on them. Serious injuries are apt to occur unnecessarily when an impatient technician carelessly gets under a vehicle without properly supporting it.

CAUTION

One should shake the vehicle to ensure it is firmly planted on the jack stands prior to getting under it.

Hydraulic Jacks and Straightening Equipment

Corrective forces as high as 20,000 pounds per square inch (psi) are not uncommon when hydraulic pulling and straightening equipment are used for repairing heavily damaged vehicles. One should stand aside from the line corrective forces travel to avoid serious injury in the event a chain or clamp should break or slip, causing the equipment to lurch or a part to rip loose and become air-borne. Never allow any hydraulic fluid leak to go without being repaired.

PROTECTION FROM VEHICLE-RELATED INJURIES

Air Bags and Hybrid Vehicles

Two areas of concern regarding the potential for serious injuries caused by a vehicle are when working around air bags and their circuitry and when working on hybrid vehicles. In both cases the manufacturer's recommendations should be consulted before proceeding with the repairs.

Hybrid Vehicles. Hybrid vehicles have high voltage circuits that, when not handled properly, could cause serious injury or even death. The manufacturer's precautions should be read and understood BEFORE initiating any work on the vehicle.

Air Bags. Air bags also present numerous opportunities for possible injury while installed in the vehicle and also when being carried and handled outside of the vehicle. Some air bags deploy in two stages and, depending on the weight of the person sitting in the seat, may have been deployed to the first stage only. Extreme caution must be used when removing a partially deployed air bag and should never be attempted without first consulting the appropriate technical service manual. A deployed air bag presents certain concerns for the technician when removing it as well. The following safety precautions should be exercised when working with any part of the system:

- Always follow the manufacturer's recommendations for disarming the system before initiating any repairs to the area where the sensors are mounted and on the module itself.
- Never probe any of the wires with a test light to test the circuit.
- Always wear gloves, safety glasses, and the appropriate respiratory protective gear when handling a deployed air bag and when cleaning the residue created from its deployment.
- When carrying a loaded or undeployed air bag, always have it facing away from your body.
- Always have an air bag face up on the bench.
- Always follow the manufacturer's recommended procedures for installing and re-arming the system.

PAINT GUN CLEANERS/ RECYCLERS

The 2008 EPA National Emission Standards for Hazardous Air Pollutants (NESHAPs) (40 CFR Part 63 HHHHHH) require the use of enclosed paint gun cleaners. This equipment significantly reduces the amount of hazardous waste generated by the shop as it allows the same solvent to be used repeatedly for cleaning the spray guns. Most shops employ a two- or three-stage approach to cleaning paint guns, which includes a coarse rinse, a second rinse, and a final cleaning which may be done by hand or in an enclosed gun cleaner. According to the NESHAPS standard, spraying solvents while cleaning a paint gun must be done inside an enclosed gun cleaner/recycler. The gun cleaner will use either a solvent for solvent base paints or an aqueous and citrus-based cleaning system for waterborne materials. Waste solvents or liquid from the two cleaning systems should never be intermixed.

Summary

Safety in the auto repair shop is usually perceived as the safe use of tools and equipment. However, because of the exposure to many hazardous chemicals, the possibility of injury is increased considerably. Therefore, the work environment of a collision repair technician is seen as very dangerous. However, through using required PPE and some common sense safe work practices, shops can make the work environment safe.

The following are some practical reminders of what must be done to keep the work environment safe and accident free:

- Always keep the work area clean.
- Clean up spills immediately.
- Ensure all tools used are working properly.
- Make sure the tool used is designed for the task.
- Use care to avoid electrical shock by frequently inspecting items for safety.
- Ensure all hydraulic and force application equipment is operating properly and is free of any leaks.
- Never use any tool or equipment you have not been trained to use.
- Never use an unfamiliar product without first consulting the MSDS for it.
- Always use the proper PPE whenever the manufacturer or circumstances warrant its use.

Despite the many health hazards and opportunities for accidents to occur, technicians have the capability of creating and promoting a work environment free of many of the perceived dangers. They must think safety, act safety, and promote safety in everything they do.

The following 29 CFR subparts are referenced most often by OSHA and the EPA for compliance in the collision repair industry:

- 29 CFR 1910 section 1200 Hazard Communication Standard
- 29 CFR 1910.134 respiratory protection standard
- 29 CFR 1910 subpart H (Hazardous Materials)
- 29 CFR 1910 subpart I (Personal Protective Equipment)
- 29 CFR 1910 subpart Z (Air Contaminants and PEL "permissible exposure limits")
- 29 CFR 1910.147 The control of hazardous energy—Lockout/Tagout
- 40 CFR Part 63 National Emission Standards for Hazardous Air Pollutants Standard (NESHAPs)

Key Terms

acute exposure
blood-borne pathogens
chronic exposure
Code of Federal Regulations (CFR)
Department of Transportation (DOT)
electrolyte
Environmental Protection Agency (EPA)
flash point
hazard communication program
hazardous material
hazardous waste
hepatitis B (HBV)
human immunodeficiency virus (HIV)
isocyanates
lockout/tagout
manifest
material safety data sheet (MSDS)
NFPA placard
Occupational Safety and Health Administration (OSHA)
personal protective equipment (PPE)
right to know
route of entry
secondary container label
volatile organic compounds (VOCs)

Review

ASE Review Questions

1. Technician A says the third prong on an electrical plug provides extra current during a power surge. Technician B says the third prong is a safety ground. Who is correct?
   A. Technician A only
   B. Technician B only
   C. Both Technicians A and B
   D. Neither Technician A nor B
2. Technician A says solvent-soaked towels should be disposed of in an airtight container. Technician B says floor dry should be disposed of in the same container used for contaminated sweepings. Who is correct?
   A. Technician A only
   B. Technician B only
   C. Both Technicians A and B
   D. Neither Technician A nor B
3. Technician A says some fumes are heavier than air. Technician B says paint storage cabinets should be vented at the top and bottom. Who is correct?
   A. Technician A only
   B. Technician B only
   C. Both Technicians A and B
   D. Neither Technician A nor B
4. Technician A says one must be trained in how to use an air purifying respirator before using it. Technician B says with proper ventilation, a dust/mist respirator can be used for painting operations. Who is correct?
   A. Technician A only
   B. Technician B only
   C. Both Technicians A and B
   D. Neither Technician A nor B
5. Technician A says the quickest way to distinguish the difference between fire extinguishers is to look at the outlet nozzle. Technician B says a class ABC fire extinguisher is filled with carbon dioxide. Who is correct?
   A. Technician A only
   B. Technician B only
   C. Both Technicians A and B
   D. Neither Technician A nor B
6. Solvents are accidentally spilled on the floor and are on fire. Technician A says a class A fire extinguisher should be used to extinguish the blaze. Technician B says a class ABC extinguisher should be used. Who is correct?
   A. Technician A only
   B. Technician B only
   C. Both Technicians A and B
   D. Neither Technician A nor B

7. Technician A says a secondary container label must be attached to any container that does not have the manufacturer's label on it. Technician B says a secondary container label is not necessary so long as everyone in the shop is aware of the contents in each specified container and it remains in the same location at all times. Who is correct?
   A. Technician A only
   B. Technician B only
   C. Both Technicians A and B
   D. Neither Technician A nor B
8. Technician A says a solvent-filled container brought into the shop area must have a label properly identifying its contents. Technician B says a secondary container label is necessary to mark a container used to store hazardous waste materials. Who is correct?
   A. Technician A only
   B. Technician B only
   C. Both Technicians A and B
   D. Neither Technician A nor B
9. Technician A says OSHA is responsible for protecting the worker in the workplace. Technician B says the EPA and OSHA's jurisdictions frequently overlap each other. Who is correct?
   A. Technician A only
   B. Technician B only
   C. Both Technicians A and B are correct
   D. Neither Technician A nor B
10. Technician A says before using a chemical with which you are not familiar, you should first read the product usage bulletin. Technician B says you should first consult the product MSDS. Who is correct?
    A. Technician A only
    B. Technician B only
    C. Both Technicians A and B
    D. Neither Technician A nor B
11. Technician A says the CFR is a set of regulations and guidelines used to create a safe work environment. Technician B says the rules and regulations outlined in the CFR come from many of the legislative acts. Who is correct?
    A. Technician A only
    B. Technician B only
    C. Both Technicians A and B
    D. Neither Technician A nor B
12. Technician A says the shop that generates hazardous waste is responsible for it until it reaches the TSDF. Technician B says the shop is responsible for hazardous waste until it is picked up by the waste hauler. Who is correct?
    A. Technician A only
    B. Technician B only
    C. Both Technicians A and B
    D. Neither Technician A nor B
13. Technician A says the manifest is a list of all the hazardous materials used in the shop. Technician B says the manifest is a list of regulations that must be followed by any facility using hazardous materials. Who is correct?
    A. Technician A only
    B. Technician B only
    C. Both Technicians A and B
    D. Neither Technician A nor B
14. Technician A says professional work habits can help create a safer work environment. Technician B says everyone in the shop must play a role in making it a safer workplace. Who is correct?
    A. Technician A only
    B. Technician B only
    C. Both Technicians A and B
    D. Neither Technician A nor B
15. Technician A says a hazardous waste must exhibit four different characteristics to be classified as such. Technician B says it is a hazardous waste if it is considered toxic. Who is correct?
    A. Technician A only
    B. Technician B only
    C. Both Technicians A and B
    D. Neither Technician A nor B
16. Technician A says a chronic exposure occurs at a low concentration for an extended period of time. Technician B says the effects of an acute exposure are usually permanent. Who is correct?
    A. Technician A only
    B. Technician B only
    C. Both Technicians A and B
    D. Neither Technician A nor B
17. Technician A says the MSDS will tell what PPE should be used when handling a hazardous material. Technician B says each time hazardous waste is picked up from the shop another document should be added to the manifest. Who is correct?
    A. Technician A only
    B. Technician B only
    C. Both Technicians A and B
    D. Neither Technician A nor B
18. The NFPA placard will:
    A. Alert the emergency responders to the potential hazards inside an area.
    B. List the hazardous materials contained inside the area.
    C. Identify the PPE required by the technicians working in the area.
    D. Identify the media required for fire fighters to extinguish a blaze.

19. Technician A says a hazardous substance with a higher PEL is less dangerous than one with a lower PEL. Technician B says the STEL of a hazardous substance or material is for a 15-minute period of time. Who is correct?
    A. Technician A only
    B. Technician B only
    C. Both Technicians A and B
    D. Neither Technician A nor B
20. Which of the following is not true of an ABC fire extinguisher?
    A. It can be used to extinguish burning solvents.
    B. It is filled with dry powder for the extinguishing media.
    C. It is not effective to extinguish a fire on metals.
    D. It is effective to extinguish a fire on metals.

Essay Questions

1. What is the manifest and what are the documents included in it?
2. What is the purpose of a secondary container label and what information should be included on it?
3. How many sections are included in a typical material safety data sheet? With which sections should the collision repair technician become most familiar?
4. Discuss the difference between a hazardous material and a hazardous waste.
5. What is the hazard communication? What should be included in it?

Critical Thinking Questions

1. With a group of other students, identify several tasks a collision repair technician performs and identify the required PPE requirement for each.
2. Identify several ways in which a collision repair technician could experience a chronic exposure if the proper PPE were not used.

Lab Activities

1. Pick five products commonly used by a collision repair technician, locate the correct MSDS, and determine the hazards each product presents and the PPE required for its use.
2. Among the materials commonly used in the shop area, make a list of at least two products that fall into each of the four hazardous waste categories.
3. Draw a scale model floor plan of the shop area, identify the location of all the fire extinguishers, and identify the class of each extinguisher. Also indicate the location of the emergency fire pulls to sound the alarm in an emergency.

Chapter 3

Hand Tools

OBJECTIVES

Upon completing this chapter, you should be able to:

- Identify general purpose hand tools used in collision repair.
- Explain the correct and safe use of general purpose hand tools.
- Identify common collision repair hand tools.
- Explain the correct and safe use of collision repair tools.
- Select the correct tools for the job being performed.
- Properly maintain hand tools.
- Understand how to safely use hand tools.

INTRODUCTION

In this chapter you will learn about the hand tools professional collision repair technicians use to perform body and refinishing repairs. Knowing how to properly select and use hand tools is necessary to understand the procedures for repairing today's vehicles. Some type of tool is required for every straightening and preparation operation. Good quality tools will increase productivity and improve the quality of the repairs. Low-cost, poor quality tools will break more easily, are usually heavier, and may feel uncomfortable.

Although it is not practical to describe all hand tools used in collision repair, this chapter will provide students with the background and understanding needed when studying other chapters in this text.

CAUTION

Injury can result from improper use of hand tools. Tools should only be used for their intended purpose. Always keep tools in proper working condition.

GENERAL PURPOSE TOOLS

Collision repair technicians use general purpose hand tools for disassembly and repair operations every day. Examples of these tools include wrenches, pliers, screwdrivers, and hammers.

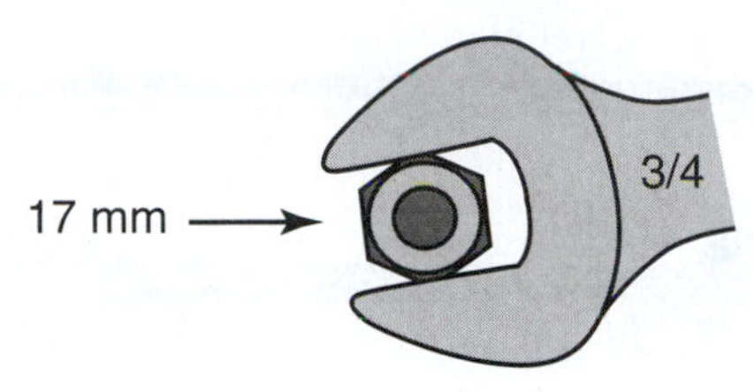

FIGURE 3-1 Use the correct size wrench to avoid damaging a fastener.

WRENCHES

Several types of wrenches are used to turn and/or hold nut and bolt heads. The word *wrench* means to turn or twist. The overall length of a wrench depends on the size of the nut or bolt it will be used on. Larger wrench sizes will be longer. The extra length will give the technician the additional leverage needed to loosen or tighten larger nuts and bolts.

The wrench size is determined by the distance across the wrench jaws. For example, a 1/2-inch wrench has a jaw opening from face to face of approximately 1/2 inch. The actual size is slightly larger so that the wrench can fit around a 1/2-inch nut or bolt head.

The **Society of Automotive Engineers (SAE)** is a nonprofit educational and scientific organization dedicated to advancing mobility technology to better serve humanity. Most standard or SAE wrench sets used in collision repair include sizes from 3/16 to 1 inch. Metric sets usually include 6- to 19-millimeter wrenches.

SAE and metric size wrenches are not interchangeable. An SAE wrench used on a metric nut will probably slip and round off the points of the nut. See **Figure 3-1**.

CAUTION

Never extend the length of the wrench with a pipe or strike it with another tool.

TECH TIP When loosening or tightening a nut or bolt, pull on the wrench—do not push. If the wrench slips off, there is less chance of injury to the hand.

WRENCH TYPES

There are various wrench types available. See **Figure 3-2**.

Open-End Wrenches

An **open-end wrench** has two flat sides that grip the flat sides of square head (four-cornered) and hex head (six-cornered) nuts. See **Figure 3-3**. They are used to disassemble bolts or nuts in confined areas when there is insufficient clearance to use a box-end wrench or socket.

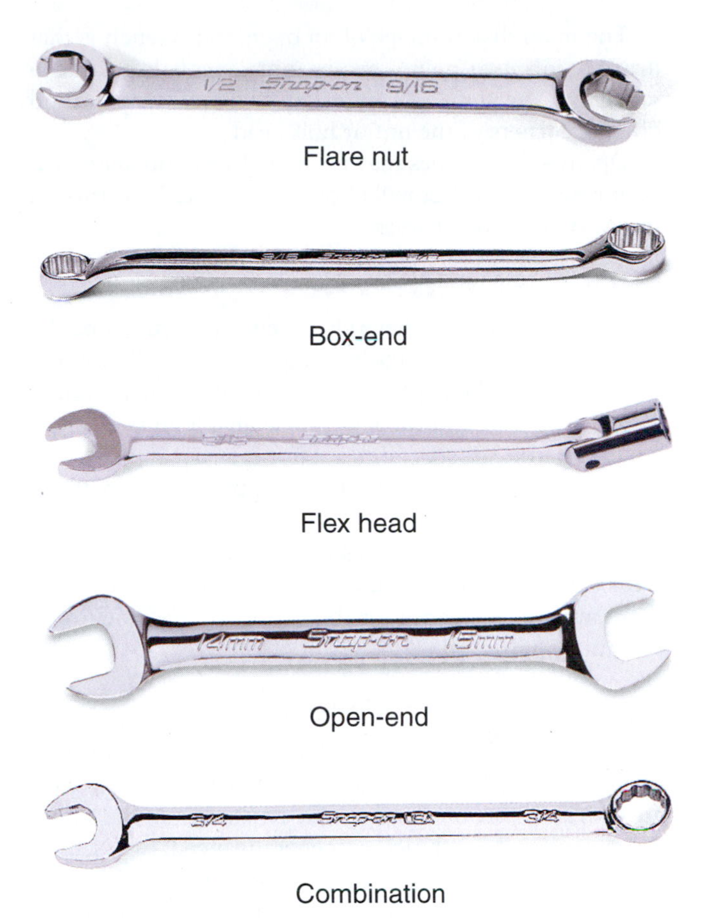

FIGURE 3-2 Several types of wrenches are available, each used for different applications.

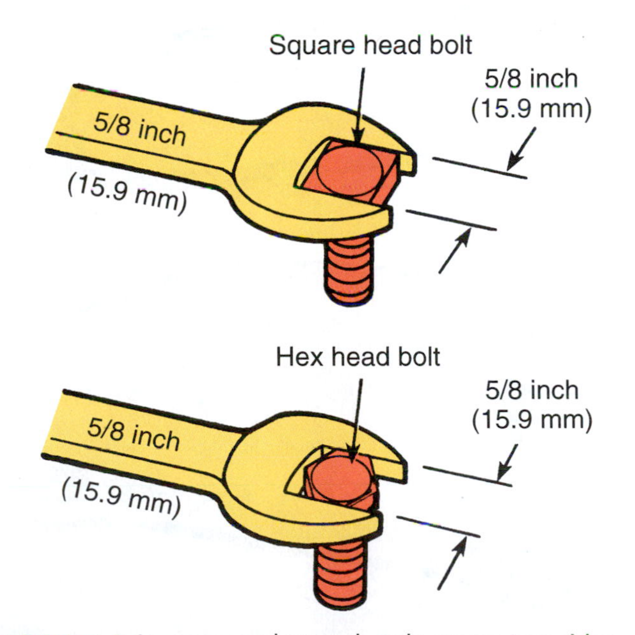

FIGURE 3-3 An open-end wrench only grasps two sides of a nut or bolt.

The main disadvantage of an open-end wrench is that it grips only two points on the nut or bolt head. If too much turning force is applied, the wrench may "round off" the corners of the nut or bolt head.

Open-end wrenches may be angled 15 to 60 degrees at both ends. This offset will allow fasteners to be turned in recessed or confined areas.

Box-End Wrenches

Box-end wrenches have closed ends that surround the bolt head or nut for better holding power. See **Figure 3-4**. More force can be applied without slipping or rounding off the nut or bolt head. They are available in 6, 8, or 12 contact points. A six point is strongest because it completely surrounds the hex nut and applies force to all six sides and points. The handle of a box-end wrench may be offset to more easily reach fasteners which are in confined areas. The main disadvantage of a box-end wrench is that there must be clearance for the end to go over the top of the nut or bolt head.

Combination Wrenches

Combination wrenches offer the advantages of both open- and box-end wrenches. See **Figure 3-5**. They have an open jaw on one end and a box end on the other. Collision repair technicians generally need two sets of wrenches while doing repair work. The combination wrench provides the necessary second set of wrenches.

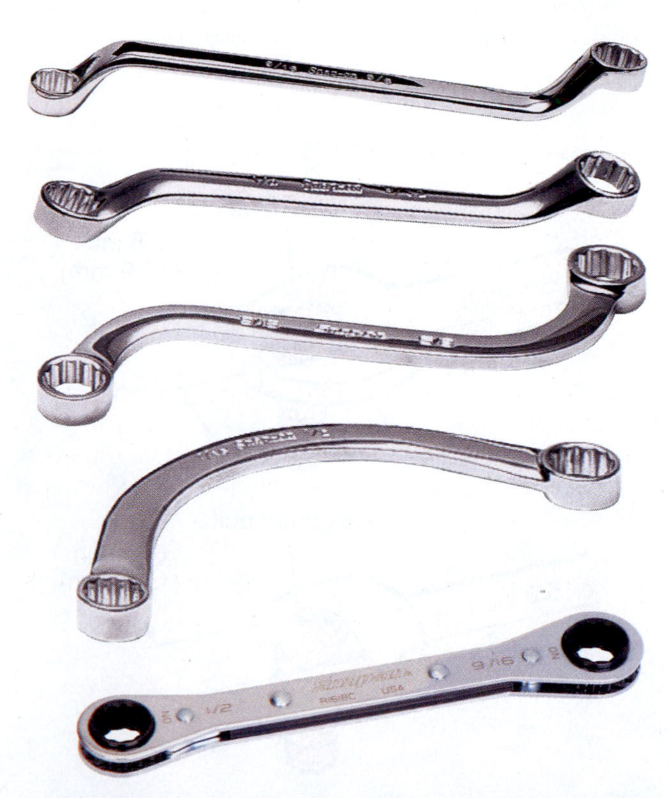

FIGURE 3-4 Box-end wrenches have closed ends that surround the bolt head.

FIGURE 3-5 Open-end combination wrenches offer the advantages of both open-end and box-end wrenches.

Flare Nut Wrenches

A **flare nut wrench** has small splits in its jaw to fit over lines and tubing. See **Figure 3-6**. It can be slipped over fuel, brake, or power steering lines. Using the flare nut wrench will reduce the chance of "rounding off" the tubing nut.

Adjustable Wrenches

An **adjustable wrench** has one movable jaw and one fixed jaw. See **Figure 3-7**. This allows the technician to adjust the wrench opening to fit different sized fasteners. This adjustable wrench has the same advantages and disadvantages as the open-end wrench. Collision repair probably requires a set of size 4-, 6-, 8-, 10-, and 12-inch adjustable wrenches.

Pipe Wrench

A **pipe wrench** is another type of adjustable wrench used for holding and turning round objects. See **Figure 3-8**. The adjusting nut allows the jaw opening to increase and decrease in size. A pipe wrench has jaws which will damage part surfaces. Therefore, the technician should only use this tool on damaged parts.

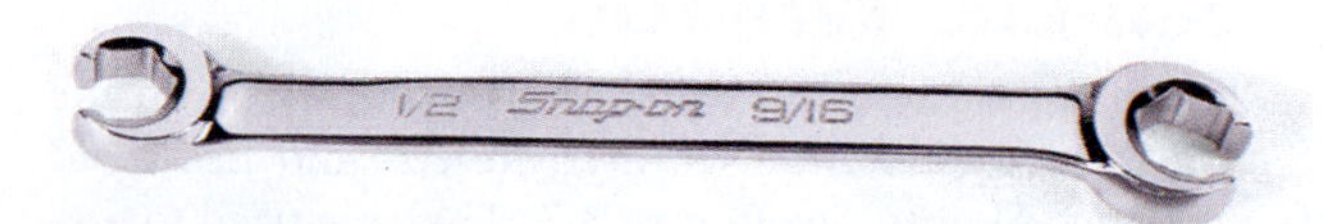

FIGURE 3-6 A double-end flare nut wrench has small splits in its jaw to fit over lines and tubing.

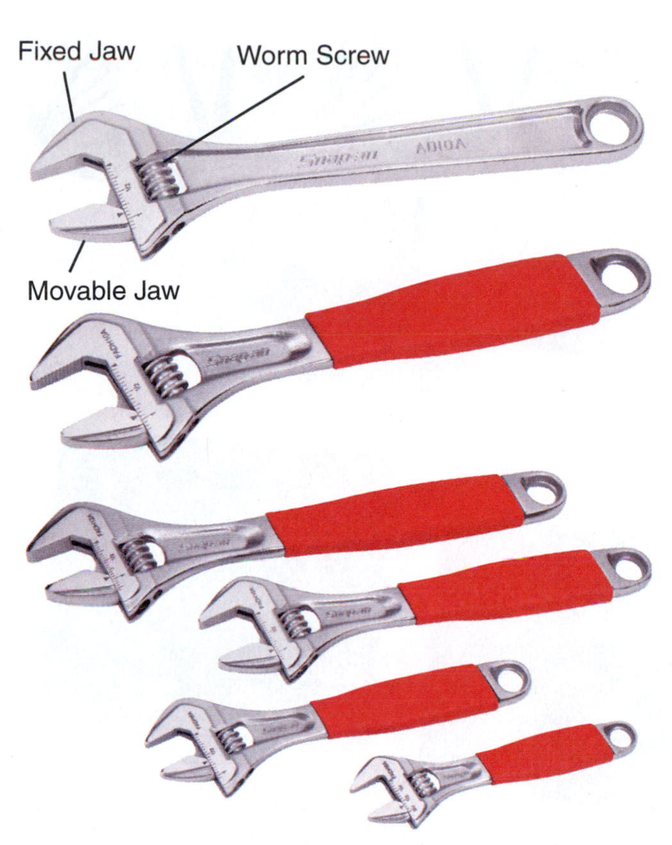

FIGURE 3-7 Adjustable wrenches have one movable jaw and one fixed jaw. Used with permission of Snap-on Tools Company, www.snapon.com.

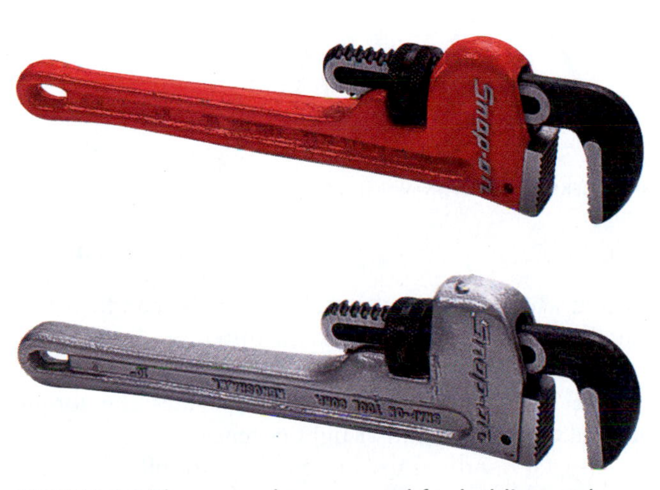

FIGURE 3-8 Pipe wrenches are used for holding and turning round objects. Used with permission of Snap-on Tools Company, www.snapon.com.

Hex and TORX Wrenches

An **Allen wrench** is a hexagon or six-sided wrench used to remove and tighten fasteners with six-point socket heads. See **Figure 3-9**. A set of SAE and metric sizes are needed for collision repair.

A **TORX wrench** or **driver** is designed to remove a six-point socket head fastener. TORX fasteners are available as a set. See **Figure 3-10**.

Tamper-resistant TORX drivers have a small hole in the end of their heads which enables the driver to fit over the peg in the center of the tamper-resistant screw.

Socket Wrenches

A **socket wrench** removes fasteners more rapidly than an open-end or box-end wrench. A **socket wrench set** consists of a socket or drive handle and detachable sockets.

Sockets are cylinder-shaped, box-end wrenches for removing bolts and nuts by rapid turning motion.

Deep well sockets are longer in length for reaching over a stud bolt. See **Figure 3-11**.

Swivel sockets have a universal joint between the drive end and socket body. See **Figure 3-12**.

Impact sockets are thicker and case hardened to withstand the forces of an air-powered impact wrench.

Sockets are available in 6-point, 8-point, and 12-point configurations. The face-to-face contact of a six-point socket provides the most holding power on a hex head. See **Figure 3-13**. The socket has one end closed with a square hold which accepts the square drive of the socket handle.

Socket Sizes. The size of a socket is determined by the drive size and the size of the fastener head it fits on. Collision repair requires both standard SAE and metric socket wrench sets. The technician will need a 1/4-inch drive set (socket sizes 3/16 to 1/2 inches or 4 to 13 mm), 3/8-inch drive set (socket sizes 3/8 to 3/4 inches or 9 to 25 mm), and a 1/2-inch drive set (socket sizes 7/16 to 1 1/4 inches or 11 to 30 mm).

TECH TIP Do not use a deep well socket when a standard size socket will do the job. The deep socket will have a greater tendency to slip off the fastener.

Drive Sizes. Common **drive** or lug sizes needed in collision repair are 1/4 inch, 3/8 inch, and 1/2 inch. The small drive sizes are designed to withstand small amounts of torque which will remove small fasteners (example: interior trim and emblems).

The larger drives will withstand more torque, which is needed to remove larger fasteners (example: bolt on body panels; bumper attachment bolts). See **Figure 3-14**.

CAUTION

Never extend the length of the drive handle with a pipe to increase leverage. Do not use the drive handle or ratchet to strike another tool.

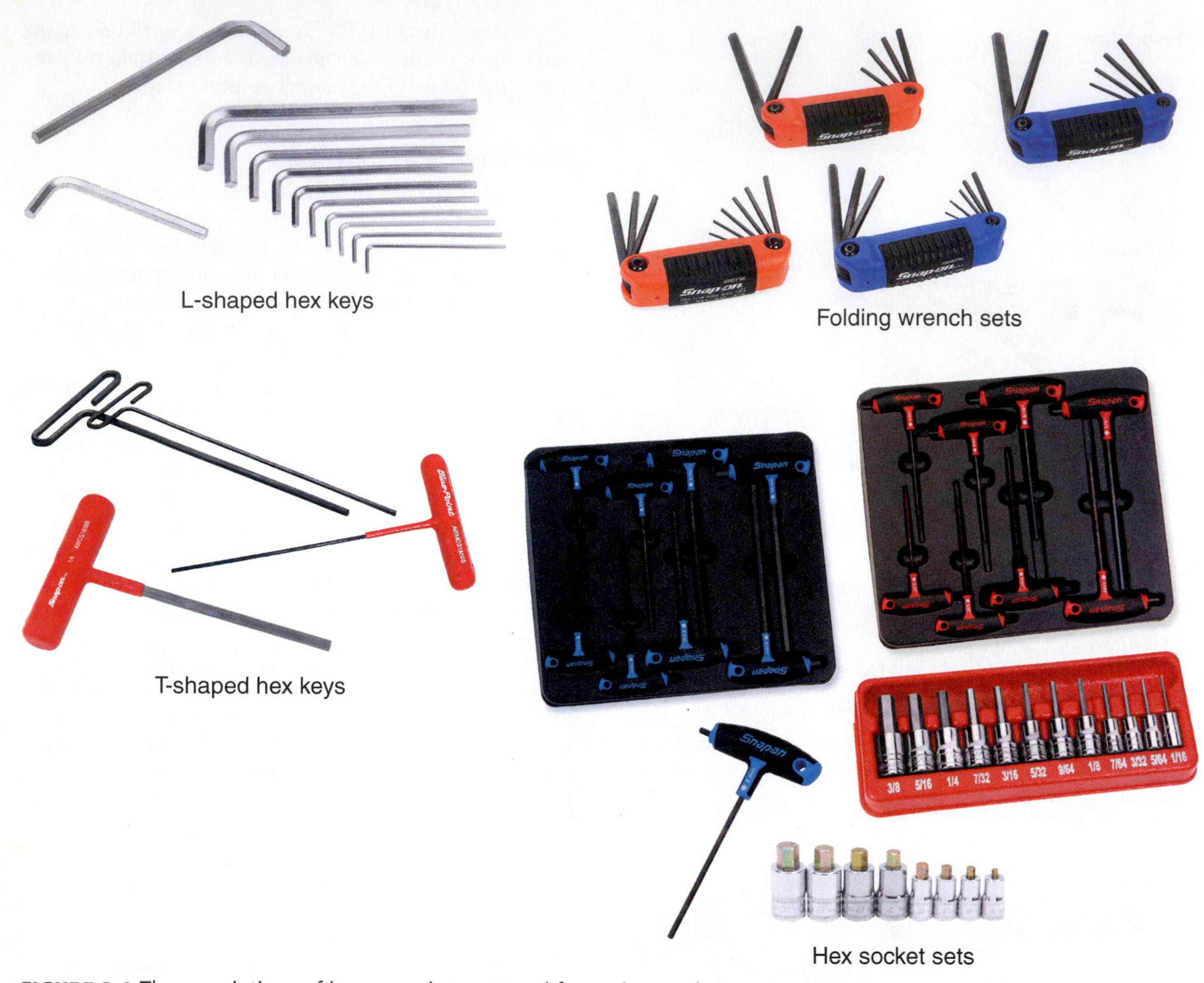

FIGURE 3-9 These variations of hex wrenches are used for various tasks. Used with permission of Snap-on Tools Company, www.snapon.com.

Socket Wrench Accessories. **Socket wrench accessories** will increase the usefulness of this wrench. See **Figure 3-15**. The following accessories are needed for collision repair:

- Breaker bar
- Sliding T-handle
- Speed handle
- Drive adapter
- Extension bars
- Spinner
- Rachet adapter
- Universal joint

Handles. A **ratchet handle** enables the removal or tightening of a fastener without removing the socket from the fastener. The reversing lever allows the turning direction to be changed to loosen or tighten nuts and bolts. Some ratchet handles have a quick release push button which unlocks the socket from the handle drive. Ratchet handles are available in different lengths, offsets, and as flex heads.

The **breaker bar** is long and provides the torque needed to loosen seized or tight fasteners.

A **speed handle** is used to quickly turn off a loosened nut or bolt.

A **sliding T-handle** can be used for a fastener with limited access.

Extensions can reach around obstacles or gain access to otherwise inaccessible fasteners.

An **adapter** allows a larger drive handle to be attached to a smaller drive socket.

CAUTION

Do not use hand-operated sockets, extensions, and accessories with air-operated impact wrenches.

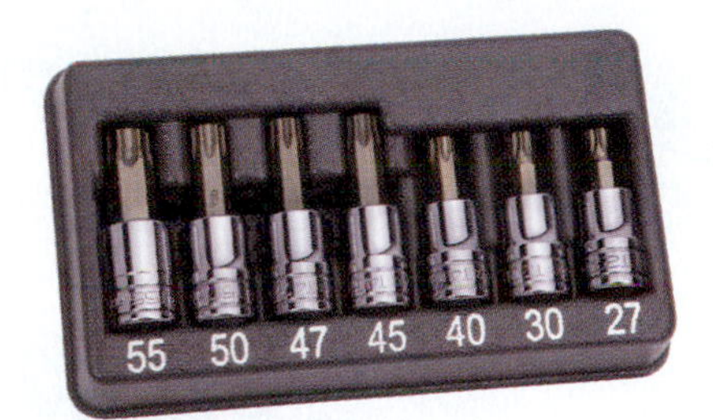

TORX®
hole drilled in the center of a tamper-resistant TORX® driver fits over the tamper-resistant pin in the fastener recess.

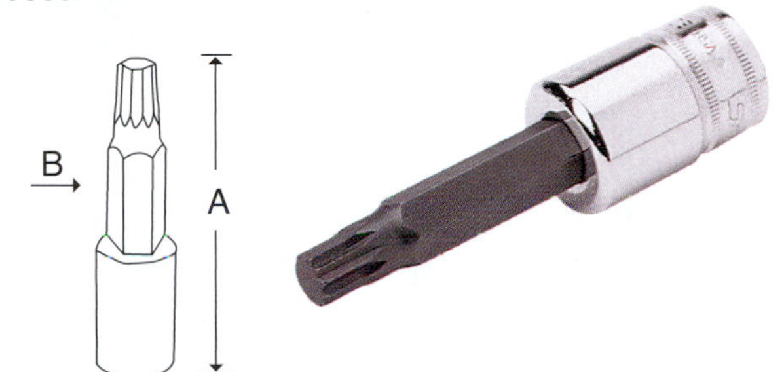

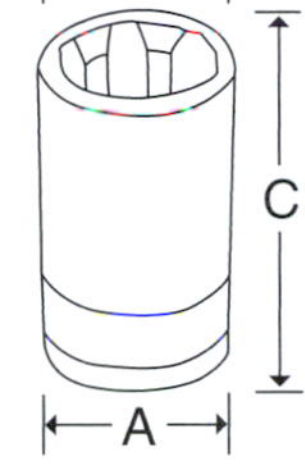

Shallow TORX®

FIGURE 3-10 Variations of TORX drivers, such as tamper-resistant and inverted TORX drivers. Used with permission of Snap-on Tools Company, www.snapon.com. All rights reserved.

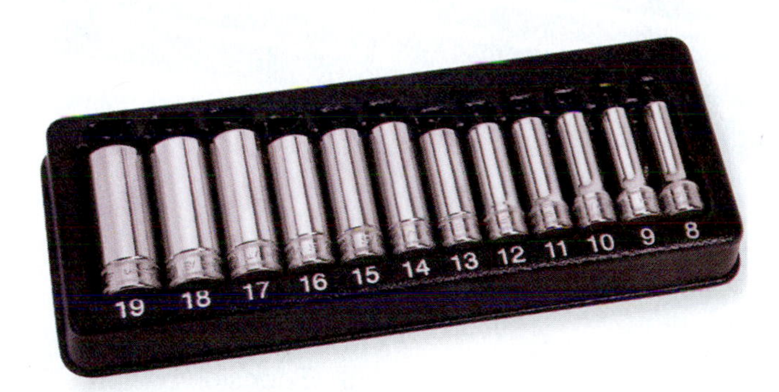

FIGURE 3-11 Deep well sockets are used for reaching over a stud bolt. Used with permission of Snap-on Tools Company, www.snapon.com. All rights reserved.

A **torque wrench** is used to measure the amount of force required to properly tighten a nut or bolt to an exact torque specification. Torque is a measure of twisting force applied to a fastener in inch-pounds, foot-pounds, and in metric measurement of Newton-meters. Fasteners which are tightened too much can strip the threads or break the

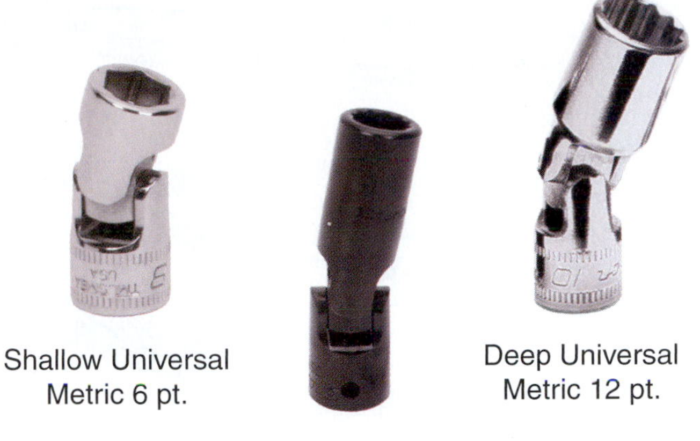

FIGURE 3-12 Swivel sockets have a universal joint between the drive end and socket body. Used with permission of Snap-on Tools Company, www.snapon.com. All rights reserved.

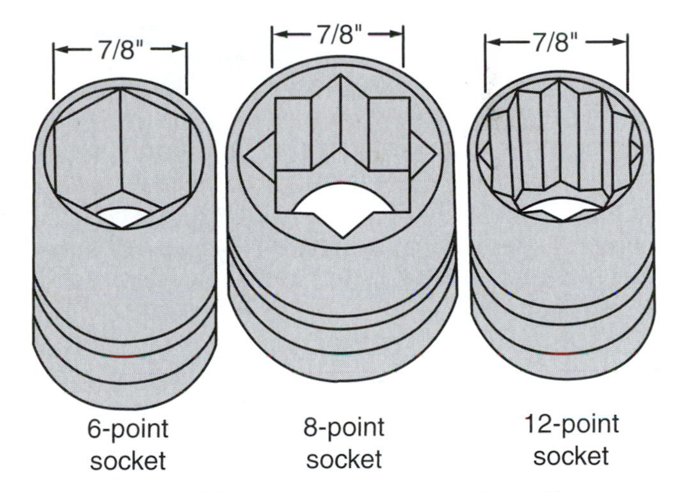

FIGURE 3-13 Different sockets are used for different nut and bolt designs.

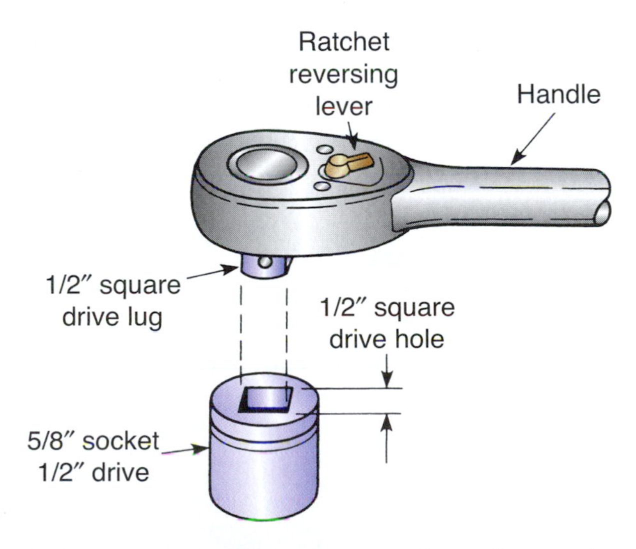

FIGURE 3-14 Socket drive sizes vary depending on the amount of torque required.

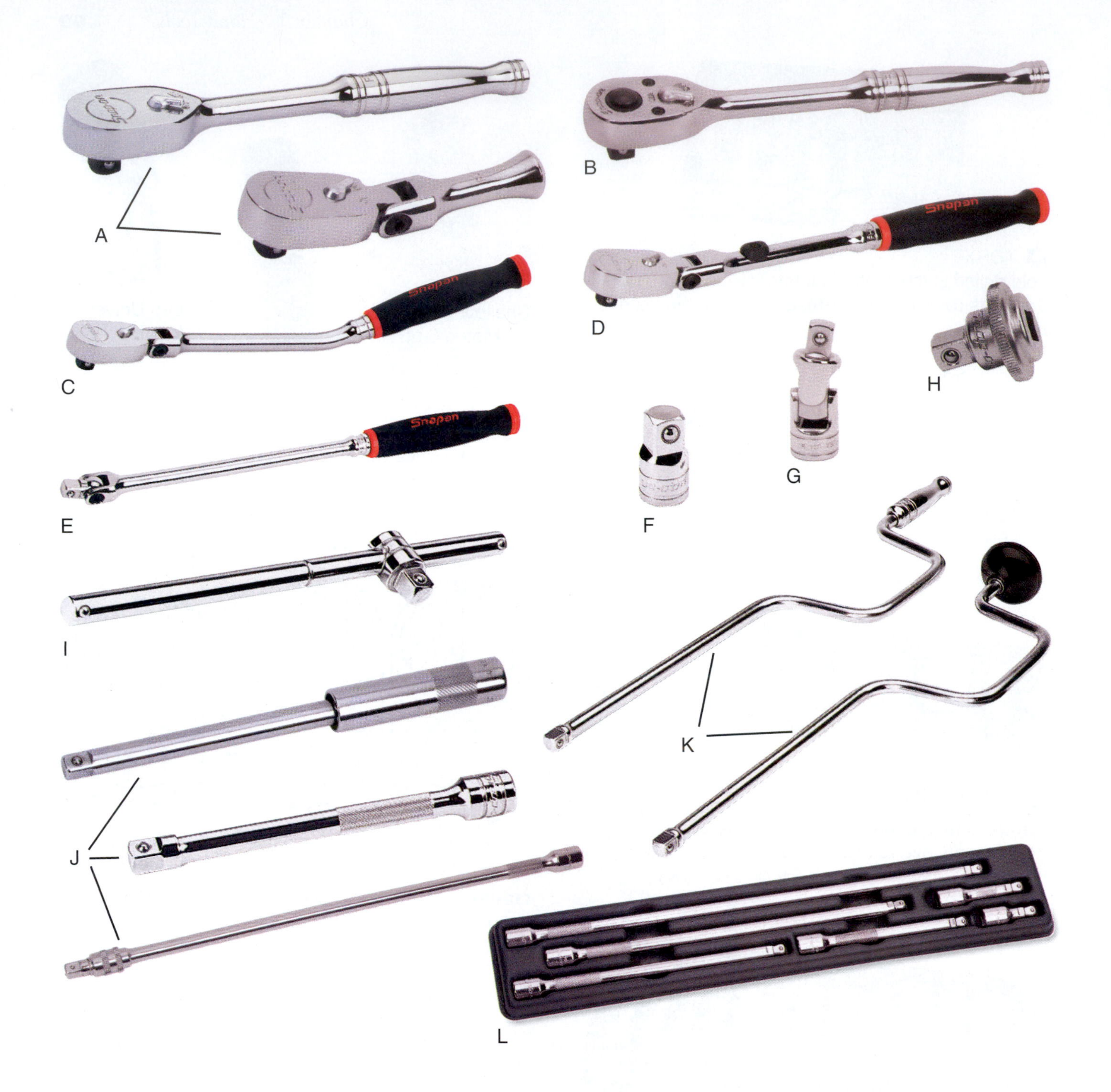

A. Sealed head	E. Breaker bar	I. Sliding T-handle
B. Quick release head	F. Ratchet adapter	J. Extension bars
C. Bent flex head	G. Friction ball universal joint	K. Speeders
D. Locking flex head	H. Ratchet spinner	L. Wobble extension set

FIGURE 3-15 Socket wrench accessories increase the usefulness of the wrench. Used with permission of Snap-on Tools Company, www.snapon.com. All rights reserved.

fastener. Vibration can cause the fastener to loosen if tightened less than the torque specification.

TECH TIP Vehicle fasteners must always be tightened to manufacturer's specifications, which are available in the vehicle service information.

SCREWDRIVERS

Many threaded fasteners used in automotive assembly are installed by some type of screwdriver. Screwdrivers come in many sizes and designs. Each type and size of screwdriver is made for a specific type and size of screw. A collision repair technician will need several types and sizes of screwdrivers.

The size of the screwdriver is determined by the length and diameter of the shank or blade. A well-designed handle will provide a better grip; this enables the technician to generate more torque to loosen fasteners.

Standard Tip Screwdrivers

A **standard tip screwdriver** has a tip with a single flat blade which fits a screw with a slotted head. See **Figure 3-16**. The blade tip thickness and width must perfectly match the slot in the screw head. A technician will need a tool set containing six to eight standard tip screwdrivers.

Phillips Screwdrivers

The tip of a **Phillips screwdriver** has two tapered crossing blades which fit the four slots in a Phillips screw. See **Figure 3-17**. This type of screw is used extensively on vehicles because of appearance and ease of installation. The four surfaces of the screw head make it less likely for the screwdriver to slip off of the fastener.

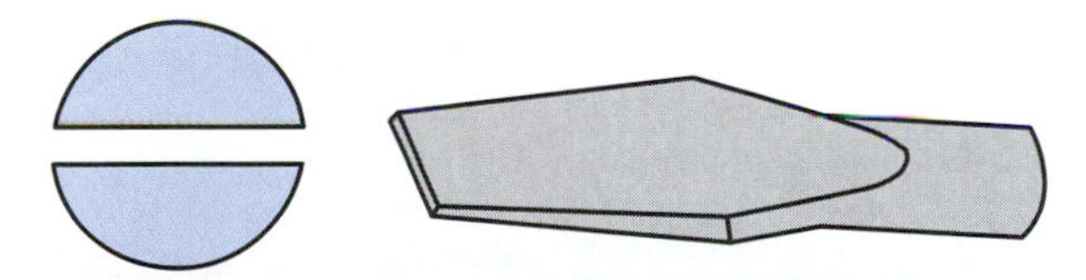

FIGURE 3-16 A standard tip screwdriver fits a screw with a slotted head.

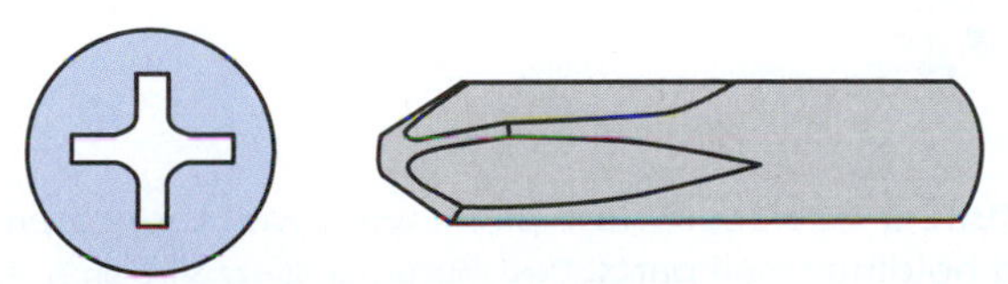

FIGURE 3-17 A Phillips screwdriver uses two tapered crossing blades.

The disadvantage of Phillips screwdrivers is that the tips tend to wear out more quickly than other screwdrivers. A collision repair technician will need a set of Phillips screwdrivers with a number 1, number 2, and number 3 tip.

Pozidrive Screwdriver

A **Pozidrive screwdriver** is similar to a Phillips screwdriver but provides a more positive connection between the driver and the screw. See **Figure 3-18**. The tip is flatter and blunter than a Phillips screwdriver.

Clutch Head Screwdriver

The **clutch head screwdriver** has four sides applying pressure which provides a more positive grip with less chance of slippage. See **Figure 3-19**.

Do not use a screwdriver as a prybar, chisel, or punch. Hammering on or prying with a screwdriver will damage its blade and tip.

FIGURE 3-18 A Pozidrive screwdriver provides a positive connection between the driver and screw.

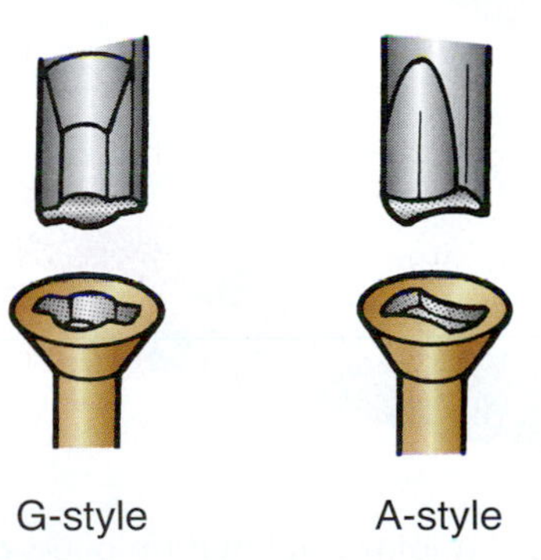

FIGURE 3-19 A clutch head screwdriver provides a more positive grip with less chance of slippage.

PLIERS

A **pliers** is designed for gripping, holding, and cutting objects. The collision repair technician will need a variety of pliers when working with wires, clips, pins, and sheet metal.

Combination pliers, or slip-joint pliers, are the most common design used by the collision repair technician. See **Figure 3-20**. The jaws are round or flat to securely grasp round or flat objects. The slip joint allows the jaws to be adjusted for two different opening sizes.

Adjustable pliers (Channel-lock®) have long handles for gripping leverage. See **Figure 3-21**. An interlocking channel allows for multiple jaw opening size adjustments. They are used for bending sheet metal, turning large nuts and bolts, and gripping round objects.

Needle nose pliers have long, thin, tapered jaws for reaching in and grasping small parts such as pins, clips, and wires. See **Figure 3-22**. They are also available with a 90-degree bend in the jaws. A technician should not twist a needle nose pliers, as it will spring the jaws out of alignment.

Locking pliers (Vise-Grip®) have a mechanism which firmly locks the jaws closed on an object. See **Figure 3-23**. They will remain in the locked position until the release lever is squeezed. These locking pliers are used for holding sheet metal pieces together during the riveting or

FIGURE 3-21 Adjustable pliers have a wide range of adjustment. Used with permission of Snap-on Tools Company, www.snapon.com. All rights reserved.

FIGURE 3-20 Slip-joint combination pliers can be adjusted for two different opening sizes. Used with permission of Snap-on Tools Company, www.snapon.com. All rights reserved.

FIGURE 3-22 Needle nose pliers work well for grasping and holding small parts. Used with permission of Snap-on Tools Company, www.snapon.com. All rights reserved.

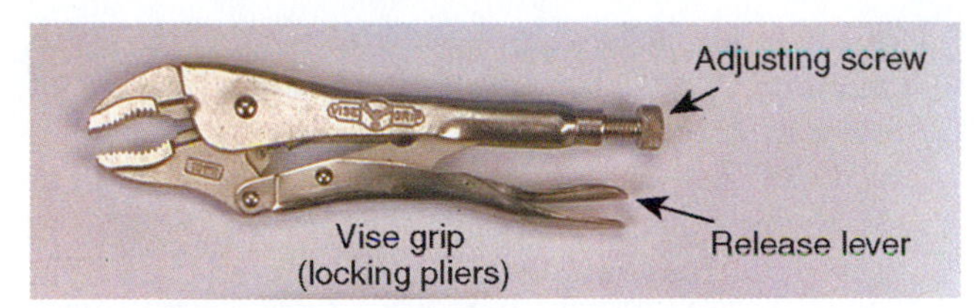

FIGURE 3-23 Locking pliers remain locked until the release lever is squeezed.

FIGURE 3-24 Diagonal cutting pliers have jaws designed to cut off wires.

welding process. They are also used for firmly gripping rounded fasteners which cannot be held by a wrench or socket.

Diagonal cutting pliers or side-cutter pliers have cutting jaws designed to cut off wires. See **Figure 3-24**.

CAUTION

Do not substitute a pliers for a wrench. They will damage nuts and bolts.

MISCELLANEOUS GENERAL PURPOSE TOOLS

The collision repair technician will need a large assortment of miscellaneous hand tools. The following discussion will include tools which, while useful, are not regularly used.

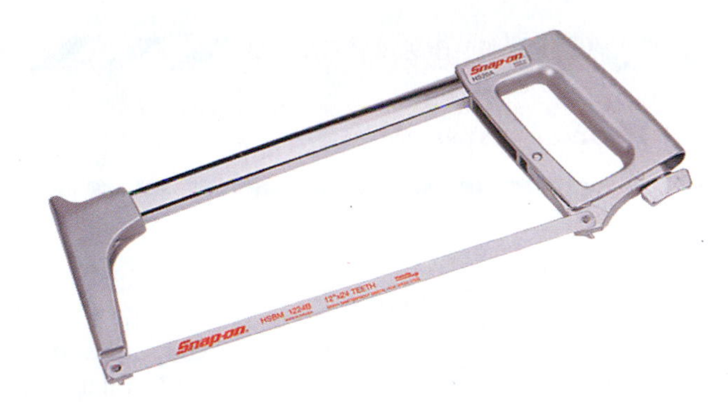

FIGURE 3-25 A hacksaw is used to cut small metal parts.

Hacksaw

A **hacksaw** is used in collision repair to cut small metal parts. See **Figure 3-25**. An adjustable frame accepts different blade lengths. When installing a hacksaw blade, the blade teeth should be pointed away from the handle. When choosing a blade, select one with the tooth configuration where at least two teeth contact the material at the same time.

Files

Files are generally used to remove burrs and sharp edges, and smooth out imperfections in metal. The three general purpose files commonly used in collision repair are flat, half round, and round. They are available in different lengths and four different cuts or filing surface configurations. **Figure 3-26**.

CAUTION

A technician should never use a file without a handle secured to the tang. One end of the file has a sharp tang which can easily puncture the technician's hand or wrist.

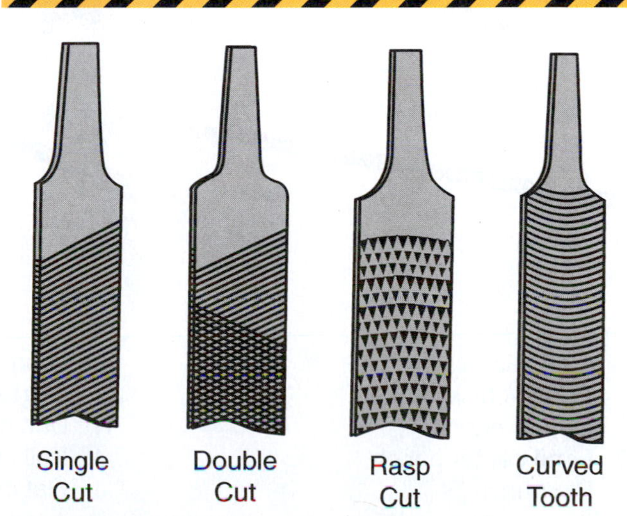

FIGURE 3-26 Files have various cutting services.

CAUTION

Never hammer on—or pry with—a file. It is very brittle and can shatter.

Chisels and Punches

Chisels are used for cutting sheet metal, separating welds, and cutting bolt heads. See **Figure 3-27**. A flat cold chisel is most commonly used in collision repair.

Punches are used for driving out pins and bolts, as well as aligning holes during component assembly. See Figure 3-27 and **Figure 3-28**.

A **center punch** has a pointed end which creates an indentation to prevent a drill bit from wandering out of position.

A **drift** or **starting punch** has a fully tapered shank to drive pins and bolts partially out of holes.

A **pin punch** has a straight shank which can push a pin or bolt completely out of a hole.

CAUTION

As the striking end of a punch or chisel becomes mushroomed, metal fragments may dislodge and fly into the technician's face. Keep the end properly shaped by grinding off the enlarged area.

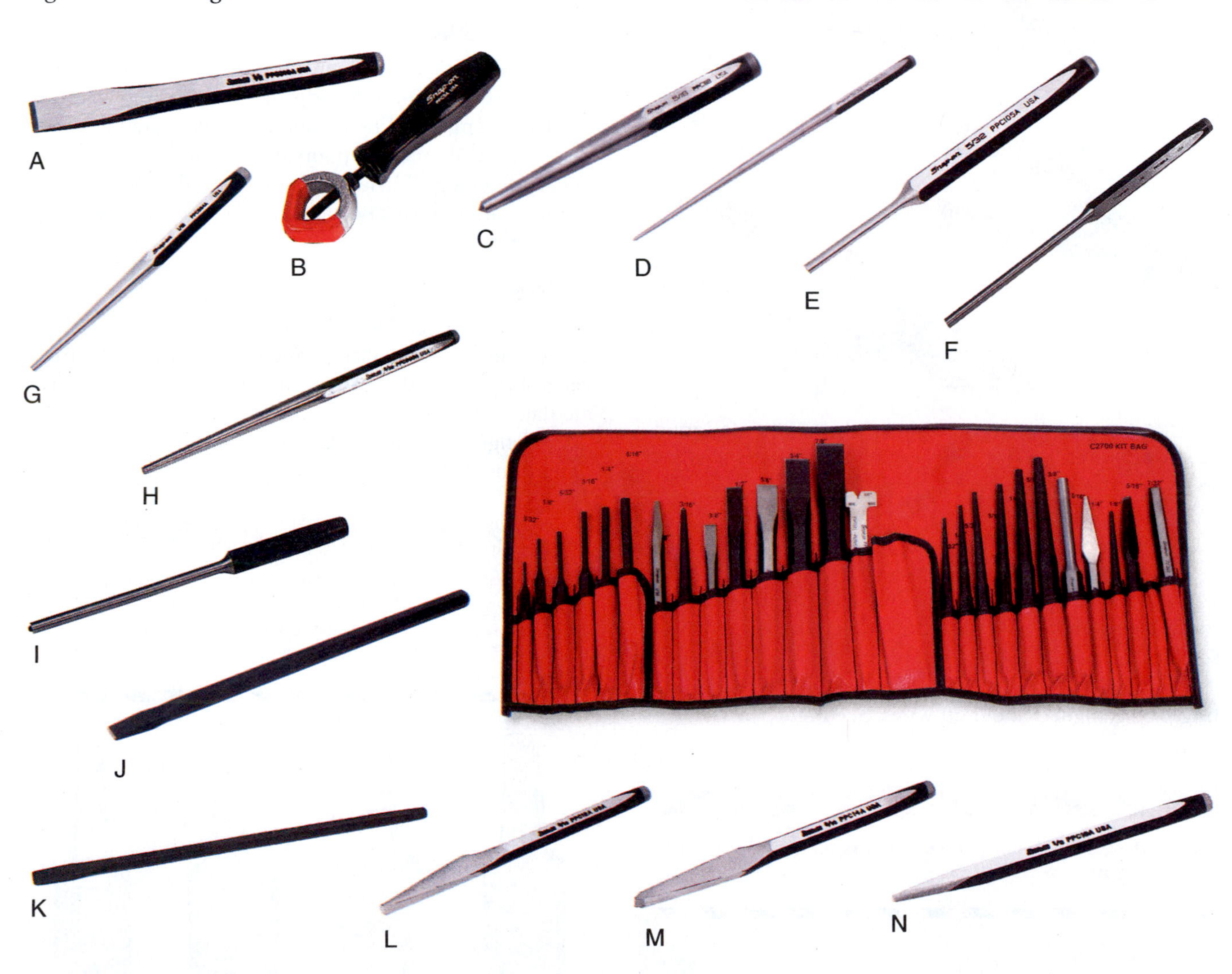

A. Flat chisel
B. Punch/chisel holder
C. Center punch
D. Long center punch
E. Pin punch
F. Long pin punch
G. Starter punch
H. Drift pin punch
I. Roll pin punch
J. Flat chisel
K. Long flat chisel
L. Half round nose cape chisel
M. Cape chisel
N. Diamond point chisel

FIGURE 3-27 Chisels and punches are available in various shapes and sizes. Used with permission of Snap-on Tools Company, www.snapon.com. All rights reserved.

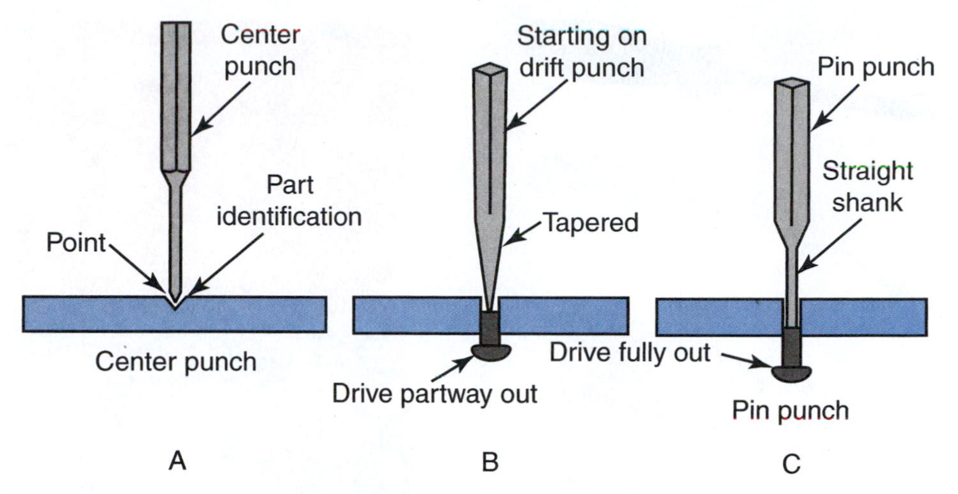

FIGURE 3-28 Many applications are available for different types of punches.

Metal Snips

Metal snips are commonly used to cut straight or curved shapes in sheet metal. See **Figure 3-29**. The jaws may be serrated to assist in cutting hard or thicker sheet metal.

Scrapers

A **scraper** is used to remove softened paint or adhesive. See **Figure 3-30**.

CAUTION

Technicians should never scrape toward themselves.

FIGURE 3-29 Different types of metal snips can be used to cut shapes in sheet metal. Used with permission of Snap-on Tools Company, www.snapon.com. All rights reserved.

Wire Brush

A **wire brush** is used to remove corrosion, dirt, and clean welds. A wire brush should be used sparingly on bare metal because it can leave scratches.

Tape Measure

A **tape measure** with a retractable blade is used to measure repair areas and check frame or unibody alignment. The tape measure should have both SAE and metric dimensions. See **Figure 3-31**.

Tap and Die Set

A **tap and die set** work in tandem. A tap cleans and cuts internal threads in holes. A die cuts external threads on bolts or cleans damaged threads on bolts. See **Figure 3-32**.

Utility Knife

A **utility knife** has a retractable blade and is used for general cutting and trimming.

HAND TOOL SAFETY

Collision repair shop safety is extremely important for the prevention of serious injury. The technician must wear safety glasses at all times in the shop. Many hand tool operations require wearing leather gloves or mechanic's gloves. Some operations will require hearing protection.

Good quality tools will not only be more durable, but also will lower the risk of injury.

The technician must know how to select the correct tool for the job and understand how to properly use the tool. Manufacturers provide specific instructions on how to use tools and what specific operation the tool is designed to perform.

Properly cleaned and maintained tools will lessen the chance of injury. All tools should be maintained as close to their original condition as possible. If a tool is damaged or broken, repair or replace it.

FIGURE 3-30 Various types of scrapers and putty knives are used to remove paint and adhesive.

FIGURE 3-31 Tape measures are used to check frame and unibody alignment.

FIGURE 3-32 A tap and die set cleans and cuts threads.

METAL WORKING TOOLS

Collision repair metal working tools can be classified as general purpose and specialized. They apply force by one or more of three ways: (1) striking a direct blow on the metal surface, (2) applying resistance to a direct blow struck on the opposite side of the metal, and (3) as a lever used to pry against the surface, usually on the underside. Technicians need to understand these tools and practice their use to develop the manual skill needed to become highly skilled. The following description will include the most common metal working tools used by the collision repair technician.

HAMMERS

Various types of hammers are designed for specific repair operations.

Ball Peen Hammer

The **ball peen hammer** is a general purpose tool. See Figure 3-33. It may be used to rough out sheet metal or drive punches and chisels. Ball peen hammers are available in several different weights.

Mallets

Mallets are available in a variety of types intended for specific uses. A **rubber mallet** can gently bump sheet

FIGURE 3-33 Ball peen hammers are available in several different weights. Used with permission of Snap-on Tools Company, www.snapon.com.

metal or be used to position trim and delicate parts without damaging the finish. Another type of mallet which is used in collision repair is a steel hammer with a rubber or plastic tip, which can be used for similar operations.

Dead Blow Hammer

A **dead blow hammer** is specifically designed to resist bouncing back after striking. See **Figure 3-34**. It is used in driving operations where marring of parts must be avoided.

Sledge Hammer

A **sledge hammer** is used for heavy driving and reforming thicker gauge sheet metal parts such as structural components. See **Figure 3-35**.

Body Hammer

There are many types of **body hammers**. See **Figure 3-36**. Collision repair technicians use body hammers to work sheet metal back into shape. Each hammer is designed for a specific repair operation. To avoid stretching and marking soft sheet metal, a hammer with a large, nearly flat face should be used. The face of the body hammer should have a dead flat center with a slight crown on the outer edge to compensate for errors when striking a blow. A small face will usually have a higher crown than a large face. Much more detail in the use of body hammers will be discussed in the following chapters.

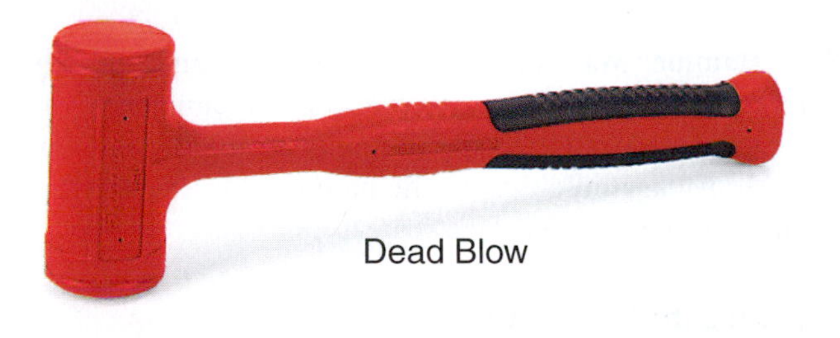

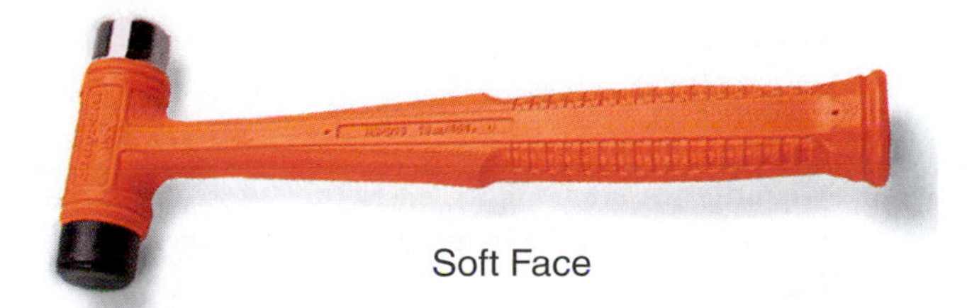

FIGURE 3-34 Types of dead blow hammers used where marring of parts must be avoided. Used with permission of Snap-on Tools Company, www.snapon.com.

FIGURE 3-35 Sledge hammers are used for heavy driving. Used with permission of Snap-on Tools Company, www.snapon.com.

Bumping Hammer. **Bumping hammers** are commonly used to "bump" out large dents during the initial straightening process. These hammers may have round or square faces and flat or crowned striking surfaces. Their faces are large so the striking force is spread over a larger area.

TECH TIP To avoid stretching the metal, the contour of the hammer face must be smaller than the contour of the panel.

Pick Hammer. A **pick hammer** is used to remove small dents. See Figure 3-36. It has a pointed tip on one end and a round smooth face on the other end.

The smooth face is used to bring down high spots on a damaged panel. The pointed (pick) end is used to gently raise the low spots from the underside of the damaged panel. Due to the variety of sizes and shapes, the technician will need to select the one best suited to the repair.

Pick Hammer Tip Maintenance. A pick hammer with a tip that is too sharp will pierce and severely stretch sheet metal. The tip can be filed or ground to a rounded or somewhat blunt tip. Excessive or repeated filing or grinding of the tip will shorten the length of the tip.

Finishing Hammer. The **finishing hammer** is used to achieve the final panel shape and contour. The finishing hammer is lighter with a smaller, more crowned face than the bumping hammer.

FIGURE 3-36 Body hammers work sheet metal back into shape. Used with permission of Snap-on Tools Company, www.snapon.com.

Shrinking Hammer. A **shrinking hammer** is similar to a finishing hammer but it has a serrated or cross-grooved face. See Figure 3-36. This hammer is used to help shrink stretched sheet metal.

TECH TIP The small serrations on the hammer's face cause the metal to gather as it is struck, creating a shrinking effect. Correct hammer blow force is critical so additional stretching does not occur.

TECH TIP To select the correct body hammer for the straightening process, consider the panel shape, type of damage, and accessibility to the damaged area.

Body Hammer Maintenance. Proper maintenance of body hammers will extend their life and prevent damage to the work surface. The face of the hammer should be flat and smooth. Nicks and chips on the hammer face will be imprinted on the work surface struck.

Repairing the Face of a Body Hammer.

1. Clean-up the hammer with the face up in a vise.
2. Use a hand file to smooth out the nicks.
3. File in all directions to keep the surface even. See **Figure 3-37**.
4. Chamfer the edge of the face to prevent sheet metal distortion when the hammer blow is delivered. See **Figure 3-38**.
5. Smooth and polish with fine sandpaper or oil stone. See **Figure 3-39**.
6. Check finish and shape by applying a marking paint to the face of the hammer.

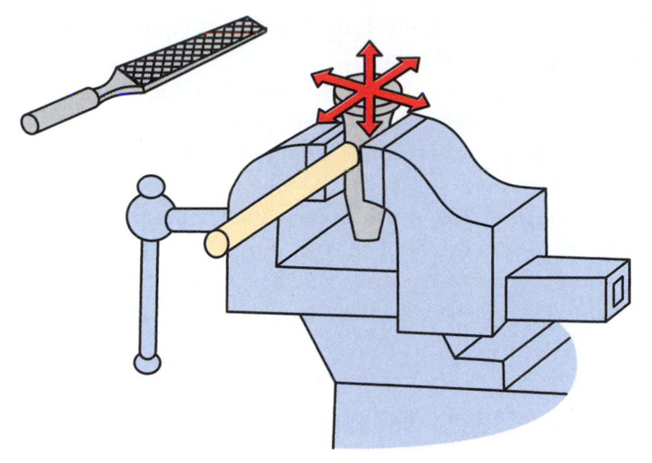

FIGURE 3-37 Repairing the face of a body hammer.

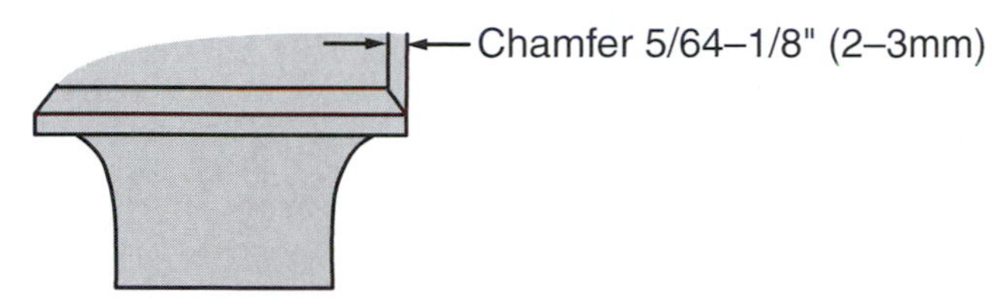

FIGURE 3-38 Chamfer the edge of the body hammer face.

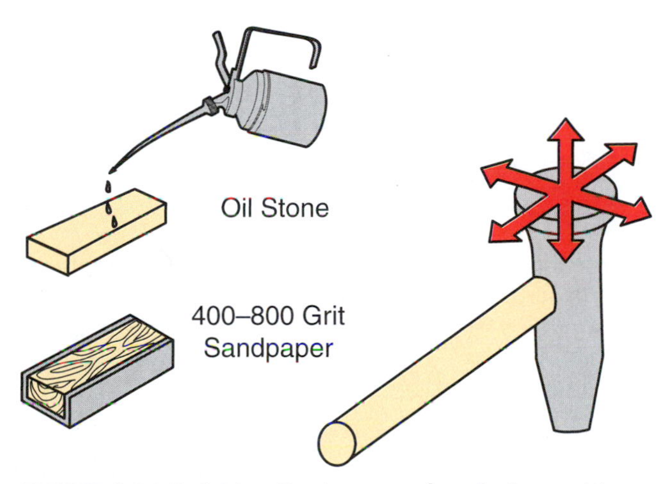

FIGURE 3-39 Polishing the hammer face is done with sandpaper or oil stone.

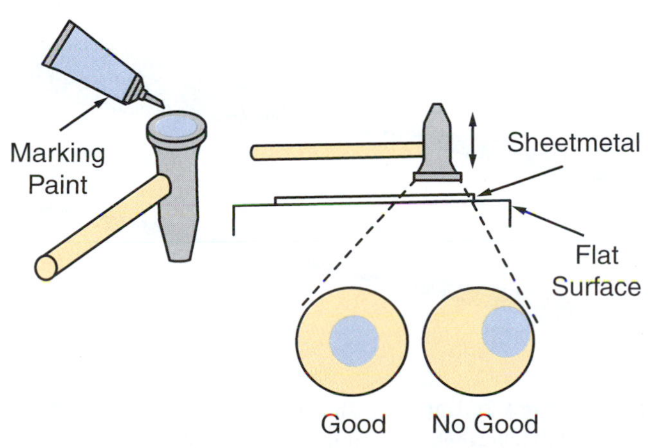

FIGURE 3-40 Checking the face of the hammer.

7. Strike a flat sheet metal surface.
8. The paint should come off the center if the face is properly shaped. See **Figure 3-40**.

Hammer Handle Maintenance

When replacing a hammer handle, insert the new handle into the smaller hole in the hammer head. It may be necessary to shave away excess wood to properly fit the handle. Lightly tap the end of the hammer handle with another hammer until it is firmly in place. Cut off any excess handle protruding through the hammer head. Insert a wood wedge into the split of the handle end. Cut off excess wood from the wedge. Then drive a steel wedge crossways to the direction of the wood wedge to complete the installation.

DOLLIES

Dollies, often called dolly blocks, are available in a variety of shapes and sizes. See **Figure 3-41**. Each dolly has curves and angles that can be used for specific dents and panel contours (high crowns, low crowns, flanges). They can be used as a striking tool or backing tool for body hammers. Other types of dolly blocks, which are not discussed here, are equally suited to the metal-straightening process. A good quality dolly block will be well-balanced so the effect of its weight is not lost when the hammer blow is struck.

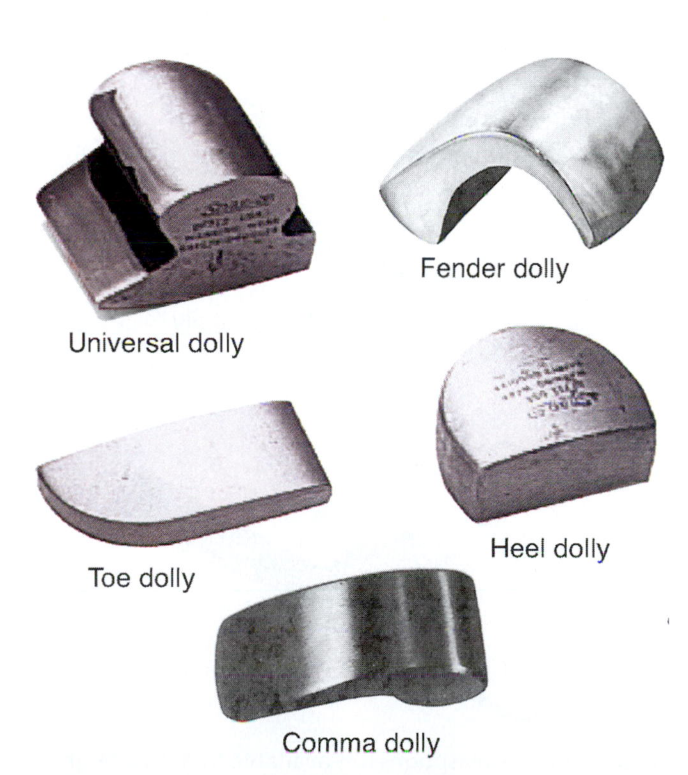

FIGURE 3-41 Dolly blocks are available in a variety of shapes and sizes.

Specific techniques for using dollies will be discussed in Chapter 7.

General purpose dollies contain many crown shapes and can be used for most general metal straightening.

Heel dollies are used in sharp corners and tight places.

Toe dollies are used for dinging flat surfaces with low crowns.

A **comma** or **wedge dolly** has a thin body to help force dents from behind narrow, curved areas.

DOLLY BLOCK MAINTENANCE

Proper maintenance of dolly blocks will prevent imprinting surface imperfections onto the sheet metal surface. To repair a damaged dolly block surface:

1. Clamp the dolly block securely in a vise with the face up.
2. Use a disc grinder with a fine disc, if desired, to shape and level the surface. Only remove enough metal to return the dolly to a usable condition. Caution: Good quality dolly blocks are hardened steel. Overheating them while grinding may soften the metal.
3. Smooth and polish the block with fine sandpaper.

SPOONS

Body spoons are designed to reach into restricted areas that conventional hammers and dollies cannot access. See **Figure 3-42**. Spoons are available in a variety of sizes and shapes, which are designed to match various panel shapes. They can be used as a driving or prying tool to remove dents. The large, flat faces are designed to spread out hammer blow force over a large area.

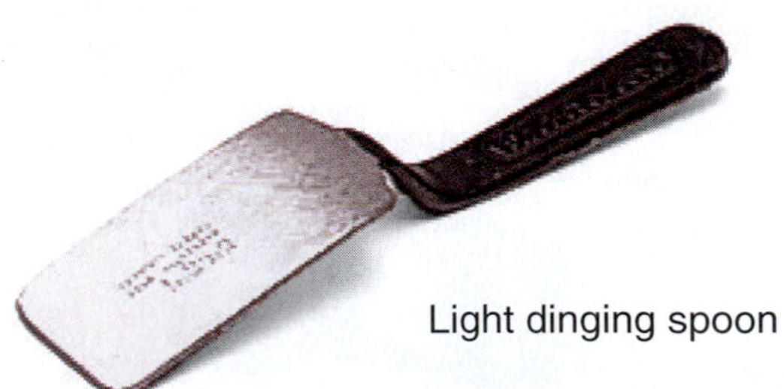

FIGURE 3-42 Body spoons are available in a variety of shapes and sizes.

Spoon dollies have long handles designed to reach into deep, recessed areas.

Dinging spoons are used to smooth and level ridges by striking a hammer on top of the spoon.

Surfacing spoons are used for spring hammering, prying, or slapping.

A **bumping file** has a spoon-like shape and serrated surfaces to slap metal back to its original shape.

Rough service spoons have an offset shape which is designed to be struck with a hammer to rough out damaged sheet metal.

CAULKING IRON

Caulking irons are special-purpose tools used to reshape bends; narrow, flat surfaces; and bead sections on fenders. See **Figure 3-43**. They are designed to be struck with a hammer.

PRYBARS

Prybars are used to reach into restricted areas. See **Figure 3-44**. They are available in different lengths and shapes. Prybars can raise low spots in enclosed components, such as doors, quarter panels, and hoods. No holes should be made to access the damaged area.

FIGURE 3-43 Caulking irons are special-purpose tools used for reshaping metal.

FIGURE 3-44 Prybars are used to reach into restricted areas.

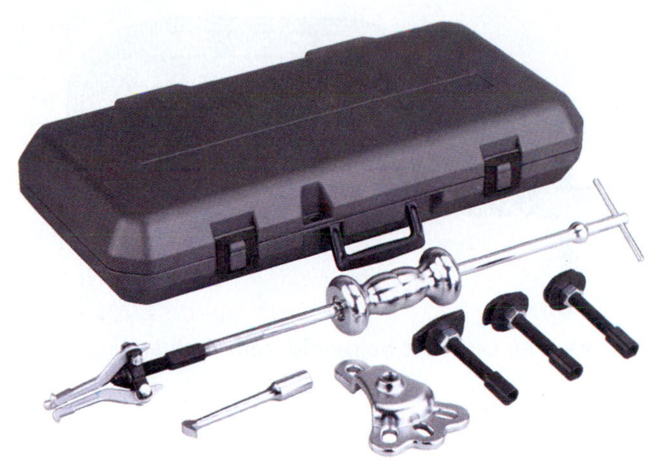

FIGURE 3-45 A dent puller can pull out dents in enclosed body panels. Courtesy of SPX/OTC Service Solutions

DENT PULLERS AND PULL RODS

A **dent puller** can pull out dents in enclosed body panels which cannot be accessed from the back side. See Figure 3-45.

Holes are punched—or drilled—in the dent's low areas. Either a hook or threaded tip is inserted into the holes. See Figure 3-46. Repeated blows of the slide hammer will pull the dent out. After the damage is removed the holes must be closed by welding.

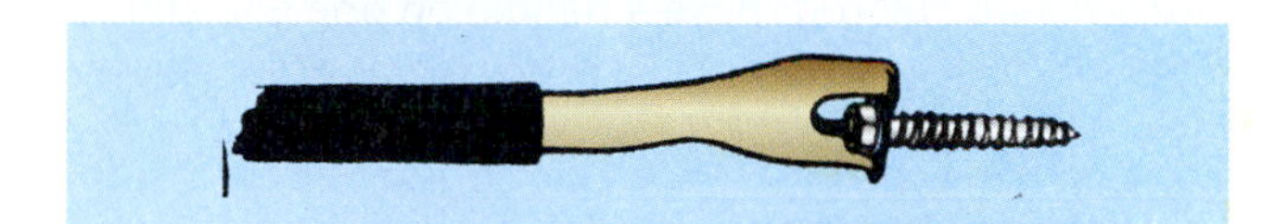

FIGURE 3-46 There are many dent puller applications.

Holes are not necessary if specifically designed pulling attachments are welded or glued to the damaged area. See Figure 3-47.

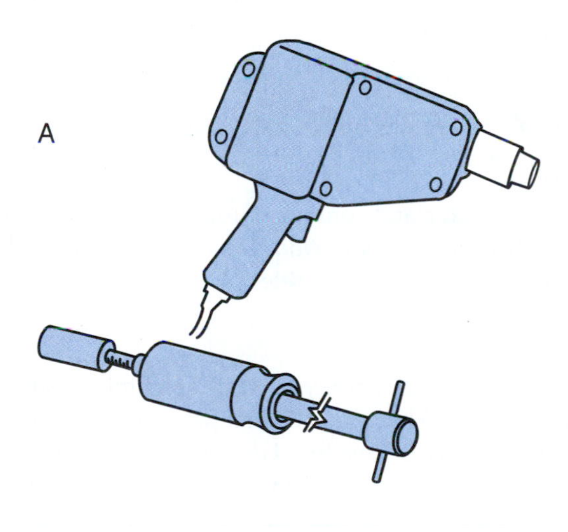

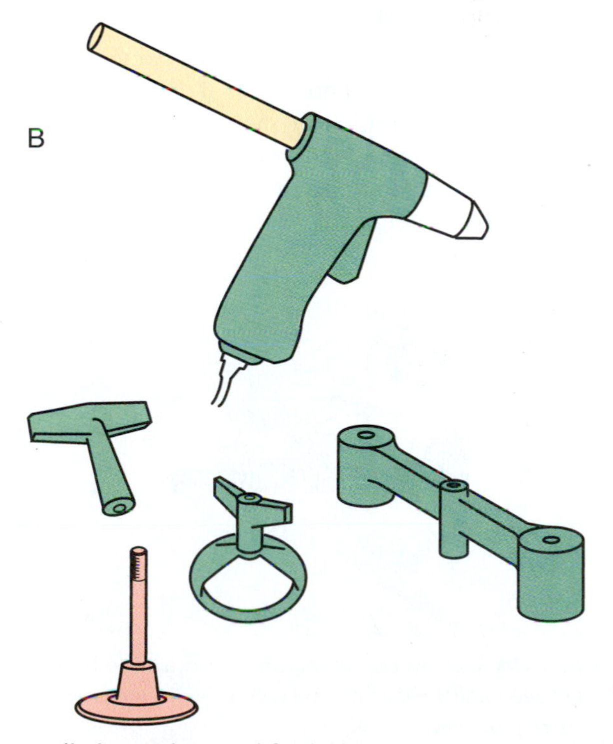

FIGURE 3-47 Weld-on (A) and glue-on (B) dent removal tools may eliminate the need for holes.

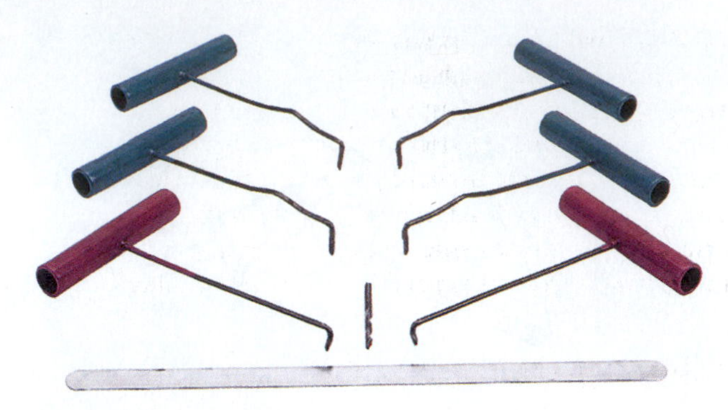

FIGURE 3-48 Pull rods have a handle on one end and a hook on the other.

Pull rods are used in a similar manner as dent pullers. See **Figure 3-48**. There is a handle on one end of the rod and a hook on the other end. Holes are drilled in the dented area. The pull rod hook is inserted into the hole and pulling force is applied while lightly tapping around the area with a hammer.

TECH TIP Dent pullers requiring holes are not frequently used. Panel distortion and warpage can occur when welding the holes closed. Proper corrosion protection is also required. Weld-on pins or studs are extensively used to avoid these problems.

SUCTION CUPS

Shallow dents may be able to be removed with a **suction cup**. This process will only work if there is no crease locking the dent in the sheet metal. The suction cup should be attached to the center of the dent and pulling force is applied. The dent may be removed with no damage to the paint. See **Figure 3-49**.

FIGURE 3-49 Suction cup attachment on a slide hammer can remove shallow dents.

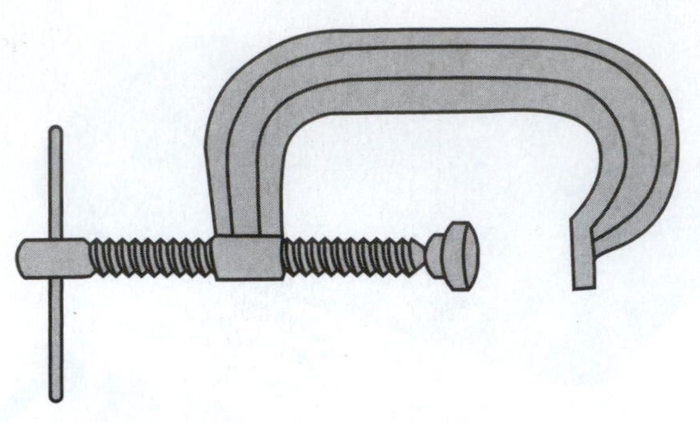

FIGURE 3-50 C-clamps help hold items in place.

C-CLAMPS

It is often necessary to hold parts in place while doing repairs. A variety of sizes of **C-clamps** can help accomplish this task. See **Figure 3-50**.

SPRING-LOADED RIVETS

Cleco clamps® can be installed to hold sheet metal parts together while welding. They are easily removed after the welding process. See **Figure 3-51**.

BLIND RIVET GUN

Blind rivets, commonly called "pop rivets," are also used to hold panels in place during the welding process. After welding, they are removed and the holes are closed by welding. A heavy-duty riveter (which uses 3/16- to 1/4-inch blind rivets) attaches mechanical assemblies such as window glass regulators and outside door handle assemblies. See **Figure 3-52**.

BODY SURFACE SHAPING TOOLS

The final shape and contour of repaired metal requires specifically designed tools. Additional tools are needed to shape and sand plastic body filler.

Body File

A **body file** is commonly used to level minor surface imperfections on sheet metal. This process is sometimes referred to as metal finishing. High and low spots in the metal surface can be identified with the body file. See **Figure 3-53**. A holder or handle is attached to the blade. File blades are available in flat or curved shapes.

Forming Tool or Surform File

The **surform file** is used to rough shape plastic body filler when the filler becomes semi-hard. This procedure reduces the amount of sanding needed to shape and level

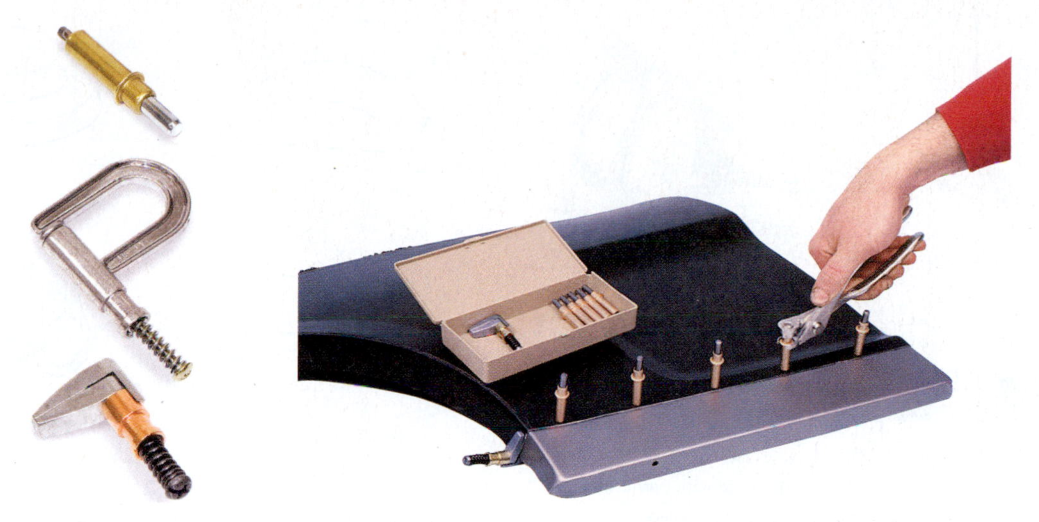

FIGURE 3-51 Spring loaded clamps (Cleco clamps) hold sheet metal together while welding. Used with permission of the Eastwood Company.

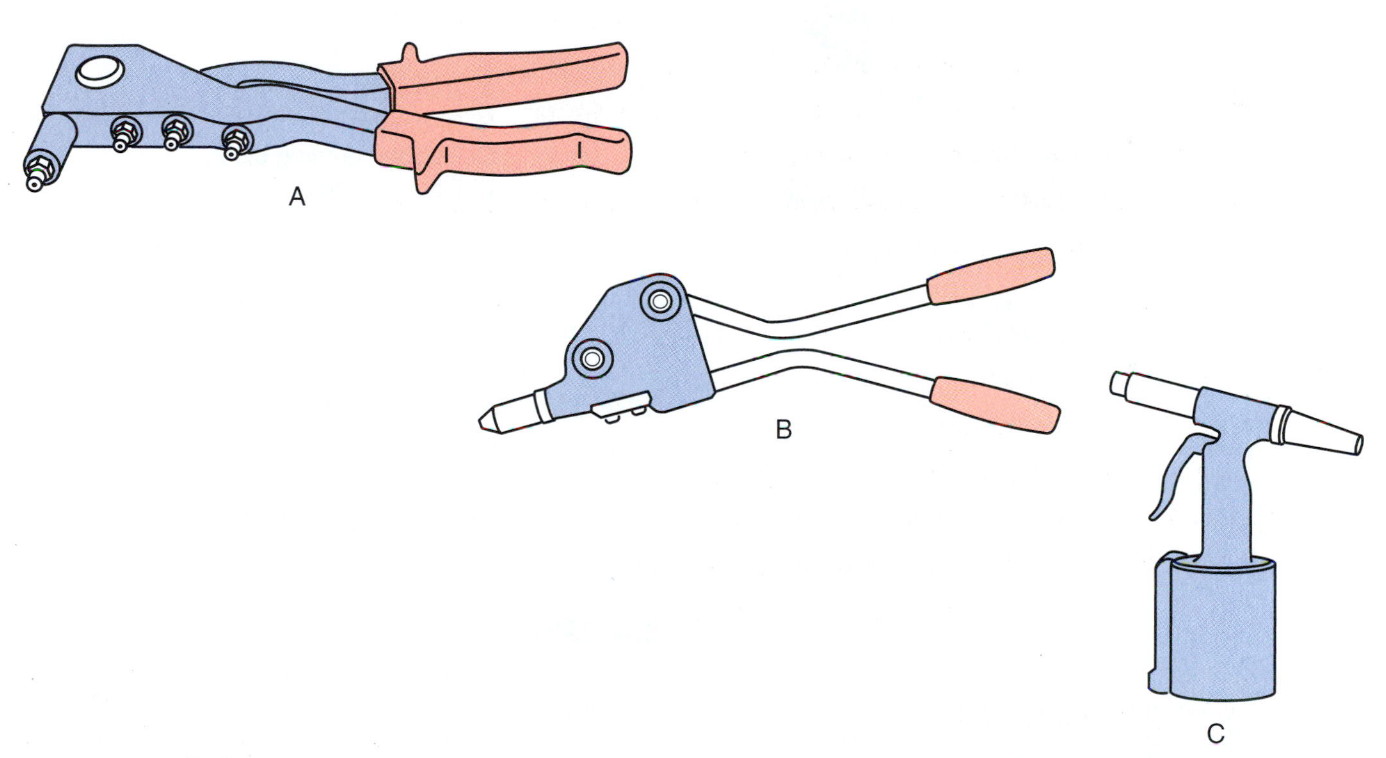

FIGURE 3-52 Blind rivet installation tools used to help hold panels in place during welding.

the body filler. See **Figure 3-54**. The surform file blade can be cleaned using a hand wire brush scrubbed in the direction of the cutting edges.

Board File

A **board file** is designed to level large flat or contoured areas of cured body filler. See **Figure 3-55**. The holder is capable of accepting different grits of sandpaper.

Sanding Block

Several sizes and types of sanding blocks are used in collision repair and refinishing. A hard rubber sanding block has a rounded top and a flat surface on the bottom. It can be used in restricted areas where a board file is too long. See **Figure 3-56**.

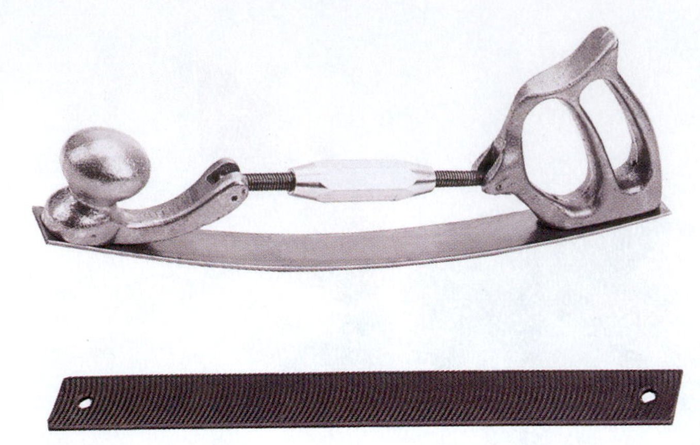

FIGURE 3-53 Body files are often used to smooth out minor imperfections on sheet metal. Used with permission of Snap-on Tools Company, www.snapon.com. All rights reserved.

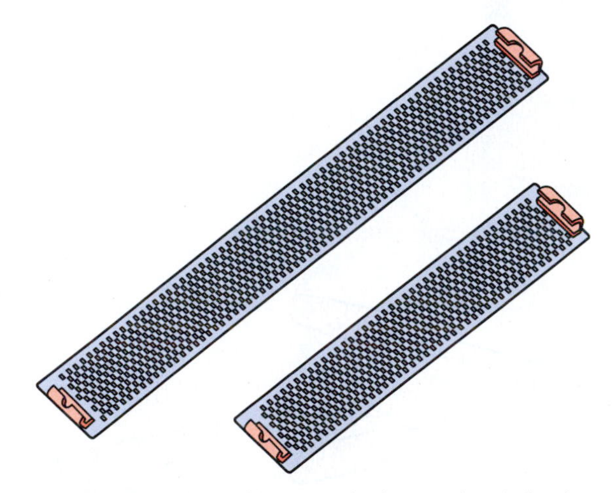

FIGURE 3-54 Surform files rough shape plastic body filler when the filler becomes semi-hard.

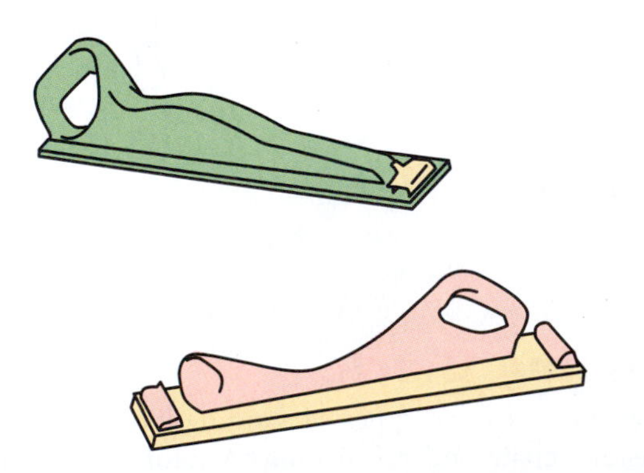

FIGURE 3-55 Sanding boards are used to level large flat or contoured areas.

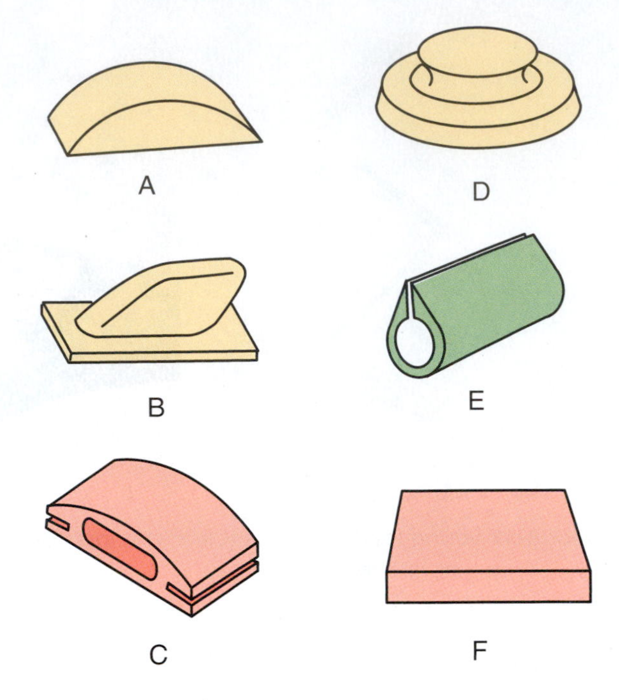

A. Half-circle
B. Rectangular sanding block
C. Rubber sanding block
D. Round hand sanding block
E. Tear drop sanding block
F. Double density sanding block

FIGURE 3-56 Sanding blocks can be used in areas where a board file is too long.

WINDOW TOOLS

A variety of special windshield and window servicing tools are used in collision repair. These tools are required to properly remove and install automotive glass.

Typical glass servicing tools are shown in **Figure 3-57**.

DOOR TOOLS

Many specialized tools are needed to service doors, specifically to remove door hardware, trim panels, and moldings. A variety of specialized mechanical clips are used to secure trim panels, door handles, and lock cylinders. Specifically designed tools are necessary for the removal of these components to prevent damage.

Typical door handle/trim pad upholstery tools are shown in **Figure 3-58**.

ANTENNA, MIRROR, AND RADIO TRIM TOOLS

This assortment of tools is used for removing antenna nuts, remote mirror control nuts, and radio retaining clips and nuts. See **Figure 3-59**.

A. Molding release tool
B. Razor blade scraper
C. Windshield knife
D. Windshield removal tool
E. Windshield locking strip installation tool
F. Windshield wiper arm removal tool
G. Window sash nut spanner socket

FIGURE 3-57 A variety of specialty door, window, and trim removal tools are available.

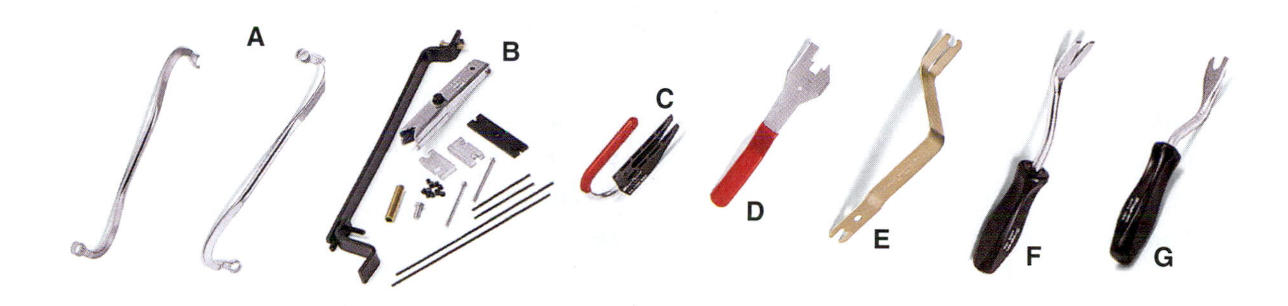

A. Door hinge bolt wrenches
B. Door removal kit
C. Door panel remover (GM and Ford)
D. Door handle tool (GM, some Fords)
E. Door handle tool (Chrysler)
F. Trim pad remover (GM, Ford, Chrysler)
G. Trim pad remover (GM)

FIGURE 3-58 A variety of specialty door, trim, and clip removal tools are available.

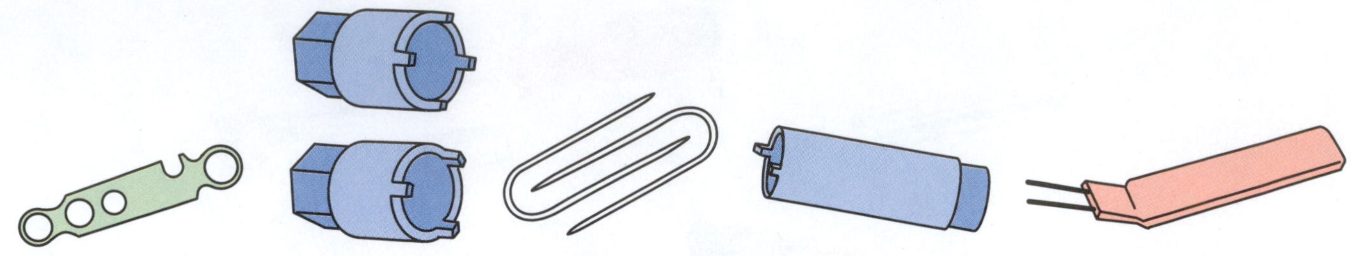

FIGURE 3-59 Specialty antenna and radio removal tools come in various styles.

Summary

- Technicians must practice all safety precautions at all times.
- Technicians should purchase good quality tools for durability and safety.
- Collision repair technicians need a complete set of wrenches (open-end, box-end, flare nut).
- Socket wrenches remove fasteners more rapidly than open-end or box-end wrenches. Select the correct socket drive size and point configuration for the job.
- The ratchet handle is a common tool for sockets when loosening or tightening automotive fasteners.
- Various types of screwdrivers are needed to remove and install many of the threaded fasteners.
- Pliers are used for gripping, holding, and cutting. A variety of pliers is necessary when working with clips, pins, and wires.
- A variety of miscellaneous tools are used to cut, separate, and drive out bolts and pins.
- Body hammers are basic sheet metal straightening tools. A variety of hammers are necessary to perform collision repair. Each hammer is designed for a specific use.
- A dolly block can be used to bump down high spots or be used as a back-up tool for hammers. Dollies have different shapes and contours to accommodate various sheet metal panel shapes.
- Body files and surface tools are used to level and shape sheet metal contours.
- A variety of specialty/miscellaneous tools are needed to perform related operations necessary to complete collision repairs.

Key Terms

adapter
adjustable pliers
adjustable wrench
Allen wrench
ball peen hammer
blind rivets
board file
body file
body hammer
body spoon
box-end wrench
breaker bar
bumping file
bumping hammer
caulking irons
C-clamps
center punch
chisels
Cleco clamps®
clutch head screwdriver
combination pliers
combination wrenches
comma/wedge dolly
dead blow hammer

deep well socket
dent puller
diagonal cutting pliers
dinging spoon
dollies
drift/starting punch
drive
extensions
files
finishing hammer
flare nut wrench
general purpose dolly
hacksaw
heel dolly
impact socket
locking pliers
mallets
metal snips
needle nose pliers
open-end wrench
Phillips screwdriver
pick hammer
pin punch
pipe wrench
pliers
Pozidrive screwdriver
prybar
pull rods
punches
ratchet handle
rough service spoons
rubber mallets
scraper
shrinking hammer
sledge hammer
sliding T-handle
Society of Automotive Engineers (SAE)
socket
socket wrench
socket wrench accessories
socket wrench set
speed handle
spoon dolly
standard tip screwdriver
suction cup
surfacing spoon
surform file
swivel socket
tap and die set
tape measure
toe dollies
torque wrench
TORX wrench/driver
utility knife
wire brush

Review

ASE Review Questions

1. Technician A says an open-end wrench provides the best grip on a hex head fastener. Technician B says a box-end wrench provides the best grip on a hex head fastener. Who is correct?
   A. Technician A only
   B. Technician B only
   C. Both Technicians A and B
   D. Neither Technician A nor B
2. Technician A says a ball peen hammer should be used to drive a chisel. Technician B says a bumping hammer should be used to shape the final contour of a damaged sheet metal panel. Who is correct?
   A. Technician A only
   B. Technician B only
   C. Both Technicians A and B
   D. Neither Technician A nor B
3. Technician A says a flare nut wrench should be used to loosen fuel and brake lines. Technician B says a box-end wrench will provide the best grip for removing fuel and brake lines. Who is correct?
   A. Technician A only
   B. Technician B only
   C. Both Technicians A and B
   D. Neither Technician A nor B

4. Technician A says a Pozidrive screwdriver should be used to remove a clutch head screw. Technician B says a Phillips screwdriver has more grip and less slippage than a Pozidrive screwdriver. Who is correct?
   A. Technician A only
   B. Technician B only
   C. Both Technicians A and B
   D. Neither Technician A nor B
5. Technician A says a TORX driver will have greater turning power and less slippage than a standard screwdriver. Technician B says an adjustable pliers is used to remove clips and wires. Who is correct?
   A. Technician A only
   B. Technician B only
   C. Both Technicians A and B
   D. Neither Technician A nor B
6. What is the best hammer to use for initial straightening on dented panels?
   A. Pick hammer
   B. Finishing hammer
   C. Bumping hammer
   D. Dead blow hammer
7. Technician A says a spoon is designed to distribute the striking force over a wide area. Technician B says holes can be drilled to provide access to the damage when using a prybar. Who is correct?
   A. Technician A only
   B. Technician B only
   C. Both Technicians A and B
   D. Neither Technician A nor B
8. Technician A says a standard screwdriver can be used as a chisel. Technician B says a chisel is used to shear off rusted bolt heads. Who is correct?
   A. Technician A only
   B. Technician B only
   C. Both Technicians A and B
   D. Neither Technician A nor B
9. Which of the following should be used to cut external threads?
   A. Tap
   B. Die
   C. Allen wrench
   D. None of the above
10. Technician A says a locking pliers will have more holding power than a box-end wrench when loosening an undamaged nut. Technician B says an open-end wrench may slip if too much force is applied. Who is correct?
    A. Technician A only
    B. Technician B only
    C. Both Technicians A and B
    D. Neither Technician A nor B
11. Technician A says a rubber mallet is an essential tool for forming the final contour on sheet metal. Technician B says a body file is used to remove a crease in sheet metal. Who is correct?
    A. Technician A only
    B. Technician B only
    C. Both Technicians A and B
    D. Neither Technician A nor B

Essay Questions

1. List the different types of body hammers used in collision repair. Then explain how you would use each body hammer for repairing sheet metal.
2. Describe the difference between a box-end wrench and an open-end wrench. Discuss the advantages of each type of wrench and how it is used in collision repair.
3. Discuss the difference between a 6-point and 12-point socket. Explain where each type would be used.
4. What is the difference between a Phillips screwdriver and a Pozidrive screwdriver?
5. Explain the difference between a hex socket and a TORX socket.

Critical Thinking Questions

1. A hex head bolt is recessed. What tool would work the best for loosening and removing this fastener?
2. A dent in a sheet metal panel is going to be removed. There is no access to the back side. How would you remove this damage?

Lab Activities

1. Tour your shop and identify the various hand tools used in collision repair. Be prepared to identify each tool and explain how it is used.
2. Obtain tool catalogs from several good quality tool suppliers. Make a list (include names and prices) which contains a complete set of tools a collision repair technician would need.
3. Do a safety inspection of all hand tools in your shop. Check for damaged or defective tools. List the tools which need repair and what repairs need to be done on each tool. Submit this to your instructor.
4. Properly repair the damaged face on a body hammer.

Chapter 4

Power Tools

OBJECTIVES

Upon completing this chapter, you should be able to:

- Name and identify pneumatic tools used in collision repair and their use.
- Name and identify electric powered tools used in collision repair and their use.
- Name and identify battery powered tools used in collision repair and their use.
- Name and identify hydraulic powered tools and equipment used in collision repair and their use.
- Cite the safe practices that must be employed whenever using power tools.
- Name and identify welders used in collision repair and their use.
- Name and identify shop equipment used in collision repair and its use.

INTRODUCTION

America moves on wheels. Our nation's economy is dependent upon the automobile and all of the other entities that comprise the transportation industry. When a vehicle is involved in an accident that takes it out of service, every effort must be made to get it back on the road as quickly as possible. This requires skilled technicians using numerous hand and power tools and a variety of equipment to restore the vehicle back to its pre-collision condition.

The purpose of this chapter is to acquaint the student with the many hand and power tools commonly used in the collision repair industry. Tool and equipment applications, recommended usage, and safety practices will be emphasized throughout. The student will be able to identify each of the tools discussed, cite its proper use, and know the applications and the recommended safety practices for each when using them.

PERSONAL SAFETY

The essence of any safety program is the awareness of the potential dangers that exist when using equipment for performing even the most routine operations. Any time a tool or piece of equipment is used, the possibility of an accident or injury exists. Many people, even though they are aware of the potential dangers, choose to ignore the possibility of injury by taking an "it won't happen to me" attitude. Every year over 200,000 work-related accidents and injuries occur unnecessarily. Many could be avoided if workers had simply used the minimum **personal protective equipment (PPE)** available.

TOOL AND EQUIPMENT SAFETY

The minimum PPE requirement in any shop or lab environment, even for someone not working there, is approved safety glasses. Safety glasses must be stamped with the numbers Z87.1 in the temple or some other part of the glasses in order to be considered approved. Other PPE that should be used include a face shield when performing grinding operations, respiratory protective equipment when performing grinding and sanding operations, gloves appropriate for the operation being performed, and a welding helmet for any welding operations. In addition, hearing protection is recommended in many situations. All protective guards and shields should be in place and function properly on any machine used by an employee. Wearing improper clothing and jewelry has also contributed to numerous injuries. Shop employees must use discretion in selecting the proper attire and PPE to help prevent the obvious accidents.

HAND AND POWER TOOL CATEGORIES

All hand and **power tools** can be divided into the following four categories based on their energy source: electric, battery, air or pneumatic, and hydraulic.

Pneumatic and Electric Tools

The majority of the cosmetic repairs are effected with the use of pneumatic and electric tools such as grinders, sanders, and cutting and shaving tools, to mention just a few. Numerous other battery or electrically energized tools such as inline screwdrivers, drills, buffers, scanners, and so on, are also used throughout the repair process.

Battery Powered Tools

With the advent of the nickel cadmium (NiCad) battery, the popularity of battery operated tools has increased considerably. The added power and extended longevity of the battery life have made these tools considerably more attractive for daily use as recharging is not required as often. The added mobility and portability of the tools often makes them a primary choice for technicians, sometimes even over a pneumatic tool. Their built-in power supply gives them an added portability and versatility and averts the restrictions of too short a cord or air hose. See **Figure 4-1**. However, battery operated tools also have their shortcomings; for example, they still require recharging, and for heavier operations the battery drain occurs quite rapidly. Their use in wet conditions is also somewhat limited.

Hydraulic Equipment

Most structural repairs are performed with the aid of hydraulic equipment such as porta-power jacks, as well as other hydraulic equipment commonly used on the frame and undercarriage pulling equipment.

FIGURE 4-1 Battery operated power tools are available with a variety of power supplies. They are very popular because they offer a great deal of portability and do not require a cord or air hose to be attached to them.

AIR TOOL DESIGN, SELECTION, AND MAINTENANCE

Technicians have many decisions to make when selecting and purchasing tools. Each tool manufacturer produces their own version of many of the tools used by collision repair technicians; each offers its own unique features, advantages, and attributes. Technicians must decide why the tool is needed and which manufacturer's product will best fulfill their personal needs. For example, must all of a technician's tools be top of the line quality? Sometimes it is better to purchase a less expensive tool if it will rarely be used or only used for short periods of time. Although a more economically priced tool may be less user friendly, one can more easily justify buying the less expensive model if it is rarely used. Ultimately, the well-advised buyer will purchase everything possible with the dollars available to spend.

Quality tool suppliers have become increasingly more concerned about what the user hopes to accomplish with the tool as opposed to the manufacturer's perceived or intended use. The power-to-weight ratio is a major concern to the manufacturers in their quest to minimize the tool user's fatigue. Reducing tool weight, eliminating tool vibration, and improving ergonomic designs are the primary motivation for engineering controls by tool manufacturers today. Ultimately, the technician's key considerations in selecting tools are the tool's comfort, performance, and price.

TOOL MAINTENANCE

Technicians in the auto collision repair industry most often choose **air tools** for a variety of reasons. All air tools have one common denominator in terms of their maintenance requirements: They must be lubricated at regular intervals with an approved air tool oil. See **Figure 4-2**.

Most air tools require little more than two or three drops of oil each time they are used, or a minimal application at the beginning of a workday when the tool will be

FIGURE 4-2 Air tools should be lubricated regularly with air tool oil only. Substituting other lubricants will likely cause premature wear and unnecessary damage to the internal parts.

used. By always using an appropriate air tool oil to keep the veins moving freely, the tools will operate smoothly and remain viable longer. Contrary to popular belief, air tool oil does not function like a lubricant in an internal combustion engine. Air tool oil's principal function is to displace moisture created by the air compressor that is pumped through the lines along with the air as it enters the tool. Using other lubricants such as engine oil may cause sludge and dirt from the air lines to collect on the drive motor and the veins, resulting in very sluggish operation. Contrary to what seems to be an accepted practice or theory, internal bearings of air tools do not require additional lubrication. All these bearings are sealed and do not require any additional lubrication other than air tool oil. One must be cautious to not overlubricate the tool as it may result in contaminating the surface being repaired. Excess oil may blow out of the tool as the air is exhausted through the air outlet, frequently directing it onto the repair area.

Lubrication Ports

Some air tools that have external moving parts and gears are designed with an external lubrication port. This port is usually covered with a removable plug, which serves as a continuous oiler for the moving parts that may not be sealed. One should check and follow the manufacturer's maintenance recommendations to ensure the tool is properly lubricated at the recommended intervals.

AIR RATCHETS AND IMPACT WRENCHES

The first power tools typically used in the repair of a damaged vehicle are the **air ratchets and impact wrenches** as they are used to remove bolted-on parts. See **Figure 4-3**. These parts must be removed to gain access to the damaged areas underneath them and usually cannot be used for pulling purposes during the roughing phase of the repairs. Pulling operations will be discussed in Chapters 14, 15, and 16.

Ratchet Drive Size

Air ratchets are generally rated by their drive shaft size and are commonly referred to as a 1/4-inch drive or 3/8-inch drive. The size of the ratchet will dictate the applications for which each can be used based on the torque they generate. See **Figure 4-4**. A 1/4-inch drive ratchet, as one would expect, has a drive shaft lug size of the stated capacity and is used for applications where less torque is required. The reduced size of these ratchets makes them an ideal tool for removing and installing smaller bolts and fasteners typically found on vehicle trim and in areas where access with larger tools is limited. The 1/4-inch ratchet is intended to be used with sockets or extensions with a matching shoulder opening. See **Figure 4-5**. With the use of a step-up adapter, one may use a larger capacity socket. However, the results are frequently less than desirable as the tool does not generate a sufficient amount of torque to warrant its use. This can also cause an undue

FIGURE 4-3 Air ratchets and impact tools are usually among the first tools used on a damaged vehicle to remove bolts and fasteners in order to gain access to the vehicle's internal parts.

FIGURE 4-4 Air ratchets are available in many sizes, shapes, and torque values and are manufactured by many different manufacturers.

FIGURE 4-5 Air ratchets are available in different sizes and, whenever possible, should be used with the sockets and extensions made for their respective size.

amount of stress on the tool; if done repeatedly, the tool may prematurely wear out.

CAUTION

When using a socket or step-up or step-down adapter with a longer handle breaker bar or ratchet, caution must be exercised to avoid exceeding its safe load capacity to prevent breakage.

The 3/8-inch drive ratchet has a drive shaft lug size of the stated capacity and is designed for applications where more torque is required, such as removing bolts and fasteners from door hinges, hoods, and other areas where larger size fasteners are used. As with its smaller counterpart, the size of the drive shaft lug is designed to accommodate sockets and extensions with a 3/8-inch square opening or shoulder.

Tool Torque Ratings

Air ratchets are usually rated by the amount of torque or foot-pounds they are capable of generating with a specified amount of air pressure. Some lesser quality ratchets may generate a maximum of 30 foot-pounds, while a better quality tool may be capable of generating up to 65 or more pounds of torque. The difference in the amount of torque these tools are capable of generating is based on the gearing mechanism inside the tool. Most air tools are capable of generating a similar amount of horsepower. In order to increase the torque value of the tool a lower gear ratio is used. This increases the power of the tool but will usually reduce its rpm (rotations per minute) rating as well.

Personal Safety

Technicians using an air ratchet must be aware of the potential danger associated with it. Even though the air ratchet is a power tool and normally is not used as a hand ratchet, it functions in the same manner as the hand ratchet. In fact, the technician is encouraged to break the bolt or nut loose by hand before triggering the tool to remove it completely. Failure to do this may result in the technician unexpectedly having a hand slammed against the edge of whatever panel the bolts are being removed from when the trigger is initially pushed. When the trigger is pushed, the tool exerts pressure against the fastener being removed, pushing the tool in the opposite direction. The more torque capacity the tool has, the harder it will push the tool in the opposite direction of the fastener being loosened—and the greater the possibility of injury to the user.

When using the air ratchet to install fasteners, the technician must use the same precautions as when removing a fastener. One must also be cautious to not hold the trigger in the on position until the tool stalls out to avoid overtightening the bolts or fasteners being installed. The preferred method for tightening a bolt or nut is to hold the trigger until the tool starts to "bog down" or to labor somewhat, and then release the trigger and finish tightening the fastener by hand. In instances where limited access or "swing room" doesn't allow the technician to do this, it may be necessary to give the trigger several short bursts of air until pressure is felt against the tool. This will eliminate the possibility of overtightening and damaging or stripping the threads on the bolt, nut, or clip being used.

Socket Selection for Air Ratchet

Contrary to popular belief and practice, it is not acceptable to use chrome-plated hand sockets with the air ratchet. The torque from the steady driving pressure exerted onto the socket shoulder will cause an undue stress that may cause the chrome plating to split or the socket to crack or split. It is advisable to use unplated impact sockets, particularly since the normal tendency is to encircle the socket or extension with one's index finger and thumb to guide and help support the tool. Any undetected peeling or cracking in the socket's chrome plating is razor sharp and may result in a severe cut on the user's fingers.

Air Impact Wrench/Gun

The air impact gun is another tool commonly used to remove and install bolts and nuts from various parts of the vehicle. This tool is available in either a 3/8- or 1/2-inch drive. The 1/2-inch drive is used more frequently than the 3/8-inch drive, as the air ratchet of the same size will perform most of the same operations. The most common application for the impact gun in the collision repair shop may be removing lug nuts from wheels and larger bolts such as those used on suspension parts, as well as some bumpers and their attaching brackets. They may also be used to install wheel lug nuts, but only when used with specially designed torque stick sockets. As their air ratchet counterpart, the impact wrench has a reversing switch to accommodate both removing and tightening bolts and nuts alike. As with the air ratchet, when using the impact wrench a special heat-treated socket, designed specifically for this tool, should be used. The air impact gun is designed to operate in such a way that it tightens the bolt or nut until it stalls out. When it stalls out it will continue tightening the fastener by a continuous hammering effect until the trigger is released. Standard hand sockets are not designed to withstand the stress from this hammering effect and, with repeated application, may likely crack, split, or even shatter. The possibility of personal injury is inevitable. One can control the amount of force or torque applied by the tool with an air metering valve built into the tool and also by varying the amount the trigger is depressed.

Electric Impact Gun

Electric impact guns are also available in both a 3/8- and 1/2-inch drive capacity. They, however, are not as versatile as their pneumatic counterparts. The most significant shortcoming of the electric impact wrench is its inability to control the amount of torque applied to the fastener when tightening it. The pneumatic impact allows the operator to control the torque by the amount of air pressure metered into the tool by the air inlet valve and by trigger control. An electric impact gun generally does not have any provisions for controlling the torque other than by triggering it or releasing the trigger entirely.

SANDERS AND GRINDERS

Most tool manufacturers use the terms *grinder* and *sander* interchangeably, which often causes confusion to the reader or user. For the sake of eliminating some of the confusion, this text will use the term *grinder* to identify tools normally used for surface removal operations, such as removing old painted finishes. Sanders are used for sanding fillers; they are also used for rough and finish paint preparation operations. See **Figure 4-6**.

FIGURE 4-6 Disc grinders are typically used for removing surface coatings and are available in several different sizes, each designed for a different application.

Although primarily designed for surface removal, grinders are used in several phases of the collision repair process. They are used to remove paint and sometimes fillers from a previous repair, remove surface rust and weld beads, and maybe even shave or form a pulling template to match the contour or shape of a surface to be pulled. Some basic precautions must be followed when using a grinder regardless of why it is being used. When using an abrasive wheel or grinding disc for surface removal with a grinder, the result is always flying sparks. These sparks, which are actually small bits of red hot metal being removed from the surface, can cause serious injury to the technician as well as damage to glass and other parts on nearby vehicles.

Personal Safety Using Grinding Tools

Technicians must be concerned about their own well-being and use the necessary personal protective equipment (PPE) when grinding. Safety glasses and a protective face shield are the minimum PPE required for such operations. In addition, respiratory protective equipment should also be worn when performing a grinding or sanding operation. A school's or shop's respiratory protection program should specify the specific respiratory protective equipment to use.

Proper Tool Setup

When using a grinder, a backing pad must be used to support the grinding disc. This step enhances the tool's effectiveness and also makes it safer to use. The tool's efficiency is also affected by this backing pad, as it prevents the abrasive disc from flexing or bending and holds it flat against the surface being ground. A variety of backing pads—made of rubber, phenolic, and urethane-reinforced

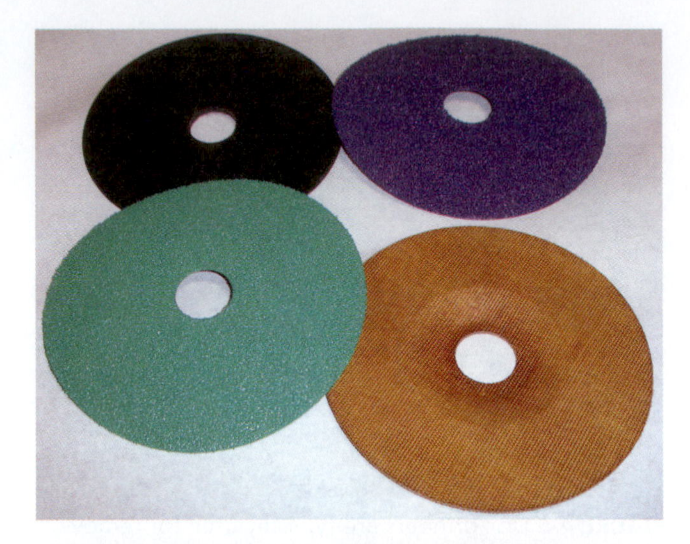

FIGURE 4-7 The backing pads are made using three primary materials: rubber, fiber-reinforced phenolic, and urethane-reinforced composition.

material—are available. See **Figure 4-7**. The backing pad selected should be smaller in diameter than the grinding disc yet large enough to ensure proper support of the abrasive disc. A general rule of thumb is the grinding disc should be approximately 10 to 12 mm (1/2 inch) larger in diameter than the backing pad, leaving no more than 5 to 6 mm (1/4 inch) of the disc exposed beyond the edge of the backing. See **Figure 4-8**. A backing pad that is too large will extend over the edge of the abrasive disc, resulting in damage and reducing the effectiveness of the grinding operation. Conversely, a backing pad that is too small will allow the unsupported edge of the grinding disc to flex excessively, which can cause metal warpage because of the excess heat generated on the surface. Very little surface removal will occur when the tool is used in this manner. The disc's excessive flexing may also cause it to tear, fracture, and eventually disintegrate, throwing fragments in all directions at a very high rate of speed.

FIGURE 4-8 The size of the backing pad selected should be closely matched to the grinding disc size. The disc should extend approximately 1/4 inch beyond the edge of the backing pad to maximize its operating efficiency.

Grinding Wheel Rotation

The user must always be aware of the grinding wheel or abrasive's direction of rotation. Unless the tool has a directional switch built into it or is specifically designed to turn in the opposite direction, all tools are designed to rotate in a clockwise motion. When performing any surface removal operations, the abrasive disc should be held so it turns off the edge, thus minimizing the possibility of removing excess material from the edge of the panel. One should never hold the grinder so the abrasive wheel turns into the edge of the panel or an exposed surface. Doing so will invariably remove excess material. See **Figure 4-9**. A more significant problem is the possibility of the disc catching or snagging on the edge, causing it to tear and destroy the backing pad in the process.

Application and Usage

Grinders are available in a variety of designs, which to some degree dictate their specific applications. A heavy-duty right angle grinder, though not as commonly used as it once was, is typically used for heavy surface removal. See **Figure 4-10**. This tool is designed to remove paint and heavy deposits of filler from a larger area as well as weldment created when making a repair where an elevated weld bead may exist. The tool is meant to be used with larger grinding discs available in 7- or 9-inch diameters. When using this heavy-duty grinder, the operator must be cautious as the tool's speed and torque capability may remove more metal than an inexperienced operator may realize. Burn through or excessive metal removal is a significant possibility with this piece of equipment.

FIGURE 4-9 When removing material on the edge of a panel, the grinding disc should always be held so the grinding surface is turning off the edge and not into it.

FIGURE 4-10 The heavy-duty right angle grinder is commonly used to remove heavy deposits of filler and grinding welds, as well as areas where surface coatings are removed from a large surface.

Surface Removal

The right angle electric grinder, once said to be the industry standard, is still a popular choice among collision repair technicians for certain surface removal operations. The smaller version of the tool is often used for weld removal when used with a hard face carbide grinding wheel. However, the sheer size of the heavy-duty grinder may limit its applications to those areas that are unobstructed, such as the vehicle's exterior.

CAUTION

One should never turn the grinder on and allow it to free spin with a disc attached to it. The grinding disc may fracture and disintegrate, causing pieces to be thrown about like shrapnel.

One must be cautious to never turn the grinder on and allow it to "free wheel" or to run without holding the grinding disc to the surface. A tear, crack, or flaw in the abrasive disc increases the possibility that the disc will disintegrate, throwing fragments over a considerable distance. These fragments could injure bystanders, as well as the operator.

Power Sources

Grinders are available that use either electricity or air as an energy source. The major difference between the electric and pneumatic grinder is its energy source and tool weight. The electric grinder invariably is heavier than the pneumatic version. To some this offers a more positive feeling of control. One of the major advantages of using electric tools over air-powered tools is that they maintain a constant speed under load. Many of the electric tools are designed with a variable speed option which allows the user to operate the tool at a specific speed. This is adjusted by the setting on the built-in adjustable rheostat or by how far the adjustable trigger mechanism is depressed. Whatever speed is chosen, the tool will maintain that rpm under load without an appreciable loss of tool speed, power, or torque. Typically the weight of electric tools is considerably heavier than air tools of similar design and application. Unlike air tools, which are constantly cooled with air flowing through them, electric power tools generate significantly more heat, especially when used for an extended period of time. This sometimes makes them uncomfortable to use. The tradeoff is that, other than an occasional armature brush replacement and blowing out the air inlet vents, these tools require little or no routine maintenance.

Variable Speed Buffer/Polisher

An electric tool that has gained popularity in recent years is the buffer, which is a variable speed version of the heavy-duty grinder. Buffers are designed to operate at

FIGURE 4-11 A variable speed buffer is often used for compounding and polishing the finish to duplicate the surface texture and remove minor flaws and imperfections after painting.

speeds ranging from 1,000 to 2,500 rpm, which is the optimum speed for buffing and polishing operations. After repainting the repair area, it frequently becomes necessary to sand and buff the surface in order to duplicate the surface texture and remove minute flaws. See **Figure 4-11**. Most vehicles require some sanding and buffing after refinishing as it is extremely difficult to duplicate the factory finish when applying the topcoat with conventional refinishing equipment.

PROTECTING AGAINST ELECTRIC SHOCK

One of the major concerns when using any electric tool is the possibility of electric shock. Many of the collision repair operations, such as the pre-repair cleaning, sanding operations, and detailing, require the use of water. Wet or damp floors are very common, particularly in the refinishing and detailing areas. In a smaller shop, where space is limited, the refinishing and detailing operations may overlap with the general repair area. In that case, the presence of water on the floor is a constant. To minimize the possibility of electric shock, a **ground fault circuit interrupter (GFCI)** becomes a critical part of the power supply equation. The GFCI should be used any time an electric tool is used on a floor, even if it is only slightly damp. See **Figure 4-12**. The GFCI may be available as an inline device added to the extension cord or it may be hard wired into the electric circuit at the breaker panel or at the outlet on the wall. OSHA regulations mandate that whenever an electrical outlet is within 4 feet of a water source, the electric outlet must be GFCI protected.

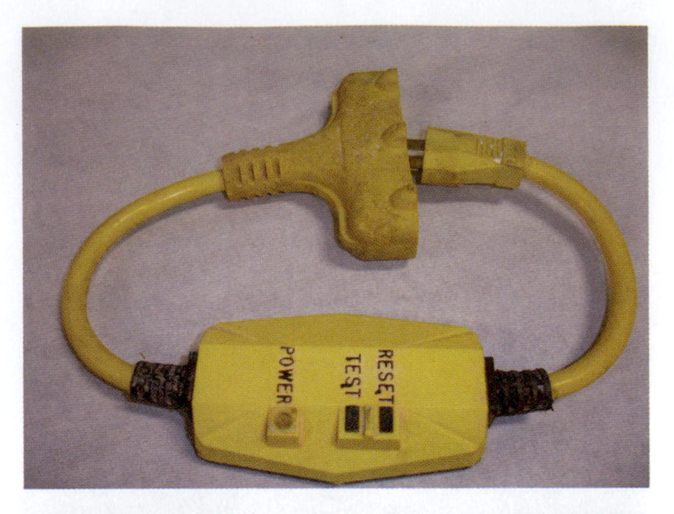

FIGURE 4-12 A ground fault circuit interrupter (GFCI) should always be part of the circuit whenever using electric tools on wet or damp floors, or when a water outlet is near the work area.

PNEUMATIC GRINDING TOOLS

Pistol Grip Grinder

The smaller pistol grip grinder is used when a smaller-sized surface must be ground and indentations left after using the large surface removal grinding operations. The smaller diameter disc is ideal when used for surface removal on minor indentations which cannot be reached without tipping the larger grinder on its edge. See **Figure 4-13**. A variety of disc sizes and backing configurations are available for use with this grinder.

The operator must be keenly aware of the dangers associated with this grinder as certain manufacturers' tools are capable of generating a disc speed up to 20,000 revo-

FIGURE 4-13 The pistol grip grinder is a popular tool for surface removal as it offers numerous sizes of sanding discs and backing pads, thus allowing its use on numerous small indentations which cannot be reached with the larger grinding disc.

lutions per minute (rpm). In order to minimize the possibility of injury, the operator should ensure the disc is held against the surface prior to engaging the trigger and not remove it from the surface until the disc comes to a complete stop. One should never pull the trigger and allow the disc to free spin. The grinding disc may fracture and disintegrate, throwing shrapnel-like fragments that could likely cause serious injury to the operator and any bystanders. Suffice to say, if a disc were to disintegrate at this speed, the possibility of personal injury is very significant.

Die Grinders

Another group of tools frequently used for surface removal are die grinders. These are available in a straightline or right angle design. Their most common application is enlarging and elongating holes, removing metal and weld from tight corners that can't be reached with a grinding disc, and beveling with various designed burring tools. See **Figure 4-14**. As with any tool used to remove metal, it is vitally important the die grinder user wear the necessary eye protection, as very small metal particles are thrown about whenever this tool is used. One typically does not even realize this is happening because the usual spark stream is not present as when using a grinding disc for similar applications. Leather gloves are also recommended PPE as the small metal slivers or shavings can embed themselves into the skin of the user's hands. When using the die grinder for enlarging and elongating holes, the operator must not try to force the burring tool into a small hole and attempt to enlarge it with the tool point. This will cause the tool to bind up in the hole or chatter and wander across the surface, often unnecessarily damaging the surrounding surfaces. This will likely cause some of the cutting teeth to break loose from the cutter. See **Figure 4-15**. Instead, the operator should drill a hole large enough to insert the cutter bit deep enough so

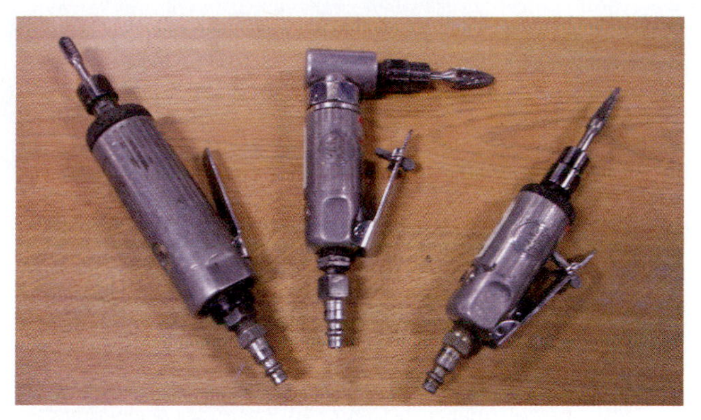

FIGURE 4-14 The die grinders are available in both a straightline and right angle body. The different tool shapes allow them to be used for enlarging and elongating holes, as well as removing metal and weldment in areas where a grinding disc cannot reach.

FIGURE 4-15 One must be cautious to not force the tool or cause it to bind up in a hole, as that action will cause the cutting teeth to break off, leaving the tool useless.

the cutting is done by the teeth on the sides of the cutting surface. It may also be used for a variety of other applications in which a buffing or other abrading attachment is used for removing various surface coatings.

METAL CUTTING TOOLS

Nibblers

Another very handy tool to have available is the metal nibbler. The nibbler can make an infinite array of cuts ranging from straight lines to very sophisticated shapes and patterns on light gauge sheet metal. See **Figure 4-16**. This can turn an otherwise painstakingly slow process of cutting metal with a hand shear into a quick and easy project. It is also an excellent means for cutting out an access window or opening in the center of a panel; for example, a door or quarter panel to expose the parts that are to be removed. See **Figure 4-17**. The only requirement for its

FIGURE 4-16 The nibbler can be used for quickly cutting an array of lines and shapes on light gauge sheet metal.

(A)

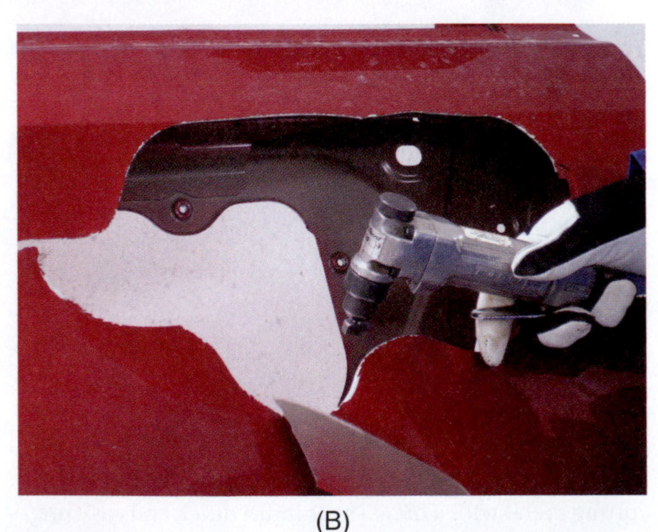

(B)

FIGURE 4-17 The metal nibbler can be used for cutting out access windows from damaged panels to expose areas inside the panels (A). It can also be used to cut intricate patterns out of sheet metal, if needed (B).

use is having enough clearance on the bottom side so as not to obstruct the shoe of the tool. Baffles, multiple metal thickness, and other reinforcements on the underside may limit its use for removing certain panels from a vehicle. On a flat piece of metal, the cuts can be initiated on the edge of the panel. However, when used somewhere besides the edge—such as in the center of a panel—a pilot hole large enough for the shoe to pass through is required. Once this is accomplished, the rest, as they say, is history.

Cutoff Tool

The cutoff tool, which was originally designed for cutting exhaust pipes and other similar parts, has become a popular choice as a **metal cutting tool** in the collision repair industry. The ability to make a clean cut through one layer of a multiple layered part makes it an ideal tool for many panel replacement operations. The tool can readily be used for cutting through layered sheet metal parts without damaging the layers below the top surface. See **Figure 4-18**. It can also be used for making cuts across the length of a single thickness panel when a sectional replacement is being performed. A real advantage of this tool is the ability to cut through the outer layer of metal without concern about damaging or accidentally severing braces on inaccessible cavities that are not visible from the top side. The tool employs a miniature carbide cutting wheel comparable to that used on a cutoff or chop saw.

As with any cutting tool that employs a spinning or rotating cutting medium for surface removal, users must employ all the appropriate personal safety precautions for themselves and others around them, the vehicle itself, and other vehicles in the adjoining areas. This tool's cutting disc operates at approximately 18,000 to 20,000 rpm and creates an array of sparks which may embed themselves into a vehicle's windows or plastic parts when directed against them. The tool offers a built-in cover shield that can be

FIGURE 4-18 The cutoff tool is a portable mini-chop saw which can be used for cutting through a single layer of a multiple layered laminated section of a panel without damaging the layers underneath it.

turned in such a way to catch or trap the hot metal being removed by rotating the tool. When cutting a single layer of metal, one must be cautious to not bind the cutting wheel in the slot being cut as it may cause the wheel to crack and disintegrate, throwing shrapnel in all directions.

Air Hammer/Chisel

One of the most popular tools used for disassembling damaged sheet metal and removing corroded and frozen fasteners is the air hammer (sometimes called the air chisel). This tool is used for cutting metal, shearing frozen bolts and nuts, loosening pins and fasteners that cannot be freed with hand tools, and for a variety of other applications. See **Figure 4-19**.

CAUTION

To avoid personal injury, leather gloves should always be worn when using the air chisel.

Despite the tool's popularity, it is one of the most dangerous pieces of equipment to use, particularly when cutting sheet metal. The hammering effect of this tool makes it difficult to make a clean or straight cut, often leaving a jagged edge on the metal surface where the cut is being made. When using the air chisel, the operator naturally tends to get somewhat aggressive or overzealous when pushing the tool to make it cut faster. An all too common result is a personal injury to the operator. The cutting chisel or tool has a tendency to disengage or slip from the groove or the surface being cut. When this happens, the operator forcibly pushes or slides a hand over the sharp and sometimes jagged edge being created by the cutter. These two factors, when working in tandem, often result in a personal injury to the user. A technician should NEVER operate this tool without wearing protective leather gloves.

Injuries can also occur when the chisel is triggered without engaging the cutting tool on any surface. This is especially true if the tool is used with a retaining spring that has become weak with use and abuse. One should never direct or playfully point the tool in the direction of someone else and pull the trigger.

The air hammer is available in a variety of power ranges from light to heavy-duty. The duty range refers to the power packed by the hammering piston within the tool. The medium-duty tool is the preferred choice for most collision repair technicians, as the heavy-duty tool is far too aggressive for typical collision repair operations. Conversely, a light-duty air chisel may be used with certain trim removal tools such as those used for adhesive bonded moldings.

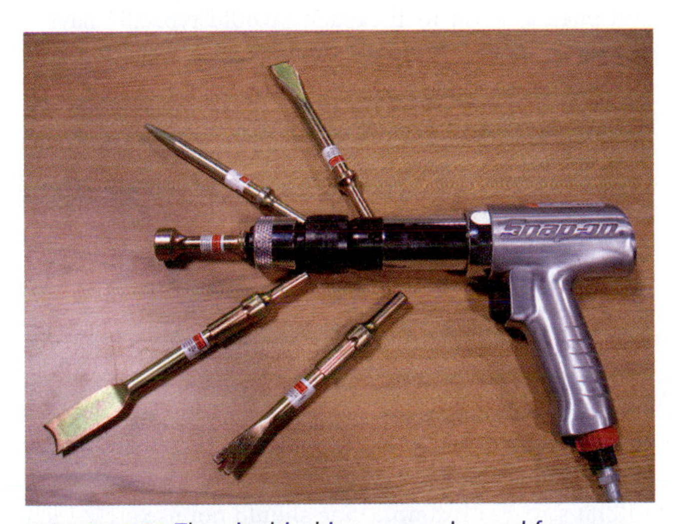

FIGURE 4-19 The air chisel is commonly used for many cutting and chiseling applications. For example, it can cut damaged sheet metal parts away from the vehicle when they are too distorted to be removed any other way. That allows the user to remove the fasteners beneath these parts. It is also used for shearing bolts and corroded fasteners.

RECIPROCATING SAW

Despite its seemingly limited use, the reciprocating saw occasionally plays an important role in the collision repair shop. The saw typically is used to cut through both the inner and outer layers of the vehicle's panel sections, especially when an entire section or recycled parts are used. For example, the user may be unable to make a clean cut at the precise required location on an inaccessible inner panel, necessitating that it be cut from outside the vehicle. The long cutting blade allows the user to cut through the two layers of the panel even when an open cavity lies between the two layers, as is found on the sail or rocker panel. See **Figure 4-20**. This is done to ensure the panels are cut at the precise location for proper fit-up when installing the replacement parts.

FIGURE 4-20 The reciprocating saw is commonly used for cutting through multi-layered panels such as the A or B pillar, the sail or rocker panel, and others that may have a cavity or large opening between the two surfaces.

Cutting blades for the saw are available in various lengths. The blade selected should be long enough to easily reach through both layers of metal yet not "stab" adjoining surfaces on the back side. This will cause the saw to "buck" each time the blade strikes the inside panel, making it impossible to control as the operator tries to make a cut at the precise desired location. The cutting blade may have to be cut shorter and trimmed—even though the manufacturer does not endorse this. The foot of the saw should always be held firmly against the surface on the side from where the cuts are made. This will also help to stabilize the saw and eliminate much of the vibration caused by the reciprocating action of the saw.

CAUTION

Steps should be taken to protect electrical wiring, as well as gas and high pressure lines, that may be routed through the cavities that are cut with the saw.

Vehicle Protection

Vehicle manufacturers sometimes route electrical wires and various other lines through the sail and rocker panel cavities. The gas and brake lines are nearly always attached to the bottom side of the rocker panel. One must check to ensure the location of these lines and move them prior to initiating any cuts to prevent damaging or severing them during these operations. This becomes even more critical with alternative fuel vehicles, as the high voltage lines are often routed in this area from the vehicle's front to the rear. The necessary precautions should be taken to protect them while cutting through the section or maybe even remove them until the vehicle is ready for reassembly.

Many tool manufacturers also offer a miniature pneumatic reciprocating saw, which is used for many of the same operations as the heavy-duty version—only on a smaller scale. This tool has a variety of uses such as cutting through a single layer of a multiple-layered part when making an offset sectional replacement. See **Figure 4-21**.

FIGURE 4-21 The pneumatic hacksaw is used for similar functions as the reciprocating saw but is used on a smaller scale. It can be used for cutting through a single layer of a multi-layered panel without damaging the metal underneath the cut area.

This will be discussed in detail in a later chapter. It can also be used for cutting out sections in the center of a panel or on the edges of it. The only restriction for using this tool is having enough clearance for the reciprocating blade to move back and forth while cutting.

DRILLS

The collision repair technician uses **drills** for a variety of operations, ranging from drilling holes to install fasteners to drilling spot welds. They are also used in conjunction with a number of other operations such as paint and surface removal with a specially designed abrasive wheel. See **Figure 4-22**. They may be electrically or battery energized, or pneumatically powered. Their drive mechanism may be a one direction single speed or may be variable speed and reversible, wherein they are referred to as VSR drills. The technician may use the battery powered drill when re-assembling small trim and more fragile items, as the amount of torque exerted can be controlled more easily with the built-in clutch on the drive shaft. The electrically energized drill may be more desirable when the rpm must be limited to a constant lower speed. The air or pneumatic drill may be preferred when the operation being performed requires a higher rpm or when the operation is a lengthy one in which the tool will remain cool and more comfortable to hold than its electric counterpart.

Size Ratings

Drills are sized and rated by their chuck capacity and are usually identified as a 1/4-, 3/8-, or 1/2-inch capacity. The drill size rating is generally determined by one or two of their features: the size of the output shaft to which the chuck is attached and the capacity of the chuck itself. A drill rated as a 1/4-inch capacity would typically have a chuck capacity large enough to hold a 1/4-inch drill bit or a tool with a similar size arbor. Likewise, a 3/8-inch drill will have the capacity to hold an arbor up to 3/8 inch in diameter, and so forth. The capacity rating of the drill is also used as means of identifying the tool's torque capability. Although the applications for a 1/2-inch drill are very limited in the collision repair shop, there is an occasional need for them.

Drill Bit Selection

As with any other power tool, the drill's user must be aware of the potential dangers associated with using the tool. One should always use common sense and a little discretion when selecting and matching the drill to the drill bit size. For example, one should not use a 1/2-inch drill to drill a hole with an 1/8-inch drill bit. Conversely, the opposite holds true—don't use a 1/4-inch drill to make a 1/2-inch diameter hole. Oversizing and undersizing the drill bit to the drill will invariably result in damaging the tool, breaking the bits, or injuring the operator. A burr forming at the edge of the hole—causing the drill to

FIGURE 4-22 The drill is perhaps the most versatile and widely used power tool in the technician's arsenal. Drills are used for drilling holes, removing surface coatings, and a variety of other operations and functions.

freeze or lock up—is a typical example of what might happen when using a drill bit that is too large for the drill size. The tool is underpowered and unable to turn the drill bit; as a result, the drill stalls out. In order to finish drilling the hole, the trigger must be activated with the bit turning before engaging it into the hole, making it difficult to control the tool. The usual result is the bit wanders around the surrounding area, marring it. It may also stall out again as it catches the burr in the hole.

Properly Securing the Tool

Another important part of using the drill properly is tightening the bit or arbor of the tool into the drill chuck. This can only be accomplished by using a chuck key, which must be matched to the gearing on the drill chuck itself. An unmatched chuck key will damage the gears on the drill chuck to the point where it will slip when attempting to engage the two. Once this happens, it becomes nearly impossible to properly tighten the drill bit or the arbor of any other tool into the chuck any longer. Once the drill bit or tool being used is tightly chucked into the drill, the chuck key should be removed from the tool prior to activating the trigger. Attempting to run the tool with the chuck key still on the tool will certainly result in injury to the user.

Keyless Chuck

Some drills are equipped with a keyless chuck. With this design the chuck is loosened and tightened simply by holding the forward part of the chuck and actuating the trigger. To tighten the chuck one simply moves the directional switch to the forward position and, while holding the chuck with one hand, actuates the trigger. This will tighten the collar until the arbor of the tool is held in place by the jaws of the chuck. Conversely, to loosen the jaws, one reverses the directional switch and, while holding the chuck with one hand, pulls the trigger until the jaws open up, thus releasing the drill bit or the arbor of the tool being used. The time required to change from one tool to another is shortened considerably with this type of chuck. One need not use a chuck key to tighten the jaws. This eliminates a time-consuming task, especially when performing operations wherein the bits require frequent changing—such as when drilling holes, inserting fasteners, and so on. Another unique feature of the keyless chuck is the drill typically has an adjustable clutch built into the drive shaft which permits the operator to adjust the amount of torque output. This helps to minimize the possibility of stripping the fastener threads by allowing the jaw to slip when a specified torque level is achieved by the output shaft. See **Figure 4-23**.

One of the biggest disadvantages of the keyless chuck is the user's trouble in sufficiently tightening it when using tools for operations requiring a significant torque, such as a drill bit used for drilling a large hole. This often causes the cutting tool to stop, jam, or freeze midway through the drilling operation. Meanwhile, the drill chuck continues to spin, often causing burrs to be formed on the end of the tool arbor. Some tools and drill bits are available with only a round shaft while others have either an octagonal or a three-sided arbor. In addition to drilling holes, some of the most common operations performed by the drill are spot weld removal with specially designed bits, surface removal with special abrasive discs designed to be used in a drill, and molding adhesive removal from the surface.

Drill Accessories

Some surface removal tools are designed to mechanically strip paint and other blemishes from the surface with a drill. The result is few or none of the sand scratches in the surface that are normally left when using a grinding disc. When an adhesive bonded molding is removed, traces of the adhesive tape frequently remain on the surface. This

FIGURE 4-23 The spot weld drill is specifically designed for drilling out spot welds using special spot weld drilling bits. The arm clamps and applies pressure to the drill bit to prevent it from wandering as the drilling operation is initiated.

FIGURE 4-24 The molding adhesive removal wheel and numerous other tools are used with the drill to perform a variety of operations.

adhesive is easily removed with a rotary eraser designed to be used on the drill. See **Figure 4-24**. Another popular use of the rotary eraser is to remove vinyl tape stripes.

SCREWDRIVERS

Battery Powered

Battery operated **screwdrivers** have become increasingly popular in recent years. These are essentially motorized screwdrivers that are primarily used for removing and installing trim and hardware. This hardware may be secured with a variety of screw heads such as a Phillips, clutch, Torx, or another popular design used for trim installations. See **Figure 4-25**. The advantage of using these tools is they have less torque than most drills, thus reducing the possibility of overtightening a screw type fastener.

FIGURE 4-25 A variety of different battery operated screwdrivers are available. Some are articulated and hinge in the middle while most are a straight body.

While they increase technician efficiency, their use also reduces operator fatigue. Many have an articulated design so they can be bent at nearly 90 degrees in the middle, allowing them to be held like a drill. This further reduces operator fatigue. This is particularly true in instances where numerous fasteners must be removed or re-installed, such as on the interior trim of a large motor home.

They may also be equipped with the necessary attachments for installing small bolts and threaded fasteners with hex heads, as well as for drilling holes. However, their slow rpm rating renders them very ineffective for the drilling operations, particularly when used on hard surfaces such as steel.

STRAIGHTLINE, OSCILLATING, AND ORBITAL SANDERS

After the damaged panels have been straightened, plastic filler is commonly used to finish smoothing them. A variety of sanders are used to finish smoothing the fillers in preparation to applying the primer and finish coats of paint. See **Figure 4-26**. The size and the surface shape or configuration often dictates which of the sanders is used for the sanding operation.

Straightline Sanders

The two most popular tools used for plastic sanding operations are the straightline and oscillating sanders. These are both equipped with a long flat shoe to which sandpaper is attached using one of the popular securing methods available from various abrasive manufacturers. With the straightline sander the shoe moves straight back and forth at approximately 3,000 strokes per minute. The

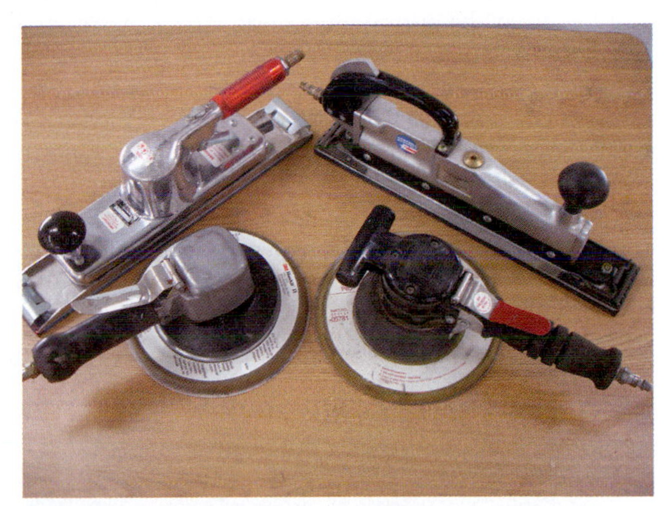

FIGURE 4-26 Plastic filler is sanded smooth using a variety of different sanders. The most popular power sanders for smoothing plastic filler are the straightline and oscillating long board sanders and the 9-inch diameter geared oscillating sander.

rapid back and forth movement of the shoe simulates the motion used when hand sanding. The operator slowly moves the sander in a criss-cross pattern over the plastic filler in much the same manner as when using a body file, being careful to follow the contour of the panel. One of the chief disadvantages of using this tool is the vibration it causes due to the rapid and frequent direction reversals made by the tool. The vibration may actually cause some of the low spots in the metal to move or shift, causing surface irregularities—especially on large flat panels. When using this tool the operator must be careful to keep the tool moving lest a low spot may be sanded into the filler, destroying the very purpose for using it.

Oscillating Sander

The **oscillating** or **orbital sander** is used in much the same manner as is the straightline. However, the principals of operation differ somewhat with this **circular sander** in that it rotates in a circular or oscillating pattern. The oscillation pattern is smaller than the **reciprocating** movement of the straightline sander. Therefore, it does not remove the filler as aggressively as its counterpart. The constant circular or oscillating motion of the tool results in a much smoother and vibration-free tool operation. It lends itself to be used more readily on larger flat panels such as a hood or roof where vibration caused by sanding may be a factor. One of the dangers of using this tool is that the handle design does not allow for a positive grip. This results in a tendency to tip slightly, thus sanding with the edge of the shoe. This may cut grooves into the filler.

Orbital and Geared Sanders

Orbital and geared rotary sanders are also used for finish sanding plastics and featheredging scratches and irregularities from the surrounding areas. The larger geared rotary sander utilizes a gearing mechanism which results in more power transfer to the disc, allowing for a larger oscillation pattern. The size of the disc used on the geared rotary sanders is usually 8 or 9 inches in diameter, which—together with larger oscillation patterns—requires more torque to drive. The larger disc size and the added power make this tool a good choice for finish sanding plastic filler. These fillers are typically sanded with a coarser abrasive grade, adding yet another dimension to the power requirements.

Lubrication

Proper maintenance—such as oiling these tools—becomes a serious matter. Many tools are designed to exhaust the spent air down onto the surface where the sanding is being performed. One must be careful to not overlubricate these tools as any excess tool oil will be blown onto the surface being sanded. This can cause problems with surface contamination, resulting in numerous paint finish problems.

DUAL ACTION, RANDOM ORBITAL, AND FINISH SANDERS

Dual Action

The **dual action (DA) sander** and the random orbital sanders look and operate on much the same principle as the larger geared sanders. The dual action (DA) sanders are generally used during the final phases of the repairs for featheredging the course scratches left in the paint immediately adjacent to the repair area. See **Figure 4-27**. Paint chips and other flaws in the existing finish are typically removed with the DA as well. Although one frequently sees advertisements using the DA for sanding plastic and other fillers, this tool is not a good choice for these operations as it often results in surface irregularities, requiring additional filler. The disc size used is typically either a 5-inch or 6-inch diameter which for all practical purposes is not large enough for sanding on a large plastic filled area.

Quite often the repair technician will rough sand the scratches and surface flaws with a coarser abrasive grade such as an 80 or 100 grit and stage up to a 320 or 400 grade for the final sanding. The purpose of the staging up is to remove the scratches left from each of the previous abrasive grades used. Using a 400 grade abrasive to remove the coarse scratches left in the old finish from sanding the plastic filler would be very time-consuming and waste a considerable amount of sandpaper. Progressively staging up to finer grades of abrasives is more time efficient and reduces excessive material usage.

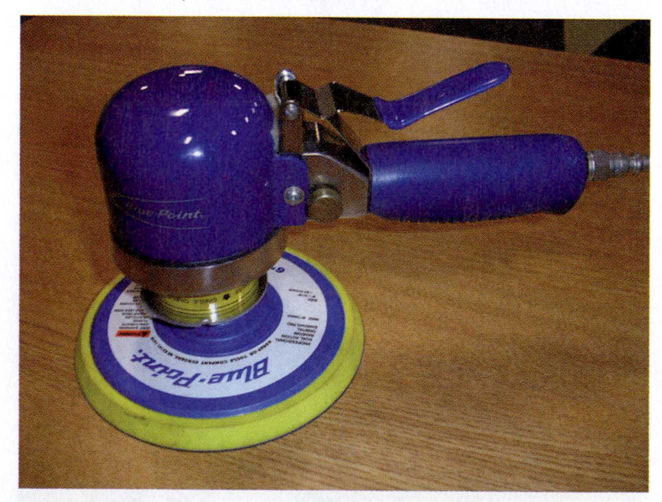

FIGURE 4-27 The dual action (DA) sander is used for featheredging the remaining scratches left in the painted surfaces surrounding the area where plastic filler was sanded with a course grade abrasive. They are also used for removing chips and other irregularities in the painted surfaces.

Random Orbital

The **random orbital sander**, while often used for the same operations as the standard DA, is actually designed for finer sanding operations. The tool's oscillation pattern is smaller, resulting in fewer and smaller sand scratches. The smaller oscillations also render the tool less aggressive; therefore, it is somewhat ineffective for removing the coarse sand scratches left in the finish from the plastic repairs. This tool is best suited for featheredging the perimeter of the repair area, **finish sanding** the adjacent panels, and scuff sanding new replacement parts prior to refinishing. They are also a popular choice for color sanding the clearcoats prior to buffing after the topcoat has been sprayed. See **Figure 4-28**. Color sanding and buffing will be further discussed in a later chapter on refinishing.

Micro DA

The micro DA is somewhat unique from its counterparts due to the size of the backing disc used for holding the abrasive in place and its pistol grip shape. The reduced disc size makes it ideal for feathering operations in smaller contoured areas where the larger 5- or 6-inch discs are not accessible. See **Figure 4-29**. It can also be used with each of the abrasive grades as the larger DAs, allowing for the same staging operations or steps used on a larger surface.

FIGURE 4-29 The micro DA sander is used for feathering in confined areas where a larger DA is not accessible and also for spot sanding individual imperfections on a newly painted surface prior to buffing and polishing.

WELDED-ON DENT PULLING SYSTEMS

Straightening damaged sheet metal requires yet another group of tools completely unique to those operations. When straightening and reshaping damaged metal, the technician must raise the surfaces that are lower than the adjoining areas by either accessing the back side and bumping, pushing, or prying them out or pulling them from the top side. Quite often it is not practical or feasible to access the back side of the panel to push or pry the low areas out. A variety of tools are available to pull minor damage from the top side.

Stud Gun

The stud gun is used to attach a specially designed nail or stud to the surface, which is then pulled with a slide hammer. The stud is permanently fused or welded to the surface, making it possible to pull the low areas out with the **dent pullers** specifically designed for this operation. See **Figure 4-30**. After the pulling operation has been completed, a twisting or rotating motion is used to remove

FIGURE 4-28 The finishing DA is used for color sanding clearcoats prior to buffing and polishing. It functions in much the same manner as the standard DA sander except the oscillations are much smaller, leaving fewer and smaller scratches in the surface.

FIGURE 4-30 A stud nail gun attaches pulling studs to the surface of a panel. This is used for pulling the damage out from the front side if the back side is not accessible to use straightening tools.

the stud with pliers. The surface is ground clean, leaving no sign that it was ever there.

Stud Gun with Carbon Electrode Used for Heat Shrinking

The stud gun includes a non-consumable carbon electrode which can be used for heating stretched surfaces for minor shrinking operations. See **Figure 4-31**. This heating tip also offers the advantage of heating small corroded fasteners which would otherwise be destroyed when removed without the use of heat.

Stinger Dent Puller

The dent puller can also be a part of the external electrical circuitry. The "stinger," as it is commonly called, functions by temporarily fusing a non-consumable tip—which is attached to the slide hammer—to the surface, which is then used for making a single pull. Each time the pull is completed, the tip must be removed and reattached to the next low area where a pull is to be made. A rotating or twisting motion on the handle of the dent puller will easily remove the tip. See **Figure 4-32**. When done, this also helps to generate a slight amount of tension in the metal to help hold it in place. This tool works particularly well for pulling long shallow indentations where the metal is "spongy" or overly flexible and access to the back side is limited. One must be careful to not use an excess amount of heat as burnthrough will result in a hole left in the surface that will require an additional welding operation. One must routinely check and re-dress the pulling tip, as a disfigured or improperly shaped tip may also cause burnthrough.

Stress Relieving When Using Pulling Equipment. When using these dent pulling devices, novice repair technicians commonly fail to stress relieve the surrounding areas. No matter how large or small the affected area being pulled, the technician must always stress relieve the adjoining areas to ensure the metal stays in its proper position. The essence of stress relieving is to cause the metallurgical structures to be set into a free state and to re-lock, causing the metal to take on a permanent set in the position where it is moved. Stress relieving will be discussed in more detail in another chapter on damage repair.

Disadvantages of Using Welded Pulling Techniques. The primary disadvantage of using any of the equipment where a pulling device must be attached to the bare steel surface is the paint must all be removed from the area where the device is to be used. The light reflection or "glare" on the surface is often used to help visually identify the location of subtle irregularities and surface

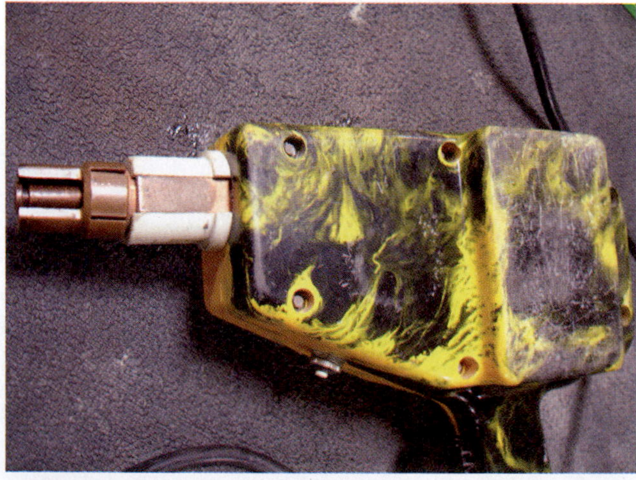

FIGURE 4-31 A carbon electrode can be attached to the stud gun and used to apply heat for small shrinking operations. It can also be used for heating corroded fasteners to accommodate removal.

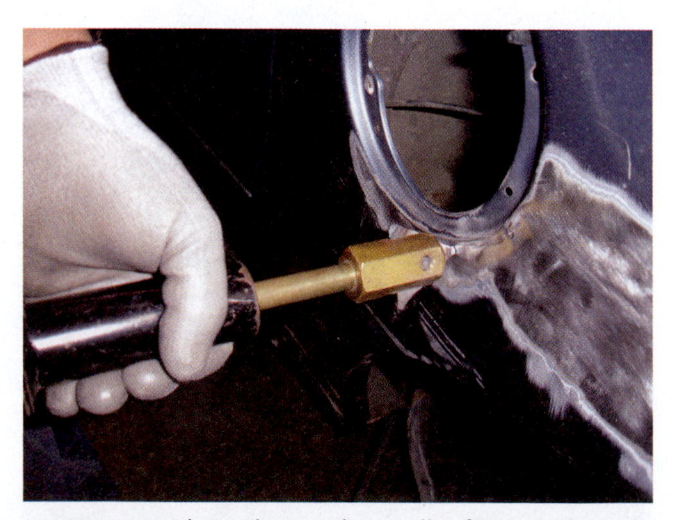

FIGURE 4-32 The "stinger" dent puller fuses a non-consumable pulling tip to the surface of the metal and then uses the dent puller to which it is attached to pull the damaged surface out. A rotating or twisting motion at the handle will quickly remove the pulling tip from the metal.

imperfections. When the paint is removed from the surface this visual advantage is no longer available and other less desirable techniques may need to be employed. To minimize the amount of metal being removed from the surface and to avoid creating grinding scratches which tend to reduce the sensitivity of the fingers when checking for surface irregularities, the technician should use a non-invasive plastic paint removal wheel for removing the paint. After the straightening and pulling operations are completed, the back side of the panel must be covered with corrosion protection material. If not properly treated, each spot where a nail or the stinger is attached becomes a corrosion hot spot.

WELDERS

Many of the repairs made on collision damaged vehicles require one or more of the many types of **welders** commonly used in repair shops today. When repairing tears or rips or when replacing sections of damaged panels, the technician may depend on the gas metal arc welding or MIG welding equipment. This process may also be used when replacing a structural rail using a series of plug welds which simulate the factory spot welds. Replacing aluminum parts requires yet another form of the MIG welding process. See **Figure 4-33**.

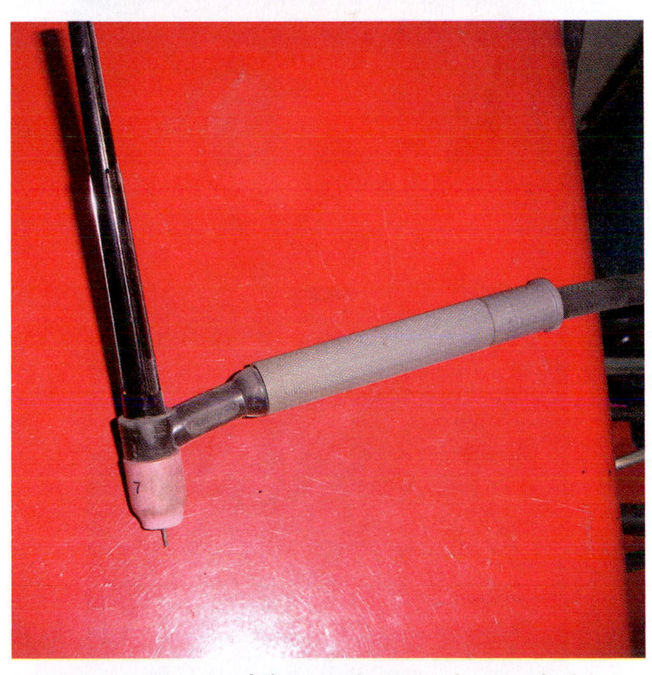

FIGURE 4-33 Many of the repair operations technicians perform require the use of some type of welding equipment. The GMAW MIG and the STRSW systems are used for most operations. The spool gun and the TIG torch are commonly used for aluminum welding operations.

Squeeze Type Resistance Spot Welder (STRSW)

When replacing large, severely damaged sheet metal parts the squeeze type resistance spot welder (STRSW) may be used to attach the new part. Although it is not as popular as it once was, a same side resistance spot welder may be used to install a small sectional panel. Suffice to say, numerous welding processes are used in the repair of an automobile. Welding processes and welding equipment will be covered in more detail in a later chapter.

TECH TIP Corrosion protection must be restored after any operations with tools or equipment using heat or welding operations are performed to protect the surface from corroding.

INDUCTION HEATER

The **induction heater**, which is a relatively new development, has become a very widely accepted piece of equipment because of its wide range of applications. The unit generates heat through an electromagnetic process when used on or around any conductive metallic surface. It may be used for removing adhesive bonded glass such as windshields, backlites, and side glass. The use of a special heating pad also enables the technician to use it for removing adhesive bonded moldings and trim without damaging any of the fragile accessories sometimes found on vehicles. It also works well for heating frozen or corroded hardware such as bolts, nuts, and other fasteners. See **Figure 4-34**. The chief advantage for its use in the latter applications is that no open flame is used, thus eliminating the danger of working around the fuel cell and connecting lines.

It can also be used for softening inaccessible or hard to reach adhesives when removing a damaged sheet metal part such as a quarter panel or outer door panel. Many times the outer panels are sealed or attached to adjoining parts with a heavy bead of adhesive to seal out moisture, dust, and the outside elements. This is sometimes found around the gas filler pocket and on the lower quarter panel to trunk floor pockets. Otherwise, to soften these sealants for panel removal, an open flame may have to be used around the area. This creates a potential explosion hazard due to the presence of gas fumes.

PERMANENT BENCH-MOUNTED TOOLS

Despite their seemingly limited use, some tools and equipment are necessary for those "once in a while" applications that would otherwise be very difficult to perform or complete. They may also be too large or cumbersome to be part of a technician's personal toolbox. Although this **permanent shop equipment** may be part of the technician's personal tool inventory, most are generally furnished by the shop. These include the chop/cutoff saw and the bench or pedestal grinder.

Chop/Cutoff Saw

Though infrequently used, the cutoff saw—or chop saw, as it is sometimes called—is a considerable time saver for certain applications, such as cutting bar stock, flat strap steel, angle iron, and a variety of other applications. See **Figure 4-35**. The most significant advantage it offers

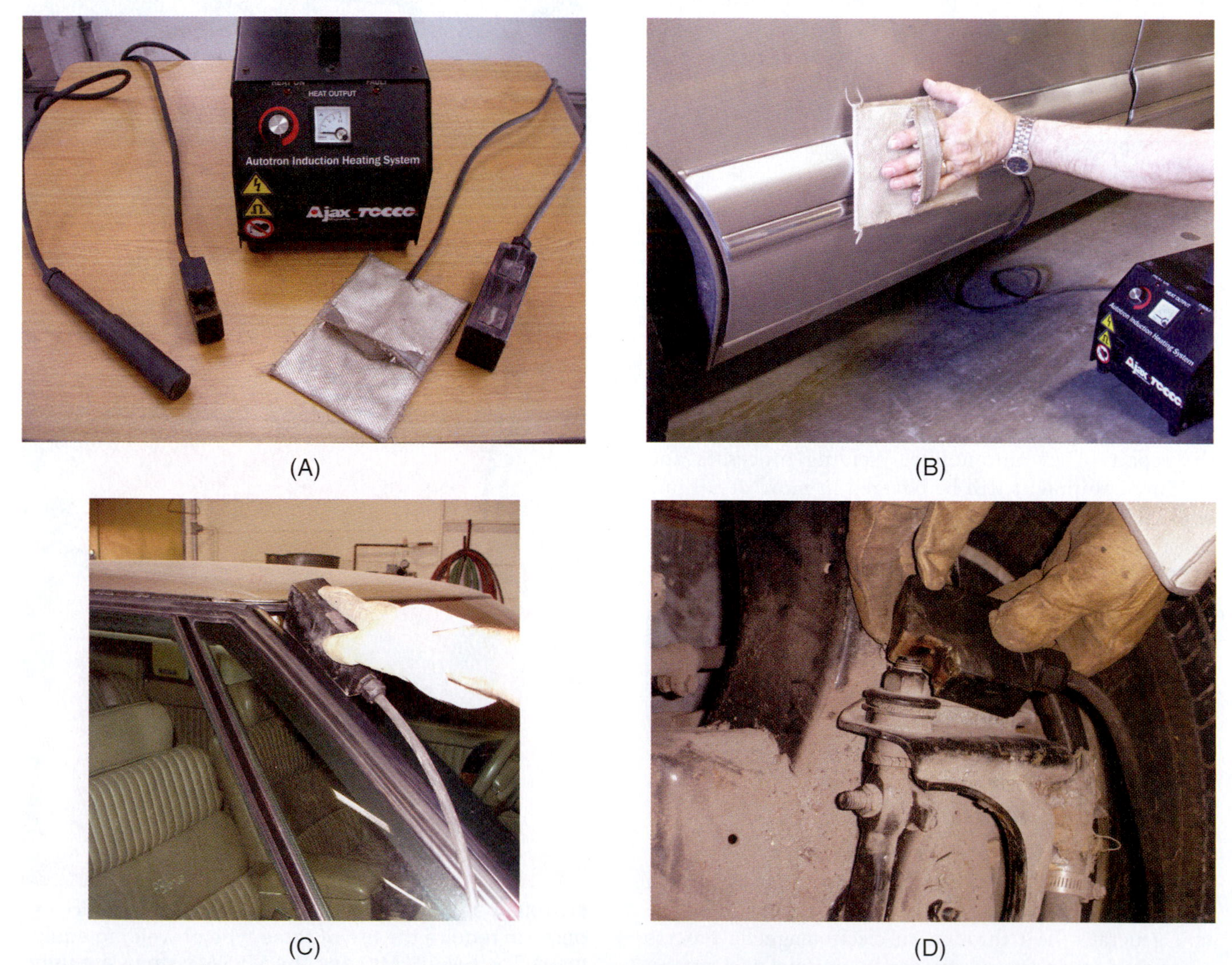

FIGURE 4-34 The induction heater (A) is used for numerous operations where heat application is required, including removing adhesive bonded moldings (B), removing adhesive bonded windshields (C), heating corroded bolts and fasteners (D), and so on.

FIGURE 4-35 The chop/cutoff saw can be used for cutting any metal item—including bar stock, angle iron, and nearly any other application—where a square cut is necessary.

FIGURE 4-36 The minimum PPE requirement for using a bench or pedestal grinder is safety glasses, a face shield, and gloves.

is the precision and accuracy of the cut that it makes. The abrasive wheel used to perform the cutting makes it an ideal media for cutting nearly any type of metal. Some common sense safety precautions must be employed when using this saw. The minimum PPE required should be safety glasses and a grinding face shield, as a significant spark pattern will emerge from the metal as it is being cut. Leather gloves should also be worn as the cutting surface and the adjoining areas become very hot when being cut. One should always inspect the cutting abrasive wheel to ensure it has not inadvertently been damaged during a prior use. A chip on the blade's cutting surface or missing fragment as small as 3 mm (1/8 inch)—or a crack—can cause the saw to vibrate quite severely and even cause the blade to disintegrate. The saw's high rpm rate will cause shrapnel to be thrown throughout the area, possibly causing serious personal injury.

To prevent "jamming" the cutting wheel onto the surface being cut, the operator should always pull the saw down to touch the blade onto the cutting material and let up on it slightly before engaging the trigger. Once the blade is turning up to speed, the saw should slowly be pulled down on the material and held there until the cut is completed. Care must be taken when picking up the pieces that were just cut as they will be very hot and cause burns to the hands.

One must also make sure the item being cut is firmly secured onto the bed of the saw body either with the built-in vise or by clamping it down with pliers or other holding tools. Technicians should NEVER try to hold the item with their hands while attempting to make a cut. A slight shifting of the item can cause the blade to bind and—ultimately—disintegrate.

Bench Grinder

The bench or pedestal **grinder** is used for a variety of applications in the repair shop, ranging from sharpening tools, shaping and forming pieces of metal, removing excess material, and truing up the end of a piece of metal. Whatever the application, some common sense things must be kept in mind. Even though the grinding shield is in place on the grinder itself, the user must always wear safety glasses along with a full face grinding shield whenever using this equipment. See **Figure 4-36**. Small hot metal shavings or particles are always created when a technician performs a grinding operation. One such hot stray spark in the eye may cause a technician to permanently lose eyesight.

Tool Rest. The placement of the tool rest is yet another safety issue the user must observe. The tool rest is a small platform attached to the grinder immediately in front of the grinding wheel. The technician lays the tool or whatever item is being ground on this platform to stabilize and support it. This tool rest must be adjusted so that the space between it and the grinding wheel does not exceed 1/8 inch. See **Figure 4-37**. A gap in excess of 1/8 inch will

FIGURE 4-37 The placement of the tool rest on the grinder is critical for the safe operation of the equipment. It should be positioned approximately 1/8 inch from the face of the grinding wheel.

allow the force of the grinding wheel to pull the object inward and wedge it between the grinding wheel and the grinder's housing, possibly causing the wheel to shatter. Flying debris and shrapnel may occur as the grinding wheel disintegrates, possibly injuring the operator and anyone standing or walking nearby.

Proper Grinding Techniques. A technician should never use the side of the wheel for a grinding surface. The side pressure exerted against the wheel can cause an undue strain, causing the wheel to disintegrate. Also, the grade or coarseness of the grinding wheel should match the task being performed. For example, when sharpening a drill bit, a fine grade abrasive wheel should be used. Conversely, when truing up the end of a piece of 1/4-inch thick metal, a coarse grade wheel should be used. Since the wheel is not easily changed out each time a different grinding operation is performed, the shop will select one that is applicable for a variety of operations. Unfortunately, in doing so the tool operates less efficiently than it should on most of the operations for which it is used. Some shops have grinders set up for specific operations. This becomes increasingly critical when performing aluminum repairs in the shop when cross metal contamination could become an issue. Even though the shop may be separated and the operations are isolated, the tools and equipment may inadvertently be used interchangeably.

Truing Tool. Another common mistake made when a technician uses the bench grinder is only using the middle of the wheel to grind a small or narrow object, thereby wearing a groove into the middle in the wheel. A grinding wheel dressing tool must be used to true up the wheel once it has become grooved. See **Figure 4-38**. To avoid grooving the wheel, the technician should move the object being ground side to side, being careful not to lose

FIGURE 4-38 The grinding wheel periodically requires truing or dressing with a special truing tool to remove grooves that may have been made through improper grinding.

FIGURE 4-39 Many times the grinder is set up with a grinding wheel on one side of the shaft and a wire wheel on the other. This setup is often used to clean corrosion or other flakey deposits on various metal surfaces.

contact with the wheel. Thus, the technician uses the entire face of the wheel.

Wire Wheel. Quite often the grinder will be set up with a wire wheel on one side of the tool and the abrasive wheel discussed previously on the other. The wire wheel is frequently used for cleaning rust and corrosion off surfaces in instances where using a **hand grinder** or another grinder may not be desirable, as using it will mar the surface. See **Figure 4-39**. The wire wheel can also be used to clean the threads on reusable bolts which may be covered with corrosion. A quick pass over the wire wheel will clean these up, making them much easier to re-install. The most significant problem encountered when using the wire wheel in this manner is holding the item firmly enough to prevent it from shifting when pushed against the wheel. A locking pliers should always be used to hold the item securely. Never try to hold a small item against the wheel—not even with a gloved hand—as it will invariably result in some lost skin on the fingers.

The biggest danger in using the wire wheel is the operator becomes impatient and starts to push the object against the wire wheel a little too hard, causing the wire bristles to bend. The harder the object is pushed against the wire bristles, the more they bend. Before long they pull the object out of the operator's hand and wedge it between the wheel and its housing. This can cause damage to the tool, the item being dressed with the wheel, or both. The same PPE must be used with a wire wheel as when using the grinding wheel, as the wire bristles occasionally dislodge from the wheel and become airborne. These can be even more dangerous than the sparks created when grinding metal as they can penetrate and lodge more deeply in the skin or—worse yet—into the eye.

Summary

- Personal safety must be PRIORITY ONE for the auto collision repair technician.
- Developing and habitually practicing proper safety procedures, using the recommended PPE, and never taking the attitude "it won't happen to me" will help to minimize the possibility of accidents and personal injury.
- To ignore any of the recommended safety procedures is almost a guarantee that an injury will eventually occur.
- The collision repair technician uses a wide variety of hand and power tools during the repair of even the simplest repair procedure.
- Many tools, simply by virtue of their incorrect use, can cause serious injuries to the individual using them, as well as innocent bystanders.
- Technicians should familiarize themselves with any tool or piece of equipment—noting do's and don'ts and the recommended safe use by the manufacturer—before using it.
- Proper maintenance of a tool will result in trouble-free use by the owner and extend its life expectancy.
- Tools have historically revolutionized the manufacturing and service industry, and that will certainly continue in the future.

Key Terms

air ratchets and impact wrenches
air tools
circular sanders
dent pullers
drills
dual action (DA) sanders
finish sanders
grinders
ground fault circuit interrupter (GFCI)
hand grinders
induction heater
metal cutting tools
oscillating/orbital sanders
permanent shop equipment
personal protective equipment (PPE)
power tools
random orbital sanders
reciprocating
screwdrivers
welders

Review

ASE Review Questions

1. Technician A says that electric tools are the technician's tool of choice because they are lighter weight than air tools. Technician B says a right angle grinder that has a disc speed of 2,500 rpm is the preferred choice for paint and filler removal. Who is correct?
   A. Technician A only
   B. Technician B only
   C. Both Technicians A and B
   D. Neither Technician A nor B

2. Technician A says chrome-plated hand sockets can safely be used with an air ratchet. Technician B says the power of an air ratchet is determined by the number of veins on the drive rotor. Who is correct?
   A. Technician A only
   B. Technician B only
   C. Both Technicians A and B
   D. Neither Technician A nor B
3. Technician A says the capacity of a drill is determined by the largest size arbor the chuck is able to hold. Technician B says the nibbler is used to break spot welds for panel removal. Who is correct?
   A. Technician A only
   B. Technician B only
   C. Both Technicians A and B
   D. Neither Technician A nor B
4. Technician A says, when using an air ratchet, the trigger should be depressed until the tool stalls out and then finish tightening the fastener by hand. Technician B says the tool rest on the bench grinder should be adjusted to within 1/8 inch of the grinding wheel. Who is correct?
   A. Technician A only
   B. Technician B only
   C. Both Technicians A and B
   D. Neither Technician A nor B
5. Technician A says the air hammer or chisel is frequently used for removing corroded bolts and nuts. Technician B says the die grinder is commonly used to dress damaged and corroded bolt threads. Who is correct?
   A. Technician A only
   B. Technician B only
   C. Both Technicians A and B
   D. Neither Technician A nor B
6. Technician A says using a GFCI will help an electric tool to run cooler. Technician B says the purpose for using a GFCI is to help avoid electric shock. Who is correct?
   A. Technician A only
   B. Technician B only
   C. Both Technicians A and B
   D. Neither Technician A nor B
7. Technician A says an impact wrench continues to tighten the fastener by a hammering effect after it has stalled out. Technician B says an air ratchet turns the bolt or nut with a continuous driving motion until it stalls out and then it stops. Who is correct?
   A. Technician A only
   B. Technician B only
   C. Both Technicians A and B
   D. Neither Technician A nor B
8. Technician A says the heavy-duty right angle grinder is the best choice for removing excess weldment that may occur during a repair process. Technician B says most pistol grip grinders rotate at a rate of approximately 10,000 rpm. Who is correct?
   A. Technician A only
   B. Technician B only
   C. Both Technicians A and B
   D. Neither Technician A nor B
9. Technician A says the induction heater is a flameless operation making it ideal to use around gas lines, fuel cells, and so on. Technician B says the impact gun may be used for installing wheel lugnuts so long as they are used with impact socket sticks. Who is correct?
   A. Technician A only
   B. Technician B only
   C. Both Technicians A and B
   D. Neither Technician A nor B
10. Technician A says an air tool should be lubricated with at least six drops of oil before putting it away after each time it is used. Technician B says the cutoff saw blade should be touching the material being cut before it is turned on. Who is correct?
    A. Technician A only
    B. Technician B only
    C. Both Technicians A and B
    D. Neither Technician A nor B

Essay Questions

1. What is the purpose of using the backing pad behind the abrasive grinding disc when using the grinder for surface removal? What is likely to occur if the backing pad is not used?
2. What are the minimum PPE requirements any time a tool or equipment is used if metal is removed from the surface?
3. Discuss the importance of the tool rest on a bench grinder being properly adjusted.
4. How does the "stinger" type dent puller differ from the stud gun regarding how they work?
5. Discuss and compare the advantages and disadvantages of electric and pneumatic tools.

Topic Related Math Questions

1. The cost of an air drill is $75.00. The cost for a comparable size electric drill is $47.00. The extension cord and the GFCI needed are $14.97 and $23.75, respectively. What is the more economical purchase and what is the difference in cost?

2. The cost to commercially bead blast the paint off a surface is $145.00. Manually removing the paint from the surface requires three man hours at $45.00 per hour, three abrasive grinding discs at $3.00 each, and six finish sanding discs at $1.50 each. Which option would be most cost effective? What would be the cost savings?

Critical Thinking Questions

1. Why should the impact wrench not be used with hand sockets?
2. What are the potential consequences of using an engine oil in an air tool?
3. What are the potential problems or hazards associated with using a 1/2-inch drill to drill a 1/8-inch hole?

Lab Activities

1. Install a 9-inch 36 grit grinding disc onto the right angle grinder with a 7-inch backing disc and attempt to remove paint from a painted surface. Now install the correct size backing pad and attempt the same grinding operation. Which system worked more efficiently? Discuss why. Discuss the potential problems and dangers that could be created with using the grinder incorrectly.
2. Obtain a battery powered 3/8-inch capacity drill and a pneumatic drill of the same capacity and secure a 3/8-inch drill bit into each. Drill a hole into a piece of metal of equal thickness. Which drill operates more smoothly, which will turn a higher rpm, and which can be used more easily away from the technician's work stall area?

Chapter 5

Welding Procedures and Equipment

OBJECTIVES

Upon completing this chapter, you should be able to:

- List the safety precautions needed to perform welding tasks.
- Explain the process for gas metal arc welding (GMAW).
- Describe the steps needed to set up a GMAW welder.
- Identify the parts of a GMAW welder.
- List three metal transfer processes for GMAW welding.
- Describe what shielding gas does during GMAW welding.
- Summarize why electrode stick-out is important.
- List the four welding positions.
- List six types of welds performed in collision repair.
- Compare and contrast weld defects and discontinuity.
- List at least five welding defects.
- Define and explain tungsten inert gas welding.
- Summarize how squeeze-type spot welding works.
- Explain the two roles of compressed gas in plasma cutting.
- Describe how a heat inducer tool works.

INTRODUCTION

Vehicle parts are joined together in three ways. Mechanically, parts are joined with fasteners such as nuts and bolts, rivets, and cotter pins. They can also be welded together with processes such as arc welding, resistance spot welding, gas welding, brazing, and soldering. In addition, parts are sometimes joined together using chemical methods such as adhesive bonding. Some parts of a vehicle are assembled with a combination of these methods; for example, with weld bonding or using adhesives with rivets or other types of mechanical fasteners.

As manufacturers continue to develop new manufacturing methods, the manner in which vehicles are joined together will also evolve. Technicians must adapt by learning and mastering these new methods as the industry changes. Within the past 30 years, technicians have needed to adapt from being skilled brazier and solder technicians working with steel, to technicians skilled in the welding and bonding of vastly different steels. Along the way, they have also learned about the use of structural aluminum, plastics, and composite materials.

Technicians today must be able to either re-join parts as they were joined during manufacturing, or be skilled at joining parts as recommended by the manufacturer. One of the most common ways of joining steel, aluminum, and plastic is through welding.

To clarify what will be discussed in the chapter, we must start by defining *welding*. Often the term is defined as the process of joining two metal pieces together using heat and a filler metal. However, this chapter will use the broader definition of welding given by the American Welding Society, which states that a weld is formed when separate pieces of material are fused together through the application of heat. This definition can also apply to the welding of plastic, which will be included in this discussion of welding. (The last category of fastening parts together, adhesive bonding, will be covered in Chapter 18, Plastic Repair.)

This chapter will present the automotive repair applications of:

- Gas metal arc welding
  - Equipment
  - Steel
  - Aluminum
- Gas metal arc welding operations
- Gas metal arc welding techniques
- Gas metal arc weld inspection
- Gas metal arc weld testing
- Squeeze-resistant spot welding
- Tungsten inert gas welding
- Induction heating
- Plasma arc cutting

SAFETY

While the collision repair technician should always be mindful of safety, the process of welding presents a large array of potential hazards that the technician must continually guard against. These safety concerns include:

- Burns
- Ultraviolet light
- Face and eye protection
- Ear protection
- Respiratory protection
- Electrical shock
- High-pressure gas cylinders
- Fire protection

All of these safety concerns for collision-welding technicians, their fellow workers, and the environment will be discussed in this chapter.

Burns

Burns are among the most frequent accidents to happen in the collision repair shop; they are also among the most painful. The process of welding produces large amounts of heat and light. Some of the light that is produced comes in the form of highly dangerous and burn-producing ultraviolet light. Burns are categorized into three types: first-, second-, and third-degree burns.

First-degree burns, the least severe of the three categories, are identified as those by which the surface of the skin reddens and becomes tender or painful, but no skin is broken. **Second-degree burns** cause the surface of the skin to be severely damaged with blisters and possibly with breaks in the skin.

First aid treatment for first-degree burns involves placing the affected area under cold water (but not ice) to reduce the heat. This treatment should be continued until the pain subsides, after which the area is patted dry and covered with a clean cloth. All second-degree burns should be treated by a medical professional to avoid the possibility of infection. Medical attention should also be sought for first-degree burns if the burned area is large or the pain persists.

Third-degree burns are the most severe of the burns; the damage extends through the top layer (**epidermis**) of the skin into the second (**dermis**) layer, and even beyond. Immediate medical attention by a medical professional is necessary for third-degree burns.

CAUTION

Burns of all degrees destroy skin, which is one of the body's first defenses against infection. Therefore, when a burn occurs, even a small one of only first-degree severity, the wound should be continually monitored until it has healed, in order to prevent infection.

To protect against burns, a technician must wear protective eyewear: both safety glasses with side shields and a welding helmet with a protective filter lens to protect the eyes from ultraviolet light. See **Figure 5-1** and **Figure 5-2**. These filter lenses, sometimes called filter plates, are rated from 4 to 12. Technicians should also protect their skin from ultraviolet (sunburn) rays by wearing fire-resistant clothing with long sleeves, as well as protective welder's gloves with long protective cuffs (gauntlets). See **Figure 5-3** and **Figure 5-4**. Trousers worn while welding should not have cuffs, as they may catch molten slag. In addition, shoes with protective covers over the laces should be worn to keep slag from getting into the shoes. See **Figure 5-5**. Lastly, the top button of the technician's shirt should be buttoned, so the skin under the chin will not become burned by ultraviolet light. See **Figure 5-6**.

Ultraviolet Light

Ultraviolet (UV) light is produced when welding occurs. It can exist without the technician feeling or seeing it.

FIGURE 5-1 Safety glasses should be worn at all times when welding, even while using a welding helmet.

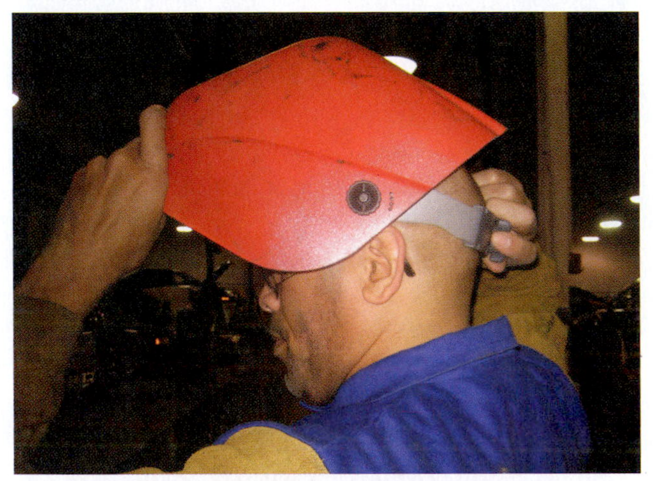

FIGURE 5-2 A welding helmet protects the eyes from ultraviolet light.

FIGURE 5-3 A welding coat protects the technician from sparks and potential UV burns.

FIGURE 5-4 Welding gloves with gauntlets protect the technician's arms. Note the gauntlet extends over the coat sleeve for added protection.

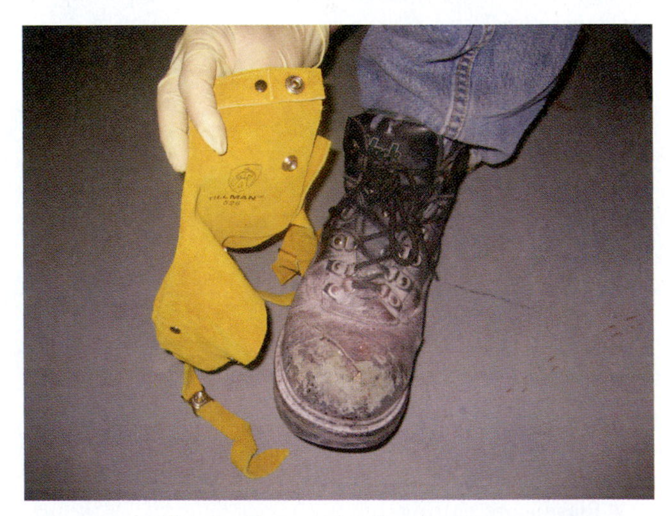

FIGURE 5-5 This leather protective lace cover prevents sparks from getting into the technician's boots.

FIGURE 5-6 If the top button of the technician's protective coat is not closed, the technician may have an exposed neck when welding overhead.

This light produces sunburn, and can produce both first- and second-degree burns. In fact, the grainy, sand-like feeling that occurs when the whites of the eyes are burned by light is a second-degree burn of the skin of the eye.

The welding technician should use eye protection (safety glasses and a welding helmet with protective lenses), as well as gauntlet gloves and a welder's coat buttoned to the top, so that no skin is exposed to the ultraviolet light produced by welding.

Face and Eye Protection

Eye protection does not mean just wearing Occupational Safety and Health Administration (OSHA) approved safety glasses with side shields, which should be worn at all times regardless of regulations. See **Figure 5-7**. It also means wearing safety goggles for protection against splashes when chemicals such as cleaners are used. See **Figure 5-8**. Welding technicians should wear face shields with safety glasses underneath them. They should also wear welding goggles for brazing and cutting, as well as welding helmets. See **Figure 5-9** and **Figure 5-10**.

The eye is an exposed organ, and is therefore easily vulnerable to injury. Materials such as flying debris from grinders and other spinning tools can injure the unprotected eye. Eyes can also become burned on both the outer white and the inner retina, so technicians should protect themselves at all times by wearing the necessary protective equipment.

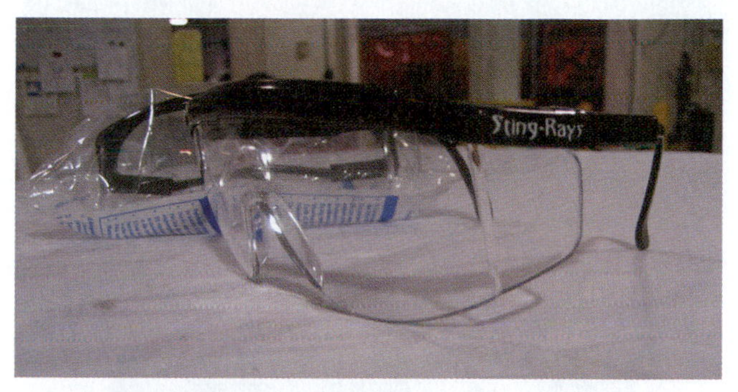

FIGURE 5-7 Safety glasses with side shields protect the technician from chemical splashes.

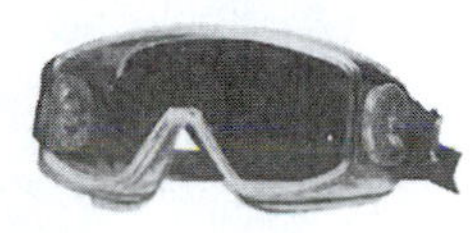

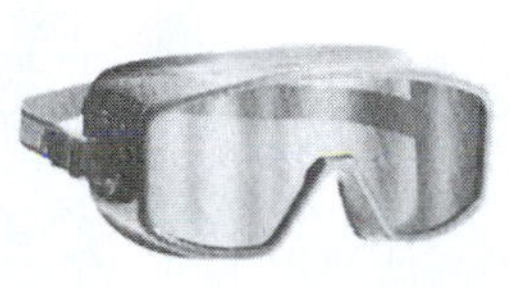

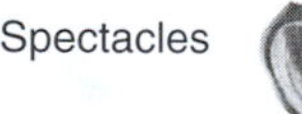

**Non-sideshield spectacles are available for limited hazard use requiring only frontal protection.*

FIGURE 5-8 Protective eyewear comes in many different types for many different applications.

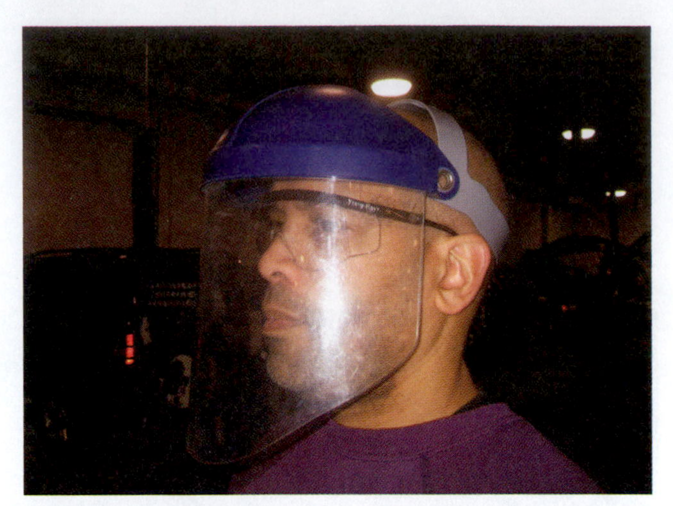

FIGURE 5-9 Welding technicians should wear a full face shield with safety glasses.

FIGURE 5-10 Welding helmet safety glasses are worn underneath the helmet.

Ear Protection

Ear protection is also necessary during welding repairs. The processes of grinding and welding produce flying and dropping embers, which can fall into the open ear. Additionally, noise in the body shop can reach high levels, and even low levels of sustained noise can cause permanent hearing loss. Technicians should protect their ears and hearing. Ear protection comes in several forms, such as earmuffs and earplugs. See **Figure 5-11** and **Figure 5-12**. When noise is particularly high, both earplugs and earmuffs can be worn together.

Respiratory Protection

Respiratory protection must also be used when welding. Technicians should protect themselves from these fumes by using a respirator. To know which type of respirator to use, welding technicians should read the material safety data sheet (MSDS) for the material being welded, which will tell them what type of respirator should be worn during the welding procedure.

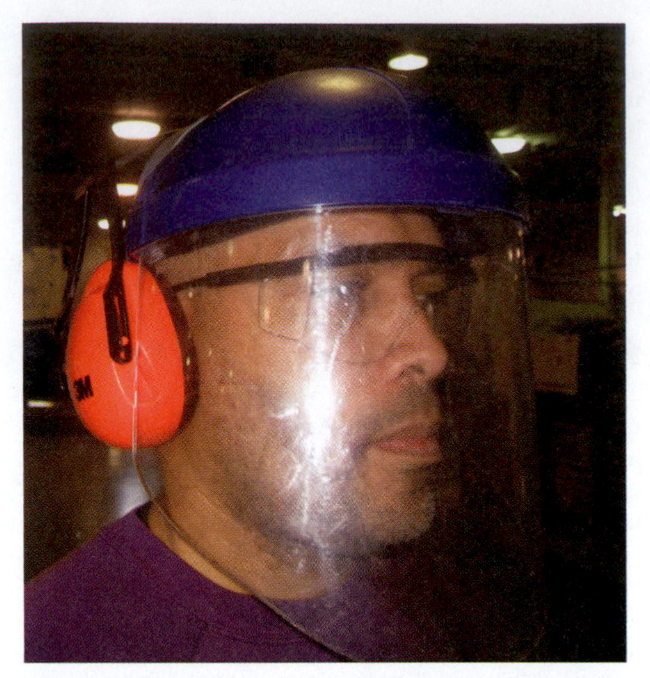

FIGURE 5-11 Ear protection can be used for both dampening noises and keeping hot sparks and slag from getting into the ear canal.

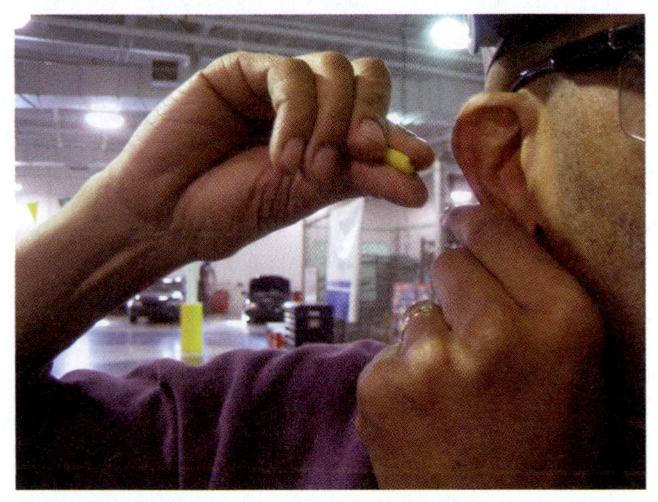

FIGURE 5-12 Properly inserted earplugs can be used with an earmuff for added protection.

Electrical Shock

Electrical shock is another safety hazard technicians face during welding. A weld is produced with electricity; therefore, the following normal electrical precautions should be adopted: Do not weld in a wet area or in standing water. Make sure that the electrical ground is working properly before welding. Make sure that all electrical connections are secured properly (power and work cable). If any type of extension cord is used, make sure it is rated for the amount of electrical current that will be flowing through it, and that it is in good working order.

High-Pressure Gas Cylinders

High-pressure gas cylinders, even the ones that are not flammable, are still a potential danger in the welding process. Because the gas in the tank is stored under pressure, if the tank were to become damaged the sudden release of the gas could cause an unsecured cylinder to explode or become a missile. To guard against explosions, flammable gas should be stored separately from nonflammable, and inert gas should be stored with the nonflammable gas. In addition, the tanks should be secured so they will not fall over. See **Figure 5-13**. The most vulnerable spot on a gas cylinder is the top, where the valve attaches. See **Figure 5-14**. When not in use, gas cylinders must be stored with their safety caps screwed in place. See **Figure 5-15**. Gas cylinders, whether flammable or nonflammable, should never be stored on their sides!

Fire Protection

Technicians must also actively consider fire protection. Fire in a body shop is a constant fear, and every precaution should be taken—requiring knowledge of both how to avoid a fire and, if necessary, extinguish one if it occurs. When welding on a vehicle, it is best for welding technicians to ask someone to be their fire watch. While a technician is welding with protective gear on and the welding helmet down, a fire can start and go undetected by the welder. If another person is providing a fire watch with a fire extinguisher close at hand, any small fire that might start can be quickly extinguished before it becomes a serious hazard.

FIGURE 5-13 To prevent them from falling over, cylinders should be secured with safety straps and protective caps.

Fire extinguishers come in four types: Type A fire extinguishers are used to control combustible solids such as paper, wood, and clothing. The symbol for a type A fire extinguisher is a green triangle with the letter A in its center. See **Figure 5-16**. Type B fire extinguishers are used to

FIGURE 5-14 This cylinder neck safety cap is being installed for cylinder storage.

FIGURE 5-15 The safety cap protects the cylinder's neck, the weakest and most easily damaged part of the cylinder.

FIGURE 5-16 A type A fire extinguisher—used for burning paper and wood—is symbolized by a green triangle with an A in the middle.

put out fires caused by oil, gas, paint thinners, and similar substances. Their symbol is a red square with the letter B in the middle. See **Figure 5-17**. Type C fire extinguishers are used to put out electrical fires such as those in fuse boxes and welding machines. The symbol for type C fire extinguishers is a blue circle with the letter C in the middle. See **Figure 5-18**. Type D fire extinguishers are used to put out fires involving combustible metals such as magnesium, zinc, and titanium. The symbol for a type D fire extinguisher is a yellow star with the letter D in the center. See **Figure 5-19**.

All four types of fire extinguishers should be readily available in the collision repair shop. Vehicles are now manufactured with flammable metals; therefore, even a type D fire extinguisher could be necessary, and at least one type D fire extinguisher should always be ready for use.

FIGURE 5-17 A type B fire extinguisher—used for flammable liquids—is symbolized by a red square with a B in the middle.

FIGURE 5-18 A type C fire extinguisher—used on electrical fires—is symbolized by a blue circle with a C in the middle.

GAS METAL ARC WELDING

Gas metal arc welding (GMAW), commonly referred to as **metal inert gas (MIG)** welding, is the most commonly used form of welding in the automotive collision repair industry. See **Figure 5-20**. It was originally developed in 1940 as a way of fusing aluminum and other nonferrous metals together, and it was quickly adapted to steel because it was faster than other welding processes of the day.

Further developments in the 1950s and 1960s gave the process more versatility, which included the adaptation to robotic welding machines. These welding robots were soon widely involved in the manufacturing process. Though robotic resistance spot welding is the most common method of welding during manufacturing, robotic GMAW is a close second.

In the repair facility, GMAW (or MIG) welding is most often performed by technicians. However, there are differences in MIG welding equipment and their setup procedures. A MIG welder can be set up to weld aluminum; however, with a different configuration, a MIG welder can be set up to weld steel. Still another MIG welder setup can be configured to weld silicon bronze. A closer look at the equipment used to MIG weld will further explain the processes.

FIGURE 5-20 This technician (with personal protective gear) is using a GMAW (MIG) welder.

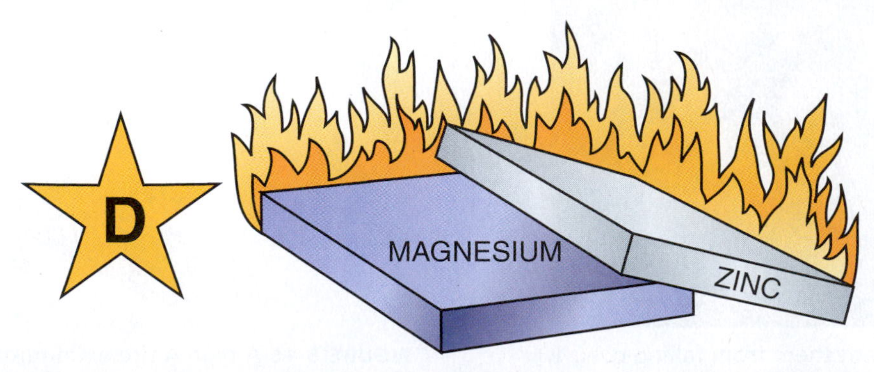

FIGURE 5-19 A type D fire extinguisher—used for flammable metals—is symbolized by a yellow star with a D in the middle.

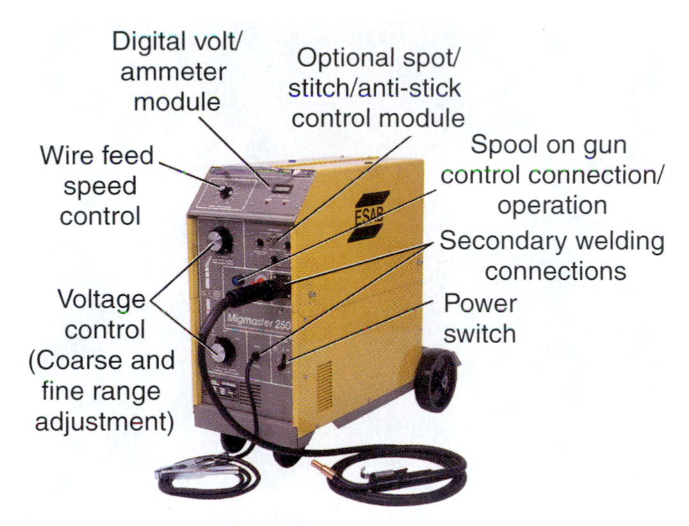

FIGURE 5-21 A GMAW (MIG) welder indicating its major components.

Equipment: Welding Steel

The following equipment is needed to gas metal arc weld (GMAW) or metal inert gas (MIG) weld:

1. Welding gun and wire feed unit
2. Power supply
3. Electrode
4. Shielding gas
5. Gas flow meters

See **Figure 5-21**.

Welding Gun. The **welding gun** has a switch which the operator activates when ready to weld. See **Figure 5-22**. The contact tip is what the welding electrode or wire passes through. A power cable transmits the power from its supply in the machine to the electrode and finally to the workplace, where the weld is formed. The nozzle directs shielding gas to the workplace, where it protects the weld from atmospheric contaminants as the weld is being formed. The wire electrode is driven through the inside of the power cable by a drive wheel, which can push the wire at rates as fast as 1,200 inches per minute (although the more common rate is from 2 to 10 inches per minute).

Power Supply. Though the power supplied to the welding machine is from alternating current, the welder uses direct current from a **power supply** to produce the welding current. See **Figure 5-23**. The electrode wire is generally positive. It tends to have the greater heat, allowing for faster wire melting and better weld penetration into the base metals. Though not generally done, one can reverse the polarity, with the electrode wire being the negative and the work clamp being positive. This technique reduces the risk of burnthrough when welding thin sheet metal.

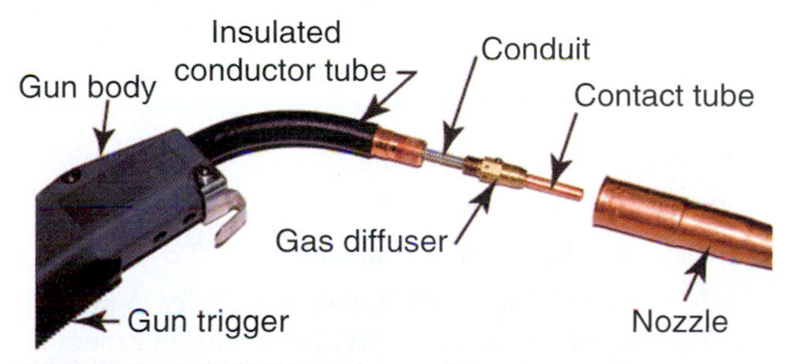

FIGURE 5-22 GMAW torch nozzle components.

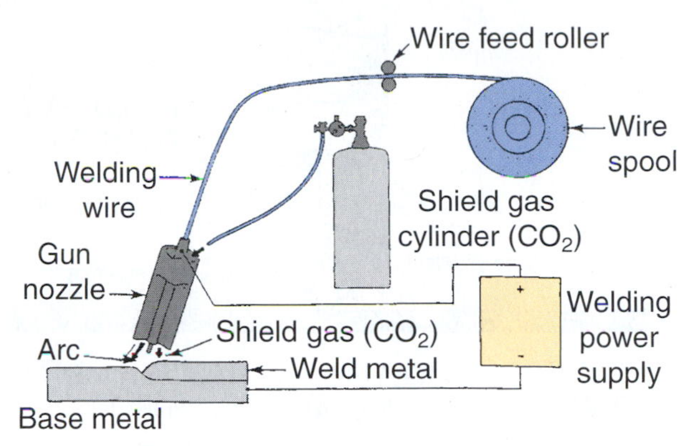

FIGURE 5-23 The principal parts of a GMAW (MIG) welder include the power supply and electrode wire.

One of the advantages to using such small electrode wire in collision repair work (0.023 to 0.035 inches) is the relatively low amperages that are needed. Therefore, welding units powered with 110-volt electricity can be used.

Electrode. The **electrode** in MIG welding is the small wire that comes out of the contact tip. This wire will vary in size and material makeup, depending on what is being welded—and, to a lesser degree, what type of joint is being welded and the position the joint is in when being welded. **Electrode wire** can run from 0.7 to 2.4 mm (0.023 to 0.035 inches) but can be as large as 4 mm (0.16 inch). The three most common electrode sizes in collision steel repair are 0.023 inch, 0.030 inch, and 0.035 inch. The specific electrode to choose for each type of application will be covered later in this chapter.

Shielding Gas. **Shielding gas** is used in GMAW to shield the weld and the surrounding areas, so that normal atmospheric gases such as nitrogen and oxygen do not interfere and contaminate the weld as it is formed. See **Figure 5-24**. If allowed to contaminate the weld as it is being formed, these naturally occurring gases will cause such defects as poor fusion, porosity (small gas bubbles in the weld), or weld metal embrittlement. See **Figure 5-25**. The gases used originally were inert (gases that form no chemical compounds), thus resulting in the more common name of metal inert gas welding, or MIG. MIG welding was first developed for welding nonferrous metals such as aluminum. When aluminum is GMAW welded today, technicians use an inert gas (argon). When GMAW welding was adapted for steel welding, gas mixtures such as 75 percent argon and 25 percent carbon dioxide were recommended. Mixtures of 70 percent argon, 28 percent carbon dioxide,

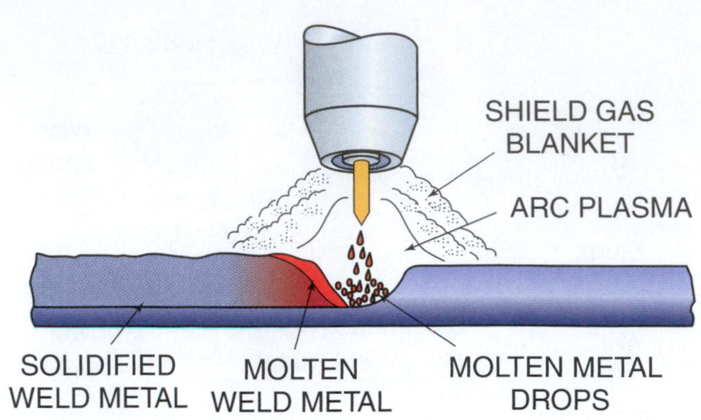

FIGURE 5-24 Shielding gas surrounds weld metal.

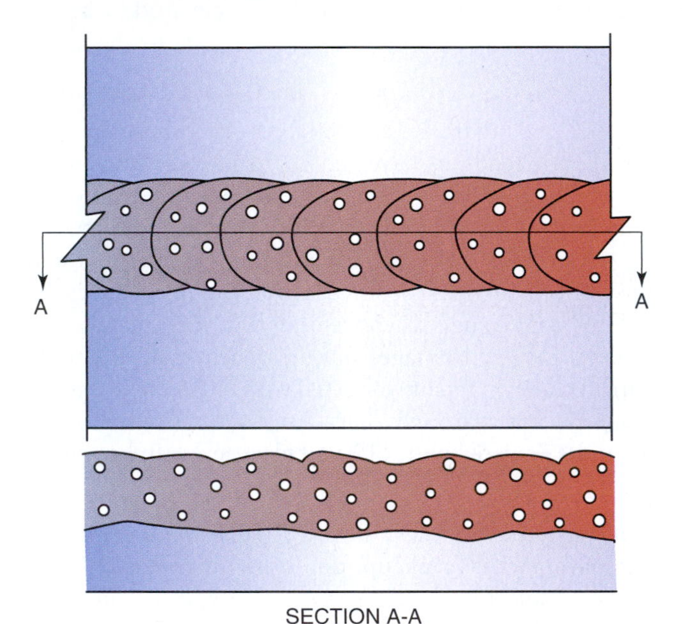

FIGURE 5-25 Porosity is defined as small gas bubbles in the weld.

FIGURE 5-26 An ultra mix of argon, carbon dioxide, and oxygen is very popular in GMAW welding today.

FIGURE 5-27 In this two-stage gas meter and regulator, the gauge on the right shows pressure in the tank, while the gauge on the left shows pressure in the hose.

and 2 percent oxygen have now become very popular. See **Figure 5-26**. Adding gases such as carbon dioxide and oxygen, which are active gases, will cause these two gases to mix with the weld, thus improving its quality.

Gas Flow Meter. The **gas flow meter** is a device which takes high-pressure gas stored in the gas cylinder and reduces it to a usable level and flow rate to supply the GMAW welder. See **Figure 5-27**. However, there is an electrically regulated open and shut valve in the welder. When the trigger is pulled, not only does it activate the wire electrode feed rollers, but it also turns on the flow of gas from the cylinder to the welding gun. The flow regulator controls the rate of flow that surrounds the weld area, enabling the proper amount of shielding gas to shield it from outside atmospheric gases. The proper rate of flow depends on the position of the weld being performed, the type of electrode being used, and the condition of the air movement around the weld area. Generally, the flow meter is set at 25 to 30 CFH (cubic feet per hour). See **Figure 5-28**. Performing GMAW welding outside with the wind blowing does not lend itself to good welding conditions, because the shielding gas will be blown away from the weld area.

Electrode Wire Type. GMAW welding uses the same equipment that was previously listed. The electrode wire that is to be used will be chosen relative to the material being welded. The American Welding Society, or AWS, has special identity codes for the different types of welding electrode wire. A common AWS code is AWS-ER70S-6, which is the wire that is recommended for welding most automotive steel. See **Figure 5-29**.

The code means:

- AWS = American Welding Society
- ER = Electrode rod
- 70 = Tensile strength of at least 70,000 psi
- S = Chemical makeup of wire and ability to keep oxygen out of the weld (Weld deoxidizers generally include manganese and silicon.)

FIGURE 5-28 In this flow meter with gas level 25 CFM, the red ball indicates the amount of gas flow when the trigger is pulled.

- The last number in the AWS code indicates the deoxidizers in the wire. Each number indicates a different element that acts as the deoxidizer during welding. The general recommendation is to use either 6 or 7, which will provide a good quality weld with reduced spatter.

Wire Size. Though wire size is chosen relative to the thickness of the metals being joined, wire size recommendations are often described with an amp range, such as 0.023 when welding from 30 to 90 amps. The reason for this is that if a welder were to be welding a piece of metal that is 0.8 mm thick, the welder manufacturer may recommend a welding setting of 30 amps.

Wire diameter recommendations:

- 0.6 mm (0.023 inch) wire is recommended for welding at 30 to 90 amps.
- 0.8 mm (0.030 inch) wire is recommended for welding at 40 to 125 amps.
- 0.9 mm (0.035 inch) wire is recommended for welding at 50 to 180 amps.

Transfer Process. In the GMAW welding of steel, the three metal transfer methods that technicians use are spray arc, short circuit, and globular. **Globular metal transfer** is the least desirable method of metal transfer. In the globular transfer method, as the weld is struck a ball of molten

WELDING GUIDE
Settings are approximate. Adjust as required.

Material	Thickness	Process	Wire		Gas		Polarity	Stickout	Welding Voltage	Wire Speed Control
			Class	Size	Type	Flow				
Carbon Steel	24 ga	GMAW	ER-70S-6 (HB 28)	0.024	CO_2 or C_{25}	20 CFH	DCEP	1/4	1	5.5–6
	18 ga							1/4–5/16	1	6–7
	16 ga							5/16–1/2	2	6.5–7
	10 ga							5/16–1/2	3 (4)	7 (8)
	3/16″							1/2	4	8
	18 ga			0.030				5/16	2	5.5–6.5
	16 ga							5/16–1/2	2	6–6.5
	10 ga							1/2	3	6.5
	3/16″							1/2	4	7–7.5
	18 ga			0.035				5/16	2	5–5.5
	16 ga							5/16	2	5.5
	10 ga							1/2	3	6
	3/16″							1/2	4	6.5
Stainless Steel	10 ga		ER-308L 308L Stainless	0.030	C_{25}			1/2	4	7.5
Carbon Steel	18 ga	FCAW	E-717-11 Fabshield 21B	0.045	None		DCEN	1/2–3/4	2	5
	16 ga							1/2–3/4	2	5.5
	10 ga							3/4	3	6
	3/16″							3/4	4	6

24 ga = 0.022″, 18 ga = 3/64″, 16 ga = 1/16″, 10 ga = 1/8″
CO_2 = Carbon dioxide
CO_{25} = 25% Carbon dioxide + 75% Argon
DCEP = DC Volts Wire Positive
DCEN = DC Volts Wire Negative

FIGURE 5-29 This typical welding guide is used for determining welding electrode wire.

metal forms on the electrode wire. This ball is larger in diameter than the electrode wire. This ball is also often irregular in shape, and when it transfers to the base metal either from gravity or short circuiting, it leaves an uneven surface on the weld and often causes spatter. When tuning a welder, the technician might note that the weld sound is like popping corn. This uneven sound is the irregular forming of the weld glob and the noise it causes as it falls to the workplace.

Short Circuit Metal Transfer. As wire feed is increased, the transfer method becomes a **short circuit metal transfer** arc. See **Figure 5-30**. The arc is struck, and when the wire electrode burns back, the gap that is developed during the globular method is bridged. This transfer produces no large irregular buildup of molten metal, since the molten filler wire is added to the base metal and thus produces a more even weld pool and a better weld with less spatter. When a GMAW welding machine is properly adjusted, the sound is more like frying eggs than popping corn.

TECH TIP When setting up and tuning a GMAW welder for steel welding of collision repairs, the technician should select wire of either 0.023, 0.030, or 0.035 electrode size, using AWS-ER70-6 wire with the short circuit metal transfer method, which sounds like frying eggs, not like popcorn popping.

Spray Arc Transfer. The **spray arc metal transfer** method of GMAW welding is accomplished only at relatively high voltages and amperages. It deposits small droplets of molten electrode to the weld surface at a high rate, with the size of the droplets being a smaller diameter than the electrode wire. Once the spray arc is accomplished, the weld is said to be "on" at all times, and the sound of the arc is a spray sound or humming/buzzing sound.

Spray arc metal transfer is generally done on base metals which have a thickness of about 6 mm or greater in the flat or horizontal positions. Because collision repair does not involve repairing steel with that thickness, the spray arc transfer method is not recommended.

Equipment: Welding Aluminum

Though aluminum may be welded using many different methods, GMAW or MIG welding have some definite advantages in the collision repair industry. One advantage is that technicians already have familiarity with the equipment, which is the same or very similar to that used with steel MIG welding. The use of aluminum MIG welding is widely accepted, and most MIG welding equipment that presently exists in the collision repair facility can be converted to aluminum MIG welding. (However, if the facility chooses **pulsed-spray metal transfer**, it will need equipment capable of this method. More information on this concept will be presented later.)

MIG aluminum welding offers good production rates, reaches into tighter areas than other welding methods, and does not have the problems of high-frequency arc, as with the **TIG (tungsten inert gas)** welding process, which can be harmful to electronic equipment.

There are a few differences between the equipment used for aluminum MIG welding and that used for steel. These differences will be discussed in the following text.

Welding Gun. Because the wire electrode is softer and more difficult to push through the long gun liner as in steel, often aluminum welders are equipped either with a spool gun or a push-pull feeder gun. See **Figure 5-31** and **Figure 5-32**. The advantages to spool guns are that they have a short gun liner, the wire spools are quick and easy to change if wire size or type must be changed, and they have a drive wheel close to the weld source, thus presenting fewer weld wire problems. The disadvantage of using a spool gun is that, because the wire spool (though much

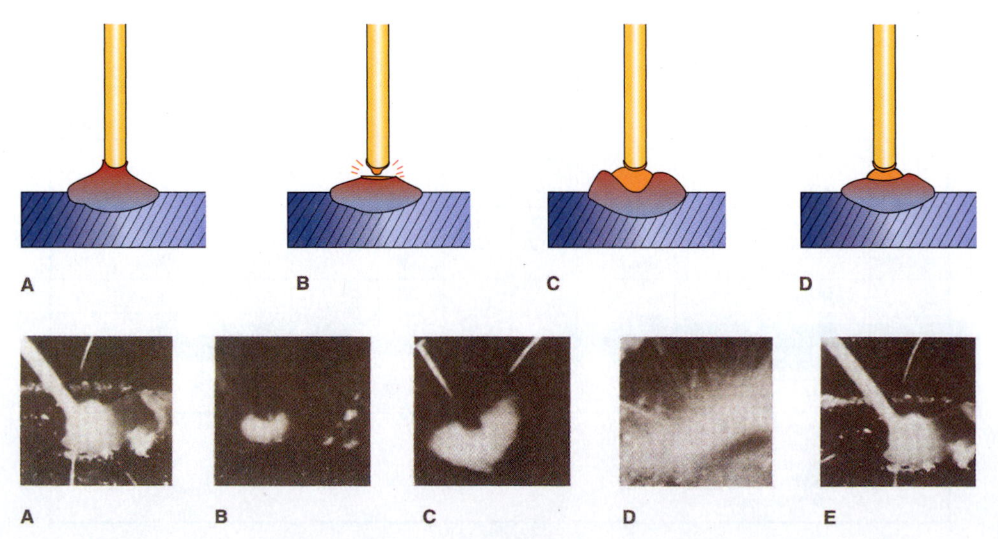

FIGURE 5-30 A short circuit metal transfer arc occurs as wire feed is increased.

FIGURE 5-31 This spool gun has a small spool of wire on the right and a push type drive wheel just ahead.

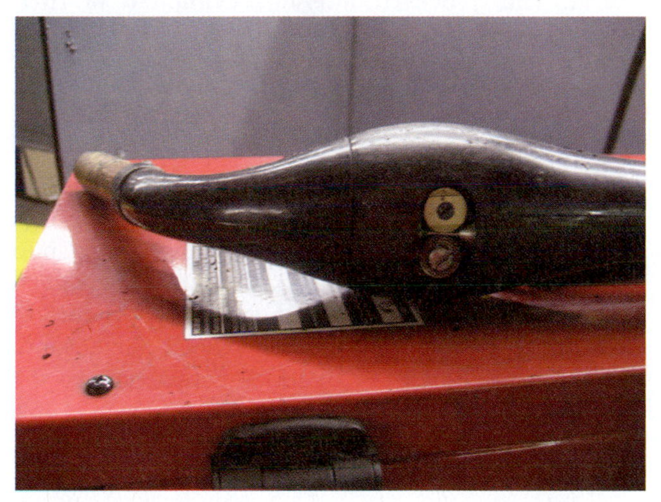

FIGURE 5-32 This push-pull gun is equipped with a common push wheel inside the welder and a second pull wheel at the gun end.

smaller) is located in the gun, it makes a spool gun much larger than a standard push wire feed or push-pull feed gun. This increased size of the gun limits access to the weld area. Push-pull feed guns have two wire feed motors, one in the welder close to the wire spool, and a second in the gun close to the weld source. The liner is much longer and more subject to dust contamination, but the addition of a second "pull" drive motor lessens the problems normally seen with a push-only drive wheel. Though there is a second drive motor in the push-pull feed gun, it requires much less space than what is needed for a drive motor and wire spool, as in the spool feed gun, thus eliminating the access problems seen with the spool gun.

Both the spool end and push-pull guns have a trigger, nozzle, contact tip, and electrode wire, as do steel GMAW welders. Both have gun liners, with the aluminum liners being made from nylon, Teflon, or coated steel, which allow for ease of passing the soft aluminum wire through the liner.

Power Supply. Since aluminum GMAW (MIG) welding equipment uses direct current, the supplied AC (alternating current) must be converted to DC, as with steel welding. The amount of current (both amperage and voltage) that is required to weld aluminum is higher than steel, and the use of 115-volt equipment is often inadequate. Thus, 208- or 220-volt machines are used. Also, when welding aluminum, the spray arc or pulsed-spray arc metal transfer method is used, and the welding machine's power source must be capable of these types of metal transfer welding.

Electrode. Aluminum welding wire electrodes are labeled as either 1000, 2000, 4000, or 5000—with the 4000 and 5000 series recommended most often for collision repair welding purposes. The numbers that come after the first one indicate the type of filler alloy. To determine the electrode wire to be used, a technician must know what type of aluminum alloy the vehicle was manufactured with. This information can be found in the vehicle collision repair service manual. The service manual often will recommend the type of wire electrode that should be used to weld a particular area of a vehicle; 4043 and 5356 wire is often recommended. The wire size is also relevant to the thickness of the material being repaired, with 0.030, 0.035, and 0.047 often being called for. Technicians should check and follow the vehicle makers' recommendations.

Aluminum welding wire is subject to oxidation. If left on a machine for long periods of disuse, the wire will deteriorate. The wire should be removed and stored in a cool dry place in an airtight bag, to reduce the amount of oxidation that will occur on the spooled wire. Because of the possibility of wire oxidizing, some collision shops prefer a spool gun because the wire spool is small, and if wire becomes contaminated they will not have lost a great deal of wire.

Shielding Gas. The gas used to shield aluminum GMAW welding is argon. Because argon is an inert gas, aluminum GMAW welding is truly a MIG (metal inert gas) welding process.

Gas Flow Meter. The gas is passed through a flow meter as with steel welding, though the flow rate is dependent on the transfer method, position of weld, thickness of the material being welded, and the nozzle size.

Transfer Process. The three transfer methods recommended for aluminum welding are short circuit, spray arc, and pulsed spray. Globular transfer, which is a type of metal transfer in GMAW welding, is not recommended for aluminum welding.

Short circuit metal transfer occurs when the wire touches the base metal being welded and electricity short circuits, thus melting both the base metal and the electrode wire. The molten wire passes to the base metal, and thus fusion of the wire and base metal occurs. This short circuit occurs at relatively low voltages, so it produces less penetration than either spray-arc or pulsed spray-arc transfer. When a properly adjusted, short-circuit metal transfer method is being used, the sound produced is a steady crackling.

The spray-arc transfer method uses a higher voltage and current than short circuit, and thus produces a larger, more fluid weld puddle with better penetration. A faster weld speed can be used in all positions. The weld "sprays" tiny streams of molten wire across the arc, with the molten drops being smaller than the electrode wire. This method of metal transfer makes a steady hum or spray sound as it welds.

Pulsed-spray arc is a spray arc transfer that pulses between high and low current as the spray arc occurs. This pulse occurs from 60 to 200 times per second, and transfers molten wire only during the high current pulse. Because of the pulsing there is less chance of weld burn-through. This method can be used in any **weld position**, and larger wire can be used than with non-pulsing spray transfer. The welding machine must be specially equipped with the ability to pulse the current, and thus the method requires sophisticated machines specifically designed for aluminum MIG welding.

Silicon Bronze or MIG Brazing

As the use of ultra high strength steel became more prevalent in the manufacturing of vehicles, new methods of bonding were introduced into the manufacturing world. Though on the manufactured vehicle it may look like old style gas brazing, the bronze was applied with a new method of **silicon bronze welding**, or **MIG brazing**. Using wire such as $CuSi^3$ or $CuAI^9$ when welding advanced high strength steel works well with pure argon as the shielding gas. The gun liner will need to be Teflon lined to ease the passing of the copper wire. MIG brazing heat is much lower than the temperatures at which high strength steel must be melted for GMAW. The temperatures that are too hot will cause the molecules of the steel to become brittle and make the steel susceptible to rust, so manufacturers have started to MIG braze them together.

Though this new type of MIG brazing has been used for some time in manufacturing, repair technicians should only use this method if the repair process is recommended by the manufacturer (and if the technician has the necessary equipment).

GAS METAL ARC WELDING OPERATIONS

This section will take the learner through the general setup of a GMAC welder and show how it operates. The following topics will be discussed:

- Welder setup
- Current
- Voltage
- Work clamp
- Gas flow
- Tip to base metal distance
- Gun angle
- Wire feed
- Welding speed
- Nozzle adjustment
- Welding positions
  - Flat
  - Horizontal
  - Vertical
  - Overhead

CAUTION

When setting up a new welder for the first time, the technician should read, understand, and follow the directions as provided by the manufacturer. The general guidelines explained in the following text are intended to give the reader only a general understanding of a GMAW welder and its operation.

A GMAW welder operates as the power enters the machine through its power cord. See **Figure 5-33**. This power could be 115 volts, 208 volts, 220 volts, or up to 440 volts on some larger industrial machines, depending on the machine and the power available. AC current passes through the power controller, which converts it to direct current. Direct current (DC) is then sent to the electrode as either **direct current electrode positive (DCEP)** or as **direct current electrode negative (DCEN)** current. Electrode positive (DCEP), sometimes called reverse polarity, is the current most often used by collision repair welders because it provides the best fusion, with less weld on the surface, and a more stable arc. Direct current electrode negative directs most of the heat to the workplace, and is used if flux-core wire is used. The volt-

FIGURE 5-33 The power cord (110 power cord) is the source of power for a GMAW welder.

FIGURE 5-34 Some GMAC welders use an interior push wire feeder.

FIGURE 5-35 Some GMAC welders use an exterior push wire feeder.

age controller also directs current to the wire feed motor. This motor supplies a steady speed of wire to the gun. As the wire feed is increased, the current is slightly increased. The voltage controller also activates an electric switch, so that when the gun is turned on, the gas valve is opened. Then shielding gas will flow through the gun cable to the gun and around the contact tip. The gas is controlled and directed to the weld site by the nozzle. Some GMAC welders have the **wire feeder** inside of the machine while others have the wire feeder outside the welder case. See **Figure 5-34** and **Figure 5-35**.

Welder Setup

The setup of a new welder is best done according to manufacturer directions. In general, the welder is set and secured on the mobile cart, and then the correct sized gas cylinder is attached to the cart. See **Figure 5-36**.

After the technician has secured the cylinder to the cart with a chain, the gas hose is then connected to the gas

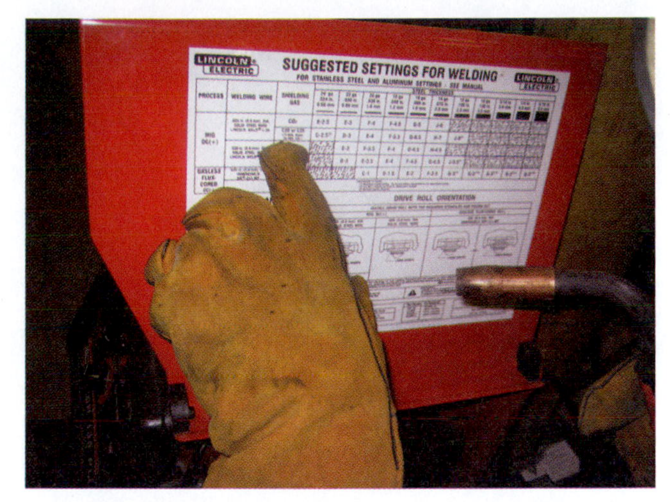

FIGURE 5-36 This technician is reading a setup chart.

TECH TIP An oversized gas cylinder attached on a welding cart will be very unstable, and therefore easy to tip over when moving the cart in the shop. See **Figure 5-37**. Gas cylinders are under extremely high pressure, and if the cylinder's top is damaged the cylinder will become a missile, which can obviously be very dangerous. When cylinders are transported, their protective tops must be in place. See **Figure 5-38**. If the cylinder is mounted on a welding cart, it should be securely fastened to the cart with a chain, and the cylinder should be of the correct size so the cart will not become unsteady. See **Figure 5-39**.

FIGURE 5-37 A GMAW (MIG) welder shown with cart.

FIGURE 5-38 A cylinder's protective top must be in place during transport.

FIGURE 5-40 Two-stage gas meter and regulator used with a cylinder.

FIGURE 5-39 This welding cart cylinder safety chain secures the cylinder to the cart.

FIGURE 5-41 Cylinder with flow meter to regulate gas flow.

inlet connector on the welding machine. The other end of the gas hose should be connected to the **gas regulator**, either a pressure regulator or a flow meter. See **Figure 5-40** and **Figure 5-41**. The gas regulator has two gauges: One shows the pressure in the gas cylinder, measured in pounds per square inch (psi), which can be very high; and the second has an adjustment valve which adjusts the gas flow. A gas flow meter also has a gauge that shows the pressure remaining in the tank. However, the second one—which may or may not be adjustable—regulates and shows the rate of gas flow in **cubic feet per hour**, or **CFH**. Many welders believe that flow of the shielding gas is very critical, and therefore prefer to use the flow meter. Some flow meters come preset to a specific flow rate, while other ones have an adjustable device so the rate of flow can be increased or decreased relative to changing welding conditions.

Once the technician has attached the tubing to the welding machine and the regulator or flow meter, the next steps are to remove the cylinder's safety cap, then open the tank valve for a short time to clear out any debris, then close the valve again. Finally, the technician inserts the gauge into the cylinder, tightening it with a wrench. See **Figure 5-42**.

Welding Gun and Cable. The welding gun is attached to the welder, according to manufacturer's recommendation, with the shielding gas line secured to the gun cable. Inside the gun cable, the shielding gas passes through and the welding wire is fed. When the gun's trigger is pulled, the gas on-off electric valve is activated, causing the gas to flow through the gun cable to the nozzle. There

FIGURE 5-42 This regulator is being attached to a cylinder.

FIGURE 5-44 The wire feed spring tension must be adjusted carefully by the technician.

it is directed around the contact tip by the nozzle, thus creating an envelope of shielding gas as the arc is struck. The metal transfer is therefore begun. When the gun's trigger is released, the current is shut off, the arc stops, the gas turns off, and finally the wire feed turns off. The wire feed stops after the current is turned off so that the weld arc does not allow the wire to burn back into the tip. This sequence allows for the proper amount of stick-out for the next strike of the arc. Some of the more sophisticated machines have an adjustment for setting the time that the machine will continue to feed after the current is turned off.

Wire and Wire Feed. Properly putting the wire spool on the machine and adjusting the wire feed tension are critical for smooth welding. The feed rollers come as either V-groove shaped, U-groove shaped, or V-knurled shaped. See **Figure 5-43**. With steel wire, a V-shaped driver roller is used. With larger wire, the V-knurled rollers work well. With soft wire such as aluminum or copper, though, the V shape and the knurling can be damaging to the wire and cause problems as the wire is passed through the lining.

The technician must also take care when adjusting the wire feed tension. See **Figure 5-44**. If too much tension is applied, the rollers can misshape the wire, making it difficult to drive through the lining. A common problem that occurs if the wire is difficult to pass through the lining is a bird's nest. See **Figure 5-45**. Also, with too little wire feed roller tension, the wire will not feed at all.

FIGURE 5-45 A bird's nest occurs if the wire is difficult to pass through the lining.

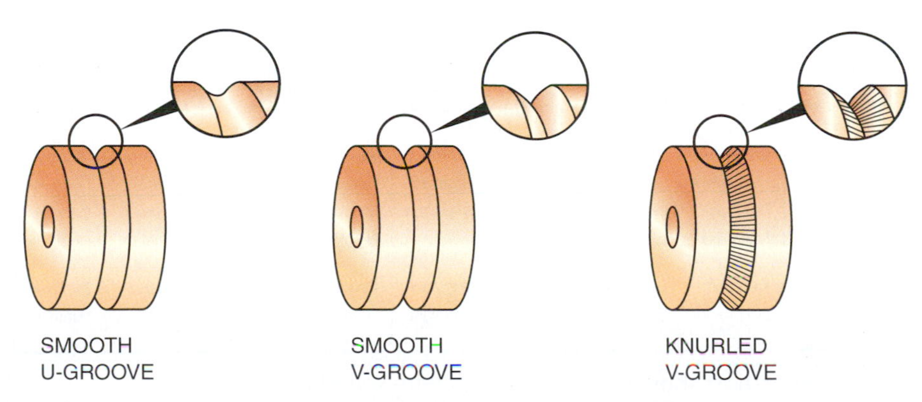

FIGURE 5-43 The V-groove and V-knurled groove drive wheel are examples of feed rollers.

Current

Current is the flow of electricity through a conductor (wire), and is measured in amperes (amps). The proper amp settings will differ depending on the type of metal transfer, wire size, shielding gas being used, and the base metal being welded. However, there are charts to help a technician find a "starting point." These charts are often found on the inside of the welder door. See **Figure 5-46**. By looking at the chart, the technician can find a welding amperage and voltage to start with. (More details on this process will be discussed later in the chapter.)

The amount of **amperage** (current) required to melt the base metal and the electrode for good welding differs with the wire feed setting. When the "wire feed" setting on the welding machine is increased, the welding machine also increases the amperage; if the wire feed is turned down, the amperage is also turned down. A technician must "tune" the welder to achieve the best weld for the type of metal being joined, and for the conditions of the weld. After finding the starting point settings on the chart, the technician then completes this "tuning."

FIGURE 5-46 This is the select amp range on a GMAW (MIG) welder.

When the current and wire speed are changed, several weld factors are affected. The depth of penetration, deposition rate, and the bead size will be affected accordingly. That is, when the wire speed (amps) is increased, the penetration, deposition, and bead size will also increase, and when the wire speed is turned down, they will decrease. See **Figure 5-47**.

As the previous information has demonstrated, the proper fine-tuning of a welder is very critical and is part of the learning curve that students must perfect to be a quality welder.

Voltage

Voltage is the force which pushes electricity through the wire. It is measured in volts; the higher the number of volts, the greater force the potential electricity has. When welding, voltage affects the arc length. When the length of the arc is set properly, and the wire feed (amps) is also correct, the arc metal transfer—along with the fusion of the metal being welded—will be exact.

To set the voltage before welding, the technician can use a chart (such as the one found in Figure 5-47) to set the machine. Then the amperage can be adjusted to tune in the wekder precisely.

Work Clamp

The **work clamp** is often called the ground, but that term is not correct. See **Figure 5-48**. The ground for the welder is one of the prongs of the welding tool that grounds the machine when plugged into the electrical supply. In contrast, the work clamp is used to complete the welding circuit. It should be fastened to the metal being welded as close as possible, so that the electrical flow of the welding circuit will not need to travel too far. In addition, when welding on a vehicle, the technician should locate the work clamp as close as possible, because the body of a vehicle is used as the path for vehicle electricity back to the battery. Strong currents such as welding currents can, if allowed to pass through the whole vehicle, cause damage to electronic components.

Welding Variables to Change	Desired Changes							
	Penetration		Deposition Rate		Bead Size		Bead Width	
	Increase	Decrease	Increase	Decrease	Increase	Decrease	Increase	Decrease
Current and Wire Feed Speed	Increase	Decrease	Increase	Decrease	Increase	Decrease	No effect	No effect
Voltage	Little effect	Little effect	No effect	No effect	No effect	No effect	Increase	Decrease
Travel Speed	Little effect	Little effect	No effect	No effect	Decrease	Increase	Increase	Decrease
Stickout	Decrease	Increase	Increase	Decrease	Increase	Decrease	Decrease	Increase
Wire Diameter	Decrease	Increase	Decrease	Increase	No effect	No effect	No effect	No effect
Shield Gas Percent CO_2	Increase	Decrease	No effect	No effect	No effect	No effect	Increase	Decrease
Torch Angle	Backhand to 25°	Forehand	No effect	No effect	No effect	No effect	Backhand	Forehand

FIGURE 5-47 A weld adjustment chart showing the effects of wire speed.

FIGURE 5-48 A work clamp (ground) is used to complete the welding circuit.

FIGURE 5-49 Electrode stick-out is the amount of electrode wire that must stick out of the contact tip before the technician starts welding.

Stick-Out

Stick-out is the amount of electrode wire that must stick out of the contact tip before the technician starts welding. See **Figure 5-49**. The wire "stick-out" length should range from 3/8 of an inch to 1/2 inch, without a large ball at the end of the welding wire. If a ball is noted on the wire, the technician should cut it off before attempting to start a weld.

Gun Angle

Gun angle refers to the angle at which a welding technician holds the gun in relation to the work. See **Figure 5-50**. As the gun angle changes, the shielding gas and the quality of weld are affected. Gun angle is also relative to the direction in which the weld is traveling. With the backhand method of welding, the technician drags the gun in the direction away from the weld pool. See **Figure 5-51**. The forehand method is used when the technician pushes the gun into the weld pool. See **Figure 5-52**. Different welding positions such as overhead, vertical, and horizontal each require different gun angles and direction.

Welding Speed

The **welding speed** is the rate at which the gun travels along the face of the weld. If the welding speed is too fast, the depth of penetration and the weld bead width decrease. If the weld travel speed is too fast, weld bead undercutting may occur. See **Figure 5-53**.

Contact Tip

The **contact tip** is a consumable item on a GMAC welder, which means that it wears out as electrode wire passes through it. The inside diameter of the contact tip is precisely sized so that specific-size wire can pass through it. That is, if a technician is welding with wire of 0.023 diameter, a 0.023-sized tip should be used. It is possible to pass smaller wire through a larger wire tip; however, the

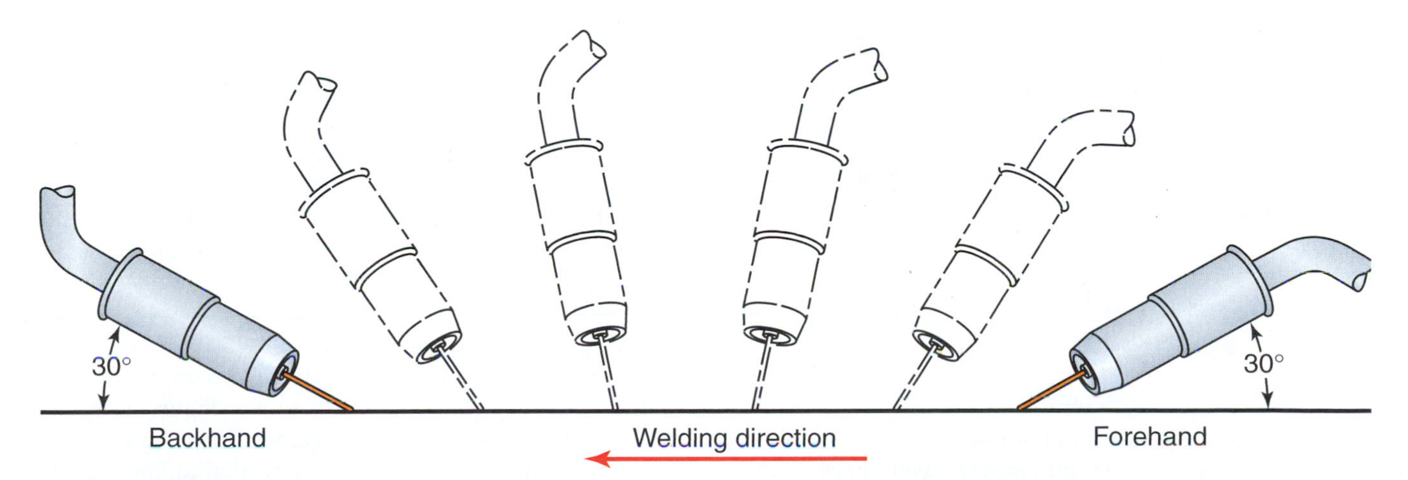

FIGURE 5-50 The push-pull welding gun angle is one of many options for the welder.

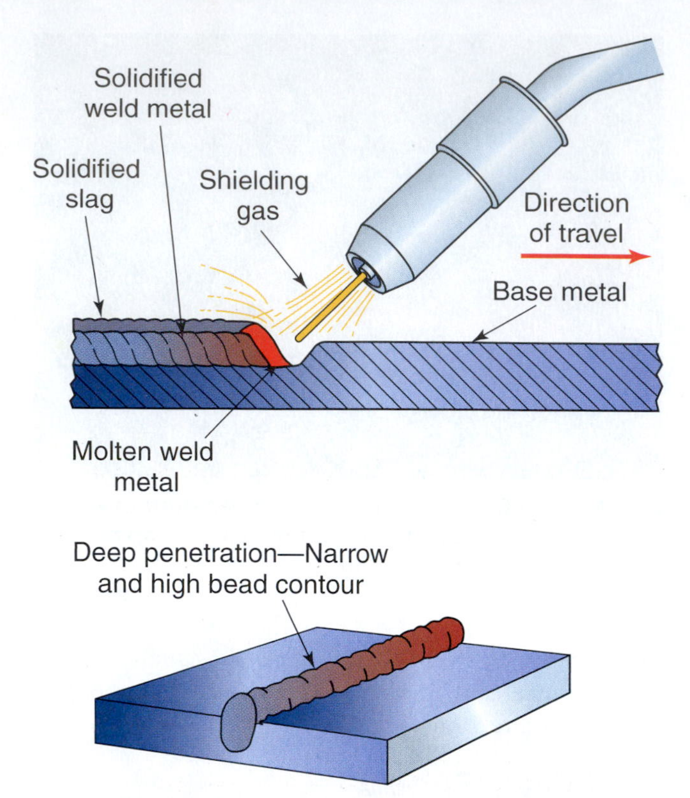

FIGURE 5-51 In backhand or drag angle welding, the technician drags the gun in the direction away from the weld pool.

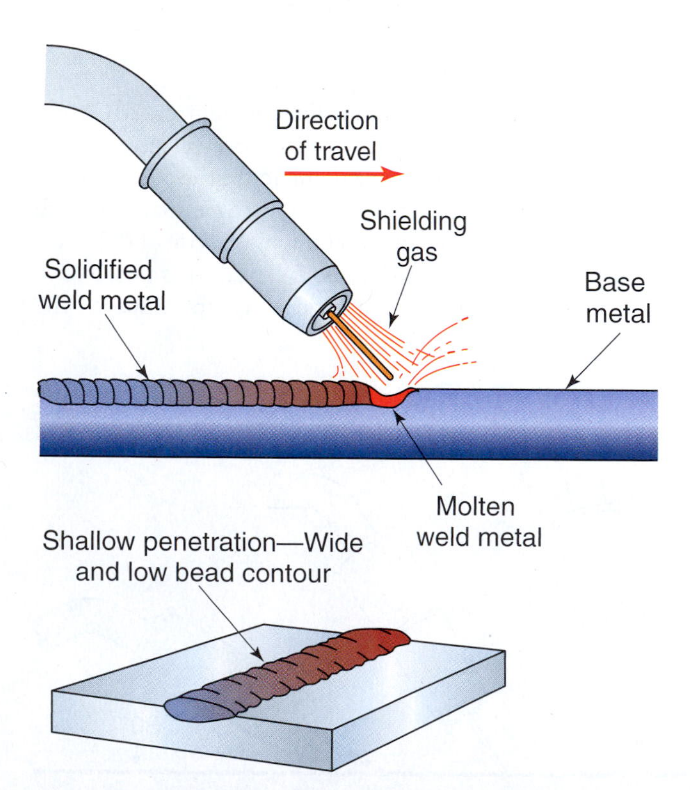

FIGURE 5-52 In the forehand or push welding angle, the technician pushes the gun into the weld pool.

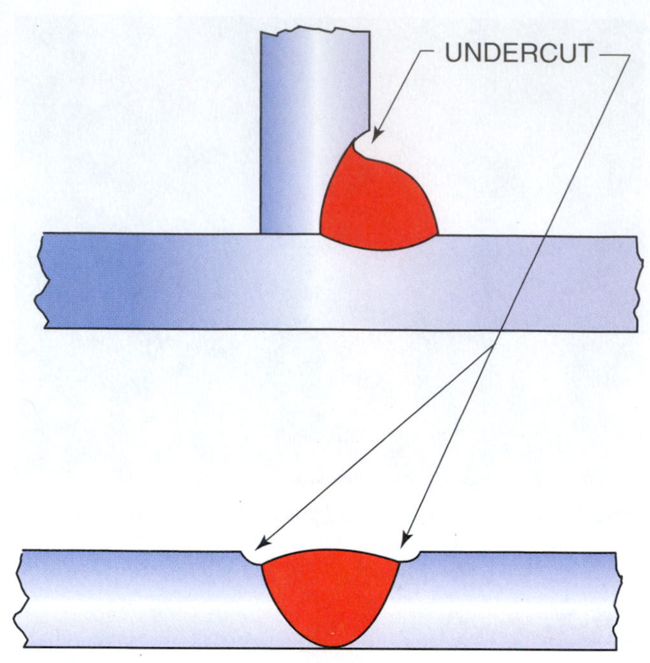

FIGURE 5-53 Undercut welding defect caused by a too-fast welding speed.

wire will wobble as it comes out, and the quality of weld will be affected. As contact tips wear out, the opening becomes larger, and again the wire may wobble. Contact tips should be checked from time to time to assess the need to change them.

Nozzle Adjustment

The **nozzles** on most welders can be adjusted, ideally so it is even with the contact tip, or with the contact tip about 1/8 inch below the level of the nozzle. The nozzle should be kept clear of spatter at all times. If a spatter is noted it should be immediately cleaned off; otherwise, the shielding gas may not flow properly.

Nozzles can be protected from spatter with the use of either a spray or dip spatter compound. In the case of an anti-spatter spray, it is applied after the tip is properly cleaned and adjusted by spraying a light coat in both the contact tip and the inside and outside of the nozzle. See **Figure 5-54**. In the case of dip, first the contact tip is cleaned, adjusted, and warmed by welding for a short time. Then the nozzle is dipped into a container of anti-spatter dip. See **Figure 5-55**. If the tip is not warm when dipped into the anti-spatter, it will not melt from the heat, and the nozzle and contact tip may become clogged.

Welding Positions

The physical position in which a collision repair technician must weld is most often dictated by where on the vehicle the weld is being performed. See **Figure 5-56**. Because of this, collision repair welders must be adept at welding in all of the various positions that follow.

FIGURE 5-54 Anti-spatter spray protects the nozzle.

FIGURE 5-55 Anti-spatter dip protects the nozzle.

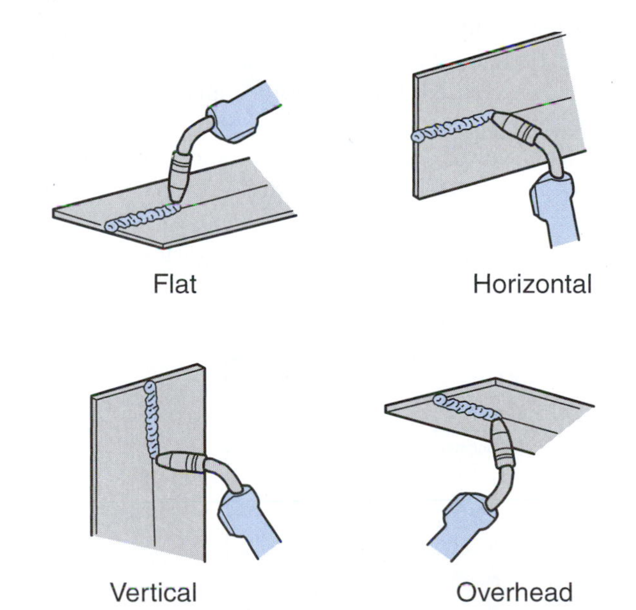

FIGURE 5-56 Welding positions are often dictated by where on the vehicle the weld is being performed.

Flat position welding is used when the pieces being welded are parallel with the shop floor. Though some parts on a vehicle will be in the flat position, often there is an access limitation, and the welding gun may not be able to be placed in an ideal position.

Horizontal position, where the piece is parallel to the shop floor, also presents itself from time to time. When the technician is called upon to weld in this position, gravity tends to pull the molten pool downward. When welding horizontally, the gun should be tilted upward to counteract the effect of gravity.

Vertical welding position means the piece is perpendicular from the shop floor, and in this case gravity strongly affects the weld pool. The effects of gravity can be used in the technician's favor by starting the weld at the top and pulling it downward, letting gravity help flow the weld pool.

An overhead weld is one of the most difficult positions to weld in, because the weld is upside down and above the technician's head. Gravity, which pulls the molten pool of metal downward, can often cause it to fall into the gun nozzle, where it will cause trouble with the shielding gas. To combat the effects of gravity in this position, the technician should keep the arc as short as possible. Often the stitch weld method, which will be discussed later, is used to counteract the effects of gravity. Overhead welds present themselves often in collision repair, and technicians should strive to become proficient when welding in this position.

CAUTION

When a technician is welding in an overhead position, slag from the molten pool can drop not only into the welder nozzle, but also onto the technician. Technicians working in this position and trying to see the weld pool have had this slag drop into their ears. If this happens, the red-hot slag can roll down the ear canal and burn a hole in the eardrum. Technicians should always wear ear protection—either the earmuff type or fire resistant earplugs—to prevent ear burns.

GAS METAL ARC WELDING TECHNIQUES

It has been said that GMAW—or MIG—welding is the easiest method to learn. It may be true that MIG welding technicians often become proficient in a shorter time than it takes to become proficient at TIG, or even stick welding. However, a technician should still not think that becoming a good MIG welder will come easily. It will take not only good hand-eye coordination, but also a good understanding of what goes on during the weld procedure, and the ability to set up and tune in the weld machine so a good weld can take place. See **Figure 5-57**. Another condition to consider is that collision repair technicians

FIGURE 5-57 This technician is welding with a vehicle on the frame rack.

FIGURE 5-58 This technician is fine-tuning the wire speed.

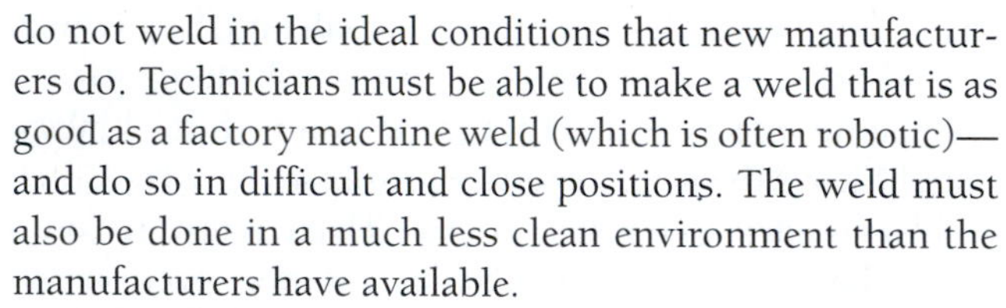

do not weld in the ideal conditions that new manufacturers do. Technicians must be able to make a weld that is as good as a factory machine weld (which is often robotic)—and do so in difficult and close positions. The weld must also be done in a much less clean environment than the manufacturers have available.

One of the most important tasks collision repair welders will do is clean the area the best that they can before welding. But with paint, undercoating, and even packed-in dirt from an accident to remove before the weld can take place, the job is much more difficult for repair technicians than for manufacturers.

Because the weld affects the structural integrity, crashworthiness, and safety of the vehicle, welds that a repair technician produce are doubly important. It is said that "B-quality paint can be applied to a vehicle for repair, but only A-quality welds are acceptable." Therefore the welds made by collision repair technicians must be top quality!

SETUP

This section will examine six important setup tasks for a weld technician to do before beginning a welding procedure. Some of the tasks are general and will be virtually the same for all six different techniques of welding. Others involve a starting point setup that will then need to be changed as the welding conditions change. See **Figure 5-58**. As an example, suppose a technician is to repair a front frame rail which will require a butt weld in the horizontal position (for the top of the rail), vertical welds (on the sides of the rail), and overhead welding (on the underside). The welding setup parameters may need to be adjusted when the technician changes positions. The chart in **Figure 5-59** shows different components that can be adjusted to provide that perfect weld. It also shows how adjusting them affects the weld quality.

Weld Current

Current (amperage) is regulated by adjusting the wire feed. As the wire feed is increased, the amperage is simultaneously increased. It takes more amperage to melt wire that is feeding out faster. Amperage adjustment also affects penetration depth, arc stability, and weld spatter. As can be seen by studying the Figure 5-59 chart, amperage affects all the categories except bead width. The only other variable that affects as many categories is stick-out.

Welding Variables to Change	Desired Changes							
	Penetration		Deposition Rate		Bead Size		Bead Width	
	Increase	Decrease	Increase	Decrease	Increase	Decrease	Increase	Decrease
Current and Wire Feed Speed	Increase	Decrease	Increase	Decrease	Increase	Decrease	No effect	No effect
Voltage	Little effect	Little effect	No effect	No effect	No effect	No effect	Increase	Decrease
Travel Speed	Little effect	Little effect	No effect	No effect	Decrease	Increase	Increase	Decrease
Stickout	Decrease	Increase	Increase	Decrease	Increase	Decrease	Decrease	Increase
Wire Diameter	Decrease	Increase	Decrease	Increase	No effect	No effect	No effect	No effect
Shield Gas Percent CO_2	Increase	Decrease	No effect	No effect	No effect	No effect	Increase	Decrease
Torch Angle	Backhand to 25°	Forehand	No effect	No effect	No effect	No effect	Backhand	Forehand

FIGURE 5-59 A technician may use a weld adjustment chart to help provide a perfect weld.

TECH TIP When adjusting voltage and amperage (wire feed), it should be noted that voltage should be changed only when the welding trigger is off. Most welding machines have voltage switches that will only move from one voltage setting to the next with no variables. Amperage switches are infinity switches that can be adjusted from 1 to 10 with an infinite amount of stopping points between the settings.

The amperage switch may be adjusted while the technician is welding to achieve the best setting for a condition. This "tuning-in" can first be done on similar scrap pieces of metal, so the finished weld will be perfect.

Voltage

Voltage is often set by referring to a chart on the inside of the welding machine cover, which suggests the proper voltage relative to the base metal thickness and the wire diameter. Once a technician sets this and the amperage is tuned in on a scrap piece of metal, welding can take place. Voltage affects the weld arc length, which in turn affects the bead width. The bead width can also be affected by travel speed. Therefore, before technicians decide to change the recommended voltage when tuning in a welder, they should explore the changes in bead width that will occur when the travel speed changes. If the proper bead width cannot be obtained by altering the travel speed, a technician could consider changing from the recommended voltage.

Travel Speed

Travel speed affects the bead width and size. Each of the six types of welds has differing requirements for both size (height) and width of weld bead. As technicians tune in a welder on scrap metal, they may need to alter both the travel speed and the amperage until they obtain an acceptable weld.

TECH TIP When tuning in a welder, technicians often must use their senses to judge when the weld is performing properly. With a welding helmet down, a technician must rely on the sense of hearing to judge when the weld speed and amperage are set correctly. If a continuous light hissing or crackling sound is heard, it is likely that the weld transfer is correct. For this correct sound, the amperage and travel speed must also be correct. If a spattering sound with no hissing is heard, the welder is likely not performing properly. This type of weld transfer will also have a larger amount of spatter. One weld technician explains the difference this way: "If it sounds like popping corn the weld is not good, but if it sounds like frying eggs the weld is OK!"

Stick-Out

Stick-out, or determining the length that the electrode sticks out past the contact tip, seems like a simple task. See **Figure 5-60**. Nevertheless, this parameter affects the weld variables greatly. The standard stick-out is from 1/4 to 5/8 inch (6 mm to 14 mm). As one can see from the chart in Figure 5-59, by increasing the stick-out length the weld will decrease the penetration and bead width, while increasing the deposition and bead size. By decreasing the stick-out, the opposite will happen.

TECH TIP Welder's pliers, which are a very useful tool, can be purchased at most welding supply stores. The long nose can be used to clean out slag from the nozzle, and it has a clamping place for contact tips as well as the nozzle. A wire cutting spot, when rested on the top of a properly adjusted nozzle, will cut the wire off at 1/4 of an inch. The long nose ends can also be used to clear wire from a bird's nest. With countless other uses also, it is an excellent tool for the technician to have when welding.

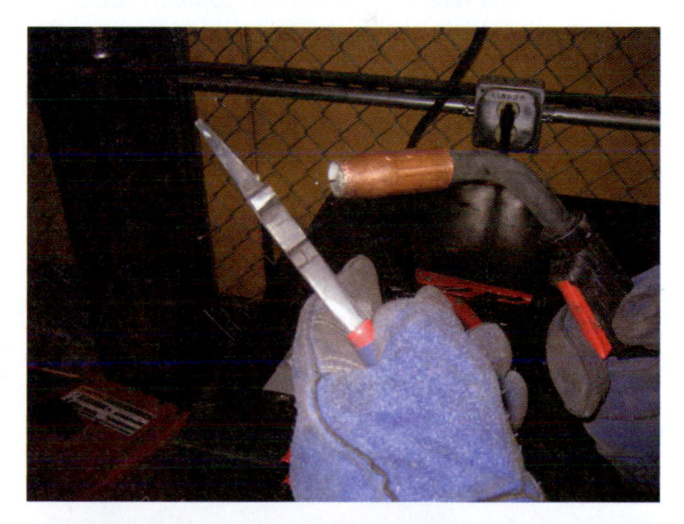

FIGURE 5-60 Welders use a 4-in-1 cutting tool to determine proper stick-out.

Wire Diameter

Wire thickness or diameter affects penetration and deposit rates; it has little or no other effect on the quality of a weld. A technician must be careful when choosing the proper electrode diameter that it is in the amperage range that the thickness of base metal to be welded requires. To determine the proper heat range, a technician could either look up the vehicle manufacturer's recommended wire size or could measure the metal thickness and find the heat range that the welder recommends. A chart showing this is normally found on the inside of the welder access door. See **Figure 5-61**. Wire size 0.023, 0.030, and 0.035 (0.6 mm, 0.8 mm, and 0.9 mm) are the most commonly recommended wire sizes, with 0.023 (0.6 mm) being recommended for light unibody sheet metal; 0.030 (0.8 mm) for medium-thickness hydroformed frames, frame horns, and full perimeter frames; and 0.035 (0.9 mm) for welding body-over-frame members. The required amperage to weld base metal and the welder's ability to produce that heat range will determine which type of welder will be able to be set up for what size wire. The table in **Figure 5-62** shows the heat range that the different types of wire diameters require. A quick glance shows that a small 115-volt, 130-amp welder would be best suited to weld with 0.023 wires. If the shop would need to weld a heavy body-over-frame that requires a 0.035 wire, it would be necessary to have a welder that could reach the 180-amp range, which would require a larger (normally 220-volt) machine.

Shielding Gas

Though changing the flow of the shielding gas could affect the penetration and bead width, what is most often noted when shielding gas is insufficient is a welding defect such as porosity. Shielding gas protects the weld from other gas contaminants such as oxygen, nitrogen, and hydrogen, which are all present in the atmosphere. Generally, the shielding gas is set at the recommended flow rate (25 to 30 CFH) for collision repair work. This flow rate is normally sufficient unless there is a high amount of air-

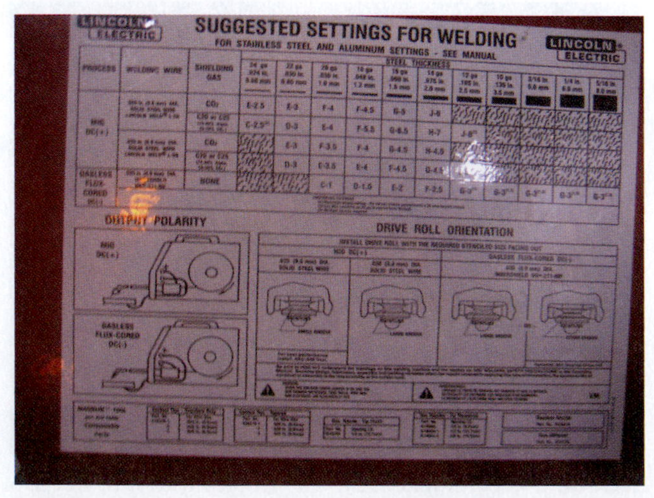

FIGURE 5-61 Manufacturer suggested welding settings.

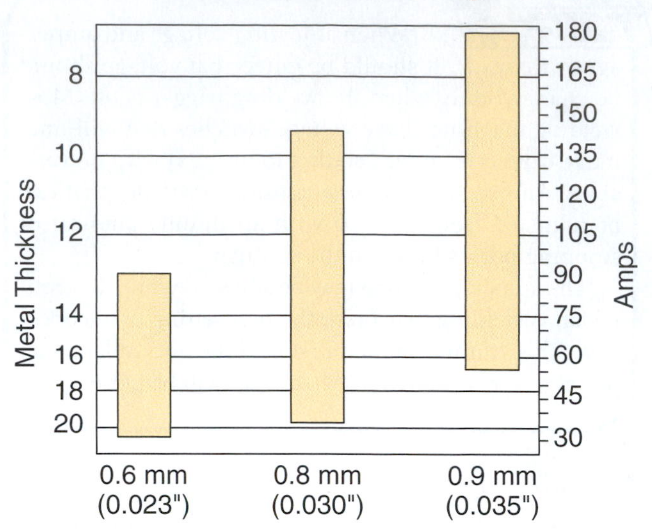

FIGURE 5-62 Comparing wire diameter vs. amperage helps a technician determine which type of welder to use.

flow, such as from a fan, which may blow the shielding gas away from the weld area.

Gun Angles

The gun angles can help shape the degree of penetration and weld bead. These angles are direction of travel such as drag or push; the travel angle is the angle in relation to perpendicular to the work; and the work angle is the direction in which the electrode is pointed when it meets the work. See **Figure 5-63**.

Direction of Travel. **Direction of travel**, though called by many names, is relative to how the welding technician moves when the weld is put down. Sometimes it is dictated by the position of the weld. If a weld is being made in the vertical position, the weld technician often chooses to weld from the top to the bottom, using the pull angle of travel to take advantage of gravity. Other times a weld technician may choose to use a push direction of travel on thin metal, because the heat of the arch will not preheat the thin metal as much as the pull method, and thus will not be as likely to create a burnthrough problem.

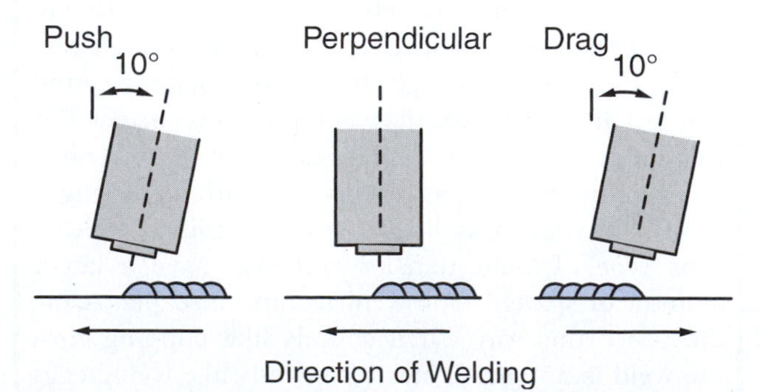

FIGURE 5-63 Gun technique can help shape the degree of penetration and weld bead.

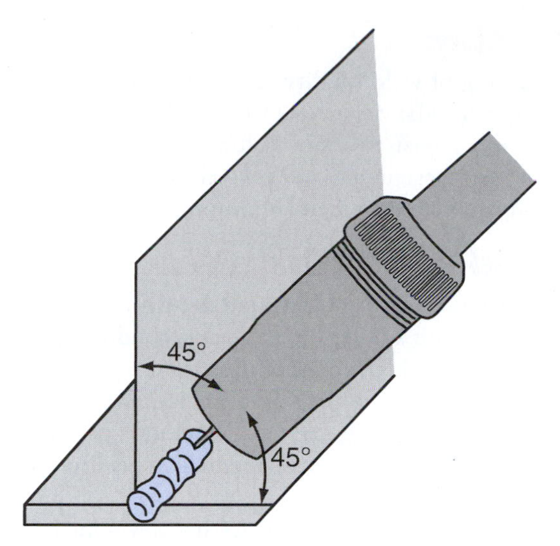

FIGURE 5-64 Work angle shows the angle of the gun in relation to perpendicular.

Travel Angle. Travel angle shows the angle of the gun in relation to perpendicular, with the gun held at true perpendicular to the weld puddle. See **Figure 5-64**. Either 90 degrees or no more than 5 to 15 degrees for either push or pull direction of travel will keep the shielding gas in place, thus lowering the possibility of defects and spatter. If the travel angle is increased beyond 20 to 25 degrees, the weld is likely to have more spatter, less penetration, and greater arc instability.

Work Angle. Work angle is the angle at which the electrode hits the pieces to be welded. A work angle may be 90 degrees when a flat butt weld is being made, and the work angle for a fillet weld may be 45 degrees to either of the two pieces being joined.

TECHNIQUES

There are six basic weld techniques. See **Figure 5-65**. They are:

- *Continuous*

  In a **continuous weld**, the arc is struck and a smooth uninterrupted bead is applied in a steady, ongoing movement. The push, or forward, direction of travel is used, and a 10 to 15 degree travel of angle should be maintained, with a 90 degree work angle. When thin body panels are being welded, the continuous weld may cause warpage, which should be guarded against.

- *Plug*

  A MIG **plug weld** is a weld that is placed through the drilled or punched piece or pieces. The direction of travel is push or forward due to the work angle being 90 degrees to the work at the bottom of the hole where the base metal is thinnest, with the travel angle being 10 to 15 degrees. The weld is started at the 2 o'clock position and pushed to the 10 o'clock position, or until the hole is filled. See **Figure 5-66**.

- *Stitch*

  A **stitch weld** is a series of short overlapping MIG spot welds which, when finished, create a continuous stream. The direction of travel, work angle, and travel angle are the same as for a continuous weld. This weld is often used on thin metal where warpage and burnthrough must be controlled.

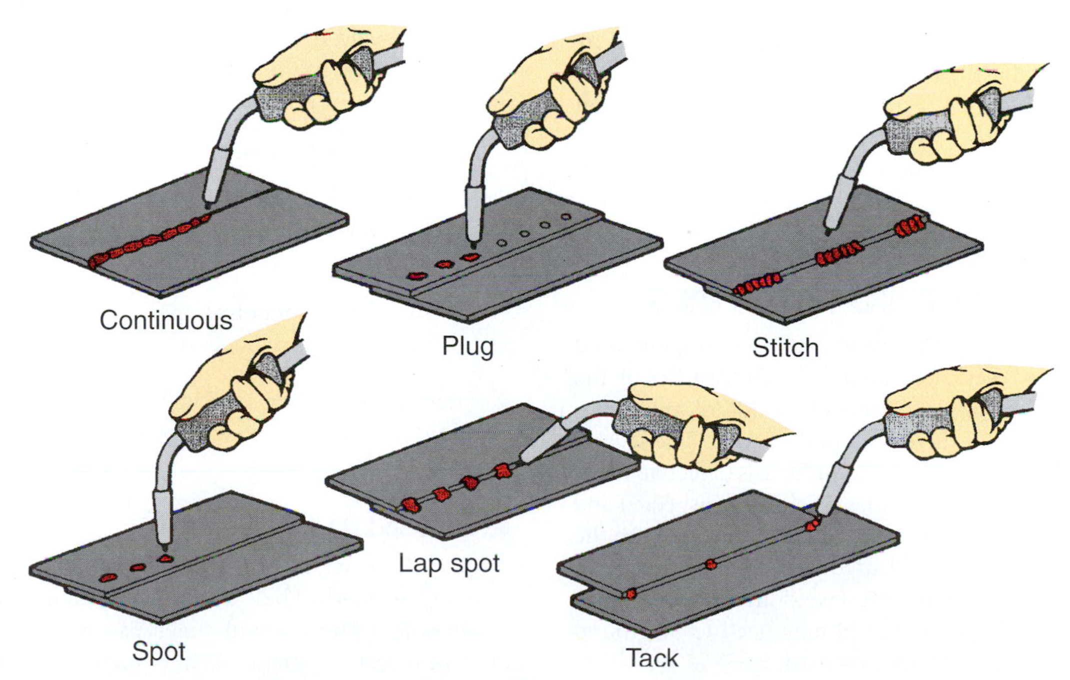

FIGURE 5-65 There are six basic GMAW (MIG) welding techniques.

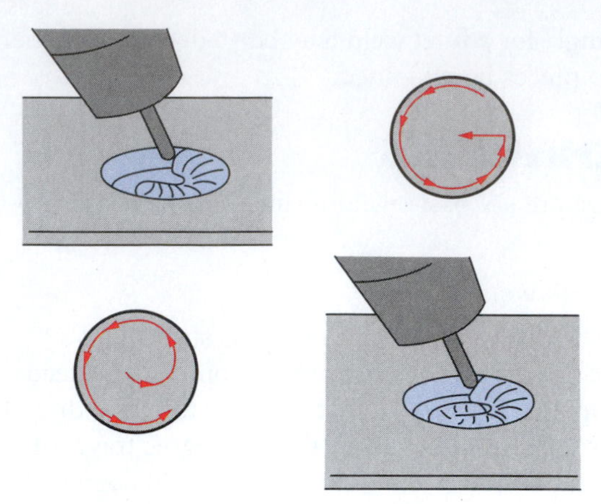

FIGURE 5-66 The plug welding technique places a weld through drilled or punched pieces.

- *Spot*

  In a MIG **spot weld**, the weld does not move. Therefore, no travel angle exists, and the work angle is 90 degrees. The trigger is pulled and the arc is directed to penetrate both pieces of metal. This type of weld is not often used in collision repair.

- *Lap*

  The MIG **lap weld** is used where two pieces of metal are overlapping themselves and the weld bead is placed to penetrate them both. The direction of travel could be either push or pull, depending on the position the weld is in. The travel angle is 10 to 15 degrees and the work angle is 45 degrees, placing the weld bead directly between the two base metal pieces.

- *Tack*

  A MIG **tack weld** is a temporary weld, used to hold fit-up pieces in place so that a permanent weld can be applied. The length of the weld is generally 15 to 30 times wider than the base metal that it is holding. There is little to no travel, with a 90 degree work angle. *A tack weld is only a temporary weld, and is not intended to be permanent.*

GAS METAL ARC WELD INSPECTION AND TESTING

In relation to evaluating welds, two commonly used terms bear some discussion. First is **discontinuity**, or the interruption of the typical structure of a weld. All welds have some discontinuity, for there is no perfect weld. The second term is **weld defect**; a weld may be judged as defective when there are too many discontinuities. Welds are inspected and judged for defects by differing standards, depending on the application of a weld. A weld used for cosmetic reasons is judged differently than a structural weld. In structural welding, if a weld is defective the weld may need to be ground out and re-applied to meet the standard.

Testing

Testing of weld quality can also be done by many different methods: visual, destructive, dye penetrant, and radiographic, just to name a few. The most common are visual and destructive when steel welds are being tested, and dye penetrant when aluminum is tested.

Welder Qualification Testing

During welding qualification testing, each welder must meet a certain set of standards and conditions to be certified as a welder. This certification may have a time limit, which means that the technician must re-test at a specific interval. The test will include specific types of welds that must be performed and specific conditions that must be met.

The collision repair industry has developed a welding qualification test required for certification. To maintain certification, a technician must re-take the test every five years. The technician must perform a plug weld, a fillet weld, a butt weld with a backer, and a butt weld without a backer. All of these welds must be performed in both the vertical and overhead positions, and the resulting welds must pass both visual and destructive testing. The visual testing is evaluated by height, width, diameter, and length. Once the welds pass the visual inspection, they are sent to destructive testing where the welds are destroyed and then judged for the quality and strength.

There are separate qualification tests for steel and aluminum welders, with some vehicle manufacturers requiring that their certified repair technicians must be weld-certified to maintain their manufacturer certification.

WELD DEFECTS

Weld defects (or the presence of an unacceptable amount of weld discontinuity) come in many forms and are given many different names. This section will discuss some of the most common weld defects:

1. Pores/pits (Porosity)
2. Undercut
3. Improper fusion
4. Overlap
5. Insufficient penetration
6. Excess weld spatter

The chart in **Figure 5-67** will help the reader to understand these common defects, their descriptions, and some possible causes of them.

Porosity

Porosity is when gas becomes trapped in the weld as it is being formed. That trapped gas then develops into small holes. These small holes can be either spherical or cylindrical in shape. While some porosity in small

Defect	Defect Condition	Remarks	Possible Causes
Pores/Pits	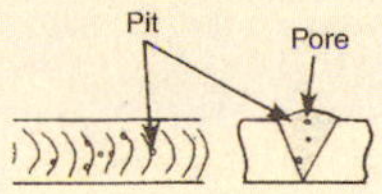 Pit Pore	There is a hole made when gas is trapped in the weld metal.	There is rust or dirt on the base metal. There is rust or moisture adhering to the wire. Improper shielding action (the nozzle is blocked or wind or the gas flow volume is low). Weld is cooling off too fast. Arc length is too long. Wrong wire is selected. Gas is sealed improperly. Weld joint surface is not clean.
Undercut	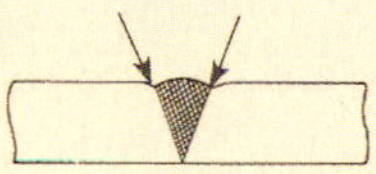	Undercut is a condition in which the over-melted base metal has made grooves or an indentation. The base metal's section is made smaller and, therefore, the weld zone's strength is severely lowered.	Arc length is too long. Gun angle is improper. Welding speed is too fast. Current is too large. Torch feed is too fast. Torch angle is tilted.
Improper Fusion	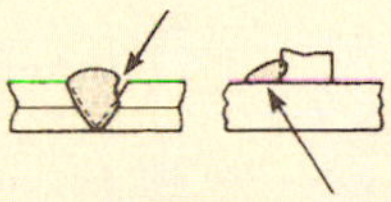	This is an unfused condition between weld metal and base metal or between deposited metals.	Check torch feed operation. Is voltage lowered? Weld area is not clean.
Overlap	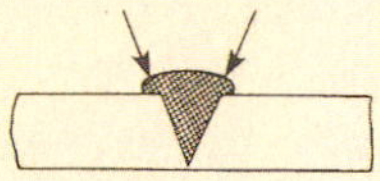	Overlap is apt to occur in fillet weld rather than in butt weld. Overlap causes stress concentration and results in premature corrosion.	Welding speed is too slow. Arc length is too short. Torch feed is too slow. Current is too low.
Insufficient Penetration	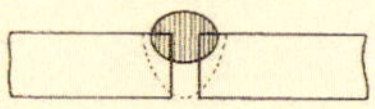	This is a condition in which there is insufficient deposition made under the panel.	Welding current is too low. Arc length is too long. The end of the wire is not aligned with the butted portion of the panels. Groove face is too small.
Excess Weld Spatter	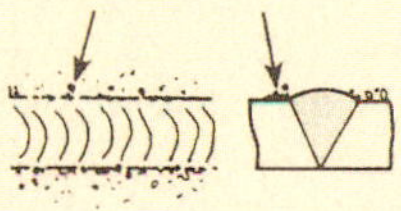	Excess weld spatter occurs as speckles and bumps along either side of the weld bead.	Arc length is too long. Rust is on the base metal. Gun angle is too severe.
Spatter (short throat)	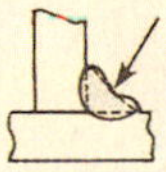	Spatter is prone to occur in fillet welds.	Current is too great. Wrong wire is selected.

FIGURE 5-67 This chart shows GMAW (MIG) welding defects and their causes.

amounts may be considered an acceptable amount of discontinuity porosity, it is a defect that must be kept to a minimum. Porosity can be caused from many different conditions, but one of those conditions that presents itself often in collision repair is the forming of hydrogen gas during the welding process because of contamination. In a new construction setting, there is much less chance of contamination during welding. However, the collision repair technician must clean all the old paint, undercoating, and dirt from the area to be welded and the surrounding area so that contamination porosity does not occur. If hydrogen gas is formed while welding, this gas may diffuse into the heat affect zone and cause cracking.

Different types of porosity are:

- **Uniform porosity**, which is scattered throughout the weld and most often caused by poor welding techniques or dirty materials. See **Figure 5-68**.
- **Cluster porosity**, which is often caused by improper starting and stopping and is clustered in these areas. See **Figure 5-69**.
- **Linear porosity**, which is caused by contamination within the joint, root, or interbead areas. See **Figure 5-70**.
- **Piping porosity**, which is also caused by contamination and forms (depending on the gas type) and escapes at the same rate as the solidity of the weld pool. See **Figure 5-71**.

Undercutting

Undercutting is generally considered a discontinuity instead of a defect, though if not controlled it can cause a reduction in the weld's cross-section. When this occurs, it may cause weld failure, and thus must be considered a defect. Undercutting is located at the junction of the weld

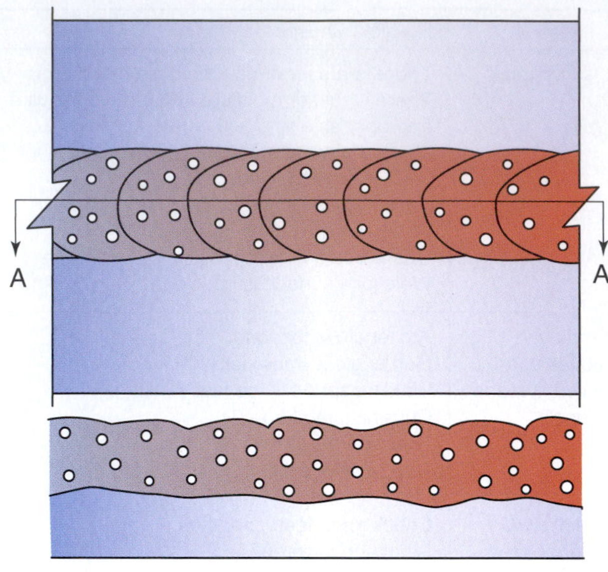

FIGURE 5-68 Uniform porosity is most often caused by poor welding techniques or dirty materials.

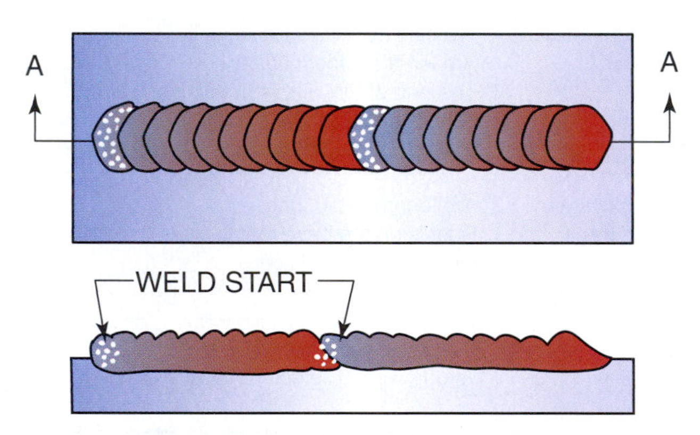

FIGURE 5-69 Cluster porosity is often caused by improper starting and stopping a weld.

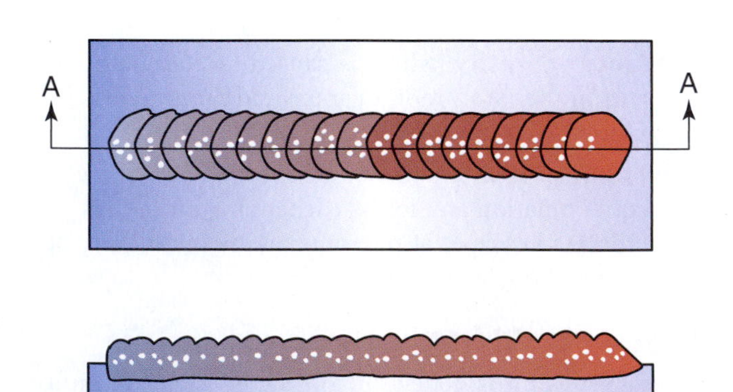

FIGURE 5-70 Linear porosity is caused by contamination within the joint, root, or interbead areas.

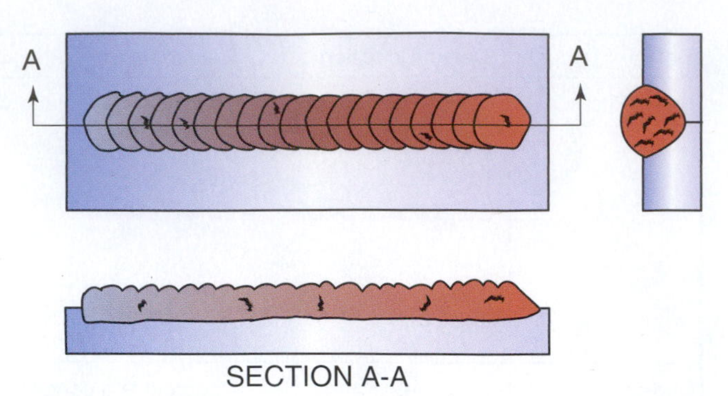

FIGURE 5-71 Piping porosity is also caused by contamination and forms.

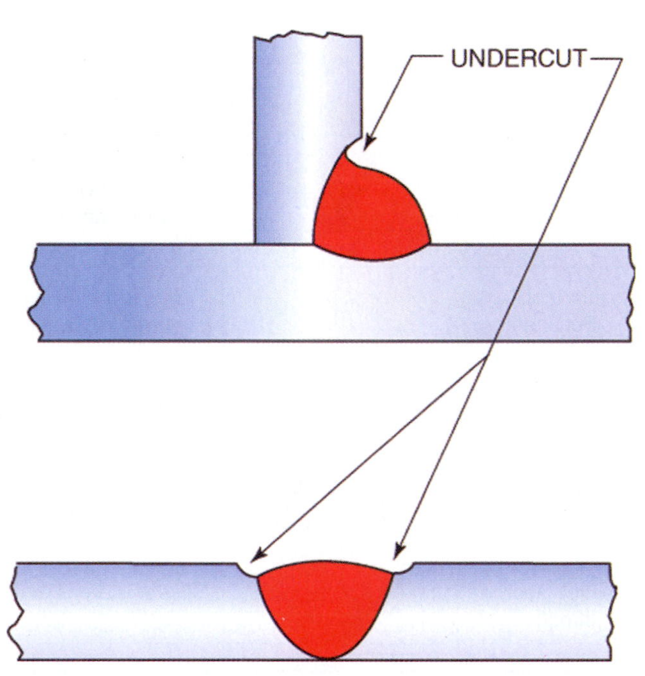

FIGURE 5-72 Undercutting is generally considered a discontinuity instead of a defect.

and the base metal (toe or root) and, as the name implies, is a small furrowing of the weld as it meets the base metal. See **Figure 5-72**. It can be caused by such conditions as an arc length that is too long, an improper gun angle, and a weld speed that is too fast. See Figure 5-67 for a more complete list.

Improper Fusion

Improper fusion is, as it sounds, the failure of the base metal and the filler metal to combine properly. Even if the base metal does melt, a thin layer of oxide may form, preventing the two from fusing. This failure of fusion can be caused by low or inadequate heat, poor weld technique, insufficient gap, or improper edge setup.

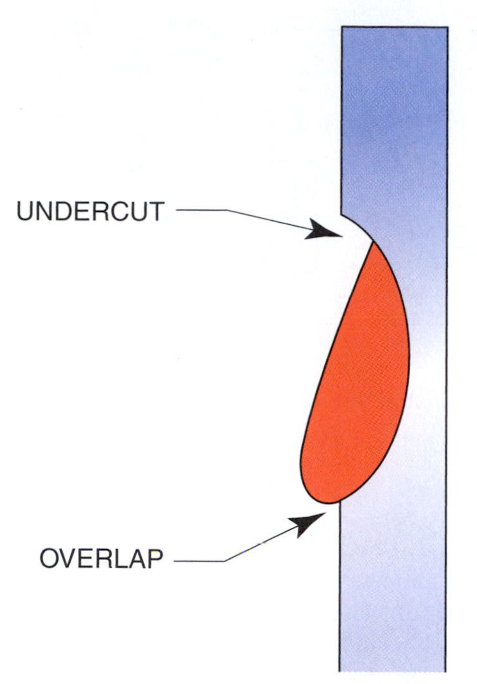

FIGURE 5-73 Overlap is the protrusion beyond the toe, face, or root of the weld without fusion.

Overlap

Overlap is the protrusion beyond the toe, face, or root of the weld without fusion. See **Figure 5-73**. This condition is also sometimes called rollover.

Poor or Insufficient Penetration

Insufficient penetration is a common defect among new welding technicians. The condition occurs when the weld fusion does not go deep enough into the base metal to produce a strong weld. See **Figure 5-74**. Different types of weld qualification requirements have penetration standards, with inadequate penetration a common cause for considering a weld deficient. When welds with inadequate weld penetration are destructively tested, they fail. The causes of poor penetration include a welding current that is too low, an arc length that is too long, an improper weld technique, and an improper joint fit-up.

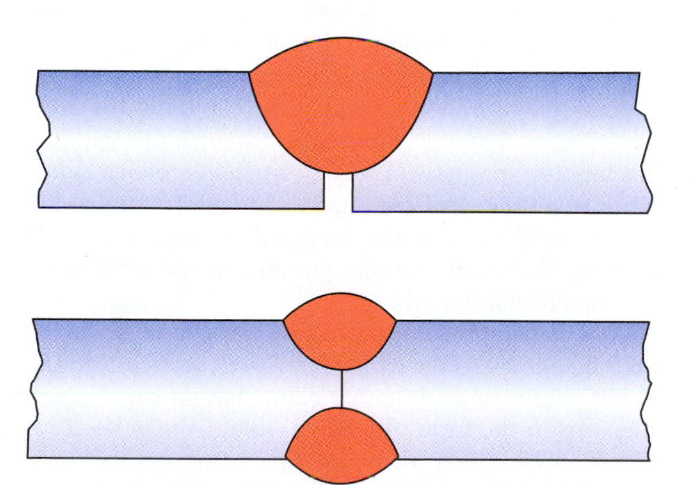

FIGURE 5-74 Poor penetration is a common defect among new welding technicians.

Excess Weld Spatter

Spatter in collision repair welding happens routinely, and is not considered a defect unless it becomes excessive. It appears as speckles of small deposits of filler electrode along the sides of the weld. These hot molten pieces of metal will cause damage to surrounding areas such as glass and painted surfaces of a vehicle; therefore, these areas must be protected against damage during the repair process. Most spatter around welds can be cleaned up when the welds are ground during preparation for refinishing.

TIG Welding

Tungsten inert gas (or TIG) welding is another type of GMAW welding that is widely used in other industries, especially for the welding of aluminum. In the collision repair industry, though, it is limited in its use. The process uses a tungsten rod on the welding gun to produce the melting arc. The tungsten rod melts only at a very high temperature, about 6,900°F; therefore, during welding the tungsten electrode is not consumed. To form the weld, a filler rod is manually fed into the molten pool as the weld is formed. TIG welding is much slower than other GMAW welding and allows more control of the weld pool, thus forming a better appearance. TIG welding does not transfer metal across the weld arc and therefore does not produce spatter or sparks; when operating properly, there is no noise. TIG uses either argon or helium for a shielding gas; therefore, it is a truly inert form of GMAW welding.

SQUEEZE RESISTANT SPOT WELDING

Squeeze-type resistance spot welding (STRSW) is the most common type of weld used by automotive manufacturers. In fact, 90 to 95 percent of all the welds used in steel unibody construction are done by using resistance spot welding.

Resistance spot welding relies on the heat which is produced by resistance when low-voltage current is passed through metal held together by pressure. See **Figure 5-75**. Resistance spot welding relies on three elements: pressure, current flow, and holding.

1. **Pressure**: Pressure is the mechanical force that holds together the items to be welded, and offers

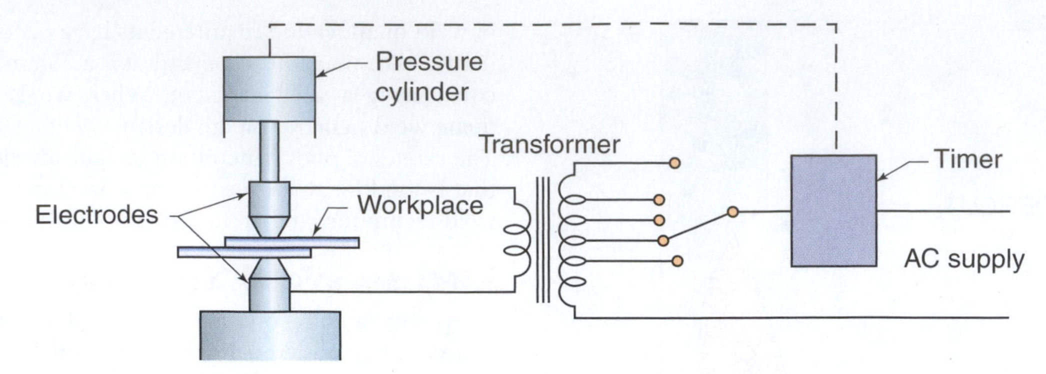

FIGURE 5-75 Resistant spot weld machine circuits rely on heat produced when low-voltage current is passed through metal held together by pressure.

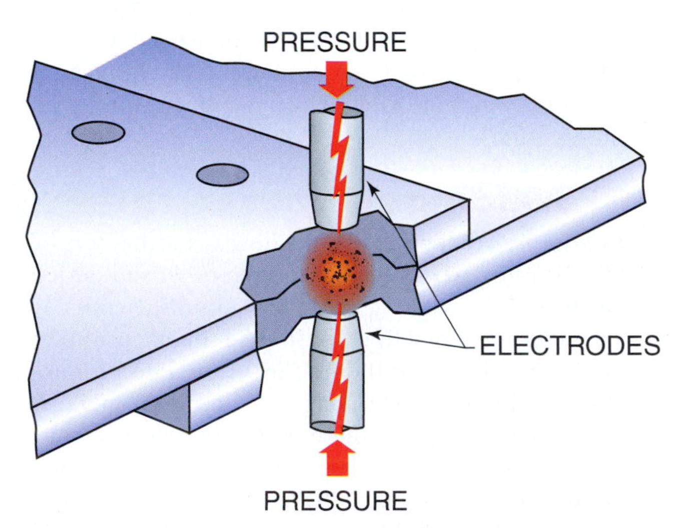

FIGURE 5-76 A pressure type resistance spot welder can hold pressure of 300 to 550 pounds on the pieces to be welded.

the electrical resistance. See **Figure 5-76**. Two electrodes, one on either side, hold pressure of about 300 to 550 pounds on the pieces to be welded.

2. **Current flow**: After pressure is applied, low-voltage high current (5 volts with 8,000 to 10,000 amps) is passed through the copper tips of the resistance spot welder. The low resistance of the copper tips do not have much resistance, but as the current passes from one copper tip through the conductive steel, the steel, which has higher resistance, heats up, melts, and—along with the pressure—forces the molten metal together. The current passes through the steel very rapidly (in 1/2 second or less).
3. **Holding time**: After the current flows through the metal and the heat caused by the resistance melts the metal between the electrode tips, the pressure is held on the weld area for 1 to 2 seconds. This allows the metal to cool, resolidify, and—because of the pressure—form a weld that is the same size as the electrode tips. This spot weld is called a weld nugget.

Squeeze-type resistance spot welders are now being recommended by manufacturers for collision repair. Because the process allows repairs to be made faster than other types of welding, body shops can now make repairs more quickly, and also in a more factory-like manner. These welders pass normal available voltage such as 230 volts/50 amps through the welding machine's transformer, converting the current to 5 volts/8,000 to 10,000 amps, and passing it through connecting cables through a set of arms which must have access to both sides of the steel to be welded. These arms are often powered by air pressure to force the two or three pieces of metal together. Welders have electronic control devices that help set the pressure, the proper current, and timers for both the current flow time and holding time. Most machines are set in relation to the thickness of the pieces being welded together. Some have settings for all three (pressure, current, and time), while others have been pre-calculated at the factory and only have one setting.

Squeeze-type resistance welders have two arms that come together to produce the pressure. These welders often come with a variety of arms for different configurations so that they can reach both sides. These arms can be changed according to the access needed, then adjusted according to the manufacturer's recommendations so the two tips align properly, forming the needed 1/4-inch (6 mm) nugget. Properly adjusted and aligned tips with the machine set properly normally do not produce much spark when the welds are made. Through continued use the tips will become mushroomed, though, and need dressing. The manufacturer's recommendations for dressing should be followed.

A considerable amount of heat will build up at the tip when in use, so machines provide a form of cooling either by air, or by liquid coolant that passes through the arms in the form of 50/50 automotive coolant. See **Figure 5-77**. Even with the tips being cooled, care should be taken while using a STRSW and the tips should be monitored. Hot tips deteriorate faster than properly cooled ones.

FIGURE 5-77 Coolant in spot welder dissipates the heat from welding.

FIGURE 5-78 Plasma arc cutting super-heats a gas, which then carries an electrical charge.

PLASMA ARC CUTTING

Plasma arc cutting is a process that super-heats a gas, which then can carry an electrical charge. See **Figure 5-78**. This narrow stream of electrically charged air is forced out a small hole in the cutting head, which melts the metal and blows it away from the cut area. The heat produced is both very high and very concentrated. With these qualities, the resulting heat affect zone is very narrow and the surrounding metal is not affected as much as it is by oxyacetylene cutting. Because of its smaller heat affect zone, the plasma cutting process has become popular.

Though it was developed in the 1950s, plasma arc cutting required refinement and further development before it became widespread. Now that common compressed air is used with plasma machines, most collision shops use plasma arc cutting.

The components that make up a plasma gun are the electrode, cutting nozzle, and shield cap, with gas (compressed air) passing in two areas. See **Figure 5-79**. The gas that passes between the shield cap and the cutting nozzle acts as the shielding gas, cooling the cutting nozzle so it does not overheat. This same compressed gas, when passed inside the gun between the electrode and the cutting heat, swirls around the electrode and becomes superheated to 60,000°F, which ionizes the gas to become plasma. Plasma, which can conduct electricity, is forced through the narrow hole in the cutting nozzle. Once on the other side, the charged plasma gas rapidly expands against the metal. The force of the superheated charged gas both melts the molten steel and forces it away from the cut area. This produces a very clean cut with little-to-no residual heat in the remaining metal.

Plasma produces heat, ultraviolet light, and infrared light. Therefore, technicians should read, understand, and follow the safety recommendations for using this tool.

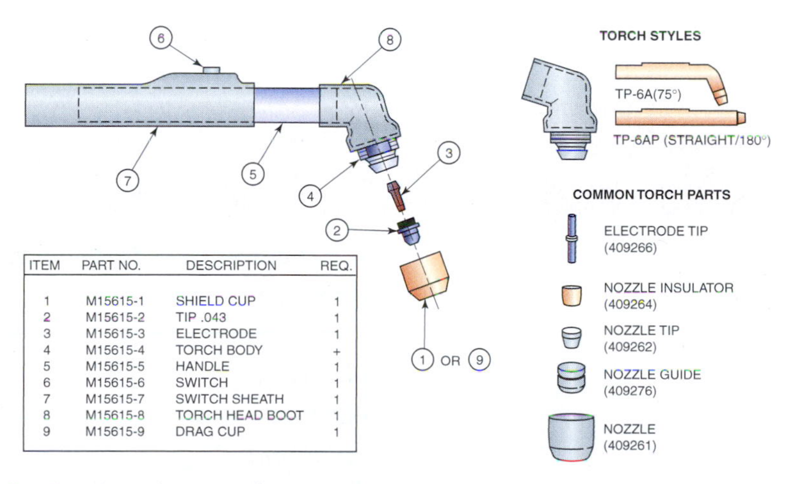

ITEM	PART NO.	DESCRIPTION	REQ.
1	M15615-1	SHIELD CUP	1
2	M15615-2	TIP .043	1
3	M15615-3	ELECTRODE	1
4	M15615-4	TORCH BODY	+
5	M15615-5	HANDLE	1
6	M15615-6	SWITCH	1
7	M15615-7	SWITCH SHEATH	1
8	M15615-8	TORCH HEAD BOOT	1
9	M15615-9	DRAG CUP	1

FIGURE 5-79 Plasma cutting involves the use of various components.

HEAT INDUCTION TOOL

Heat induction tools are devices that act like a transformer, setting up a magnetic field between the inducer tool and the metal object. See **Figure 5-80**. A magnetic field is created and heat is produced at the metal area. (The tool does not get hot.) This heat can be used to remove moldings and glass, and even to heat up rusted-on bolts for removal. See **Figure 5-81**. Induction tools are now also being used to heat metal for small dent removal.

FIGURE 5-80 Heat induction tools are devices that act like a transformer.

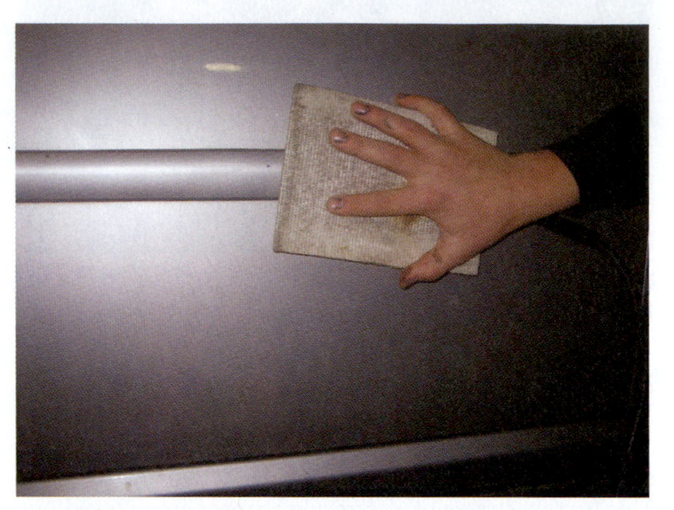

FIGURE 5-81 Induction tools can be used to remove molding.

Summary

- Welding in all its forms, and the skill to create those welds, are among the most exacting and challenging tasks a collision repair technician must learn.
- Manufacturers and insurance companies require technicians to be certified welders to perform repairs.
- Welding takes good hand-eye coordination, knowledge of the theories of the process, and hours of practice.
- A technician can become "rusty" if welding is not done on a regular basis, so constant practice to maintain acquired skills is a must.
- New technologies and changes in the present technologies force technicians to constantly adapt.

Key Terms

amperage
cluster porosity
contact tip
continous weld
cubic feet per hour (CFH)
current
dermis
direct current electrode negative (DCEN)
direct current electrode positive (DCEP)
direction of travel
discontinuity
electrode
electrode wire
epidermis
first-degree burns
gas flow meter

gas metal arc welding (GMAW)
gas regulator
globular metal transfer
gun angle
heat induction
improper fusion
insufficient penetration
lap weld
linear porosity
metal inert gas (MIG)
MIG brazing
nozzle
overlap
piping porosity
plasma arc cutting
plug weld
porosity
power supply
pulsed-spray metal transfer
second-degree burns
shielding gas
short circuit metal transfer
silicon bronze welding
spot weld
spray arc metal transfer
squeeze-type resistance spot welding (STRSW)
stick-out
stitch weld
tack weld
third-degree burns
tungsten inert gas (TIG)
ultraviolet (UV) light
undercutting
uniform porosity
voltage
weld defects
weld position
welding gun
welding speed
wire feeder
work clamp

Review

ASE Review Questions

1. Technician A says that a first-degree burn occurs when the skin is severely damaged and that it should have ice placed on it for relief. Technician B says that first-degree burns can be caused by ultraviolet light. Who is correct?
   A. Technician A only
   B. Technician B only
   C. Both Technicians A and B
   D. Neither Technician A nor B
2. Technician A says that the epidermis is the top layer of the skin. Technician B says that skin is the body's first defense against infection. Who is correct?
   A. Technician A only
   B. Technician B only
   C. Both Technicians A and B
   D. Neither Technician A nor B
3. Technician A says that glasses protect eyes against flying debris. Technician B says that goggles protect eyes against splashes. Who is correct?
   A. Technician A only
   B. Technician B only
   C. Both Technicians A and B
   D. Neither Technician A nor B
4. Technician A says that ultraviolet light is bright, and very hot. Technician B says that ear protection does not need to be used when welding. Who is correct?
   A. Technician A only
   B. Technician B only
   C. Both Technicians A and B
   D. Neither Technician A nor B
5. Technician A says that when welding galvanized steel, a welder needs to have respiratory protection. Technician B says that MSDS sheets have the local fire marshal's phone number on them. Who is correct?
   A. Technician A only
   B. Technician B only
   C. Both Technicians A and B
   D. Neither Technician A nor B
6. Technician A says that high-pressure gas cylinders can be hauled laying down in a pickup truck, as long as they are secured with chains. Technician B says that a welder is grounded, and because of this

technicians can weld even if the floor is wet. Who is correct?
A. Technician A only
B. Technician B only
C. Both Technicians A and B
D. Neither Technician A nor B

7. Technician A says that fire extinguishers come graded as A, B, C, and D. Technician B says that a C grade fire extinguisher is for putting out normal combustibles like paper. Who is correct?
A. Technician A only
B. Technician B only
C. Both Technicians A and B
D. Neither Technician A nor B

8. Technician A says that MIG stands for metal ignitable gas. Technician B says that gas metal arc welding (GMAW) is a more accurate term than MIG. Who is correct?
A. Technician A only
B. Technician B only
C. Both Technicians A and B
D. Neither Technician A nor B

9. Technician A says that much of the lighter GMAW welding work in a body shop can be done with a 115-volt welder. Technician B says that a 115-volt welder can easily weld with wire of 0.095 diameter. Who is correct?
A. Technician A only
B. Technician B only
C. Both Technicians A and B
D. Neither Technician A nor B

10. Technician A says that GMAW in the collision repair industry uses pure argon shielding gas. Technician B says that GMAW in the collision repair industry uses pure helium shielding gas. Who is correct?
A. Technician A only
B. Technician B only
C. Both Technicians A and B
D. Neither Technician A nor B

11. Technician A says that a shielding gas flow of 25 to 30 CFH is the general recommendation for shielding gas for collision repair GMAW. Technician B says that AWS-ER70S-6 is the type of electrode recommended for collision repair welding. Who is correct?
A. Technician A only
B. Technician B only
C. Both Technicians A and B
D. Neither Technician A nor B

12. Technician A says that globular metal transfer is the type of metal transfer recommended for GMAW aluminum welding. Technician B says that aluminum welding wire of the 3000 series is the wire type recommended for automotive welding. Who is correct?
A. Technician A only
B. Technician B only
C. Both Technicians A and B
D. Neither Technician A nor B

13. Technician A says that GMAW aluminum welding uses argon and is therefore truly a MIG welding process. Technician B says that when silicon bronze welding, a TIG welding is used. Who is correct?
A. Technician A only
B. Technician B only
C. Both Technicians A and B
D. Neither Technician A nor B

14. Technician A says that current is measured as volts. Technician B says that the voltage setting on a welder will depend on where the work clamp is. Who is correct?
A. Technician A only
B. Technician B only
C. Both Technicians A and B
D. Neither Technician A nor B

15. Technician A says that wiggling electrode wire from a GMAW welder has no effect on a weld's quality. Technician B says that an overhead welding is a difficult welding position in which to produce a good weld. Who is correct?
A. Technician A only
B. Technician B only
C. Both Technicians A and B
D. Neither Technician A nor B

16. Technician A says that there are six basic weld techniques in collision repair. Technician B says that a tack weld is a strong and permanent weld. Who is correct?
A. Technician A only
B. Technician B only
C. Both Technicians A and B
D. Neither Technician A nor B

17. Technician A says that if a discontinuity is found in a weld, it is considered an unacceptable weld. Technician B says that a good weld will have no spatter. Who is correct?
A. Technician A only
B. Technician B only
C. Both Technicians A and B
D. Neither Technician A nor B

18. Technician A says that TIG welding is the only recommended method of welding aluminum. Technician B says that TIG stands for tungsten inert gas. Who is correct?
A. Technician A only
B. Technician B only
C. Both Technicians A and B
D. Neither Technician A nor B

19. Technician A says that most manufacturing welds of steel unibody vehicles are squeeze-type resistance

spot welds (STRSW). Technician B says that STRSW are not recommended for collision repair use. Who is correct?

A. Technician A only
B. Technician B only
C. Both Technicians A and B
D. Neither Technician A nor B

20. Technician A says that plasma arc cutting uses acetylene as the shielding gas. Technician B says that heat induction tools produce a magnetic field, which produces the heat. Who is correct?

A. Technician A only
B. Technician B only
C. Both Technicians A and B
D. Neither Technician A nor B

Essay Questions

1. Summarize why shielding gas is needed in GMAW welding.
2. Describe the overhead GMAW welding position.
3. List five welding defects in GMAW welding and explain what could cause each of them.
4. Describe the two roles of shielding gas in plasma arc cutting.
5. Explain the role that sound plays when a welder is tuning in a GMAW machine.
6. Explain how ultraviolet light could affect exposed skin.
7. Describe how wire stick-out affects the quality of a GMAW weld.

Topic Related Math Questions

1. If a collision repair shop used 1.5 cylinders of shielding gas per week, how many cylinders would the shop need for a year?
2. A plasma cutter needs a minimum of 35 pounds of compressed air to operate. The air regulator is set to 15 pounds. The plasma machine is connected with a 25-foot-long rubber hose that loses 5 pounds of pressure per every 5 feet of line. At how many pounds should the regulator be set?
3. A plasma cutter needs a minimum of 35 pounds of compressed air to operate. The air regulator is set to 15 pounds. The plasma machine is connected with a 25-foot-long rubber hose that loses 5 pounds of pressure per 5 feet of line. How many pounds should the regulator be increased?
4. If a GMAW welder is set to feed wire at 17 feet per minute, how many minutes would it take for the welder to feed out 1,523 feet of wire?

Critical Thinking Questions

1. What could cause porosity in a GMAW weld?
2. Why is overhead welding more difficult than welding flat?
3. Why is structural weld quality so important?
4. How does sound affect how a weld technician adjusts his travel speed?
5. Why is ultraviolet light dangerous to the unprotected eye?

Laboratory Activities

1. Determine among different types of fire extinguishers which type is best suited for each type of fire.
2. Locate and adjust the proper voltage for welding 18-gauge high strength steel.
3. In the flat position, practice welding a plug weld.
4. In the overhead position, practice welding a lap weld.
5. In the vertical position, practice butt-welds with a backer.
6. Inspect welds for defects.

Chapter 6

Vehicle Construction

OBJECTIVES

Upon completing this chapter, you should be able to:

- Distinguish between body over frame and unibody designs.
- Identify the role of frame on a full frame vehicle.
- Identify the different frame designs utilized on the modern day vehicle and the advantages of each.
- Identify the structural design and role of the undercarriage on a unibody design.
- Identify the difference between a space frame design and a true unibody vehicle.
- Identify the difference in how the suspension and steering parts are attached on a unibody and body over frame vehicle.
- Identify the difference in the design characteristics of a front- and rear-wheel-drive vehicle.
- Identify the difference between the front engine, mid engine, and rear engine design.
- Discuss the concepts of collision energy management to safeguard the vehicle occupants.
- Review the emerging trends and materials used in vehicle construction.

INTRODUCTION

The automobile as we know it today has evolved and emerged from a crude and noisy open air vehicle, offering little or no protection for the driver, to the safe, luxurious, and environmentally controllable vehicle we enjoy today. The car has evolved from an open air vehicle to one offering numerous options such as air conditioning, power steering, and so on. Despite the introduction of numerous on-board computer-controlled creature comforts and electronic technology, the automobile's drive mechanism and basic design have remained very much the same as those used in the original automobile invented by the Duryea brothers in 1895. The engineering and design features that have changed most significantly are those intended to ensure passenger safety. Most notably, these changes involve the vehicle's undercarriage and superstructure, which includes the frame on the body over frame (BOF) vehicles and the undercarriage, structural pillars, and reinforcing members of the unibody.

RESEARCH AND DEVELOPMENT

Consumers expect an aesthetically appealing automobile. In addition, auto manufacturers have been driven by various private and governmental agencies such as the National Highway Transportation Safety Administration (NHTSA), the Insurance Institute for Highway Safety (IIHS), and the **Environmental Protection Agency (EPA)** to produce vehicles that ensure improved passenger safety and increased fuel economy. Despite manufacturers' efforts to reduce the automobile's weight, many of the safety initiatives, emission controls, and electrical systems required to operate all the on-board systems have made some of the current model vehicles heavier than their early model predecessors. For example, the curb weight of the 1965 through 1967 Ford Mustang was approximately 2,800 pounds compared to 3,500 pounds for the 2006 Mustang. In their quest to ensure passenger safety, auto manufacturers have developed numerous innovative design features that have resulted in a significant reduction of accident-related injuries and deaths. Many of the innovations used on the automobile have been modified from the findings of other organizations, most significantly the aircraft and space industries. A number of these innovations were originally introduced into the automobile industry on the unibody vehicle, and many have since been integrated into the design of the body over frame vehicles as well.

MANUFACTURING AND MATERIALS USED

Steel is still the most common material used in the manufacture of both domestic and commercial vehicles. However, the type of steel used on late model cars, **sport utility vehicles (SUVs)**, and trucks has been modified to impart a number of different characteristics depending on where they are used on the vehicle. Most of the rails and structural members on a **full frame vehicle** are relatively thick and are made of a standard grade, hot rolled **low carbon steel**. The steel used for making the structural reinforcements on the unibody vehicle are made of either a high strength steel (HSS), high strength low alloy (HSLA), or ultra high strength steel (UHSS). The **tensile strength** of these metals ranges from 60,000 to 120,000 pounds per square inch (psi), compared to 30,000 psi offered by the standard low carbon steel. Most of the **panels** on unibody vehicles are also **die stamped** which further increases their strength characteristics.

The increased strength value of the high strength steels has enabled manufacturers to reduce the thickness of the material used without sacrificing or compromising the vehicle's structural integrity. In areas where additional strength is required, a lamination of multiple layers of steel is often robotically welded together using a **squeeze-type resistance spot welding (STRSW)** process. See Figure 6-1. This allows the manufacturer to sometimes combine two or more thicknesses of metal with varying tensile strength values into one structure. An exacting number of welds that are precisely gapped and placed at very strategic locations are used to ensure the structure will respond in a precise manner during a collision. It is paramount that the repair technician understands and follows the procedures recommended by the manufacturer when repairing or replacing any of these parts, as they are very heat sensitive. (The recommended repair procedures will be covered in Chapters 14 and 15.) Corrosion protection between the layers of metal is another concern to both the manufacturer and the repair technician.

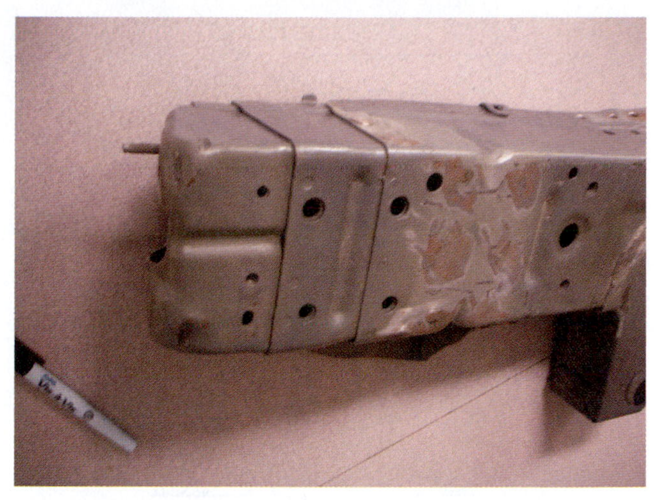

FIGURE 6-1 Manufacturers use layers with varying thicknesses of steel and laminated steel to control the collapsing of the rails during a collision.

STRENGTH CHARACTERISTICS

Shape Affects Strength

The degree of strength and the subsequent resistance to bending and distortion of a panel is determined by a number of factors unrelated to the metal's tensile strength. The more irregular the surface is made when it is formed, such as stamping crowns into it, the more rigid and resistant to bending it will be. Conversely, a large flat surface such as a roof panel offers very little resistance to bending and distortion from outside forces. The difference between these two surfaces is a result of **work hardening** that occurs on the crowned panel when it is formed. Any time metal is bent or permanently changed in shape below the temperature of red hot, it becomes work hardened. See Figure 6-2.

Work Hardening

When metal is formed the **metallurgical structures** or **lattices** are arranged in a very precise and systematic formation with random spacing between each structure. This spacing—together with the natural elasticity of the lattices—allows the metal to bend and flex, yet spring back

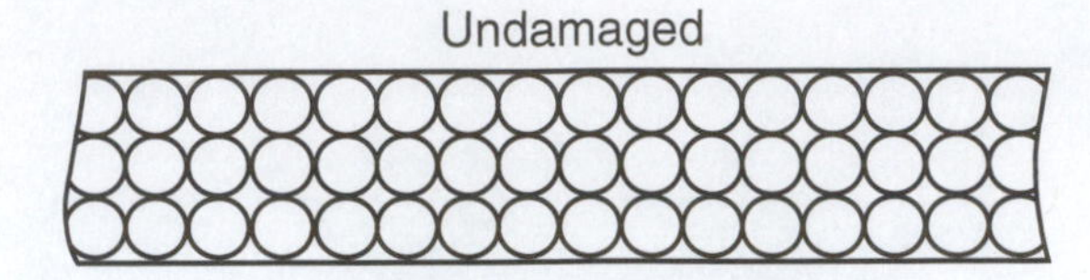

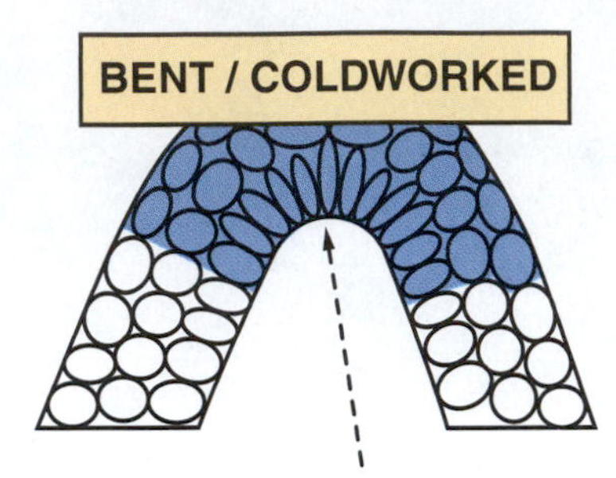

FIGURE 6-2 The metallurgical structures become permanently distorted when the metal is bent beyond its yield point below the temperature of red hot. Examples of the metal becoming work hardened are during die stamping operation and when the panels get damaged in an accident.

to its original configuration when the force causing it to change shape is removed. However, when a sufficient amount of force is applied, causing the metallurgical structures to become permanently distorted as in Figure 6-2, the metal will become work hardened in that area. It will become permanently distorted from its original shape, thus taking on a new set or shape. Any time metal becomes work hardened, it becomes stronger and increasingly more resistant to bending and flexing. It should also be noted: the sharper the formation of a bend (or, in the case of a crown, on the surface of a panel), the more severe the work hardening becomes. As one would expect, the metal loses its flexibility and becomes brittle, eventually fatiguing, cracking, and breaking if subjected to repeated bending in the same area.

The auto manufacturers frequently use work hardening as a tool for strengthening the parts when manufacturing **cosmetic panels** and structural panels. Depending on the strength requirement of a part, one or more shapes are formed into a panel when they are die stamped. The severity of a crown or bend can dramatically affect the resulting amount of strength a panel will have. A surface with a sharp 90-degree bend, for example, offers more strength and resistance to flexing than one that has rounded corners. A soft or gentle crowned surface offers even less than either of these two.

Common Shapes Used

The most common configurations utilized for reinforcement purposes by auto manufacturers are a simple 90-degree L-shaped surface, a three-sided "U" channel, and—in areas where substantially more strength is required—a four-sided box-like structure. See **Figure 6-3**. The L-shaped bend is commonly used on the mating flanges of nonstructural panels where they are spot welded together. The three-sided U-shaped channel is used on frame rails, undercarriage channels, mounting brackets, suspension mounting pockets, and other areas where additional structural support is required. The four-sided box-like structure is commonly used on the front and rear rails, in the cowl pillar area, and other locations where the maximum possible strength is required. Many times these box-shaped panels are comprised of one U-shaped piece of metal or rail with a flat piece welded into place making the fourth side. This is commonly referred to as a **hat channel**. Any panel becomes exponentially stronger each time another side is added or enclosed because each surface supports and helps the others resist becoming distorted. This is the essence of a unibody design: As each additional panel is welded into place it further adds strength and integrity to all the others until all the sides are completely connected, making a completely enclosed structure.

FIGURE 6-3 Automobile manufacturers use essentially four different shapes or configurations to strengthen the metal: A four-sided enclosed box structure, a "U" or "C" channel, a crowned surface, and a 90-degree bend or angle for a mating surface on the edge of a panel.

NEW ENGINEERING CONCEPTS

Many of the improvements for vehicle safety are not so much the result of using a higher quality steel and other materials, but rather are the result of improved engineering and design. Through these new engineering concepts the auto manufacturers have been able to develop a systematic means of reducing and re-directing the collision energy around the passenger compartment, instead of through it. This protects the occupants from a significant amount of the impact forces. The use of crush zones, stress risers, stress concentrators, and kickup areas in the frames, undercarriages, and various structural members has been instrumental in the development of an automobile that will react and respond very predictably in a collision situation.

Sacrificial Parts

Crush zones are designed into certain areas of a structural member to absorb and deflect much of the collision energy. A variety of techniques are used to create the crush zones including the use of dimples, slots, and piercing holes at very precise locations in a rail. Dimensionally reducing the size of the rail and the metal thickness in the area intended to collapse, or increasing the distance between spot welds to create a weaker area to control the bending location, are other design principles used to manage the collision forces. See **Figure 6-4**.

Convoluted Rails. Body over frame vehicle designs sometimes place a convoluted section at the front of the rail which is also designed to collapse in a collision. See **Figure 6-5**. The use of offsets and kickups in the rails are some other methods used to create a crush zone. Whatever design may be used, the reinforcement or structural members are designed to bend, fold, or collapse in a very specific location when a specified amount of force is applied. This force causes them to move in a predetermined direction. See **Figure 6-6**. For example, a front rail will collapse and move in a downward direction, causing the engine to move under the passenger compartment rather than moving straight back into it. Crush zones are incorporated at strategic locations throughout the vehicle, thus reducing the collision forces regardless of the angle or the point of impact on the automobile.

Predictable Damage. The common location of the crush zones in most car and truck models offers a significant advantage to the repair technician as the type and location of the damage sustained in a collision is much easier to predict, identify, and isolate. Most vehicles will sustain a similar type of damage when subjected to a given collision situation. For example, if one were to take four different vehicles and subject them to similar impact forces, each would sustain similar damage in the crush zones on the front rails and adjacent areas. Regardless of the manufacturer, each of the vehicles would predictably sustain similar damage, thus making it easier to trace and identify the flow of the damage. Based on the individual's past experience, when certain damage conditions exist, the technician can assume that other specific areas will have been affected or damaged as well. The need to check deeper into the vehicle should be self-evident by these tipoffs. If

FIGURE 6-4 A variety of techniques—such as dimples stamped into the rails, piercing holes, and making slots in critical areas—are commonly used to create crush zones on the vehicle's structural rails.

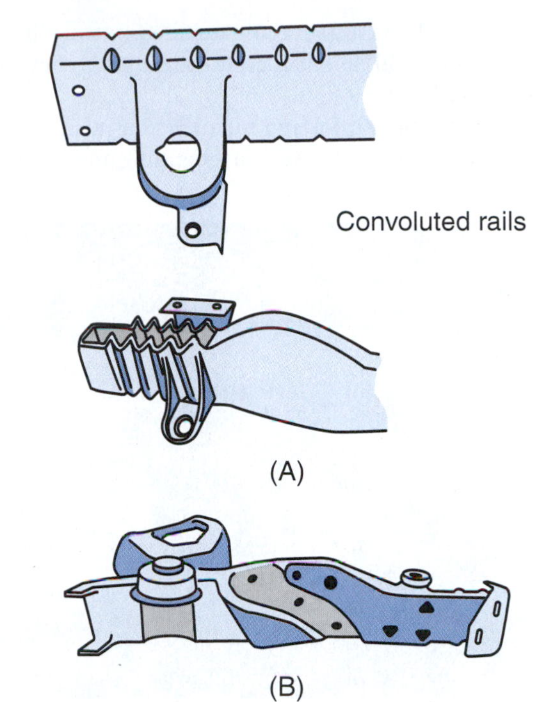

FIGURE 6-5 On body over frame (BOF) vehicles, kickup areas, convolutions, shape variations, irregularities, and offsets (A) are other methods used to create crush zones (B) to dissipate the collision energy.

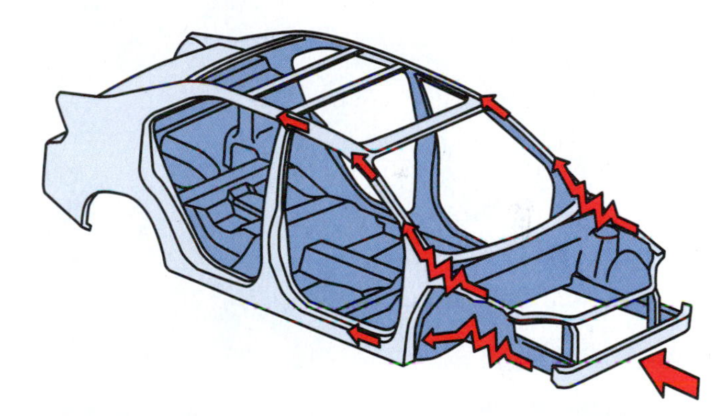

FIGURE 6-6 The placement of the crush zones and the collapsing of the structural parts of the vehicle help to direct the collision energy above, below, and around the passenger compartment to reduce its effect on the occupants.

some of the mechanical parts are attached in the affected area, they too should be checked for damage as they may require replacement.

The shape of the reinforcing members has a significant effect on the vehicle's ability to resist bending and buckling when outside or hostile forces are applied. See **Figure 6-7**. A smooth, flat surface—such as those commonly used for the roof and trunk lid of the vehicle—offer little or no resistance to bending. Any strength factor they impart is largely dependent on the reinforced areas to which they are attached.

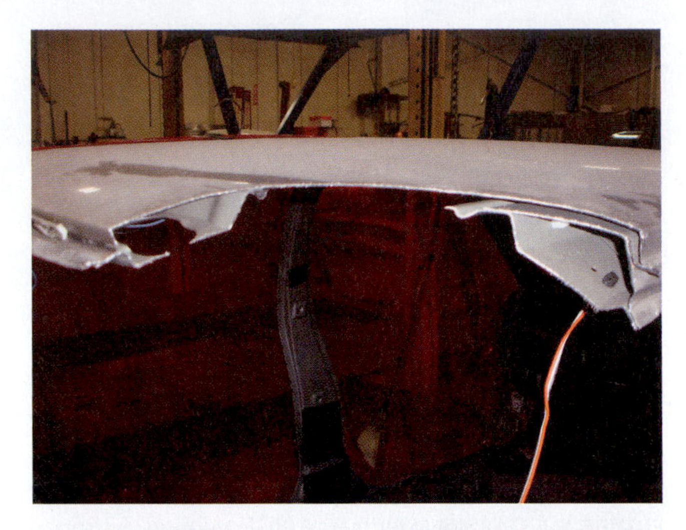

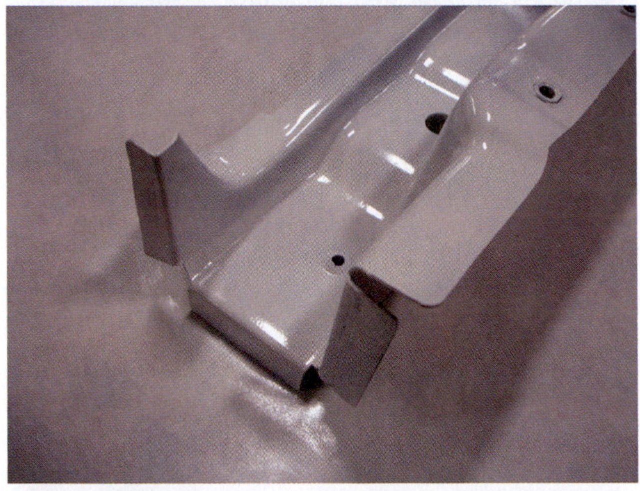

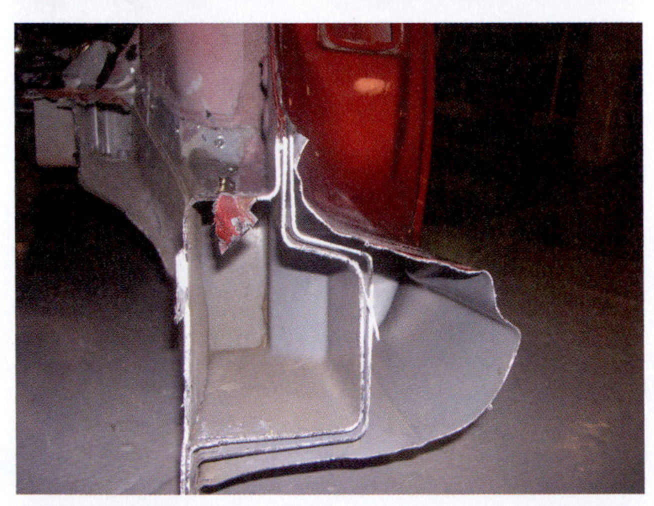

FIGURE 6-7 The shape of the exterior panel and the reinforcing members determines the degree of strength and resistance to bending by collision and other outside forces.

MANUFACTURING MATERIALS USED

Steel vs. Aluminum

Since the beginning of time, steel has been the metal of choice for U.S. car manufacturers because it has been abundantly available and it is thought to be more repairable than any other metal. Aluminum, which for many years was extensively used by many European manufacturers, was shunned by the domestic car manufacturers because it created numerous repair issues. Special tooling, straightening and pulling machines, and welding equipment are needed and a great deal more skill is required of the technician to repair the aluminum vehicles. However, the weight reduction and the corresponding increase in fuel efficiency achieved with the use of aluminum has been a major incentive for U.S. manufacturers to give serious consideration to its use. See **Figure 6-8**. The increased number of imported **aluminum-intensive automobiles** has forced equipment manufacturers to respond to the changing needs of the collision repair industry. Repair technicians have become increasingly more knowledgeable and adept at repairing these vehicles. Aluminum is likely to be the new frontier for car manufacturers because, with new recycling and refining methods and techniques, it is cheaper to produce and more abundant than steel.

FIGURE 6-8 The increased use of aluminum has been a significant factor in reducing the weight of the automobile. Aluminum has become a popular choice of metal for many Asian and European imports, as well as American-made sports cars.

Plastics

Plastics have been used in the manufacture of automobiles for many years for two primary reasons; They are lightweight and easily molded into very dramatic shapes to fit and mate up to adjoining panels. A variety of other **composite materials** have also been used in an effort to reduce the weight of the vehicles, resulting in an increase in fuel efficiency.

CHANGING VEHICLE DESIGN

Body Over Frame vs. Unibody

The **body over frame (BOF)** was the standard design for nearly all vehicles manufactured in the United States from the beginning of the automotive age until the 1970s, when surging gas prices caused by an oil embargo by Arab nations forced the American auto manufacturers to re-evaluate their products. Consumers demanded smaller and more fuel efficient vehicles than the current offerings allowed. In order to comply with consumer demands, the size and weight of the vehicles had to be reduced significantly. This lead to a significant series of changes in the design and size of the cars produced, the materials used in their production, and the manner in which the vehicles were manufactured. Ultimately whole new concepts in **domestic vehicle** design and manufacturing were introduced. The **unibody**, which had been widely used for many years by European auto manufacturers, became the design of choice, nearly replacing the BOF system. This section will review the differences between the body over frame and the unibody design and evaluate the advantages and disadvantages of each system.

Body Over Frame

The body over frame design is commonly used to describe those vehicles that have a separate frame and body, both of which are independent of each other. The vehicle may have either a full or **partial frame**. On the full frame, the rails span the entire length of the vehicle, extending from the front completely to the rear end. On the partial frame or **stub frame**, the frame is a shortened section under the front of the vehicle only. The frame is essentially the vehicle's backbone and is responsible for the majority of the body's strength characteristics and support. The body, engine, and drivetrain, as well as the steering and suspension parts, are all attached to flanges that are welded or riveted to the frame. Special body mounts with isolators or bushings made of rubber or neoprene are used to hold the body in place and prevent it from shifting or moving during normal use. See **Figure 6-9**. This bushing is used to dampen the shock effects and isolate the road noise and vibration from being transmitted into the passenger compartment through the floor pan. Without the use of these **sound deadening materials**, the noise level and vibration would be very uncomfortable for the passengers.

Frame Designs. The two principal frame designs commonly used on today's cars and trucks are the ladder and perimeter design. The ladder frame gets its name from its likeness to a conventional ladder. See **Figure 6-10**. It is typically used for larger and heavier applications such as trucks, passenger and cargo vans, and four-wheel-drive vehicles where added strength is required. Two relatively straight side rails that run parallel to the body—with several cross-members welded or riveted at strategic locations—are generally employed to complete the assembly. The side rails are typically a three-sided "U" channel design except in areas where additional strength is required, such as under the engine and transmission mounting locations, where they may be a closed box design. The wheels and much of the steering mechanism—such as the spindle, control arms, steering arms, and so

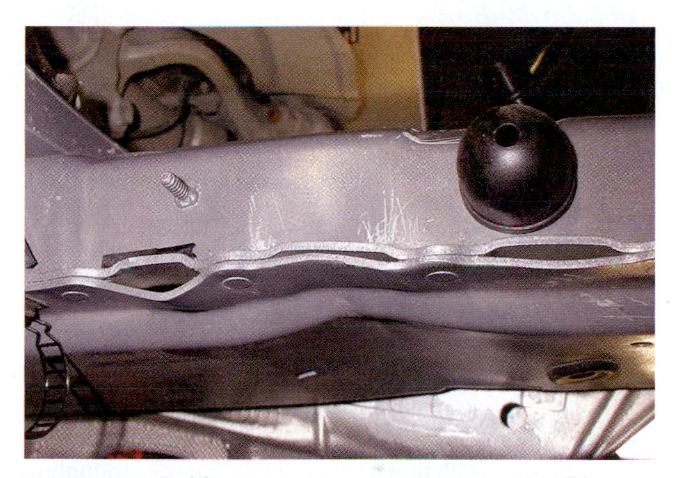

FIGURE 6-9 Rubber or neoprene spacers and bushings are placed between the body and the frame and between the frame and the core support to eliminate the road noise and vibration that would otherwise be transmitted into the body through the frame.

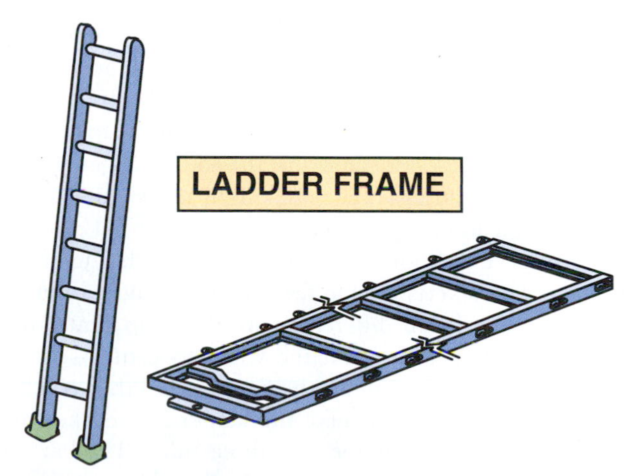

FIGURE 6-10 The ladder frame holds a close likeness to a conventional ladder, hence its name. Its use is limited to larger and heavier trucks and a limited number of four-wheel-drive vehicles, all of which require additional strength and rigidity.

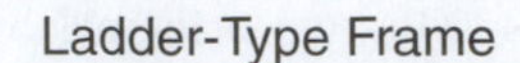

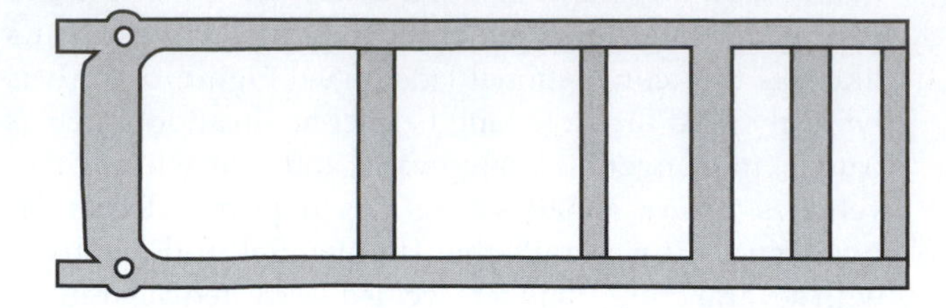

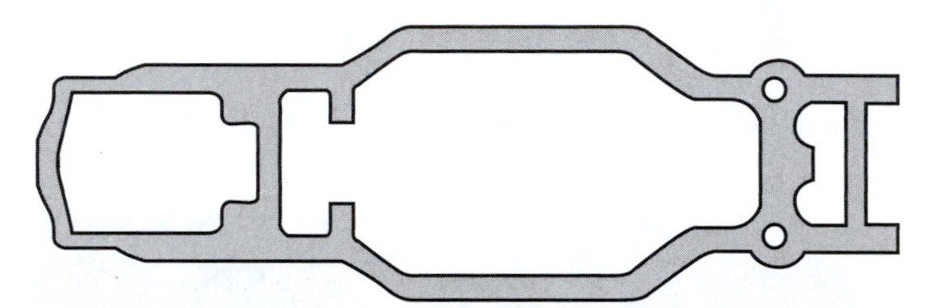

FIGURE 6-11 The steering mechanism, suspension, and wheels appear to extend out further on a ladder frame than on a perimeter because the rails are in a straight line from front to rear.

on—nearly always extend to the outside of the frame side rails. See **Figure 6-11**.

The cross-members are used to reinforce the frame and for mounting most of the body, steering, and suspension parts. They are strategically placed in areas where mounting the cab and box is most easily accomplished, to simplify removal and installation, and to maximize the strength of each. They are also used to attach and support the steering and suspension, the transmission, and other drivetrain parts. The area immediately under the engine is generally quite heavily reinforced with additional supports and metal thickness to accommodate the additional weight.

Vehicles using the ladder frame typically have a higher center of gravity since the cab and the remaining body components are attached above the frame and most of the steering and suspension parts are attached below it. This gives the vehicle added height, making it prone to tipping over more easily. At the same time the added height results in additional road clearance, making it more suitable for use in rough and irregular terrains.

With few exceptions, the perimeter frame has many of the same characteristics and features as the ladder frame. The most significant differences are the drop center design, the kickup areas, and the widened center section used to protect and reinforce the perimeter of the vehicle's passenger compartment. Like the ladder frame, the side rails on the perimeter frame are a three-sided "U" channel design with several areas where an enclosed box design is used for added strength, reinforcement, and rigidity. The closed sections are typically used in areas such as where the engine, transmission, and suspension parts are mounted, and in the torque box area.

Torque Boxes. The torque boxes are nearly always an enclosed rail design. They are located under the four corners of the passenger compartment where the front and rear rails are attached to the center section rails, forming a fully enclosed box-like structure. This area—which is very rigid because of the numerous layers of steel that are joined together—is commonly used for many other applications such as attaching the body to the frame, lifting the vehicle off the ground, anchoring the vehicle for repairs, measuring reference points, and so on.

The torque boxes serve several purposes both structurally and to improve ride quality. One such function is to absorb and dissipate the twisting effect that occurs on both the front and rear frame rails upon acceleration, particularly on lateral mounted engines and on a rear-wheel-drive vehicle. Upon acceleration, the engine causes a downward thrust on one side of the frame, causing the other rail to raise up. At the same time, at the rear of the vehicle, the transfer case causes an uplifting force upon acceleration. The torque boxes absorb and dissipate this twisting effect, thus preventing it from being transferred rearward from the engine or forward from the rear axle. Without them, the entire body of the vehicle might twist uncontrollably each time engine acceleration occurs.

The torque boxes are used as one of the primary measuring reference points for diagnosing damage and affecting the necessary repairs. The sturdiness of this area makes it less apt to become damaged in a collision, thus rendering it a reliable and frequently used reference and measuring site for damage analysis and diagnosis. This same strength feature makes it a choice area for anchoring the vehicle during repairs. It's also used for lifting the vehicle off the ground, whether with a floor jack or a four point shop lift.

While not true in all cases, several vehicles equipped with the perimeter frame design also incorporate the body for reinforcement of the total unit. Cross-members are strategically placed near the front of the passenger compartment and near the rear of the vehicle. This allows the body to be attached to the frame rails at more locations, making the entire unit more solid. The front cross-member is also utilized to support the tail mount of the transmission and other mechanical parts commonly found on four-wheel-drive models. The body is mounted to the frame using a special bolt which holds a neoprene or rubber bushing in place between the body and the frame.

Drop Center Design. The drop center design allows the overall body to be lowered, giving it a lower center of gravity. This makes the vehicle more stable, especially on turns and on rough and irregular driving surfaces. In this design, the **rocker panels** are placed immediately to the outside of the frame rails, providing the occupants additional protection in a side collision.

As part of the drop center design, a kickup effect of the side rails in front and behind the passenger compartment are utilized to attach the suspension and steering parts below the rails. This is a significant safety feature in both a frontal and rear collision. The rails are designed to collapse and fold in such a way that it causes the engine and drivetrain to go under the passenger compartment, rather than be pushed straight back into the occupants. As another advantage, it allows for easier access and egress from the vehicle because it is built lower to the ground. Until the late 1970s and early 1980s—when the unibody nearly became the universal or exclusive design offered—the perimeter frame undercarriage was the most popular frame used by all U.S. auto manufacturers. See **Figure 6-12**.

A relatively recent development in undercarriage and other structural design is a process called hydroforming. See **Figure 6-13**. Both body and structural parts are fabricated using this process. After rough bending and forming the part, which is usually a tubular structure, the hydroforming process is performed to create the final shape of the part. The process involves the use of a combination of extremely high air, water, or hydraulic pressures ranging upward to 10,000 metric tons (over 10 million kilograms or over 20 million pounds). The metal part is placed into a mold and the pressure is applied to achieve the final shape, often involving smoothing and rounding the edges and corners when the pressure is applied. The final product may require some simple drilling operations for attaching various parts and for mounting them to the undercarriage. This process offers many advantages over conventional frame fabrication, as little or no welding is required to attach mounting flanges and the strength factors exceed any other system used up to this point. The simplicity of the process will certainly find its way into the fabrication of many other parts of the automobile than what is currently used.

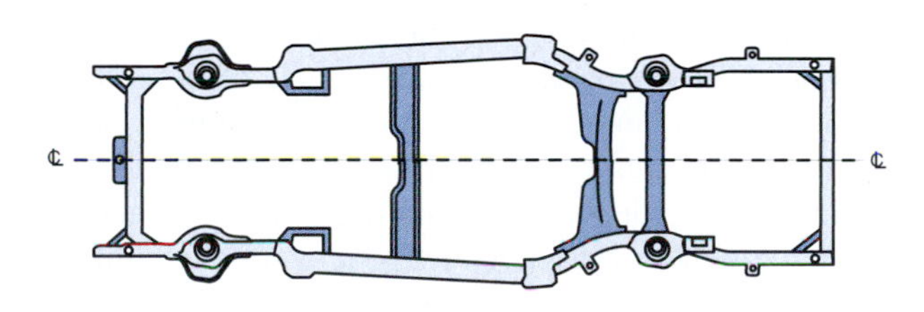

FIGURE 6-12 The perimeter frame is primarily used on passenger cars, with some modifications on SUVs. It usually incorporates a drop center section for easier entrance and egress from the vehicle.

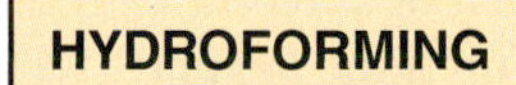

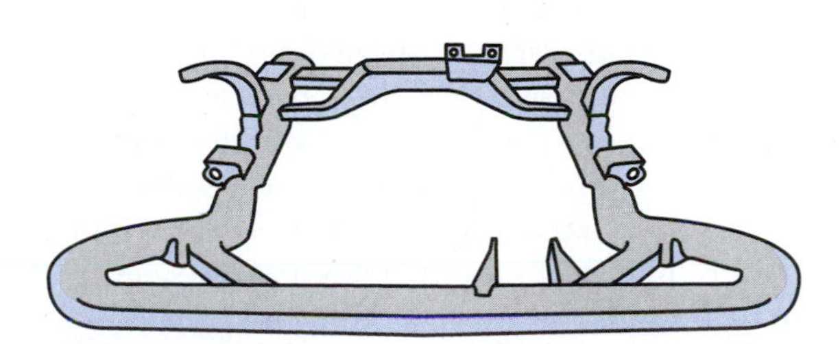

FIGURE 6-13 Hydroformed frames are manufactured as stub frames only and in small sections. They are rough formed first before being subjected to extremely high pressure to round out the corners and edges. They have no flanges or seams, and any mounting brackets are welded on after the fact.

Unibody Vehicles

The unibody design became the most popular choice for nearly all automobiles manufactured after 1980. The sudden increase in crude oil prices and the public's demands for more fuel efficient cars forced the automobile manufacturers to re-think what they offered to consumers. For many years the unibody had been used very successfully by European manufacturers, and with certain modifications was embraced by the U.S. auto manufacturers as well. It offered an immediate reduction of nearly one-third of the curb weight of many cars, which of course resulted in a significant increase in fuel economy. However, although the consumer greeted the unibody with a great deal of enthusiasm, that acceptance wasn't shared by the repair industry. This was a completely new design for most repair facilities, except those who specialized in repairing imported vehicles prior to the late 1970s and early 1980s. Very few shops were equipped with the anchoring, pulling, and measuring systems required to make repairs to these automobiles. Very few of the technicians were knowledgeable or trained in the use of the specialty welding equipment and techniques the manufacturers specified. Many repair shops simply avoided scheduling any of these vehicles that required any structural repairs into their shop. Quite often the repairs that were made to vehicles were done incorrectly, sometimes leaving them unsafe and unable to withstand the collision forces in a subsequent accident. The entire collision repair industry was in turmoil as they attempted to respond—with little success—to the changes that had been implemented overnight. The Insurance Institute of America, together with the big three automobile manufacturers as well as many of the equipment manufacturers, ultimately came to the rescue by creating a coalition committed to training the American auto collision repair workforce. This was called the **Inter-Industry Conference on Auto Collision Repair**, an organization known and recognized throughout the industry today as **I-CAR**.

The Unibody Design. What is it about the unibody vehicle that makes it so unique from all of the body over frame vehicles? Perhaps the most significant difference of the unibody from the BOF vehicle is the undercarriage design and materials used. The heavy frame that is constructed of nearly 3/8-inch thick steel is replaced with multiple layers of thinner gauge steel in some areas on the unibody. These thin multi-layered sections are made of special **high strength steel** that are considerably thinner but stronger than the old standard frame rails, often resulting in a considerable weight reduction. This steel is also very heat sensitive, and special care must be taken when heating and welding on any part where this material is used. Overheating that may occur from using incorrect welding techniques or even minor heat applications will result in destroying its structural characteristics. The result is a vehicle that is no longer able to withstand the forces of a subsequent collision, nor the wear and tear from normal driving conditions.

Steering and Suspension Mountings. Unlike on a BOF vehicle, the steering and suspension parts on a Unibody are attached directly to the front side rails, which are part of the body on a unibody design. The design, attachment, and placement of the engine and drivetrain have also been modified considerably. A transverse mounted engine and drivetrain have replaced the lateral placement historically used in the BOF vehicle. The front-wheel-drivetrain and transmission used on most unibody designs has eliminated the drive shaft running to the rear axle where the drive mechanism was once located. The subsequent elimination of the center hump in the passenger compartment is a significant benefit for passenger comfort and increased cabin space. In light of all these changes, one can readily understand the need for the special equipment and techniques that must be employed and why different rules, procedures, and repair guidelines must be applied when repairing such a vehicle. Numerous new design features have been incorporated into the unibody with the ultimate goal of protecting the vehicle's occupants.

In recent years, consumers and consumer advocacy groups have become increasingly more concerned and verbal about the safety and well-being of the public in nearly all facets of society. In response to these public demands, the automobile manufacturing industry has committed itself to producing products that offer more occupant protection than ever before. Overall, the automotive industry is responsible for instituting more advances and changes for consumer protection than any other area in our society. The research and developments engineered into today's automobile, particularly the unibody, have made the car a great deal safer than it has ever been before.

UNIBODY CONSTRUCTION AND DESIGN FEATURES

The unibody vehicle is constructed of a combination of structural and semi-structural panels that are spot welded together, forming a single unit. Each panel is dependent upon all the other adjacent and non-adjacent panels to give it strength and rigidity. When all the parts are welded together, three distinguishable—yet interdependent—sections of the car are formed: the front section, the passenger compartment, and the rear end of the car. While each section of the vehicle can be isolated and is identified as somewhat of a separate unit for measuring and dimensioning purposes, they are all dependent upon each other to ensure the vehicle's structural integrity.

In a unibody, the structural parts are used in harmony with the cosmetic parts, further defining to the interdependence of all the parts to make a solid unibody structure. An example of this is the **quarter panel**, which—although not considered a structural panel—is used to reinforce the pillars and other structural members around it. Due to its interdependent construction, any stress—whether it be a damaging or corrective force—will be transmitted through the entire length of the vehicle, affecting it throughout. Depending on the nature and severity of the damaging force, the unibody vehicle is designed to absorb, dissipate, and transmit the damage around the passenger compartment, thus protecting the occupants. In order to protect the vehicle occupants, many engineered provisions are made to route the damage around—instead of through—the passenger compartment.

Space Frame

A close relative of the unibody vehicle is the **space frame** design. See **Figure 6-14**. The undercarriage on this vehicle is very similar to that of the unibody in that the steering and suspension parts are attached to the body. The principal difference between the space frame and the unibody is the main structure: It is a self-contained, crash worthy frame made of lightweight high strength steel with composition panels fastened to it. Unlike the true unibody, it is not dependent upon the exterior panels for strength and rigidity and will perform the same with or without the exterior panels attached. Sheet molded compound (SMC) exterior panels are attached by applying adhesive bonding, attaching them with fasteners, or using a combination of both.

Cross-Over Vehicles

Over the years the unibody has been modified, evolving into somewhat of a hybrid with regard to the undercarriage and structural support system. As discussed previously, on a true unibody, the mechanical, suspension, and steering parts are attached directly to the body. However, a number of models have been available where a partial frame section is bolted to the bottom of the front rails, forming a modified body over frame vehicle. These are commonly referred to as stub frames or sub-frame vehicles, meaning the frame only covers a small section of the front of the vehicle. In some instances a similar frame section may be attached on the rear of the vehicle as well. The upper body still resembles that of a true unibody in that the vehicle's shell is combined to form a complete assembly. The primary difference between these and the unibody is the drivetrain, steering, and suspension parts are all attached to this sub-frame.

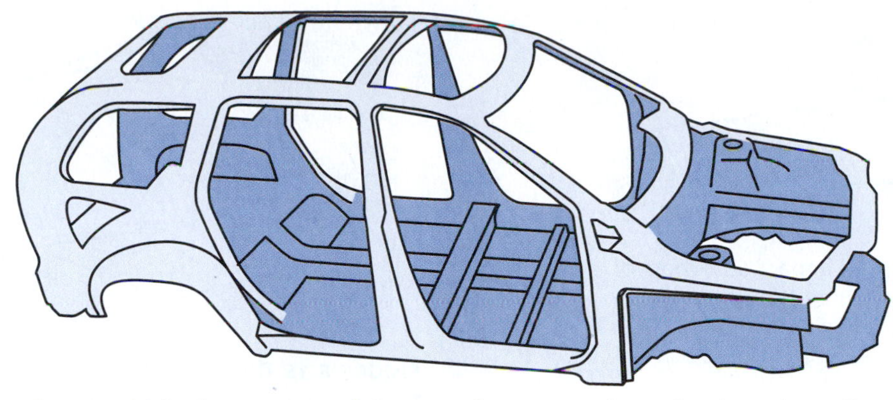

FIGURE 6-14 A space frame vehicle shares many of the same features as the unibody and usually consists of a crash worthy, self-contained framework made of lightweight high strength steel. The outer panels are commonly made of plastic and are adhesive bonded to the frame.

MANAGING COLLISION FORCES

In order to understand how the collision energy is diverted around the passenger compartment one must review and understand the features and techniques used to absorb and dissipate these forces. One must also understand that a nearly simultaneous reaction occurs throughout the vehicle as any significant force is applied. It must also be understood that because all the panels are interconnected, any force applied to the vehicle will affect all the other areas throughout the car. The intensity of the collision forces are reduced by employing a variety of collision energy management principles and engineering techniques.

Bumper Assembly and Fascia

The bumper at the front of the vehicle is said to be the vehicle's first line of defense. See Figure 6-15. Many vehicles utilize a plastic front cover that is integrated into the design with the hood and front fenders. A steel reinforcement bar together with some form of energy absorbing material such as a **structural foam** pad is used on most late-model vehicles. In some instances an **impact absorber** or **isolator** is used to attach the bumper to the front side rail. They are designed to protect the vehicle in most "parking lot" impacts by absorbing 5 miles per hour or slower impacts without sustaining any damage on the front of the vehicle. This absorber can be spring loaded, gas charged, or it may be filled with hydraulic fluid or another form of compressible material. They are designed to collapse or compress during a low speed collision and retract or extend back out when the pressure is removed. See Figure 6-16. Many are designed with slotted attaching holes to afford up and down and side-to-side adjustment to align the bumper with the adjacent front parts.

Front Rails

Immediately behind the front bumper are the core support and the front side rails. The front side rails together with the radiator or core support form the structural framework for the entire front end of the vehicle. The rails are usually a combination of two or three die stamped pieces of metal that are welded together to form an enclosed box-shaped assembly. See Figure 6-17. This four-sided configuration is considerably sturdier and more resistant to bending, buckling, or collapsing inappropriately in a crash situation. Many times these parts are made

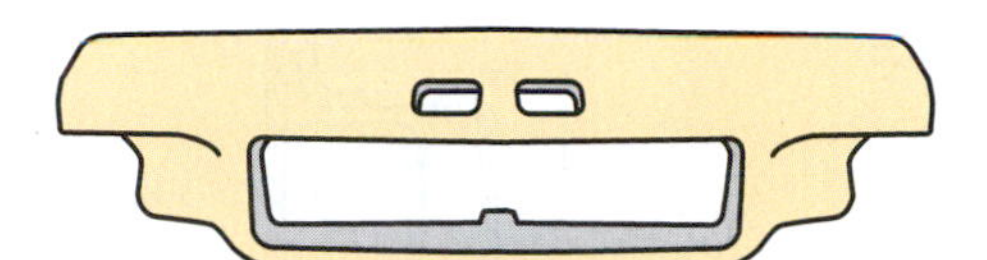

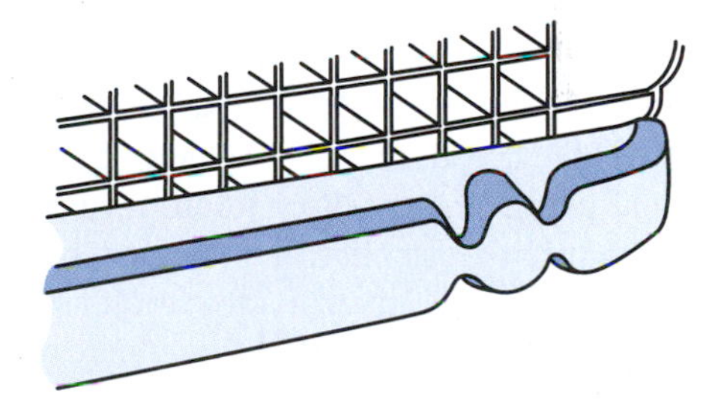

FIGURE 6-15 On unibody vehicles the energy absorbing devices used on the front bumper to dissipate the collision energy include a high density structural foam pad, an eggcrate honeycomb plastic energy absorber, stranded fiberglass reinforcements, and steel bumper reinforcements.

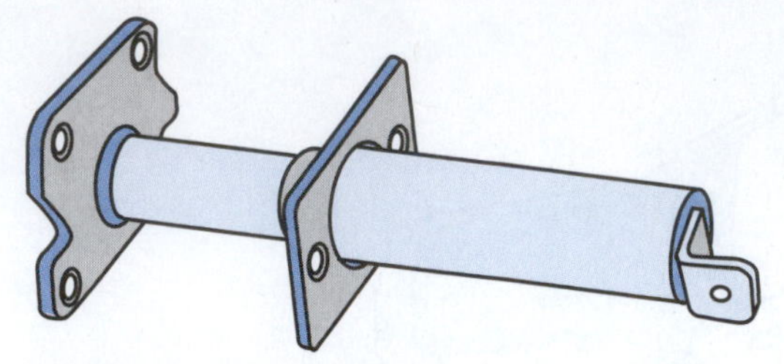

FIGURE 6-16 Mechanical impact absorbers or isolators may also be used to cushion the impact forces. They are either gas charged or they may be filled with hydraulic fluid, which is displaced by a piston upon impact.

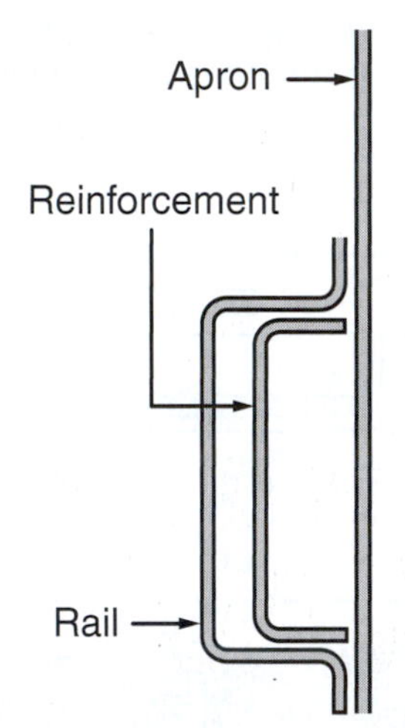

FIGURE 6-17 The front side rails are usually a laminated series of two or more die stamped panels made of conventional low carbon and high strength steel which are welded together to form an enclosed box structure or assembly.

of high strength steel and require special attention when repairs are made. In order to achieve a specific strength factor, the manufacturer may incorporate a combination of two or more layers of standard low carbon steel together with high strength steel to form a front rail assembly. The front rails extend from the front of the vehicle to below the passenger compartment where they are spot welded into place, often becoming a part of the floor pan reinforcement. The **strut tower** or strut housing, which is used to attach part of the steering and suspension parts, may be an integrated part of the front rails or they may be a separate piece welded onto the assembly. See **Figure 6-18**.

Core Support

On many models the core or radiator support is a one-piece assembly that is welded between the two front rails to further reinforce the front assembly. This part is frequently made of high strength steel and should be treated accordingly when making repairs near or on it. On some models the radiator support may be two or more individual sections welded or bolted into place.

The engine cradle, which is usually a heavily reinforced support member, may be bolted to the bottom of the front

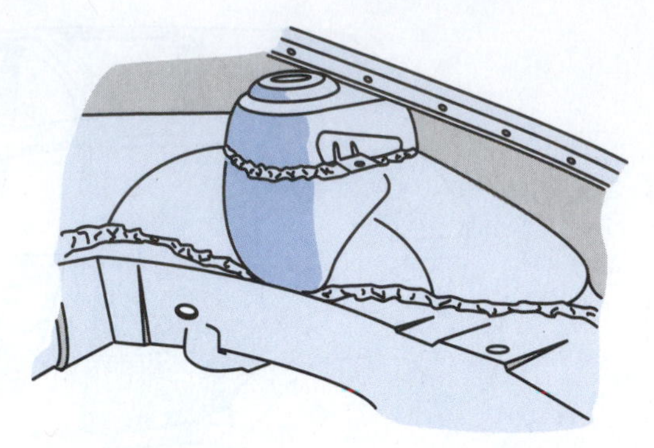

FIGURE 6-18 The strut tower or strut housing may be integrated as part of the front rail assembly or it may be serviced as a separate piece that is welded into place. In some instances, it may even be a casting that is bolted on.

rails or it may be welded into place. The engine mounts, which also hold and stabilize the transmission, may be attached to a number of places including the cradle, the core support, or the side rails.

DISSIPATING COLLISION ENERGY

Auto manufacturers have employed a number of different systems and techniques to reduce and dissipate the collision forces which ultimately reduce the effects on the vehicle's occupants. Things such as pierced holes, convolutions, offset designs, and kickups placed in the front side rails to enhance the absorption and dissipation of the collision energy are commonly used on both the unibody and full frame vehicles as well. See **Figure 6-19**. The convolutions cause the metal to accordion or compress together at a specific instant when the collision energy forces are exerted upon the area. See **Figure 6-20**. The pierced holes are a means of weakening the rail at a precise location so it will bend in a designated area as opposed to kinking or buckling from the collision force. Other methods commonly used to direct the collision forces include decreasing the size of the front rail at specific locations and increasing the distance between welds and the use of tailor welded blanks.

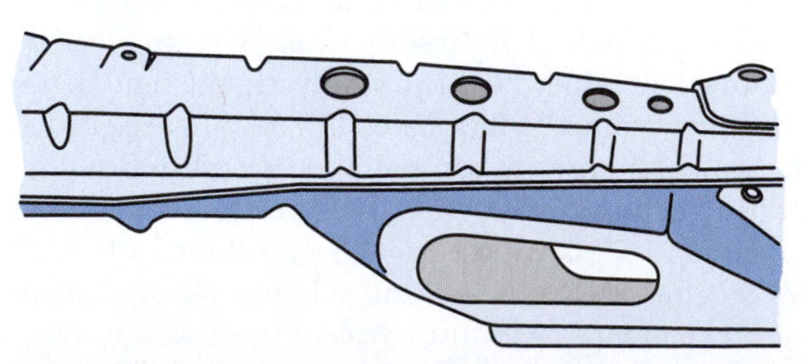

FIGURE 6-19 Auto manufacturers employ a number of techniques such as kickups, convolutions, offset designs, and piercing holes into the rails to create crush zones to reduce the effects of the collision energy.

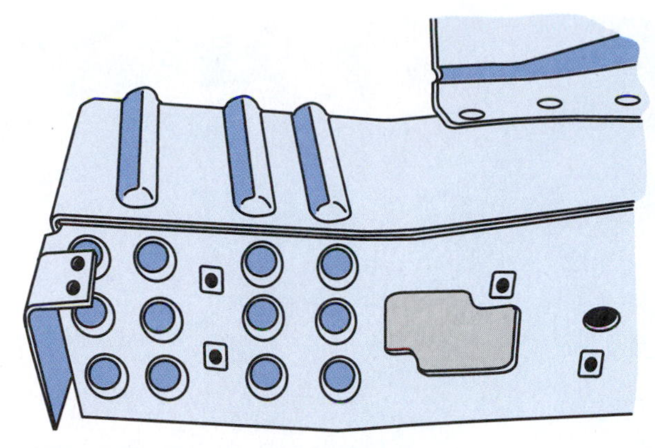

FIGURE 6-20 Convolutions, indentations, and dimples, as well as efforts to reduce the size of the rail, are methods commonly used to create crush zones in the front rails.

Tailor Welded Blanks

The manufacturers use a laser welding technique called **tailor welded blanks** to join together sheets of steel made of different strengths and thicknesses. A section of thinner gauge high strength steel is laser welded to a piece of thicker material, forming a sheet which is then die stamped to form a part. The part—for example, a front rail—is welded into place on the vehicle with the thinner and weaker area at the front and the thicker and stronger part at the rear end of the rail. See **Figure 6-21**. Sometimes two and three different thicknesses of metal or laminations are fused together, creating a similar number of collapsing locations. Thus, in a collision situation, the front of the rail will collapse before the center or rear section does. Tailor welded blanks are also used in other sections of the vehicle such as the **uniside panel** which encloses the doors. The use of tailor welded blanks often results in the elimination of flanges and frequently eliminates much of the required spot welding during the assembly process.

FIGURE 6-21 Tailor welded blanks are sheets of two or three different thicknesses of metal that are laser welded together, which are then used to fabricate parts for areas such as the front rails, center pillars, and hinge pillars where crush zones are required.

Kickups

Kickup areas are also placed at strategic locations in the front and rear rails to help manage the collision forces. The kickup areas are typically located above and slightly behind the suspension mounting areas in both the front and rear side rails of the vehicle. On the front rail they are designed to bend upward in a collision situation, thus starting the engine to move in a downward path. Ultimately it causes the engine to move under the passenger compartment in a severe frontal crash. The possibility of occupant injury is reduced considerably as the engine and drivetrain are moved away and underneath the passenger compartment as opposed to being pushed directly into it. Both the front and rear rails are designed to predictably react in a similar manner. When assessing the collision damage, the technician must be aware of this fact.

Passenger Compartment Cage

As was stated previously, a major concern for automobile manufacturers is protecting the passengers by reducing the collision energy and directing the impact forces around the passenger compartment. To that end, the entire perimeter of the passenger compartment is encased with box-like structural members that essentially form a protective cage. See **Figure 6-22**. All of the structural members are made of high strength steel—or, at a minimum, are reinforced with it—to ensure the safety of the occupants. The rocker panels, together with the floor pan, form the foundation of the vehicle, much like the frame does on a body over frame vehicle. The areas where the front and rear rails are joined to the floor and the rocker panel form the equivalent of the torque box on the full frame vehicle. The rocker panel is the focal point of strength as it consists of multiple overlapping layers of metal welded

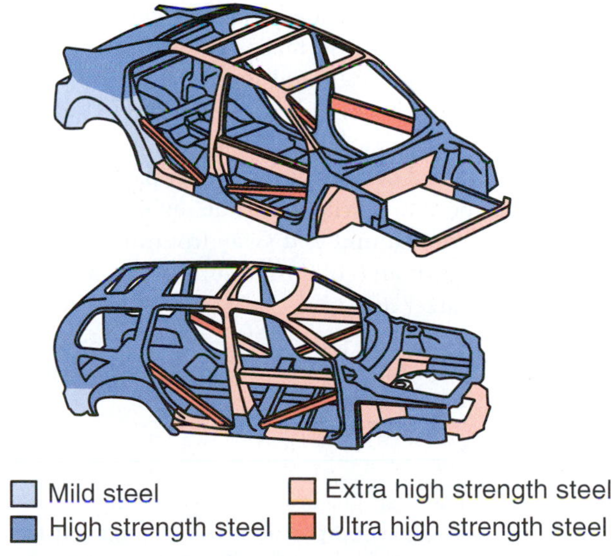

FIGURE 6-22 The passenger compartment on the typical unibody automobile is a cage-like structure designed to transmit the collision forces around the outer perimeter of the vehicle, thus reducing the possibility of passenger injury.

FIGURE 6-23 Lateral stiffeners are commonly welded into the inside of enclosed cavities in areas such as the rocker panels where additional strength and rigidity are required.

FIGURE 6-24 The front door hinge pillar is the most heavily reinforced pillar on the vehicle, as it must support the front of the roof and the front doors as well as help route the collision forces around the passenger compartment.

together to form a completely enclosed structure. See **Figure 6-23**. Depending on the model of the vehicle, it may be additionally reinforced with lateral stiffeners, which are layers of a thicker gauge metal with stamped ribs running through them. These reinforcements are welded inside the enclosed cavity and are visible only after removing the outer layer of the rocker panel. This area of the vehicle is least likely to be affected or become distorted in a collision, therefore making it a logical reference point from which measurements are made for damage diagnosis. It is also the area commonly used for anchoring the vehicle during repairs as well as the lifting points for raising the vehicle off the ground.

Cowl Assembly

The **cowl assembly**, located at the forward part of the passenger compartment, is made up of the front door hinge pillars, the fire wall, and the upper plenum area. As this is the forward part of the passenger compartment and the primary line of defense in a frontal crash situation, it is reinforced much like the rocker panel. In a collision it must be able to withstand and deflect or channel the impact energy around and away from the passenger compartment without collapsing under those forces. High strength steel and multiple layers of seamless or welded overlapping pieces are often used to form a nearly completely enclosed box design for the hinge pillars. See **Figure 6-24**.

Firewall

The firewall is a single layer of metal located at the front of the passenger compartment that protects it from the heat, noise, and fumes coming from the engine compartment. It is commonly used to attach certain mechanical parts, heater and air conditioning assemblies, and various bulky electrical components that would be difficult to place inside the vehicle.

Pillars

All the pillars used to support the roof are designed to withstand a considerable amount of force and minimally must be able to support at least two and a half times the weight of the vehicle to resist collapsing in an accidental rollover. See **Figure 6-25**.

A Pillars. The **A pillar**, which is also commonly known as the windshield post, is a fully enclosed tube with a built-in flange to which the windshield is attached. It may be made of two pieces of metal welded together to form an enclosed structure. Several models—including the Cadillac Deville and the Buick Park Avenue A pillars—are formed by using the hydroforming process. The windshield is bonded to the A pillar with a special urethane adhesive to give it added strength and rigidity. To further

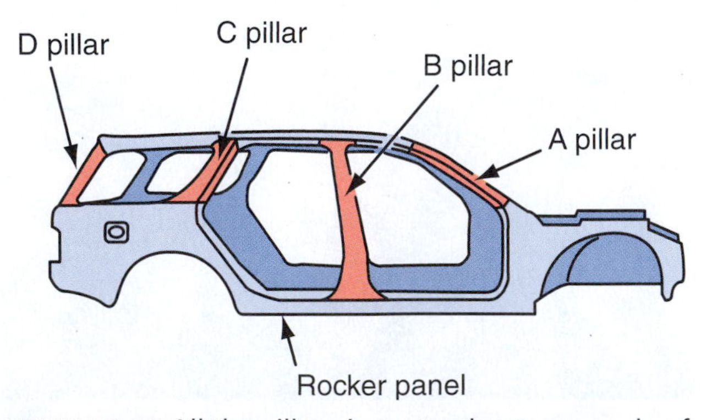

FIGURE 6-25 All the pillars, in part at least, are made of high strength steel because they must be able to withstand at least 2 1/2 times the weight of the vehicle in a rollover situation. Some are further reinforced with structural foam.

reinforce the pillars, some manufacturers partially or completely fill the A pillar and some of the other pillars with a structural foam. When cured, this foam will add a considerable amount of structural integrity to the pillars as it will help to resist flexing, bending, and buckling if the vehicle were to roll over in an accident. (Structural foams will be covered in greater detail in Chapter 8.)

B Pillars. The center pillar, or **B pillar** as it is often called, is located between the two doors at the center of the vehicle on a four-door model or at the rear door on a two door. One of its primary functions is to support the roof, especially during a rollover. It is also the attaching point for the rear door hinges on the four-door model vehicles. This assembly is a multiple piece construction, sometimes with lateral stiffeners added for additional strength. In recent years some manufacturers have incorporated the use of structural foams to add strength and rigidity to this and certain other structural components and pillars. Structural foams are covered in more detail in Chapter 8.

The center pillar is incorporated into the roof and the rocker panels with a very precise number of strategically placed spot welds at both locations to ensure the proper transfer of the collision energy around the passenger compartment. The number and placement of the welds has significant ramifications for the repair technician when replacing this assembly. This is the only roof support pillar with a crush zone built into it to ensure that it flexes and bends at the correct location during a collision. See **Figure 6-26**.

C Pillars. The **C pillar** is also known as the sail panel on most passenger cars or it may be the support pillar behind the rear door on a sport utility vehicle. On passenger cars this is the area where the front upper section of the quarter panel and the roof are joined together. The rear window package tray, along with several overlapping layers of reinforcements from adjacent interior panels, are integrated and welded together to help reinforce this area of the roof.

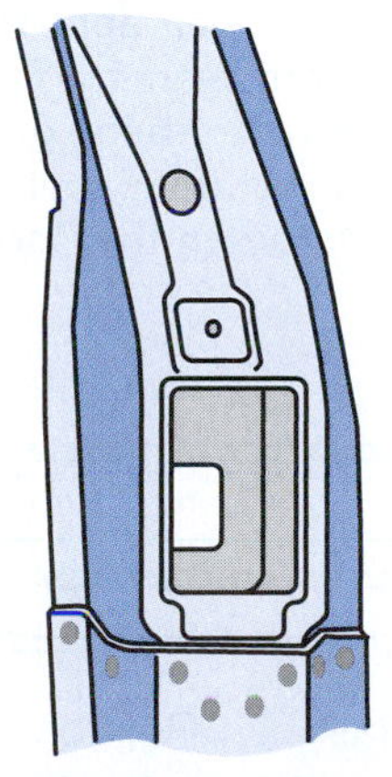

FIGURE 6-26 The B pillar is the only pillar that has provisions for a crush zone built into it, presumably to ensure it flexes properly to transmit the collision energy around the passenger compartment.

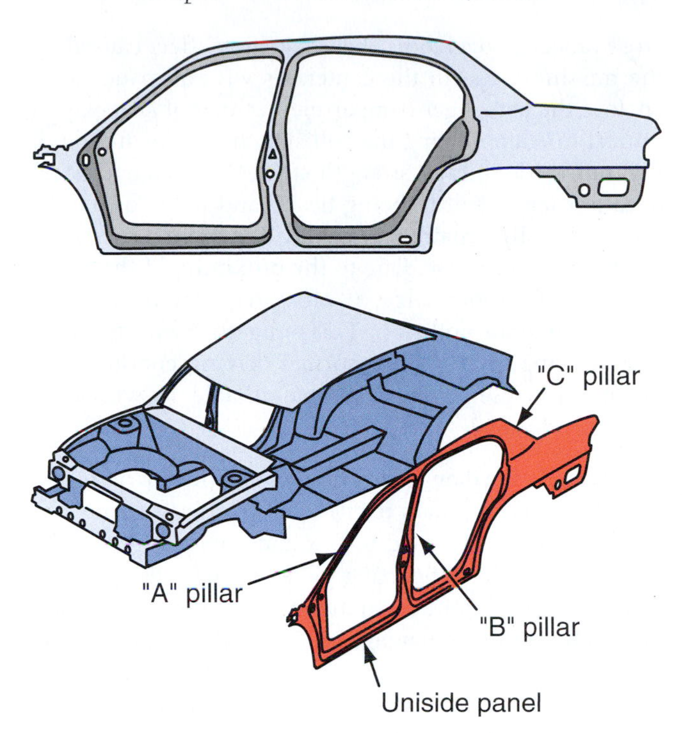

FIGURE 6-27 The uniside panel is one continuous panel extending from the quarter panel to the front door hinge pillar and incorporates all three of the roof support pillars.

In recent years passenger car manufacturers have introduced the uniside panel, which is a revolutionary new design in that it incorporates the outer panels of nearly the entire side of the vehicle into one continuous piece. See **Figure 6-27**. This panel is another example of where the tailor welded blanks are used. The quarter panel section of the uniside, which is made of standard low carbon steel, is attached to the A, B, and C pillars, which are made of a heavier gauge high strength steel. This creates a continuous, uninterrupted panel. The advantage of this design is there are no welded seams and the likelihood of corrosion forming in the weld areas is eliminated. Significantly fewer welds have to be made in the assembly process, resulting in the entire panel being stronger because it is a continuous, seamless structure. The use of the uniside panels further helps to reduce the vehicle's weight. Repairs are rarely made which include replacing the entire uniside panel. Instead, only a section of the panel, such as the quarter and sail panel, may be replaced. This reduces the invasion of undamaged sections and minimizes disruption of the manufacturer's corrosion protection.

The roof panel may be a single or a two-layered design with the outer and inner panels reinforced by a boxed framework that is integrated with all the support pillars and extends around the entire perimeter. In some instances the inner layer of metal is a solid piece that stretches across the entire roof, while on other models it is only a partial panel or reinforcing cross brace extending from one side across to the other. Vehicles equipped with a sunroof are apt to be more heavily reinforced with cross braces to overcome the weakening effect caused by the missing metal in the center. As with all panels that enclose the passenger compartment, the roof also plays a critical role in reducing the collision energy even though it is not made of high strength steel. High strength steel does not lend itself to being flexed and bent constantly, nor is it easily repaired as would be required if it were used for this purpose. Due to the proximity of the crush zones at the roof's edge, it often requires repair after even a moderate collision. This—together with the constant flexing, even under normal driving conditions—dictates the use of material that is able to withstand these conditions.

D Pillars. As the damage travels through the top of the vehicle from the A pillars to the C or **D pillars** at the rear end, the roof is designed to flex and give along with the movement of the supporting pillars to help dissipate the collision forces. Quite often in the aftermath of a collision in which significant damage is sustained, a characteristic buckle will be visible immediately above the center pillar. This is the result of a crush zone designed into the roof and surrounding reinforcements designed to deflect the collision energy away from the passenger compartment.

TECH TIP A buckle in the center of the roof directly above the B pillar should be a tipoff to the estimator and the technician that additional damage was sustained to the roof's inner structure as well.

The rear end of the vehicle is comprised of the quarter panels along with the C pillar or sail panels which are an integrated part of them; the D pillars; the trunk, which includes the floor pan; and the rear bumper assembly. Depending on the model of the vehicle, the trunk lid, rear deck lid, or rear hatch is used to enclose this area. A rear deck lid made of steel or aluminum is used in most sedan and hardtop models, and on a hatchback the rear glass is typically incorporated into the rear hatch. The rear hatch is usually attached to the rear of the roof panel. On sedan and hardtop models, the upper rear body panel is placed between the two quarter panels and between the trunk lid and the rear window. This part is used to attach the hinges for the trunk lid and is integrated into the rear window package tray which, when welded together, forms the **pinchweld** for the rear window. The lower rear panel located between the two quarter panels is typically used to attach the trunk latching mechanism and the taillight assembly. It is also integrated into the trunk floor pan, forming the area used to secure the rear bumper attaching components. On most SUVs, vans, and mini-vans the D pillar attaches the rear door and taillight assemblies. This area is typically fabricated using at least one layer of high strength steel and, in some models, is further reinforced by filling the pillar with structural foam.

Floor Pan

The trunk floor pan, like the floor under the passenger compartment, is not a structural part. See **Figure 6-28**. However, the numerous ridges and convolutions stamped into it would give one the impression that it is indeed a structural panel. It is, however, reinforced by the rear side rails, which are the primary structural members on the rear of the vehicle. Spot welding the trunk floor to the "U"-shaped rear rails—which are made of high strength steel—creates an enclosed hat channel. This results in a very sturdy structural assembly. See **Figure 6-29**. The rails extend from the outermost part of the rear end—where

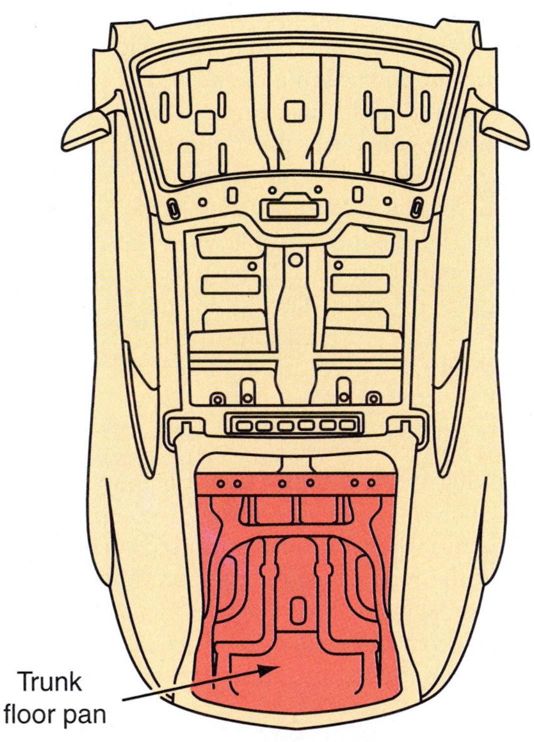

FIGURE 6-28 Contrary to their appearance, both the trunk and passenger compartment floor pans have numerous ridges stamped into them, giving them the appearance of being a structural panel. They are welded to U-shaped rails underneath the floor to obtain the required strength.

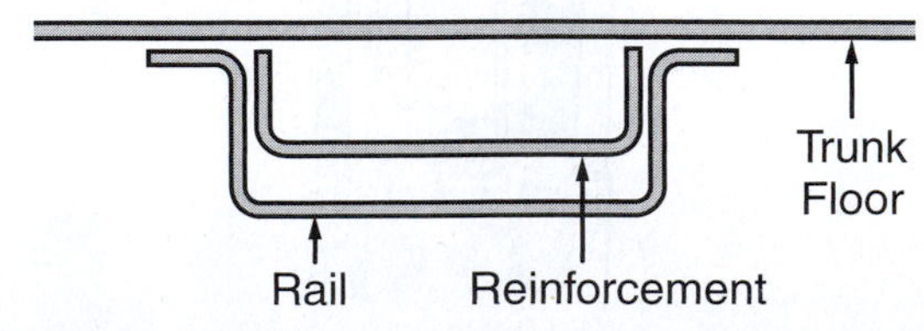

FIGURE 6-29 The trunk floor is reinforced by welding it to the U-shaped high strength steel rails, forming an enclosed hat channel to create a very strong structural assembly.

the bumper assembly is attached—forward to the passenger compartment, where they are attached at the rear torque box area. Additional lateral rails may also be placed between the two side rails, both in front and behind the spare tire pocket, to help reinforce the area and create a structural barrier capable of diffusing the collision energy.

NONSTRUCTURAL AND SEMI-STRUCTURAL PARTS

Welded Panels

To this point our discussions have included a study of the makeup, design, and function of the structural parts only. Other panels may be classified as semi-structural and nonstructural, depending on where they are used. Certain cosmetic panels also play a crucial role in the overall structural integrity of the vehicle, even though they are not designed in the tradition of a structural part. The quarter panel, for example, is not made of high strength steel nor is it reinforced with any of the typical reinforcing members. Yet it reinforces the entire rear section of the vehicle. It is often used to close out a structural panel such as the trunk floor and the rear door pillars; as such it is commonly referred to as a semi-structural part. Without the quarter panel the entire body shell would be very weak and unable to withstand any outside forces like those in a collision.

Bolted Panels

Doors. The doors are not part of the body's structural makeup as they are bolted with a hinge at the front in most cases and are held in a closed position by the engagement of the latch mechanism. However, the doors are a highly reinforced independent unit which help to direct the collision forces around the passenger compartment. The doors are individually constructed similar to the overall unibody. See **Figure 6-30**. They are made up of an inner and outer panel which are formed, flanged, and welded or bonded together around the entire perimeter to form a solid unit. An intrusion beam constructed of ultra high strength steel extends from the front to the rear of the door and is welded to the door frame at each end. The intrusion beam is a protective barrier for the passenger compartment during a side collision, commonly called a "T-bone" impact. Due to the ultra high strength characteristics of the intrusion beam, it must be replaced when bent in an accident. Attempting to repair such a structure will likely cause the metal to crack at the point where it is bent. The part will eventually fatigue, leaving that area of the vehicle unprotected in a subsequent collision.

The doors also house the mechanism used to raise and lower the side windows. The window regulators and other associated parts are either individually attached to the door's inner frame or they are mounted on a removable modular panel that houses all the movable parts of the side glass.

FIGURE 6-30 The door assemblies are made of two individual panels held together by lapping the outer panel over the inner panel using a hemming flange around the outer perimeter. The assemblies are reinforced by welding or adhesive bonding them together, and an intrusion beam made of ultra high strength steel is welded to the inside.

Hood and Trunk Lid

Other frequently overlooked panels that play a semi-structural function are the hood and trunk or deck lid. While the function of these panels is largely aesthetic, they are also designed with the intent of deflecting some of the collision energy. Stress risers, in the shape of slight indentations on the reinforcement rails, are incorporated into the underside of both the hood and deck lid that cause them to buckle and bend at specific locations rather than be driven straight into the passenger compartment when hit from the front or the rear. The hood hinges are made of the weaker low carbon steel, causing them to bend and buckle more easily, further causing the hood to bend at strategic locations. Likewise, the hinges holding the deck lid are bowed to make them bend at a specific location, causing the lid to follow suit.

Retractable Sun Roof

The nonstructural parts include items such as the outer retractable cover of a sun roof. They are attached to a motor-driven regulator with light cables, small chains, or some other means used to move the metal or glass cover back and forth to open and close the roof. Since these panels offer no strength or integrity to the roof, the framework around the outer perimeter must be reinforced to overcome that lost by the hole in the roof.

MODERN DAY VEHICLE DESIGNS

Numerous vehicle designs are commonly used on present day cars and trucks, and most are identified by their roof design and their drivetrain. Some of the shapes used

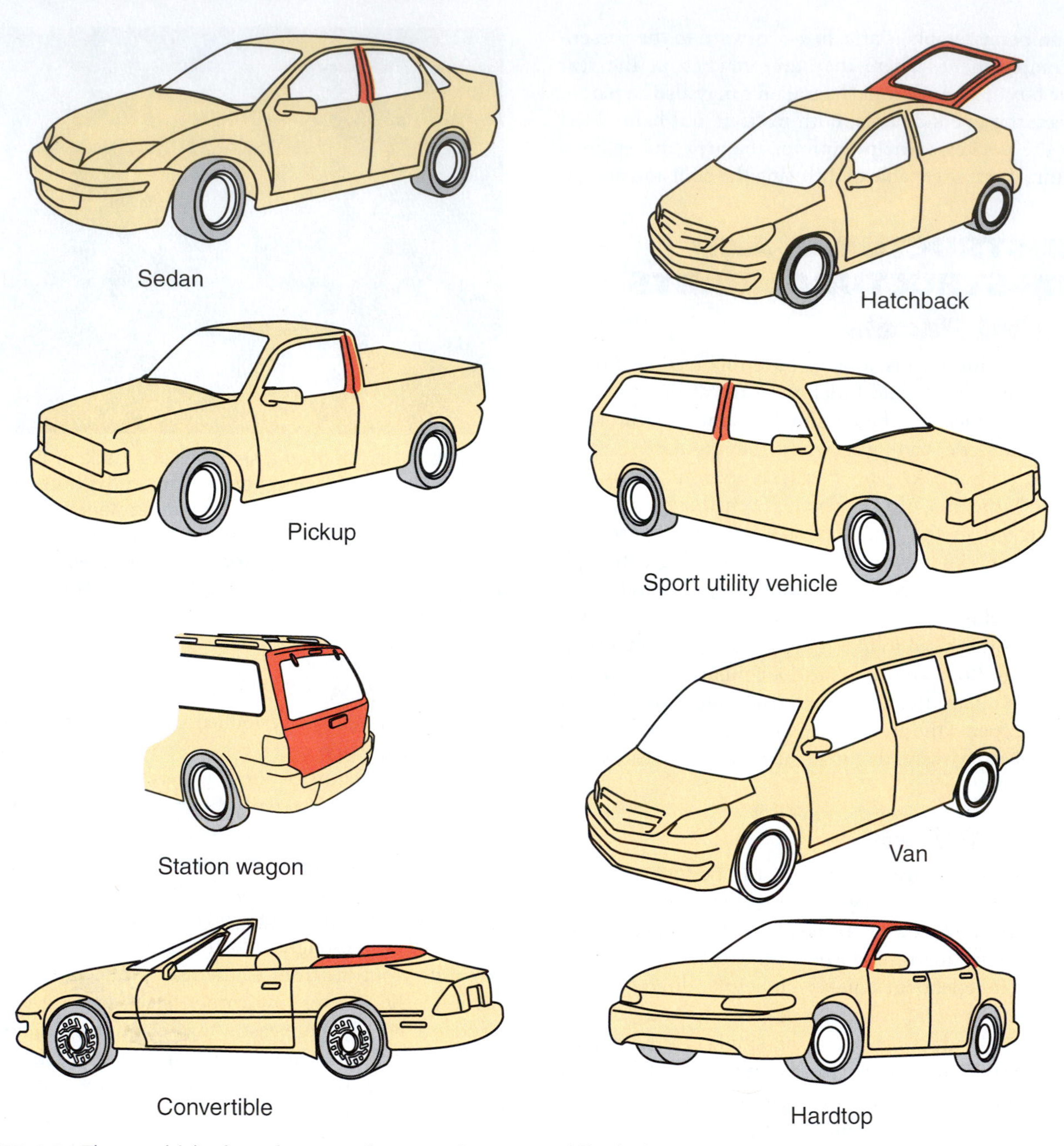

FIGURE 6-31 These vehicles have become the norm for automobile designs in recent years.

are reminiscent of the 1960s and 1970s designs, while others are more contemporary drawing board creations introduced within the past decade. The following designs have become the norm in recent years. See **Figure 6-31**.

- Sedan. This vehicle may be a two-door or four-door model with a center post used to support the roof. On the four-door model the rear door hinges are attached to the center pillar.
- Hardtop. A hardtop may also be a two-door or four-door model. Typically a hardtop does not have a center pillar; therefore, the roof must be reinforced to support the roof in a rollover situation. A four-door pickup may also be an example of a hardtop design.
- Hatchback. The hatchback is usually a two-door model with a large lid, called the hatch, attached at the rear of the roof panel. The rear glass is normally installed into the hatch; when lifted, the entire rear of the car opens up. This allows for more storage and cargo space in the rear of the compact cars on which it is used most often.
- Station wagon. This vehicle offers the most storage and cargo space as the passenger compartment extends completely to the rear of the vehicle. The tailgate, which swings open to gain easy access to

the rear of the vehicle, also houses the rear window. The tailgate often offers a two-way operation in that it can be swung open like a side door or it can be laid down like a tailgate on a pickup truck.

- Convertible. This vehicle is usually thought of as having a retractable canvas roof attached to a mechanized tubular steel frame, which is driven by an electric motor. Some models use a removable formed hardtop which may be made of metal or plastic. Certain other models use a retractable hardtop which stores in the trunk area of the car when it is down.
- Van. The van is a large box-like vehicle offering a large amount of interior storage and cargo room. The mini-van is typically a front-wheel-drive unibody construction and in recent years has been the most popular multi-seat family vehicle. The larger model is a rear-wheel-drive vehicle with a full frame design and is most popular as a work or utility vehicle.
- Sport utility vehicle (SUV). The SUV offers a multitude of options to the consumer. It is typically a full frame design, although some hybrid vehicles sport a unibody design. They may be either a two- or four-wheel-drive model. These vehicles are usually built higher off the ground than cars to accommodate the four-wheel-drive mechanism, as well as for additional clearance for off-roading and other rough terrain applications for which they are very well suited.
- Pickup truck. Pickup trucks typically have a separate cab and box which are bolted to the frame as individual sections. They may be a two-door, four-door, or any one of a number of designs in between. In recent years an extended cab with reduced size doors has been available as well. They may be two-wheel-drive or four-wheel-drive models, generally with a front engine and a rear-wheel-drive.

DRIVETRAIN DESIGN

While all unibody vehicles are characteristically the same unit construction, several different design variations are used to accommodate a variety of drivetrains. The **drivetrains** (sometimes called **powertrains**), which consist of the motor, transmission, and drive shafts, are classified as either a front engine with a front-wheel-drive, a front engine with rear-wheel-drive, a mid-engine with rear-wheel-drive, or a rear engine with rear-wheel-drive.

Front Engine

The front engine front-wheel-drive—the most popular design used on unibody vehicles—is considered the safest design in a frontal collision. The majority of the vehicle's mass weight is in front of the passengers; thus, there is little chance of it slamming forward into the passenger compartment as is likely to occur with other designs. The engine—which may be mounted either **longitudinally** or **transversely**—appears to be an integral part of the transmission. In many instances the engine and transmission mounts are used to support the two units as one. Transaxles are used to transfer the power to the front wheels via a **constant velocity joint (CV joint)**. Since the drive shafts (transaxles) are nearly in a direct line with the front drive wheels there is no need for the center hump, which was necessary to accommodate the drive shaft on rear-wheel-drive vehicles. This, of course, increases the size of the passenger compartment and enhances the passenger comfort. The weight distribution of this vehicle is approximately 65/35 with approximately 60 to 65 percent of the weight in the front one-third of the vehicle and 35 to 40 percent throughout the rest of the vehicle. The weight distribution of this vehicle makes it ideal for driving in snow and other wet driving conditions.

The front engine rear-wheel-drive vehicle is a more traditional design for vehicles of the past. The engine located at the front is mounted longitudinally and the drive shaft runs in line with it to the rear cross-member that houses the differential and drive axle. The placement of the transmission, which is bolted to the rear of the engine—the floor tunnel—is necessary for the routing of the drive shaft to the rear differential. The weight distribution of this vehicle is relatively uniform from front to the rear.

Mid-Engine

The mid-engine design places the engine immediately behind the passenger compartment, generally directly over the rear wheels. This design is typically used on high performance sports cars. Because the engine is immediately behind the passenger compartment, it is usually limited to two passengers. It is generally a lower profile vehicle and has a lower center of gravity, making it ideal for cornering and responsiveness when driving. Due to the lowered body design, it is also a very aerodynamically efficient vehicle. See **Figure 6-32**. Because of the engine's placement, the entire undercarriage—particularly the center and rear sections—must be constructed more heavily as much of the vehicle weight is confined to that part of the car. It offers the least amount of protection to the passenger compartment in a frontal collision. Even though the center section of the body is more heavily reinforced than any of the other Unibody sections, there is very little space between the engine and passenger compartment to absorb and deflect the collision energy.

Due to the engine placement, other mechanical functions become somewhat more problematic on the vehicle. The cooling system is less efficient because the cooling radiator and air conditioning condenser must be placed at the front of the car, which requires routing the coolant lines over a longer distance to get to and from the engine.

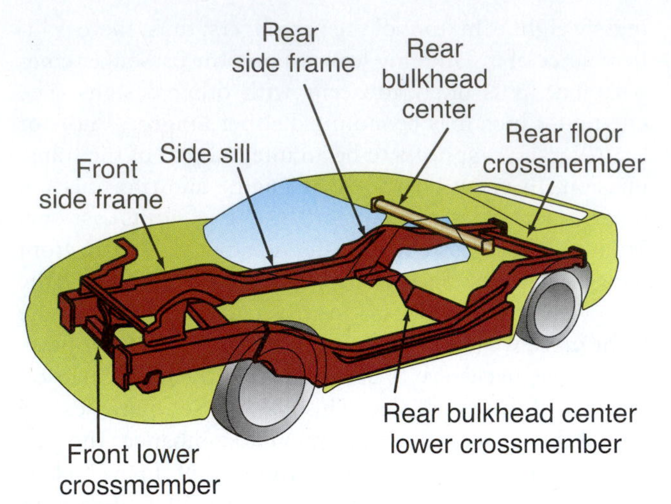

FIGURE 6-32 The mid-engine design vehicle requires the entire undercarriage to be constructed more sturdily throughout the undercarriage because of the proximity of the engine and drivetrain support these components require.

The radiator may also be mounted under the vehicle, which makes it infinitely less efficient. Access to the engine compartment is somewhat limited and much of the service work must be completed from underneath the car.

Rear Engine

The rear engine rear-wheel-drive, with few exceptions, has mostly been used on European performance sports cars and an occasional Asian manufactured model. See **Figure 6-33**. The transmission is typically placed forward of the engine to accommodate and minimize the angle of the transaxles to the rear drive wheels. Due to the placement of the engine and transmission it is still less safe for the passengers than the front engine vehicles, but it does offer more protection than the mid-engine design. It offers traction comparable to the front-wheel-drive. However, similar mechanical issues exist with this design as they do with the mid-engine design, because the radiator and air conditioning condenser are also located at the front of the vehicle.

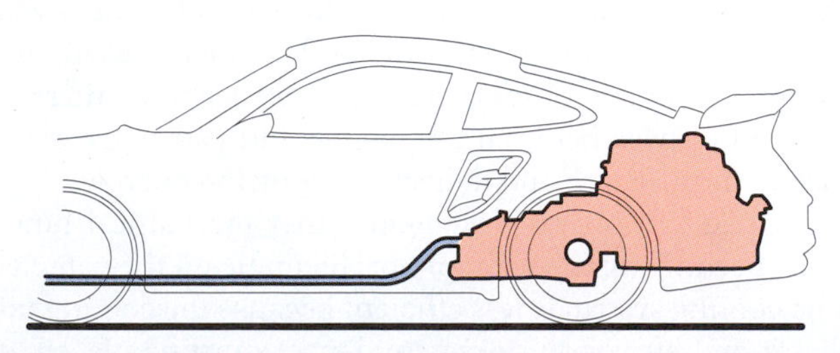

FIGURE 6-33 The rear engine drive design has largely been limited to European sports cars and an occasional American and Asian model vehicle.

Some vehicles offer an all-wheel-drive option. This can be found on vehicles with either front- or rear-wheel-drive with numerous variations on the manner in which they function. Although they are called an all-wheel-drive, all the wheels are not necessarily driving at all times. However, they are potentially engaged at any time when one or more wheels begins to slip, as is the case when driving on ice, snow, or in rainy conditions and on wet terrain. Most have a primary drive axle that may be either front- or rear-wheel-drive. The other end becomes an assist axle. On vehicles with rear-wheel-drive, the front wheels become a front wheel assist axle. Conversely, the rear wheels become the assist axle on a front wheel drive. The power ratio split may be 60:40 and in some cases 80:20. This suggests the primary drive shaft generates 60 percent of the driving power and the secondary or assist shaft produces 40 percent when slippage starts to occur. Likewise, on an 80:20 split, 80 percent is the primary drive and 20 percent is the assist drive when needed.

AERODYNAMIC CONSIDERATIONS

Another area to which the automobile designers and manufacturers have devoted a great deal of effort is the study and application of **aerodynamics**. Like the modern day jet aircraft designers attempting to make the aircraft sleeker and more economical to operate, automobile manufacturers are constantly testing new ideas to make the car more economical to operate. In order to appreciate the benefits of improved aerodynamics on the modern day automobile, it is necessary to understand the rules and application of theory on the modern day vehicle design.

As the vehicle moves down the road the air exerts a pressure-like force against it, which increases as the speed of the vehicle increases. The force, known as drag, is caused by friction or the resistance between the air and the surface of the automobile. The smoother and less disrupted the surface of the automobile, such as by protruding objects like mirrors, headlamp openings, and so on, the less the amount of drag that is created. As the vehicle's speed increases, drag becomes significantly more important as it can inhibit the vehicle's performance and efficiency.

Automobile designers commonly use wind tunnels to simulate the effects of the air flowing over the car while driving down the road. They are able to alternate the angle at which the air passes over the vehicle, thus simulating the wind blowing against the vehicle on the highway. Smoke trails passing over the car are used to enable the engineers to visualize the air movement over, under, and around the vehicle.

The shape and design of the vehicle has a very significant effect on the vehicle's efficiency of operation. Therefore, the designers have incorporated numerous methods to smooth and round out the shape of the vehicle. The front ends have been lowered and a raised bubble has been incorporated into the hood to accommodate the engine rather than building the entire front end high enough to accomplish this. The lines of the adjacent panels have been smoothed out so the air flows rearward, smoothly passing around the soft uninterrupted edges of the windshield and roof and on beyond, passing off over the trunk lid. In some models an aerodynamic spoiler or wing has been incorporated into the trunk lid to enhance the benefits of the air flowing over the vehicle. The use of advanced design ground effects are also used to route disruptive air from under the vehicle, thus creating less pressure against the irregular surfaces under the car. This reduces the amount of drag in that area.

Areas in which the manufacturers have made the most visible changes are the shape and placement of the headlights. The car of the 1980s and 1990s had headlights that were placed at the very front of the vehicle with square edges and provided very little effort to divert the air around them. Some vehicles had concealed headlamps molded into the hood or front end of the vehicle. However, when the lights were turned on, the entire mechanism raised up from out of the hood, totally disrupting the aerodynamic efficiency the automobile's design offered. However, many of the headlamps on the modern day vehicle are a capsule-shaped composite unit that houses all the front lamps that once were installed as separate or individual units. They are usually integrated and molded into the front fenders and hood, becoming a vital part of the overall aerodynamics for the front end of the vehicle.

GLASS AND WINDOWS

Laminated Windshield

The windshield and other windows in most vehicles play a more vital role in today's automobile than ever before. The windshield, which is made of laminated safety glass, is used for a structural function to help support the roof in a collision. Even though the glass may crack from the impact and the twisting that occurs, a tough plastic laminate sandwiched between the layers of glass holds it in place to support the roof and prevent occupant ejection. See **Figure 6-34**. Separate exterior moldings, which once were used to cover or disguise the unsightly gap and edges between the windshield and the pinchweld or fence, have been replaced with a more aesthetically appealing modular design. This modular design includes a plastic molding that is permanently attached to the windshield or side glass. When installed onto the vehicle, it looks like it is a part of the body. See **Figure 6-35**. This design has been implemented to help improve the aerodynamics of the vehicle as well.

Tempered Glass Windows

The vehicle's side and rear glass are made of tempered safety glass. This glass is designed to implode or break out into many small pieces with rounded edges to protect the occupants from cuts in the event of breakage.

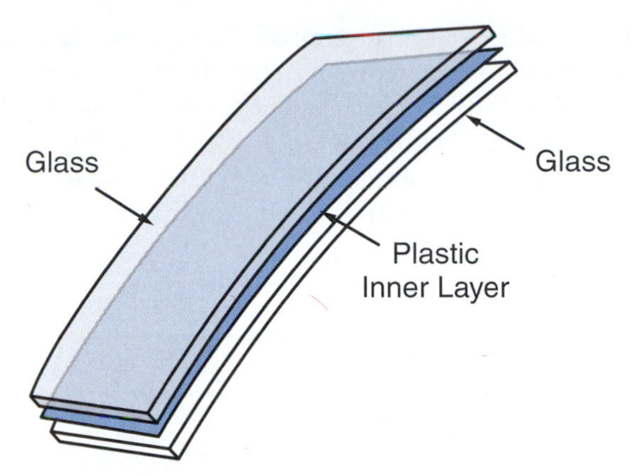

FIGURE 6-34 The National Highway Transportation Safety Board requires all automobile windshields to be made of laminated safety glass which serves a structural function by helping to support the roof in the event of an accident.

FIGURE 6-35 The use of the modular glass has been instituted primarily to improve the aerodynamics of the vehicle.

EMERGING MANUFACTURING CONCEPTS

Along with improved aerodynamic designs, efforts to reduce the vehicle weight have also been a driving force behind automobile manufacturers experimenting with new and lighter weight materials. Obviously any effort to reduce the vehicle's weight will result in increased fuel efficiency. A number of new products, along with modifications to traditionally used materials and manufacturing procedures, have been introduced in recent years. As a result, manufacturers have significantly reduced the weight of most cars and trucks.

Plastics

For years, plastics have been a common ingredient in the overall weight reduction equation. New and improved fiber-reinforced plastics offering improved performance and durability have substantially increased their use. Some examples are replacing steel bumper impact absorbers and reinforcements, suspension parts such as leaf springs, and reinforcing members which previously were made of steel. See **Figure 6-36**. Egg crate, composite, and structural foam impact absorbers have largely replaced the heavy steel front and rear bumper beams, and multi-layered leaf springs have been replaced by one layer of fiber-reinforced plastic. Fenders, hoods, and other bolted-on nonstructural parts have also been used to replace steel panels as well.

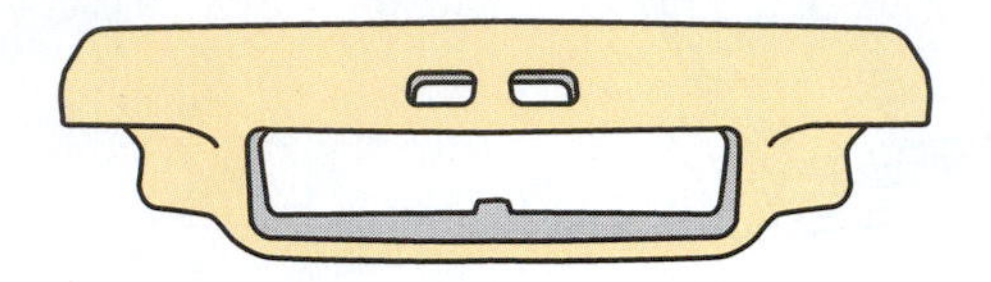

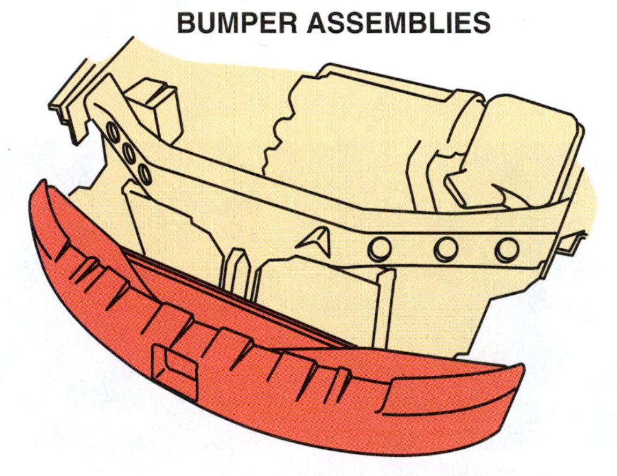

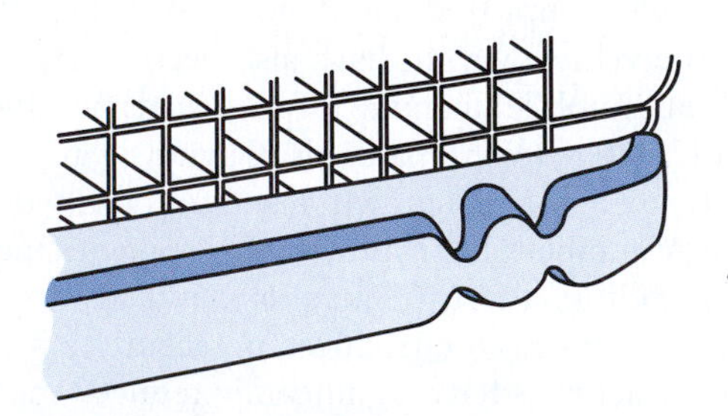

FIGURE 6-36 Front and rear impact energy absorbers—once a heavy steel assembly—have been replaced with one of several lighter weight materials such as structural foam, an egg crate honeycomb plastic design, or a composite fiber material.

Sandwiching

Perhaps one of the most innovative applications for plastic is sandwiching it between two layers of steel in areas such as the trunk floor of certain vehicles. See **Figure 6-37**. These **steel sandwiched parts** offer an excellent means of increasing strength, reducing road noises, and reducing the weight of a thicker single layer of steel than would otherwise be required—by as much as 50 percent. Balsa wood is another means of steel sandwiching for weight reduction, strength enhancement, and elimination of vibration and road noise.

Carbon Fiber

Carbon fiber reinforced plastic (CFP), a close relative of plastic, is currently being used extensively by the military on their jets and helicopters. It is credited with much of their improved performance, as the weight reduction from using this material instead of steel is said to be over 50 percent—and nearly 30 percent over aluminum. The material is a woven carbon fiber and resin that offers com-

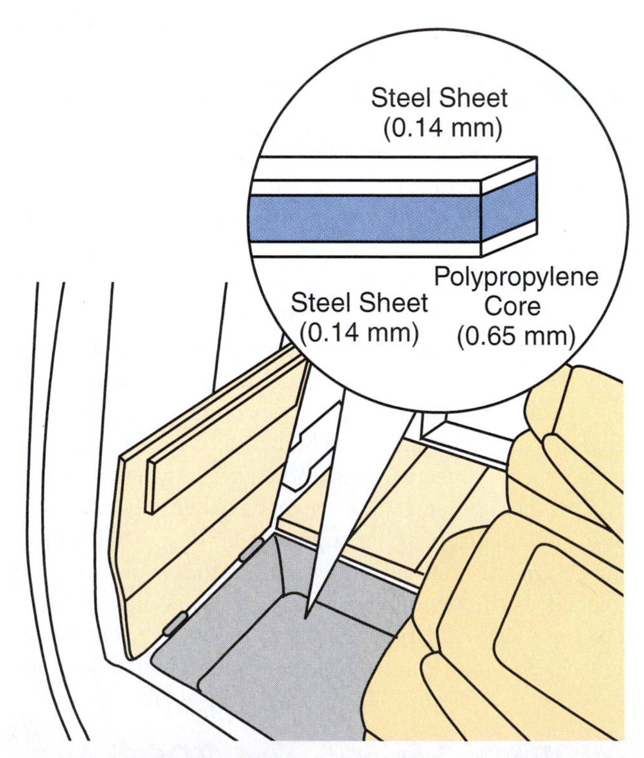

FIGURE 6-37 In an effort to reduce vehicle weight, manufacturers use steel and plastic sandwiching in the vehicle floors. This results in increased strength, reduced road noises, and decreased weight by approximately 50 percent.

parable strength characteristics as that of both steel and aluminum. While much of the efforts at integrating CFP have been directed toward performance vehicles, these findings will certainly be applied to the domestic vehicles as well. (For example, CFP is used on the hood of the 2004 Corvette Z06, the front structure of the Mercedes Benz 2003 SLR Mclaren, and the roof of the 2003 BMW M3 CLS.) According to Tim Sramcik, contributing editor of ABRN, researchers predict a weight reduction as high as 68 percent can be achieved in the automobile when CFP is used in place of steel and aluminum, which will lead to an increase of 40 percent in fuel efficiency.

Aluminum

Aluminum is often not considered a viable replacement for steel. However, in recent years it has been viewed much more favorably by U.S. automakers. Aluminum offers the same strength factors as does steel while at the same time allowing at least a 20 to 30 percent weight reduction for all parts used to replace steel. The weight reduction will result in a marked improvement in fuel efficiency. The raw materials for manufacturing aluminum are also more plentiful and cheaper to produce.

Thus far, the use of aluminum has been very insignificant compared to steel. Although not widely used other than for sport car models and a very small percentage of family and commercial models, its use is expected to increase dramatically in the next few years.

Other metals offering increased performance and strength have been introduced on a small but expanding scale. Magnesium, because of its ultra light weight and outstanding strength characteristics, has been used in areas where increased strength is required without adding additional weight from adding multiple layers of steel or aluminum. The 2000 and later Ford F150 pickup radiator support and some of the Chevrolet Corvette suspension components are examples of areas where magnesium has been used.

RESTRAINT SYSTEMS

The restraint systems used in all automobiles are of two types or systems: the active and the passive systems.

Active Restraint System

The active restraint system requires the passenger to engage the seat belt, or to "buckle up" so to speak.

Passive System

The passive restraint system, which is also called the supplemental restraint system (SRS), is an automatic system which includes the air bags and seat belts with motorized tensioners. The seat belt tensioners are tied into the same system as the air bags and deploy simultaneously with them to assist in restraining the front seat passengers. The joint deployment of the air bag and the seat belt tensioners is no doubt the single most effective passenger safety feature to be introduced in recent years. When used in conjunction with the seat belts, the air bags have had a more significant impact on reducing the number of accident fatalities than any other safety device on record. The air bags are designed to deploy when an abrupt stopping or deceleration occurs and the impact angle is head on to approximately a 30 degree angle from the vehicle's front center section. See **Figure 6-38**. A vehicle that travels at approximately 15 miles per hour (approximately 20 kilometers per hour) and crashes into a fixed barrier will cause the air bags to deploy. However, two vehicles colliding head-on must travel at a higher speed than a cumulative 15 miles per hour because both will absorb some of the collision energy, therefore cushioning the impact to less than that required to deploy the air bags.

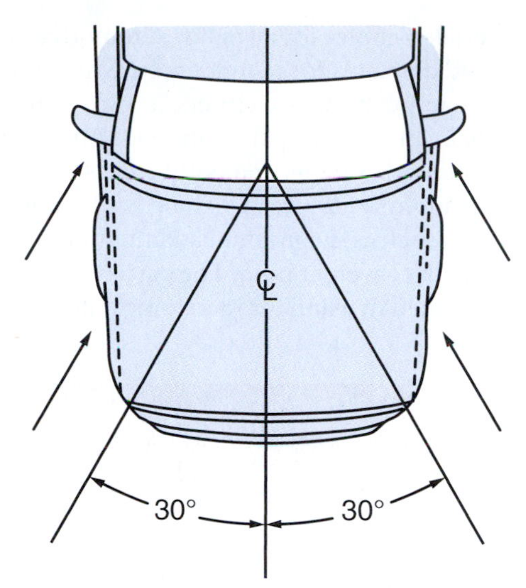

FIGURE 6-38 When an abrupt stopping or deceleration occurs and the impact angle is head on to approximately a 30 degree angle from the vehicle's front center section, the vehicle's air bags will deploy.

Side air bags can also be found on an increasing number of higher end cars and SUVs. Like those found at the front of the passenger compartment, they are designed to deploy on side collisions when impacted by a vehicle traveling at a predetermined speed. On some vehicles they are designed to deploy when the vehicle tilts at a predetermined angle in anticipation of it rolling over or onto its side.

FUTURE MANUFACTURING CONCEPTS AND TRENDS

New features and design concepts are offered with each new model year of cars that is introduced. Currently there are over 2,000 different vehicles available to the consumer, and this number is certain to increase in the future. What does the automobile of the future have to offer? What changes can we expect to see in cars and trucks 10 years from now? The American Plastics Council indicated in an issue of *Automotive Engineering* that by 2020 plastics will be the material of choice for all major automotive systems. The increased demands for petroleum base materials such as plastics may push manufacturers to use more carbon fiber materials for exterior panels instead. Aluminum use for both structural systems and body panels will become increasingly more widespread—along with other lightweight metals such as magnesium and titanium—because of their strength to weight ratio. The current fuel shortage and the continued demand for increasingly more fuel efficient vehicles virtually guarantees that the automobile will continue to be downsized. As crude oil prices continue to skyrocket, the use of plastic parts will become increasingly more costly, as plastic is an oil derivative. As the technology continues to develop and improvements are made on carbon fiber, it will continue to replace the use of plastics. Likewise, hybrid vehicles will become increasingly more prevalent.

Summary

- Consumer expectations, changes mandated by private and government agencies, and a reduced fuel supply have all been instruments of change in the automobile design in recent years.
- The unibody design has been a very significant contributor to the downsizing of the automobile.
- The body over frame, once the primary design both cars and trucks, has continued to play a significant role in vehicle design and innovation.
- High strength and advanced high strength steels, an increased use of plastics and other composition materials have all played a significant role in reducing the vehicle weight to improve fuel economy.
- New engineering design concepts using sacrificial parts such as crush zones, taylor welded blanks, hydroformed and collapsible structures have been incorporated into the vehicle design to make the car safer by diverting the collision energy around the passenger compartment instead of through it.
- Cavities in structural parts are filled with foams to increase their strength and structural integrity.
- The vehicle of the future will include an increased use of aluminum, plastics, and other more exotic materials such as composition and carbon fiber materials.

Key Terms

A pillar
aerodynamics
aluminum-intensive automobile
B pillar
body over frame (BOF)
C pillar
carbon fiber reinforced plastics (CFP)
composite material/composites
constant velocity joint (CV joint)
cosmetic panels
cowl assembly
crush zones
D pillar
die stamped
domestic vehicle
drivetrain/powertrain
Environmental Protection Agency (EPA)
full frame vehicle
hat channel
high strength steel
I-CAR (Inter-Industry Conference on Auto Collision Repair)
impact absorber/isolator
longitudinal engine
low carbon steel
metallurgical structures/lattices
panel
partial frame vehicle
pinchweld
quarter panel
rocker panel
sound deadening materials
space frame
sport utility vehicle (SUV)

squeeze-type resistance spot welding (STRSW)
steel sandwiched parts
structural foam
strut tower
stub frame
tailor welded blank
tensile strength
transverse engine
unibody structure
uniside panel
work hardening

Review

ASE Review Questions

1. Technician A says that the body over frame vehicle was the most popular design in the United States for both cars and trucks prior to the introduction of the unibody. Technician B says the vehicle weight of a unibody is heavier than the body over frame. Who is correct?
   A. Technician A only
   B. Technician B only
   C. Both Technicians A and B
   D. Neither Technician A nor B
2. Technician A says much of the technology for automobile safety design concepts comes from the aircraft industry. Technician B says the effects of collision energy travel throughout the vehicle when it is struck. Who is correct?
   A. Technician A only
   B. Technician B only
   C. Both Technicians A and B
   D. Neither Technician A nor B
3. Technician A says all unibody vehicles utilize a stub frame under the front of the vehicle to attach the drivetrain. Technician B says the enclosed box type construction is typically used throughout the length of the frame rails. Who is correct?
   A. Technician A only
   B. Technician B only
   C. Both Technicians A and B
   D. Neither Technician A nor B
4. Technician A says the manufacturers use plastic sandwiching as a means to enhance the strength of the floor pan. Technician B says wood is also used for sandwiching. Who is correct?
   A. Technician A only
   B. Technician B only
   C. Both Technicians A and B
   D. Neither Technician A nor B
5. Technician A says the weight distribution of a unibody is approximately 60 percent at the front and 40 percent at the rear of the vehicle. Technician B says the manufacturers use foam to strengthen and reinforce some pillars. Who is correct?
   A. Technician A only
   B. Technician B only
   C. Both Technicians A and B
   D. Neither Technician A nor B
6. In a front-wheel-drive vehicle the engine and transaxle are mounted:
   A. Longitudinally.
   B. Transversely.
   C. Laterally.
   D. Inline.
7. Technician A says the suspension parts for a unibody with a sub frame are mounted to the body. Technician B says a hat channel is commonly used on the rear floor pan for reinforcement. Who is correct?
   A. Technician A only
   B. Technician B only
   C. Both Technicians A and B
   D. Neither Technician A nor B
8. Technician A says the ladder frame is best suited for larger and heavier vehicles. Technician B says the perimeter frame is most often used for larger vehicles. Who is correct?
   A. Technician A only
   B. Technician B only
   C. Both Technicians A and B
   D. Neither Technician A nor B
9. Technician A says stress risers are used to cause the metal to bend at a predetermined location. Technician B says an enclosed box-like structure is used in areas where maximum strength is required. Who is correct?
   A. Technician A only
   B. Technician B only
   C. Both Technicians A and B
   D. Neither Technician A nor B

10. Which of the following is NOT used to create a crush zone?
    A. Increasing the distance between spot welds
    B. Piercing slots or holes into the metal
    C. Using heavier gauge metal at the front of a rail section
    D. Using tailor welded blanks
11. Technician A says a full frame vehicle is usually a rear-wheel-drive model. Technician B says the aerodynamics of a vehicle will not affect its fuel efficiency. Who is correct?
    A. Technician A only
    B. Technician B only
    C. Both Technicians A and B
    D. Neither Technician A nor B
12. Which of the following is true of hydroforming?
    A. It is used to reinforce the perimeter frame.
    B. It is used for forming large truck frames.
    C. It is done using high pressure.
    D. It is used to make the B pillars.
13. Technician A says the pillars supporting the roof of the car must be able to withstand at least one and a half times the weight of the car. Technician B says the A pillar is sometimes hydroformed. Who is correct?
    A. Technician A only
    B. Technician B only
    C. Both Technicians A and B.
    D. Neither Technician A nor B
14. Technician A says lateral stiffeners are sometimes welded into the rocker panel for reinforcement. Technician B says the sail panel is also called the C pillar. Who is correct?
    A. Technician A only
    B. Technician B only
    C. Both Technicians A and B
    D. Neither Technician A nor B
15. Technician A says convolutions are formed into front rail members to reinforce them. Technician B says additional ridges are stamped into the passenger compartment and trunk floor to increase their strength. Who is correct?
    A. Technician A only
    B. Technician B only
    C. Both Technicians A and B
    D. Neither Technician A nor B

Essay Questions

1. Discuss what the auto manufacturers do to increase the strength factor of a die stamped panel and what happens to accomplish this.
2. Discuss the most significant things that have been changed in automobile design to reduce their weight in the past 10 years.
3. Discuss what effect die stamping has on the strength value of a panel.
4. What is the correlation between the shape of a part and the resistance it offers to bending?

Topic Related Math Questions

1. The use of aluminum and plastics will significantly reduce the weight of the vehicle. A vehicle with a curb weight of 3,100 pounds will be reduced by one-third if it is made with aluminum and an additional 10 percent if plastic is used to replace certain parts. What will be the final weight of the vehicle?
2. The curb weight of a body over frame vehicle is 4,200 pounds. The curb weight of a unibody vehicle is 3,100 pounds. How much more fuel efficient (miles per gallon) would the unibody be if, for every 200 pounds of weight reduction, a savings of 3 miles per gallon in fuel consumption would be realized?

Critical Thinking Questions

1. Organize a group discussion and identify as many areas of our society—including our jobs and work locations—and the world economy that have evolved over the years as a result of the invention of the automobile.
2. Cite as many of the safety features built into the automobile that you can remember. Discuss where they are found and how they work to safeguard the vehicle's occupants.
3. Tracing the patterns of the consumer's changing demands for automobiles in the past two decades, discuss what changes have taken place as far as the type of vehicles that have been most popular and what type of design best describes those vehicles.

Lab Activities

1. Locate one body over frame (BOF) vehicle and one unibody vehicle, place them side by side, and elevate them. Identify the differences in their undercarriage designs, noting the methods of mounting the drivetrain, suspensions, and the difference in the thickness of the metal used on each vehicle.
2. Using the same vehicles, identify the lifting points that should be used when raising the vehicle off the ground to avoid damaging any part of the vehicle.

Chapter 7

Straightening Steel and Aluminum

OBJECTIVES

Upon completing this chapter, you should be able to:

- Identify types of sheet metal used in vehicle construction.
- Describe characteristics of automotive sheet metal.
- Analyze sheet metal damage.
- Identify and describe sheet metal repair tools and techniques.
- Discuss stretched metal and shrinking.
- Describe paintless dent repair.
- List uses of aluminum in automotive manufacturing.
- Compare and contrast types of aluminum used in vehicle construction.
- Describe the characteristics of automotive aluminum.

INTRODUCTION

This chapter will explore the art of straightening steel and aluminum. It will identify the types of steel used to construct vehicles, the characteristics of each of the types of metals used, and how to analyze damage that may occur in a collision. It will identify the varying tools used to straighten both steel and aluminum. In this chapter the student will explore spring hammering, hammer-on-dolly and hammer-off-dolly techniques, and the use of pick tools, spoons (including slapping spoons and slapping files), and welded-on and adhesive dent removing tools and techniques. The chapter will also discuss the use of heat for shrinking stretched steel, how heat is used to remove dents from aluminum, and how steel can be shrunk without using heat.

In this chapter, students will also examine the characteristics of aluminum alloys and how each alloy differs in strength and reparability. The chapter will also examine the different challenges posed by straightening aluminum, and how galvanic corrosion can be prevented.

SAFETY

Technicians who practice the techniques outlined in this chapter should use appropriate personal safety equipment such as—but not limited to—safety glasses, gloves, protective clothing, safety work boots, and ear protection. They should read, understand, and follow all safety recommendations when working with tools. They should also find, read, understand, and follow all safety precautions outlined by MSDS documents when working with chemicals.

Remember that working in a collision repair shop can be hazardous, and technicians should be ever vigilant to protect themselves, their fellow workers, and the environment.

TYPES OF SHEET METAL USED IN VEHICLE CONSTRUCTION

For technicians to become skilled at repairing damaged vehicles, they must understand how the different types of materials that they are working on respond to both the collision damage and the repair process. This chapter will discuss repairing steel and aluminum. Though all steel basically looks the same, there are very different varieties. These various steels respond differently to collision forces. They also respond differently to repair forces. In fact, steel can become stronger and more difficult to correct because of the type of damage forces the vehicle incurs during a collision. Technicians might waste a great deal of time if they do not understand these changes and know what to do to correct them. If technicians do not use proper repair techniques when repairing steel, more damage may be created during the repair process, In that case, more work will be needed to correct the "new" damage.

Two steel manufacturing processes are utilized to make the different types of steel used in automobile manufacturing: hot-rolled and cold-rolled processes. **Hot-rolled steel** is created through a process that rolls newly manufactured steel at a temperature of 1,472°F (792°C) to a thickness of 5/16 to 1/16 (7.9 to 1.6 mm) thick. See **Figure 7-1**. Parts such as frame cross-members and other thicker parts are often made from hot-rolled steel. **Cold-rolled steel** is a process that uses previously hot-rolled steel, washes it with an acid rinse (cooling it), then rolls it even thinner. See **Figure 7-2**. After it is cold rolled, the steel is annealed to make it more flexible for press-forming into body parts such as fenders and quarter panels. Annealing is a process of heating steel and then allowing it to cool at a controlled rate, thus preventing it from becoming brittle.

FIGURE 7-1 Hot-rolled steel is created when newly manufactured steel at a high temperature is rolled to a desired thinness.

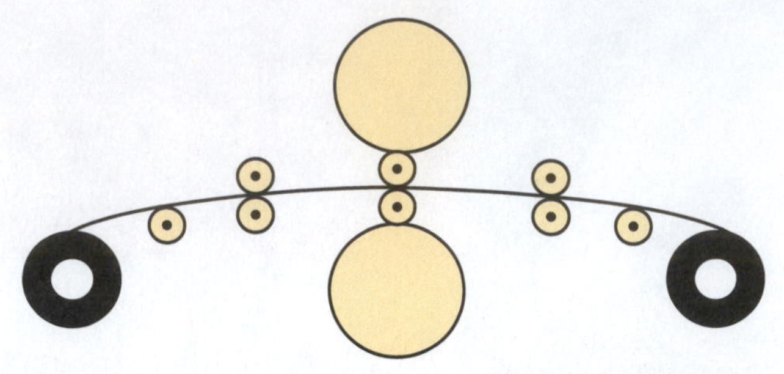

FIGURE 7-2 To make cold-rolled steel, hot-rolled steel is washed in an acid rinse and then rolled even thinner.

Steel used in the manufacture of automobiles also comes in different types such as low carbon or mild steel, high strength steel, ultra high strength steel, and high strength low alloy steel. Each of these steel types has different characteristics and will respond differently in a collision. Automotive engineers specifically choose steels with these different characteristics to perform in different areas of a vehicle. By using the different steels appropriately, the engineers can predict how a vehicle will respond in a collision and thus add to the vehicle's ability to either resist collision forces or channel those forces away from the passenger compartment, thus increasing the vehicle's "crash worthiness," which protects passengers.

As forces are applied to steel, they are defined as "tensile strengths," the amount of pulling force that a type of steel can withstand before it deforms or cracks. Tensile strength can be measured as "yield strength"—the pulling force that can be applied before the steel will deform—and as "ultimate strength"—the force that can be applied before the steel breaks.

Other forces are measured as "**compressive strength**," which is the steel's ability to resist crushing forces; "**shear strength**," its ability to withstand cutting or slicing forces; and "**torsional strength**," its ability to withstand twisting forces.

Mild Steel

Low carbon or **mild steel** has a low level of carbon and is the softest of the steels used in auto manufacturing. It can withstand yield stresses of up to 30,000 psi (pounds per square inch). Though mild steel is the easiest to form during manufacturing (and repair), it is also thicker and thus heavier. Manufacturers have switched to other types of steels to reduce the weight of vehicles. Steel, as it is being manufactured, forms a grain structure: the larger the steel grain structure, the more pliable it is. The larger the grain structure, the less effect heat has on the steel's strength. Mild steel has a fairly large grain structure. See **Figure 7-3**.

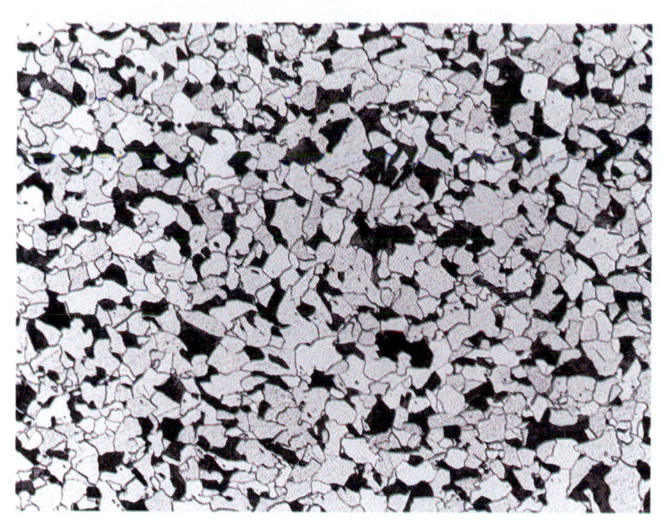

FIGURE 7-3 This microscopic view of mild steel grain shows it is fairly large.

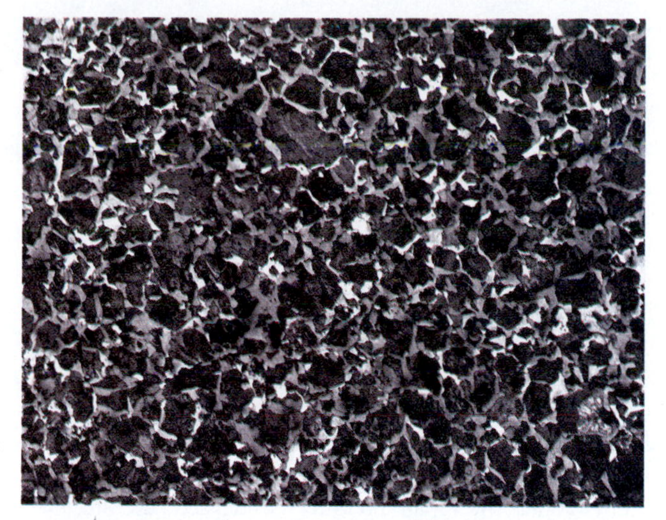

FIGURE 7-4 This microscopic view of high strength steel (HSS) grain shows the structure is smaller than that of mild steel.

High Strength Steel

High strength steels (HSS) have a yield strength of 60,000 psi, which means that parts that are manufactured from this type of steel can be made thinner and will resist the same amount of yield stress as thicker mild steel parts. See **Figure 7-4**. Reducing the weight of a vehicle will make it become more fuel efficient. Though high strength steel is lighter and has a higher yield level, it is sensitive to heat after manufacturing. This limits the amount of heat that can be applied to HSS during the repair process.

The practice of heating steel to make it more pliable during repair can only be used sparingly with the repair of HSS. Each manufacturer has its own heat recommendations or limitations, and technicians should know these recommendations before they repair high strength steel. High strength steel's grain structure is smaller than that of mild steel, so it becomes more sensitive to heat.

High Strength Low Alloy Steels

High strength low alloy (HSLA) steels are between mild and high strength steels in yield strength and heat sensitivity, and are thus used to manufacture parts such as windshield frames and some moldings, where strength is needed and a reduced weight is also desired.

Ultra High Strength Steel

Ultra high strength steel (UHSS) is the strongest of the steels, with yield strength of over 100,000 psi. It is also the most sensitive to heat, and no recommendations for heating while repairing are given. If a part made of UHSS is bent and requires repair, it must be repaired cold. This type of steel is also very sensitive to cracking and fracturing; therefore, UHSS parts that are damaged in a collision are often replaced rather than repaired. Ultra high strength steel is very sensitive to heat, and most manufacturers do not recommend that parts made with it are repaired unless it is done cold.

Boron Steel

Boron steel could be called the most ultra of the ultra high strength steels. This steel alloy's yield point is from 196,000 to 203,000 psi, which is 100 percent higher than ultra high strength steel. This extremely strong yield strength also comes with extreme sensitivity to heat, so much so that even the process of galvanizing steel harms the steel.

Boron steel's increased strength provides some advantages and disadvantages. An advantage to very strong steel is that—because of its strength—a part can be made much thinner and thus much lighter than other steels with the same strength. However, a disadvantage is that the steel cannot be repaired following a collision. If a boron steel part is damaged, it must be removed and replaced with a new part. When boron parts are bent from their original shape, they are likely to have large visible cracks due to the material's brittleness. Or, if cracks are not visible, they may still exist in very small form that a technician may not notice. If a technician attempts to straighten the part, it will crack even further. Therefore, all boron parts that become bent must be replaced.

CHARACTERISTICS OF AUTOMOTIVE SHEET METAL

Steel also possesses different characteristics that a technician must be familiar with to understand and plan for a vehicle's repair. Steel can bend, and when the forces are released it may return to its original shape. However, if the steel is bent past a certain point it will return some, but not to its original state. Steel's ability to bend and form into different shapes is called its "**plasticity**," and steel's ability to bend and stretch but then return to its original shape is called its "**elasticity**." A technician must understand both properties to successfully repair steel following a collision.

The last of steel's characteristics that must be understood by repair technicians is "**work hardening**," which occurs both during a collision and during repair. This concept can become very troublesome to a technician if not properly understood.

Elasticity

Steel has the ability to bend and stretch to a certain point. After that point, when the pressure is released, it will return to its original shape without permanent deformation. This elasticity is continually evident in the repair process. As forces are applied by hammering or by hydraulic straightening equipment, a vehicle will seem to be in perfect alignment. When the repair pressure is released, the steel will return. If new technicians do not understand steel's elasticity, they might start a repair process in an unnecessary area. That is, simply relieving the stress on the steel may return it to its original shape without any work by the technician. Relieving stress in one area may also let another return to its previous shape without any additional work.

Steel will bend with force, then return. The process simulates a tree in the wind: As the wind applies force, the tree bends, but when the wind stops the tree will return to an upright position. See **Figure 7-5**. If the amount of force that is applied to steel goes past its point of elasticity or its yield point, the grain structure will be changed and the steel will be permanently bent. See **Figure 7-6**.

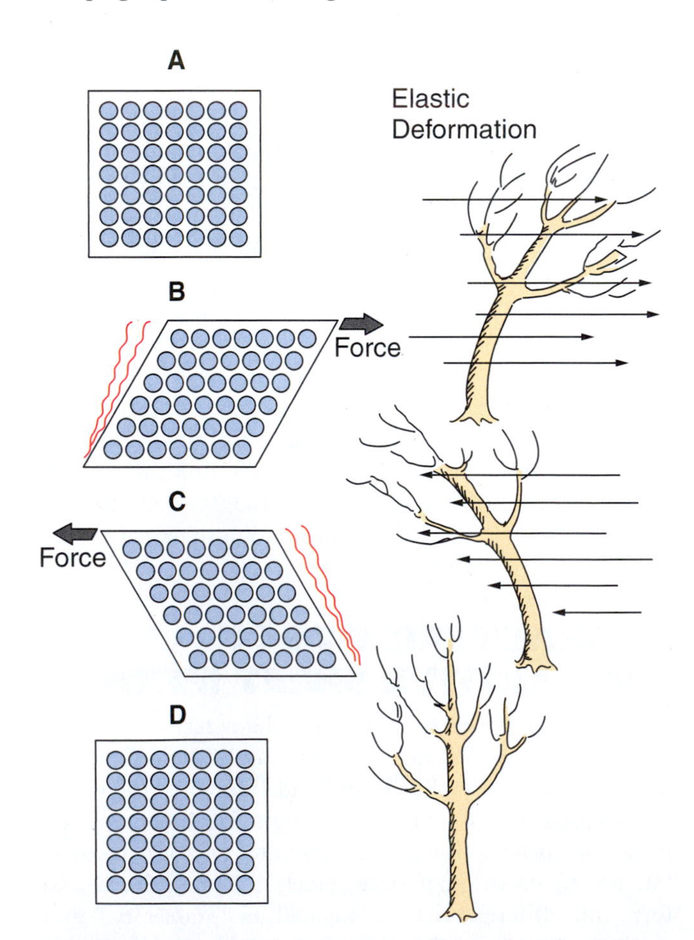

FIGURE 7-5 Steel under pressure will bend and return, similar to a tree in the wind.

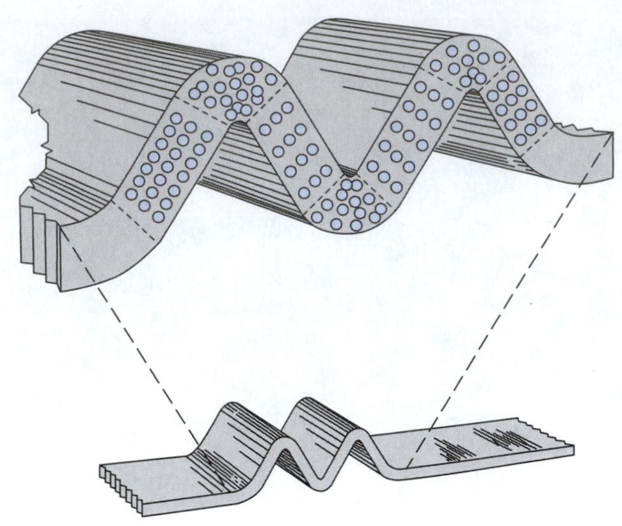

FIGURE 7-6 Deformed grain is found in steel bent past its point of elasticity.

Plasticity

Plasticity, or the ability of steel to take on new shapes, is one of the most utilized characteristics of steel. A flat steel sheet with uniform thickness is placed into a press where it is shaped into a vehicle's body parts, such as doors and fenders. As the flat steel is forced beyond its yield point into shapes, the grain structure is changed. This granular change causes the steel to become stronger, a process called work hardening. Relatively weak, thin steel can be pressed into shapes. As each shape is formed the steel becomes stronger. Even the gentle low curves (**crowns**) of a hood or roof strengthen steel because of the forming of the steel.

Work Hardening

Work hardening is the process that occurs each time steel is bent. The changing of the grain structure of steel causes the outer molecules to stretch and become thin-

TECH TIP Work hardening and increased brittleness can be seen by the bending of a mild steel coat hanger. Take a straight length of coat hanger and bend it into a sharp bend, then straighten it and rebend it several times. After a few bends it will become so brittle, due to work hardening, that the coat hanger will break. Some believe that the speed at which the hanger is bent causes it to break, but even if the hanger is bent and straightened, then rebent at 5-minute intervals, it will still become so work hardened and brittle that it will break with the same number of bends as if done rapidly.

ner, and the inner molecules to compress and become stronger. See **Figure 7-7**. Work hardening also is used by engineers to make relatively thin and weak metal considerably stronger. See **Figure 7-8**. Each bend, shape, and flange work hardens the metal as it is stamped, thus making it stronger and more able to withstand the rigors of everyday operation. Every time steel is work hardened, it is also made a bit more brittle and more likely to fracture.

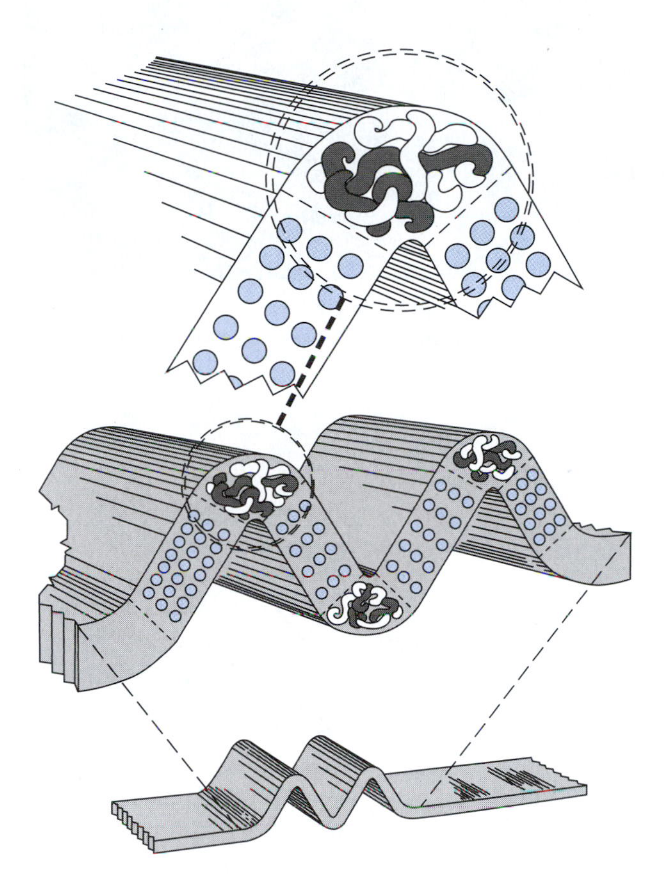

FIGURE 7-7 Work hardened and stretched metal is created through a change in grain structure.

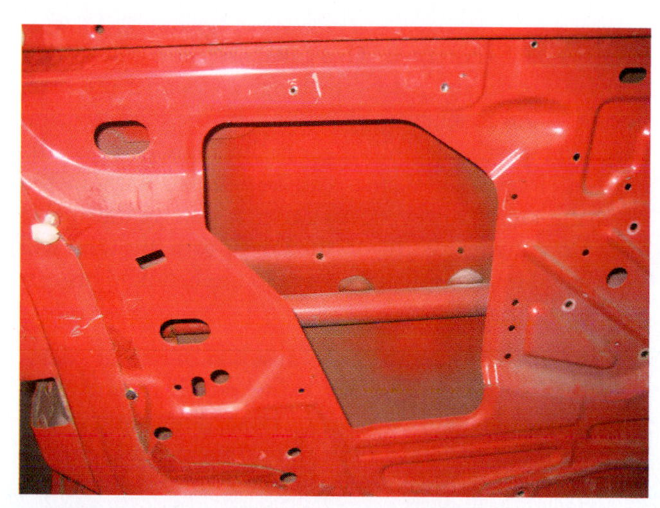

FIGURE 7-8 This intrusion beam is designed to strengthen a door, protecting occupants during a collision.

How Steel Responds to a Collision and Repair. Work hardening occurs in collisions as well. As the steel is bent past its yield point from a collision, it hardens and makes the area more brittle. As a technician works to return the steel to its original shape, the act of bending the steel will work harden the steel as well. If the part to be repaired is made of mild steel, it can be worked back into shape with normal bumping techniques such as hammer-and-dolly straightening. If the part is made from some of the stronger metals more sensitive to work hardening (such as high strength steel or high strength low alloy steel), care should be taken not to overwork the metal, thus causing brittleness from work hardening. Overworked high strength steels can develop cracks and fractures.

SHEET METAL DAMAGE ANALYSIS

Before a technician starts any work, he or she should make a repair plan. The plan includes thinking through what the best method is to repair the job, and how it could be done most efficiently. To make these decisions, the technician should perform the following mental steps: imagine the direction of impact; identify the direct damage and the indirect damage; note where the high and low crowns should be in the body panel; determine if access to both sides of the damage is possible; and determine whether it is to just be "roughed in" and then replaced, or whether the part is to be completely finish-repaired.

Direction of Impact

To restore a damaged vehicle to its pre-accident condition, a technician must be able to visualize the direction of travel that caused the damage. Simply put, to repair the damage a technician must reverse the force and direction of the original damage. As a collision happens, two types of damage occur. First is **direct damage** (sometimes called primary damage), that area where the vehicle was hit directly. Next is **indirect damage** (sometimes called secondary damage), which can be a much larger part of the damage to the vehicle.

Direct Damage

Direct damage is the area that actually made contact with whatever the vehicle collided with. The impact will often leave an imprint of that hit. Direct damage is often small in comparison to the overall damage, averaging 20 to 25 percent. However, the area of the direct damage will be one of the more difficult areas to repair, due to the work hardening in that area. Direct damage is normally easy to identify, which helps determine the direction of the damage.

Indirect Damage

Indirect damage is more difficult to assess than direct damage. There is no imprinting of what hit the vehicle. Furthermore, indirect damage may appear in an area away from the direct damage. See **Figure 7-9**. Indirect damage is the result of force traveling through the panel, causing damage in areas different from the original damage. For example, in the drawing, the indirect damage is the ripple in the roof of a vehicle that was hit in the front.

Repair efforts should never begin with the indirect damage. As the stresses are relieved from the direct damage, the indirect damage may return to its original point. Indirect damage often has not exceeded the panel's yield point; when stress is relieved, the steel's elasticity may allow it to return to its original shape. To straighten damaged areas, a technician must identify areas of the vehicle that are normal, such as high and low crowns and body lines, as well as types of damage in a panel, such as hinge buckles, collapsed hinge buckles, and rolled buckles.

Crowns. **Crowns** are common shapes that manufacturers put into panels to strengthen them. The gentle curves of a hood and roof are crowns, and they strengthen the panel through work hardening. A high crown, or a more curved area of a panel, will afford the panel more strength than a low crown or more gently curved part of the panel. Damage often occurs in crowned areas. Depending on how and what type of damage has occurred, a technician must restore the crown to its original shape.

Hinge Buckle. A **hinge buckle** often occurs in a crowned area of a panel where the direct damage pushes the surrounding metal. The pressure causes the metal to bend either up or down, and the resulting **simple buckle** resembles a hinge on a door. See **Figure 7-10**. Often the area has both pressure and tension forces on the buckle. Repairs can be made by pushing or hammering up from the lowest area of the hinge buckle, working from outer edges of the hinge inward to the lowest areas. See **Figure 7-11**. Most simple hinge buckles have little to no metal that is stretched or shrunk, so normally only simple repair techniques are needed.

Collapsed Hinge Buckle. A **collapsed hinge buckle** is a hinge buckle that extends through a stamped-in reinforced flange, bead, or ridge on a panel. As the hinge buckle passes through a ridge or bead, the overall length of the panel is shortened. This causes tension in the outer edges of a panel and places the hinge area under pressure. If the tension is relieved by pulling on the outer tension areas and gently hammering on the high parts of the hinge, the center low areas can be relieved without too much difficulty. However, if one tries to remove the buckle by hammering or passing on the lowest area, the steel will stretch and it will eventually need to be shrunk to bring it back to its original shape.

Rolled Buckle. **Rolled buckles** result as secondary damage. They occur either from being pulled inward (collapsed rolled buckle) or by being pushed upward (simple rolled buckle) by forces from the primary damage. Pressure on the panel from being pushed inward forces the panel down. At the other end of the roll, the panel is pushed upward. By relieving the pressure, the rolled buckle will be relieved.

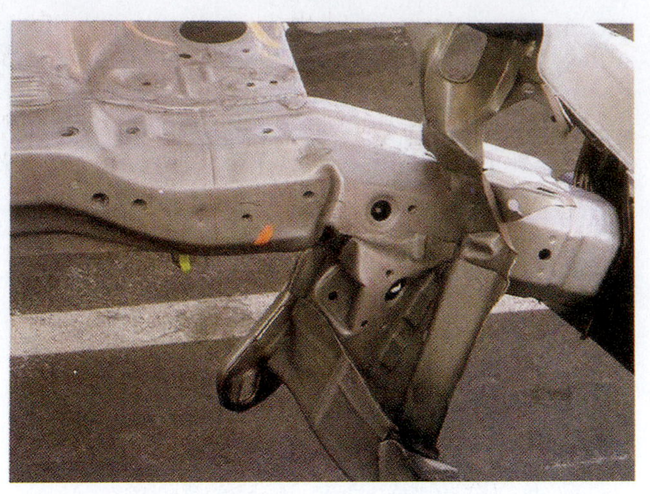

FIGURE 7-10 A hinge buckle often occurs in a crowned area of a panel where direct damage pushes the surrounding metal.

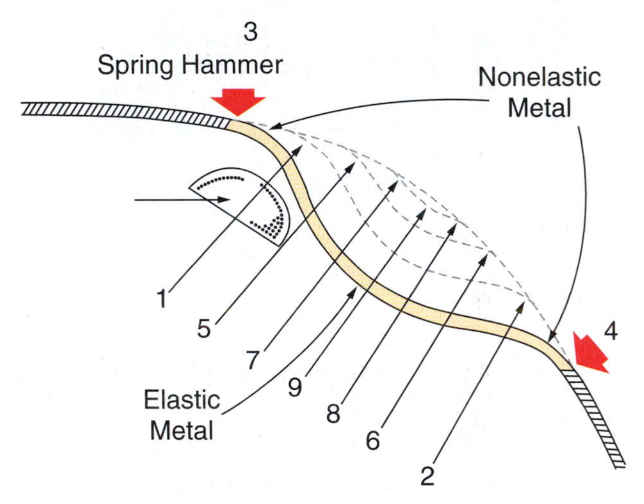

FIGURE 7-11 Repairs can be made by hammering from the lowest area of the hinge buckle, working from the outside inward.

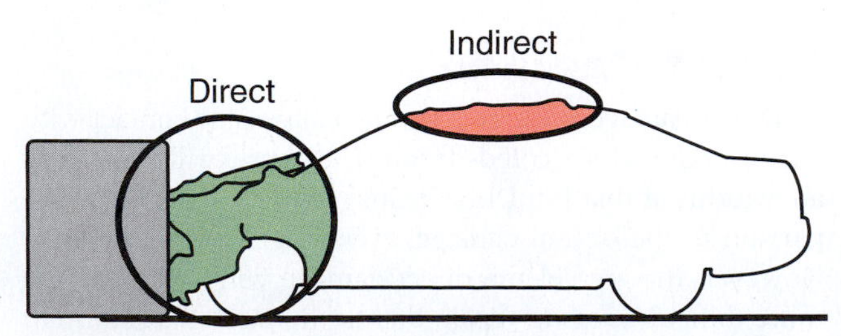

FIGURE 7-9 A technician needs to identify both direct and indirect damage from a collision.

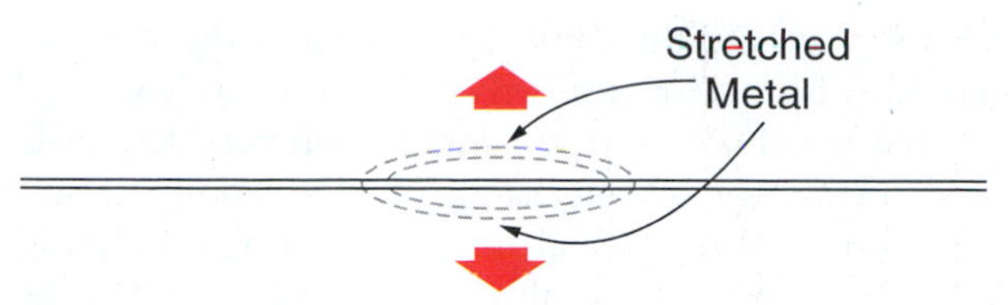

FIGURE 7-12 Stretched metal will move up or down under stress, a process called "oil canning."

TECH TIP Try this demonstration to visualize the rolled buckle damages and their repair. Take an empty soda can, and with your thumb on the side in the center of the can, press inward. The pressure from your thumb will not only push the dent inward but will also pull the outer edges of the soda can inward (forming a collapsed roll). At the upper and lower edges of the can, high areas will roll upward (producing a simple rolled buckle). Notice that the overall length of the can has been shortened. To repair this type of damage, pull gently and gradually on both sides of the shortened can, bringing the collapsed roll upward. Gentle hammering on the compound roll will also help restore the can's original shape. See **Figure 7-12**.

Access to Damage

When technicians assess an area needing repair, they must determine if both sides of the damaged area are accessible. With access to both sides, technicians will have a variety of repair technique options, the most common being the hammer-and-dolly techniques. If technicians do not have access to both sides, more creative techniques such as the stud gun, "wiggle" wire, and glue-on techniques may be the only alternatives available. A third situation may arise if technicians have only limited access to both sides; in these situations, a prybar may be used instead of a dolly or pick hammer.

Repair with Access to Both Sides. The most common "bumping" technique technicians use when they have access to both sides of a vehicle is the **hammer-on-dolly** technique. In this technique, the dolly is held on one side of the damaged panel and the hammer is swung against the opposite side. Both tools will act to either raise or lower the damaged sheet metal. A technician can choose to use either the hammer-on-dolly or the hammer-off-dolly methods. See **Figure 7-13**. Alternately, the pick hammer alone can be used to raise dents in damaged areas.

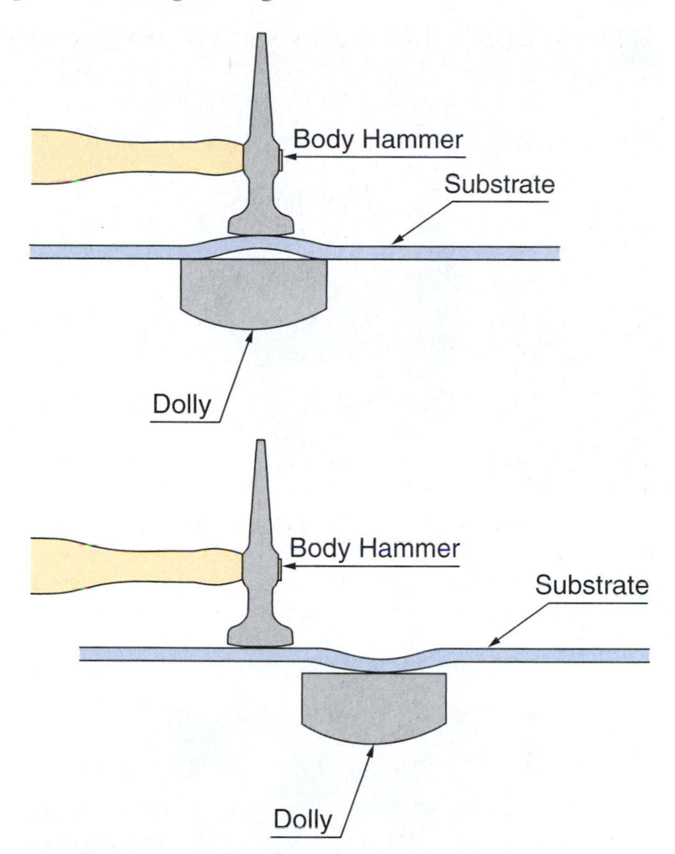

FIGURE 7-13 A technician may use the hammer-on-dolly or hammer-off-dolly techniques when they have access to both sides of a vehicle.

Repair with Access to One Side. When access is only available on one side, a technician must use tools that will raise the low spots and lower any spots that may be too high. A variety of tools can help. The **stud welder** gun can temporarily weld a stud to the outer surface of a panel, allowing a technician to attach a slide hammer and pull up the low areas. See **Figure 7-14**. A variation of this tool is a welder that attaches a "**wiggle wire**." See **Figure 7-15**. After being attached with a special tool, the wiggle wire is pulled upward to help raise low areas with the force being placed over a larger area. Both the studs and the wiggle wire can be removed following the repair, so that the repair area can be restored to its original shape. Other tools

FIGURE 7-14 Stud gun and pulling tools are used to remove dents when a technician does not have access to both sides of a vehicle.

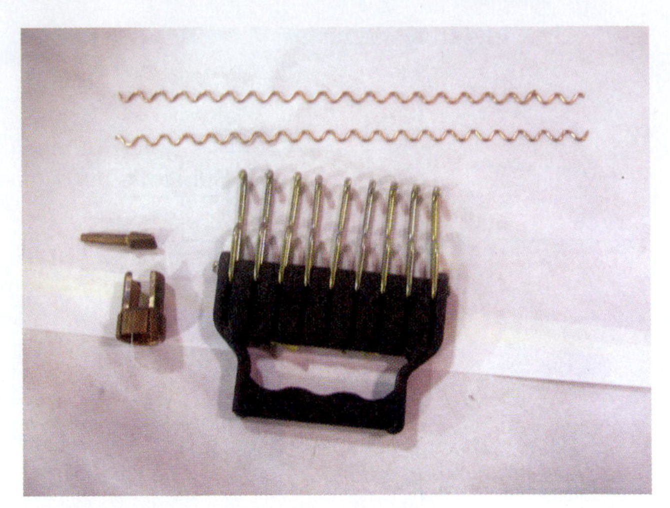

FIGURE 7-15 A wiggle wire and pulling tool are used as a single-sided dent removing tool. Wiggle wire is temporarily welded on for pulling and later removed.

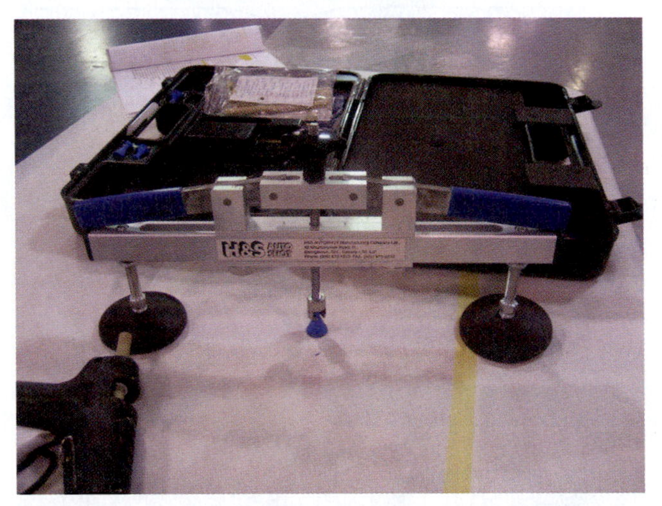

FIGURE 7-16 A glue-on dent pulling tool is glued on temporarily, then removed after the dent has been removed.

have been developed which accomplish the repair in a similar manner. The **glue-on dent puller** glues on a device that is used to pull up low areas (instead of welding). See Figure 7-16.

Limited or Partial Access. If access is only partially available to technicians, they can slide in a prybar and use it to raise up low areas while simultaneously using a hammer to "ding down" high areas. Though technicians have used this basic method for years, more recently it has developed into a complete method of dent removal, called **paintless dent removal (PDR)**.

When to Replace and When to Repair

In the past most body repair technicians have prided themselves on their bumping ability. Although it may seem that many panels can be repaired, sometimes it is better to rough out the damaged areas until the vehicle's dimensions have been restored, and then remove and replace that panel with a new panel. When vehicle panels were all made with mild steel and work hardening was not a large consideration, metal finishing through bumping was the best and most profitable method. However, as more and more panels are being constructed with high strength steel and high strength low alloy steels, each time that the panel is bent either in the collision or through bumping techniques, the metal becomes too brittle and subject to developing cracks, even through normal usage.

With these factors in mind, how does a technician or estimator make the decision whether to replace or to repair a panel? If a panel has been bent in a tight radius, one that is less than 1/8 inch (3.2 mm); or if the bend is greater than 90 degrees when the part is straightened, it will crack. A cracked part should be replaced. Even mild steel will crack under the described bend conditions. Therefore, if a part is bent and can be straightened cold (without the use of heat), it should be straightened. Otherwise, if a part is kinked, it should be replaced.

SHEET METAL REPAIR TOOLS AND TECHNIQUES

Though many tools and techniques can be used to repair damaged sheet metal, this section will discuss hammers (including their maintenance) and hammering, dollies and their usage, spoons, picks, sappers, and pulling devices. It will also explore hammering techniques, hammer-on-dolly and hammer-off-dolly techniques, spring hammering, using picks, and pulling with tools such as stud welders, suction cups, and glue-on pulling devices.

Each new job presents a technician with the challenge of knowing which of these tools and techniques is best suited to a particular repair situation. Therefore, technicians should become skilled at them all so that they are able to use the appropriate technique.

Hammers

Body hammers come in many different types and designs, each for a specific use. They are divided into four classes. **Aligning** and **roughing hammers** are usually large and heavy, and are used to align a panel, bringing it back to its original dimensions. **Bumping** and **dinging hammers** are used to straighten and flatten already roughed-out sheet metal, further restoring it to its original shape. **Finishing hammers**, which include pick hammers, are used to smooth low and high areas and produce a perfectly smooth surface. The fourth and last category of hammers is **specialty hammers**, which are each designed to do a specific straightening job.

Alignment and Roughing Hammers. These large, heavy hammers are used often with the assistance of hydraulic pulling devices to bring the damaged metal back to its original shape and alignment. Sledge hammers have two heads that are shaped the same; they can vary in size and weight from a 10-pound, long-handled hammer that may require two hands to operate, to smaller short-handled sledges. See **Figure 7-17**.

This category also includes hammers such as the ball peen hammer. This tool is used similarly to the sledge hammer but for smaller jobs, with the ball end of the hammer used to rough out metal. See **Figure 7-18**. The ball peen hammer is particularly used when bucking (mushrooming) the end of a rivet. Cross peen hammers are often favorites of body technicians for aligning and roughing out damaged metal. See **Figure 7-19**. A cross peen, often called a blacksmith's hammer because of its chisel-like cross peen side, will raise up body lines in the rough-out stage of repair.

Bumping and Dinging Hammers. These hammers are used following the rough-out step to restore the surface to its original shape and smoothness. They come in many different types, shapes, and weights. The one chosen for use depends on the job at hand, relative to a specific need. Collision repair technicians often have a large assortment of hammers to use in the bumping and dinging stage of repair.

Heavy-duty bumping hammers, which may also be called shrinking hammers, have a round face and a square face and are used for badly creased metal or heavily gouged areas. Though they may not be larger than their counterparts, the light bumping hammers, they have a heavy body between the two faces and thus have more mass for the heavier jobs. See **Figure 7-20**.

The light bumping hammer has a shape and size similar to the heavy-duty bumping hammer, but the body is lighter and is used for lighter smoothing operations. See **Figure 7-21**.

The balanced dinging hammer often comes with two different sized round heads, or with a round head and a square head. It is used for bumping in areas where there is more room, or where a flair—such as on a fender—needs to be reshaped. A balanced dinging hammer that comes with a square head can also be used as a light dolly when straightening metal.

Finishing Hammers. Often called pick hammers, finishing hammers are distinguished by having one end with a round head and the opposite end with a pick of some type or length. Their use in the finish process is to pick up the small low spots; then, if needed, with the aid of a dolly and the round head of the hammer, the tool lowers the picked area back to the proper shape of the panel. Pick hammers come as short pick hammers, long pick hammers, straight chisel hammers, and curved chisel hammers. See **Figure 7-22**, **Figure 7-23**, **Figure 7-24**, and **Figure 7-25**. Chisel hammers are used to straighten and lift the body lines in panels.

Special Hammers. The category of special hammers could be endless, for new types of hammers for specific job

FIGURE 7-17 A sledge hammer can vary in size from a 10-pound, long-handled hammer to a smaller, shorter hammer.

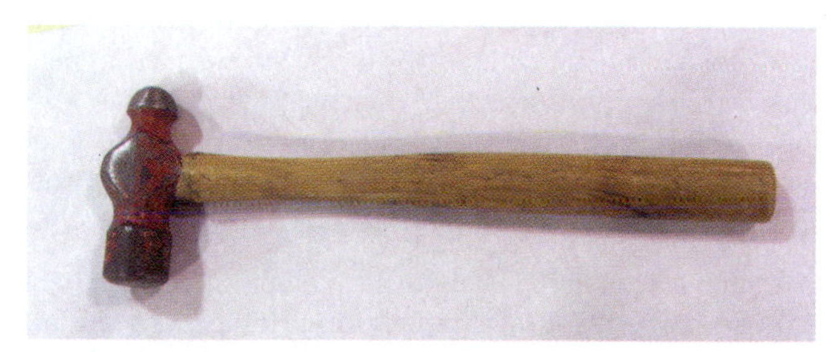

FIGURE 7-18 A ball peen hammer is often used when bucking the end of a rivet.

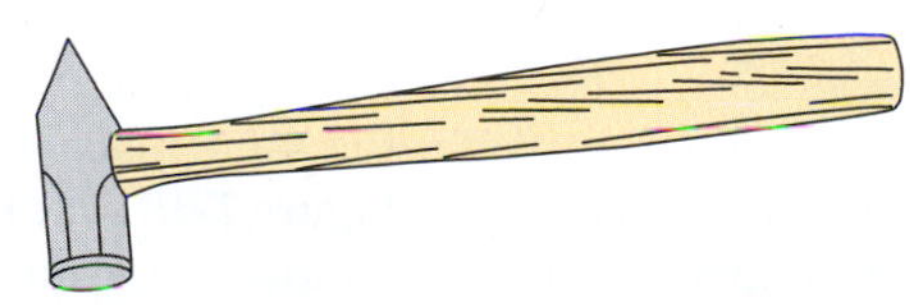

FIGURE 7-19 A cross peen hammer will raise up body lines in the rough-out stage of repair.

FIGURE 7-20 Heavy-duty bumping hammers are used for badly creased metal or heavily gouged areas.

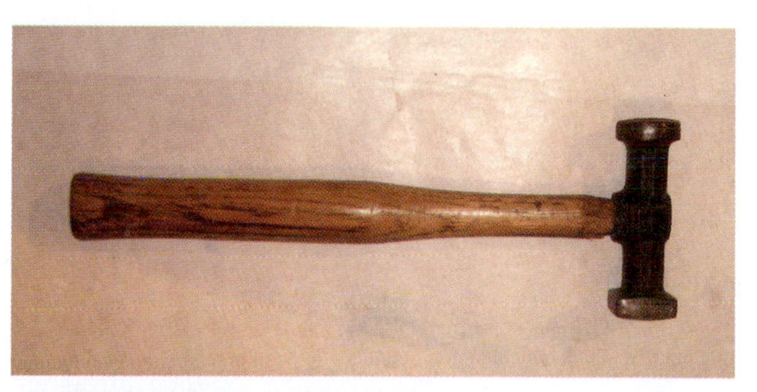

FIGURE 7-21 Light bumping hammers are similar to heavy-duty bumping hammers, but they are used for lighter smoothing operations.

FIGURE 7-22 A short pick hammer is a type of pick hammer.

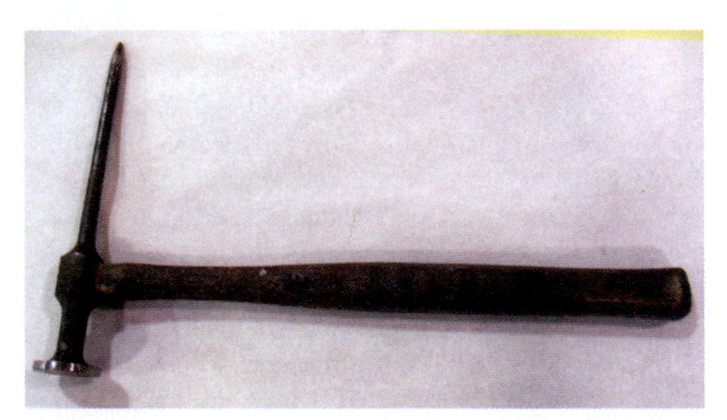

FIGURE 7-23 A long pick hammer is a type of pick hammer.

FIGURE 7-24 A straight chisel hammer is a type of pick hammer.

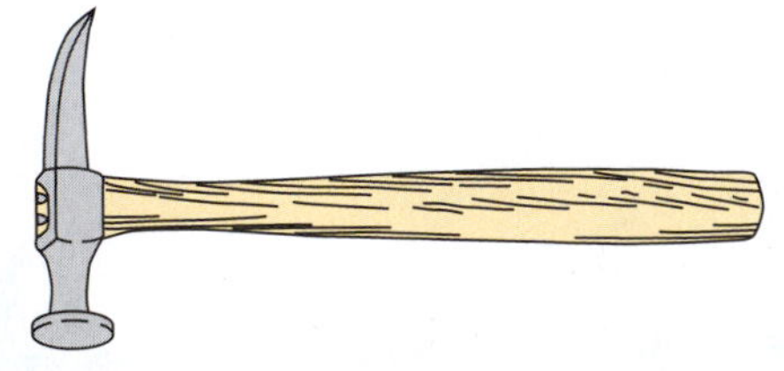

FIGURE 7-25 A curved chisel hammer is a type of pick hammer.

needs are developed each year. This category includes the reverse curved light bumping hammers, used for aligning and shaping crowns. See **Figure 7-26**. This specialty hammer features two curved heads that are mounted in opposite directions from each other. Another type, the

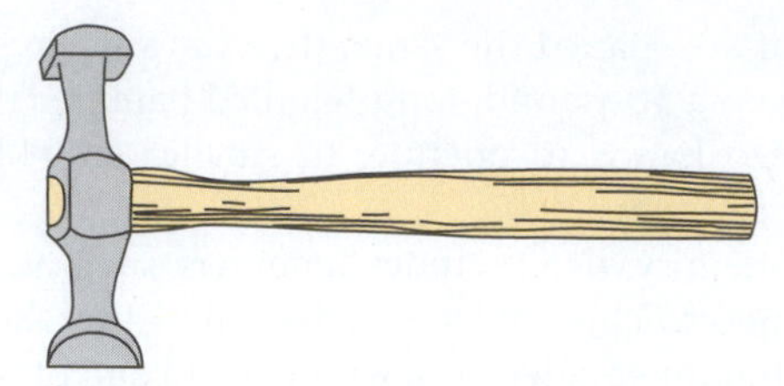

FIGURE 7-26 Reverse curved light bumping hammers are used for aligning and shaping crowns.

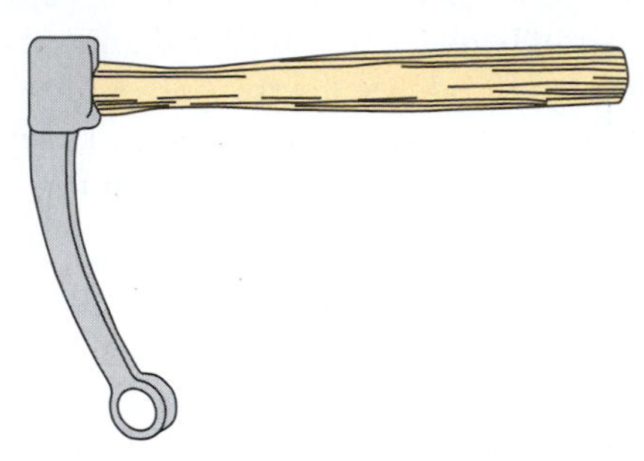

FIGURE 7-27 Heavy-duty fender-bumping hammers have a crowned face on one end and a heavy face on the other.

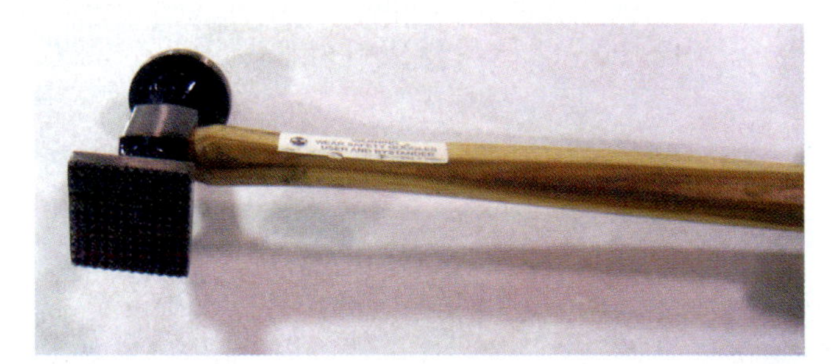

FIGURE 7-28 Shrinking hammers are equipped with a knurled head for shrinking.

heavy-duty fender-bumping hammer, has a crowned face on one end and a heavy face on the other, which can be struck with a heavy roughing hammer such as a cross peen hammer for difficult areas. See **Figure 7-27**. Other specialty hammers include shrinking hammers, equipped with a knurled head for shrinking. See **Figure 7-28**. A European shrinking hammer has a spiral knurled head that is equipped with a ratcheting mechanism that turns slightly when struck, aiding in the shrinking process.

Hammer Use. Using a hammer for body work is much different than using a hammer for driving nails or for masonry work. The technician should grasp the body hammer at about one-fourth of the length of the handle up from the bottom. See **Figure 7-29**. The thumb, first, and second fingers hold the hammer, while the third and the little finger guide the hammer blows. Some technicians choose to grasp the hammer at the same place, but with the first finger in the handle. See **Figure 7-30**. The first finger is used to direct the hammer blows, or as an aiming technique as the technician bumps the repair area. The hammer should be swung in a circular fashion, using the

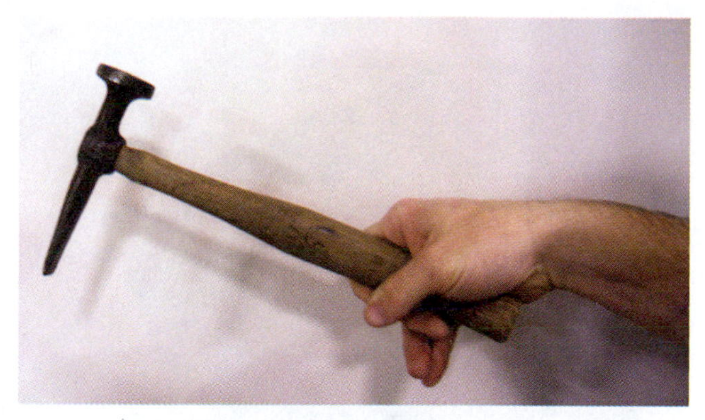

FIGURE 7-29 A body hammer is driven mostly by wrist movement.

FIGURE 7-30 Some technicians place their finger on the hammer's handle to help direct its blows.

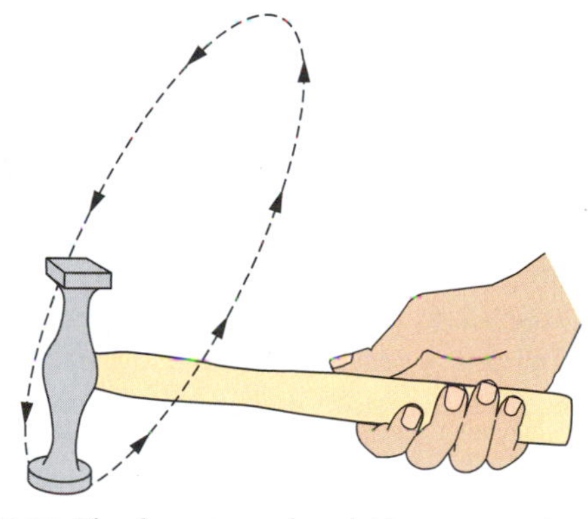

FIGURE 7-31 The hammer should be swung in a circular fashion, using the wrist instead of the arm.

wrist instead of the arm. See **Figure 7-31**. The hammer should be allowed to hit the surface and then bounce back, which helps the wrist with the next blow. The blows should be light and should be aimed about 3/8 inch apart. This technique is needed because the hammer head, with its slightly curved surface, has only about 1/2 inch surface space when it makes contact with the surface. Therefore, many light blows will move the surface with more control than would a few heavy blows.

Hammer Maintenance. Hammers will need to be maintained from time to time. First, the handle should be inspected; it should always be in good shape, without nicks or splits. The hammer head should be firmly attached without slop, since without the head being firmly attached the hammer cannot be guided accurately enough to be an effective tool. A sloppy head may come off during the dinging process, possibly breaking something or injuring someone.

The face of the hammer should be kept clean of nicks and mars. A marred hammer cannot smooth out the metal's surface as it should. To remove mars, the technician should clamp the hammer in a vise and use a file to remove the mars, taking care to maintain the same curve that was originally in the face. See **Figure 7-32**. Once the face is clear of imperfections, the edges should be beveled slightly. See **Figure 7-33**. Once the head is cleaned and beveled, the face should be finish sanded, first with P80 grit, then with P180, and so on, until the face is smoothed with P400. See **Figure 7-34**. A smooth and mar-free hammer will produce the desired finish.

FIGURE 7-32 A technician will file the face of a hammer to keep it smooth.

FIGURE 7-33 Once the hammer face is clear of imperfections, the edges are beveled slightly.

FIGURE 7-34 The hammer is finish sanded with 80 grit sandpaper to complete its restoration.

TECH TIP Wooden hammer handles, which are preferred by some technicians (because the wood absorbs some of the shock), will sometimes need to be replaced. The grain of a good replacement handle runs in the opposite direction of the hammer handle. This grain pattern will make the handle stronger and more likely to withstand missed blows, which is one of the main causes of hammer handle damage. When replacing the handle, there should be no gaps between the handle and the head where it is inserted into the steel part of the hammer. Though both wooden and metal wedges are normally provided and should be used, the additional use of epoxy cement will fill any voids and attach the head as firmly as it should be. Epoxy can be drilled out with the wood if it ever needs further replacement.

Dollies

Dollies are heavy metal devices that are used either by themselves or with a hammer to rough and finish metal. They come in many different types and shapes, each suited for a particular purpose. Technicians often have an assortment of dollies available for different uses. The most common of the dollies are general purpose; low crown; wedge dollies; heel, both covered and non-covered; toe; and shot bag. Though many other dollies are available, including special purpose dollies, these are the most common.

General Purpose Dolly. As the name implies, the general purpose dolly can be used for many different applications. See **Figure 7-35**. Because it has many different curved surfaces, along with sharp corners, the general purpose dolly can be fitted to many different surfaces. It is sometimes referred to a railroad dolly because of its resemblance to a railroad track. This dolly is easy to hold, and—because of its mass—it is efficient when roughing out severely damaged surfaces. If a technician were only able to have one dolly, this would be the one to choose.

FIGURE 7-35 A general purpose dolly can be used for many different applications.

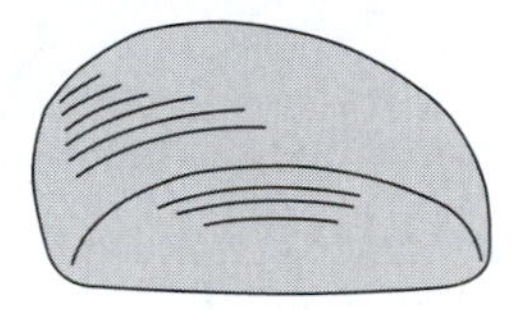

FIGURE 7-36 A low crown dolly is often used to shrink thin metal.

Low Crown Dolly. A low crown dolly is often used to shrink thin metal. See **Figure 7-36**. It has enough mass to be grasped easily and held in place as the technician performs the shrinking process, either cold or hot. As the name implies, the low dolly's shape does limit its use to low crowns.

Wedge Dolly. Wedge dollies are designed for working crowned areas such as fender flares. The pointed end on this dolly can get into areas that are very tight, such as between bracings of doors that cannot be reached by any other type of dolly.

Heel Dolly. Heel dollies get their name from their shape. With their curved surfaces, they are handy for the shaping of both low and high crowns. Their flat surfaces can be used to straighten flanges, and the near flat surface can be used to support the flat outer surface of a door skin as the flange is being hemmed. This dolly is the one most often chosen for door skin hemming. It is available as a standard steel heel dolly and as a plastic covered dolly, so it is less likely to mar the outer skin of a door as it is being hemmed. See **Figure 7-37**.

FIGURE 7-37 A heel dolly gets its name from its shape.

Toe Dolly. This dolly, which also gets its name from its shape, is probably the second most-used dolly in a metal technician's toolbox. The toe dolly has flat and gently crowned surfaces and is ideal for the shrinking of flat and low crowned areas. One caution with this and other finish dollies: Technicians should avoid making hard blows between the hammer and the dolly when using the hammer-on-dolly technique; they will cause the metal to stretch and may cause "oil-canning."

Spoons

Spoons come in different shapes and sizes, all for different jobs. They come in three categories: surfacing, driving, and backup spoons. Each type is designed for a different job during the repair process. During the initial rough-out stage, technicians use spoons to lift and pry areas that a pick hammer may not be able to get to. They can be used like a dolly to back up hammering; when used with a hammer, a surfacing spoon is used with spring hammering to final-smooth the surface of a body panel.

Surfacing Spoons. **Surfacing spoons**, such as light spoons, are nearly flat and are used to finish bumping a panel. See **Figure 7-38**. A light spoon can be placed on the high area of a panel, then hit lightly as it is dragged over an unwanted crown to lower it to the desired contour. The technician holds the hammer loosely; when the hammer bounces or springs off the spoon, the hammer is lifted for the next blow. The spoon distributes the hammer blow over a greater surface area than the hammer by itself would produce, thus making a mar-free or hammer mark-free surface. See **Figure 7-39**.

Backup Spoons. **Backup spoons** are much heavier and larger than surfacing spoons, and can be used in the place of a dolly for finish work to a panel. They also can be placed in areas that a dolly cannot access. Because they are heavier, they are often used to pry out damaged areas in the roughing stage of panel repair.

FIGURE 7-38 Light surfacing spoons are used to finish bumping a panel.

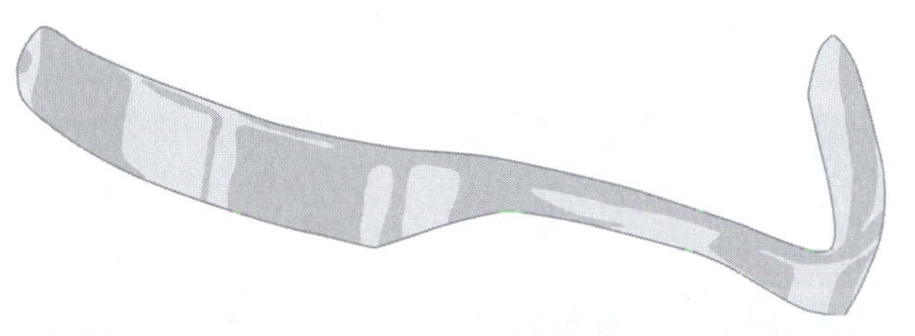

FIGURE 7-39 A double-ended backup spoon helps distribute a hammer's blow over a greater surface area than a hammer would produce by itself.

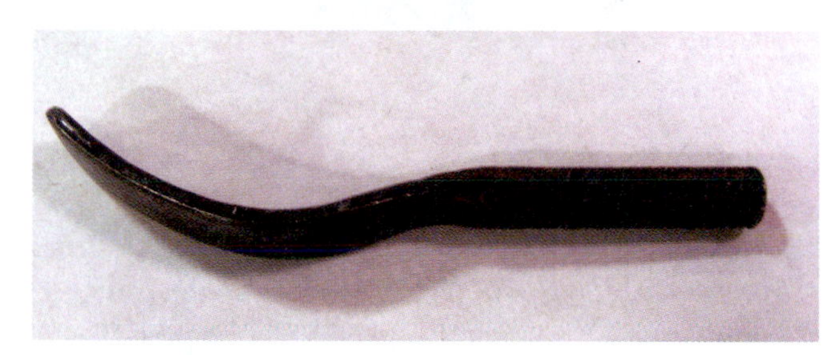

FIGURE 7-40 Driving spoons are used in the roughing stage of metal finishing.

Driving Spoons. **Driving spoons** are the heaviest of the spoons and are, as the name implies, used in the roughing stage of metal finishing. See **Figure 7-40**. They are used to bring up low areas on the back side of fenders and in quarter panels.

Dolly Spoons. **Dolly spoons** are often used to get to places that a technician might not otherwise be able to reach. See **Figure 7-41**. A dolly spoon is heavy and is positioned on a long steel handle, so it can be held firmly against the back side of a panel as a hammer is used on the front side. It can also be used as a driver to lift low areas that a technician may not be able to reach with a hammer.

Slapping Spoons. A **slapping spoon** (also called a bumping file) is a misunderstood tool, but an excellent tool to use for final finishing when metal finishing is performed. See **Figure 7-42**. It can be used with a dolly or backup spoon on the opposite side of the panel. In addition, when slapped and slightly dragged against a nearly finished surface, it can eliminate most if not all of the

FIGURE 7-41 Dolly spoons are often used to get to places that a technician might not otherwise be able to reach.

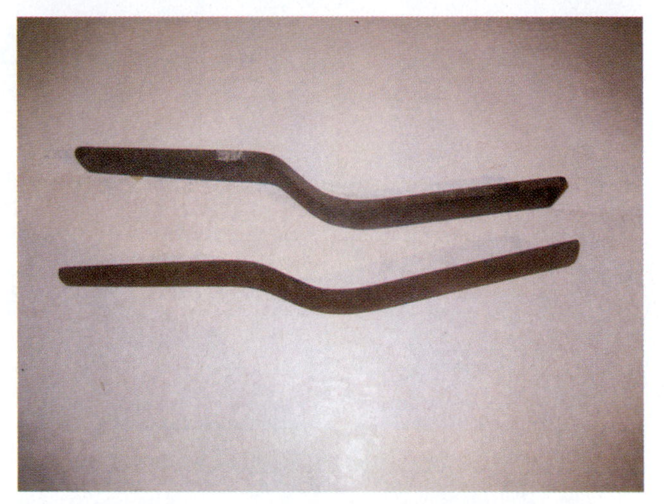

FIGURE 7-42 A bumping file or slapping spoon can eliminate most, if not all, of the hammer marks left from the metal finishing process.

hammer marks left from the metal finishing process. This is the last step before the technician sands and primes the metal for painting.

Picks

Picks are tools used when access to the back side of a vehicle is not possible. They are often inserted through holes in a door, after the taillight has been removed, or under the hood. In the past, and even now sometimes when "paintless dent removal" (PDR) techniques are used, a hole is placed in an area where it is not visible. A plug is then inserted into the hole after the work is completed. This practice remains very controversial among both auto manufacturers and technicians. Some believe that the creation of an access hole is acceptable. Others, however, believe that once the hole is placed in the vehicle's body it causes irreversible damage, and therefore it should never be done.

Picks come in an assortment of shapes and lengths: short and long picks, chisel bit, as well as many tools specifically designed for paintless dent removal. See **Figure** 7-43, **Figure** 7-44, **Figure** 7-45, and **Figure** 7-46.

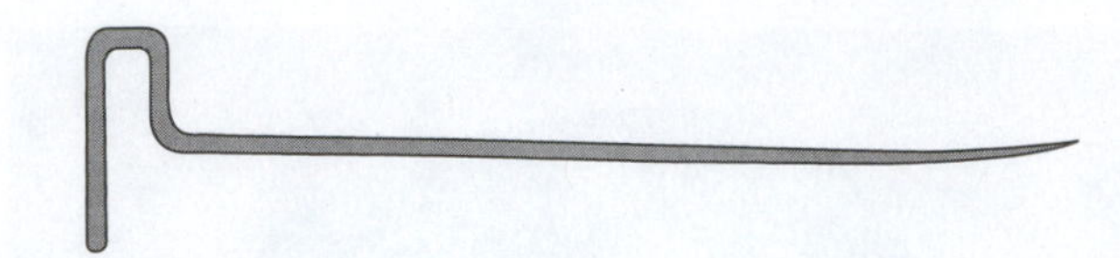

FIGURE 7-43 Short curved pick.

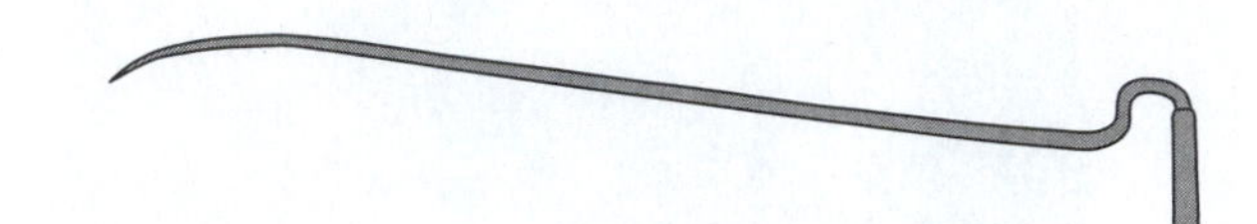

FIGURE 7-44 Long curved pick.

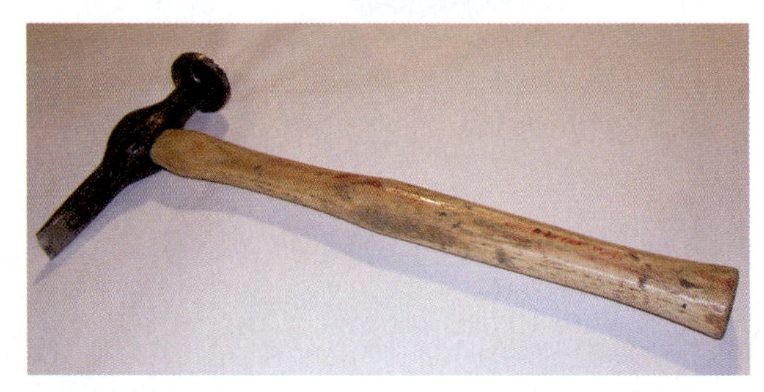

FIGURE 7-45 Chisel bit pick.

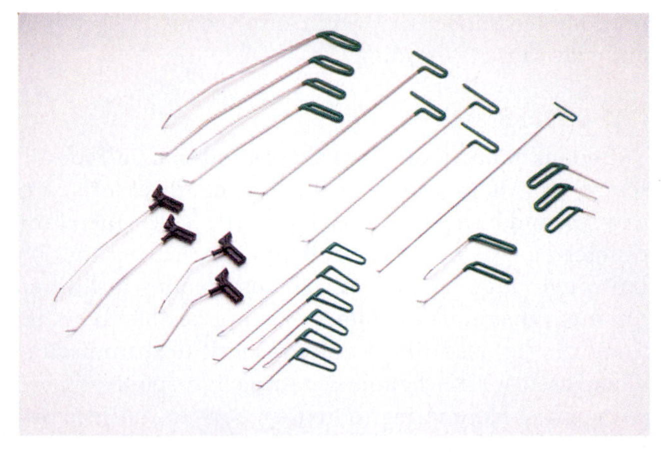

FIGURE 7-46 Paintless dent repair (PDR) tools.

TECH TIP When repairing a vehicle with either a pick hammer or a pick tool, the technician should keep in mind that each time the factory corrosion protection is disturbed, it will create a corrosion hot spot which will cause that damaged area to corrode. Because a technician is not always able to see these damaged areas due to close or limited access, corrosion protection should always be re-applied following the repair. Epoxy paint followed with petroleum or wax base corrosion protection is best.

FIGURE 7-47 A slapping file is a necessary tool when metal finishing dents.

Slappers

Although nearly forgotten when doing modern collision repair work, the slapper is a necessary tool when metal finishing dents. See **Figure 7-47**. It works in two ways. When used with a dolly, the serrated surface shrinks the stretched metal, while revealing low areas that may need more hammer work. It can also be used to remove metal as it is being slapped on the surface of worked metal, thus removing hammer marks.

Pulling Tools

Pulling tools are used often in collision repair shops when two-side access is not available. They can be used with weld-in tabs, a stud welder and slide hammer, or with glue-on tabs. See **Figure 7-48**, **Figure 7-49**, and **Figure 7-50**. All of these tools are used to raise areas of low metal or to stabilize metal while hammering on high areas.

Techniques

Specific techniques are used with each of the previously discussed tools. Likewise, different techniques are used with different situations. The way a technician holds

FIGURE 7-48 A weld-on pulling tool is used to raise areas of low metal or to stabilize metal while hammering on high areas.

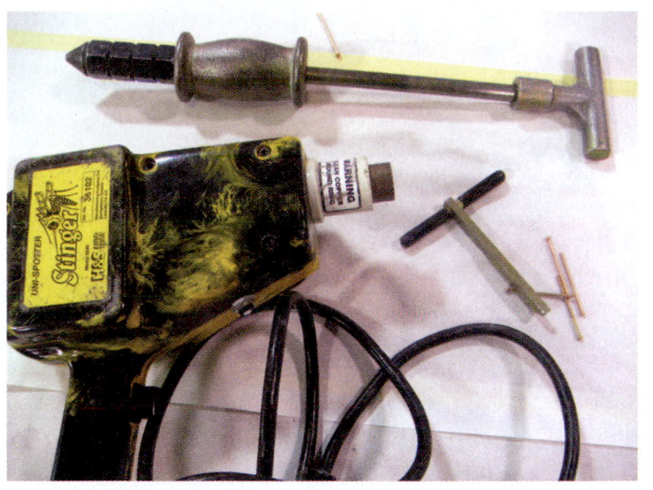

FIGURE 7-49 A stud welder with slide hammer is used to raise areas of low metal or to stabilize metal while hammering on high areas.

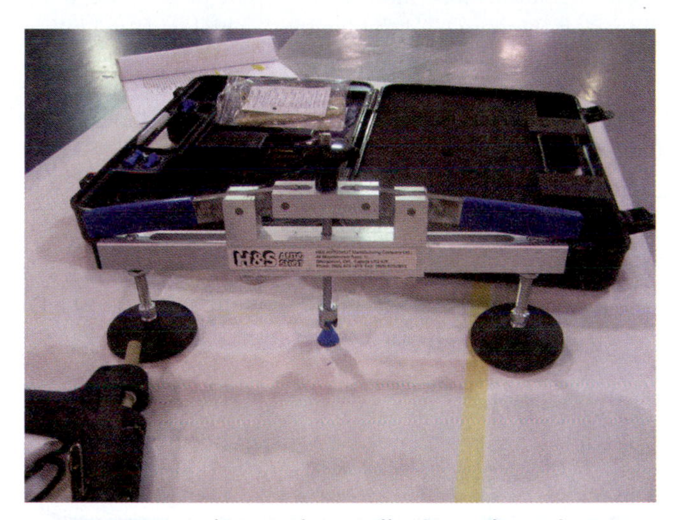

FIGURE 7-50 A glue-on dent puller is used to raise areas of low metal or to stabilize metal while hammering on high areas.

a body hammer differs depending on what the technician is trying to do with it; hammer-and-dolly techniques also differ depending on what the technician is trying to accomplish. Pick tools must be used in a precise way so they will not cause more damage than they repair. Slapping files, when used properly, will remove hammer marks and other fine damage when finishing metal, and pulling tools have been developed for specific purposes as well.

Hammering Techniques. Though hammering may seem like a simple, almost natural process, the way a technician grasps a hammer and applies the hammer blows to the work is very critical; in some cases, the techniques take a great deal of practice. The technician holds the hammer with the thumb and the first two fingers. See **Figure 7-51**. The third and fourth fingers are used to lightly grasp the handle (or in the case of spring hammering, to help produce the bounce needed for that technique). In a technician's performance of

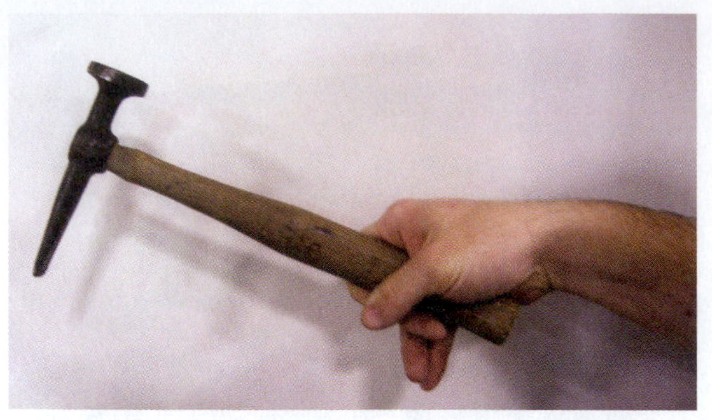

FIGURE 7-51 In the bump hammer grip, the technician holds the hammer with the thumb and first two fingers.

body work, the force of the blow is produced mostly using the wrist, not the arm and shoulder as when driving nails. New technicians may modify the grip by placing the index finger on the handle of the hammer, which aids in aiming the hammer blow. See **Figure 7-52**.

TECH TIP Anthropologists believe that a hammer or a striking tool (at first a rock) was the first tool that ancient people used and then fashioned. See **Figure 7-53**. They believe that hammers, then axes, were developed to make shelters, and also to hunt with even before knives were used.

The fact that our ancestors with opposable thumbs (the mechanism that allows us to grasp tools) fashioned tools was one of the most significant proofs that humans are superior to other species.

The amount of hammer force is also very critical when doing body work, and is often misunderstood by technicians. Though hard and forceful blows are sometimes very useful when repairing frames and performing heavy work, they are often not needed when repairing body metal. In

FIGURE 7-52 Technicians may modify their grip on a hammer by placing the index finger on the handle, which aids in aiming the hammer blow.

FIGURE 7-53 This flint ax is from Papua, New Guinea.

fact, multiple light blows will repair a dent in sheet metal at a much faster rate than heavy blows.

Spring Hammering. **Spring hammering** is used when a technician needs to give a series of light blows to metal, causing it to move or relax without causing the metal to stretch.

The technician grasps the hammer as described before, but when the head hits the surface of the metal and the steel gives or springs (from the elastic nature of the steel), the wrist is lifted. Then the technician adds a subsequent blow. The last two fingers aid in this repetitive spring hammering.

Spring hammering is often used with a light body spoon so the force of the hammer blow will be spread over a larger surface area. By spreading the hammer force over a larger area, the process is less likely to cause marks from the hammering.

Hammer-on-Dolly Technique. The hammer-on-dolly technique means a dolly is held directly underneath where the hammer blows are placed. See **Figure 7-54**. The

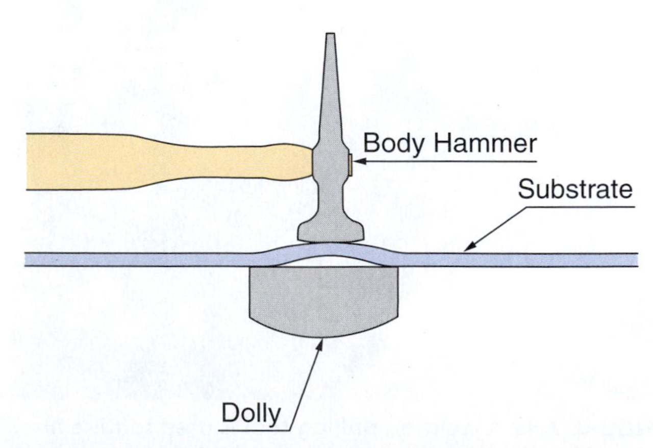

FIGURE 7-54 In the hammer-on-dolly method, a dolly is held directly underneath where the hammer blows are placed.

hammer blows remove the high areas of the metal as it is trapped between the dolly beneath and the hammer blow above. Placing the hammer blows exactly where they will do the most good is critical. The technician must strike the high spot, forcing it down. If the hammer blow misses the spot, it will cause further damage. Because the metal is between the dolly and the hammer blows, the distinct possibility of stretching the metal exists. The technician must be careful when hammering so stretching does not occur. Shrinking, either hot or cold, is a difficult process and should be avoided when possible.

TECH TIP When using the hammer-on-dolly method and a blow is placed precisely, a dinging sound is produced, much like a hammer hitting an anvil. If the blow misses, a dull sound instead results.

Because a body hammer's head is slightly curved, the area that hits the metal can be quite small. Therefore, when using a hammer-on-dolly method the technician must continually move the head of the hammer and the dolly beneath it. Often high areas are worked down by placing blows from the outside inward. See **Figure 7-55**.

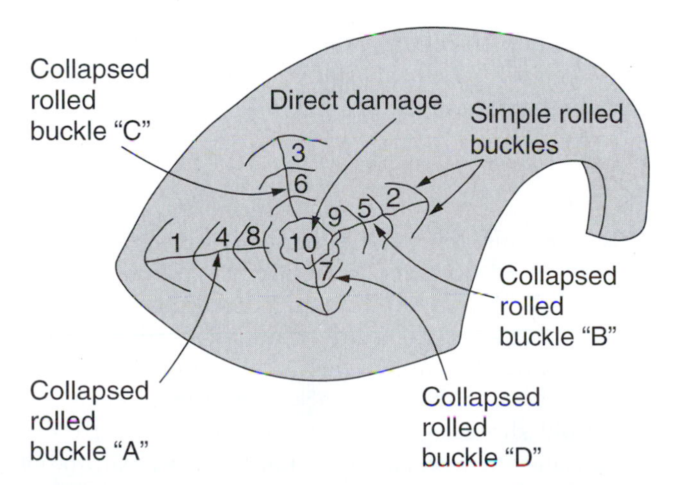

FIGURE 7-55 When using a hammer-on-dolly method, high areas are often worked down by placing blows from the outside inward.

Hammer-off-Dolly Technique. The **hammer-off-dolly** technique is used to raise low areas of metal. The dolly is placed beneath the metal where it is low. The dolly is held firmly as the technician places hammer blows near where the dolly is located, using the spring method of hammering. The dolly will rebound off the metal and spring back, causing the metal to rise. See **Figure 7-56**. When using the hammer-off-dolly technique, the technician works the dent from the outside in, thus raising the dent slowly.

The hammer-off-dolly method is used after the dent has been roughed out by hammering or picking up the dent. Unlike the hammer-on-dolly method, there is little fear of stretching the metal from the hammering.

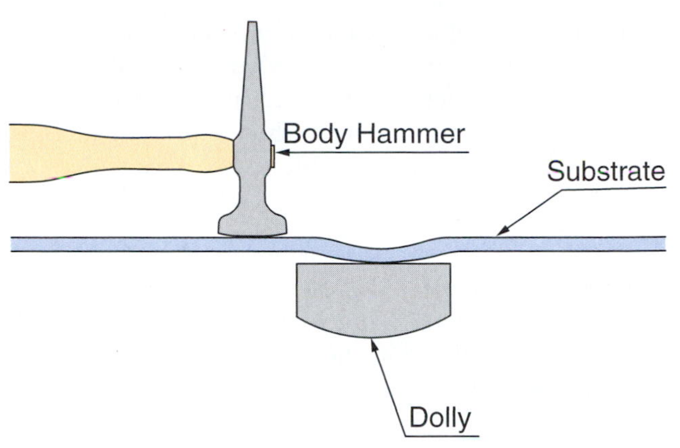

FIGURE 7-56 When using the hammer-off-dolly technique, the technician works the dent from the outside in, thus raising the dent slowly.

Pick Tool Techniques. Technicians can pick up a low spot—either when roughing out the dent or when metal finishing—using the force from either a hammer blow or a pick bar. See **Figure 7-57** and **Figure 7-58**. The point where the pick hits the metal is important. If the pick is placed in the correct spot, the dent can be raised with less force and a lower number of impacts. If the point where the pick hits the metal is incorrect, instead of repairing the damage, more damage will be created. The technician must be familiar with these **pick tool techniques**.

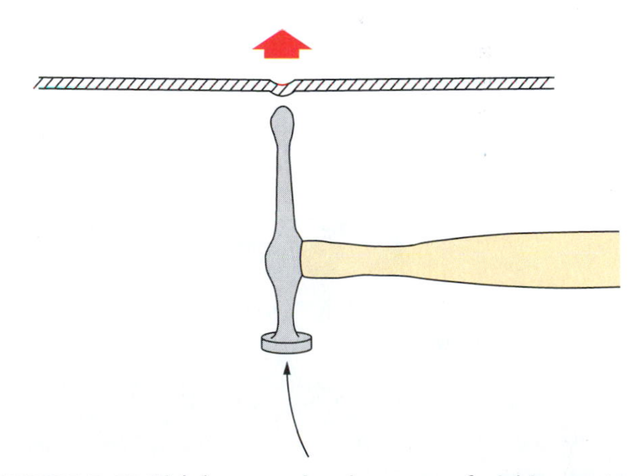

FIGURE 7-57 Pick hammering is a way of picking out low spots when metal finishing.

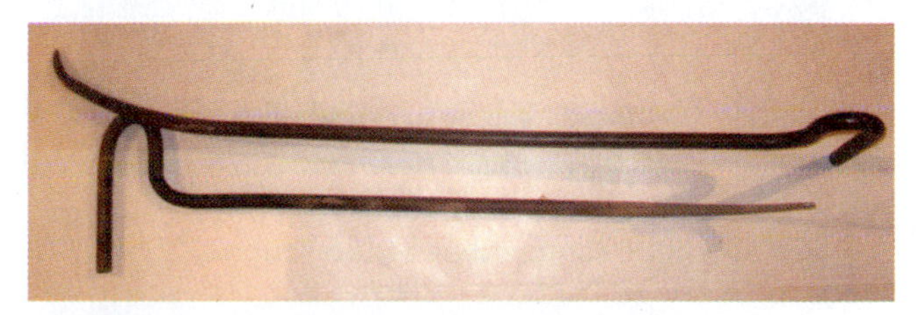

FIGURE 7-58 Pick bars need to be placed in the correct position, or more damage can occur.

Because aiming a pick hammer is so critical, a technician should practice on steel where both sides can be seen. Through practice, a technician will be able to see where the blows hit the metal, thus improving their accuracy. Often it is not possible to see both sides of a dented piece of metal. Therefore, the technician must be very accurate.

Slapping File. The **slapping file** is a finish tool that many young technicians have not learned to use. As a result, many do not use the tool when finishing metal work. The slapping file looks like a bent steel file. When it is used with a dolly, the slapping file is brought down directly over the top of the dolly. As it hits the dolly, it is dragged slightly before the next slap is applied. In this way, the area of high metal is both pushed down and filed off at the same time. The areas that are not smoothed out by the filing action are low and very easily seen.

The technician can use the slapping file without the dolly when small imperfections such as hammer marks need to be removed before priming.

Because finish metalworking is often more labor intensive and thus more costly than replacing parts, it is not often done in modern collision repair. However, the ability to metal finish when needed often is the dividing mark between a technician and a journeyman.

Weld-On Dent Removal Tools. **Weld-on dent pullers** are tools, often tabs, which are welded onto an area of metal that will be removed after the rough repairs are made. These tabs provide a convenient area on which to attach a hydraulic pulling tool, in order to apply force in the opposite direction from the collision direction. See **Figure 7-59**. With this force, corrections can be made to bring a body structure back into pre-accident dimensions. After the dimensional corrections have been made, the tab and the damaged sheet metal is removed, and the part is replaced with a new part, such as a quarter panel.

FIGURE 7-59 Weld-on tabs are welded onto an area of metal that will be removed after the rough repairs are made.

Adhesive Dent Removal Tools. Adhesive dent removers work similarly to the weld-on tabs, except that when these are removed they do not damage the panel that they were connected to. The area to which the tab will be attached is cleaned and sanded; then a special hot gun glue is used to attach the tab. When the glue has cooled, the technician attaches the pulling tool and lifts the dent. When the repair is completed, the glued-on tabs can be removed.

STRETCHED METAL AND SHRINKING

When a vehicle is involved in a collision, the force of the object that strikes the vehicle often stretches the metal that it hits. The repair technician must shrink stretched metal back to its original shape and size. Stretched metal most often occurs at the point of direct damage. Metal shrinking—both cold (hammer and dolly) and hot (with a torch or other heat source)—can be tricky, and will require much practice for a new technician to master. Technicians who are able to shrink metal will find that they become very productive as collision repair masters.

Metal stretches because, when impacted, it is forced beyond its point of elasticity and the metal is re-shaped. As the metal is yielding, some of the grains are compressed while others are stretched. It is the technician's responsibility to force this stretched metal back to its original size and shape. Metal can be moved by either hammering it back or by heating it. In both methods, the metal grains are forced from their stretched condition to the original thickness. When the metal—through hammering—has been returned to its original size and shape, the process is considered cold shrinking. Although it is less likely to damage the metal, it is also the slower way to shrink metal. In heat shrinking, heat is applied to the stretched area. Then, with the metal still hot, the area is hammered back into shape and rapidly cooled.

Cold Shrinking

Cold shrinking is done with a shrinking hammer and dolly. The tools come with a gridded face; the grid allows the molecules to shrink as the metal is hammered. See **Figure 7-60**. When cold shrinking, the technician should take care not to use the hammer-on-dolly method, which is likely to further stretch the metal. Light blows, however, gently move the metal back to its original shape and size. Cold shrinking can be a slow process, and many technicians often decide to use the faster heat shrinking method.

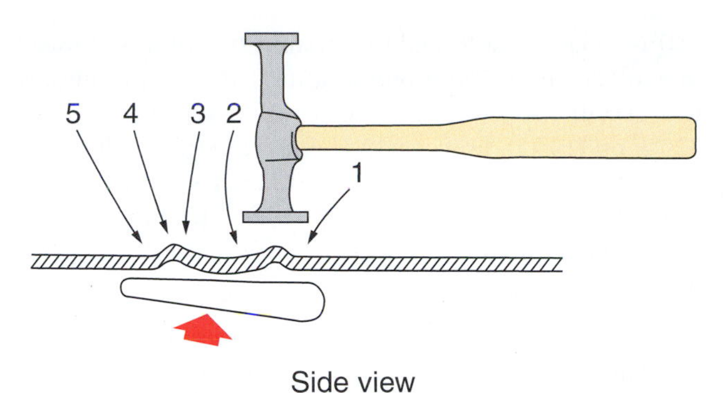

FIGURE 7-60 Cold shrinking is performed with a hammer and dolly.

Heat Shrinking

Heat shrinking is a process that heats the stretched metal with a torch—or, more often, with a carbine tip on a welding machine—to a red-hot state. Then, before the metal has a chance to cool, it is quenched with either a wet rag or sponge, which causes the metal to shrink during the rapid cooling. If the metal does not shrink back completely, the technician can repeat the process.

The technician must be cautious when heat-shrinking metal, especially the newer high strength steels. Heat damages the metal; thus, when a technician is applying heat, it should be applied only to the stretched metal. To do this, the technician often uses a carbon electrode and a welding machine. The tip can be applied to the exact area that needs to be shrunk, thus not affecting the surrounding metal.

PAINTLESS DENT REPAIR

Paintless dent repair (PDR) is a method of repairing dents that does not damage the painted surface. Pick tools are placed on the back side of the dented panel. Using a light to show where the pick is pressing up on the back side of the damage, the technician raises the low area. This method is very effective for repairing hail damage, door dings, and other light damage. However, the method is rarely used with heavy collision damage. Though the picks used for PDR look similar to normal body picks, they have blunter tips and come in various sizes. Because of their size and shape, the picks allow the technician to gain access to the underside with more ease, causing less damage. With their blunter ends, the picks are also less likely to cause damage as the technician gently pries up on the underside of the damaged area.

Because this type of repair is completed without needing to repair the vehicle finish, the work must be gentle. As the technician pries up with the tool, little movement is noted on the outside of the panel. With the aid of a light, the technician watches as the work progresses so that no high spots are created.

Though PDR is commonly used in the repair collision industry, some vehicle manufacturers still do not recommend the process. They note the corrosion protection damage caused by the tool and the possible damage to weatherstripping and other parts in areas where the tool is placed into the vehicle's body. As mentioned before, many believe that drilling access holes and the resulting corrosion damage afterwards do not equate to restoring the vehicle to its pre-accident condition, and so they feel the procedure should not be done.

ALUMINUM AND ITS USE IN AUTOMOTIVE MANUFACTURING

Although the use of aluminum in automotive manufacturing is not new to the industry, the demands for decreased automotive emissions and increased fuel economy not only have encouraged the use of aluminum outer body parts, but also have brought about the use of aluminum structural parts.

Originally, aluminum was limited to stamped sheet aluminum parts such as fenders, hoods, and deck lids, which were fastened to structural steel parts. Because these parts could be easily removed and replaced, technicians did not concentrate on the differences in repair procedures that were necessary for aluminum repair. The differences between the repair of aluminum and steel are not precisely in technique difference, but rather are procedure or tool differences. Aluminum is repaired with hammer and dolly, just like steel. However, the hammer may be plastic or wood instead of steel, and the dolly may be coated with a plastic covering so that steel particles will not be transferred to the aluminum being repaired. See **Figure 7-61**. Heat is used to repair aluminum as it is with steel; however, the heat ranges are much lower than for steel. Even paint removal from aluminum is necessary as it is with steel, but again the process is different than for steel, due to the different characteristics of aluminum. Therefore, though the techniques are similar, technicians must take care to use the proper process and tool for aluminum repair. Otherwise, the repair quality and longevity will not be acceptable.

Another change in aluminum usage has come with the introduction of "**aluminum-intensive**" **vehicles**. Manufacturers have introduced the use of cast and extruded aluminum for structural parts. These parts are attached to other structural parts of the vehicle with rivets, clinches, and adhesive bonding. Though not new to the aircraft industry, these techniques are new to automotive repair, and technicians need to adapt to these new methods. To be successful at aluminum repair, technicians should also understand the differences between the types of aluminum used in automobile manufacturing.

FIGURE 7-61 Aluminum repair tools may include a hammer and dolly, just like when repairing steel.

TYPES OF ALUMINUM USED IN VEHICLE CONSTRUCTION

Aluminum in its pure form is not found in nature. Rather, an ore called **bauxite** is mined. Bauxite ore (a type of clay), a combination of oxygen and aluminum hydroxide, is processed into pure aluminum. Aluminum in its pure form is a corrosion-resistant metal that conducts both heat and electricity well, but it is very soft and weak. To make it stronger and more resistant to bending, other elements are mixed with it in a process called alloying.

Alloys

Aluminum is grouped into two types of **alloys**: heat-treatable and non-heat-treatable. **Heat-treatable** alloys will gain strength by heating, whereas **non-heat-treatable** alloys gain strength by work hardening. These two groups of alloys are classified according to how strength is gained during manufacturing. Both these types of alloys, though, can be heated during the repair process to help complete the repair.

Heat-Treatable Alloys. Heat-treatable alloys are generally used for exterior body panels such as hoods, door skins, and quarter panels. They are 2000-, 6000-, and 7000-series aluminum alloys. The 2000 series alloys are alloyed with copper, 6000 series are alloyed with magnesium, and 7000 series are alloyed with zinc and magnesium or copper.

Heat-treatable alloy 2000 series aluminum is often used to manufacture both outer and inner hood panels. During the manufacturing process, the material is heated to gain strength. During the repair process the material will gain strength by work hardening. If heated within a specific repair threshold (400°F to 570°F, or 200°C to 300°C) it can be softened for repair. Then, if it is allowed to cool slowly to room temperature, the aluminum part will regain its pre-heat strength. If overheated (above 755°F or 415°C), however, its cooled strength will be decreased significantly. It loses significant strength in the heat affect zone when welded; since preheating aluminum before welding increases the heat affect zone, preheating should be avoided.

Non-Heat-Treatable Alloys. Non-heat-treatable alloys do not gain strength by heating, but through cold forming. The inner panels are generally built with this class of alloys. As the sheet aluminum is stamped into its needed forms, the non-heat-treatable alloys gain strength. This group of alloys will also continue to gain strength as it is deformed by a collision, and will also continue to become stronger and more brittle as it is work hardened during the repair process. It can be heated to soften during the repair process, as long as the repair threshold of 400°F to 570°F is not exceeded, and that the surface is allowed to cool naturally. Quenching with water or a blowgun may cause the alloy to crystallize, which will change the mechanical properties. Consequently, the practice is not recommended.

The non-heat-treatable alloys include the 1000-, 3000-, 4000-, and 5000 series. The 1000 series is 99 percent or greater aluminum and is not alloyed with other elements. The 3000 series is alloyed with magnesium, the 4000 series is alloyed with silicon, and the 5000 series is alloyed with magnesium.

Non-heat-treatable alloys are strengthened by work hardening, and are thus subject to cracking during repair. Although such a material can be heated to soften for repair, specific repair heat thresholds should not be exceeded. Though it is subject to cracking due to work hardening, a non-heat-treatable alloy is less subject to cracking during welding, and has excellent formability in weld areas.

CHARACTERISTICS OF AUTOMOTIVE ALUMINUM

Aluminum has significant differences from steel. Understanding these differences will help a technician be better equipped to return aluminum to its pre-accident condition.

Aluminum differs significantly from steel in its:

- Elasticity
- Plasticity
- Work hardening
- Collapsing response in a collision
- Manner of corroding

Elasticity

Elasticity is a structure's ability to return to its original shape. Steel and plastic formed body parts have what is known as "memory." That is, the part seems to "remember" after a collision what its manufactured shape was and, during the repair process, tends to naturally return to that shape. Aluminum does not have that advantage. Following a collision, aluminum panels will require more force and/or more work during the repair process to return to their original shape and state. Because of this required effort, aluminum repair often uses low heat during the repair process. The repair heat threshold is between 400°F and 570°F. This low heat will not adversely affect either heat-treatable or non-heat-treatable alloys, but will allow them to soften enough so that the part can be worked back to its original shape. Then, when allowed to cool naturally (that is, not forced by quenching with water or air), the part will maintain its original strength.

Plasticity

Plasticity is the ability of a material to be formed into a particular shape. Because aluminum parts are generally 1 1/2 to 2 times thicker than steel, while they have the same or more strength than steel, it can be harder to shape them into stamped parts. Because of their plastic abilities, certain aluminum alloys can be formed through extrusion and casting. Aluminum can also be formed through hydroforming, thus making its plasticity as versatile as its steel counterpart, while maintaining both its strength and its light weight. Through the use of alloying, aluminum can be as strong and formable (plastic) as steel, while being much lighter and less corrosive than steel.

Work Hardening

Work hardening—the hardening of a material when cold worked—can be significant with non-heat-treated alloys. Because of this, the material can have significant cracking. However, work hardening can be lessened by heating aluminum to soften it for straightening. To make sure that cracking did not occur during repairs, technicians should afterwards test the area by using a dye penetrant.

The process of checking for cracks has three steps: first, the area to be tested should be cleaned of contaminants. Some dye penetrant kits come with a cleaning agent, but if one is not provided a wax and grease remover can be used. After the area has been cleaned and allowed to dry, the dye is sprayed on the area to be tested. The technician should follow the recommended waiting time, which will allow the dye to seep into any cracks or defects that may have formed during the repair process. The dye is then wiped off. Following the application of dye, a developer is applied to the test area. The developer will activate the dye that is in small cracks, thus allowing a technician to see any imperfections. If weld repairs are to be made, the area must be cleaned again, because dye penetrant can cause weld zone contamination. However, repairing cracks in aluminum is usually not recommended; the part is most often removed and replaced by a new part. The technician should seek out and follow the manufacturer's recommendations in this matter.

How Aluminum Responds to a Collision

Some technicians believe that the main advantage to using aluminum is its resistance to corrosion. Others believe that its lightness and strength are its main advantage, which is also a valid viewpoint. Though aluminum is 1.5 to 2 times thicker than steel of comparable strength, it is considerably lighter, which is also a big advantage. Design engineers also appreciate aluminum's predictability in a collision. Unlike steel, aluminum collapses in a predictable manner; thus, aluminum parts can be designed to contain collision forces. By knowing how a material can and will collapse under differing conditions, engineers can design safety collapse zones into frame rails to protect passengers in collision emergencies. With a substance such as aluminum, which is corrosion-resistant, strong, light, and predictable when it collapses, it is no wonder that design engineers are using more in automotive production. Another seldom-thought-of advantage to using aluminum is its recyclability. Aluminum costs less to recycle than it does to produce in the first place, which cannot be said of steel and plastic. Nearly 90 percent of the aluminum in use today is recycled.

Aluminum and Corrosion

Though it has been stated that aluminum is corrosion-resistant, when exposed to air it does develop an oxidative coating almost immediately. Obviously, when **oxides** are mentioned in regard to steel, technicians think immediately about rust, which in the case of steel is correct. With aluminum, though, this oxide coating is "**self-limiting**." That is, it will form an oxide coating on aluminum exposed to oxygen until this oxide coating covers the aluminum's surface, and thus stops more oxygen from getting to the bare aluminum. The oxidation process then stops or does not let corrosion continue. Steel does not have this self-limiting property, and thus continues to corrode.

Aluminum, though, is quite susceptible to **galvanic corrosion**, sometimes called **contact corrosion**. This reaction occurs when two dissimilar metals such as steel and aluminum are exposed to each other. The more active of the two, in this case steel, transfers its electrons to the less active one. The less active one (aluminum) corrodes and the active one (steel) remains intact. To prevent galvanic corrosion from developing, the two dissimilar metals must be kept separated. This is done in the industry by placing nonconductive materials between them. Examples of these separation materials are protective coatings on fasteners, plastic, or rubber hardware, as well as specially designed adhesives which form a layer between them, thus separating the two metals.

Cross-Contamination. A constant concern when repairing aluminum is **cross-contamination**, whereby steel particles are spread to aluminum by way of sanding dust on tools. Small particles from tools such as hammers and dollies may transfer if the tools are used to repair first steel, then aluminum. Even drill shavings that are allowed to fall on aluminum parts can cause cross-contamination. That is why manufacturers of aluminum-intensive vehicles recommend that all tools that are used to repair steel should be keep separated from those tools used to repair aluminum. For hammer-and-dolly work, a wooden or plastic hammer is often used with the hammer-off-dolly method, and the dolly is coated with a plastic or rubber coating. See **Figure 7-62**. As discussed previously, when working with aluminum it is better to use lighter blows (with the panel heated to its working threshold temperature to return it to the proper state and shape) than when repairing steel. Therefore, such lighter tools can be used, and kept solely for aluminum repair.

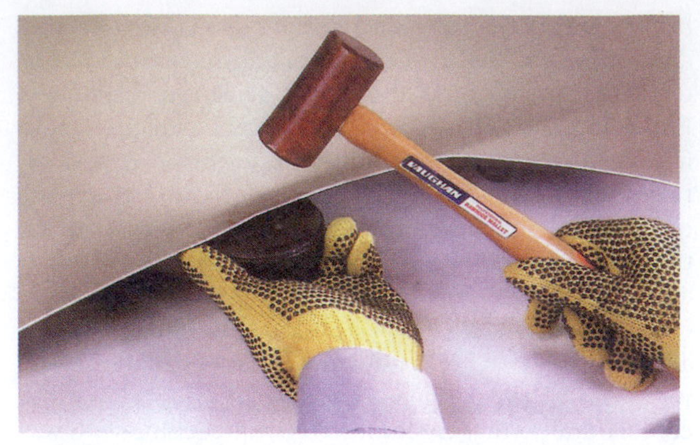

FIGURE 7-62 A wooden hammer is often used for aluminum dent repair.

Summary

- Straightening steel and aluminum are basic skills required by collision repair technicians today.
- Although more panels are replaced than repaired, technicians who are skilled in straightening both steel and aluminum are highly sought after.
- Straightening techniques are highly desirable skills to possess, and technicians who have mastered them have done so through long hours of practice and experience.
- Even talented technicians continue to hone their straightening skills throughout their careers.
- Metal finishing is a skill worth the continued practice and development it requires.

Key Terms

aligning/roughing hammers
alloys
aluminum-intensive vehicles
backup spoons
bauxite
boron steel
bumping/dinging hammers
cold-rolled steel
cold shrinking
collapsed hinge buckle
compressive strength
contact corrosion
cross-contamination
crowns
direct damage
dollies
dolly spoon
driving spoons
elasticity
finishing hammers
galvanic corrosion
glue-on dent pullers
hammer-off-dolly
hammer-on-dolly
heat shrinking
heat-treatable
high strength low alloy (HSLA) steel
high strength steel (HSS)
hinge buckle
hot-rolled steel

indirect damage
low carbon/mild steel
non-heat-treatable
oxides
paintless dent removal (PDR)
pick tool technique
plasticity
rolled buckle
self-limiting
shear strength
simple buckle
slapping file
slapping spoon
specialty hammers
spring hammering
stud welder
surfacing spoons
torsional strength
ultra high strength steel (UHSS)
weld-on dent pullers
wiggle wire
work hardening

Review

ASE Review Questions

1. Technician A says that hot-rolled steel is steel manufactured and rolled with very hot rollers. Technician B says that hot rolling is the process of rolling steel while it is still hot from manufacturing. Who is correct?
   A. Technician A only
   B. Technician B only
   C. Both Technicians A and B
   D. Neither Technician A nor B
2. Technician A says that Boron steel is a type of ultra high strength steel. Technician B says that low carbon steel is another term for mild steel. Who is correct?
   A. Technician A only
   B. Technician B only
   C. Both Technicians A and B
   D. Neither Technician A nor B
3. Technician A says that plasticity is the ability of steel to be bent and then return to its original shape. Technician B says that elasticity is steel's ability to be formed into different shapes. Who is correct?
   A. Technician A only
   B. Technician B only
   C. Both Technicians A and B
   D. Neither Technician A nor B
4. Technician A says that work hardening is the process during which metal becomes harder when cold worked during manufacturing. Technician B says that work hardening is the process during which metal becomes harder when cold worked during repair. Who is correct?
   A. Technician A only
   B. Technician B only
   C. Both Technicians A and B
   D. Neither Technician A nor B
5. Technician A says that direct damage refers to the area that made contact with whatever hit the vehicle. Technician B says that direct damage is often the largest amount of damage that occurs in a collision. Who is correct?
   A. Technician A only
   B. Technician B only
   C. Both Technicians A and B
   D. Neither Technician A nor B
6. Technician A says that indirect damage is often the largest amount of damage seen. Technician B says that crowns are shapes that manufacturers put in panels to strengthen them. Who is correct?
   A. Technician A only
   B. Technician B only
   C. Both Technicians A and B
   D. Neither Technician A nor B
7. Technician A says that roll buckles are secondary damage. Technician B says that the ability to access damage for repair will strongly influence the repair technique used by a technician. Who is correct?
   A. Technician A only
   B. Technician B only
   C. Both Technicians A and B
   D. Neither Technician A nor B
8. Technician A says that the hammer-on-dolly technique is used when the technician has access to only one side of a panel. Technician B says that the hammer-off-dolly technique is used when the technician has access to only one side of a panel. Who is correct?
   A. Technician A only
   B. Technician B only

C. Both Technicians A and B
D. Neither Technician A nor B

9. Technician A says that HHS stands for huge strength steel. Technician B says that hammers come in a variety of shapes and sizes, which allows them to be used for specific types of dent removal. Who is correct?
   A. Technician A only
   B. Technician B only
   C. Both Technicians A and B
   D. Neither Technician A nor B
10. Technician A says that hammers must be faced when scarring occurs. Technician B says that a smooth and mar-free hammer will produce the desired finish. Who is correct?
   A. Technician A only
   B. Technician B only
   C. Both Technicians A and B
   D. Neither Technician A nor B
11. Technician A says that a general purpose dolly is sometimes called a railroad dolly. Technician B says that a surfacing spoon is nearly flat and is used for finishing work. Who is correct?
   A. Technician A only
   B. Technician B only
   C. Both Technicians A and B
   D. Neither Technician A nor B
12. Technician A says that a slapping spoon is used for raising up low spots in metal surfaces. Technician B says that pick hammers come in different lengths and sizes. Who is correct?
   A. Technician A only
   B. Technician B only
   C. Both Technicians A and B
   D. Neither Technician A nor B
13. Technician A says that pick hammers and pick tools cause corrosion hot spots that must be repaired with corrosion protection. Technician B says that pulling tools are often used when two-side access is not possible. Who is correct?
   A. Technician A only
   B. Technician B only
   C. Both Technicians A and B
   D. Neither Technician A nor B
14. Technician A says that spring hammering means a spring-like hammer is used. Technician B says that a slapping file is a rough-out tool. Who is correct?
   A. Technician A only
   B. Technician B only
   C. Both Technicians A and B
   D. Neither Technician A nor B
15. Technician A says that series 990 aluminum alloy is nearly 100 percent pure and used for electrical wire. Technician B says that heat should not be used when straightening aluminum alloys. Who is correct?
   A. Technician A only
   B. Technician B only
   C. Both Technicians A and B
   D. Neither Technician A nor B
16. Technician A says that aluminum is both thinner and stronger than its comparable steel counterpart. Technician B says that aluminum's corrosion resistance, light weight, and crash predictability make engineers prefer to use it instead of steel. Who is correct?
   A. Technician A only
   B. Technician B only
   C. Both Technicians A and B
   D. Neither Technician A nor B
17. Technician A says that aluminum does not work harden when repaired. Technician B says that galvanic corrosion occurs with galvanized steel only. Who is correct?
   A. Technician A only
   B. Technician B only
   C. Both Technicians A and B
   D. Neither Technician A nor B

Essay Questions

1. Analyze and describe the differences between aluminum and steel in regards to corrosion.
2. Explain what advantages aluminum has for use in manufacturing automobiles.
3. Explain work hardening.
4. Describe the hammer-off-dolly repair technique.
5. Explain how a pick tool is used and when it is the best tool for a particular repair problem.
6. List the steps needed to properly face a bumping hammer.
7. Describe a collapsed hinge buckle.

Topic Related Math Questions

1. If a technician can repair 15 fenders in 1 hour, how many fenders will the technician repair in 27.5 hours?
2. If a technician can repair 15 fenders in 1 hour, how long will it take the technician to repair 626 fenders?
3. A shop had 26 hammers with broken handles. If it takes a technician in the shop 15 minutes to repair

each hammer, how many hours will it take the technician to repair them all?

4. A shop has 26 hammers with broken handles. If it costs $3.26 to repair each one, how much will it cost to repair them all?
5. If hammer handles cost $17.00 per dozen, how much would it cost to buy three handles? (Round to the nearest penny.)

Critical Thinking Questions

1. Explain why it is important for new technicians to master straightening techniques.
2. Explain why and when a technician would use the hammer-on-dolly repair technique.
3. After using the hammer-on-dolly technique, a technician notices that there is a bulge in the steel panel that was worked on. What has happened to the panel to cause the problem?

Laboratory Activities

1. On a fender with free access to both sides, place a 4-inch long, 1/4-inch deep dent, which the student will practice removing using the hammer-on-dolly bumping technique.
2. On a door shell with access to only one side but room for a pick bar, have students practice removing dents using a pick bar.
3. On a fender with light damage and access to both sides, have students practice dent removal using the hammer-off-dolly technique.
4. On an aluminum fender with light damage and access to both sides, have students practice removing light damage using heat and the hammer-off-dolly techniques.
5. On a lightly damaged aluminum hood, have students practice dent removal using an adhesive pulling tool.

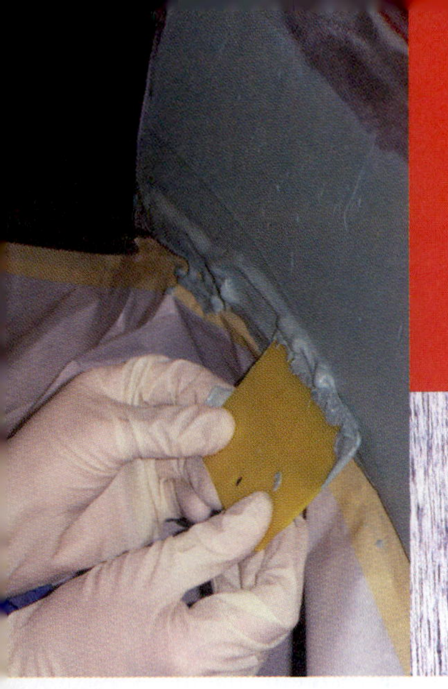

Chapter 8

Fillers

OBJECTIVES

Upon completing this chapter, you should be able to:

- Understand the history of fillers in this industry from their inception to modern day applications.
- Be acquainted with the many fillers commonly used in automobile manufacturing as well as those used when a vehicle becomes involved in an accident.
- Select fillers for effecting repairs on specific substrates.
- Understand the composition of fillers and their effect on the longevity of a repair.
- Understand surface preparation and proper application issues when applying fillers.
- Identify health and safety concerns and PPE requirements relating to the use of fillers and catalysts.
- Identify tools and equipment typically used for working with plastic and other fillers.
- Locate and identify specific hazards on MSDS pertaining to filler materials.

INTRODUCTION

Various fillers are used in both the manufacture and repair of automobiles after they sustain collision damage. Until recently, automobile manufacturers have used fillers exclusively for cosmetic functions to disguise and cover seams and fill unsightly surfaces between panels. In the collision repair industry, the word *fillers* has usually implied plastic fillers used to finish repairing collision damaged panels. Most recently fillers have taken on a whole new function and purpose for the manufacturer: structural reinforcement in body panels and increased noise reduction, as well as their traditional functions of preventing wind, water, and dust leaks into the passenger compartment.

HISTORY OF FILLERS

Plastic fillers—as we know them today—were originally introduced to the automobile collision repair industry in the mid to late 1950s as a more user friendly replacement for solder. **Solder**, a mixture of lead and tin, had been the industry standard for both automobile manufacturers and the collision repair industry. Although auto manufacturers continued to use solder for many years after the introduction of plastic fillers, the collision repair industry gradually shifted from solder to plastic fillers, and eventually embraced their use industry wide.

The widespread acceptance of plastic fillers by the repair industry came about for a number of reasons, the principal ones being the expertise required of the technician and the reduction of the metal thickness on the newer style cars. The increased awareness of potential health problems associated with constant exposure to lead was yet another contributing factor to the shift to plastic fillers.

FILLER TYPES AND APPLICATIONS

Historically, fillers have only been used for surface repairs and for putting the finishing touches on collision damaged panels. However, in recent years auto manufacturers have incorporated fillers into the structural makeup of a number of vehicles. For the purpose of discussion in the remainder of this chapter the fillers will be referred to as either plastic fillers or structural fillers, which include noise reduction filler materials.

Filler Types

All plastic fillers are categorized into four general groups: body fillers, special reinforced fillers, putties, and special glazes. The products in each group are designed to perform various functions for differing applications and circumstances. The substrate or the surface being repaired (i.e., steel, plastic, aluminum, etc.) and the location of the repair on the vehicle may dictate the specific filler the repair technician may select and use. Some fillers, for example, are designed specifically for use on a surface that is subject to numerous impacts, abrasion, and jarring, while others are designed for use where preventing moisture penetration is a significant concern. A thorough knowledge and understanding of their function and specific application recommendations is critical to ensure a long lasting repair.

Plastic Fillers. Body fillers offer the largest variety of products from which one may choose. Some are identified as **premium fillers**, others are referred to as **lightweight** products, while yet another group is called a **conventional** or **standard filler**. To the untrained eye they all offer similar characteristics such as excellent adhesion, sanding and **feathering** properties, fast curing, sag resistance on vertical surfaces, and so on. Despite these seeming similarities, however, many significant differences exist in the preparation, application, and sanding and feathering characteristics of each. Although the premium fillers are generally more costly than any of their counterparts, they are made of better quality ingredients and offer many favorable working characteristics over their less expense counterparts. Their real advantage is their compatibility to the coated metals found on many modern day vehicles. Another advantage lies in their usability on certain urethanes, plastics, and sheet molded compounds (SMC). The user is well advised to specifically determine on which plastics these fillers may be used. They may also be applied to an abraded original equipment manufacturer's (**OEM**) finish, thus eliminating the need to completely remove the finish and leave the substrate more corrosion resistant. This procedure is a significant departure from the standard recommendations, as most other fillers require the complete removal of the entire surface coating. Perhaps the most significant advantages are the fillers' **tack free** and **stain resistant** qualities. The possibility of redo's or comebacks due to a staining problem, which will be discussed later in this chapter, is virtually eliminated with their use. As was mentioned previously, premium fillers are generally more costly to purchase than any of their counterparts. However, the ease at which they can be sanded compared to other products—not to mention eliminating the possibility of bleaching and staining in most cases—is worth the additional cost.

Lightweight Fillers. Lightweight plastic fillers are also commonly used by repair shops. They are usually referred to as lightweight fillers because, like their premium counterparts, they use a combination of talc and microscopic spheres for the filling ingredients. These **microspheres** are considerably less dense and lighter weight than talc; consequently, the weight of the filler is reduced significantly compared to the conventional fillers, which use talc exclusively. Because these two fillers are made with a higher concentration of talc, they are less moisture resistant as the talc will absorb moisture when exposed to the atmosphere. This can become a significant issue when performing a **rust** repair. In the event the repaired rustout is not completely sealed, moisture will penetrate through the back side and be absorbed by the filler, resulting in a premature repair failure.

Another very important consideration is neither the lightweight nor the standard products are tack free or are guaranteed to be stain free like the premium fillers. This can mean the technician will need to make additional applications of material and perform more sanding to finish the repair. The most significant difference between the premium and the lightweight and conventional plastic fillers, however, is the quality and refinement of the resin and other ingredients used. Despite the shortcomings of the lightweight and conventional fillers compared to the premium fillers, many shops will choose these options over the premium fillers because they are slightly less costly to purchase. It is difficult to justify the artificial cost savings when one repair re-do can cost many times the savings of buying the less expensive product.

Two Part Polyester Glazing Putty. Many times the repair technician will choose to apply a thin coat of a specially formulated two-part polyester glazing putty over the entire repair area. This putty ensures the elimination of any minute flaws left in the plastic filler and covers any unfilled sandscratches left in the substrate. This extra thin material—very similar to the plastic filler—is applied in a micro-thin application. Unlike plastic filler, it should not be used to fill surface indentations, as it lacks the required filling ingredients to accomplish such a task. However, it can be mixed with most of the plastic fillers to enhance a smoother application and for easier sanding.

FILLER INGREDIENTS

Although numerous types of plastic fillers are available, most are made from a combination of similar ingredients. With few exceptions fillers are a two-part material which, when mixed together, create a chemical reaction causing the material to cure or set up. The main body or the filler part of the plastic filler essentially consists of a **polyester resin**, talc, and—in some of the fillers—a combination of talc and microspheres. **Talc** is a very fine and soft powdery-like mineral consisting of magnesium silicate. One might liken or compare it to baking soda or corn starch as its texture and characteristics are similar to these common kitchen ingredients. It is much heavier than the newer microspheres, which are commonly used—in part at least—in both the premium and lightweight plastic fillers. One might compare these minute, round, and hollow microspheres to miniature Ping-Pong balls.

Resin

Resin—a syrup-like material—is responsible for uniformly suspending all the filler ingredients, making it possible to mix and smoothly apply the resin to the desired surface. Once the filler has been catalyzed with a chemically matched hardener, it is applied to the surface, whereupon the resin cures or hardens. Thus, it becomes the binding agent that secures and permanently holds the plastic filler to the surface.

Catalyst

Another principal ingredient of the plastic fillers is the **catalyst** or hardener, which must be mixed into the filler in specified quantities to form a hardened mass of material when cured. The principal ingredient of the catalyst is **butyl benzoyl peroxide** suspended in a variety of inert liquids and solvents. A more detailed discussion of its function and purpose will be covered later in the chapter.

FILLER STORAGE

Before using plastic filler, a technician must thoroughly understand how to use the material and what problems may develop, starting with the storage of the material. Plastic fillers are available in a variety of quantities, from pint, quart, gallon, and 5-gallon cans to plastic bladders encased in cardboard containers. Whatever size container is purchased, the material must be rotated—or the container must be inverted or flipped over—at least monthly if it is not used within that time period. The talc in the plastic will settle to the bottom of the container, forcing the resin to the surface if it is left on the shelf for an extended period of time. This separation of the ingredients can cause numerous application and adhesion problems. Therefore, the technician should thoroughly agitate the container to uniformly re-suspend the ingredients within the resin prior to removing any of the material from the container. Many times when plastic has been warehoused or stored for an extended period of time without being rotated, the user will find an accumulation of resin on the top of the can when first opening it. Before removing any of the material from the container, this resin must be thoroughly mixed into the rest of the material in the container.

TECH TIP Never pour off any of the resin material that may collect at the top of a newly opened container of plastic filler, as it will destroy the balance of the ingredients. This will result in numerous adhesion and curing problems. The material should be thoroughly mixed prior to removing any of it.

Agitating resin to uniformly re-suspend all the materials should be done with a combination of an up-and-down motion and a slow deliberate stirring motion using a stirring paddle or stick. The container should never be placed in a paint shaker to agitate as it will form air pockets, which may cause problems later when applying the material. Many shops purchase plastic fillers in large plastic bladders contained in cardboard boxes. It becomes increasingly important to routinely rotate the stock as it is not possible to agitate the filler when it becomes completely separated. In extreme cases it may be necessary to dispose of the material.

SURFACE PREPARATION

No one step in filler preparation and application is more or less important than any of the others. Proper surface preparation is as critical to the longevity of the repair as is the selection, application, and finishing of the material being used. One must first make sure the metal—or other substrate being repaired—is correctly straightened, repaired, and ready for the application of the plastic filler. An inexperienced repair technician may be tempted to apply filler over an improperly repaired metal surface—one that has not been sufficiently restored to the state of tension required to maintain the correct contour once filler has been applied. This may result in an oil-canning effect once the filler has been applied. This effect is created by the vibration and flexing action that occurs during the grating and sanding operations. (Proper metal straightening techniques are covered in Chapter 7.) The heat generated from the chemical reaction during the curing process may also cause the metal to lose some of its tension once the filler has been applied.

REPAIR DAMAGE FIRST

Inexperienced technicians commonly think that it is faster and easier to fill a damaged surface than it is to repair it. However, filling a damaged area is never an acceptable substitute for properly straightening and repairing the metal. In most cases the time required to fill a damaged surface without first restoring it to its proper contour is considerably longer than may be expected. It requires several applications of filler to fill the damaged area, with grating and sanding required between each application. The result is typically disappointing and considerably less than desirable as the filler rarely can be contoured properly and the repair area is noticeably different even to the untrained eye.

Once the damaged surface has been properly repaired, it may be advisable to wipe down the entire area to be filled with a wax and grease remover—particularly if hydraulic equipment has been used—as fluid leakage is an imminent possibility. Hand tools used on a previous job that have not been properly cleaned may be another source of some surface contamination that may go undetected. Any contamination that may have been introduced to the surface from hand tools and other equipment may result in a variety of problems at a later time in the repair process. The importance of cleanliness cannot be overlooked at any time during the repair, as even the slightest contaminant may be cause for additional work at a later time in the repair process.

CAUTION

All the proper PPE—including respiratory protective equipment, safety glasses, and face shields, as well as gloves—must be worn when performing any grinding and sanding operations. Consult the school's or shop's respiratory protection and safety program.

SURFACE REMOVAL

Once the surface has been properly cleaned the technician needs to remove all the old finish down to the bare substrate. This may be accomplished with any of several surface removal tools such as a high speed sander, a dual action sander, or even a drill with a special mandrel designed to hold an abrasive wheel. See **Figure 8-1**. The size of the area to be covered with the plastic filler will dictate which surface removal tool is used. The surface coating should be removed to approximately 3 or 4 inches beyond the edge of the damage repaired. This will allow a sufficient amount of space to properly feather the filler into the adjacent areas. Sometimes when the damage is near the edge of a panel or a crowned surface on the panel, it may

FIGURE 8-1 A number of tools are available to remove the surface coating prior to using a filler. The size of the area and the amount of surface to be removed will dictate which tool to use.

be advisable to remove the coating all the way to the end of that panel or the crown. Finishing the filler will be completed more easily when it can be feathered into the edge of the panel.

Protect Adjoining Areas

When grinding or sanding near the edge of a panel or a crowned surface that is to be used for a natural break line for the repair, one must take extra precautions to protect the adjoining areas. The smart repair technician will cover the edge of the adjoining panel or crown with protective tape prior to initiating any of the surface removal with the sander or grinder. See **Figure 8-2**. A special extra thick protective tape is available for this application; multiple layers of masking or duct tape may also be used as well. One accidental grinder mark or sandscratch on the surface of an adjacent panel will require additional finishing and painting time, as well as additional costs for the shop and extended down time for the vehicle.

Abrasive Selection

Historically all surface removal operations were accomplished by using a 24 or 36 grade abrasive disc on a high speed pistol grip or a right angle grinder. A 36 grade abrasive disc was recommended for aluminum panels. Technicians thought the additional abrasive scratches left in the surface would increase the bonding surface and actually promote better adhesion of the plastic filler. However, this often required additional finishing steps to fill and eliminate these scratches in the outer perimeter of the repaired surface and often resulted in sandscratches showing through once the topcoat was applied.

Most plastic filler manufacturers currently recommend a different approach to removing the surface than the recommended past practices. The metal thickness on most modern day vehicles has been reduced considerably.

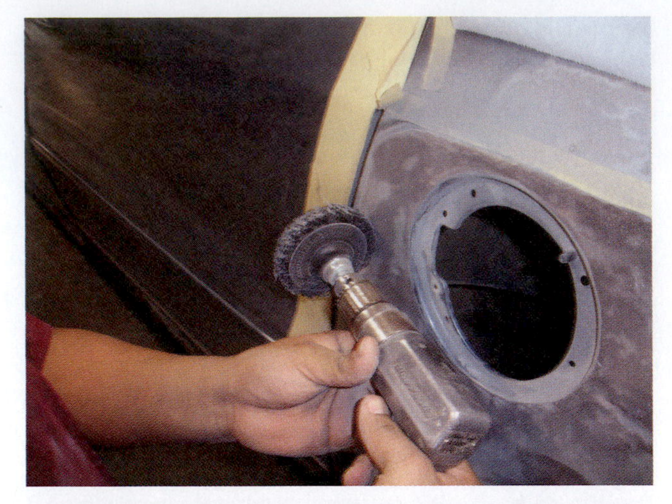

FIGURE 8-2 Care must be taken to protect the adjoining areas from possible damage when removing the surface coating and when sanding the fillers.

Therefore, using the same coarse and aggressive abrasive grades from the past would certainly remove an excess amount of material from the substrate. In this situation, grinding or burning through the metal is a real possibility. Refer to the chart in **Figure 8-3** to select the recommended abrasive grade for the specific surface removal task being performed.

A. METALLIC SURFACES	
OEM Finishes	80D
Paint (Previous Repair)	80D (36D optional if excessive)
Plastic Filler Removal	36D
Scale Rust	36D
Welds	50D/36D
B. PLASTIC/COMPOSITION PARTS AND MATERIAL	80D

FIGURE 8-3 This chart lists the recommended abrasive grades to use for removing the finish from the repaired area in preparation to applying filler.

As indicated in Figure 8-3, the current recommended procedure is to not exceed an 80 grade abrasive to remove the original primer and topcoat to minimize the amount of substrate removal while taking off the surface coating. This will also minimize the depth of the grinding scratches left in the surface, which may not get filled with the plastic or other fillers. Many modern day automobile manufacturers apply a micro-thin layer of **zinc** on the surface of the panels to maximize the metal's corrosion protection. See **Figure 8-4**. The use of an excessively harsh abrasive will most likely remove this protective layer, leaving the surface somewhat compromised to corrosion.

Many times a vehicle that comes into a shop for repair of a surface has had prior repairs made on the same damaged surface. The technician may find several layers of material consisting of fillers, primers, and additional topcoat and clearcoat. In this instance a more aggressive abrasive than the 80 grade may be necessary, as the task will be overly time-consuming. A high speed or right angle grinder with a 36 grade abrasive may be necessary in order to accomplish the task in a reasonable amount of

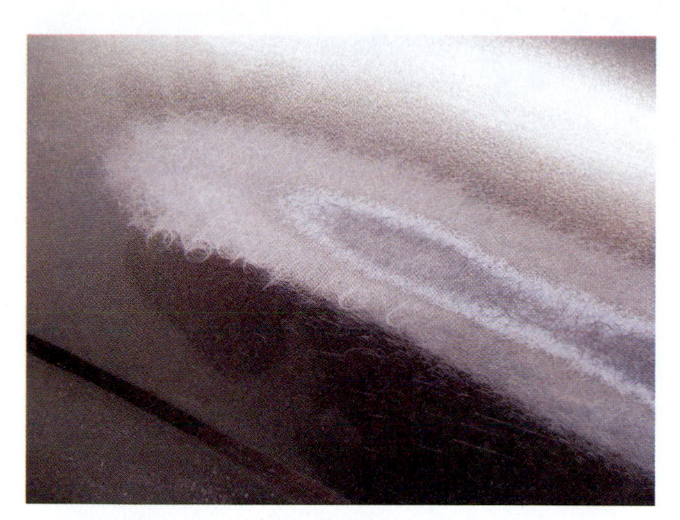

FIGURE 8-4 When removing the old finish, the technician must exercise care to minimize disturbing the protective zinc coating on the surface.

time with minimum material usage. In extreme cases wherein the filler may be excessively thick, a 24 grade grinding disc may even be recommended. To minimize the amount and the depth of the grinder marks left in the surface when using an aggressive grinding disc, it may be advisable to remove the majority of the material and leave a thin layer which may be removed with the 80 grade abrasive. This will require less finish time and material when preparing the surface for priming and finish coating.

CHEMICAL CLEANING SURFACE

After all the surface coatings have been removed down to the bare substrate, it is advisable to solvent wipe the surface with acetone. This ensures the surface is free of any contaminants that may be left behind from the air tool exhaust, air-borne particles, or even from wiping the surface with one's bare hands. Acetone is a fast drying solvent capable of dissolving most petroleum-based contaminants as well as any that may come from the technician's skin. A wax and grease remover is not recommended as they are much slower drying. If the surface is over-wetted, it may be difficult to remove all the residue from the scratches when wiping it off. Any solvent that does not get completely wiped dry and off the surface or out of the sandscratches and grinder marks may affect the adhesion of the plastic filler. Certain wax and grease removers may also leave an invisible film or residue on the surface that can cause similar adhesion problems. Once the surface coatings have been removed and the surface has been solvent wiped, the technician should apply the plastic filler as soon as possible to avoid prolonged exposure, the possibility of surface corrosion from moisture in the air, and contamination from air-borne contaminants in the shop.

PLASTIC APPLICATION TOOLS

Applicators

Prior to removing the filler from the container and preparing it for application, the technician must select the appropriate applicator as well as the mixing board. Despite the critical role these items play in successfully applying filler materials, their importance is often overlooked. The application of plastic filler is usually accomplished using one or more of a variety of plastic application tools or spatulas. The two most popular types of applicators are a urethane-type plastic and a steel putty knife. They range in size from approximately 2 inches to 6 inches in width. The applicator selected should be representative of the size of the area being coated. The urethane applicator can be bent slightly by the technician to conform to the surface to which the filler is being applied. Some technicians prefer to use a thin flexible steel putty knife. Its chief advantage is the edge stays straight and the filler is applied in a more uniform thickness. As one becomes more familiar with the use of the many fillers available, selecting the appropriate applicator for each type of filler will become second nature. Each individual has preferences as to which applicator works best in each situation.

One of the keys to successfully applying the filler is to ensure the applicators used are properly cleaned after each use and free of nicks or divots in the application surface. The applicator's surface should be free of any hardened material, particularly on the edge of the surface being used to apply the filler. If any dried material is left on this surface, each pass of the tool will result in lines or streaks in the plastic. This creates extra work as additional fills will be required to cover the surface, and each additional fill requires shaping, sanding, and finishing. See **Figure 8-5**.

FIGURE 8-5 Using a dirty plastic applicator will inevitably create additional work for the technician. Any hardened material left on the surface will leave streaks when applying the new fill, causing additional work to sand them out.

FIGURE 8-6 A urethane plastic applicator can be redressed to remove nicks in the application edge and to remove small bits of hardened filler by simply running it back and forth over a fine grade abrasive on a flat surface.

Occasionally a urethane applicator gets a nick in the edge used for applying the filler. However, this is not a reason to dispose of the applicator, as it can be redressed and used again. To redress the applicator, the technician holds a piece of fine grade sandpaper on a flat surface, holds the applicator in a vertical position, and runs the edge of it back and forth over the abrasive surface, smoothing it in the process. See **Figure 8-6**. Likewise, a technician can redress a steel putty knife simply by holding the blade's edge on a hard smooth surface and using a mill file to smooth the surface. See **Figure 8-7**.

Plastic Mixing Board

The mixing board should be made of a material that will not absorb any of the ingredients from the filler and the hardener. Many times an unknowing individual will use a discarded cardboard box as a mixing board. However, most cardboard boxes have a wax coating to make them somewhat water repellent. The wax on the card-

FIGURE 8-7 The edge of a metal plastic applicator can be redressed by using a mill file to remove the rough edge and any scratches found in the blade.

board surface can contaminate the material, as it may be scraped off the surface and incorporated into the plastic as it is being mixed on the board. This may cause some serious adhesion problems as the unwanted waxy residue will likely compromise the plastic's ability to stick to the surface. In addition, some of the chemicals from the plastic or the catalyst will be absorbed by the paper, leaving an unbalanced mixture of material. This improperly balanced mixture can result in a variety of problems, ranging from the plastic not curing properly and affecting its workability to adhesion problems at a later time.

A variety of commercially produced mixing boards are available from any jobber supply house. The most popular items are made of an acrylic plastic which can readily be cleaned and reused over and over again. Another available type has tear-off sheets of paper, similar to a bound tablet, which are mounted on a board. After each use the technician simply tears off the used sheet, exposing a new or clean one for the next batch to be mixed. See **Figure 8-8**. One may want to check local regulations to ensure the disposal of the used sheets is allowed in the shop's **solid waste stream**.

Another way to easily fabricate a mixing board is to cut a piece of flat metal from a replaced panel and remove the surface coating.

CAUTION

When working with metal, it may be advisable to cover the edges with tape to avoid getting cut by the sharp edges created when cutting the metal out.

This method can be used repeatedly without concern for contamination being introduced from the mixing surface. To maintain successful use of the same mixing

FIGURE 8-8 The plastic mixing board should be made of a non-absorbent material to protect against absorbing any of the chemicals from the filler.

board, the technican must clean it after each use. This can be accomplished simply by scraping the excess filler from the mixing board that may be left over and disposing of it with the other shop solid waste materials. Wiping the board with a mild solvent such as acetone will remove all the remaining residue, leaving a clean surface for the next time filler is to be mixed. One should never attempt to mix filler on a board that has dried or hardened material left over from a previous mix. There is always a possibility of the old material breaking loose from the board and getting mixed into the new material, which will make it impossible to apply the material smoothly and free of streaks. Likewise, when using a container of filler—such as a gallon or quart can—that the material must be dipped out from, one should always ensure the tool being used is clean and free of any hardened and dried plastic.

CAUTION

When mixing and applying plastic filler, the minimum PPE the technician should use are safety glasses and either latex or nitrile gloves. Consult the product MSDS to obtain the specific recommendations for the product being used.

SELECTING AND MIXING THE FILLER

At this point the plastic filler should be mixed and applied to the surface using all the safety precautions and the required PPE. The final decision the technician must make is which specific filler is best suited for the type of repair, the type of surface being repaired, and the area of the vehicle where it is to be applied. (This will be covered later in the chapter.) Along with determining the type of filler to use, the repair technician must also be knowledgeable about the type and amount of catalyst that must be used with the material. As mentioned previously, the catalyst is the agent that must be mixed into the plastic to cause it to cure. Using the correct amount of catalyst is vital to ensure that the repair will last for the life of the vehicle. Using an excessive or inadequate amount of catalyst will be a great detriment to the longevity of even the smallest repair. Therefore, it is important to understand and practice proper mixing of the material at all times.

Catalyzing Filler

Determining the exact amount of catalyst or hardener that must be added to the material is one of the most difficult things the untrained individual must learn to do, especially when mixing the materials manually. Despite the importance of adding the proper amount of catalyst added to the mix, the directions for doing so are often very vague, nonscientific, and hard for the beginner to understand. Therefore, the technician may need to experiment in order to achieve a workable solution for the prevailing shop conditions. Most manufacturers recommend adding the catalyst in an amount equal to 2 percent of the plastic filler. See **Figure 8-9**. Some relatively sophisticated equipment is required in order to accomplish this, and it is extremely time-consuming to measure the filler this precisely each time a batch is mixed. Therefore, one must devise a more expedient procedure that may require mixing one or more experimental batches of the material and timing the cure time, sanding time, and so on. Since the catalyst is pigmented, a large chip of the batch that produces the most desirable results can be saved to compare the coloration on future batches mixed. One way to determine the amount of hardener required is to divide the material into golf ball sized parts and add approximately an inch long ribbon of hardener for the equivalent of each imaginary ball. See **Figure 8-10**. Another method that may be utilized is placing approximately a 4-inch diameter pad of filler on the mixing board and squeezing a ribbon of hardener across the middle of the material. Regardless of the system used, it must be applicable for mixing batches of filler ranging in an amount for filling a 1-inch diameter spot up to a very large area.

Mixing Filler and Catalyst

After adding the catalyst to the filler, one must thoroughly mix the entire batch of material together to ensure the hardener is uniformly distributed throughout the entire mass of material. The material should be the same color throughout and be free of any streaks of color resembling that of the catalyst. Any discoloration in the mix indicates additional mixing is required. When mixing the catalyst into the plastic, one should avoid stirring the material in a circular motion; doing so may form air pockets in the mix, which may be difficult to eliminate. Instead,

FIGURE 8-9 Plastic manufacturers recommend a 2 percent mixing ratio of hardener to the amount of plastic being mixed. The proportions shown are a precise ratio of the two ingredients.

one should mix the material by using a folding motion. See **Figure 8-11**. This method works more effectively than stirring and reduces the possibility of air pockets forming in the material.

One must always ensure the catalyst is properly mixed or kneaded before squeezing any of it out onto the batch of filler to be applied. This is particularly true if the tube being used was just recently taken off the shelf or put into use. The ingredients are known to separate while in stor-

FIGURE 8-10 Filler manufacturers frequently cite a ratio of 1-inch ribbon of hardener to a golf ball size part of filler. The pad of filler in this figure is approximately that size once it has flowed out.

age and must be mixed to uniformly re-suspend them to ensure a proper balance of material is being dispensed. A tube of hardener that has exceeded its normal shelf life will not re-suspend uniformly and should be disposed of, as it will cause numerous unnecessary problems. Proper catalyzing of the plastic filler must be accomplished each time a batch of filler is mixed. This is not possible if one element of the entire mix is not fit for use. Numerous immediate and long-term problems affecting adhesion, workability, and longevity of the repair may likely occur if the material is overcatalyzed or undercatalyzed.

Figure 8-12 will help one understand the curing process for plastic filler and the importance of through-curing of the filler mass. The sketch in Figure 8-12 is a very simple example of what may occur when the resin in the plastic (receptors) and the catalyst (reactors) are mixed together. A chemical reaction occurs, causing a cross-linking process which bonds and fuses all the material together. As the chemical reaction occurs, heat is generated. This promotes the cross-linking to occur, causing a wider range of networking. This ultimately links the entire mass of material together. In order for this scenario to occur, however, the correct proportions of catalyst must be present for the amount of filler being catalyzed. When the plastic is overcatalyzed (too much hardener is present), the chemical reaction occurs too quickly, causing too much heat. Therefore, the cross-linking of the receptors cannot fully take place. The result is a mass of material that is brittle, poorly anchored to the surface, and likely will compromise the longevity of the repair. In the event a batch of the filler is inadvertently overcatalyzed, one can add additional filler to the mix until an acceptable mix is achieved. However, in the event a catalyst-rich mixture is applied to the surface, it should be removed to avoid any potential problems.

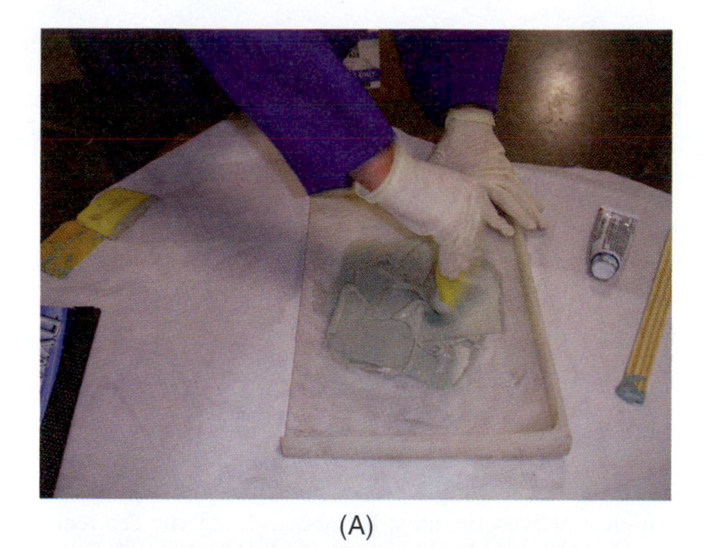

(A)

(B)

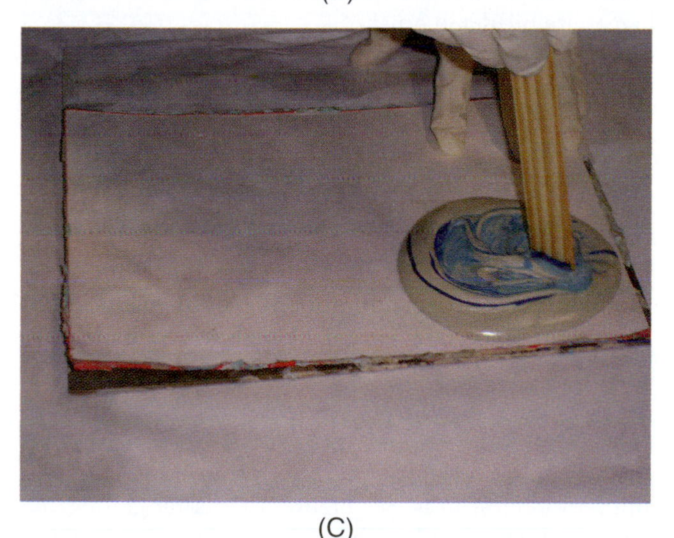

(C)

FIGURE 8-11 When mixing the filler and catalyst, the technician should use a folding and sweeping motion to blend the catalyst into the plastic, as in views (A) and (B). Stirring the mixture, as shown in view (C), may cause air pockets.

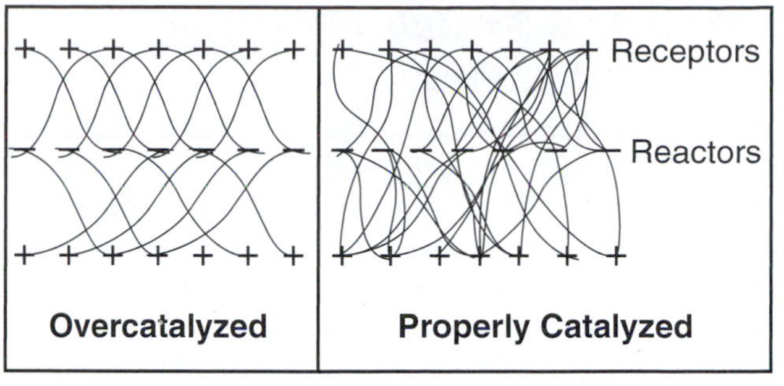

FIGURE 8-12 A cross-linking process occurs as the catalyst and the resin in the filler start to chemically react with each other. Overcatalyzed filler will react too quickly, causing excessive heat. This prevents the filler from completely cross-linking, therefore leaving the material brittle and lacking sufficient adhesion.

PROBLEMS WITH IMPROPER CATALYZING

The following are some of the most common problems likely to occur if the filler is *over-catalyzed.*

Pinholes and Air Pockets

When two chemicals are mixed together, such as the plastic and catalyst, a chemical reaction occurs. Whenever a chemical reaction occurs, heat is generated and gases are given off, which under normal circumstances will escape through the surface. Overcatalyzing the mixture causes excessive heat and gases which cannot escape through the mass of material, resulting in numerous **pinholes and air pockets** throughout the layer of material. These pinholes and air pockets can cause serious problems, especially when the surface and topcoats are applied.

Adhesion Loss

Plastic fillers are chemically formulated to adhere to the substrate through a chemical bonding process. However, the chemicals must have ample time for the chemical bonding to occur. An overcatalyzed mix will cure too quickly, resulting in **adhesion loss** because the mix is not allowed to link to the substrate. The material will become very brittle and chip very easily if bumped, such as when a door is opened against it.

Shortened Working Time

The presence of excess hardener will cause the filler to cure too quickly. In many instances this happens before the filler can be applied to the surface. This results in difficulty applying the material smoothly, and frequently requires a subsequent fill to smooth the surface. More waste is also generated as the unused, prematurely hardened material must be thrown away.

Extended Finish Sanding and Feathering Time

Excess hardener causes a "short circuiting" effect between the reactors and receptors. Some of the material never fully cures, nor does it bond to the substrate. The uncured filler around the outer perimeter never fully hardens or attaches to the surface. When it is sanded, the edges merely break away rather than feathering to a fine taper. In some rare instances, extending the cure time will allow for proper feathering. However, if this happens additional problems may be discovered later in the repair.

Repair Mapping

Occasionally after the plastic has been finish sanded and the primers, topcoat, and clearcoats have been applied over the repair area, a slight texture outlining the areas covered with plastic appears in the surface. This situation, known as **repair mapping**, is usually the result of the solvents in the refinishing materials penetrating through the thin edge of the plastic that may be marginally bonded to the surface, softening it and causing it to lift. In extreme cases the repair may need to be redone.

Staining/Bleaching

Staining is a slight to moderate discoloration that may appear in the surface after the topcoat and clearcoat have been applied. This may occur immediately upon completion of the refinishing or it may not occur until after a period of time passes, possibly as long as 6 months. Staining is caused when peroxide in the catalyst is not fully used up to cure the plastic reacting with the solvents and other ingredients in the paint materials, which causes a discoloration.

Likewise, similar problems such as those outlined below may occur if the plastic is undercatalyzed. In order for the plastic to through-cure, a sufficient amount of catalyst must be present to react with all of resin in the mix. In order for the material to properly adhere to the substrate, it must cure and become hardened. Unless there is enough catalyst to react with all of the plastic filler and resin, it will not properly cure and become hard. Any attempt to recoat with additional applications of filler over the top of the improperly cured layer will eventually result in the entire surface delaminating and peeling. If this **delamination** and peeling should happen, the uncured layer of filler must be removed by whatever means is available.

PLASTIC APPLICATION

The technician now must determine the amount of material required to cover the repair area with the first coat. This requires practice—one should not expect to expertly dispense the required amount of material during the first several attempts made. For the first application, one should dispense enough filler material to cover all the indented and low areas and to apply a thin glaze coat over most of the surface from which the surface coating has been removed. The preferred procedure is to fill the low areas first, then follow immediately with a glaze coating over the remaining surface. The technician must exercise caution when applying the glaze coat to not remove the filler from the low areas, as the material is still soft and can easily be removed with the applicator if too much pressure is applied. Depending on the size and shape of the surface, two or more applications of filler may be required to properly fill the area and correctly shape and sand the surface smooth.

Flat Surfaces

Applying plastic filler is said to be an art form that requires considerable practice. Occasionally it will try the patience of even the most learned and experienced technician. However, modern day vehicles are manufactured using many different-shaped surfaces with a combination of crowns and configurations that require skill and adaptability to duplicate. The most common surfaces encountered include a flat or low crowned surface, a reverse and high crown, and a diminishing crown. In many cases a combination of these crowns are adjacent to each other and must be repaired and finished with filler. The experienced technician will formulate a plan of attack for applying the filler before even dispensing and mixing the material onto the mixing board or palette.

TECH TIP When applying plastic filler to any surface, the shape and configuration of the peak crown must always be established before any of the adjoining areas are filled and finished.

Establish Crowned Surfaces

A basic understanding of the techniques and sequence used for applying filler will help eliminate many of the pitfalls commonly encountered by the novice technician. As is the case when straightening and repairing collision damaged sheet metal, the more defined crown surfaces such as the high and reverse crowns must be restored first. Likewise, when applying filler, particularly over a larger area, the most vividly defined crowns should be established first. A good rule of thumb is to start by restoring the surface which extends the farthest outward and then address all the adjoining areas. See **Figure 8-13**. After grating and rough and finish sanding the crowned surface, the adjoining areas may be covered with filler. To protect and maintain the shape of the crowned surface, one can cover its peak with tape along the formed edge until the next application is completed. The tape is then pulled off while it is still creamy or before it hardens. See **Figure 8-14**. Once the filler hardens it can be grated, sanded, and finished. Care

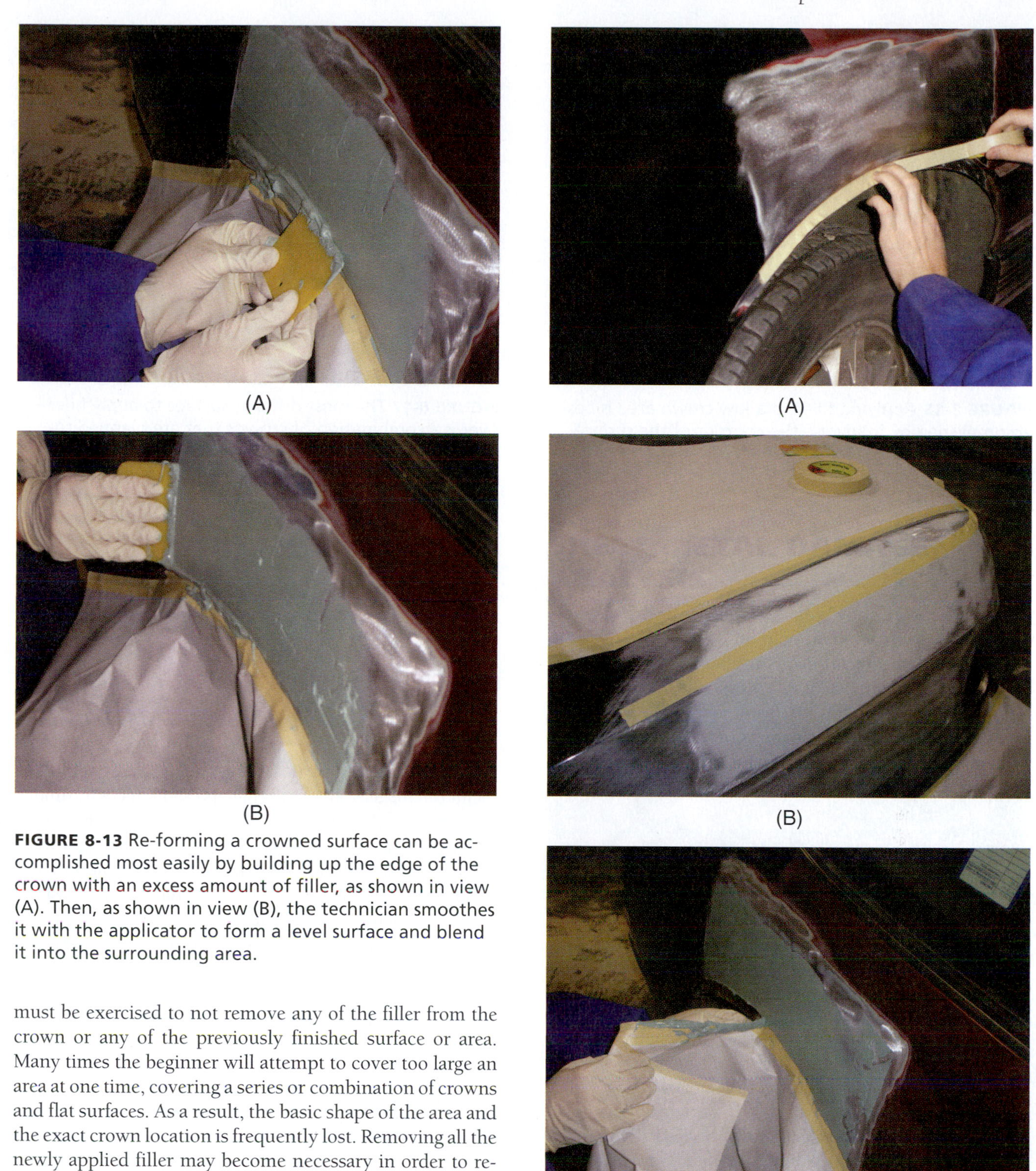

FIGURE 8-13 Re-forming a crowned surface can be accomplished most easily by building up the edge of the crown with an excess amount of filler, as shown in view (A). Then, as shown in view (B), the technician smoothes it with the applicator to form a level surface and blend it into the surrounding area.

FIGURE 8-14 Re-creating a high or sharply formed crown can be accomplished more easily by taping along the edge to be formed as in views (A) and (B). The filler should be applied by building it up along the edge of the tape line and blending it into the adjoining area and, as in view (C), removing the tape before the plastic cures and hardens.

must be exercised to not remove any of the filler from the crown or any of the previously finished surface or area. Many times the beginner will attempt to cover too large an area at one time, covering a series or combination of crowns and flat surfaces. As a result, the basic shape of the area and the exact crown location is frequently lost. Removing all the newly applied filler may become necessary in order to re-establish proper configuration and the crown locations.

Low Crown Surfaces

When applying filler to a low crowned panel it is advisable to follow the contour moving from either the top to the bottom of the area being filled or from the bottom to the top with the applicator. See Figure 8-15. This will ensure the applicator will make contact with the panel across its entire width, thus depositing a uniform thickness of material with

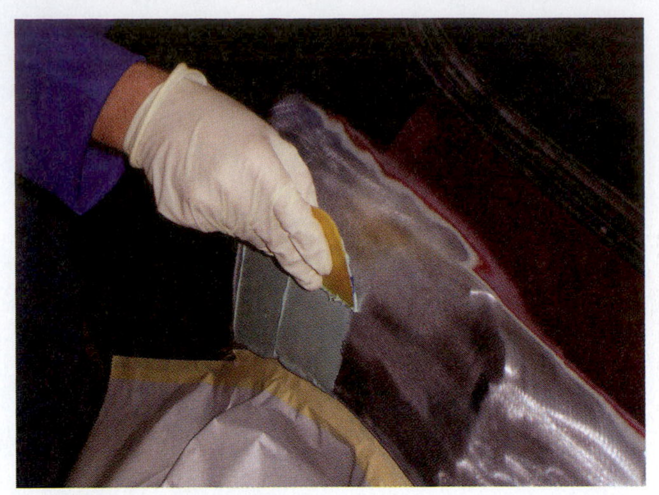

FIGURE 8-15 Applying filler to a low crown area is best accomplished by following the contour of the surface with the applicator moving from the bottom up or from the top down—whichever method is most easily accomplished.

FIGURE 8-16 Attempting to apply filler to a low crown surface by following the wrong contour will result in numerous lines and flat spots that will require additional applications and sanding in between each fill.

each pass. The technician should not attempt to apply the material by moving the applicator laterally along the length of the panel, as only a narrow section of the applicator will contact the surface. This will result in narrow lines of unevenly applied material on the surface and require unnecessary future fills. See **Figure 8-16**.

Combination and Reverse Crown Surfaces

Many times a combination of a slight reverse crown and a low or high crown will be found immediately adjacent to each other. It may be advisable to fill and finish these surfaces simultaneously as they flow and are blended into

FIGURE 8-17 The most difficult surface to apply filler over is a combination of crowns that are blended together. One example is the surface above where a high crown at the top blends into a reverse surface and ends in a low crown at the bottom. An irregular surface such as this may require three or more filler applications to accomplish the task.

each other. However, if the reverse crown is relatively sharply formed, it may be wise to repair it independently of the adjoining areas. See **Figure 8-17**.

Most experienced technicians still resort to some trial and error approaches and experiment with different methods when they are confronted with a new or unique surface design. There is no sacred or foolproof method or technique to use, as each repair presents its own unique challenges to the repair technician. The approach or technique that was successful on the last repair may not work on the next vehicle. Technicians must be flexible and able to adapt their approach and plan of attack to successfully overcome each unique challenge.

The application of the plastic filler is only the first in a series of many steps that must be followed to ultimately restore the damaged surface to where the repair is undetectable by even a professional. See **Figure 8-18**.

Filler Removal

PLASTIC, ETC.

Cheese Grate	
Rough Sand	40D
Finish Sand	80D

GLAZE COATS (Optional)

Rough Sand	80D
Final Sand	150

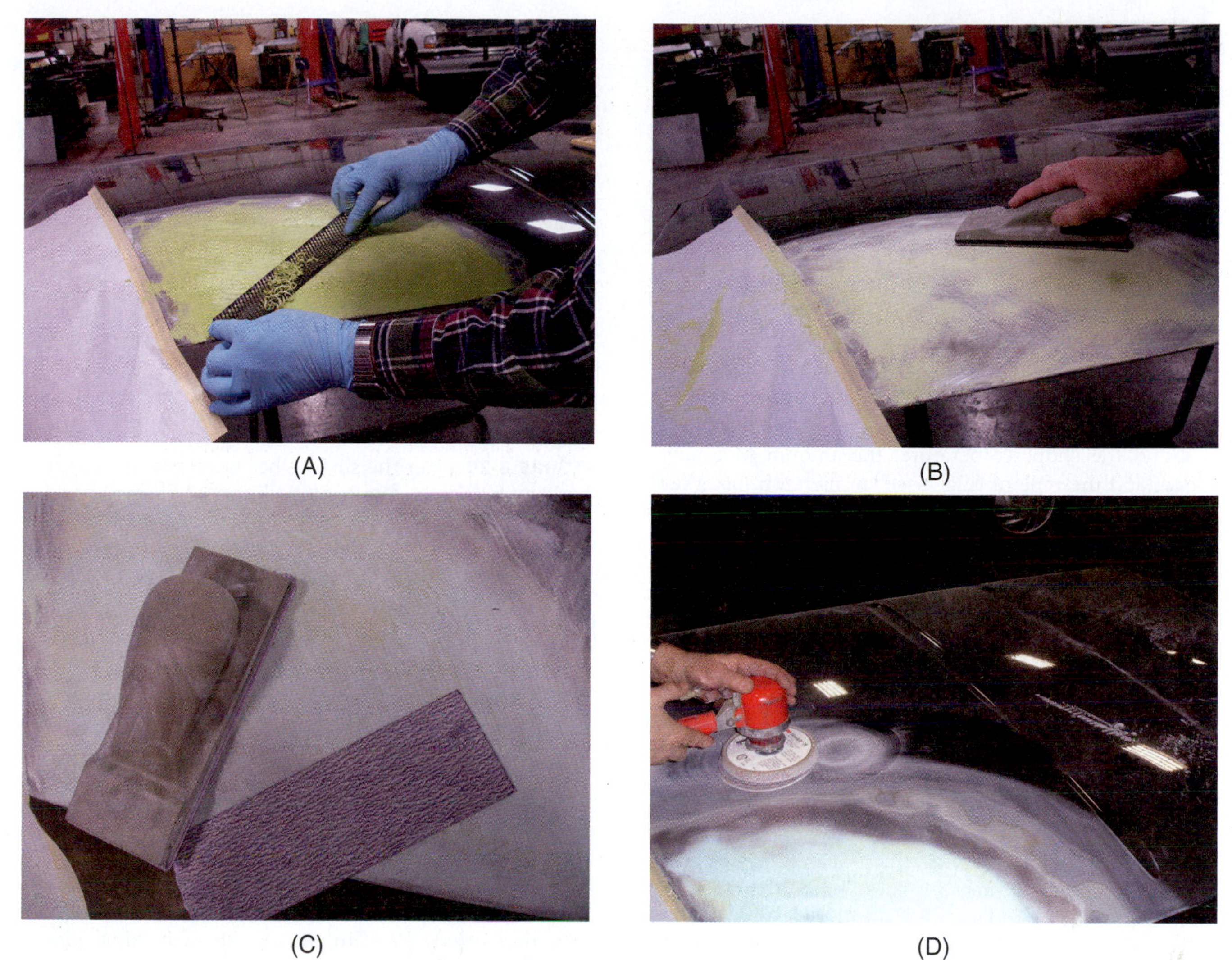

(A) (B) (C) (D)

FIGURE 8-18 Finishing the plastic filled surface is accomplished in several stages. The first step is to grate the filler to remove the heavy deposits of excess material (A). This is followed by sanding with a coarse grade abrasive and staging from a 40 grade up to an 80 grade abrasive (B and C). The glaze coat is finished using an 80 grade followed by a 150 grade abrasive (D).

FILLER REMOVAL, SHAPING, AND SANDING

Grating

After the technician applies the filler, the finishing sequence incorporates a series of several steps. The first is to "cheese grate" the filler, which is commonly called grating. It gets this name because a surform plane, the tool commonly used to remove the first layer of the partially cured filler, removes the filler in strands similar to when a person is grating cheese. See **Figure 8-19**. The grating step removes the surface irregularities left by the applicator and excess material, which helps to level the filler to a smooth plane.

FIGURE 8-19 Grating the filler is done with the use of a "cheese grater" or surform file to remove the excess filler and start shaping the material on the surface.

TECH TIP When **grating plastic**, one must not become overzealous and remove too much material from the surface. The plastic is soft and is removed very easily. An inexperienced technician is apt to quickly remove too much material if the proper precautions are not taken.

Grating Time

The length of time one must allow the filler to cure before the grating can be started is dependent on a number of variables, including the temperature of the repair surface and the shop temperature, the amount of catalyst added, and the type of filler used, to mention just a few. When using a premium filler, these materials are tack free and sand easily enough that grating to remove the excess material may not be necessary. However, most beginners have a difficult time applying the material smoothly enough to eliminate this step. In the interest of expedience, it may be advisable and necessary to grate the material to minimize the time required to finish it properly.

A constant rule while working with plastic is this: The warmer the temperature of the shop and the surface, the faster the cure rate. To determine if the material has cured sufficiently for grating, the technician should try digging a fingernail into it. If the surface is formed and semi-hard but can still be scratched with a fingernail, it may be ready to grate. However, if grating is started prematurely it will literally pull the material off the substrate, as the material isn't properly anchored to the surface.

TECH TIP One must not start grating the filler prematurely as it will pull the material off the substrate, as the material isn't properly anchored to the surface until it is thoroughly cured.

After grating the surface, any low areas requiring additional filler will become evident as the grater will have skipped over them, leaving a different color and texture. See **Figure 8-20**. One should stop the grating at precisely the point when the first signs of the substrate start to become visible. Further removal of additional material will necessitate re-filling again. Rough sanding the filler to feather the outer edges prematurely may prevent an excessive material buildup, but it will require additional sanding at a later time. Sanding may further identify areas where additional filler is required. When the next layer of plastic is applied, the low areas identified by the grating and sanding should be filled first and then the entire surface glaze coated once again. It may seem like a waste of material and effort to recoat the entire area rather than just the low areas. Rest assured, it is in the technician's best interest to do so because sanding the filler off what were the low areas will also remove filler off the adjacent areas. The result will be a low area on either side of the area where the filler was applied and sanded most recently. See **Figure 8-21**.

FIGURE 8-20 After the surface has been grated, any remaining low areas requiring additional filler will be revealed, as they will have been skipped over by the grater, leaving a crater-like sunken area.

Rough Sanding

The technician should use a coarse grade of **production sandpaper** to remove any uncured filler which was not removed with the grater from the low lying areas. This ensures proper adhesion of the next coat of plastic applied over these areas. Paraffin, an ingredient in most plastic fillers, is added to protect the uncured surface from oxygen in the atmosphere while it is curing. As the filler cures, the paraffin migrates to the surface and remains there until it is either grated or sanded off. Applying a subsequent layer of plastic without first removing this layer of uncured material would compromise its adhesion to the previous coat or layer.

Restoring the damaged surface to its pre-collision condition may require two, three, or even more applications of plastic filler. Each application should be followed with the appropriate amount of grating and finish sanding until the surface is restored to its proper configuration.

PLASTIC FILLER SANDING TOOLS

The final shaping and finishing of the plastic is usually accomplished using a variety of mechanical air sanders, hand-held sanding boards, and backing pads covered with various grades of production sandpaper. See **Figure 8-22**. Because the plastic is relatively hard once cured, it is highly advisable to use a firm to hard backing to support the sandpaper whenever attempting to sand an unfinished surface smooth. Using a firm backing pad ensures

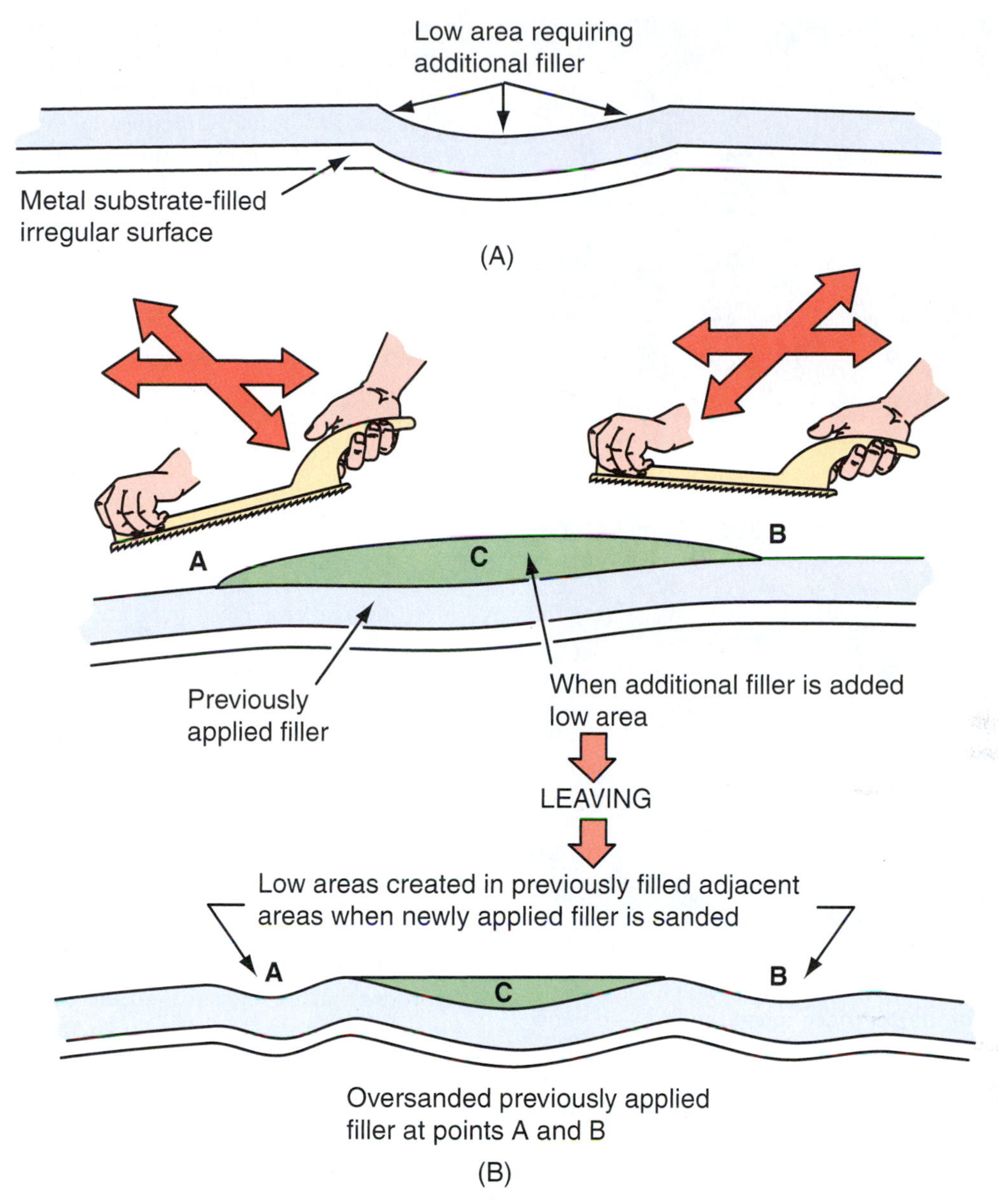

FIGURE 8-21 When filling the low areas identified while grating, the entire previously filled surface area should also be covered with a skim coat. In view A, sanding the filler if applied over the low areas only will have some disappointing results. As shown in view B, filler material will also be sanded away from the adjacent areas, creating low areas at points A and B while attempting to remove the newly applied filler at point C.

that the sandpaper is held on a firm, flat surface, thus removing the filler off the high areas and leaving it in the low areas until the surface is completely flat or matches the intended contour of the panel. Any attempt to smooth the surface without using a firm backing, such as holding the sandpaper by hand or using a soft or flexible object in its place, will yield some pretty undesirable and unsatisfactory results. When a soft or flexible backing pad is used, the sandpaper will conform to the surface irregularities and, as a result, the surface will never be sanded smooth.

The rough and final forming of the plastic filler is commonly accomplished with production grades of sandpaper. Selecting the abrasive grade, or grit as it is commonly called, is often based on personal preference.

Rough Sanding and Forming Filler

Most plastic filler and abrasive manufacturers recommend the use of either a 36 or 40 grade production sandpaper for the initial stages of sanding. (Refer to the abrasive grade chart in the Tech Tip on page 210 for specific recommendations.) Sandpaper should be used only after the surface has been grated, as the paraffin in the filler causes the outermost layer of the plastic to remain

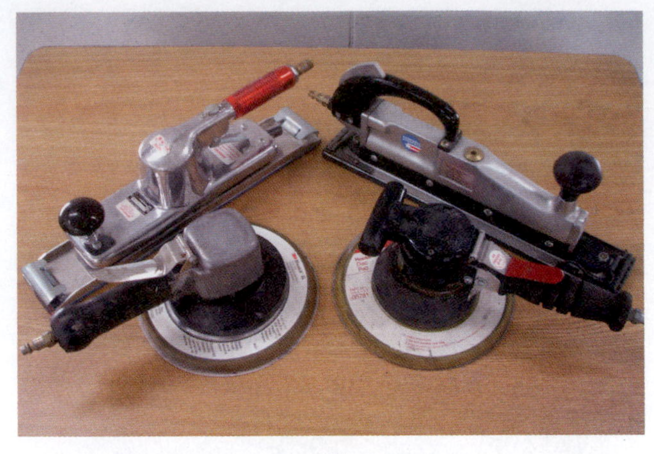

(A)

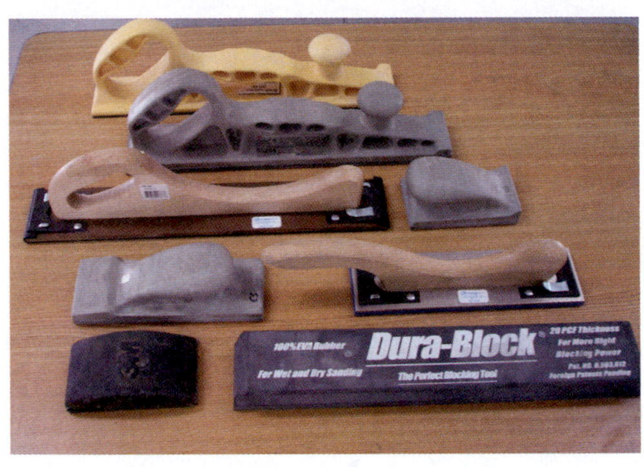

(B)

FIGURE 8-22 A number of mechanical sanders and hand sanding boards are available to finish sanding the filler. View A shows several oscillating straightline and circular sanders, and view B shows a number of hand sanding boards of varying lengths.

uncured and soft. Sanding this outer layer of material will load up the sandpaper, rendering it useless.

After the filler has been shaped to match the contour of the adjacent areas, it is advisable to change to perhaps an 80 grade sandpaper. This will help eliminate the coarse sandscratches from the production paper and also help to create a smoother featheredge at the outer perimeter of the plastic filler. The point where a technician makes the switch to the finer grade paper is subject to personal preference. Keep in mind that making the change prematurely will require additional time to remove the remaining excess material. However, waiting too long will likely result in oversanding the filler, which may necessitate an additional application of filler to bring it up to the level of the adjoining surface. See **Figure 8-23**.

Finish Sanding

Many times minor irregularities and slight voids remain in the surface of the plastic after it would otherwise be finished, feathered, and ready for priming. At this point the technician may select one of several available glazing fillers. See **Figure 8-24**. A two part glazing putty, a product very similar to plastic filler, may be used to fill these minor imperfections and any remaining coarse grade sandscratches. This product should be applied in a micro-thin layer, cover the entire plastic repair area, and extend to nearly the outer perimeter where the topcoat was first removed. This will ensure that any minute flaws and surface imperfections that may have been overlooked, as well as slightly oversanded edges, are filled and ready for priming once the material has been finish sanded. Grating the glaze coat is not recommended as too much material is apt to be removed. The initial sanding should be done with 80 grit. A staging up to a 150 or finer grade is recommended for finish sanding. A finer grade paper, such as 180 or 220 grade, may be used if so desired. However, keep this in mind: The smaller the sandscratches that are left in the surface, the less likelihood of them showing through the finished repair. Using a firm backing for sanding this material—such as a sanding board—is recommended as well.

The ultimate goal when finished with the plastic is to restore the exact shape and configuration of the surface so it looks the same as before it became damaged. Several attempts at filling may be required to accomplish what may seem to be a simple task. However, even the most experienced technician occasionally has difficulty accomplishing this task with only one or two applications of filler. The beginner should not become discouraged if it takes several attempts to restore the original contours.

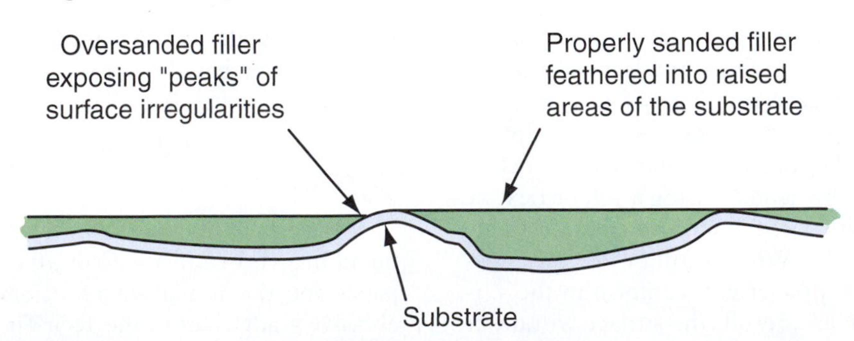

FIGURE 8-23 On the left above, the oversanded filler appears as a slightly low area next to a bare metal spot showing through the filler. On the right, the properly feathered filler ends at the precise point where the peak of the crown appears. The filler blends into it, forming an uninterrupted smooth surface.

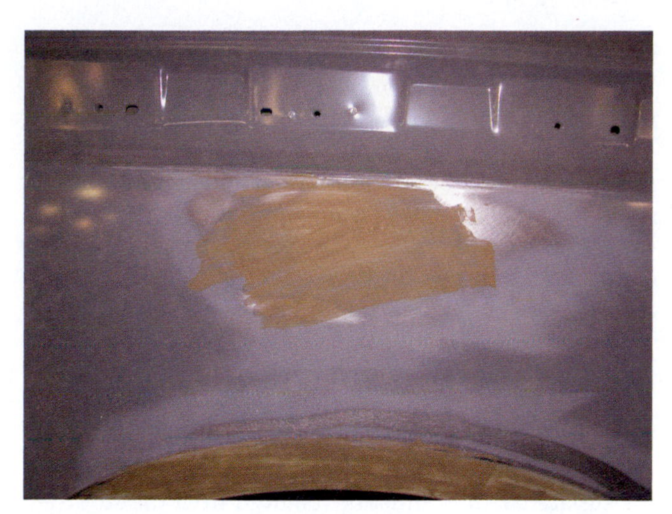

FIGURE 8-24 Even after finish sanding the plastic filler, some minor imperfections and deep sandscratches may remain in the surface. A micro-thin layer of polyester glazing compound or putty can be applied over the entire repair area to fill any surface flaws and imperfections that may still be in the surface.

Perimeter Sanding

After the plastic filler and the finish glaze coats have been finish sanded, it is necessary to featheredge all the sandscratches in the painted surface around the entire perimeter of the repaired area. See **Figure 8-25**.

The most expedient way to accomplish this is to use an 80 grade sandpaper on a dual action sander to remove the most severe scratches, which are usually found immediately on the outer edge of the repair. After the coarse scratches have been sanded out, the technician should select a finer grade such as a 180 or 220 grade, followed by a 320 grade on the dual action sander to finish the feathering process. When feathering, at least a 3/8-inch taper should be visible for every layer of paint material on the surface. This will eliminate any possibility that visible signs of the repair will remain after the basecoats are all sanded and the topcoat has been applied. A more in-depth study of abrasives and surface preparation will be covered in Chapter 25.

Rough feathering	80D on DA Sander
Fine feathering	150 on DA Sander
Finish feathering	220 on DA Sander
Perimeter sanding	320 on DA or Hand Sander

FIGURE 8-25 After all the filler has been finished, the final step is featheredging the sandscratches from the painted surface surrounding the repair area. This is done in a series of steps starting with an 80 grade abrasive and following the stages shown in the chart above.

DO'S AND DON'TS ABOUT PLASTIC FILLER

Many times an individual will unknowingly use plastic filler in a manner not recommended by the manufacturer. This often results in a variety of problems and material failures that are inappropriately blamed on the product. Knowing the manufacturer's recommendations for the safe use of any product should be a prerequisite to its use. The following are several do's and don'ts that one should follow to avoid any unnecessary problems.

- Surface temperature. Always ensure the surface to which plastic is being applied is at least 70°F. It will react best between the temperatures of 70°F to 90°F. It may be advisable to warm the surface with a heat lamp prior to applying the filler, as it will not cure properly if the surface is too cold. Caution should be taken to not overheat the surface, however, as it may affect the adhesion of the material as well.
- Proper catalyzing. Never overcatalyze the plastic to ensure a proper cure when working in cold shop conditions. On the surface, the filler may seem to cure properly, but numerous problems such as staining, bleaching, or adhesion loss may result at a later time.
- Rotate stock. Always rotate the stock in storage by inverting the cans and storage containers at least monthly. This helps to prevent the resin from separating from the filling ingredients.
- Ensure material is uniformly mixed. When dispensing the filler out of a can, always mix the material thoroughly before removing any of its contents to ensure a proper mix of material. Once the container has been opened, mix or stir the plastic in the storage container at routine intervals and never allow the resin to separate from the talc and other materials.
- Ensure proper filler consistency. Frequently the material in the bottom of a nearly empty container appears to be dry and lacking enough resin to properly mix or spread smoothly. One can add plastic honey, a special resin extender, to revitalize

the remaining filler. One should never compensate for a dry mix by adding additional catalyst to liquefy the material when mixing it. Never use the same tool that has been used for mixing plastic to remove additional filler from the container unless it has been properly cleaned. The presence of unused catalyst on the tool may cause clumps of filler to form in the container and contaminate the remaining contents.

- Keep the container tightly sealed. When dispensing the filler from a can, replace the cover on the container immediately to prevent the evaporative solvents from escaping into the atmosphere. The evaporation of these solvents will result in an unbalanced mix in the material that is left in the can. This can cause many of the problems—such as adhesion loss, improper curing, and so on—discussed earlier in the chapter. In extreme cases, the material may dry out to the point where it must be thrown away.
- Select the proper material for the substrate. One should not automatically assume that all products are made to be used on any substrate. Because a technician uses one product made for a specific application doesn't automatically mean similar products can be used for the same purpose. Some products are made specifically for use on galvanized coated steel and aluminum while others bearing the same name are not intended for these applications.
- Read label instructions. All manufacturers include printed instructions on the label of the can. IT IS NOT AN INSULT TO ONE'S INTELLIGENCE TO READ THE DIRECTIONS FOR USING A PRODUCT!

SPECIAL REINFORCED FILLERS

Despite their widespread use for finishing collision damaged panels, many of the plastic fillers have an inherent weakness for absorbing moisture. Many times repairs must be made on areas of the vehicle that have deteriorated due to rust and long-term exposure to the elements. These areas are often suspect for the repair shop because microscopic holes not visible to the eye may exist throughout the entire area. Sometimes panels that have rusted away or been completely perforated by rust or **oxidation** must also be repaired. Unless these microscopic pores are completely sealed during the repair, moisture penetrates through the metal and is absorbed by the plastic filler, ultimately causing the repair to fail prematurely. Even if only one small hole is left unnoticed when a technician finishes welding a replacement panel into place, it can still result in a repair failure. A waterproof filler should be used for repairs where these conditions may exist.

Moisture Barrier

Most filler manufacturers offer a variety of fiberglass-reinforced and aluminum-filled products that are specifically designed for these types of repairs. These materials are water resistant because they do not incorporate talc or any other moisture-absorbing ingredients. This makes them a popular filler choice in areas where rust and oxidation are suspected. Even though many of these products are not as easily finished and feathered as is plastic filler, most can be used interchangeably. Most plastic fillers, glazing compounds, and putties can be applied directly over them. This makes finishing the repair much quicker, as the finish sanding can be done on the glazing putty as opposed to the fiberglass. It should also be noted that all the surface coatings must be removed from the substrate prior to applying any of these materials.

Fiberglass-Reinforced Filler

The fiberglass-reinforced fillers are made with various lengths of fiberglass strands suspended in a polyester resin. See **Figure 8-26**.

Short Stranded Reinforced Filler. The short stranded filler—or chopped filler, as it is sometimes called—is used in the same manner as plastic filler. It is a creamy mixture that can be applied as smoothly as plastic filler and should be used to cover the entire repair area. However, unlike plastic filler, this product should not be grated when it starts to gel. It should be allowed to cure or harden completely and then sanded and feathered with a sanding board using the appropriate grade of production paper. After the surface has been sanded and featheredged properly, it can be finished with a glaze coat of a two part poly-

FIGURE 8-26 The long stranded reinforced fiberglass filler is somewhat water resistant and suitable for bridging small holes and gaps in the surface, making it a popular choice for making temporary rust repairs. The chopped reinforced fiberglass filler has a consistency similar to plastic and is also water resistant, making it a popular choice for waterproofing welded seams.

ester putty. This surface is then ready for the primers and finish coats.

Long Stranded Reinforced Filler. A long stranded fiberglass product known industry-wide as Tiger Hair is widely used for repairing vehicles with panels that have completely rusted through. This product is best known for its ability to bridge across a hole between the edges of an area that has completely rusted away. While this is not the recommended procedure for repairing rust-perforated panels, it is occasionally used as a "quick fix" by used car lots and do- it-yourselfers. Even though the fiberglass offers a moisture barrier to prevent further deterioration, if the rust is not removed from the surrounding areas, it will continue to grow. The repair rarely lasts for longer than a year.

Thixotropic Resin

Using a fiberglass cloth soaked with a **thixotropic resin** is yet another system technicians use for bridging and repairing areas that may have rusted out. One should not assume that since manufacturers make these products available that this is the recommended or an acceptable procedure to use. There is no substitute for removing the rust or oxidation from the affected areas and replacing it with fresh new metal. This is the only way one can accomplish a legitimate repair that will last the vehicle's lifetime.

Metallic Clad Fillers

The aluminum clad body repair fillers are also very effective in areas where moisture penetration and absorption are a concern. See **Figure 8-27**. These filler materials are formulated with a very fine aluminum powder or flakes suspended in a polyester resin instead of talc. The aluminum powder is waterproof, corrosion resistant, and has become a very popular choice for filler with classic car enthusiasts, as it is known to be the next best thing to lead-based solder. In the event the paint or topcoat should get scratched down to the substrate, it appears to be solder covering the surface rather than plastic and it won't turn the rust color as is often the case after prolonged exposure. The use of this product in areas of the vehicle that are subjected to constant abrasion and road splash—such as wheel openings and rocker panels—is a wise choice. The paint is frequently chipped or abrated off these areas of the vehicle because of the constant harsh road exposure. Using this product for a filler to repair these areas offers the metal additional protection, as it is harder than standard plastic filler. It will also appear to be metal exposed rather than the unsightly rust color that would otherwise appear.

FIGURE 8-27 Fillers containing metal particles for their filling ingredients as opposed to talc are a good choice to use in areas where surface abrasion from road splash and other intruding objects are likely to occur. These materials are also relatively water resistant, making them a good choice for filling on nearly any part of the vehicle.

This product is unique from most of the other fillers in that a liquid hardener is used instead of the cream hardener/catalyst that is used with plastic fillers. The manufacturer's recommendations for catalyzing must be followed precisely as the liquid is not visible once it is dispensed onto the filler and the quantity added is difficult to judge. It may be advisable to dispense the filler into an unwaxed paper cup, then add the catalyst and stir it in this container to avoid losing or splashing any of it out of the mix. Once the material is completely mixed, paddle it out of the container onto the mixing board—folding it to eliminate any bubbles that may have formed when stirring it in the cup—and apply it to the surface.

With any repairs where oxidation, moisture penetration, or absorption is a concern, the surface must be properly cleaned prior to applying the filler. In addition, the entire surface area that is suspected of being perforated must be coated with the waterproof filler. In order to prevent a leaching effect underneath the repaired area, it is necessary for the technician to apply a thick enough layer of material to ensure the moisture cannot penetrate through it after it has been finish sanded. In some instances it may even be necessary to slightly indent the surface to ensure a thick enough layer of material is applied.

FOAM FILLERS

For many years auto manufacturers have used foams and other fillers for preventing water leaks, eliminating wind noises, filling seams where panels overlap, and filling unsightly gaps. In recent years the use of foam fillers has taken on a whole new function in the automobile design and manufacturing industry, as they are being used to enhance and improve the vehicle's structural integrity. While foams are still being used for nonstructural functions—such as improving the aesthetics of the vehicle and to reduce road noise—their structural functions include reinforcing certain parts of the vehicle to strengthen and improve their crash worthiness. The foam fillers currently used by the manufacturers are identified as nonstructural and structural. According to ICAR, these foam fillers are used for the following reasons:

- Reduce **noise, vibration, and harshness (NVH)**
- Eliminate water, dust, and wind leaks or air noise
- Stiffen the body structure and control flexing of the vehicle
- Provide collision energy management
- Provide additional crash protection and help maximize occupant safety

CAUTION

Many of the chemicals used in the automotive foams are known hazardous materials. It is imperative to consult the MSDS for each product prior to its use to determine the PPE required for its safe handling.

Structural and Nonstructural Characteristics

It is sometimes difficult to differentiate or identify the specific foam used by its outward appearance. However, one can differentiate between a structural and nonstructural NVH foam by checking the material's flexibility, density, and hardness. The NVH foams are identified as nonstructural material. Depending where they are used, they may be either very flexible or relatively rigid. The **NVH flexible foam** is usually black in color and very porous or open celled. When squeezed it will not deform permanently but will return to near its original shape. The **NVH rigid foam** is less porous but will deform permanently when depressed. The **structural foams** are commonly used to enhance the vehicle's structural integrity and crash worthiness. They are very dense, closed cell epoxy base materials and are very brittle and compression resistant. They are likely to break or crack when flexed or bent. The rigid NVH and structural foams exhibit many similar outward characteristics. Therefore, one should be cautious to ensure that the rigid NVH foam is never substituted for structural foam as it could seriously compromise the vehicle's crash worthiness.

Nonstructural Foams

A large percentage of the American workforce spends a significant amount of time in their automobiles. As a result, auto manufacturers have tried to enhance and improve the creature comforts of these automobiles by incorporating a variety of features aimed at the driver's and passenger's well-being. One of the most significant features a would-be car buyer is concerned about is reduced NVH levels when driving down the road. The NVH is caused by a number of factors, including the texture of the roadway or driving surface, airflow around the cabin of the car, vehicle design, and sometimes normal wear and tear on the vehicle. By filling select cavities and hollow voids between panels with a soft expandable urethane foam, auto manufacturers are able to reduce panel vibration and absorb and block the sound waves from entering the passenger compartment. The location and placement of these foam materials vary with different manufacturers and vehicle models. As indicated in **Figure 8-28**, the NVH foams may be used between roof skin and the cross-beams, between the door skin and intrusion beam, and in the quarter panel area to fill any large voids and pockets, as well as areas where water and dust might enter the vehicle. They may also be found in the dog leg area, rocker panel, and inside the upper rails and pillars. One can never be certain where they are located. Therefore, it is advisable to consult the vehicle maker service manual or technical service bulletin for their exact location prior to attempting to disassemble a vehicle, ap-

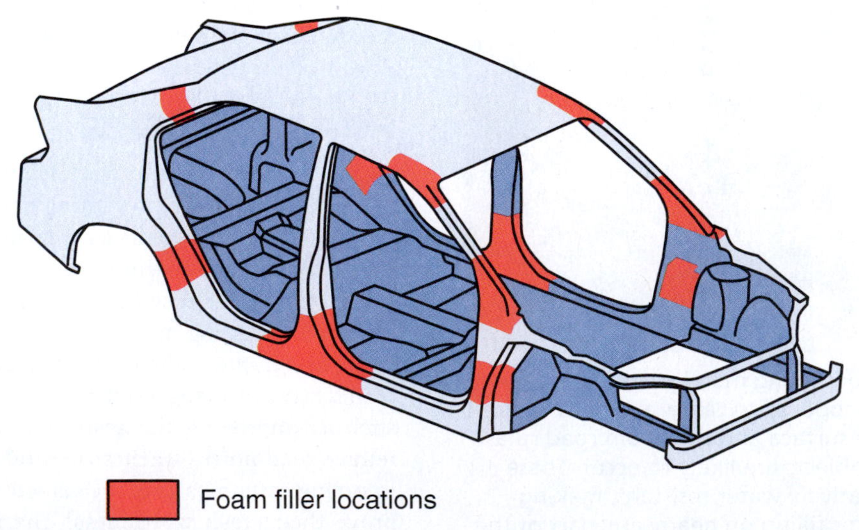

FIGURE 8-28 Nonstructural foams are used in various parts of the vehicle to reduce noise, vibration, and harshness (NVH) and to help keep water and dust out as well.

ply any heat, or weld on a damaged section, as their use may differ with each vehicle model.

CAUTION

One should never apply an open flame to any area where foam may be used as it is very flammable and will give off toxic fumes, including carbon monoxide and cyanide gas.

These materials are very heat sensitive and will melt or burn at temperatures well below the **sensitive heat range** of all steel or aluminum parts used on the vehicle.

Soft Flexible Foam. Two distinct different methods or materials are used for reducing the NVH in a vehicle. The most prevalent is the soft flexible foam, which—because of its open cell construction—is the most effective material for absorbing, blocking, and reducing sound waves. This material is extremely flexible but offers no strength or reinforcing features and is most commonly used to fill larger gaps such as in the areas between the quarter panel and gas filler pocket, and the roof rails where the most flex occurs in the vehicle's body. When removing a panel where foam is present, caution should be exercised to avoid destroying the material as it can be re-used by adding an adequate amount to re-fill the gap when the replacement panel is installed. A hot air gun or induction heater should be used to soften and release the foam from the damaged part, making removal much quicker and easier. It also minimizes the damage and distortion of it.

Rigid Foam. The rigid NVH foam is not only used to absorb and block sound waves, but it is also used to cushion and control the movement of adjacent parts and body panels by preventing them from contacting each other, causing vibrations. They are not used to improve the structural integrity of the vehicle, but they are used to stiffen the vehicle by reducing the twisting and flexing as it goes down the road. Depending on the model of the vehicle, they are used in many of the same areas as is the flexible foam. (See Figure 8-28.) In order to determine the exact location of application, it is necessary to consult the vehicle maker service manual or technical service bulletin. The same precautions must be taken with the rigid NVH foam when applying heat in and around the repair area because—like the flexible foams—it is very flammable and easily melted or ignited with a direct flame. All foams should be removed from the repair area before attempting to make any repairs.

NVH rigid foam is brittle, inflexible, and will deform permanently when it is struck or compressed by some means. It has a very limited ability to be deformed and will crack or break when bent or distorted. NVH rigid foam can be differentiated from the structural foam because when cut one will see a fine cellular structure not characteristic of the others. It is typically white or yellow in color. See **Figure 8-29**.

FIGURE 8-29 The purpose of structural and rigid foams should not be confused. The structural foam on the left is usually a hard, smooth, almost glossy and nonporous material. The rigid foam on the right is hard yet very porous material.

Foam Composition. The NVH rigid foam is a two-part urethane material which is applied or installed with a dispensing gun specifically designed by the product manufacturer. Both the flexible and rigid NVH foams require special dispensing equipment and **static mixing tubes** for their application. Before attempting to use the material, one should consult the technical service bulletin to determine the specifics for its use, as the flow and expansion rate vary from one product to another. Installing the foam at the incorrect location may result in only partially filling the cavity, resulting in noises, vibrations, and compromised vehicle integrity.

Most manufacturers also utilize formed foam parts to absorb collision energy. The best example is the foam impact absorber on both the front and rear bumper of many vehicles. Many of these parts can be repaired; however, the manufacturer's specific recommendations and requirements must be ascertained prior to attempting to do so. Some manufacturers also install preformed rigid and flexible foam blocks called stuffers into strategic locations. These are removable and replaceable service parts that are bolted or clipped into position or they may simply be stuffed into a cavity inside the door assembly, quarter panel, or nearly any other part of the car.

Structural Foams

Structural foam is currently used by many automobile manufacturing companies to improve the crash worthiness of the vehicle by reinforcing certain structural parts and redirecting the collision energy around the passenger compartment. The structural foams are most commonly used for reinforcing the A and D pillars, but are also used to strengthen the front lower rails and full frame torque box areas on select vehicles. One must be cognizant of their location prior to initiating a repair, especially when heat is applied. Most of the foams will melt or start to burn at temperatures well below the safe heating range of even the most temperature-sensitive steel parts. The automobile manufacturers have found that structural foam is very effective for reducing the twisting and flexing of an area of the automobile. The foam also significantly enhances the vehicle's strength and rigidity with the addition of considerably less weight than with the use of additional steel. Selecting and re-installing the correct foam is vital to restoring the vehicle's structural integrity to its pre-collision condition. Learning to identify them and recognizing their characteristics is a must for the successful restoration of collision damaged vehicles.

Characteristics of Structural Foam. While not totally unique from other foams commonly used by auto manufacturers, structural foams exhibit some characteristics that are distinctly different from the others. The structural foams are typically gray or orange in color whereas the others are black, yellow, or white. Their cell structure is either very small or non-existent, and the material looks like a hard, solid mass of plastic. The foam is very hard and brittle and will break when a technician attempts to pierce it or when it is bent too far. The technician should consult the vehicle maker service manual, technical service bulletin, the service procedures guide, or perhaps even the product makers hotline if not certain about the type of material to use. The location of its use on the vehicle should also be a tipoff to the technician to determine its composition. See **Figure 8-30**. In the interest of saving time and eliminating a lot of guess work each time a vehicle is repaired that requires foam re-installation, the shop might obtain a kit of all the adhesives most commonly used and make a sample of each. This could be accomplished by dispensing some of each foam material into a paper cup, letting it cure, and cutting it in half with a saw. The samples could then be properly identified and mounted on a board with the cut side facing out. Whenever a foam has to be matched, the technician simply compares the material found in the vehicle to the samples on the board. One can quickly compare the compression characteristics, the texture, the cell structure, and the density of the material on the vehicle against the material on the board and select the one which best matches the original.

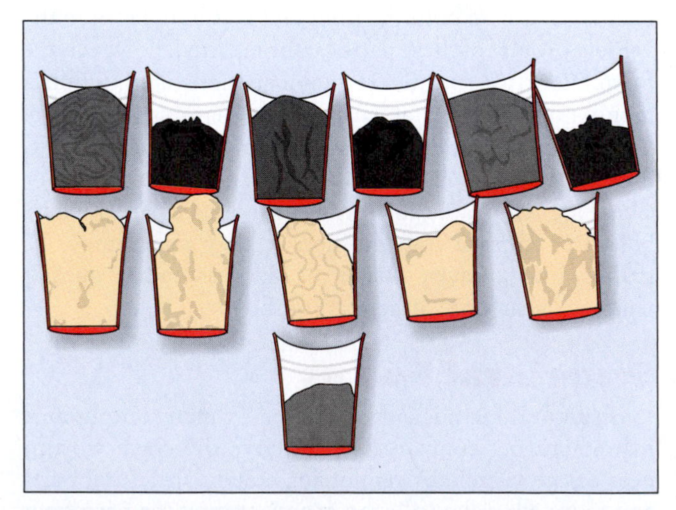

FIGURE 8-30 In this figure a sample of each of the structural and nonstructural foams has been mixed and dispensed into a glass that has been cut in half. This will help the user select the correct foam by comparing the foam in the vehicle against the foam in the sample.

SELECTING THE CORRECT FOAM

After selecting the best foam product for a particular application, the technician must consider several other factors. The amount of material required can be determined only after calculating the length and size of the cavity. Certain foams will expand as much as 10 times their liquid volume, while others have no expansion characteristics. This must be determined to ensure enough material is on hand to complete the job once started. The flow rate and foam time must also be considered. If the foam expands too quickly it will not flow to the bottom of the cavity being filled, thus leaving the area unprotected. See **Figure 8-31**. The location of the opening through which the material is injected and how far it must flow to the lowest point to be filled must also be considered. A lengthy cavity that must be filled from the top to the bottom would best be filled with a material having a fast flow rate and a slower foam time. This ensures the material will be allowed to flow completely to the bottom before it foams up, thus assuring the cavity is filled from bottom to top. An extension hose may also be attached to the static mixing tube to ensure the material reaches the furthest point that needs to be filled. However, if the cavity must only be filled from the middle to the top, the technician should use a slower flow rate material with a faster foam time.

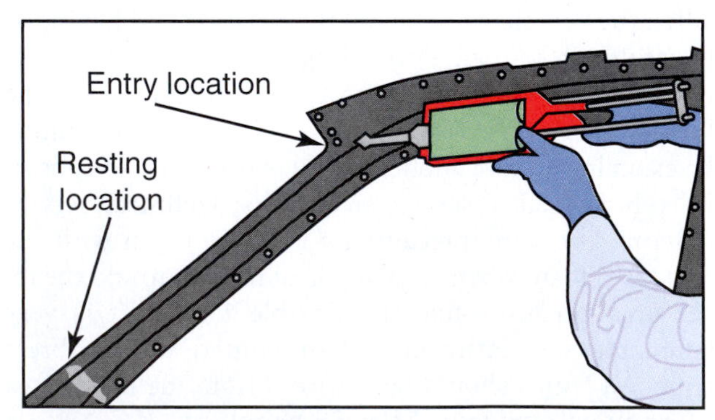

FIGURE 8-31 A technician must consider several things when selecting the foam used to replace what was removed. The flow rate must be determined to ensure adequate time is allowed to reach the bottom of the area to completely fill the cavity. The expansion rate must also be calculated to determine the precise amount of material to prevent either over- or underfilling.

SURFACE PREPARATION FOR FOAM APPLICATION

The final considerations that one must exercise before installing the foam materials is to ensure all the bare metal areas are clean, dry, and primed. An epoxy or urethane primer is preferred as it will provide the best corrosion protection. Structural foam is epoxy based, therefore reducing the chances of incompatibility between the products. While acid etch primers are not the best choice, they can be used provided they are fully dried or cured before the foam is applied. The possibility of corrosion forming at a later date is also possible, as acid etch primers tend to attract moisture which can cause problems at a later date.

FOAM INSTALLATION

Installing the foam, simple as it may seem, may require a little home grown ingenuity. Placing the foam at the precise location may require some forethought prior to even installing the replacement panels. The undamaged foam left on the vehicle after the repairs are completed may be left in place and new foam added to replenish what was removed during the repair. One should continue to remove the loose material to the point where whatever is left is firmly attached to the surface. Any foam that is not firmly attached to the surface or has been affected by any welding operations should be removed. Sometimes when welding, the foam may need to be removed before installing the new panel and then re-installed after the fact as the adhesion is apt to be affected by the heat from the welding operation. Structural and rigid foam should be removed with a sander; any flexible material may be scraped off with a chisel or other appropriate tool.

Planning

Planning ahead and determining the location, method of installation, and type of material to use frequently are not given the necessary consideration until after the replacement panel has been installed. For some applications, this may be too late. Unfortunately, in a desperate attempt to comply with the requirements for re-installing the material, the technician may resort to techniques that are not approved by either the product or the auto manufacturer. One cannot simply drill a hole at a convenient location in the part to install the foam. Sometimes **foam dams** have to be put into place prior to installing the replacement panel to ensure the foam flow is limited to a specific area. The dams may be made of re-used parts left in place on the vehicle when the damaged parts were removed. They can also be fabricated from foam blocks, plastic, compressible foam, properly corrosion protected steel, or even a balloon. The primary requisite for material used to make a dam is it must be corrosion proof and should not absorb moisture.

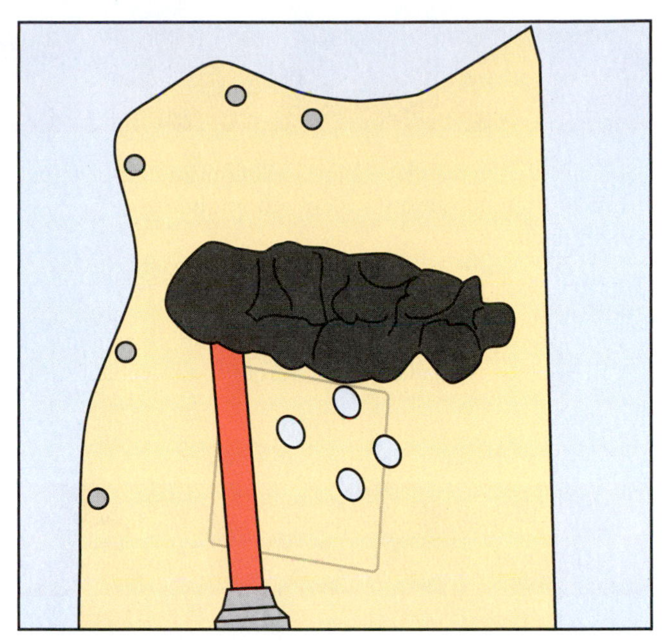

FIGURE 8-32 Occasionally it becomes necessary to create a shelf or dam with a fast foaming material to hold the structural foam in place until it cures inside an inaccessible area.

Foam Dam

Occasionally a dam must be made by forming a shelf or ledge with foam that is suspended between two inaccessible panels on a vertical surface. See **Figure 8-32**. However, access may be too limited to install a fabricated dam, making it necessary to resort to another technique. A fast foaming material dispensed very slowly will enable the technician to accomplish this. When dispensed slowly, the material will foam as it leaves the mixing tube, thus clinging to the side of the panel without dripping off as it comes out of the tube opening. Patience is required when doing this, as the process may require two or three passes—with a cure time between each—in order to build the ledge wide enough to fill the width of the cavity. Additional dry or cure time may be required prior to installing additional foam above or below this area.

Install Proper Type and Amount

Occasionally one might be tempted to conserve on the cost of the materials used by using a consumer foam such as one might find at a hardware store instead of the automotive foams. This temptation should be avoided at all cost. Consumer foams should never be used. They are a one part aerosol product that lacks the strength and integrity required of the NVH foams, resulting in inadequate NVH control. They also require moisture to cure. Moisture is absorbed into them as they cure and they continue to absorb it, causing a significant corrosion problem.

Summary

- Foam fillers have played, and will continue to play, an increasingly important role in automobile manufacturing.
- Using foam fillers to enhance the vehicle's structural integrity will no doubt become increasingly widespread, because it offers a considerable benefit with a minimum addition of weight to the vehicle.
- When technicians understand the makeup and function of the various types of foams, it ensures that the repairs are made with the correct materials that are equivalent to those used by the manufacturer.
- Using the correct foam fillers and tools can help assure the repaired vehicle has been restored to its original integrity and crash worthiness.

Key Terms

adhesion loss
butyl benzoyl peroxide
catalyst
conventional/standard fillers
delamination
feathering
foam dam
grating plastic
lightweight fillers
microspheres
noise, vibration, and harshness (NVH)
NVH flexible foam
NVH rigid foam
original equipment manufacturer (OEM)
oxidation
pinholes and air pockets
polyester resin
premium fillers
production sandpaper
repair mapping
rust
sensitive heat range
solder
solid waste stream
stain resistant
staining
static mixing tube
structural foam
tack free
talc
thixotropic resin
zinc

Review

ASE Review Questions

1. Technician A says that all plastic fillers and fiberglass fillers are catalyzed using the same butyl benzoyl peroxide hardener. Technician B says most plastic fillers use an epoxy base resin. Who is correct?
   A. Technician A only
   B. Technician B only
   C. Both Technicians A and B
   D. Neither Technician A nor B
2. Technician A says clean cardboard can be used for a plastic mixing surface. Technician B says plastic fillers can readily be applied over a rust-perforated surface without consequence so long as the scale has been removed. Who is correct?
   A. Technician A only
   B. Technician B only
   C. Both Technicians A and B
   D. Neither Technician A nor B

3. Technician A says a can of plastic filler can be put into a paint shaker to thoroughly agitate it. Technician B says the liquid-like material that collects on top of the can of filler is responsible for its adhesion to the surface. Who is correct?
   A. Technician A only
   B. Technician B only
   C. Both Technicians A and B
   D. Neither Technician A nor B
4. Technician A says structural foam is used to reduce the NVH on the vehicle. Technician B says repair mapping may occur if plastic filler is overcatalyzed. Who is correct?
   A. Technician A only
   B. Technician B only
   C. Both Technicians A and B
   D. Neither Technician A nor B
5. Technician A says grating the plastic filler will reveal the low spots on the surface when the excess material that was applied has been removed. Technician B follows up the grating step with 80 grade sandpaper. Who is correct?
   A. Technician A only
   B. Technician B only
   C. Both Technicians A and B
   D. Neither Technician A nor B
6. Technician A says the aluminum clad body fillers require the use of liquid hardener. Technician B says the premium grade fillers are less apt to stain than the conventional fillers. Who is correct?
   A. Technician A only
   B. Technician B only
   C. Both Technicians A and B
   D. Neither Technician A nor B
7. Technician A says a two part glaze coat should be used as a final finish coat over the entire repair area. Technician B says the paint should be removed approximately 3 to 4 inches beyond the edge of the repaired surface when using conventional or lightweight fillers. Who is correct?
   A. Technician A only
   B. Technician B only
   C. Both Technicians A and B
   D. Neither Technician A nor B
8. Technician A says automobile manufacturers use stuffers inside doors and other cavities to help reduce the NVH. Technician B says most manufacturers use preshaped foam to help absorb collision energy. Who is correct?
   A. Technician A only
   B. Technician B only
   C. Both Technicians A and B
   D. Neither Technician A nor B
9. Technician A says structural foams are used in the areas of the vehicle's A pillars and D pillars. Technician B says the only place structural foam is used is on the lower rails and in the torque box area of some frame rails. Who is correct?
   A. Technician A only
   B. Technician B only
   C. Both Technicians A and B
   D. Neither Technician A nor B
10. Technician A says visible streaks of catalyst or hardener left in the plastic when applying it can cause staining problems. Technician B says most plastic filler manufacturers recommend a ratio of approximately 2 percent hardener to the total plastic batch. Who is correct?
    A. Technician A only
    B. Technician B only
    C. Both Technicians A and B
    D. Neither Technician A nor B

Essay Questions

1. List the four primary types of fillers commonly used in the auto collision repair industry today. Briefly discuss the difference in their content or makeup.
2. Discuss what causes staining problems in plastic filler and discuss three things a technician can do to prevent them.
3. Discuss the difference in the texture of flexible and rigid NVH and structural foam characteristics.
4. What must be done with the foam left on the vehicle after the repairs have been completed prior to installing the replacement foam?
5. What primer(s) or basecoat(s) should be used on the surface where foams are to be applied?

Topic Related Math Questions

1. A cavity that is 26 inches long and 4 inches wide by 6 inches high must be filled with flexible foam. Each foam kit will fill the equivalent of 200 square inches. How many kits will the technician need to fill the cavity?
2. A cavity that is 36 inches long, 1 1/2 inches high, and 4 inches wide must be filled. Brand "X" flexible foam kit, when full, holds 9 square inches of foam but will expand to 10 times its size when dispensed. If the kits cost $32.00 each, how many kits must the technician purchase and what will be the cost of buying enough kits to fill the cavity?

Critical Thinking Questions

1. Locate the MSDS for each of the three randomly selected products that will be used for this exercise and do the following: (A) Determine the PPE requirements for each. (B) Determine the health hazards of each. (C) Identify the primary routes of entry into the body by each material. (D) Determine

the fire extinguishing media to use in the event the material catches on fire.

2. What would the potential consequences be if one were to install NVH foam in place of structural foam after completing the collision work?

Lab Activities

1. Using two different grades of plastic filler to perform each of the following steps, mix three batches of plastic filler using the following ratios or proportions of catalyst for each:
   A. 1 percent catalyzation rate—one golf ball size with a 1/2-inch ribbon of hardener.
   B. 2 percent catalyzation rate—one golf ball size with a 1-inch ribbon of hardener.
   C. 4 percent catalyzation rate—one golf ball size with a 4-inch ribbon of hardener. Thoroughly mix the three batches of material and place each on a clean separate piece of 12-inch square masking paper in as large a glob of material as possible. Observe and record the length of time it takes for each batch of material to reach the gel stage, the grating stage, and the sanding stage. Also observe the color variations and the temperature change of the material as it progresses through the curing stages.
2. After the pieces of plastic have cured and cooled, using a hacksaw, completely cut through a cross-section of each in three different locations and observe any pinholes that may have occurred.
3. Using three paper cups of the same size, dispense approximately 10 ounces of structural foam into one cup, approximately 10 ounces of rigid NVH foam into another cup, and approximately 1 ounce of flexible expandable NVH foam into a third cup. Allow the foams to cure and harden and then cut the cups in half from top to bottom. (A) Note the difference in the expansion characteristics of each. (B) Note the difference in the cell structure of each. (C) Note the compression characteristics of each. (D) Learn to identify each foam by its physical characteristics.
4. Place each of the foams above into warm water to raise their temperature approximately 10°F, and dispense approximately 1 ounce of each. Observe to see if there is any difference in how the material responds when it is dispensed.

Chapter 9

Trim and Hardware

OBJECTIVES

Upon completing this chapter, you should be able to:

- Label and store removed trim, moldings, hardware, and other parts for reinstallation following the repair of a vehicle.
- Identify and properly remove or reinstall the more common bolts, fasteners, screws, and retainers.
- Identify and use specialty trim and hardware tools.
- Remove, store, prepare for reinstallation, and reinstall in the proper location all trim, moldings, and emblems.
- Remove, store, and reinstall common automotive hardware.
- Store and organize for reinstallation parts which were removed for repair or refinish.
- Remove and reinstall common interior trim, molding, and weatherstripping.
- Test for wind and water leakage of the reinstalled weatherstripping.
- Remove and reinstall or replace exterior trim and body cladding.
- Remove and reinstall both painted-on and vinyl pinstriping or decals.

INTRODUCTION

Trim and hardware are often viewed as rather unimportant elements when studying collision repair, even though many vehicles have failed to be delivered back to a customer on time because of something as simple as not having the correct fastener or clip. Even worse than not ordering the proper clip, however, is having the clip when the vehicle was disassembled, but losing it, simply because the clip was not labeled and stored properly. In fact, knowledge and care in trim and hardware removal, storage, selection, and reinstallation are essential for the technician.

This chapter starts with a few points on safety, and then goes on to discuss the proper labeling and storage of hardware removed from a vehicle before repair and refinishing. The most common fasteners, bolts, screws, and retainers will be identified and discussed. The chapter will also discuss the need for collision repair technicians to stay efficient in the collision repair industry by learning about the new and different fasteners developed each model year. A section of the chapter is devoted to the removal, saving, and reinstalling of trim and emblems, with a special section on how to make templates from the original vehicle to help in the proper location of a part following repair and refinishing. This chapter will also examine the removal of items such as light assemblies, mirrors, and body moldings to aid in either the repair or refinish of a vehicle, and the reattachment of these items at a later date. The body shop term for this process is **R&I (remove and install)**, and the items that are removed as an assembly are often referred to as hardware.

Though the intention of a collision repair facility is to get a vehicle repaired as quickly as possible, some severely damaged vehicles may be in the repair facility for some time. Therefore, it is critical to properly store, label, and organize parts, fasteners, and hardware for each vehicle. They must be organized and stored until the vehicle is put back together so an invisible repair will be accomplished. This chapter will devote a section to different types of storage equipment and ways to organize parts and hardware when they are removed from the vehicle.

Other areas that will be studied are interior trim and weatherstripping. Openings such as doors and trunks must be repaired so that they look the same as before the collision occurred; however, they must also operate properly. Not only must the fit and finish be correct, but no water and/or wind leaks should exist following the repair. Tests for water and wind leaks will be presented in this chapter, along with Tech Tips to help the technician when inspecting for them.

Some vehicles are heavily equipped with trim, molding, and body cladding. Each of these materials has its own special concerns when being removed and reinstalled in vehicles. Some of these moldings are single-use only; they must be removed and then, even if handled with care, replaced. (The body shop term for this is **R&R**, for **remove and replace**.)

The final section will examine the removal of vinyl pinstriping and decals, along with the removal of painted-on pinstriping. Techniques for both removal and installation of new decals will also be examined. The method of reinstalling painted-on pinstripes will be examined, too, with tips on how to make masked and painted stripes look like a freehand pinstripe.

FIGURE 9-1 Gloves are designed to keep the technician's hands safe while working in hot or sharp areas.

FIGURE 9-2 These gloves are designed without covering on the thumb and first two fingers so technicians can manipulate tool and fasteners.

SAFETY

Though trim and hardware, at first glance, do not seem to be topics that require much discussion regarding safety, many technicians have been seriously injured during the seemingly simple operations of removing trim. Therefore, some safety tips should be kept in mind.

First, one of the easiest things to forget during a simple operation in trim and hardware removal is wearing safety glasses. Technicians often think that because the task will take only a short time and no power tools will be used during the operation that they can leave their safety glasses on the toolbox. But accidents and emergencies can happen at any time. Even if removing only a small clip that might take less than a minute, the technician should always wear safety glasses.

Next, technicians should wear gloves for safety—but not bulky work gloves that don't allow them to feel anything, or flimsy latex gloves; technicians should choose gloves designed for the job at hand. When working on the frame machine, technicians need gloves that will protect their hands from sharp or heavy steel. See **Figure 9-1**. When fasteners must be taken off, special fingerless mechanics gloves may be used. See **Figure 9-2**.

If the task requires the use of sanders or other machines that produce dust, the technician should also use a particle mask or respirator. See **Figure 9-3**.

When technicians use any product that requires safety precautions, either for them or the environment, it must be accompanied by an MSDS. The MSDS includes sections on:

- Hazardous ingredients
- Physical data

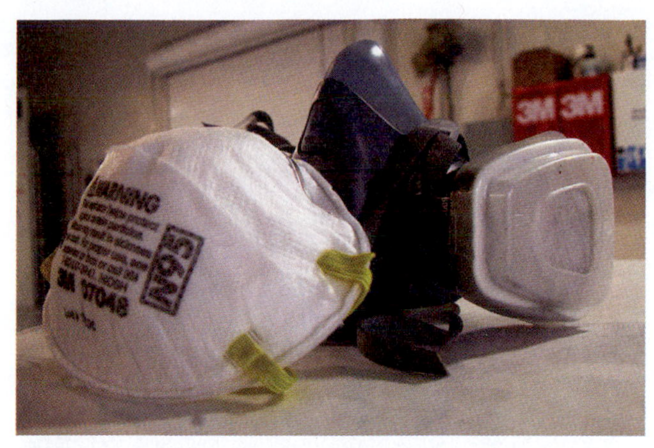

FIGURE 9-3 Safety equipment used around sanders includes a particle mask for dust and respirators for vapors.

- Fire and explosion hazards
- Reactive data
- Health hazard data, or toxic properties
- Preventive measures
- First aid measures
- MSDS preparation information

Technicians must read, understand, and follow the recommendations in the MSDS for the protection of themselves, their fellow workers, and the environment.

SCREWS, BOLTS, NUTS, WASHERS, RIVETS, AND RETAINERS

Automobiles are constructed or assembled by using many different types of fastening devices that hold two or more parts together. They come in a staggering variety and are ever-changing as newly engineered fasteners are developed. This section will deal with the major categories of screws, bolts, nuts, washers, rivets, and retainers.

Screws and **bolts** are designed with many different types of driving heads. See **Figure 9-4**. Each requires a specific tool and has specific advantages and disadvantages. Engineers choose different types of driving heads for different purposes. For example, engineers need four different types of tools to remove a 2005 Ford air bag. Engineers do this to make an air bag difficult for thieves to remove. If only one tool were needed, the air bag could be removed quickly; the need for multiple tools for the removal, therefore, becomes a deterrent to thieves.

Some of the terms used to describe fasteners can also be confusing. For example, a fastener that passes through a part and threads into another is often called a bolt, but technically it is a "cap screw," while a bolt uses a nut to fasten it together. As other technical terms arise within this chapter, Tech Tips will be inserted to help explain or clarify items.

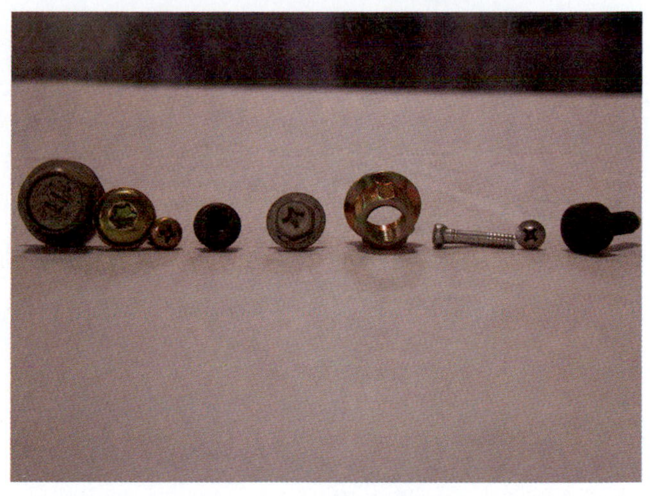

FIGURE 9-4 Fasteners have different types and sizes of heads which require different tools.

FIGURE 9-5 Machine screws come with different types of driving heads.

Screws

Screws come as either **machine screws** or **self-tapping (sheet metal) screws**. Machine screws are threaded into one or more parts, which draws them together and holds them in firm contact. Machine screws can have different heads designed for different types of fastening work. They also come with different types of driving heads. See **Figure 9-5**. Because screws do not use a nut to fasten together, they are much quicker to use during assembly, and thus may be used to increase production efficiency.

Sheet metal or self-tapping screws are generally used to attach pieces of thin sheet metal together, or to attach molding or trim to sheet metal. They come in varying types of head designs and can be "driven" or tightened with different types of tools. See **Figure 9-6**.

Sheet metal or self-tapping screws are best inserted into a punched hole that is slightly smaller than the screw. When the screw is driven into the punched hole, the metal that is driven through when punched is pulled back by the

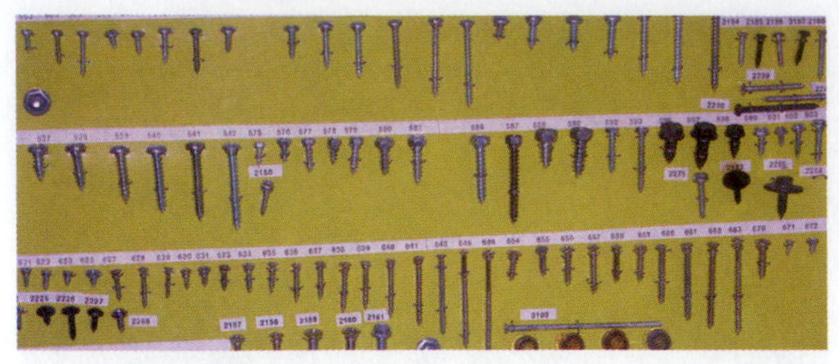

FIGURE 9-6 Sheet metal and self-tapping screws come in different sizes and lengths as well as different heads.

metal screw. This makes the bond tighter and stronger than if the hole is drilled.

Bolts

A bolt is a metal rod or pin, normally threaded and having a head or socket easily gripped by a wrench, which is used with a nut to fasten things together. Bolts are passed through two or more parts. Then a threaded nut is attached and tightened with two wrenches, often to a specific torque specification, causing the parts to hold tightly together.

TECH TIP Torque specification is a specific twisting force that is placed on fasteners, which causes them to be held in place under adverse conditions without loosening. All fasteners have an ideal torque; that is, when tightened to that point, the fastener or the parts being fastened will not be damaged, and they will stay at that torque (or tightness) without loosening. Different types of torque specifications and torque techniques will be covered later in this chapter.

Do not confuse **torque** (a twisting motion) with **TORX** (a type of fastener head)!

Bolts are often used to connect two or more parts together that are thicker than sheet metal, even up to large parts on military tanks and steel beams in large multi-floor buildings. In the 1920s, steel beams in buildings were held together with "hot" rivets that were pounded together, and then would tighten as they cooled. In the 1940s, rivets were replaced with high-grade bolts and nuts, which are stronger.

Bolt Classifications. When ordering or describing an **SAE (Society of Automotive Engineers)** bolt, many things must be taken into account: the bolt size (overall diameter), the pitch of the threads (number of threads per inch), the type of thread (such as UN, or unified screw thread standard), depth of the thread (such as C for coarse), the fit (or how tightly the fastener fits on the bolt), and the length of the bolt. While all of this information is occasionally needed to get the precise replacement bolt, normally all that would be needed is the designation 1/4-20 × 3″; this describes a bolt 1/4 inch in diameter with 20 threads per inch that is 3 inches long. When describing a metric bolt, the higher the pitch number, the coarser the thread pitch, or number of threads. To determine the pitch of a fastener a technician should use a thread-pitch gauge. The pitch gauge, either SAE or metric, is inserted into the threads, checking each one until the precise fit is found. See **Figure 9-7**. The marking on the gauge will reveal the proper pitch for the fastener.

FIGURE 9-7 This technician is using a thread-pitch gauge.

Bolt threads are generally right-handed threads that are tightened by turning them clockwise, as one can remember by the phrase *"righty tighty, lefty loosey."* Some threads, though, are left-handed threads and will tighten by turning the bolt counterclockwise. Left-handed threads are usually found on parts that rotate, and often have head markings that identify them. If a technician suspects that a bolt is a left-handed thread but identification marks are not found on the head, the technician should check a service manual to avoid damaging the fastener by overtightening.

Grading and Class. Both nuts and bolts are classified by grade (SAE) and class (metric)—the larger the grade or class number, the stronger the bolt or nut.

Metric classes are:

- 4.6
- 4.8
- 5.8
- 8.8
- 9.8
- 10.9

Classification numbers will be stamped on the head of the bolt.

SAE grades are generally:

- 1 or 2
- 5
- 8

To determine the grade of an SAE bolt, count the number of slash marks on it, and then add 2. That is, if there are no slashes, the grade is 2; if there are three slashes, the grade is 5; and if there are 6 slashes the grade is 8.

Care should be taken to replace fasteners with fasteners of the same grade or higher. Never replace a fastener with a lower grade fastener, or it may fail under normal operating conditions.

Nuts

Nuts are the devices that thread onto a bolt and, when tightened, hold the parts being fastened together. Nuts come in many different types and serve many different purposes. See **Figure 9-8**. By having different shapes, nuts can be fastened either by hand (wing nut) or with special driving tools such as torque drivers or 12-point sockets. They can also come as a **castle** or **slotted nut**, where a **cotter pin** can be placed through them, ensuring that the nut will not loosen. See **Figure 9-9**. **Self-locking nuts** have special plastic inside them; as the bolt is tightened the plastic presses against the threads, holding the bolt in place. See **Figure 9-10**. Self-locking nuts are a single-use fastener; once tightened and then removed, the nut should not be reused.

FIGURE 9-8 Nuts of different size, thread pitch, and head shape are used for different types of applications.

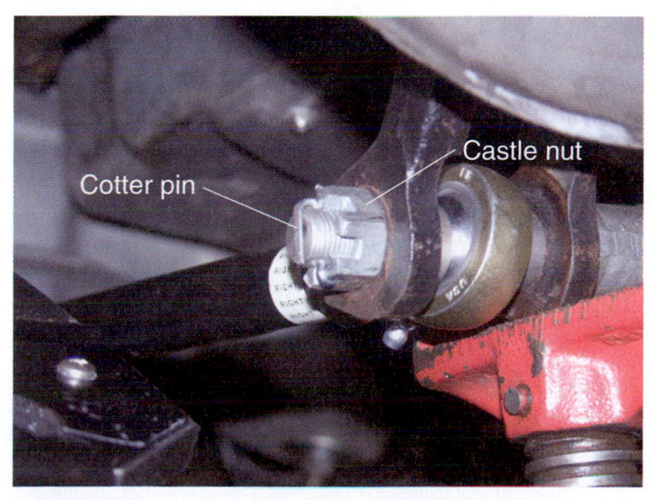

FIGURE 9-9 A castle nut with a cotter pin protects against the nut loosening when in use.

FIGURE 9-10 On this self-locking nut, note the plastic which applies pressure when tightened. Often these nuts are considered a single-use fastener.

TECH TIP Cotter pins, self-locking nuts, and other nuts can be "single-use" fasteners. That means that they can be used only once. If they are loosened, for whatever reason, they must be replaced. That also means that if a technician puts a self-locking nut on a part and tightens it down, then finds that the part must come off for some reason, the self-locking plastic is ruined and a new nut must be used.

A **speed nut** is also a fastener, one that does not fit cleanly into either the bolt or nut category. Though it was stated earlier that nuts are generally only used on bolts, speed nuts are an exception. Speed nuts are types of fasteners developed in the automotive industry to speed up the assembly process. See **Figure 9-11**. They come in different types such as flat, barrel, j-nut, and u-nuts. These nuts are placed on parts before assembly for places where a technician has little or no access to use a wrench on the other side of the part. Speed nuts are made for use with both bolts and screws; then they are tightened from the side that has access.

Washers

Washers come in many different types, all designed for specific tasks. See **Figure 9-12**. The flat washer comes as a common washer that fits between a nut and the clamping surface to spread out the clamping surface and hold the parts tighter. Though flat washers generally are used on the nut side of a bolt and nut combination, a flat washer

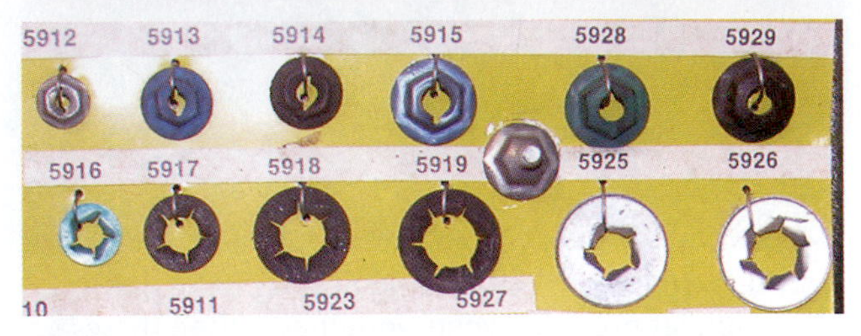

FIGURE 9-11 Speed fasteners are used to make assembly quicker and are often single-use fasteners that must be replaced with new after disassembly.

FIGURE 9-12 Different types of washers. Washers help spread the force of a fastener over a larger area, making them more effective at holding.

can be used on the head side. This spreads out the clamping power of the fastener. Some automotive bolt fasteners come with an enlarged area on the bolt head which acts like a flat washer and spreads out the load. See **Figure 9-13**. Flat washers are also commonly used with a locking washer to prevent marring the parts. If the parts are being clamped, first the flat washer, then the lock washer, and finally the nut is attached. Lastly, specialty flat washers called "fender" washers are made. They are larger than standard flat washers, thus spreading out the load in an even greater area than standard washers. See **Figure 9-14**. In the 1950s and 1960s, some automobiles were manufactured using these fender washers; however, they have not been used by OEMs for some time.

Another type of washer, trim washers, are often used on interiors and other places where appearance is a concern. See **Figure 9-15**.

FIGURE 9-13 This bolt has a large flat washer, called a fender washer. The large surface area of the fender washer applies the force of the fastener to a larger area.

FIGURE 9-14 Flat washers with different sizes (same size inside) used to apply more force when needed.

FIGURE 9-15 Trim washers are designed to have a screw head below the surface with a decorative rim.

FIGURE 9-16 Locking washers come in split-spring (left), external and internal (center), and countersunk (right). When pressure is applied the spring or barbs keep fasteners from loosening.

TECH TIP Collision repairs should be as undetectable as possible, and this applies to the use of fasteners as well. All automotive fasteners that are replaced on a vehicle should be identical to the fasteners used in the original manufacturing.

Copper or fiber washers, which are used in mechanical applications, are soft. They will deform to match the surface they are being clamped against. Copper and fiber washers are often used between two parts to assure a snug fit and to stop leakage, such as on brakes. These are single-use washers; once used and removed, they must be replaced.

Locking washers also come in many different types and configurations. They can be a simple wave or spring washer, or a more aggressive split-lock washer. See **Figure 9-16**. There are also external, internal, external/internal, and countersunk locking washers; all are designed to grab the part being clamped and the fastener, so vibration will not cause them to loosen.

Rivets

Rivets—which have been used to hold metal parts together since before the invention of welding—are still in use today. In the past, two pieces of metal with holes in them were joined by passing a piece of hot rivet through them. One end of the rivet had a head on it, and the other end was pounded or "bucked down" until the joint was tight and a second head was formed. This was done with the rivets heated to "red hot" so the bucking could be done with less force. Then, the hot rivet would tighten as it cooled.

With the increased use of aluminum parts, especially where structural aluminum parts are joined to steel, the union is held in place with adhesive bonding and rivets because the two dissimilar metals cannot be welded.

Common rivets such as blind or pop rivets, **straight rivets** (rivets that need to be bucked), high-strength blind rivets, and plastic blind rivets are used for different applications. See **Figure 9-17**. **Blind rivets** come in a variety of sizes, lengths, and materials. Though 1/8-inch rivets are common for holding on molding and trim, they also come in larger sizes. For example, 1/4-inch rivets are used to hold on door handles. Aluminum shaft rivets are made for holding together aluminum parts, as steel shaft rivets do for steel parts. Steel blind rivets, which have a special coating to separate dissimilar metals, are used when two different types of material are bound together.

Rivets used for bonding aluminum are **self-piercing rivets (SPR)**, **flush-mount straight rivets**, and blind rivets. See **Figure 9-18**. All **rivet bonding** methods are made for use with structural aluminum construction and repair. No hole is needed for self-piercing rivets, as the name implies; instead a special gun is used to drive the rivet through two or more layers of aluminum. As it passes through, the rivet holds the parts together using the pressure the gun applied to the rivet as it was forced through the aluminum. Self-piercing rivets are the most

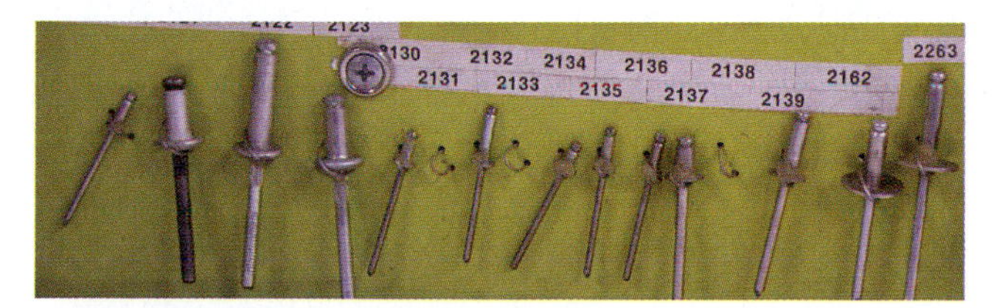

FIGURE 9-17 Pop rivets, or blind rivets, can be installed with a special tool from one side.

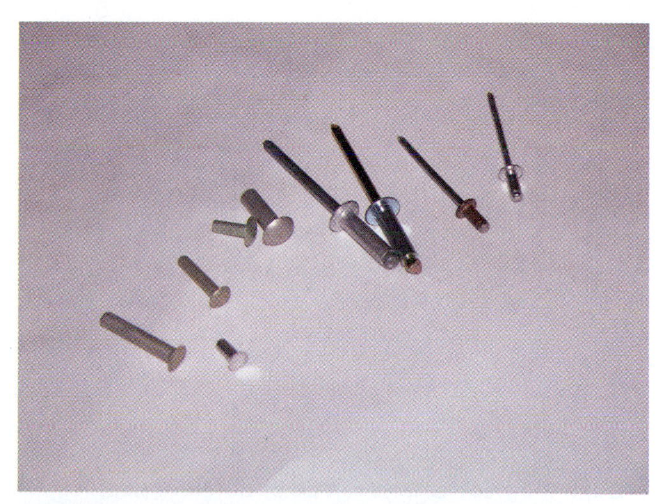

FIGURE 9-18 Self-piercing rivets (SPR) (on the left) and blind rivets (on the right) both need special tools for application.

common type of rivet used when a vehicle is manufactured. However, because they require a special gun for installation, smooth flanges without holes, and access to both sides when being reinstalled, this type of rivet is not always reinstalled when repairs are done. The manufacturer's recommendations may be to use flush-mount straight rivets or blind rivets in place of the factory SPR. For more on riveting and bonding aluminum parts, see Chapter 7.

Retainers

Retainers such as **snaprings**, pins, and keys are used to hold parts together where other types of fasteners would not be practical. See **Figure 9-19**, **Figure 9-20**, and **Figure 9-21**. Snaprings fit into grooves on the outside of parts that the snapring is forced to open slightly. When in place, snaprings squeeze around the groove, thus not letting the parts come apart. Internal snaprings are fasteners that fit into a groove on the inside of a part. When they are forced to become smaller with a special tool, then placed in the interior groove and the tool is relieved, the snapring's natural springing action holds in the groove and the parts cannot come apart. Generally, snaprings are used for parts that do not need to be held tightly, but should not come apart.

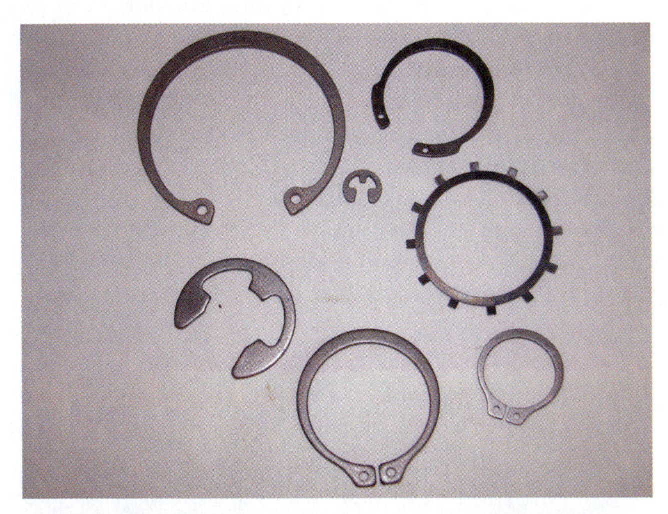

FIGURE 9-19 Snaprings use the pressure of the ring to hold parts in place. Special tools are needed to install and remove them.

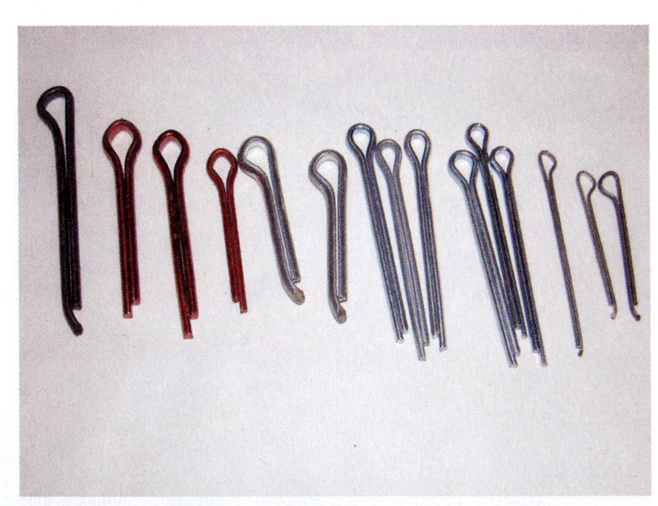

FIGURE 9-20 Cotter keys can be of different sizes, length, and protective coatings. Though often used a second time, cotter keys (or pins) are designed to be a single-use fastener.

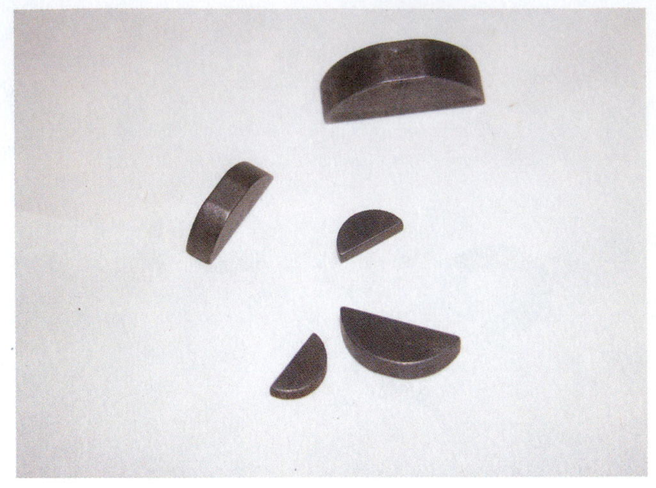

FIGURE 9-21 Keys, which slide into a matching groove in shafts with pulleys, are placed to prevent the pulley from spinning on a shaft.

Pins are fasteners used to keep loosely fitting items from coming apart. Fasteners—including cotter pins, hair pins, and clevis pins—are used on items that may need to be taken apart without damaging the part, and in some cases taken apart often. Therefore, these fasteners must be easy to use.

Some pins, such as cotter pins, are single-use fasteners and should not be reused. That is, each time a cotter pin is removed, it should be discarded and a new one should be used to reinstall the parts.

TRIM TOOLS

Because fasteners come in many types, and also because trim is often held against parts that are painted or have been upholstered, many different trim-removing tools are needed. One such group consists of door handle removing tools, which are used to remove a hidden **C-clip** attached to the door handle. See **Figure 9-22**. Windshield wiper removal tools are used to push back a retaining clip on the wiper arm and raise the arm up and off the wiper shaft. See **Figure 9-23**. Trim clip-removing tools are used to pry clips out of their holes without damaging the clip or the trim. See **Figure 9-24**. Technicians also use plastic windshield tools around glass and other trim parts so they don't mar or damage the parts. See **Figure 9-25**.

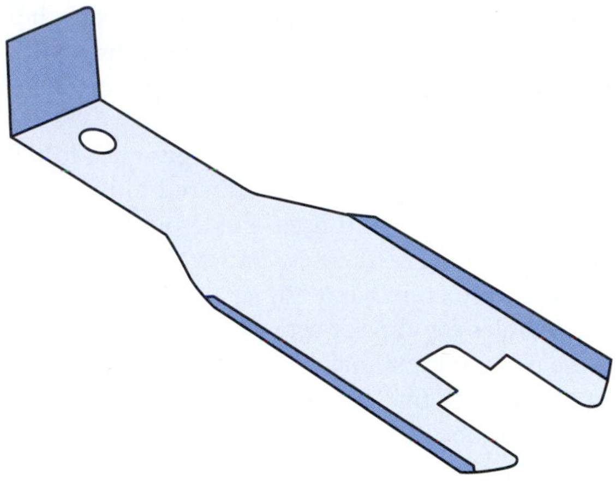

FIGURE 9-22 Door handle and window crank removing tools are used to remove C-clips attached to the door handle.

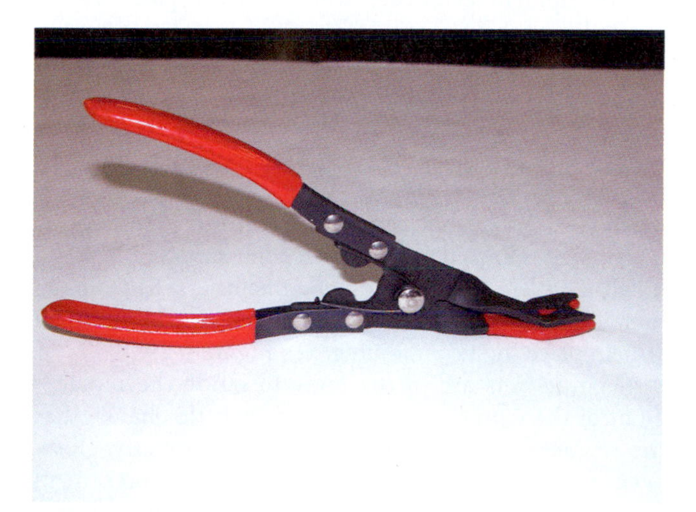

FIGURE 9-23 The windshield wiper removal tool pushes back a retaining clip on the wiper arm and raises the arm up and off the wiper shaft.

FIGURE 9-24 Trim clip-removing tools pry clips out of their holes without damaging the clip or trim.

FIGURE 9-25 Some older vehicles require a special tool, like this windshield molding removal tool, to remove chrome moldings around glass.

LABELING AND STORING

One of the first tasks performed when repairing a vehicle is to remove the damaged parts. To do this, undamaged parts must often be removed as well to gain access to the damaged area. These removed parts must be stored and labeled so the technician is able to reinstall them following repairs.

As noted earlier in this chapter, many different types of clips and fasteners are used to hold these parts in place. If these parts are not put back where they came from or placed in a container that is labeled, the technician may easily forget exactly which bolt goes where. See **Figure 9-26** and **Figure 9-27**. To repair a vehicle so that it is restored to pre-accident condition and the repair is truly undetectable, even the bolts, screws, and fasteners must be the correct type and applied in the proper place. One of the best and easiest ways to accomplish this task is to place the hardware removed from each part in an individual self-sealing plastic bag. Next, the technician should label the bag accordingly (for example, one bag might be marked "Front Door-Left") and place it on a **parts cart** or in the vehicle. See **Figure 9-28**. The parts cart should be labeled as to which vehicle the parts belong to. A running list of the clips and fasteners that must be replaced can also be kept on the

FIGURE 9-26 This part has the nuts put back in place for convenient storage.

FIGURE 9-27 This plastic bag is labeled for parts storage and quick reassembly.

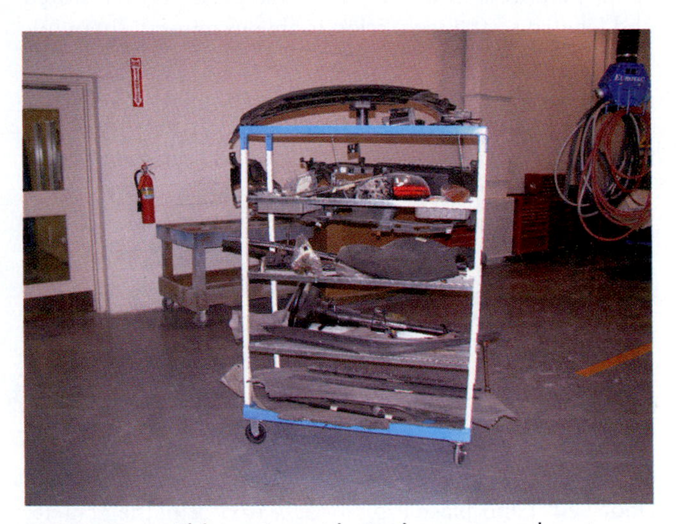

FIGURE 9-28 This parts cart is storing removed parts and labeled fastener storage bags.

parts cart. Then, by giving the list to the person responsible for ordering parts, the technician can proceed as scheduled when the vehicle is ready for reassembly.

Emblems and trim can also be stored on the parts cart, along with templates that the technician may have made as the emblems were removed.

REMOVAL OF TRIM AND EMBLEMS

Often in the course of a repair job, a technician must remove trim, molding, emblems, or nameplates—either in the paint department, or when new parts are replaced. Each item may require a different type of removal process or need a special removal tool. If the trim or molding is to be reinstalled because it is not damaged, the technician must remove it without causing damage to the vehicle or the part. The part must also be replaced precisely as it was removed. However, not all of the parts have pins or other locaters; therefore, a template can be made to help with the reinstallation of the removed part or the placement of a new one.

However, many technicians rely on experience to tell if the trim or molding part is to be adhered to the vehicle with just adhesive, or if it has a guide pin for relocation. Often by looking at the new part a technician can tell the best way to remove a trim part. For example, there may be holes for clips or screws that hold on either a clip or the part; or there may be just a guide pin on the new part and a hole that it fits into, which will alert a technician to the location for replacement while taking off the old part.

If for some reason none of these conditions indicates the way a part can be removed, a technician should then use the vehicle's service repair manual. Either vehicle-specific repair manuals or computerized vehicle service manuals can be helpful to technicians when removing parts that are unfamiliar to them.

Molding, trim, and emblems that are fastened on with clips or fasteners pose little problem for a technician to remove. However, when a part or emblem is adhered to the vehicle with adhesive, the removal is sometimes difficult. Before removing an emblem or piece of trim, technicians must be sure that they can replace it in the precise place that it was removed from. If a technician is replacing a molding where other moldings remain attached on the same line, a piece of tape can be extended between them to help place the new molding. See **Figure 9-29**.

If guide pins are in the body to guide the replacement of the trim or molding, there is little chance that the replacement part will be placed incorrectly. However, if neither bodylines nor guide pins are present to help a technician, a template should be made before a part is removed to assure its proper replacement. See **Figure 9-30**.

Making Templates to Replace Trim and Emblems

Before an emblem or nameplate is removed, technicians must be sure that they can replace the new part in the proper place. To help with this, a replacement template can be made before the old part is removed so the new one will line up with ease. To make a template, masking tape (several layers to make it strong and sturdy) is applied to the body of the vehicle, using holes or bodylines as reference points. See **Figure 9-31**. It is best to have the template tape on two sides of the part being removed, so that when replacing that part or putting on a new one, the technician can precisely accomplish the right/left and up-and-down placement. After the **tape template** is made and used to confirm the reference points, it should be carefully removed—taking care that the template shape is not altered—and stored in a safe place. See **Figure 9-32**.

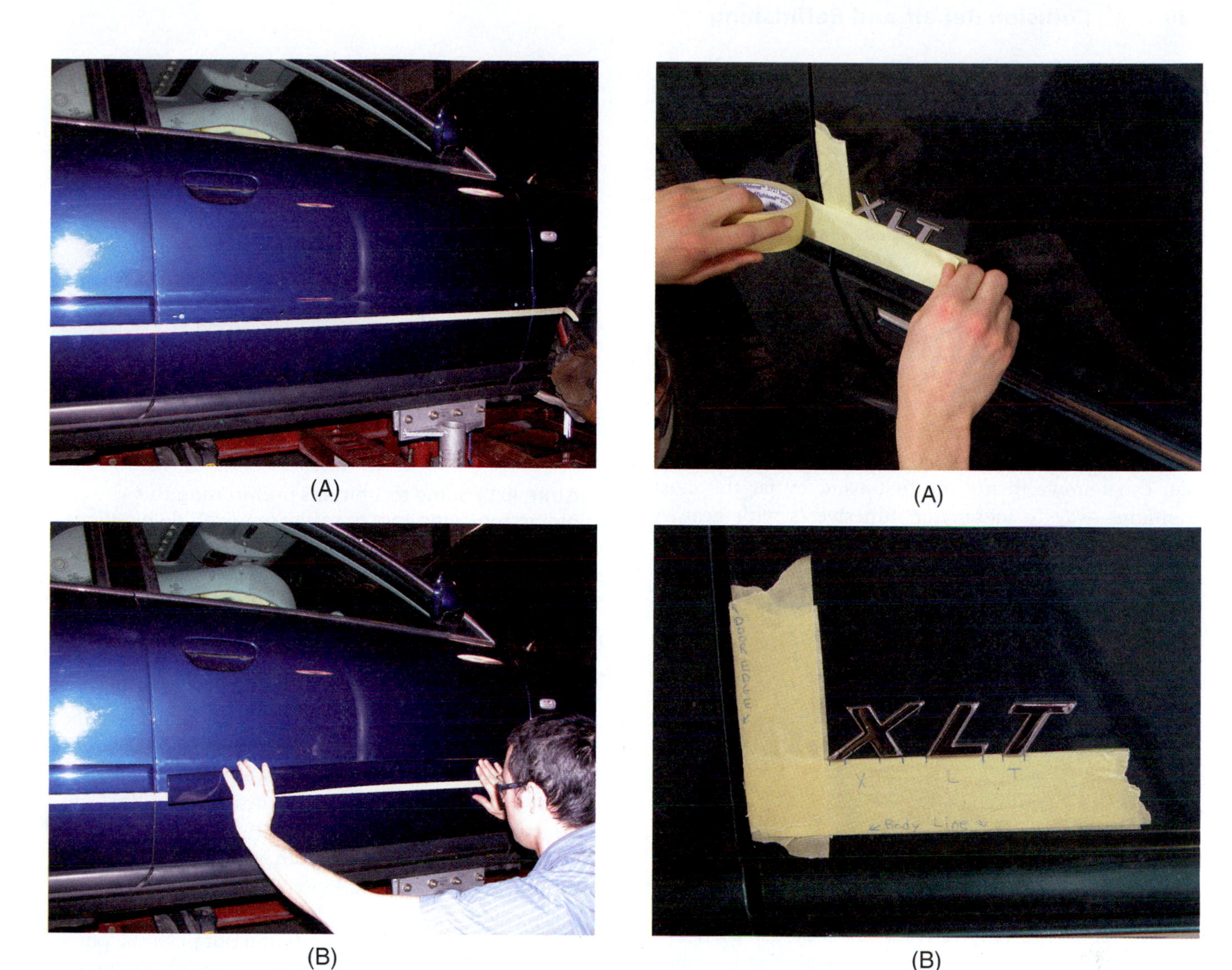

(A)

(B)

FIGURE 9-29 Tape is placed between two installed moldings (A) to provide a straightline guide for reinstalling removed moldings (B).

(A)

(B)

FIGURE 9-31 (A) This tape template is being applied for guidance. Tape is placed along the door edge and bodyline. (B) Marks are made on the tape for reapplication during reassembly.

FIGURE 9-30 This molding has a guide pin and part with plastic barrel nut. Pressure from the plastic barrel nut holds the part in place.

FIGURE 9-32 This tape template can be removed and stored on the inside glass for reapplication during assembly. Using a template assures exact placement.

TECH TIP Tape templates can be safely stored by sticking them on the inside of the section of the vehicle's glass nearest to the place where it will be replaced.

Removing Glued-On Trim and Emblems

Once the template is made so the technician can replace the new part, the old part can be removed. Typically trim, moldings, and emblems are held in place with two-sided adhesive tape. There are several ways that a technician can remove them. The first—and by far the most common—way to loosen the adhesive is with heat. A technician often uses a heat gun, and then slides a thin-bladed plastic putty knife under, thus removing the part. See **Figure 9-33**. Care should be taken when using the heat method, though, because too much heat can cause damage to either the part being removed or to the painted surface below the part.

Plastic or metal razor blades can also be used to remove moldings and emblems by sliding the blade underneath the molding and slowly cutting the adhesive. However, care must be taken when using the razor blade method to avoid personal injury! Technicians have been severely cut in cases when the razor blade slipped. For safer use when removing small emblems, technicians can use an emblem-removing tool that holds a razor blade securely. See **Figure 9-34**.

Another way to remove emblems is by using the fish line method. In this technique, protective tape is first placed around the part being removed. Then a length of high-strength plastic fishing line is cut, and the middle of the length is slipped under one corner of the emblem. The line is pulled in a seesaw method under the emblem, cutting the adhesive until it breaks free and the part is removed. See **Figure 9-35**.

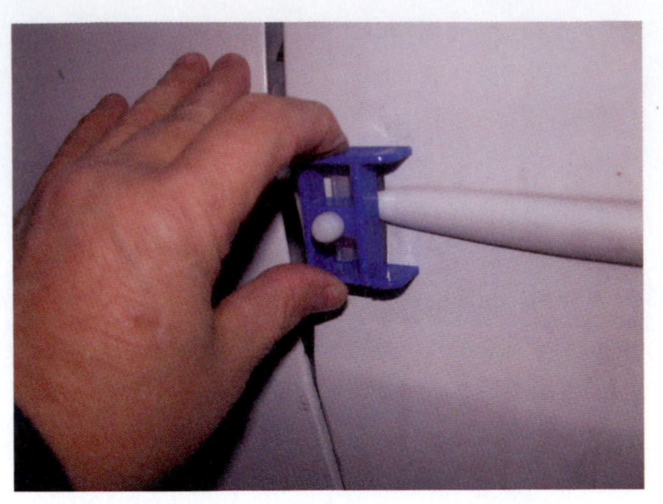

FIGURE 9-34 Some technicians prefer using this molding-removing tool to safely remove side moldings.

TECH TIP When using the heat and putty knife method, a technician should first be careful not to use too much heat. As the heat is being applied to the part, the technician should occasionally feel the part. If it is too hot to keep a hand on for a short period of time, it is too hot, and damage may occur. Secondly, the technician should not use a metal putty knife. A preferred choice is a plastic one, which can be found at most hardware stores. The plastic knife is soft, and should not cause damage to the paint below or the part being removed. Also, the technician should not push the putty knife too aggressively if the adhesive is not coming off easily. A better method is to try applying more heat. The heat-and-pry method may need to be done in steps, a little at a time.

FIGURE 9-33 A heat gun and plastic putty knife can be used to remove molding. Care should be taken not to overheat the emblem or vehicle.

Induction removal is another removal method. An **induction removal tool** has greatly simplified the removal of molding and emblems; it is a tool with attachments for different applications of heating and removal. See **Figure 9-36**. The tool, when turned on, sets up a magnetic field around the part being heated. The attachment used for moldings is a plastic mat-like tool that is lightly rubbed over the molding with the foot pedal activated. See **Figure 9-37**. This heats the metal below the molding, thus loosening the adhesive and making it easy to remove the molding. Though the pad being held by the technician does not get hot, the molding, adhesive, and the vehicle below it do. Therefore, the temperature can become high enough to damage paint if care is not taken. The induction tool can be used to remove glass and loosen bolts, as well. See **Figure 9-38**.

FIGURE 9-35 In the fish line removal technique, fishing line is moved back-and-forth under the emblem to cut the adhesive without damaging the vehicle's finish.

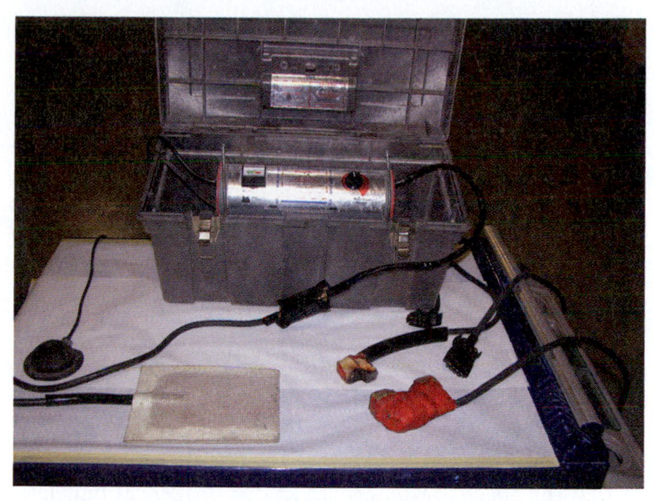

FIGURE 9-36 This induction tool with attachments is used for removing glass, nuts, and molding. The tool uses an electromagnetic field to apply heat. Care should be used not to overheat painted surfaces.

FIGURE 9-37 This induction tool is being used to remove a molding.

FIGURE 9-38 Induction tools can be used on glass as well as when heating a bolt for removal.

HARDWARE

Hardware on a vehicle comes in many forms. Unfortunately, it can have many different names as well. Hardware includes interior door trim, kick panels, moldings, cladding, appliqués, emblems, decals, and pinstriping. Each type of hardware has its own methods of removal and care once removed. Each type, including the appropriate labeling and storing, will be discussed in this section.

When hardware is removed, it must be stored along with the fasteners that are used to attach it. Though this may seem like a small and insignificant matter in the collision repair process, it can be one of the most frustrating. If the vehicle is "finished" and ready to be reassembled, but the technician finds that a hardware part is missing or broken, or that the special fasteners are lost or broken, it may take several hours or even days to get the needed parts.

This section will discuss the organization and storage of these types of hardware: exterior moldings, appliqués, cladding, emblems, decals, pinstriping, interior moldings, trim, and weatherstripping.

Organizing Removed Hardware

When trim or hardware is removed during the repair or refinish process, it should be labeled and stored. A convenient way to do this is to place the fasteners back into the piece that has been removed. See **Figure 9-39**. Alternately, the fastener could be placed into a labeled plastic locking bag for storage on a parts cart. See **Figure 9-40** and **Figure 9-41**. If removed parts, hardware, and fasteners are "bagged and tagged," then stored on a labeled cart properly, the reassembly process can be expected to go smoothly, without delays.

Another important aspect of the "**bag and tag**" process is to note and locate any needed replacements as the trim and hardware is being removed. For example, a technician might note that a fastener is broken and unusable. (Plastic fasteners, especially, may break when removed.) As the part is being bagged and tagged, the technician can either locate the new fastener, or—if it is not a stock part—order it so it is ready to be reinstalled when the rest of the vehicle is ready.

TECH TIP Proper labeling of the parts cart with the vehicle that is being worked on is important. As collision repair shops get larger and larger, the possibility of having two of the same make and model of vehicle in the shop at the same time increases. Therefore, if a shop has two blue 2004 Chevy pickups at the same time, each cart should be labeled to distinguish which Chevy pickup each set of parts goes with.

FIGURE 9-39 Technicians can place nuts back onto removed parts for storage.

FIGURE 9-40 This plastic bag (bagged and tagged) is labeled for quick reassembly.

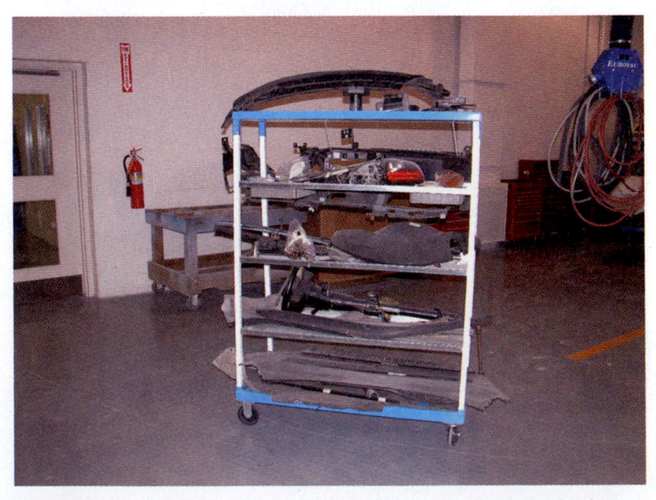

FIGURE 9-41 This parts cart has labeled "bag and tag" parts on it.

Exterior Trim

Exterior trim—such as moldings, appliqués, cladding, emblems, decals, and pinstriping—comes in many forms. Unfortunately, it is attached in many different and often-changing ways. Each new car model year, the fastening method changes, and technicians must learn the safest way to remove these current items without damaging them. They may be glued to the surface of the vehicle; there may be bolts, pins, or holding devices that go through the part that they are fastened to; and when a new part is attached to the vehicle, new holes may need to be drilled. If the part is to be removed and reinstalled (R&I), the adhesive must be removed from both the part and the vehicle. Then new adhesive or tape needs to be applied to the part to be reinstalled. New plastic parts may need to be painted to match the vehicle or painted the trim color, as noted on the identification label. Trim labeling, which is often found on the same code label as a vehicle's paint codes, can be found in different locations depending on the make and model of the vehicle being repaired. These labels are sometimes found in the vehicle's trunk—either under the trunk lid or in the tire compartment. Inside the vehicle they may be found under carpet, on the door jamb, in the console, or in the glove box. They may also be found in the engine bay. To find the location for the vehicle that is being worked on, paint manufacturers provide a label locater guide in their color chip manual. See **Figure 9-42**. Also, a technician can use a computer search to help find the paint code location for specific vehicles.

Emblems. **Emblems**, sometimes called nameplates, generally include the vehicle make, its model, and possibly other identifying titles of vehicle features. Emblems may be of different types, and there are many different ways to attach them. On some nameplates each letter is connected, and the entire plate can be removed and replaced. See **Figure 9-43**. Others may be separate individual letters. See **Figure 9-44**. In this case, it is impractical to totally remove and reinstall those nameplates.

Nameplates may be attached by several different methods, one of the most common ways being with adhesive glue. For example, a technician can use a heat gun to warm the emblem and the attaching glue to remove the glued-on nameplate. See **Figure 9-45**. Once the glue is softened, a technician can use a plastic putty knife to remove the emblem. The technician can also use fishing line to cut the glue. See **Figure 9-46**. In addition, a technician can use a steel emblem-cutting tool which attaches to an air hammer. See **Figure 9-47**. However, this method can cause damage to the paint surface and should be used with caution. Lastly, an induction tool can be used to heat and loosen the attaching glue. See **Figure 9-48**. Again, care should be taken when learning how to use this type of equipment. Too much heat can easily be applied to the molding, which may damage the painted surface under it.

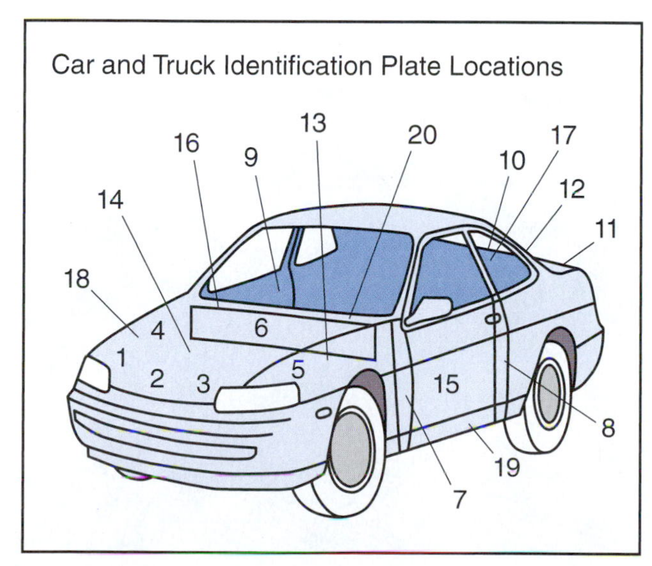

FIGURE 9-42 A paint code locater page from a color chip guide book.

FIGURE 9-43 This nameplate has all the letters connected.

FIGURE 9-44 This nameplate is made up of individual letters.

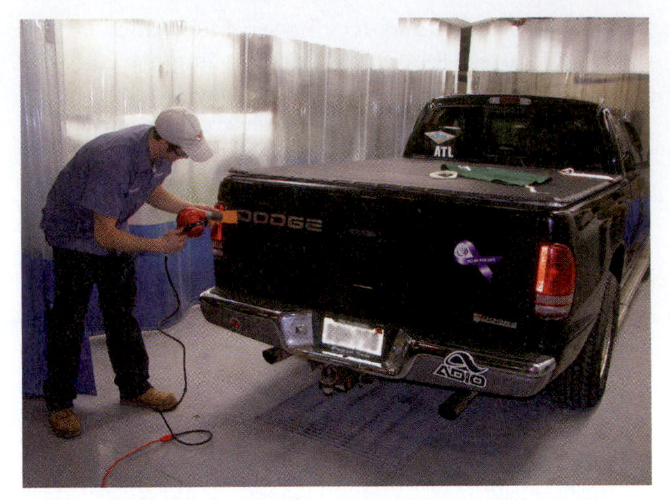

FIGURE 9-45 A technician can use a heat gun to remove a nameplate with a plastic putty knife.

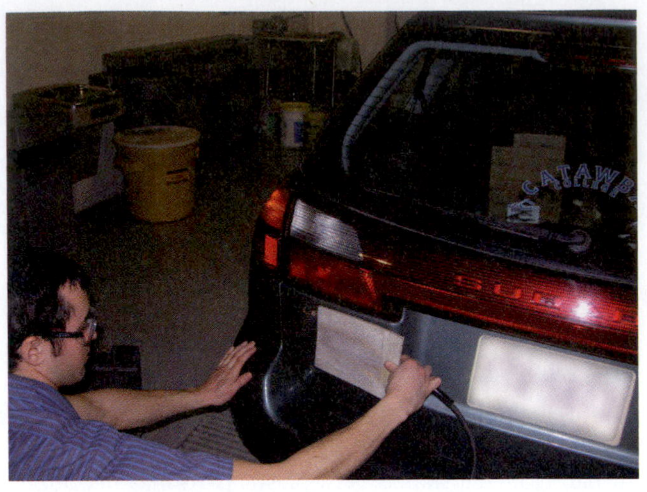

FIGURE 9-48 A technician can use a heat inductor to remove a nameplate.

FIGURE 9-46 Fishing line can be used to remove the adhesive under a name tag.

Care should be taken when removing nameplates because there may be pins or fasteners that protrude through the body part that it is mounted on. It may be necessary to gain access to the back side of the body part to remove some nameplates without harming the surface.

Each of these methods is effective and can be used to remove nameplates; it is up to the technician to choose the one that best serves the individual job being performed.

Moldings. **Moldings** come in many different types and have different names as well. There are body side moldings, which can be either OEM or aftermarket. See **Figure 9-49**. There are also belt-line moldings and even windshield-reveal moldings. See **Figure 9-50** and **Figure 9-51**. Some say that cladding, which will be discussed in the next section, is also a type of molding. See **Figure 9-52**.

Body molding should be removed before refinishing takes place; it is nearly impossible to either sand or mask a panel properly with molding still attached to the vehi-

FIGURE 9-47 When using a steel trim removing knife, care should be taken to prevent damage to the vehicle's finish.

FIGURE 9-49 Body moldings can be either OEM or aftermarket.

FIGURE 9-50 Belt-line moldings are the moldings between glass and the vehicle.

FIGURE 9-51 The windshield-reveal molding is attached to the glass and removed during windshield removal.

FIGURE 9-52 Cladding is the plastic covering at the bottom of the door and rocker.

cle. See **Figure 9-53**. To assure that the surface is prepared properly, the molding should be removed.

Molding removal can be done in much the same fashion as that used for nameplates. However, sometimes moldings are much larger. If the fishing line method is chosen the technician may need to get new line several times before the molding is removed. Often a heat gun and plastic putty knife is the most practical method. The technician's intention is to remove the molding without harming either the molding or the vehicle. Care should be taken when removing molding, since if the molding is bent it will likely need to be replaced.

Whichever removal method is used, though, there will be a residue of adhesive left behind on both the vehicle and the molding, which must be removed. Though there are many adequate adhesive removers on the market, the bulk of the adhesive should be removed before these products are used. The most common tool used for adhesive removal is the eraser disk. This is a rubber disk that is commonly installed into a drill. See **Figure 9-54**. Though a cordless drill provides mobility when using an eraser disk, the main reason for using the drill is control of how fast the eraser is spun. If the eraser is turned too fast the eraser will heat up and deteriorate faster than it should. By using a drill the speed can be slowed down and the tool will be more efficient when used.

The tool can be used on the back side of the molding as well. This will prepare it for application of new adhesive tape for re-attachment. See **Figure 9-55**.

After the bulk of the residue adhesive is removed with the eraser disk, any glue that remains can be removed with adhesive remover solvent.

Belt-line moldings are often attached with plastic clips which, when removed, may break and need replacement.

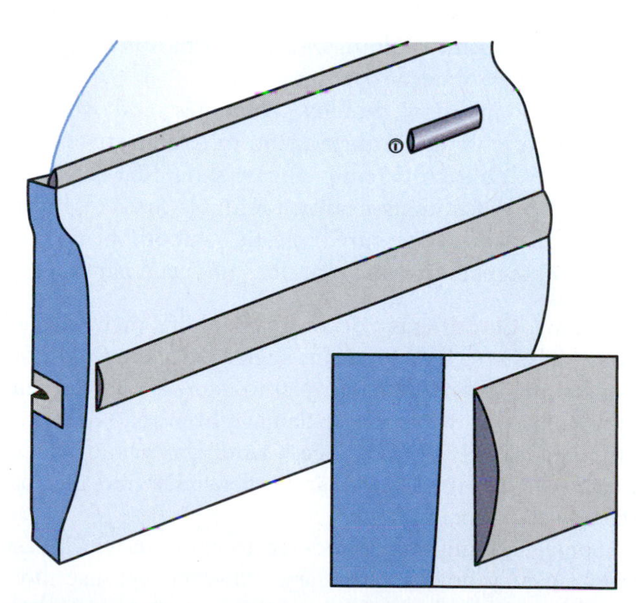

FIGURE 9-53 Body molding attached to a vehicle with adhesive is difficult to refinish. Although the molding can be masked properly, proper sanding is extremely difficult.

FIGURE 9-54 An eraser tool is used in a drill to remove adhesive after the molding has been removed.

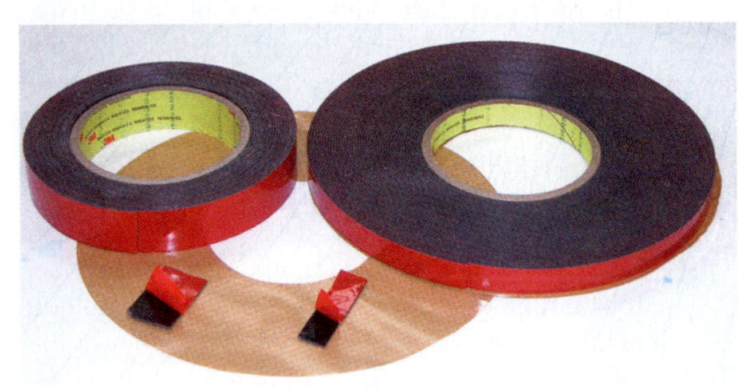

FIGURE 9-55 Two-sided tape (dabble sticky back tape) is used for molding reinstallation.

See **Figure 9-56**. Removing a belt-line molding typically requires the removal of inside trim to gain access to the molding. In fact, in some cases door glass may also need to be removed. Again, the technician should take care in removing belt-line moldings, since if bent it is likely that the molding will need to be replaced.

Belt-line molding, window appliqués, and weatherstripping are often interdependent, or closely attached to each other. As each is removed care should be taken not to damage those items nearby. Moldings that are plastic might be painted, textured plastic, smooth plastic, or color-impregnated plastic that does not need painting.

Cladding. Cladding is a decorative molding that is placed on vehicles in several locations such as rocker covers, bottoms of doors, and tailgates, just to name a few. It is generally placed in areas where damage from road debris or chipping is likely to take place. Cladding provides protection to the part underneath and is generally more durable than steel and paint alone.

Some cladding can be ordered to color, in which case painting may not be necessary. However, because most cladding is made from plastic, painting may require some special preparation. Technicians should check the paint manufacturer's recommendation.

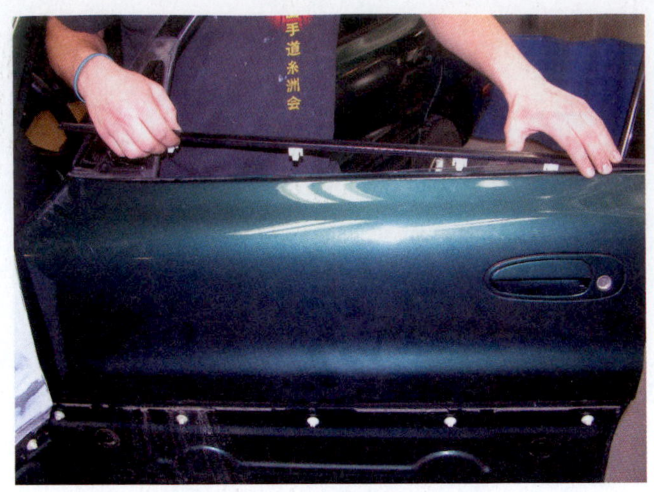

FIGURE 9-56 Belt-line molding removal (note plastic clips) often requires the removal of inside trim.

Cladding may be held in place with adhesive, such as tailgate cladding. However, it is more likely attached with clips or plastic rivets. Even when care is taken while removing cladding, sometimes the clips become damaged and must be replaced. Because of the vast array of clips that are available, it may not be practical to keep clips in stock; therefore, as the vehicle is disassembled in the body shop or de-trimmed in the paint department, the parts department should be notified to order any replacement clips that will be needed for reassembly.

Appliqué. Appliqué is decorative trim typically found near or surrounding a window. It can be painted on gloss black or dull black; it can be made of plastic and adhered with adhesive, clips, or screws which may be hidden under weatherstripping or seals. Appliqué may be a decorative designed decal which—if removed—may need to be ordered from the manufacturer for re-application. Generally appliqué is removed last, after the mirror, belt-line molding, and weatherstripping or door seals. All of the other parts generally overlap onto the appliqué, necessitating that it be removed last.

Preparing the vehicle for reinstallation of appliqué varies, depending on what type of appliqué is used. If the appliqué is painted, the surface is prepped as for any other paint preparation. If fastened with clips or fasteners, the re-attachment will be completed when all old fasteners are checked for being in good working order, then re-attached or replaced. If applied with adhesive, the old surface of both the vehicle and the appliqué are cleaned, and then new adhesive is applied and the appliqué re-attached.

Decals. Decals or vinyl trims come in a vast array of decorative applications. This trim might be as simple as a decorative decal box side application, such as a 4 × 4 decal on a pickup, to elaborate aftermarket vinyl application that is popular with import tuner customizing. See **Figure 9-57** and **Figure 9-58**. Other decals or vinyl applications could be imitation wood covering on a vehicle's body side.

FIGURE 9-57 A 4 × 4 decal bagging may be used as a decorative application.

FIGURE 9-58 Custom vinyl decoration can appear in many styles.

To remove decals, the technician can use a solvent remover, or a heat gun and plastic razor blade once the adhesive has been removed by hand. Following the removal of the decal, the residue glue can be removed with adhesive remover solvents. Once the vehicle's collision damage is repaired, the vehicle is painted and the decal is reinstalled.

Small vinyl decals can be replaced dry, but if the technician is concerned that the application of the decal may need to be moved or changed, decal application liquid should be sprayed lightly on both the vehicle and the new decal. This allows the decal to be moved or relocated if needed before the liquid evaporates.

Once the decal is in place, all excess liquid application solution should be squeezed out so the decal can adhere firmly to the vehicle. Care must be taken to make sure that all air bubbles are removed, and also that the surface that will be under the decal is free from dirt or other imperfections.

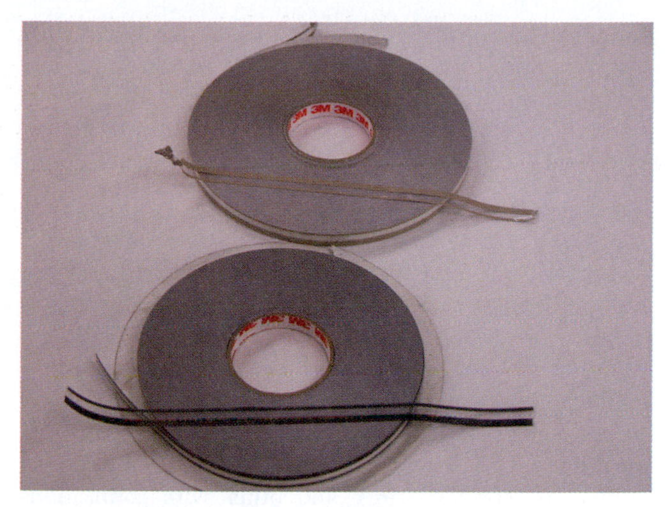

FIGURE 9-59 Vinyl pinstriping tape is available in a variety of colors.

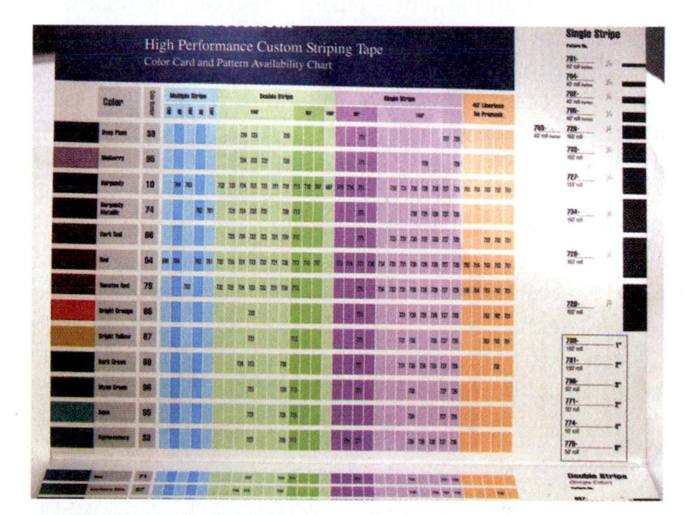

FIGURE 9-60 A pinstriping tape chart is used to match size, color, and spacing.

Pinstriping. Though **pinstriping** can take many forms—including vinyl application, painted on, and freehand pinstriping—most of the pinstriping that will be replaced in a collision repair shop is of the vinyl type. Most replacement pinstriping comes prespaced and in an assortment of colors. See **Figure 9-59**. A technician chooses the color, then the size and spacing, that matches the vehicle being repaired. See **Figure 9-60**. After the vehicle is repaired and refinished its pinstriping can be installed.

Though it is true that most vehicles have vinyl striping, some come with painted-on striping. In this case a technician can choose the proper configuration and size of pinstripe masking tape. After it is applied the paint can be applied with a sprayer or brushed on to match the original stripe.

Interior Trim

Nearly all of the interior parts are considered trim. A rear package tray, speakers, headliner, door trim pads, kick panels, interment panel, consoles, sill plates, carpet, and even things such as a third light are considered trim. We will discuss such things as removal of seats, carpet, and door, kick, and pillar trim. These items are often removed when repairing severely damaged vehicles. If these items are not removed they may be damaged by the repair process.

Trim. To access the inside of doors to remove window glass or window regulators, the interior **trim** must be removed. Each door is different—when removing a door trim for the first time a technician must proceed with care. Plastic clips are easily damaged and often screws are hidden beneath decorative covers. Plastic tools are made to help pry the clips and push pins to avoid damage to the trim piece on the vehicle and allow the reuse of some fasteners. See **Figure 9-61** and **Figure 9-62**. As the trim is being removed electrical connectors, speaker wires, and other wire connections should be carefully unplugged. Once the door trim is successfully removed, it should be stored in a safe place with all of its clips and fasteners labeled. Between the door trim and the door is a dust and moisture shield that is sometime referred to as a "vapor barrier." This barrier, when removed, must be kept intact for reinstallation. Its purpose is to keep moisture, dirt, and dust from getting into the vehicle. If it is damaged it should be repaired before reinstallation.

Other interior trim pieces, like the plastic covering over the pillars or kick panels, can be removed with the same plastic trim tools as used with the door. They should also be stored and kept away from harm until reinstallation. Though plastic seems quite flexible, it can be brittle. Especially in cold weather, plastic may be damaged while being removed.

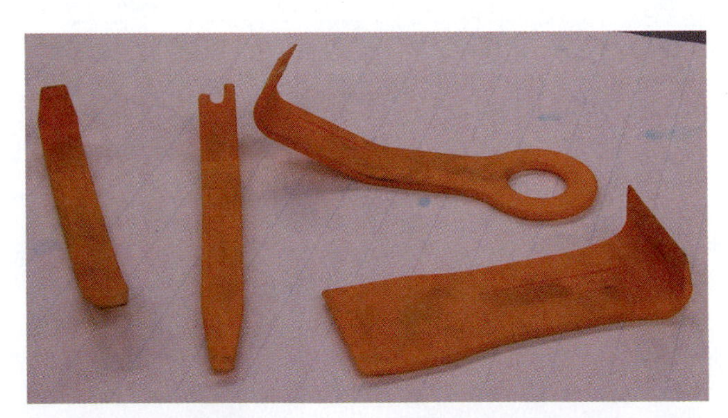

FIGURE 9-61 Plastic trim tools help to pry clips to avoid damage to the trim.

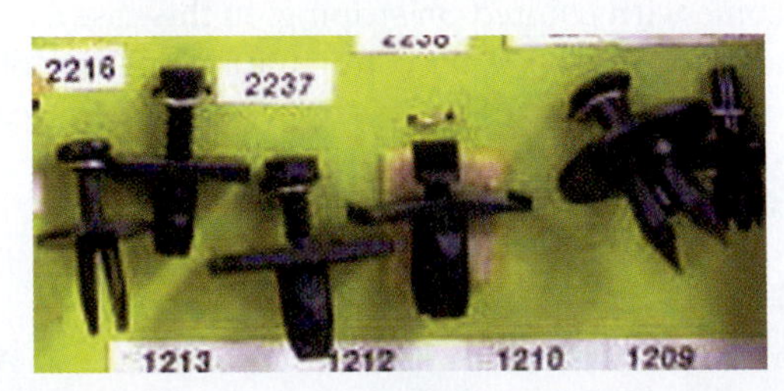

FIGURE 9-62 Plastic push pins help avoid damage to the trim.

Seats and Carpet. To remove seats and carpet, technicians should be careful to remove all the electrical connections, and then remove the bolts that may be accessed from inside or even underneath the vehicle. Often seat belts must be removed to allow for the removal of seats. To properly remove and reinstall seat restraints and other safety equipment, the technician should check the vehicle's service manual for the recommended procedure for removal and reinstallation.

Once the seats and trim have been removed, the carpet can be easily removed and stored for safety while the vehicle is being repaired. The carpet can be detailed much more easily when it is out of the vehicle, so that the technician has full access to all areas of the carpet and can clean it thoroughly.

Weatherstripping. Weatherstripping is a soft rubber attached around any opening. When the opening—like a door or window—is closed, the weatherstripping fills the gaps between the vehicle and the closure. It guarantees a tight fit, keeping wind, water, dirt, and dust from entering the vehicle. Following a collision repair, technicians must be sure not only that all closures work properly, but also that the weatherstrips are in working order and make a weather-tight seal. See **Figure 9-63**.

Weatherstripping can be held on with adhesive, in channels, with plastic pins or clips, and in some cases with screws or other fasteners. If the weatherstripping has not been damaged in the collision, the technician must be careful not to damage it during removal. The weatherstripping then should be bagged and tagged. Often, during weatherstripping reinstallation, the left and right of the same door can look very similar to the technician. However, if they are each labeled properly, they can be easily distinguished and reinstalled correctly. Then, after reinstalling the weatherstripping, the technician should perform a fit test and water leak test.

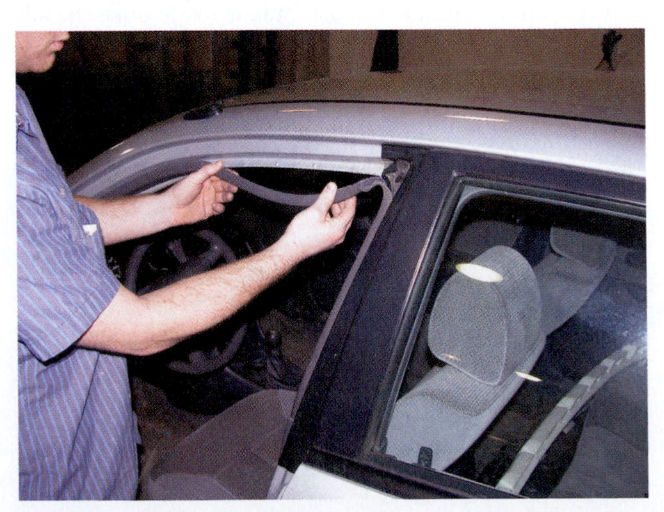

FIGURE 9-63 This weatherstripping is being removed for masking and refinishing.

Summary

- Understanding fasteners and trim removal is an ever-changing task for technicians.
- Each new model of vehicle brings new methods of attaching trim with new clips, fasteners, and often new tools to remove them.
- Technicians should note that fasteners sometimes break (or may be one-use fasteners), and that even with the best of care plastic clips often break during removal.
- Sharp tools can damage interior and other trim as it is being removed.
- The removal and reinstallation of trim is both challenging and critical.
- Proper labeling and storage of trim pieces when removed is critical, perhaps more so than sometimes realized by new technicians.
- Although the removal, labeling, storage, and reinstallation process is not one of the most glamorous jobs in the collision repair field, there are few tasks that are more important.

Key Terms

appliqué
bag and tag
blind rivets
bolts
castle/slotted nuts
C-clips
cladding
cotter pins
decals
emblems
flush-mount straight rivets
induction removal tool
machine screws
molding
nuts
parts cart
pinstriping
remove and install (R&I)
remove and replace (R&R)
rivet bonding
screws
self-locking nuts
self-piercing rivet (SPR)
self-tapping/sheet metal screws
snaprings
Society of Automotive Engineers (SAE)
speed nuts
straight rivets
tape template
torque
TORX
trim
weatherstripping

Review

ASE Review Questions

1. Technician A says that self-tapping screws are used to hold sheet metal together. Technician B says that self-tapping screws come as Phillips head only. Who is correct?
   A. Technician A only
   B. Technician B only
   C. Both Technicians A and B
   D. Neither Technician A nor B
2. Technician A says that cladding is the name of the interior trim of automobiles. Technician B says that weatherstripping is always attached with adhesives. Who is correct?
   A. Technician A only
   B. Technician B only

C. Both Technicians A and B
D. Neither Technician A nor B

3. Technician A says that bagging and tagging is the covering of a vehicle to protect it from overspray. Technician B says that the proper labeling and storing of removed parts is vitally important to the reassembly after repairs. Who is correct?
   A. Technician A only
   B. Technician B only
   C. Both Technicians A and B
   D. Neither Technician A nor B
4. Technician A says that the use of protective mechanics gloves is important when removing trim and is part of a repair technician's personal protective clothing. Technician B says safety glasses should be worn at all times when working in a collision repair shop. Who is correct?
   A. Technician A only
   B. Technician B only
   C. Both Technicians A and B
   D. Neither Technician A nor B
5. Technician A says that a bolt is typically used to hold together two removable parts. Technician B says that SAE stands for Society of American Engineers. Who is correct?
   A. Technician A only
   B. Technician B only
   C. Both Technicians A and B
   D. Neither Technician A nor B
6. Technician A says that a 1/4 -20 × 3″ is a 1/4- inch diameter fastener with 20 threads per inch that is 3 inches long. Technician B says that to determine the pitch of a fastener a thread-pitch gauge should be used. Who is correct?
   A. Technician A only
   B. Technician B only
   C. Both Technicians A and B
   D. Neither Technician A nor B
7. Technician A says that bolts come in both right- and left-handed thread. Technician B says that to loosen a right-handed thread bolt, it should be turned to the right. Who is correct?
   A. Technician A only
   B. Technician B only
   C. Both Technicians A and B
   D. Neither Technician A nor B
8. Technician A says that flat washers are used to spread out the clamping power of the fastener. Technician B says that pop rivets are also referred to as blind rivets. Who is correct?
   A. Technician A only
   B. Technician B only
   C. Both Technicians A and B
   D. Neither Technician A nor B
9. Technician A says that an SPR is a self-punching rivet. Technician B says that to "R&I" a part, a technician will repair and install a part. Who is correct?
   A. Technician A only
   B. Technician B only
   C. Both Technicians A and B
   D. Neither Technician A nor B
10. Technician A says that a castle nut uses a cotter pin to assure that it stays tight. Technician B says that snaprings are fasteners that fit into a groove on the inside of a part. Who is correct?
    A. Technician A only
    B. Technician B only
    C. Both Technicians A and B
    D. Neither Technician A nor B
11. Technician A says that if a bolt has a grade of 4.6 marked on its head, it is a metric bolt. Technician B says that grade 8 bolts can be replaced with grade 5 bolts if needed. Who is correct?
    A. Technician A only
    B. Technician B only
    C. Both Technicians A and B
    D. Neither Technician A nor B
12. Technician A says that nameplates can be removed with fishing line. Technician B says that nameplates can be removed with heat and a plastic putty knife. Who is correct?
    A. Technician A only
    B. Technician B only
    C. Both Technicians A and B
    D. Neither Technician A nor B
13. Technician A says that appliqués can be either painted on or glued on. Technician B says that an eraser disk should be turned fast to be effective. Who is correct?
    A. Technician A only
    B. Technician B only
    C. Both Technicians A and B
    D. Neither Technician A nor B
14. All EXCEPT which one of the following could be considered exterior trim?
    A. Cladding
    B. Belt molding
    C. Weatherstripping
    D. Moldings
15. Technician A says that decals should be removed with a razor blade. Technician B says that decals can be installed either wet or dry. Who is correct?
    A. Technician A only
    B. Technician B only
    C. Both Technicians A and B
    D. Neither Technician A nor B

Essay Questions

1. Explain why parts should be labeled and stored when removed from a vehicle during repair.
2. Analyze and describe how SAE bolt grading can be recognized by a technician.
3. Explain why it is important to inform the parts department if clips or fasteners are damaged when removing parts from a vehicle.
4. Describe the procedure used to remove adhesive bonded nameplates with "fishing line."
5. Describe how to make a template from tape before removing a nameplate.
6. Explain how a self-locking nut works.
7. Explain why weatherstripping is needed around a door.

Topic Related Math Questions

1. A technician is removing weatherstripping which has 34 plastic fasteners. Of these, 50 percent of the fasteners break during removal. How many fasteners must the technician purchase?
2. The parts department uses 32 fender fasteners a week. How many fasteners would the department use in a year?
3. If a bolt is 2 1/2 inches long, how many millimeters long is it?
4. If two-sided adhesive comes in a 7-yard roll that costs $26.34, how much does the adhesive cost per foot?
5. If you bought two-sided adhesive for $3.00 per foot and you marked it up 150 percent, how much would you charge for a yard of it?

Critical Thinking Questions

1. Explain the importance of the parts department in a body shop.
2. Why is a cotter key a single-use fastener?
3. Why should a collision repair be an undetectable repair?

Lab Activities

1. In the lab, set up an assortment of fasteners which students must identify and label.
2. Have students remove a vehicle's door and disassemble it, then reassemble the door and reinstall it on the vehicle.
3. Have students remove a body side molding, clean the vehicle and molding of adhesive residue, then reinstall the molding on the vehicle.
4. Have students prepare a vehicle for refinishing by detrimming the vehicle: removing side mirror, belt molding, body side molding, door handle and lock, appliqué, and any other trim that would interfere with refinishing.
5. Have students remove the interior of a vehicle (seats, carpet, and needed trim), then reinstall.

Chapter 10

Estimating Collision Damage

OBJECTIVES

Upon completing this chapter, you should be able to:

- Define the estimate and damage report types, such as:
  - Manual
  - Computer
  - Digital imaging
- See how estimates are made and what they are utilized for, such as:
  - Sales
  - Parts orders
  - Repair orders
- Understand various facets of insurance coverage and programs, such as:
  - Liability
  - Collision
  - Comprehensive
  - Deductibles
  - Direct repair
- Utilize and understand industry terms, such as:
  - Betterment
  - Depreciation
  - Appearance allowance
  - R&I (remove and install)
  - R&R (remove and replace)
  - O/H (overhaul)
  - IOH (included in overhaul)
  - O/L (overlap)
  - AT (access time)
  - PED (preexisting damage)
  - SUB (sublet)

- Identify a vehicle
  - Decoding a VIN
  - Refinish, trim, and various other codes
- Identify and analyze types of damage to a vehicle, such as:
  - Visual damage indicators
  - Measuring for damage
  - Primary and secondary damage
  - Repair or replace?
  - Plastics and other substrates
  - The different areas of a vehicle (subject to damages)
- Utilize estimating guides and understand the facets of them, such as:
  - Parts information
  - Labor information
  - Procedural explanations
  - Refinish operations
  - Price information
- Evaluate a vehicle's worth, and determine whether or not it is a:
  - Total loss
  - Repairable vehicle

INTRODUCTION

Any complex venture or significant process that one wishes to complete successfully requires *a plan*—a set of steps that outline and help guide an entire process to a goal. For example, the architect creates and works off of a blueprint, and the surgeon makes diagnoses from charts and x-rays. The collision repair process begins with—and works off of—a *damage report*. The damage report is itself a plan, and assists in the continual planning and execution of the collision repair process. You may have heard the expression "You never get a second chance to make a first impression." One of the first impressions a collision repair center will make on a prospective customer is the damage report.

Can a damaged motor vehicle be repaired without a damage report? Technically, yes. A collision repair center's technicians could take a vehicle apart and repair it "as they go," so to speak. However, confusion, wasted time, and misunderstandings would rule the entire process. The odds of completing a safe, pre-loss repair will decrease in direct proportion to the increased odds of making mistakes! The chance of operating a profitable business would be just that—a *slight* chance, at best. If a repair center wants to satisfy and retain their customer base— if not *grow* their business—all of the aforementioned points should not be left to a haphazardly run venture. Picture an architect trying to build a house without a detailed plan of action. That provides a visualization of what a collision repair technician would be like trying to restore a damaged vehicle without a damage report.

WHAT IS A DAMAGE REPORT?

A damage report—also referred to as a repair estimate or damage appraisal—is a step-by-step *detailed* account of what needs to be done to restore a collision-damaged motor vehicle to its original, factory-manufactured condition. A better question than "What is a damage report?" might be "*Why?*"—as in, why use a damage report and what does it accomplish, or help to accomplish?

Among other things, the professional damage report provides:

- A written record of estimated repair costs that can serve as an agreement between the repairer, customer, and insurer
- A tool that aids in the sale of the repair center's services
- A sound basis for preventing misunderstandings
- A plan or "road map" to performing safe, cosmetically pleasing, pre-loss condition collision repairs
- All necessary information to create accurate part and work orders
- The means to price the collision center's product (collision repairs) accurately and competitively
- A pathway to optimum shop productivity and the assurance of the fastest return delivery possible

Let's take a look at the types of damage reports that can be created.

DEFINING ESTIMATE AND DAMAGE REPORT TYPES

What *type* of report can be created? Reports can be a manual estimate (by hand), a computerized one, or a report created by the use of *imaging*. However, regardless of what kind of report is produced, the *means to feed that report* remains the same—it emanates from the report writer's technical and investigative skills, experience, and education. A computer-produced estimate still relies upon the appraiser's skills. Whether one writes and computes an estimate longhand, or feeds decisions into a software database program, the final product will only be as good as the information and methods that are utilized to produce it.

Manual Estimates

Before the advent of personal computers, collision damage reports or estimates were created "by hand." Although it is becoming a rarity, some collision centers still create estimates this way.

In order to create a manual damage report, the appraiser needs a set of *collision estimating guides*—also known as "crash books"—and blank damage report sheets, along with something to write with. The appraiser systematically goes over the vehicle and decides if the individual parts and assemblies are repairable or need to be replaced. Parts requiring replacement are located in the collision estimating guide for the specific vehicle being appraised. Guides also contain information such as the part numbers and prices, the time necessary to replace the parts, the refinish time needed to paint parts, and an array of other information that aids in the completion of an accurate, detailed report.

After the appraiser appraises all repairable parts, identifies all replacement parts, documents the various operations and materials, and writes down all the refinishing procedures, the final calculations are computed manually, and the damage report is completed.

Computer-Generated Estimates

Like many other manual documents that have moved to computers recently, the computer-generated damage report has become quite common in the collision repair industry. See **Figure 10-1**. The main advantage of using computerized estimating is it creates a more streamlined process. Once again, the appraiser's *skill* and *experience* are critically important, and the overall information inputting process remains basically the same. The estimated times to restore repairable parts, the decision to replace the unrepairable ones, and decisions such as which operations will be completed "in-house" as opposed to being subletted to a **specialty shop** are still made by the damage report's writer. However, there are functions that can be completed by the computerized system as the information is entered, thus streamlining the process by saving time and promoting added efficiency.

Some of the features that are provided by computerized estimating include:

- Automatic mathematical calculations and **cost totaling** (that eliminates human error in that regard).
- Neat, "official-looking" documents that can help make a good impression on customers.
- Automatic technical calculations such as overlap, and additional or **included operations**. For example, if a quarter panel and rear face bar are replaced, the overhaul time for the face bar includes remove and install (R&I) time for that assembly. The R&I time is also included in the *replacement* of the quarter panel. The computer can deduct the "overlapping" included operation time *automatically*.
- A list of part numbers to assist in making a parts order and obtaining parts.
- Storage of **claim-related documentation**.
- Allowing the repair shops and insurers the ability to work "online" with each other, and communicate with customers and vendors electronically.
- The ability to track productivity, workflow, and **job costing**. Shops can analyze sales trends, profits by job, and even which area or company the shop's work has been coming from during any period of time.
- The creation of management reports from a conglomerate of different claims and business transactions.

Although there are advantages to computerized estimating, the need to know and understand areas like vehicle construction, repair and replacement procedures, and **OEM** (original equipment manufacturer) repair requirements remain. Think of the difference between manual and computerized estimating this way: If one were to take away a construction crew's power tools, they would still be able to build a house with *hand tools*. It may take a bit longer and/or be a bit more difficult to accomplish but the carpentry skills, safety factors, mathematical acumen, and ongoing education and experience needed to complete the project *remains the same*. Similarly, an appraiser should be able to *manually create* a damage report, if a computer is not available to provide assistance.

It is important to understand how to write—as well as *read*—a computerized estimate. A computerized damage report uses symbols to alert one to any "overrides." An override occurs when the writer makes a manual entry in place of an entry that would normally be created by the computer's software program. Codes and symbols such as pound signs (#) and asterisks (*) typically signal things like:

- A user-entered allowance or value
- **Recycled part** and **aftermarket** part pricing
- **Judgment labor times** (never supplied by the database or crash guide for *repairs*)
- **Judgment replacement times**
- Adjustments for price changes, or additions such as access time

In any case, whether dealing with a computer or a handwritten report (when reading, or creating an estimate) *know the estimating system*, its codes, "rules," where **options** are listed, what **labor rates** are being allowed for whatever task, how allowances are generated, what—if anything—has been "overrided," and what is included and not included in the various allowances.

WE CARE INSURANCE
234 Maple Drive, Anywhere, USA 11234
555-4000

Damage Assessed By. Joseph Smith

Claim Rep: Sally Jones

* Product Type : Auto
* Date of Loss : 04/30/2009
* Deductible : 1,000.00
Days to Repair: 4

Insured: PETER PRATTI
Address: 123 Main Street, USA
Telephone: Work Phone: 555-1000 Home Phone: 555-2000

Mitchell Service: 911754

Description: 2005 Toyota Corolla S
Body Style: 4D Sed
VIN: 2T1BR32E75C100000
Mileage: 8,745
OEM/ALT: A
Color: GRAY

Vehicle Production Date: 10/04
Drive Train: 1.8L Inj 4 Cyl 4A FWD
License: DBU-1357 TN

OPTIONS: ANTI-LOCK BRAKE SYS. (ABS), ALUM/ALLOY WHEELS, AIR CONDITIONING, POWER STEERING
POWER BRAKES, POWER WINDOWS, POWER DOOR LOCKS, TILT STEERING WHEEL
CRUISE CONTROL, ELECTRIC DEFOGGER, AUTOMATIC TRANSMISSION
AM-FM STEREO/CD PLAYER (SINGLE), PASSENGER-FRONT AIR BAG, POWER REMOTE MIRROR
4 WHEEL DISC BRAKES, FRONT WHEEL DRIVE, L-4 ENGINE, 4-DOOR, DRIVER-FRONT AIR BAG

Line Item	Entry Number	Labor Type	Operation	Line Item Description	Part Type/ Part Number	Dollar Amount	Labor Units
				FRONT BUMPER			
1.	100027	BDY	OVERHAUL	FRT COVER ASSY			1.8 #
2.	102343	BDY	REPAIR	FRT BUMPER COVER	Existing		1.5*#
3.		REF	REFINISH/REPAIR	FRT BUMPER COVER		C	1.3*
4.	102367	BDY	REMOVE/REPLACE	R FRT BUMPER SPOILER	76081-02030-B1	350.10	INC
				FRONT LAMPS			
5.	102382	BDY	REMOVE/REPLACE	R FRT COMBINATION LAMP ASSEMBLY	81110-02370	214.87	INC #
6.		BDY	CHECK/ADJUST	HEADLAMPS			0.4
				FRONT FENDER			
7.	100220	BDY	REMOVE/INSTALL	R FENDER ASSY			1.4 #
8.	100224	BDY	REPAIR	R FENDER PANEL	Existing		2.0*#
9.				refinish in repair panel			
10.		REF	REFINISH/REPAIR	R FENDER PANEL		C	1.8*
				FRONT INNER STRUCTURE			
11.	100255	BDY	REPAIR	UPR FRONT BODY TIE BAR -S	Existing		1.0*
12.		REF	REFINISH	UPPER TIE BAR			0.5
13.	102445	BDY	REPAIR	R FRONT BODY FRONT APRON PANEL -S	Existing		1.0*#
14.		REF	REFINISH	R APRON			0.5
				ABS/BRAKES			
15.	102463	MCH	REMOVE/INSTALL	ABS MODULATOR UNIT -M	Existing		0.5*
16.				d & r for apron repair			
				FRONT SUSPENSION			

FIGURE 10-1 A computerized estimate is often used in collision repair shops. The entire (handwritten version of this) estimate will be analyzed beginning on page 253. (See also Figure 10-5.)

Digital Imaging

Another method of creating damage reports is through the use of *digital imaging*. In this method, the repair shop takes electronic or digital photographs (or video images) of a damaged vehicle and attaches and sends them online to insurers (and back). Some of the procedures that can be accomplished through the use of digital imaging include allowing:

- Some insurance appraisers to write damage estimates from the images without leaving their office.
- Appraisers to ascertain if damages are related to a new claim, or are from a preexisting one.
- Insurance claims handlers to view a request for a supplemental estimate without going to the repair shop.
- Reduced repair **cycle time**, and payments to be made to the repairer much faster.
- Easier storage of claim records and documentation.

Using imaging tends to reduce the time needed to accomplish certain tasks. For example, in most cases, supplements that would normally take a few days might be completed in less than an hour! Collision repair training and technical information can be viewed, stored, and shared over the Internet. In fact, it is reaching the point where a shop will have to be "electronically connected" to conduct business in many aspects—whether time conscious or not. Many direct repair relationships are now built around—and conducted over—the Internet using computerized estimating, electronic communications, and digital imaging. With a direct repair relationship (covered in the Insurance Coverages section of this chapter) an insurer may send an interested customer to a repairer, contact the shop via **electronic mail (e-mail)**, and send the claim and insurance coverage information to the shop online. In turn, the shop may write a computerized estimate, take digital images of the damages, and send the entire electronic file back to the insurer. Any discrepancies, additional (supplemental) estimates, and discussion can be handled in like manner, and payment can be sent faster, oftentimes being deposited electronically. Speed—along with a safe, quality repair—complements and brings collision repair centers and insurers together efficiently.

HOW DAMAGE ESTIMATES ARE MADE

A good damage report (estimate) begins with gathering as much relevant information as possible.

Obtaining Information Concerning the Loss

If possible, a conversation with the driver and/or owner of the vehicle in order to obtain details about the loss and *physical damage* is an excellent place to begin. Whether or not someone "ran a light" or was intoxicated may be important to the *liability* claim(s) that may arise but not necessarily to *the physical damage* to the vehicle.

Here are some of the relevant points one may wish to cover with the principals involved:

- *How fast were the vehicles involved traveling?* **Collision energy** travels through a vehicle; the more speed, the further the energy may have transversed the structure.
- *What direction were the vehicles traveling and what was the direction of impact?* Motor vehicles rarely strike each other or other objects symmetrically. Energy travels, and damages occur at different angles throughout.
- *What color were the other vehicle(s) or objects involved?* **Paint transfer** helps identify recent or old damages.
- *How much time has elapsed since the loss and has the vehicle been cleaned since?* Debris, and other visual clues in the undercarriage, wheels, and suspension parts, may now be removed.
- *Did you apply the brakes? If so, how hard?*
- *How many passengers were seated in the vehicle and which seat belts were in use?* Passengers may cause individual **inertia energy** to tweak parts of the vehicle's structure. Supplemental restraint system components in use may need to be replaced if they were being utilized during the collision.
- *Are there any unusual noises, feelings, or performances being heard, felt, or experienced that were not existent prior to the occurrence?*
- *Are there any accessories or functions no longer operating properly? Any steering or handling problems? Any dash lamps now illuminated?*
- *Was the vehicle driven after the collision or towed?* Towing and/or driving a collision-damaged vehicle can result in further damages.
- *Were temporary repairs made?*

Any other relevant facts should be gathered and noted. A thorough **damage analysis** and report can now begin.

Initial Vehicle Inspection

When beginning a physical damage analysis, appraisers will want to create a plan that allows all of the damage to be found. The following steps should be included during the planning process:

- *Imagine the direction and force of the impact(s); identify* **direct damage**.
- *Think about where—and how far—the collision energy traveled; identify* **indirect damage**.
- *Identify how and where the collision energy was absorbed.*

- *Determine if there is structural damage, and where. Raise the vehicle and check the undercarriage structure.*
- *Visualize how the repaired area(s) will be refinished.*
- *Ask yourself—Can the vehicle be safely repaired? Is it economically feasible to do so?*

A thorough, complete inspection increases the chances of avoiding supplements; utilizes measuring to identify and/or rule out structural damages; is utilized to develop a parts order, a **repair plan**, and a work order; and may require some disassembly to locate and/or rule out hidden damages.

This chapter will also cover how measuring a vehicle during damage analysis can help determine if it has (or does not have) structural damages. A tape measure, tram gauge, or center gauges can be used for this purpose. Later sections will also cover direct and indirect damages, structural parts, and the decision to repair or not repair.

Writing a Damage Report

Whether writing a damage report *longhand* or feeding the information into an estimating system's database, it takes a well-organized plan of action and ongoing education in the collision repair field to complete it properly. After the initial vehicle inspection, analysis, and complete identification, the *writing* begins. Writers can utilize the crash manual's layout as a guide or, if they have a specific method of working through the damages, they can utilize that instead. If using the manual as a guide, they should refer to the alphabetized Section Index on the first page of each specific vehicle. The manual is arranged from front-to-rear, from outside the vehicle inward, and usually has accompanying illustration(s) for each individual section. The illustrations are "exploded views" that usually present the most frequent kinds of collision damages. See **Figure 10-2**. If writers use the manual as a guide, they should work from the outside inward, and examine each damaged component and any attached parts. For each and every part, they should list the price, the labor time, and any refinish allowance necessary. If a part is deemed by the writer to be safely repairable, the estimated repair time—*which is never found in the guides*—will need to be noted. If writing the report by hand, it is a good idea to document any replacement part numbers.

Moving through each section of the guide affords the writer one of the sequential, systematic methods of writing a damage report. The writer should carefully follow all footnotes, references, and the **procedural** or **"P" pages**, which are described later in this chapter. When finished, the writer will carefully total up the entries (if writing without a computer).

For the purpose of demonstrating the *art of writing* a damage report, the following example will be completed without the aid of a *computer* database. See **Figure 10-4**.

TECH TIP A number of different sequential methods can be utilized in writing a collision damage estimate. For example, one can follow the crash guide's layout from front to back, going through each section of the guide and comparing it to the vehicle one is writing an estimate (damage report) for.

P&L Consultants, a Knoxville, Tennessee-based training and consulting firm, with an office in New York City, has a unique 10-Step Sequential Estimating Laminate® that covers all three sections of a vehicle (front, side, and rear) depending upon which area(s) of the vehicle are damaged. See **Figure 10-3**. P&L includes the laminate and the instructions for utilizing it in their Pre-Production Damage Analysis and Estimating Seminar©. Contact them at 865-288-0981 or 917-860-3588.

The information that one documents on a handwritten damage report is the same information that would need to be fed into a computerized estimating system; the basic differences being the longhand writer will have to total up the figures, as well as manually make adjustments—such as overlap, and included or **not included operations**. See **Figure 10-5**.

1. FRONT BUMPER SECTION: The bumper cover has been scraped on the right side; the attached right front plastic spoiler was torn off at the mounting tab and is not repairable. No other damage is visible to the front bumper assembly. The cover has been loosened and, when peeled back, reveals an undamaged reinforcement and absorber. See **Figure 10-6**.
   - **Line 1:** Overhaul front bumper assembly; 1.8 hours from the crash guide.
   - **Line 2:** Repair front bumper cover and refinish; 1.5 repair hours (estimated from experience) plus 2.5 hours of refinish time, and 1.0 hour of clearcoat (40 percent of first major panel of fascia) from the crash guide.
   - **Line 3:** Replace front bumper cover spoiler (right side); $350.10 with labor included in overhaul of bumper assembly. Part #76081-02030-Bl which comes (color impregnated) silver.
2. FRONT LAMPS SECTION: The right front combination (head-) lamp is broken at the tabs and is not repairable. Anytime a headlamp is removed and/or replaced, arming of both lamps should take place.
   - **Line 4:** Replace combination lamp assembly; $214.87 with labor of 0.2 hour. Overlap of 0.9—for the R&I of the cover and grille—was deducted from the full amount of replacing the headlamp since those operations are already

FRONT LAMPS

COMBINATION LAMP

Aim Lamps 4
R&J Combination Lamp Assy .... R/L #1.1

Includes R&I Front Cover Assy & R&I/R&R Grille Assy

NOTE: R&R Does Not include aim lamps.

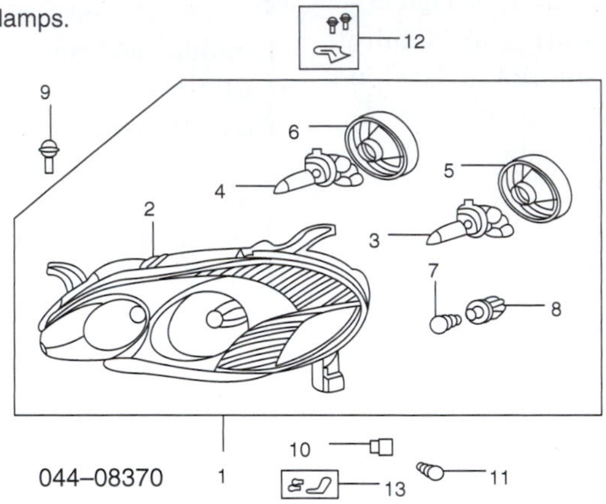

044–08370

1.	Lamp Assy, Combination				
	CE, LE Model	03-04 R	81110-02190	#1.1	214.87
		L	81150-02200	#1.1	214.87
		05 R/L	N.A.	#1.1	0.00
	S Model	03-04R	81110-02200	#1.1	214.87
		L	81150-02210	#1.1	214.87
		05 R	81110-02370	#1.1	214.87
		L	81150-02360	#1.1	214.87
		R	81110-02210	#1.1	214.87
	XRS Model	L	81150-02220	#1.1	214.87
2.	Lens & Housing¶				
	CE, LE Model	03-04 R	81130-02190	#1.1	157.51
		L	81170-02200	#1.1	157.51
		05 R/L	N.A.	#1.1	0.00
	S Model	03-04 R	81170-02200	#1.1	157.51
		L	81170-02210	#1.1	157.51
		05 R/L	N.A.	#1.1	0.00
	XRS Model	R/L	N.A	#1.1	0.00
	#Includes R&I Front Cover Assy & R&I/R&R Grille Assy				
3.	Bulb, Low Beam¶	R/L	90981-13047	#.2	19.41
4.	Bulb, High Beam¶	R/L	90080-81041	#.2	20.63
5.	Cover, Low Beam¶	R/L	90069-81008		7.80
6.	Cover, High Beam¶	R/L	90069-81008		7.80
7.	Bulb, Signal Lamp¶				
	CE, LE, S Model	R/L	90084-98027	#.2	3.94
	XRS Model	R/L	90084-98048	#.2	3.94
	#R&R One Side Complete, included in R&I/R&R Combination Lamp Assy				
8.	Socket, Signal Bulb				
	CE, LE, S Model	R/L	90075-60049		7.26
	XRS Model	R/L	90075-99106		7.34
	¶included w/Combination Lamp Assy				
9.	Balt, Upr Lamp (2/Side)	R/L	90084-10015		.63
10.	Insert, Bolt	R/L	90179-06127		.57
11.	Bolt, Lwr Lamp	R/L	90105-06246		.52
	Repair Kit, Housing				
12.	Upper	R	81193-02030	#.3	6.69
		L	81194-02030	#.3	1.74
13.	Lower	R	81195-02040	#.3	6.69
		L	81196-02030	#.3	6.69
	#w/Combination Lamp Assy Removed				

FIGURE 10-2 This exploded view of a headlamp is found in the Mitchell Crash Guide.

AUTOMOBILE PHYSICAL DAMAGE ESTIMATING SEQUENCES

FRONT SEQUENCE

1- **Structural Damage Analysis.** *Unibody/frame realignment needed? Set-Up & Measure? Number of pulls required? Create a repair plan. Determine the condition(s), extent of damages and the time needed.*

2- **Frame Repairs.** *Repair or Replace upper & lower frame rails, aprons, firewall, cowl, windshield & posts and all attached parts* after *specification restoration.*

3- **Radiator Core Support Analysis.** *Repair or Replace? Includes the support assembly and all other attached parts (i.e. radiator, condenser, information labels, air bag sensors, hoses, etc.)*

4- **Face Bar Assembly.** *Repair or Replace? Includes reinforcement, cover, absorbers and all other related attached parts.*

5- **Headlamp Mounting and/or Header.** *Repair Or Replace? Includes headlamps, grille(s), moldings & related attached parts.*

6- **Hood Assembly.** *Repair or Replace? Includes all related attached parts.*

7- **Fenders (left & right).** *Repair or Replace? Includes liners, splash shields, moldings, lamps and all other attached parts.*

8- **Steering & Suspension Analysis.** *Wheels through Alignment. Includes wheels, tires, S&S, cradle analysis and all other related parts and/or operations.*

9- **Mechanical Analysis.** *Drive Train & Mechanicals check. (Includes engine block, related and/or attached components, transmission, brakes, and related computer and/or electrical systems.*

10- **Supplemental Restraint System Analysis.** *Air Bag, Seat Belts, Dash Panel, Seats & all other related components check (includes any/all interior damages).*

FIGURE 10-3 P&L Consultants' 10-Step Sequential Estimating Laminate® covers all three sections of a vehicle. *(Continued)*

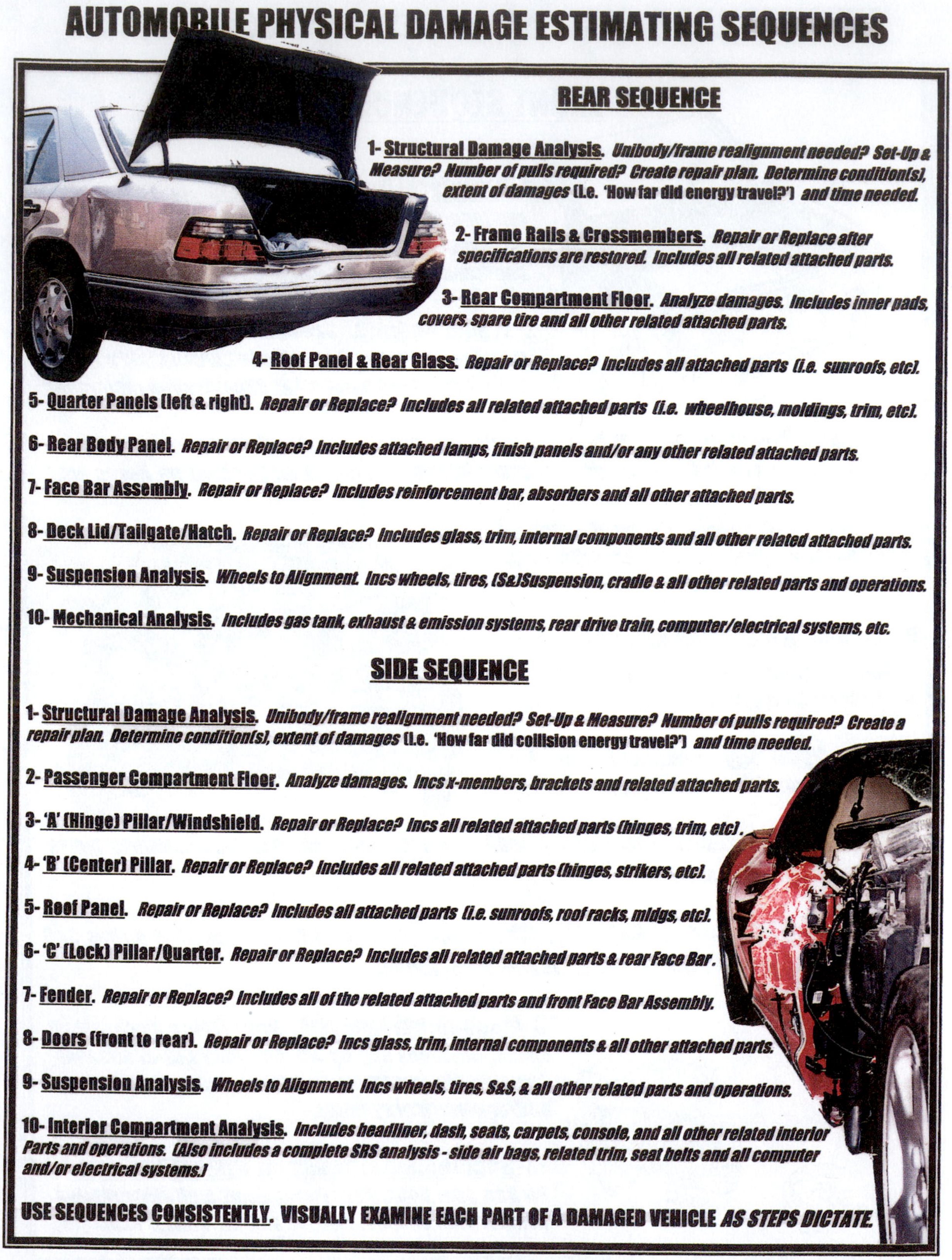

AUTOMOBILE PHYSICAL DAMAGE ESTIMATING SEQUENCES

REAR SEQUENCE

1- **Structural Damage Analysis**. *Unibody/frame realignment needed? Set-Up & Measure? Number of pulls required? Create repair plan. Determine condition(s), extent of damages* (i.e. 'How far did energy travel?') *and time needed.*

2- **Frame Rails & Crossmembers**. *Repair or Replace after specifications are restored. Includes all related attached parts.*

3- **Rear Compartment Floor**. *Analyze damages. Includes inner pads, covers, spare tire and all other related attached parts.*

4- **Roof Panel & Rear Glass**. *Repair or Replace? Includes all attached parts (i.e. sunroofs, etc).*

5- **Quarter Panels** (left & right). *Repair or Replace? Includes all related attached parts (i.e. wheelhouse, moldings, trim, etc).*

6- **Rear Body Panel**. *Repair or Replace? Includes attached lamps, finish panels and/or any other related attached parts.*

7- **Face Bar Assembly**. *Repair or Replace? Includes reinforcement bar, absorbers and all other attached parts.*

8- **Deck Lid/Tailgate/Hatch**. *Repair or Replace? Includes glass, trim, internal components and all other related attached parts.*

9- **Suspension Analysis**. *Wheels to Alignment. Incs wheels, tires, (S&)Suspension, cradle & all other related parts and operations.*

10- **Mechanical Analysis**. *Includes gas tank, exhaust & emission systems, rear drive train, computer/electrical systems, etc.*

SIDE SEQUENCE

1- **Structural Damage Analysis**. *Unibody/frame realignment needed? Set-Up & Measure? Number of pulls required? Create a repair plan. Determine condition(s), extent of damages* (i.e. 'How far did collision energy travel?') *and time needed.*

2- **Passenger Compartment Floor**. *Analyze damages. Incs x-members, brackets and related attached parts.*

3- **'A' (Hinge) Pillar/Windshield**. *Repair or Replace? Incs all related attached parts (hinges, trim, etc).*

4- **'B' (Center) Pillar**. *Repair or Replace? Includes all related attached parts (hinges, strikers, etc).*

5- **Roof Panel**. *Repair or Replace? Includes all attached parts (i.e. sunroofs, roof racks, mldgs, etc).*

6- **'C' (Lock) Pillar/Quarter**. *Repair or Replace? Includes all related attached parts & rear Face Bar.*

7- **Fender**. *Repair or Replace? Includes all of the related attached parts and front Face Bar Assembly.*

8- **Doors** (front to rear). *Repair or Replace? Incs glass, trim, internal components & all other attached parts.*

9- **Suspension Analysis**. *Wheels to Alignment. Incs wheels, tires, S&S, & all other related parts and operations.*

10- **Interior Compartment Analysis**. *Includes headliner, dash, seats, carpets, console, and all other related interior Parts and operations. (Also includes a complete SRS analysis - side air bags, related trim, seat belts and all computer and/or electrical systems.)*

USE SEQUENCES CONSISTENTLY. VISUALLY EXAMINE EACH PART OF A DAMAGED VEHICLE *AS STEPS DICTATE.*

FIGURE 10-3 Continued

FIGURE 10-4 This 2005 Toyota Corolla 'S' four-door sedan with a clearcoated silver color that was struck in the right front is drivable. The step-by-step analogy on pages 253, 257, and 259 of the writing of the damage report is illustrated in Figure 10-5.

included with bumper overhaul time in header notes. Part #81110-02370

- **Line 5**: Aim headlamps; 0.4 hour from crash guide. (Refer to Figure 10-2 for front lamp section of crash guide.)

3. HOOD SECTION: The right corner underside has a small imperfection from sliding against the fender. Rather than disturb the factory finish and clearcoat on this part, the repairer decided to touch up the (unseen) area for future corrosion protection purposes.
   - **Line 6**: Repair the hood panel and touchup (estimated from experience).
4. FRONT FENDER SECTION: The right fender is damaged but repairable; 2.0 estimated repair hours (from experience) plus full refinish of 1.8 and 0.4 hour of clearcoat (20 percent of additional panels' refinish time) with no overlap taken from base refinish time, since the adjacent bumper cover is NOT considered a major panel. See **Figure 10-7**.
   - **Line 7**: Repair the fender and refinish (inside the fender—avoid color at rear portion of the fender which is undamaged) with no need to disturb (blend) right front door for color match; 2.0 hours to repair plus 1.8 base refinish time and 0.4 hour for clearcoat.
5. FRONT INNER STRUCTURE SECTION: The right front side of the **vehicle's nose** collapsed ever so slightly toward the centerline. Measurements taken of the point-to-point specifications reveal the right side apron and tie bar to be "in" 4 millimeters. The support tie bar is bent slightly at the right end while the right fender apron is bent at the same juncture. The structure will need to be pulled back* into specification, and the bent areas repaired. Any components that could impede repairs and/or refinishing—in this case, the undamaged **ABS modulator** and the undamaged windshield washer bottle—will need to be moved out-of-the-way for access. See **Figure 10-8**.
   - **Line 8**: Repair the radiator support upper tie bar and refinish (after pulling operations); 1.0 hour (from experience) plus 0.5 hour from the crash guide to refinish with no clearcoat necessary.
   - **Line 9**: Repair the right side front section of the apron panel and refinish; 0.5 hour (from experience) to repair and 0.5 refinish time from the crash guide.
6. ABS/BRAKES SECTION
   - **Line 10**: Disconnect and reconnect (D&R) the ABS modulator to refinish the apron (time estimated, based upon what needs to be done—lift unit off the surface and mask-off area).
7. WINDSHIELD (Wiper System) SECTION
   - **Line 11**: Remove and install (R&I) undamaged washer bottle for access to repair the apron; 0.5 hour from the guide.
8. WHEEL SECTION: The right front alloy wheel was scraped in the collision. No other damages exist to the tire, suspension, or steering. However, any hit to a wheel requires that an alignment adjustment be done. In addition, the tire will need to be mounted to the replacement rim, a new valve stem installed, and a high-speed balancing accomplished. The replacement wheel will also need to be remounted on the vehicle. Note that in this case, the repairer chose to "exchange" the alloy wheel for an aftermarket company's one in order to save the cost of purchasing a new OEM wheel.
   - **Line 12**: Replace the right front alloy road wheel (aftermarket); $129.95 plus 0.3 labor (from crash guide).
   - **Line 13**: Valve stem and a high speed balancing; $15. Entered as a sublet operation since the repairer does not have the equipment for this service.
9. FRONT SUSPENSION SECTION
   - **Line 14**: Four-wheel alignment; 1.4 hours as written from the crash guide.
   - **Line 15**: Measure and pull tie bar and apron (from previous explanation above); 2.5 hours (from experience). See **Figure 10-9**.
10. MISCELLANEOUS ENTRIES: Paint and materials (P&M) costs—and in this case rust-proofing and a car cover—are entered at the end of the estimate.

**Note: See Line 15 for "pull time."*

Car Owner Peter Pratti
Business Phone 555-1000
Date 5/1/08

Address 123 Main St. USA
Home Phone 555-2000
Due Date 5/4/08

Insurance Co. WE CARE
Phone 555-4000
Retain Parts Yes ☑ No ☐ VIN

VIN 2TTBR32E75C100000
Adjuster Joe Smith
Destroy Parts Yes ☐ No ☑

Year	Make	Model	License No.	Speedometer
2005	Toyo	Corolla 'S'	DBV-1357	8,745

Customer Initial [signature]

Line No.	Repair	Replace	Details of Repair and Parts Index	Parts	Labor	Paint	Sublet/ Misc.
1	O/H	-	BUMPER ASSEMBLY (Front)		1 8		
2	✓	-	BUMPER COVER (Front) & Refinish		1 5	2 5 / 1 0c/c	
3	-	✓	BUMPER COVER (Front) SPOILER	350 10	I O/H		
⟶			PART #76081 - 02030 - Bi (Silver)				
4	-	✓	COMBINATION LAMP (R/S)	214 87	0 2 (0/2)		
⟶			PART #8110-02370				
5	✓	-	AIM HEADLAMPS		0 4		
6	✓	-	HOOD PANEL & TOUCH-UP		0 2	0 2	
7	✓	-	FENDER PANEL (R/S) & Refinish		2 0	1 8 .4ck	
8	✓		RADIATOR SUPPORT TIE BAR & ref		1 0	0 5	
9	✓	-	APRON PANEL (R/S Front) & Refinish		0 5	0 5	
10	D&R	-	ABS MODULATOR (To refinish Apron)		0 4		
11	R&I	-	WASHER TANK ASSEMBLY (for Ace)		0 5		
12	-	✓	ROAD WHEEL (R/S front) Aftermarket	129 95	0 3		
13	⟶		VALVE STEM & High Speed Bal				15 00
14	⟶		4-Wheel Alignment		1 4		
15	✓	-	MEASURE & PULL TIEBAR/APRON		2 5		
16	⟶		P & M/Clearcoat				176 24
17	⟶		RUST-Proofing				15 00
18	⟶		CAR COVER				5 00

Total Parts		$ 694.92
Total Labor	12.7 × $50.00	$ 635.00
Total Paint	6.9 × $50.00	$ 345.00
Total Sublet		211.24
Subtotal		$ 1886.16
Tax	5 %	$ 94.31
EPA/Waste Disposal Charge		$ ——
Total Charges		$ 1980.47
Revised Estimate		$ ——
Customer ______		Date 5/1/08

FIGURE 10-5 This is the complete handwritten damage report for the 2005 Toyota Corolla 'S' from Figure 10-4.

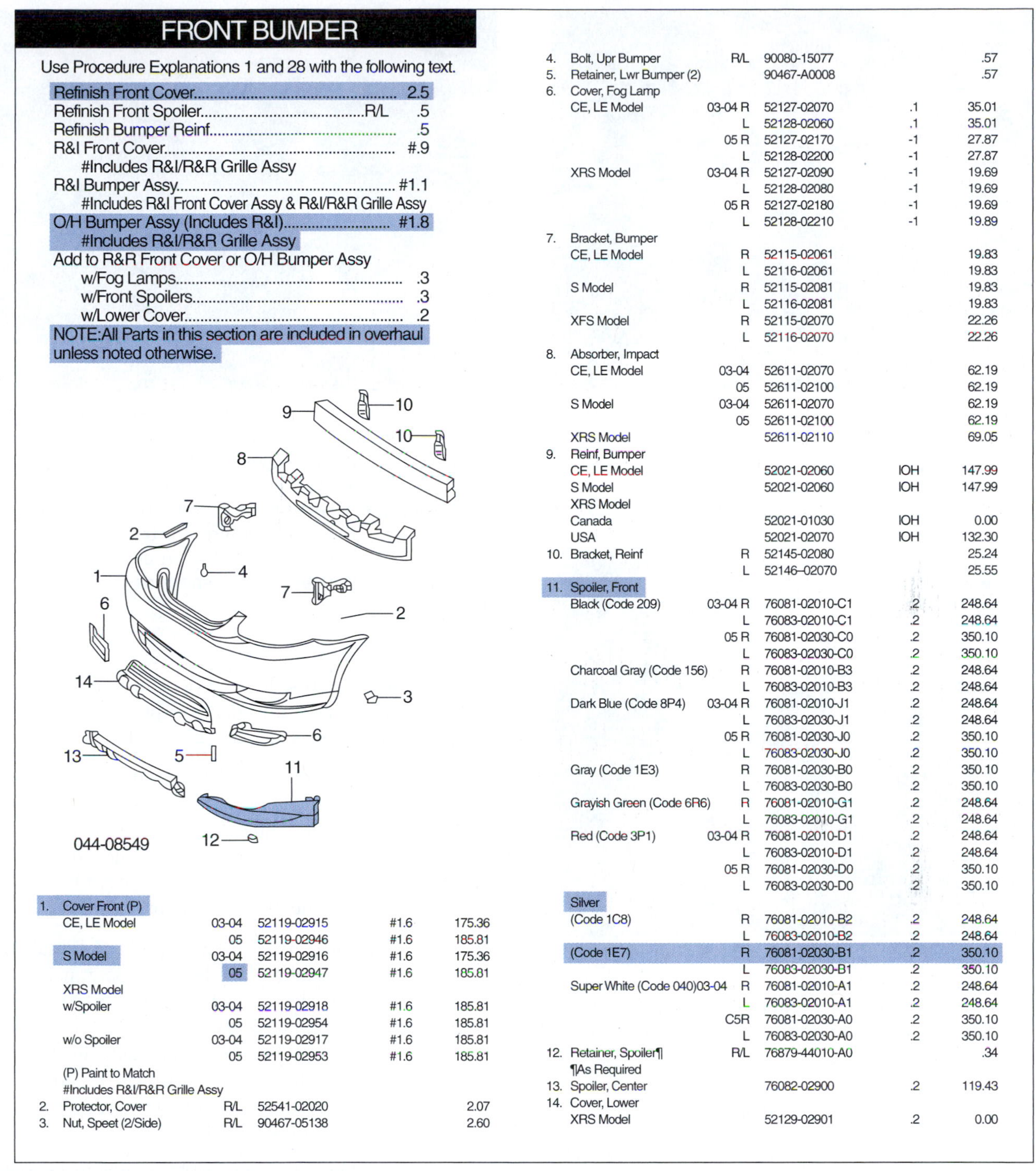

FRONT BUMPER

Use Procedure Explanations 1 and 28 with the following text.

Refinish Front Cover.. 2.5
Refinish Front Spoiler.......................................R/L .5
Refinish Bumper Reinf... .5
R&I Front Cover... #.9
#Includes R&I/R&R Grille Assy
R&I Bumper Assy... #1.1
#Includes R&I Front Cover Assy & R&I/R&R Grille Assy
O/H Bumper Assy (Includes R&I).......................... #1.8
#Includes R&I/R&R Grille Assy
Add to R&R Front Cover or O/H Bumper Assy
w/Fog Lamps.. .3
w/Front Spoilers.. .3
w/Lower Cover.. .2
NOTE: All Parts in this section are included in overhaul unless noted otherwise.

1.	Cover Front (P)				
	CE, LE Model	03-04	52119-02915	#1.6	175.36
		05	52119-02946	#1.6	185.81
	S Model	03-04	52119-02916	#1.6	175.36
		05	52119-02947	#1.6	185.81
	XRS Model				
	w/Spoiler	03-04	52119-02918	#1.6	185.81
		05	52119-02954	#1.6	185.81
	w/o Spoiler	03-04	52119-02917	#1.6	185.81
		05	52119-02953	#1.6	185.81
	(P) Paint to Match				
	#Includes R&I/R&R Grille Assy				
2.	Protector, Cover	R/L	52541-02020		2.07
3.	Nut, Speet (2/Side)	R/L	90467-05138		2.60
4.	Bolt, Upr Bumper	R/L	90080-15077		.57
5.	Retainer, Lwr Bumper (2)		90467-A0008		.57
6.	Cover, Fog Lamp				
	CE, LE Model	03-04 R	52127-02070	.1	35.01
		L	52128-02060	.1	35.01
		05 R	52127-02170	-1	27.87
		L	52128-02200	-1	27.87
	XRS Model	03-04 R	52127-02090	-1	19.69
		L	52128-02080	-1	19.69
		05 R	52127-02180	-1	19.69
		L	52128-02210	-1	19.89
7.	Bracket, Bumper				
	CE, LE Model	R	52115-02061		19.83
		L	52116-02061		19.83
	S Model	R	52115-02081		19.83
		L	52116-02081		19.83
	XFS Model	R	52115-02070		22.26
		L	52116-02070		22.26
8.	Absorber, Impact				
	CE, LE Model	03-04	52611-02070		62.19
		05	52611-02100		62.19
	S Model	03-04	52611-02070		62.19
		05	52611-02100		62.19
	XRS Model		52611-02110		69.05
9.	Reinf, Bumper				
	CE, LE Model		52021-02060	IOH	147.99
	S Model		52021-02060	IOH	147.99
	XRS Model				
	Canada		52021-01030	IOH	0.00
	USA		52021-02070	IOH	132.30
10.	Bracket, Reinf	R	52145-02080		25.24
		L	52146-02070		25.55
11.	Spoiler, Front				
	Black (Code 209)	03-04 R	76081-02010-C1	.2	248.64
		L	76083-02010-C1	.2	248.64
		05 R	76081-02030-C0	.2	350.10
		L	76083-02030-C0	.2	350.10
	Charcoal Gray (Code 156)	R	76081-02010-B3	.2	248.64
		L	76083-02010-B3	.2	248.64
	Dark Blue (Code 8P4)	03-04 R	76081-02010-J1	.2	248.64
		L	76083-02030-J1	.2	248.64
		05 R	76081-02030-J0	.2	350.10
		L	76083-02030-J0	.2	350.10
	Gray (Code 1E3)	R	76081-02030-B0	.2	350.10
		L	76083-02030-B0	.2	350.10
	Grayish Green (Code 6R6)	R	76081-02010-G1	.2	248.64
		L	76083-02010-G1	.2	248.64
	Red (Code 3P1)	03-04 R	76081-02010-D1	.2	248.64
		L	76083-02010-D1	.2	248.64
		05 R	76081-02030-D0	.2	350.10
		L	76083-02030-D0	.2	350.10
	Silver				
	(Code 1C8)	R	76081-02010-B2	.2	248.64
		L	76083-02010-B2	.2	248.64
	(Code 1E7)	R	76081-02030-B1	.2	350.10
		L	76083-02030-B1	.2	350.10
	Super White (Code 040)	03-04 R	76081-02010-A1	.2	248.64
		L	76083-02010-A1	.2	248.64
		C5R	76081-02030-A0	.2	350.10
		L	76083-02030-A0	.2	350.10
12.	Retainer, Spoiler¶	R/L	76879-44010-A0		.34
	¶As Required				
13.	Spoiler, Center		76082-02900	.2	119.43
14.	Cover, Lower				
	XRS Model		52129-02901	.2	0.00

FIGURE 10-6 The front bumper section from the crash guide. The information from this section that was utilized to create the first four lines of the estimate for the 2005 Toyota Corolla 'S' are highlighted in blue.

P&M costs were computed from the 5.3 hours of base refinish time and the 1.4 hours of clearcoat time. The actual paint code was found on the driver's door jamb (in this case).

- **Line 16:** Paint and materials total $176.24 from a refinish materials guide.
- **Line 17:** Rust-proofing is a $15 entry and is used to protect the repaired fender.
- **Line 18:** A car cover ($5.00) is utilized to protect the vehicle past the refinish time—included 36 inches.

Totaling the Estimate

After checking over all entries, calculate the totals.

Parts. Add up the prices of the parts written: in our example, the bumper spoiler, combination lamp, and the aftermarket

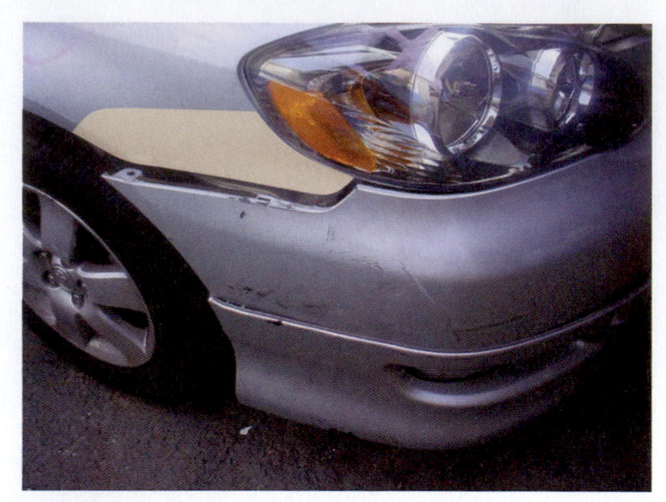

FIGURE 10-7 The damaged area of the fender (tan area) is repairable. There will be good access to the damage from the *inside* as well as the *outside* of the fender since the headlamp will be removed for replacement.

FIGURE 10-9 Point-to-point measurements should be taken during damage appraisal. The measurement between the left strut tower and the right front area of the nose is considered with a tram gauge. The right front apron section and the tie bar is collapsed 4 millimeters. Note the tan-highlighted section of the tie bar. It is the same area of the structure that "kinked" and is the exact same area of the tie bar as seen in Figure 10-8.

FIGURE 10-8 The radiator support tie bar is "kinked" as pointed out. After pulling the structure back to correct specifications, the tan-highlighted area of the tie bar will require repair and refinishing.

wheel total $694.92. Once the repairer has permission to repair the vehicle, these parts should be ordered.

Labor. Total up the labor allowances on the sheet. The 12.7 hours allowed for all labor (except refinish) is multiplied by the hourly rate that the shop operates at. (When an insurance company is involved, the shop may negotiate with them and agree upon an hourly rate.) Hourly rates can fluctuate within an estimate. There may be a designated rate for body, frame, mechanical, and/or paint labor. Thus, an estimate may be written with two or more different hourly rates. This example uses $50 per hour for *all* operations; $50 multiplied by 12.7 equals $635.00 for labor.

Paint. The refinish allowances total 6.9 hours. Multiplying 6.9 by $50—our fictional hourly rate—yields a monetary paint labor allowance of $345.00. This figure does NOT include any *materials* that may be necessary to accomplish the refinishing work. An in-depth explanation of how to calculate refinish hours appears later in this chapter.

Sublet. Sublet—and/or "miscellaneous items"—also need to be considered. (Sublet costs are covered later.) There is a total of $211.24 for "Sublet/Misc" items. This includes $15 for a valve stem and high-speed balancing for the right front tire, $176.24 for paint materials,* $15 for rust-proofing materials, and $5 for a car cover.

Tax. We are now ready to calculate a taxable subtotal. This is done by adding total parts, labor, refinish time, sublet, and miscellaneous charges. That figure ($1,886.16) is multiplied by the prevailing tax rate in the area for a final gross taxable sum. In this case, we utilized a fictional tax rate of 5 percent. Thus 0.05 multiplied by $1,886.16 equals $94.31 tax. Adding the subtotal to the tax figure gives us a gross total of $1,980.47.

After-Tax Add-Ons. After-tax add-ons would be added last. For example, if this car was towed and an independent company charged a taxable fee, it might be added on as an after-tax allowance (with no additional taxes added).

**Paint materials (P&M) allowance was calculated by utilizing a* Refinish Materials Guide *published by* Mitchell International, Inc. *Some repairers and/or insurers may prefer to use a similar guide, or calculate the materials utilizing a "multiplier," which is a figure that the total refinish hours are multiplied by. Hence, if $20 was set as the P&M multiplier, one would multiply that figure by 6.9—the total number of refinishing hours allowed. Thus, the P&M allowance would be $138.*

If an insurance deductible applies, it would be subtracted from the final figure. The "after-deductible" figure is paid by the insurer while the deductible amount—owed to the repair shop—is the vehicle owner's responsibility.

Supplements and Hidden Damage

No matter how diligent, knowledgeable, and experienced a damage appraiser is, there will be occasions when the repairer will locate "hidden damage"—damages that could not be seen, or simply went undetected during the initial inspection. These discoveries usually result in an additional report called a *supplement*.

A supplement contains additions (and/or subtractions) to an initial damage report. Supplements can also be generated by parts price differences, errors made on the initial inspection, and changes to a repair plan after work has begun—in addition to finding hidden damages. It is best to avoid supplements when at all possible since they can result in delays, additional operating costs, and a loss of customer confidence in the entire process.

Sales (Selling). It is often said that successful businesses view themselves as "a product to be sold each and every day." How does a repair shop "sell its product"? That process must *start* and revolve around a professional damage report. The damage report is one of, if not THE first impression one makes upon a customer. Only after a repairer produces a proper, thorough inspection and an accurately detailed report can the shop know *exactly* what challenges are to be faced, and what will be needed to *sell* a profitable yet competitive product. Utilizing a damage report, one can

- *Plan* the repairs properly
- Know the exact *cost* of the job and what will be needed to accomplish the work ("If you cannot measure it, you cannot manage it. . . . Your *business* that is!")
- Accurately communicate how much *time* will be needed (Shops compete for time—which *is* money)
- *Educate* customers (a smart, well-informed consumer is a shop's ally!)
- Gain and retain *control* of the situation, as well as the business
- Keep up with the competition
- "**Upsell**" the job, selling services unrelated to the loss (instead of doing various repairs free of charge!)

One study has shown that approximately 35 percent of all estimates result in a "sale." Out of that number, approximately one-third of those customers will develop other relationships, become dissatisfied, and move (or pass) away. THE REMAINING TWO-THIRDS OF THEM BECOME *INDIFFERENT!* Can collision repairers afford NOT to take the time to properly prepare themselves, thus making a tremendous and lasting impression on their customers? Those who can write a professional, highly detailed estimate can view themselves from that moment forward as *salespeople!*

Parts Order. From an accurate estimate, a technician can formulate *a parts list*. To order the parts and components needed to repair a vehicle, one could need some—or all—of the items found on a professionally prepared estimate that are listed below:

- Vehicle identification number (VIN)
- Production date
- Options and "packages"
- Color and trim codes
- Engine and/or transmission code
- Specific parts and their numbers

Repair Order. A repair or work order is prepared from the damage report. It lists all of the operations and parts needed for the technician to perform the specific repairs. How accurate the estimate is can have a direct or indirect effect on the:

- Quality of the work
- Timeliness of completion
- Proper payment for some collision technicians
- Overall profitability of the repair center

INSURANCE COVERAGE

A majority of collision-damaged vehicles are covered by an insurance policy. Insurance companies fund over 85 percent of all collision repair jobs during any given year since motor vehicle insurance is mandatory in most areas of the country, and physical damage insurance is usually mandatory if a vehicle is being paid for over *time*. The entity that loaned the driver the collateral to pay for the vehicle, or leased the vehicle to the driver, typically wants to protect their investment until the loan is satisfied. In fact, most people who purchase new vehicles usually purchase physical damage coverage to protect the vehicle itself, as well as liability insurance to protect *themselves* in the event they become responsible for damages to another person's vehicle or property while operating it.

Insurance was conceived in order to "spread the losses of a few among the masses." To simplify this concept, picture it this way: if one out of every one hundred *insured* persons were to sustain a loss of one hundred dollars, the hundred people contribute one dollar *each*—rather than that one person sustaining *the entire loss of one hundred dollars*. Naturally, it is a bit more complicated than that but it remains the basic premise of insurance coverage.

How does an insurance policy affect the collision repair center? When an insured vehicle is involved in a loss, the policy's owner usually reports the damages to the company that holds that policy. This can be done directly or, at times, through an **independent agent** that sold the policy to the vehicle's owner on behalf of the insurer. The agent and/or the claims handler establishes a file and assigns a claim number to the case. The file will contain documents and information that are vital to settling the claim and getting the vehicle repaired properly and in a timely manner. Those documents may include:

- Police reports and statements
- Any prior loss history of the vehicle and/or the driver
- Vehicular loan information (all lienholders)
- Appraisals or damage reports, and photos of the damages
- All related receipts and invoices
- Involved outside agency information such as theft bureau, motor vehicle, and insurance department reports
- Statements from drivers, passengers, witnesses, and others with any link to the claim or vehicle, and all documentation of any conversations with involved parties

Once insurance coverage has been verified to be in force on the vehicle for the date and time of the accident or damage occurrence, a **reserve** (of money) is established to pay the eventual costs of the claim, contacts are made, a damage appraiser is sent to see the vehicle, and any necessary investigations begin.

Vehicle damages are commonly appraised in the collision repair center of the vehicle owner's choice. However, depending upon (1) the circumstances of the vehicle's owner, (2) if the vehicle is still safely driveable or not, and (3) any programs or protocols that the insurer follows, the vehicle may be inspected for damages at:

- The owner's home or workplace
- An insurer's drive-in claim center
- Various repair centers
- A preferred collision repair provider recommended by the insurance company

It is a good practice to have the vehicle inspected in an environment where all damages can be safely and efficiently observed and analyzed—preferably with the repairing center's representative present for collaboration and agreement. After writing a damage report or estimate, the insurance company's representative typically reviews the damages with the repair center's representatives—who often will have created a damage report of their own. After review, discussion, and negotiation of points such as damages, repairs and replacement parts needed, labor rates and allowances, and who will be paying for what, the insurance company and repair center representatives need to come to an agreement—commonly called an agreed price (A/P)—so repairs can commence.

While attempting to reach an agreement with an insurer, many points in the process can make the experience positive and cause the least discomfort for the customer. However, if overlooked or neglected, they can *sour* the customer and all of the participants during the process. Since the majority of collision damages are paid for by insurers, it should provide at least some sense of security to the collision repair industry. However, what provides comfort can also be a source of *contention*. The following, although by no means all-encompassing, are some of the things that all collision repairers should be aware of, and pay attention to:

- Although insurers share a "mutual customer" with repairers, the two entities approach that customer from two *different viewpoints*. As long as no one has been hurt, the collision shop is happy to see the customer! They provide a service that restores the owner's property and provides themselves with revenue. The insurance company—although legally obligated to service their policyholder's needs—are *not* necessarily happy with a loss. Although they also help to restore the owner's property by providing the necessary finances and services, they make more profit by NOT having to pay losses. Thus, they are not thrilled to see a new "customer" unless they are signing up a new policy. The collision shop needs to recognize, understand, and respect these divergent approaches to a loss. In that way, they will better communicate, cooperate, and ultimately achieve their goal—a safe, pre-loss condition repair—while garnering the funds they deserve to do so.
- Insurance companies are usually larger and more "multifaceted" than repair centers. They employ more people, handle more different functions and situations, and have more volume on any given day. Therefore, constant communication, education, understanding, and efficiency are vital to the collision shop's success. For example, one should have the claim number handy when communicating, and strive to understand each company they are conducting business with. Does the insurer utilize staff appraisers, or are they employing "independents"? Are their policies sold by them "in-house," or by an independent agent that is off-site? What are their parameters around parts replacement suppliers? Is the damage appraiser also handling the injuries, coverages, and payments of the claim, or is the insurer using other personnel for these and other functions? These are a few considerations that can help the collision repairer to better cope with the challenges of doing business with an insurance company.
- Vehicular owners usually have a negative impression of insurance companies. If that impression is *not* overcome and/or is encouraged by them or the repairers, *everyone* loses! Dissatisfied customers rarely walk away from a bad experience blaming the insurer or the repairer. They usually are

soured on *everyone* involved in the experience, and/or decide that they were correct in assuming the worst before the repair process began. Collision repairers and insurers *cooperating with each other* during every step of the process is vital to the ultimate success of collision repairs, as well as the growth of the collision repair center and industry.

Following are some specific parts of motor vehicle physical damage insurance coverage that affect the collision repair industry to some extent.

Liability

Liability is the part of the policy that insures the vehicle's owner for any damages he or she may *cause* to other people and/or their property. Causing damage to another person or vehicle is commonly called a "third-party" claim. The insured and their insurance company are the first and second parties, respectively; a vehicle or person that was involved in a loss with the insured (first) party is the *third party*. To illustrate, suppose vehicle A is sitting at a traffic light, and is rear-ended by vehicle B. Vehicle B is insured by ABC Insurance. They contact ABC and notify them they have hit another vehicle. ABC Insurance agrees to inspect and pay for the damages to vehicle A since they insure vehicle B for such occurrences. Vehicle A's owner is the third party in the case. Subsequently, an appraiser and/or claims handler from ABC Insurance will inspect and pay for the damages to vehicle A—usually negotiating with the repair center of that vehicle owner's choice.

Depending upon circumstances, some third-party claims are either paid for in full, in part, or denied altogether. For example, if two vehicles backing out of parking spots collide with each other—rear bumper-to-rear bumper, in a **comparative negligence** locale (where "blame" is typically divided from 10 percent to 100 percent)—the two insurers of the vehicles may conclude that each driver is 50 percent at fault. In that case, an appraiser may write a full damage report on the third-party vehicle but pay only half (50 percent) of that report. In other cases, a third-party claimant may be found to be 100 percent at fault, and the first-party insurer may deny *all* liability in the matter. Still other third-party claims may be denied for lack of adequate coverage on the part of the first party with the policy. In any case, the collision repairer, although repairing each car properly and completely, may not be getting sufficient funds from the owner to repair the vehicle. It is imperative that the collision repair center inquire about—and understand the financial arrangements to pay for—the repairs *before* they begin. See **Figure 10-10**.

Collision

Collision covers damage to an insured owner's vehicle that occurs as a result of collision-related mishaps—regardless of fault. This coverage applies only to the insured vehicle that is listed under the policy contract and is thus considered a *first-party* claim—the insurance company that issued the policy to its policyholder has *first party* in the claim.

Collision-related mishaps are typically any damage that occurs on the first-party vehicle as a result of a motor vehicle accident with another vehicle or vehicles (a moving traffic occurrence); another vehicle, person, or object colliding with the vehicle while it was parked and unoccupied (such as a shopping cart, bicycle, another vehicle, etc.); a driver striking any object besides another vehicle (house, fence, tree, etc.); and, at times, having an object hit the vehicle as a result of a natural occurrence (tree limb, hail stones, etc.).

Again, the collision repair center should be sure of proper, in-force coverage, and where the full payment for the complete repairs will be coming from before beginning a repair job.

Comprehensive

Comprehensive coverage usually applies to damages that are willfully carried out on the insured vehicle such as vandalism (spray paint, scratches, etc.), the theft of parts or the entire vehicle, and specifically detailed (in the policy) occurrences not covered under collision such as striking a live deer or other select animals. In addition, towing, glass breakage, and, at times, "claims-related damages" (i.e., damages incurred *after* a collision but not directly related to it, such as tow truck-induced damages, or weather damages to an interior as a result of collision-related glass breakage that went unattended in a tow yard).

All of these losses can be challenging to handle and adjust for different reasons. In many instances the average consumer is upset and victimized by these losses. The insurance appraiser and/or repairer is often not aware of exactly what transpired, what is damaged, or even all of the parts that may be missing from the vehicle. In any case, as always, communicate with all parties and establish what is being covered, what is related, and what is and is not being paid for—and by whom—before beginning repairs.

Deductibles

A *deductible* is a set and agreed-upon figure that an insured driver has had written into his or her policy contract, and represents the amount of money that is *deducted* from the total physical damage repair payment on any one individual claim. Deductibles come in many different amounts from as low as $50 (a rarity) to as high as $2,000 or more, depending upon the type of coverage and/or the amount the policy coverage is costing the insured party. Generally, the *lower* the policy payments are, the *higher* the deductibles will be. Deductibles only come off first-party claims, and apply to collision as well as some comprehensive claims.

For example, if a policyholder chooses to have collision coverage on a late-model vehicle, and that coverage costs $600 per year to carry with a $500 deductible, he or she may elect to change to a *higher deductible*—for example, $1,000—thus lowering the cost of the policy premium to

PROGRESSIVE DIRECT
P.O. BOX 31260
TAMPA, FL 33631

Policy number 12345678-0
Underwritten by:
Progressive Direct Insurance Co
January 16, 2008
Policy Period: Jan 14, 2008 - Jul 14, 2008
Page 1 of 2

RYAN JONES
123 MAIN STREET
PITTSBURGH, PA 15236

progressive.com
Online Service
Make payments, check billing activity, update policy information or check status of a claim.

800-PROGRESSIVE (800-776-4737)
For customer service and claims service, 24 hours a day, 7 days a week.

Auto Insurance Coverage Summary

This is your Declarations Page

Your coverage began on January 14, 2008 at 12:01 a.m. This policy expires on July 14, 2008 at 12:01 a.m.

Your insurance policy and any policy endorsements contain a full explanation of your coverage. The policy contract is form 9610D PA (05/06).

COLLISION COVERAGE FOR RENTAL VEHICLES

IF THIS POLICY PROVIDES COLLISION COVERAGE, IT WILL APPLY TO VEHICLES YOU RENT, BUT NOT TO VEHICLES RENTED FOR 6 MONTHS OR MORE.

Underwriting Company

Progressive Direct Insurance Co
P.O. Box 31260
Tampa, FL 33631
800-888-7764

Drivers and household residents

	Additional information
RYAN JONES	Named insured

Outline of coverage

1997 Ford Mustang CP

VIN 1FALP4040VF123456	Limits	Deductible	Premium
Liability To Others			$195
Bodily Injury Liability	$50,000 each person/$100,000 each accident		
Property Damage Liability	$50,000 each accident		
Uninsured Motorist	$50,000 each person/$100,000 each accident		8
Underinsured Motorist	$50,000 each person/$100,000 each accident		3
Medical Payments	$5,000 each person		20
Comprehensive	Actual Cash Value	$250	53
Collision	Actual Cash Value	$500	113
Rental Reimbursement	$30 each day/maximum 30 days		18
Roadside Assistance			7
Total 6 month policy premium			**$417**

FIGURE 10-10 This automobile insurance policy *Declarations Page* outlines all coverages—including liability, collision, and comprehensive—and details important points such as deductibles, glass coverage, and towing limits. Courtesy of Progressive Casualty Insurance Company.

$400 per year. The good news is the policyholder is saving $200 a year on premiums; the bad news is that if a loss does occur, instead of paying the first $500 of the claim out-of-pocket, the policyholder now must pay the first $1,000. This is a trade-off that many people choose—often without really stopping to understand the consequences. In any case, the collision center is often caught in the proverbial "middle" when a loss does in fact occur, and drivers realize they will be paying at least some part of the damages themselves. Many insured parties do not even understand what a deductible is or even know they are responsible for one until they have a loss. Always define *who* is paying for *what* before beginning repairs.

Direct Repair Programs

A *direct repair program* (DRP) is a way of doing business employed by many insurance companies. A DRP is a group of collision repair centers that an insurance company aligns itself with based upon mutual understanding, goals, and the meeting of selected criteria that the insurer and repairers agree to. In turn, when their insureds and/or third-party claimants are involved in a collision, comprehensive, or liability loss, the insurer recommends a group of selected DRP shops, and the owner can choose one at which to have the vehicle repaired.

These programs are advantageous for a number of reasons, namely:

- Less paperwork and red tape, leading to a faster, more efficient repair.
- The insurer and the repairer have a good working relationship that is designed to satisfy their customers.
- The work is usually guaranteed by both the insurer and the shop—sometimes for the life of the vehicle.

Unless otherwise agreed to and stated in the policy contract, an insured party does not have to utilize the DRP shop(s) that are recommended by the insurance company.

INDUSTRY TERMS

The collision repair industry is far from a haphazard field. Precision, accuracy, and a common ground with all involved parties is essential to successful restoration of damaged vehicles. What follows are the main terms that are essential to creating, understanding, explaining, and making proper use of a professional vehicular damage report.

Betterment

The term *betterment* is defined as "the increase in value of property that occurs as a result of an improvement." Insurance companies usually apply it when a vehicular part is replaced that has had *measurable* wear; the insurance company pays for the replacement part but charges the vehicle's owner for the amount of wear or usage they got out of that part *prior to the accident* or loss. Here is an example: a tire with 6/32nd's of thread wear—that would have had 11/32nd's of thread when *new*—is damaged in a collision loss and paid for by the insurer. The insurance company might charge the customer 45 percent of the price of the replacement tire, since the customer got to use 5/32nd's (45 percent) of the usable thread before the collision loss. Thus, the insurer would be paying 55 percent of the cost of a replacement tire, while the vehicle's owner, the policyholder, pays 45 percent of the cost. See **Figure 10-11**.

Depending upon where the loss occurred and what the policy reads, betterment applications may vary. Local, regional, and even national laws may affect the application of betterment. Even the part itself could factor in.

Here is an example of a part that may be difficult to assess with regard to betterment: a vehicular radio would be difficult, if not impossible, to apply betterment to. How would one measure how much usage an owner already had? A component like a radio either works, or it does not. Unlike a tire, one cannot see or measure the life left in such a part. In fact, even if one *could* measure how much life was left, would anyone have a log of how much a particular radio had been used?

Betterment may also be affected by regulations. For example, one region or state may deem betterment to apply to the increase in value to the particular part or assembly being replaced, while another state may apply betterment to how much the value of the entire vehicle was affected (or not) by any replacement parts. In fact, some states may prohibit the application of betterment (or depreciation, which will be covered later) altogether.

In addition to tires, some other parts that may be subject to betterment include batteries, drivetrains, brake pads, and shock absorbers.

FIGURE 10-11 This tire has 4/32nds of usable thread life left. Since a new tire has 11/32nds of thread, this tire has 36 percent usable life left. An insurer may choose to have the owner pay for 64 percent of the price of the new tire—if this one requires replacement.

Depreciation

Depreciation is similar to betterment and is often utilized in place of it. Like betterment, the term *depreciation* is often used to adjust the amount of money paid for a part that requires replacement after a loss—but has been worn, and/or has a measurable and *finite* life expectancy, and/or has preexisting, unrelated (to the loss) damages on it. Similar to betterment, when depreciation is taken on a replacement part, the insurer may pay for the percentage of the replacement part that was *unused* before the loss.

Another application of depreciation can occur when a part or assembly requiring replacement had unrelated damages on it prior to the physical loss. For example, suppose a fender had a dent in it prior to a loss. That same fender, after being hit during a collision, is now rendered *unrepairable*. The insurer might calculate how much the fender's "old damage" devalued the fender prior to the loss. Subject to laws and regulations, they may decide to deduct a percentage of money from the value of the *new* fender. Thus, if the replacement fender were $200, the insurer might decide to deduct $50 from the payment—with $50 representing the amount of damage that was on the fender *prior to the loss*.

When applying depreciation or betterment, the reductions are meant to be charged to the owner of the vehicle, not the repairer. Therefore, it is important that both are applied fairly, and are explained to the owner clearly and completely before work on the vehicle begins.

Appearance Allowance

An appearance allowance is an amount that is paid directly to the vehicle's owner when the vehicle has sustained minor *cosmetic damage* that the owner agrees will go unrepaired. Instead of repairing the minor imperfection or replacing the part altogether, the owner agrees to "live with" the damage and the insurance company awards the policyholder an appropriate amount to do so. See **Figure 10-12**.

FIGURE 10-12 These minor scrapes to the edge of this roadwheel may warrant an *appearance allowance*—provided the owner is agreeable.

For example, suppose a wheel sustains a scratch during a collision. The owner agrees to leave the scratched wheel on in exchange for $100 that goes into the owner's pocket (or is applied to the deductible). The appearance allowance does not go to the repairer; it is the owner's compensation for *living with the minor damages*.

Appearance allowances should never be written on parts that affect safety! For example, if a tire sustained a minor cut in a side wall but was still holding air, it should be replaced. If that cut makes the tire deflate during high speed operation, it could affect the safety of the vehicle's occupants.

R&I (Remove and Install)

Remove and install (R&I) means that a part is removed from a vehicle as an assembly, is set aside, and is reinstalled and aligned on the vehicle at a later time. Nothing else is done to the part—no disassembly, replacements, or cleaning. The time for R&I is usually found in the appropriate crash manual.

One example of a part that commonly undergoes R&I is a bumper assembly. If a vehicle's quarter panel is struck and requires replacement, the rear bumper assembly will need to come off but may not require any repair or replacement (of the bar or its attached components). The rear bar assembly would be written for R&I, taken off the vehicle as an entire assembly, and placed aside. After the new quarter panel was installed on the vehicle, the bumper assembly would be reattached and realigned to the vehicle. (In this case, the time to remove and install the rear bumper assembly is usually included in the time to replace the quarter panel.)

R&R (Remove and Replace)

When remove and replace (R&R) is employed, an old part is removed; any necessary, reusable parts that are attached to it are transferred to the new part; and the entire assembly is installed and properly aligned to the vehicle. The time for the remove and replace operation is typically found in the collision manual crash guide.

An example of remove and replace can be described using a radiator core support. If a core support requires replacement, the old support is removed and all attached parts that are reusable are stripped off. The new core support is installed, and all undamaged reusable parts from the old support are installed on the new one. Any attached parts that are in need of replacement are then attached to the new support.

O/H (Overhaul)

Overhaul (O/H) is utilized to describe a process that involves removing an assembly; taking that assembly completely apart; inspecting, cleaning, and replacing its parts as necessary; reassembling the assembly; and installing and properly aligning it back onto the vehicle.

A bumper assembly can be written from the crash manual for O/H. The bumper is removed from the vehicle and disassembled. The various attached parts are taken off; all parts that can be reused are inspected and cleaned. The part or parts that are in need of replacement are replaced. Any parts in need of repair are repaired (the repair times are NOT included in overhaul time) and the entire assembly is reassembled and reattached to the vehicle in proper alignment.

IOH (Included in Overhaul)

Operations or labor tasks that are included in overhaul (IOH) are simply all of the operations and procedures that the overhaul of a specific part or assembly include. Some examples of operations that could fall under this category are alignment of parts, welding and cutting operations, and assembly removal and installation.

For example, if a front bumper assembly were written to be overhauled, aligning the entire assembly back onto the vehicle, replacing any damaged parts of that assembly, and reassembly of the entire bumper assembly would be included in that overhaul operation.

In order to ascertain exactly what is—and what is not—included in an overhaul operation, the procedural pages should be referenced. (Procedural explanations ["P" pages] will be covered later in this chapter.) See **Figure 10-13**. For example, suppose an SLA (short [upper control] arm, long [lower control] arm) suspension assembly was written for overhaul. The "included" and "not included" sections of the "P" pages would describe what is—and is not—included in that particular overhaul. In two different providers' crash guides, the disassembly and replacement of damaged parts are included in the overhaul of the front suspension system. However,

Front Suspension

IMPORTANT REMINDER: Due to the design of suspension on unibody vehicles, it may be necessary to perform four wheel alignment.

Front Suspension Component R&R

Included Operations	Not Included Operations
• Each operation identified in the text is considered to be a stand-alone operation unless noted otherwise	• Wheel alignment
• Remove and install wheel	• Bleed brakes if necessary

Front Suspension O/H

Included Operations	Not Included Operations
• Remove and install wheel	• Replace steerage linkage parts
• Disassemble and clean parts	• Remove and replace:
• Visual check for damage	Torsion or stabilizer bar
• Replace needed parts	Drive axle parts
• Assemble	• Bleed brakes if necessary
	• Wheel alignment

FIGURE 10-13 To understand what is included or not included in crash guide time allowances, one must refer to the procedural ("P") pages. In this example, the removal and reinstallation of the roadwheel is *included* with the front suspension R&R or O/H. Used with permission of Mitchell International, Inc. All rights reserved.

one provider specifically includes the R&I of the road wheel in its overhaul time, while the other provider lists the road wheel in the "not included" (in overhaul) section. Thus, the need to utilize the procedural explanations is apparent at all times.

O/L (Overlap)

Overlap (O/L) is a term used to describe labor operations that are duplicated when two or more parts are serviced, replaced, or refinished at the same time. A common overlap situation occurs when two parts that are joined together with a common seam or surface, or share parts that attach to both panels, are replaced.

For example, when a quarter and rear body panel are replaced during the same repair job, the rear lower portion of the quarter panel and the side of the rear body panel *share* a welded seam. Therefore, less time is required to replace these two adjoining parts *collectively* than is required to replace either the quarter or rear body panel *individually*. The common welded seam is considered overlap, and an appropriate time allowance is deducted—either by the crash manual, or an agreed upon time by all parties negotiating the loss.

Those two panels also share a mutually attached *part*—the rear face bar. Thus, if both panels were replaced during the same repair job, one of the R&I allowances that the individual replacements allow would be deducted.

Overlap can also be used during refinish operations. If a fender and an adjacent door were written for refinishing, some "overlap" would need to be deducted because of the commonality of refinishing two (or more) adjacent panels. Operations such as mixing the materials, cleaning the equipment, and adjusting the spray gun would naturally be done *once*—not before every additional panel to be refinished. Deductions for non-adjacent panels are usually relevant as well. In either refinish overlap case, the first panel *never* has overlap applied to it; only the ensuing panels can cause overlapping of allowances.

Overlap has another connotation as well—when repairs are done to one part of the vehicle that can affect other parts of it. For example, if a vehicle that has been hit in the rear suffers frame misalignment that causes distortion in the quarter panels (commonly called secondary or conjunctive damage), pulling the vehicle back to specification may relieve some of the distortion to the quarter(s). Therefore, the quarter damages that are alleviated from the pulling operations can sometimes be referred to as overlapping damages. This is NOT to be confused with the common definition as it applies to replacing and refinishing of parts.

AT (Access Time)

Access time (AT) is commonly needed and written whenever one part's removal is affected or impeded by other damages. Crash guide operational allowances are commonly determined by removing and replacing parts from *undamaged* vehicles. Therefore, a crushed radiator core support could require additional time to remove it from the vehicle. A damaged bolted-on part's removal could be longer than the standard allowance if the bolts are rusted on. The damaged hood that cannot be opened may require more time than the manual allowance if the hood must be cut and/or the latch needs to be accessed first. In all three examples, an access time allowance may be added to the published allowance to remove the parts in question.

PED (Preexisting Damage)

Preexisting damage is damage that occurred prior to the current collision, comprehensive, or theft loss. It may require repairs but is *not* a part of the claim. Examples of preexisting damages include dents, corrosion, broken glass, cracked lamps, nonfunctioning mechanical and/or worn or missing parts, ripped upholstery, poor prior repairs, scratches in the vehicle's finish, and faded or mismatched paint.

A collision repair center—as well as the insurance damage appraiser—must identify these unrelated damages as such and be sure to make them known to the vehicle's owner before repairs commence. Repair shops and insurers alike have far too often been deemed responsible for these and other damages when they fail to make them known to the customer from the outset. In fact, when making these damages known to a customer upfront, a collision repairer not only avoids being held responsible, but they sometimes are able to "sell" more services to the customer in the way of repairing these damages for an additional charge.

SUB (Sublet)

Sublet (SUB) repairs are any operations or tasks performed by any other repairer or company other than the enlisted collision repair center. Commonly sublet or "farmed out" operations are towing, **wheel alignments** and balancing, glass replacements, and plastic, vinyl and leather trim repairs. Regardless of who does what additional **sublet repairs** to a vehicle, those operations should be listed on the damage report, explained to the customer, done with the same care and attention as the rest of the repair, and backed-up by the "general contractor"—the collision repair center entrusted with the vehicle's repair.

IDENTIFYING A VEHICLE

Before attempting any repairs, a damaged motor vehicle must be properly *identified*. The following information is vital to many repairs in the collision repair environment:

- Year, make, and model of the vehicle
- Production date and assembly plant
- Optional equipment and all accessories
- Paint and trim codes
- Engine size and accessories
- Current mileage on the vehicle

- Restraint systems
- Emission system information
- Transaxle or transmission model

Perhaps the most important link to identifying the vehicle is the *Vehicle Identification Number,* commonly referred to as the "VIN." This number—which can, and will need to, be decoded—is typically found on a plate which is mounted on the top of the instrument panel at the left corner. It is visible through the windshield and occasionally may be found mounted on the A pillar at the same general location.

In addition to the VIN plate, there are various other labels on motor vehicles, and it is important to be able to read and correctly decipher the information from them. Some of these plates or labels are regulated by law and a few—such as emissions and VIN labels—*are federally* regulated. Types of vehicular labels and plates may include the:

- Vehicle identification number (VIN) plate
- Service parts identification label
- Paint code label
- Emission control label
- Supplemental restraint system (air bag) system labels
- Tire and weight labels
- Fuel label
- Anti-theft labels

TECH TIP Be careful when documenting a vehicle identification number. Some states have registration stickers that adhere to the windshield near the VIN plate. Oftentimes, an estimator will copy the VIN from *that* sticker rather than take the time to read it off of the VIN *plate*. In cases where the sticker contains an error, it can lead to problems such as errant VIN decoding, wrong parts ordered, and other mistakes that could be avoided by always taking the VIN off the VIN plate!

Decoding a VIN

The VIN plate contains a coded "number" (consisting of a combination of 17 letters and numbers since the model year 1980) that is decoded before a damage report or repairs begin, in order to ascertain the following information:

- Nation where the vehicle was built
- Vehicular manufacturer
- Vehicle model line or type
- Restraint systems
- Body style
- Model year
- Engine type
- Assembly plant
- The sequential number (unique to each vehicle)

Among other things, a VIN must be decoded to order the replacement parts that may be necessary to repair the damaged vehicle. Additionally, if the collision repair shop requires the use of electronic diagnostic equipment to diagnose and/or repair the vehicle, the VIN becomes vital.

The collision crash guides, as well as service manuals, provide charts, lists of options, and instructions to decode a VIN. When utilizing a computer database estimating system, the VIN can be automatically decoded in most instances. Here is an example of a 17-digit VIN from a Lexus vehicle that we can manually decode together:

JTHBJ46G572061889

Turning to the crash manual, we find the necessary information under "VIN Interpretation." Beginning with the first digit of the VIN, here is the decoded information for the vehicle:

1. Manufacturing Country: J = Japan
2. Manufacturer: T = Lexus
3. Vehicle Type: H = Passenger Car
4. Body Type – Passenger Cars: B = Four-Door Sedan
5. Engine: J = 3.5L V6
6. Series Model: 4 = ES350
7. Restraint System: 6 = Dual Front, Front Side, Head Air Bags w/front knee curtain air bags
8. Car Line: G = ES350
9. VIN Check Digit: 5 = Manufacturer's Internal Code
10. Vehicle Model Year: 7 = 2007
11. Assembly Plant: 2 = Japan

12–17. Serial Number: 061889 = Sequential Production Number

There is other information we may need during the course of the repairs to this vehicle. For example, suppose we need to refinish part (or all) of the vehicle. How would a technician go about ordering materials to match the color?

Refinish, Trim, and Various Other Codes

Collision estimating guides also provide refinish and trim code locations. For example, a paint "label" may provide the color and type of vehicular finish. It may also provide trim and interior color codes (if a separate label for such is not provided). The colors, trim information, and types are provided to various suppliers for the proper parts and materials to be utilized.

Paint codes may also be located on tags such as the *identification* and/or the *vehicle option* labels. A vehicle option label may contain the vehicle, engine, and transmission (or transaxle) model codes, in addition to paint and trim codes.

IDENTIFYING AND ANALYZING TYPES OF DAMAGE

To the average person, "damage is damage." That is not the case when a professional is analyzing a collision-damaged vehicle! Collision damage comes in all different "types" and presents challenges with respect to identifying and analyzing it. For example, do we have—or do we need to rule out—structural damage that may (or may not) have affected the *frame* of the vehicle? If structural misalignment exists, where and how far did it travel through the vehicle? Will individual parts, panels, and components require replacement, or can they be repaired? What kind of material(s) are the various affected parts made of? These, and other questions, will be examined in this section.

Visual Damage Indicators

As collision (inertia) energy travels through a vehicle it leaves a "path of evidence." Telltale signs can usually be seen wherever energy traveled. These signs, commonly referred to as *visual damage indicators*, are generally regarded as evidence that energy reached or exceeded the area where they are found, especially when a number of different indicators are present. See **Figure 10-14**. Where indicators are present, there is likely to be *secondary damage*.

The following visual damage indicators can suggest that structural damage has occurred:

- Buckles, bowing, and/or waves in sheet metal
- Kinks in panels
- Split seams
- Popped or stressed spot welds
- Cracked or ruptured seam sealer
- Chipped or fractured undercoating
- Chipped or peeling paint finish
- Uneven, wide, narrow, and/or closed gaps between panels
- Undamaged doors or lids that will not open (or close)
- Doors that drop or pop upward when opened

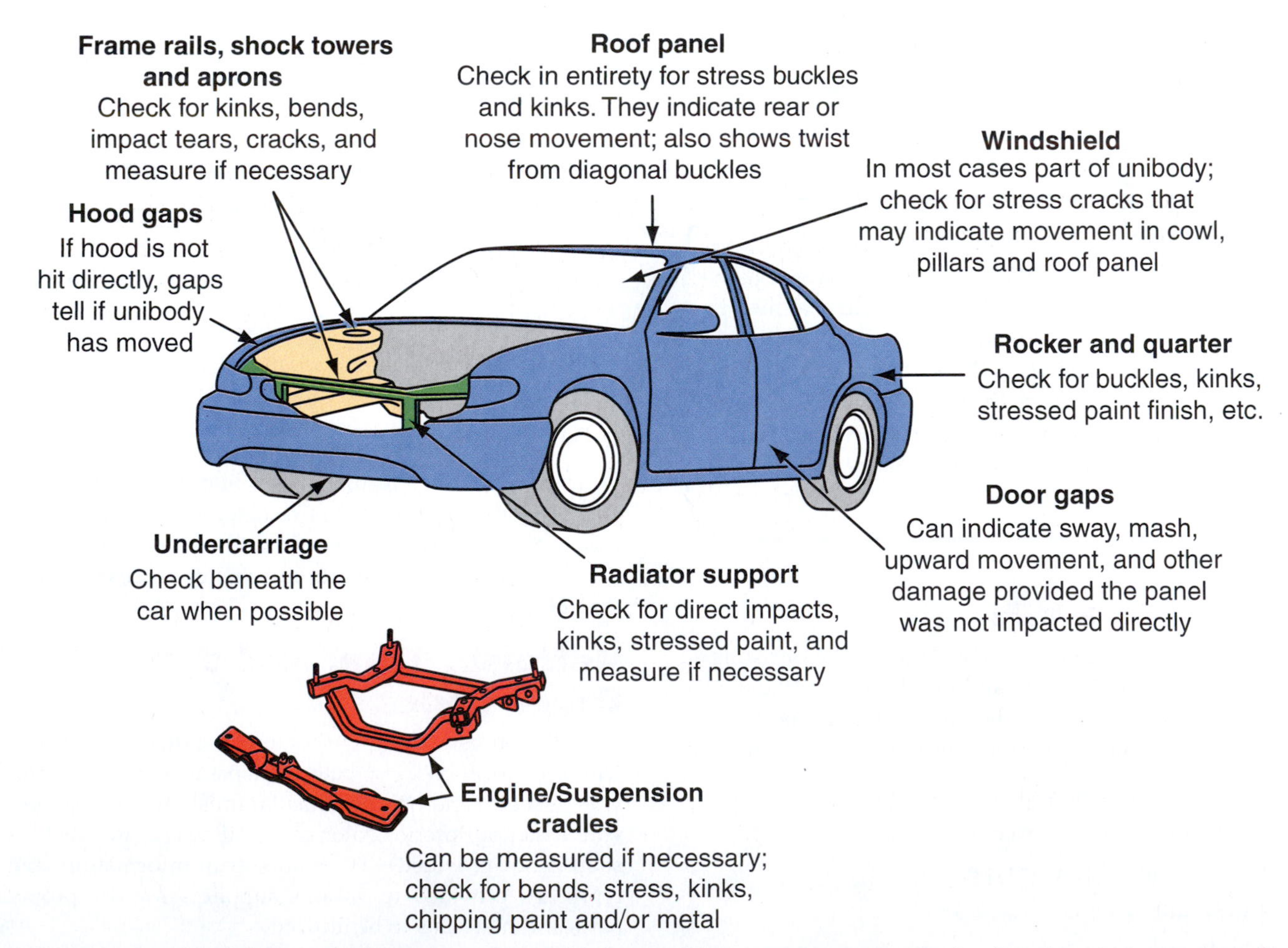

FIGURE 10-14 Visual damage indicators can ascertain that collision energy traveled to (and/or through) a section of a vehicle. Adapted from P&L Consultants LLC.

- Moved, or otherwise disturbed body parts
- Disturbed grommets
- Engine vibrations
- Cracked stationary glass
- Uneven ride height
- Mechanical (or other) parts that rub abnormally

In the damage report example involving the 2005 Toyota Corolla, there was a **buckle** found in the upper radiator tie bar. This visual indicator prompted the estimator to *suspect* that the structural specifications of the vehicle could have been affected. Further analysis (see the section Measuring for Damage) determined this to be the case.

Other visual damage indicators were searched for but NOT found on the Corolla. Examples included split seams and/or seam sealer, uneven gaps, doors that opened and/or closed improperly, cracked stationary glass, and kinks in structural components. When NOT found, it is a good indication that damages can be *ruled out*. In any case, visual indicators NEVER overrule the conclusions that are gathered by *measuring* a vehicle. Visual indicators are considered *guides* to help locate damage, while measuring supports the findings.

Measuring for Damage

Like visual indicators, simple measurements help determine if a vehicle has structural damage. Measurements are *exacting*, and always support—if not overrule—visual indicators.

Measurements taken during damage analysis verify that a structure has moved (or not). Determining the exact specifications and the *precise* amount of misalignment is usually done (and extremely important) during the repair process. A tape measure, tram gauge, and even centering gauges can be utilized for damage analysis procedures.

Tape Measure. A tape measure can make point-to-point measurements. The tape, stretched between two points being measured, must lay *flat* as bends or other abnormalities can skewer the true reading of the area. Care must be exercised with regard to where the tape measures, as well as where the tape *ends* for a true reading. If, for example, one were to measure from the outside edge of a bolt or hole, the tape would need to end at the inside edge of the opposite edge or hole—provided that the two bolts (or holes) were of the same size. See **Figure 10-15**.

TECH TIP Some damage appraisers grind the front edge of their tape measures down to a point. This allows them to sink the altered fixture into a control point, hole, or bolt head indentation to take measurements without needing another technician to hold one end of the tape measure. It also ensures more accurate point-to-point measuring.

FIGURE 10-15 A tape measure can be utilized to measure for structural damage during appraisal. (A tram gauge can also be utilized, as illustrated in Figure 10-9.)

Tram Gauge. The tram gauge is a tool that is based upon the tape measure. It produces point-to-point measurements between two control points. The tram gauge tends to be more accurate than the tape measure since it is rigid in structure and can circumvent obstacles in the way of the measurement. At each end it has pointers that allow the measurement to be taken above any obstructions that would otherwise block a tape measure. Most gauges have a graduated scale on the bar. Pointers must be set at equal heights when utilizing the scale. If the situation does not allow for the pointers to be set equally, a tape measure can be used to obtain the exact distance between the two unequally set pointers in order to obtain a true point-to-point measurement.

Symmetrical and Asymmetrical Measurements. Whether the estimator uses a tape measure, a tram gauge, or even centering gauges (covered in Chapter 11) it is important to ascertain whether the areas being measured are *symmetrically* or *asymmetrically* built. *Symmetrical* dimensions are *equal* between corresponding points on opposite sides of the vehicle's centerline for length, width, and height. If, for example, one measures the distance between the strut tower and the opposite fender's front mounting bolt on a symmetrically built nose section, the exact opposite measurement should be the *same*.

Asymmetrical measurements are *unequal* for length, width, and/or height between corresponding sides of a vehicle across the centerline. Therefore, it is imperative that the person analyzing the damages know the dimensions he or she is measuring—especially if they are asymmetrical in nature.

Critical to the use of measuring devices is at least a basic understanding of the terms that define the areas of a vehicle that may be measured. The *datum plane* is an imaginary plane or surface that sits parallel to the underbody of a vehicle, and is used to measure proper height

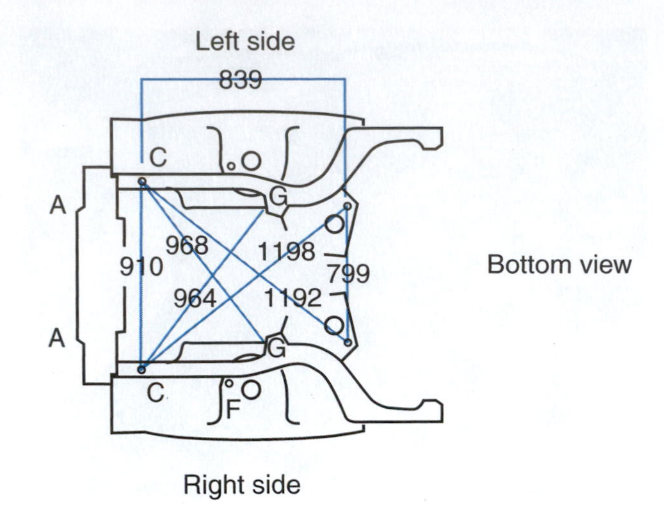

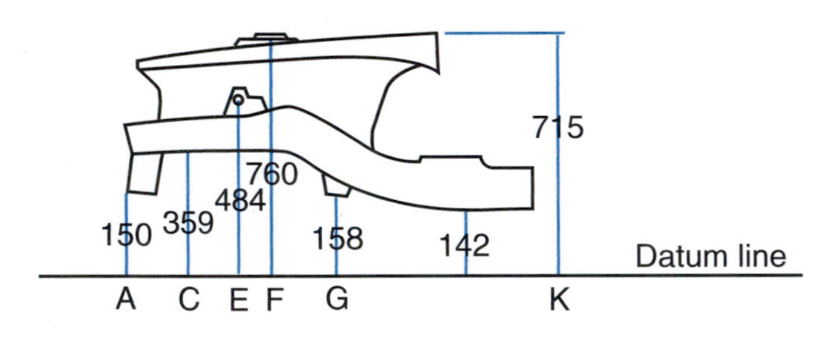

* All dimensions are metric.
* All control points are symmetrical side to side unless otherwise noted.

* All tolerances are +/– 3 mm.

FIGURE 10-16 Since two-thirds of all collision damage events involve a nose of a vehicle, the point-to-point dimensions of the nose are included in the crash guides. Note the asymmetrical dimensions of this nose.

dimensions. *Centerline* is an imaginary line that divides the vehicle into two equal halves lengthwise. On many vehicles, the centerline point is scribed on the firewall. The centerline is the basis for making comparative measurements from side-to-side, or diagonally.

In most crash guides, whether or not in hard copy or computer form, the underhood point-to-point measurements will be published and illustrated. This is useful since two-thirds of all collision losses damage the front of a vehicle. See **Figure 10-16**.

Primary (Direct) and Secondary (Indirect) Damage

While analyzing a damaged vehicle, one will locate primary and, sometimes, secondary damage. Primary (or direct) damage is often easy to locate as it is caused by the primary impact(s) to the vehicle. For example, the Toyota Corolla mentioned and written about earlier in this chapter suffered primary damage to the right side fender, bumper cover, and right front wheel. One can readily see where the primary impact(s) occurred.

Secondary (or indirect) damage is more difficult to locate, requires the search for visual indicators, and may need to be verified by measuring. Secondary damage is usually caused by inertia energy traveling through a vehicle. The split seams, cracked stationary glass, and buckles in panels are but a few of the evidences of secondary damages. On the Corolla, the buckled upper radiator support's tie bar was an indication of secondary damage to that part. This indirect damage was supported by the measurements taken during the damage analysis, as was seen in Figures 10-8 and 10-9.

Repair or Replace?

Using a crash guide or computer database offers the estimator help in a number of ways such as identifying where a particular part may be located, what it costs, how much time should be allowed to change or remove and install a component, and providing a sequential approach to locating all of the damage. However, it *cannot* tell a writer what to repair and what to replace. Those decisions are based on prior experience, education, and a case-by-case analysis of each individual vehicle, to name but a few factors.

The following questions are a *guide* to making proper repair or replace decisions:

- **Is the part repairable?** Materials such as sheet metal, aluminum, high-strength steels, fiberglass, sheet molded compounds, and plastics can technically be repaired. However, pieces that may be severely torn, stretched, punctured, or burned severely may be beyond repair.
- **Is the piece missing?** Although collision repair technicians are extremely inventive, and can fabricate pieces of a part—or a part itself—in most cases without the entire part or component, repairs are not likely to be effective.
- **Will repairs keep the vehicle safe?** Some parts can be repaired cosmetically but may not be returned to safe, pre-loss operating condition. A severe **kink** or tear in a structural crush area can be made to *look* good. However, how a part "looks" in this case is irrelevant; how it *performs* in any subsequent collision is critical.
- **Will repairs change the *substance* of the part?** Polyester or plastic filler is an acceptable means of repairing damaged sheet metal. However, a sheet metal door that is completely covered with plastic will change what the OEM intended the door to be made of. Again, pre-loss condition is what collision repairers strive for. Repairs—NOT *redesign*—is the ultimate intention.
- **Is it cost-effective to repair a repairable part?** First, understand that price NEVER undermines safety! If something is NOT repairable for *any* reason, replace

it regardless of cost. However, price can at times dictate replacement *even if a part is safely repairable.* To illustrate, a damaged fender requiring three hours of labor at $50 per hour to properly restore may be replaced if a replacement fender can be found for $85. With a variety of replacement options available (i.e., aftermarket, used, remanufactured, etc.) and competition between companies fierce, *replacing repairable parts* has become more prevalent.

Plastics and Other Substrates

Today's vehicles are built with a variety of **substrates** (materials). Plastics, aluminum, high strength and ultra high strength steels, and magnesium are but a few. The damage estimator needs to routinely consider questions such as, "What family (i.e., Thermoplastic or Thermoset) and type of plastic (i.e., Polyurethane, Xenoy, fiberglass, etc.) is damaged?" "Is it repaired by using heat, epoxy, fillers, or a combination?" "Does the plastic have a textured finish, and can it be duplicated?"

Plastics can be found virtually anywhere on a vehicle. Exterior body panels, interior and exterior trim, radiator and gas tanks, and even floor panels are being made from plastic. These parts may have to be removed to see *past* them since plastic can return to its basic shape after being struck, sometimes hiding further damages to other parts. Plastic can be stressed on the back side, requiring removal to check its reparability. Some OEMs prohibit repair to some plastic parts although the *material* itself may in fact be repairable. Plastic fuel lines and tanks, battery cases, suspension parts, and some bumper reinforcements and energy absorbers should *never* be considered for repairs.

Other substrate considerations include: Does the vehicle contain aluminum parts, or is it considered an "**aluminum-intensive vehicle**" (i.e., Jaguar, Audi) whereby the entire structure and outer panels are aluminum? What sort of heat—if any—can be utilized in the repair process? Can the shop weld aluminum (i.e., utilizing 100 percent argon, etc.)? If a vehicle contains magnesium part(s) (i.e., dash panel support), is the shop equipped to handle the repair and/or any resulting emergency because of the highly flammable nature of this substance? What sort of urethane should be utilized to adhere the stationary glass to a particular structure? These and many other considerations must permeate the thought process of a damage estimator who endeavors to create a professional, and accurate damage report.

The Different Areas of a Vehicle (Subject to Damages)

Motor vehicles are damaged at all different sections, in a variety of ways. In addition, we have discussed primary, as well as secondary, damages, whereby energy travels through a vehicle—leaving damage in its wake. Vehicles can be struck in the front, rear, side(s), and undercarriage, as well as suffer a rollover in which multiple impacts occur and multiple areas are affected. In addition to collision, vehicles occasionally catch fire, are stolen and/or vandalized, get flooded with water, are oversprayed with paint or other materials, and sustain glass breakage.

The damage appraiser should always cover an entire vehicle regardless of the severity of damages. The basic sections of a vehicle are:

- **Front (nose) section.** This section contains the hood, fenders, headlamps, grill, front bumper assembly, front wheels and tires, front suspension and steering, and the inner structure (i.e., rails, aprons, radiator support, firewall, suspension and other cross-members, etc.).
- **Center (passenger) section.** This section consists of the roof, doors, glass, center post, passenger compartment floor, and the rocker panels.
- **Rear (tail) section.** This section includes the deck lid, quarter panels, rear body panel, taillamps, rear bumper assembly, rear suspension (and steering if so equipped), and the rear inner structure (i.e., rear rails, rear compartment floor, and wheelhouses).
- **Interior (passenger compartment).** This section should always be examined after a collision. The headliner, dash panel and all warning lamps, SRS (air bags and seat belts) consoles, door pads, seats, carpeting, and garnish/trim moldings are subject to direct or indirect damages.
- **Mechanical components.** These consist of the drivetrain (engine and transmission) as well as all extremities of it (i.e., air conditioning compressor, condenser, radiator, alternator, battery, etc.), exhaust system, and gas tank.
- **Undercarriage.** This area of the vehicle can reveal up to 40 percent of the structural misalignment. Whenever possible, raise the damaged vehicle on a lift, and/or get safely under it in order to check the vehicle thoroughly.

Collision repairers and those who analyze damages should be looking to find, or rule out, damages to all structural parts (i.e., A, B, C, and D pillars, rails, aprons, wheelhouses, strut towers, roof rails, rocker panels, firewall and cowl, and rear body panels), bumper assemblies, hoods, fenders, lamps, grill, tires/wheels, cooling and A/C systems, electrical systems, drivetrains, drive axles and drive shafts, engine cradles, brake systems, steering and suspension systems, air bags, seat belts, steering columns, glass, seats, doors, floor pans, quarter panels, roof, instrument panel and frame, deck lids, taillamps, fuel tank, exhaust and systems, moldings, and interior/exterior trim.

USING ESTIMATING GUIDES

Estimating guides are manuals that contain specific information repairers and insurance appraisers can utilize to create an accurate damage report. Part diagrams and availability, specific details and pricing, labor times necessary to accomplish certain repair and replacement tasks—along with what may be included or not included in the time provided—and refinish operations, procedures, and times are some of the many details provided.

A collision estimating crash guide is just that—a *guide*. Information such as replacement and refinish times, procedural cautions and explanations, and parts prices are provided as a basis for starting, writing, and agreeing to a damage repair plan and monetary figure. Although as accurate and well-intentioned as it can be in that regard, one should never feel compelled to stick to the suggested information as though the guide were a Bible! When in doubt or disagreement, technical experience and education, additional sources, and common sense can—and should—be relied upon to complete an accurate, fair damage report.

Parts Information

Information concerning vehicular parts affected by collision losses is found in the *estimating guides* whether the guide is published in hard copy (book) or database (computerized) form. Illustrations in the form of exploded diagrams are provided. Usually, parts most likely to be affected together in a physical damage loss are grouped together and linked to the guide's text for that section, complete with the parts' corresponding number. See **Figure 10-17**.

FRONT

Disable & Enable Air Bag System ....m .3
Diagnose Air Bag System ....m #.5
 # Includes disconnect & connect scan tool & retrieve/clear trouble codes

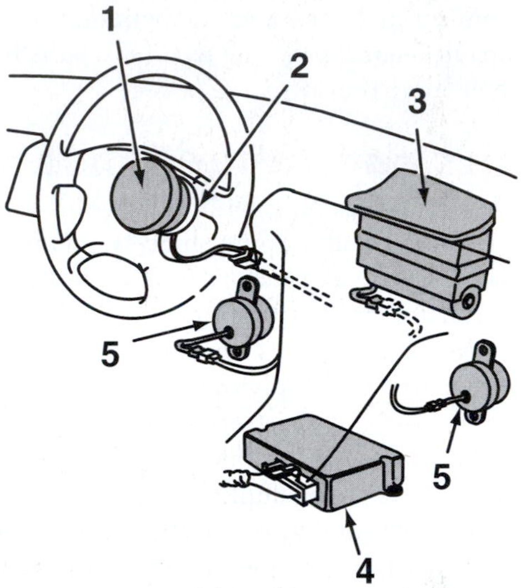

	Part	Year	Part No.		Time	Price
1	**Module, Driver Air Bag**					
	AA, BB Model					
	Bone	03-04	01270-EO	m	.3	646.14
		05	02221-EO	m	.3	648.10
	Stone	03-04	01270-BO	m	.3	646.14
		05	02221-BO	m	.3	648.10
2	**Cable, Spiral**					
			01040-EO	m	.7	202.00
3	**Module, Passenger Air Bag**					
	AA, BB Model					
	Bone	03-04	02070-EO	m	.5	751.66
		05	02021-EO	m	.5	751.66
	Stone	03-04	02070-BO	m	.5	751.66
		05	02021-BO	m	.5	751.66
4	**Sensor, Center Air Bag**					
	w/Side Air Bags					
		03-04	02040	m	.4	288.55
		05	02050	m	.4	215.42
	w/o Side Air Bags					
		03-04	02140	m	.4	288.55
		05	02151	m	.4	215.42
5	**Sensor, Front Air Bag**					
		03-04R	01270	m	.4	54.89
		L	02221	m	.4	54.89
		05 R/L	01270	m	.4	54.89

FIGURE 10-17 Be sure to choose correct parts and prices! With this air bag system, the correct model, color, options, year, and side are but a few of the choices.

At times, manufacturers may discontinue the sale of a part or sell a part that has been *remanufactured*. In either case, a symbol will alert the user so that the shop may seek other means of replacement, and/or be alert to the fact that a "core charge" (exchange of the damaged part for the remanufactured one) is expected. Parts that may be sold as remanufactured by an OEM may include alternators, air conditioning compressors, and steering racks, to name a few.

Other parts information that may be provided by the crash guides include parts that come finished from the manufacturer (or "**color-impregnated**") such as bumper covers or moldings, non-reusable parts like suspension fasteners or plastic clips, and any specially priced parts or groups of parts. Special footnotes may also be attached to a part; an example is a molding that may need to be "ordered by application" (specific to a certain model or package).

Labor Information

Published labor times—most often arrived at by doing *time studies* on undamaged vehicles and utilizing new, undamaged replacement parts—are included in crash estimating guides. These times are listed in tenths of an hour (0.1, which equals 6 minutes) to *hours* (1.0 hour and up). It is of the utmost importance that all parties involved in die handling of a collision loss—including repairers and insurers—understand that the *actual* time necessary to complete an operation may vary depending upon factors such as severity of the damages, skill of the repairer(s), and parts, materials, and type of repair equipment utilized.

Occasionally, the crash estimating guide will *not* provide a labor time for a particular operation. This may be due to a lack of a time study and/or an inability to determine a suitable time. In these cases, all involved in settling a loss should agree upon time(s) before proceeding with the damage report.

In addition to the aforementioned *industry terms* (i.e., R&I, R&R, and O/H), labor operations and times are categorized appropriately. For example, operations that are considered mechanical or structural in nature will be designated as such. Other designations—such as drill, paint, or diagnostic times—may be warranted and utilized.

Estimating crash manuals may suggest one refer to an outside testing organization's suggested repair procedures. *Tech-Cor*, ***I-CAR***, and other industry-recognized organizations do not usually provide die times to an estimating guide company. They do, however, work in harmony with them to provide procedural information such as sectioning, corrosion protection, and supplemental restraint system (SRS) repairs.

There are various "additions" to published labor times that may be necessary to add to a damage report or estimate. Collision access time, glass cleanup, used parts preparation, working with non-OEM parts, pulling a vehicle to align parts and/or structural components, pinchweld clamp repair, and/or removal and restoration of corrosion-resistant materials are but a few examples of what may be considered "additionals." Likewise, "subtractions" (see Overlap) may be in order. An example cited earlier is the quarter and rear body panels' *common surface* where two parts share a welded seam. Usually, a footnote or procedural page will explain the time alteration necessary, and provide an amount to deduct (or do so automatically, if a computerized estimating database is being utilized). See **Figure 10-18**.

Some crash guides house additional information and/or sections that further assist in the creation of a damage report and/or the repair of a vehicle. Material lists (i.e., masking tape, adhesives, and glass cleaners), plastics identification charts (i.e., TPO, SMC, PE, etc.), and special precaution sections (i.e., computer module locations and handling, SRS handling, etc.) should be referenced accordingly.

Procedure Explanations

Each section of a crash guide (i.e., bumpers, hood, suspension, etc.) has a corresponding *procedural explanation* ("P" page) section that complements and supports it. Each "P" page is a listing of "included" and "not included" operations that either fall within or outside of the time(s) listed for a particular section of a vehicle's repair service. If included, an operation has already been deemed a necessary part of the overall procedure and no additional time needs to be added. Some individual operations may or may not be necessary depending upon the individual repair job requirements. Such an operation may or may not have a published time. If needed and published, add it. If needed and not established, agree on a time and list it on the damage report.

For example, the "P" page section for the full replacement time (remove and replace or *R&R*) of a quarter panel *includes* the removal and reinstallation of the rear window glass and moldings, the door striker, and the quarter interior trim. See **Figure 10-19**. It also includes "grind, fill, and smooth welded seams (up to 150 grit sandpaper)" and the rear seat R&I. However, it may NOT include the R&R or R&I of an antenna or the replacement of stripe tape. It is fairly obvious why—not all quarter panels house an antenna nor do all cars have stripe tape decor.

Refinish Operations

Like labor times, refinish times are listed in tenths and hourly increments, and are usually for one color (single stage finish) applied to new, undamaged OEM replacement parts and components without any trim or other attached parts. Typically, the published crash guide's refinish times cover the following:

- Solvent wash
- Sanding, cleaning, priming/sealing a panel
- Masking adjacent panels (up to a fixed point, usually 36 inches; areas beyond 36 inches would require a car cover, and/or masking of interior entries and areas such as the engine compartment and trunk openings)
- Mixing paint materials (NOT the cost of the materials, however)
- Preparing, adjusting, and cleaning spray guns and equipment
- Applying color coats
- Removal of masking materials

Types of Finishes. There are different types of OEM finishes, namely single stage, two-stage, and three-stage finishes. Single stage consists of color coat finish that achieves durability and gloss without any "clearcoat" or additional paint steps added. The two-stage—also called basecoat/clearcoat—consists of a color coat and layer(s) of clearcoat. The three-stage finishes consist of three steps—the color coat, the clearcoat, and an additional step sandwiched between the two, a *mid-coat*—which is basically a suspension or color with mica or aluminum particles added. The particles, covered in the final stage by the clearcoat layer, give the finish a "frosted" appearance or "glow."

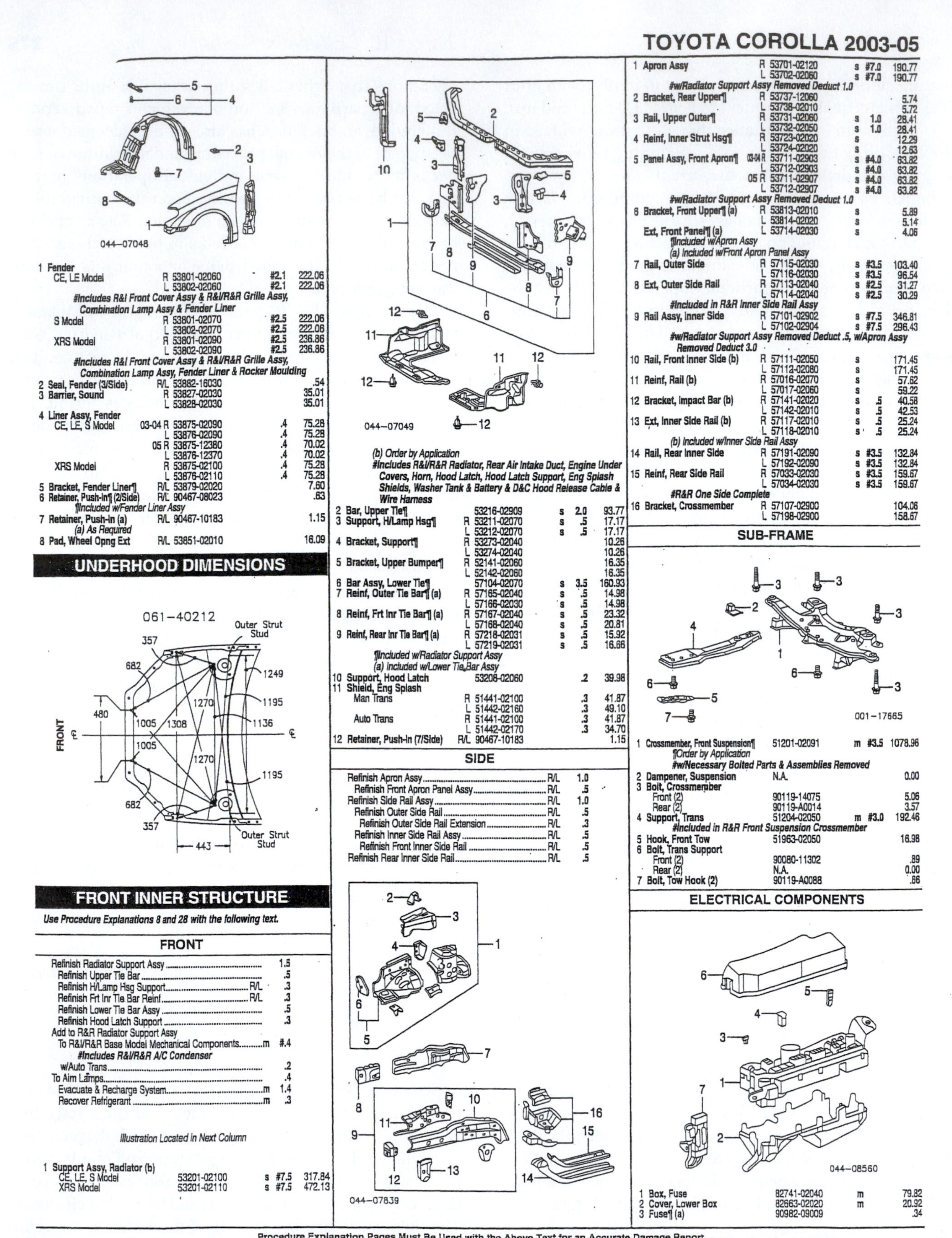

TOYOTA COROLLA 2003-05

	Part		Part No.		Labor	Price
1	Fender					
	CE, LE Model	R	53801-02060		#2.1	222.06
		L	53802-02060		#2.1	222.06
	#Includes R&I Front Cover Assy & R&I/R&R Grille Assy, Combination Lamp Assy & Fender Liner					
	S Model	R	53801-02070		#2.5	222.06
		L	53802-02070		#2.5	222.06
	XRS Model	R	53801-02090		#2.5	236.86
		L	53802-02090		#2.5	236.86
	#Includes R&I Front Cover Assy & R&I/R&R Grille Assy, Combination Lamp Assy, Fender Liner & Rocker Moulding					
2	Seal, Fender (3/Side)	R/L	53882-16030			.54
3	Barrier, Sound	R	53827-02030			35.01
		L	53828-02030			35.01
4	Liner Assy, Fender					
	CE, LE, S Model	03-04 R	53875-02090		.4	75.28
		L	53876-02090		.4	75.28
		05 R	53875-12380		.4	70.02
		L	53876-12370		.4	70.02
	XRS Model	R	53875-02100		.4	75.28
		L	53876-02110		.4	75.28
5	Bracket, Fender Liner¶	R/L	53879-02020			7.60
6	Retainer, Push-In¶ (2/Side)	R/L	90467-08023			.63
	¶Included w/Fender Liner Assy					
7	Retainer, Push-In (a)	R/L	90467-10183			1.15
	(a) As Required					
8	Pad, Wheel Opng Ext	R/L	53851-02010			16.09

UNDERHOOD DIMENSIONS

FRONT INNER STRUCTURE

Use Procedure Explanations 8 and 28 with the following text.

FRONT

Operation			Labor
Refinish Radiator Support Assy			1.5
Refinish Upper Tie Bar			.5
Refinish H/Lamp Hsg Support	R/L		.3
Refinish Frt Inr Tie Bar Reinf	R/L		.3
Refinish Lower Tie Bar Assy			.5
Refinish Hood Latch Support			.3
Add to R&R Radiator Support Assy			
To R&I/R&R Base Model Mechanical Components		m	#.4
#Includes R&I/R&R A/C Condenser			
w/Auto Trans			.2
To Aim Lamps			.4
Evacuate & Recharge System		m	1.4
Recover Refrigerant		m	.3

Illustration Located in Next Column

	Part		Part No.		Labor	Price
1	Support Assy, Radiator (b)					
	CE, LE, S Model		53201-02100	s	#7.5	317.84
	XRS Model		53201-02110	s	#7.5	472.13
	(b) Order by Application					
	#Includes R&I/R&R Radiator, Rear Air Intake Duct, Engine Under Covers, Horn, Hood Latch, Hood Latch Support, Eng Splash Shields, Washer Tank & Battery & D&C Hood Release Cable & Wire Harness					
2	Bar, Upper Tie¶		53216-02909	s	2.0	93.77
3	Support, H/Lamp Hsg¶	R	53211-02070	s	.5	17.17
		L	53212-02070	s	.5	17.17
4	Bracket, Support¶	R	53273-02040			10.26
		L	53274-02040			10.26
5	Bracket, Upper Bumper¶	R	52141-02060			16.35
		L	52142-02060			16.35
6	Bar Assy, Lower Tie¶		57104-02070	s	3.5	160.93
7	Reinf, Outer Tie Bar¶ (a)	R	57165-02040	s	.5	14.98
		L	57166-02030	s	.5	14.98
8	Reinf, Frt Inr Tie Bar¶ (a)	R	57167-02040	s	.5	23.32
		L	57168-02040	s	.5	20.81
9	Reinf, Rear Inr Tie Bar¶ (a)	R	57218-02031	s	.5	15.92
		L	57219-02031	s	.5	16.66
	¶Included w/Radiator Support Assy					
	(a) Included w/Lower Tie Bar Assy					
10	Support, Hood Latch		53208-02060		.2	39.98
11	Shield, Eng Splash					
	Man Trans	R	51441-02100		.3	41.87
		L	51442-02160		.3	49.10
	Auto Trans	R	51441-02100		.3	41.87
		L	51442-02170		.3	34.70
12	Retainer, Push-In (7/Side)	R/L	90467-10183			1.15

SIDE

Operation		Labor
Refinish Apron Assy	R/L	1.0
Refinish Front Apron Panel Assy	R/L	.5
Refinish Side Rail Assy	R/L	1.0
Refinish Outer Side Rail	R/L	.5
Refinish Outer Side Rail Extension	R/L	.3
Refinish Inner Side Rail Assy	R/L	.5
Refinish Front Inner Side Rail	R/L	.5
Refinish Rear Inner Side Rail	R/L	.5

	Part		Part No.		Labor	Price
1	Apron Assy	R	53701-02120	s	#7.0	190.77
		L	53702-02060	s	#7.0	190.77
	#w/Radiator Support Assy Removed Deduct 1.0					
2	Bracket, Rear Upper¶	R	53737-12060			5.74
		L	53738-02010			5.72
3	Rail, Upper Outer¶	R	53731-02060	s	1.0	28.41
		L	53732-02050	s	1.0	28.41
4	Reinf, Inner Strut Hsg¶	R	53723-02020	s		12.29
		L	53724-02020	s		12.63
5	Panel Assy, Front Apron¶	03-04 R	53711-02903	s	#4.0	63.82
		L	53712-02903	s	#4.0	63.82
		05 R	53711-02907	s	#4.0	63.82
		L	53712-02907	s	#4.0	63.82
	#w/Radiator Support Assy Removed Deduct 1.0					
6	Bracket, Front Upper¶ (a)	R	53813-02010	s		5.89
		L	53814-02020	s		5.14
	Ext, Front Panel¶ (a)	L	53714-02030	s		4.06
	¶Included w/Apron Assy					
	(a) Included w/Front Apron Panel Assy					
7	Rail, Outer Side	R	57115-02030	s	#3.5	103.40
		L	57116-02030	s	#3.5	96.54
8	Ext, Outer Side Rail	R	57113-02040	s	#2.5	31.27
		L	57114-02040	s	#2.5	30.29
	#Included in R&R Inner Side Rail Assy					
9	Rail Assy, Inner Side	R	57101-02902	s	#7.5	346.81
		L	57102-02904	s	#7.5	296.43
	#w/Radiator Support Assy Removed Deduct .5, w/Apron Assy Removed Deduct 3.0					
10	Rail, Front Inner Side (b)	R	57111-02050	s		171.45
		L	57112-02080	s		171.45
11	Reinf, Rail (b)	R	57016-02070	s		57.62
		L	57017-02060	s		59.22
12	Bracket, Impact Bar (b)	R	57141-02020	s	.5	40.58
		L	57142-02010	s	.5	42.53
13	Ext, Inner Side Rail (b)	R	57117-02010	s	.5	25.24
		L	57118-02010	s	.5	25.24
	(b) Included w/Inner Side Rail Assy					
14	Rail, Rear Inner Side	R	57191-02090	s	#3.5	132.84
		L	57192-02090	s	#3.5	132.84
15	Reinf, Rear Side Rail	R	57033-02030	s	#3.5	159.67
		L	57034-02030	s	#3.5	159.67
	#R&R One Side Complete					
16	Bracket, Crossmember	R	57107-02900			104.06
		L	57198-02900			158.67

SUB-FRAME

	Part	Part No.		Labor	Price
1	Crossmember, Front Suspension¶	51201-02091	m	#3.5	1078.96
	¶Order by Application				
	#w/Necessary Bolted Parts & Assemblies Removed				
2	Dampener, Suspension	N.A.			0.00
3	Bolt, Crossmember				
	Front (2)	90119-14075			5.06
	Rear (2)	90119-A0014			3.57
4	Support, Trans	51204-02050	m	#3.0	192.46
	#Included in R&R Front Suspension Crossmember				
5	Hook, Front Tow	51963-02050			16.98
6	Bolt, Trans Support				
	Front (2)	90080-11302			.89
	Rear (2)	N.A.			0.00
7	Bolt, Tow Hook (2)	90119-A0088			.66

ELECTRICAL COMPONENTS

	Part	Part No.		Price
1	Box, Fuse	82741-02040	m	79.82
2	Cover, Lower Box	82663-02020	m	20.92
3	Fuse¶ (a)	90982-09009		.34

Procedure Explanation Pages Must Be Used with the Above Text for an Accurate Damage Report.
Comments or suggestions? Need assistance? Call 1-800-854-7030 or 858-578-6550. Ext. 8220. Fax 1-888-256-7969 or 858-549-0629.

FIGURE 10-18 What does 7.5 hours of labor include to install a radiator support on a 2005 Toyota Corolla? Follow the footnotes (and the "P" pages) to be sure to be fair to everyone involved with the vehicle's repair.

GUIDE TO ESTIMATING

QUARTER PANEL - FULL PANEL R&R

LABOR PROCEDURES

SPECIAL NOTATION:

Quarter panel R&R times are generally based upon replacing at the factory seams.
On some vehicles, it is not practical to replace at the factory roof seam.
If a sectioning operation is not listed, then the R&R time on the parts line
represents the time to splice at the best area.

INCLUDED:

- Back window & reveal moldings.
- Bolt on extensions & fillers.
- Carpet & mats turn back.
- Cutting & welding as necessary.
- Door striker.
- Electrical wiring D&R.
- Fuel door.
- Fuel filler neck D&R (if necessary).
- Glass & regulator assembly.
- Grind, fill, & smooth welded seams (up to 150 grit sandpaper).
- Headliner D&R.
- High mounted stop lamp (if necessary).
- Interior molding.
- Quarter interior trim.
- Rear seat R&I.
- Rear lamp assemblies R&I (std equip).
- Rear bumper assembly and/or cover R&I (If necessary).
- Rear package tray trim.
- Seat belt D&R.
- Sill plates.
- Stationary quarter glass & reveal moldings (unless noted).
- Trunk lid, lift gate, and tall gate D&R.
- Trunk compartment trim.
- Weatherstrips, seal strips.

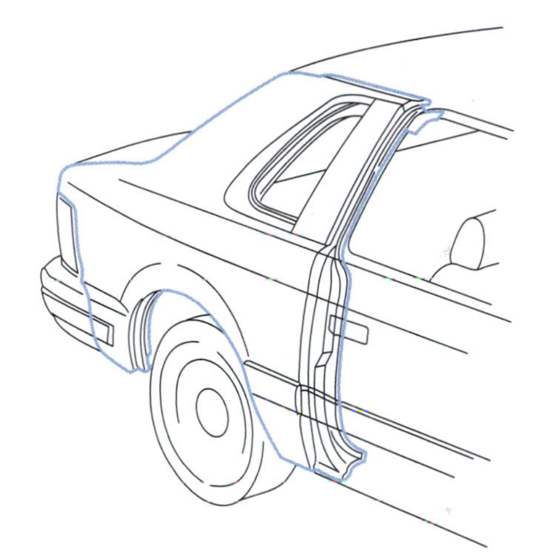

DOES NOT INCLUDE:

- Accessory items.
- Antenna.
- Emblems & nameplates.
- Foam.
- Fuel tank.
- Headliner R&I.
- Lamp R&I (optnl equip).
- Moldings & ornamentation.
- Mud guard.
- Panels inward of quarter.
- Road wheel.
- Roof cover R&I or roll back.
- Spoilers & flares (optnl equip).
- Stripe tape, decals, or overlays.
- Trunk lid, lift gate, or tail gate R&I.

REV. 10 01

Information on these pages may be superseded by information contained in footnotes attached to specific suggested labor operation times.

G33

FIGURE 10-19 The procedural ("P") page indicates operations for the replacement of a full quarter panel. Used with permission of Mitchell International, Inc. All rights reserved.

As with labor operations, there are "P" pages for refinishing operations, along with allotted times, additions, and sometimes subtractions to the published allowances.

Calculations and Operations. As stated earlier, the basic refinish time allotment is for *single stage refinishing*. If clearcoat (two-stage) needs to be calculated, the typical guide allotment is 0.4 (24 minutes) to be added to the time for the first major panel or soft bumper cover. The guides list what a "major panel" is (i.e., fender, hood, rocker panel). *Additional* panels would typically have 0.2 (12 minutes) added to each of them—after any applicable "overlap" is first deducted. The three-stage refinish also carries a formula for calculation of additional time(s); typically, the first major three-stage panel would have 0.7 (42 minutes) added to it, followed by 0.4 (24 minutes) for *additional* panels.

Overlap exists in refinishing operations as well as the other repair operations mentioned previously. The standard deduction for refinish overlap is 0.4 for each adjacent panel (to the previous panel painted NOT counting the first one, as there is no overlap taken for the first major panel refinished) and 0.2 for each non-adjacent panel refinished. See **Figure 10-20**.

Refinish times are NOT intended for complete vehicular refinishing nor are they necessarily complete when addressing used, or repaired, panels. Operations necessary to refinish repaired and/or used panels (i.e., removing or taping off of glass, door handles, and other trim, and/or featheredge-primer/surfacer application and masking procedures) are typically not included in refinish times.

Other operations that are typically not included but may need to be accomplished are tinting (adjusting the color by adding or subtracting portions of toners in a particular paint formula), blending into undamaged adjacent panels, and final buffing, polishing, and/or **color sanding** to eliminate imperfections and/or match textures. These operations—including two-tone painting (the operation in which panel[s] are painted two different colors)—carry specific formulas in order to calculate them accordingly.

FIGURE 10-20 After a completed repair, if the finish is not "perfect" the owner will take notice!

Some parts may need to be refinished *in part* before being installed on a vehicle. Examples include fender edges ("edging"), undersides of deck lids and hoods (sometimes finished with a variation of the exterior finish color that may or may not be clearcoated), and door jambs. These times are also additions to published refinish times.

Additional refinish-type operations and materials that may or may not need to be addressed on the damage report include undercoating and corrosion-protectant materials (i.e., cavity wax, weld-thru primers) and seam sealers which seal up welded areas to prevent fumes from entering a passenger compartment, as well as corrosion from attacking welds.

Blending and Spot (or Partial) Painting. With some colors and/or textures, it may be necessary to "blow-in" or blend paint into undamaged, adjacent panels in order to "fool the human eye" into thinking there is an absolutely seamless, flawless color match. Blend time formulas are provided in the guides, and usually include time to blend color and provide a second stage of a three-stage finish into an undamaged exterior surface, covered by clearcoat over the entire panel.

Not to be confused with blending, "spot refinishing" or *partial painting* a panel usually occurs over *repaired areas* of a panel. Unlike blending into an area of a panel, the spot repair requires undercoat application, as well as basecoat application "to full coverage"—plus a possible basecoat application to blend the new layers of color with the existing ones. If no formula exists for such an operation, negotiation between all parties should result in an agreement as to how to accomplish and pay for it.

Price Information

Only OEM *suggested retail* part prices are provided. An exception to this is glass pricing; since some OEMs occasionally do not provide glass price(s), *National Auto Glass Specifications* (NAGS) pricing is always listed. All prices are generally the most recent prices provided at the time of the guide's printing, or the computer database's latest update. In relatively rare cases, no manufacturer's suggested list price is provided. In cases such as this, the most accurate pricing information can be obtained by contacting the OEM dealer.

Note that published part prices *generally* do not include applicable taxes, refinishing or paint and/or body repair materials, and necessary attachment fasteners.

EVALUATING A VEHICLE'S VALUE

When a vehicle is involved in a loss, it is wise to give attention to its "**pre-loss condition**" worth. "Value" can be viewed, determined, defined, and *even felt* in many different ways. For example, the young college student who is commuting to and from school in a 15-year-old Toyota Camry that runs well, has been maintained properly, and recently had four new radial tires and a battery installed might state he or she "cannot replace" the car. In *that person's view* their vehicle is "priceless." To the little old couple who go to church, the food market, and the bank once a week in their 1971 Chevy Impala that they paid $3,685 for new, the value might be "a million bucks"!

For the collision repairer, insurer, and other "official" entities, the value of a vehicle is viewed from *objective terms*. "Sentimental" value, purpose that the vehicle is serving, or even what one "added" to the vehicle or "planned to restore" contribute little—if anything—to the vehicle's *monetary worth*.

There are a number of "official" ways to value a vehicle—to achieve its *actual cash value* or ACV—depending upon locale, specific state and local laws, the prevailing market, and even what type of insurance policy was purchased. In most cases, a "book value" (i.e., NADA, Red Book, Blue Book, etc.) is established and configured according to the year, make, and model; various options and add-ons; the prior condition and upkeep; the mileage; and even the date the loss occurred. In some locations, a "replacement vehicle" may be found (i.e., comparable year, make, and model with similar options, in a similar condition, and within a mileage range that is close to the vehicle it is replacing). If the owner does not want the replacement, he or she may be given the stated *value* of THAT vehicle.

Other laws exist that allow for a mileage deduction if the vehicle was "totaled" inside a certain number of months and/or miles. For example, a person who suffers a total loss of a vehicle who paid $19,895.00 for it from a dealership six weeks and 1,000 miles ago may be granted the purchase price, *less* a mileage deduction (i.e., 1,000 miles @ 0.35 per mile) of, say, $350.

Most insurers are bound and/or rely on an average of at least two book values. The books list the base retail value, the allowance (or deduction) for equipment (or lack of it), the mileage addition or subtraction (for high or low mileage), and the applicable taxes that would be added. There are book values for recent-to-new models, older and special interest vehicles, trucks, heavy equipment, motorcycles, campers, and boats. See **Figure 10-21**.

Is the Vehicle a "Total Loss"

When analyzing a damaged vehicle in order to create a damage report, the appraiser should pay attention to a number of ways a vehicle might be considered a total loss.

- **Is the vehicle unsafe to repair?** A collision repairer and/or insurer may deem a vehicle a total loss because it is considered unrepairable with regard to *safety*. In these cases, the value of the vehicle does not factor into the decision. It may have been submerged in saltwater, burned beyond recognition, or hit in a way that rendered it unsafe to restore.
- **Does the cost to repair exceed the value of the vehicle?** If, for example, a safely repairable 3-year-old vehicle will require $14,750 to repair but is worth $11,000, it would not be cost-effective to repair it. If one can *replace* a possession worth $11,000, why would anyone want to spend a dollar more to *repair* it?
- **Is the vehicle a "constructive total loss"?** A constructive total loss occurs when the ACV of a vehicle, *less the value of the salvage* (i.e., a vehicle in its unrepaired state), is exceeded by the cost to repair it. For example, a Honda Accord is deemed to be worth $12,475 *before* a collision. The insurance adjuster writes an estimate to repair it to pre-loss condition for $11,000. However, a salvage buyer gives the adjuster a "bid" for $2,000—for the damaged vehicle. If the insurer pays out $12,475 to their insured (ACV) and takes the car to the salvage yard where they recoup $2,000 (salvage value) they have sustained a net loss of $ 10,475 ($ 12,475 less $2,000). *Why pay a repairer $11,000 to restore a car that they are only losing $10,475 on if they total it?*
- **Is the vehicle repairable?**

If none of these situations apply, the vehicle is *repairable*. Remember that a safe, pre-loss condition repair of the vehicle must be accomplished. If the amount to repair the vehicle exceeds the ACV (pre-loss value) of the car, or the amount to repair exceeds the value less a salvage value, the vehicle is usually considered a total loss.

Autosource Valuation

Administrative Data — 2006 Dodge Durango 4WD

Default Office
Autosource - USA
Milwaukie Branch
4211 SE International Way
Suite A1
Milwaukie OR 97222

Claimant	(None)
Insured	Training
Claim	TEST FILE
Loss Date	04/21/2008
Loss Type	Collision
Policy	
Other	

VINSOURCE Analysis — 2006 Dodge Durango 4WD

VIN	1D4HB48N06Fxxxxxx
Decodes as	2006 Dodge Durango SLT 4WD 4D Wagon
Accuracy	Decodes Correctly
History	Activity was reported

- **Autosource activity: Reported by Autosource - USA in Milwaukie, OR on April 21, 2008. Call them at (503)652-3350 regarding Claim: TEST FILE (Request: 22165363 DOL: 04/21/2008 Odometer: 28,121).**
- **Autotrak activity: (NONE).**
- **Audatex/Estimating activity: (NONE)**
- **Sales history activity: (NONE)**

Valuation Summary — 2006 Dodge Durango 4WD

	Typical Vehicle	Loss Vehicle	Adjustment
Price	$19,612		$19,612
Engine	8 Cylinder 4.7 Engine	8 Cylinder 4.7 Engine	
Transmission	5 Speed Automatic	5 Speed Automatic	
Odometer	32,520 Mi(Typical)	28,121 Mi(Actual)	310
Equipment/Package Adjustment (See Valuation Detail)			1,105
Autosource Value Before Condition Adjustments			21,027
Total Condition Adjustments (See Condition Adjustment Detail)			-750
Total Condition Adjusted Market Value			**$20,277**
General Sales Tax		%	
Title Fee			
Transfer Fee			
Deductible			-
Net Adjusted Value			
Salvage/Other			-

Warning

The market value displayed may not reflect the activity detected by VINSOURCE and/or NICB research.

Vehicle Valuation Detail — 2006 Dodge Durango 4WD

FIGURE 10-21 The value of a vehicle before a loss occurs oftentimes determines whether or not a vehicle is *worth* repairing. Courtesy of Progressive Insurance Company and Audatex North America, Inc. All rights reserved.

Summary

This chapter covered:

- What a damage report is, the different types of damage reports, and why it is important to create one
- How damage estimates are made and how they are utilized for making collision repair sales, ordering necessary parts, and creating repair orders
- Insurance coverages such as liability, collision, and comprehensive, and what role the insurance company and its employees have in the collision repair process
- Various collision industry terms such as betterment, depreciation, appearance allowance, remove and install, remove and replace, overlap, access time, preexisting damage, and sublet and the role they have in the repair of a vehicle
- Identifying a vehicle utilizing the Vehicle Identification Number (VIN) and various other codes found on a vehicle
- Identifying and analyzing various types of damages by utilizing visual damage indicators and measurements
- How to make repair and replace decisions, analyzing plastic and other materials that are found on a typical motor vehicle, and the different areas of a vehicle that are subject to damages
- How and why to use estimating guides including the parts, labor, procedural, refinish, and pricing information contained within them
- How and why to evaluate a vehicle's overall worth in order to determine if the vehicle should be repaired

Key Terms

ABS modulator
access time (AT)
aftermarket
aluminum-intensive vehicles
buckles
claim-related documentation
collision energy
color-impregnated
color sanding
comparative negligence
cost totaling
cycle time
damage analysis
direct damage
electronic mail/e-mail
I-CAR (Inter-Industry Conference on Auto Collision Repair)
included operations
independent agent
indirect damage
inertia energy
job costing
judgment labor time
judgment replacement time
kink
labor rates
not included operations
options
original equipment manufacturer (OEM)
overlap
paint transfer
pre-loss condition
procedural/"P" pages
recycled parts
repair plan
reserve
specialty shop
sublet repairs
substrate
upsell
vehicle nose
wheel alignment

Review

ASE Review Questions

1. A damage report is sometimes referred to as a:
   A. Police report.
   B. Damage appraisal.
   C. Final invoice.
   D. Phone call that a shop's estimator makes to the vehicle's owner.
2. Technician A says that a vehicle cannot be repaired with a manually written damage report. Technician B says that a computerized damage report can only be created by the original equipment manufacturer. Who is correct?
   A. Technician A only
   B. Technician B only
   C. Both Technicians A and B
   D. Neither Technician A nor B
3. All of the following can be found on a typical computerized damage report EXCEPT:
   A. The policy owner's deductible.
   B. The police report file number.
   C. The OEM parts number.
   D. The vehicle's options.
4. Sheet metal repair allowances can be found in:
   A. A frame sheet.
   B. The crash manual.
   C. The OEM manual.
   D. None of the above.
5. Technician A says that a direct report program is a repair program that directly involves the insurance company, the repair shop, the original equipment manufacturer, and the agent that sold the policy. Technician B says that blending an undamaged panel is also referred to as "fooling the customer." Who is correct?
   A. Technician A only
   B. Technician B only
   C. Both Technicians A and B
   D. Neither Technician A nor B
6. All of the following are important to ascertain while gathering information concerning a collision loss EXCEPT:
   A. The color of the other vehicles involved in the accident.
   B. What temporary repairs were made to the vehicle following the loss.
   C. The mileage that the vehicle had at its last inspection.
   D. The owner's perception of what might not be working properly since the accident occurred.
7. The collision report center's estimator should primarily concern him- or herself with what question while analyzing a collision-damaged vehicle?
   A. Can this vehicle be safely repaired?
   B. How many speeding tickets does the driver have in the past 3 years?
   C. Were any of the drivers involved cited for driving while intoxicated?
   D. Is the vehicle worth less than $10,000?
8. Technician A says that a crash guide is published in book form. Technician B says that the exploded view of parts that present the most frequent kinds of collision damages are only found in computerized versions of the crash guides. Who is correct?
   A. Technician A only
   B. Technician B only
   C. Both Technicians A and B
   D. Neither Technician A nor B
9. Technician A says that "P" pages are included in crash manuals only to assist trainee damage estimators in creating a damage report. Technician B says digital imaging can be utilized by collision repairers and insurers to correspond with each other. Who is correct?
   A. Technician A only
   B. Technician B only
   C. Both Technicians A and B
   D. Neither Technician A nor B
10. All of the following are typically included in the R&R of a quarter panel EXCEPT:
    A. The antenna.
    B. The R&I of the rear bumper assembly.
    C. The drop of the headliner.
    D. The removal and installation of rear stationary glass.
11. Which of the following is NOT important when ordering parts?
    A. Included vehicle options
    B. Production date of the vehicle
    C. Insurance appraiser's code
    D. Color and trim codes of the vehicle
12. A work order was created by a repair center for a damaged vehicle. The document—also called a repair order—was created from the:
    A. Damage report.
    B. Parts order.
    C. VIN.
    D. Vehicle owner's instructions.

13. Technician A says that refinish overlap can be applicable on adjacent, as well as non-adjacent, panels. Technician B says that liability coverage is what a first-party collision claim is paid off of. Who is correct?
    A. Technician A only
    B. Technician B only
    C. Both Technicians A and B
    D. Neither Technician A nor B

14. Technician A says that comprehensive coverage may cover an insured policyholder's collision damage from striking a live deer. Technician B says a deductible is the amount of money that is deducted from a first-party collision payment. Who is correct?
    A. Technician A only
    B. Technician B only
    C. Both Technicians A and B
    D. Neither Technician A nor B

15. Betterment and depreciation charges are paid to the:
    A. Shop.
    B. The frame repairer.
    C. The insurance department.
    D. None of the above.

16. A collision repair technician is told to remove and install (R&I) a bumper assembly. Basically, the technician will be:
    A. Taking off the assembly and immediately reinstalling it.
    B. Taking off the assembly, washing it down, refinishing it, and reinstalling it later on.
    C. Replacing the old assembly with a brand new one later on.
    D. Removing the assembly and setting it aside, so as to reinstall and align it later on.

17. A collision repair technician is overhauling a bumper assembly. All of the following will be included in that process EXCEPT:
    A. Taking the bumper completely apart.
    B. The price of a new energy absorber.
    C. Inspecting each and every part of that assembly.
    D. Cleaning the reusable bolts.

18. Technician A says that overlap is a term used to describe a method of plastic repair. Technician B says that wheel alignment is one of a number of operations that a collision center may "sublet." Who is correct?
    A. Technician A only
    B. Technician B only
    C. Both Technicians A and B
    D. Neither Technician A nor B

19. A collision repair center's damage estimator is decoding a VIN. Which of the following will the estimator NOT be able to obtain from that number?
    A. The original manufacturer of the vehicle
    B. Where and when the vehicle was built
    C. The type of transmission that the vehicle is equipped with
    D. The type of restraint system that the vehicle has installed at the factory

20. Kinks in sheet metal, engine vibrations, cracked windshields, and peeling paint are considered a few of what are known as visual damage indicators, and may signal to a damage estimator that a vehicle has sustained:
    A. Electrical system damages.
    B. Primary damages to the taillamps.
    C. Secondary damages.
    D. None of the above.

21. Technician A says that measurements can be symmetrical or asymmetrical. Technician B says that collision damages can be considered primary or secondary. Who is correct?
    A. Technician A only
    B. Technician B only
    C. Both Technicians A and B
    D. Neither Technician A nor B

22. Technician A says that a production motor vehicle can include aluminum, plastic, and/or magnesium components. Technician B says that a crash manual will never include plastic identification charts. Who is correct?
    A. Technician A only
    B. Technician B only
    C. Both Technicians A and B
    D. Neither Technician A nor B

23. All of the following are typically included in a crash guide's refinish time EXCEPT:
    A. Solvent washing.
    B. Masking adjacent panels indefinitely.
    C. Mixing paint materials.
    D. Removal of all masking material.

24. After achieving an agreed price with the insurance company, a collision repair center proceeds to start repairs but finds two damaged parts that neither the shop nor the insurance appraiser saw on the initial inspection. The subsequent estimate that is written for additional damages is called a(n):
    A. Second estimate.
    B. Additional estimate.
    C. Subsequent estimate.
    D. Supplemental estimate.

25. An insurance company appraiser deems a collision repair center customer's vehicle to be a "constructive total loss." By this the insurer means the:
    A. Vehicle's construction was faulty to begin with.
    B. Insurer feels that a repairer could never restore the vehicle to a safe operating condition.
    C. OEM parts that are needed are no longer manufactured.
    D. ACV of the vehicle, less the salvage value, is exceeded by the cost to repair it.

Essay Questions

1. Discuss three or more procedures that can be used to create a damage report through the use of digital imaging.
2. Discuss the differences among the following three parts of motor-vehicle physical damage insurance coverage that affect the collision repair industry: liability, collision, and comprehensive.
3. Discuss the advantages of a direct repair program.
4. Explain what an *appearance allowance* is using a specific example.

Topic Related Math Questions

1. The subtotal of parts, labor, refinish time, sublet costs, and miscellaneous charges for a vehicle's damage is $2,341.88. The prevailing tax rate in the area is 6 percent. What is the gross total of the damage estimate?
2. A person who suffered a total loss of a vehicle bought it from a dealership five weeks ago for $17,630. At the time of the loss, the vehicle had an additional mileage of 850. The owner of the vehicle is granted a mileage deduction of $0.25 per mile. What is the actual cash value of the vehicle?

Critical Thinking Questions

1. Discuss the value of a damage report and what it accomplishes in the process of repairing a collision-damaged vehicle.
2. Cite as many visual indicators of possible structural damages to an automobile as you can remember. Discuss how a repair technician could use these visual indicators to assess underlying damages.
3. Organize a group discussion and identify the different types of information that can be found in an estimating guide. Discuss ways in which repair technicians and insurance appraisers can use this information to create an accurate damage report.

Lab Activities

1. Locate a special-interest vehicle. Using two book values, compute the pre-loss value—the actual cash value (ACV)—of the vehicle. Then write an estimate that supports your calculations.
2. Find the VIN plate on the same vehicle. Use a collision crash guide to decode the Vehicle Identification Number (VIN).

Chapter 11

Collision Damage Analysis

OBJECTIVES

Upon completing this chapter, you should be able to:

- Discuss the essentials of collision theory and the effects of force on shape and structural members.
- Discuss the various structural designs used in the manufacture of the automobile and how they are affected by the collision energy.
- Identify and distinguish the difference between direct and indirect damage including the effects of inertial forces.
- Identify the damage sustained while performing a visual inspection of the damaged vehicle.
- Identify and isolate common and discrete damage sustained by each of the three sections of the automobile.
- Trace the flow of damage caused by a collision.
- Identify types of quick check measuring methods and techniques that can be used to identify damaged areas of the vehicle.
- Identify the specification manuals and other available resources commonly utilized to identify the damage sustained by the vehicle.

INTRODUCTION

The modern day automobile has evolved into a very sophisticated combination of electronic and mechanical systems requiring very knowledgeable and learned individuals to perform what was once thought to be very routine service and maintenance. Likewise, when an automobile is involved in a collision, it requires highly skilled technicians with a broad range of knowledge and expertise to diagnose and repair numerous problems not encountered by the individuals performing the required routine maintenance. The wide scale use, expansion, and integration of the many electronic systems affected in an automobile accident create a host of problems normally not encountered from normal driving conditions. The body and structural members are subjected to shocking, twisting, and distorting forces. These forces cause damage that must be repaired and reversed and panels that must be repaired or replaced to restore the vehicle to its pre-collision condition.

DAMAGE DEFLECTION

The damage sustained by a vehicle in a collision is often very deceiving, and much of it will go undetected by an untrained individual who does not understand the principles of design on the modern day automobile. Most cars and trucks are designed, engineered, and constructed to absorb, deflect, and divert the forces of the collision energy around the vehicle's outer perimeter, thus reducing the effects of the impact on the passengers. This occurs because a series of spontaneous reactions and shock absorbing responses are triggered throughout the vehicle from the instant of initial contact with the **impact object** until the vehicle comes to a complete rest. One must understand the design principles commonly used to deflect the collision energy around the vehicle, thus reducing the shock effect that would otherwise radiate through the entire structure and the damage that occurs as a result of it.

IDENTIFY DAMAGE SUSTAINED BY THE VEHICLE

Damaged vehicles typically can be classified in one of three general ways: those with mere cosmetic damage, those that are more severely damaged but still driveable, and those that are so severely damaged they are inoperable. Regardless of the severity of the damage, one should obtain as much pertinent information about the incident as possible from the vehicle owner or driver. The more severe the vehicle's damage, the more information that is necessary to perform an in-depth inspection, formulate an accurate damage analysis, and formulate a repair plan.

Vehicle Occupants. Most damaged vehicles are still driveable and will be driven to the repair shop to obtain a damage estimate. To help with diagnosis, the estimator should ask the vehicle owner several questions concerning the vehicle's occupants and contents at the time of the accident. For example, were any other passengers in the car using seat belts? Some manufacturers recommend replacing any seat belt that was being worn by a passenger during a collision as the collision causes it to become stressed. Likewise, the seats and backs may be damaged if they were occupied at the time of the crash. Were the passengers transporting any heavy objects inside the car or trunk that may have become air-borne, causing damage to the vehicle's interior when thrown against soft interior parts? An unrestrained heavy object is apt to become air-borne and cause untold amounts of damage to the interior that could go undetected unless the estimator is made aware of it.

Accessories. While evaluating the damage, the estimator should determine if there is additional invisible damage by asking the owner if all the accessories—such as the blower motor, air conditioning, radio, turn signal indicators, horn, lights, and any other accessories—are functioning properly. A non-functional accessory can be a tipoff to check for damage to the wiring and other parts of the electrical system. Also, the estimator should determine if any of the warning lights on the dashboard are on longer than normal or at any time when the engine is running. This may be a sign that some of the sensors or mechanical parts have been affected or damaged.

Driveability of the Vehicle. One should also determine if the vehicle handles differently when driving, turning, and starting and stopping. Any change perceived in the vehicle's driving and handling characteristics should be noted. Inquiring about the road conditions may be helpful to determine whether the suspension was subjected to more than normal side thrust—especially if it skidded sideways—or whether the vehicle may have been pushed into or jumped over a curb. These conditions will greatly increase the possibility of wheel misalignment and suspension damage that must also be noted and included in the damage analysis. If the vehicle left the road, there could be further damage to the structural parts which must be carefully inspected when reviewing the undercarriage.

Vehicle Stationary or Moving During the Accident. In order to accurately identify the damage sustained by the vehicle, it is necessary to obtain as much critical information about the incident as possible. One important item to identify is the speed and direction of travel of the vehicle being assessed and, if possible, the same information about the other vehicle or impact object that caused the damage. Knowing the angle or direction of travel will be helpful to determine if the damaged parts of the vehicle will have shifted, been pushed to the side, or moved straight forward or backwards. This will help the estimator and repair technician identify areas to look for possible damage that may otherwise go unnoticed. Items such as transmission and throttle linkages may become damaged in a straightforward collision but not be affected in an angular impact. Likewise, damage to some of the steering and suspension components, mechanical parts, and the engine and transmission mounts are more likely to occur under certain collision conditions than in others. The area of a vehicle where the impact occurs will have a significant bearing on the amount and type of damage the vehicle will sustain. An angular frontal impact, for example, will likely cause a different type of damage on the vehicle's structural members than will occur if the vehicle is struck at nearly the same angle in the center or another section of the vehicle. Several control modules such as the main electronic control module (ECM)—which is the main computer or control module for all of the vehicle's electrical functions—are usually located in the right kick panel rearward of the engine compartment. The rear lamp control module is attached to the inner reinforcing panel, which is located immediately inside the outer quarter panel. During a side impact they are in the direct damage path. Knowing the location of these modules is helpful to ensure their inclusion in the damage report.

VEHICLE IDENTIFICATION NUMBER (VIN)

As part of the actual damage assessment, the estimator should obtain and record as much information about the vehicle as possible from the various vehicle identification tags. The first and foremost source of information about the vehicle is the **vehicle identification number (VIN)**. See **Figure 11-1**. The VIN is usually a 17-digit identification number that is unique to each vehicle. According to federal regulations, it must be located on the top left corner of the dashboard at the bottom or at the lower left **A pillar** and must be visible through the windshield. This identification number bears a considerable amount of information specific to that particular vehicle. See **Figure 11-2**. By using a cross reference guide to interpret the information one can determine all of the following about the vehicle:

Country where manufactured	Engine size
Make	Type of body
Manufacturer's vehicle line	Transmission
Restraint system	Year of manufacture
Manufacture plant	Vehicle serial number

It frequently becomes necessary to know the exact engine size, manufacturing plant, or the country of origin when choosing the parts to be replaced. Occasionally the month and the year the vehicle was manufactured is necessary to select the correct part for the vehicle being evaluated. Therefore, having ready access to this information—which is obtained from the VIN and other identification tags—will save a considerable amount of time when the damage report is actually written.

FIGURE 11-1 The VIN provides unique information about each specific vehicle.

VEHICLE INFORMATION LABELS

Other vehicle option labels—located in a variety of locations—can aid the estimator in obtaining information needed to make a definitive decision about a part or assembly. See **Figure 11-3**. The air conditioning specifications are often located at the top of the **radiator** support or on the bottom side of the hood. Other informational labels—such as the belt routing directions, the emission control label, tire pressure recommendations, manufacturing conformity labels, and paint and trim code labels—are all placed at strategic points unique to each manufacturer. If these labels are damaged or are located on damaged parts that are to be replaced, they should also be included in the list of parts to be replaced. One should not attempt to peel these off the original part and transfer them to the new replacement part. Transferring or tampering with a VIN label is considered a felony. If the label is a serviceable item, a service part number is usually located at the label's bottom left or right corner.

ANTI-THEFT LABELS/DOT LABELS

Many late-model vehicles also display **anti-theft labels** on most of the exterior body panels. See **Figure 11-4**. These labels bear the VIN of the vehicle and are designed to deter the indiscriminate theft of these parts. They are also used for tracking when they become salvage parts for other vehicles. Any time a part is replaced with a new one, a replacement anti-theft number should be installed as well. The replacement label will not be an exact match to that of the original as a number must be created. Otherwise, the manufacturer's service part number—which is preceded with the letter "R"—is typically used. This indicates that the part is a replacement for the original one installed by the manufacturer. If the anti-theft number on a part does not match the VIN of the vehicle being inspected and assessed for damage, it is an indication that the vehicle has likely been repaired previously. This should be cause for closer scrutiny while making the overall damage assessment. A considerable amount of time can be saved by ensuring all the right questions have been asked of the vehicle owner and all the specifics about the vehicle have been documented. While this is being done, the estimator is also able to accomplish another very important part of the overall repair scenario: gaining the vehicle owner's confidence that the vehicle will be repaired properly.

DIRECT AND INDIRECT DAMAGE

In order to do a thorough and detailed analysis of the overall damage sustained by the vehicle, one must mentally re-create the series of events, actions, and reactions that occurred within the total framework of the vehicle during

VIN INTERPRETATION

CHRYSLER LLC
Jeep
1 J C U N 7 7 1 X G T 0 0 0 0 0 1
① ② ③ ④ ⑤ ⑥ ⑦ ⑧ ⑨ ⑩ ⑪ ⑫ ⑬ ⑭ ⑮ ⑯ ⑰

1 Manufacturing Country
1 • U.S.A.
2 • Canada
2 Manufacturer
B • Jeep (Canada)
J • Jeep (U.S.A.)
3 Type (1983-88)
C • Multi-Purpose Vehicle
D • Incomplete Vehicle
T • Truck
3 Type (1989-06)
1 • Multi-Purpose Vehicle
2 • Pickup
4 • Multi-Purpose Vehicle (w/o Side Air Bags)
6 • Incomplete
7 • Pickup
8 • Multi-Purpose Vehicle (w/Side Air Bags)
4 Engine Type (1983-88)
B • 2.1L 4Cyl. Turbo Diesel
B • 2.5L 4Cyl. 2BC (GM Built)
C • 4.2L 6Cyl. 2BC
E • 2.5L 4Cyl.
H • 2.5L 4Cyl. TBI
M • 4.0L 6Cyl. EFI
N • 5.9L V8 2BC
U • 2.5L 4Cyl. (AMC Built)
W • 2.8L 6 Cyl. 2BC
Y • 2.5L 4Cyl.
4 GVWR (1989-07)
E • 3001-4000 lbs. GVW
F • 4001-5000 lbs. GVW
G • 5001-6000 lbs GVW
H • 6001-7000 lbs. GVB
5 Transmission/Drive (1983-88)
A • Auto. Column Shift (Full-Time 4WD)
B • Auto. Floor Shift (Part-Time 4WD)
C • Auto. Floor Shift (Full-Time 4WD)
D • Auto. Floor Shift (2WD)
5 Transmission/Drive (1983-88)
E • Auto, Column Shift (Part-Time 4WD)
F • 5-Speed Manual (Part-Time 4WD)
G • 4-Speed Manual (Part-Time 4WD)
H • 4-Speed Manual, Floor Shift (4WD)
J • Auto, Column Shift (4WD)
K • Auto, Floor Shift (4WD)
L • 5-Speed Manual (Part-Time 4WD)
M • 4-Speed Manual (Part-Time 4WD)
N • 5-Speed Manual (Part-Time 4WD)
P • 5-Speed Manual (Full-Time 4WD)
P • Auto, Column Shift (Part-Time 4WD)
R • Auto, Floor Shift (Part-Time 4WD)
S • 4-Speed Manual, Floor Shift (2WD)
T • Auto, Floor Shift (Full-Time 4WD)
U • Auto, Floor Shift (2WD)
V • 5-Speed Manual, Floor Shift
W • 5-Speed Manual (2WD)
X • 4-Speed Manual (Part-Time 4WD)
Y • 4-Speed Manual, Floor Shift
Z • Auto (NP231)
5 Model Line (1989-92)
J • Cherokee (4WD)
J • Comanche (4WD)
N • Wagoneer
T • Cherokee (2WD)
T • Comanche (2WD)
S • Grand Wagoneer
Y • Wrangler
5 Model Line (1993-07)
A • TJ (Canada)
A • Wrangler (U.S.A.)
B • Cherokee (2WD)
F • Cherokee (4WD)
G • Commander (4WD)
H • Commander (2WD)
J • Cherokee (4WD)
K • Liberty (2WD)
L • Liberty (4WD)
N • Cherokee (4WD)
R • Grand Cherokee (4WD)
S • Grand Cherokee (2WD)
S • Grand Cherokee (4WD)
T • Cherokee (2WD)
W • Grand Cherokee (4WD)
W • Wrangler (Mexico)
X • Grand Cherokee (2WD)
Y • TJ (Canada)
Y • Wrangler (U.S.A.)
Z • Grand Cherokee (4WD)
Z • Grand Cherokee (2WD)
6 Series (1989-91)
1 • S (E)
2 • Base
3 • Islander
3 • Pioneer
4 • Sahara
5 • Laredo, Grand, Renegade
6 • Eliminator, Sport
7 • Limited
7 • Briarwood
6 Series (1992-07)
1 • S (YJ)
1 • Sport (Wrangler)
2 • Base
2 • SE (Wrangler)
2 • X
3 • Islander
3 • Pioneer
3 • Renegade (Liberty)
3 • Sport
3 • X
4 • Base (Commander)
4 • Laredo
4 • Sahara
4 • Sport
4 • Unlimited (Wrangler)
5 • Classic
5 • Sahara (Wrangler)
5 • Laredo
5 • Limited (Grand Cherokee, Liberty)
5 • Limited (Commander)
5 • Renegade
6 • Limited (Cherokee)
6 • Overland
6 • Renegade, Sport, SE
6 • Rubicon (Wrangler)
7 • Briarwood, Limited, Country
8 • Sport
8 • 5.9 Limited
6-7 Body Type (1983-88)
15 • Grand Wagoneer
25 • J10 Pickup (119" WB)
26 • J10 Pickup (131" WB)
27 • J20 Pickup (131" WB)
63 • Comanche 6' Bed (4WD)
64 • Comanche 6' Bed (2WD)
65 • Comanche 7' Bed (4WD)
66 • Comanche 7' Bed (2WD)
73 • Cherokee 2-Door (2WD)
74 • Cherokee 4-Door (2WD)
75 • Wagoneer
77 • Cherokee 2-Door (4WD)
78 • Cherokee 4-Door (4WD)
81 • Wrangler
87 • CJ7
88 • Scrambler
7 Body Type (1989-07)
4 • 2-Door Open Body (LWB)
6 • 2-Door Pickup
7 • 2-Door Sport Utility
8 • 4-Door Sport Utility
9 • 2-Door Open Body
8 GVWR (1983-88)
Grand Wagoneer, J10 & J20 Pickup
N • 5975 lbs.
S • 7001-8000 lbs.
Grand Wagoneer, J10 & J20 Pickup
U • 6001-7000 lbs.
Y • 8001-9000 lbs.
CJ7
A • 3750 lbs.
E • 4150 lbs.
Cherokee, Wrangler Base
J • 3001-4000 lbs.
1 • 4001-5000 lbs.
Cherokee Pioneer
K • 3001-4000 lbs.
2 • 4001-5000 lbs.
Wrangler Sahara
M • 3001-4000 lbs.
2 • 4001-5000 lbs.
Cherokee, Wrangler Laredo
L • 3001-4000 lbs.
4 • 4001-5000 lbs.
Comanche Base
F • 5001-6000 lbs.
1 • 4001-5000 lbs.
Comanche Pioneer
P • 4001-5000 lbs.
R • 5001-6000 lbs.
Comanche Laredo
T • 4001-5000 lbs.
U • 5001-6000 lbs.
8 Engine Type (1989-07)
B • 2.4L 4Cyl.
D • 2.5L 4Cyl.
E • 2.5L 4Cyl.
F • 2.5L 4Cyl.
G • 2.5L 4Cyl. TBI
H • 2.5L 4Cyl. MPI
J • 4.7L V8 (High Output)
K • 3.7L 6Cyl.
L • 4.0L 6Cyl.
M • 2.5L 4Cyl. Turbo Diesel
M • 4.2L 6Cyl.
N • 4.7L V8
P • 2.4L 4Cyl.
P • 2.5L 4Cyl.
P • 4.7L V8 Cyl Flex Fuel
P • 5.2L V8
S • 4.0L 6Cyl.
T • 4.2L 6Cyl.
U • 4.0L 6Cyl.
V • 4.0L 6Cyl.
Y • 5.2L V8 MPI
Y • 5.9L V8
Z • 5.9L V8
1 • 2.4L 4 Cyl (Wrangler)
1 • 3.8L V6 Cyl.
2 • 5.7L V8 Hemi
3 • 6.1L V8 Hemi
5 • 2.8L 4 Cyl Turbo Diesel (Liberty)
7 • 5.9L V8
9 VIN Check Digit
• Manufacturer's Internal Code
10 Vehicle Model Year
D • 1983
E • 1984
F • 1985
G • 1986
H • 1987
J • 1988
K • 1989
L • 1990
M • 1991
N • 1992
P • 1993
R • 1994
S • 1995
T • 1996
V • 1997
W • 1998
X • 1999
Y • 2000
1 • 2001
2 • 2002
3 • 2003
4 • 2004
5 • 2005
6 • 2006
7 • 2007
11 Assembly Plant
C • Detroit, MI
L • Toledo, OH
P • Toledo, OH
W • Toledo, OH
12-17 Serial Number
• Production Sequence Number

FIGURE 11-2 Each digit of the VIN is representative of a feature or option for that particular vehicle and is available in sources such as the crash estimating guide. Courtesy of Mitchell International, Inc.

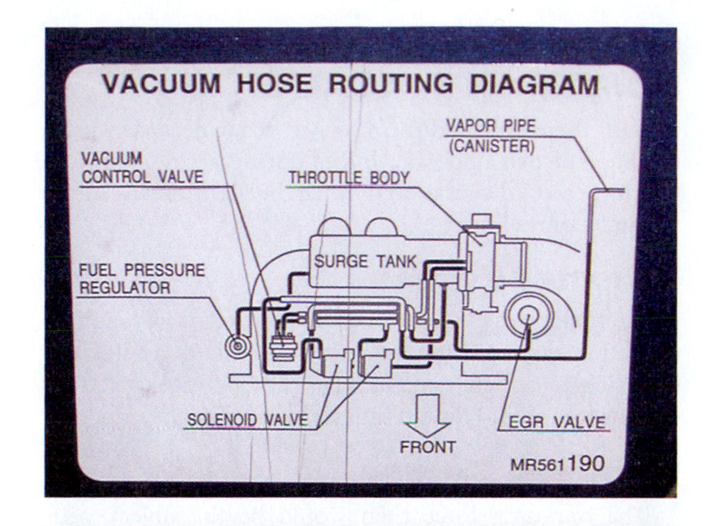

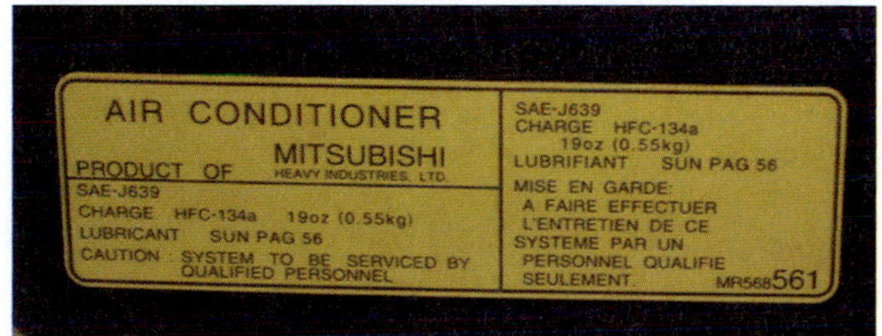

FIGURE 11-3 Various vehicle option labels can help the estimator obtain needed information about parts or assemblies.

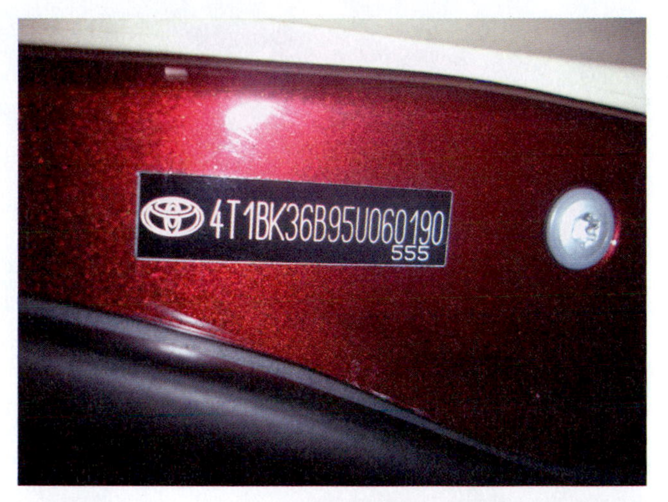

FIGURE 11-4 Anti-theft labels deter the theft of parts and also aid in tracking when parts are used on other vehicles.

the actual collision. One must understand the opposing forces imposed upon the structure and how these forces affect and create the damages sustained at the **point of impact (POI)** and points away from this area. No matter how minor or severe the impact, there are two types of damage that occur as a result of any collision: direct damage and indirect damage (sometimes called related damage).

Direct Damage

Direct damage is always more easily identified than indirect damage as it appears in the immediate area where the impact object came into contact with the vehicle. Direct damage is the obvious damage that even the untrained eye recognizes. It may affect only one panel or it may affect several panels, depending on the size and shape of the impact object and the vehicle's speed at the time of the impact. In a more severe frontal impact, the direct damage typically affects a number of exterior panels such as the front fascia, grill, hood, fenders, and any other panel that came into contact with the impact object. These panels often show varying amounts of damage, which frequently necessitates replacement of them. See **Figure 11-5**.

By carefully inspecting and analyzing the point of impact and the adjacent areas, the estimator can obtain a vast amount of information about the circumstances surrounding the accident. For example, it may be possible to determine the angle at which the impact object struck the vehicle, where the impact object initially made contact with the vehicle, and the order in which all the subsequent damage occurred.

Indirect Damage

Indirect damage is often more difficult to identify than direct damage because it is usually located in areas away from the immediate point of impact. Unlike direct damage, it has numerous characteristics. Much of the indirect damage happens as a result of the energy deflection that occurs during a collision. As the collision forces travel through the vehicle and around the passenger compartment, the panels and structures in its path twist,

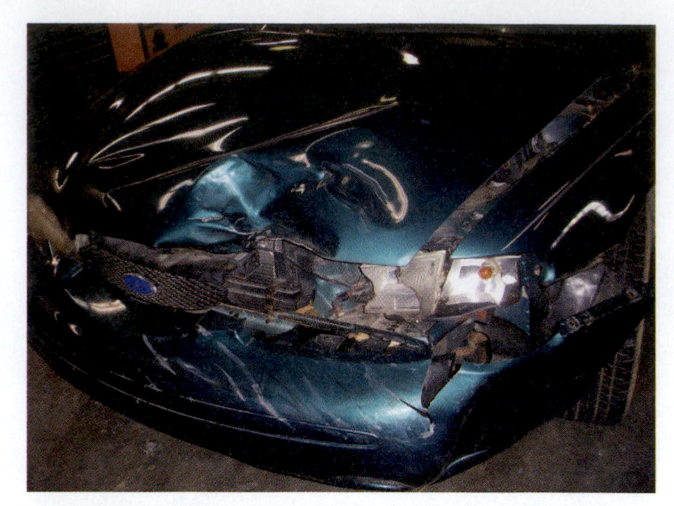

FIGURE 11-5 This picture shows the point of impact at the front of the vehicle, an example of direct damage.

shift, and move to absorb and deflect its energy, often causing damage that is very subtle and difficult to identify. Many areas of a vehicle are designed to react in this manner when they are subjected to collision forces of a certain magnitude, ultimately to reduce the effects of the collision forces in the passenger compartment. In order to isolate and identify these damaged areas, it may become necessary to remove some of the trim on the vehicle's interior or splash shields on the exterior to ascertain that the damage exists. These areas are often thought of as sacrificial structures common to most vehicles as they are found on nearly all vehicles and can be expected to react in a very like manner, no matter who the vehicle manufacturer may be. These are areas one can expect to be damaged in even a minor collision because of their design and intended purpose in the overall scheme of crash energy management; for example, the front rails in the area where the bumper is attached. In even a low speed collision, these rails typically bend and slightly collapse as a result of the impact forces. It takes a knowledgeable and trained estimator to know where to look when assessing the overall damage of the vehicle.

INTERNAL AND EXTERNAL FORCES IN A COLLISION

Both direct and indirect damage occur because of the presence of two opposing forces during a collision. For the purpose of discussion, they can be identified as internal and external forces.

Internal Forces

Internal forces exist because the vehicle's mass weight is in motion and tends to remain in motion until a sufficient opposing force acts upon it from the opposite direction, causing it to stop. This force is commonly known as inertia.

External Forces

The opposing force then would be the object with which the vehicle collides, causing it to abruptly come to a halt. In a typical frontal collision, the part of the vehicle that strikes the impact object stops first while the remaining parts of the vehicle tend to remain in motion momentarily longer or until the opposing force catches up with them. See **Figure 11-6**. During this short moment in time, while the upper part of the vehicle is still in motion and the lower part of the vehicle has stopped, a considerable amount of damage may occur in areas totally removed from the impact area. The roof support pillars may buckle at the location of a crush zone, the roof panel may flex and twist violently, the undercarriage could bend or buckle at the location of a stress riser—all of which may return to near their original location when the vehicle comes to a complete stop. This is the damage known as indirect damage—because there are no visible signs of contact with the impact object, it often goes undetected or overlooked. To put things into perspective, the roof of a vehicle that crashes into a brick wall at 40 miles per hour tends to keep moving at 40 miles per hour for a fraction of a second longer than the front end of the car. Since the roof is attached to the lower part of the body by the A, B, and C pillars and can't become separated from the lower body, it will continue to move forward until the opposite force stops its forward movement. This perpetual effect of motion, or

FIGURE 11-6 This sketch of a car moving forward shows arrows of the moving vehicle and arrows pointing in the opposite direction, labeled "opposing and inertial forces."

inertial forces, must be identified and reversed during the repair procedure. Many times more damage is caused by these inertial forces than by the impact object itself.

MASS WEIGHT MOVEMENT AND ENGINE AND TRANSMISSION CHECKS

While all of this is taking place, damage caused by **mass weight movement** is occurring simultaneously in the engine compartment. Mass weight shifting commonly occurs because the engine and transmission, which are very heavy, are subjected to the same abrupt opposing forces as is the body. Since they are not part of the body, but in fact are attached indirectly by the engine and transmission mounts, they are subject to a third force, known as a backlash effect. The mounts are commonly made of steel-reinforced rubber parts that sometimes have cavities filled with hydraulic fluid for stability. These create more opportunity to spring and shift when subjected to abrupt forces, such as those present in a collision. Because of their weight and the flexibility of the mounts, the engine and transmission will shift forward initially until the forward motion stops, at which time they will move backwards again, and then forward one more time to find their own resting spot. See **Figure 11-7**.

Engine and Transmission Checks

It is very common for the engine and transmission mounts to sustain damage which cannot be detected unless specific tests are performed while doing the damage analysis.

Transmission Mount. One of the things that can be done to determine if one or more engine or transmission mounts have been damaged is to engage the transmission and—while stepping on the brake—push down on the accelerator. The engine normally rocks or shifts slightly under normal operating conditions. However, an excessive amount of back and forth movement on a transverse mounted or side-to-side movement on an inline mounted engine is an indication one or more of the mounts has been damaged and needs to be replaced.

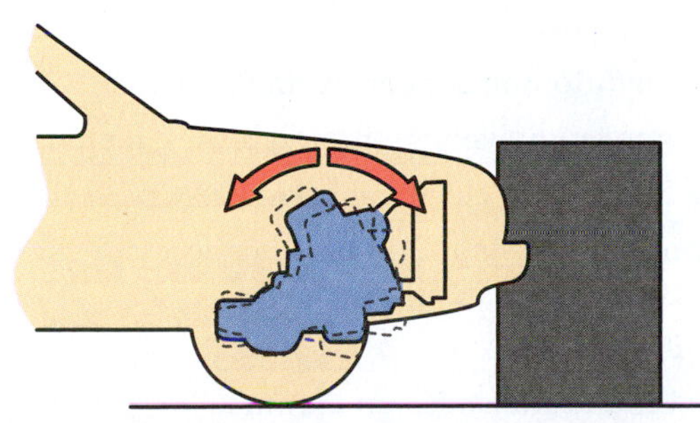

FIGURE 11-7 This sketch shows an engine rocking forward, backward, and forward to stop at its own resting spot.

Engine Mounts. With the assistance of someone else to manipulate the shifting and accelerator, one can confirm damage to these mounts by observing the movement at the lower engine mounts and the dog bone, an engine mount sometimes used where the engine is attached to the upper core support.

While performing this test, one should also operate the transmission shifting lever by alternately shifting the vehicle through all the forward gears and back to reverse to ensure no damage was sustained by the shifting linkage. Damage to the engine and transmission mounts is frequently accompanied with damage to the shifting linkage. The shifting lever should operate smoothly without any signs of binding or difficulty when moving through the entire throw pattern.

Shifting Linkage. While checking the shifting linkage, also note the gear indicator on the dash or steering column to ensure it is properly adjusted and correctly identifies the gear the transmission is in. The throttle linkage may also be located and attached in nearly the same location as the shifting linkage. Damage to one nearly always signals that damage has occurred to the other as well. Also check to see if the accelerator pedal works smoothly and easily both when stepped on and released.

Heater and Air Conditioning Plenum. Another area that automatically becomes suspect whenever engine mount damage occurs is the heater and air conditioning plenum which is attached to the fire wall immediately behind the engine. A minor crack in the housing can result in exhaust and other fumes from the engine compartment entering the passenger compartment. Many times the manufacturer places a cover over most of the engine. This cover is also made of plastic and may become cracked and have pieces broken out of it. Damage to the air conditioning lines and other components within the system may also occur because of their proximity to the engine.

A/C and Cooling Lines. The accumulator—a relatively large and usually silver-colored canister normally located in front of the plenum—should be checked for signs of physical damage such as indentations and questionable alignment with areas to which it is attached and kinks in the lines entering and exiting from it.

As was discussed in Chapter 6, Vehicle Construction, most automobiles are designed and constructed in such a way that one can predict the existence of collision damage when certain areas of the vehicle become damaged. Both the estimator and the technician must be aware that the collision forces affect not only the immediate impact areas, but also radiate throughout the entire vehicle. As these **hostile forces** pass through the vehicle, they affect many areas, leaving little more than very subtle signs of damage—unlike the obvious signs found at the impact area. An example may be a slightly wider gap between the fender and door or between the door and the A pillar. These conditions are apt to occur without the usual signs of damaged moldings, a cracked windshield, or other obvious things one would expect to see.

VEHICLE DESIGN FACTORS

Other things the estimator and the technician must consider prior to making a damage assessment is the vehicle's design. They must determine if the vehicle is a body over frame (BOF) or a unibody design.

Body Over Frame Vehicle

The type of damage incurred by the vehicle will vary considerably depending on its design or vehicle type. The damage to the exterior or cosmetic panels on a BOF vehicle tends to remain somewhat concentrated or more localized to the immediate point of impact. This is because the collision energy is largely absorbed by the frame and the mounts used to attach the cab to it. The damage to the undercarriage or the frame is apt to be more concentrated and less widespread since the frame is rather heavily constructed and it absorbs the brunt of the collision forces at the point of impact.

Unibody

In contrast to the BOF design, one can expect to see more widespread damage on the unibody vehicle, as the collision forces typically affect the entire vehicle due to its integrated design.

Knowing the approximate speed of the automobile at the time of the accident will also be beneficial to making a more accurate assessment of the overall damage. The higher the rate of speed at the time of the collision, the deeper the collision forces will penetrate into the vehicle, increasing the likelihood of hidden damage. Hidden damage often occurs in areas that are covered by carpeting on the floor and flexible trim parts on the sides and on the ceiling. The plastic parts tend to spring back to their original shape if the surface to which they are attached is not distorted too severely. This is also true of the flexible exterior panels such as those commonly used on a space frame design. Knowing and understanding how the vehicle will react and respond under certain collision conditions and what areas are affected when these circumstances are present will help to ensure none of the damage will go undetected. Having a thorough understanding of the vehicle's design and construction will also help to ensure that none of the potentially damaged areas are overlooked.

SYSTEMATIC DAMAGE ASSESSMENT

When assessing the damage, one should approach the task as systematically and methodically as is possible.

Re-Creating the Damage

Re-creating the damage can best be accomplished by mentally re-creating the sequence in which the damage occurred, starting from the point of impact and moving through to the damaged area furthermost from the starting point. The estimator should walk around the vehicle several times to visualize any obvious damage incurred, not only at the point of impact, but the secondary damage from the inertial forces and other areas that may have become damaged from a secondary impact into another object. Excessive gaps or no space between panels, disrupted bodylines from one panel to the next, panels that look to be out of their normal position, and buckles in non-adjacent parts are all signs of damage that must be addressed. One must also remember that on a unibody vehicle especially, the damage is apt to radiate throughout the entire length of the vehicle. Therefore, damage may occur some distance from the point of impact without actually coming into contact with the impact object. This may also be the result of the inertial forces present at the moment of the collision or the effects of a loose mass weight object in the vehicle. After doing an overall visual analysis to obtain the "big picture," one should then work through each of the three sections of the vehicle, one at a time, to ensure no damaged item is overlooked.

Inspection Sequence

During an inspection, the estimator should work from the front exterior of the vehicle and move inward to include all the areas under the hood, followed by the hood fenders, and so on. The following sequence will minimize the possibility of inadvertently excluding or overlooking any damaged area or part in the front section of the vehicle:

- Front bumper cover including all lights, accessories, and so on
- Bumper reinforcement and energy absorbing devices
- Mounting brackets and hardware
- Front rails behind bumper
- Core support
- Air conditioning condenser and radiator
- Cooling fan housings, shroud, blades, clutch
- Cooling system electrical wiring
- Air intake housings and all other accessories attached to core support
- Front side rail
- Engine and transmission mounts
- Hood, hinges, latches, gas shock support
- Fenders
- Suspension and steering parts that will be inspected with the undercarriage

Using Vehicle Specification Manuals

Nearly every domestic and commercial vehicle ever assembled has manufacturer's dimensions or specifications available that were used for manufacturing specifications. These can also be used to determine the existence of damage, as well as to ascertain the restoration and removal of the damage at the time of the repairs. For purposes of measuring, these three sections are arbitrarily separated into the front section, which is located under the engine compartment; the center section, located under the passenger compartment; and the section found in the trunk area. See **Figure 11-8**. These specifications show the length, width, height, and diagonal measurements of the vehicle's undercarriage, door, and window openings and are used as references or a guide to determine when the vehicle is returned to the same specifications used by the manufacturer to assemble it. The manufacturer's undercarriage specifications for most vehicles are laid out in three sections reminiscent of the three sections of the unibody design: the front, the center or passenger compartment, and the rear section. In addition to the undercarriage dimensions, a separate set of underhood dimensions are also available due to the many critical tolerances that must be maintained in this area.

Comparative X-Checking

Depending on the degree of accuracy required for the task at hand, two different methods of measuring are employed. One can make **comparative quick check measurements** to ascertain whether damage actually did occur. See **Figure 11-9**. If it is necessary to pinpoint the exact location and identify the extent of the damage more accurately, one can perform several measurements and compare them to the published dimensions. The published dimensions list the exact measurements used by the manufacturer when the vehicle was assembled. These are the dimensions to which the vehicle must be restored when it is repaired.

Comparative Quick Checks

A comparative quick check measurement can be accomplished by simply measuring from a fixture hole located near the front on the right fender or core support to the rear mounting or alignment hole on the left rear fender, for example. Record these measurements and repeat the same

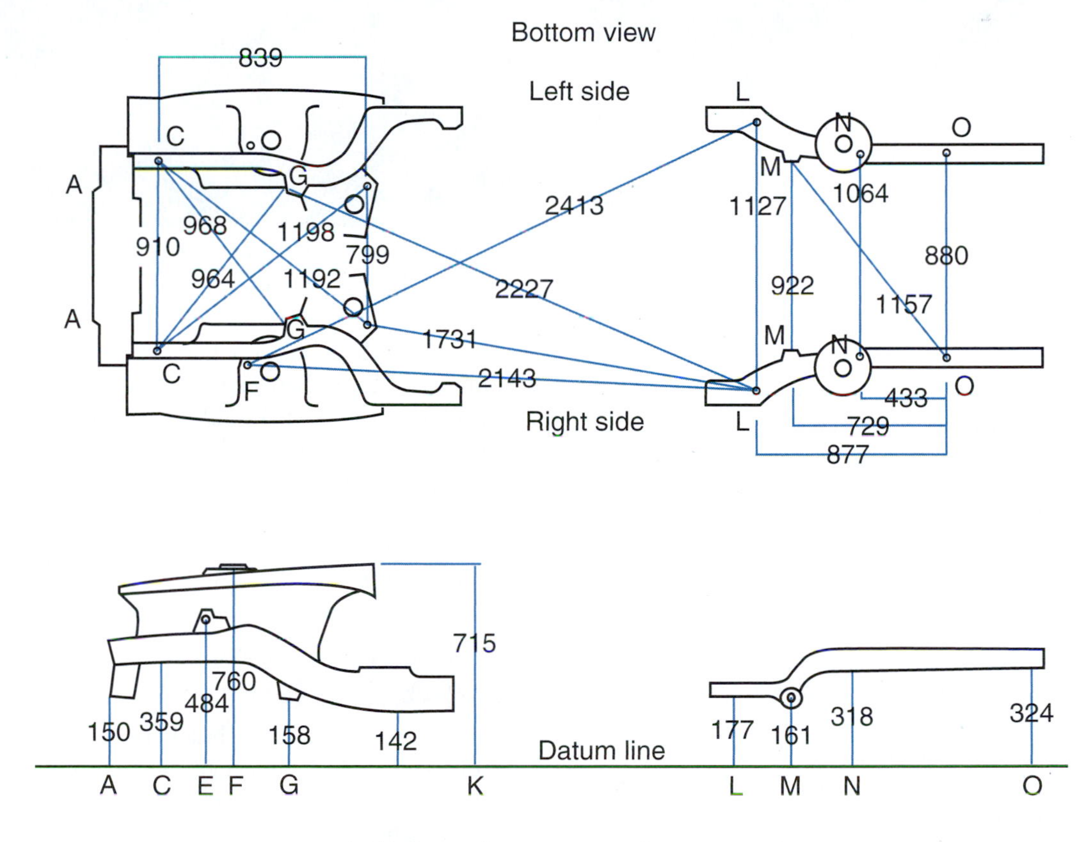

FIGURE 11-8 The undercarriage specs of a vehicle show the length, width, height, and diagonal measurements of the vehicle's undercarriage. Underhood, door, and window opening dimensions are also available.

FIGURE 11-9 These pictures show making comparative underhood X-checking measurements.

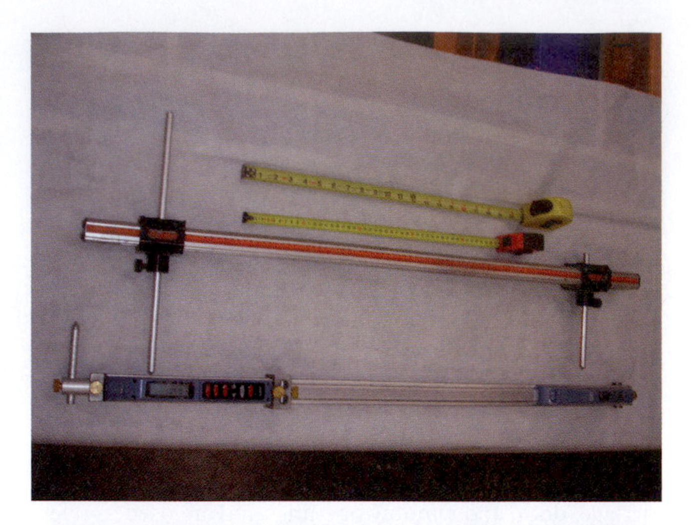

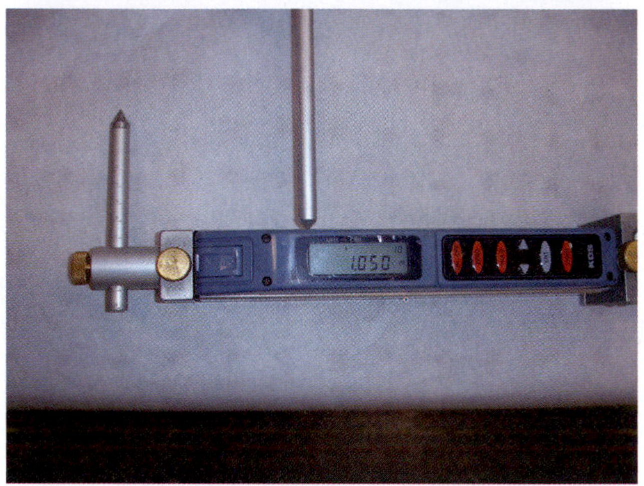

FIGURE 11-10 A tram bar or tape measure can be used for X-checking measurements.

process on the opposite sides of the vehicle. These measurements should be recorded and checked to see if there is any difference in them. In the event the measurements are not alike, one can determine approximately how much the front area of the vehicle has shifted or moved sideways or rearward by subtracting the difference of the shorter measurement from the longer one. X-checking measurements can be performed using the tape measure or a **tram bar**. See **Figure 11-10**. At this time when the damage is being analyzed and identified, the measurements are taken to confirm that damage did indeed occur, but not to determine the location or the exact amount that may have occurred. It will become necessary to take more precise measurements at a later time when the repairs are actually being made, using the printed underhood dimensions. This will be covered in Chapter 13, Measuring Structural Damage.

INSPECTION SEQUENCE

Front Fascia

In a front collision one would analyze the damage to the front bumper assembly, followed by the grill and all the lighting accessories, the hood, and the front fenders, including all the items attached thereto. The section of the front rails located between the bumper attaching points and the core support should be carefully inspected from above and below the vehicle. This is one of the sacrificial areas that nearly always sustains some damage in even a minor collision. Crush zones are frequently incorporated into the rails at this point to absorb and dissipate some of the initial impact forces.

Bumper Mounts and Absorbers

The bumper mounting brackets and impact absorbers should be carefully checked to ensure they were not shifted sideways. These areas are not always readily visible without removing some of the trim covering them. Therefore, in order to ascertain the existence of damage at this point in the analysis, it may become necessary to check for some characteristic damage deeper into the vehicle—such as a misaligned core support—and work backwards. After inspecting these areas, one would then move to the area under the hood to inspect the core support, the heating ventilation and air conditioning systems, and all the me-

FIGURE 11-11 The front bumper assembly on a BOF vehicle consists of a heavy steel bar that runs the width of the vehicle and is bolted to the frame.

FIGURE 11-12 The fascia on a unibody vehicle is usually a large flexible plastic panel that covers the entire front of the vehicle.

chanical parts that were likely in the direct path of the collision forces. By using a systematic process such as this, one is less apt to overlook any of the damaged areas regardless of how large or small.

Knowing whether the vehicle's structural design is body over frame, unibody, or space frame will help the estimator identify more accurately the damage incurred. While each design will sustain a certain amount of damage, the parts affected will be somewhat different. On a body over frame vehicle, for example, the front bumper is usually more heavily constructed with the use of a steel reinforcement and is less likely to reveal the damage in this area as obviously as can be noted on the unibody. The bumper assembly is typically attached to a mounting bracket, which in turn is attached to the front side rails. In some instances, a mechanical impact absorber is also used to absorb the shock effect. See **Figure 11-11**. The damage in these areas is typically more obvious and viewable than on the unibody or space frame. The unibody and space frame frequently use a larger plastic **fascia** which covers the front bumper reinforcement. It is sometimes accompanied by a high density Styrofoam energy absorbing pad or an egg crate plastic reinforcement to fill any gap between the outer cover and the reinforcement. The reinforcement may be welded to the front rails, making it a part of the front structural makeup, or it may be bolted to the front rails. Realizing the differences in the design used by the manufacturer will be beneficial as each system is made to absorb and sustain the damage by a different part of the structure. Many times the fascia is attached to the lower grill mounting panel, the front fenders, and other areas with various fasteners. The mounting tabs and holes should be carefully inspected to determine if they were damaged and, if so, if the fascia should be replaced or can be repaired. Knowing where to look for the damage and understanding the type that is likely to have occurred will help create a speedy and accurate analysis and repair. See **Figure 11-12**.

FRONT LIGHTS AND ELECTRONIC SENSORS

As one moves inward from the front of the vehicle and toward the back, knowing the characteristics of the vehicle being evaluated and the equipment options typically located and attached in this area will be helpful in making a more accurate assessment. In addition to the standard equipment such as the headlights, a considerable number of highly sophisticated sensors and other electronic devices are attached to this part of the vehicle. One must be knowledgeable of their existence and be able to recognize them by their appearance in order to ensure they are not omitted from the damage assessment.

Headlights

The headlights may be the first items that one must consider as they are frequently incorporated into the front bumper and fascia assembly. Many times the entire lighting system—including the high and low beam headlight, directional signal lamps, the side turning lights, and the emergency flasher—are all incorporated into one composite assembly. Due to their critical function, one minor damage flaw may necessitate the replacement of the entire capsule. Other vehicle lighting systems may be designed with each light as an individual unit totally independent of the others. See **Figure 11-13**.

Mechanized Front Lamps

The lighting systems on several manufacturers' vehicles utilize a very sophisticated mechanical system in addition to their assumed function of illuminating the roadway. The 2005 Jaguar XK and the Lexus IS 350 are equipped with auto leveling headlamps which adjust themselves automatically to compensate for a shift in the ride height and levelness of the vehicle. The 2006 BMW 330 and the Volkswagen Passat are equipped with turning headlamps. The

FIGURE 11-13 The headlight capsule on some vehicles—which includes the headlight, turn signal, parking lamps, and their mounting surfaces—are frequently damaged in a frontal collision.

headlights follow the steering wheel of the car and adjust or turn in the same direction as the front wheels are turned. These systems are usually self-checking when the vehicle is first started. An inactive or non-moving system is a sign of a damaged or binding motor. The wiring, motors, and all sensors must be checked very carefully to ensure they are all functional and undamaged.

Headlight Mounting Brackets

One must also carefully inspect the headlight mounting panel and brackets because if the headlight assembly or any of the attaching parts are damaged, these parts typically require replacement as well. The wiring for the entire lighting system is often routed through holes and openings in the radiator support and is often pinched and damaged whenever the front of the vehicle is involved in a collision. Other vehicles utilize **fiber-optic cables** to illuminate some optional items at the front of the vehicle. These fiber-optic cables are very fragile; when cut, kinked, or damaged in any way, they won't properly project light through them. These must be carefully inspected for possible cuts and crushing damage and replaced as needed.

Many times the grill assembly is integrated into the headlamp assembly and may have been damaged by the impact object as well. The mounting panels and housings are often covered by baffles, gussets, and other trim, making it difficult to determine if they sustained damage. Seals and baffles that direct the airflow through the cooling system are also located in this area and—because of their location—may be damaged as well. These parts may seem rather insignificant to the uninformed person, but they perform a critical function to the cooling of the engine and must be replaced when damaged. One cannot assume that any part of the vehicle is insignificant.

At this point the estimator may follow through with inspecting the structural panels and the engine and mechanical parts under the hood, or continue with the inspection of the exterior front body panels. Since the drivetrain and the suspension systems are closely tied together, they will be evaluated together after assessing the exterior panels.

Hood and Fender

When inspecting the hood and fenders, one should look for any obvious differences in the side-to-side gaps between the fender and hood as well as any differences in spacing between the panels from the front to the rear. Any differences between the height of the two panels should also be noted.

Hinges. The hood hinges should be checked for damage or possible misalignment by raising and lowering the hood slowly to ensure it is properly aligned with the fenders. See **Figure 11-14**. The primary and secondary latch should be checked for possible misalignment and proper latching and release. See **Figure 11-15**. The fenders should also be carefully inspected for irregular gaps between them and the doors. Many times tape stripes and decals run parallel to each other from one panel to another. Adhesive bonded moldings are often installed parallel to each other from one panel to the next. These can be another quick visual indicator to see if they are in line with those on the adjacent panel.

Tires and Wheels

The tires and wheels should be checked for damage and any sign of unlikely material such as grass, weeds, or dirt that may be stuck between the tire and wheel. This is a definite indicator of the vehicle leaving the road, which could be a setup for additional damage. This can be substantiated by tire rub marks on the fender liner or splash shields. Any sign of rub marks would indicate the suspension moved back far enough during the collision to warrant a closer inspection when that area is assessed. One should also look for cracked undercoating, pulled spot welds, dirt that may have been rubbed off, and any other signs that lead one to believe movement may have occurred. See **Figure 11-16**. Any discrepancy in the gaps between the fender and hood or doors may be cause for additional damage and may warrant X-checking this section of the vehicle to confirm the suspected damage.

At this point one would commence evaluating any possible damage under the hood. Moving inward to the radiator or core support, one must be aware of the many parts attached at this location. Any misalignment of the

FIGURE 11-14 Any notable difference in the fender to hood gap might indicate damage to the hood and hinges as well as the fenders.

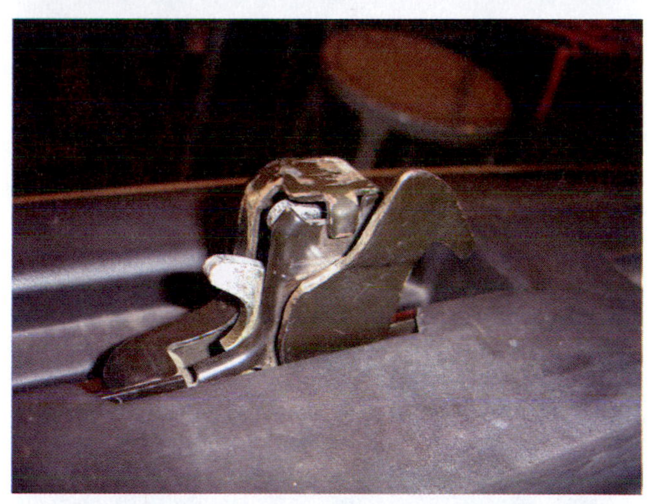

FIGURE 11-15 The primary and secondary latch should be checked for possible misalignment and proper latching and release.

fenders or damage, however slight, to any of the core support panels is a clue to look for damage to the heating and ventilation system—particularly in a frontal collision. These parts are located in the immediate area of the direct impact and consequently are subjected to most of the collision energy. All the cooling and many of the air conditioning system parts are attached directly to the radiator support or core support at the front of the vehicle. Since the air conditioning condenser and the radiator have little or no protection from intruding objects they are highly susceptible to becoming damaged. Both of these parts are relatively susceptible to punctures; when the vehicle flexes during the impact, they also bend and flex a considerable amount—in some cases to the point that the radiator core becomes separated from the fluid reservoirs or side tanks attached to it. Some vehicles have an auxiliary transmission fluid or engine oil cooler integrated inside the tanks of the radiator. One should remove the radiator cap to check for signs of oil mixed with the coolant as this could be a sign of a broken oil reservoir inside the radiator. The radiator cooling fins should be carefully inspected to ensure adequate airflow is possible for proper cooling and that none of the coolant circulation veins have been punctured. Combing the fins may be possible if they have not been damaged excessively. See **Figure 11-17**, **Figure 11-18**, and **Figure 11-19**.

FIGURE 11-16 One should look for cracked undercoating, pulled spot welds, dirt that may have been rubbed off, and any other signs that lead one to believe movement may have occurred.

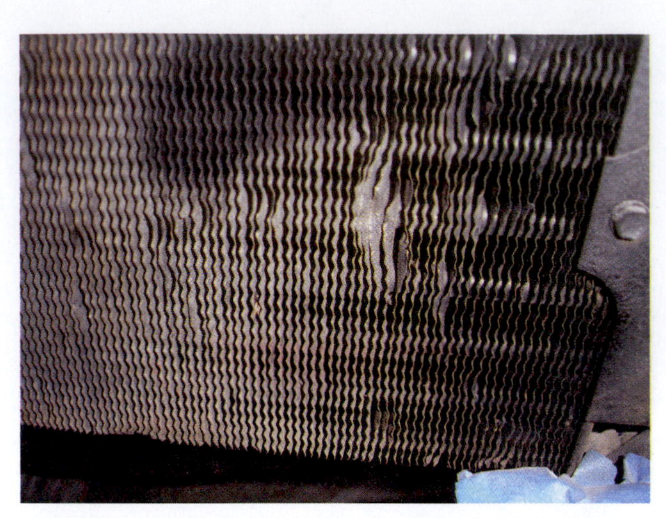

FIGURE 11-18 The radiator coolant circulation veins should be checked for puncture holes and to ensure they have not collapsed or been crushed.

FIGURE 11-17 Since the radiator has little or no protection from intruding objects it is highly susceptible to becoming damaged.

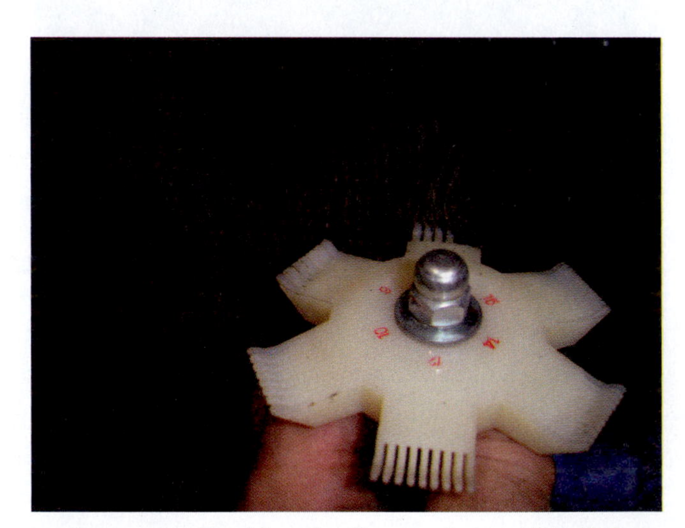

FIGURE 11-19 Combing the fins may be possible if they have not been excessively damaged.

Air Conditioning Components

In most cases the air conditioning condenser is subjected to even more of the initial impact than the radiator, as the condenser is located in front of radiator. Many times it may be punctured by some of the breaking fragments of the grill or by the impact object. A condenser that has been punctured—resulting in a loss of the refrigerant—is easily recognized, as an oily looking spot will appear where the refrigerant and oil escaped. This should be a tipoff to the estimator and technician to include replacing the receiver dehydrator-dryer or accumulator, as this is the recommended procedure whenever the system has been exposed to the air for even a short period of time.

Occasionally, in the aftermath of a collision, the radiator and **A/C condenser** remain slightly twisted but still intact and holding the coolant and maintaining pressure. See **Figure 11-20**. The estimator and technician must de-

FIGURE 11-20 A twisted A/C condenser should be replaced even though it remains pressurized.

cide if they must be replaced or if they will be able to withstand a second twisting or flexing when straightened along with the core support. It is advisable under these circumstances to replace the part rather than taking a chance of the part failing at a later time.

ENGINE COOLING SYSTEM

Inspecting and pinpointing damage to the cooling and heating and ventilation (HVAC) system is sometimes difficult to do as many of the parts are closely crowded together, making it difficult to see each individual item. This is further complicated by the presence of **gussets**, covers, insulators, and seals at the top of the core support that are often used to cover and disguise many of the otherwise unsightly parts on the front of the vehicle. However, knowing and understanding where the critical parts are located and knowing which of them are apt to become damaged under certain circumstances will help the estimator to identify any potential damage. The engine cooling hoses should be checked for cuts and a note should be made of where on the radiator the supply and return ports are located as this information is necessary when a new part is ordered. Even though the coolant intake and outlets may be located in the same location on vehicles using a common engine by a given manufacturer, a different engine size and transmission type may change where on the radiator tanks they are placed. The heater hoses are frequently routed around the sides and sometimes the back side of the engine. They should also be inspected to ensure they are free of any rub marks from a pulley and they have no cuts or puncture holes that would result in coolant loss. The transmission and auxiliary engine oil cooling lines may require replacement even if they sustain what appears to be only a very slight kinking. Any obstruction, however minor it may appear, can affect their ability to properly circulate the oil, thus causing a problem with overheating. This is particularly true in lines made of metal.

Radiator Shroud and Fans

On a unibody vehicle, the radiator shroud and cooling fans are attached to the core support immediately behind the radiator. Even though the radiator may not be damaged, the cooling fans, housing, and shroud often are damaged as a result of the shifting mass weight from the engine and the movement of the front structure. The cooling fan housing, the wiring, and all other parts of the assembly must be carefully inspected for damage or misalignment. The cooling fan blade may have chips of plastic broken out of it, causing the shaft of the motor to get bent. If not replaced, it will cause a severe vibration problem. The motor housing may get cracked as a result of the engine shifting or moving forward during the impact. See **Figure 11-21**.

Body over frame vehicles are typically equipped with a belt-driven cooling fan which is attached to the water pump at the front of the engine. The blades and the **fan clutch** should be carefully inspected to ensure they are not misaligned or twisted. See **Figure 11-22**. A scraping or

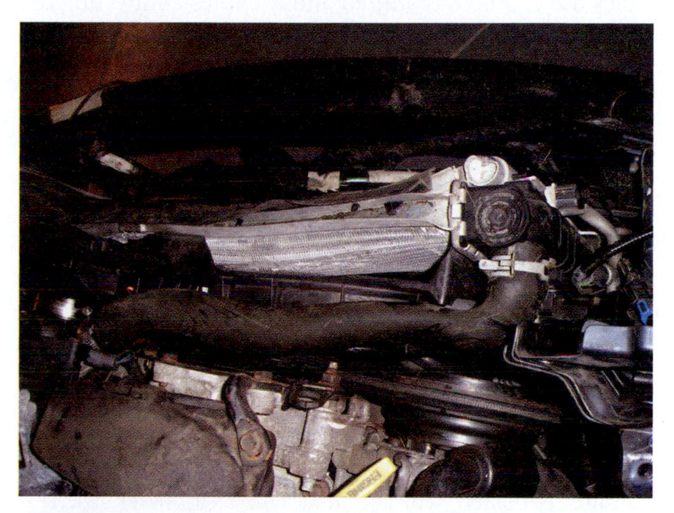

FIGURE 11-21 Any damage to the cooling fan blades or the fan housing will likely cause a severe vibration when the engine is run.

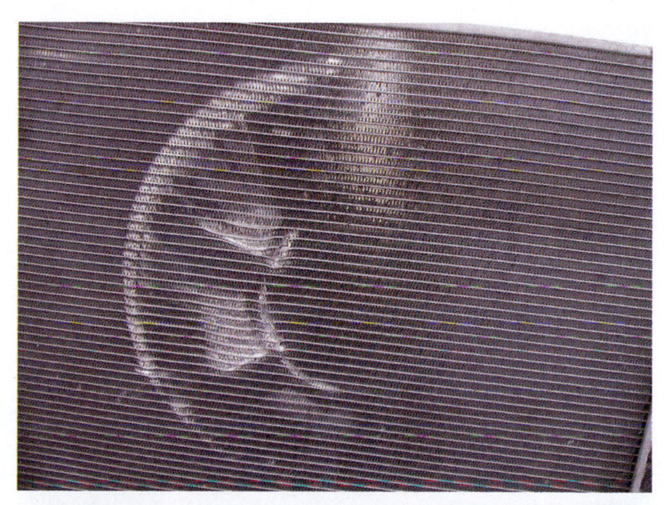

FIGURE 11-22 A scrape or grooved indentation on the radiator core resembling the shape of the fan blade will result in damage to both items.

grooving effect identical to the size and shape of the fan blade on the back side of the radiator should be a tipoff to carefully inspect all of these parts. A bent fan blade can cause severe vibration problems when the engine is operated at even low speeds. Any damage to the radiator shroud or any of the baffles designed to direct the airflow across the engine should be replaced.

Heater Core

Even though the heater core is not located anywhere near the front of the vehicle, like the radiator it is part of the heating and cooling system. Hot coolant circulated from the engine through the heater core is used to heat the passenger compartment. The heater motor blows air over the heater core, picking up heat from the fins and distributing it throughout the passenger compartment. The coolant is eventually circulated back to the radiator where it is cooled and re-circulated once again. Much like the radiator, the heater core is subject to damage—particularly when the engine has shifted and the lines connecting the core have been moved as a result. A film covering the windshield inside the car, a sweet and pungent aroma inside the car, and a wet spot on the passenger side floor are signs of a leaking heater core.

Accessory and Control Modules

On some vehicles an auxiliary electrical fuse/control box is also attached immediately behind the radiator support. On certain other vehicles vital mechanical parts such as the antilock brake modulator assembly may be attached in this area as well. Separate lines running to each wheel are attached to this modulator and may be susceptible to damage from the impact. The engine air intake plenum is commonly placed near this area. Due to its sheer size, it is often crushed or cracked as a result of the moving mass weight of the engine. See **Figure 11-23**. An item by item inspection must be performed to ensure that no damaged parts are overlooked.

Engine and Transmission Mounts

The engine and mounts should be inspected for obvious signs of damage and movement. See **Figure 11-24**. To check for damaged or broken mounts, the engine should be started, if possible. The vehicle should then be put into gear. While stepping on the brake, the technician should push on the accelerator to see if the engine moves or rocks excessively. A slight amount of movement is normal, but any excessive shifting or keeling is a sign of a damaged engine mount. In the event the engine cannot be started, a prybar may be used against the engine to determine if any movement is possible. Being able to move the engine with minimal effort is an indication the mounts are damaged. The technician should try to isolate which one—or ones—is damaged. The transmission shifting and throttle linkages may be affected by the rocking of the engine, thus causing damage to the mounting brackets and possibly even the cables themselves. The water pump, power steering pump, and their respective mounting brackets are attached in areas where they are susceptible to damage by the engine moving. Their respective pulleys should be checked to ensure they did not become bent or misaligned.

Valve and Engine Covers

The valve cover and the timing chain cover should also be checked for possible damage. The valve covers on many late-model cars are made of plastic—if any indentations are evident they should be replaced. Likewise, if any damage is visible on the timing chain covers they should also be replaced as dirt infiltration and contamination is apt to occur if left exposed by a crack or pieces missing. Occasionally an informational label with pertinent details about the engine is placed on the timing chain cover. This infor-

FIGURE 11-23 The engine air intake plenum is often crushed or cracked as a result of the moving mass weight of the engine.

FIGURE 11-24 The engine and mounts should be inspected for obvious signs of damage and movement.

mation should be recorded along with the part number so it can also be ordered and replaced along with the cover.

Exhaust System

The exhaust system should be checked to ensure that no cracks or other damage to the system is evident. If the engine starts, any damage to the system is readily evident because of the noise level. However, if the vehicle cannot be started it may be difficult to determine if any damage exists. Any sign that the engine may have moved or any indication of damage to the engine mounts should be a clue of possible damage to this area.

Fuel Pump

Attempting to start the engine may require more than the normal amount of cranking for a variety of reasons. The possibility of damage to the fuel pump from the abrupt collision forces is always present. On many vehicles, the fuel pump is located inside the fuel tank. This may necessitate the removal of the cell to determine any damage. One can make a quick, easy check to determine if the fuel pump is functioning properly by listening for it to run when the ignition key is first turned on. A short audible noise can be heard coming from the fuel pump inside the tank as it pressurizes the system each time the ignition is turned on. One should stop attempting to start the engine immediately at the first sign of any gas leaks from broken lines or fittings when the fuel pump is running or when attempting to start the vehicle. Even if the engine starts running, the fuel pump will attempt to maintain pressure in the lines, which typically exceeds that needed to keep the engine running. Excess fuel escaping out of lines is apt to cause a fire or possible explosion.

Fuel Cutoff Switch

Some vehicles have an **inertia cutoff switch** that is activated when the vehicle is stopped abruptly, such as in a

FIGURE 11-25 An inertia cutoff switch may be activated on some vehicles when the vehicle is stopped abruptly, such as in a collision.

collision. See **Figure 11-25**. This switch is activated to prevent the fuel pump from pumping gas out of possible broken fuel lines and to stop the engine from running under extreme conditions. This switch can be found in a variety of locations inside the trunk. One should consult the vehicle service manual to determine its exact location as it must be reset in order to start the engine.

ANALYZING A ROLLOVER VEHICLE

Two or three additional checks need to be made as part of the routine inspection performed under the hood prior to startup if a vehicle has been rolled over.

Check Fluids

This is especially important if the vehicle was allowed to set in an inverted position for any length of time. The first thing that should be done is to check all the fluids to ensure they are clean and not contaminated. The coolant should not have any oil floating in it, nor should the engine oil show any signs of stray liquids. Also check to ensure the oil level is not higher on the dipstick than the recommended markings suggest. This condition could signal additional problems exist. The transmission fluid should not show any higher on the dipstick than at the very bottom or the tip of it. This is normal as this fluid is normally checked after the engine has been running for a period of time.

Check Engine Cylinders for Fluid

Before starting to crank the engine, the spark plugs should be removed and the engine should be rotated manually to ensure no fluids are in the cylinder. This should be continued until at least two or three full extensions of the pistons have occurred. Check for fluid coming out of the plug holes and wipe the area clean of any liquid that may have run out during this test. Before attempting to start the engine, the fluids should be topped off to ensure against damage when it is started. The spark plugs should now be reinstalled, the plug wires connected, and the engine cranked over with the starter. When the motor starts some discolored smoke may be noticed during the first two or three minutes the engine is running. This is a normal phenomenon provided it doesn't last for longer than two or three minutes. A white exhaust smoke is indicative of coolant burning, which will have been left in the combustion chamber. A blue exhaust smoke is indicative of burning oil that was left in the engine. The colored smoke should stop after a short time of running the engine. Cranking the engine before performing these steps could result in **hydraulic lockup** and cause additional damage. If the engine has not been started prior to this time the shifting linkage and the transmission shifting can be checked at this time.

Check Doors to Opening Fit

After completing the inspection under the hood, the next areas that should be carefully inspected are the doors and their respective components. The placement of the doors within the openings can reveal a wealth of information concerning the overall body alignment. An undamaged door, for example, with gaps that are not the same at the front as on the rear would indicate that the hinge pillar may have been affected. Also, the possibility that the fender, rear door, or quarter panel may have shifted should not be ruled out. An uneven gap between the bottom of the door and the rocker from the front to back of the panel might also be an indication that the entire body has been affected by a twisting condition. This is especially true if the vehicle has left the road by going into a ditch or jumping a curb or a road obstacle. When these signs are present, it should be an automatic signal of the possibility of damage to the undercarriage.

In the event the door has sustained some obvious damage, one should also check for uniform gaps between it and all the adjacent panels. Uneven gaps between it and the adjacent panels is an indication the hinges may again have been affected. If the door sustained a hard hit in the lower section, it may have bent the intrusion beam, causing a shortening effect which would require replacement of the entire panel. In a less severe damage scenario, the area where the door intrusion beam is welded to the door should also be checked for pulled spot welds or any distortion of the inner door frame. Any sign of damage to this area of the door warrants replacement, as no allowance is made for these to be repaired.

Check Doors and Hinges for Damage

One should open and close the doors to ensure they operate smoothly and there is no rubbing or binding at any point while opening or closing them. Any time more than a normal effort is required to open or close the door, it is an indication the hinges may have been affected. While opening and closing the door, the latch mechanism should also be checked for proper latching and releasing. The windows should be fully lowered and raised to rule out any possible damage to the tracks, or the wiring on electric lifts.

Check Locks and Latch

If the vehicle is equipped with electric locks, they should also be checked for proper operation. Keyless entry optional equipment is available in a variety of systems. A number or digital push pad locking and unlocking system may require the use of the combination programmed into the vehicle to operate it. This code should be obtained from the vehicle owner during the time when all the other pertinent information is obtained. A keyless fob, if available, should also be used to check for proper locking and unlocking from outside the vehicle. A vehicle equipped with a remote start option should become part of the checklist as well.

The side glass, moldings and trim, mirrors, global positioning system, antennas, and radio masts and their respective mounting surfaces should be inspected for any damage. These items are nearly always replaced when they become damaged.

ANALYZING SIDE IMPACT DAMAGE

A side impact differs from a frontal impact in a number of ways that offer numerous implications for repair versus replacement that must be considered when evaluating the damage and deciding what must be done to best repair the vehicle. Many times decisions must be made that impact not only the repair issues being dealt with but also the vehicle's longevity. One must decide whether the proposed repairs will in fact restore the vehicle to its pre-collision condition. Many times inner structural panels must be removed to gain access to the parts that require replacing. Would using an entire section of recycled parts be a more prudent choice to ensure the restoration of the structural integrity? At the same time one must decide if using recycled parts creates any potential consequences of compromising the overall structure and the integrity of the vehicle in a subsequent collision and for normal driving as opposed to using new parts?

When analyzing the damage on a side impact, especially in the center section of the car, one must be aware that this is perhaps the most critical section of the vehicle because it is the passenger compartment. This area of the vehicle includes more disguised or hidden structural members that are apt to be affected in a collision than any other part of the vehicle. Typically the entire interior is covered with upholstery and trim, therefore making damage assessment more difficult than on other parts of the vehicle because the damage frequently cannot be seen until the trim pieces are removed. One must know and understand the vehicle's structure and be able to anticipate many of the damages that are apt to occur in a collision.

BEND VS. KINK

One must also determine whether the parts are repairable or if they will require replacement. The principles of bend vs. kink must be exercised whenever these decisions are made. The rules of **bend vs. kink** suggest that when a body panel, particularly a structural part, is bent it can be straightened, but when it is kinked it must be replaced. While there is no description or rule of what constitutes a kink, one might say any time a sharply formed bend is formed within a short distance or radius it is considered a kink. See **Figure 11-26**.

Kinked Area

One is apt to find a kink in or near a crush zone in the immediate impact area; however, kinks may also be found in areas away from the point of impact as well. See **Figure 11-27**. A gradual bend without any sharply formed edges or ridges spread over any area—however long or short—may be considered a bend. An area that is bent is apt to be restored to its original shape by roughing and removing the severely damaged parts around the area or by applying the principles and techniques for normal metal straightening. A rule of thumb that should be exercised is any structural part that is kinked should always be replaced. Many times the area where metal is kinked is not completely restorable and the area where the actual bend occurred is apt to fatigue, eventually becoming the area

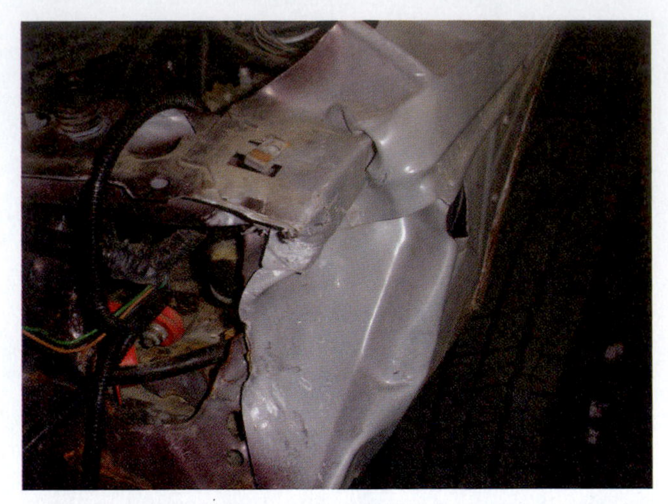

FIGURE 11-26 Any time a sharply formed bend is formed within a short distance or radius it is considered a kink.

FIGURE 11-27 Kinks may be found in or near a crush zone in the immediate impact area, as well as in areas away from the point of impact.

where the breaking down of the structural integrity begins. An uninformed repair technician might think reinforcing the area by welding an additional layer of metal over the top of it would be a proper repair. However, doing this would be **re-engineering** that section of the vehicle, causing it to respond and react unpredictably in a subsequent collision.

In a side collision the areas that are nearly always affected are the front and rear doors and the hinge pillars, also known as the A, B, and C pillars. Because of their strength and the manner in which they are attached to the roof and rocker panel, these two areas of the vehicle are also subject to damage as well even though they did not come into contact with the impact object. The effects of the inertial forces must be taken into consideration when analyzing the damage that occurred. The roof panel, the door opening frames, and the area where the center pillar is attached to the roof must also be checked for distortion, pulled or damaged spot welds, and seams that may have been stressed and pulled apart from the collision forces. The same detailed inspection must be done at the bottom of the **B pillar** where it is attached to the rocker panel.

To determine the exact extent of the damage it may be necessary to remove the interior and perform a series of comparative diagonal measurements from the roof to the floor on the opposite side of the vehicle, and then repeat the same measurement for both sides. This may have to be done in several locations in order to pinpoint the exact location of the damage. The measurements should also be compared to the available published dimensions. See **Figure 11-28**.

ROOF, B PILLARS, DOOR HINGES, AND SIDE AIR BAG SENSORS

Roof

The roof, being less reinforced than a rocker panel, frequently sustains more damage in a side collision. A pulling down effect on the roof frequently occurs, particularly if the impact is located more to the center of the car and directly on the B pillar. This may result in the roof becoming damaged to the point of requiring the replacement of the outer panel. The decision that must be made is whether it is more practical or feasible to repair the roof as opposed to replacing it. Replacing it means the entire perimeter of the roof will have to be exposed and the possibility of leaks, squeaks, rattles, and corrosion are potential issues that must be dealt with. If the vehicle is equipped with a sun roof, the process becomes even more complicated. Another consideration is whether the roof is made of aluminum, steel, or another composite material. If the roof is made of steel, knowing whether the manufacturer endorses the **weld-bonding** procedure will be helpful in making the decision to repair or replace it. Weld-bonding uses a combination of special panel bonding adhesives to bond the outer panel to the mating surfaces along with plug welding for added strength and repair integrity. Weld-bonding helps to reduce the possibility of **noise, vibration, and harshness (NVH)** and also offers added

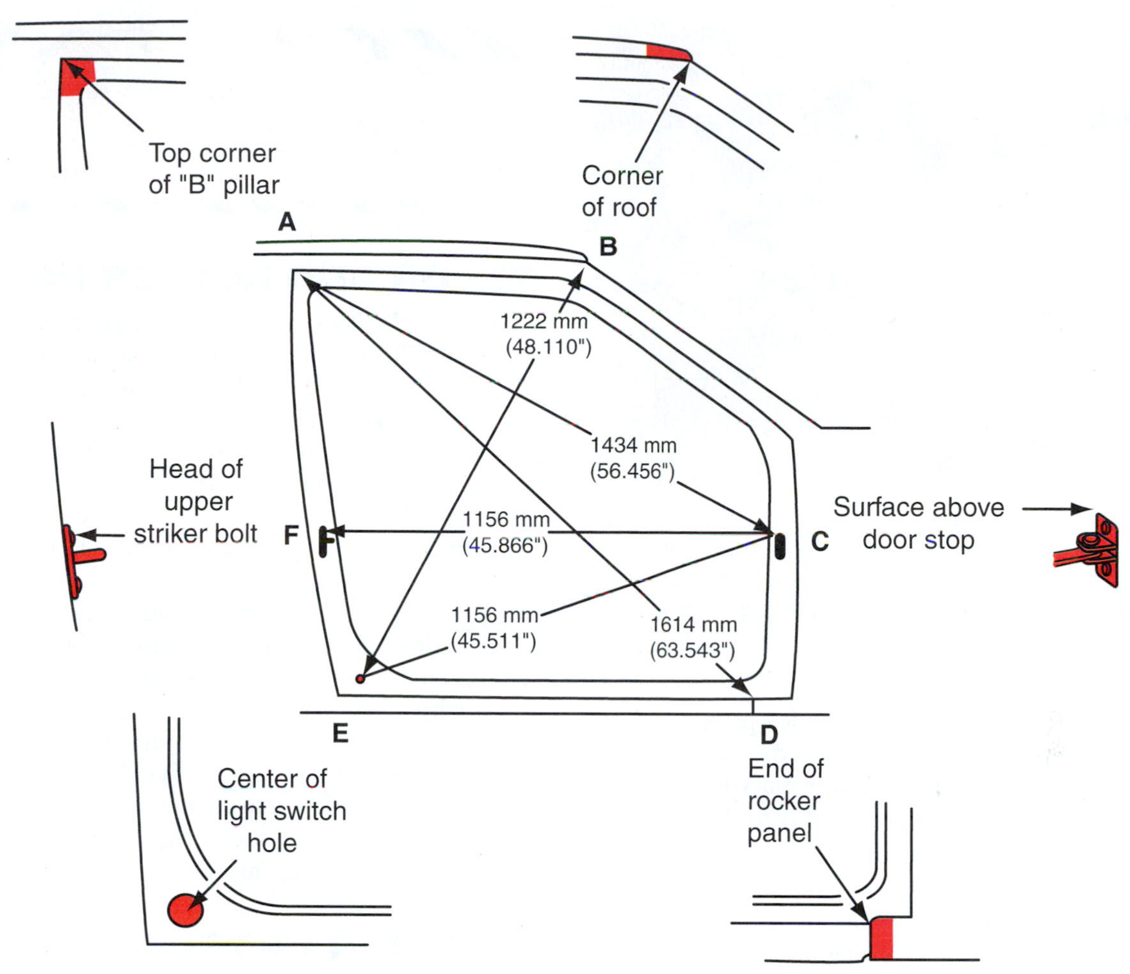

FIGURE 11-28 To determine the exact extent of the damage it may be necessary to perform a series of comparative diagonal measurements in several locations in order to pinpoint the exact location of the damage.

corrosion protection in the seam areas where it is applied for bonding the panel into place.

B Pillar Replacement

As in the case of the roof, a number of considerations enter into the decision of repairing versus replacing the center or B pillar. One must determine if it can be removed without compromising the roof, as it is often integrated into several layers of the roof's perimeter reinforcements. The decision may be made more easily if the roof and the B pillar both require replacement, as disrupting the mating surfaces is no longer an issue. The technician must decide if it is possible to access all the mounting surfaces where it is attached to the rocker panel or if it will be necessary to remove the outer rocker panel in order to gain access to the areas where it is welded into place. Another consideration is whether it is incorporated into a uniside panel, in which case it may require replacing the outer layer of the lower A pillar as well. Does the manufacturer allow a sectional replacement in the damaged area or is it necessary to replace the entire damaged panel at the factory seams? See **Figure 11-29**. One must determine whether the construction of the pillars and structural reinforcements will allow for properly affected sectional replacement. In some cases replacing an

FIGURE 11-29 The manufacturer of a uniside panel may allow a sectional replacement in the damaged area. However, in some cases, it may be necessary to replace the entire damaged panel.

FIGURE 11-30 Replacing an entire section of the vehicle with recycled parts may be more feasible than using new parts, as fewer factory seams are disrupted and the original manufacturer's corrosion protection will remain intact.

entire section of the vehicle with recycled parts may be a more feasible repair than using new parts. Fewer factory seams are disrupted and the original manufacturer's corrosion protection will remain intact. See **Figure 11-30**.

Door Hinges

While inspecting the center pillar, one should also check the rear door hinges to ensure they did not sustain any damage. The hinges may either be bolted or welded to the B pillar. The bolted hinges are readily adjustable in at least three directions to ensure a proper fit for the door into the opening. However, the welded hinges are not adjustable; therefore, any misalignment must be correctable by either repairing or replacing the center pillar. Lexus, in particular, requires that the center pillar be replaced if the door hinges have been damaged.

Side Air Bag Sensors

The side air bag sensors are located in the B pillar and rocker panel area on vehicles with this equipment. Any damage or misalignment to the mounting surface for these sensors should be noted as well. It is critical that the sensors be positioned and mounted level as the air bag deployment can be affected in a subsequent collision if they are not. The air bags invariably are covered by trim. When they are deployed the trim is nearly always damaged or destroyed and requires replacement. Before starting to pull, tug, or make any type of repair in the area where they are located, one must take the necessary precautions to protect against accidental deployment by removing the battery cable and allowing the system to discharge.

The technical service manual should be consulted to confirm the recommended procedure to follow for the vehicle in question.

CAUTION

The air bags are activated by a capacitor which stores energy for varying lengths of time. One may have to wait up to 30 minutes after disconnecting the battery for the system to fully discharge.

FLOOR AND ROCKER PANEL

Because the floor is almost always covered with carpeting, it is difficult to determine if it has sustained any damage. Even if one looks at the suspected damage area from the bottom side of the vehicle, damage is still difficult to determine as it may be disguised by the multi-layering created by the reinforcements in the area where the rocker panel is attached. The numerous stamped convolutions and the reinforcements attached to the underside of the floor further increase the difficulty to observe any existing damage. Even once the carpeting is removed from the floor, it is still difficult to observe any damage because of the surface irregularities built into the floor for strength and rigidity. Many times it becomes necessary to resort to making comparative diagonal measurements of the entire passenger compartment to determine the existence of damage as published dimensions for the floor pan are not available. Many times a technician's critical and discerning eyes are the best tool for identifying the existing damage.

QUARTER PANEL

The quarter panel and sail panel (**C pillar**) are frequently affected by a side impact. Some manufacturers consider the quarter panel a structural part and it should be replaced at the factory seams when replacement becomes necessary. Most, however, do not consider it a part of the vehicle's structural makeup. Whenever a sectional replacement is allowed by the manufacturer, technical service bulletins should be consulted to determine where acceptable cutting and seam areas are located. Many times the collision forces radiate through the door via the intrusion beam, causing a pulling effect on the latch area in the dog leg and damaging it. Unless it comes into direct contact with the impact object—whether it be an intruding vehicle or the post it was slammed into—the damage in this area is usually less severe than that sustained in the A pillar area. This is due in part to the unsupported area and the openness of the wheel well, which will move more easily to accommodate the shifting mass of metal. The wheel well housing and the surrounding areas should be carefully inspected to ensure any damage sustained in these areas is included in the damage report. One must be aware of the presence of the rear light control module and the fuel inertia switch. See **Figure 11-31**. Even if not broken by the impact, the quarter glass and **backlite**, or rear glass, may have to be removed to accommodate the repairs in that area of the vehicle. The shop may perform the re-

FIGURE 11-31 The rear light control module and fuel inertia switch must be inspected carefully for damage.

moval and reinstallation of the glass in-house or sublet these functions to another entity.

CAUTION

Broken glass should be vacuumed up prior to getting inside the vehicle to protect both the upholstery and the technician from getting cut when getting inside.

STRUCTURAL FOAMS

Another issue that must be addressed is the use of structural foam in the pillars of the areas repaired. Many vehicles, particularly sport utility vehicles, use structural foam for reinforcing the roof and the A pillar, B pillar, and the **D pillar**, all of which support the roof. Some manufacturers also use the structural foam in the front door hinge pillars and in the lower section of the B pillar. Many times this material is not visible until the outer skin is removed. When roughing out the damage or removing the outer skin one must exercise caution not to apply heat to the pillars in the areas filled with the material. When installing a new roof, one must ensure that any of the foam removed with the damaged panel—along with the additional material that must be removed for welding the new part into place—is reinstalled after the replacement panel is welded into place. It is imperative the correct type and amount be installed to ensure the restoration of the vehicle's integrity. The vehicle service manual should be consulted for specific locations where these structural foams are used.

REAR SECTION INSPECTION

The damage sustained by the rear of the vehicle is usually less extensive than in any other section of the vehicle because very few fragile or delicate parts are located in this area. The taillights and brake lights are perhaps the most critical parts that are found in this section. When inspecting for damage, one should start at the rear bumper, checking for damage to the outer cover, the reinforcement, and any attached impact absorbing device. From that point one should continue looking deeper into the vehicle to include all the rear lights, the trunk floor, and lid. When inspecting the trunk lid for possible damage, it should be checked for proper and uniform gaps and to ensure the height is correct between it and the adjacent panels. The hinges and latching mechanism should be checked to ensure they are functioning properly. When unlatched, the trunk lid should pop open and rise slowly to a full open position. Some models may use a system of crossover torsion springs to raise and hold the trunk lid open while others with larger lids may use gas shocks to assist with holding them open. See **Figure 11-32**. When closed, the trunk lid should latch without someone having to slam it shut. Some manufacturer's more luxury vehicles have automatic pull-down closers that draw the lid down the last 4 or 6 inches when the latch mechanism is engaged. This should be checked for proper and smooth operation without pausing or requiring outside assistance.

Trunk Floor

The trunk floor is most easily inspected for damage when looking at it from the top side or inside the trunk. The carpet and the spare tire cover should be removed to allow for an overall inspection and side-to-side comparison. One should check for any unusual buckles on the flat areas on either side of the spare tire housing and also in the spare tire pocket itself. Many times the manufacturers include engineered shapes into the floor for added

FIGURE 11-32 Some vehicles use a system of crossover torsion springs to raise and hold the trunk lid open.

FIGURE 11-33 The trunk floor has numerous convolutions and ribs stamped into it for reinforcement. These should not be confused with damage such as buckles.

reinforcement. See **Figure 11-33**. The estimator and repair technician should be cautioned to not confuse these as suspected damage areas.

Hat Channel Design. Many times the trunk floor becomes the fourth side of an enclosed hat channel because it is welded to a "U" channel underneath it. See **Figure 11-34**. The rails used to form the hat channel usually form a border of reinforcement around the entire perimeter of the trunk floor. This serves as the primary structural reinforcement for the rear end of the vehicle. The rear spring and suspension parts are all attached to these rails—any damage that causes them to bend, shift, or move out of their normal position will affect the vehicle's driveability. It is imperative that any damage—however major or minor—be addressed during the repairs as this may also compromise the suspension alignment as well. Any suspected damage sustained on the rear end of the vehicle must be restored to the proper dimensions to ensure the trunk lid and the rear lamps all seal properly. The suspension must be restored to the manufacturer's dimensions to ensure proper tracking by the rear wheels.

Fuel Tank. The gas tank or fuel cell is normally mounted under the trunk floor, usually immediately forward of the spare tire compartment between the rear rails. Damage to the trunk floor may cause secondary damage to the fuel cell, and occasionally the hoses and lines leading to the tank may become kinked or damaged. One should ensure that no cracks occurred in the gas line from the jarring and flexing that took place during the impact. Even the slightest crack can result in a significant amount of gas being pumped out onto the floor when the engine is cranked or even by merely turning the ignition on. The fuel pump located in the fuel tank will start and continue to pump gas until it reaches a certain pressure. A leak in the line will not allow the pressure to build up; therefore, the pump will continue to pump gas until the ignition is turned off.

BRAKES, STEERING, AND SUSPENSION SYSTEM

While checking the suspension system for possible damage, one should also check the braking parts for any

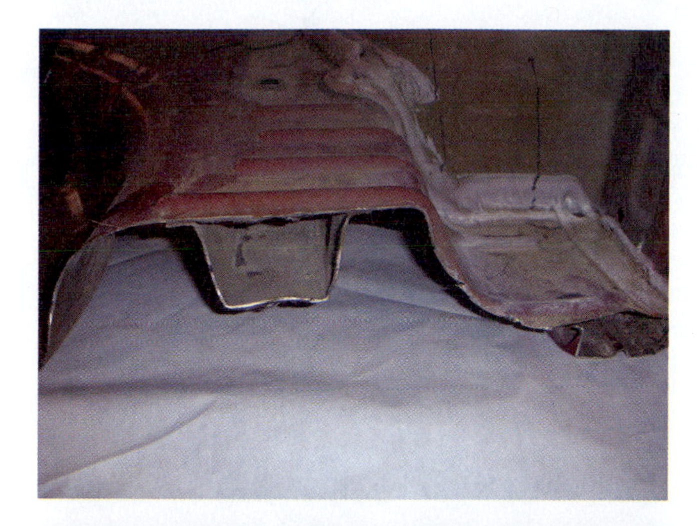

FIGURE 11-34 The trunk floor is welded to a series of interconnecting "U" channels to form the closed hat channels that reinforce the entire rear end of the vehicle.

visible signs of obvious damage. If there are any signs of damage to the wheel, rotor, caliper, or any part of the brake lines or flexible hoses, one should also check for possible damage to the wiring for the ABS system. Other wiring harnesses located in this same area that can be affected are the traction control system (TCS), electronic stability control (ESC), and the tire pressure monitoring system (TPMS). These are all safety control systems for the safe operation of the vehicle. Overlooking them could cause a subsequent driving problem.

Popular Steering and Suspension Systems

There are essentially four distinctly different suspension system designs currently being used by the automobile manufacturers worldwide: the **MacPherson strut**, parallelogram, solid axle, and the **double wishbone suspension**. The parallelogram system is often referred to as the short-arm/long-arm (SLA) and the wishbone is a combination of features or principles from the MacPherson strut and the parallelogram systems. A fifth and very popular system used by the Ford Motor Company on all their trucks until the late 1990s was the **twin I-beam suspension** system. This system is no longer being used on any of their vehicles, however. Each of these systems offers its own unique advantages and disadvantages. See **Figure 11-35**.

Steering System Damage

The steering and suspension systems frequently sustain varying amounts of damage depending on a variety of conditions including the type of road surface, weather conditions, and if the vehicle slid sideways or left the road. An asphalt or concrete road may result in a vehicle sustaining more damage than would occur on loose dirt or a gravel road, especially if it was pushed or slid sideways. The possibility of damage is nearly inevitable whenever a vehicle leaves the road by jumping a curb or running into a ditch or median. The first step in determining these conditions is to question the driver or vehicle owner about various circumstances about the accident. One should find out as much about the circumstances as possible by asking exactly what happened, how the vehicle reacted, if other vehicles were involved, or if the vehicle left the road, slammed into a curb sideways, and so on. One should also attempt to ascertain any unusual driving characteristics that have become evident since the collision. Does the vehicle exhibit any different or unusual steering symptoms than it previously had? After this is determined, the next step is to visually inspect the wheels for the obvious. Do the wheels show any visible signs of taking a direct hit from the impact object or a secondary impacting force such as the curb? Are there any signs—such as scratches, scuff marks, dents, or missing pieces—that would be a tipoff to look for additional damage?

Types of Suspension Damage Sustained

The front suspension and steering systems include several components that may become damaged in a collision. Among the damage conditions that commonly occur are wheel setback, a misaligned or bent strut, and bent steering and lower control arms. In more severe collisions the mounts for these parts may become damaged as well. Understanding the principles of operation of the suspension systems can also help to identify the type of damage that may be unique to their design. Despite knowing each system, it is often difficult to pinpoint the exact part or location that may have become damaged by a visual inspection alone. Most of the affected parts are hidden or visually inaccessible because of the wheel and splash shields. Any suspicion about damaged or misaligned parts can quickly be confirmed or rejected by performing one or a series of simple diagnostic tests. These tests can be performed with a minimum amount of

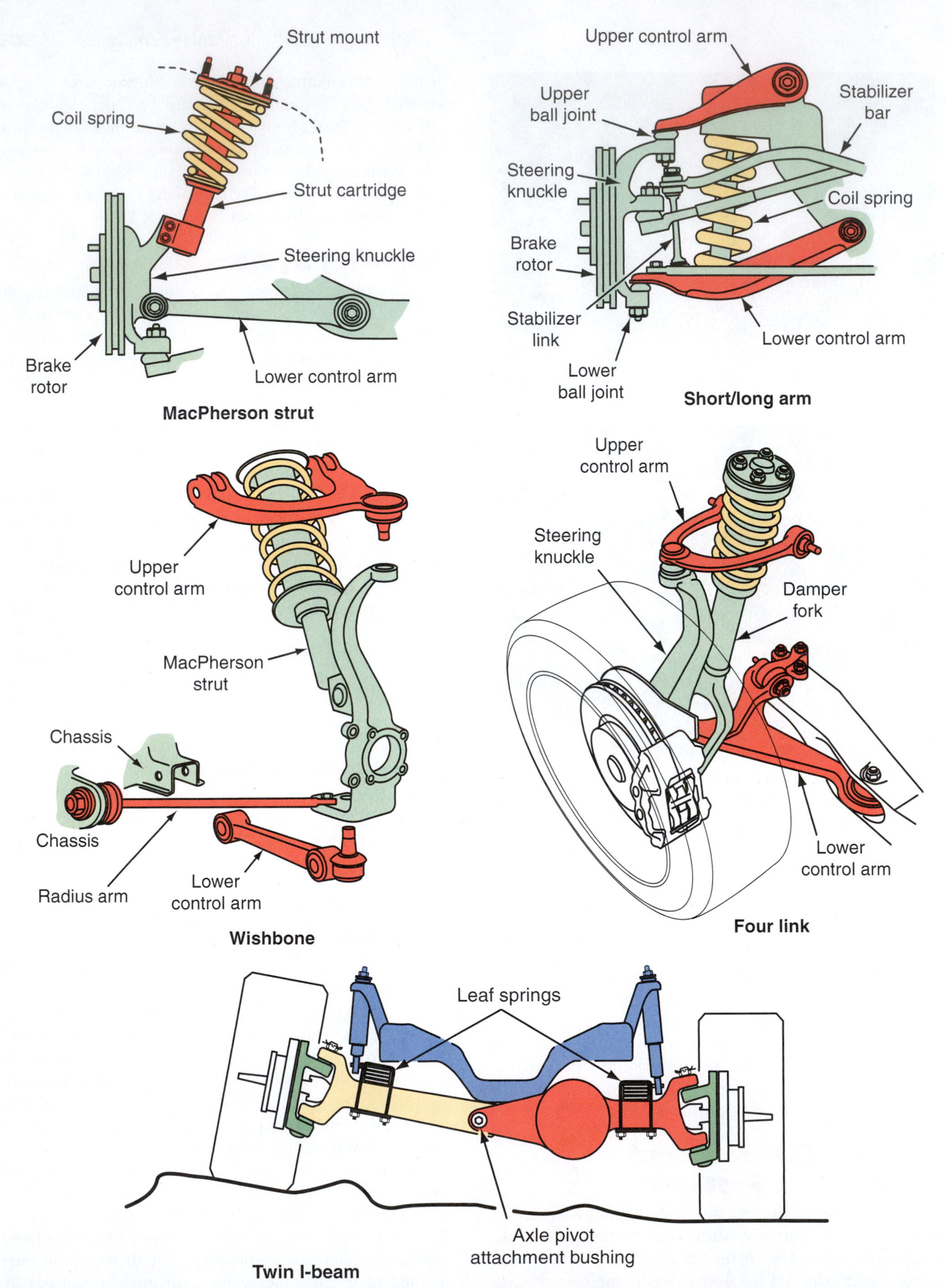

FIGURE 11-35 The four systems currently being used by automobile manufacturers worldwide are the MacPherson strut, parallelogram, solid axle, and double wishbone suspension. The Ford Motor Company used the twin I-beam suspension system until the late 1990s.

required equipment which ranges from a tape measure to a full scale undercarriage measuring system.

Wheel Setback. Checking for wheel setback is done by simply measuring and comparing the two sides of the vehicle's wheelbase. See **Figure 11-36**. This will quickly reveal if the front or rear wheels have been shifted forward or rearward. When performing this operation, one must ensure the front wheels are turned straight forward as accurately as possible. The technician then takes a measurement from the center of the front wheel to the center of the rear wheel on both sides and compares the two. Any discrepancy beyond 1/16th of an inch is an indication of a possible damaged lower control arm in the front or a damaged trailing arm or an axle that has been shifted. Another quick and simple check that can be performed without requiring any disassembly is to encircle the four wheels with a string. When the front wheels are pointing straight forward, the string should contact each tire on both the front and rear side wall. A gap between the tire and the string may indicate an axle that has shifted or a **front wheel setback**. If either condition is revealed by these simple tests, it warrants further analysis at a later time. See **Figure 11-37**.

Strut Damage. A wheel that is leaning inward or outward excessively at the top may have an obvious camber misalignment problem. This can be caused by a bent strut housing, a strut rod, or a misaligned strut tower. To check for a bent strut housing, one should remove the wheels and, using a ruler or tape measure, compare the distance from a common surface on either side such as the rotor to the housing and two or three other locations. Additional measurements may be made, including from the housing to the inner rail. Any difference exceeding 1/16th inch would indicate a bent strut housing. See **Figure 11-38**.

A bent strut rod can be revealed by performing a strut rotation check. This is accomplished by loosening the upper strut locknut and rotating the strut rod at least one full revolution with a wrench. While the rod is being turned one should watch to see if there is any inward or outward movement at the top of the wheel. The wheel moving outward when turned in one direction and inward when turned in the opposite direction may be an indication of a bent strut rod. One should ensure the locknut is properly torqued again after the test has been completed. Oil leaking from the housing where the strut rod goes into it may also be a symptom of a bent strut rod.

Steering Arm Damage. Observing any discrepancy in the angle or the amount of the inboard or outboard movement of the wheels when they are fully turned to one side or the other reveals a bent steering arm. A bent steering arm or shaft on the left side, for example, will turn the wheel harder or inward further on the left turn than on the right because the arm has been shortened. Conversely, it will turn the left wheel outward a lesser amount on a right turn than will the right side wheel. A misaligned rack-and-pinion steering

FIGURE 11-36 Checking for wheel setback is done by simply measuring and comparing the two sides of the vehicle's wheelbase.

FIGURE 11-37 A quick and simple check for setback that can be performed without requiring any disassembly is to encircle the four wheels with a string.

FIGURE 11-38 To check for a bent strut housing, one should remove the wheels and, using a ruler or tape measure, compare the distance from a common surface on either side such as the rotor to the housing and two or three other locations.

gear or the surface to which it is mounted can also be the cause of this and similar problems.

A bent or misaligned lower control arm can be the cause of both a camber and caster problem. When bent by an impact from the side—such as the vehicle slamming into the curb or getting struck directly on the lower part of the wheel by the impact object—the control arm will be pushed inward, causing it to collapse slightly or to become shortened. This will pull the wheel inward at the bottom, forcing it out at the top and causing a positive camber angle. Likewise, on a **parallelogram/short-arm/long-arm (SLA) system**, if the upper control arm becomes bent by a side impact it may also result in a shortening effect. This may cause the upper portion of the wheel to be pulled inward, causing a negative camber effect.

IDENTIFYING DAMAGE

Camber Quick Check

A camber quick check can be made very quickly by using one of several gauges, such as a camber/caster quick check gauge. Some are designed to be attached to the wheel magnetically, while others are used by attaching them to the chime of the wheel. One may even use an angle locator—provided it can be placed flat against the wheel and it is graduated in degrees so a relatively accurate comparison can be made from side to side. See **Figure 11-39**. Prior to attaching the gauge to the wheel, one must make certain the vehicle is setting on a relatively level surface and the tires are correctly inflated. The vehicle does not need to be placed on an alignment table to conduct this test. Remember, the technician is merely attempting to verify the existence of damage at this point, not repair it.

FIGURE 11-39 One may use an angle locator—provided it can be placed flat against the wheel and it is graduated in degrees so a relatively accurate comparison can be made from side to side.

Misaligned Strut

A possible strut tower misalignment can be checked simply by taking diagonal measurements from side to side and to a common reference point behind the strut location and comparing them to the published underhood specification charts. It is important that the common reference point used be located behind the strut tower as any side movement, however slight, in front of the tower will skew the measurements. The likelihood of sidesway movement is infinitely less behind the strut tower; therefore, the measurements are more precise and accurate. This can be done with the use of a tape measure or a tram bar.

Caster Quick Check

A caster misalignment can be caused by the lower control arm being pushed forward or to the rear. This can happen either from hitting the impact object or from striking a hole or some other impression when the vehicle leaves the road. The lower ball joint is the most common measuring site used when checking for potential caster misalignment, as it is the attaching point of the steering knuckle which holds the wheel in position. To check for a caster misalignment problem, one needs only a tape measure in most cases. A tram bar may be used in the event a straightline measurement is not possible. One should measure from two or three different locations to ascertain the correct location of the lower ball joint. Some of the measurements that can be made include side-to-side comparison measurements to a reference point that is common to both sides, a forward measurement from a common reference point on each side behind the ball joint, and a diagonal measurement using common locations.

REAR SUSPENSION DAMAGE

The rear suspension is apt to incur similar damage as that sustained by the front of the vehicle. In some instances it may be more likely to become misaligned as it is not as heavily reinforced as is the front suspension—particularly to withstand damaging forces from the side.

Rear Suspension Designs

Automobile manufacturers use one of four principal designs for rear suspension systems: **solid axle**, trailing arm, short-arm/long-arm (REAR SLA), and independent suspension. See **Figure 11-40**. Prior to the widespread use and popularity of the unibody design, the solid rear axle suspension was the most popular system for domestic vehicles. Most cars and trucks were rear drive, which is the system best suited for this design. The third member, as it is sometimes called, may be held in place by the rear leaf springs that are bolted to the frame in front and behind the axle. Alternatively, a trailing arm and coil spring design may be used. In either case the axle is subject to becoming misaligned, causing a steering problem called dogtracking. This may come to

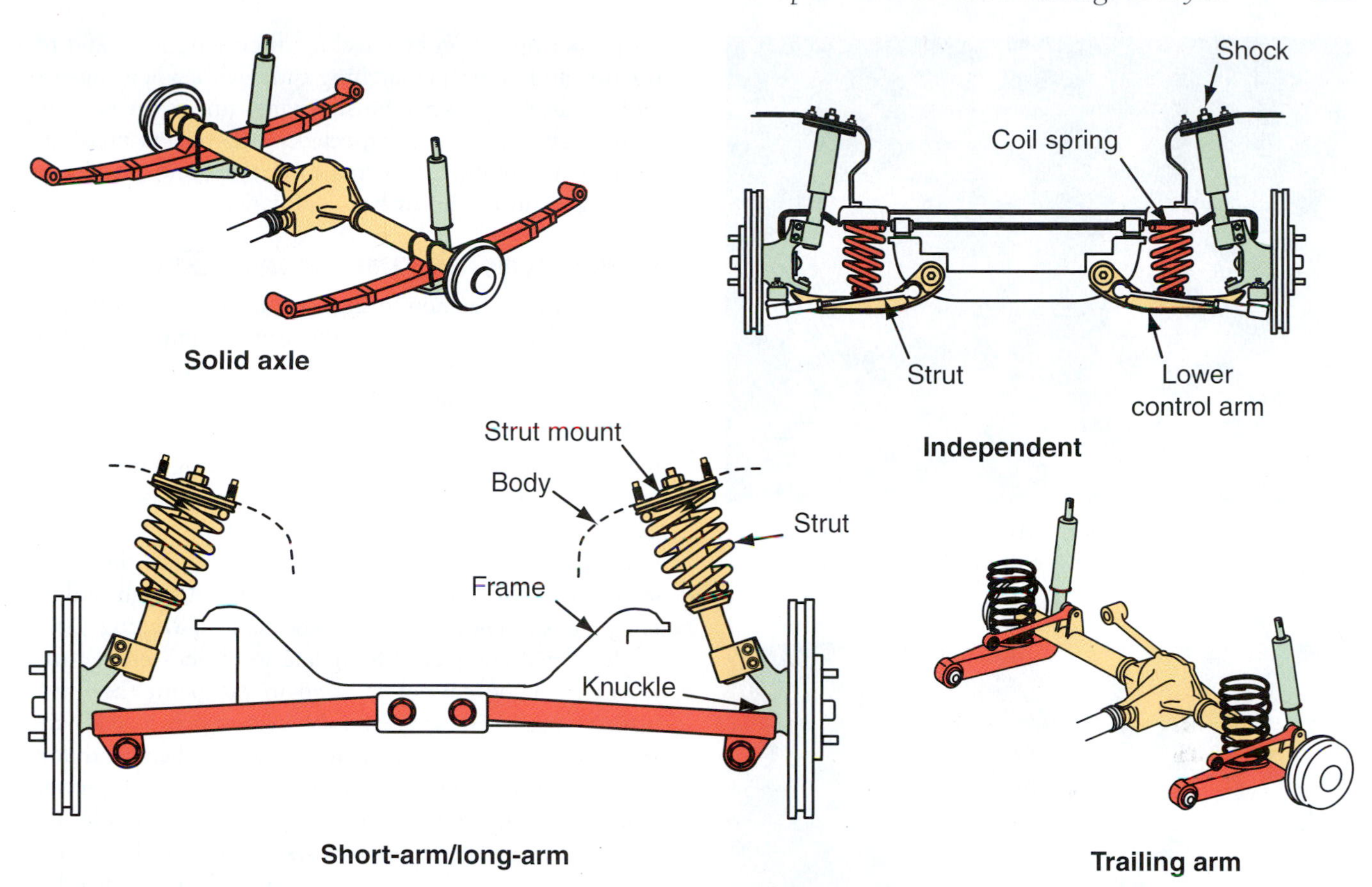

FIGURE 11-40 Automobile manufacturers use one of four principal designs for rear suspension systems: solid axle, trailing arm, short-arm/long-arm (REAR SLA), and independent suspension.

light when a wheelbase measurement is made as was discussed when checking for a wheel setback condition on the front suspension. A camber misalignment typically does not become an issue with this system because the rear axle is very heavily designed; therefore, wheel mounting parts are less likely to become misaligned.

With the advent of the unibody vehicle design, the trailing arm rear suspension became increasingly more popular because it was lighter weight and more adaptable to the less bulky mounting systems required of the unibody rails. However, like other rear end designs, this system is prone to damage from side forces. The wheels may or may not have a camber adjustment, so this must be taken into consideration when inspecting for damage.

Rear SLA. Both the REAR SLA and the independent rear suspensions are similar to the SLA and MacPherson strut front suspension as they both utilize very similar parts, construction, and adjustments. The REAR SLA—such as that used on the 1997 and earlier through present Chevrolet Corvette—utilizes both an upper and lower control arm, knuckle, shock absorbers, coil springs, and adjustable tie rods. It makes allowances for camber and toe adjustments much like those on the front suspension. The independent rear suspension system likewise incorporates the use of a strut-spring-shock configuration with a detachable knuckle or spindle that is adjustable, as well as a lower control arm.

Trailing Arm. When assessing the rear end for damage on the trailing arm suspension design, the areas that are most often affected in a collision are the axle, trailing arms and their mounts, and the stabilizer bar. The rear wheel hub usually bolts directly to the axle end with no provisions for adjustments. Any sign of wheel misalignment warrants replacing the entire axle assembly. The remaining three systems (i.e., the solid axle, independent, and the REAR SLA) have provisions for adjustments to overcome misalignment problems by either shimming or by bending the axle housing. The same quick checks can be made on these systems as were described for the front end systems to check for camber, strut, and setback problems.

VEHICLE INTERIOR AND OPTION SENSORS

The interior of the vehicle, unless it is part of a damaged outer panel, is usually the last area to be inspected as the type of damage it sustains is usually completely removed from the mainstream of the damage on the vehicle's exterior. Whenever an air bag is deployed, it gives an untrained observer the impression that the vehicle's

FIGURE 11-41 A deployed airbag often gives the impression that the interior is completely destroyed.

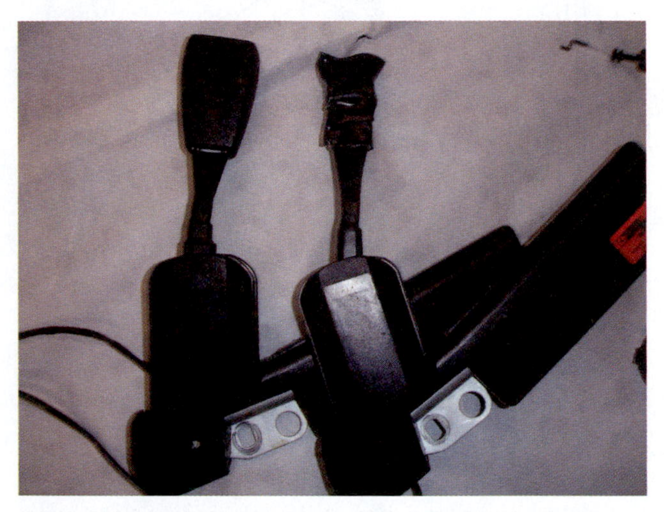

FIGURE 11-42 If a vehicle is equipped with automatic seat belt tensioners, they are deployed simultaneously with the air bags.

interior is damaged beyond repair. The dash looks to be ripped apart, the steering wheel looks like a large shopping bag is protruding from a large rip in it, and the windshield is often cracked from side to side. See **Figure 11-41**. Simultaneous to the air bag deployment, if a vehicle is equipped with automatic seat belt tensioners, they are also activated at the same time and require replacement as well. See **Figure 11-42**.

Seats and Backs

While checking for damage to seats and backs, one should check all the seats for proper operation and ensure all are positioned at a proper seating angle. Any seat with a ratcheting back should be moved completely up and down to ensure it works freely and locks into place at all positions. Likewise, any electric seats should be checked to ensure the motors are all functioning properly and the tracks are not damaged. The seat belts should be checked for proper operation by checking their extension and retraction and for proper latching and release when engaged and released. The knee bolster panel under the steering column should also be inspected for dents and misalignment that may have occurred from the occupant's knees striking it during the air bag deployment.

Dash and Monitoring Screens

Many times the monitoring screens, warning lights, and auditory devices used to alert the driver of impending danger or the screen used to give the driver directions are placed in highly exposed areas. As a result, they are often damaged in a collision. Many cars now have a global positioning system (GPS) built-in to assist drivers in finding a destination when driving in areas with which they are not familiar. For example, when driving in a large city, the address of the destination is programmed into the system. Once the driver starts driving, it gives constant updates through audible commands whenever a turn is to be made. If a wrong turn is made, it will automatically update itself and re-direct the driver from the current location to the point the driver wants to reach. The screen for this system may be built into the dash in the same area where the radio and air conditioning controls are typically located or it may be an adjustable external screen that is attached to the dash or console. The non-adjustable antenna for this system is typically attached to the roof of the vehicle, many times near the rear window area. Because they are located in such a vulnerable part of the interior, one must carefully inspect all the monitors and warning devices as they are apt to be struck by loose and unrestrained cargo that may become air-borne during a collision. See **Figure 11-43**. The directional system can be checked by turning on the ignition switch, activating the system, and entering any destination in the region. Many times these systems are programmed for a particular region of the country. If the vehicle is from outside a given region, it may appear to be non-functional.

FIGURE 11-43 Because of the vulnerable location of the GPS screen, it is a likely target for loose or unrestrained cargo during a collision.

Backup Cameras

Backup cameras, together with parking sensors, have become increasingly popular in many larger vehicles, such as sport utility vehicles (SUVs), because of the blind areas that commonly exist when backing up. The lenses for a backup camera are typically located above the license plate or above the rear window. See **Figure 11-44.** A damaged rear bumper may automatically require replacement of the lenses as repairs to these items is not recommended because visibility will be reduced. The backup camera is automatically activated when the transmission is shifted into reverse, much as the backup lights are on all vehicles. The screen for these cameras is also located in the dash.

Satellite Communication Systems

The OnStar® Telecommunication system is an option that has been made available predominantly by General Motors Manufacturing Corporation on many of their models. The vehicle owner has the option of activating the system by subscribing to a monthly or annual service. In addition to being integrated with the GPS system, in the event the vehicle is involved in an accident wherein the air bag is deployed the system automatically alerts a national clearing house of a deployed air bag on a vehicle in which the system has been activated. A law enforcement agency in the area will then be alerted automatically and emergency assistance will be dispatched to the scene of the accident. Another popular feature of the system is, in the event one inadvertently locks the keys into the vehicle, a specified 800 number can be called and the vehicle doors can be unlocked via satellite. In the event of an emergency, the OnStar system can also be used by activating a button typically located on the rear view mirror. A voice operator will respond and a description of any emergency can be communicated to the operator, who in turn can give the exact location to a law enforcement agency in the area.

Head Up Display (HUD)

The head up display (HUD) located in the vehicle's dash projects information such as the driving speed and the information normally displayed on the gauges in the dash up onto the windshield. This is done so drivers do not have to look down or take their eyes off the road to read any of the information. The projector for this system is located inside the dashboard and instrument panel in front of the driver so it can be read from that position only. In the event the windshield is damaged, a special windshield must be reinstalled in order to read the projected images. See **Figure 11-45.** A limited number of service parts are available for the camera and projector and may require replacement if they become damaged. In order to diagnose and determine if the system is functioning properly, the technician must start the vehicle to activate it. One should check for proper focusing and ensure it does not project a doubled image.

Adaptive Cruise Control

Another group of very sophisticated electronic equipment sensors that should be checked at this time is the **adaptive cruise control** system sensors. These sensors are located behind the grill at the front of the radiator support or in the front bumper or fascia. See **Figure 11-46.** The wiring for them is routed through the radiator support in the same general area as those for the automatic headlight adjusting motors and other options. These wires are susceptible to cutting and scraping damage and may even be severed. This equipment is apt to be found on vehicles like the Lexus, Jaguar, Mercedes Benz, Infinity, and certain high-end General Motors products.

If estimators do not recognize the adaptive cruise control sensors or know where to look for them during the inspection, they will often overlook their existence. The

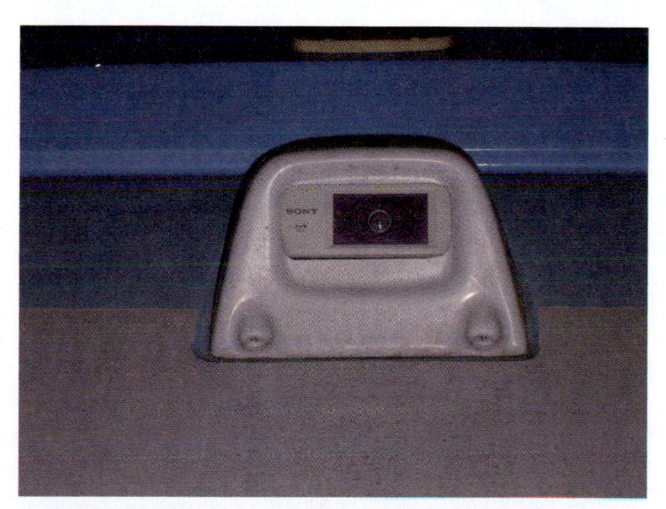

FIGURE 11-44 The lenses for a backup camera are typically located above the license plate or above the rear window.

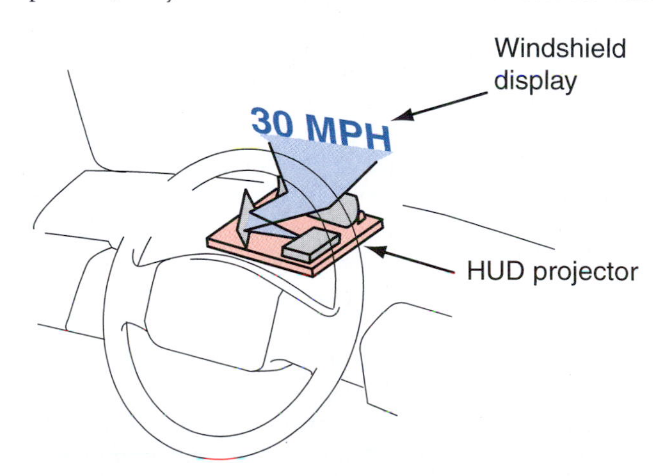

FIGURE 11-45 The head up display (HUD) projects information normally displayed on the gauges in the dash up onto the windshield.

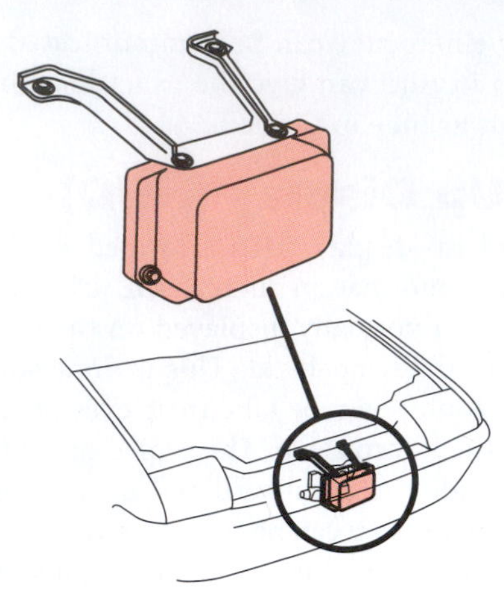

FIGURE 11-46 The adaptive cruise control sensors are usually located in the front of the vehicle near the center of the core support.

system will alert the driver and respond automatically—for example, if approaching another vehicle from the rear and coming too close—by decelerating and slowing down the vehicle. The system utilizes either radar or laser sensors, which are usually located behind the grill in the center part of the vehicle or in the front bumper. Unless the estimator or technician recognizes them, they will likely be overlooked.

TECH TIP The laser or radar sensors for the adaptive cruise control may not function properly if the front suspension and wheel alignment is not properly set.

Parking Sensors

The parking sensor is yet another device which, due to its mounting location, is very susceptible to becoming damaged. Parking sensors are located in the front fascia and are somewhat sensitive to impact damage. These can be checked by starting the engine, placing a large object near the front of the vehicle, and listening for an audible warning indicating the presence of said object. These are also located in the rear bumper cover and can be checked in the same manner.

Lane Departure System

In addition to the usual functions of the inside rearview mirror, the lane departure system is used to house the controls for several optional systems. In addition to the On-Star activation button, the sensors or camera used for the lane departure warning system (LDWS) may be installed in or near the rearview mirror as well as the rain sensors. See **Figure 11-47**. The LDWS warns the driver if the vehicle is veering off the normal course of the roadway by either sounding an audible sound system or by flashing lights. One should check the cameras or sensors for damage if the rearview mirror sustained any.

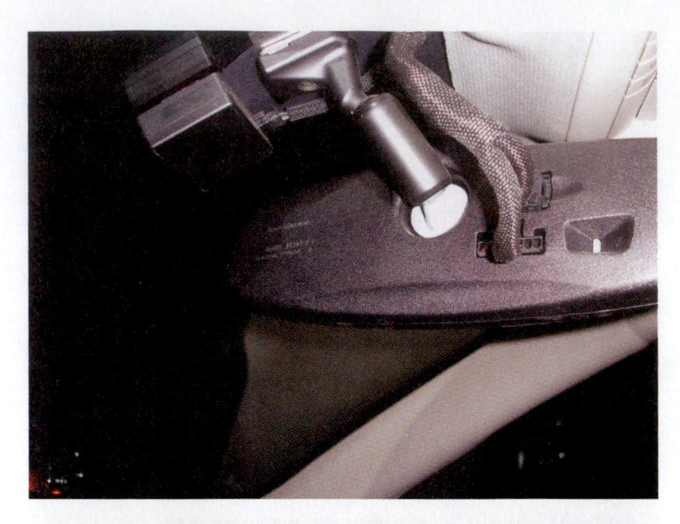

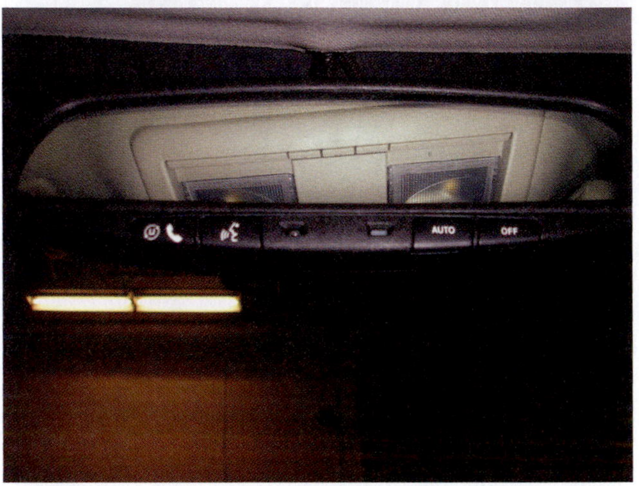

FIGURE 11-47 Numerous sensors and activation switches are located on or near the interior rearview mirror, such as the automatic windshield wiper rain sensor and the lane departure warning system (LDWS).

Rain Sensors

Other systems sometimes integrated into the mirror include the rain sensors, which automatically turn on the windshield wipers when a specified number of water droplets are on the windshield. Automatic headlight dimming sensors will dim the headlights when they sense the light of oncoming vehicles. The mirror will also transition to a lower or darker reflection automatically to reduce the glare of vehicles approaching from the rear. Other items commonly displayed on the mirror are the outside temperature, a compass reading that indicates the direction of travel, and—on some vehicles—the air bag status. To check for the proper function of these options, the technician may simply shine a flashlight on the front and back side of the mirror to check for headlight dimming and transitioning for glare reduction. Spraying water on the windshield is all that needs to be done to check for rain sensing by the activation of the windshield wipers. Many times map lights can be checked for proper operation simply by activating the switch.

As with the vehicle's exterior, a systematic and methodical analysis will often result in a plan to restore the vehicle to look new and undamaged once again.

ALUMINUM

The use of aluminum has gained in popularity in recent years, particularly among European auto manufacturers. Even though it offers numerous advantages over steel—such as being very light weight, recyclable, and very corrosion resistant, it also has shortcomings as well. Many shops are not equipped to repair aluminum because it requires tools, equipment, and specialized fixtures that may be dedicated to one vehicle use only. The training and certification required by technicians may limit them to repairing a specific manufacturer's vehicles. The lack of training on the intricacies of aluminum repair may result in a disastrous outcome if an untrained individual attempts to perform repairs for which that person is not trained. Most auto manufacturers specify that before a shop is allowed to make any structural aluminum repairs on their vehicles, they must become a certified repair facility. This mandates that technicians have been trained specifically on repairing their vehicles, the shop has an area dedicated exclusively to aluminum repair, and they have the equipment measuring and pulling necessary to properly effect the repairs. Therefore, when diagnosing the repairs to be made, it may become necessary to sublet all or part of the work to be done.

Aluminum Repair Considerations

Understanding the differences in the methods and techniques used in manufacturing vehicles will help technicians to understand some of the uniqueness of the aluminum vehicle from its steel counterparts. Most of the exterior aluminum body parts and their reinforcing pieces are die stamped and resistance spot welded, much like those of steel. Unlike steel, however, the repairability of aluminum is very limited as it becomes work hardened and unworkable with a limited amount of flexing and bending, such as typically occurs in a collision. The loss of flexibility and the work hardening that occurs on even a minor damaged surface often makes repairing an outer body panel a questionable task Many times the panel must be replaced as opposed to being straightened, which results in higher repair costs. Small dents and minor dings are normally repairable so long as access is available from the back side. However, because aluminum does not have a "memory," damage such as hail dents is not as easily repaired as when it occurs on steel panels. One must also

FIGURE 11-48 The outer perimeter of an aluminum panel is usually less repairable due to the increased work hardening that occurs as a result of die stamping.

understand that certain areas, such as the center of a panel, are more easily repaired than the outer edges, because of the additional work hardening that occurs when they are die stamped. What may appear to be a very repairable crease on the edge of a hood or a door, for example, may result in the metal becoming cracked before the damaged surface is smoothed out. See **Figure 11-48**.

Repair vs. Replacing Aluminum Parts

When deciding whether to repair or replace an exterior panel, one must understand the technology and equipment that must be employed to install the replacement part. The manufacturers utilize several methods to secure the panels together during the vehicle's manufacture. In addition to GMA MIG welding and squeeze-type resistance spot welding (STRSW), blind and self-piercing rivets (SPR), adhesives, rope hemming flanges, and clinches are the most common methods used in assembly of the vehicles. When repairing and replacing parts or sections, it is important to duplicate as closely as possible the process used by the manufacturer. Special equipment used exclusively for these operations may be required to accomplish this. In addition, the matchup of exterior panels to the existing parts on the vehicle may be far from exact, thus requiring that modifications be made to accomplish this. The replacement parts are frequently made by outside vendors, not the original manufacturer.

Aluminum Parts Manufacturing Methods

Many of the structural and reinforcement parts may be either castings or extruded, and frequently require replacement if restoring the shape and state of the part is even remotely questionable. BMW, for example, specifically forbids repairing any structural damage on the 2004 and 2005 series 5 and 6 cars and specifies these parts be replaced instead. Due to the unique design, their often intricate shapes, and their limited reparability, most casting parts are nearly always replaced when they become damaged. These parts are usually dedicated structural parts used for reinforcements in the undercarriage, in the vehicle's load bearing areas, and—in some instances—in the strut tower area. Most of these parts are welded together as an assembly, and occasionally they are bolted into place and used as a sacrificial part on the front and rear rails of the vehicle. These are considered "throwaway" parts and typically are not repairable; thus, they are designed to be unbolted and replaced with a new one. Knowing and recognizing which parts are repairable, which ones require replacing, and which require special attaching techniques will help to streamline the damage analysis process.

Extruded Parts

Extruded parts, which are typically used for reinforcement purposes, may be a continuous seamless boxed construction or they may be formed in two sections and welded together to form an assembly. An example of this may be the lower front rails of a vehicle, in which varying degrees of thickness may be required to ensure the vehicle collapses at a precise location in an accident. Another example may be the lower A pillar or the center pillar. Most extruded parts must be replaced if they are damaged with any degree of kinking or when there is any question of being able to restore both the shape and state of the structure. Any cracks in the structure resulting from the collision will also require replacing the part—or, at a minimum, a section of it—provided the manufacturer makes provisions for it. Repairing a cracked area may result in altering its structural integrity. In some cases, a cracked area may be prone to further cracking after the repairs have been made and the vehicle put back into service. An unwritten rule often cited regarding aluminum repair is any time a crack or tear appears around the area where a structural part is attached to another part, the parts in question should be replaced.

Many times performing a sectional replacement may seem like a more logical repair than performing a complete panel replacement. Before committing to this approach, however, one should determine if the OEM recommended procedures are available.

Measuring System Requirements

The final considerations that must be weighed when attempting to determine the feasibility of repairing the vehicle are the measuring system and the repair equipment required to accomplish the task. Most aluminum-intense auto manufacturers require the use of a dedicated fixture bench system for anchoring and measuring the vehicle while it is being repaired. Repair equipment manufacturers must obtain certification for their equipment from the auto manufacturers prior to becoming a designated repair facility. Due to the amount of flexing aluminum undergoes and its spring-back characteristics, a system of multiple anchoring fixtures is required to hold the vehicle in place while applying the necessary corrective forces to restore the vehicle. Attempting to repair a damaged undercarriage without the correct equipment will undoubtedly bear disastrous results and leave the owner with a vehicle that may not be safe to drive again.

ALTERNATIVE FUEL VEHICLES

The most recent newcomers to the automotive industry are the alternative fuel or hybrid vehicles. Depending on their propulsion system, these vehicles may be classified as either a full hybrid or a mild hybrid. A full hybrid vehicle is powered exclusively by electricity without the aid of any form of internal combustion power source. A mild hybrid is one that incorporates the use of an electric energy source along with a gasoline internal combustion engine. An electric motor may be used to power or drive some of the accessory systems such as the power steering and brakes. Typically, the electric motor is initially used from a dead stop—such as when waiting for a red light to change. When the vehicle reaches a certain speed the gasoline engine will take over. The electric side of the propulsion system is powered by a high energy **nickel metal hydride battery (NiMH)**. This battery is continually being recharged as the vehicle is being driven and even when it is being braked to a stop.

CAUTION

Before doing anything on the vehicle the technician should take the necessary steps to become familiar with how the vehicle and its systems work and what precautions one must take for protection from possible electrocution and injury to those nearby.

Before attempting to disassemble or remove any part of the vehicle for inspection or to make any repairs, no matter how minor they may seem, it is critical that technicians attempt to familiarize themselves with how the vehicle operates and the safety precautions that must be exercised when working in and around the vehicle. Extreme caution must be used whenever working around one of these vehicles as the electric system uses extremely high voltage as a propulsion source. Failure to act cautiously could possibly result in electrocution or serious injury not only to the technician but also to those nearby. General Motors uses a 42-volt battery system on its vehicles, and most other manufacturers use a similarly high voltage system.

Hybrid Critical Parts

The two most critical parts of the system are the battery and the **inverter**, which is used to convert direct current (DC) from the battery to alternating current (AC). This is used to operate some of the accessories and convert AC back to DC for storage purposes. Depending on the repairs to be made, the battery may need to be removed from the vehicle if any work is to be performed within 12 inches of it. This may include something so simple as hammering for minor stress relieving as well as any type of welding or cutting operations.

CAUTION

Lineman's high voltage gloves should be worn before attempting to disable the electrical system on a hybrid vehicle. No jewelry should be worn when performing any disabling steps on the vehicle.

Disable the Electrical System

A series of steps should be followed when attempting to disable an electrical system. The first is to remove the key from the ignition and put it someplace where it won't inadvertently be put back in the car until it is ready to start back up once again. Vehicles with a keyless remote start can be started without the key in the ignition. One should follow this step by disconnecting the 12-volt battery—usually located under the hood—which is used to energize the sensors and relays. After this step has been taken, much of the danger of an electrical shock will have been eliminated. The next step to disabling or deactivating the system is to disengage the high voltage battery module switch, which is usually found on the top of the battery pack in the trunk. The single most important piece of personal protective equipment required when attempting to disable the high voltage energy system or remove the battery from the vehicle is a pair of electrical lineman's gloves that are rated to protect a person against up to 1,000 volts of power.

CAUTION

Before a technician puts on lineman's gloves they should be checked to ensure they are free of any pinholes or tears, as even the smallest opening may allow current to pass through them when the technician is attempting to disable the system or remove the battery.

Discharge/Drain off Time. After the high voltage battery has been disconnected, one should wait a minimum of 5 minutes for the system to completely discharge before attempting to remove or detach any item from the vehicle to inspect it. This is the minimum amount of time required to allow the energy to drain off the system. Even after waiting the specified amount of time one should check the system with a DVOM meter to ensure it has fully discharged by placing the probes of the meter on each of the cables coming off the battery. If the digital display reads 0 volts, the system has been discharged, and it is thus safe to work on the vehicle.

Moving the Vehicle. One final precaution that must be taken is to ensure any time the vehicle is moved in the shop, it is done with the wheels off the ground. A charge can build up in the system once again simply by moving the car, even a short distance.

TECH TIP All hybrid vehicle manufacturers have vehicle-specific disabling instructions available online at no cost to the user. The following websites can be visited to obtain this information

- General Motors: *www.gmspc.com* and go to the *Emergency Personnel Training* link
- Ford Motor Co: *www.motorcraftserv.com* and go to the *Quick Guides* link followed by the *ESCAPE emergency response guide* link
- Toyota and Lexus: *www.techinfo.toyota.com* and go to the *Toyota and Industry* link followed by the *Hybrid and Alternate Fuel Vehicle emergency information* link
- Honda and Acura: *www.techinfo.Honda.com* and go to *Hybrid emergency response* link

In addition to the aforementioned instructions, each manufacturer offers at least one alternate method of disabling the electrical system.

Alternate Disabling Method

Each system has a main fuse, plug, relay, or a switch that can be disconnected and removed from the vehicle. Most also have a switch under the steering wheel in the steering column that can be used for this same purpose. The technician can also disconnect the ground cable from the 12-volt battery. The high voltage system is not able to function if the 12-volt battery is disconnected or discharged. After disconnecting the battery cables one can remove the ignition circuit relay, which is located in a box attached at the front cowl panel. The technician can also remove the high voltage fuse from the underhood fuse box, which is located right behind the core support at the front of the engine compartment. High voltage gloves should be worn to perform this operation. These two methods are not the preferred ways but are usable in the event the high voltage battery is not accessible. Whichever method is used, it is necessary to wait the specified length of time to allow the energy to be drained from the sensors and condensers in the system.

Inspecting and Identifying Damage

At this point one can commence with the inspection of the vehicle. While making the initial inspection, one should look to ensure that the battery pack has not been damaged, resulting in any electrolyte leakage. An electrolyte leak that is not coming from the high voltage battery will not have a wet look as in a wet cell battery, because the electrolyte is gel-like material instead. The area should be adequately ventilated prior to removing any parts, as the NiMH battery electrolyte will react with the aluminum used for the body. This causes hydrogen gas to form and accumulate under the hood or trunk lid. This material is also very caustic; litmus paper should be used to confirm it is electrolyte, after which it should be neutralized with a mixture of water and boric acid or ammonia and water. A spill should not be wiped up without first checking to make sure it is electrolyte and neutralizing it. In the event a spill should occur from a wet cell 12-volt battery, it should be neutralized with a mix of water and baking soda.

The inspection for collision damage from this point should be performed in the same manner as is done with any other damaged vehicle, with the exception of the high voltage system and the wiring. Any signs of leakage from the high voltage battery—such as those discussed previously—warrant its replacement. The wiring is always identifiable by the bright orange-colored corrugated cable which is run inside another sheath of plastic on the underside of the vehicle or on the floor of the passenger compartment. Any sign of a tear or break in the outer sheath or the corrugated cable, resulting in exposing any part of the cable, warrants replacement of the part. Any exposed wire may result in arcing or short circuiting; thus, replacement is warranted.

Reinstalling the Battery

Once the repairs have been completed, the battery must be reinstalled as soon as possible. Completely discharging the battery may shorten its life. In some cases, allowing a battery to completely discharge will require that it be replaced, as there are no provisions for recharging it other than by driving the vehicle. Driving the vehicle under battery power is the only means to start the gasoline engine; if the battery is completely discharged, there is no provision for propulsion. Replacing the battery will then be the only means of starting the vehicle. Likewise the 12-volt battery should be recharged using the vehicle-specific methods outlined by the manufacturer. One cannot assume the battery can be recharged in the same manner as when a battery has been discharged from normal use. For example, the 2001 to 2003 Prius must be charged at a rate not to exceed 3.5 amps, and the Honda Insight must be recharged at a rate of 40 amps.

Provisions should also be made for the proper disposal of the damaged battery. Local ordinances may dictate how the battery must be treated for disposal.

INSPECTING THE UNDERCARRIAGE FOR DAMAGE

Inspecting the undercarriage for damage is usually part of the last inspection phase of the vehicle's exterior. It has been said that nearly 40 percent of the damage will be overlooked if the undercarriage is not included in the overall damage analysis. By the time the upper body inspection has been completed there will have been many signs or hints of possible damage that may exist under the vehicle. The first area that will be inspected is the direct damage area, for a number of reasons. The entire impact forces of the damage were encountered at this point and little or none of the damage deflection occurred until the damage passed through this area. The initial shock effect was felt at this point and the damage will reflect this. An impact at the front of the vehicle will result in the sacrificial rail or rails being buckled or collapsed as the impact forces pass through them. The damage may continue to radiate deeper into the vehicle, leaving buckles and other forms of damage in its path. Depending on the severity of the impact and the intensity of the impact forces as they pass through the vehicle, the damage may be vividly visible in the form of buckles and kinks in the side rails. In some cases the damage will be considerably more subtle and obviously more difficult to identify. Instead of being able to see any distinct buckles as a sign of damage, one may have to look for gradual curves and bends in the rails. Other signs of possible damage include cracks in the undercoating or accumulated dirt and surface rust, scrub marks where something may have rubbed the surface, and so on. The deeper into the vehicle the damage penetrates, the more subtle the signs become.

Identifying Damage Location

During this phase of the entire repair scenario, we are not necessarily concerned with identifying the severity of the damage but are more concerned with locating all the areas that have been affected by the impact forces. We are interested in obtaining a view of the "big picture," which will help us to formulate a more accurate repair plan. Several methods and equipment options can be used to ascertain the location of the damage.

Quick Checks with Tram Bar

The first—and perhaps the simplest—quick check method is by making length, width, and comparative diagonal measurements using the tram bar or a tape measure. See **Figure 11-49**. The comparative diagonal measurements will perhaps result in more immediate results than the length and width because the latter two usually require

FIGURE 11-49 A series of length, width, and diagonal quick checks can be made easily with a tram gauge or tape measure to identify any damage to the undercarriage or suspension parts.

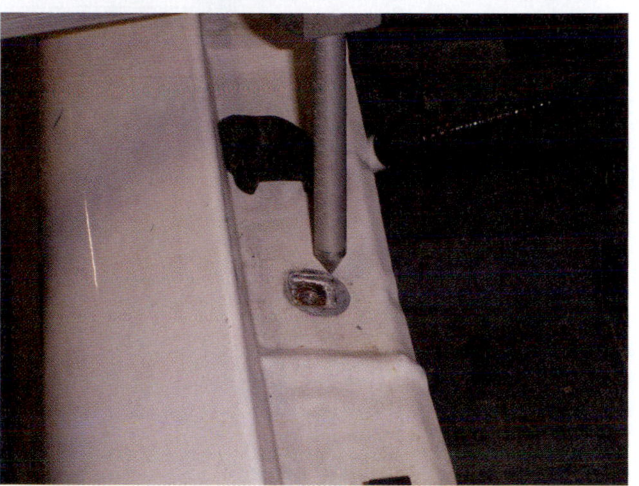

FIGURE 11-50 If the pointers on the tram bar don't align identically to the opposite side of the vehicle, the section of the undercarriage being checked has sustained some damage.

published dimensions to determine what the dimensions should be according to the manufacturer. When making comparative diagonal measurements, one must locate at least two reference points at the same location on both the left and right sides of the vehicle from which the measurements can be made; for example, a hole in the exact same locations, front and rear, on both the right and left rails. These reference points may be a hole in the bottom of the rail, a rivet or bolt used to attach a bracket, or a mechanical part or even a notch or seam in the bottom of the rocker panel. The tram bar is held so the pointers can be placed at the precise location of the two reference points, such as on the left front corner diagonally back to the right rear corner. The bar is then turned and held to the same locations—only on the opposite sides of the vehicle. If there is no damage, the pointers will align perfectly to the opposite reference points. However, if the pointers don't align identically to the opposite side of the vehicle, the section of the undercarriage being checked has sustained some damage. See **Figure 11-50**. It is advisable to make comparative reference measurements from at least two or three different locations to confirm or dispell the existence of damage.

The straight or linear measurements will not be as revealing for identifying any damaged sections of the vehicle as the diagonal checks. Without the use of the published undercarriage dimensions, it is difficult to make accurate comparisons of the vehicle's dimensions compared to what the manufacturer's specifications may be.

Self-Centering Gauges

Another technique that can be used to quickly identify a damaged undercarriage is the use of mechanical self-centering gauges. Four of these gauges are typically hung at predetermined locations from the undercarriage. By sighting down through the center pins one can quickly determine if any part of the vehicle has been shifted or moved to one side or the other, as indicated by pins that are not aligned properly. See **Figure 11-51**. After sighting down the gauges from the rear to the front of the vehicle and determining that damage is present on the front section of the vehicle only, one of them can be moved to that part of the undercarriage to get a more accurate indication of the exact location of the sidesway damage. See **Figure 11-52**. In other words, taking the number 4 gauge from the far rear of the vehicle and placing it between the number one and two gauges at the front will help to isolate the location of the damage more precisely. One can also determine if any part of the undercarriage has been kicked up or pulled down by whether the horizontal bars are hanging parallel to each other. See **Figure 11-53**.

Portable Laser System

Portable laser systems are also available. These systems are attached directly to either the vehicle's undercarriage or

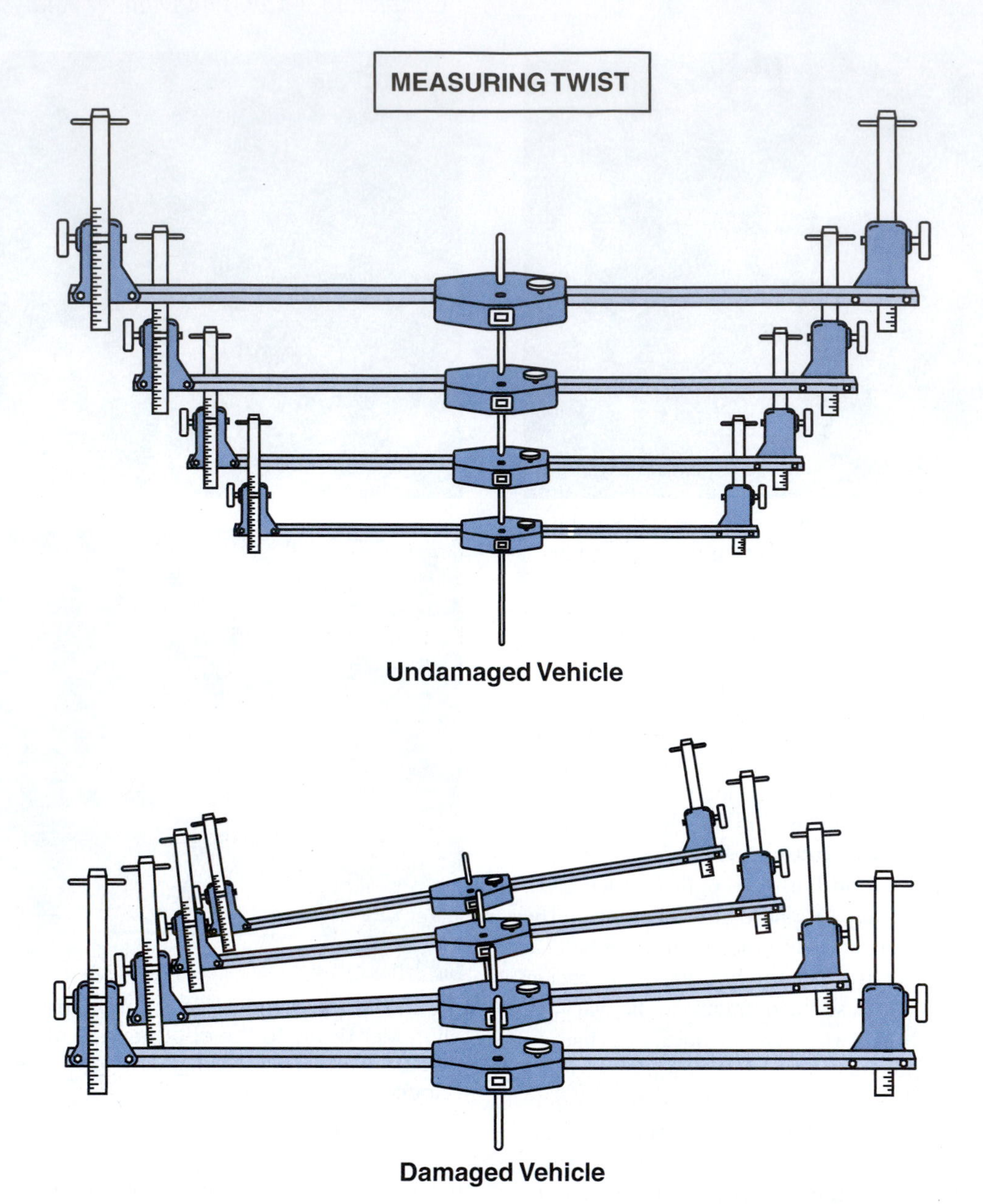

FIGURE 11-51 By using centering gauges, a technician can tell by the way pins align (or don't) if the undercarriage needs repair.

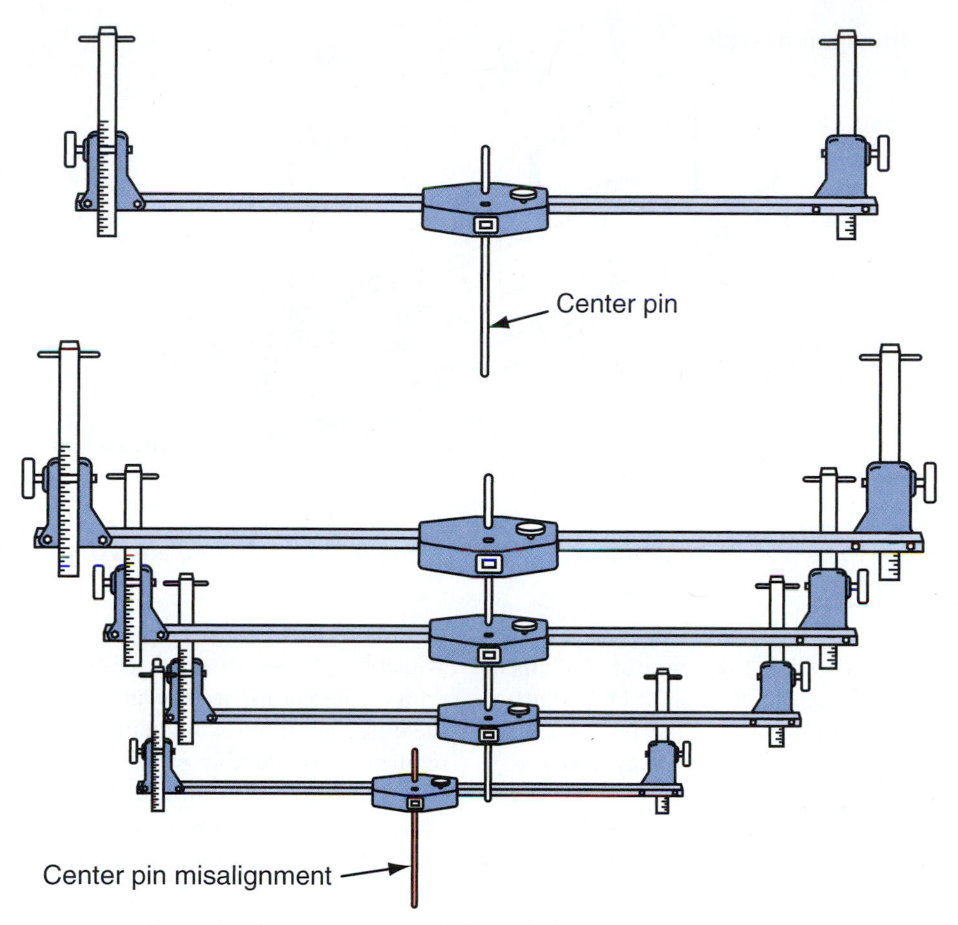

FIGURE 11-52 Taking a gauge from the far rear of the vehicle and placing it between the gauges at the front will help to isolate the location of the damage more precisely.

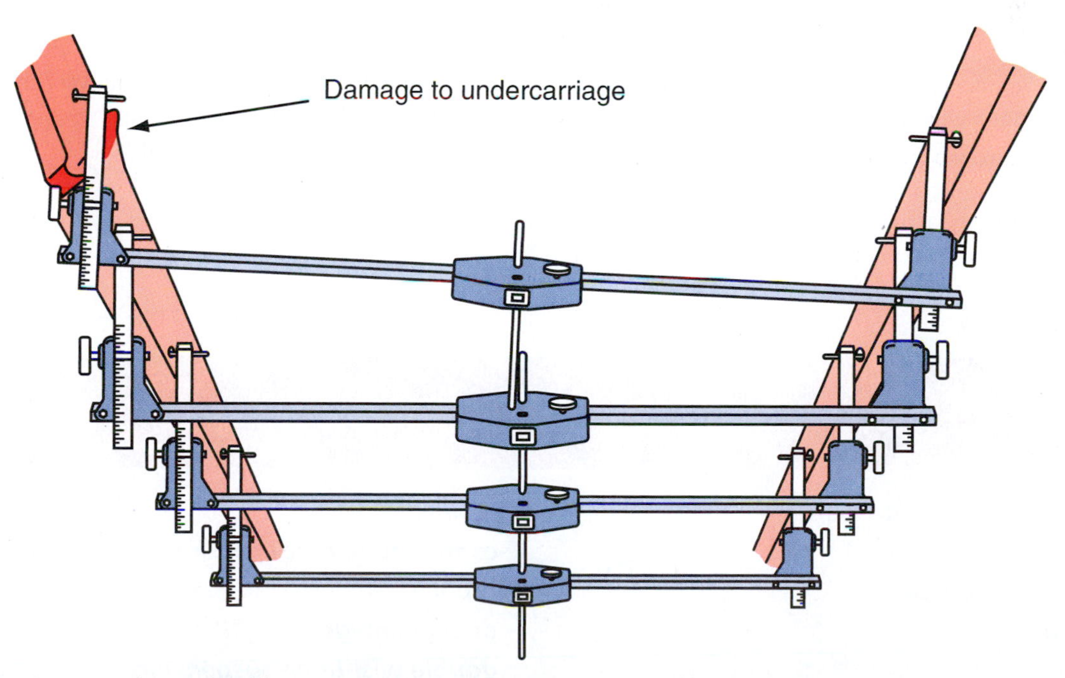

FIGURE 11-53 One can determine if any part of the undercarriage has been kicked up or pulled down by whether the horizontal bars are hanging parallel to each other.

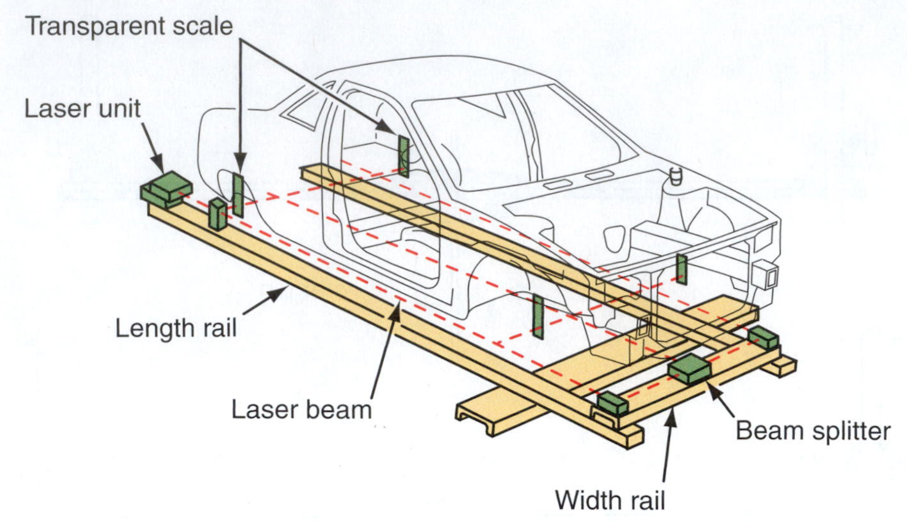

FIGURE 11-54 This Laser Mate is attached to the undercarriage and projects a beam to areas of the vehicle to check for damage.

the rocker panel and are used to sight to a variety of locations to determine the presence of damage. See **Figure 11-54**. One can quickly and easily identify not only the presence of damage, but the location of it as well with this type of system. The most significant disadvantage of these systems is the need for a lift to raise the vehicle off the ground to get under it. A number of other computerized systems are available that use either a laser or sonar, and one uses a simultaneous comparison of an LED readback to determine the difference in measurements. Since these systems often require a considerably more involved setup and calibration process, they are not likely used for a quick damage analysis. These systems are addressed in greater detail in Chapter 13, Measuring Structural Damage.

Summary

- Making a systematic and detailed analysis of the damage to a vehicle is necessary no matter how minor or severe the damage appears to be.
- In order to ensure that none of the damage is overlooked, it is sometimes necessary to obtain information from the driver or vehicle owner, from manufacturers, and other sources of information.
- The more intent one is to include all the resources available, the more precise the analysis and the less likelihood of any unexpected surprises once the repairs are initiated on the vehicle.

Key Terms

A pillar
A/C condenser
adaptive cruise control
anti-theft label
B pillar
backlite
bend vs. kink
C pillar
comparative quick check measurements
D pillar
direct damage
double wishbone suspension
fascia
fan clutch

fiber-optic cable
front wheel setback
gusset
hostile forces
hydraulic lockup
impact object
indirect damage
inertia cutoff switch
inertial forces
inverter
MacPherson strut
mass weight movement
nickel metal hydride battery (NiMH)
noise, vibration, and harshness (NVH)
parallelogram/short-arm/long-arm (SLA) system
point of impact (POI)
radiator
re-engineering
solid axle
tram bar
twin I-beam suspension
vehicle identification number (VIN)
weld-bonding

Review

ASE Review Questions

1. Technician A says inertial forces frequently cause damage immediately around the point of impact. Technician B says the point of impact and the direct damage are usually found in the same area of the vehicle. Who is correct?
   A. Technician A only
   B. Technician B only
   C. Both Technicians A and B
   D. Neither Technician A nor B
2. Technician A says many vehicle manufacturers use sacrificial structures to deflect the effects of some of the impact forces. Technician B says mass weight movement can cause damage to the engine and transmission mounts. Who is correct?
   A. Technician A only
   B. Technician B only
   C. Both Technicians A and B
   D. Neither Technician A nor B
3. Technician A says hybrid vehicles are either a mild hybrid or a full hybrid. Technician B says aluminum parts lack the flexibility of steel and become work hardened more easily than steel. Who is correct?
   A. Technician A only
   B. Technician B only
   C. Both Technicians A and B
   D. Neither Technician A nor B
4. Technician A says many of the sensitive warning and informational sensors are placed at the front of the vehicle in the area of the grill and radiator support. Technician B says some vehicles have an inertial shutoff switch that automatically turns the ignition off under certain operating conditions. Who is correct?
   A. Technician A only
   B. Technician B only
   C. Both Technicians A and B
   D. Neither Technician A nor B
5. Technician A says the VIN is a 17-digit number which provides vehicle-specific information to the technician. Technician B says the anti-theft labels have the same number on them as does the VIN. Who is correct?
   A. Technician A only
   B. Technician B only
   C. Both Technicians A and B
   D. Neither Technician A nor B
6. Which of the following would NOT normally be used to make comparative measurements on the undercarriage when checking for damage?
   A. Tram bar
   B. Self-centering gauges
   C. Tape measure
   D. Tram gauge
7. Technician A says crush zones are built into the A pillar for added reinforcement. Technician B says reinforcement foam need not be replaced if some of it comes off when removing the damaged panel. Who is correct?
   A. Technician A only
   B. Technician B only
   C. Both Technicians A and B
   D. Neither Technician A nor B
8. Technician A says the collision damage radiates throughout the vehicle, many times damaging panels a considerable distance from the point of impact. Technician B says the inertial forces are frequently the cause of damage that occurs in a part

of the vehicle away from the impact area. Who is correct?
A. Technician A only
B. Technician B only
C. Both Technicians A and B
D. Neither Technician A nor B

9. Technician A says knowing the speed of the vehicle and the angle of impact are helpful to make a more accurate damage assessment. Technician B says using a recycled part may reduce some of the corrosion problems. Who is correct?
A. Technician A only
B. Technician B only
C. Both Technicians A and B
D. Neither Technician A nor B

10. Technician A says a vehicle skidding sideways during a collision can affect the front suspension parts. Technician B says the damage sustained by a body over frame vehicle tends to remain more localized than on a unibody. Who is correct?
A. Technician A only
B. Technician B only
C. Both Technicians A and B
D. Neither Technician A nor B

Essay Questions

1. What safety precautions should be exercised by a technician and others in the shop when working around a hybrid vehicle?
2. What are the advantages of using weld-bonding as an installation method as opposed to merely welding the panel into place?
3. List the four primary suspension systems commonly used on modern day vehicles and briefly describe how they differ from each other.
4. Discuss the differences in the three principal types of aluminum parts commonly used.
5. Discuss how one would go about X-checking the undercarriage of a damaged vehicle without the use of published dimensions.

Topic Related Math Questions

1. The new quarter panel that must be replaced on the customer's car costs $450.00 The allowable labor time to replace the part is 12.5 hours. You have the opportunity to purchase a recycled part for $275.00 and it will take 3.0 hours to trim and prepare the recycled part for installation. The shop labor rate is $42.00 per flat rate hour. Will it be more economical to repair the vehicle with the new or recycled part?
2. The estimator must decide whether to replace or repair a part. The labor time required to repair the part is 9.0 hours plus an additional material cost of $65.00. The new part will cost $625.00 and the allowable flat rate time is 2.0 hour to replace the new part. The labor rate is $45.00 per flat rate hour. Which is the more cost effective repair?

Critical Thinking Questions

1. A vehicle is two years old. To repair the part on the vehicle in question will compromise the manufacturer's corrosion protection considerably. Is repairing the part more legitimate than replacing it?
2. The cost of repairing a vehicle is $10,590. The retail value of the vehicle is $16,000. The totaling threshold by the insurance company is 70 percent. What are the long-term implications if the vehicle is repaired vs. totaled out?
3. The door and quarter panel on a vehicle sustained some minor damage prior to the accident for which the current estimate is being prepared. Both panels are now being replaced. Should the current party at fault's insurance company pay for the total cost of repairing the vehicle? Why or why not?

Lab Activities

1. Using a tram bar, make a series of comparative quick checks under the hood of a damaged vehicle and identify and record the extent of the misalignment.
2. Locate a damaged unibody and a BOF vehicle that sustained damage on the front rails. Locate and identify the damage sustained in the crush zones and compare the amount and type of damage sustained by both vehicles.
3. On the same two vehicles, identify the direct damage and the indirect damage and compare the distance the damage radiated into the vehicle.

Chapter 12

Bolted Exterior Panel Replacement

OBJECTIVES

Upon completing this chapter, you should be able to:

- Discuss the purpose for attaching parts using bolts or other types of fasteners commonly used to secure exterior parts.
- Discuss the different types of materials used for manufacturing the component parts.
- Discuss the structural functions that component parts offer.
- Describe the attaching methods used and the basic concepts used for installing and adjusting techniques.
- Explain the types of fasteners and fastening commonly used on these panels.

INTRODUCTION

All automobiles are constructed with a variety of different-shaped panels that are attached in numerous ways using a wide selection of accessories and fastening methods. These panels range from the purely cosmetic ones that enhance the vehicle's aesthetics to those functional panels in the doors, hood, trunk lid, and semi-structural and structural parts. The structural parts, as one might surmise, are designed to ensure the vehicle's structural integrity and to protect the occupants during a crash. Other parts, such as the doors, are constructed by using a combination of two or more sheet metal parts to make up an assembly which in turn is held onto the vehicle with bolts. Other panels that require less structural enhancement or integrity are made of plastics and composite materials which offer little or no strength. They are often held in place with plastic or other types of removable fasteners. For the purpose of discussion in this chapter, all parts that can be removed and reinstalled using mechanical fasteners such as bolts, rivets, and so on, will be referred to as "component" panels.

COMPONENT PANELS

All motor vehicles are constructed of a combination of structural parts welded together to form the vehicle's framework and foundation. In addition to the structural framework, a number of component or adjustable panels are used to finish enclosing the vehicle. Many of these parts—such as the hood, door, deck lid, and so on—serve a functional role. Others are merely used for convenience. Their use varies depending on the manufacturer, but in many cases they are utilized for ease of access when servicing mechanical and other functional parts that are frequently installed behind or underneath them. Many times the outer part must be removed in order to replace a damaged or non-operational item. An example is the blower motor on some vehicles, which is accessible only after removing the fender. On other late-model cars, a splash shield must be removed to gain access to service or replace the battery. Replacing and adjusting the headlights may require removing a part to gain access, and servicing the windshield motor and wiper mechanism requires the removal of the upper cowl panel. In some instances even the hood has to be removed.

Replacing

The largest percentage of accidents involve the front end of the vehicle. Removing and replacing items such as the fenders, hood, and grill are accomplished much more easily by unbolting them and installing a replacement part as opposed to removing a welded panel such as a quarter panel. Consequently, the repairs requiring part replacement are considerably less involved and are accomplished much more quickly, thus returning the vehicle to the owner in less time.

Most collision-damaged automobiles require a variety of repairs including straightening damaged panels that sustained minor damage, replacing unrepairable component panels, and replacing welded structural parts, panels, and even complete sections of the vehicle. As one would expect, the panels that are welded by the manufacturer must be replaced using welding equipment. The methods, procedures, and techniques are standard for most of these operations regardless of the manufacturer or the vehicle being repaired. However, when replacing any component parts one is apt to discover numerous different styles and types of fasteners used. The most common ones found on today's vehicles are bolts, nuts, screws, clips, and adhesives.

Securing

The methods used for securing parts on vehicles have changed significantly in recent years as manufacturers seek ways to become more efficient and cost effective in manufacturing their vehicles. Many parts that were once attached with bolts and nuts are now installed with adhesives or plastic clips. Some of the fasteners are used exclusively for one model year while others are generic to the point of using them on nearly all vehicles manufactured. On some model vehicles a plastic pop rivet is used to hold the grill or special cladding and ground effects in place. Some of the fasteners—such as a blind or pop rivet—are a one-time use fastener and must be replaced whenever the part or panel for which they are used has been removed. One must make provisions to order these items along with the parts that are being replaced. See **Figure 12-1**.

FIGURE 12-1 Plastic clips, aluminum and plastic pop rivets, plunger expanded rivets, adhesives, and one-time use hardware have replaced the bolts, nuts, and what once were reusable clips for installing trim on today's automobiles.

The type of fastener used is often linked to the likelihood of having to remove and replace a panel and whether it is apt to require adjusting to fit to the adjacent parts. The amount of road exposure a panel receives is also a determining factor in the type of fastener used. Plastic fasteners do not rust or corrode so they are frequently used to attach the bottom of the bumper cover to the reinforcement. In contrast, plastic pop rivets are frequently used on splash shields inside the fender wells where constant road spray is present. Aluminum pop rivets are frequently used to secure the inner bumper liners in high visibility areas because they will not discolor from oxidation as does steel. Whichever fasteners are used, one must become familiar enough with them to recognize those that require replacing and those that can be reused.

FENDERS

The front fenders are perhaps the most frequently replaced front panel. They may be made of steel, aluminum, or composite materials. Regardless of the material used, the methods are usually the same for removing and installing them, with the exception of the mounting hardware. Steel and aluminum fenders are typically installed with generically designed, reusable shoulder bolts. Plastic and composite fenders require a specifically designed bolt which will allow for **thermal expansion and contraction**. Using an incorrect fastener may likely cause problems with the fender shifting and binding when the door is opened and closed. During the first attempt at removing a fender or other component panel—and until one becomes familiar with the type and location of the hardware commonly used—it may be advisable to label the parts or immediately reinstall them into the same area from where they were removed once the part has been taken off the vehicle. This will help to ensure their proper placement and use when reinstalling the new part. See **Figure 12-2**. A variety of different bolt lengths and head designs are used, and reinstalling one in the wrong location will leave tell-tale signs that repairs were made to the vehicle. See **Figure 12-3**. Damage to attaching and adjacent parts may also be caused by using the incorrect size and length of bolt.

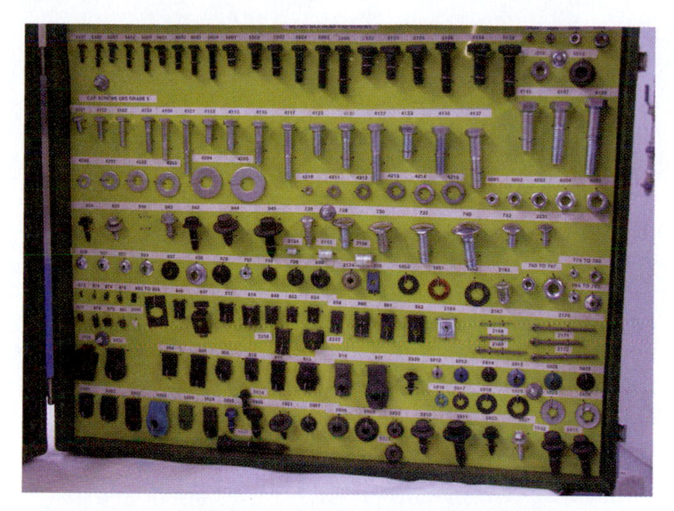

FIGURE 12-2 A variety of different bolt lengths and head designs are used for installing parts.

FIGURE 12-3 Reinstalling the wrong bolt in the wrong location will leave tell-tale signs that repairs were made to the vehicle.

Adjusting Fenders

Fenders may require adjusting for a variety of reasons. With few exceptions, when replacing the front fenders they must be adjusted or shifted to accommodate their fit to the front edge of the door and the edges of the hood. Sometimes as the hinges on the door and hood start to wear, they tend to allow the part to shift, resulting in a change in the gaps between them and the adjacent parts such as the fender and hood. Therefore, it becomes necessary to adjust the parts to establish the correct gaps once again. In other instances a vehicle may have been repaired and the shape of the door may not have been restored to its exact original configuration. The headlights, grill, front bumper cover, and many other adjustable parts are often attached to the front fenders. In order to obtain a uniform gap between all these parts it becomes doubly important that these parts are adjustable to accomplish an exact fit.

Core Support

On most body over frame (BOF) vehicles the entire front end assembly is attached via an adjustable radiator support (also referred to as the core support), which is bolted to the front of the frame rail. The radiator support can be adjusted approximately 1/2 inch in any direction at the bottom because the bushings or isolators used to secure it are placed over an enlarged hole to accommodate this movement. See **Figure 12-4**. The fenders, front grill, headlights, and hood latch in turn are attached to the radiator support. In order to properly align them to all their adjacent panels, each is designed to be adjusted. Without this movement, the front end assembly would be extremely difficult to align properly. See **Figure 12-5**. Manufacturers use several methods to make allowances for adjusting the fenders including enlarging and elongating the bolt holes, thus allowing a significant amount of movement. They may also use an adjustable **cage nut** or **U-clip**, which is designed to slide back and forth inside

FIGURE 12-4 The radiator support on a BOF vehicle is designed with a considerable amount of adjusting capacity to accommodate the necessary movement of all the front panels.

FIGURE 12-5 Many times a cage nut or J-nut is used in conjunction with an elongated hole on the front fenders and other panels to accommodate the adjustment required to properly align all the panels to each other.

the hole through which the bolt passes. The fenders are usually designed to be moved forward and rearward and in and out from the engine compartment. In order to obtain any upward movement, **shims** must be placed between the bottom of the fender lip and the mounting surface—such as the top of the radiator support and the upper cowl panel where it is attached. Shims, however, should never be used to compensate for incomplete or inaccurate repairs made to the structural members. As a rule, there is a sufficient amount of movement in each fender to adjust it individually to properly fit on each side.

In order to obtain the proper gaps between the fender and all the adjacent panels and components, a recommended sequence should be followed. After all the bolts and fasteners have been finger started, the area between the fender and the front of the door should be addressed next when permanently securing the panel. This area is most visible and is also the area where function and appearance must be used in harmony. The rear of the fender must be adjusted so that the door opens and closes repeatedly without coming into contact with the fender. Therefore, establishing the proper gap is very important as the consequences could be quite serious if the two panels rub together when the door is opened. Any further shifting may cause the two parts to bind, resulting in some serious damage as the door may inadvertently be forcibly opened. After the rear has been properly secured, one should then move forward to the top of the fender where it is bolted to the rail. The splash shield on the bottom side should be adjusted and tightened at the same time if it is the type that is bolted into place. Otherwise, it should be left until the entire fender is installed. The associated components at the front should be aligned and secured at this point. After all the fasteners have been tightened, the hood should be lowered slowly to ensure it is still fitting properly.

TECH TIP One must always close the hood slowly and observe the gaps between it and the fenders when closing it the first time after installation. This ensures the hood closes down between the two fenders and does not ride on top of them.

ADJUSTING FRONT ASSEMBLY

Occasionally it becomes necessary to shift the entire front assembly from one side to the other to obtain proper gaps between the panels. This may be the result of an improperly repaired or aligned frame rail during a previous repair or it may be from the front assembly coming loose and shifting on its own. When this happens the entire core support is shifted or moves off center, which in turn makes the fenders, hood, and all the other parts—including the front bumper—appear to be improperly placed. When examining the panel gaps, it appears that the only one that is correct is between the rear of the hood and the cowl panel. In order to correct the problem, it becomes necessary to move the entire core support, fenders, grill, and all other components attached to it. Before jumping into shifting the parts, however, it is advisable to X-check the front end assembly with a tram bar. This will confirm whether the front assembly is improperly attached or if the frame rails are misaligned. In extreme cases it may become necessary to repair the front rails to achieve a proper alignment.

Before attempting to move the core support from side to side, one must ensure that the bolts into the cowl at the rear of the fender have been loosened to allow for the movement of the metal at that area. As the front is shifted the rear of the fender must be allowed to move accordingly or a binding effect will occur, which will preclude any of the movement necessary to complete the operation. See **Figure 12-6**. It may also be necessary to move the fenders forward or rearward to achieve the proper gap between

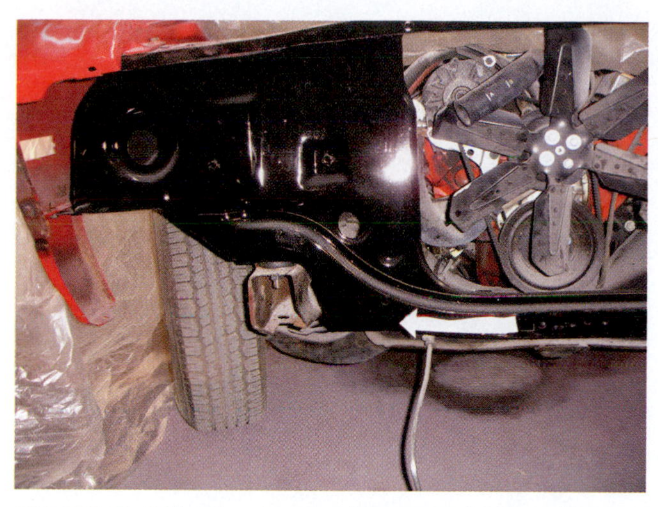

FIGURE 12-6 The fenders can become misaligned in numerous ways and may require shifting the core support from side to side in order to realign them.

them and the front of the door once the core support has been shifted.

ADJUSTING REACTION

Let us look at how and why this adjustment becomes necessary. The trainee must first understand that only the front one-half of the fender is adjusted by moving the core support from one side to the other. The rear of the fender must be adjusted individually with the bolts holding it to the cowl panel. It must also be understood that when the front of the fender is moved inboard or outboard, a sympathetic reaction occurs and a certain amount of movement in the opposite direction occurs at the rear of the fender. This often results in either opening or closing the gap between it and the front edge of the door, thus requiring additional adjusting at that location. See **Figure 12-7** and **Figure 12-8**.

FIGURE 12-7 Adjusting the core support may require the use of a hydraulic or friction jack to shift the entire front assembly in one direction or another.

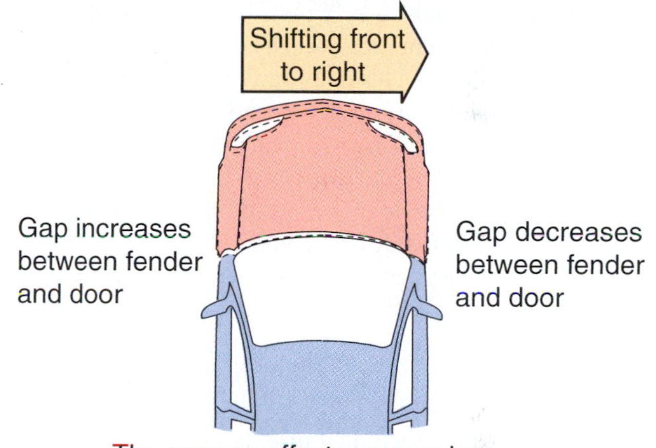

FIGURE 12-8 When the core support is adjusted, the rear fender will shift in the opposite direction that the front fender is being moved. The gap at the rear must be adjusted accordingly.

RESTORING ATTACHING SURFACES

The procedure for adjusting the fenders on a unibody vehicle is considerably more restrictive by comparison to doing so on BOF vehicles. Because the restoration of the structural members on a unibody must be more precise, fewer allowances are made for adjustments, which rarely exceed 3 mm (1/8 inch). The attaching surfaces must be restored to the precise location used by the manufacturer. Many times the only movement or adjustment allowed is raising the fender slightly by using shims between it and the inner fender rail to which it is attached. The bolt holes used to attach the fender are usually only a small fractional size larger than the bolts used for securing the fender. Therefore, very little adjustment is possible for fitting the fender

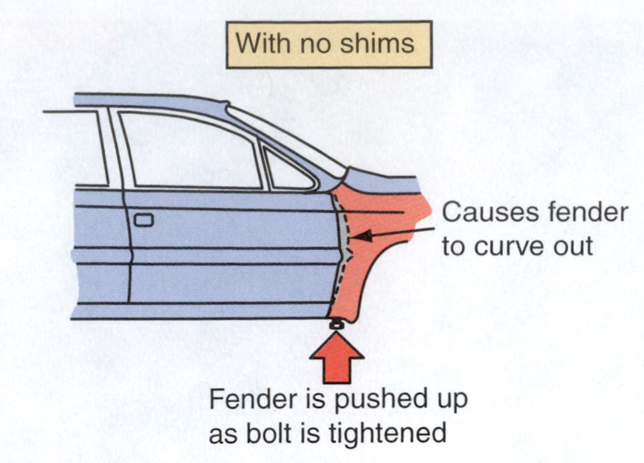

FIGURE 12-9 The curvature of the rear edge of the fender can be increased or decreased by removing or inserting shims at the lower rear edge of the fender where it is secured to the body.

FIGURE 12-10 Occasionally it is necessary to reshape slightly the opening edge of the fender using a hammer and a block of wood and a hydraulic jack.

to the adjacent panels. See **Figure 12-9**. Occasionally the inward or outward curvature of the fender's rear edge can be altered by adding shims or removing them from the bottom of the fender to match the leading edge of the door. This may become necessary, especially on those vehicle models where the fenders must be adjusted to a **positive flushness**. When the door is opened, it is designed to slip or move behind or inside the fender's rear edge. A positive flush fender adjustment positions the rear edge of the fender outward beyond the front edge of the door. If the fender has a negative or flush adjustment, the rear edge of the fender is even or slightly inboard of the door. In this case it will not allow the door to clear the rear edge, causing the two panels to bump into each other. When the door is forced open, it will inevitably cause one or both to become damaged.

Fender to Hood Adjustments

Occasionally it becomes necessary to make some minor adjustments on a fender even though the bolt holes do not make allowances for it. This may become necessary when the gap between the fender and the hood are still too tight even after all the normal provisions allowed by the slots and holes in the mounting surface have been used. This can be accomplished in two ways. See **Figure 12-10**. One is to take a block of wood at least 150 to 200 mm (6 to 8 inches) long and pad it with two or three layers of a heavy towel. While holding the wood block against the peak of the crown, strike several firm blows against the edge of the block using a dead blow hammer. Make sure to move the block back and forth several inches after each hammer blow is struck. This is usually enough to move or roll the fender's top edge out just far enough to give it the proper gap. This can be repeated several times until the desired results are achieved.

Another method involves placing a hydraulic jack with a properly padded flat push plate against the surface and applying pressure from the opposite side of the front

CAUTION

One must exercise extreme caution to avoid applying an excessive amount of force, thus damaging the edge of the crown on the fender.

fender. While pressure is being applied the flat surface in the adjacent area can be massaged using a hammer and a dinging spoon protected with tape to prevent marring the surface. One must use considerable caution and discretion to avoid damaging or "dinging" the surface with either tool being used.

Patience is a virtue when using either of these techniques. Being in a hurry to either hammer or push too hard can lead to disastrous results. After making two or three passes with either setup, one should stop and check the progress made.

INSTALLING SEQUENCE

When installing an entire front clip—which includes the hood, fenders, and grill—one should start by installing the hood first. All the adjoining parts must be installed around it and aligned to it. The center part of the cowl assembly is usually the least likely part of the vehicle to become damaged in a collision. Therefore, assuming the hinges, their mounting surface, and the hood are not damaged, one should install and align it to the upper cowl

TECH TIP

Caution must be exercised when loosening and tightening the bolts, as this operation is done repeatedly while adjusting the hood. Either a hand-held ratchet or a 1/4 drive air ratchet should be used for this operation. The bolts are small with fine threads; overtightening them may result in stripping the threads on the hood, thus requiring drilling and tapping new ones.

panel first—especially if it is the type that is welded into place. If the upper cowl filler panel is the removable type, one can take measurements from the lower windshield or other common surfaces to the back edge of the hood to ensure a uniform distance and then mount it to the hinges accordingly by **snugging the bolts**.

It is not necessary to fully tighten the bolts at this time as they will have to be loosened again to fine tune them when the fenders are installed. The height of the hood cannot be determined until the fenders have been installed.

Installing Front Panels

The front panels usually aren't installed until after they match the color of the vehicle, which is done in a process called **edging** or **edge painting**. This is necessary because many of the visible surfaces are not accessible for painting after they have been installed onto the vehicle. Because of its size and weight and to avoid causing any damage to the adjacent areas, attaching the hood should be done only with the assistance of a coworker. The bolts should be installed and tightened by merely snugging them down to hold them in place—they should not be torqued at this point. Many times the hood hinges function in conjunction with a hood prop to help hold it in an open position. These may be gas-charged pistons that are attached alongside the hinge or they may even be a part of the hinge assembly. They may also be a coiled spring that is put under tension to hold up the hood when it is opened. Likewise, a more conventional prop rod that is attached to the fenders or radiator support may be used.

TECH TIP The gas-charged piston and the spring-loaded props must be removed prior to removing the hood from the hinges. They should not be reinstalled until after the hood has been secured to the hinges.

The prop rods should be installed at this time to help hold the hood open while all the parts are being installed and adjusted. Once the hood has been attached to the hinges, the fenders can be installed by placing a bolt in each of the holes used to secure it. At this time the technician is interested in merely getting all the fasteners into place and started, by finger tightening them. They will be torqued down to specifications once they all have been installed and the final adjustments are made for permanent installation.

Checking Hood Alignment

At this point extreme caution must be used because the hood is being closed for the first time after the fenders have been installed. The hood should be closed very slowly, checking constantly from side to side to ensure the fenders are spread wide enough for the hood to fit between them. Dropping or slamming the hood before the opening has been checked for proper clearance may result in a significant amount of damage to the hood, the fenders, or both. The height of the hood must be matched to the fender edges. The forward or rearward adjustment and the side-to-side gaps must also be established between the panels at this time. If the fenders are made of a composite material, their position or location must be determined at this time and the bolts torqued to specification.

Checking Fender Alignment

Permanently installing the fender is usually accomplished in a series of sequential steps. The areas that are most visible—such as the area between the fender and the door and the fender and the hood—should be addressed first. One should start by establishing the gap between the front edge of the door and the rear edge of the fender—this is usually the most visible line around the fender, as it can be seen from nearly every angle around the vehicle. Most manufacturers recommend using a specific gap between stationary panels such as the fender and a functional panel such as the door. In the event these recommendations are not available, a simple rule of thumb to follow is to leave approximately a 4 to 5 mm (3/16 to 1/4 inch) gap between the panels in question. A gap that may be totally acceptable between two stationary panels may not be tolerable in a case such as the door and fender, where the door is opened and closed constantly and the possibility of rubbing or brushing against the rear edge of the panel is possible. This may become more prevalent with plastic parts, as they expand when the temperature rises. See **Figure 12-11**. A positive flushness may be preferred or required because it places the fender's rear edge slightly outboard of the leading edge of the door, allowing it to slide behind the

FIGURE 12-11 A positive adjustment between the rear edge of the fender and the front of the door may be necessary to keep the door from scraping against the rear edge of the fender and to eliminate wind and turbulence noises.

fender's rear edge without making contact. This may also be recommended to eliminate any wind noise while driving at highway speeds. The contour or curvature of the fender's rear edge must also be matched to that of the door. This can be accomplished by adding or removing shims at the bottom of the fender or by the slotted holes at the mounting surface.

Fine Tuning the Hood

The height of the hood can be adjusted in one of two ways: by shifting it up or down on the hinges at the rear or by adjusting the bumpers that are attached to the front of the hood or the upper core support. The hinge design is usually one of three types; a vertical mount, a horizontal mount, or one with enlarged holes on the hinge body. See **Figure 12-12**. With the vertical mounted hinge, the rear of the hood can be raised or lowered using the enlarged slots in the hinge strap. This is accomplished by loosening either the front or rear bolt and pivoting the hood on the bolt that is left snug. This will allow the hood to move either up or down at the front or rear, whichever needs to be adjusted. This may require two or more attempts before the proper height is achieved. Raising or lowering the hood using the horizontal hinge is accomplished by placing shims or spacers between the hinge strap and the bottom of the hood attaching surface. Both designs require the use of the adjustable bumpers located at the front of the hood or core support to raise and lower it to match the height of the fenders.

Adjusting Leveling Bumpers

Leveling bumpers help level the hood from side to side as well as eliminate the "**hood flutter**" that commonly occurs when the vehicle is driven at highway speeds or on rough terrain. They also exert a slight amount of upward pressure against the hood so when the latch release is actuated it will cause the hood to spring up slightly, separating the catch from the latch.

FIGURE 12-12 A variety of hood hinges are used to accommodate the adjustment required to set the height of the hood to match the fender.

HOOD LATCH ASSEMBLY

The hood latch system is responsible for holding the hood in a closed position once the latch has engaged the **striker**. See **Figure 12-13**. Most vehicles are equipped with a primary and secondary (or emergency) latch. The secondary latch is not used to hold the hood in place except during a failure of the primary latch, at which time it will prevent the hood from popping open while the vehicle is being driven. The primary and secondary latch may both be incorporated into one mechanism or they may be two independently functioning mechanisms. The latching mechanism—which is usually attached to the top or the front of the radiator support—is equipped with either enlarged or slotted holes. These holes accommodate the necessary adjustments to align the latch to the striker on the hood. To open the hood the latch is released by pulling a release handle inside the car, which is attached to the latch via a cable at the front of the radiator support or by pushing on a release lever under the front hood area outside the car. The secondary latch is usually attached to the bottom side of the hood. Adjusting the hood and the fenders should be done with the latch removed because it can give the technician incorrect signals concerning its alignment. When the latch is engaged, it will pull an improperly aligned hood into its proper alignment position, giving the technician a false reading concerning its location between the fenders. When left like this, both the latch and hinges are put under an undue stress, causing them to wear prematurely. This may cause the latch to not release properly.

(A)

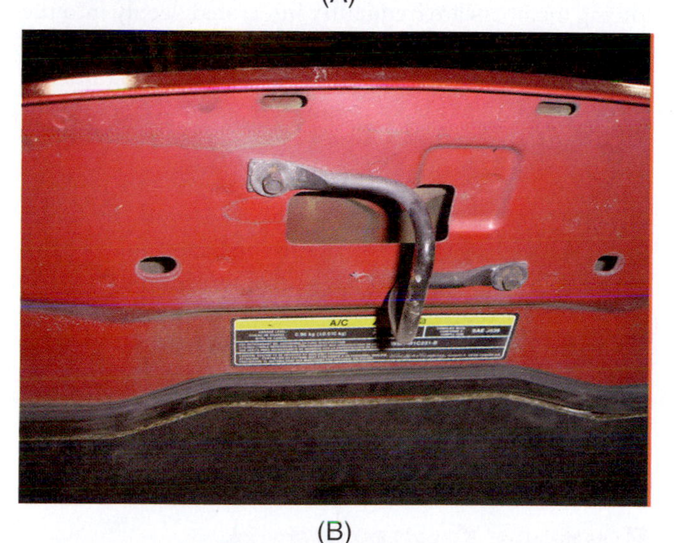

(B)

FIGURE 12-13 (A) This hood latch consists of a primary and secondary latching mechanism incorporated into one assembly, typically attached at the top of the radiator core support. (B) The catch part of this type of latching mechanism is usually attached at the front of the hood.

TRANSFERRING UNDERHOOD ITEMS

Several items that must be transferred from the damaged hood to the new one are attached to the bottom side of it.

Accessories

An insulation pad that covers the majority of the underside is attached with special clips. This pad insulates the hood from the tremendous heat generated by the engine and protects it from warping. All of the clips are reusable provided care is taken to not break them when they are removed. Other items that must be transferred are the windshield washer squirter nozzles and hoses and the underhood emergency light. The wire for the light is usually fed through the underhood rails and wired into a disconnect or splice joint at the front of the fire wall. See **Figure 12-14**. To expedite installing the light before installing the hood onto the vehicle, the technician can feed a separate piece of small gauge wire through the cavity in which the wire is routed and leave it in place. Once the hood is installed the connector end of the wire can be attached to the wire left in the cavity and used to pull the connector through the routing path for the wire. Once the connector end is pulled through the routing path, the pulling wire can be clipped and the connector can be plugged into the wiring receptacle.

Vehicle Informational Stickers

Several vehicle informational stickers are also attached to the underside of the hood. These cannot be transferred, however, and must be ordered new and installed after the fact. A part number is located at the bottom of each sticker, This number should be recorded on the estimate sheet to ensure the correct item is obtained when ordering the replacement parts. Occasionally the adjustable hood stops or bumpers are attached to the bottom of the hood. Some vehicles may also have hood scoops, hood scoop screens, and other original or add-on accessories that must be transferred as well. One last item that must be checked is the grounding strap that is used on some vehicles wherein an aluminum hood is used. This strap—usually either a copper or brass stranded cable with a formed bolt eyelet at each end—is

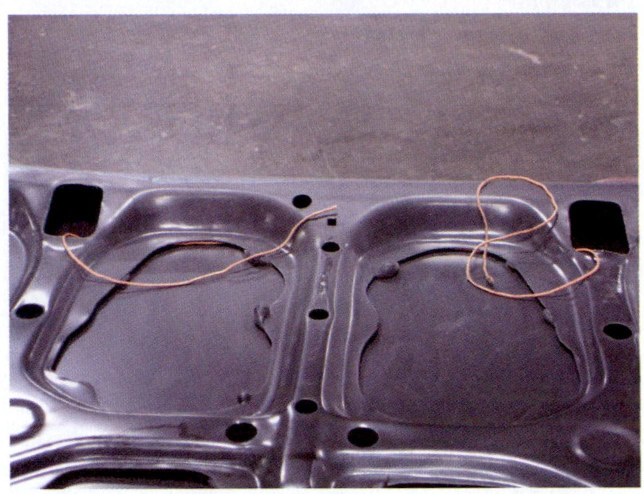

FIGURE 12-14 Pulling a retrieving wire or string through the routing path of an electrical wire when removing it will make the reinstallation much easier.

used to prevent oxidation from occurring on the hinge mounting surfaces. It is also used to ground certain electrical circuits. When installing this strap one must ensure a good clean contact surface is made on both the hood and on the area of the body to which it is attached. The best time to transfer these items is strictly a judgment call for the technician, because some straps may be more easily installed off the car on a work bench while others are more easily done while the part is on the vehicle. Only after the fenders and hood have been attached and properly aligned can the grill, headlights, and bumper assembly be installed.

TRANSFERRING FENDER ACCESSORIES

As with the hood, several removable parts must be transferred from the damaged fender to the replacement.

Radio Antenna

The radio antenna is usually held in place with a special painted or decorative nut and housing which must be installed to hold the mast in place. If the antenna and lead make up a one-piece unit, the lead must be re-routed through the cowl side panel and plugged into the radio under the dashboard. Many vehicles have a motorized retracting antenna which is usually installed prior to attaching the fender to the vehicle. The electrical connector must then be reconnected.

Side Marker Lamps

The side marker lamps may be attached to the fender or to the front fascia. If attached to the fender, they must be reinstalled prior to attaching the splash shield under the fender. Occasionally moldings and emblems are attached with mechanical fasteners; if the holes are not predrilled this must be done to accommodate them.

Hood Cushions

Hood cushions are commonly located on the inner lip of the fenders on both sides. Their function is to support the center of the hood and to protect the finish by keeping the hood from rubbing against the fender.

FRONT BUMPERS AND FASCIA

The federal government has mandated that bumper designs on today's automobile must be able to withstand a minor collision, such as those that occur in a parking lot, without sustaining any damage. Initially, the requirements were a 5 mph crash, but these regulations have since been reduced to approximately 2 1/2 mph. By incorporating energy absorbers and other reinforcing members, the front and rear bumper assemblies are able to assist in managing the collision energy in a minor to moderate collision. These energy absorbing devices and reinforcing members are frequently integrated deeply into the bumper and hidden behind the bumper cover, making it difficult to complete a thorough damage analysis until the entire assembly has been removed from the vehicle.

Bumper Designs

Three uniquely different bumper designs are commonly used on today's cars, trucks, and sport utility vehicles (SUVs). A steel bumper—such as the kind that has been used for many years—is primarily used on BOF vehicles, SUVs, certain mini trucks, and nearly all medium and large size trucks. Another design commonly used is a combination of steel and plastic wherein a fascia is attached to the steel facebar and another uses a complete fascia or bumper cover.

Bumper Components

The bumper assembly is usually comprised of three primary parts: the bumper reinforcement, an energy absorbing device, and the fascia or facebar. See **Figure 12-15**. The reinforcement is usually made of ultra high strength steel (UHSS), aluminum, or some type of reinforced com-

FIGURE 12-15 The front bumper on a unibody vehicle consists of three parts: the fascia or cover, an energy absorbing device, and the reinforcement, which is a solid steel assembly spanning across the entire front.

FIGURE 12-16 A variety of different collapsible rail end cap designs may be incorporated into the front of the rail that will collapse or accordion during a collision.

posite material. On some vehicles it is welded directly to the front rails, making it a part of the front structural members. On some vehicles the reinforcement is bolted to the front rails, making it a replaceable OEM serviced part. Due to the critical safety function of the bumper reinforcement, manufacturers warn against attempting to make any repairs to them. Both the aluminum and steel reinforcements are often made in a box-like design and restoring their shape is not possible. The steel reinforcement is made of UHSS and should not be heated. Cold straightening is virtually impossible. Likewise, the aluminum reinforcement is apt to crack if any attempts at straightening it are made.

ENERGY ABSORBERS

Most unibody vehicles place an energy absorbing device made of a high density composite material or an egg crate or honeycomb plastic immediately in front of the reinforcement to help dissipate the collision energy and to hold the cover to the proper contour. On a body over frame vehicle a collapsible mechanical impact absorber or a collapsible end cap may be used as well. See **Figure 12-16**. Although not as popular as it once was, the retracting piston energy absorber is still used on limited applications. The piston type energy absorber is filled with either a hydraulic fluid, gas, or a heavy bodied composite material. When disposing of a damaged unit a small hole should be drilled into the side of the piston to prevent the possibility of personal injury in the event it is exposed to high temperatures or extreme conditions. Nearly all auto manufacturers warn against repairing any energy absorbers.

STEEL BUMPERS

The steel bumpers may be chrome plated or painted to match or contrast the vehicle's color. They are typically used on pickup trucks and other vehicles that are normally used for heavier duty applications and industrial settings. This design is customarily attached to the vehicle with a collapsible frame horn which is designed to collapse or retract during a moderate impact. See **Figure 12-17**. These are typically bolted in place and are made to be adjusted by using slotted holes in the bumper itself, the reinforcement, and the front of the frame rail. Most manufacturers advise against straightening or repairing bumper mounting brackets when they become damaged during a collision. Improper repairs may cause improper air bag deployment timing.

Replacing Steel Bumpers

Due to their size and weight, removal and installation of bumpers should not be attempted without the assistance of a helper. All the painted surfaces adjacent to the bumper edges should be protected by covering them—or

FIGURE 12-17 Once a very popular design used by manufacturers, the mechanical impact absorber is used on a limited basis today.

the edges of the bumper—with tape to avoid scratching the surfaces should they come into contact with each other when the bumper is installed. It may also be advisable to use a floor jack to help support the bumper while it is being installed to avoid personal injury. The floor jack can also be used to raise or lower the bumper to ensure uniform gaps between it and the body sheet metal are achieved on both sides by using the slotted holes in the impact absorber. In some rare cases it may be necessary to shift the entire front sheet metal assembly by moving the core support to center the bumper and to obtain a uniform gap between all parts. The bolts holding the bumper to the frame rails and to the reinforcement should be torqued to manufacturer's specifications.

Steel Bumper with Fascia

Some bumper designs may incorporate a plastic cover over either or both the lower and upper part of a center **facebar**, forming an **air dam** to direct airflow to ensure proper engine cooling. This design is commonly used on SUVs and on a limited number of large domestic BOF cars. The facebar may be chrome plated or painted to match or contrast the vehicle's color, and the plastic fascia is usually painted to match the vehicle's color. With this design the fascia is attached to the facebar using special plastic fasteners or bolts designed specifically for this application. The bumper and reinforcement are usually attached to the vehicle using similarly designed mounting brackets as those used on trucks.

Full Fascia

The majority of the cars manufactured today are equipped with a full fascia or front cover made of either plastic or other composite materials. Even though they are equipped with a reinforcement and energy absorbing pad, their function is entirely cosmetic. They usually have openings to allow airflow to the engine compartment and to house the front lights. Some of the fascia may be equipped with a textured or dull-colored rub or impact strip incorporated into the center of them. This is designed to withstand a mild impact—such as would occur in a parking lot—without scratching or leaving signs of damage. A high density energy absorbing pad is commonly placed between the fascia and bumper reinforcement attached to the front rails. This is designed to help maintain the shape of the cover and also to help manage the collision energy. This energy absorbing pad may be bolted, riveted, or glued to the reinforcement or it may be left unattached and held in place by pressure from the bumper cover pushing against it.

The fascia is attached to the front of the vehicle in a variety of ways. It may be riveted with one-time use aluminum or plastic rivets, attached with plastic push pins, or bolted to the upper core support with specially designed shoulder bolts. It is usually attached to the front of the fenders with studded mounting brackets molded to the shape of the front edge of the fender and the fascia.

Fascia-Mounted Grill and Lights

The headlights, turn signal lights, cornering and fog lamps, and the grill are all incorporated into the front bumper cover. Even though the headlights are incorporated into the front cover, they are attached to the core support by a separate reinforced mounting bracket. The lights perform a critical safety function, and the flexing of the soft fascia would not hold them steadily enough. The mounting surface to which they are attached on the core support must be properly shaped to fit the contour and shape of the mounting bracket. Even though the headlights are designed to be adjusted with special adjusting screws and levelers built into the assembly, all the usable adjustment may not be enough to compensate for an improperly shaped mounting surface.

Grill

The grill is yet another component that is attached to the front cover. Some grill panels are inserts attached to the fascia with bolts or rivets while others may be molded directly into the bumper cover. Others are snugly centered into the opening and attached or held in place with mounting brackets that are bolted to the front of the core support. Many times the grill and **headlight covers** or **bezels** overlap each other, making it necessary to remove the headlight moldings in order to remove the grill. They may be chrome plated, painted plastic, and in many instances are a textured surface of a contrasting color. The grill opening is also used to allow airflow into the engine compartment.

REAR BUMPER ASSEMBLY

The rear bumper assembly is designed in much the same way as the front bumper assembly and cover. Like the front bumper assembly, the rear bumper reinforcement is attached to the rail ends as was done on the front of the vehicle. The high density energy absorbing pad is also placed between the cover and the reinforcing pad.

Lighting and Accessories

While lighting is not normally attached to the rear bumper, other accessories may be secured to the fascia or the rear facebar—whichever may be used. The backup sensors that warn the driver about blind obstacles—or identify when the driver is about to back into a large object—are usually attached to the rear bumper cover. The license plate lights are also attached to the rear bumper cover. Whenever removing the rear bumper cover or bumper assembly, one must be careful to first remove the wiring and the light attaching brackets to avoid damaging any part of the loom sensors. After reinstalling or replacing the bumper, one must ensure all the lights are functional once again. Failure to do so may result in a vehicle that is unsafe to drive.

Diagnosing Damage

Damage to both the front and rear bumper reinforcement, the energy absorbing pad, and the cover itself are all difficult to diagnose because the soft plastic fascia readily returns to its original contour when it becomes damaged. None of the damage may become evident until the outer cover has been removed. Suffice to say, one must inspect them very carefully so as not to overlook any existing damage.

DOORS

The doors are the most frequently utilized component on the vehicle as they are opened and closed each time one gets in and out of the vehicle.

Hinge Mounting and Location

The doors may be attached to the vehicle with adjustable hinges that are bolted to the **hinge pillar**, or the hinges may be welded to both the door and the body. The most common mounting location for the hinges is at the front of the door. However, with the increased popularity of four-door trucks they are apt to be located at the rear of the door as well. Other hinge designs and locations are found on SUVs and hatchbacks, where the hinges are located at the top of the rear gate or hatch and on sliding doors. Due to the frequency of being opened and closed, doors are subject to becoming misaligned due to wearing of the hinge mechanism or because of the hinge shifting or slipping at one of the attaching points.

Adjusting Slots. The bolted hinges are designed with enlarged or slotted holes to accommodate adjusting or they may be attached to a **floating hinge plate** on the inside of the hinge pillar. See **Figure 12-18**. In either case, they may be shifted approximately 12 mm (1/2 inch) in either direction fore and aft, up or down, and in and out. The hinges that are permanently welded to the body utilize a replaceable hinge pin and bushing design, which—when they become worn—cause the door to sag and thus become misaligned.

Removing and Installing Doors

Removing the Door. Although removing and replacing a door may seem like a simple task, several problems can occur if the technician follows an incorrect sequence when removing it. The first step is to remove the interior trim panel. This is done by removing any friction clips around the outer edges and screws that may be used to hold it in place, particularly in areas where closing pull straps are attached. The trim panel must be carefully pryed away from the inner door frame with special door trim removal tools designed to remove the clips without destroying them. All the lighting and accessory electrical connectors must be disconnected to avoid damaging them and the wires. All the electrical lighting and accessories are connected to the

FIGURE 12-18 Most door hinges are adjustable by using an encased floating plate that allows the door to move approximately 12 mm (1/2 inch) in either direction—fore and aft, up or down, and in and out.

main wiring loom with an individual pigtail which must be disconnected. The main wire loom must then be removed from the door panel. It may be advisable to sketch a diagram of the wiring route and mark each pigtail with the name and location of the accessory from which it was removed to help during reassembly. Each connector and the wiring attached to it should be inspected to ensure they are not damaged or that there are no wires pulled out of the sockets or connectors. Any damaged items should be noted at this time so they can be reordered, leaving enough time to secure them for when the vehicle is reassembled. Underneath the trim panel one will find an insulator or barrier shield adhesively attached to the inner door panel. This must be carefully removed to prevent tearing or damaging it, because the shield must be reinstalled when reassembling the door. At this point the wiring loom should be disconnected from the inner panel and removed through the access hole at the front of the door jamb. One can then start removing the bolts from the bottom hinge first, followed by those from the top hinge, which will allow the door to be pulled away. Removing the door from the hinges prior to removing the trim panel and pulling the wiring out will likely result in the door hanging by the wiring loom, almost guaranteeing some unnecessary damage will occur.

Installing the Door. When installing the door onto the vehicle, it is advisable to have an assistant help lift the door into place because it is very heavy and cumbersome and difficult to control. A portable door stand may also be used to hold the door steady until all the bolts have been started into the hinge pillar.

Before attempting to install the door, the hinges should be attached to either the door jamb or to the hinge pillar, whichever is more difficult to access while the door is being installed. Because the hinges have a significant amount of free movement for adjusting purposes, it is advisable to attach them at the midpoint of their adjustment extremes and snugly attach the bolts. This will serve as a good starting point to fine tune the door adjustments when it is fully installed. Before attaching the door or any of the hardware, the front, bottom, rear edges, inner panel, and any other surfaces that are not accessible once they are mounted onto the vehicle must be edged or painted to match the vehicle's color. The hinges and any other attaching hardware—such as the door check straps and the striker assembly—must also be painted prior to installing.

TECH TIP The first bolt is always the most difficult one to get started when hanging a door, because the hinge plate is difficult to align with the holes in the hinge and, as such, it is hard to get the first threads started. A simple solution to help overcome this problem is to cut the head off a bolt with matching threads and screw it into the hinge plate part way. By inserting this bolt through the corresponding hole on the hinge, the door can now be put into place and rested on it while all the others are aligned and started to be threaded into the hinge plate. Once all the other bolts have been started, the "headless bolt" can be removed and the permanent one put into place.

Adjusting the Door

Adjusting the doors may become necessary for a variety of reasons. Any time a door is replaced or has to be removed to accommodate repairs to it or an adjacent panel, it will require adjusting to ensure uniform gaps are achieved. It may also require adjusting to compensate for normal wear and tear on the hinges or even to correct an improper alignment made during a previous repair. Occasionally a hinge may slip and move on the mounting surface because a bolt came loose. Any visible disparity in the door gaps is an indication of a potential problem that must be addressed because it is certain to become a more serious issue at a later time. On the other hand, a uniform gap around the entire perimeter of the door is an indication that it fits into the opening properly and it is functioning in harmony with all adjacent surfaces.

Door adjustments can be a very frustrating experience for the beginner. However, when a few simple rules and guiding principles are applied, it can be much less difficult and frustrating. The most logical starting point is to first align the door with the rocker panel and the quarter panel as both are permanently welded into place and very unlikely to be moved out of their proper place.

Four-Door Model. On a four-door model vehicle, it may be necessary to adjust both the front and rear doors, starting at the rear and moving forward. As part of the inspection sequence, one must analyze how—and at what part of the hinge—the desired adjustment must be made. See **Figure 12-19**. Most bolted hinges are designed with a combination of slotted, elongated, or enlarged holes

(A)

(B)

FIGURE 12-19 The door fore and aft adjustments are made on the surface where the hinge is mounted to the hinge pillar (A). The up and down and inboard and outboard adjustments are made where the hinge is mounted to the door (B).

which allows the hinge to move in a combination of three distinctly different directions; fore and aft, up and down, and in and out. As a rule, the fore and aft adjustments are made on the surface where the hinge is mounted to the hinge pillar. To move the door forward or rearward, simply loosen the bolts and push the door forward or pull it back. The in and out adjustment is made where the hinge is mounted to the door, and the up and down adjustment may be made at either location depending on the manufacturer. Simply stated, the door is adjusted by loosening the bolts on the hinge, sliding it in the intended direction, and then retightening them. The simplest way to determine which part of the hinge the adjustment is made on is to close the door and observe in which direction each part of the hinge will allow movement to occur.

Occasionally the door hinges may be frozen or corroded into place and may require some additional persuading. There are several ways of breaking the bond loose between two surfaces. After loosening all the bolts, one can raise or yank the rear of the door upwards sharply by hand or place a jack under the door's rear and raise it up. It can also be accomplished by placing the floor jack under the rear of the partially opened door and bouncing on the door opening sill plate. One may also place a heavy duty spoon against the back of the hinge and strike it several times with a large hammer, thus breaking the bond.

Coarse and Fine Tuning Adjustments

Door adjustments are commonly made in two phases or stages: coarse adjusting and detail adjusting or fine tuning. **Coarse adjusting a door** is commonly used when the door is initially installed. It is roughly placed into position in the opening and attached. When coarse adjusting, the entire door is moved in one or more of the three directions allowed and bolted into place. At this point the technician is not particularly concerned about the gaps and flushness of the fit. **Detail adjusting** or **fine tuning a door** is used to precisely position the door so the gaps are uniform around the perimeter and the flushness is correct.

Fine Tuning Adjustments. Fine tuning or detail adjusting is usually most difficult for a novice to understand and complete. This process is best understood if one can visualize the door rotating in circular motion around an axis or pivot point (the hinges) with all corners moving in a uniform direction and distance. In **Figure 12-20**, using the top hinge as an **axis point**, any shifting or movement made causes the door to move in a circular motion. The most significant amount of movement occurs at the corners, each moving a corresponding amount equal to each other. In order to effect this adjustment one should loosen all the hinge to pillar bolts except one on the top hinge, which is used for the axis point.

For example, to move the entire door forward slightly, the following scenario should be followed: to use the top hinge as a pivot point, loosen two of the three bolts holding the top hinge to the pillar and loosen all the hinge to pillar bolts on the bottom hinge. This single bolt on the top hinge will serve as the axis point for the first stage of the adjustment. Close the door nearly completely and pull down on the upper rear corner of the door, causing it to move downward at the rear. One bolt on the bottom hinge should now be snugly tightened. To check its progress, close the door. One will observe that the most obvious movement will be seen at the corners of the door in the following way: The top corner will have moved back closer to the adjacent panel and down slightly, while the lower corner moved down and forward away from the adjacent panel. In addition, the lower front corner will have moved forward and up slightly. A larger gap will now exist between the upper rear fender edge and the front upper door edge.

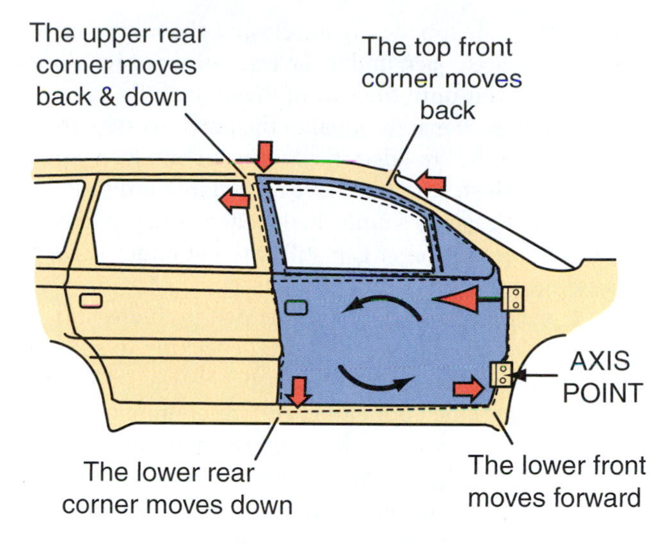

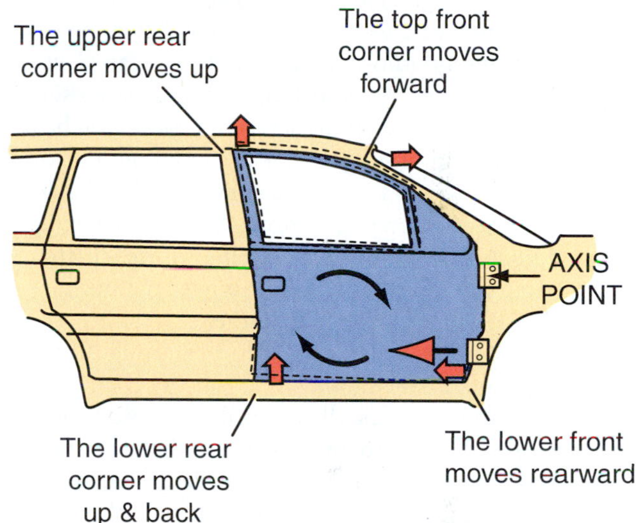

FIGURE 12-20 By using the top or bottom hinge as an axis point, any shifting or movement made causes the door to move in a circular motion.

To equalize the gaps between the door and fender and the movement around the door's perimeter, the bolt on the top hinge should now be loosened. The one bolt on the bottom hinge now becomes the axis point for the next phase of the adjustment. Again closing the door nearly completely, place a jack under the rear corner of the door and raise up on it until the rear of the door shifts upward an amount approximately equal to the previous step, then retighten the bolt. Provided the door was shifted an equal amount in both steps, it will have moved forward slightly, thus closing the gap to a uniform distance from top to bottom. In the event a larger gap still exists at either the top or bottom, one should repeat the steps that were performed by moving the door to reduce or increase the gaps—whichever is necessary. When checking to see if a proper alignment with the adjacent panels was achieved, one should close the door and check the alignment of the most visible crown surfaces between the door and the adjoining panels. If the crown runs without interruption or deflection from one panel to the next, one can assume the door is properly aligned to the panels next to it. The flushness should also be checked at this time. After the door has been properly adjusted, it should be opened completely. The bolts on the top hinge should be loosened first, retightened, and then the same process should be repeated on the bottom hinge. Prior to loosening the bolts, a floor jack or door jack should be placed under the door to support it and to prevent it from shifting, thereby losing the adjustment just completed. This is necessary to relieve any binding or stressing of the hinges that may have occurred while they were being adjusted. Failure to do this may cause the hinges to wear prematurely because they are left under stress.

The procedure just described can also be used to adjust the inward and outward movement of the door to the body. Remember, this adjustment is made on the hinge to door mount and is used to set the door flush to the surface of the adjoining panel. Occasionally the front door may have to be placed with a **negative flushness** to the rear of the front fender. A negative flushness involves placing the front edge of the door inboard of the rear edge of the fender a very slight amount, perhaps no more than 1 mm (1/16 inch). This may be necessary to eliminate **wind leaks and turbulence noises**. Unless otherwise specified, the rear edge of the front door and the front edge of the rear door should be mounted flush to each other. Likewise the rear edge of the door should also be mounted flush to the leading edge of the quarter panel. These same principles can be applied when adjusting any part that is attached with adjustable hinges.

Installing Striker

After the door is attached and fully adjusted one should install the striker, which should be removed before the installation and adjusting procedure is initiated. Many times the striker will give the technician a false impression of the door's adjustment as it may raise or lower the door slightly when it is engaged, giving an incorrect reading that it is aligned properly. To avoid this, the striker should be removed until the door is fine tuned and then it should be installed to fit into the location where it matches the door latch. The purpose of the striker is to hold the rear of the doors inward or outboard to achieve flushness with the adjacent panels. Even though it can be shifted upward/downward and in/out, it is not designed to raise or lower the door to align it with the adjacent panels. It is movable to make it match the location of the door latch. On some vehicles the striker may not be adjustable, thus necessitating some additional shifting of the door to properly align it to the stationary striker.

TECH TIP All door weatherstripping should be fully in place before making final door adjustments.

Even though it may have to be removed again prior to painting, all of the weatherstripping should be installed while fine tuning the door adjustments. Attaching them after the door has been fully adjusted may alter how the door opens and closes and fits into the opening. The door latch to striker depth should also be adjusted at this time. The striker should be adjusted so that when the door lock is released, it should move out a very slight amount—only enough to clear the latch to prevent it from re-latching. The door should then swing open with a minimum effort. One should also be able to close the door with a minimum amount of effort without having to slam it to engage the latch mechanism. The **latch mechanism** has two catches: the pre-latch, which holds the door in position and lines it up with the striker, and the latch position, which holds it in a tightly closed position. If any excessive slamming is required to close the door, it is an indication that the door is not properly aligned.

Latch Assembly

Replacing the door requires transferring or replacing all of the bolted accessories such as the latch mechanism, the inner and outer door handle, the lock release mechanism, and occasionally the locking button. These accessories may be bolted, riveted, or even held in place with screws or plastic clips. See **Figure 12-21**. The latch mechanism, which is attached at the rear of the door, is held in place with machine screws where an exacting cutout is made to accommodate the latch and striker engagement when the door is closed. Opening and closing the door is accomplished by releasing the latch via the remote connecting rods that connect it to the door handle. Installing the outside door handle is commonly done after the outer door panel has been painted so the surface under it will be the same color as the vehicle.

Installing and Adjusting Composite Panels

On some vehicles the doors are constructed of a combination of a steel frame with a **composite outer panel**. Adjusting these doors basically requires the same procedure

FIGURE 12-21 The door latch mechanism is disengaged by moving either the inside or outside release handles, which are connected to the latch with the remote control connecting rods.

FIGURE 12-22 Adjusting the doors on a vehicle with sheet molded compound (SMC) or other composite outer door panels sometimes requires the removal of the outer panel in order to obtain the correct door frame shape.

used for adjusting the hinges. However, most of the detail adjusting is accomplished with the **outer skin**, which is attached with special bolts. See **Figure 12-22**. Once the door shell or frame has been positioned and secured into the opening, all the fasteners should be installed onto the door and snugly tightened to hold the outer skin in position. The skin must be positioned with very specific gaps between it and adjacent panels to allow for thermal expansion and contraction. In some instances it may need to be adjusted and positioned using the current **ambient temperature** as a guide. The fasteners then must be torqued to very specific recommendations to allow for the proper amount and direction of the expansion to occur. The front door and fender are usually fastened and torqued to allow the thermal expansion to move forward and the rear door and quarter panel to move rearward. Improper tightening will likely cause the panel to buckle or warp under extreme temperature changes if the correct torque amount is not used. The manufacturer's vehicle service manual should be consulted for specific information and recommendations.

Most vehicles have movable windows attached to the inside of the doors. The windows are attached to the bottom of the door with either a manually operated or electrically activated regulator. The regulator is responsible for moving the window up and down as the motor is activated or the handle or crank is turned by hand.

Rear Sliding Doors

The rear sliding doors are commonly found on minivans and full size vans on either the passenger side or on both sides of the vehicle. They may be operated manually, by remote control, or by a switch usually located near the front of the door, on the dash or console, or the control panel located in the ceiling. With very few exceptions, the glass in the sliding doors is stationary or fixed. See **Figure 12-23**.

Door Tracks and Accessories. The door is attached to the vehicle with movable hardware that rolls or slides on a track on both the upper and lower part of the door. The tracks are usually curved in such a way that they pull the door inward into place against the body and the weatherstripping when the door is pulled forward. Because the doors do not swing out, little or no adjustment is required to fit them into place and seal the opening. The striker and latching mechanism both operate in the same manner as on other doors. While some manufacturers still use conventional wiring systems to connect the electrical accessories in the sliding doors, some utilize a matching **electrical plunger** mechanism—one on the door and one on the body—to connect the electrical accessories inside the door to the vehicle's wiring system. See **Figure 12-24**. The two components of the plunger assembly must be precisely aligned to each other to ensure the accessories—such as the door opening and closing motor, the door locks, and so

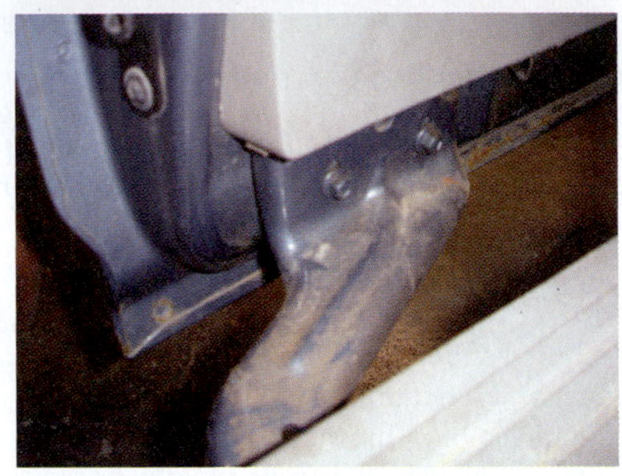

FIGURE 12-23 The sliding side doors on most minivans roll open and shut on a fixed roller track and require little more than an occasionally inboard and outboard or up and down adjustment.

FIGURE 12-24 Electronic contact plungers are used in place of electrical wire connectors to activate the door lock, alarms, and so on, on the sliding doors.

on—will function properly. The manufacturer's service manual must be consulted to determine which of the plunger contacts operates any given accessory.

Door Hardware and Accessories. A number of additional accessories that are attached to the door's interior or exterior must be removed and transferred from the damaged door to the replacement panel. These items include the inside and outside door handles, which are connected to the latch by remote connecting rods or cables; the lock cylinder assembly, which is also attached to the latch with connecting rods; as well as the lock button connector.

Replacing the lock cylinder will necessitate re-coding the lock tumbler to ensure its operation with the keys used on the old locks and ignition. If the vehicle is equipped with electric locks, a vacuum or electric solenoid is likely to be attached to the lower rear section inside the door. This mechanism is responsible for locking and unlocking all the doors when the remote control or the manual switch is activated.

The outside door handle may be attached to the outer panel with threaded studs that are a part of the handle body, bolts, rivets, or screws and clips. Many times the outside door handle is painted to match the vehicle's color. This requires that it—as well as the area underneath the handle where it is attached on the door—be painted prior to installation. Therefore, the outside door handle is typically installed after it has been refinished. Covering the surface immediately around the door handle location with tape will help to ensure the area will not be damaged or scratched during its installation.

DECK LID AND HATCH

The **deck lid and hatch** are the closures for the rear of the vehicle spanning across the entire width between the two quarter panels and the lower rear panel. While the deck lid and hatch both perform the same basic function, they are considerably different from each other in their design and how they are attached to the vehicle. The deck lid is commonly used on all standard car models while the hatch is utilized on most hatchback model cars and on some SUVs. The deck lid is typically a solid panel that spans across the width of the opening between the two quarter panels. It may be made of steel, aluminum, or a composite material, depending on the type of vehicle and the manufacturer. The hinges are attached to the vehicle by welding or bolting them to the underside of the package tray at the top of the quarter panels inside the trunk compartment. On some model vehicles, such as the automatic convertible, the deck lid may have dual openings as the deck lid is designed to open in the front to accommodate the top when it is opened. See **Figure 12-25**.When the top is up, the trunk may be accessed from the rear for storage. The rear window is commonly installed into the rear hatch and it opens and closes along with the hatch whenever it is raised or lowered.

FIGURE 12-25 The deck lid is held open by either putting the hinges under a state of tension with a spring, by pressure with two interconnecting torsion rods, or with gas-filled shocks.

FIGURE 12-26 A deck lid attached to a vertically mounted hinge allows for a significant amount of up and down adjustment of the deck lid.

There are three primary deck lid hinge designs used on vehicles: the torsion bar type, which is most commonly utilized; the piston type, which is similar to those used on the hood; and the spring type. The purpose of the torsion bars, the piston, and the spring is to raise the deck lid once the latch has been disengaged and hold it in an open position until it is pulled down and closed. Unlike the other designs, the torsion bar type allows one to adjust the spring tension to raise and hold the deck lid with more or less tension as needed to compensate for wear. On the other hand, the piston must be replaced when it is no longer able to hold the deck lid in an open position. Although rarely done, the spring is replaced to overcome losing the tension from wear and fatiguing.

Adjusting the Deck Lid

Adjusting the deck lid is accomplished by using many of the same principles discussed for door adjustments. Using one of the hinges for the rotation or axis point and shifting the lid around it is the same concept as is used when adjusting the door on its hinges. The hinges may be attached either vertically, horizontally, or—on some models—they may be installed into the cavity which is part of the inner reinforcing panel. See **Figure 12-26**. Vertically mounted hinges and those mounted to the inside of the deck lid reinforcement cavity usually offer more adjusting latitude than the horizontal types. In addition to the up and down adjustments, one can also adjust the **pitch of the deck lid** to match the contour of the quarter panels from front to rear. See **Figure 12-27**. This is also possible on the cavity inserted hinge by simply raising and lowering the deck lid at the bolt locations.

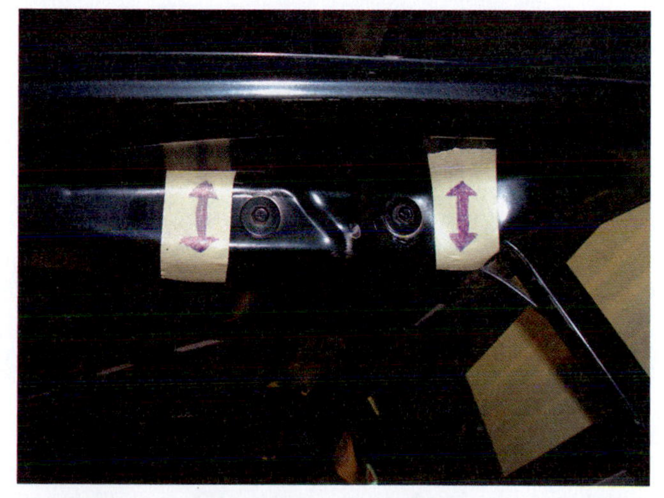

FIGURE 12-27 Adjusting the deck lid up and down to match the quarter panel height is done in the same way as adjusting the doors by using either the front or rear bolt as an axis point and pivoting the panel up or down around it.

Any up and down adjustment on the horizontal mounted hinge must be accomplished by placing shims between the hinge straps and the deck lid. The deck lid can be adjusted fore and aft with all the hinge designs by simply moving it within the enlarged holes provided on the hinge arms or the deck lid mounting surface, whichever area is made to be moved. On some vehicles the hinge assembly is bolted to the rear package tray, thereby allowing additional adjusting latitude.

Adjusting Deck Lid Flushness

The **deck lid tension** is typically adjusted with the striker. It should be positioned to hold the edges of the deck lid flush or even from front to rear with the matching edges of the quarter panel. It is designed to be raised, lowered, and moved from side to side to match the location of the striker, which is normally attached to the lid itself. As with the hood latch, the striker should not be used to pull the deck lid into alignment by moving it from side to side. It should be adjusted so that when the lid is closed down, the striker scissors directly through the middle of it, not causing it to shift.

Striker

The striker can be adjusted from side to side and up and down to properly hold the trunk lid in alignment with the opening once it has been aligned. On some high end vehicles the deck lid striker is designed to pull the deck lid down into a closed position automatically. The deck lid must be closed far enough to engage or contact the latch, at which time it will automatically pull it down into a tightly closed position. One must be careful to not slam the lid as is normally done on a manual design as it may damage the automatic closer.

FIGURE 12-28 The rear liftgate on an SUV functions in several different ways, such as opening the rear window separately or opening the gate completely.

REAR HATCH

The rear hatch on the hatchback model vehicles and the rear **liftgate** on the SUVs serve the same function as the trunk lid does on sedan and other model vehicles. They are the closures for the rear of the passenger compartment. Their design, however, is somewhat unique from the trunk lid in several respects. They are typically attached to the rear of the roof with hinges that allow them to be opened and closed. A gas-charged prop rod is typically used to hold them in an open position once they are raised up. The rear glass is attached to the rear liftgate with mechanical fasteners, or the glass may be permanently installed by adhesive bonding it to the framework of the assembly. See **Figure 12-28**. On the SUVs the rear glass is nearly always attached to the liftgate with hinges so it can be opened without opening the entire gate assembly. Conversely, on a hatchback, the glass is usually permanently bonded to the hatch framework; when the hatch is lifted, the glass is raised with it as well. Many times a rear window wiper assembly is attached to the rear liftgate or hatch. This includes the wiper motor, window washers, and all the additional functional parts found on the front windshield wiper assembly. When replacing the rear hatch or liftgate, one must transfer all these items from the damaged part to the replacement assembly.

One of the most significant differences between the sedan model vehicles and the hatchback or SUVs is the entire interior of the vehicle is exposed when the rear gate on either of them is opened. There is no dividing wall between the passenger compartment and the trunk, such as is found on the sedan models.

Split Door Design

Another design that has been used for many years on vans and SUVs is the vertically mounted split door design. This design uses two doors split in the center of the opening that are attached to the D pillar using an upper and lower hinge. Some limited applications use a 1/3 to 2/3 split design—wherein one of the rear doors covers approximately 2/3 the width of the opening and the other door covers the remaining area. The rear windows may be hinged to open a small amount at the bottom for ventilation or they are permanently installed. As with the hatch, the rear doors are adjustable to fit tightly enough to seal the opening.

REAR LIGHTS

Regardless of the vehicle design, the rear lights are attached to multiple locations on the rear of the vehicle. They may be attached to either the rear panel, to the hatch, the D or rear hinge pillar, the deck lid, or a combination of two or more of them. The wiring for the license plate lights, the rear windshield wiper, the rear heated glass, and the latch release are often routed through the cavities between the inner and outer layers of the panel to which they are attached. See **Figure 12-29**. Before attempting to remove the bolts or fasteners holding the rear liftgate in place, the wires must be disconnected at the bulkhead which is usually located somewhere near the top of the panel. Failure to do so will inevitably result in damaging the wiring, because once the bolts have been removed from the hinges the wiring is the only thing with which the panel will be suspended or attached to the body. The taillamp and rear lamp assembly may be a continuous modular unit which spans across the entire width of the deck, extending onto the rear panel. It may also be made up of several individual pieces attached separately to make up the assembly. On some of the base model vehicles, the taillamps are small individual units attached to the rear of the quarter panel only. Whichever design is used, they are usually attached with studs molded to the lamp assembly that are held in place with nuts on the inside of the rear panel to which they are attached. A foam rubber padding or a ribbon of sealing caulk is placed between the lamp assembly and the body to seal out water and dust.

FIGURE 12-29 A common source for the rear vehicle wiring is a wiring module that may be located in the trunk, somewhere on the quarter panel, on the rear package tray, or on an SUV near the rear door opening. This module may be covered by the headliner.

COMPOSITE PANELS

Many of the space frame design vehicles and a limited number of others, such as the Buick Park Avenue, use composite fenders that are attached to the vehicle with mechanical fasteners such as bolts, rivets, and other special application hardware. Many of the exterior parts of the Saturn "S" series, VUE, and the ION—including the quarter panel, doors, and fenders—are made of a composite material. These parts are attached with special application fasteners that allow for thermal expansion and contraction as the temperature rises or drops.

Installing Panels

When installing these panels, one must consult the manufacturer's service manual to determine the recommended torque and tightening sequence that must be followed. Overtightening the fasteners or tightening them in the incorrect sequence may cause buckling and warping of the panel as the temperatures vary.

Adjust for Expansion

A rule of thumb to remember is the front fender and the outer door panel are torqued to move forward as they expand and the rear door and quarter panel are attached in such a way the expansion will shift them to the rear. In addition to tightening to the proper torque specifications, one must also tighten them in a specific sequence. Failure to do so may result in the panels becoming warped during normal but dramatic temperature changes.

PICKUP BOXES

Most pickup boxes are made of steel. However, some have bolted side panels made of a composite material such as a sheet molded compound (SMC). These are commonly found on pickup trucks with step side boxes and they are installed with bolts, while screws may be used in the wheelhouse areas—especially when a flare opening is used. Removing the side panel may require the removal of the bolts holding the box to the frame to slide it rearward and access the bolts. On some trucks—such as the 2001 and newer Ford Explorer SportTrac—the entire box is made using SMC.

Many times heat must be applied to loosen bolts, particularly those that are exposed to road spray and other corroding elements. One must exercise extreme caution when applying this heat—especially if using an open flame—as the fuel tank and fuel lines are usually in the immediate area where this is done. An induction type heater should be used instead as no exposed or open flame is used and the heat can be concentrated in the exact area where it is needed.

CAUTION

Caution must be exercised when applying heat to the bolts to avoid damaging or melting the SMC material in the box.

Removing the Pickup Box

Removing the pickup box off the frame requires access from the bottom side of the vehicle as the attaching bolts are run through the frame rail and nutted from the bottom side. This may necessitate raising the vehicle off the ground to gain access to them. Many times heating the nuts may become necessary if the vehicle has been in service for any period of time, as they are constantly exposed to road spray and debris which promotes corrosion. Caution must be exercised to avoid overheating them, especially when a composite material box is used, as it may cause the material to melt or become disfigured from the heat.

Preparation for Removal

Several items must be disconnected or removed prior to lifting the box off the frame. The fuel filler connecting hose and the connecting ground strap must be removed from the box side panel or the fuel filler pocket. The wiring for all the rear lamps must be disconnected and any wiring that may be attached to the box must also be removed. Many times an aftermarket trailer hitch attached to the vehicle's frame must be removed in order to gain access to the fasteners holding the box to the frame. When the box is lifted off the vehicle it is critical to make note of any additional shims that may be placed on any of the body mounts to ensure they are reinstalled in the same location when the box is put back into place. Many times shims or spacers are used to ensure the box is set on a level plane and to eliminate any twisting that may occur if they are not placed properly. Removing the box from the frame will require the assistance of several people to lift it up high enough to clear the rear tires and move it back away from the frame. A **box lifter** or a series of planks with chains can also be used to lift the box off the rear of the vehicle. The box lifter should be set on a table or a cart capable of supporting its weight rather than setting it on the floor as the lower panels may collapse from the weight if the lifter is set down on them.

GLASS

Among the least noted—yet most functional—parts of a vehicle are the windows that surround the passenger compartment. The use of glass windows permits visibility into and out of the passenger compartment, security for the occupants, and protection from the elements and flying debris when driving down the road. By being able to open and close them, windows also allow for controlled ventilation of the passenger compartment.

Laminated and Tempered Glass

There are two primary types of glass used for windows on all motor vehicles: laminated safety glass (LSG) and tempered—sometimes called safety—glass.

Laminated Safety Glass. The LSG is typically used for windshields but rarely for movable glass applications. It is made by bonding two layers of glass together using a clear plastic layer as a bonding material between the layers. The laminated glass is extremely durable in protecting against foreign objects penetrating the glass and entering the passenger compartment. Likewise, it protects the passengers from being ejected from the vehicle in an accidental situation. Even after the glass has become cracked, the plastic laminate layer prevents the glass from disintegrating. This—together with the urethane adhesive bonding material used to attach the windshield to the pinchweld—allows the glass to be used as part of the vehicle's structural makeup even in a rollover situation. Even though it is extremely durable, it imparts a considerable amount of flexing characteristics. Consequently it is less resistant to cracking and is an undesirable choice for areas that are constantly exposed to jarring, such as the type door glass must sustain.

Tempered Glass. Tempered safety glass is commonly used for the backlite or the rear vehicle window, nearly all the adhesive bonded side glass, and in the doors and other movable glass on most motor vehicles. When it is manufactured, tempered safety glass is heated to a very high temperature and then cooled very rapidly. This tempering process forms a very hard outer shell, making the glass as much as ten times stronger than other standard glass types. The outer shell is extremely durable against outside forces, blunt objects, and the repeated jarring door glass must be able to sustain. The most significant shortcomings of the glass occur when the outer layer of the glass is penetrated, such as by a stone or some another sharp object. The glass will literally disintegrate, thus leaving the area it is intended to protect completely compromised or exposed. The repairability of the glass is limited to removing very small or light scratches and scuff marks.

Glass Coloration

The glass used in automobiles is available in a variety of tints or colors—depending on the vehicle's type and style—and may be determined by the vehicle's interior and exterior color. Tinted glass is commonly used in all the windows for aesthetic purposes and to reduce eye strain for the driver and passenger by reducing the ultraviolet ray exposure from the sun. All vehicles equipped with OEM air conditioning have tinted glass as well. The shades or tints that are available include green, bronze, blue, and gray. A clear or non-tinted glass is also available. When ordering a replacement glass it is necessary to determine the correct coloration or tinting of the glass. This is best accomplished by holding a piece of white paper on the inside of the glass and viewing it from the opposite side. Privacy glass, which is considerably darker and less transparent, is commonly used in SUVs, vans, and other special applications. The privacy glass can only be used on the windows behind the windshield and the front door windows, as it may reduce the driver's visibility and ability to see into the vehicle from the outside.

CAUTION

DISARM THE AIR BAGS PRIOR TO ATTEMPTING TO SERVICE ANY INTERIOR PART OF THE DOOR. With the increased popularity of side air bags and curtains, one must use extreme caution when attempting to service any of the door and window mechanisms. The air bag sensors are frequently located inside the doors, and an innocent bumping of a wire could cause them to deploy on the spot, or perhaps prevent them from deploying in a future accident.

Door Glass

The door glass is designed to move up and down in the vast majority of vehicles. The only known exceptions to this are the windows on the sliding and swing doors on most vans. This allows the driver and occupants to open and close the windows as needed to suit their particular ventilation needs. The vehicle model and the door style usually dictate the style of door glass and the regulating mechanism used for this purpose.

Sedan and Hardtop Design

The doors on present day vehicles are designed with either a frame around the top that encloses the glass or they are frameless. See **Figure 12-30**. The framed design encompasses the entire edge of the glass even when the door is opened. For many years a vehicle using this design was called a sedan model. A frameless design leaves the window completely exposed when it is rolled up and the door is opened. This design is commonly referred to as a hardtop model vehicle. The latter design typically incorporates a number of adjustments into the regulating assembly to ensure the glass fits properly into the opening. The sedan model vehicle typically offers few adjustments as the glass is forced to move up and down within the framework at the top of the door. See **Figure 12-31**.

FIGURE 12-30 Some vehicles have a frame around the edge of the glass while others, such as hardtop models and convertibles, leave the edge exposed, giving the vehicle a more sporty look.

Regulator Mechanisms

The glass is attached to a regulator inside the door which is used to raise and lower the window. The regulators are either manually operated with a hand crank or they may be powered electrically. The three most common **window regulator** designs used on today's automobiles, SUVs, and trucks are the twin-arm scissoring design, a cable and pulley arrangement, and a plastic ribbon or strap design.

Twin-Arm Scissor System. On the twin-arm scissor system, the primary arm—which is attached to the regulator gearing system—is the drive arm that raises and lowers the glass while a secondary arm balances and holds the

FIGURE 12-31 The principal parts of the window mechanism are the regulator, primary and secondary lift arms, lower window/glass sash front and rear guides, and the inner panel cam.

FIGURE 12-32 The twin-arm scissor system raises and lowers the glass using the primary arm as the drive mechanism and the secondary arm to balance the window and keep it level. The inner panel cam is used to adjust the pitch of the glass (raising or lowering it) at the front.

glass level as it moves up and down in the tracks. See **Figure 12-32**. The scissoring arm arrangement has been used successfully for many years and is still the most popular choice for vehicles where a larger than normal size door glass is used. The twin-arm arrangement is the most effective of all the designs to balance and bear the weight of larger and heavier glass. Unlike the other designs, this regulator system makes some allowances for adjusting the glass and the guides and tracks to compensate for wearing within the mechanism.

FIGURE 12-33 The cable style regulator uses a cable routed through a tubular enclosure to pull the window up or down depending on the direction the drive motor or the handle are being moved.

Cable and Pulley System. The cable and pulley system works on a principle of actually pulling the sash to which the glass is attached up and down, depending on the direction in which the regulator is actuated by the handle or the motor. See **Figure 12-33**. The cable is routed along a specific path through tubular enclosure and wrapped several times around the drive wheel. When the regulator is turned in one direction the cable pulls the sash and glass upward; when turned in the opposite direction, it pulls the glass down. The glass is made to move within the confines of the guides located at the front and rear of the door, leaving little or no room to adjust the glass when it becomes misaligned.

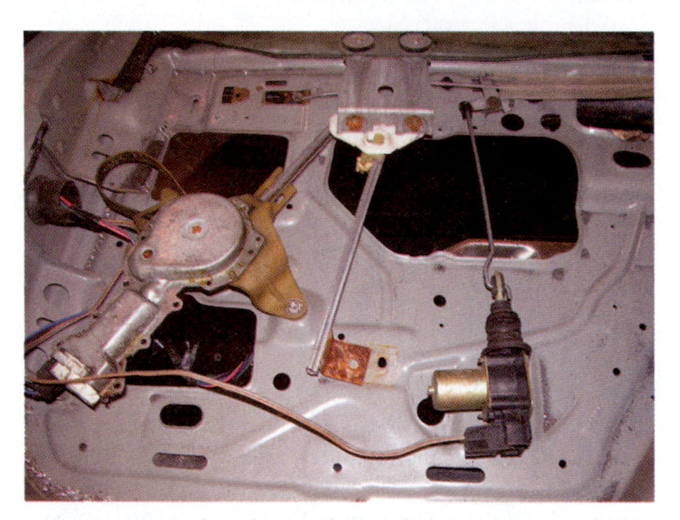

FIGURE 12-34 The plastic ribbon design pushes or pulls a semi-rigid plastic ribbon inside an enclosed track guide to move the window up or down.

Gear-Driven System. The gear-driven plastic ribbon design operates by a drive wheel with lugs that mesh with a heavy duty plastic ribbon that is routed inside a guide track, much like the cable and pulley arrangement. See **Figure 12-34**. When the regulator is activated the drive wheel turns, causing the meshing with the plastic ribbon which raises or lowers the window.

The glass is indirectly connected to the regulator in a variety of ways. The glass is first fastened to the lower sash rail with either adhesives, bolts and nuts, rivets, or plastic retainer discs and clips. The sash rail or channel is then connected to the regulator arms or cables using specially designed connecting hardware for this purpose. As the regulator gear or drive mechanism is activated, it moves the arms, cable pulley, or plastic ribbon, causing the window to move up or down as desired. The recommended procedures for removing and installing the window and regulator from the door assembly can be found in the manufacturer's repair service manual.

Hinged and Mechanized Side Windows. Many vehicles are equipped with movable side glass systems such as the quarter windows. The most common style of movable glass is designed with a hinge on the front of it, allowing the glass to be opened by pivoting it out at the rear. These are usually found on the rear quarter panel of some two-door models, minivans, and extended SUVs. They can be manually operated or operated by an electric-driven motor with the controls usually located somewhere on the dashboard. When opened while the vehicle is in motion, they create somewhat of a vacuum, thus promoting a positive airflow through the cabin. They are not intended to be opened out fully and their outward movement is usually restricted by a check strap on the manually operated type and by limiting the amount of movement allowed in the motor on the electric opening and closing motor.

SLIDING GLASS WINDOWS

Sliding window assemblies are very popular on the rear of pickup trucks. The main framework is attached to the vehicle by permanently bonding it with adhesives or it may be set in a removable gasket They are designed with a fixed glass on each side and a sliding glass in the middle that moves in either direction, allowing the window to be opened.

Summary

- The body of an automobile is constructed of two types of panels: stationary and adjustable.
- The stationary panels are usually considered structural parts responsible for holding the vehicle together and to protect the occupants in the event of an accident.
- In order to be able to get into the vehicle, it is necessary to incorporate the movable panels such as the doors, hood, and deck lid.
- In order to make panels movable, it becomes necessary to provide adjustments which allow one to shift and move them to fit into their respective openings and yet seal out the outside elements.
- Without adjustments the vehicle's outward appearance and functionality would be marginal at best.

Key Terms

air dam
ambient temperature
axis point
box lifter
cage nut/U-clip
coarse adjusting a door
composite outer panel
deck lid and hatch
deck lid tension
detail adjusting/fine tuning a door
edging/edge painting
electrical plunger
facebar
floating hinge plate
headlight covers/bezels
hinge pillar
hood flutter
latch mechanism
liftgate
negative flushness
outer skin
pitch of the deck lid
positive flushness
shims
snugging the bolts
striker
thermal expansion and contraction
wind leaks and turbulence noises
window regulator

Review

ASE Review Questions

1. Technician A says the air bag sensors for the side curtains are located at the front of the vehicle next to the front air bag sensors. Technician B says the air bags should be disabled before attempting to service the door if the vehicle is equipped with side air bags. Who is correct?
   A. Technician A only
   B. Technician B only
   C. Both Technicians A and B
   D. Neither Technician A nor B
2. Technician A says manufacturers use component panels to make access to mechanical parts easier. Technician B says bolts, nuts, and clips are the most common methods used to secure component parts. Who is correct?
   A. Technician A only
   B. Technician B only
   C. Both Technicians A and B
   D. Neither Technician A nor B
3. Technician A says some fasteners are one-time use and must be replaced when a part is replaced. Technician B says composite panels may require a different style hardware for securing them than does steel. Who is correct?
   A. Technician A only
   B. Technician B only
   C. Both Technicians A and B
   D. Neither Technician A nor B
4. Technician A says adjusting front sheet metal is more restrictive on unibody than on BOF vehicles. Technician B says one may X-check the front end to determine if it is out of proper alignment. Who is correct?
   A. Technician A only
   B. Technician B only
   C. Both Technicians A and B
   D. Neither Technician A nor B
5. Technician A says the hood has two catches, the primary and secondary for emergencies. Technician B says the hood latch should be the first thing to attach when adjusting the hood. Who is correct?
   A. Technician A only
   B. Technician B only
   C. Both Technicians A and B
   D. Neither Technician A nor B
6. Technician A says adjusting doors should be done by loosening all bolts on both hinges and moving the door. Technician B says the fore and aft adjustments are made on the hinge to pillar location. Who is correct?
   A. Technician A only
   B. Technician B only
   C. Both Technicians A and B
   D. Neither Technician A nor B
7. Technician A says the hatch on a hatchback model vehicle is attached to the rear of the roof with hinges. Technician B says LSG is typically used for windshields. Who is correct?
   A. Technician A only
   B. Technician B only
   C. Both Technicians A and B
   D. Neither Technician A nor B
8. Technician A says one must torque the fasteners in a specific sequence on composite panels. Technician B says electrical plungers are sometimes used in place of hard electric wires. Who is correct?
   A. Technician A only
   B. Technician B only
   C. Both Technicians A and B
   D. Neither Technician A nor B
9. Technician A says the bumper usually has either a reinforcement or an energy absorber. Technician B says the reinforcement is made of low carbon to bend so it can absorb some of the impact forces. Who is correct?
   A. Technician A only
   B. Technician B only
   C. Both Technicians A and B
   D. Neither Technician A nor B
10. Technician A says one should start to torque the fenders down first at the front and work back toward the door. Technician B says a positive flushness may be required between the rear of the front fender and the door. Who is correct?
    A. Technician A only
    B. Technician B only
    C. Both Technicians A and B
    D. Neither Technician A nor B
11. Which of the following statements is correct?
    A. On a body over frame vehicle the front panels are attached to the frame via an adjustable radiator core support.
    B. The front fenders should be shimmed to raise them up to the correct height to compensate for insufficient structural repair.
    C. The hood should be installed last when replacing the entire front sheet metal assembly.
    D. The rear glass is nearly always attached to the rear hatch with hinges on a hatchback model.
12. Technician A says tempered glass is used on nearly all side windows of a car. Technician B says the

rubber bumpers on top of the core support are used to eliminate the flutter on the hood. Who is correct?
A. Technician A only
B. Technician B only
C. Both Technicians A and B
D. Neither Technician A nor B

13. Technician A says an egg crate design, a high density foam, and a collapsible mechanical piston are all designs used for energy absorbers. Technician B says thermal expansion occurs on all exterior panels. Who is correct?
A. Technician A only
B. Technician B only
C. Both Technicians A and B
D. Neither Technician A nor B

14. Technician A says the hinges on some vehicles are welded to both the door and the hinge pillar. Technician B says even though they are welded to the body, the door hinges can still be adjusted. Who is correct?
A. Technician A only
B. Technician B only
C. Both Technicians A and B
D. Neither Technician A nor B

15. Technician A says the radiator support is attached to the frame rail with a large bushing or isolator between the two. Technician B says the radiator support can be adjusted 1/2 inch in either direction to accommodate properly aligning the panels. Who is correct?
A. Technician A only
B. Technician B only
C. Both Technicians A and B
D. Neither Technician A nor B

Essay Questions

1. Discuss the sequence that should be followed when permanently attaching the fender to the vehicle.
2. Discuss the sequence that should be followed when installing all the panels in the front of the vehicle and discuss why this should be done.
3. Discuss the three different adjustments that can be made when adjusting the door on the hinges and what part of the hinge is used to affect each adjustment.
4. What is meant by a positive flushness and a negative flushness adjustment? Where and why would one use this method?
5. Discuss why one must consider the ambient temperature when installing a composite panel.

Topic Related Math Questions

1. A normal gap between two composite panels is 1/4 inch when they are installed at normal room temperature. When installing the part below that temperature, an additional 1/16th inch gap should be left between the panels for every five degrees. The shop temperature is 60°F when installing them. How much gap should be left between the panels?
2. By substituting a plastic fascia for a steel bumper assembly, the manufacturer can reduce the overall vehicle weight by 30 percent. The steel bumper, brackets, and reinforcement weigh 110 pounds. (A) What is the weight of the fascia and mounting attachments? (B) How many pounds less did the fascia assembly weigh than the steel bumper assembly?

Critical Thinking Questions

1. Prior to the vehicle leaving the shop, all the door to fender gaps were correct. However, after driving the vehicle for a period of time, one of the doors rubs the rear edge of the fender whenever it is opened. What may be the cause of the problem?
2. Today's automobiles are constructed using a wide variety of materials. What are the potential problems and consequences if the repair shop doesn't use the correct hardware when reinstalling either old or new parts?

Lab Activities

1. Locate a vehicle with an SMC door skin and note the method for securing it and the provisions made for placement and location on the door's inner frame when permanently attaching it.
2. On a BOF vehicle loosen the bolts holding the fender in place at the front and along the length of it, leaving the bolts only snug at the top and bottom of the fender. Observe the gap between the fender and the door and along the length of the fender when it is shifted inboard and outward.
3. Locate a misaligned door and determine the cause of misalignment, then—using the shifting axis point technique—realign the door to properly fit into the opening.

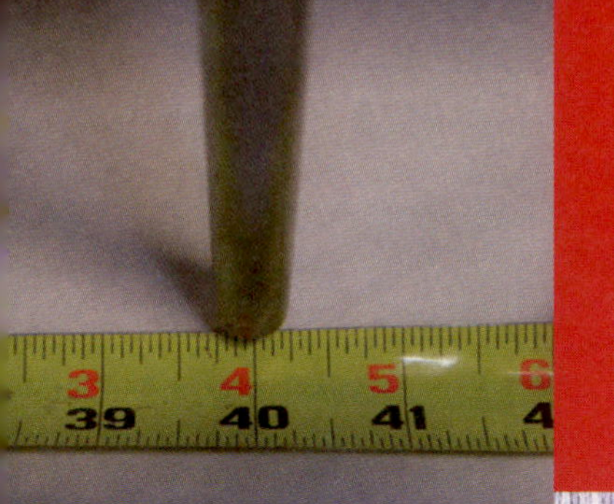

Chapter 13

Measuring Structural Damage

OBJECTIVES

Upon completing this chapter, you should be able to:

- Apply personal and vehicular safety measures.
- Identify and describe the use of manual measuring equipment such as the tram bar, centering gauges, and other measuring devices.
- Discuss the unibody's three section design and its measuring implications.
- Identify, define, and discuss the significance of the zero point, centerline, and datum plane.
- Identify and discuss the three-dimensional measuring systems.

INTRODUCTION

The essence of an accurate and expedient repair is frequent and precise damage analysis. Today's collision-damaged vehicle must be restored to within closer tolerances than has ever been required before. In order to ensure these tolerances are maintained, a finite measuring system is required to first identify the damage and then—with the use of pulling and straightening equipment—apply the precise amount of corrective forces at the exact angle necessary to restore the vehicle to its pre-collision condition. All unibody vehicles must be straightened to within a maximum allowable tolerance of 3 mm; on some select models, they must be restored to precisely the same dimensions used by the manufacturer. This requires a combination of a very precise measuring system and a highly skilled technician to operate the systems.

Although measuring for damage analysis does not seem like a dangerous task, numerous opportunities exist for accidents to occur. Many of the steps require that the individual get under the vehicle to inspect the damage, locate target-hanging locations, and perform various other tasks that could put one at risk. One must be certain the vehicle's wheels are properly chocked to ensure against it rolling off the frame rack or causing any injury. The vehicle should be placed in park if it has an automatic transmission, or the transmission should be put into reverse if it is a standard shift. One must be cautious about using the emergency brake—if it is not used routinely by the owner, corrosion will have formed on the cables and other moving parts, making it apt to lock up when used. In that case, additional work must be performed to free the brakes before the vehicle can be moved.

SECURING THE VEHICLE

Many of the operations performed during this phase of the repair are tasks the trainee has not done before. As a result, the trainee will not be familiar with the equipment being used, which creates additional opportunities for personal injuries to occur. Most equipment used to measure the undercarriage and perform the repairs requires the vehicle to be lifted off the ground and mounted onto the holding clamps to secure it in place.

CAUTION

Technicians should NEVER place their hands between the pinchweld of the vehicle and the pinchweld clamps.

Care must be taken when securing the vehicle into the clamps that hands and fingers are not placed between the vehicle pinchwelds and the holding devices. See **Figure 13-1**.

If the vehicle inadvertently slips on the jack and drops, it could easily pinch one's hands between the holding clamp and the pinchweld, resulting in not only serious personal injury but damage to the vehicle as well. The area of the vehicle used to clamp the jack into place should be inspected to ensure it is structurally sound and free of any signs of corrosion or rust perforation. Once clamped into place, the entire weight of the vehicle will be resting on this area. Therefore, any evidence of corrosion—even if it isn't perforated—should be cause for concern as the rocker panels may collapse from the weight. This could cause serious damage and require significant unplanned repairs.

Before using any equipment one must become sufficiently familiar with it in order to anticipate and avoid any potential safety issues. Each equipment manufacturer has their own unique requirements regarding the safe use of their equipment. It is paramount those safety requirements are met and practiced. Ignoring the manufacturer's recommendations will inevitably lead to disastrous results. Safety glasses are always a must, even when performing the most mundane tasks. As one gets under the vehicle, debris, dirt, and rocks that may have been packed into the pockets during the collision may become dislodged and get into the eyes. Furthermore, immediately after installing the measuring system and performing the initial analysis, the repairs are initiated. At that point, the appropriate safety precautions must be taken.

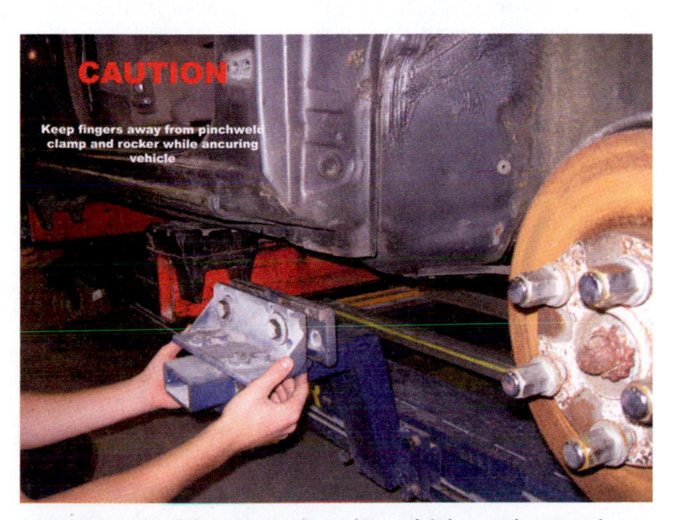

FIGURE 13-1 When securing the vehicle to the repair bench, do not place hands and fingers between the vehicle pinchweld and the holding clamps.

MAKING A REPAIR PLAN

One might ask, "Why even measure?" or "Why is measuring so important?" There are numerous reasons for measuring the vehicle. The essence of any repair—regardless of how severe or minor the damage may be to the structural system, the undercarriage, and other areas of the vehicle—is to locate, identify, and confirm the exact amount of damage. Identifying the primary damage is usually very easily done. These are the areas where the exterior panels are obviously damaged as a result of coming into contact with the impact object. However, in the vast majority of cases, the damage radiates well into the surrounding areas, frequently with no readily visible signs of its presence. Many times the damage migrates into areas that are covered with trim or other overlapping parts. This damage is frequently not evident unless the parts are removed or the area is measured and checked for damage. While these damage conditions are often very subtle, they must become a part of the overall repair plan which ultimately is formulated from the results of the measuring operations. Usually the further the damage radiates out, the less severe it becomes in the outer fringes at least. As the repair plan is formulated, these are the areas that must be corrected first. The sequence of the repairs typically affected is from the outer fringes—where the damage is least severe—to the center or point of impact, where it is most severe.

ASSESSING COLLISION DAMAGE

Primary and Secondary Damage

In collision repair, whenever damage occurs to a vehicle, it is identified as being either primary (direct) damage or secondary (indirect) damage.

Primary Damage. When assessing the damage, the first area that one should evaluate is the primary or direct damage. This damage is sustained by the vehicle because it came into direct contact with another object or vehicle. For this reason, it is also known as the **point-of-impact (POI)** damage. POI damage is any type of damage that affects the vehicle—whether it is structural or merely cosmetic—at or near the point of impact. Depending on the severity of the impact, this area of the damage may manifest itself as very sharply creased or folded and bent metal damage that may require cutting away to remove it. However, removing it may expose additional damage to the area underneath that may require repair or replacement as well.

Secondary Damage. Secondary or indirect damage occurs away from the point of impact. Any type of structural damage or form of misalignment—including even cosmetic damage anywhere on the vehicle other than the POI—is considered secondary damage.

MISALIGNMENT OF VEHICLE SECTIONS

A very good example of secondary damage occurs when a stationary vehicle is struck from behind. The vehicle may show an uneven door to quarter panel gap, a downward buckle in the roof, misaligned doors, and possibly a larger than normal fender to door gap at the bottom of the front fender. This damage is not caused by direct contact with the other vehicle or any of the other forces at the point of impact. It is in fact caused by the misalignment of vehicle sections. A typical unibody is built in three interconnecting sections that make up the whole body or vehicle structure. While they are connected and joined together to create the vehicle's structure, they function and react individually in a collision. When assessing the damage, one should think of how the collision forces affected not only the immediate impact area, but how the damage may have traveled or radiated from one section to the next, throughout the vehicle, and the overall damage that may have occurred as a result.

COMMON COLLISION DAMAGE AREAS AND CONDITIONS

Four distinctly different types of collision damage conditions are directly dependent upon the area of the vehicle that was struck or impacted. They are a front end collision, a rear end collision, a side hit (on either side), and finally a vehicle rollover, the latter of which is usually the most difficult to repair.

In a typical rollover the damage is often sustained from several different directions as the forces push or move the metal more than once as the vehicle rolls first onto its side and then onto its top. A rollover is essentially a combination of all three of the previously noted damaged areas thrown into one.

Walk-Around Vehicle Inspection

Before making any attempts to diagnose the damage by measuring any part of the vehicle, one can obtain a considerable amount of information about the type and location of the damage simply by walking around the vehicle and making several very quick and simple observations. Studying and analyzing the damaged panels that came into direct contact with the impact object can be very revealing as to what happened in the initial phases of the collision. The initial point of impact, the angle or direction of movement by the vehicle, and the impact object can frequently be determined by the effects and impressions left on them. At this point one can re-create an imaginary or visual replay of what likely happened as the impact forces continued to move deeper and deeper into the vehicle and affected the structural members within. Once beyond the direct damage areas and the obvious mangled and distorted panels, the next thing one must observe are the gaps between the panels. The technician might consider these questions:

- Are the gaps even or are they larger at the top than at the bottom?
- Are the doors still properly aligned in the opening or is part of one or more overlapping an adjacent panel?
- Are the doors still movable without any excessive pressure or do they bind when someone tries to open and close them?
- Are the gaps between the deck lid and the upper rear panel and the quarter panels even? If not, is there a consistent pattern of misalignment between these panels that might be a tip as to what may have been shoved or moved by the effects of inertia?

One should also check the gaps between the hood and the cowl ventilator panel to determine if there is any damage.

The next area to check is the wheels and their position in the wheel opening areas. Is there any sign of damage to the wheels that might be a tipoff of possible suspension damage? Is there any debris stuck between the tire and the wheel that would be a sign that the vehicle slid sideways, possibly bumping a curb and causing some suspension damage that would otherwise not be noticed? One should also check for any scrapes or scrub marks in the wheel housing and on the fender liner that would be an indication of possible wheel setback or other steering angle suspension parts damage. While checking in these areas one should also check to see if there is any evidence of cracked caulking and seams or pulled spot welds. These are all very visible signs of damage that may have occurred and should serve as a reminder to check all critical suspension parts as these are all signs of possible damage.

Front End Damage

The front end collision usually involves more individual items to consider than any other part of the vehicle—not because the damage is more severe, but because of the many mechanical parts and vehicle functions that are housed in this area. To further complicate matters, most of the exterior panels on the front are bolted into place, sometimes making locating and identifying the damage to the inner reinforcements and structural members more difficult.

Damage Transmission. Many times after a vehicle collides head-on with another vehicle or some type of barrier or obstacle, the damage will be transmitted throughout the entire vehicle. This is because the front of the vehicle will have been stopped by the object it collided with. However, due to the momentum, the rear end of the vehicle will continue traveling forward a short time, causing damage throughout the entire vehicle.

Suspension, Drivetrain, and Exhaust. Front end damage involves not only the upper body, but includes areas underneath the vehicle such as the driveshaft, exhaust, and even hard parts such as the suspension and drivetrain components. Depending on the speed of travel at the time of the collision, the rear end may even become misaligned upward. This is because the natural tendency of the rear of the vehicle is to rise up because of the forward momentum. Since the rear of the car is much lighter weight, a kickup reaction is a natural phenomenon. The front section of nearly all modern unibody vehicles is designed to dissipate energy by buckling, collapsing, and folding up to protect the vehicle's occupants. In the process of redirecting the collision energy, damage to parts that one wouldn't necessarily associate with the collision is created because of the collapsing and buckling effect that routinely occurs.

Therefore, when measuring the vehicle one must be intent to check and evaluate every possible area—such as the rails, reinforcements, and structural members—that could possibly have been affected. One must expect the unexpected when measuring this part of the vehicle. A learned technician will always measure the entire vehicle to not only assess the obvious damage but to locate the subtle and inconspicuous damage as well.

Rear End Damage

A rear end collision frequently causes the back of the vehicle to collapse and move forward somewhat. However, because the bulk of the vehicle's weight (approximately 60 percent) is in the front, that area tends to resist the force of the collision. That, however, does not mean that the front and center sections are exempt from the effects of the collision energy forces.

Localized Damage. Like the front of the vehicle, the rear is also designed to fold up and collapse in a collision, thus dissipating some of the effects of the collision energy. Because the rear of the vehicle is not reinforced as heavily as the front end, it will typically sustain more localized damage than the front when subjected to a comparable impact force. The damage usually looks more severe, but it will usually not radiate into the vehicle quite as deeply as it would in a front end collision. So even though the damage looks more serious or severe, it is usually limited to the deck lid, lower rear panel, quarter panels, and the trunk floor section. In a more severe rear end collision, the lower structure of the center section can be pushed forward and the back of the vehicle may be pushed so hard that it pushes upward or downward, with the possibility of the roof buckling and the door openings becoming misaligned. Once again, the technician should visually inspect and measure the entire vehicle for damage. Do not be fooled by focusing only on the damaged section.

Side Impact

During a collision where the vehicle is impacted on the side, the center section of the vehicle being struck will move in the same direction as the external force. However, the rest of the vehicle being struck, from approximately the beltline and above, and the area on either side of the direct point of impact, will not be affected to the same degree as the immediate impact area. In fact, this area will have a tendency to resist the forces of the striking object and seemingly even move in the opposite direction of the force being exerted against the adjoining areas by the impact object. As the center section of the vehicle being struck moves in the direction of the impact object, causing that area to collapse, a shortening effect of the overall length will occur as the center section (of the struck vehicle) continues to collapse from the collision force. When making length measurements on a vehicle experiencing this type of impact, the side that was struck will be significantly shorter in length than the opposite side. The vehicle assumes a bowed effect along the entire length of the body that causes it to be shorter. The upper part of the body may likely display a significant leaning effect in the opposite direction of the impact object's movement. This effect is caused by inertia and the effects of the inertial forces.

Rollover

A rollover can be described as a series of several rapid succession crashes from several different directions. Each time the vehicle makes one complete revolution or rolls 360 degrees it has been exposed to a minimum of at least four different crashes from at least four different directions. This is further compounded by forces thrust upon the structural members that push them in the opposite direction each time a different surface or side hits the ground as the vehicle rolls until it comes to a rest. There are numerous points of impact in a rollover, and attempting to isolate or trace the damage flow is very difficult. Even on a "soft" rollover, in which the vehicle may have rolled while moving at a slow speed and perhaps didn't make a complete revolution, predicting the amount of damage sustained will be equally difficult even for the most learned technician. The most common type of damage sustained in a rollover will result in the upper body being pushed to one side, making the vehicle look lopsided or out of square. One must be keenly aware of the inevitable misalignment of the lower body and undercarriage, which will likely affect the vehicle's structural integrity.

INERTIA

Newton's first law of motion is often paraphrased in the following way: An object at rest tends to stay at rest and an object in motion tends to stay in motion with the same speed and in the same direction *unless acted upon by an external force*. The dynamics of inertia are comprised of two elements: the behavior of stationary objects and the prediction of objects that are in motion. Any moving vehicle involved in a collision is an excellent example of the effects of inertia and why technicians must be concerned with its effects when they attempt to repair a vehicle that has been damaged. The study of a vehicle involved in a rollover accident is an even more classic example of how the effects of inertia impact the measuring and repair procedure. Imagine a vehicle becoming airborne and coming back down at a slight angle in such a way that the A pillar is the first area to come into contact with the ground. That pillar has contacted the ground and is no longer moving downward other than to bend, buckle, and collapse. At this same exact moment the ground is actually pushing up against the pillar. The rest of the vehicle has not made contact with the ground yet and so it is still moving downward toward the road surface. At this precise moment we have two opposing forces on the verge of acting upon each other because one pillar on the corner of the vehicle is actually moving up while the other three pillars are still moving down. The opposing forces that are about to interact will likely cause some severe buckling and a twisting effect on the roof and the entire center section of the vehicle. It is important to realize that until the vehicle finally comes to rest, damages such as this will be repeated over and over, causing additional damage each time it occurs.

Opposing Forces Collide

The external forces are apt to cause each of the three sections of the vehicle to respond and move in different directions, thus totally disrupting the body dimensions. In an accident in which a forward moving vehicle strikes a barrier and stops abruptly, the same effect would be repeated. In this case the entire roof shifts forward until the pillars stop the movement, causing it to remain in that position until another force—such as the corrective forces applied to reverse the effects of it. Another example of the effects of inertia is if one were driving in a vehicle without wearing a seat belt. In the event of an abrupt stop, the driver's body reacts exactly like a loose package would behave on the rear package shelf. As the driver begins to brake the vehicle to a stop, the loose package continues to move in the same direction and at the same rate of speed as the vehicle until it strikes something to stop the forward motion. Keep in mind that *objects in motion tend to stay in motion*. When a seat belt is fastened, it will hold the person in place and protect the individual from striking objects in the passenger compartment like the steering wheel, dashboard, or the windshield.

Inertia Causes Indirect Damage

The damages caused by the inertial forces are usually said to be indirect damage because they are usually a distance away from the point of impact and do not come into contact with the impact object. The more subtle the damages, the more difficult they are to locate and identify, which further reinforces why measuring is such a critical step in the overall repair process. Any attempt to repair the direct damaged area without first addressing the indirect area will inevitably result in additional work to correct subsequent alignment problems.

CONFIRM THE CONDITION OF THE CENTER SECTION

One of the most important tasks accomplished by measuring is to determine the trueness of the vehicle's center section. Structurally and dimensionally, this is the single-most important part of the vehicle. No repairs should be attempted or even considered until the exact location and amount of the damage is identified. Making an incorrect pull on a damaged vehicle may result in creating permanent irreversible damage and require replacing a part that would otherwise have been salvageable and repairable.

While the center section is being checked for dimensional accuracy, all the dimensions for the steering and suspension attaching locations should be checked as well. Other items that should be included in the damage analysis are all of the attaching locations and the components that affect the vehicle's driveability. These areas must also be monitored constantly while the repairs are progressing. Adjustments and alignment provisions to compensate for body misalignment are rarely allowed on any of the steering, suspension, and drivetrain components and their attaching points on a unibody vehicle. Therefore, it is critically important that any misalignment in these areas be addressed as part of the repair plan and procedures.

Typically most of the attention is given to the collision damage sustained by the upper body because this is most visible and any obvious departure from the vehicle's normal shape tends to grab our attention. While one should not diminish the significance of this damage, many more important issues must be addressed on the undercarriage. It is often said that approximately 40 percent of the damage exists on the undercarriage, suspension, and the drivetrain. This is the area of the vehicle where all the repairs are initiated. Unless the undercarriage, the vehicle's foundation, is squared first and returned to its dimensional tolerances, the upper body cannot be restored to the manufacturer's specifications.

UNDERBODY DAMAGE CONDITIONS

There are essentially five different damage conditions typically sustained by the undercarriage depending on the vehicle design and the circumstances of the collision or impact causing the damage.

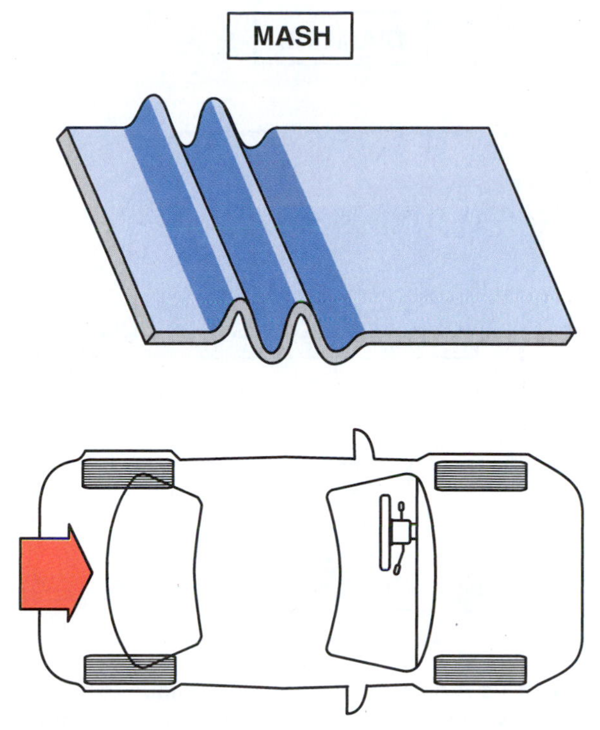

FIGURE 13-2 A mash condition usually occurs when a vehicle is involved in a direct frontal or rear impact, causing the rail to become compressed and the length to become shortened.

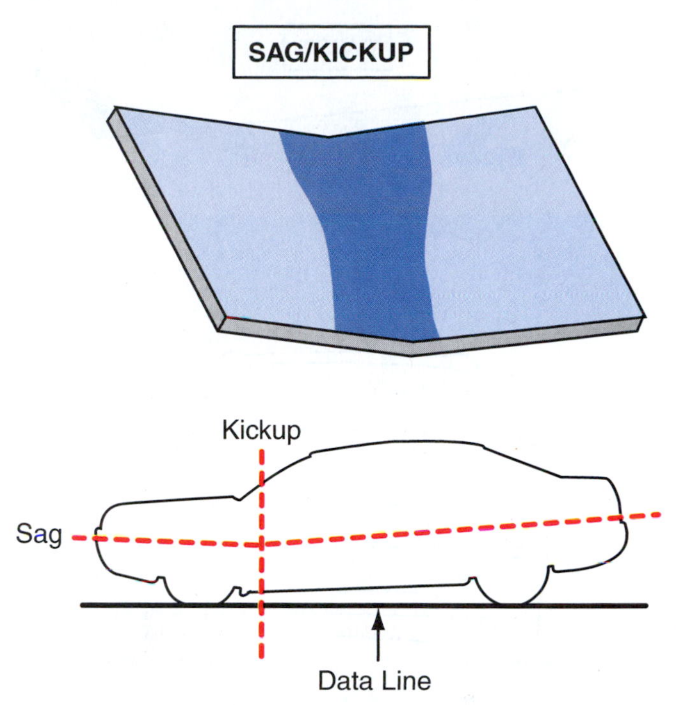

FIGURE 13-3 A sag or kickup condition usually is secondary damage caused by the rail being pushed down or kicked up, such as when a vehicle hits an object, resulting in an upward projection.

Mash

In a mash condition, one or both of the rails are shorter than the manufacturer's specifications. See **Figure 13-2**. This is usually the result of a direct hit on the end of the rail, causing it to collapse or buckle, thus reducing it in length. It is also referred to as short rail or collapse. It is apt to be found on both unibody and body over frame damaged vehicles.

Sag

In the sag condition, either the front or the rear of the vehicle height measurement may be out of the proper height specification. See **Figure 13-3**. It may affect either one or both rails on either the front or the rear of the vehicle. It gives the impression the vehicle is not level. This is very commonly found on either a damaged unibody or body over frame vehicle.

Another common condition very similar to the sag is a condition referred to as kickup. In the kickup condition, the rail end may have been struck from the bottom, causing it to be pushed upward. Both conditions give the impression that one side is higher or lower than the other. The repair procedures for the sag and kickup conditions are very similar in that the application of the corrective forces applied must merely be reversed.

Sidesway

In a sidesway condition, a section of the undercarriage is moved or shoved to one side or the other. See **Figure 13-4**.

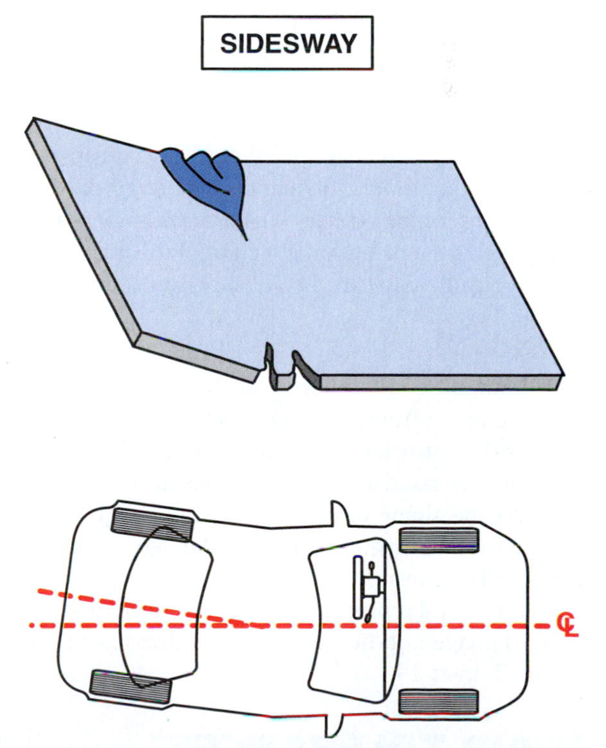

FIGURE 13-4 Because most accidental impacts occur at an angle, a sidesway is one of the most common damage conditions found on a vehicle.

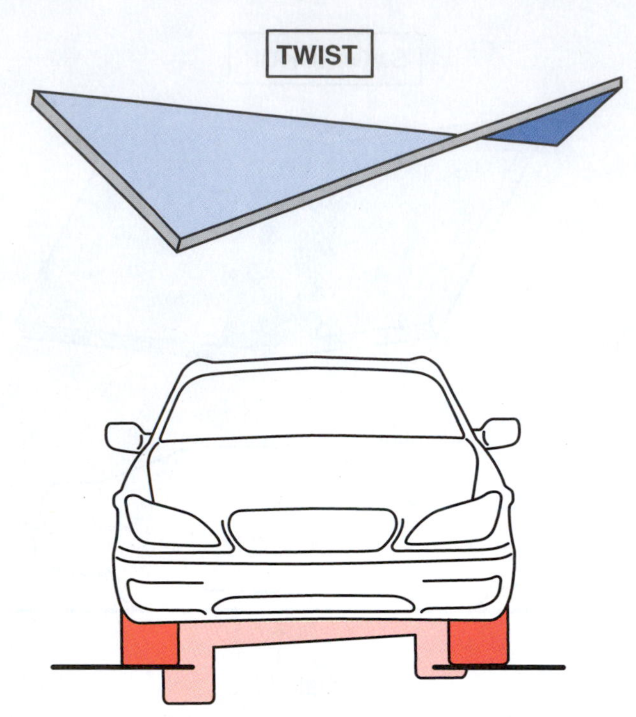

FIGURE 13-5 A twist condition is unique to a BOF vehicle and usually occurs when a vehicle becomes airborne. When it lands, the vehicle first suffers a hard impact on one corner, causing the rail to be kicked up enough to push the rear of the rail downward.

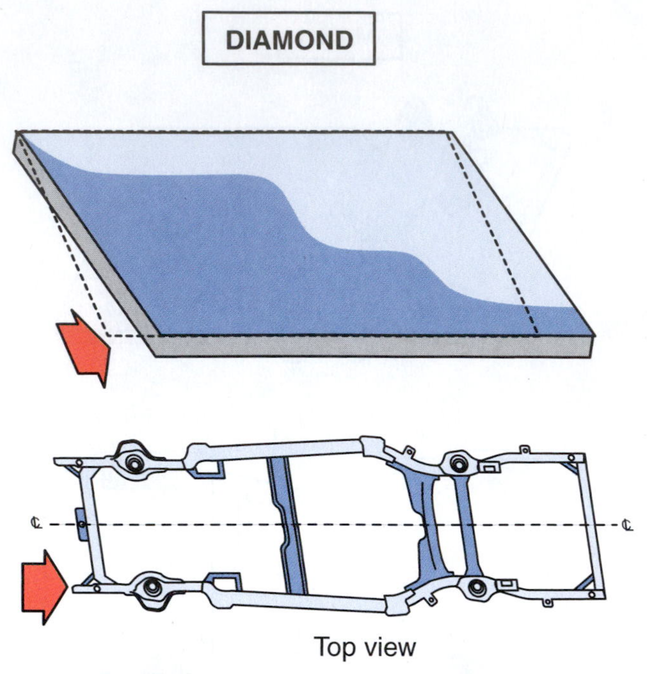

FIGURE 13-6 A diamond condition is unique to a BOF vehicle and is caused by the vehicle striking a solid impact object with one corner while moving straight forward.

It is usually found on the ends of either the front or rear rail and, depending on the nature of the impact, may move either one side or both. This may also be found on either a unibody or body over frame vehicle.

Twist

A twist is a type of damage that typically occurs in the center section of the vehicle, frequently radiating out to either the front or rear of the vehicle. See **Figure 13-5**. It is a condition that affects the height on a section of the rail. This condition is typically only found on a body over frame vehicle.

Diamond

In a diamond condition, one of the side rails has been struck hard at the front, causing it to be moved back many times over the entire length of the vehicle. See **Figure 13-6**. This is characterized by the center section being out of square and the shape of the rails actually slightly resembling a diamond shape. This type of damage is limited exclusively to body over frame vehicles.

The causes and repair methods and techniques for each of these damage conditions will be addressed in greater detail in Chapter 14.

REPAIR TOLERANCES

A collision-damaged vehicle must be restored to within very specific tolerances of the manufacturer's original specifications for several reasons. The structural components must be able to withstand the collision energy and react in exactly the same manner as originally intended in a subsequent collision. The vehicle's steering and driveability are very dependent upon the proper alignment of all the mechanisms associated with driving and directing the vehicle down the road. A steering gear that is misaligned 3 mm (1/8 inch) from its original location may seem like a very insignificant issue, but it is sufficient to cause the vehicle to steer very erratically as it is being driven. Consumers place their family's trust and lives into the hands of those who repair and correct collision-damaged vehicles. Collision repair technicians have a legal and moral obligation to restore a damaged vehicle to pre-accident factory specifications. Accurate measuring is a critical part of uncovering unseen damage remaining after the visual damage analysis, and makes properly repairing the damage easier. In this section we will investigate and identify a variety of different measuring systems commonly used by the collision repair technician to accomplish the precise measuring required to restore a collision-damaged vehicle back to its pre-collision condition.

MEASURING TECHNIQUES

Different measuring techniques using a variety of different equipment and information are used during the repair

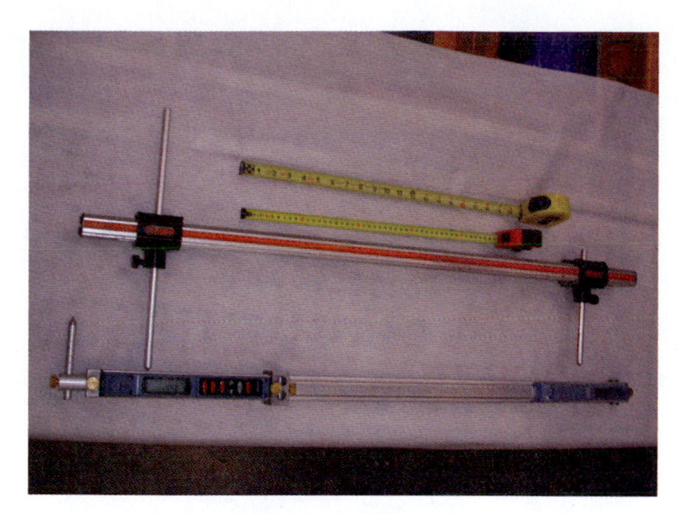

FIGURE 13-7 A number of measurements are taken during the repair process, some of which are made using a tape measure and tram bars and gauges.

FIGURE 13-8 When making point-to-point measurements, the pointers must be parallel to each other on both ends of the bar.

of collision-damaged vehicles. See **Figure 13-7**. Some of these measurements are made manually using a tram bar and tape measure, while the bulk of them are made using very elaborate equipment capable of simultaneously monitoring a number of **reference points** under the vehicle. All the measurements taken are based on (compared to) printed specifications that are developed to be used with a specific manufacturer's measuring equipment. The equipment made to take multiple measurement readings simultaneously may use a **laser** system, it may require mounting the vehicle onto a fixture system, or it may also use other measuring devices in conjunction with multiple targets attached at strategic locations on the vehicle's undercarriage. Whatever equipment is used, the outcome should result in restoring the vehicle to the precise dimensions specified by the manufacturer.

Point-to-Point Measuring

As the name implies, **point-to-point measurements** are derived by measuring the distances between two points on the vehicle. See **Figure 13-8**. The dimensions may be compared to known or fixed measurements published in specification manuals or a database used for this purpose. Alternatively, the dimensions can be taken and compared to the same location on opposite sides of the vehicle. Occasionally the data are not available from preprinted sources, in which case it may become necessary to comparatively check side-to-side measurements on the vehicle or even compare the measurements to an identical undamaged vehicle. Point-to-point measuring is an excellent tool or means to identify collision damage with a minimum effort using little more than a tape measure and a tram bar or digital tram gauges.

Tape Measure. When using the tape measure for making point-to-point measurements one must remember that the tape must run in a straight line or remain flat to achieve an accurate measurement. Any bowing or dipping will result in an inaccurate measurement because the point-to-point measurements are to be made in a straight line. For example, when checking the underhood dimensions various components may be in the way, causing the tape to bow, bend, or buckle in the middle. This will show the measurements to be longer than the actual distance being measured. The tram bar can be used to bridge over the obstacles between the two locations because the pointers can be adjusted so the bar will level, thus maintaining a straight line between the two reference points.

Checking for Squareness. Point-to-point measurements can be used to check the squareness of virtually any opening or area of the vehicle. These measurements can be done very quickly and easily by comparing the dimensions of the vehicle opening to those in print, or they may be made by comparing the dimensions from the damaged side of the vehicle to the undamaged side. See **Figure 13-9**. Seams, welds, defined edges such as the corner of a sharply formed crown, and hinge and striker mounting locations can all be used as measuring reference points. It is not necessary for a reference point to be included in the published dimensions. The only criteria required is that the point be identical on both sides of the vehicle.

Measuring Scales Used. Even though the majority of the printed dimensions are listed in both **metric and inch increments**, the student will find that—with a little effort—the metric dimensions are easier to use. See **Figure 13-10**. Tape measures with both scales printed on them are available, but it may be advisable to obtain one that has only the metric scale printed on it as it is easier to read. Since the future of the industry points to the continued and expanded use of the metric system, one is well advised to become better versed in the metric system by using it for all applications.

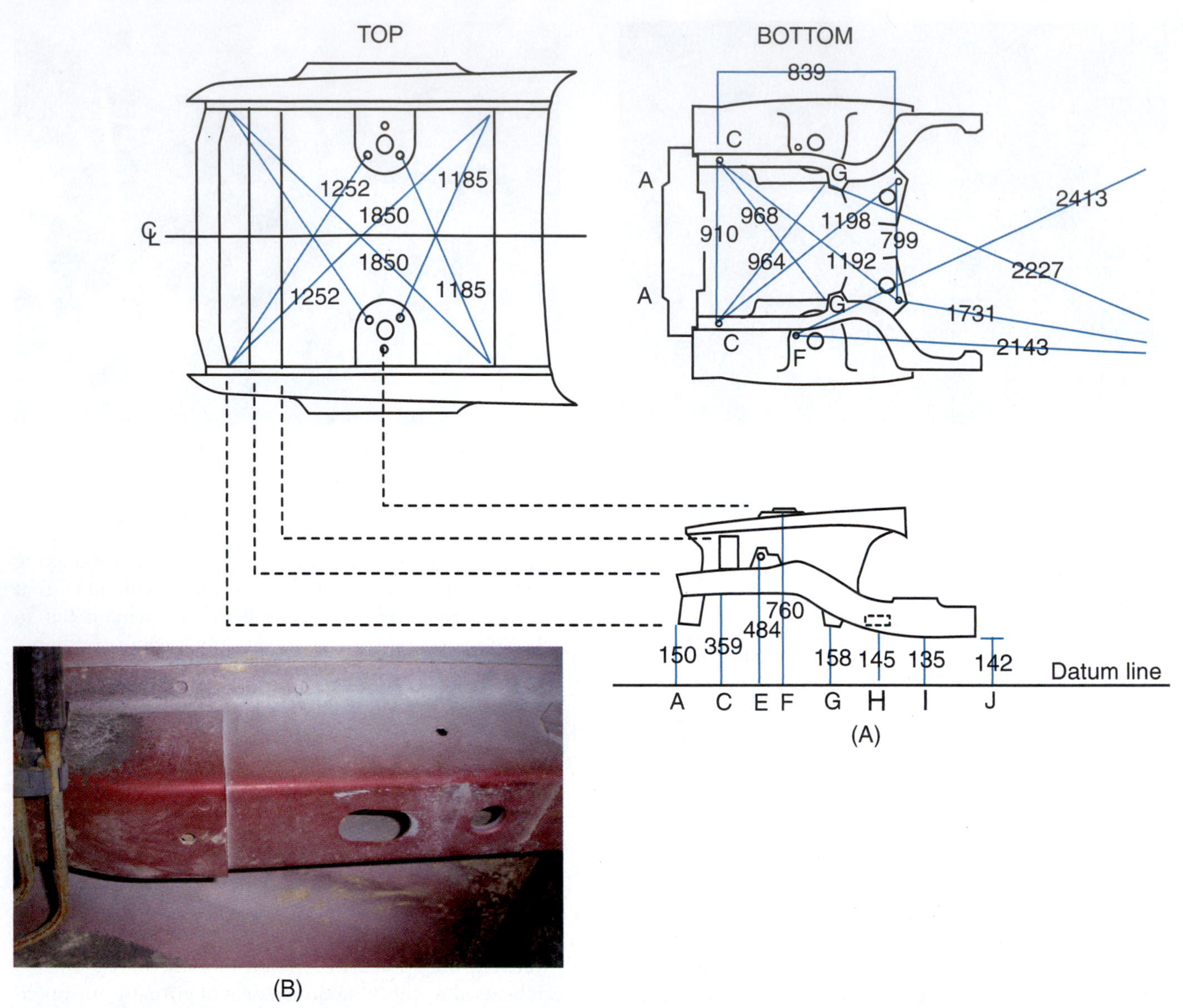

FIGURE 13-9 In (A), comparative measurements are made using reference points that are not listed in the specification manual. In (B), surfaces such as reinforcement overplating, seams, and jig holes are good reference points to use.

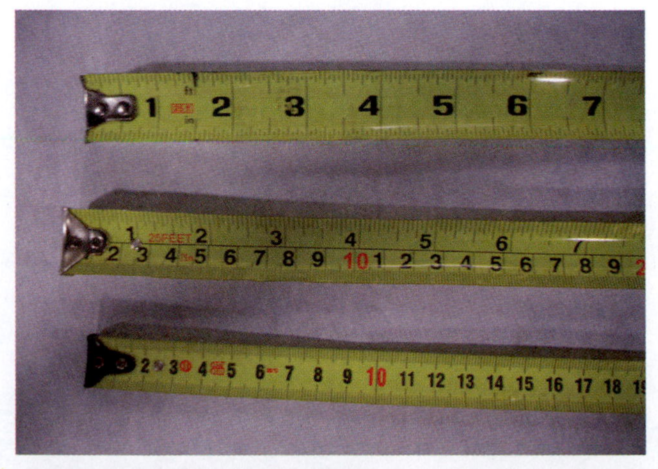

FIGURE 13-10 The dimensions given in the dimensional specification manual are given using both the metric and U.S. standard SAE measuring systems.

Regardless of the source of printed dimensions used, some common applications and rules apply with regard to measuring points. Whenever possible, one should use the reference points used by the publishers for measuring purposes as they are identified and recognized for dimensional accuracy. See Figure 13-9. The technician should be aware that most measurement reference points are bolt heads or nuts, fixed holes or slots, and occasionally the edge of a welded section that has a surface common to both sides. It is assumed that whenever using any bolts, holes, or other fasteners as reference points, one should measure to the center of each unless otherwise specified by the publisher.

Occasionally it becomes necessary or desirable to measure to reference points that are not included in the published manuals. In other words, that one elusive dimension needed is simply not included in the specification manual. In this instance a comparative measurement must be taken

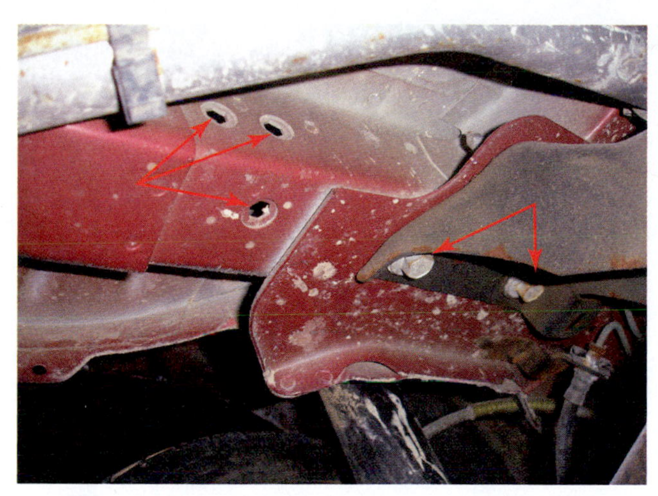

FIGURE 13-11 Bolts and fasteners, holes and slots are common elements used as reference points from which measurements are taken.

comparing the dimension of the damaged side to the undamaged side of the vehicle. While the reference points one can use are not identified in the manual, one can quickly and easily find reference points from which these measurements can be taken. See **Figure 13-11**. The automobile manufacturer uses **net building** as a tool to properly align the sheet metal parts to be welded together. These are common to both sides and can easily be used for a reference point when measuring. See **Figure 13-12**. Other locations—such as seam joints and overlapping areas—also offer a relatively high degree of accuracy for measuring reference points. The most critical factor to ensure accuracy is measuring to the precise same location on both sides of the vehicle.

FIGURE 13-12 Net building and lancing locations and seams and overlapping areas can be used as measuring reference points because they are usually common to both sides of the vehicle.

Tram Bar and Tram Gauge

A tool that is frequently used for point-to-point measuring is a tram gauge or tram bar. See **Figure 13-13**.

Tram Bar. A **tram bar** can be an unspecified length of continuous tube or a telescoping measuring device with adjustable pointers on each end. A tram bar can be a relatively accurate tool to use for making point-to-point measurements. One need only have a tape measure to check and confirm the distances between the pointers to create relatively accurate references. Many of the older style tram bars require the use of a tape measure to determine the distance between the two pointers. Manual tram bars may also be equipped with a sliding tape measuring scale, as the movable pointers are extended for measuring and the pointers may also be marked with incremental markings.

Digital Tram Gauge. A more modern version of the tram bar is the **digital tram gauge**. It is capable of measuring as each of the telescoping sliding bars are extended. They may extend from as short a distance as 300 mm to in excess of 2,500 mm (12 inches to in excess of 10 feet). Some may have a built-in ruled scale that shows the length of the bar being pulled out while the more sophisticated ones are equipped with a liquid crystal display (LCD) readout that shows the changing dimensions. Some may be equipped with an electronic memory bank, capable of storing up to as many as 10 different measurements from the same vehicle. In this way the operator need not constantly refer back to the specification manual to determine the correct measurement.

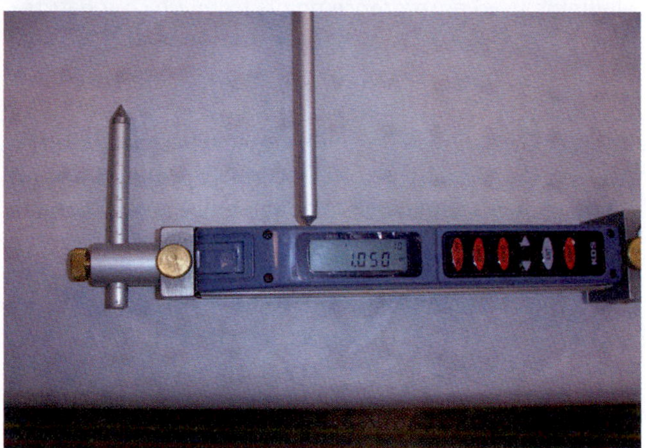

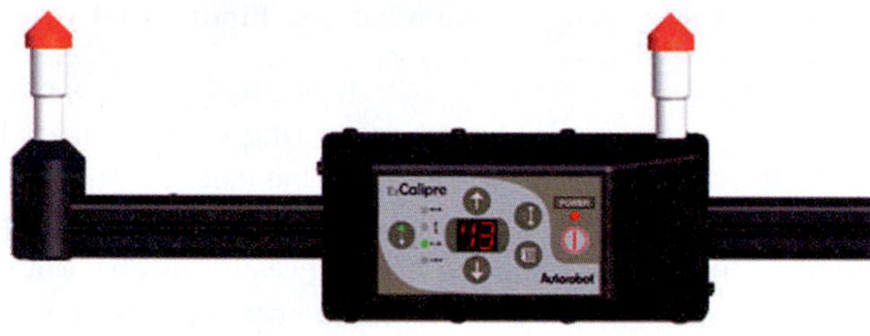

FIGURE 13-13 Tram bars and tram gauges with linear scales, LCD readouts, and wireless remote measuring systems are used for point-to-point measurements.

FIGURE 13-14 Enlarged cones and tips are put over the ends of the pointers for measuring to the center of the enlarged holes in the undercarriage with the tram bar or gauge.

A wireless remote system that uses a computer database is also available. These systems are capable of storing numerous measurements on the computer and printing them out for future reference.

Tram Gauge Accessories. Tram gauges may also be available with various accessories such as different-sized tips or cones to make measuring to the center of larger holes easier and more precise. See **Figure 13-14**. Various pointer lengths and bar extensions are also available, allowing for extended measurements to be taken. The various pointer lengths are designed to help avoid or clear obstacles, and the tips and cones help to position the tram gauge securely in reference holes. One must understand that when performing point-to-point measurements with a tram gauge using published dimensions, it is necessary to have the tram gauge pointers set at the same height or distance from the bar. When they are in an offset position or at different heights, a different measurement will result than if the pointers are set at the same level. See **Figure 13-15**.

(A)

(B)

(C)

FIGURE 13-15 It is imperative that the pointers on the tram bar/gauge are placed even or parallel to each other when making point-to-point measurements as in (A). In (B) and (C), a 12 mm (1/2-inch) discrepancy can occur over a 1,000 mm (40-inch) span by not placing the pointers parallel to each other.

FIGURE 13-16 It may be necessary to offset the pointers in order to make a straight line measurement when an obstacle, such as the rear end, is in the way.

Occasionally it becomes necessary to adjust the pointers to different heights or perform **offset measuring** at various heights to clear obstacles, such as the engine and transmission or suspension parts that may be located between the two reference points. See **Figure 13-16**. When placing the pointers in an offset position becomes necessary, one should use a tape measure to measure between the two pointers to achieve an accurate measurement. Always make certain the pointers are securely fastened into the holder as it may otherwise allow the pointer to be tilted, causing an inaccurate measurement. One must also make certain that the tram gauge pointers are positioned securely in the recommended reference holes cited in the vehicle's specification manual. See **Figure 13-17**. As was pointed out previously, when using holes, bolt heads, nuts, or rivets as a reference point, one should measure from center-to-center of the fasteners.

MEASURING TO LARGE AND DISSIMILAR SIZE LOCATIONS

Hole Size and Shape Dictates the Technique Used

When using the tram gauge to measure to a larger-than-normal hole, it is difficult to ensure the tram pointer is being placed at the center of both holes. Three different techniques can be used when measuring large and irregular-shaped holes: near-edge to far-edge, inside-to-inside, and outside-to-outside.

When two large holes being used are the same size, measurements can be taken by using the edges of the holes. Holding one pointer at the edge of the opening on the near side of one hole and the other pointer at the far edge of the other hole will automatically make a center-to-center measurement. See **Figure 13-18**.

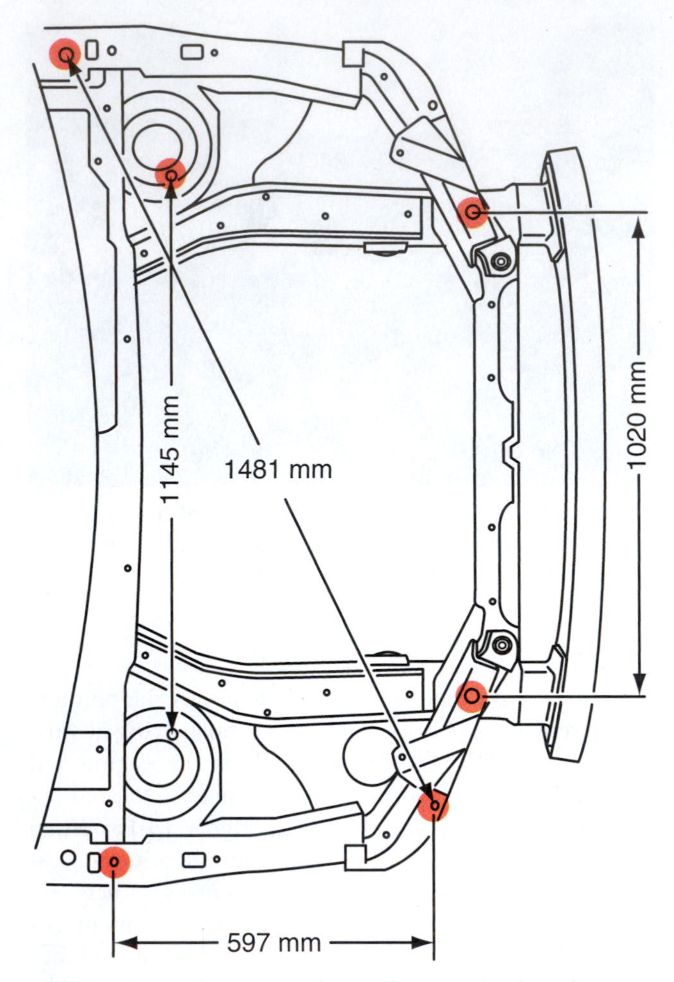

FIGURE 13-17 One must also make certain that the tram gauge pointers are positioned securely in the recommended reference holes cited in the vehicle's specification manual.

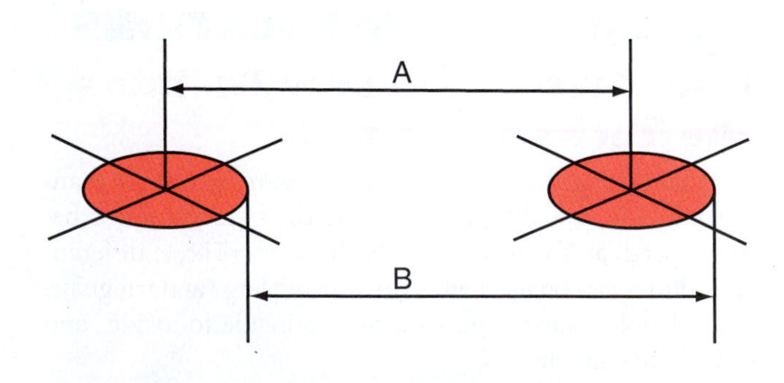

FIGURE 13-18 When using bolt heads, nuts, or rivets as a reference point, one should measure from center-to-center of the fasteners.

Measuring Outside-to-Outside or Inside-to-Inside

To determine the dimension when the holes are a different size, one simply totals the diameter of the two holes, divides that total by 2, and then adds or subtracts that amount to the **center-to-center measurement** of the holes in question. Subtracting that amount or half of the combined diameter of the two holes will provide the dimension for the inside-to-inside distance between the two holes. See **Figure 13-19**. To determine the outside-to-outside dimension, one must add half of the combined diameter—or 50 percent of the combined hole diameters—to the center-to-center dimension. See **Figure 13-20**. These **edge-to-edge measurements** can be quite accurate.

Taking measurements on the undercarriage usually involves a little different approach than when measuring underhood, door opening, or other dimensions. These measurements are usually taken from under the vehicle,

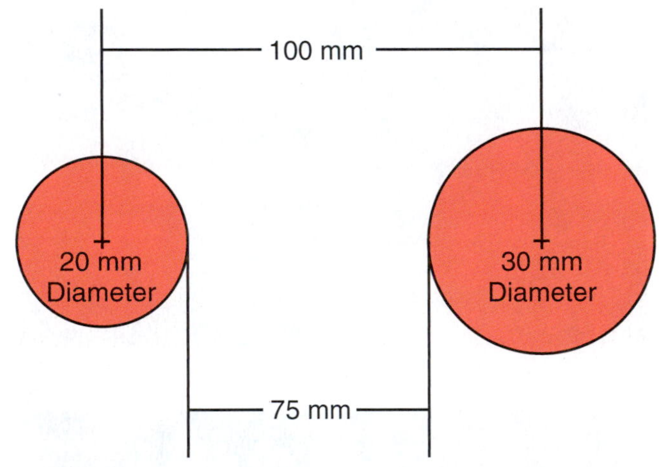

FIGURE 13-19 Measuring from edge-to-edge using the inside-to-inside measurements offers a more precise reference in many cases.

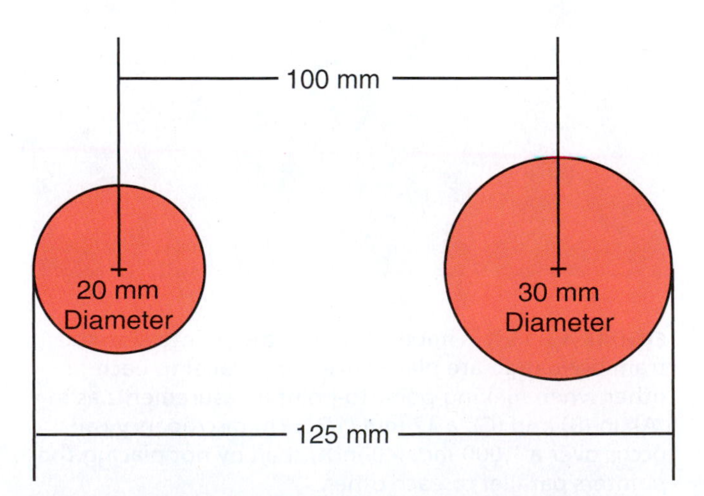

FIGURE 13-20 Measuring from edge-to-edge can also be accomplished using the outside-to-outside measurements.

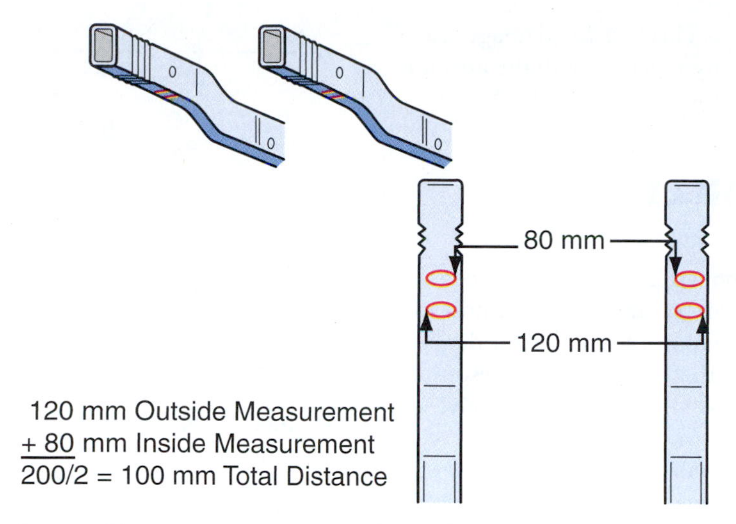

FIGURE 13-21 Many times when making measurements on the undercarriage the holes used for reference points may not be the same diameter, or they may be oblong or some other configuration.

which requires one to look at things from an unfamiliar perspective. In addition, these measurements often cover a considerably longer span than underhood or door openings, for example. When making undercarriage measurements, the holes used for reference points may not be the same diameter or they may be oblong or some other configuration. When the holes are not the same diameter or one is elongated, the same principles for mathematically calculating the dimensions discussed previously can be used to determine the distances between the pointers. See **Figure 13-21**. One should make the measurements as long as possible with the tram gauge as they are more accurate than shorter ones. The possibility of creating an error is compounded with each additional measurement made. Therefore, it is best to make one long cumulative measurement consisting of two or three short ones as opposed to two or more short lengths.

THREE-SECTION DESIGN

One must keep in mind that while the unibody appears to be a solid contiguous structure, in reality it is three somewhat interdependent yet independent units consisting of the front, center, and rear sections welded together to form the whole of the vehicle. For the purpose of construction and for diagnostic measuring, each section is assigned its own set of identifiable dimensions together with those of the overall vehicle. Likewise, when diagnosing the vehicle's damage, each individual section—along with the overall length of the vehicle—must be monitored and compared to the published dimensions. See **Figure 13-22**.

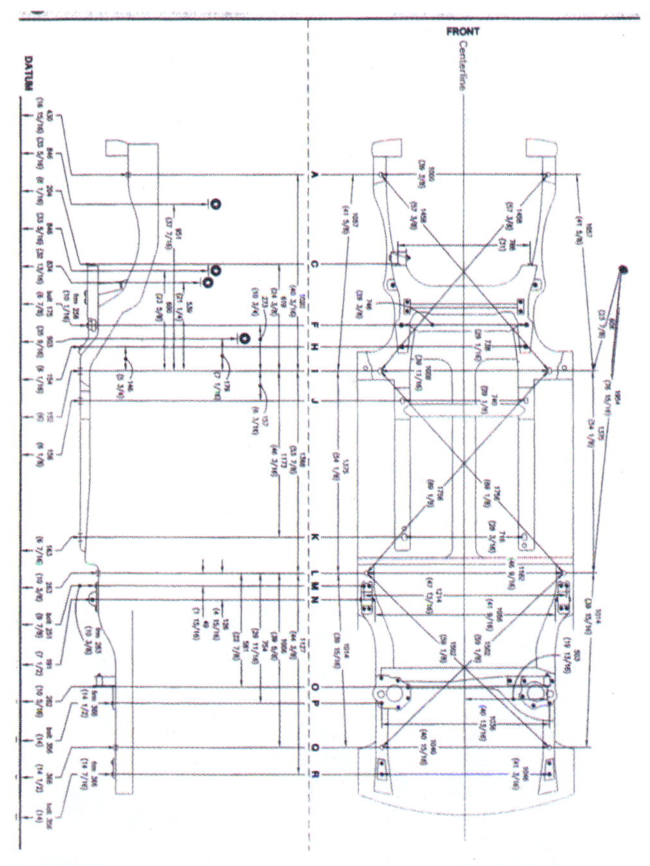

FIGURE 13-22 Each vehicle, model year, and option package has its own unique set of dimensions; therefore, it is necessary to match the chart being used to the vehicle being measured.

Quick Check Measurements

As one might expect, in a less severe collision the vehicle's damage is frequently contained or isolated to the immediate area or section that sustained the impact. Therefore, measuring beyond that point is usually not necessary. However, in a more severe crash, by design of the vehicle, the damage radiates from one section to another—and eventually throughout the entire vehicle—making it necessary to know the dimensions of each section as well as that of the overall vehicle. Therefore, when the vehicle's undercarriage is checked for damage, each section is checked individually, together with the area immediately adjacent to it and the overall vehicle as well.

CHECKING VEHICLE DIMENSIONS

How then does one go about determining a vehicle's dimensions? The collision repair technician has a number of tools available to accurately assess the collision damage sustained by the vehicle. Perhaps the simplest resources are the individual **body dimensional charts** that give dimensional specifications for the **undercarriage dimensions**, **underhood dimensions**, and a variety of other critical areas. These are frequently used when manual gauges such as the **self-centering gauges** and the tram gauge are used to locate and identify the damaged areas. See **Figure 13-23**. Each vehicle, model year, and option package has its own unique set of dimensions; therefore, it is necessary to match the chart being used to the vehicle being measured. The specifications include a variety of parallel, linear, and diagonal measurements. When using the body dimension charts, the tram bar is commonly used by placing the pointers at the specific reference points—for example, the 1,756 mm diagonal dimension between points I and L in Figure 13-22—and comparing the dimensions of the chart to those taken from the vehicle. The first comparative measurements made should be those in the vehicle's center section. The four corners under the passenger compartment in the torque box areas are the primary control points from which the majority of the measurements originate. For this reason, this area is also referred to as the **zero plane area**. See **Figure 13-24**. These are the same areas the manufacturer uses for the fixture locations and to securely hold the vehicle while it is being manufactured. Because it is the most rigid area of the vehicle it is least likely to become damaged or distorted in a collision. This makes the area an ideal location from which to originate most of the dimensions.

FIGURE 13-23 When using the body dimensional charts, the tram bar is commonly used by placing the pointers at specific reference points and comparing the dimensions of the chart to those taken from the vehicle.

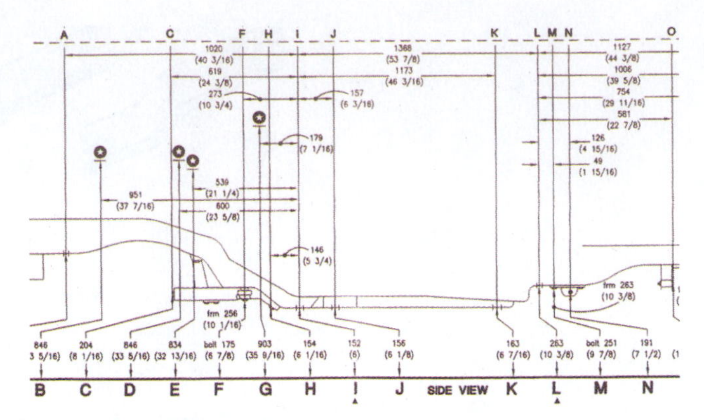

FIGURE 13-24 The darkened triangle under the letters I and L are indicators that these are the point zero locations on the undercarriage. Most measuring systems require three undamaged areas be used for this purpose.

Three Reference Points Required as Starting Point

All body dimensional charts cite length, width, and diagonal measurements to the base areas because they are the most critical locations to establish an accurate measuring foundation. Ideally four undamaged locations are desirable; however, using three locations is sufficient until the fourth area can be rough pulled into position to restore it dimensionally. If three correctly positioned reference points do not exist, it may become necessary to move to the front or rear section of the vehicle until at least three reference points can be identified for a starting point. Remember that in order to accurately measure a vehicle, one must start with at least three accurate reference points.

In order to confirm the dimensional trueness of the section, diagonal, length, and width measurements should be taken and compared to the specification manual. When this has been determined, both the front and rear sections of the vehicle should be checked individually for dimensional trueness in the same manner as the center section. To further confirm the accuracy of the dimensions, one may take measurements that span the entire center section and carry into either the front or rear section as well. One long measurement spanning an area is considered to be more accurate than two or three short ones covering the same distance. As part of diagnosing the front and rear sections, these measurements must also be checked and compared to the overall vehicle length dimensions to determine if the overall length measurements are correct.

Specification Manual Linear Measurements

A common misconception of the dimensions shown on nearly all body dimensional charts is the intended use of the linear dimensions shown on the vehicle's side view. Contrary to common belief, they are not point-to-point measurements. In fact, these datum line dimensions are straightline measurements which run parallel to the vehicle's centerline. See **Figure 13-25**. In order to better understand why these dimensions cannot be used for point-to-point measurements, refer to the undercarriage specifications in Figure 13-25. There is a significant angle between points A and L, making the dimension notably longer than when the distance is measured in a straight line that runs parallel to the centerline. Any attempt to use these dimensions to make point-to-point comparisons will certainly create a great deal of confusion. Any attempt to pull the vehicle to conform to these dimensions will result in a vehicle that has been "repaired" to totally incorrect dimensions. Although these dimensions can be used to make point-to-point measurements, it is necessary to change how the tram gauge is set up. One must perform some simple mathematical calculations first. See **Figure 13-26**.

Since the dimensions used to make these measurements are based on the **datum plane** which is located at specified distances from the vehicle's undercarriage, one must adjust the tram gauge pointers to an offset position that will replicate the datum line under the vehicle. As shown in Figure 13-26, the datum line is 450 mm below the rail at the front and 300 mm at the rear. Because the tram gauge must be held level to simulate the datum plane, the tram bar pointers are adjusted 100 mm above the known datum line to simulate it and still leave enough distance from the floor to move the bar about freely while measuring. Once the tram bar has been set up, one must perform some simple calculations (which will be discussed in the following text) to determine the exact distance between the two points identified as measurement A in **Figure 13-27**.

The formula used for this is $A^2 + B^2 = C^2$. Using the diagram of the undercarriage in Figure 13-27, one can carry out the calculations in the following way:

A = 200 millimeters (linear dimension on undercarriage chart)

C = 100 millimeters (distance from the centerline to the reference point)

D = 60 millimeters (distance from the centerline to the reference point)

B = 40 millimeters (the difference between C and D)

The calculations are performed as follows:

- Dimension "A" × "A" (200 × 200) = 40,000
- Dimension "B" × "B" (40 × 40) = 1,600
- 40,000 + 1,600 = 41,600

Using a calculator, determine the square root of 41,600.

$$\left(\begin{array}{c}203.96\\ \sqrt{41{,}600}\end{array}\right)$$

The correct figure is 203.96 or 204 mm, the distance the pointers should be placed apart on the tram bar. This is not a point-to-point measurement. Therefore, it should not be measured from tip to tip on the pointers, but should be taken parallel to the tram bar itself.

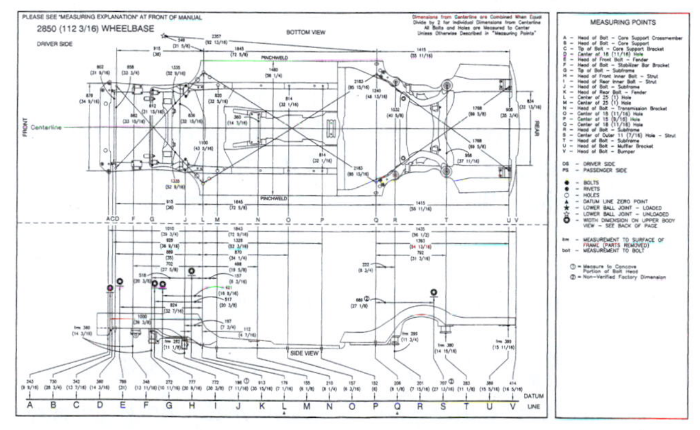

FIGURE 13-25 The tram gauge pointers are adjusted to an offset position to simulate the datum plane when using datum dimensions for point-to-point measuring.

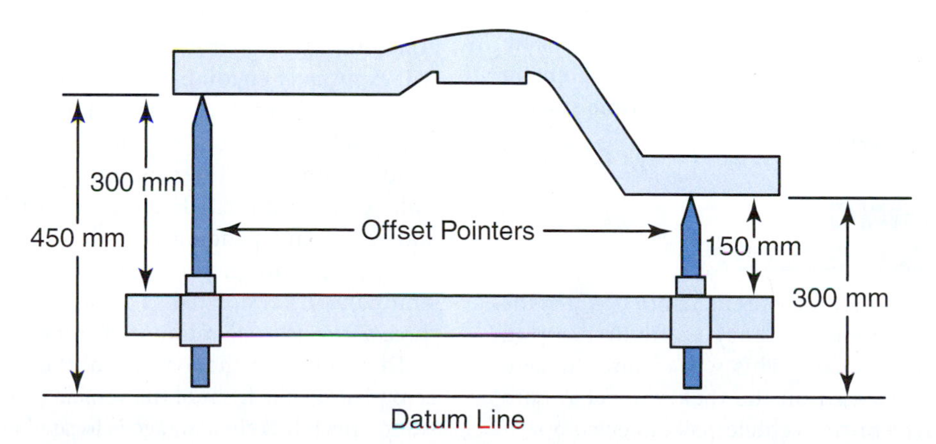

FIGURE 13-26 Since the dimensions used are based on the datum plane which is located at specified distances from the vehicle's undercarriage, one must adjust the tram gauge pointers to an offset position that will replicate the datum line under the vehicle.

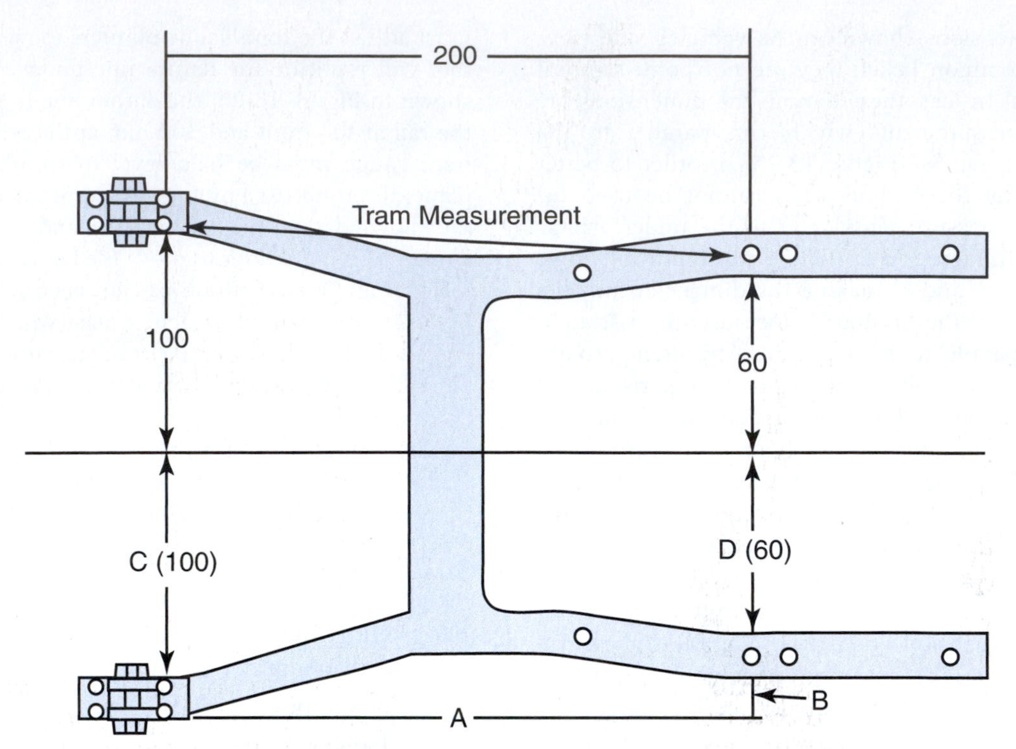

FIGURE 13-27 Once the tram bar has been set up, one must perform some simple calculations to determine the exact distance between the two points identified as measurement A.

Occasionally when diagnosing damage, it becomes necessary to make some quick comparative diagonal measurements to determine the exact location of a misaligned section of a vehicle. This can be done with a tram bar or tram gauge without the aid of published dimensions. See **Figure 13-28**. It is accomplished simply by making a comparative measurement from the damaged section on one side of the vehicle and comparing it to the same undamaged section on the opposite side of the vehicle. In the event both sides of the vehicle have been damaged, comparative measurements can be made using an undamaged vehicle that is of the same vintage and body style. One need not be as concerned about the placement or positioning of the tram gauge pointer's height when making these measurements as they are merely comparative and do not require the use of published dimensions. One does, however, need to ensure the same pointer is used on the common reference point on the opposite side of the vehicle to ensure the same angle is maintained for both measurements.

ESTABLISHING THE VEHICLE BASE

Performing the diagonal measurements to confirm the undercarriage's dimensional accuracy is the first step in establishing the vehicle's base. This will ensure the zero plane of the vehicle, located on the rocker panel at both the front and the rear of the vehicle passenger compartment, can be used for all the required baseline measurements. Until this has been accomplished, the technician cannot begin attaching the measuring system to the undercarriage. The measuring system must be installed on the vehicle in order to identify the vehicle's centerline and datum plane. The centerline and datum plane are the basis from which all the length, width, and height measurements are monitored while the vehicle is being repaired. One cannot identify the precise location and extent of the damage until these planes are identified.

Identifying the Zero Plane

Because the rocker panel section of the undercarriage beneath the passenger compartment is the basis for all measuring, most measuring system manufacturers and undercarriage dimension publications refer to this area as the zero plane or the body zero point. In addition to being the basis for all measurements, it is also the starting point for attaching the numerous measuring system targets which will eventually encompass the entire undercarriage, spanning from the front to the rear of the vehicle. When repairing a unibody vehicle, one must remember the vehicle is divided into three interconnecting units or sections: the front, center, and rear. Even though each section responds and reacts independently of the others as the collision forces pass through them during a collision, they must be simultaneously repaired as a single unit. This requires all three sections to be monitored and repaired as one unit.

Depending on the location of the primary damage, the zero plane at the front of the vehicle is used for length measurements when the damage is located at the rear of the vehicle. Conversely, the rear zero plane is used when the damage is located at the front of the vehicle. Nearly all of the measuring systems in use today require that all four control

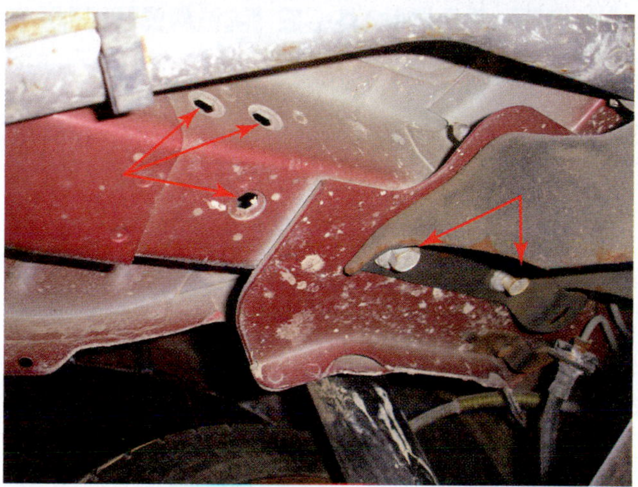

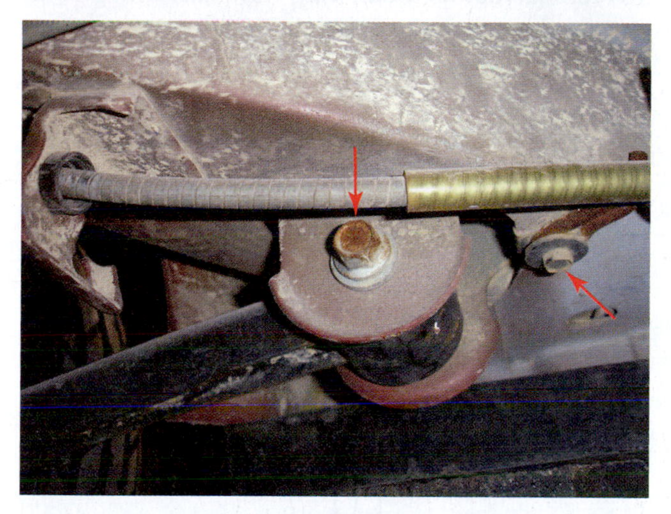

FIGURE 13-28 These reference points are being measured to locations that would not necessarily be published.

points be within the specified tolerance of each other. However, on most systems, a measuring foundation can be established with three reference points while the fourth is being realigned by rough repairing the damage sustained in the area. Some equipment requires only two undamaged points from which a measuring base can be established.

TECH TIP The Snap-On Tool Company's Car-O-Tronic measuring system requires only two undamaged points to establish a foundation for measuring. One must keep in mind that this only establishes length and width dimensions until a third point can be established.

IDENTIFYING REFERENCE PLANES USED FOR MEASURING

A vehicle involved in a side impact may likely sustain damage to one—or possibly both—of the control points on the damaged side, thus requiring re-establishing at least one of the affected control points before the measuring system can be installed. Occasionally it becomes necessary to remove some of the more severely damaged parts and to rough straighten at least the center section of the vehicle before the measuring system components can be attached. However, when both the control points on one side of the vehicle are damaged or distorted, the vehicle is usually totaled or rendered unrepairable.

TECH TIP Whenever a vehicle is raised up, taken off its wheels, and placed on the pinchweld clamps, the weight of the engine and drivetrain on the front-wheel-drive vehicles causes the front section to droop down. This area must be supported to ensure the entire undercarriage is level to eliminate any inaccuracy in the measurements taken.

Identifying the Centerline

Once the targets or reference indicators have been hung at their designated locations, a set of reference lines can be established. The first such reference line that becomes evident is the **centerline plane**. This is an imaginary line that runs through the middle of the vehicle from the front to rear and from the floor to the roof. See **Figure 13-29**. The centerline is used as a reference point to measure and monitor all of the vehicle's width measurements and to determine any side-to-side movement or deviation from the vehicle specifications.

Identifying the Datum Plane

The datum plane—an invisible line that establishes an imaginary horizontal plane parallel to the bottom

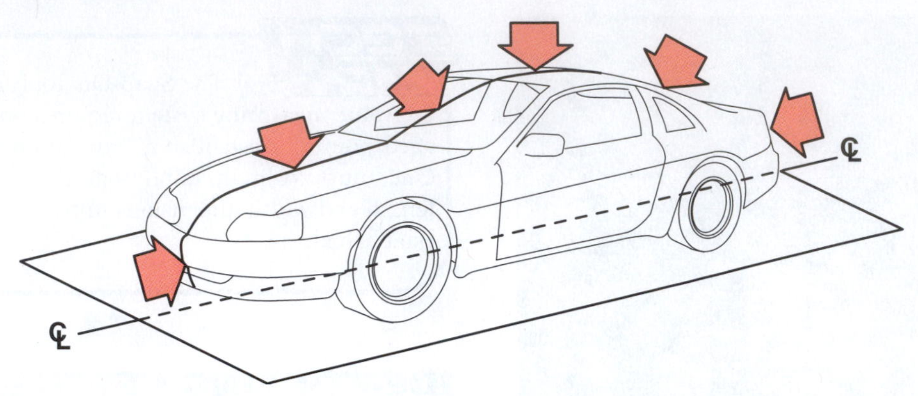

FIGURE 13-29 The centerline is an imaginary line that runs down the middle of the vehicle from the front to rear and from the floor to the roof.

of the vehicle—is used to measure and monitor all height dimensions. See **Figure 13-30**. The datum plane is established differently depending on the type of measuring system used. The most common concept is using the datum plane as a reference from which perpendicular measurements or lines are projected upwards to specific locations on the vehicle's undercarriage. This establishes height dimensions. Another concept is measuring down from specific locations of the undercarriage to the datum line. However the plane is determined, the datum line is used to determine any upward or downward deviation of the undercarriage. Perpendicular lines are projected forward or rearward to designated reference points such as engine cradle bolts, suspension mount locations, and other critical points to establish the precise length measurements required.

UNDERCARRIAGE MEASURING SYSTEMS

Once the diagonal checking has been completed to determine the state of the vehicle's base, one of a variety of measuring systems may be employed to check the undercarriage's overall alignment. One such system may be the manual mechanical system, which monitors the undercarriage two-dimensionally—length and width. A more sophisticated **three-dimensional measuring system**—which simultaneously monitors the length, width, and height dimensions—may also be used. The purpose for using either system is to determine the degree of misalignment that occurred. No matter which measuring system is used, the technician must establish two reference indicators that are key to monitoring the undercarriage—the centerline and the datum plane.

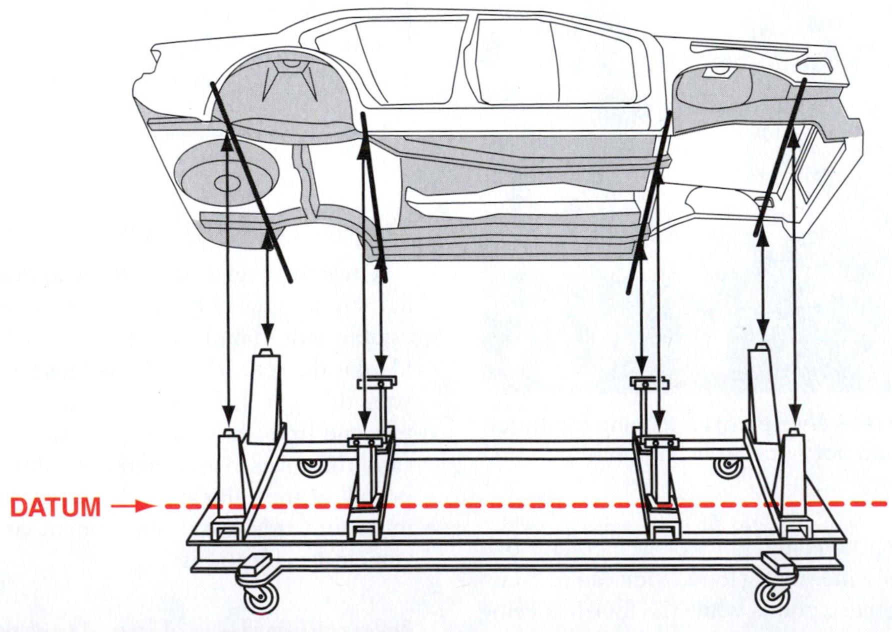

FIGURE 13-30 The datum line is an imaginary line created when a series of measurements are projected down a specified distance from the bottom of the vehicle.

Manual Mechanical Measuring Systems

The **manual mechanical measuring systems** typically include the self-centering gauges and the **strut tower gauges**, which may or may not be used in conjunction with each other. See **Figure 13-31**. In cases where damage to the strut tower is suspected, the strut tower gauge is used with the centering gauges to check for misalignment. If the diagonal measurements are made and no misalignment exists or the vehicle being repaired is a body over frame, the centering gauges are used alone.

Self-Centering Gauges. When compared to the more modern and sophisticated **laser measuring systems**, the centering gauges are sometimes thought to be outdated technology. However, they are still very useful as a quick reference to determine whether a vehicle undercarriage is within specs. See **Figure 13-32**. These gauges must be

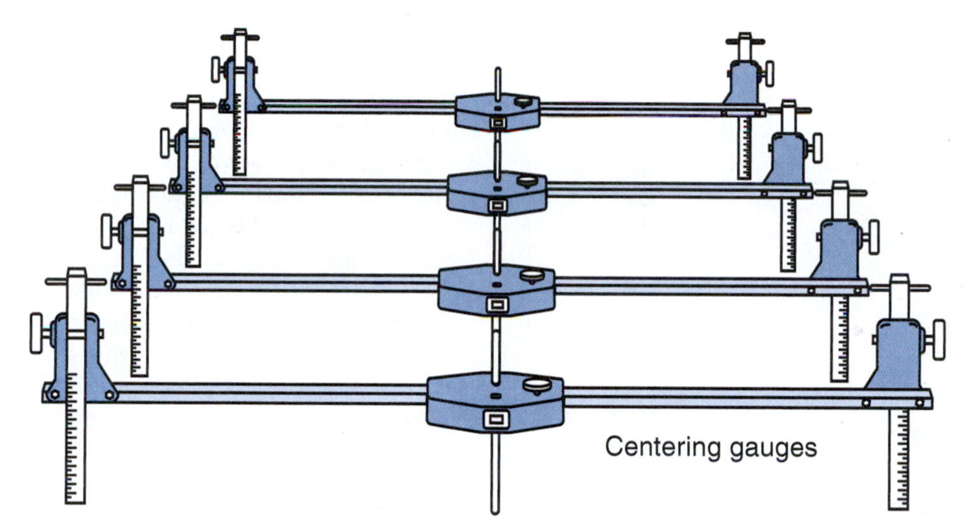

FIGURE 13-31 A minimum of four centering gauges must be hung at the four control locations to determine whether the undercarriage is misaligned.

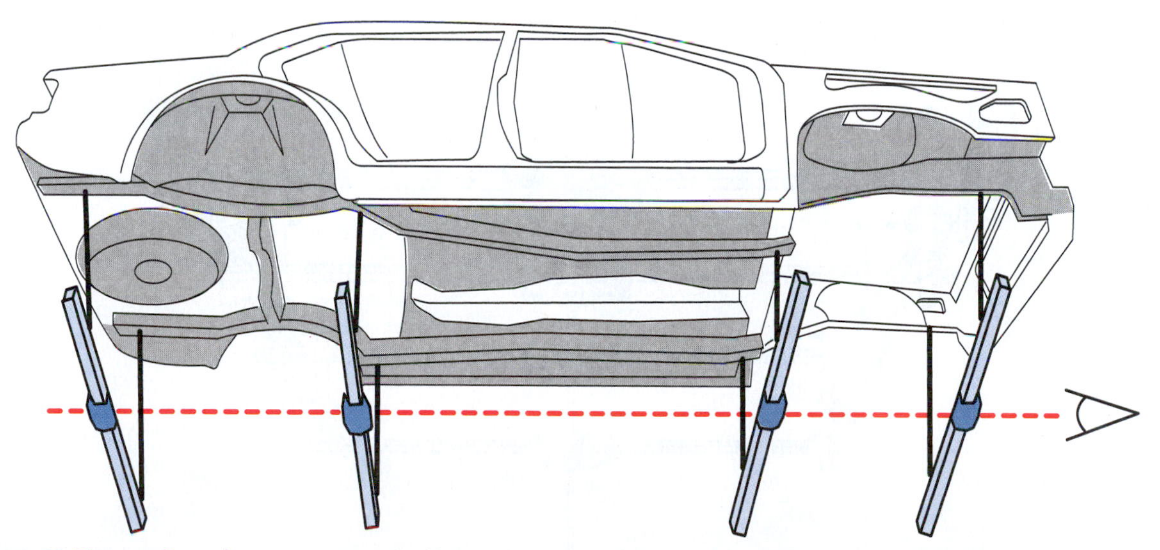

FIGURE 13-32 The preferred or recommended location to hang the four centering gauges is at the four control locations under the vehicle.

used in a group, with a minimum of four used, each placed at a strategic location on the undercarriage. The preferred or recommended location to hang them is at the four control locations under the vehicle. These control points are located at the front immediately behind the bumper mounting area, at the four torque box areas, and in the area immediately in front of the rear bumper mounting location. The body dimensional charts are used to determine the specific location where the gauges are to be hung and the distance they should be suspended from the undercarriage. See **Figure 13-33**. When hanging them, it is extremely important that both sides of the gauge are attached to the undercarriage at the same precise location and in the exact same way on both sides of the vehicle. Provided they are hung according to the specifications on the undercarriage dimensional chart, the centerline and the datum plane immediately become visually identifiable. The centerline—which otherwise is invisible—is identifiable by the location of the centering pins located on the center of the horizontal bars of the centering gauge hanging below the vehicle. All width dimensions are referenced from this location. If the vehicle is symmetrical, the distance from the centerline is the same to a common location on both sides. An asymmetrical vehicle will have different dimensions from the centerline to the same location on the opposite sides. When measuring or checking an asymmetrically designed vehicle, it is imperative the undercarriage dimensional chart be consulted because the dimensions are not the same from the centerline to either side. This can result in a considerable amount of confusion when attempting to determine the amount and the location of damage that exists.

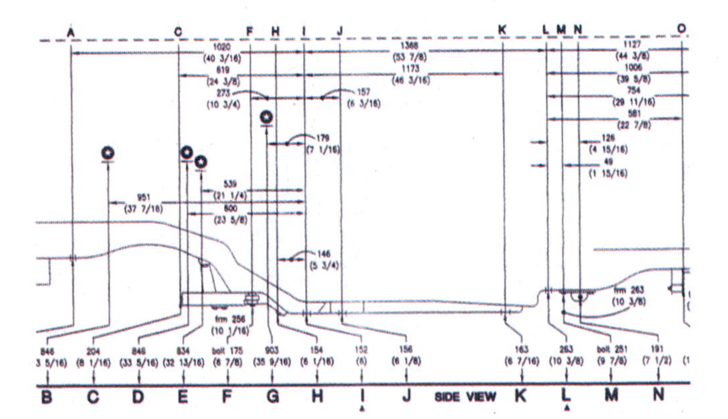

FIGURE 13-33 The body dimensional charts are used to determine the specific location where the gauges are to be hung and the distance they should be suspended from the undercarriage.

Hanging and Reading Centering Gauge. These gauges are initially hung at the four control locations to give the repair technician a "bird's eye" view of the overall damage condition. This will also identify both the centerline and the datum plane. See **Figure 13-34**. When sighting down the gauges, one should always use the undamaged end of the vehicle as the focal point from where to start the sighting. By sighting down the center pins from the vehicle's undamaged end to the damaged end, one can visualize if any side-to-side movement took place and the extent to which it occurred on the undercarriage during the collision.

When performing this visual inspection, the center pin on the vehicle's damaged end should be compared to those in the center and the section on the end of the vehicle from where the sighting is being done to determine if any undercarriage misalignment exists. See **Figure 13-35**. Any misaligned center pin on the gauge is visual evidence that the rails shifted to one side or the other. In order to more accu-

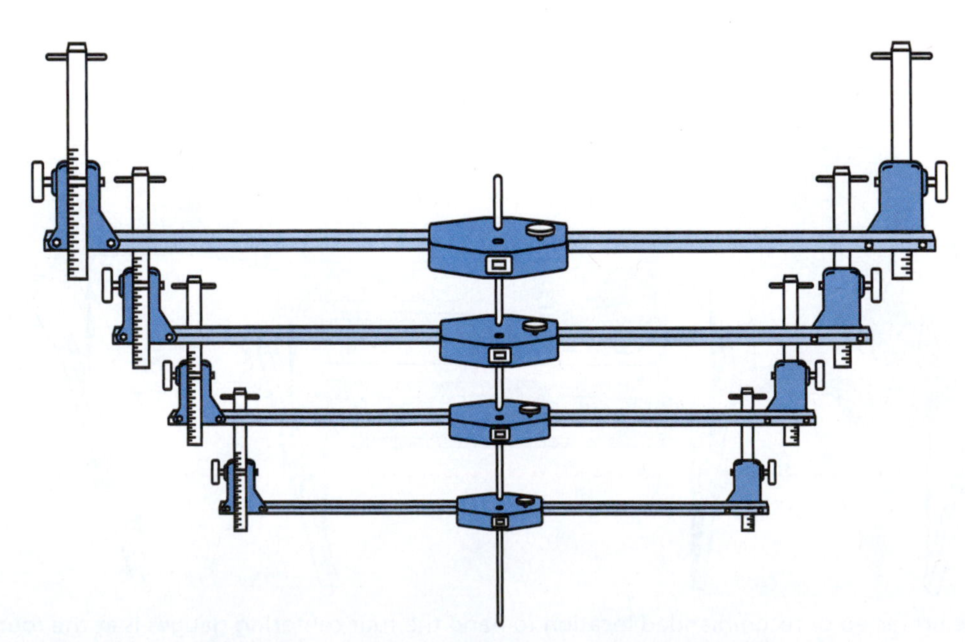

FIGURE 13-34 The centering gauges are initially hung at the four control locations to give the repair technician a "birds eye" view of the overall damage condition.

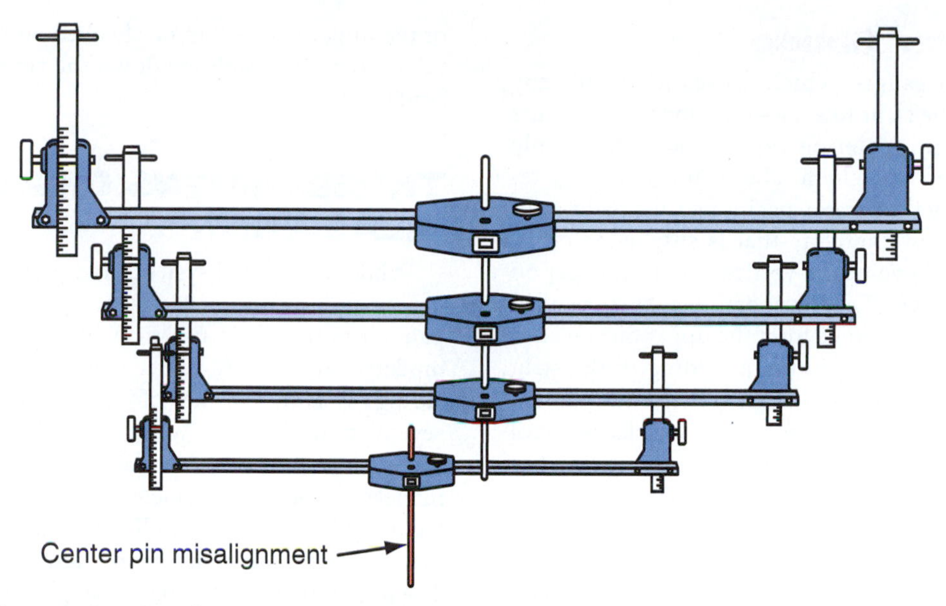

FIGURE 13-35 When performing this visual inspection, the center pin on the vehicle's damaged end should be compared to those in the center and the section on the end of the vehicle from where the sighting is being done to determine if any undercarriage misalignment exists.

rately isolate the location of the damage, one can place additional gauges between those already hung—provided extra ones are available. In this way the exact location of the damage becomes increasingly more visible and can be tracked from the starting location to the end of it.

The datum plane—an invisible line that establishes an imaginary horizontal plane parallel to the bottom of the vehicle at a given distance below it—is used to measure height dimensions. See **Figure 13-36**. The datum plane is established only after the gauges are suspended from the undercarriage at specified locations and distances identified on the body dimensional chart. When hung correctly the horizontal cross bars of the centering gauges visually identify the datum plane. Any existing damage that has caused a section of the undercarriage to move up or down is quickly identified when sighting along the parallel bars because the misaligned side will be either higher or lower than all the other gauges.

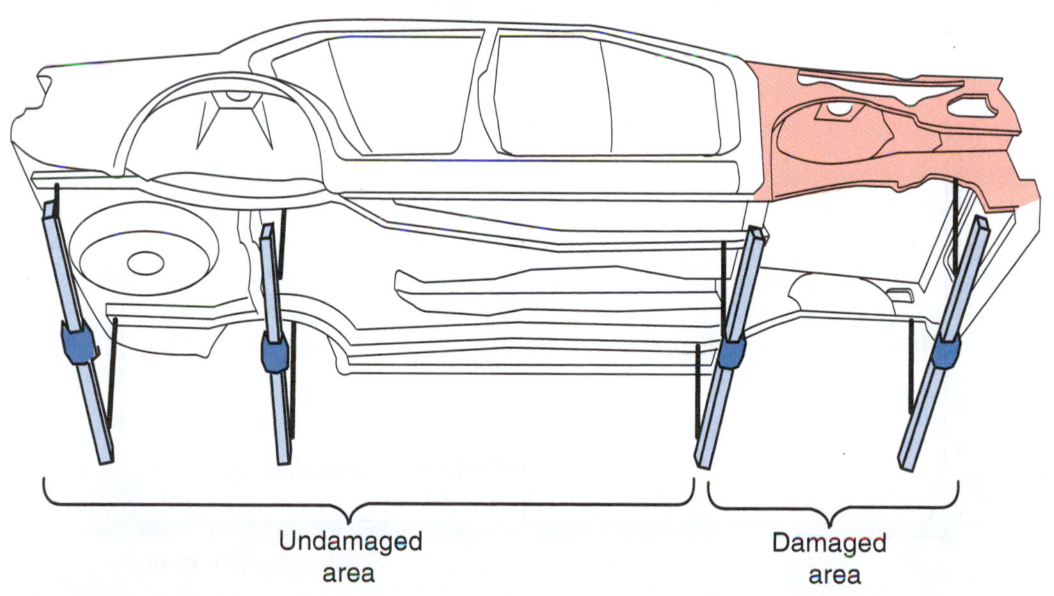

FIGURE 13-36 The datum plane is established only after the gauges are suspended from the undercarriage at specified locations and distances identified on the body dimensional chart.

Strut Tower Gauge

The strut tower gauge—which is used to identify any misalignment of the strut towers—functions in much the same manner as the centering gauges and is commonly used in conjunction with them. The strut tower gauge requires two horizontal bars: one positioned above the strut tower and one below the body that is suspended or attached to the one above with a vertical connecting bar on each side of the vehicle. The lower bar must be suspended at precisely the same distance from the upper one on each side in order to obtain an accurate reading of the strut tower height. The technician must use a body dimensional chart to determine the dimensions for placement of the lower horizontal bar in relation to the upper one. See **Figure 13-37**. Like the centering gauge, the strut tower gauge readings are taken from the lower horizontal bar. By comparing the strut tower gauge to the horizontal bars on the centering gauge, the technician can determine the location of any height damage. Like the centering gauges, the lower horizontal bar features a center pin that can also be used to identify the centerline and to determine if any side-to-side movement has occurred. This is accomplished by using the alignment relationship of the center pins on the centering gauges hanging from the undercarriage to the center pin on the horizontal bar of the strut tower gauge. While the strut tower gauge is typically used to detect any misalignment of the strut towers, it can also be used for checking the radiator support, center pillar, cowl, quarter panel, and even the roof line. Any misalignment of the strut tower area—or, for that matter, any area of the upper body—can be checked and compared to the undercarriage by sighting down the center pins and horizontal bar.

THREE-DIMENSIONAL MEASURING

While the manual systems and point-to-point measuring are still relatively effective tools for diagnostic and repair purposes, their use may be inadequate on certain modern day unibody vehicles. A three-dimensional measuring system capable of simultaneously monitoring several critical length, width, and height locations has become the industry standard. See **Figure 13-38**. The modern day unibody vehicle's center section is the manufacturing foundation, as well as the base from which all dimensions are taken and monitored for repair purposes. The four corners under the passenger compartment, commonly referred to as the vehicle base, become the **principal control points** from which all measurements originate and are taken. The vehicle's dependency on this area is due in part to the vehicle's structural design, making this area the strongest part of the undercarriage and the least likely to become damaged in a collision. Before any of the measuring equipment can be attached or set up on the vehicle, it is necessary to verify the trueness of the dimensions on at least two points or locations. On most equipment, three locations are required. This is accomplished by the point-to-point linear and diagonal measurements discussed previously.

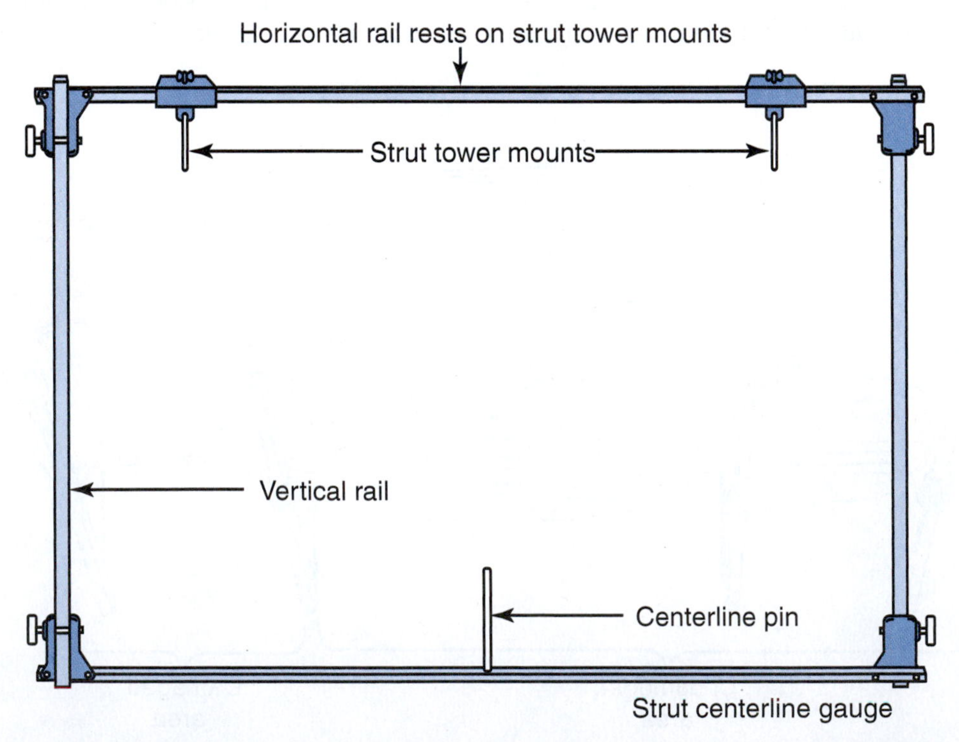

FIGURE 13-37 Like the centering gauge, the strut tower gauge readings are taken from the lower horizontal bar by comparing it to the horizontal bars on the centering gauge to determine the location of any height damage.

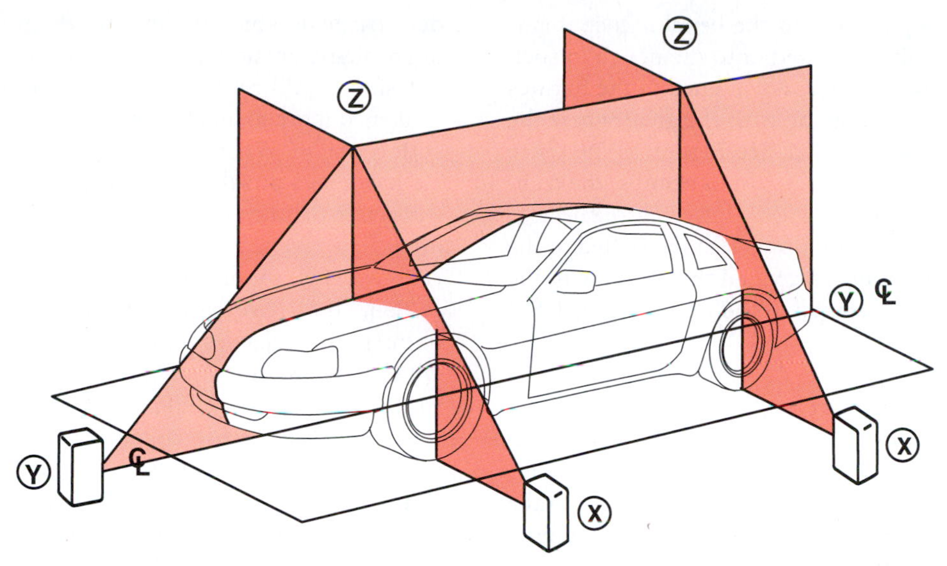

FIGURE 13-38 A three-dimensional measuring system that simultaneously monitors a number of critical length, width, and height locations has become the industry standard.

The measuring systems are categorically divided into two general groupings or types depending on their use. They may be a dedicated system, which is dependent upon a bench or a rack that the vehicle is clamped against to secure it while it is being measured and straightened, or they may be a universal system that can be used with a variety of straightening systems. The dedicated system requires the vehicle to be clamped securely to the bench, which essentially forms a box or perimeter around the vehicle for measuring purposes. Once this is accomplished, the vehicle is said to be "squared to the bench." After the vehicle is secured to the bench all the measurements are taken from known points on the bench to specified locations on the undercarriage. These dedicated systems may be a **dedicated fixture measuring system**, a **universal mechanical measuring system**, or an electromechanical system. The dedicated fixture system uses special fixtures which—when bolted to the bench at specified reference points—will match a given surface on the undercarriage, whereas the universal mechanical system uses pointers to locate the reference points. The electromechanical system uses a highly sophisticated articulated arm resembling that of a robot to take measurements.

The systems may also be either manual or computerized. Manual systems require the technician to maneuver and manipulate the measuring apparatus to identify damage and to determine the dimensional accuracy of the repair. Computerized systems are dependent on a database which monitors the undercarriage dimensions by hanging targets at predetermined locations. A computer uses the database to analyze the location and the amount of the damage on the vehicle and compares this to the correct dimensions. The system then displays a sketch of the undercarriage with both the correct dimensions and the current misaligned dimensions on the computer monitor. These systems may use a laser or ultrasonic sound waves directed at targets as a guiding system.

Dedicated Fixture Measuring System

The dedicated fixture system has passed the test of time, as it was among the first bench systems to be introduced in the United States and, to date, is the only system certified to be used for many of the high-end aluminum-intensive vehicles. The Celette system, for example, is the most predominant equipment used in Europe for repairing many of the high-end luxury vehicles such as the Audi, BMW, Jaguar, Mercedes-Benz, Porsche, Volkswagen, and Volvo. The Celette and the Car-O-Aligner fixture system are the only equipment meeting the manufacturer's certification requirements in North America.

When properly used, these systems are extremely accurate. This makes them the equipment of choice for vehicles with little or no dimensional tolerance. It allows unparalleled precision for returning a vehicle to its pre-collision condition because each point that is monitored on the vehicle must match the fixture to very precise tolerances. When they don't match, the sections that can be straightened are straightened, and those that cannot are removed and replaced with a new section.

The dedicated bench system uses an extremely strong, portable steel work bench with a perfectly flat working surface to which transverse beams are attached at strategic locations to match vehicle-specific tie-down locations identified by the manufacturer's reference manuals. The pinchweld or sill clamping fixtures are positioned and attached to the transverse beams matching the location of each of the four torque box areas for the specific vehicle being repaired.

Prior to attaching the vehicle to the bench, it must be raised on a two-post hoist to accommodate the removal of certain components such as the suspension. This will also make other critical dimensional locations accessible to the fixture mounts. This must be done in order to secure the vehicle once it is set down onto the bench. Once this is done, the vehicle is then lowered onto the bench in such a way that the torque box areas are aligned to the fixtures on the bench and secured along with the areas where the suspension parts were removed. See **Figure 13-39**. Once the vehicle is secured to the pinchweld clamps, the technician utilizes a vehicle-specific **fixture setup sheet** to add a number of additional fixtures that are bolted to the bench at specified locations. Provided they align with the undercarriage, they are secured to the vehicle as well. See **Figure 13-40**.

A Go or No-Go System. Checking the vehicle's condition or dimensional accuracy is determined by whether the fixture locations match their mating surfaces on the undercarriage. Provided the fixtures properly conform with the vehicle's contact surfaces, it is said to be within the manufacturer's accepted tolerances. The fixtures are said to mimic the factory platform fixtures used in manufacturing the vehicle. One can assume that any area of the undercarriage in which the fixtures do not align properly is out of alignment and needs to be restored to the proper dimensions. Unlike most others systems, the fixture system is likened more to an alignment procedure than to a measuring system. When all the fixtures for the specific vehicle have been attached and all properly align with their respective mating surface, one can be assured the undercarriage is precisely aligned to the manufacturer's specifications. This arrangement is said to be a "go or no-go system." If the reference points precisely line up with the fixtures, "the repair is a go." If they don't properly line up with each other, it's a "no go." In other words, the undercarriage must be restored more precisely. In order to maximize the use of the system, a designated minimum number of the fixtures must be used.

In addition to establishing the undercarriage dimensions, the fixtures can also be used to perform several other functions as part of the repair. For example, they can be used for holding replacement body panels in their correct position while they are being welded into place.

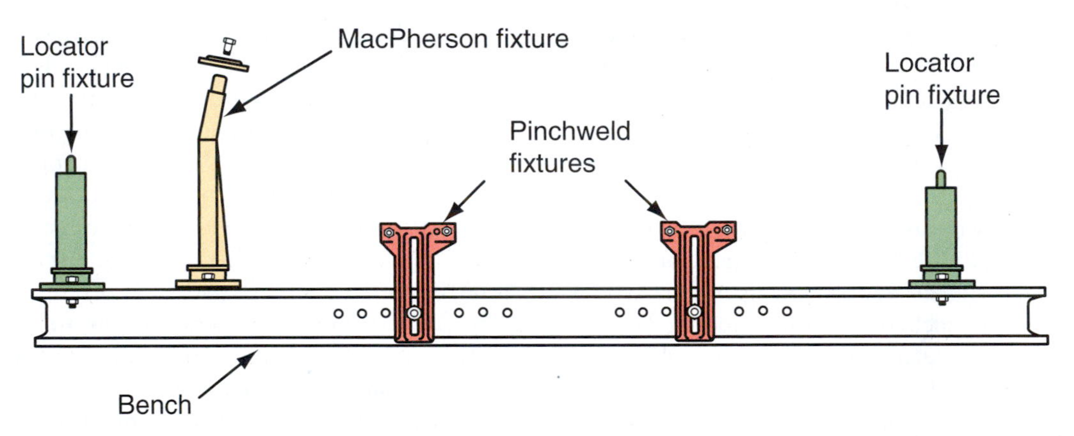

FIGURE 13-39 The dedicated fixture system may require removal of some suspension parts prior to mounting the vehicle to the bench.

FIGURE 13-40 Once the vehicle is secured to the pinchweld clamps, the technician utilizes a vehicle-specific fixture setup sheet to add a number of additional fixtures which are bolted to the bench at specified locations.

This will help to ensure they are welded at the precise location and ensure a proper fit with all the adjacent panels. A significant disadvantage, however, is that a set of vehicle-specific fixtures is required for nearly every model that is being repaired. In most cases they must be rented from a central clearing house because the cost of purchasing every set would be prohibitive. Another disadvantage of the system is upper body dimensioning and monitoring is limited to those areas where the fixtures are designed to go. Other than for under the engine compartment, very few fixtures for checking any area above the undercarriage are available with the system. The need to remove various suspension parts and other components in order to secure the vehicle to the bench also makes this system very labor intensive. The inability to furnish any written documentation showing proof of the repair's dimensional accuracy, such as is common on many of the computerized systems, also makes this equipment less desirable in some repair facilities.

Universal Mechanical System

A universal mechanical system utilizes a bridge or ladder to which the vehicle is attached at the pinchwelds. See **Figure 13-41**. The universal mechanical measuring system also uses the anchoring system to calibrate the measuring system to the specific vehicle. Once the vehicle is secured to the clamps, width gauges are attached perpendicular to the ladder at predetermined, vehicle-specific locations. Adjustable height tubes are attached to the inward and outward adjustable arms on each side of the width gauges that are then used to check the vehicle height.

The vehicle height is measured by using the calibrated scales on the height tube by raising them up to make contact with the undercarriage at strategic points or locations. This is done to establish the vehicle's datum plane. When the tubes match the height of the undercarriage, the datum plane is at the correct or designated level or height. If the tubes do not make proper contact with the designated area, it indicates the undercarriage is misaligned and requires straightening to restore the datum plane. To make the measurements more precise and exacting, a variety of cones and other attachments are used to match the holes, openings, and other measuring locations on the undercarriage.

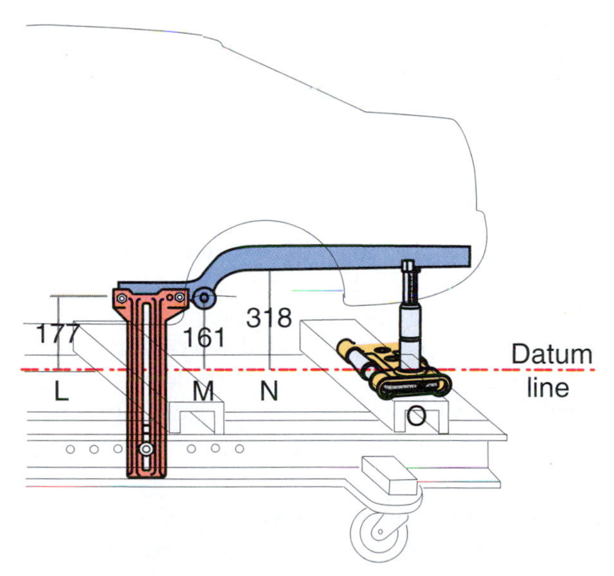

FIGURE 13-41 A mechanical universal measuring system uses a portable bench much like that which is used with the fixture system. The vehicle is attached to it at the pinchwelds, and the undercarriage is measured using height tubes and length and width scales.

The vehicle width is also measured by using the scale on the width gauge and by installing height tubes at predetermined locations. This is another means to establish the center plane, even though pointers are usually not placed under the center section of the vehicle. See **Figure 13-42**. The placement of the measuring tubes or pointers is determined by the manufacturer's specifications for each individual vehicle's body style. These dimensions are available in either a printed manual or on CD-ROM. The degree of accuracy in using this measuring system is largely dependent on the precision of the measurements used to locate the placement of the pointers.

An advantage to this system is that a technician can obtain immediate visual results by comparing reference points on the vehicle to where they are supposed to be according to the measuring system. A technician will know that if the points do not match up then the car is misaligned and needs to be properly straightened and aligned. However precise or accurate the alignment, it is up to the technician to record this information as there are no electronic links or means of communicating this information and transferring it onto a paper document. One of the chief disadvantages of this system is the measuring system must be calibrated from the bench to the vehicle. In the event the vehicle should shift or move even slightly during the pulling or straightening operation, the entire calibration process must be repeated because it is no longer in the same position from where the measurements were initially taken.

Occasionally the pointer tubes get in the way of the pulling equipment or prevent the technician from gaining access to an area where they must perform some stress relieving while the pulling is being done. This may require moving or partially disassembling the measuring rack in order to get to the areas in question. Simultaneously monitoring the reference points while the pulls are being made is sometimes difficult because the pointers must be dropped down and away from the undercarriage to avoid damaging them. They must be moved back into position on the undercarriage each time a check is made. Therefore, one must pull, stop, and reposition the reference tubes, check their position and the status of the undercarriage, and then lower them again when more pulling is required. The advantage is the status of the undercarriage is very visible when the monitoring tubes are raised up and placed at their designated location.

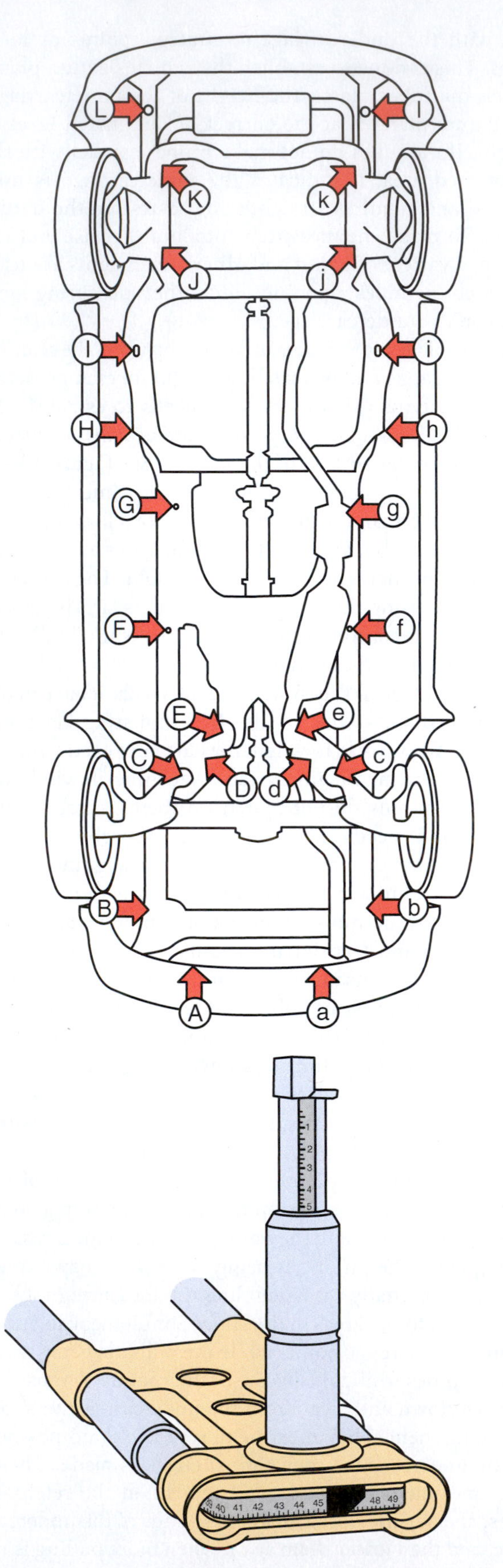

FIGURE 13-42 A legend is published by the manufacturer for every vehicle to locate the placement of the tubes and the scales used to monitor the dimensions.

Simultaneous Universal Laser Measuring Systems

Simultaneous universal measuring systems, as the name implies, can be used with a variety of different bench and straightening systems. They may be manual systems or they may work from a computerized database that allows the technician to monitor the undercarriage dimensions with targets that are hung at predetermined locations. The manual system usually requires the technician to manually take readings using either a projected laser beam or a digital tram gauge that is built into the mounting bridge or hardware. Whichever system is utilized, the undercarriage's length, width, and height dimensions are simultaneously monitored and observed.

Manual Laser Measuring Systems. Laser systems are infinitely accurate as a measuring system because they typically use one or more laser beams which are projected onto a series of targets which presumably are attached at the correct location. See **Figure 13-43**. The laser beam is directed through one or a series of transparent targets or it may be projected onto an opaque scale attached at one or a number of strategic locations to monitor the dimensions. This system is designed to project either one or two laser beams depending on which part of the system is being used. A universal mounting bracket or bridge can be attached in a variety of locations and is calibrated to the vehicle. Once attached, it can be used to monitor reference points in a number of locations on either the upper part of the vehicle, the undercarriage, or both simultaneously.

Computerized Laser Measuring Systems. Most of the laser measuring systems currently in use work from a computerized database which monitors the undercarriage dimensions by hanging targets at predetermined locations. A computer analyzes the data of the target locations on the damaged vehicle, compares them to the correct dimensions in the database, and then displays an illustration of the undercarriage with both the correct dimensions and the current misaligned dimensions in a contrasting color on the computer monitor. These systems may use a variation of a laser or ultrasonic sound waves directed at the targets as a guiding system. Most of the laser systems can be used with a variety of different straightening and repair bench systems and frame machines.

Advantages of Computerized Systems. One of the most significant advantages of any computerized system is the ability to print out a document that shows the exact amount and location of the damage prior to initiating the repairs. When the repairs have been completed a document can also be produced showing the exact dimensional accuracy of the completed repairs. As our society has be-

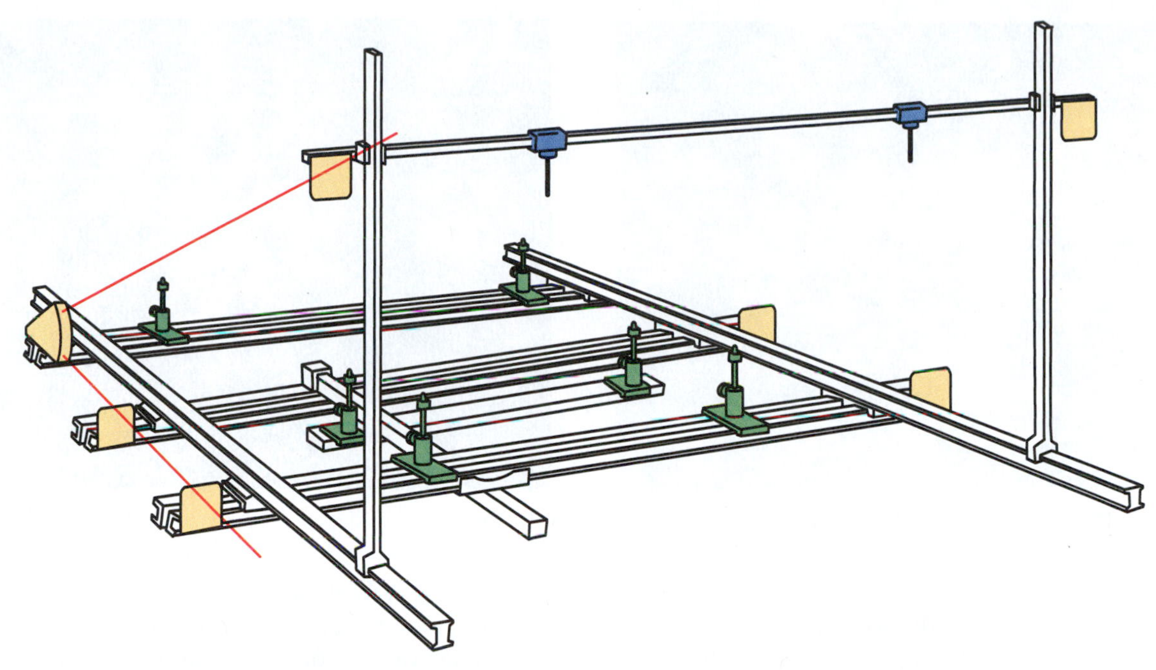

FIGURE 13-43 Laser systems typically use one or more laser beams, which are projected onto a series of targets which presumably are attached at the correct location.

come increasingly more litigious, it has become increasingly more important to document and communicate all the procedures and repair steps to the vehicle owner and the insurance company. This document is adequate proof that the repairs have actually been made and the vehicle has been restored to the manufacturer's specifications.

Another notable advantage of most computerized measuring systems is the technician's ability to observe the undercarriage's movement as it is being repaired. With many systems an image of a properly shaped undercarriage is displayed in one color and another image of the damaged undercarriage is shown in a contrasting color. Most systems update themselves approximately three times every second, so any movement resulting from the application of the corrective forces is readily visible. Their versatility is further recognized because they can be used to measure nearly anywhere ranging from underneath the vehicle to the top of the roof. Since the targets are attached to the bottom of the vehicle and the system monitors the movement from target to target, recalibrating is not necessary in the event the vehicle shifts from the corrective forces.

Most of the laser measuring systems in use today are computerized. They provide more convenience, and are very accurate and expedient to operate. The computerized laser systems function off of vehicle-specific data that are preloaded onto a computer program. Information to properly identify the vehicle must be entered into the computer, which will then show a diagram of the vehicle's undercarriage on the screen. A body scanning unit—or the hub, as it is sometimes called—must be placed in a centrally located area under the vehicle and connected to the computer as well. Once the measuring system has been calibrated by selecting the zero plane location on the computer screen, the technician must hang targets at the zero plane and several other critical measuring locations identified on the screen. See **Figure 13-44**. Once this is completed, the computer—through a simultaneous triangulation process with each of the targets—measures, records, and calculates the exact location of each target. Using this data, the computer then creates and displays a second image of the undercarriage in a contrasting color to show where any misalignment may occur. As was pointed out previously, some systems will display the correct dimension in one color and the current vehicle dimensions in another color to show the degree of misalignment that exists.

FIGURE 13-44 Each target is a reference point from which measurements are taken and interpreted by a series of bouncing light beams emitted by the scanning unit updating the image on the monitor several times a second.

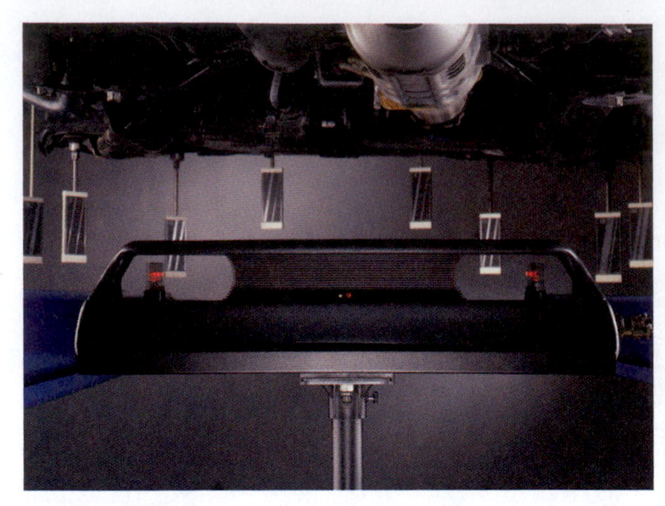

FIGURE 13-45 A solid laser beam is not projected from the hub to the target.

The targets used are system specific and their appearance and reflective surface will vary depending on the system and sensing unit used. Unlike what one might expect to find on a laser measuring system, a solid laser beam is not projected from the hub to the target. See **Figure 13-45**. Instead a series of bouncing or moving light beams are steadily projected from the scanning unit onto a bar-coded target with angled stripes on the surface. The location or level of the angle on the target where the laser light strikes the surface is used to monitor and update the datum plane. The computer updates the image on the monitor several times each second. Because of this simultaneous updating, one can observe the progress of the repair on the screen as the damaged area is pulled and straightened.

As with any system, one must be aware of some potential problems or pitfalls to avoid. One must be very careful to ensure the reference points used match those specified by the dimension publisher or the measuring system manufacturer. An incorrectly placed reference target can produce some serious consequences and the results could be disastrous if the vehicle is mistakenly overpulled while attempts are made to stretch it to an improperly mounted target. Many times the reference locations may require attaching the target to the underside of a rail that may be slightly deformed or possibly coated with dirt or undercoating. One must ensure the mounting surfaces are properly shaped and free of any surface coatings other than paint. Excessive air movement in the shop can also be a hindrance when this system is being used. Any movement of the targets can affect the accuracy of the measuring system and may even cause it to freeze up or stop monitoring and updating the progress being made.

Ultrasonic Measuring System

The **ultrasonic measuring system** uses high-frequency sound waves to measure the vehicle's dimensions. The ultrasonic system consists of a computer to operate the system and store data, the reference probes that are emitter targets, and a unit that receives the incoming sound waves. A database of the vehicle's dimensions is programmed into the computer which then displays an image of the undercarriage on the monitor when the vehicle has been identified. The reference points to be used are clearly identified to ensure the correct areas are being measured.

The emitter probes send out inaudible sound waves or clicking noises, and the receiving unit translates the sounds into electrical signals which the computer system receives to interpret the information. The ultrasonic measuring system determines how long it takes the sound to reach the emitter target, and if the times are not equal for the targets hung at common locations, the computer will determine that measurements are off. Therefore, it is imperative that the targets or emitters are attached at the precise location specified by the system. As with most other computer-driven systems, the measurements are constantly upgraded by the target locations and are constantly displayed on the monitor as the vehicle is being straightened.

As with the laser measuring system, base targets to identify the datum and the center planes must be hung under the vehicle base followed by others at specified locations on the undercarriage. From that point the remaining targets are hung one at a time to completely identify the dimensions of the undercarriage.

Electromechanical Measuring System

The **electromechanical measuring system** is available and used only on a limited basis. See **Figure 13-46**. Like all computerized measuring systems, this system uses a database to display the undercarriage and its corresponding dimensions on the computer monitor. The measuring system consists of a programmable robot-like device that relies on a movable arm to take and record measurements. The probe at the end of a very mobile articulated arm is placed into position-designated reference points. With the push of a button, the dimension at that location is entered into the computer, stored, and used for comparing the vehicle to the correct dimensions.

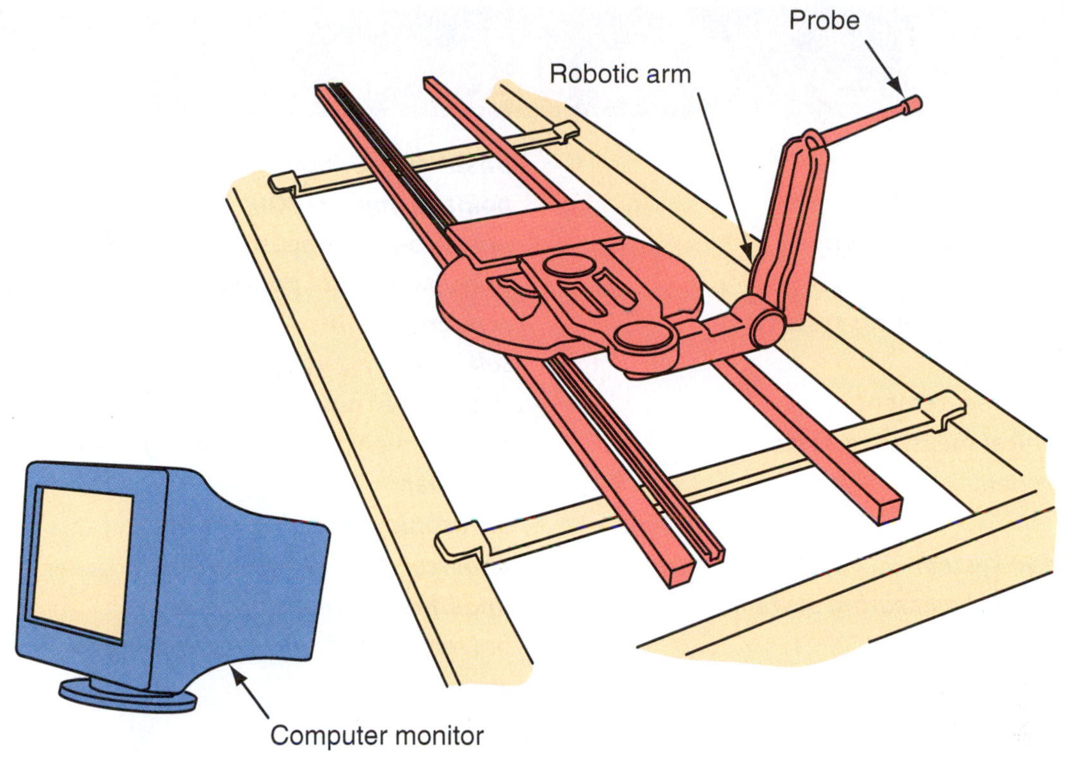

FIGURE 13-46 The electromechanical measuring system consists of a programmable robot-like device that relies on a movable arm to take and record measurements.

The robotic arm rests upon and glides along two rails (known as a measuring bridge) that reaches from front to rear of most modern unibody vehicles. The measuring slide communicates wirelessly with the computer through sensors in each joint and the measuring head. This system delivers spontaneous measuring data which again is refreshed or updated several times a second. Once the entire vehicle has been measured and the data entered, the computer can print an image of the undercarriage showing both the correct dimensions and those of the damaged area as well. However precise and accurate it may be, monitoring the repair progress is limited with this system as it can track only one location at a time by leaving the measuring probe intact with the damaged area while pulling and making the corrections. In order to observe the progress in the remaining areas of the vehicle, the probe must be moved to each individual location and the data re-entered each time a check is made.

Summary

- Knowing how to properly use a measuring system is the essence of a safe, accurate, and expedient repair.
- The industry standard allows a maximum of 3 millimeters tolerance for dimensional accuracy when restoring the damaged vehicle.
- Although there are literally dozens of different measuring systems available, no one system is the best one in the market, nor is it able to fulfill everyone's needs.
- The most current, sophisticated, and up-to-date equipment—when not used correctly—will likely provide some very disappointing outcomes.
- The success of a measuring system lies more with the individual programming and operating the equipment than with the machine itself.
- The system selected for use by technicians should be the one with which they feel most comfortable.
- Each repair facility has its own unique requirements based on the available space in the shop, the repair technician's knowledge base for computers, as well as the degree or level of the technician's talent, competence, and expertise.

Key Terms

body dimensional chart
centerline plane
center-to-center measurement
datum plane
dedicated fixture measuring system
digital tram gauge
edge-to-edge measurement
electromechanical measuring system
fixture setup sheet
laser
laser measuring system
manual mechanical measuring system
metric and inch increments
net building
offset measuring
point of impact (POI)
point-to-point measurement
principal control points
reference points
self-centering gauges
strut tower gauges
three-dimensional measuring system
tram bar
ultrasonic measuring system
undercarriage dimensions
underhood dimensions
universal mechanical measuring system
zero plane area

Review

ASE Review Questions

1. Technician A says the point zero plane is the area from which most of the undercarriage measurements are taken. Technician B says diagonal measurements should extend from one section of the vehicle into the next. Who is correct?
   A. Technician A only
   B. Technician B only
   C. Both Technicians A and B
   D. Neither Technician A nor B
2. Which of the following is NOT considered part of the three-dimensional measuring process?
   A. Length
   B. Width
   C. Height
   D. Depth
3. Which of the following is NOT a computerized measuring system?
   A. Laser system
   B. Jig and fixture system
   C. Electromechanical system
   D. Sonic measuring system
4. Technician A says the centering gauges can be used to determine the datum plane. Technician B says the height measurements are taken from the datum plane. Who is correct?
   A. Technician A only
   B. Technician B only
   C. Both Technicians A and B
   D. Neither Technician A nor B
5. Technician A says the same areas used as the control points for repair purposes are used by the manufacturer as holding fixture locations. Technician B says the center plane is used to take side-to-side measurements. Who is correct?
   A. Technician A only
   B. Technician B only
   C. Both Technicians A and B
   D. Neither Technician A nor B
6. Technician A says the jig and fixture system is usually a "go or no go" measuring system. Technician B says the universal mechanical measuring system frequently requires removing

several suspension parts prior to securing it to the bench. Who is correct?
A. Technician A only
B. Technician B only
C. Both Technicians A and B
D. Neither Technician A nor B

7. Technician A says the damage will radiate deeper into the unibody on a front hit than it will on a rear impact with a comparable force. Technician B says in a collision, each section of the unibody responds in a like manner to the section in front or behind it. Who is correct?
A. Technician A only
B. Technician B only
C. Both Technicians A and B
D. Neither Technician A nor B

8. Technician A says when making point-to-point measurements with a tram gauge, the pins should always be adjusted so the bar can be held parallel to the datum plane. Technician B says one long diagonal measurement is more accurate than three shorter ones. Who is correct?
A. Technician A only
B. Technician B only
C. Both Technicians A and B
D. Neither Technician A nor B

9. Technician A says most of the indirect damage is caused by the inertial forces in a more severe collision. Technician B says most three-dimensional measuring systems require two correct dimensional locations to establish the system. Who is correct?
A. Technician A only
B. Technician B only
C. Both Technicians A and B
D. Neither Technician A nor B

10. Which of the following would be the most usable for taking point-to-point measurements on the undercarriage?
A. Tram bar
B. Self-centering gauge
C. Tape measure
D. Strut tower gauge

11. Technician A says the electromechanical measuring system is able to monitor multiple locations while the repairs are being made. Technician B says most of the computerized measuring systems are dedicated to only one specific pulling or straightening system. Who is correct?
A. Technician A only
B. Technician B only
C. Both Technicians A and B
D. Neither Technician A nor B

12. Technician A says a diamond damage condition exists when the frame rail on one side of the vehicle has been raised at the front and lowered at the rear in a collision. Technician B says the dimensions from the center to the outside of the vehicle are equal on both sides on an asymmetrical vehicle. Who is correct?
A. Technician A only
B. Technician B only
C. Both Technicians A and B
D. Neither Technician A nor B

13. Technician A says the roof of a vehicle that is struck hard in the center of the side will lean toward the side of the vehicle that was struck. Technician B says since the indirect damage occurred last, it should not be removed until after the direct damage is roughed out. Who is correct?
A. Technician A only
B. Technician B only
C. Both Technicians A and B
D. Neither Technician A nor B

14. Technician A says when making point-to-point measurements where a large hole is used as a reference point, it may be advisable to measure edge to edge for more accuracy. Technician B says all the damaged exterior parts should be removed before anything else is done to gain access to any areas where damage may exist. Who is correct?
A. Technician A only
B. Technician B only
C. Both Technicians A and B
D. Neither Technician A nor B

15. Technician A says most measuring reference points are taken to the center of bolts, bolt heads, holes, and rivets. Technician B says occasionally it may be necessary to make comparative side-to-side measurements at reference points that are not used in the undercarriage specification manual. Who is correct?
A. Technician A only
B. Technician B only
C. Both Technicians A and B
D. Neither Technician A nor B

Essay Questions

1. Discuss the difference between where one is apt to find the direct damage and the indirect damage on a vehicle.
2. Discuss why the collision forces may penetrate deep into the front of the vehicle rather than into the rear provided it is subjected to a comparable impact.

3. List four areas one can visually inspect that can be used as tipoffs for damage identification.
4. Discuss the difference between primary and secondary damage.
5. List and describe four different three-dimensional measuring systems.

Topic Related Math Question

1. A technician is measuring a distance between two points and is using two different-sized holes on the undercarriage for his reference points. One of the holes is 48 mm in diameter and the other is 36 mm in diameter. The distance between them is 1,480 mm. What would be the distance between them when measuring inside-to-inside and outside-to-outside?

Critical Thinking Questions

1. Imagine a vehicle being struck by another vehicle on the left front corner in a near head-on collision. Discuss how the front section of the vehicle would react and trace the damage flow into the center section.
2. Two people took a diagonal measurement in the same area of the vehicle with a tram bar. They obtained significantly different dimensions. What might they have done differently to arrive at their respective measurements?
3. Why, if the pointers on the tram gauge are set the same distance apart but are offset, does one arrive at a different dimension than when the pointers are set even when measuring between two points with them?
4. How and why did the metric measurement system come about?

Lab Activities

1. Select several random measuring locations on a unibody vehicle. Make some quick check comparative measurements to those locations and create a measurement legend by documenting the dimensions.
2. On the undercarriage, locate two different-sized oversize holes (12 mm or larger) that are common to both sides of the vehicle. Using the averaging technique described in the chapter, determine the center-to-center dimensions for each location. Using the same technique, determine the dimensions of the near side to far side of these holes and record your findings.
3. Locate two possible alternate zero plane locations on the front and rear of the vehicle that may be used in the event the designated area is damaged and another must be used as a starting point for measuring the undercarriage.

Chapter 14

Straightening and Repairing Structural Damage

OBJECTIVES

Upon completing this chapter, you should be able to:

- Discuss the precautions that must be taken to avoid personal injury and prevent further vehicular damage during the repair procedures.
- Discuss the types of damage commonly sustained by a vehicle in a collision.
- Discuss the types of equipment commonly used for repairing damaged vehicles.
- Discuss the repair concepts commonly employed to restore a damaged vehicle to its pre-collision condition.

INTRODUCTION

We have all no doubt seen the television commercials that show a car crashing into a barrier wall to convince the consuming public of the vehicle's collision worthiness. Every microsecond it takes for the actual crash to occur from start to finish represents numerous hours to repair and restore the damage. Before any repairs can be made, all of the damage must be identified and isolated in all of the areas affected in the collision. This is usually accomplished with the aid of a three-dimensional measuring system and other hand-held measuring tools and equipment. Once all the damage sustained by the vehicle has been identified and pinpointed, a technician formulates a repair plan and maps out the sequence and procedures to be used. The ensuing procedures of repairing, straightening, and replacing the reinforcement members are perhaps the most critical part of restoring the vehicle's structural integrity. One must keep in mind the vehicle must be restored to its pre-collision condition or—as it is sometimes called—maintain its **re-collision integrity**. This means if the vehicle is involved in a subsequent collision it must be able to withstand the same degree of collision forces and each structural member must respond and react in the same identical manner the manufacturer designed it to do. In order to ensure this will happen, any part of the structural system affected in the collision—however large or small—must be restored to withstand the same forces, bend and deform in the same manner, and be able to deflect and dissipate the effects of the collision energy in the same identical manner it did originally.

SAFETY AND PPE REQUIREMENTS

Repairing the vehicle's structural damage presents more opportunities for personal injury than any other part of the repairs. Many times the exterior sheet metal panels are ripped and torn, leaving dangerously sharp and jagged edges exposed. These create many opportunities for personal injuries. In order to gain access to damaged areas deeper inside the vehicle, parts and sections must be removed. This is frequently accomplished by removing and cutting away sections of a panel with air chisels or metal cutting equipment, which often creates additional possibilities for injury by exposing sharp and unprotected surfaces. The roughing and straightening procedures frequently require the use of corrective forces which sometimes approach 10,000 pounds of pressure per square inch on the hydraulic system.

The possibility of personal injury is imminent if the repair technician becomes careless around this equipment. Gloves and safety glasses are the minimum PPE equipment that should be used, along with a welding face shield any time heating or welding operations are performed. The sheet metal in the attachment area may tear, releasing an undirected 5- to 10-pound projectile with a considerable amount of force. Pulling clamps might slip off the surface to which they are attached, creating a safety hazard, Anti-recoil cables should always be attached to the pulling clamp to prevent it—or whatever pulling device is being used—from becoming air-borne should it tear loose or slip from the surface being pulled. An unattentive technician or an innocent passerby could easily fall victim to serious injury because of the corrective forces required to straighten a damaged area of the vehicle. The possibility of further damage always exists to areas of the vehicle that are adjacent to the repair area—particularly when working around or near the glass. The necessary precautions must be taken to protect the windows by either removing the windshield or **backlite** before initiating the repairs. The door glass should also be rolled down into the door or covered with a protective cover if any grinding or sanding operations are performed near it.

STRAIGHTENING THE METAL

During a collision numerous internal and external forces react to each other, causing the body of the vehicle to respond and behave in ways it would never do under normal usage.

Conflicting Forces Collide

The internal forces—those inside the vehicle that are usually referred to as inertial forces—have to do with the vehicle's motion and momentum as it is either moving down the road or standing still.

Hostile or External Forces

The external forces, sometimes called hostile forces, are the opposing forces thrust upon the vehicle by the impact object. As these hostile forces move through the vehicle they cause the metal to become bent, buckled, and torn at the primary or direct damage area. As the damage radiates out and away from the **point of impact (POI)**, the panels tend to resist the forces thrust upon them. In the process they become stretched out of position as they strain to remain attached to their adjacent panels, resulting in those panels being put under a strain as well. As the damaging forces continue to travel through the vehicle, the indirect or secondary damage becomes broader and more widespread until the impact forces stop. As a rule, the evidence of damage is greater near the point of impact because the extent of damage is reduced as the energy is dissipated into the adjacent structure. However, in some cases, the energy is passed through the impact point with little evidence of visible damage and is propagated to a point that is deep within the body.

Identifying the Path of Damage

In the aftermath of all these forces resisting one another, a significant amount of damage has occurred. A unibody vehicle is designed so that the energy thrust upon it during impact travels along a predetermined path. The collision energy flow starts at the point of impact and flows through the structure until all the energy has been dissipated. In the process, metal has been buckled, ripped, and torn in the immediate impact area. The metal in the adjacent panels has become strained as the structural reinforcing members route the collision energy around the passenger compartment. Many times this causes welded mating surfaces to become stressed and—depending on the shape of the adjacent panels—puts them under either a state of tension or pressure. As the collision energy is routed through the structural members, they bend and buckle at their designated collapse zones to help absorb, deflect, and dissipate the effects of the damaging forces. These same structural members often become deformed in areas away from the designated collapse zones as well. As a rule, as the damage travels deeper into the vehicle the path of damage and destruction becomes more widespread but less severe until it stops. This is the point at which the repairs must be initiated to reverse the effects of the collision energy.

Damage Analysis

To the untrained observer, removing the damage simply involves reversing the effects of the collision energy. In reality, although reversing the effects of the damaging forces is exactly what must occur, a considerable amount of analyzing and planning must first be done. One must determine where the damage ended or where the effects of the collision energy stopped. This is done through a series of steps that involves looking for visual tell-tale signs that are frequently tipoffs for locating damage. See **Figure 14-1**.

For example, a larger gap between the top and bottom of the fenders and the doors is a sure sign that the front clip has been kicked up in the front end. See **Figure 14-2**.

FIGURE 14-1 An uneven gap between the fender and the door is an indication of damage to the front section of the vehicle where the fenders are attached.

FIGURE 14-2 The rear of the front door overlapping either the rear door or the rocker panel is an indication the hinges or the hinge pillar have become misaligned.

FIGURE 14-3 A buckle over the area of the B pillar in the roof is an indication the A and B pillars have been affected in a frontal collision, and the sail panel has been affected in a rear collision.

Tell-Tale Signs of Damage

A sagging door or one that overlaps the rocker panel or the quarter panel may indicate a misaligned hinge pillar. Likewise, a non-uniform gap between the front and rear door or between the door and the quarter panel are a sure sign that the vehicle's rear section has shifted around the center section. See **Figure 14-3**. A characteristic buckle in the roof above the vehicle's B pillar indicates the A pillars have been affected, as the damage was transmitted through them, causing them to be pushed back at the bottom and upward at the roofline. This may also create an excessive gap between the upper frame of the front door and the roof line. The obvious cracking of the windshield may likely have occurred as the pillars shifted to accommodate the transmission of the damaging forces around the passenger compartment.

Other Subtle Signs. A number of additional, more subtle signs may be evident that can help one identify the damaged areas. They may include broken or missing parts, unusual gaps in strengthening materials such as reinforcements or patches, shifted part-to-part joints, and misaligned corners and edges of components. Other symptoms may be torn, pulled, or strained spot welds in any of the direct or indirect damage areas. These welds may indicate a strained panel or structural member. Broken undercoating in the wheel housing or on the undercarriage, tire scrub or scuff marks on the wheel housing or splash shield, a damaged wheel, or debris stuck between the tire and the wheel all indicate potential damage to the wheel and suspension areas. Fresh rust on areas that are typically painted or coated, as well as cracked or flaked paint surfaces, are all suspect areas where buckles may have occurred. Extensive measurements must be made by using any one or a combination of tramming, X-checking, or a three-dimensional measuring system to identify the location and extent of the damage.

Revealing Damaged Areas. In many cases it becomes necessary to remove some of the more severely damaged panels to gain access to additional damage that would otherwise not be visible or accessible. This may include unbolting panels, chiseling and cutting away remaining sections of severely damaged panels, or cutting openings or windows into the outer layers of damaged panels. The structural members—or a part of them—should not be removed or separated from their mounting location until an adequate amount of stress relieving has been effected. At this time it may also become necessary to rough repair the undercarriage to ensure the location of the zero plane for attaching the measuring system, as discussed in Chapter 13. Once the measuring system has been put into place, the process of applying the corrective forces may begin.

STRAIGHTENING EQUIPMENT

The type of straightening equipment used will dictate the entire repair process and the procedures employed to complete the job. The first thing that must be determined is whether the available equipment is capable of accommodating the required repairs. Some vehicles, especially the 4 × 4s, are larger, heavier, and considerably more heavily reinforced than the conventional framed and unibody vehicles. Many times they are considerably longer and, because of their large tires, are wider than most of the domestic vehicles. Some of the pulling machines are not long enough to properly pull some of these vehicles, especially when an end-to-end application is necessary. Attempting to repair a vehicle on a machine that is too short or too small to properly accommodate it will inevitably result in personal injury to the technician or additional damage to the vehicle. The capacity of the hydraulic system used with the repair equipment will also be a significant consideration in determining whether the vehicle is repairable.

Avoid Overextending Equipment Capability

Safety issues, for both the technician and the vehicle, become significantly more profound when repairing a body over frame vehicle than when repairing a unibody. Because they are more heavily reinforced, the pulling equipment may require a pulling capacity up to as much as 10 tons. More and more vehicles manufactured today are equipped with frame sections made of high strength steel and hydroformed steel. Because of the increased amount of force required to pull a damaged body over frame (BOF) vehicle, the possibility of clamps breaking, blockage slipping, and chains breaking becomes more of a concern. Each in their own way increases the possibility of injury. Additionally, simultaneous multiple pulls are necessary to minimize the amount of stress put on the metal and to allow for localizing the pull. That is, it becomes necessary to attach the pulling clamps as close to the damaged area as possible to maximize the corrective energy application. The longer the distance from the pulling location to the point that must be affected, the less effective the pulling forces become and the more imminent the possibility of equipment breakage and vehicle slippage and shifting.

TYPES OF EQUIPMENT

The equipment most commonly used in today's repair facilities is one of three primary types. a floor-mounted system, a portable system such as the bench used with a **dedicated fixture system** or with a simultaneous **universal system**, and a bench or frame rack, as they are commonly called.

Floor-Mounted System

A **floor-mounted system** may consist of a series of interconnecting rails installed flush to the floor that are used with special hydraulic rams that fit into the rails via special adapters or a series of steel pots that are encapsulated in the concrete floor. A power post equipped with a hydraulic ram and pulling chain is anchored to these pots with another chain and is used to apply the corrective forces. See **Figure 14-4**. An adaptable or universal clamping system that can be anchored to the floor with the remaining pots in the floor is typically used to secure a unibody vehicle. When making repairs on a BOF vehicle, it is typically secured with a series of clamps and chains attached to the frame rails and secured at the floor pots. The advantage of this type of system is a stall is not necessarily dedicated to performing only frame and undercarriage repairs as is the case with a **frame rack** or **bench system**. One can also have more than one vehicle set up for pulling at any given time, which frees up some critical shop floor space. This eliminates the "wait in line until this one is finished" dilemma. The most significant disadvantage of this system is the work height is limited to floor height only; therefore, access to the underside is somewhat limited and relatively restrictive. The type of measuring systems that can be used with this system is also somewhat limited because of the height at which the vehicle is raised off the ground.

Portable Bench System

The portable bench systems usually consist of an extremely strong portable steel work bench on wheels with a perfectly flat working surface to which the vehicle is attached at the pinchwelds with pinchweld clamps or sill clamping fixtures. See **Figure 14-5**. Transverse beams are installed onto the bench to hold the vehicle in place, as well as to attach the vehicle-specific measuring system.

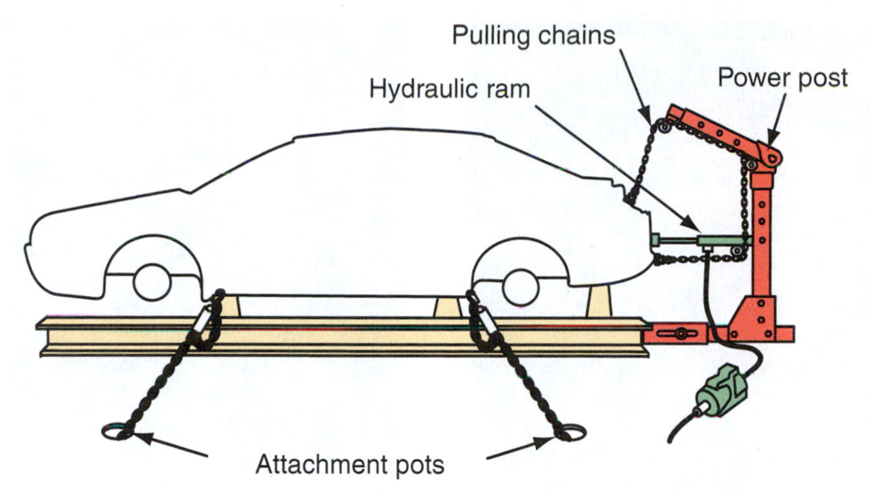

FIGURE 14-4 A floor-mounted pulling system may be a series of flush-mounted rails in the floor or a series of pots in the floor used to secure the vehicle and the power pulling posts that are used to pull the damaged vehicle.

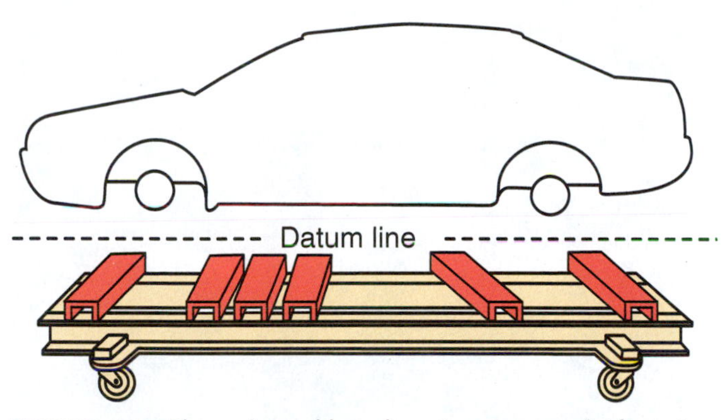

FIGURE 14-5 The universal bench system uses a platform to secure the vehicle at the pinchweld with special clamps attached to the framework of the bench. Adjustable vertical tubes—together with a measuring system created by attaching transverse beams to the platform—are used to monitor the vehicle's undercarriage while repairs are made.

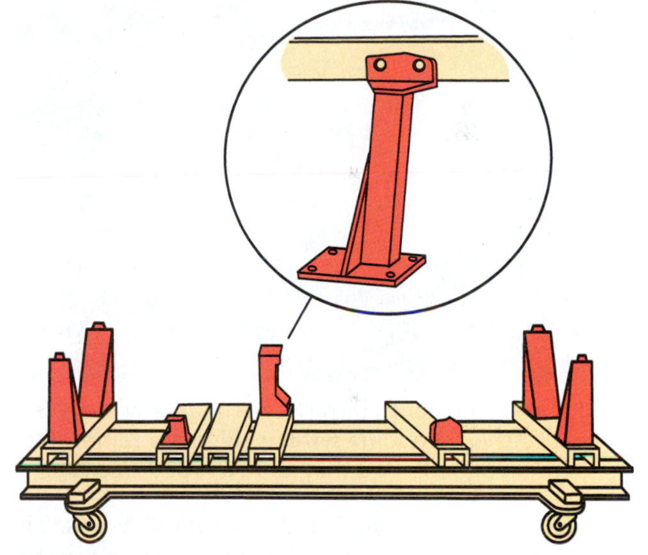

FIGURE 14-6 The dedicated fixture system is often called a "go or no go system" because fixtures are attached to the platform and, after removing specified undercarriage and steering parts from the vehicle, it is lowered onto the fixtures. If the undercarriage lines up with the fixtures it is considered dimensionally true. If not, it is out of alignment and must be straightened.

Dedicated Fixture System. On a dedicated fixture system, in addition to the vehicle being attached to the pinchweld clamps, a series of vehicle-specific fixtures are attached to the bench that must align with designated locations on the vehicle's undercarriage. See **Figure 14-6**. Designated suspension and other mechanical parts must be removed from the vehicle before it can be mounted onto the fixtures on the bench, as these locations are monitored by the measuring system.

Dozer System. A **dozer pulling system** that is adapted to the specific bench being used is commonly employed as a straightening system with these benches. These dozers are essentially articulated arms that are energized by a hydraulic jack pushing against them to pull the damage out. A chain is attached to the arm, which in turn is attached to the damaged area of the vehicle. In this way correction forces are applied. Depending on the manufacturer, one—or as many as three—of the dozers are attached to the bench at any given location fully 360 degrees around the vehicle. When multiple pulling dozers are available, multiple pulls can be made simultaneously in addition to a complete end-to-end pull, giving the equipment the flexibility that other models do not offer.

Rack System

The rack systems are usually a hybrid carried over from the traditional frame repair systems with some modifications. See **Figure 14-7**. They are usually equipped so they can provide both pulling and pushing forces by changing some of the tooling and fixtures. The size of their working surface is usually fairly substantial so that in addition to using them for straightening operations, they may also be used

FIGURE 14-7 The bench or rack system usually has a relatively large deck or working system to which the pulling towers are attached. On some systems these are removable. They can usually be lowered to ground level for ease of moving them on and off, and they can be raised to a comfortable working height.

for such tasks as alignment and suspension repairs. The alignments can be accomplished by using portable turntables that are placed on the bed of the machine. The tie-down system may be a dedicated clamping system that clamps directly to the bed of the machine, or it may use a portable universal system. Vehicles may be winched onto the bed or they may be driven or pushed onto them. The bed of the machine is usually elevated from the floor and the vehicles are loaded via ramps by tilting the machine bed. Likewise, the vehicle can be driven straight onto the bed of the machine and then raised and lowered using a built-in hydraulic system.

Multiple Pull Towers. These machines are commonly equipped with three pulling towers that are attached directly to the bed of the machine. On some machines the pull towers roll along an integrated track at the front, sides, and rear of the machine bed, thus enabling them to pull a full 360 degrees around the vehicle. On some machines, the towers may swing from a central location at the front of the rack, enabling them to pull from an approximate 270-degree radius around the vehicle. Since many of these machines are a hybrid of the original frame machines, they are constructed heavily enough so they can be used for straightening BOF vehicles. The anchoring system used makes them very useful for repairing unibody vehicles as well. Several different three-dimensional measuring systems are adaptable to this equipment as well.

Bench/Rack System

The **bench/rack system** is yet another hybrid system with characteristics that are similar to both the portable bench and the rack. While the bench is usually not portable or movable, it is designed to be elevated off the ground for access underneath while the corrective pulls are being made. The pull towers can be attached to any area of the machine's perimeter, affording a full 360-degree pulling capability around the vehicle. The Car-O-Liner system is designed in such a way that the pulling towers can be attached to the bench, thus raising the entire tower—along with the bench—to a more convenient working position or height. The anchoring system is usually a positive or solid system wherein the vehicle is attached to the clamps. These, in turn, are solidly bolted to the bench. This anchoring system can be used on both BOF and unibody designed vehicles.

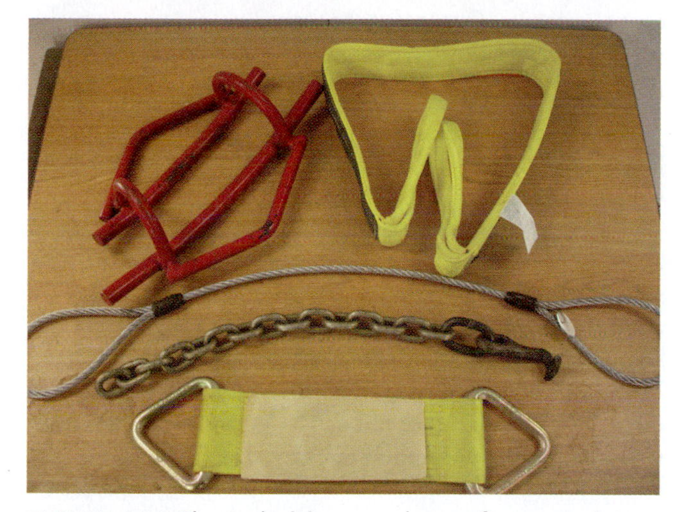

FIGURE 14-8 The technician can choose from a variety of pulling clamps, as well as hooks, chains, pull straps, cables, and other accessories, to enhance the repair process.

PULLING ACCESSORIES

The accessories available for use with the straightening and pulling equipment are as varied and diverse as are the vehicles being repaired. The technician has the option of choosing from a wide array of pulling and holding clamps, pulling straps and cables, hooks, and other oddly shaped devices. See **Figure 14-8**. Some of the accessories are very vehicle specific, but most tooling is quite universally applicable. Each equipment manufacturer has their own unique set of accessories they recommend using with their equipment. However, despite the equipment manufacturers' recommendations, the accessories selected for use by the technician are based more on personal preference, not because they match the equipment manufacturer's system.

Attaching and Pulling Techniques

Pull clamps may be attached directly to the damaged metal that will be removed once the roughing part of the repair is completed. See **Figure 14-9**. They can be attached directly to an exposed edge of the panel or one can cut a window or hole into the metal and install the clamp onto the edge of the opening. Another technique commonly used is to weld a **sweat tab** onto the surface and attach a pull clamp to it. This allows the technician to apply a corrective force in the immediate area where it is needed. One should always make the first series of pulls before removing any of the damaged metal—especially if it is a welded panel. See **Figure 14-10**.

Pull Clamps. As a rule, when pulling on single-layered nonstructural parts a light-duty single bolt expandable jaw clamp is used because it is lightweight and can be

FIGURE 14-9 Pulling clamps can be used in a number of different ways. In this illustration, a hole has been cut into the panel immediately in front of the damaged area. This allows the technician to apply corrective force directly on the damaged area, thus maximizing the pulling effects.

FIGURE 14-10 Because the part will be replaced after roughing it out, a pull tab may be welded directly to the damaged area to make a corrective pull.

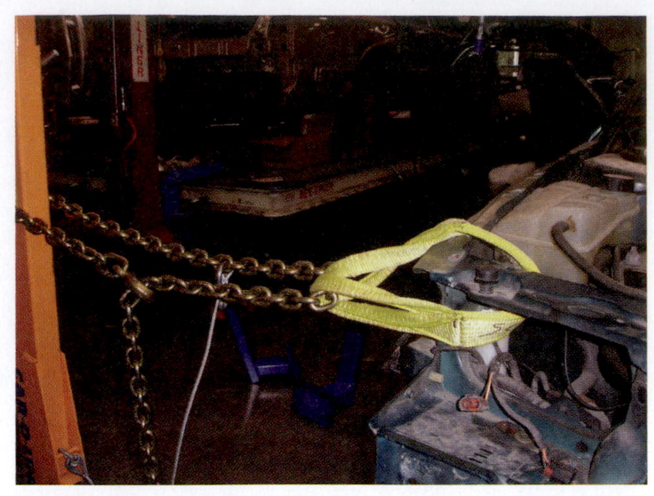

FIGURE 14-11 A flexible nylon reinforced strap can be used for making corrective pulls, particularly in instances where an entire section of the vehicle has to be moved simultaneously.

installed and removed quickly and easily. As the metal becomes thicker due to multi-layering or because of increasing thickness, the pull clamps used must be heavier duty. These clamps are often secured with two or more bolts. Thicker or multi-layered metal usually indicates that the area is a structural part, which would suggest that high strength steel may be used. Accordingly, it will require additional force to move the metal. In addition, the amount of force applied requires that the "teeth" on the clamp's biting surface be able to hold more aggressively. When a greater than normal amount of tension is to be applied, a weld bead can be made near the edge of the pull tab to keep the clamp from slipping off.

CAUTION

When making pulls using more force than normal, one should stop once the slack has been taken up and retighten the clamp to keep it from slipping.

One must take the necessary precautions to prevent the pulling clamp from slipping off the surface being used.

Reinforced Nylon Straps. Reinforced nylon straps can also be used in areas where an entire section of a vehicle must be pulled or moved. See **Figure 14-11**. The use of these straps allows the technician to exert a significant amount of tension or force to a flat surface without causing any damage to the area because of the strap's flexing characteristics. One must be cautious to not use the pulling strap in areas where sharp edges of metal are exposed, as they may cut or damage the straps.

Using Damaged Parts as Pulling Points. Many times it becomes necessary to re-align openings and areas of the vehicle that have become misaligned in the collision. The door openings, for example, are likely targets to become misaligned or be pushed out of square on either a side hit or an angular impact. In order to restore the opening to dimensional accuracy, special clamps designed for a particular shape or area of the vehicle may be required. Instead of using a specialty clamp, one may choose to use the door hinges as a pulling attachment since they will most likely be replaced. This places the corrective force in the exact location or area that caused the misalignment.

TECH TIP The anchor points must be able to withstand the total of the cumulative forces applied when repairing the damage. No one part of the vehicle can withstand the total amount of the corrective forces that are applied.

PRINCIPLES OF ANCHORING

None of the aforementioned operations of rough repairing and attaching the three-dimensional measuring system can be accomplished without first anchoring the vehicle. A variety of anchoring systems are available, some of which are frame rack or bench specific while others are universally adaptable. See **Figure 14-12**.

Solid Anchor System

A unibody vehicle is typically anchored to the bench by clamping it at the pinchweld with special pinchweld clamps. This is commonly called a **positive** or **solid anchoring system**. Once the vehicle is attached to the pinchweld clamps, they are bolted to securely lock them down onto the bench. This holds the vehicle and prevents it from moving as the corrective forces are applied.

Since the unibody does not have a frame that can hold the vehicle against the corrective forces used during the repairs, a facsimile of the frame is created by clamping the vehicle to the pinchweld clamps. Anchoring the vehicle to the bench or frame rack is the only means of securing the vehicle to prevent it from shifting or moving as the corrective forces are applied. Therefore, one might say the pinchweld clamps temporarily become the vehicle's frame.

Anchoring at the Pinchweld

Correct anchoring is critical to protect the vehicle from further damage which can be caused by the corrective forces. The pinchweld clamps are typically installed or attached at each of the four torque box areas under the corners of the passenger compartment. This area of the vehicle is used because these are the most heavily reinforced areas of the undercarriage, thus minimizing the possibility of creating any damage with the corrective forces. This also distributes the pulling force uniformly throughout the entire vehicle, which further helps to stabilize the vehicle and allows for force application in any direction without causing any distortion or damage to any one of the anchoring points. Attempting to repair the vehicle without properly securing it—or by attempting to hold it at fewer locations than is recommended—will inevitably result in creating additional damage and also increases the possibility of personal injury.

FIGURE 14-12 Many benches and racks have a dedicated vehicle clamping system which is designed to secure nearly all vehicle models.

Preparing the Anchoring Area

Before attempting to secure any part of the vehicle to the pinchweld clamps, several preparatory steps must first be performed. One must carefully inspect the pinchweld area that will be used for supporting the vehicle's weight to ensure it is free of any corrosion perforation and that it is structurally sound and able to support the weight without collapsing. No signs of any previous repairs should appear in the area. Any corrosion protection or chip guard material that has been applied to the area must be scraped off the surfaces where the clamps are to be attached. The jaws of the holding clamps should also be checked to ensure the teeth are not filled with filler material from a previous operation. Failure to do so will prevent the clamps from properly "biting into the metal," thus allowing the vehicle to move or shift as the corrective forces are being applied. Depending on the measuring system being used, recalibrating the machine may be required any time the vehicle shifts, even if only a very slight amount.

Protecting Brake and Fuel Vapor Recovery Lines

On many vehicles, electrical wiring, gas and brake lines, and other high pressure hydraulic lines are routed through the area where the pinchweld clamps are attached and must be moved out of the way. These lines are typically routed along the rocker panel immediately to the inside of the pinchweld. They can easily be relocated simply by removing the hardware holding the lines in place and pushing them out of the way. Failure to do so may result in crushed, cut, or damaged hoses and lines that may go undetected until the vehicle is ready to deliver to the customer. This failure could easily cause the customer to doubt the quality of the repairs on the rest of the vehicle. Once the corrective forces have been initiated, one should always retighten the clamps to ensure they are tight enough to securely hold the vehicle, because they tend to grab and dig into the metal once the pulling is initiated. One must also keep an eye on the pinchweld areas during the repair process to ensure they do not collapse or become distorted from the effects of the corrective forces.

TECH TIP It is imperative that the pinchweld areas used for clamping the vehicle are redressed, cleaned, and refinished completely to restore the corrosion protection and to ensure no visible signs of previous repairs are left behind.

ANCHORING THE BODY OVER FRAME VEHICLE

Anchoring a body over frame (BOF) vehicle requires a completely different method for securing the vehicle to the frame rack or bench than what is used for the unibody. See **Figure 14-13**.

Holding Devices

A series of chains, tensioners, stabilizing beams, blocking devices, and occasionally steel or wood blocks may be used for bridging to prevent the vehicle from shifting or moving as the repairs are performed. One must be cautious about the chains selected to ensure they are proof tested to withstand the amount of the corrective forces that are applied. A dollar saved by purchasing a lesser quality chain will be quickly negated the first time it breaks, causing additional damage to the vehicle and/or to the technician.

Some frame machine manufacturers offer machine-specific anchoring clamps and tooling similar to that used for unibody repair. In these instances, clamps and hold-down fixtures are manufactured to fit the exact dimensions of the bench or frame rack. A variety of adapters may be required to accommodate and securely tie down the numerous different-sized vehicles that may be encountered. Another manufacturer offers a universal adjustable blocking system that is adjusted to fit the inside perimeter of the frame. Once it is set up for the specific frame that is being repaired, it is bolted to the vehicle's frame as well as the frame rack or the bench, thus preventing the vehicle from shifting as the corrective forces are applied. The corrective forces used to repair the vehicle can be applied from nearly any conceivable angle. Regardless of the equipment used or the method employed to secure the vehicle to the rack or bench, it must be able to hold the vehicle securely regardless of the angle from which the pull is being made.

FIGURE 14-13 Unlike the unibody, on which the vehicle is secured to the bed with pinchweld clamps, the BOF vehicle is generally secured by using chains and other blocking attachments.

STRAIGHTENING CONCEPTS

The essence of an accurate and expedient repair is analysis. Therefore, the first order of business for straightening and restoring any collision-damaged vehicle is to thoroughly analyze the damage and formulate a repair plan based on that analysis. The repair plan must include a thorough understanding of how—and the order in which—the damage occurred. Using the visual tips and indicators for locating damage discussed previously, one must mentally re-create the accident and visualize the order and sequence in which the damage occurred and where it stopped. One can nearly always assume that the point where the damage flow stopped will most likely be a reinforced area such as a crowned surface, an area where a structural member is welded to another, or an area immediately adjacent to a crush zone. That is the location where the damage reversal must begin.

LIFO

Regardless of the severity of the damage, when restoring a collision-damaged vehicle, the concept of "last-in first-out" (**LIFO**) must be practiced. In other words, the damage that occurred last must be removed first. In order to accomplish this, not only must the damage sequence be reversed, but the direction of the damage flow must be reversed as well. Failure to do so will inevitably result in creating additional damage which may require replacing parts that otherwise would have been repairable.

Choosing Pulling Locations

Several considerations must be weighed when deciding the how, when, and where of the repair procedure. For example, where should the first pull be made? At what angle should the pull be made? Where and how should one attach the pulling clamps? All these questions may be thought through and answered, but one won't know the exact sequence of events that will occur until the pulling and straightening procedure begins. A "roll with the flow" sequence may have to be followed, as it is often difficult to predict how the damaged areas will react and move once the corrective forces are applied. Ultimately, the goal is to move the metal in a reverse direction of how it was formed or created.

FIGURE 14-14 Using a multiple pull setup will expedite the removal of the damaged areas a great deal.

Reversing Damage Flow

In order to reverse the damage flow most effectively, a series of multiple pulls should be used with the initial setup. See **Figure 14-14**. This will allow the technician to pull in two or three different directions by alternating back and forth from one pulling point to another as the metal starts to unfold. It might be noted that the surface is left under tension and is not released when switching from one pulling point to another. Many times as the damage flow is reversed, the metal "locks up." In order to get it to move or to unlock it again, it becomes necessary to pull from another angle or from another adjacent area. If the equipment used is not able to accommodate a multiple pull setup, a considerable amount of time will be lost by repeatedly having to release the corrective force and switching from one pulling point to another. A significant amount of the spontaneous corrective forces are also lost in the process by constantly having to move back and forth from one area to another. In many cases the metal at the POI is collapsed and crumpled so severely that it may be necessary to pull the more severely damaged sections of metal out of the way and remove them to make room for the damage deeper into the vehicle that needs to be pulled out. Once this is accomplished, one can get down to the finite pulling techniques that will ultimately remove the stressed area, particularly on the panels that are to be salvaged.

PULLING AND ATTACHING LOCATIONS

One of the most critical decisions that must be made is where on the vehicle the clamps, straps, or other pulling devices should be attached to maximize the effort and obtain the best results. One must remember that most of the area that is being considered as a potential pulling surface has been damaged in the collision and is likely going to be replaced. Therefore, nearly any area in and around the POI is "fair game" to attach pulling clamps, straps, cables, and so on, to complete the removal of the damaged sections in order to gain access to the deeper damaged areas. One need only remember that nearly any area is a potential pulling location so long as it is able to withstand the pulling forces exerted in the area.

Attach at End of Damage Flow

As the direct damage is removed and the corrective forces move inward, other areas that can be used for attaching points include the bumper reinforcements, their mounts, and the bolt holes used to attach them to the vehicle. Steering and suspension parts, as well as mounting bolts and locations, may also be used because they are usually the end of the damage flow as it is transmitted through the parts. These areas are usually more heavily reinforced. In addition, since the bolts and hardware used to attach the parts to the vehicle are likely to be replaced, they make excellent fastening locations because of their added strength. A word of caution, however: Do not attach clamps to any of the mechanical, steering, or suspension parts that are not going to be replaced—use only their attachment areas or locations. See **Figure 14-15**.

Use Damaged Exterior Panels for Pulling

Another area that may be used for pulling attachment points is the damaged sheet metal, particularly if it is going to be replaced. In some cases it may be necessary to pull on the damaged exterior panels in order to relieve some of the collision stresses and also to gain access to the seams and attaching points to remove the panel in preparation for replacement. A very quick and easy way to pull on the outer panels is to install weld-on tabs—or sweat tabs as they are often called—onto the surface and attach a pulling clamp to it. This offers a significant advantage in that the corrective force can be applied at the precise location where it will be most effective. When the pulling is completed, the panel will be removed and no further cleanup is necessary in the pulling area. As was discussed previously, a window can be cut out of any area of the panel and a pull clamp may be attached in that area as well.

STRESS RELIEVING

Another element of the repair process that is a vital link to successfully straightening and repairing the damage is stress relieving. The three most common methods used for stress relieving are (1) hammering or shocking the metal; (2) pulling the buckled and distorted metal to unlock the stressed areas, allowing it to reflow; and (3) applying controlled amounts of heat.

Shocking the Metal

The old theory of using a heavy sledge hammer and beating on the surface of the frame rail no longer applies,

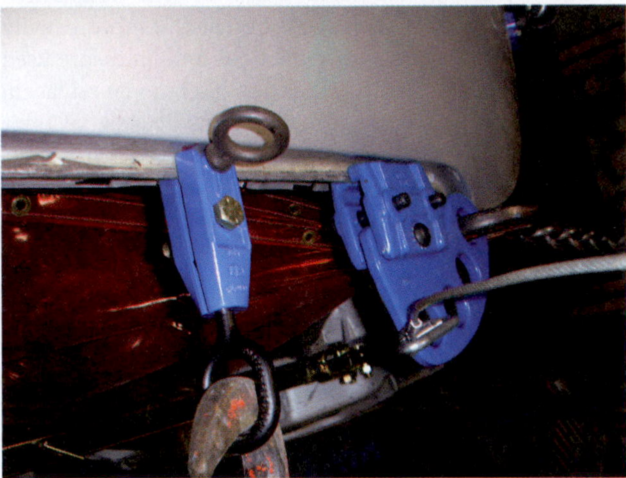

FIGURE 14-15 The corrective forces can be applied using a variety of different pulling clamps and other accessories by attaching them to the damaged sheet metal and many other reinforced areas.

especially on the undercarriage of a unibody vehicle. This would do little more than create a significant amount of additional damage by the crushed surface where each hammer blow is struck. When hammering to stress relieve around the repair area, one must remember that with high strength steels a much larger surface must be addressed. One must remember that when hammering to relieve stress on a unibody vehicle, it must be done with more finesse than when stress relieving on the frame rails of a body over frame vehicle. It is best to use a lighter weight hammer to avoid making numerous indentations on the surface that are likely to occur when too large a hammer is utilized. See **Figure 14-16**. Stress relieving is best accomplished with lighter and less massive hammers and tools as opposed to the old tried and true 2-pound ball peen hammer. The principle of stress relieving is to shock the metal, momentarily setting the metallurgical structures into a free state that relieves the stresses and allows the metal to move. Numerous hammer blows with a lighter weight tool are much more effective than a few with a heavier hammer. Fewer divots and less surface marring will occur with a lighter weight hammer, thus requiring less cleanup and repair to ensure no tell-tale signs of repair are left behind.

Pulling Damaged Areas

Pulling the damaged rails is another very effective way to relieve stress. Many times the panels and damaged surfaces are collapsed and buckled so severely that the only way to unlock the damage is to forcibly pull the crushed metal apart. As the damaged structural parts are pulled, allowing the damaged bolted-on parts to be removed, a considerable amount of stress relieving is accomplished. This sets the stage for the technician to gain access to the damage deeper into the vehicle, as well as to further release and remove some of the effects of the collision energy. See **Figure 14-17**. The rails and structural members closest to the outside of the vehicle—such as the front or rear rails—occasionally have to be sacrificed in order to remove damage deeper into the vehicle. In most cases the damage sustained will justify replacing the part anyway. Pulling on it will help to relieve the stresses in the surrounding areas prior to removing it.

Heat Application

The third, and perhaps the least desirable, method of stress relieving is to use controlled heat. Using heat is the most difficult method to administer, and when employed should be done as a last resort only. Heat is said to be high strength steel's worst enemy because the steel is very temperature sensitive—to the point where it cannot be heated to temperatures above 1,200°F for 2 minutes. However, some forms of high strength steel have a lower **temperature threshold**, allowing for temperatures up to a maximum of 700°F. The safest method for applying heat is using

FIGURE 14-16 Stress relieving on a damaged rail doesn't necessarily require a heavy sledge hammer, as was the custom on older vehicles with heavier gauge frame rails. The same effect can be accomplished using a lighter weight body hammer and directing more light blows with a spoon to distribute the force of each blow over a larger area.

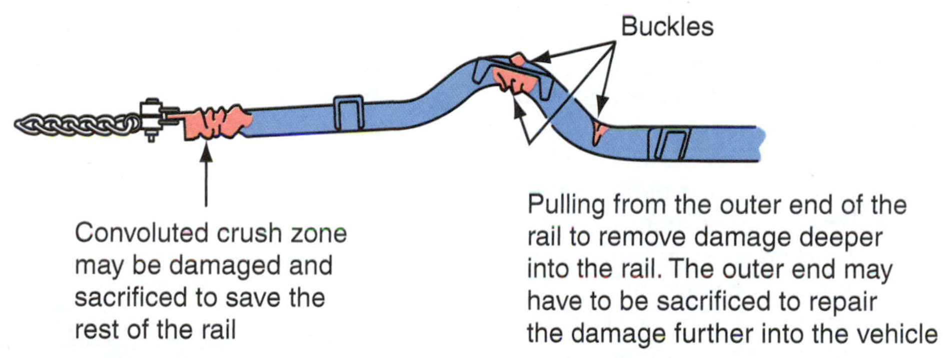

FIGURE 14-17 A damaged rail end may have to be sacrificed in order to remove the damage and restore the dimensions deeper into the rail. A convoluted crush zone that sustained damage in the collision must be replaced; therefore, pulling on it is acceptable.

an induction heater. An open flame can be used but is less desirable because of the possibility of undercoating and other sealants catching on fire. In order to ensure the surface is not overheated, the temperature must be monitored by using a non-contact thermometer, heat sensitive crayons, or other temperature-sensing materials. See **Figure 14-18**. Some manufacturers do not allow the use of any heat. Toyota Manufacturing Company, for example, prohibits the use of any heat on the Lexus and certain other models except for welding panels into place. Other manufacturers specify very limited amounts of heat be used. Perhaps the safest rule of thumb to follow is this: Heat may be applied only to those structural members that are going to be replaced. When in doubt about the use of heat and the amount that can safely be administered, one should consult the manufacturer's service manual for the specific vehicle being repaired. In addition to destroying the physical and structural characteristics of high strength steel, it can also destroy the corrosion-resistant coatings on the surface.

FORCE APPLICATION CONCEPTS

Tension and Pressure

The two force application methods or techniques commonly used for correcting collision damage are **tension** and **pressure**.

Tension. With few exceptions, nearly all the straightening equipment manufactured today uses tension as the principal

FIGURE 14-18 Applying heat may be necessary to help relax the frame rail or inner structure for straightening purposes. An induction heater is best to apply the heat, and a non-contact thermometer should be used to monitor the temperature to prevent overheating.

source for applying the corrective forces. The reason is quite simple: Almost every aspect of force application can be controlled when using tension. In order to successfully apply a corrective force, four key elements or factors must be controllable: the source or base of the corrective force, the angle at which the corrective force is applied, the direction of movement, and the routing of the energy (from anchor point to pulling point). In order to successfully direct any of the latter three elements, one must be able to select the exact location from which the pulling energy will originate. The technician has almost complete control of this by the placement of the chain on the pulling post. See **Figure 14-19**. The corrective energy radiates in a straight line from the energy source—the power post—through the chain to the pulling clamp at the attachment point, and on through the body to the anchor points. As it passes beyond the pull clamp attaching point, it becomes somewhat diffused as it travels through the entire body and is distributed equally to all the anchoring locations. See **Figure 14-20**. To ensure the stresses caused by the collision forces are relieved from throughout the entire vehicle, it must be anchored to the bench or frame machine at all four of the control points.

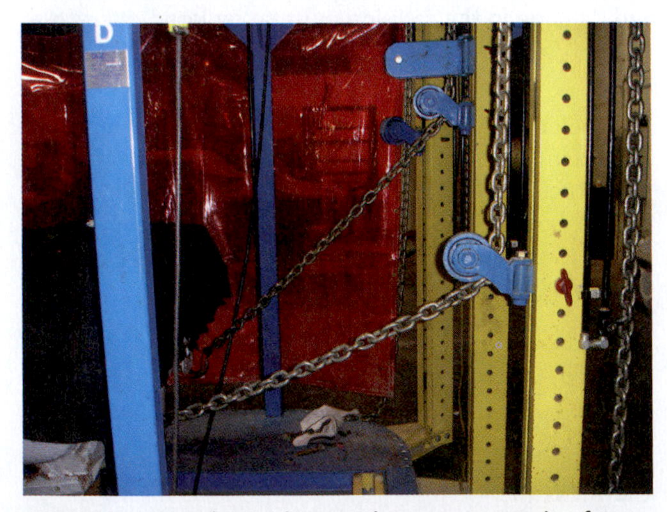

FIGURE 14-19 When using tension as a corrective force, the angle of the corrective force can be regulated and controlled through the angle of the chain in relation to the source of the force to the point where the pull is being made.

Pulling Angle. The angle of the pulling chain leading from the power post to the pulling clamp will largely dictate the direction of movement the metal will follow as it is pulled. In this way the technician can control and dictate the exact angle at which the metal moves. As the damage starts to unfold, it may become necessary to change the angle of the pull. One need only raise or lower the chain or swing the pulling post from one angle to another, thus ensuring the metal will follow the correct path.

Pressure. The use of pressure as a corrective force is considerably less effective than tension. See **Figure 14-21**. Because the force of pressure tends to follow the path of least resistance, it is virtually impossible to control the direction of movement by the metal. As the buckled panels and structural members unfold the metal will shift, changing direction. The corrective force—the pushing pads on the jack extensions—will follow the metal as it moves. To further reduce the effectiveness of pressure as a corrective force, being able to push at the exact angle required is very difficult. A flat surface—which is required to push or

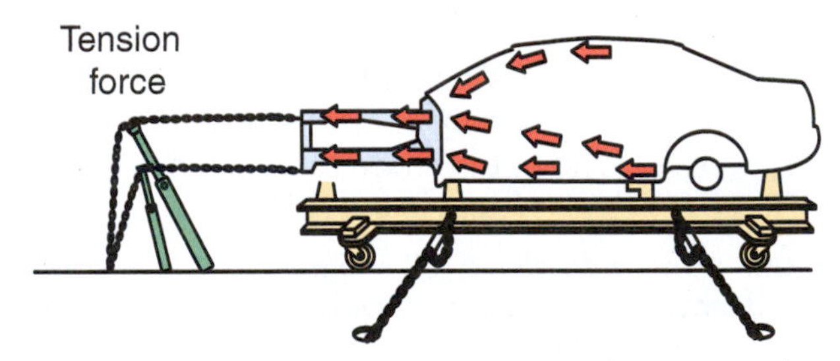

FIGURE 14-20 The corrective energy uniformly travels through the entire area of the vehicle from the point where it is attached to the vehicle to all the anchor points no matter how close or far from the attaching point.

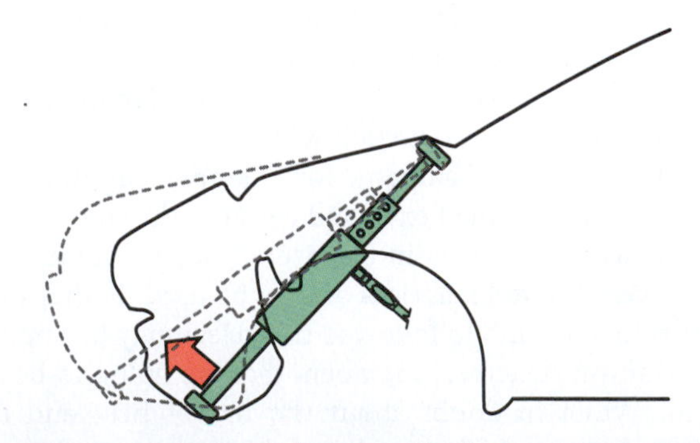

FIGURE 14-21 Pressure has very limited use as a corrective force because it tends to follow the path of least resistance or movement of the metal as it unfolds.

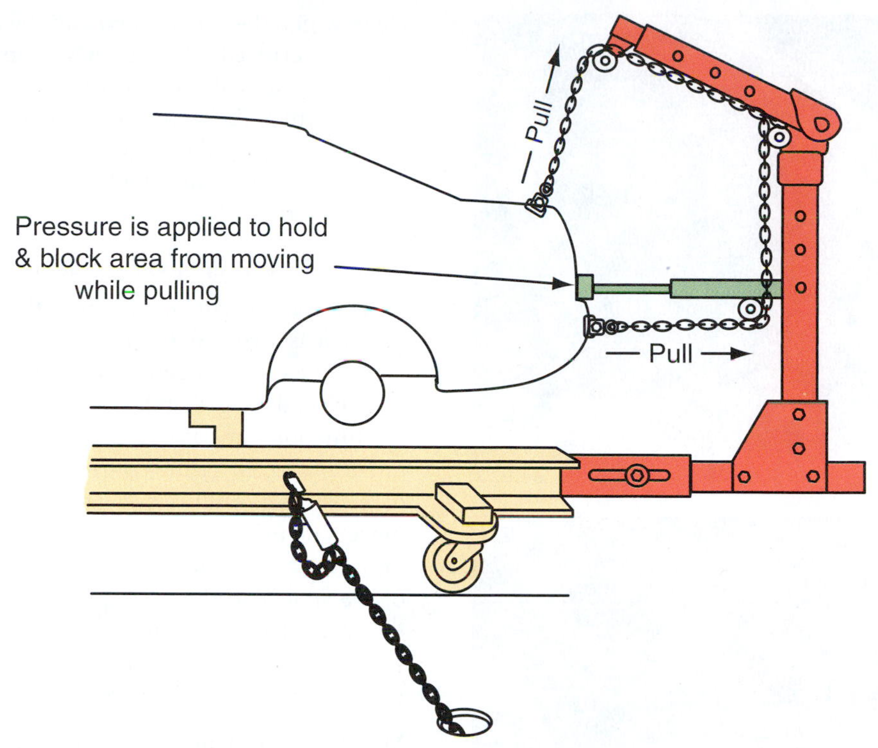

FIGURE 14-22 Pressure is commonly used for holding or blocking an area that must remain in a given location while a corrective force is applied in an adjacent area. It may also be used in tandem with tension by pushing against the back side of an area being pulled. (Adapted from Autorobot Finland, www.autorobot.com.)

block against—is rarely available on both ends of the hydraulic jack. Even if there are two surfaces parallel to each other that can be used to push from, matters are further complicated because the surfaces on either end of the jack often become damaged because of the small pads on the hydraulic jack extensions. Additional repair work will be needed to remove the damage caused by collapsing the metal at those locations.

Pressure as Supplemental Force. Pressure is frequently used to supplement the tension forces. It is commonly used for holding and blocking sections that should not move while an adjacent area is being pulled, thus increasing the effectiveness of the corrective forces. See **Figure 14-22.** It may also be used to assist or increase the amount of the force being applied by pushing against the back side of an area being pulled.

Turnbuckles

Another device commonly used when repairing damaged structural panels is **holding turnbuckle**. This is a stationary device that neither pushes nor pulls, but instead firmly holds an area in a given position. See **Figure 14-23.** It is used to hold an opening such as a door or sunroof, preventing it from either becoming larger or smaller and

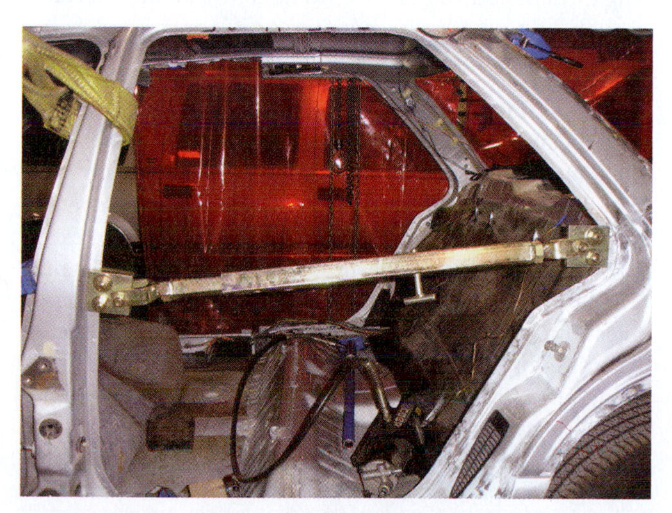

FIGURE 14-23 The turnbuckle is used to prevent an unsupported opening—such as the sunroof or door—from being pulled out of alignment by the corrective forces passing through the area as the corrective forces are applied.

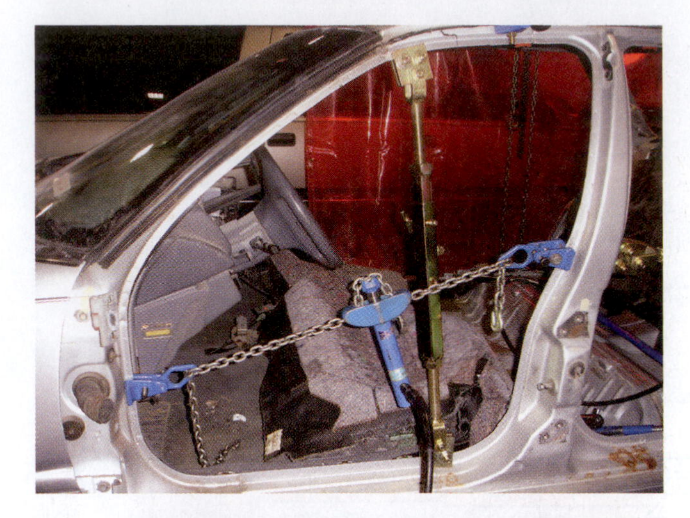

FIGURE 14-24 A hydraulic pull ram may be used to hold the door from spreading as the corrective forces are applied in the top photo, and a ram exerting pressure can be used to prevent the opening from collapsing or changing shape.

losing its shape as the corrective forces around the opening are being applied. Even though there is no metal in the area where the sunroof or door openings are located, the corrective forces pass across and through these openings, affecting them as though the surface were covered with a layer of metal. See **Figure 14-24**. This may cause the opening to become out of square unless it is held or blocked with a turnbuckle or another force application device such as a hydraulic push or pull ram. One might say it exerts a force of both tension and pressure, depending on the corrective force that passes through it. After the repairs have been completed, one must remember to dress up the areas where the jaws were engaged to remove any signs of damage they may have caused.

COLLISION DAMAGE AREAS

There are essentially four types of damage encountered by the repair technician. They are not so much identified as damage conditions but instead by the area of the vehicle in which they are located and the manner in which the damage occurred. These damage conditions are the result of a front end collision, a rear end collision, a side collision in which the vehicle may be hit on either side, and a complete or partial rollover. The latter condition presents more challenges and usually is far more difficult to repair than the others.

Front End Collision

There are usually many more things to consider when a vehicle is involved in a frontal collision than in any other part of the vehicle. In addition to the exterior of the vehicle becoming damaged and distorted, many of the mechanical functions are also located in this area and are frequently affected by the collision. Sometimes they become damaged as a result of being struck directly by the impact object and many times because they have been crushed or impacted by the sheet metal, rails, and structural sections of the vehicle as they collapse under the collision forces. Many times the exterior panels must be removed, and the structural parts must be pulled and roughed out in order to gain access to these parts to remove them.

Mass Inertia Damage. In addition to the damage sustained by the mechanical parts in the immediate area of the impact, the damage frequently flows rearward to include other sections of the vehicle. The transmission, drive shaft, and rear end of the vehicle may also be affected as the damaging forces radiate through these components. Due to the effects of mass inertia movement, damage to the suspension and undercarriage are likely to occur as well. The mass weight of the engine, transmission, differential, and other suspension parts tends to continue moving when the vehicle comes to an abrupt stop. The sudden transfer of force or direction of the impact energy as the vehicle comes to a stop causes the engine and transmission mounts to break or causes their mounting locations to become severely misaligned. The inertial forces are known to cause damage as simple as a small ding or dent and can escalate to affecting an area severely enough to tear a cross-member or a convoluted frame member loose from its attaching points.

Forces of Mass Inertia. The majority of the frontal collisions occur while the vehicle is moving forward. One must remember that as the vehicle is moving forward there is a significant amount of momentum or mass movement. There are also significant amounts of hostile external forces created by the impact object whether it is a vehicle moving toward our subject or if it is a stationary object. In either case, essentially the impact object is moving in the opposite direction. These two opposing forces acting upon each other perpetuate the theory that is commonly referred to as inertia, which ultimately causes the inertial or indirect damage on the vehicle. As was discussed in Chapter 11, Collision Damage Analysis, the front of the vehicle comes to a halt at the precise moment it comes into contact with the impact object. The entire area and

FIGURE 14-25 The rear of the vehicle remains in motion for a microsecond longer than the front when it crashes into a stationary object. Before the rear of the vehicle comes to rest, the rear wheels raise off the ground momentarily and the rear end moves forward. This causes a chain reaction of movement through the roof, causing the A and B pillars to shift outward.

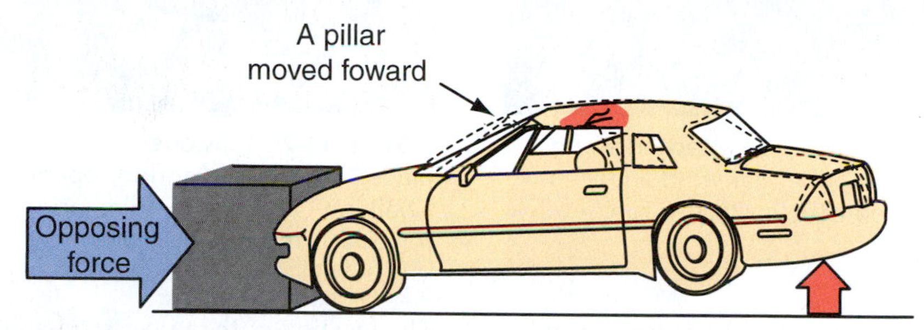

FIGURE 14-26 While no part of the roof came into contact with the impact object, a significant amount of indirect damage radiated all the way through to the rear of the vehicle and must be addressed. A series of pulls that will reverse the damage flow to the upper part of the vehicle while holding the lower part is required to remove the effects of the indirect damage.

mass behind this section continues to move forward for a microsecond until a sufficient amount of resistance is created or met to stop this forward motion. See **Figure 14-25**. When this forward motion is stopped, the rear end of the vehicle may be raised off the ground momentarily. At the same time the roof panel above the center pillars becomes buckled in response to the bending effect from the motion in the rear end of the vehicle and the opposing forces from the front of the vehicle. In a more severe collision, the A pillars may be pushed up and out to dissipate the damaging forces around the passenger compartment.

Indirect Damage. All things considered, most of the panels are not in their normal position when the vehicle and all the opposing forces finally come to a stop. The panels, however, maintain the shape they took on at the precise moment they stopped moving forward. As a result, several parts of the vehicle that didn't even come into contact with the impact object are misaligned and in need of repair. One must understand that when the vehicle is damaged to the point where the A pillars have shifted and moved up and out, the air bags will also have been deployed. Together with the cumulative damage to the rest of the vehicle, this damage is usually so severe it is generally rendered a total loss. However, in the interest of discussing the normal sequence of repairs, we will assume the vehicle is repairable in this instance. See **Figure 14-26**.

To repair these damages, the technician will require corrective forces that will move the sheet metal in the reverse direction it moved while the vehicle was in its forward motion. See **Figure 14-27**. The LIFO principle obviously must be applied at this point in the repair. The indirect damage, the damaged areas away from the POI, must be corrected first.

Removing Outer Perimeter Damage. Assuming the vehicle is properly anchored and the measuring system has identified the misaligned areas, the first thing that needs to be done is to start moving the rear section back and down. This reduces the stress and moves some of the

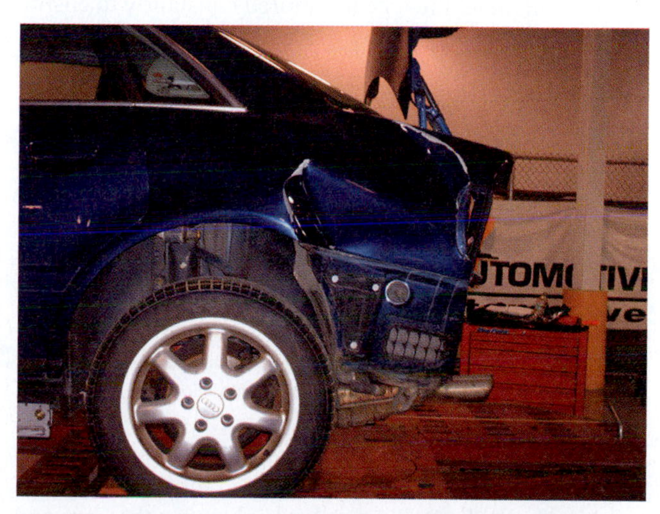

FIGURE 14-27 The rear of the vehicle is pulled down and back simultaneously with the roof. A "bunching effect" often occurs that necessitates moving the outermost displaced metal first in order to make room for the deeper damage to move back.

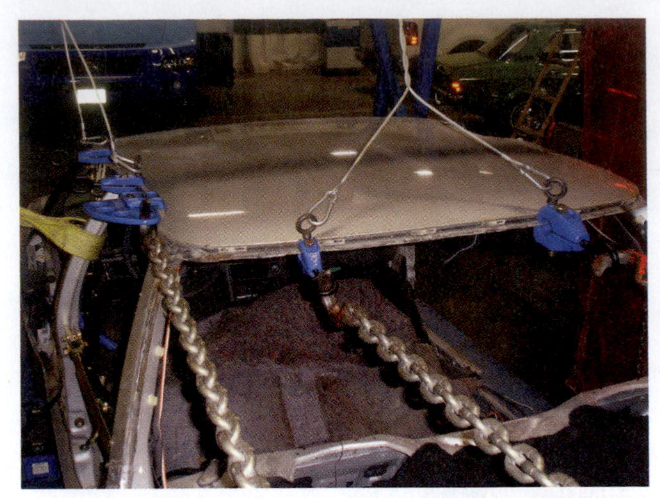

FIGURE 14-28 Pulling clamps may be attached along the side of the roof and onto the rear window pinchweld to simultaneously apply the corrective forces to the entire roof.

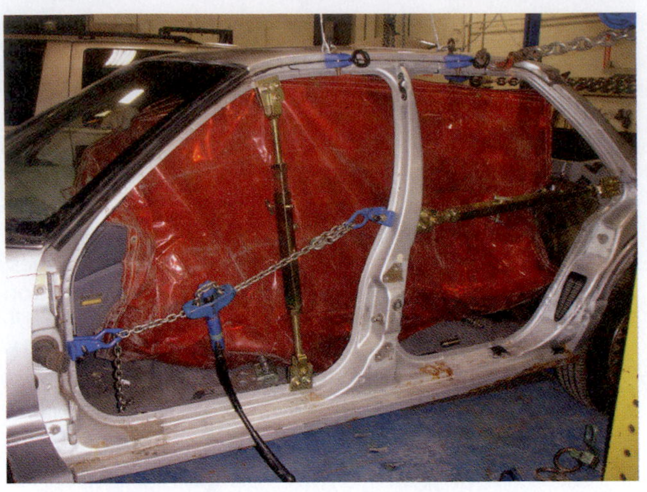

FIGURE 14-29 Tensioners and turnbuckle braces are placed into the lower door opening to prevent it from "walking" or changing dimensionally as the corrective forces are applied.

bunching or compressed metal out of the way to make room for the metal that shifted forward to be pulled back again. As the pull continues, a point is reached where an excessive amount of resistance is felt and the metal stops moving. See **Figure 14-28**. At this point the clamps should be attached to the roof sides and rear to initiate pulling the roof back to its proper location once again. As the damage is pulled back, any buckles and other damage in the roof must be removed. The A pillars, B pillars, and the C pillar (**sail panel**) must be addressed inside and out to ensure they are properly stress relieved to allow the proper movement of the metal. Any twisted or bent areas must be stress relieved while the corrective forces are being applied.

Monitoring Pillar Openings. The A pillars, the B pillar from top to bottom, and the attachment areas to the roof and the sail panel must be monitored constantly to ensure they are stress relieved and that the openings are squared up once again. See **Figure 14-29**. One must monitor the openings often, readjust the pulling location as they are pulled into shape, and place tensioners or jacks into the opening to prevent them from growing or collapsing.

After the indirect damage has been restored, the front of the vehicle will be addressed. It may be advantageous—or even necessary—to initiate the pulling and straightening operation on the front of the vehicle while the rear is being pulled back into position. Before any of this can occur, however, the damaged parts—such as the bumper assembly, front fenders, hood, and any other bolted-on parts—must be removed to gain access to the damaged rails. Because they were in the immediate area of the direct impact, the damage to the rails and inner panels is apt to be very visible. Pulling and straightening the bent and kinked rails is usually done as a prerequisite to replacing them or a section of the front end of the vehicle. The front suspension angles are also checked for proper placement at this time and their mounting locations are addressed. The location of the upper strut and the placement of the strut tower are major factors to consider when pulling and straightening the front of the vehicle.

Rear End Collision

A rear end collision usually has a less significant impact on the overall structure than just about any other part of the vehicle. Like any other part of the vehicle, the rear panel, trunk floor quarter panels, and upper rear panels are all designed to collapse—and thereby absorb and dissipate the collision energy. However, unless it is a mid-engine or rear-wheel-drive automobile, the number of mechanical functions and parts located in this part of the vehicle are minimal. The rear suspension is typically located inboard a fair distance from the end of the vehicle. The most significant damage the vehicle will sustain in a rear end collision is in the area of the trunk floor and the rails that support it.

Localized Damage. As a rule, when a vehicle is impacted in the rear the damage largely remains localized. The reinforcements are designed to route the damage around the passenger compartment to protect the occupants, but there is a minimum amount of mass weight in this area compared to the front of the vehicle. While the rails under the trunk floor are made of high strength steel like most others on the undercarriage, they are less substantial in construction and fewer in number and frequency than in other parts of the vehicle. Therefore, in a rear end collision, the rear section of the vehicle will likely move forward or be kicked up or pushed down. However, because so much of the weight displacement is in the front, the rest of the vehicle tends to absorb and dissipate the collision energy forces. While the damage remains more localized, it tends to be more evident and much more severe looking. Because of the reduced amount of inner reinforcement, more overall bending and crumbling or distortion of the outer panels will occur. One, however, should not

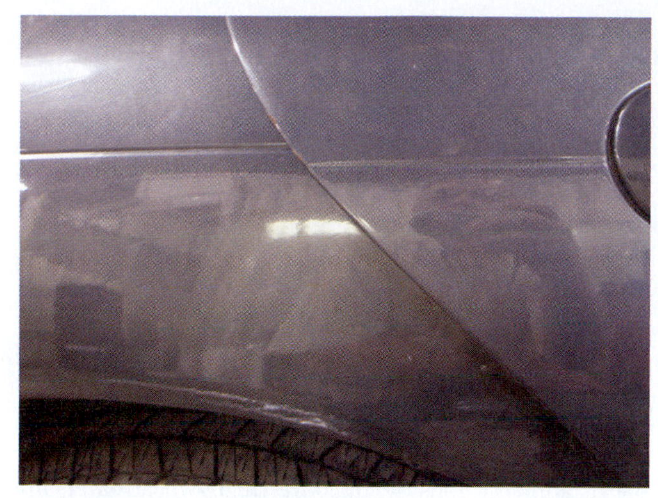

FIGURE 14-30 In a severe rear end collision, the quarter panel may be pushed forward even to the point of causing the rear of the door to overlap the front edge of the quarter panel.

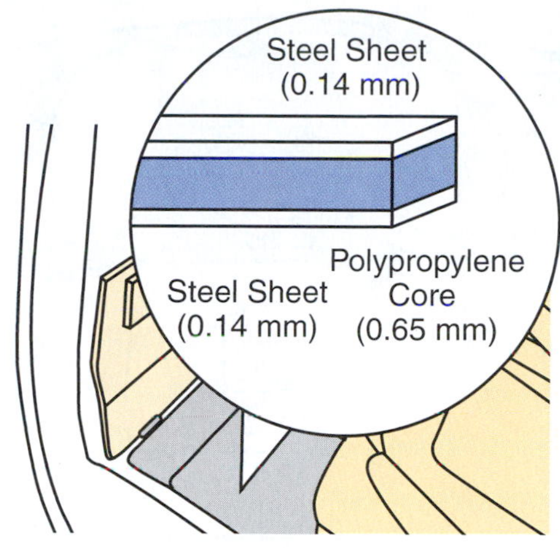

FIGURE 14-31 Laminated steel is used in designated areas of the passenger compartment floor on a limited number of vehicles to reduce road noise. It also enhances the strength characteristics.

assume that because the damage flows more easily in this section of the vehicle that the front or center section will not sustain any damage during a rear end collision. See **Figure 14-30**. In a more severe rear collision the lower structure of the center section can be pushed forward and the roof may also become buckled, causing the doors to become misaligned as well. Depending on the severity of the collision, the quarter panels may have moved forward enough so the rear door gaps have disappeared completely. Alternatively, the doors may even be overlapping the front edge of the quarter panel.

Pulling Sequence. The repairs in this area are affected in the same way as in any other part of the vehicle. No matter what type of damage exists, the reinforcing members must be pulled first to start relieving the damage. In this case the rails under the trunk floor must be addressed and pulled first because of the overall shortening effect from the collapsing and buckling of the metal. In a sense, the outer layers of metal at the POI must be pulled out of the way to make room for the deeper damaged area to move as it is pulled back into its proper location. An example of this includes the spare tire pocket which often sustains a significant amount of damage in a rear collision, frequently trapping the spare tire and shoving the metal in front of it forward. Little can be done to remove the damage that is forward of the spare tire until the damage at the POI has been roughed out and the spare tire is removed from its storage area. Only then can the deeper damage be restored.

Laminated Steel. In recent years laminated steel has been introduced and used in the trunk floor, pockets for fold-down seats, and other areas of the vehicle undercarriage. See **Figure 14-31**. The use of **laminated steel** involves sandwiching a layer of either soft wood or a plastic film between two thin layers of metal to reduce road noise and vehicle weight, while at the same time enhancing the strength of the metal considerably. As a rule, when the laminated floor pan sustains any damage beyond a simple bending or bowing it must be replaced because it is impossible to determine whether the inner laminate has been cracked or broken. The consequences of applying heat to any part of the laminated metal surfaces is also obvious. Therefore, any heating or welding operations performed must be done to exacting manufacturer's specifications.

Side Collision

In a side collision the affected areas are often damaged more severely by the indirect damage than in any other type of collision. By nature the vehicle's design lends itself to the assemblies moving more significantly when the vehicle gets impacted. See **Figure 14-32**. The B pillar,

FIGURE 14-32 The indirect damage resulting from a side collision is usually more severe than from any other angle on the vehicle because the side of the vehicle has little to reinforce it other than the B pillar.

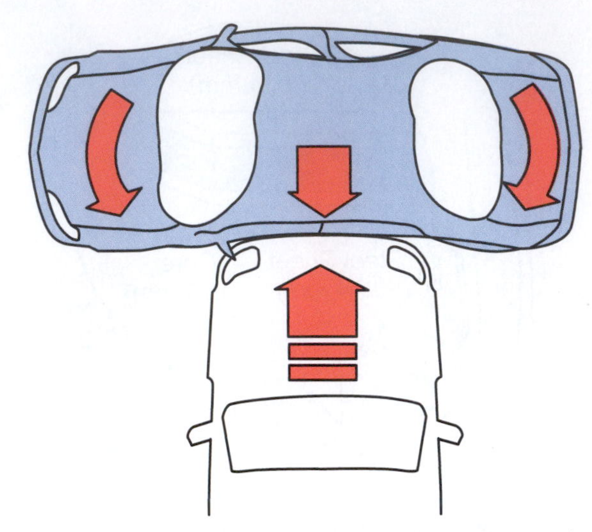

FIGURE 14-33 During a side collision the entire roof assembly tends to shift in the opposite direction the impact forces moved during the collision.

for example, is a long single structural unit that has no additional support on either side because there is no metal structure other than the doors to support it. The large open areas between it and the A and C pillars, respectively, offer no support for either pillar. See **Figure 14-33**. Therefore, as the lower section of the vehicle moves in the direction the impact object is pushing it, the inertial forces will tend to move the upper part of the vehicle in the opposite direction. In addition to the leaning effect on the upper portion of the vehicle, the front and rear sections of the vehicle may also have moved in the opposite direction of the forces from the impact object. This could cause the areas on either side of the impact area to "wrap around" the POI, causing a shortening effect on the vehicle's overall length. Therefore, in the aftermath of the collision the roof may be leaning in the opposite direction than the impact forces were moving and the overall vehicle may also be shortened. The vehicle essentially assumes an exaggerated "C" shape from the front to the rear on the side of the impact. The same bowing effect—only to a lesser degree—may also occur on the opposite side of the vehicle.

The door assemblies will also have sustained a significant amount of direct damage and, in turn, may have inflicted some indirect damage to the areas where they are attached. The front door hinge pillar may have been affected by the pulling effect to the door as it is pushed inward. Although perhaps not as severe, a similar effect may have occurred on the rear door lock pillar at the front of the quarter panel.

TECH TIP Due to the limited ability of some straightening equipment, hydraulic jacks used to apply pressure at strategic locations may need to be incorporated into the repair equation to supplement the pulling efforts being performed.

Repair Sequence. The repair sequence will require simultaneously reversing the effects of the collision energy in several areas. See **Figure 14-34**. An end-to-end pull will be necessary to reverse the overall shortening effect of the vehicle while the center section at the POI is pulled in the opposite direction than the impact object was moving. At the same time, the roof and upper body must be moved in the opposite direction the inertial forces moved it. This will require a series of pulling setups that simultaneously apply a corrective force on both sides of the vehicle. See **Figure 14-35**. Some straightening equipment is not capable of applying tension in more than two areas at any given time. In this instance, porta powers and other hydraulic jacks can be employed to augment the pulling equipment by pushing against the edge of the roof and B pillar from the floor at the opposite side of the vehicle. Caution must be exercised to ensure the pushing forces

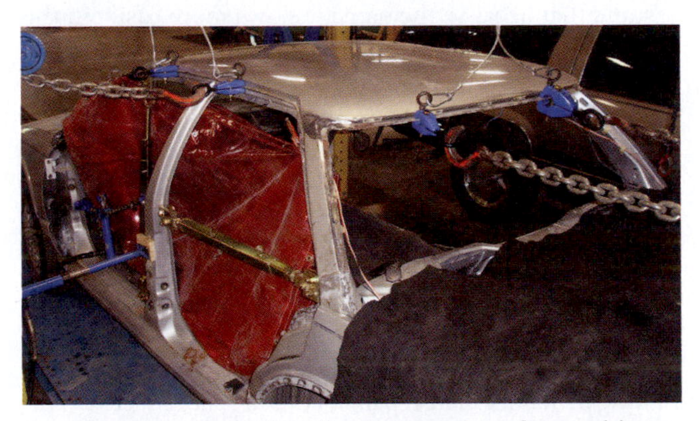

FIGURE 14-34 Restoring the damaged roof assembly requires a multidirectional application of corrective forces simultaneously.

FIGURE 14-35 The simultaneous use of both tension on the outside of the vehicle and pressure on the inside of the vehicle is necessary to reverse the shifting effect of the roof that occurred in the collision.

don't become misdirected as any pressure application will generally follow the path of least resistance. Since the force application is already being applied at an angle from the floor on the opposite side of the vehicle, it may start to move in an incorrect direction as the straightening gets under way.

Angular Impacts. Most of the frontal impacts are angular as opposed to a head-on collision. This is because drivers often attempt to steer away from the oncoming vehicle or impact object to avoid a collision. Typically only one corner of the vehicle comes into contact with the impact object. However, in so doing it usually affects a larger area because it moves further down the side of the vehicle before the turning arch steers the vehicle away from the impact object. As one would expect, most of the same panels and reinforcement members are affected as in a straight frontal collision. The difference is the damage may move the metal in two or three different directions—backwards, inward, and sideways—at the same time. The repair process will require more pulls that must be made from more different angles than would be required if it were a direct head-on collision.

Rollover

When a vehicle rolls over in an accident a whole series of events occurs that undermines all the logic, activity, and analysis that goes into any of the repairs discussed previously. When a vehicle rolls over the damage that was sustained a fraction of a second earlier will be further complicated as another part of the vehicle pounds into the ground almost simultaneously as it rolls. The vehicle undergoes several impacts, creating a like number of POIs—each at seemingly a different angle. It is often difficult to determine which part was damaged first and which was the last to become distorted. Many of the typical signs used to visually pinpoint damage are destroyed or are not applicable because they are manifested by subsequent damage as the vehicle continues to roll. Knowledge based on past experience is the only tool the technician has to draw upon to anticipate how the damaged areas will respond when the corrective forces are applied. Provided the damage didn't penetrate too deeply below the upper section of the A, B, and C pillars, a vehicle that has rolled over lightly at a low speed is often repairable with a sectional roof replacement and a select group of other structural parts. However, if the vehicle rolled more than once, the damage is usually too severe to warrant any kind of repair consideration.

UNDERCARRIAGE DAMAGE

Thus far all the discussion has centered around the damage that occurs in the upper body area. However, during a collision in which the upper body sustains structural damage, the lower body and undercarriage rarely escape incurring a predictable amount of damage as well.

Damage Conditions

The most common damages that occur are mash, collapse, sidesway, twist, and diamond. Each has its own unique characteristics and identifiable appearances and each requires a unique approach or method for straightening and restoring the vehicle to the manufacturer's specifications. The diamond and twist damage conditions are usually limited to occurring on full frame, body over frame vehicles. Only in rare instances do they occur on a unibody vehicle. The mash, collapse, and sidesway, however, routinely occur on these vehicles.

Anchoring BOF Vehicles for Repair

Since most of these damage conditions are apt to occur on both a unibody design and the BOF, we will discuss the unique corrective concepts that must be applied to each. One must understand that while the damage that occurs may have similar characteristics on both a unibody and BOF vehicle, the repair procedures are often unique to each body design. For example, unlike the unibody design, the BOF vehicle is not always clamped or attached to the repair bench with the pinchweld clamps. Therefore, each repair made will require different anchoring concepts. A variety of different methods or techniques can be used to anchor the BOF vehicle for repair purposes. In the interest of simplicity, we will discuss the blocking system previously discussed in this chapter, augmented by chaining the vehicle where necessary to ensure the vehicle does not shift or move about as the corrective forces are applied.

One of the things that must be overcome when applying a corrective force to a BOF vehicle is the sympathetic reaction that occurs. This phenomenon creates a tendency for the vehicle to rotate or move in the opposite direction of the corrective forces on the far end of the vehicle. In order to overcome this phenomenon the vehicle must be blocked or chained down at strategic points to prevent this shifting effect. This is accomplished in a variety of ways and is largely dependent on the equipment that is being used. Because the unibody vehicle is secured by clamping it at the torque box areas with the pinchweld clamps, one rarely has to be concerned with overcoming these sympathetic reactions.

REPAIRING UNDERCARRIAGE DAMAGE

In order to restore the undercarriage to the manufacturer's specifications, the order or sequence that should be followed are to first repair the diamond condition, followed by the twist, mash, sidesway, and collapse, respectively. The sag/kickup is a secondary condition or part of the collapse condition and are the last repairs to be made, but they are addressed along with collapsing as it is often a residual damage condition. It is usually addressed or repaired last. Rarely would all of these damages occur on the same vehicle from the same collision. However, whichever of these damage conditions are evident after a collision, the sequence specified should be followed for the damages that did occur.

When pulling to straighten a frame and any other structural member, it is generally necessary to pull the rail or the section of the undercarriage in question beyond the normal position to allow for **springback**. Recall that the elasticity in steel gives metal the ability to flex and bend yet return to its original position. The same principle applies in repairing a damaged frame and undercarriage, but on a larger scale because the damage typically encompasses a larger area than when repairing sheet metal on a cosmetic panel. Therefore, the movement will be more obvious. Depending on the size and shape of the area being pulled or repaired, it may be necessary to move the area as much as 10 to 12 mm (1/2 inch) and, in extreme cases, as much as 18 to 20 mm (3/4 inch) beyond its correct location to allow for the springback.

Diamond

The **diamond** condition occurs on a body over frame vehicle only. Normally it occurs in a frontal impact in which the impact object strikes the frame rail directly on one side of the vehicle only. See **Figure 14-36**. Recalling the effects of the inertial forces, the rail on the opposite side of the vehicle continues to move forward some distance before it comes to a stop. The effect is further enhanced by the mass weight of the engine and transmission that are bolted to the frame rails as they continue to move forward on the unrestrained side, essentially pulling the rail along with them until it all comes to a halt. As a result, the center section of the vehicle is usually out of square, often assuming an exaggerated diamond shape. The damage may be manifested to affecting the centerline readings in the front and rear areas of the vehicle.

A diamond damage condition is often difficult to identify because it may be very subtle in appearance and can be even more difficult to repair. The engine cross-member, the area of the frame that is most heavily reinforced, is usually responsible for causing the diamond condition. It is reinforced heavily enough to absorb the collision energy and transfer it on through the entire length of the rail on the side of the impact. A diagonal measuring quick check with a tape measure or tram gauge in the center section of the vehicle will quickly reveal the existence of the diamond condition. Conversely, a lateral measurement on the frame rails will not identify the existence of the damage condition because the rail has been pushed straight back. The damage condition on the frame in **Figure 14-37** is a diamond condition which likely occurred when the right front rail of the vehicle collided head-on with a stationary object. The inertial forces caused the entire left rail to shift forward. This damage condition will become evident when X-checking the center section of the frame. The dimensions between point B and E will be shorter than between points C and D. To confirm the diamond condition one should also X-check the areas between points B and C and the rear end of the rail and between points D and E and the front of the rail, provided the rail at point A is not distorted too badly. In order to straighten this condition, the vehicle must be tied down and possibly pull on the left rear end of the rail to prevent it from moving forward when the corrective forces are applied at point A. In addition, the frame must be blocked at points B, C, D, and E to overcome the sympathetic reaction when pulling at point A. One must constantly monitor the progress of the pull to ensure that overpulling does not occur.

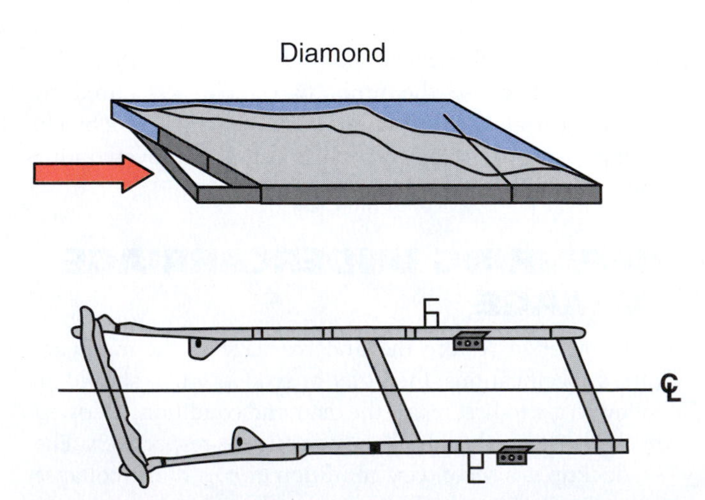

FIGURE 14-36 A diamond condition is usually the result of a BOF vehicle striking a solid object on one side of the frame while moving in a straight line that causes the entire rail to shift rearward on that side.

Twist

Many times after the vehicle has struck the impact object—and before it comes to a complete halt—it will have slid up over the impact object, which will likely add an additional dimension of damage to consider. That damage is usually a **twist** condition. This is another damage condition that rarely ever occurs on a unibody vehicle. Whenever a twist condition occurs, it should be the second in the series of repairs made whether the vehicle is a unibody or a BOF. See **Figure 14-38**. It usually results in the center section of the vehicle becoming out of level with the rest of the undercarriage and is usually characterized by the rail forward of the center section being pushed up and the rear of the rail being pushed down due to the sympathetic reaction that occurs. A typical dimensional difference in the rail height may show the front of the rail to be pushed up 8 to 10 mm above the normal height and down an equal amount on the rear of the rail. This is quickly evidenced by both the front and rear doors overlapping the openings an equal amount at the top or bottom on the same side of the vehicle.

Occasionally when the vehicle becomes air-borne and hits the ground on one front wheel first, the twisting may affect the rail on the opposite side of the vehicle. When this occurs the twisting will be likened to the "wringing effect," as is shown in **Figure 14-39**. The corner at point A is kicked up and the opposite rear corner

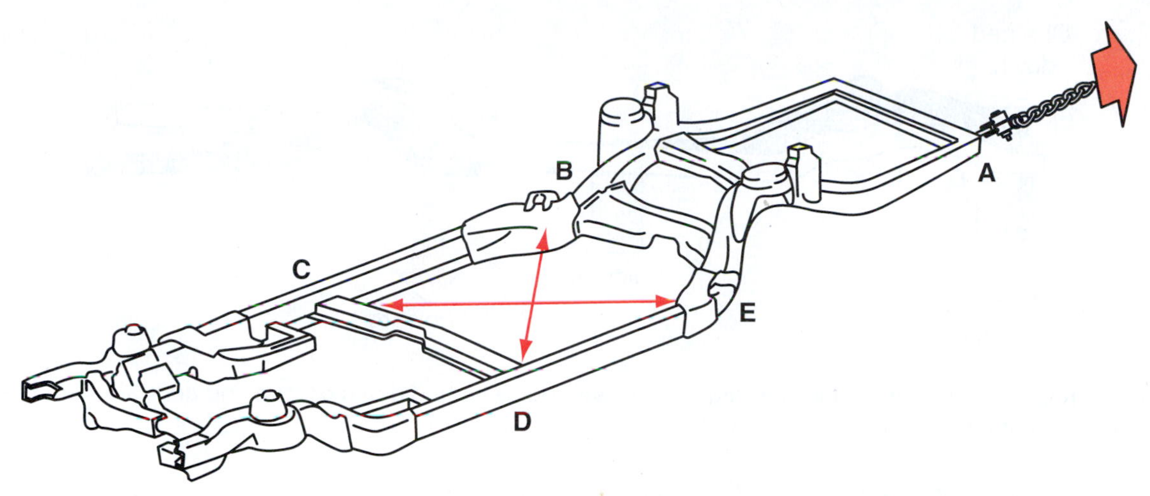

FIGURE 14-37 The frame must be blocked from the side at points C and D to overcome the effects of the sympathetic reaction, which causes the rear to shift to the right.

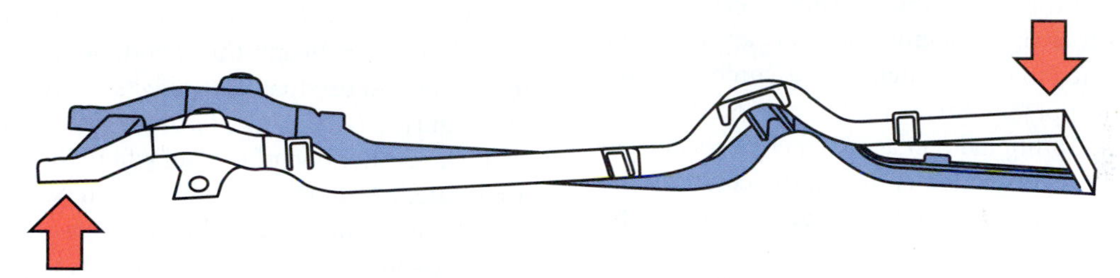

FIGURE 14-38 A twist condition causes one rail to be completely out of level with the rest of the undercarriage. The front of the rail is kicked and the rear is shifted down.

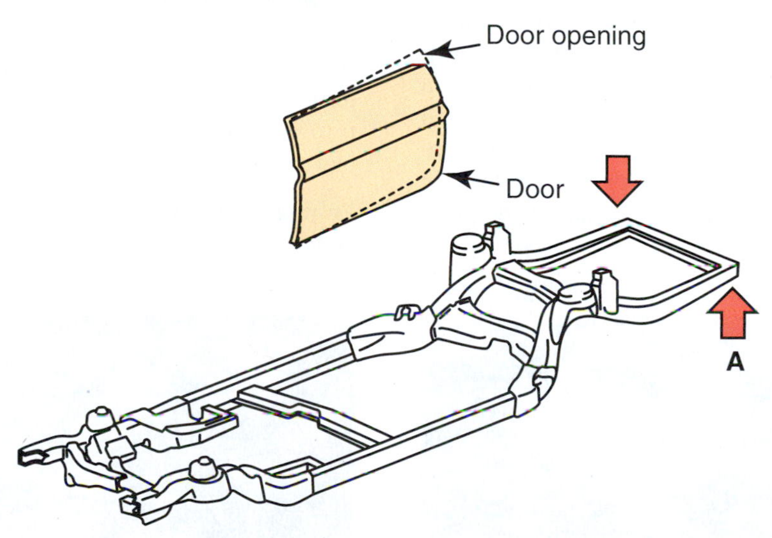

FIGURE 14-39 A twisting effect may occur throughout the body when a vehicle that becomes air-borne comes down hitting the ground on one corner first. The doors will show a sagging effect on one side and a kicked-up appearance on the opposite side of the vehicle.

may be kicked up due to the reaction by the inertial forces. Generally the opposite corner will be kicked up and the rear corner on the same side may be pushed down. This is characterized by the doors overlapping their respective openings in an opposite manner. The doors on the right side of the vehicle may overlap their openings at the top and conversely at the bottom on the left side of the vehicle.

Repairing the Twist Condition. Repairing the twist condition typically involves blocking the center section, using it as a **fulcrum**, and moving the ends of the misaligned rail in the opposite directions of each other. See **Figure 14-40**. This will involve pulling down on the end of the rail that is kicked up at point A while pushing up on the end that is pushed down at point B. A downward pull is also used at point D to remove the kickup resulting from inertial forces.

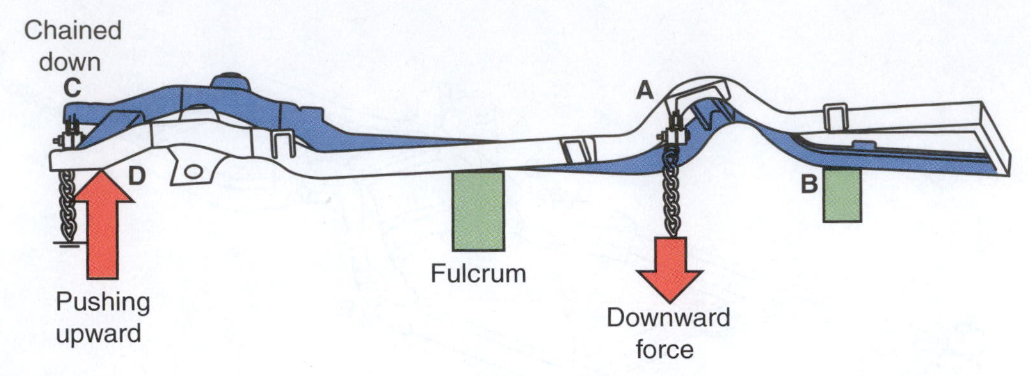

FIGURE 14-40 Repairing the twist condition requires the vehicle to be tied down on one end and raised on the opposite while blocking or creating a fulcrum in the center section of the vehicle.

At the same time, the opposite corner at point D is chained down to overcome the sympathetic reaction and keep it from pushing up. Most likely the center section will require anchoring in the area of the fulcrum to keep it from shifting or moving as the corrective forces are applied. One should be cautioned to not automatically assume the most effective place to apply the corrective forces is on the outer ends of the rails. Additional damage is likely to occur because the distance between the ends and the center of the frame where the damage exists are too great to place the effects of the corrective forces where they will be most effective. See **Figure 14-41**. Therefore, the most effective placement of the corrective forces will be approximately midway between the end of the rail, preferably near a reinforced area, and the point where it is blocked. As is always the case when repairing damage on a structural member, the rails must be moved beyond their normal location to overcome the elastic factor and the entire length of the rail must be stress relieved while the corrective forces are applied. Ultimately the goal here is to restore the levelness of the entire undercarriage from side to side and throughout the length of the vehicle.

Mash

The third damage condition to repair is the mash condition. A **mash** condition most often results in a shortening of either the front or rear rail length because in a frontal or rear collision they bend, buckle, and fold under to deflect and dissipate the effects of the collision forces. See **Figure 14-42**. This shortening effect usually occurs first at the outer end of the rail where the **crush initiators** are located. After they collapse, the next surfaces to be affected are the offsets and kickup areas, respectively. The rails get shortened more and more as the damage travels beyond the crush zones, moving deeper into the vehicle. Depending on the severity of the impact, the damage travels deeper into the rails, causing damage ranging from simple issues like a mild bending or bowing to a severe buckling effect. After the vehicle has been measured to determine the precise location of the damage, the next step is to pull and straighten the structure even though it may be replaced as part of the overall repair plan.

Pulling Sacrificial Parts. The damaged crush initiators and any sections in the crush zone that sustained any

FIGURE 14-41 Improper placement of the corrective forces can result in creating additional damage that may require more effort than was originally required to remove the collision damage.

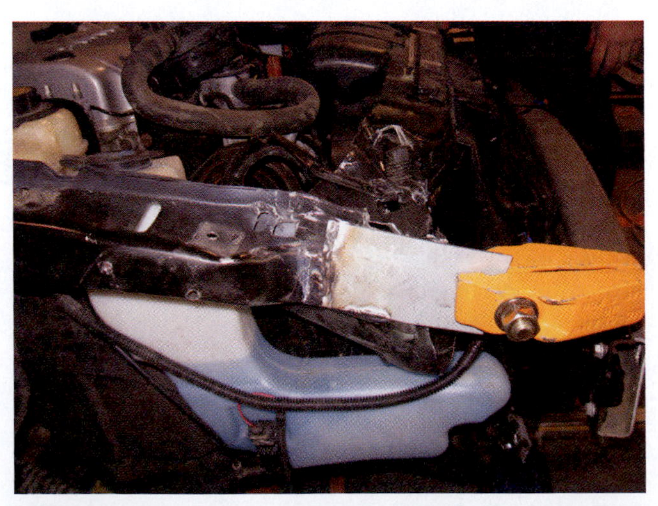

FIGURE 14-42 The rail is shortened only a short amount when only the end of it is affected, or it may affect an even larger area when the damaging forces travel deeper beyond the first crush zones.

damage must be replaced. Since these are all considered sacrificial collision structures, they are also logical locations to attach the pull clamps for making the first series of multiple pulls. See **Figure 14-43**. This is frequently necessary to make room for the damaged areas deeper into the rail to be able to move back to their proper location. One cannot assume all the damage can be removed by pulling in this one area alone. See **Figure 14-44**. Occasionally it becomes necessary to sacrifice a section of a rail to avoid having to replace an entire section of the vehicle. An example of this is demonstrated in Figure 14-44 where the pulling attachment is located in an undamaged area in such a way to enhance pulling the damage deeper into the vehicle. This may also become necessary to overcome the effects of the work hardening resulting from the collision. The metal at point D in the illustration has become work hardened at the buckled area, and attempting to pull it straight by attaching to the end of the rail may cause the entire rail to bend improperly in the entire area between points A and D. To overcome this, a pulling device may have to be attached at point C and the corrective force applied from that point. When the damaged area is restored to the manufacturer's specifications, the damaged sacrificial sections can be removed and the replacement parts installed using the manufacturer's procedures or the uniform procedures for collision repair (UPCR) guidelines for replacing the parts.

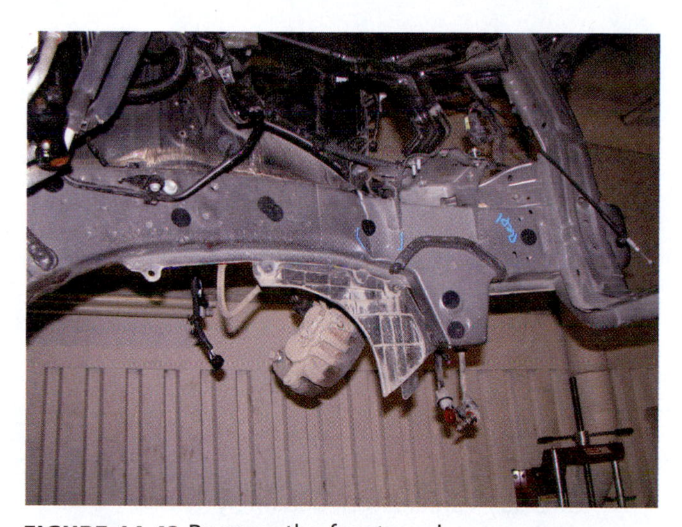

FIGURE 14-43 Because the front crush zone areas are going to be replaced, areas like this convoluted rail end are a logical point to make the initial pulls to restore the overall damage back to dimensional correctness.

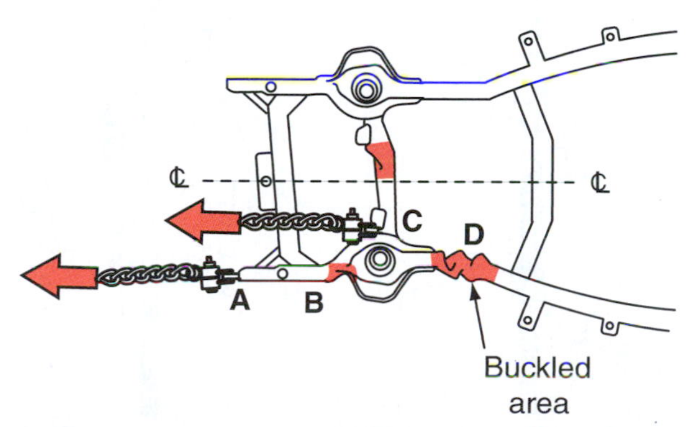

FIGURE 14-44 To ensure the rail moves at the correct location when pulling it, the pulling clamp may need to be attached deeper into the rail at a point nearer the location of the buckling.

Sidesway

Any time a vehicle is struck at an angle with a sufficient force for the collision energy to penetrate beyond the cosmetic panels and into the inner fender rail, a **sidesway** will almost certainly occur. See **Figure 14-45**. The sidesway is the fourth condition in the sequence that should be repaired. It usually occurs when one side of the front rail is shifted toward the center of the vehicle or, if the collision forces are sufficient, it will shift the entire front section of the vehicle—including the radiator support—off to one side. The same phenomenon can occur on the rear of the vehicle as well when the rear floor section of the trunk and the support rails are impacted. Due to the structure of the rear section on a unibody vehicle, the entire rear section is more apt to shift than just one side moving toward the center section. Depending on the type of impact, the damage will move in the same direction as the impact object was moving. However, if the vehicle struck a stationary object it will move in the opposite direction than the vehicle was moving.

Use Multiple Pulls. Repairing a sidesway condition should always be done using a multiple pull setup. As the damage was formed it invariably moved or shifted the metal in several directions in a near simultaneous motion. See **Figure 14-46**. In order to reverse the effects of the collision forces, a series of pulls must be made alternating from one clamp to another to enhance reversing the damage. As with the mash condition, the pull clamps may be attached to the end of the rail, being careful to not damage the crush zones while pulling the damaged section out. In instances where the metal may be damaged more severely, it may become necessary to attach to an area deeper on the rail away from the end. Occasionally it may be necessary to support the areas between the rails by placing a hydraulic jack between them to prevent the rails from collapsing from the corrective force. See **Figure 14-47**.

Sag/Kickup

The last form of undercarriage damage to be repaired is the **sag/kickup** condition. See **Figure 14-48**. The sag is characterized by the outer end of the rail on one side of the vehicle being pulled down. Even though it is referred to as a sag condition, many times the opposite—a kickup condition—will exist on a vehicle. Even though they are on opposite ends of the spectrum of one another, they will give the same visual damage appearance on the vehicle. The only difference is the kickup causes the front fascia to

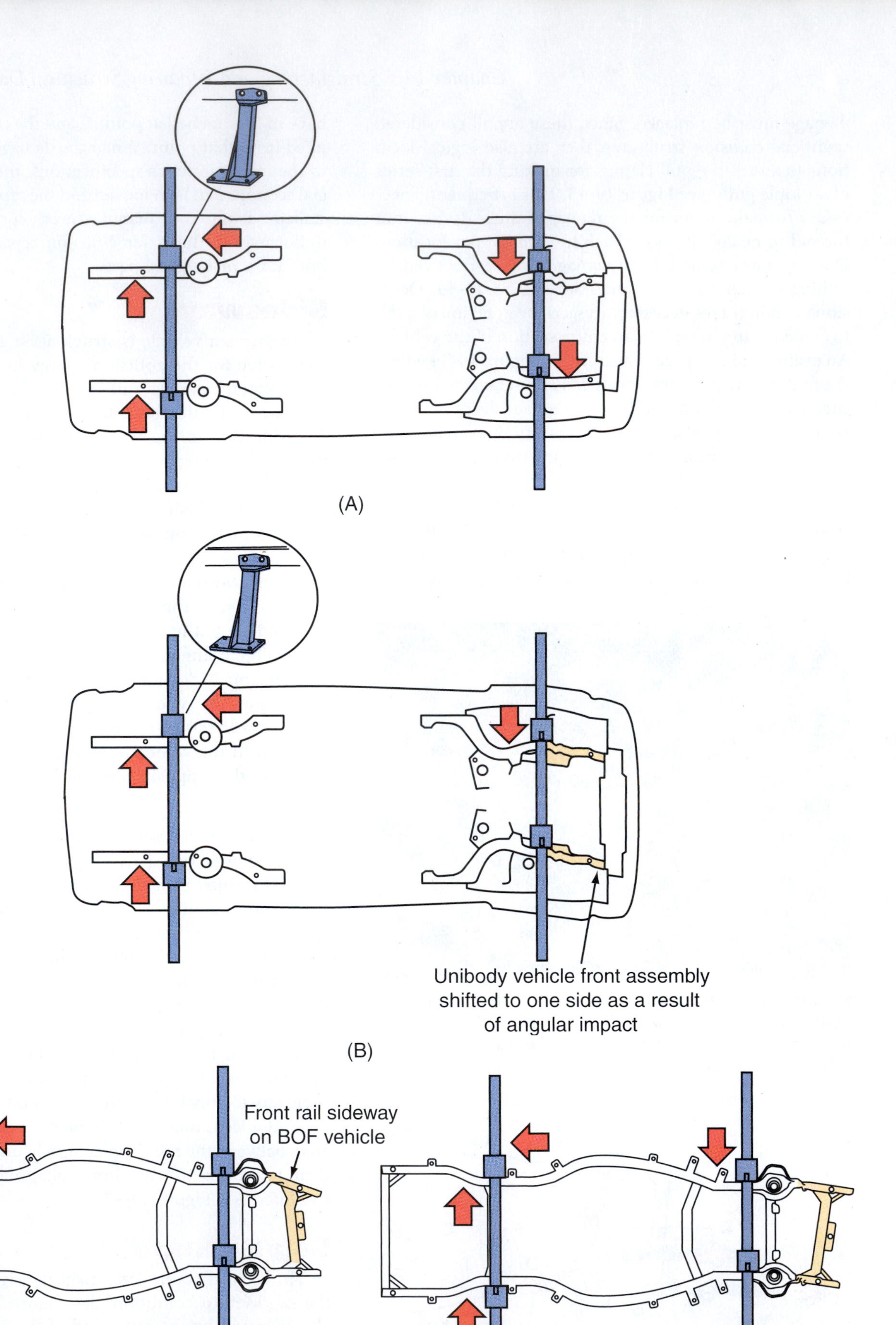

FIGURE 14-45 The sidesway conditions in (A) and (B) are representative of what occurs in a unibody design on a side impact. In (C) on a BOF vehicle, only the left rail is shifted slightly in a minor collision. Or both front rails may be shifted if a sufficient force is present as is shown in (D).

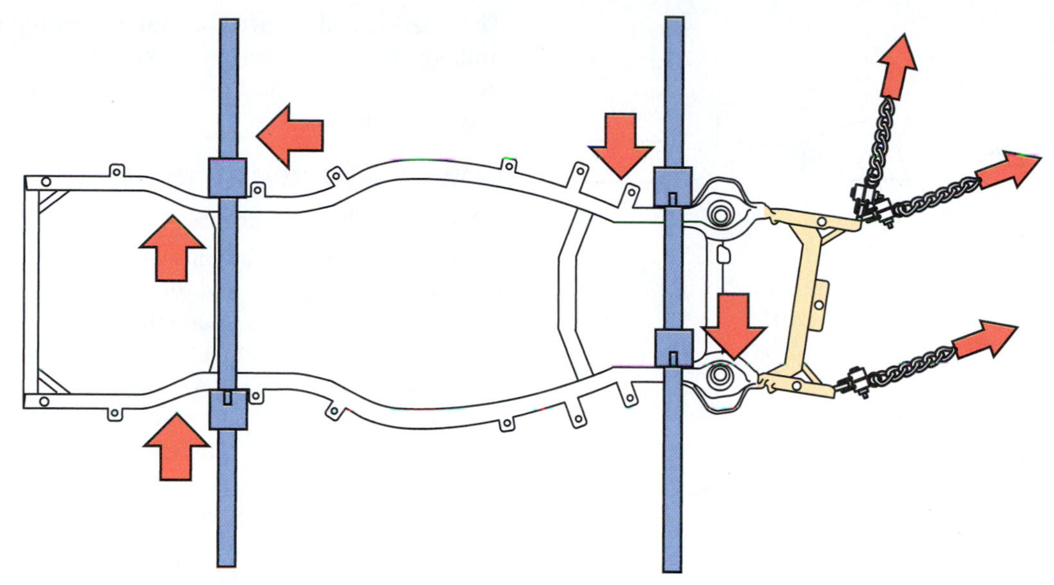

FIGURE 14-46 A multiple pull setup is most effective to move the entire front assembly back to its proper location.

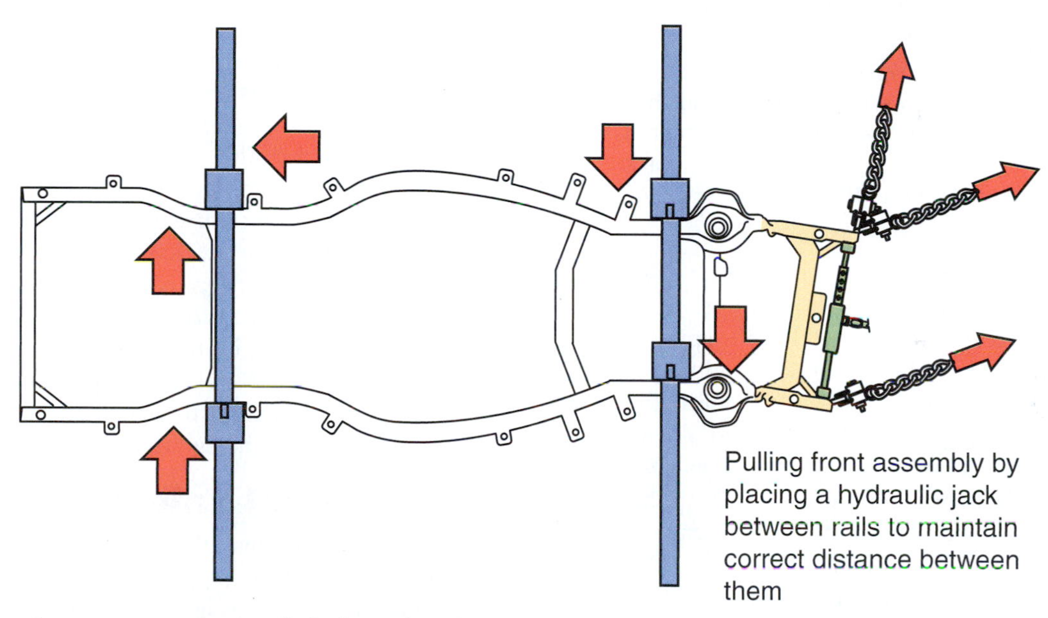

FIGURE 14-47 A porta power hydraulic jack can be placed between the front rails, enabling the technician to simultaneously move the entire front assembly by pulling from the far side of the front structure without being concerned of crushing or tearing the panels between the two rails.

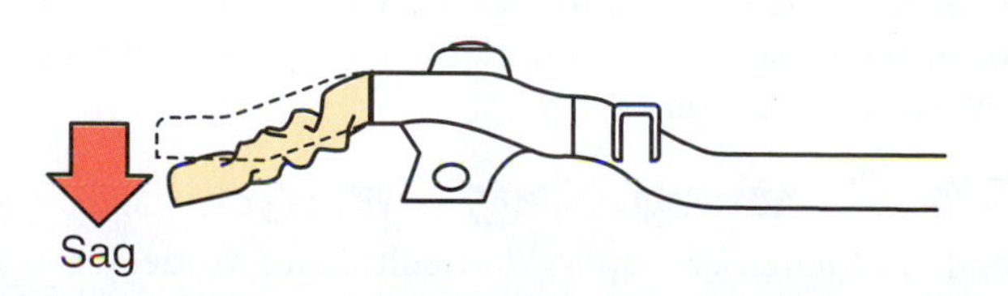

FIGURE 14-48 On the sag condition, the end of one side rail is lower than the opposite side of the vehicle or lower than the specifications indicate.

ride high on one side while the sag will give it a drooping effect. When the bumper reinforcement and the attaching brackets are not attached to the front rails, it is difficult to distinguish which condition exists until a measurement is taken. Sometimes—to complicate matters—both sag and kickup damage exists. Restoring the correct height from the datum plane can be accomplished only after a measuring system that identifies the datum plane is set up or attached to the vehicle.

Repairing the Sag Condition. The sag/kickup condition is the easiest of all the damage conditions to repair, particularly if it affects the outer extremes of the front or rear rails. See

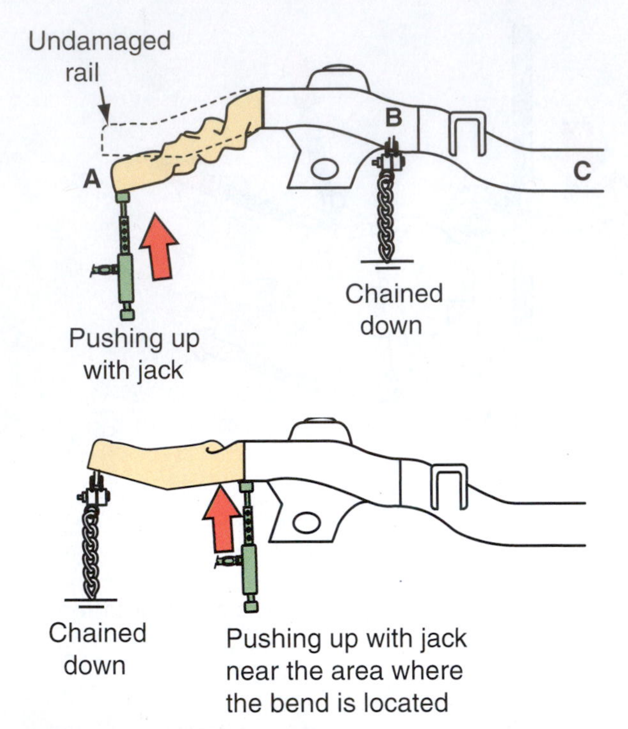

FIGURE 14-49 Correcting the sag condition is being accomplished by tying down the side rail behind the buckled area and raising up on the outer end of the rail. The rail is tied down at the end and is being raised up behind the buckled area to pull the kickup area down.

Figure 14-48 and **Figure 14-49**. In the case of a sag, one must anchor the vehicle down securely to prevent it from raising up when corrective upward force is applied at the end of the rail. In Figure 14-49, the sag condition is being corrected by first holding down the area behind the damage and then raising up on the front of the rail with a hydraulic jack. The tie down is placed as close as possible to maximize the effects of the corrective force. In the event the damage is more gradual and covers a larger area, it may be necessary to tie the vehicle down in the torque box area to obtain a more widespread movement of the metal. In the case of a unibody, it may not be necessary to further anchor the vehicle as it is already being held in place at the torque box with the pinchweld clamp. Changing the location of the strut tower, however, should be a primary concern when making the correction using this setup. Even though the unibody vehicle and the BOF vehicle may require a completely different anchoring system to maximize the corrective forces, the same basic concepts and laws of tension apply on both vehicle designs.

STEERING/SUSPENSION ALIGNMENT

While all the undercarriage repairs are being performed by the technician, the suspension—another very critical part of the repair—must also be monitored constantly. The steering and suspension systems on the modern day unibody and body over frame vehicles demand that the two independent—yet interrelated—systems be repaired at the same time.

Parallelogram and Rack-and-Pinion Systems

Historically there have been—and still are—two primary steering systems used on the automobile and small trucks: **parallelogram**, sometimes referred to as the **short-arm/long-arm (SLA)**, and the **rack-and-pinion**. Several variations of the SLA system have been introduced in recent years, but their principles of operation essentially remain the same. Therefore, the same close tolerances and specifications must be maintained throughout. The rack-and-pinion is used primarily on the unibody vehicles and on some select BOF vehicles, whereas the SLA system is used on most BOF vehicles.

Steering and Suspension Relationship

The close relationship between the steering system and the body to which it is mounted, especially on the unibody design, necessitate repairing both at the same time because if one or the other is not restored to the manufacturer's specifications the vehicle will not be considered safe to drive when it is returned to the vehicle owner. The use of an automated or mechanical measuring system is the only means of ensuring that all the suspension parts and their mounting points have been restored to the precise location specified by the manufacturer. Only when this is accomplished can all the suspension and steering angles be properly placed to ensure the vehicle will respond correctly when being driven.

Dimensional Tolerances

Most automobile manufacturers specify that a maximum tolerance of 3 mm (approximately 1/8 inch) must be maintained when restoring the collision-damaged sheet metal and "0" tolerances on steering and suspension parts. The dimensional tolerances must be maintained to this degree because most unibody vehicles make no allowances for movement or adjustments on any part of the steering or suspension system. The correct placement of all steering angles is achieved exclusively by the correct placement of the strut tower, the mounting surfaces for the steering and suspension parts, and the use of the correct and undamaged parts. Not only does the part have to be in good shape, but it must also be precisely mounted in the correct location.

The Unforgiving Design

It is incumbent upon the technician to decide if a misalignment issue is due to a damaged part or if the problem is the result of incorrectly aligned sheet metal.

It is no longer feasible to sublet the undercarriage repairs to another repair facility because one can no longer differentiate between where the alignment operations end and the body work begins. Both are totally interdependent upon the other to correctly restore the vehicle to its pre-collision condition. This is why the unibody vehicle is sometimes referred to as "The Unforgiving Design."

AFFECTED PARAMETERS

There are essentially three parameters that are affected with the adjustments, settings, and placement of the suspension angles: tire wear, vehicle stability, and diagnostics. On a unibody vehicle the correct placement of each of these is almost totally dependent upon the correct alignment of the sheet metal, the structural members, and the mounting surfaces for suspension parts on the undercarriage.

Camber and Caster Positioning

On a vehicle equipped with the MacPherson strut suspension, the correct positioning of both the camber and caster angles are largely dependent upon the placement and alignment of the strut tower. Because the top of the strut is attached to the upper part of the tower, when moved inboard or outboard it will result in a more negative or positive camber reading. In the event it is moved too far forward or rearward, either too much or too little caster adjustment will result. While these angles may be adjusted, not all vehicles are manufactured with provisions for moving these parts to ensure the proper alignment of these angles.

DEGREES VS. LINEAR DIMENSIONS

Strut Tower Placement

Many people don't understand why the dimensional accuracy of the strut tower placement is so critical. It is because the camber and caster angles are always cited in degrees and the dimensions for the sheet metal parts are always given in linear dimensions. It is difficult for many people to make the connection and to understand the correlation between these two differing descriptors. The following illustration should help to put the two forms of measuring into perspective. See **Figure 14-50**.

SLA Suspension

On the SLA suspension system, the distance from the lower ball joint to the upper ball joint is 305 mm (12 inches). In that short distance 1 degree of misalignment equals 5.3 mm of misalignment. This represents nearly 1/4 inch of shimming that must be done to overcome the misalignment or the frame rail must be pulled out that amount to correct the incorrect angle.

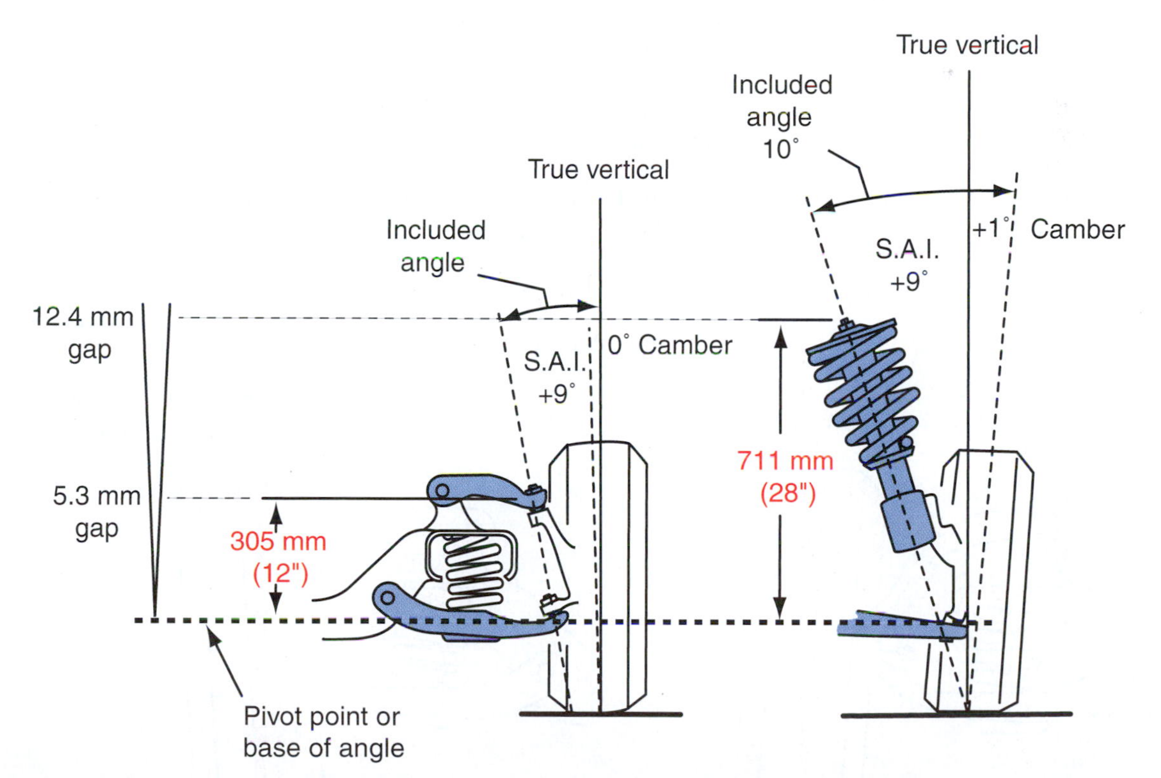

FIGURE 14-50 Many technicians have difficulty understanding the significance of an alignment angle that is off as little as 1 degree. In reality this seemingly small amount is over 5 mm (nearly 1/4 inch) of misalignment at the upper control arm on a SLA system and 12.4 mm (1/2 inch) at the strut tower mount on a unibody design.

MacPherson Strut Suspension

On the MacPherson strut suspension the typical strut is approximately 711 mm (28 inches) from the lower ball joint to the top of the strut. In that distance 1 degree represents 12.4 mm of misalignment. In other words, the strut tower must be shifted nearly 12.4 mm (1/2 inch) to obtain the correct alignment angle. In either case the camber setting is sufficiently incorrect to cause a significant amount of tire wear.

CALCULATING MISALIGNMENT

With the help of some basic assumptions, one can determine the degree of misalignment on nearly any suspension system by using the following formula. One must understand that any circle—regardless of its size—has 360 degrees. All calculations are made using the circumference (the distance around the outside of the circle) and the radius (1/2 of the diameter) of the circle. Since most misalignment problems become evident during the alignment process, the number of degrees of suspension misalignment become known. With this information, the degree of misalignment can easily be converted to distances and the amount of misalignment can easily be determined with some simple calculations using the following formula:

$$\frac{\text{\# of degrees of misalignment} \times \text{circumference of pivot distance}}{\text{360 degrees}} = \text{Distance for correction}$$

Assuming 1 degree of misalignment exists in the example on the right in Figure 14-50, one must do the following:

- Calculate the length of the pivot axis = (711 mm or a 28 inch radius)
- Calculate the circumference of the circle (711 × 2π) = 4,465.08 mm
- Divide the pivot axis by 360 degrees (4,465.08 ÷ 360) = 12.4 mm
- The distance the strut must be moved if off 1 degree is 12.4 mm.
- If the amount of misalignment were 2 degrees, the tower would need to be moved 24.8 mm.

One should not automatically assume the strut tower is out of alignment and start to pull without first checking to ensure other areas are not causing the problem. For example, the strut could be bent or the subframe may also have shifted, causing a misalignment problem.

WHAT IS CAMBER?

Camber is described as the inward or outward tilt at the top of the wheel when viewing it from the front of the vehicle. It is commonly identified as negative camber, positive camber, or "0" camber. See **Figure 14-51**. As is shown in Figure 14-51, positive camber exists when the wheel is tilted out at the top or away from the engine compartment. Negative camber exists when the top of the wheel is tilted inward toward the engine compartment. The camber angle is measured in degrees from true vertical.

Wear Angle

Camber is commonly referred to as a wear angle because if the angle is not correct either the inside or the outside of the tire is contacting the road surface exclusively, thus placing the entire weight of the vehicle on only a small part of the tire. On a positive camber setting the primary road contact area on the tire is on the outside edge, therefore causing a significant amount of wear on that surface.

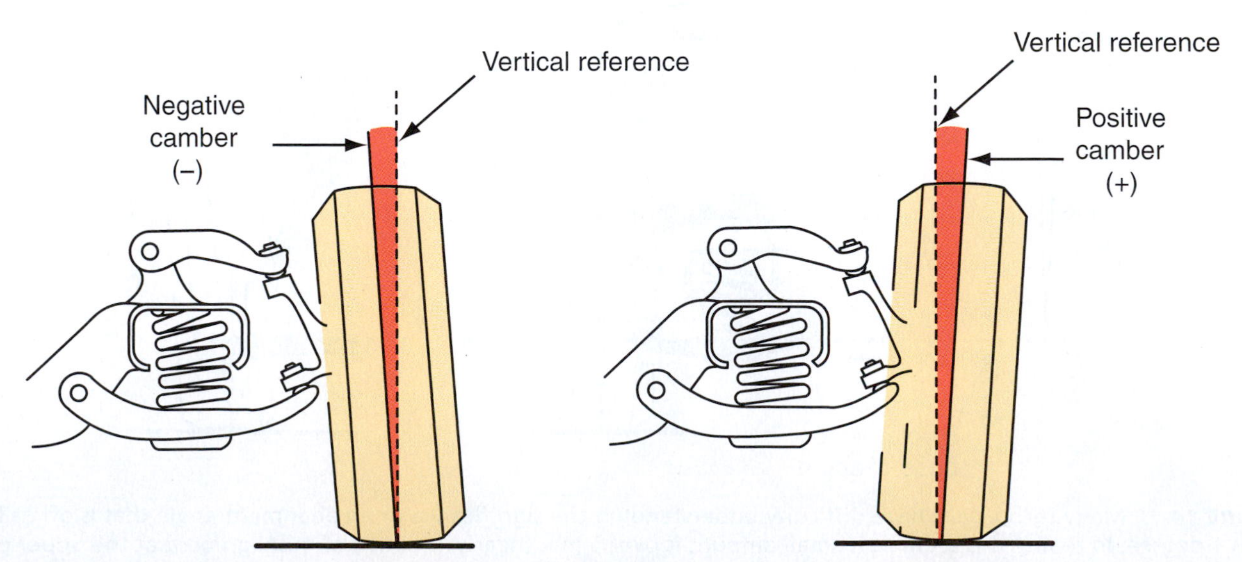

FIGURE 14-51 Camber is an alignment angle that refers to the tilt of the wheels and is identified as negative (leaning inboard at the top), positive (leaning outward at the top), or "0" camber when the wheels are perfectly vertical.

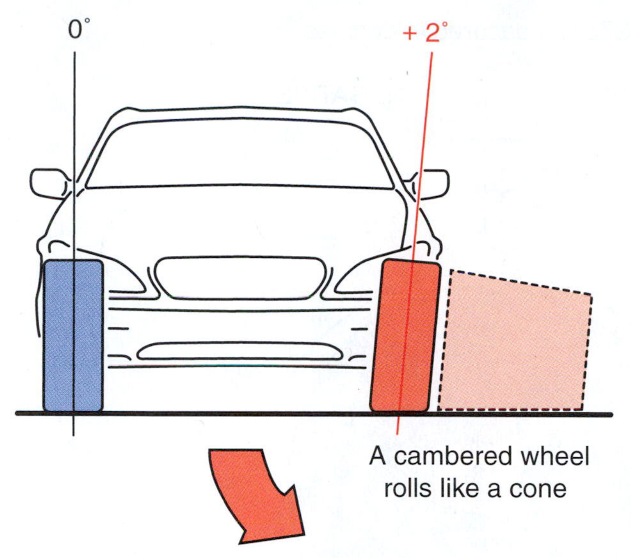

FIGURE 14-52 When an uneven camber angle between the two sides of the vehicle exists, the vehicle will always "pull" or drift to the side where the positive angle exists, much like a cone effect.

Pulling and Drifting

On a negative camber setting, the contact surface is on the inside of the tire, again causing accelerated tire wear. In addition to causing an excessive amount of tire wear, an improper camber setting will also cause the vehicle to "pull" or drift to one side when driving down the road. Whenever a pulling condition caused by a camber misalignment exists it will always pull to the side of the vehicle that has the most positive camber adjustment. See **Figure 14-52**. One might think of camber as a cone because it never rolls in a straight line. The large side of the cone gives it a positive angle; therefore, it has the same effect as does the steering force moving it in the direction of the tip. Unlike a cylinder, it will move toward the direction in which it is tipped.

Because the camber settings place a steering force on the wheels—causing the vehicle to pull toward the side of the vehicle where the most positive camber exists—both wheels need to be set at an angle nearly identical to each other to eliminate this effect. Therefore, the camber setting on both wheels should be set near "0" with perhaps 1/4 degree more positive on the left side.

CAMBER

Troubleshooting Camber-Related Problems

The following troubleshooting tips may be helpful to achieve a properly aligned suspension system.

Recommended Settings

- "0" DEGREES + 1/4 to 1/2 degrees more positive on left side

Results of Excessive Positive or Negative Camber Settings

- Poor steering and handling
- Pulling or drifting to one side (to side with greatest positive setting)
- Excessive tire wear
- Excessive wheel bearing wear

Causes of Incorrect/Out of Specification Camber Settings

- Unibody not within specification (strut tower pulled inboard or pushed out from specification)
- Ball joint pulled inboard or out from centerline
- Bent spindle/knuckle
- Bent strut
- Bent lower control arm (MacPherson strut suspension)
- Bent upper or lower control arm (SLA suspension)

Adjustments/Corrections

Adjustments will vary depending on suspension type.

Unibody.

- Straighten/move strut tower to specification
- Adjusting slots available on strut tower
- Adjustment on strut to steering knuckle
- Adjust cams on lower control arm

Parallelogram.

- Cams
- Shims
- Slots

WHAT IS CASTER?

Caster is described as the forward or rearward tilt of the front wheels on the steering axis, which is a straight line between the top of the strut through the lower ball joint. It is usually described or viewed from the side of the vehicle. See **Figure 14-53**. Caster is commonly referred to as either positive or negative, depending on whether the tilt of the wheel is facing the front or the rear of the vehicle. If the top of the steering axis is behind the wheel or the tilt of the wheel is facing toward the front of the vehicle, it is said to be positive caster. Conversely, if the top of the steering axis is in front of the wheel or the tilt is facing toward the rear of the vehicle, it is said to be negative caster. It is commonly considered a stability angle and may be used to adjust the firmness of the steering wheel when the vehicle is being driven. Historically, European vehicle manufacturers recommend more negative caster

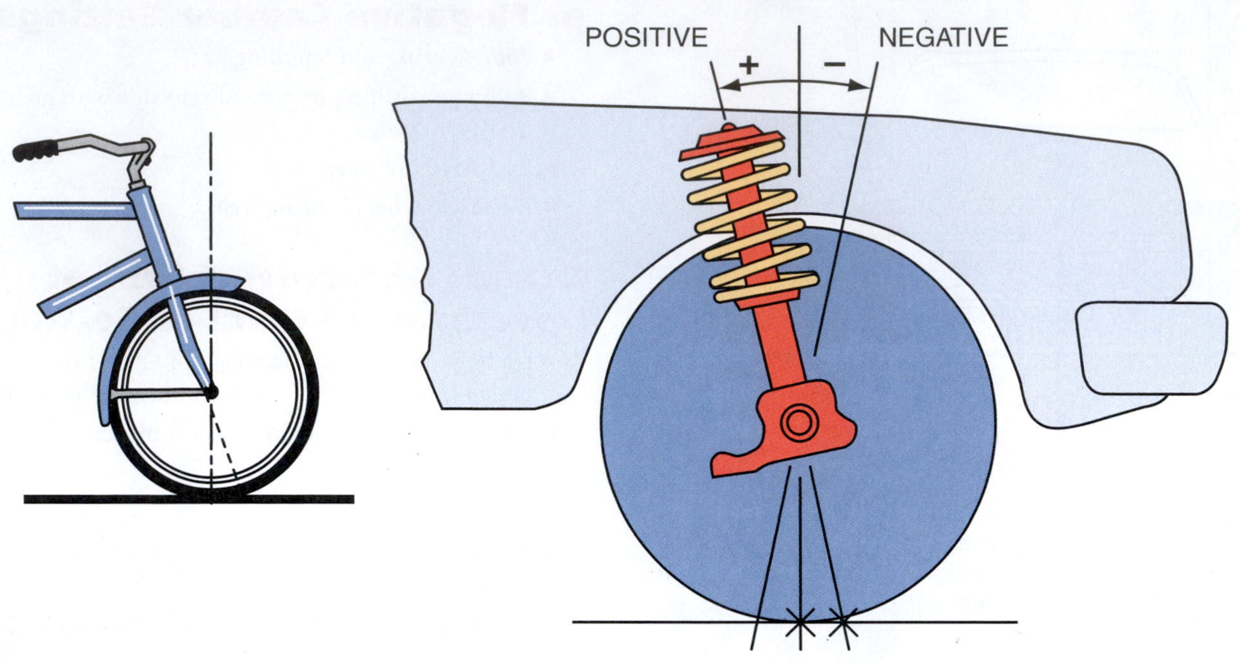

FIGURE 14-53 Caster is the forward or rearward tilt of the front wheel steering axis, a straight line between the top of the strut or upper ball joint and the lower ball joint. It is measured in negative and positive degrees.

setting than do domestic vehicles to give the steering a firmer feel.

Caster may also be used to create more or less directional stability and **wheel returnability**. A vehicle with a high degree of positive caster tends to force the wheels to roll straight ahead. It also tends to help the wheel return to a straight-ahead angle An example of this is the angle on the front fork of a two-wheel bicycle. The rider can let go of the handlebars and the bike will continue to move straight ahead. When the wheel or the handlebar is turned to negotiate a turn it will return to a straightforward position when the steering wheel or handlebar is released to find its own center. The caster angle should be set equal on both wheels with a possible variance of 1/4 degree more positive on the right wheel. Vehicles with a steering problem that pulls from caster misadjustment will drift to the side with the least positive setting.

CASTER

The following guide may be used for troubleshooting caster problems and may be helpful to achieve a properly aligned suspension system.

Positive Caster

- Backward tilt at the top of the strut or pivot point
- Enhances directional control
- Always pushes the wheel toward the center of the vehicle—the high point of the arc
- Increases the wheel returnability to the straight-ahead position when the steering wheel is released after cornering

Excessive Positive Caster

- Hard steering
- Causes shimmy in the steering
- Excessive road shock felt in the steering wheel

Negative Caster

- A forward tilt at the top of the strut or pivot point
- Reduces steering effort
- Reduces steering and vehicle stability

Excessive Negative Caster

- Handling instability
- Tendency to cause oversteering

Possible Causes of Caster Out of Spec

- Top pivot of strut forward or rearward
- Bottom pivot point forward or rearward
- Bent strut
- Bent upper or lower control arm
- Strut tower misaligned
- Unibody not within spec

Adjustments

- Vary from vehicle to vehicle
- Some vehicles have no adjustment
- Slots
- Shims
- Cams
- Strut bar

TOE—TOE-IN/TOE-OUT

Toe is the single most critical angle that is set during an alignment. When not set correctly, it can cause extreme tire wear in a very short distance. Toe is the dimensional difference between the front of the front wheels/tires and the rear of the front wheels. Another way to think of it is the distance from center to center between the front of the left and right tires compared to the center-to-center measurement on the rear of the same tires. It can be measured in degrees or millimeters, minutes and fractions. See **Figure 14-54**.

Causes Steering Instability

In addition to causing excessive tire wear, an incorrect toe setting can cause steering and handling problems such as the vehicle wandering and steering instability. In Figure 14-54, the wheels should be pointed straight ahead when the vehicle is in motion. Excessive **toe-in/toe-out** setting is also shown in Figure 14-54, demonstrating the "scrubbing" of the tires that occurs if it is not set correctly. Automobile manufacturers publish toe recommendations as do most undercarriage dimension manuals in their vehicle-specific specifications. The vehicle manufacturer's and the undercarriage dimension manual specs are not always correct or they may suggest such a wide range of specifications that when followed, it may not result in the best possible steering characteristics.

Toe Settings

A broad range of settings is recommended ranging from a 21.5 mm (toed-in) to 14 mm (toed-out) on some vehicles when in a static condition or not moving. In the

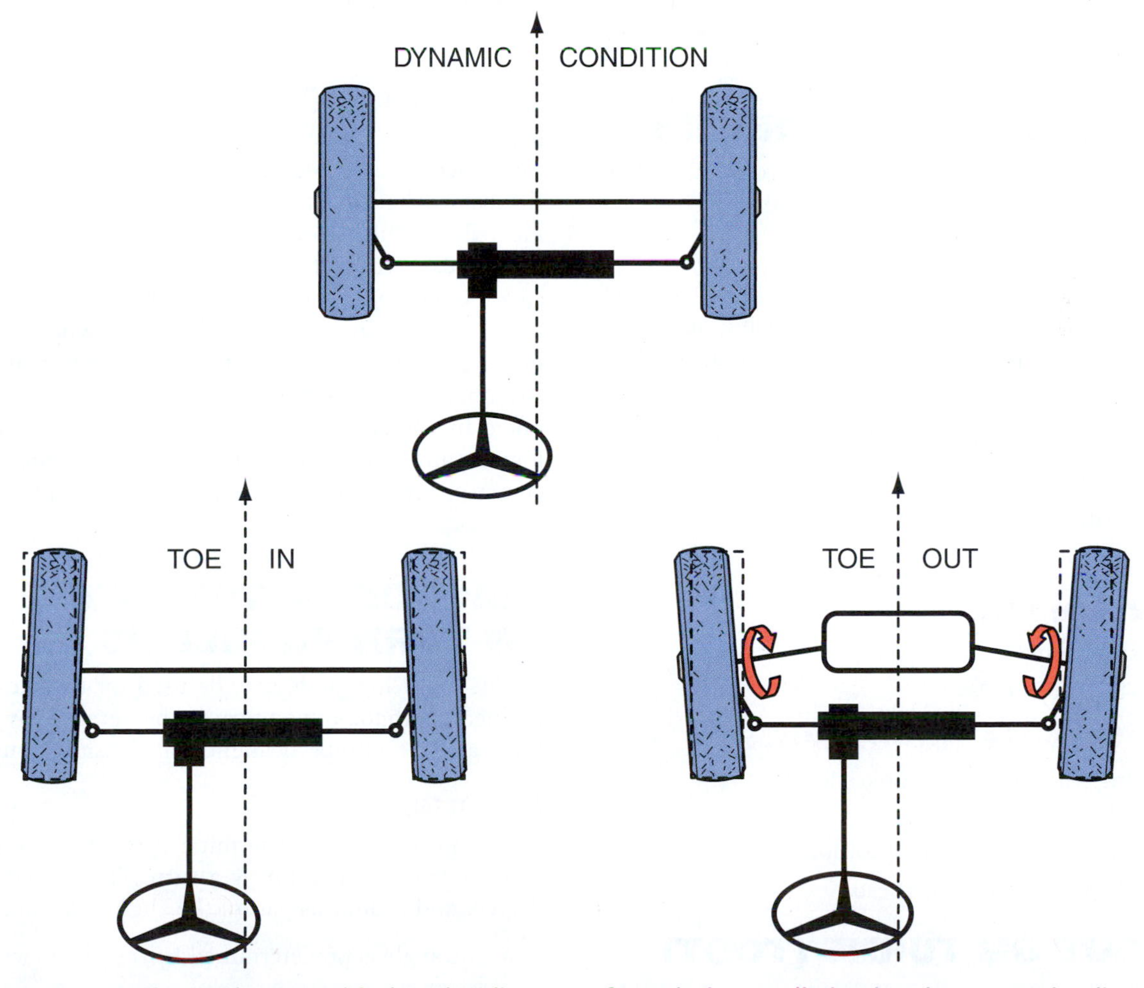

FIGURE 14-54 Toe is the single most critical angle adjustment for reducing or eliminating tire wear. It is adjusted negative, positive, or as "0" dynamic toe when the vehicle is in motion.

event access to vehicle-specific recommendations are not available, a "rule of thumb" is that most front-wheel-drive vehicles are adjusted to 2 to 3 mm of toe-out (1/8 inch) and rear-wheel-drive vehicles are usually set for approximately 2 to 3 mm (1/8 inch) of toe-in.

Purpose for Toe Settings

In a static condition (not moving) the wheels may be pointed either inboard or outboard slightly, but as the vehicle starts to move, the tires tend to shift to a straightforward attitude. The purpose of the toe-out is so that—under dynamic conditions—the vehicle moving forward will have the tire and wheel attain a **zero running toe**. The toe setting is made for the tendency to overcome the torque of the drive axles on a front-wheel-drive vehicle and the friction against the road on a rear-wheel-drive vehicle. A zero running toe is most desirable to minimize tire wear. A vehicle with 2 mm (1/16 inch) of toe misalignment will produce approximately 9 feet of sideways scuff for every mile driven—in other words, the equivalent of dragging the tires sideways or perpendicular to their normal rolling direction for 9 feet every mile driven. The life of the tires can be reduced by as much as 50 percent if toe setting is not adjusted properly. The toe setting is adjusted after all other settings have been corrected during a wheel alignment.

DIAGNOSING TOE PROBLEMS

The following guide may be used for troubleshooting toe-related problems and may be helpful to achieve a properly aligned suspension system.

Purpose

Toe compensates for the toeing-in and toeing-out of the front wheels when the vehicle is moving down the road to achieve "0" running toe.

Excessive Toe-In (positive toe)

- Tires scrubbing
- Tires feather in

Excessive Toe-Out (negative toe)

- Tires scrubbing
- Tires feather out

Adjustments Made with

- Steering wheel straight (spokes leveled horizontally)
- Steering gap in high point position
- Wheels pointed straight ahead

TOE-OUT ON TURNS (TOOT)

Toe-out on turns is a non-adjustable angle that is affected by the steering arms only. It can be corrected only by replacing one or both of the steering arms that may have become bent in a collision. This is also a tire wear angle but affects the vehicle most significantly during turns. See **Figure 14-55**. When a vehicle turns a corner the inside wheel (in the direction of the turn being made) must always turn more than the outside wheel. In reality, this increases the distance between the wheels at the front so a toe-out condition exists. The wheel to the inside of the turn is always ahead of the outer wheel whenever making a turn. Therefore, a toe-out condition always exists on a turn which prevents excessive tire wear. As noted by the arrows in the illustration, each wheel follows a path of its own but all the paths are circles with a common center point. This helps to prevent scrubbing the tires.

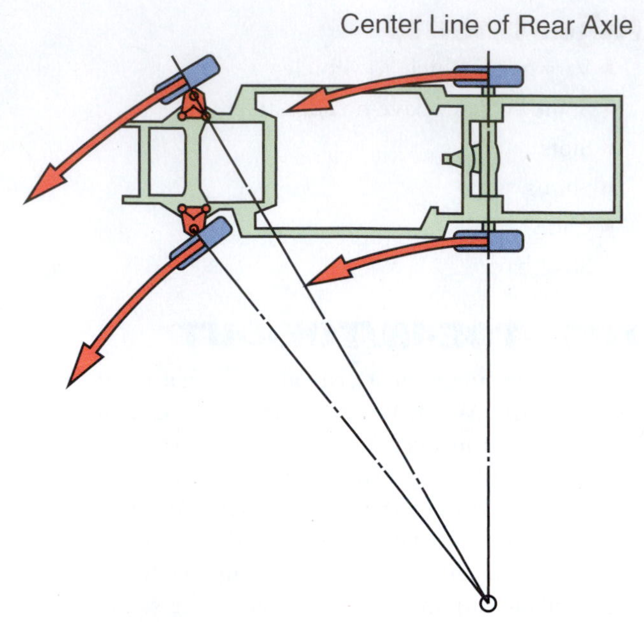

FIGURE 14-55 Toe-out on turns is a non-adjustable angle that is determined by the steering arms. To change the angle, one must replace the steering arms. Each wheel takes a slightly different path on a turn that is affected by the steering arms on the vehicle.

DIAGNOSING TOE-OUT ON TURN PROBLEMS

The following guide may be used for troubleshooting problems with toe-out on turn problems and may be helpful to achieve a properly aligned suspension system.

Purpose

The purpose is to determine if the wheels turn out enough or too much during a turn. This is measured in degrees and is done as part of the alignment checks.

- The angle is predetermined by the manufacturer's design.
- The angle allows the inside wheel to turn a narrower radius than the outside wheel on a turn.

- It minimizes tire wear that occurs on turns.
- It improves handling during and immediately after a turn.

Turn Angle Out of Spec

- Poor response on a turn
- Squealing tires
- Excessive front tire wear

Adjustments

No adjustment possible. Replace the steering arm(s).

STEERING AXIS INCLINATION (SAI)

Steering axis inclination (SAI) is the degree or amount of inward tilt of the upper pivot point from a true vertical line on the steering assembly. This inward tilt tends to keep the wheel pointed straight ahead. It is also largely responsible for returning the wheels to a straightforward position after a turn is completed. When the wheels are turned the spindle tends to rotate forward and down, causing the vehicle to rise up. See **Figure 14-56**. On a MacPherson strut suspension it is usually identified as the angle that is formed by the lines which run down through the length of the strut to the ball joint which intersects with a true vertical line on the ground. On a short-arm/long-arm (SLA) suspension, it is the angle formed by the straight line that runs through the upper and lower ball joint which intersects on the ground with a true vertical line. This intersecting point is the precise location where the spindle exerts the downward pressure when the steering wheel is turned, causing the vehicle to rise up slightly. When the steering wheel is released the downward pressure from the weight of the vehicle causes the wheels to return to a straight-ahead position once again when the turn is completed. In reality the weight of the vehicle brings the wheels back to a straightforward position after the turn is completed and also helps to give directional stability. SAI, is largely a diagnostic angle for tire wear. It will help determine whether the upper strut towers are properly located, if the control arms are damaged, or if the center cross-member or subframe has shifted. It is also known as the king pin inclination (KPI).

STEERING AXIS INCLINATION PROBLEMS AND DIAGNOSIS

The following guide may be used for troubleshooting SAI problems and may be helpful to achieve a properly aligned suspension system.

Purpose

SAI is the inward tilt or the upper pivot point from true vertical.

It provides a pivot point near the center of the tire.

- It promotes ease of steering.
- It provides even braking.
- It adds to the directional control of the vehicle.
- It returns the wheel to the straight-ahead position.
- It holds the wheels in the straight-ahead position.
- It helps distribute the vehicle weight evenly to the tires.

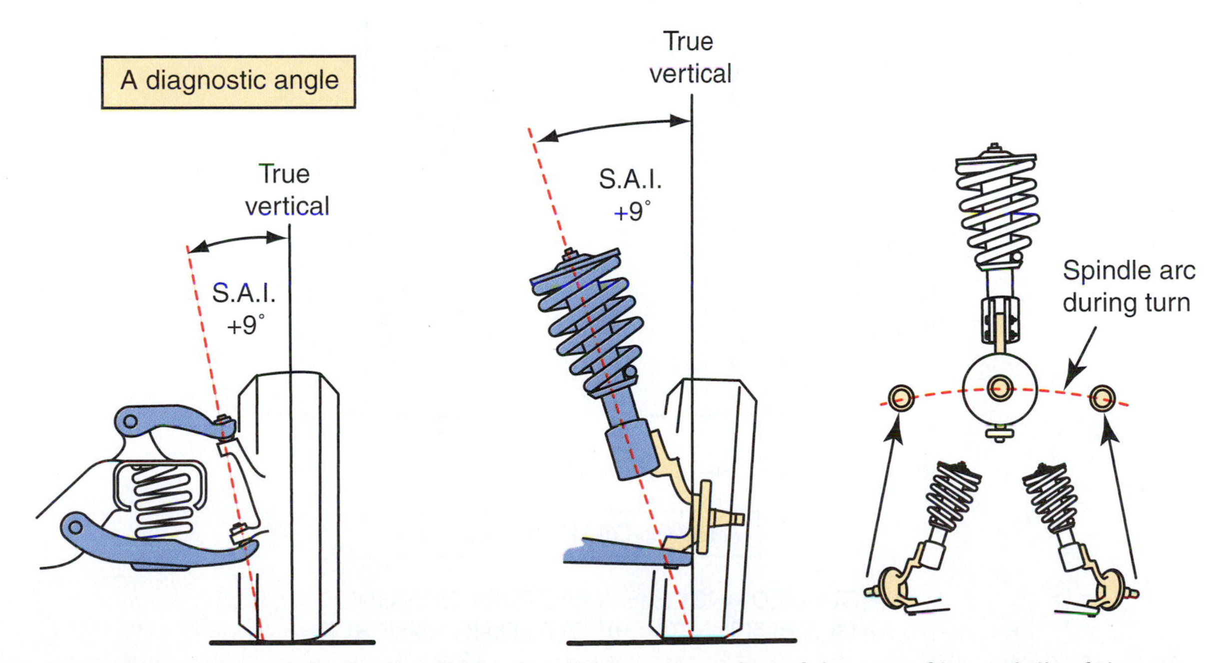

FIGURE 14-56 The steering axis inclination (SAI) is the distance or number of degrees of inward tilt of the upper steering axis compared to a true vertical line that intersects at the bottom of the tire and runs straight up vertically.

Possible Causes for Out of Spec SAI

- Top pivot point moved in or out
- Bottom of pivot point in or out
- Unibody structure out of spec

Adjustments/Corrections

- Align frame/undercarriage
- Replace damaged parts
- Shift subframe

INCLUDED ANGLE

The term **included angle** is the sum total of SAI and camber. When the camber is positive the two are added together. See **Figure 14-57**. If the camber is negative, the degrees of negative camber are subtracted from the total angle degrees. The included angle is a factory designed angle built into the steering knuckle and is used as a diagnostic angle if damage to the suspension parts is suspected. It tells the technician only one thing—if it is out of spec, on a MacPherson strut suspension, there is a bent part somewhere between the strut tower and the lower ball joint or between the upper ball joint and lower ball joint on a SLA suspension system.

THRUST ANGLE

The **thrust angle** is the direction all four wheels move in relation to the vehicle's centerline. It is determined by the position of the rear axle and the toe of the rear wheels. See **Figure 14-58**. If the rear wheels are aimed straight forward and are parallel to the centerline, the vehicle will move straight forward. This is referred to as "0" thrust angle or no thrust angle, which is the ideal effect. On the other hand, if one of the rear wheels is toed-out and the other is toed-in, the rear of the vehicle will tend to run out toward the side with the toed-out wheel. This is commonly referred to as **dog tracking**. The rear wheels tend to steer the vehicle as it is moving along. All four wheels must be pointed straight forward and be parallel to the centerline of the vehicle in order for it to move straight down the road.

Troubleshooting

The following guide may be used for troubleshooting thrust angle or tracking problems and may be helpful to achieve a properly aligned suspension system.

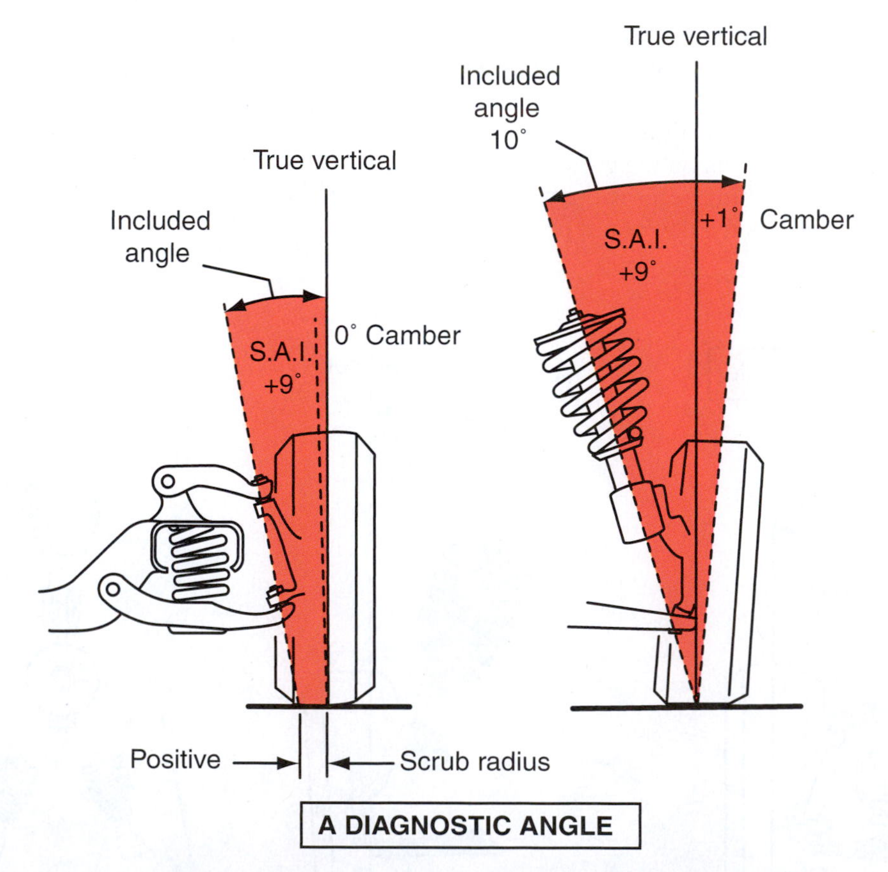

FIGURE 14-57 The included angle is designed into the steering system by the manufacturer and is used as a diagnostic angle when damaged suspension parts are suspected.

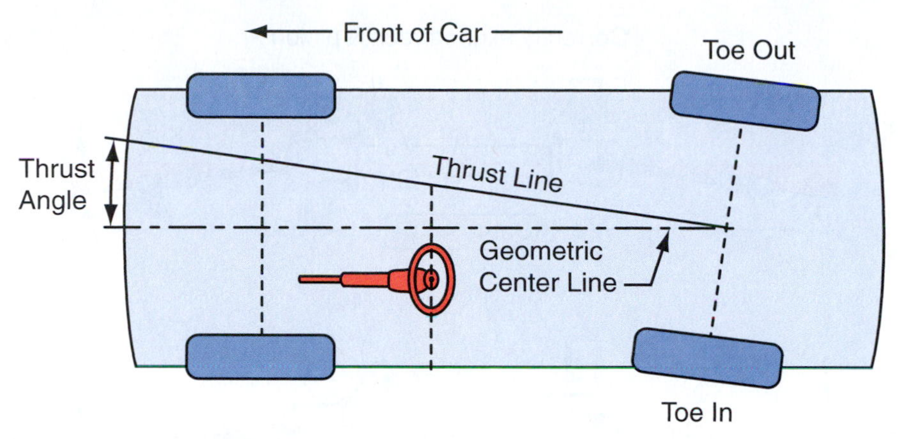

FIGURE 14-58 The thrust angle is the direction the four wheels move with respect to the vehicle's centerline. Any time one side of the axle is toed-out and the opposite side is toed-in a similar amount, the vehicle will have a tendency to run out toward the side with the toed-out wheel.

Thrust Angle

This is the direction all four wheels move in relation to the vehicle's centerline.

- This is determined by the position of the rear axle and rear wheel toe.
- On a misaligned solid axle, one wheel is toed-in and one is toed-out.

Thrust Angle Out of Specification Causes

- Dog tracking
- Unbalanced steering on turns
- Oversteering or understeering
- Fast tire wear

Adjustments

Adjustable Suspension.

- Adjust to specs for toe and thrust angles

Non-adjustable Suspension.

- Check for misaligned rear suspension mounts
- Replace damaged or worn parts

BUMP STEER

Bump steer is not considered one of the alignment geometry angles or one of the adjustments usually made as part of a wheel alignment, but it is a condition that can become an issue after a frontal collision. Bump steer describes a condition in which, after a vehicle hits a bump, it wants to go in one direction and then another or it darts and dives, without moving the steering wheel. See **Figure 14-59**.

Causes of Bump Steering

On a vehicle with a rack-and-pinion steering system a misaligned rack is usually the cause of bump steering. A variance as little as 3 mm (1/8 inch) is sufficient to cause the vehicle to dart and dive unpredictably. When the rack is out of alignment, it causes the tie-rod ends to be positioned at different angles. This causes the tie rods to move in a different arc radius, one longer than the other, causing the toe to change differently on each wheel during the jounce and rebound cycle when a bump is hit. See **Figure 14-60**.

Bump steer can also be experienced on vehicles that have a standard gear box and idler arm. On this design the cause for the bump steer response is the intermediate tie rod not being parallel with the vehicle. This will in turn cause the tie rods to be at a different angle to each other. There again, a different arc radius causes the vehicle to dart and dive.

BUMP STEER TROUBLESHOOTING

The following guide may be used for troubleshooting bump steering problems and may be helpful to achieve a properly aligned suspension system, resulting in a smoother ride.

Bump steer is a condition that causes a vehicle to shift direction very unpredictably or to dart and dive when it hits a bump without the steering wheel being moved or turned.

Causes

- Misaligned rack-and-pinion gear (rack-and-pinion steering)
- Improperly aligned intermediate tie rod (on standard gear box)

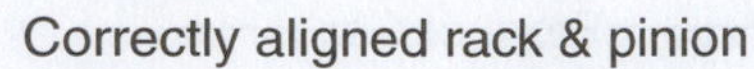

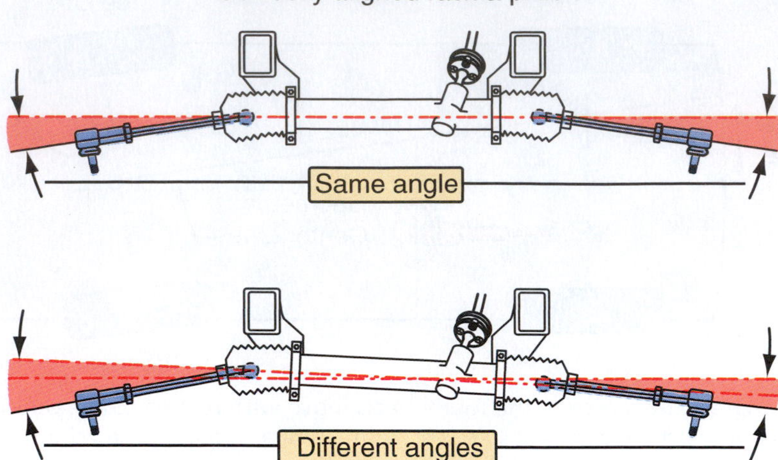

Which in turn causes the tie rod to swing in different arcs.
This causes the toe to change.

(A)

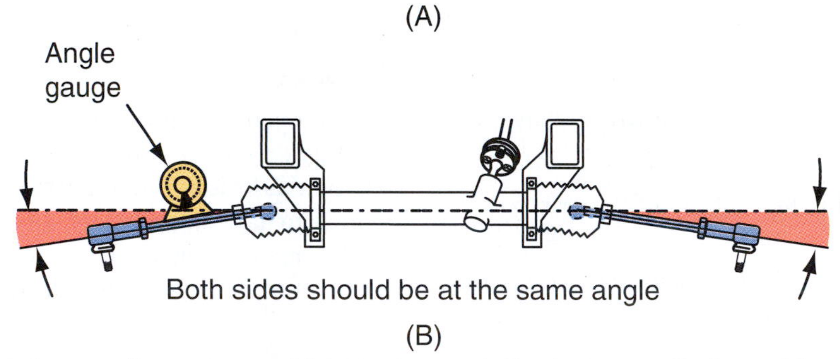

(B)

FIGURE 14-59 Bump steering usually causes a vehicle to dart and dive involuntarily from side to side whenever a bump is hit in the road. (A) shows the difference in the steering arm angles between a properly aligned steering gear and one that isn't. (B) shows the correct alignment of the steering rack.

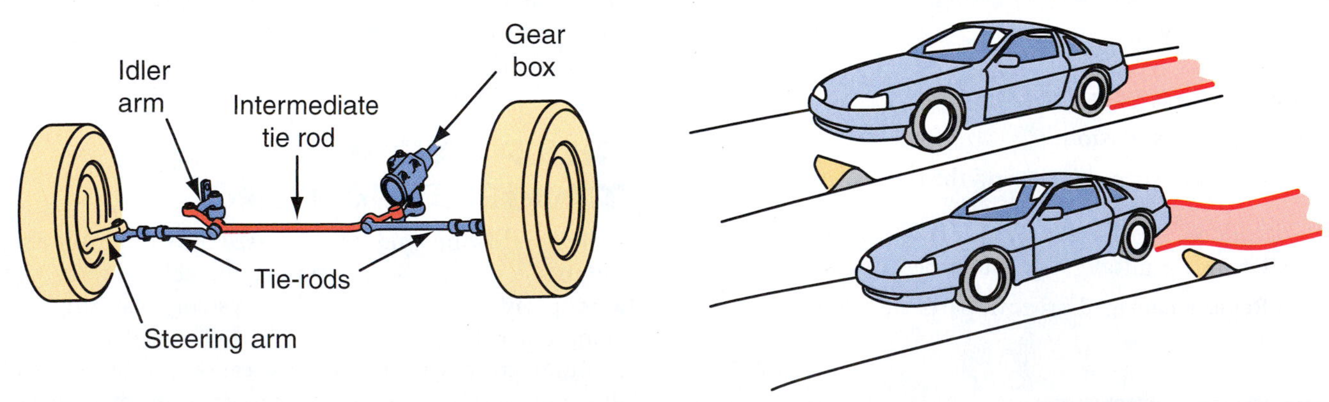

FIGURE 14-60 Bump steering on an SLA steering system is caused by the intermediate tie rod not being level with the vehicle or a bent mounting surface.

- Improperly repaired strut tower (too high on one side)
- Misaligned lower control arm mounting location (inner pivot point too high or too low)

Diagnosing Bump Steer

Rack-and-Pinion Steering System.

- Visually check levelness of steering rack on vehicle
- Check tie rods for like angles (use simple hardware store protractor)
- Check height of inner lower control arm mount from floor or compare to opposite side–they should be the same

Corrections

- Level steering rack on vehicle
- Repair strut tower to proper height
- Align lower control arm mount location

Standard Gear Box Steering System.

- Check levelness of intermediate (center) tie rod
- Check idler arm bracket (If adjustable, raise or lower to level of the intermediate tie rod)
- Check height of inner lower control arm mounting location from floor and compare to opposite side

Corrections

- Adjust idler arm to proper position if movable
- Replace idler arm if bent
- Replace lower control arm if bent at the mounting location
- Repair lower control arm mounting surface if bent or deformed

MEMORY STEERING

Memory steering is another steering or driving condition that is not considered to be part of the alignment procedure. In most cases it won't appear until after the vehicle has been repaired, the wheel alignment is completed, and the vehicle is returned back to the owner. Road testing the vehicle is the only way this condition would come to the repair shop's attention—unless the customer brings the vehicle back to complain about it. Memory steer is a condition where the vehicle wants to continue to drift or pull in the same direction as a turn that was just executed. This can be caused by a damaged or bad strut bearing or one that was installed incorrectly. In either case, it will not allow the steering mechanism to completely return to a straightforward latitude or position.

It can also occur as a result of defective or faulty rubber bonded socket (RBS) tie-rod ends, which are used largely by Ford Motor Company. These are preloaded torque elastic joints that provide as much as 20 percent of the effort required to return the wheels to a straightforward position. If the RBS is installed incorrectly or is damaged, the steering wheel would not have the necessary assist required to return the wheels to a straightforward position.

Summary

- Repairing the modern day automobile requires more knowledge, skill, and intuitive reasoning than has ever been required.
- Technicians rely on their knowledge, experience, and intuition to guide them as they go about repairing a damaged vehicle.
- For many years the auto collision repair industry has been in a state of transition as the automobile manufacturers have attempted to meet consumer demands for a more economical and fuel-efficient automobile.
- The materials used in manufacturing fuel-efficient vehicles have changed markedly in recent years as manufacturers have attempted to find a balance between safety, efficiency, and functionability.
- The introduction of higher strength steels has allowed manufacturers to reduce the thickness and weight, making it necessary to develop new repair techniques.
- The introduction of boron and laminated steels requires additional technical expertise to repair these vehicles.
- As the use of aluminum increases, new techniques will be needed to repair those vehicles.
- The future of the collision repair industry is grounded in the constant need for change and upgraded skills to ensure technicians will have the knowledge and expertise to repair the automobile of tomorrow.

Key Terms

backlite
bench/rack system
bump steer
camber
caster
crush initiators
dedicated fixture system
diamond
dog tracking
dozer pulling system
floor-mounted system
frame rack/bench system
fulcrum
holding turnbuckle
included angle
laminated steel
LIFO
mash
memory steering
parallelogram/short-arm/long-arm (SLA)
point of impact (POI)
positive/solid anchoring system
rack-and-pinion
re-collision integrity
sag/kickup
sail panel
sidesway
springback
steering axis inclination (SAI)
sweat tab
temperature threshold
tension/pressure
thrust angle
toe-in/toe-out
twist
universal system
wheel returnability
zero running toe

Review

ASE Review Questions

1. Technician A says a unibody should be anchored at the end of the front and rear rails to ensure the entire vehicle is uniformly affected by the corrective forces. Technician B says the area where the pinchweld clamps are attached should be finished, primed, and painted after removing them. Who is correct?
   A. Technician A only
   B. Technician B only
   C. Both Technicians A and B
   D. Neither Technician A nor B
2. Technician A says that pressure as a corrective force is more easily controlled than any other corrective force. Technician B says the use of tension allows one to pull in a more precise direction than pressure. Who is correct?
   A. Technician A only
   B. Technician B only
   C. Both Technicians A and B
   D. Neither Technician A nor B
3. All of the following may be used as an attaching point when making corrective pulls EXCEPT:
   A. A flange.
   B. A damaged suspension part.
   C. An undamaged impact absorber.
   D. Damaged sheet metal.
4. Technician A says whenever using tension as a corrective force the entire area between the attaching point and the anchor point is affected equally. Technician B says when using tension the pulling device will always attempt to seek its own center of the pulling line of force. Who is correct?
   A. Technician A only
   B. Technician B only
   C. Both Technicians A and B
   D. Neither Technician A nor B
5. During a collision the inertial forces generally:
   A. Cause damage away from the point of impact.
   B. Affect only areas immediately adjacent to the point of impact.

C. Affect only non-structural parts.
D. Create damage that is readily visible and identifiable.

6. Technician A says the tie-down/anchoring points on a unibody vehicle are the same as for a BOF vehicle. Technician B says the anchoring points are usually found under the four corners of the passenger compartment. Who is correct?
   A. Technician A only
   B. Technician B only
   C. Both Technicians A and B
   D. Neither Technician A nor B
7. Which of the following is true of using pressure as a corrective force?
   A. It is rarely used in conjunction with tension.
   B. It is readily directed to correct damage.
   C. It follows the path of least resistance.
   D. The direction it follows can easily be controlled.
8. Technician A says a unibody vehicle must be tied down to a bench to form a frame as the vehicle has none. Technician B says when making corrective pulls on a unibody one must pull in only one direction at a time to avoid overpulling it, causing additional damage. Who is correct?
   A. Technician A only
   B. Technician B only
   C. Both Technicians A and B
   D. Neither Technician A nor B
9. The anchoring points used on the unibody vehicle must:
   A. Be parallel to the direction of any pulls made when repairing the vehicle.
   B. Be able to withstand the total of all the combined corrective forces used.
   C. Be at the outermost part of the vehicle, front and back.
   D. Be located at the front and rear of the passenger compartment.
10. Technician A says when stress relieving on high strength steel surfaces it must be done over a larger area than on standard or mild steel. Technician B says a limited amount of heat can be used so long as the safe temperature time threshold is not exceeded. Who is correct?
    A. Technician A only
    B. Technician B only
    C. Both Technicians A and B
    D. Neither Technician A nor B
11. Technician A says an incorrect camber setting will cause a vehicle to wander in the direction of the least positive setting. Technician B says caster is a stability angle. Who is correct?
    A. Technician A only
    B. Technician B only
    C. Both Technicians A and B
    D. Neither Technician A nor B
12. All of the following can be used as an attachment location for performing corrective pulls EXCEPT:
    A. Damaged sheet metal that will be replaced.
    B. The steering gear bolted to the frame rail.
    C. A damaged inner fender rail that will be salvaged.
    D. A collapsed B pillar.
13. Which of the following is NOT considered a diagnostic angle?
    A. SAI
    B. Included angle
    C. Toe
    D. Camber
14. Technician A says a certain degree of thrust angle is desirable to keep the vehicle running straight. Technician B says "0" running toe is the desired setting to achieve. Who is correct?
    A. Technician A only
    B. Technician B only
    C. Both Technicians A and B
    D. Neither Technician A nor B
15. Technician A says when pulling the damaged area to restore it to the correct dimension, it may need to be pulled beyond its normal position to overcome the elastic factor. Technician B says the point of impact will likely incur a considerable amount of inertial damage. Who is correct?
    A. Technician A only
    B. Technician B only
    C. Both Technicians A and B
    D. Neither Technician A nor B

Essay Questions

1. Why is tension the predominant corrective force used for collision repair?
2. What are at least five different pulling methods and accessories that can be used for removing damage on a vehicle?
3. If all five of the primary undercarriage damage conditions existed in a single vehicle, in what sequence should they be repaired?
4. What are the most common undercarriage repair systems (i.e., frame racks, etc.) used? What are their primary advantages?
5. Which suspension angles will cause tire wear when they are not properly adjusted? Why will the tire wear occur?
6. How does stress relieving differ on a high strength steel rail compared with a rail made out of mild steel?

Topic Related Math Questions

1. Determine the included angle from the following set of parameters: The left wheel has a 1.5 degree negative camber reading and the SAI is 9 degrees. The right wheel has a 0.5 degree positive camber reading and the SAI is 9.5 degrees.
   A. What is the included angle on the left side?
   B. What is the included angle on the right side?
   C. How much difference is there in the included angle?

Critical Thinking Questions

1. Why is a unibody vehicle not likely to sustain a diamond undercarriage damage condition?
2. During the repair process, the technician overlooked a buckle in the floor that prevented the seat from setting level until after the rocker panel was pulled straight. What might the consequences be for such an oversight and what should be done to repair the overlooked damage?
3. Not until the repairs have been initiated does the technician discover a high strength steel rail has been repaired previously and from all indications heat has been used on it. What are the possible consequences and what should be done at this point?
4. Why does the type of damage sustained by a unibody automobile body differ from that sustained by a body over frame vehicle?

Lab Activities

1. Looking at the undercarriage of a unibody vehicle, identify where the front and rear pinchweld clamps should be attached to properly secure the vehicle for pulling. Do the same for a BOF vehicle.
2. Locate a BOF and unibody vehicle and identify which vehicle uses a parallelogram steering system and which uses a rack-and-pinion system. Observe and record the difference between the two.

Chapter 15

Structural Parts Replacement

OBJECTIVES

Upon completing this chapter, you should be able to:

- Identify safety practices that protect both the vehicle and the technicians performing the repairs.
- Remove damaged cosmetic exterior parts, identify them, and catalog them, and do the same for other peripheral parts and pieces.
- Rough repair damaged cosmetic and structural parts prior to removing them and replacing them with new panels.
- Locate and utilize recommended vehicle-specific repair procedures from a variety of sources.
- Identify the welding sequences and techniques used for permanent panel installation.
- Compare adhesive bonding versus welding and note a combination of each method.
- Reinstall foam and corrosion protection.

INTRODUCTION

The structural parts on an automobile might be compared to the foundation of a house or any other structure that is built to withstand repeated exposure to stress, damaging outside forces, and the environment. They are the load-bearing members for the automobile and are largely responsible for protecting the passenger compartment, especially when subjected to outside forces such as those experienced in a collision. The importance of performing technically correct repairs is never more critical than it is on an automobile because an incorrect repair can greatly affect the vehicle's reaction and response—as well as the degree of responsiveness in the event of a collision. Improper repairs are also apt to have a very adverse effect on the vehicle's drivability and road performance after it has been repaired. As an industry, we are committed to restore a collision-damaged vehicle to its pre-collision condition, if not better. To accomplish this, we must strictly adhere to the rules and guidelines established by the manufacturers and various other research organizations whose primary concern is to protect the motoring public. Therefore, technicians must commit to performing each repair they make to the best of their ability, using sound tested and proven methods to ensure the vehicle is as safe as it was when manufactured. That means every panel that is repaired or replaced—and every weld made in the process—must meet or exceed the minimum industry standards established.

SAFETY IN THE REPAIR SHOP

Many times repair technicians may become somewhat lax and almost uncaring or unconcerned about basic safety issues. They seemingly perform the same—or at least similar—operations repeatedly without consequence to themselves and those who work around them. However, trainees are constantly exposed to new tools and equipment, unfamiliar materials, and new operations and procedures which they haven't performed before. One should always be cognizant of the need to protect not only oneself, but also the others in adjacent areas.

Personal Safety

For protection from injury, technicians should wear, at a minimum, safety glasses, gloves, and protective clothing. In addition to safety glasses, technicians should wear a protective face shield whenever performing any drilling, grinding, or surface removal operations. Welding helmets are an absolute must when performing any of the many cutting and welding operations. When welding on any zinc-coated metals, technicians should use the appropriate respiratory protective equipment. Fellow workers in the surrounding work area should also be made aware of any potential hazards any time materials are used that may "spill over" into their work area. This may include fumes, overspray, dust, and any other form of potentially hazardous exposure.

Vehicle Safety

Likewise, one must be aware that additional damage may occur to the vehicle while the technician is performing the repairs. One of the first things that should be done when a disabled vehicle is brought into the shop for repairs is to ensure the battery is disconnected. Although the battery may have been left connected without consequence before being brought into the shop, damaged and exposed wires may suddenly start to cause problems during parts removal, pulling, and straightening operations. Fluids such as windshield washer solvent and coolant leaking out of damaged reservoirs and overflow tanks may run or drip into electrical components, causing short circuits and permanent damage. At this time it may also be advisable to disconnect any fluid lines and their storage reservoir to prevent leaking when the initial pulls and repair operations are started. Many times the storage reservoirs for the windshield wiper solvent and the transmission coolant lines are not readily accessible. Therefore, disconnecting and removing those lines must wait until some of the damaged parts have been moved out of the way.

Protecting the Glass. Even though they may seem to be a safe distance from the damaged areas being repaired, any glass or windows that may be affected during the pulling and roughing operations should be removed to protect them from breakage. The appropriate precautions must be exercised when removing the glass to avoid breakage. Many times the windshield and side glass are put under a severe strain as a result of the twisted body panels to which the glass is attached. Whenever there are any signs of this, extreme caution must be exercised when removing the glass because it is already under an undue stress and breakage is increasingly more imminent.

Protecting Trim Parts. Many of the vehicle's exterior moldings, ornaments, and trim are very fragile and sensitive to heat and abrasion. These items should always be removed, especially if they are to be transferred and reinstalled onto the replacement part on the vehicle. Most of the moldings and trim are attached with two-sided tape or other adhesives, and care must be exercised when removing them to avoid bending or twisting these pieces. Once they are removed, care should be exercised to store them flat or in a position similar to the configuration of the panel to which they are permanently attached when on the vehicle. Failure to remove the parts before initiating the repairs may result in damaging them, thus causing additional expenses for the shop because these parts must then be replaced. The most effective means of removing these parts without damaging them is with the use of an induction heater or hot air gun which heats the surface, warming and softening the adhesive. This makes it possible to remove the molding with a minimum effort.

CREATING A REPAIR PLAN

There is always an air of excitement each time another damaged vehicle is brought into the shop, especially if it is a little more heavily damaged than the norm. The first reaction is to jump into it and start removing parts to get to the structural damage where the actual repairs must be initiated. However, the technician should be cautioned and advised to resist this temptation and instead make a detailed analysis of the damage. Studying the damage from the point of impact and tracing it throughout the vehicle can be helpful to mentally formulate a repair plan. The obvious damage—such as the misaligned panels in the areas adjacent to the impact area—should be used as directives to further investigate the possibility of damage deeper into the vehicle. As the damage is traced from the point of impact and followed throughout the vehicle, the repair sequence can be mentally laid out. Removing exterior parts prematurely will likely "spoil the evidence" and make the repairs more difficult to effect.

INVENTORYING DAMAGED PARTS

Once the visual analysis is completed and the repair plan has been formulated, the next step is to look at the damage report to determine if all the damage has been included in the estimate. The technician should make a detailed inventory of the parts listed on the damage report and compare it to the parts delivered. If any items are overlooked or missing, they should be noted immediately so they can be acquired—if available—as quickly as possible. Failure to do this may waste a considerable amount of time and consume much needed shop space if a part is not available and the vehicle must be stored in the work area while waiting for the part's arrival. Once the vehicle is partially or completely dismantled to initiate the repairs, moving it back outdoors is not a viable option as exposure to the elements may be detrimental to most of the interior and other delicate parts.

TECH TIP All fasteners and items used to secure the component panels to the vehicle should be properly marked and identified as they are removed from the vehicle to ensure the correct ones are reinstalled when installing the replacement parts. Using a bolt that is too long may cause considerable damage to a new hood, for example, if used instead of the correct-length one.

MARK AND INVENTORY HARDWARE

Most of the large exterior **component panels** are secured to the vehicle with bolts and other removable fasteners. Due to their size, removing them will require assistance from a coworker. Prior to starting with the dismantling procedure, provisions should be made to inventory, identify, and store both the damaged and undamaged parts along with the hardware that will be used in the vehicle's re-assembly. The technician should "**bag and tag**" all the removable and reusable hardware—that is, it must be properly marked and identified so when the replacement parts are installed the correct fasteners will be used for each part. These large damaged parts will need to be set aside until the vehicle is reassembled, as many of the functional peripheral parts—such as the hinges, latches, and specialty fasteners and accessories—must be transferred to the replacement parts after **jamming and edging** and prior to installation. Reinstalling some of the clips and fasteners is more easily done when they can be removed from the damaged parts and immediately installed onto the new ones, because remembering exactly how they were installed is sometimes difficult after a significant lapse of time.

REMOVING DAMAGED PANELS

As the large component panels are unbolted and removed, much of the damage to the structural parts becomes more evident. Many times on a front collision, it is necessary to rough out or pull some of these damaged structural panels—such as the core support and grill and headlight mounting brackets—away from the mechanical parts to accommodate their removal. In a side collision the same may have to be done with the B pillar to accommodate the removal of the seats and other interior parts. Likewise, in a rear collision it may become necessary to remove the taillights, trunk liners, and spare tire, if necessary. This is accomplished by a variety of methods including the use of pulling clamps, pull straps, and—in some instances—even welding a **sweat tab** or **pull tab** onto the surface to which a pull clamp is attached for pulling. At this point, the purpose or intent of pulling the damaged sheet metal parts is to make room to remove the mechanical parts. It is also used to start the roughing out and re-alignment of the damaged rails. Prior to initiating any of the pulling sequences, the vehicle must be anchored to the bench or the frame machine used to restore the damaged structural parts. This will be covered in greater detail in a later chapter.

RESTORING DIMENSIONS AND STRESS RELIEVING

Most structural panels are a combination of two or more layers of metal welded together, with each side supporting all of the others to resist bending or becoming distorted. Whenever any such panel or structural rail becomes damaged in any way, numerous conflicting stresses result within the members and on the surrounding areas. Before removing any of the damaged rails or any part of them, it is necessary to rough repair them and **square the door and window openings** as much as possible. This is necessary to relieve the stresses around the area while at the same time restoring the damaged openings to their proper dimensions. In addition to restoring the framework of the openings and any damaged rails, efforts must be made to restore any outer sheet metal panels attached to them as closely as possible to their proper dimension before removing any part of them. Removing any damaged rail, panel, or section thereof without first truing the openings may make the task extremely difficult, as there is no telling how the metal in the adjacent areas will shift, move, or respond when a section under stress is removed.

ROUGH REPAIRING DAMAGED AREAS

As was pointed out previously, the primary purpose of pulling is to roughly restore the damaged rails and openings as nearly as possible to their factory specifications before removing them. One of the advantages of pulling them first is the damaged parts can be used to apply corrective forces at the precise location and angle where the most effective results may occur. Many times a pull clamp, cable, or strap will be attached to such an area and used to move the damaged metal without regard to damaging it, as

FIGURE 15-1 A pull tab can be welded directly to the damaged area to pull and rough out the damage prior to removing it for replacement.

the metal eventually will be removed and replaced. See **Figure 15-1**. An example is the pull tab that has been welded to the upper rail of the core support to which a pull clamp will be attached. This will allow the technician to pull in the exact direction required to restore the front dimensions prior to removing the damaged parts. This same approach can be used in any section of the vehicle regardless of the location or severity of the damage.

The methods required to repair each damage condition are as unique as the vehicles to which the repairs are made. In some instances the panels may be repaired using conventional metal bumping and straightening techniques. With the aid of various filler materials, they can be restored to look like new. Any panels not considered to be repairable are replaced. However, before removing them a series of similar steps must be performed to restore the repair to its pre-collision condition. These steps ultimately include measuring, pulling, removing, and replacing the damaged parts. Panels that are not considered repairable are replaced by using either **OEM**, **LKQ (like, kind, and quality)**, or recycled parts. Replacing the parts may be accomplished by using one of three methods commonly employed: replacing them at the factory seams, performing a sectional replacement with new parts, or performing a sectional replacement with recycled parts. Whichever method is utilized, the final result must ensure that the vehicle's original structural integrity is restored to protect the vehicle's occupants in the event of a subsequent collision.

LOCATING AND DRILLING SPOT WELDS

After the exterior bolted panels have been removed, the damaged area has been roughed out, and the openings have been restored, the next step in replacing structural parts is to locate and drill the spot welds. Locating the spot welds can sometimes be difficult because some of the protective coatings applied by the manufacturer either fill or disguise their location. Once the approximate whereabouts of the welds have been established, several techniques can be used to reveal their exact location. One of the quickest methods is using a coarse piece of sandpaper that has been doubled over to scrub back and forth across the area where the welds are placed. See **Figure 15-2**. The firmness of the paper will quickly locate the otherwise difficult to see welds. A wire brush or abrasive wheel on a drill is another effective way to accomplish this task. See **Figure 15-3**. Occassionally the spot welds are covered with heavy-duty undercoats and seam sealants which require scraping for removal. See **Figure 15-4**. While these materials are often flexible, they are relatively firm and hard and sometimes difficult to remove without first warming them up to soften them. A hot air gun is a very safe and effective way to accomplish this. An acetylene welding torch or a small butane torch may also be used; however, extreme caution must be exercised any time an open flame is used on or near any part of the vehicle.

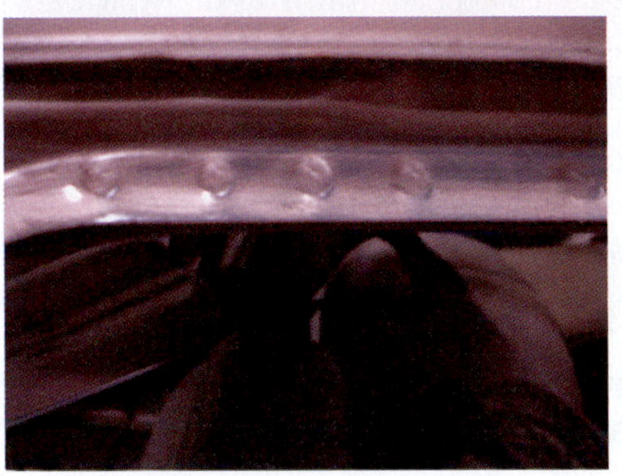

FIGURE 15-2 The exact location of the spot welds can be determined by folding a piece of production grade sandpaper and lightly sanding over the area to highlight them.

FIGURE 15-3 The abrasive stripping wheel is being used to identify the exact location of the spot welds.

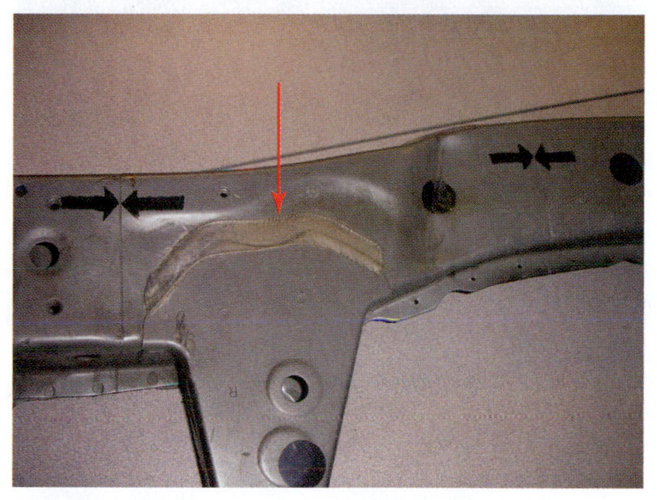

FIGURE 15-4 Many times the welded seam is covered with a heavy layer of seam sealer to disguise and waterproof it—as well as for aesthetic purposes.

(A)

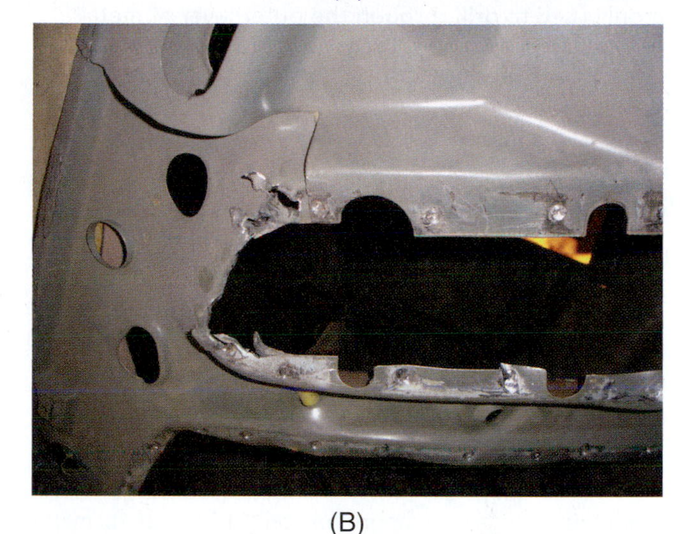

(B)

FIGURE 15-5 Once the spot welds are properly drilled out (A), the layers of metal can easily be separated and the panel removed from the vehicle. (B) shows what happens if the welds are not drilled properly.

SEPARATING THE SPOT WELDS

Once the welds have been located the often tedious task of drilling them out must be completed. A number of different spot weld drills are available. The location of the spot welds and their degree of accessibility sometimes dictates the tool used to remove them. The preferred method of removing the spot welds is drilling them out with a specially made spot weld cutting drill which is operated with either an electric or air-operated motor drill. The idea of drilling spot welds is to cut around the outer perimeter of the actual spot weld on the outer layer of metal without disturbing the layer to which it is attached. Once the spot welded layer has been cut through it is no longer attached by the spot weld and it can be removed or separated from the piece below it. See **Figure 15-5**. A chisel may be used to help in separating the layers of metal. If the welds are accurately drilled, a minimum amount of effort is all that will be needed. An effort should be made to drill through the center of the weld as the chisel tends to tear the welds apart when not properly drilled, leaving a jagged mess on the edge of the metal. This will have to be smoothed, welded, ground, and repaired before the new panel can be re-attached.

CAUTION

One should be cautious to avoid drilling through both layers of metal when removing the spot welds as the welds for the replacement panel should be placed in the same area. If a hole appears in the weld placement area, it will need to be filled and smoothed before the replacement part can be installed. This may result in weakening the area due to the heat affect zone.

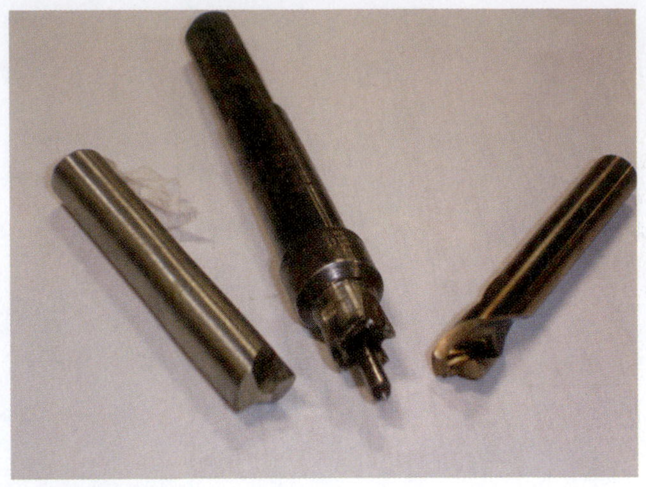

FIGURE 15-6 Spot weld drills such as these are commonly used to drill through the outer layer of metal, thus separating it from the layer to which it is welded. Others have graduated shoulders which are not as desirable to use.

Spot Weld Cutters

There are three specific designs of spot weld removal tools. Some can be used with a regular drill, while others are specifically designed for use in special drills. See **Figure 15-6**. Some of the cutters are available in a specific size, with the most popular being the 5/16 or 3/8 inch or 8 mm sizes because that is the approximate size of the spot welds to be removed. Another style is the graduated bit, which has a series of tapered shoulders that allow its use for several different size holes starting from 1/8 inch and graduating up to an approximately 1/2 inch diameter hole. This bit cannot be used if there is another layer of metal close behind the surface where the spot welds are being drilled. The drill bit must be able to penetrate through the outer layer without any obstruction for at least 1 to 1 1/2 inches on the back side. The hole saw design bit offers various sizes of removable cutters that are attached to a spring-loaded shank, which is used as a guide to keep the bit from wandering while drilling. For best results it is advisable to center punch or mark each of the spot welds before attempting to drill them to keep the cutter bit from wandering. A standard drill bit can be used to make a small pilot hole; however, it is not recommended for removing the entire spot weld as it will bore a hole through both layers, leaving a void in the area where the welds for the replacement panel are to be placed.

Spot Weld Drill

A specially designed drill with a clamping lever is also available. This drill allows the technician to clamp the drill to the precise location where the weld is to be removed. The cutter bit has a small centering pin which can be placed into the center of the weld. The tool is then clamped into place, thus eliminating any possibility of the bit wandering about while drilling. This method, however, will only work for drilling welds on an open edge. Access to the back side is necessary in order to engage the clamping device.

FIGURE 15-7 When drilling the spot welds one must be careful not to drill through both layers of metal.

When access with the spot weld removing drills is not possible, the use of a grinding wheel such as on the cutoff tool can be used to remove the surface of the metal where the weld is located. This usually results in the removal of metal in a larger area, which makes cutting too deep a possibility. See **Figure 15-7**. Caution should be used to ensure the vehicle's windows are protected from flying sparks created in the process. Once the spot welds have all been drilled, a large flat chisel on an air hammer can be used to finish separating the layers of metal. Provided all the spot welds have been properly drilled, a minimum amount of effort will be needed to finish removing the damaged parts.

PANEL ASSEMBLY METHODS

While some panels on a vehicle are not considered structural parts as such, they do serve a vital role in the overall function of maintaining the vehicle's integrity and passenger safety. These panels are the quarter panel, roof panel, and the doors. None of these panels are made of the same steel quality as is used for structural parts; however, each is attached to a structural part and enhances the parts to which they are attached. The roof, for example, is attached to each of the pillars on the vehicle which are used to deflect and distribute some of the impact energy around the passenger compartment. Likewise, the quarter panel is attached to the C pillar, which also supports the roof. In a rollover situation the combined strength of these pillars must be able to support the vehicle.

Doors

The doors are reinforced with intrusion beams made of ultra high strength steel designed to prevent passenger compartment penetration in a side collision. Boron steel, a form of steel that the repair industry knows little about at this time, has been introduced on a limited basis for outer door skins to further reinforce the side of the vehicle. All of these parts serve a functional role in the vehicle's overall structural design. When replacing any part of them, they should be treated and repaired as though they were a structural part.

Quarter Panel and Roof Seams

While most panels are either spot or fusion welded to each other, occasionally a soft, more flexible weld filler material is used to attach a panel to the vehicle. The high strength steels used to structurally reinforce vehicle parts lack the flexibility required to withstand the constant flexing and vibration that occurs in certain sections of the vehicle as it moves down the road. In some instances, brass or bronze is used to join panels together and finish out the weld joint on a large flat panel, such as is found between some of the pillars and the roof. A continuous **silicon bronze** joint is frequently used where the sail panel and the roof panel are joined together. See **Figure 15-8**. The silica bronze is used for a filler as it offers adequate strength to the structure yet is flexible enough to withstand the constant vibration and flexing that occurs in that area of the vehicle. Steel weld joints are not used in this area as they are apt to fatigue, crack, and eventually start to rust. The result is a very unsightly surface in a highly visible part of the vehicle.

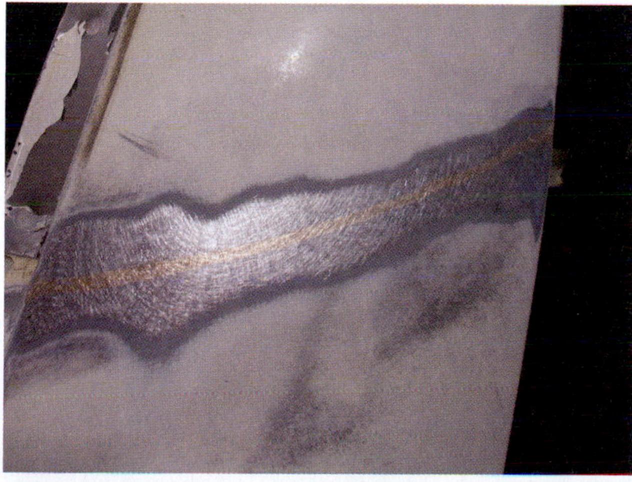

FIGURE 15-8 A silicon bronze joint is commonly used to weld the roof panel and the quarter panel on the C pillar.

When removing the quarter panel one can either cut through the silicon bronze joint with an abrasive wheel or cut the panel off below the joint. In the event it is removed by cutting through the silicon bronze weld joint, it must be reinstalled using the same welding process to secure the new panel and ensure adequate flexibility is restored to the area. If the panel is removed by cutting below the weld joint, the replacement panel would then be installed with a lap joint below the silicon bronze seam, leaving it in place to continue to function as a natural flexing point.

QUARTER PANEL REPLACEMENT METHODS

The quarter panel may be replaced using one of three different methods depending on the area damaged, the age of the vehicle, and the availability of the parts.

Factory Seams

Replacing the quarter panel at the factory seams is perhaps the most common way to perform this task. However, this method is the least often used because of the amount of effort required and the areas of the vehicle that are disturbed when removing and replacing the panel in this manner. To replace the panel at the factory seams, it must be separated from the roof. This usually requires the removal of the backlite, quarter glass, and all the upholstery and interior trim from that part of the interior. It may also require removing all—or at least part—of the headliner attached to that area of the interior.

To remove the quarter panel, the spot welds must be drilled around the entire perimeter of the panel. This includes the water draining trough inside the trunk opening, along the backlite pinchweld, the quarter window opening, and around the door and wheel opening—as well as the lower trunk pockets, dog leg, and rocker panel area. The silicon bronze seam on the sail panel must be cut to separate the quarter panel from the roof panel. See **Figure 15-9**. Cutting below the bronze joint is a more desirable area to make the cut and lap the replacement panel over the remaining layer of the sail panel. No matter where the panel is cut for removal, numerous welds and the original corrosion protection is disturbed unnecessarily, thus increasing the possibility of corrosion forming at a later time. As a rule, the factory replacement option is most commonly used on automobiles that are less than 2 years old and is done to preserve the vehicle's vintage characteristics.

FIGURE 15-9 Removing the quarter panel requires that the spot welds be drilled out around the entire perimeter of the panel or, in the case of a uniside panel, it may need to be cut at the location where it is seamed for a sectional replacement.

Abbreviated Replacement

The abbreviated method is a more common approach as it requires considerably less intrusion into the vehicle and results in disturbing considerably less of the corrosion protection. Except in rare cases, it also eliminates the need to remove the rear window and interior upholstery. Much like the factory seam replacement method, the spot welds must still be removed from the perimeter of the panel. The damaged quarter panel is then cut and seamed on the lower part of the sail panel, usually on a crowned surface where the replacement panel seam is most easily disguised. Many times the vehicle manufacturer will make recommendations as to where on the panel the damaged part should be cut and how the new panel should be trimmed and prepared to be attached. See **Figure 15-10**. If uncertain about the best location to make the cut, it may be advisable to consult the **I-CAR Uniform Procedures for Collision Repair (UPCR)** to determine the recommended or preferred procedures for replacing the panel.

INSTALLING THE PANEL

Trial Fitting the Panel

Installing the new panel can be accomplished by using the adhesive bonded method, plug welding, or a combination of both. Whichever method is used, the panel must be temporarily installed to trial fit it. During this phase of the installation, the panel will be secured with a combination of sheet metal screws, blind rivets, or clamping pliers. This is done to ensure all the surfaces and openings will align properly when they are fitted for the final installation. Any inconsistent gaps with the adjacent door and trunk openings, improper alignment of the crowns with adjacent panels, and the correct gaps between panels are corrected at this time. The panel will be removed once again before permanently installing it to remove any burrs on the contact surfaces, coat all the bare areas with weld-through coating, and apply the adhesive materials if they are to be used.

Installing the Panel

Provided all the seams and joints are properly marked, the reassembly and alignment will be considerably faster and less involved than the first time when it was fitted to align with all the adjacent panels and surfaces. Once the panel has been bonded together, it should be allowed to stand undisturbed for at least 4 hours or the minimum time recommended by the adhesive manufacturer. In the event the panel is to be welded into place, one must ensure all the holes for the plug welds have been pierced into the panel's edge and weld-through coating is sprayed over any area that has been bared in preparation for installation.

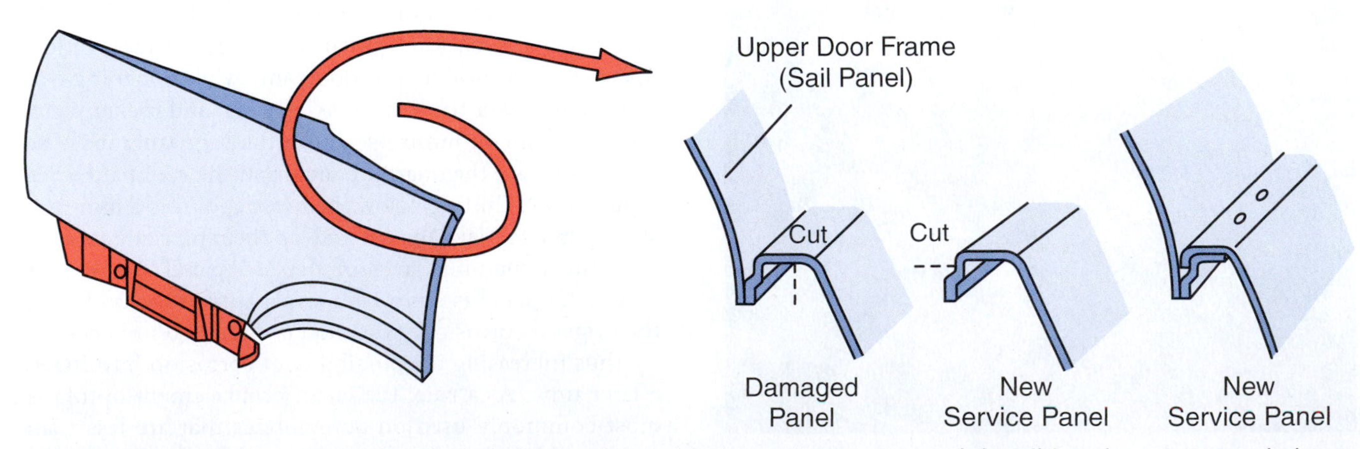

FIGURE 15-10 Manufacturers often include instructions with the replacement panel describing the recommended replacement method for a specific part, as in this quarter panel.

ROOF PANEL REPLACEMENT

Removal

The roof panel is perhaps the most difficult and involved panel to replace on the vehicle. On some models, removing the roof may include disturbing and removing the spot welds that hold together the B pillar, the inner roof reinforcement panel, and the outer roof. It may also require cutting into the A and C pillars as well. See **Figure 15-11**. On some uniside models, the seam connecting the roof to the side panel is very accessible as it is exposed with a trim molding covering it. See **Figure 15-12**.

CAUTION

Proper care must be exercised to protect the interior upholstery and trim from possible damage when removing and replacing the outer panel. Any damage that occurs during these procedures becomes the shop's responsibility. All items that can readily be removed should be taken out and those that can't should be covered with protective covers.

Whatever means is used to install the roof, nearly all of the vehicle's interior must be removed to protect it from damage and to gain access to all the surfaces. Any upholstery—such as the seats and carpeting—that is not removed must be covered to protect it from falling debris during the removal and installation of the new panel.

The repair technician may find that the damaged roof panel has been installed by **weld-bonding** (i.e., using both spot welds and adhesive bonding). Therefore, it may require more than merely drilling the spot welds to remove it. A heat gun will also need to be used to soften the adhesive to release it, allowing the panel to be separated from the mating surface. This is best accomplished by inserting a putty knife or a flat chisel between the two surfaces to separate them once the glue has been softened with the heat gun. The spot welds must be drilled carefully and precisely to minimize the amount of chiseling necessary to completely separate the mating surfaces of the two panels. Any excessive chiseling will increase the possibility of damaging the attaching surfaces, thus requiring additional work to install the replacement part. It will also result in the need for a thicker layer of adhesive to fill the gap between the two surfaces, which can cause the joint to become weakened.

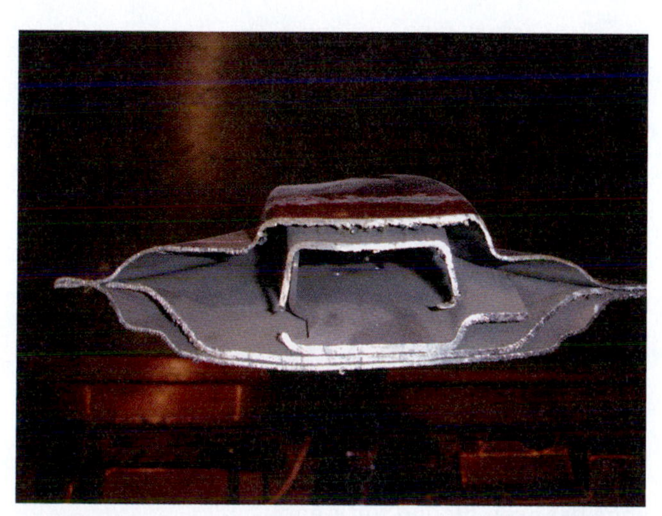

FIGURE 15-11 Separating the roof from the B pillar may be as easy as drilling the spot welds holding it and the roof together. On other models, separating the roof panel from the B pillar is a very invasive procedure as it frequently requires disturbing the welds at the A, B, and C pillars.

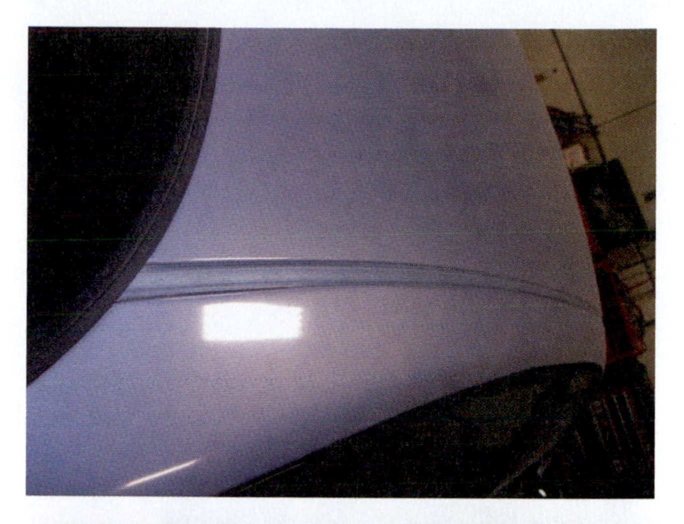

FIGURE 15-12 On vehicles using a uniside panel, the seam connecting the roof panel to all the side panels is exposed and very accessible from the exterior of the vehicle.

Installing the Roof

After the roof has been set into place and clamped, either a plug weld or an STRSW should be made in each corner of the roof to ensure the structural integrity has been restored. See **Figure 15-13**. The use of adhesive to bond the panel into place offers two distinct advantages in that the

FIGURE 15-13 Spot welds in select locations may be required to ensure the structural integrity is restored when using the adhesive bonded panel replacement method.

corrosion protection step has been eliminated and it also helps to eliminate the possibility of noise, vibration, and harshness (NVH).

DOOR ASSEMBLY REPLACEMENT AND REPAIR

The door panel is made up of two separate layers of metal that are held together by crimping the edges of the outer panel over the inner panel to form a hemming flange that is either glued or spot welded into place. In some cases, both methods may be used. Adhesive bonding offers a more complete corrosion protected panel, particularly at the bottom of the door where the possibility of rusting is most apt to occur. The door intrusion beam, which is used to protect the occupants from intrusion by another vehicle in a side impact, is welded to the frame of the inner panel. This beam is made of ultra high strength steel (UHSS) and cannot be heated to repair. Therefore, when it becomes damaged it should not be repaired. Replacement of the entire door is mandatory whenever the intrusion beam becomes bent or damaged, as it usually has a shortening effect on the overall door length.

Outer Panel Removal

Many times when a door becomes damaged, replacing the outer panel is a more economical and desirable repair option than replacing the entire door assembly. When replacing the outer panel, the recommended procedure is to grind through the edge of the metal on the hemmed flange, thus separating the outer layer of metal from the hemmed flange. See **Figure 15-14** and **Figure 15-15**. The

FIGURE 15-14 Removing the outer door panel requires grinding through the edge of the outer layer of metal, exposing the layers of the hemming flange.

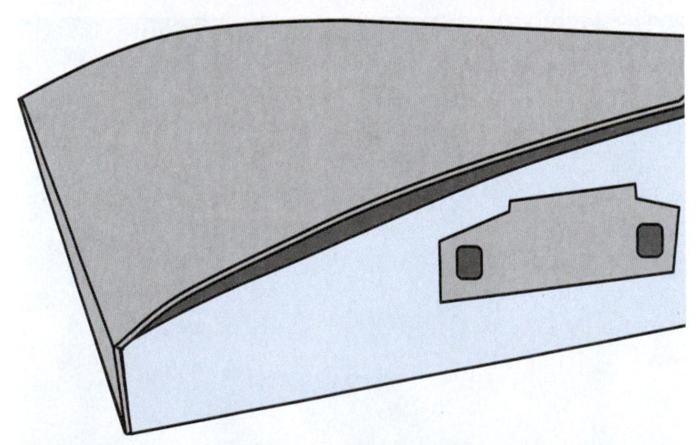

FIGURE 15-15 The edge of three layers of metal that make up the hemming flange will be exposed when the edge has been ground sufficiently.

outer panel can then be removed from the door frame, thus exposing the entire inside mechanisms and the sides of the door frame. At this time one must remove the remaining inner layer of the hemmed flange from around

FIGURE 15-16 A chisel may be required to separate the spot welds holding the outer panel from the inner panel after grinding through the hemmed flange.

the outer edge of the door by heating it to soften the glue and separating it, and by using a chisel with a wide flat bit to finish cutting through the spot welds placed around the edges by the manufacturer. See **Figure 15-16**. When chiseling the spot welds apart, care must be used to prevent tearing or ripping the door flange. Any damage to the door frame should be repaired at this time, and tears must be welded up and smoothed. All surfaces must be smooth and free of imperfections prior to the installation of the new outer replacement panel.

Installing the Outer Door Panel

Prior to installing the replacement panel, one should place the new part face down with the flanged edges facing up and place the door frame into it to ensure a proper fit. The outer walls of the door frame should be properly positioned to promote a tight fit and any minor surface damage and irregularities should be repaired at this time. The door frame should fit snugly inside the outer repair panel and any area where a gap larger than 3 mm (1/8 inch) exists should be corrected. In addition, any signs of corrosion that may have started on the mating surface of the door frame should be repaired prior to installation. Assuming all surfaces are properly aligned, the panel is ready to be installed.

Crimping vs. Bonding

Prior to this time the technician should have decided whether the door outer repair panel is to be installed using the adhesive bonded method or if the outer panel is to be crimped over the inner panel, forming the hemming flange and placing spot welds at strategic locations to prevent it from shifting. The most significant advantage of using the adhesive bonded method, in addition to bonding the two surfaces together, is that the adhesive

FIGURE 15-17 When installing the replacement outer door panel using the adhesive bonding method, both contact surfaces must be thoroughly cleaned before installing the adhesive. The manufacturer's instructions must be followed as each company may be slightly different.

serves as an excellent corrosion-inhibiting material, thus minimizing the possibility of rust forming at an otherwise corrosive hot spot.

Adhesive Bonding. Provided the panel is to be installed using the adhesive bonded method, some manufacturers recommend that the inside edge of the flanged surface of the replacement panel and both sides of the door frame are to be cleaned and free of any paint, primer, corrosion, and so on, before fitting the door and the outer panel together. See **Figure 15-17**. A special adhesion-promoting primer is to be applied over the bare steel surface, followed with the application of bead of the sealer to all of the primed surfaces. See **Figure 15-18**. One must also reinstall the expansion foam pads between the door intrusion beam and the outer repair panel prior to the final installation. The manufacturer's recommendations should be consulted regarding the type of foam that is recommended at this location.

FIGURE 15-18 Some manufacturers recommend putting the adhesive onto both surfaces while others recommend it be put on only one. The technician should follow the manufacturer's instructions.

FIGURE 15-19 A four-sided hat channel is commonly used for reinforcing many structural parts of the vehicle.

Once the adhesive has been installed and all of the mating surfaces are aligned, the next step is to roll the flange over the outer door frame around the entire perimeter of the door, using a specially designed hammer with a rubber-coated dolly on the bottom side to back up the hammer blows. A door edge crimping tool designed to automatically roll the flange over—thus forming the hemming flange—may also be used. Whichever crimping method is used, it must be completed within a time frame of approximately an hour as that is usually the working window of the adhesive materials before they start to dry or cure.

Crimping. In the event the door outer repair panel is not adhesive bonded, the same procedure should be used to roll the flange over the edge of the door frame as when using the adhesive bonding system. After the flange is rolled over the door edge, small spot welds are placed at strategic locations to reinforce the entire assembly and to prevent the outer door panel from shifting. The door is now ready to be installed and adjusted to fit.

STRUCTURAL PARTS DESIGN

All structural parts are used for one primary purpose; to reinforce and support any part to which they are attached. They invariably are a three-sided rail, although frequently a fourth side is welded to the rail to form an enclosed box. See **Figure 15-19**. Some are extruded while most are stamped into a U-shaped channel first, and the fourth side is welded into place to completely enclose all the sides. This is commonly referred to as a hat channel design. To further strengthen their reinforcement capabilities, and depending on where on the vehicle they are located, these parts may be comprised exclusively of high strength steel or they may be a laminated combination of both high strength and standard grade mild steel. Some structural parts are formed by **hydroforming**. Whichever method is used to make the rail or reinforcement member, they are all made with one central purpose: to reinforce and strengthen the vehicle's overall structure.

Replacement Options

When parts become damaged in a collision, a set of nearly universal rules can be used when replacing them.

One can categorically identify four different options or methods that are used for replacement of all structural panels. The following methods are used industry-wide to affect these repairs:

- Replace the part or portion of it at factory seams using original manufacturer's parts or those manufactured by aftermarket suppliers. This is commonly referred to as performing a factory seam replacement.
- Section with new or recycled parts. In this method, only a section of the part is replaced with either OEM parts or by using a recycled part.
- Replace an entire assembly or a section with recycled parts. A section of the vehicle which includes an area larger than the damaged part alone is installed.
- In some rare cases, an entire vehicle may even be "clipped" by using a full body section replacement. In this procedure, the rear half of a vehicle with a severely damaged front end may be installed onto a vehicle with a good front end that sustained irreparable damage to the rear end.

Depending on the area of the vehicle where the damage is located, the location and the amount of damage sustained on the part or assembly and the structure's shape will determine the method that is used to replace the part. Another factor that will enter into the decision is the type of joint design that must be used to join the replacement part to the vehicle.

FACTORY SEAM REPLACEMENT

Replacing the part at the factory seams invariably becomes more involved than any other replacement method. When this method is used, all the spot welds must be drilled out so the part can be completely separated from the rest of the body. This frequently requires the complete—or at least partial—removal of overlapping braces and reinforcements from adjacent body parts. Because these same parts must be reinstalled, the technician must be careful not to damage the mating surfaces during the removal to minimize the effort required when reinstalling it. Accessibility to the spot welds is often very difficult to achieve and drilling them out without damaging adjacent areas sometimes becomes a real challenge. Fitting the new panel and the overlapping braces into place is equally difficult; therefore, a sectional replacement becomes a great deal more desirable.

SECTIONAL REPLACEMENT

A **sectional** or **partial replacement** is often more desirable, as it leaves the vehicle as crashworthy in a subsequent collision as it originally was by the manufacturer. When performing a sectional replacement, only a part of the structural member is replaced. It usually involves cutting and installing a section of a replacement panel at locations other than at factory welded seams, thereby disturbing fewer of the OEM welds. The corrosion protection remains intact in this method as well. This also requires less disassembly and removal of the interior and exterior upholstery and trim in the repair area. The placement of the seam or the area where the part is sectioned must be placed at strategic locations using specific procedural guidelines. These guidelines are established and made available to the industry to ensure that technically correct and structurally sound repairs are consistently performed by the repair shops. They are available from a variety of different sources, some of which require a service fee while others are available at no charge to the user. The following are the most accessible by the shop at little or no cost to them:

- Manufacturer's Body Repair Service Manual
- Tech-Cor @ tech-cor.com
- I-CAR Advantage @ I-car.com
- Uniform Procedures for Collision Repair (UPCR) @ I-car.com

If there is any doubt or uncertainty about the correct location or method of removal of the old part or installation of the replacement part, the technician should consult one or all of the aforementioned sources.

Sectional Replacement Using Recycled Parts

Replacing the damaged parts with a recycled part or section is a very common practice in the auto collision repair industry—so much so that the automobile repair industry is recognized as the world's single largest recycling organization. Using recycled parts offers numerous advantages to the repair technician. One of the most significant advantages is when the recycled section is removed from the parts vehicle, it is cut large enough to include most of the surrounding attaching surfaces. This allows the technician to choose whether to replace only the damaged part or to replace a larger section at a location away from the damaged area, making it easier for alignment, fitting, and permanent installation. When properly executed, this procedure can be accomplished without compromising the vehicle's structural integrity. Another significant advantage of this method is the corrosion protection in some of the critical areas is not affected as this reduces the invasiveness of the factory spot welds. When the structural part must be replaced separately, the excess materials that are attached to the recycled part must be removed by first drilling all the spot welds and separating the parts with a chisel. As is always the case when using the air chisel to separate panels that are spot welded together, care must be used to not tear or damage the mating surfaces.

Welding or securing a structural part at the factory seam usually involves placing a series of plug welds at the same precise locations as the manufacturer. If both sides are accessible to allow the use of the STRSW welder arms, it may also be used. When MIG plug welding is used, the holes for the welds must be pierced into the replacement panel prior to installing it. The required corrosion protection steps must be taken accordingly on both surfaces regardless of which welding option is utilized. Because the replacement part is shaped to match the surface to which it is attached, installing it involves securely clamping or holding it in place with screws or blind rivets while the welding operation is performed. The technician should be advised to resist the temptation of placing more welds than were used by the manufacturer. This would alter the design of the structural part and re-engineer it.

WELDING REQUIREMENTS

One must be careful to not re-engineer or alter any of the structural characteristics in the repair area by either overwelding or underwelding the part being installed. When using plug welds to install the replacement part, it is critical that the structural part be welded into place using the same number of welds used by the manufacturer and to place them in the same precise location as they were on the original part. Some manufacturers recommend that when using the STRSW method the number of welds should be increased by 20 to 30 percent—provided the flange surface offers enough space to accomplish this.

Some manufacturers may recommend placing the welds at a different location than the originals were located to minimize the size of the heat affect zone.

Weld Placement

The weld placement must be performed with a relative degree of accuracy to ensure the structural part collapses and reacts correctly in a subsequent collision. An overwelded part may not bend at the precise moment it is designed to transmit the damage elsewhere, or collapse at the right location. An underwelded part may rip or tear away from its attaching location, again not responding at the correct time or location. The number of welds and their location can readily be determined by studying the damaged part that was removed.

Recommended Welding Procedures

When performing a sectional replacement, several different joint designs are used depending on the location of the seam, the shape of the structure, and the intended function of the part. Whichever joint design is used, the technician must adhere to some common rules or procedural guidelines concerning the welding techniques used. For example, one must be careful to avoid welding a bead longer than 19 mm (3/4 inch) at a time. One should also avoid making plug welds closer than within 144 mm (6 inches) of each other without allowing adequate time for the metal to cool completely. The main seam of the replacement panel should be tack welded into place, first using a stitch welding procedure and then welded solid by filling the areas between the spots made when stitching. Patience is required when welding these parts to ensure the metal is allowed to cool properly between welds. One should never attempt to quench the metal or cool it by blowing forced air onto it, as doing so will cause the metal to become crystallized. This will cause the metal to fatigue, crack, and break at a later time.

CAUTION

Many of the structural parts are made of high strength steel. Allowing the metal to cool between welds is critical to maintain the steel's structural integrity. Overheating will destroy the metal's strength characteristics, causing it to revert to the same quality and strength of a common steel.

JOINT PLACEMENT AND LOCATIONS

Numerous other guidelines must be followed concerning joint design, placement of the seams, and joint reinforcement techniques. Failure to comply with certain guidelines may result in the parts not responding correctly in the event of a subsequent collision. The following are considered the most critical areas to avoid when choosing the location of the replacement seam.

Avoid welding in the following areas:

- Any restraint system mounting location such as the D ring anchor point
- Any multi-layered reinforced area where inner layers are inaccessible for welding and corrosion protection cannot be accomplished
- Areas where the B pillar and the roof intersect
- Areas where the rocker panel and B pillar intersect
- In laser-welded/tailor blank areas
- At any silicon bronze seams
- On mounting surfaces such as door hinges and striker plate areas
- At mounting locations such as door hinges
- Through holes that would otherwise be used for measuring reference points (usually larger than 3 mm [1/8 inch])
- In collapse or crush zone areas
- In areas where mechanical parts are attached and their locations
- On compound-shaped surfaces such as a convoluted rail

Nearly all of these areas represent surfaces on which a solid weld cannot be assured because they are inaccessible from both sides. Some of the areas are laminated and multi-layered and also spot welded together, making it difficult to ensure accomplishing a sound weld joint. The obvious concern with sectioning in any of these areas is restoring the vehicle's collision worthiness and to ensure passenger safety in a subsequent collision.

ACCEPTABLE SEAM LOCATIONS

Despite the limitations cited where section joints are not allowed, there are a number of surfaces and areas where one can safely effect a repair without consequence or compromising the structural integrity to any part of the vehicle. The following guidelines can be used to effect a proper repair using the joint design best suited for the area in question.

- Areas that are uniform in shape and metal thickness
- Areas with adequate clearance where accessibility can be gained for welding on both sides
- Areas where access can be gained with corrosion protection equipment
- Areas recommended by vehicle manufacturers
- Using general guidelines where recommendations are not available

JOINT FABRICATION CRITERIA

A number of joint designs are commonly used when performing a sectional replacement with either new or recycled parts. The joint design that is used will be dictated somewhat by the location of the damage on the vehicle and the shape of the part itself. Many times a combination of one or more of the joints is incorporated into the replacement of certain parts.

TECH TIP Prior to installing any replacement panel or joining any area together, it is of critical importance that all the necessary steps are taken to ensure that the corrosion protection has been properly restored.

CAUTION

Caution must be exercised to allow the metal to cool properly between welds or to space them sufficiently to prevent overheating the surface and minimizing the heat affect zone.

MAKING PROPER WELDS IS CRITICAL

Avoid Overheating the Weld Area

When welding the butt joint together, particularly if it is a lengthy one, it is advisable to tack weld (stitch weld) the surfaces together at routine intervals to prevent warpage and distortion when the joint is joined together permanently. If the structure includes any defined crowns, the crown surfaces should be welded first to ensure their alignment, followed by skip welding the remaining seam. The first weld should be placed in approximately the center of the seam and alternate from side to side from the first weld, placing another weld approximately 26 mm (1 inch) from the previous one completed until reaching the end of the seam. A second pass of stitch welds should be made until a maximum gap of 13 mm (1/2 inch) exists between the tack welds so that when the seam is welded solid, that is the maximum bead length that must be made. Another consideration is the thinner the metal, the shorter the bead length that can be made without the danger of burning through. The space between tack welds may need to be even shorter in these instances. Adequate time must be allowed between welds to properly cool, helping to minimize the possibility of warpage. See **Figure 15-20**. Before filling the space between the stitch welds, it is necessary to grind the excess **weldment** from each individual bead. The added thickness of the excess weldment could result in inadequate fusion when welds are affected. A uniform metal thickness on the surface will ensure a sound and uniform weld bead will be made.

FIGURE 15-20 When replacing a panel with a distinctly shaped definition crown, the crowned surface should be aligned and tack welded before any other welds are made on the seam.

TECH TIP Before attempting to make a weld on the part to be welded to the vehicle, the repair technician must ensure the welding equipment is properly adjusted for maximum performance. One should never attempt to weld on the vehicle without first practicing making the exact welds on like materials and duplicating the joint design to be used. Damaged parts removed from the vehicle are ideal materials to use for this purpose.

Setting Up the Welder

When practicing and setting up the welding equipment one should experiment to determine which of the many variables will most effectively achieve the desired results. The direction of travel, the angle of address with the welding gun, the amount of electrode stickout, and the resulting gun distance will dramatically affect the quality and integrity of weld that is accomplished. If a weld-through coating is to be used in the weld joint, it should also be used when practicing. One must remember that when welding on thinner gauge metal the drag or reverse method should be used to minimize the amount of burn away from the seam area and reduce the amount of burning through. Every possible variable should be attempted and considered. When the right formula is achieved, that exacting technique should be duplicated for welding the part onto the vehicle.

Establishing Root Gap

A **root gap** is sometimes left between the panels being joined together to ensure sufficient penetration when welding and also to eliminate problems associated with expansion and contraction forces. The amount of root gap that is left is often dictated by the thickness of the metal. Where the metal thickness at the joint is approximately 1.6 mm (1/16 inch) thick or less, no root gap should be necessary and the metal should be butted tightly together. When working with a thicker gauge metal—such as a front rail or any structural member—the gap should increase accordingly with the thickness of the metal being welded. A rule of thumb to follow when making a butt weld on thicker materials is the root gap should be at least the equivalent of the metal thickness. In instances where it is not practical to leave a root gap, the edges should be beveled or V-grooved to ensure adequate penetration is achieved.

JOINT DESIGNS

Open Butt Joint

The open butt joint is made by fitting or butting the edges of two adjacent panels together and welding them with a solid seam. The contact surfaces on both must be precisely matched to each other and the face level must be carefully aligned to the other to avoid creating an irregular or uneven surface when the two pieces are welded together. See **Figure 15-21**. The thickness or gauge of the metal will determine the size of the allowable gap between the two surfaces. The thinner the gauge of metal, the narrower the gap that is allowed. An excessive gap will result in frequent burn-throughs as the welding operation is performed, causing the joint to be weak, thus compromising the structure's integrity.

FIGURE 15-21 Proper alignment of the parts is critical to attaching the sectional replacement part to the vehicle. An excessive or uneven root gap requires additional welding to properly fuse the seam together.

Offset Butt Joint

The offset butt joint is very similar to the open or straight butt joint except that it is used on a two-sided structure with a hollow cavity such as an A pillar. This structure is often obtained by welding two U- or similar shaped panels together, forming the hollow cavity in the middle. When making an offset butt joint, the seam is staggered by placing the cut line above or below the one made on the opposite side of the pillar instead of being in a straight line. See **Figure 15-22**. This joint design is used where additional strength and structural integrity are required. The added strength and resistance to bending is obtained because the routine stresses and the damaging forces are spread over a larger area of the welded seam. The breakover or bending that occurs from the collision forces is not in a straight line as it would be in a straight butt joint. All the same alignment and preparation required for the straight butt is required of this joint as well. All welding procedures discussed previously should also be incorporated in this repair.

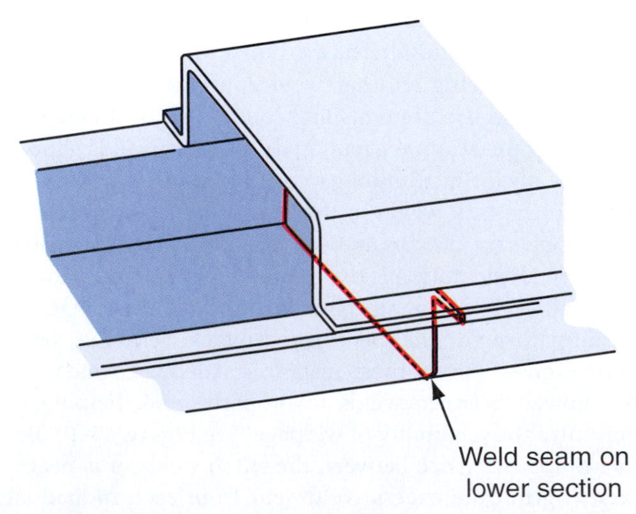

FIGURE 15-22 The offset butt joint is used in areas such as the A pillar when making a sectional replacement.

Butt Joint with Insert

The butt joint with insert—sometimes called butt joint with backer or backing—is used where restoring structural integrity is critical. It is frequently utilized where

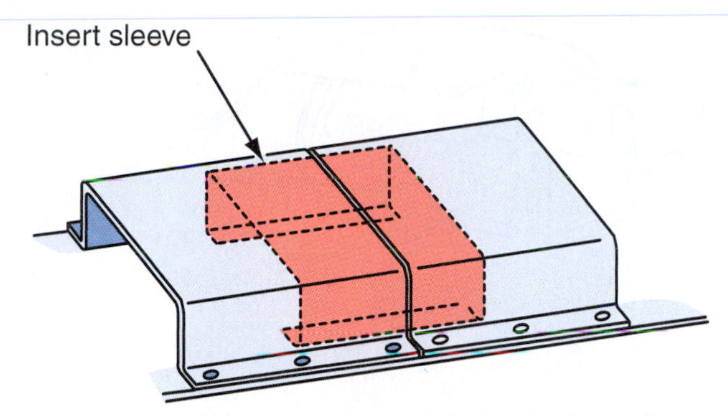

FIGURE 15-23 The insert is used to strengthen and reinforce the welded joint area, leaving it with the same strength as the vehicle originally possessed.

high strength steel parts are used for both the inner and outer layer of the structure. Examples of this design are the A and B pillars and also areas where a combination of high strength and standard grade steel are used together, such as in the vehicle's rocker panel section. The insert is used to strengthen and reinforce the welded joint area, leaving it with the same strength as the vehicle originally possessed. See **Figure 15-23**. When fitting the rails together, the two surfaces are cut and fitted to match, leaving approximately a 3 mm (1/8 inch) gap for the root gap when welding them together. A short piece, approximately 75 to 100 mm (3 to 4 inches) in length, that is to be used as the insert can be cut from the excess of the replacement part or from the damaged part. This piece is then fitted and plug welded to the inside of the cavity where the weld joint is to be made. This insert can be very helpful to align the two rails and ensure a proper fit between the two surfaces. It will also be beneficial in controlling the heat and ensuring that a sound structural joint will be created because the insert can be used to absorb some of the heat necessary to join the two edges together. When welding the edges together a root gap equal to 2 to 3 times the metal's thickness should be left between the two edges to ensure proper penetration. All the preparations, setup of the equipment, and practicing discussed for previous joints should be adhered to in this operation as well. Before the insert is permanently installed, it should be coated with a weld-through coating to ensure no corrosion will occur in the repaired area.

Lap Weld

The lap joint is perhaps the most frequently utilized option as it is frequently incorporated with several other joint designs. There are essentially three variations of the lap joint: the flat, flanged or recessed, and the tapered. The flat lap simply involves placing one layer of metal over the top of the other and securely welding the edge of it to the piece below it. See **Figure 15-24**. This joint design may be used where a long seam is required to attach the replacement part to the existing part of the vehicle and an open butt joint would not be practical due to warpage and other structural

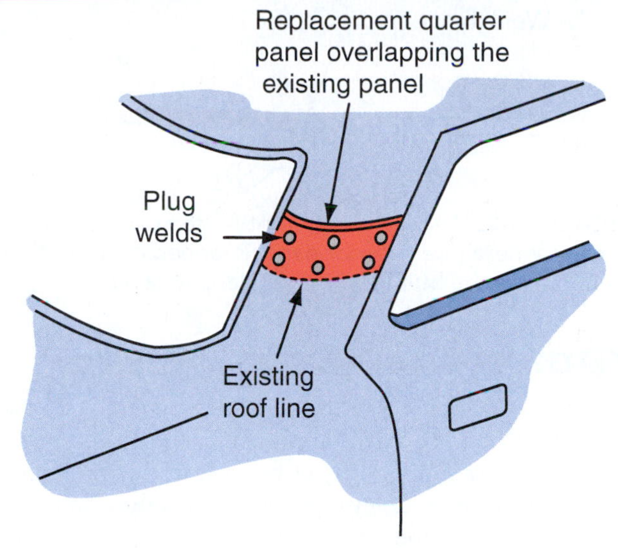

FIGURE 15-24 A recessed or flat lap joint can be used on the sail panel where the quarter panel is attached to the roof. Plug welds are placed below the seam area to secure the edges underneath the lapped section and the seam is welded solid.

implications. The vehicle manufacturers commonly use a lap weld on rear rails, in the floor pans where the passenger compartment floor and the trunk floor overlap, and on the B pillars. The panel may be welded into place at the seam only, by using plug welds, or using a combination of both.

Tapered Lap Joint

The tapered lap joint is commonly used where a section of an enclosed box or rail is replaced. In this procedure a V slot double the width of the metal thickness is cut out of each corner of the rail, allowing each side to be bent inward slightly. This allows the rail to be butted together by slightly inserting the rail end with the reduced size into the other. When the two rail ends are welded together and the excess weldment is ground from the seam, the rail will be flat and not show any visible signs of a joint having been made.

Flanged or Recessed Lap

This joint is more apt to be used on exterior sheet metal panels than on structural panels. When a panel is lapped together using a standard lap the seam area will be raised up at least the equivalent of the metal thickness. See **Figure 15-25**. For cosmetic reasons this is not acceptable because finishing and disguising the repaired area is very difficult. A special recessing or flanging tool is used to make a sharply formed flange or recess—usually the equivalent of the metal thickness. To disguise the seam area, one of the edges of the metal is recessed, forming a trough or a lowered ledge or surface. The other edge is then placed into the lowered surface and welded. When the excess weldment is ground off, the surfaces are flush with each other.

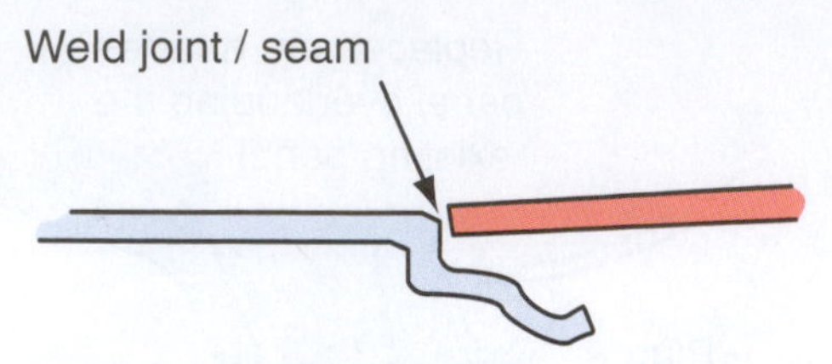

FIGURE 15-25 A recessed lap joint is used on the flat surface where one exterior panel is lapped over another to minimize the buildup from the lapped layer.

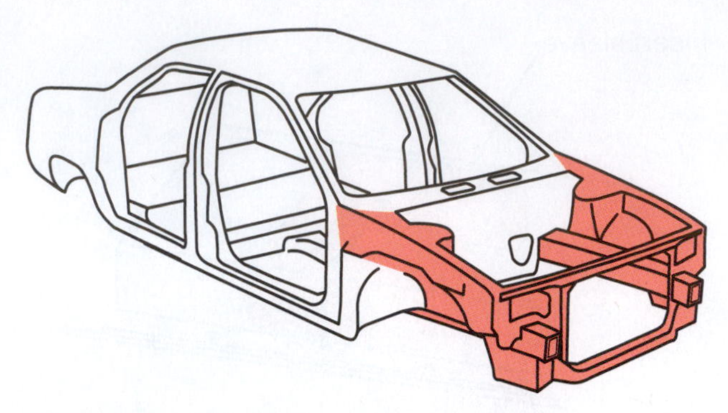

FIGURE 15-26 The front structural assembly on a unibody may consist of an enclosed assembly where all the parts are welded into one solid assembly. It may resemble a modular design where the front upper tie bar may be bolted into place.

FRONT RAIL REPLACEMENT

The front assembly of the vehicle is a combination of several uniquely different and interdependent structural units. The core support may be one of a variety of designs. It may be made up of multiple pieces including an upper tie bar, two side panels, and a lower tie bar, each of which may be serviced as individual parts, or the entire assembly may be obtained. This design allows the replacement of individual pieces by simply drilling the spot welds to separate them from adjoining surfaces and welding the replacement part back into place. A sectional replacement may also be made without necessitating the removal of the undamaged fender and associated parts on the opposite side of the vehicle.

TECH TIP Regardless of the design of the core support or the option used to replace it, the damaged assembly must be pulled and roughed out to restore its dimensional tolerances as closely as possible to the manufacturer's specifications.

When replacement of the welded core support is necessary, it may be performed by replacing it at the factory seams or by performing a sectional replacement. In addition to weighing the cost effectiveness of performing a sectional replacement, one must also consider the practical side of it as well. Remember that many of the underhood dimensions are based on the core support location; therefore, it is imperative that numerous measurements are made prior to welding it into place to ensure it is precisely located. In many instances it is wiser and more practical to replace the entire unit as an assembly.

Another design may be an assembly that is more heavily constructed but, instead of being welded into place, it may be bolted directly to the front rails. Whichever design is used, the core support performs a structural function and closes out the two side rail assemblies to which it is attached to help reinforce the entire front structure. See **Figure 15-26**.

FRONT SIDE RAILS

The sides of the front structure are made up of the front upper rails, the strut tower assemblies, and the lower rails which are welded together as a unit to form the apron assembly. The apron assembly is welded to the sides of the cowl, the fire wall, and the lower body. Together this assemblage of panels and structures is commonly referred to as the front vehicle structural assembly.

Replacing Front Rails

When replacing either of the front rails, the sectional replacement option can only be used if the cut is made in front of the strut tower. The tower is the load-bearing area for all the weight of the front of the vehicle, and sectioning it in any part of the panel may compromise its structural integrity. However, a sectional replacement can be performed in several locations in front of the tower assembly. The side rails are usually constructed of one or more hat channels that are attached to the apron assembly, enclosing them to form a closed hat channel. The upper rail and apron assembly can be replaced as a factory seam installation or it may be sectioned. See **Figure 15-27**. The factory seam replacement is accomplished by removing the assembly at the appropriate factory seams. The sectional replacement is performed by separating the front of the apron assembly at the seam where it is attached to the strut tower and using a butt joint with backing in an area of the rail where a sectional joint can be installed, using the guidelines prescribed for such repairs on the rail. See **Figure 15-28**. Replacing the lower rail is more difficult and involved because it is frequently reinforced with additional layers of metal welded to the outside along with various stamped ridges and recesses. Lateral stiffeners may also be used at critical points on the inside of the rail to add strength and rigidity. Locating a surface that meets all the criteria for a sectional replacement is difficult on this structure, particularly when an insert is necessary to reinforce the joint area. The technician should consult the vehicle manufacturer's recommendation for performing a sectional replacement at the chosen area or consult the I-CAR UPCR. The number of locations where a sectional replacement can be made are limited. Therefore, a

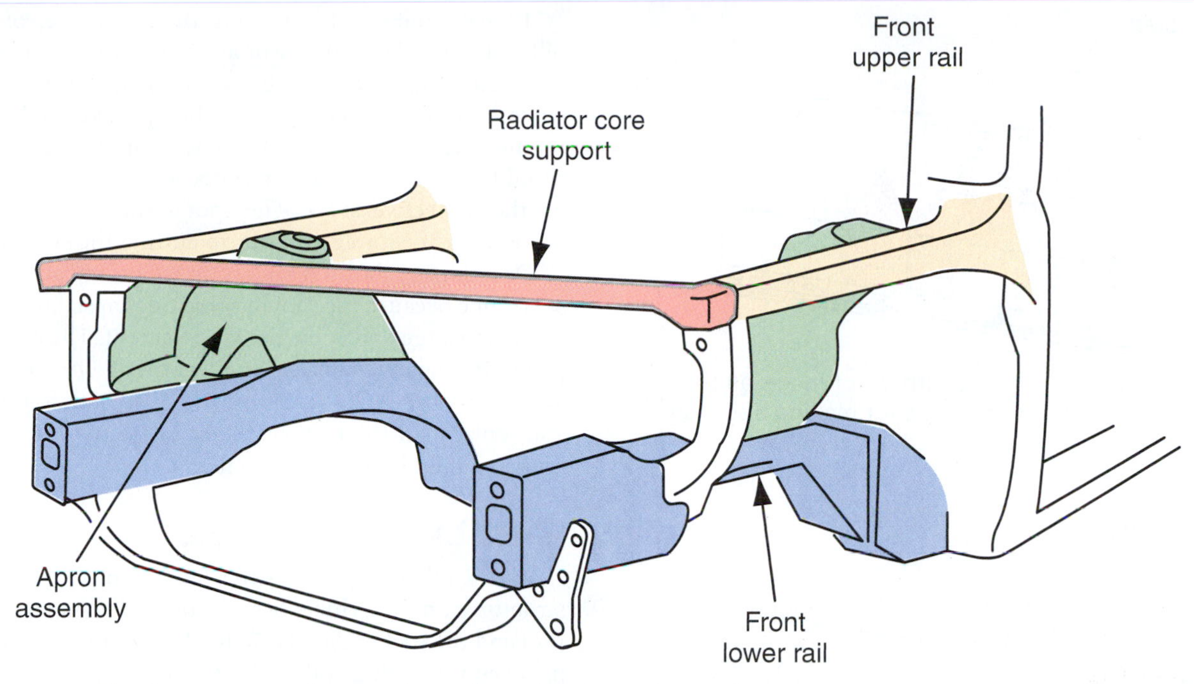

FIGURE 15-27 The front structural assembly on the unibody consists of the core support, the upper and lower side rails, and the strut tower, which are all welded together to form a complete assembly.

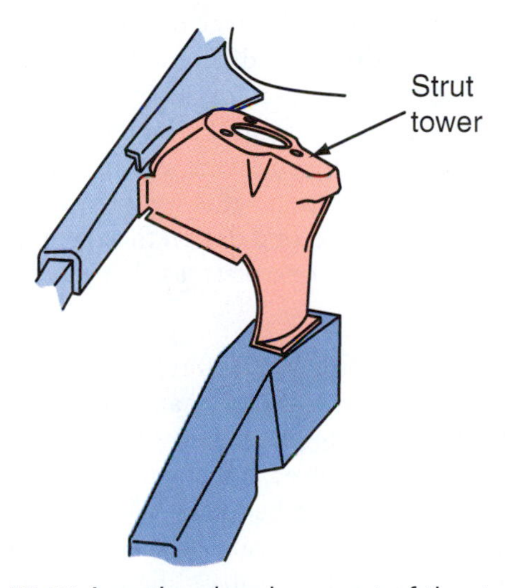

FIGURE 15-28 A sectional replacement of the strut tower can be used only if the seam is made in front of the tower or at the factory seams on the tower.

complete replacement at the factory seams may be a more viable option.

PILLAR REPLACEMENT

Depending on the design or vehicle model, the vehicle will use either three or four vertical pillars to support the roof—commonly referred to as the **A, B, C, and D pillars**. These structural members must be constructed sturdily enough to protect the passenger compartment and support the vehicle in the event of a rollover. Nearly all manufacturers utilize some variation of an enclosed box design using from two up to four layers of metal in some isolated sections of the pillar. Depending on the type of accidental involvement, the extent of the damage may affect one—or possibly all—of the pillars. Each pillar offers some structural characteristics that are somewhat unique from the others. Consequently, it may require differing approaches when replacing them.

A Pillar

The A pillar, located at the front of the vehicle, extends parallel with the windshield edges from the roof down to the top of the cowl area. It may be made up of two or three separate pieces of metal welded together in such a way to form a lip or fence upon which the windshield rests and is glued into place. The A pillar is usually reinforced at both the top and bottom where they are attached to their respective mating surfaces. These reinforcements limit the size of the area in which a sectional replacement can be made to approximately the center one-half of the pillar. Sectioning an A pillar requires some planning and thought before even attempting to cut the damaged one out of the vehicle. One can use either a butt joint with insert, although an offset butt may be more desirable in some instances. The type of joint design that will be used will determine the exact location of the cut. See **Figure 15-29** and **Figure 15-30**.

When the butt joint with insert is used, it is done to ensure the restoration of the strength and structural integrity. Therefore, one must fashion an insert to resemble the configuration of the cavity, use a piece of the damaged

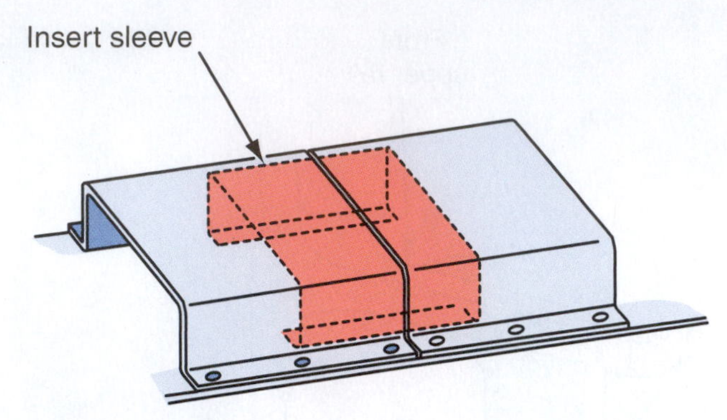

FIGURE 15-29 A butt joint with insert meets all the specified strength requirements for making a sectional rail replacement.

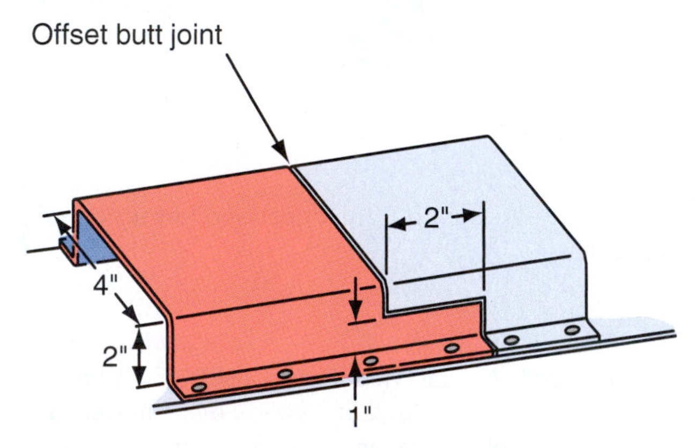

FIGURE 15-30 Another variation of the offset butt joint can be made either with or without an insert.

section of the pillar, or cut one from the excess material of the replacement part. The insert should be cut approximately 75 to 80 mm (3 inches) long and inserted inside the cavity so the length is split equally between the two sides of the cut. The insert should be plug welded onto the original part of the pillar before slipping the replacement section over the protruding end. See **Figure 15-31**. The insert should be plug welded on the replacement section and the seam welded solid using the techniques discussed previously. The excess weldment should be removed, being cautious not to remove any of the surface from around the area.

To make the offset butt joint, the outer layer of the A pillar should be cut higher or above the inner layer rather than in a straight line cut, thus creating the offset. The distance between the cuts should be approximately 75 to 80 mm (approximately 3 inches) apart. Technicians should be careful to make the cuts between the spot welds on the respective sides. The spot welds may be drilled more easily if completed prior to cutting the seams.

Occasionally both A pillars must be replaced at the same time because the roof may have collapsed from a collision or a large object having fallen onto it. The outer roof panel is usually replaced under these circumstances. When that becomes necessary, the A pillars are removed and replaced at the factory seams where they are welded at the top and bottom of the pillar.

B Pillar Replacement

The B pillar is perhaps the most complex of all the structures on the vehicle because of how it is tied into the roof and the rocker panels. To further complicate matters, it serves as the hinge pillars for the rear door and the front door latch, thus requiring an infinite degree of dimensional accuracy when installing the replacement part. Each of these functions requires that additional reinforcements be added to the pillar which, in turn, creates additional obstacles that must be overcome when replacing it. There are also two distinctly different designs for the center pillar. One design is used on many of the body over frame (BOF) vehicles wherein the center pillar is a separate assembly with a visible seam where it overlaps the rocker panel, at which point it is welded into place. See **Figure 15-32**. On the second design the center pillar and rocker panel are all in one piece as they are part of the **uniside panel**.

B Pillar Full Replacement. A full replacement frequently involves separating both the outer and inner layers of the B pillar where it is integrated and welded into the roof. This becomes very involved and one must exercise a great deal of care and patience because—when separating the pillar from the roof—the possibility of damaging the outer roof skin is imminently possible. The pillar's inner reinforcement is usually welded to the roof header on the in-

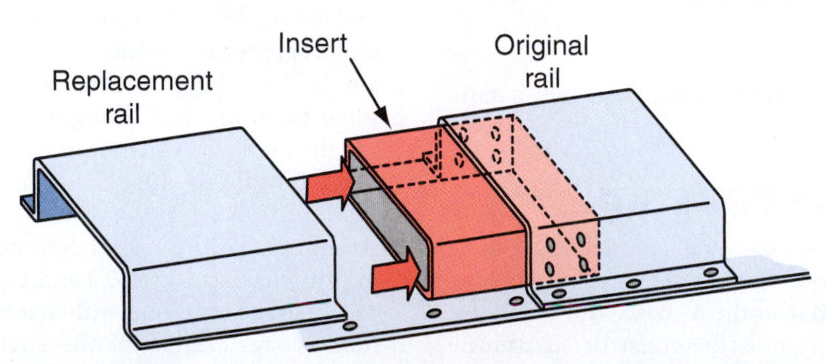

FIGURE 15-31 The insert should be plug welded into the original rail on the vehicle before the replacement part is installed.

(A)

(B)

FIGURE 15-32 (A) shows how on some models the center or B pillar is an overlay that is welded over the top of the rocker panel. (B) shows how the center pillar on a uniside design is incorporated into the rocker panels as one continuous piece.

side of the vehicle and the spot welds are usually quite accessible. However, the outer section of the pillar is integrated and welded together with the outer roof skin, the header, and the drip rail if the vehicle in question has one. A vehicle using a uniside panel frequently requires separating the seam where the uniside and roof are joined together. The most effective means of removing the B pillar assembly from the roof is to sever it near the roof line first so it can be cut apart and removed in pieces. Ultimately the ideal solution is to replace both the outer roof skin and the B pillar.

B Pillar Sectional Replacement. Depending on the location and severity of the damage, the center pillar may be replaced completely or a sectional replacement option may be performed. The sectional replacement option is more popular because the complete replacement method requires a great deal more disassembly, disturbs many more factory welds and the original corrosion protection, and increases the possibility of squeaks and other such sounds. The sectional replacement is best accomplished by cutting the pillar above the **belt line** which is generally the narrowest area of the B pillar. See **Figure 15-33**. The area approximately midway between the belt line and the **D ring anchor location** near the top of the pillar is usually reinforced with only one additional reinforcement layer, therefore making it an ideal location to section the pillar because one can gain access to all sides of the pillar for welding purposes. The D ring anchor is the area where the shoulder seat belt is attached. This area must be avoided when making a cut because it is relatively heavily reinforced, making it difficult to duplicate a solid joint when welding the area back together.

The area of the center pillar just described is usually designed as a closed hat channel with the U-shaped channel facing the outside of the vehicle and the enclosing flat piece on the passenger side of the structure. An additional reinforcement layer is usually welded between these two layers. The cut should be made in such a way that a staggered offset butt joint can be made, allowing access to all surfaces for welding. A lap joint on the back side will allow access to all the surfaces to accommodate the welding operation. See **Figure 15-34**. The lapped piece on the back

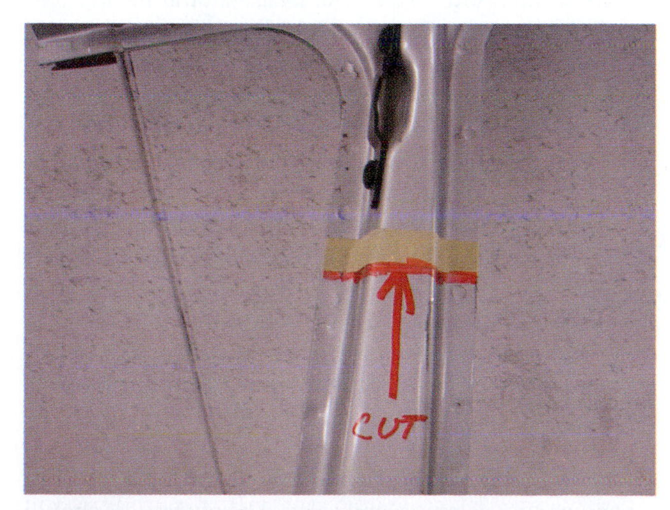

FIGURE 15-33 Making a sectional replacement on the B pillar is done most easily approximately midway between the roof and the belt line, as this is the narrowest point of the panel and it is well below the D ring anchor point.

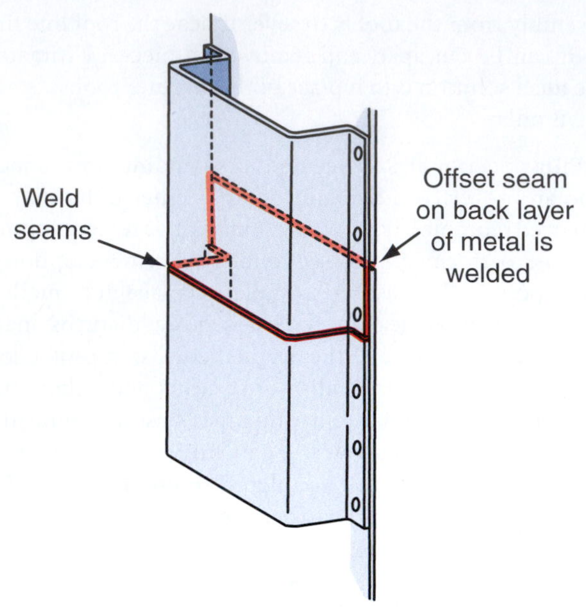

FIGURE 15-34 An offset butt joint with a lapped piece on the back side is used when access is required to weld a reinforcement on the inside of the pillar.

side should be offset approximately 100 mm (4 inches) above the cut line on the outer surface. One must drill and separate the spot welds that hold the flat layer on the back side to the "U" channel. A straight line cut should then be made on the back side where the two pieces will be welded together once again. The spot weld nuggets must be removed, the surface dressed up, and a coat of weld-through coating should be sprayed over the entire contact surface prior to welding them together.

An insert or sleeve should be made to fit inside the "U" channel to reinforce that section of the pillar and ensure the structural integrity has been restored when the pillar is fully welded into place. The insert should be approximately 75 mm (3 inches) long and can be cut from part of the damaged old pillar from the immediate area where it was removed. It may need to be modified somewhat to ensure it will fit inside of the front section of the pillar.

Prior to installing the insert several steps must be taken. Holes must be put into the upper pillar for plug welding it into place, any burrs that may have formed on the inside of the pillar at the cut line must be removed, and the insert should also be prepped by removing the paint from it. It should also be re-sprayed with weld-through coating to ensure the corrosion protection has been restored. Only after all these steps have been performed can the sleeve be clamped into place and welded by using holes made for the plug welds. When the replacement pillar is installed, it should be aligned and temporarily held in place with screws, welding clamps, and so on. The plug welds should then be welded first, followed by welding the seam solid using a skip welding technique to avoid overheating the metal as would likely occur if the seam were welded with a continuous non-stop bead. Remember, any distinct or definition crown should be welded first, followed by filling the areas between using a skip weld technique.

Lower B Pillar Section. The lower section of the center pillar where it is attached to the rocker panel usually requires drilling the spot welds to remove the section. The B pillar is usually the last layer installed by the manufacturer; therefore, it is the top layer of metal at that location. When drilling the welds to separate the panel, one must be cautious to not drill too deeply to avoid drilling holes into the rocker panel reinforcement. The replacement welds must be placed in very nearly the exact location of the original ones made by the manufacturer. Any holes inadvertently drilled into the mounting pad must be welded shut to ensure there is a metal surface to which the replacement part can be welded. Removing the lower section of the B pillar frequently requires chiseling some of the spot welds to finish separating the two layers where the spot welds were too large to catch with the drill. Caution should be exercised to avoid damaging the mounting surface or pad to which the replacement will be welded.

CAUTION

Caution must be exercised whenever using a grinder in an area where the metal is ripped or where the surface is uneven with multiple layers of material. The possibility of injury is infinitely increased due to the grinding disc catching on an edge and tearing when working in an area such as this.

Prior to installing the replacement part the entire attaching surface must be cleaned and properly prepped to accept the new part. All the **spot weld nuggets** must be removed, and any rips or tears on the attaching surface must be welded and smoothed. The surface must be coated with a weld-through coating to ensure no corrosion hot spots exist between the layers of metal once it is assembled.

Preparing the replacement part is no less important than performing any other part of the total procedure. Due to the location and shape of the structure, it will be permanently installed using GMAW MIG plug welds. The placement of the welds must precisely simulate the original weld locations and also be precise in the number made. One must avoid welding along the edge on any part of the overlapping seam as this would result in **re-engineering the vehicle**. This may alter the reaction and response to the collision forces in the event of a subsequent collision.

To ensure the welds are affected correctly, it is advisable to practice making the same welds that will be used on the vehicle. One can be assured the same type and gauge of steel is being used by practicing on some of the excess material that was trimmed off from the replacement part or from the damaged part removed from the vehicle. This will help to ensure the welding machine is set up and adjusted properly so nothing is left to chance when the replacement part is permanently installed onto the vehicle. After the welding operation is completed, the excess weldment must be sanded/ground off the part, being careful to not remove any more of the surface layer than is necessary. The seams should be caulked where recommended, and the area should be prepped for painting.

Uniside Panel B Pillar Replacement. Replacing the B pillar on a vehicle that utilizes a uniside panel design requires a considerably different approach. Here again, the pillar is a combination of at least an inner and outer layer spot welded together and may be replaced at the factory seams. Alternatively, the sectional option may be used. See **Figure 15-35**. Unlike the pillar described previously, wherein the outer layer is welded over the top of the rocker panel, on the uniside design it is integrated into the rocker panel and the entire side of the vehicle. When removing the damaged part, the **wind laces** from around the door opening and other trim, seat belts, and all the necessary interior trim must be removed. The spot welds that hold the inner and outer panels together and those used to attach the pillar to the roof will be exposed. If only the outer layer of the pillar is to be replaced, one needs to only drill the spot welds along the front and rear door openings slightly higher than the damage extends, making a sectional replacement more practical. Since the bottom of the pillar is incorporated into the rocker panel, it will eventually become necessary to cut through the outer layer of the rocker panel somewhere near the middle of both the front and rear door openings. See **Figure 15-36**. However, before making these cuts, it may be advisable to cut out a window just below the pillar, allowing access to determine the location of any reinforcements that are not visible from the outside. However, these windows will have to be dealt with when removing the remainder of the panel. The replacement seams on the rocker panel are normally placed near the middle of the front and rear door openings because fewer reinforcements are used in this

FIGURE 15-35 To make a sectional replacement on a uniside design, the panel must be spliced into the rocker panel.

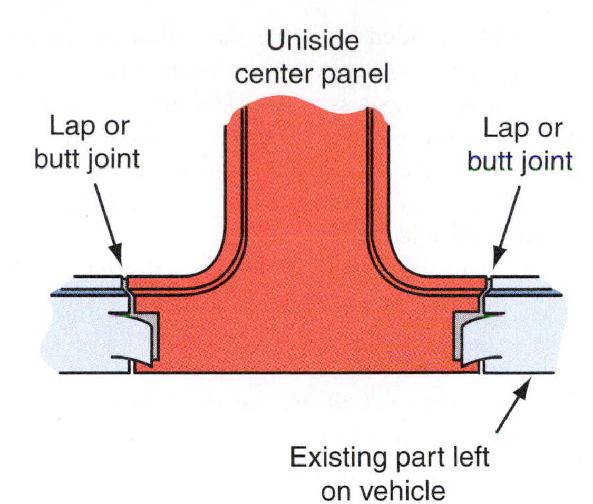

FIGURE 15-36 A window may need to be cut into the rocker panel to gain access to weld the inner reinforcement when making a sectional replacement. The windows will be welded shut and finished after all the structural welding has been completed.

section of the rocker, making the installation of the replacement part less difficult. Prior to cutting into the rocker panel, it is advisable to place some type of support member—such as a jack stand, floor jack, and so on—under the area immediately in front of and behind the area where the cut is to be made. This will support the vehicle and prevent the cutting tool from binding as the body tends to collapse inward when the cuts are being made.

Before starting to drill the spot welds, one should carefully analyze the manner in which the pillar is assembled to determine which welds will best be drilled and which ones need not be addressed. In the event the entire pillar requires replacing, the inner and outer members will be removed as a unit by drilling out the welds where the assembly is attached to the roof header panel and the outer skin.

C Pillars and D Pillars

The C pillar, commonly referred to as the sail panel, is the rear roof support pillar on most cars which is incorporated into the upper part of the quarter panel. It also houses the ledge to which the rear window is attached on nearly all model cars except the hatchback. The D pillar is the rearmost pillar on most SUVs and minivans. The procedure for sectioning the C pillar and the D pillar are very similar to sectioning the upper B pillar. One must decide if the operation will best be accomplished by using new parts and piecing it together or using recycled parts and replacing an entire section of the vehicle in one piece. Perhaps the most logical and expedient way to accomplish this is to perform a complete quarter sectional replacement using recycled parts. The principal criteria to successfully completing the operation is taking accurate measurements, recording these measurements, and carefully laying out the plan of attack.

When recycled parts are removed from the parts vehicle at the salvage yard, pieces of the adjoining parts are removed and included with the parts that are to be replaced. Therefore, the recycled part usually requires trimming to remove the excess sections left intact when it was removed at the salvage yard. The spot welds should be located, marked, and carefully drilled to avoid damaging the contact surfaces more than necessary as this will affect the fit when it is installed.

As with the other two pillars discussed previously, the C and D pillars are made of an inner and outer layer of die-stamped metal parts that are welded together with a relatively large cavity separating them. One must carefully measure from at least two or three locations to the area where the seam is to be made because the same cut must be duplicated on the replacement part. As with any sectional replacement, the joint location on the inner reinforcement should not be in a straight line with that of the outer sail panel. An offset butt joint with an insert should be used to weld the two panels together. The insert is to be installed in the same manner as described on the A and B pillars. When welding the inner and outer panels together, the technician should use the prescribed precautions to prevent overheating the metal by making short skip welds and allowing sufficient time for the metal to cool between welds. The first welds should be made at the location of any definition crowns to ensure their alignment, followed by welding the rest of the seams, using a skip welding technique alternating from the inside surfaces to the outside panel. Alternating with the welding from the inner reinforcement to the outside panel will help prevent overheating the metal's surface. If there is any doubt about any part of the replacement procedure, the recommendations furnished by the manufacturers or the I-CAR Uniform Procedures for Collision Repair (UPCR) should be consulted.

UNISIDE PANEL INSTALLATION

When installing the uniside center pillar, many of the same steps must be performed as were discussed previously. The steps required will differ slightly depending on whether a sectional or full panel replacement is to be performed. A full replacement requires cleaning and preparing the area around the roof header for installation and welding the new part into place. See **Figure 15-37**. All spot weld nuggets from the old welds must be removed; any tears, rips, and surface flaws resulting from removing the damaged part must be smoothed; and a weld-through coating must be sprayed over the area. In both the sectional and full replacement option, a sleeve must be fabricated to be inserted at both seams on the rocker panel, observing all the necessary corrosion protection steps required.

The insert should measure approximately 100 mm (4 inches) in length and should be inserted so the length is split equally between the two sides of the seam. Prior to installing the insert, the paint should be removed and the surface sprayed with weld-through coating. Several plug

FIGURE 15-37 The spot weld nuggets left from the drilled out welds and any tears in the surface must be welded before the replacement part can be installed.

weld holes should be punched into both the rocker panel and the replacement part, being careful to leave at least a 25 mm (1 inch) space between each weld. The insert should be welded onto the rocker panel on the vehicle first using the plug holes made earlier. The new part can then be put into place and welded. One should leave approximately a 3 mm (1/8 inch) root gap between the two butted edges of the existing and the replacement panels to ensure that proper penetration and fusion occurs when welding the seam solid. When welding the seam, one must observe all the precautions and techniques discussed previously to ensure the metal is not overheated.

UNISIDE PANEL SECTIONAL REPLACEMENT

For a sectional replacement, a sleeve must also be fabricated and prepared for installation at the seam in the upper rail section in the same manner as discussed previously. The placement of the seam should be approximately midway between the belt line and the D ring anchor mount to avoid cutting into its mounting reinforcement. An offset staggered butt joint should be used to weld the reinforcement and the layer on the back side of the pillar.

REAR RAILS

The rear rail on a vehicle is typically a U-shaped rail with the trunk floor welded over the top of it to finish enclosing the fourth side, making an enclosed hat channel. The rails usually run from the rear of the vehicle forward to the rear torque box under the passenger compartment. Additional lateral rails are located at the rear of the vehicle. Another is located in the area in front of the spare tire pocket, which usually intersects with the side rails to form the reinforcements of the trunk floor. Like the front rails, the rear rails represent a variety of shapes and configurations. Some of them are multi-layered, while others are convoluted or notched and are designed to absorb and dissipate collision energy. The end of the rear rails are designed with end caps welded into place to form pockets to accept the impact absorbers. The rear suspension parts, axle housing, trailing arms, and most other parts associated with the rear wheels are usually attached to the rails as well. By their very design of being a closed box construction, when these rails become damaged in a collision, a kinked area often occurs. Many times the repairs include replacing a section of the rail or, in some instances, the complete rear rail. However, prior to removing the damaged rail, the area should be pulled and roughed out first to relieve all the stresses left in the area.

Full Replacement

When replacing the entire rail, it usually becomes necessary to remove the suspension parts attached to them. The spot welds on the damaged rail should be drilled out from whichever side is most accessible in that area. Sometimes access is not possible from inside the vehicle, making it necessary to drill some out from underneath the vehicle and some from inside. All the nuggets must be removed, any tears or rips that may have occurred during the removal of the rail must be repaired, and weld-through coatings must be applied prior to installing the new part. The plug weld holes must be pierced into the edges of the replacement part at the same intervals used in the original rail and then it should be prepared to install. Most generally a screw or some type of removable fastener is used to temporarily secure the rail until it is permanently welded into place. Welding the rail will require the technician to weld in an overhead position in some areas. Practicing these welding procedures is recommended prior to attempting to perform them on the vehicle.

Sectional Replacement

Replacing a section of the rail requires a similar amount of preparation except the damaged rail is cut off at an area that meets the criteria for making a sectional replacement. A sufficient amount of distance away from any holes, convolutions, and reinforcements—when they are used—should be allowed to install an insert into the rail prior to welding. All the necessary preparations for installing the insert should be performed in the same manner as when using them in any other area of the vehicle. In the event the configuration of the rail does not allow for an insert, a smooth section of the rail away from the kinked area should be selected to make the seam and prepare to install the replacement part by making a tapered lap joint. All the necessary corrosion protection precautions should be performed. Occasionally the manufacturer utilizes structural foam to reinforce sections of the rail. Any of these materials originally used by the manufacturer should be reinstalled after the corrosion protection steps have been performed. Corrosion protection in this area is critical and all surface coatings used by the manufacturer should be reinstalled as well because this part of the vehicle is highly exposed to road spray and surface abrasion.

SECTIONING FLOOR PANS AND TRUNK FLOORS

Although the passenger and trunk floor pans are not considered part of the vehicle's structural unit, they occasionally require replacing. Even though most of the surfaces are very irregular and single layered, many of the same principles and techniques used when replacing pillars are also utilized when performing a floor pan or a trunk floor sectional replacement. Both the trunk floor and the passenger compartment floor have many stamped convolutions and compound shapes limiting the areas where a sectional replacement can be used. Since a reinforcing insert is not required as part of the replacement procedure, almost any flat area that measures more than 25 mm (1 inch) wide, extends across the width of the floor pan, and will not require cutting through any reinforcement or undercarriage rail is a suitable area for sectioning.

Measure Before Cutting

Prior to cutting the floor pan, several measurements must be taken and recorded to ensure the overall dimensions are not altered when reassembled. After measuring, the cut line should be marked in an area that is away from any reinforcements and suspension mounting locations. One must be careful not to cut through any part of the reinforcement rails that run underneath the floor. The cut should be made with a reciprocating saw or other cutting tool that will make a straight smooth cut without leaving any jagged edges because the seam for the replacement part will be placed at that exact location.

Installing Floor Pan Section

When installing the replacement part, an overlapping joint is used with the edge of the top piece always pointing forward. By overlapping the metal in this way, the road splash and debris will stream past or away from the joint seam, minimizing the possibility of exhaust gases and foreign materials entering the vehicle. The two pieces of metal should overlap each other by at least 25 mm (1 inch), allowing adequate space for the plug welds that are used to weld the layers of metal together. The plug holes should be pierced at approximately 25 mm (1 inch) intervals into the top piece, thus eliminating the need to weld overhead. To ensure proper alignment during the welding procedure, the two layers of metal should be temporarily secured with removable fasteners until the welding is completed. Upon completion of the welding, these items should be removed and the holes welded shut.

RESTORE CORROSION PROTECTION

Prior to welding the two sections together, all the necessary steps should be taken to ensure all the corrosion protection has been restored. All the paint should be removed from both sides of the metal in the joint seam area and a weld-through coating should be sprayed over the entire mating surface. After the welding has been completed, any excess primer that extends beyond the seam area should be removed as it does not offer good adhesion to any materials applied over it. A urethane primer or a self-etching primer and an appropriate coating should be applied over the seam area. It should then be caulked on both the top and bottom sides with a non-hardening flexible caulking material. The repair area should be finished with a topcoat, simulating the original finish used by the manufacturer.

Sectioning the Trunk Floor

Sectioning the trunk floor entails essentially the same procedures used when sectioning the floor pan. The most significant difference is the number of reinforcing cross-members that will be found under the trunk floor. Most of these reinforcements are welded to the trunk floor, forming a closed hat channel. When possible, one should section the trunk floor over the top of the rear flange of the cross-member. Plug welds should be used to secure the floor to the rails. As was discussed previously, the plug weld holes should be put into the trunk floor piece so the welds can be made from the top side. All the corrosion protection precautions and finishing should be completed in the exact same manner as was discussed previously for the floor pan. Whenever possible, the I-CAR UPCR or the manufacturer's recommendations should be consulted to ensure the correct procedures are followed.

COMPLETE REAR CLIP REPLACEMENT

The most dramatic repair that one can make is the full body sectioning. This involves making one vehicle out of two otherwise totaled vehicles. Two vehicles that are the same year and model that sustained irreparable damage to opposite ends of each other are cut into two and the two undamaged ends are re-attached to each other, making a fully functional vehicle. Before attempting to perform a repair so dramatic as this, one should consult the manufacturer's and I-CAR UPCR recommendations to determine if this procedure is recommended on this particular vehicle. Nearly every joint design and replacement option discussed in this chapter is exercised in this operation. It is highly recommended that an experienced technician with a knowledge and understanding of all the operations to be performed be involved throughout the procedure.

The procedure involves removing the good halves from both vehicles by cutting both A pillars, B pillars, and through the floor and rocker panels on both sides. During the initial cuts made, both the A pillar and B pillar on the parts vehicle should be cut approximately 75 mm (3 inches) longer than where they will be welded together on the repaired vehicle. This will allow enough material for making the offset butt joint, the final trimming, and precisely fitting the pillars to the vehicle.

Measuring Is a Critical Step

Prior to making any cuts, all the door and window openings must be carefully measured and meticulously recorded to ensure the information and dimensions required to reassemble the vehicle are available and dimensionally correct. This includes diagonal, linear, and vertical measurements made to identical reference points on both vehicles. The location of the cuts should be identified and marked, and measurements should be rechecked. The location of the seam should be checked to confirm that it complies with all the sectioning guidelines. Remember that nothing should be left to chance. The seats, seat belts, carpeting, and any other interior parts that will be affected should be removed from both vehicles to ensure the needed access.

Support the Undercarriage

Before making any cuts, provisions should be made to support the undercarriage with jack stands or by clamping it to a bench. This will minimize any binding or pinching at the rocker panel and both pillars when they are cut. All fuel and vapor recovery lines—along with the brake lines—should be separated at their mechanical connectors and, if possible, removed from the vehicle. Any electrical wire looms that are routed through any of the cavities from the front to the rear of the vehicle should be disconnected and marked for easy reinstallation. The fuel tanks should be removed as well from both vehicles to prevent any inadvertent spillage and also eliminate the possibility of fire. Any unnecessary chance taken by taking a shortcut or omitting a step can increase the possibility of an injury to the technician or unnecessary damage to the vehicle. No matter how routine or mundane a repair procedure may seem at the outset, anything less than a total commitment to quality will result in compromising the repair's overall effectiveness.

Summary

- To reduce the weight of automobiles and subsequently improve their fuel economy, automobile manufacturers have introduced numerous dramatic design and structural changes on nearly all cars.
- The changes have included reducing the gauge of metals commonly used, the size and shape of reinforcements, their installation methods, and the materials used for these structures.
- Through the combined efforts of the auto manufacturers, equipment manufacturers, and the collision repair industry, the challenge to stay abreast with the changes presented to the auto collision repair industry have been met with enthusiasm and a spirit of cooperation.
- The future of the industry will continue to present a whole new series of challenges as the use of aluminum and various composition materials becomes more widespread.
- The future of the automobile repair industry lies in the hands of many, and a continued joint effort will certainly result in an industry filled with promise and many additional challenges.

Key Terms

A, B, C, and D pillars
bag and tag
belt line
component panels
D ring anchor location
heat affect zone
hydroforming
I-CAR Uniform Procedures for Collision Repair (UPCR)
jamming and edging
like, kind, and quality (LKQ)
OEM
re-engineering the vehicle
root gap
sectional/partial replacement
silicon bronze
spot weld nuggets
squaring the door and window openings
sweat tab/pull tab
uniside panel
weld-bonding
weldment
wind laces

Review

ASE Review Questions

1. Technician A says a straight cut open butt joint can be used to resect a C pillar when making a sectional replacement. Technician B says an insert will help to strengthen the area where the joint is welded. Who is correct?
   A. Technician A only
   B. Technician B only
   C. Both Technicians A and B
   D. Neither Technician A nor B
2. Technician A says a tapered lap joint may be used in certain areas if there is not enough of a smooth and uninterrupted surface to insert a sleeve. Technician B says welding the trunk floor to the open side of a "U" channel creates a hat channel. Who is correct?
   A. Technician A only
   B. Technician B only
   C. Both Technicians A and B
   D. Neither Technician A nor B

3. Technician A says using recycled parts can be advantageous because the recycler may include parts that may require replacing that are not necessarily included with new parts. Technician B says a uniside panel may include the quarter panel, the center pillar, and door openings. Who is correct?
   A. Technician A only
   B. Technician B only
   C. Both Technicians A and B
   D. Neither Technician A nor B
4. Technician A says the flanged or recessed lap joint is commonly used when performing a sectional rail replacement. Technician B says the abbreviated quarter panel replacement method usually includes replacing the sail panel. Who is correct?
   A. Technician A only
   B. Technician B only
   C. Both Technicians A and B
   D. Neither Technician A nor B
5. Technician A says when installing a replacement center or B pillar the technician should add an additional 30 percent more MIG plug welds than the manufacturer originally used. Technician B says increasing the number of MIG plug welds over the amount used by the manufacturer may alter the vehicle's ability to distribute the impact forces in a subsequent collision. Who is correct?
   A. Technician A only
   B. Technician B only
   C. Both Technicians A and B
   D. Neither Technician A nor B
6. Technician A says before any part of the damaged structural part is removed it must be roughed out to relieve the strain from the area. Technician B says all door and window openings must be roughly restored before removing any part of the structural member. Who is correct?
   A. Technician A only
   B. Technician B only
   C. Both Technicians A and B
   D. Neither Technician A nor B
7. Technician A says when welding the replacement structure into place, the first welds should be made on the definition crowns. Technician B says a stitch weld technique should be used when welding the seam to minimize the possibility of warpage. Who is correct?
   A. Technician A only
   B. Technician B only
   C. Both Technicians A and B
   D. Neither Technician A nor B
8. Technician A says a sectional replacement can be made by cutting through the center of the strut tower and overlapping the replacement part over the old tower. Technician B says it is advantageous to use an insert for reinforcement any time it is possible. Who is correct?
   A. Technician A only
   B. Technician B only
   C. Both Technicians A and B
   D. Neither Technician A nor B
9. Technician A says when drilling out spot welds, one should drill through both surfaces leaving a hole where the weld once was located. Technician B says using the air hammer will separate the metal surfaces more quickly and cleanly than drilling the spot welds first. Who is correct?
   A. Technician A only
   B. Technician B only
   C. Both Technicians A and B
   D. Neither Technician A nor B
10. Technician A says when performing a full factory replacement of the B pillar, the outer skin of the roof must also be replaced. Technician B says the best location for the seam is between the D ring anchor point and the belt line when performing a sectional replacement on the B pillar. Who is correct?
    A. Technician A only
    B. Technician B only
    C. Both Technicians A and B
    D. Neither Technician A nor B
11. Technician A says when making a sectional replacement of the A pillar, a staggered butt may be used. Technician B says a butt joint with insert may be used. Who is correct?
    A. Technician A only
    B. Technician B only
    C. Both Technicians A and B
    D. Neither Technician A nor B
12. Technician A says weld bonding can be used on all structural panels provided adequate curing time is allowed. Technician B says weld bonding affords better corrosion protection in the seam area than other approved methods. Who is correct?
    A. Technician A only
    B. Technician B only
    C. Both Technicians A and B
    D. Neither Technician A nor B
13. Technician A says any time the door intrusion beam is bent the door assembly should be replaced. Technician B says the door outer repair panel can be installed using either adhesives and crimping the hemming flange or by crimping the edge and using small spot welds around the outer edge to hold the skin in place. Who is correct?
    A. Technician A only
    B. Technician B only
    C. Both Technicians A and B
    D. Neither Technician A nor B

14. Technician A says silicon bronze is used in some seams to increase the flexibility of the joint without compromising strength. Technician B says when using STRSW one should increase the number of welds by approximately 30 percent over the amount the manufacturer used. Who is correct?
    A. Technician A only
    B. Technician B only
    C. Both Technicians A and B
    D. Neither Technician A nor B
15. Technician A says a full body section involves cutting through the A, B, and C pillars and across the floor and rocker panels. Technician B says an open butt weld joint should be made on both the B pillar and the C pillar when welding them together during a full body section. Who is correct?
    A. Technician A only
    B. Technician B only
    C. Both Technicians A and B
    D. Neither Technician A nor B
16. Technician A says a sectional replacement can be made on any rail so long as the seam location or the insert do not overlap a hole larger than 12 mm (1/2 inch). Technician B says a sectional replacement can be made on a tailor welded area so long as a tapered lap joint is used. Who is correct?
    A. Technician A only
    B. Technician B only
    C. Both Technicians A and B
    D. Neither Technician A nor B
17. Technician A says an insert or a sleeve used in a sectional replacement will reinforce the seam area so it is as strong as the original rail. Technician B says the size of the root gap is determined by the metal thickness where the weld is being made. Who is correct?
    A. Technician A only
    B. Technician B only
    C. Both Technicians A and B
    D. Neither Technician A nor B
18. Technician A says the location of the seam for a sectional replacement may be limited due to inaccessibility for welding. Technician B says some manufacturers reinforce structural parts with structural foams. Who is correct?
    A. Technician A only
    B. Technician B only
    C. Both Technicians A and B
    D. Neither Technician A nor B
19. Technician A says the door and quarter panel are not designed like most structural parts, but they reinforce all areas adjacent to them. Technician B says weld-bonding requires less corrosion protection and will help to reduce vehicle NVH. Who is correct?
    A. Technician A only
    B. Technician B only
    C. Both Technicians A and B
    D. Neither Technician A nor B
20. Technician A says when making a sectional replacement on the passenger compartment floor the rear section of the floor should be lapped over the front to avoid carbon monoxide intrusion into the passenger compartment. Technician B says the I-CAR UPCR is a set of guidelines one can use when not certain about the recommended procedures for a specific repair. Who is correct?
    A. Technician A only
    B. Technician B only
    C. Both Technicians A and B
    D. Neither Technician A nor B

Essay Questions

1. List at least four different types of welding joints used when making a sectional replacement. Discuss how each of them is made and where they are most likely to be used.
2. What are the advantages of using recycled parts as opposed to OEM parts?
3. Discuss what advantages an insert or sleeve offers when it is used in a sectional replacement.
4. What are three things that will likely be included in the I-CAR UPCR for a sectional replacement of a trunk/rear floor rail?
5. Discuss why it should be necessary to rough repair a rail section even though it will be replaced.

Topic Related Math Question

1. The cost of the OEM replacement rail and reinforcements is $790 and a 7 hour labor time is allowed. A recycled sectional assembly is available for $450—it will take 5 hours to trim it up and an additional 6 hours to install it. With a prevailing labor rate of $45 per hour, which is the more economical repair route to follow?

Critical Thinking Questions

1. The rear rail is estimated to be replaced on the vehicle, but when the technician starts to rough repair it prior to removal, it is discovered that it can easily be repaired. Should the rail still be replaced, or should the technician go ahead and straighten it and receive payment for the replacement, keeping in mind the repair is a very legitimate one and complies with all the guidelines for straightening?
2. The estimate indicates the quarter panel is to be replaced at the factory seams, but the technician feels by spending an additional 2 hours rough

repairing the damaged one he can use the abbreviated method without leaving any signs of damage. The additional time required to repair the damaged panel is equal to the additional amount of time allowed to replace the panel at the factory seams. What should the technician do?

3. Discuss what the potential consequences are of re-engineering the vehicle by placing additional welds and reinforcements in the repair area when making repairs.

Lab Activities

1. Locate a vehicle with a uniside panel design and one without and observe the difference in how the two designs differ.
2. Locate a unibody design automobile and locate at least four different areas where the reinforcing members—such as hat channels, multi-layered surfaces, and other forms of reinforcements—are used. Identify what procedure may be required to replace these parts.
3. Locate a BOF vehicle. Identify and compare the structural reinforcement methods used on the undercarriage of these vehicles and compare them to a unibody design.

Chapter 16

Full Frame Sectioning and Replacement

OBJECTIVES

Upon completing this chapter, you should be able to:

- Identify the various frame designs.
- Discuss the most common applications for each frame design.
- Identify the types of metals used and the repair and replacement variables for mild, HSS, HSLA, and UHSS steel.
- Follow the recommendations for proper and correct heat applications.
- Review considerations for sectional replacement.
- Review considerations for full frame replacement.
- Develop a repair plan for replacing the damaged section or the entire frame assembly.
- Identify the variables and considerations for repairing and replacing frames utilizing recently introduced technology.
- Distinguish different repair techniques used for repairing and replacing aluminum structural and frame sections.
- Follow safety procedures that must be practiced to protect both the vehicle and the technicians performing the repairs.

INTRODUCTION

The full frame design has been used on vehicles since the invention and introduction of the first car by the Duryea brothers in the late 1800s. Shortly after its introduction, the first automobile accident occurred. Since that time repairs involving replacing sections of—or the entire—frame have been effected to restore collision-damaged vehicles and make them roadworthy once again. In the past, this meant forging and replacing sections of the frame or the entire assembly since the equipment to repair the damaged frame and undercarriage did not exist at the time.

PERSONAL SAFETY

The repair and replacement of frame and undercarriage parts often presents more opportunities for serious personal injury to the technician than any other phase of the repairs carried out in the shop. The novice technician may not be familiar with the equipment and many of

TECH TIP PERSONAL SAFETY SHOULD ALWAYS BE A PRIORITY!! One must never diminish the importance of personal safety when making any repairs that require the use of force application equipment. The proper PPE—such as safety glasses and gloves—are minimal requirements and the proper face shield should always be used whenever performing any metal cutting and grinding operations. A welding helmet should also be used when performing any welding operations.

the tools, which could lead to personal injury—or damage to the vehicle. Even though the frame section or rail may be replaced, it is still necessary to rough repair the section in order to relieve the stresses and strain in the surrounding areas. This often involves the use of force application equipment by exerting extremely high levels of pulling and pushing forces. The possibility of pulling clamps coming loose, chains breaking, and metal tearing is very high. Care must be taken to prevent personal injury and additional damage to the vehicle by securing the pull clamps with anti-recoil straps. These straps limit the distance the clamp will travel in the event the metal tears away from where the clamp is attached or if the clamp were to slip and come loose from the attaching point. The possibility of damage to the adjacent areas of the vehicle from the flying clamp is also averted.

Recoil Straps

Whenever using any corrective force one should be careful to not stand directly behind or near the line of pull. Even though the anti-recoil straps are used, the clamp will still travel some distance. Standing near or behind the pulling device will not preclude the technician from being struck by the clamp. A chain or pull strap breaking is always a possibility whenever an elevated amount of force is used. It is safer to monitor the progress of the pull by standing off to the side of the pulling clamp or strap as near to the vehicle as possible. In this way the projectile consisting of the clamp, chain, and ripped metal will be pulled away from the technician, thus reducing the possibility of being struck by the mass of material—resulting in personal injury.

Vehicle Safety

The use of any equipment with which one is not familiar always presents new problems and potential hazards. Whether driving the vehicle onto the equipment, chocking the wheels, or so on, one must become properly versed and familiar with the equipment to ensure the recommended procedures and safety principles are practiced. Raising the vehicle and securing it to the frame machine are all unique to each particular type of machine being used. The frame bench manufacturer's warnings and recommended operating procedures must be followed whenever using the equipment. Ignoring their recommendations will inevitably lead to disastrous results.

FRAME DESIGNS AND APPLICATIONS

The full frame has been in use as a support mechanism for the vehicle since the beginning of the automotive industry. The first automobile required a variation of a full frame in order to attach and hold the engine and drivetrain to the vehicle's body. Even today, the full frame is utilized in the design of many different models of vehicles that are used for many different applications. The full frame can be found on many passenger and sports cars, as well as sport utility vehicles (SUVs) and pickup and large industrial application trucks. The vehicles that utilize a separate frame and body design are commonly referred to as either body over frame (BOF) or a full frame vehicle if in fact a full frame is used.

Frame Types

Over the years numerous changes have been made to strengthen and adapt the frame to fit the needs of a variety of different vehicle designs. However, the same basic concept and function of the frame is still used today. The modern day vehicles may utilize one of the following three basic designs: ladder frame, perimeter frame, or a stub frame. See **Figure 16-1**. A modified version that incorporates concepts of both the ladder and perimeter frame has been introduced in some vehicles, resulting in somewhat of a hybrid version.

Ladder Frame. By virtue of the strength characteristics it offers, the ladder frame is most commonly utilized on trucks used for large and heavy industrial applications where increased road clearance is important. The ladder frame has several cross-members welded or riveted at strategic locations along its entire length for added strength and rigidity. The result is a higher center of gravity, making it less than desirable for domestic use. However, some SUVs and smaller pickup trucks have also used the ladder design because running boards are an option commonly used on them and easy access to the vehicle is not a perceived inconvenience.

Perimeter Frame. The perimeter frame is most commonly utilized on full frame passenger and sport cars. The perimeter frame incorporates several unique features including a drop center design which allows the passenger compartment to be lowered, making for easier entry and egress from the vehicle. A torque box is located at each corner under the passenger compartment to which offsets or kickups are attached. These kickups are used to attach the front and rear rails to the center section. They are also used to ensure for proper suspension mounting heights and—most importantly—for collision force management and energy dissipation.

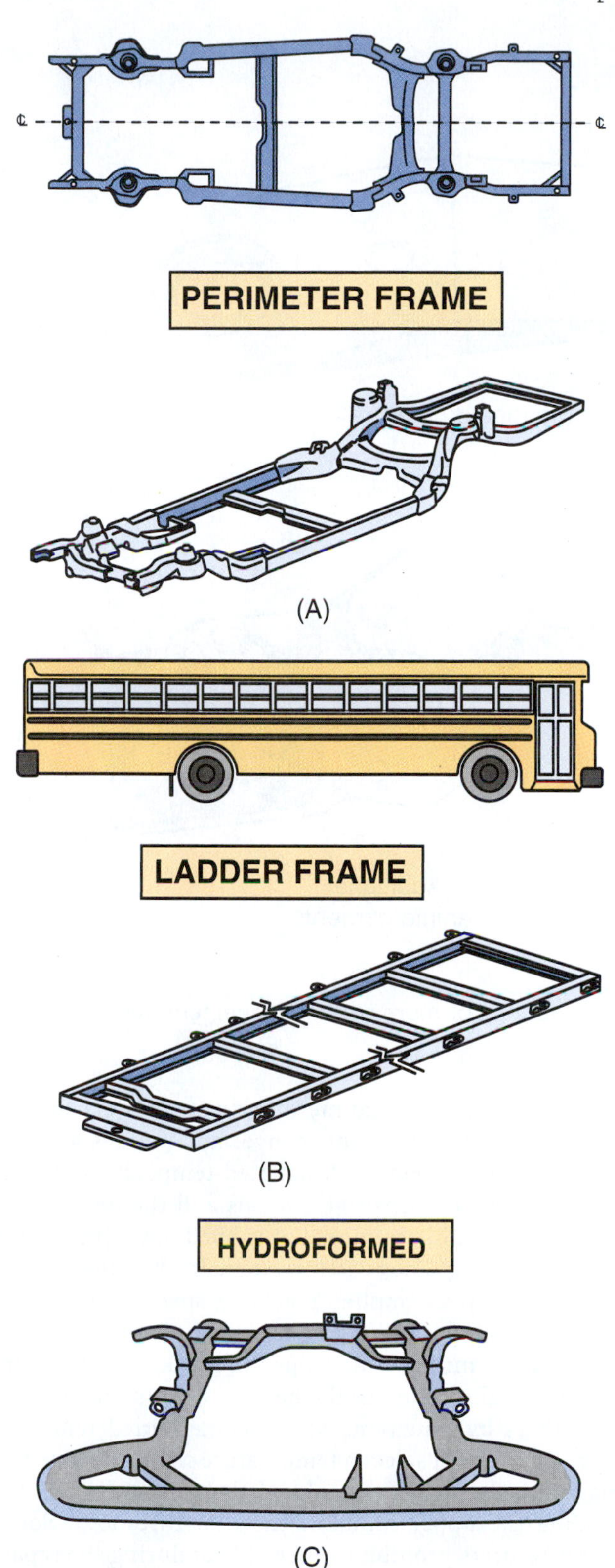

FIGURE 16-1 The perimeter (A), ladder (B), and stub or sub frame (C) are commonly used on the modern day body over frame automobiles, trucks, and SUVs.

Stub Frame. Stub frame vehicles are a combination of a unibody and BOF vehicle. The stub frame—which is usually designed as heavily as that of a full frame on a BOF vehicle—is attached to the bottom of the core support at the front and the underbody of the passenger compartment approximately one-third of the way rearward on the vehicle. The engine, transmission, and suspension parts are attached to it, much like on a full frame vehicle. It is used on select unibody designs only.

TECH TIP Even though the damaged section of the frame will be replaced, rough straightening is required to relieve the stress to the adjacent areas prior to removing it. Therefore, many of the same repair concepts that are used when repairing a frame, such as heating, stress relieving, and general repair concepts recommended by the manufacturer, must be employed as part of the pre-removal steps.

REPAIR CONSIDERATIONS AND IMPLICATIONS

Before making any attempt to repair or replace any part of the damaged frame, the technician must become familiar with the numerous design concepts, types of steel and materials used by the manufacturer, and the recommended repair procedures that must be practiced and avoided. The technician must also become keenly aware of the energy management concepts used by the manufacturer. An awareness of the designs and methods used for collision energy management and the location of each energy absorbing device is crucial to the correct repair of the damage. See **Figure 16-2**. Manufacturers may use one or a combination of any of the following methods to manage the collision forces: kickup areas, collapse zones, elongated holes and slots, incrementally reduced surfaces, tapering the rail size, and placing additional reinforcements in strategic areas. Other design characteristics may include convolutions at the outermost end of the rails and surface indentations to accommodate and control the bending and collapsing of the rail at a specific location when a predetermined amount of force is applied.

Heat Applications

An age old attitude about repairing damaged frames is that heat must be used for stress relieving and softening the metal to help make it movable. Today's technician must be keenly aware of numerous facts about the type of steel used on the modern automobile and the parameters for repairing them. Certain types of steel used by the manufacturer are extremely heat sensitive and cannot be heated, while others may be heated but to very low threshold limits—whether the frame is fabricated by conventional means such as **hot rolling**, or by **die stamping**, or if it is **hydroformed**. Some

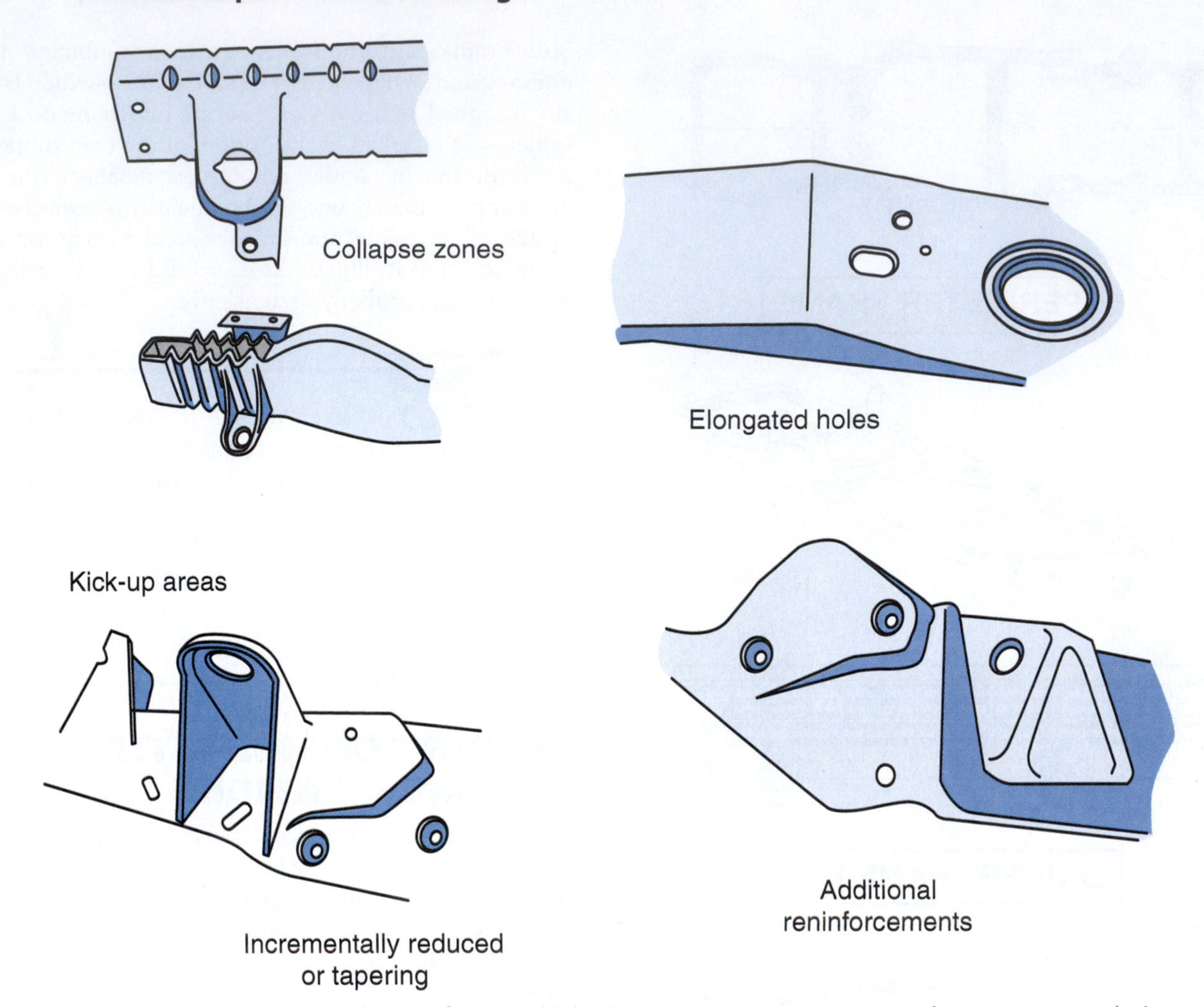

FIGURE 16-2 The frames on most body over frame vehicles incorporate one or more crash management designs to reduce the impact of the collision energy into the passenger compartment.

frames are made of mild steel while others may be made of either **high strength steel (HSS)**, **high strength low alloy (HSLA)**, or **ultra high strength steel (UHSS)**. General Motors' small and mid-size pickups are made of a combination of hydroformed mild steel and high strength steel. Dodge Ram pickup frames are made of a combination of both mild and high strength steel as well. All the frames made of high strength steel are very heat sensitive, thus limiting the temperature to which they can safely be heated without affecting their strength or integrity. Most manufacturers have very specific "safe" temperature ranges or upper temperature thresholds which must never be exceeded. These temperatures may be as low as 700°F (371°C) while others may be as high as 1,200°F (648°C). Before applying heat to any part of the frame rail, it is highly recommended that the technician consult the vehicle manufacturer's manual to avoid inadvertently causing any irreversible damage from overheating.

Monitoring Heat. To ensure that overheating does not occur, one should always use temperature-sensitive crayons, heat detection paint, heat monitoring strips, or a non-contact digital thermometer to monitor the temperature whenever heat is applied. Each of the crayons, paint, and heat monitoring strips are available in a variety of temperature ranges and are designed to melt at their respective designated temperature range. The temperature-sensitive crayons and the heat detection paint should be painted or rubbed onto the surface prior to heating. They will start to melt when the surface to which they are applied reaches a specified temperature, at which point the heat application must stop. The temperature monitoring strips are stuck onto the surface and will respond to the heat in the same manner as the others by liquefying at a predetermined temperature. One should select a temperature slightly below the maximum heating threshold of the steel to avoid exceeding the upper limits of the heat threshold. Some manufacturers prohibit the use of heat during the repair process. The technician is well advised to reference the manufacturer's recommendations or consult the I-CAR website for a comprehensive chart which lists the heating recommendations for most manufacturers before initiating any heat application.

Heating Duration. In addition to the maximum allowable temperature, most manufacturers also limit the length of time heat may be applied to a surface. The length of time that heat may be applied to any frame made of HSS, HSLA, and UHSS is cumulative. Some manufacturers specify that heat may be applied but for a maximum of two minutes cumulatively, while others may recommend a shorter period of time.

TECH TIP The cumulative time is the total length of time that elapses during the application of heat—whether it is done only once or more often.

Temperature and Duration. One must be extremely cautious to not apply heat above the specified maximum temperature or for any length of time in excess of the manufacturer's recommendation at the risk of destroying its strength and structural integrity. In a subsequent collision the improperly repaired rail may fail, resulting in serious injuries or a possible loss of life. When repairing frames made of mild steel, heat may be used to help relieve the stress while pulling, but it must be used with a considerable amount of discretion. Heating will not affect the strength or integrity of mild steel so long as it is not applied excessively. When repairing a frame regardless of whether it is made of high strength or mild steel, the goal should always be to restore it to its proper original dimensions and strength.

Composition of Frame Steel

How then does one differentiate between whether the steel used is one of the high strength classes or if it is mild steel? Once a vehicle has been put into service, there is no definitive way to identify material or composition of any frame as it becomes covered with dirt, road film, and rust. However, the surest way to determine the type of steel used is to consult the auto manufacturer's service manual. Information included in the manual will describe the material used as well as any other data concerning the recommended temperature ranges for heating along with a variety of other information pertinent to the specific vehicle. See **Figure 16-3**.

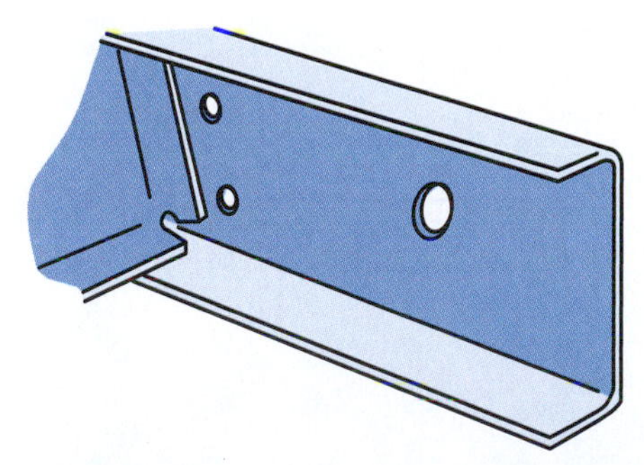

FIGURE 16-3 Most frame rails made of standard low carbon steel are U-shaped and are formed by the steel manufacturer while they are at or near the temperature of red hot.

Conventional Mild Steel. A conventional frame made with mild steel is readily recognizable as the rails are usually formed into a "U" or "C" channel shape and are rarely ever painted. One should, however, not automatically assume this to be the case. If in doubt, the safest approach is to treat all steel as though it is high strength.

HSS Frame Rails. Most rails made of high strength steel characteristically are painted and are frequently a double "C" shape welded together to form an enclosed box. See **Figure 16-4**.

Hydroformed Frame. Another design concept that has become increasingly more popular with recent production years is the use of hydroformed frames—or at least sections of the frame that are formed by that means. The Chevrolet Z06 Corvette, for example, utilizes a hydroformed aluminum frame. Chrysler Corporation's Dodge Ram pickup truck frames and front inner fender rails are also hydroformed. The hydroformed frame is commonly made of high strength steel and must be treated as such. There is no guarantee of being able to visually identify a hydroformed frame as there are no common visible characteristics that differentiate it from any other type of frame design. The only recognizable feature typically present on a hydroformed design is a solid, seamless tube that may be bent to shape with continuous, smooth, and rounded corners made to conform to the adjacent areas. See **Figure 16-5**. The only welding done on these frames is used to attach brackets and body mounting tabs to the rails. The corners or edges at the top and bottom of the rail are also more rounded as opposed to sharp bends typically found on a stamped or hot rolled rail. One, however, should not assume this to apply universally. Some manufacturers may use only a small hydroformed section that is welded to the other parts of the frame. The learned technician will consult the manufacturer's vehicle manual to determine the total design of the frame to ensure the correct repair techniques are affected.

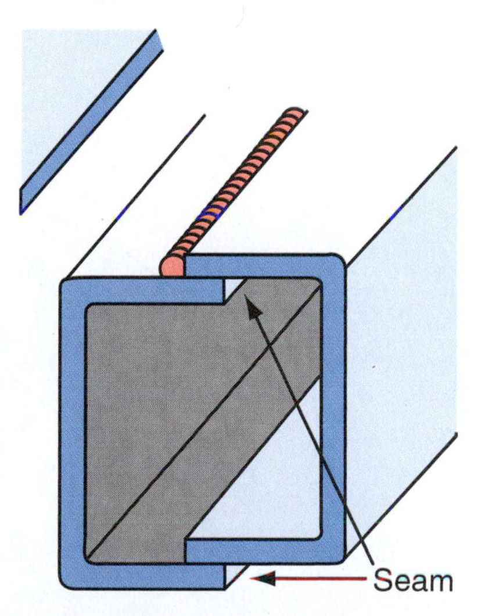

FIGURE 16-4 Many times the frame rails or sections made of high strength steel are made in the shape or form of an enclosed box and are commonly painted.

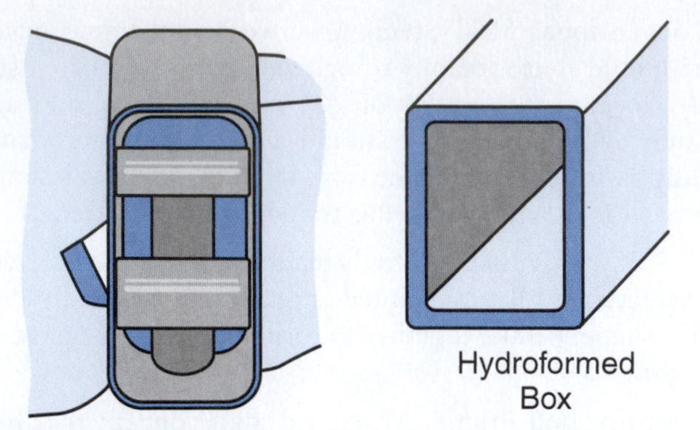

FIGURE 16-5 Hydroformed frame rails are usually tubular shaped and have more gentle rounded corners as opposed to the sharply formed ones found on most other frames.

KINK VS. BEND

Another consideration that will determine whether a frame section is repaired or replaced is the **kink vs. bend** rule or concept. A bend is defined as a change in the shape of a rail or part that is smooth, continuous, and perceived to be repairable without leaving any signs of permanent deformation. It should be repairable without the use of heat. See **Figure 16-6**. However, provided the manufacturer specifies that heat may be applied, it may be acceptable to use within the specified time and temperature guidelines. A kink is a relatively sharp bend, usually 90 degrees or more, within a short radius. It is not likely to be straightened without the use of heat. A kink usually results in tears and cracks in the area where it was bent when it is straightened. One should not attempt to weld these tears and cracks as the metal has been permanently deformed and weakened. The area will likely not be restorable to its original state or

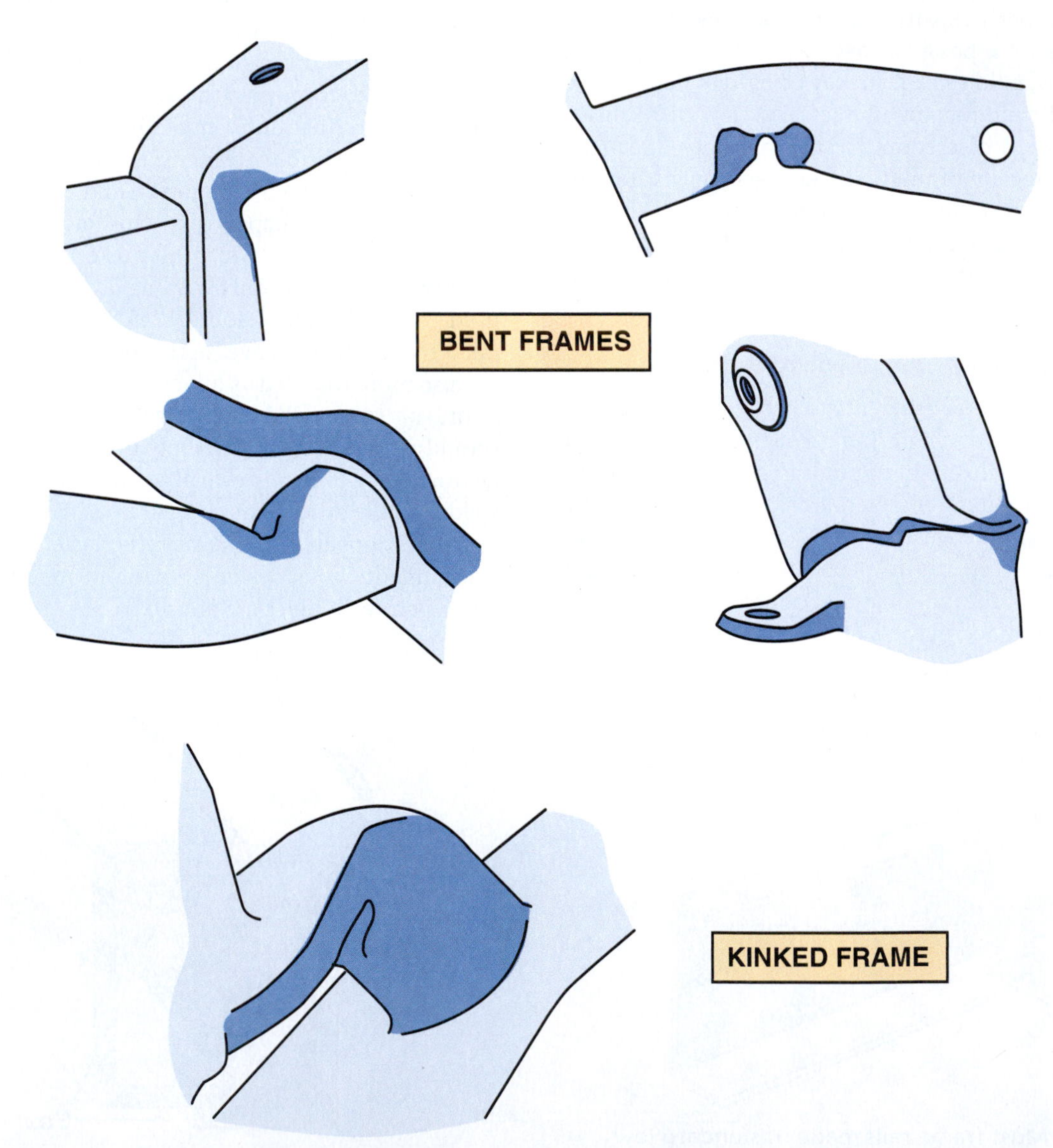

FIGURE 16-6 Most manufacturers recommend the frame section be replaced when it becomes sharply bent or kinked. These are some examples of kinked frames that cannot be repaired.

shape nor its proper dimension. Therefore, it is not capable of withstanding the forces of a subsequent collision. After roughly pulling and restoring the damage, one should remove the rail or section and replace it with a new part.

STRUCTURAL COMPONENT REPLACEMENT METHODS

The three most common methods used for replacing damaged frames and their component parts include a complete frame replacement, a sectional replacement, and a component replacement. Each has certain advantages and disadvantages.

Vehicle Preparation

The full frame replacement is perhaps the least common method of repairing damage because it involves the removal and transfer of the engine, suspension, drivetrain, mechanical components, and the body from the damaged frame to the new replacement. While this may seem like an extremely labor intensive operation, realistically it is less involved than one might think as it mostly involves unbolting a series of components and transferring them to the new frame. Transferring the shell of the body is no doubt the most delicate part of the total operation. With some serious analysis and preparation, this can be accomplished with less effort than one might think.

Removing Damaged Components

Whether replacing the entire frame, a section, or just a single component, a series of operations must be performed prior to the removal of the damaged part. The first thing that must be done is to remove the damaged exterior body parts and trim to expose the area that is to be repaired. All fluids must be drained from the radiator and other reservoirs located in the repair area, and the parts removed and placed in a protected storage area if they are to be reinstalled. The damaged parts that are to be replaced should also be stored because many times mounting brackets and other parts must be transferred to the replacement part prior to installation. The air conditioning refrigerant must be recovered and the lines capped off to prevent them from becoming contaminated and collecting excessive moisture and dust. Any of the fuel and vapor recovery lines, brake lines, and any other items that may become damaged during the repair should be removed from their mounting brackets so they can be moved out of the way. Most of these steps are performed prior to moving the vehicle onto the frame rack.

TECH TIP The single most important anchoring consideration is this: The area of the vehicle used for securing must be able to withstand the total of the cumulative corrective forces applied.

Roughing Damaged Frame

One of the most important parts of the overall repair is to rough repair the frame prior to removing any part or section of it. To the uninformed individual this may seem like a total waste of time and energy. However, this must be done to relieve the stresses that resulted from the twisting throughout the frame and structural members of the vehicle during the collision. Removing a section of—or cutting into—the damaged rail without first taking the necessary stress relieving steps will likely result in the entire frame unwinding like a loose spring. Regaining the dimensional trueness will be extremely difficult.

Anchoring

In order to rough repair the vehicle it must be properly anchored to the bench or frame rack used for pulling it. In some cases, securing the vehicle to the bench or frame rack must be done exclusively within the frame straightening equipment manufacturer's specifications. In other cases it is accomplished with a combination of chains, clamps, pulling straps, bracing apparatuses, and other hydraulic assisted jacks. During this phase of the repair operation the use of multiple pulls moving in several directions simultaneously are commonly used to reverse the flow of damage created in the collision. Stress relieving by massaging or hammering on the rails in the immediate damaged areas and on the adjacent areas while they are under tension is also necessary. The combination of pulling the rails multi-directionally and stress relieving will help restore the metal to its original shape and state, as discussed in a previous chapter. Whatever means is used for rough restoring rails, the area of the vehicle used to anchor it to the rack must be strong enough to support the corrective forces applied.

Certain securing systems may be required for specific models of vehicles as some automobile manufacturers are very direct about the location and type of clamping system that must be used on their vehicles. Improperly anchoring a vehicle may have some devastating results when the corrective forces are applied; for example, a tear may occur in the metal of the undercarriage or rail if it is anchored in an area not constructed heavily enough to withstand the corrective forces. One can easily create more damage to the vehicle than what may already exist if the correct approach is not used.

After the repairs have all been completed, one last very important step must be taken to ensure no tell-tale signs of a repair are left behind. The teeth used to keep the tie-down clamp from slipping when the corrective forces are applied bite into the metal, often leaving a very obvious disfigured surface. Care must be taken to smooth these areas out, restore the corrosion protection, and finish them just as is done on any other part of the vehicle—even though it is on the underside.

FULL FRAME REPLACEMENT

The increased value of many of the high-end vehicles makes it a more feasible repair alternative than has been the case in recent years. Many insurance companies consider the repairing versus totaling threshold at approximately 70 percent of the vehicle's current retail value. Suffice to say a vehicle with a market value of $50,000 is an excellent candidate for replacing the entire frame in order to keep from totaling it out. One might wonder what circumstances would require replacing the entire frame. Some manufacturers do not allow welding on certain sections of the frame, or the frame may be severely kinked in areas where it cannot be reached to repair it and the manufacturer prohibits sectioning at that specific location. Another set of circumstances that would require replacing an entire frame is that cross-members are frequently not serviced individually; if recycled parts are not available, the only option is for a full frame replacement. Another thing that may cause one to consider replacing the entire frame is damage in multiple locations throughout the structure where dimensional restoration is simply not practical or feasible.

Transferring Mounts and Brackets

When this repair option is exercised, new parts are more apt to be used than recycled ones provided they are available from the manufacturer. A number of factors must be taken into consideration even though the frame is an OEM part. Most generally the frames available to the repair aftermarket are for the **baseline/standard model vehicle** only. Therefore, some of the mounts may have to be transferred from the old frame to the new one because the engine size and transmission type are different or some of the optional equipment on the damaged vehicle is not part of the standard models. Some of the damaged parts on the existing frame may not be part of the new one; therefore, they must either be repaired and transferred or obtained from an auto recycler. Sometimes the manufacturer will make a mid-year production modification to the frame which will require some additional retrofitting and modification in order to properly match it to the body or vice versa. These considerations help determine whether to use new or recycled parts.

Using Recycled Parts Considerations

The use of recycled replacement parts offer yet another whole new set of problems to overcome that may result in a less than desirable outcome. For example, the vehicle from which the recycled frame is used may have been involved in a collision, making the frame less than dimensionally perfect. Also, one does not know whether the repairs made by an unknown shop were performed correctly.

Previous Repairs. Any incorrect repairs made by the unknown repair facility are likely to become the technician's responsibility if they fail in the future. The vehicle may have had a different engine, transmission, and drivetrain; therefore, some of the necessary mounting brackets may not be attached or may have to be moved. The recycled frame may have been one that underwent the mid-production year change, thus altering it in some manner.

Corrosion. Excessive corrosion may be yet another issue with which the shop, insurance company, and the vehicle owner must be concerned.

Previous Modifications. One other important consideration is modifications that may have been made to the replacement part—such as adding reinforcements for attaching a snow plow or other equipment such as a trailer hitch or other options. One must determine whether the replacement frame is adequately sturdy or structurally sound enough to return the vehicle to the same function level it once had, or if it was intended for use as a lighter duty vehicle. Whether the replacement frame is OEM equipment or recycled, one must ensure that it is the same model and design. This is best determined by placing the two frames side by side and taking numerous comparative measurements. Any significant dimensional differences should be cause for rejecting the part.

SECTIONAL REPLACEMENT CONSIDERATIONS

A sectional replacement is perhaps the most common method used because an area as little as a front rail end cap may be replaced up to a large section of a rail or even the entire front or rear section of the frame.

Disassembly Minimized

A sectional replacement usually limits the amount of disassembly required of the vehicle to only the area in which the repair is to be made. The use of recycled parts is a common practice as parts are generally abundantly available, and obtaining recycled parts from an identically matched or similar donor vehicle is usually not a problem.

Removal Methods

Another very important factor one must remember is the recycler frequently uses a flame cutting torch to remove the parts from the donor vehicle. When ordering the part one should request the replacement part be cut at least 6 inches longer than what is actually needed. This will allow for enough excess to remove the heat affected zone caused by the heat created for removal.

OEM Replacement

In recent years most manufacturers have begun to service partial frame sections for those areas that are commonly damaged in a collision. General Motors, for example, offers three different frame sections for the Corvette as well as sections for their ST model SUVs and the CK 1500 model truck. Likewise, Ford Motor Company services similar parts for the Expedition and the Lincoln Navigator while DaimlerChrysler also offers three sections for the Dodge Ram Pickup. When available, the use of OEM replacement parts may be advisable and certainly more desirable as opposed to a recycled part. The OEM part typically includes most of the mounting brackets for the specific vehicle being repaired. More importantly, one can be assured that it will be dimensionally accurate.

MODULE SECTION REPLACEMENT

Module replacement of frame sections has become increasingly more prevalent as the auto manufacturers have begun to offer **frame replacement modules**. A module typically consists of a section of the frame that includes both side rails and the respective cross-member(s) that are normally attached at either end of the frame. The module is designed to replace a section of the frame that, by design, may collapse during a collision. These modules are designed in such a way that the damaged part is cut out at specified locations and the replacement part is welded into place. The seam for the replacement part may be in front of the collapse zone on the rear of the vehicle and behind the collapse zone on the front of the vehicle. By replacing these modules, the damaged collapse zone is replaced, thus restoring the manufacturer's original collision management system along with the manufacturer's recommended reinforcements.

COMPONENT REPLACEMENT

Many times it becomes necessary to replace only certain components of the frame.

Collapsible End Caps

Most vehicle manufacturers offer a variety of components such as collapsible frame end caps and bumper mounting brackets, core support and suspension mounting brackets, and in some instances the frame rail ends. Ford Motor Company offers the **convoluted front frame rail** end for the Windstar minivan—as does DaimlerChrysler, who services the frame tip for the Ram pickup truck.

Sacrificial Parts

Many times these are sacrificial parts that are designed to collapse in a collision as both an energy management device and also to protect the rails in front or behind them from damage. Rail end caps are not intended to be straightened or repaired but replaced instead. Straightening them will result in altering their structural makeup, which could possibly cause them to not react correctly in the event of a subsequent collision.

ATTACHING CRITERIA

The frame replacement components may be welded, riveted, or bolted into place by the manufacturer. If they have been riveted the replacement part is usually welded or bolted as riveting is not a common practice in the industry and most shops are not set up to perform such tasks. When bolting the replacement part into place, a specific bolt hardness is commonly recommended by the manufacturer. When the parts have been welded by the manufacturer, the replacement should be welded using a GMAW MIG welder that is capable of welding heavier gauge steel.

FRAME SECTIONING CRITERIA

When replacing a section of the frame, whether it be an end cap, a module, or an OEM or recycled part, certain rules and guidelines must be followed regarding the area where the joint or seam is to be made and areas that must be avoided. Vehicle-specific guidelines for frame sectioning are available from several sources.

Information Sources

These information sources include the manufacturer's body repair and service manuals, the vehicle maker's compilation manuals, and other vehicle-specific procedure bulletins that are frequently included with the OEM replacement parts. Tech-Cor is yet another source for vehicle-specific and generic information concerning the required procedures.

I-CAR GENERAL SECTIONING GUIDELINES

The I-CAR General Sectioning Guidelines—which are a series of generically applicable procedures—have been developed and made available by I-CAR. These guidelines are not vehicle specific, but the procedures have been adequately researched and tested to make them readily usable for nearly any vehicle so long as the manufacturer does not prohibit sectioning or the design of the damaged area lends itself to being repaired in this manner. These guidelines are specific to the point of advising the repair technician of any areas to avoid when making a cut or installing a new part. More accurately, they are more of a list of what not to do and what areas to avoid. The following is a brief and not all inclusive summary of the areas to avoid when selecting a location for frame parts replacement:

- Areas that should be avoided for making a seam include engine, suspension, and drivetrain mounting locations and brackets. The dimensional tolerances and the strength and rigidity required in these areas do not allow for any cutting and welding in or near them.

- One should also avoid cutting through any hole that is over 3 mm (1/8 inch) in diameter as it may be a measuring reference location.
- All crush zone areas are to be avoided as welding and installation of inserts will alter their strength characteristics, causing them to collapse incorrectly. The deployment timing of the air bags in the front or side of the vehicle may be affected as well.
- Surfaces with compound shapes, convolutions, and areas where the width and height diminish or become a tapered shape should be avoided as matching the replacement part and properly reinforcing it will be difficult.
- The area selected for the seam should be readily accessible from all sides for welding both the inserts and the replacement part.
- The area should also be relatively uniform in shape for ease of installing the insert required for reinforcing the joint area.
- The area must be accessible to install and restore the required corrosion protection.

TECH TIP One of the most critical steps that must be taken during the replacement of any frame rail section is to take and accurately record the measurements so there is no guessing about them when re-assembling the vehicle. These measurements may be taken from the undamaged section on the opposite side of the vehicle or they may be obtained from undercarriage dimension specification manuals—or even from similar vehicles.

SECTIONAL REPLACEMENT JOINTS

When performing a sectional replacement, several options are available for joint design, some of which are mandated by the manufacturer while others are preferred procedures developed by Tech-Cor and I-CAR. Although some manufacturers prohibit their use for certain applications, the butt joint with insert is the most common joint design when using recycled parts. As discussed previously, the sectioning joint should be made in an area that meets the criteria specified by the general sectioning guidelines to ensure the restoration of the vehicle's structural integrity. When an OEM replacement part, section or module is used, the manufacturer may offer specific guidelines to follow regarding fitting the parts to the vehicle and the preferred joint design.

TECH TIP Unless specified otherwise, all welding operations performed on any frame repair or replacement are to be done using the gas metal arc welding (GMAW) metallic inert gas (MIG)—a.k.a. **GMAW MIG welding** process. Occasionally the manufacturer will require the shielded metal arc welding **(SMAW) MIG** or **flux core welding** process be used.

Inserts

To make the insert, one may use a section of the damaged rail that has been cut off or a piece of the excess material that was removed from the replacement part. See **Figure 16-7**. If the section is an enclosed box design it will be necessary to cut the insert on all four sides to reduce the dimensions for fitting it inside the rails. It will then need to be welded together once again. See **Figure 16-8**. Another option is for the technician to fabricate the insert from material that is of comparable thickness. However, whenever possible, the insert should be made from either the original or replacement part to ensure the same strength quality of the steel is used.

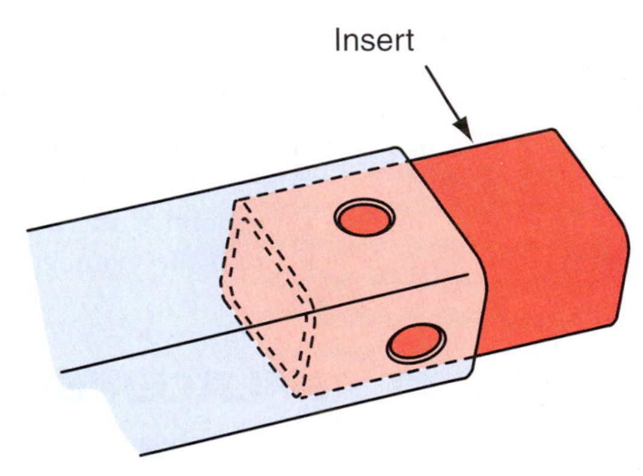

FIGURE 16-7 One of the most popular options for performing a sectional replacement is to use a butt joint with insert.

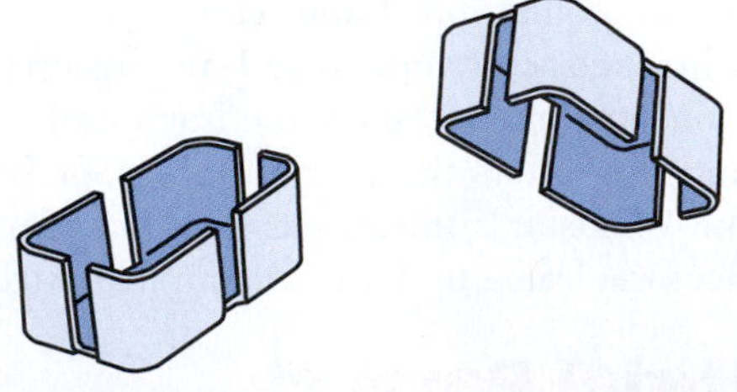

FIGURE 16-8 An insert can be fabricated from a piece of either the replacement or damaged rail. The section used for the insert must be cut into 4 pieces and rewelded so it will fit into the cavity of the frame rail being repaired.

Installing Insert

Once the insert has been fabricated, holes should be drilled into the existing and replacement rails approximately 12 mm (1/2 inch) from the end at approximately 25 mm (1 inch) intervals to hold the insert in place while fitting the two parts together. They will also become part of the welds required to hold the assembly into place permanently. See **Figure 16-9**. The insert should be approximately 50 to 75 mm (2 to 3 inches) in length and fitted as snugly as possible to the inside of the box, then welded through the plug holes made. The replacement rail section should then be fitted over the insert and the measurements should be taken and checked to ensure the dimensional accuracy of the rail has been restored.

A **root gap**—which should be the equivalent of approximately two thicknesses of the metal—should be left between the two mating edges to ensure adequate penetration when the two pieces are welded together. Once the correct dimensions have been determined, the part should be clamped into place, tack welded at the seams, and plug holes made in the replacement part. Once the tack and plug welds have been accomplished, the measurements should be rechecked prior to finish welding a solid bead around the rail. Depending on the thickness of the rail, it may be necessary to weld the surfaces together by making two passes. See **Figure 16-10**. Figure 16-10 is an example of a preformed insert which is formed into this crush cap assembly for a Chevrolet pickup and SUV. See **Figure 16-11**.

TECH TIP The purpose for the insert is to help control the heat and to ensure that proper fusion occurs between the two edges when welding the rail together. If the backing is not fitted snugly against the back side of the rails, the welding heat may burn the edges away, leaving a weak joint or seam. In order to prevent the burning back, it will be necessary to reduce the heat, again resulting in a weakened joint.

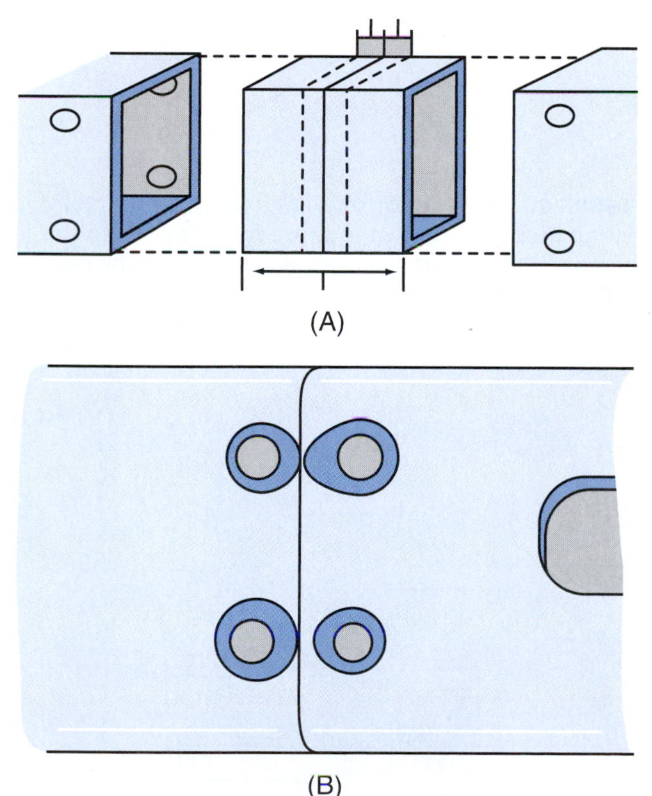

FIGURE 16-9 (A) After the insert has been re-welded together, the excess weldment must be removed, in order to insert it into the rail cavity. It should then be plug welded on one side to ensure it doesn't move when fitting the other section of the rail over it. (B) When the rails have been aligned to the proper dimensions, the remaining plug welds should be completed and the seam should be welded with a solid seam weld using the skip welding technique to avoid overheating the rail.

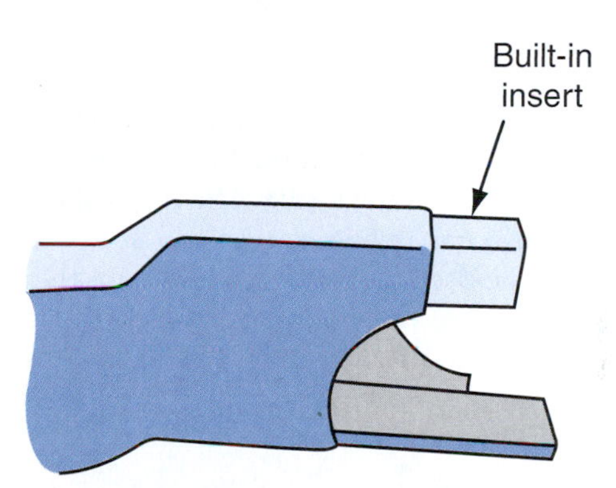

FIGURE 16-10 Some manufacturers' frame rail ends are designed so preformed inserts can be used for replacement parts. This requires cutting and removing the damaged part, inserting the new one, and welding it into place.

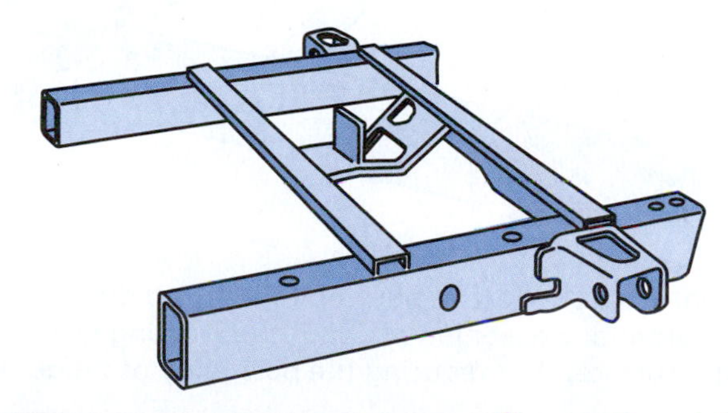

FIGURE 16-11 Some manufacturers offer a rail module or section as a replacement part, making it a great deal more economically feasible to repair the vehicle because purchasing the entire frame is not required.

Offset Butt

An offset butt joint is sometimes recommended by the manufacturer when welding a replacement rail section into place. This is commonly used to distribute any potential weakening effect resulting from the heat affect zone over a larger area, thus reducing the risk of the metal fatiguing, breaking, or tearing in the event of a subsequent collision. See **Figure 16-12**. An example of this is the recommended joint design for replacing the rear module H-section of a Dodge Ram pickup. The mating surface or edge on the manufacturer's replacement part is cut at an angle across the **frame web/fence** to ensure that an open offset butt joint is used when installing the replacement part. The placement of the seam is staggered on the other side of the rail so the seams are not directly in line or across from each other. The manufacturer further recommends that the joint not be reinforced by **fish plating** the back side.

Offset Fillet or Lap Joint

An offset fillet or lap joint may be used for several applications. One may be where a box rail is slid inside another and welded at the opening edge of the receiving rail. The seam may be welded with a solid bead or it may be skip welded according to the manufacturer's recommendations. Another example may be where one layer is placed over the top of another or overlapping the edges and both edges are welded, one from the front or outside and one from the back side or the inside. The edges of the entire overlapped surface may be seam welded or a skip weld technique may be used. See **Figure 16-13**. Another example is the area where the front bumper mounting brackets are installed onto the frame rails. They are slid down over the outside of the rail and are welded solid along the seams. See **Figure 16-14**.

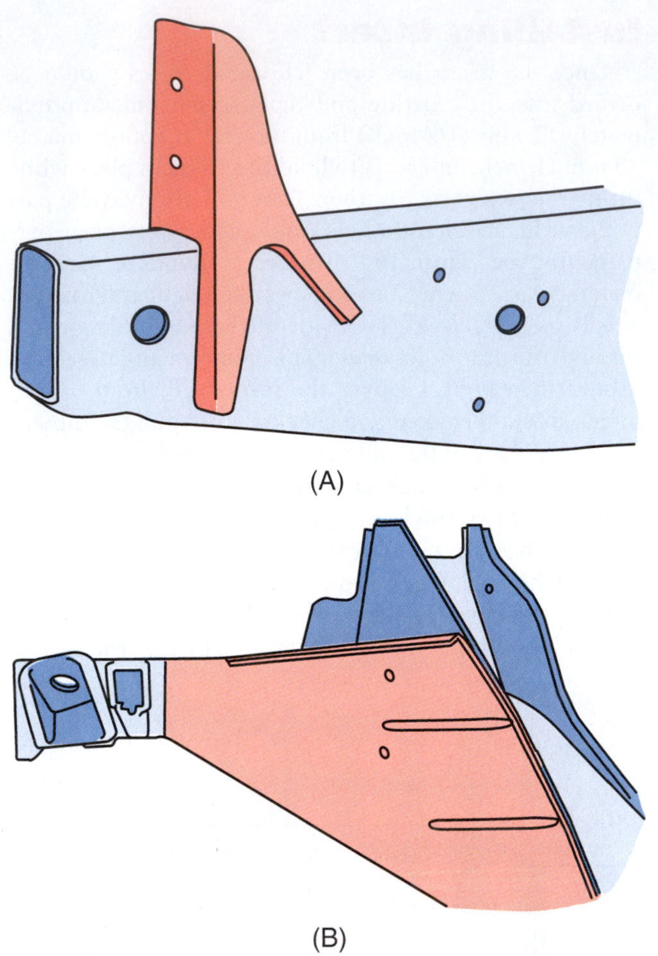

FIGURE 16-13 (A) An offset fillet joint is another joint design used to secure various pre-formed parts to the frame rails. This part simply slips into or over the rail and is welded at the specified location. (B) Offset fillet weld showing an overlapping seam on a Corvette rail replacement.

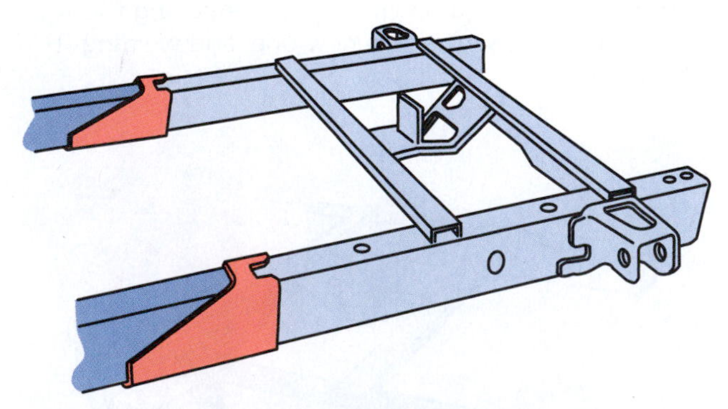

FIGURE 16-12 An offset butt joint is used to distribute the potential weakening effect of the welding over a larger surface, thus reducing the possibility of fatiguing.

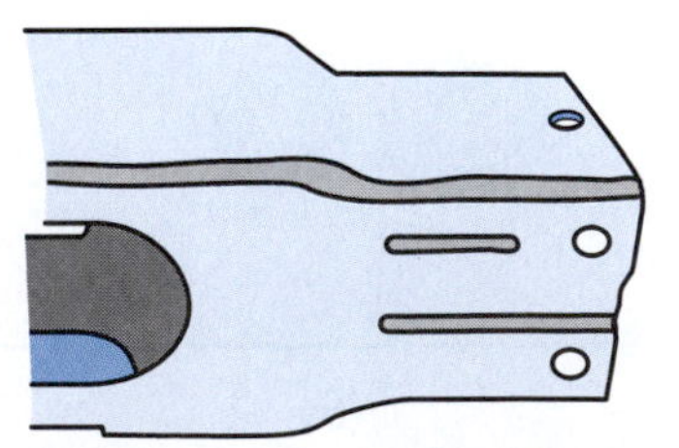

FIGURE 16-14 The tapered butt joint is another joint design sometimes used for making a sectional replacement near a collapse zone or to install the bumper mounting end seen here.

STRESS AND FATIGUE CRACKS

Sometimes frame repairs must be made that are not the result of an accidental or crash situation. Occasionally a frame will start to crack from normal wear and tear and fatiguing. It may also start to show signs of weakening in previously repaired areas because it wasn't welded properly or from unnecessary stresses caused by incorrect heating and repairs. The technician must be knowledgeable about the types of damage that can be repaired and which circumstances will require the replacement of an affected part. One must also recognize that not all parts of the frame can safely be welded and repaired. It may be necessary to replace the entire frame—or at least a section of it—to correctly fix the problem. The issue then becomes one of repair or replace. Where does one go to determine whether the manufacturer prohibits repairs or if they will allow repairing it? If fixing is allowable, how must one proceed? This is all information that must be determined before making a final decision.

Stress cracks are not a common occurrence on a frame rail but they do occur from time to time. When this happens one must determine if the crack is the result of fatiguing and what is causing the weakening effect on the metal—or if it is the result of an improper or incomplete previous repair. An improper or incomplete previous repair will usually result in cracks forming in a single area. However, stress cracking that results from broken welds or a missing fastener will result in a spider web appearance in more than one area. When one set of cracks is found, one must become a detective to locate any other areas showing signs of cracking and not only repair them—provided they are in a repairable area—but also determine the cause and repair or replace that part as well.

One cannot assume that cracks in a frame can automatically be welded. Even if they can be welded, the equipment normally used in the shop may not be suitable for the repair. Some manufacturers prohibit the use of heat on any part of their frame. This includes heat generated by welding. In this instance, the entire frame—or at least a section of it—must be replaced. DaimlerChrysler specifies the use of the SMAW MIG or flux core welding process to weld on their Dodge Ram truck frames. This is a departure from the norm with the type of welding equipment and electrode commonly used for making repairs.

Allowable Crack Repair Locations

The location of the cracks on the frame rail will also determine whether they can be repaired. According to the general procedure guidelines, cracks in the frame fence or web can be repaired, but they should not be repaired if they are located in the **frame flange** area. See **Figure 16-15**. Before attempting to weld a crack in the rail two or three things must be done. The surface should be cleaned sufficiently to be able to see the exact location and direction of the crack(s) and clearly mark them with chalk or soap stone. A 3 to 6 mm (1/8 to 1/4 inch) diameter hole should be drilled at the end of the crack to ensure the location of where it stops. The crack should then be V-grooved or beveled to approximately 50 percent of the metal's thickness. The crack can then be welded using the equipment specified by the manufacturer. Reinforcing the back side or fish plating may be advisable if it is allowed by the manufacturer.

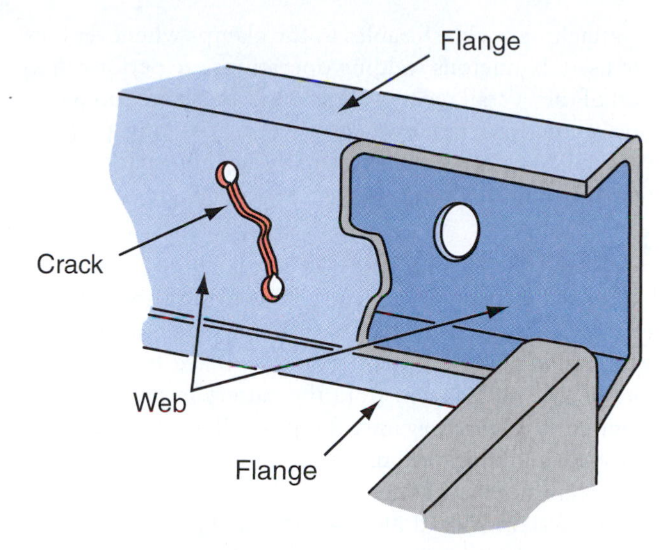

FIGURE 16-15 Cracks can be repaired on the web of the frame rail but are not recommended on the flange. Manufacturer's recommendations must be followed to ensure additional damage or weakening will not occur.

MEASURE FOR ACCURACY

Regardless of the simplicity or complexity of the operations being performed, certain protocols must be adhered to at all times to ensure an accurate and lasting repair. Precise measurements must be made, recorded, and referenced routinely any time a part, component, or assembly is removed as part of the overall repair sequence. The cliché "measure twice and cut once" is never more applicable than it is in this type of repair. In addition to referencing and using commercially produced specifications, it may also be advisable to record additional dimensions that may not be included in the prepared specifications. Many times, that one dimension that would be handy to have or be able to use is not available. One can never have too much data to use when reassembling the vehicle.

PROTECTING THE VEHICLE

Care must be taken to protect the vehicle from further damage that may occur during the repairs.

Pulling Clamps

Pulling clamps may recoil if they tear loose while performing various pulling operations. The necessary precautions should be taken to prevent this from happening by attaching backlash cables to the clamps whenever they are used. Numerous welding operations are performed as part of the overall repair. All the glass and other soft moldings should be removed or—if not possible—properly covered to protect them from flying hot sparks when grinding and welding.

Electrical System

Several steps must be taken to safeguard the vehicle's electrical system. The first precautionary step that must be taken is to ensure the battery cables have been disconnected and pulled away from the battery to prevent them from creeping back against the posts. They have a memory and sometimes will move back into a similar position they were in when bolted to the battery. The passive restraint system should also be disarmed according to the manufacturer's recommendations.

Electronic Control Module

One must also be cognizant of the location of the **electronic control module (ECM)** and take the necessary measures to protect it when welding on the vehicle. It may be advisable to remove it from the vehicle prior to initiating any repairs. If it is not removed, care must be exercised to keep the welding cables at least 300 mm (12 inches) away from it when welding. Never allow the welder cables to be draped across the ECM when welding. This is a very expensive part to replace, and an arc spike from the welder is all that is necessary to destroy it. When welding in areas where electrical wires are knowingly located, the welder cables should be run perpendicular to them if possible.

PRACTICE WELDING FOR PERFECTION

Performing structurally sound welds consistently is the single most important factor to ensuring longevity of the repair. Regardless of how proficient the technician may be at welding, not a single bead should be attempted on the vehicle without first practicing the same weld on scrap pieces. These pieces may be material or damaged parts that were removed from the vehicle or from the excess materials cut off the replacement part. A single weld that may not have penetrated sufficiently may be the starting point for the repair to eventually fail and conceivably cause an accident. To ensure this does not happen, the welding machine should be adjusted and practice beads made on like materials and in the same welding positions that will be used when installing the parts. The practicing should all be done off the vehicle.

Welding Equipment Needs

The welding equipment must also be capable of welding at 125 to 130 amps consistently. Even though the popular 110 to 115V welders are capable of welding at this level, their duty cycle is not adequate for welding at that level. They typically lack an adequate duty cycle to weld at that level for a long enough period of time. The welding electrode must also meet the tensile strength standard and the alloy structure prescribed by the manufacturer.

Repair Objectives

Many times the repair decisions made are based on very subjective criteria made available from a variety of sources including the manufacturer, I-CAR, and Tech-Cor, as well as the technician's past experiences. The manner in which the information is utilized will have a significant effect on the outcome and the longevity of the repairs. Ultimately the principal objective of the repairs rendered is to restore the vehicle to the same level of road worthiness it offered prior to the collision. Every step and operation performed by the repair technician in the total scheme of the repairs should be devoted to that end. No part of any repair should be made with an expected outcome of anything less than perfection. Only when these principles are practiced can we assume the vehicle's road worthiness will have been restored.

ALUMINUM

For many years the principal users of aluminum for manufacturing automobiles have been those located in the European Continent. Only on rare occasions did the collision repair shops have to concern themselves with repairing these vehicles because they were usually sublet to shops specializing in these repairs. However, in recent years the U.S. automobile manufacturers have shown an increased interest in aluminum because it offers weight reduction features. This has created a whole new frontier for the collision repair industry as a whole new technology has begun to emerge. Because of this, the age old techniques used for repairing steel are no longer usable for repairing the aluminum vehicles.

Shop Certification

To a large degree, the collision repair industry is not yet ready to meet the challenges put before it. Much of the repair work is thought to be so critical that the manufacturers producing these vehicles have stipulated very specific mandates for the shops who perform structural repairs on their collision damaged vehicles. Most of these manufacturers mandate that the facility be fully certified, which includes extensive technician training, specialized measuring and pulling equipment, and a facility dedicated exclusively to repairing aluminum. As part of the certification process, technicians must be fully trained on each of the specific vehicles they are attempting to repair. All hand and power tools must be dedicated exclusively for aluminum use as well. Currently, only a very elite group of shops—which are usually located in larger metropolitan areas and specialize in this area—have the credentials required to per-

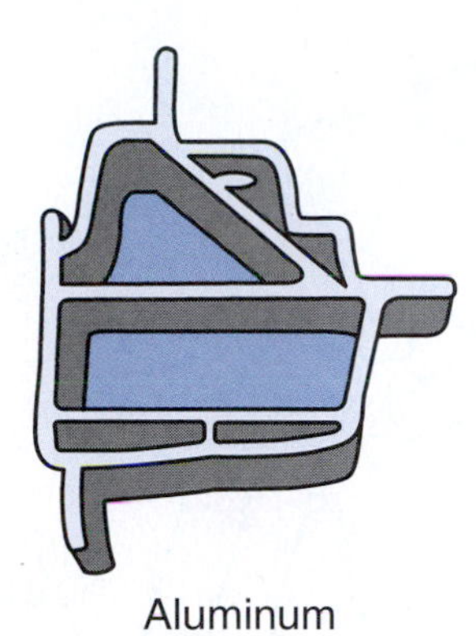

FIGURE 16-16 Extruded aluminum parts are nearly always replaced when they become damaged because their insides are made up of numerous reinforcing cells making repairs impossible.

form these repairs. However, as the use of aluminum for automobile manufacture becomes more widespread, the collision repair industry will be compelled to make the necessary adjustments.

Aluminum Repairs vs. Steel

There are numerous differences in how aluminum must be treated during damage repairs. Until these differences are understood and the techniques mastered, one should not attempt to repair a damaged aluminum-intensive vehicle. Many of the rules to determine if a part is to be repaired or replaced are no longer applicable when working with aluminum. Unlike steel, aluminum has a very limited flexing or motion life. Steel can be bent and straightened several times without it showing signs of fatiguing or breaking down. It has a "memory" of its previous shape and, with some effort, it can be restored to that same configuration. Aluminum, on the other hand, cannot withstand more than two or perhaps three bending motions that change its shape before it becomes hardened to the point where it will crack when continued bending occurs. Therefore, a whole new set of guidelines and procedures must be adhered to when attempting to repair it. When they become damaged, structural aluminum parts are more apt to be replaced than not. See **Figure 16-16**. The shape or configuration of the part is a significant contributing factor to this because they are frequently multi-layered and many times have one or more internal cells such as those found in extruded parts.

FORMS OF ALUMINUM PARTS

Aluminum parts used on automobiles are in the form of sheet metal, extrusions, and castings.

Sheet Metal Parts

Sheet metal is commonly used for the production of the outer cosmetic panels. Unless one were to lift it or perform a magnetic test on the panel, one could not differentiate sheet metal from a hood made of steel or a composite material. The underside of the hood, on the other hand, is usually another complete layer of metal with numerous die-stamped ridges and indentations that span across the entire length and width of it. Even with this added layer of material, the hood is considerably lighter in weight than one made of steel. See **Figure 16-17**.

FIGURE 16-17 The underside of an aluminum hood frequently includes another layer of metal stamped with reinforcements and spans the entire back side.

Extruded

Extrusions or **extruded parts** and assemblies are commonly used for structural members such as frame rail ends. These parts are frequently bolted or riveted into place to accommodate both their installation and replacement in the event of a collision. They may also be used for outer rocker panels, sill members, and a variety of other applications. They are formed by pressing or forcing heated aluminum through a die and then cut to exacting lengths to ensure their fit for proper installation. See **Figure 16-18**. Since the extruded parts are formed in this manner they are a continuous and seamless tube or rail, frequently having multiple layers or inner cells for strength and rigidity. They are also relatively uniform in thickness. Parts that are used for structural purposes are usually approximately 50 percent thicker than those made of steel.

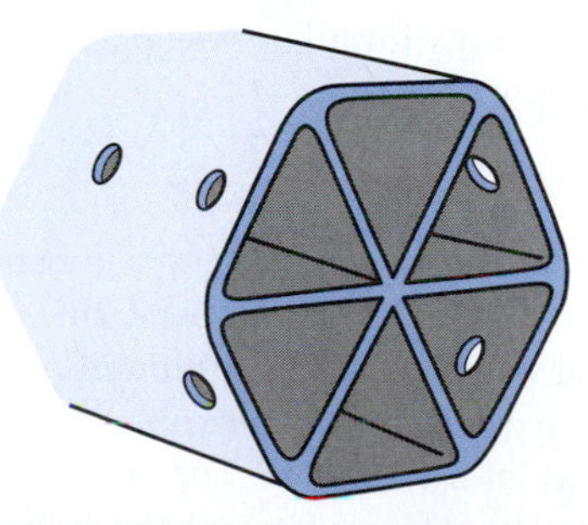

FIGURE 16-18 Extruded aluminum parts are usually a continuous seamless tube or rail with multiple cells throughout the center section for added strength and rigidity.

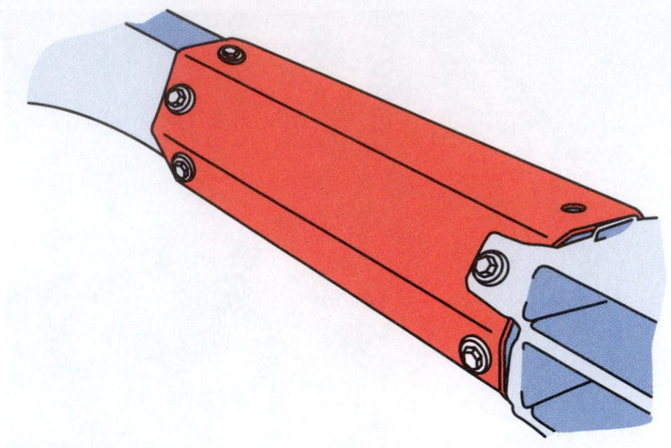

FIGURE 16-19 Some aluminum parts are made by casting, which enables the manufacturer to vary their thickness—allowing a bolt boss to be incorporated into them. These items are used to fasten extruded parts to them with bolts, as shown.

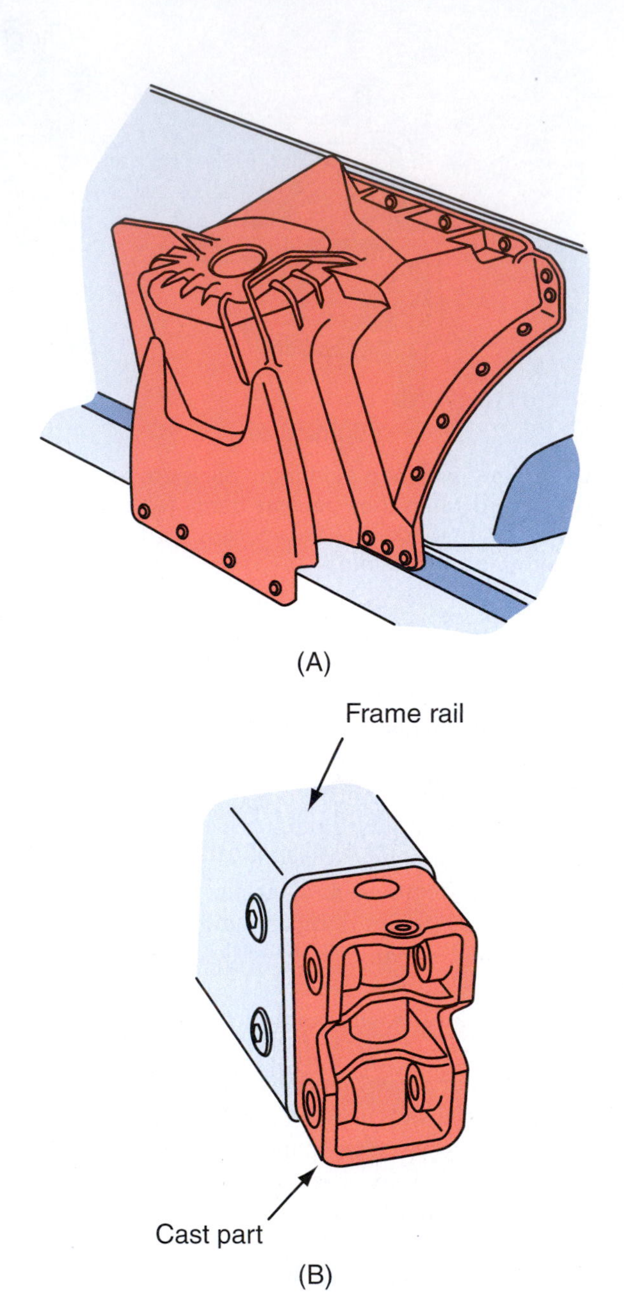

FIGURE 16-20 (A) Some parts that are normally welded together on steel bodied vehicles such as this cast strut tower are bolted or riveted to the rails on an aluminum-intensive vehicle. (B) Some of the parts may be short nodes, such as these that are used as adapters and reducers between two different size parts.

Cast

Cast parts are formed by pouring molten aluminum into molds and allowing it to cool. Extruded parts are sometimes used as an intermediate part to attach the front bumper mounting hardware to the front or rear frame rail, for example. They are also used in areas where another panel or structural member is bolted to the vehicle, as well as for mounting brackets. See **Figure 16-19**. The casting of the part makes it possible to vary its wall thickness, thus allowing for a **bolt boss** to be incorporated into it. Machining, drilling, and threading operations are commonly performed on these parts after they have been formed to accommodate bolting attaching parts to them. See **Figure 16-20**.

DEVELOPING A WORK PLAN

A whole new work plan must be developed and many different approaches must be used when attempting to repair aluminum compared to steel. Rarely is one able to use the same repair techniques on aluminum as those used on steel. The aluminum responds very differently and becomes hardened to the point of cracking when it is bent or damaged. Because it lacks many of the workable characteristics commonly found in steel parts, they must be replaced. See **Figure 16-21**.

Repair vs. Replace

The common practice of replacing or repairing a damaged steel part based on the kink vs. bend rule does not apply when working with structural aluminum parts. One can no longer assume a panel or structure can be repaired by the outside appearance of the damage. The visual indicators of the damage severity used for repairs on steel may be overruled by the manufacturer's recommendations or requirements to replace an aluminum part when it has been damaged.

Sacrificial Parts

The damaged structural aluminum rail extensions often become a sacrificial item—especially on the front rails—because they are used for attaching the front bumper to the rails. They are frequently used to attach pulling clamps, and so on, to straighten or repair the adjacent indirect damage. When completed, that part is then replaced. The alloy structure and the working characteristics of certain aluminum parts requires that once damaged, the part must be replaced because it no longer has the required structural

(A)

(B)

FIGURE 16-21 Aluminum parts frequently crack and tear when damaged, such as in (A) and (B), due to the work hardening that occurs during manufacture and again during the collision, making it necessary to replace them.

characteristics. The norm is to replace any structural parts with any visual signs of damage.

Rarely are any of the aluminum-intensive vehicle undercarriages repaired when they become seriously damaged in areas other than on the front of the vehicle. As a rule, all of the affected parts that make up the structural assembly are replaced with new OEM parts. Likewise, if the vehicle has a separate frame, the entire frame assembly is replaced with an OEM frame assembly. The affected areas become highly stressed in a collision and much of the damage may be overlooked or invisible—even when using a **die penetrant**, such as was discussed in a previous chapter. Welding on these parts can further add to these stresses, eventually causing metal fatigue and ultimately a repair failure.

REPAIR GUIDELINES

All aluminum-intensive vehicle manufacturers have very specific guidelines and procedures that must be followed for any structural parts that are replaced. One should never assume that one manufacturer's required procedures are the same as those of another. Some may require a welded part to be replaced with rivets, while others may specify replacing the same part by rivet bonding, while yet another may require rewelding the part. The adhesives used may also be completely different from one manufacturer to another. The vehicle manufacturer may even recommend the surface preparation for the adhesive be done differently than what the adhesive manufacturer may suggest. In order to properly effect the repair, the vehicle service manual must be consulted each time a repair is made, because one part of the vehicle may require a completely different approach than another.

ATTACHING ALUMINUM STRUCTURAL PARTS

Aluminum parts are attached to the vehicle by any one or a combination of bolts, welding, riveting, and rivet bonding. Each manufacturer offers very specific recommendations to follow when replacing any of the serviceable parts. Therefore, before attempting to replace any of the serviced parts, one must consult the manufacturer's specific service manual to determine their recommended procedures.

Bolted

Some vehicles—such as the 2004 and later Jaguar—bolt the front rail to the frame as a sacrificial part. It is designed to collapse in a collision, thus sparing the rest of the front structural members from becoming damaged. The bolts are made of aluminum or coated steel. When installing them, they should be secured using hand tools and a torque wrench only. The use of pneumatic tools are apt to overtighten them, thus stretching the threads. Most manufacturers recommend that new bolts be used to replace the part because most are a one-time use fastener. In addition, galvanic corrosion should be avoided when coated steel is used.

Welded

Welding is the most common means of attaching structural parts to the vehicle by the manufacturer. The parts are welded using one or a combination of either spot welding, slot welds, fillet welds, and plug welds. Spot welding is commonly used in areas where thinner metals are used. However, one will rarely find this type of weld to be used on structural members, although it is very commonly used on body parts and where lighter gauge materials are attached to heavier gauge materials. **Slot welds** are commonly used where a removable structural part is welded to the main frame member and is often done in conjunction with fillet welds on the edges of overlapped areas. Plug welds are used only occasionally in areas where structural requirements are not as critical. The incidence of fillet welds is considerably more significant than is used on vehicles manufactured out of steel.

Welding Equipment. The welding operations performed during the repair of vehicles are primarily done with the **GMAW MIG pulse** process with a spool gun or other conventional welding equipment converted for this application. See **Figure 16-22**. The most common welds used when repairing these vehicles are the slot and fillet. Certain manufacturers even warn against using the plug weld. The GMAW MIG pulse is the most closely matched system available in the repair industry to duplicate those welds. See **Figure 16-23**. Spot welding is not possible in the repair industry because it requires too much current and the possibility of injury to the technician and the vehicle's electrical system is very high. Even though some parts are

FIGURE 16-22 Most welding operations performed on aluminum are done with the MIG pulse welding process using either a spool gun—such as the one shown above—or they can also be made with a standard MIG gun set up for aluminum welding.

welded into place by the manufacturer, they may recommend they be replaced using rivets or rivet bonding. This is yet another example of where the technician must follow the manufacturer's guidelines.

Rivets

Some of the parts near the very front of the frame are installed with rivets. These again are usually sacrificial members and are riveted to more easily accommodate their replacement. The manufacturers use several different types of rivets, including the self-piercing, solid, blind, flush-mounted, and the protruding head. They may be made of aluminum or coated steel.

Self-Piercing Rivet. A riveted part is generally replaced using the same type of rivet. However, the manufacturer will occasionally recommend something other than what was originally used during the manufacture. For example, a solid protruding head coated steel rivet or a flush-mounted blind rivet should be used to replace a self-piercing rivet (SPR). The manufacturers usually recommend placing the SPRs in an alternate location used by the manufacturer and to not use the same area used during manufacture.

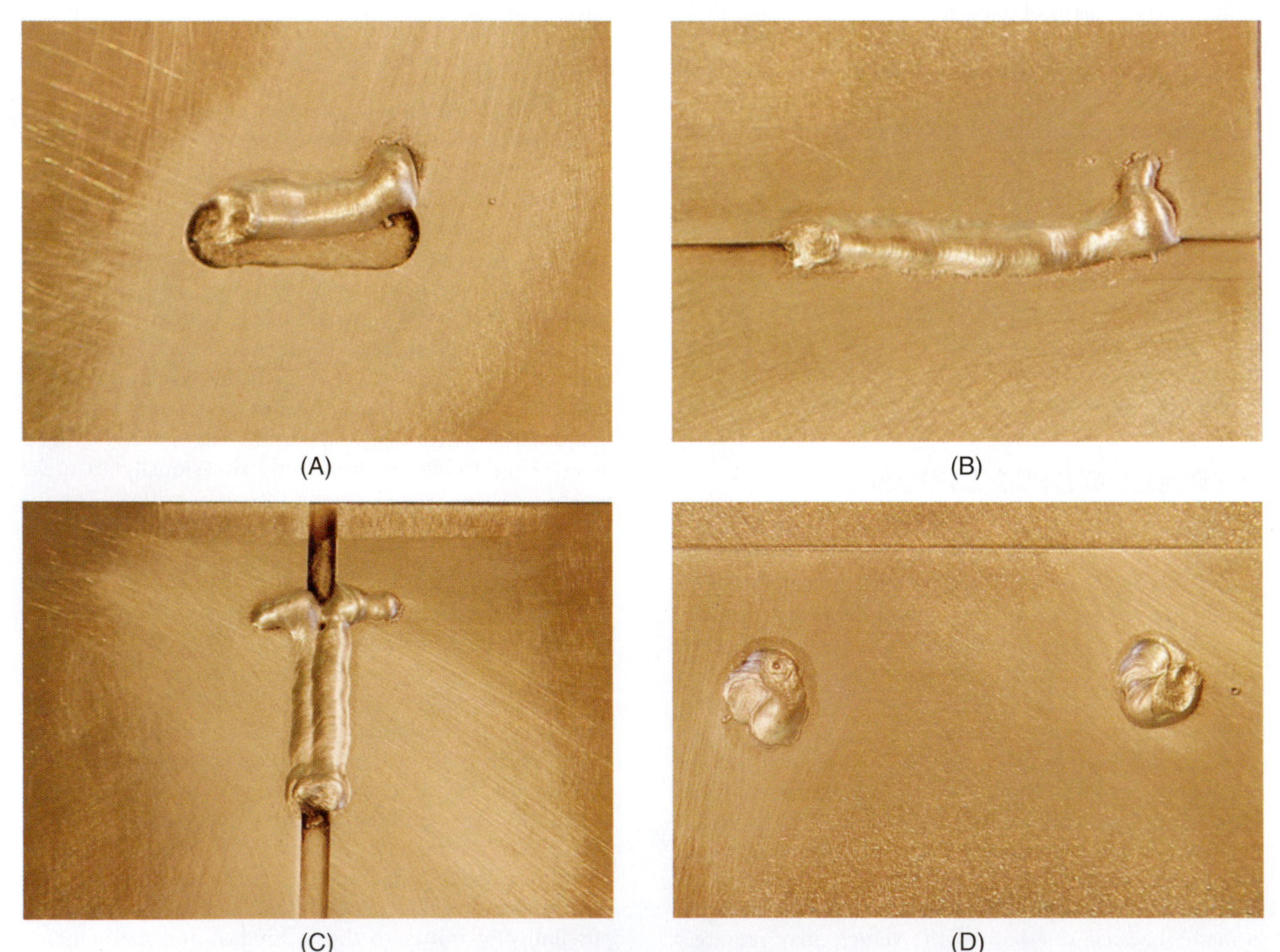

FIGURE 16-23 The type of weld bead most commonly used when welding on aluminum is either a slot weld (A) or a variation of a fillet joint (B). Other types include a butt joint with backing (C) and a plug weld (D).

TECH TIP Adhesives play a vital role in replacing certain parts of the vehicle. One must be extremely vigilant to follow the vehicle manufacturer's recommendations whenever this procedure is used. Failure to follow these recommendations may result in a premature repair failure.

Rivet Bonded

Rivet bonding aluminum is very similar to replacing a steel panel by weld bonding. The process includes the use of a combination of adhesives and rivets instead of welding to attach a replacement part to the existing rail or structural member. This same procedure may be used for making a sectional replacement on a rocker panel or similar areas where the manufacturer may make allowances for such repairs. The repair technician must be cautioned to use the adhesives recommended by the manufacturer. One must be further advised to use the vehicle manufacturer's recommendations for applying and using the adhesives when they differ from the adhesive manufacturer's recommended use. Jaguar, for example, recommends flame treating the metal prior to applying a primer, followed by the adhesive application. Another manufacturer (Audi) recommends using a grinding stone to abrate the adhesive contact surface prior to applying a primer, followed by the adhesive application. When the vehicle manufacturer recommendations are not available, one should use those furnished with the adhesives. Other manufacturers may recommend using adhesives when replacing a part with rivets even though they were not used in the original manufacture of the vehicle.

Expandable Insert

Another form of bonding aluminum parts is to use an expandable insert when performing a replacement with insert. See **Figure 16-24**. In this instance, the insert is fabricated by the auto manufacturer to perform this repair on a specific vehicle's rail. Both the existing and replacement rails are prepared to accept the insert, which will be held in place using a combination of adhesive and the pressure of the bolt against the insert until the adhesive cures. A hole is drilled at the butting edges of both rails to allow the bolt to fit through them. When it is tightened, it causes the insert to expand and fill the cavity. The adhesive must be allowed to thoroughly cure before it is put into service.

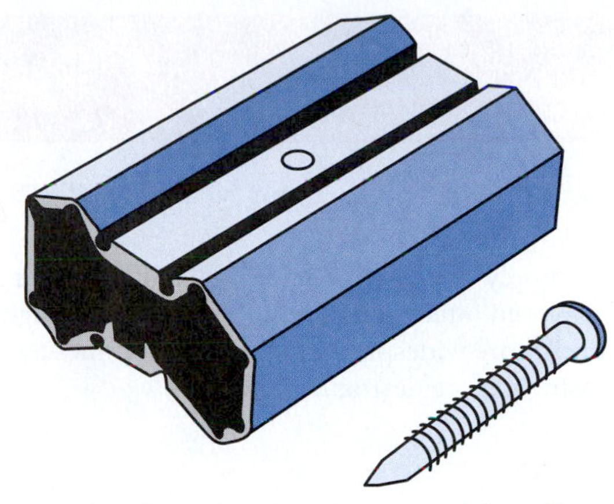

FIGURE 16-24 A sectional replacement of a rail can also be accomplished with aluminum by using a combination of an expandable insert and adhesive to bond it to the inside of the rail.

RIVET REMOVAL

Rivet removal is as critical a part of the repair as any other part of the job. Incorrect removal may result in additional damage such as ripping and tearing the surface to the degree that the mating surfaces are damaged beyond repair. The type of rivet used must be identified before it can be properly removed. Likewise, the removal methods used may vary depending on the type of rivet used. An SPR, for example, may be removed by inserting a different set of dies into the same tool used for installing them, thus pressing the rivet out in the reverse of how it was installed. The SPRs can also be drilled out or removed by grinding them down and punching them out. One must be very cautious if the grinding method is used as some of the surface immediately around the rivet head will also be removed, resulting in the area lacking in strength. Some vehicle manufacturers warn against this method of removal for this reason. Some manufacturers recommend grinding off the heads and punching out the remaining mandrel and body, leaving a hole to use for reinstallation of the new part. The holes must be drilled into the replacement part to match the corresponding locations of those left when removing the rivet. Whichever method is used for repairing damage, one must consult the vehicle manufacturer's guidelines because each vehicle requires uniquely different repairs—even within the same family of cars. One must never take for granted that the techniques used on one type of vehicle will be allowed on another.

Summary

- The continued use of aluminum by automobile manufacturers is inevitable.
- As the supply of raw materials for steel diminishes and the increased interest in reducing the vehicle weight becomes more widespread, aluminum becomes an increasingly more desirable alternative.
- As more and more aluminum-intensive vehicles are produced by the auto manufacturers, it will become increasingly more important for the technicians in the industry to rise to the occasion to continually upgrade themselves to meet future challenges.

Key Terms

baseline/standard model vehicle
bolt boss
cast parts
convoluted front frame rail
die penetrant
die stamping
electronic control module (ECM)
extrusions/extruded parts
fish plating
frame flange
frame replacement module
frame web/fence
GMAW MIG pulse
GMAW MIG welding
high strength low alloy (HSLA)
high strength steel (HSS)
hot rolling
hydroformed
kink vs. bend
rivet bonding
root gap
slot welds
SMAW MIG/flux core welding
ultra high strength steel (UHSS)

Review

ASE Review Questions

1. Technician A says all frame rails are made of common low carbon mild steel. Technician B says the frame rails are made from several steel grades. Who is correct?
   A. Technician A only
   B. Technician B only
   C. Both Technicians A and B
   D. Neither Technician A nor B
2. Technician A says that heat can safely be applied to high strength steel so long as the upper temperature threshold of 1,500°F is not exceeded. Technician B says heat can be applied for 3 minutes cumulatively so long as it is not heated to more than 1,200°F. Who is correct?
   A. Technician A only
   B. Technician B only
   C. Both Technicians A and B
   D. Neither Technician A nor B
3. Technician A says the ladder frame is typically used on large heavy-duty trucks. Technician B says the ladder frame is used because it is stronger than any other type of frame design. Who is correct?
   A. Technician A only
   B. Technician B only
   C. Both Technicians A and B
   D. Neither Technician A nor B
4. Technician A says that before a damaged frame section can be removed from the rest of the frame, it must be roughly repaired to relieve the stresses.

Technician B says a butt joint with insert is commonly used when an enclosed box rail section is replaced. Who is correct?
A. Technician A only
B. Technician B only
C. Both Technicians A and B
D. Neither Technician A nor B

5. Technician A says the manufacturer uses crush zones to help manage the collision energy. Technician B says if vehicle-specific repair procedures are not available one should use the general sectioning guidelines provided by Tech-Cor and I-CAR. Who is correct?
A. Technician A only
B. Technician B only
C. Both Technicians A and B
D. Neither Technician A nor B

6. Technician A says the root gap should be approximately the equivalent of two thicknesses of the metal in the frame rail when making a butt joint with an insert. Technician B says two welding passes may have to be made when welding the new rail onto the existing one. Who is correct?
A. Technician A only
B. Technician B only
C. Both Technicians A and B
D. Neither Technician A nor B

7. Technician A says the SMAW MIG or flux core welding process must be used for welding on nearly all frame rails. Technician B says ideally the electronic control module should be removed from the vehicle prior to making any MIG welds. Who is correct?
A. Technician A only
B. Technician B only
C. Both Technicians A and B
D. Neither Technician A nor B

8. Technician A says the back side of the rail must always be reinforced with another layer of metal whenever a butt joint is used to secure a rail section. Technician B says a 110 to 115V welder is adequate for performing all the required welds when replacing frame sections. Who is correct?
A. Technician A only
B. Technician B only
C. Both Technicians A and B
D. Neither Technician A nor B

9. Technician A says an offset fillet weld is typically used when one section of the rail is slid inside another one. Technician B says high strength steel and mild steel may be used for parts that are hydroformed. Who is correct?
A. Technician A only
B. Technician B only
C. Both Technicians A and B
D. Neither Technician A nor B

10. Technician A says some manufacturers use sacrificial parts at the outer end of the rail to protect the rail deeper into the vehicle and to help absorb some of the collision energy. Technician B says disconnecting the battery will help to protect the electrical system. Who is correct?
A. Technician A only
B. Technician B only
C. Both Technicians A and B
D. Neither Technician A nor B

11. Technician A says an advantage of making a sectional replacement is that it requires less disassembling of the vehicle. Technician B says most OEM frame assemblies sold as replacement parts for repairing damaged vehicles are usually made for the standard model. Who is correct?
A. Technician A only
B. Technician B only
C. Both Technicians A and B
D. Neither Technician A nor B

12. Technician A says a hydroformed part is usually recognizable because it has sharply formed square corners. Technician B says a kinked frame rail should be replaced because it cannot be repaired without the use of heat. Who is correct?
A. Technician A only
B. Technician B only
C. Both Technicians A and B
D. Neither Technician A nor B

13. Technician A says a non-contact thermometer is the only means used to monitor the temperature of the metal when applying heat. Technician B says plug welds are used to hold the insert into place when the two rails are aligned. Who is correct?
A. Technician A only
B. Technician B only
C. Both Technicians A and B
D. Neither Technician A nor B

14. Technician A says the perimeter frame uses a drop center design to assist in managing the collision energy. Technician B says placing seams at suspension mounting locations should be avoided when making a sectional replacement. Who is correct?
A. Technician A only
B. Technician B only
C. Both Technicians A and B
D. Neither Technician A nor B

15. Technician A says even though the general sectioning guidelines are not vehicle specific they can be used for making most repairs safely.

Technician B says placing a seam in a crush zone will not affect the timing of an air bag deployment. Who is correct?
A. Technician A only
B. Technician B only
C. Both Technicians A and B
D. Neither Technician A nor B

16. Technician A says that structural aluminum replacement parts can be welded into place using the resistance spot welder. Technician B says aluminum parts must all be flame treated before bonding them. Who is correct?
A. Technician A only
B. Technician B only
C. Both Technicians A and B
D. Neither Technician A nor B

17. Technician A says the SPRs can be removed using the same tool used to install them. Technician B says the same type of rivet used by the manufacturer must be used when replacing an aluminum part. Who is correct?
A. Technician A only
B. Technician B only
C. Both Technicians A and B
D. Neither Technician A nor B

18. Technician A says aluminum can be flexed or bent more frequently without consequence than steel. Technician B says aluminum becomes work hardened more easily than steel. Who is correct?
A. Technician A only
B. Technician B only
C. Both Technicians A and B
D. Neither Technician A nor B

19. Technician A says extruded parts usually have a single cell cavity. Technician B says extruded parts are usually seamless and the wall thickness is very uniform. Who is correct?
A. Technician A only
B. Technician B only
C. Both Technicians A and B
D. Neither Technician A nor B

20. Technician A says rivet bonding may be recommended by the manufacturer even though it was not originally used when the vehicle was manufactured. Technician B says the use of rivet bonding is usually left to the discretion of the technician making the repairs. Who is correct?
A. Technician A only
B. Technician B only
C. Both Technicians A and B
D. Neither Technician A nor B

21. Technician A says aluminum structural parts may be 50 percent thicker than those made of steel. Technician B says a cast aluminum part may be drilled and machined to specifications. Who is correct?
A. Technician A only
B. Technician B only
C. Both Technicians A and B
D. Neither Technician A nor B

22. Technician A says cast aluminum parts are formed in a mold. Technician B says extruded parts may have multiple cells in the center of them. Who is correct?
A. Technician A only
B. Technician B only
C. Both Technicians A and B
D. Neither Technician A nor B

23. Technician A says the bend vs. kink rule used for steel repairs rarely applies when working with aluminum parts. Technician B says aluminum parts are approximately 50 percent heavier in weight than those made of steel. Who is correct?
A. Technician A only
B. Technician B only
C. Both Technicians A and B
D. Neither Technician A nor B

24. Technician A says some aluminum-intensive vehicle manufacturers recommend replacing damaged parts using rivets. Technician B says resistance spot welding is a common practice when replacing aluminum parts. Who is correct?
A. Technician A only
B. Technician B only
C. Both Technicians A and B
D. Neither Technician A nor B

25. Technician A says coated steel bolts are used for attaching parts to aluminum. Technician B says the bolts should be installed with a pneumatic tool to ensure they are properly tightened. Who is correct?
A. Technician A only
B. Technician B only
C. Both Technicians A and B
D. Neither Technician A nor B

Essay Questions

1. List at least three different joint designs that could be used when performing a frame sectional replacement and briefly describe how each is made.
2. Cite at least four things that must be considered when trying to determine whether to use a re-cycled or new unit for performing a frame sectional replacement.

3. List the most common frame designs used on today's vehicles and describe how they differ from each other.
4. Cite the principle differences between the undercarriage of a unibody and BOF vehicle.

Topic Related Math Questions

The following formula is used to convert degrees FAHRENHEIT to degrees CELSIUS:

$$°C = \frac{5}{9} \times (°F - 32)$$

The following formula is used to convert degrees CELSIUS to degrees FAHRENHEIT:

$$°F = (\frac{9}{5} \times °C) + 32$$

1. Your non-contact thermometer does not have a Celsius scale and you need to convert the Fahrenheit temperature to Celsius. The Fahrenheit temperature is 750 degrees. What is the Celsius temperature?
2. The temperature on the surface is 1,150°C. Calculate the Fahrenheit temperature.
3. A steel frame on a comparable size vehicle weighs 985 pounds. Aluminum is approximately 65 percent of the weight of steel. What would be the approximate weight difference between the aluminum frame and the steel frame vehicles?
4. An automobile's fuel economy is improved by three miles per gallon for every 100 pounds that is removed. On the two vehicles above, which offers better fuel economy and by how many miles per gallon?

Critical Thinking Questions

1. What is the likelihood of aluminum becoming the metal of choice over steel for car manufacturers in the future? Why will this happen? Why will it likely not happen?
2. Why are most manufacturers so intent on requiring the repair shops to have dedicated tools, equipment, and work areas for repairing aluminum-intensive vehicles?
3. Develop a list of as many differences as you can cite between aluminum and steel.

Lab Activities

1. Locate and describe at least three different crash management areas on the front frame section of a BOF vehicle in the shop and discuss how they function.
2. Locate three different areas where a sectional replacement could be performed on a BOF vehicle in the shop.
3. Using a BOF vehicle in the shop, locate the different methods used to attach parts, brackets, and reinforcements to it.

Chapter 17

Welded Exterior Panel Replacement

OBJECTIVES

Upon completing this chapter, you should be able to:

- Identify the personal safety equipment that is needed to protect yourself and others when performing the described repair procedures.
- Explain the differences between kinked parts and bent parts.
- Identify the structural and nonstructural parts on a vehicle and know where to find the information that identifies structural parts for specific vehicles.
- Explain why precise measurements are vital when analyzing structural damage.
- Describe what items must be taken into consideration when deciding to replace or repair a structural part.
- Explain when replacing a part at factory seams is the best method of repair.
- Explain when sectioning a part is indicated.
- List the four main types of joints used in sectioning.
- Describe how each of the following is made: a lap joint, offset butt joint, open butt joint, and butt joint with a backer.
- List the steps in the general procedures for the removal of: spot welds, welded joints, and bonded joints.
- List the steps in the general procedures for welded panel replacement.
- Describe where to find specific or general vehicle specifications for replacing the following items: door skins, quarter panels, a vehicle roof, and rear body panels.

INTRODUCTION

Vehicles involved in a collision commonly need to have welded exterior panels either repaired or replaced. While many other vehicle parts add to the vehicle's **structural integrity**, welded exterior parts contribute substantially to a vehicle's structural integrity and its crashworthiness. A repair technician is responsible not only for returning a vehicle to its **cosmetic** pre-collision condition, but also for returning it to its pre-accident crashworthiness status. That is to say, if the vehicle should be involved in another collision following its repair, it must perform during that collision as it was originally designed to perform by the manufacturer. If the vehicle does not collapse as designed, the occupants may sustain injuries that they might not have otherwise. To return a vehicle to its proper crashworthiness, manufacturer recommendations must be followed precisely. If no manufacturer's recommendations are available, the technician should follow tested and published repair procedures.

Restoring corrosion protection is a necessary part of all structural repairs, because the repair process itself can cause corrosion which, if left unchecked, can compromise the vehicle's structural integrity. Manufacturers vary in their recommendations for restoring corrosion protection. In fact, some manufacturers have changed their recommendations for items such as weld-through primer over the years. Although one manufacturer recommends the use of one type of corrosion protection, not all manufacturers may use the same recommendation. Technicians must read, understand, and follow the most current recommendation.

Also, recommendations may have changed since the most current repair manual was printed, These manuals are updated by the publication of a subsequent "technical bulletin" (sometimes called a **technical service bulletin** or **TSB**), which **supersedes** the repair manual. Technicians can best find this information on the manufacturer's service website. The most current information is usually posted there. These websites may be accessed for free or by subscription, depending on the manufacturer.

This chapter will explore general non-specific procedures for repairing a vehicle's structural components. For specific vehicle recommendations, the manufacturer's repair procedures should be followed.

SAFETY

Technicians who practice the techniques outlined in this chapter should use appropriate personal safety equipment such as—but not limited to—safety glasses, gloves, protective clothing, safety work boots, and ear protection. They should read, understand, and follow all safety recommendations when working with tools. They should also find, read, understand, and follow all safety precautions outlined by MSDS documents when working with chemicals.

Remember that working in a collision repair shop can be hazardous, and technicians should be ever vigilant to protect themselves, their fellow workers, and the environment.

STRUCTURAL PANEL REPAIRS

Not all welded-on panels are considered structural. That is, some do not add to the strength of the vehicle, nor do they add to the safety of the passengers within the vehicle. Panels such as door skins and the outer portion of **deck lids** are not considered structural. On the other hand, roof panels and sometimes quarter panels *are* considered structural. When replacing a welded-on structural panel, its precise placement and way that the panel will react in a subsequent collision are quite critical. Therefore, following the recommended procedures for replacement is also very critical.

Replacement procedures have been tested by the related agency before the procedures become recommendations. That is, the intended replacement procedure has first been followed by company technicians, and then the vehicle has been put through a controlled crash to test the procedure's integrity. If the repair does not perform as well or better in crashworthiness than a non-repaired vehicle, that procedure will not be published by the testing agency as the recommended procedure.

Analyzing Damage

Welded-on exterior parts, when replaced, must be measured and returned to their pre-accident measurements. There is little to no tolerance for the vehicle to vary from the manufacturer's dimensions in all three planes (length, width, and height). The vehicle must be measured and returned to factory **tolerances** before the damaged part is removed. If the part is removed before it is rough straightened, it becomes extremely difficult or impossible to return the vehicle to its pre-accident conditions. If the vehicle is repaired while even slightly out of tolerances when welded, there is not room for adjustment afterwards. Therefore, the vehicle should be measured, fit, adjusted, and re-measured multiple times to assure that it will be restored to its pre-accident conditions.

"Repair or Replace" Decisions

When evaluating a damaged vehicle, the technician who must decide whether to repair or replace a panel needs to assess many different elements of the damaged panel.

Some of the questions the technician must answer include: Is the part bent or kinked? Will it cost more to repair the panel than it would to replace it? (Or, in some cases: Will it cost more than a certain percentage of the replacement cost to repair the panel instead?) Will the replacement require more removal of adjacent parts or disturb more factory corrosion protection than repair would require? Does the manufacturer recommend replacement over repair?

TECH TIP

A part is "kinked" if it has a sharp bend, usually over 90 degrees over a short area, or if after repair there are visible cracks, tears, or deformation. See **Figure 17-1**.

A part is "bent" if the deformation of the part is smooth and consistent, or if by straightening the part can be restored to its pre-collision condition. See **Figure 17-2**.

The decision to repair versus replace might be based on the cost of replacement. Though the procedure may vary

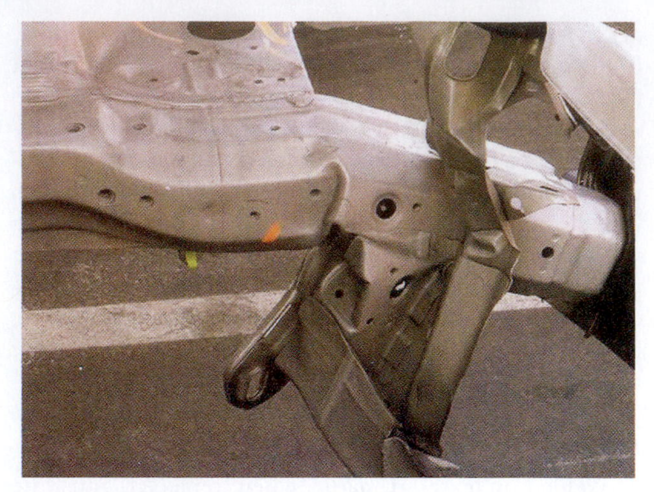

FIGURE 17-1 Kinked parts are parts that have been bent past their point of elasticity and will not return when stresses are relieved.

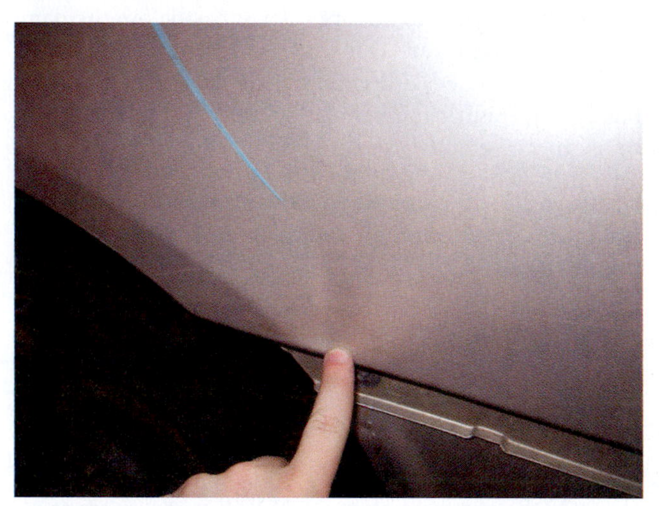

FIGURE 17-2 Bent parts may return to their original state and shape when pressure is relieved or can be easily metalworked back into shape.

from business to business, the comparison is generally determined by calculating a set percentage of the cost to repair as compared to the cost to replace. As an example, if the percentage allowed is 75 percent and the repair would cost $80, versus a cost of replacement of $100, the business would replace the part since this repair cost would be 80 percent of the cost of replacement.

In addition, damage may be repaired rather than replaced if the replacement of the part at factory seams will be more intrusive than the repair. That is, if the replacement requires the removal of intact welds, adjacent parts, or the destruction of otherwise undisturbed factory corrosion protection, the repair procedure may be to section a new part to the undamaged old part.

Additionally, the decision to replace versus repair must always be based on the vehicle manufacturer's recommendation. As an example, a manufacturer may not recommend the replacement of a door skin, but rather recommend that the entire door shell be replaced, believing that when a skin is replaced (even with the most careful replacement of corrosion protection), a skinned door may be more likely to corrode than will a replacement door shell.

Panel replacement can be done either by replacing a new panel at the original factory seams or by sectioning (partial replacement) of the new part.

Replacing at Factory Seams

Replacement parts as produced during manufacturing come as whole parts. These parts, such as a uniside, are replaced at the original factory seams. See **Figure 17-3**. The technician must follow the manufacturer's recommendation as to the type and number of replacement welds. Technicians must also follow the manufacturer recommendations for the use of adhesion bonding and/or special fasteners—as with aluminum repairs—as well as the careful replacement of corrosion protection. By following the recommended repair procedures, a technician can not only produce an undetectable repair but also can maintain the vehicle's pre-accident crashworthiness.

TECH TIP One of the technician's responsibilities is to restore a damaged vehicle to its pre-accident condition. That includes restoring the vehicle's ability to perform as designed by the manufacturer in an additional accident, known as the vehicle's **crashworthiness**. Those parts that are designed to absorb energy during a collision—thus protecting the vehicle's occupants—must remain intact following the vehicle's repair.

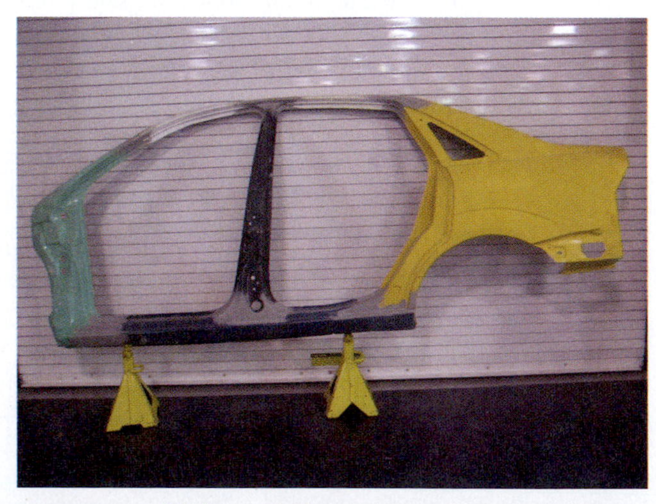

FIGURE 17-3 With a uniside replacement part, the whole side of the vehicle may be manufactured as one piece.

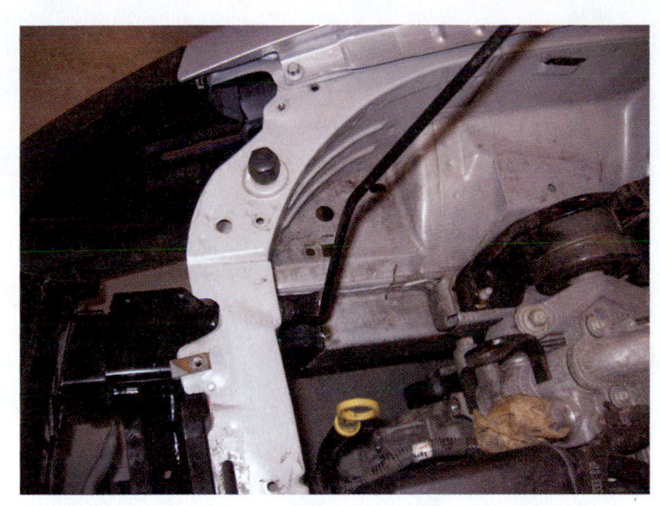

FIGURE 17-4 When sectioning a panel, follow the vehicle-specific sectioning location recommendations.

SECTIONING

Sectioning of a factory replacement panel is the partial replacement of that panel. The new part is cut in at an area other than the factory seam, thus replacing only the damaged area of the vehicle. Replacement parts that are commonly sectioned, such as door skins and quarter panels, are done so as not to disturb factory corrosion protection or require the need to remove intact factory welds. See **Figure 17-4**.

Sectioning is also done in such areas as A pillars, trunk floors, B pillars, quarter panels, rear rails, floor panels, rocker panels, and front rails. When choosing to section a structural or nonstructural part, the *manufacturer's recommendation must be followed*. In some cases the manufacturer specifically recommends against sectioning, while in others the manufacturer will recommend not only that sectioning is acceptable, but also describe the precise area where the part should be cut and the type of joint and welds to use. Manufacturer's recommendations can be found in the vehicle repair manuals. If there are no published manufacturer's recommendations for or against sectioning, a technician should consult industry recommendations such as **I-CAR's Uniform Procedures for Collision Repair (UPCR)**, or insurance company recommendations. Technicians should remember that the vehicle manufacturer's recommendation takes precedence over all other recommendations when choosing a repair procedure.

TECH TIP It is the repair technician's job to be current on all the vehicle manufacturers' recommendations. It is possible that a vehicle repair manual has a published recommendation for sectioning that is later retracted.

Manufacturers publish repair bulletins—commonly called "technical service bulletins," or TSB—which may retract previously recommended procedures. Keeping current on the repair bulletins is critical in regards to sectioning a vehicle.

SECTIONING JOINTS

There are generally four types of sectioning joints:

1. Lap joint
2. Offset butt joint
3. Open butt joint
4. Butt weld with a backer

The specific joint that is used will depend on the manufacturer's recommendation, the location of the cut line, and/or the configuration of the part being joined.

A butt joint with or without an insert is commonly used to close sections such as pillars, rails, and rocker panels. When inserts are used, the recommendation may call for the use of plug welds attaching the insert. Inserts help the technician with fit-up and weld ease, although some manufacturers do not recommend their use.

Offset butt joints—or staggered butt joints, as they are sometimes called—are used without an insert and can be recommended for use on pillars and rails.

Lap joints can be recommended on rear rails and trunk and floor pans. They may also be recommended on B pillars.

Use of Lap Joints

A lap joint is used when repairing trunk floors, floor pans, rear rails, hat channels, and flat closures. See **Figure 17-5**. The parts should be overlapped according to manufacturer's recommendations or approximately overlapped 1/4 to 1 inch. Plug welds are often recommended to help with fit-up. See **Figure 17-6**. After the joint is positioned correctly and measurements have been taken to verify, the bottom or outside joint is welded using a continuous weld. The inside or top of the lap weld is sealed to prevent exhaust and water leakage into the passenger compartment. Proper corrosion protection measurements should be taken with all welds. Recommendations vary among manufacturers, so technicians should read, understand, and follow along according to the specific manufacturer's guidelines.

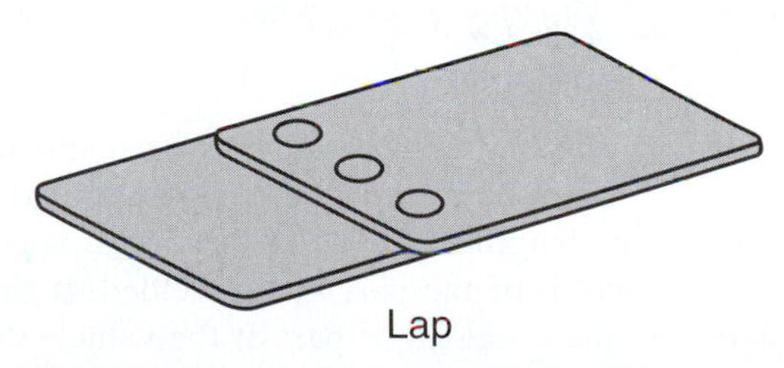

FIGURE 17-5 A lap joint is used when repairing trunk floors, floor pans, rear rails, hat channels, and flat closures.

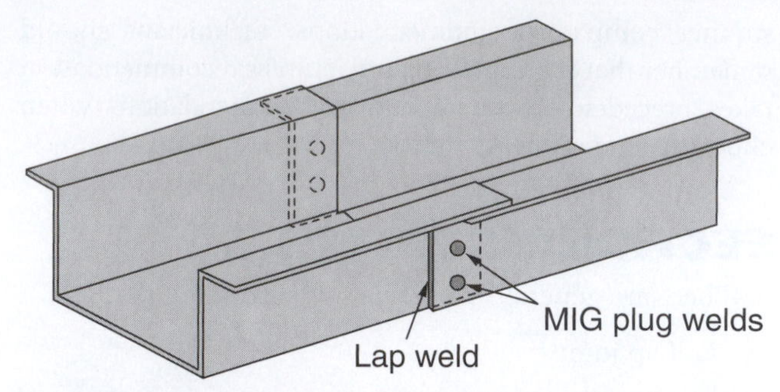

FIGURE 17-6 Plug welds are often recommended on lap joints to help with fit-up.

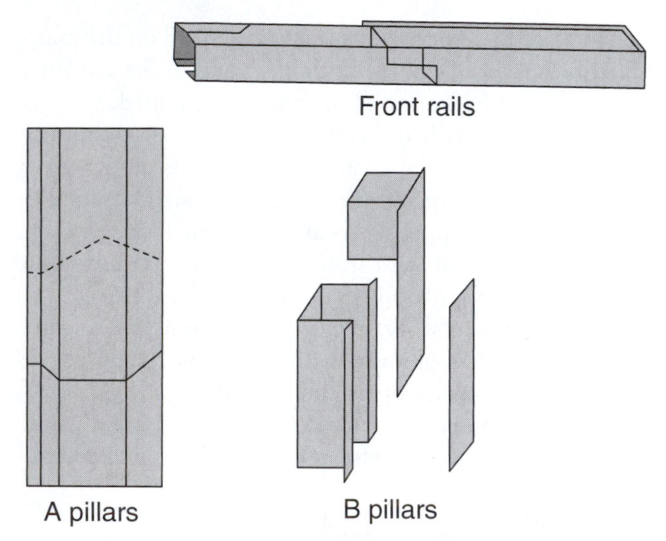

FIGURE 17-7 This offset butt joint has an insert.

FIGURE 17-8 This offset butt joint does not have an insert.

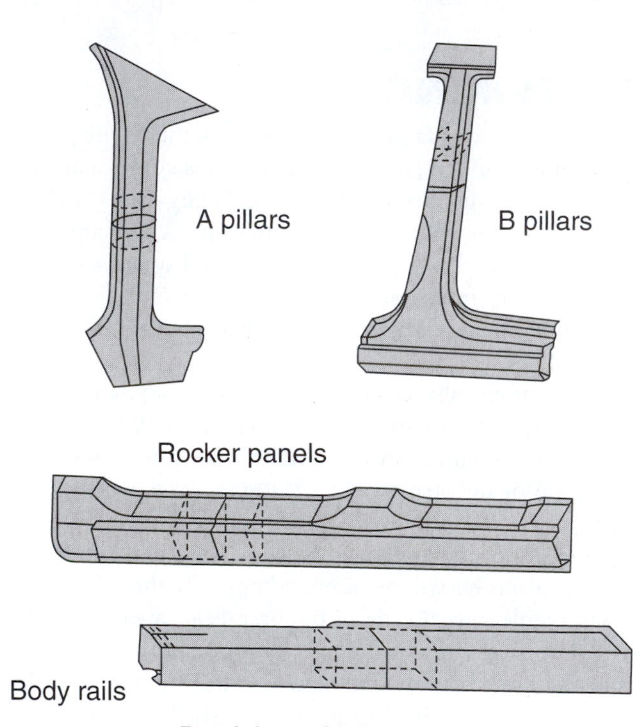

FIGURE 17-9 Butt joints with a backer are used by some manufacturers when sectioning the front rail or A and B pillars.

Use of an Offset Butt Joint

The offset butt joint, sometimes called a staggered butt joint, can be used with or without an insert. See **Figure 17-7** and **Figure 17-8**. It is often used for closed sections such as A and B pillars, as well as to keep the original D ring intact when repairing a B pillar. Plug welds are often recommended for fit-up, with the joint closed with a continuous weld. The proper corrosion protection recommended should be applied. This is an ideal joint where there are three or more pieces within the joint, thus making it impossible to weld without using an offset joint.

Use of a Butt Joint with a Backer

Butt joints with a backer are used by some manufacturers when sectioning the front rail or A and B pillars. See **Figure 17-9**. The length of the backer or insert is generally twice the width of the part being welded. It can be harvested from the undamaged part of the vehicle that is being repaired or from the unused part of the replacement part. This will assure that the insert is the same thickness and type as the vehicle. Flanges are removed, the insert is placed inside of the repair part, and 5/16 (8 mm) plug weld holes are drilled, taking care that the **heat affect zone** of the plug welds and the butt weld will not overlap. By following proper corrosion protection recommendations, the insert is plug welded in place on one side. The part is fit-up, measured to verify proper fit, and the second plug welds are welded on the new part. The fit is checked again, and then a continuous weld is used to close the butt weld. When an insert is used to join a part, it is strengthened, thus changing its crashworthiness. Some manufacturers do not recommend using an insert with a backer for front rails; therefore, technicians should read, understand,

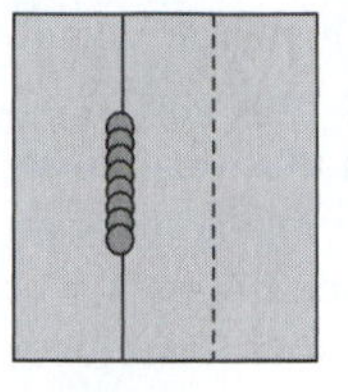

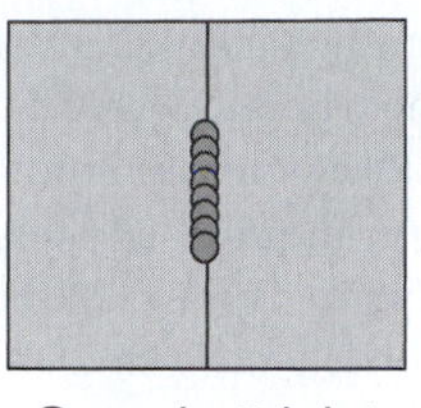

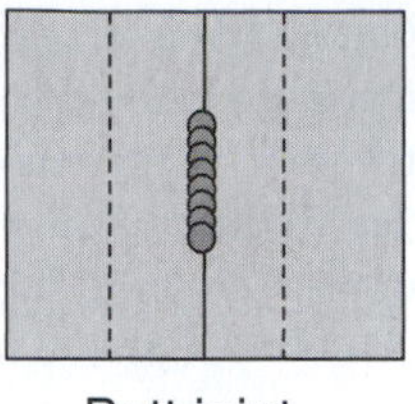

FIGURE 17-10 For sectioning front rails, some manufacturers recommend an open butt weld or a butt weld without a backer.

and follow the vehicle manufacturer's recommendations for repairing specific parts.

Use of an Open Butt Joint

For sectioning front rails, some manufacturers recommend an open butt weld or a butt weld without a backer. See **Figure 17-10**. The use of these welds will ensure that the repair does not strengthen the rail, thus affecting the way it will collapse in a subsequent collision. The use of an open butt weld is more difficult to fit-up than one with a backer, because there are no plug welds to hold the parts in place for welding. The new part must be placed in position with the proper gap between the new part and the vehicle. This gap allows for the needed weld penetration. The parts are tacked in place, then measured to assure that the parts are aligned correctly, then welded with a continuous weld. **Tack welding** produces a temporary weld; thus, a continuous weld is used to join and strengthen the weld for closure. Since this type of weld is more likely to move during the welding process, the technician should afterwards verify that the part has not moved during the welding process, thus assuring that the proper fit-up is maintained.

PANEL REMOVAL PROCEDURES

Welded-on panels are removed either at the factory seams, for complete replacement, or sectioned as outlined previously. A technician could attach these panels with spot welds, continuous welds, or adhesive bonding to replace that which was applied at the factory. On some vehicles, a technician will encounter a continuous weld in a material other than steel, such as bronze, which should not be mistaken for older, heat-intense brazing. Modern MIG brazing using silicon bronze is applied with a MIG welder, which is faster than older methods. More importantly, the heat affect zone of bronze is lower in melting temperature than steel MIG welding, making it better for high strength steels. When replacing factory-installed MIG brazing, a technician should follow manufacturer's recommendations. The technician first must identify the factory type of bonding, which most often will be a factory-installed resistance spot weld. Manufacturer repair manuals sometimes identify the precise number and location of **original equipment (OE)** spot welds, but others do not. In these cases, a technician must locate them. Continuous welds including MIG brazing, as well as adhesive bonding removal, also require specific methods of removal so new panels can be replaced with little repair to the mating flanges. Proper removal of the OE panels will help with new part fit-up and installation.

Spot Weld Removal

As mentioned earlier, some repair manuals give the exact location of OE spot welds and others do not. Consequently, finding spot welds can sometimes pose a challenge. If the new panel is available, its mating flanges will often give a technician a good idea where spot welds could be located. Spot welds are often covered, thus masking their location. Paint sealers, undercoating, and a variety of other coverings can also make them difficult to locate. These coverings should be removed, using an abrasive wheel, scraping, or sanding. See **Figure 17-11**. Sometimes soft undercoating can be removed with a wax and grease remover. Generally these methods will reveal most spot weld locations. In the past, some technicians used heat to burn off coatings that covered spot weld locations. However, heat can damage high strength steels. If used, the technician should proceed carefully. Tools which produce low heat, such as a heat gun rather than a torch, are best used for removing or softening coatings for quicker removal, thus not damaging the remaining underlying mating surfaces.

If the spot welds are not revealed after the technician has removed the coatings from mating surfaces, a hand chisel can be carefully driven between the mating surfaces to reveal them. Care should be used with this method, because when a chisel is driven into the mating surfaces, it will deform one of the panels. The technician should make sure that the deformed panel is the one that will be removed, not the remaining mating surface. There are many tools used to remove spot welds: cut-off wheel, hole saw, compound drill, conventional drill, and spot weld

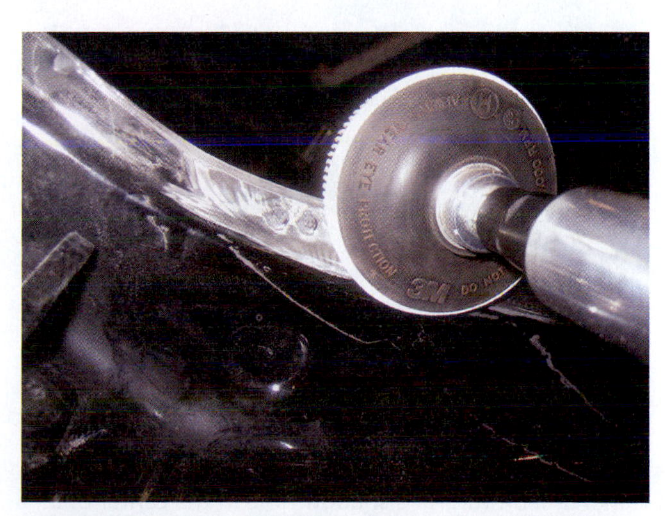

FIGURE 17-11 To locate spot welds, coverings should be removed by using an abrasive wheel, scraping, or sanding.

cutting tool. See **Figure 17-12**. Regardless of the removal method chosen, the technician should take care to keep the remaining mating surface as undamaged as possible.

TECH TIP If a spot weld will be removed with a hole saw or compound bit, technicians will often use a small (1/8 inch) conventional drill bit to drill a shallow hole in the center of a spot weld in order to keep the spot weld removing tool from wandering when in use. See **Figure 17-13**.

FIGURE 17-12 There are many tools used to remove spot welds: cut-off wheel, hole saw, compound drill, conventional drill, and spot weld cutting tool.

FIGURE 17-13 Technicians will often use a small conventional drill bit to drill a shallow hole in the center of a spot weld to keep the spot weld removing tool from wandering when in use.

Cut-Off Wheel. When the technician uses a cut-off wheel, it should be held so the wheel grinds off the spot weld from the side of the panel that is to be removed and discarded. See **Figure 17-14**. Care should be taken to grind through one layer only. See **Figure 17-15**. Often technicians will grind through the weld, leaving a small amount of the layer to be removed; then they use a separating tool or hand chisel to remove the remaining part. See **Figure 17-16**.

FIGURE 17-14 When the technician uses a cut-off wheel, it should be held so the wheel grinds off the spot weld from the side of the panel that is to be removed and discarded.

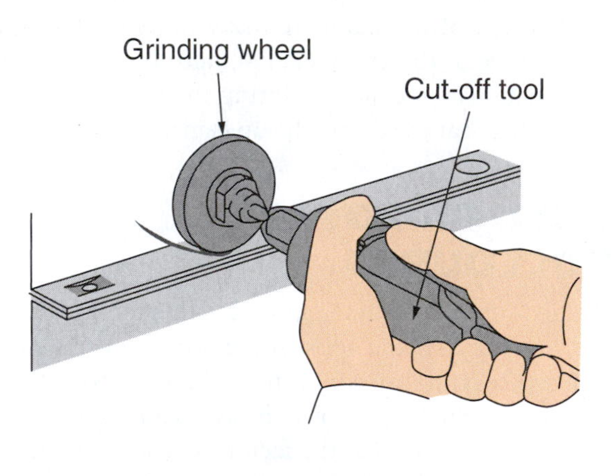

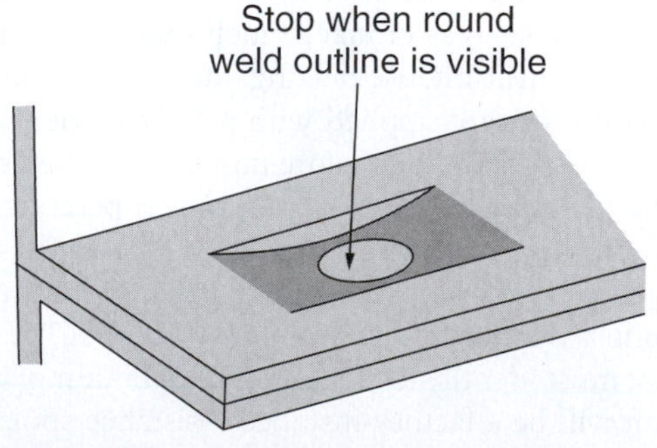

FIGURE 17-15 Care should be taken to grind through one layer only when grinding a spot weld.

FIGURE 17-16 Often technicians will grind through the weld, leaving a small amount of the layer to be removed; then they use a separating tool or hand chisel to remove the remaining part.

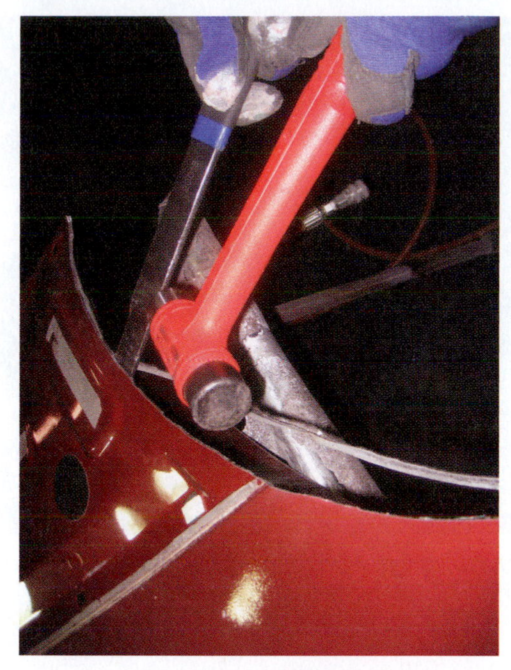

FIGURE 17-17 In spot weld removal, the panel should be drilled nearly completely through, leaving a small amount remaining which can be easily removed with a panel-separating tool.

Grinding off a weld with a cut-off wheel generates heat, which could cause damage to the underlying surface, so care should be taken not to damage it. Sparks are also generated during the use of a cut-off wheel; thus, the vehicle, particularly glass surfaces which can be easily pitted by sparks, should be protected from these sparks.

Hole Saw, Compound Bit, and Conventional Bit. All of these methods of spot weld removal involve drilling or cut-out on the weld or near the weld on the surface of the panel that will be removed, while being careful not to damage the mating surface that remains. The panel should be drilled nearly completely through, leaving a small amount remaining which can be easily removed with a panel-separating tool. See **Figure 17-17**. When this method is used, a small "nugget" of the spot weld will remain on the mating surface which can be "dressed," or ground smooth, for a smooth fit-up and re-weld later in the process. See **Figure 17-18**.

FIGURE 17-18 A small "nugget" of the spot weld will remain on the mating surface which can be "dressed," or ground smooth, for a smooth fit-up and re-weld later in the process.

Continuous Weld Removal

Some of the OE bonds are continuous MIG welds, whether with steel or bronze. Because the welds are longer, they should be removed with a grinding wheel. The size of the grinding wheel should be taken into consideration, because if too large a wheel is used it may damage areas which should not be ground. A tool that fits the job should be used, such as a small angle grinder. See **Figure 17-19**. When removing a continuous MIG bronze weld, the weld will grind much faster than a steel weld. Therefore, the technician should be especially careful not to damage underlying areas.

Adhesive Bonding Removal

Adhesive comes in many different forms; some types need to be completely removed before new adhesive is attached, while others, such as one-part urethanes, may need to have a level layer of the original adhesive left intact before applying new adhesive. Technicians must read, understand, and follow the vehicle manufacturer's recommendation. If no vehicle maker's recommendation is available, the recommendations provided by the bonding manufacturer should be followed.

Adhesive bonding that must be removed can be mechanically removed with a grinder. As with spot welds,

FIGURE 17-19 Because continuous welds are longer, they should be removed with a grinding wheel.

the mating surface that will remain should be kept as undamaged as possible while removing the old panel and bonding agent. Some bonding agents can best be removed with heat.

The first step when removing adhesive bonded panels is to remove all the mechanical fasteners that may be used, along with all of the trim and hardware. The technician can roughly remove the outer edge with a grinder, leaving the flange areas where the adhesive has been placed. A flameless heat device such as a heat gun that will reach 400°F is effective; as the adhesive warms it will soften, thus breaking the bond. The parts can then be pried off without damaging the mating surface. Once the part has cooled, any remaining adhesive can be removed.

PANEL REPLACEMENT

As with any repair process, one of the first steps when replacing a panel is preparing the mating surface. The mating surfaces must be smooth, straight, and in proper alignment to receive the new panel. Old spot welds should be ground smooth, removing all burrs or tears that may occur from the removal of the damaged panel. Then the mating surface should be straightened and measured to make sure that the dimensions are proper to receive the new part. All paint and undercoating should be removed from the mating surfaces, and corrosion protection should be applied following the manufacturer's recommendations. The vehicle will then be ready for the next step, the positioning and fitting of the new panel.

Positioning and Fitting New Panels

The replacement panel should be put in place on the vehicle and held in place with locking pliers, or fastened with screws where pliers cannot be used. The vehicle should be assembled as much as needed to assure proper fit and alignment. In order to verify their visual check for alignment, technicians should then use a measuring system. Visual alignment is a good first step when aligning a replacement part, but accuracy should always be verified by specific measurements. New OE parts may come with alignment marks or holes, which should be used when installing new parts, but they too should be verified by measuring. Even bolt-on parts should be temporarily attached before welding, to verify that the new part will be exactly in position.

Panel Installation Procedures

Once again, the manufacturer's repair manual should be checked for the recommended weld locations and the number and type of welds. This information should be marked on the new part. Once all of these details are checked and within satisfactory recommendations, those parts that would get in the way when welding should be removed. With the panel in correct position and corrosion protection precautions taken, the panel can then be welded in place. Large panels that require a large number of welds should have a few welds applied, and then the technician should check the panel to be sure that it has not shifted during the welding process. Though the processes of welding, then checking the panel position, then welding more, may seem time consuming, these actions will assure that the panel remains in position throughout the process.

When the welds have been completed they should be dressed cosmetically, sealed, and caulked for leakage as recommended to complete the process.

REPLACEMENT PROCEDURES

Several commonly replaced welded-on panels include door skins, quarter panels, pickup box sides, body uniside, roof panels, and rear body panels. Though many of the replacement steps for each of these parts are similar, each part will be covered individually. The steps detailed will be presented in a generic manner here. To obtain more model-specific procedures, a technician should consult the vehicle manufacturer's repair manual for precise procedural recommendations.

Though many of these parts are welded in place, they may also be attached with a bonding adhesive or a combination of welding and bonding. The part may be manufactured from steel in one of its many forms, from mild steel to very hard ultra high strength steels sometimes called boron. Parts could also be aluminum, plastic, or other types of composites. If so, the replacement techniques will vary. The three most common types of door skins that a technician will replace are those manufactured from steel, aluminum, and plastic. Therefore, procedures for each of these three materials will be covered in some depth in the following discussion.

Door Skins

Replacements for damaged doors may be accomplished by replacing the entire **door shell**, which is an intact door as manufactured from the factory (the inside frame, including the **intrusion beam** and outer shell). See **Figure 17-20** and **Figure 17-21**. On the other hand, one might also replace a **door skin** (the outer portion of the door shell), which is installed on the door frame after the damaged outer shell has been removed. The choice of replacement procedure will be dictated by the extent of damage or by recommendations of the vehicle manufacturer. Both replacement methods are commonly performed during the normal course of automotive collision repair.

FIGURE 17-20 Replacements for damaged doors may be accomplished by replacing the entire door shell.

The following steps outline the general procedure and order for the technician to correctly install a replacement door skin. See **Figure 17-22**.

1. Disconnect the negative terminal of the battery to isolate the vehicle's electrical components.
2. Protect the interior and adjacent panels to the door that is to be removed.
3. Remove all exterior parts such as door handles, trim, mirrors, and moldings.
4. Remove glass to ensure that no damage will occur during door skin replacement. Some types of vehicles can have the door skin replaced with the glass intact, but it is often removed to assure that it does not become damaged.

TECH TIP Glass is extremely sensitive to heat, especially in the form of grinding sparks and welding slag. Even the slightest spark will melt into the glass, causing irreparable damage. Glass must be protected from sparks, and the best protection is often to remove it from the vehicle before grinding and welding.

5. Door removal: Though it may be possible to replace a door skin with the door remaining on the vehicle, it is often necessary to remove it for better and free access to all edges during replacement.
6. Measure and straighten the door to assure proper door alignment after repair.
7. Cut and remove the center portion of the door skin for access, taking care not to cut or damage the interior parts, intrusion beam, or door frame. Remove the center section. (Adhesive bonding may be attached to the outer door beam and intrusion beam. A heat gun can be used to help soften the adhesive.)

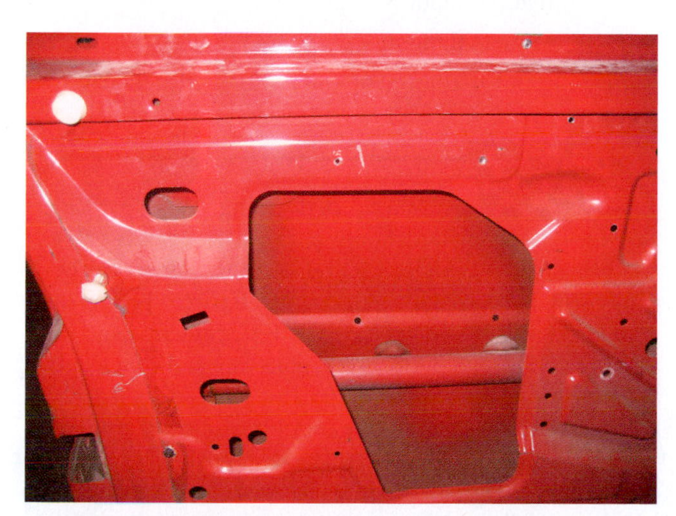

FIGURE 17-21 The intrusion beam keeps passengers safe during a side collision.

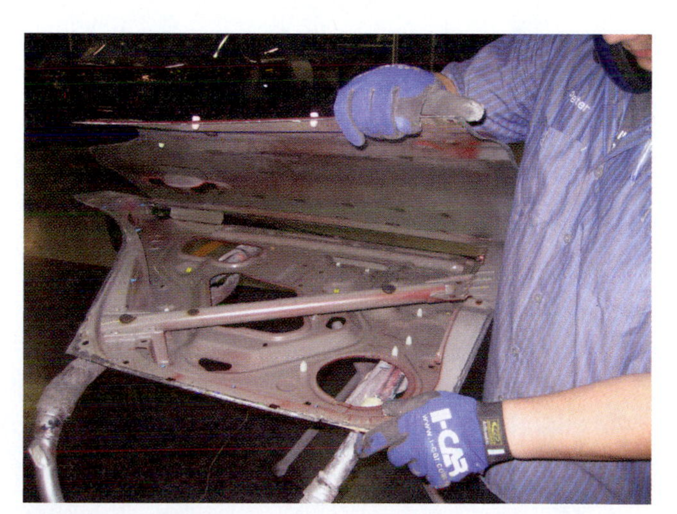

FIGURE 17-22 This door skin is being removed after grinding.

8. Remove sealant, foam, and/or sound deadening material from inside the door to assure proper fit-up and attachment of replacement door skin.
9. While some door skins will be replaced without sectioning, some may be sectioned according to the manufacturer's recommendations. Verify the manufacturer's recommendations for sectioning. Then measure, mark, and cut accordingly.
10. Remove the remainder of the damaged door skin.

TECH TIP There are many tools available to aid in the removal of an old door skin. Commonly a technician will grind the door flange. See **Figure 17-23**. Carefully grind the flange only until the two sides have been revealed. See **Figure 17-24**. If grinding is not done with care, the door frame can be damaged, thus severely complicating the repair!

FIGURE 17-23 To aid in the removal of an old door skin, a technician will commonly grind the door flange.

FIGURE 17-24 The technician should carefully grind the flange only until the two sides have been revealed.

11. Remove the inner welded-on door skin flange with pliers, carefully grinding the spot welds off the door frame without damaging the door frame.
12. Using a heat gun, remove any remaining sealer and/or adhesive.
13. Straighten any door frame damage.
14. Check the door frame for proper fit.
15. Trial-fit the new door skin to the vehicle's repaired door frame.
16. Transfer any parts. This step must be done with the door skin off the frame, because the technician may not have open access upon attachment of the door skin.
17. Apply corrosion protection to bare metal parts.
18. Cut the door skin if sectioning is recommended.
19. Prepare the new door skin for installation. Apply sound deadening material if needed, as access is unavailable after installation. Apply **weld-through primer** as recommended. Drill or punch holes as recommended for welding. If possible, partially flange the door skin before installation.
20. Test-fit and mark the door skin before removal to ensure identical location during final fit-up.
21. Remove skin for final preparation.
22. Apply seam sealer or foam on the intrusion beam as recommended.
23. Install the door for final fit-up. Assure proper fit-up by aligning the marks from prior fit-up.
24. Partially flange the door skin so a final fit-up can be done.
25. Verify proper fit and alignment by installing the door, and adjust the skin if necessary. *This step can be time-sensitive due to the* **open time** *of the adhesive.* The door must be bonded, fit, installed, and adjusted as needed before the adhesive has set up and adjustment is impossible.
26. Complete flanging.
27. Check for alignment and adjust as necessary.

TECH TIP When performing any welding process, a technician should perform test welds before welding on the vehicle. The welder should be set up. Using metal harvested from the removed part, which is the same thickness and type as the replacement part, the technician should perform test plug, lap, and spot welds. The test welds should undergo a **weld destructive test**, and the welder or technique should be adjusted until acceptable high-quality welds are accomplished. The technician can then with confidence perform welds of the highest quality on the vehicle.

28. Weld the recommended type and number of welds.
29. Clean up welds as needed, and apply additional corrosion protection as needed.
30. Apply seam sealer to match pre-accident conditions.
31. Verify fit, refinish, and reassemble.

Quarter Panels

Quarter panel replacement is a common welded-on replacement procedure in a collision repair shop. Though a damaged quarter panel can often be straightened, the procedure to replace one is often the more economical repair. See **Figure 17-25**.

Before a quarter panel is replaced, its dimensions (length, height, and width) must be restored; otherwise, the replacement panel cannot be properly placed. Specific make and model replacement procedures are best obtained from a manufacturer's vehicle-specific repair manual. However, the procedure outlined in the following generic list will give the student an idea as to the steps and precautions that must be followed for the procedure. Remember also that personal, environmental, and fellow worker safety precautions must be followed at all times.

1. Disconnect the negative battery cable and disarm the passive restraint system.
2. Remove adjacent interior parts which may be damaged, and protect the adjacent parts by removing them also.
3. Remove **quarter glass** and **backlite** (rear window).
4. Remove the fuel tank if needed. (Gas filler neck may be within the damaged quarter panel, and thus the neck and tank must be removed.) Store the removed gas tank according to fire codes.
5. Remove the rear bumper, deck lid, and/or **fascia** as needed.

FIGURE 17-25 Though a damaged quarter panel can often be straightened, the procedure to replace one is often the more economical repair.

6. Measure the vehicle to verify that it has been returned to its proper dimensions; repair as needed.
7. **De-trim** the vehicle by removing lights and moldings.
8. Remove any electrical wiring and relocate for protection.
9. Locate and mark spot welds and/or bronze welded areas for removal.
10. Remove any sealers and/or foams prior to weld cutting, to avoid possible toxic smoke.
11. Verify, measure, and mark the sectioning area if recommended.
12. Drill, cut, or otherwise remove spot welds and welded areas.
13. Cut the quarter panel slightly larger than required at the recommended section area so it will have enough excess for later fit-up and trimming.
14. Remove the damaged quarter panel.

TECH TIP If the technician chooses to lift the rear of the vehicle for better access to the lower areas, the vehicle should be supported in such a way that it does not twist the unibody. If the vehicle is supported improperly when the quarter panel is removed and becomes less stable, it will twist, making it very difficult to fit-up the new part. In addition, if a new part is welded in place with the vehicle twisted, it will remain that way when lowered. It is convenient to lift the vehicle while removing the old damaged quarter panel. The vehicle should be placed back in its normal resting position when fitting the new panel to assure that it will be aligned properly!

15. Trim, straighten, and de-burr the mating surfaces of the vehicle.
16. Trial-fit the replacement quarter panel. When proper alignment is obtained, mark the location so the replacement panel can be positioned in that exact location when final alignment is performed.
17. Clean the mating surfaces and mark the recommended areas to replace resistance spot welds if recommended by the manufacturer.
18. Trim and fit the replacement part until an exact fit is obtained.
19. Remove the replacement panel and prepare the mating surfaces for final installation.
20. Apply weld-through primer and corrosion protection according to the vehicle maker's recommendations.
21. Verify the recommendation for the use of adhesion bonding, and apply as necessary.

22. Final-fit the replacement panel and align according to previously made marks. Verify fit by measuring the adjacent panel alignment.
23. Tack the panel in place.
24. Measure and re-check for proper fit prior to final welding. Verify that proper gaps have been met and maintained during the fitting process.
25. Once the final fit has been verified, welding can be completed.
26. Re-check the fit during the welding process to ensure that the panel does not move.
27. Dress and finish the welds, apply corrosion protection as needed, then seal according to recommendations. Sealer should appear the same as the factory-installed sealer. Apply the required sound deadener and foams as needed.
28. Refinish as needed.
29. Reassemble and inspect. Along with inspecting for fit and finish, also inspect the vehicle for wind, water, and dust leaks.

Roof

Roof panels are often adhesive bonded when replaced. See **Figure 17-26**. Therefore, the technician should check the manufacturer's recommendation when replacing a roof panel. The generic replacement procedure that follows is for a welded replacement of a roof panel.

1. Disconnect the negative battery cable to isolate electrical parts while performing roof replacement.
2. Disarm passive restraint while replacing roof panels.
3. Read, understand, and follow all personal, environmental, and fellow worker safety precautions while performing this procedure.
4. De-trim parts on and adjacent to the replacement part.

FIGURE 17-26 The technician should check the manufacturer's recommendation when replacing a roof panel.

5. Remove the windshield, backlite, roof-mounted glass, and quarter glass as needed.
6. Remove interior parts, including headliner, and protect from damage.
7. Remove or relocate wiring, drain tubes, and **side curtain air bags** as needed.
8. Measure and verify that the vehicle is within dimensional specifications. Repair as needed.
9. Cut out the center section of the roof for access, making sure that cross-members and edges of mating surfaces are not damaged.
10. Remove any bonded areas by using a heat gun to soften the bonding agent, making it easier to separate the roof from the adhesive.
11. Identify and remove welds at corners of the roof if needed; then separate the remaining roof panel at factory seams.
12. Straighten the mating surfaces as needed.
13. Clean and dress the mating surfaces.
14. Trial-fit the replacement panel.
15. When proper fit is achieved, mark the roof and mating area for later reference.
16. Verify the recommended replacement attachments. Mark where and how many resistance spot welds will be used. If weld bronze should be applied, mark where, and where adhesive bond will be used.
17. Prepare the surfaces for the appropriate attachments, including weld-through primer and corrosion protection as recommended.
18. Apply adhesive and foam on all mating surfaces—including cross-members—as needed.
19. Apply new replacement parts and position as previously marked.
20. Verify the height, fit, and gaps are correct.
21. Weld as necessary, or secure and hold until the bonding agent has cured.
22. **Dress welds** as needed.
23. Apply seam sealers and sound deadener as needed.
24. Refinish as required.
25. Reassemble and inspect. Along with inspecting for fit and finish, inspect the vehicle for wind, water, and dust leaks.

Rear Body

Replacing a rear body panel more often than not requires that the vehicle be brought back to pre-accident dimensional tolerances. The procedure for replacing a rear body panel begins after the technician has repaired and measured to verify that the vehicle's dimensions are within tolerances. See **Figure 17-27**. Students should review the procedure for straightening a vehicle, covered

FIGURE 17-27 Replacing a rear body panel begins after the technician has repaired and measured to verify that the vehicle's dimensions are within tolerances.

in Chapter 14, as needed. The technician then follows these steps:

1. Remove or protect adjacent parts that may become damaged during the repair process.
2. Make sure that the negative terminal of the battery is disconnected.
3. Re-route any electrical wires to protect them from damage.
4. Inspect and locate spot weld locations.
5. Remove spot welds, taking care not to damage mating surfaces.
6. Remove damaged parts. Either note or keep damaged parts to verify locations of labels that must be replaced when repairs have been completed.
7. Grind the remaining **weld nuggets** on mating surfaces, and then clean off old sealer.
8. Inspect and straighten mating surfaces as needed.
9. Apply the recommended weld-through primer or corrosion protection as needed.
10. Trial-fit the replacement part.
11. Measure the temporarily attached replacement part to assure that it is located in proper position. This should include having proper gaps on the deck lid.
12. Refer to the service manual to verify the proper number and locations for spot welds, and mark locations on the replacement part.
13. Remove the part and prepare the mating flange on the replacement part. Prepare the marked locations for spot weld, apply weld-through primer, and apply corrosion protections according to recommendations.
14. Re-attach the rear body panel and test-fit all parts—including the taillight, deck lid, floor pan, and so on—to be certain that all are positioned properly.
15. Clamp in place and tack weld.
16. Remove clamps and verify that the fit has remained in place.
17. Complete all required welds.
18. While welding, periodically check the position of the replacement part to verify that it remains in the proper position.
19. After welding, measure and fit all openings and parts, and check that the proper gaps have been maintained.
20. Dress welds as needed.
21. Apply corrosion protection that may have been damaged during the welding process and apply sound deadener and sealers as recommended.
22. Refinish.
23. Reassemble and inspect. Along with inspecting for fit and finish, also inspect the vehicle for wind, water, and dust leaks.

Summary

- Replacing welded exterior parts requires skills in welding, measuring, and grinding, along with many other skills.
- One of the most valuable skills, however, is preciseness.
- The proper positioning is critical because once a part is welded, no further adjustment can be made.
- Fit-up is further complicated by the joining of two or more parts, with their position in length, width, and height being critical.
- Once welded in place, most of these parts become part of the vehicle's structural integrity and must perform as designed prior to being damaged.
- Maintaining the vehicle's original crashworthiness is critical to the safety of passengers if the vehicle is involved in a subsequent collision.
- The precise checking and rechecking is to verify that a replacement part is placed in proper position and remains there as the work is completed, and also that the recommended procedures are followed—in order

for a vehicle's structural integrity and safety to be maintained.

- In addition to the importance of positioning, it is critical that all welds the technician completes be precise and of the highest quality.
- Customer satisfaction is generally judged by the fit and finish of the vehicle.
- In the case of replacement of welded-on parts, often the more critical aspects of the repair are areas such as safety, durability, and structural integrity.
- These qualities cannot be easily inspected for; to be assured, they depend upon the repair technician's thorough knowledge and precise performance of replacement procedures.

Key Terms

backlite
cosmetic
crashworthiness
deck lid
de-trim
door shell
door skin
dress welds
fascia
heat affect zone
intrusion beam
open time
original equipment (OE)
quarter glass
sectioning
side curtain air bags
structural integrity
supersedes
tack weld
technical service bulletin (TSB)
tolerance
I-CAR Uniform Procedures for Collision Repair (UPCR)
weld destructive test
weld nugget
weld-through primer

Review

ASE Review Questions

1. Technician A says that all welded-on panels are structural panels. Technician B says that although welded-on panels are often structural parts, not all are. Who is correct?
   A. Technician A only
   B. Technician B only
   C. Both Technicians A and B
   D. Neither Technician A nor B
2. Technician A says that if a part is bent, it should be repaired. Technician B says that if a part is kinked, it should be replaced. Who is correct?
   A. Technician A only
   B. Technician B only
   C. Both Technicians A and B
   D. Neither Technician A nor B
3. Technician A says that the decision about when to replace a part can be based on the manufacturer's recommendation. Technician B says that the cost of repair is often used to judge if a part should be replaced instead of being repaired. Who is correct?
   A. Technician A only
   B. Technician B only
   C. Both Technicians A and B
   D. Neither Technician A nor B

4. Technician A says that all welded replacement parts are installed at factory seams. Technician B says that sectioning cut lines are always left to the technician's discretion. Who is correct?
   A. Technician A only
   B. Technician B only
   C. Both Technicians A and B
   D. Neither Technician A nor B
5. Technician A says that crashworthiness is how a vehicle is designed to perform in a collision. Technician B says that crashworthiness is how a vehicle is designed to absorb energy during a collision. Who is correct?
   A. Technician A only
   B. Technician B only
   C. Both Technicians A and B
   D. Neither Technician A nor B
6. Technician A says that a lap joint is the type of joint used in sectioning all welded-on parts. Technician B says that a lap joint is often used to repair trunk floors. Who is correct?
   A. Technician A only
   B. Technician B only
   C. Both Technicians A and B
   D. Neither Technician A nor B
7. Technician A says a butt joint can be either used with a backer or without. Technician B says that to make a strong butt weld a backer must be used. Who is correct?
   A. Technician A only
   B. Technician B only
   C. Both Technicians A and B
   D. Neither Technician A nor B
8. Technician A says that spot welds are removed using a hole punch. Technician B says that an air chisel is the best tool for removing factory spot welds. Who is correct?
   A. Technician A only
   B. Technician B only
   C. Both Technicians A and B
   D. Neither Technician A nor B
9. Technician A says that the manufacturer recommendation for application of adhesive bonding should be followed precisely. Technician B says that all adhesive bonding is mixed at the same ratio. Who is correct?
   A. Technician A only
   B. Technician B only
   C. Both Technicians A and B
   D. Neither Technician A nor B
10. Technician A says that measuring and fit-up are critical when replacing welded-on replacement parts. Technician B says that when welding-on a replacement part, the technician should check the fit often because the act of welding may move the part slightly due to heat. Who is correct?
    A. Technician A only
    B. Technician B only
    C. Both Technicians A and B
    D. Neither Technician A nor B
11. Technician A says that all replacement door skins are adhesive-bonded in place without the use of welds. Technician B says that all replacement door skins are welded in place without the use of adhesive bonding. Who is correct?
    A. Technician A only
    B. Technician B only
    C. Both Technicians A and B
    D. Neither Technician A nor B
12. Technician A says that a technician may use a combination of welds and adhesive bonding when replacing a quarter panel. Technician B says that before a quarter panel is replaced it should be measured for length, width, and height. Who is correct?
    A. Technician A only
    B. Technician B only
    C. Both Technicians A and B
    D. Neither Technician A nor B
13. Technician A says that side curtain air bags may accidentally deploy if not disabled when replacing a roof panel. Technician B says that the back light should be removed when replacing a roof panel. Who is correct?
    A. Technician A only
    B. Technician B only
    C. Both Technicians A and B
    D. Neither Technician A nor B
14. Technician A says that factory spot welds are easily found. Technician B says that weld-through primer is always used when welding-on a replacement panel. Who is correct?
    A. Technician A only
    B. Technician B only
    C. Both Technicians A and B
    D. Neither Technician A nor B

Essay Questions

1. Describe what should be considered in the decision to replace vs. repair a part.
2. Explain what a kinked part is.
3. In your own words, explain crashworthiness.
4. Where would a technician find a recommended procedure for sectioning a replacement part?
5. What is the best tool for removing a factory spot weld, and why?
6. What is destructive testing of a weld?

Topic Related Math Questions

1. If a factory quarter panel had 15 spot welds and the recommendation was to increase the factory welds by 33 percent, how many total welds should a technician put on the replacement part?
2. A quarter panel is 54 inches long with a recommendation to place a spot weld every 1.5 inches. How many spot welds are needed?
3. A 39-inch long replacement rail must have 13.5 inches removed. What will be the length of the remaining rail?
4. A replacement roof is 54 inches long and 32.5 inches wide. One tube of adhesive will cover 76 linear inches. How many tubes are needed?

Critical Thinking Questions

1. The replacement part cost and the labor cost for the part's replacement total $250.00. The cost of repair is $197.50. Should the part be replaced or repaired, and why?
2. Which is a better repair, welding or adhesive bonding, and why?
3. Which is a better repair, a sectioning or replacing at factory seams?
4. Why is crashworthiness important?
5. Why should a technician perform destructive testing on practice welds before welding in a replacement part?

Lab Activities

1. Using a door that has been removed from a vehicle, remove the outer door skin.
2. Measure and fit a new door skin to the above door.
3. Set up and practice an open butt weld.
4. Set up and practice a butt weld with a backer.
5. Set up and practice a lap weld.
6. Set up and practice plug welds.

Chapter 18

Plastic Repair

OBJECTIVES

Upon completing this chapter, you should be able to:

- Explain the personal, shop, and environmental safety precautions a technician must take when working with plastics.
- Compare and contrast the differences between thermoset plastics and thermoplastics.
- Explain how to identify the different types of plastics.
- List the different types of plastics which are considered polyolefin.
- Describe the processes for the plastic float test and sand test, and explain the results.
- Explain what "plastic memory" is and how technicians can use it to their advantage.
- Describe the process of hot air plastic welding.
- List the steps for airless plastic welding.
- Describe the differences between a two-sided repair and a single-sided repair.
- Distinguish between sheet moldable compound (SMC) and fiber-reinforced plastic (FRP).

INTRODUCTION

Plastic automobile parts first appeared in the late 1930s and the 1940s in the form of fiberglass, the material we now refer to as fiber-reinforced plastic. Plastic parts interested automotive engineers back then because they were extremely light compared to their steel counterparts, and also because they would not rust. By the 1950s, fiberglass production was simplified, changing from a difficult and time-consuming hand lay-up to a much faster production process of chop-gun application. In 1953, the first all-fiberglass car, the Corvette, went into production, and automotive plastic (which is now called composite plastic) began to come into use.

Composite plastics have changed greatly since that time, though. In the 1960s, plastic bumpers were first developed. Then, in the 1980s, some fiberglass items were replaced with a more stable material—sheet moldable compound (SMC)—and technicians were faced with a major change in repair procedures. The SMC repair methods were different from those used for fiberglass; when technicians naively used old fiberglass repair methods, the repair failed. Plastic bumpers were soon manufactured from a staggering array of plastics, many with unpronounceable names. So many plastics became available—each with their own differing repair recommendations—that repair technicians were unable to remain up-to-date on recommended repair procedures for them all. As a result, plastic parts that could have been repaired were instead often replaced.

Plastic or composite parts continued to improve, though. As gas prices rose, engineers developed more and more plastic into automobile parts to lower the vehicle's weight, thus improving its gas mileage. Repairers began to ask the question, "How can we repair these parts successfully?" With the improvement of plastic composite repair materials, technicians were able to make plastic repairs without needing to keep a staggering array of differing repair materials on hand.

At one point repairers—before they could begin the repair—had to identify what materials the part was made from, then pick the specific repair for that type of plastic. Now, plastic repair procedures have become greatly simplified. Although repairers still need to know which method is best for the type of plastic being repaired, the process is faster and more severely damaged parts have become candidates for repair.

This chapter will discuss how to identify plastic, the safety precautions needed to work with plastic, welding and adhesive repairs, repair materials, and repair procedures for flexible plastic repair and reinforced plastic. The chapter will also include a discussion of adhesive bonding.

SAFETY

Repairing plastic parts is no different than working with other potentially hazardous materials, whether the parts are rigid or flexible, and whether they are made from a thermoplastic or a thermoset plastic. Therefore, as with other materials, the collision repair technician should be constantly concerned with assuring personal, environmental, and workplace safety. Common safety precautions include the following:

1. Read, understand, and follow all safety precautions set by the manufacturer.
2. Read, understand, and follow the MSDS related to the materials being used.
3. Wear safety glasses at all times.
4. When using liquid chemicals that could splash, wear safety goggles.
5. Wear the recommended respirator, gloves, and other safety clothing.
6. Always work in a well-ventilated area.
7. If resin or other catalysts come in contact with your skin, follow the first aid treatment set out in the MSDS.

Collision repair technicians should think about safety at all times.

PLASTIC CLASSIFICATION

Plastics or composite materials refer to a wide range of manmade synthetic components created from substances such as oil, coal, natural gas, soybeans, and corn, just to name a few. These plastics are grouped into two large categories: thermoplastics and thermoset plastics.

Thermoplastics can be repeatedly softened and shaped when heated without changing their chemical makeup. These plastics soften when heated; then, when placed into a mold, they can be shaped. When the molded plastic is allowed to cool, it will harden. This group of plastics is repaired by plastic welding and with adhesives.

TECH TIP Thermoset plastics, which will not soften when heated, cannot be plastic welded; they must be repaired with a chemical bond. Thermoset plastics—when in a soft form—will harden either by heating them, by adding a catalyst, or by applying ultraviolet light. As they harden, their chemical properties change, and they will not re-soften. Therefore, they are not repairable by the plastic welding process.

TECH TIP Plastic welding is a process in which the technician melts plastic with the use of heat, then adds a filler rod and allows the repair to cool. See **Figure 18-1**. The repair is accomplished by the fusion of the part and the filler rod, and is thus called welding. This process is similar to gas welding on steel. It can only be performed on plastics that are capable of softening with heat and then re-hardening when cooled, or on thermoplastic parts.

FIGURE 18-1 In plastic welding, the technician melts plastic with the use of heat, then adds a filler rod and allows the repair to cool.

Thermoset plastics will permanently take the shape of a part by heating, using a catalyst, or applying ultraviolet light. They will not re-soften by heating or through the reapplication of a catalyst or light. These are repaired with chemical adhesives, not welded.

Composite plastics are combinations of two or more plastics formulated to achieve specific performance characteristics.

TYPES OF PLASTIC

Because of the differences in types of repair methods a technician may use, the levels of identification of plastic will differ. If the repair must be accomplished by the welding of a thermoplastic part, the repair technician will need to know specifically from what type of thermoplastic the part was made, so that the proper filler rod can be chosen. If the repair method will be adhesive bonding and filler, the level of plastic identification will be determined by reading the recommendation of the repair material manufacturer. The repair material manufacturer will also determine the method used to identify the type of plastic.

The most specific way to identify a plastic is by its **International Organization for Standardization (ISO)** code. See **Figure 18-2**. An ISO code is a standard set of letter codes that identify either the type of plastic the part was made from, or the manufacturing process the part was made from. The ISO code may be molded into the back side of the part. See **Figure 18-3**. However, the location of this code is not standardized, and often the part must be removed in order to find the code. Furthermore, because the placement of the code on plastic parts is not required, sometimes the code will not be found even when the part is removed.

The repair technician can also locate the type of plastic a part is made from in a service repair manual, though a mid-year change of plastic by the manufacturer may mean that the information in a repair manual will be obsolete. One of the most reliable ways to identify which plastics a specific repair material can repair is by seeking out the recommendation of that repair material maker. If an ISO code cannot be found, the plastic must be tested to identify which category it falls into. The three most reliable tests for identification are the float test, sanding test, and flexibility comparison.

TECH TIP **Polyolefins** are defined as any of the polymers and copolymers of the ethylene family of hydrocarbons. This group of thermoplastic plastics is often used to make flexible bumpers called fascias. See **Figure 18-4**.

Float Test

The **float test** is used to identify thermoplastics. A small sliver is trimmed from the plastic to be repaired and placed in a clear glass filled with water. If the plastic sliver floats, it is a thermoplastic. If the sliver sinks, the plastic may not be a thermoplastic. The reason a piece of plastic that sinks is not conclusively a thermoset plastic is that many other things may also cause it to sink. For example, perhaps not all of the refinish was removed, and the weight of that contaminant caused the sliver to sink. Another reason is that many plastics found on vehicles today could be a plastic blend, which could either float or sink. See **Figure 18-5** and **Figure 18-6**. This test is only conclusive when a sliver floats.

Sand Test

The **sand test** may also help identify whether a plastic is a thermoplastic polyolefin. See **Figure 18-7**. On the back side of a plastic part, the technician sands an area with an angle grinder, using a 36-grit sanding disk and medium speed (about 5,000 rpm). If the plastic smears, melts, and is waxy, the test indicates that the plastic is a polyolefin. If the part sands dry, producing dust, the plastic part may not be a polyolefin. This test, like the float test, does not conclusively show if the part is a polyolefin or not. Mixtures may sand either by smearing or by producing dust.

Flexibility Test

Some plastic part repair kits use a **flexibility test** to classify plastic parts by their flexibility. See **Figure 18-8**. Plastics might be classified as:

1. Flexible
2. Semi-flexible
3. Semi-rigid
4. Rigid

The filler or adhesive being used may rely either on the part's ability to flex, or on where it is mounted on the vehicle.

PLASTIC MEMORY

Plastic has memory; in other words, if deformed by a collision, plastic remembers what its original shape was and can return to that molding. This **plastic memory** quality helps technicians as they reshape the plastic, and heat helps plastic return to its original shape. By heating a plastic part and gently reshaping it by hand (wearing gloves), the technician can return the part to its original—or near-original—shape. The part should be warmed with gentle heat—either by heat gun or ultraviolet light—or baked in the spray booth. The objective is not to melt the plastic but to raise the temperature so the plastic can be manipulated. Once the part is back to its original shape, it should be allowed to cool naturally. Though heat alone cannot remove all the defects, especially if the area is stretched, it is a useful tool in the initial repair process.

Identifying Symbol	Chemical Composition	Typical Usage	Common or Trade Names	Suggested Repair Method	Repair Tips
PUR (RIM, RRIM)	Thermoset Polyurethane	Bumper covers, front and rear body panels, filler panels	Elastoflex, Bayflex, Specflex (Reaction Injection Molding)	Weld with urethane rod (5003R1) or Uni-Weld (5003R8)	Don't try to melt the base material; just melt rod into the V-groove.
TPU (TPUR)	Thermoplastic Polyurethane	Bumper covers, soft filler panels, gravel deflectors, rocker panel covers	Pellethane, Estane, Roylar, Texin, Desmopan	Weld with urethane rod (5003R1) or Uni-Weld (5003R8)	
TPO, EPM, TEO	Polypropylene + Ethylene Propylene rubber (at least 20%) (polyolefin)	Bumper covers, valence panels, fascias, air dams, dashboards, grilles	TPO (Thermoplastic Olefin), TPR (Thermoplastic Rubber), EPI, EPII, PTO, HiFax	Weld with Uni-Weld (5003R8) or TPO Blended Gray rod (5003R5)	Use adhesion promoter before applying filler or coating.
PP	Polypropylene (polyolefin)	Bumper covers, deflector panels, interior moldings, radiator shrouds, inner fenders	Profax, Oreflo, Marlex, Novolen, Carlona	Weld with Uni-Weld (5003R8) or Polypropylene Black rod (5003R2)	Use adhesion promoter before applying filler or coating.
PC + PBT	Polycarbonate + Polybutylene Terephthalate	Bumper covers (Ford)	Xenoy (GE)	Weld with Polycarbonate Clear rod (5003R7) or Uni-Weld (5003R8)	Preheat groove before welding with polycarbon-ate rod.
PPE + PA (PPO+PA)	Polyphenylene Ether + Polyamide	Fenders (Saturn, GM), exterior trim	Noryl GTX (GE)	Weld with Nylon (5003R6), Uni-Weld Ribbon (5003R8), two-part epoxy system, or instant adhesive.	Preheat groove before welding with nylon rod. Use fiberglass mat with instant adhesive.
ABS	Acrylonitrile Butadiene Styrene	Instrument clusters, trim moldings, consoles, armrest supports, grilles	Cycolac (GE), Magnum (Dow), Lustran (Monsanto)	Weld with ABS White rod (5003R3) or repair with Insta-Weld instant adhesive	Instant adhesive works great on ABS.
PC + ABS	Polycarbonate + Acrylonitrile Butadiene Styrene	Door skins (Saturn), instrument panels	Pulse (Dow), Bayblend (Bayer), Cycoloy (GE)	Weld with Polycarbonate rod (5003R7), ABS rod (5003R3), two-part epoxy, or instant adhesive.	Preheat groove before welding. Use fiberglass mat with instant adhesive.
UP, EP	Unsaturated Polyester, Epoxy (Thermoset)	Fender extensions, hoods, roofs, decklids, instrument housings	SMC, Fiberglass, FRP	Repair with two-part epoxy system (2020 or 2021) or polyester resin and glass cloth.	This material cannot be repaired with the welder.
PE	Polyethylene (polyolefin)	Inner fender panels, valences, spoilers, interior trim panels	Lacqtene, Lupolen, Dowlex, Hostalen	Weld with Polyethylene Opaque White rod (5003R4)	Use adhesion promoter before applying filler or coating.
PC	Polycarbonate	Interior rigid trim panels, valence panels	Lexan, Merlon, Calibre	Weld with Polycarbonate Clear rod (5003R7)	Preheat groove before welding.
PA	Polyamide	Radiator tanks, headlamp bezels, quarter panel extensions, exterior trim finish parts	Nylon, Capron, Celanese, Zytel, Rilsan, Orgamide, Vydyne, Minion	Weld with Nylon Opaque White rod (5003R6) or Uni-Weld (5003R8)	Preheat groove before welding, especially on radiator tanks.
TEEE	Thermoplastic Ether, Ester Elastomer	Bumper fascias (Bonneville SSE, Park Ave., '91-'96 Vette front), rocker panel covers (Camaro & Firebird)	Bexloy V (DuPont)	Weld with Uni-Weld (5003R8) or two-part epoxy system (2000 or 2020)	
PET	Polyethylene Terephthalate + Polyester	Fenders (Chrysler LH)	Bexloy K (DuPont), Vander (Hoechst)	Weld with Uni-Weld (5003R8) or two-part epoxy system (2020 or 2021)	
EEBC	Ether Ester Block Copolymer	Rocker cover moldings, fender extensions ('91-'96 DeVille)	Lomod (GE)	Weld with Uni-Weld (5003R8) or two-part epoxy system (2010)	
EMA	Ethylene/Methacrylic Acid	Bumper covers (Dodge Neon)	Bexloy W (DuPont)	Weld with Uni-Weld (5003R8)	

FIGURE 18-2 An ISO code is a standard set of letter codes that identify either the type of plastic the part was made from, or the manufacturing process the part was made from.

FIGURE 18-3 The location of the ISO code is not standardized, and often the part must be removed in order to find the code.

FIGURE 18-4 The group of thermoplastic plastics called polyolefins is often used to make flexible bumpers called fascias.

FIGURE 18-5 If the plastic sliver floats during a float test, it is a thermoplastic.

FIGURE 18-6 If the sliver sinks during a float test, the plastic may not be a thermoplastic, depending on other factors.

FIGURE 18-7 The sand test may also help identify whether a plastic is a thermoplastic polyolefin based on the presence of smears or dust.

PLASTIC WELDING

Plastic welding uses heat—either hot air or airless welding—and often a plastic filler rod to fuse the broken plastic together. The fusing of plastic takes place when the base plastic is melted by the heat source and a molten puddle is formed. The correct plastic filler rod is melted and then joins with the molten base plastic. When the weld is complete, the plastic is allowed to cool naturally.

Hot Air Welding

Hot air welders are usually connected to a compressed air supply such as the shop's air compressors. The welder has an air orifice that heats that base plastic in the area that will be repaired, and a tube that a plastic welding rod is passed through which preheats the rod. As the base plastic is heated, the welding tube is pulled over the repair

FIGURE 18-8 This plastic part is bent past 90 degrees, showing its flexibility.

FIGURE 18-10 A tacking tip helps hold the part in proper alignment while welding.

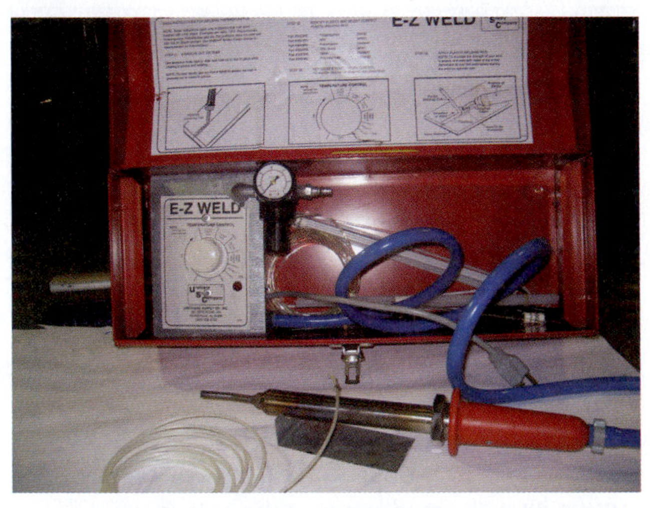

FIGURE 18-9 Hot air welders are usually connected to a compressed air supply such as the shop's air compressors.

FIGURE 18-11 A round tip is used for smoothing plastic welds.

area as the rod is pushed down into the weld area. This forms the weld. See **Figure 18-9**.

Hot air welders sometimes come with different tips, such as a tacking tip which does not use a filler rod. See **Figure 18-10**. A plastic tack is like other tacks, except that it is only temporary and can be taken apart later if needed. Another type of tip—a round one—is used to fill holes in plastic. See **Figure 18-11**. Still another tip—a production tip—is designed to weld longer and larger areas.

Hot air plastic welders require more technical skill to use than airless plastic welders. Care should be taken when operating them, because the hot air may be as high as 1,100°F.

Airless Plastic Welding

Airless plastic welders are more common than hot air welders, and they require a little less practice to develop skill in their use. See **Figure 18-12**. They are powered by 110-volt shop electricity, have adjustable heat controls, and are rated at about 80 watts at the tip. They may have additional tips, specifically without a melt tube. Kits often come with a welder that includes welding rods (assorted) and extra tips.

Airless Welding Procedure

Before welding plastic, the technician must choose the proper welding rod. The most precise way to do this is to find the ISO identification code, then consult the welding rod manufacturer for the recommended welding filler that is best for that type of plastic. If the ISO code cannot be found, the technician must perform a welding rod test. To do this, first clean the back of the plastic part or an area that will not be seen. Prepare the airless welder by cleaning the melt tube with a 3 mm (1/8″) drill, and also clean

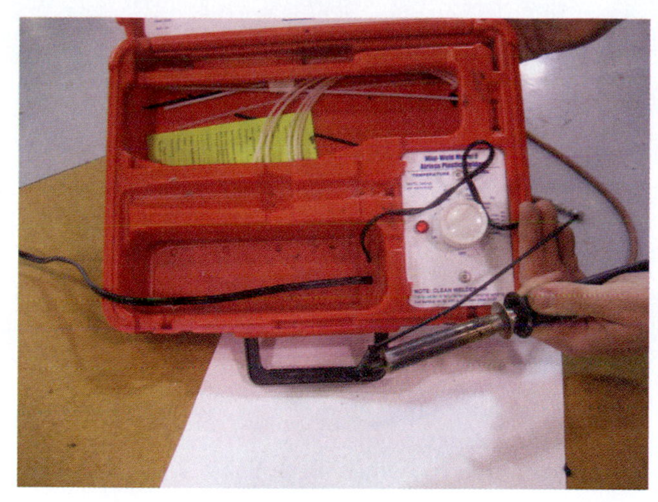

FIGURE 18-12 An airless plastic welder may have different settings for different types of plastic.

the welding shoe by sanding all the previous welding debris with a P220-grit sandpaper. Then weld a rod to the plastic, let it cool, and then pull it off. Do this with a number of rods to find the one that holds the best. This weld rod test can—and should—be performed for choosing the proper rod for hot air welding as well, though it is not the most common method of plastic welding.

Cleaning. The first step when preparing a plastic part for repair (and in fact, for refinishing also) is cleaning. It is a three-step method: soap and water, isopropyl alcohol, and wax and grease remover. See **Figure 18-13**. Plastic bumpers are manufactured using a mold release agent, which helps when removing the part from the die. This release agent is injected into the mold as a pellet mixed with the plastic pellets during manufacturing, and will continue to rise to the surface long after manufacturing. Because of this, plastic parts must be washed with soap and hot water to dissolve any water contamination, even if the bumper being repaired is older. These parts should be washed both inside and out. After they are dry (by forcing them dry with air or using another method), they should be washed with an isopropyl alcohol. The alcohol will act as an anti-static agent and will remove any remaining contaminants that are not water-soluble. The third step is to clean with a mild wax and grease remover, which ensures the removal of any remaining substance. Strong solvents used on open raw plastic may cause swelling and could contaminate the repair area. To avoid such contamination, the technician should never clean plastic with lacquer thinner or strong reducers.

Tapering. Plastic welding is successful when a large enough surface is involved so that a welding rod is fused into the old base plastic. The welder's shoe cannot heat plastic that it does not touch. Therefore, to make sure that the weld will be strong, the repair area must be V'ed out—that is, the edges tapered out in a V shape so filler material will have greater contact with the original plastic. See **Figure 18-14**. This V-shaped cut should be made on both sides, if both sides have access. It should be wide enough that the welding shoe can comfortably touch the lower part of the V. A rotary tool works well for this. If there is no access to both sides, the V should extend to the lower level of the plastic part so that the weld can fully penetrate the weld area. To reinforce a weld, wire screen is imbedded on the back of the welded area. See **Figure 18-15**.

FIGURE 18-13 This plastic bumper is being washed with soap and water to remove water-soluble contaminants.

Following the welding, the surface can be sanded to level. This can be done by using first a low speed pistol grip grinder, then a dual action sander (DA). The repair area may need filling to complete the repair process (which will be covered in the adhesive repair section).

Though both types of plastic welding—airless and hot air—can do an acceptable repair job, these methods have been overshadowed by chemical adhesive repairs. The skill level, supplies, testing, and time needed for plastic welding are all disadvantages over adhesive bonding. Welding plastic repairs have not enjoyed the popularity found with adhesive bonding.

FIGURE 18-14 To make sure that the weld will be strong, the repair area must be V'ed out—that is, the edges tapered out in a V shape so filler material will have greater contact with the original plastic.

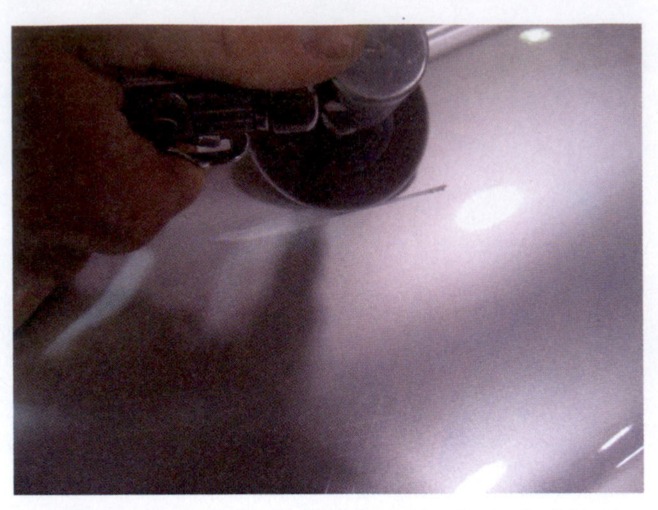

FIGURE 18-16 In this one-sided repair, the technician is preparing a V for repair.

FIGURE 18-15 To reinforce a weld, wire screen is imbedded on the back of the two-sided repair area.

FIGURE 18-17 In this two-sided repair, the back side has a temporary backer in place.

ADHESIVE REPAIR

Chemical adhesive bonding has become very popular with collision repair technicians due to the advances that have been made in ease of application, sandability, and product durability. These adhesives can be epoxy, urethane, acrylic, or even cyanoacrylate ("super glue"), though cyanoacrylate may not be strong enough to be used alone. They come as bonding agents and filling agents, which sand better than those that are purely bonding adhesives.

Many competing adhesive bonding products are available to choose from, each with a slightly different repair method that should be followed for maximum strength of repair. Their cleaning, straightening (heat), and setup methods are similar. The types of repairs are also similar; they include straightening or stress-relieving with heat, performing one-sided gouge repairs with primarily filling agents, and using two-sided repairs in cases where the plastic part has been damaged completely through the plastic. See **Figure 18-16**. **Two-sided repairs** generally require that the part be removed from the vehicle so access to the back is possible. See **Figure 18-17**.

REPAIR MATERIAL

Adhesive plastic repair materials are two-part: the first material is the resin and the second one is the catalyst.

When the catalyst is mixed with resin, they react and become hard. Plastic repair material comes in differing amounts of flexibility to match the plastic that is to be repaired. If a repair adhesive sets up with less flexibility than the original plastic, the repair will become visible and will not perform as it should. Catalysts should not be mixed with resins that they are not intended for. Because of this, **adhesive repair** companies supply the resin and catalyst in separate tubes that are packaged together. See **Figure 18-18**. Special application guns are also supplied which dispense the resin and catalyst together. See **Figure 18-19**. Each resin and catalyst combination has a very specific rate at which they should be mixed; some require a 1:1 ratio, while others may be 1:1/2 as the plunger of the dispensing gun is depressed. See **Figure 18-20** and **Figure 18-21**. Because the diameter of each tube is different, the precise amount of resin to catalyst is dispensed. The precise combination of the correct amounts is critical to the repair's success. In earlier times, when catalyst and resin were shipped in separate containers, technicians did not always combine them as directed, which often caused repair failures.

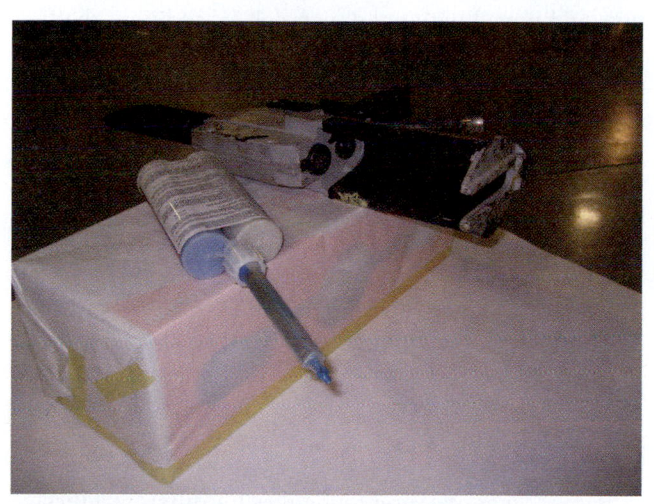

FIGURE 18-18 Adhesive repair companies supply the resin and catalyst in separate tubes that are packaged together.

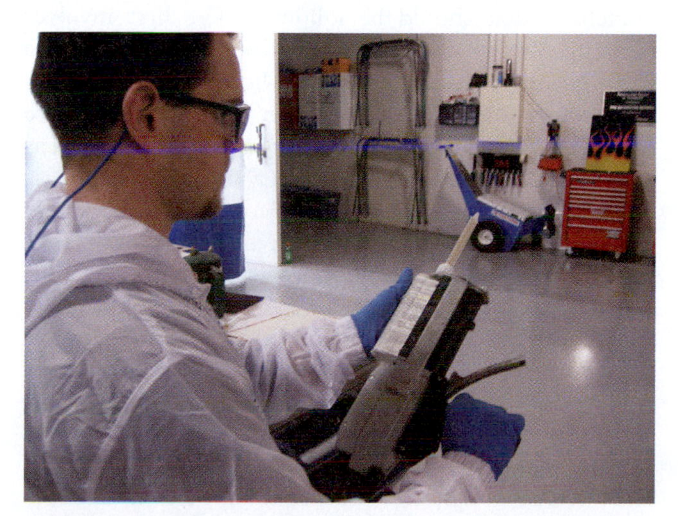

FIGURE 18-19 Special application guns dispense the resin and catalyst together.

FIGURE 18-20 Some resin and catalyst combinations require a 1:1 ratio, in which equal amounts are dispensed.

FIGURE 18-21 Some resin and catalyst combinations use a 1:1/2 ratio, in which different amounts are dispensed from the different tubes.

Mixing is also critical. Both agents—catalyst and resin—must be thoroughly mixed so that the repair product will perform correctly. In the past, hand mixing proved to be insufficient to adequately combine the two materials. To aid technicians, a mixing tube was developed. See **Figure 18-22**. This tube attaches to the end of the combined catalyst and resin tubes. As the plunger dispenses the two agents, they are mixed in the dispensing tube as they pass through. See **Figure 18-23**.

The only precaution that technicians must take before attaching a mixing tube is to squeeze out a small amount of both catalyst and resin—about 1 inch—to assure that neither has reacted to either water or air. See **Figure 18-24**. Due to the expense of plastic repair material, some technicians are reluctant to do this step, but it is necessary to assure that the material used for the repair is fresh. This step should also be done before the mixing tube is attached so that no contaminated non-fresh material is placed into the

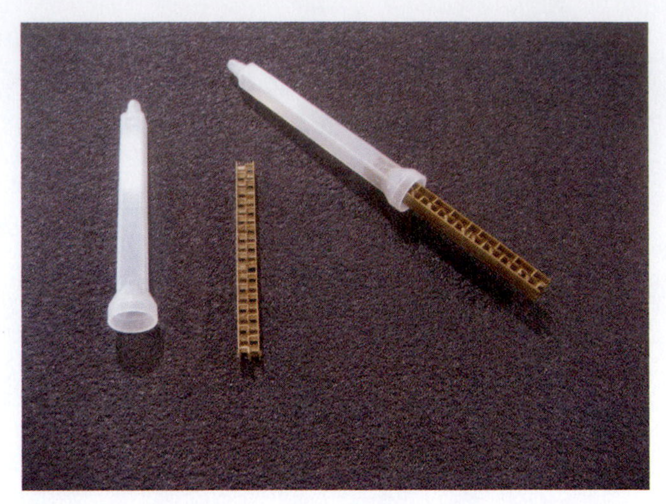

FIGURE 18-22 To aid technicians, a mixing tube was developed that attaches to the end of the combined catalyst and resin tubes.

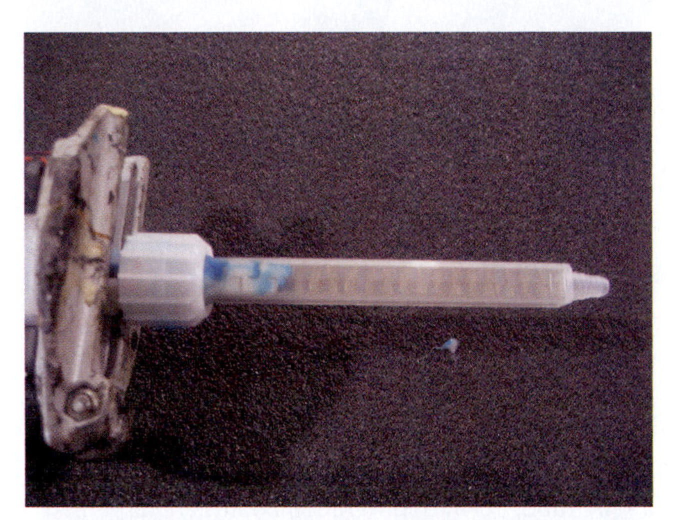

FIGURE 18-23 As the plunger dispenses resin and catalyst, they are mixed in the dispensing tube as they pass through.

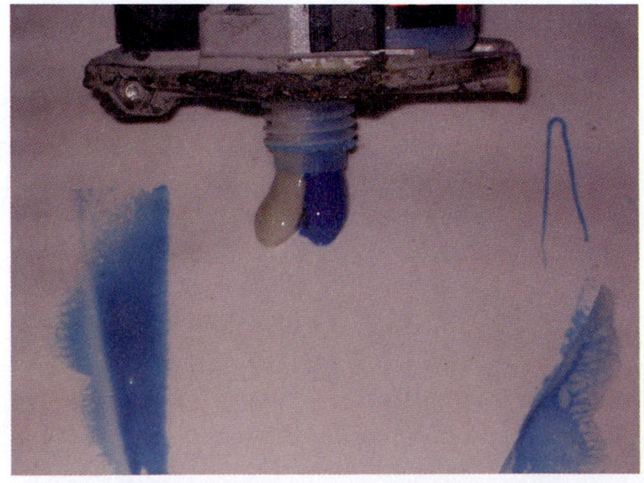

FIGURE 18-24 By dispensing 1 inch of plastic repair material before attaching the mixing tube, the technician ensures that no dry or contaminated material is used.

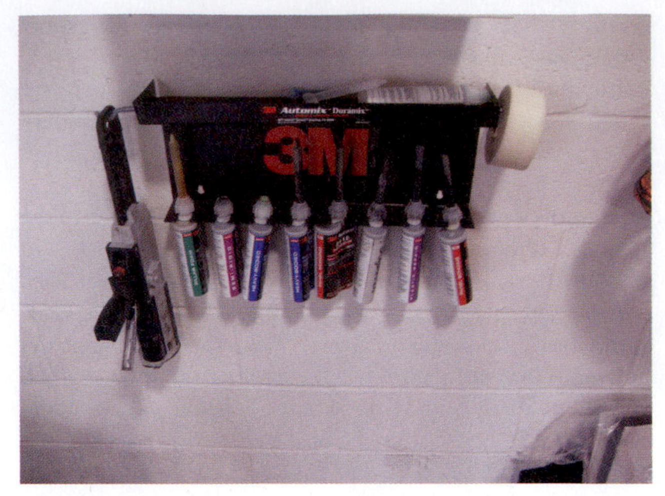

FIGURE 18-25 Leaving the dispensing tube on half-used tubes of material will keep the unused portion from catalyzing or drying.

tube and mixed with good material. In addition, the first 3 to 4 inches of mixed material should be run out and discarded to ensure that the adhesive is completely mixed.

If there is material remaining in the container when the repair is finished, the dispensing tube is left in place for storage. See **Figure 18-25**. The combined material in the dispensing tube will catalyze and keep air and moisture from the remaining material. Then, when the repair material is needed again, the technician removes the old dispensing tube and squeezes out and discards a small amount of material. Once a new tube is installed, the material is ready to use again.

FLEXIBLE PLASTIC REPAIR

Because manufacturer guidelines for different adhesive repair materials may be different, each manufacturer's specific recommendations for repairing **flexible plastic** should be followed. However, there are also some general guidelines that should be followed. The first involves cleaning. As in plastic welding and refinishing, it uses a triple-clean process: first soap and hot water wash, to remove water-soluble materials; then isopropyl alcohol to clean and remove alcohol-soluble contaminants as well as static electricity from the part; then a mild wax and grease remover, to assure that the part is clean. After the cleaning process, the technician should stress-relieve the plastic with heat to return the part to its original shape. Once the shaping is completed, the technician must decide whether a single-sided repair is sufficient or a two-sided repair is necessary.

Single-Sided Repair

After the surface to be repaired has been triple-cleaned as described earlier, the surface should be stress-relieved with heat to return it to its original shape before repair. See **Figure 18-26**. The area to be repaired must then be tapered

FIGURE 18-26 In a single-sided repair, the surface should be stress-relieved with heat after cleaning to return it to its original shape.

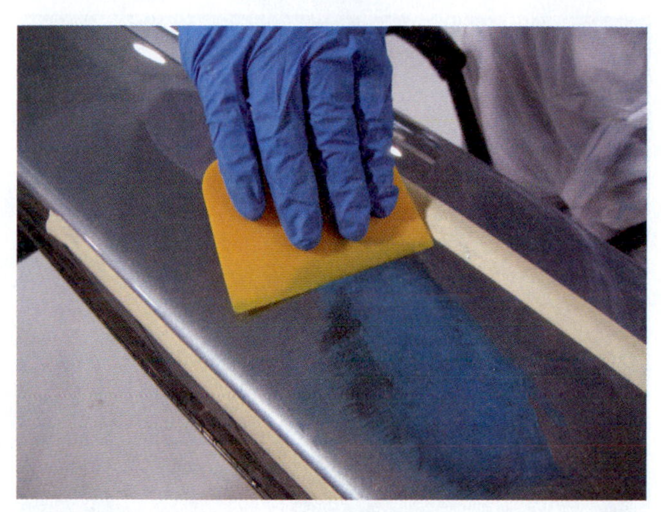

FIGURE 18-27 The taper in a single-sided repair should go from the surface of the base plastic to at least the depth of the deepest scratch or gouge, without going through the plastic.

so the repair material can be placed as deep as the damage has gone into the base plastic. Most repair material recommendations call for sanding with between 24- to 80-grit sandpaper, then tapering the repair area to about 1 1/2 inches around the area to be repaired. See **Figure 18-27**. This taper should go from the surface of the base plastic to at least the depth of the deepest scratch or gouge, without going through the plastic. Next, the repair area should be sanded either by hand or with a DA. Manufacturers will recommend sandpaper ranging from 40- to P180-grit. The repair material should not be placed over paint finish, and therefore the repair area should be feathered. The technician should take care when sanding that the plastic is not melted. If melting occurs, the melted material must be removed to assure good repair adhesion. After sanding, the area should be cleaned with a recommended cleaner such as isopropyl alcohol.

FIGURE 18-28 Most adhesion promoters are applied as an aerosol from a spray can, though the manufacturer may call for flame treatment.

Different types of **adhesion promoters** may be called for by the bonding repair manufacturer. See **Figure 18-28**. Most are applied as an aerosol from a spray can, though the manufacturer may call for flame treatment, which changes the plastic's molecular structure so it will accept the plastic repair material better. If the manufacturer calls for flame treatment, a butane or propane torch can be lightly passed over the plastic repair area without heating up the plastic.

The majority of the adhesion promoters are sprayed on the repair areas just before the repair mixture is applied. The manufacturer of the repair material will offer a specific adhesion promoter, which is typically applied in one to two light coats, with the technician allowing 10 to 20 minutes of dry time prior to application of the repair material. When repairing plastic, it is important for a repair technician to follow all of the repair material recommendations to ensure a successful repair.

Application. Plastic repair material is applied similarly to plastic body filler. The application gun allows the technician to apply the mixed material directly onto the area to be repaired. It can then be spread with a plastic body spreader, until it is smooth and just slightly above the area that is to be repaired. Once the material has cured, it can be sanded smooth—either by hand or with a DA—and primed with a plastic primer filler for final sanding. See **Figure 18-29**.

Two-Sided Repair

Two-sided repairs are generally done with the part removed from the vehicle. The technician will first do most of the structural repairing from the back side. This type of repair is done on damage that has penetrated the plastic part, and at times some of the plastic may even be missing. Often when a two-sided repair is needed, the plastic has been forced out of shape. In this case, some may need to be removed so the plastic can be returned to its original

FIGURE 18-29 Once material has cured, it can be sanded smooth either by hand or with a DA.

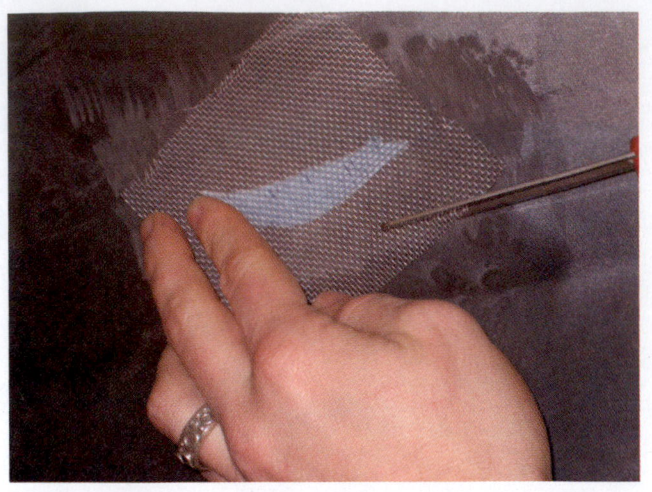

FIGURE 18-31 Instructions for applying the adhesive on the back side of the repair recommend that it be reinforced.

shape without bulges. In some cases, the damage is so severe that proper alignment is not possible, and technicians should take care to heat stress-relieve the part. When the part is realigned, the surface should be triple-cleaned as described earlier.

The edges of the plastic should be gradually tapered from the surface of the front of the part to the rear. See **Figure 18-30**. Repair manufacturers often call for a taper of from 1/2 to 1 1/2 inches wide. Next, the front and back side of the repair should be sanded with the recommended grit (normally 40- to 80-grit) and cleaned in preparation for application of the repair material. The repair area is next treated with an adhesion promoter. Typically, instructions for applying the adhesive on the back side of the repair recommend that it be reinforced. See **Figure 18-31**. Though reinforcement recommendations vary, with some suggesting a peel-and-stick type of reinforcement and others recommending the application of one or more layers of mesh, the integrity of the repair is

FIGURE 18-30 The edges of the plastic in a two-sided repair should be gradually tapered from the surface of the front of the part to the rear.

FIGURE 18-32 Some technicians suggest using a peel-and-stick type of reinforcement.

likewise assured with the application of back reinforcement. See **Figure 18-32** and **Figure 18-33**. On the front tapered area, the technician applies repair material in the same way as described in the discussion of one-sided repairs. Then, after the plastic has cured, the front is sanded smooth, primed with a plastic primer filler, and refinished.

REINFORCED PLASTIC REPAIR

Reinforced plastic comes in two basic categories. One is **fiber-reinforced plastic (FRP)**, which for years was called fiberglass. It uses catalyzed resin, which is spread onto a fiberglass mat while in a mold. This process hardens the part, which is then removed from the mold,

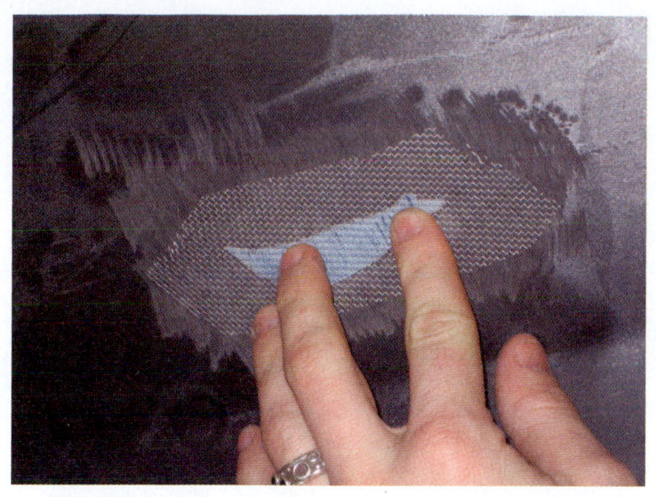

FIGURE 18-33 Some technicians recommend applying one or more layers of mesh as reinforcement.

trimmed, and used. Fiber-reinforced plastic has a smooth front and has visible fibers on the back.

The second category of reinforced plastic repair is **sheet moldable compound (SMC)**. Sheet moldable compound is a product that combines both fiber reinforcement and resin that is placed into a die and heated. The heat activates the SMC, which hardens into the desired part. Manufacturing using SMC is more economical than using fiber-reinforced materials, and thus it has become the material of choice. Sheet moldable compound is smooth on both sides and requires the use of specific SMC repair materials. The Corvette, which was well-known for using "fiberglass" (fiber-reinforced plastic) in its manufacture, started using SMC in the 1980s.

Fiber-Reinforced Plastic (FRP) Repair

Some manufacturers of FRP may recommend against their material being repaired, while others offer specific repair procedures. The damage can, like other plastics, require only shallow, one-sided cosmetic repairs. If so, two-part epoxy can be used. More commonly, however, polyester filler is used. A typical **single-side repair** process for FRP is as follows:

1. Wash with soap and water to remove water-soluble contaminants.
2. Clean with a mild wax and grease remover to remove non-water-soluble contaminants.
3. Remove paint from the surrounding area with DA and 80-grit sandpaper.
4. Bevel the damaged area, allowing for sufficient bonding of filler agent.
5. Mix filler, according to manufacturer directions, and apply.
6. After the repaired area has cured, level it by sanding, finishing to P220.
7. Apply primer filler as needed and block sand for proper contour.

Two-Sided Fiber-Reinforced Plastic Repair. More severe damages, ones that have penetrated the plastic, will require a two-sided repair. This repair usually involves correcting damage of the fiber reinforcement. When FRP is broken, it may have damaged fibers sticking above the surface. See **Figure 18-34**. These must be removed before completing the repair. A die grinder cutting tool can be used to cut away the damaged FRP, revealing a repairable hole.

The technician should follow this sequence for two-sided FRP repair:

1. Wash the part with soap and water.
2. Clean with a wax and grease remover.
3. Remove paint with a grinder and 80-grit sandpaper. The paint should be removed to approximately 3 inches beyond the repair area on both sides.
4. Bevel the edges of the repair area to a 30 degree bevel on both sides.
5. Remove sanding dust.
6. Cut enough fiberglass matting to make three layers that are each slightly larger than the area to be repaired.
7. Prepare and mix the resin and hardener, according to manufacturer's directions.
8. With a paintbrush, saturate the first layer of fiberglass reinforcement with the resin. Also brush resin on the inside of the FRP area to be repaired. Place the saturated fiberglass on the back side of the repair area and tap into place with the bristles of the paintbrush. Make sure that the saturated fiberglass makes a tight contact with the FRP.
9. Repeat the process in step #8 with saturated fiberglass until three layers of fiberglass have been

FIGURE 18-34 When fiber-reinforced plastic (FRP) is broken, it may have damaged fibers sticking above the surface.

placed on the back of the repair area. Allow this to catalyze.

10. After the back has cured, grind the front of the repair with 50-grit sandpaper, then clean.
11. The front can then be repaired with polyester plastic body filler, sanded and primed as needed.

This repair can also be completed with chemical adhesives, which are designed for FRP. The adhesive is substituted for the resin, but the process to follow is the same as previously explained.

Sheet Moldable Compound (SMC) Repair

Sheet moldable compound (SMC) can be distinguished from FRP by looking at the back side. For example, SMC will have a smooth back, while FRP will have visible fibers. If the plastic to be repaired is SMC, the repair process differs from FRP in that resin cannot be used. Two-part adhesive repair designed for SMC is the repair medium of choice. SMC may have light single-sided repairs and deeper two-sided repairs performed, as in the other types of plastics.

Single-Sided Repair of SMC. Small cosmetic gouges or scratches in SMC should be repaired using this process:

1. Wash with soap and water.
2. Clean with wax and grease remover.
3. Sand with 80-grit sandpaper to remove paint.
4. Taper the repair area to remove defects, without penetrating the part.
5. Remove sanding dust.
6. Apply the recommended SMC repair filler with an application gun, according to manufacturer's recommendation. Allow repair to cure.
7. Sand initially with 80-grit or P120-grit paper. Then finish-sand to P320.

Two-Sided SMC Repair. Though SMC does not have visible fibers on its back side, it does have reinforcing fibers on its inside. When damage extends through the material, some fibers will be damaged and—like with damaged FRP—they must be removed. This can be done by using a die grinder or a small reciprocating saw. See **Figure 18-35**. The damaged SMC must be removed before the repair is completed. Most manufacturers of SMC repair material recommend using a scrap piece of SMC for the backing, though there are also backing reinforcement tapes and other materials that can be used for the backing repair.

FIGURE 18-35 The damaged SMC must be removed with a die grinder or small reciprocating saw before the repair is completed.

Though the recommendations of the adhesive material manufacturer should be followed, a general repair procedure to follow is as follows:

1. Wash with soap and water.
2. Clean with a wax and grease remover.
3. Remove paint with a grinder and 80-grit sandpaper. The paint should be removed to approximately 3 inches beyond the repair area on both sides.
4. Bevel the edges of the repair area to a 30 degree bevel on both sides.
5. Remove sanding dust.
6. Prepare a SMC backing from a scrap piece of SMC. It should have the same contour as the repair area. The repair piece should be washed, cleaned, and have the paint removed with 80-grit paper.
7. Prepare the recommended product used to bond the backer. This may or may not be a different product than the one used to repair the front. Some SMC adhesives require the use of heat to catalyze the adhesive. Therefore, the manufacturer's recommendation for time or heat should be followed. Allow to cure.
8. After the backing has been applied and cured, clean the front and apply the repair material to the front. Some manufacturers may recommend the use of a "pyramid" front repair, which means that repair material and reinforcement tape are stacked upon each other until the level of the repair reaches the surface of the defect.
9. Sand the repaired area to contour. Then apply primer filler, and block.

Summary

- The use of plastic in the manufacture of automobiles has steadily increased as more fuel-efficient vehicles have been in demand by consumers.
- There is no reason to believe that the use of plastics will not continue to develop.
- The demand for technicians who have the skills to repair plastic will also continue to be in demand.
- The development of new plastics is also inevitable, and the repair procedures will change with it.
- Because of this inevitable change, technicians will need to stay abreast of the changing repair techniques for plastics.

Key Terms

adhesion promoter
adhesive repair
airless plastic welding
chemical adhesive bonding
fiber-reinforced plastic (FRP)
flexibility test
flexible plastic
float test
hot air plastic welding
International Organization for Standardization (ISO)
plastic memory
plastic welding
polyolefin
sand test
sheet moldable compound (SMC)
single-side repair
thermoplastic
thermoset plastics
two-sided repair

Review

ASE Review Questions

1. Technician A says that a thermoplastic will soften when heated. Technician B says that a thermoplastic will not soften when heated. Who is correct?
   A. Technician A only
   B. Technician B only
   C. Both Technicians A and B
   D. Neither Technician A nor B
2. Technician A says that a thermoset plastic gets hard when it is heated. Technician B says that you can identify a thermoset plastic because it doesn't have visible fibers on its back. Who is correct?
   A. Technician A only
   B. Technician B only
   C. Both Technicians A and B
   D. Neither Technician A nor B
3. Technician A says that ISO stands for International Society of Operators. Technician B says the FRP stands for filament-reinforced plastic. Who is correct?
   A. Technician A only
   B. Technician B only
   C. Both Technicians A and B
   D. Neither Technician A nor B

4. Technician A says that polyolefin plastics are all thermoset plastics. Technician B says that polyolefin plastics are all thermoplastic plastics. Who is correct?
   A. Technician A only
   B. Technician B only
   C. Both Technicians A and B
   D. Neither Technician A nor B
5. Technician A says that a float test can identify a thermoplastic plastic. Technician B says that a sand test can identify a thermoplastic plastic. Who is correct?
   A. Technician A only
   B. Technician B only
   C. Both Technicians A and B
   D. Neither Technician A nor B
6. Technician A says that plastic memory helps a technician to re-form damaged plastic. Technician B says that plastic reshaping is aided by heating because of plastic memory. Who is correct?
   A. Technician A only
   B. Technician B only
   C. Both Technicians A and B
   D. Neither Technician A nor B
7. Technician A says that hot air welding takes more skill than airless plastic welding. Technician B says that polyolefin plastics cannot be repaired with chemical plastic adhesives. Who is correct?
   A. Technician A only
   B. Technician B only
   C. Both Technicians A and B
   D. Neither Technician A nor B
8. Technician A says that the correct adhesive repair material mixing ratio is 1:1. Technician B says that when starting a tube of adhesive repair material, the technician should extract and discard a small amount before placing the mixing tube on the dispenser. Who is correct?
   A. Technician A only
   B. Technician B only
   C. Both Technicians A and B
   D. Neither Technician A nor B
9. Technician A says that adhesive promoter needs to be used with an SMC repair. Technician B says that two-sided repair is needed when plastic is scratched, in order to assure strength. Who is correct?
   A. Technician A only
   B. Technician B only
   C. Both Technicians A and B
   D. Neither Technician A nor B
10. Technician A says that FRP has visible fibers on its back. Technician B says that fiberglass rosin can be used to repair SMC. Who is correct?
   A. Technician A only
   B. Technician B only
   C. Both Technicians A and B
   D. Neither Technician A nor B

Essay Questions

1. Describe how to perform an airless plastic weld.
2. Explain the differences between thermoplastic and thermoset plastics.
3. How would a technician do a two-sided SMC plastic repair?
4. Describe why plastic must be cleaned with both isopropyl alcohol and wax and grease remover.

Topic Related Math Question

1. If a manufacturer recommends removing the paint off a part to 50 mm, what is that measurement in inches?

Critical Thinking Questions

1. How could you identify a type of plastic?
2. If you have a puncture in a rigid plastic part, which repair method should be used?
3. If you need to identify whether or not a part is a thermoplastic, what is the best way to identify it? How do you do this action?

Lab Activities

1. Make a one-sided repair on a scratched plastic bumper.
2. Re-form a plastic bumper using a heat gun.
3. Make a two-sided repair on a damaged SMC part.
4. Distinguish between a part that is FRP and one that is SMC.

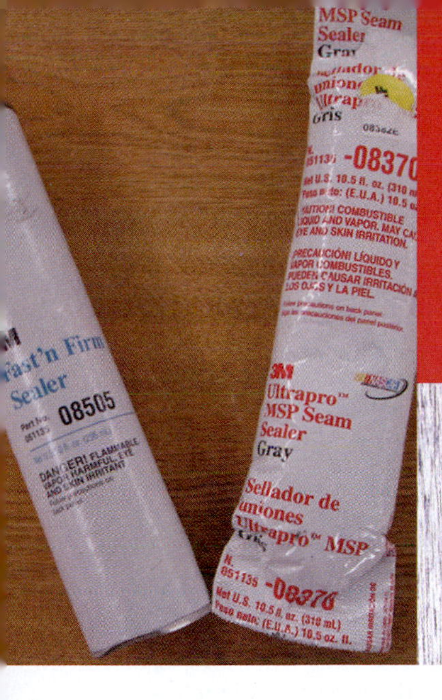

Chapter 19

Corrosion Protection

OBJECTIVES

Upon completing this chapter, you should be able to:

- Discuss the methods employed by the automobile manufacturing industry to prevent oxidation and corrosion.
- Define corrosion and identify the elements and environment that promote formation and growth of oxidation.
- Identify environments that may promote oxidation on vehicle bodies and undercarriages.
- Define corrosion hot spots and how they impact the collision repair industry.
- Discuss metal treatment techniques to retard oxidation on automobiles.
- Discuss the types of anti-corrosion materials used to prevent or inhibit corrosion.
- Identify the corrosion protective basecoat materials commonly used in repairing collision-damaged vehicles.
- Identify various seam sealing materials and discuss their intended uses.

INTRODUCTION

Never before in automotive history has so much research and technology been put into the ongoing development of the automobile as is being done by the current automobile manufacturers. Many of the recent developments have been driven by the need to increase fuel economy. Great strides have been made in the development of stronger, yet lighter weight materials for manufacturing modern day cars and trucks. One of the most significant achievements has been reducing the thickness and much of the mass of steel required for each vehicle produced. Frames and undercarriages once made of 5 to 7 mm (3/16 to 1/4 inch) of thick steel have been replaced with laminated rails half that thickness and weight, which nonetheless offer 100 percent more strength. The results are very attractive to the consumer because they generally translate into a considerably more fuel-efficient vehicle.

However, a significant trade-off concerning the vehicle's life expectancy was made in the process. The once thick and bulky structural frame rails were expected to last at least 15 to 20 years before rust and corrosion would become a structural concern. Reducing the metal's thickness by this amount meant a considerably shorter vehicle service life unless something else was done. Numerous alloy materials and special treatments and coatings were introduced to enhance the corrosion resistance of the metals used to manufacture automobiles. This created a whole new array of problems for the repair industry because the same corrosion resistant characteristics could not be duplicated when a vehicle was repaired. A whole new series of products and materials had to be developed to help the repair industry respond to this added dimension for corrosion protection.

PERSONAL PROTECTIVE REQUIREMENTS

Because most of the anti-corrosion materials are petroleum derivatives, they contain materials that are known to be hazardous to one's health. Many of the materials used are applied using either spraying equipment or other application tools such as brushes, spatulas, or other similar devices. Whatever means is used, the technician is likely to be exposed to the fumes emitted or to the overspray particles coming off the material as it is sprayed. The minimum PPE used should include safety glasses, protective gloves, respiratory protective equipment, and any other equipment deemed appropriate for the operation. The technician should consult the manufacturer's material safety data sheet (MSDS) for each product used to identify the required PPE and make the appropriate provisions when any of these materials are being used.

CORROSION

In its simplest form, **corrosion** is the result of a chemical reaction that occurs when exposed metal such as steel reacts to oxygen and an electrolyte. An electrolyte, which must be present for oxidation to occur, can be in the form of an acid or moisture even when in as small a quantity as the humidity in the atmosphere. The result is a gradual deterioration (**oxidation**) of the steel surface. The residual left behind is a loosely attached accumulation of flaking layers of **iron oxide** commonly called rust. The presence of some elements like road salt, **acid rain**, and certain other pollutants promote or accelerate oxidation. Because the iron oxide that collects on the steel's surface is an unattached, very loosely layered material, the oxidation process is allowed to continue to occur unchecked so long as the three elements required for it to occur are present. This process will continue until all the steel is gone or an external protective coating is applied to the surface to seal out the elements.

Self-Healing Metals

Other metallic substances such as aluminum, magnesium, stainless steel, and titanium also corrode or oxidize in much the same way. However, with these metals, the oxides that form on the surface create a very tightly bonded, protective, and corrosion-resistant coating that protects the surface from further deterioration. This is an almost invisible micro-thin oxide or nitride layer of material which is only a few atoms thick and is capable of regenerating itself. When the surface becomes scratched or the natural protective coating is compromised in any way, it will quickly regenerate itself, thus eliminating the heavy buildup of oxidized material such as is the case with steel. This process is commonly referred to as **passivation**. Despite the degree of effectiveness against corrosion from the passivation process, little or no protection is realized when the metal is exposed to a saltwater spray environment. Unless aluminum has a protective coating, it will turn white, corrode, and even dissipate completely in a span of only a few years in such an environment.

The term *corrosion* automatically gives one a negative impression or thoughts because it usually confirms our preconceived notion of a vehicle "rusting away" with age. For most of us, corrosion implies a deterioration of metal—most commonly steel—resulting in rust. However, like many forms of energy, when directed in a positive manner it can be used to our advantage. Corrosion occurs or takes on numerous forms. When redirected, we can actually use it to extend the life of steel or other metals. To understand this we must first realize that when different metals come into contact with each other in the presence of oxygen they react to each other. Because some metals are more reactive than others, the surface of the more reactive metals breaks down or deteriorates, causing an oxide layer likened to its matter to form on the surface. This prevents further exposure to oxygen, thereby slowing down the oxidation process. Because zinc is a more aggressively reactive material than steel, it will sacrifice itself and form a sacrificial barrier coating of zinc oxide over the steel surface. This barrier coating virtually eliminates further exposure to oxygen by the steel, thus nearly stopping the corrosion process. This is commonly known as sacrificial corrosion.

Protective Coatings

Protecting the steel surfaces from oxidizing can be accomplished by several different measures or techniques. The automobile manufacturing industry utilizes three such methods to varying degrees to reduce or prevent oxidation of metal: **cathodic coatings** and galvanizing, chemical treating and etching metal, and applying sealants and surface coatings for protection and decorative purposes. Galvanizing is the most prevalent metal treatment system used because it is relatively inexpensive to do and it affords more flexibility in the type and amount of coating applied.

Galvanizing. **Galvanization** is a process that consists of coating metal with a thin layer of another metal such as zinc. See **Figure 19-1**. As zinc is a more chemically active metal than steel, it oxidizes and forms a tight barrier coating over the steel, thus protecting it from further exposure to oxygen. See **Figure 19-2**. This is known as **sacrificial corrosion** because the zinc breaks down before the steel, forming a zinc oxide coating over the surface of the steel which protects it from further oxygen exposure. Once the surface is coated, the oxygen and moisture will react with

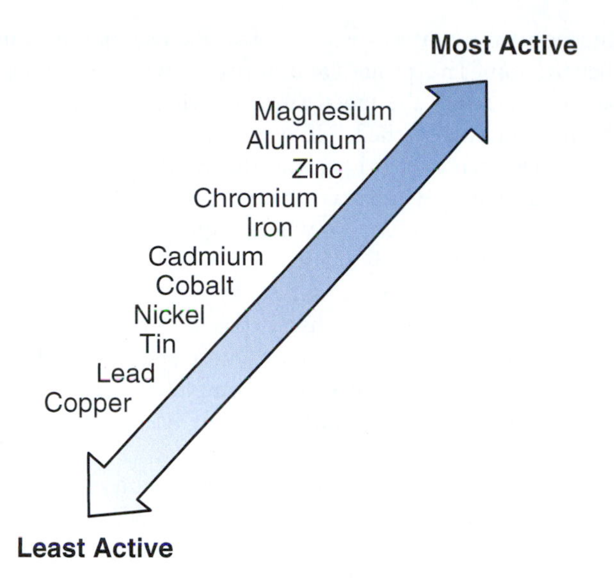

FIGURE 19-1 Some metals such as zinc are significantly more chemically active than others. Therefore, these metals are used as a protective coating on automotive steel.

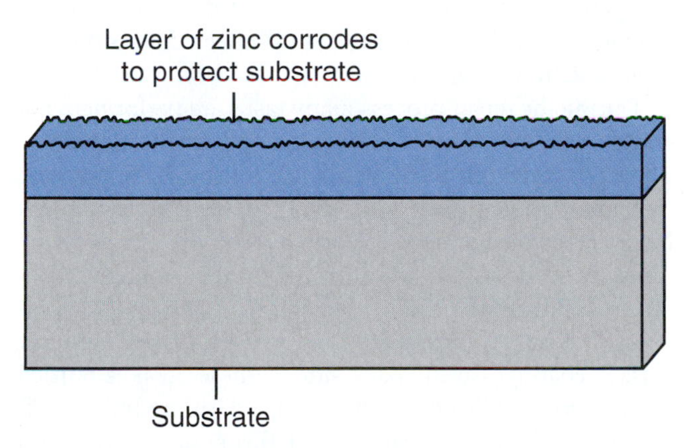

FIGURE 19-2 The zinc forms a protective layer over the steel, creating a sacrificial coating to protect it from corroding.

the zinc coating instead of the steel underneath it. Thus, the metal's life expectancy is considerably extended. Once the zinc has formed an airtight barrier on the surface, the oxidizing process will nearly be stopped until such a time as the surface becomes scratched or abrated. These actions will remove the barrier coat. At that time the entire reactive process starts all over and continues until the surface is recoated with another airtight coating. This process is repeated each time the surface is attacked or exposed by something breaking through or penetrating the protective layer. In a sense, the zinc passivates the steel in much the same manner as aluminum coats and protects itself. Depending on whether the coating is required on one or both sides of the metal, the coating may be applied by hot dipping or completely submersing the metal or by **electrogalvanizing** or spraying the coating onto one side of the metal. In this way just one side or both sides are galvanized. The longer the steel is left exposed to the zinc bath or the spray application, the heavier the coating that forms on the surface.

Despite the widespread use of galvanizing for corrosion protection on the automobile, it is not totally effective against corrosion. Vehicles that are constantly exposed to a saltwater environment—such as that found along the ocean shore lines—have a considerably shorter life expectancy than in less arid states such as Arizona where the climate is not saturated with metal's two worst enemies: salt and moisture. It is inevitable that all automobiles eventually start to rust or corrode—in fact, it is the single most significant factor responsible for diminishing the automobile's value. Each year consumers lose a significant amount of their automobile's value because of corrosion and the deterioration resulting from it. Therefore, the auto collision repair technician has a moral and legal obligation to take every step possible to retard and eliminate the corrosion process whenever a vehicle is repaired.

E-Coating. Despite all the efforts made by the manufacturer to protect the vehicle from the corrosion process, there are numerous ways it can start on a vehicle without anyone's knowledge of its presence. In their quest to eliminate or at least minimize the number of **corrosion hot spots**, the majority of the automobiles produced today are primed using an electrostatic deposition process commonly called **e-coating**. After the vehicle is completely assembled, it is subjected to a series of cleaning and prepping stages in preparation to priming. It is dipped in vats filled with special cleaning and etching chemical agents. Exposure to these chemicals results in the entire vehicle being covered with a phosphate and a zinc phosphate coating, which is an ideal surface for corrosion protection and primer adhesion. After the vehicle has been cleaned, it is submersed into the priming vats, which are electrically charged to maximize the priming process. In this process the vehicle is charged with a negative electrical charge and the primer vat is given a positive charge. Because the negative charge in the vehicle will attract the positively charged primer, the process ensures that every possible void and exposed surface of the vehicle is completely coated with a uniformly thick coating of primer. This is the single-most effective method of coating any surface and cannot be duplicated in the aftermarket or repair industry. This primer serves multiple functions. In addition to being a basecoat to which all the finish coats adhere, it is also used to create a uniformly smooth metal surface, thus creating a flawless surface for the topcoats to go over. It also gives the finish added chip resistance to help prevent chipping from incidental contacts and bumps.

In all, the vehicle is subjected to five specific steps to ensure that the necessary corrosion protection is in place: galvanizing, cleaning, phosphate treating, rinsing, and priming. While these steps cannot be duplicated in their exact form when repairing a vehicle, every effort must be made to use the materials and methods available to ensure the original corrosion protection has been restored as closely as possible. See **Figure 19-3**. Even though not all of the auto manufacturers use the exact same process or procedures, a similar variation is exercised to ensure the maximum corrosion protection steps possible are taken.

Seam Sealers Applied to Overlapping Joints

After the primer and basecoats are dried and before the finish coat of paint is applied, every seam, welded flange, and all of the overlapping and exposed panel seams are coated with a protective **seam sealer** to minimize exposure to the elements. The vehicle is then covered with the finish paint coat to seal out any exposure to moisture or other corrosion-promoting elements such as salty and gritty road splash. This protects the metal surfaces from exposure to water and oxygen. The sealing is commonly done using an electrostatic process, which again ensures the entire vehicle's surface becomes coated and protected. The thoroughness of this process cannot be duplicated by the repair industry; many times the method used to repair a vehicle is selected to safeguard this corrosion protection.

The Corrosion Hot Spots

Despite the effectiveness of the coating process obtained by submersing the vehicle, corrosion hot spots will eventually occur. As the vehicle is driven over roads and rough terrain, the body panels move and shift. This causes them to rub against each other, rubbing through the primer and paint coats in the process. Sealants begin to be pulled and stretched, causing bare spots to occur. These areas all become potential hot spots for corrosion to get started, many times in areas that are not visible until corrosion starts to spread and becomes visible on the vehicle's exterior. By this time it may be too late to stop it other than by replacing the affected panels.

Collision and Repair Related Hot Spots. Corrosion hot spots are areas that become suspect for corrosion to get started because they lose or have their protective coatings removed by abrasion, vibration, heat, or other means. This may also occur as the result of a collision—which frequently causes seams and welded areas to become stretched and pulled apart—or from the subsequent collision repairs that were made. In some instances they are the result of routine wear and tear on the vehicle.

Taking a closer look at what happens during a collision, one can readily see what may happen to give corrosion a foothold for a starting point. When the vehicle is subjected to collision forces the panels shift and move as the damage radiates through the body. Spot welds and other welded areas are pulled and severely stretched. Many times they even return to their original position, thus hiding the stretched welds from the technician's naked eye. Anticorrosion materials covering seams and welded joints are pulled and torn loose from where they are attached, leaving numerous surfaces exposed for microscopic moisture penetration. The primer and topcoat may form microscopic cracks on the metal surfaces where the metal has been severely stretched. Notwithstanding the areas that may have been exposed prior to the accident, rust may already have formed on areas where the paint was rubbed off from a previous incident or where improper or incomplete repairs were previously made. Unless the panels with the affected seams and spot welds are replaced, the pulled welds will remain somewhat exposed. Even though attempts are made to seal and paint these areas when the repairs are made, it is impossible to completely penetrate and seal the affected areas by spray painting basecoats and topcoats. When the vehicle is put back into service, the newly exposed bare areas will have ample opportunity to develop into new corroding hot spots despite the efforts made to stop them. Because these areas are usually covered by trim and are located deep in the recesses of the joints, the corrosion goes undetected until the trim and other panels are removed for a subsequent repair or to service other parts—or it becomes widespread to the point of perforating the exterior panels. At this point it is too late to do anything other than to replace the panel or to cut the affected areas out and replace them with new metal.

During the repair process many tasks are performed that could result in hot spots developing. Whenever heat is applied—such as in cutting and welding operations—the metal surface is exposed and left somewhat compromised for a corrosion hot spot to form. See **Figure 19-4**. Many times welded pulling studs or a nonconsumable heat-attached puller are used for repairing the metal. A considerable amount of heat is required to attach them, and any surface coatings on the back side of these areas is burned away or forms blisters on the surface. Drilling holes is often required to temporarily hold the replacement panel during installation and occasionally to attach the trim back onto the vehicle. In addition to the exposed edges of the holes becoming a potential hot spot, the shavings that are created from the drilling operation often fall into crevices and seams at the bottom of the panel being drilled. Rarely are these shavings removed, thus leaving another spot for rust to start forming. Bare or uncoated steel left lying against even a coated surface is an open invitation for oxidation to eventually occur. Hammering to stress-relieve the area while pulling and repairing the metal may also cause the corrosion protection and surface coatings to become cracked and compromised. The use of spoons, pry picks, and other hand tools invariably scratches the paint and protective coatings. Therefore, any area that is part of the repair procedure must be addressed prior to returning the vehicle to the owner.

Atmosphere-Induced Hot Spots. Numerous additional possibilities exist for hot spots to occur during a vehicle's lifetime. **Marine atmospheres** found along the coast lines of the oceans have a significant effect on the life expectancy of automobile bodies even when they have been

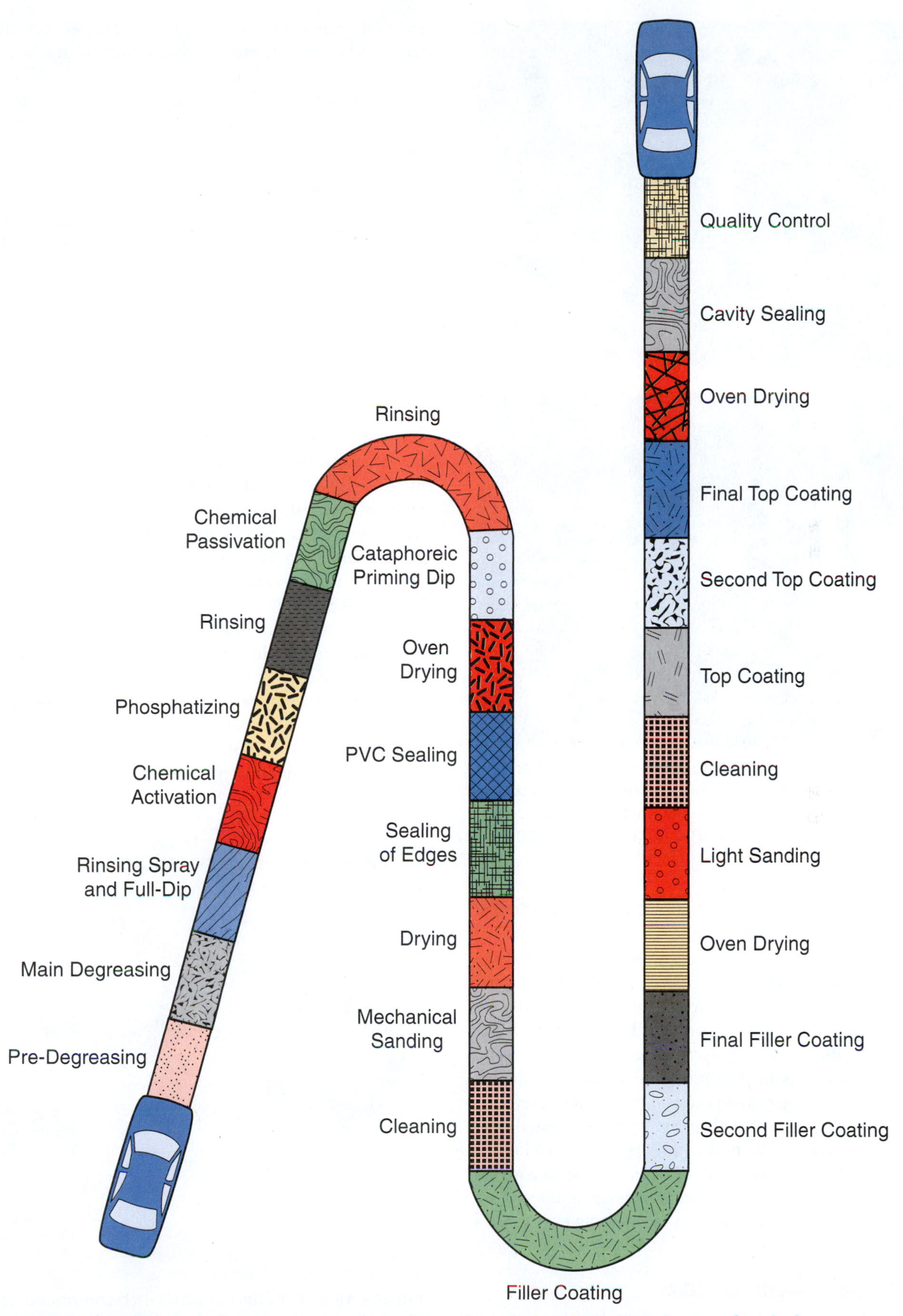

FIGURE 19-3 Automobile manufacturers use a series of steps for treating the metal's bare surface before applying the corrosion protective coatings and the finish coats.

FIGURE 19-4 Each time a welded-on pulling stud or a similar pulling device is attached to the surface, a corrosion hot spot is created on the back side of the panel. The surface must be treated and re-coated to protect it from corroding.

properly undercoated by the manufacturer. A vehicle that is exposed to constant salt spray—such as that used for deicing roads in winter—or driven in certain industrial climates will deteriorate much more rapidly than would otherwise occur. The atmosphere in many areas of the nation is considered relatively corrosive due to industrial pollution and fallout such as may occur near chemical and metallurgical processing facilities. **Sulfur dioxide**—which is the essence of acid rain—is considerably more prevalent in large metropolitan areas because of the increased concentration of vehicles using fossil fuels. It is also a known cause of oxidation and premature deterioration of a metallic surface. Although it may seem very insignificant, an abrasion or paint chip that results in exposing even a small spot of metal can be the source of an oxidation hot spot.

Galvanic Corrosion

Sometimes optional items are attached to the automobile using screws, rivets, or other types of metallic fasteners. **Galvanic corrosion** is apt to occur between the fastener and the surface to which the item is attached, especially if they are made of dissimilar materials. All of these sources are an invitation for corrosion to get started and—if left unchecked—will continue to spread to the point of requiring an extensive repair to stop it.

Corrosion Insulators

The best solution for protecting any vehicle from corroding is preventing corrosion from getting started. This may seem like a near impossible task considering all the possible ways it can get started. However, when followed correctly, a few simple rules will certainly eliminate the vast majority of opportunities for corrosion. See **Figure 19-5**. Whenever drilling, always ensure the edges of the holes or the fastener threads are covered with a protective coating that will prevent corrosion. Whenever possible, use the fasteners that are compatible with the surface. If available, the best preventative measure is to duplicate the original fastener. When parts made of dissimilar metals are used, always insulate them from each other. See **Figure 19-6** and **Figure 19-7**. Always prime any bare metal spots before coating them with paint or covering them with a piece of trim. See **Figure 19-8**.

Unlike steel, aluminum has the ability to protect itself from most of the elements previously mentioned by the formation of the oxide coating it naturally generates. While the importance of taking all the necessary corrosion protection steps is not lessened with aluminum, oxidation does tend to not be as serious a problem as with steel under normal driving conditions. Unfortunately, this

FIGURE 19-5 A coating of a zinc-rich compound can be applied to the fasteners prior to installing them to provide additional corrosion protection.

FIGURE 19-6 An insulator is placed between the outside door handle and lock cylinder and the door panel to prevent corrosion from occurring between the two dissimilar metals used.

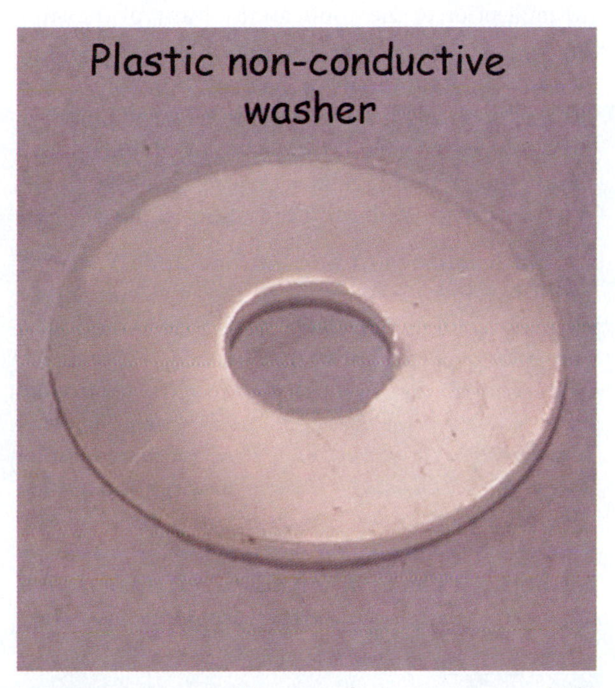

FIGURE 19-7 Automobile manufacturers may use bolts and other fasteners with a non-conductive insulator integrated into the washer to prevent the possibility of galvanic corrosion.

naturally occurring oxide offers no protection from a continuous exposure to saltwater spray environments—such as occurs in a marine atmosphere or from highway deicing applications. Aluminum will turn white as it deteriorates in such caustic environments; if left unprotected, it will eventually disintegrate. In order to protect it from these environments, various alloys must be added or it must be coated with special coatings. **Alodine iridite** is a chemical coating commonly used to help protect the surface from premature oxidation.

FIGURE 19-8 One should ensure that even the smallest bare metal spot is coated with primer prior to topcoating to prevent corrosion that results from a metal molding or trim piece contacting the area.

RESTORING CORROSION PROTECTION

Each paint manufacturing company produces the metal cleaning and etching chemicals required to prepare the bare metal surface to match their primers and basecoats. Their respective products are chemically formulated to ensure the surface created by their etching procedures are completely compatible with each other. No matter how insignificant or major the repair being performed, a systems approach should be used. This suggests that whenever possible, all the products and materials used for this part of the repair should be selected from the same manufacturer's product line. This will ensure all the materials are compatible with one another and eliminate the possibility of product failure due to improper bonding and cross-linking.

Surface Preparation

Restoring the corrosion protection to the repaired areas requires many of the same steps performed by the manufacturer with minor variations. Many of these steps are also the initial stages for preparing a surface for refinishing. The first—and perhaps the single-most important—thing one must do is to ensure the surface is clean before starting to apply any of the basecoat material. This suggests that the area to be corrosion protected must be thoroughly washed to remove any water-soluble contaminants that would otherwise affect the adhesion and proper coverage of the affected area. After washing the surface, it should be thoroughly cleaned with a wax and grease cleaner to remove any petroleum base contaminants. These steps may already have been performed before the repairs were initiated.

However, the surface is exposed to dust and dirt from the grinding operations or oil contamination from the hydraulic equipment. In addition, air tool exhaust may have splashed onto the surface and cross-contamination from other tools may have been deposited onto the surface. Unless the surface is properly cleaned, the adhesion of the corrosion protection materials can be compromised—much as when refinishing. Regardless if the affected area is on a cosmetic panel or a structural reinforcing member that is not visible when the vehicle is in service, it must be prepped for painting and for restoring the corrosion protection with a product that is compatible with the paint manufacturer's system.

Chemical Etching

Any bare metal surfaces that are not coated with fillers, or any surface that is not coated with a basecoat, must be abrated before chemical etching to ensure both **mechanical** and **chemical adhesion** will occur. The abrating should be done with an abrasive ranging from 80 to 120 grade and should be done thoroughly enough so the effect is visible. Even though the primer coats are designed to adhere to the coatings left by the chemical cleaning and etching steps, they lack the same adhesion the manufacturers obtain by the baking process. The adhesion quality will be compromised considerably if this step is not performed. It must then be chemically treated and prepared for basecoating.

CAUTION

Regardless of the severity or mildness of any chemical being used, one should use the manufacturer's recommended PPE for the operations performed. Many of the metal cleaning and etching chemicals are an acid base; therefore, the minimum PPE should be latex gloves and safety glasses. Failure to use these may result in chemical burns wherever the material contacts exposed skin.

Application of Etching Materials. The first step in chemically treating the metal is to saturate any bare or uncoated exterior surface with a phosphoric base metal conditioning material. This is the equivalent of the phosphate treatment used by the manufacturer. The metal etching chemical—metal conditioner, as it is commonly called—should be applied liberally and the solution must be allowed to **dwell on the surface** for at least 1 minute. While the surface is wet, it should be scrubbed with a plastic scouring pad, being careful to not let the surface dry prematurely. While the etching solvents dissolve the corrosion on the surface, the scrubbing action will help to break the corrosion loose and remove it from the surface. While the material is still wet it may be necessary to rinse it off with water, wipe it dry, or simply let it air dry, depending on the manufacturer's recommended procedures. The primary purpose for using metal etching chemicals is to remove and neutralize rust and corrosion from the surface and chemically clean the bare metal, which ultimately will promote adhesion and longevity of the repair. Improperly using these chemicals may completely negate the repair's longevity by destroying much of the adhesion that would otherwise have been achieved.

Metal cleaning or etching materials and conversion coatings should be used on the exterior surfaces only. Because they contain acid, allowing them to set inside a cavity for an extended period of time may cause the metal to deteriorate and weaken. Overetching the metal may result in preventing proper adhesion. Many of the etching chemicals require dilution with water in varying proportions, although others are intended to be used as packaged. One should consult and follow the manufacturer's label instructions to ensure the correct use of the product. Incorrectly using the material may likely nullify the desired effects. Under no circumstances should the metal etching chemicals be used prior to the application of plastic fillers. Contrary to the improved adhesion characteristics these chemicals provide for primers and other basecoats, plastic fillers are not compatible with them and adhesion loss will occur if the surface is etched prior to its application. Therefore, any repairs requiring the use of fillers must be performed prior to the application of any of the surface treating materials. See **Figure 19-9**.

FIGURE 19-9 Paint manufacturers recommend the use of a metal etching chemical followed with a conversion coating to maximize protection for the substrate.

CONVERSION COATINGS

Some manufacturers recommend following a two-step procedure with their product to maximize its effectiveness. The second step involves what is commonly called a conversion coating. This material is also a **phosphoric acid** base material with additional additives such as mild detergents that help to clean contaminants from the surface, along with certain other ingredients to convert the surface to a **zinc phosphate coating**. This material—the second stage of the etching and surface conversion and coating process—creates an airtight protective surface similar to what the manufacturer accomplishes in their zinc bath. A combination of the metal etching chemical and the conversion coating form a nearly perfect airtight coating on the surface that will preclude corrosion from creeping under the edges when the paint surface has been broken.

A wet-on-wet application is commonly used when applying the metal conditioning and conversion coatings. This means the surface is left wet after rinsing the metal etching material and following it immediately with the conversion coating material without first drying it. Dwell time again is relatively important as the material must be left wet on the surface for one to three minutes. As always, the manufacturer's recommendations must be followed to ensure the correct usage of the product. However effective the corrosion barrier achieved may be with the use of these materials, they are completely ineffective if they are not covered with a protective coating of primer and topcoat. Even though the combination of the two steps is an ideal approach and the zinc phosphate coating produces nearly optimum results of cleaning and prepping the surface, it is rarely used. The widespread use of certain acid etch primers has essentially eliminated the need for the additional step of conversion coating the surface, as the primer will accomplish all of what the conversion coatings accomplish.

BASECOATS

After the surfaces have been chemically cleaned and etched, they must be coated with one of several available basecoats, commonly called primer. These coatings fall into one of four major categories: weld-through coating, self-etch primer, wash primer, and two-part epoxy primer. One cannot overemphasize the importance of using these products correctly as some are completely incompatible with the others. For example, spraying an epoxy primer over an acid-laden product such as self-etch primer is guaranteed to make the surface blister and delaminate. The best intentions of doing the best job possible may backfire completely, resulting in the technician having to re-do everything because the finish fell off.

Weld-Through Coatings

One product that stands out from the other primers is weld-through coatings. See **Figure 19-10**. These are special zinc-rich primers designed to protect the microscopic voids that form from the welding heat on the inside of the flanges and mating surfaces. These surfaces cannot be reached to properly treat after the panels have been welded together; therefore, they must be treated with the weld-through coating prior to assembling them. These coatings should be applied to any bare surfaces of the flanges, seams, and mating surfaces where the welding is to be done before joining them together. See **Figure 19-11**. Despite the added corrosion protection obtained in these areas from the weld-through coatings, optimum results will be achieved if all the original e-coat primer is left in place on all the flange mating surfaces except in the immediate areas where the welds are to be made. Coating any area from which the e-coat has been removed in the weld area and on the insert with weld-through coating will maximize the protection to the area.

FIGURE 19-10 A zinc-rich weld-through coating should be sprayed over the bare substrate prior to welding the panels together.

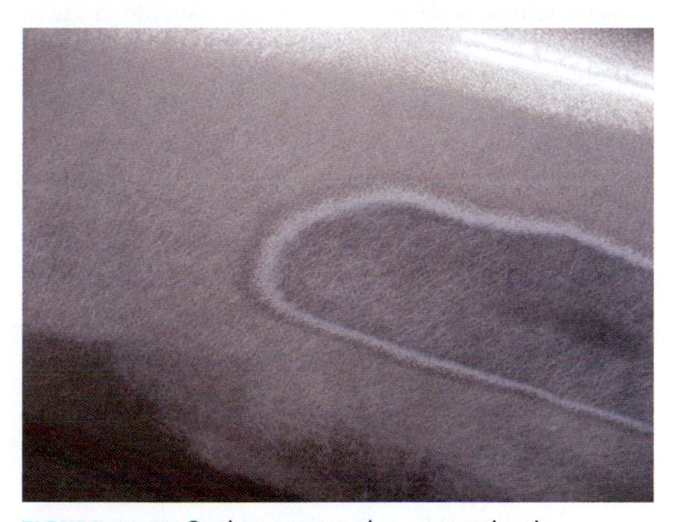

FIGURE 19-11 Optimum corrosion protection is achieved by leaving the original e-coat on the surface except in the areas where the welds are to be made. The welded areas are to be bared and coated with weld-through coating before assembly.

The protection is achieved as the zinc in the weld-through coating liquefies from the welding heat, allowing it to flow into and around the weld nugget, filling any microscopic voids and pinholes left as a result of the welding operation. These pinholes and microscopic voids, if left unprotected, quickly become a major source of corrosion at the weld site. When the surface cools, the zinc becomes a solid coating once again. Any remaining remnants of the weld-through coating outside of the immediate weld site must be removed after the welding operation has been completed. Failure to do so will result in the primers and other coatings flaking and peeling off because the active zinc will attack and destroy the binder of any material applied over the top of it. The purpose of the weld-through coating is to protect the immediate area of the weld site. It is not intended to be used for any other form of surface coating or protection.

Self-Etching Primers

Self-etching primers may be more desirable to use in hard-to-access areas where it is difficult to perform the scrubbing and cleansing tasks that precede the material's application. They contain phosphoric etching material that neutralizes and etches the surface as it dries, thus eliminating the need for using the metal etching chemicals and conversion coatings prior to its application. They are also a zinc-rich material which affords excellent corrosion resistance. While most manufacturers still recommend pre-cleaning and etching the bare metal surface prior to applying the self-etch primers, those steps are not as critical with this product as they are with other base coating materials. A near perfect environment for corrosion resistance and adhesion is created when the two steps are used together. One of the most significant shortcomings of this product is that it cannot be used over or under plastic fillers. Many of the paint manufacturers and those that manufacture plastic fillers recommend that these materials not be sprayed over plastic fillers nor should they be applied to any surface over which plastic will be used. The acid in the primer will react with the polyester resin and the catalyst, causing an adverse reaction which will result in the filler becoming stained and also losing adhesion.

Because many of the self-etch primers are **hydrophilic**, they must be coated with a sealer or another form of a topcoat to finish sealing it off from the atmosphere. Leaving the primer exposed to the atmosphere for any length of time is almost a guarantee that rust will form and quickly spread throughout the bare surface that was coated with it. One can apply this primer on the inside of an enclosed cavity such as a frame rail or similar structural member. However, caution must be used to not overwet the inside surface, as this will cause the dry time to be extended significantly and possibly cause the metal to weaken. Overwetting can also cause adhesion failure between it and the topcoat applied if sufficient dry time is not allowed between coats. One must be sure the surface is thoroughly dried to avoid a reaction between the acid and the material sprayed over it. Due to the product's nature, one must also ensure the inside surface is coated with a topcoat or another corrosion protecting material to prevent moisture absorption and the subsequent formation or corrosion on the surfaces.

A common misconception about the use of a self-etching primer is that it can be sprayed over any surface without first abrating it. As with any paint product, no matter what surface it is applied over—whether painted or bare steel—it must first be sanded or abrated to ensure proper adhesion.

Wash Primers

Wash primers impart some of the same features as the self-etching primers in that they have phosphoric acid in them and they are also zinc rich. They also offer excellent surface cleaning and corrosion resistance once they are cured or dried. One of the chief advantages of these materials is there is no need to chemically etch the surface prior to applying it. In fact, the product warranty may be voided if the surface is chemically etched prior to its application. It must also be coated with a compatible topcoat within a reasonable period of time. Unlike the self-etch primers, the wash primer is designed to be sprayed over a bare substrate only. Therefore, it lacks some of the versatility offered by the self-etching products available.

Epoxy Primers

The most universal primer available is the two-part **epoxy primer**, commonly referred to as a 2-k product. It can be applied over a properly prepared or etched bare steel and aluminum substrate and nearly every imaginable surface. When applied over a painted surface it need only be cleaned and abrated first. When applying it over steel or aluminum, the surface must first be abrated and chemically treated and etched. Under no circumstances, however, should it be applied over an acid-laden primer such as self-etching or wash primers as one of the ingredients is phosphoric acid. The chemical reaction between the acid and the binder will result in blistering and ultimately adhesion loss.

Unlike most of the other surface coatings which dry or harden by the evaporation of the solvents, the epoxy cures as a result of a chemical reaction between the base material and the catalyst. It offers outstanding adhesion and corrosion protecting characteristics when the surface is properly cleaned and prepped prior to its application. The use of this product over a properly prepared surface is as close to duplicating the OEM protection as is possible. Because preparation requirements differ from one manufacturer to another, the manufacturer's instructions must be consulted and followed to ensure the proper recommendations and procedures are followed. Failure to do so will likely result in a product failure or the product prematurely breaking down and deteriorating. Like all other basecoats, the manufacturers recommend that this product be coated with a sealer or topcoat to ensure complete protection. However, in a number of lab tests conducted, this product offered better adhesion and corrosion protection than any other product of its kind—even in the absence of a top or finish coat.

ANTI-CORROSION COMPOUNDS

Anti-corrosion compounds are a heavy bodied material that may be either a petroleum or wax base.

Petroleum Base

The automobile manufacturers or the dealership selling the vehicle commonly apply a coating of petroleum base product over the exposed underbody surfaces such as wheel housings and the undercarriage.

Wax Base

The wax base compounds are commonly applied to the inside surfaces of the doors and the quarter panel, to floor pockets in the trunk area, and occasionally in other inaccessible areas such as the lower section of the B pillar. Many are designed to creep and flow into joints, seams, skips, and into otherwise inaccessible areas that are hard to reach with other spray equipment and materials. This flowing characteristic promotes **self-healing** and reflowing in the event the surface becomes scraped, scratched, or burned during repairs or from exposure to any other elements. They are normally applied after the finish coat has been applied to avoid surface contamination and the possibility of adhesion loss.

In order for these materials to be effective, they must be able to resist water and the constant exposure to atmospheric moisture and the caustic chemical residues carried by it. Examples of these chemical residues are the salt residue typically found in a marine environment and the sulfur dioxide found in the atmosphere of densely populated areas. They must also be able to withstand oil and solvent exposure because they are occasionally sprayed with misdirected lubricants when lubricating the inner door and window mechanisms. If they were to break down from this exposure, the affected areas could be compromised unknowingly. A good quality material will remain flexible as it ages and continue to be self-healing when the outer coating becomes scratched during repairs or is intruded upon. Anti-corrosion compounds are one of the most effective corrosion protecting materials available for the repair industry. When used in conjunction with an epoxy primer, they are almost totally effective in preventing corrosion.

Anti-Corrosion Compound Use

Most manufacturers selectively coat areas of the vehicle that are susceptible to oxidizing with a corrosion protection compound. Areas such as the interior of the door panels are typically coated with a wax base compound. The interior of the quarter panels and the trunk floor seams are frequently coated as well. These areas are subject to water intrusion from runoff and leaks and because of poor ventilation in these areas. In addition, the moisture is slow to evaporate, thus creating a haven for corrosion to occur. Any repairs made on a vehicle that invade or disrupt these coatings must be finished by recoating the back side or inside of the repair area with a product similar to that used by the manufacturer. In some instances the vehicle manufacturer will recommend taking corrosion protection steps that were not part of the original manufacturing process. For example, after performing a sectional replacement of a front or rear rail, they may recommend coating the entire interior of the rail with an anti-corrosion compound. This is done to prevent corrosion at the repair area hot spots and also to ensure the continuation of the manufacturer's original corrosion warranty protection plan. Likewise, when any repairs are made in which heat is used or welding operations are performed, additional steps should be taken to clean and prepare the underside of the affected area. Before attempting to re-apply any type of coating, the surface should be lightly scrubbed with a scour type pad or mild abrasive to break the ash from the burned paint and remove any blisters, as they will otherwise trap moisture between them and the metal surface. This creates an ideal environment for the corrosion to grow rampantly.

Anti-Corrosion Compound

Corrosion compounds are available from several manufacturers in a variety of forms. The three principal forms or bases of these compounds are a wax base, a petroleum base, and one that uses recycled rubber products.

Wax Base. The wax base materials are typically used on the insides of the doors and areas subject to moisture entrapment, much like what is done by the manufacturer.

Petroleum Base. The petroleum base products are to be used on exterior panels such as on the underbody, undercarriages, wheel housings, and the exposed surfaces of the rocker panels seen from underneath the vehicle. This product should be used when the frame of a BOF vehicle is to be coated to protect a repair area.

Rubberized Materials. The rubberized materials are designed for use on surfaces that are constantly exposed to road splash and abrasion, such as the rear wheel housing. This material is more flexible and resilient and capable of withstanding more severe road exposure than the others.

Equipment Requirements

The equipment requirements for applying the non-aerosol materials are quite minimal. They usually include a custom designed low pressure spray gun with a variety of extension tubes and hoses and atomizing nozzles. The extension tubes and hoses are required to reach deep inside the rails and channels to direct the spray into every part of the cavities. A variety of nozzles are required to ensure the material is atomized correctly while it is being applied, particularly to the insides of the rails, channels, and other enclosed surfaces. The same equipment can be used to apply each of the three different material types. One must ensure the equipment is properly cleaned after each use, particularly if the same gun and hoses are used to apply all of the different types of compounds. Material left inside the gun, hose, and extension tubes may result in the materials chemically reacting to each other and causing a breakdown of the product or adhesion loss.

SURFACE PREPARATION FOR APPLICATION

The surface preparation requirements vary as widely as do the manufacturers and the products they make. One should not assume all manufacturers require the same surface preparation. Some manufacturers may require a primer coat be applied and sanded prior to their application. Others may allow the technician to apply it directly over bare steel that has been chemically treated and etched. Yet another manufacturer may require the surface be sanded without chemical etching. All of these application options are possible but the most effective results are achieved when they are applied over a surface that is first coated with an epoxy primer.

Epoxy Primer

Near optimum results are achieved when an epoxy primer is used. In order to further maximize the results and ensure against oxidation forming in the repair area, the entire back side of the repair area—where picks, hammers, pulling studs, and other tools were used—should be completely coated to form a protective coating.

Remove Ash

In areas where pulling studs or any other heat application was used in the repair process, the ash from any previous surface coating should be removed by lightly abrating the area with an abrasive pad. The material should not be applied excessively, yet should be added heavily enough to "heal itself" when it is scraped or when it is brushed against. Another important consideration to remember when installing it is not to apply it so heavily that it will plug the drain holes in the areas where it is being sprayed. As a rule, these materials are applied after the finish coat has been applied to prevent intrusion and possible contamination of the substrate before the topcoat application.

Application of Material

These materials are usually applied using spray equipment; therefore, they require a certain amount of masking to protect adjacent surfaces and areas from inadvertently being sprayed in the process. Some common sense precautions should be exercised to keep the spray away from any parts or surfaces that conduct heat, electrical parts and components, and vehicle informational and identification tags and numbers. One must pay particular attention not to spray any moving parts such as the drive shaft and wheels because it could cause an out-of-balance condition, resulting in a drivetrain vibration when the vehicle is driven. Caution must also be exercised to prevent inadvertently spraying any of the material onto any of the following parts:

- Power window motor, regulator, guides, gears, and pulleys and cables
- Power antenna
- Engine accessories and exhaust parts
- Shock absorbers
- Transmission, drive shaft, and rear axle
- Shifting linkages
- Brake parts
- Interior parts
- Interior upholstery
- Seat belt retractors
- Door locks, cylinders, door latches, and lock remote control connecting rods

Any material accidentally sprayed onto any of these surfaces should be thoroughly wiped and cleaned off as it may permanently stain interior parts and affect the function of other parts.

Accessing Areas for Application

Gaining access to the areas where the anti-corrosion compounds have to be sprayed or applied is sometimes difficult to achieve. However, access may be gained through holes, slots, and other voids commonly used for crush zones on the front and rear rails.

Fenders. The fenders are occasionally covered with a corrosion-resistant coating that can be applied prior to installing the fender onto the vehicle. It may be necessary to add additional material once the fender is permanently mounted on the vehicle to match the amount used by some manufacturers.

Doors. The doors can usually be accessed through drain holes at the bottom of the door or through the removable vent cover plates at the back of the door. Alternatively, they should be sprayed prior to reinstalling the trim panel if it has been removed for the repairs. One should ensure the windows are in their full up position to ensure the entire area up to the belt line is accessible to be coated and the window is not covered with the corrosion protection material. One must ensure the bottom of the door is thoroughly covered by making more than one pass over the area from the front to the rear and making a point of spraying both the front and rear corners where access is somewhat restricted.

A Pillars. In the area of the A pillar, access can usually be gained around the area where the roof and pillar are joined together.

Crush Zones. Crush zones and other openings designed for accessing hinge bolts and other components on the B pillar, C pillar, and D pillar may be used to insert the spray wand to coat the interiors of those areas. Occasionally it becomes necessary to drill a hole into the pillar for access. However, one must check the vehicle manufacturer's endorsement or recommendations before drilling any holes because doing so may possibly weaken the pillar area.

Removable Plugs. Many times the rocker panels have removable plugs through which access may be gained. Occasionally it becomes necessary to drill an access hole into the bottom of the rocker panel near the vehicle's center to gain access. One can then insert a flexible cone spray wand and spray in either direction until the area is properly coated.

Quarter Panel. Spraying the quarter panel and the inside of the trunk is done with a flexible spray wand. By spraying from above, the materials are sprayed downward, allowing them to flow into the recesses between the inner wheel house and the quarter panel and the trunk floor.

The Dog Leg and C Pillar. The dog leg and C pillar may also be coated while inserting the wand into openings that are usually accessible from this vantage point.

Inside Trunk Lid. Spraying the inside surfaces of the trunk lid can be done at the same time as it is usually coated with the same material. Care must again be taken to ensure the material is sprayed in a downward direction to ensure it flows into the recesses in the area of the hemming flange where the outer and inner layers are crimped together. These materials should be applied in any area or surface that one can imagine to be affected by corrosion even though they may have been coated previously. This is one instance where a "little extra" attention is certainly warranted.

STRUCTURAL FOAM COVERED AREA

Before starting to spray any corrosion protection, one must determine if any of the area in question is to be filled with a structural foam. Covering the surface with any coating may affect the adhesion of the foam as it will no longer have a clean, dry surface to which it can anchor itself. There may also be issues of compatibility between the two different materials where one or both may cause the other to deteriorate. It is better to investigate these things before starting to apply the material rather than waiting until the task is completed and then discover one material must be removed before the other can be applied.

CHIP RESISTANT COATINGS

The design of some vehicle models leaves the lower sections—such as the areas behind the wheel openings, the lower door sections, and the rocker panels—exposed to road abrasion more than others. To protect these areas, the manufacturers may apply a 6- to 8-inch high band of a soft, flexible **chip resistant coating** that has a soft spongy or rubber-like consistency along the entire lower section of the vehicle. These coatings are sprayed to a considerably thicker film build than normal, giving them the ability to absorb the shock effect of any rocks and debris without chipping the paint. The coating may be made of a pigmented material or it may be clear. The pigmented materials are usually applied prior to the topcoat so the surface will match the vehicle's color after painting. The clear or unpigmented products can be sprayed either before or after the finish topcoat has been sprayed. While not all vehicles have a chip resistant coating applied in this area, it is a part of the vehicle that requires some special attention—especially if it has been mounted to the pinchweld clamps commonly used when repairing any structural damage. The paint and surface coatings invariably are broken, exposing the bare steel surface underneath. All the necessary steps must be taken to restore the finish in this area. This is done not only to protect against corrosion, but also because it is one of the primary areas of evidence that the vehicle has sustained collision damage.

SEAM SEALERS

Whenever two panels are joined together—whether by welding or with mechanical fasteners—a seam exists at the overlapping areas. The edges and contact surfaces of these panels are potential corrosion hot spots and the necessary steps must be taken to protect them. Moisture and contamination must be kept out of these joints as they can become a haven for corrosion to occur undetected, because the seams are usually in an obscure location.

Moisture and Vapor Barriers

Seam sealers are commonly used to protect these areas by covering and filling them to prevent the moisture from getting into them. They are also used to prevent dangerous fumes from entering the passenger compartment, particularly in the areas behind the engine compartment and underneath the vehicle where exhaust fumes could escape from their normal routing.

Because they are often used in highly visible areas and in locations where there is a significant amount of vibration and flexing, these barriers must remain flexible to prevent cracking and splitting. To provide protection, they must also be resistant to chemical exposure. Their use in areas such as the strut tower, joints and seams around the cowl, fire wall, rocker panels, and so on, requires that they be paintable. Tooling and painting them to match the color of the vehicle frequently disguises their location and they become much less obtrusive to the viewer.

SURFACE PREPARATION FOR SEAM SEALERS

Prior to applying any type of seam sealer, the surface should be properly prepared and coated with at least a coat of primer. Seam sealers have no corrosion protection characteristics or capabilities; therefore, they should never be applied over a bare metal surface. Their adhesion is dependent on the primer or surface coating over which they are applied. Applying these products over a bare uncoated seam will create a corrosion hot spot that will go undetected until it is too late. The first sign of an existing problem will become evident when the sealant starts to bubble or when the rust starts to leach out from under the sealer. At this point the damage is nearly irreversible.

Seam Sealer Consistencies

Seam sealers—available in several different forms or consistencies—are used for a variety of different applications. Some sealers are heavy bodied, others are a thinner consistency for self-leveling, and some are even made to be sprayed. Solid ribbon-like sealers are also available for special applications. They may be a one component or a two-part material that is applied directly to the surface with a special nozzle attached to their dispensing tube. Some are a latex or water base, while others are a petroleum base material which in some instances may dictate how they are applied and smoothed and where on the vehicle they may be used. See **Figure 19-12** and **Figure 19-13**.

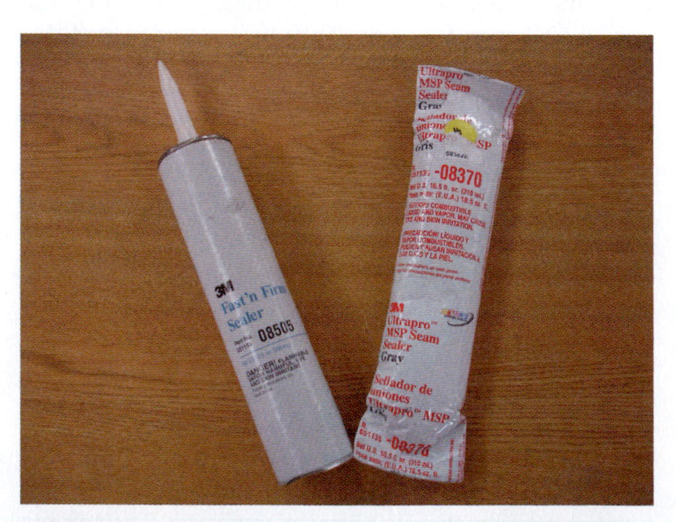

FIGURE 19-12 The single-component caulking and sealers are either a latex or solvent base and require "open time," which allows them to dry prior to coating them with topcoat.

FIGURE 19-13 The two-part sealants are catalyzed and generally dry or cure more quickly.

Single and Two-Part Component

The single component sealers cure by evaporation of the solvents or the vehicle in them. The two-part products consist of the primary filler material—which is the main body—and a second ingredient, which is the catalyst required for the material to cure. See **Figure 19-14**. They are typically dispensed with a special application gun or dispenser that mixes them as they are being squeezed from the tubes through a special **static mixing tube**. The mixing tubes are available in varying lengths for different products; the length selected is somewhat dependent upon the amount of mixing required and the working time allowed.

Leveling Characteristics

Some sealers are self-leveling, which means they are to be applied to the surface and left to flow and level out,

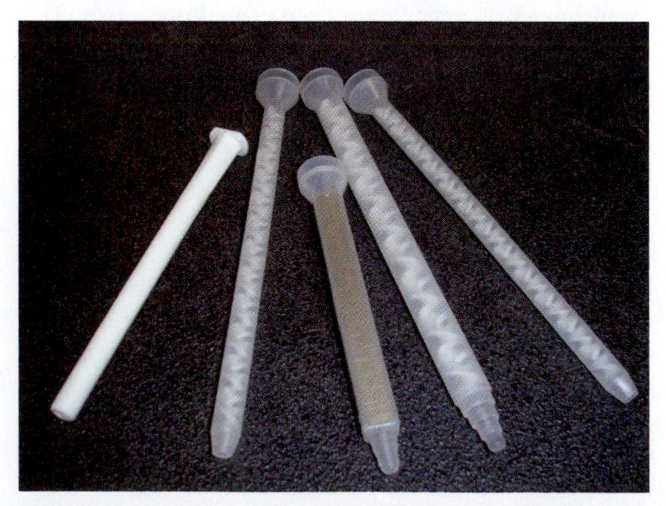

FIGURE 19-14 A static mixing tube that completely mixes the two ingredients typically attaches to the dispensing gun used for applying the two-part sealants and caulking materials.

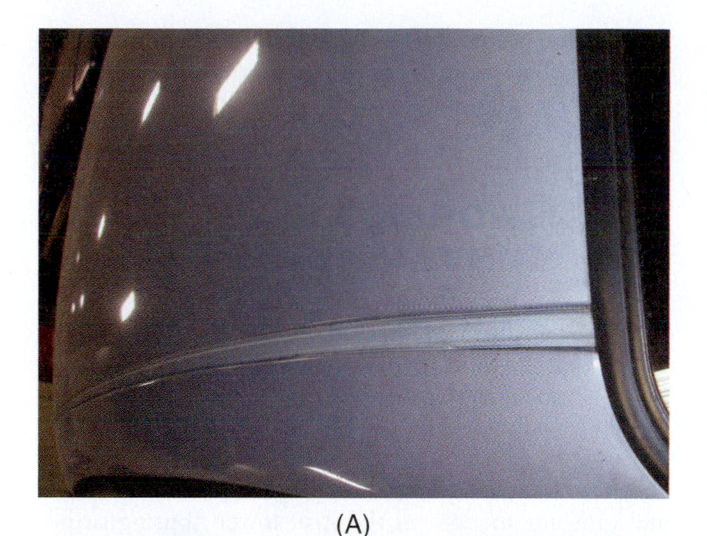

(A)

(B)

FIGURE 19-15 The seam between the roof and side panel on uniside designs is either covered with a removable molding (A) or filled with a sealant (B).

while others may require paddling or feathering to achieve the smoothness required. The self-leveling materials are commonly used in areas where the seam between two panels forms a cavity that must be filled with a special sealing material. See **Figure 19-15**. An example of this is the seam between the roof and a uniside panel on certain late model vehicles. The sealant is applied to a specified level in the cavity to ensure it is filled to the proper level and left to cure. Smoothing or tooling is usually not required, but one must ensure a sufficient amount is applied in one application. The materials may not flow together undetected once it has started to cure and additional sealant is added.

Thin-Bodied Sealers

The thin-bodied seam sealers may also be used to fill small narrow voids that are less than a 3 mm (1/8 inch) gap. When applied, they should be allowed to flow into the seam, adding additional material as needed to fill any

FIGURE 19-16 Many times the manufacturer covers the overlapping edges at the seam to protect against corrosion and water and dust infiltration.

skips in the event it runs away to fill a void. They will shrink slightly as the solvent or vehicle evaporates from them, and a sufficient amount of material should be added to ensure no skips occur once the material has dried. See **Figure 19-16**.

Heavy-Bodied Sealers

The heavy-bodied seam sealers are commonly used in areas where the overlapping seams may not fit tightly together. They can safely be used to fill seams ranging from 3 to 7 mm (1/8 to 1/4 inch). Seams with gaps this large may be found in areas such as the passenger compartment floor to the rocker panel, on the cowl side panel, and in the pockets between the quarter panel and the trunk floor. These sealers may be either a single component or a two-part material typically dispensed from a single tube. The sealers can also be dispensed from tandem tubes from which the two parts are squeezed out through a static mixing tube and mixed simultaneously in the process.

Tooling Sealers

Depending on the location where they are applied, the seam sealers may require some tooling or paddling to feather the edges to a smooth taper—especially when used in areas that are visible. The tooling can be done with an applicator similar to those used for applying plastic fillers, or one may find using fingers covered with a latex glove to be the most effective method. One will find the use of a lubricant—such as a mild solvent or water—will enhance the smoothing and feathering considerably. If the sealer is a latex base, water must be used. If it is a petroleum base, a mild solvent—such as a wax and grease remover or similar product—will be most effective. One should avoid using strong solvents such as lacquer thinner as it will soften and damage the painted or primered surface to which they are applied. It can also dissolve the sealer and cause it to become runny. A brushable seam

FIGURE 19-17 Auto manufacturers sometimes use an expandable product to fill hollow cavities and pockets in areas of the rear quarter panel, door bottoms, and other areas for added strength; to eliminate noise, vibration, and harshness; and to prevent water and dust infiltration.

sealer may also be applied over the top of them to give them a more finished appearance.

Solid Seam Sealers

Sometimes a larger void or pocket may have to be filled in the upper rear recesses of the quarter panel area or other parts of the vehicle. Other than to cover and smooth the overlapping areas of the seam where the panels are joined together, this is not a job for the heavy-bodied seam sealers. Instead, a solid seam sealer or an expandable foam should be used after the seam sealer has dried or cured. The use of foams is discussed in Chapter 8. See **Figure 19-17**.

Manufacturers have varying drying time or curing requirements before the sealers can be topcoated with a primer or finish coat of paint. One should consult and follow the sealant manufacturer's dry time recommendations because premature coating may cause adhesion problems. In addition, the material may never completely cure; therefore, it won't properly anchor itself to the surface. Some sealants may require priming and sealing prior to the application of the topcoat to ensure proper adhesion of the finish. It is not advisable to use silicone seam sealers commonly distributed by hardware stores, because they are usually not paintable with automotive paints.

Brushable Sealers

Brushable seam sealers are commonly used in conjunction with the heavier-bodied seam sealers or they may be used independently as a finish coating. As the name implies, brushable seam sealers are applied with a brush, often over the top of panel seams and in areas where a heavy-bodied sealer is used to give the panels a finished look. Since they are a thinner material, they flow and level out, giving the surface a more finished appearance even though brush streaks are still visible in them. When used as a finish coating over a heavy-bodied sealer, one must allow adequate flash time for the heavier bodied material before its application or solvent trapping may occur, which could result in adhesion loss and incomplete curing.

Application Equipment

In order to duplicate the pattern and texture used by the manufacturer, some seam sealers must be sprayed using a special air compressed applicator gun. See **Figure 19-18**. These textured surfaces are commonly found on the door skin hemming flanges, quarter panel seams, trunk floors, and on some models on the strut tower housing seams. The applicator gun is equipped with adjustments that allow the user to vary the application pattern to duplicate a variety of textures with the material. The materials applied with this equipment are generally single-component seal-

FIGURE 19-18 Automobile manufacturers frequently use a special textured seam sealer and their likeness must be duplicated to prevent tell-tale signs of repairs in areas such as the strut tower, door hemming flange, and other relatively high visibility areas.

FIGURE 19-19 Ribbon and other solid seam sealers that must remain flexible over time are used to cover and fill large gaps. These sealers must withstand drastic temperature changes and exposure to the elements as they are not paintable.

ers used to duplicate special textured effects used by the manufacturer. Before attempting to spray the material onto the vehicle, it is advisable to practice on a piece of scrap material until the desired texture is achieved. An incorrect or unmatched texture of the seam sealer may be a tell-tale sign, to even the casual observer, that the vehicle has undergone collision damage repairs.

Solid Ribbon Seam Sealers

Solid ribbon seam sealers are used for filling gaps and open joints in areas where they may require removal and reinstallation. See **Figure 19-19**. Large gaps such as the areas where the air conditioning lines enter the plenum are an example where these sealants must be used. They must remain soft and pliable, be able to withstand the flexing of the pipes leading into the housing, and must accommodate their removal and reinstallation when servicing any components inside.

PANEL BONDING ADHESIVES

Panel bonding adhesives are not normally thought of as being a corrosion protecting sealant. However, they can be the most effective defense available to the repair technician for protecting an otherwise vulnerable area. When replacing a door outer repair panel with the adhesive bonding material, for example, the adhesive serves a dual purpose by bonding the metal panels together and completely sealing off and protecting the seam area from possible corrosion. The adhesive characteristics make the material even more effective than other seam sealers. Complemented by a coating of corrosion protecting material sprayed on the inside of the door panel, this nearly eliminates any possibility of corrosion getting started in the seam area at the bottom or sides of the door.

FOAM SEALANTS

In recent years the use of foams has begun to play an increasingly more critical role to enhance the automobile's structural integrity. Because of their appearance, the foams used—and the location in which they are used—could be mistaken for corrosion protection devices. One must ensure they are properly reinstalled if removed to restore the vehicle's structural integrity. Furthermore, to avoid the possibility of a chemical reaction that may destroy the foam's structural characteristics, the automobile manufacturer may specify the type of corrosion protective materials that can be used in the same areas with them. They may also specify the sequence in which the materials must be installed to maximize the vehicle's corrosion protection and to protect the foam's structural integrity. The vehicle manufacturer service manual must be consulted to ensure the correct restoration of both these materials.

Many times the process of restoring the vehicle's corrosion protection after the repairs are completed is thought of as a necessary evil. Therefore, the entire process is treated with little or no enthusiasm and dignity. The process should instead focus on protecting and preserving the vehicle by constantly maintaining a clean repair site free of contaminants and making sure that no steps are inadvertently omitted. A concerted effort should be made to chemically treat all surfaces to ensure they will not deteriorate under the materials that will be applied over them. Ultimately, the surface preparation and the manner in which the corrosion protection materials and the finish coats are applied should be of equal importance as those applied to the vehicle's exterior.

Even though most of the areas that must be treated and protected against corrosion are not visible at the time of the repairs, ignoring them will soon become a costly oversight. Once the corrosion gets started on the back side or inside of the repaired areas, it will continue to spread, eating its way through the unprotected metal from behind and eventually causing the paint to flake off, exposing an unsightly rusted surface.

WARRANTY PROTECTION

The repair shop has a moral and legal obligation to restore any remaining manufacturer's corrosion protection warranty. Many auto manufacturers warranty their vehicles against rust through for 10 years. When a collision repair shop performs any collision repairs, they assume the balance of the remaining warranty. Therefore, the corrosion warranty on a new vehicle repaired in any shop becomes their responsibility for the next 10 years. This should not be cause for concern provided the recommended safeguards are performed. However, failure to comply will certainly result in the vehicle being brought back to the repair facility for repairs—this time at a very high cost not only for labor and materials, but also to your reputation.

Summary

- When dealing with anti-corrision materials, technicians must protect themselves by using the appropriate personal protective equipment.
- Corrosion is a chemical reaction that occurs when certain metals are exposed to oxygen or electrolytes.
- Three methods that can reduce or prevent oxidation are cathodic coatings and galvanizing, chemical treating and etching metal, and applying sealants and surface coatings.
- Hot spots are areas where corrosion is likely to occur, often from panels rubbing against each other.
- Basecoats fall into four categories: weld-through coating, self-etch primer, wash primers, and two-part epoxy primer.
- Anti-corrosion compounds can be petroleum base or wax base.
- Seam sealers are available as either a single component or two component material some of which are applied with a special dispensing gun that mixes the two components as required as they come out of the dispensing tube.

Key Terms

acid rain
alodine iridite
anti-corrosion compounds
cathodic coatings
chemical adhesion
chip resistant coatings
corrosion
corrosion hot spots
dwell on the surface
e-coating
electro-galvanizing
epoxy primer
galvanic corrosion
galvanization
hydrophilic
iron oxide
marine atmosphere
mechanical adhesion
oxidation
passivation
phosphoric acid
sacrificial corrosion
seam sealers
self-etching primer
self-healing
static mixing tube
sulfur dioxide
wash primer
zinc phosphate coating

Review

ASE Review Questions

1. Technician A says a corrosion hot spot can be the result of not coating pry pick marks on the back side after repairing a damaged panel. Technician B says corrosion hot spots only occur when heat is used for repairing damage. Who is correct?
   A. Technician A only
   B. Technician B only
   C. Both Technicians A and B
   D. Neither Technician A nor B
2. Which of the following is NOT one of the corrosion protective steps automobile manufacturers use?
   A. E-coating
   B. Passivation
   C. Galvanizing
   D. Wet-on-wet cleansing and etching
3. Technician A says a properly applied anti-corrosion compound applied to the interior of a door should be self-healing. Technician B says before applying any corrosion protecting materials several steps

required for refinishing operations should be followed as well. Who is correct?
A. Technician A only
B. Technician B only
C. Both Technicians A and B
D. Neither Technician A nor B

4. All of the following are links for corrosion to occur EXCEPT:
A. Aluminum.
B. Moisture.
C. Oxygen.
D. Heat.

5. Technician A says steel oxidizing is a form of passivation. Technician B says galvanized steel is a form of sacrificial corrosion. Who is correct?
A. Technician A only
B. Technician B only
C. Both Technicians A and B
D. Neither Technician A nor B

6. Which of the following is the recommended procedure when plastic filler is part of the repair scenario?
A. Etching the metal prior to applying plastic filler
B. Spraying self-etch primer onto the surface prior to applying plastic filler
C. Applying anti-corrosion compound prior to applying plastic filler
D. Cleaning the metal and applying filler directly over it

7. Technician A says the zinc phosphate coating is nearly an airtight coating over the bare metal surface. Technician B says the zinc phosphate coating must be covered with a topcoat in order to preserve its corrosion protecting capabilities. Who is correct?
A. Technician A only
B. Technician B only
C. Both Technicians A and B
D. Neither Technician A nor B

8. Which of the following is best suited for corrosion protection on the inside of the doors?
A. Self-etch primer
B. Wash primer
C. Wax base anti-corrosion compound
D. Petroleum base anti-corrosion compound

9. Technician A says the dwell time for phosphoric acid on bare metal is approximately 1 minute. Technician B says it is advisable to scrub the metal surface with a scuffing pad after the phosphoric acid has dried from the surface to remove the excess coating. Who is correct?
A. Technician A only
B. Technician B only
C. Both Technicians A and B
D. Neither Technician A nor B

10. Technician A says weld-through coating is designed to liquefy and reflow around the weld nugget to protect the weld joint. Technician B says prior to applying the two-part epoxy primer, the surface should be chemically etched to obtain optimum results. Who is correct?
A. Technician A only
B. Technician B only
C. Both Technicians A and B
D. Neither Technician A nor B

11. Technician A says most corrosion hot spots can be slowed by etching, priming, and sealing them from the elements. Technician B says some anti-corrosion compounds are made of recycled rubber. Who is correct?
A. Technician A only
B. Technician B only
C. Both Technicians A and B
D. Neither Technician A nor B

12. Technician A says seam sealers can be applied over bare metal surfaces. Technician B says aluminum creates a self-protecting anti-corrosion coating as it ages. Who is correct?
A. Technician A only
B. Technician B only
C. Both Technicians A and B
D. Neither Technician A nor B

13. Technician A says the zinc coating on the galvanized steel corrodes to protect the steel surface. Technician B says aluminum is more corrosion resistant than uncoated steel. Who is correct?
A. Technician A only
B. Technician B only
C. Both Technicians A and B
D. Neither Technician A nor B

14. Technician A says galvanic corrosion occurs when two dissimilar metals come into contact with each other. Technician B says an insulated steel bolt or one with a special coating can be used to hold two aluminum parts together. Who is correct?
A. Technician A only
B. Technician B only
C. Both Technicians A and B
D. Neither Technician A nor B

15. Technician A says sulfur dioxide is one of the key elements causing acid rain. Technician B says a marine environment commonly results in an accelerated oxidation process. Who is correct?
A. Technician A only
B. Technician B only
C. Both Technicians A and B
D. Neither Technician A nor B

Essay Questions

1. List at least four sources of corrosion hot spots and cite how they can be avoided.
2. List four steps in sequence that must be followed when corrosion protection must be put into place on a bare steel surface that hasn't been previously coated.
3. List the three primary types of anti-corrosion compound and where each is designed to be used.
4. Discuss the difference between sacrificial corrosion and galvanic corrosion.
5. List three of the primers that are used to prevent corrosion and where they are to be used.

Topic Related Math Questions

1. The inside of a rear wheel well has a surface area of approximately 4 × 8 feet. The vehicle owner wants both sides coated with rubberized undercoating. A quart of rubberized undercoating will cover approximately 40 square feet of surface with each coat. How many quarts of undercoating/corrosion protection material will be needed to complete the job if two coats are sprayed over the entire area?
2. A 22 oz. can of aerosol corrosion protection material costs $22.75. The can contains enough material to cover the inside of 1 1/3 doors. How many cans will it take to cover all four doors and what will be the cost?

Critical Thinking Questions

1. In a group discussion, discuss the different sources of oxidation on a typical automobile. What can the manufacturer do to eliminate each of them?
2. Discuss the economic impact of corrosion on the overall automobile industry.

Lab Activities

1. Using an old discarded steel body part from a vehicle, remove the paint from the front side and apply several pulling studs commonly used to remove damage from outer panels. Now observe the back side of the panel to see the corrosion hot spots created by their use.
2. Cut out three pieces of metal approximately 12 inches square and prepare them to be coated with various corrosion protection materials. Coat each of them with one of the three commonly used anti-corrosion compounds. Observe the difference in the amount of build that can be accomplished with each, their dry times, and any other unique characteristics about them. Record the required preparation and the materials that can be used to coat each.

Chapter 20

Paint Spray Guns

OBJECTIVES

Upon completing this chapter, you should be able to:

- Operate a spray gun using proper:
    - Personal safety
    - Environmental safety
    - Fellow worker protection safety
- Describe spray gun types.
- Identify the different gun designs.
- List proper air supply and regulation needs for refinish equipment.
- Describe the function of spray gun components.
- Demonstrate the proper spray gun adjustment.
- Demonstrate correct cleaning and maintenance of a spray gun.

INTRODUCTION

This chapter will be devoted to the study of spray paint application and the spray gun. Although coatings are mostly applied through spraying today, that's not how coatings were applied in the early years of automobile manufacturing. Before the spray gun was invented, automotive coatings were applied by brush. See **Figure 20-1**. This was a very time-consuming and labor intensive operation. Coats of finish, varnish, and India enamels were initially brushed on. Then the vehicle was allowed to dry, after which the brush strokes were sanded smooth. Subsequent coats of finish were applied until the manufacturer reached the desired thickness of finish. After the final finish was polished, the vehicle was assembled and prepared for delivery.

The invention of the spray gun, which occurred in the early 1900s, is generally credited to two different persons, Joseph Binks and Alan DeVilbiss, a physician. See **Figure 20-2**. Dr. DeVilbiss invented a bulb spray device in 1888 that would spray medicine to the back of a patient's throat. See **Figure 20-3**. In 1890, Binks—a painter who painted the basement's storage areas for the Marshall Field & Company department store—figured out how to spray the walls instead of brushing them. He developed his spray finishing equipment, and by 1893 it was in use outside the automotive refinishing industries.

In 1907, Thomas DeVilbiss, son of Alan DeVilbiss and an inventor in his own right, experimented with adapting the original atomizer to create a spray gun to meet the challenges of spray finishing. See **Figure 20-4**. Soon after, spray technology was adapted to the repair

FIGURE 20-1 Before the spray gun was invented, automotive coatings were applied by brush. Used with permission of PPG Industries.

FIGURE 20-3 Dr. DeVilbiss invented a bulb spray device in 1888 that would spray medicine to the back of a patient's throat.

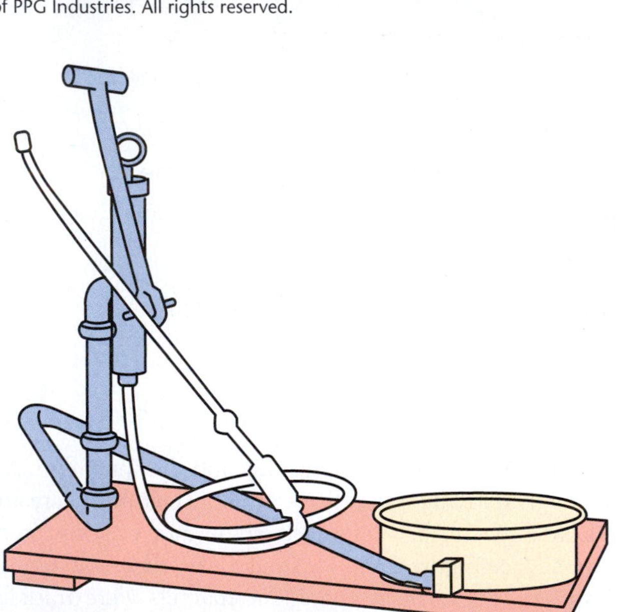

FIGURE 20-2 Binks developed his spray finishing equipment, and by 1893 it was in use outside the automotive refinishing industries.

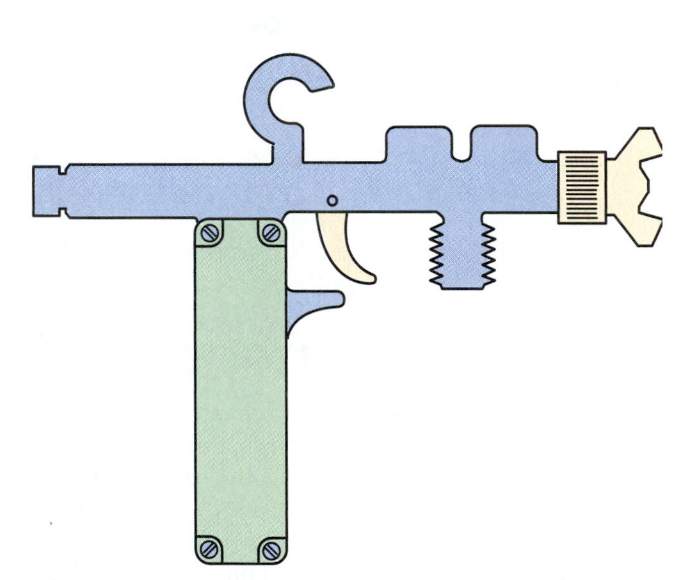

FIGURE 20-4 Thomas DeVilbiss (Dr. DeVilbiss' son) developed one of the first spray guns.

industry, which also increased the efficiency of automobile repair. In 1923, nitrocellulose-based enamels were introduced with their much-reduced dry times; this also reduced the amount of time needed to refinish a vehicle. Before spray painting and nitrocellulose-based enamels were available, it could take as long as 40 days to finish a vehicle; with these two innovations, the time was reduced to only days.

From the early 1900s until 1971, when the high volume, low presure (HVLP) spray gun was patented, there were only minor changes in spray gun technology. And while HVLP spray gun technology has significant advantages over the normally aspirated spray gun, it is not used exclusively in the automotive refinish industry today. Many states do mandate the use of HVLP spray guns, however, because they provide very high **transfer efficiency (TE)**. An HVLP gun can have as high as 80 percent transfer efficiency, as compared to a conventional gun which may have as little as 25 percent transfer efficiency. It is easy to see that with a high transfer efficiency of paint material to the vehicle, the cost of materials is greatly reduced and the amount of **volatile organic compounds (VOCs)**, or the pollutants released, are also greatly reduced as well.

The latest adaptation of coating application is the rolling of undercoatings. Though used in Europe for years, the technique has only recently been adopted in America. The main reason for this adaptation, like others in the past, is to speed up the repair process.

This chapter will be devoted to the spray gun: its types, setup, usage, and maintenance. For vehicle refinishers, the spray gun is one of their most valued tools.

SAFETY

Though a spray gun itself is not dangerous, the products sprayed through it are potentially harmful. Paint spray gun operators must protect themselves, their fellow workers, and the environment from the damages that these substances can cause. See **Figure 20-5**. The refinish technician not only needs to know how to use a spray gun safely, but must also know how to safely use the coatings being sprayed through it. Personal protective equipment that must be used includes:

- Safety glasses
- Safety goggles
- Chemical-resistant gloves
- Paint suit
- **Respirator**
    - Supplied air for isocyanine finishes
    - Charcoal for non-isocyanine finishes
- **Ear protection** in areas of noise levels greater than 90 db

See **Figure 20-6** and **Figure 20-7**.

FIGURE 20-5 To maintain a long and healthy career, a technician should always wear the correct personal protective equipment.

FIGURE 20-6 Respirators and paint suits are key personal safety items.

Read, Understand, and Follow Directions

To safely use spray equipment and the coatings which are sprayed through them, a technician should fully read, understand, and follow the safety instructions provided by the manufacturer. Spray gun manufacturers provide safety instructions in their product instruction manuals, as do manufacturers for coatings. **Material safety data sheets (MSDS)** are, by law, provided to each purchaser of hazardous products. Technicians should also read, understand, and follow their directions, as well as follow basic safety rules.

See **Figure 20-8**.

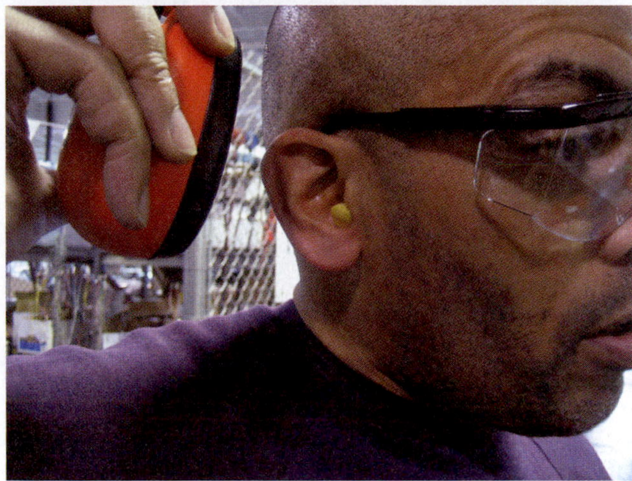

FIGURE 20-7 Ear protection should be worn, as even moderate noise over long periods of time can cause hearing damage.

CAUTION

Master Safety Rules

For safety, keep these suggestions in mind:

- **All accidents can be prevented.**
- **All operating exposures can be controlled.**
- **Regular safety audits are essential.**
- **All deficiencies must be corrected promptly.**
- **All injuries must be reported to the supervisor.**
- **All "horse-play" should be avoided.**
- **Strictly avoid smoking in the shop.**
- **Do not consume food or beverages in the shop.**
- **Safety equipment must be worn.**
- **Report all equipment failures.**
- **Report all spills at once.**
- **Know the facility's emergency exit plan.**
- **Recognize and protect the safety of others.**

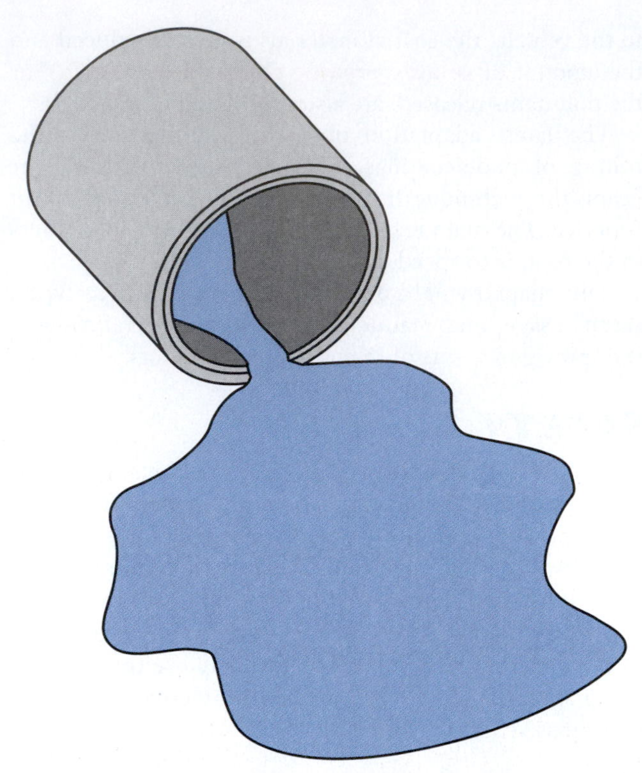

FIGURE 20-8 Paint spills are considered toxic and should be contained and removed immediately.

Inspection

To operate under safe conditions, a technician should inspect equipment to assure that it is in proper operating condition. Personal safety devices should be closely inspected to be sure that they are safe to use. If defects are found, the equipment should be repaired or replaced immediately. **Personal protective devices (PPDs)** should be cleaned and checked for proper fit on a regular basis. Spray guns should be inspected for proper and safe operations, and if found to be defective they should be replaced. Hoses, booths, and the working area should be inspected for proper and safe operations. During equipment inspection, the work area should be inspected for cleanliness; any clutter found should be cleared up to ensure a safe work area. Remember: "*All accidents can be prevented.*" Keeping the work area clean and free from clutter is one of the steps needed for accident prevention.

SPRAY GUN TYPES

Spraying in the automotive refinish business generally uses either siphon feed, gravity feed, or pressure feed spray equipment. These types of equipment can come in as high pressure, low volume (standard) or high volume, low pressure (HVLP) spray equipment. Different types of equipment each have their advantages and disadvantages, which will be discussed within this chapter.

Other types of spray equipment used for coatings application, though not generally for automotive refinishing, include rotating bell, electrostatic, powder coating, and airless application equipment. These are normally used in manufacturing and not generally in collision refinishing.

Siphon

The siphon spray gun has been in use from the early part of the 1900s. It is designed with its paint cup mounted below the main part of the gun. This type of spray gun relies on high pressure air which must pass over a tube extending into the paint; that air pressure creates a vacuum, thus lifting the paint into an air stream and then out through the gun's tip. Though this type of spray gun does very well in atomizing coatings, it needs so much air pressure to lift the paint into the body of the gun that the gun's transfer efficiency is very low. In this process the paint travels at such a high rate when it leaves the gun that it hits the target and bounces off, only depositing about 25 percent of the coating onto the vehicle. Overspray is abundant, causing large volumes of waste and environmental contamination. The use of this type of gun has been outlawed in many states.

Siphon Feed Operation. As its name implies, the **siphon feed gun** operates by creating a siphon over a tube, which extends into the paint cup. See **Figure 20-9**. As high pressure forces air over this tube, a siphon, or **venturi** is created. Because of the low pressure created in the tube, the paint is lifted into the airstream. See **Figure 20-10**. The paint now in the airstream passes over the needle, through the nozzle, and out the air cap. Up to this point, the paint is not atomized; it has only been drawn out of the paint cup and to the paint cap. Automotive paint guns are generally external-mix paint guns, meaning that the paint and air mix to cause atomization outside of the air cap. See **Figure 20-11**. Outside the air cap a large amount of air bombards the coating, and it is forced into small droplets or atomization. See **Figure 20-12**. There are, on the air cap, two air horns, which also have air passing through them. These columns of air force the spray pattern from a

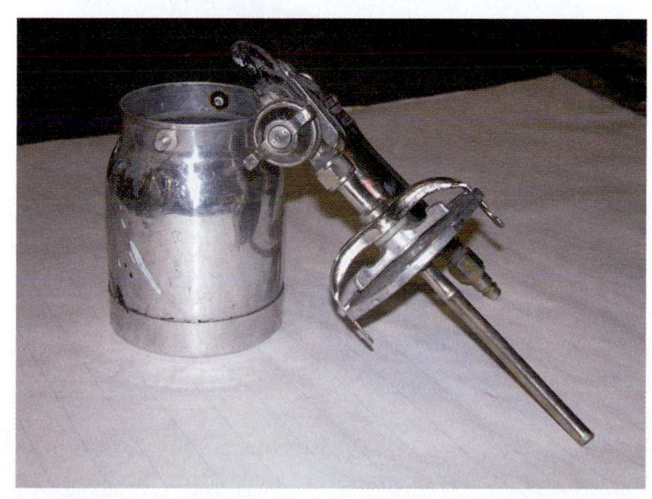

FIGURE 20-9 A siphon feed gun draws paint up the paint tube through the venturi or vacuum effect.

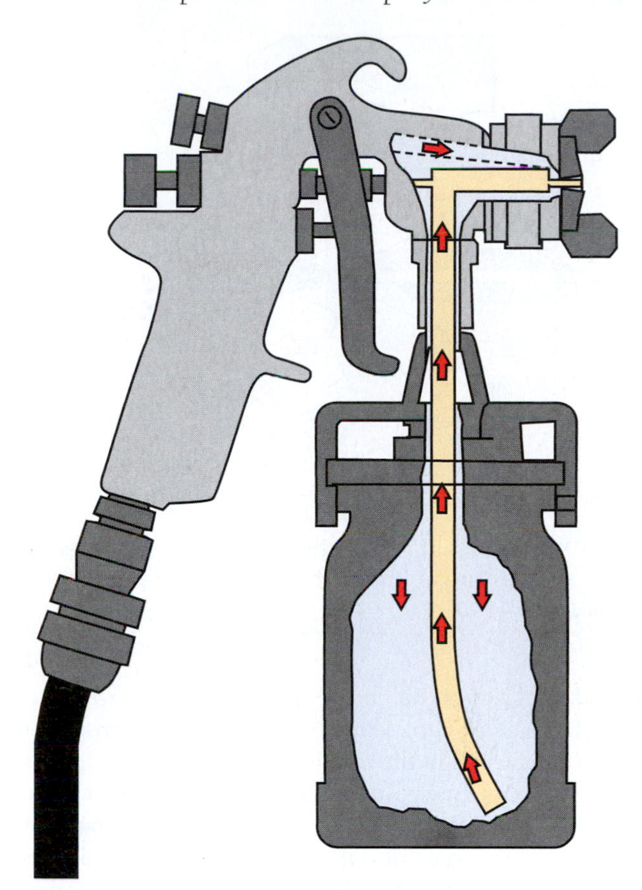

FIGURE 20-10 Siphon is caused by air passing over the paint tube, drawing the paint out of the cup into the gun. Adapted from Binks/DeVilbiss.

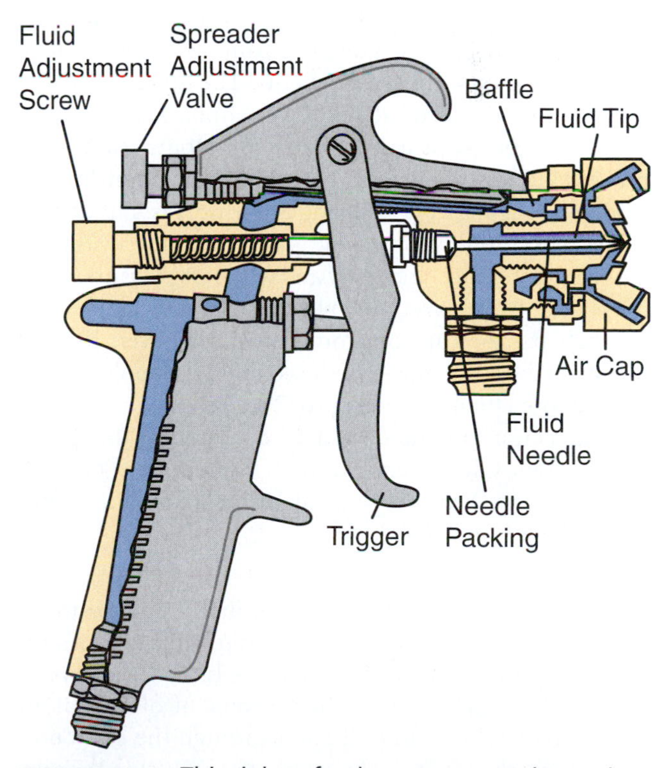

FIGURE 20-11 This siphon feed gun cutaway shows the needle, nozzle, cap with horns, and atomization. Adapted from Binks/DeVilbiss.

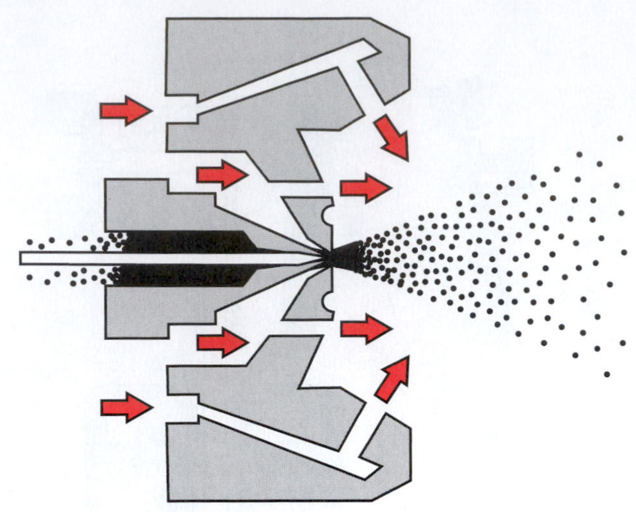

FIGURE 20-12 Paint passes through the air cap where more air pressure atomizes the paint outside the gun. Adapted from Binks/DeVilbiss.

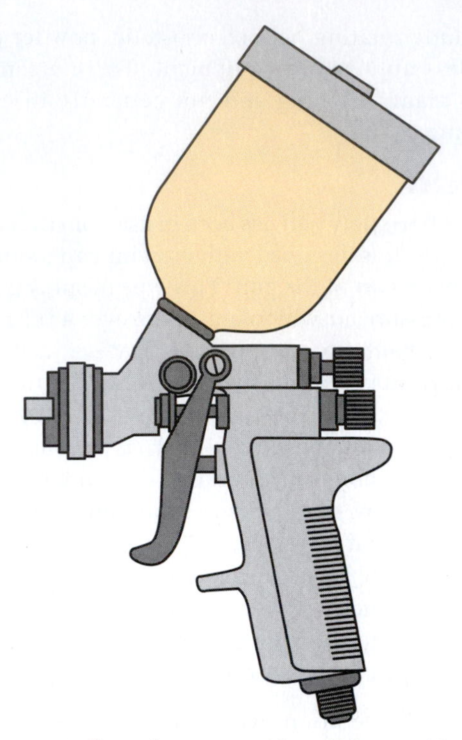

FIGURE 20-13 Though a venturi is still formed in a gravity feed gun, less pressure is needed due to the effects of gravity. Adapted from Binks/DeVilbiss.

round circle to an oblong spray pattern, which is better suited to spray application. For the siphon process to work, an air hole in the cap must be present and open, so that as paint is drawn out of the gun, outside air can come in to equalize the cup pressure. If the vent hole is not open and air cannot come into the cup, a pressure will develop as paint is removed through the siphon process. This pressure will not allow paint to be moved up the paint tube.

Gravity Feed

On gravity feed spray guns, the paint cup is mounted on the top of the spray gun. The paint can flow through the gun with only the aid of gravity. The gravity provides the air pressure needed to move paints, and therefore all the air pressure can be used for atomization. This process allows this type of gun to be more transfer efficient. See **Figure 20-13**. Though a venturi is still created as air pressure flows past the paint tube, the coating is moved much more efficiently and with less pressure. Paint must still travel over the needle, out the nozzle, and through the air cap for atomization. However, because gravity is the main mover of the coating, the application is much more efficient. See **Figure 20-14**.

The technician should not be confused at this point: *not all* gravity feed guns are high volume, low pressure (HVLP) guns. Various types of HVLP guns will be discussed later in this chapter.

Gravity Feed Operation. A **gravity feed gun** works in the following manner. Coating passes from the cup, on top of the spray gun, and down a tube to the airstream. As the trigger is pulled, the needle moves back, which allows air to pass through the gun. This movement of air—along with gravity—helps to pull paint through the gun's nozzle and out the air cap. The paint, when outside the gun, is mixed with air, atomizing the paint. The atomization process is the same for siphon and gravity feed guns. The only difference is that in using the gravity feed gun, a large amount of pressure is not needed to lift the paint from the cup to the gun, thus making it more efficient than the siphon feed gun. Though the process does not

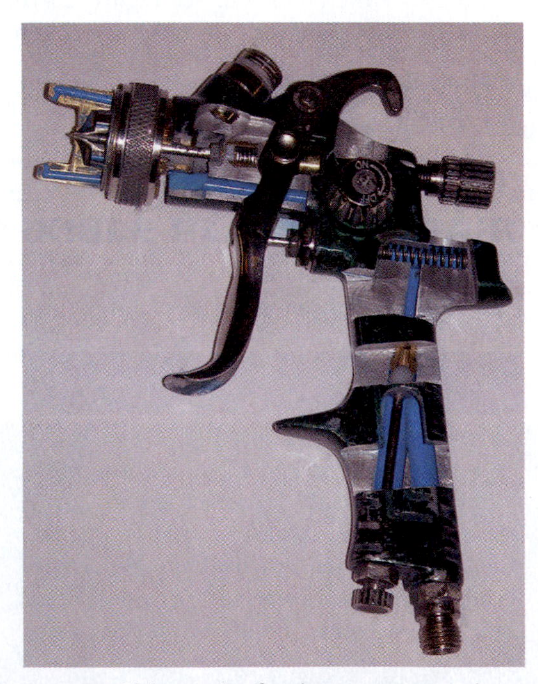

FIGURE 20-14 This gravity feed gun cutaway shows the needle, nozzle, cap with horns, and atomization. The blue areas are air passages.

completely rely on a siphon to pull paint out of the gun, the vent hole in the top of the cup must be kept open and free, allowing air to come in as paint is moved out. This eliminates cup pressure, which would prevent the paint from moving out of the cup.

Pressure Feed

A **pressure feed gun**, though similar to the types described previously in the way it atomizes paint, is quite different from both the siphon feed and gravity feed guns in the way it moves paint from the cup or pot. With the pressure feed gun, air is diverted from the gun to the pot (paint vessel), where pressure is applied. Pressure forces paint through a hose to the gun. See **Figure 20-15**. This pressure delivery system allows the gun to be even more transfer efficient than either the gravity feed or the siphon feed gun. A vacuum is not created in the gun as the paint passes through. Instead, the paint is forced over the needle, through the nozzle, and out the cap, where—for the first time—it meets with air, causing atomization. A pressure feed gun is normally used in manufacturing tasks where larger volumes of paint of a single color are used. The gun has a much larger spray pattern, making it most suitable for production painting. It is more difficult to clean than either the siphon or gravity feed types, for the gun, pot, and paint supply hose must all be cleaned. This gun can also be either a high pressure, low volume or a high volume, low pressure gun.

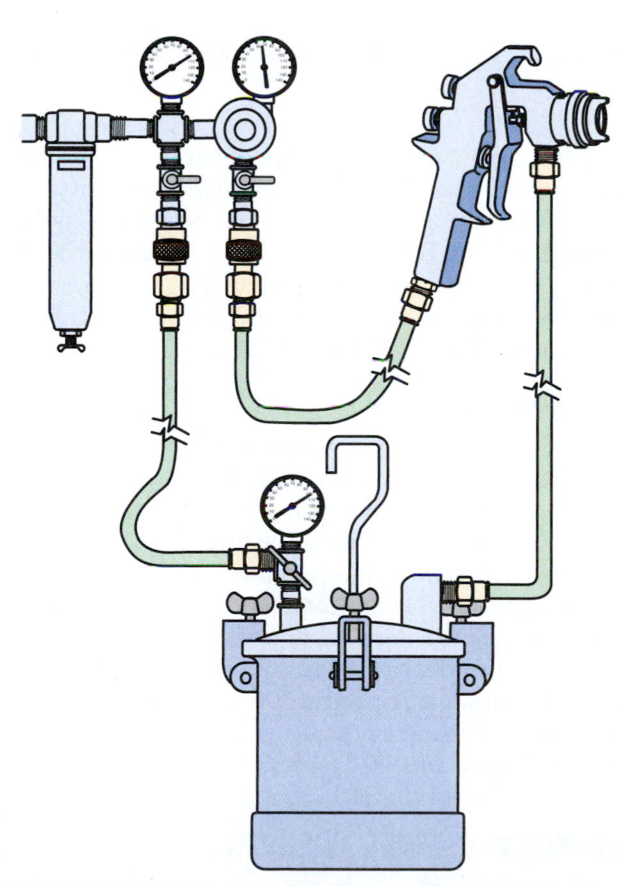

FIGURE 20-15 The left line in this pressure feed gun applies pressure to the paint, forcing it up the right line. The center line supplies atomization pressure. Adapted from Binks/DeVilbiss.

Rotating Bell

The **rotating bell** spray application is normally used in manufacturing conditions. It is often found either (1) mounted stationary, while the vehicle is moved under it; or (2) on the arm of a robot which travels in a set pattern to apply coatings to a stationary vehicle. See **Figure 20-16**. The paint is atomized using **centrifugal** force. The paint is released onto the bell, which rotates at a very high speed; this centrifugal force breaks even the most **viscous** liquid into very small, mist-size droplets. This type of atomization provides large patterns, allowing the finishes to be applied rapidly. Because no air is used to propel the finish, this method offers the benefit of very high transfer efficiency.

Electrostatic

Electrostatic painting does not use air pressure to propel the finish from the gun to the vehicle being painted. It uses the physics law of "opposites attract." To cause this attraction, the atomized paint is passed through an electrostatic spray gun where the paint is given a negative charge. See **Figure 20-17**. The object to be painted is given a negative charge, causing the finish to be drawn to the charged object. The finish will coat all exposed areas of the charged object. This means that even the back side will be coated, and that there will be little overspray. Though this type of finishing can be used for small objects, it is not generally used in the automotive refinishing industry.

Electrostatic application of finishes can be and is applied through the dipping of the vehicle or part into a large vat of finish. The vat is charged with the negative ions, and the object being dipped into the vat is charged with a positive charge. The charged part is coated with the finish as it is dipped; then it is baked to harden the finish before other topcoats are applied.

TECH TIP Electrostatic painting should not be confused with e-coat painted parts. Though the new part primer is applied by electrostatic methods, the term *e-coat* has come to mean that the primer is a thermoset corrosion resistant coating, which is one of the best corrosion protections available.

Powder Coating

Powder coating is a process which uses dry finish that turns to liquid when heated. Then the liquid flows over the entire part. As it cools, the finish becomes a solid, providing a very hard and durable surface. The finish is propelled—using low air pressure—over an

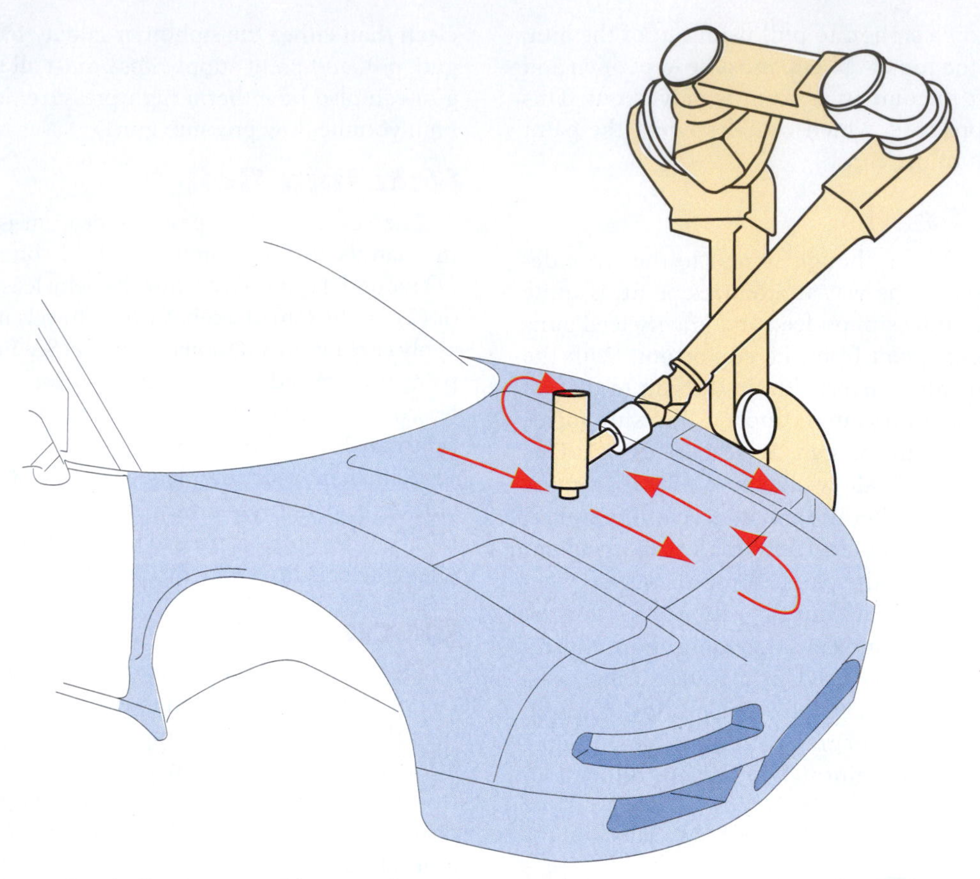

FIGURE 20-16 A rotating bell robot, used in manufacturing, travels in a set pattern to apply coatings to a stationary vehicle.

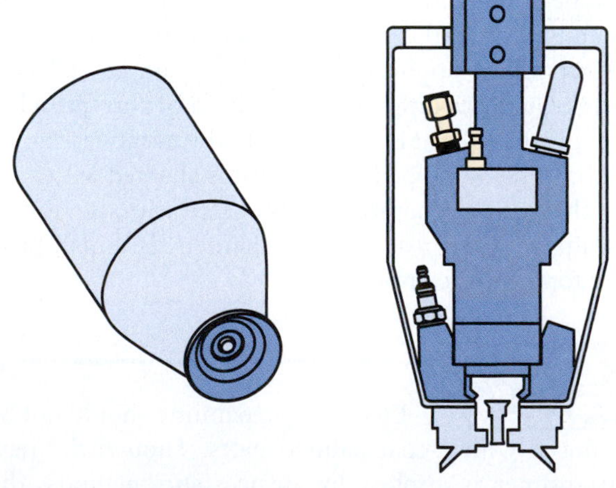

FIGURE 20-17 An electrostatic spray gun is used in powder coating.

electrode, where it is given a negative charge. The object to be painted is given a positive charge, and—as in electrostatic painting—the charged finish is attracted to the charged object. The dry paint particles cling to the charged object being painted. The sprayed objects are then taken to the baking area. The baking process causes the dry finish to become liquid and flow over the surface of the charged object being painted. When the part is cooled, a very hard and durable finish results. Because the paint is dry and does not contain the pollutants normally found in solvent-borne paints, this process is very environmentally friendly. Particles not attracted to the charged part can be recovered and used again, which also makes this type of finishing nearly 100 percent transfer efficient. See **Figure 20-18**.

Airless

Airless application is a process in which no air is used either to atomize the finish or to propel it to the object being painted. It uses high hydraulic pressure to force the finish through the spray gun. The pressure both atomizes the finish and propels it to the object being painted. This type of application is not commonly used in automotive refinishing, though it is sometimes used to finish large objects such as trucks or construction equipment. It is more difficult to clean and maintain the equipment in this application, and is therefore not often seen in collision repair. See **Figure 20-19**.

SPRAY GUN DESIGNS

Though spray guns come in many types and designs, the remainder of this chapter will confine itself to the spray gun designs used commonly in collision repair and refinishing. High pressure guns were the first to gain pop-

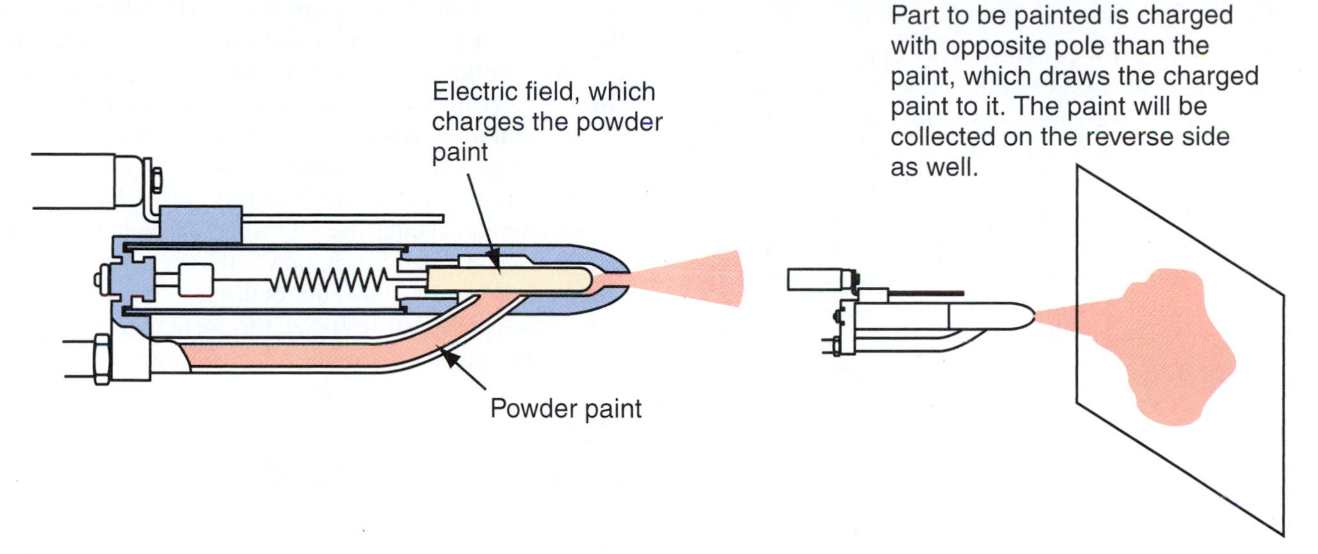

FIGURE 20-18 Powder coating uses dry finish that turns to liquid when heated. Then the liquid flows over the entire part.

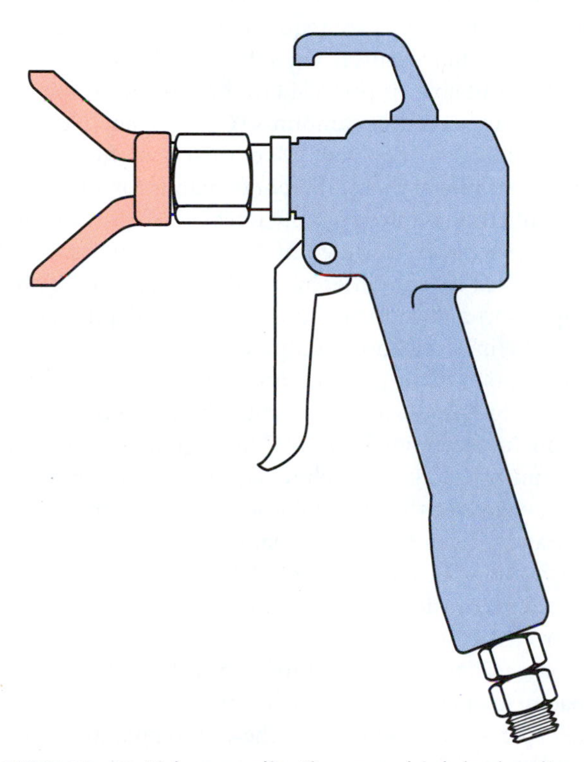

FIGURE 20-19 Airless application uses high hydraulic pressure to force the finish through the spray gun.

ularity, and they worked efficiently with the types of finishes being used at that time. High volume, low pressure guns became popular starting in the middle 1970s, as paint viscosity and environmental laws changed. About that same time low pressure, low volume (LPLV) guns were also developed, but never became popular with automotive refinishers.

More recently, guns that claim to be high transfer efficient and high pressure have become popular with refinish technicians, and have even been allowed in states that have restrictions on air-borne pollutants. The makers claim these guns have the qualities of a high pressure gun with the transfer efficiency of the HVLP gun.

High Pressure Guns

High pressure, low volume (HPLV) guns—sometimes referred to as **high pressure guns**—were the most common type of spray gun used from the early 1900s until the late 1970s. They provided a smooth finish when spraying low viscosity coatings, but generated a large amount of overspray and low transfer efficiency (25 to 30 percent). They would operate at pressures of 50 to 70 psi, which propelled the finish at such velocity that it would hit the object being finished, and then bounce off. See **Figure 20-20**. During this time, paint was relatively inexpensive, so paint compensated for the large volumes of material needed to cover the vehicles being refinished with high pressure guns.

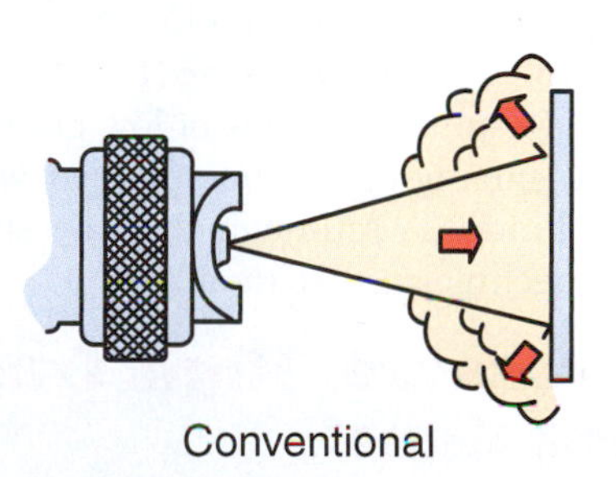

FIGURE 20-20 Finish sprayed under high pressure can bounce off a part as overspray.

HVLP

High volume, low pressure (HVLP) guns, though invented early in the 1900s and sold as an attachment to vacuum cleaners, were not used in automotive refinishing until the late 1970s to the early 1980s. These guns use very low pressures (8 to 10 psi) at the cap, but must rely on large volumes of air (CFM) to atomize and propel the finish. Because of the low pressures used by HVLP guns, the amount of overspray was greatly reduced, and the transfer efficiency was increased (65 to 80 percent).

In the 1970s and 1980s, the population became more concerned about the environmental damages caused by the release of volatile organic compounds (VOCs) contained in solvents. Some states legally required that HVLP guns—with their higher transfer efficiency—be used. The new laws also mandated that finishes be manufactured with less solvent in them, and that finishes be applied at a thicker, more viscous state than was used in the past. These new finishes were called "high solids paints." However, due to this high solid content, high solid paints were more difficult to atomize. When HVLP guns first became commonly used, they did not atomize finishes as well as the older high pressure guns had. The lower level of atomization caused the paint to have more texture (orange peel) than when high pressure guns were in use. However, HVLP guns have changed over the years. Today, even with the high solid paints used, they can atomize finishes well, providing a smooth finish that is acceptable to both the refinish technician and customers.

TECH TIP Transfer efficiency (TE) is most affected by the technician's ability to spray efficiently. Though spray guns have the capability to spray at 65 to 80 percent, national studies of transfer efficiency have shown that an experienced technician who has not been trained in TE spray techniques routinely sprays at less than 65 percent. It is believed that if all automotive technicians were to spray at a *minimum* of 65 percent, low-level ozone pollutants (smog) would be significantly reduced.

LVLP

At about the same time that HVLP guns were being developed, **low volume, low pressure (LVLP) guns** were being marketed. Despite the claims of low pressure and low volume, these guns never became popular with automotive spray technicians. Although they are still available, few—if any—technicians use them.

High Pressure, High Transfer Efficient Guns

Spray gun manufacturers continually develop new designs of spray equipment, and even though some states mandate the use of low pressure spray guns (less then 10 psi at the cap), what they are concerned with is the gun's transfer efficiency. In the past guns that sprayed with higher pressures (greater than 10 psi at the cap) have been less than 65 percent transfer efficient. Spray gun manufacturers have now developed guns that spray at the desired 65 percent transfer efficency or better, with pressures which exceed 10 psi at the cap. Though these guns spray at higher pressures than HVLP guns, they do not spray at the pressures used with the so-called high pressure guns of the past. Therefore, some states which previously restricted the use of non-HVLP guns are now allowing the use of these high-TE guns.

AIR SUPPLY AND REGULATION

The shop's air supply and regulation plant is responsible for having the proper volume (cubic feet per minute, or CFM) at the needed pressure (pounds per square inch, psi). The needs of both volume (CFM) and pressure (psi) in the vehicle repair part of the business may be greatly different than in the refinish shop. In the refinish shop, larger amounts of air (CFM) are required at lower pressures than in the body shop section. The refinish side may also need class D breathable air provided to the fresh-air respirators, which again need larger amounts (CFM) of air at lower psi than in the body shop side. Modern shops often separate the two because of their different demands for air. The size and configuration of air supply lines is also critical within the two shops. Even different compressor types may be better suited for the different needs. A full discussion on compressors and shop layout is covered in depth in Chapter 22, Refinishing Shop Equipment.

Regulation of air is also critical. The size of air supply lines, including the air hoses, must be taken into consideration in the refinish shop. The regulators, lines, and hoses must be large enough to supply the amount (CFM) of air needed for the high volume spray guns and fresh air respirators. Even the quick change couplings need to be high volume. See **Figure 20-21**. All regulators are not alike and careful consideration should be taken when choosing the proper regulators.

Some wall mount regulators do not have a high enough capacity to operate both an HVLP spray gun and **supplied air respirators**. To know that the wall regulator has sufficient capacity, the technician must know the requirements of the supplied air respirators and the HVLP gun. Knowing these needs, the technician must choose a wall regulator that exceeds those minimums.

Gun inlet regulators also need to be high capacity, sufficient to allow the needed amount of air to pass through. If the gun is starved for air at the inlet regulator—even though the inlet pressures are reading correctly—the gun may not operate properly due to not getting the needed volume (CFM) of air. Some gun regulators use a gate valve to restrict pressure into the gun. See **Figure 20-22**. As the valve is turned to reduce the pressure (psi), it also greatly

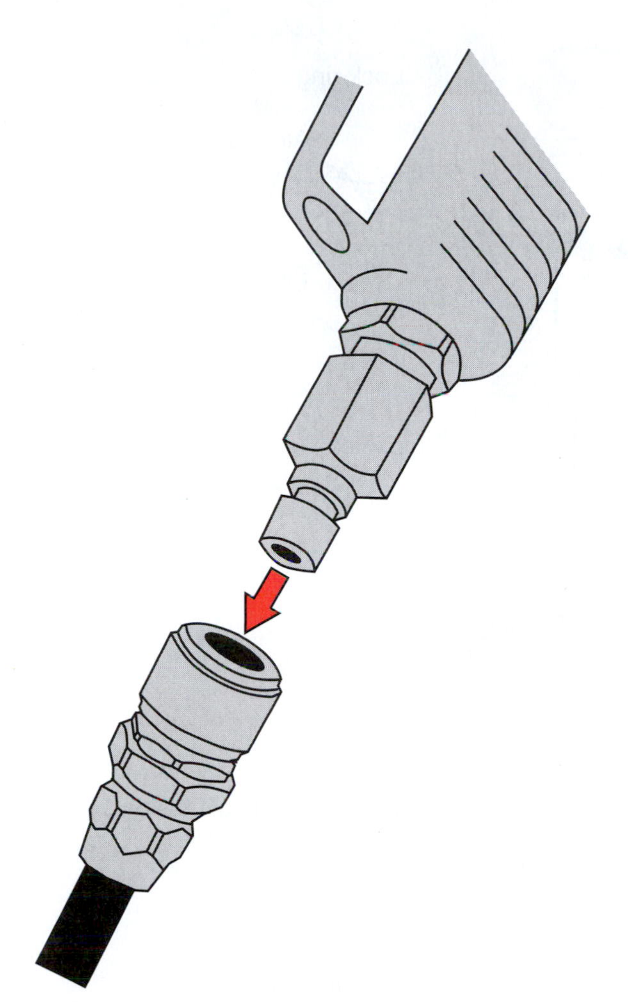

FIGURE 20-21 High volume quick change couplings are needed to supply higher volumes of air for HVLP spray guns.

FIGURE 20-22 A gate valve or "cheater" valve reduces the volume of air as it reduces the pressure.

FIGURE 20-23 This high volume (right) quick change coupler has the standard lower volume on the left.

reduces the volume (CFM) of air. Therefore, it doesn't allow the gun to operate properly. High capacity gun regulators are needed. Those chosen are commonly diaphragm regulators, which can restrict pressure without restricting the needed volume.

Though most quick change cupping devices use 1/4-inch connectors, they should also be high volume in the paint shop. As can be seen in **Figure 20-23**, the high volume style is much larger than the standard.

Hoses should also be considered when using HVLP guns. Again, even though the guns use 1/4-inch threaded connectors, a 1/4-inch hose line has a much lower capacity than the 3/8-inch hose. The large hose is recommended for use with higher volume equipment.

SPRAY GUN COMPONENTS

Though there are many different manufacturers of spray guns, they all operate similarly. **Figure 20-24** shows a breakdown of gun parts. A technician should understand the importance of these parts and how the gun operates. This section will explain the main parts of a spray gun and how the gun should be set up and adjusted.

Air Cap

The **air cap** at the front of the gun directs the airflow from the air horns against the paint, which comes out of the center orifice. See **Figure 20-25**. This airflow helps shape the spray pattern from a round to an oblong pattern. Oblong patterns provide the technician with a controllable wide stroke. This pattern also provides an even distribution of atomized finish over the vehicle as it is being painted. The auxiliary air nozzles help break the fluid finish into small droplets for proper atomization. This type of spray gun is called an external-mix spray gun, because even though air is used to bring the paint to the gun tip, the paint and air mixing for atomization don't occur until the fluid leaves the gun. Though some air caps have different configurations of air cap orifice, they basically have the same three locations: center, side, and auxiliary air nozzle. See **Figure 20-26**.

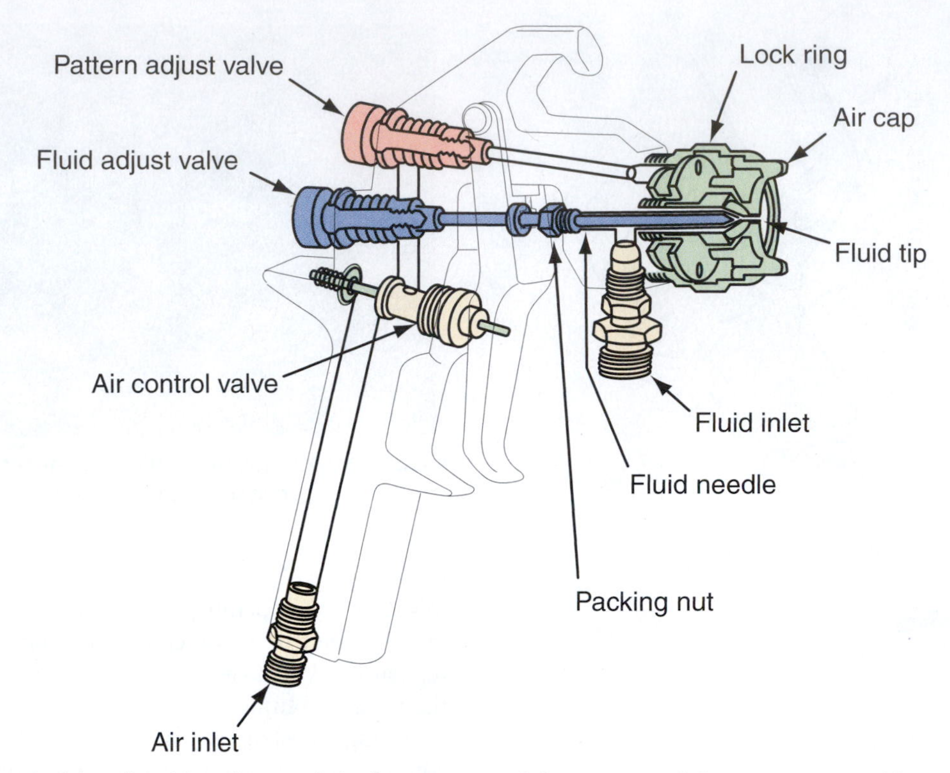

FIGURE 20-24 A technician should understand the importance of these parts of the spray gun and how the gun operates.

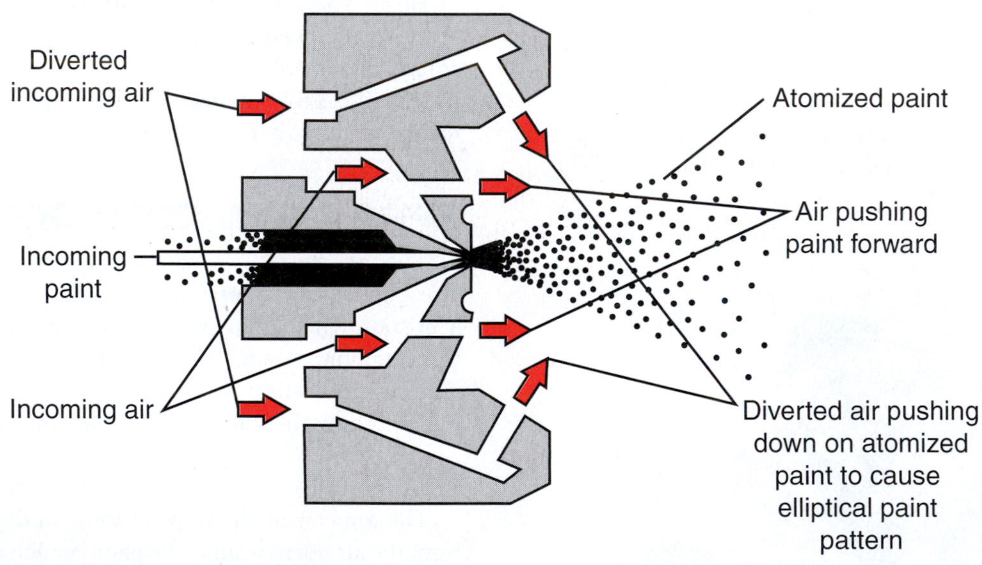

FIGURE 20-25 The air cap sprays in an oblong pattern, which provides an even distribution of atomized finish over the vehicle as it is being painted. Adapted from Binks/DeVilbiss.

Atomization. **Atomization** is the process which breaks the surface tension of a liquid into small droplets, allowing it to be deposited as a spray. The smaller the droplets are, the smoother the painted surface will be. In a spray gun, the paint is distributed in three stages. See **Figure 20-27**. In stage one, the finish is released around the fluid needle and passes through the fluid nozzle center. The paint does not start to atomize until it leaves the gun. The needle, nozzle, and cap combination—and the sizes of their openings—are critical to how well the gun will atomize a finish product.

In stage two of the paint application process, openings around the outer ring of the fluid nozzle allow air to bombard the liquid, thus breaking down the finish into droplets. See **Figure 20-28**. As it passes further outward, the round stream of partially atomized droplets are hit with streams of air from the air cap.

FIGURE 20-26 An air cap can have different orifice configurations, although they basically have the same three locations: center, side, and auxiliary air nozzle.

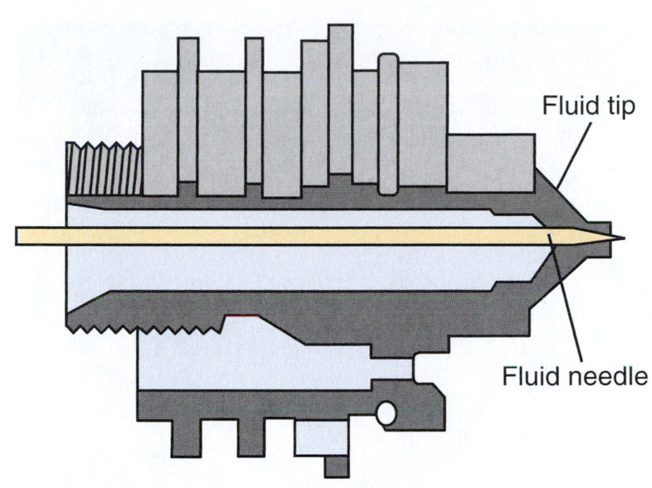

FIGURE 20-28 In stage two of the paint application process, openings around the outer ring of the fluid nozzle allow air to bombard the liquid, thus breaking down the finish into droplets. Adapted from Binks/DeVilbiss.

Stage three involves the streams of air coming from the air cap center orifice, which help further break the finish into smaller atomized droplets. At the same time, larger streams of air from the air cap horns bombard the round spray pattern, changing it into the needed elliptical spray pattern.

Because spray guns are external-mix and the three stages of atomization described previously change liquid finish into the needed small droplets, the full atomization does not take place until the finish has traveled out from the gun. It now becomes easy to see why holding the gun at the proper distance is so critical. (Distance, angle, and techniques will be covered in Chapter 21.)

Air Cap Pressure Gauge. Though air pressure can be measured at different locations when spraying (compressor wall, gun inlet, and air cap), the most accurate location is at the air cap. See **Figure 20-29**. When measured at any of the other locations, it is possible to have variables that may affect the true measurement. It should also be noted that most recommendations for air cap pressure are taken at the center outlet, not the air horn. Many **air cap pressure gauges** have both locations for measurement. The recommendation for air cap pressures can be found in information from either the paint gun manufacturer or the paint manufacturers. Most paint manufacturers will recommend the proper paint needle cap and pressure setup for their paint in specific guns. It is best to locate these recommendations and follow them.

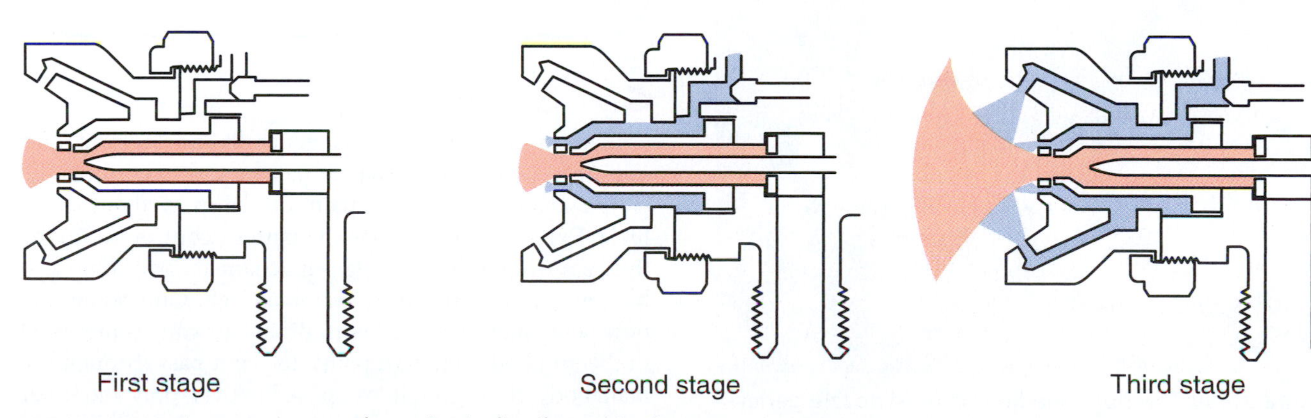

FIGURE 20-27 In a spray gun, the paint is distributed in three stages. Adapted from Binks/DeVilbiss.

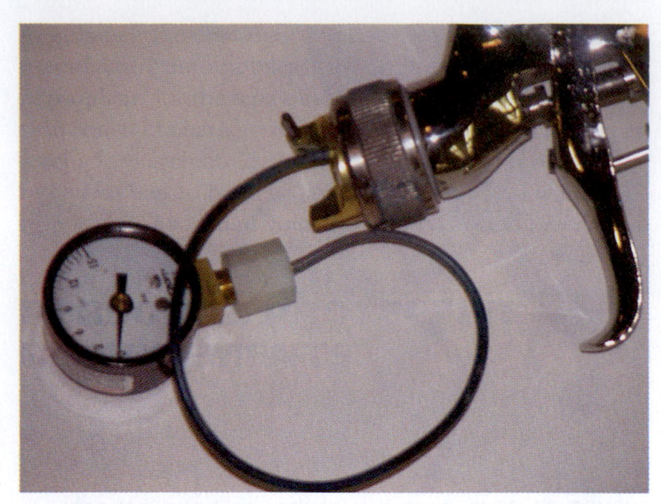

FIGURE 20-29 The air cap gauge measures the final air cap pressure.

Fluid Needle and Nozzle

The **fluid needle** and **fluid nozzle** generally come as a set; they regulate the amount of fluid that can pass from the cup to the spray cap for atomization. The recommendation for fluid nozzle size can be acquired from either the paint gun manufacturer or from the paint manufacturer. See **Figure 20-30** and **Figure 20-31**. It generally uses a larger needle and nozzle setup such as a 1.4 to 1.6 for more viscous material such as clearcoats or primer-surfacer, and smaller 1.3 to 1.2 for less viscous materials such as basecoats. These sizes allow the spray gun to operate at its best when properly set up, and it is critical for a technician to have the proper needle and nozzle setup for each specific spray application. The setting of the proper needle and nozzle is commonly referred to as the gun setup.

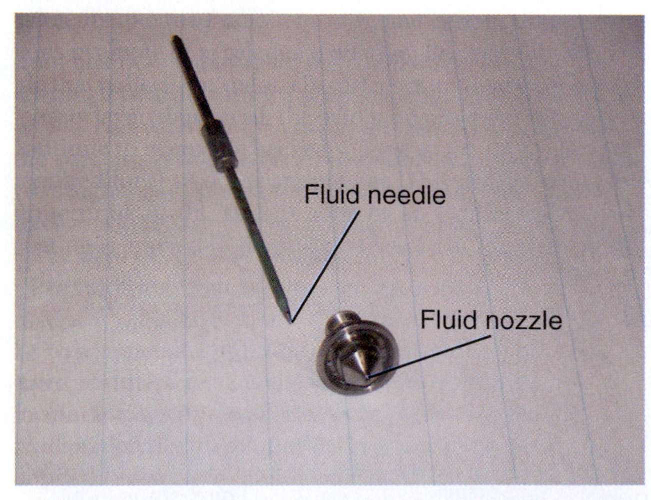

FIGURE 20-30 The fluid needle and fluid nozzle generally come as a set.

Cup

The spray **gun cup**, which may seem very simple, also requires the technician to give some thought to choosing the proper one. In fact a technician may not have just one for a gun. The cups come in different sizes, are made from different materials, and are now available with cup liners; some cups themselves are disposable. Different cup sizes are available for the different amounts needed for refinishing. A 1,000 ml cup will hold a lot of material, but may be too big when the technician is doing a spot repair. It may also be too heavy for some technicians. See **Figure 20-32**. A technician must use a non-metallic cup when applying acid etch material to avoid damaging the cup. See **Figure 20-33**. Some technicians prefer a smaller plastic cup with a twist-off cap. See **Figure 20-34**. Still other technicians may prefer a gun with a plastic liner or throwaway cup system for easy cleaning. See **Figure 20-35**. Therefore, one can see that although the cup may appear to be a simple part of the spray **gun setup**, the technician has many cup choices for different applications.

Spray Gun Cap

Spray gun caps come in many very different styles and are made from various types of material. They can be of a screw-on type or a plastic rubber type. Other than keeping them clean and in good operating shape, they require little maintenance. The critical part of the spray gun cap is the breather hole, which must be kept open and free from debris. If the hole becomes clogged, no air will come into the cup as the paint is taken out while spraying. This will cause pressure inside the cup, and paint flow will stop. When cleaning the cap, use a toothpick or other soft object to clean the breather hole, and make sure it is in good working order.

TECH TIP When cleaning a plastic paint gun cup cap, the technician should not soak it in solvent. Plastic, when soaked, will swell, and the cap will no longer fit the cup; even after drying out, it may no longer fit the cup opening.

Gun Body

The spray gun's body is generally made from aluminum. It contains internal air passages, which propel the paint and atomize it so the paint can be sprayed onto a vehicle. Externally, it also has the pattern control knob, the fluid control knob, the trigger mechanism, and some also have an air volume control knob as well. Gun bodies are now being manufactured with digital air gauges mounted in the gun body. Though paint does not pass through the main body of the gun, it is exposed to overspray and other contaminants. Therefore, keeping the main body clean,

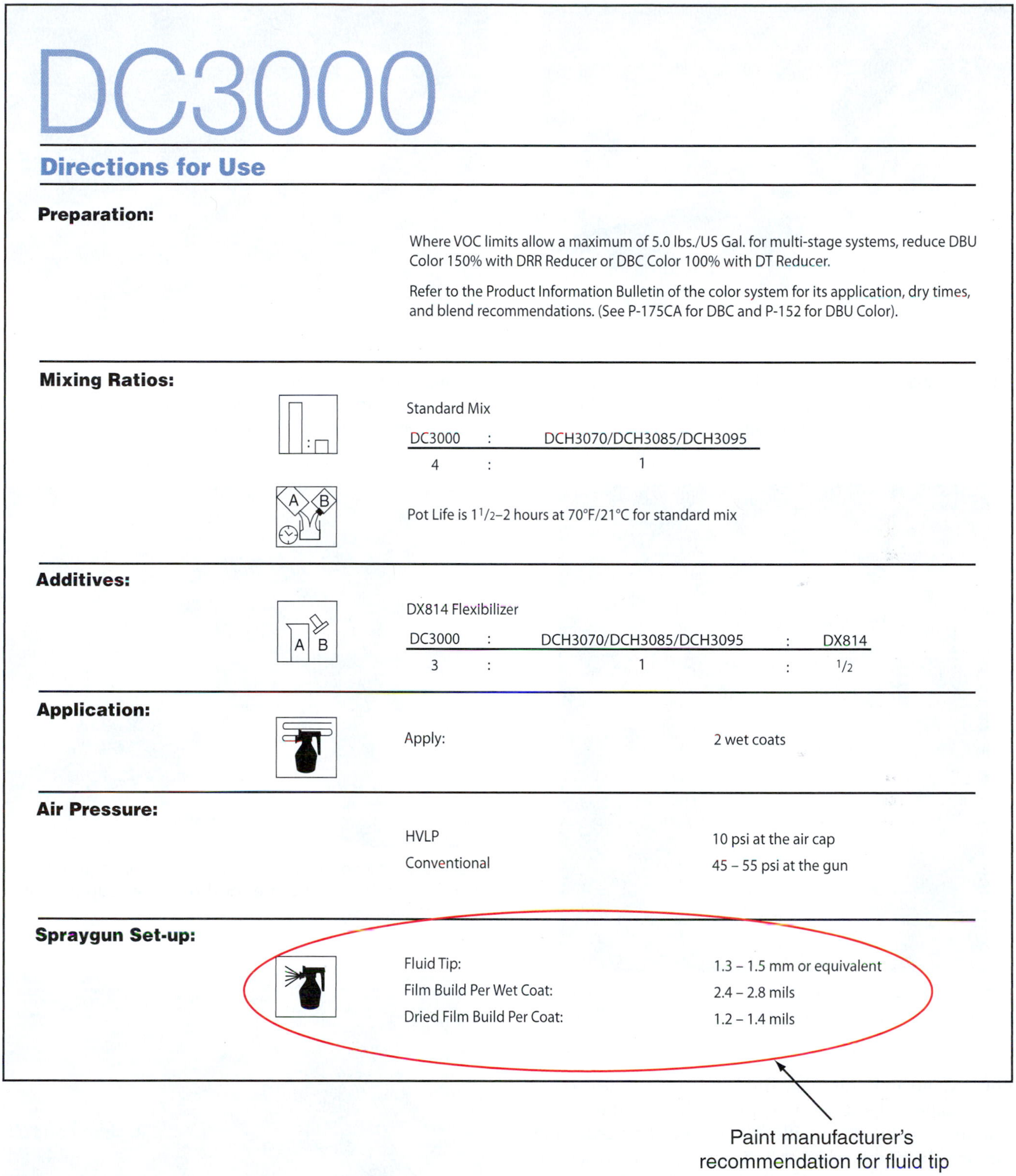

DC3000

Directions for Use

Preparation:

Where VOC limits allow a maximum of 5.0 lbs./US Gal. for multi-stage systems, reduce DBU Color 150% with DRR Reducer or DBC Color 100% with DT Reducer.

Refer to the Product Information Bulletin of the color system for its application, dry times, and blend recommendations. (See P-175CA for DBC and P-152 for DBU Color).

Mixing Ratios:

Standard Mix

DC3000	:	DCH3070/DCH3085/DCH3095
4	:	1

Pot Life is 1 1/2–2 hours at 70°F/21°C for standard mix

Additives:

DX814 Flexibilizer

DC3000	:	DCH3070/DCH3085/DCH3095	:	DX814
3	:	1	:	1/2

Application:

Apply:	2 wet coats

Air Pressure:

HVLP	10 psi at the air cap
Conventional	45 – 55 psi at the gun

Spraygun Set-up:

Fluid Tip:	1.3 – 1.5 mm or equivalent
Film Build Per Wet Coat:	2.4 – 2.8 mils
Dried Film Build Per Coat:	1.2 – 1.4 mils

Paint manufacturer's recommendation for fluid tip

FIGURE 20-31 This PPG "P" sheet provides recommendations for the needle and nozzle setup. Used with permission of PPG Industries.

lubricated, and in good working shape is very critical to proper operation.

The two—and sometimes three—control knobs located in the main body are the points at which the paint gun—once it is set up (proper needle and nozzle set)—must be fine-tuned so it will operate at its best. These control knobs are the fluid control knob (generally located in back), the pattern control knob (located either in back or on the side), and the air control knob, located either in the back or on the bottom. See **Figure 20-36**.

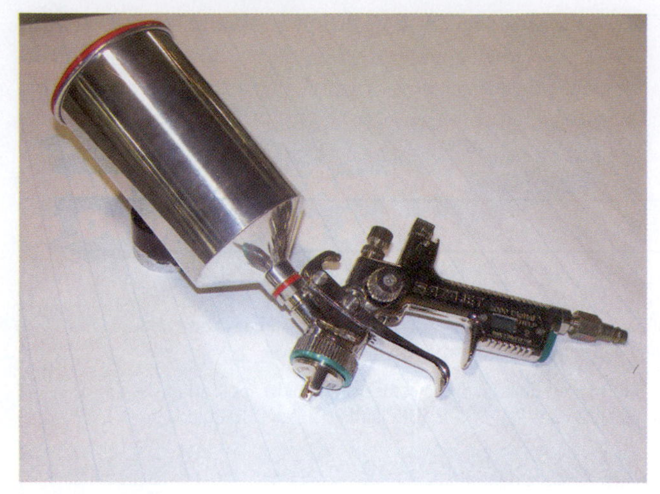

FIGURE 20-32 A 1,000 ml cup in a gravity feed cup will hold a lot of material, but may be too big when the technician is doing a spot repair.

FIGURE 20-33 A technician must use a non-metallic cup with the gravity feed gun when applying acid etch material to avoid damaging the cup.

FIGURE 20-34 Some technicians prefer a gravity feed gun that has a smaller plastic cup with a twist-off cap.

FIGURE 20-35 Some technicians may prefer a gravity feed gun with a plastic liner or throwaway cup system for easy cleaning.

Fluid Control Knob. The **fluid control knob**—sometimes called the **trigger control knob**—adjusts the amount of paint that is allowed around the needle and through the nozzle when the trigger is pulled. When adjusted further out, more fluid is allowed past the needle. When adjusted further in, less fluid is allowed out. This mechanism is sometimes called the trigger adjustment knob because the further out the knob is adjusted, the further back the trigger can be pulled. This adjustment is very critical; after the gun is properly adjusted, the technician should pull the trigger back until it stops each time the gun is triggered. Doing this assures that each stroke has the same amount of paint each time.

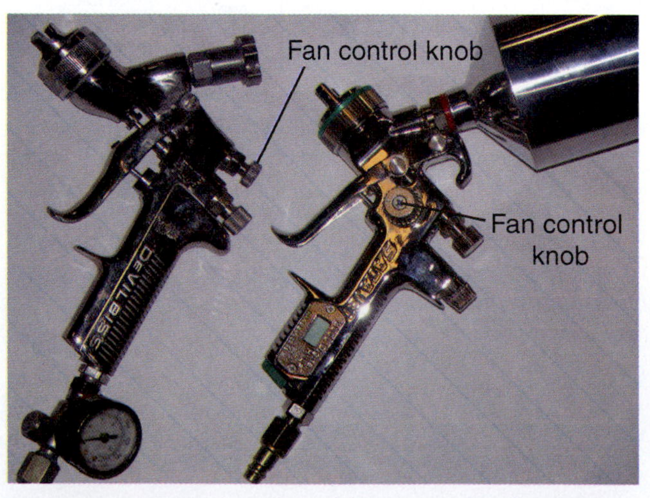

FIGURE 20-36 The location of fan control knobs can vary from gun to gun. Courtesy of Binks/DeVilbiss.

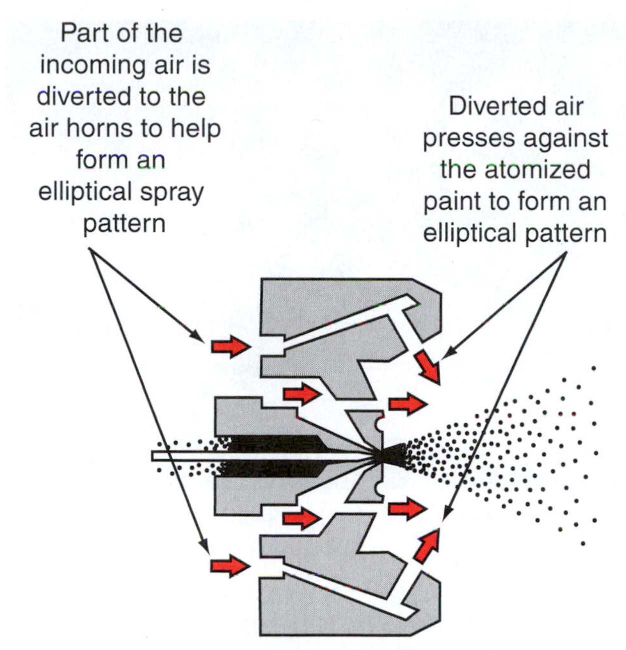

FIGURE 20-37 The amount of air allowed to pass through the air horns will adjust the spray pattern from round to elliptical. Adapted from Binks/DeVilbiss.

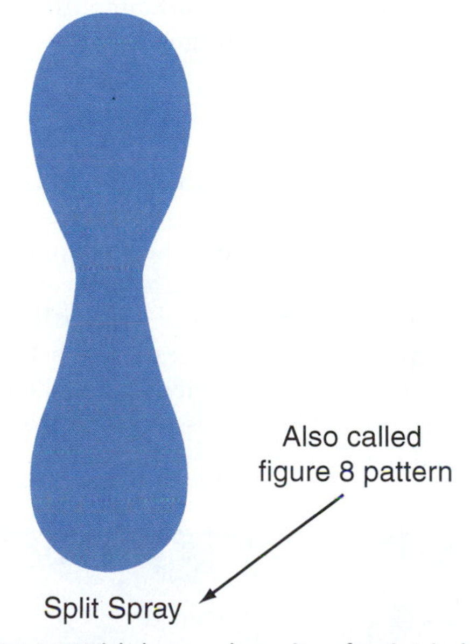

FIGURE 20-38 With heavy deposits of paint in both the top and bottom, and the center light on fluid, the figure 8 pattern causes streaking when sprayed.

When the fluid adjustment knob is adjusted fully back or fully open, the gun's air needs will change. In fact, each adjustment (fluid, air, and pattern) will affect the others. Many technicians will adjust their guns by setting the fluid adjustment first, then the air, and finally the pattern. Then they will recheck them all, to be assured that each setting is exactly adjusted.

Pattern Control Knob. The pattern, or fan, adjustment knob adjusts the amount of air to be sent through the air cap horns. See **Figure 20-37**. The amount of air allowed to pass through the air horns will adjust the spray pattern from round to elliptical. When small amounts of air are allowed through, the elliptical pattern will be smaller and the fluid heavier. If the **pattern** or **fan control knob** is turned open, the spray pattern will get larger. Though many technicians operate their guns at "full fan," that does not mean the pattern control knob is fully open. When adjusting the spray pattern to "full open," the technician should watch the pattern while turning the knob slowly. When the pattern stops getting wider, the technician should stop turning the control knob. If the knob is turned fully open and the gun air pressure is set at high, the air stream from the horns may be at such high pressure that the elliptical pattern the technician is trying to achieve may become what is called a **figure 8 pattern**: With heavy deposits of paint in both the top and bottom, and the center light on fluid, this pattern causes streaking when sprayed. See **Figure 20-38**.

ADJUSTMENT

Spray gun adjustment is one of the most important activities that a technician will perform to enable quality refinishing. However, even though this activity is critical, many technicians do not use a routine method to adjust their guns. This section will describe a set way of adjusting a spray gun that has been used by a large number of technicians over the years. Though the method described is not the only way to obtain a good spray pattern, it will provide a repeatable method that can consistently result in proper adjustment.

When adjusting a spray gun, the technician must keep in mind that each of the adjustable controls affects the others. For example, a control like the air pressure will affect the controls more than one like the pattern/fan control.

This necessary, consistent method of gun adjustments starts with air pressure. The technician then adjusts the fluid (trigger) control, then pattern (fan), then checks the air pressure again, using an air cap pressure gauge to check that it remains as it should. Finally, the technician should spray a small pattern on a test panel to confirm that the gun is adjusted properly. This final check should be performed before either the vehicle is refinished or a complete test panel is sprayed.

Though a new technician must master many skill sets, few are more important than adjusting the spray gun.

Once the gun is adjusted properly, the spray technique becomes critical to a quality refinish. This task will be covered in depth in Chapter 21.

Air Pressure

When adjusting a spray gun, **air pressure (psi)** is the best place for the technician to start. Of all the adjustments the technician will perform, setting the proper spray gun air pressure affects the others most. If air pressure is set too low, the gun will never atomize the finish properly; if set too high, the gun's pattern will be streaky. Even if the finish is properly atomized, the resulting product may have

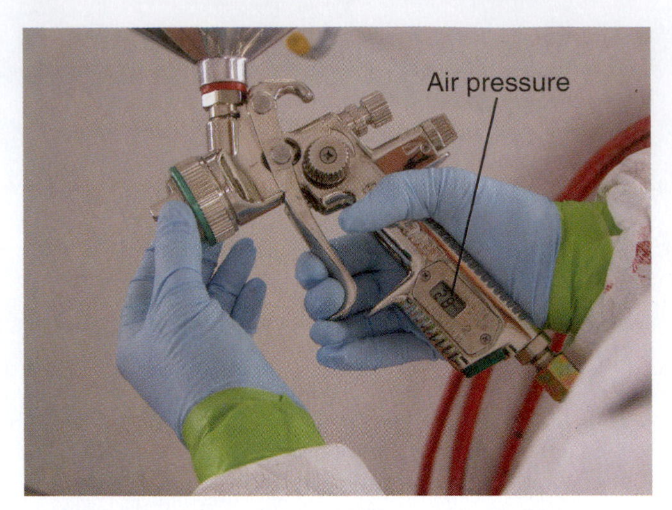

FIGURE 20-39 A spray gun may have a pressure gauge in its handle.

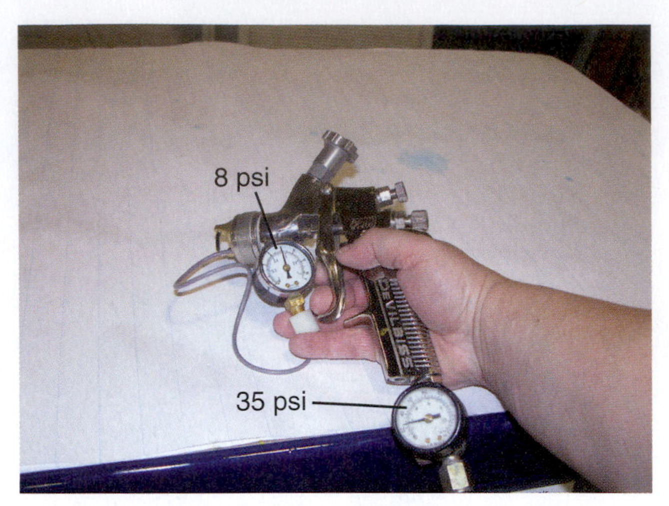

FIGURE 20-40 Pressure changes significantly from the inlet to the cap.

visible streaks of light and dark finish. Technicians will refer to this type of defect in the paint as streaks, striped, or zebra paint. Repairing this defect can become difficult and untimely. The defect may require sanding and re-shooting the affected area. (Paint defects and their repair will be discussed fully in Chapter 23.)

Each gun, depending upon its setup and its make, will need a different amount of air pressure and volume to operate properly. There are no gauges to check the amount of **air volume (CFM)** that is coming through the air line. The business facility must calculate that the compressor, the air lines, and the **quick connect couplers** are sufficient for the CFM needs of the equipment being used. If equipment needs change, the air supply system must be re-evaluated to confirm that it will continue to supply the needed air.

As mentioned earlier, each tool has a specific air pressure need to operate properly. Some tools, such as an air impact tool, may have a range of 65 to 90 psi within which the tool will operate properly. These types of tools may be set at the wall regulator, with little concern for fine air pressure regulation. A spray gun, on the other hand, needs to have very precise regulation. A spray gun inlet regulator, though more accurate than the wall regulator, still may not provide proper air regulation. See **Figure 20-39**. Spray gun air pressure requirements may state that the gun be set at 8 to 10 psi. The most accurate place to check this pressure to assure precise adjustment is at the cap. See **Figure 20-40**.

Air hoses and their flexible lines provide the spray technician the opportunity to move freely while refinishing a vehicle. However, this freedom comes at a cost: As this flexible line moves, some of the air pressure (pushing in all directions with equal force) is used up by expanding the air hose's line. Air lines can lose up to 10 psi for every 25 feet of line the air must pass through. With this in mind, the technician should use the most accurate way to adjust a spray gun: by having the gauge attached to the gun cap and adjusting the air pressure at the wall. This eliminates the need for the gun inlet gauge. If the gun inlet gauge remains on the gun, it should be set at full open, and the air should be adjusted at the wall. The technician should be careful to avoid stepping on the air hose while refinishing a vehicle. To do so will change the air pressure while he or she is standing on the hose, causing the gun to work improperly.

With all of these critical facts in mind, the technician will be able to accurately set the gun at the recommended pressure.

Fluid Control

The fluid or trigger control is set next. See **Figure 20-41**. This controls the amount of fluid that passes by the needle, through the nozzle, and out the air cap of the spray gun. It is often called the trigger control, because the farther a technician opens the control, the farther back the trigger can be pulled. As the trigger comes back, more finish is allowed to pass through the nozzle. The larger the nozzle is, the more liquid can pass through it or the more viscous a fluid that can pass through.

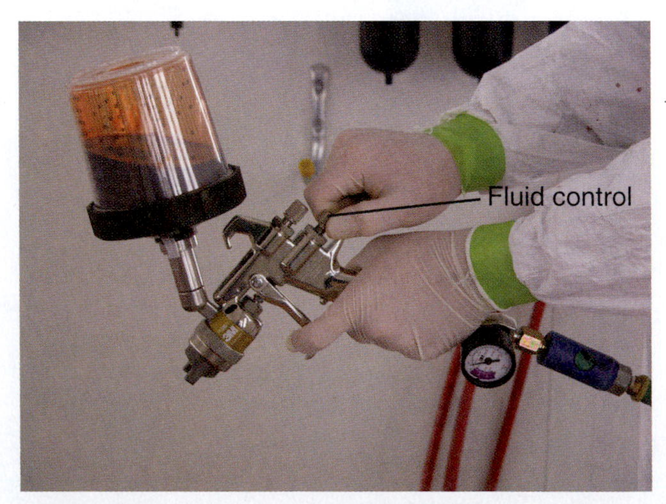

FIGURE 20-41 The trigger controls the amount of fluid that passes by the needle, through the nozzle, and out the air cap of the spray gun.

TECH TIP To better see how the nozzle size affects how fluid passes through it, set up a funnel on a paint gun strainer stand. Then measure 4 ounces of corn syrup into one container, and 4 ounces of water into another. Using a stopwatch, time how long it takes for water to pass through the funnel, and how long it takes for the corn syrup to pass through.

The thicker the liquid, the longer it takes to pass through the nozzle, or the higher the air pressure it will take to force the same amount of liquid through the same size opening. Gun manufacturers give general recommendations for certain viscosities of liquids as they pass through each of the gun models of that manufacturer. Finish manufacturers not only give specific recommendations for the paints they make, but they also give recommendations for that paint in specific guns, with specific needle and nozzle setups. So it can be seen that when setting up a gun and adjusting the proper air pressure, it is best to follow the paint manufacturer's recommendations for the specific finish and gun.

Pattern Adjustment

Pattern, or fan, control is next. See **Figure 20-42**. The pattern control adjusts the width of the gun's elliptical pattern. When the pattern control knob is turned in the streams of air, which comes out the air cap horns, it causes the gun's pattern to elongate into an elliptical spray pattern. See **Figure 20-43**. The further back the knob is turned, the more air is allowed through, and the wider the pattern becomes. The further in the knob is adjusted, the less air is allowed through the air horn orifice, and the smaller the fan or pattern becomes. When no air is allowed through (pattern control knob turned completely in), the gun's pattern will be round. See **Figure 20-44**. A technician should, when opening the pattern control, be cautious to avoid opening it too far. As the pattern control knob is opened, the spray pattern will become larger and elliptical in shape. At some point the size of the spray pattern will stop enlarging even though the pattern control knob can be opened further. When opened further, the air stream will have higher pressure, and will force the atomized finish to be heavier at both the top and bottom of the pattern and lighter in the center. This will result in a figure 8 pattern or "**zebra striping**" finish, which may be difficult to repair.

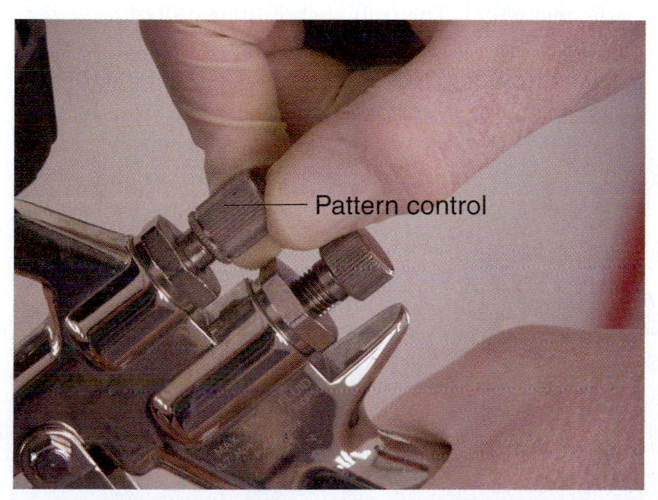

FIGURE 20-42 The pattern control adjusts the width of the gun's elliptical pattern.

FIGURE 20-43 The further in the knob is adjusted, the less air is allowed through the air horn orifice, and the smaller the fan or pattern becomes.

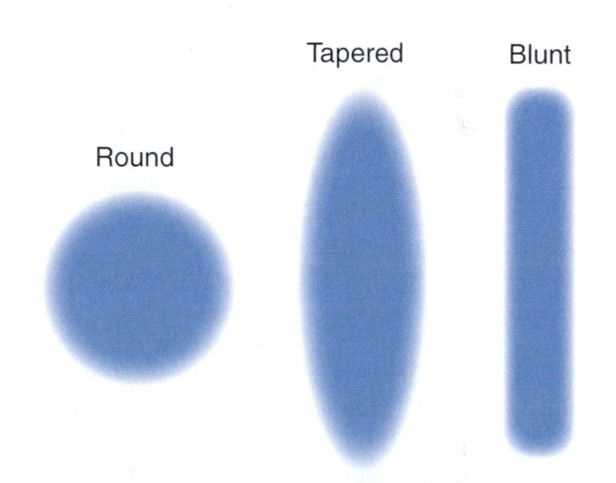

FIGURE 20-44 Spray patterns can be adjusted from round (fully in) to elliptical (fully open). On some guns, because of their design, they produce an elliptical pattern with blunted ends.

From time to time, when the object to be painted is small, the technician will adjust the fluid control and pattern to spray a smaller pattern. Having the proper pattern size for the object being painted will increase the technician's transfer efficiency, because it better controls the overspray. The better a technician is at controlling overspray, the higher transfer efficiency will be, thus reducing waste, pollution, and cost.

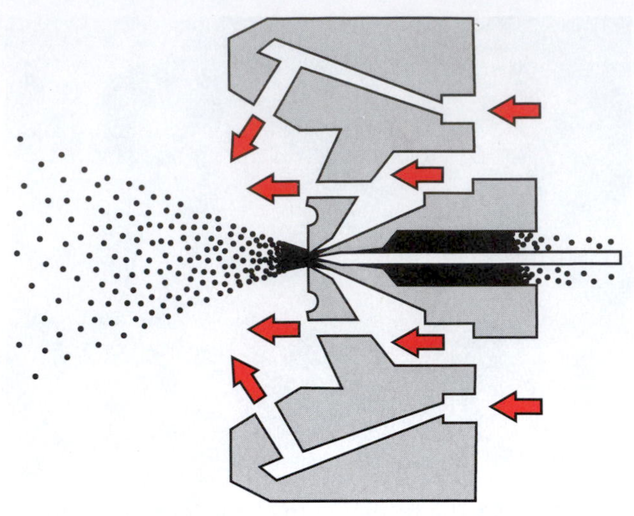

FIGURE 20-45 The elliptical pattern, a result of air streams coming from the air cap horn orifice, also controls the directions of the gun's ellipse pattern.

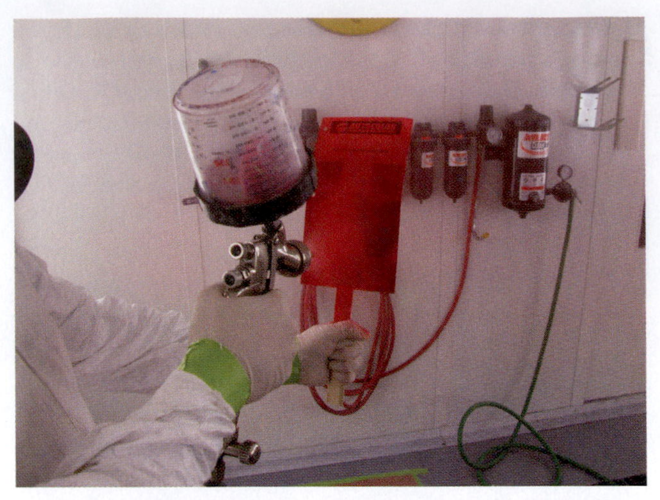

FIGURE 20-46 Once all the adjustments have been re-checked, a spray pattern test should be sprayed out.

Cap

Though the cap may seem to be a minimal part of the spray gun setup since there is little to adjust, the cap can in fact play a large role in finish quality and speed of application. The elliptical pattern, a result of air streams coming from the air cap horn orifice, also controls the directions of the gun's ellipse pattern. The ellipse will be in the opposite direction of the cap horns. See **Figure 20-45**. If the horns are placed in a horizontal position to the gun body, the spray ellipse will be vertical (the common position for spraying a vehicle). If the horns are placed vertical to the gun body, the spray ellipse will be horizontal. When planning their spray techniques, technicians should decide the position that the cap horns should be placed for each individual job.

Test

Once the gun has been fully adjusted, the technician should test that the gun is correctly adjusted. This test should start with a re-check of all of the adjustments explained previously, to confirm that they have remained correct. (As mentioned before, each adjustment affects the others, and when adjusting the pattern control knob, the air pressure may have been affected, and so on). It is a good idea to double-check all the adjustments before testing the pattern.

Once all the adjustments have been re-checked, a **spray pattern test** should be sprayed out. See **Figure 20-46**. To spray a test pattern, first check that atomization is correct and that the droplets are small, to aid in flow-out and leveling. Second, check that fluid is not coming out too fast or too slow. When the fluid releases too fast, the pattern will deposit too much paint, resulting in runs, orange peel, or too high a film thickness. When fluid is too slow, the pattern will result in light spray coverage, poor film build, or dry spray.

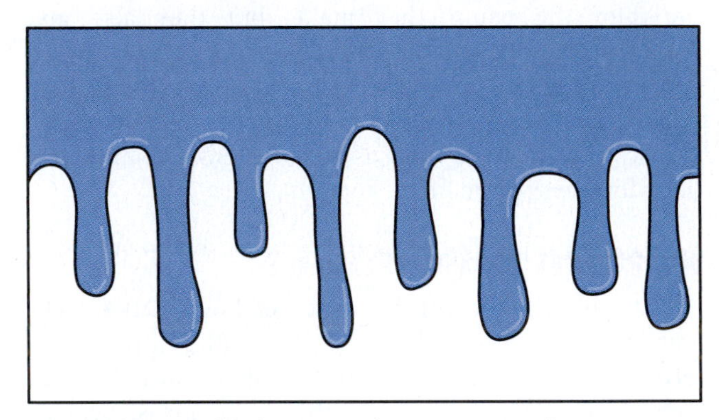

FIGURE 20-47 This horizontal test pattern shows an even run throughout.

If the spray pattern is not correct, the technician should further adjust the gun, correcting to the proper spray pattern.

To check that the pattern is even throughout, the cap horns can be adjusted to the vertical position, with the gun held still at the correct distance and perpendicular to the test panel. The trigger should be pulled for a 2- to 3-second period, which will produce a run. The resulting run pattern should be even over the length of the test pattern, showing that the fluid deposit is even through the ellipse. If the run shows a larger run pattern at each end, the gun pattern adjustment is uneven and further adjustment is needed. See **Figure 20-47**.

TECH TIP Though it may initially seem that there are many steps to adjusting a spray gun and that it would take considerable time to do so, the process becomes fast—and even automatic—to the experienced refinish technician. When protected correctly, a properly adjusted spray gun helps make the difficult task of refinishing a vehicle easier.

CAUTION

Technicians should use the personal protection devices recommended in the specific MSDS when working with chemicals.

CLEANING AND MAINTAINING SPRAY GUNS

Like adjusting a spray gun, proper cleaning and maintenance also aids greatly in a professional quality finish. Aside from needing to be cleaned after each color or finish, a spray gun also has parts that wear and must be replaced. The spray gun needs to be lubricated on a regular basis for smooth operation and to avoid premature part wear. In fact, if not cleaned properly, the gun adjustment may become impossible to adjust. Many painters not only clean their gun after each use, but also break it down, clean each part thoroughly, and reassemble and lubricate it on a regular basis. The frequency of this thorough **spray gun cleaning** depends on the amount the gun is used. For a busy technician, it could be as often as each week. The better a gun is cleaned and maintained, the better it will operate.

Manual Cleaning

Manual cleaning may be all that is available to some technicians. It could be stated that manual cleaning may be better than using a gun washer, because each part is inspected and cleaned thoroughly. For this purpose, some technicians continue to manually clean their guns even if they have a gun washing machine available to them. There are even those who say that a gun should only hold paint in a small portion, and should not have harsh solvents on the gun's packing and springs. See **Figure 20-48**. Soaking the disassembled gun can be very harsh on packings, causing them to prematurely fail. Even though they know that solvents are harsh on packings, many technicians nonetheless soak their disassembled guns in cleaning solvents, which allow guns to be completely cleaned more easily. See **Figure 20-49**.

CAUTION

Some states and local governments may require that all vessels with solvents in them must be covered, in order to prevent volatile organic compounds from evaporating into the atmosphere. Technicians should know, understand, and follow their federal, state, and local laws regarding use of volatile organic compounds.

When one is cleaning a spray gun, the small orifices of the gun may become clogged with paint debris. They should never be cleaned with steel or other hard objects, however. Cleaning with such devices may alter the size or shape of a spray gun, which in turn may alter the way that the gun operates. It may also cause the gun's parts to need replacement before their natural wear cycle. Using a plastic bristle brush or wooden toothpicks for cleaning instead will not harm the small holes.

One gun-cleaning process is *not recommended,* however. Backpressure cleaning is a method in which solvent is placed in the gun. Then the gun is sprayed, forcing clean solvent through the gun. A cloth is then placed over the air cap, causing the solvents to be forced back through the gun. This process is repeated several times until the gun is clean. Those in the refinish field should know that this method is *very dangerous to both the technician and the environment.* In fact, backpressure cleaning has been outlawed in some states.

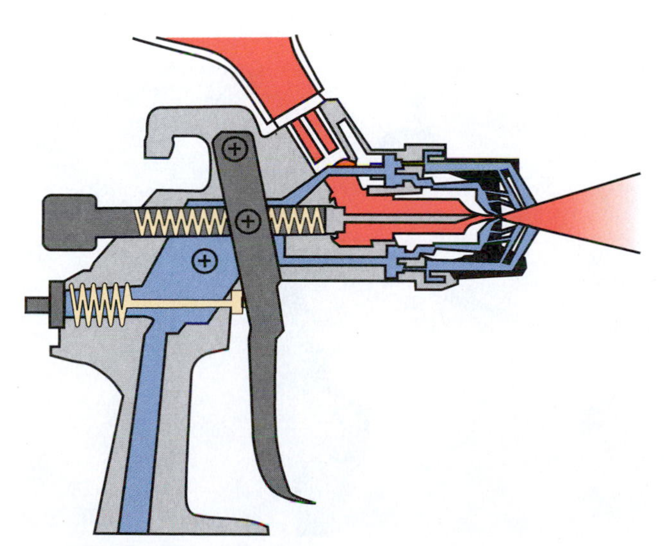

FIGURE 20-48 This figure shows the path of paint (red) and air (blue) in a gravity gun.

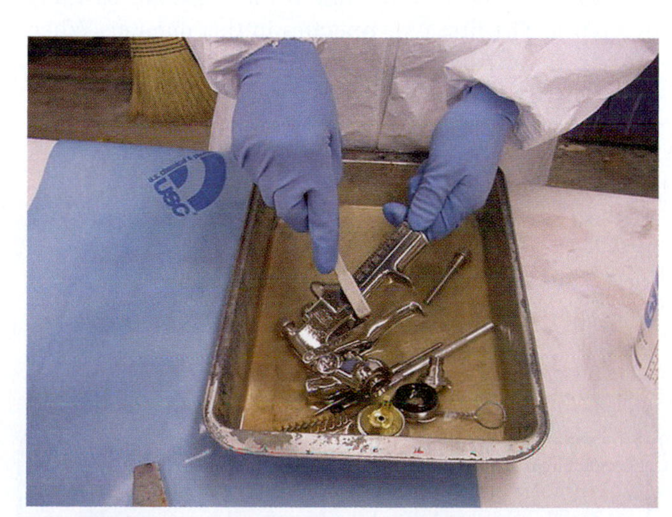

FIGURE 20-49 Guns must be kept clean for optimal operation.

FIGURE 20-50 A gun washer helps keep equipment clean and in good operating order.

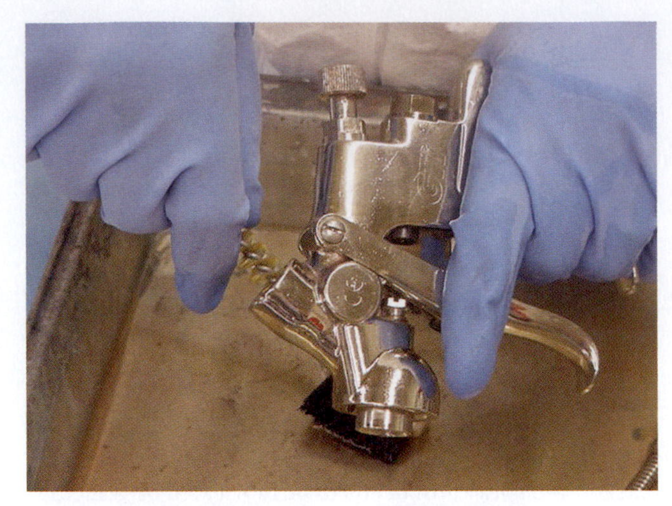

FIGURE 20-51 Using a manual cleaner assures cleanliness.

Cleaning with a Gun Washer

Several gun washers are available for use, and though they vary slightly in how they work, there are basically two types. The first type operates in this manner: When the gun is partially broken down, the parts are placed in an enclosed vat, where solvent is sprayed over them like what happens in a dishwasher. See **Figure 20-50**. The other is a part-washer type, where the gun is cleaned with a brush which has solvents pumped through it, aiding in the cleaning of the gun. See **Figure 20-51**. Some states have required that both these types of **gun cleaning** machines be covered when in use, to reduce the amount of VOCs that are released into the atmosphere.

TECH TIP Approximately 20 percent of all the volatile organic compounds (VOCs) released into the atmosphere are released during the cleaning operation.

Lubrication

While oils and silicones are not compatible with most finishes and care must be taken when using petroleum products in the refinishing department, spray guns must be lubricated following cleaning. When the adjustment knobs are under pressure from their springs, they may become difficult to turn if not lubricated. The trigger connecting screw should also be lubricated to keep it working smoothly. **Figure 20-52** shows the areas to which a light spray **gun lubrication** should be applied after cleaning. This lubrication should be applied thoroughly but sparingly, to avoid contact with the finish in the gun. The spring for the fluid control knob should be lightly coated with petroleum jelly before reinstalling it after cleaning.

Rebuilding

Depending on how much it is used, a spray gun may need rebuilding from time to time. The process can be

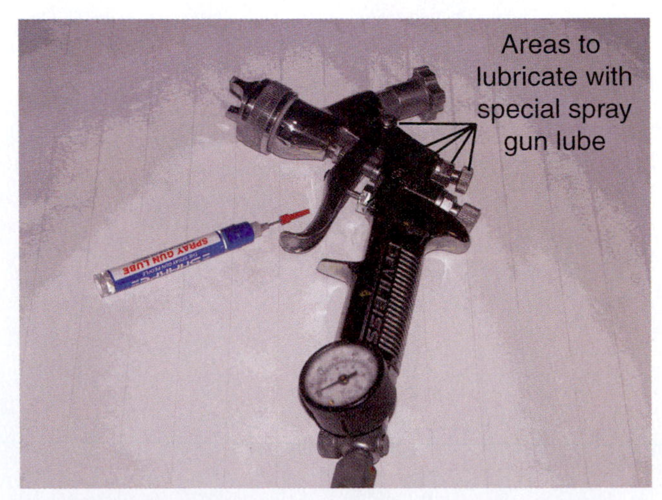

FIGURE 20-52 Occasional lubrication is needed using special spray gun lubrication.

difficult because when a gun is in need of rebuilding, it is generally not working well and will be difficult to disassemble. It is wise to rebuild a gun on a regular basis, before it no longer operates properly. A **gun rebuilding** kit can be obtained from the spray gun makers. In rebuilding, all the parts that come in the kit should be replaced on the gun. See **Figure 20-53**. Spray guns need to be completely disassembled when rebuilding. The process may require the use of special tools, which are generally supplied by the manufacturer when the gun is purchased. It may be as long as 2 years after the purchase of the gun before rebuilding is required, and the technician should keep the tools that are included to disassemble the gun available. Most gun kits come with rebuild instructions, which should be followed closely. After the gun is rebuilt, it should be lubricated and tested for proper operation.

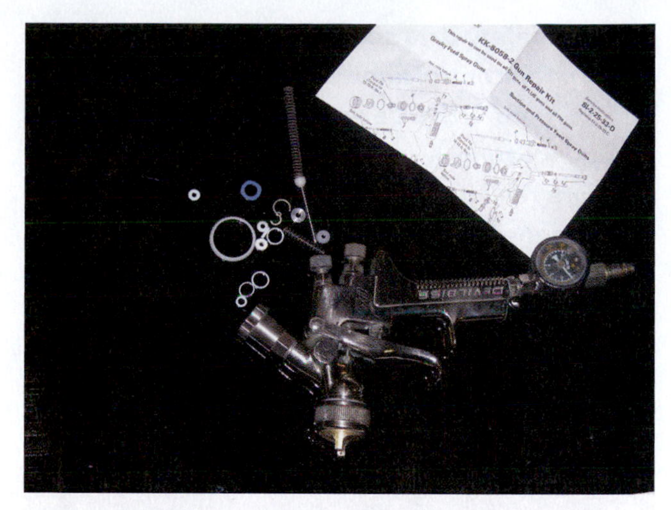

FIGURE 20-53 Spray guns may require occasional rebuilding. Gun rebuild kits supply the needed replacement parts.

Summary

- The spray gun is a precision tool which, when used by a skilled technician, can provide years of service.
- When properly set up, adjusted, cleaned, and maintained, a spray gun is a reliable and profitable tool.
- One of the most important tasks a refinish technician must learn is the proper setup and adjustment of the gun.
- Though this process may at first seem difficult and time consuming, it soon becomes second nature.
- The new technician should practice, and if not satisfied with the way the gun is operating, should seek help from the instructor or supervisor, because a gun that is not operating well will not give satisfactory results.
- Consequently, a poorly operating gun can cause much more work to repair the flaws that will occur.
- Applying a professional quality refinish on a vehicle can only come with instruction, practice, and skill.

Key Terms

air cap
air cap pressure gauge
air pressure (psi)
air volume (CFM)
atomization
centrifugal
ear protection
electrostatic
figure 8 pattern
fluid needle
fluid nozzle
fluid/trigger control knob
gravity feed gun
gun cleaning
gun cup
gun lubrication
gun rebuilding
gun setup
high pressure guns
high volume, low pressure (HVLP) guns
low volume, low pressure (LVLP) guns
manual cleaning

material safety data sheet (MSDS)
pattern/fan control knob
personal protective device (PPD)
powder coating
pressure feed gun
quick connect couplers
respirator
rotating bell
siphon feed gun
spray gun adjustment
spray gun cleaning
spray pattern test
supplied air respirator
transfer efficiency (TE)
venturi
viscous
volatile organic compounds (VOCs)
zebra striping

Review

ASE Review Questions

1. Technician A says that a charcoal vapor respirator is the proper respirator to use when refinishing a vehicle. Technician B says that the way to find the correct respirator to use when finishing a vehicle is to check the MSDS for the coating being used. Who is correct?
   A. Technician A only
   B. Technician B only
   C. Both Technicians A and B
   D. Neither Technician A nor B
2. Technician A says that 20 percent of the volatile organic compounds (VOCs) that are released into the atmosphere in the collision repair industry are done so during cleaning operations. Technician B says that gloves do not need to be worn during gun cleaning. Who is correct?
   A. Technician A only
   B. Technician B only
   C. Both Technicians A and B
   D. Neither Technician A nor B
3. Technician A says that HVLP stands for high velocity light pressure. Technician B says that HVLP stands for high viscosity low pressure. Who is correct?
   A. Technician A only
   B. Technician B only
   C. Both Technicians A and B
   D. Neither Technician A nor B
4. Technician A says that HVLP spray guns not only cut material costs, but that they are also required in some areas. Technician B says that only HVLP spray guns can have a transfer efficiency of above 65 percent or greater. Who is correct?
   A. Technician A only
   B. Technician B only
   C. Both Technicians A and B
   D. Neither Technician A nor B
5. Technician A says that a rotating bell spray gun is commonly used in collision repair refinishing. Technician B says that electrostatic painting uses positive and negative charges on the paint and object being painted to attract paint to the part. Who is correct?
   A. Technician A only
   B. Technician B only
   C. Both Technicians A and B
   D. Neither Technician A nor B
6. Technician A says that when cleaning a spray gun manually, safety glasses are not necessary. Technician B says that gloves are not necessary when cleaning spray guns. Who is correct?
   A. Technician A only
   B. Technician B only
   C. Both Technicians A and B
   D. Neither Technician A nor B
7. Technician A says that cleaning the small orifices on the spray gun cap is best done with a small piece of steel. Technician B says that cleaning the small orifices on the spray gun cap is best done with a soft object such as a toothpick, so no damage will be done to the small hole. Who is correct?
   A. Technician A only
   B. Technician B only
   C. Both Technicians A and B
   D. Neither Technician A nor B
8. Technician A says that adjusting a spray gun is a simple task, and a technician does not need to check it often. Technician B says that an air cap pressure gauge is not necessary, since air regulation can be done by listening to the gun as it sprays. Who is correct?

A. Technician A only
B. Technician B only
C. Both Technicians A and B
D. Neither Technician A nor B

9. Technician A says that the pattern control knob is always located on the back of the spray gun. Technician B says that the fluid control knob may also be called the trigger control knob. Who is correct?
A. Technician A only
B. Technician B only
C. Both Technicians A and B
D. Neither Technician A nor B

10. Technician A says that the pattern width should be set to the size of the part being painted. Technician B says that the pattern control knob should be set at fully open at all times. Who is correct?
A. Technician A only
B. Technician B only
C. Both Technicians A and B
D. Neither Technician A nor B

11. Technician A says that the needle and nozzle setup directions are found in the MSDS for the paint being sprayed. Technician B says that needle and nozzle setup is the same for all sprayable materials. Who is correct?
A. Technician A only
B. Technician B only
C. Both Technicians A and B
D. Neither Technician A nor B

12. Technician A says that atomization starts inside the gun at the needle and nozzle. Technician B says that atomization starts outside the gun as the paint leaves the nozzle. Who is correct?
A. Technician A only
B. Technician B only
C. Both Technicians A and B
D. Neither Technician A nor B

13. Technician A says that if the air cap horns are set at the horizontal position, the spray pattern will be vertical. Technician B says that if the air cap horns are set at the vertical position, the spray pattern will be horizontal. Who is correct?
A. Technician A only
B. Technician B only
C. Both Technicians A and B
D. Neither Technician A nor B

14. Technician A says that testing the spray gun for proper adjustment should be done on a test panel before spraying on a vehicle. Technician B says that just looking at the pattern when the product is sprayed into the air will tell the technician if the gun is adjusted properly. Who is correct?
A. Technician A only
B. Technician B only
C. Both Technicians A and B
D. Neither Technician A nor B

15. Technician A says that a spray gun should not be lubricated because it will contaminate the finish when spraying. Technician B says that spray guns should be lubricated after cleaning to assure smooth operation. Who is correct?
A. Technician A only
B. Technician B only
C. Both Technicians A and B
D. Neither Technician A nor B

Essay Questions

1. Write the description for the steps necessary to adjust a spray gun.
2. Explain the directions for hand cleaning a spray gun.
3. Compare and contrast the accuracy of spray gun pressures at the paint booth wall, the spray gun inlet, and the cap pressure.
4. Write the description of the three stages of spray gun atomization.
5. Explain the places that need to be lubricated on a spray gun.
6. Compare and contrast cleaning a spray gun using a gun cleaning machine and cleaning it by hand.

Topic Related Math Questions

1. If a gun has a fluid flow of 4 ounces per minute, how long will it take for a pint of fluid to flow through this gun?
2. A technician has two fluids: one with a viscosity of 17 seconds in viscosity cup, and the other with viscosity that is double the first one's viscosity. How long will the second fluid take to flow through the same viscosity cup?
3. While hand cleaning paint guns, the solvent you are using evaporates at the rate of 1/4 ounce per minute, and it takes you one-half hour to clean the gun. You started with 32 ounces of solvent. How much solvent will be left when three guns have been cleaned?
4. A shop uses 3.5 quarts of solvent per month in a spray gun cleaner to clean the refinish department's spray guns. How much solvent is needed for a year?
5. The refinish department must change their respirators' pre-filters every 6 weeks. There are 13 employees using respirators. How many respirator pre-filters are needed per year?

Critical Thinking Questions

1. A technician has been spraying basecoat with a needle nozzle setup of 1.3 mm. The next task will be spraying clearcoat. Will the needle need to be larger or smaller than 1.3 mm?
2. Which type of gun cleaning would release the least amount of volatile organic compounds into the atmosphere? (Explain your reasoning.)
   a. Pressurized back flushing
   b. Enclosed gun washing machine
   c. Open hand cleaning
3. Explain what transfer efficiency is and how it reduces environmental impact.
4. List and describe the steps needed to adjust a spray gun.
5. List and explain the advantages and disadvantages of using an HVLP spray gun.

Lab Activities

1. Demonstrate the proper use of these personal protective devices:
   a. Safety glasses
   b. Ear protection
   c. Gloves
   d. Respirator (air supply)
   e. Paint suit
2. Demonstrate the steps needed to adjust a gravity-feed HVLP spray gun.
3. Test a spray gun to confirm that the gun's air pressure, fluid needle, and pattern are properly adjusted.
4. Demonstrate the disassembly, cleaning, and lubrication of a spray gun.
5. Demonstrate the proper use of a gun washing machine.
6. Demonstrate the rebuilding of a spray gun.

Chapter 21

Spray Techniques

OBJECTIVES

Upon completing this chapter, you should be able to:

- Demonstrate how to adjust a spray gun for proper:
  - Atomization
  - Fluid
  - Spray width
  - Distance
- Explain the proper stroke/movement overlap and rate of movement for proper spraying.
- Outline the proper spray sequence, including the plan of attack for large, small, tall, and short vehicles.
- Demonstrate the proper spray gun positions.
- List the proper steps for different spray techniques, including:
  - Cutting in
  - Parts painting
  - Single-stage solid color
  - Single-stage metallic
  - Basecoat
  - Clearcoats
  - Panel painting
  - Basecoat blending
  - Standard blending
  - Reverse blending
  - Wet bedding blending
  - Clearcoat blending
  - Roller application
- Explain transfer efficiency and its importance to the environment and to health.

INTRODUCTION

Spray techniques are among the most critical areas of refinishing, yet they are among the most difficult to master. While this chapter will help you understand the many different and varied aspects of spray techniques, in the end the only way to become a quality refinish technician is to practice, then practice some more.

This chapter will discuss the adjustments of spray equipment, help the technician develop the correct stroke and smooth movement while painting, and assist the technician to understand the need for a well-thought-out spray sequence. It will also discuss the importance of the different aspects of spray gun positioning and how gun position affects the quality of the finish. This chapter will help the reader understand the different techniques of application in relation to the object being painted and the effect that each technique has on the quality of the finished product, as well as the environment.

The only way to perfect the application of finishes is to practice. With practice and coaching while applying finishes, new technicians can understand and perfect the techniques discussed in this chapter.

SPRAY GUN ADJUSTMENTS

Spray gun adjustments are essential if technicians want to obtain a professional refinish application. Air pressure must be high enough to atomize the coating and move it from the spray gun to the surface that is being refinished. However, the air pressure must not be so high that it wastes the product being sprayed. Air pressure requirements will be different according to the type of equipment being used and the type of coating that is being applied. The technician must also be aware of—and adjust—the fluid needle according to the type of coating, equipment used, and the application spray method or techniques. The spray width must also be adjusted for each specific application. If these adjustments are not set properly, the overall quality of the finish will suffer and defects may occur which will then need to be repaired. Each of these adjustments must be set properly to obtain a quality finish.

TECH TIP Though travel, speed, and distance are not considered "adjustments," they are critical for the overall quality of the coatings being applied. As such, they must be taken into consideration while setting adjustments to obtain a quality spray finish.

CAUTION

When applying refinish, the technician should use all the recommended personal safety equipment.

Air Pressure

Air pressure is one of the most critical settings in relation to proper atomization. It must be set so that the paint is broken into small driblets, which will allow the fluid paint to be transferred from the spray gun cup through the gun body and out the air cap. See **Figure 21-1**. The three locations where air pressure is measured are at the wall, at the gun, and at the gun's air cap. Air regulator pressure is normally found on the spray booth wall. See **Figure 21-2**. Spray gun inlet pressure, found on the lower inlet of the spray gun, is a more accurate measurement of true air pressure. See **Figure 21-3**. However, air cap pressure, though the least often checked, is the most accurate reading of true air pressure. See **Figure 21-4**.

Regulator Pressure. The regulator is most often the first—and sometimes only—location at which a technician will adjust the air pressure. It is also the least accu-

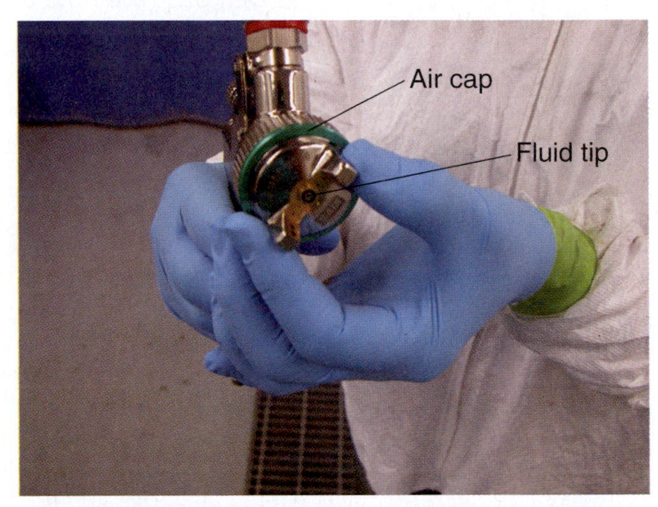

FIGURE 21-1 Air cap and fluid tip design are the key factors in atomization and transfer efficiency.

FIGURE 21-2 When using a wall air regulator, pressure from the wall to the gun will drop due to the hose and its length.

FIGURE 21-3 A spray gun inlet pressure reading is a more accurate reading than one taken at the wall.

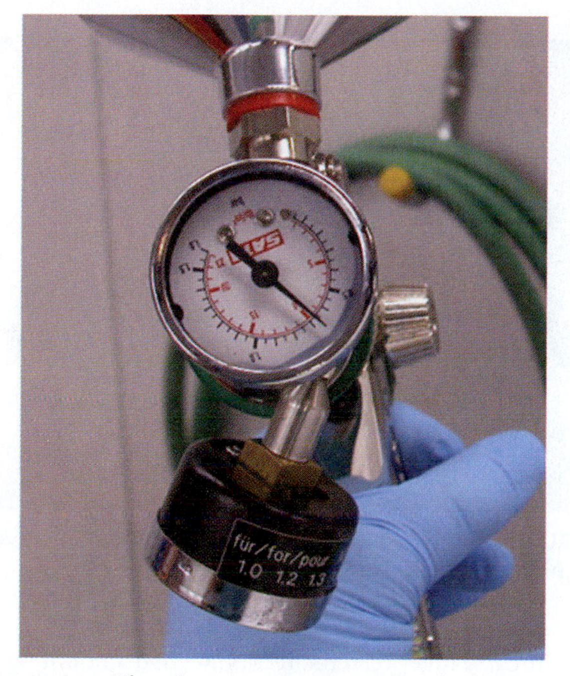

FIGURE 21-4 The air cap pressure reading is the most precise reading of air pressure.

FIGURE 21-5 If air inlets and quick connectors are not the proper size, they will restrict airflow to the gun, and therefore reduce its ability to spray properly.

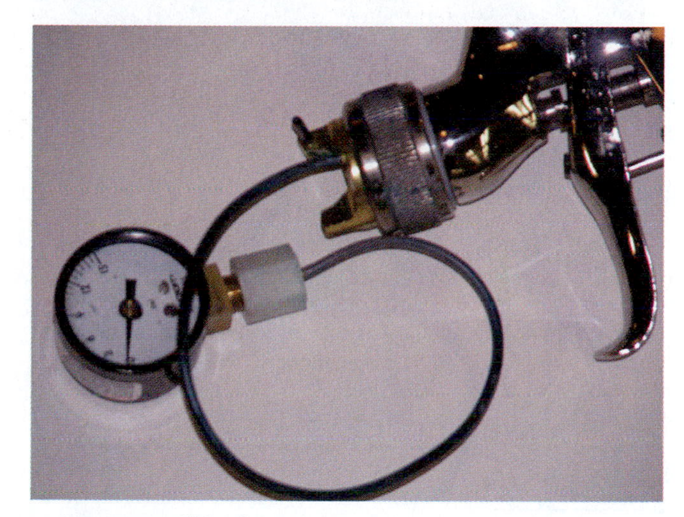

FIGURE 21-6 This air cap pressure gauge is measuring air pressure at the cap.

rate of the locations, because airflow is influenced by many variables prior to reaching the place where it must atomize the finish coating. If air inlets and quick connectors are not the proper size, they will restrict airflow to the gun, and therefore reduce its ability to spray properly. See **Figure 21-5**. The length and size of air hoses used to connect the spray gun to the booth regulator, if not chosen properly, may also reduce the spray gun's efficiency. The proper selection of air hose fittings and other equipment that affects spray gun efficiency is covered in Chapter 22.

Gun Pressure. Setting the spray gun pressure is initially done from the booth regulator. The pressure should be set according to the spray gun manufacturer's recommendations. The booth pressure setting is the starting place for fine-tuning the spray pressure. To regulate the pressure precisely, the refinish technician should fit the spray gun with a cap pressure gauge. See **Figure 21-6**.

Air Cap. Pressure gauges come in different configurations, depending on the manufacturer, and may have both air cap and air horn gauges. See **Figure 21-7**. Paint manufacturers have in their production sheets the recommended air cap pressure for their material. If the paint manufacturer calls for 9 **pounds per square inch (psi)** at the cap for their material, the spray gun can now (with the air cap gauge installed) be set properly. First, if the spray gun is equipped with a gun air regulator, adjust it to the fully open position. See **Figure 21-8**.

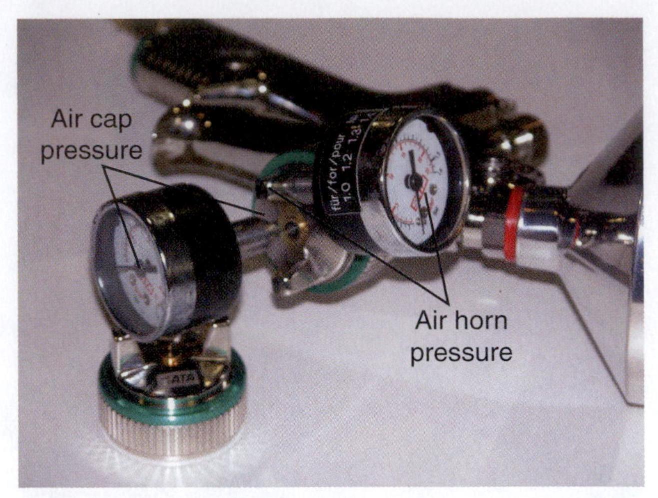

FIGURE 21-7 Some manufacturer's gauges measure both air cap and air horn pressures.

FIGURE 21-9 A gate or cheater valve (left) will restrict volume as it reduces pressure while the high flow (right) regulators will allow for full air volume at any pressure.

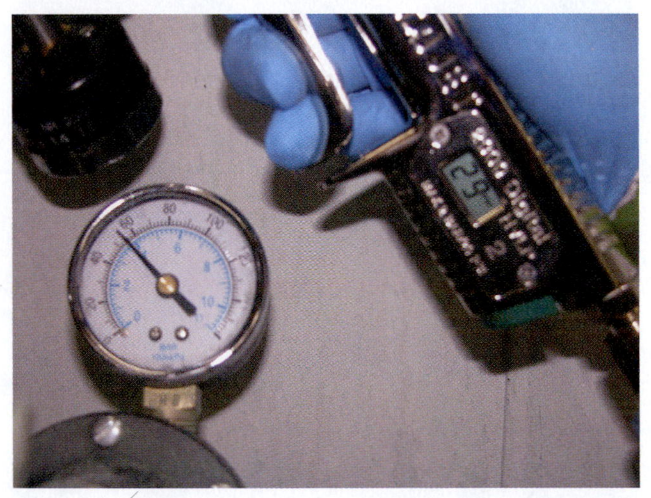

FIGURE 21-8 This wall gauge is at 55 psi with an air inlet reading of 29 psi (26 lb psi are lost in the 30-foot hose).

FIGURE 21-10 The wall gauge and air cap gauge have significant differences.

TECH TIP Many different types of aftermarket air gauges and regulators are available for spray gun monitoring. Some of them are very accurate and can be used with confidence. Others, though, are not so accurate. If used, they can restrict the airflow to the gun so much that the gun cannot operate as it was designed. If the technician chooses to use a gun end regulator—or even has a gun with a regulator as part of the handle—it should be of the proper inlet size to allow the needed amount of airflow through the gun (measured in **cubic feet per minute**, or **CFM**). See **Figure 21-9**. Even when the technician is sure that the CFM airflow is sufficient, he or she should check the cap pressure to verify that the gun reading is correct. See **Figure 21-10**.

Then, with the trigger pulled to allow airflow only (no paint should be coming out), adjust the air pressure so the air cap gauges read the recommended amount by adjusting the booth regulator. See **Figure 21-11**. As can be seen, the three pressures do not match each other. See **Figure 21-12**. The air cap reading, which is the most accurate, should be the gauge used to set the manufacturer's regulated setting.

TECH TIP When a technician, after multiple checks, becomes convinced that the wall setting will be the required setting to meet the manufacturer's cap recommendations, it may not be necessary to install the cap gauge each time the gun air pressure is set. However, it is still wise to periodically check that the wall setting routinely used is accurate by attaching the air cap gauge from time to time to confirm its accuracy.

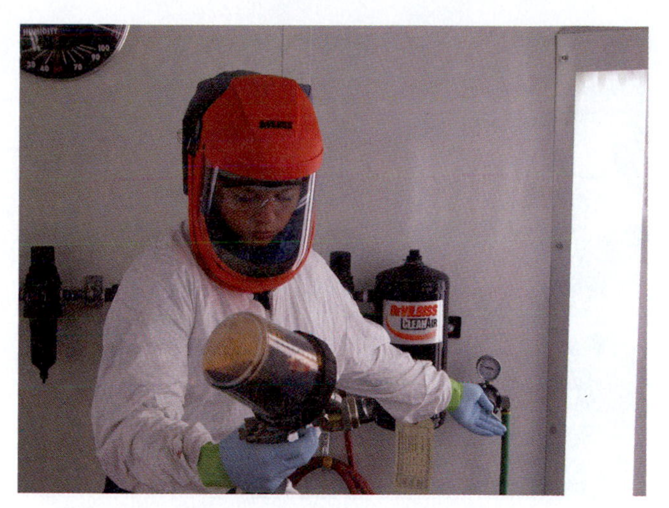

FIGURE 21-11 This technician is adjusting a gun at the wall while reading the gun air cap gauge.

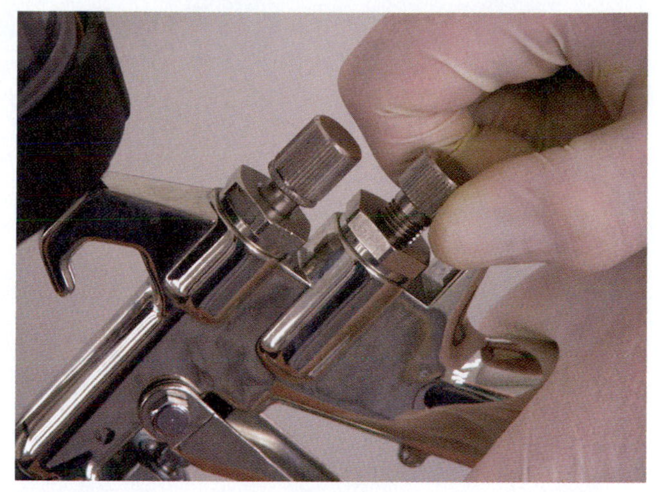

FIGURE 21-13 One of the first steps in atomization is adjusting the fluid needle.

FIGURE 21-12 Notice that all three gauges are reading a different pressure.

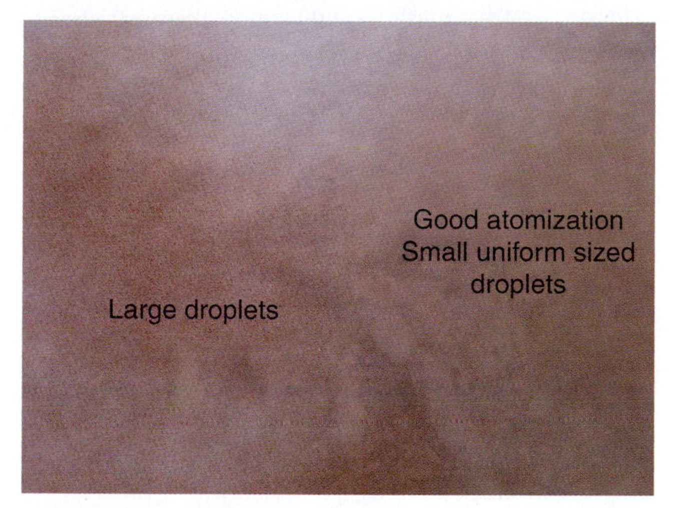

FIGURE 21-14 Technicians should spray a test pattern to check for proper atomization before spraying a vehicle.

Fluid Adjustment

By turning the fluid adjustment knob, the technician regulates all of the following: the atomization of the coatings emitted from the gun, the texture of the coating on the vehicle to which it is being applied, the film thickness of the coating on the vehicle, and the speed at which the technician must spray the coating on the vehicle.

Atomization. In atomization, the technician should first adjust the fluid needle. See **Figure 21-13**. This will allow coating to flow out of the gun at a low or high rate, depending on how far back the needle is adjusted. If the needle is adjusted in, the flow will be slowed. As the needle is adjusted outward, the flow of fluid will increase. The needle should be adjusted so that when the coating is sprayed, the air coming from the gun will properly atomize the coating. When the coating is atomized properly the droplets hitting a test panel will be small, allowing the coating to flow together smoothly. If the droplets hitting the test panel are large, they will not flow together and the surface will have an "orange peel" texture (described below). See **Figure 21-14**.

Texture. The most common texture that appears while spraying is called orange peel. (Other textures and defects will be covered in Chapter 31.) **Orange peel** can be caused from many different conditions, but in regard to adjustments, orange peel may occur due to an improperly adjusted fluid needle. See **Figure 21-15**. If the needle is adjusted out too far and high volumes of fluid come out of the gun, the air pressure available to the spray gun cannot atomize the fluid properly and the atomized droplets will be large. These large droplets do not "flow out," or connect with each other and lay flat and smooth on the

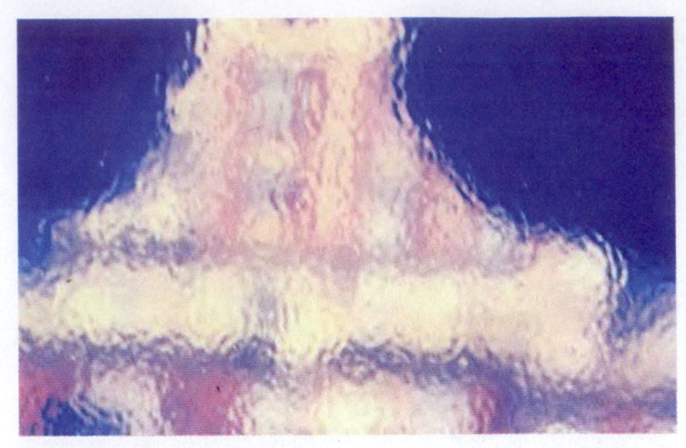

FIGURE 21-15 When a panel has orange peel, painters must match its color and texture to create a proper undetectable repair.

vehicle's surface. (However, if this rough texture is already present on a vehicle being refinished, it is the technician's responsibility to match the texture of the vehicle being repaired to restore it to its pre-accident condition.) Orange peel can also appear if the gun is held too close to the vehicle's surface while being sprayed, if the coating is improperly mixed, and as a result of many other conditions which will be discussed later within this text.

Film Thickness. The coating applied to the vehicle should be thick enough to properly cover the surface, which will provide the needed protection and the required color (**hue**). If the film thickness is too little, the coating will not protect the surface from corrosion and ultraviolet (UV) damage, and the surface will rapidly deteriorate. If the film layer applied is too thick, the coating film may fail. In addition, applying excessive film thickness is costly both to the repair facility and to the environment, as more **volatile organic compounds (VOCs)** are released into the atmosphere. (Methods to guard against this hazard will be covered in the transfer efficiency section, later in this chapter.)

Spray Width

Spray width is the width of the spray "fan" occurring when coatings are applied to a vehicle. The spray can be adjusted to a small round pattern, or to a wide fan-like pattern for coverage of larger areas. See **Figure 21-16**. Normally the spray width is adjusted to what is called "full fan," which means that the fan is adjusted as much as possible and still has proper coverage within the pattern. Some guns, when adjusted to the fully open setting on the fan width adjustment knob, can cause the spray pattern to have heavier deposits of coating at the top and bottom of the spray gun pattern. This is commonly called a figure 8 pattern, and spraying the paint in this manner can cause streaks in the vehicle's finish.

When the trigger of a spray gun is pulled, air flows through the air horns and hits the paint stream, forcing it into an elongated pattern. The amount of air coming from the air horns is adjusted using the fan adjustment knob on the spray gun. This knob may be located at different places on different spray guns. See **Figure 21-17**. The position of the air horns can be adjusted by slightly loosening the air cap, then rotating the air horns into the desired position, then re-tightening the air horns.

Spray width can be adjusted to match the size of the object being sprayed. If the object being sprayed is small, the spray width should be adjusted so as to properly cover the

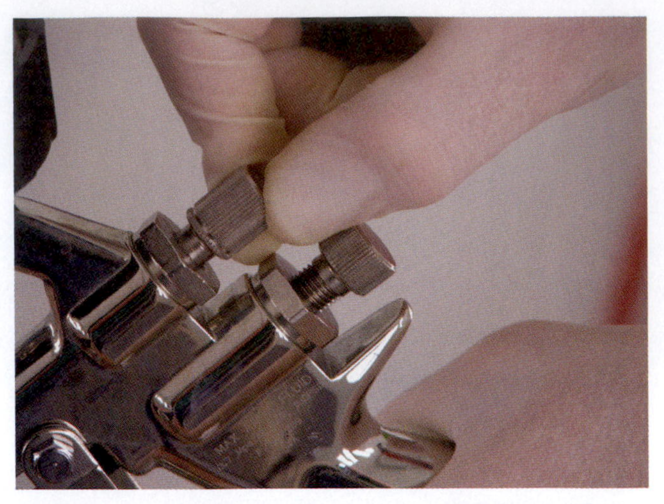

FIGURE 21-16 The fan width pattern adjustment knob may be either on the side—as in the photo—or at the back of the gun.

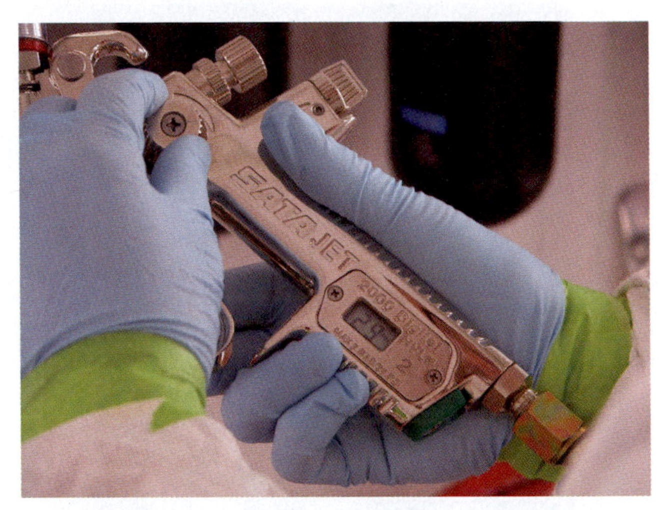

FIGURE 21-17 Fan adjustment knobs can be found in different locations.

TECH TIP Stripes of alternating light and dark lines in a paint job are commonly called "**zebra striping**." Though the problem can be caused by poor fan adjustment, other causes may also produce the same effect. Zebra striping will be covered in Chapter 31, Paint Problems and Prevention.

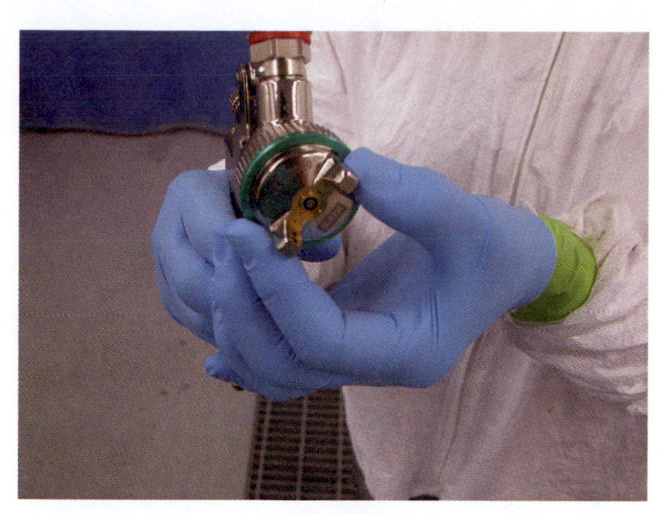

FIGURE 21-18 In a spray horn adjustment, rotation of the air horns changes the spray pattern.

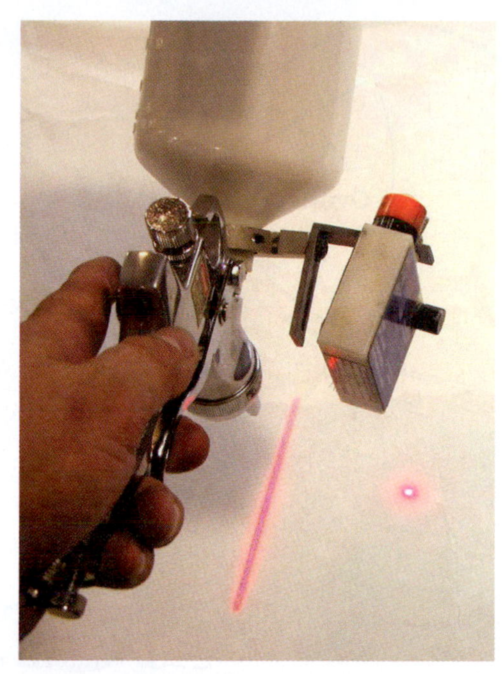

FIGURE 21-19 When using a laser guidance practice gun, a long laser beam simulates the width of the pattern, while dots indicate distance. When two dots are visible, the gun is either too far back or too close to the surface.

object. However, it should not be so wide that the spray overshoots the object, causing large amounts of coating to be sprayed into the atmosphere as overspray.

The fan direction can also be adjusted according to the object being sprayed. If the desired spray direction is horizontal, the spray pattern should be vertical so the panel can be covered as quickly and efficiently as possible. To adjust the gun to spray a vertical pattern, the technician should adjust the air horns to a horizontal position. See **Figure 21-18**. If a vertical spray pattern is needed, the spray horns should be adjusted to a horizontal position.

Distance

The distance at which the gun is held from the object being coated is very critical. This distance will be different in various procedures, depending on the type of gun used—and its manufacturer. The technician should check with the manufacturer for their gun's recommended distance. Paint manufacturers may also have recommendations for the proper fluid needle and seat to be used, as well as distance recommendations that the technician should follow.

Proper distance has traditionally been very difficult for technicians to learn. However, it's been made easier with the invention of the laser guidance tool. See **Figure 21-19**. This tool, when attached to a spray gun and adjusted to the recommended distance, will show two laser dots when the gun is either too close or too far away from the target. When the distance is correct, only one dot will be seen.

This tool helps technicians know that they are operating at the proper distance at all times. As the vehicle's surface changes, technicians can adjust their distance accordingly. Keeping the proper distance makes the technician more efficient, and less of the coating is lost as overspray.

Stroke/Movement

The technician's stroke and movement while covering the vehicle is one of the most valuable—and sometimes most difficult—operations to learn. It involves such factors as overlap of each stroke on the panel being painted to achieve the proper coverage without defects or streaking. It includes the rate of movement, so the vehicle will be covered efficiently and to the proper film thickness to achieve the protection needed. Stroke and movement also include the use of the proper spray pattern while covering the vehicle.

Overlap

Overlap is the amount of fan that is sprayed over the previous lap to obtain good coverage. When spraying coatings on a vehicle, the technician should assure that each path that the spray gun travels covers the previous one. The amount of coverage varies according to the type of finish being sprayed, but generally the technician should spray at a 50 to 75 percent overlap. That is, as the technician returns following the first path over the vehicle, the next one should be halfway on the old spray path and halfway beyond the old path. See **Figure 21-20**. This 50 to 75 percent overlap can be achieved by aiming the spray horns at the bottom of the previous spray pattern, while keeping in mind to maintain the proper rate of speed (1 foot per second) and the proper distance (one laser dot).

If the overlap is not maintained (that is, usually if it is less than 50 percent), the finish will appear streaked and will not be satisfactory. If the overlap is too great, the finish may develop excessive orange peel, sags, or runs, and the film thickness will be too great.

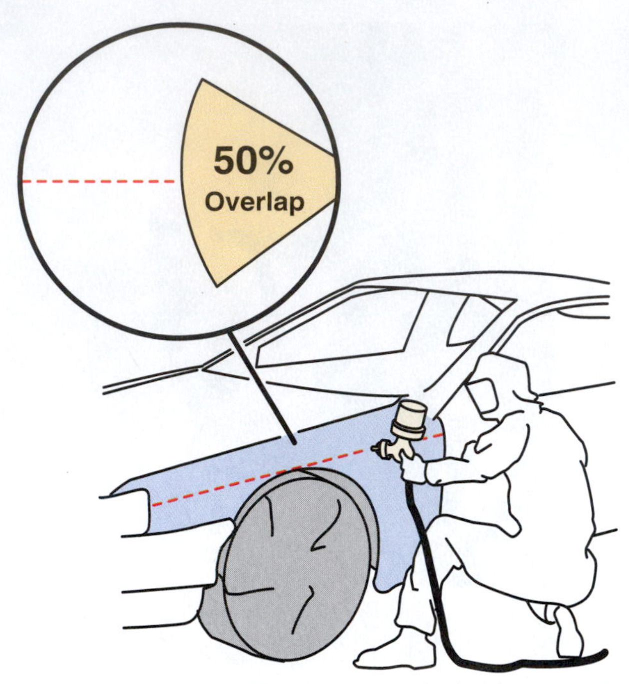

FIGURE 21-20 Each paint path should overlap the previous lap from 50 to 75 percent.

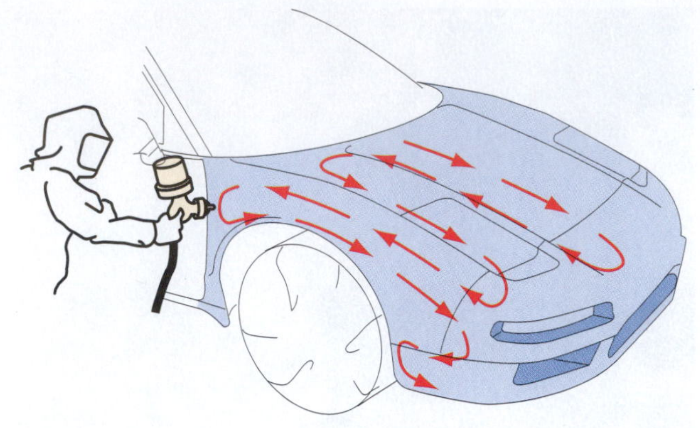

FIGURE 21-21 The paint pattern should move from one side down, keeping a wet edge from the previous stroke.

Rate of Movement. Though gun speed is not an adjustment that can be set on the gun, it is affected greatly by the other adjustments mentioned previously. If the gun is not set to atomize the coating well, there is little that the technician can do by adjusting the gun speed. Assuming that the atomization is properly adjusted, at its lowest amount the gun speed will need to be very slow. If the atomization is adjusted properly at the higher side of fluid flow, the technician can have a greater rate of flow. Spray guns, depending on their type, may or may not be able to spray at a higher rate of gun speed.

When a gun is properly adjusted, the rate of application speed is approximately 1 foot per second. Though travel speed will become natural with experience (and there are experienced technicians who paint at higher rates than 1 foot per second), new technicians should try to adjust the speed to meet the 1-foot-per-second standard until they become more comfortable with applying coatings.

SPRAY SEQUENCE

Each vehicle to be refinished will require a specific spray sequence. The technician will need to plan a starting place as well as determine where the sequence will end. The plan will also include how the gun spray pattern will be set up; the direction of movement according to the type of spray environment; whether the object will be refinished using horizontal streaks or vertical ones; whether the vehicle has a large surface (because of this requirement, the technician may need to mix the coating with different reducers); and if what is being painted is short (and therefore will need to be raised up) or tall (in which case the technician will need something to stand on).

Plan of Attack. The plan of attack refers to how the technician will decide on the spray sequence for a certain vehicle. What will be sprayed first and where will the spraying process end? If the job involves painting mostly on the front of the vehicle, the technician may choose to back the vehicle into the spray booth. This will allow the technician more ease in moving around the vehicle while painting. Knowing that one should spray in the direction of air movement, the technician should plan to spray from the top down in a downdraft booth, and from the front to the rear in a cross-draft or semi-downdraft spray booth. See **Figure 21-21**. Keeping a "**wet line**," though less important when spraying basecoat, becomes very important when spraying single-stage paint and clearcoat. The technician should not start or stop in the middle of a panel such as the hood or roof, then return to that area to continue spraying. By doing so, the area may have dried enough that the new spray line may not melt into the old spray line, leaving an obvious area of **dry spray** that may or may not be repairable when **detailing** or buffing later.

Though most vehicle spraying will be done with the gun set up to spray in the horizontal position, some objects may be best sprayed while suspended in the booth—or, if tall, sprayed in the vertical position. If the technician believes it would be best to spray the object being refinished in the vertical position, the spray gun's air horns should be turned horizontally to allow for vertical spraying. See **Figure 21-22**.

Painting Large or Low Objects. The technician must also consider the object's size when making the plan of attack. Large objects—such as the side of a van—will need to be sprayed from the front to the rear in one long motion. In such a case, the technician should plan how to move while in the booth so that the spray hose will reach far enough, and determine where to place the hose so the technician will not step on it while walking. The techni-

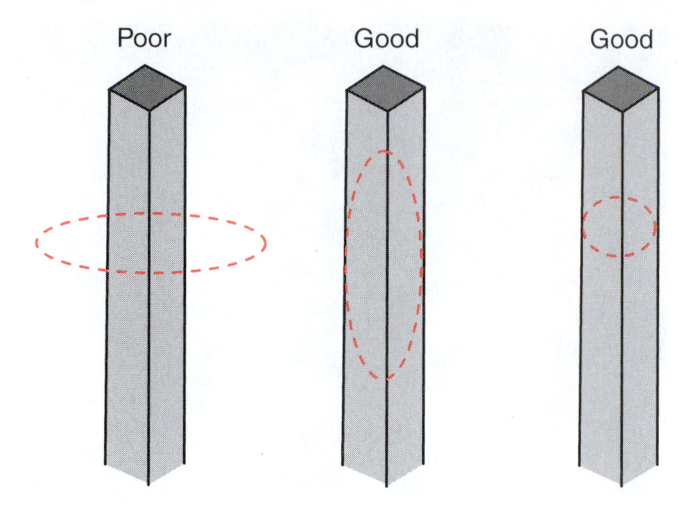

FIGURE 21-22 The spray pattern width should match the size of the object being sprayed. Overspray represents wasted paint and is also dangerous to the environment.

FIGURE 21-23 The technician may need to use a platform to reach a vehicle's roof.

cian will also need to have something to stand on to reach the sides or top. See **Figure 21-23**. Some vehicles that may be too tall when in the spray booth can be reached if air is let out of the tires, thus lowering the vehicle.

> **TECH TIP** The plan of attack can be made while the technician does the final cleaning and blowing off, just outside the booth. See **Figure 21-24**. Like most procedures, the planning stage may not be a separate time spent just thinking about that one thing, but can be developed while doing other steps in the process. Doing so reduces the time it will take to refinish a vehicle, thus speeding up the vehicle's **cycle time**.

FIGURE 21-24 An efficient technician will make a plan of attack before starting to refinish a vehicle.

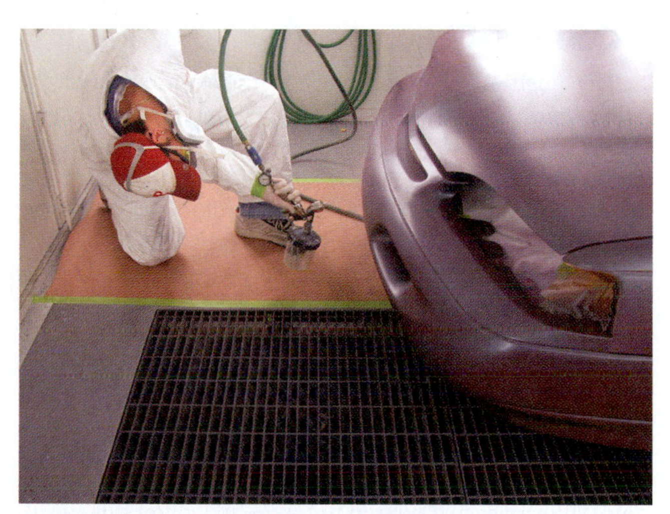

FIGURE 21-25 Some disposable spray cups and liners allow a technician to spray upside down.

Other vehicles may need to be raised so the technician will be able to spray low objects or portions such as the rocker panels, low bumper covers, and **ground effects**. Some booths come with lifts installed in them to raise vehicles, allowing the technician better access to these low objects. In other cases, the object can be raised and placed on jack stands, which also allow better access. Some spray guns are able to spray in all positions. A gun capable of doing so allows the technician to spray with the gun upside down. This type of gun may be needed to reach very low objects. See **Figure 21-25**.

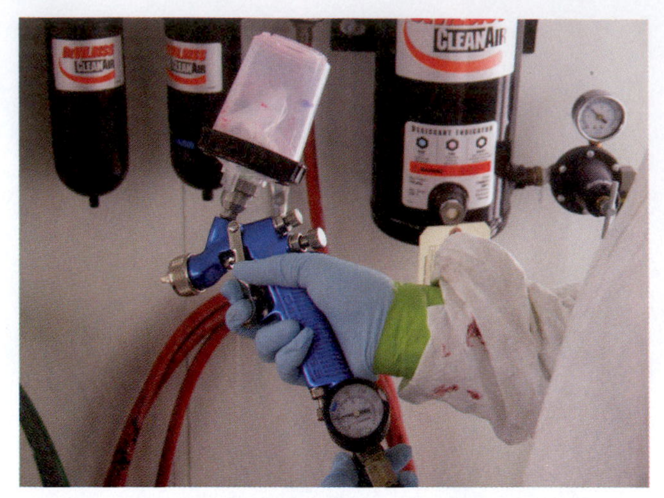

FIGURE 21-26 Small spray guns, sometimes called detail guns, are ideal for smaller tight work.

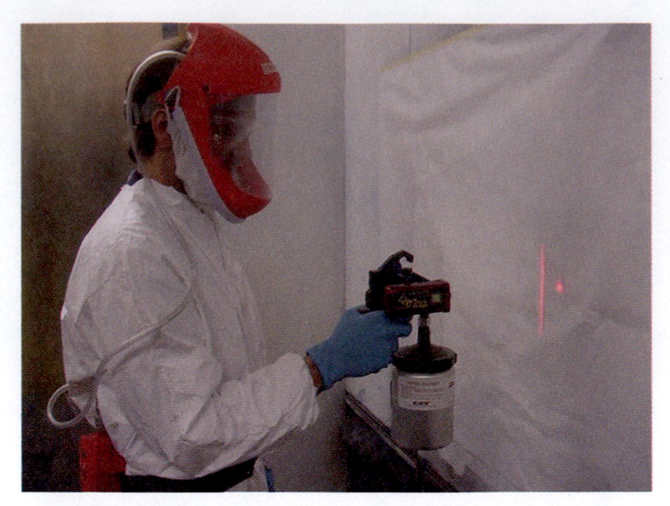

FIGURE 21-27 A new technician may practice with a laser-guided gun to help hold the gun perpendicular during service work.

The technician may choose to use a smaller spray gun with small paint cups when spraying small objects. See **Figure 21-26**. Small spray guns are also very useful when spraying small areas on a larger panel. Sometimes areas to be primed are small, and the primer may be sprayed best with a smaller gun.

POSITION

The position at which the technician must hold the gun is very critical to the quality of the work. As mentioned earlier, whether technicians move horizontally or vertically, they *must* keep the spray gun perpendicular to the varying surfaces of the vehicle. In addition to holding the gun at the proper distance in the proper position, the technician must also avoid bad spraying habits such as arching, pitching, improper rotation, and large lead and lag distances.

Perpendicular Movement

Technicians should keep the spray gun perpendicular to the vehicle being sprayed at all times, Though this may sound like an easy direction to follow, it becomes more difficult when spraying vehicles of particular shapes. On flat and long panels such as a quarter panel or a pickup truck's box sides, the perpendicular movements are long flowing and easy to do. But when the technician needs to refinish areas such as a vehicle's door fender or the hood, it becomes more difficult. The technician must take care to follow the vehicle's contour as the spray gun passes over it. Tools such as laser-guided practice guns help train students to hold the gun perpendicular. Both the fan laser and the distance laser help with this practice as well. See **Figure 21-27**.

Distance

The distance between the spray and surface is critical to the quality of the work. It is also critical to the technician's transfer efficiency, which will be covered later in this chapter. To produce a high quality, properly atomized coating with the proper film thickness and a smooth finish, the technician needs to spray at the proper distance and maintain that correct distance. Before the development of the laser-guidance device, keeping this proper distance was very difficult to learn. Technicians used a lot of trial and error to perfect their methods. See Figure 21-27. Today, the new technician can learn first with the laser-guided practice gun, then attach a laser distance device to a gun while spraying and be able to perfect the technique much faster. By simply keeping the two laser dots together as one, the technician can be assured that the proper distance is maintained. Though the laser may be kept on the gun, most technicians will remove it when they have built confidence in their ability to maintain proper distance.

Arching

Arching is an improper distance pitfall that some technicians practice when spraying. When the gun is moved in an arching movement instead of by keeping the proper distance throughout the stroke, numerous problems occur. See **Figure 21-28**. The finish will have uneven film thickness, may need more coats to provide proper coverage, and may have uneven surface texture, all because the technician is not holding the gun at the proper distance through the stroke. When a highly metallic coating is sprayed in an arching pattern, there may be uneven **metallic lie**, causing improper color match. Technicians can determine if they are arching during their stroke by videotaping themselves while spraying.

Pitch

Pitch occurs when the technician is not holding the gun perpendicular to the surface of the vehicle being painted. The gun may be held so the **gun fan** is closer at its bottom, mak-

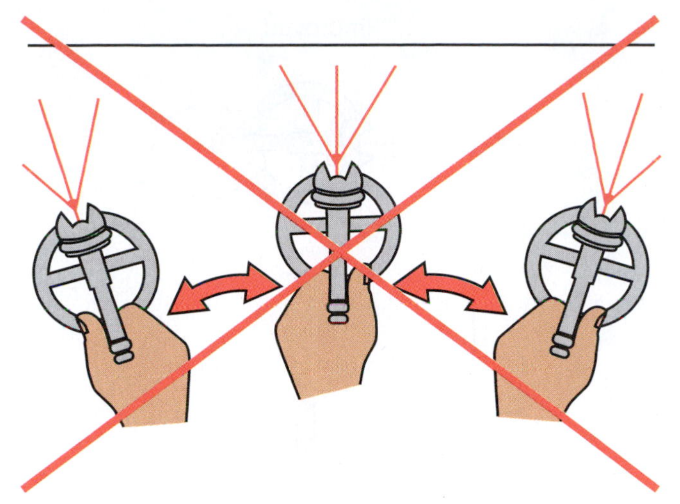

FIGURE 21-28 Arching is a common gun pattern mistake. The gun should be kept perpendicular and the same distance from the surface throughout the stroke.

ing the finish deposit heavier, or held closer at its top (causing the paint to be heavier at the top). See **Figure 21-29**. Both of these pitch problems will cause the paint to be unevenly deposited, and may result in streaks.

FIGURE 21-29 Pitch occurs when the gun is tilted either to the top or bottom, causing an uneven spray pattern.

Rotation

Rotation occurs when technicians turn the wrist either clockwise or counterclockwise as they pass over the surface being painted, thus shortening the fan width. See **Figure 21-30**. Though rotation will not cause paint application defects if all other angles and the distance are correct, it will require more passes over the vehicle being

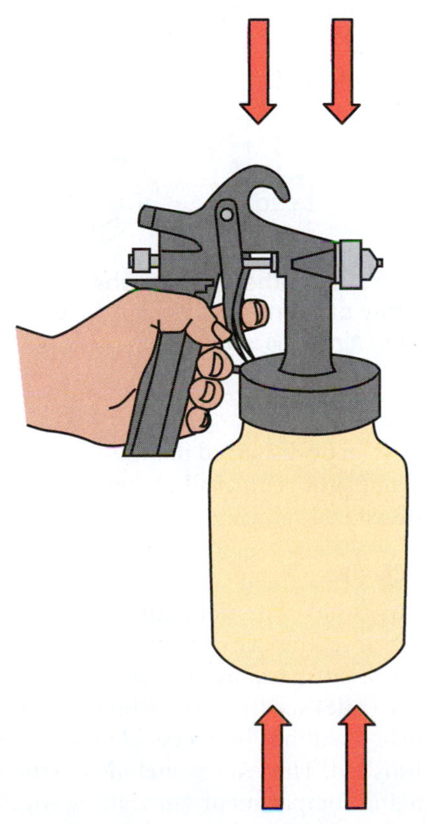

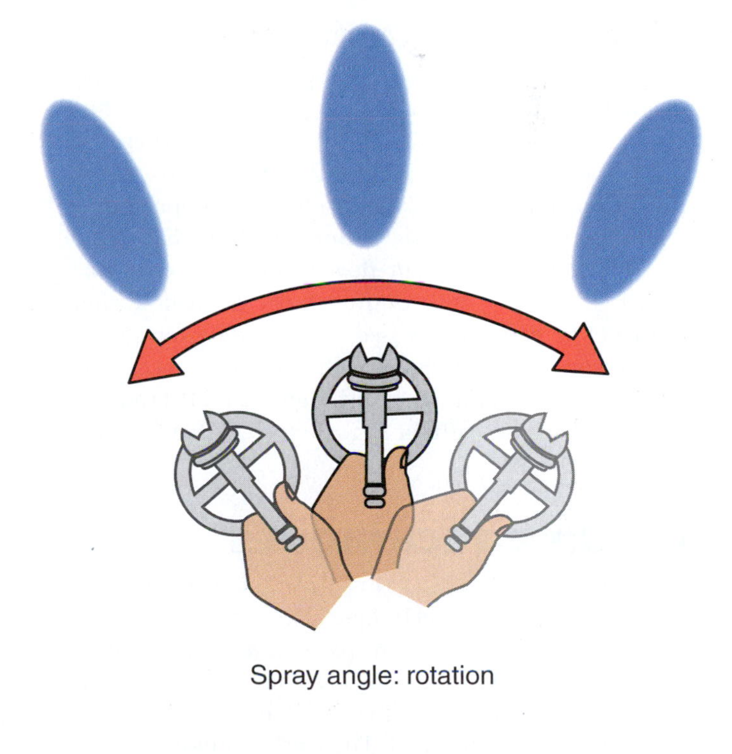

FIGURE 21-30 Rotation occurs where the pattern is rotated either right or left, which will lower the stroke's coverage pattern.

refinished. In fact, rotating the gun to cause the fan to be smaller is beneficial at certain times. Using rotation *to a technician's advantage* will be discussed in this chapter when transfer efficiency and **banding** are explained.

Gun Triggering

Spray guns have two trigger positions. The first one, which is from trigger at rest to about halfway pulled, will allow air—but no coating—to escape from the gun tip. The second position, from halfway to fully pulled, will allow both air and coating to escape from the gun tip. Proper triggering of the gun means while the technician is spraying, the air will constantly be escaping from the gun tip while the paint flow is turned on and off, depending on the gun's position and the object being painted.

The technician should start the airflow when the gun is at the proper distance and perpendicular to the object being painted. Just before the object is reached, the paint flow is turned on, and the trigger is continually pulled through the stroke while over the object being painted. After the target area is passed, the paint flow is stopped. However, the airflow should remain on as the technician changes direction for the return pass. Just before the target object is reached, the trigger is pulled completely and the paint flow is again started. This process continues until the object is coated. Keeping the airflow on and the paint flow triggered on and off will increase the quality of the work.

Lead and Lag Distances

Lead is the distance between the point where the gun is triggered and the point where the gun pattern spray hits the part. Lag is the distance between the point where the pattern leaves the part and where the gun is untriggered. As mentioned, the technician must learn to properly trigger a spray gun to properly perform high quality work. Likewise, starting the paint flow before the area to be painted is reached—and continuing the paint flow after the area to be painted is passed—is necessary for quality refinish work. The successful technician will keep these two distances to a minimum. See **Figure 21-31**. Lead and lag will be discussed further in the transfer efficiency section of this chapter.

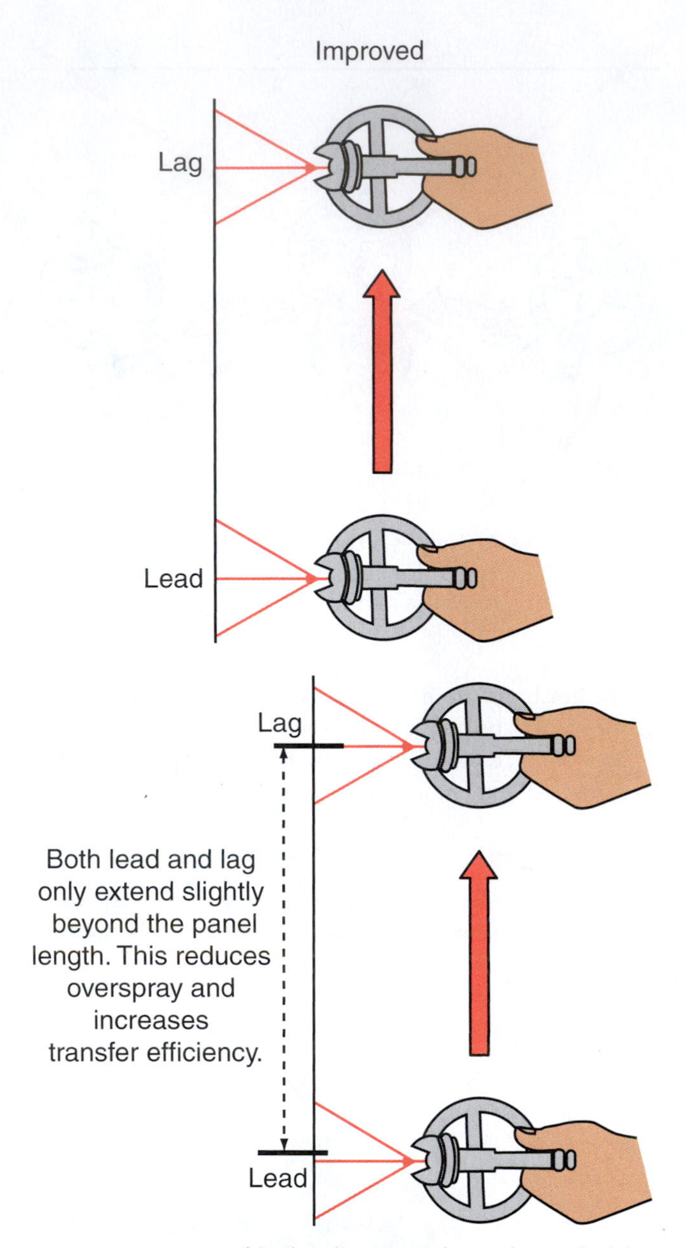

FIGURE 21-31 Lead is the distance where the technician starts the spray pattern before it hits the object. Lag is the amount of time the spray is kept on when passing the object.

REFINISH TECHNIQUES

Refinish technicians need to perfect many different refinish techniques. One technique, called "**cutting in**," is used for painting replacement parts in areas that would not be accessible when attached to the vehicle. Another technique involves attaching completely refinished parts after the vehicle is refinished. There are techniques for refinishing with single-stage solid colors and single-stage metallic colors, for applying basecoat, applying clearcoats, and even a technique for applying undercoats with a roller. Each of these application techniques—and their differences—will be discussed in this section. Technicians need to become proficient at all of these procedures to become a professional refinisher.

Cutting In

Cutting in is the process of outlining a new or replacement part before the part is attached to the vehicle. These replacement parts need to be refinished with the proper color (hue) on the areas which—when attached to the vehicle—will not be accessible but would be visible if not finished. These areas include behind lights, inside the engine compartment, the trailing edge of finders (which is seen when the door is opened), and other areas similar to these. These areas would be visible with the hood or door open or could be seen behind the light. Though a technician may be able to apply paint following careful masking and creative gun work, it is much easier to prepare the parts and paint all the hard to reach areas before the part is attached. Some technicians will prepare the part to be refinished by masking off the areas where they don't want paint, while others will prepare and cut in the part to be painted without masking. See **Figure 21-32** and **Figure 21-33**. The choice to use one of these techniques over the other one is generally determined by each particular repair facility.

After the part's surface is inspected, cleaned, and sanded according to the shop's standards (see Chapter 25, Surface Preparation), the technician places the part in the spray booth and carefully sprays the areas that are cut in. The part should be finished to the same specifications as the original equipment manufacturer (**OEM**) part. That is, if the OEM part was finished with basecoat clearcoat, the replacement part should also have the same finish.

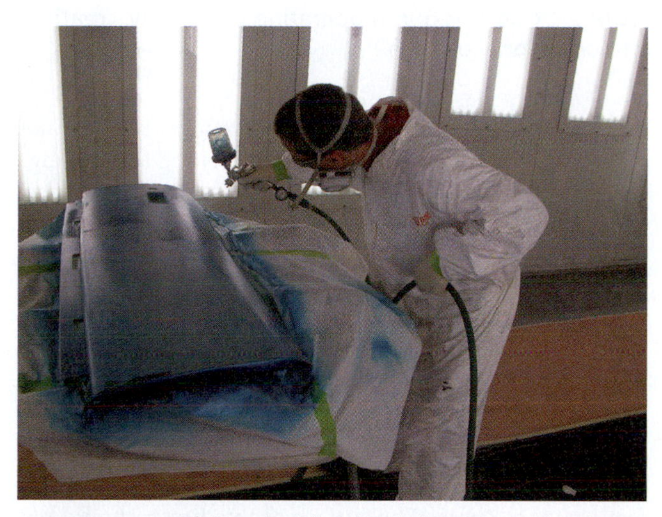

FIGURE 21-32 This technician is edging a part that has been masked to protect other areas from overspray.

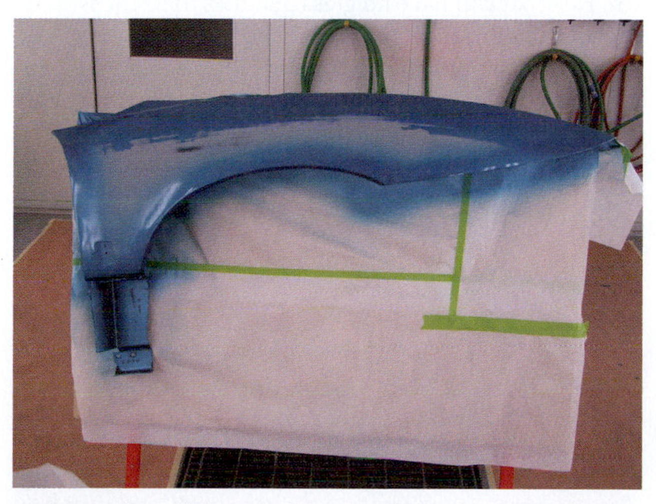

FIGURE 21-33 This technician is edging in a part that has not been masked.

Part Painting

Some parts—such as outside mirrors, cladding, ground effects, and door handles—may need to be sprayed while off the vehicle, then attached later. These parts should be sprayed using the same color and finish as the vehicle, and they should be sprayed in the same position as they will be when mounted on the vehicle. When sprayed in the same position as they will be on the vehicle, the metallic orientation will be the same as the remainder of the vehicle. By using this method, the technician does not risk a difference in color and metallic reflection between the parts. Bumper covers or **fascia** stands—along with mirror holding devices—are made to hold these parts in the correct orientation to maintain proper metallic lie. See **Figure 21-34**. Care should especially be taken while refinishing smaller parts. As they are difficult to detail or polish, every effort should be taken to prevent defects.

Single-Stage Solid Colors

Though some refinishers believe that it is one of the techniques easiest to practice and perfect, single-stage solid color spray techniques have their difficulties. Although black and other single-stage solid colors are rather easy, single-stage metallics can prove quite challenging. There is little margin for error and the refinish technician must take care that the "wet line" is maintained.

The refinisher should apply single-stage spray according to the paint manufacturer's recommendations. Often this recommendation is a medium to **full wet coat**. That is, the amount of paint applied in a single stroke is wet in appearance with full gloss. Spraying a full wet coat is one of the more difficult techniques, for the finish must be so wet that it is dangerously close to running. When applying a finish this wet, care must be taken not to rush the **flash time**. Maintaining proper flash time will allow the solvents time to evaporate before the next coat is applied.

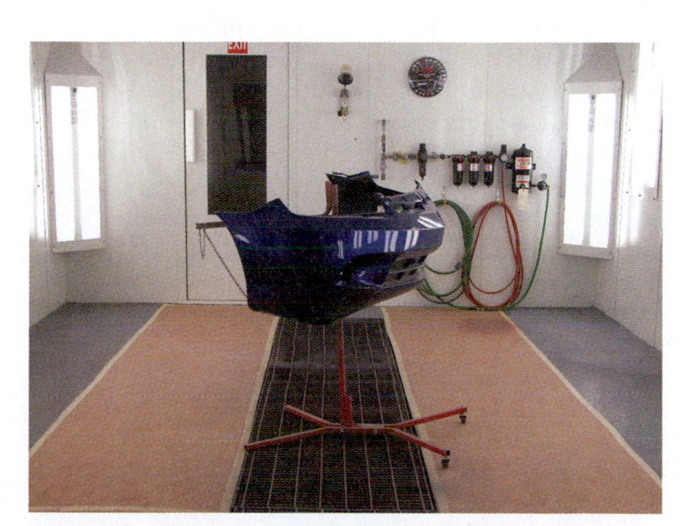

FIGURE 21-34 Fascia on a paint stand made to hold parts in the correct orientation to maintain proper metallic lie.

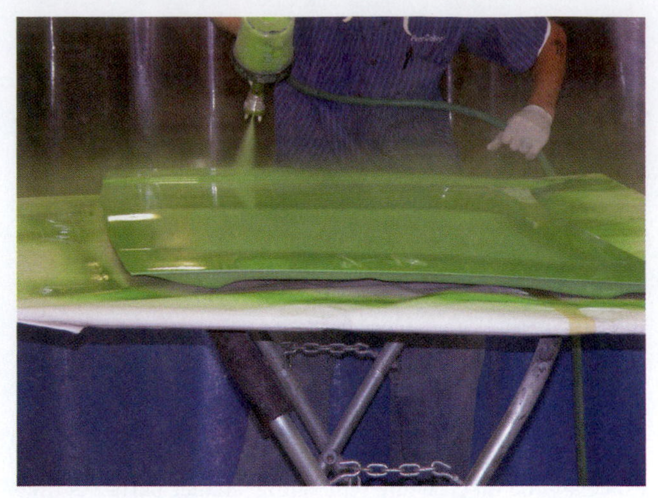

FIGURE 21-35 While spraying some colors, the technician must look at the area being sprayed at a transverse angle.

Some colors—such as white in a brightly lit booth—may be difficult to see when spraying the spray (wet) line. The technician must look at the area being sprayed at a transverse, or 45-degree, angle. See **Figure 21-35**. This angle helps the technician to view the wet line when it otherwise would not be visible.

Though solid colors will not normally pose a streaking problem, the surface must be kept wet to ensure the proper gloss as it is being sprayed. Distance (as always) should be maintained and the gun kept perpendicular to the surface. The number of coats should be applied according to the manufacturer's recommendations along with the proper film thickness.

Single-Stage Metallic

Though surface preparation and cleanliness would be the same as with solid colors, a metallic single stage is more likely to have **paint striping**. Gun adjustment is critical, including the fan adjustment. If the fan adjustment is set fully open, the spray application may be slightly heavier at the top and bottom of the spray streak. See **Figure 21-36**. Even if proper distance is maintained and the gun is held perpendicular to the surface, this figure 8 spray pattern will streak the finish. Paint manufacturers' recommendations for single stage are often to spray a medium wet to full wet coat. Even if this guideline is followed, however, streaking may still become a problem.

If streaking is noticed, the technician may need to let it flash for a longer period of time, then apply a **mist coat** in a "cross-checked" stroke pattern. A cross-check stroke pattern means the technician sprays in the opposite direction from the previous stroke pattern. If a technician is spraying a hood in a front-to-rear stroke when striping occurs, first check the gun for proper adjustment and check the fan for even coverage on a test panel, to eliminate the possibility that the gun is causing the problem. Next, apply a light mist stroke, spraying across the hood in the opposite direction. Mist coats are applied (with the correct flash time between) until the striping is no longer seen. Care must be taken not to have excessive film build while eliminating the striping.

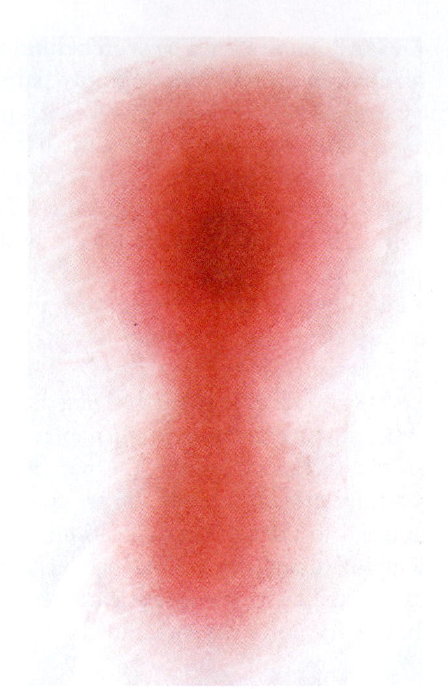

FIGURE 21-36 This gun is operating incorrectly, creating a figure 8 spray pattern.

The best practice is not to get striping in the first place, which means that the technician should be confident that the gun is adjusted properly and that the distance and angles are maintained to avoid striping.

Basecoat

Basecoat—the first half of a basecoat/clearcoat refinish application—is the most common spray application in collision repair.

The recommendation for applying basecoat is often in **medium wet coats**. The surface of this finish should not be applied to a gloss finish. In fact, the finish will be dull either as it is being applied or shortly after the solvents **gas out**. Basecoat will have no gloss qualities to it. Gloss is obtained when the clearcoat is applied.

Basecoat application recommendations may not identify how many coats are required. Some may only state that basecoat should be applied to "**full coverage**," which means that when the basecoat no longer can be seen through, it has sufficient coverage. However, technicians must be careful when judging "full coverage." What looks good in the booth may not be full coverage in bright sunlight. Technicians should use coverage stickers or a **sprayout panel** with a coverage sticker on it. See **Figure 21-37**. This will ensure not only the proper color (hue) of the mixed paint, but will also show how many coats it will take to achieve full coverage. The sprayout panel should be sprayed in the same manner as the vehicle will be sprayed, with the same flash time between each coat. The coats are applied until the black and white squares can no longer be seen; the same num-

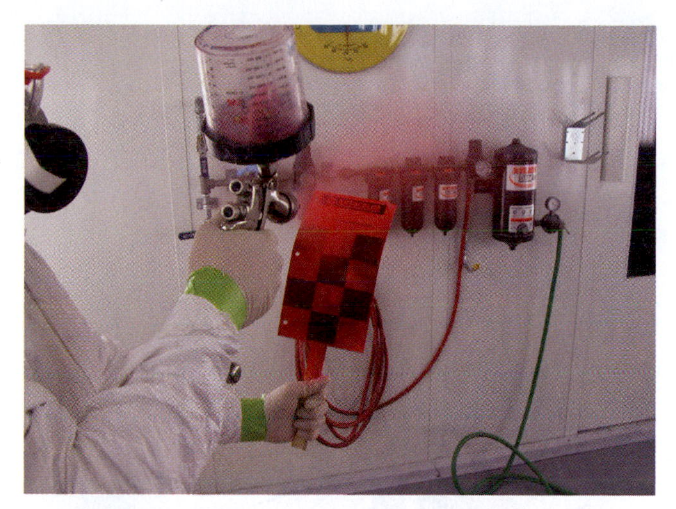

FIGURE 21-37 Technicians should use coverage stickers or a sprayout panel with a coverage sticker on it to check for full coverage.

ber of coats should then be applied when the vehicle is refinished.

If imperfections are noted and repairs are needed, basecoat can be repaired with light sanding and tacking. Following the repair, however, a full coat should be applied to the repair area so the sanding will not be visible.

Basecoat must be clearcoated. It has no resistance to the elements and will soon fail if not covered with a clearcoat. Although there may be a time period the technician must wait before the clearcoat can be applied, there also is a time limit by which the technician must apply the clearcoat. Otherwise, the surface must be scuffed before the clearcoat is applied. (This process is discussed in more detail in Chapter 24.)

Basecoat may be applied to the full panel, applied on a portion of the panel (spot repair), or applied to the repaired area and then sprayed into the next panel to assure proper color match (blending), which will be discussed later in this chapter. The one thing that all these basecoat techniques have in common is that they *must be covered with clearcoat* to be able to stand up to the elements.

Clearcoats

Clearcoats are the second type of coating applied during the basecoat/clearcoat two-stage spray application. Though a clearcoat does not contain color, metallic, pearls, or other additives, which add to the glamour of the finished product, it does provide all the necessary additives to help the finish hold up to the elements. It also adds the shine (**depth of reflection**), and generally enhances the basecoat below it.

Because clearcoats are formulated to provide different qualities, they may require differences in the way a technician will apply them. Some clearcoats are called hyper-curing clears. Because of their very fast curing, they will need to be applied in a particular way. Technicians will need to make basic decisions when using hyper-curing clears and know their limitations. Because of their hyper-curing quality, these clears will not stay wet long enough to spray a large repair. These very fast-curing clears are ideal, though, for cutting in new parts, as well as for small repairs where only one or two panels are being refinished. Other clearcoats are designed to cure much more slowly; when using one of these, the technician can take longer with the application without fear that the finish will be cured before the first coat is completed. The more "forgiving" curing time helps the technician avoid leaving dry spots where the end meets the beginning.

The technician's technique when applying different clearcoats will vary, depending on the size of the repair and the clearcoat chosen for the repair. The most important consideration when the technician is making the plan of attack is keeping a wet line, to avoid having dry areas when the finish is applied. The direction of airflow is also important when applying clearcoats, but this does not vary from job to job. Therefore, the technique will not vary because of it.

If the area on the vehicle to be refinished is a hood and fender, in a downdraft booth, the technician can choose a hyper-cure clear. The plan of attack would be to begin at the edge of the hood, spray to the center of the hood, then go to the opposite side of the vehicle and continue spraying from the center to the opposite edge of the hood, then continue to spray down the fender until the application is completed, keeping a wet edge throughout the procedure. See **Figure 21-38**. As with other spray operations, the gun must be kept perpendicular to the part being sprayed and

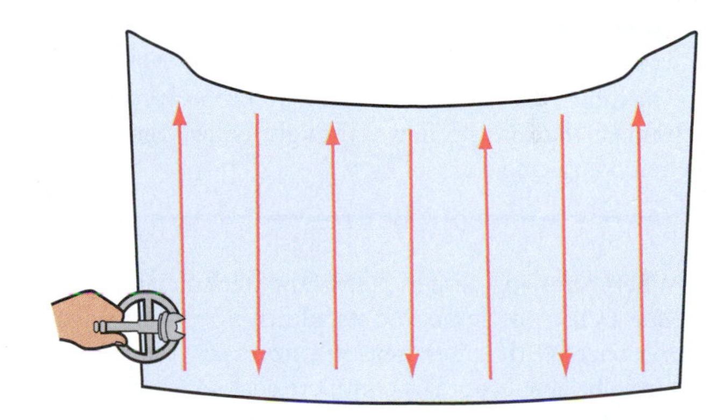

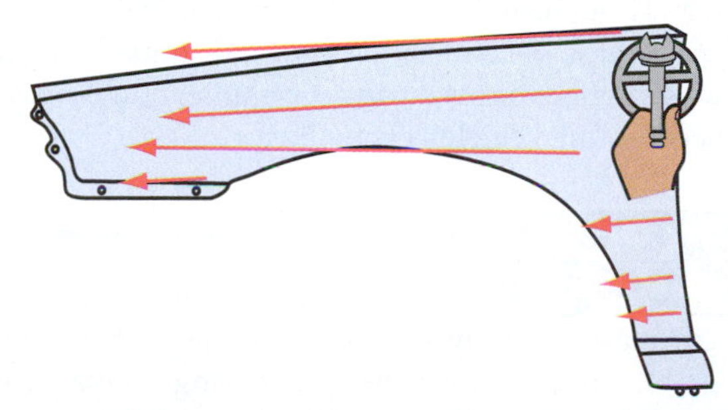

FIGURE 21-38 If the area on the vehicle to be refinished is a hood and fender, in a downdraft booth, the technician can choose a hyper-cure clear.

the gun should be kept at the manufacturer's recommended distance throughout applications.

The amount of clear (number of coats) applied will vary, depending on the manufacturers' recommendations. Some paint manufacturers recommend that the first coat of clear be applied just beyond the last coat of color over the repair area, with the second coat being applied over the entire panel. Other manufacturers will recommend that the clear be applied to all the area being covered. The technician should know, understand, and follow the paint maker's recommendations for the specific clear being applied. It is recommended that most clears be applied as full wet coats. This means that the clear being applied must be sprayed so it looks fully wet following each gun pass. The appearance should look somewhat like a puddle of water, and each 50 percent overlap pass should push this wet puddle further along the panel being sprayed. It is sometimes difficult to view this, and the technician should view the "wet edge" using "**transverse viewing**." This means the technician looks at the panel being sprayed at a 45-degree angle. See **Figure 21-39**. This angle allows the technician to note the reflection while spraying. If the panel is viewed from straight down or from a 90-degree angle, it is very difficult to see the wet edge adequately. For some colors, such as white in a very brightly lit booth, it is difficult to watch the wet edge even when the technician views it at a transverse angle.

TECH TIP The technician should be warned that when spraying coatings to full wet, the possibility of having runs becomes greater. A properly sprayed clearcoat that has no flaws, though, is truly gorgeous when completed.

When spraying larger areas, the technician should choose a clear (or sometime its additives) that cures at a slower pace. With larger areas it is necessary for the technician to be able to spray a complete coat before the finish cures to the point where the last area meets the first area sprayed. If it is not sufficiently wet, the meeting place will not "melt into" the existing finish and will show a dry area. Though sometimes dry areas can be repaired in the detail phase of refinishing, they still must be repaired. If flaws can be avoided during the application process, it will speed up the vehicle's cycle time.

TECH TIP Some finishes are formulated to cure fast and only use one hardener, while other finishes can have their speed of curing regulated by the hardener that is chosen.

FIGURE 21-39 The technician should view the area being sprayed at a 45-degree angle. This aids with viewing the spray overlap area or "wet edge."

Paint manufacturers provide detailed application directions for each product that they produce. These production sheets should be fully read, understood, and followed for a more trouble-free application of the product.

Panel Painting

Panel painting is a technique used when the color is known to be a perfect match (which is not often obtained) to the existing finish. Solid colors are sometimes applied where the entire panel is refinished. That is to say, the panel is properly cleaned, the surface is prepared, and finish is applied to all areas of the panel.

The technician should know that when the panel painting method is used, even a slight mismatch is more likely to show. Where the newly painted panel lies next to the old finish (such as the fender/door area), the color difference may be more easily seen. In fact, with today's glamour colors, this technique is often not even discussed during manufacturer training classes.

When applying color using the panel painting technique, gun angles must be kept perpendicular to the panel, with the proper distance maintained at all times. The technician also should know, understand, and follow the paint manufacturer's recommendations.

Basecoat Blending

Basecoat blending is a process in which the basecoat is applied to a repair area that is smaller than the complete panel being repaired. It differs from spot repair, where no attempt is made to gradually reduce the film thickness around the edges of the repair. In blending, the technician applies coating at a reduced thickness slightly beyond the repair area. Though there are different techniques used to accomplish this (such as standard blending, reverse blending, and wet bedding, all of which will be discussed in this section), all of these processes have been designed to help in the color match, which panel painting does not accomplish. The repair area will be covered with sufficient amounts of basecoat to provide complete coverage, and the areas surrounding the repair will have lesser amounts of coverage, letting the color beneath it show through the new color. This combination of colors will trick the eye into not noticing slight differences in the two colors.

TECH TIP Blending should only be attempted with colors that are only slightly unmatched in color. If the color is more than slightly different from the original color, it no longer is considered a "**blendable match**" color. No matter how well the color is blended, it will still be visible as a color mismatch.

To determine if a color is a "blendable match," the refinish technician should prepare a sprayout panel to compare it to the original color. (Sprayout panels are covered in Chapter 30.) Once the newly mixed color is confirmed to be a blendable match, the technician can choose which spray technique to use to blend the color.

Blending. In the standard blending process, the first coat of basecoat is applied to cover an area just slightly larger than the repaired area. This procedure is performed with the trigger of the spray gun being gradually let off to reduce the amount of color film thickness that is applied. This first coat fully covers the repair area, with thinner amounts of color being applied just beyond the repair area. The second coat is applied in a similar manner, but with just a slightly larger area being covered. Subsequent coats are applied to progressively larger areas until the repair area is completed to full coverage. See **Figure 21-40**.

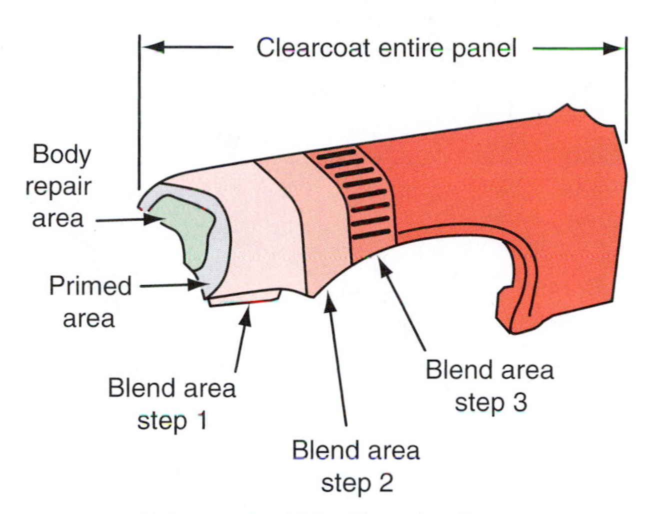

FIGURE 21-40 In standard blending, the first coat extends just past the repaired area, the second coat extends slightly past the first, and the third extends beyond the second.

CAUTION

Be careful not to develop a fanning motion to accomplish blended edges. Instead of slowly letting the trigger off and keeping the gun perpendicular to the surface as they blend, technicians sometimes arch—or fan—the gun, and in doing so they increase the distance. This may create a color difference, or the refinished surface may have a different metallic lie, because the increased distance does not accomplish a blendable match.

The standard blending technique can be used when the paint technician is blending colors that are very close in appearance when the sprayout panel is compared to the vehicle. The technique can also be used if the color is not highly metallic in nature. However, if the technician suspects that the color may be difficult to match, reverse blending or wet bedding should be used instead.

Reverse Blending. **Reverse blending** is a process similar to standard blending, but applied in reverse. The first coat is applied to the anticipated largest area with the outer edge being off-triggered to apply a lighter amount of color. The second coat is applied to an area smaller than the first, but with full coverage over the repair area. See **Figure 21-41**. The last coat should be applied over the repair area and just slightly beyond, for **full hiding**. The sprayout panel not only tells the technician how closely the new mixed color will match the vehicle, but will also tell the technician how many coats will be needed for full coverage. For example,

REVERSE BLEND

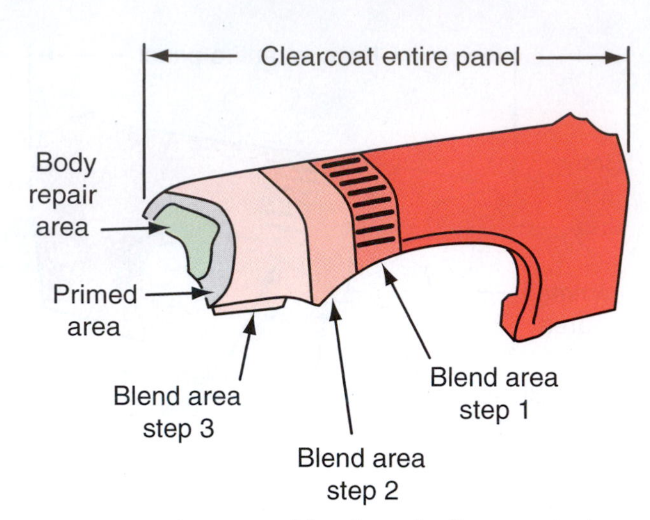

FIGURE 21-41 In reverse blending, the first coat is the largest, with the second coat slightly smaller, and the third and subsequent coats becoming progressively smaller.

if the technician finds that the sprayout panel will take four coats to achieve full coverage, the previous process should be adjusted to accommodate four steps.

Reverse blending is the suggested procedure to use when the newly mixed color is considered to be "a difficult one." If metallic lie problems are anticipated, reverse blending will allow each successive coat being applied to fall on a previous coat, and thus will not result in dry edge problems. With better metallic lye, the color match will be less likely to show a mismatch.

Wet Bedding Blending. Reverse blending does not always completely eliminate some of the more difficult match problems. When a color is thought to be a blendable match, but is also known to be difficult to apply, **wet bedding** blending may be the best technique to use. Though wet bedding is similar to reverse blending in that the first coat is applied to the largest anticipated area, the difference is that the first application of color (ready to spray) is mixed with clearcoat (ready-to-spray) at a 1:1 ratio and is applied to the vehicle as the wet bed. The clear chosen to mix with the color should be a slow curing type, to allow the wet bed to remain wet as the other coats are applied over it. The second coat is applied without being mixed with clear and applied over the wet bed. This procedure allows the metallic to lie in the wet bed and orient properly, because the wet bed allows it to **melt in** properly. The subsequent color coats (also without clear added) are applied according to the number indicated by the sprayout panel. See **Figure 21-42**. The only coat with clear added is the first one, creating the wet bed, with the remainder of the coats applied in a reverse manner.

Blending Option #3 Using basecoat color blenders as a foundation:

Step #1. Prep panel as normal for basecoat blend. Reduce **basecoat color** as normal.

Step #2. Reduce ***basecoat color blender*** exactly like the basecoat color. Pour into a separate basecoat spray gun. Take both guns into the spray area.

Step #3. Apply one coat of RTS ***basecoat color blender*** to the repair area. Applying the blender first may reveal small imperfections such as find sandscratches that were missed in final preparation steps. These sandscratches can cause metallic "streaks" in high metallic colors, they should be removed.

Step #4. Apply basecoat color as normal keeping the blend edge well within the ***basecoat color blender*** from Step #3.

WET BED

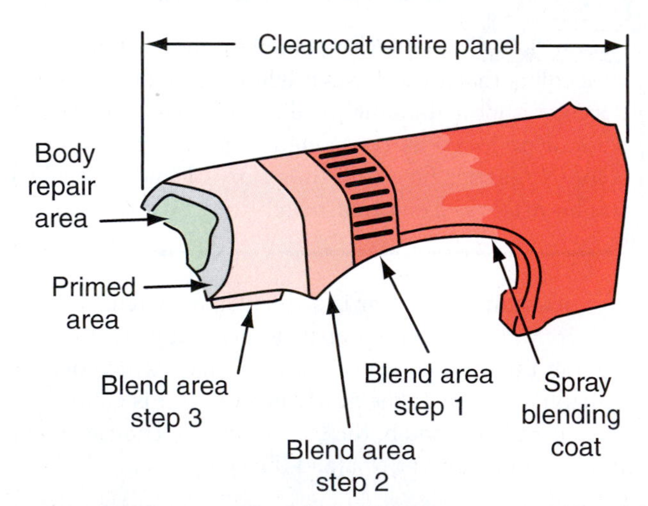

FIGURE 21-42 In wet bedding, the first coat is blending clear to the complete panel, while the reverse blend is done with basecoat.

Clearcoat Blending

Clearcoat blending is a technique used to blend the clearcoat into an area where masking cannot be applied to a specific separating line. Examples of such areas are the lower dog leg of a quarter panel where it meets the rocker, as well as where the quarter panel meets the roofline. In each case, there is no separating line and no definitive line to mask to. The procedure in such situations is to blend the clear into the old finish. However, paint makers do not often recommend this process due to its limitations. Clear must have a minimum of two mils of coverage, or the area with less will be affected by ultraviolet (UV) light and may fail. Though not often recommended by paint companies, the paint technician may still need to use the technique from time to time.

To achieve clearcoat blending, the basecoat is applied using the techniques the technician feels will best accomplish an invisible repair. The first coat is applied to full wet, to the complete panel *except* the area to be blended. In that area, the clearcoat is tapered off by slowly **off-triggering** the gun in the blend area, thus causing less clear to be deposited on that area. Where the clear is blended, the technician will use a second gun to spray a light coat of slow reducer over the blend area. This operation does not cause the new paint to blend into the old paint (see Chapter 27, Solvents for Refinishing), but it allows the new clear to flow out in a smoother manner, and therefore produces a more invisible blend line. The second and subsequent coats of clear are sprayed just slightly larger than the previous coat, with a light coat of slow reducer applied to aid in flow out. See **Figure 21-43**. *Caution should be taken when applying reducer to the blend area, for it will cause the clear to run easily.*

When the vehicle has cured, the blended repair should be polished by hand until the blend area is invisible.

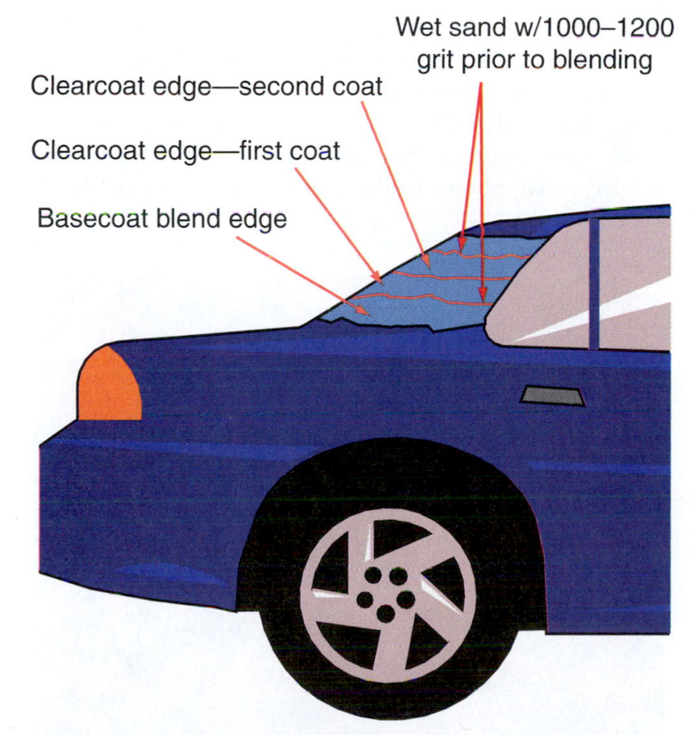

FIGURE 21-43 The second and subsequent coats of clear in clearcoat blending are sprayed just slightly larger than the previous coat, with a light coat of slow reducer applied to aid in flow out.

Rolled Application of Undercoats

Though rolling of undercoating has been a common practice in Europe for years, the technique has not been widely adopted in North America. The process can save time when the application is performed properly; it does take some time, though, for technicians to learn the proper technique. The process also requires the proper equipment, which appears in the following list:

- Roller handle
- Roller covers (light and heavy rollers)
- Coating container
- Masking tape
- Personal protective equipment.
  - Glasses/goggles
  - Gloves
  - Respirator
  - Ear plugs, if work area noise level measures over 100 decibels (db)

See **Figure 21-44**.

The surface to be covered should be prepared for normal undercoating application (see Chapter 25). The panels adjacent to the areas being primed should be masked with 1 1/2-inch tape to protect surfaces from accidental coverage. Then the technician should mix the roll primer in the application tray or cup. A non-waxed paper cup or small plastic mixing cup can be used for roll application. See **Figure 21-45**. Then the proper application roller must be chosen. One with thicker mat will apply the undercoat with a higher film thickness, while one made from foam will apply lighter film build. Rollers are also available with longer handles for applications in hard-to-reach areas.

The roller is loaded by dipping it into the tray, then working some undercoating into the applicator. The applicator should be full, but not so full that it is dripping when

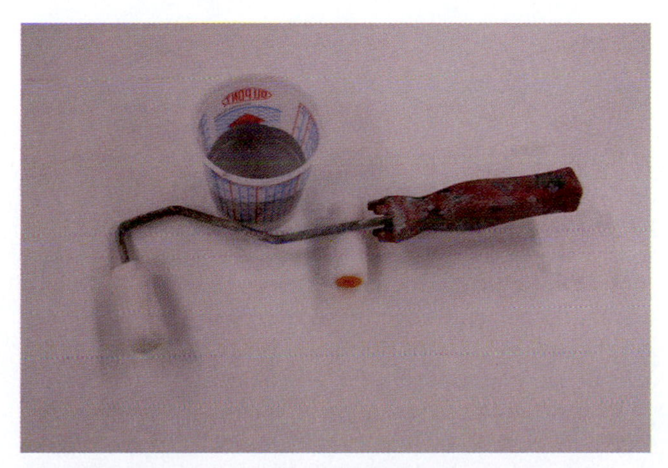

FIGURE 21-44 Roll priming equipment includes the technician's personal protective equipment.

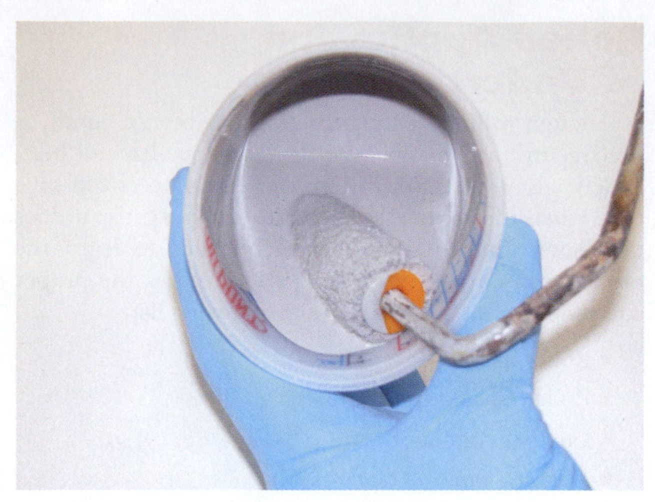

FIGURE 21-45 Roll primer is mixed in a cup using the edge of the cup to load primer on the roll.

moved to the panel for application. Apply the undercoating starting in the middle of the repair and moving outward. The stroke should be applied with light pressure in the center. As the roller reaches the outer edge the pressure should be reduced, feathering the end of the stroke. The technician should continue covering from the center out, with the outer edges feathered for lighter coverage, until the entire area is covered. Flash times should be observed the same as with spray applications, and the number of coats should be applied according to paint maker recommendations. *The technician should be warned, though, that if undercoats are roller-applied in a box-like manner, without feathering the outer edges as just explained, the dried undercoat will be very difficult to sand without having edges show.*

Following the application of primer using a roller, cleanup is also a very quick process. If a plastic cup is used when the undercoat cures, the cup can be removed and reused. The roller is removed and not reused, with little to no cleanup needed for the application handle. When technicians learn to apply undercoats with a roller, it is easy to see that the process can take less time and use less material.

TRANSFER EFFICIENCY

Transfer efficiency (TE) is the measure of how much of the finish material that is placed in the gun is transferred to the target surface. Though the ideal amount is a range of about 62 to 65 percent, the average operator sprays at the 40 to 45 percent level, even when using a spray gun capable of much better transfer. In some areas of the United States, refinish technicians are now required to use spray guns that operate at low tip pressures in order to increase the transfer efficiency of the operation. Without training, however, most technicians still spray at much lower rates of efficiency than the low-pressure guns are designed for. If all spray technicians were to spray at the 65 percent transfer efficiency level, the savings in costs of coatings alone would be 10 to 20 percent.

Transfer Efficiency and the Spray Gun

There are some spray guns that spray at 65 percent transfer efficiency; however, most guns spraying at this level or higher are **high volume, low pressure (HVLP) guns**. These guns are ones whose cap pressure does not exceed 10 psi. This measurement is taken with a special spray gun cap pressure gauge. See **Figure 21-46**. Though some of these gauges are capable of measuring both cap and horn pressures, this discussion will use the cap pressure measurement.

In their procedure sheets, paint makers have recommended spray pressures ranging from 8 to 10 psi at the cap for applying coating. The only truly accurate way to measure this pressure is to attach a cap pressure gauge to verify that the pressure set at the wall regulator will not exceed the recommended cap pressure. The wall regulator will be set higher than 8 to 10 psi due to the connectors, air hose, and **valves** that the air must pass through. Measuring pressure at regular occasional intervals will ensure the accuracy of the settings.

Because the HVLP guns are set at a lower pressure (psi), they will require much higher volumes of air. The wall regulator must be able to meet or exceed the required cubic feet per minute (CFM) needed in the booth. That is, if two spray guns requiring 25 CFM and a fresh air respirator requiring 30 CFM are being used, the wall regulator must be capable of at least 80 CFM, if they are all to be operated at once. In addition, the air hoses and quick connections should be capable of delivering the needed 25 CFM to that gun. Large capacity quick connections with a minimum of 3/8-inch hose are needed to supply air to the gun to operate as an HVLP.

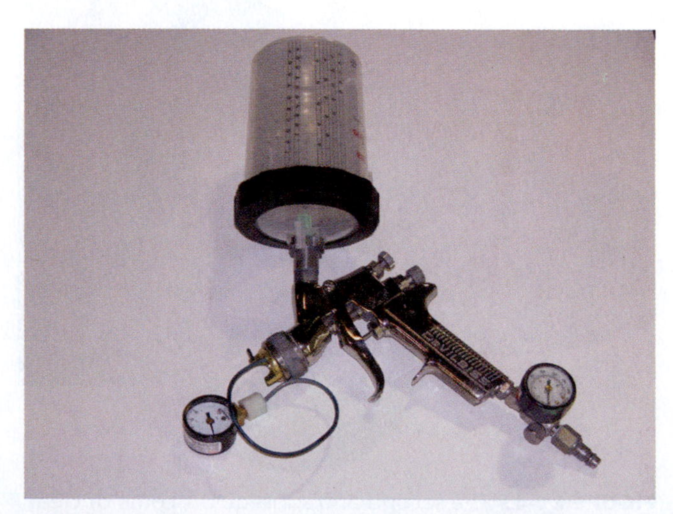

FIGURE 21-46 This gun has tip and inlet pressure gauges.

Unfortunately, when the wall regulator hoses and quick connector are not capable of the needed CFM capacity, the gun still operates, but not properly, and may not be able to be adjusted to eliminate excessive spray texture (orange peel).

There are also spray guns which will spray at the 65 percent transfer efficiency rate though they are not truly HVLP guns. It should also be noted that not all gravity-feed guns are HVLP guns, nor are all suction feed guns high pressure guns.

Transfer Efficiency and the Environment

It has been stated that HVLP guns have the highest transfer efficiency of all spray guns produced for applying coatings, and that some states have mandated their use. The reason is that HVLP guns emit far lower amounts of toxic substances into the atmosphere than non-HVLP guns. The national rule (see Chapter 27) limits the amount of volatile organic compounds (VOCs) that can be present in reducers, and it also limits the amount of solvents that can be added to a coating to prepare it for spraying. With these limitations and the use of HVLP spray guns, the VOCs emitted into the environment are lessened. As fewer VOCs are emitted into the atmosphere, the ground level ozone pollution will be reduced. High levels of atmospheric ozone or smog affect people with lung diseases. Reducing the VOCs emitted into the atmosphere will reduce smog. Transfer efficiency can be increased by technique as well. By banding the part to be painted (spraying the perimeter of the part first), a technician can reduce the amount of material needed to obtain full coverage. Banding directs the paint at the target with less overspray going into the air, thus increasing the painter's transfer efficiency and reducing environmental load. Banding reduces cost as well by using less material.

Summary

- Spray techniques are among the most critical areas of refinishing.
- Spray gun adjustments are essential if technicians want to obtain a professional refinish application.
- Air pressure is one of the most critical settings in relation to proper atomization.
- Other important settings include fluid adjustment, spray width, distance, and stroke/movement.
- Technicians should keep the spray gun perpendicular to the vehicle at all times.
- Distance, arching, pitch, rotation, gun triggering, and lead and lag distances are other positional concerns when spraying a vehicle.
- Refinish techniques include applying single-stage solid colors and single-stage metallic colors, basecoats, clearcoats, and undercoats.
- Transfer efficiency is the measure of how much of the finish material placed in the spray gun is transferred to the target surface.

Key Terms

banding
basecoat blending
blendable match
cubic feet per minute (CFM)
cutting in
cycle time
depth of reflection
detailing
dry spray
fascia
flash time
full coverage
full hiding
full wet coat
gas out
ground effects
gun fan
high volume, low presssure (HVLP) guns
hue
medium wet coat

melt in
metallic lie
mist coat
OEM
off-triggering
orange peel
paint striping
pounds per square inch (psi)
reverse blending
sprayout panel
transfer efficiency (TE)
transverse viewing
valve
volatile organic compounds (VOCs)
wet bedding
wet line
zebra striping

Review

ASE Review Questions

1. Technician A says that when adjusting a spray gun, the most accurate place to read pressure is at the wall regulator. Technician B says that when adjusting a spray gun, the most accurate place to read pressure is at the air cap. Who is correct?
   A. Technician A only
   B. Technician B only
   C. Both Technicians A and B
   D. Neither Technician A nor B
2. All of the following are places where spray gun air pressure can be measured EXCEPT:
   A. Spray booth gauge (wall regulator).
   B. Gun gauge.
   C. Compressor tank gauge.
   D. Air cap gauge.
3. Technician A says that when adjusting a spray gun, the atomized coating coming out of the gun cap should be a fine mist with small droplets of spray. Technician B says that when refinishing a vehicle, its original orange peel texture should be matched. Who is correct?
   A. Technician A only
   B. Technician B only
   C. Both Technicians A and B
   D. Neither Technician A nor B
4. Technician A says that when spraying, a 50 percent overlap stroke should be used. Technician B says that when spraying, a 75 percent overlap stroke should be used. Who is correct?
   A. Technician A only
   B. Technician B only
   C. Both Technicians A and B
   D. Neither Technician A nor B
5. Technician A says that refinish technicians will always spray finish to a full wet coat. Technician B says that some spray application techniques will call for medium wet coats. Who is correct?
   A. Technician A only
   B. Technician B only
   C. Both Technicians A and B
   D. Neither Technician A nor B
6. Technician A says that basecoats do not need to be covered with clearcoat. Technician B says that single-stage finish cannot be covered with clearcoat. Who is correct?
   A. Technician A only
   B. Technician B only
   C. Both Technicians A and B
   D. Neither Technician A nor B
7. Technician A says that when spraying larger areas the technician should choose a clear (or sometimes its additives) that cures at a slower pace. Technician B says that dry spray may not be repairable with detailing. Who is correct?
   A. Technician A only
   B. Technician B only
   C. Both Technicians A and B
   D. Neither Technician A nor B
8. All of the following are blending techniques EXCEPT:
   A. Wet bedding.
   B. Reverse blending.
   C. Two gun blending.
   D. Standard blending.
9. Technician A says that some paint manufacturers do not recommend clearcoat blending. Technician B says that some automotive manufacturers do not recommend clearcoat blending. Who is correct?
   A. Technician A only
   B. Technician B only
   C. Both Technicians A and B
   D. Neither Technician A nor B

10. Technician A says that applying primers by rolling can reduce both cycle time and coats. Technician B says that rolling is a very easy technique to use and no special application techniques are needed. Who is correct?
    A. Technician A only
    B. Technician B only
    C. Both Technicians A and B
    D. Neither Technician A nor B
11. Technician A says that transfer efficiency is how efficiently technicians complete each vehicle. Technician B says that transfer efficiency is how efficiently the paint is transferred from the gun to the part being coated. Who is correct?
    A. Technician A only
    B. Technician B only
    C. Both Technicians A and B
    D. Neither Technician A nor B
12. All of the following are spray techniques EXCEPT:
    A. Wet bedding.
    B. Single stage.
    C. Blending.
    D. Banding.
13. Technician A says that even colors that are severely mismatched can be successfully blended. Technician B says that a color that is considered a blendable match is only slightly different from the original color. Who is correct?
    A. Technician A only
    B. Technician B only
    C. Both Technicians A and B
    D. Neither Technician A nor B
14. Technician A says that panel painting is a technique used when the color is known to be a perfect match to the existing finish. Technician B says that when applying color using the panel painting technique, gun angles must be kept perpendicular to the panel, with the proper distance maintained at all times. Who is correct?
    A. Technician A only
    B. Technician B only
    C. Both Technicians A and B
    D. Neither Technician A nor B
15. Technician A says that when spraying coatings to full wet, the possibility of developing runs becomes greater. Technician B says that when spraying larger areas, the technician should choose a clear that cures at a slower pace. Who is correct?
    A. Technician A only
    B. Technician B only
    C. Both Technicians A and B
    D. Neither Technician A nor B

Essay Questions

1. A vehicle is about to be refinished. The left fender, which is new, will be finished. The hood and front door will be blended. Describe the plan of attack needed for refinishing this vehicle with basecoat/clearcoat finish.
2. Describe the steps and order needed for proper spray gun adjustment.
3. Define atomization.
4. Name the most common type of unwanted texture, and explain the main causes and remedies for it.
5. What is gun triggering, and why should a technician use it?

Topic Related Math Questions

1. A shop has three paint booths with air supplied respirators and spray gun hookups. Their HVLP guns require 21.5 CFM, and their fresh air supplied respirators require 9.75 CFM. Three painters are using their supplied air respirators, but only two are spraying with their HVLP guns. What would be the total required CFM for the three painters?
2. A technician is spraying a vehicle with three coats of basecoat and two coats of clearcoat. There is a 12-minute flash time between coats, and it takes the painter 9.5 minutes to complete each coat. How much time will it take the painter to complete the refinish task?
3. A spray gun needs 22 psi of pressure; the air must travel through a flexible air hose 50 feet long, and the air hose loses 12 psi every 25 feet of length. At what pressure must the wall regulator be set?

Critical Thinking Question

1. A technician is about to blend a silver color vehicle. The technician believes the blend to be difficult. Though it appears to be a blendable match when compared to the sprayout panel, it is not perfect. The air temperature is 95°F with 98 percent humidity. The vehicle is masked to refinish the left front fender and blend into the hood and door. The normal flash time is 12 minutes between each coat and 30 minutes dry time before clearcoating. The clearcoat will normally need 15 minutes between coats. The sprayout panel shows that it will need three coats of basecoat and two coats of clearcoat. The time is now 4 o'clock, and the shop closes at 5 o'clock. Explain what the technician must take into consideration in planning, and determine if the job will be completed by closing time.

Lab Activities

1. Air pressure adjustment:

   With the following items, the student should set up and adjust an HVLP spray gun to10 psi at the cap.

   - HVLP gravity spray gun
   - Cap spray gauge
   - Gun inlet air gauge
   - Hose with connectors

2. Fluid adjustment:

   With the following items, the student should set up and adjust an HVLP spray gun to full width, proper atomization, and 10 psi at the cap.

   - HVLP gravity spray gun
   - Cap spray gauge
   - Gun inlet air gauge
   - Hose with connectors
   - Liquid mask reduced with water to a viscosity of 17 to 19 seconds, and tinted with food coloring

3. Spray technique:

   With the previously adjusted gun loaded with a spray mask, the trainee should spray a fender using good gun positioning and techniques such as:

   - Distance
   - Perpendicular to the surface
   - Rate of travel
   - Triggering
   - Proper lead and lag
   - Banding
   - Flow
   - Overlap

4. Spray Technique:

   After the student has practiced and become comfortable with the spray techniques in item 3, the student should be video-recorded to help critique himself or herself.

Chapter 22

Refinishing Shop Equipment

OBJECTIVES

Upon completing this chapter, you should be able to:

- Know, understand, and use the safety equipment necessary for the task.
- Identify the components of an uncompressed air system.
- Identify and explain why air supply size and configuration is critical.
- Calculate compressed air needs for a shop.
- Identify and explain the components of an air cleaning system.

INTRODUCTION

The refinish department of a typical collision repair facility has many large pieces of equipment which need to be maintained on a regular basis. Some equipment allows complex air handling: air compressors, breathable air supply compressors, air monitoring equipment, air-drying equipment, and air handling and control equipment, along with air lines and hoses. The facility must also maintain spray booths, prep decks, and baking and curing equipment such as ultraviolet (UV), infrared IR, convection hot air, and venturi blowers for waterborne curing.

Compressed air needs in the refinish department are complex, and three different types of equipment supply air for these various needs. Clean, breathable air for the air supply respirators is needed for the technicians. Compressed air is needed for operating power tools which require higher air pressure but less volume, and high volume, low pressure air is needed for HVLP spray guns. Because three different types of air are needed—in some cases requiring three different types of air compressors—there may be three different maintenance schedules for the compressors. The air must be cooled, dried, and cleaned differently for all three needs. All of these variables require a technician to have knowledge of an array of equipment information and maintenance requirements.

Within the spray environment, the technician must maintain the equipment properly so it will control the dust, dirt, and debris; airflow; temperature; light; and heat (and with some spray booths, coolness and humidity). Technicians should keep and follow a maintenance schedule to assure that the equipment is working at its best and will not prematurely break down. For example, the air pressure inside the booth must be kept at a slightly positive pressure, in order to keep the dirt outside the booth from getting in, thus keeping the finish cleaner. This is one of the many maintenance checks to be performed regularly.

In short, the refinish technician must not only be knowledgeable about how to prepare and refinish vehicles, but also must know, understand, and maintain a vast amount of equipment.

This chapter will explore the different types of refinish equipment, their care and maintenance, and ways to properly test and operate each type.

SAFETY

In the refinish department, safety often refers to personal safety: the proper use of personal protective devices and environmental safety when working with chemicals. Refinish materials—and the chemicals they are made from—pose personal and environmental protection concerns. The refinish technician is governed by—and must comply with—many local, state, and federal laws concerning protection.

Each product must be accompanied with a material safety data sheet (MSDS). The MSDS includes sections on:

- Hazardous ingredients
- Physical data
- Fire and explosion hazards
- Reactive data
- Health hazard data, or toxic properties
- Preventive measures
- First aid measures
- MSDS preparation information

Technicians must read, understand, and follow the recommendations in the MSDS for the protection of themselves, their fellow workers, and the environment.

In addition to being aware of chemical hazards, technicians must also be careful when maintaining equipment. When cleaning areas that may have paint residue, for example, the technician should use the same personal protective devices as outlined in the original MSDS for the paint product.

Technicians must use care and caution to ensure personal safety when working on electrical equipment. For example, the reminder "Lock out, tag out" indicates that circuit breakers for the equipment are to be disconnected, and the breaker should be locked so it cannot be reconnected while someone is working on it. Then, the electrical box is tagged to let others know that it is being worked on and should not be reconnected.

Each employee should be trained to carry out equipment maintenance. If not properly trained, the technician should not perform it. Technicians should always read, understand, and follow all safety directions provided by equipment manufacturers.

All safety guards for each piece of equipment should be properly installed and kept in good working order. Technicians should not remove or disable safety guards.

AIR COMPRESSORS

Air compressors come in different styles, operate under different conditions, and provide different types of air pressures or volumes. Each style compresses air differently and can supply air of different qualities. Each of these compressors has advantages which best suit a particular air supply need. For example, piston compressors supply higher pressure at lower volumes, while rotary compressors supply larger volumes at a lower pressure. Rotor vane or diaphragm compressors can provide the **class D breathable air** needed for **air supply respirators**. A typical repair facility may have all three compressor types within the working environment.

Piston Compressors

Piston compressors are by far the most common type of air compressors in the collision repair facility; in fact, they are the most common type of air compressor in general. See **Figure 22-1**. They work much like an automobile engine, with valves and a piston that travels up and down. When the piston travels down, the intake valve opens. Outside air is drawn into the compressor. As the piston starts to travel up, the intake valve closes and the exhaust valve opens. The air that was taken in is then compressed into a smaller area. See **Figure 22-2**. The compressed air is released into either a holding tank or directly into the air lines for use.

Air compressors have rings around the compressor's pistons. These rings work to create a seal so that the air being compressed will not escape down into the crankcase. In the bottom of the crankcase, a piston air compressor has

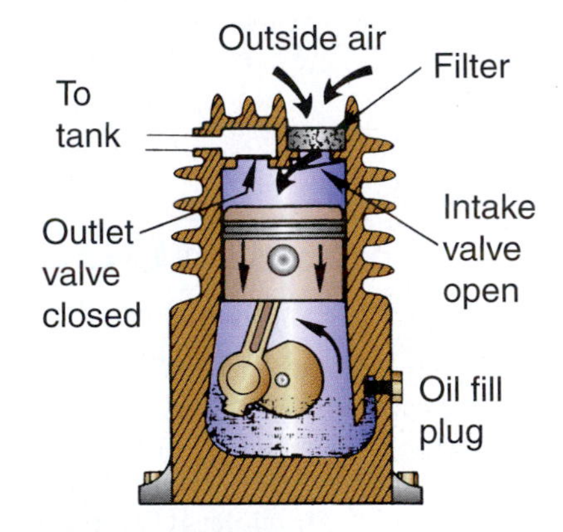

Downstroke—air being drawn into compression chamber

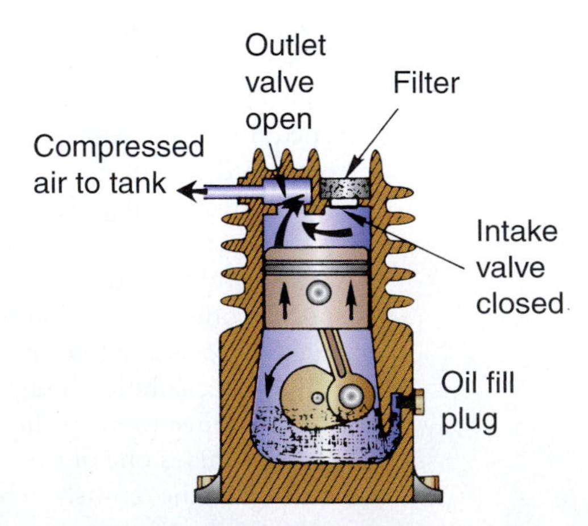

Upstroke—air being compressed and forced to tank

FIGURE 22-1 Two-stage piston air compressors are by far the most common type found in collision repair facilities.

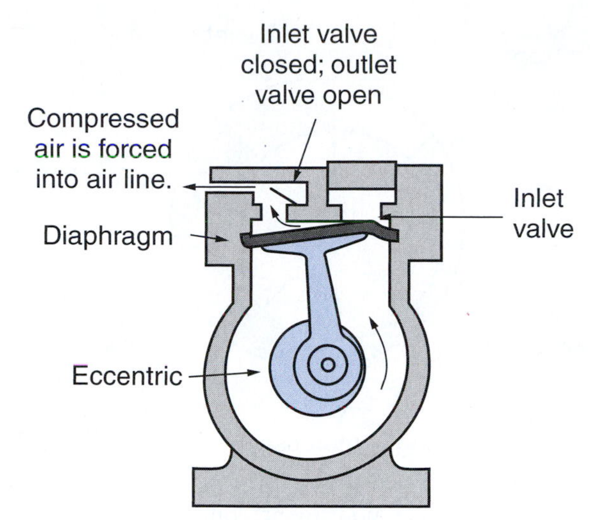

FIGURE 22-2 A single stage compressor brings in atmospheric air and compresses it to the desired pressure. Then it is either stored in a tank or used in the shop.

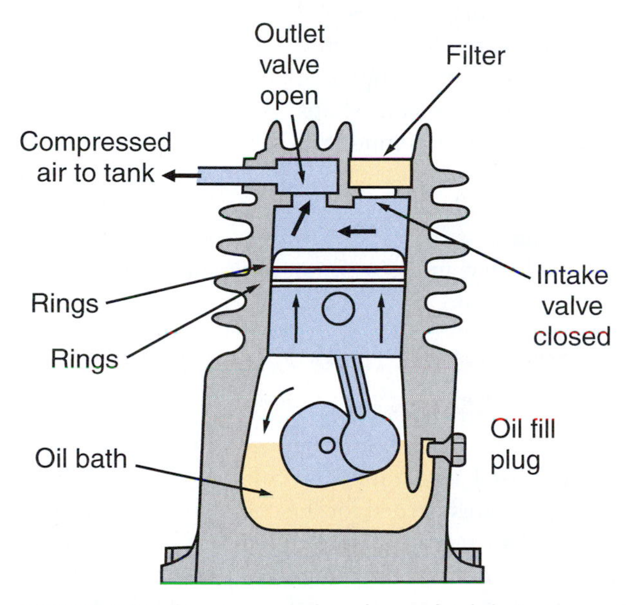

FIGURE 22-3 Compressor rings keep the lubrication oil in the bottom from getting into the compressed air. As the rings wear, oil may "blow by" the rings, thus contaminating the compressed air and causing paint defects.

an oil bath that lubricates the compressor as it runs. See **Figure 22-3**. As the crank and piston travel up and down, the oil splashes on the cylinder walls. The compressor pistons also have "oil rings" which help remove the oil from the cylinder wall so it does not get into the compressed air. If either the compression or oil ring becomes worn, however, the compressor may not work to its capacity, or it may develop oil contamination.

As the outside air is taken in, it also contains some water as humidity. Although the air can be compressed, the water will not be compressed. On very humid days, as the compressor provides air, it will have water as a by-product. This water will find its way into the air tank where most of it can be drained. Sometimes, though, water may find its way into the air lines, and must be regularly drained out of the lines. This is why most air supply lines have a drain. This draining will need to be done occasionally, but more often in hot and humid weather. Since water that is left in the compressed air can cause finish problems, it should be eliminated.

Piston air compressors come as either a single-stage compressor or a two-stage compressor.

Single-Stage Compressors. **Single-stage compressors** are generally smaller than other compressors. Although they may have more than one piston, the compressing process only occurs once. They are generally rated from one to five horsepower, though horsepower is not the best way to judge the amount of air pressure that is being produced. A single-stage air compressor which uses a five horsepower engine—either gas or electric—may not produce many cubic feet of air per minute. **Cubic feet per minute**, or **CFM**, measures the volume of air being produced. The amount of CFM will change with the amount of pressure (**pounds per square inch**, or **psi**) that the compressor produces. These two measurements of pressure to volume are "inversely related." That means that if one goes up, the other goes down. The best way to judge an air compressor is to match the air requirements to the amount of air being produced. If the shop requirements are 72 CFM at 100 psi, then the compressor must provide the needed pressure and volume. Generally, a two-stage compressor can produce larger amounts of psi and CFM than a single-stage compressor with the same size (horsepower) power supply.

Two-Stage Compressors. A **two-stage compressor** is similar to a single-stage air compressor in that it has pistons with rings and oil in the crankcase. What is different is that there are two different-sized pistons. The first one—the larger—compresses the air the same way a single-stage compressor does. However, instead of venting air into a holding tank or to the supply lines, it sends air to another small piston to be compressed again in a second stage. See **Figure 22-4**. This second stage compresses air at a higher pressure using less energy.

Two-stage air compressors can be very large with multiple pairs of cylinders. In some factory settings, compressors can be 200 to 300 horsepower—and larger, if needed. If a large amount of high pressure air is needed, a piston compressor is suited for the job. If large volumes (CFM) at lower pressures are needed, the best compressor is a rotary screw or turbine compressor.

Rotary Screw Compressors

The **rotary screw compressors** have long been the compressor of choice for industrial applications. They provide pulsation-free, continuous air, and are energy efficient at full load. They can operate at a 100 percent **duty cycle**, which means that they do not need to shut down.

Rotary screw compressors operate by having air enter a sealed chamber, where it is trapped by two contra-rotating

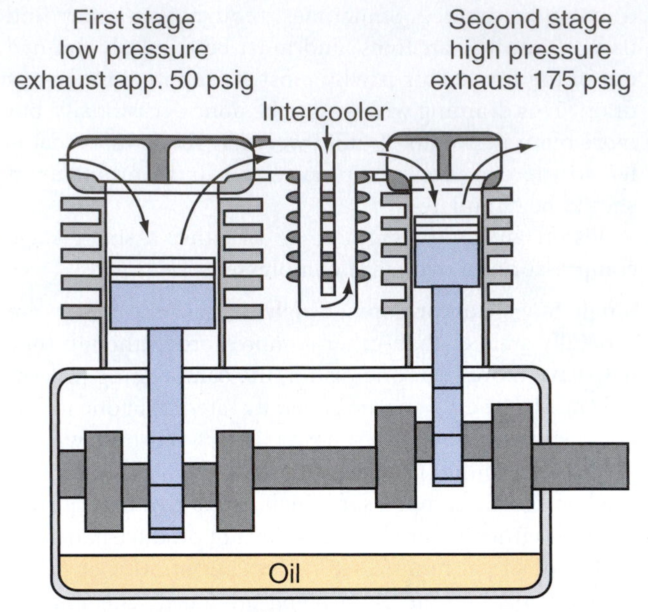

FIGURE 22-4 In a two-stage compressor, the first stage compresses the air partially, and then passes through a cooler before going to the second and final compression cycle. Adapted from Binks/DeVilbiss.

rotors. With a screwing motion, the rotors compress the air as it passes through. See **Figure 22-5**. This simple mechanism allows air to be compressed at about one-half the temperature of piston-compressed air. With lower operation temperatures and fewer moving parts, this type of air compressor can operate at full-load, continuous operation for years without maintenance problems. Rotary compressors are also significantly quieter than piston compressors.

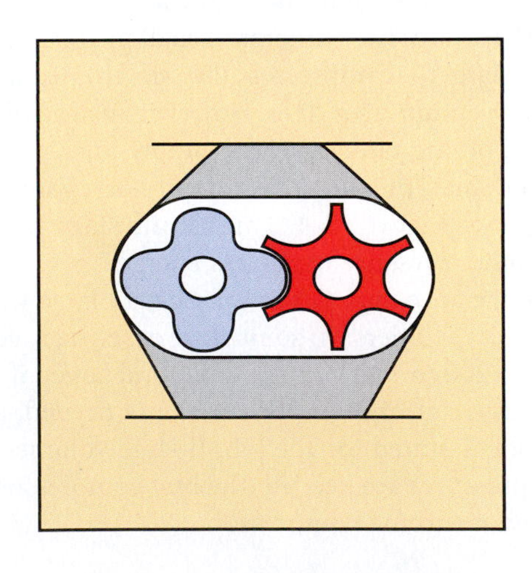

FIGURE 22-5 More efficient than piston compressors, screw compressors produce more volume of air with less noise.

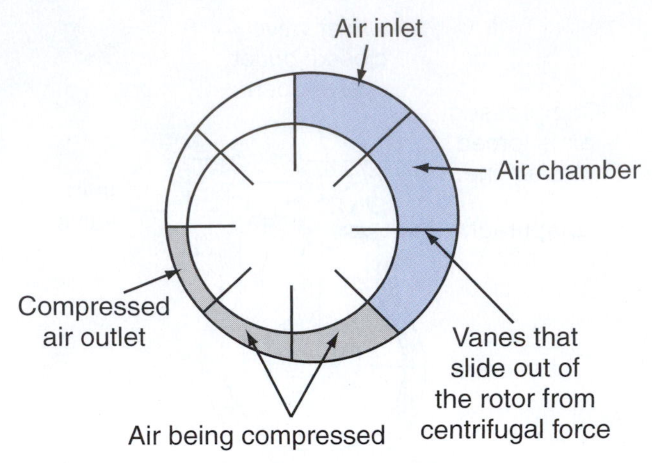

FIGURE 22-6 Rotary vane compressors rotate in a decreasing radius compartment, compressing the air before discharge.

As collision repair facilities' compressed air needs have changed to higher volumes of air per worker—and with the increased use of **high volume, low pressure (HVLP) spray guns**—rotary compressors have become the compressor of choice in many shops.

In the past, rotary compressors were not used because of a preserved oil contamination problem. Thanks to design improvements, oil contamination is no longer a problem. As these compressors are able to provide maintenance-free, clean air at high volumes with quiet operation, they are rapidly becoming more popular.

Rotary Vane and Diaphragm Compressors

Rotary vane and **diaphragm compressors** are not commonly used to produce the volumes of high (above 100 psi) pressures needed for the collision repair facility's compressed air needs. They are often used in producing class D **breathable air** for supplied air respirators. These small machines can operate at low temperatures and with a continuous duty cycle that eliminates the need for an **air holding tank**. See **Figure 22-6**. They are generally connected directly to the air supply lines, switched on, are run continually when needed, and then are switched off. The volume and pressure of air is sufficient for air supply. Because of the clean, contamination-free operation, the resulting air can be used for a breathable air supply.

AIR SUPPLY REQUIREMENTS

The air needs of a collision repair facility can be complex: How many cubic feet of air (CFM) are needed to supply all the tools that will be in operation? At what pressure must each worker operate, no matter where that worker is in the supply line? And what quality of air is needed for the work being done?

Because of all these varying needs, air supply and delivery has become very complex.

Breathable Air Requirements

As refinish material uses isocyanine as a catalyst, air supply respirators are now needed. These positive airflow respirators supply clean class "D" breathable air to the technician's hood, eliminating the need to clean or filter the contaminated air from within the spray booth.

Though class "D" breathable air can be produced from the shop's compressed air, through filtration for contaminants and the use of a carbon monoxide monitor, breathable air is often generated from a different compressor.

Nine grades of air are currently listed by the U.S. Compressed Gas Association. A, B, and C are grades used in the medical facilities and are not suited for supplied air respirators. The fourth, grade D, must contain 19.5 to 23.5 percent oxygen, which is the same as atmospheric oxygen levels. The carbon monoxide levels must be less than 10 parts per million (ppm), carbon dioxide must be less than 1,000 ppm, and oil contamination levels must be less than 5 milligrams per meter cubed. Due to these strict regulations, instead of monitoring the large amounts of compressed air needed for the repair facility, businesses produce breathable air in smaller quantities in a system that can be closely monitored and maintained.

How Much Air Is Needed?

To determine what volume of air will be needed, one must determine which air tools—and how many of them—may be operating at the same time. If the air supply is not sufficient to operate all the tools at the same time, the air reserves in the compressed air tank may become depleted. Piston compressors have a specific duty cycle rating, which means that they should not run continually. The air reserve tank serves to hold air when the compressor shuts off. When the supply is depleted, a pressure switch signals the compressor, which starts again and will refill the tank.

The amount and pressure of air needs must also be determined. As is the case for air tools, will there be a need for air of higher pressures (80 to 100 psi) but with lower volume per tools? Or will the needs be, as with spray guns, for lower pressure (30 to 45 psi), but at significantly higher volumes per tool?

To calculate the air supply needs, the number of tools and their air volume needs must be calculated; then these needs must be matched with an air compressor and type to meet the needs.

Because of the introduction of high volume, low pressure spray guns in the refinish department and their specific needs for high volumes of air, some collision repair facilities have divided the air supply into two zones. In the body shop, a piston compressor is used to supply high pressure. In the refinish area, a rotary compressor is used to supply the higher volume needs of that section. In the refinish department, cleaner and moisture-free air is critical. With the two departments separated, the air requirements can be monitored more closely. In the refinish department, a third system may be provided which also produces breathable air.

Air Holding Tank

So far, this chapter has concerned itself with discussing the different ways to produce compressed air. The next sections will explore the storage and distribution of air. When air is compressed, depending on the type of compressor, it will be stored in a holding tank. The size of the holding tank will differ, depending on how many tools will be operated at one time. Piston air compressors produce air in cycles. Each time the piston compresses air, a large amount of highly compressed air is driven into the storage tank; then, as the piston goes down, drawing in air for another compression, it delivers no air. This is why piston compressors are sometimes called precipitating compressors. They deposit air into the holding tank in cycles of air and no-air bursts.

Piston compressors also have a duty cycle, which means that they can work for a certain percentage of time. Then they must stop and rest to cool. After that, they can start up again to compress air for a period of time. If a compressor has a duty cycle of 65 percent, it can work for 65 percent of the time and rest 35 percent of the time. Stated another way, such a compressor will provide 39 minutes of work per hour, and will require 21 minutes of rest per hour.

As can be seen, the holding tank for a piston compressor must be large enough to have reserve air to supply the working tools while the compressor is in its rest cycle. It also needs to be large enough to cancel out the cycles of air and no-air that the compressor provides.

With a rotary screw compressor, the air supply is continuous, and it has a 100 percent duty cycle. Therefore, the only need for a holding tank is to have enough reserve air so that tools will have sufficient compressed air at all times. If the compressor has the capacity to run 10 tools, and—for a short period of time—11 are hooked up to it, the air in the holding tank will be sufficient to operate all 11 tools.

Rotary vane or diaphragm compressors, which are used primarily to supply breathable air to air supply respirators in collision repair shops, are "**on-demand compressors**." This means that because the air supply is not cyclic and they can run at a 100 percent duty cycle, they come on and stay on as long as they are needed, then shut off. Therefore, they do not generally have a holding tank.

Though illustrations often show the holding tank with the compressor attached to it, it can also be configured in such a way that the holding tank is located away from the compressor. See **Figure 22-7**. In industrial or large collision repair shops, the compressor may be in a compressor room (to contain the noise), while the storage tank is located in another room. See **Figure 22-8**.

With the air compressed and stored, the next step is to deliver the air to the work area in the amounts (CFM) and pressure (psi) needed.

FIGURE 22-7 This compressor is configured to have the holding tank below.

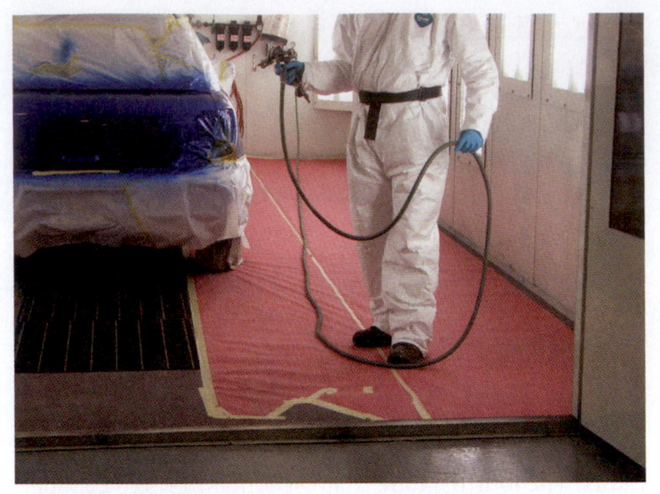

FIGURE 22-9 This 25-foot paint booth has 35 feet of air hose, a typical size for a spray booth.

FIGURE 22-8 Some compressors use a remote storage tank.

Line Size and Layout

The amount of air that can travel through an air line is limited by the **air line size**. If the air line is only 1/4 inch, it stands to reason that less air can travel through this air line than one that is 1 inch or 1 1/2 inches in size. Therefore, if the minimum line size needed is 3/8 inch, the supply lines delivering it must be that size or larger.

The material used to supply air must also be taken into consideration. If the supply line is all hard, such as steel, there will be minimum air pressure (psi) lost. However, if the air supply line is flexible, as with a rubber air hose, the flexibility of that hose will cause the air pressure to drop. Specifically, a rubber spray hose with a 3/8-inch inside diameter will lose 10 pounds of pressure for every 25 feet of hose. Continuing this example, typically a spray booth air hose is 35 feet long; therefore, it will lose 16 psi. See **Figure 22-9**. This loss of air pressure occurs when the pressure inside the hose pushes on the flexible air line and the line expands a little, thus losing pressure.

TECH TIP If technicians hold a flexible air hose firmly in their hand when it is not hooked up to air, and someone plugs it into an air source, they can feel the line get slightly bigger as the air rushes into the air hose. This movement in a flexible air line is where some air pressure is lost.

The air supply line leading to the workplace that requires a minimum of 3/8-inch inside diameter means that the line leading up to the 3/8-inch hose must be bigger. Typically, a collision repair shop will start with a 1 1/2-inch line as the main supply line, then reduce to a 3/4-inch line for the "drops" (the line from the main run to the wall regulator), then to 1/2 inch into the regulator, and 3/8 inch out of the regulator to the hose. See **Figure 22-10**.

Though little air pressure is lost using hard air supply lines, in large facilities where air must travel long distances, *some* air pressure is lost, and air supply lines may need to be larger than 1 1/2 inches. When designing an air supply line layout, it is best to consult a pipe-sizing chart to calculate the proper pipe sizes. See **Figure 22-11**.

Though air should be dry to operate tools properly, and in the refinish area air must not have any water in it (air drying will be covered later in this chapter), some moisture may occasionally get into the air supply lines. To be able to remove this from the air supply lines, an air drainage line—which is the lowest point in the supply line—should be available for draining occasionally. See Figure 22-10. The air "drop" which comes off the main supply line should not be plumbed to drop down as the name implies, but should be sent up, then elbowed downward to the air regulators. This configuration will minimize water being sent into the work regulator, and will send it instead to the more downhill "dependent" air drain line, or the lowest part of the air supply system.

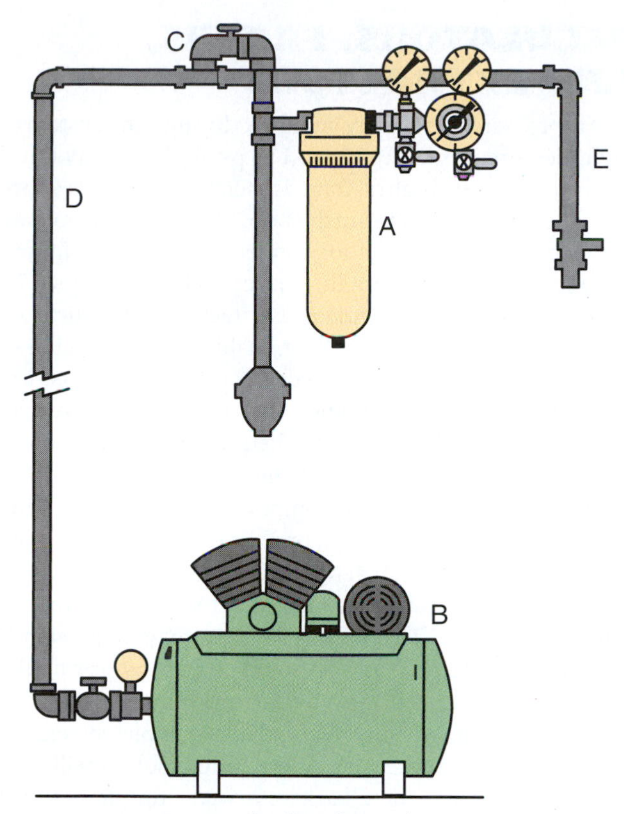

FIGURE 22-10 Drops, the line from the main run to the wall regulator, are part of this air line configuration. Adapted from Binks/DeVilbiss.

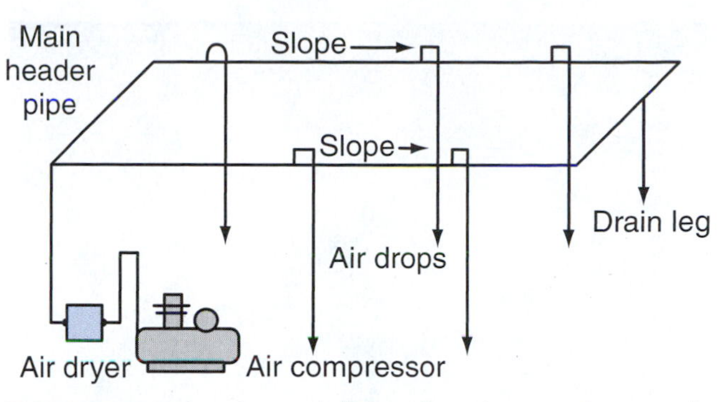

FIGURE 22-12 In a loop air line, all outlets receive equal air pressure. Adapted from Binks/DeVilbiss.

In many collision shops, the air supply delivery system is configured in what is called a terminal-end design. See Figure 22-10. This means that the air which comes from the compressor and reservoir tank will run in one direction only. If a large demand is placed on the first air outlet, the air outlets which are "downstream" from the first one will have a decrease in air pressure and volume. To keep this from occurring, the better design is a "piping loop" system. See **Figure 22-12**.

A piping loop is a system that is connected in a loop fashion. The air leaves the holding tank and enters the line in the center; at this point, the air can travel in either direction, depending on the demand. If a large demand is placed in one air outlet, the air can travel in another direction to reach other air outlets. With this system, there are no "downstream" air outlets, and no **air line pressure drop** is noted.

In some systems, as in the drawing in Figure 22-12, a second tank is placed in the line, which assures that there is no air pressure drop in the line. With this setup, when air demands are placed on the system in any location, air can flow freely in any direction to supply the demand.

Types of Air Lines

Air lines are generally made of black steel threaded piping. The installation of this type of line is plumbed into the building on the outside of the wall, or sometimes extended down from the ceiling on retractable hose lines. See **Figure 22-13** and **Figure 22-14**.

Length of pipe lines in feet

CFM	25	50	75	100	200	300
5	1/2"	1/2"	1/2"	1/2"	1/2"	1/2"
10	1/2"	1/2"	1/2"	3/4"	3/4"	3/4"
15	1/2"	3/4"	3/4"	3/4"	3/4"	3/4"
20	3/4"	3/4"	3/4"	3/4"	3/4"	3/4"
30	3/4"	3/4"	3/4"	3/4"	1"	1"
50	1"	1"	1"	1"	1"	1"
60	1"	1"	1"	1"	1 1/4"	1 1/4"
100	1 1/4"	1 1/4"	1 1/4"	1 1/4"	1 1/2"	1 1/2"

FIGURE 22-11 This air line size chart has air pressure drop indicated per foot and diameter of line. Courtesy of Binks/DeVilbiss.

FIGURE 22-13 Air line run with one pressure regulated outlet, one that is not regulated, and a drain outlet.

FIGURE 22-14 Retractable hose lines can extend from the ceiling.

Though air pressure of 150 psi is well below the air rating of schedule 40 PVC plastic piping (200 psi), manufacturers of PVC plastic piping do not recommend the use of their product for air supply. Plastic, if used for air pressure, would break the pipe. It could shatter into small shards of plastic that could cause serious injury or damage.

Although aluminum piping has been developed for air supply, it is generally used in small home shop installations, and is not generally seen in collision repair shops at this time.

REGULATORS, FILTERS, AND EXTRACTORS

Air pressure is typically supplied through the system at high pressure, generally 150 to 175 psi in the supply lines. This pressure is higher than is needed for **pneumatic tools** in the collision refinish shop. To efficiently operate air tools such as a dual action sander (DA) or an air buffer, the recommended operating air pressure is 90 psi. To lower the pressure, a regulator is attached to the airdrop. See **Figure 22-15**. The regulator enables the technician to adjust the air pressure as needed for different tools. Because some air contamination can be in the air supply lines, the air regulators are attached to both filters and extractors. **Air filters** remove elements such as debris that may not have been cleaned completely from the incoming air at the compressor, or debris which may come from older air lines. **Air extractors** are devices which will remove oil or moisture that may find its way into the air lines. See **Figure 22-16**. Oil in air lines may be present if aging air compressor rings let some oil pass by during the compressing of the air. A well-designed system will have a cooler/dryer in it, but may not be able to completely remove all the moisture from the air. As the water cools, it will condense from mist or dew back into water droplets. If not forced to the terminal end for drainage, it should be eliminated at the workstation airdrop.

Air Transformers

Air regulators and **air transformers** are devices used to help transform high air pressure into lower air pressure. Though such a device is often called a regulator, one of the

FIGURE 22-15 An air regulator with water trap will help remove moisture from the lines.

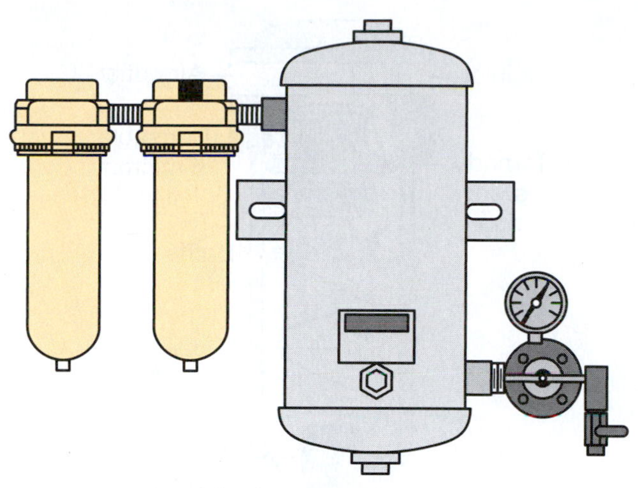

FIGURE 22-16 Methods to prevent contamination in the air lines include the air filter (left), water extractor (center), and desiccant moisture filter (right).

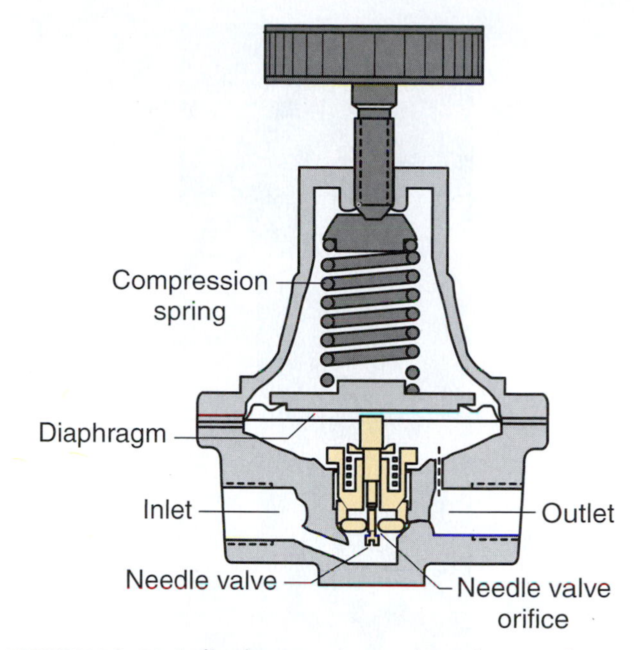

FIGURE 22-18 A diaphragm air regulator is sometimes called a "transformer." Adapted from Binks/DeVilbiss.

key functions of the transformer is to change psi without changing or affecting the CFM of the air from the supply line. A gate valve will regulate the air pressure, but while doing this, it also restricts the amount of CFM that can pass through the opening that it has made smaller. See **Figure 22-17**. Also, when using a gate valve type regulator when the air pressure is not in use (with the tool stopped), the pressure will return to line (high) pressure. When the tool is restarted, it will—after a short period—go back to the lower pressure, because of the line size restriction.

A transformer works with a diaphragm and a spring. See **Figure 22-18**. Air on one side of the diaphragm pushes against the other. Depending on the amount of pressure the spring exerts on the other side, the air pressure is regulated without restricting volume flow. Also, because the pressure exerted by the spring is constant, the pressure will not fluctuate to the higher line pressure.

When choosing a device to regulate air pressure, the transformer style is more consistent and will supply a more regulated pressure than a gate valve regulator.

Debris Filters

Though some debris filters and moisture extractors are packaged in the same device, they may also come separately. See **Figure 22-19**. The debris filter is the final cleaning device used to eliminate any dirt, rust flakes, or other foreign objects in compressed air. It can be a device which looks like a copper filter—as would be seen in an automobile fuel filter—or one made of dry paper with fine holes in it designed to filter out objects, similar to an automobile oil filter. See **Figure 22-20**. Each of these types of

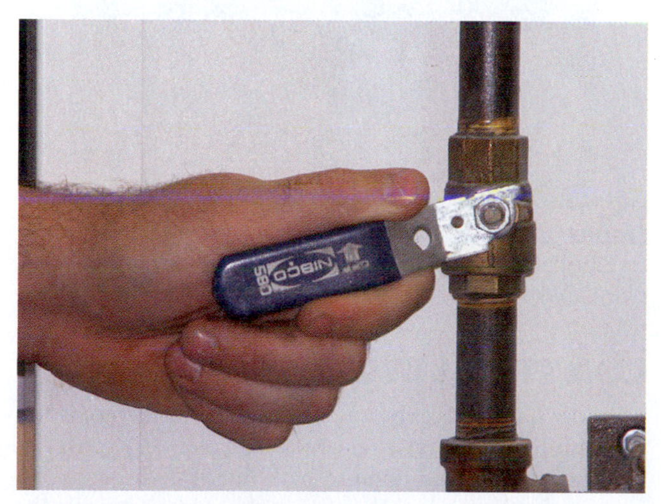

FIGURE 22-17 An air line shutoff gate valve will regulate the air pressure. However, while doing this, it also restricts the amount of CFM that can pass through.

FIGURE 22-19 Debris filters can include an air filter, extractor, dryer, and regulator.

FIGURE 22-20 Inline air filters may resemble automobile inline air filters.

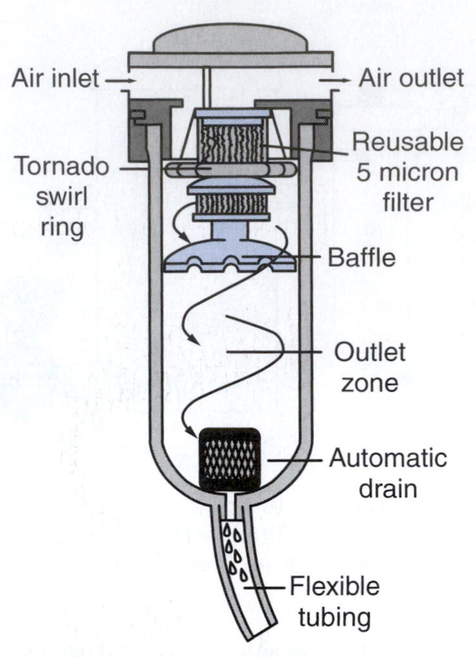

FIGURE 22-21 A centrifugal partial extractor removes oil and water that may get by a filtering device. Adapted from Binks/DeVilbiss.

debris filters must be inspected on a regular basis and replaced as needed.

Extractor

An extractor is a device that extracts oil and water, both of which may be small enough to pass through a filtering device; it generally does this through centrifugal force. As the air is passed through an extractor, it is directed to flow in increasingly smaller circles by passing over a deflector, which looks like a screw. The heavier water and oil are thrown to the outside walls of the extractor, where gravity pulls them to the bottom. This pooled oil and water collection can be forced out the bottom through a drain. See **Figure 22-21**.

Lubricators

Air line lubricators put oil into the air line to lubricate the tools, keeping them in good working condition. See **Figure 22-22**. Although the compressed air has just been sent through an extractor to remove oil and water, the lubricator puts back just oil—the correct type of oil used to lubricate tools.

Though air line lubricators are good devices to have on compressed air lines to help lubricate tools, they should not be used in the refinish department. Spray guns need to be lubricated, but they should not have oil in the compressed air. Any oil in compressed air should not be allowed to get to the paint gun.

Many modern collision repair facilities have two separate air compressors and supply lines for this reason. In the body repair section of the business, the air lines may have an inline air oiler, but in the refinish portion of the shop, the air lines do not have an oiler.

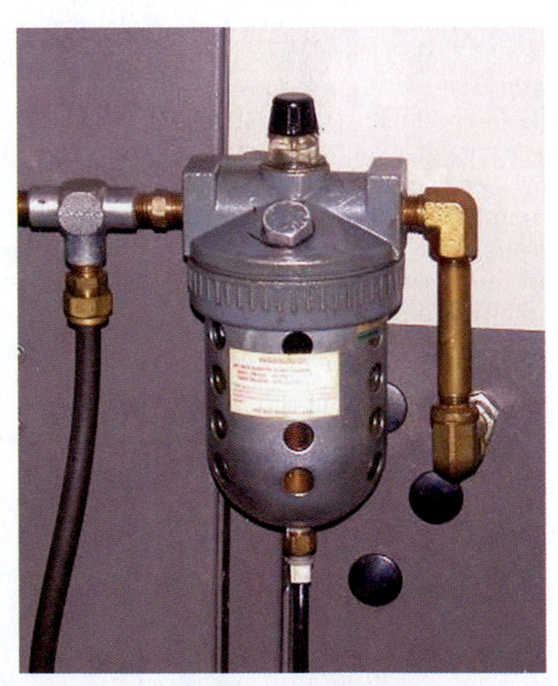

FIGURE 22-22 Air line lubricators put oil into the air line to lubricate the tools. However, compressed air lines for paint shops should not have oilers installed.

CONTROLLING MOISTURE

Though it has been stressed to this point that moisture in compressed air must be controlled, the devices (other than extractors) used to do so have not been discussed. In a compressed air system, moisture should be controlled at many stages of the system or the moisture could overwhelm a device, and the system would end up with water in it. The first place moisture is controlled is on the air compressor through its "after-cooler"; then in the tank with a "dump trap valve"; then on to an air dryer which may be a **desiccator** or an air conditioner-style dehumidifier. Finally, hopefully with the majority of the air cleaned of water, it will pass through an extractor at the air drop, and in the paint booth in a second dryer, generally a desiccator.

After-Coolers

As air is compressed from a large volume to a smaller volume, the air becomes hotter. This air contains water, debris, and sometimes oil along with heat. All of these products should be removed from the air before it is distributed into the air delivery system. An **after-cooler** is a device—often a line—that is installed between the compressor and the storage tank. See **Figure 22-23**. This line is covered with fins similar to the fins on a radiator, which help dissipate the heat from the compressed air as it passes through. In larger systems it could be a device that uses water circulating over the top of the air line to cool the air before it is stored in the air holding tank.

As the air cools, the water in the air condenses and is collected at the bottom of the air tank, where it is drained off either through an automatic air drain or manually during normal air system maintenance. Also, as the air cools, the oil and debris will fall to the bottom of the tank and will be removed with the water.

Dump Trap

A **dump trap** is a device that automatically opens to allow the condensed water and oil to be removed from a compressed system. The dump trap is generally placed at the bottom of the compressed air storage tank, which is sometimes called a recover tank; this is often the lowest point in a compressed air system. The automatic emptying of the collected water helps the system to remain in good operating order.

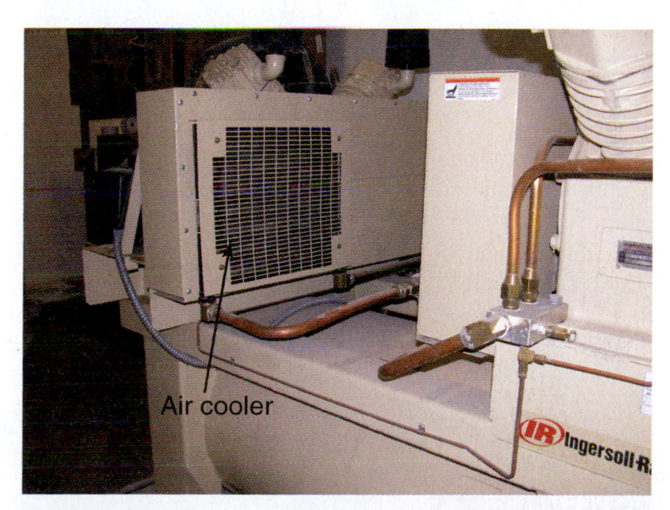

FIGURE 22-23 This air compressor has an after-cooler (in center) to help dissipate heat as compressed air passes through. Air compression produces large amounts of heat that must be removed.

Most automatic dump traps work on a set time of opening and closing, which means that most of the time all the water is extracted from the **air receiver** tank. However, in high humidity times or in areas where there is more water collected in certain seasons, the timed dump trap may be unable to handle all the water, and some may not be drained. As a normal maintenance, therefore, the dump trap should be checked for proper operation and to see that the water has been completely drained.

Air Dryers

Compressed air can be sufficiently dried for most applications with the use of extractors and after-coolers. In the refinish department, though, the air must be completely dry, without any moisture, for painting applications. To remove the remaining water, different types of equipment are used. The most common equipment is a refrigerant type, which runs the air from the receiver tank through an air conditioning system to remove all the water that may remain in the system. Another type is a desiccant style dryer. Desiccant is a dry chemical (such as calcium oxide) that has a high affinity for water; this chemical removes the remainder of the water as it passes over the chemical beads. The desiccant beads must be removed and replaced with fresh ones when they become water-soaked.

HOSE

Though hoses are generally referred to as air hoses, there are two different types of hoses used in the collision repair department. The first is a flexible hose made from rubber that can be either single braided or double braided. **Single braided hose** is an inner rubber hose covered with a braided mesh, then with a second covering of rubber over the braiding. See **Figure 22-24A**. The single braided type is used for general shop work, while **double braided hose** can be used for higher pressures. See **Figure 22-24B**.

The second type of hose is fluid hose, which is made to withstand not only pressure, but also the chemicals that are within them. When a "pressure pot" paint application is used, a large "pot" of mixed paint with liquid paint hoses running to the paint gun supply the paint. In businesses where large amounts of paint are applied, such as in bus and truck painting facilities, pressure pot paint application is common. See **Figure 22-25**.

Over time, as baking paint booths have become more commonly used in collision repair facilities, it became evident that hoses were breaking down faster than they had in the past. This premature hose deterioration was found to be due to the constant heating and cooling of the hoses as they remained in the booth during the paint curing cycle. Varied recommendations have been made to combat this determination. First, the hoses could be quickly disconnected and removed from the booth before the **bake cycle**. Another recommendation was that the hoses be replaced on a regular basis to combat the deterioration

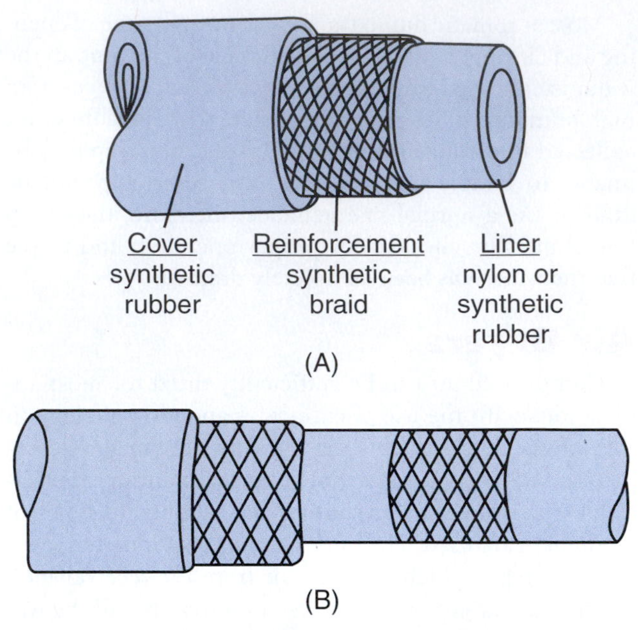

FIGURE 22-24 Multi-layered, reinforced braided flexible air hose can be single braided (A) or doubled braided (B).

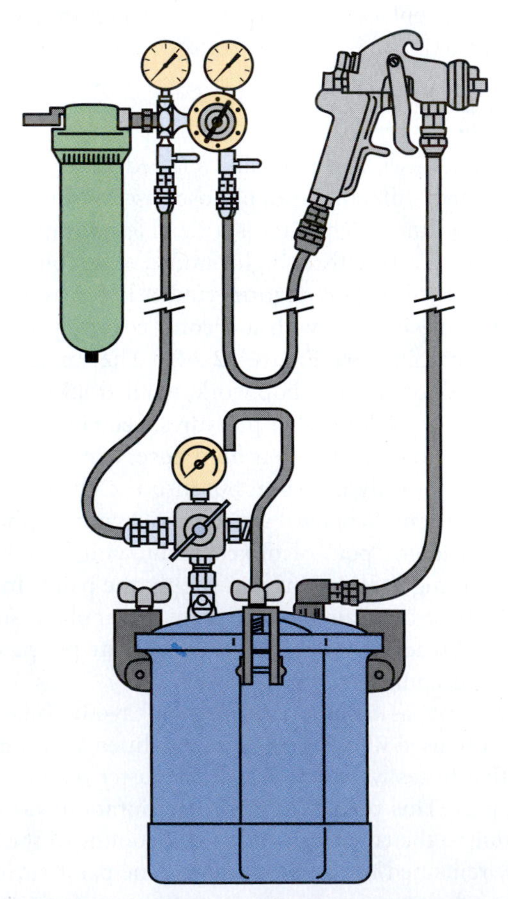

FIGURE 22-25 In businesses where large amounts of paint are applied, such as in bus and truck painting facilities, pressure pot paint application is common. Adapted from Binks/DeVilbiss.

caused by the heat. Hose manufacturers are now making hoses that can withstand the repeated heating and cooling cycle that they experience in paint booths.

Other considerations concerning hose are size, length, connectors, condition, and normal hose maintenance.

TECH TIP As a hose deteriorates from heating and cooling, the inside lining becomes cracked and brittle. Small pieces of hose debris become dislodged and are then spread from the hose to the paint surface. The wise technician will check booth hose conditions regularly for possible deterioration in order to prevent paint contamination.

Sizes

The size of a hose is important for the proper use of air tools, but is especially critical when using spray equipment such as high volume, low pressure spray guns. To deliver the needed air volume, or CFM, to the spray equipment, a sufficient **inside diameter (ID)** must be present. A 3/8-inch inside diameter (ID) air hose is the minimum size to deliver the 25 to 30 CFM required. Unfortunately, when a tool such as a spray gun is supplied with the proper pressure and not the proper CFM, it may still appear and "sound" as if it is working properly. However, the technician may experience lower quality results, such as an increased amount of orange peel. It may be difficult to track the cause back to hose size or connectors that are unable to accommodate the volumes needed.

When choosing an air line for a specific workstation, be sure that the size (ID) will be sufficient for the tool it is supplying.

Length

Air hose length is also a factor to be considered. Because they are flexible, air hoses will get larger as the pressure increases inside. Consequently, the air pressure will drop after the air passes through the hoses. Hoses are rated for maximum air pressures, so choosing an air hose which has a maximum pressure rating of four times the operating pressure will minimize the pressure drop. Even with the proper psi-rated hose, pressure will drop with length. See **Figure 22-26**. Within a 3/8-inch ID air hose that is 50 feet long, air pressure will drop 6 psi. Though this is not a significant pressure drop to affect an air impact wrench, which is run at pressures from 90 to 100 psi, it is sufficient to affect an HVLP spray gun, which requires 28 to 30 psi for proper operation.

Connectors

Quick release connectors can also be a source of reduction in CFM supply. As can be seen in **Figure 22-27**, the

	15 CFM	18 CFM	20 CFM	25 CFM
1/4" × 20'	20 psi	26 psi	28 psi	34 psi
5/16" × 20'	7 psi	10 psi	12 psi	20 psi
3/8" × 20'	2.8 psi	4 psi	4.8 psi	7 psi

FIGURE 22-26 This air hose pressure drop chart shows diameter and length at the left, psi and CFM in the columns.
Courtesy of Binks/DeVilbiss.

FIGURE 22-27 Hose connected downward relieves pressure on the flexible hose wall, extending its usable life.

two different connectors can both screw into 3/8-inch lines. However, as also seen, the inside diameters are quite different. The larger "high flow" connectors will allow a larger amount of air to flow through. When standard or non-high flow connectors are used, the air supply available becomes the lowest capacity of volume that air can pass from that point on. If a technician has chosen and sized the proper air hose, but used the incorrect connectors, the tool may starve for the proper CFM that it needs.

Condition

Though hoses last a long time, they will eventually deteriorate with age and use. Therefore, they should be checked from time to time to make sure they are in good operating condition. In air hose lines, whether single or double braided, the braiding reinforcement can break down. When this braiding is damaged, air can enter between the laminations of rubber and cause a bulge. Delamination bulges can occur in any part of the hose, but often occur at end connections. If the bulge breaks, the hose can become dangerous as it flies around the workplace.

High flow quick connectors for air hoses should be arranged so they face down from a connecting point. See Figure 22-27. If hoses are connected so they must bend downward after the connection, the hose may have premature delamination and failure. See **Figure 22-28**.

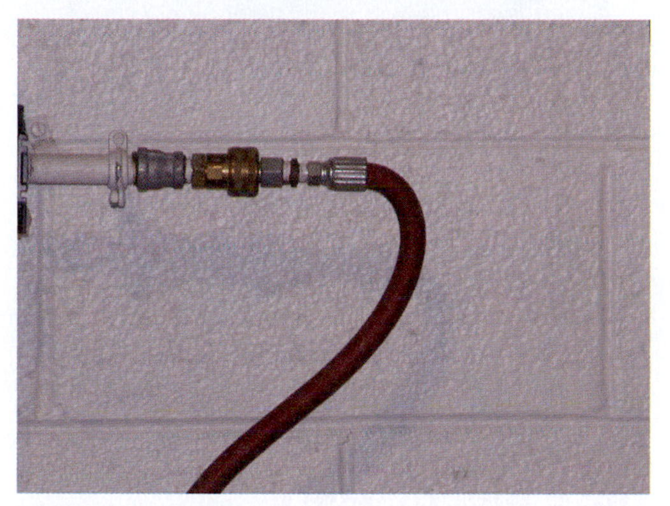

FIGURE 22-28 This hose is connected on an angle. The downward pressure will shorten the hose's durability.

FIGURE 22-29 Hoses should not be subject to frequent abuse such as being pulled, run over, or twisted. In poorly supported hose, even one hook shortens the hose life.

Hoses should not be subject to frequent abuse such as pulling them to free them from around tires, running over them with vehicles, or even leaving them in a twisted bundle where they are forced into small areas. Air hoses are made to be flexible, but cannot withstand being constantly forced into tight bundles. See **Figure 22-29**. When not in use, hoses should be loosely coiled and stored up off the floor.

FIGURE 22-30 Using two hooks will support the hose and lengthen its working life.

Hose Maintenance

Hose maintenance is remarkably straightforward. Hoses should not be abused during their use, as mentioned previously, and they should be stored properly as well. When not in use, they should be loosely coiled and hung on supports that will not force them into too tight a radius. See **Figure 22-30**. If hung on a single hook, the hose—even when loosely coiled—may be forced into too small a radius and damage the inside braiding, causing hose malfunction. Instead, a storage rack with two hooks will allow the hose to have a gentle curve in the coil, so that even if the hose is stored for longer periods of time, the hanging position will not prematurely deteriorate the hose.

The outside of the hose should also be cleaned occasionally. If it is used where it has been exposed to overspray, it should be wiped down with a mild solvent to remove any overspray that might otherwise become air-borne contaminants when the hose is next used for spraying.

COMPRESSOR MAINTENANCE

Compressors can and do last for years, and in some cases for decades. If a compressor is sized properly, not run outside the manufacturer's recommended duty cycle, and properly maintained, a compressor can be expected to operate without failure for long periods.

The best compressor maintenance is preventive maintenance. If a facility checks the compressor daily before starting it, and maintains a monthly and yearly compressor maintenance plan, the compressor should be relatively failure-free.

Daily Compressor Maintenance

Each day, before starting the compressor, a technician should:

- Check that the oil levels are within the recommended operating levels.
- Add the recommended oil type if needed, taking care not to overfill.
- Check to make sure that the dump valve is operating properly, and if not automatic, drain any water from the air storage tank.
- Inspect that all parts are in good operating condition before starting.
- Listen for any unusual noises when the compressor is started following the inspection and checks. If any are noted, stop the compressor to investigate.

Weekly Compressor Maintenance

Each week, a technician should:

- Check that the pressure safety valve is working properly.
- Check for air leaks at the compressor, and repair any if found.
- Clean by blowing off the after-cooler fins, the compressor fins, and any accumulation of dust.
- Check that the belt guard is secure, and tighten if necessary.
- Clean the air inlet filter.
- Keep the compressor area clean and free from clutter.

Monthly Compressor Maintenance

On a monthly basis, the technician should:

- Have the oil checked, topped off, or changed as recommended by the manufacturer.
- Check that the pressure switch for start pressures and stop pressures is working properly.
- Check for oil leaks.
- Check for loose bolts and fasteners, and tighten if needed.
- Check drive belts for wear and proper fit, and tighten if needed.

To perform a system check, the technician should:

- Check and repair all air supply system leaks.
- Drain and clean all extractors.

- Check for proper operation of all regulators/transformers.
- Drain all line outlets of water.
- Check hoses for cracks or bulges.
- Check hose ends for bulges and check that quick connectors are working properly.
- Check and change (if necessary) desiccant air dryers.
- Check referent dryer.

SPRAY BOOTHS

Spray booths provide many advantages for the spray technician; they provide a clean spray area, with good temperature control, controlled airflow, proper lighting, fire safety, spill control, curing, and—in areas where it is needed—sometimes cooling. With all of these advantages, technicians seldom remember that one of the best advantages is that it protects other employees in the refinish department from hazardous paint vapors and overspray.

Spray booths come in different configurations and styles. They can be "front load," where the vehicle is driven into the booth and must be taken out the same way, or booths with openings at both ends that can be driven through for loading and unloading.

They can be cross-draft booths, semi-downdraft booths, or downdraft booths. Each type has distinct advantages and disadvantages.

Cross-Draft Spray Booths

Cross-draft spray booths draw the incoming air from the shop, generally through the booth doors. The air travels over the length of the vehicle, is filtered again at the opposite end of the booth, and then is vented to the outside of the building. See **Figure 22-31**.

Though cross-draft booths protect other employees from hazardous overspray, their dirt control is poor. If dirt is in the outside area and is not filtered out through the door filters, it will contaminate the finish, as it must pass by the entire length of the vehicle before it is vented outside.

Though the cross-draft booth has filters, this style of booth has the least filtering ability among the various styles of booths. The booth has only one exit fan; therefore, the booth's ventilation is a negative one, and when the doors are open, dirt from the shop is drawn in.

Cross-draft spray booths, though the least expensive to purchase and operate, represent one of the earlier designs. Today, they are generally not the booth of choice when new facilities are constructed.

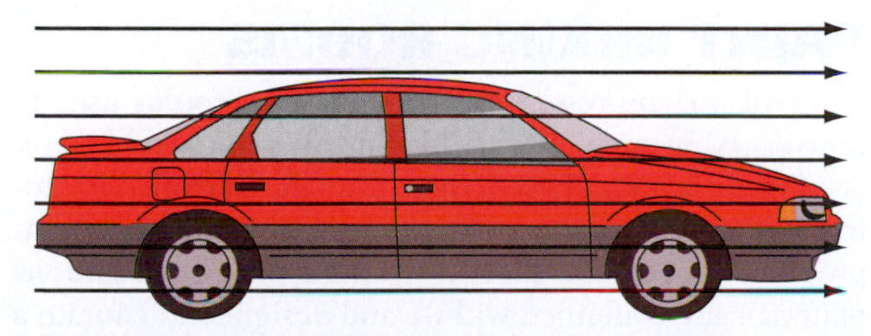

FIGURE 22-31 Cross-draft spray booths draw the incoming air from the shop.

Semi-Downdraft Spray Booths

Semi-downdraft spray booths, though introduced at approximately the same time as downdraft booths, are somewhat cheaper to purchase. However, they are less efficient and less clean than downdraft booths. See **Figure 22-32**.

An advantage of this style of booth is that air is brought in through the rear roof of the booth. This allows the air to be collected from outside the building, where it is cleaner and freer of debris. It is also forced in through a fan system, allowing the regulation of how much cleaned air enters the booth. The air blows to the rear of the booth at the top and is drawn out from the lower front of the booth. The air is filtered before it is vented outside.

Because the air is brought in from outside, in cooler weather an "**air-makeup" system** must heat the air to spray temperatures before entering the booth. This booth design can—with its fan system bringing air into the booth and a second fan system taking air out of the booth—have the airflow regulated so slightly more air is coming into the booth than is going out. This positive pressure regulation helps keep the dirt from outside the booth from being drawn in, as it would be with a booth having only an exhaust fan.

Downdraft Spray Booths

Downdraft spray booths draw air in from outside, where the air is filtered, heated to the proper spray temperature, and filtered again before entering the booth from on top of the vehicle. The air is drawn around and under the vehicle, where it is filtered, then exits through floor vents. See **Figure 22-33**.

FIGURE 22-32 Semi-downdraft booths are cheaper than downdraft booths, but are also less efficient and less clean.

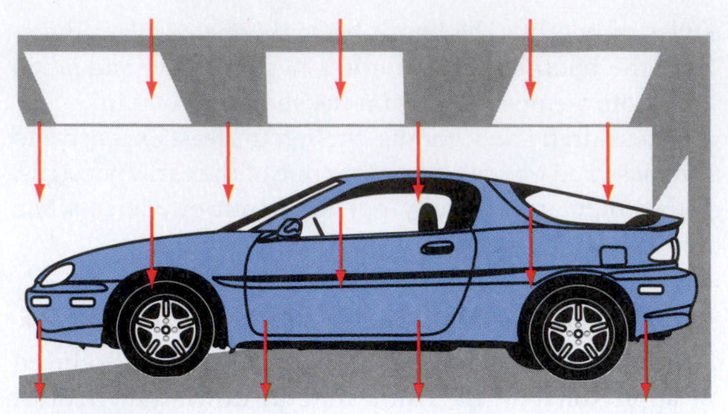

FIGURE 22-33 Downdraft spray booths provide the best air and dirt control of all the spray booth styles.

Downdraft spray booths provide the best air and dirt control of all the spray booth styles. Because the overspray is drawn to the floor quickly, it protects the spray technician best. Dirt does not need to travel over the length of the vehicle to be vented out the front. Because the booth has both incoming fans and exhaust fans, the airflow can be controlled.

In addition, temperature control can be closely monitored during the spray cycle. Also, when the booth is switched to bake mode, the vented air is stopped and the heated air is recirculated from the heat source to the booth, reducing the baking cost.

Downdraft spray booths are the most common new booth installation, for good reasons. The computer controls have become so sophisticated that spray temperature can be controlled to within one degree. Also, paint-specific purge and bake times can be programmed for different products to automatically purge, bake, and stop after spraying.

Spray Booth Balancing

Spray booth balancing helps control the air movement inside the spray environment to assure that outside dirt does not come into the booth. The amount of air coming into the spray booth is controlled along with the amount of air flowing out of the booths. This control can create a slightly larger amount of airflow coming in, or "positive airflow," or a larger amount of airflow going out of the booth, or "negative airflow."

Positive Airflow. With more air coming into the spray environment than is leaving, a **positive airflow** will keep dirt that is outside the booth outside. If a door is opened, air will come out of the booths instead of drawing air that has not been filtered—and may be dirty—into the booth. Positive airflow is monitored by using a manometer or a magnehelic gauge. See Figure 22-34.

The booth balance should be checked often, because as the filters become dirty and air movement is decreased either into or out of the booth, the balance will change. The booth can be adjusted to accommodate the filter to a point, but when the booth can no longer be adjusted to positive airflow, the filters must be changed and the booth readjusted.

FIGURE 22-34 A manometer can be used to balance the spray booth's pressure and monitor air filter life.

Negative Airflow. If the booth airflow is restricted and there is more air leaving the booth than is coming in, the booth is said to be in a **negative airflow**. This will draw air in, and dirt with it. If the booth cannot be adjusted to a positive airflow, the booth filters should be inspected and changed as needed.

PREP DECKS

Prep decks are areas that contain the vehicle as it is being sanded; they serve to protect other technicians from dust contamination as the vehicle is being prepared for paint. Though a prep deck's main function is to protect others, it also provides an environment that extracts dust rapidly and offers a brightly lit environment to work in. Some prep decks also control the temperature for spraying primer, while others can have a bake mode to speed-up curing time.

Prep decks can be made of steel walls with curtain dividers, and may have single or multiple vehicle capacity. See Figure 22-35. They may be downdraft or semi-downdraft style, and can be positive airflow adjusted.

PAINT MIXING ROOMS

Paint mixing rooms are self-contained rooms used to house mixing and cleaning equipment, and to store paint. See Figure 22-36. These rooms, though controlled by different state and local regulations, are generally explosion-proof and spillproof. Most—if not all—of the hazardous materials are contained within, and designed to contain a spill if one occurs. In this clean environment, paint mixing banks and computer mixing systems are housed and

FIGURE 22-35 A two-car prep deck can be made of steel walls with curtain dividers.

FIGURE 22-36 Paint mixing rooms are self-contained rooms used to house mixing and cleaning equipment.

used. Paint gun washers and solvent recyclers are also found in the mixing room. These rooms are often conveniently located with doors into the spray booth so that—after a technician has mixed and reduced the proper paint—the technician can enter the booth without going into a dirty environment.

Paint mixing rooms help the refinish department to consistently produce clean and professional-quality refinished vehicles.

CURING EQUIPMENT

As paint technology has changed from paints that dry through evaporation to paints that cure through oxidation and canalization, the curing time has been reduced. Changing the environment around the finish accelerates the cure, and this has reduced vehicle repair times. Paint booths were first adapted with forced air heaters, which will "bake" a finish using hot forced air or **convection curing**. With the introduction of infrared lights, the object that was finished was heated—instead of the air around it—to speed up the curing process. As paint chemistry developed, finishes were developed with reduced cure cycle times resulting from exposing the finish to ultraviolet light. Other ways to speed up the curing of waterborne paints have also been developed, through use of venturi blowers.

Each of these curing methods has advantages and disadvantages, and one may be better than another for a specific shop's conditions. Additionally, a refinish facility may choose to use one type of curing accelerator for a certain situation and another for a second situation, depending on the place in the shop and the specific needs of the situation.

Forced Hot Air Curing

Forced hot air is the most common spray booth curing accelerator. With this method, air is drawn in from either outside the paint environment or the booth itself, heated, then moved into the booth area. See **Figure 22-37**. The hot air, which is recirculated back into the heating burner, heats the air further until the set temperature is reached; then the temperature is held at the set level for a specific time. The booth then opens the outside air vents, and exhausts the heat. At that point, workers can re-enter the booth when the cool-down cycle has been completed.

Infrared Curing

Infrared is a classification of light waves. **Infrared (IR) curing** devices use either short infrared or medium infrared waves to heat solid objects instead of the air around the object. Because the infrared waves pass through the

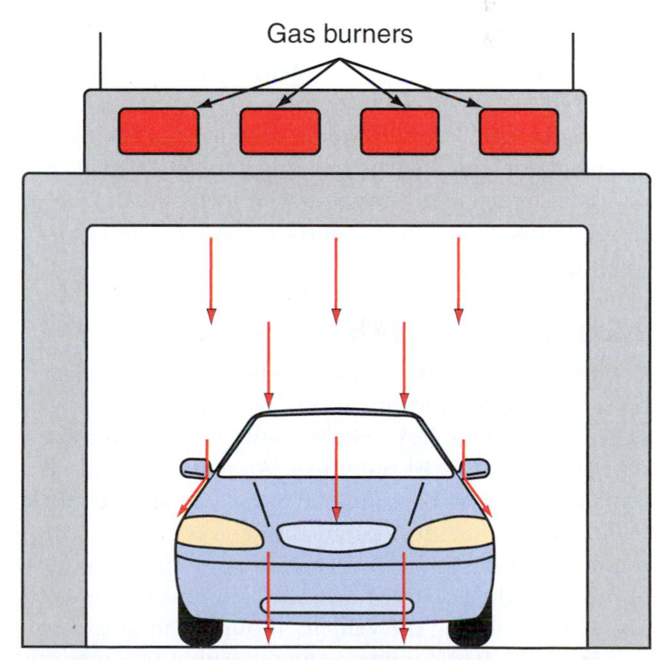

FIGURE 22-37 In the booth burner configuration, the hot air—which is recirculated back into the heating burner—heats the air further until the set temperature is reached.

FIGURE 22-38 Portable IR curing lights are commonly used because they can be moved to the vehicle as needed.

finish and heat the solid object below the finish, many technicians refer to this type of curing as inside-out curing. Infrared devices can be mounted in paint booths, or they can be portable. See **Figure 22-38**. These portable IR machines are more commonly used, because they can be moved to the vehicle as needed.

Infrared heating is fast. Therefore, the technician operating the IR curing machine should be careful when placing the machine for curing. If the machine is placed too close to the vehicle, the heat can be so great as to bubble the paint off the vehicle.

Because the heat is directed onto the solid object and not the air around it, sensing that the vehicle is getting too warm may be difficult until it is too late. Infrared curing machines supply recommended distances that machines should be placed from the vehicle, and the technician using the IR device should take care to stay within the recommended distances. Some sophisticated IR machines also have heat-sensing devices which will turn off the IR lights if the vehicle's surface temperature becomes too high.

Ultraviolet (UV) Curing

Ultraviolet, or UV, is also a light wave. By using UV light with special formulations of finishes—a process called **ultraviolet (UV) curing**—the technician can accelerate the curing of the finish. After the specially formulated finish has been applied to the vehicle, a UV light is directed at the finish for curing. The needed exposure to the UV light is a very short amount of time (2 minutes) in order for the finish to be cured. The next step is sanding. Before sanding the vehicle, though, the finish must be wiped down to eliminate any overspray that may still be uncured. However, following that procedure, the surface can be sanded.

Ultraviolet finishes are presently only available in primer surfacers, but further development is expected.

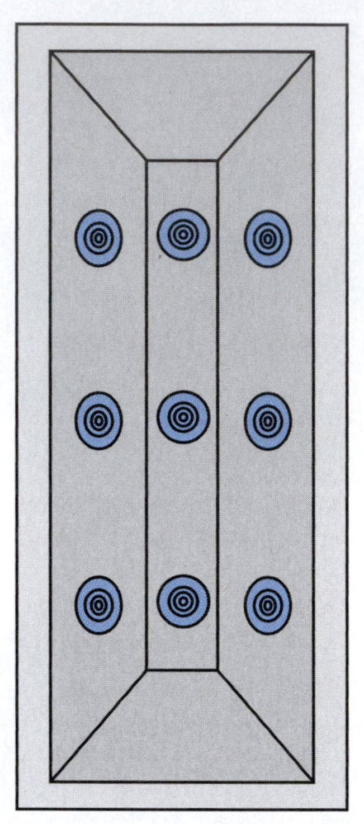

FIGURE 22-39 A venturi cone blower in a spray booth is used for drying waterborne finishes.

Venturi Blowers

Waterborne finishes need air movement to evaporate the water from the finishes. To accelerate the drying of waterborne finishes, unheated dry air is blown across the finish through small **venturi blowers** to remove moisture. These devices can be either mounted permanently in a spray booth, or may be portable machines, which can be moved to the vehicle. See **Figure 22-39**.

SPRAY BOOTH OPERATIONS

Though spray booths come in many different configurations and with several different types of operating controls, there are three standard phases of the spray booth operation. The first is the spray cycle, during which the booth's environment is set for the best application of finishes. Next, the purge cycle occurs. In this cycle the booth operates for a time period at slightly higher temperatures than spray temperatures. This process allows the small amount of solvents to be removed. Finally, the bake or cure cycle is operated, during which the temperatures are much higher than in the spray or purge cycles, which accelerates the curing time.

Spray Cycle

When a spray booth has its spray temperature set at the ideal spray setting, that temperature will be 70° to 80°F, depending on the manufacturer's recommenda-

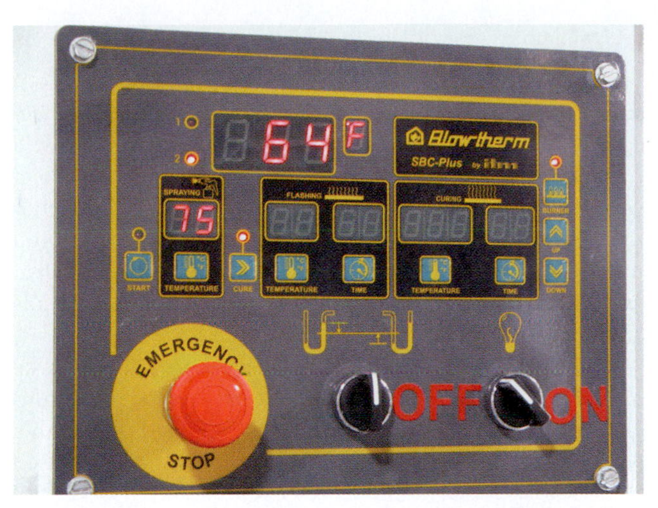

FIGURE 22-40 Booth controls in the spray cycle booth set the temperature to the ideal spray setting.

tions. See **Figure 22-40**. In the **spray cycle**, the air enters the booth and is vented outside. Then fresh air at the set temperature is passed through the booth. The movement of air draws overspray from the booth during finish application.

The spray cycle is not timed, and will run continually until changed to another cycle. The booth's lights are on and air pressure is available to the wall-mounted transformer.

Purge Cycle

During the **purge cycle**, the booth will draw in fresh air and maintain the proper temperature, which is often slightly higher than spray cycle temperatures. This allows the booth to run in at this setting for a specific set amount of time. See **Figure 22-41**. The air is vented outside the booth to eliminate any solvent accumulation as it evaporates from the freshly applied paint. The booth lights are generally on, and in some booths, the air supply to the wall transformer has been turned off.

Bake Cycle

The bake cycle can be set to come on after the purge time has elapsed. Its purpose is to change the booth environment to accelerate the cure cycle. Generally, the lights are switched off (because heat shortens the light bulbs' life expectancy). The temperature is raised to the manufacturer-recommended cure temperature, approximately 120°F to 140°F, for a set amount of time. This again depends on the paint manufacturer's recommendations (which may be from 8 to 40 minutes). During the curing cycle, the air is recirculated over the heat source to rapidly raise the temperature and reduce the cost of heating the air. Following the set cure time, the booth will cool down. The lights will relight when the booth is cooled to the point that workers can re-enter. Some lights will automatically switch off after a set amount of time, unless the process is started again.

Surface Temperature. Although the temperature is set and measured by the booth, the true critical temperature is the vehicle's **surface temperature**. A vehicle has a large mass that will change temperature more slowly than the air around it. The booth measures the air in the booth, not the vehicle. Depending on the time of year and the weather conditions, the vehicle surface temperature may be different than the booth reading.

The best way to monitor the surface temperature is with a magnetic surface monitor. See **Figure 22-42**. This device has two temperature indicators: a free-floating red one, and a black temperature indicator, which reads the temperature of the steel. As the vehicle is being placed into the booth, the monitor is set by moving the red movable indicator to rest on the temp indicator. It is then placed on the vehicle surface, away from where it is to be painted but close enough to get a true reading. As the temperature rises, the black indicator will push the movable indicator. When the bake cycle has been completed, the location of the red indicator will be the highest temperature that the vehicle's surface

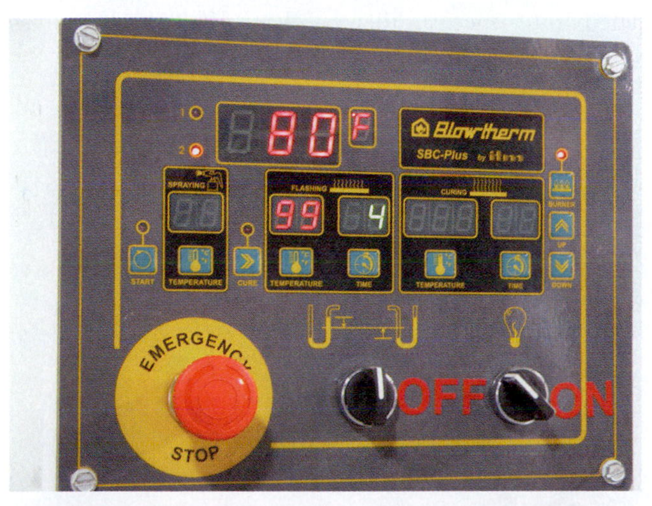

FIGURE 22-41 Booth controls in the purge, or sometimes called flash, cycle are used to set the proper temperature, which is often slightly higher than spray cycle temperatures.

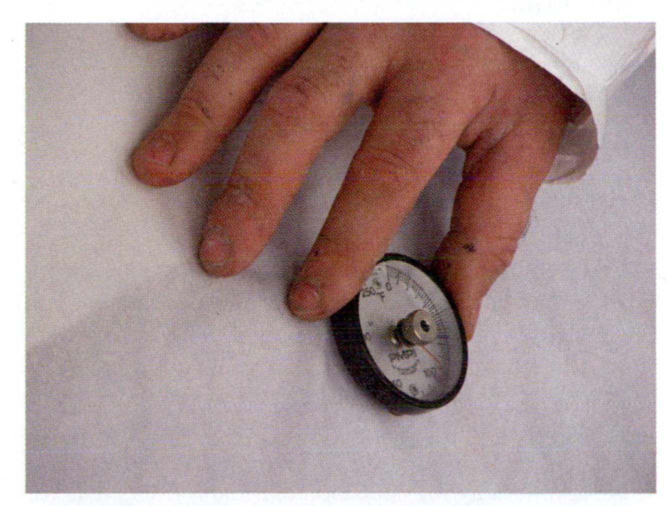

FIGURE 22-42 This surface temperature monitor, placed on the vehicle in a booth, gives two readings: One indicates the highest temperature (near 100) with the second indicating the surface temperature (near 70) now.

got. If the difference between the booth temperature and the surface temperature is significant, such as if the booth temp was 140°F and the highest surface temperature of the vehicle was 100°F, the booth temperature can be changed to have the surface temperature reach the needed 140°F.

Lighting

Lighting in a spray booth is significantly different than lighting in general. To properly match colors, the booth light must be as close to sunlight as possible. The light bulbs must have an 85 to 92 percent **color rendering index**, or **CRI**, which is a measurement that compares artificial light to sunlight (100 percent CRI). The amount of light, which is measured in foot **candlepower**, should be from 5 to 10k foot candlepower in all areas of the booth for the technician to see properly.

TECH TIP In a spray booth, light can sometimes be *too* bright. When the candlepower exceeds 10k foot candlepower when spraying clear on a white vehicle, it becomes difficult to see the spray line. As clear is being sprayed, the best way to observe the wet line is to look at an angle, instead of straight down.

Booth lights also lose both their CRI and their candlepower with age. If they have been operated during the baking cycle, they will prematurely lose both. Booth lights should be changed on a regular basis, not simply when they stop working.

Booths can lose their candlepower rapidly even when the bulbs are in good condition and relatively new. This loss may be due to the glass covering over the lights having overspray accumulate on them. To minimize the loss, the booth—including the lights—should be cleaned on a regular basis.

BOOTH CLEANING

The refinish spray booth should be cleaned on a regular basis. Not only should it be cleaned of debris and vacuumed of dirt between each sprayed vehicle, but it should also be thoroughly cleaned on a regular basis. The walls should be cleaned and booth liquid mask should be applied. The light lenses need to be cleaned of overspray and the bulbs tested for proper operation. The **floor filters** should be changed often, and the grates covering them should be cleaned—or even sandblasted—to remove overspray before it accumulates. Spray booth transformers, air filters, and desiccators should be checked for proper operation; additionally, the condition of the air hoses should be inspected for wear.

The cleaner the booth is kept, the cleaner the refinish work will be.

Walls

Booth walls accumulate overspray readily but may not show that they are becoming dirty. To combat the walls becoming contaminated, they should be wiped down on a regular basis. Booth liquid mask is available to protect walls from dirt. After application and use, the old liquid mask can be easily washed off with water. As the liquid mask is washed off, the dirt and overspray come with it and can be cleaned up from the floor. After the clean walls are dry, the liquid mask can be reapplied.

When applying liquid mask, the booth lights can be masked off to keep them clean. Alternatively, the technician can spray the entire booth and wipe the lights with a damp cloth to clean off the liquid masking as needed.

Light Lenses

As mentioned before, light lenses can also become covered with overspray, which changes their candlepower and CRI. To keep them in good operating condition they must be cleaned often. Overspray may need to be removed using a scraper; then the lenses should be washed with glass cleaner. See **Figure 22-43**. In fact, they may need to be washed and scraped several times to get them properly cleaned. When the lens is clean, the light should be tested for proper operation, and any bulbs replaced if needed to continue proper operation.

Booths lights are sealed to keep spray vapors out. The light gasket sealers should be checked as the lights are being cleaned, and repaired if found to be faulty.

Floor

Paint booth floors not only should be vacuumed on a regular basis, but also should be mopped to clean off any dirt and dust that may not have been completely removed during vacuuming. To keep booth floors clean from the overspray that is drawn toward them, they can be covered with booth spray masking or rosin paper. See **Figure 22-44**. The booth paper can then be changed when it becomes dirty to avoid contaminants from accumulating on the floor and making cleaning more difficult.

FIGURE 22-43 Booth lights will need to be cleaned often to remove overspray.

FIGURE 22-44 Paper on the booth floor keeps dirt from getting into paint work.

Filters

Spray booths have from two to four different filters that must be checked and changed on a regular basis. The floor filters will become contaminated the fastest, and may need changing often because the overspray is drawn toward them. See **Figure 22-45**. After the booth's floor grates are removed, the old filters should be taken out, disposed of according to state and local requirements, then replaced with new filters designed for the booth.

Some booths have incoming air profilers, which should be checked and changed as needed. See **Figure 22-46**. Although **ceiling filters** may not need changing often, they can be time consuming to replace when the need arises. Some booths are equipped with **pre-filters**, which are filters that clean incoming air and extend the life of the more expensive ceiling filter. See **Figure 22-47**. Lastly, **exhaust filters**, if equipped, should be checked and changed as needed. See **Figure 22-48**.

FIGURE 22-45 The floor filters in the spray booth are the fastest item to be contaminated and the first to be checked for change out.

FIGURE 22-46 Some booths are equipped with pre-filters to extend the ceiling filter life.

FIGURE 22-47 Ceiling filters clean the air before it enters a downdraft booth.

FIGURE 22-48 Booth floor exhaust filters should be checked and changed as needed.

SPRAY BOOTH TESTING

To keep a booth in good operating condition, it should be tested for proper operation. Though some refinish departments may have and use the testing equipment, they may also have a booth testing service perform the test for them. They will test the booth for proper air volume, light, air movement, and dirt.

Velometer

A **velometer** tests the amount of air volume moving in a spray booth. Each booth is designed to move a specific amount of air for proper operating conditions. Spray booths will have different volumes of air moving when they are empty than when they have a vehicle in them. The booth should be tested in both these conditions—first with the booth empty, to see if the booth meets the required air movement specifications. See **Figure 22-49**. Then, with a vehicle in the booth, the test should be performed at the vehicle's side. See **Figure 22-50**. When the booth is tested with a vehicle in it and the readings are compared from side to side and front to rear, the tester can see if the booth has "dead spots," or areas where the air does not flow properly over the vehicle. If dead spots are found, a smoke generator should be used to see how the flow is affected.

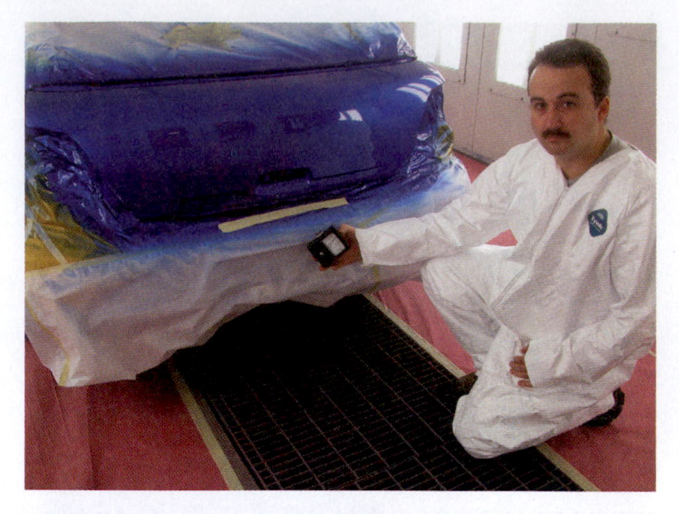

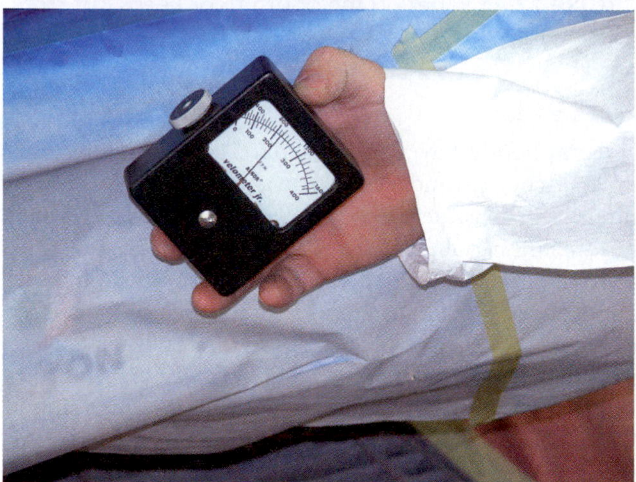

FIGURE 22-50 This velometer is monitoring airflow with a vehicle in the booth. A filled booth will have more airflow than an empty one.

FIGURE 22-49 This velometer is monitoring airflow with the booth empty.

Smoke Generators

A **smoke generator** is a device that generates a nontoxic but visible smoke near the vehicle. The tester can watch as the generator moves smoke around the vehicle and into the vents for exhausting out of the booth. The observation of the smoke will indicate if the booth has proper airflow, or where the airflow is the best and where the airflow is the least. By knowing where the air movement is best, a vehicle can be better placed in the booth. It can also help the technician determine if the spray booth is in need of repair.

Light Meter

The paint technician may often have use for a **light meter** in the booth. A light meter is used from time to time to test the intensity (brightness) of the booth lights; it is also used to test the color rendering index (CRI). If the lights are tested on a regular basis, the facility can change light bulbs as they age and lose their CRI and candlepower. See **Figure 22-51**.

Particle Counter

A **particle counter** measures the amount of debris in the spray booth's air. If the booth is tested and is found to have an excessive amount of dirt, the facility can either use a different type of filter or find and eliminate the source of the dirt.

FIGURE 22-51 This light meter is monitoring the light intensity in the booth.

Summary

- The operation and maintenance of paint department equipment is extensive.
- To operate at its peak performance, equipment should be cleaned and inspected on a regular basis.
- Preventive maintenance is the first and best line of defense to avoid costly and unexpected breakdowns.
- Compressors, whether piston or rotary, supply the working power to not only spray equipment, but also to breathing devices and other pneumatic tools.
- Breathable air regulations require special monitoring, and the refinish shop may have a separate system designed specifically for air supply generation.
- Air distribution and regulation control the smooth delivery of clean air to the worker through regulators/transformers, filters, extractors, and dryers.
- The better the air distribution system is maintained, the more clean, dry air that is supplied to tools and workers, leading to a long working life for them both.
- Prep decks, mixing rooms, and spray booths provide clean and efficient spray and mixing environments, which make for better safety and a cleaner working area, both inside and out.
- As can be seen, the proper cleaning, testing, and safe operation of all refinish department equipment will lend itself to a smooth, efficient, and profitable operation.

Key Terms

after-cooler
air extractor
air filter
air holding tank
air line lubricator
air line pressure drop
air line size
air-makeup system
air receiver
air supply respirators
air transformer
bake cycle

breathable air
candlepower
ceiling filters
class D breathable air
color rendering index (CRI)
convection curing
cross-draft spray booths
cubic feet per minute (CFM)
desiccators
diaphragm compressor
double braided hose
downdraft spray booths
dump trap
duty cycle
exhaust filters
floor filters
high flow quick connectors
high volume, low pressure (HVLP) spray guns
infrared (IR) curing
inside diameter (ID)
light meter
negative airflow
on-demand compressor
paint mixing rooms
particle counter
piston compressor
pneumatic tools
positive airflow
pounds per square inch (psi)
pre-filters
prep deck
purge cycle
rotary screw compressor
rotary vane compressor
semi-downdraft spray booths
single braided hose
single-stage compressor
smoke generators
spray booth balancing
spray cycle
surface temperature
two-stage compressor
ultraviolet (UV) curing
velometer
venturi blowers

Review

ASE Review Questions

1. Technician A says that compressed air is used to drive pneumatic tools in the collision repair industry. Technician B says that all compressors used in collision repair are piston-type compressors.

   Who is correct?
   A. Technician A only
   B. Technician B only
   C. Both Technicians A and B
   D. Neither Technician A nor B

2. Technician A says that a two-stage compressor is a compressor with two compressor motors. Technician B says that in two-stage compressors, both pistons are the same size.

   Who is correct?
   A. Technician A only
   B. Technician B only
   C. Both Technicians A and B
   D. Neither Technician A nor B

3. Technician A says that CFM is a measurement of air pressure. Technician B says that psi is a measurement of air pressure.

   Who is correct?
   A. Technician A only
   B. Technician B only
   C. Both Technicians A and B
   D. Neither Technician A nor B

4. Technician A says that a rotary screw compressor has a 100 percent duty cycle. Technician B says that a rotary screw compressor can produce air at about one-half the temperature of piston-compressed air.

   Who is correct?
   A. Technician A only
   B. Technician B only
   C. Both Technicians A and B
   D. Neither Technician A nor B

5. Technician A says that breathable air used for air supply respirators must be class A air. Technician B

says that breathable air used for air supply respirators must be class D air.

Who is correct?
A. Technician A only
B. Technician B only
C. Both Technicians A and B
D. Neither Technician A nor B

6. Technician A says that the air hose used for HVLP spray guns must be 3/8 inch inside diameter. Technician B says that quick-change air connectors must be high flow for HVLP spray guns.

   Who is correct?
   A. Technician A only
   B. Technician B only
   C. Both Technicians A and B
   D. Neither Technician A nor B

7. Technician A says that a piping loop air supply delivery system assures that all air leads will have the same pressure. Technician B says that with a piping loop system air can flow in both directions.

   Who is correct?
   A. Technician A only
   B. Technician B only
   C. Both Technicians A and B
   D. Neither Technician A nor B

8. Technician A says that an air transformer is a device used to regulate air pressure. Technician B says that an air extractor is used to oil the air lines.

   Who is correct?
   A. Technician A only
   B. Technician B only
   C. Both Technicians A and B
   D. Neither Technician A nor B

9. Technician A says that moisture control in air lines can be controlled with delaminates. Technician B says that moisture control in air lines can be controlled with desiccators.

   Who is correct?
   A. Technician A only
   B. Technician B only
   C. Both Technicians A and B
   D. Neither Technician A nor B

10. Technician A says that a semi-downdraft spray booth controls air best. Technician B says that a paint booth can be temperature-controlled, including cooling.

    Who is correct?
    A. Technician A only
    B. Technician B only
    C. Both Technicians A and B
    D. Neither Technician A nor B

11. Technician A says that the most common spray booth installed is a downdraft spray booth. Technician B says that the most commonly installed spray booth today is the side draft spray booth.

    Who is correct?
    A. Technician A only
    B. Technician B only
    C. Both Technicians A and B
    D. Neither Technician A nor B

12. Technician A says that exposing primer to infrared light can speed up curing time. Technician B says that exposing some primers to ultraviolet (UV) light can speed up their cure times.

    Who is correct?
    A. Technician A only
    B. Technician B only
    C. Both Technicians A and B
    D. Neither Technician A nor B

13. Technician A says that during the purge cycle, the booth will recirculate the heated air to cut down on the cost of heating air. Technician B says that during the purge cycle, heated air is vented outside the booth to eliminate any solvent accumulation.

    Who is correct?
    A. Technician A only
    B. Technician B only
    C. Both Technicians A and B
    D. Neither Technician A nor B

14. Technician A says that the bake cycle heats the air in the booth to drive off the solvents, thus speeding up the curing time. Technician B says that during the curing cycle, the air is recirculated over the heat source to rapidly raise the temperature.

    Who is correct?
    A. Technician A only
    B. Technician B only
    C. Both Technicians A and B
    D. Neither Technician A nor B

15. Technician A says that the critical temperature called for by paint manufacturers is the temperature of the vehicle's surface. Technician B says that the temperature that is needed to speed up the curing time is the temperature of the air around the vehicle.

    Who is correct?
    A. Technician A only
    B. Technician B only
    C. Both Technicians A and B
    D. Neither Technician A nor B

Essay Questions

1. Describe the differences between a cross-draft booth and a downdraft spray booth.
2. Explain how the three spray settings on a spray booth (spray, purge, and cure) differ in temperature, and why.

3. List the three weekly spray booth maintenance items, and explain how they should be completed.
4. Define cubic feet per minute (CFM) and pounds per square inch (psi).

Topic Related Math Questions

1. If a shop has 15 technicians using dual action (DA) sanders, each requiring 4.5 CFM; three painters using spray guns, each requiring 12 CFM; and two of those painters are using air supply respirators, each requiring 15 CFM, how many total CFM of air are required?
2. It has been determined that booth filters should be changed after 36 hours of spray time. It takes an average of 47.25 minutes to spray a vehicle. How many vehicles can be sprayed within each 36-hour period?
3. If a compressor has a 65 percent duty cycle, how many minutes per hour can that compressor be run continually, and how much down time must it have in each hour?
4. If a spray hose loses 10 pounds of pressure for every 25 feet, how much pressure would a 250-foot air hose lose?
5. A shop has five desiccant bead air dryers that need to have their desiccant changed. Each one requires 3.12 pounds of desiccant. How many pounds of desiccant must be purchased to change these filters? (Desiccant can only be purchased in 1-pound increments.)

Critical Thinking Questions

1. A shop owner is designing a new large collision repair shop. Explain how that shop owner would determine the air requirements for the shop, and also explain which type of compressor would provide the best air supply for the new facility.
2. The spray department is experiencing dirt in their paintwork. How would you, as manager, go about eliminating this problem?
3. A shop that has been in business for a long time (35 years) starts to experience fisheyes in the paintwork. As shop foreman, how would you go about solving this problem? (What would you check?)

Lab Activities

1. Have students perform weekly and monthly booth maintenance.
2. Have students start and check a spray booth for proper operation.
3. Have students perform monthly compressor maintenance.
4. Have students spray primer and properly set up infrared lights for curing it.
5. Have students start and set up a spray booth for proper spray setting (including balancing), purge setting, and curing time.

Chapter 23

Detailing

OBJECTIVES

Upon completing this chapter, you should be able to:

- Outline the methods of detailing a vehicle when new, following a repair, and on older finishes to maintain the vehicle's protection and luster.
- Explain the detailing tools that are commonly used and outline the techniques needed to use them.
- Identify the common supplies used in detailing, the cautions regarding detailing, and how to avoid or correct the common mistakes that may be encountered when detailing a vehicle.
- Perform the following tasks: prepare a new vehicle's finish for sale; detail a newly refinished vehicle, providing an invisible repair; and detail the finish on an older vehicle.
- Maintain personal safety as well as environmental safety.
- Perform cleanup following detailing, and identify the common types of defects that can be repaired and those that cannot.
- Explain the theory of least aggressive first methods and the advantage of using them.

INTRODUCTION

In the automotive industry, **detailing** is defined as the deep cleaning of a vehicle. Detailing is done at many stages in an automobile's life. For example, it is often done at the factory to remove minor defects that were caused during the manufacturing process.

> **TECH TIP** The author says: "I started my career in the automobile industry as an online paint repair person for Ford Motor Company. It was there that I learned how to make an automobile finish look its best. It was also there that I contracted the 'curse of Detroit' (the attraction to all things automotive!); I have been working in the automobile industry ever since, enjoying the basic pride of producing beautiful vehicles."

Every new vehicle is prepared for sale after it is received from the manufacturer. After a vehicle is delivered to the dealership, it is inspected, prepared for delivery, and undergoes a new vehicle detailing. Because consumers now often keep their vehicles for longer periods of time, the vehicle may undergo additional detailing throughout its lifetime to maintain its luster and new car feel.

A thorough vehicle detailing is labor-intensive and requires specialty equipment and products; therefore, an independent industry has developed to fill this growing need. Some detailers will even come to a customer's workplace to detail the vehicle in the parking lot while the customer is at work. Dealerships also provide detailing services for customers while the vehicle is being serviced. In fact, every vehicle that goes through the collision repair process will receive some degree of detailing.

Detailing is often the service area where people start their careers as collision repair technicians. This task will help a new technician realize what quality truly is, and that in the collision repair industry the difference between a good job and a great job is often in the details—or in this case, in the detailing.

SAFETY

The technician must consider many safety issues while using detailing products and equipment. Some technicians do not follow personal safety precautions when using detail equipment and materials, even though some of the most harmful materials in the collision repair shop are found in the detail shop! As an example, silicone in the rubbing compound is dispersed into the air while buffing. Therefore, the technician needs to wear a particle mask and safety glasses during this process for protection. When using chemical cleaners, the technician should wear protective gloves. Even while washing a vehicle, the technician should wear safety glasses to protect the eyes from soap and other cleaning chemicals. Whitewall cleaners often have corrosive chemicals in them; a technician should use personal safety equipment while using these products as well.

Although it would be impossible to list all the safety precautions needed while performing detailing operations, the best source for personal, fire, environmental, and other protections is the product's material safety data sheet (MSDS). By law, the manufacturer must provide an MSDS for all hazardous products. It is the technician's responsibility to read, understand, and follow all safety precautions within the MSDS.

TOOLS AND EQUIPMENT

The tools that are needed to detail a vehicle come in many forms; they can be powered by electricity or air, depending on the operator's preference. Specialty tools are developed each year. The following equipment list, though extensive, may not include the most recently introduced items.

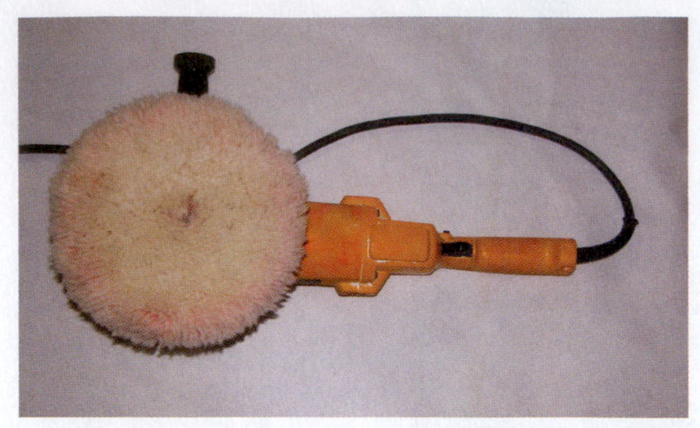

FIGURE 23-1 A large electric powered buffer should not be used in wet areas.

FIGURE 23-2 A pneumatic powered buffer can be used in wet areas, but ear protection should be worn when it is being used.

Buffers can be large, electrically powered machines or pneumatically (air) powered ones. See **Figure 23-1** and **Figure 23-2**. They can have buffing pads or bonnets, which are different sizes ranging from 9 inches to 3 inches in size. See **Figure 23-3**. They can use one speed or have adjustable speed controls. All of these different buffers and their bonnets have advantages and disadvantages.

Buffing should be performed at different speeds, depending on the type of finish that is being polished. For lacquers, the buffer should be run at a high rate of speed. This high speed will produce heat and the lacquer finish will soften and reflow, causing the finish to level out; this type of buffing will increase the shine. One of the cautions when buffing lacquer at this fast speed, however, is that it could remove all the finish very quickly, causing a problem called burn-through. Therefore, be careful when polishing at a high speed. To polish enamels or two-part urethanes, the buffer should be run at a much slower speed. The buffer should never exceed 2,000 **rpm (revolutions per minute)** during this procedure. In fact, some

FIGURE 23-3 Buffing pads come in a variety of sizes and are made from different materials. Choosing the proper buffing bonnet is as critical as choosing the correct polishing compound.

manufacturers of buffing bonnets and compounds recommend that the buffer be run at 800 rpm. Although enamels and urethanes do not reflow—and burn-through is not as common with these finishes as with lacquers—it could still happen if incorrect techniques such as high-speed polishing are used.

Modern electric buffers have become smaller and lighter than their predecessors, which is also an advantage when buffing two-part finishes. Light pressure is best when buffing these materials, and the older electric buffers are too heavy. The extra weight not only tires the technician out rapidly, but the machine's weight also applies too much pressure while buffing today's finishes. Although **pneumatic** polishers can be speed controlled, some operators dislike the constant and loud noise produced by them. Many prefer the quieter electric buffers.

CAUTION

Technicians should be aware that long-term hearing loss will occur when loud noises are present, but also that exposure to noises of as little as 100 db (or decibels) *for long periods of time* can be just as detrimental. If one must use a raised voice to be heard when talking to someone close by, that indicates that the ambient (surrounding) noise is above 100 db. In this case, hearing protection should be used.

The different bonnet sizes also affect the buffer's speed. If the buffer shaft is turning at 1,000 rpm and a 9-inch buffer pad is used, the outer edge of that buffing pad will be turning much faster than if the pad was 6 inches in size. Because the outer edge of a buffing pad is where most of the pad's work is done, a smaller (thus slower) pad should be used. The smaller pads are also easier to use in tighter or smaller places. The small 3-inch buffing pad is very good for polishing around antennas and door handles. See **Figure 23-4**. Electric buffers are made that can be adjusted from very low speed, such as 500 to 600 rpm, to 3,000 or 4,000 rpm. See **Figure 23-5**. Most detailers prefer these lighter, adjustable electric buffers today. See **Figure 23-6**.

FIGURE 23-4 A small 3-inch buffing tool and assorted bonnets are used for small, hard-to-reach areas.

FIGURE 23-5 Buffer speed is critical when polishing a vehicle.

CAUTION

Although electric buffers are preferred by detailers, caution should be taken when operating electric equipment near water. Unfortunately, water is often present in the collision repair department.

Sanding blocks or pads are among the tools commonly used while detailing. Some finishes will contain runs, sags, orange peel, or dirt that will necessitate sanding. When sanding, the technician should place the sanding paper on a hard sanding block or a sanding pad. See **Figure 23-7**. Sanding pads range from very soft, for use over areas where the paper should not dig into the surface—such as louvers in front of windows; to medium pads, used when the technician needs the sandpaper to be more aggressive, as when

FIGURE 23-6 Electric buffers, although heavier than pneumatic buffers, have become lighter in recent years and thus easier to handle, and often the preferred tool of detailers.

FIGURE 23-7 Hard sanding blocks help level a paint surface (such as orange peel), but caution should be taken around high crown ridges to avoid a "cut through."

removing orange peel; to stiff pads, for use when very aggressive sanding is required, as when removing dirt nibs. See **Figure 23-8**, **Figure 23-9**, and **Figure 23-10**. Specialty sanding pads are used to hold small sandpaper for spot sanding such as de-nibbing dirt in the new finish. See **Figure 23-11**. Special detailing sanding stones are manufactured for sanding operations ranging from run removing to de-nibbing. These stones are made from volcanic pumice pressed into small, easily handled stones of different grit, which can be used in small areas. See **Figure 23-12**.

Sandpaper used in the detailing process is not much different than that used in the refinish department, but it is much finer. This finer paper, ranging from P400 to P3000 grit, is needed to perform the least aggressive process to get the repair completed. Using the least aggressive sanding possible prevents scratches from forming that would take much effort to remove after the sanding process. Most sanding in the refinish department is done using the wet sanding method. This method uses water combined with a slight amount of car wash soap, which acts as a lubricant. Although a hose with running water may be used, the preferred method is to use a spray bottle to squirt clean soapy water onto the area being repaired. This method results in less water being used and less of a mess in the shop. See **Figure 23-13**. When using the older method of applying water with a rag or sponge from a bucket of water, the water in the bucket often became contaminated from the sanding debris. If the water has even small amounts of grit in it—from sandpaper as fine as P2000 grit—the debris in the water can be caught under the sandpaper. Sanding with the debris-containing paper will cause scratches that will damage the finish and may not be removed with sanding.

Dual action (DA) sanders have evolved from the original sanders. They are now sanders with large sanding orbits—3/8 inch—which are aggressive and remove mate-

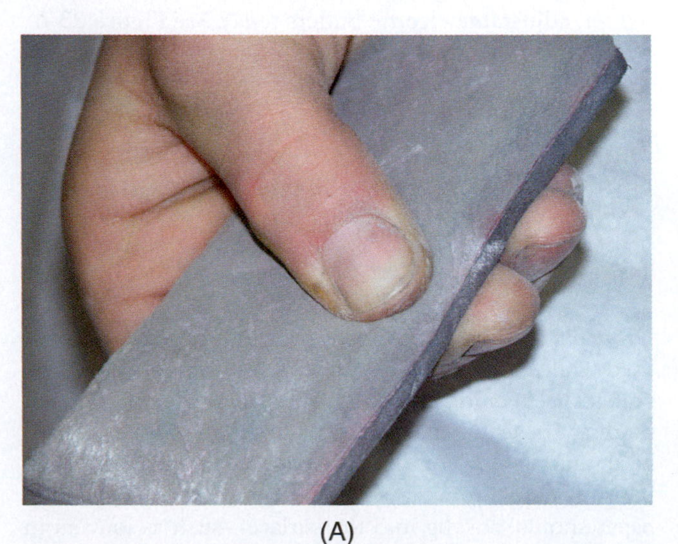

(A)

(B)

FIGURE 23-8 (A) Soft pads are less aggressive and can be used on uneven surfaces though care should still be taken. (B) This note pad has one soft side (top) and a harder (bottom) side.

(A)

(B)

FIGURE 23-9 (A) Medium soft paper can be used to remove orange peel prior to polishing. (B) This shows the harder side of a sanding pad.

(A)

(B)

FIGURE 23-10 (A) A hard pad is used to remove dirt nibs. (B) This pad has little to no give when squeezed, which makes it excellent for leveling a paint surface evenly.

rials very fast. There are also general-purpose sanders with an orbit of 3/16 inch, as well as fine DA sanders, which have small sanding orbits (3/23 inch) for very fine, lightly aggressive sanding. These small-orbit or fine sanders do not remove material fast, but can be used with finer grit detailing sandpaper that will not leave "DA signs" when the sanding is finished. See **Figure 23-14**. To further aid in the mechanical sanding without being overly aggressive, a foam pad is added to the sanding surface to keep the sandpaper cooler, and also allow for wet sanding with a standard DA sander. This fine DA sander is now commonly used to de-nib or remove orange peel prior to buffing or polishing.

Polishing pads come in many different sizes and types. The size differences were explained previously, but the technician must also know the different types as well. Pads might be wool or a wool blend mix, or might be foam

FIGURE 23-11 Sanding "stones" are used for de-nibbing defects from paint; note the P-grit on the stone.

(A)

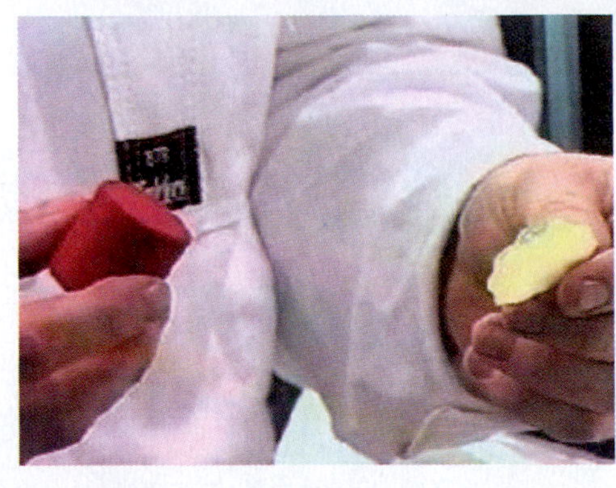

(B)

FIGURE 23-12 (A) Small detailing sanding blocks are also available. (B) Assorted fine grit sanding disks are available for detailing blocks.

FIGURE 23-13 Clean debris-free soap and water is squirted on the area being detailed for sanding lubrication.

FIGURE 23-14 Less aggressive "painters" dual action (DA) sanders are available with a small 3/32-inch stroke.

TECH TIP If the orbit of the DA sander is not known, a technician can perform this simple test to find out its orbit. First, turn the DA sander over. With sandpaper on the device and using a felt tip marker, run the DA sander while lightly touching the marker to the sandpaper. When an orbit outline has been made with the marker, stop the DA sander and measure the size of the orbit. An orbit of 3/8 inch can be used with aggressive paper of P150 or coarser. The 3/16-inch orbit can be used with above P150 to P400 grit paper; and the 3/32-inch orbit can be used with finer than P400 paper. Paper that is finer than P400 should use a foam intermediate pad on the DA sander. See **Figure 23-15**.

pads. See **Figure 23-16**. Foam pads can also be different in the cell size (the size of those small air holes in the pads) or in the amount of aggression they have. The ones with large cells (holes) in them are generally stiffer and more aggressive, while pads with smaller cell size are less aggressive and should be used with finer polishing compounds. See **Figure 23-17**. Pads also may be flat or waffled (a lumpy uneven surface). See **Figure 23-18**. This waffling allows the pad to run at a cooler rate, thus making the pad less aggressive than ones without waffling.

Orbital polishing machines are used for applying very fine compounds and **glazes** used for finish application while detailing. These tools have small 6-inch highly waffled pads, and they run very coolly and slowly. See **Figure 23-19**. This is exactly what is needed to apply the glazing compound and bring about the perfect finish that the detailer is

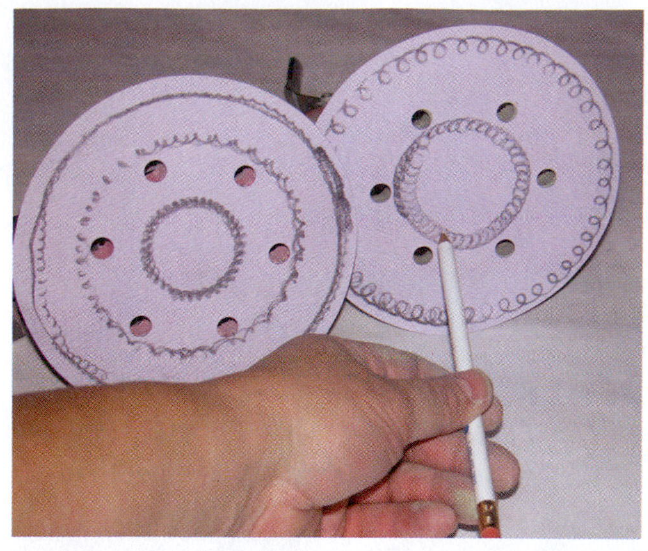

FIGURE 23-15 When checking a sander's stroke pattern, note the smaller pattern on the left.

FIGURE 23-16 Wool pads come as all wool (more aggressive) and wool blends (less aggressive) pads. The proper choice of pads is critical.

FIGURE 23-17 Foam pads also come in differing degrees of aggression. Larger cell size generally indicates a more aggressive pad.

FIGURE 23-18 Smaller and softer waffled pads are used for less aggressive, more delicate work.

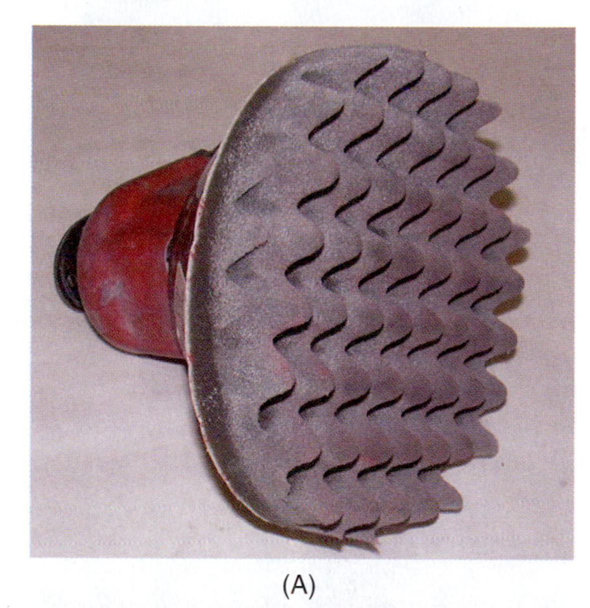

(A)

(B)

FIGURE 23-19 (A) A large waffling pad with very soft foam is used for fine detail work. (B) A polishing pad is used on a dual action sander for removing fine defects and high gloss polishing.

FIGURE 23-20 A cleaning spur is used for wool bonnets only.

FIGURE 23-21 A cleaning brush is used for foam pads.

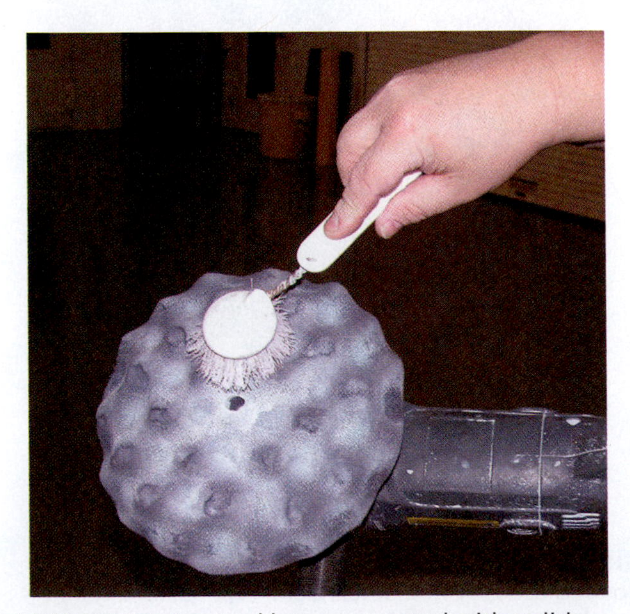

FIGURE 23-22 As a pad becomes coated with polish, gentle cleaning with a brush is needed.

(A)

(B)

FIGURE 23-23 (A) After extended use, both foam and wool bonnets can be washed and used again. (B) A pad cleaning machine can also be used to clean pads.

seeking. An orbital polishing machine is less likely to leave swirl marks than a standard buffing machine.

Pad cleaning equipment is also needed during detailing jobs. Although pads can—and should—be cleaned as they get filled with compound and debris, they can be temporarily cleaned while being used. Wool and wool blend pads should be cleaned with a *spur.* See **Figure 23-20**. However, foam pads should not be cleaned with a spur, but rather with a *cleaning brush.* See **Figure 23-21**.

Even though the pads are cleaned from time to time with a brush or spur, they should also be washed with soap and water occasionally to remove imbedded debris. This can be done by hand in a sink or in a *pad-cleaning machine.* See **Figure 23-22** and **Figure 23-23**.

FIGURE 23-24 A heat gun and plastic razor blade are used to help remove stickers.

Power washers are also used for many detailing operations, from cleaning engines to cleaning fully cured painted surfaces. (On new paint, though, it is recommended that the vehicle be *hand washed* with the **two-bucket method of washing**.) A power washer can operate at pressures as high as 3,000 psi, and should therefore be used with caution around delicate engine electrical and computer parts.

Vacuum cleaners are another necessity in the detailing process; no vehicle is thoroughly cleaned without the use of a vacuum. They are used for cleaning headliners, carpet, trunks, floor mats, pickup boxes, and for many other duties in a collision repair shop. Though vacuums are available in what are called wet/dry models, it is best to have separate vacuums: one set up specifically for wet use, and another for dry.

Paint thickness gauges are used to determine the thickness of the paint film before the vehicle is finished, and then again after the vehicle is refinished. With this procedure the detailer can determine just how much finish can be removed during the detailing process. Most paint makers require a minimum thickness of 4 **mils** (a mil is the standard unit of paint thickness) of finish to remain, in order to sufficiently protect the vehicle from harm. Without a **film thickness gauge**, a detailer will not know how much finish has been removed.

Heat guns are often used to heat up unwanted stickers on vehicles, which then can be removed easily with a plastic razor blade. See **Figure 23-24**.

DETAILING MATERIALS

Detailing materials come in many different types and names. In fact, the names of the different materials can be confusing. One material may be called a "polishing compound" even though the label or product directions indicate that the material is equal to P800 grit. This is an aggressive material and should only be used when something this aggressive is needed. It is far more aggressive than the "**polish**" name implies. Another material may be called a "**compound**," which implies that the material is aggressive and needed only when a defect or texture needs to be removed; but in fact, some polishing agents are named compounds. In short, a technician must thoroughly study a material's properties before it is used. When new products are introduced into a detailing system, they should be used with caution until the technician fully understands how they work.

A staggering array of materials can be chosen to detail a vehicle. Although it is not as critical to use the same brand products throughout the detailing process, as with finishes, it is still a good practice to do so. If the same company's materials are used through the detailing process and a problem is encountered, that company will be able to help identify how the problem can be eliminated.

> **TECH TIP** Technicians should use the "**least aggressive method first**." This means that the technician initially uses the product or technique that is the least damaging to the finish when removing a defect. Many techniques for removing defects will damage the finish slightly in the process, by leaving scratches behind after their removal. That situation requires another procedure to remove those scratches, until the finish is restored. It is always easier to move to a more aggressive method if needed than to be disappointed with the amount of work needed to "repair the repair."

BUFFING/POLISHING TECHNIQUE

In order to become both skilled and quick at buffing and polishing, technicians must practice. These two traits—speed and quality—will earn a technician both a favorable reputation and increased income. Many new technicians to the field start by mimicking a technique that has been developed and perfected by others. Technicians eventually develop the individual techniques that work well for them, but to do so they must both practice and observe, then practice more. With this observation, they will develop their own specific techniques.

While there are a number of techniques that must be developed to become a master (sanding, nibbing, vacuuming, washing, glass cleaning, odor removal, tire cleaning, waxing, and many others), the technique of most importance is buffing and polishing.

Buffing and polishing are not the same things. In fact, they require two different techniques, when done correctly. Buffing is generally more aggressive, and is done with the more aggressive wool or wool blend bonnet. See **Figure 23-25**. It also uses a different, coarser, and more aggressive compound, with more pressure applied to the buffing machine as the work is done. Buffing is not needed as often on basecoat/clearcoat; when such buffing is done,

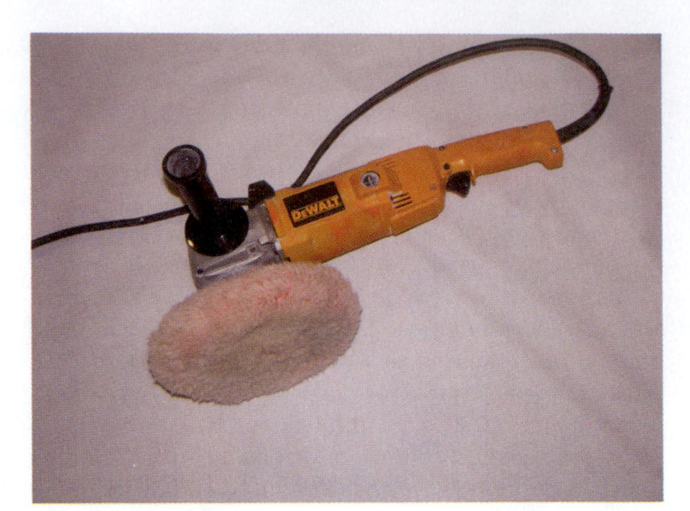

FIGURE 23-25 Aggressive buffing equipment includes large heavy buffers with a wool blend bonnet.

the technician must be careful to avoid being too aggressive with a new finish. Swirl marks are more likely to occur when buffing.

TECH TIP **Swirl marks** are circular marks left by aggressive buffing and, less often, during polishing. They are *defects* that should be removed. These defects should not be covered with glazing compound, as that will cause the swirl marks to reappear after several washings.

Polishing, on the other hand, is more often done with foam pads, which are less aggressive and manipulated with less pressure.

TECH TIP It is a common misbelief that the more aggressive a technician is when detailing, the faster the job will get done. This could not be further from the truth. Often an aggressive detailing process will cause more damage, which will take longer to correct than if the detailer would have used the least aggressive method first!

Once the service has been inspected, and the repair completed and cleaned, it is time to either polish or buff.

To buff, the technician should choose a wool pad. A pure wool pad is used to perform the most aggressive buffing. If the pad is new, it should first be cleaned with a spur to remove the loose wool hairs. See **Figure 23-26**. Then compound is either applied to the bonnet in an X pattern, or applied to the area to be buffed. See **Figure 23-27**. The buffer is turned on at a very slow speed, and the compound is evenly applied to the area to be buffed.

FIGURE 23-26 This wool bonnet is being cleaned with a spur.

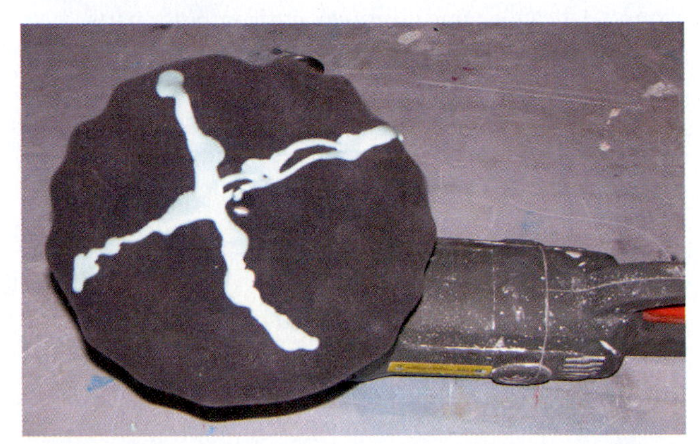

FIGURE 23-27 Compounds can be applied to the pad in an X pattern, or applied directly to the area to be buffed.

When the compound is spread on the vehicle instead of the bonnet, the operator can move the machine from the vehicle up into the line of compound with a left to right movement. Buffing machines generally turn in a clockwise manner. If the buffing bonnet is turning to the right and the line of compound is hit with the bottom of the bonnet as it spins to the right in a slow manner, the compound will be applied more evenly, without too much loss of the compound. The machine should be turning at a slow rate until the compound is spread.

Buffers that are made to start up gradually when triggered are better than buffers which come on at high speed immediately. With gradual power-up, the buffer allows the compound to stay on the area to be buffed, rather than being slung off by the rotating speed.

The area to be buffed or polished should be kept to about a 1 1/2 square foot section. This is about the width of person's shoulders. Compound should not dry out until its work has been done. When it is applied to too large an area the desired results will not be realized.

TECH TIP Detail technicians often keep a squirt bottle with water handy, especially during hot days, to keep the compound moist until the buffing or polishing process is complete.

Buffing and polishing can generally be done in four passes—first in one direction then the opposite—which are repeated until the compound is used up or the buffing is complete. On the first pass with the buffer, light to moderate pressure is applied while moving the buffer. Each back and forth stroke should overlap the previous one slightly, to make sure that the complete area is covered. The second set of passes is done in the opposite direction with lighter pressure. This lighter pressure will allow the machine to spin slightly faster again, overlapping the previous streak slightly. The third pass uses even lighter pressure, and for the fourth, only the weight of the buffing machine is on the bonnet. To summarize the technique, each subsequent pass has lighter pressure and will allow the machine to spin faster. Keep in mind that when the last strokes are being performed, the machine will be turning the fastest. That speed should not exceed the compound's maximum recommended speed. At this point the technician needs to decide whether to move on to the next step or repeat the compounding step again.

What a technician wants to accomplish is called "**depth of reflection**." This is the clarity of the reflection of objects seen in the finish. If technicians can see their face reflected in the finish clearly while standing upright, and there are no swirl marks, the detailer is ready to proceed to the next step. After compounding comes polishing, which may be a two- or three-step process.

First, the compound residue should be removed from the area that has just been compounded with a clean towel. One of the most common mistakes that a technician will make when compounding is to apply too much compound. When compound is overused, not only will it be slung all over the technician and in unwanted areas on the vehicle, but it may also "soak" the bonnet, leading to the bonnet becoming "loaded." This will require frequent cleaning with a spur or washing.

TECH TIP When compounding, be careful around open body lines. Compound could get into them and become very hard to remove. Many technicians cover these openings with tape while buffing to keep them clean. See **Figure 23-28**.

Polishing is very similar to compounding in the way the technician uses the tool. What is different is the polishing material—the polishing bonnet (usually foam) and the smoother, more scratch-free finish with even more depth of reflection. The bonnet is generally foam, the recommended speed may be slower, the bonnet is cleaned with a brush instead of a spur, and the compound is much less aggressive. See **Figure 23-29**.

Polishing may be the first step in the detailing process, even on a new finish. However, if what is needed to produce a "like new" finish is only nibbing and light polishing, the compounding step is completely skipped.

Polishing may be a two-step process, with the technician first using a light aggressive polish—sometimes called a cleaner/**polisher.** This light polish will remove

FIGURE 23-28 This body opening is covered with masking tape to keep compound out.

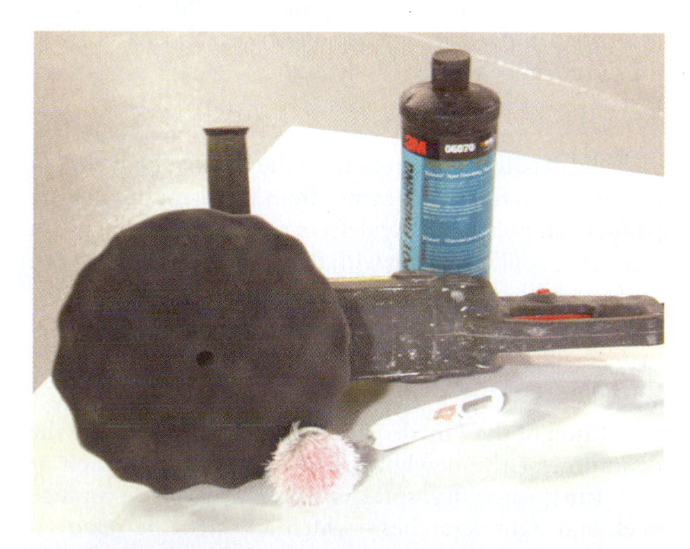

FIGURE 23-29 Polishing equipment includes foam bonnets, compound, and a cleaning brush.

only very small scratches, sometimes referred to as **micro-scratches**. These micro-scratches are so fine that they may appear like a haze on the surface, causing reduced depth of reflection. If the haze or "micros" don't come out after polishing the surface a couple of times, the technician may need to compound the surface to remove them, then progress to polishing, then move to glazing.

If the polishing to be done is a two-step process, not only is the polish material different, but the foam bonnet is also changed when the next step is started. The new bonnet is less aggressive, as is the material.

Foam bonnets, like wool, need to be cleaned from time to time. However, they are cleaned with a brush instead of a spur. A spur would destroy the foam, whereas the brush can be applied lightly to remove the excess material.

TECH TIP If a bonnet becomes filled with either compound or polish, it should be washed. Washing can be done by hand, or by using a special machine.

DETAILING NEW FINISH

Detailing a new finish is the most common detailing task in a collision repair facility. It differs from other detailing processes because the finish has not fully cured. New finish can take as long as 90 days to fully cure; if detailing is needed, it must be performed before that time.

Each paint manufacturer has different recommendations for when paint can be detailed. Some new hyper cure paints are baked for as little as 8 minutes at 120°F. Then, when the vehicle is at room temperature, it can be sanded and polished as needed. It is easy to see that a different, less aggressive method must be used when a finish is this new.

Many decisions must be made before the vehicle is detailed. The technician should inspect the vehicle, determine which repair (detailing) method should be used, color sand as needed, buff or polish, then apply a sealer to finish the process.

Inspection

If the detailer is different from the painter, both should inspect the vehicle as it comes from the paint booth. The painter can point out any defects noticed while the finish was being applied. Along with the detailer, the painter can decide whether the defect can be repaired with detailing, or whether it needs to be refinished. The detailer should inspect the condition of the finish, determine the film thickness, and make a detailing plan.

Condition of the Finish. The detailer should evaluate the condition of the newly applied finish. Defects such as dirt, runs, sags, dry spray, overspray, excessive orange peel, and light scratches—which all might be removed through detailing—should be noted. The finish should also be inspected for conditions which cannot be removed by detailing, such as color mismatch, bleeding, fish eye, and contour mapping, just to name a few. After the surface is inspected the detailer can make a repair plan. How aggressive should the repair be? Does it need buffing, or will just polishing be required? Will sanding need to be done, and—if so—how much? By having a repair plan, the technician can place the car in the proper area of the shop for the repairs needed; then the tools and materials may be gathered, and the proper amount of masking will be removed before the detailing process is started.

If extensive detailing is needed, the detail technician will not remove all the masking. In fact, much of it may be left on to protect the areas that will not be detailed from being contaminated with the detailing materials. Doing so will make it much easier to clean up the vehicle following detailing.

Film Thickness. As mentioned previously, film thickness gauges are used to determine the amount of finish that has been placed on the vehicle. Although most new vehicle manufacturers recommend that no more than 0.5 mils of finish be removed during detailing, most paint manufacturers recommend that a minimum of 2 mils of clearcoat remain following detailing of new finish. Two mils of clearcoat is the minimum amount that must be present to protect the finish from **ultraviolet (UV) rays**, which are damaging to automotive finishes.

How Much Film Thickness Can Be Removed? The only way that a detail technician will know how much clearcoat is on the vehicle is to take film thickness readings. See **Figure 23-30**. Most paint makers recommend that the basecoat film thickness should be from 1.5 to 2 mils thick. If the technician knows the film thickness before the refinish work is started, and what the thickness is following the repair, the before-repair film thickness reading and the 2 mils of basecoat can be subtracted from the end film thickness reading. The result determines the

FIGURE 23-30 Film thickness readings are critical when detailing a vehicle. Sufficient clearcoat must remain after detailing to protect the vehicle's finish.

FIGURE 23-31 This panel has film thickness readings as follows:

P800 grit sanding—6.8 remain
P1000 grit sanding—6.3 remain
P2000 grit sanding—6.1 remain
Compounding—5.9 remain
Polishing—5.7 remain

amount of clearcoat that can be removed during detailing to still have the minimum of 2 mils clearcoat remaining to protect the vehicle.

Detailing technicians should be aware of just how much finish is removed after each step of the detailing process. **Figure 23-31** shows a panel painted with basecoat; its film thickness is recorded at the top. Different types of detailing processes are performed on the panel, with the reading of the film thickness recorded on the area that was detailed. The amount of finish removed is also listed. Once a plan is made and the technician knows what processes are needed to repair the area, the detailer can determine if enough finish will remain after the necessary detailing has been completed.

Least Aggressive Methods First

As stated earlier, the least aggressive method that will complete the job should be chosen first. If the technician believes that overspray can be removed either by using detail clay or by using polishing compound, the clay should be used because it is the less aggressive of the two choices. If it is believed that washing alone will remove water spots on a finish, the technician should not start with polishing compound.

Many low or minimally aggressive methods can be used to remove defects, such as deep cleaning, detail clay, waxing, and glazing—just to name a few. These less aggressive methods should be considered before progressing to a harsher method.

Deep Cleaning. **Deep cleaning** differs from soap and water washing in that chemical cleaning usually follows it. Soap and water will remove dirt, salt, bugs, and some tree sap. Such debris is a large part of what is deposited onto a vehicle by normal use, and for this reason soap and water is used first. It does not, however, remove oil-based materials and waxes. Road tar, oils, greases, waxes, and many other products, which are also commonly found on vehicles, will not be removed by soap and water, and therefore need a wax and grease remover. Technicians are often surprised to see how dirty their cloth becomes when using wax and grease remover, even after the vehicle has been cleaned with soap and water.

Although a vehicle has been cleaned with soap and water *and* with wax and grease remover, there may still be debris that remains on the vehicle. To remove this, detail clay will be needed.

Detail Clay. Although it looks like children's modeling clay, **detail clay** is not the same. See **Figure 23-32**. It is a product designed to clean off those products that stick to the vehicle that are not removed through washing.

To use detail clay properly, the technician should take it from its airtight container and pull a small amount from the block. Then it should be kneaded to soften it. When the clay becomes soft and pliable, the area to be detailed should be lubricated with either the recommended spray, or with a solution of water with a light amount of

FIGURE 23-32 Detail clay is used to remove minor defects such as overspray.

FIGURE 23-33 Before using detail clay, warm it in the microwave or use a heat gun. This will soften the clay for ease of detailing.

soap in it. CLAY SHOULD NOT BE USED WITHOUT LUBRICANT!

The clay is rubbed over the area to be cleaned with light pressure; as the clay passes over the debris, it will stick to the clay and be removed. As the clay gathers debris, the detailer should occasionally knead the clay to provide a clean surface. When the clay becomes laden with debris and can no longer be kneaded clean, the piece should be discarded and a new one prepared for use.

If the clay is dropped it should be kneaded before being used again, for it will have picked up all the dirt that it hit on the floor.

TECH TIP To make the detail clay more pliable, it can be placed in a microwave for a few seconds. Don't heat it too much, though, as that will make it too hot to handle. Just warm it enough to become more pliable. The clay will then be easier to knead. See **Figure 23-33**.

Polishes. Polishes can range from cleaning polishes, which have only a slight amount of abrasion, to those that are very aggressive. Just because the product is called a polish does not mean that it has no grit. With this in mind, the technician should carefully read a product's application instructions to see how much damage it may cause. Polish may be applied with a **random/orbital buffer**/polisher; however, using an orbital applicator reduces the possibility of swirl marks. In fact, polishes and a DA-type applicator may be the precise tools needed to remove swirl marks. See **Figure 23-34**.

Although polishes are often the last item applied in the detail process, glazes may be used following polishes to provide a light layer of protection.

FIGURE 23-34 DA polishers are used to remove "swirl marks."

Glazes. New finishes will continue to "gas out" for up to 90 days, depending on the type and manufacturer of the finish. Because of the need to let the finish continue to gas out until all the solvents have evaporated, the new finish should not be waxed earlier, which would cover the surface and stop the evaporation.

Many technicians like to provide a degree of protection to the new finish and will finish the job by using a glazing compound. Not all glazing compounds are alike, though. Some are called "fill and glaze" compounds. Although they are very good at covering up swirl marks, this glaze will wash off in just a few washings and again reveal the swirl marks.

CAUTION

Swirl marks are *defects* and should be removed, not just covered up with "fill and glaze." If the fix is only temporary, a customer's satisfaction will only be temporary!

Other glazes may not be of the fill and glaze type; they may be just a breathable covering that will provide shine similar to wax but still allow the finish to completely cure. Their protective ability will only last a short time, and these glazes should be followed up with wax after the proper time has elapsed, to better protect the vehicle.

Color Sanding

Color sanding, the most aggressive process, should only be chosen when the surface of the finish is in need of leveling due to texture, dirt, or runs, all of which must be in the clearcoat. If the defect is in the basecoat below the clearcoat, it will not be repairable through detailing.

Color sanding can be done by hand with a block, or with fine detailing DA sanders. Sandpaper from P800 grit to P3000 can be used by hand with sanding blocks to level large areas as in removing orange peel. Alternatively, small nibbing sanding blocks with sandpaper in very fine grits can be used for spot-removing dirt or runs. Detail blocks are also available for spot detailing.

Hand Sanding (Blocking). Hand sanding, or blocking, is generally done on large areas where the finish needs to be leveled, then buffed and polished. This procedure may be necessary to remove dry spray, light sand scratches, or long runs. Because the sanding direction is back and forth, the scratches that result never truly go away: they just get so fine that the eye does not see them. Block sanding should be done in what is called "**line of sight**" directions. This means that the technician sands in the direction that the vehicle is normally viewed in. For example, a hood should be sanded from the front to the rear, not across the hood. Likewise, a roof is not sanded across and down the sides; instead, it is sanded from front to rear.

Sanding is done in a progression from the coarsest to the finest sandpaper, and most detailers will sand to a grit of P2000 before starting buffing or polishing. By sanding to this fine level, the vehicle will buff or polish faster.

Although starting with P1500 grit paper and progressing to P2000 can repair most vehicles, some may require the technician to start with P1000 grit, then go to P1500, and finally use P2000 grit paper before buffing/polishing.

Depending on the finishing system being used, a vehicle that has been finished to P2000 may only need to be polished to reach the desired depth of reflection.

Mechanical Sanding. As stated earlier in the tool section of this chapter, a finish or detail DA sander is one with a small orbital pattern of usually about 3/32 inch. Finish DA sanders with a foam pad attached using P300 sandpaper can produce a nondirectional finish that will polish out very quickly.

Although a DA sander with a foam pad will not level the finish much, neither will it leave behind marks that may be hard to polish out. Often detail technicians will first level the surface to below the defect by hand, which is accomplished very quickly. By then using a finish DA sander with P3000 grit paper before a light polish, the job is done as quickly as possible. Although many different companies produce finish DA sanding products and their recommendations should be followed, most of these products can be used either dry or wet. The area to be sanded is lightly sprayed with a squirt bottle and the DA pad is moistened; then the area is quickly sanded. The DA sander should be kept moving and only light pressure should be applied. The surface, if being wet sanded, should be kept damp with the squirt bottle. See **Figure 23-35**. Squeegee off the area being sanded occasionally to monitor the progress. As soon as the desired repair has been reached, the panel can be cleaned and prepared for polishing.

FIGURE 23-35 Fine sanding before polishing is done with a finish DA sander and a squirt bottle to moisten the surface.

Buffing/Polishing

As explained earlier, buffing with a wool bonnet may be too aggressive for new clearcoat paint. In fact, if the vehicle has been prepared properly with sanding to P2000 grit sandpaper, it will not need to be buffed. Polishing with a foam bonnet and then a polishing DA sander with glaze may be all that is needed to obtain the like-new depth of reflection. This polishing should be done in the same manner that was explained in the technique section of this chapter.

Glazing. When the desired depth of reflection is reached and all defects—including swirl marks—have been removed, the finish should be glazed with a product designed for fresh clearcoat finishes. Using a product of this type will allow the finish to completely cure and will also provide a temporary protective coating.

DETAILING FULLY CURED FINISH

Detailing a fully cured finish (such as OEM), or a refinish which has cured for 90 days or longer, requires a different approach than detailing fresh finish. The finish that has cured is "harder" than fresh finish. It may have been "faded" by the sun, and it will certainly have been bombarded by road and environmental debris; therefore, the way it is detailed is much different.

Washing

The fully cured vehicle must be washed before it is closely inspected. First the detail technician should power

FIGURE 23-36 The two-bucket washing method keeps debris picked up by the mitt from damaging new finishes.

FIGURE 23-37 Dragging a chamois or terry cloth towel will help to dry the surface without causing streaks.

wash the vehicle to remove as much of the loose debris as possible. Then it should be soaped with a wash mitt to remove the dirt that was not removed by power washing. The washing should be done using the two-bucket method. See **Figure 23-36**. The first bucket is filled with warm soapy water; the second bucket is filled with clean, clear water. The wash mitt is first soaked in the soap and water; then a panel of the vehicle is washed. The wash mitt is then cleaned off in the clear water to get rid of the dirt that it picked up while washing the panel. Then the mitt is soaked in the clean soapy water again, and the process is repeated on the next area of the vehicle. This method reduces the chance that debris picked up by the washing mitt will cause damage to the surface while the vehicle is being cleaned.

TECH TIP The soap used for cleaning a vehicle should be car wash soap. It is designed to be **pH** neutral (7) and free from other additives that could contaminate the surface following cleaning.

After the vehicle is hand cleaned, it should be dried prior to using wax and grease remover.

Drying

Although drying may seem like a simple process, there are several cautions that should be followed. When drying a vehicle, the synthetic **chamois** or terry cloth towel used should be dragged across the surface, not scrubbed. See **Figure 23-37**. Natural chamois cloths should not be used, because they attract dirt and debris, and will become abrasive with age. Cotton terry cloth is an excellent material to use to dry a vehicle because of its ability to pick up dirt, and then to release it when it is rinsed or washed. Terry cloth also can be easily dried by wringing it by hand or by using a wall-mounted wringer. See **Figure 23-38**.

FIGURE 23-38 A wall-mounted ringer helps make a drying chamois or towel last longer.

Microfiber cloths, or **detail cloths**, have become popular for drying, compound cleaning, and wax removal. Their ability to pick up things and then be cleaned again has led to their popularity.

Chemical Cleaning

Chemical cleaning is used to remove substances that are not dissolved with soap and water. Some of these substances are tar, grease, oils, and waxes. This deep chemical cleaning removes items that may be stuck to the surface that, if picked up during buffing, would cause damage. This method also cleans the surface so it can be properly inspected. **Chemical cleaners** work by dissolving the substance, then causing it to float to the top of the chemical. With this in mind, the technician should realize that when doing chemical cleaning, small areas should be cleaned, and then wiped off. If the surface is not wiped before it has a chance to dry, the floating debris may be redeposited on the finish.

When considering chemical cleaners, the technician should be aware that some areas of the country no longer allow the use of solvent-based wax and grease removers. In such cases, water-based wax and grease removers should be used instead.

Inspection

To properly inspect the vehicle, it first must be washed—that is, deep cleaned. The cleaning will require power washing with an automotive soap, followed by cleaning with wax and grease remover. Then, with the dirt and road grime no longer imbedded into the surface, the finish can be fully inspected.

To begin the inspection, first take a film thickness reading to determine how much paint is on the vehicle. See **Figure 23-39**. Several readings should be taken from each panel to obtain an average thickness.

TECH TIP There are several different types of film thickness gauges. See **Figure 23-40**. The less expensive slide types, which use a magnet and sliding gauge, are the least accurate.

The more accurate—and also more expensive—digital gauges are very accurate. Since the recommendation from at least one manufacturer is to not remove more than 0.5 mils when detailing, the meter must be able to read accurately to a fraction of a mil.

1 mil equals 1/1000 of an inch.

Next the vehicle must be inspected for any debris that may have stuck to the surface and was not removed when washed and chemically cleaned. Using a magnifying glass—either one with its own light or one that allows light to get in—will allow the detailer to closely inspect the surface. See **Figure 23-41**. Scratches (gouges into the finish) can be inspected to determine if they go through the clearcoat into the basecoat below. If so, they will not be able to be removed with detailing. Debris stuck into the film, such as rail dust or cinder flakes, must be removed before buffing. If not removed, such materials cause scratches when caught in the buffing bonnet. Other defects that might be noted are **acid rain**, which appears to be shallow craters etched into the finish, and **hard water marks**, which appear to be spots not etched into the finish. See **Figure 23-42** and **Figure 23-43**.

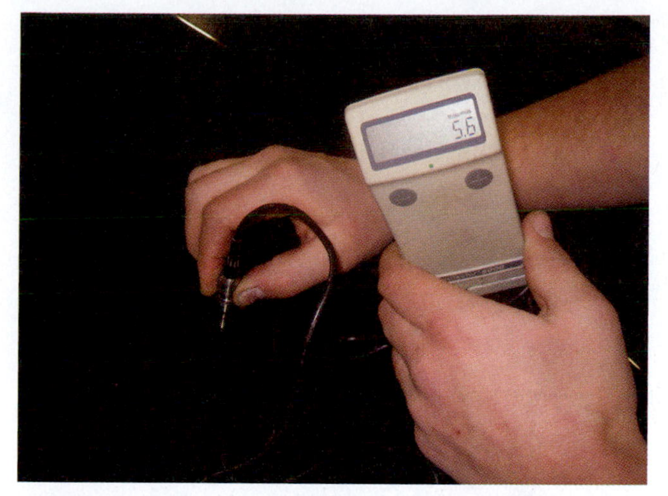

FIGURE 23-40 Electronic film thickness gauges are the most accurate gauges available.

TECH TIP A scratch that can be felt when a fingernail is dragged across it is most likely too deep for simply detailing without repair. See **Figure 23-44**.

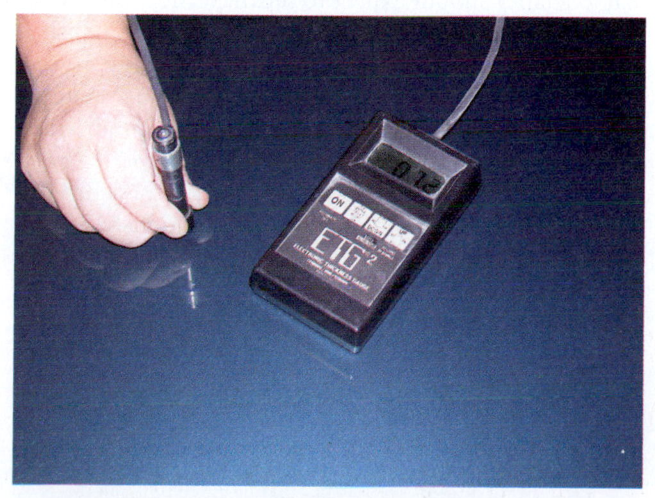

FIGURE 23-39 A final film thickness reading is taken to determine how much protective clearcoat remains.

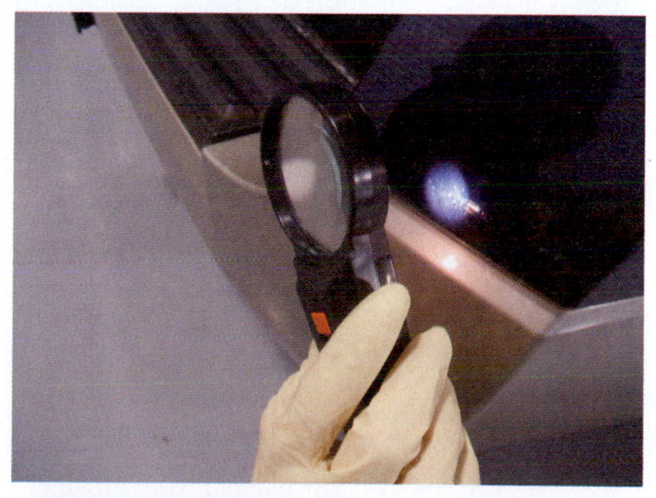

FIGURE 23-41 A magnifying glass with light can be used for close inspection.

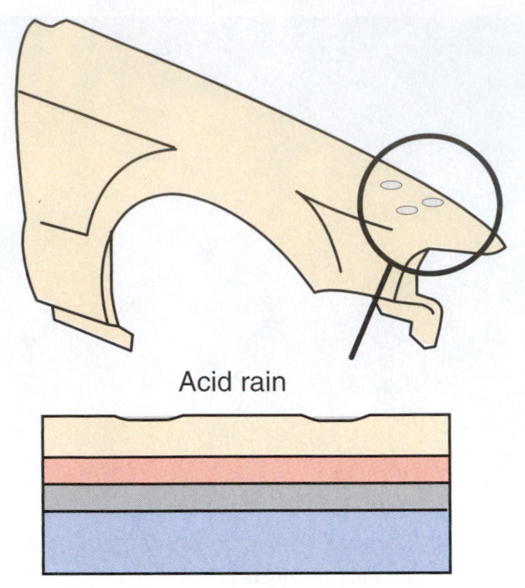

FIGURE 23-42 Acid rain defects appear to be shallow craters etched into the finish.

FIGURE 23-43 Hard water marks appear to be spots not etched into the finish.

FIGURE 23-44 When performing the thumb nail scratch test across a surface, if the scratch can be felt with a fingernail, it cannot be polished out!

If debris stuck to the finish is noted on inspection, it must be removed before buffing. The technician can do this by using detail clay. In the case of rail dust, though, to remove the unwanted substance the technician instead may need an acid that dissolves the metal. (This technique will be explained later in this chapter.) With the surface inspected and cleaned, the technician can make a plan to detail the vehicle's finish.

INDEPENDENT DETAILING SHOP

Detailing has become an often-needed process in several situations: when a new car is being prepped, when detailing a vehicle for owners who don't have the needed equipment, following a collision repair, or before a used car is placed on the lot for sale. The independent detailing shop has become a very lucrative business. One only needs to do an Internet search to find detail shops, suppliers, or even franchises with schools that will teach technicians how to develop their own businesses. Although independent detailers don't often need to detail fresh paint, they do need to have the skills to repair light damage caused from abuse, normal wear, and natural disasters such as floods and accidents. Although the same techniques that have been outlined earlier in this chapter apply to these technicians, some conditions that the technician in a collision repair shop or new car dealership would not often see may also need to be learned.

Repairing Finish Damage

One of the main reasons for a customer to bring a vehicle to a detail shop is to have damage repaired. Among these damages are light scratches, water spots, acid rain, rail dust, and **paint degradation** (fading). After the damage is removed, customers then want their vehicle to shine again and be protected. Independent detailers are then often called upon to apply waxes for this purpose.

Scratches. Scratches are one of the most common defects that a detailer should expect to repair. If the scratch does not extend into the basecoat, it can often be polished away. Even if the scratch cannot be completely removed, it can be made to look much less obvious. Some scratches are just buffed away using polishing compound, while others may need to be removed with compound followed by polish. Scratches that are too deep to be polished out should be removed as much as possible, polished, then covered with a good protective wax. Remember that most vehicle manufacturers recommend that a minimum of 2 mils of clearcoat must remain on the surface to protect the vehicle from UV rays. This normally means that no more than 0.5 mils can be removed from an OEM finish.

Vehicles that have been repeatedly brush-washed at automated car washes can develop light scratches; however, if these scratches are not too deep or severe, they can be removed with polishing.

Water Spotting. Water spotting, also called hard water spotting, is caused from contaminants in the water used while a vehicle is being washed. Water spots can also be caused from rain, though actually rain often contains a higher acid content, which will in fact cause acid rain spots.

Although water from municipal water supplies is normally treated or filtered to remove contaminants such as iron, its pH is also tested and corrected to a neutral (pH 7) before being distributed to its customers. Water used to detail vehicles should be free of iron or other minerals, have a neutral pH of 7, and have a low salt content. The better the water, the easier it will be to clean a vehicle.

Water spotting on vehicles can often be removed by deep cleaning as explained earlier in this chapter. If the water spots still remain after the vehicle has been soap and water washed, properly dried, then chemically cleaned, the technician should reexamine the vehicle to see if the spotting is caused from acid rain. Acid rain will appear to be etched into the paint surface, and if the spots are suspected to be acid rain, they should be neutralized before polishing. If the technician is sure it is not acid rain, the vehicle should be polished with a very light cleaning polish, then waxed, which will normally remove the hard water spotting.

Acid Rain Damage. Acid rain is created when rain falls through polluted air. The sulfur dioxide in the air combines with the water to form a very light mixture of sulfuric acid. Although the mixture of sulfuric acid is not strong enough to damage the vehicle's finish when the rain first lands on the vehicle, the mixture becomes stronger as it evaporates. This stronger acidic mixture will etch into the vehicle's finish. When the vehicle surface has dried completely, most of the damage has already been done, although the water used when washing may reconstitute the sulfuric acid. To get rid of the acid, it must be neutralized by mixing bicarbonate of soda (baking soda) with water at a rate of 1 tablespoon per gallon and rinsing the vehicle with the solution. Once the vehicle is neutralized, it can then be washed in the normal way, dried, and chemically cleaned. It must then be compounded to remove the etched-in acid rain marks. Following compounding, the vehicle should be polished, and then waxed for protection.

Rail Dust. **Rail dust** is a deposit of metal fragments into the paint on a vehicle. This occurs when the car is transported by rail or parked near railroads. Other types of metal deposits from industrial fallout can also contaminate a vehicle. Rail dust or other metal dust is most often noted on the horizontal vehicle surfaces such as roofs, hoods, and deck lids. It can be recognized on visual inspection by noting small rust particles on the surface of the finish. See **Figure 23-45**. Alternatively, it can be noted by feeling small sharp particles when a hand is slowly dragged over the finish.

To remove rail dust, detail clay may be sufficient, if the concentration is light. For more severe problems, though, a solution of sulfuric acid may be needed. Several companies market such rail dust removing solutions. The technician applies the product by soaking a towel in the solution, then laying the towel on the affected surface for a prescribed amount of time. Then the vehicle is rinsed with a neutralizing solution (baking soda and water). Be careful when using rail dust solutions, and follow all the recommendations and cautions. Technicians must read, understand, and follow the personal safety and environmental safety cautions contained in the product's MSDS at all times.

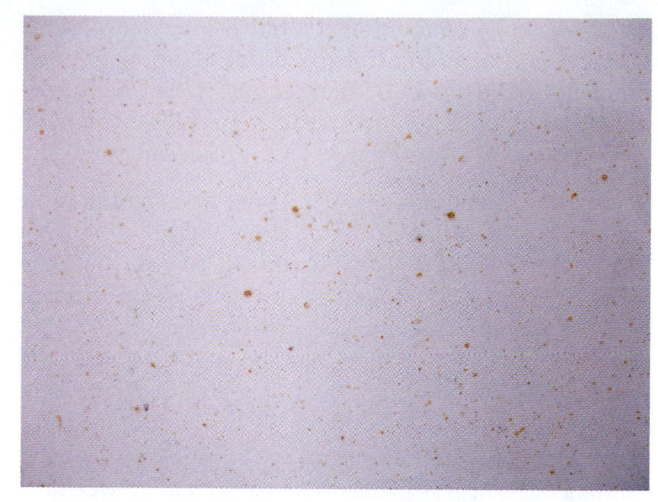

FIGURE 23-45 Rail dust with rust can be recognized on visual inspection by noting small rust particles on the surface of the finish.

Clearcoat Degradation. Clearcoat degradation means the clearcoat has faded over time due to exposure to the elements. Any finish that has been fully cured will start to break down with time. Its surface will dull as it breaks down; on some colors, as with white, it will look dirty; on black, red, and other colors, the finish will appear "milky." The degraded finish must be deep cleaned, checked for imbedded particles, clay cleaned as needed, then polished back to its original shine. Following the cleaning and polishing, the technician should apply a layer of wax to protect the finish from UV rays. Although all finish will eventually degrade, this degradation can be slowed with protective wax. Many vehicles—especially older finishes that have been detailed several times—will benefit from wax protection.

Interior

Independent detailing shops often detail interiors. Northern winter conditions can play particular havoc on interiors, as snow and water are tracked into the vehicle; then the heater runs, causing vapors to be deposited on nearly all parts of the vehicle's interior. This mist carries with it salt, oils, and other contaminants that not only dirty the vehicle's interior but also stain and cause fabric wear and corrosion on other parts of the vehicle and its electrical equipment.

Interiors should be vacuumed, then deep cleaned (if cloth interiors) or cleaned and dressed (for leather and vinyl). The shop should also clean painted surfaces, as well as the glass and mirrors. Interiors also present other problems that may not be found in other areas of the vehicle: stains and odors that result from spills, people becoming sick, and the deposit of films from smoking.

Vacuuming. Vacuuming of a vehicle, though seemingly simple, is made difficult by the degree of grinding in of dirt on a vehicle's carpet. The areas that need—but may infrequently receive—vacuuming are also nearly unreachable. To properly vacuum a vehicle, the technician first should remove all the larger and loose objects, and then do a quick vacuuming. Then, using a blowgun and wearing safety glasses and a face shield, the operator should blow the particles from the hard-to-get-to places out to a more accessible area. Areas such as between the seats and the console can be blown to either the back seat foot area or to the front. Debris from under the seat can be blown forward, and then vacuumed up with ease. Once the large particles have been removed, the carpet should be vacuumed with the aid of a stiff bristle brush in front of the vacuum. See **Figure 23-46**. The brush loosens the dirt from deep in the carpet and then the vacuum can remove it. The carpet can then be washed and the water and dirt removed with a wet-dry vacuum or **carpet extractor**.

Fabric and Carpet Cleaning. Fabric cleaning can also be somewhat tricky. Stains in a cloth interior or carpet may need to be spot treated to remove them. Generally the stains are acidic (or below a pH of 7). To neutralize them, most general purpose cleaners are slightly alkaline. To remove the stain, the technician should wet the area with all-purpose cleaner, and then blot it with a clean towel. See **Figure 23-47**. Never scrub the stain, because the scrubbing action is likely to spread it. When the blotting is finished, check to see if the stain has been removed, and repeat the process if necessary. When the stain has been removed, all the water should be removed from the carpet or cloth with a vacuum or extractor to speed the drying.

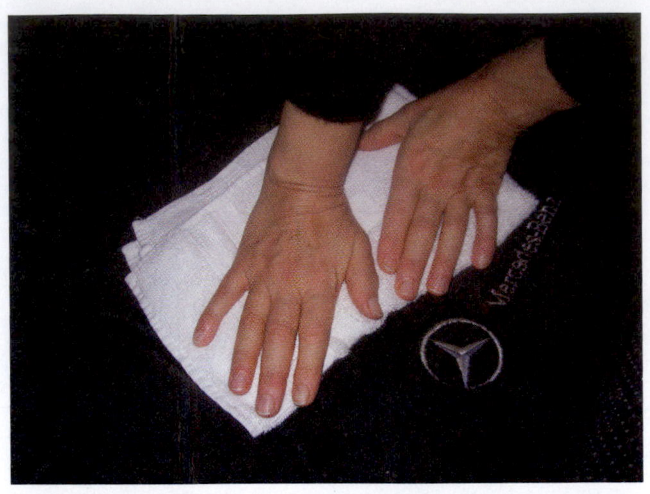

FIGURE 23-47 Blotting with a towel removes stains and spills without spreading them.

Headliner Cleaning. Headliners are fragile and should not be vacuumed or washed, or they may be harmed. A clean cloth dampened with all-purpose cleaner should be used instead to wipe them down. See **Figure 23-48**.

Dust Removal. Dust is a common contaminant in a vehicle's interior. It can be removed with a cloth dampened with all-purpose cleaner, with a vacuum by itself, or even with a vacuum and a detail brush combination. See **Figure 23-49**. A **detail brush** can be made by cutting the bristles of an inexpensive painter's brush down to about 1/2 inch and wrapping the metal part with tape to prevent scratching. See **Figure 23-50**. Dashboard dressing or interior dressing is also sometimes used to shine the vehicle's interior. However, be careful when using dressings, since some contain silicone oils that may contaminate a collision repair

FIGURE 23-46 A brush loosens the dirt so the vacuum can remove it.

FIGURE 23-48 All-purpose cleaner is sprayed on, then wiped off, a headliner to clean it. Scrubbing may damage the headliner.

FIGURE 23-49 A vacuum and detail brush combination can be used to clean hard-to-get-to areas.

FIGURE 23-50 To build a detail brush, cut the bristles of an inexpensive brush shorter to make them firmer and tape over the metal part to protect the vehicle.

shop; they can also easily get onto glass and be very hard to remove.

Odor Removal. Odor removal is often needed in vehicles for varying reasons, and the best way to control an odor is to clean the vehicle and eliminate it. If after a thorough cleaning an odor remains, there are several ways to combat it. First, DO NOT use scents, which merely cover the odor, for they will only cover it for a short time. Although odor eliminator liquids are available that can be used, eliminating the odor by cleaning is best.

A common place for odor to originate is in the air conditioning vents. Mold and fungus can grow in the air conditioner **plenum**, though the air-conditioning water weep hose being plugged commonly causes this growth. Spraying the central dash air conditioning vent with an antifungal disinfectant can eliminate the odor. The vent hole should be cleaned as well.

Odor can also come from doors where the trim pad was removed and the **vapor barrier** was removed or damaged. If the vapor barrier is not in place or is damaged, it will allow moisture into the interior, where fungus and mold can grow. This causes unpleasant odors. The vapor barrier should be replaced or repaired, the door trim back side sprayed with an anti-fungicide disinfectant, then the door reinstalled.

If the source of the odor cannot be found, an ozone machine can be placed in the vehicle with all doors closed and the windows rolled up for the recommended time, which will—in most cases—eliminate the odor.

Cleaning Glass

Glass cleaning should be one of the last—if not the last—interior detailing steps to be completed. By leaving it for last, the technician does not have to be concerned about dirtying the glass during the other cleaning steps.

General glass cleaners are good products to use for cleaning both interior and exterior vehicle glass. Interior glass on a vehicle in which the occupants have been smoking may be hard to clean. In fact, it may take more than one application of the cleaner to complete the job.

Because both the inside and outside of the glass must be washed, it is best to wipe the cleaner off one side in a vertical direction and off the other side in a horizontal direction. In that way, if streaks are noted when the job is completed, the technician will know which side of the glass still has streaks by their direction.

TECH TIP When cleaning glass that has had window-tinting film installed, do NOT use cleaners containing ammonia. Ammonia can damage window-tinting film!

Cleaning Convertible Windows

Convertible windows may be made from glass or they may be plastic. Although convertible glass windows can be cleaned in the same way other glass windows in vehicles are cleaned, plastic convertible windows must be treated with care. They are fragile and will scratch and damage quite easily, especially with harsh cleaning chemicals. Special plastic cleaners and polishes should be used so no damage will come to the window.

Cleaning Convertible and Vinyl Tops

Convertible and vinyl tops need special care both when washing and in dressing them following washing. Although they are cleaned with simple all-purpose cleaners, they often must be scrubbed with a brush to remove the grime from within the textures of the materials. The material also contains moisture that is removed when exposed to the UV rays and strong cleaning detergents. This means that after being cleaned, they should be dressed with a vinyl top conditioner or dressing.

Cleaning Wheels

Vehicle wheels are no longer just the plain steel rims of past generations that can be washed with a brush and soap. Today they are often made of chrome or metal alloy, and in fact they are the part most commonly damaged from detailing. They also become contaminated with very difficult debris that may even cause staining, such as brake dust on front mag wheels.

Wheel cleaners are often caustic and should only be used on wheels on which the technician knows the cleaner will not cause damage. The wheel should be cool when being cleaned, and care should be taken not to spray the wheel cleaner on surfaces that it will damage. Only the two wheels on the same side should be cleaned at one time. The technician should rinse the cleaner off with large amounts of clear water before the cleaner is able to dry. Then the opposite side can be cleaned. Wheels can be waxed following cleaning, which will protect the wheel surface from brake dust, at least for a short time.

Cleaning Engine Compartments

Although engine bays are not always cleaned when a vehicle is detailed, they are often included with a detailing service. They can be cleaned with a power washer, as long as the power washer is not pointed directly at sensitive electrical and electronic devices. Wrapping these sensitive parts in plastic before spraying the engine bay with a power washer should protect them.

The protected engine bay should be wet down first, sprayed in the bay with a degreaser, and then left to allow the degreaser to work. Following the recommended soaking time, the technician can rinse off the engine bay. The detailer should not let the engine cleaner dry before it is sprayed off, or the process will need to be repeated. Also, one should not forget to clean the underside of the hood when washing the engine bay.

Waxing/Sealing

For some detailers, the waxing step is one of the most enjoyable steps in detailing the vehicle. The wax often smells good; it glides over the surface of a thoroughly cleaned vehicle and produces a smooth, shiny, great looking finish. The depth of reflection is great, and all the previous hard work of detailing a vehicle is realized. There are many different types and configurations of wax on the market. Some contain carnauba; others are soft, hard, paste, liquid, and non-hazing, just to name a few. Most detailers find a particular wax that fits their style of application and provides the required amount of protection. One thing to keep in mind is that if a wax goes on with little application trouble, it may come off just as quickly. Waxes that are applied through a spray hose at a pressure car wash will come off just as quickly when the vehicle is rinsed. A good paste carnauba wax may take a little more time to apply, but in the long run the results are very hard to beat.

Summary

- Detailing is defined as the deep cleaning of a vehicle, and is done at many stages in a vehicle's life.
- Tools used in detailing include buffers, sanding blocks, polishing pads, paint thickness gauges, and power washers.
- The two-bucket method of washing involves using one bucket of soapy water to clean and one bucket of clean water to remove debris from the cleaning mitt.
- Detailing materials may be classified as polishes or compounds.
- The "least aggressive first" method means the technician initially uses the product that is least damaging to the vehicle's finish when repairing a defect.
- Film thickness, measured in mils, indicates how much finish is on the vehicle.
- Detail clay is used to clean off products that are stuck to the vehicle that can't be removed through washing alone.
- Finish damage can occur from scratches, water spotting, acid rain, or rail dust.

Key Terms

acid rain
ambient
buffer
carpet extractor
chamois
chemical cleaner
color sanding
compound
decibels (db)
deep cleaning
depth of reflection
detail brush
detail clay
detail cloth
detailing
detrimental
film thickness gauge
glaze
hard water marks
least aggressive method first
line of sight
micro-scratches
mil
paint degradation
pH
plenum
pneumatic
polish
polisher
rail dust
random/orbital buffer
revolutions per minute (rpm)
swirl marks
two-bucket method of washing
ultraviolet (UV) rays
vapor barrier

Review

ASE Review Questions

1. Technician A says that UV rays stand for ultravolatile rays. Technician B says that UV rays stand for ultraviolet rays. Who is correct?
   A. Technician A only
   B. Technician B only
   C. Both Technicians A and B
   D. Neither Technician A nor B
2. Technician A says that a buffer should not be run above 2,000 rpm when buffing urethane basecoat/clearcoat. Technician B says that larger buffing pads have a higher edge speed than the smaller ones. Who is correct?
   A. Technician A only
   B. Technician B only
   C. Both Technicians A and B
   D. Neither Technician A nor B
3. Technician A says that fine sandpaper for color sanding is used to smooth out orange peel. Technician B says that when hand sanding to remove orange peel, a sanding pad should be used. Who is correct?
   A. Technician A only
   B. Technician B only
   C. Both Technicians A and B
   D. Neither Technician A nor B
4. Technician A says that rpm stands for revolutions per meter. Technician B says that ROM stands for right turns per minute. Who is correct?
   A. Technician A only
   B. Technician B only
   C. Both Technicians A and B
   D. Neither Technician A nor B

5. Technician A says that polishing pads can come as wool or wool blend. Technician B says that polishing pads come as foam of differing aggressive qualities. Who is correct?
   A. Technician A only
   B. Technician B only
   C. Both Technicians A and B
   D. Neither Technician A nor B
6. Technician A says that an orbital polishing machine is a DA-like machine that is used to apply polish and will not leave swirl marks. Technician B says that DA stands for dual action. Who is correct?
   A. Technician A only
   B. Technician B only
   C. Both Technicians A and B
   D. Neither Technician A nor B
7. Technician A says that a film thickness gauge is used before detailing to find out how much film is on the vehicle. Technician B says that a film thickness gauge is not needed when detailing a vehicle. Who is correct?
   A. Technician A only
   B. Technician B only
   C. Both Technicians A and B
   D. Neither Technician A nor B
8. Technician A says that a buffing pad is cleaned with a brush. Technician B says that a polishing foam pad is cleaned with a spur. Who is correct?
   A. Technician A only
   B. Technician B only
   C. Both Technicians A and B
   D. Neither Technician A nor B
9. Technician A says that new finish is fully cured if it is baked. Technician B says that new finish is not fully cured for as long as 90 days after it is painted. Who is correct?
   A. Technician A only
   B. Technician B only
   C. Both Technicians A and B
   D. Neither Technician A nor B
10. Technician A says detail clay and children's modeling clay are the same. Technician B says that detail clay must be used with a lubricant. Who is correct?
    A. Technician A only
    B. Technician B only
    C. Both Technicians A and B
    D. Neither Technician A nor B
11. Technician A says that the best thing with which to dry a vehicle is a natural chamois. Technician B says that a synthetic chamois can become abrasive with use. Who is correct?
    A. Technician A only
    B. Technician B only
    C. Both Technicians A and B
    D. Neither Technician A nor B
12. Technician A says that a neutral pH is 10. Technician B says that a neutral pH is 7. Who is correct?
    A. Technician A only
    B. Technician B only
    C. Both Technicians A and B
    D. Neither Technician A nor B
13. Technician A says that swirl marks are natural and unavoidable. Technician B says that swirl marks can be eliminated with "fill and glaze" compounds. Who is correct?
    A. Technician A only
    B. Technician B only
    C. Both Technicians A and B
    D. Neither Technician A nor B
14. Technician A says that acid rain residue must be neutralized. Technician B says that a neutralizing solution is made by mixing 1 tablespoon of baking soda for every gallon of water. Who is correct?
    A. Technician A only
    B. Technician B only
    C. Both Technicians A and B
    D. Neither Technician A nor B

Essay Questions

1. In your own words, explain a buffing technique.
2. What is the "least aggressive first" theory?
3. List the steps used in the two-bucket washing method.
4. Describe how to use detail clay.

Topic Related Math Questions

1. A vehicle has a film thickness of 3.5 mils before refinishing. After refinishing, it has an average film thickness of 9 mils. The average amount of basecoat is 2 mils. How much of the 9 mils of film thickness can be removed to maintain 2 mils of new clearcoat?
2. A shop uses 17 sheets a day of P1000 grit sandpaper. The shop works five days a week, 50 weeks a year. How much P1000 grit paper will they need for a year of business?
3. A detail technician can complete seven vehicles a day for which she gets paid $37.50 for each vehicle. The detailer has worked 35 weeks. How much money has she earned?
4. A detail shop can buy a 5-gallon pail of polishing compound for $240.00. The shop is currently buying the polishing compound for $12.34 per quart. How much money will be saved by buying the compound in the 5-gallon container?

Critical Thinking Questions

1. As a shop manager, it is your job to purchase supplies at the most economical manner possible. You have the opportunity to purchase a 5-gallon container of polishing compound at an overall savings of $7.00. Taking all things into consideration (time taken to transfer it into a smaller container, spillage, and others), is it most economical to buy the larger container? Explain your answer.
2. Your daily schedule is full when a new customer comes in, requesting to have a vehicle detailed immediately. If you add the vehicle to today's schedule, one of the vehicles that has already been promised for delivery today may not get done. What factors in customer service must you consider to arrange the schedule?

Lab Activities

1. Have students deep clean a vehicle with the two-bucket method followed by chemical cleaning.
2. Have students block sand to remove orange peel.
3. Have students remove minor runs.
4. Have students buff a finish.
5. Have students polish a finish.
6. Have students detail the interior of a vehicle.

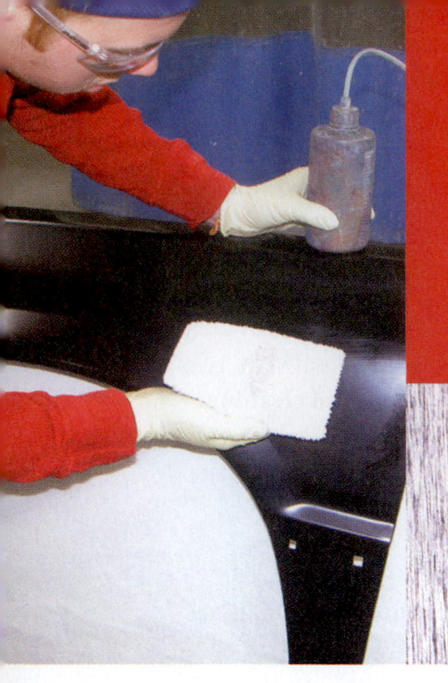

Chapter 24

Understanding Refinishing

OBJECTIVES

Upon completing this chapter, you should be able to:

- Describe the operations of different types of refinish shops, including the customer expectations in each instance.
- List the functions of a topcoat.
- Explain how paint chemistry affects a finish.
- List the different aspects of a finish's appearance.
- Describe how to protect both old and new finishes after application.
- Explain how to find a vehicle's paint code and how to convert it to a paint manufacturer's code.
- List the different types of color coats and their components.
- Explain the different types of refinish systems, their component layers, and how each layer affects the finish.

INTRODUCTION

The process and chemistry of automotive finishes and refinishes has evolved over the automotive industry's short 100-year history. Automotive finishes and aftermarket refinishes have developed and changed at a rapid pace throughout this history. In fact, the refinish market continues to be one of the most rapidly changing segments of the collision repair industry. This chapter will explore the different aspects of the refinish industry.

Collision repair refinishing involves many variables that the painter has no control over. When they occur, the technician must be able to recognize them and have a strategy to continue to apply a quality finish in spite of them. The technician should also be aware that many variables associated with refinishing *can* be controlled. Knowing what can be controlled and taking the necessary steps to do so helps the refinish technician to continually produce a high quality finish with short **cycle time**.

Industry experts believe that the better a technician understands all aspects of the finish being applied, the easier it becomes for the technician to consistently produce a quality finish.

SAFETY

Refinish materials and the chemicals they are made from pose personal and environmental protection concerns. The refinish technician is governed by—and must comply with—many local, state, and federal laws concerning protection.

Each product must be accompanied with a material safety data sheet (MSDS). The MSDS includes sections on:

- Product information
- Hazardous ingredients
- Physical data
- Fire and explosion hazards
- Reactive data
- Health hazard data, or toxic properties
- Preventive measures
- First aid measures
- MSDS preparation information

Technicians must read, understand, and follow the recommendations in the MSDS for the protection of themselves, their fellow workers, and the environment.

TYPES OF FACILITIES

Although this textbook is largely devoted to the collision repair industry, many other types of facilities exist that also refinish vehicles, either as their entire operation or as a part of their customer services. These different types of businesses may have varying primary goals. The facility's goal—or goals—will govern how the refinish technicians perform their tasks. In the economy refinish shop, for example, the technician's main goal may be to refinish a vehicle in the fastest and most economical way possible. Although speed and economy are of concern to the collision repair industry as well, other goals—such as returning a vehicle to its pre-accident condition—are key considerations. The key goal of a custom painting shop may be to produce a flawless, one-of-a-kind, highly creative finish. Each of these different refinish businesses will not only have different goals for their technicians, but the customer expectations in each case will vary as well. This chapter will explore the differences within these different facilities.

Collision

The collision repair refinishing industry is challenging. See **Figure 24-1**. The technicians must be able to restore a vehicle that has been involved in a collision to its pre-accident condition, and do so in an economical and efficient manner. The repaired vehicle must be as durable—and have the same life expectancy—as before its repair. The repair must be done in a timely fashion, and when completed, the repaired area should have the same life expectancy as the rest of the finish. The collision repair facility is responsible not only to the vehicle's owner, as their customer, but also often to the customer's insurance company, which may be paying the repair costs.

The repair must also meet or exceed the vehicle manufacturer's repair standards, as well as federal, state, and industry standards. A refinish technician must match the finish to the vehicle's **hue, value, chroma, and texture**,

FIGURE 24-1 Collision repair technicians must adapt to the ever-changing developments of vehicle construction.

and must meet these challenges under differing conditions while maintaining consistent quality standards. The refinish department is also obligated to follow the ever-changing environmental and clean air standards, while continuing to meet the customer's expectations of quality, speed, and economy.

Custom

The custom paint industry is also very challenging, but its challenges appear in different areas than in the collision repair industry. See **Figure 24-2**. Their finishes are original and seldom expected to match an existing finish. In addition, the beginning **substrate** is often not in need of repair prior to the custom paint being applied. Although excessive **film build** is a concern to the custom refinisher, custom painting often exceeds the film thickness of the paint maker's recommendation. Finally, though time is still a concern for the custom painter, it is not as critical as within a collision repair facility.

A custom painter's challenges lie mostly within the creativity and expertise of technique. Although most customers of the custom paint industry have a specific scheme in mind, the input and skill of the painter/artist

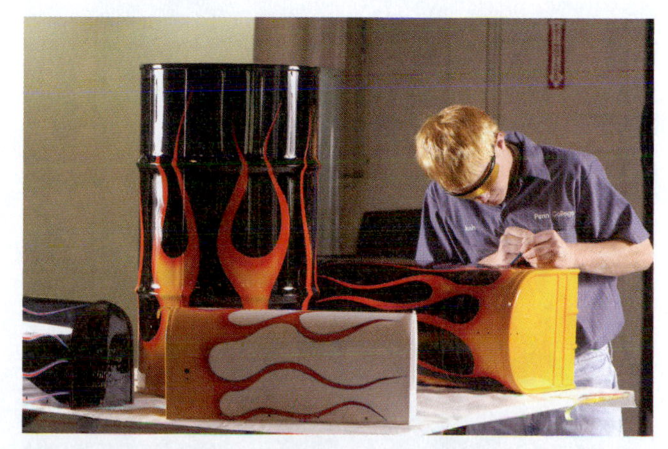

FIGURE 24-2 Custom painting, such as hand pinstriping, remains popular and in demand among vehicle owners.

are often sought. This means that the person applying the custom paint must not only be very good at applying a designed finish, but must also be able to create unique, one-of-a-kind designs for each customer. Custom painters with good reputations who have developed notable skills are often sought after for their work, for which they are able to receive attractive fees.

Economy

The economy refinish industry has developed to support those customers who are keeping their vehicles longer. These older vehicles may or may not have been involved in a collision. Often a vehicle in good working order is brought in just to freshen up the vehicle's exterior. These vehicles generally need to be completely refinished, rather than refinishing only a repaired area. While economy repair and refinish businesses do some collision repair work, generally their main repairs are to the normal wear and rust that vehicles may encounter during the aging process.

Because economy refinishing businesses do not often perform spot repairs, the technician does not need to be as skilled in matching color and texture as with the collision repair industry. The technician, though, must be skilled in performing the task as quickly and economically as possible. In this way, these businesses meet the market's need for fast, economical refinishing.

Customer Expectations

As in any business, the customer's needs and expectations must be recognized and met. See **Figure 24-3**. This can be done by thoroughly explaining what the technician will do while repairing the vehicle. Customers who own luxury vehicles may expect a higher level of repair quality than those customers who seek repairs in the economy market.

Collision repairs are those needed when a vehicle has been involved in an unexpected accident. Statistically, the average customer is involved in a collision once in 10 years. Due to the time that may elapse between each repair need, most customers are not fully aware of how to proceed with getting their vehicles repaired. Some will call their insurance agent before going to a collision repair center, while others will go directly to a favored repair center.

FIGURE 24-3 Customers expect high-quality, professional, and efficient collision repairs.

Many consumers do not have spare vehicles to drive while one is being repaired, and with some accidents an extended time period may be needed to gather the parts and repair the vehicle. Collision repair centers know the inconvenience accidents cause their customers. Shops that strive for customer satisfaction not only provide a fast, high-quality repair, but also may offer services such as rental vehicles available for pick-up and drop-off at the collision center. Such collision shops become direct repair facilities for as many insurers as possible to expedite the repair process. They have convenient and comfortable waiting areas for customers who may need to stay at the facility during the process. Employees also need to be aware that customers have been greatly inconvenienced by their vehicles being out of commission, and should do everything possible to assure that the repair process is done quickly.

Customers expect to have a vehicle **repaired to its pre-accident condition** in appearance and to be as roadworthy in operation as before. They expect the repair to be undetectable when completed. Some customers with newer luxury vehicles will expect to have their vehicle returned in showroom condition with no **diminished value**. Those who have had an accident and seek repairs through the economy repair market may only wish to have their vehicle returned to a safe, operable condition, and will accept some cosmetic flaws.

All customers expect to have their vehicles returned to them in a clean condition, with their personal items intact. These expectations include the radio station presets and the car seat being in the same position as when the customer dropped off the vehicle. Technicians should not play the vehicle's radio, look through the glove box when not necessary, or use any of the customer's personal items without permission while the vehicle is being repaired.

FUNCTIONS OF COATINGS

Coatings serve many purposes. They may be undercoats, to provide adhesion to both the substrate below them and the next coating that will be applied over them. See **Figure 24-4**. They also provide corrosion protection by covering metal and eliminating oxidation. They could be sealers, which act as barrier coatings between other undercoatings and the topcoats placed over them. Sealers also provide a smooth, flat surface for topcoat application, enhancing **flow out** and **finish holdout**. Basecoats provide the pigment and special additives—such as metallic, mica, and prismatic effect—which add to the finish's glamour and appeal.

The final coating is the clearcoating. Although this coat does not add to the color of the finish, it does have many

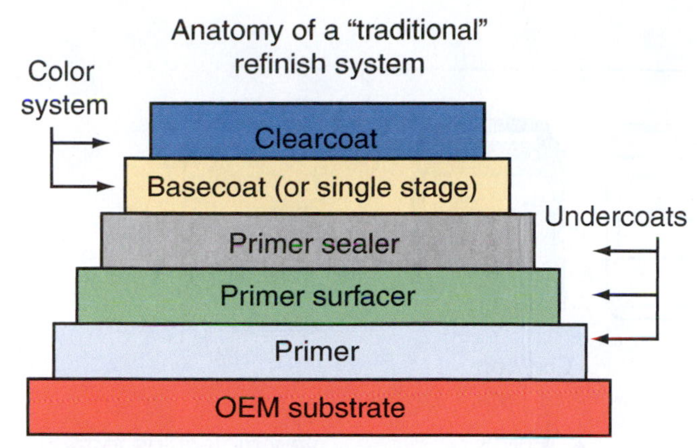

FIGURE 24-4 Coatings serve many purposes: protection (primers), surface leveling (primer surfacer), barriers (sealers), color coats (basecoat), and protection from long-term deterioration (gloss and clearcoats).

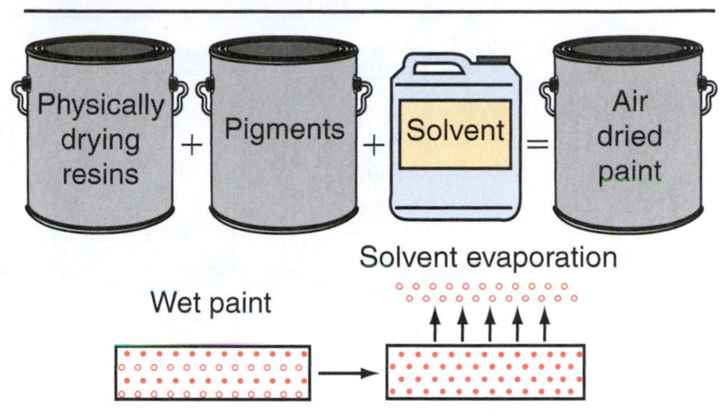

FIGURE 24-5 Some coatings primarily cure by evaporation (lacquer).

qualities that protect the underlying coatings and enhance their appearance. Clearcoats adhere to the basecoat under them, protecting them from environmental elements such as acid rain, industrial fallout, bird droppings, insects, tree sap, road debris, and the many other items that bombard vehicles in their daily operation. Even the sun's **ultraviolet (UV) rays** can be very destructive to vehicle finishes; if not for the protection in the clearcoat, these coatings would fail very rapidly.

There are two main categories of finish lacquers: (1) solvent evaporating finishes and (2) enamels. Enamels also evaporate, but they oxidize as well to form a lasting finish. In addition, there are waterborne and urethane types of enamel finishes, which will also be discussed in this chapter.

Lacquers

Lacquers, first introduced by the Dupont Corporation, are paints which are suspended in thinners. When sprayed onto a vehicle, lacquer will evaporate into the atmosphere at room temperature. The chemicals used in thinners and reducers will change from a liquid to a gas at room temperature, and are therefore considered "volatile" compounds. Lacquers are thinned to a more liquid state and sprayed onto the surface to be coated. As the coating reaches the vehicle's surface, the "thinner" continues to change from a liquid to a gas as the coating "dries." See **Figure 24-5**. Lacquer paints use large amounts of thinners, sometimes reduced 300 percent (that is, three parts of thinner to one part paint). When lacquer is fully dried, the coating is still subject to the surface re-dissolving if it comes in contact with thinners again. This reflowing with thinners will continue throughout the life of the vehicle's finish. Following application, a lacquer finish must be polished to obtain a gloss. Still, lacquer finishes are not very durable and are adversely affected by environmental elements such as sunlight and acid rain.

By the mid-1980s, vehicle manufacturers stopped using lacquer finishes. In fact, environmental regulations have banned the use of lacquers in vehicle refinishing in many areas of the country. Because of the vast amounts of thinners needed, the process releases large amounts of harmful **volatile organic compounds (VOCs)** into the atmosphere.

Enamels

Although these finishes also have agents that reduce their viscosity, enamels are generally called reducers rather than thinners. The chemical agents in enamels aid in the transfer of the coating from the paint gun to the object being coated. Once on the target area, the reducers evaporate, but the remainder of the coating cures through oxidation. See **Figure 24-6**. The result is a harder, more durable finish which—after fully curing—will not reflow and become liquid again. This application process takes longer than the evaporation of lacquers, which makes the fresh paint more prone to having contaminants land in the finish as it is oxidizing. Although applying heat to the freshly painted surfaces could speed up the oxidation, it

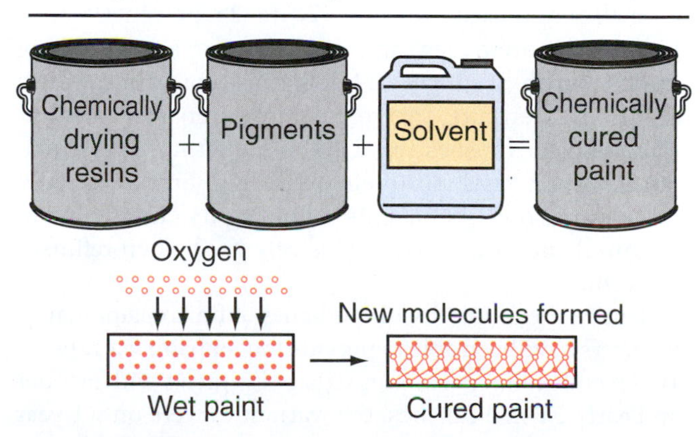

FIGURE 24-6 Some coatings, such as enamels, primarily cure by oxidation.

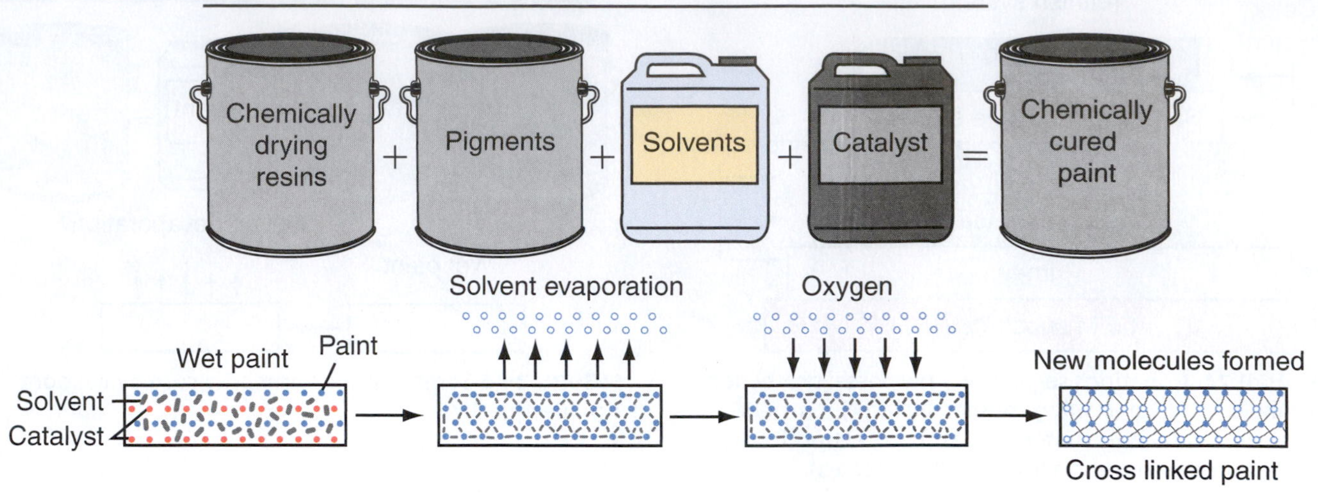

FIGURE 24-7 Some coatings, such as urethanes, primarily cure by chemical cross-linking.

was still a slow process. Therefore, the vehicle being refinished may need to remain in a spray booth for long periods of time. Although enamels ensured a more durable finish, the long process made them even less practical than lacquer finishes.

Early in the 1970s, refinish companies added hardeners or catalyst to enamels, which caused them to cure faster and maintain their hard solvent-resistant finishes.

TECH TIP A **catalyst** is a chemical which, when added to two other chemicals, causes a reaction. The reaction would not take place—or would occur much more slowly—if that chemical were not present. Paint hardeners are catalysts, which cause the cross-linking—or binding—of polyurethanes.

Urethanes. In the 1980s, catalyzed enamels gave way to urethanes and polyurethanes, which cure by evaporating a small amount of solvents (20 to 25 percent), cross-linking the urethanes, and oxidizing for a long lasting durable finish. See **Figure 24-7**. Because these finishes still oxidize as they cure, they remain in the enamel category. Polyurethane finishes are a durable thermoset product, which is very resistant to environmental breakdown and will not dissolve in solvents following its curing. Today, polyurethanes are nearly exclusively used when refinishing vehicles.

Due to the durability of urethanes, refinish paint manufacturers have been able to provide longer paint warranties. When collision repair shops repaired vehicles with lacquer and early enamel finishes, the warranties were only 1 year in length. With the use of urethanes and polyurethanes, warranties have been significantly extended.

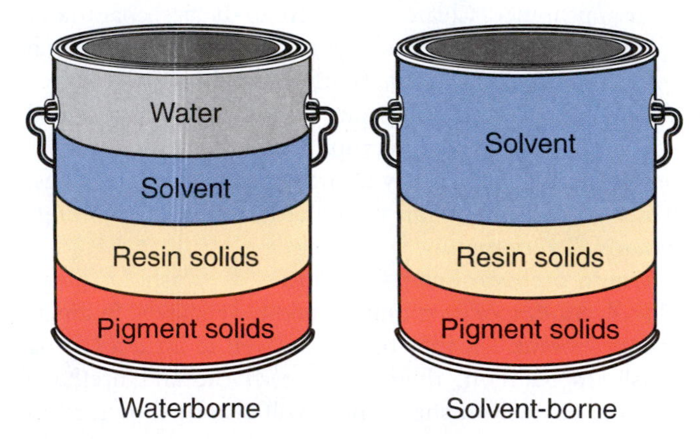

FIGURE 24-8 Water-base finishes that cure through evaporation can have as much as 25 percent solvents which must evaporate as well.

Water-Base Finishes. Although water-base finishes have been used since the 1980s (and their use has been mandated in some areas of the United States to help reduce VOCs), they largely remain unpopular with most painters.

Although the name would lead one to believe that these paints have no solvent or VOCs in them, that is not the case. See **Figure 24-8**. When compared to solvent-borne finishes, the water only replaces about 25 percent of the solvents present in other urethane finishes. Furthermore, water-base finishes are only available in basecoats, and so must be covered with a solvent-based topcoat.

This type of finish is also costly to the refinish business. Large amounts of air must be circulated in the paint booth to speed the evaporation; this is accomplished by fitting spray booths with special drying venturis. See **Figure 24-9**. This added expense and the relatively small savings in solvent contamination of the environment have as yet discouraged the widespread use of waterborne finishes.

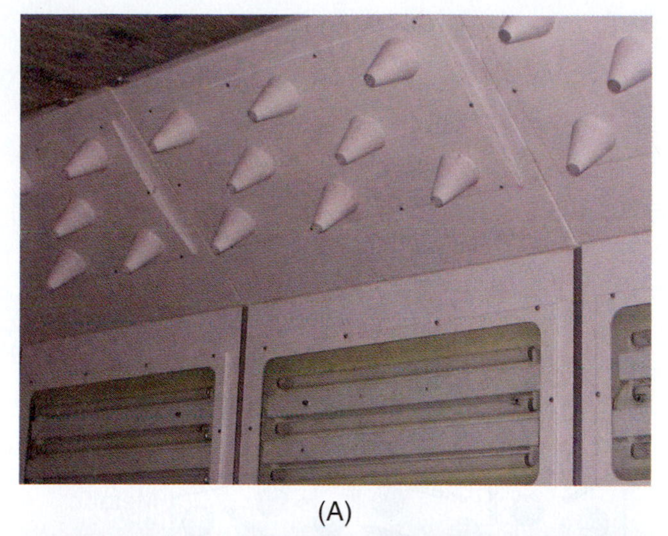

(A)

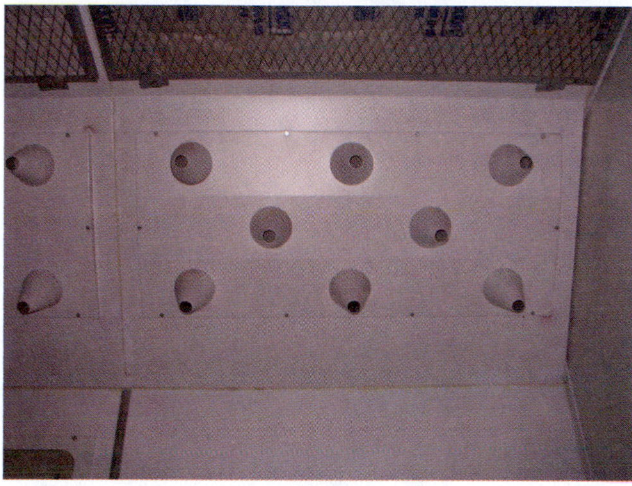

(B)

FIGURE 24-9 Paint booths used with waterborne coatings must circulate larger amounts of air to cure the water-base finishes (A). One method to circulate air is with the use of venturi nozzles mounted in the ceiling (B).

PAINT CHEMISTRY

The student can see from the brief history just presented regarding the changes in coatings and the way they work that paint chemistry has evolved dramatically. Lacquers were pigments dissolved in a liquid solvent that, after application, would evaporate back into a solid coating. Today's coatings involve sophisticated chemical reactions using solvents, catalysts, and a curing process to produce both a very durable solid surface to protect the vehicle beneath, and also a very attractive finish.

Because paint finishes result from sophisticated chemical reactions, the end product is influenced by environmental conditions. The paint product's responses to heat, humidity, airflow, air pressure, and the spray technique used all make vehicle refinishing a complicated process. The better technicians understand these variables, the better they will be able to manage their results. Managing all of these variables to consistently offer a finish that meets the customer's needs and expectations is at best a difficult process.

The refinish technician who hopes to provide a quality product should strive to learn the variables of paint chemistry. By studying and understanding the ingredients in coatings, how they react when mixed, and how they react to different spray conditions, a technician can provide a better product. Technicians with knowledge of these factors will also provide a safer and healthier working environment for themselves and their fellow workers and community.

Paint Composition

Paint is made by mixing different substances together: binder, pigments, solvents, and special additives. See **Figure 24-10**. Although the amounts or percentages of each item added and what each category is made from have changed over automotive refinish history, the four main categories of ingredients have remained the same. Binders do as their name implies—they help the pigment bind to the surface they are being sprayed upon. They also help provide the **holdout** qualities of finishes (help finishes resist marring). Pigments are the substances that produce the color, or hue, of the finish, while solvents help dissolve the mixture into a liquid that can be applied to the vehicle. Finally, the additives make up the smallest percentage of the mixture. Additives help with durability, leveling, adhesion, corrosion resistance, and other needed qualities.

Although the four categories have not changed over the years, the amounts used to produce the paint product have. In the early finishes used by the collision repair industry, solvents made up over half the total volume of each gallon of paint sprayed. At that time it was not known how damaging these volatile solvents were. When technicians in those earlier years mixed paint, they might have added as much as 150 percent solvents (1 part paint to 1 1/2 parts solvents) to the mixture, then sprayed it at 25 percent transfer efficiency or less. Around 75 percent of the paint sprayed did not stay on the vehicle, meaning that the majority of the solvents and paint solids were wasted. Paint coatings have evolved to the point where now reductions are as little as 20 percent (5 parts paint to

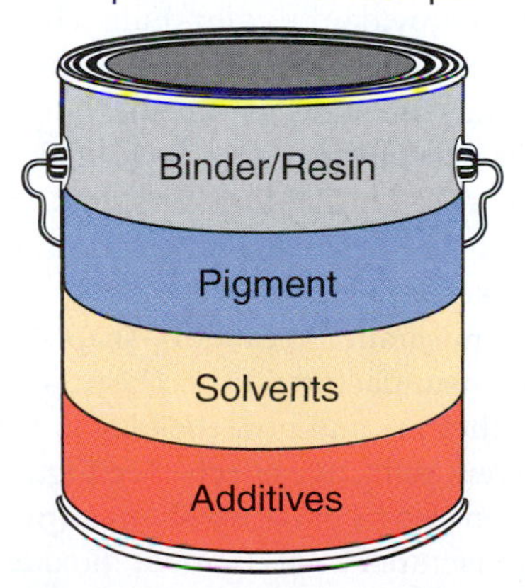

FIGURE 24-10 The principal ingredients in solvent-borne paint are: binders/resins, pigments, solvents (as much as 85 percent), and additives.

1 part solvent). Also now the coatings are sprayed with guns which are 60 to 75 percent transfer efficient. These advancements make the high solid paints used today more efficient and better for the environment.

TECH TIP "**High solids**" is a term commonly used regarding finishes; it refers to the ingredients in finishes that do not evaporate following application and curing. When paints were first developed, they contained only 25 percent solids. The remaining 75 percent of ingredients were solvents, which dissipated during the curing process. High solid paints have 50 percent or more solids, which makes them less harmful to the environment.

The four categories of materials used in coatings, though they have changed over the years, still provide the needed qualities in refinish industry coatings. Details of each category are discussed next.

Binders. Although binders—sometimes called resins—have been formulated from different chemicals over the years, their purpose has not changed. The binder is one of the portions of the paint mixture that does not evaporate, and is therefore considered part of the solids. In finishes, as the name implies, the binder helps bind the pigments to the substrate.

The type of binder or resin a paint is made from will determine what type of paint it is. Binders can be made from natural resins such as linseed or cottonseed oil, as older finishes were, or from plastics such as methyl methacrylate, urethanes, polyurethanes, and polyesters, to name just a few. As binders have changed from using natural resins to manufactured plastics, the durability of the finishes has vastly improved. Finishes using these new binders provide better luster, durability, strength, and protection for the vehicle.

Pigments. Pigments are also part of the solids in finishes. These additives provide the color, hide what is under the finish, and help provide a small part of the finish's durability. Before they are added to finishes, pigments can be either powder-like substances or larger flake-like materials (metallic). When in the finish, each of these pigment types will reflect the desired color or effect.

Metallic flakes range from objects that are round and saucer-like in appearance to random-shaped objects. They are commonly manufactured from plastic and, depending on how deep they are and how they lay in the finish, will produce different reflective effects. See **Figure 24-11**.

Pigments can also be made up of **mica (pearl)** material, which is manufactured from titanium dioxide, chromium, and iron oxide. These pigment flakes are semi-transparent and will both reflect light back and allow some light to pass through. This changes the appearance of the pigments under it, reflecting different colors depending on the angle that it is viewed from. See **Figure 24-12**.

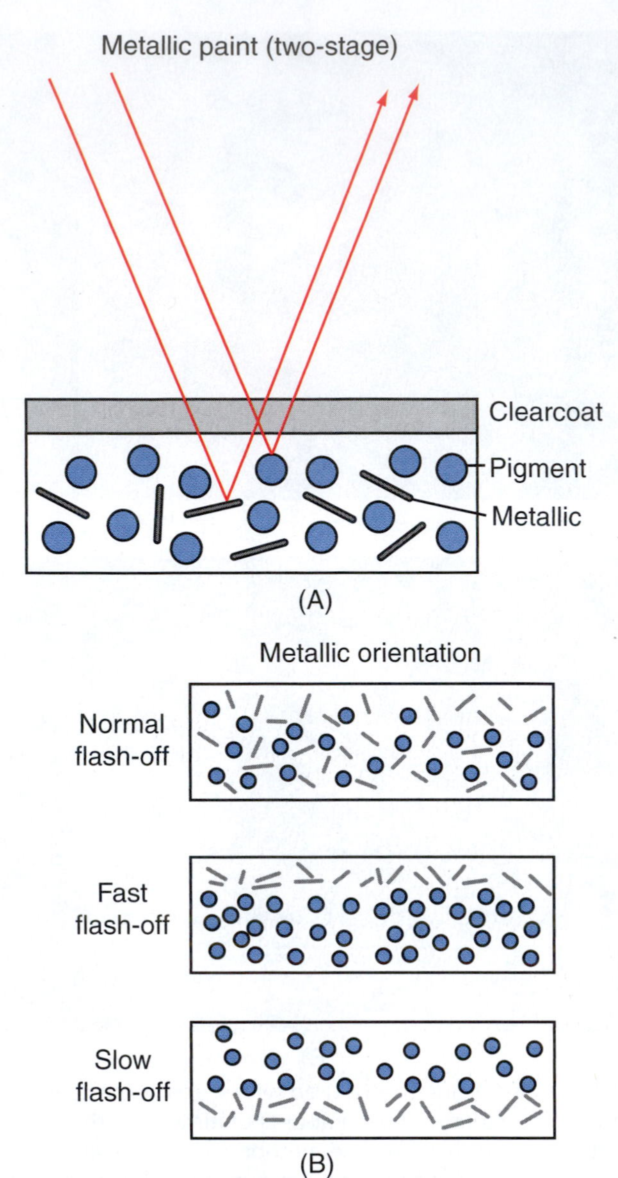

FIGURE 24-11 Metallic (small reflective materials) in paint reflects light, which enhances the vehicle's colors (A). The precise orientation (how metallic flakes lie in the paint) is critical to color matching (B).

Other pigments are **prism-effect flakes**. These flakes are also semi-transparent flakes and, depending on the angle that they are viewed from, will shift color and reflect and refract different colors. Some prismatic flake pigments can shift as much as five different colors as one moves around the vehicle and view it from different angles.

Solvents. Solvent is the material added to the binder and pigments that dissolves the solids and allows them to be applied to the surface. Solvents are designed to evaporate, and are therefore considered a volatile organic compound.

These agents that liquefy the solid pigments and binders are sometimes referred to as the "vehicle," or the part of the paint which helps transport its solids. Finishes need these agents in the paint. Even though finishes may be sprayed,

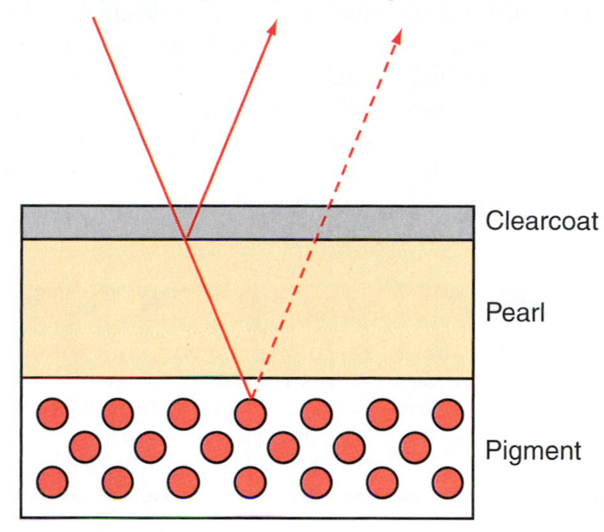

FIGURE 24-12 Mica, also called pearl, both reflects (like a mirror) and refracts (allows light to pass through), thus producing glamorous paint colors.

TECH TIP All matter on Earth is in one of three states: solid, liquid, or gas. A substance may change from a solid to a liquid, then to a gas, depending on temperature and pressure. Water (at sea level) is a solid when its temperature is below 32°F. When heated above 32°F, it changes into a liquid; when heated beyond 112°F, this liquid changes into a gas. Though all substances will be in one of these three states of matter, depending on their temperature and pressure, some substances change from liquid to a gas with only little temperature or pressure change. Chemicals that change from a liquid to a gas at room temperature—such as the solvents used in automotive refinishing—are considered **volatile chemicals**. They evaporate easily, and therefore easily pass into the atmosphere.

brushed, or rolled, they must be liquefied to help transfer them from the paint can to the object being painted.

Different pigments—such as metallics, mica, and prismatic flakes—need to be oriented at different levels in the finish to accomplish their effect. Without solvents to liquefy these pigments, their effect would be lost. The liquefaction of the binders also aids in their ability to bind to the surface below. If binders didn't have this property, they would not flow over the surface in a manner allowing them to level and provide a smooth surface.

Flow out is the motion of pigments and binders as they are suspended in the solvents. With the aid of gravity, the liquid spreads, allowing the surface to become very smooth and level. This action reduces the texture in the finish and increases the gloss as the surface becomes level. **Leveling** can only be accomplished when the finish is in its liquid state. As the pigments and binders are deposited on the vehicle's surface, gravity pulls the finish to a flat and smooth surface. When a vehicle is sprayed and texture is present, it will be greater on the horizontal surfaces than on the vertical surfaces, due to the gravitational pull being greater on the vertical sides of the vehicle.

Additives. Additives are special chemicals added to the paint mixture (containing pigment, binders, and solvent) either when the finish is manufactured or when the technician reduces the mixture to prepare it for application. Some additives aid the finish with leveling, binding, adhesion, and metallic or mica orientation. Others may provide special qualities that the paint would not have otherwise, qualities such as:

- Improving flexibility of the paint
- Improving adhesion to plastic
- Speeding up or slowing down of the curing time
- Lowering the freezing point
- Reducing the amount of bubbles when the finish is shaken
- Reducing foaming as paint is sprayed
- Increasing durability
- Increasing the amount of flatness or gloss
- Preventing rankling, blistering, and improving chemical resistance
- Providing ultraviolet protection

One of the most striking changes in paint technology has come from additives. Paint can now be protected from the sun's ultraviolet rays. Clearcoats used now cure at a high rate of speed, and the life of the finish has increased from 1 year in length (which was typical when lacquer was used) to 5 to 10 years of durability.

Reducers and Thinners

Reducers and thinners are solvents that are used to liquefy finishes and transport them to the surfaces of the object being painted. In the past, when a technician was working with lacquer finishes or using a solvent to clean the equipment, he or she would use a product called thinner. When technicians were using enamel topcoatings, they would instead use a reducer to make the finish less viscous, or thick. Both reducers and thinners are a mixture of chemicals designed to make the finish liquid, to transport it to the object being finished, and to evaporate at different speeds to accomplish the needed result. Thinners and reducers are designed for specific finishes and to do specific jobs as the liquid in the finish is evaporating.

Reducers can be manufactured with slightly different amounts of these solvents to either speed up or slow down the evaporation rates. Although the amount of time that finishes need to be liquid before they dry or evaporate is the same, no matter what the temperature is, the evaporation process is strongly influenced by the temperature. As temperature rises, some chemicals will evaporate faster; or as the temperature is reduced, they will evaporate more slowly. Reducers may be formulated to evaporate at a specific speed, through different mixtures of solvents used in different temperatures or conditions. The technician will choose the reducer that will provide the best results for the conditions encountered.

Although evaporation is affected by factors other than temperature, temperature is the factor most often used to choose or formulate a reducer. Other factors that will affect evaporation are airflow and humidity. If the airflow is high, as in a downdraft paint booth, some reducers will evaporate at too fast a rate to have the desired effects (gloss, texture, flow, and so on), and the next slower reducer may be best. If the humidity is high, it will also affect evaporation, and a faster reducer may be needed. The technician must also bear in mind that reducers used in colder weather will evaporate at a faster rate than reducers formulated for hotter weather.

Some experienced technicians are able to mix the different temperature reducers for different spray conditions; this practice is commonly referred to as "cocktailing." This method requires a keen understanding of reducers and spray conditions, and should not be attempted until the technician is familiar with the paint product and the way different conditions affect finishes. Some manufacturers do not recommend this practice at all.

TECH TIP When humidity is low, gases and mist evaporate readily into the atmosphere. However, when humidity is high, the air is saturated with liquid (water), and evaporation is markedly reduced. When humidity exceeds 90 percent, the evaporation is markedly slowed and a reducer, which is a faster evaporator, may be better suited for some refinish tasks.

Hardeners

Hardener, *catalyst*, and *activator* are all terms used by manufacturers for an additive placed in the finish during the reduction process that causes the finish to cross-link. Cross-linking is a chemical reaction that joins the urethanes in the binder together. This joining causes a thermoset reaction. Not only does the finish become hard and able to resist the elements, but it also becomes impervious to solvents. Even though the solvents will evaporate and some oxidation will occur, the finish will not fully cure, or cross-link, if the correct amount of catalyst is not present. Any finish having a catalyst added is referred to as a 2K. Not only are the topcoats catalyzed, but primers and other undercoatings are also 2K. Catalyzing the undercoatings causes them to become stronger, more durable, and less likely to have problems such as shrinkage, surface mapping, sand-scratch swelling, and other imperfections that are more prevalent in non-catalyzed undercoatings.

The chemical that is commonly used to catalyze automotive finishes, isocyanine, is not only a carcinogen, but also is irritating to the skin, eyes, and lungs. This chemical can be absorbed through the skin and eyes if the technician does not protect these membranes as directed by the product's MSDS. When spraying finishes with isocyanine added, the technician should wear an air-supplied respirator that uses class D breathable air.

CAUTION

Technicians should read, understand, and follow all the recommendations found in the product's MSDS—particularly those found in the sections on health hazards and protective measures—for their personal protection when using the product.

Thermoplastic/Thermoset

Plastics such as urethanes come in two categories: **thermoplastics**, which are plastics that will reflow or become liquid again when solvents are introduced; and **thermoset plastics**, which will not reflow or become liquid, following their cure process, when solvents are applied to them. It is easy to understand that because of its properties, a thermoplastic finish is less durable and unable to hold up to the environment or road conditions. On the other hand, a thermoset finish uses a hardener to cross-link as the finish cures, causing it to become impervious to thinners and other solvents.

Once the cross-linking has completed, the thermoset finish is very durable and strong. However, there are different phases in this finish process that the technician should be aware of. In the first phase, when the finish is initially applied, it can be recoated without fear of lifting. Little or no cross-linking has occurred, and the finish will flow properly as the technician applies the subsequent coats. In the second phase of the process, the finish is partially cross-linked. This means that some of the finish will reflow as solvents are applied to it, and some will not. If a coating comes in contact with solvents through spraying additional coats or by any other means, some of the coating will not flow, while some will, causing a "wrinkling" or "lifting" effect. See **Figure 24-13**. The third phase begins

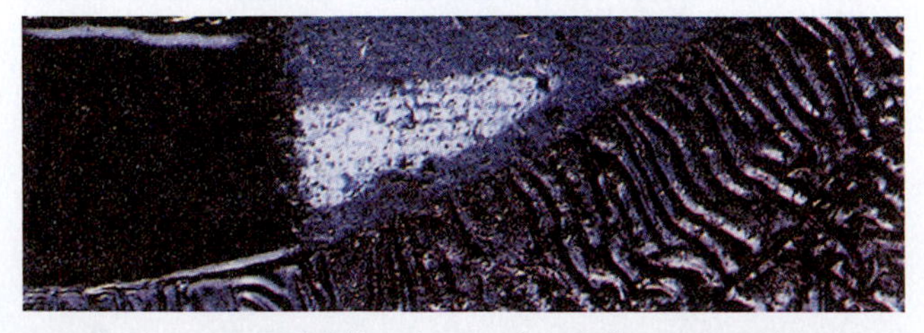

FIGURE 24-13 Wrinkling is a paint defect caused by a chemical reaction that occurs during application. Used with permission of PPG Industries.

when the finish has completely cross-linked and is no longer subject to solvent flowing; in this phase, the technician can apply solvents without fear of the finish lifting.

Thermoplastic finishes are less durable. If covered with a thermoset coating, the duality of the resulting finish will still be only as durable as the weakest finish, in this case the thermoplastic. If a vehicle is repaired with a thermoplastic primer (such as lacquer primer), then topcoated with a thermoset (such as 2K polyurethane), the overall finish is only as strong as the lacquer undercoating. If a technician will be finishing a vehicle with a thermoplastic product such as lacquer, he or she should continue with a thermoplastic topcoat, keeping in mind that such a finish is less durable than a thermoset finish. If the technician is finishing a vehicle with the more durable thermoset finishes, he or she should be certain that the old finish is a thermoset finish as well, by performing a rosin or solvent test.

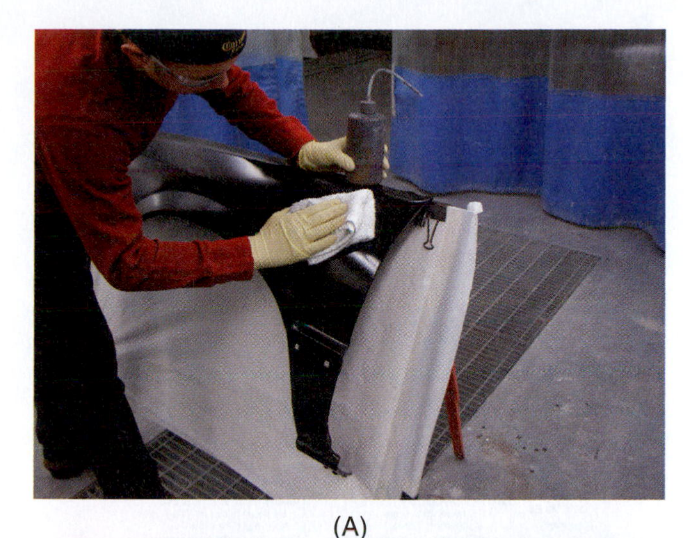

(A)

(B)

FIGURE 24-14 (A) In the solvent test, solvent is applied on the coating to be tested. (B) Shortly after application, the solvent is wiped off. If the coating can be wiped off with the clean cloth, it is not a thermoset coating.

Rosin Test

Sometimes called the solvent test, the rosin test is performed to see if the old finish on the vehicle will reflow when a strong solvent is applied. To test a painted surface, a technician should apply thinner to a shop towel and rub it onto the surface in question. If the coating comes off on the application rag, or if the surface becomes soft following the test when scratched, the coating is a thermoplastic and will not be durable enough to accept thermoset coatings over it. See **Figure 24-14**.

The rosin/solvent test is useful when purchasing LKU (used) parts, or parts from an **aftermarket** vendor. In such cases a technician may not be sure whether the finish on the replacement part is thermoset or not. Once the test is completed, the technician will know if it will be necessary to strip off the offending thermoplastic finish and apply a thermoset undercoating before applying a durable top finish.

CAUTION

If a coating is found to be thermoplastic, it must be removed or it will remain the weakest link in the finish. If a thermoset undercoat is applied over the old finish, it is in danger of failure, and most paint manufacturers will not cover this failure under their warranty.

TECH TIP Although the four main factors which affect finishes are temperature, humidity, airflow, and time, the film thickness is also a factor, and technicians should try to apply film within the recommended limits.

Standard Conditions

The standard conditions are the starting point for measuring how paint will react to various factors. The four factors which affect finishes the most are temperature, humidity, airflow, and time.

Paint manufacturers have created product procedure sheets, which outline the proper way to use each product. See **Figure 24-15**. These sheets state the amount of time the technician should wait between applying coats, how long after spraying basecoat the technician should wait before clearcoating or re-masking for two-toning, and the amount of time needed to force-dry the finish with heat. These times are based on standard conditions, which are 50 percent humidity at 70°F, with "sufficient airflow to readily remove the overspray." If conditions in which the technician is spraying vary from this standard, the times in the "P" sheets (procedure sheets) must be adjusted accordingly. If the humidity rises above 50 percent, the flash and purge times must be extended. Percent of humidity is a measurement of how much moisture (most of the time water) the

PROCESS		
MIX RATIO	PRIMER-SURFACER	HIGH BUILD
	P565-510/511 5 P210-796/798 1 P850-16XX 1	P565-510/511 5 P210-796/798 1 P850-16XX 0.5
	Note: Only P210-796 or P210-798 MS Hardeners should be used in HS Primer Surfacers.	
VISCOSITY & POTLIFE	**Viscosity**: 19-23 seconds Din#4 @ 70°F (21°C) **Pot Life:** 1 hour at 70°F (21°C)	**Viscosity**: 30-35 seconds Din#4 @ 70°F (21°C) **Pot Life:** 30 Minutes at 70°F (21°C)
SPRAYGUN & AIR PRESSURE	**Siphon:** 1.6-1.9mm (.063"-.075") 45-55 psi at the gun **Gravity:** 1.7-1.9mm (.069"-.075") 45-55 psi at the gun **HVLP:** 1.7-2.2mm (.069"-.087") Max 10 psi cap press. 3-8 fluid press. (pressurized cup)	**Siphon:** 1.8-2.2mm (.070"-.086") 45-55 psi at the gun **Gravity:** 1.8-2.2mm (.071"-.086") 45-55 psi at the gun **HVLP:** 1.7-2.2mm (.069"-.087") Max 10 psi cap press. 3-8 fluid press. (pressurized cup)
	(**HVLP:** Refer to gun manufacturer's recommendations for required inlet pressure.)	
APPLICATION	**Apply 3 coats** (approx. 3.0-5.0 mils) **Note:** Film build will depend on fluid tip selection.	**Apply 3 coats** (approx. 7.5-10 mils) **Note:** Film build will depend on fluid tip selection.
FLASH TIME	**5 minutes** between coats Allow 5-minute flash-off before force dry or Infra Red.	**5 minutes** between coats Allow 5-minute flash-of before force dry or Infra Red.
DRY TIMES	**Air-dry times at 70°F (21°C)** 1.5-2 hours @ 3-4 mils 2.5-3 hours @ 5 mils **Force Dry (metal temperature):** 20 minutes at 140°F (60°C) **Infra-Red (short wave):** 8-12 minutes after 5 minute flash	**Air-dry times at 70°F (21°C)** 4-6 hours minimum @ 3-4 mils *Overnight Preferred **Force Dry (metal temperature):** 30 minutes at 140°F (60°C) **Infra-Red (short wave):** 8-12 minutes after 5 minute flash
SANDING	**Wet Sanding:** P400 for Single Layer Color P600 for Basecoat color **Machine Sanding:** P320 or finer for Single Layer colors P360 or finer for Basecoats	**Wet Sanding:** P400 for Single Layer Color P600 for Basecoat color **Machine Sanding:** P320 or finer for Single Layer colors P360 or finer for Basecoats

FIGURE 24-15 Paint companies provide a procedure to follow when using specific products. Courtesy of Standox Paint.

The 15°F Rule

This rule has two parts and applies to **thermoset** products.....

#1. For every 15°F increase in temperature above 70°F, a refinish product's dry time may be reduced by as much as 1/2. The same rule applies to the product's pot life.

#2. For every 15°F decrease in temperature below 70°F, a refinish product's dry time may almost double! The same rule applies to the product's pot life.

NOTE: All product crosslinking and curing in 2K products slows significantly below 60°F/16°C. Thermoset paint will not "set" properly if subjected to cool temperatures during initial curing stages. Such conditions can result in a finish that may eventually **dry** but will exhibit reduced durability, gloss, and repairability. This loss of performance is due to the film **never reaching a fully cured state.**

FIGURE 24-16 Knowing the 15° rule will help technicians judge how solvent paint cures under differing conditions. Used with permission of PPG Industries.

air has absorbed; if the humidity is 60 percent, that means that 60 percent of the air's capacity for absorbing mist is full, with only 40 percent remaining. As can be seen, if the air is at 90 percent humidity, it only has 10 percent of its ability to absorb vapors or mist (the solvents) remaining, and this will markedly slow the evaporation of the solvents.

TECH TIP Evaporation of solvents into the atmosphere is sometimes called "**gassing out**." That is, the solvents which have turned from liquid to gas are being released from the finish.

Time is the other condition that affects the evaporation and curing times of finishes. As the temperature rises from the standard 70°F, the curing time decreases, and as the temperature decreases from the standard, the curing and evaporation times increase. When temperatures vary from standard, the times in the product sheets must be adjusted using the 15° rule.

15° Rule

The **15° rule** states that for every 15° the temperature rises over the standard of 70°F, the time to cure is doubled; and for every 15° the temperature drops below the standard, the cure time is cut in half. This means that if the product procedure sheet states that a finish will cure to the touch in 4 hours at 70°F and the actual temperature is 55°F, it will actually take 8 hours to cure. Accordingly, if the actual temperature is only 40°F, the finish will take 16 hours to cure. On the other hand, if this same product is sprayed in an 85°F shop, it will take only 2 hours to cure, and only 1 hour if the temperature is 100°F. The 15° rule applies to finishes as they are being forced dry as well, and it is easy to see why forcing a finish to 140°F will cure it in 40 minutes. See **Figure 24-16**.

TECH TIP Although the 15° rule applies to a finish's curing times in both directions, it should be noted that all catalyzation stops below 40°F. That means if the spray environment drops below 40°F, the finish will no longer cure, and may remain wet until heat is applied.

Solvents and What They Do

Although it is commonly known that solvents are used to liquefy the finish's solids, and then transport them to the surface being coated, it is not always understood just how the solvents accomplish these tasks. A purchased finish—or one mixed by a spray technician—is too thick, or viscous, for spraying and must be reduced. To reduce or thin the finish, reducers are added. As mentioned earlier, the reducers gas out at different speeds to suit the spray environment. In fact, there are three different stages at which reducers must gas that affect the end result of the finish. These stages are "in-flight" loss of solvents, purge

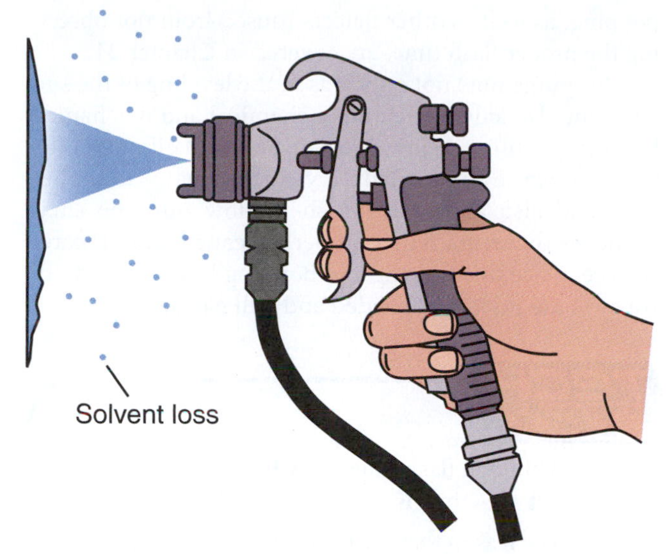

FIGURE 24-17 The first stages of solvent evaporation occur when the paint leaves the gun and before it hits what is being painted. This is called "in flight loss of solvent."

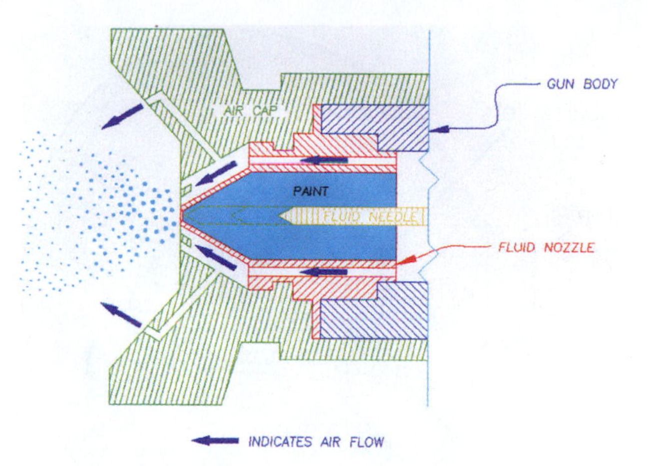

FIGURE 24-18 In external mix guns, atomization (the breaking down of paint to small particles) occurs outside the paint gun's fluid nozzle. Courtesy of Binks/DeVilbiss. All rights reserved.

time, and final curing. All of these stages are necessary for a finish to do the following: atomize properly, level and adhere as designed, polish properly, resist marring, and provide long-term durability. If technicians know how solvents work, they will be better equipped to solve problems as they arise.

TECH TIP The difference between a good painter and a great painter is the great painter knows how to prevent or correct problems as they arise.

In-Flight Loss of Solvent. In-flight loss of solvent is, as the name implies, the portion of the solvent which is designed to both reduce the paint's viscosity and evaporate "in flight," or between the time it leaves the gun and when it hits the object being finished. See **Figure 24-17**. Refinish guns are external mix, meaning the paint mixes with air outside the gun, both to propel the finish to the vehicle and to atomize the finish. Although some air is used to cause a vacuum and transport the finish from the cup to the air cap of the gun, this air only minimally aids in the atomization. See **Figure 24-18**. The air from around the paint gun nozzle—and, to a lesser degree, the air from the cap's air horns—provides the necessary power to break the finish into the small atomized droplets needed for proper finishing. See **Figure 24-19** and **Figure 24-20**. After the in-flight solvent has aided in the atomization process, it is no longer needed. If it remained, it would slow down the gassing out and curing process, since the chemicals used for this process are chemicals that change from liquid to mist, then to gas, very rapidly. The remaining solvents are carried

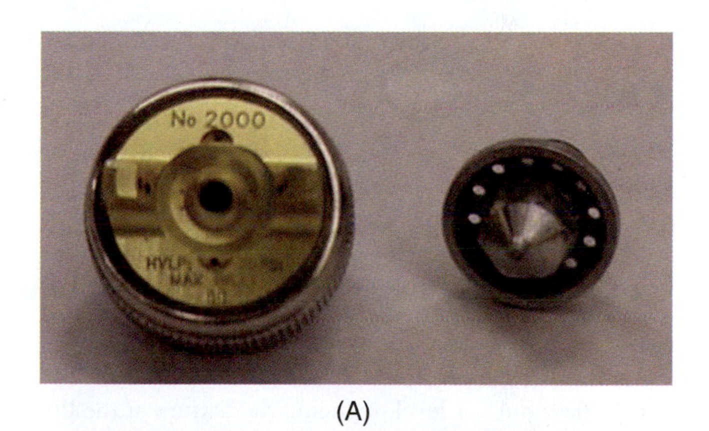

(A)

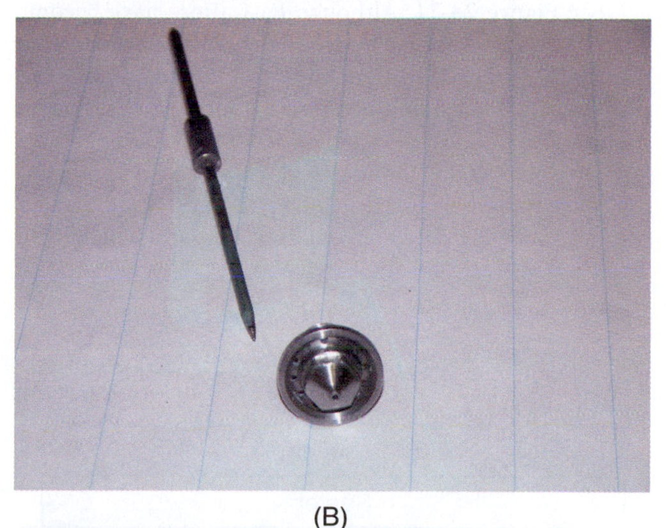

(B)

FIGURE 24-19 The paint gun cap (left) and nozzle (right) help atomize paint for application (A). The needle (left) and nozzle (right) also aid with atomization and must be precisely matched to the type of paint being applied (B).

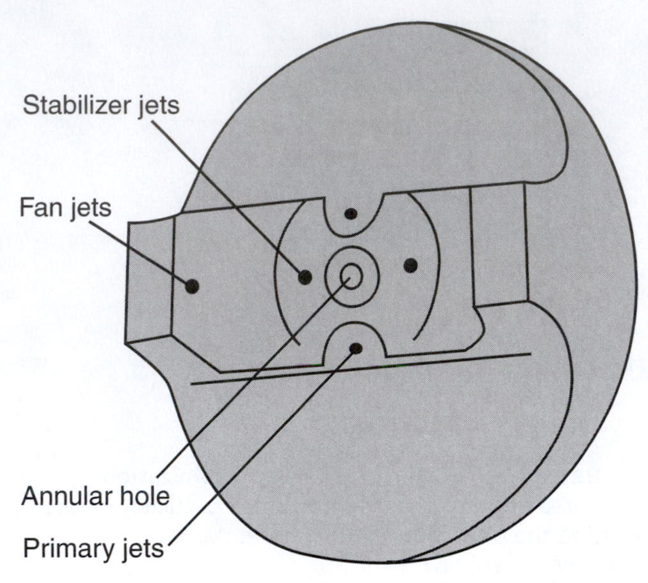

FIGURE 24-20 As air from the paint gun horns hits the paint stream coming from the nozzle, it starts to break paint down from large droplets to smaller ones.

from the spray gun to the vehicle's surface, where they can perform the second stage of this process, which is the purge—or leveling—stage.

Purge or Leveling Stage. The **purge time** or leveling time is the stage during which solvents are gassed out of the finish more slowly than in the first stage. This amount of time is sometimes called the **flash time**. Solvents will evaporate between the first and second coats of the finish. This time in between coats allows the small droplets formed during atomization to reconnect, creating a continuous film over the surface of the painted object. This also allows the finish to "flow out" or level, reducing the texture of the finish. See **Figure 24-21**. Although flash times have become shorter as new finishes have been developed, a technician must be sure that the flash time recommendations supplied by the paint manufacturer are followed. If sufficient flash time is not observed and a coat of paint is applied before the solvents are completely gassed out, these remaining solvents will be large enough to cause bubbles and result in a paint defect called solvent popping. Solvent popping, as well as other defects caused from not observing the proper flash time, are covered in Chapter 31.

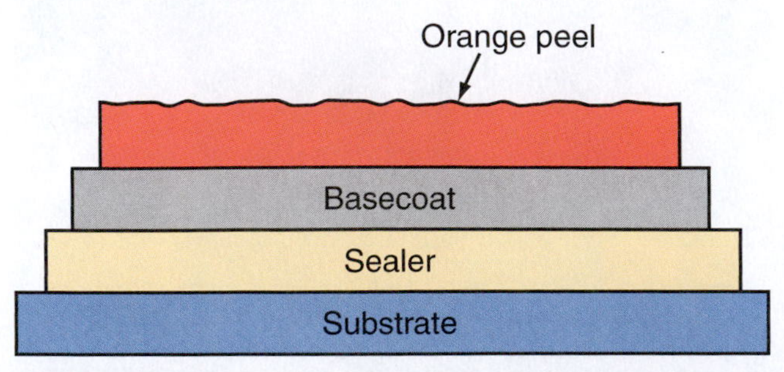

FIGURE 24-21 Coatings, when first applied, may have a slight "orange peel" appearance during the flash time. When the secondary solvent evaporates the coating levels or "flows out," leaving a smooth appearance.

This purge time not only aids in the leveling of the surface, but also aids in creating a chemical and mechanical bond to the finish below. The solvents will allow the finish to attach to the partially dry surface that was applied first, and also allow the finish to flow into the small scratches that remain in a properly cleaned and prepared surface. Once this flowing and bonding has occurred, the solvents are no longer needed and will gas out.

TECH TIP

1. Although flash times vary from product to product, this time is relativity fast: 8 to 15 minutes.
2. The reason that the purge time following application of the finish is longer than flash time is that there are multiple coats on the surface, and the solvents must have enough time to gas out, before baking or applying another topcoat is done.

Final Curing. Final cure time is the amount of time that a finish needs to release all of its solvents, and to have full catalyzation through cross-linking. The majority of the solvents are released during the previous two stages, though some will remain. The remaining solvents may take up to 90 days to be completely gassed out. Most paint manufacturers include recommendations to not polish the finish for a certain amount of time following the air-dry cure time or force-dried time. See Figure 24-15. They often have a time recommendation requiring the finish to cure for 60 to 90 days before waxing, since waxing sooner seals the finish and would not allow this final gassing out of the solvents. This slower final gassing out allows the finish to reach a hard, durable, and mar-resistant state.

TECH TIP Some finishes will have a recommendation that directs the technician to let the finish cure for a period of time before buffing and polishing, if needed. It may also have a maximum time that the finish can be left unbuffed before it becomes so hard that rubbing is no longer possible.

APPEARANCE

Whether a customer buys a new vehicle or picks up a refinished vehicle from a collision repair business, the first thing the customer will look at is the finish. "Fit and finish," or its appearance, is what is evaluated first. Although fit is important, finish is nearly always the first observation made by the customer. In the case of a repaired vehicle, the customer will evaluate the new gloss or depth of reflection, the color match, and the texture; and will check for defects in the new finish.

TECH TIP The only standard to compare the new finish against is the vehicle's factory applied finish.

Although customers evaluate the vehicle's fit, cleanliness, and operation, their first impression will rest on the quality of the finish.

Gloss

Gloss is how shiny a vehicle is. This term is not to be confused with **depth of reflection**, which is how clear an image looks when viewed in the reflection of a finish. See **Figure 24-22**. Vehicles may be very glossy, but not have a clear depth of reflection, due to texture or other defects. It is the refinish technician's obligation to match the gloss as closely as possible. In fact, it may be possible to apply a finish that has a better gloss or depth of reflection than the OEM finish. However, this is not what is required of a quality repair. The repair should be as undetectable as possible.

TECH TIP A common discussion between technicians is that a clearcoat finish will be glossier when it has not been polished, and that the depth of reflection will be greater when polished. The reason is that a polished finish will have less texture than an unpolished finish. When the surface of a clearcoat is sanded and polished, it can be polished to a high gloss, but may never return to its pre-sanded gloss. "Show finishes" are often cleared, then sanded to remove the texture, then re-cleared to have both the depth of reflection and the desired gloss.

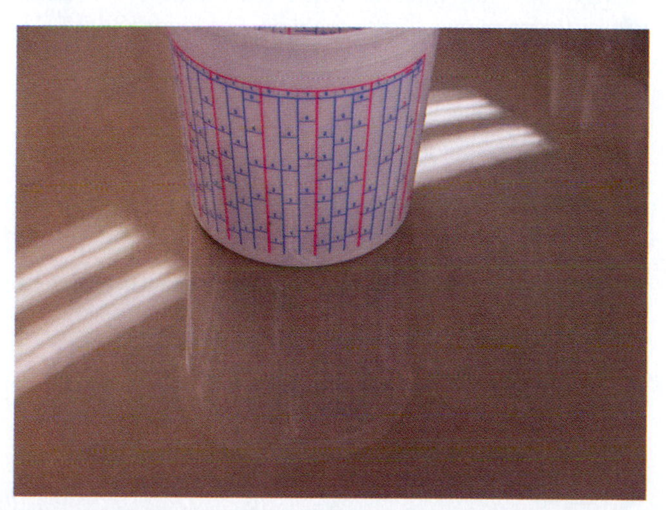

FIGURE 24-22 Depth of reflection, as seen in the reflection of the paint mixing cup and lights on the paint surface, is one way to identify a quality repair (it should match the vehicle).

It may be necessary to polish the vehicle's adjacent panels to match the new finish so the whole repair becomes undetectable.

Color Match

Color matching is one of the most difficult aspects of creating an undetectable repair, a task that can from time to time be very challenging and even frustrating. Color matching needs to compare the vehicle to the new value (lightness or darkness), hue (color, tint, or shade), and chroma (intensity, richness, or muddiness).

When viewing a vehicle to evaluate its color match, one must first understand how light affects color and the viewer. Natural or white light has all the colors of the rainbow: red, orange, yellow, green, blue, indigo, and violet. See **Figure 24-23**. These colors are easily learned by remembering the acronym "ROY G BIV," which sounds like a person's name.

We see color when light with the previous spectrum reflects off a surface. A color or combination of colors is reflected off the surface while the others are absorbed. That is, a blue car will reflect blue while absorbing the remaining colors in the spectrum. See **Figure 24-24**.

TECH TIP Black and white are not considered colors. White is the presence of all colors. That is, all colors are reflected at once, and the person perceives white. Black is the absence of color. That is, no color is reflected, and the person perceives black.

If a vehicle is viewed in a different type of light than sunlight, it may affect the way the color appears. If viewed in florescent light, the vehicle will appear to have more violets and reds than in natural light. If evaluated in incandescent light, it will appear to have more yellows, oranges, and reds.

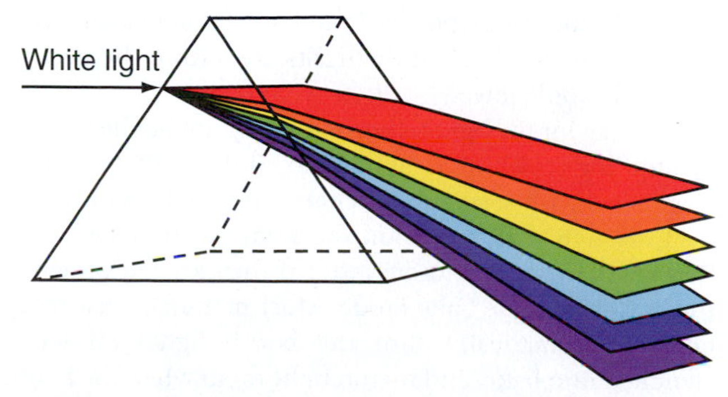

FIGURE 24-23 Light passing through a prism breaks white light (sunlight) into its individual colors.

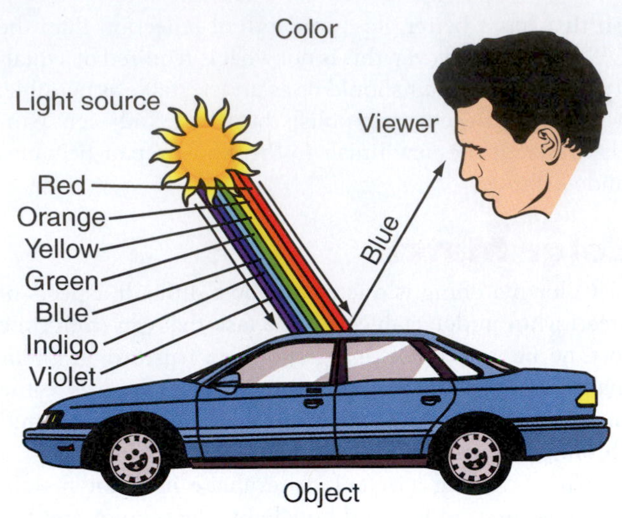

FIGURE 24-24 The light of a color will reflect only its color (blue in this case) and absorb the others.

FIGURE 24-25 Having the correct type (color corrected) and intensity is critical in paint booths.

When colors appear to be different shades (hues) when viewed under different lights, the effect is called metamerism. To avoid metamerism, the technician should view finishes either in direct sunlight or in an environment that has color-corrected lights.

Most modern manufactured spray booths have color-corrected light. Otherwise, the refinish area can have color-corrected lights installed to avoid metamerism. To be assured that the lights are color-corrected, check the light bulb for the marking **CRI (color rendering index)**. If the bulbs are from 85 to 92 percent CRI, they are considered to have sufficient color correction to evaluate the color.

TECH TIP When in doubt about the color match, use natural sunlight to evaluate the match.

The spray area should also be bright enough for the technician to view the vehicle being sprayed. Brightness is measured in candlepower, and the spray area should be from 5 to 10k foot candles bright. The brightness can be measured with a light meter, which should be able to read both CRI and candlepower. Using a light meter on a regular basis will indicate if the lights are within the needed CRI and candlepower.

Waiting for the light to burn out may not be the best indicator for replacing a light bulb. Bulbs will lose their brightness with age, and can change their CRI with age as well, though they may look the same to the technician. Bulbs will even deteriorate faster if they are left on when the booth is in the bake mode. Most manufactured spray booths automatically shut the booth lights off when switched into bake, and then relight them when the booth has cooled down following the baking process. To assure that the light is sufficiently bright (5 to 10k foot candles) and of the proper CRI (85 to 92 percent), a light meter should be used on a regular basis.

Besides light, many other conditions can influence the way a person sees color. Drugs (prescription and illegal varieties), caffeine, alcohol, and even carbon monoxide levels can also affect the way a person sees color. One may also experience "color fatigue" when the same color is viewed for a long period of time. When evaluating a color, look at the color for 10 seconds at a time, and then look away for a short time before reviewing it, to prevent perception to be altered by color fatigue.

Color blindness is also a factor that may affect technicians. Color blindness does not always mean that a technician will see only in black and white or gray shades. Instead, an individual may have trouble with certain colors in the color wheel (typically blue, red, yellow, and green). As blue changes to red passing through purple, the technician may have trouble distinguishing the subtle changes. Or the interval when red changes to yellow through orange may be the problem. See **Figure 24-25**. Although color blindness occurs in both males and females, it has a higher incidence in males. Color blindness can also range from very mild, nearly undetectable, to severe. Therefore, having a degree of color blindness does not necessarily mean that a technician will not be able to refinish vehicles. If a technician is aware of the deficiency and can ask for another technician to help with specific colors, quality workmanship can still be accomplished. See **Figure 24-26**.

Texture

Textures in a finish can occur in any level, from the undercoating through the topcoat. See **Figure 24-27**. Although unwanted texture is often removed when primer filler is sanded, it may still occur when the sealer, basecoat, or clearcoat is applied. If the texture appears under the clearcoat, the only way to repair it is to sand, smooth, and repaint the surface. If it occurs in the

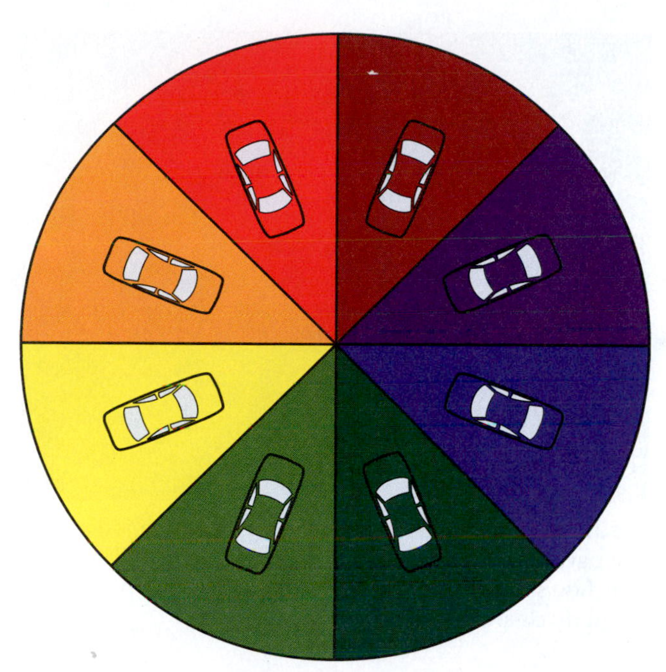

FIGURE 24-26 This color wheel shows primary and secondary colors. Adapted from DuPont.

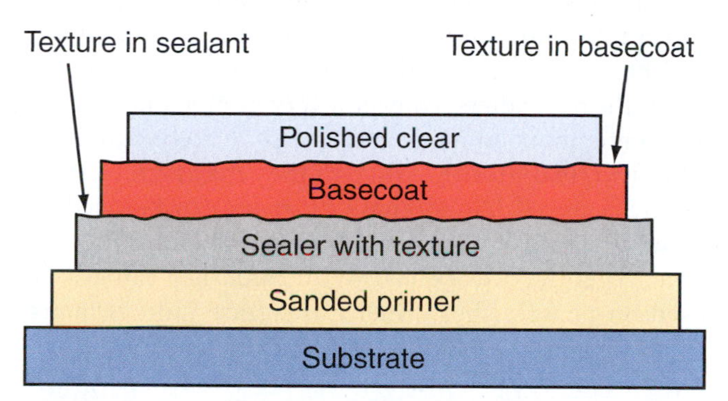

FIGURE 24-27 Although the clearcoat has been sanded, polished, and is smooth, the underlying texture shows through.

clearcoat, however, the technician may be able to sand out the texture and polish (taking care not to remove too much clearcoat in the process).

As mentioned before, texture is generally an unwanted defect. However, if some exists in the original finish, the refinish technician must match the original texture with the new finish. This is accomplished by either slowing the rate of travel when spraying, or by adjusting the fluid adjustment knob on the spray gun to spray at a higher flow rate.

PROTECTION

Although protection of a fresh finish (such as wax) may not be recommended until 60 to 90 days following the application of new finish, the spray technician should be able to advise customers on the proper care of the finish. There are generally three types or steps to protect a finish. These protection steps are cleaning, glazing, and waxing.

> **TECH TIP** Some detail technicians will wax a vehicle then **glaze** it, while others will glaze it first, then wax.

Cleaning

Although cleaning is generally thought of as soap and water cleaning, finishes can also be cleaned by other means such as chemical cleaning and detail clay cleaning. Soap and water cleaning is not as simple as it appears. First, neither fresh nor cured finishes should be washed using dish soap or laundry detergents. A soap which is pH balanced (7 pH) is best. In addition, flake soaps may not dissolve well and may scratch the surface. Dish soap is very concentrated, and most technicians use too much, making it very difficult to wash off even with large amounts of water. Dish soap also has many additives which, although they help clean dishes and protect the dish washer's hands, are unneeded when washing vehicles. See **Figure 24-28**.

Chemical cleaners are also very useful for cleaning off the debris that is not water-soluble, such as road tars, oils, and greases. Debris can come from newly repaired blacktop highways as well as from concrete dust created by the normal action of the vehicle driving on cement roads. Besides these, coolant, oil, air-conditioning water, acid rain, bird droppings, and bugs—just to name a few sources—also bombard a vehicle's finish with many harmful substances on a regular basis.

Cleaning off the water-soluble contaminants is generally the first step. If the finish is fully cured, a pressure washer can be used, although its pressure *is* abrasive (which is how it removes the dirt!). Hose end or low-pressure water is the best; alternately, the use of the two-bucket method is

FIGURE 24-28 Car wash soap (right) has been specifically designed for automotive refinishing use. The use of other soaps may cause defects in the paint being applied.

advisable. The two-bucket method consists of one bucket of warm to hot soapy water, and the second with warm to hot clear water. To wash the vehicle, the technician will first rinse the vehicle with low-pressure water (warm, if possible), then load the washcloth or mitt with the clean soapy water and wash a section of the vehicle. When the mitt needs to be reloaded for a new section, the technician will rinse the wash mitt in the second bucket to release the debris that has been taken off the vehicle. Once the mitt is clean and wrung out, it is reloaded with clean soapy water, and another section of the vehicle is cleaned. This process is repeated until the whole vehicle is cleaned. The technician should not let the soapy water dry on the vehicle before it is rinsed off. If the cleaning process is taking too long, the vehicle's clean sections should be rinsed. A vehicle is considered to be clean when the rinse water sheets off from the cleaned section. See **Figure 24-29**. If it beads up on the surface, the vehicle may need to be re-washed until it is clean.

Chemical cleaning uses a product generally called "wax and grease remover," though some waxes are water-soluble and will be removed with soap and water wash. Wax and grease removers are made with solvents, which dissolve road tars, oils (including silicones), and greases, along with other non-water-soluble **hydrocarbons** (products made from crude oil), which remain following soap and water wash. If these contaminants are not removed before work is done on a vehicle, they will cause contamination in the new finish. Also, if they are not cleaned off before work is done, sanding and other repair work will spread them, at least, and at worst may drive them into the repair area, making them much more difficult to remove. When a new part is being cleaned, all sides should be cleaned so the contaminants will not be spread when handled.

The area being chemically cleaned should be kept to an area of 2 square feet (about shoulder width). See **Figure 24-30**. The chemical cleaner is designed to lift

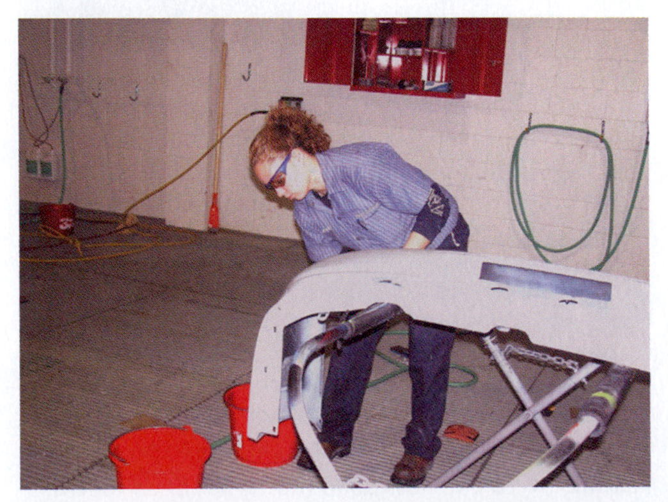

FIGURE 24-29 The two-bucket method of cleaning uses two buckets: one with clean, soapy water, the other with clean, clear water.

FIGURE 24-30 After a chemical cleaner is applied and the part is properly cleaned, the rinse water will sheet as it flows off the part. If beading occurs, the part is not properly cleaned.

contaminants off the vehicle and float them on top of the cleaner. See **Figure 24-31**. Then, before the solvent can dry, it is wiped up with a clean cloth. If the cleaning solvent is allowed to dry, the floating contaminant will be redeposited on the vehicle. There are different strengths of cleaners; some are very strong and used for very contaminated areas and on completely cured finishes (original finish), although they may be too strong for fresh paint. Mid-strength cleaners can be used on cured finishes, but may need to be used more than one time to get the finish clean. With care, they can also be used on new finishes. Finally, there are mild cleaners, which can be used on fresh finishes (primer, sealers, and other newly applied finishes). These cleaners may have names such as "final wipe." Some companies are now presoaking cleaning cloths with final cleaning solvents for convenience.

Although chemical cleaners may be applied by soaking a cleaning towel straight from the container, spraying on the cleaner instead from a refillable spray container reduces the amount of waste. See **Figure 24-32**.

Chemical cleaners contain the largest amount of VOCs, and the technician should be aware of all the state, federal,

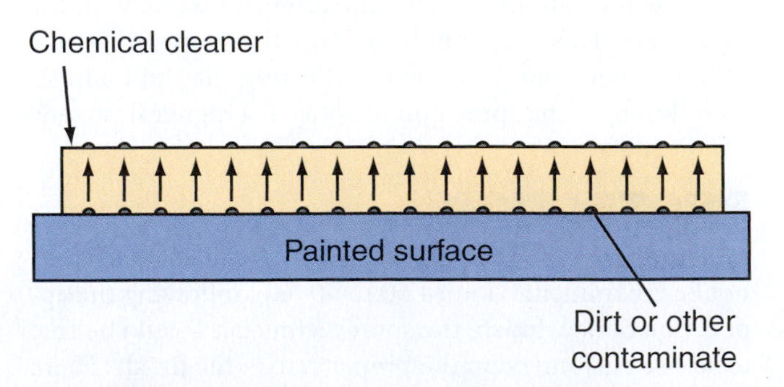

FIGURE 24-31 Chemical cleaners float contaminates to the surface of the cleaner to be wiped up.

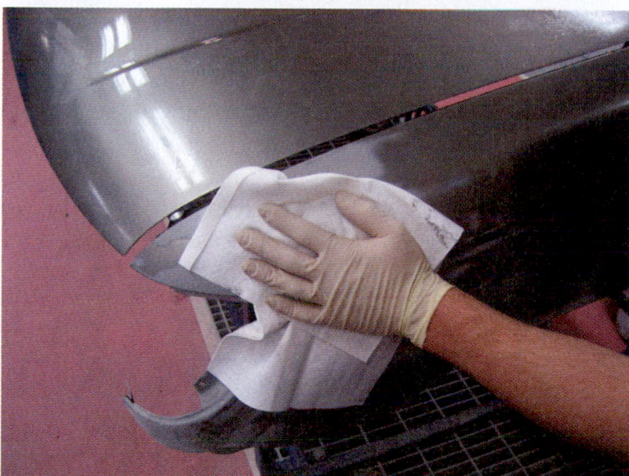

FIGURE 24-32 Chemical cleaner is most efficiently applied by spraying it onto the surface. Less cleaner is used; therefore, less air pollutants are produced. Following spray application, the chemical cleaner is wiped off.

and local restrictions on their use. Some locales restrict the use of solvent wax and grease removers; therefore, most paint manufacturers provide waterborne wax and grease removers, which work in the same way as the solvent-borne ones.

FIGURE 24-33 Detail clay must be kneaded prior to use to bring a clean surface to the top. Clay that will not clean through kneading should be replaced.

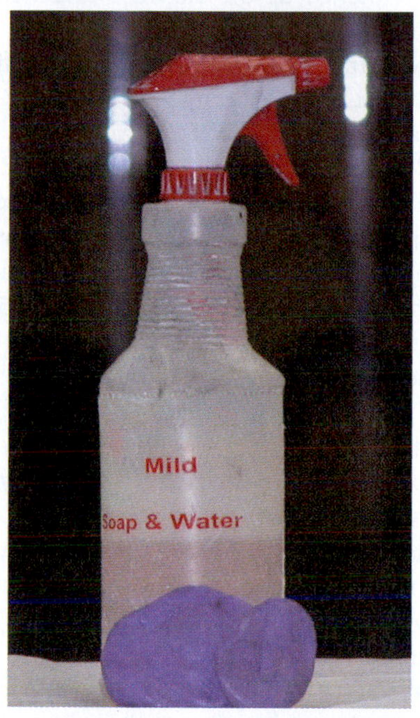

FIGURE 24-34 Detail clay must be used with a lubricant such as soap and water.

Detail clay is a cleaning substance which looks like children's modeling clay. It is designed to pick up contaminants on the surface of a finish. Although it is not designed to clean such contaminants as solvent-soluble substances—like road tar and grease—or water-soluble substances, it will clean such substances as overspray, light rail dust, bugs, and other contaminants.

Before using this clay, it must first be kneaded into a pliable clean object that can be rubbed over the vehicle, using a lubricant such as soap and water. See **Figure 24-33** and **Figure 24-34**. As it picks up contaminants, it needs to be re-kneaded at intervals to reveal a clean surface before continuing over the vehicle. If kneading does not bury the

FIGURE 24-35 To soften detail clay for kneading, it can be warmed in a microwave.

contaminants picked up off the floor under the clay surface, the piece should be discarded.

Some clay manufacturers also make a specific lubricant to be used with their product. Whether the technician chooses the manufacturer's product or soap and water, clay should never be used without a lubricant.

Detail clay may be hard to knead in cold weather. To soften it, the clay can be placed in a microwave oven for 10-second intervals until it becomes soft enough to work into a clean and sticky ball. See **Figure 24-35**. However, the technician should be careful that the clay doesn't become too hot to handle.

Glazing

Glazings are products designed to quickly remove light surface imperfections, swirl marks, and light oxidation or minor stains. They also produce a deep rich, swirl-free finish. Glazes may or may not contain light abrasives to help remove swirl marks left from polishing. Technicians should be careful in choosing a glaze, because some products are called glazes and have "filling" agents, which cover swirl marks without removing them. When such products are used, the swirl marks will return following rain or washings. Swirl marks are a defect and should be removed before final protection is applied.

Waxing

Waxes are protective coatings designed to clean finishes, then to shine and protect the finish. They may have light abrasives to clean, or they may have only wax to shine and protect. Wax will provide a thin protective coating, which with time will need to be replaced. Some waxes are designed to wash off each time the vehicle is cleaned, and therefore they must be replaced. Others are designed to last up to three months before they must be replaced. Some wax companies recommend that wax removers be used before applying a new coat of wax to reduce buildup. The care and protection of vehicle finishes is such a vast field that an industry of "automotive detailers" has arisen, whose sole purpose is to clean and protect vehicle finishes.

On fresh finishes, the technician should be aware that most paint manufacturers recommend a period of final curing of 60 to 90 days before waxing. The technician should always read, understand, and follow the recommendations on waxing.

PAINT FINISH IDENTIFICATION

One of the main reasons that a newly mixed finish sometimes does not match is that the technician did not get the proper vehicle manufacturer's paint code. Once the **vehicle paint code** is obtained, the technician must convert it into the **paint manufacturer's code** to retrieve the formula needed to mix the color. Recently, aids for identifying vehicle manufacturer's codes have been devised, such as the "vindicator" software. One using this software places the vehicle's VIN (vehicle identification number) into the computer to identify the vehicle's color. See **Figure 24-36**. However, these systems are not always available. The technician should also know how to find the vehicle paint code.

OEM Vehicle Codes

Most paint manufacturers will provide their customers with an information guide to find a vehicle's paint code information. See **Figure 24-37**. These codes may be on the inside of a vehicle's door or in the trunk. See **Figure 24-38** and **Figure 24-39**. The location of the vehicle's paint code may vary from manufacturer to manufacturer; further-

TECH TIP If the vehicle's paint code can't be found after searching all the likely locations, try doing a computer search. Using a typical search engine, type in the vehicle's make and the phrase "paint code." Once the technician reaches the site, the paint code may be listed by VIN number, or by the vehicle's model and year. The technician will probably find an illustration that looks something like **Figure 24-40**.

FIGURE 24-36 The VIN (vehicle identification number) can be found on the driver's side lower windshield.

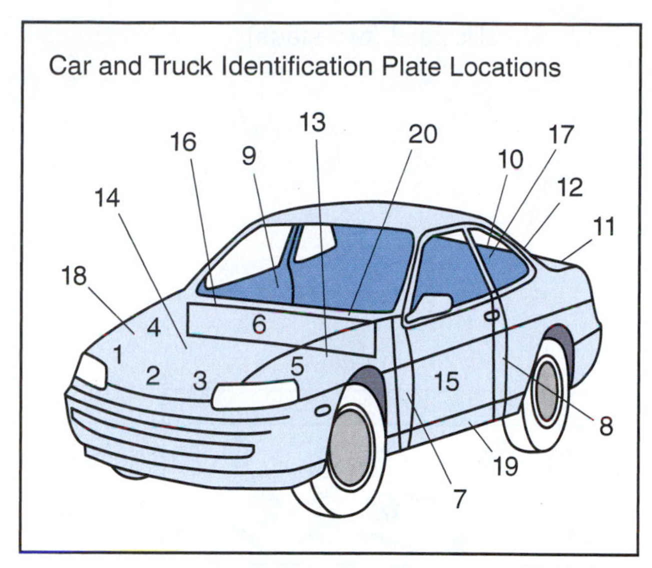

FIGURE 24-37 Paint codes are found in differing locations depending on the vehicle's make and model. A paint code guide can help the technician find locations for specific vehicles.

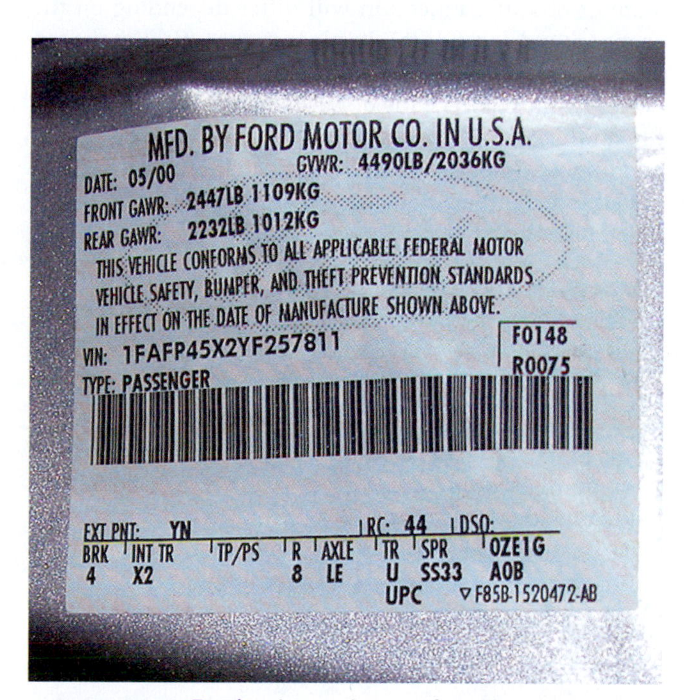

FIGURE 24-38 Ford paint codes are found on the driver's side door.

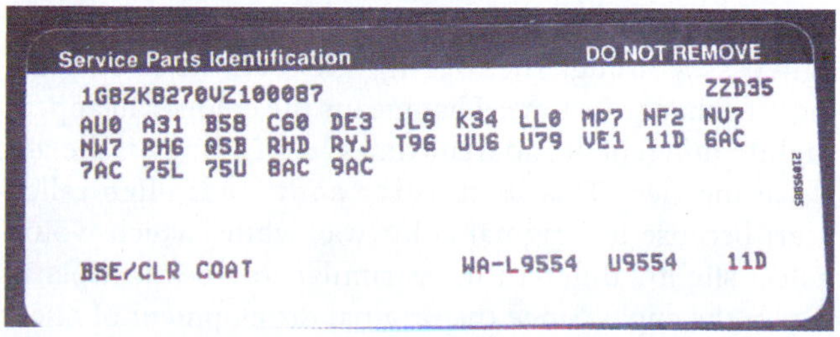

Service Parts Identification DO NOT REMOVE

1GBZK8270VZ100087 ZZD35
AU0 A31 B58 C60 DE3 JL9 K34 LL0 MP7 NF2 NU7
NW7 PH6 QSB RHD RYJ T96 UU6 U79 VE1 11D 6AC
7AC 75L 75U 8AC 9AC

BSE/CLR COAT WA-L9554 U9554 11D

FIGURE 24-39 GM paint codes might be found in the trunk on some models.

FIGURE 24-40 Paint codes can be retrieved through Internet libraries. Courtesy of Google.

more, the location may be different within manufacturers, depending on either the vehicle's model or where it was assembled. Finding the location of the vehicle's paint code may be difficult with all these variables. If a technician cannot locate it from the information provided in the paint manufacturer's paint code location guide, a call to the jobber who supplied the paint will generally provide the needed information.

Paint Maker Codes

Once the vehicle's paint code has been retrieved, it must be converted into the proper code for the paint system being used. To assist with this conversion, the paint manufacturer may provide a paint chip color book. See Figure 24-41. This guidebook not only has the locator guide for the different vehicle makers' paint code

FIGURE 24-41 A color chip book provides visual verification of color matches. Courtesy of DuPont

locations, but it also lists the vehicle code and the paint company code. In addition, it provides a color chip which can verify that the chosen code number is in fact the correct one. Then the formula can be retrieved and mixed. Care must be taken to properly pour the paint tints into the formula, since not doing so is a chief reason for color mismatch while refinishing a vehicle (see Chapter 29).

TYPES OF COLOR COATS

Topcoat finishes can come in different types such as solid color, metallic, pearls/mica, and prismatic. By adding different substances, light will be reflected back with different qualities or effects. Either changing their position in the paint film or the position in which they lay (or are oriented) will also change how light reacts to them. Other substances will let light pass through them, changing the effect of the color. The light may be reflected, as in metallics, refracted, as in prismatics, which allow light to pass through to change it; or refracted, as a prism would. See **Figure 24-42**. Pearl (mica) allows *some* light to pass through, refracting it and reflecting some back.

Solid

Solid colors contain no special light-changing additives such as metallics, pearls, or prismatics. The light that hits a solid color will absorb the entire color spectrum except the desired color, and will reflect that color back. At first, one might think that these would be the easiest to match when refinishing, but even solid color formulas are composed of many different colors. Even white or black shades may have several other colors within the mixture, and may therefore present some challenges for matching.

Metallic

Metallics are flakes of plastics that are added to a finish to provide a glittery appearance. See **Figure 24-43**. The light is not only reflected back, but is reflected back at

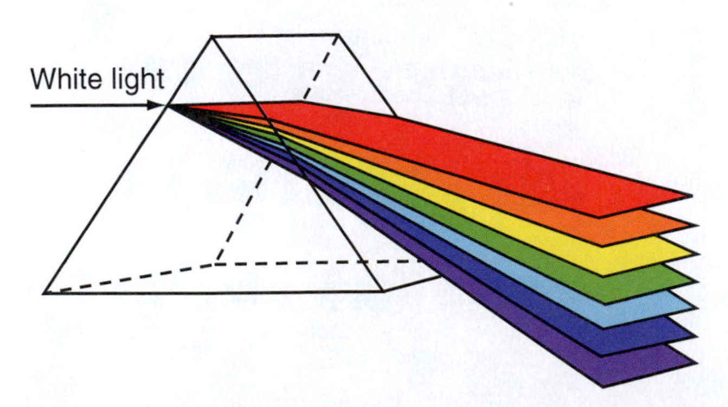

FIGURE 24-42 Light refraction breaks light into its color spectrum as with a prism.

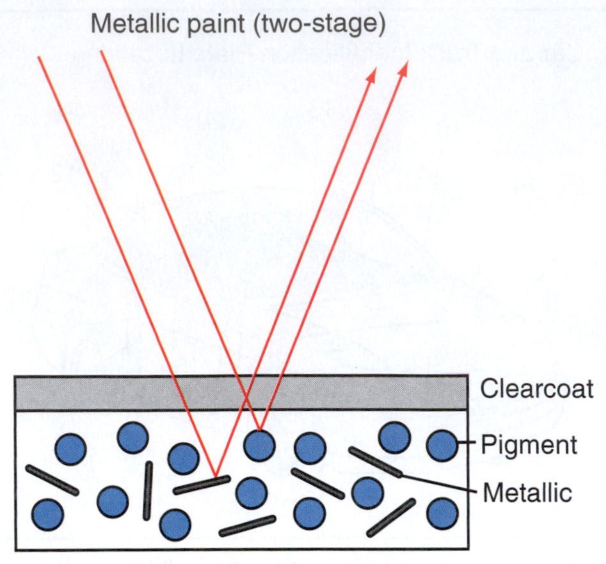

FIGURE 24-43 Metallics reflect light as with a mirror.

slightly different angles to provide the glitter effects. The amount of glitter reflection will differ depending on the shape (round or random), its place in the finish (at the top or driven to the bottom of the film), and its orientation (whether the flake is lying down or standing up within the film of the finish). Consequently, there are many different conditions within the finish's film that alter the effect that metallics will produce, making them challenging to spray when matching a vehicle.

TECH TIP Most collision repair shops do little or no complete refinish work. The majority, if not all, of the work requires a technician to match an existing finish. Matching an existing finish is much more challenging than doing a complete refinish. Considerable skill and knowledge is needed to spray a refinish repair that matches the vehicle's existing color (value, hue, and chroma), metallic orientation, texture, and gloss.

Pearls/Mica

Mica is mined mineral flake that has been coated with titanium dioxide, chromium, and iron oxide powders. It is semi-transparent, which allows some of the light to pass through and be reflected back. See **Figure 24-44**. As the light passes through the coating, it refracts the color, causing it to change its tone. That means the colors will appear slightly different at different times, depending on the angle of the view. This **mica color additive** is often called pearl because its original color was white, which would reflect slightly different tones similar to a pearl, depending on the angle. Since the original development of mica, though, many other colors have been developed.

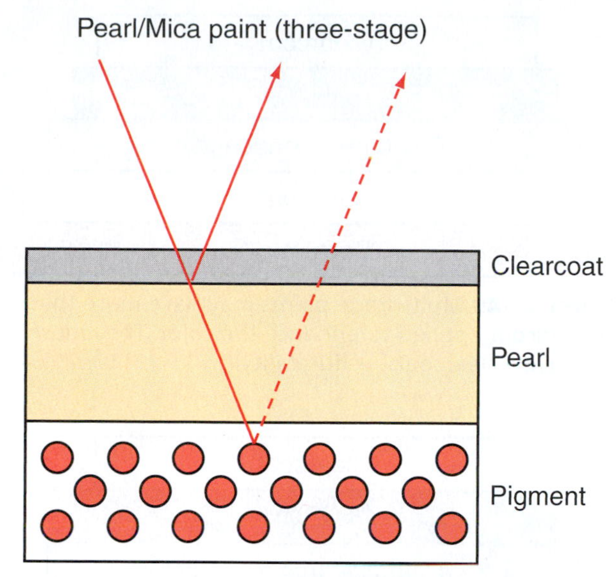

FIGURE 24-44 Mica can both reflect and refract light.

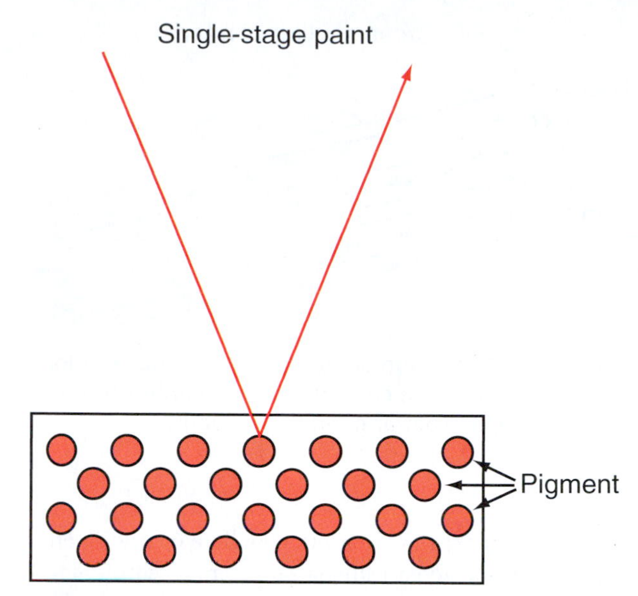

FIGURE 24-45 Single-stage paints do not need to have clearcoat applied over them for durability.

Prismatic

Prismatic color additive, which is manufactured from specific color-shifting products, will change as much as five color shifts depending on the added material. It also has both reflecting and refracting effects, and will have different colors depending on the angle at which it is viewed. Prismatics are not affected by orientation, and therefore do not present the same challenges as do metallics and mica.

REFINISH SYSTEMS

Refinish systems consist of single-stage colors, which have all the components of the finish in a single product. They are sprayed in one step and do not require a clear protective coating to be placed over them. Basecoat/clearcoat requires two steps: a basic color coat, followed by a clearcoat. Multi-coat products will have three or more coats: a basecoat, which is the color; intermediate coats, which may have mica, flake, tinted clear, or another special effect; and protective clearcoat, which will be applied last.

Single Stage

Single-stage finishes were the original method used in vehicle refinishing. They have the color and any additives—such as metallics and mica—contained within. They also have the gloss and protective portion, all of which is applied in a single process. See **Figure 24-45**. These types of finishes are often no longer sprayed in the collision repair industry. When they are used, some paint manufacturers recommend that the repair be covered with a clearcoat to aid in blending.

Basecoat/Clearcoat

Basecoat/clearcoat refinishing is what the name implies. The base color—which may have glamour additives

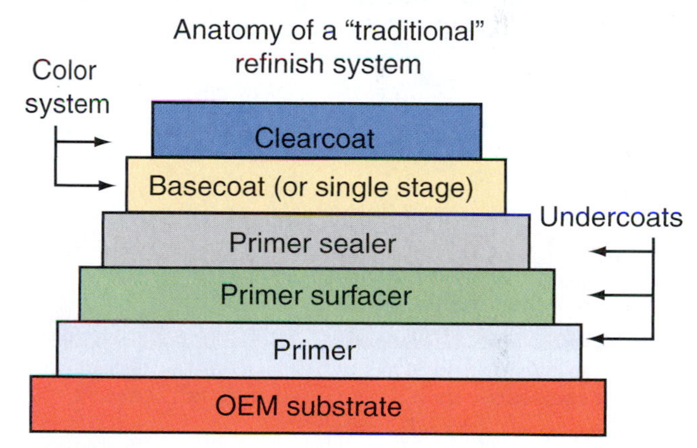

FIGURE 24-46 A basecoat/clearcoat finish must have clearcoat applied over it for durability.

such as metallic and/or mica in it—is applied first. See **Figure 24-46**. This basecoat will not have any gloss, and needs only to be sprayed sufficiently to cover the undercoat beneath. A clearcoat, which is required over it, has the necessary gloss, depth of reflection, and protective components of the finish. This coating must have a minimum amount of film thickness to provide the needed protection. However, if that film thickness is exceeded it may shorten the finished life.

Multi-Coats

Multi-stage finishes have a basecoat similar to a basecoat/clearcoat finish. They also have the color coat, which may have glimmer additives such as metallic and mica. In that case, the surface is covered with an intermediate coating. This step could be as little as one coat (tri-stage, see Figure 24-46) with mica or prismatic mixed in,

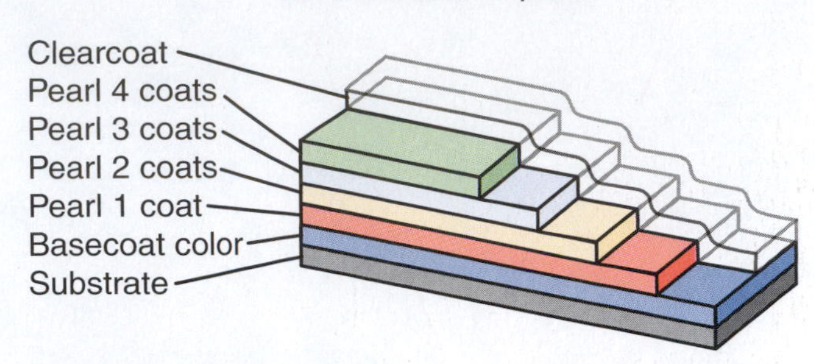

FIGURE 24-47 Tri-stage paints consist of a base (color) coat, an intermediate coat (midcoat) to enhance the color, and a clearcoat applied for durability.

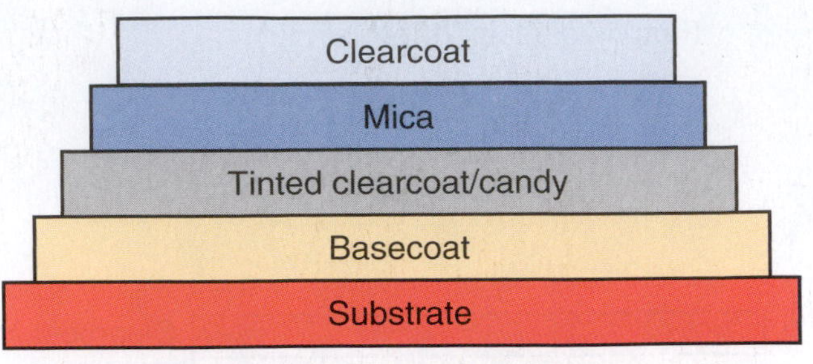

FIGURE 24-48 Multi-stage paints may have more than one midcoat applied to enhance the color. They must also have a clearcoat for durability.

or may have more than one intermediate coat with effects in it, such as clearcoat transparently tinted with color. Finally, the one or more intermediate coats are followed with the protective clearcoat. See **Figure 24-47**.

The combination of intermediate coats are endless, and can cause many different challenges for the refinish technician. See **Figure 24-48** and **Figure 24-49**.

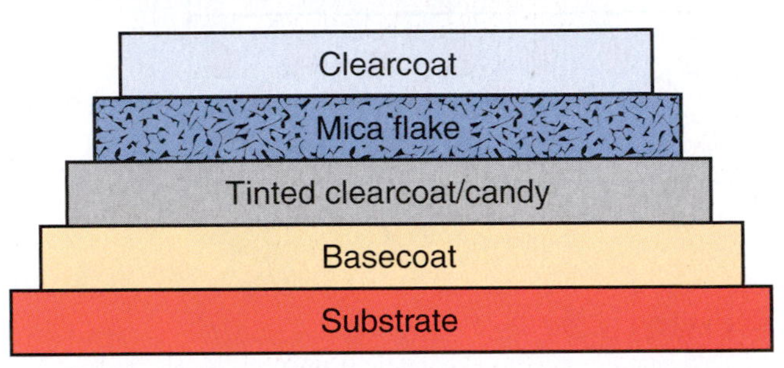

FIGURE 24-49 Candy colors are an example of multi-stage coatings with a basecoat, tinted (translucent) midcoat, possibly a second midcoat with pearls, and then a clearcoat for gloss and durability.

Summary

- Many facilities—including collision repair, custom, and economy—do refinish work.
- Coatings serve many purposes: to serve as undercoats, to provide adhesion, to offer corrosion protection, and to act as sealers.
- The two main categories of finish lacquers are solvent evaporating finishes and enamels.
- Paint is made by mixing different substances together: binder, pigment, solvents, and special additives.
- Reducers and thinners are solvents that are used to liquefy finishes and transport them to the surfaces of the object being painted.
- Plastics such as urethanes come in two categories: thermoplastics and thermoset plastics.
- The four main factors that affect finishes are temperature, humidity, airflow, and time.
- Gloss, color match, and texture all are part of a finish's appearance.
- Three ways to protect a finish are cleaning, glazing, and waxing.
- Paint finish identification can come from OEM vehicle codes or paint maker codes.
- Color coats may be solid, metallic, pearls/mica, or prismatic.

Key Terms

aftermarket
catalyst
color rendering index (CRI)
cycle time
depth of reflection
detail clay

diminished value
15° rule
film build
finish holdout
flash time
flow out
gassing out
glaze
gloss
high solids
holdout
hue, value, chroma, and texture
hydrocarbons
leveling
metallics
mica color additive
mica/pearl
paint manufacturer's code
prism-effect flakes
prismatic color additive
purge time
repaired to pre-accident condition
substrate
texture
thermoplastic
thermoset plastics
ultraviolet (UV) rays
vehicle paint codes
volatile chemical
volatile organic compounds (VOCs)
wax

Review

ASE Review Questions

1. Technician A says that lacquer paint was used for automotive refinishing before urethane paint. Technician B says that urethane paint was used for automotive refinishing before lacquer paint. Who is correct?
   A. Technician A only
   B. Technician B only
   C. Both Technicians A and B
   D. Neither Technician A nor B
2. Technician A says that customer expectations may be different for a collision repair business than for a custom paint company. Technician B says that a collision repair business returns vehicles to their pre-accident condition. Who is correct?
   A. Technician A only
   B. Technician B only
   C. Both Technicians A and B
   D. Neither Technician A nor B
3. Technician A says that a vehicle's paint provides corrosion protection. Technician B says a sealer acts as a barrier coat between the repair below and the coating above. Who is correct?
   A. Technician A only
   B. Technician B only
   C. Both Technicians A and B
   D. Neither Technician A nor B
4. Technician A says that clearcoat only adds gloss and shine to the finish, and that the protection is in the basecoat. Technician B says that clearcoat has ultraviolet (UV) protection in it. Who is correct?
   A. Technician A only
   B. Technician B only
   C. Both Technicians A and B
   D. Neither Technician A nor B
5. Technician A says enamel paint cures through light evaporation and oxidation. Technician B says that urethane paint dries by evaporation. Who is correct?
   A. Technician A only
   B. Technician B only
   C. Both Technicians A and B
   D. Neither Technician A nor B
6. Technician A says that urethane paint cures through evaporation and cross-linking. Technician B says that urethane is not very durable and must be protected with wax. Who is correct?
   A. Technician A only
   B. Technician B only
   C. Both Technicians A and B
   D. Neither Technician A nor B
7. Technician A says that water-base paint has no VOCs. Technician B says that water-base clearcoats help reduce VOC release. Who is correct?
   A. Technician A only
   B. Technician B only
   C. Both Technicians A and B
   D. Neither Technician A nor B

8. Technician A says that thinners and reducers are a single solvent with a set reaction time. Technician B says that thinners and reducers are a combination of solvents mixed to react in precise ways. Who is correct?
   A. Technician A only
   B. Technician B only
   C. Both Technicians A and B
   D. Neither Technician A nor B
9. Technician A says that paint is made from binders, solvents, pigments, and additives. Technician B says that paint is made from pigments and solvents. Who is correct?
   A. Technician A only
   B. Technician B only
   C. Both Technicians A and B
   D. Neither Technician A nor B
10. Technician A says that a binder is sometimes called rosin. Technician B says that binders can be either natural or manufactured. Who is correct?
   A. Technician A only
   B. Technician B only
   C. Both Technicians A and B
   D. Neither Technician A nor B
11. Technician A says that pigments are the portions of paints that supply its color. Technician B says that additives give paint special qualities such as antifoaming. Who is correct?
   A. Technician A only
   B. Technician B only
   C. Both Technicians A and B
   D. Neither Technician A nor B
12. Technician A says that reducers act as a vehicle for the finish. Technician B says that reducers come in only one type, and can be used in any temperature. Who is correct?
   A. Technician A only
   B. Technician B only
   C. Both Technicians A and B
   D. Neither Technician A nor B
13. Technician A says that reducers evaporate all at once. Technician B says that reducers are designed to have in-flight loss of solvent, with some of the reducer evaporating before it reaches the vehicle. Who is correct?
   A. Technician A only
   B. Technician B only
   C. Both Technicians A and B
   D. Neither Technician A nor B
14. Technician A says that final solvent evaporation is done when the vehicle has been baked at 140°F for 40 minutes. Technician B says that a vehicle can be polished as soon as the vehicle has been baked at 140°F for 40 minutes. Who is correct?
   A. Technician A only
   B. Technician B only
   C. Both Technicians A and B
   D. Neither Technician A nor B
15. Technician A says that a 2K paint is one that uses a hardener. Technician B says that an isocyanine is a carcinogen as well as a skin, lung, and eye irritant. Who is correct?
   A. Technician A only
   B. Technician B only
   C. Both Technicians A and B
   D. Neither Technician A nor B
16. Technician A says that a thermoset paint, when fully cured, will get soft when solvents are applied. Technician B says that a thermoplastic paint, when fully cured, will *not* get soft when solvents are applied. Who is correct?
   A. Technician A only
   B. Technician B only
   C. Both Technicians A and B
   D. Neither Technician A nor B
17. Technician A says that hue refers to the paint's color. Technician B says that value is the intensity of the color. Who is correct?
   A. Technician A only
   B. Technician B only
   C. Both Technicians A and B
   D. Neither Technician A nor B
18. Technician A says that the 15° rule states that for every 15°F the temperature rises, it will take longer for a urethane finish to cure. Technician B says that cold does not affect the way urethane finishes cure. Who is correct?
   A. Technician A only
   B. Technician B only
   C. Both Technicians A and B
   D. Neither Technician A nor B
19. Technician A says that orange peel will not affect the finish's depth of reflection. Technician B says that swirl marks are a defect and should be removed, not covered. Who is correct?
   A. Technician A only
   B. Technician B only
   C. Both Technicians A and B
   D. Neither Technician A nor B
20. Technician A says that CRI stands for color rendering index. Technician B says that the vehicle paint code is found in the trunk on all vehicles. Who is correct?
   A. Technician A only
   B. Technician B only
   C. Both Technicians A and B
   D. Neither Technician A nor B

Essay Questions

1. Describe the function of "in-flight loss of solvent" when transporting a reduced finish from the paint gun to the target object.
2. Describe the "two-bucket" method of automotive soap and water washing.
3. Explain the functions of automotive finish components (binders, pigments, solvents, and additives).
4. Explain how humidity in the spray environment affects flash times of finishes.
5. How does the 15° rule work?
6. Describe how to locate the vehicle paint code and how to convert it to a paint maker's paint code.
7. Describe how to use detail clay.

Topic Related Math Questions

1. If a 1,000 ml container has 68.5 percent solids, 17.3 percent solvents, and 12.3 percent binder, what is the percentage of additives?
2. How many ml of additives are present in the item 1 situation?
3. If paint is to be reduced 3:1:1 (the three parts are paint, and the remainder are reducer and hardener), and you need 762 ml of paint, how much reducer do you need?
4. The paint department uses 3 quarts of DT 885 reducer each week, and the boss wants to take advantage of a sale and buy a year's supply of DT 885. How many gallons should the boss order?

Critical Thinking Questions

1. A refinish technician arrives at work on a very hot day (100°F) with high humidity (95 percent), and has a very busy schedule (10 cars). The paint booth is new and in very good condition, but the spray conditions remain 95°F, with 90 percent humidity in the booth, and good airflow. What are the variables that the technician must consider when deciding on the *spray technique* and the *choice of reducer* to use on this day?
2. A 7-year-old silver vehicle is to be refinished on its left fender with blending into the hood, fascia, and left front door. Explain the different steps that should be taken to prepare this vehicle for refinishing.
3. What can be done when choosing reducers and hardeners for a urethane finish in order to speed up a vehicle's cycle time?

Lab Activities

1. Fill four pint jars: two with rock salt and two with road tar. Have the students fill one of the salt jars with warm water and soap, and the other with wax and grease remover, and shake them both. In the jars that have road tar, have the students fill one with warm water and soap and the other with wax and grease remover, and shake them both. Have the students record the results.
2. With a sensitive scale that can weigh to 0.001 grams, mix and reduce a urethane finish to be ready to spray (RTS). Weigh a 4-ounce paper cup when it is clean and empty, and record the weight. With a paint stick, drip 5 drops of reduced RTS finish into the cup, and quickly reweigh the cup. With the cup open to the air, let the solvents evaporate. The next day, reweigh the cup. Record its current weight and calculate the difference from its original weight. What percentage of the RTS finish is solvent, and what percentage is made up of solids?
3. With a light meter, have the students check the following locations for both candlepower and CRI (color rendering index): booth, prep area, masking area, and body shop lights. Determine which areas of the shop will cause metamerism.
4. With five panels (one with lacquer, one with enamel, one with urethane, and two with canned spray paints), have the students perform a thermoset solvent test.
5. On two painted panels that have overspray on them, have students prepare detail clay and use it both with and without lubricant. Have students explain their findings.

Chapter 25

Surface Preparation

OBJECTIVES

Upon completing this chapter, you should be able to:

- Inspect the vehicle for surface preparation.
- Prepare a repair plan.
- Clean a vehicle's surface.
- Be able to perform tasks such as:
  - Removing the old finish when necessary
  - Feather edging
  - Priming
  - Blocking
  - Sanding
  - Scuffing
- Prepare aluminum substrates.
- Prepare steel substrates.
- Prepare plastic substrates.
- Prepare e-coated substrates.
- Prepare the vehicle for blending.

INTRODUCTION

This chapter will discuss what most refinishers believe to be the most important part of the whole refinishing process: preparing the surface for refinishing. This involves examining the repaired surface—or the surface that is to be finished or refinished—and preparing it to accept the topcoats. Other chapters examined aspects of achieving an invisible repair, such as matching the color, texture, gloss, and depth of reflection, which are very important parts of the refinish technician's responsibility. This chapter will examine how to correctly prepare the vehicle's surface to provide the lay and adhesion of the undercoats to the vehicle.

If the surface to be refinished is not properly prepared, the repairs underneath may be visible in the new finish. See **Figure 25-1**. In other cases, even if the surface is not properly prepared, the repair may not be visible. However, the new finish may not have proper adhesion to the old surface and may separate, causing a condition called peeling. See **Figure 25-2**. If the undercoating is not smooth when topcoat is applied, the surface can never be polished smooth, and the finish will need to be re-prepared for topcoating.

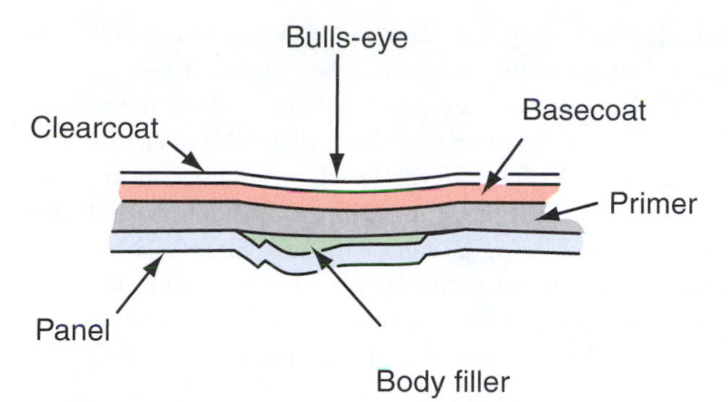

FIGURE 25-1 Even the slightest imperfection in preparation could show through the new finish, causing a condition called "couture mapping" or "bull's eye."

FIGURE 25-2 Improperly prepared surfaces can cause poor adhesion and "peeling."

Surface preparation is often the first task that a new technician is required to learn, do, and become proficient at. The vehicle is prepared for the lead painter to apply the topcoat, though some would argue the preparation steps are as difficult to learn and surely as important as the topcoat. Most times, when the surface preparation is not done properly, the only repair which will remedy the defect is sanding and re-preparation of the surface for topcoating, which is a costly and time-consuming process.

While each vehicle in a collision repair facility is similar to every other vehicle, each is nevertheless different. Therefore, each vehicle must have its surface prepared slightly differently than other vehicles. A technician who can quickly prepare a plan for the needs of each different vehicle contributes to a more profitable refinish department.

Successful refinish technicians are able to judge each surface preparation situation as it arises, and to efficiently prepare the vehicle for topcoating. It is easy to see how important surface preparation is, and how important the technician responsible for the task is to the overall collision repair facility.

SAFETY

Refinish materials and the chemicals they are made from pose personal and environmental protection concerns. The refinish technician is governed by—and must comply with—many local, state, and federal laws concerning protection.

Each product must be accompanied with a material safety data sheet (MSDS). The MSDS includes sections on:

- Hazardous ingredients
- Physical data
- Fire and explosion hazards
- Reactive data
- Health hazard data, or toxic properties
- Preventive measures
- First aid measures
- MSDS preparation information

Technicians must read, understand, and follow the recommendations in the MSDS for the protection of themselves, their fellow workers, and the environment.

INSPECTION

Collision repair shops are generally divided into two departments. One performs the vehicle's structural and cosmetic repairs, returning the vehicle to its pre-accident condition. The other—the refinish department—restores the vehicle's finish to its pre-accident condition. Each vehicle will have a work order outlining the agreed-upon repair. Using this work order, the technician will do a preliminary inspection of the vehicle. As vehicles pass through the repair process, undiscovered or unreported repairs often become evident. If that happens, the original work order or job estimate needs to change to accommodate the newly discovered damage. The refinish technician should inspect the vehicle to determine if all the refinish needs were addressed in the vehicle's repair order. If the technician inspecting the vehicle for refinishing is not the lead painter, the two should agree on the refinish plan. This eliminates confusion or lost time when preparing the vehicle. Next, the vehicle should be inspected for quality and completion of repairs. Has everything on the repair order been completed and is it repaired to the quality needed for refinishing? If items have been missed, or if the repair quality is such that the paint department cannot restore the finish to its pre-accident condition, the vehicle should be returned to the repair department for completion. Then, the technician in the refinish department should inspect and identify the type of finish on the vehicle, identify the type of **substrate**, evaluate the film thickness, make a repair plan, and clean the vehicle.

This chapter will examine each of the variables for preparing a vehicle for refinishing, and will discuss ways to prepare the surface for many different conditions, giving the reader a better understanding of surface preparation.

TYPES OF FINISH

Although the different types of finishes were discussed at length in Chapter 24, it is important for technicians to know what type of finish they will be applying. By correctly covering the existing finish, the new finish will perform as originally designed. Each paint manufacturer provides production recommendations in their P-sheets, or sometimes called technical service bulletins (TSB). See **Figure 25-3**. These sheets identify the type of finish it is, what other finishes with which they are compatible, and preparation instructions. The instructions will also outline the proper surface preparation steps to take for optimum results when using the new finish.

New finishes are generally designed to be applied over OEM finishes; that is, over a finish applied at the manufacturer's facility. These finishes were baked at a high temperature before the vehicle was fully assembled. Other finishes are designed to be placed over previously refinished surfaces such as **lacquers**, **enamels**, **urethanes**, or epoxies. Although many **aftermarket** refinishes can be applied over different finishes, some may not be compatible to the finish below them. In this case, special steps need to be taken to avoid mishaps.

Original Manufacturer Finish

The finishes originally applied at the factory are not the same as the finishes applied in the collision repair facility. The original equipment manufacturer (OEM) finishes need to be heated to properly cure. Although a refinish aftermarket repair facility may heat finishes, they are not able to heat the vehicle to the 450°F (or higher) needed to cure OEM finish. During the manufacturing of a vehicle, the undercoats and topcoats are applied before the final assembly is completed, and the required heat does not destroy heat-sensitive equipment added during assembly. In fact, the finish applied at the factory is one of the best substrates over which to apply new finish. When preparing a vehicle for blending over old undamaged finish, the technician only needs to properly clean and sand the vehicle to prepare it for new finish.

Previously Finished Areas

When inspecting the finish of a repaired vehicle, the technician must decide whether it is the original finish or an area that has been refinished. One of the ways to discover if the vehicle has been refinished is to take a film thickness reading. If the film is thicker than what should be expected for a factory finish, it should be assumed that the vehicle has been previously painted. In this case, the technician should perform the surface preparation steps for this type of surface. If the technician does not have the equipment needed to measure the film thickness, looking for the tell-tale signs of previous work can aid the technician. Such signs include overspray on moldings or lights; evidence of moldings that have been removed and reattached, which may not fit properly when re-applied; surface defects; or color mismatch. See **Figure 25-4**.

If the vehicle has been previously refinished, the technician needs to determine whether the finish is a lacquer finish. Lacquer finishes will refold when solvents are applied. The **solvent test** should be performed when the old finish is suspected of being lacquer. (Review Chapter 24 for the solvent test.) Once the finish is determined not to be a lacquer finish, and if it does not exceed the film thickness limits, it should be inspected to determine its substrate.

TYPES OF SUBSTRATE

Vehicles at one time were generally manufactured from steel; today, they come to the repair facility with many different types of substrates. They could be mostly **steel substrates**, or have some **aluminum substrate** parts. Some are aluminum-intensive vehicles, made of mostly aluminum; while others may have a mixture of steel and aluminum for the structural portions, with some or all of their exterior panels manufactured from plastic. These are called **plastic/composite substrates**. Each different substrate needs to be prepared differently. A technician must identify the different parts to make a surface preparation plan for each vehicle's particular needs. Aluminum, steel, and plastics all need to be prepared slightly differently, as the following sections will show.

Aluminum Preparation

Aluminum comes in many different types and strengths. However, the material's strength in the way it responds in a collision and how it collapses under the force of an accident are somewhat different than how the cosmetic panels respond to **sanding** and refinishing. When sanded, aluminum will scar more easily than steel. Therefore, an aluminum surface should never be sanded with sanding grits more aggressive than 80-grit. Aluminum expands with normal day-to-day heat, and must have larger gaps between panel openings. Although aluminum does not "rust" as steel does, it will corrode; therefore, the technician needs to take special anti-corrosion protection measures.

Sometimes it may not be obvious to the technician performing an inspection that a panel is made from aluminum. Some technicians will test the surface by trying to apply a magnet. While a magnet will not stick to aluminum, it will not stick to plastics, either. The technician should be sure that the panel in question is in fact aluminum and not plastic, if a magnet is used. The best way to determine if a panel is manufactured from aluminum is to check the collision repair service manual for the vehicle. If a repair manual is not available for identification, and the panel is suspected of being either aluminum or plastic, the technician may use some film thickness gauges that read the aluminum and not the plastic. See **Figure 25-5**.

Steel, which still is the most common component of the automobile, is easily identified with the use of a magnet.

OC-1

Plastic Prep System

SU4901
Clean and Scuff Sponge

SU4902
Plastic Adhesion Wipe

SU4903
Advance Plastic Bond (Quart)

SUA4903
Advance Plastic Bond (Aerosol)

PPG *One Choice®* brand *Plastic Prep System* is designed to simplify the plastic refinish prep process and deliver superior adhesion to all common automotive plastic substrates.

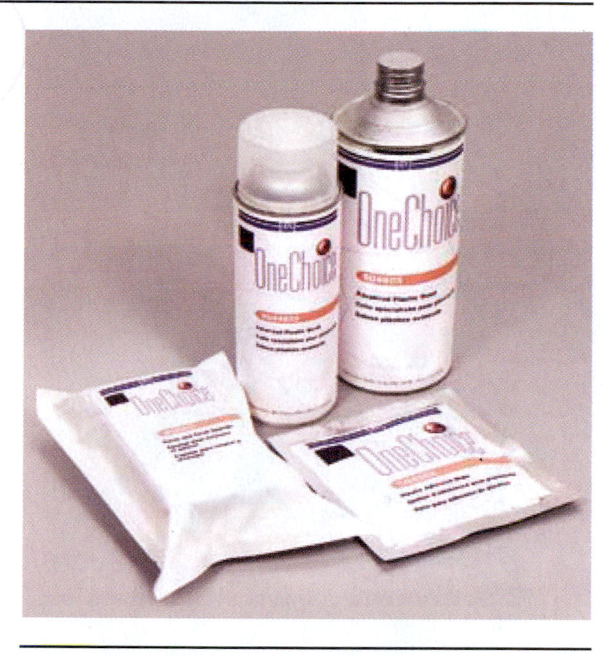

Features

- Convenient ready-to-apply packaging
- Fast application and dry times

Advantages

- Can be used on all common automotive plastic substrates
- Significantly reduces the potential for warranty repairs

Benefits

- Simple, easy-to-use-process
- Shorter refinish cycle time
- Superior adhesion

Required Products

- SU4901 Clean and Scuff Sponge
- SU4902 Plastic Adhesion Wipe
- SU4903 Advance Plastic Bond
- SUA4903 Advance Plastic Bond (Aerosol)

Related Products

- SXA103 Multi-Prep (Aerosol)
- DX103 Multi-Prep

Compatible Products

The *One Choice Plastic Prep System* is for universal use with PPG Brand Topcoats and Undercoats.

Note: When applying PPG topcoats and undercoats over plastic substrates, please refer to the products specific technical bulletin for proper application.

Compatible Surfaces

All common primed and unprimed automotive plastic substrates

Application Data

Process for Pre-Primed Plastic Substrates

Step 1:

Using the SU4901 Clean and Scuff Pad...

Tear open SU4901 and clean the substrate thoroughly using the scuff pad side of the pre-saturated sponge, then rinse with water. Blow dry or wipe with a clean cloth. Entire surface must be totally de-glossed. Make sure surface is thoroughly dry before proceeding.

FIGURE 25-3 Most paint manufacturers provide a production information ("P") sheet. Used with permission of PPG Industries.

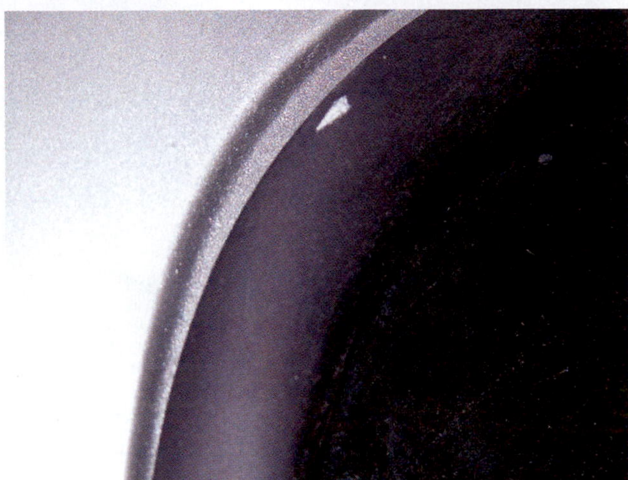

FIGURE 25-4 Overspray caused by poor masking is a tell-tale sign that the vehicle has been refinished.

FIGURE 25-5 This film thickness gauge can measure both steel (black end) and aluminum (gray end) film thickness.

TECH TIP Although some technicians believe that the use of aluminum is a new change in the manufacturing of automobiles, it is not. Some Ford Model T's were made with aluminum hoods to reduce the weight of the "Tin Lizzy."

Film thickness gauges also are able to identify steel, as is a service manual. The surface preparation of steel is the most common type of preparation, and many products have been developed for the proper adhesion, filling quality, and corrosion protection of steel. Although there are many different types of steel used in automobile manufacturing, the preparation of steel substrates are the same.

Plastic/Fiberglass/Composite

Plastic is a large category that could generally be placed into two large categories. Thermoset and thermoplastics do have different surface preparations to be considered. There have been many different ways used to identify which category plastics fit into. Methods include the burn test, the sanding test, and finding the manufacturer's identification mark to identify the type of plastic used. All of these types of identification are limited, and may not properly identify the plastic. Using the manufacturer's collision repair service manual to identify the plastic is the most accurate way to identify different types of plastic. The vehicle manufacturer's recommendation for surface preparation remains the best guide, and should be followed if available. If the vehicle manufacturer's recommendation is not available, the next best recommendation to follow is that of the paint maker. Each paint manufacturer will publish recommendations on how to identify plastic and the surface preparation steps needed to apply its refinish product. These steps should be followed for best results. See **Figure 25-6**.

PAINT FILM THICKNESS

Should All or Some of the Finish Be Removed?

Film thickness is the amount of coating film that is placed on the vehicle when it is painted. It is measured in one-thousandths of an inch, normally referred to as mils (1 mil = 1/1000 inch). An OEM finish will range from 4 to 6 mils, with refinish thickness ranging from 6 mils upward, depending on the methods of repair used. Most paint manufacturers recommend that the film thickness not exceed 10 to 12 mils total following refinishing. However, if a vehicle has already been refinished once, and a second application of paints is required, it may exceed the recommended total film thickness limits. If these film thicknesses are exceeded, the refinish manufacturer may not place a warranty on its product.

To stay within the recommended film thickness maximums, the technician must know the vehicle's film thickness before refinishing. Taking multiple film thickness readings of a panel will determine its average thickness.

P-194

Universal Plastics Adhesion Promoter

DPX801

PPG's Universal Plastics Adhesion Promoter DPX801 is a red tinted transparent product specifically designed for use on automotive bumpers and various plastic parts to provide adhesion on these substrates for subsequent topcoats.

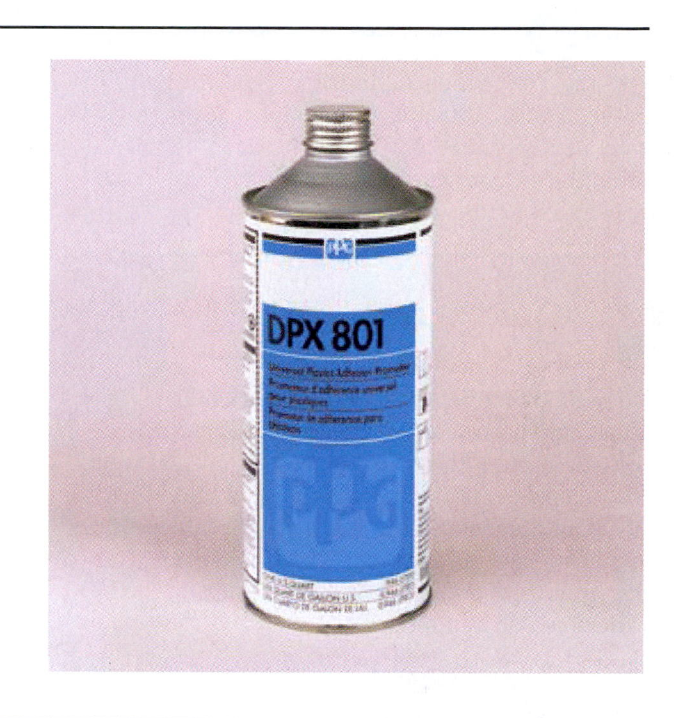

Compatible Surfaces

DPX801 may be applied over:

- Properly cleaned & sanded plastic and rubber parts

Directions for Use

Surface Preparation:

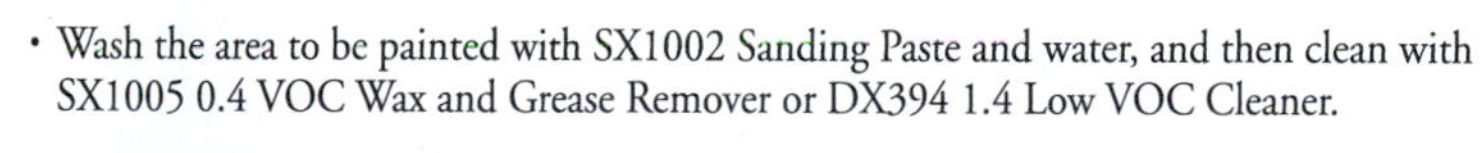

- Wash the area to be painted with SX1002 Sanding Paste and water, and then clean with SX1005 0.4 VOC Wax and Grease Remover or DX394 1.4 Low VOC Cleaner.
- Lightly scuff with 400 – 600 grit sandpaper or a gray/white Scotchbrite.
- Re-clean with SX1005 0.4 VOC Wax and Grease Remover or DX394 1.4 Low VOC Cleaner.
- Final wipe with DX103 MULTI-PREP™.

Mix Ratio:

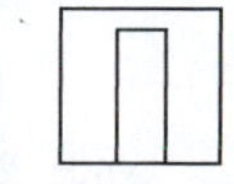

DPX801 is ready to spray.

No reduction is necessary. Thoroughly shake or stir DPX 801 before use.

Pot Life: N/A

FIGURE 25-6 Plastic preparation requires specific preparation steps to ensure good adhesion. Follow the paint manufacturer's recommendations.

With this average, the technician can determine if the surface must be stripped, or if a **partial film removal** is needed to remain within the required film thickness after refinishing is complete.

It is not unusual to find that the vehicle's horizontal surfaces have a thicker reading than the vertical or side surfaces. If a panel's average film thickness is 8 mils or above, the technician will need to remove part of the finish before applying new coatings.

> **TECH TIP** When taking film thickness readings for an average, if the reading jumps to a very high film thickness (from 8 or 10 to 23 or even 40), it is likely that the area has plastic body filler under it. If so, the area should not be included in the average.

Defects and How to Repair Them

When refinishing vehicles following a collision, the technician may discover that the original finish has defects in it due not to the collision, but to normal wear. Common wear defects are stone chips, scratches, cracking, and oxidation, just to name a few. Paint problems and their repairs are covered thoroughly in Chapter 31. Complete descriptions of their causes and the steps needed to correct them can be found there.

Repair Plan Made

After inspecting the vehicle, the repair technician will create a refinish plan using all the information found when performing the inspection. This is done even though the vehicle already has both an estimate and a repair order. This plan will outline which steps must be taken to prepare the vehicle, which products will be used to prepare the surface and ensure its proper corrosion protection, and which techniques of topcoat application should be used to ensure an invisible repair and restore the vehicle to its pre-accident condition. Based on the answers to these questions, the plan will ensure that the technician will perform a speedy, efficient, and cost-effective refinishing job.

CLEANING

Cleaning is done at many stages of the repair process. The vehicle is soap and water washed, then chemical cleaned, before the repairs are even started. It also gets a **cleaning in the refinish department**, both with soap and water and chemically, when it arrives. In fact, it will be cleaned once again just before the vehicle is masked for spraying.

It is important to clean the vehicle when it arrives from the body repair area to eliminate any contaminants that may have been deposited during the repair. Soap and water washing will eliminate the water-soluble contaminants (salt, body filler dust, smoke residue from welding, and other such contaminants). Then the area to be refinished and blended should be cleaned with chemical cleaner (wax and grease remover) to remove the contaminants produced from working with frame machines and their hydraulic oil, or other oil contaminants that may have gotten onto the vehicle during collision-related mechanical repairs.

Soap and Water

When **soap and water cleaning** a vehicle, the technician should use a pH balanced (pH 7), liquid car washing soap. Dish soap, an alkaline mixture that is highly concentrated and does not wash off well, is not suitable for vehicle washing. Powdered soap is another less-than-ideal choice for cleaning a vehicle, because it may not dissolve properly and the flakes may become aggressive on the washing mitt, causing scratches. When mixing the liquid car washing soap in water to wash a vehicle, the technician should use warm to hot water for best results.

To assure that the washing mitt is clean each time it is rinsed and new soap is picked up by it, the technician should use the **two-bucket method of washing**. One bucket is filled with a mixture of hot water and car washing soap. A clean washing mitt is dipped in the clean soap and water bucket, and then is applied to the vehicle surface. When a panel is cleaned, the mitt is rinsed in a second bucket of clean water to remove the dirt that was picked up by washing. When the mitt is clean, it is then re-soaped, a new section of the vehicle is cleaned, and the process repeated until the entire vehicle has been washed. See **Figure 25-7**. The vehicle should be rinsed with clear, low-pressure, warmed water, making sure all the soap has been rinsed off thoroughly. If the washing takes a long time or is interrupted for some reason, the soaped areas should be rinsed before the soap has a chance to dry on the surface; otherwise, the panel will need to be washed again. If the water in the rinse bucket becomes too dirty to properly clean the wash mitt, it should be discarded and replaced with fresh water in the bucket.

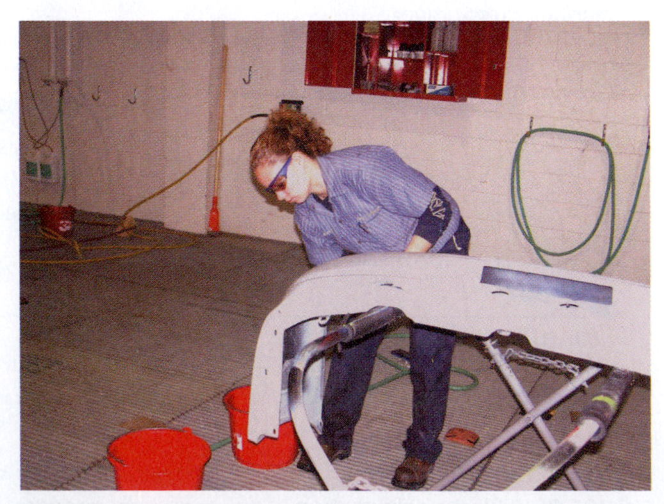

FIGURE 25-7 Cleaning plastic parts is critical. The two-bucket cleaning method keeps debris off the cleaning mitt, thus resulting in better cleaning.

When a technician washes a vehicle that has been repaired, the cleaning process should be thorough. This means that the doors should be opened, and the inside door jambs should be cleaned of any dirt and dust that may have entered during repairs. The trunk and the area under the hood should also be cleaned, to prevent any dirt from blowing onto the vehicle during spray operations.

While washing the vehicle, the technician also should check the finish for any damage that may need repair before it is refinished.

Chemical/Solvent

Washing the areas that will be refinished with a wax and grease remover is a form of **chemical cleaning**. It is intended to remove any wax, grease, silicones, and other contaminants that cannot be dissolved with soap and water. The chemical cleaner should either be applied with a clean lint-free cloth or sprayed on with a spray bottle. See **Figure 25-8**. The cleaner will dissolve the contaminants and float them to the surface, where the technician can remove them by using a clean, lint-free cloth or a paper towel. Because the contaminants are floated to the surface of the cleaner, the cleaner cannot be allowed to dry before it is wiped off. Therefore, the technician should apply the remover to only approximately 2 square feet at a time. Then the technician should wipe the cleaner off while it is still wet, and continue to wet up another portion and wipe it off, until the area to clean is completed.

If the vehicle's surface is heavily contaminated with road tar and grease, it may take a second washing to remove the contaminants. Most paint manufacturers are now making different strengths of solvents—some for use with heavily contaminated vehicles, and others for use with newly applied coatings that need a lighter cleaning. The technician should pick the cleaning solvent best suited to the specific application.

FIGURE 25-8 The chemical cleaner in this pressurized sprayer is used to remove the non-water-soluble contaminants.

After completing a chemical cleaning, the technician should be certain that the surface is clean by feeling it. A gloved hand on a clean surface will have resistance as it is being pulled over the surface; however, if the surface is still contaminated with wax or grease, it will smear and have little resistance as the finger is moved over the surface.

PAINT REMOVAL

If the surface in need of refinishing has a high build of finish on it, the technician may need to strip or partially remove the finish before applying anything new. If the film thickness is 8 to 10 mils and a new finish needs to be applied, the addition of 4 to 6 mils of new paint would mean that the finished product would have 12 to 16 mils of paint. Most paint manufacturers recommend that the paint film not exceed 10 to 12 mils; therefore, such a case requires **paint removal**.

The finish may also be damaged and in need of complete removal, which is sometimes done with sanding using a dual action (DA) sander. It can also be done with chemical removers or blasting, either using sand blasting or other less aggressive means such as peanut shells or plastic beads. This aggressive process—called media blasting—may need to be done by professional media blasters.

Partial Paint Removal

Partial removal of paint is generally done mechanically using a DA sander. See **Figure 25-9**. The technician must carefully choose a sandpaper that is aggressive enough to take off the needed finish in a reasonable amount of time, but not so aggressive that it leaves deep scratches in the substrate below. As mentioned previously, aluminum should never be sanded with paper that is more aggressive than 80-grit. Plastic will also scar easily, and should not be sanded with paper more aggressive than 80-grit, either. Steel, however, may be sanded with more aggressive sandpaper, because it will not leave the deep scratch grooves seen with aluminum and plastic.

FIGURE 25-9 This dual action (DA) sander is used to remove excess finish.

When partially removing the finish, the technician should take care not to sand through to bare metal in large spots. When fresh steel becomes exposed, it needs to be treated with a corrosion-resistant paint such as acid etch or epoxy primer to keep the surface from corroding.

If bare metal is exposed, primer should be applied to the surface as soon as possible. Corrosion will start within an hour of sanding on steel, and in less than 30 minutes with aluminum. If the bare surface is exposed to air for longer than these times, it should be lightly re-sanded and then immediately coated with a corrosion-resistant coating.

Chemical Stripping

Chemical stripping is generally done when the finish must be completely removed. If the part can be removed and stripped easily while off the car, it will help control the spread of the chemicals. See **Figure 25-10**. To begin the chemical stripping procedure, the outer edge of the part is covered with tape to keep the chemical stripper from creeping under the edge. The technician should use personal protective equipment as instructed within the MSDS that accompanies the chemical stripper.

The technician should apply the chemical stripper to a section that is small enough to be easily managed, such as about half of a hood. The stripper is brushed on in one direction and allowed to bubble up the finish. When the finish has lifted as much as is likely, it should be scraped off using a plastic spreader. If some stubborn spots of finish are still adhering to the part, a second application of the stripper should be applied, allowed to work, and then removed. The part is then rinsed off with water and dried. When the chemical stripping is completed and the part is rinsed off, the outer edge of masking tape is removed. Then the part is DA sanded, after which the correct corrosion protection is applied.

(A)

(B)

FIGURE 25-10 (A) Chemical stripping can be done with the part off the vehicle. Note that chemical stripping may be restricted in some areas! (B) Chemical stripper may need to be applied more than once to remove heavy paint film.

Media Blasting

Sand blasting has been used for years to clean rusty steel; it removes the oxidation along with any finish that may remain. Sandblasting is done with silica sand blasted with high-pressure air against the steel. The fast blasting causes the steel's surface to heat up, often warping sheet metal in the process. Other blasting media have been developed to reduce the warpage, such as peanut shells, glass and plastic beads, and even soda and water. All of these materials strip and clean the metal. However, the steel or aluminum must be treated with corrosion-resistant primers shortly after the blasting, or the surface will rapidly start to corrode.

Most of the equipment needed for blasting is large and expensive, and may not be available at collision repair shops. The process of **media blasting** is also quite dirty. Consequently, most shops will locate the blasting area away from the shop, or will take their parts to a professional blaster to be stripped. See **Figure 25-11**.

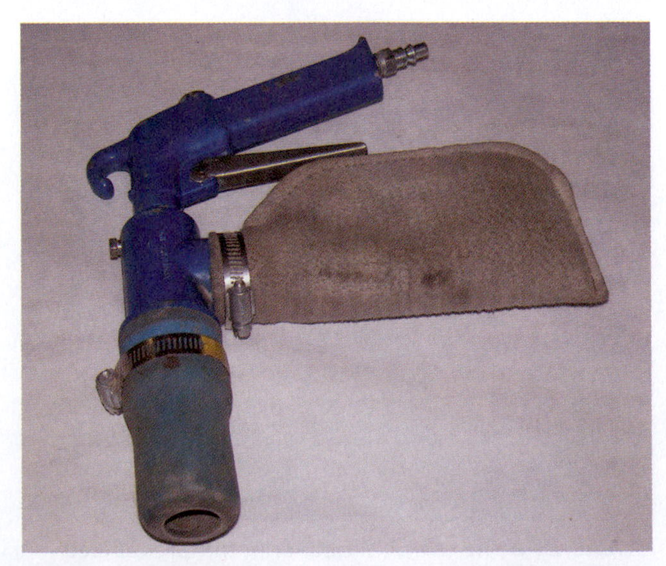

FIGURE 25-11 A spot sand blaster can be used in small areas.

BARE STEEL PREPARATION

In bare **steel preparation**, it is best not to strip off the factory-installed corrosion protection. Corrosion protection is applied to the steel at two specific times and places. First, it is applied over the steel when it is manufactured at the steel mill. The steel may be coated with zinc or other coatings during a process called galvanizing. These applied coatings protect the steel but can be destroyed or disturbed during manufacturing. After the steel has been stamped into parts and welded together, the bare vehicle body is covered with an electro-deposition coating commonly called "**e-coat**." During the stripping process, one or both of these protective coatings can be destroyed. If so, they will need to be re-applied.

In some collision repair facilities, restoring corrosion protection may be part of the structural technician's responsibility. Regardless, a refinish technician should know how to perform these procedures in the refinish area, in order to do so as needed.

There are different ways by which a technician can restore corrosion protection. The technician can choose from three different types of corrosion protection, depending on the needs of the part being protected. These three material choices are metal cleaners and conversion coatings, acid etch primers, and epoxy primers.

Cleaners and Conversion Coating

If during the stripping process the metal is exposed, it is best to use the most aggressive form of corrosion protection, which is chemical cleaning, and then apply a conversion coating, followed with a corrosion protective primer. To be assured that the corrosion protection has been replaced as closely as possible to factory corrosion protection, the best way is to clean freshly stripped and sanded bare metal with chemical cleaning, and then apply conversion coating. This procedure is explained in detail in Chapter 19. In review, the steps are:

1. Clean the surface with wax and grease remover.
2. Sand with P80 to 150-grit sandpaper. (Do not sand with paper finer than P150 grit, or adhesion will be lost.)
3. Re-clean with wax and grease remover.
4. Using a scuffing pad, scrub the bare metal with the chemical cleaner.
5. Rinse with water and dry.
6. Scrub in the conversion coating, and keep it wet for the recommended time. Then rinse with water. Water should sheet off. Otherwise, reapply the conversion coating.
7. Dry the surface, and apply corrosion protective primer.

This type of aggressive corrosion protection should be applied to exterior vehicle surfaces that have been either chemically or mechanically stripped. If new patch panels or rust repair has been performed, this is the procedure of choice. However, if a new part comes in with no bare metal, or if the repair procedure has not removed the e-coat, this cleaning and conversion coating may not be necessary.

Acid Etch Primer

Acid etch primers were developed to cover freshly exposed steel in order to reduce the lengthy cleaning and conversion coating process outlined previously. The acid etch (the steel cleaning step) and the zinc coating (the conversion coating step) are both contained in one sprayable application. For this process to work properly, the surface being sprayed needs ample airflow to allow the acid to gas out during the flash time. The coating will not gas out if it is sprayed on the vehicle too thickly, either. Additionally, the recommended film thickness should not be exceeded, in order to achieve the proper results. If sprayed in enclosed rails, the product will not properly neutralize, and will not protect the steel from corrosion. For these reasons, acid etch primers are not recommended for use in enclosed areas.

TECH TIP As the name indicates, acid etch primers contain a substance designed to etch into metal. When sprayed through a spray gun, the primer will etch into a vehicle's surface. Plastic cups or a plastic cup liner is needed to protect the metal surface. The acid does not etch into the metal of the spray gun because it passes through it so rapidly, unlike the cup which is in full contact during the time the acid is in the gun.

Epoxy Primers

Epoxy primers are highly recommended as the first step when applying corrosion protection to a repaired area. They can be applied on enclosed interior surfaces, exposed interior surfaces, exposed exterior surfaces, or exposed joints. They have excellent adhesion, do not need to neutralize, do not need to be topcoated, and are not adversely affected by heavy application. Some manufacturers allow plastic body filler to be applied over their epoxy primer product, which protects the steel below it from corrosion. Epoxy primers can be sprayed, rolled, or even brushed onto areas that have been properly cleaned and prepared, allowing a technician the ability to protect corrosive hot spots caused from the repair process. See Chapter 19.

ALUMINUM PREPARATION

Although the corrosion protection process for bare aluminum is similar to the procedure for steel, there are subtle differences. First, aluminum should never be sanded with sandpaper more aggressive than P80. Using a coarser paper will cause the surface to have deep sand scratches that may not cover properly. Also, there are cleaning and conversion coating products specifically designed for aluminum; therefore, products intended for steel should not be used. While the steps used for application are similar, the technician should read, understand, and follow the specific recommendations for the aluminum application.

As with steel, the application of conversion coating followed with a corrosion protection primer is also recommended for aluminum.

ABRASIVES

One of the most commonly used groups of products in the refinish area—and particularly in the surface preparation steps—is abrasives. This category of tools contains items such as sandpaper, grinding disks, compounds, and polishes, along with other products. Abrasive materials are glued to a backing paper, then used to scrape or cut off the unwanted surface. The technician can determine how aggressive the abrasives are and how deep the scratches are they leave behind through **grit numbering**. This system assigns a number to indicate aggressiveness. The more aggressive or coarse group uses smaller numbers, such as 24-grit grinding pads, whereas the finer group uses larger numbers, like the very fine-textured and less aggressive P3000 paper. The abrasive particles are arranged on the paper very closely, as a closed coat, or farther apart, as an open coat. They can be attached to very thick backing paper, to enhance the cutting power, or to very light backing paper, to keep the cutting power of the paper to a minimum. The collision repair technician must know which type of paper, the proper grit, and which backing pad should be used for each application in order to enhance each job. The incorrect choice of sanding paper may make more work for the technician than is needed.

Open/Closed Coat

Sanding papers and grinding disks are manufactured as either **closed** or **open coat paper**. The closed coat paper has the abrasive material glued to the backing pad very tightly. Around 90 percent or more of the paper is covered with abrasive material. This type of product works best when a technician is removing hard materials, such as steel that does not get stuck in the closely arranged abrasive products. Open coat paper has only 50 to 70 percent of the backing covered with abrasive material. See **Figure 25-12**. This allows the material being removed to spin off with the centrifugal force of the sanding machine without clogging the sandpaper as much. Therefore, it allows the paper to last longer. Soft substances such as paint will clog closed coated material faster.

(A)

(B)

FIGURE 25-12 Open coat paper has loose or "open" spacing of the abrasive (A). With closed coat, the abrasive is tightly packed with no open areas (B).

Sandpaper Backing

Sandpaper backing may be light paper, heavier stiff paper, or even fiber for very aggressive and stiff surfaces. The different types of backings help each cutting material do its job better. On light sanding paper, the backing is rated A for the lightest with C-, D-, or E-weights being progressively heavier and stiffer. The lighter the paper, the better it can conform to the surface that it is sanding. Light paper lasts the least length of time, and will need to be replaced more often than other backings. The heavier and stiffer C- through E-weight paper is more aggressive, and can be used with heavier grit material. The backing paper and the glue used to attach the sanding material to the paper may either be waterproof (used for wet sanding) or paper that will get soft in water (for dry sanding). With wet/dry P400-grit paper, for example, the technician would use A-weight backing paper that does not break down in water, and use waterproof glue to attach the P400-grit abrasive.

Coarser sanding abrasive, which the technician needs to remove material faster, will use cloth backing rated with letters such as J (for the lighter) through X (for the heavier and stiffer backing).

For the coarser, very aggressive sanding, fiber backing is used. These very tough and semi-rigid backings are used for heavy work where the pressures placed on the sander may be high and the speed may be high. These heavy backing disks need a strong backing so they aren't torn or broken during their operation, thus sending parts of the disk flying, and possibly injuring the operator or others in the work area.

1. When using fiber-grinding disks, always use a backing pad with no more than 1/4 inch overhang.
2. Never use another fiber-grinding disk as a backing pad.
3. Never exceed the recommended speed.
4. If the edge becomes damaged, it should be trimmed to eliminate the defect.
5. If improperly stored, sandpaper may curl, becoming unusable. It should be stored in a dry, moderate temperature area in the original container.
6. Some sandpaper manufacturers recommend soaking fine grit paper in water before use. Even if this recommendation is not made, P1500 or finer paper may still need to be softened in water to prevent heavy scratching.

Type of Attachment

Grinding disks and sandpaper can be attached to the sanding machine in different ways. Cloth-backed sanding paper can be attached with manual clips, it can be attached with **hook and loop** type attachments, or it can be glued in place (either self-adhesive or using glue).

When glued in place, the sandpaper can come with self-adhesive **glue-on sandpaper**, which allows the paper to be attached and removed with little difficulty. It is generally less expensive than the hook and loop type, but can be attached only once. Some sandpaper comes with no glue on the paper. A tube with glue is placed on the pad; then the paper is attached. A hook and loop is attached by placing light pressure on the sandpaper when attached to the backing pad, allowing the hooks and loops to attach. Hook and loop paper can be removed and reattached later, if needed.

SANDING

Sanding in the refinish department is the most common task performed; some technicians believe that it is the most important step in surface preparation. When properly done, sanding will level, smooth, help make the transition from repaired area to OEM finish, promote adhesion, and generally make the repair undetectable. Sanding can be done by hand, or added by power equipment. Many tools have been developed to help technicians become faster and better while making the job easier to perform.

Still, the sanding step is complex. A technician must learn to detect and remove imperfections as small as 1/2000th of an inch; otherwise, they will be visible in the finish. A technician must master wet and dry sanding, sanding with a machine and by hand, blocking, and scuffing. The technician must learn when and how to use sanding paste, guide coats, sanding pads, and abrasive pads. The technician must master which direction must be used when sanding. Most of all, when using any of these tools or methods, the technician must learn how to make the work performed undetectable.

Hand Sanding

Although **hand sanding** may be more time consuming and physically difficult than machine sanding, the technician has, by far, more control over the end result. Most refinish technicians will at some point do some sanding (or scuffing) by hand, to assure themselves that the surface is prepared correctly. Hand sanding can be done either wet or dry; it can be done with a sanding block or a sanding pad. See **Figure 25-13** and **Figure 25-14**. Generally, the intention of hand sanding is to make the surfaces of the repair match the substrate and the old finish, leaving all three level, without waves, and free from defects. See **Figure 25-15**. A technician could feather by hand, but it would be very time-consuming and labor intensive.

When hand sanding, the technician should use a half sheet of paper folded in thirds and wrapped around a sanding pad. See **Figure 25-16**. Sanding pads are used to even out the pressure applied to the paper, and to assure that no finger marks are transferred to the vehicle's surface. See **Figure 25-17**. Sanding pads can be hard, or may have holes

FIGURE 25-13 Hand sanding can be done manually with a sanding block.

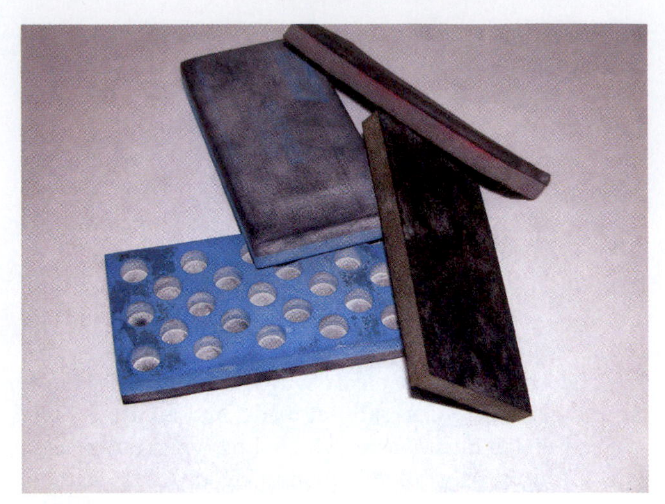

FIGURE 25-14 Assorted soft and hard sanding pads are available for the technician.

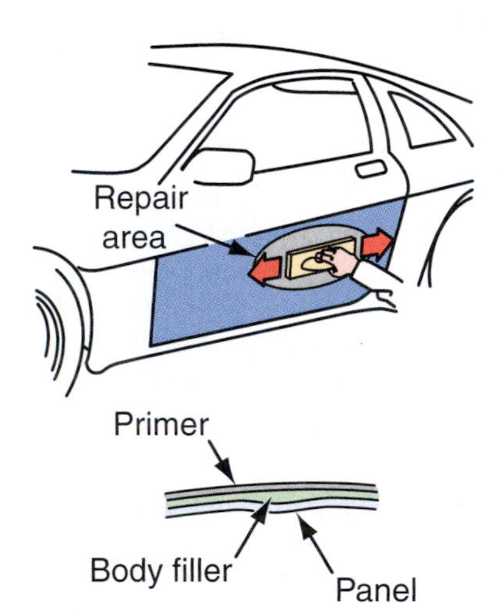

FIGURE 25-15 Blocking is done to level the area of repair, making it undetectable.

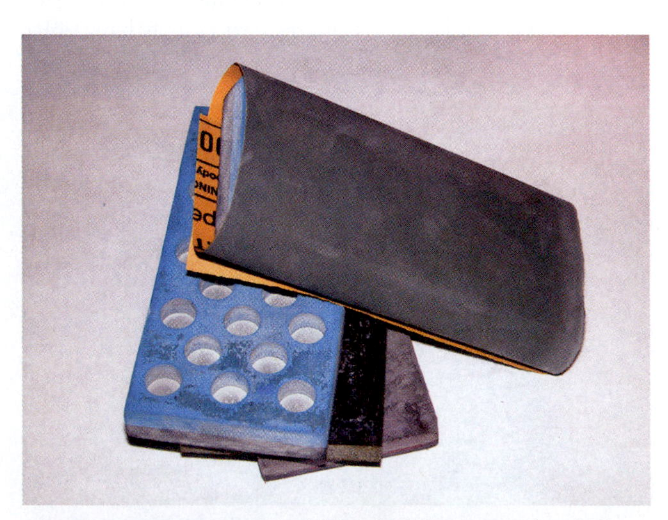

FIGURE 25-16 Paper is held in place by wrapping it around a sanding pad.

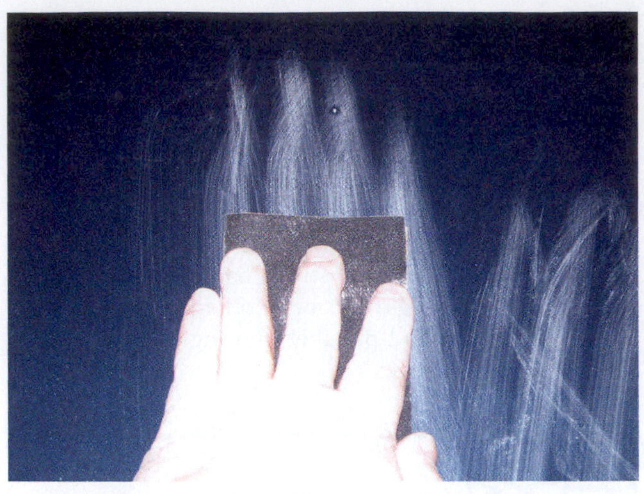

FIGURE 25-17 Sanding without a pad can create visible finger marks.

FIGURE 25-18 This sanding pad has a blue, aggressive side and black, less aggressive side.

in them for aggressive sanding. See **Figure 25-18**. Or they may be softer, for a less aggressive sanding pressure.

Blocking. **Blocking** is a process that levels an area in order to blend in the uneven areas created by the repair process. Following the repair, a technician may have the substrate at the lowest level, the body filler slightly higher than that, and the old finish (although feathered) at the highest surface. Primer filler is added to the area to fill and level these three heights. The technician must then use a sanding block to level or plane off the three different areas. The block should be large enough to bridge the repair area. As the block travels over the high spots, the technician sands them very aggressively—and with little pressure on the low spots, to have little effect on them. As all the high spots get lower, the surface becomes flat. The process is stopped when all the areas are level, or until there is no more primer on the high spots. If the technician sees the substrate showing ("breaking through"), the

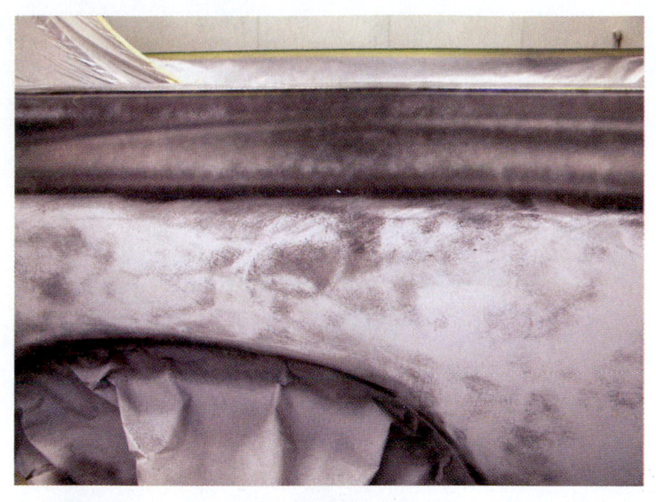

(A)

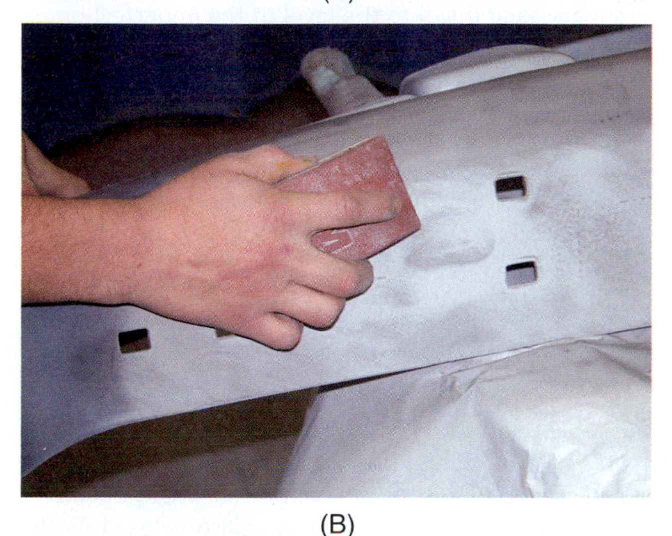

(B)

FIGURE 25-19 A guide coat and long standing block show imperfections while leveling the repaired area (A). This guide coat reveals low spots during the blocking process (B).

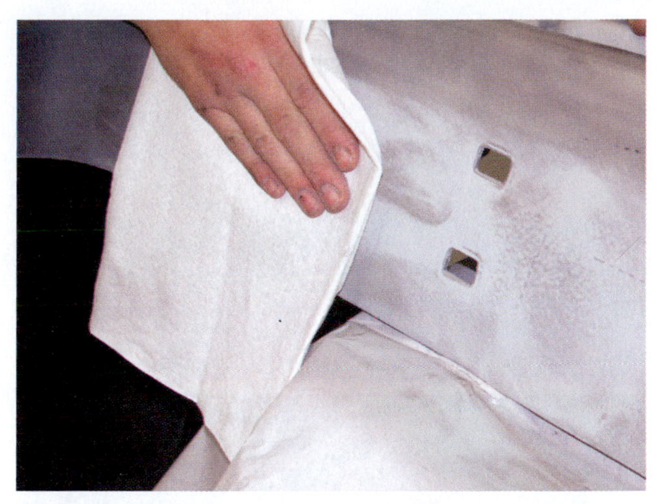

FIGURE 25-20 By using a cloth between the vehicle and hand, a technician will be better able to feel for low spots.

blocking should be stopped. The surface is then cleaned and evaluated. It may need to have primer filler applied again, followed by another round of blocking.

To help a technician see the imperfections better, a guide coat can be applied over the surface to be blocked. The guide coat should be a contrasting color so that when the surface is sanded, the high areas will be removed, leaving the guide coat in the low areas. The remaining guide coat's contrasting color will reveal the scratches or other imperfections. A guide coat can be a sprayed-on color, or a dry powder that adheres to the surface. See **Figure 25-19**.

For a surface to be completely prepared, all defects must be removed, and the surface should be flat and straight. This desired condition, though, is not always easy to see or feel! To help the technician feel the surface better, a shop towel can be placed between the surface and the technician's hand. By rubbing the surface with the towel slowly, the technician can better feel and attend to the imperfections. See **Figure 25-20**.

Wet Sanding. To use the **wet sanding** technique, the technician wets the paper and the surface with a water bottle. A small amount of soap is added to the water to lubricate the sanding action. The technician uses a back and forth motion, holding the sandpaper firmly against the surface to be sanded.

The sanding should be done along the bodylines in straight—not circular—motions. If done in circles, the scratches may be visible under the finish. In hand wet sanding, the technician should use either a water bottle or hose to rinse off the debris as the sanding is done. (If the sandpaper were to be dipped into a bucket of water, the water in the bucket would get contaminated with the sanding dirt, and could cause scratches deeper than the intended abrasion caused by the grit of the paper. This is particularly important when using sanding grits finer than P600.) As the sanding process progresses, a squeegee should be used to remove any sanding debris and water. The surface is then checked, and the sanding stopped as soon as the repair is completed.

Dry Sanding. Sanding can also be done dry. The technique is similar to wet sanding in the way the paper is prepared and wrapped around a block or sanding pad. However, instead of using water to lubricate the surface, the technician will dry-sand the area, using a shop towel to wipe the dust away. As with dry blocking, the technician uses a guide coat to see the high and low spots.

Dry sanding produces a large amount of dry, dusty debris. For this reason, the technician should wear a dust respirator, even if the operation is being performed in a prep station and vacuum assist is used to remove the dust. Dust gets into every opening it is near. Because it is difficult to remove all the dust, many technicians prefer the wet sanding method to assure a clean finish.

(A)

(B)

FIGURE 25-21 (A) Dual action (DA) sanders have different stroke sizes. The larger the stroke, the more aggressive the sanding action. (B) This is a standard orbit vacuum DA sander 3/8 stroke.

Feather Edging

Although feathering, or **feather edging**, can be done by hand, it is most often done with a sanding machine called a dual action (DA) sander. See **Figure 25-21**. The machine, which is normally air operated, has a random pattern as it sands. When held flat, the DA sander is able to taper the surface of the old finish down to the bare substrate, or taper the body filler to the substrate. A well-feathered area will expose all the levels gradually, tapering it from the highest to the lowest, then back again. See **Figure 25-22**. With the area feathered well, the primer filler can be applied with little blocking (leveling) needed. Any area of different heights must be feathered before priming, to allow the primer filler to work.

Scuffing

Scuffing is done to provide a mechanical adhesion for the next coating that will be applied over it. Scuffing is not intended to level any imperfections in the surface or to feather out scratches or chips. It will only take the shine off, or "deglaze," the surface of the old finish. This deglazing breaks the surface of the old finish, causing small scratches. This promotes a larger surface area for the next finish to adhere to. It also helps clean off the old oxidized surface contaminants, and provides a freshly abraded substrate for the new finish. Scuffing is generally done with a plastic abrasive pad. These pads come in different grit equivalents, although they do not use a numbering system. Depending on the manufacturer, there may be three different abrasive qualities. They are generally color coded relative to their aggressiveness.

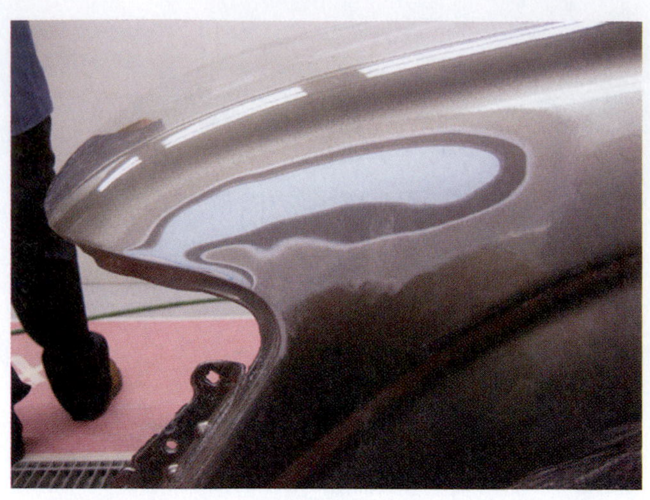

FIGURE 25-22 In this feathered area, the technician will gradually sand down to the level of the imperfection, making the repair nearly undetectable.

Sanding Paste

Sanding paste is a substance added to a plastic abrasive pad to aid in the pad's scuffing (deglazing) ability. Scuffing can be done either wet or dry, and the added paste will help get into close areas that are otherwise difficult for the pad by itself to reach. The paste also helps clean the surface and prepare it for topcoat. Sanding paste is particularly helpful when preparing plastic parts for refinishing. The paste's cleaning quality works especially well on a residual mold release agent, which is sometimes difficult to remove from these parts.

TECH TIP Some technicians will use household cleaners for sanding paste. Although this product will do a good job of cleaning, due to the bleach which is added to it, it is very hard to adequately rinse off. Even after rinsing with large amounts of water, a dry, gritty film can be felt on the plastic's surface. Bathroom cleaners are not the best choice for sanding paste substitute.

NEW E-COAT PANEL PREPARATION

Although new OEM replacement parts are one of the simpler surfaces to prepare, some care should be given to assure the necessary steps are taken. New OEM parts arrive with an electro-deposition coating. If this coating is a true e-coat—not just a black covering with little adhesion—the finish can be cleaned, scuffed, and painted. If it is not e-coat, it needs to be stripped, and the bare metal treated as outlined earlier in this chapter.

To ensure that a finish is e-coat, a technician should perfom a solvent test. Once the test confirms that the coating is thermoset, the normal preparation can be completed:

1. Clean with hot water and soap.
2. Clean with a wax and grease remover.
3. Scuff with a red sanding pad, using a sanding paste. Deglaze the entire surface to be refinished.
4. Seal the topcoat as needed for edging.

Some paint manufacturers now provide a sealer that can be applied directly over properly cleaned and degreased new parts, without first scuffing them. These new products speed up the refinishing process considerably. The technician should read, understand, and follow the recommended instructions provided by the paint manufacturer for preparing new e-coated parts.

PLASTIC SURFACE PREPARATION

When preparing plastic parts, the technician should identify which category the plastic parts fit into. The first of two major categories of plastic parts is **rigid plastic**, which includes polycarbonate, sheet moldable compound (SMC), and fiberglass. The second category is **flexible plastics**, which include polyurethane, polyamide, and thermo polyurethane, among others.

A technician should know that special procedures must be taken when working with these plastics to assure a quality refinish. Some of these plastics have small hairs, which may wick in solvents during cleaning. For these, a special plastic cleaner should be used. Some plastics will need to have adhesion promoters applied before priming to supply proper adhesion. They may have mold release agent on them, which will inhibit adhesion and must be taken off before surface preparation is attempted. This section will cover the surface preparation of bare plastic parts, primed plastic parts, and repaired plastic parts.

Bare New Plastic Surface Preparation

Although new plastic parts may come primed, not all do. In **new plastic preparation**, a new unprimed flexible plastic part should be inspected to check for damage, and to be sure that the part is the correct one. The part should then be cleaned to assure that all mold release agents have been removed, and scuffed to provide proper adhesion. Then the part should be cleaned again to remove the scuffing debris, and next have adhesion promoter applied if needed. The part should then be primed and the primer scuffed. Finally, the part is sealed and topcoated. Each of the steps will be discussed in detail next.

Inspection. Even though the inspection process may have been done when the part was delivered to the collision repair facility, it is still wise to make sure that the part delivered is the correct one. It should also be inspected for damage or deformity. If the new part is damaged, the facility must decide whether to return the part for a new one, or to repair it.

If the part is deformed from improper storage or from shipping, it may be possible to bring it back to its original shape. A plastic bumper, for example, has "memory." This means that if placed on a bumper tree and heated, it may return to its original shape without needing repairs. See **Figure 25-23**. Although this baking is not a foolproof method, it is easily done and should be tried. The bumper tree holds the bumper in the position that it will assume when on the vehicle. When the bumper is heated and it becomes soft, the bumper will try to return to its original shape.

Once the technician knows from inspection that the part is the correct one, and that it has no damage, the next step is the cleaning process.

Cleaning. The thorough cleaning of a plastic part is extremely important. The first step in the process is to clean the part with soap and water. In order to remove contaminants, the water used should be as hot as the technician can use without getting burned. The hot water should contain pH-balanced car washing soap, as cleaning should follow the two-bucket method. The part should be washed both inside and out, to remove the mold agent on the back side of the part. This procedure will keep the oil from contaminating the front of the part when it is being worked on. Following the washing step, the part should

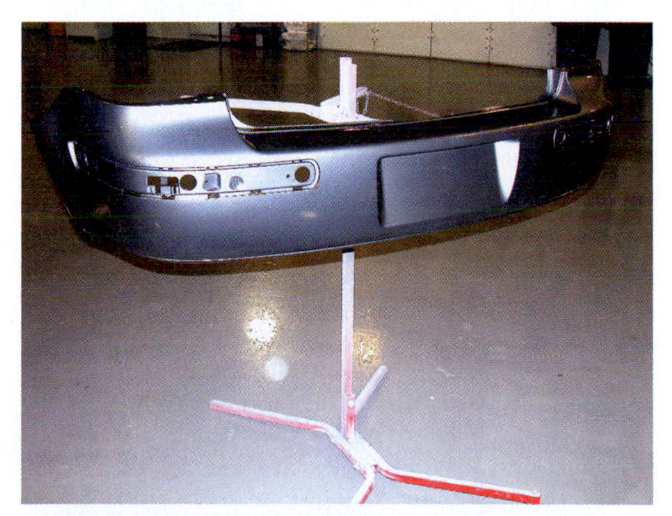

FIGURE 25-23 A plastic bumper tree holds a bumper in position to regain its proper shape.

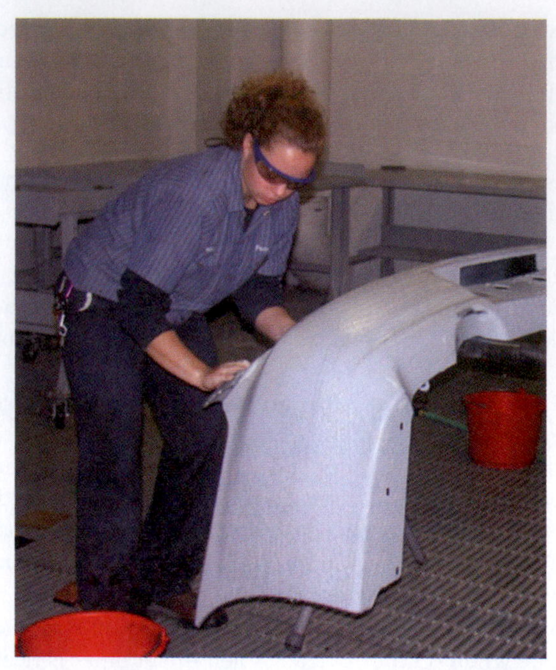

FIGURE 25-24 Plastic cleaning is critical, starting with the soap and water stage.

FIGURE 25-25 Scuffing with a gray abrasive pad and sanding paste helps provide a mechanical adhesion bond.

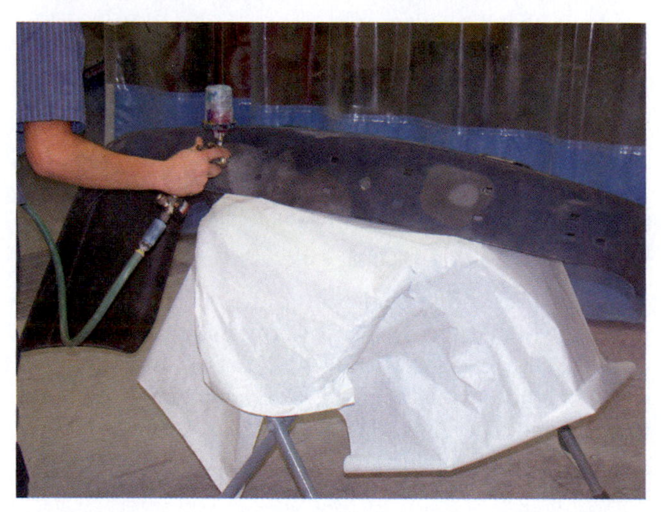

FIGURE 25-26 Certain types of plastics require an adhesion promoter to be applied before the undercoats.

be rinsed with clear water. During this step, the technician should note that if the water does not sheet off, the part is not completely clean, and should be cleaned again. See **Figure 25-24**.

The second cleaning should be done with an anti-static plastic cleaner. The action of rubbing plastic with a shop towel will charge the part with static electricity, which will attract dust and dirt. Anti-static plastic cleaner, on the other hand, will remove wax and grease, and decharge the part as it is being cleaned.

The oily contaminants on some parts are difficult to remove. The main contaminant is a mold release agent used in manufacturing. It is sometimes injected into the mold with the plastic as it is being manufactured, making it difficult to completely remove. To help release these agents, the part should be baked at 140°F for 30 minutes, then should be re-cleaned. The mounting of the part on a holding device will assure that the part will not deform when it softens from the heat. Once the part is heat-treated, or "tempered," it can be cleaned and prepared for scuffing.

Scuff. After cleaning, the part should be scuffed to provide a mechanical adhesion bond. The technician should scuff the part with a medium to fine plastic abrasive pad. Using the coarse abrasive pad would be too aggressive, and may leave scratches that will need to be blocked out. Therefore, a medium pad is preferred. Sanding pastes have been manufactured to aid in the scuffing process. This paste is applied to the abrasive pad and can be used either wet or dry. With the paste on the pad, the technician is able to get into small areas that the pad does not easily reach, thus deglazing the entire part well. If a paste is used, it must be rinsed off with water. Most sanding pastes are also cleaning agents, and are formulated to rinse clean easily. See **Figure 25-25**.

Cleaning. The now freshly scuffed part should be cleaned again with a plastic anti-static cleaner to prepare it for the application of plastic adhesion promoter. The technician should read, understand, and follow the paint manufacturer's recommendation for cleaning of the parts prior to applying adhesion promoters.

Adhesion Promoter. The application of adhesion promoter prepares the cleaned and scuffed part to accept the primer undercoats. Some adhesion promoter is applied in more than one coat. In this case, specific flash times are recommended, and should be closely adhered to. The flash time before applying the primer or sealer is also often critical, and should be adhered to as well. See **Figure 25-26**.

Primer/Sealer Application. At this point, the technician should apply the primer/sealer recommended by the paint

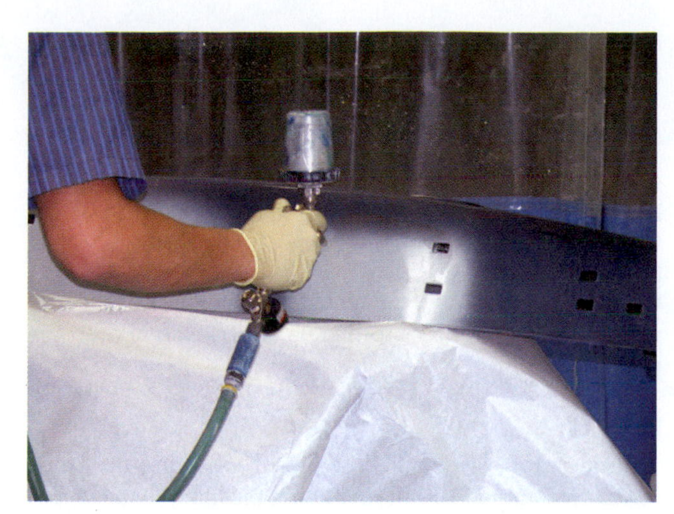

FIGURE 25-27 The technician applies primer over the adhesion promoter.

maker to the cleaned and prepared part. The specific recommendations for the products used should be followed to ensure the desired results are achieved. See **Figure 25-27**.

Final Cleaning. Following the recommended flash time, the part is given a final cleaning and is tacked, readying it for topcoat application. The instructions provided by the paint maker may call for flex agent in the sealer and/or the basecoat, or only in the clearcoat. The technician should read, understand, and follow these specific recommendations.

Primed New Plastic Surface Preparation

The **primed plastic preparation** steps for new plastic parts that are delivered in primer include inspection, cleaning, sanding and scuffing, sealing, final cleaning, then topcoating.

Inspection. New part inspection is the same for primed parts as it was for unprimed parts. The part's correctness and its condition should be checked before starting any preparation.

Cleaning. The initial cleaning for new parts is identical for primed and unprimed parts. The contaminants need to be removed both inside and out prior to scuffing. Even though mold release agents are not as prevalent as with unprimed parts, the technician must still thoroughly clean the new part.

Sanding/Scuffing. Sanding should be done by hand with a P600-grit paper. If the primer is found to be particularly hard, the technician may choose to start with P320 grit, followed by P600 grit. The surface should be completely deglazed and appear to be uniformly dull in appearance. If the part is primed but has a textured surface, the part should be scuffed with a coarse plastic abrasive pad, also until deglazed and uniformly dull.

Re-Cleaning. The sanded and scuffed part should be re-cleaned with anti-static plastic cleaner and allowed to dry. The part should be inspected to confirm that the entire part is deglazed and that no unsanded areas are visible. It should also be checked for any primer sand-through spots. If bare primer is found, the spots should be re-primed and scuffed according to the paint maker's recommendations.

Topcoating. The part is now prepared for topcoating. An adhesion promoter should be applied, if required, and the topcoats applied as directed by the paint maker's recommendations.

Repaired Plastic Surface Preparation

Repaired plastic preparation is more complicated, due to the potentially different levels of the surface, which will need to be made level through using primer filler and blocking. The steps are as follows.

Cleaning. The repaired part is first soap and water washed as described earlier, then cleaned with an anti-static plastic cleaner.

Feather Edging. The repaired area is feathered with a dual action (DA) sander using P240-grit paper. See **Figure 25-28**.

Sanding. The area to be primed should be sanded using P320-grit paper.

Re-Cleaning. Following the sanding, the parts should be re-cleaned with anti-static plastic cleaner, and then tacked.

Adhesion Promoter. Apply adhesion promoter if required.

Priming. The recommended plastic primer filler should be applied according to the paint maker's recommendations, which may call for a flex additive.

FIGURE 25-28 Feather edging can be accomplished using a DA sander and P240-grit paper.

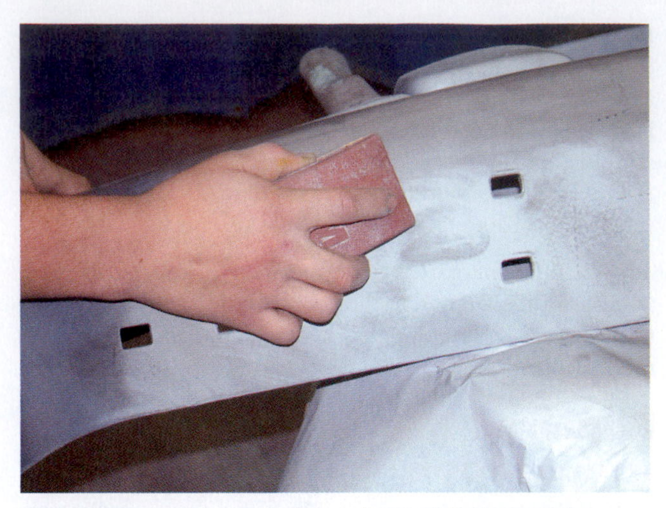

FIGURE 25-29 The technician should block the primer with a hand block, using P240-grit paper.

Blocking. Following the paint maker's recommended flash and cure time, the technician should block the primer with a hand block, using P240-grit paper. A guide coat will help the technician identify high and low spots, indicating when the surfaces of the repair area are no longer visible. Then the part should again be cleaned with anti-static cleaner. See **Figure 25-29**.

Sanding/Scuffing. The primed area should then be sanded with P320-grit paper, followed by sanding the entire part with P400- to P500-grit paper. The next step is scuffing the part's surface with an ultrafine scuffing pad. The area to be finished should be deglazed and have a uniformly dull area.

Final Cleaning. The area to be refinished is given a final cleaning with an anti-static plastic cleaner, then tacked for topcoat.

PREPARING BARE STEEL

Surface preparation on a steel substrate requires its own special considerations. The proper corrosion protection steps must be taken, and the blocking needed may be a larger area. The steps needed to prepare the surface on a steel substrate are as follows.

Cleaning

Clean the repair area and adjacent panel with a wax and grease remover.

Sanding

The area should be sanded with P240-grit paper to assure proper adhesion and that the bare steel substrate is free of oxidation contamination. Even if oxidation is not visible, it can start to form within a short time, and should be sanded to assure that the steel is clean. Do not remove the galvanized coating by using aggressive DA sanding or coarse grit paper.

Apply Corrosion Protection

At this point, the refinish technician should apply the proper corrosion protection. If the shop's policy is to use chemical cleaner and conversion coating, it should be applied as outlined in the paint maker's recommendations. If the shop policy is to use an acid etch, epoxy, or other corrosion protection direct-to-metal primer, it should be applied at this time.

Priming

Primer/surfacer should be applied following the paint maker's recommendations, including the film thickness recommendation. This may take two to three coats of primer filler. The finish should be allowed to flash as directed and cure either by force or normal curing times as directed.

Blocking

Following the needed curing time, the primed area should be blocked using P240-grit paper, guide coat, and a hand block. It should be sanded until an acceptable surface is achieved, or until the primer is sanded through. This would need an additional application of primer and additional blocking until an acceptable surface is achieved. The area should be cleaned with a wax and grease remover to clean off the sanding debris.

Sanding/Scuffing

The blocked area should be sanded with P320-grit paper, followed by scuffing with a fine or ultrafine scuffing pad.

Final Cleaning

The prepared surface should be given a final cleaning with a wax and grease remover, then tacked to prepare it for topcoating.

BLENDING PREPARATION

In **blending preparation**, the steps to prepare any repair should be taken as outlined previously. The surrounding panels that will be color blended and clearcoated should be prepared for topcoating.

Cleaning

If the finish technician does not know if the vehicle has been soap and water washed, it should be done, followed by cleaning the areas to be finished with a wax and grease remover.

Sanding/Scuffing

The area that will be color blended should be sanded with a P400- to P500-grit paper or equivalent plastic scuff pad. The areas to be clearcoated should be sanded with P1000- to P1200-grit sandpaper or equivalent plastic scuffing pad. The area should be completely deglazed and, when dry, should be uniformly dull with no shiny spots visible. See **Figure 25-30**.

Final Cleaning

The final cleaning is important to the overall cleanliness of the vehicle's finish. Many technicians will at this point strip all the masking off the vehicle and rewash it with soap and water, making sure that all the door jambs, under hood, the trunk area, and the wheel wells are clean. The cleaner the vehicle, the cleaner the final finish. After the final washing, the vehicle is dried and masked, then blown off and placed in the booth to be chemically cleaned with wax and grease remover, then tacked for topcoating.

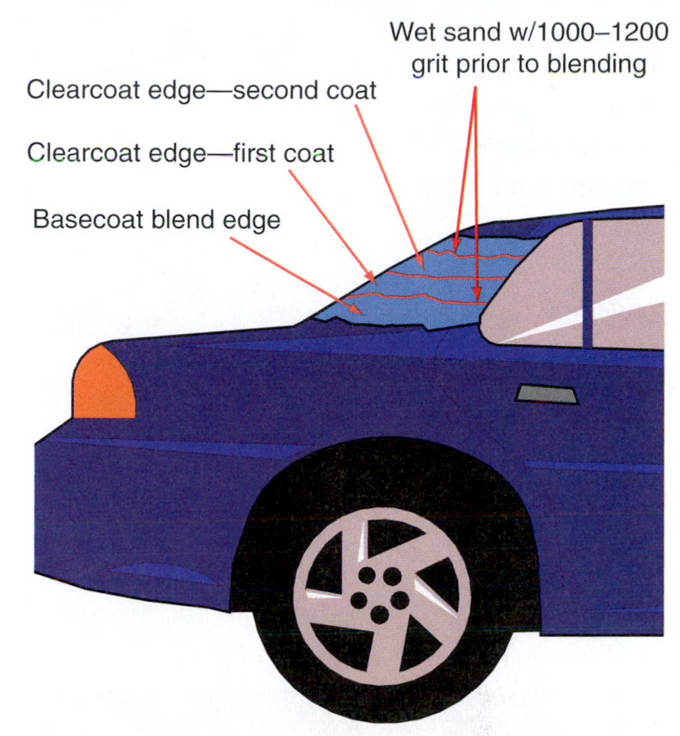

FIGURE 25-30 Many steps are required to prepare a surface for clearcoat blending.

Summary

- Surface preparation is often the first task a new technician is required to learn, do, and become proficient at.
- A good and thorough vehicle inspection is needed prior to beginning work, which reduces the cycle time of the repair.
- A well thought-out repair plan makes the preparation process smoother, as does having a clean refinishing area.
- Types of finishes include lacquers, enamels, urethanes, and epoxies.
- Cleaning is done using both soap/water and chemicals/solvents.
- Techniques for old finish removal include feather edging, blocking, sanding, and scuffing.
- A variety of steps are needed for preparing surfaces of aluminum, steel, and plastic, as well as the preparation steps for e-coat and the remainder of the vehicle for blending.
- Surface preparation remains one of the key processes that a refinish technician must learn to master in a highly skilled and challenging career.

Key Terms

aftermarket
aluminum substrate
blending preparation
blocking
chemical cleaning
chemical stripping
cleaning in the refinish department
closed/open coat paper
dry sanding
e-coat

enamel
feather edging
film thickness
flexible plastic
glue-on sandpaper
grit numbering
hand sanding
hook and loop
lacquer
media blasting
new plastic preparation
paint removal
partial film removal
plastic/composite substrate
primed plastic preparation
repaired plastic preparation
rigid plastic
sanding
sanding paste
sandpaper backing
scuffing
soap and water cleaning
solvent test
steel preparation
steel substrate
substrates
two-bucket method of washing
urethane
wet sanding

Review

ASE Review Questions

1. Technician A says that lacquer finish cures through oxidation. Technician B says that enamel finish cures through oxidation. Who is correct?
   A. Technician A only
   B. Technician B only
   C. Both Technicians A and B
   D. Neither Technician A nor B
2. Technician A says that 40-grit sandpaper is the best to use when stripping old finish from aluminum hoods. Technician B says that aluminum is used in automobile manufacturing because it doesn't corrode. Who is correct?
   A. Technician A only
   B. Technician B only
   C. Both Technicians A and B
   D. Neither Technician A nor B
3. Technician A says that plastic is refinished the same way, no matter which type of plastic is used. Technician B says that some plastics require that an adhesion promoter be applied to keep the new finish from failing. Who is correct?
   A. Technician A only
   B. Technician B only
   C. Both Technicians A and B
   D. Neither Technician A nor B
4. Technician A says that if a vehicle's film thickness is too high, part of the paint film must be removed to assure that the new finish will not fail. Technician B says that paint manufacturers have different film thickness limits. Who is correct?
   A. Technician A only
   B. Technician B only
   C. Both Technicians A and B
   D. Neither Technician A nor B
5. Technician A says that each vehicle will have a slightly different refinish plan. Technician B says that the first step in cleaning a vehicle for refinishing is to use a wax and grease remover. Who is correct?
   A. Technician A only
   B. Technician B only
   C. Both Technicians A and B
   D. Neither Technician A nor B
6. Technician A says that automotive soap is pH neutral (pH 7). Technician B says that dish soap is pH neutral (pH 7). Who is correct?
   A. Technician A only
   B. Technician B only
   C. Both Technicians A and B
   D. Neither Technician A nor B

7. Technician A says that media blasting is always done with peanut shells. Technician B says that sand blasting can cause warpage. Who is correct?
   A. Technician A only
   B. Technician B only
   C. Both Technicians A and B
   D. Neither Technician A nor B
8. Technician A says the bare metal should be cleaned with wax and grease remover before applying conversion coating to prevent rust. Technician B says that a plastic paint cup or paint cup liner should be used to prevent acid etch primer from harming a steel paint cup. Who is correct?
   A. Technician A only
   B. Technician B only
   C. Both Technicians A and B
   D. Neither Technician A nor B
9. Technician A says that an open coat sandpaper is 50 to 70 percent covered with abrasive grit. Technician B says that closed coat sandpaper is 90 percent covered with abrasive grit. Who is correct?
   A. Technician A only
   B. Technician B only
   C. Both Technicians A and B
   D. Neither Technician A nor B
10. Technician A says that sanding will remove and level the surface being sanded. Technician B says that the purpose of scuffing is primarily to deglaze the surface. Who is correct?
    A. Technician A only
    B. Technician B only
    C. Both Technicians A and B
    D. Neither Technician A nor B
11. Technician A says that blocking is used to remove the shine from a paint surface before refinishing. Technician B says that blocking is done to level the high spots in an uneven repaired surface. Who is correct?
    A. Technician A only
    B. Technician B only
    C. Both Technicians A and B
    D. Neither Technician A nor B
12. Technician A says that feather edging is another term for blocking. Technician B says that sanding paste is used to attach sandpaper to a DA sander. Who is correct?
    A. Technician A only
    B. Technician B only
    C. Both Technicians A and B
    D. Neither Technician A nor B

Essay Questions

1. Describe how to block sand a repaired and primed area.
2. Compare and contrast the differences between sanding and scuffing.
3. Describe the process for solvent testing a new part.
4. Write a step-by-step procedure sheet for preparing a bare steel repair area for refinishing.
5. Write a step-by-step procedure sheet for preparing an aluminum substrate for refinishing.
6. Write a step-by-step procedure sheet for preparing an e-coated part for refinishing.
7. Write a step-by-step procedure sheet for preparing a plastic area (with adhesion promoter) for refinishing.

Topic Related Math Questions

1. A technician is preparing an aluminum fender for refinishing. One quart of diluted conversion coating is required. The conversion coating must be diluted 1 to 2 with water. How many ounces of conversion coating are needed, and how many ounces of water are needed?
2. Another technician is preparing an aluminum fender for refinishing. One liter of diluted conversion coating is required. The conversion coating must be diluted 1 to 2 with water. How many milliliters of conversion coating are needed, and how many milliliters of water are needed?
3. A vehicle's fender has been measured and found to have thicknesses of 8.5 mil, 8.9 mil, 9.0 mil, 8.2 mil, and 8.8 mil, in five different locations on the fender. What is the average film thickness?
4. A technician is preparing to spray primer filler on three panels, each requiring 6 ounces of primer. The primer is mixed 5 parts primer to 1 part hardener. How much hardener is needed? How much primer is needed?
5. Another technician will be mixing soap and water to clean a vehicle. The soap is diluted 1 part soap to 50 parts water. Four gallons of water is in the bucket. How much soap should be put in the water?

Critical Thinking Questions

1. A vehicle has entered the refinish department. Through inspection, the technician finds that the

surface for painting needs to be re-primed and blocked. If the technician sends it back to the body shop, it will not be completed on time. If they fix it in the refinish department, it can be delivered on time. What should be done?

2. Three vehicles arrive in the refinish department at the same time, after there are already three other vehicles—in varying stages of completeness—in the department. The vehicles in the department now are all on time. In fact, two are ahead of schedule. The technician learns that one of the newly arriving vehicles is late, and if the vehicle is repaired in the order it has arrived, it will not be delivered on time. As the lead painter, what are the technician's options for repairing these vehicles?

Lab Activities

1. Have the student clean, feather edge, prime, and block a part for final prep.
2. Have the student inspect, clean, prime, block, and prepare for topcoat a part with a steel substrate.
3. Have the student inspect, clean, prime, block, and prepare for topcoat a part with an aluminum substrate.
4. Have the student inspect, clean, prime, block, and prepare for topcoat a part with a plastic substrate.
5. Have a student inspect, clean, and prepare a part for blending.

Chapter 26

Masking Materials and Procedures

OBJECTIVES

Upon completing this chapter, you should be able to:

- Inspect a vehicle to determine how it should be masked for its specific repair.
- Decide which parts should be removed, which parts should be masked, and which method should be used to mask them.
- Clean the vehicle for masking.
- Know which masking material to choose for the technique being done.
- Understand what each type of material is used for and how it is best applied.
- Properly handle and maintain the tools used in masking.
- Identify the different techniques used for masking.
- Know how and when to remove masking.

INTRODUCTION

This chapter will examine the many and varied decisions that a refinish technician will need to make on each refinish project. Although many of the tasks will generally be done in the same manner, the way they are applied will vary from project to project. The chapter will discuss decisions about which parts to remove and which ones to mask, as well as cleaning the vehicle for masking. Cleaning preparations for masking include inside the door jambs and door openings, under the hood, and in the engine compartment, trunk, and wheel wells to prevent dirt from getting in the new finish.

Other topics discussed in this chapter include the different types of masking material, the advantages and disadvantages of each of these materials, when one should choose one type over another, and the equipment used with these materials. Masking materials include:

- Coverings
  - Paper
  - Liquid
  - Plastic
- Tape
  - Beige
  - Green
  - Blue

- Fine line tape
  - Blue plastic
  - Beige plastic
- Aperture tape

This chapter will also explain the different methods of masking application and identify why one may be chosen over another. Masking techniques include:

- Masking for protection
- Complete refinish masking
- Color blending masking
- Clear blending masking
- Back masking
- Reverse masking
- Masking for multi-color

The chapter will then examine the methods of masking removal, and note when and how much masking material should be removed prior to detailing the vehicle.

SAFETY

Refinish materials and the chemicals that they are made from pose personal and environmental protection concerns. The refinish technician is governed by—and must comply with—many local, state, and federal laws concerning protection. Each product must be accompanied with a material safety data sheet (MSDS). Technicians must read, understand, and follow the recommendations in the MSDS to protect themselves, their fellow workers, and the environment.

INSPECTION

Performing a vehicle inspection before proceeding with masking will help the technician determine which masking technique will be needed and what areas on the specific vehicle will need to be masked. It will also help the technician decide which parts should be removed and which parts will remain on the vehicle. See **Figure 26-1**.

While inspecting the vehicle, the technician may find that special cleaning tasks need to be performed before the masking can be properly done. The inspection process also provides a good opportunity to decide if specialty masking materials will be used for this vehicle, and if so, what type of special materials to use. With the inspection completed, the technician can gather the needed materials and proceed with the cleaning.

As with other tasks in the collision repair industry, the technician must balance doing the task quickly against doing it well. To that end, masking can be done fastest if the technician has a good "**plan of attack**" and has gathered the needed materials for the task. See **Figure 26-2**. With the proper tools and plan, the vehicle can be masked quickly with no lost steps or wasted time.

One other masking technique is masking for priming. Although similar to other types of masking, this procedure is meant to protect the areas of the vehicle that are not being refinished from unwanted overspray. The main difference is the nature of primers. Primers are made to adhere or stick to everything they touch; they are heavy and have high solids. This means that if primers are in the

FIGURE 26-1 This technician is inspecting a vehicle with an estimator prior to refinishing.

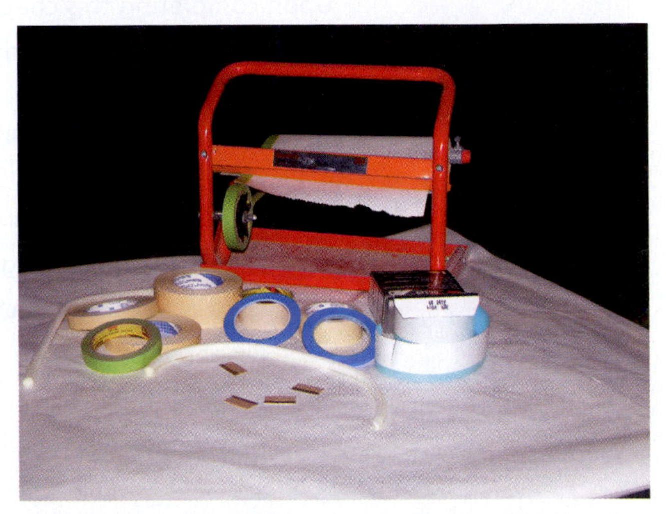

FIGURE 26-2 A plan of attack may include choosing the proper masking machine and the needed materials.

TECH TIP Masking to a break line is placing masking tape over a body line with a break for edge that occurs in the body design. By doing this there will not be a visible "hard line" that shows following the refinish application. See **Figure 26-3**. The more common break lines are the door, trunk, fender, and other gaps where the masking is applied to the next panel after the gap. See **Figure 26-4**. Other break line areas can be the roof channel, window, door, and other moldings. See **Figure 26-5**. All of these areas can be masked to break the refinish without leaving a hard tape line.

FIGURE 26-3 A tape "hard line" may occur when masking is applied to a normally occurring area.

FIGURE 26-5 Other break line areas include the roof channel, window, door, and other moldings.

air and they fall on any unprotected surface, they may stick to areas where a technician doesn't want them. They can also be very difficult to remove. A vehicle is often masked for priming, then blocked, sanded, and scuffed. The masking will then be removed to clean the vehicle, after which it is re-masked for refinishing. Many of the techniques used when masking for priming are unique and cannot be used for the other refinishing techniques.

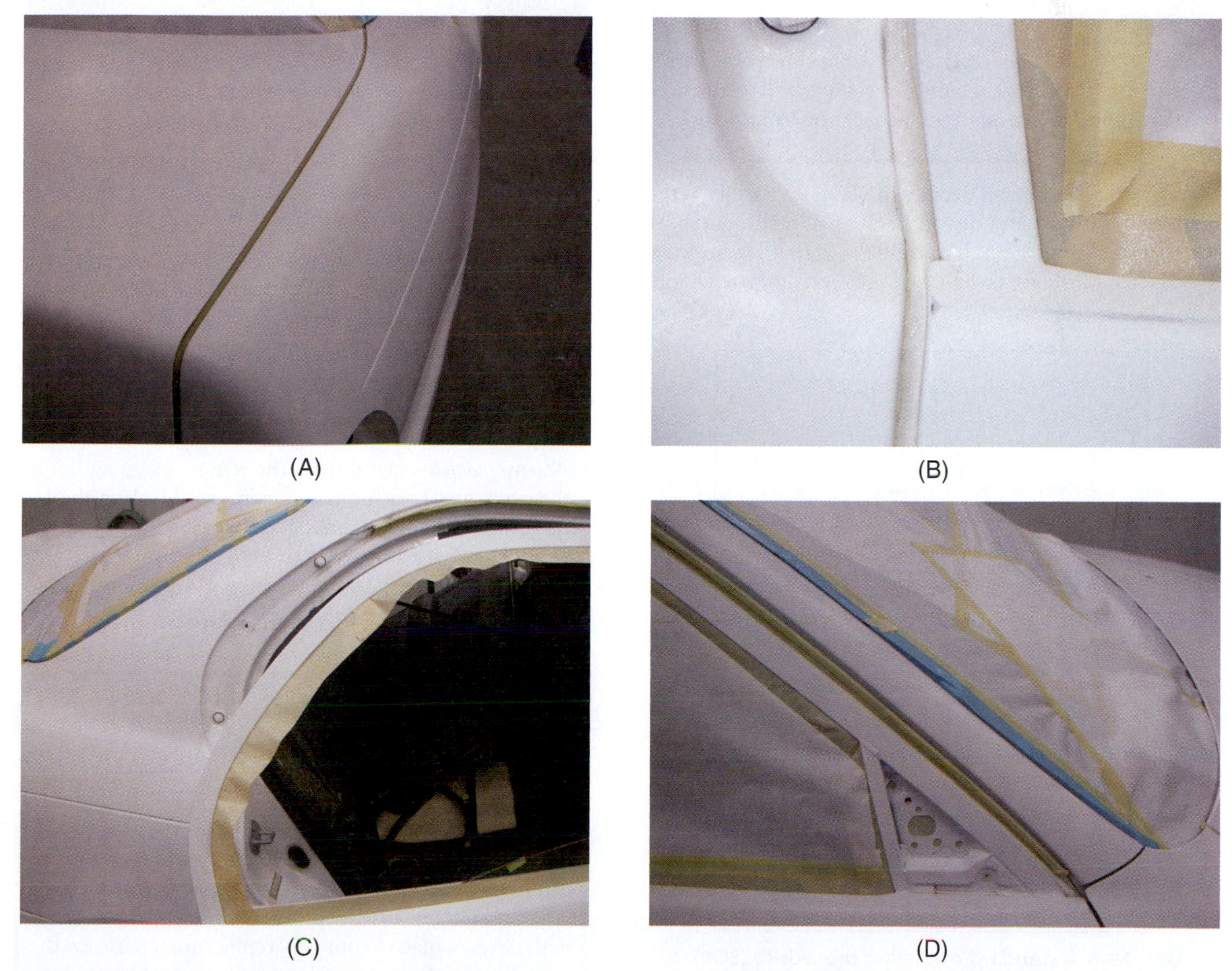

(A) (B) (C) (D)

FIGURE 26-4 Common break lines include the trunk gap (A), door and quarter (B), masked door and pillar (C), and back masked window (D).

A refinish technician—as well as a "**painter's helper**"—must know and be proficient at all of these techniques. They must be able to assess each vehicle quickly and make a masking "plan of attack," then execute it quickly so as not to slow the vehicle's cycle time.

Complete Refinish

When masking for a complete refinish, the technician needs to decide what parts should be removed and what can be masked. Each shop may have a standard operating procedure (SOP), a set of standards compiled to help a technician with the many varied decisions that must be made throughout the workday. SOPs may be written and posted in the facility. See **Figure 26-6**. Some manufacturers have recommendations for **parts removal** when refinishing. In addition, some paint manufacturers also may have recommendations for what should be removed when refinishing to maintain the paint company's guarantee. See **Figure 26-7**.

Parts Removal

Moldings, door handles, lock cylinders, etc.

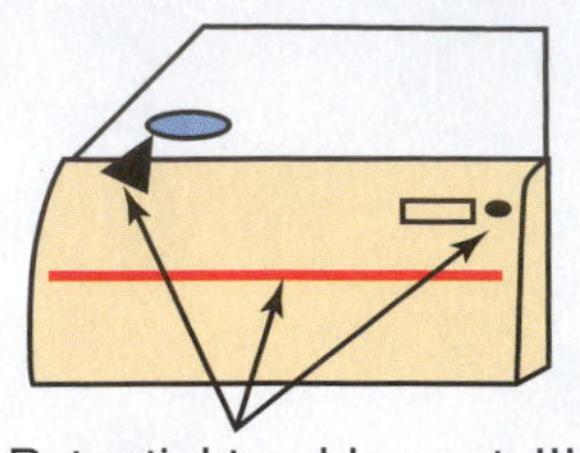

To remove or not to remove? That is the question!

Potential trouble spots!!!

If the part fits very "snugly," remove it to avoid peeling problems later!

FIGURE 26-7 Paint manufacturers sometimes offer recommendations for parts removal.

TECH TIP With tight-fitting window moldings, some stationary quarter glass and backlite (rear windows) may need to be removed to comply with the manufacturer's recommendations. See **Figure 26-8**.

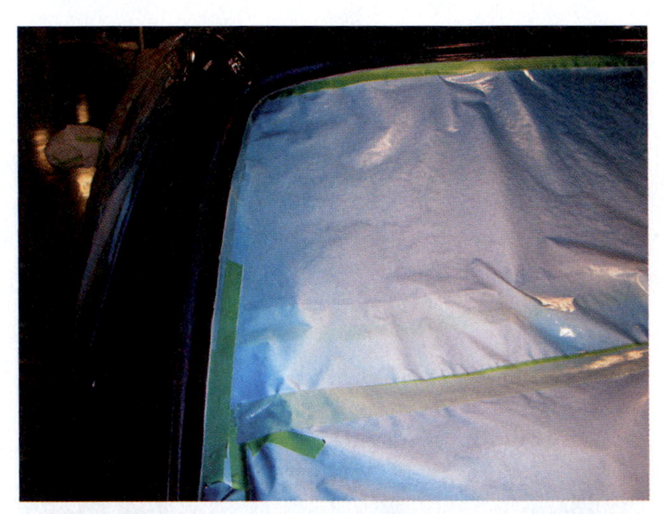

FIGURE 26-8 When masking, some stationary quarter glass and backlite may need to be removed to comply with the manufacturer's recommendations.

When the technician decides to remove a part from the vehicle, it is important to properly store it. In addition, the fasteners holding the part onto the vehicle may be damaged when the part is removed. The technician should note any parts broken during the removal step and order replacements. Then the technician should place the fasteners in a labeled container. That way, the technician will know which fasteners to use in each location on the vehicle. See **Figure 26-9**. The removed parts and their fasteners should then be stored in a safe location.

Masking and Paint Prep
SOP

- Read the work order: formulate a plan.
- Soap and water wash vehicle.
- Chemically clean areas to be refinished.
- Place vehicle outside spray booth for masking.
- De-trim the vehicle. (Remove all trim that will hinder an undetectable paint job.)
- Cover any holes produced by de-trimming.
- Apply foam tape to gaps and other areas as needed.
- Outline the area to be masked with $\frac{3}{4}$-inch green tape.
- Fill in the outlined area with as few pieces of paper as possible.
    - Keep paper as flat and wrinkle-free as possible.
    - Any folds must be taped down.
    - Mask down loose paper.
- Clean booth.
- High-pressure blow the vehicle off.
- Place vehicle in booth.
- "Bag" the vehicle. (Apply plastic covering to vehicle, masking down as needed.)
    - Don't let masking touch the floor, which will impede air flow.
- Cover any exposed wheels.
    - First cover with a plastic wheel cover (to protect brakes from overspray).
    - Then mask wheel opening with paper (to protect against dirt.)
- Blow off the vehicle again.
- Clean area to be painted with final wax and grease remover.
- Tack vehicle for painting.

FIGURE 26-6 A standard operating procedure (SOP), a set of standards to help the technician, may be written and posted in the facility.

TECH TIP Although many technicians will store removed parts in the vehicle, keeping them there may damage or soil the vehicle's interior. Instead, **parts storage** racks are preferred for the safe storage of these removed parts. See **Figure 26-10**. The parts and their labeled bags of fasteners are then easily at hand for reinstallation when the vehicle is being reassembled for delivery. The racks also provide the parts department with an ideal place to put new replacement parts for the vehicle until they are used. See **Figure 26-11**.

FIGURE 26-9 Fasteners can be stored in a labeled zip-lock bag to help ensure the right piece is returned to the right part of the vehicle.

FIGURE 26-10 Parts and fasteners should be stored on a parts cart.

FIGURE 26-11 By putting items in labeled bags on a parts cart, they are easily at hand when the technician needs them.

After the parts are removed, the technician should mask any areas that overspray may get into for protection. Such areas include gaps at the doors, hood, trunk, and gas filler door. If not blocked by masking, overspray will adhere to unwanted areas and may be very difficult to remove without sanding and refinishing. These areas can be masked with narrow paper, back taped, or have aperture tape applied. See **Figure 26-12**, **Figure 26-13**, and **Figure 26-14**. Then the openings are closed and the remainder of the vehicle is masked.

TECH TIP A vehicle is generally masked and sealed outside the spray booth, leaving the driver's door unfinished so the vehicle can be blown off, taken into the booth (where the masking is finished), then final cleaned, tacked, and sprayed. See **Figure 26-15**. The entire vehicle is generally not masked in the booth, as this helps keep the booth clean.

FIGURE 26-12 Masking with 6 in. paper in the trunk can prevent overspray from getting into undesirable areas.

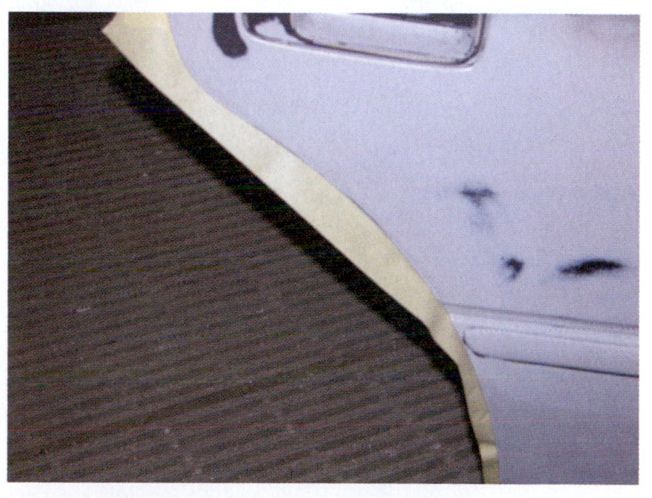

FIGURE 26-13 This door is back taped to prevent overspray from getting into unwanted areas.

FIGURE 26-14 Aperture tape can also be used to mask against overspray, as is done on this gas filler door.

FIGURE 26-15 Vehicles are generally masked outside the booth to help keep the booth clean.

Wheel opening masking can be done with a wheel-masking device called a "wheel cover." It can come as a reusable cover, a one-use plastic **disposable wheel cover**, or a paper cover. See **Figure 26-16** and **Figure 26-17**. All

FIGURE 26-16 A reusable wheel cover is fast and easy to apply.

FIGURE 26-17 Plastic wheel covers are discarded after a single use.

of these methods have advantages and disadvantages. The reusable wheel cover is fast and easy to apply, although the overspray that sticks to it when used often becomes contaminated with debris, which can—and does—fly off when the vehicle is sprayed and may contaminate the adjacent painted surface. In addition, this device does not protect the inner wheel well from overspray.

Plastic single-use covers are also fast. Because they are used only once—then discarded—they do not contaminate the surrounding painted area with flying debris. Like other wheel covers, though, they only protect the wheel—not the wheel well and its parts—from overspray.

When a wheel well is masked with paper, the wheel—as well as all of the parts inside the well—are protected from overspray. Because it is a single-use masking method, using a paper **wheel well cover** is also cleaner. See **Figure 26-18**.

Wheel wells are generally dirty from normal use. Although they should be cleaned with soap and water prior to refinishing, some debris may still remain. By paper

FIGURE 26-18 A paper-masked wheel well is quick to apply, clean, and protects all wheel well parts.

FIGURE 26-19 Using a plastic cover to protect the wheel along with masking paper to eliminate dirt from entering the wheel opening provides double the protection.

masking, the vehicle's surface is protected from debris being dislodged and contaminating it during the spray application.

TECH TIP When a plastic wheel cover is applied to the wheel and the wheel well is paper masked off afterwards, the vehicle's brake rotor is doubly protected from overspray. See **Figure 26-19**.

Panel Painting

To prepare a vehicle for panel refinishing, a technician blocks and sands the repair area to ensure an undetectable repair. The technician then prepares the entire panel for refinishing. See **Figure 26-20**. That is, if a hood and fender were replaced or repaired, both the entire hood and fender will be surface prepared; then the vehicle will be masked at the hood and fender line and at the fender-door line. Although most refinish repairs require a blend, from time to time the color match is so good that it will not need to be blended. In that case, the vehicle may be panel-blended instead.

FIGURE 26-20 This technician is preparing a vehicle for refinishing in the production shop.

Blending

Although blending starts out with the vehicle being masked like in a panel blend, there are some minor differences. The technician will panel mask the vehicle, exposing every panel to be painted and cleared. Then the technician will cover the remainder of the vehicle, followed by the adjacent panels that will only be blended onto—then cleared—later. These panels are covered in a way that will allow the technician to remove this second masking after the repaired panels have had basecoat applied to full coverage. See **Figure 26-21**. The technician

FIGURE 26-21 This blending masking is being removed following basecoat application.

FIGURE 26-22 Blending within a single panel can be done if no new panels require complete coverage.

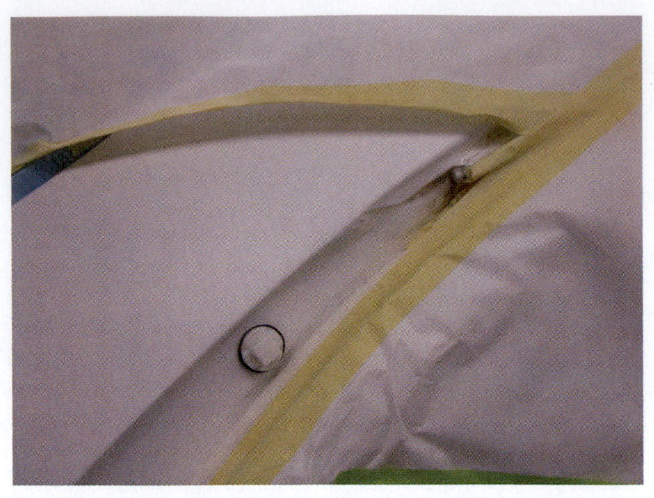

FIGURE 26-23 Bridge masking of a sail panel allows some of the finish to go under the masking without forming a hard line.

then removes the second masking, lightly blends the adjacent panels, and—after the needed flash and cure time—clears all the panels. This technique for **blend masking** works well when new parts have been placed on the vehicle. The lead painter can spray the new parts until coverage is acquired, without fear of getting overspray on the adjacent panels, which can cause match problems later.

If there are no new panels requiring complete coverage, then blending into an adjacent panel is done. The vehicle is masked the same as for panel painting. See **Figure 26-22**.

TECH TIP All masking should lie flat, without open creases or folds. If the paper is not flat it should be folded, then the fold taped to close it. By doing this, the paper will not catch dust that could be sprayed out later, causing dirt to contaminate the finish.

Blending the Clearcoat. Although blending of clearcoat is rarely recommended by either paint manufacturers or vehicle manufacturers, there is occasionally a need to blend the clearcoat into the lower part of the quarter panel or the A pillar. To do this, the area can be bridge masked, then sprayed. See **Figure 26-23**. The bridge allows some of the finish to go under the masking without forming a hard line. The bridge is then removed, and blending solvent is applied to help the clearcoat flow out. Some technicians will mask beyond the needed spray area, spray the clearcoat well below the masked line, and use blending solvent to help the clearcoat flow out. After the finish has cured, the technician removes the bridge masking and hand-details away any overspray that may have accumulated.

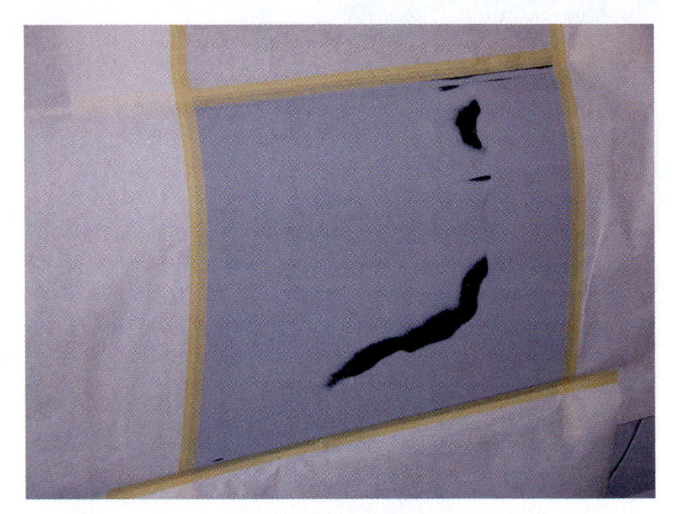

FIGURE 26-24 Square masking produces tape edges, and is a common mistake when masking for priming.

Masking for Priming. Masking for priming is done mostly to prevent primer from getting on unintended parts of the vehicle. The **primer masking** process is completed very much like a panel masking. The panel to be primed is exposed; all other areas are covered (especially openings like door cracks, so the primer will not get inside the door jambs). If a door handle or lock has been removed, that area must be covered to keep overspray off the glass below it, as well as off the door regulator and lock mechanism.

A common mistake when masking for priming is square masking. See **Figure 26-24**. If a technician square masks a panel, then blocks it, it becomes nearly impossible to feather the edge that has been created by the masking hard line. The primer builds up at the masking tape edge, causing a ridge that is actually higher than the tape.

See **Figure 26-25**. When the masking is removed, a sharp line is left behind which will require considerable feathering to make it undetectable.

TECH TIP All liquids have a phenomenon called surface tension. In this phenomenon, the liquid's molecules press equally in all directions. When another molecule pushes back against them, these forces cancel themselves out, allowing the molecules to lay flat. On the surface, they have no molecules to press against, and therefore they cause the surface to arc. See **Figure 26-26**. At the edge of the liquid in a jar, the surface tension will force the liquid to rise up. See **Figure 26-27**. Against tape, the surface tension will cause the liquid to rise up higher than the tape edge; then, when the tape is removed once the liquid has dried, it will leave behind a ridge which is actually higher than the surrounding surface. See **Figure 26-28** and **Figure 26-29**.

FIGURE 26-25 Masking tape edges are difficult and time consuming to feather out.

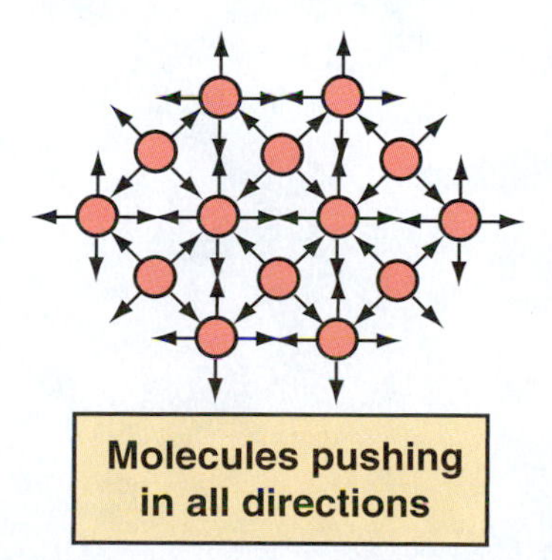

FIGURE 26-26 When molecules can't press against each other, they may create a curve on the surface due to surface tension.

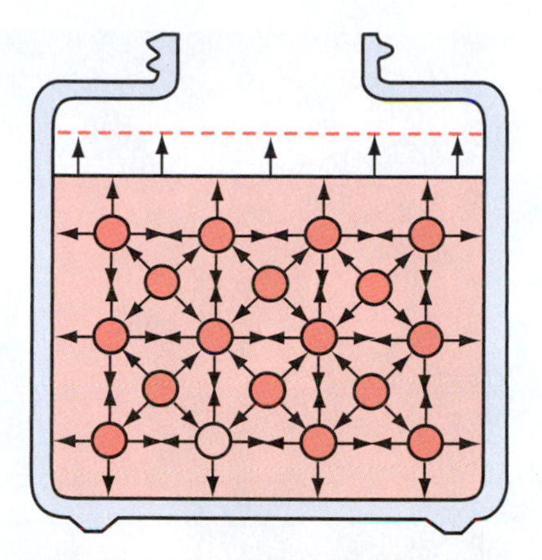

FIGURE 26-27 Surface tension in a jar will force the liquid to rise up.

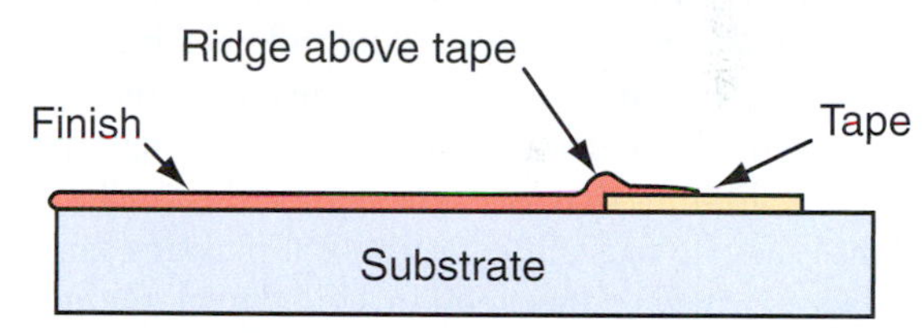

FIGURE 26-28 Surface tension against tape will cause liquid to rise higher than the tape edge.

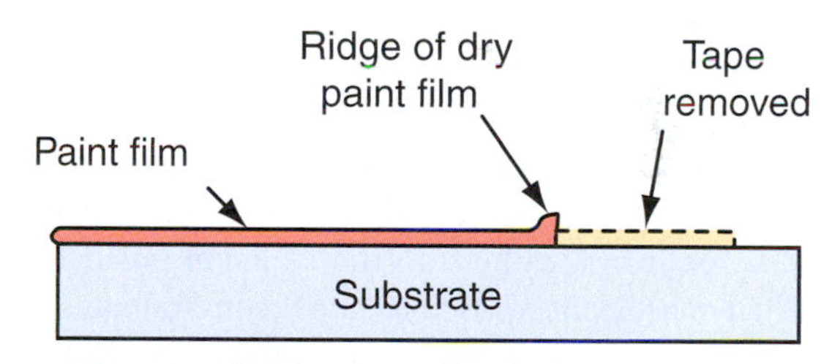

FIGURE 26-29 When tape is removed, the paint ridge is higher than the paint level.

MASK OR REMOVAL OF PARTS

The refinisher's goal is to closely match the original manufacturing finish to make the vehicle's refinish undetectable. Therefore, the more parts that can be removed from the vehicle, the more likely that will be. The first rule of mask vs. removal is that if it will take less time to remove the part than mask it, then it should be removed. Many parts are not only difficult to mask, but their presence makes it difficult to properly prepare the surface. The second rule to follow is that if the surrounding surface cannot be properly prepared, then the part should be removed.

This second rule is the difficult one. For example, can a door lock be properly masked? The metal portion of the lock may be maskable, but the little gasket under it

FIGURE 26-30 Back glass with tight molding is difficult to both mask and sand correctly without removal.

may not. Furthermore, the area close to the gasket may not be able to be scuffed properly, and thus may be at risk for peeling later. In such cases, even if time consuming, the part should be removed to assure an undetectable refinish.

Some vehicle manufacturers may also have recommendations regarding removal for masking. Although quarter glass with moldings that are tight against the finish (reveal molding) may be maskable with reveal masking tape, the manufacturer may recommend instead that all the stationary glass—such as the front windshield, backlite, and quarter glass—be removed for refinish. See **Figure 26-30**.

CLEANING

By this stage in the repair process, the vehicle has been washed several times. It must be washed again. Although the vehicle may not need the thorough cleaning that it received when it first came into the collision repair facility, the vehicle has been exposed to many contaminants in the facility that may cause problems during its refinishing. Greases might have contaminated the vehicle while the structural repair personnel handled it. Every crack may have been exposed to dust during any nonstructural repairs (plastic body filler). Even the dust and debris generated by the refinish department during the blocking and sanding steps for surface preparation must now be removed.

The technician should keep in mind that performing a **pre-masking cleaning** on the vehicle is the last chance to clean, and that the majority of the dirt that gets into a finish comes from the vehicle.

Soap and Water

First, the refinisher will perform a **soap and water cleaning**. Everything will need to be cleaned using the two-bucket system. The refinisher will clean the vehicle's exterior, being careful to keep the washing device clean in clear water. It may be wise to switch to a used washing mitt to finish the job for areas like the inside of the door jambs, the trunk opening, under the hood, and of course the wheel wells. If these areas are particularly dirty or contaminated with grease, it may be necessary to clean the area with wax and grease remover, then re-clean the area with soap and water. Even though the vehicle will be masked tightly—and hopefully the inside area of the vehicle will not contaminate the remainder of the vehicle—*time spent cleaning the vehicle is never wasted time*.

Doors. Doors present some difficult cleaning problems, since they present areas where dust and debris commonly collect: on the sill plate from passengers' feet, on the hinges from lubricant, and on the weatherstripping from silicone. Wax and grease remover works well to clean the hinges, the door striker, and normal dirt on weatherstripping. To remove silicone from the plastic (to help the tape stick), isopropyl alcohol also works well; it is inexpensive and will not add to the vehicle's static.

Under Hoods. The under-hood area is another place where contaminants are commonly found. Although it may not be possible to completely clean this area, it should be cleaned as well as possible. The under-hood area should be sprayed off (low hose pressure), soaped, and sprayed again. Then the engine compartment can be cleaned. If a high pressure washer is used, the refinish technician should cover any electrical components that may be damaged from the pressure washer with plastic. See **Figure 26-31**. Although engine degreasers are available to clean the engine compartment completely, the main purpose for this cleaning is merely to dislodge and remove those contaminants that might otherwise reach the finish during the spraying process. The engine compartment will eventually be sealed with aperture tape at the hood-fender line, or it will be covered with paper. See **Figure 26-32** and **Figure 26-33**.

Some shops use liquid mask for the engine compartment and the fender wells. When it is applied, liquid mask sticks to the dirt and holds it in place so it doesn't become air-borne and contaminate the finish. Some liquid masks even become soap to aid in washing the vehicle as it is being rinsed off.

FIGURE 26-31 Areas that may be damaged by water should be covered with plastic before power washing.

FIGURE 26-32 This hood is covered with aperture tape to keep out overspray.

FIGURE 26-33 This engine compartment is covered with paper to keep out contaminants.

Trunks. Although the inside of the trunk doesn't need to be soap and water cleaned, the drip rail around the edge and the weatherstripping do need to be cleaned and degreased. Plastic body filler and other dust can easily accumulate there and must be washed out. If not cleaned away when sprayed, the dust will become air-borne and contaminate the vehicle.

Wheel Wells. As stated previously, these vehicle areas need special considerations, not just because of dirt from normal use, but because they will be covered with masking paper, and the tape must be able to stick to the surface. Tape will come loose if a wheel well area is not cleaned well before the tape is applied. See **Figure 26-34**.

Chemical

Chemical cleaning uses (1) wax and grease remover and (2) isopropyl alcohol to clean the vehicle for masking and refinishing. Wax and grease remover should be applied to the areas that will be refinished to remove contaminants after the vehicle has been washed with soap and water and allowed to dry. Doing so will eliminate defects such as fish eyes.

FIGURE 26-34 A wheel masked with paper will not only keep out overspray but will also keep dirt from the wheel area off the paint.

TECH TIP It is not necessary to wear protective gloves while masking a vehicle. In fact, it is difficult to mask with gloves, because the tape sticks to the gloves and will not release easily. The technician should put on gloves after the masking is done and before the wax and grease remover is applied.

The vehicle will be cleaned with wax and grease remover before it is masked, and then again after the vehicle is masked, just before it is sprayed. This is because contaminants from the technician's bare hands can transfer salt and oils to the vehicle's surface during the masking process.

The second cleaner, isopropyl alcohol, can be used not only to clean the vehicle's surface, but also to remove the vehicle's static charge that builds up while it is being worked on. Static electricity can attract dust from the air in the paint department and hold it on the vehicle. A quick spray of alcohol will discharge the static from the vehicle. Unfortunately, as soon as the vehicle is again tacked or cleaned, the act of passing a cleaning cloth over the vehicle will start up the static charge again.

Weatherstripping. Weatherstripping may be contaminated with silicone from the products used by the vehicle's owner. Many types of interior cleaners and dressings are made from silicone and are hard to remove. In fact, it may take several wax and grease remover or alcohol cleanings to adequately remove it all. However, to repeat, if the weatherstripping is not cleaned properly, tape will not stick to it. If the weatherstripping does not come clean the first or second time wax and grease remover is used, try alcohol and then wax and grease remover again. Do not use lacquer thinner to clean weatherstripping, as it is too aggressive and may cause damage.

MASKING MATERIALS

Masking materials come in many different forms, such as different sizes and types of paper for different applications. There are also liquid masks and plastic drapes for protecting larger areas. Aperture tape in different sizes is used for filling gaps to keep overspray from getting in. Tape comes in a staggering amount of sizes, with different types of backing (beige, green, and blue). Tape may also be made of plastic (fine-line tape for masking small areas where the tape needs to be applied in tight curves). Each of these masking materials has its advantages and disadvantages. The technician will need to choose which is best suited for the particular job, and have all of them at hand for quick access when they are needed. The technician will need to know when to fold masking material and when to cut the material. Although the masking machine will help with getting the rough amount of paper and tape for a particular application, the technician still may need to fine-tune the measurement when on the vehicle, and this generally requires cutting. The way that a technician chooses to cut masking material is important, and will be covered in this section.

Paper

Masking paper is used to cover areas of the vehicle that the paint technician does not want paint to reach. Paper comes in varied widths, including 3, 6, 9, 12, 15, 18, 24, and 36 inches. However, the most commonly used are 6, 12, 18, and 36 inches wide. See **Figure 26-35**. They also come in different types of paper, which can be resistant to solvents that are mild as well as those that are very aggressive. Some types of paper have polycoated backing to prevent solvent bleed-through when aggressive solvents are used. Paper also comes in different thicknesses, with the thinner being easier to conform to the vehicle's many curves, and the thicker suited for covering large areas to be protected. Paper may come in varied lengths as well. A refinish technician may use paper rolls with small amounts on hand masking machines, a 450-foot roll of general-purpose utility grade "green" paper, or the longer 750- and 1,000-foot rolls. It is important for the refinish technician to know which paper is appropriate for the situation—and be able to identify each paper's advantages and disadvantages.

FIGURE 26-35 Masking paper comes in different sizes and types.

Mild Solvent-Resistant General Purpose Paper. Mild solvent-resistant paper is made for utility masking where the solvents are not sprayed directly onto the masking paper. These types of paper are generally used for masking large, flat areas that need protection from overspray. They are cheaper and generally thicker, with less conformability than thinner paper. Paper comes in different colors so technicians can tell—at a glance—which type of paper they are grabbing. General purpose paper may also come in different lengths: 450 feet, 750 feet, or 1,000 feet per roll.

Solvent-Resistant Paper. Solvent-resistant paper is used for areas where the paint will come in contact with the masking paper. Some of these types of papers resist solvent penetration due to their thickness. This thickness, though, will make it more difficult to conform to the vehicle's body shapes. Others are thinner but use a tighter bonded paper, which allows them to conform to the vehicle more easily and also be resistant to solvents. However, if maximum solvent resistance is the main criterion for a particular application, a thin paper with a polycoated backing will be the best choice.

TECH TIP Technicians will often set up a multiple-dispensing paper machine with different sizes and types of paper on it: for example, 6-inch thin solvent-resistant paper to mask small curved areas; 12- or 18-inch thin solvent-resistant paper for the areas where paint will be sprayed directly onto the paper and solvent resistance is most needed; then a second 18-inch, thicker paper with less solvent resistance, for masking areas (such as windows) where paint will not come in contact with the masking paper. The dispenser will also provide 18-inch low solvent-resistant paper for covering large areas, as well as plastic masking film for quickly covering large areas of the vehicle. See **Figure 26-36**.

Sizes. The paper width should be chosen with two things in mind. First, it should be wide enough to cover the object being masked in one application or, if that is not pos-

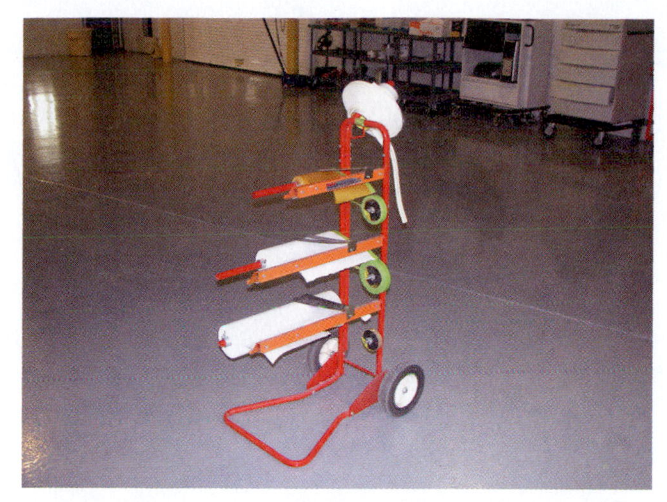

(A)

(B)

FIGURE 26-36 A multiple-dispensing paper machine may include 6-, 12-, and 18-inch paper (A). Plastic is used to protect the areas not being painted from overspray (bagging the vehicle) (B).

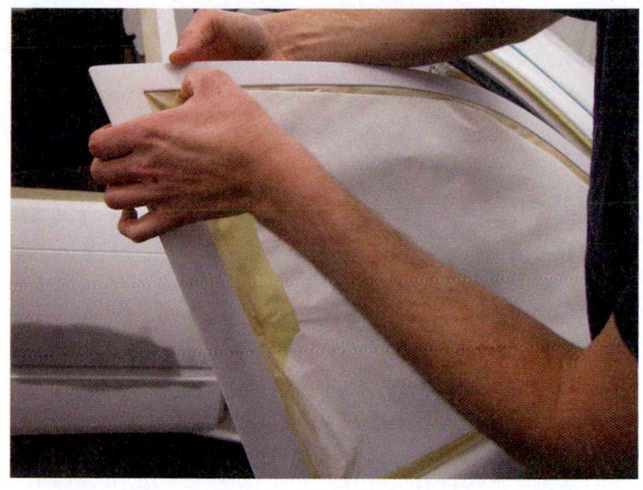

FIGURE 26-37 Some technicians will try to cover an area, like this window, with one sheet of paper if possible.

FIGURE 26-38 All folds should be masked to prevent dirt from being trapped by them.

sible, it should be wide enough to cover the area with as few pieces as possible. Second, it should not be so wide that much of the excess will be cut or folded under.

When applying paper, it is best (for less dust) if the paper lies flat, with no untaped folds. Some technicians will try to cover an area with one sheet of paper if possible. See **Figure 26-37**. If that is not possible, and the paper needs to be either folded or cut to size, any fold should be sealed with tape to prevent overspray from gathering under it as dust, as this dust could later become air-borne and contaminate the finish. See **Figure 26-38**.

TECH TIP If masking paper needs to be folded, it should be folded under if possible. If folded under, the outside perimeter can be taped and sealed and will not attract dust. See **Figure 26-39**. If it must be folded, all the edges must be sealed with tape so dust will not collect.

FIGURE 26-39 This masking paper is being folded under and taped so it will not attract dust.

Liquid Mask

Liquid mask is a material used to cover large areas of the vehicle, or areas that technicians feel cannot be cleaned as well as they would like. The vehicle is first masked to protect the area that will be finished, exposing the remainder of the vehicle. This can be done very rapidly and with the least expensive masking material. The cleaned vehicle is then sprayed with liquid mask over the remainder of the vehicle. See **Figure 26-40**. Depending on the manufacturer, liquid mask may need to be applied in two or more coats. When the mask dries, the masking can be removed. The area to be finished is then exposed using normal masking materials. Although liquid mask can withstand solvents from direct spraying, it is best to outline the vehicle with 12 to 18 inches of solvent-resistant paper, using the normal fold-sealing precautions as with standard masking.

To remove liquid mask, apply clear warm water and allow it to soak in for a short time; then perform a second rinsing to remove the remaining liquid mask. Some liquid mask will turn into a biodegradable soap when rinsed off the vehicle, leaving the surface behind clean and ready for delivery.

Liquid mask is available to do a variety of tasks for surface protection, dust control, and dirt/debris control, and may take several forms. Some are designed for spraying on paint booths and prep decks; they will withstand heat, last for longer periods of time, and still rinse off easily. Others are designed for dust control and are sprayed on paint booth and prep deck floors to trap overspray and make sweeping up easier. These can also be rinsed off and reapplied. The overspray liquid masking also comes in different types: a dry application and a tacky one. The dry application liquid mask will form a continuous film over the entire area to which it is applied. The tacky style, some technicians believe, will trap air-borne dust particles and hold them in place. It is also used to hold contaminants in place in dirty, hard-to-clean areas such as wheel wells and undercarriages.

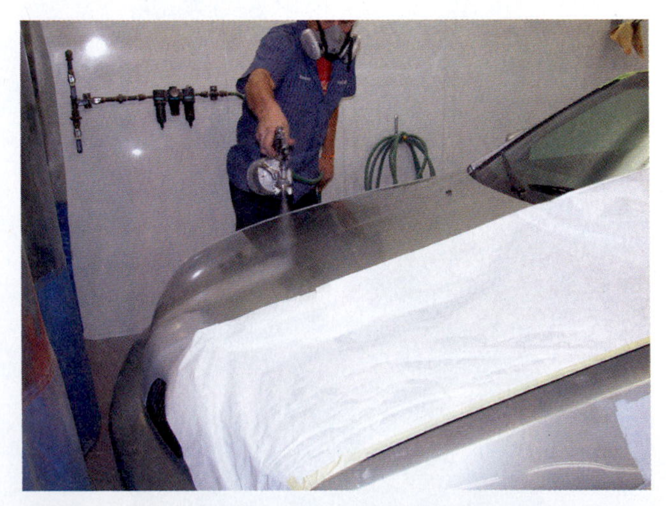

FIGURE 26-40 Liquid mask covers large areas of a vehicle, especially ones that cannot be cleaned as well as desired.

FIGURE 26-41 This vehicle is draped (bagged) with plastic film.

Plastic Drape Film

Plastic drape film also comes in different sizes and qualities. The better drape films are made of a nonporous material that resists both paint penetration and dried paint film flaking. It comes in rolls that may only be 9 inches wide, but when cut to length will unfold to 72 inches wide. The film is see-through to aid in application, and two widths cut to the proper length can cover an entire vehicle. See **Figure 26-41**.

TAPES

Tapes come in a staggering variety of types and widths for many different purposes. A refinish technician should keep many at hand to be used for different tasks. Some are thicker (beige) masking tape for use on paper machines (3/4 in. or 19 mm) and masking large areas (1 1/2 to 2 in. or 38 mm to 50 mm). The thinner and more conformable green tape, often seen in 3/4-inch or 19-mm width, will be used for outlining a part to be masked with paper. The "safe-release" long-term tape is designed to stay on vehicles for longer periods and still be removed without residue or breaking apart.

Although these and other tapes can look similar, all "**masking tape**" is not the same. First, the adhesive on automotive masking tape is "pressure sensitive." This means that once applied and pressed into place, it will adhere to a grease-free and silicone-free surface until it is released or pulled off. When it is removed, the tape will not leave an adhesive residue. Some non-automotive tapes may look similar to automotive tape but do not have these qualities. Besides not sticking well, they may contaminate the surfaces they were applied to, resulting in a time- and labor-intensive removal and cleanup.

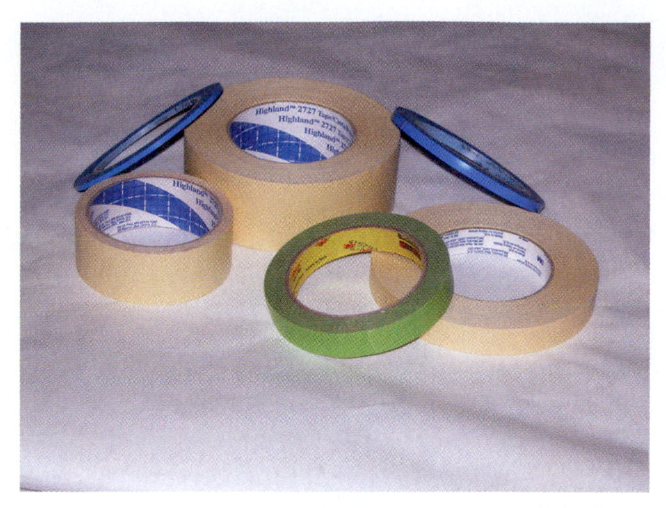

FIGURE 26-42 Although available in many sizes, beige tape commonly comes in 3/4, 1 1/2, and 2 inch widths.

Beige Tape

Beige tape has been used in the automotive industry for decades and has served it well. It comes in many widths: 1/8, 1/4, 1/2, 7/8, 3/4, 1, 1 1/2, and 2 inches—though it is generally used in the 3/4 and 1 1/2 to 2 inch widths. See **Figure 26-42**. It is thicker than the thin green tape designed for bending around curves, but thinner than the blue tape designed to withstand long-term placement. Generally the beige tape is seen on masking machines and used for sealing down masking paper. It is a good general purpose tape with many versatile uses.

Performance ("Green") Tape

Crepe masking tape comes in many forms and is continually being improved. Automotive masking tape has gone through some changes with the introduction of a thinner tape (6.7 mils), which will lay tighter to the surface and slope under objects being masked more easily than the thicker beige tape. It is highly conformable and provides thin paint lines. It will also turn in tighter radii than the thicker tape, and is a very good tape to border parts being masked, before the beige tape is used with masking paper. Its pressure-sensitive adhesive sticks with a touch, and will release without leaving an adhesive residue.

Safe-Release ("Blue") Tape

Safe-release tape is designed for vehicles that will have masking tape left on for longer periods. It has proven to be safe on the vehicle finish for 7 to 14 days in direct sunlight, and even longer when the vehicle is out of direct sunlight. It uses a pressure-sensitive synthetic adhesive system that will not leave adhesive transfer. The tape is ideal for applying masking over freshly cured paint for multi-color application.

TECH TIP Before safe-release tape arrived in the refinish industry, technicians used beige and "green" tape for multi-color masking. It was common practice for the technician to pull off a line of tape, then pull the glue side over a pants leg, to lessen the amount of adhesive and make the tape stick less (therefore reducing the possibility of leaving a tape line in the fresh paint). Safe-release masking tape can be applied directly without dulling the adhesive.

Specialty Tapes

Specialty tapes, also called fine line tapes, are plastic tapes designed to adhere more flatly against the surface, leaving little or no possibility of the finish bleeding under the tape. Therefore, it makes a sharp dividing line. All masking tapes are made of a crepe paper material with adhesive applied to one side. The wrinkly surface of the crepe backing can allow a finish to creep under the tape, producing an uneven line. Specialty tape generally consists of polypropylene film backing and vinyl backed tape.

Green Tape. The polypropylene film tape can be applied over freshly painted finishes sooner than crepe masking paper with less fear of leaving either tape tracking or adhesive residue. It comes in an assortment of sizes, from as small as 3/32 inch to as large as 3/4 inch wide. It turns well and can easily be applied in straight lines.

Blue Tape. Vinyl tape is also a specialty tape used for two-tone color application and fine masking when precision is needed. It will not allow finish to creep under it, and can be applied to fresh paint earlier than masking tape. It turns with ease and can be stretched around even the tightest corners. It comes in widths from 1/8 to 3/4 inch; these different widths are sufficient to tackle even the most challenging of fine masking requirements.

Aperture Tape

Soft-edge masking tape comes in the round 12 mm and 19 mm sizes used to mask off door openings. It also comes in the wider 28 mm soft-edge foam masking tape used for masking A pillars and other wide gaps. See **Figure 26-43**. These specialty masking forms have been designed to apply quickly to gaps, which otherwise may allow finish to get into areas where it is not wanted and would be difficult to remove.

Aperture tape was developed to replace traditional paper and tape masking; the two round aperture tapes have an adhesive on one side that can be attached to the door jamb near the edge. Then the door is closed touching the tape, which keeps the overspray out. Aperture tape is also ideal for keeping polishing and compounding sludge out of the door jamb area. The wider and flatter 28 mm tape was designed for larger gaps, though it works in a similar way.

FIGURE 26-43 Wide foam tape or aperture tape comes in different sizes for different application needs.

Trim Masking Tape

Trim tape is a masking tape with a hard band strip which is slipped under moldings such as flush mount windshields, side lights, and backlite windows. The protective covering is then removed to expose the adhesive, which is rolled over, lifting the molding slightly. This allows the paint to flow under the molding. The hard strip protects the underside of the molding and comes in different widths, such as 5, 7, 10, and 15 mm. If the decision is made to refinish a vehicle without removing the glass with flush molding, this tape will allow the technician to perform a clean and professional masking.

MASKING MACHINES

Masking machines have been developed to help the technician apply masking paper more quickly. The machines apply masking tape to the paper as it is being drawn from the roll. Masking machines come in different types and configurations. Some hold only one size of paper, while others are masking stations that not only hold many sizes of paper, but also the equipment commonly needed for masking. See **Figure 26-44** and **Figure 26-45**. Although the variety of masking machines is already vast, technicians are continually developing different masking machines to speed the process of masking a vehicle.

Use of the Masking Machine

The main purpose of a masking machine is to make masking paper with pre-loaded tape readily available to the refinish technician. To accomplish this, the paper is loaded on the machine and adjusted to a precise stationary position. The tape is also loaded on a dispensing wheel. This wheel is then adjusted so that the tape, when applied, is 50 percent on the

FIGURE 26-44 This portable masking machine contains only one size of paper.

FIGURE 26-45 This masking station has all the equipment needed at hand.

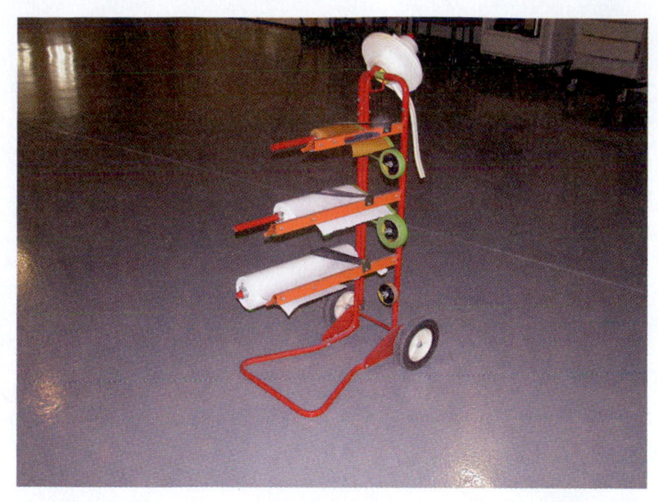

FIGURE 26-46 This masking machine (tree) has different types and sizes of paper.

FIGURE 26-47 This masking machine applies tape to paper automatically. The tape is 50 percent on the paper and 50 percent off, allowing it to stick to the vehicle as well.

paper and 50 percent over the paper edge as the paper is pulled off the roll. See **Figure 26-46**. The taped paper then passes through a spring and under a cutting bar, where it can be cut to length. The dispensed paper can then be applied to the vehicle quickly and easily. See **Figure 26-47**.

To cut a length of paper, the technician will pull off the length that is needed at an angle, so as not to engage the cutting bar. When the proper length has been pulled off, the technician will raise the paper against the cutting bar, tape side first, to cut the paper.

Adjusting the Masking Machine

To operate properly, masking machines need to be adjusted properly. Most masking machines have adjustments for the paper roll. Some of these adjustments have spring tensioners on them. There should be enough tension on the paper so that when the technician stops pulling, the paper will not roll out freely. However, it should not have so much tension that it is difficult to pull off the paper. As previously noted, the masking tape should be adjusted so it dispenses the tape at about 50 percent on the paper and 50 percent off the roll. There should be sufficient clearance so that neither the paper nor the tape is crushed as it pulls off.

A masking machine that is operating well will help the technician mask a vehicle, almost without notice. However, a machine that is constantly malfunctioning can become a point of great frustration. Keeping the machine loaded and adjusted is an easy way to lower the amount of stress in the workplace.

TECH TIP Although the tension on the masking machine is generally adjusted at the paper mounting location with a spring, this is not the only place that tensions can occur. Tape that is old may have become quite tacky and will not unroll readily. This may cause the masking machine to have so much tension that it does not work well. In such a case, the tape should be replaced with fresh, easily unrolled masking tape.

TECHNIQUES

Masking is an art. Each vehicle, with its specific masking needs, is a unique challenge for the technician. Although some basic masking techniques are used often with little modification, each new task is different. While it would be impossible to list all the different challenges possible in masking a vehicle, the explanations that follow suggest masking techniques for overcoming some of the challenges. Hopefully the demonstrated methods will aid new technicians in developing quick and efficient masking methods of their own.

TYPES OF MASKING TO BE PERFORMED

Although many different types of maskings can be performed, they generally fall into four **masking techniques**. The first technique is **masking of complete finish** or masking the "overall" paint job. With this type of masking technique, the vehicle may have parts removed, after which all the remaining finish of the vehicle will be prepared and exposed to be painted. Some parts may be best refinished off the vehicle and reattached following the finish, or some may be partly finished, then attached, with the remainder of the parts being finished as the "overall" refinishing is completed. This technique is not often done in the collision repair industry, however. See **Figure 26-48**.

In the second masking technique, **masking for panel painting**, one or more panels—such as a hood, or hood and fender—are prepared and masked so as to only expose those panels, after which they are completely finished.

FIGURE 26-48 Complete painting is not often done in collision repair, although some near total rebuilds will come close to needing complete refinishing.

Parts in the areas that need to be protected from paint are removed. The remaining parts of the vehicle are then masked for protection. See **Figure 26-49**.

TECH TIP Masking for protection protects the vehicle from overspray. Overspray can settle on a vehicle from any spray process in a shop. It may settle on one vehicle when another vehicle is being sprayed close to it. Some shops protect the areas of a vehicle that are not to be refinished as soon as the vehicle arrives in the refinish shop. Liquid mask and other complete vehicle coverings will protect the vehicle from unwanted overspray falling on the vehicle. The use of spray booths and prep decks eliminate much of the problem of random overspray. See **Figure 26-50**.

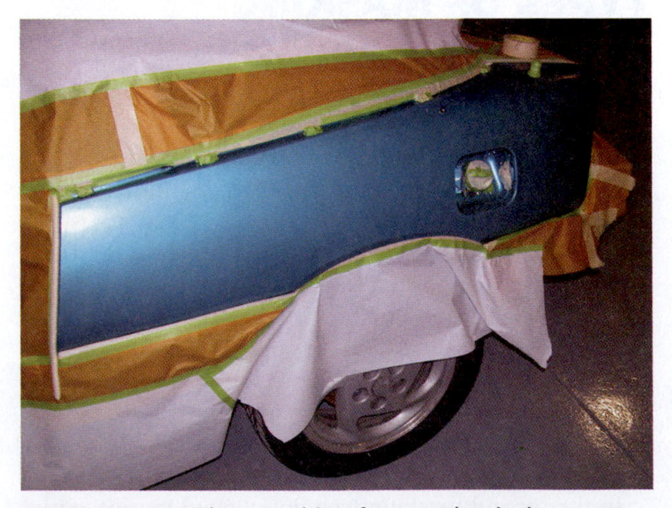

FIGURE 26-49 When masking for panel painting, complete panels are exposed and painted. The adjacent panels do not receive paint, therefore they are covered with masking to protect them from overspray.

(A)

(B)

FIGURE 26-50 A double prep deck (A) and spray booths (B) eliminate much of the problem of random overspray in a shop.

The next technique is **masking for blending**, in which the panel is prepared, basecoat is applied to the repaired area of a panel, and then the basecoat is blended gradually out from the repair. The remainder of the panel—or the repaired panel and adjacent panels—are then covered with clearcoat to render the blend undetectable. See **Figure 26-51**.

Blending of either panel—or single stage within a panel—is not generally recommended. It is, however, done from time to time if the area to be blended is small. Technicians should understand how to mask for this process in case they are ever called upon to do so. **Masking for clear blending** within the panel can be accomplished by preparing and exposing the area which needs to be refinished, then finding an area suitable for this type of blending process. See **Figure 26-52**.

Masking for Protection

The purpose of masking for protection is to cover the area of a vehicle from overspray. There are three possible

FIGURE 26-51 This vehicle has been masked for blending.

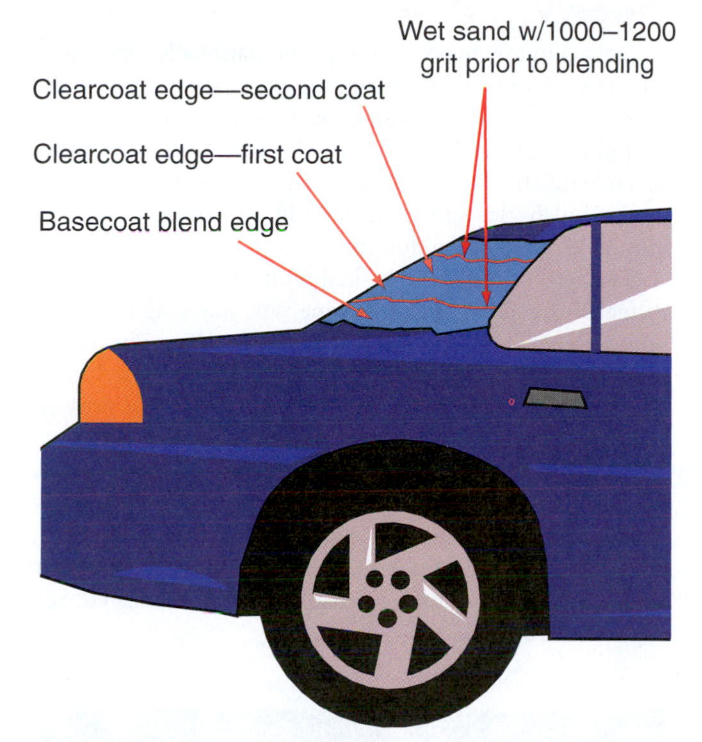

FIGURE 26-52 This vehicle is being prepared for clear blending.

alternatives that will be explained: liquid mask, plastic drape, and 36-inch paper. All of these methods will do a sufficient job, but may not suit each situation. All three types of masking for protection will be covered with their advantages and cautions.

Liquid Mask. Liquid mask, as its name implies, involves a liquid that is sprayed on the vehicle where protection is needed to provide a barrier between the vehicle and any overspray or contaminant that may damage or dirty the vehicle. It comes in two types: one dries creating a non-sticky film, and the other dries while remaining somewhat sticky. This type will attract air-borne contaminants. The dry type has a protective film that some technicians like, while other technicians may use the sticky type in particularly dirty areas—like wheel wells and undercarriages—to combat flying debris.

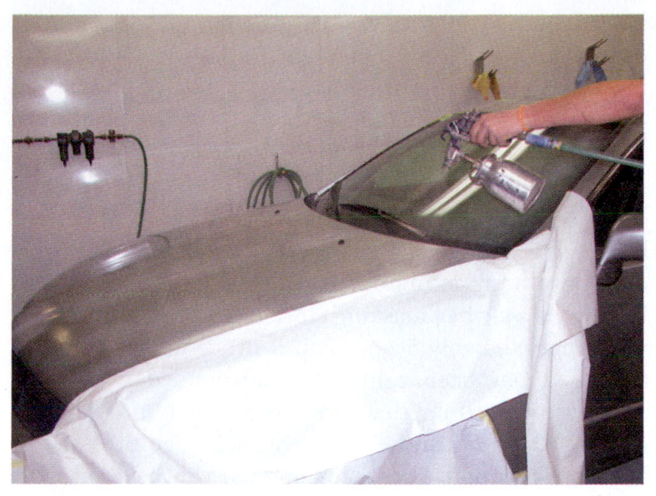

FIGURE 26-53 Before applying liquid mask, the area to be painted must be masked off, exposing the remainder of the vehicle.

To apply liquid mask, the area to be painted must be masked off first, exposing the remainder of the vehicle. See **Figure 26-53**. With this masking done, the liquid mask can be sprayed on the remainder of the vehicle. When the liquid mask has been applied and the needed flash time has elapsed, the vehicle can then be masked for refinishing. Only an outside border of paper, 12 to 18 inches in width, is needed, although the remainder of the areas should be masked normally. Aperture gaps between the doors, fenders, and other openings should be sealed to eliminate overspray. Moldings should either be removed or masked, and all areas where finish is unwanted should be protected.

Plastic Draping. Plastic can cover a vehicle rapidly; it comes on rolls, allowing a technician to pull it over a vehicle, cut it to length, then unfold it to make an envelope over the vehicle. It can be applied after the area to be painted has been masked by then cutting an opening to expose the needed area and sealing the edges of the cut area. Alternatively, it can be applied before the area is masked, then opened and taped down when the area is masked. Either way, the vehicle can be covered rapidly.

Paper Draping. When using paper coverings for protection, 36-inch paper is generally used and is applied until

TECH TIP When masking for protection using a drape, such as plastic film or paper, *do not let the masking material hang to the floor in a downdraft paint booth.* Doing so will interrupt the airflow, decreasing the booth's efficiency.

the entire vehicle is covered. Although an experienced technician may learn to complete this process rapidly, it is the slowest of the three methods.

Back Masking

Back masking has many purposes. **Figure 26-54**, for example, features a hood that has been removed from a vehicle and is being prepared for priming. To keep paint and overspray from getting to the underside, the technician has placed a strip of masking tape on the back of the hood, with approximately 50 percent of it overhanging. When the hood has been completely surrounded with tape, masking paper is applied and either folded under, or left hanging down. The hood will have a soft edge and will not have a visible tape line. This method can be used with many masking applications, either on or off the vehicle.

Reverse Masking

Reverse masking is done to help eliminate a hard tape line. In a reverse masking method, the masking paper is applied in the opposite direction as it would normally. Then it is brought back upon itself until a slight amount of tape is exposed. The exposed tape should not appear to be folded flat against itself, but should have a gradual curve. This masking method eliminates a hard line when the finish is sprayed onto this area.

Masking for Multi-Color

To mask a vehicle for multi-colors, the vehicle is painted with the first color beyond the area that it is required. Then, when the recommended flash time has passed, the area is masked for the second color. When applying masking tape to the freshly painted area, it is best to soften the adhesive by dragging the glue side over the pant leg. This reduces the amount of adhesive that is on the tape, and will help reduce the possibility of tape tracking residue being left behind.

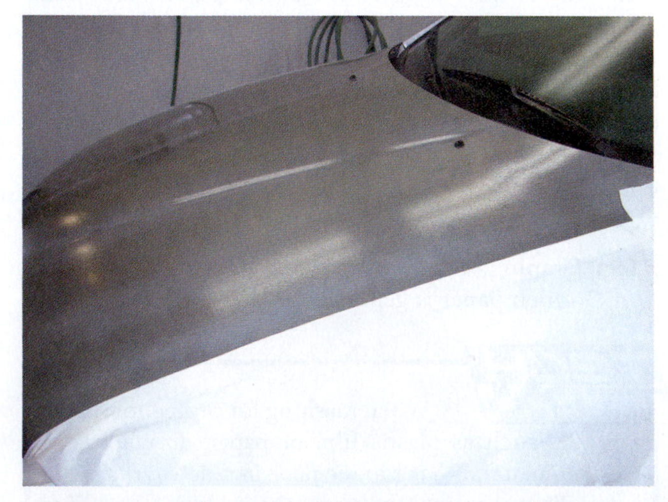

FIGURE 26-54 The technician, using back masking, places a strip of masking tape on the back of the hood, with approximately 50 percent of it overhanging.

MASKING REMOVAL

Masking removal should be done soon after the vehicle has cured, although some technicians—if the vehicle requires sanding and buffing—will leave some masking on until those procedures are completed. Leaving the masking on the vehicle during the detailing process will keep detailing sludge from getting into areas that would not need cleaning otherwise, and prevent material from entering cracks that would be hard to clean.

Masking paper and coverings can be removed and folded against itself in large pieces, which makes their disposal easier. If the masking is taken off in small parts, it can make a significant mess that will take additional time to clean up later.

APPLICATIONS

Door/Window

Door/window masking has particular challenges. First, all the decisions about removing parts should be made. The removal of door handles, key locks, window weatherstripping, and moldings may produce a better, more undetectable finish. If they are left on the vehicle, the deglazing of the finish close to them should be carefully done to eliminate any peeling due to poor adhesion.

These parts must be masked with care, and the choice of masking tape is critical. Crepe type masking tape may allow creeping of the finish under its edge. The better choice would be the smooth-surfaced plastic vinyl tape, which can be stretched around a small object and pressed tightly against it. See **Figure 26-55**.

The window should be first surrounded with 3/4-inch masking tape, followed by an application of masking paper. The paper should be applied in one piece if possible, with the excess either trimmed with a razor blade or folded under with all the open edges and folds sealed. See **Figure 26-56**.

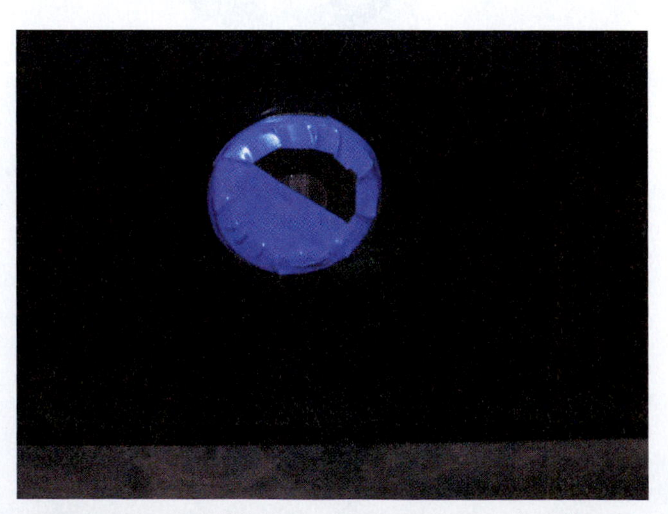

FIGURE 26-55 Masking a key lock with blue fine line protects these parts from contamination.

FIGURE 26-56 This window of a door is masked with a single piece of paper to reduce dirt.

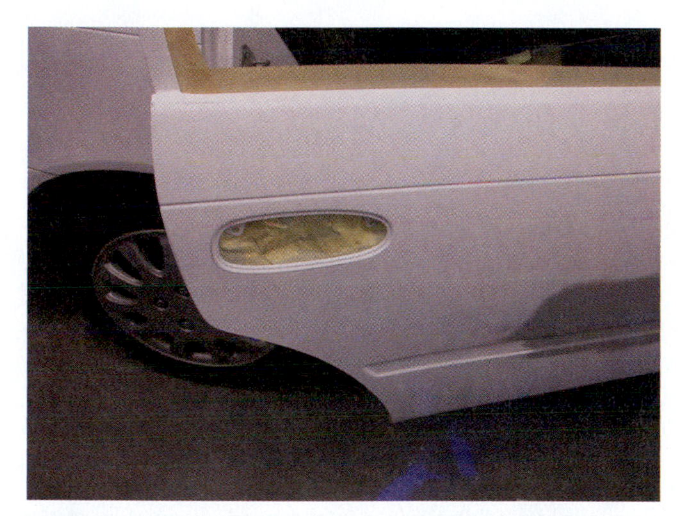

FIGURE 26-57 Door handle holes can be sealed from the back side with tape to prevent contamination.

If there are any openings from the removal of parts (such as door locks and handles), they should be sealed from the inside to prevent contamination from the inside of the door coming out, and from overspray going into the door. See **Figure 26-57**.

Windshield

To mask a windshield, the technician should surround the windshield with 3/4-inch tape first, if the molding has been removed. If the moldings have not been removed—as with flush mount—trim masking tape should be applied.

Trim masking tape has a rigid side without adhesive that comes in different sizes (either 5, 7, 10, or 15 mm wide) and must be matched to the width of the molding being masked. The rigid side is slid under the molding; then the protective covering is removed to expose the adhesive glue. The tape is pulled so the molding is lifted off the vehicle's surface slightly, to allow finish to be applied slightly under the molding as it is sprayed. See **Figure 26-58**.

To stretch around corners, the technician should cut from the outside edge to the rigid plastic, which will allow the tape to make the curve. See **Figure 26-59**. If any of the molding is exposed due to the cuts, it must be protected with a small amount of masking tape.

Once the window has either been surrounded with 3/4-inch tape or trim masking, the paper can be applied. Although it is best to apply masking paper in one sheet if possible, in cases such as front and back glass it may be impractical to apply 36-inch paper. Two or more rows of paper may be necessary instead. Automobile glass is generally curved, and having the paper lie flat is best, so the first row should be placed at the lower edge. See **Figure 26-60**. The ends should be either folded or cut to accommodate the curves and sealed with tape. It may be best to open the hood and extend the first row of paper under it into the engine compartment.

FIGURE 26-58 Trim masking tape has a rigid side without adhesive that comes in different sizes.

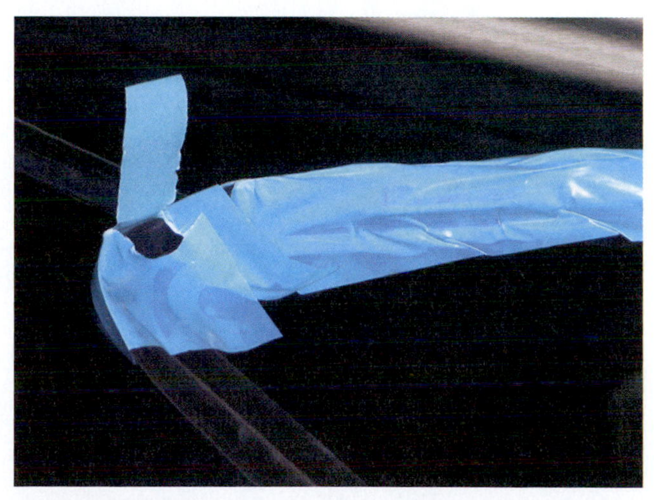

FIGURE 26-59 Trim tape must be cut and overlapped to go around curves.

FIGURE 26-60 Paper is first placed at the lower edge of a window, then the upper, for a cleaner masking application.

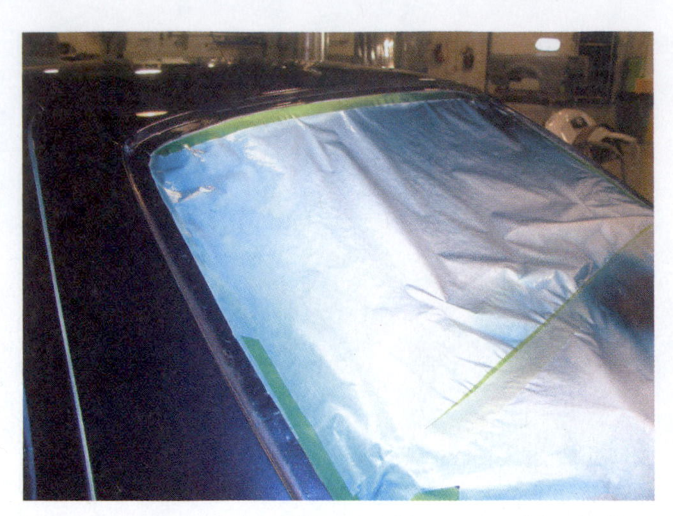

FIGURE 26-62 This backlite (glass) is removed for better masking.

The second and third rows, if needed, can be applied and sealed, going from the lower to the top of the windshield. When applied in this manner, the masking paper can be made to lie flat, and will not have edges to catch dust.

TECH TIP When cutting masking, care should be taken when using a razor blade or other sharp object. Sharp cutting tools can damage chrome, plastic, glass, and just about anything they come in contact with. The best way to cut tape on a vehicle is to place the razor where the cut is to be made, then lift the tape against it. See **Figure 26-61**.

If cutting paper, the paper should be lifted slightly off the vehicle, then cut; otherwise, fold the excess under, then seal it in place.

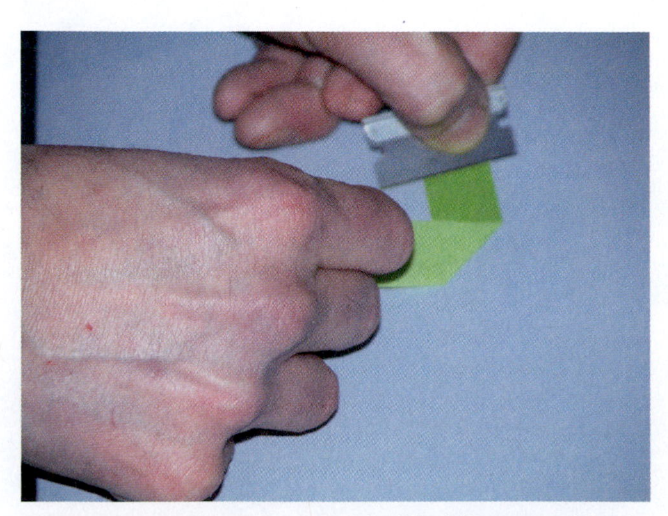

FIGURE 26-61 This razor is held still while tape is lifted to it for cutting.

Backlite or Windshield Removed

It is often necessary to mask up openings such as the windshield and backlites with the glass out. To do so, special care should be taken to protect not only the inside of the vehicle, but also the adhesive application shelf of the glass. Stationary glass adhesives should not be placed over topcoats. An adhesive primer is applied prior to applying the glass adhesive, and it must not be placed over finish.

First, apply masking tape to cover the adhesive shelf; then apply masking paper in the needed rows, from the bottom to the top. See **Figure 26-62**. Each row should be sealed with masking tape; to do this, the technician may need to reach inside the vehicle to have one hand under the paper being sealed and the other pressing down the masking tape. See **Figure 26-63**. All gaps and openings should be sealed.

Engine Compartment

The engine compartment should be protected from overspray, and it should be sealed from any contaminants that become air-borne.

First, remove parts such as hood bumpers, and apply a strip of 6-inch paper to the inside shelf of the hood. The front grill can be partially covered with paper at this time as well. Apply a strip of aperture tape on top of the paper in the inside shelf. See **Figure 26-64**. The paper and aperture tape are placed high enough to seal the compartment (when the hood is closed), but not so high that it will leave a line when removed.

With the outside openings sealed, the inside of the engine compartment can be covered with paper to seal the engine area.

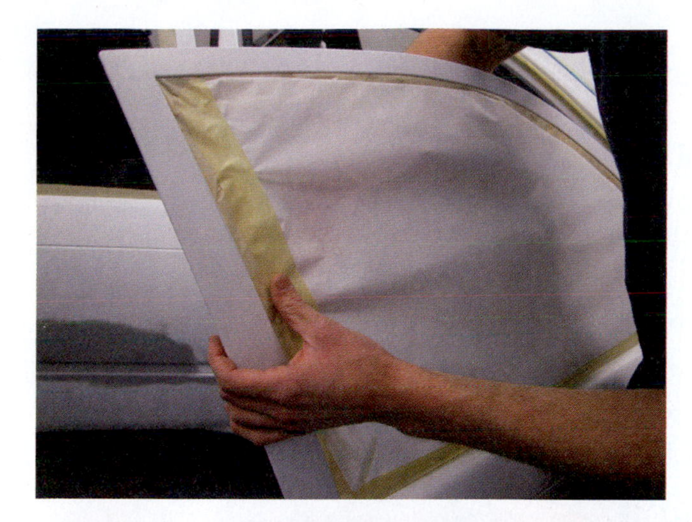

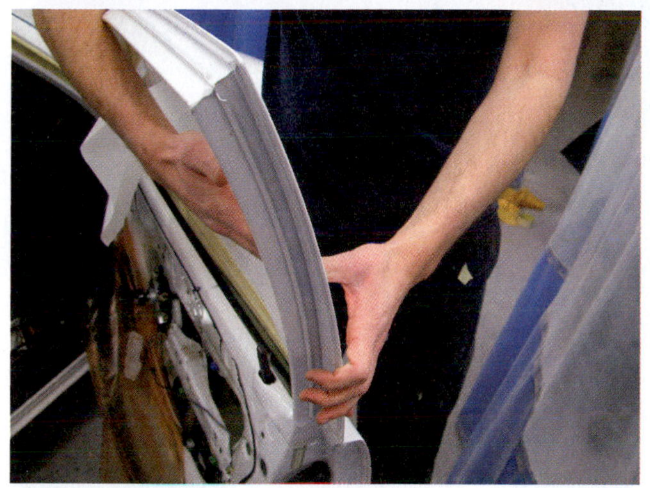

FIGURE 26-63 To mask some areas, access to both sides is necessary.

FIGURE 26-64 The engine compartment should be protected from overspray, and it should be sealed from any contaminants that become air-borne.

FIGURE 26-65 To protect a door jamb, the door may be masked with paper and aperture tape.

Door Jambs

To mask a door jamb that will be sprayed with the door closed, the main concern is to seal the door aperture to prevent overspray from getting in. This can be done in a manner similar to masking a hood. The door is opened and a row of 6-inch paper is placed on the inside of the door, then sealed with aperture tape. See **Figure 26-65**. The process continues by closing the door against the aperture tape, and checking it for proper seal and height. (The tape should not be too high.)

Doors may sometimes need to be masked in order for the technician to spray inside the door jamb. To do this, the weatherstripping should be removed and masking tape applied to the pinchweld area where the weatherstripping is mounted. It can be placed using the back mask method, where the first row of 1 1/2 inch masking tape is placed on the back of the pinchweld; then the paper is applied over it, sealing the inside completely. See **Figure 26-66**. Then the remainder of the inside of the door must be masked as well, exposing the area to be finished and protecting the areas that do not get finished.

Gas Filler Door

It may be just as easy to remove the gas filler door and spray it off the vehicle as it is to mask it. If the part is to remain on the vehicle, though, mask the opening as with the inside of the hood and door, placing aperture tape on the inside, then close the gas filler door against the tape. See **Figure 26-67**.

TECH TIP When applying aperture tape, care should be taken not to stretch it as it is being applied. The foam is soft and easily stretched, but will try to shrink back to its original size and may pull loose when shrinking.

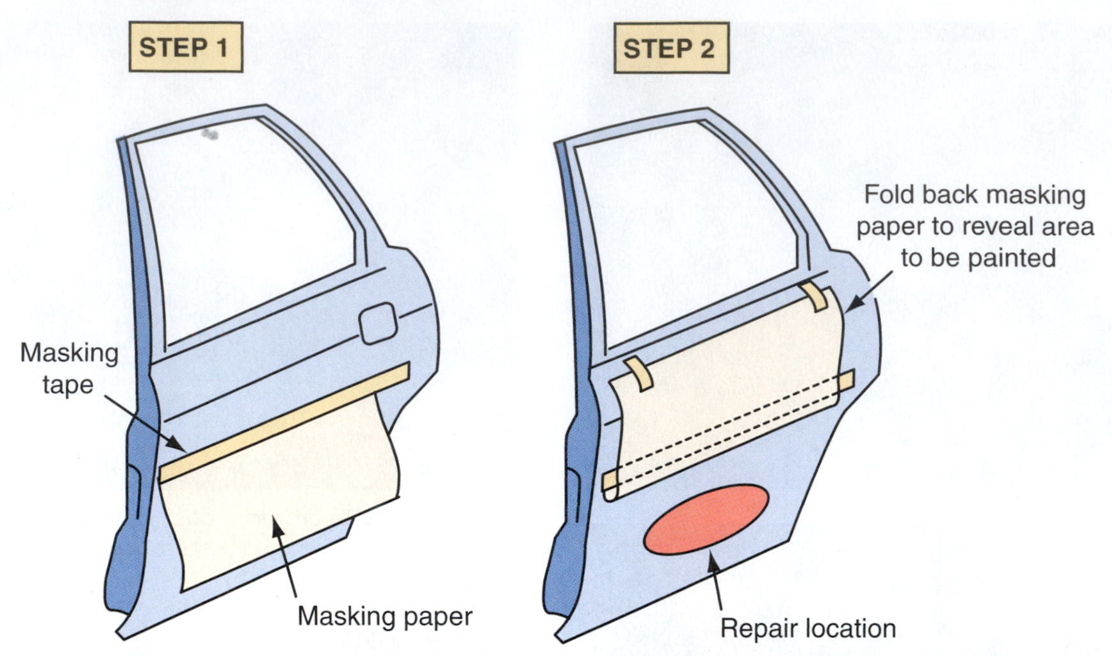

FIGURE 26-66 A technician can back mask the inside of the door with masking tape and paper.

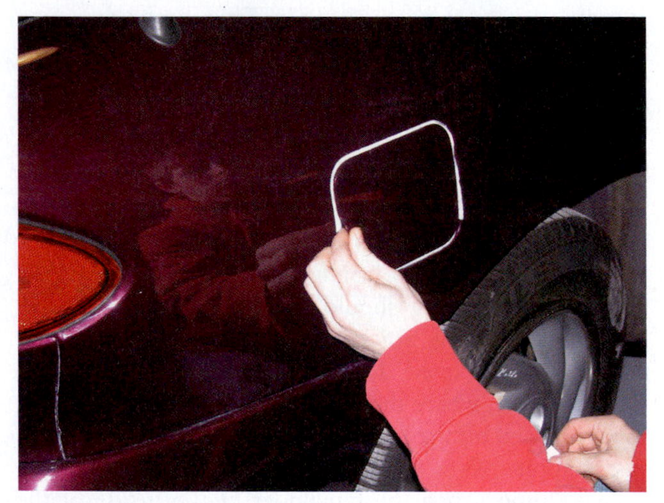

FIGURE 26-67 The gas filler door may be as easy to remove as it is to mask.

FIGURE 26-68 The trunk may need to be masked with aperture tape to keep overspray out of the gaps.

Trunk

The trunk may need to be masked to keep overspray out of the gaps, using 6-inch masking tape and aperture tape. See **Figure 26-68**. Or it may need to be completely sealed from overspray. See **Figure 26-69**. Note that the hinges have been covered with masking tape to protect them from overspray.

A method used to rapidly mask off objects, such as a trunk hinge or an antenna that cannot be easily removed, is to make an envelope with paper, then seal the top and bottom with tape. See **Figure 26-70**.

TECH TIP When taping an object that is long and round, like an antenna or door handle, do not wrap masking tape around it. Although this process covers the object rapidly and may seem like the thing to do, it is very difficult to remove the masking tape later. Instead, apply the tape in long runs. See **Figure 26-71**.

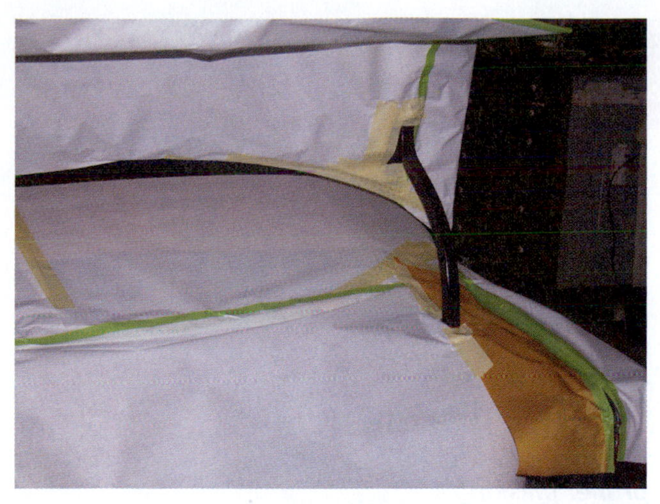

FIGURE 26-69 In some cases, the entire trunk may need to be completely sealed from overspray.

FIGURE 26-70 To rapidly mask off objects, make an envelope with paper, then seal the top and bottom with tape.

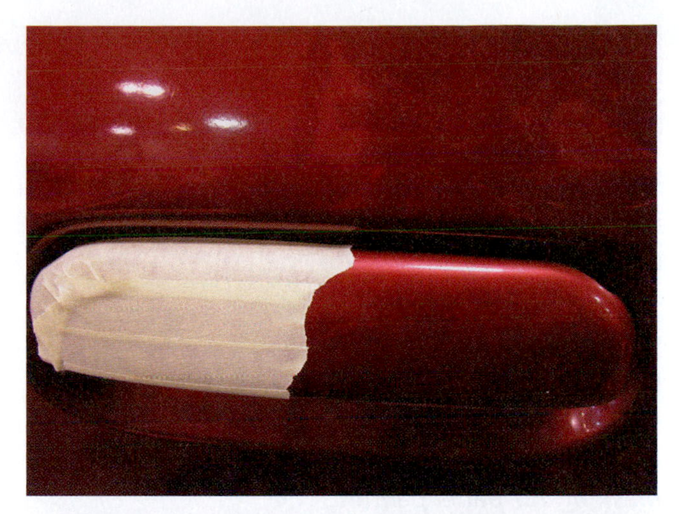

FIGURE 26-71 When taping a long, round object, like a door handle, use long runs of tape instead of simply wrapping masking tape around.

Summary

- Although some may believe masking to be a simple, low skill level task in the paint department, it is actually quite complicated.
- The vehicle needs to be properly soap and water washed and chemically cleaned before refinishing; then the masking technician needs to make a plan specifically for that vehicle, and execute the plan.
- The technician must know when to use liquid mask or plastic film drape for protection, and how to mask each area of a vehicle as quickly as possible, eliminating the vehicle's exposure to overspray that would otherwise need to be removed later.
- The vehicle needs to be masked in such a way that it does not hinder the operation of the spray booth, and also does not trap and then redeposit dirt on the vehicle.
- The vehicle must be masked so that it can be unmasked rapidly, and without damaging the fresh finish.
- The technician must master all these masking skills to accomplish a professional quality, speedy refinish product in the modern collision repair/refinish department.

Key Terms

aperture tape
back masking
blend masking
chemical cleaning
disposable wheel covers
liquid mask
masking for blending
masking for clear blending
masking for panel painting
masking machines
masking materials
masking of complete finish
masking paper
masking removal
masking tapes
masking techniques
painter's helper
parts removal
parts storage
plan of attack
plastic drape film
pre-masking cleaning
primer masking
reverse masking
soap and water cleaning
specialty tapes
trim tape
wheel well covers

Review

ASE Review Questions

1. Technician A says that when a new product is introduced into the refinish department, it should be accompanied with an MSDS provided by the manufacturer. Technician B says that an MSDS will list the personal protective devices that must be used when working with a product. Who is correct?
   A. Technician A only
   B. Technician B only
   C. Both Technicians A and B
   D. Neither Technician A nor B
2. Technician A says a vehicle that has been repaired in the body shop has been soap and water washed before it was worked on, and therefore it is not necessary to re-wash the vehicle when it comes to the refinish department. Technician B says that when washing a vehicle it should be soap and water washed using a two-bucket method. Who is correct?
   A. Technician A only
   B. Technician B only
   C. Both Technicians A and B
   D. Neither Technician A nor B
3. An aftermarket part has arrived in the refinish department. Two technicians have been assured that the black coating is an e-coat primer, but they disagree about how to proceed. Technician A says that it is OK, and they should refinish it as they normally do. Technician B says that they should perform a wipe-down test to be assured that it is a thermoset product. Who is correct?
   A. Technician A only
   B. Technician B only
   C. Both Technicians A and B
   D. Neither Technician A nor B
4. Technician A says that masking is done to protect the parts of the vehicle where new finish is unwanted. Technician B says that overspray is not a problem and won't travel far, so covering a vehicle in the back when painting in the front is not necessary. Who is correct?
   A. Technician A only
   B. Technician B only
   C. Both Technicians A and B
   D. Neither Technician A nor B
5. Technician A says that a good painter can mask most parts, and not many need to be taken off for refinishing. Technician B says that the removal of parts will help the refinish technician to refinish the vehicle so the repairs will be undetectable. Who is correct?
   A. Technician A only
   B. Technician B only
   C. Both Technicians A and B
   D. Neither Technician A nor B

6. Technician A says that when a vehicle is being masked for panel painting, the finish will be sprayed on the entire panel. Technician B says that when masking for blending, the adjacent panels are masked so the color can be blended over the repair and the clear applied to all the exposed panels. Who is correct?
   A. Technician A only
   B. Technician B only
   C. Both Technicians A and B
   D. Neither Technician A nor B
7. Technician A says that when preparing a panel for blending of clearcoat, it is masked the same as it would be for blending. Technician B says that reverse masking or bridge masking is used to blend clear, to eliminate a hard line. Who is correct?
   A. Technician A only
   B. Technician B only
   C. Both Technicians A and B
   D. Neither Technician A nor B
8. Technician A says that when masking for priming, overspray is not a problem and 12 inches of paper will be enough to protect the vehicle. Technician B says that when masking for priming, it is not necessary to use aperture tape. If primer gets in the cracks, it can be wiped off at any time. Who is correct?
   A. Technician A only
   B. Technician B only
   C. Both Technicians A and B
   D. Neither Technician A nor B
9. Technician A says that besides a liquid mask for vehicles, one is made for spray booths as well. Technician B says that there is even a liquid mask made specifically for booths' floors. Who is correct?
   A. Technician A only
   B. Technician B only
   C. Both Technicians A and B
   D. Neither Technician A nor B
10. Technician A says that when doing the final washing of a vehicle before masking, it should be cleaned inside all the openings where repair dust may have accumulated. Technician B says that the entire vehicle should be cleaned with wax and grease remover before masking as well. Who is correct?
   A. Technician A only
   B. Technician B only
   C. Both Technicians A and B
   D. Neither Technician A nor B
11. Technician A says that it matters very little which type of paper is used for masking, so get the cheapest. Technician B says that paper comes in 36-inch rolls that can be cut down to smaller sizes if needed. Who is correct?
   A. Technician A only
   B. Technician B only
   C. Both Technicians A and B
   D. Neither Technician A nor B
12. Technician A says that some papers are resistant to solvents, due to their thickness. Technician B says that the thinner the paper is, the more conformable it will be. Who is correct?
   A. Technician A only
   B. Technician B only
   C. Both Technicians A and B
   D. Neither Technician A nor B
13. Technician A says that most paper machines are very difficult to load, adjust, and work with; therefore, they are not helpful with masking. Technician B says that paper machines come in many different sizes and configurations and greatly speed up the process of masking. Who is correct?
   A. Technician A only
   B. Technician B only
   C. Both Technicians A and B
   D. Neither Technician A nor B
14. Technician B says that masking tape backing is made from a paper called crepe. Technician B says that tape made from plastic or vinyl helps to eliminate paint from creeping under the tape. Who is correct?
   A. Technician A only
   B. Technician B only
   C. Both Technicians A and B
   D. Neither Technician A nor B
15. Technician A says that aperture tapes come in four different sizes to match different sizes of openings. Technician B says that trim masking tape is made to lift flush mount moldings slightly, to allow paint to flow underneath the moldings during refinishing. Who is correct?
   A. Technician A only
   B. Technician B only
   C. Both Technicians A and B
   D. Neither Technician A nor B

Essay Questions

1. Describe the process for masking off a quarter panel with a flush mount molding on the stationary side glass.
2. Compare and contrast liquid masking with paper masking.
3. It will take 25 minutes to remove all the parts on a door that is being blended, but only 25 minutes to mask them. Which method would you recommend, and why?
4. Some technicians trim off extra paper by cutting it with a razor blade, while others fold the excess

under. What are the advantages and disadvantages of each method?

5. Why is it best to leave some of the masking paper on a vehicle that needs to be polished?

Topic Related Math Questions

1. Five identical vehicles come into the shop. The first one requires 625 feet of 12-inch paper to mask it. How many rolls of 12-inch paper will be needed if each roll has 750 feet?
2. In a store, 3/4-inch masking tape comes 12 rolls to a sleeve and 4 sleeves to a case. How many rolls are in 12 cases?
3. Liquid mask will cover 653 square feet per gallon. An average vehicle has 231 square feet, and the department paints 750 vehicles per year. How much liquid mask will be needed for a year?
4. An average vehicle requires 32 feet of plastic drape film. Each month the shop covers 46 vehicles. How many rolls of drape will be needed per month, if each roll has 450 feet of film?
5. If the average technician is able to mask 10 vehicles a day and Susan, the new masking technician, is able to mask 17 vehicles per day, what is Susan's percentage of efficiency?

Critical Thinking Questions

1. Two vehicles come into the paint shop at the same time. The one that is scheduled to leave later is the vehicle that can be finished in the least amount of time. Do you move it up and mask it first, or attend to the vehicle that is scheduled to leave first, and why?
2. You have just finished masking the first vehicle of the day, and you notice that some of your masking supplies are low. There are only two other vehicles in the shop now for masking. What must you do to prepare an order to have enough supplies delivered to mask the vehicles that will be painted today?

Lab Activities

1. Have students apply liquid mask to a spray booth.
2. Have students back mask a hood that is off the vehicle.
3. Have students mask a vehicle for final priming.
4. Have students clean a vehicle for masking.
5. Have students mask a vehicle for a complete repaint.
6. Have students mask a vehicle for color blending and clearing.
7. Have students reverse mask a vehicle to blend clearcoat into a sail panel.

Chapter 27

Solvents for Refinishing

OBJECTIVES

Upon completing this chapter, you should be able to:

- Understand how solvents are used in the refinish industry.
- Control the variables that affect the quality and speed of refinishing.
- Choose the proper solvents to use when changes occur in factors such as temperature, humidity, and air movement.
- Understand why certain spray techniques, such as keeping the proper distance from the object being sprayed, are important.
- Understand why different blending techniques work better with certain colors.

INTRODUCTION

Although solvents were previously discussed in Chapter 24, this chapter covers them in more detail. The chapter will explain why and how solvents are used during the manufacturing, shipping, and application of finishes. It will also explore the decisions that a refinish technician must make every day when using solvents. The chapter will discuss the following areas: how solvents are used when spraying a coating, regarding how they react to the temperature, humidity, thickness, time, and air movement; how they are used in cleaning spray equipment, surface preparation, plastic preparation, and general cleaning in the body shop; and how they affect both drying and curing times, their relation to hardeners, and how they affect gloss, flow out, hardness, and long-term holdout.

In addition, the chapter will explore how solvents must evaporate at different times to produce a high-quality professional refinish, and how they affect the environment by emitting volatile organic compounds (VOCs) into the atmosphere. See **Figure 27-1**. Because of solvents' effects on the environment and the cost of disposing of them once used, this chapter will also explain the advantages of recycling.

This chapter is designed to help a technician become thoroughly familiar with solvents and how they work in the collision repair environment.

SAFETY

Safety is very important to the technician when using solvents! Solvents are harmful to the health and safety of people and therefore are considered a hazardous material that is toxic to not only spray technicians, but also to the environment.

Like many toxic and hazardous materials, safety precautions and equipment have been developed to keep persons using solvents safe from harm. **Material safety data sheets (MSDS)**—which

FIGURE 27-1 Volatile organic compounds, when released into the atmosphere, are a leading cause of smog.

SECTION 8 - Exposure Controls or Personal Protection

Engineering controls and work practices:

Ventilation:

Provide sufficient ventilation in volume and pattern to keep contaminants below applicable exposure limits.

Respiratory:

Do not breathe vapors or mists. Wear a properly fitted air-purifying respirator with organic vapor cartridges (NIOSH approved TC-23C) and particulate filter (NIOSH TC-84A) during application and until all vapors and spray mists are exhausted. In confined spaces, or in situations where continuous spray operations are typical, or if proper air-purifying respirator fit is not possible, wear a positive pressure, supplied-air respirator (NIOSH TC-19C). In all cases, follow respirator manufacturer's directions for respirator use. Do not permit anyone without protection in the painting area.

Protective clothing:

Neoprene gloves and coveralls are recommended.

Eye protection:

Desirable in all industrial situations. Goggles are preferred to prevent eye irritation. If safety glasses are substituted, include splash guard or side shields.

FIGURE 27-2 Material safety data sheets (MSDS) tell technicians which personal safety equipment should be used when working with a particular chemical.

FIGURE 27-3 Solvents are added to paint to ensure it is "carried" from the spray gun to the object being painted.

are developed for each product—have all the needed information for the safe use and handling, storage, cleanup, and safety equipment that must be used when handling the material, as well as emergency information in case of an accident or exposure to people and the environment. Law requires that facilities keep MSDS readily available for persons using toxic and hazardous substances.

It would be nearly impossible—and certainly impractical—to try to list the safety requirements for all the solvents that spray technicians may encounter. However, it is important for all technicians to know that information explaining how to use solvents safely is readily available.

Technicians should read, understand, and follow the safety information contained in the MSDS for the materials that they are using. See **Figure 27-2**.

COMPONENTS

Solvent is a general word for any substance that dissolves and carries another chemical. For example, water is a solvent when it dissolves and carries salt in human sweat. Many solvents are completely harmless to technicians and the environment. However, most of the solvents used in the collision repair industry are both hazardous to the environment and harmful to the technician using them. Solvents are used to dissolve paint solids and "carry" them from the manufacturer to the collision repair shop. Once there, additional solvents are added to ensure that the paint is "carried" from the spray gun to the object being painted. See **Figure 27-3**. At this point, the solvents must evaporate under very controlled or predictable conditions. The better paint technicians are at controlling or predicting the way these solvents evaporate, the better they can control the quality for their work and the profitability of the shop.

Many solvents are grouped in the category of **volatile organic compounds (VOCs)**. These chemicals are "**volatile**," which means they vaporize at room temperature; in fact, some of these chemicals are more volatile—or vaporize more rapidly—than others. This volatility is important when formulating solvents (reducers and thinners) so that the solvent will perform in a certain manner under certain conditions. This volatility is also affected by humidity, heat, time, and temperature. Manufacturers of paint reducers and thinners can, by mixing different solvents in their reducers and thinners, cause them to perform under different conditions. Refinish technicians can also choose the proper reducers for the conditions at hand when refinishing a vehicle by knowing what affects the evaporation (volatility) of a certain reducer and which one to use in certain conditions.

The term "**organic**," when referring to a solvent which is a VOC, means chemicals containing the element **carbon**. Organic chemicals are basic chemicals found in natural things such as coal and its cousin, oil. Many other organic chemicals are not found in nature, but are constructed by chemists in laboratories. Reducer, thinners, and other solvents used in the collision repair industry result from crude oil and its distillation. These chemicals are called **hydrocarbons** (chemicals containing hydrogen and carbon). This means that they are carbons, which are derived from crude oil. The term **compound** refers to a substance which has combined elements in a fixed proportion. These components cannot—and do not—separate through physical means.

It should be evident that a VOC can be made up of many combinations of organic compounds that rapidly evaporate. With this in mind, the technician should know that most of the reducers, thinners, and other solvents and cleaners fall into the category of VOCs, and are strictly regulated by state and local governments. They must be disposed of according to state and local laws when they are no longer of use to the collision repair facility. However, because they evaporate so rapidly, they can be separated from the solids that they suspend by heating the waste product, thus providing solvent vapors that can be captured, condensed into a liquid, and reused as a solvent. The recycling of used solvents will be covered later in this chapter.

Solvents

The category of solvents is very large; it includes reducers, thinners, adhesion promoters, cleaners, performance additives, and other specialty solvents.

The use of solvents starts with the manufacture of paint. The four major components of paint are: pigments, resin or binders, solvents, and special performance additives. See **Figure 27-4**. When solvents are added to paint during the manufacturing process, they dissolve the resins and pigments, forming a liquid that is easily transportable from the factory to the point of use in the collision repair shop. During manufacturing, solvents also dissolve the special additives that are put in paint to enhance its performance when applied.

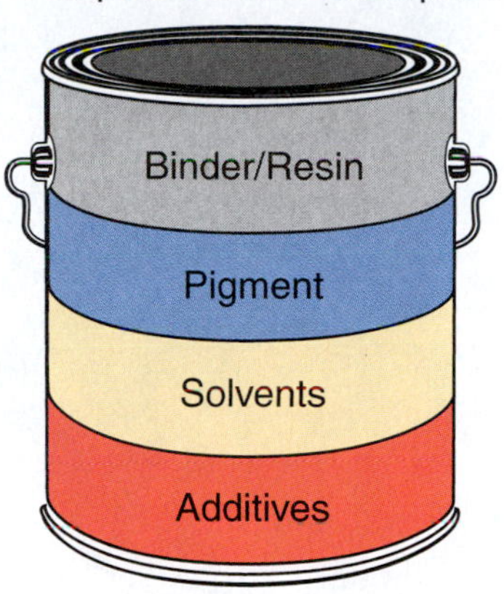

FIGURE 27-4 The four components of paint are resin or binders, pigments, solvents, and additives.

Once the manufactured paint is transported to the shop, it will be further thinned or reduced, so that it can be placed in a spray gun and applied to the vehicle. The reducer or thinner that is used in the shop differs from the one added during manufacturing. The reducer used in the repair facility is a mixture of chemicals that have been combined to evaporate at three different times during the repair process. A portion of the solvent evaporates in the time between when it leaves the gun to when it reaches the object being painted. This evaporation process is called in-flight loss of solvent. The next portion evaporates after the paint lays on the painted surface for a short time. This allows the coating to settle and the paint to level before it starts to become thicker. These chemicals, called leveling solvents, evaporate during the flash time. Because these solvents are predictable, paint manufacturers can reasonably predict when solvents will be gassed out, and when a second or subsequent coat can be applied. This time is generally referred to as flash time.

The last, and much slower, solvent evaporation allows the coating to harden slowly and become much more durable because of this slow release of solvents. Often paint manufacturers do not recommend the application of wax for as long as 90 days, to allow the freshly painted surface to completely gas out, or evaporate, before wax is applied. Doing so eliminates—or severely limits—any further evaporation. These three facets of evaporation—and the way a paint technician chooses the proper reducer and thinners to accomplish the process—will be covered in depth later.

Reducers and Thinners

Although **reducers** and **thinners** have similarities, they perform different tasks for different types of finishes. There are generally only two types of paint used in collision repair: lacquers and enamels. The paint in both of these categories must be thinned from the original thickness that comes from the manufacturer. Viscosity indicates the thickness of paint. Thick paint has a high viscosity, and a very thin paint has a low viscosity. Viscosity is measured with a **viscosity cup**. See **Figure 27-5**. This cup will hold a precise amount of paint, and it has a very specific size hole in its bottom. Viscosity is measured by how many seconds it takes for the paint to drain from the cup through the hole. A paint with 15 seconds' viscosity is thinner than one with 25 seconds' viscosity.

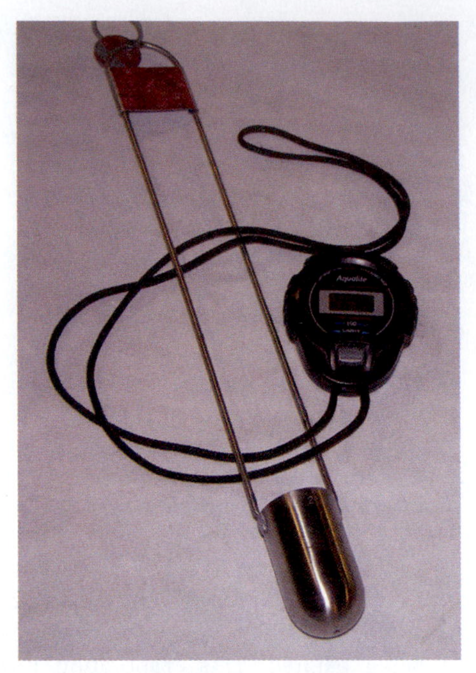

FIGURE 27-5 To test for correct reduction, use a viscosity cup and stopwatch to time the flow.

Most coatings that come from the manufacturer must be thinned. Generally, reducers are used when enamel is to be thinned; however, thinners are used for lacquer paints. Although this may be somewhat confusing to a new painter, it is not generally a problem for technicians. Lacquer is not often used in modern refinishing, and so thinners are mostly used for purposes of cleaning equipment.

In-Flight Loss of Solvent. Because this portion of the added reducer or thinner is designed to evaporate from the time it leaves the gun and before it reaches the object being sprayed, the chemical used must be very volatile. Most paint guns used in the aftermarket painting industry are external mixing guns. This means that paint and atomizing air do not mix until the paint travels out of the gun's tip. Because paint and air are mixed outside the gun, the paint's viscosity or thickness must allow the paint to be broken into small enough particles to allow the paint to lie smoothly on the vehicle.

TECH TIP All liquids have surface tension. This means the molecular components of the liquid want to stick together. The surface of a liquid acts as a thin elastic film under tension, and when paint (a liquid) is sprayed out of the end of a spray gun into the air, the paint forms drops. To make these drops small, the surface tension must be overcome, causing the liquid to form very small droplets. This is called atomization, and allows paint to lie on the surfaces flatly.

If the droplets are too large, the paint will not lie on the vehicle smoothly, and will cause a rough surface called "orange peel." See **Figure 27-6**.

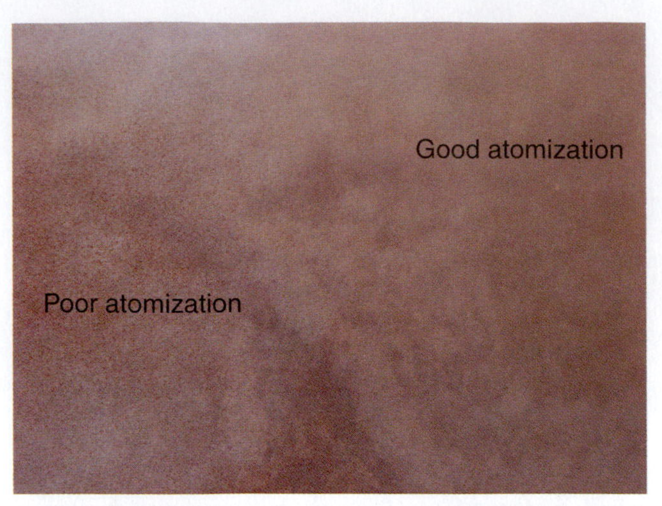

FIGURE 27-6 Poor atomization will result in the finish having poor flow out and is more likely to have orange peel. With good atomization the finish will lie on the surface more smoothly resulting in a better flow out and less texture when cured.

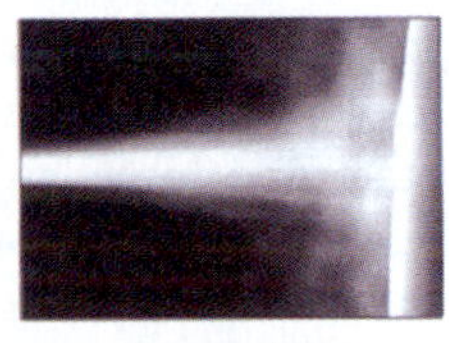

FIGURE 27-7 A conventional high pressure gun produces large amounts of overspray.

Two ways to increase atomization are to increase the pressure sprayed from a gun, or to make the paint being sprayed less thick. Although increasing air pressure is an efficient way to break paint into smaller droplets, this increased pressure propels the paint so rapidly that when it hits the surface it bounces off as overspray, and doesn't stick to what is being painted. See **Figure 27-7**. Many states now have laws about the type of gun that can be used to spray paint. These states require the use of a high volume, low pressure (HVLP) gun. By definition, HVLP guns have an air cap pressure that is 10 psi or below. Another way that paints can be atomized is to reduce the paint's viscosity, as it will need less pressure to atomize then (since thicker liquids have higher surface tension).

To reduce paint, a technician must add more solvents to it. The amounts of solvents that can be put in paints have been regulated by the National VOC Regulations, and some states require the recording and reporting of solvents that are used during automotive refinishing. Therefore, it may be illegal to add more than the recommended amount of solvents. The options remaining for a technician are to reduce the amount of paint coming from the gun, and to keep the gun distance correct.

With the proper solvent chosen and the gun adjusted properly—especially the fluid adjustment—even thick, high-solids materials can be atomized and sprayed to achieve a smooth surface which does not require buffing to reduce the orange peel.

TECH TIP Not all "orange peel" textures are bad. In fact, a good paint technician must become expert at matching not only color (hue), shade (value), and metallics (chroma) and how they lie on the surface, but also must be able to match the texture of the original vehicle's surface. By adjusting the gun, a technician can match the orange peel that surrounds the area being refinished.

Leveling Solvents. Leveling solvents are the solvents that remain in the paint after it hits the surface of the vehicle being painted. They allow atomized droplets to reconnect and "**flow out**." The solvent must remain in the coating, allowing gravity and other factors such as temperature, humidity, and airflow to help the paint surface become flat and smooth.

Although this portion of solvent is less volatile (evaporates more slowly) than the in-flight loss solvents, it must also evaporate fast enough to allow the technician to apply additional coats within a reasonable time. The time that must pass before the technician applies additional coats of a coating—or before the paint can be baked—is referred to as flash time. If sufficient flash time is not allowed and a subsequent coat is applied, a condition called "solvent popping" will occur. See **Figure 27-8**. Solvent popping is a condition in which solvents from below new coats of paint try to escape (**gas out**), and get trapped under a paint film that is starting to harden. As the bubble of gas solvent breaks through the surface, a small crater is formed, causing an imperfection in the paint surface.

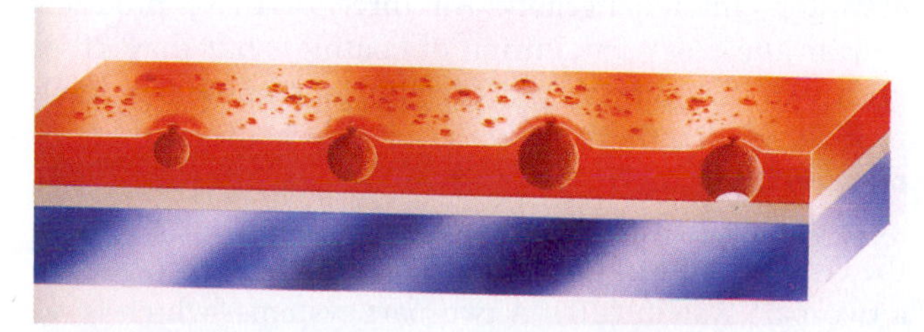

FIGURE 27-8 If flash time is not long enough for leveling solvent to evaporate before a second coat is applied, the finish may develop "solvent popping."

Although this leveling time is very critical, the refinish industry continually asks paint manufacturers to reduce the time needed to spray a vehicle. One of the ways to accomplish this request is to reduce the flash time by using a more volatile solvent during flash time.

Final Solvents. The final solvent evaporation can take up to 90 days, in some cases, to completely gas out. Solvents that remain in the coating allow the second **curing** process to take place.

Thus far, we have been exploring the evaporation process of paint. However, evaporation is not how modern urethanes, polyurethanes, and epoxies cure. In fact, only 25 percent of the curing process is truly **drying** or evaporating. Paints cure by a process called **cross-linking**, which forms chemical bonds between two separate molecular chains, generally **polymers**. This cross-linking is added, and a much harder, more durable bond is formed if some of the solvent remains in the paint as the cross-linking takes place. As this three-dimensional molecular structure is formed, the resulting coating is more shear-resistant, has a higher temperature resistance, and is oil- and solvent-resistant. This process takes an extended period, and is aided when small amounts of solvent remain in the coating.

Although cross-linking takes some time, the paint does not remain wet. Cross-linking can be sped up by applying heat or radiation, and most manufacturers recommend "baking" polymer paints after spraying to speed up the curing process. By applying heat of 100°F to 140°F for a specific amount of time (generally 40 minutes, or with some hyper-curing clears only 8 minutes), the cross-linking can be sped up. Still, even after the coating has been baked and the surface of the paint reaches ambient (room) temperature, it is not fully cured. The surface is mar-resistant; it can be sanded and polished, and can then be delivered back to the customer. However, most paint manufacturers recommend that the vehicle not be waxed for up to 90 days. Waxing applies a sealant to the coating's surface and does not allow the final gas out of the remaining solvents, which needs to occur for the coating to completely cure.

Adhesion Promoters

Technically an **adhesion promoter** is any material used to improve adhesion between two materials; thus, primers are adhesion promoters. However, in the automotive refinishing industry, adhesion promoters are solvents that increase the adhesion between a substrate, providing a better mechanical bond with what is sprayed over it.

Most currently used adhesion promoters are sprayed on a polyolefin plastic part; then coatings are applied over it. The adhesion promoter will act as a binding agent of the two products.

Before urethane, polyurethane, and epoxy paints were developed, lacquer paints would reflow when solvents were applied to them, no matter what their age. Because of this, coatings were marketed as "adhesion promoters." These products were used to soften the old lacquer paint; when new paint was added, the two would "flow together."

Paints that use a hardener (2K) for cross-linking will not reflow when solvent is added. Products marketed as adhesion-promoting coatings do not soften the old paint, but provide a better flow out of the paint as it is applied. Fully cured urethane, polyurethane, or epoxy paints will not reflow from the application of solvents. These thermoset coatings are durable and resistant to marring, environmental attack, and solvents.

Cleaners

Soap and water and chemical cleaning are done at several stages when repainting a vehicle in the collision repair industry. Soap and water cleaning (a type of solvent) is the first step performed on a vehicle, sometimes even before the vehicle is examined for a repair estimate.

Some contaminants are only soluble in water, or in soap (a wetting agent) and water. Other contaminants are only dissolved with a wax and grease remover (solvent). The two containers in **Figure 27-9** each started out with rock salt in them, a common de-icer used to clear roads. As the picture shows, the salt in the container with soap and water has dissolved, while the salt in the container with wax and grease remover has not dissolved at all.

In **Figure 27-10**, the two containers have road tar in them. The road tar in the container with soap and water has not dissolved, but the road tar in the container with wax and grease remover has dissolved. These illustrations demonstrate the need for various cleaning techniques in the refinish shop.

Vehicles which enter a repair shop should be soap and water washed before any work is started. Following the soap and water wash, the vehicle should be chemically cleaned (with wax and grease remover). Although the best practice is to chemically clean the entire vehicle, the technician should—at a minimum—clean the panels that will be repaired.

FIGURE 27-9 Chemical cleaner did not dissolve the rock salt (left) but soap and water did (right).

FIGURE 27-10 Road tar did not dissolve in soap and water (left) but did with chemical cleaner (right).

During the repair process, the vehicle may become contaminated as the technician performs the necessary work. For example, if workers who are not wearing gloves touch the vehicle, the oils and salt from the workers' hands will be deposited on the vehicle.

TECH TIP Body repairpersons often use either work gloves or latex gloves while working on a vehicle in the repair shop. These gloves primarily protect the worker from chemical contaminants, as they often become contaminated with both water-soluble and petroleum contaminants. A vehicle entering the refinish area should be first soap and water washed, and then should be chemically cleaned to remove any contaminants deposited during the repair.

Hardeners

Although **hardeners** are not truly a solvent, they are included here because—often after using a reducer—a hardener, catalyst, or activator is added to reduce a coating's viscosity.

Paint technicians often use hardeners, catalysts, and activators interchangeably. All three of these products help in the cross-link curing of coatings, but they differ greatly in how they work. If technicians know how each product works, they can better choose between them to meet differing conditions.

Hardener is a product added to polyurethane to start the cross-linking or curing process. It is the second part of a two-part system (2K). A two-part system—which is reduced, then sprayed onto a vehicle—will dry. That is, the solvents will gas out of the film and a dried coating will remain; however, this coating will not be cured and no

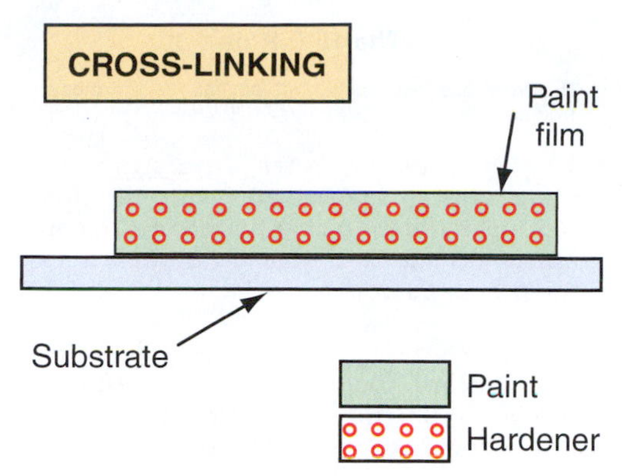

FIGURE 27-11 Urethane paints cure by chemically cross-linking.

cross-linking will have occurred, thus leaving the coating weak and fragile. It can be easily marred, scratched, and even removed by solvents. When the hardener is added, it causes the paint to cross-link. (Cross-linking ties polymer molecules together into a network, forming larger molecules.) See **Figure 27-11**. This cross-linking of polymers binds them together to make a large single sheet of polymer (plastic) over the area being coated.

Although the term *hardener* is used often in coating manufacturers' recommendations, as are the terms *catalyst* and *activator*, hardener is a general term which does not explain how it works. However, activators and catalysts do explain their actions, and both could be in paint's hardener.

An **activator** is a "chemical or other form of energy that causes another substance to become reactive, or that induces a chemical reaction." In other words, it is a chemical that, when added to paint, causes it to react or cross-link.

A **catalyst** is a substance, typically used in small amounts, that modifies and increases the rate of a reaction without being consumed in the process. This product will speed up the cross-linking that was started by the activator.

Suppose a technician applying a coating has three different hardeners to choose from. Hardener A will cause the coating to cure very fast, and is therefore good for small jobs. Hardeners B and C each allow the coating to harden at progressively slower rates, therefore making them better for larger jobs or complete paintwork. Hardeners B and C contain less catalyst.

When catalyst is added to a polymer, it causes cross-linking as explained previously.

TECH TIP **Isocyanates** are the main chemicals used to cross-link polyurethane paints. A large group of chemicals with many substances fall into the isocyanates category. They are added, in differing amounts and types, to paint hardeners, and altering these amounts and types can change the time needed for the completion of cross-linking when the paint is fully cured.

PAINT SCIENCE

The area of paints and coatings is a complete science in the chemistry field. Some colleges devote separate departments to their study. Other colleges have full paint chemistry research areas. See **Figure 27-12**. Paint chemistry involves the way coatings react. Although this is a chapter about solvents, the following paint science explanations will help a technician better understand the application of paint and how paint reacts to its environment when it is applied.

Thermoset

Coatings that are thermoset are hard, cross-linked materials that will not soften with solvent or heat. Thermoset polymer coatings do not stretch. This quality is what gives urethane, polyurethane, and epoxy coatings their highly durable qualities.

It is important to keep in mind that a coating that will eventually become a thermoset coating is not in that thermoset state before it is cured or cross-linked. Prior to cross-linking, the coating is in its thermoplastic state, and can be reflowed by solvent. This property is necessary so that multiple coats can be applied to reach the proper color hiding needed.

While in this thermoplastic state, the solvents in the coating allow it to "flow"—or level—as it is applied to the object being painted.

Thermoplastic

A thermoplastic is defined as a solid that can be reformed or can be reshaped when heated. When in a thermoplastic state, a thermoplastic polymer resin can be reflowed when a solvent is applied to it. This state only occurs when the resin has not yet cross-linked. From the time that an activator is placed in the coating, it starts to cross-link, and there are very specific times when the

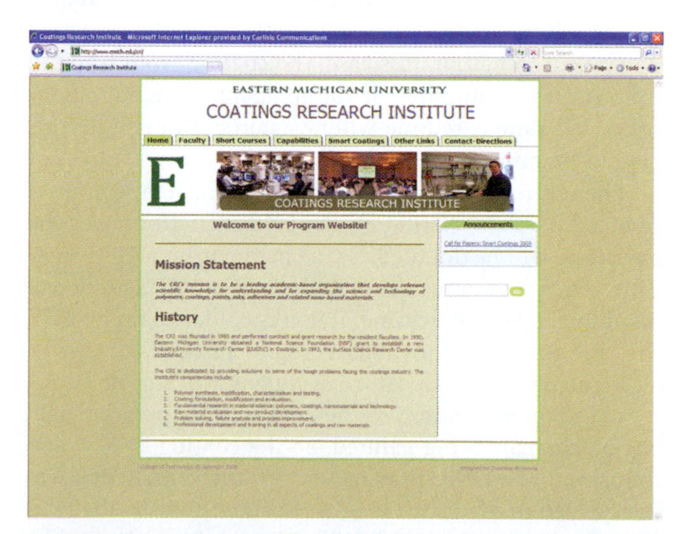

FIGURE 27-12 Some colleges have specified areas or departments devoted to paints and coatings, as shown by the Coatings Research Institute EMU website.

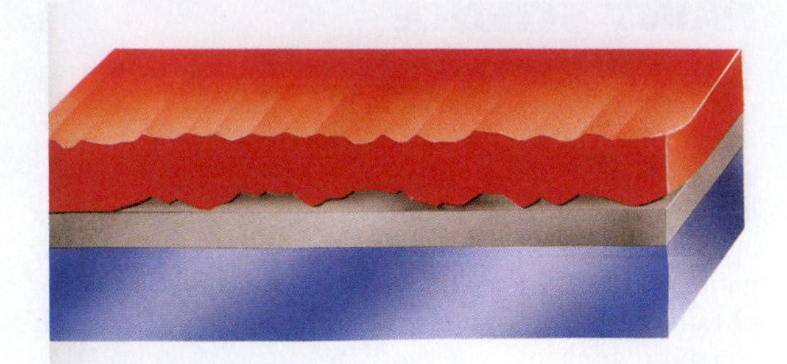

FIGURE 27-13 There are very specific times when an additional coating can be applied. If the times are not followed closely, the coating could lift or wrinkle, causing a defect.

coating can be recoated, or a second or subsequent coat can be applied. If the times are not followed closely, the coating could lift or wrinkle. See **Figure 27-13**. These windows of opportunity for successful coating will be explored more fully later in this chapter.

Elastomers

Elastomers are pliable plastic materials that are good at withstanding deformation. They are added to a coating with an activator and catalyst, which cross-link with the coating. This makes the coating more flexible. These additives are often called flex agents. Some paint manufacturers recommend that they be added to flexible parts, such as bumper fascia, when they are painted so they can withstand the flexing these parts endure.

15° Rule

If a coating recommendation states that the coating in a certain temperature setting will cure at a certain time, raising the temperature by 15°F will generally cut the cure time in half. For example, a certain coating in a 70°F environment will cure in a certain time—for example, 3 hours. If the temperature is raised by 15°F, the cure time will be cut in half. In this case, the coating would cure in 1 hour and 30 minutes. If the temperature is raised another 15°F, the cure time will be again cut in half, in this case to 45 minutes. See **Figure 27-14**.

The reverse of the rule applies when the temperature is reduced by 15°F. That is, in the colder environment, the cure time will be doubled. So, a change from a 70°F environment, with a cure time of 3 hours, to a 55°F setting will result in a 6 hour cure time. If it gets colder again by 15°F, the cure time needed will lengthen to 12 hours.

Remember, though: *All cross-linking will stop at 40°F.*

These temperatures are the coating's temperature as it sits on the part being painted. So if a coating is 70°F and is sprayed in a 70°F spray booth, but the vehicle that was just placed in the booth is only 40°F, the coating that was 70°F when it was sprayed will become colder when it hits the colder vehicle. Until the coating temperature reaches 70°F, it will not respond at the times indicated.

The 15°F Rule

This rule has two parts and applies to **thermoset** products.....

#1. For every 15°F increase in temperature above 70°F, a refinish product's dry time may be reduced by as much as 1/2. The same rule applies to the product's pot life.

#2. For every 15°F decrease in temperature below 70°F, a refinish product's dry time may almost double! The same rule applies to the product's pot life.

NOTE: All product crosslinking and curing in 2K products slows significantly below 60°F/16°C. Thermoset paint will not "set" properly if subjected to cool temperatures during initial curing stages. Such conditions can result in a finish that may eventually **dry** but will exhibit reduced durability, gloss, and repairability. This loss of performance is due to the film **never reaching a fully cured state.**

FIGURE 27-14 The 15° rule states by raising the temperature by that amount, the curing time will generally be cut in half. Used with permission of PPG Industries.

Window Rules

There are three windows of time to be aware of as the coating lies on the vehicle cross-linking. The first is the "window of opportunity," the time when the coating is soft enough (because not much cross-linking has taken place) that subsequent coats can be placed over it. This time period spans the time after the leveling solvent has gassed out, but before much cross-linking has taken place. Because little or no cross-linking has occurred, the next coat of paint with its solvents will not lift or wrinkle the previous coat.

The second window is the "window of danger." This is the time period when the leveling solvent has gassed out (beyond the flash time), and cross-linking has begun, but is not complete. If a subsequent coat is applied during this time, the solvent in the new coat will reflow the part of the previous coat that has not cross-linked, but will not reflow the parts of the coat that have cross-linked. This causes what is called lifting or wrinkling, creating an unacceptable finish.

The third window is the "window of stability." This refers to the time frame when the freshly applied coating has cross-linked enough that it becomes stable and, if needed, can be topcoated for either two-toning or repair.

Percentage of Solids

Although the percentage of solids is not a solvent, this topic does underscore how paint has changed, specifically how it now contains more solids. As stated earlier, paint is made up of pigments, resins, solvents, and additives. The part of the paint remaining after it becomes completely cured is the solids. Solids generally consist of the pigments (color), resin (which binds the pigments to the surface), and some of the additives. The solvents evaporate and are not counted as a solid.

Coatings produced in the 1980s had very low percentages of solids (12 to 19 percent). However, as the environmental laws have changed, the solid content has increased significantly, to 45 to 60 percent. Obviously, coatings with these amounts of solids (and significantly lower concentrations of solvents, as delivered from the manufacturer) would be considerably more viscous (thicker).

These thicker coatings—coupled with laws that restrict the amount of reducer that can be added to a coating before it is applied to a vehicle and the regulation which requires the use of high volume, low pressure (HVLP) guns—has made the job of a paint technician much more complicated.

FIGURE 27-15 Hyper-curing clearcoats are good for small jobs, such as blending the fender, door, and hood of a vehicle.

CONDITIONS THAT AFFECT COATING PERFORMANCE

Many conditions may affect coating performance, including the amount of time that is available to refinish a vehicle, the temperature and humidity of the day when a vehicle is being refinished, the thickness of the coating being applied, and the amount of airflow around the vehicle as it is being refinished. A professional refinisher must first control as many of these variables as possible by having good equipment, which can control the airflow, temperature, and—to a lesser degree—humidity. For the conditions that cannot be controlled, either by the technician or the equipment, the technician should know how the coating will perform by knowing how possible solvents and additives will react to the conditions. This section of the chapter will explore how these conditions affect coatings, and how a paint technician can adapt to conditions through the selection of solvents as well as hardeners.

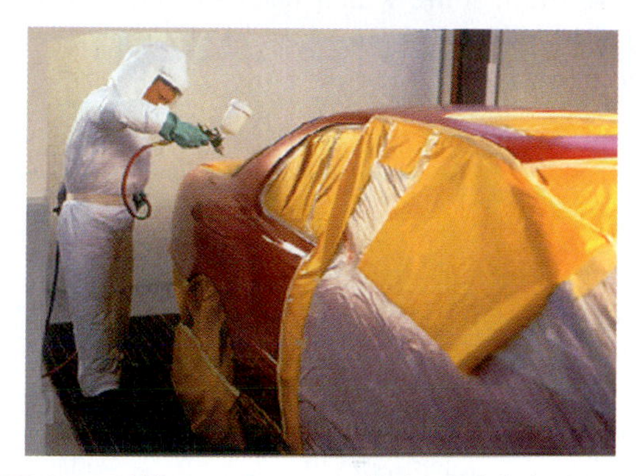

FIGURE 27-16 When a larger refinish job is being completed, slower clearcoats are needed, so that a refinisher can apply a coat of paint completely around the vehicle before it starts to cure.

Time

Time is one of the most valuable commodities in a production shop. The faster a vehicle can be finished, the more vehicles can be done in a day; thus, the more profit that will be realized by the shop. Paint managers are always looking for ways to complete vehicles faster. Seldom does a paint technician have the luxury of painting a vehicle, then letting it sit in the paint booth for 8 to 10 hours to slowly cure. Paint companies have developed coatings, particularly clearcoats, with alternate curing capabilities. Some are formulated to cure in **ambient temperature**. They have also developed others, such as hyper-cure clearcoats that can be forced to cure in as little as 8 minutes at low (120°F) temperatures. Manufacturers continue to develop products that accelerate the speed of curing. For example, **ultraviolet (UV) curing**, a technology that is still in its infant stages, permits a finish to be cured in just 2 minutes.

Although these fast-curing coatings are used often in production shops, there are times when a production painter will choose to use a slower-curing finish. Hyper-curing clearcoats are good for small jobs, such as blending the fender, door, and hood of a vehicle. See **Figure 27-15**. However, on larger applications, where from half to a complete paint job is being done, the hyper-cure coatings may not be appropriate. When a larger refinish job is being completed, slower clearcoats are needed, so that a refinisher can apply a coat of paint completely around the vehicle before it starts to cure. Otherwise, dry spots can become a problem. See **Figure 27-16**.

Carefully choosing the proper curing coating, to match the vehicle being painted and the time allotted, is critical to the shop's productivity. Determining the size of the job and when it fits into the schedule are also important elements of a refinish shop's production and profit.

Temperature

Temperature affects both the drying (evaporating) and curing (cross-linking) of coatings. As the temperature increases, the reaction time of hardeners speeds up. In fact,

TECH TIP As a technician is planning the production for a day, vehicles that may take longer times in a booth may best be completed just before the technician is going to take a break. If a larger paint job is to be completed, it may be best to schedule it close to lunch, when it can be left in a booth for a longer time while the paint team goes to lunch. For even larger paint work that may take a longer time to complete, then cure, it may work best to do it at the end of the day, when the vehicle can stay in the booth until the next workday.

if the recommended temperature is increased by as little as 15°F, the cure time can be decreased by as much as one-half. To help technicians cope with varying temperature conditions, paint companies now offer different hardeners to choose from: some which react slower, and therefore can be used in hotter weather; and faster ones, which can be used in colder weather.

TECH TIP Most paint companies that make differing hardeners market them with numbers that correlate to the temperatures for which they have been developed. So, if a hardener is numbered "3070," for example, the last two numbers indicate that it is suited for painting in temperatures of 70°F, plus or minus 5°F—that is, at 65°F to 75°F.

If a hardener is labeled 3095, its temperature range may be from 90°F to 100°F. To know precisely what the range is for the brand and formulation that a technician is using, the manufacturer's production recommendation sheets should be read.

Read, follow, and understand both the safety precautions and the production recommendations of the manufacturer.

Humidity

Although humidity does not affect the drying and curing time as much as temperature does, it is still a factor, and should be taken into consideration when painting.

When the day's humidity goes up, the amount of vapors that can be held in the air decreases; therefore, solvent does not escape as quickly as it should from the finish that has just been sprayed. If a day has 40 percent humidity, the solvents will evaporate much faster than on a day with 90 percent humidity, when the air around the vehicle being sprayed is already saturated with vapors.

Although most technicians make their choice of solvent solely on the temperature, the humidity should also be taken into consideration. If a day's temperature is 85°F at 90 percent humidity, it may be wise to choose a solvent that evaporates faster than the temperature requires, to compensate for the high humidity of the day. If given a choice of 70°F or 85°F or 95°F solvents, the technician should remember that the warmer the day, the faster the solvent will evaporate, so slower-evaporating solvents are formulated for hot days. A technician who needs to compensate for the humidity should choose a slower-evaporating reducer, or one for a lower temperature. In this case, the best choice would be 70°F reducer.

TECH TIP When paint manufacturers make recommendations for paint curing and flash times, the times indicated are for what are considered normal conditions: 50 percent humidity, 70°F air temperature, and 60 to 100 linear feet per second of airflow. If the conditions under which a vehicle is being painted differ from these, the spray technician must make the appropriate compensations.

Although humidity does affect some hardeners, it does so much less than it affects the evaporation. Spray booths have helped technicians to control the spray environment. In conditions where the temperature and humidity are high, the booth can be equipped with air conditioning, which not only lowers the temperature but also lowers the humidity in a booth.

Thickness

Film thickness is critical to the way a coating will perform as it is being applied, and to how it will hold up to long-term use. If a finish is applied too thickly, it will not expand and contract with the substrate it is sprayed on as the day's temperature changes. The thicker a coating is, the more likely it is to fail by cracking.

Recent studies show that most painters exceed the recommended film thickness guidelines, commonly by as much as 100 percent. This extra thickness slows the evaporation of solvents and extends flash times. When a coating is sprayed on too thickly, solvent popping can occur, even if the recommended flash time has been followed. Most painters do not measure their film thickness, though it is very important to know.

One of the ways to avoid exceeding the recommended film thickness is to prepare a sprayout panel. See **Figure 27-17**. The technique measures how many coats are required for full coverage. The technician can gauge a wet film thickness taken from the sprayout panel. See **Figure 27-18**.

Airflow

As air passes over a freshly painted vehicle, the solvents evaporate. If the rate of airflow is increased, the evapora-

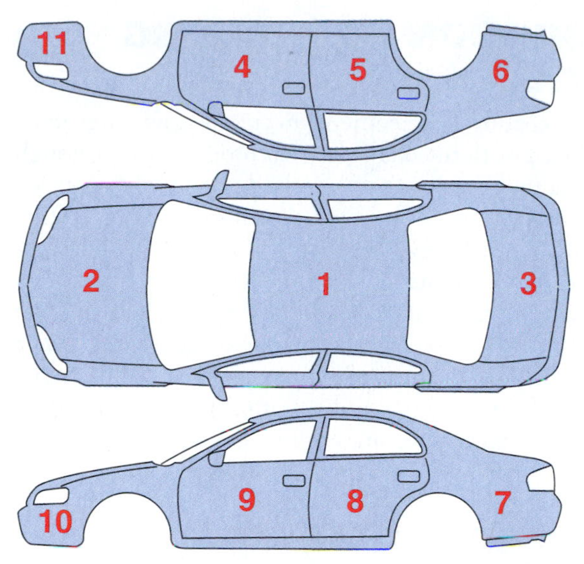

FIGURE 27-17 One way to avoid exceeding the recommended film thickness is to prepare a sprayout panel.

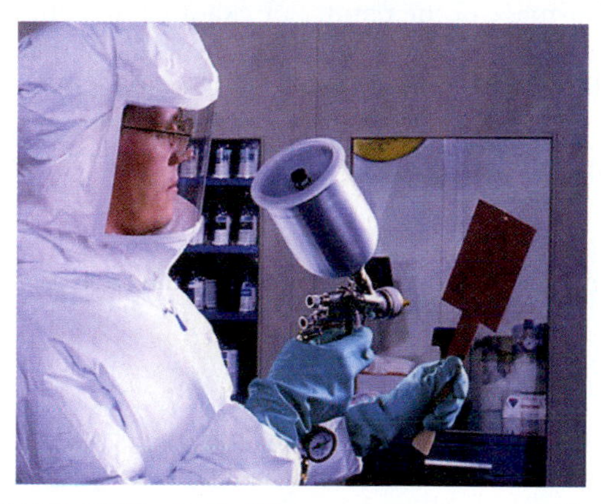

FIGURE 27-18 The technician can gauge a wet film thickness taken from the sprayout panel.

TECH TIP When a technician takes a wet film thickness reading, small marks remain in the paint film. Therefore, it is impractical to do this to a newly refinished vehicle. See **Figure 27-19**. However, when the reading is taken on a sprayout panel, the panel tells a technician how many coats are required for full coverage, and if the color is a blendable match. In addition, a wet film thickness test assures that the technician's application is not exceeding the recommended film thickness.

tion will also increase. In an area with low airflow, such as a side-draft booth or a semi-downdraft booth, no solvent compensation needs to be taken. However, if a vehicle is in a high airflow booth—such as a downdraft booth where the airflow typically exceeds 60 to 100 linear feet per minute—the paint technician may need to compensate for

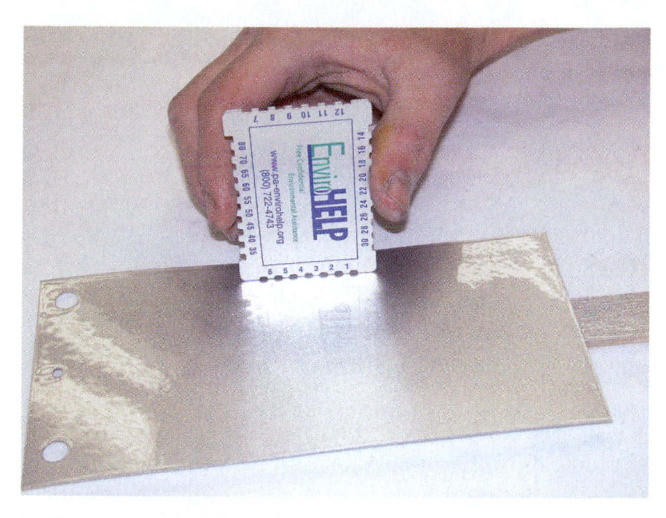

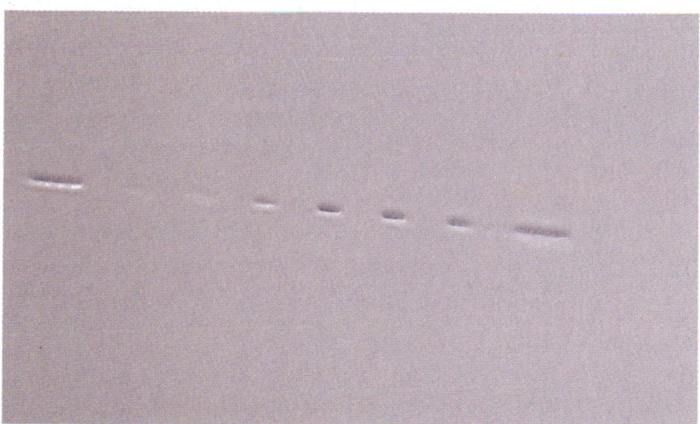

FIGURE 27-19 When a technician takes a wet film thickness reading, small marks remain in the paint film, which is why these readings should be done on a sprayout panel.

the higher airflow and subsequent faster evaporation. Many paint companies recommend that the next higher temperature solvent be used to compensate for the higher airflow.

TECH TIP Some paint manufacturers market reducers and thinners under the descriptions slow, medium, and fast. These names may be confusing to some technicians, because the fast solvents are used in the colder conditions where volatility is slowed down and the slow solvent is used in warmer conditions where the evaporation is increased by the heat.

SPECIALTY PRODUCTS

Specialty products are used to accomplish specific tasks. Most of them contain solvents. Color blenders or blending solvents are coatings containing a large amount of solvent, to help blend subsequent coats of color together. Other specialty products include accelerator and retarder. These products are either applied to the vehicle before painting it, or they are added to the paint to accomplish a specific task.

Color Blenders

Color blenders, or blending solvents, are substances used to "flow out" either color or clearcoat in areas where the new paint and **OEM (original equipment manufacturer)** finish meet. See **Figure 27-20**. The blending solvent is sprayed on the blend area to allow all the layers to flow together.

Accelerator

Accelerators are additives used in paint to speed up the evaporation process. Although products like this have been around for a long time, they are seldom used and seldom recommended in production paint application. Some custom paint applications use accelerators to speed up the process when multiple layers of finish need to be applied.

Retarder

Retarders are additives that slow down the evaporation and curing process of coatings. This type of product has also been around for many years, but is seldom recommended by the manufacturer for production collision repair work. When custom paint applications require "marbleizing" coatings, retarders are added to extend the evaporation, allowing the technician more time to perform all the steps in the process.

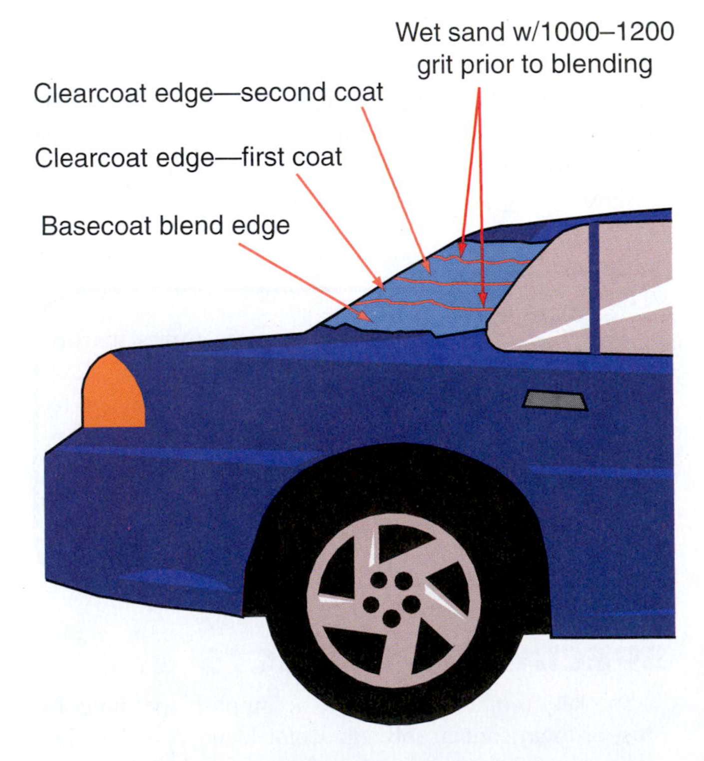

FIGURE 27-20 Color blenders, or blending solvents, are substances used to "flow out" either color or clearcoat in areas where the new paint and OEM finish meet.

REDUCTION AND MIXING RATIOS

Paint coatings that come from the manufacturer must be reduced or thinned so they will properly pass through a spray gun, resulting in the desired quality of finish. See **Figure 27-21**. Some paint manufacturers recommend that the paint be reduced by percentage; others will recommend that parts reduce the paint; and still others may recommend that the coating be reduced by weight. Although all of these methods are acceptable, as coatings become more sophisticated, the reduction process becomes more critical as well. Stirring and agitation also become more critical as paints become more sophisticated and as more special additives are used.

Percentage Reduction

Reduction by percentages is a method of reducing where the amount of paint to be reduced is measured; then the amount of reducer that is placed in the mixture is a percentage of the paint. For example, if paint is to be reduced at a 100 percent reduction and the amount of paint to be reduced is 500 ml, then 500 ml of reducer will be added. If that same amount of paint is to be reduced by 50 percent, then only 250 ml of reducer (or if reduced by 200 percent, then 1,000 ml) should be added.

Although manufacturers rarely call for percentage reduction now, a paint technician should be able to use this method.

Parts Reduction

Reduction by parts is a common recommendation when two or more additives are called for. If paint requires

FIGURE 27-21 A paint graduated mixing cup can be used to get the desired ratio of paint and thinner.

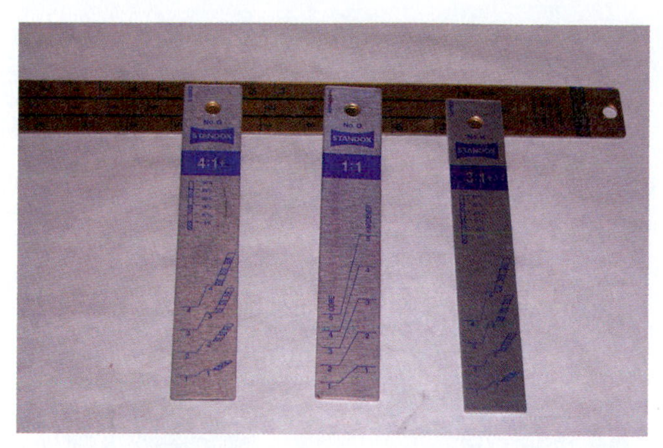

FIGURE 27-22 Paint manufacturers often provide part reduction sticks, which can be used in any size container with parallel sides.

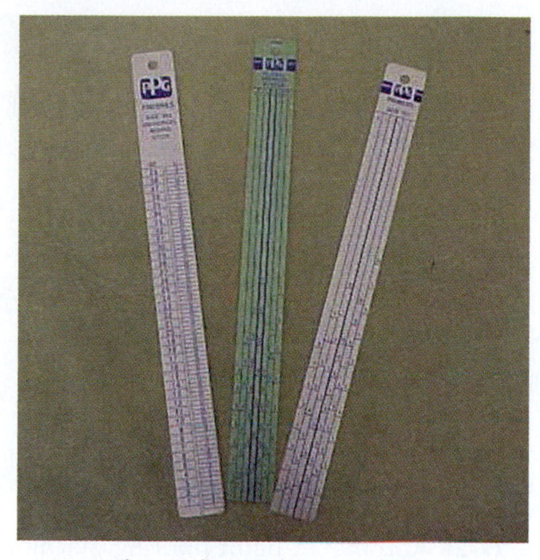

FIGURE 27-23 The stick may be specifically designed for a product with the proper ratio on it, like this non-coating-specific one.

both a reducer and a hardener, the reduction requirement may be something like a 2:2:1 reduction. That means two parts of paint has two parts of reducer added, then 1 part of hardener, followed by stirring or agitation to accomplish a proper reduction for application. When the technician is reducing using the parts technique, it does not matter what size the part is, as long as all other parts are the same. If the reduction is 2:2:1 and the painter adds 2 gallons of paint, then the two parts of reducer must be 2 gallons as well, and the 1 part of hardener must also be measured in gallons.

Paint manufacturers often provide part reduction sticks, which can be used in any size container with parallel sides. See **Figure 27-22**. The stick may be a generic one, may be divided into equal parts, or may be specifically designed for a product with the proper ratio on it. See **Figure 27-23**.

Mixing cups are available with reductions printed on their face, to eliminate the need for a stick. See **Figure 27-24**. Disposable lining paint cups are also available with graduated measurements on them for parts reduction, which can also speed up the paint process. See **Figure 27-25**.

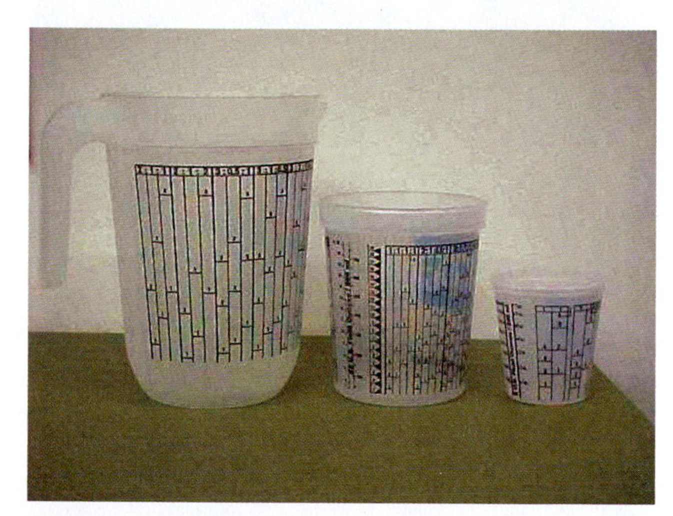

FIGURE 27-24 Mixing cups are available with reductions printed on their face, to eliminate the need for a stick.

FIGURE 27-25 Disposable lining paint cups are also available with graduated measurement on them for parts reduction, which can also speed up the paint process.

TECH TIP When measuring for reductions, no matter which method is used, the technician should calculate parts, percentages, or weight as accurately as possible. Even slight over- or under-measurements can cause problems.

Weight Reduction

Reduction by weight is the most accurate method of measurement, and often is the method recommended by the paint manufacturer. In this method, reduction is done on the same scale by which the paint formula is

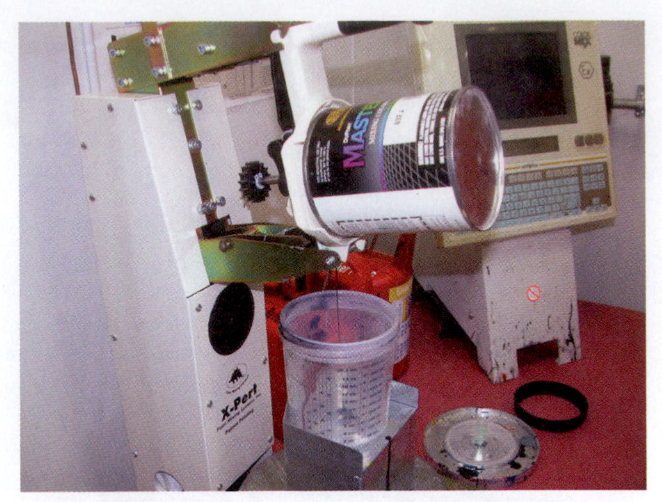

FIGURE 27-26 Automated paint pouring machines can provide greater accuracy in measuring and reducing waste.

FIGURE 27-27 Paint sticks are available as wooden or metal.

measured. Most paint manufacturers now have the paint formula and the reduction formula together, so that the paint technicians who are mixing the color formula can continue to reduce the paint for application on the scale. With the computerization of this process, the accuracy has improved greatly, and waste has been reduced as well. See **Figure 27-26**.

FIGURE 27-28 A paint shaker makes sure all components are thoroughly mixed.

Stirring or Agitating

When the reducer, hardener, and other additives are combined, they must be thoroughly stirred or agitated in a paint shaker. This can be accomplished by stirring it with a paint stick, or by placing the combined paint in a paint shaker for a short time to make sure that the components are fully combined. See **Figure 27-27** and **Figure 27-28**. This stirring or agitation is critical to the overall paint quality, and care should be taken to assure the products are fully combined.

SOLVENTS AND THE ENVIRONMENT

Most solvents are considered volatile organic compounds (VOCs), as mentioned earlier. Their volatility or evaporation qualities are critical to the painting process, but this same evaporative quality is harmful to the environment. See **Figure 27-29**. When VOCs are released into the atmosphere, especially on warm days, they contribute to the formation of lower level ozone gases. These gases are a major factor in the formation of smog, and are very harmful to persons with respiratory diseases such as asthma and emphysema. Paint technicians should be very aware of the amount of VOCs that they release into the atmosphere, as

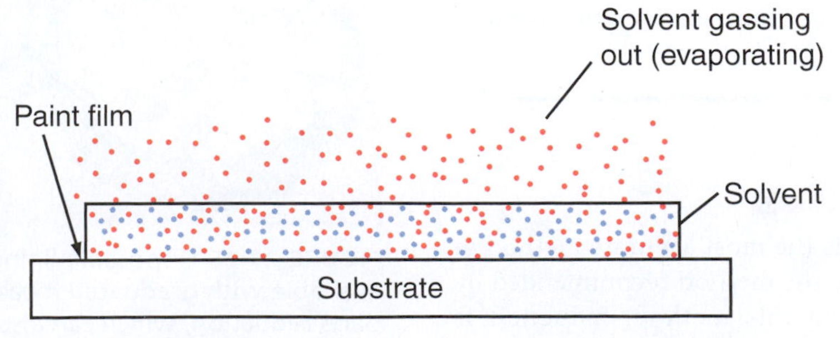

FIGURE 27-29 Solvents evaporate at different rates.

well as other **harmful air pollutants (HAPs)** which are released during collision repair and refinishing. The topics of reducing air pollutants during spray painting will be discussed in Chapter 28.

SOLVENT RECYCLING

Many solvents are used in the collision repair industry. They are used for reduction of paints as well as for many cleaning operations. In fact, 20 percent of all solvents are used in the cleaning of spray guns. To reduce the amount of toxic air pollutants produced and to reduce the amount of toxic waste from the gun cleaning process, paint recyclers have been developed. Gun cleaners and solvent recyclers have been produced specifically for this purpose. See **Figure 27-30** and **Figure 27-31**.

Solvent recyclers are stills, which heat the waste to a point that the solvents boil and become vapor. Because the solids in the waste boil at a higher point than solvent, the solvent vapors can be collected, cooled, and reused, thus separating the solids from the solvent. This process recycles 80 to 90 percent of the solvents while reducing the volume of waste by 75 to 90 percent, thus significantly reducing the cost of toxic waste disposal. See **Figure 27-32**.

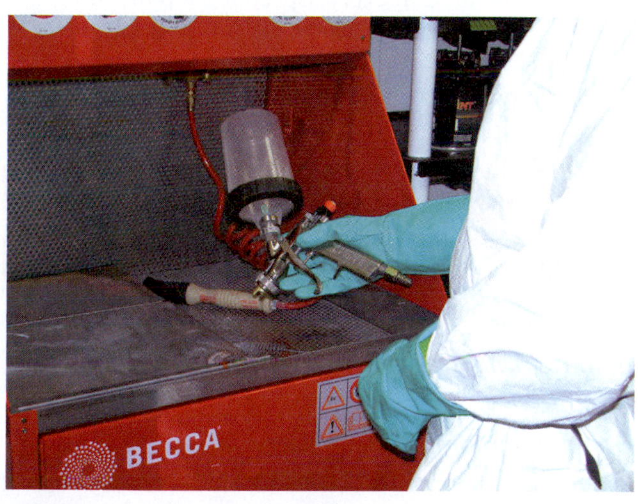

FIGURE 27-30 An automated gun cleaner helps reduce the amount of toxic waste produced in the gun cleaning process.

FIGURE 27-31 An automated solvent recycler is specifically designed to reduce the amount of toxic air pollutants produced.

FIGURE 27-32 This figure shows solvent before recycling and the remains after solvent is taken off. The residue (still bottom) has been reduced by up to 90 percent.

Summary

- When using solvents, safety is of the utmost importance.
- A solvent dissolves and carries another chemical.
- Reducers and thinners reduce the viscosity of finishes.
- A thermoset coating is a hard, cross-linked material that won't soften with heat, whereas a thermoplastic can be reshaped when heated.
- The 15° rule means, in certain conditions, raising the temperature by 15°F will cut the curing time in half.
- Conditions that affect coating performance include time, temperature, humidity, thickness, and airflow.
- Specialty products include color blenders, accelerators, and retarders.
- Popular reduction methods include reduction by percentages, reduction by parts, and reduction by weight.
- By choosing the proper solvent, a painter can control the outcome, even when spraying conditions are not ideal.
- Solvents are used in every aspect of the refinish industry—the better a technician understands them, the easier choices become and the more consistent the finished product can be.

Key Terms

activator
adhesion promoter
ambient temperature
carbon
catalyst
compound
cross-linking
curing
drying
elastomers
flow out
gas out
hardener
harmful air pollutants (HAPs)
hydrocarbon
isocyanates
material safety data sheet (MSDS)
organic
original equipment manufacturer (OEM)
polymers
reducer
reduction by parts
reduction by percentages
reduction by weight
thinner
ultraviolet (UV) curing
viscosity cup
volatile
volatile organic compounds (VOCs)

Review

ASE Review Questions

1. Technician A says that the personal safety equipment used when handling solvents is the same as when using engine coolant. Technician B says that to find the personal safety equipment needed to handle solvents, a technician should read, understand, and follow the MSDS instructions. Who is correct?
   A. Technician A only
   B. Technician B only
   C. Both Technicians A and B
   D. Neither Technician A nor B
2. Technician A says that the first cleaning done to a vehicle before working on it is a chemical cleaning with a wax and grease remover. Technician B says that a solvent is a general word for many substances

that dissolve and carry another chemical. Who is correct?
A. Technician A only
B. Technician B only
C. Both Technicians A and B
D. Neither Technician A nor B

3. Technician A says that the term *volatile*, when talking about VOCs, refers to chemicals which evaporate easily. Technician B says that organic, when talking about VOCs, refers to chemicals which contain carbon. Who is correct?
A. Technician A only
B. Technician B only
C. Both Technicians A and B
D. Neither Technician A nor B

4. Technician A says that many of the products that paint technicians use contain solvents. Technician B says that paints coming from the manufacturer do not have solvents in them, so a paint technician must add them so the coating can properly pass through the spray gun. Who is correct?
A. Technician A only
B. Technician B only
C. Both Technicians A and B
D. Neither Technician A nor B

5. Technician A says that lacquer paints are thinned. Technician B says that enamel and urethane paints are reduced. Who is correct?
A. Technician A only
B. Technician B only
C. Both Technicians A and B
D. Neither Technician A nor B

6. Technician A says that leveling solvents evaporate from paint last, sometimes as long as 90 days after spraying. Technician B says that the first solvents to evaporate are leveling solvents. Who is correct?
A. Technician A only
B. Technician B only
C. Both Technicians A and B
D. Neither Technician A nor B

7. Technician A says that the surface tension of cured paint makes it scratch-resistant. Technician B says that surface tension only exists on liquids. Who is correct?
A. Technician A only
B. Technician B only
C. Both Technicians A and B
D. Neither Technician A nor B

8. Technician A says that before blending new paint onto OEM paint, an adhesion promoter should be applied so the old finish will soften and let the new finish blend into it. Technician B says that an adhesion promoter is used to help with the adhesion of a coating on polyolefin plastics. Who is correct?
A. Technician A only
B. Technician B only
C. Both Technicians A and B
D. Neither Technician A nor B

9. Technician A says that wax and grease remover is a chemical cleaner. Technician B says that wax and grease remover is a solvent. Who is correct?
A. Technician A only
B. Technician B only
C. Both Technicians A and B
D. Neither Technician A nor B

10. Technician A says that lacquer paint is a two-part paint, and is sometimes called 2K paint. Technician B says that isocyanates are chemicals used to reduce paints. Who is correct?
A. Technician A only
B. Technician B only
C. Both Technicians A and B
D. Neither Technician A nor B

11. Technician A says that when paint reaches its cross-linked and thermoset stage, it will no longer reflow when solvents are applied to it. Technician B says that a thermoplastic paint will soften when solvents are applied to it. Who is correct?
A. Technician A only
B. Technician B only
C. Both Technicians A and B
D. Neither Technician A nor B

12. Technician A says that the four components found in paint shipped from the manufacturer are resins, pigments, solvents, and additives. Technician B says that the way to measure the thickness of a coating is with a viscosity cup. Who is correct?
A. Technician A only
B. Technician B only
C. Both Technicians A and B
D. Neither Technician A nor B

13. Technician A says that when referring to a spray gun, HVLP stands for high viscosity, low pressure. Technician B says that temperature, humidity, and airflow will affect the way a coating dries and cures. Who is correct?
A. Technician A only
B. Technician B only
C. Both Technicians A and B
D. Neither Technician A nor B

14. Technician A says that the most accurate way to reduce paint is in the paint mixing scale, by weight. Technician B says that if 1 quart of paint is to be reduced by 150 percent, then 1 quart and 1 pint of reducer must be added to it. Who is correct?
A. Technician A only
B. Technician B only
C. Both Technicians A and B
D. Neither Technician A nor B

15. Technician A says that although most technicians stir paint, the step is not that important and could be skipped. Technician B says that once paint and solvents are mixed together, they cannot be separated again. Who is correct?
    A. Technician A only
    B. Technician B only
    C. Both Technicians A and B
    D. Neither Technician A nor B

Essay Questions

1. Explain the three times at which solvents evaporate when spraying, and why.
2. Describe why a paint technician must wash a vehicle with both soap and water and with wax and grease remover.
3. In a paragraph, explain how temperature affects the curing of paint.
4. Describe what thermoset is, in regard to a polyurethane paint when it is curing.
5. Explain how humidity affects solvent evaporation.

Topic Related Math Questions

1. A painter must reduce 1,525 ml of paint at a 125 percent ratio. How much reducer should be added?
2. If a paint department, working 5 days a week, uses 1/2 pint of wax and grease remover a day, how many gallons of wax and grease remover would that paint department need each year?
3. A paint department's lead painter has derived a method of production that will reduce the use of solvent by 25 percent. That department now purchases 64.5 gallons of solvent at $12.34 per gallon. How much will the new method save the company per year?
4. A vehicle is sprayed with a clearcoat, which normally can be sanded and polished after curing for 5 hours at 70°F. The temperature on the day that the vehicle is painted is 85°F. How long will it be before that vehicle can be sanded and polished, using the 15° rule?
5. A painter is mixing 25 gallons of paint, which is reduced at a 4:1:1 ratio. How many quarts of reducer must be added to the paint?

Critical Thinking Questions

1. A vehicle is being painted in an environment which is 90°F, 90 percent humidity, in a downdraft booth which has 130 linear feet per minute of airflow. The paint technician has three temperature reducers to choose from (70°F, 85°F, and 98°F). Which reducer should she use, and why?
2. On a hot and humid day, a paint technician is making a test panel for a silver vehicle. All is going well, except that when the clearcoat is being sprayed, it has an unacceptable amount of orange peel. What could the technician do to correct this problem?

Lab Activities

1. Have students reduce 1 quart of paint using 150 percent reduction; then agitate, strain, and place in a spray gun prepared for spraying, explaining how to properly do each step.
2. Have students properly clean an incoming vehicle to remove both dirt and road tar, explaining how to perform each step.
3. Have students properly clean and prepare a polyolefin bumper for topcoating, explaining each step.
4. Have students make a sprayout panel for a basecoat/clearcoat vehicle to determine the needed number of coats to achieve full coverage.
5. Using a wet film thickness gauge, have students determine the wet film thickness of the paint sprayed on a test panel.

Chapter 28

Application of Undercoats

OBJECTIVES

Upon completing this chapter, you should be able to:

- Become familiar with the different types of undercoatings and their usage.
- Understand the safety precautions needed when applying specific undercoats.
- Perform the different application techniques needed for undercoatings, and determine their advantages and disadvantages.
- Learn which techniques will speed up the production of the repair, and know when to use specialty undercoats such as aerosol spray primers.
- Learn how to use the tools needed for application of undercoats, and how to properly clean those tools.
- Identify the specific surface preparation necessary for each undercoat before the topcoat can be applied.
- Note when to choose the different undercoats.

INTRODUCTION

Although undercoats and their usage have been covered in other chapters of this book, this chapter will deal with the classification, the specific usages, the application and curing techniques, and cautions to follow when using undercoatings.

There are advanced ways of applying undercoats. Although the methods may seem simple at first glance, it does take a good amount of instruction and practice to master these application techniques.

With the introduction of undercoats for specific applications, and the confusion that some technicians have when it comes to undercoatings, the topic warrants a separate chapter. This chapter will define each undercoat as noted in the U.S. national VOC regulation of September 11, 1998, including:

- Primer, pretreatment, and wash primers
- Primer surfacers
- Primer sealers
- Sealers
- Specialty coatings

It will also explain what each undercoat does, what precautions should be taken when using them, and various application techniques and their advantages and disadvantages.

This chapter will also outline techniques used in collision repair that increase productivity while maintaining a high standard of quality and durability. Finally, the chapter will examine new products such as UV curing undercoats and water-base undercoats.

SAFETY

In the refinish department, safety refers to personal safety, the proper use of personal protective devices, and environmental safety when working with chemicals. When technicians work with refinish materials, the chemicals that they are made from pose personal and environmental protection concerns. The refinish technician is governed by—and must comply with—many local, state, and federal laws concerning protection.

Each product must be accompanied with a material safety data sheet (MSDS). An MSDS includes sections on:

- Hazardous ingredients
- Physical data
- Fire and explosion hazards
- Reactive data
- Health hazard data or toxic properties
- Preventive measures
- First aid measures
- MSDS preparation information

Technicians must read, understand, and follow the recommendations in the MSDS for the protection of themselves, their fellow workers, and the environment.

In addition to these cautions, other safety precautions must be taken when maintaining equipment. When cleaning areas that may contain paint residue, the technician should use the same personal protective devices as outlined in the original MSDS.

Technicians should use care and caution measures for personal safety when working on electrical equipment, such as the "lock out, tag out" procedure. (In this precautionary measure, the circuit breakers for the equipment are disconnected and the breaker is locked, so that it cannot be re-connected while someone is working on it, and the electrical box is tagged to let others know that it is being worked on and should not be reconnected.) Each employee should be trained how to safely perform equipment maintenance procedures. If technicians are not trained in the proper safety measures, they should not perform the maintenance.

All safety guards for each piece of equipment should be properly installed and in good working order. Removing or disabling safety guards should not be tolerated.

Technicians should read, understand, and follow all safety directions provided by equipment manufacturers.

TOOLS AND EQUIPMENT

When applying undercoatings, technicians must use personal protective equipment, including:

- Safety glasses
- Particle mask (used during sanding process)
- Charcoal respirator (used when spraying non-isocyanate-laden materials)
- Supply air respirator (used when spraying isocyanate materials)
- Solvent resistant gloves
- Spray suit
- Steel-toed boots

Other tools that will be needed are:

- Spray guns—such as gravity-fed HVLP primer guns and mini primer guns—for the spray application of primer. See **Figure 28-1**.
- Roller handle, cup, and roller cover for the application of rolled primer. See **Figure 28-2**.

FIGURE 28-1 HVLP gravity guns come in different sizes such as standard (left) for typical applications and smaller detail (right) guns for smaller applications.

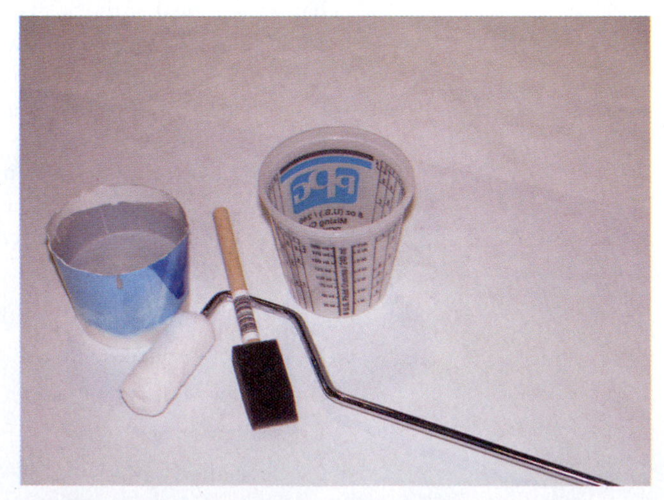

FIGURE 28-2 Undercoats are sometimes applied by rolling and foam brush application.

- Brushes, both bristle and foam applicator. See Figure 28-3.
- Masking material, guide coat, film thickness gauge, wax and grease remover, sanding blocks (soft and hard), and clean paper towels. See Figure 28-4.

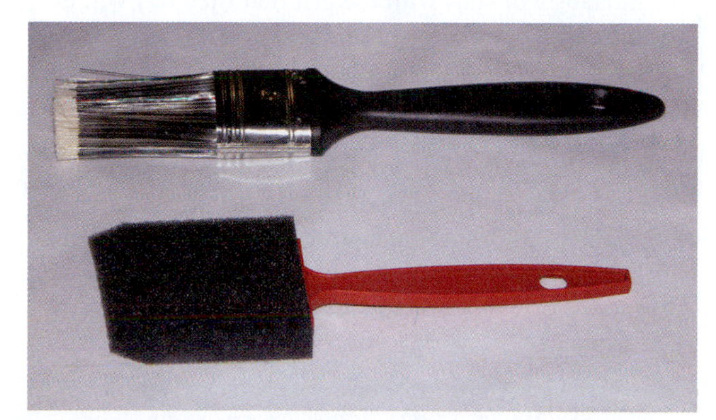

FIGURE 28-3 Undercoating can be applied by brush through either bristle or foam application.

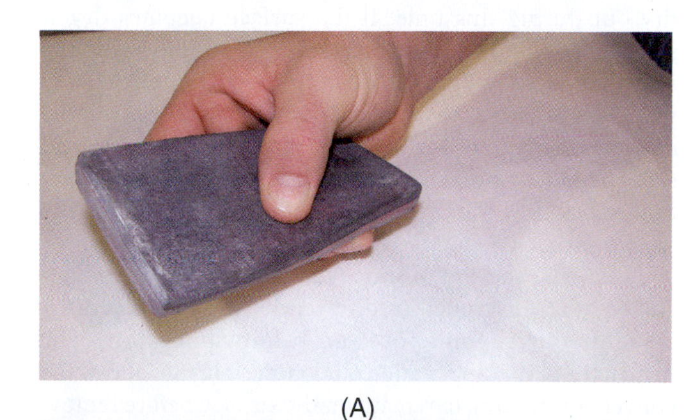

(A)

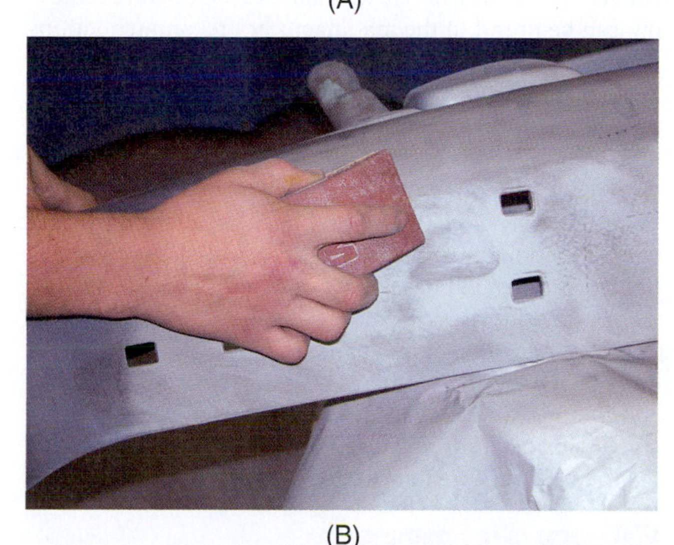

(B)

FIGURE 28-4 A soft hand-sanding block is often used on uneven surfaces (A). A hard hand-sanding block is often used on flat surfaces (B).

UNDERCOATS

Although undercoats were used for years before the national volatile organic compound (VOC) regulation of 1998, the materials were not clearly defined. Generally, technicians referred to all undercoats as "primer," and before the 1970s the undercoat commonly used was a one-part lacquer primer surfacer.

Two-part coating (2K) was introduced in the early 1960s, but primers did not become two-part (2K) until later in the 1960s, and were not generally used until the 1970s.

Sealers are an undercoating that is used to "seal," or create a barrier, between the old finish and repairs that may be under the new finish. This barrier coat was especially needed when lacquer finishes were used and the solvents in the new finish could reflow the old finish and cause defects to show through. A sealer was placed on the repaired area to form a barrier, stopping the solvent from passing. Sealers became popular in the 1970s, although most spray technicians then referred to them as primers. Undercoatings can be listed into categories such as pretreatment or as primers, primer surfacers, sealers and primer sealers, and specialty coatings. These categories cover all the undercoats, including such things as rock guard or chip resistant coatings.

Although we now have the categories developed by the national volatile organic compound (VOC) regulation, products still change continually and product definitions become blurred. For example, products such as multipurpose undercoatings, depending on how they are mixed, can be used either as a sealer or a primer surfacer. When new coatings are developed, and old products are improved, the technician needs to be aware of the changes. One product may do different tasks, depending on how it is reduced or what type of catalyst is added to it. Single products can serve as a sealer when reduced in a certain manner. The same product, if reduced differently, can be used as a primer surfacer. The one certain fact about coatings, especially undercoatings, is that they will continue to change.

Because some technicians use the word *primer* as the single word to describe undercoatings, this section will start with definitions of undercoat products and a description of what each is intended to do.

The common categories of undercoatings are:

- Metal cleaner/conversion coatings
- Epoxy primers
- Self-etching primers
- Wash primers
- Primer
- Primer surfacers
- Primer sealers
- Adhesion-promoting primers

Metal Cleaners and Conversion Coatings

Under the national VOC regulations, **metal cleaners** and **conversion coatings** must have 6.5 pounds of VOC per gallon of coatings or less.

TECH TIP Epoxy primers, metal cleaners, conversion coatings, and self-etching primers are listed in the national rule as "pretreatment/wash primers," and are defined as a "coating applied to bare metal to deactivate the metal surface for corrosion resistance."

This metal-treating undercoating is used to help replace the factory corrosion protection, which may have been removed with repairs. It consists of two parts, a metal treatment followed by a conversion coating. This corrosion protective undercoating is applied in two steps. The first step is application of the metal treatment. Manufacturer's recommendations for dilution should be followed. It is applied to bare metal that has been properly cleaned and sanded as recommended by the product's manufacturer. Each **substrate** or surface that is being sanded will have recommended sand grit provided by the paint manufacturer. By looking at the paint coatings' recommendations, the technician will know what grit to finish the substrate to, and how much time can elapse before the undercoating must be applied. See **Figure 28-5**.

The diluted metal conditioner is normally applied with a spray bottle and worked into the prepared metal with an abrasive pad. The conditioner is then allowed to dwell on the surface for the recommended amount of time. Most metal conditioner directions recommend that they do not dry out during this time. If the surface becomes dry, it should be re-wet with the metal treatment. Following the recommended working time, the metal conditioner is washed off with water and dried in preparation for the conversion coating. Metal treatment is a solution of acid that chemically prepares the substrate for the conditioner that will follow. The technician should always wear the proper protective devices while using these products.

The following tables are a guideline for preparing the various substrates appearing on the previous page. The processes listed explain basic and necessary steps to prepare the substrate prior to applying undercoats.

Steel
When steel has been thoroughly cleaned it is **very prone to corrosion/rust** and should be primed quickly to prevent oxidation.

Type	Process
Rusty Steel	Clean surface thoroughly with Wax and Grease Remover DA sand surface using P80 - 120 grit discs Clean off dust, reclean with Wax and Grease Remover Apply appropriate Primer within one hour
Sanded Steel	Clean surface thoroughly with Wax and Grease Remover Apply appropriate Primer within one hour
Shot Blasted Steel	DA sand surface using P80 - 120 grit discs Clean off dust, reclean with Wax and Grease Remover Apply appropriate Primer within one hour
Stainless Steel	Clean surface thoroughly with Wax and Grease Remover DA sand surface using P120 - 240 grit discs Clean off dust, reclean with Wax and Grease Remover Apply appropriate Primer within one hour
Chromed Steel	Clean surface thoroughly with Wax and Grease Remover DA sand surface using P120 - 240 grit discs Clean off dust, reclean with Wax and Grease Remover Apply appropriate Primer within one hour
Galvanized/Zinc Coated Steel	Clean surface by applying Wax and Grease Remover with a fine Scotchbrite pad Reclean with Wax and Grease Remover Apply appropriate Primer within one hour

FIGURE 28-5 Different types of surfaces require different surface preparation. Paint manufacturers often publish recommendations for surface preparation.

TECH TIP When manufacturer information recommends that the technician sand a surface with P400-grit sandpaper, it does not mean that technicians will start with that grit paper, but instead that they will end with that grit paper. When sanding certain surfaces, technicians will start with the grit that they feel will do the job quickly, and then progress to the next higher grit as the job nears completion, finishing with the recommended grit. Technicians should not jump more than 100 grit numbers when progressing to the finished product. An example might be that a technician starting with P80-grit sandpaper would progress to P180, then P220, then P320, and finally to P400. There are six important points (six "P's") to remember regarding surface preparation: "Proper Preparation Prevents Poor Paint Performance."

Following the use of the metal conditioner, a conversion coating is applied as directed by the manufacturer. Some products require dilution before application. The instructions for application and how long to allow the conversion coating to remain on the surface before removing can be found in the manufacturer's recommendations and should be strictly followed.

Metal cleaner/conversion coating is an excellent way to protect new metal patch panels, or the larger exposed metal portions that restoration sometimes requires. Although it is not often used in collision repair, it should be used whenever a technician is concerned about corrosion contamination during repairs. This two-step process is only recommended on metal surfaces that are exposed to the air, such as quarter panels or other outer body parts. Interior parts, or enclosed parts such as inside frame rails, do not have the airflow that is necessary when using metal conditioners and conversion coatings. Metal cleaners and conversion coating must be covered with an additional undercoating before topcoating.

Epoxy Primers

Epoxy primers are primers with excellent adhesion to properly cleaned and treated bare metal. In tests they have proven to possess excellent corrosion protection qualities. Manufacturers often recommended that they be applied to bare metal before the application of plastic body filler, to an adhesion for the body filler to the metal, and to provide corrosion protection as well.

Epoxy primers are two-part primers with a coating and a hardener. Depending on the manufacturer, epoxy primers may require an "induction period" following reduction. An **induction period** is the amount of time that the reduced and hardened primer must set prior to application on the vehicle. This time period allows the paint coating and hardener to react to each other, thus providing the product's full potential for both adhesion and corrosion protection.

Epoxy primers can be applied onto exterior, interior, and even enclosed areas such as inside frame rails. Although they may cure slowly when inside a frame rail, they will perform as well as if they were applied on an exterior surface.

Epoxy primers are not intended to fill any irregularities or imperfections, and though they are normally applied with a spray gun, they will also perform satisfactorily if applied with a roller or foam brush. Areas of bare metal exposed during the repair process can be primed with epoxy primer before being coated with primer surfacer; or unexposed areas can be brushed with epoxy primer, and then have a petroleum- or wax-based corrosion protection applied over it.

Self-Etching Primer

Self-etching primers come under the national VOC regulations, and are allowed to contain a maximum of 6.5 pounds of VOCs per gallon of coating. They are used instead of metal cleaner and conversion coatings. Self-etching primers are generally used on bare metals that are smaller in size and are less likely to develop corrosion. They also contain an acid, which etches into the properly prepared metal surface, but the acid that is contained in them will stop by itself if the proper airflow is available.

When applying self-etching undercoatings, the technician must understand that—to work properly—the product must be applied to the recommended amount of **film thickness**. However, in the case of self-etching primers, it should also not be applied too thickly. See **Figure 28-6**. A recommendation of 0.2 to 0.4 mils dry film build is not uncommon; in fact, some products will have a maximum film build recommendation. If the film build is too great, the acid in it will not properly "**gas out**," or evaporate. Self-etching primer is not recommended for application on interior surfaces, nor on enclosed areas. See **Figure 28-7**. Acid etch undercoatings must have active airflow to neutralize the acid in them.

Acid etch primers cannot be directly topcoated. A primer, primer sealer, or other suitable undercoat must be applied over self-etching primer. See **Figure 28-8**.

Self-etching primers do not need to be sanded before priming, unless they have exceeded the recommended flash time. Some self-etch primers that are not covered within a prescribed amount of time will need to be scuffed before applying anything over them. The technician should read, understand, and follow the recommendations given by the manufacturer.

Wash Primers

Wash primers are very similar to acid etch primers in that they contain an acid that etches into the metal surface they are applied to and they must be thinly applied. What is different is that they must be covered with a primer or sealer before topcoating with color. Their main purpose is to provide corrosion protection to exposed substrate.

TECH TIP Because they both contain an acid, acid etch and wash primers should not be sprayed using a metal paint gun cup. The acid contact with the cup will pit the metal; thus, a plastic or lined cup should be used. See **Figure 28-9**.

Primer Surfacers

Primer surfacers are undercoats that provide adhesion to the surface they are sprayed on and adhesion to that which is sprayed over them. They also provide light filling qualities, which means that when applied to an area they will provide a film build that can be sanded to remove small imperfections. They are allowed by the national VOC regulations to contain no more than 4.8 pounds of VOCs per gallon. Primer surfacers are sometimes referred to as primer fillers, and generally have a recommendation of applying two to four coats with a dry film build of 2 to 2.25 mils per coat. This means a film build can be as much as 8 to 10 mils of fill before sanding. However, that is not to say that primer surfacer should be sprayed and left that thick. In fact, most paint makers recommend a 2-mil thickness film build after the surface has been block sanded. As seen in **Figure 28-10** and **Figure 28-11**, if an imperfection of 3 mils has 5 mils of primer surfacer applied over it, then 2 mils are sanded off where the imperfection will no longer be seen. In this way, the primer surfacer has "leveled" the substrate, and the imperfection is removed.

TECH TIP A primer/primer surfacer is a coating applied before the topcoat; it is designed to promote adhesion to both the substrate and the topcoat that will follow. A primer surfacer is used to fill "minor" imperfections, and is used for corrosion protection and topcoat adhesion.

The "art" of surface preparation and blocking is extremely important to the refinish technician. As can be seen, imperfections of as little as 1/2000 of an inch can be visible in a painted surface and must be removed before topcoating is applied. It may take some time and experience to develop the feel needed to spot and remove imperfections this small.

To help find surface imperfections, the technician often applies a guide coat to the primed surface to help properly

DPX170/171

Directions for Use

Surface Preparation:

- Wash the area to be painted with soap and water, then clean with DX330 ACRYLI–CLEAN®, Wax and Grease Remover, SX1005 0.4 VOC Cleaner Wax and Grease Remover, DX394 1.4 Low VOC Cleaner or SX1004 Plastic Cleaner & Prep.
- Sand the bare metal areas completely with 80–180 grit abrasive.
- Re-clean with DX320, DX330, DX394, or SX1005. Final wipe with a clean damp cloth to remove any cleaner residue.
- DX103 may be used as a final wipe.
- Prime aluminum substrate as soon as possible and no later than 8 hours after cleaning steps.
- Prime carbon steel immediately after cleaning.

Mix Ratio:

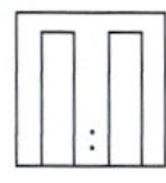

DPX170 or DPX171	:	DPX Catalyst
1	:	1

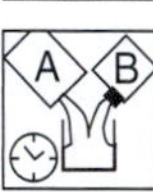

Pot Life is 24 hours at 70°F (21°C)
Pot Life is shortened as temperatures increase.

Spraygun Set-up:

Apply:	1 coat for small areas 2 coats for large areas DO NOT apply excessive film builds in order to avoid poor adhesion or drying characteristics.
Fluid tip:	1.3–1.5 mm or equivalent
Air pressure:	5–10 PSI at the air cap for HVLP guns 40–50 PSI at the gun for conventional guns

Dry Times:

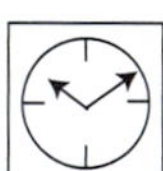

Between coats:	10–15 minutes
Dry to topcoat:	20 minutes at 70° (21°C) for 1 coat applications 30 minutes at 70° (21°C) for 2 coat applications

Note: After 24 hours, lightly scuff DPX Primer. Maintain 0.5 mil minimum. Recoat with additional DPX Primer, if necessary.

DPX170 contains chromium compounds.

DO NOT inhale sanding dusts from DPX170. See the warnings on the label and MSDS for additional information.

FIGURE 28-6 Film thickness is critical when applying undercoatings. Some undercoatings require that a minimum of 2 mils must be applied, while other undercoatings recommend that more than a certain amount be applied. Used with permission of PPG Industries.

identify these small imperfections. A guide coat is a contrasting color applied over a surface that is to be sanded. When the technician sands the surface, the high spots will be removed first, exposing areas that are lower. The lower areas will still have the contrasting color, and—if the technician is unable to feel the imperfection or low spot—the guide coat reveals it. Places where the guide coat remains indicate a low spot. If the primer surfacer has enough build, the technician can continue to sand until the entire guide coat is removed, thus leveling the entire area. If in the process of sanding, the primer filler is completely removed and metal is revealed, the technician must stop sanding and may need to apply more undercoating. Most manufacturers recommend that film thickness of 2 mils of primer remain on the surface for the needed protection and adhesion of the subsequent topcoatings.

Although most primer filler is applied by spray painting, another more rapid application method has been

FIGURE 28-7 Self-etching primer should not be applied too thickly. Note that the yellow colored primer can be seen through slightly when properly applied.

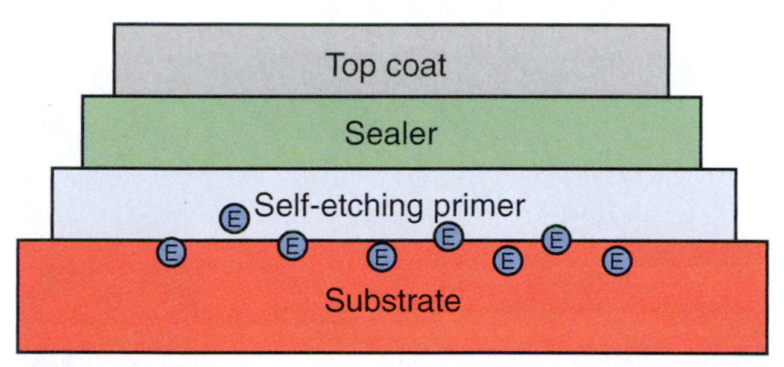

FIGURE 28-8 Acid etch primer (sometimes called self-etching or wash primer) is applied over substrate for corrosion protection.

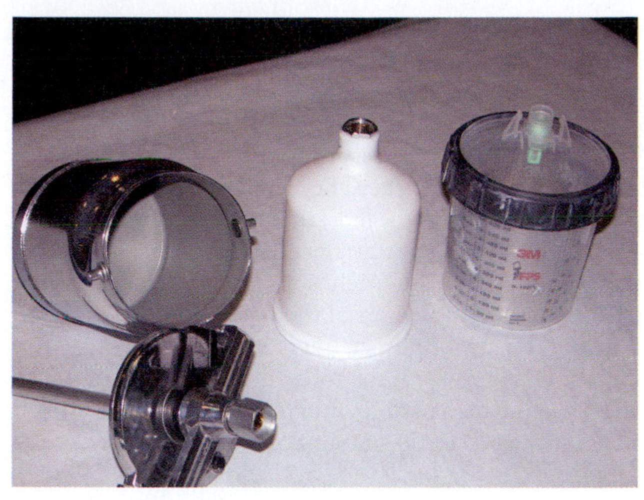

FIGURE 28-9 Acid etch primer can damage some paint cups; the cup on the left is coated with a protective coating inside and the two other plastic cups are not affected by the acid.

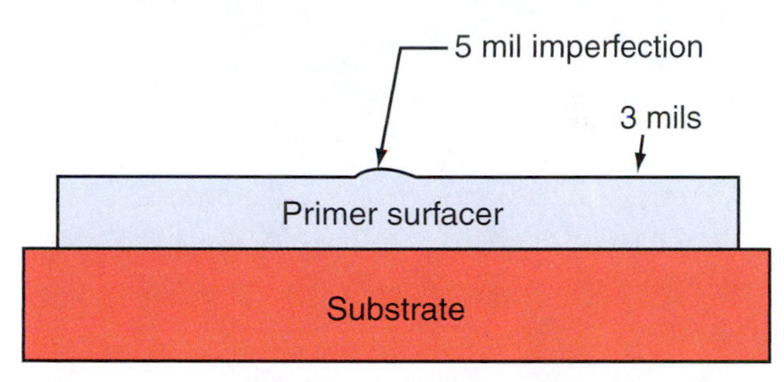

FIGURE 28-10 Primer filler, though applied thickly, is block sanded to remove imperfections. Most recommend that a 2 to 3 mil film remain after sanding.

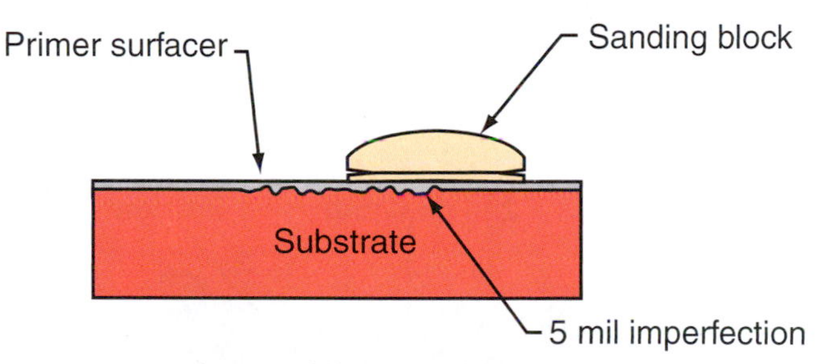

FIGURE 28-11 Primer filler is applied to repair minor imperfections. If the recommended amount of primer is not present after the needed sanding has been completed, more must be applied.

adapted to automotive refinishing. Primer filler applied using the roll method has proven to speed up the application process. When the material is applied with a roller, little or no masking is needed. Although the technician should use a respirator when rolling, the substance is not forced by air pressure into the remaining portions of the work area. When an undercoating is applied by the spray process, the paint booth needs to be used to protect other workers from hazardous gasses. Instead, roller application of primer filler can be done outside of a booth. Although there are still VOC emissions when rolling primer filler, comparatively very little VOC gasses out into the workplace. Another benefit is the technician does not need to move or mask the vehicle, which cuts down the cycle time the vehicle needs for its repair. Many repair facilities roll primer surfacer on in the body shop without masking it or moving it from the metal repair area. Then, when sent to the paint shop for refinishing, the vehicle is block sanded and refinished.

Roll priming is also much more efficient when mixing, since small amounts can be mixed and applied with little or no waste. Small amounts can be mixed in a small container, and then applied. The roller can be disposed of, the roller handle wiped down, and the mixing container disposed of. All these features provide significant time savings. Technicians can continue on to other repairs, and their time is not wasted with menial cleanup. Roll priming occurs at nearly 100 percent **transfer efficiency**, as well. The only waste is the material left on the roll after the application and what remains in the mix container.

Primer surfacer is one of the few coatings that leaves the finished product slightly rough, because it must be sanded. As the technician rolls primer filler, much can be done to control the smoothness of the application. The pressure applied as the product is rolled on the vehicle can control priming film thickness. The more pressure that is applied to the roller as it is being applied, the thinner the coating is spread. Therefore, if applied from the outside in—with the greatest pressure applied at the start and less as the technician reaches the center—the center of the primed area will have the most film build. See **Figure 28-12**. Also, if the technician starts with the first coat applied to the largest area, with each subsequent coat becoming smaller in area, the center of the repair area will have the thickest amount of primer filler. See **Figure 28-13**. After the last coat is applied, the roller is passed over the entire area as lightly as possible to smooth or level the primed area, thus reducing the amount of blocking needed.

FIGURE 28-12 When rolling primer, it should be applied from the outside in. More pressure is applied on the outside (applying less primer) and less pressure is applied over the repaired area where more primer is needed.

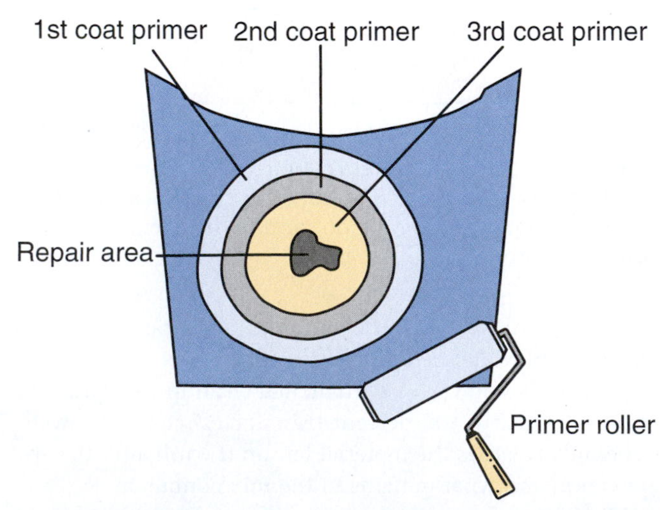

FIGURE 28-13 Rolled primer should be applied with the largest area covered first and the subsequent coats becoming progressively smaller. The last coat should be the same size as the repaired area.

TECH TIP Although roll priming sounds difficult, and it does take a technician some practice to master these skills, it saves significant time. When mastered, roll application can significantly speed up a technician's repair times. The six "P's" of painting—Proper Preparation Prevents Poor Paint Performance—is important with all surface preparation, including roll priming.

Primer surfacer is often applied using spray application methods. Because primer surfacer will need to be applied to small areas of the vehicle, surface masking for primer is much different than masking for topcoating. To keep overspray contained to the smallest area possible, back masking is used. In the **back masking** technique, the masking paper is applied to cover the area to be primed, then folded over on itself. See **Figure 28-14** and **Figure 28-15**. This

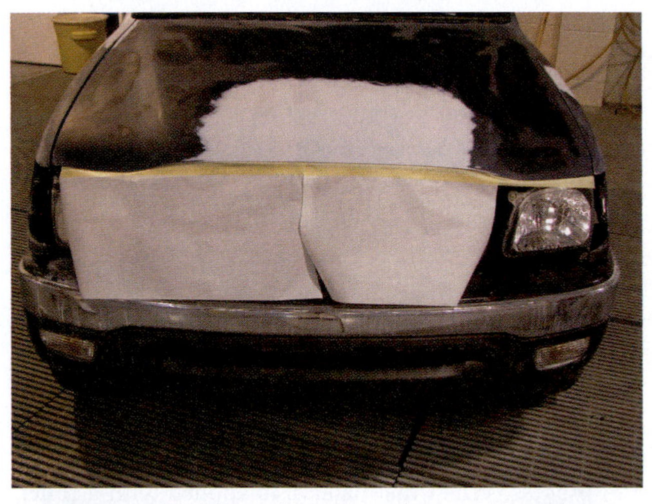

FIGURE 28-14 Back masking is used to avoid a primer hard line.

FIGURE 28-15 Masking paper is applied backwards, then rolled, exposing where the primer should go but covering the area to be protected.

causes the paper to make a slight roll, which eliminates a **hard tape line** that would be seen if the masking tape with paper were applied and the material were sprayed directly over the tape. See **Figure 28-16** and **Figure 28-17**. This back masking is completed until the area to be primed is enclosed with paper, protecting areas where primer is unwanted. See **Figure 28-18**.

With the masking complete, the repaired area can be primed. The repaired and feathered area will need two to four coats to provide enough primer surfacer to be blocked, with priming extending beyond the feathered area. The first coat should be the largest, to cover the repaired area, the feathered area, and slightly beyond. The second and subsequent primed areas are each smaller than the first, with the final coat being applied just on the area of the repair. See **Figure 28-19**.

This "reverse" method of priming should be done so that each subsequent coat of primer surfacer is applied over a "wet" or previously coated area. The edge of a primer coat is thin, and may not have sufficient adhesion if a second coat is applied over it; thus, it may be at risk of failing. When primed with the first coat, the smallest coat—and each progressively larger coat applied over it—will have these areas of potentially poor adhesion, and the possibility of failure will be multiplied. See **Figure 28-20**.

Although primer surfacer should not be thicker than 2 to 3 mils when topcoated, application of primer surfacer by either rolling or spraying can cause significant film build. When block sanded to level any imperfections in the repaired area, however, most of the primer surfacer is removed. The remaining surface will have the required 2 to 3 mils of primer protection.

FIGURE 28-16 The roll is not pulled tightly so the paint will not create a primer hard line as is present with masking tape.

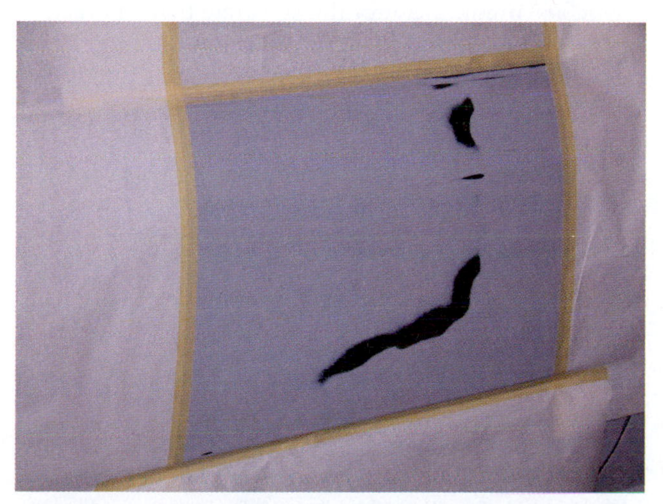

FIGURE 28-17 If taped in a "block," the masking tape would produce a hard to sand line (hard line) that must be feathered or it will be visible when painted.

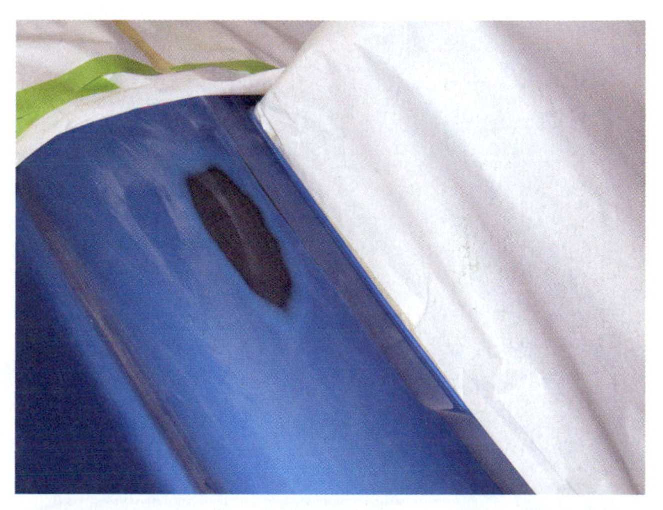

FIGURE 28-18 Back masking is when masking tape is applied half on and half off the back surface of a part. Then masking paper is applied to the exposed tape. This allows paint to be applied to the edge without a hard line on the part.

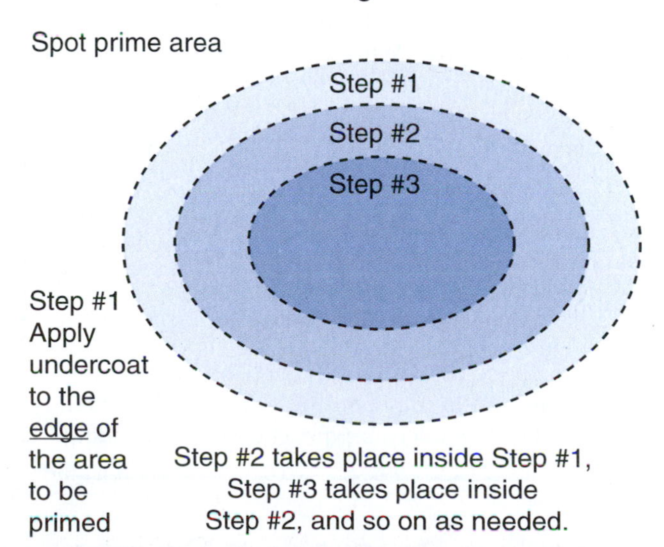

FIGURE 28-19 Sprayed primer should be applied in reverse order, that is: the first coat should be the largest with subsequent coats becoming smaller.

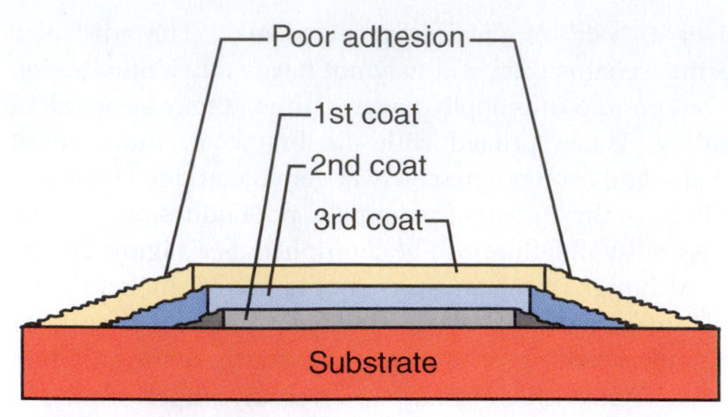

FIGURE 28-20 Priming from the smallest area to the largest has subsequent coats covering areas of poor adhesion, which may fail due to poor adhesion.

Primer Sealers

A **primer sealer** is a coating applied following a primer, or after a previously painted surface has been prepared for paint, but before the topcoat is applied. It acts as a barrier coat, stopping the solvents in the topcoat from penetrating its surface, which could possibly cause adverse reactions. Primer sealer promotes adhesion to the coatings below it, as well as adhesion to the coatings above it. The sealer also provides "holdout," or the topcoat's ability to shine and maintain its luster for many years after being applied. By providing a smooth and consistent surface for topcoats to be applied upon, primer sealers allow the topcoat and clearcoat to perform to the best of their ability.

TECH TIP There is an ongoing discussion among paint technicians as to whether or not a sealer should be used on every paint application. The short answer is that not every job requires it. However, there is a lengthier answer, which follows.

If sealer is used to eliminate paint defects and to create the best possible surface for topcoat application, when is it best used? When properly cured, 2K primers and their thermoset curing ability will not allow solvents to penetrate their surface. Since the ability to be a barrier coat is one of the main reasons for using primer sealer, that reason is eliminated in this case. If the surface has many different colors, a sealer may be used to provide an even-colored surface for the topcoat. However, if the repair has been performed properly using 2K primer filler, and the surface is properly cleaned and sanded, OEM substrate sealer is not required.

As can be seen, the decision to use or not use sealer is dependent on the conditions of the surface being refinished and the technician's experience with refinishing.

Adhesion-Promoting Primers

Adhesion-promoting primers are required when painting unprimed or raw polyolefin plastics. To prepare unpainted polyolefin plastic, it is cleaned and sanded. Then adhesion-promoting primer is applied. The soap and water cleaning removes the water-soluble contaminants such as mold release agents, and the adhesion promoters provide a chemical bond of the next coating to the raw plastic. Any subsequent coatings that will be added to the part will not need adhesion promoters. See **Figure 28-21**.

The term *adhesion promoter* can be confusing. These are products sold to help promote adhesion of topcoats over OEM clearcoat finishes. There are also blender solvents that some technicians use, falsely believing they promote adhesion to the surface of catalyzed and cured OEM or aftermarket finishes. Some also believe that these blending solvents soften the old finish, allowing a new topcoat to "blend" or "melt" into the old finish. One of the characteristics of polyurethane finishes is that when the finish has cured, it will not reflow when chemicals are applied over them. Chemicals such as paint stripper can cause them to lift off the metal, but thinners, reducers, or "blending solvents" do not soften the old finish.

Blending solvents do, however, help with metallic orientation, which will be covered in Chapter 29.

Specialty Coatings

Specialty coatings are a category of undercoats containing items such as plastic adhesion promoters, rock guard, chip resistant undercoats, and other coatings used to perform specific tasks.

Rock guard is a coating that can be applied either before topcoating or following the application of the finish. It protects the areas of potential impact with a coating which, because of its flexibility, can resist impact. These coatings can appear to be textured; depending on the make of vehicle, this coating can be lightly to heavily textured. See **Figure 28-22**.

SURFACE PREPARATION

Surface preparation of the substrate that will receive topcoats is one of the most critical operations in refinishing. If the surface is not properly prepared, all of the coating applied afterwards will fail. Each undercoat that follows will require a different preparation. Depending on the type of substrate, an undercoat may also have its own specific requirements.

This section will cover the surface preparation of each type of undercoat and will list the specifics of metal, aluminum, and plastic preparation, keeping in mind the six "P's" of painting: "Proper Preparation Prevents Poor Paint Performance."

Metal Conditioner/Conversion Coatings

Surfaces that will need preparing for metal conditioner and conversion coating are generally bare steel, either new metal or newly repaired surfaces. The surface may have

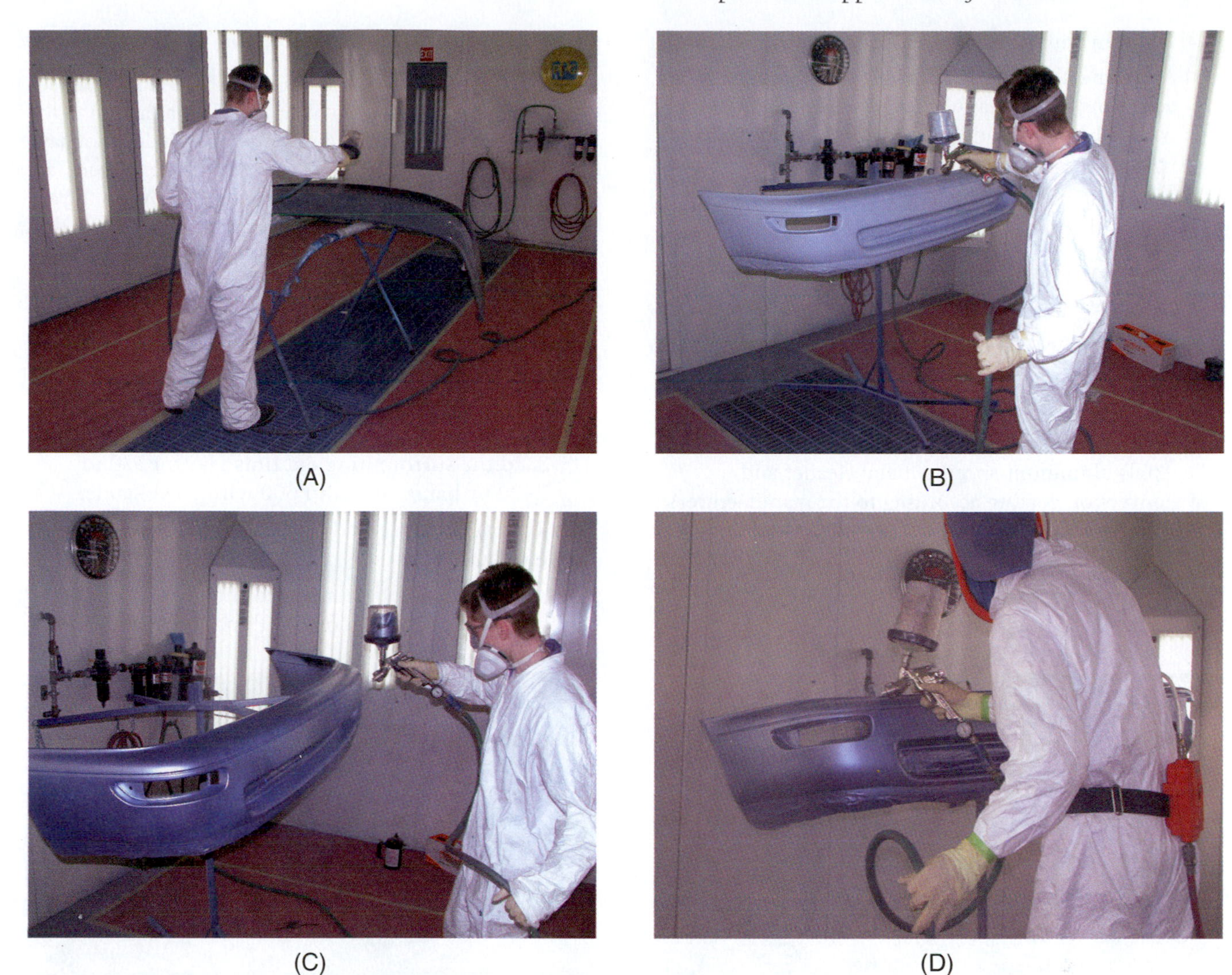

FIGURE 28-21 Plastic parts must be painted using adhesion-promoting primers (A). This step is followed with a flexible sealer (B). Then basecoat is applied (flex additive is added as recommended by the paint manufacturer) (C). This is followed with clearcoat, once again following the paint manufacturer's recommendations for flex additive (D).

FIGURE 28-22 Some special coatings—such as a rock guard—are textured and should be applied to match the vehicle's original texture.

been media blasted or ground while being prepared, or could be aluminum which has been stripped of any protective topcoating. This process also removes any factory applied corrosion protection, which will be replaced when metal cleaner and conversion coating is applied.

The first step in preparing a surface for application of undercoats is to soap and water wash. This should be done with all types of undercoats and with all types of substrates.

Steel. The steps in preparing steel are:

- Perform a soap and water wash.
- Clean the surface with wax and grease remover.
- Sand the surface with a dual action (DA) sander with P80- to P180-grit sandpaper. Do not touch the freshly sanded bare steel with a bare hand (gloves should be worn), as oils and salts from skin will contaminate the surface.
- Remove dust (with compressed air or a clean, lint-free towel).

- Re-clean with wax and grease remover.
- Apply metal cleaner and conversion coating according to the manufacturer's recommendations within 1 hour of sanding, or the surface must be resanded.
- Apply the needed undercoating (primer surfacer, sealer, or epoxy primer). See **Figure 28-23**.

Aluminum. The steps in aluminum preparation are:

- Perform a soap and water wash.
- Clean the surface with wax and grease remover.
- Sand the surface with a dual action (DA) sander with P120- to P240-grit sandpaper. Do not sand aluminum with grit coarser than P80.
- Apply aluminum-specific metal cleaner and conversion coating according to the manufacturer's recommendations, within a half hour of sanding, or the surface must be resanded.
- Apply the needed undercoating (primer surfacer, sealer, or epoxy primer).

Epoxy Primers

Epoxy primer is used on bare steel and aluminum for corrosion protection. It also offers excellent adhesion to the bare steel, aluminum, plastic, and fiberglass. It does not have any fill quality, however, and if surface imperfections are noted, a primer surfacer must be applied following the application of epoxy primer. Epoxy primer is applied to a thin film thickness of 0.75 mils to 1.5 mils. If the surface has no imperfections, it can be topcoated directly with a basecoat, and then clearcoated. The steps for epoxy preparation are:

- Perform a soap and water wash.
- Clean the surface with wax and grease remover.
- Sand bare metal with P80, then P180 sandpaper. Sand the surrounding old finish with P320 to P400 by hand, or with P600 with a DA sander.
- Remove the dust, and re-clean with wax and grease remover.
- Apply the epoxy primer according to the manufacturer's recommendations, within 1 hour of sanding.

(A)

(B)

(C)

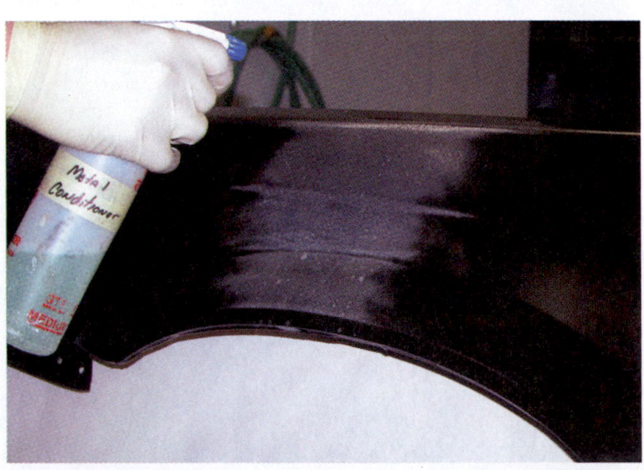

(D)

FIGURE 28-23 Metal chemical treatment is used to prevent corrosion. The first chemical used is the cleaner (A). The second chemical used is the conditioner (B). The metal cleaner is first sprayed on, then scrubbed into the surface with a coarse fiber pad. After the recommended waiting period it is rinsed off (C). Then metal conditioner is applied according to label directions to complete the corrosion protection (D).

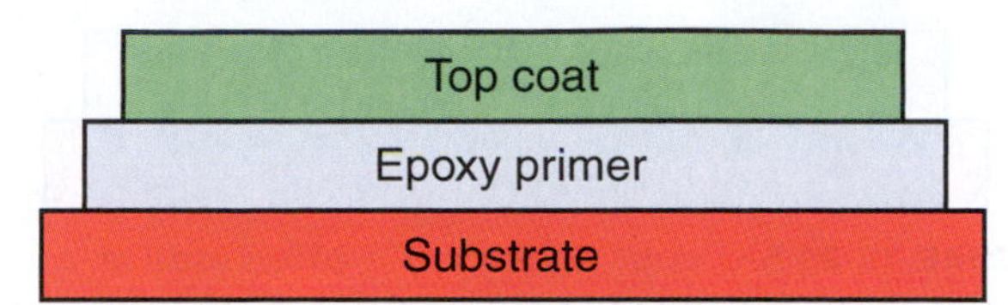

FIGURE 28-24 Epoxy primer is also used for corrosion protection. A single application is sprayed on, which makes the process more time efficient.

- Apply the needed undercoating (primer surfacer, sealer, or epoxy primer), or if it has no imperfections, it can be topcoated. See **Figure 28-24**.

Self-Etching Primers

Self-etching primer is used for corrosion protection. Most products of this type must be topcoated with a sealer or primer surfacer prior to topcoating. Self-etching primers must be applied as directed by the manufacturer. Many of them are thickness sensitive; if applied too thickly, the acid within them will not deactivate properly and may cause failure. The technician should read, understand, and follow the manufacturer's recommendations. The steps for self-etching primer preparation are:

- Perform a soap and water wash.
- Clean the surface with wax and grease remover.
- Sand the surface with P180 grit with a DA sander, followed with P240 grit.
- Remove the dust, and re-clean with wax and grease remover.
- Apply self-etching primer according to the manufacturer's recommendations, within 1 hour of sanding.
- Apply the needed undercoating (primer surfacer, sealer, or epoxy primer). If the surface has no imperfections, it can be topcoated. See **Figure 28-25**.

Primer Surfacers

Primer surfacer is applied to a substrate that has minor imperfections in need of repairing. Although some primer surfacers are **direct-to-metal** and have some corrosion protection, this type of primer may not be recommended for application over bare metal. Therefore, the surface may need to have an epoxy or acid etch primer applied first. Primer fillers are used as spot panel coatings and overall coatings. When applied to the area that has minor imperfections, the primer filler is applied, and then sanded smooth, leveling the surface and thus removing the imperfections. See **Figure 28-26**.

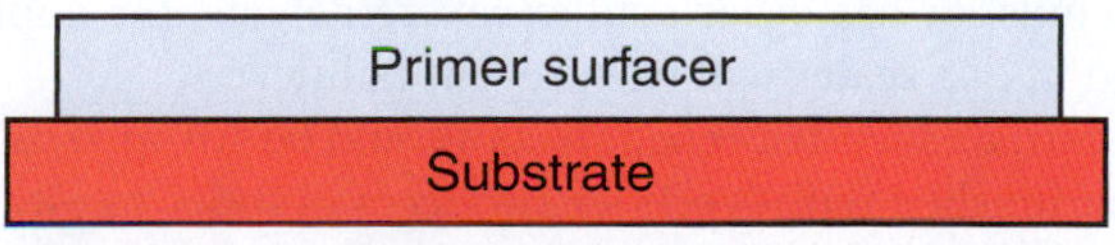

FIGURE 28-25 Self-etching primer is also sprayed on. This corrosion protection coating is fast and effective.

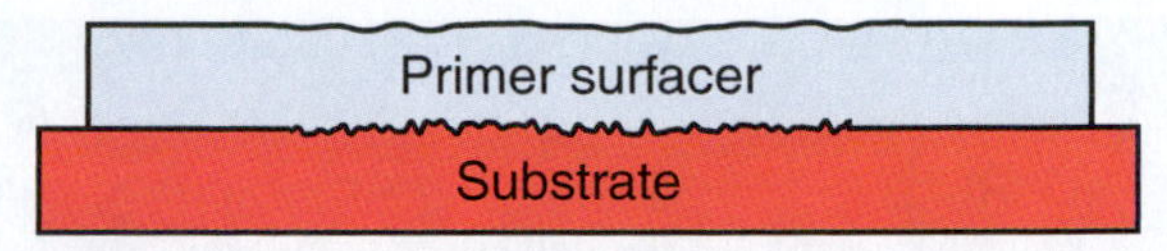

FIGURE 28-26 Primer surfacer is sprayed on to fill minor imperfections. It must be sanded before application of topcoat.

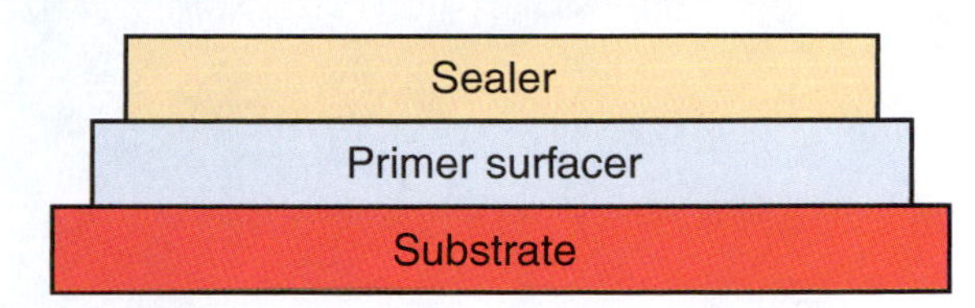

FIGURE 28-27 Primer surfacer may need to be covered with a sealer prior to topcoating.

The steps in primer surface preparation are:

- Perform a soap and water wash.
- Clean the surface with wax and grease remover.
- Sand the surface with P80 grit with a DA sander, followed with P180 grit.
- Remove the dust, and re-clean with wax and grease remover.
- Apply self-primer surfacer according to the manufacturer's recommendations within 1 hour of sanding.
- Properly sanded, cleaned, and prepared primer surfacer can be directly topcoated, although it also may need a sealer applied before being topcoated. See **Figure 28-27**.

Adhesion-Promoting Primers

Adhesion-promoting primers are required when the technician is painting unprimed, or raw, polyolefin plastics. The proper preparation of plastics is specific to the type of plastic substrate that is being coated. Plastic parts that have been primed or that have OEM finish on them do not need to be coated with an adhesion promoter. See **Figure 28-28**. However, if the part that is being finished is a new, unprimed plastic part, adhesion promoters should be used. Unprimed plastic parts should be:

- Washed thoroughly with hot, soapy water, both inside and out, then rinsed well, also on both sides. The part should be forced-dried with compressed air or allowed to air dry.
- A plastic cleaner should be used to clean the soap-and-water-washed part to remove any contaminants that might not have been removed when washed. Many of these plastic cleaners contain an anti-static component and will remove the static charge that builds up as they are wiped.

FIGURE 28-28 Plastic adhesion promoter may be clear or only slightly colored.

- Sand the part with a medium abrasive pad and sanding paste to provide the necessary adhesion. If the plastic is particularly soft, a fine abrasive pad should be used. A coarser (red) abrasive pad is too aggressive for plastic, and may leave "sand scratches" if used. The abrasive pad and paste should be used wet; then, following the scuffing, the part should be rinsed with a large volume of clear water. The rinse water should sheet off the part when rinsed; if it beads up, the part must be re-scuffed, then rinsed again until the water sheets off the entire part.
- The part should be re-cleaned with a plastic-specific wax and grease remover, then wiped in one direction with a clean disposable wipe. The part should not be handled with bare hands from this stage on. If it must be handled at all after the final cleaning, re-cleaning should be done.
- The part should be primed with an adhesion promoter according to the manufacturer's recommendations, followed by application of topcoat, then clearcoating.

Primer Sealers

Primer sealers have been developed with four distinct purposes. See **Figure 28-29**.

1. To provide adhesion to the substrate that it is being placed over, and to provide adhesion to the topcoat that is being applied over it.
2. To provide **uniform color** over substrates that may have a different color or colors within the area being refinished on the vehicle.
3. To provide **uniform holdout** over a substrate that may not be the same as the remainder of the vehicle.
4. To act as a barrier coat, preventing or minimizing the penetration of refinish solvents into the coatings below it.

These functions of primer sealers are needed in specific situations. When the spray technician notes one of these situations in which a sealer is needed, it should be applied to prevent potential problems.

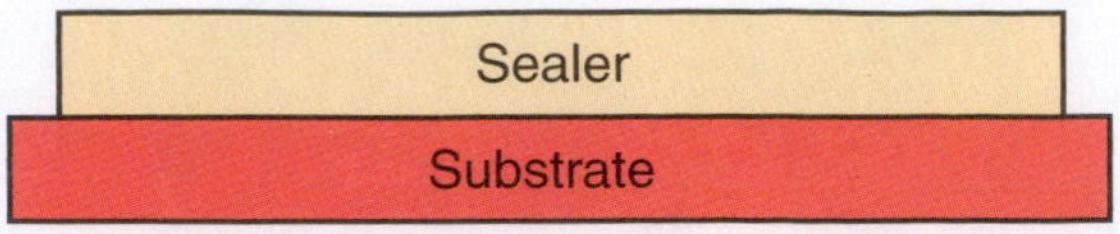

Sealer
Substrate

FIGURE 28-29 Sealer or undercoat is often used as a barrier coat between old painted surfaces and the new coating being applied.

When the adhesion of a topcoat to the part being refinished is in question, a sealer should be used. When a new e-coated part is being installed, a sealer should also be applied—after the new part has been properly prepared—to ensure that the topcoat will not delaminate from the part later. It should also be used on an **LKQ (like, kind, and quality)** used part, which may have a good finish although its adhesion may be in question.

Sealers are also used to provide uniform holdout. **Holdout** is a term used to describe the long lasting of the color and depth of reflection (shine) that a new finish has. With uniform holdout, the finish will have no **bleed-through**, **dieback** (dulling of the surface overall or on repair spots), **bleaching** or staining (color change over areas of plastic body filler repairs), **contour mapping** (lifting or visible areas of rough edges around the feathered area), or other defects sometimes caused by poor substrate conditions.

Sealer is commonly used over LKQ parts to provide a uniform or neutral color under a replacement part, which is often a different color than the vehicle. It can also be chosen when repairs have been completed, resulting in different colors on the parts to be refinished. Colors that may have a "**poor hiding**" quality if sealed will have a uniform substrate color, and the possibility of poor **hiding** or repairs showing through the new finish will be eliminated.

When a vehicle is being refinished, or when a used LKQ part is being installed on a vehicle, the technician may be concerned that the old finish may not be of top quality. If the finish is in poor condition or questionable in any way, a sealer can be applied to provide the barrier between it and the new finish.

When sealers are applied and cured according to the manufacturer's recommendations, they create a barrier coat, which prevents or restricts solvents from the topcoat

TECH TIP Sealers *cannot* prevent failure over severely damaged or poor substrate. If the old surface is cracked, soft, has a film thickness of greater than 8 to 10 mils, or has other similar severe defects, the application of a sealer will not prevent failure.

Surfaces with severe defects should be stripped to—and finished with—primer surfacers in order to establish a good substrate for the application of topcoat.

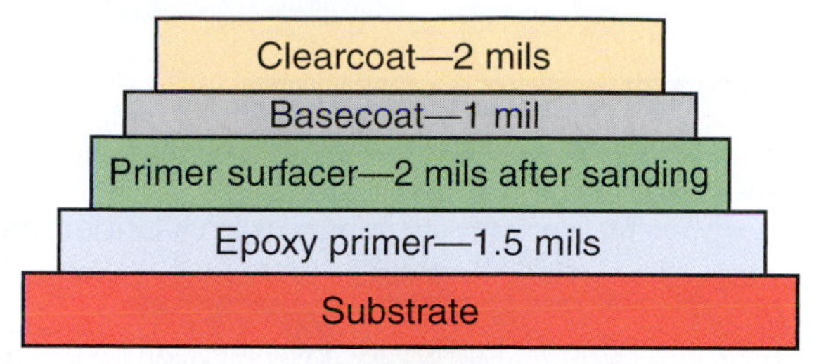

FIGURE 28-30 This shows a cross-section of the coating needed to follow a panel repair.

or clearcoat from penetrating into the substrate below. When solvents do penetrate old, unstable, or questionable finishes, a multitude of problems can result. However, the use of sealers as a barrier coat can prevent such problems. See **Figure 28-30**.

Preparation for Topcoat

Although each undercoat has its purpose and should be chosen and applied as designed, each has some slight surface preparation and application requirements. In fact, each manufacturer may have slightly differing surface application preparation procedures. Each manufacturer's product recommendations should be read, understood, and followed for the product being applied. Some will recommend that the surface be cleaned with soap and water, then with a wax and grease remover, followed by sanding to P400 to P600, followed with wax and grease remover cleaning again before application of the specific undercoating. See **Figure 28-31**. Others may instruct that the sanding be performed through P800. See **Figure 28-32**.

PROCESS	PRIMER-SURFACER	HIGH BUILD
MIX RATIO	P565-510/511 5 P210-796/798 1 P850-16XX 1	P565-510/511 5 P210-796/798 1 P850-16XX 0.5
	Note: Only P210-796 or P210-798 MS Hardeners should be used in HS Primer Surfacers.	
VISCOSITY & POTLIFE	**Viscosity**: 19-23 seconds Din#4 @ 70°F (21°C) **Pot Life:** 1 hour at 70°F (21°C)	**Viscosity**: 30-35 seconds Din#4 @ 70°F (21°C) **Pot Life:** 30 Minutes at 70°F (21°C)
SPRAYGUN & AIR PRESSURE	**Siphon:** 1.6-1.9mm (.063"-.075") 45-55 psi at the gun **Gravity:** 1.7-1.9mm (.069"-.075") 45-55 psi at the gun **HVLP:** 1.7-2.2mm (.069"-.087") Max 10 psi cap press. 3-8 fluid press. (pressurized cup)	**Siphon:** 1.8-2.2mm (.070"-.086") 45-55 psi at the gun **Gravity:** 1.8-2.2mm (.071"-.086") 45-55 psi at the gun **HVLP:** 1.7-2.2mm (.069"-.087") Max 10 psi cap press. 3-8 fluid press. (pressurized cup)
	(**HVLP:** Refer to gun manufacturer's recommendations for required inlet pressure.)	
APPLICATION	**Apply 3 coats** (approx. 3.0-5.0 mils) **Note:** Film build will depend on fluid tip selection.	**Apply 3 coats** (approx. 7.5-10 mils) **Note:** Film build will depend on fluid tip selection.
FLASH TIME	**5 minutes** between coats Allow 5-minute flash-off before force dry or Infra Red.	**5 minutes** between coats Allow 5-minute flash-of before force dry or Infra Red.
DRY TIMES	**Air-dry times at 70°F (21°C)** 1.5-2 hours @ 3-4 mils 2.5-3 hours @ 5 mils **Force Dry (metal temperature):** 20 minutes at 140°F (60°C) **Infra-Red (short wave):** 8-12 minutes after 5 minute flash	**Air-dry times at 70°F (21°C)** 4-6 hours minimum @ 3-4 mils *Overnight Preferred **Force Dry (metal temperature):** 30 minutes at 140°F (60°C) **Infra-Red (short wave):** 8-12 minutes after 5 minute flash
SANDING	**Wet Sanding:** P400 for Single Layer Color P600 for Basecoat color **Machine Sanding:** P320 or finer for Single Layer colors P360 or finer for Basecoats	**Wet Sanding:** P400 for Single Layer Color P600 for Basecoat color **Machine Sanding:** P320 or finer for Single Layer colors P360 or finer for Basecoats

FIGURE 28-31 Note the recommendation at the bottom of P400 for single layer colors and P600 for basecoat colors. Also note the different recommendations for machine sanding. Courtesy of Standox Paint.

Standox® Basecoat

WORKING PROCESS:	Blending-in (droplet technique)
SUBSTRATE:	- **Repair area: Standox Fillers** - **Original paintwork: Sanded with P800-P1000 for Basecoat** - **Sand the area for application of basecoat with P800-P1000. Prep the rest of the panel with P1200-P1500 and finish with a gold scuff pad used with Standohyd® Sanding Paste.**
PRETREATMENT / CLEANING:	For substrate preparation information see **Standox** Painting System S1!
APPLICATION: Use air fed respirator. Refer to relevant Health and Safety Data sheets.	If necessary, spray 1-2 coats of **Standox** 1K Basecoat Colorless onto the entire area. 2-5 min/68°F (20°C) Apply 2 coats of **Standox** Basecoat to cover the repair area, taper each coat. Approx. 10 min/68°F (20°C) Final blend and droplet coat - thin the ready to spray basecoat again approx. 30% 17 s/DIN 4 MM @ 68°F (20°C) Finely spray border area 15-25 psi Droplet technique **Standocryl®** 2K Clears

FIGURE 28-32 Different companies recommend different sanding grits and do not distinguish between single-sand and basecoat clearcoat applications. Courtesy of Standox Paint.

Although these are slight differences, the specific paint manufacturer's recommendations should be followed. The technician who does so risks little chance of unforeseen defects arising.

TINTED PRIMERS

Before the 1980s, topcoats were more opaque, and would cover the undercoats without fear of shadows of the color underneath showing through. Undercoats were typically red or gray, and hiding was not a concern. When "glamour colors" were added, however, the basecoats became more **translucent**. As these translucent, poor hiding colors became more popular, the standard black, green, or red undercoats would often show through the new colors. To combat this poor hiding condition, undercoats were formulated to be tintable. When mixed, they could have color similar to the topcoat added to aid with poor hiding topcoats.

After the introduction of tintable undercoats, most paint manufacturers had recommendations as to how much of the color could be added to the undercoat to help with hiding. Two manufacturers decided to pursue the problem in a different way, by experimenting with each color, then recommending an undercoat that could be tinted specifically for the topcoat being applied. They did not color the undercoat similar to the topcoat, but instead had varying shades of gray, which changed the "**value**" of the undercoat from white through gray to black. Determining the proper value for the topcoat eliminated the need for applying a high number of basecoats, which otherwise were needed to prevent poor hiding.

MULTIPURPOSE UNDERCOATS

With all these specific undercoats to choose from, it was only logical that paint manufacturers would come up with a multipurpose undercoat, one that would have the best qualities of all the other undercoats in a single product. A multipurpose undercoat would provide the following features:

- Fast, reliable coverage that would eliminate the multiple applications of traditional undercoats
- Reduction in the amount of hazardous waste produced with traditional undercoat processes
- Reduction of the amount of coatings needed, thus reducing the cost of refinishing
- Corrosion protection in an undercoat that could be sprayed directly on bare metal (direct-to-metal)
- Fast-sanding primer surfacer
- Tintable
- Primer surfacer
- Excellent leveling and **flow** characteristics

With these goals in mind, multipurpose undercoats were developed. The result is single coating that can be used as either primer filler or a sealer, depending on how it is reduced. Using a multipurpose undercoat eliminates the need for application of primer and sealer, thus reducing the material costs, VOCs escaping into the atmosphere, and the labor needed to complete the vehicle's refinish. See **Figure 28-33** and **Figure 28-34**.

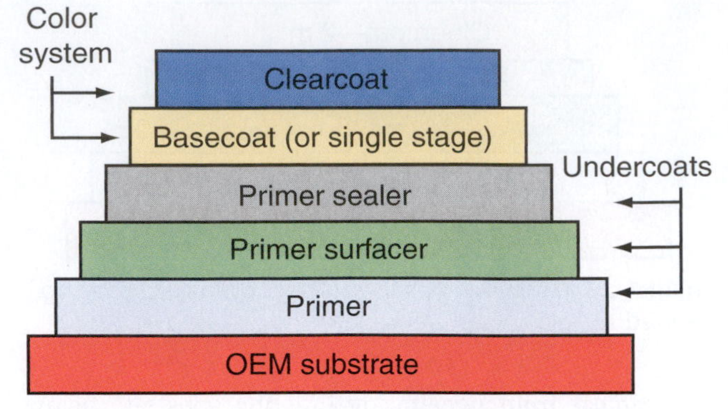

FIGURE 28-33 Cross-section of a repaired finish using primer (acid etch), primer (primer surfacer), sealer basecoat, and clearcoat.

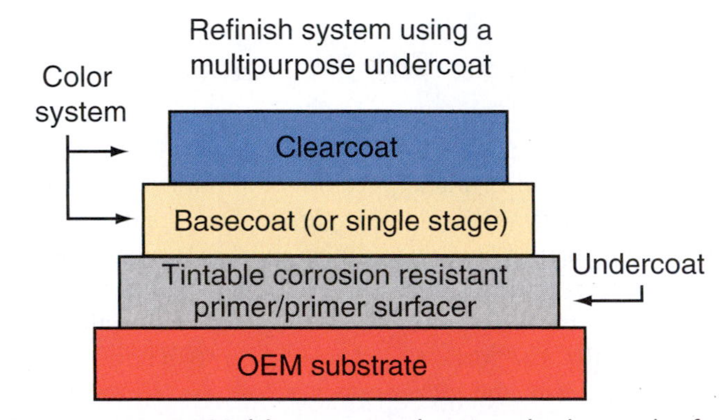

FIGURE 28-34 Multipurpose undercoats do the work of all the undercoats combined and often can be applied directly on bare metal.

Summary

- Undercoats are provided to the refinish technician to be used for varying and specific purposes.
- Undercoats can provide specific outcomes, such as corrosion protection.
- Some undercoats are designed to do more than one task, such as to provide adhesion, corrosion protection, and the filling of imperfections.
- Other undercoats are designed to provide a barrier from what is under them, allowing the topcoat to perform without problems.
- Still other undercoats are designed to perform many—or even all—of the tasks needed by a refinish technician.
- The proper use of undercoats, the understanding of what they are designed for, and the proper surface preparation and application are critical to quality refinishing of vehicles.
- Each manufacturer provides the recommendations for surface preparation, reduction, and application of their products.
- To use these products as they were intended, refinish technicians should read, understand, and follow the manufacturer's recommendations.
- Although different manufacturers have different recommendations for surface preparations, the technician will benefit from remembering the "six P's of surface prep" slogan shown below:
  1. Proper
  2. Preparation
  3. Prevents
  4. Poor
  5. Paint
  6. Performance
- Although many varied undercoatings have been developed for many different conditions, the surface preparation is still the single most important step in refinishing a vehicle.

Key Terms

adhesion-promoting primers
back masking
bleaching
bleed-through
contour mapping
dieback
direct-to-metal
film thickness
flow
gas out
hard tape line
hiding
holdout
induction period
like, kind, and quality (LKQ)
metal cleaner/conversion coating
poor hiding
primer sealers
primer surfacer
rock guard
roll priming
self-etching primers
specialty coatings
substrate
surface preparation
transfer efficiency
translucent
two-part coating (2K)
uniform color
uniform holdout
value
wash primer

Review

ASE Review Questions

1. Technician A says that the national VOC rule was enacted in 1998. Technician B says that 2K paint products require a hardener. Who is correct?
   A. Technician A only
   B. Technician B only
   C. Both Technicians A and B
   D. Neither Technician A nor B
2. Technician A says that metal cleaning and conversion coating is no longer required. Technician B says that some paint manufacturers may recommend metal cleaning and conversion coating before using their epoxy primers. Who is correct?
   A. Technician A only
   B. Technician B only
   C. Both Technicians A and B
   D. Neither Technician A nor B
3. Technician A says that epoxy paint is a primer used as a sealer before spraying topcoat. Technician B says that epoxy paint is a primer used to fill minor imperfections. Who is correct?
   A. Technician A only
   B. Technician B only
   C. Both Technicians A and B
   D. Neither Technician A nor B
4. Technician A says that some paints may have an "induction period," which is the length of time that the coating must rest after adding the catalyst before spraying. Technician B says that primer surfacer does not require sanding before topcoating. Who is correct?
   A. Technician A only
   B. Technician B only
   C. Both Technicians A and B
   D. Neither Technician A nor B
5. Technician A says that VOC stands for volatile organic components. Technician B says that an acid etch primer must be sanded before topcoating. Who is correct?
   A. Technician A only
   B. Technician B only
   C. Both Technicians A and B
   D. Neither Technician A nor B

6. Technician A says that wash primers contain acid. Technician B says that primer surfacer must be sanded before topcoating. Who is correct?
   A. Technician A only
   B. Technician B only
   C. Both Technicians A and B
   D. Neither Technician A nor B
7. Technician A says that paint thickness is measured in mils. Technician B says that paint booths protect workers who are outside the booth from hazardous overspray. Who is correct?
   A. Technician A only
   B. Technician B only
   C. Both Technicians A and B
   D. Neither Technician A nor B
8. Technician A says that roll primer reduces the amount of VOCs vented into the atmosphere. Technician B says that roll primer has 75 percent transfer efficiency. Who is correct?
   A. Technician A only
   B. Technician B only
   C. Both Technicians A and B
   D. Neither Technician A nor B
9. Technician A says that adhesion promoters are primers used on polyolefin plastics. Technician B says that adhesion promoters are primers used to promote adhesion to OEM paints. Who is correct?
   A. Technician A only
   B. Technician B only
   C. Both Technicians A and B
   D. Neither Technician A nor B
10. Technician A says that blending solvents help the new paint melt into the OEM finish. Technician B says that blending solvents help with metallic orientation. Who is correct?
    A. Technician A only
    B. Technician B only
    C. Both Technicians A and B
    D. Neither Technician A nor B
11. Technician A says that the first step in surface preparation is soap and water washing. Technician B says that wax and grease remover is the first step in surface preparation. Who is correct?
    A. Technician A only
    B. Technician B only
    C. Both Technicians A and B
    D. Neither Technician A nor B
12. Technician A says that aluminum should never be sanded with sandpaper coarser than P180 grit. Technician B says that sanded aluminum should be coated within a half hour of sanding, or the surface will need to be resanded. Who is correct?
    A. Technician A only
    B. Technician B only
    C. Both Technicians A and B
    D. Neither Technician A nor B
13. Technician A says that primer sealers provide adhesion to the substrate they are applied to and to the topcoat sprayed on them. Technician B says that sealers act as a barrier coat between the old finish under them and the new topcoat being applied to them. Who is correct?
    A. Technician A only
    B. Technician B only
    C. Both Technicians A and B
    D. Neither Technician A nor B
14. Technician A says that some primers are tintable. Technician B says that some paint manufacturers provide primers that come in different colors, which can be mixed to create an undercoat with the best value for topcoating. Who is correct?
    A. Technician A only
    B. Technician B only
    C. Both Technicians A and B
    D. Neither Technician A nor B
15. Technician A says that multipurpose undercoats have been developed to perform like a sealer, a primer filler, or a corrosion protector. Technician B says that multipurpose undercoats are primer fillers only. Who is correct?
    A. Technician A only
    B. Technician B only
    C. Both Technicians A and B
    D. Neither Technician A nor B

Essay Questions

1. List the steps needed to prepare an OEM finish for refinishing.
2. How does an acid etch primer adhere to the surface of bare metal substrate?
3. What are the differences between a primer and a primer surfacer?
4. What functions does a primer sealer perform?
5. Explain why a paint technician would choose to tint primer.

Topic Related Math Questions

1. Five fenders need to be sealed, each requiring 12.5 ounces of sealer. How much sealer should be prepared?
2. If a body shop uses an average of 28.5 ounces of primer per job and the shop completes 300 jobs per month, how many gallons of primer will be needed in 1 year?

3. If a technician applies 3.5 mils per coat of clearcoat and applies three coats, then removes 1.5 mils of that clearcoat by sanding, how many mils of clearcoat will remain?
4. A part has just been stripped to bare metal. Conditioner will need to be mixed at a ratio of 5 parts water to 3 parts conditioner concentrate. If 67 ounces of the solution are needed, how many ounces of water and how many ounces of conditioner are needed?
5. Primer surfacer is being mixed at a ratio of 4 parts primer to 2 parts reducer and 1 part hardener. If 54 ounces are needed for the job, how many ounces of primer are needed, how many ounces of reducer are needed, and how many ounces of hardener are needed?

Critical Thinking Questions

1. As the shop owner, you are considering switching to a multipurpose direct-to-metal rollable primer. No one in the shop has been trained to use the product, and it will take a bit of time to train the technicians and still maintain the shop's productivity. However, by switching to the rollable primer, shop profits will increase by 8 percent over a year's production. How can you implement this new process and not suffer production loss during the training period?
2. The paint representative is promoting a new primer that he claims will increase production by 5 percent. What must you consider before changing to this new product?

Lab Activities

1. Have students reduce and mix primer filler/sealer/epoxy primer and acid etch primer.
2. Have students block sand, clean, and prepare a primed fender for topcoating.
3. Have students prepare a hood with some areas of bare steel and others with OEM finish for refinishing.
4. Have students prepare a hood with some areas of bare aluminum and others with OEM finish for refinishing.
5. Have students prepare a polyolefin bumper with adhesion promoter for refinishing.
6. Have students seal a prepared part for topcoating.

Chapter 29

Advanced Refinishing Procedures

OBJECTIVES

Upon completing this chapter, you should be able to:

- Calculate the amount of coating needed for specific applications.
- Demonstrate and be able to paint at a high transfer efficiency level.
- Demonstrate techniques of application that will reduce time and material:
  - Basecoat application
  - Clearcoat application
  - Single-stage application
  - Engine bay application
  - Edging parts application
  - Wet-on-wet application
- Demonstrate blending techniques such as:
  - Standard blending
  - Reverse blending
  - Wet-bedding blending
  - Zone blending
  - Single-stage blending
- Repair multi-stage paints.
- Efficiently perform overall vehicle refinishing.
- Be cognizant of the technician's role in sales and profitability.

INTRODUCTION

Since basic painting techniques were introduced in Chapter 21, it may seem to some that this chapter is unnecessary! However, by the time students study this chapter, the textbook authors hope they will have perfected the steps of basic spray techniques. Although spraying is a skill that requires good hand-eye coordination and hours of practice to become proficient, it is never too early to learn about advanced techniques. So, even if technicians have not yet mastered all the finer points of basic spray techniques, learning—or at least looking at—the more advanced spray techniques will extend their knowledge base.

This chapter will study methods that a lead painter must know, such as being able to calculate the amount of paint needed for a given job. By knowing just how much paint is needed, a technician is able to mix a paint formula with little or no waste, which is a very useful skill (especially since "**wet product**" is one of the most costly items in a collision repair business).

A lead technician should also know what spray transfer efficiency is, and should be able to spray at 62 to 65 percent transfer efficiency. Some states now require paint technicians to be trained and tested for transfer efficiency. Other states, believing that training is very valuable, arrange for their state environmental protection agencies to offer free training to paint technicians in their state.

Lead painters should know that some undercoatings can be applied by methods other than spraying that speed up the repair time and do not influence quality. For example, time can be saved by rolling or applying undercoats with a foam brush. In addition, VOC emissions can be greatly reduced, and cleanup time and cost can be markedly reduced as well.

By properly following basic techniques like triggering, overlap, and lead and lag distances (along with not picking up bad habits, such as fanning or yawing the gun), the advanced paint technician will perform more efficiently and will keep the cost of painting a vehicle down. Paint guns should be held perpendicular to the object being painted. If the gun is twisted either right or left of perpendicular, it is considered yawing, which narrows the spray pattern.

A lead painter who can recognize and efficiently use blending techniques such as standard blending, reverse blending, and wet-bedding will produce significant cost and time reductions. Although single-stage blending and clearcoat blending have their drawbacks, advanced painters should be able to perform these techniques, for they are sometimes needed.

As multi-stage paints become increasingly popular, some paint manufacturers predict that they will become commonly used; therefore, a lead painter should be comfortable with blending them as well.

In addition, even though complete or overall paintwork is not often done in collision repair shops, lead painters should be able to quickly and efficiently do this type of paint job as well. Not only should they be able to produce a high-quality overall paint job, but they should know how to fit it into their paint line schedules. That way, even with its special time requirements, the job does not disrupt the efficient flow of a profitable shop.

A final advanced skill for a lead painter is to consider the shop's profitability. Not all painters—lead or otherwise—always consider themselves as salespeople. However, they often serve in that role for the business, since the customer's satisfaction depends on the finished product that the painter provides. Painters must be mindful of the unique position that they hold, since customers pay considerable attention to the finish when evaluating the shop's quality. Also, paint shops that move vehicles through their departments in a timely way also significantly impact the business's profits and customer satisfaction. Lastly, if a lead painter uses the needed paint product efficiently, the profit of the business will increase as well.

As can be seen, painters who become efficient, fast, and profitable can be among the most valuable assets in a collision repair business.

SAFETY

Although a spray gun itself is not dangerous, the products sprayed through it are dangerous. See **Figure 29-1**. Paint spray gun operators must protect themselves, their fellow workers, and the environment from the damages that these substances can cause. The refinish technician not only needs to know how to use a spray gun safely, but must also know how to safely use the coatings being sprayed through it. See **Figure 29-2**. Personal protective equipment must be used, such as:

- Safety glasses
- Safety goggles
- Chemical-resistant gloves
- Paint suit
- Respirator
  - Supplied air for isocyanine finishes
  - Charcoal for non-isocyanine finishes
- Ear protection in areas of noise levels greater than 90 db. See **Figure 29-3**.

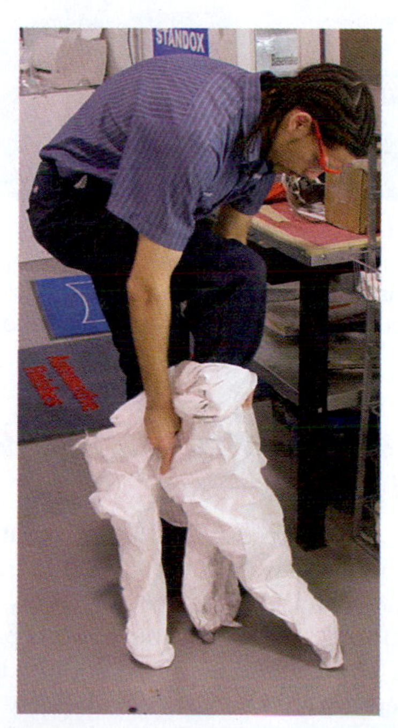

FIGURE 29-1 Painters must wear personal protective gear when painting.

FIGURE 29-2 Paint suits protect the painter's skin, hair, and clothing.

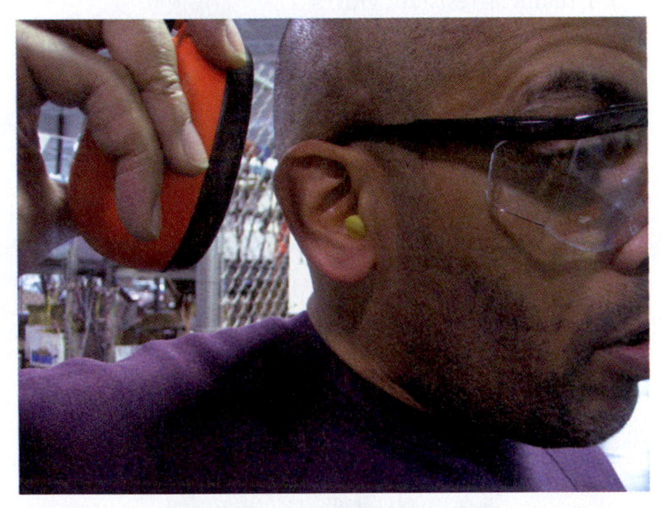

FIGURE 29-3 Although paint areas are not as loud as some working areas in a collision repair shop, the noise is nearly constant and can cause hearing loss over time. Painters should protect their ears from prolonged noise damage.

Read, Understand, and Follow Directions

To safely use spray equipment and the coatings that are sprayed through them, a technician should fully read, understand, and follow the safety instructions provided by the manufacturer. Spray gun manufacturers provide safety instructions in their product instruction manuals, as do manufacturers for coatings. Material safety data sheets (MSDS) are, by law, provided to each purchaser of hazardous products. Technicians should also read, understand, and follow their directions.

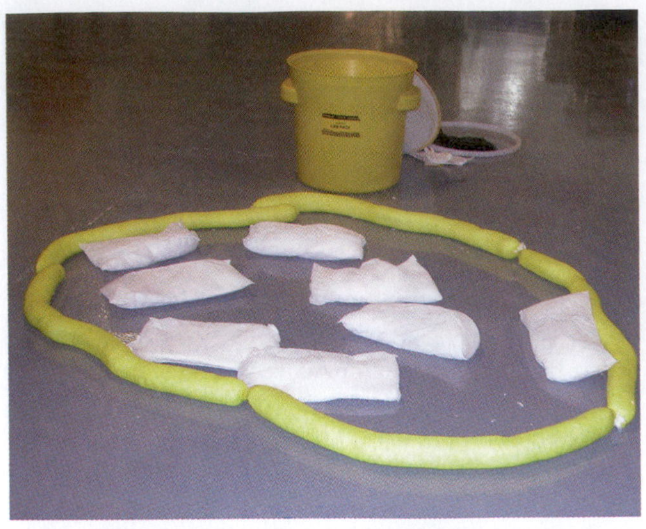

FIGURE 29-4 Spills must be both reported to a supervisor and cleaned up immediately.

Technicians should also follow basic safety rules. These master safety rules include:

- Prevent all accidents.
- Control all operating exposures.
- Perform regular safety audits.
- Promptly correct all deficiencies.
- Report all injuries to the supervisor.
- Avoid all "horse-play."
- Strictly avoid smoking in the shop.
- Do not consume food or beverages in the shop.
- Always wear appropriate safety equipment.
- Report all equipment failures.
- Report all spills at once. See **Figure 29-4**.
- Know the emergency exit plan.
- Recognize and protect the safety of others.

CALCULATING THE AMOUNT OF COATING NEEDED

Controlling cost in the collision repair paint department is important to the business's overall profitability. One of the most costly areas is the paint material, or what is often called "wet product." If a paint department controls the paint it uses—keeping it to just what is needed—it not only cuts cost when paint coatings are purchased, but it also cuts down on waste disposal and hazardous and toxic waste costs.

A common area of waste in the paint department occurs in the mixing of paint formulas, which often are not completely used. If a quart of paint is mixed for a repair and only half of it is used, not only is there the cost of the unused paint that is wasted, but also the cost to dispose of it. Clearcoat or 2K undercoats that are mixed, reduced,

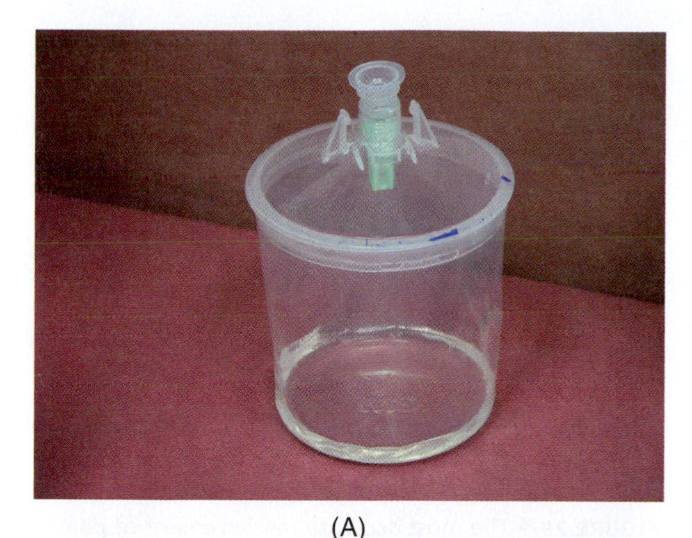

(A)

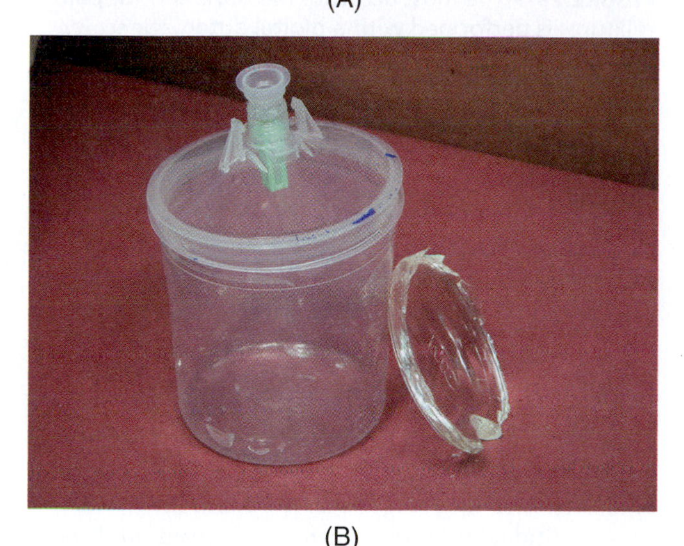

(B)

FIGURE 29-5 (A) Pot-life is the amount of time that a paint product will remain usable (i.e., in a suitable condition for application). (B) After a coating's pot-life is expired, it may look sprayable before it hardens (clear in a cup). However, it should not be used.

and hardened cannot be used later, because at the end of their **pot-life** (the time that they are sprayable before they harden into a solid or semi-solid), they need to be disposed of. See **Figure 29-5**. As can be seen, the cost of overmixing is not just in the initial cost; the added disposal cost may be as much as the cost of the paint itself.

It is not difficult to calculate the amount of each part that must be mixed if the total amount of coating needed is known. Paint companies have created calculation charts to tell approximately how much paint, clearcoat, or undercoatings—such as sealer—are needed for a specific task. See **Figure 29-6**. However, this chart is only approximate, and painters must adjust it according to their experience. Nonetheless, with some attention, a painter can become good at estimating very close to the exact amount that is needed.

Once the exact amount of product needed is known, that amount can be mixed from a formula, then reduced

Sealer Quality Needed

Formula: 3 oz. per hood, roof, and deck lid.
2 oz. per side panel.
3 oz. for a bumper.
1-2 oz. for under hoods.
1-2 oz. for jamming a fender.

Product is figured Ready To Spray!

Note: This information is for mid-sized vehicles.

Paint Quality Needed

Formula: 4 oz. per New Panel.
2 oz. for Blend Panel.
6 oz. per Bumper, Hood, and Roof.
4 oz. for Under Hoods, Deck Lids etc
2 oz. for jamming a Fender, Quarter.

Product is NOT Ready To Spray! This is amount to mix.

Exceptions to the rules; Mix whites in pints and quarts.
Mix difficult mixes per computer recommendation.

Note: This information is mid-sized vehicles.

Clear Coat Quality Needed

Clear needed per panel ready to spray:

Clear Coat Hood	6- 8 oz.
Clear Coat Fenders	2- 4 oz.
Clear Coat Bumper	4- 6 oz.
Clear Coat Quarter	3- 5 oz.

Product is figured Ready To Spray!

1.3-1.4 needle and fluid tip recommended

Note: This information is for mid-sized vehicles.

Examples of Common Job

Front Clip, Blend Doors (New Hood, Fender and bumper)

Jamb hood	4 oz.
Jamb two fenders	4 oz.
Paint hood	6 oz.
Paint two fender	8 oz.
Paint bumper	6 oz.
Blend both doors	4 oz.
Paint Needed	32 oz

to mix on scale

Note: This information is for mid-sized vehicles.

FIGURE 29-6 The precise amount of paint coating needed can be calculated by referring to the paint manufacturer's amount recommendations.

on the scale with the aid of the computer. If a computer with reduction capability is not available, it can be calculated. For example, suppose 32 ounces of **ready to spray (RTS)** coating is needed, and the reduction ratio is 2:1:1. The amounts of each material can be calculated by adding the reduction parts together (2 + 1 + 1 = 4), and then dividing the amount of RTS material (in this case 32 ounces) by 4. The technician would know that each part of the formula is equal to 8 oz. Then, in a graduated mixing cup with ounce measurements, 16 ounces of paint (8 × 2 = 16), 8 ounces of reducer, and 8 ounces of hardener are added, for a total of 32 ounces. See **Figure 29-7**.

When using this method, the paint can be mixed in the same container—such as a reduction cup or a disposable paint cup—which will also save on the cost of containers and cleanup. See **Figure 29-8**.

FIGURE 29-7 Paint measurement can be done with the aid of a graduated cup.

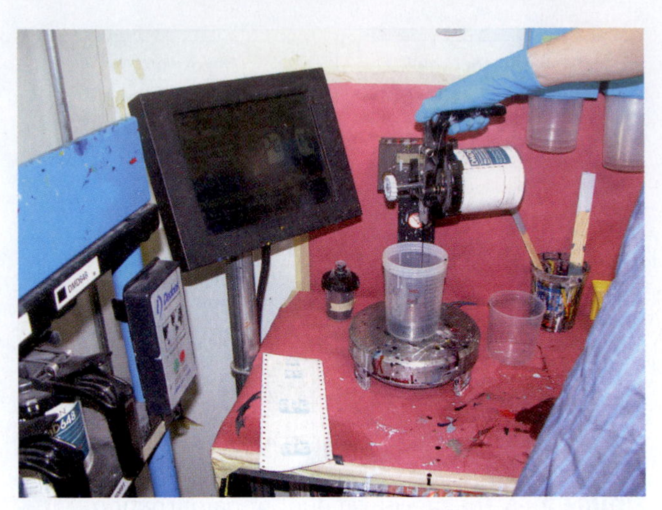

FIGURE 29-8 Paint measurement is most accurate when a scale is used.

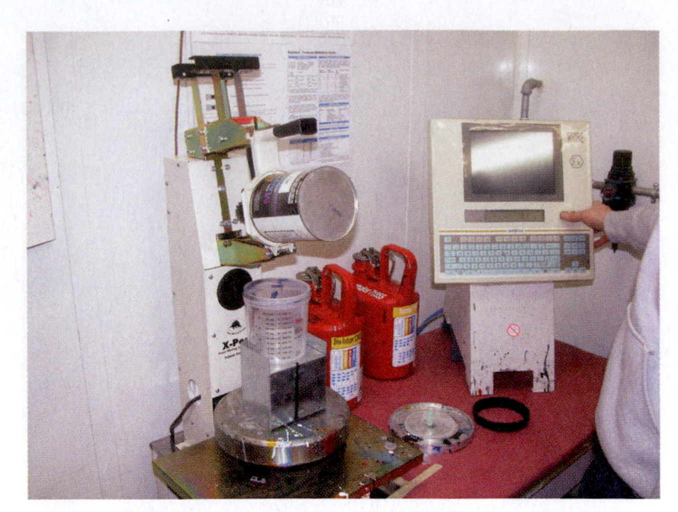

FIGURE 29-9 The most accurate measurement of paint mixtures is performed with a digital automatic scale.

By being able to mix the precise amount of paint needed, the technician will save the department considerable money in the end.

32 oz of reduced ready to spray needed.

Needs to be mixed 2:1:1, or 2 parts paint, 1 part reducer, and 1 part hardener.

2+1+1 = 4 32/4 = 8 or 8 ounces per part =
16 oz paint
8 oz reducer
8 oz hardener

Total: 32 ounces of ready-to-spray reduced paint

With some computerized pouring systems, a formula is retrieved by using the vehicle's **vehicle identification number (VIN)**. See **Figure 29-9**. The paint can be quickly and accurately retrieved and mixed without fear of over- or underpouring. The exact amount needed can also be mixed and reduced on the scale, making it more efficient and cost effective than ever.

Automated clearcoat mixing systems are also quite convenient and efficient. Clearcoat, the precise amount of hardener, and reducer (if called for) are mixed mechanically as needed. If painters need 20 ounces to clearcoat a vehicle, they could mix 20 ounces. Or if the exact amount needed is not known, painters can get the amount they think is needed for one coat, say 12 ounces. They can then spray the coat, measure what is left, then calculate the exact amount needed for the second coat. As can be seen, this type of "on-demand" clearcoat mixing can be very efficient.

By using these methods, a lead painter can make the paint department much more efficient, which will translate into a higher profit margin.

TRANSFER EFFICIENCY

Paint **transfer efficiency (TE)** is a measurement of how efficiently a painter and the equipment can transfer the liquid paint from the spraying equipment to the object being painted. The higher transfer efficiency that a painter can develop, the more cost efficient the paint department will become. The higher transfer efficiency a spray painter can develop, the less paint that is needed for each application. Besides saving the business the cost of wasted paint, high paint transfer efficiency also results in less paint being emitted into overspray in the booth. The less overspray that is emitted, the longer the booth filters will last; thus, additional business expenses are saved on disposal and filter replacement.

For some time now, the Department of Environmental Protection has looked for ways to reduce the production of waste VOCs emitted into the atmosphere. To this end, the National VOC Regulations of 1998 were made, restricting how much solvent could be shipped in the manufactured paint products. In many states, laws have been enacted that require the repair facility to track all of the paint products and solvents as they are being used. Other states have enacted laws requiring that painters use spray guns that spray with high volume and low pressure. See **Figure 29-10**.

Paint gun companies have developed paint guns which now have the ability to spray at 70 percent or higher transfer efficiency. Also, today's paints have fewer solvents in them when they come from the manufacturer. However, extensive tests in waste reduction research at the University of Northern Iowa show that the average painter still sprays in the 40 to 45 percent transfer efficiency range.

However, tests have also proven that—with the proper training—technicians can improve their transfer efficiency to an average of 65 percent. This improvement can be accomplished with as little as 12 hours of training. If spray technicians everywhere could be trained to spray at

FIGURE 29-10 As HVLP guns may look the same as non-HVLP guns, most HVLP guns will be clearly marked.

65 percent transfer efficiency, the amount of VOCs released into the lower ozone could be reduced significantly.

The techniques that a technician must perform to become more transfer efficient are not difficult to learn, nor are they particularly new. They are things such as full-fan banding, where none of the spray pattern is sprayed over the edge while edging a part. See **Figure 29-11**. Once full-fan banding is learned, the technician can rapidly band parts that are about to be sprayed, like the fender shown in **Figure 29-12**. The illustration shows the fender being banded, or the outer edges being painted around the circumference of the part. When parts are properly banded, fewer lead and lag distances are used. This increases transfer efficiency (lead is when a spray is started before hitting the target and lag is when the spray is stopped after moving past the target). See **Figure 29-13** and **Figure 29-14**.

FIGURE 29-11 Full-fan banding sprays a coat of color on a panel without having overspray extend over the panel.

Lastly, painters should know how much paint they are applying. Many painters apply from 25 to 50 percent more paint than the manufacturer recommends. Although this does not represent wasted overspray emissions, it does result in more paint than is needed, and therefore emits more VOCs as they evaporate off the vehicle. If a technician makes a sprayout panel to determine both the color of the mix and number of coats needed for coverage, a wet film thickness gauge can be used so technicians will know how much paint they have applied.

As the student may recognize, none of these "advanced spray techniques" are particularly difficult to master, and when mastered they do not represent an increase of time spent finishing a vehicle. What they do represent is a significant reduction in paint materials and cost, and a significant reduction in VOCs emitted into the atmosphere. This VOC reduction represents a significant reduction in cost (filters, waste material, and cleanup).

APPLICATION PROCEDURES

Since Joseph Binks and Dr. Alan DeVilbiss invented the spray gun at the turn of the 20th century—and since DeVilbiss's son, Thomas, refined it and developed it for wide use in the automotive industry—paint is rarely brushed on anymore. In fact, if anyone were to even suggest to most technicians that coatings—or more specifically undercoatings—be applied by means other than spraying, they would think that person was a quack, or at best a bit ignorant about automotive coatings application. However, in the automotive finish and refinish business, many different ways of application are used regularly. If a lead painter understands when these various methods can each be used to the best advantage, the decisions can speed up production significantly.

Brush Application of Undercoats

While this text is not suggesting that a technician should "paint a car with a brush," it is suggesting that some undercoats can effectively be applied by brush—more specifically, with a foam brush or foam roller. See **Figure 29-15**. Epoxy undercoats (those with a long pot-life, perhaps 72 hours, for example) can be mixed at the beginning of the week, and then applied to bare metal with a foam brush before the application of plastic body filler. Then, following the smoothing and feathering of that body filler, a second application could be applied to

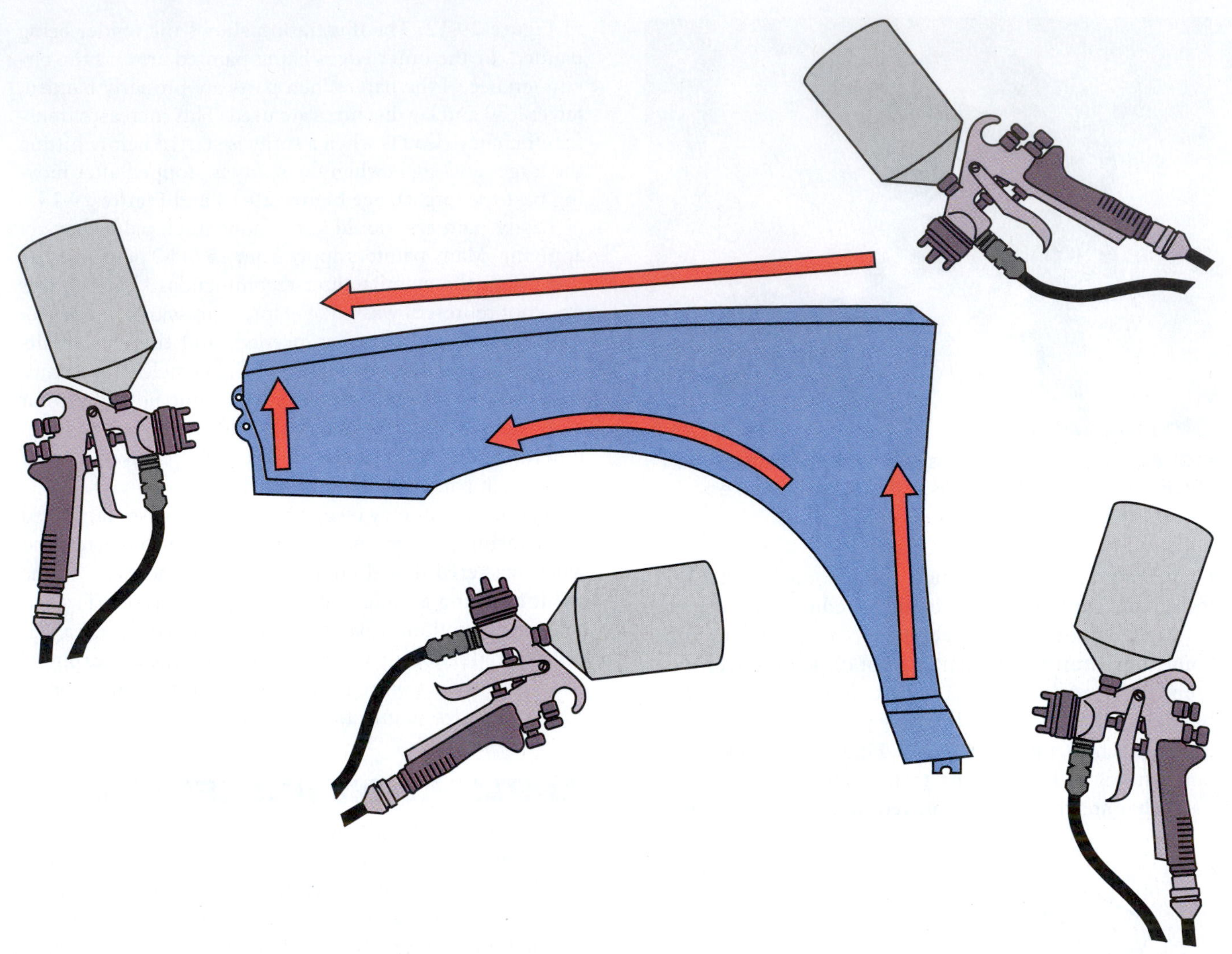

FIGURE 29-12 Fender banding is completed by spraying the perimeter of a fender, then returning to fill in the body of the fender. This practice cuts down on overspray waste and increases transfer efficiency.

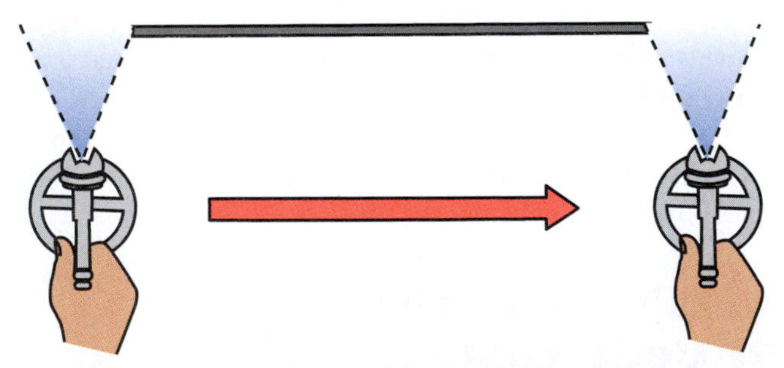

FIGURE 29-13 Improper lead and lag is when a technician starts spraying well before the part is approached and holds the trigger far beyond the part. This causes excessive overspray, waste, and pollution.

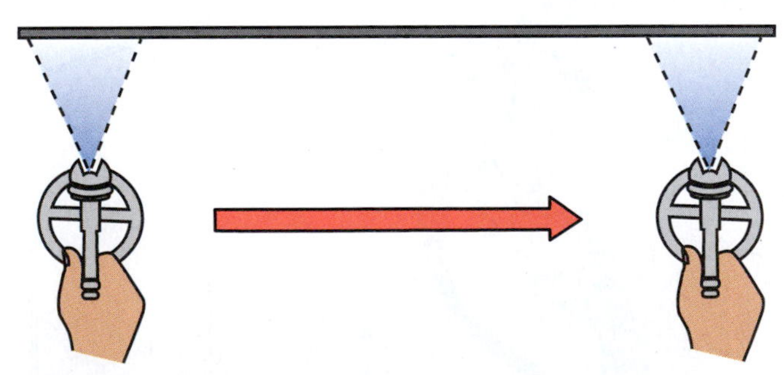

FIGURE 29-14 Proper lead and lag keeps overspray, waste, and pollution to a minimum.

the areas of exposed steel without getting it on the filler, as recommended for corrosion protection. This process is fast, since it involves no masking of the vehicle or moving it to the prep area. Also, the brush is inexpensive and can be thrown away instead of cleaned. See **Figure 29-16**. The vehicle would then be ready for primer filler as needed.

TECH TIP Unfortunately, acid etch primers usually have a film thickness limit that cannot be exceeded and still allow the coating to perform properly. Therefore, unlike epoxy, this material would not be a candidate for foam brush or roller application.

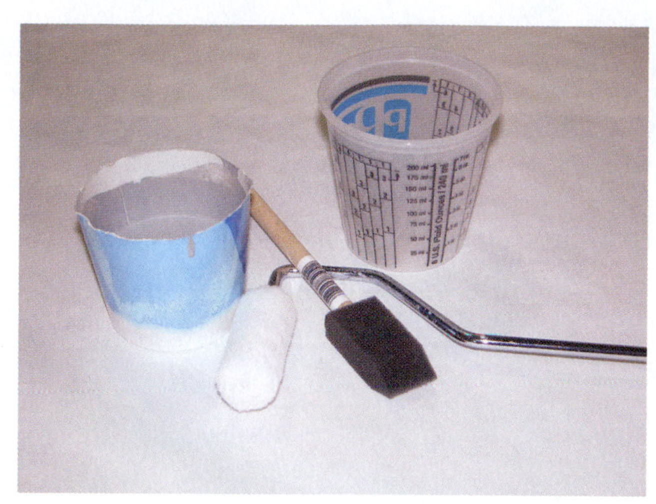

FIGURE 29-15 Brush painting vehicles—using a foam brush or a foam roller—during early manufacturing can be a fast and relatively inexpensive process.

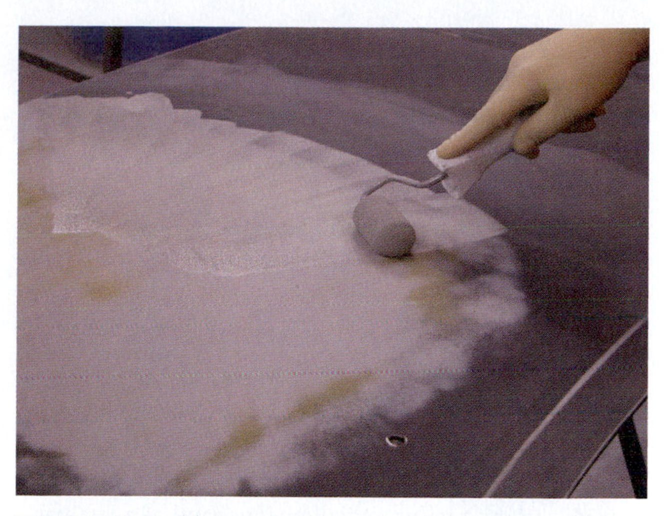

FIGURE 29-17 The application of undercoating with a roller or foam brush will significantly decrease waste and pollution.

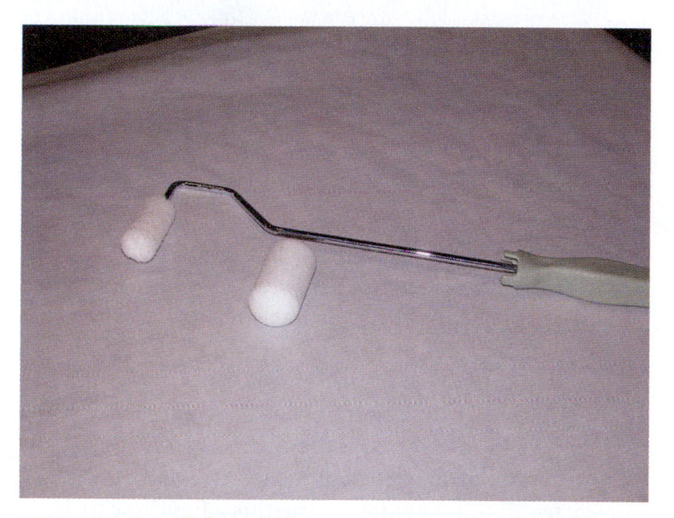

FIGURE 29-16 Primer rollers are inexpensive and are discarded instead of cleaned after use. This makes the process more time efficient.

FIGURE 29-18 The more pressure applied when rolling on the undercoat, the thinner the coating. The less pressure applied, the thicker the coating.

Not only can the epoxy paints be applied by foam brush or roller, but some primer fillers also are ideal—or even designed specifically—for roller application. Rollers come in different surface configurations. Some are made of cloth and have a rough texture. These apply more primer per coat than the smooth foam rollers, which can be a bit more controllable. See **Figure 29-17**.

To apply rolled primer, a technician should start applying primer from the center of the repair, moving outward. The application starts with light pressure, which diminishes as it rolls toward the outer edge. At or near the edge, little to no pressure is applied, and the roller is lifted off the area being primed. This will feather the application of primer from the thickest in the center to the thinnest in the outer parts of the application. See **Figure 29-18**. Rolled application will need several coats, just as sprayed applications would. Proper flash time should be observed between coats. The technician should apply the number of coats needed to fill imperfections.

If a foam applicator is used, it will leave a smoother coating which will block out easily for topcoating. As can be seen, this technique may take some practice before it is perfected. However, when perfected, much time, money, and material will be saved. Roll priming is fast—no moving the

TECH TIP Although **foam brush application** is not practical with acid etch primer, aerosol spray can application sometimes is. (Aerosol spray acid etch primer, while readily available, is still prohibitively expensive for many projects.) A foam-tipped applicator is available to apply acid etch primers.

vehicle, little to no masking, little to no material waste, no cleanup—and if the primed area is heated with **infrared lights**, the primed area can be block sanded in a short time (some in as little as 8 minutes or less).

Topcoats

While no alternative method for the application of topcoats is available to the aftermarket at this time, new types of paint and application processes are being explored. Although **powder coating** has been adopted by many of the manufacturers for finishing vehicles before shipping them to a dealer, a practical aftermarket finish has not yet been developed for the collision repair industry—although it may be, in the future.

If anything about the collision repair and refinish business is certain, it is that it will change, and at a fast pace. Working in the collision repair industry is without argument a life-long learning process, because as new methods and products become available, technicians must learn and adapt.

ENGINE BAY

Although finish application of engine bays does not involve new methods of application, the process does have some differences that must be taken into consideration. Many engine bays can be sprayed a different color than the outer part of the vehicle. Because they are not exposed to ultraviolet light, they may also be sprayed with a different type of clearcoat than is on the outer part of the vehicle. In fact, engine bays may be sprayed with a less glossy clearcoat, or no clearcoat at all. Some engine bays are sprayed with single-stage paint, and this should be taken into consideration.

Most engine bays are sprayed with only part of the engine removed and the remainder masked off; however, the complete removal of an engine may be necessary, and large portions of the bay might be sprayed before the engine is reinstalled. See **Figure 29-19** and **Figure 29-20**.

FIGURE 29-19 Careful masking must be performed when engine parts are not removed.

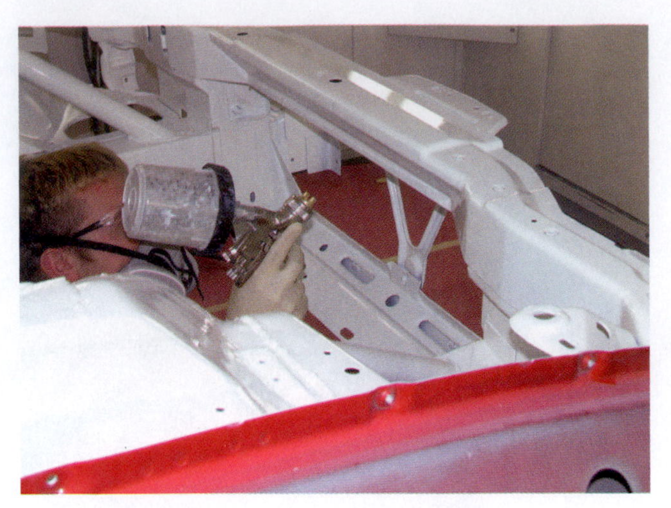

FIGURE 29-20 With engine bay components, coating application with the engine removed is more convenient and speedy.

Therefore, a lead spray technician will need to work engine bay painting into the day's schedule. Then the job will be sent back to the repair shop for engine installation and vehicle reassembly, before it returns to the paint shop for completion.

EDGING-IN PARTS

The process of **edging-in parts**, sometimes called **cutting-in parts**, is normally reserved for the painter's helper or the apprentice in the paint department. It is a perfect task to help the trainee to learn gun control and develop into a good painter. It is also a great place for the new painter to develop speed and confidence. All of the basic techniques taught in Chapter 21, as well as advanced techniques explained in this chapter, can be—and often are—developed while an apprentice works at edging-in parts.

Techniques and methods that can be used to make a paint department more efficient and cost effective while edging-in parts include the use of no-sand sealers and wet-on-wet sealers, as well as using old paint for first coats.

No-Sand Sealers

The use of **no-sand sealers** can speed up the edging-in process. No-sand sealers are sealers for new e-coated parts without imperfections. After being cleaned (with both soap and water, and chemically) such a part can be coated with a sealer, then topcoated. One can see that this would significantly speed up the preparation process.

Wet-on-Wet Sealers

Wet-on-wet sealers are sealers that have no dry time, and just a short flash time before the topcoat can be applied. Many sealers must dry for around 30 minutes before they can be topcoated. With wet-on-wet sealers, that dry time is eliminated. A part that comes into a shop that uses a wet-on-wet non-sanding sealer could have parts edged-in and ready for installation in less than an hour, or maybe even 30 minutes—the amount of time a technician needed to wait for just dry time before wet-on-wet was developed.

Using Old Colors for First Coat

The use of an old, unused but similar color for the first coat of edging-in is not new to the industry, but still a good way to reduce the amount of waste.

For example, perhaps the part that needs to be edged-in is metallic blue, and the technician has some waste metallic blue that is close to the color of the vehicle that is being painted. This surplus paint color can be applied for the first and even second coats; then the correct color of the vehicle can be sprayed over for the last coat, or until the edged-in part has full coverage.

On the other hand, if the technician already mixes just what is needed for each job, and mixes and reduces on a scale to minimize waste, this additional cost-saving practice may not be necessary.

BLENDING PROCEDURES

Blending is by far the most common paint procedure done in the collision repair department. Blending methods are techniques used to trick the eye into not noticing if a color does not perfectly match the surrounding panels.

Color has changed greatly over the past years and will continue to become more vibrant and attractive to the customers. However, the reflective additives that have been added to paint to get these colors, while attractive, make the paint more difficult to match. Consequently, paint technicians have needed to come up with methods to accommodate this increased difficulty with matching.

Blending provides for full coverage with the new paint in the area of the repair, with each subsequent coat covering an area larger than one with full coverage, resulting in some of the old color showing through. As the blended area becomes larger and has less coverage, more of the old paint shows through, until the new color (which may have been slightly different than the original) is no longer noticeable, due to the gradual fading a blend provides. See **Figure 29-21**.

Although blending works well, some problems became evident as it was first being developed. To eliminate these problems, some modifications have been developed to correct the shortcomings of the standard blending method. Three types of blending are commonly in use, each designed for a specific situation. The three blending methods are:

- Standard blending
  - For non-metallics, and for colors which cover well and lie down easily.

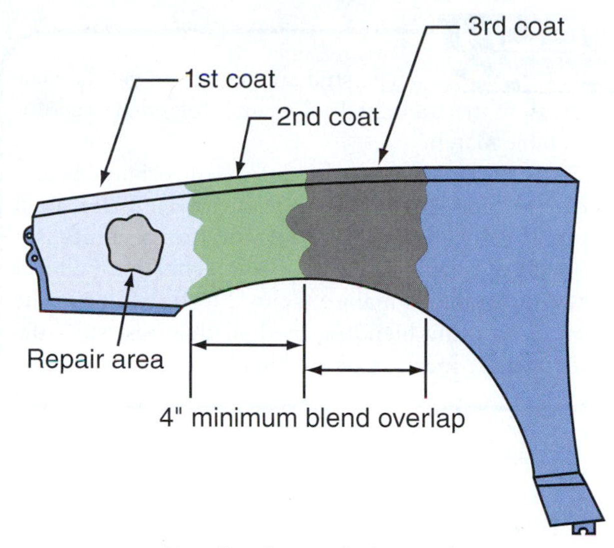

FIGURE 29-21 Blending is a technique where consecutive coats are applied. Each coat is extended approximately 4 inches larger than the next, with the last coat being the largest.

- Reverse blending
  - For most metallic colors, poor hiding colors, and colors that may have metallic orientation difficulties.
- Wet-bedding
  - For metallic colors such as silver, gold, and bronze, and for paint with high amounts of mica in it.

Blending can be done on a single panel. If the repair is small enough to allow sufficient room to extend the blend, blending generally requires 12 inches to properly complete. If the area being blended is closer than 12 inches to the next panel, the blend area will need to be extended into the adjacent panel and therefore it should be prepared for blending. Do not limit the size of a blend to accommodate its closeness to the next panel. The only reason blends work is that they "trick the eye" as some of the old paint shows through. If the blend is kept too small, the small blend will *not* trick the eye.

Just how large should a blend be? It all depends on the color, and on which blending method is best for a vehicle. Generally, each new coat will need to be applied to a minimum area about 4 inches larger than the previous coat. See Figure 29-21. Again, the precise amount by which the area gets larger depends on the color, as well as where on the vehicle the blend is being applied. Consider an example: A blend in the lower part of a vehicle's door with a color that has little metallic in it will not need to be too large. On the other hand, a blend on the hood of a silver or bronze car will need to be larger. Painters sometimes avoid blending from a horizontal to a vertical panel, believing that the blend would not be undetectable on the flat (horizontal) surface. Actually, though, many times when a horizontal panel is not blended into, that is the place where it becomes noticeable that the color does not match.

TECH TIP The student should not get the false impression that all mixed color can be blended to an undetectable match.

The mixed color must be close to the original before it can be blended. In fact, after mixing a color, then comparing it to the vehicle, the question becomes is it a "blendable" color or not. If the technician believes that it is possible to make a **blendable match**, he or she must choose the blending method that best suits the color and the location of the blend.

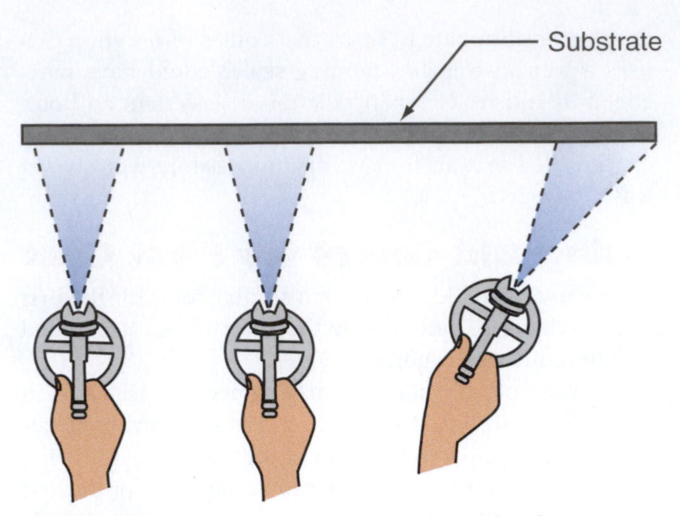

FIGURE 29-22 Fanning when applying a blend may cause color mismatch with poor metallic lie.

Standard Blending

Standard blending was one of the first methods developed. A technician using this method will make a plan to refinish it by first looking at the panel or panels to determine the farthest area that will receive paint. This area will be cleaned, sanded to P400/P500 grit, and prepared to receive color. The remaining area of the panel or panels—which will be cleared after the color is applied—will be cleaned and sanded to P1000 grit or equivalent.

The repaired area should be sealed if the painter believes that it is necessary for an undetectable repair. Sealer should be kept to as small an area as possible, to keep the overall blend small.

The paint should be mixed and reduced according to the manufacturer's specification. The technician should choose the best solvent for the spray conditions. The first coat of paint is applied to cover the repaired and sealed area, but not much beyond it.

The second coat should be applied, following the proper amount of flash time, but extended further than the first by about 4 inches. The third and subsequent coats should be extended a little larger each time, until full coverage is achieved. See Figure 29-21.

When the recommended amount of flash time between the basecoat and clearcoat has passed, the technician should apply the first coat of clearcoat to the entire panel or panels. Then following the flash time, the technician applies a second coat, again to the entire panel.

A common mistake made by technicians when they blend with color is that they fan their strokes when painting. The painter may believe that this action feathers out the paint, making a better blend. In actuality, it may cause metallic orientation problems. When fanned, a paint stroke does not maintain the proper distance. Thus, the metallic orientation changes, and a halo or visible edge may become apparent. See **Figure 29-22**. To avoid dry spray and still have the paint applied lighter at the edges of a blend, the technician should slowly let off the trigger at the outside of a blend coat. This is called "**off-triggering**."

TECH TIP Some paint manufacturers recommend that when applying clearcoat to a blended area, the technician should apply the first coat to an area just beyond the last color coat but not to the entire panel, with the second coat being applied to the full panel.

This method is helpful, especially when applying clearcoat over a yellow vehicle. Clearcoats are not all *completely* clear, and may have a yellowing effect when two coats are applied over an already yellow vehicle. The painter should consider this effect when determining whether to use two full coats of clearcoat on a repair. See **Figure 29-23**.

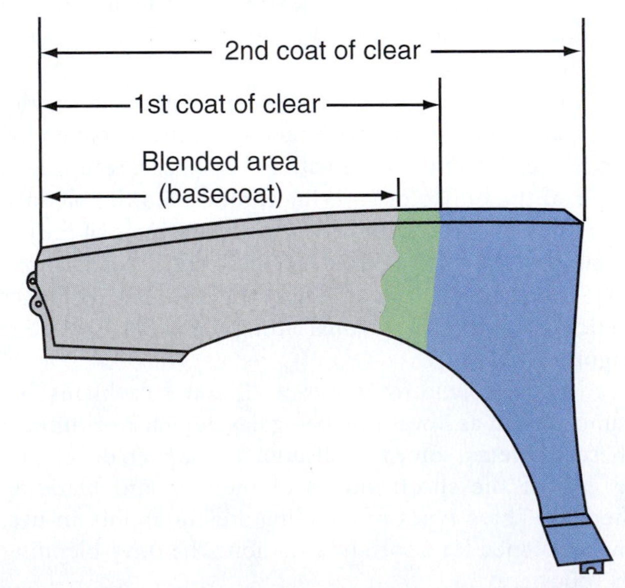

FIGURE 29-23 Clearcoat application with 1 1/2 coats may prevent a yellowing effect from appearing on the vehicle.

Reverse Blending

Not all colors are easy to blend. In the past painters sometimes confronted problem colors such as highly metallic ones, or colors with different-sized metallics that may have an orientation problem when being sprayed. Even with colors that have poor hiding when sprayed, technicians knew that the standard blend technique needed to be modified.

To solve these problems—as well as the problem of each subsequent coat being sprayed over the dry area of the previous coat (which may cause an adhesion problem)—the **reverse blending** method was developed. See **Figure 29-24**.

In the reverse blend method, the panel or panels are prepared in the same manner as for standard blending, but the first coat is applied to the largest area that will receive paint. Following the recommended first coat time, the second coat—which is smaller—is applied. Then the third, when applied, covers the smallest area, which is approximately the size of the repaired area. See **Figure 29-25**.

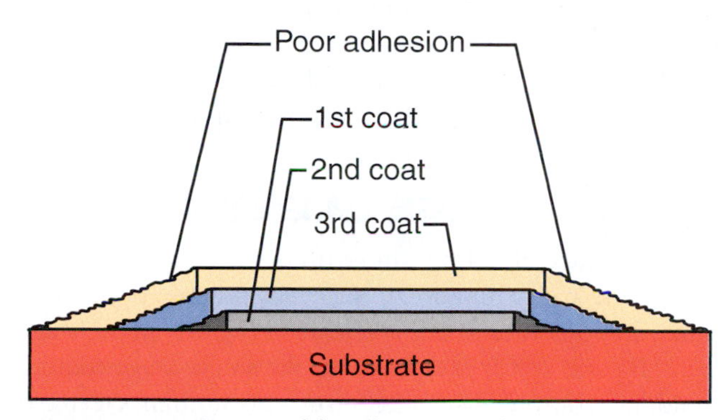

FIGURE 29-24 Reverse blending is used to ensure that dry spray under each coat of a blend will be eliminated.

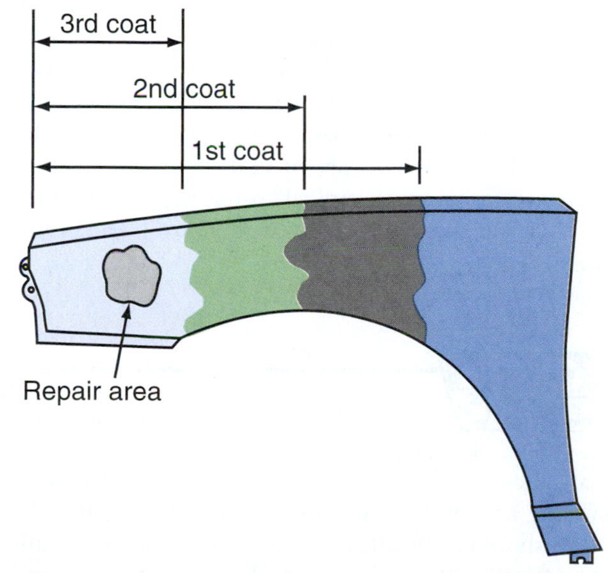

FIGURE 29-25 Reverse blending is a process where the largest coat is applied first and each consecutive coat becomes smaller.

This method improved the blending process greatly, and is now the method recommended by some of the leading paint manufacturers. Unfortunately, it did not solve all the color, or orientation, problems with some of the more difficult colors such as silver, gold, and bronze. It also didn't solve problems with colors containing large amounts of mica or pearl. These products will dry very fast, and sometimes they show at the edge of a blend.

Wet-Bedding Blending

To solve these problems with difficult colors, a blending method was needed which would allow the newly applied colors to lay on a wet surface, so that they could be held in place properly when sprayed. Also, if the color were to fall into a wet area, the additives like mica would not show as a dry area at the end of the blend.

To accomplish these goals, the **wet-bedding blending** method is now used. A technician sprays a coating of blending clearcoat on the panel to be blended before the color is applied. Then the color can be applied in the reverse blending method. Thus, all of the color that hits the vehicle will land in a wet "bed." See **Figure 29-26**.

If a blending clearcoat is not available, the technician can mix the clearcoat that will be used following the application of color, over-reducing it by 10 to 20 percent with a slow reducer. By doing so, the clearcoat will stay wet longer. The technician should apply a light coat of this blending clearcoat to the complete panel, and then apply the color using the reverse blending method.

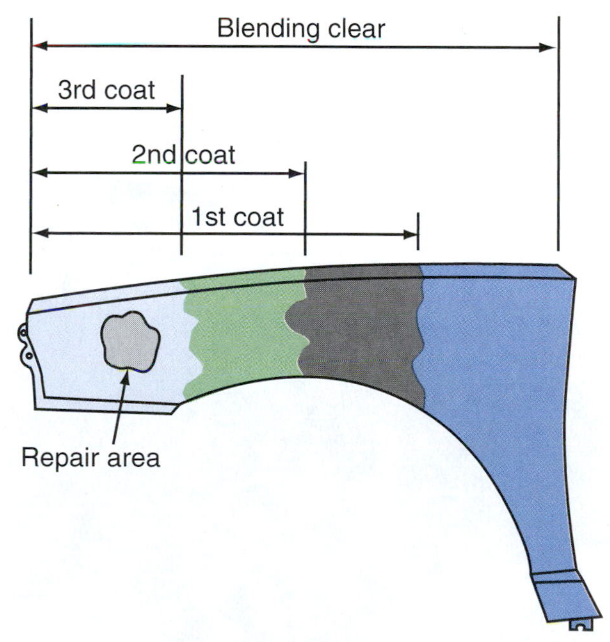

FIGURE 29-26 "Wet bedding" is a process where blending clearcoat is first applied to the entire panel before reverse blending. This method is best used for difficult color matches. Metallic lie is best achieved with this method.

CLEARCOAT BLENDING

At times, a technician may be asked to blend the clearcoat that has been put over a repair in an area where there is no convenient tape line as such. These areas might include a sail panel and the lower portion of a quarter panel. See **Figure 29-27**.

> **TECH TIP** Many automotive manufacturers or paint companies do not recommend clearcoat blending, because the thin blend area (that area where the clearcoat feathers out onto the OEM finish) will often fail in a short period of time. Clearcoat must have a minimum of 2 mils of thickness to withstand the abuses of ultraviolet (UV) light, and at the blend area there is much less than 2 mils with this method. The area will either fade, showing a repair line, or it will delaminate.

When attempting to clearcoat blend, the technician should prepare the surface in the same way as with clearcoat application. It should be sanded to P1000 or P1500 in the area where the clearcoat is to be applied. The vehicle should have the area where the clearcoat is to be applied sanded slightly beyond the spot where the clearcoat will stop.

After the color has been applied and the proper flash time has been observed, the first coat of clearcoat should be applied over the entire panel and into the area where the blend will take place. At the point of blend, the technician should off-trigger the spray gun, thus reducing the amount of paint coming out of the gun. However, the technician should also maintain the proper distance, so that the last part of the clearcoat to hit the vehicle will not have dried, as it would if the gun were fanned at the end of the stroke. The second coat of clearcoat is applied in the same manner as the first, but the blend is off-triggered lower than the first. If a third coat is required, it is also applied like the previous ones, with the third stopping below the second.

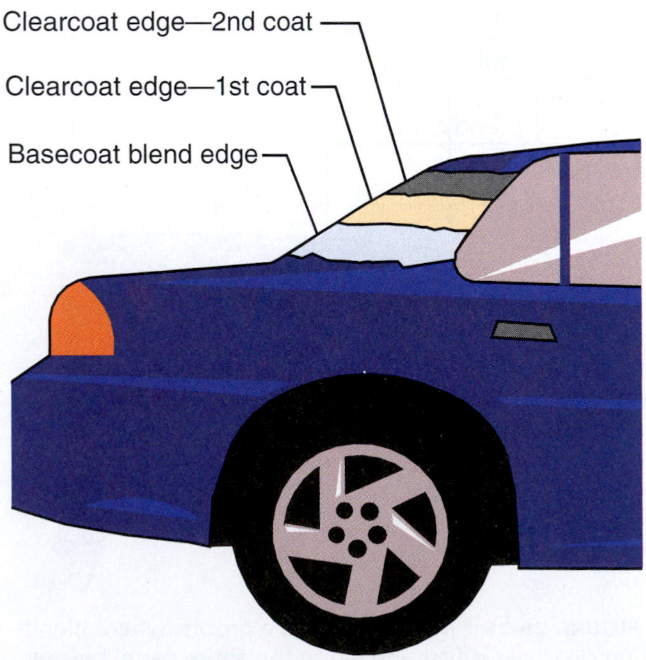

After the clearcoat has been applied, the technician applies a blending solvent to the blend edge. Some manufacturers market blending solvent, but if it is not available a technician could pour out most of the clearcoat that was used to blend, add solvent, and apply this mixture of mostly solvent with a little clearcoat (75 percent solvent to 25 percent clearcoat) to the blended area. Care must be taken to apply just enough of this mixture so the blend area will flow out without causing a run.

Some experienced technicians use pure solvent in a gun to flow out the blended area. However, much care should be taken with this method, because it is very easy to run the blend area with pure thinner in the gun.

FIGURE 29-27 Clearcoat blending on a sail panel, though not recommended by most paint companies, can be achieved if care is taken. However, its longevity may not be sufficient.

SINGLE-STAGE BLENDING

Single-stage blending can be done sometimes, depending on the paint manufacturer's recommendation. Most paint manufacturers allow clearcoating of single-stage paint. If paint repair is to be done, the single-stage coating could be sprayed over the repair area, then clearcoat sprayed over the entire panel to make the blend undetectable. Clearcoat can be blended using the off-trigger method (as explained in the section on clearcoat blending). In the area where the paint is to be blended, the first coat should be the largest and the gun should be off-triggered, not fanned. Each subsequent coat should be smaller and off-triggered as well. In the area where the paint coats meet the non-painted portion of the vehicle, a mixture of 75 percent solvent and 25 percent color is placed in the gun and applied to the edge of the blend area. This will allow any dry spray to flow out and become undetectable.

> **TECH TIP** Imagine that a technician has clearcoat blended or single-stage blended an area and finds that, after the area has cured, it must be detailed by sanding and buffing. In this situation, the area of the blend should be *detailed by hand* (not by machine buffing). Because this area is very thin and easily burned through, hand polishing with the least aggressive method first is the necessary technique.

MULTI-STAGE BLENDING

Multi-stage repairs are very difficult. When completing such a repair, the technician should not only make a sprayout panel for the basecoat, but should also make a **letdown panel** for the intermediate coat.

Multi-stage paint consists of a ground coat with color (and sometimes a pearl), and a **midcoat**, which could be clearcoat with pearl or prismatic color, or even a tinted clearcoat. Applying the proper number of midcoats needed to match the vehicle is critical. Then a clearcoat is sprayed over both coats to protect the finish. See **Figure 29-28**.

To repair a multi-stage vehicle, the technician first prepares it like any other blend repair. The bodywork repair (where it is primed and where the sealer and ground coat will be applied) should be finished to P400/P500 or equivalent. The remainder of the panel—where the midcoat, then clearcoat, will be applied—should be sanded to P1000/P1500 or equivalent.

The paint codes (both ground coat and midcoat) should be retrieved, and the technician should confirm they are the correct ones for the vehicle. After this, a **sprayout panel** should be made. See **Figure 29-29**. (For information on sprayout panels, see Chapter 21.) The sprayout panel can be compared to the vehicle to determine if the ground coat is correct and help the technician decide how many coats of midcoat are necessary for a proper match. The sprayout panel's ground coat color can be checked to the ground coat color on the vehicle.

To find a ground coat-only area on a vehicle for comparison, the technician must hunt for it. Sometimes a ground coat-only area can be found in the trunk under carpet, or under a sill plate in the door or even in the engine bay area. What is important is that the technician is sure that the ground coat is a good match before applying the other two coats.

After a sprayout panel has been made and the number of coats required for full coverage is known, a letdown panel for the midcoat should be made.

Because midcoats are coatings that either have mica, prismatic, or even tinting added to clearcoat, the technician needs to know how many coats of midcoat must be sprayed over the ground coat to match the original finish.

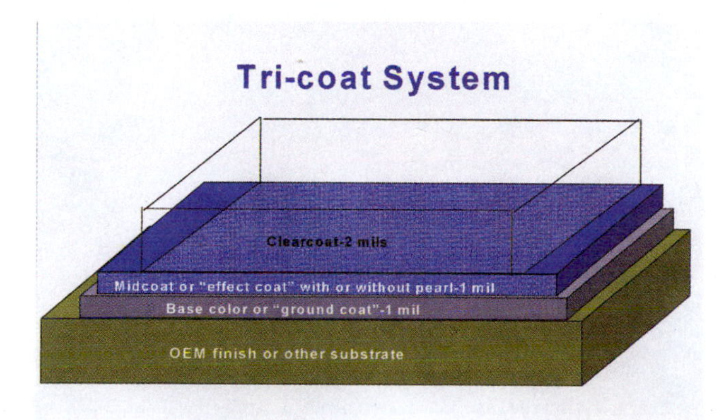

FIGURE 29-28 Multi-coats or tri-coats have a color coat (ground coat) applied, after which an effect coat or midcoat is applied before clearing.

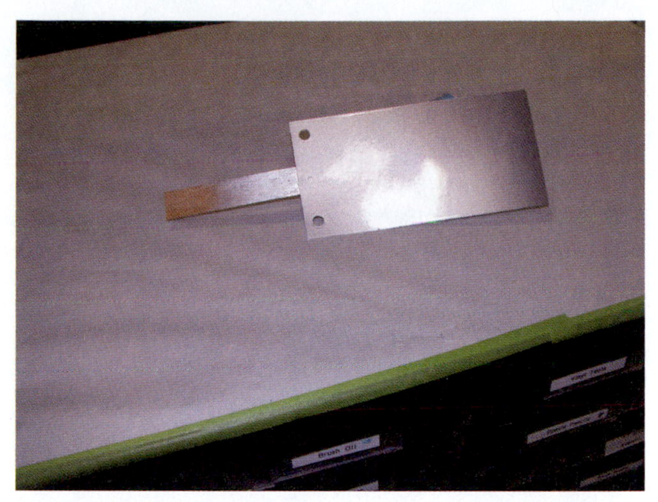

FIGURE 29-29 A sprayout panel is a small panel that helps the technician check for color match and application needs.

> **TECH TIP** Many paint technicians believe that the midcoat is the most important layer, and that a sprayout panel of the ground coat is not necessary. They are very careful to make a very accurate letdown panel of the midcoat, then spray the resulting blend on the vehicle, only to find that the blend is not the right color.
>
> Either the ground coat or the midcoat can be the culprit in a color mismatch. To avoid the problem, both a sprayout panel and a letdown panel should be made.

A letdown panel will let the technician know this. To make a letdown panel, the technician prepares a panel with the same amount of coats of ground coat that the sprayout panel revealed earlier. The letdown panel must be larger than the panel used for a sprayout test. See **Figure 29-30** and **Figure 29-31**.

The panel is then masked off with paper, starting from the bottom so it has five to six panels that can be opened later. See **Figure 29-32**. The last panel is left open, is cleaned, and then the first coat of midcoat is applied to the panel. When the proper flash time has been observed, the next panel is opened and a second coat is applied. This is continued until the panel is completely coated. The letdown panel now has only one coat on the bottom of the panel, two on the next, and so on until the top one, which has the most coats. When the proper flash time is observed, the panel is masked off in the opposite way from the midcoat application, and two coats of clearcoat are applied to half of the panel. See **Figure 29-33**. The panel now has ground coat on it with five to six progressive coats of midcoat, and finally two coats of clearcoat. The same amount of coatings is also used on the vehicle.

FIGURE 29-30 A letdown panel is made to determine the number of midcoats that will be needed to color match when applying multi-stage paints.

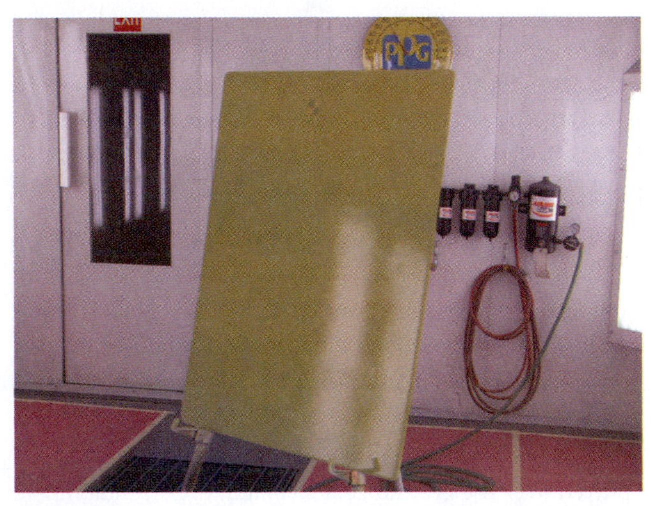

FIGURE 29-31 A panel is initially sprayed with the ground coat for hiding. Note the check hiding sticker.

The letdown panel can now be checked against the vehicle so the technician knows how many coats of midcoat must be applied.

Application

Ground coat is applied in the same manner as normal blending color is applied, in either the standard blend method or the reverse method. The same choices should be observed with ground coat as with basecoat. If the ground coat is highly metallic or has pearl in it, it should be applied in a reverse blend method. If the ground coat is non-metallic and contains only color pigment, a straight blend should be used.

Blending

Following the proper flash time, the midcoat should be applied in the reverse blend method, spraying the largest

(A)

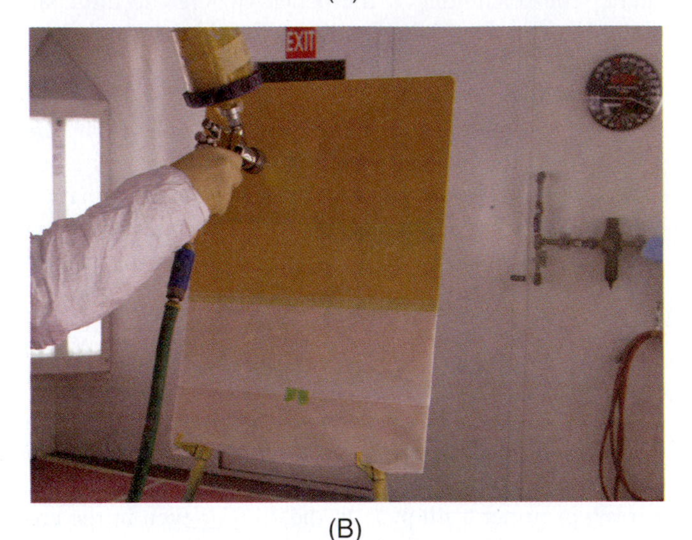

(B)

FIGURE 29-32 The letdown panel is masked for a minimum of five midcoat applications (A). Each time masking is removed, another midcoat is applied (B).

FIGURE 29-33 The panel is masked and clearcoat is applied to half the panel.

of the coats first and each subsequent coat smaller, until the ground coat is covered with the number of midcoats the letdown panel revealed would be a match.

Care must be taken to observe the proper flash time for each coat of midcoat. If the coats of midcoat are rushed and all the solvents don't evaporate properly, the midcoat will have a wet halo look to it when the clearcoat is applied.

As can be seen, the repair area for a multi-stage paint is much larger than for a standard blend. Care must be taken for each step performed in the process.

Flash times must be observed, or the repair will fail by being visible. However, if care is taken to apply the ground coat—then the proper number of midcoats—as indicated by a sprayout panel, and final clearcoat is properly applied, the result will be a professional and undetectable repair.

OVERALL VEHICLE REFINISHING

In collision repair, **overall** or **complete paint jobs** are not often done, but are required from time-to-time. In the paint shop, preparing for an overall job follows the basic steps as one would follow for a blend. The vehicle is washed with soap and water; it is chemically cleaned; parts are removed as needed; and the area to be painted is sanded with P400 or P500 grit sandpaper or equivalent. A plan of attack (or sequence) is developed for the vehicle. See **Figure 29-34**. After being placed in the booth, the vehicle is tacked off, and the basecoat is applied. To this point, the choices of solvent and application have been very similar to blend; only the area being sprayed is much larger.

When the clearcoat is to be applied, things change. The painter must be able to start clearing the vehicle at one point, then move on to the remainder of the panels, following the plan that was made, until the complete vehicle is cleared. When the vehicle is finished, the clearcoat where the painter started must still be wet enough so the last panel sprayed will melt into the first panel sprayed and there is no "dry line." To do this, the reducer used must be one that will not evaporate too quickly, and the hardener and/or catalyst must be ones that will not cure too quickly, either.

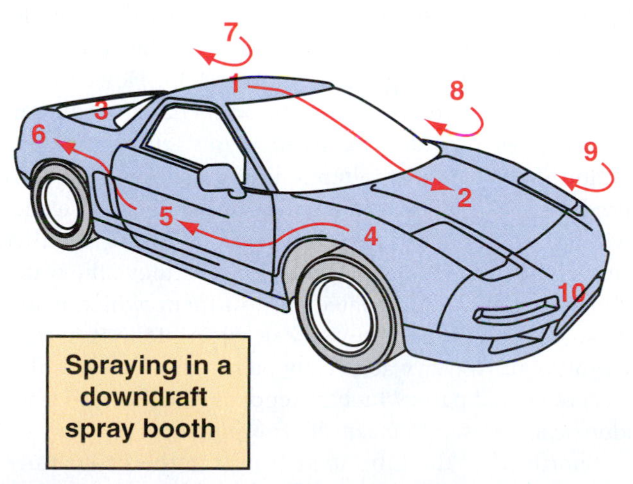

FIGURE 29-34 Overall or complete paint work must have a plan of attack.

As can be seen, complete or overall paintwork takes much longer than routine jobs, and the solvents and hardeners will take longer to cure, so most lead painters will save this type of work for the last vehicle of the day. This way, when the vehicle is completed it can be left in the spray booth to cure overnight.

WATERBORNE PAINTS

Paints that use water as their *primary solvent* have been used successfully in Europe for years. Note the emphasis on the words *primary solvent*. A common misconception is that waterborne paints have no solvents other than water in them. This is not true. Water—not tap water—is the primary solvent, but the paints still contain as much as 15 to 20 percent traditional VOC (volatile organic compounds)-laden solvents, depending on the brand of paint. See **Figure 29-35**.

By reducing the amount of VOC solvents in basecoat paint, a technician can reduce the amount of smog-causing emissions from automotive aftermarket refinishing by 75 percent or more. Some states have not met their current clean air emissions goals; others have, or are awaiting new regulations with a goal of reducing the VOC emissions by 70 percent or more.

Waterborne paints are not new technology. In fact, even before the introduction of waterborne basecoats in the

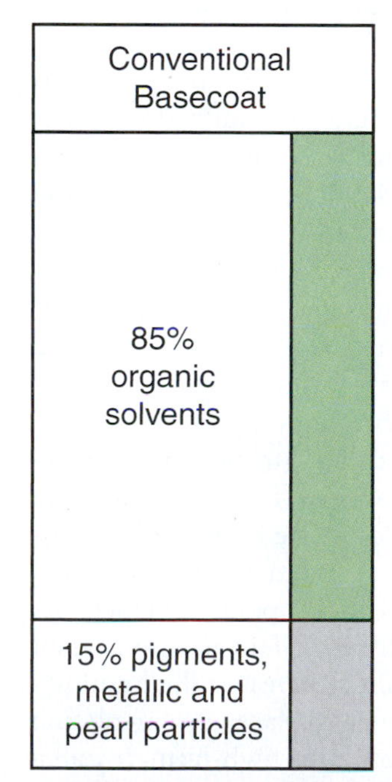

FIGURE 29-35 In waterborne paints, water is the primary solvent, but the paint can still contain as much as 15 to 20 percent VOC-laden solvents as well.

1990s, waterborne undercoat was available as a barrier coat between troublesome lacquer substrates and new finish. Although waterborne technology has advanced since its general availability in the 1990s, a suitable waterborne clearcoat is not yet available. While waterborne finishes for undercoats and basecoats are common, basecoat must still be protected with a solvent-laden, high-solids clearcoat.

There are other advantages to waterborne paint as well. First, cure time is reduced—if the booth has the proper airflow. Additionally, color match, metallic orientation, and blending can be enhanced with the use of waterborne basecoats. In addition, technician safety is markedly increased when waterborne coatings are used. Although some believe technicians need to use special respirators when applying water finishes, this fact is not supported in the product MSDS literature.

So what must be done differently when using waterborne coatings? Some people claim that waterborne coatings can be applied using the same equipment and techniques as with solvent paint. However, in order to benefit from all the advantages of fast curing and application, some new or additional equipment will be needed. Storage and cleanup are different, and the shop may need to make spray gun changes.

Equipment

Booths. Although new spray booths will not be needed, some changes in drying/curing equipment may be required. Most paint manufacturers recommend that a downdraft spray booth, which has the capacity to move large amounts of air, is the best type of booth to use; some booths may not move the needed amount of air for fast drying of waterborne finishes.

Traditional solvent-borne basecoats dry through evaporation. Even though atmospheric humidity does affect the drying time of these finishes, temperature has a stronger influence on their curing time. Baking-type spray booths can significantly increase the cost of using solvent-borne paints. When waterborne paints are used, the cost of baking is markedly reduced. When using solvent paint to speed up the cure time, a technician can choose a "faster" reducer and/or increase the temperature of the booth to reduce the coating's cure time.

When using waterborne finishes, air movement influences dry time the most, because water must evaporate from the finish. Although atmospheric humidity does influence a waterborne coating's ability to evaporate, if air is moved rapidly over the surface of the painted vehicle, its evaporation occurs more rapidly as well. In addition, when water evaporates out of a finish, it increases the humidity of the air just above the painted surface. By circulating large amounts of air in the booth, this high-humidity area just over the evaporating surface is moved and lower-humidity air takes its place, thus increasing drying.

Booths designed for solvent curing may not move the needed amount of air. Therefore, to increase booth air movement, special blowers called venturis may need to be added. Booths designed for waterborne finishes have venturis permanently installed. See **Figure 29-36**. For existing booths, a movable venturi or venturis can be placed in the booth, circulating air around the area on the vehicle which has been painted. See **Figure 29-37**.

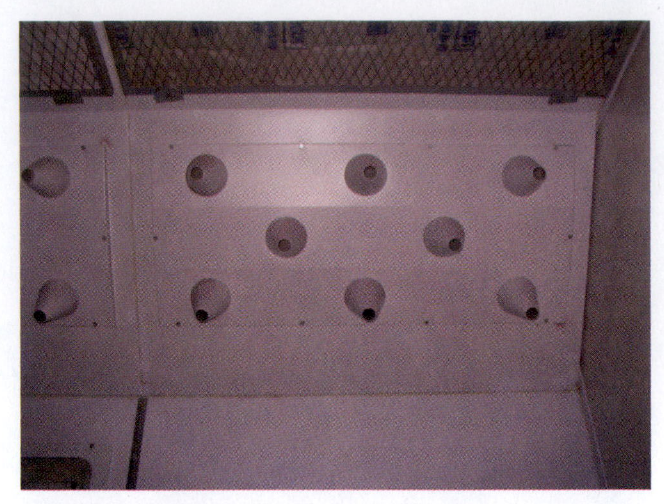

FIGURE 29-36 This paint booth has a roof-mounted venturi, which speeds up the airflow to help evaporate waterborne basecoats.

The natural question is, "If all this air must be moved past fresh paint, isn't it likely that the paint will become dirty?" The answer is yes. Dirt is among the painter's biggest enemies. The air that is passed through the venturis must be scrupulously clean. Instead of directing the airflow at the painted surface, the air should be directed so it passes over or by the painted surface. This moves the air past the evaporating paint without driving dirt into the paint.

Another consideration that may also need addressing is filters. Although paper booth filters are rarely used anymore, some booths still have them. Because of the water used with waterborne paints, paper filters can be damaged when used in conjunction with waterborne coatings.

Spray Guns. Most spray guns being used today for solvent-based paints can also be used for the application of waterborne coatings. See **Figure 29-38**. However, the technician must consider the recommended needle and nozzle sizes provided by paint manufacturers, as gun cleanup may need to be altered. When traditional guns are used with waterborne paints, the guns should be cleaned with water first to remove the waterborne paint, and then rinsed with either solvent or alcohol to remove the water. Older guns have non-coated steel in them, which would be susceptible to corrosion when water alone is used to clean them. Newer water-ready paint guns have stainless steel or coated parts which water does not affect, so water alone can be used to clean the equipment.

Shortly after the introduction of waterborne coatings, specially designed guns were developed for water application. These guns not only addressed the corrosion problem but also addressed the gun's ability to atomize waterborne

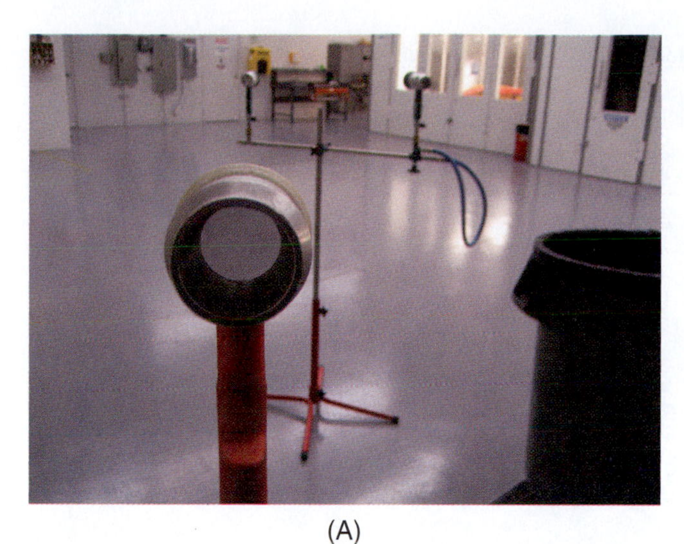

(A)

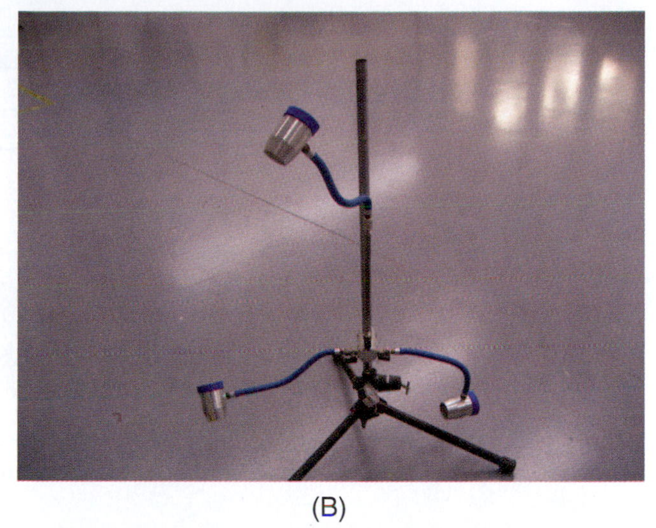

(B)

FIGURE 29-37 (A) A mobile venturi (for air augmentation) is shown in the foreground, with two more on stands in the background. They help meet the needs of air movement in an existing paint booth. (B) These mobile venturis stand with adjustable mounts. This allows the precise aiming of air movement when evaporating waterborne basecoat.

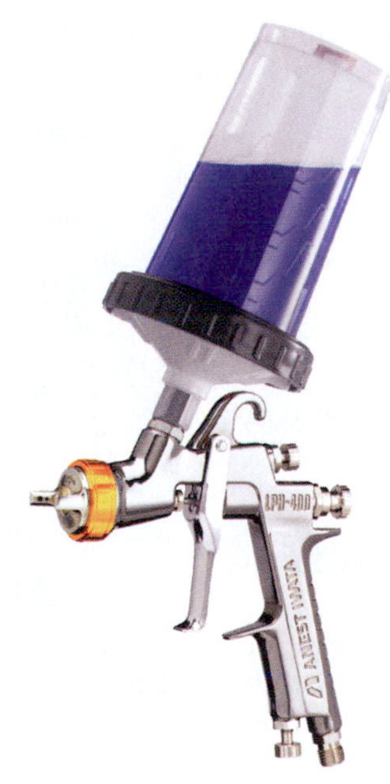

FIGURE 29-38 This spray gun is designed for waterborne basecoat application.

FIGURE 29-39 On this waterborne tint bank, note the green labels indicating waterborne product.

paint. As mentioned earlier, when applying any coating the technician should follow the paint gun setup recommendations for the coating being applied. If a tip size of 2.5 is called for, that's the size that should be used.

Storage

When using waterborne coatings, freezing or even long-time cold (below 40°F) may harm the paint, since water is the principal component. See **Figure 29-39**. When transporting waterborne paint products in the winter, vehicles must have the ability to keep the paint product from freezing. When stored at paint stores and in body shops, the paint must also be in heated areas to avoid freezing. Even when there is no danger of freezing, the paints need to be applied at suitable temperatures to avoid atomization and flow problems. Waterborne paint can be damaged by high temperatures as well, which leads some paint manufacturers to recommend that waterborne paint be stored in areas that do not go below 46°F and not above 75°F. The ideal temperature is a constant 68°F. See **Figure 29-40**.

Reduction

A common misconception with waterborne coatings is that a paint shop can just take water from the tap to reduce the coating. In reality, the water used to reduce waterborne

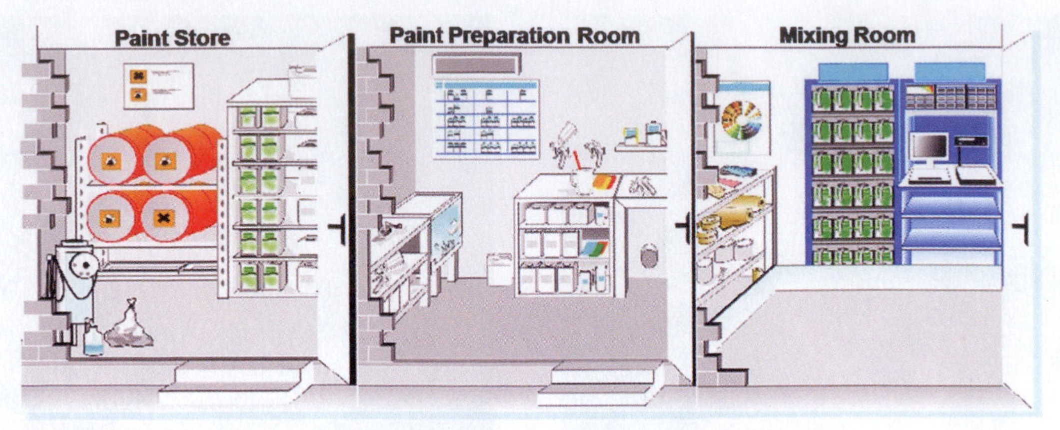

FIGURE 29-40 Waterborne paint product should ideally be stored at a constant 68°F. The recommended temperature parameters are between 46°F and 75°F. Rapid temperature changes should be avoided.

coatings must be de-ionized. Thus, most companies recommend and supply the precise type of reducing water needed for their finishes. See **Figure 29-41**. Contaminants such as iron or other minerals that appear in common tap water would cause undesirable side effects. The way the waterborne paint is reduced and the choices of reducers are significantly different than for solvent paint. Waterborne paints commonly have only one reducer to choose from. This makes reducer selection much simpler than with solvent paints. Also, some companies recommend that a non-metallic paint be reduced only 10 percent, and metallic/pearl paints be reduced 20 percent. This is a significant change from a 1:1 reduction, which is often recommended with solvent paint.

One paint line has introduced a micro-gel technology where their latex core swells, developing small finger-like hairs that help keep the paint from settling. See **Figure 29-42** and **Figure 29-43**. This anti-settling technology allows the paint to be stored without agitators or a mixing rack. The only mixing requirement before pouring the paint into a mix is a few gentle rocks back and forth to make sure that it is in suspension. See **Figure 29-44**.

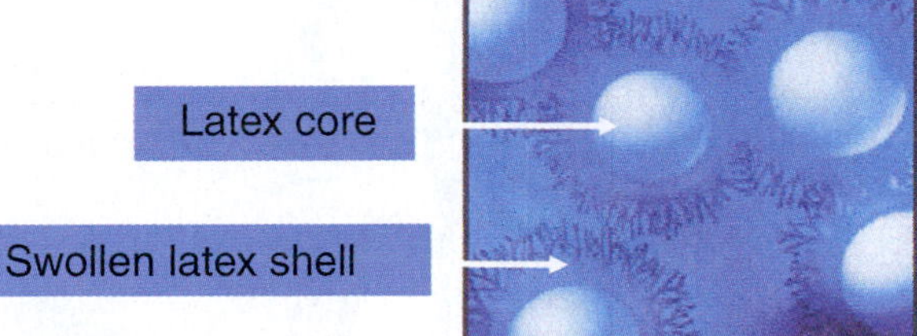

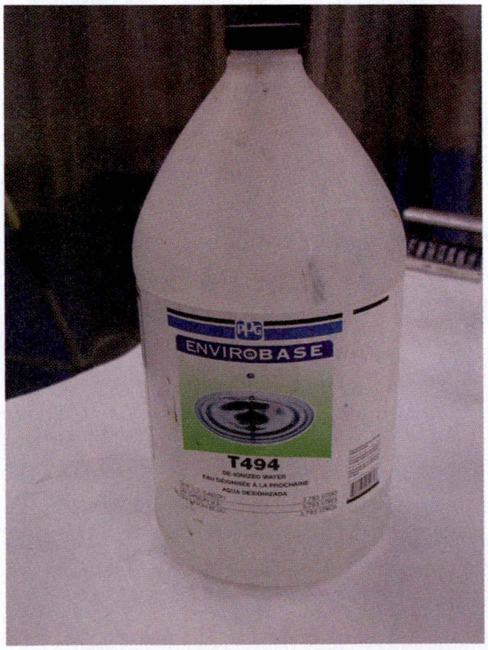

FIGURE 29-42 Some products have developed non-settling technology, which swells the latex shell. This develops "fingers" which keep paint in suspension, thus eliminating the need for internal agitators or a mixing bank.

FIGURE 29-41 Waterborne paint must be reduced with specially formulated water.

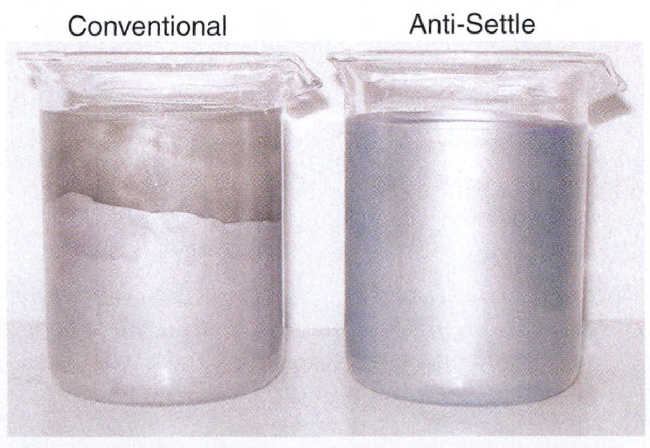

FIGURE 29-43 Conventional paint is shown on the left, anti-settle technology on the right.

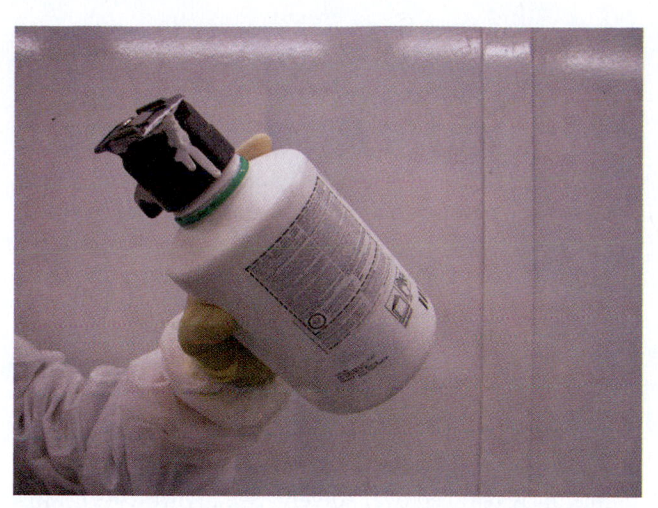

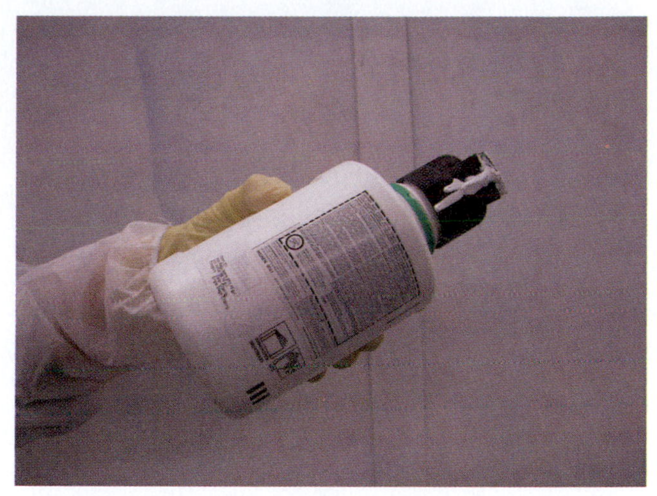

FIGURE 29-44 Anti-settle paint only needs a gentle rock to the left and right prior to mixing.

FIGURE 29-45 Although waterborne coatings have reduced VOCs, technicians still need to use personal protective gear.

FIGURE 29-46 During waterborne final cleaning, a wax and grease remover is used just prior to the application of a waterborne basecoat.

Safety

As mentioned earlier, waterborne coatings are much safer for both the environment and the technician applying the coating. However, personal protective equipment will still be needed. See **Figure 29-45**.Technicians must still read, understand, and follow all the safety precautions listed in the product MSDS. Paint manufacturers, in their quest to help paint shops reduce throughput cycle times, work continuously on developing safer coatings. Nonetheless, considering the amount of hazardous ingredients they are routinely exposed to, paint technicians must be diligent in their pursuit of personal safety.

PREPARATION

Surface preparation for using waterborne coatings is similar to that used for solvent paints, where the surface must be cleaned and de-greased, sanded and scuffed, masked, and then placed in the booth for refinishing. However, some minor differences do exist. Most paint manufacturers recommend that the surface to be painted be finish sanded to a finer grit than used for solvent paints—such as P-800 by hand or P-1000 by DA—to avoid metallic tracking into the sandpaper scratches. The cleaning process is also very similar, although a specific solvent-reduced waterbase wax and grease remover may be recommended for use by the paint manufacturer or mandated by local law before painting. See **Figure 29-46**.

Masking and Tape

Masking with a coated paper that is water resistant will avoid the problem of paper becoming saturated with water and thus allowing paint to bleed through. Most high-grade paper contains a solvent resistance coating which is likely to resist water as well. However, it is prudent to confirm this with the paper manufacturer or your jobber. If saturated by the water process, tape may also become difficult to remove. Because it is applied in light coats then forced to evaporate before the next coat is applied, waterborne spraying is unlikely to cause difficulty with the masking or tape.

WATERBORNE APPLICATION

The application of water finishes is different than the application of solvent finishes. Those who are familiar with solvent application will need to develop new application techniques. Although waterborne paints have significantly reduced the use of organic solvents in their mix, approximately 15 percent still remains. Therefore, an NIOSH-approved vapor respirator is necessary, along with gloves and glasses for its application. The first two coats—or the number needed until hiding is accomplished—are applied as with solvent paint.

TECH TIP As with solvent paints, the manufacturers' recommendations for paint gun setup and operations—such as needle and nozzle recommendations as well as air pressure—should be followed for the coating that you are painting. Each paint company's recommendations can be different, yet are critical to follow. To obtain the best results, a paint technician should read, know, and follow the recommendations for the products being used.

Blending

Most paint manufacturers recommend two or three blending techniques, all of which use a product that may be called a blend-in or colorless coat. This product is used to produce an undetectable repair.

Standard, reverse, and wet-bed blending with waterborne coatings is similar to the process used for solvent paint for the first two coats or until near coverage is accomplished. The last coat is sometimes called either a "control coat" or a "drop coat." It is critically important that this coat is applied correctly.

Control/Drop Coat

To apply a control/drop coat, the gun is adjusted to a lower air pressure (reducing the pressure by 10 to 20 percent). The gun distance is also lengthened and the travel time is increased. By applying a light coat to the blended area, the metallic orientation can be smoothed out, providing an undetectable blend. This technique has been used for years by some painters with the application of solvent metallics. When a metallic paint becomes meddled or streaky, it is necessary to apply every time a waterborne basecoat is applied. It may not be easy to detect streaks in the finish prior to applying clearcoat; therefore, the final coat (or two) should be control/drop coated.

Evaporation/Flash Time

Flash time recommendations have, in the past, been printed in the technician support bulletins provided by the paint manufacturers. With waterborne paint, these recommendations may not be available. The time that a waterborne coating takes to evaporate prior to applying the subsequent coat will vary with every application. Depending on the dryness of that day's air and the air movement in the booth, flash/evaporation varies significantly. A paint technician could—with the choices of reducers—control the flash time with the paint. With waterborne, air movement becomes the controlling factor. After a coat has been applied and air movement for waterborne coating is initiated, either by venturis or fans, the flash time can be significantly reduced over solvent. See **Figure 29-47**. Most paint recommendations state that the basecoat should dry to a uniformly dull coat with no visible "wet" areas prior to the application of the next coat.

Color Matching

Color matching for waterborne paint is similar to the process used with solvent paint. The formula retrieval must be precise, including checking for the color variant that will provide a blendable match. The formulas must be mixed precisely and a sprayout panel made to assure that the mix is correct.

Clearcoat is applied according to the paint makers' recommendations after the basecoat has sufficiently cured.

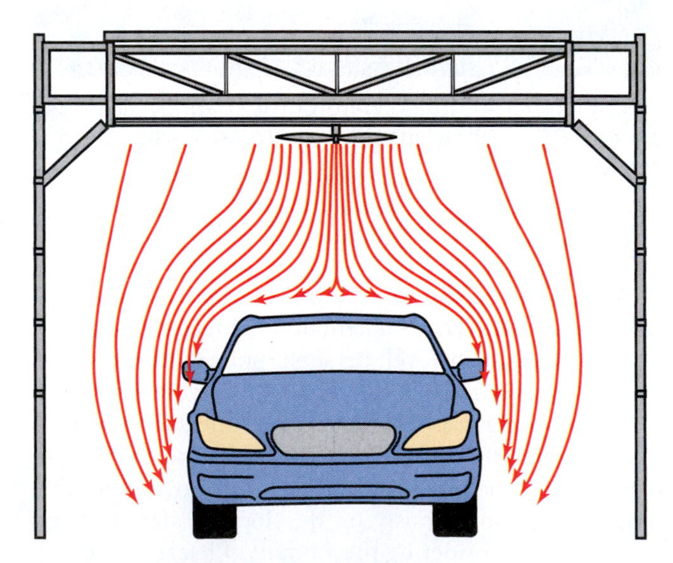

- Additional air velocity
- Waterborne flash cycle
- Shortens bake times for clears
- Retrofit to existing booths
- Easy installation

FIGURE 29-47 One of the ways to increase air movement in an existing spray booth is to install a ceiling fan.

FIGURE 29-48 A waterborne paint gun cleaner. Waterborne waste must be kept separated from solvent waste. Used with permission of Hedson Technologies N.A., Inc. All rights reserved.

Waterborne gun cleaning equipment is available that cleans the painting equipment and then, employing a filter system, can periodically clean the cleaning water. See **Figure 29-48**.

CLEANUP AND DISPOSAL

Waterborne guns must be cleaned separately from those used to apply solvent coatings. The solution used, if not specifically recommended by the manufacturer, must be de-ionized and de-mineralized to avoid residual deposits. The water used for cleaning is considered a toxic waste after cleaning in most areas and must be disposed of accordingly. The water can be cleaned of the paint residue by adding a powdery substance called a "flocculating agent." When added to waterborne contaminated cleaning fluid and agitated, it will cause the paint particles to clump together. The particles can then be strained out, and the cleaning water used up another 10 times as recommended by some manufacturers. This gun-cleaner waste, like any paint waste, must be disposed of separately from the solvent waste in most areas.

Summary

- As the student is learning, the job of a lead painter is complex, as it requires mastering each of the blending techniques, and also being able to recognize which one would work best on which vehicle as it arrives in the paint department.
- Lead painters must be able to accomplish these techniques as rapidly as possible, while still producing high quality and durable work.
- To accomplish these outcomes, lead painters must use all the tools and resources they have, including:
  - A sprayout panel to check the color match and to determine how many coats of basecoat will be needed to get full coverage.
  - Letdown panels to determine how many coats of midcoat will be needed to obtain a quality match.
  - Spray techniques which will increase the painters' transfer efficiency, thus reducing the amount of overspray and of refinish coat waste.
  - Above all, correct safety techniques. Successful painters master the principles of working in a safe manner to protect themselves, their coworkers, and the environment.
- It may take a new technician years to perfect all of the techniques and skills outlined in this chapter; however, with practice, study, and continued training, a technician can progress in a paint department to become a lead painter.
- By reducing the amount of VOC solvents in waterborne basecoat paint, a technician can reduce the amount of smog-causing emissions from automotive aftermarket refinishing by 75 percent or more.
- In order to benefit from all the advantages of fast curing and application with waterborne coatings, new or additional equipment will be needed and storage and cleanup are different than with solvent paints.

Key Terms

blendable match
complete paint job
cutting-in parts
edging-in parts
foam brush application
infrared light
letdown panel
midcoat
no-sand sealers
off-triggering
overall paint job
pot-life
powder coating
ready to spray (RTS)
reverse blending
sprayout panel
standard blending
transfer efficiency (TE)
vehicle identification number (VIN)
wet-bedding blending
wet-on-wet
wet product

Review

ASE Review Questions

1. Technician A says that painters who have been painting for years can develop an eye for paint mixing and no longer need to measure when they reduce paint. Technician B says that paint reduction is critical, and paint must be measured each time it is mixed, no matter what the painter's experience. Who is correct?
   A. Technician A only
   B. Technician B only
   C. Both Technicians A and B
   D. Neither Technician A nor B
2. Technician A says that mixing by scale is the most accurate method. Technician B says that a paint's pot-life refers to how long it can stay in a can after it is shipped from the factory to the paint shop. Who is correct?
   A. Technician A only
   B. Technician B only
   C. Both Technicians A and B
   D. Neither Technician A nor B
3. Technician A says that a 2K product refers to epoxy paint products only. Technician B says that RTS stands for Reactive Trisulfide. Who is correct?
   A. Technician A only
   B. Technician B only
   C. Both Technicians A and B
   D. Neither Technician A nor B
4. Technician A says that in a paint ratio of 4:1:1, the 4 stands for the amount of reducer that is added to the mixture. Technician B says that a painter with a high transfer efficiency is a painter who will save a paint department money by using less materials. Who is correct?
   A. Technician A only
   B. Technician B only
   C. Both Technicians A and B
   D. Neither Technician A nor B
5. Technician A says that some undercoats can be efficiently brushed or rolled on to speed up production. Technician B says that engine bays are always the same color as the vehicle's outer color. Who is correct?
   A. Technician A only
   B. Technician B only
   C. Both Technicians A and B
   D. Neither Technician A nor B
6. Technician A says that edging-in of parts is often the paint apprentice's job. Technician B says that wet-on-wet sealers are a type of sealer that allows the color to be applied with only a short flash time, thus speeding up paintwork. Who is correct?
   A. Technician A only
   B. Technician B only
   C. Both Technicians A and B
   D. Neither Technician A nor B

7. Technician A says that color that was mixed for another car should never be used to edge-in parts, even when it is a close match. Technician B says that blending, though used occasionally in paint shops, is too costly to use on a regular basis. Who is correct?
   A. Technician A only
   B. Technician B only
   C. Both Technicians A and B
   D. Neither Technician A nor B

8. Technician A says that standard blending means a technician sprays the largest area coat first, with each coat thereafter being smaller. Technician B says that the wet-bedding blending spray technique is used for solid colors only. Who is correct?
   A. Technician A only
   B. Technician B only
   C. Both Technicians A and B
   D. Neither Technician A nor B

9. Technician A says that when preparing a panel for blending, one should sand the entire panel to P400/P500 or equivalent. Technician B says that when a panel is being prepared for blending, the area that will receive color is sanded to P400/P500 or equivalent. Who is correct?
   A. Technician A only
   B. Technician B only
   C. Both Technicians A and B
   D. Neither Technician A nor B

10. Technician A says that when preparing a panel for blending the area, it should be sanded with P1000 or equivalent. Technician B says that clearcoat can be used over some single-stage blends. Who is correct?
    A. Technician A only
    B. Technician B only
    C. Both Technicians A and B
    D. Neither Technician A nor B

11. Technician A says that a multi-stage paint's midcoat could contain pearl, prismatic color, or even be tinted with color. Technician B says that a sprayout panel is made to find out how many midcoats are needed with a multi-stage paint. Who is correct?
    A. Technician A only
    B. Technician B only
    C. Both Technicians A and B
    D. Neither Technician A nor B

12. Technician A says that the ground coat of a multi-stage paint job should have a letdown panel made, so the technician knows how many coats are needed. Technician B says that a multi-stage paint job will take up less room on a panel than a standard blend. Who is correct?
    A. Technician A only
    B. Technician B only
    C. Both Technicians A and B
    D. Neither Technician A nor B

13. Technician A says that because there are so many coats that must be applied to a multi-coat paint job, the flash times recommended can be cut in half to speed up the work. Technician B says that both the ground coat and the midcoat need to be checked for color match before application, because either being off could cause the new finish to be an unacceptable match. Who is correct?
    A. Technician A only
    B. Technician B only
    C. Both Technicians A and B
    D. Neither Technician A nor B

14. Technician A says that the plan of attack for a complete paint job is not as critical as a blend. Technician B says that the type of reducer and hardener used for a complete paintwork may be one that causes the evaporation and curing to be slower than a blend. Who is correct?
    A. Technician A only
    B. Technician B only
    C. Both Technicians A and B
    D. Neither Technician A nor B

15. Technician A says that after blending the color into a panel, the clearcoat should be applied to the entire panel. Technician B says that when blending clearcoat, even if it looks good after it cures, it may fail later from exposure to the elements. Who is correct?
    A. Technician A only
    B. Technician B only
    C. Both Technicians A and B
    D. Neither Technician A nor B

Essay Questions

1. Explain how a painter's transfer efficiency affects the paint cost of a paint department.
2. Describe the steps that must be taken to prepare a panel or panels for blending.
3. List the steps needed to make a letdown panel.
4. Explain why making a sprayout panel is necessary.
5. Describe the method of wet-bedding blending.

Topic Related Math Questions

1. A painter needs 64 ounces of paint to complete a job. The paint is mixed at a 4:2:1 ratio. How many ounces of paint are needed?
2. A paint department uses 8 gallons of paint per week. How many gallons of paint are needed for a year?
3. If a painter needs to over-reduce a liter of paint by 15 percent, how many milliliters are needed?

4. If a painter is mixing paint for a job with a mixing ratio of 4:1:1 and she adds 12 ounces of paint, how much hardener must be added?
5. A painter's starting transfer efficiency is 46 percent. After training, testing shows that the painter's transfer efficiency has improved 25 percent. What is the new transfer efficiency level?

Critical Thinking Questions

1. A silver metallic vehicle arrives in the shop for painting. Explain which method of blending should be used, and why.
2. Three vehicles come to the paint department at the same time. They all must be completed within 24 hours. Two are blends, and one is an overall paint job. Which should be painted first and which last? Explain why.

Lab Activities

1. Have students prepare a vehicle for blending.
2. Have students apply a blend using the standard method.
3. Have students apply a blend using the reverse method.
4. Have students apply a blend using the wet-bedding method.
5. Have students blend a multi-stage repair.

Chapter 30

Color Evaluation and Adjustment

OBJECTIVES

Upon completing this chapter, you should be able to:

- Explain Munsell's color theory.
- List the definitions and terms used to explain and understand color.
- Explain the effect of light on color, in terms of how it reflects and how it refracts.
- Explain what metamerism is.
- Define color deficiencies, and how a technician can correct for them.
- Explain how color variances happen during manufacturing, how the paint manufacturers make adjustments for them, and the tools at a painter's disposal to correct them.
- Outline how pigments, metallic flakes, flop, mica, and prismatic colors affect a mixed and sprayed color.
- Describe why a color effect panel helps when tinting a mismatched color.
- Demonstrate the steps needed to color plot and tint a color, thus making it a blendable match.

INTRODUCTION

This chapter will explain the topics of color theory, how people perceive color, and the definitions of terms used in explaining color. The chapter will discuss the effects that different types of light have on color, and how different colors both reflect and refract light. It will introduce the students to *metamerism* and how it affects the way a color appears in different types of light. Students will explore color deficiencies, and discover how some persons see colors differently than others.

This section will also introduce the way pigments, metallics, mica, and prismatics affect color, and how the technician can evaluate color and adjust it regarding each material. The paint property of "flop," and how it can be recognized and adjusted, will be covered in this chapter as well.

By studying and understanding these concepts, a student will be able to understand and apply color plotting, and therefore be able to tint a color which is not a blendable match so that it can become one.

Color plotting is the science and art of adjusting a mixed color to create a blendable match. Even if a technician does everything correctly to match a color, from time to time the result may still be an unblendable match. That is, the technician may have correctly found the paint code on the vehicle by converting it from the paint manufacturer's code; correctly retrieved the manufacturer's formula, mixed without mistakes (no over- or underpours), and prepared a color effect panel. In spite of doing all of these steps correctly, the color may still be an unblendable match. To correct an unblendable match, the technician should add toner or toners, a process commonly called "tinting."

Although some technicians are able to look at the color and adjust it using tints from the tinting bank, most are not able to do so. Over the years, a process of color plotting has been developed so the technician can first evaluate the mismatched color, and then determine which of the colors in the formula can be added to it in order to adjust the color, thus making it a blendable match.

Color plotting will allow even those less experienced technicians to adjust or tint a mismatch color. To become proficient at tinting, a technician must first understand the steps needed. Then, through practice, one can become a proficient tinter.

This chapter will first introduce students to the terms and theory of tinting, then take them through each step of the process, thus preparing them for actual tinting.

SAFETY

Chemicals used in the refinish business can be hazardous to both personnel and the environment. The refinish technician is governed by—and must comply with—many local, state, and federal laws concerning protection.

All hazardous products must be accompanied with a material safety data sheet (MSDS). The MSDS includes sections on:

- Product information
- Hazardous ingredients
- Physical data
- Fire and explosion hazards
- Reactive data
- Health hazard data, or toxic properties
- Preventive measures
- First aid measures
- MSDS preparation information

Technicians must read, understand, and follow the recommendations in the MSDS for the protection of themselves, their fellow workers, and the environment.

COLOR THEORY

Color theory dates back to the work of Isaac Newton (1642–1727) and his experiments with prisms in the 17th century. By shining sunlight through a glass prism, Newton noted that seven colors would appear: red, orange, yellow, green, blue, indigo, and violet. See **Figure 30-1**.

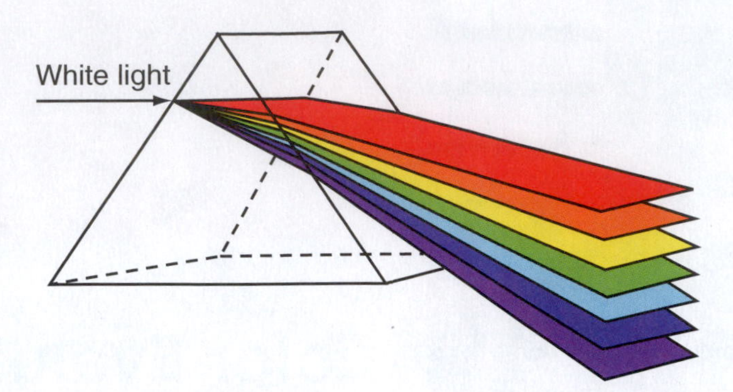

FIGURE 30-1 Prisms will show seven colors—red, orange, yellow, green, blue, indigo, and violet—as sunlight passes through.

TECH TIP Students of color theory have sometimes found it difficult to remember which colors appear—and in what order—when white light passes through a prism. To assist them, a work trick, or mnemonic device, was developed: The "name" **ROY G BIV** stands for red, orange, yellow, green, blue, indigo, and violet. This helps one to remember both the color names and the order in which they occur.

When Newton observed sunlight as it passed through a prism, it was thought that the glass had seven colors (ROY G BIV) within it, and the light passing through the glass reflected the seven colors like light shining through a stained glass window. However, Thomas Young (1773–1829) discovered that sunlight, or what would come to be known as white light, consisted of electromagnetic radiation. As the light passed through the prism, it was divided into the seven different wavelengths, with purple and blue being the shortest, green and yellow being medium lengths, and orange and red being the longest.

To help explain the various colors, three colors—red, yellow, and blue—became known as **primary colors**. Mixing different combinations of these three colors produced all the other colors. For example, to get green, one mixes together equal parts of yellow and blue.

Another aid created to help art students better understand colors and how they affect each other was the **color wheel**. See **Figure 30-2**. This wheel was divided into the three primary colors (red, blue, and yellow), then further divided into three **secondary colors**: orange, green, and violet. It was divided again into six additional colors: red-orange, red-violet, yellow-green, yellow-orange, blue-green, and blue-violet. Black and white are not considered colors at all. Black is considered the presence of all the col-

FIGURE 30-2 A color wheel is divided into the three primary colors (red, blue, and yellow), as well as the three secondary colors: orange, green, and violet.

ors (equal parts of yellow, blue, and red mixed together will produce black), and white is considered the absence of all color.

TECH TIP

The color wheel used for automotive tinting (which will be discussed later in this chapter) uses four divisions of color: red, yellow, blue, and green. Although green is not a primary color, it is used as the fourth spoke of the automotive color wheel to help visualize the colors and how they affect each other. See **Figure 30-3**.

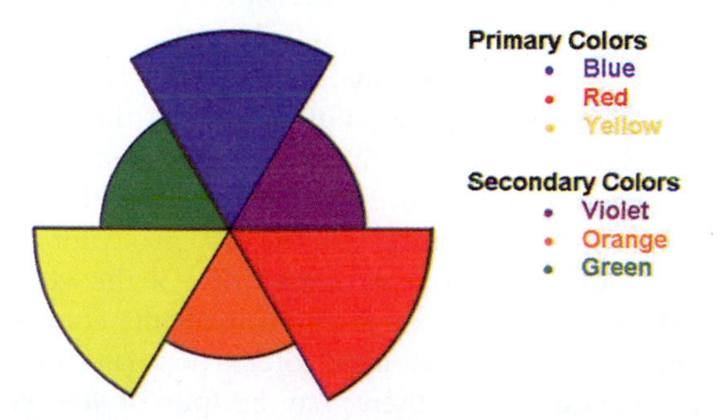

FIGURE 30-3 The color wheel used for automotive tinting uses four divisions of color: red, yellow, blue, and green.

Albert H. Munsell (1858–1915), an American art instructor, came up with a way of describing color and all of its variables. He also developed a visual tool that not only described the three dimensions of color, but also arranged them in a three-dimensional wheel that was easier to understand. See **Figure 30-4**.

Color is described as having three dimensions: hue, value, and chroma (sometimes called intensity).

Hue is the quality or dimension of color, such as red, blue, or yellow. When sunlight, which has all the wavelengths of all colors, shines on a red car, all the colors are absorbed except red, which is reflected back. See **Figure 30-5**. Therefore, a car's color can be described as having a red hue. Hue is illustrated on the outside spokes of Munsell's color wheel.

Another dimension of color is its value. **Value** is the lightness or darkness of color. On Munsell's color wheel, this is the center column, with white (the absence of color) at the top and black (the presence of all color) at

FIGURE 30-4 Munsell color theory is described as having three dimensions: hue (the outside spoke of the color wheel), value (from outer to the center axes), and chroma (the center axes).

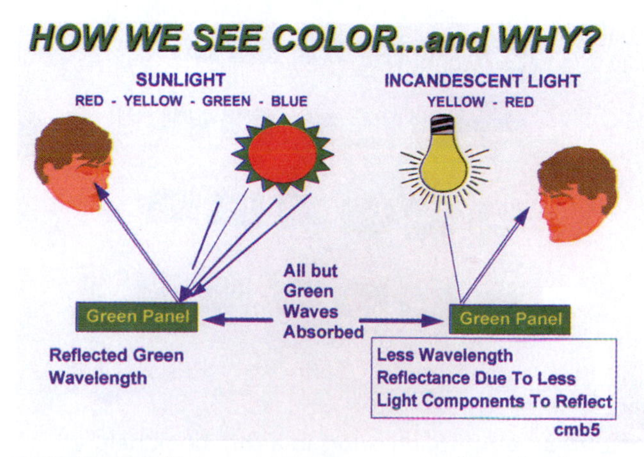

FIGURE 30-5 The eye perceives green after all other color waves are reflected.

the bottom. If red were to be illustrated at its different values, the top would be light pink and the bottom would be a very dark red.

The last dimension of color is chroma, sometimes called intensity. **Chroma** is the color's brightness. If a vehicle is said to be *bright* red, its hue is red and its chroma is bright, or pure intense red. On the Munsell color wheel, chroma is represented as the spokes that go through the center of the wheel.

FIGURE 30-7 Along with color chip books, variance decks provide common variance of particular formulas.

EVALUATING THE NEWLY MIXED COLOR

There are many ways to describe color, depending on *why* one needs to describe the color. In the automotive refinish industry, when trying to match a color, the newly mixed color is always compared to the vehicle that it must match. That is to say that the standard is always the vehicle being matched, not the **color chip book** or the color variance book or deck. See **Figure 30-6** and **Figure 30-7**. The newly mixed paint should not be compared to the vehicle while still in the can. Neither is it accurate to compare paint on a paint stick to the vehicle by looking at them. See **Figure 30-8**. Newly mixed paint is best compared when it is applied to a sprayout panel or color effect panel. This panel has color hiding devices (the black and white squares). When the new color is sprayed until one cannot see the difference between the two colors, the panel has been sprayed to "hiding," and the technician will know the number of coats that the new paint needs to reach "complete hiding." Then, when clearcoat is applied, the sprayout panel can be compared to the vehicle. By using a sprayout panel, the metallic orientation will be the same as when the vehicle is sprayed, and it is the truest way to compare the new paint to the vehicle to decide if it is a blendable match.

FIGURE 30-8 Comparing paint from a paint stick against a car is not accurate enough for color match.

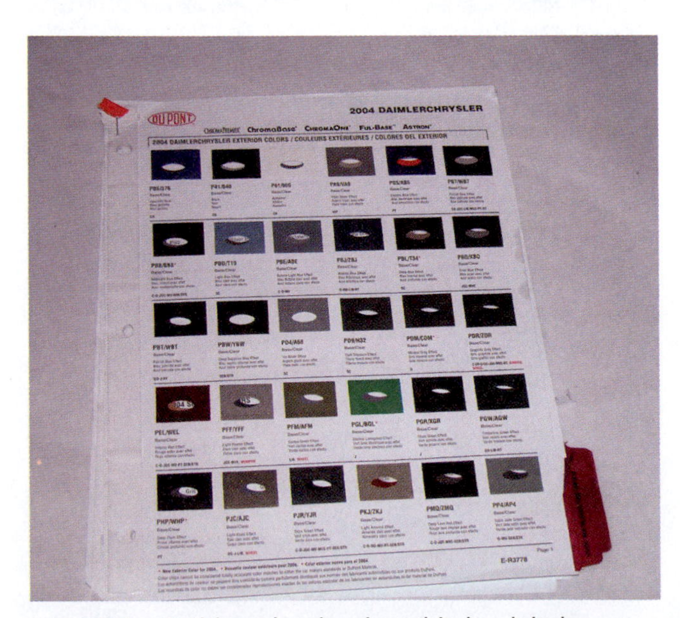

FIGURE 30-6 Although colors in a chip book help painters, the vehicle being refinished is the standard against which to match the paint being mixed.

Tinting

Tinting is the last resort when matching a vehicle, as it can be costly and time consuming. However, when an experienced painter feels the newly mixed color is not a blendable match—even when sure the correct formula has been mixed correctly—then the decision to tint should be considered.

When a color is a mismatch, its code should be rechecked by retrieving the paint code from the vehicle and the color retrieval system. A common mistake that new technicians often fall prey to is trying to get the vehicle's color code by comparing it to the chip in a color code book. This method can sometimes work, if the vehicle manufacturer does not offer many of the same color options for that year. However, if the color is popular—for example, a metallic blue—there may be four or five very close blue chips in one paint book. The small chip in the chip book does not give the technician a large enough color surface to make a proper match. The second mistake that a new painter can make is to try to retrieve the color by its name. The name of the color will change from year to year and from vehicle to vehicle, and is never a good method of retrieving a paint code.

The factory paint code should always be retrieved from the vehicle. See **Figure 30-9**. This is not always an easy task, since the paint code is not in the same place on different makes of cars. Even same-make vehicles may have the paint code in different places from year to year and model to model.

The best way to find the paint code location is to go to the paint manufacturer's paint code book, which contains a directory of paint code locations. Another way to find the paint code is to look up its location in the vehicle's repair manual. See **Figure 30-10**.

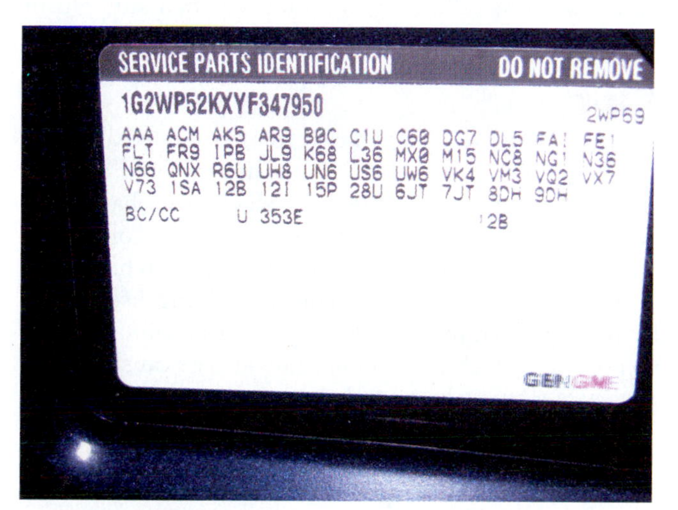

FIGURE 30-9 All vehicles have a manufacturer's paint code. This GM vehicle has the code U353E in the trunk.

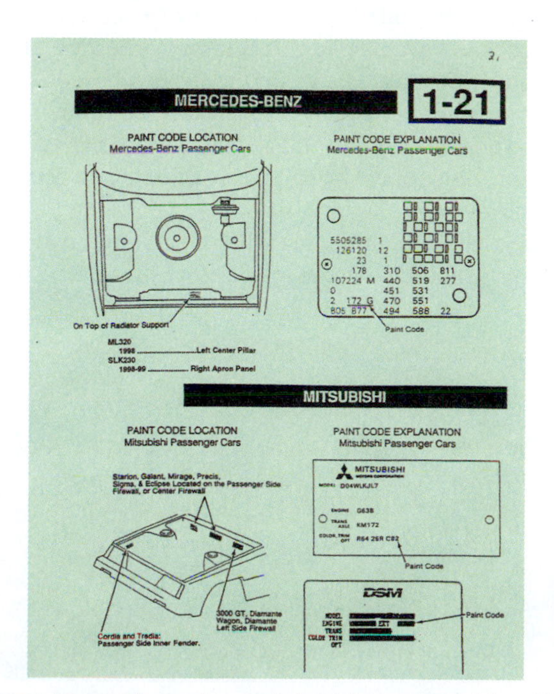

FIGURE 30-10 The paint code location can be identified using the manufacturer's repair manual.

FIGURE 30-11 Computerized paint pouring machines will pour even small amounts accurately.

When the paint code is confirmed, if the paint is still a mismatch, the next step is to confirm the mixing accuracy. If the paint, while being mixed, was either overpoured or underpoured, the paint will not be correct.

The development of paint pouring machines has greatly diminished the likelihood of a paint formula being poured incorrectly. See **Figure 30-11**.

If all the previous variables have been confirmed and no mistakes have been made in the preparation of the new color, the technician may decide to tint.

Describing Color

Although there are other methods of tinting colors, the plotting method is most commonly used.

When the plotting method is used, the dimensions of color are described with value first. Is the newly mixed paint lighter or darker than the vehicle it is being compared to? Second, the hue is compared to the vehicle. Here the technician only has a choice of the colors on either side of the vehicle's color on the color wheel. Say, for example, that the color being compared is blue. The vehicle could only be either a "redder" blue or "greener" blue than the mixed paint. Red is to the right of blue on the color wheel, and green is to the left of blue on the color wheel. The blue color could never be more yellow than the vehicle. Yellow is opposite blue on the color wheel, and crossing through any of the four colors on a color wheel would dirty the color, which would change its value, not its chroma. When matching a color and describing hue, only the color to the right or left of the color to be matched can be compared against it.

The last color dimension that is compared and described is chroma (sometimes called intensity, saturation, or richness). A color's chroma increases as it moves away from the center of the Munsell color wheel, and it will decrease in chroma as it moves into the center of the wheel. See **Figure 30-12**.

Plotting color description will be covered in the color section of this chapter.

FIGURE 30-12 Chroma on the Munsell color wheel increases as it moves away from the center, and it will decrease as it moves into the center of the wheel.

EFFECTS OF LIGHT ON COLOR

The type of light used to compare one color to another greatly affects the appearance of color. As mentioned earlier, the type of light and how the surfaces are viewed can greatly affect how a person perceives color. When light is applied to a color, some of the light rays are absorbed, while others are reflected. Others can pass through either the mica or a prismatic component of the finish and be refracted back to the viewer.

Depending on the light source and whether it contains a full color spectrum of wavelengths, the paint may look different in different types of light, or have *metamerism*. Another variance is *flop*, which means that a paint color looks different when viewed at different angles.

Light Reflection

Absorption and **reflection** are the most common occurrences to light or to the visible waves in regard to color. White light of natural sunlight has a complete spectrum of light waves. As light shines on a colored surface, all the wavelengths that are in white light are absorbed, except for the wave of that color. For example, when light hits a red vehicle, all the wavelengths—except for red—are absorbed. The red wavelengths are reflected, and the vehicle is perceived as red to persons looking at it. These reflected waves are bounced back without changing their wavelength or changing their direction. The same amount of light that shines on a surface is reflected back, and if viewed from different angles, it will look the same color. Even if viewed under less intense light, the vehicle will still be red, even if shadows are cast on it.

Light Refraction

Refraction, on the other hand, occurs when light waves pass through an object and are not reflected back, but instead reflect a different color wave back to the eye. Refraction happens when light passes through prismatic colors, or what are sometimes called color-shifting paints.

Metamerism

Metamerism, or the condition of paint colors looking different in different light sources, can happen when paint is matched using light that is not color corrected. Most spray booths have light that is color corrected, which means that the booths have the same wavelengths as sun or white light. Color-corrected light is measured by a **color rendering index (CRI)**, and must be at least 85 to 92 percent CRI to properly match colors. If paint colors are compared under fluorescent light, which has more violet and reds in it, the colors will not match when viewed outside. This effect occurs because incandescent light contains more yellow, orange, and red than sunlight or CRI-corrected light.

Flop

Flop occurs when a vehicle's color is viewed at different angles. The flop or change could appear to be lightness or darkness change, or the color shade can change. Although flop is sometimes a desired aspect of a color, most times it is considered a color defect or mismatch.

Flop is usually the result of metallics being disoriented, or what is called the paint having "poor metallic lie," due to poor spray technique. Flop could also be caused by the incorrect metallic being added.

Metallics come in a staggering array of sizes and shapes. In fact, metallics commonly make up the single largest category of tints on a mixing bank. Although it may be tempting to think that a tint should be added if the color is a mismatch, only the same metallics used in the original formula should be added, or flop is likely to occur.

To properly evaluate flop, a color must be viewed from different directions. The color should be viewed straight on at a right angle, which is called face or straight viewing, and at a 45-degree or less angle, which is called pitch angle (or "side tone," by some paint manufacturers). See **Figure 30-13** and **Figure 30-14**.

When checking for flop with a sprayout panel, the technician should compare the panel to the vehicle first at

FIGURE 30-13 Straight-on viewing helps determine color match.

FIGURE 30-14 Side tone viewing helps evaluate flop for metallic orientation.

FIGURE 30-15 A technician can check for flop with a sprayout panel and a color-corrected light.

face angles, then by moving the sprayout panel to the side tones (45 degree or greater) to compare. See **Figure 30-15**.

TECH TIP Care should be taken when evaluating a vehicle for color match. If the color is viewed for a prolonged time, the eye becomes fatigued and an accurate comparison is no longer possible. A technician should compare for 10 seconds, then look away to view something else, then reexamine the colors. If the technician is having difficulty, a coworker with good color acuity can be asked to help.

COLOR DEFICIENCY

A person with color deficiency or **color blindness** does not just see things in black and white. In fact, one of the reasons that the condition is now called color deficiency is because when a person does have it (and it is more common in men than in women), it is usually a deficiency in a color group. Color deficiencies are a condition in which certain colors cannot be distinguished. Among people with color deficiency, the most common color deficiency is with red/green (99 percent). While blue/yellow deficiency also exists, it is rare, and there is no commonly available test for it. Total color deficiency, or true color blindness, where a person sees only shades of gray is extremely rare.

Color deficiency occurs in 8 to 25 percent of men, and much less frequently in women (in only about 0.5 percent).

In about 1917, **Dr. Shinobu Ishihara** at the University of Tokyo developed one of the best-known tests for color deficiencies. It is a series of colored dots with visible numbers contained within them that test for different types of color deficiencies. See **Figure 30-16**.

Although color deficiency is generally an inherited condition, other factors can also cause color deficiencies, such as age. Temporary color deficiencies caused by things such as drugs (alcohol, caffeine, nicotine, and some prescriptions) can also occur. Another common phenomenon is color fatigue, which comes from staring at one color for too long a time.

TECH TIP Looking at a color for less than a 10-second interval can help avoid color fatigue. When evaluating a color, a technician should look at the color for 8 to 10 seconds, then look away for 10 to 12 seconds or more, then look back at the color that is being evaluated for another 8 to 10 seconds.

Although color deficiency can be troublesome, it does not need to eliminate a person from the refinish trade. If technicians believe that they are color deficient, the first step is to be tested for it. Although there are quick Ishihara tests available online, it is better to be evaluated by a professional. Once tested, a technician will know which color—or colors—are troublesome and can properly compensate.

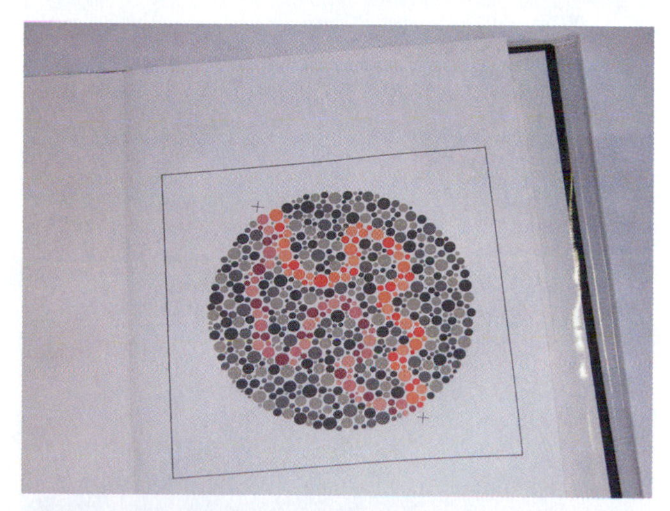

FIGURE 30-16 Dr. Shinobu Ishihara at the University of Tokyo used a series of colored dots with visible numbers contained within them to test for different types of color deficiencies.

COLOR VARIANCES

When a color is chosen for a new vehicle, a sample of that color is sent to the paint manufacturer. The paint manufacturer is allowed a 5 percent **color variance**, + or −, from the paint sample. Some vehicles are manufactured at different locations, which may mean the paint can vary from plant to plant. At each of these plants, or within a plant, the paint may vary from application differences. So, as can be seen, the same color paint can vary from vehicle to vehicle.

When a color variance is noted, paint manufacturers will develop a variant color formula. This could show up on paint formula in the computer, or a variance deck may be developed. See **Figure 30-17** and **Figure 30-18**.

> **TECH TIP** Even if a technician is sure that the variant color is perfect, a sprayout panel should be made to help decide how to blend the color, and to be sure that the color is correct.

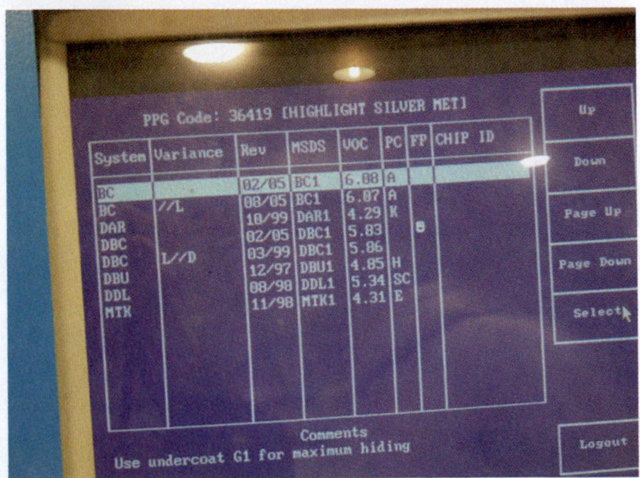

FIGURE 30-17 Color variance formulas can be retrieved electronically.

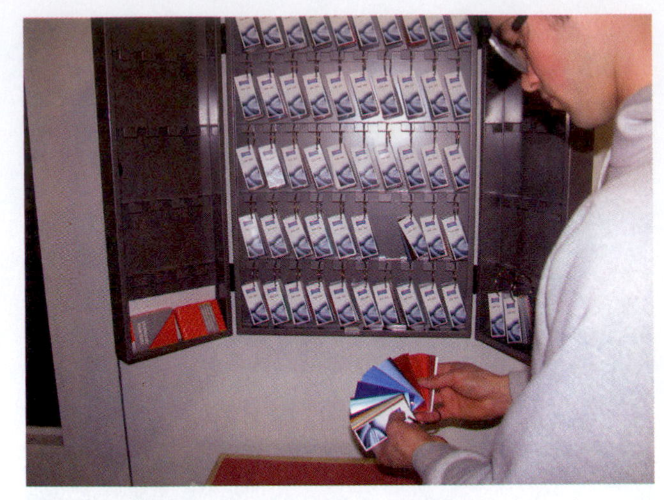

FIGURE 30-18 Variance decks with actual sprayed colors of the variation have been developed for visual comparison.

With a color variant deck, the technician can check the color that has varied from the standard against the variance deck and choose a blendable match.

MIXING TONERS

Color toners or the cans of paint on a mixing bank that hold the varying color pigments and metallics can lead to color mismatch. See **Figure 30-19**. Even if a formula is followed perfectly, if a **mixing toner** or mixing bank is not handled properly, it can lead to every formula being mismatched.

Mixing systems have lids with paddles that are inserted into a can of toner. See **Figure 30-20**. These paddles are designed to bring the toner from the bottom of the can to the top and mix it together so that the toner is consistent throughout the can. Mixing banks are designed to mix all the toners at once and for a specific amount of time. Most mixing banks will mix the paint for

FIGURE 30-19 Mixing toners are used to mix the needed formulas.

FIGURE 30-20 A mixing paddle helps keep toners from settling.

15 minutes at a time, and often paint manufacturers recommend that the bank be mixed twice a day. (Follow the paint maker's recommendations.)

If a toner is not mixed properly, the heavy portions of the toner will settle to the bottom. If that happens, the toner poured off into a formula will be lighter than it should be. However, even more unfortunately, the remainder of the toner will no longer be correct, even when properly mixed. As a result, every formula using that toner may be a mismatch. This is sometimes referred to as a "corrupt rack."

TECH TIP If only one toner is compromised and if that toner is replaced, it's possible to remedy the problem. However, if two or more toners are compromised—which could easily be the case—it may be nearly impossible to locate the responsible toners; thus, the entire toner deck may need to be replaced. To determine which toners are corrupted, the technician must track any mismatched formula and determine any common toners which should be replaced. Identifying and eliminating the toners which have been corrupted may be a long and time-consuming process.

METALLIC FLAKE TONERS

Metallic flakes—one of the largest groups of toner on a mixing bank—are classified as extra fine, fine, medium, medium coarse, coarse, and extra coarse. Although most flakes come in the silver color, they can also come as gold flake.

To keep a mixing bank properly mixed, each company will recommend how often and for how long the toners should be mixed. For the most part, toners cannot be overmixed, although metallics can be. A paint company may recommend that metallics should not be overmixed, especially the extra-large-flaked metallics. If the **metallic toner** is overmixed, the action of the mixing paddles could break the flakes down into a smaller size than indicated, and thus cause a mismatch within a formula.

Technicians should know, understand, and follow the paint manufacturer's recommendations for toner bank mixing, for both under- and overmixing.

MICA

Mica is a semi-transparent crystalline material that absorbs and both reflects and refracts light in a prismatic fashion. When mica is suspended in a paint film and light passes through it, some light will continue through to the color below (semi-transparent) and is reflected back, revealing the hue. Other light is refracted back in a prismatic fashion, giving the coating a back color, such as a silver or white shimmer.

Mica may be contained in the basecoat, where the effect is less dominant, or in a midcoat, where the mica is suspended over the basecoat and thus has a more dominant back color effect.

Mica's back color effect will change depending on what angle it is viewed from, and thus may cause flop.

Any color, either basecoat or midcoat, that contains mica—and, to a lesser degree, that contains metallic—can produce flop. When comparing a sprayout panel to the vehicle, the technician should evaluate it in sunlight or by a light with a CRI of 85 percent or better. It also should be viewed in both a head-on and side view to evaluate the color for flop.

TECH TIP Mica comes in powder and liquid forms, which are added to color formulas. Because it is a mined mineral and does not dissolve in solvents or the paint mixture, it can clump and be difficult to suspend in paint formulas. It is often one of the heaviest components of a paint formula. When a formula sits for a time after being mixed, it has a tendency to settle to the bottom of a container. Both of these conditions should alert a technician to the fact that mica-containing paints should be agitated well before being added to a paint formula, and a mixed color should also be agitated well before using, to combat settling.

PRISMATIC COLORS

Prismatic color paints—which are sold under many brand names such as "Chromalusion by Dupont," "Kameleon by House of Kolor," and many others—are paints which will shift color, depending on the angle at which the sun hits the surface and on where the viewer is located. This color has layers of reflective and semi-transparent (refracting) metals suspended in it. See **Figure 30-21**. Some of these colors can shift as many as five different colors, depending on how the paint is mixed and what is mixed in it.

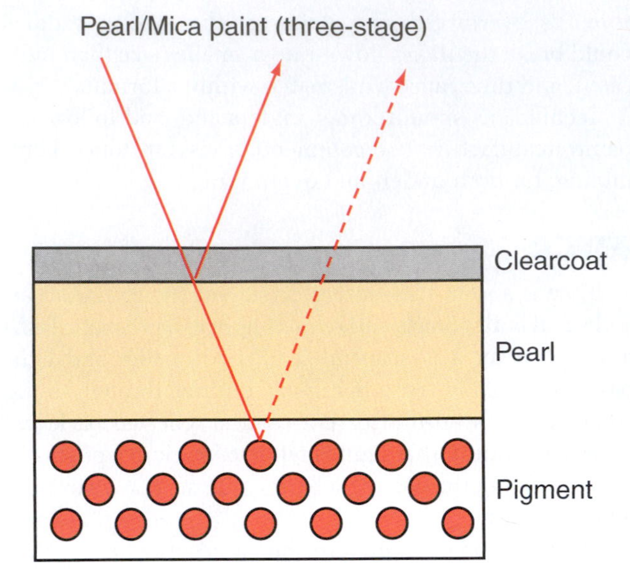

FIGURE 30-21 Mica both reflects and refracts light.

Because this coating depends on the light and how that light hits it, color match does not pose a problem when blended. In fact, prismatic paint can be blended without concern for either metamerism or flop.

TEST PANELS

Test panels have been discussed in other chapters with regard to the amount of coats that will be needed to reach full coverage, which is an important part of automotive refinishing. This section will discuss use of the two types of test panels in regard to color match. **Sprayout panels**, sometimes called **color effect panels**, and letdown panels are used to not only compare the number of coats of mica containing midcoat, but also to examine the ground coat match and final finish match with clearcoat applied. The proper use of these two types of panels not only will increase the technician's matching ability, but will also speed the painting process, thus making a painter more productive. It will also reduce the amount of materials used and reduce toxic waste and emissions, making the painter more cost efficient.

Sprayout or Color Effect Panel

A color effect panel, when prepared properly, will give a technician much needed information. Because it is prepared ahead of time, it will allow the technician time to make adjustments in either the paint application (such as spray technique, solvent changes, or number of coats) or changes in paint formula (tinting) as necessary, and still maintain a reduced cycle time for the vehicle.

Number of Coats. A sprayout panel is generally made of paper with the spray out front having large black and white checks on it. See **Figure 30-22**. The back has an area for information regarding the panel and the vehicle that it is being made for. See **Figure 30-23**. A technician will often tape the panel to a paint stick. After mixing a coating precisely as it would be when the technician sprays the vehicle, the technician will apply the basecoat to full coverage, or until the black and white checks on the sprayout panel can no longer be distinguished from each other. The technician should observe the recommended flash time between coats, the same as when spraying the vehicle. When full coverage is reached, the technician will know the number of coats that will be required for the vehicle. Next, the painter will need to check the sprayout panel for proper color match.

Color Match. To check color match against the vehicle, the sprayout panel should be clearcoated to view it against the vehicle. The sprayout panel, after it has flashed and can be taped, should have half of the panel covered with masking paper and the recommended number of coats of

FIGURE 30-22 This sprayout panel is being used with a color-corrected light to assess color match before painting.

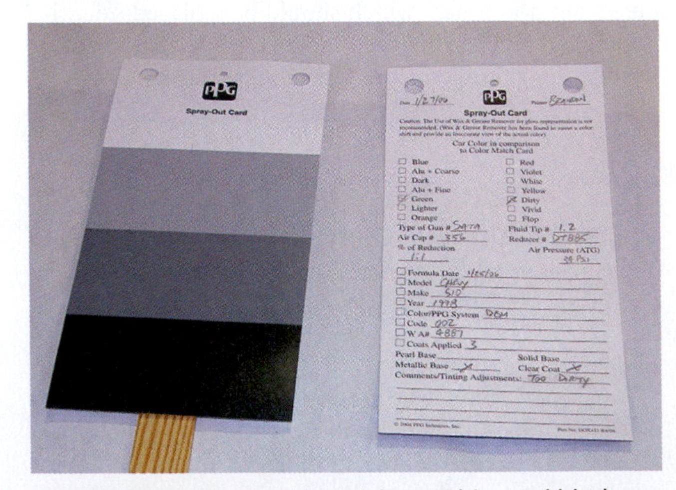

FIGURE 30-23 Sprayout panels have white and black areas to help check for proper coverage on the front, with areas for vital information to be recorded on the back.

clearcoat applied. See **Figure 30-24**. After the panel has cured, it should be compared to the vehicle.

The completed basecoat and clearcoat panel is the only true comparison of the new finish to the vehicle. The panel should be viewed from head on and side view angles, checking for flop. See **Figure 30-25** and **Figure 30-26**. Remember that the viewing of any color for match should be in sunlight or color-corrected lighting with a CRI of 85 percent or better. The sprayout panel should be checked for film thickness while it is still wet, so the paint technician will know the final film thickness on the vehicle following refinishing.

Wet Film Thickness. A sprayout panel will also tell the refinish technician the film thickness. A wet film thickness gauge will leave small marks in the finish. Because of this, the gauge cannot be used on a visible part of the vehicle's finish. Since this thickness is important information for a refinish technician, the wet film thickness reading should be taken on a sprayout panel. See **Figure 30-27**.

FIGURE 30-24 This sprayout panel is masked for clearing.

FIGURE 30-25 This technician is viewing the panel with a head-on view.

FIGURE 30-26 This technician is viewing the panel with a side angle view.

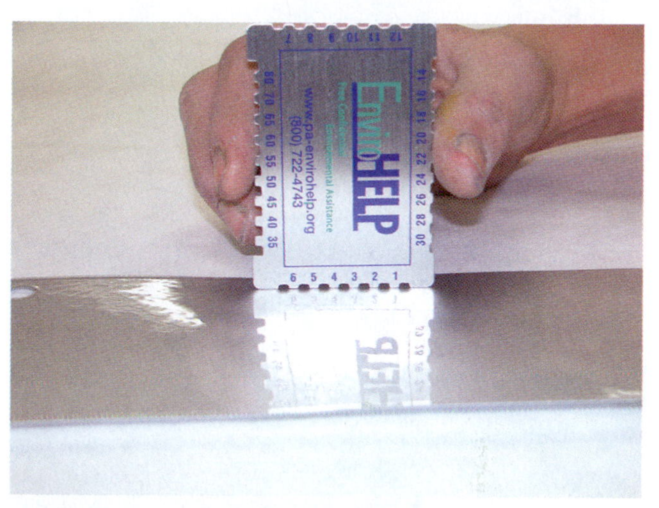

FIGURE 30-27 A wet film thickness gauge should be used on a sprayout panel instead of the vehicle, as it will leave small marks in the finish.

Letdown Panel

A **letdown panel** is used to match a multi-stage finish. It is generally larger than a sprayout panel and should have a hiding sticker applied. See **Figure 30-28**. When the ground coat is applied, the technician will know how many coats of ground coat are needed for full coverage. Then the panel should be divided into seven sections by masking off sections from the bottom to the top, as seen in **Figure 30-29**. The panel is then sprayed with the midcoat. After the recommended flash time, the first section of masking paper is removed, and a second coat is applied to the panel. Now the first section has two coats and the second has only one. This is repeated until the six panels have been removed and a coat is applied to each newly unmasked section. With the last section on the panel unmasked, it will have one section with six coats, then five, then four, until the last panel section, which has only ground coat. See **Figure 30-30**. The panel is then masked

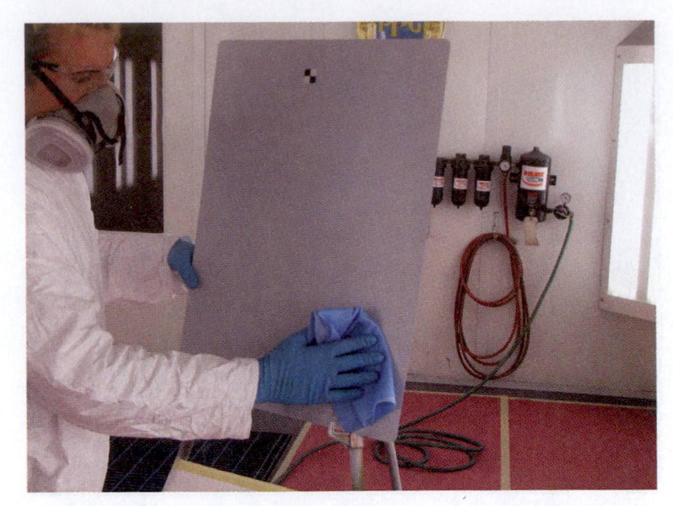

FIGURE 30-28 A letdown panel is generally larger than a sprayout panel and should have a hiding sticker applied.

vertically, and the recommended numbers of coats are applied. See Figure 30-31.

Once the letdown panel is completed, it can be used to check for color match. First the ground coat at the bottom of the panel is compared to the vehicle in an area where no midcoat or clearcoat is applied; this can be found sometimes in the trunk behind a trim panel, or under the kick panel in a door. See Figure 30-32. If the ground coat does not match, the overall finish will not match.

Next, the letdown panel should be compared to an undamaged and cleaned part of the vehicle to determine which of the six coats of midcoat is the best match to the vehicle. When this comparison is completed, the painter can then prepare a refinish plan.

The letdown panel will have revealed how many coats of ground coat are needed (the number of coats it took to cover the hiding sticker), the number of midcoats needed

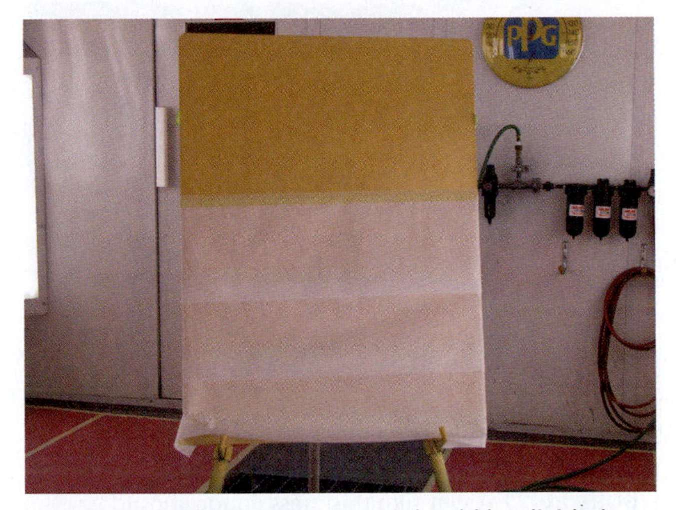

FIGURE 30-29 The letdown panel should be divided into seven sections by masking off sections from the bottom to the top, then sprayed with the midcoat.

FIGURE 30-31 The letdown panel is then masked vertically, and the recommended number of clearcoats is applied.

FIGURE 30-30 The letdown panel will have five coats of midcoat for color comparison.

FIGURE 30-32 Exposed basecoat under the door sill panel can be used for ground coat comparison.

to match the vehicle (the best comparison to the midcoat sections), and the final film thickness (wet film thickness reading). All these pieces of information allow the technician to make adjustments as needed to have a plan for applying a sprayable match for a multi-coat finish.

COLOR PLOTTING

Although there are other methods of evaluating and adjusting paint when tinting, the **color plotting** method is the one most often taught in paint manufacturers' schools. To color plot, a technician will need the formula, a manufacturer's color wheel, and a color plotting chart.

When tinting, one of the technician's most common mistakes is to look at the new color, say, "As compared to the vehicle, it needs more blue," go to the tinting bank, grab a blue, and add it to the formula. Paint manufacturers have gone to a great deal of trouble to make a formula that is as close to the standard as possible. If all the other variables have been checked and the painter has confirmed that they are not the cause of the color mismatch, then the painter should tint. If tinting is chosen, the only colors that should be considered when tinting are the colors within the formula!

TECH TIP Things that can cause paint to mismatch, other than the formula that has been mixed, are:

- Incorrect manufacturer's paint code
- Incorrect conversion of manufacturer's paint code to paint company code
- Variance deck not consulted
- Over- or underpouring
- Poor or no agitation
- Incorrect air pressure
- Incorrect reduction
- Incorrect spray technique
- Poor hiding (not enough coats)
- Incorrect choice of reducers

If all the paint color mismatch possibilities have been evaluated and found not to be the cause, the painter should then consider tinting.

To tint, the painter must first write out the formula and the amounts. See **Figure 30-33**. Then the lightness and darkness should be evaluated by looking at the formula.

When plotting the formula, first evaluate its lightness or darkness, or its value. If the formula has more white than color (by adding up the amounts) then its value should be plotted above the centerline on the value bar.

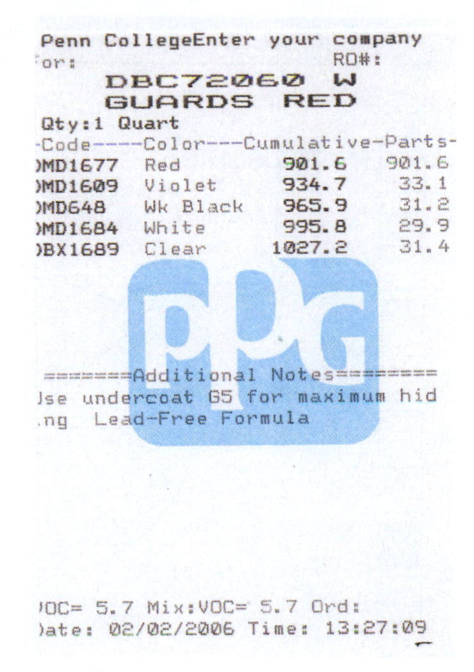

Penn CollegeEnter your company
'or: RO#:
DBC72060 W
GUARDS RED
Qty:1 Quart

-Code-----	Color---	Cumulative-	Parts-
)MD1677	Red	901.6	901.6
)MD1609	Violet	934.7	33.1
)MD648	Wk Black	965.9	31.2
)MD1684	White	995.8	29.9
)BX1689	Clear	1027.2	31.4

========Additional Notes========
Jse undercoat G5 for maximum hid
.ng Lead-Free Formula

'OC= 5.7 Mix:VOC= 5.7 Ord:
)ate: 02/02/2006 Time: 13:27:09

FIGURE 30-33 The painter will evaluate the color formula, like this one for Guards Red.

Black should be considered a color when evaluating value. Use an F, signifying formula, when plotting the formula on the value bar.

If there is more color (including black) than white, the value should be plotted below the center line on the value bar. See **Figure 30-34**.

Next, the hue should be plotted. By looking at the formula, in the case of Porsche red, the formula is red, or red with slight blue cast, or red with slight yellowish cast.

Remember that the color can only move right or left from the dimension (red in this case). It cannot move across the wheel, which would muddy the color.

When plotting the formula's value, hue, or chroma, use an F to indicate where the formula should be represented on the three bars. See **Figure 30-35**.

Next, the chroma or intensity should be plotted. When plotting the formula's chroma at this point, plot it in the middle, which will allow for change later.

When a formula and vehicle has metallic or mica in it, the flop value and flop hue should be evaluated, and the results should be entered at the bottom of the chart.

After the formula has been evaluated, the sprayout panel (formula) should be evaluated. That is, it should be compared to the vehicle in the order of value, hue, then chroma, and finally flop.

If the vehicle is lighter than the sprayout panel, the vehicle should be plotted above the F in the value bar (use a V for vehicle). The painter should then evaluate the vehicle's color or hue, place a V where it should be, then identify the chroma. (Is the vehicle's red, in this case, more intense than the formula? If so, the V should be placed right of the formula on the chroma bar.) See **Figure 30-36**.

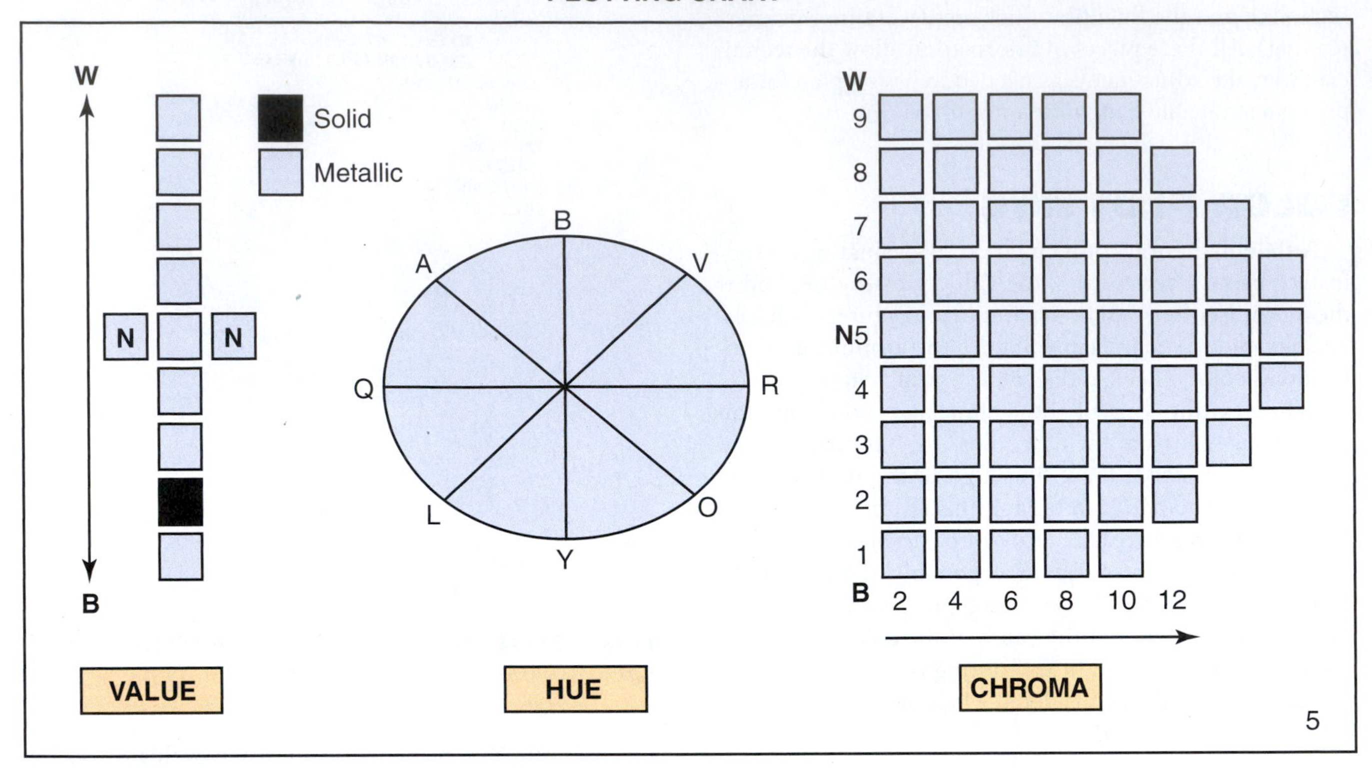

FIGURE 30-34 A technician can use a plotting chart for color tinting. Adapted from PPG.

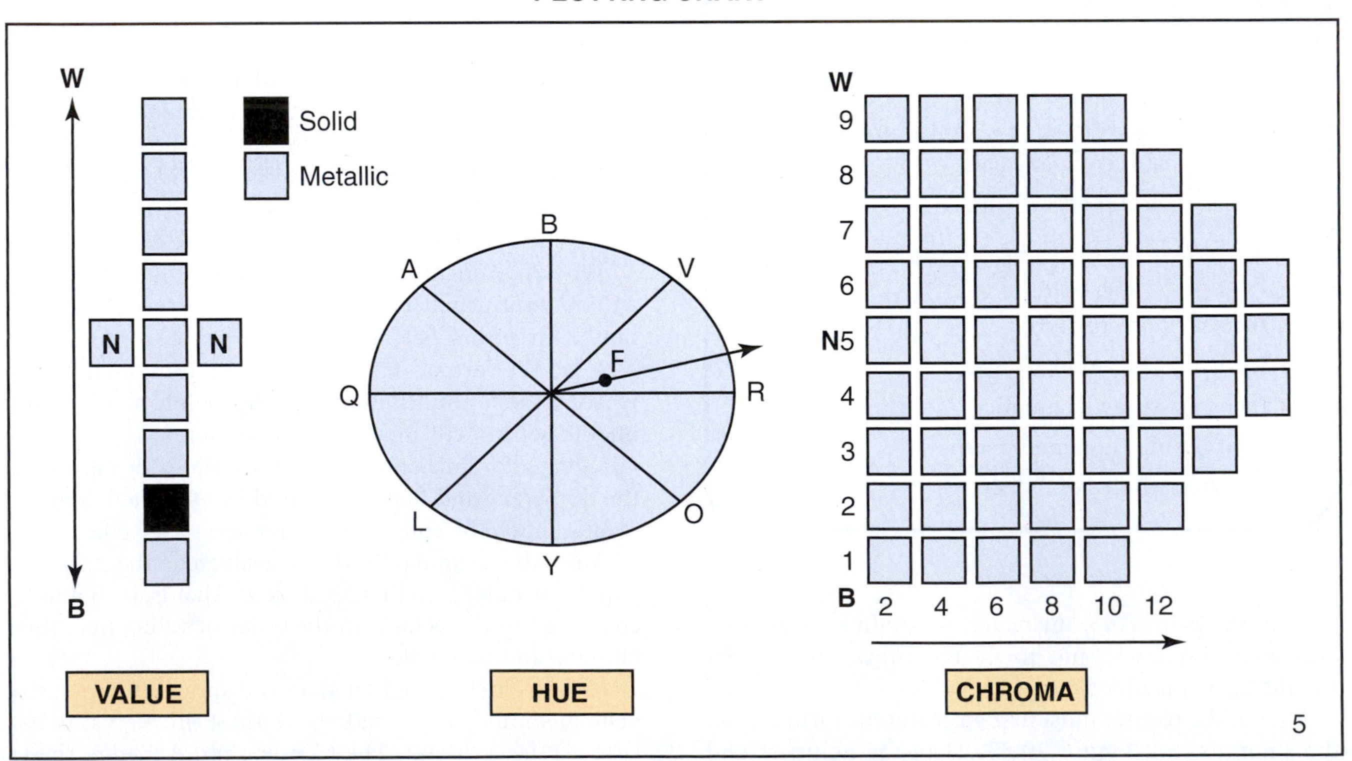

FIGURE 30-35 Red plotted with an F used to indicate the formula's color placement and a V to indicate the vehicle's color placement. Adapted from PPG.

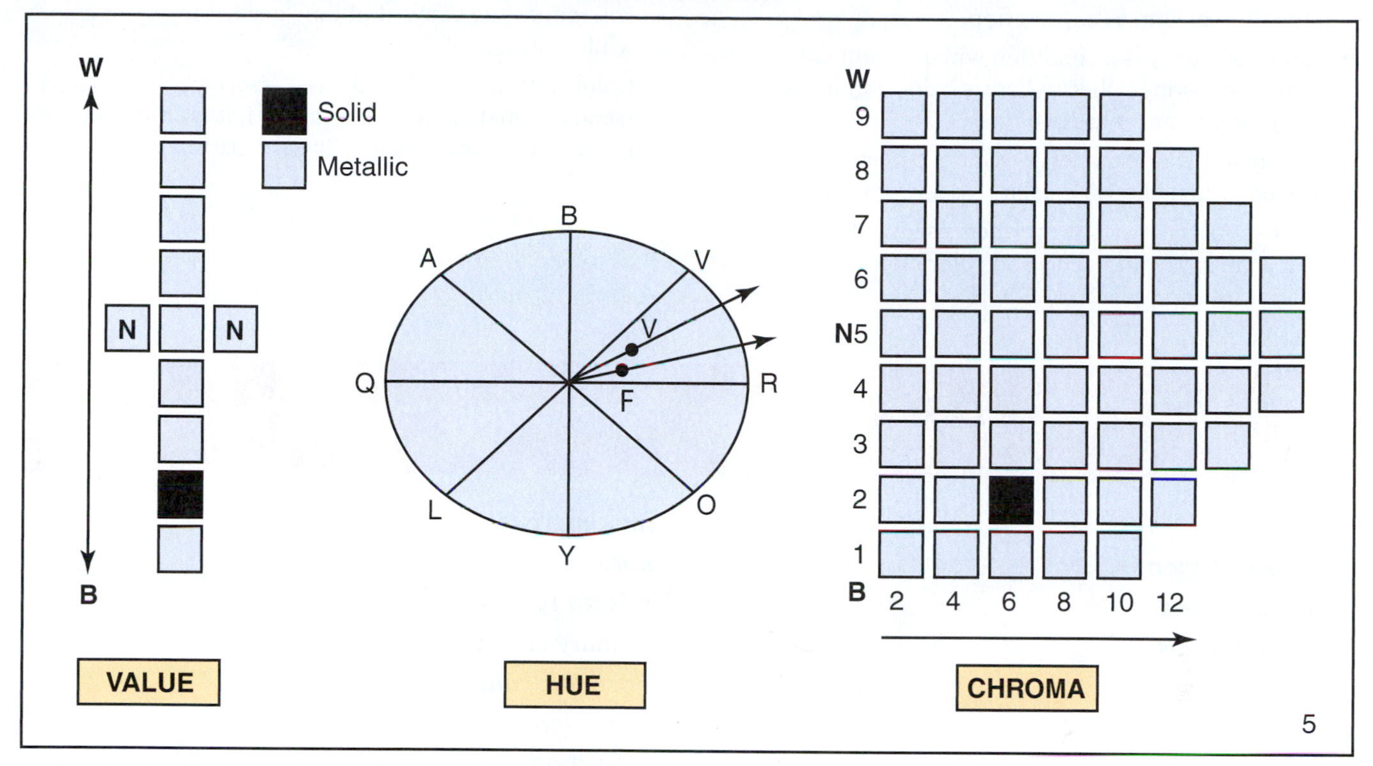

FIGURE 30-36 This completed plotting chart shows the value, hue, and chroma. Adapted from PPG.

If the vehicle has metallic or mica in it, the bottom section should be filled in after comparing the vehicle to the formula at front and side. This viewing should be done in color-corrected light.

With this information gathered, the tinting adjustments can begin.

Color Adjustment

When performing **color adjustment** to match the vehicle, first look at the manufacturer's color wheel. It will show where on the color wheel all the colors lay in the tint bank. By knowing which direction the formula must move to match the vehicle's color, and where the tint bases are on the color wheel, the technician will know which color must be added to have the formulas move toward the vehicle's color.

When tinting a mismatched color, it is best to divide the color to be tinted into a smaller batch, record what the small batch weighs, and note how much tint is added to that small batch. This allows the technician to tint without fear. If an unsuccessful first attempt at tinting is discarded, some of the original will still be available for tinting.

Tints should be added in small and measured amounts, recorded, then the paint agitated and checked for correction. When the formula is correct, a sprayout panel is prepared, checked against the vehicle, and the remainder of the batch can be tinted identically to the first one, using the recorded amounts gathered when tinting.

Tinting should be the last resort. However, on those occasions when it is necessary, if approached with care and order using the previous plotting method, a blendable match can be reached in a reasonable amount of time.

Summary

- The seven colors that make up ROY G BIV (the colors that appear when sunlight shines through a glass prism) are red, orange, yellow, green, blue, indigo, and violet.
- The primary colors are red, yellow, and blue, whereas the secondary colors are orange, green, and violet.
- Albert Munsell developed a three-dimensional wheel that helped describe color and all its variables.
- The three dimensions of color are hue, value, and chroma.

- Effects of light on color can include reflection, refraction, metamerism, and flop.
- Color deficiency is a condition when certain colors cannot be distinguished, whereas color blindness means seeing only shades of gray.
- Paint manufacturers are allowed a 5 percent color variance when choosing colors for a new vehicle.
- The two main types of test panels used with color match are sprayout panels and color effect panels.
- Color plotting, when understood and used properly, will reduce the amount of time needed to adjust colors while tinting.
- Color tinting should be the last resort when making a blendable match, but when needed, it is an invaluable tool in a collision repair refinish department.

Key Terms

chroma
color adjustment
color blindness
color chip book
color plotting
color rendering index (CRI)
color variance
color wheel
Dr. Shinobu Ishihara
flop
hue
letdown panel
metallic toners
metamerism
mica
mixing toners
primary colors
prismatic colors
reflection
refraction
ROY G BIV
secondary colors
sprayout/color effect panel
test panels
value

Review

ASE Review Questions

1. Technician A says that ROY G BIV is a mnemonic (memory device) for red, orange, yellow, green, blue, indigo, and violet. Technician B says that primary colors are yellow, blue, red, and green. Who is correct?
   A. Technician A only
   B. Technician B only
   C. Both Technicians A and B
   D. Neither Technician A nor B
2. Technician A says that Albert H. Munsell developed a test for color deficiency in 1917. Technician B says that Albert H. Munsell developed a three-dimensional color tree used to identify and describe color. Who is correct?
   A. Technician A only
   B. Technician B only
   C. Both Technicians A and B
   D. Neither Technician A nor B
3. Technician A says that hue is the dimension of color that describes its color. Technician B says that chroma is the dimension of color that describes its lightness or darkness. Who is correct?
   A. Technician A only
   B. Technician B only
   C. Both Technicians A and B
   D. Neither Technician A nor B
4. Technician A says that value describes how intense a color is. Technician B says that chroma is sometimes called intensity. Who is correct?

A. Technician A only
B. Technician B only
C. Both Technicians A and B
D. Neither Technician A nor B

5. Technician A says that if a color is found to be a non-blendable color, the first consideration should be tinting. Technician B says that there are few things that can cause a mismatch other than the tints added to a formula. Who is correct?
A. Technician A only
B. Technician B only
C. Both Technicians A and B
D. Neither Technician A nor B

6. Technician A says that factory color codes are found on the driver's door of vehicles. Technician B says that to find a vehicle manufacturer's paint code, check the vehicle's repair manual for the location. Who is correct?
A. Technician A only
B. Technician B only
C. Both Technicians A and B
D. Neither Technician A nor B

7. Technician A says that color can only move either right or left of itself on a color wheel when describing a color. Technician B says that a vehicle viewed under fluorescent light will look the same when it is compared to sunlight. Who is correct?
A. Technician A only
B. Technician B only
C. Both Technicians A and B
D. Neither Technician A nor B

8. Technician A says that CRI is a measurement of color-corrected light. Technician B says that CRI stands for color rendering index. Who is correct?
A. Technician A only
B. Technician B only
C. Both Technicians A and B
D. Neither Technician A nor B

9. Technician A says that reflection means a light passes through something and a different color comes out the other side. Technician B says that reflection means light bounces off of a surface and a color is viewed. Who is correct?
A. Technician A only
B. Technician B only
C. Both Technicians A and B
D. Neither Technician A nor B

10. Technician A says that metamerism is a condition where a color appears different under different types of light. Technician B says that flop is a condition where a color containing metallic and/or mica looks different at different angles. Who is correct?
A. Technician A only
B. Technician B only
C. Both Technicians A and B
D. Neither Technician A nor B

11. Technician A says that a side tone angle means a vehicle's color is evaluated at a 45-degree to 60-degree angle. Technician B says that metamerism should be judged by viewing it at a side tone angle. Who is correct?
A. Technician A only
B. Technician B only
C. Both Technicians A and B
D. Neither Technician A nor B

12. Technician A says that color deficiency means a person sees no color at all. Technician B says that color deficiency affects women more than men. Who is correct?
A. Technician A only
B. Technician B only
C. Both Technicians A and B
D. Neither Technician A nor B

13. Technician A says that color fatigue can occur after staring at a color for only 5 seconds. Technician B says that a color variance on a vehicle can occur because a model of a car could be painted at different factories. Who is correct?
A. Technician A only
B. Technician B only
C. Both Technicians A and B
D. Neither Technician A nor B

14. Technician A says that "Chromalusion" is a brand of prismatic color. Technician B says that a sprayout or color effect panel is a test panel used to check a multi-stage color. Who is correct?
A. Technician A only
B. Technician B only
C. Both Technicians A and B
D. Neither Technician A nor B

15. Technician A says that to know the number of coats needed for a multi-stage color, a sprayout panel should be made. Technician B says that when plotting a color, it should be evaluated in the order of value, hue, and chroma. Who is correct?
A. Technician A only
B. Technician B only
C. Both Technicians A and B
D. Neither Technician A nor B

Essay Questions

1. Describe the three dimensions of color according to Munsell.
2. Explain how different types of light affect the way we see colors.
3. Describe how light from the sun is absorbed and reflected so the eye can perceive color.

4. Explain what metamerism is.
5. List the steps of color plotting and explain the process.

Topic Related Math Questions

1. If a quart of paint weighs 2.43 pounds, how much would 5 gallons weigh?
2. A technician has 2,936 ml of paint that he divides into five equal containers. How many liters are in each container?
3. If a technician uses 54 grams of tint to adjust one of the divided colors in item 2, how much would be needed to adjust the remaining four containers?
4. If 12 percent of the male population has color deficiency and there are 23 male students in a class, what percentage of the men in the class could have color deficiency?
5. In the paint department, a white toner is used at the rate of 247 ml per week. How many liters of that toner will be used in one calendar year?

Critical Thinking Questions

1. In the paint department of a large collision repair shop, multiple painters mix paint off of the same mixing station. They consistently need to tint only 5 percent of the paint jobs each month. Suddenly the percentage jumps to nearly 80 percent. What could be the problem?
2. A lead painter is faced with a paint job for which the sprayout panel shows that the mixed color is too dark to be a blendable color. What should the painter do, and why?

Lab Activities

1. Have the students prepare a color effect panel.
2. Have the students prepare a letdown panel.
3. Have the students plot a formula.
4. Have the students compare a color variance deck to a vehicle and choose a blendable match.
5. Have the students view a color effect panel against a vehicle and describe the difference.

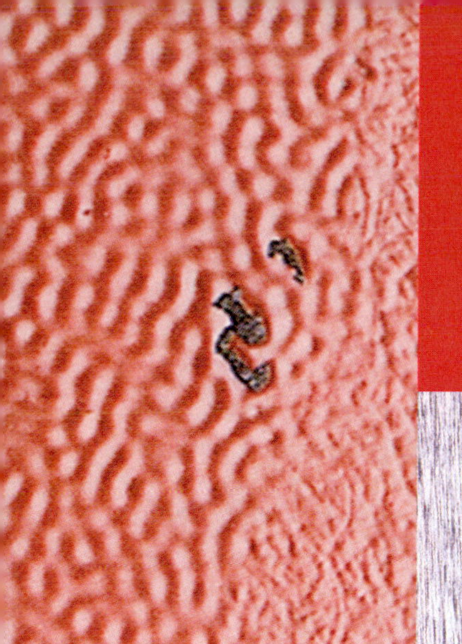

Chapter 31

Paint Problems and Prevention

OBJECTIVES

Upon completing this chapter, you should be able to:

- Identify and know the likely causes and corrective measures to repair:
  - Defects caused by poor preparation.
  - Defects caused by poor spray techniques.
  - Defects caused by contamination.
  - Defects caused by poor drying or curing.

INTRODUCTION

Many variables in the refinish environment can cause defects. This section will deal with defects that might be caused by poor preparation or spray technique; or by contamination, drying, or curing. For each of these categories, this chapter will identify the defects that may occur, explain the cause of the problem, discuss how to prevent the problem from occurring, and explain the repair of the problem if it does occur.

All paint defects can be prevented. The technician should take great care when preparing the vehicle for painting—and during the painting process—to minimize the possibility of defects. Most can be avoided when proper precautions are taken.

SAFETY

Safety must be emphasized and reviewed in the area of preventing and correcting paint defects, as it is in other areas of vehicle repair and refinish. In the refinish department, safety refers to personal safety, the proper use of personal protective devices, and environmental safety when working with chemicals. Refinish materials—and the chemicals they are made from—pose personal and environmental protection concerns. The refinish technician is governed by—and must comply with—many local, state, and federal laws concerning protection.

Each product must be accompanied with a material safety data sheet (MSDS). The MSDS includes sections on:

- Hazardous ingredients
- Physical data
- Fire and explosion hazards
- Reactive data
- Health hazard data, or toxic properties
- Preventive measures

- First aid measures
- MSDS preparation information

Technicians must read, understand, and follow the recommendations in the MSDS for the protection of themselves, their fellow workers, and the environment.

In addition to this information, other safety precautions should be taken when maintaining equipment. When cleaning areas that may have paint residue, the technician should use the same personal protective devices as outlined in the original MSDS.

There are care and caution procedures for personal safety when working on electrical equipment, too, such as "lock out, tag out." This means the circuit breakers for the equipment are disconnected, and the breaker is locked so it cannot be reconnected while someone is working on it. Also, the electrical box is tagged to let others know that it is being worked on and should not be reconnected.

Each employee should be trained to carry out equipment maintenance. If technicians do not have that training, they should not perform that procedure.

Technicians should read, understand, and follow all safety directions provided by equipment manufacturers.

All safety guards for each piece of equipment should be properly installed and in good working order. Removing or disabling safety guards should not be tolerated.

DEFECTS CAUSED BY POOR PREPARATION (IDENTIFICATION, CAUSES, PREVENTION, AND REPAIRS)

If caution is not taken during the preparation phase of refinishing, defects will occur. Although dirt can contaminate a finish after it has been painted, dirt can and does also occur from not properly cleaning the vehicle before it is painted. Sand scratch swelling can occur if the proper primer is not used or the proper sealer is not applied before the vehicle is refinished. Overspray results from poor masking. Bleeding may be caused by not tinting undercoats, or by poor sealing. If feather edges are not properly primed or sealed, they can show through with featheredge splitting. Finally, if an incompatible ground coat is used, the finish may lift. All of these defects will add considerable time and cost to the completion of a vehicle's repair. They should be avoided if possible; however, if they do occur, the technician should know the proper repair.

> **TECH TIP** A wise old painter once told a young apprentice that the difference between a good painter and a great painter *isn't* that the great painter never has problems. The difference is that when the occasional problem arises, the great painter knows how to quickly repair the problem.

Dirt

Identification. **Dirt contamination** is the most common defect problem in refinishing. See **Figure 31-1**. Some shops have such a severe dirt problem that they have resigned themselves to the fact that they will polish each vehicle before it is returned to the customer. Other shops believe that dirt can be avoided and that polishing can be eliminated altogether. Practically speaking, dirt is avoidable, and all precautions to prevent its harmful effects on the refinish procedure should be taken. Still, from time to time a finish will need to be polished and the dirt removed to maintain the needed quality.

Dirt will show in the paint surface as different-sized particles. The distribution will be random. Sometimes solvent popping may look similar to dirt particles, except when solvent popping is examined with a 10X magnifying glass and a light, it will appear as small craters—or indentations—rather than particles. See **Figure 31-2**. Solvent popping is also generally distributed evenly over the entire panel, while dirt can have a concentration of particles near the source.

Causes. The causes of dirt contamination include:

- The vehicle was not blown off before being placed in the booth.
- Wax and grease remover was not used, or not used properly.
- The vehicle was not tacked properly before painting or before each coat.
- Poor masking techniques allowed dirt to build up on folds or on areas that were not taped down.

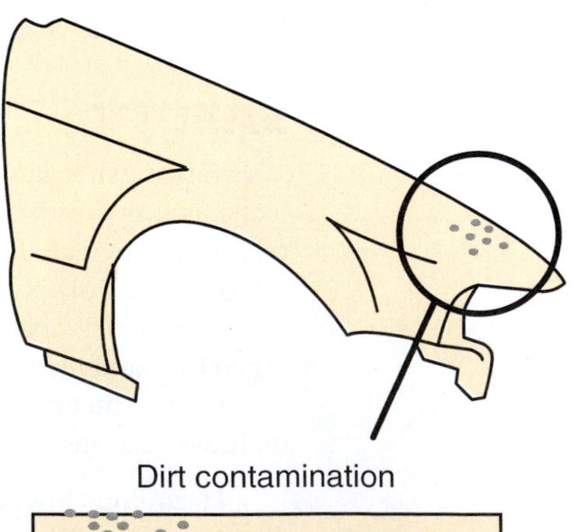

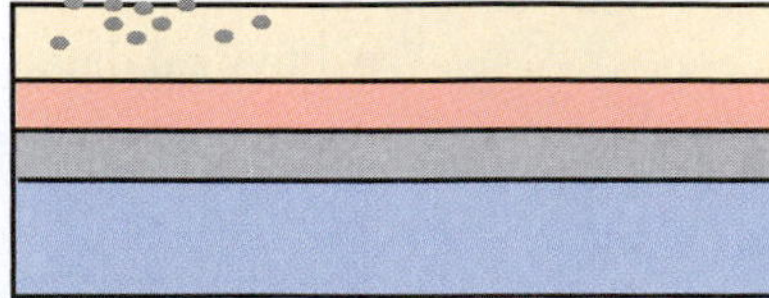

FIGURE 31-1 Dirt contamination is the most common defect problem in refinishing.

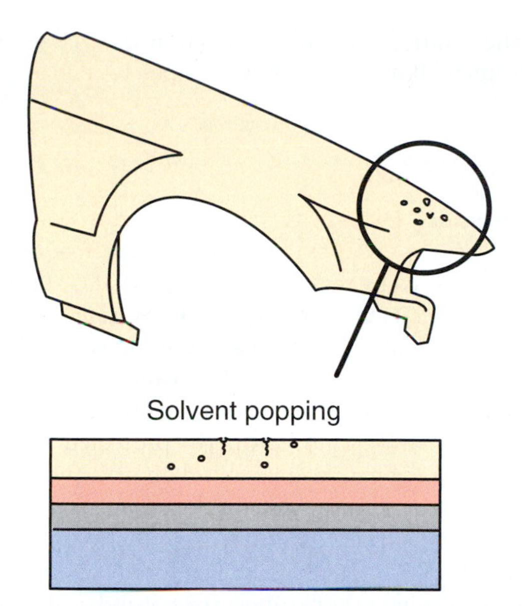

FIGURE 31-2 When solvent popping is examined with a 10X magnifying glass and a light, it will appear as small craters—or indentations—rather than particles.

- People other than the paint technician were allowed in the booth.
- Paint was not strained.
- Dry spray dust settled on wet paint.
- Contamination from the air supply settled in the finish.
- Contamination from an old air hose caused dirt.
- Poor airflow in the booth (clogged filters) spread dirt.
- Static electricity drew dirt to the vehicle.
- Paint formula mixed in a dirty container transferred dirt.
- The painter's clothing or paint suit was not clean.

Prevention. To prevent dirt contamination, follow these steps:

- Thoroughly blow off the vehicle before it enters the paint booth.
- Use wax and grease remover before painting the vehicle.
- Tack the vehicle with a new tack cloth before the vehicle is painted; tack down the complete vehicle, even the masking paper covering areas not to be painted. Lightly tack the area to be painted between coats with a clean spot on the tack cloth.
- Check that good quality masking paper is used, and that there are no areas of the vehicle left unmasked or with loose masking.
- Keep all other personnel out of the booth until the vehicle is complete and tack free (people bring in dirt).
- Mix, stir, and strain paint before painting.
- Keep overspray down and safeguard against dry spray falling into wet paint.
- Check the air supply for contamination on a regular basis.
- Change air hoses on a regular basis to eliminate hose debris.
- Change filters on a regular basis to keep airflow optimum.
- Ground the vehicle before painting, or use an anti-static spray.
- Mix, store, and reduce paint in clean containers.
- Wear a clean paint suit, not only for individual safety but also for the vehicle's cleanliness.

How to Repair. If the finish has a dirt contamination problem, it can be corrected by doing the following:

- If the dirt problem is not too severe, the finish may be able to be fine sanded and polished.
- If the dirt is severe or in the basecoat, it may need to be sanded and resprayed.
- If the technician notices dirt in the sealer, the sealer can be baked or air-dried, then sanded with fine sandpaper to remove the dirt. Then the surface is resealed and painted.

Sand Scratch Swelling

Identification. **Sand scratch swelling** is a defect that appears following application of topcoat. See **Figure 31-3**. Coarse scratches that have not been sanded smooth swell as the solvents from the basecoat or topcoat soak into them. The scratches in the paint or in the repair underneath the new paint become visible by reflecting light differently than the new topcoat. The scratches, which run in the direction of the sanding of the repair below, may not

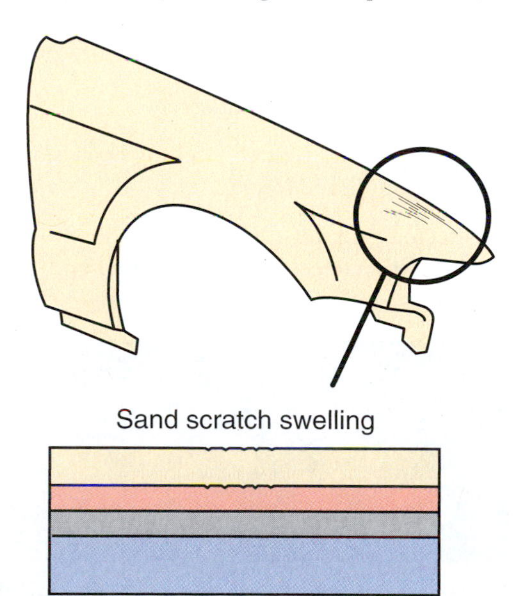

FIGURE 31-3 Sand scratch swelling is a defect that appears following application of topcoat.

appear until the paint is cured. These sand scratch swelling marks may only appear on the outer edge of the repair, if sealer is not applied out far enough.

Causes. The causes of sand scratch swelling include:

- Preparing the repair using sandpaper that is too coarse.
- Primer applied too heavily, bridging deep scratches. See **Figure 31-4**. The bridged primer sinks into the scratch when strong solvents are applied.
- Incomplete sanding of the repair to the old finish area.
- Not using guide coat when preparing the surface. Guide coats not only help technicians see high and low spots, but also help them see when coarse sand scratches have been sanded smooth with finer grit paper.
- Not sealing the repair area, or not applying the sealer out far enough to cover the underlying repair.
- Using too slow a solvent in topcoat.
- Applying topcoating too soon after applying sealers.
- Allowing too little flash time between coats.
- Applying topcoats before primer is dry.
- Applying topcoat too heavily.

Prevention. To prevent sand scratch swelling, follow these steps:

- Prepare the surface using the proper grit sandpaper.
- Use a guide coat when preparing the surface.
- Properly reduce undercoating.
- Use sealer.
- Use the correct grade and quality of solvents.
- Follow the paint manufacturer's flash time recommendations.
- Do not apply primer too heavily.
- Use proper cure time of undercoats.

How to Repair. If sand scratch swelling occurs, it can be corrected by doing the following:

- The affected area must have the finish removed to the base of the defect, then be repaired and refinished.

FIGURE 31-4 The bridged primer sinks into the scratch when strong solvents are applied. Used with permission of PPG Industries. All rights reserved.

- The splitting cannot be bridged or covered with 2K primers. It must be removed, repaired, and re-topcoated.

Overspray

Identification. Overspray can take many forms, but generally it is spray dust that settles on areas of the vehicle not protected by masking. See **Figure 31-5**. It can settle on horizontal surfaces such as an unprotected roof or deck lid. It can also find its way under masking that has not been properly taped closed. Some technicians include as overspray any finish that has settled on molding, lights, or chrome that was not properly protected by mask. Primer overspray may also be found on tires and inside of wheel openings. If a technician does not take the same precautions when masking for primer as when masking for painting, primer overspray may find its way into areas such as jambs, wheel openings, tires, hoods, trunks, and gas filler doors. Overspray may appear as a dull or dirty film on unprotected finish. It may look like a dull rough area on glass; it will have a rough feeling that will not wipe off with window cleaners or wax and grease removers. Overspray that is allowed to get on the interior parts may be difficult to remove.

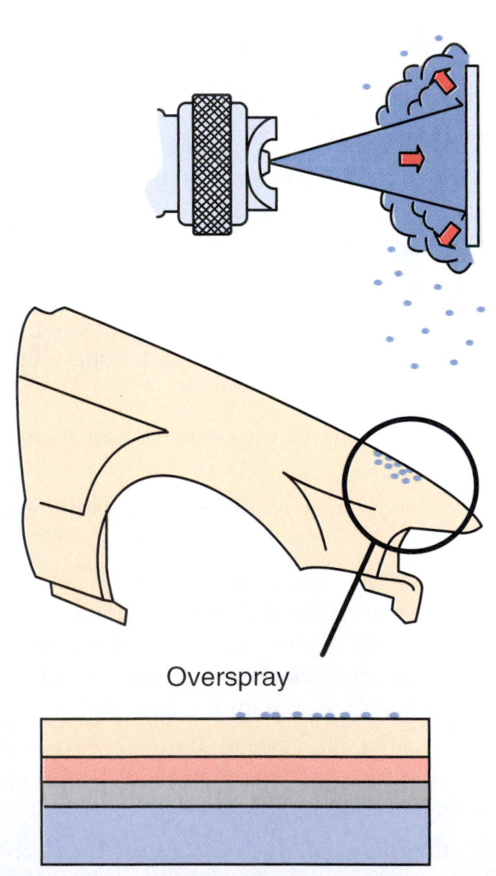

FIGURE 31-5 Overspray can take many forms, but generally it is spray dust that settles on areas of the vehicle not protected by masking.

Causes. The causes of overspray include:

- Poor masking methods.
- Areas of the vehicle not covered with masking, especially horizontal surfaces, which are more susceptible to overspray.
- Too high an air pressure when applying topcoats.
- Not protecting areas such as wheel wells and jamb openings.
- Use of poor or low grade masking paper.
- Spraying without using the recommended air pressure.

Prevention. To prevent overspray, follow these steps:

- Cover all areas to be painted.
- Check that all masking is fully taped.
- Spray at recommended pressures.
- Cover areas such as wheels and jams.

How to Repair. If overspray occurs, it can be corrected by doing the following:

- Severe overspray may need to be sanded and refinished
- Light overspray may be removed with:
  - Normal washing
  - Wax and grease remover
  - Wiping with a solvent rag
  - Polishing
  - Sanding and polishing

Bleeding

Identification. **Bleeding** refers to a color underneath the topcoat, usually darker, which shows through the finish. See **Figure 31-6**. It may appear as a stain of a darker color, and it could take the shape of the repair under the refinished area. This bleeding may only show through when in bright light.

Causes. The causes of bleeding include:

- The repair was not well sealed before topcoating.
- Dark sealer or primer was used under the topcoat.
- Plastic body filler has been overhardened.
- Silhouettes are left from removed lettering on the old (often red) finish.
- Sealer or primer has not been tinted.

Prevention. To prevent bleeding, follow these steps:

- Use proper sealing of old finish.
- Tint undercoats to help with hiding.
- Use the correct amount of hardener in plastic body fillers.
- Seal removed lettering thoroughly before topcoating.

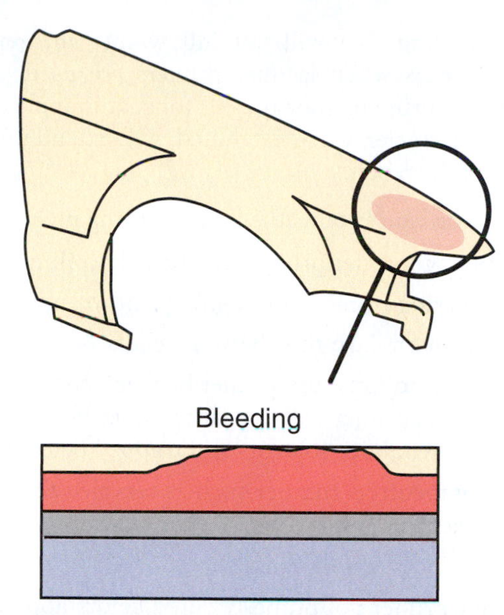

FIGURE 31-6 Bleeding refers to a color underneath the topcoat, usually darker, which shows through the finish.

How to Repair. If bleeding occurs, it can be corrected by doing the following:

- Sand the area smooth.
- Apply sealer as needed.
- Topcoat.

Featheredge Splitting (Contour Mapping)

Identification. **Featheredge splitting (contour mapping)** appears as sand scratches or splitting around the repair area. See **Figure 31-7**. This defect can look similar to sand

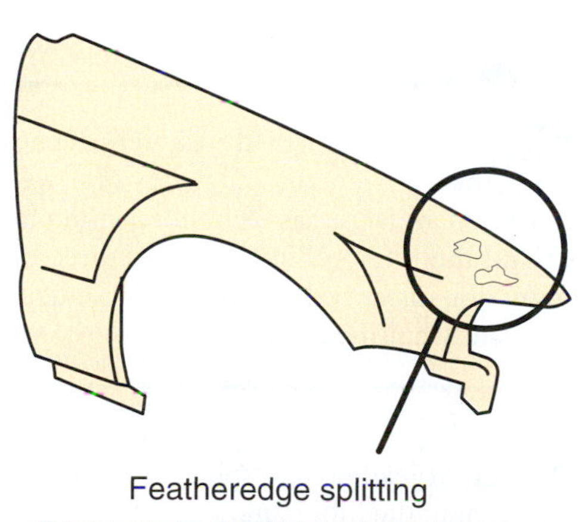

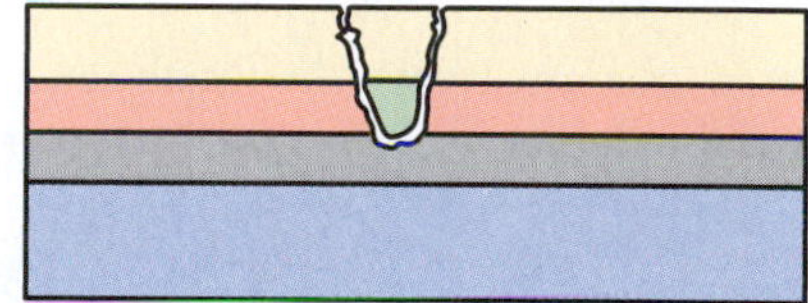

FIGURE 31-7 Featheredge splitting (contour mapping) appears as sand scratches or splitting around the repair area.

scratch swelling, but will not follow into the repaired areas. It occurs when lacquer primers are used, and it will appear during or shortly after topcoat is applied. The scratches or cracks, when examined, will extend down to the base of the repair.

Causes. The causes of featheredge splitting include:

- Applying lacquer primers too heavily (piling).
- Not mixing primer thoroughly prior to application.
- Not enough flash time between coats.
- Using air to force-dry primer between coats. (The top of the primer will appear to be flashed off by the air, but will have solvents still trapped below.)
- Use of incorrect thinner such as recycled thinner to thin lacquer primer.
- Use of lacquer spot putty.
- Heavy primer not properly cured before applying topcoat.

Prevention. To prevent featheredge splitting, follow these steps:

- Use 2K-urethane primer.
- Apply lacquer primer according to the paint manufacturer's recommendation.
- Do not use compressed air to force-dry primer between coats.
- Do not pile on primer.
- Properly mix and reduce primers with quality solvents.
- Allow the proper flash time between coats of primer.
- Properly cure primer before sanding or topcoating.

TECH TIP If primer dry time needs to be speeded up, the use of an infrared light works well. Follow the paint manufacturer's recommendation for time per coat and the infrared light manufacturer's recommendations for distance back from the vehicle. This can significantly cut the dry time.

How to Repair. If featheredge splitting occurs, it can be corrected by doing the following:

- The affected area must have the finish removed to the base of the defect, then it must be repaired and refinished.
- The splitting cannot be bridged or covered with 2K primer. It must be removed, repaired, and re-topcoated.

Lifting

Identification. **Lifting** appears as a small wrinkling of the finish below the topcoat. See **Figure 31-8**. It may cover the entire sprayed area or may occur only around the perimeter of the repair. This defect happens when strong solvents attack the underlying finish before it has a chance to fully cross-link. The finish below will lose adhesion as the new solvents are applied. The part will swell when the solvent is soaked up and lift, while the other part will not be affected by the new solvent.

Causes. The causes of lifting include:

- Recoating fresh paint before it is properly cured.
- Topcoats applied too heavily or "wet."
- Improper flash time.
- Sanding through fresh finish and not sealing before refinishing.
- Lacquer substrate.
- Too aggressive solvents being used.

Prevention. To prevent lifting, follow these steps:

- Follow the paint manufacturer's recommendation for repair and refinishing fresh topcoat.
- Perform proper surface preparation, priming, and sealing.
- Identify the old finish before repairs are attempted.
- Use the proper reducing solvents for the conditions.
- Use quality solvents.

How to Repair. If lifting occurs, it can be corrected by doing the following:

- The lifting must be completely removed.
- Feather edge the repaired area well before priming.
- Prime using a 2K primer reduced with the proper reducer.
- Apply primer using lighter, drier coats.
- Seal the area beyond the repair.
- Refinish using light coats of topcoats.

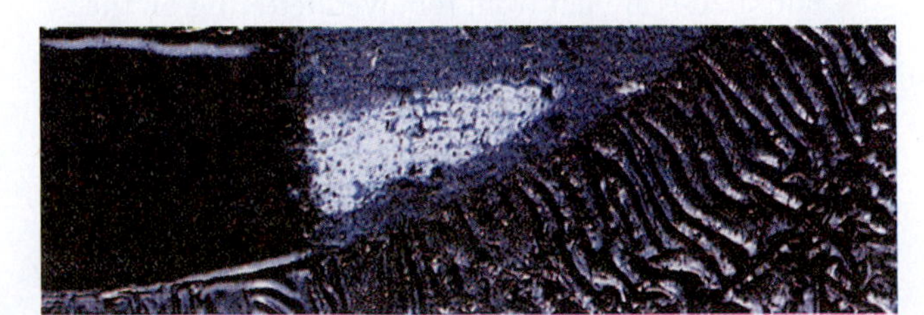

FIGURE 31-8 Lifting appears as a small wrinkling of the finish below the topcoat. Used with permission of PPG Industries.

FIGURE 31-9 Peeling occurs when a topcoat is placed on a vehicle and the adhesion—both mechanical and chemical—fails. Used with permission of PPG Industries.

Peeling

Identification. **Peeling** occurs when a topcoat is placed on a vehicle and the adhesion—both mechanical and chemical—fails. See **Figure 31-9**. If the surface is not abraded or cleaned properly, the new paint will not adhere to it and will peel. The finish may:

- Separate between the old finish and the new topcoat.
- Separate near door handles or moldings because of poor adhesion.
- Come off in large sheets (of new finish).

Causes. The causes of peeling include:

- Surface not properly cleaned or abraded before application of new topcoat.
- Improper flash time with excessive film build.
- Applying topcoat dry.
- No or poor sanding.
- Silicone contamination.

Prevention. To prevent peeling, follow these steps:

- Perform proper and thorough cleaning (with both soap and water and wax and grease remover).
- Perform proper sanding preparation.
- Follow the manufacturer's recommendation for application.

How to Repair. If peeling occurs, it can be corrected by doing the following:

- Alkalize for film thickness.
- Remove excess film thickness as needed.
- Feather the peeled area back until a good feather edge is obtained.
- Prime and block.
- Seal.
- Topcoat.

Chips

Identification. **Chips** refer to the breaking away of a topcoat, leaving small exposed areas. See **Figure 31-10**. The chipping may be present only in the topcoat or it could extend down to the substrate. Chipping occurs due to the topcoat's inability to flex when impacted with road debris, or because of thermal expansion or contamination on the surface under the topcoat. This contamination has no adhesion to the substrate and will cause the finish to fail with little abuse.

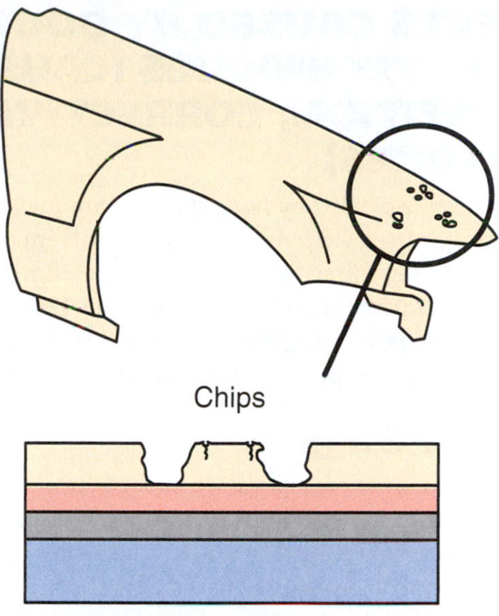

FIGURE 31-10 Chips refer to the breaking away of a topcoat, leaving small exposed areas.

Causes. The causes of chipping include:

- Use of low chip resistant coatings.
- Use of low chip resistant undercoatings.
- Driving frequently on gravel roads.
- Having either too thick or too thin a film build.

Prevention. To prevent chipping, follow these steps:

- Use chip resistant coatings.
- Apply coatings according to the paint manufacturer's film thickness recommendations.
- Use the recommended catalyst or activators.

How to Repair. If chipping occurs, it can be corrected by doing the following:

- Clean frequently with soap and water and wax and grease remover during surface preparation.
- Fill small chips, if possible, with paint, then sand and buff.
- Feather larger chips.
- Properly treat and clean bare metal.
- Use proper undercoats.
- Seal and topcoat.

DEFECTS CAUSED BY POOR SPRAY TECHNIQUES (CAUSES, PREVENTION, CORRECTIVE MEASURES)

Some defects caused by poor spray techniques are runs or sags, orange peel, striping, poor hiding, solvent popping, mottling, and dry spray. All of these can be eliminated by proper gun adjustment and spray technique. Distance, rate of travel, overlap, keeping a wet edge, and lead and lag are all important to the application of a defect-free topcoating.

Runs or Sags

Identification. **Runs** or **sags** are caused from heavy application finishes. See **Figure 31-11** and **Figure 31-12**. Paint is pulled downward by gravity to form slight sags, heavier runs, or even drips when the run is severe. Runs or sags form shortly after the finish has been applied and before it dries. When the surface has become dry, the runs can be felt above the surface of the paint. A metallic gather may form when spraying metallic paint. See **Figure 31-13**. A slight run where the metallic bunches up in the sag line may be more visible, due to the metallic, than it is noticeable by feeling the surface.

Causes. The causes of runs or sags include:

- Heavy application
- Cold spray environment
- Over-reduction
- Wrong solvents for reducing (too slow)
- Incorrect air pressure (low)
- Too-close spray technique
- Incorrect overlap
- Paint leaking from the gun
- Poor gun setup
- Vehicle surface temperature too cold
- Poor lighting in the spray area
- Too slow rate of travel
- Stepping on air hose

FIGURE 31-11 Paint is pulled downward by gravity to form slight sags. Used with permission of PPG Industries.

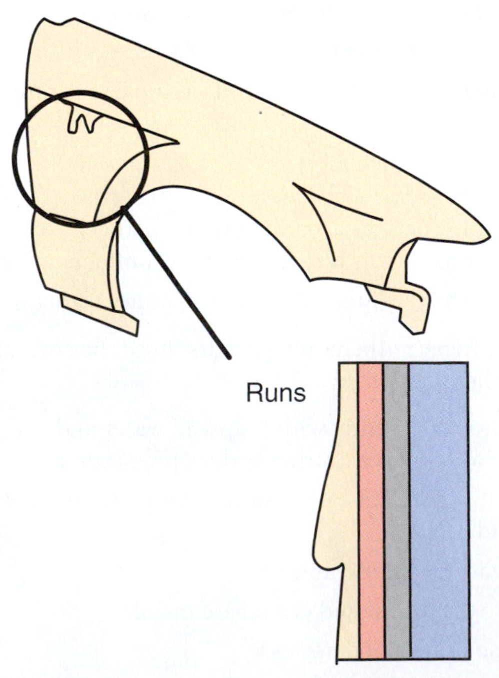

FIGURE 31-12 Paint is pulled downward by gravity to form heavier runs.

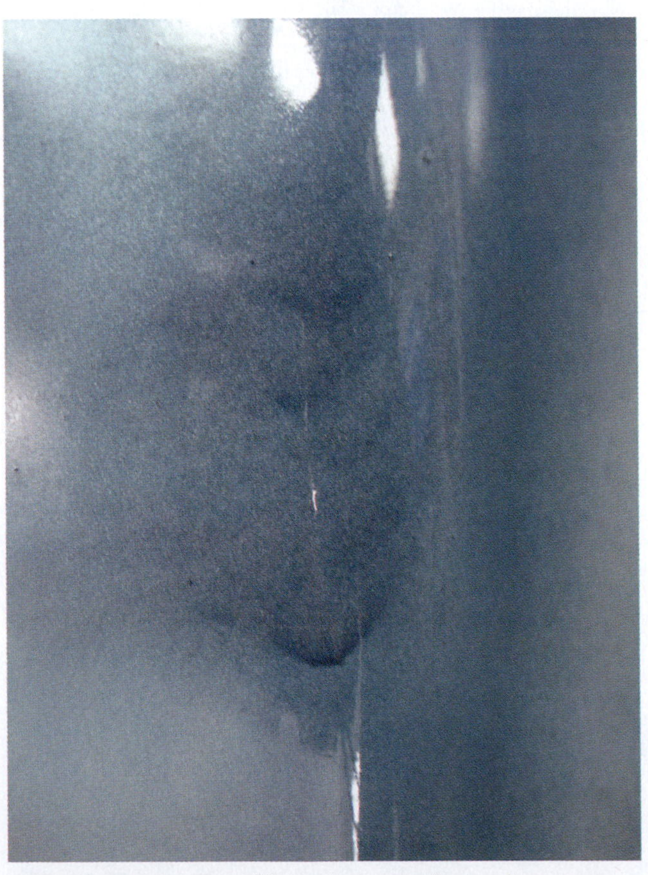

FIGURE 31-13 A metallic gather may form when spraying metallic paint.

Prevention. To prevent runs and sags, follow these steps:

- Use proper gun adjustment.
- Use proper distance.
- Use proper rate of travel.
- Use proper overlap.
- Use proper air pressure.
- Use correct temperature for vehicle and booth.
- Follow paint manufacturer's recommendations for reductions.
- Use proper choice of reducers for spray conditions.
- Follow the paint gun's recommendations for setup and air pressure.
- Repair gun leaks.
- Have sufficient lighting in the spray area.

How to Repair. If runs and sags occur, they can be corrected by doing the following:

- If the run is in the topcoat:
  - Remove the run by sanding using a hard block, which will remove the run only. Then sand the panel with ultra fine sandpaper to remove scratches, and polish.
- If the run is in the basecoat:
  - If the run is noticed before clearcoat is applied, sand out the run, then recoat with basecoat and clearcoat.
  - If the run is not caught until after the clearcoat has been removed, the run defect should be sanded smooth. This may be accomplished before the clearcoat is removed depending on the severity of the run. Then reapply basecoat and clearcoat.
- If the run is in the undercoat:
  - If the run is found to be in the undercoat, it will need to be completely sanded to the undercoat, then resealed, basecoated, and cleared.
- If a metallic gather occurs:
  - Metallic gathers can be sanded until the defect is removed and recoated. The sanding should be stopped when the run's height has been removed, even though it can still be seen. If it is not detectable by touch, it can be recoated.

Orange Peel

Identification. **Orange peel** appears as an uneven surface of the topcoat. See **Figure 31-14**. It resembles the peel of an orange, hence the name. Although it looks rough, it may feel smooth. The paint that has not atomized well is not able to "flow out" or "level" following spraying, and will appear rough. The appearance will generally be uniform throughout the panel, although it may be slightly rougher on horizontal surfaces (hoods, roofs, and deck lids) where gravity has not been able to help the paint flow out.

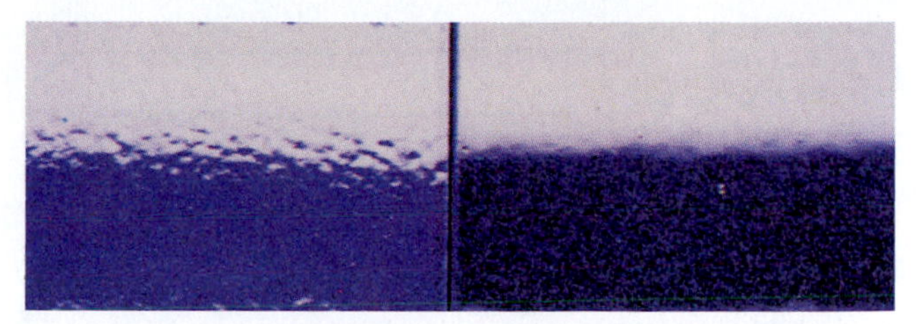

FIGURE 31-14 Orange peel appears as an uneven surface of the topcoat. Used with permission of PPG Industries.

Causes. The causes of orange peel include:

- Poor gun adjustment
  - Fluid needle adjusted out too far.
  - Fan adjusted too small.
- Poor spray technique
  - Spray distance too close.
- Extreme temperature
  - If the temperature is extremely high, the solvents are evaporated from the finish before it reaches the paint surfaces, causing poor flow out.
- Incorrect reduction of finish
  - If the finish is not reduced to the proper viscosity, the paint will not be atomized properly.
- Air pressure
  - If air pressure is too low, paint will not be atomized well and flow out will be poor.
- Gun setup
  - If the proper needle and tip are not used, atomization will not occur below 10 pounds of pressure at the tip. Some areas are regulated to use spray guns that do not exceed 10 pounds of pressure at the tip.
- Under-reduced
  - If the finish is not fully reduced, the viscosity will be too slow for the gun to atomize, causing orange peel.
- Material not fully mixed
  - If the finish is not thoroughly mixed before application, the mixture may have a lighter portion at the top, which will atomize well, and a thicker one at the bottom, which will not atomize well.
- Sealer applied with orange peel
  - If the sealer is applied with an orange peel texture, all other coatings applied will also have that appearance.
- Substrate not smooth
  - The surface must be prepared without texture; if any texture is in the substrate, all other coatings that are applied after will have texture.

TECH TIP Although orange peel is considered a defect, many vehicles will arrive from the manufacturer with texture in the finish. The refinish technician is obligated not only to match the vehicle's hue, cast, and chroma, but its texture as well. This means that if a vehicle that is being repaired has texture from the manufacturer, the refinisher must match it in the repaired area. Otherwise, the vehicle still cannot be considered "returned to its pre-accident state."

Prevention. To prevent orange peel, follow these steps:

- Proper gun adjustment
  - Gun should be adjusted to the recommended air pressure, fan, and fluid for the type of gun being used and the type of paint being applied.
- Proper gun setup
  - Fluid needle, fluid tip, and air cap should be set up according to recommendations for the finish being applied.
- Use of proper spray technique
  - The technician should use proper overlap, distance, rate of travel, and approach.
- Proper temperatures of booth and vehicle
  - The vehicle's surface temperature should be within the paint manufacturer's recommendations.
  - Paint booth temperature should be as recommended by the paint maker.
- Follow recommended reduction
  - The airflow, type of finish, temperature, humidity, type of booth, and the paint maker's recommendations should all be taken into consideration when choosing a reducer.
- Follow recommended air pressure
  - Different types of guns take different air pressures for proper atomization of paint. The technician should know and follow the recommendation of air pressure for the gun being used.
- Stir, mix, and agitate the finish completely
  - Whether the technician stirs the paint with a paint paddle, mixes it on a machine, or uses a paint shaker for agitation, it should be done thoroughly.
- Apply sealer without texture
  - Sealer should be treated as a topcoat when each coat is applied. If texture is applied in the sealer, all other finishes to follow will have texture. The previously noted precautions should be followed when sealer is applied.
- Prepare substrate properly without texture
  - Texture in the substrate will produce texture that cannot be polished out in the finishes.

How to Repair. If orange peel occurs, it can be corrected by doing the following:

- Orange peel in clearcoat only
  - If the orange peel is only in the clearcoat, fine sanding and polishing may remove it.
- Orange peel in basecoat
  - If the orange peel is not detected before the clearcoat is applied, the surface will need to be sanded, recoated with basecoat, and re-cleared.
  - If the texture is noted in basecoat before the clearcoat is applied, the basecoat can be sanded, then recoated with basecoat, then coated with clearcoat.
- Orange peel in sealer
  - The texture will need to be removed and re-finished.
- Orange peel in substrate
  - The texture will need to be removed and refinished.

Striping

Identification. **Striping** or "zebra striping" appears as uneven color or stripes of light and uneven color in the finish. See **Figure 31-15**. The stripes occur in metallic, mica, and impregnated (clear with color) colors, and appear as though the metallic particles are gathering in the stripe. It occurs when the finish is applied using poor spray technique or the finish is applied too wet. It usually appears in the first coat and does not correct itself with additional coats.

FIGURE 31-15 Striping or "zebra striping" appears as uneven color or stripes of light and uneven color in the finish.

Causes. The causes of striping include:

- Dirty spray gun
  - Spray guns that do not have a proper spray pattern due to dirt may cause striping with application.
  - Spray guns should be thoroughly cleaned after each paint application.
- Cold spray area or vehicle surface
  - Temperatures of the spray environment and vehicle surface should be within the paint manufacturer's recommendations.
- Spray distance too close
  - Spray distance is critical, and when incorrect it can contribute to many different defects.
- Poor overlap technique
  - A 50 percent overlap is a critically important aspect of spray technique. If correct overlap is not maintained, poor finish application will result.
- Not reduced according to recommendations for the situation
  - The airflow, type of finish, temperature, humidity, type of booth, and the paint maker's recommendations should all be taken into consideration when choosing a reducer.
- Paint applied too wet
  - When finish is applied at too wet a rate, the solvents can cause many different defects, including striping.

TECH TIP The use of a laser-aiming device helps the technician keep proper distance, no matter the vehicle's shape. See **Figure 31-16**. When attached and adjusted properly, the laser will guide the technician at the recommended distance.

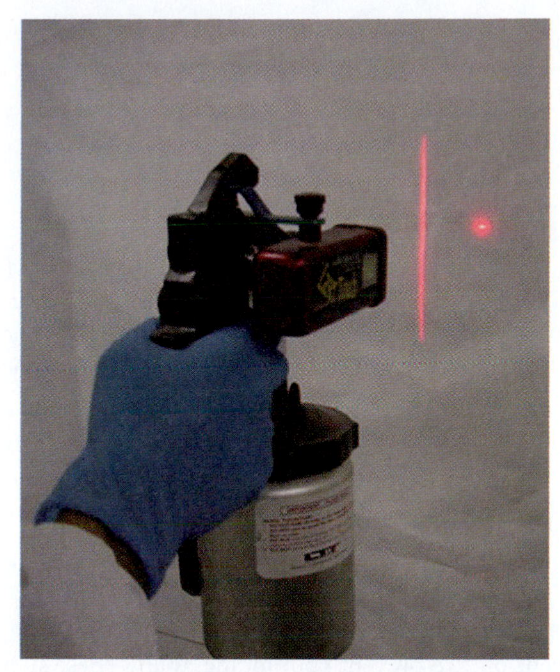

FIGURE 31-16 The use of a laser-aiming device helps the technician keep proper distance, no matter the vehicle's shape.

Prevention. To prevent striping, follow these steps:

- Clean gun
  - Paint technicians should keep their spray guns clean at all times. Cleaning after each application is necessary; if not properly done, the gun will not properly operate.
- Use of proper solvents
  - The airflow, type of finish, temperature, humidity, type of booth, and the paint maker's recommendations should all be taken into consideration when choosing a reducer.
- Proper mixing
  - Whether the technician stirs the paint with a paint paddle, mixes it on a machine, or uses a paint shaker for agitation, mixing should be done thoroughly.
- Testing the spray pattern
  - When the technician has completed the preparations for spraying, a test pattern should be sprayed on to confirm that all adjustments have been properly made.
- Proper spray gun technique
  - The technician should use proper overlap, distance, rate of travel, and approach.
- Not applying finish too wet
  - The finish should be applied according to the paint manufacturer's recommendations.
  - The technician should avoid spraying finish in too wet a manner.

How to Repair. If striping occurs, it can be corrected by doing the following:

- Allow correct flash time between coats
  - If a coat is too wet, the technician should allow for the finish to flash before applying a second coat. Corrections in the spray gun adjustment and setup can be made before the next coat.
- Cross-checking
  - Cross-checking means the technician will spray the panel in the opposite direction of the striped one. The technician should allow for proper flash time, after corrections to gun setup have been

made and a test pattern confirms that the spray pattern and adjustment is correct.

- Sand and refinish
  - If striping is severe, the finish will need to be reapplied.

Poor Hiding

Identification. **Poor hiding** appears when too little film thickness is applied over the repair, and in bright light the thin topcoat can be seen through. See **Figure 31-17**. Incorrect film thickness will contribute to poor durability, holdout, and adhesion.

Causes. The causes of poor hiding include:

- Too few coats of finish
- Poor equipment setup
- Poor lighting in the spray environment
- Over-reduction

Prevention. To prevent poor hiding, follow these steps:

- Follow the paint manufacturer's reduction recommendations.
- Use hiding stickers to check for full hiding in the booth.
- Have sufficient lighting in the spray environment.
- Set up equipment according to the paint manufacturer's directions.

How to Repair. If poor hiding occurs, it can be corrected by doing the following:

- Sand and refinish.

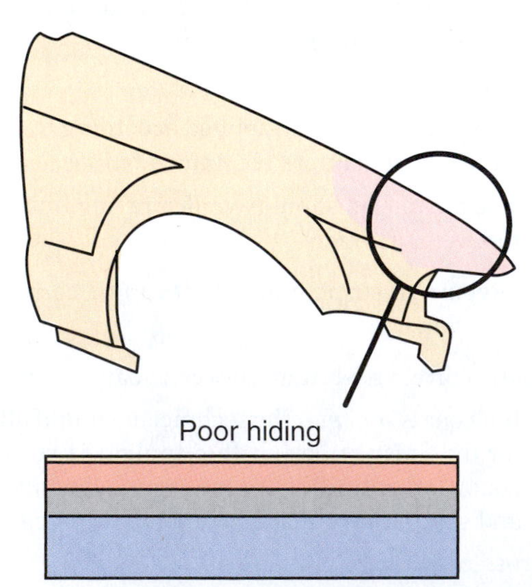

FIGURE 31-17 Poor hiding appears when too little film thickness is applied over the repair, and in bright light the thin topcoat can be seen through.

Solvent Popping

Identification. **Solvent popping** is a condition that occurs when trapped solvents rise to the top of a finish and burst, causing small craters in the finish's surface. See **Figure 31-18**. The problem may be made worse from force-drying or uneven drying with infrared lights. When the surface is examined with a magnifying glass, small blisters can be seen. See **Figure 31-19**.

This condition can be mistaken for dirt in the finish, if not examined carefully.

Causes. The causes of solvent popping include:

- Temperature during baking too high
- Not enough flash time between coats
- Infrared lights not placed properly
- High film thickness
- Improper surface cleaning
- Improper air pressure

Prevention. To prevent solvent popping, follow these steps:

- Follow the recommendations of the paint maker for baking temperatures.
- Follow the recommendations of the paint maker for flash time and reduction.
- Place infrared lights at the proper distance.
- Do not exceed the film build recommendations.
- When cleaning the surface with a wax and grease remover, remove the cleaner thoroughly before applying topcoats.

FIGURE 31-18 Solvent popping is a condition that occurs when trapped solvents rise to the top of a finish and burst, causing small craters in the finish's surface.

FIGURE 31-19 When the surface is examined with a magnifying glass, small blisters can be seen.

FIGURE 31-20 Mottling is a defect that can occur in metallic paint applications.

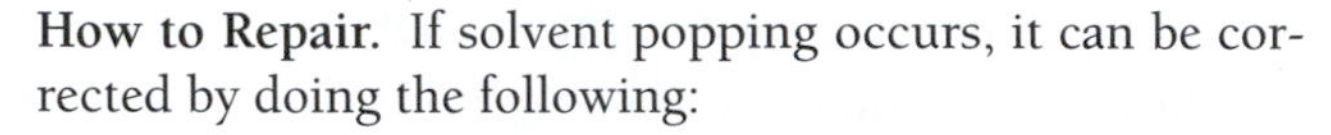

How to Repair. If solvent popping occurs, it can be corrected by doing the following:

- Severe solvent popping
  - When the defect is severe, the topcoat will need to be removed down to below the level of the damage. If the popping is in the sealer, the finish will need to be removed down to that level; then the area can be refinished.
- Light solvent popping
  - Sand out the defects and polish (if very light), or if needed, recoat.

Mottling

Identification. **Mottling** is a defect that can occur in metallic paint applications. See **Figure 31-20**. The metallic appears inconsistent in its flow, or it has a random—not smooth—appearance to the metallic lie. The appearance sometimes will look uneven, blotchy, or cloudy.

Causes. The causes of mottling include:

- Poor gun setup
- Incorrect reducer
- Heavy application
- Low shop or vehicle temperature
- Insufficient gas-out time before clearcoating
- Spray distance too close
- Wet application of color coat

Prevention. To prevent mottling, follow these steps:

- Follow the recommended gun setup.
- Follow the paint maker's recommendations for reducer.
- Apply medium wet coats, not heavy wet coats.
- Maintain correct shop and vehicle temperatures.
  - Allow the proper amount of time for flashing of solvents (gas out) before applying topcoat.
  - Maintain proper spray gun distance.

How to Repair. If mottling occurs, it can be corrected by doing the following:

- Sand and refinish.

Dry Spray

Identification. In an application resulting in **dry spray**, the finish will have lost all gloss, appear rough, and have a granular texture. See **Figure 31-21**. Although dry spray commonly affects small areas, poorly applied coats may affect complete panels.

Causes. The causes of dry spray include:

- Poor gun setting
- Too fast a reducer for conditions
- Gun distance too far
- Rate of travel too fast
- Over-reduction
- High air pressure

Prevention. To prevent dry spray, follow these steps:

- Follow the recommended gun settings.
- Use the recommended reducer for the conditions.
- Follow the recommended distances for the gun.
- Follow the recommended air pressure.

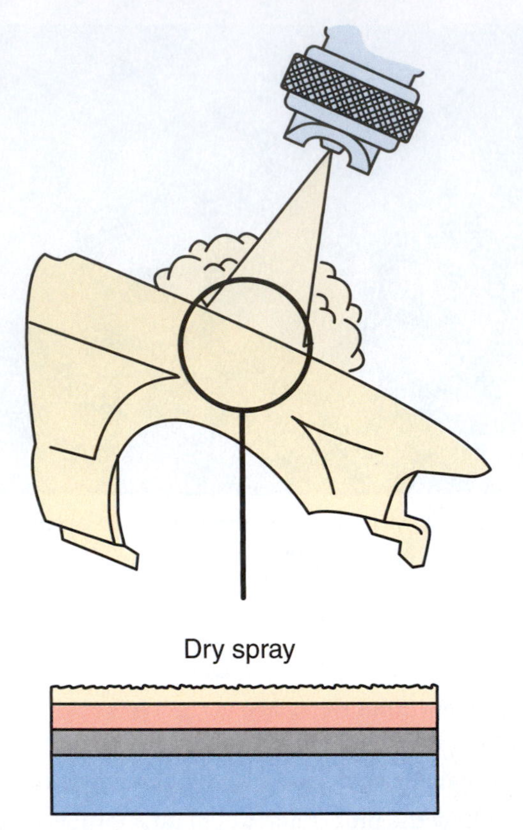

FIGURE 31-21 In an application resulting in dry spray, the finish will have lost all gloss, appear rough, and have a granular texture.

How to Repair. If dry spray occurs, it can be corrected by doing the following:

- Severe
  - Sand and refinish.
- Light
  - If dry spray is very light, the surface should be sanded with ultra fine sandpaper and polished.

DEFECTS CAUSED BY CONTAMINATION (CAUSES, PREVENTION, CORRECTIVE MEASURES)

Defects caused by contamination are common and often overlooked. When refinishing a vehicle, a technician should try to control as many of the conditions that will affect the finish's outcome as possible. Contaminations that cause defects such as fish eyes, blistering, craters from acid rain, staining, and marks from bird droppings can be controlled. This section will help the technician spot these conditions and eliminate possible defects before they happen. It will also examine the solutions to repairing the defects if they should happen.

Fish Eye

Identification. **Fish eye** can appear at any time in the painting process: in the ground coat, basecoat, and even in the clearcoat. See **Figure 31-22**. Fish eyes may appear near the end of a paint project from air-borne contaminants. They look like an area on the wet paint that will not allow the paint to cover the surface, a crater that may or may not show through to the bottom of the coat. It may even take on a distinguishable shape, such as a fingerprint. Fish eyes are caused by silicone, a very light oil found as an ingredient in many products. Silicone is light and may easily become airborne. The size of fish eyes may range from those large and easily seen on the surface to those so small they require magnification to be identified. They may be spread over the entire surface or concentrated on a certain spot. Fish eyes are the most common of the contamination defects.

Causes. The causes of fish eye include:

- Poorly cleaned surfaces with contamination such as hand prints, wax, oil, or grease remaining
- Compressed air with oil contaminants
- Incorrect cleaning solvents
- Diesel fuel or exhaust in the air
- Excessive use of fish eye eliminators
- Silicone contamination brought in from outside air

Prevention. To prevent fish eye, follow these steps:

- Clean the surface of the vehicle with a good quality wax and grease remover.
- As a paint technician, remember to wear solvent resistant gloves while working on a vehicle.

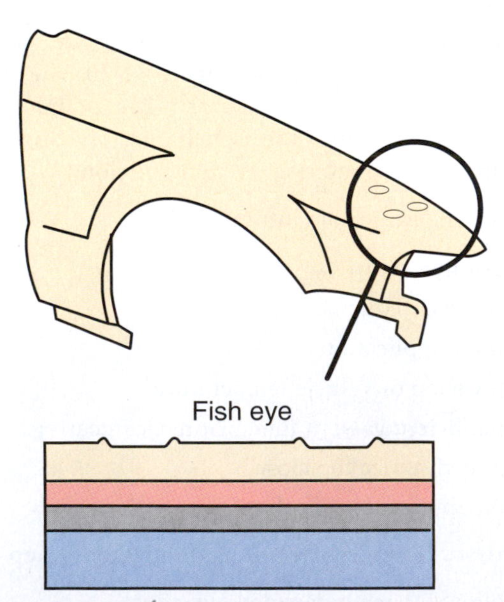

FIGURE 31-22 Fish eye can appear at any time in the painting process: in the ground coat, basecoat, and even in the clearcoat.

- Regularly maintain the cleanliness and performance of the air compressor.
- Keep the shop clean.
- Wear a clean paint suit.

How to Repair. If fish eye occurs, it can be corrected by doing the following:

- Stop spraying as soon as fish eye appears.
- Find the contamination source and eliminate it.
- If the fish eye is laminated to a small area, let the paint flash; then spray using light, medium wet coats.
- If on second coat the fish eye is not better, stop spraying and allow drying. Clean with wax and grease remover, sand to eliminate the craters, re-clean, and refinish.

Blistering

Identification. **Blistering** refers to bubbles, or what appear to be water blisters, under the paint film. See **Figure 31-23**. They may range in size from small (0.5 mm) to large (2 mm), and they may not always be round, but may take on the shape of a repair. They generally do not require magnification to identify, but can sometimes be mistaken for dirt. If the bubble is broken, it may have a liquid in it or just vapor. Blisters are most often seen on horizontal areas. They are not always caused from underlying rust, but rust may form if the defect is not repaired.

Causes. The causes of blistering include:

- High moisture in air lines which is deposited while spraying
- Vehicle not thoroughly dried after washing and before finishing
- Vehicle left in a high-moisture area for a long time
- Covered and stored vehicles

Prevention. To prevent blistering, follow these steps:

- Dry the vehicle completely before finishing.
- Eliminate water from the compressed air.
- During storage, use a breathable cover.

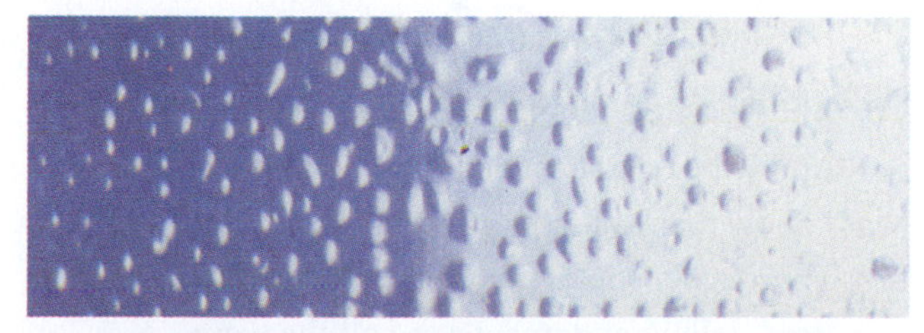

FIGURE 31-23 Blistering refers to bubbles, or what appear to be water blisters, under the paint film. Used with permission of PPG Industries.

How to Repair. If blistering occurs, it can be corrected by doing the following:

- Sand away the blister.
- Eliminate rust if present.
- Prime, seal as needed, and refinish.

Acid Rain

Identification. **Acid rain** appears as an etch mark that cannot be washed away. See **Figure 31-24**. It is noticeable as a discolored area with low gloss, and can even be felt with a fingernail. Although usually seen on the horizontal surfaces of a vehicle, it can also occur on the vertical areas. Acid rain is generally seen more often in urban areas, especially manufacturing areas. The etched areas on an affected vehicle may have a gray color to them.

Causes. The causes of acid rain include:

- Industrial contamination of the surface of a vehicle

Prevention. To prevent acid rain, follow these steps:

- Wash the vehicle often, especially if the vehicle is often in industrial areas.
- Clean and wax regularly to help protect the finish.
- Use urethane clearcoats, as they are less susceptible to acid rain. (However, even they will show signs of decay if exposed for long periods of time.)

How to Repair. If acid rain occurs, it can be corrected by doing the following:

- Wash with automotive soap.
- Neutralize by sponging on a mixture of water and baking soda (one teaspoon baking soda to quart of water).

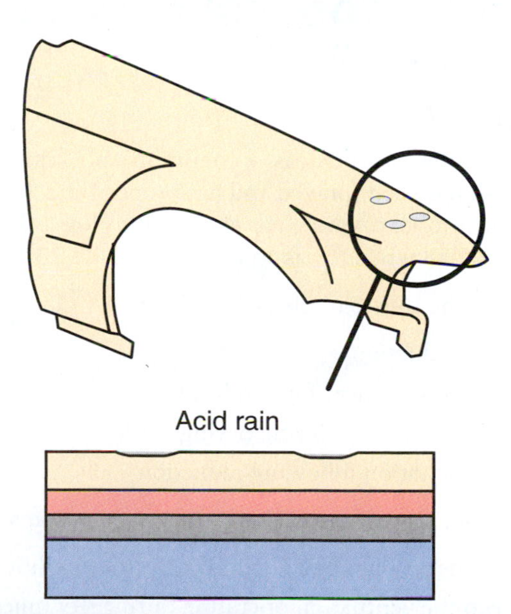

FIGURE 31-24 Acid rain appears as an etch mark that cannot be washed away.

- Rinse with clear water.
- Polish to remove etching.
- In severe cases, sand and refinish.

Bird Droppings

Identification. **Bird droppings** can look similar to acid rain and will etch the surface of the paint, but they will be more commonly found on horizontal surfaces. Also, red cars seem to attract bird droppings more than other colors. Droppings often discolor or stain the paint surface, and if tested may have an acid pH.

Causes. The causes of bird droppings include:

- The seasons. Bird droppings are seasonal. They can be worse in the fall, when berries are eaten by birds and the droppings stain the finish more.

Prevention. To prevent bird droppings, follow these steps:

- Wash vehicles in high-risk areas frequently. Bird droppings that are left on the painted surface will etch into the finish each time it rains and the droppings are re-moistened.
- Maintain the vehicle's finish with wax to help combat bird dropping etching.

How to Repair. If bird droppings occur, it can be corrected by doing the following:

- Wash and neutralize the area.
- Clean the area with mild polishing compound.
- Wax after cleaning to protect the finish.

DEFECTS CAUSED BY POOR DRYING OR CURING (CAUSES, PREVENTION, CORRECTIVE MEASURES)

Dieback

Identification. **Dieback** is a condition appearing when the finish has been sprayed and has proper gloss, but then when the proper dry time has expired the gloss is less satisfactory than when it was wet. See **Figure 31-25**.

FIGURE 31-25 Dieback is a condition in which, as the dry time expires, the gloss becomes less satisfactory.

Causes. The causes of dieback include:

- Over-reduction
- Too fast a reducer for conditions
- Poor ventilation during drying
- Poor agitation following reduction

Prevention. To prevent dieback, follow these steps:

- Use the recommended reducer for the conditions.
- Keep the ventilation operating during dry times.
- Follow the recommended mixing ratios.
- Stir the reduced coatings well before spraying.

How to Repair. If dieback occurs, it can be corrected by doing the following:

- Severe
  - Sand and refinish.
- Light
  - If very light, the surface should be sanded with ultra fine sandpaper and polished.

Cracking

Identification. The **cracking** or splitting of the paint film commonly appears as straight lines and extends through the entire film thickness. See **Figure 31-26**. It can occur in the topcoat or undercoats, and is a result of thermal expansion and contraction. It can also be caused by application of too heavy a film thickness.

Causes. The causes of cracking include:

- Coating not thoroughly mixed
- Too little activator or catalyst in 2K products
- Applying topcoat over cracked surface
- Temperature of surface too hot or cold
- Exceeding recommended film thickness
- Fanning, or using compressed air to speed flash time
- Incorrect catalyst

Prevention. To prevent cracking, follow these steps:

- Apply the recommended film thickness.
- Follow the surface preparation recommendations.
- Thoroughly stir all mixtures before application.
- Read, follow, and understand all the paint manufacturer's recommendations for application.
- Do not dry with compressed air or fan the coating to speed flash time.

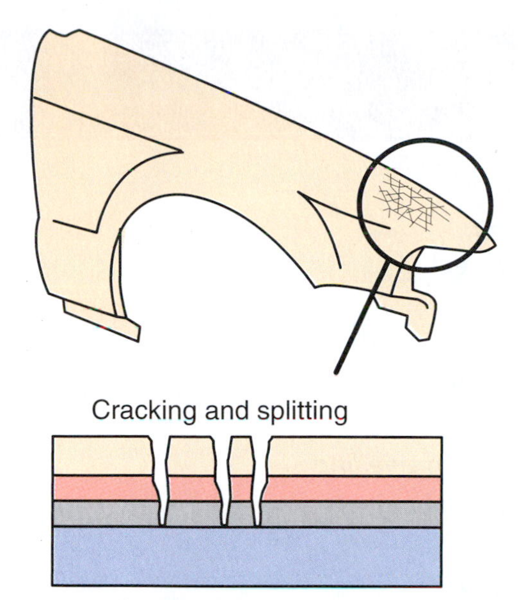

FIGURE 31-26 The cracking or splitting of the paint film commonly appears as straight lines and extends through the entire film thickness.

How to Repair. If cracking occurs, it can be corrected by doing the following:

- Remove all cracks in the affected area.
- Prepare the surface with undercoats and sealers according to the manufacturer's recommendations.
- Reapply topcoat.

Wrinkling

Identification. **Wrinkling** is a condition on the surface of the paint film that occurs when still liquid absorbs liquids,

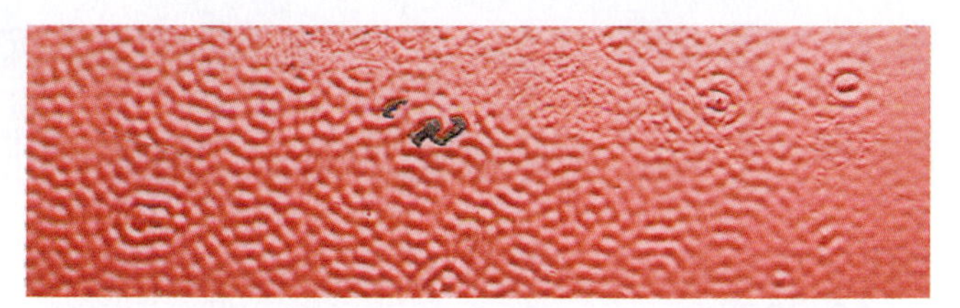

FIGURE 31-27 Wrinkling is a condition on the surface of the paint film that occurs when still liquid absorbs liquids, causing the surface to swell.

causing the surface to swell. See **Figure 31-27**. The surface looks leathery and prunelike.

Causes. The causes of wrinkling include:

- Wrong or inadequate activator or catalyst
- Flooding of the topcoat, exceeding film thickness recommendations
- Paint film not properly drying and curing

Prevention. To prevent wrinkling, follow these steps:

- Use the recommended catalyst.
- Apply the recommended film thickness.
- Stir the coating thoroughly before application.
- Use the recommended amount of catalyst.

How to Repair. If wrinkling occurs, it can be corrected by doing the following:

- Remove all cracks in the affected area.
- Prepare the surface with undercoats and sealers according to the manufacturer's recommendations.
- Reapply the topcoat.

Summary

- Paint problems or defects can be caused by many varying conditions including poor preparation, poor spray techniques, contamination, and poor drying or curing.
- A skilled refinish technician has specific corrections to use to correct each specific defect.
- Some corrections or preventions can be as simple as cleaning and polishing, while others may be so involved that the vehicle needs to be refinished.
- All of the defects discussed in this chapter can be prevented or minimized by using proper cleaning techniques, and by knowing and executing the precise spray techniques required for different applications.
- The experienced technician has seen firsthand the benefits of eliminating contamination on the vehicle, in the shop, and on equipment; and of proper mixing, drying, and curing of coatings.
- In the paint shop, the best cure for paint defects is prevention.

Key Terms

acid rain
bird droppings
bleeding
blistering
chips
cracking
dieback
dirt contamination
dry spray
featheredge splitting/contour mapping
fish eye
lifting
mottling
orange peel
overspray
peeling
poor hiding
runs/sags
sand scratch swelling
solvent popping
striping
wrinkling

Review

ASE Review Questions

1. Technician A says that dirt contamination occurs from not properly cleaning the vehicle before it is painted. Technician B says that dirt contamination can sometimes be mistaken for solvent popping, but on close examination dirt is found to be more random in its distribution than in solvent popping. Who is correct?
   A. Technician A only
   B. Technician B only
   C. Both Technicians A and B
   D. Neither Technician A nor B
2. Technician A says that sand scratch swelling is a defect that occurs due to air-borne contamination. Technician B says that sand scratch swelling occurs due to poor surface preparation. Who is correct?
   A. Technician A only
   B. Technician B only
   C. Both Technicians A and B
   D. Neither Technician A nor B
3. Technician A says contour mapping appears as sand scratches or splitting around the repair area. Technician B says that contour mapping is a condition in which the vehicle finish flakes off around areas that were not properly sanded. Who is correct?
   A. Technician A only
   B. Technician B only
   C. Both Technicians A and B
   D. Neither Technician A nor B
4. Technician A says that lifting occurs due to poor spray technique. Technician B says that poor curing and drying cause lifting. Who is correct?
   A. Technician A only
   B. Technician B only
   C. Both Technicians A and B
   D. Neither Technician A nor B
5. Technician A says that runs or sags are caused by poor surface preparation. Technician B says that poor spraying with too slow a travel speed causes runs and sags. Who is correct?
   A. Technician A only
   B. Technician B only
   C. Both Technicians A and B
   D. Neither Technician A nor B
6. Technician A says that if a gun is not adjusted properly, orange peel may occur. Technician B says that if too fast a gun travel speed is used, orange peel will occur. Who is correct?
   A. Technician A only
   B. Technician B only
   C. Both Technicians A and B
   D. Neither Technician A nor B
7. Technician A says that striping can be caused by poor overlap technique. Technician B says that striping can be caused by a dirty spray gun. Who is correct?
   A. Technician A only
   B. Technician B only
   C. Both Technicians A and B
   D. Neither Technician A nor B

8. Technician A says that solvent popping is caused by poor finish mixing. Technician B says that solvent popping is caused by improper flash time between coats. Who is correct?
   A. Technician A only
   B. Technician B only
   C. Both Technicians A and B
   D. Neither Technician A nor B
9. Technician A says that fish eye is caused by surface contamination before painting. Technician B says that dieback is caused by over-reduction of the finish. Who is correct?
   A. Technician A only
   B. Technician B only
   C. Both Technicians A and B
   D. Neither Technician A nor B
10. Technician A says that cracking can be prevented by applying the recommended film thickness. Technician B says that wrinkling is cured by removing all cracks in the affected area, preparing the surface with undercoats and sealers according to the manufacturer's recommendations, and reapplying topcoat. Who is correct?
    A. Technician A only
    B. Technician B only
    C. Both Technicians A and B
    D. Neither Technician A nor B

Essay Questions

1. After a vehicle has been refinished, the technician notices that dirt and runs have occurred during the application of the finish. What corrective measures must the refinish technician take to prepare the vehicle for delivery?
2. While painting a vehicle, the technician notices that striping is occurring. What corrective measures must be taken to cure this condition?
3. How can a technician prevent overspray from occurring on a vehicle to be refinished?
4. List four defect conditions that can be cured by detailing.

Topic Related Math Questions

1. A shop details 35 cars a week using a total of 2.5 quarts of polishing compound per week. How many gallons of polishing compound must the shop buy to last a year?
2. A shop averages two defect-refinishing repairs per week, at an average cost to the shop of $37.45 per repair. How much profit is lost per year due to defect repairs?
3. It takes an average of 32 minutes to remove dirt from each vehicle. The shop paints 47 vehicles per week, at a shop rate of $38.00/hour. What is the yearly cost to the company of dirt removal?

Critical Thinking Questions

1. A painter in a refinish department paints, on average, three more vehicles per week than another painter in the shop. The faster painter, however, has two to three vehicles each week that he must repair due to defects in the finish. How would you determine which painter is the more cost-efficient worker? Explain your answer.
2. Describe how the manager of a busy paint shop could decide on the acceptable percentage of defect repairs in the paint department. Explain how the manager could find out how much defect repairs were costing the business. Explain your answer.

Lab Activities

1. Have students properly identify and demonstrate the proper sanding and buffing techniques needed to remove dirt from a finish.
2. Have students properly identify and demonstrate the proper sanding and buffing techniques needed to remove runs and sags from a finish.
3. Have students properly identify and demonstrate the proper sanding and buffing techniques needed to remove overspray from a finish.
4. Have students properly identify and demonstrate the proper sanding and buffing techniques needed to remove orange peel from a finish.
5. Have students properly identify and demonstrate the proper sanding and buffing techniques needed to remove solvent popping from a finish.

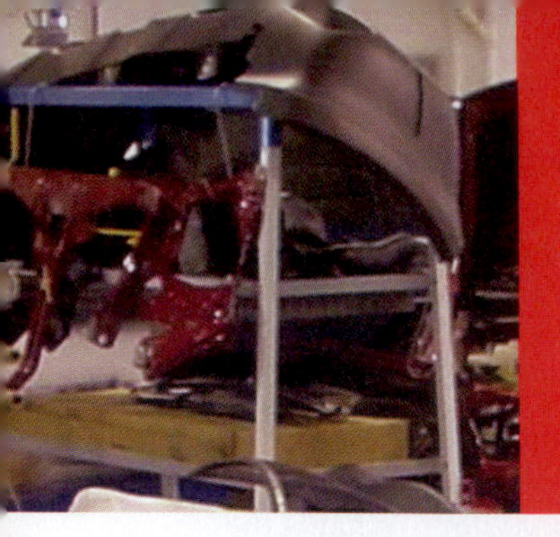

Glossary

Note: Terms are highlighted in color, followed by Spanish translation in bold.

A

A pillar. The pillar, sometimes called a windshield post, that extends down from the front of the roof to the top of the cowl assembly.

pilar A. El pilar, a veces llamado poste de parabrisas, se extiende hacia abajo desde el frente del techo hasta el ensamblaje de la capucha.

ABS modulator. A device that balances the braking force in an automobile with anti-lock brakes.

modulador ABS. Dispositivo que equilibra la fuerza de frenado en un coche con frenos antibloqueo.

A/C condenser. Part of the air conditioning system located in front of the radiator, which is used to dissipate the heat from the cabin or passenger compartment.

condensador de aire acondicionado (A/C por su acrónimo en inglés). Parte del sistema de aire acondicionado localizado delante del radiador, el cual es usado para disipar el calor de la cabina o área de pasajeros.

access time (AT). Extra time needed when part removal or other repair operations may be impeded by other damage to the vehicle, such as a damaged door being difficult to remove because of structural damage to the vehicle.

tiempo de acceso. Tiempo adicional requerido cuando la remoción de partes u otras operaciones de reparación pueden ser impedidas por causa de otro daño al vehículo, como una puerta dañada difícil de quitar debido al daño estructural del vehículo.

acid rain. A form of precipitation which contains pollutants that are a result of sulfur dioxide and nitrogen oxide emitted from burning fossil fuels, which combine and chemically react with each other in the upper atmosphere, forming an acidic by-product.

lluvia ácida. Forma de precipitación que contiene contaminantes que son resultado del dióxido de azufre y óxido de nitrógeno emitido al quemar combustibles fósiles; éstos se combinan y reaccionan químicamente uno con el otro en la atmósfera superior, formando un subproducto ácido.

activator. A chemical or other form of energy that causes another substance to become reactive, or that induces a chemical reaction. In other words, it is a chemical that, when added to paint, causes it to react or cross-link.

activador. Forma de energía química o de otro tipo que hace que otra sustancia se haga reactiva, o que induzca una reacción química. En otras palabras, es un químico que, cuando se añade a la pintura, la hace reaccionar o produce un enlace cruzado.

acute exposure. A high level exposure to a hazardous environment or to a chemical for a short period of time, usually without resulting in any permanent injury or complications.

exposición aguda. Exposición de alto nivel a un medioambiente peligroso o a un químico, por un período de tiempo corto, usualmente sin daños o complicaciones permanentes.

adapter. An accessory that allows a larger drive handle to be attached to a smaller drive socket.

adaptador. Accesorio que permite que una manija de propulsión grande sea unida a un cubo de propulsión menor.

adaptive cruise control. A cruise control option available mostly in high end vehicles which will automatically reduce the vehicle's speed when approaching and getting too near the rear end of another vehicle.

control de velocidad crucero adaptable. Opción de control de velocidad crucero disponible en vehículos de última generación que reduce automáticamente la velocidad del vehículo cuando se aproxima y se acerca demasiado a la parte trasera de otro vehículo.

adhesion loss. A reduction in bonding properties on a vehicle that may result in chipping, peeling, or cracking; may be caused by overcatalyzing or undercatalyzing material when mixed for application.

pérdida de adherencia. Reducción de las propiedades de enlace en un vehículo que puede causar desconchado, peladuras o agrietamientos; puede ser causada por materiales sobrecatalizadores o subcatalizadores cuando son mezclados para su aplicación.

adhesion promoter. Any material used to improve adhesion between two materials; thus, primers are adhesion promoters. However, in the automotive refinishing industry, adhesion promoters are solvents that increase the adhesion between a substrate, providing a better mechanical bond with what is sprayed over it.

promotor de adherencia. Cualquier material usado para mejorar la adherencia entre dos materiales; de esa forma, las imprimaciones son promotoras de adherencia. Sin embargo, en la industria de barnizado automotriz, los promotores de adherencia son solventes que aumentan la adherencia entre un sustrato y lo que es rociado sobre él, proporcionando un mejor enlace mecánico.

adhesion-promoting primers. Primers required when painting unprimed or raw polyolefin plastics. After unpainted polyolefin plastic is cleaned and sanded, adhesion-promoting primer is applied. The adhesion promoters provide a chemical bond of the next coating to the raw plastic. Any subsequent coatings that will be added to the part will not need adhesion promoters.

imprimaciones promotoras de adherencia. Imprimaciones necesarias cuando se pintan plásticos de *polyolefin* no imprimados o crudos. Luego de que el plástico de *polyolefin* pintado es limpiado y lijado, la imprimación promotora de adhesión es aplicada. Los promotores de adherencia proporcionan un enlace químico de la siguiente capa con el plástico crudo. Cualquier capa subsiguiente que sea aplicada a la parte no necesitará promotores de adherencia.

adhesive repair. A two-part mixture used as a repair method: the first material is the resin and the second one is the catalyst. When the catalyst is mixed with resin, they react and become hard. Plastic repair material comes in differing amounts of flexibility to match the plastic that is to be repaired.

reparación adhesiva. Mezcla de dos materiales usada como un método de reparación: el primer material es la resina y el segundo es el catalizador. Cuando el catalizador es mezclado con la resina, éstos reaccionan y se endurecen. El material de reparación plástico provee varios niveles de flexibilidad que permiten emparejar el plástico que debe ser reparado.

adjustable pliers. Pliers that have long handles for gripping leverage and an interlocking channel that allows for multiple jaw opening size adjustments. They are used for bending sheet metal, turning large nuts and bolts, and gripping round objects.

pinzas ajustables. Pinzas que tienen mangos largos para apalancar y un canal entrelazado que permite múltiples aberturas de mordaza. Son usadas para doblar láminas metálicas, ajustar tuercas y tornillos grandes, y agarrar objetos redondos.

adjustable wrench. A wrench with one movable jaw and one fixed jaw, which allows the technician to adjust the wrench opening to fit different-sized fasteners.

llave ajustable. Llave con una mordaza móvil y otra fija que permite al técnico ajustar la abertura para encajar sujetadores de diferentes tamaños.

aerodynamics. The.study of the effects of wind resistance on the shape of a vehicle as it travels down a road.

aerodinámica. Estudio de los efectos de la resistencia del viento de acuerdo a la forma de un vehículo cuando éste viaja en un camino.

after-cooler. A device—often a line—that is installed between the compressor and the storage tank. This line is covered with fins similar to the fins on a radiator, which help dissipate the heat from the compressed air as it passes through. In larger systems it could be a device that uses water circulating over the top of the air line to cool the air before it is stored in the air holding tank.

post-enfriador. Dispositivo, a menudo una línea, que es instalado entre el compresor y el tanque de almacenaje. Esta línea está cubierta de aletas similares a las de un radiador, las cuales ayudan a disipar el calor del aire comprimido a medida que éste pasa a través de ellas. En sistemas más grandes, éste podría ser un dispositivo que usa el agua que circula por encima de la línea de aire para enfriar el aire antes de que sea almacenado en el tanque contenedor.

aftermarket. Replacement parts for a motor vehicle manufactured by a company other than the original equipment manufacturer.

mercado alternativo. Repuestos para un automóvil fabricados por una compañía diferente del fabricante de equipo original.

air cap. A device at the front of the spray gun that directs the airflow from the air horns against the paint, which comes out of the center orifice.

Tapa de aire. Dispositivo localizado en frente de una pistola rociadora que dirige el flujo de aire desde los canales de aire hacia la pintura, la cual sale del orificio central.

air cap pressure gauge. A tool used to measure air pressure. Although air pressure can be measured at different locations when spraying (compressor wall, gun inlet, and air cap), the most accurate location is at the air cap. When measured at any of the other locations, it is possible to have variables that may affect the true measurement.

medidor de presión de la tapa de aire. Instrumento usado para medir la presión de aire. Aunque ésta puede ser medida en diferentes sitios cuando se está rociando (pared del compresor, entrada de la pistola y tapa de aire), el sitio más preciso es la tapa de aire. Cuando la presión es medida en cualquiera de los otros sitios, es posible tener variables que pueden afectar la medida verdadera.

air dam. A hole or opening in the lower front bumper or grill designed to funnel or direct air over the radiator and into the engine compartment to promote cooling the engine.

represa de aire. Agujero o abertura en el parachoques o rejilla delantera inferior diseñado para canalizar o dirigir el aire sobre el radiador y dentro del compartimiento del motor para facilitar el enfriamiento de éste.

air extractor. A device which will remove oil or moisture that may find its way into the air lines.

extractor de aire. Dispositivo que remueve el aceite o la humedad que se puede encontrar en las líneas de aire.

air filter. A device that removes elements such as debris that may not have been cleaned completely from the incoming air at the compressor, or debris which may come from older air lines.

filtro de aire. Dispositivo que remueve elementos tales como residuos que no pudieron ser limpiados completamente del aire que entra en el compresor, o residuos que pudieron llegar a través de líneas de aire más viejas.

air holding tank. A storage container for compressed air.

tanque contenedor de aire. Contenedor de almacenaje de aire comprimido.

air line lubricator. A device that puts oil into the air line to lubricate the tools, keeping them in good working condition. Although the compressed air is sent through an extractor to remove oil and water, the lubricator puts back just oil—the correct type of oil used to lubricate tools.

lubricador de línea de aire. Dispositivo que pone aceite en la línea de aire para lubricar los instrumentos manteniéndolos en buena condición. Aunque el aire comprimido es enviado a través de un extractor para remover aceite y agua, el lubricador sólo regresa aceite: el tipo correcto de aceite usado para lubricar instrumentos.

air line pressure drop. A decrease in air pressure in air lines.

gota de presión de la línea de aire. Disminución en la presión de aire en las líneas de aire.

air line size. The width of an air line hose, which determines the amount of air that can travel through. Less air can travel through a 1/4 inch air line than a 1 inch air line.

tamaño de la línea de aire. Anchura de una manguera de línea de aire, la cual determina la cantidad de aire que puede ser transportada. Menos aire puede viajar por una línea de aire de 1/4 de pulgada que por una línea de aire de 1 pulgada.

air-makeup system. A system that will heat to spray temperatures before it enters the spray booth. This booth design can have the airflow regulated so slightly more air is coming into the booth than is going out. This positive pressure regulation helps keep the dirt from outside the booth from being drawn in, as it would be with a booth having only an exhaust fan.

sistema de composición de aire. Sistema que se calienta para dispersar la temperatura antes de entrar en la cabina de dispersión. Este diseño de cabina puede regular la corriente de aire de forma que ligeramente más aire entra en la cabina que el que sale. Esta regulación de presión positiva ayuda a mantener la suciedad del exterior de la cabina de entrar, tal como lo sería con una cabina de dispersión teniendo un solo ventilador de escape.

air pressure (psi). Measured in psi, this is the force exerted by air over a certain location.

presión de aire (psi). Medida en psi, es la fuerza ejercida por el aire en un determinado lugar.

air ratchets and impact wrenches. These tools, available as air-driven or electric, should only be used with sockets and attachments designed for the stresses created by impact tools.

trinquetes de aire y llaves de impacto. Estos instrumentos, que pueden ser de aire comprimido o eléctricos, sólo deben ser usados con enchufes y accesorios diseñados para las tensiones creadas por instrumentos de impacto.

air receiver. A tank designed to hold the air in the compressed air process.

receptor de aire. Tanque diseñado para mantener el aire durante el proceso de compresión.

air supply respirators. *See* supplied air respirator.

respiradores de suministro de aire. Ver "respiradores de aire suministrado."

air tools. Tools which run significantly cooler than electric tools and require little more than a daily lubrication for their maintenance requirements.

instrumentos de aire. Instrumentos que funcionan considerablemente más fríos que instrumentos eléctricos, y que requieren poco menos que una lubricación diaria para sus requerimientos de mantenimiento.

air transformer. Sometimes called air regulators; these are devices used to help transform high air pressure into lower air pressure. Although such a device is often called a regulator, one of the key functions of the transformer is to change air pressure without changing or affecting the volume of the air from the supply line.

transformador de aire. Algunas veces llamado regulador de aire; éste es un dispositivo usados para ayudar a transformar presión de aire alta en presión de aire baja. Aunque este dispositivo es a menudo llamado regulador, una de las funciones claves del transformador es cambiar la presión de aire sin cambiar o afectar el volumen del aire de la línea de abastecimiento.

air volume (CFM). Measured in cfm, it is the space taken up by air.

volumen de aire (CFM). Medida en pies cúbicos por minuto (ft^3/m), es el espacio ocupado por el aire.

airless plastic welding. More common than hot air welders, and requiring a little less practice to develop skill in their use, these welders are powered by 110-volt shop electricity, have adjustable heat controls, and are rated at about 80 watts at the tip.

soldador plástico sin aire. Más común que los soldadores de aire calientes, y que requiere un poco menos de práctica para desarrollar la habilidad de su uso, estos soldadores son impulsados por electricidad de 110 voltios, tienen mandos de calor ajustables, y tienen una medida de aproximadamente 80 vatios en la punta.

aligning/roughing hammers. Large and heavy hammers used to align a panel, bringing it back to its original dimensions.

martillos de alineado/desalineado. Martillos grandes y pesados usados para alinear paneles y regresarlos a sus dimensiones originales.

Allen wrench. A hexagon or six-sided wrench used to remove and tighten fasteners with six-point socket heads.

llave Allen. Llave hexagonal o de seis lados usada para remover y apretar sujetadores de cabeza de enchufe de seis puntos.

alloys. A mixture of elements designed to make the final material stronger and more resistant to bending.

aleaciones. Mezcla de elementos diseñada para hacer que el material final sea más fuerte y más resistente a la flexión.

alodine iridite. A chemical treatment process or coating used for treating aluminum to enhance its corrosion protection and form a surface that will promote adhesion for paint and other coatings applied over it.

alodine iridite. Proceso de tratamiento o recubrimiento químico para tratar el aluminio y mejorar su protección contra la corrosión, y formar una superficie que mejorará la adherencia para la pintura y otras capas aplicadas sobre él.

aluminum substrate. Any material, made of aluminum, that makes up part of a motor vehicle's structure or outer panels.

sustrato de aluminio. Cualquier material, hecho de aluminio, que hace parte de la estructura o paneles exteriores de un automóvil.

aluminum-intensive vehicles/automobiles. Motor vehicles constructed primarily from aluminum, including the structure and exterior panels.

vehículos de aluminio intensivo. Automóviles construidos principalmente de aluminio, incluyendo la estructura y los paneles exteriores.

ambient. Surrounding.

ambiente. Alrededor.

ambient temperature. The temperature of the atmosphere in the immediate environment; for example, the ambient temperature of a room may be 72°F and the ambient temperature outdoors may be 95°F.

temperatura ambiental. Temperatura de la atmósfera en el medio ambiente cercano; por ejemplo, la temperatura ambiental de un cuarto puede ser 72°F, y la temperatura ambiental al aire libre puede ser 95°F.

amperage. The measuring unit of current.

amperaje. Unidad de medida de la corriente eléctrica.

anti-corrosion compounds. A heavy bodied compound that is sprayed or brushed onto enclosed areas and on the back side of many panels that are susceptible to corrosion.

compuestos anticorrosivos. Compuesto pesado que es rociado o cepillado en áreas encerradas y sobre el lado trasero de muchos paneles que son susceptibles a la corrosión.

anti-lock brake system (ABS). A system that prevents wheel lockup during braking.

sistema anti-bloqueo de frenos (ABS). Sistema que previene que las ruedas se bloqueen durante el frenado.

anti-theft label. A label bearing the vehicle's VIN attached to most of the exterior body panels on select vehicles; used for tracking and identification purposes.

etiqueta antirrobo. Etiqueta que lleva la identificación (VIN por su acrónimo en inglés) del vehículo y que es colocada en la mayoría de los paneles exteriores de determinados vehículos; esta es usada con objetivos de identificación y rastreo.

aperture tape. Developed to replace traditional paper and tape masking, the two round aperture tapes have an adhesive on one side that can be attached to the door jamb near the edge. Then the door is closed touching the tape, which keeps the overspray out. Aperture tape is also ideal for keeping polishing and compounding sludge out of the door jamb area.

cinta de abertura. Desarrollada para sustituir el papel tradicional y la cinta de enmascarar, las dos cintas de abertura redondas tienen un pegamento en un lado que puede ser adherido a la jamba de la puerta cerca del borde. La puerta entonces es cerrada tocando la cinta, la cual mantiene los excesos de rociado alejados. La cinta de abertura también es ideal para mantener los residuos de pulido y de compuestos alejados del área de la jamba de la puerta.

appliqué. A decorative trim typically found near or surrounding a window. It can be painted a gloss black or dull black; it can be made of plastic and adhered with adhesive, clips, or screws which may be hidden under weatherstripping or seals.

appliqué. Moldura decorativa elegante típicamente encontrada cerca o bordeando una ventana que puede ser pintada con negro brillante o mate; y que puede ser hecha de plástico y adherida con adhesivos, clips, o tornillos, los cuales puedes estar ocultos bajo coberturas o sellos a prueba de intemperie.

atomization. The process which breaks the surface tension of a liquid into small droplets, allowing it to be deposited as a spray. The smaller the droplets, the smoother the painted surface.

atomización. Proceso que rompe la tensión superficial de un líquido en pequeñas gotitas, permitiéndole ser esparcido. Entre más pequeñas sean las gotitas, más lisa quedará la superficie pintada.

axis point. The alternating pivot point used to adjust the door to effectively move it in a circular motion until it is properly centered with uniform gaps between all adjacent panels.

puntos de eje. Punto de pivote alterno usado para ajustar la puerta para efectivamente moverla en un movimiento circular hasta que esté apropiadamente centrada con espacios uniformes entre todos los paneles adyacentes.

B

B pillar. The pillar or post, sometimes called the center pillar, which extends from the center of the roof down to the top of the rocker panel.

pilar B. Pilar o poste, a veces llamado pilar central, que se extiende desde el centro del techo hacia abajo hasta el panel del estribo.

back masking. Technique in which a technician places a strip of masking tape on the back of the hood, with approximately 50 percent of it overhanging. When the hood is completely surrounded with tape, masking paper is applied and either folded under, or left hanging down.

enmascarado trasero. Técnica en la cual un técnico coloca una tira de cinta adhesiva en la parte trasera del capó, con aproximadamente el 50 por ciento de la cinta sobresaliendo. Cuando el capó está completamente rodeado por la cinta, el papel de enmascarado es colocado, y doblado hacia bajo o dejado colgando.

backlite. A term commonly used to describe the rear window of an automobile.

luces traseras. Término comúnmente usado para describir las luces de la parte trasera de un vehículo.

backup spoons. Much heavier and larger than surfacing spoons, these can be used in the place of a dolly for finish work to a panel. They also can be placed in areas that a dolly cannot access. Because they are heavier, they are often used to pry out damaged areas in the roughing stage of panel repair.

cucharas traseras. Mucho más pesadas y más grandes que las cucharas de moldeado, éstas pueden ser usadas en lugar de una carretilla para realizar trabajos en un panel. Éstas también pueden ser usadas en áreas en donde una carretilla no puede entrar. A causa de su mayor peso, éstas son a menudo usadas para abrir áreas dañadas en la etapa ruda de la reparación de paneles.

bag and tag. The identification and inventorying of fasteners, hardware, and misc. parts while a vehicle is being disassembled to ensure they are put back into the correct place on the right panel when the vehicle is reassembled.

bolsa y etiqueta. Identificación e inventariado de sujetadores, herrajes y partes accesorias, mientras un vehículo está siendo desensamblado para asegurar que ellos sean colocados en el lugar correcto cuando el vehículo sea reensamblado.

bake cycle. The amount of time a vehicle being finished is placed in heat to dry the paint.

ciclo de horneado. Cantidad de tiempo que un vehículo terminado es expuesto al calor para secar la pintura.

ball peen hammer. A general purpose tool available in several different weights that is used to rough out sheet metal or drive punches and chisels.

martillo de bola. Herramienta de uso general disponible en varios pesos que es usada para golpear láminas metálicas o dirigir golpes y cinceles.

banding. Spraying first the perimeter of the part to be painted.

estratificación. Rociado del perímetro de la parte que va a ser pintada antes de pintarla.

basecoat blending. A process in which the basecoat is applied to a repair area that is smaller than the complete panel being repaired. In blending, the technician applies coating at a reduced thickness slightly beyond the repair area.

mezclado de recubrimiento base. Proceso en el cual el recubrimiento base es aplicado a un área de reparación que es más pequeña que el panel que está siendo reparado. En la mezcla, el técnico aplica la capa con un grosor reducido ligeramente más allá del área de reparación.

baseline/standard model vehicle. A vehicle equipped with the minimum options all vehicles in the same family line offer; the vehicle purchaser may choose to add any additional options and features available for said vehicle at an additional cost.

vehículo modelo de línea base o estándar. Vehículo equipado con las mínimas opciones que todos los vehículos de la misma familia tienen; el comprador de vehículo puede añadir cualquier opción y características adicionales disponibles para el vehículo por un costo adicional.

bauxite. An ore, a combination of oxygen and aluminum hydroxide, which is processed into pure aluminum.

bauxita. Mineral que es una combinación de oxígeno e hidróxido de aluminio, el cual es procesado en aluminio puro.

belt line. An arbitrary, non-descript line approximately two-thirds the way up the door where the metal for the door panel ends and the window opening begins.

línea de correa. Línea arbitraria indescriptible localizada a aproximadamente dos terceras partes hacia arriba en la puerta, en donde el metal del panel de la puerta termina y la apertura de la ventana comienza.

bench/rack system. A hybrid frame repair system with characteristics that are similar to both the portable bench and the frame rack systems. While the bench is usually not portable or movable, it is designed to be elevated off the ground for access underneath while the corrective pulls are being made.

sistema de elevado de banco. Sistema híbrido de reparación de estructuras con características que son similares tanto al banco portátil como a los sistemas de elevado de estructuras. Mientras que el banco por lo general no es portátil o movible, éste está diseñado para ser elevado desde el suelo para acceder por debajo, mientras que los jalones correctivos están siendo aplicados.

bend vs. kink. A guideline commonly used to determine whether a steel frame rail or structural part should be replaced or if it is repairable (i.e., can it be corrected without leaving any signs of damage in order to "restore it to its natural state"); a rail that is bent 90 degrees or more in a short radius must be replaced, whereas one that is gradually bent over a longer area may be repaired.

doblado versus curvado. Guía comúnmente usada para determinar si un riel de marco o parte estructural de acero debe ser sustituido o si es reparable (es decir, éste puede ser corregido sin dejar signos de daño para "restaurarlo a su estado natural"); un riel que es doblado 90 grados o más con un radio pequeño debe ser reemplazado, mientras que uno que es gradualmente doblado sobre un área más larga puede ser reparado.

bird droppings. Fecal matter of birds which can cause damage to a finish similar to acid rain if not removed.

deposición de pájaros. Materia fecal de pájaros que puede causar daño a un terminado, similar al causado por la lluvia ácida, si no es removida.

bleaching. Color change over areas of plastic body filler repairs.

blanqueamiento. Cambio de color sobre reparaciones de relleno de partes plásticas.

bleeding. Situation in which color underneath the topcoat, which is usually darker than the topcoat, shows through the finish.

traspaso (de color). Situación en la cual el color que está debajo del recubrimiento exterior es visible a través del terminado. El color suele ser más oscuro que el recubrimiento exterior.

bleed-through. Condition where the previous color or coating shows through the newly applied finish.

traspaso de color. Condición en la cual el color o el recubrimiento previo es visible a través de terminado aplicado recientemente.

blend masking. Masking technique in which, after exposing every panel to be painted and cleared, the technician will cover the remainder of the vehicle, followed by the adjacent panels that will only be blended onto—then cleared—later. These panels are covered in a way that will allow the technician to remove this second masking after the repaired panels have had basecoat applied to full coverage. The technician then removes the second masking, lightly blends the adjacent panels, and—after the needed flash and cure time—clears all the panels.

enmascarando combinado. Técnica de enmascarado en la cual después de exponer cada panel a los procesos de pintado y limpieza, el técnico cubre el resto del vehículo, junto con los paneles adyacentes que sólo serán combinados, y limpiados, después. Estos paneles son cubiertos de forma que permita al técnico quitar este segundo enmascarando después de que a los paneles reparados se les haya aplicado el recubrimiento base con cobertura total. El técnico entonces quita el segundo enmascaramiento, ligeramente combina los paneles adyacentes, y, después de los tiempos de destello y de curado necesarios, limpia todos los paneles.

blendable match (color). A color that is close enough in appearance to another color that blending will be successful. This is often determined by comparing possible color matches on a sprayout panel.

color combinable. Color que es bastante cercano a otro color en aspecto, que la combinación tendrá éxito. Esto es a menudo determinado al comparar posibles colores combinables en un panel de pruebas de rociado.

blending preparation. A preparation process similar to that used to prepare any other repair. The surrounding panels that will be color blended and clearcoated should be prepared for topcoating.

preparación de la combinación. Proceso de organización similar al hecho para preparar cualquier otra reparación. Los paneles circundantes que serán cubiertos de colores combinados y transparentes deben ser preparados para la aplicación de la capa final.

blind rivets. Commonly called "pop rivets," these are used to hold panels in place during the welding process. After welding, they are removed and the holes are closed by welding.

remaches ciegos. Comúnmente llamados "remaches *pop*", éstos son usados para sostener paneles en su sitio durante el proceso de soldado. Después de la soldadura, éstos son quitados y los agujeros son cerrados por soldadura.

blistering. Refers to bubbles, or what appear to be water blisters, under the paint film.

ampollado. Se refiere a burbujas, o lo que parece ser ampollas acuosas, bajo la capa de pintura.

blocking. A process that levels an area in order to blend in the uneven areas created by the repair process. Following the repair, a technician may have the substrate at the lowest level, the body filler slightly higher than that, and the old finish (although feathered) at the highest surface.

bloqueado. Proceso que nivela un área a fin de combinar las áreas desniveladas por el proceso de reparación. Después de la reparación, el técnico puede poner el sustrato al nivel más bajo, el material de relleno ligeramente más alto que éste, y el terminado final (aunque seccionado) en la superficie más alta.

blood-borne pathogens. Microorganisms or viruses that are carried in the blood or other body fluids that are transmitted from one person to another by exposure through cuts, abrasions, sores, and so on.

patógenos transportados por sangre. Microorganismos o virus que son llevados en la sangre u otros fluidos corporales, los cuales son transmitidos de una persona a otra por la exposición a través de cortaduras, abrasiones, llagas, etcétera.

board file. A tool designed to level large flat or contoured areas of cured body filler. The holder is capable of accepting different grits of sandpaper, held in place by manual clips as opposed to hook and loops or glue.

sujetador de tablero. Herramienta diseñada para nivelar áreas grandes o perfiladas de rellenos curados. El sostenedor es capaz de aceptar lijas de diferentes tamaños de grano, sostenidas en su lugar por sujetadores manuales; a diferencia del colgado y enlazado, o pegado.

body dimensional chart. A specification chart that gives dimensional specifications for the undercarriage dimensions, underhood dimensions, and a variety of other critical areas. These are frequently used when manual gauges such as the self-centering gauges and the tram gauge are used to locate and identify the damaged areas.

gráfica de dimensiones vehiculares. Gráfica que provee especificaciones dimensionales para las medidas del tren de aterrizaje, dimensiones interiores del capó y una variedad de otras áreas críticas. Estás gráficas son frecuentemente usadas cuando medidores manuales, tales como medidores de auto-centrado y medidores de precisión, son usados para localizar e identificar las áreas dañadas.

body file. A tool commonly used to level minor surface imperfections on sheet metal, a process sometimes referred to as metal finishing. High and low spots in the metal surface can be identified with the body file.

limadora de superficies. Herramienta comúnmente usada para nivelar imperfecciones superficiales menores en láminas metálicas; este proceso a veces se denomina acabado metálico. Los puntos altos y bajos en la superficie metálica pueden ser identificados con la limadora de superficies.

body hammer. A variety of hammers used to work sheet metal back into shape. Each hammer is designed for a specific repair operation. To avoid stretching and marking soft sheet metal, a hammer with a large, nearly flat face should be used. The face of the body hammer should have a dead flat center with a slight crown on the outer edge to compensate for errors when striking a blow. A small face will usually have a higher crown than a large face.

martillo de cuerpo. Variedad de martillos usados para trabajar láminas metálicas y darles su forma original. Cada martillo está diseñado para una operación de reparación específica. Para evitar estiramientos y marcas en láminas metálicas blandas, un martillo con una cara grande y casi plana debe ser usado. La cara del martillo de cuerpo debe tener un centro plano muerto con una corona ligera en el borde exterior para compensar posibles errores al dar un golpe. Una cara pequeña tendrá por lo general una corona más alta que una cara grande.

body over frame (BOF). A vehicle design wherein the frame and body operate independently of each other.

cuerpo sobre marco. Diseño de vehículo en donde el marco y el cuerpo funcionan independientemente el uno del otro.

body spoon. A tool designed to reach into restricted areas that conventional hammers and dollies cannot access. Spoons are available in a variety of sizes and shapes, which are designed to match various panel shapes. They can be used as a driving or prying tool to remove dents. The large, flat faces are designed to spread out hammer blow force over a large area.

cuchara de cuerpo. Herramienta diseñada para llegar a áreas difíciles en las que martillos convencionales y carretillas no pueden tener acceso. Estas cucharas, disponibles en una variedad de tamaños y formas, están diseñadas para emparejar varias formas de paneles. Éstas pueden ser usadas como una herramienta de manipulación o intromisión para remover abolladuras. Las caras grandes y llanas están diseñadas para extender la fuerza de un martillazo sobre un área grande.

bolt. A connecting device often combined with a washer and nut that is inserted into a hole to connect two materials together.

perno. Dispositivo de conexión que a menudo es combinado con una arandela y una tuerca, y que es insertado en un agujero para unir dos materiales.

bolt boss. An area or a surface on a cast part made thicker—often as much as twice as thick as the surrounding walls—to accommodate drilling holes to tap threads into it, allowing bolts to attach parts.

área de pernos. Área o superficie de una parte fundida que es más gruesa, a menudo dos veces más gruesa que las paredes circundantes, y que se usa para alojar agujeros para le paso de roscas, permitiendo que los pernos unan las partes.

boron steel. Steel with a yield point from 196,000 to 203,000 psi, which is 100 percent higher than ultra high strength steel. This extremely strong yield strength also comes with extreme sensitivity to heat, so much so that even the process of galvanizing steel harms the steel.

acero de boro. Acero con punto de fluencia de 196.000 a 203.000 psi, que es 100 por ciento más alto que el acero de alta resistencia. Esta resistencia extrema a la fluencia viene también acompañada de una extrema sensibilidad al calor; tanto que aún el proceso de galvanizado del acero, lo afecta.

box-end wrench. A wrench with closed ends that surround the bolt head or nut for better holding power. More force can be applied without slipping or rounding off the nut or bolt head.

llave de cubo cerrado. Llave con extremos cerrados que rodean la cabeza de un perno o tuerca para mejor poder de agarre. Más fuerza puede ser aplicada sin bajar o desgastar la tuerca o la cabeza del perno.

box lifter. A lifting device that, with the aid of an engine hoist, can be used to lift a pickup box off the vehicle's frame.

levantador de caja. Dispositivo de levante que, con la ayuda de una grúa de motor, puede ser usado para levantar la caja de una camioneta y separarla de la estructura del vehículo.

breaker bar. A long bar that provides the torque needed to loosen seized or tight fasteners.

barra de quiebre. Barra larga que proporciona la torsión necesaria para aflojar sujetadores agarrados o muy apretados.

breathable air. *See* class D breathable air.

aire respirable. Ver "aire respirable clase D."

buckles. Distortions in sheet metal and/or frame structures that occur as a result of collision energy forces traveling through that area of a vehicle.

combas. Deformaciones en láminas metálicas y/o estructuras del marco vehicular que ocurren como resultado del viaje de las fuerzas producidas por la energía de una colisión a través de aquella área del vehículo.

buffer. A power tool that usually offers a variable speed RPM rating used to polish an existing or newly applied finish to remove minor surface flaws and texture in the newly applied finish.

pulidora. Herramienta eléctrica que por lo general gira a varios niveles de revoluciones por minuto; se usa para pulir un acabado existente o recientemente aplicado para remover defectos menores superficiales y texturas en el acabado recientemente aplicado.

bump steer. A condition in which, after a vehicle hits a bump, it involuntarily and uncontrollably darts and dives, from side to side, without moving the steering wheel. It can become an issue after a frontal collision.

conducción en choque. Condición en la cual, después de que un vehículo choca, éste involuntaria e incontrolablemente se lanza y salta de lado a lado, sin mover el timón de conducción. Esto puede ser importante luego de una colisión frontal.

bumping/dinging hammer. Commonly used to "bump" out large dents during the initial straightening process, these hammers may have round or square faces and flat or crowned striking surfaces. They are often used to straighten and flatten already roughed-out sheet metal, further restoring it to its original shape.

martillos de choque/abolladura. Comúnmente usados para "sacar" abolladuras grandes durante el proceso de alineamiento inicial, estos martillos pueden tener caras planas redondeadas o cuadradas y superficies de golpeo con coronas. Éstos son a menudo usados para alinear y aplanar láminas metálicas que están previamente deformadas, para luego darles su forma original.

bumping file. A tool with a spoon-like shape and serrated surfaces to slap metal back to its original shape.

lima para choques. Herramienta con una forma parecida a una cuchara y superficies aserradas que sirve para golpear un objeto metálico y regresarlo a su forma original.

butyl benzoyl peroxide. The principal ingredient in the catalyst or plastic hardener that is responsible for creating the chemical reaction that causes plastic to harden.

peróxido de Butil-Benceno. Principal ingrediente en el catalizador o endurecedor de plástico que es responsable de crear la reacción química que hace que el plástico se endurezca.

C

C pillar. The support panel for the rear of the roof, often called the sail panel, that extends upward from the quarter panel and ties the roof and quarter panel together.

pilar C. Panel de soporte para la parte de atrás del techo, a menudo llamado panel de navegación, y que se extiende hacia arriba desde la cuarta parte del panel; éste une el techo con la cuarta parte del panel.

cage nut/U-clip. A special fastener /clip attached to an oversized hole, allowing it to move around to enable movement of the panel being attached; commonly used in conjunction with a bolt to secure an adjustable panel such as a fender, hood.

tuerca de jaula o clip en U. Sujetador especial o clip unido a un agujero de gran tamaño que lo deja moverse para permitir el movimiento del panel que está siendo sujetado; éste es comúnmente usado en conjunto con un perno para asegurar un panel ajustable, tal como un guardabarros, o capó.

camber. The inward or outward tilt at the top of the wheel when viewing it from the front of the vehicle. It is commonly identified as negative camber, positive camber, or "0" camber.

comba. Inclinación hacia adentro o hacia fuera en la parte superior de la rueda cuando es vista desde el frente del vehículo. Ésta es comúnmente identificada como comba negativa, comba positiva, o comba "0".

candlepower. The measurement of the amount of light.

bujía. Medida de la cantidad de luz.

carbon. An element found in coal, petroleum, asphalt, and some organic compounds.

carbono. Elemento encontrado en carbón, petróleo, asfalto, y algunos compuestos orgánicos.

carbon fiber reinforced plastics (CFP). A very durable material made of woven carbon fibers and a special resin extensively used on military aircraft and for limited applications on automobiles.

plásticos reforzados con fibra de carbono (CFP). Material muy duradero hecho de fibras de carbono tejidas y una resina especial extensamente usada en aviones militares y en usos limitados en automóviles.

carpet extractor. A machine that removes dirt and water from a carpet during cleaning.

extractor de alfombra. Máquina que quita la suciedad y el agua de una alfombra durante su limpieza.

cast parts. Parts usually made with thicker walls than the extruded parts which are obtained by pouring melted aluminum into a form and allowing it to solidify. Once the metal is solid, it will be machined, drilled, and threads will be cut into the openings to accommodate the use of bolts threaded fasteners.

partes de fundición. Partes usualmente hechas con un espesor mayor que las partes extrudidas, las cuales son obtenidas al verter aluminio derretido en un molde dejando que se solidifique. Una vez el metal está sólido, esté será trabajado, perforado, y los hilos serán cortados dentro de las aberturas para alojar pernos y sujetadores enhebrados.

caster. The forward or rearward tilt of the front wheels on the steering axis. It is commonly referred to as either positive or negative, depending on whether the tilt of the wheel is facing the front or the rear of the vehicle.

ángulo caster. Inclinación hacia delante o hacia atrás de las ruedas frontales sobre el eje de conducción. Éste es comúnmente definido como positivo o negativo dependiendo de si la inclinación de la rueda está hacia el frente o hacia atrás del vehículo, respectivamente.

castle or slotted nuts. A nut, sometimes called a slotted nut, used with a cotter pin to ensure it will not loosen.

tuerca tipo castillo. Tipo de tuerca, a veces llamada tuerca de seguridad, que es usada con un alfiler de chaveta para asegurar que no se afloje.

catalyst. A chemical which, when added to two other chemicals, causes a reaction. The reaction would not take place—or would occur much more slowly—if that chemical were not present.

catalizador. Sustancia química que, cuando se agrega a otras dos sustancias, causa una reacción. La reacción no ocurrirá, u ocurrirá mucho más lentamente, si ese químico no estuviera presente.

cathodic coatings. A form of electrochemical coating similar to galvanizing.

recubrimiento catódico. Forma de recubrimiento electroquímico similar a la galvanización.

caulking irons. Special-purpose tools designed to be struck with a hammer that are used to reshape bends; narrow, flat surfaces; and bead sections on fenders.

hierro de calafateado. Herramientas de uso especial, diseñadas para ser golpeadas con un martillo, que son usadas para dar forma a zonas curvas, superficies estrechas y planas, y secciones de cuenta en guardabarros.

C-clamps. A tool used to hold parts in place while doing repairs.

abrazaderas tipo C. Herramienta usada para mantener las partes en su lugar mientras se hacen reparaciones.

C-clips. A part of the door handle assembly.

gancho tipo C. Parte del ensamblaje de la manija de la puerta.

ceiling filters. Air filters in the ceiling of a spray booth that screen out contaminants which, although not changed regularly, are time consuming to change when needed.

filtros de tope. Filtros de aire en el tope de una cabina de rociado que mantiene los contaminantes apartados. Estos filtros, aunque no sean cambiados regularmente, requieren mucho tiempo cuando es necesario su reemplazo.

center punch. A punch with a pointed end which creates an indentation to prevent a drill bit from wandering out of position.

perforadora de centro. Perforadora con un extremo puntiagudo que crea una mella para impedir que una parte una broca se mueva de su posición.

centerline plane. An imaginary line that runs through the middle of the vehicle from the front to rear and from the floor to the roof. The centerline is used as a reference to measure and monitor all of the vehicle's width measurements and to determine any side-to-side movement.

plano de línea central. Línea imaginaria que pasa por en medio del vehículo desde el frente hasta la parte de atrás y desde el piso hasta el techo. La línea central es usada como referencia para medir y monitorear todas las medidas de anchura del vehículo y para determinar cualquier movimiento de lado a lado.

center-to-center measurement. When using reference holes, the policy to measure from the center of one hole to the center of another.

medida de centro a centro. Cuando se usan agujeros de referencia, es la regla de medir del centro de un agujero al centro del otro.

centrifugal. An outward force associated with rotation.

centrífuga. Fuerza externa asociada con la rotación.

chamois. A type of cloth frequently used for drying vehicles. Only synthetic chamois should be used, as the natural chamois clothes attract dirt and debris, and will become abrasive with age.

gamuza. Tipo de tela frecuentemente usada para secar vehículos. Sólo gamuza sintética debe ser usada, ya que las telas de gamuza naturales atraen suciedad y residuos, y se vuelven abrasivos con el tiempo.

chemical adhesion. A process which occurs through chemical cross-linking when chemically matched paint products are applied over the top of each other. An example is applying a clearcoat over the color coat made by the same manufacturer. Cross use of products usually negates the cross-linking process.

adherencia química. Proceso que ocurre a través de un enlace cruzado cuando productos de pintura químicamente compatibles son aplicados uno sobre el otro. Un ejemplo es aplicar un recubrimiento transparente sobre el recubrimiento de color hecho por el mismo fabricante. El uso cruzado de productos usualmente anula el proceso de enlace cruzado.

chemical adhesive bonding. A bonding method that's very popular with collision repair technicians due to the advances that have been made in ease of application, sandability, and product durability. These adhesives can be epoxy, urethane, acrylic, or even cyanoacrylate ("super glue"), though cyanoacrylate may not be strong enough to be used alone. They come as bonding agents and filling agents, which sand better than those that are purely bonding adhesives.

enlace de adhesivo químico. Método de enlace que es muy popular en los técnicos de reparación de colisiones debido a los avances que han sido logrados para facilitar su aplicación, para el lijado y la durabilidad del producto. Estos adhesivos pueden ser epóxicos, uretanos, acrílicos o aún cianoacrilatos ("superpegante"); aunque el cianoacrilato puede no ser suficientemente fuerte para ser usado solo. Éstos vienen como agentes de enlace o de relleno, los cuales lijan mejor que aquellos que son puramente adhesivos de enlace.

chemical cleaner. A substance which removes contaminants that are not dissolved with soap and water. They work by dissolving the contaminant, then causing it to float to the top of the chemical.

limpiador químico. Sustancia que remueve contaminantes que no se disuelven con jabón y agua. Éstos trabajan disolviendo el contaminante, luego hacen que éste flote sobre las sustancias químicas.

chemical cleaning. A process used to remove substances that are not dissolved with soap and water, such as tar, grease, oils, and waxes. This deep chemical cleaning removes items that may be stuck to the surface that, if picked up during buffing, would cause damage. This method also cleans the surface so it can be properly inspected.

limpieza química. Proceso usado para remover sustancias que no se disuelven con jabón y agua, tales como alquitrán, grasa, aceites y ceras. Esta limpieza química profunda remueve partículas que pueden adherirse a la superficie que, si se atrapan durante el brillado, pueden causar daño. Este método también limpia la superficie de forma que pueda ser apropiadamente inspeccionada.

chemical stripping. A process generally done when the finish must be completely removed. To begin the chemical stripping procedure, the technician applies the chemical stripper to a section, allows it to bubble up the finish, then scrapes off the lifted finish with a plastic spreader. The procedure is repeated as needed.

limpieza química. Proceso generalmente hecho cuando el acabado debe ser completamente removido. Para comenzar el proceso de limpieza química, el técnico aplica el limpiador químico a una sección, permitiéndole que reaccione con el acabado, luego quita el acabado raspándolo con una paleta plástica. El procedimiento se repite como sea necesario.

chip resistant coatings. A special coating of a soft, heavy bodied material applied to the lower section of a car—either over or under the topcoat—to protect the paint from chipping by cushioning the effects of rocks and other road abrasion materials.

recubrimiento resistente al desconchado. Recubrimiento especial de un material robusto y suave aplicado a la sección inferior de un vehículo, sea sobre o por debajo del recubrimiento superior, para proteger la pintura del desconchado al amortiguar los efectos de rocas y otros materiales abrasivos de carreteras.

chips. The breaking away of a topcoat, leaving small exposed areas. Chipping occurs due to the topcoat's inability to flex when impacted with road debris, or because of thermal expansion or contamination on the surface under the topcoat.

desconchado. Ruptura del recubrimiento exterior que deja pequeñas áreas expuestas. El desconchado ocurre debido a la incapacidad de la capa exterior de flexionarse cuando es impactada con escombros de caminos, o a causa de expansión térmica, o por contaminación de la superficie bajo el recubrimiento exterior.

chisels. A tool used for cutting sheet metal, separating welds, and cutting bolt heads. A flat cold chisel is most commonly used in collision repair.

cinceles. Herramienta usada para cortar láminas metálicas, separar soldaduras y cortar cabezas de pernos. Los cinceles planos son más comúnmente usados en la reparación de colisiones.

chroma. Sometimes called intensity; it is the color's brightness.

chroma. Algunas veces llamada intensidad; es la brillantez del color.

chronic exposure. A low level exposure to a harmful substance that occurs over an extended period of time, frequently resulting in irreversible damage affecting vital organs and the central nervous system.

exposición crónica. Nivel bajo de exposición a una sustancia dañina que ocurre durante un periodo extenso de tiempo, resultando frecuentemente en daño irreversible, lo que afecta los órganos vitales y el sistema nervioso central.

circular sanders. These tools, which operate at a lower speed than grinders, are used for polishing and buffing the new finish after completion of the refinishing operations.

lijadora circular. Estas herramientas, que operan a una velocidad más baja que los discos abrasivos, son usadas para pulir y brillar el nuevo acabado después de completar las operaciones de barnizado.

cladding. A decorative molding that is placed on vehicles in several locations such as rocker covers, bottoms of doors, and tailgates, just to name a few. It is generally placed in areas where damage from road debris or chipping is likely to take place. Cladding provides protection to the part underneath and is generally more durable than steel and paint alone.

revestimiento. Moldeado decorativo que es colocado en vehículos en varios sitios, como tapas oscilantes, fondos de puertas, y puertas posteriores; por mencionar algunos. Éstos son generalmente colocados en áreas en donde daños por escombros de carretera o desconchamiento pueden ocurrir. El revestimiento proporciona la protección a la parte de abajo y es generalmente más durable que el acero y la pintura solos.

claim-related documentation. Anything of significance that can have a direct or indirect effect on a physical damage insurance claim such as police reports and statements, any prior loss history of the vehicle and/or the driver, vehicular loan information, appraisals or damage reports, and photos of the damages, all related receipts and invoices, involved outside agency information such as theft bureau, motor vehicle, and insurance department reports, all statements from drivers, passengers, witnesses, and others with any link to the claim or vehicle, and all documentation of any conversations with involved parties.

documentación relacionada con reclamos. Cualquier cosa significante que pueda tener efecto directo o indirecto en un reclamo de seguro de daño físico tales como reportes y declaraciones de policía; cualquier historia previa de pérdida del vehículo y/o del conductor; información del crédito del vehículo, reportes de valoración o daño, y fotos de los daños; todos los recibos relacionados y facturas; información de las agencias exteriores involucradas tales como oficina de robo, vehículo motorizado y reportes de los departamentos de la compañía de seguros; todas las declaraciones de los conductores, pasajeros, testigos, y otras personas que estén relacionados con el reclamo o el vehículo; y toda la documentación de cualquier conversación con las partes involucradas.

class D breathable air. Air needed for supplied air respirators, provided by rotor vane or diaphragm compressors.

aire respirable clase D. Aire necesario para respiradores de aire suministrados, proveídos por una veleta de rotor o compresores de diafragma.

cleaning in the refinish department. The soap and water and chemical cleaning performed when a vehicle enters the refinish department, which will be performed even if it has been washed in previous departments.

limpieza en el departamento de acabados. Limpieza con jabón, agua y químicos, realizada cuando un vehículo entra en el departamento de terminado, la cual será realizado aun si el vehículo ha sido lavado en departamentos anteriores.

Cleco clamps®. Spring-loaded tools installed to hold sheet metal parts together while welding. They are easily removed after the welding process.

abrazaderas tipo *Cleco*. Herramientas con resortes instaladas para mantener partes de láminas metálicas juntas mientras son soldadas. Éstas son fácilmente removidas después del proceso de soldadura.

closed coat paper. Paper with the abrasive material glued to the backing pad very tightly. Around 90 percent or more of the paper is covered with abrasive material. This type of product works best when a technician is removing hard materials, such as steel that does not get stuck in the closely arranged abrasive products.

papel de recubrimiento cerrado. Papel con el material abrasivo adherido fuertemente al respaldo. Cerca del 90% o más del papel está cubierto con material abrasivo. Este tipo de producto trabaja mejor cuando un técnico está removiendo materiales duros, tales como el acero que no se adhiere a los productos abrasivos fuertemente adheridos.

cluster porosity. Porosity often caused by improper starting and stopping which is clustered in specific areas.

porosidad de racimo. Porosidad que a menudo causada por inicios y paradas impropios, la cual se acumula en áreas específicas.

clutch head screwdriver. A screwdriver that has four sides applying pressure, which provides a more positive grip with less chance of slippage.

destornillador de cabeza de embrague. Destornillador que tiene cuatro lados que aplican presión, lo cual provee un más efectivo apretón con menos oportunidad al deslizamiento.

coarse adjusting a door. Placing the door into its opening and fastening the bolts to hold it into place as a stop-gap method, while not being concerned about the gaps or spacing with adjacent panels.

ajuste burdo de puerta. Colocación de la puerta en su lugar ajustando los pernos para mantenerla fija, como un método de limitación de holgura, mientras que no se estén alterando las holguras o espaciados de paneles adyacentes.

Code of Federal Regulation (CFR). A collection of general and permanent rules and regulations published by the federal government that serve as guidelines for employers to follow, as well as an enforcing tool to ensure compliance.

Código de Regulaciones Federales (CFR por su acrónimo en inglés). Colección de reglas generales y permanentes, y regulaciones publicadas por el gobierno federal, que sirven como guías a seguir para empleadores, así como una herramienta para hacer cumplir estas provisiones.

cold-rolled steel. Steel made from a process that uses previously hot-rolled steel, washes it with an acid rinse (cooling it), then rolls it even thinner. After it is cold rolled, the steel is annealed to make it more flexible for press-forming into body parts such as fenders and quarter panels.

acero laminado en frío. Acero hecho a partir de un proceso que usa acero previamente laminado en caliente, el cual es lavado con un enjuague ácido (enfriándolo), y luego es laminado de nuevo más delgadamente. Después es laminado en frío, el acero luego es templado para hacerlo más flexible y poder darle forma de partes tales como guardabarros y paneles de cuartos.

cold shrinking. A technique done with a shrinking hammer and dolly. The tools come with a gridded face; the grid allows the molecules to shrink as the metal is hammered. Cold shrinking can be a slow process, and many technicians often decide to use the faster heat shrinking method.

encogido en frío. Técnica realizada con un martillo de encogido y carretilla. Las herramientas tienen una cara enmallada; la malla permite que las moléculas se encojan a medida que el metal es martillado. El encogido en frío puede ser un proceso lento, y muchos técnicos a menudo deciden usar el método de encogido en caliente, el cual es más rápido.

collapsed hinge buckle. A hinge buckle that extends through a stamped-in reinforced flange, bead, or ridge on a panel. As the hinge buckle passes through a ridge or bead, the panel's overall length is shortened, which causes tension in the outer edges of a panel and places the hinge area under pressure.

broche de bisagra colapsada. Broche de bisagra que se extiende a través de un reborde, cuenta o canto reesforzado sellado en un panel. A medida que el broche de bisagra pasa a través de un reborde o cuenta, la longitud total del panel es reducida, lo cual causa tensión en los bordes exteriores de un panel y ubica el área de la bisagra bajo presión.

collision avoidance systems. Vehicular systems that assist the driver in maintaining control of a vehicle and/or avoiding objects that may cross its path.

sistemas anticolisión. Sistemas vehicular que ayudan al conductor a mantener el control de un vehículo y/o evadir objetos que puedan estar en su camino.

collision energy. The forces created by a motor vehicle impacting with another vehicle, object, person, and/or the ground; inertia energy that is transmitted throughout the vehicle subsequent to an impact, usually leaving a wake of damage in its path.

energía de colisión. Fuerzas creadas por un automóvil al impactar otro vehículo, objeto, persona, y/o el suelo; la energía de inercia que es transmitida en todas las partes del vehículo luego del impacto usualmente deja una estela de daño en su camino.

color adjustment. Making a change to a color which is not quite yet an exact match to what's needed. This is often aided by using color wheels and tints.

ajuste de color. Hacer un cambio a un color que no es exactamente idéntico al color necesitado. Esto es a menudo hecho con la ayuda de ruedas de color y tintes.

color blindness. A condition in which a person can only see varying shades of gray. This is much rarer than color deficiency, which is simply the inability to distinguish certain colors.

daltonismo. Condición en la cual una persona puede solo ver escala de grises. Ésta condición es mucho más rara que la deficiencia de color, la cual es simplemente la incapacidad de reconocer ciertos colores.

color chip book. A guide to various manufactured colors used for vehicles.

catálogo de colores. Guía de colores fabricados usados en vehículos.

color effect panel. *See* sprayout panel.

panel de efecto de color. Ver "panel de prueba de rociado."

color plotting. A method of evaluating and adjusting paint when tinting. To color plot, a technician needs the formula, a manufacturer's color wheel, and a color plotting chart.

trazo de colores. Método de evaluar y ajustar una pintura cuando se le da tono. Para trazar un color, el técnico necesita la fórmula, la rueda de colores de un fabricante, y una gráfica de trazado de colores.

color rendering index (CRI). A measurement that compares artificial light to sunlight (100 percent CRI); a means to measure color-corrected light.

índice de interpretación de color (CRI por su acrónimo en inglés). Medida que compara la luz artificial con la luz solar (CRI del 100%); esta es una forma de medir la luz modificada por los colores.

color sanding. The most aggressive sanding process, which should only be chosen when the surface of the finish is in need of leveling due to texture, dirt, or runs, all of which must be in the clearcoat. Color sanding can be done by hand with a block, or with fine detailing DA sanders.

lijado de color. Este es el proceso de lijado más agresivo, que sólo debe ser elegido cuando la superficie del acabado necesita nivelación debido a textura, suciedad o corridas, todos éstos deben estar en el recubrimiento transparente. El lijado de color puede ser hecho a mano con un bloque, o con lijas de detallado fino tipo DA.

color variance. The amount a paint manufacturer is allowed to shift within a color range. Generally a 5 percent color variance from a paint sample is allowed.

variación de color. Cantidad que un fabricante de pinturas puede cambiar dentro de un rango de color. Generalmente es permitida una varianza del 5% a partir de una muestra de color.

color wheel. An aid created to help art students better understand colors and how they affect each other. The wheel is divided into the three primary colors, then further divided into three secondary colors, then divided again into six additional colors: red-orange, red-violet, yellow-green, yellow-orange, blue-green, and blue-violet.

rueda de color. Guía creada para ayudar a los estudiantes de arte a entender mejor los colores y cómo éstos se afectan entre sí.

La rueda está dividida en los tres colores primarios, luego está más adelante dividida en tres colores secundarios, luego en seis colores adicionales: rojo-anaranjado, rojo-violeta, amarillo-verde, amarillo-anaranjado, azul-verde y azul-violeta.

color-impregnated. A part that is colored during the manufacturing process that will usually not require refinishing at the collision repair facility. Common examples include color-impregnated bumper covers, dash panels, and garnish (interior) moldings.

coloreado. Parte que es coloreada durante el proceso de fabricación que usualmente no requerirá barnizado en las instalaciones de reparación de colisiones. Algunos ejemplos comunes incluyen tapas de parachoques, paneles de carrera, y molduras de guarnición (interiores) coloreadas.

combination pliers. Also called slip-joint pliers; these are the most common design used by the collision repair technician. The jaws are round or flat to securely grasp round or flat objects. The slip joint allows the jaws to be adjusted for two different opening sizes.

tenazas de combinación. También llamadas tenazas de unión deslizante; éstas son las que tienen el diseño más común usado por los técnicos de reparación de colisiones. Las mordazas son redondas o planas para agarrar en forma segura objetos redondos o planos, respectivamente. La unión deslizante permite que las mordazas sean ajustadas a dos tamaños de abertura diferentes.

combination wrenches. A wrench with an open jaw on one end and a box end on the other that offers the advantages of both open- and box-end wrenches. As collision repair technicians generally need two sets of wrenches while doing repair work, the combination wrench provides the necessary second set of wrenches.

llaves combinadas. Llave con una mordaza abierta en un extremo y un cubo en el otro que ofrece las ventajas de dos tipos de llaves: las de extremo abierto y las de cubo. Ya que los técnicos de reparación generalmente necesitan dos juegos de llaves mientras hacen el trabajo de reparación, las llaves combinadas proveen el segundo juego de llaves necesitado.

combo repairer. A collision repair technician that handles a variety of repair tasks and is usually responsible for the vehicle's repair from start to finish.

reparador en general. Técnico en reparación de colisiones que maneja una variedad de tareas de reparación y es por lo general responsable de la reparación del vehículo desde el principio hasta el final.

comma/wedge dolly. A dolly with a thin body to help force dents from behind narrow, curved areas.

carretilla tipo coma o cuña. Carretilla con cuerpo delgado que ayuda a forzar las abolladuras de áreas estrechas y curvas por la parte de atrás.

comparative negligence. The determination of who is at fault and to what percentage in any individual collision claim event.

negligencia relativa. Determinación de quién tiene la culpa y a que porcentaje corresponde cualquier evento de reclamo por colisión individual.

comparative quick check measurements. Comparative measurements of a vehicle's openings—such as under the hood or a door—that are made with either a tape measure or tram bar to compare the dimensions of the vehicle's damaged side to the undamaged side; to determine if a section of the vehicle shifted as a result of the collision forces.

medidas de verificación rápida relativas. Medidas relativas de los espacios de un vehículo, tal como bajo el capó o en una puerta, que son hechas con una cinta métrica o con un calibrador, para comparar las dimensiones del lado dañado del vehículo con las del lado intacto y así determinar si una sección del vehículo cambió a consecuencia de las fuerzas de la colisión.

complete paint job. A repainting of the entire vehicle, which is rarely done in the collision repair industry.

trabajo de pintura completo. Pintar completamente el vehículo de nuevo; esto es raramente hecho en la industria de reparación de colisiones.

component panels. A panel that is typically attached to the vehicle with removable fasteners such as bolts, clips, or screws, allowing it to be adjusted to match the crowns and gaps to the stationary panels adjacent to it.

paneles componentes. Panel que es típicamente unido al vehículo con ajustadores removibles tales como pernos, ajustadores o tornillos, que le permite ser ajustado para emparejar las coronas y holguras de los paneles inmóviles adyacentes a él.

composite material/composites. Various plastics used in the manufacture of vehicles; may include sheet molded compounds, thermoplastics, thermoset plastics, urethanes, and a variety of other synthetic materials used on exterior and interior parts of the vehicle.

compuestos o material compuesto. Varios plásticos usados en la fabricación de vehículos; éstos puede incluir compuestos laminados, termoplásticos, plásticos tipo *thermoset*, *urethanes*, y otros materiales sintéticos usados en partes exteriores e interiores del vehículo.

composite outer panel. The exterior or outer panels on a space frame vehicle. They are frequently made of sheet molded compound (SMC), which is commonly referred to as a composite material.

panel externo compuesto. Paneles exteriores o externos en un vehículo de marco espacial. Éstos son con frecuencia hechos de compuestos moldeados de láminas (SMC por su acrónimo en inglés), que son comúnmente referidos como un material compuesto.

compound. A substance which has combined elements in a fixed proportion. These components cannot—and do not—separate through physical means.

compuesto. Sustancia en la cual se combinan elementos en una proporción fija. Dichos componentes no pueden, ni no se, separan a través de medios físicos.

compressive strength. Steel's ability to resist crushing forces.

fuerza de compresión. Capacidad del acero para resistir fuerzas aplastantes.

constant velocity joint (CV joint). A special universal joint attached to the outer end of the drift shafts used to accommodate the slight off angle between the drive shaft and the drive angle.

unión de velocidad constante (CV por su acrónimo en inglés). Unión universal especial ligada al extremo externo de los ejes de deriva usada para acomodar el ligero ángulo entre el árbol de leva y el ángulo de conducción.

contact corrosion. A reaction that occurs when two dissimilar metals such as steel and aluminum are exposed to each other. The more active of the two, in this case steel, transfers its electrons to the less active one. The less active one (aluminum) corrodes and the active one (steel) remains intact. To prevent galvanic corrosion from developing, the two dissimilar metals must be kept separated. *See also* **galvanic corrosion.**

corrosión por contacto. Reacción que ocurre cuando dos metales distintos, como acero y aluminio son expuestos el uno al otro. El más activo de los dos, en este caso el acero, transfiere sus electrones al menos activo. El menos activo (aluminio) se corroe y el activo (acero) permanece intacto. Para impedir el desarrollo de la corrosión galvánica, los dos metales deben ser mantenidos separados. Ver también "corrosión galvánica."

contact tip. A consumable item on a GMAC welder, which means that it wears out as electrode wire passes through it. As contact tips wear out, the opening becomes larger, and again the wire may wobble. Contact tips should be checked from time to time to assess the need to change them.

punta de contacto. Artículo consumible en un soldador tipo GMAC; lo que significa que éste se desgasta a medida que el alambre del electrodo pasa a través de él. Cuando las puntas de contacto se desgastan, la apertura se hace más grande, y el alambre puede tambalearse. Las puntas de contacto deben ser revisadas de vez en cuando para evaluar la necesidad de cambiarlas.

continuous weld. Welding method in which the arc is struck and a smooth uninterrupted bead is applied in a steady, ongoing movement. The push, or forward, direction of travel is used.

soldadura continua. Método de soldadura en el cual el arco es golpeado y una cuenta lisa ininterrumpida es aplicada en un movimiento estable y en curso. La dirección de empuje o avance de viaje usada.

contour mapping. Lifting or visible areas of rough edges around a feathered area. *See also* **featheredge splitting.**

mapeo de contorno. Levantamiento o áreas visibles de bordes ásperos alrededor de un área seccionada. Ver también "agrietado en secciones."

convection curing. A method of drying the finish on a vehicle using heat of the air around the vehicle to "bake" the coating, as opposed to heating the vehicle itself.

curado por convección. Método de secar el terminado de un vehículo usando el calor del aire que circula alrededor de él, para "hornear" la capa, a diferencia de calentar el vehículo mismo.

conventional/standard fillers. *See* **premium fillers, lightweight, conventional-standard fillers.**

rellenos convencionales o estándar. Véase rellenos premium, ligeros, rellenos convencionales o estándar.

convoluted front frame rail. A frame rail end usually located at the front of the vehicle that has a convoluted or rippled surface as opposed to the usual flat shape; commonly used in the collision energy management design to help dissipate the effects of the collision forces.

riel del chasis delantero en forma de acordeón. Extremo del riel de un marco, que usualmente es colocado en la parte frontal del vehículo, y que tiene una superficie doblada o tensada, a diferencia de la forma plana habitual. Éste es comúnmente usado en el diseño de administración de la energía de colisión para ayudar a disipar los efectos de las fuerzas de colisión.

corrosion. A form of degradation of steel aluminum or nearly any other metal or form of matter, wherein the material reacts with oxygen together with an electrolyte—usually water—causing the material to break down and eventually dissipate.

corrosión. Forma de degradación del aluminio de acero, o casi cualquier otro metal, o forma de materia, en donde el material reacciona con el oxígeno junto a un electrolito, usualmente agua, causando que el material se rompa y eventualmente se disipe.

corrosion hot spots. Any surface or area that offers an environment conducive for corrosion to occur. It may be an area of a seam, a welded area, an area where repairs have been made, or an area where the metal has become exposed.

puntos calientes de la corrosión. Cualquier superficie o área que ofrece un ambiente apropiado para que la corrosión ocurra. Puede ser el área de una costura, un área soldada, un área en donde las reparaciones han sido hechas, o un área en donde el metal ha estado expuesto.

cosmetic. Items that relate to the vehicle's appearance rather than its structural integrity, such as decals, pinstriping, upholstery, and so on.

artículos cosméticos. Artículos que están relacionados con el aspecto del vehículo, en lugar de relacionarse con su integridad estructural; tales como etiquetas, *pinstripping*, tapicería, etcétera.

cosmetic panels. The exterior body panels which are used to enhance the vehicle's aerodynamics, style, and beauty.

paneles cosméticos. Paneles de cuerpo exteriores que son usados para realzar la aerodinámica del vehículo, el estilo, y la belleza.

cost totaling. The step in the estimating process in which the various subdivisions of a damage estimate are calculated mathematically and added up to produce the bottom line cost to repair a vehicle.

estimado total. Paso en el proceso de estimado en el cual las varias subdivisiones de una estimación de daños son matemáticamente calculadas y sumadas para producir el costo total de la reparación un vehículo.

cotter pins. A small piece of metal placed through a hole in a nut or bolt to keep it from loosening.

alfileres de chaveta. Pequeña pieza de metal colocada a través de un agujero en una tuerca, o un perno, que le impide aflojarse.

cowl assembly. An assemblage of panels located at the front of the passenger compartment which includes the front door hinge pillars, fire wall, and upper cowl panel.

ensamblaje de la capucha. Ensamblaje de paneles localizados en el frente de un compartimiento de pasajeros el cual incluye los pilares de gozne de la puerta principal, la pared contrafuegos y un panel de la capucha superior.

cracking. Splitting of the paint film that commonly appears as straight lines and extends through the entire film thickness. It can occur in the topcoat or undercoats, and is a result of thermal expansion and contraction. It can also be caused by application of too heavy a film thickness.

agrietado. Rompimiento de la película de pintura que comúnmente se manifiesta con líneas rectas y se extiende por completo a través del grosor de la película. Esto puede ocurrir en el recubrimiento exterior o en las primeras manos de pintura, y es resultado de la expansión y contracción térmica. También puede ser causado por la aplicación de una capa de película demasiado gruesa.

crashworthiness. The vehicle's ability to remain structurally sound in the event of a collision. It is a goal of the repair shop to get the vehicle back to its original crashworthiness after a repair.

resistencia a colisiones. Capacidad de un vehículo de permanecer estructuralmente estable en caso de una colisión. Es un objetivo de la tienda de reparaciones, el regresar al vehículo a su capacidad de resistencia a colisiones original después de una reparación.

cross-contamination. A process whereby materials contact another material, leading to reactions, corrosion, or other breakdown. For example, steel particles are spread to aluminum by way of sanding dust on tools. That is why manufacturers of aluminum-intensive vehicles recommend that all tools that are used to repair steel should be keep separated from those tools used to repair aluminum.

contaminación cruzada. Proceso en el cual unos materiales se ponen en contacto con otros, produciendo reacciones, corrosión, u otro daño. Por ejemplo, las partículas de acero son transportadas al aluminio a través del polvo resultante del limado sobre herramientas. Esta es la razón por la cual los fabricantes de vehículos, principalmente hechos con aluminio, recomiendan que todas las herramientas usadas para reparar el acero sean separadas de las herramientas usadas para reparar el aluminio.

cross-draft spray booths. A spray booth that draws the incoming air from the shop, generally through the booth doors. The air travels over the length of the vehicle, is filtered again at the opposite end of the booth, and then is vented to the outside of the building.

cabinas de corriente cruzada. Cabina de rociamiento que dirige el aire entrante de la tienda generalmente a través de las puertas de la cabina. El aire que viaja a través del vehículo es filtrado de nuevo en el otro extremo de la cámara, y luego es expulsado al exterior del edificio.

cross-linking. The process through which paints cure by forming chemical bonds between two separate molecular chains.

enlace de cruz. Proceso por el cual las pinturas se curan formando enlaces químicos entre dos cadenas moleculares separadas.

crowns. The gentle low curves of a hood or roof that strengthen steel because of the forming of the steel.

coronas. Curvas suaves de un capó o techo que refuerzan el acero a causa del trabajado que ha recibido.

crush initiators. The area at the outer end of the rail where the shortening effect usually occurs first.

iniciadores de acortamiento. Área del extremo externo de un riel en donde el efecto de acortamiento generalmente ocurre primero.

crush zones. An area of reinforcement that is designed and engineered to bend and buckle a certain way at a specific location to help absorb and dissipate the effects of the collision energy.

zonas de acortamiento. Área de refuerzo que es diseñada y trabajada con determinados dobleces y curvas para ayudar a absorber y disipar los efectos de la energía de colisión.

cubic feet per hour (CFH). A measure used to show the rate of gas flow.

pies cúbicos por hora (CFH por su acrónimo en inglés). Medida usada para medir el flujo de gas.

cubic feet per minute (CFM). The measure of air's volume.

pies cúbicos por minuto (CFM por su acrónimo en inglés). Medida de volumen del aire.

curing. Paint process in which a component is placed into a temperature-controlled environment for a specified period of time to ensure the paint tightly bonds to the painted metal.

curado. Proceso del pintado en el cual un componente es colocado en un medioambiente de temperatura controlada, durante un periodo de tiempo, para asegurar que la pintura se fije fuertemente al metal pintado.

current. The flow of electricity through a conductor (wire), which is measured in amperes (amps).

corriente. Flujo de electricidad a través de un conductor (alambre), que es medido en amperios (amps).

cutting in/cutting-in parts. A refinish technique used for painting replacement parts in areas that would not be accessible when attached to the vehicle.

cortado en partes. Técnica de terminado usada para pintar repuestos de partes en áreas que no serían accesibles cuando éstos están puestos en el vehículo.

cycle time. The amount of time required to repair a damaged vehicle, starting with pre-production and ending with its delivery to the customer in pre-accident condition.

tiempo de ciclo. Cantidad de tiempo requerida para reparar un vehículo dañado, el cual comienza con la preproducción y que termina con su entrega al cliente en la condición que tenía antes del accidente.

D

D pillar. The rear roof support pillar on SUVs and vans which is also used to attach the rear doors on these vehicles.

pilar D. Pilar de apoyo de la parte trasera del techo en vehículos deportivos (SUV por su acrónimo en inglés) y camioneta, el cual es también usado para las puertas traseras en esos vehículos.

D ring anchor location. The reinforced attaching point for the shoulder seat belt located on the B pillar on any car and pickup truck.

ubicación del ancla de anillo D. Punto de unión reforzado para el cinturón de seguridad de hombro localizado en el pilar B en cualquier coche y camioneta.

damage analysis. The process by which an estimator, damage appraiser, and/or collision repair technician inspects a vehicle to create a damage report and repair plan.

análisis de daño. Proceso por el cual un perito, estima el daño, y/o un técnico de reparación de colisiones inspecciona un vehículo para crear un reporte de daño y un plan de reparación.

datum plane. An invisible line that establishes an imaginary horizontal plane parallel to the bottom of the vehicle at a given distance below it, which is used to measure height dimensions.

plano de referencia. Línea invisible que establece un plano horizontal imaginario paralelo al fondo del vehículo, a una distancia dada debajo de éste, el cual es usado para medir dimensiones de altura.

dead blow hammer. A specifically designed hammer that will resist bouncing back after striking an object. It is used in driving operations where marring of parts must be avoided.

martillo de golpe muerto. Martillo específicamente diseñado para rebotar luego de golpear un objeto. Éste es usado en operaciones de manejo de partes cuando se deben evitar deformaciones.

decals. Sometimes called vinyl trims, these come in a vast array of decorative applications. This trim might be as simple as a decorative decal box side application, or as elaborate as aftermarket vinyl applications that are popular with import tuner customizing.

etiquetas. Algunas veces llamadas molduras de vinilo, éstas tienen una gran variedad de usos decorativos. Estas molduras pueden ser tan simples como para ser usados en aplicaciones exteriores decorativas, o muy elaboradas tales como las etiquetas de vinilo personalizadas disponibles en el mercado alternativo.

decibels (db). A measure of loudness. Technicians should be aware that long-term hearing loss will occur when loud noises are present, but also that exposure to noises of as little as 100 decibels for long periods of time can be detrimental.

decibeles (db). Medida de la intensidad del sonido. Los técnicos deben estar concientes de que la pérdida de audición a largo plazo ocurre por la exposición a ruidos fuertes; y también, de que la exposición a ruidos de baja intensidad, de tan solo 100 decibeles, durante períodos largos de tiempo puede ser perjudicial.

deck lid and hatch. The trunk lid or the hatch lid found on the rear of the vehicle which encloses and protect the vehicle's rear end from the elements, much like the doors do in the passenger compartment. The deck lid is usually a stamped piece of steel, aluminum, or a composite material, whereas the hatch usually includes the rear window mounted in the framework of the hatch lid.

tapa de cubierta y escotilla. Tapa de cajuela o tapa de escotilla encontrada en la parte trasera del vehículo, la cual cierra y protege el vehículo de elementos, tal como lo hacen las puertas en el compartimiento de pasajeros. La tapa de cubierta es usualmente una pieza sellada de acero, aluminio o un material compuesto; mientras que la escotilla por lo general incluye la ventana trasera montada en le marco de la tapa de la escotilla.

deck lid tension. The amount of tension exerted against the deck lid by the automatic closer or by the latch when it engages the striker; the degree of tension will help cause the lid to "pop" open when the lock is released.

tensión de la tapa de la cubierta. Cantidad de tensión ejercida contra la tapa de la cubierta por el sistema de cerrado automático o por el pestillo cuando éste se engancha al ariete; el grado de tensión ayudará a hacer que la tapa se abra "saltando" cuando la cerradura es liberada.

dedicated fixture (measuring) system. A measuring system that uses special fixtures which, when bolted to the bench at specified reference points, will match a given surface on the undercarriage.

sistema de partes integrantes. Sistema de medición que usa partes integrantes especiales las cuales, cuando son remachadas en el banco en puntos de referencia específicos, coincidirán con una superficie dada en el tren de aterrizaje.

deductible. The dollar amount designated in an insurance policy that the policy owner must pay before an insurer pays the rest.

deducible. Cantidad de dinero designada en una póliza de seguros que el dueño de la póliza debe pagar antes de que la aseguradora pague el resto.

deep cleaning. A thorough soap and water cleaning, followed by a chemical cleaning.

limpieza profunda. Limpieza completa con jabón y agua, seguida de una limpieza química.

deep well socket. A socket which is longer in length that standard sockets; used for reaching over a stud bolt.

enchufe de entrada profunda. Enchufe que es más largo que los enchufes comunes; éste es usado para alcanzar un espárrago (tipo de tornillo sin cabeza).

delamination. The peeling of plastic filler from the surface to which it is applied; may be one or more layers thick depending on the amount of additional material applied over an undercatalyzed batch.

deshojado. Desconchado del relleno plástico de la superficie en la cual éste es aplicado; pueden ser una o varias capas gruesas según la cantidad de material adicional aplicado sobre una pila subcatalizada.

dent puller. A tool commonly used to pull out dents in enclosed body panels which cannot be accessed from the back side. Holes are punched—or drilled—in the dent's low areas. Either a hook or threaded tip is inserted into the holes. Repeated blows of the slide hammer will pull the dent out. After the damage is removed; the holes must be closed by welding.

sacador de abolladuras. Herramienta comúnmente usada para sacar abolladuras in paneles cerrados los cuales no pueden ser alcanzados desde la parte de atrás. Agujeros son perforados, o taladrados, en las áreas bajas de la abolladura. Ya sea un gancho o la punta de un hilo es insertado en los agujeros. Los golpes repetidos del martillo de moldeado sacan la abolladura. Después de que el daño es eliminado; los agujeros deben ser cerrados con soldadura.

Department of Transportation (DOT). The agency responsible for regulating the transportation of a variety of consumer and industrial goods and merchandise.

Departamento de Transporte (DOT por su acrónimo en inglés). Agencia responsable de regular el transporte de una variedad de bienes y mercancías, de consumidores y de la industria.

depth of reflection. The clarity of the reflection of objects seen in the finish. If technicians can see their face reflected in the finish clearly while standing upright, and there are no swirl marks, the detailer is ready to proceed to the next step.

profundidad de reflexión. Claridad de la reflexión de objetos vistos en el acabado. Si los técnicos pueden ver su cara reflejada en el acabado claramente estando de pie, y no hay señales de remolinos, el detallado está listo para seguir al siguiente paso.

dermis. The second layer of skin.

dermis. Segunda capa de piel.

desiccator. A device used for preserving moisture-sensitive items. It is one of the last steps used in controlling moisture in a compressed air system.

desecador. Aparato usado para conservar artículos sensibles a la humedad. Este es uno de los últimos pasos usados en el control de la humedad en un sistema de aire comprimido.

detail adjusting/fine tuning a door. The process of adjusting the door to obtain a uniform gap between it and all adjacent panels, as well as a flush level with all surrounding surfaces.

ajuste de detalle o afinación fina de puerta. Proceso de ajuste de una puerta para obtener una holgura uniforme entre ella y todos los paneles adyacentes, así como una nivelación con todas las superficies circundantes.

detail brush. A cleaning tool which can be made by cutting the bristles of an inexpensive painter's brush down to about 1/2 inch and wrapping the metal part with tape to prevent scratching.

cepillo de detallado. Herramienta de limpieza que puede ser hecha cortando las cerdas de una brocha barata a aproximadamente 1/2 pulgada y envolviendo la parte metálica con cinta para prevenir rasguños.

detail clay. A product designed to clean off those products that stick to the vehicle that are not removed through washing.

arcilla de detallado. Material diseñado para remover productos que se adhieren al vehículo que no son quitados por el lavado.

detail cloth. Microfiber cloths which have become popular for drying, compound cleaning, and wax removal due to their ability to pick up things and then be cleaned again.

tela de detallado. Telas de microfibra que se han hecho populares para secar, limpiar compuestos, y remover cera, debido a su capacidad de adherir partículas y luego ser limpiadas otra vez.

detailing. The deep cleaning of a vehicle, done at many stages in an automobile's life. For example, it is often done at the factory to remove minor defects that were caused during the manufacturing process.

detallado. Limpieza profunda de un vehículo, hecha en muchas etapas de la vida de un coche. Por ejemplo, a menudo es hecho en la fábrica para quitar defectos menores que fueron causados durante el proceso industrial.

de-trim. Removal of the vehicle's lights and trim.

desvestido. Remoción de las luces y la tapicería del vehículo.

detrimental. Hazardous.

perjudicial. Peligroso.

diagonal cutting pliers. Sometimes called a side-cutter pliers, this tool has cutting jaws designed to cut off wires.

tenazas de corte diagonal. A veces llamadas tenazas de corte lateral, éstas herramientas tiene mandíbulas cortantes diseñadas para cortar alambres.

diamond. A condition which occurs in a frontal impact in body over frame vehicles in which the impact object strikes the frame rail directly on one side of the vehicle only. Recalling the effects of the inertial forces, the rail on the opposite side of the vehicle continues to move forward some distance before it comes to a stop. As a result, the center section of the vehicle is usually out of square, often assuming an exaggerated diamond shape.

diamante. Condición que ocurre en un impacto frontal, en el cuerpo sobre el marco de un vehículo, en el cual el objeto del impacto golpea el riel del marco directamente a un solo lado del vehículo. Recordando los efectos de las fuerzas de inercia, el carril en el lado opuesto del vehículo sigue adelantándose alguna distancia antes de que esto se detenga. Como resultado, la sección central del vehículo está por lo general fuera del cuadrado, a menudo asumiendo una forma exagerada de diamante.

diaphragm compressor. Although not commonly used to produce the volumes of high (above 100 psi) pressures needed for the collision repair facility's compressed air needs, these compressors are often used in producing class D breathable air for supplied air respirators.

compresor de diafragma. Aunque no es comúnmente usado para producir los volúmenes altos de presión (arriba de 100 psi) requeridos en la reparación de colisiones, estos compresores a menudo son usados en la producción de aire clase D para respiradores de aire suministrado.

die penetrant. A special spray-on material used to locate cracks in an aluminum part that may not be visible without magnification. A developer is used in conjunction with the die penetrant, which helps to make the die in the cracks more visible.

penetrante muerto. Material que se aplica rociándolo usado para localizar grietas en una parte de aluminio que pueden no ser visibles sin aumento. Un revelador es usado con el penetrante muerto, lo cual ayuda a hacer que las grietas sean más visibles.

die stamping. The process used to form all outer metal panels for cars and trucks by placing a piece of thin, flat sheet metal between the two dies of a press, which are stamped together to form the part.

sellado muerto. Proceso usado para formar todos los paneles metálicos externos para coches y camiones; se realiza al colocar una pieza de lámina metálica delgada entre los dos dados de una prensa, los cuales son sellados juntos para formar la parte.

dieback. A condition appearing when the finish has been sprayed and has proper gloss, but then when the proper dry time has expired the gloss is less satisfactory than when it was wet; dulling of the surface overall or on repair spots.

opacado. Condición que aparece cuando el acabado ha sido aplicado y tiene un brillo apropiado, pero entonces cuando el tiempo de secado apropiado ha pasado el brillo es menos satisfactorio que cuando estaba húmedo; opacando la superficie en general o en puntos de reparación.

digital tram gauge. A more modern version of the tram bar with a digital readout which shows the length of a dimension without the use of a tape measure.

calibrador digital. Versión más moderna de la barra de medición con una lectura digital que muestra la longitud de una dimensión sin el uso de una cinta métrica.

diminished value. A reduction in a vehicle's worth, either because of age or damage.

valor disminuido. Reducción del valor de un vehículo, debido a su edad o al daño que presente.

dinging spoon. A tool used to smooth and level ridges by striking a hammer on top of the spoon.

cuchara de suavizado. Herramienta usada para alisar y nivelar crestas al golpear con martillo su parte superior.

direct current electrode negative (DCEN). Current that directs most of the heat to the workplace, and is used if flux-core wire is used.

electrodo negativo de corriente directa (DCEN por su acrónimo en inglés). Corriente que dirige la mayor parte del calor al lugar de trabajo, y que es usada si el alambre principal está siendo también usado.

direct current electrode positive (DCEP). Sometimes called reverse polarity, this is the current most often used by collision repair welders because it provides the best fusion, with less weld on the surface, and a more stable arc.

electrodo positivo de corriente directa (DCEP por su acrónimo en inglés). Algunas veces llamada polaridad inversa, esta es la corriente más a menudo usada por soldadores de reparación de colisiones, porque ella proporciona la mejor fusión, con menos soldadura en la superficie, y un arco más estable.

direct damage. The damage created at the point where the vehicle impacts another vehicle or object; also called primary damage.

daño directo. Daño creado en el punto en donde el vehículo impacta a otro vehículo u objeto; éste es también llamado daño primario.

direct repair program (DRP). A business arrangement in which an insurance company recommends a particular collision repair business for repairs on their claims.

programa de reparación directa (DRP por su acrónimo en inglés). Arreglo comercial en el cual una compañía de seguros recomienda una agencia de reparación de colisiones específica para realizar las reparaciones de sus reclamos.

direction of travel. The decision made by the technician of how the weld is put down, sometimes dictated by the position of the weld. If a weld is being made in the vertical position, the weld technician often chooses to weld from the top to the bottom, using the pull angle of travel to take advantage of gravity. Other times a weld technician may choose to use a push direction of travel on thin metal, because the heat of the arch will not preheat the thin metal as much as the pull method, and thus will not be as likely to create a burnthrough problem.

dirección de viaje. Decisión tomada por el técnico de cómo la soldadura es aplicada, algunas veces es determinada por la posición de la soldadura. Si una soldadura está siendo colocada en posición vertical, el técnico de soldadura a menudo decide soldar de arriba hacia abajo, usando el ángulo de jalón de aplicación para aprovechar la acción de la gravedad. Otras veces, los técnicos de soldadura pueden decidir usar una dirección de empuje de aplicación en metales delgados, porque el calor del arco no precalentará el metal delgado tanto como con el método de jalón, y así probablemente no se creará un problema de sobrecalentamiento.

direct-to-metal. Applying a material on metal without first coating with an epoxy or acid etch primer.

directo al metal. Aplicación de un material sobre un metal sin la aplicación previa de una imprimación de fijación epóxica o ácida.

dirt contamination. The most common defect problem in refinishing. Although many shops have a dirt problem, dirt is avoidable, and all precautions to prevent its harmful effects on the refinish procedure should be taken.

contaminación por suciedad. Éste es el problema de defecto más común en barnizado. Aunque muchas tiendas tengan un problema de suciedad, la suciedad es evitable, y todas las precauciones para prevenir sus efectos dañinos en el procedimiento de barnizado deben ser tomadas.

discontinuity. The interruption of the typical structure of a weld. All welds have some discontinuity, for there is no perfect weld.

discontinuidad. Interrupción de la estructura típica de una soldadura. Todas las soldaduras tienen algunas discontinuidades, ya que no hay soldadura perfecta.

disposable wheel covers. Plastic single-use covers that are only used once—then discarded. Because they are disposable, they do not contaminate the surrounding painted area with flying debris. Like other wheel covers, though, they only protect the wheel—not the wheel well and its parts—from overspray.

tapas de rueda desechables. Cubiertas plásticas de un solo uso que luego de usarse son desechadas. Como son desechables, éstas no contaminan el área pintada de los alrededores con escombros volantes. Tal como con otras tapas de rueda, sin embargo, éstas sólo protegen la rueda, no el pozo de la rueda y sus partes, de excesos de rociado.

dog tracking. An incorrect tracking condition caused by one of the rear wheels being toed-out and the opposite side being toed-in.

descuadrado. Condición de pisado incorrecto en la cual las ruedas traseras pisan hacia fuera, mientras que las delanteras hacia dentro.

dollies. Often called dolly blocks; these tools are available in a variety of shapes and sizes. Each dolly has curves and angles that can be used for specific dents and panel contours (high crowns, low crowns, flanges). They can be used as a striking tool or backing tool for body hammers. A good quality dolly block will be well-balanced so the effect of its weight is not lost when the hammer blow is struck.

carretillas. A menudo llamadas bloques de carretilla; estos instrumentos están disponibles en una variedad de formas y tamaños. Cada carretilla tiene curvas y ángulos que pueden ser usados para abolladuras específicas y contornos de paneles (coronas altas, bajo coronas, rebordes). Éstas pueden estar usadas como un instrumento de golpeo o instrumento de apoyo para martillos de cuerpo. Un bloque de carro de buena calidad estará bien balanceado de forma que el efecto de su peso no se pierde cuando el martillazo es dado.

dolly spoon. Often used to get to places that a technician might not otherwise be able to reach, a dolly spoon is heavy and positioned on a long steel handle, so it can be held firmly against the back side of a panel as a hammer is used on the front side. It can also be used as a driver to lift low areas that a technician may not be able to reach with a hammer.

cuchara de carretilla. A menudo son usada para llegar a sitios en donde un técnico de otra forma no podría llegar; una cuchara de carretilla es pesada y posicionada en un mango de acero largo, entonces puede ser sostenida firmemente contra el lado trasero de un panel cuando un martillo está siendo usado en el lado delantero. También puede estar usado como un impulsor para levantar áreas bajas que un técnico podría no alcanzar con un martillo.

domestic vehicle. A vehicle manufactured in the United States.

vehículo doméstico. Vehículo fabricado en los Estados Unidos.

door shell. An intact door as manufactured from the factory, including the inside frame, the intrusion beam, and the outer shell.

cascarón de puerta. Puerta intacta como es producida por la fábrica, incluyendo el marco interior, la viga de intrusión, y el cascarón externo.

door skin. The outer portion or covering of the door shell.

piel de puerta. La parte externa o cobertura del cascarón de la puerta.

double braided hose. An inner rubber hose covered with two layers of braided mesh, often used for high pressure work in the repair facility.

manguera de trenzado doble. Manguera de interior de hule cubierta de dos capas de malla trenzada; ésta es a menudo usada para trabajos de alta presión en los sitios de reparación.

double wishbone suspension. An SLA steering system that incorporates a combination of the features of a parallelogram and MacPherson strut suspension system; utilizes both a separate upper and lower control arm along with a strut design very similar to that of the MacPherson strut.

suspensión de espoleta doble. Sistema de manejo tipo SLA que incorpora una combinación de los rasgos de un paralelogramo y un sistema de suspensión de puntales tipo *MacPherson*. Este sistema utiliza un brazo de controles separados superior e inferior, junto con un diseño de puntales muy similar al del puntal tipo *MacPherson*.

downdraft spray booths. These booths draw air in from outside, where the air is filtered, heated to the proper spray temperature, and filtered again before entering the booth from on top of the vehicle. The air is drawn around and under the vehicle, where it is filtered, then exits through floor vents.

cámara de rociado descendente. Éstas cámaras dirigen el aire desde adentro hacia fuera, en donde el aire es filtrado, calentado a una temperatura adecuada y filtrado de nuevo antes de entrar en la cámara por la parte de arriba del vehículo. El aire es dirigido alrededor y bajo el vehículo, en donde es filtrado, y luego sale a través de las aberturas del piso.

dozer pulling system. A straightening system that uses articulated arms energized by a hydraulic jack pushing against them to pull the damage out. A chain is attached to the arm, which in turn is attached to the damaged area of the vehicle. In this way correction forces are applied.

sistema de empuje corregido. Sistema de enderezamiento que usa brazos articulados accionados por un gato hidráulico que los empuja para sacar el daño. Una cadena es atada al brazo, que a su vez es atado al área dañada del vehículo. De esta manera las fuerzas de corrección son aplicadas.

Dr. Shinobu Ishihara. Developer of one of the best-known tests for color deficiencies—a series of colored dots with visible numbers contained within them that test for different types of color deficiencies.

Doctor Ishihara. El es el desarrollador de una de las más famosas pruebas para deficiencias de color; consiste en una serie de puntos coloreados, con números visibles entre ellos, que prueban las diferentes deficiencias de color.

dress welds. The process of cleaning up the welded metal to ensure structural stability and cosmetic appearance.

vestido de soldaduras. Es el proceso de limpiar el metal soldado para asegurar su estabilidad estructural y aspecto cosmético.

drift (starting) punch. A punch with a fully tapered shank to drive pins and bolts partially out of holes.

golpe de perforado (inicial). Golpe con un perno en forma de cuña para sacar parcialmente clavijas y pernos fuera de sus orificios.

drills. Tools which come in a variety of sizes designed to bore holes into a material to assist with connecting items, removing surface material, and cleaning.

taladros. Herramientas que vienen en una variedad de tamaños diseñados para perforar agujeros en un material para facilitar su unión con otros objetos, removiendo el material superficial y limpiándolo.

drive. The drive sizes needed in collision repair are 1/4 inch, 3/8 inch, and 1/2 inch. The small drive sizes are designed to withstand a lesser amount of torque and is used to remove small fasteners (example: interior trim and emblems), while the larger drives will withstand more torque, required to remove larger fasteners (example: bolt on body panels; bumper attachment bolts).

dado. Los tamaños necesarios de dados en la reparación de colisiones son 1/4, 3/8 y 1/2 pulgadas. Los tamaños más pequeños de dados están diseñados para soportar una menor cantidad de torque y son usados para quitar ajustadores pequeños (por ejemplo: molduras y emblemas interiores), mientras que los dados más grandes soportarán más torque, el cual es requerido para quitar ajustadores más grandes (por ejemplo: pernos en paneles, o pernos de accesorios de parachoques).

drivetrain/powertrain. The entire system used to energize the vehicle; includes the engine, transmission, drive shafts, and—if the vehicle utilizes it—the rear differential.

tren de transmisión o tren de potencia. Es el sistema completo que mueve el vehículo; incluye el motor, la transmisión, los árboles de leva, y, si el vehículo lo utiliza, el diferencial trasero.

driving spoons. The heaviest of the spoons, these tools are, as the name implies, used in the roughing stage of metal finishing. They are used to bring up low areas on the back side of fenders and in quarter panels.

cucharas de control. Éstas son las más pesadas de las cucharas, y son usadas en la etapa de manejo pesado de metales. Estas cucharas son usadas para levantar áreas bajas sobre la parte de atrás de guardabarros y en cuartos de paneles.

dry sanding. A sanding technique in which the technician chooses not to lubricate the surface to be sanded and rubs the sandpaper directly against it, brushing away dust as the process continues.

lijado en seco. Técnica de lijado en la cual el técnico decide no lubricar la superficie a ser lijada, y frota el papel de lija directamente contra la superficie, quitando el polvo a medida que el proceso avanza.

dry spray. When paint spraying, a condition where a line of paint dries before the adjoining line of paint is applied, creating a defect in the finish. The finish will have lost all gloss, appear rough, and have a granular texture. Although dry spray commonly affects small areas, poorly applied coats may affect complete panels.

rociado seco. Es una condición que se presenta al pintar en donde una línea de pintura se seca antes de que la línea siguiente de pintura sea aplicada, creando un defecto en el acabado. El acabado habrá perdido todo el brillo y parecerá áspero, y tendrá una textura granular. Aunque el rociado seco afecta comúnmente áreas pequeñas, recubrimientos pobremente aplicados pueden afectar paneles completos.

drying. Evaporation that takes place in the curing process.

secado. Evaporación que ocurre en el proceso de curación.

dual action (DA) sander. This tool is used to perform a variety of sanding operations, ranging from fine detail sanding to rough feathering sand scratches, chips, and other flaws from the painted surface after it has once been repaired in preparation to topcoating.

lijadora de operación dual (tipo DA, por su acrónimo en inglés). Esta herramienta es usada para realizar una variedad de operaciones de lijado, que van desde lijado fino de detalles hasta lijado rudo de rayones, desportillados y otros defectos de la superficie pintada después de que esta ha sido reparada como preparación para la aplicación del recubrimiento exterior.

dump trap. A device that automatically opens to allow the condensed water and oil to be removed from a compressed system. The dump trap is generally placed at the bottom of the compressed air storage tank, which is sometimes called a recover tank.

trampa de vaciado. Dispositivo que automáticamente se abre para permitir que el agua y el aceite condensados sean removidos de un sistema comprimido. La trampa de vaciado es generalmente colocada en el fondo del tanque de almacenaje de aire comprimido, que es a veces llamado un tanque de recuperación.

duty cycle. The amount of time a piece of equipment runs before needing to be shut down.

ciclo de uso. Cantidad de tiempo que un equipo funciona antes de tener que ser desconectado.

dwell on the surface. The length of time an etching material—such as a metal etching acid—is left wet on the surface to ensure adequate time for the etching process to occur.

permanencia en superficie. Tiempo en que un material de grabado, tal como un ácido de grabado metálico, es dejado húmedo sobre la superficie para asegurar que el proceso de grabado ocurra.

E

ear protection. Plugs or other PPE worn to reduce the decibel amounts to which the technician is exposed.

protección de oído. Tapones u otros equipos de protección personal (PPE por su acrónimo en inglés) que son usados para reducir la cantidad de decibeles a los que los técnicos están expuestos.

e-coating. The process of dipping or submersing automobiles in a primer-filled vat using an electrical charge to attract the primer to the exposed metal on the vehicle; applying an electrodeposition coating.

capa electrónica. Proceso de sumergido o bañado de vehículos en una tina llena de imprimación usando una carga eléctrica para atrae la imprimación al metal expuesto del vehículo; esto es aplicar una capa de electro-deposición.

edge-to-edge measurement. When measuring using reference holes, a technique where measurements are made from the outside edge of one hole to the outside edge of the other, or the inside to the inside.

medidas de borde a borde. Cuando se hacen medidas de agujeros de referencia, es una técnica en donde las medidas son hechas desde el borde exterior de un agujero hasta el borde exterior del otro, o del interior al interior.

edging/edge painting. Painting or color coating the edges of a panel that is inaccessible and cannot be painted once the part is installed onto the vehicle.

bordeado o pintado de bordes. Pintar o colorear los bordes de un panel que es inaccesible y que no puede ser pintado una vez la parte es instalada en el vehículo.

edging-in parts. *See* cutting in/cutting-in parts.

partes bordeadas. Ver "cortado en partes."

elasticity. Steel's ability to bend and stretch but then return to its original shape.

elasticidad. Capacidad del acero de doblarse y estirarse para luego volver a su forma original.

elastomers. Pliable plastic materials that are good at withstanding deformation. They are added to a coating with an activator and catalyst, which cross-link with the coating. This makes the coating more flexible.

elastómeros. materiales plásticos flexibles que son resistentes a la deformación. Éstos son añadidos a una capa con un activador y un catalizador, lo cual hace un enlace cruzado con el recubrimiento. Esto hace que la capa sea más flexible.

electrical plunger. An electronic sensing device equipped with a number of contacts and mated receptacles that perform various functions such as opening or locking the door when activated or energized.

émbolos eléctricos. Dispositivo electrónico de detección equipado con varios contactos y receptáculos emparejados que realizan varias funciones, como la apertura o el cierre de la puerta cuando éste es activados o energizado.

electrode. In MIG welding, it is the small wire that comes out of the contact tip. This wire will vary in size and material makeup, depending on what is being welded—and, to a lesser degree, what type of joint is being welded and the position the joint is in when being welded.

electrodo. En soldadura tipo MIG, es el alambre pequeño que sale de la punta de contacto. Este alambre varia de tamaño y material, según lo que se esté soldando, y, a un grado menor, el tipo de unión que está siendo soldada, y la posición de la unión de la soldadura.

electrode wire. Wire used in MIG welding, which can run from 0.7 to 2.4 mm (0.023 to 0.035 inches) but can be as large as 4 mm (0.16 inch). The three most common electrode sizes in collision steel repair are 0.023 inch, 0.030 inch, and 0.035 inch.

alambre de electrodo. Alambre usado en soldadura tipo MIG, que puede medir entre 0.7 y 2.4 mm (de 0.023 a 0.035 pulgadas), pero que puede ser tan grande como 4 mm (0.16 pulgadas). Los tres tamaños de electrodo más comunes en la reparación de acero de colisión son 0.023, 0.030 y 0.035 pulgadas.

electro-galvanizing. A form of galvanizing in which the application of the zinc coating is accomplished by spraying it onto one side of the steel surface only.

electro-galvanizado. Forma de galvanización en la cual la aplicación de la capa de zinc es llevada a cabo rociándola sólo en un lado de la superficie de acero.

electrolyte. A very strong acid-based liquid found in batteries which is usually laden with lead from exposure to the lead plates inside the battery.

electrolito. Líquido de base ácida muy fuerte encontrado en baterías que por lo general está cargado de plomo, por la exposición a las placas de plomo dentro de la batería.

electromechanical measuring system. A measuring system that uses a programmable robot-like movable arm to take and record measurements that are input into a computer data and then displays the undercarriage dimensions on a computer monitor.

sistema de medición electromecánica. Sistema de medición que usa un brazo movible parecido a un robot programable para tomar y registrar medidas, las cuales son introducidas como datos de computador, y que luego despliega las dimensiones del tren de aterrizaje en un monitor de computador.

electronic control module (ECM). The central control unit or vehicle computer that literally controls every electrical function directly, as well as many of the mechanical functions.

módulo de control electrónico (ECM por su acrónimo en inglés). Unidad de control central u computador de vehículo que literalmente controla cada función eléctrica directamente, así como muchas de las funciones mecánicas.

electronic mail (e-mail). Correspondence and data transmitted via the World Wide Web (Internet).

correo electrónico (e-mail). Correspondencia y datos transmitidos a través del World Wide Web (Internet).

electrostatic. A painting technique that uses the physics law of "opposites attract." To cause this attraction, the atomized paint is passed through an electrostatic spray gun where the paint is given a negative charge. The object to be painted is given a negative charge, causing the finish to be drawn to the charged object. The finish will coat all exposed areas of the charged object.

electrostática. Técnica de aplicación de pinturas que usa la ley de la física "de la atracción de partes opuestas". Para causar esta atracción, la pintura atomizada es pasada por una pistola rociadora electrostática donde se le da carga negativa a la pintura. También se le da carga negativa al objeto a ser pintado, haciendo que el acabado se dirija hacia él. El acabado cubrirá todas las áreas expuestas del objeto cargado.

emblems. Sometimes called nameplates; these generally include the vehicle make, its model, and possibly other identifying titles of vehicle features. Emblems may be of different types, and there are many different ways to attach them.

emblemas. A veces llamados letreros de nombres; éstos generalmente incluyen la marca del vehículo, su modelo, y posiblemente otros títulos de identificación de las características del vehículo. Los emblemas pueden ser de tipos diferentes, y hay muchas formas de unirlos.

enamel. A finish designed to be placed over a previously refinished surface, which will flow smoothly and provide a glossy appearance once dried.

esmaltes. Acabado diseñado para ser aplicado sobre una superficie previamente refinada, el cual fluye suavemente y proporciona una apariencia brillante una vez se seca.

Environmental Protection Agency (EPA). The independent government agency charged with safeguarding the environment, including land, air, and water, from pollution.

Agencia de Protección del Medio Ambiente (EPA por su acrónimo en inglés). Agencia independiente del gobierno encargada de salvaguardar el medioambiente (incluyendo tierra, aire, y agua) de la contaminación.

epidermis. The top layer of skin.

epidermis. Capa superior de piel.

epoxy primer. A highly corrosion-resistant two component primer, hence the name 2K product, which uses a catalyst to enhance the curing of the material.

imprimación epóxica. Imprimación de dos componentes altamente resistentes a la corrosión, de ahí el nombre del producto 2K, la cual usa un catalizador para mejorar el curado del material.

estimate/damage report. A document created by a damage appraiser or estimator during the vehicle's damage inspection designed to determine the cost of restoring a damaged vehicle to pre-accident condition.

reporte o estimación de daño. Documento creado por un estimador de daños o perito, durante la inspección de daño del vehículo, diseñado para determinar el costo de la restauración de un vehículo dañado y regresarlo a la condición que tenía antes del accidente.

exhaust filters. Air filters that screen the air leaving the spray booth.

filtros de escape. Filtros de aire que protegen el aire que sale de la cámara de rociamiento.

extensions. An accessory that can reach around obstacles or gain access to otherwise inaccessible fasteners.

extensiones. Accesorio que puede pasar alrededor de obstáculos o ganar acceso a sujetadores que de otra forma son inaccesibles.

extrusions or extruded parts. A part made by heating the aluminum to a temperature high enough to soften the metal so it can be forced through a form or die to obtain its shape. Extruded parts frequently have one or several individual cells in the center, each separated from the others by a wall.

extrusiones o partes extruidas. Parte hecha al calentar el aluminio a una temperatura bastante alta hasta ablandarlo y forzarlo a través de un molde para darle forma. Las partes extruidas con frecuencia tienen una o varias celdas individuales en el centro, cada una separada de las demás por una pared.

facebar. Often used synonymously with the front bumper, it is the steel bar that is either chrome plated or painted on the front bumper assembly.

barra delantera. A menudo se usa como sinónimo de parachoques delantero; ésta es la barra de acero que es cromada o pintada en el ensamblaje del parachoques delantero.

fan clutch. A mechanism attached to the fan which engages and disengages as needed, allowing the fan blade to "freewheel" once the vehicle reaches a certain speed.

embrague del ventilador. Mecanismo unido al ventilador que se engrana o desengrana como sea necesario, permitiendo que el ventilador gire libremente una vez el vehículo alcanza una cierta velocidad.

fascia. The large plastic covering over the vehicle's front and rear areas where the bumper is typically attached; in many cases it houses all the signal, marker, and headlights in the front and the taillights, as well as the backup and license plate lights, on the rear of the vehicle.

facia. Cobertura plástica grande que cubre las áreas delanteras y traseras del vehículo en donde el parachoques es típicamente unido; en muchos casos la *facia* aloja toda la señal, marcador, y luces frontales y traseras; así como las luces de placa vehicular y de reserva, en la parte trasera del vehículo.

feather edging. A method of paint application where the paint is spread out thinner near its edges as it blends into the existing surface finish.

bordeado fino. Método de aplicación de pintura en donde la capa de pintura extendida es más delgada cerca de los bordes, ya que ésta se combina con el acabado de la superficie existente.

featheredge splitting. A defect that appears as sand scratches or splitting around the repair area. Although it can look similar to sand scratch swelling, it will not follow into the repaired areas. Instead, it occurs when lacquer primers are used, and it will appear during or shortly after topcoat is applied.

agrietamiento en secciones. Defecto que luce como rayones de arena o divisiones alrededor del área de reparación. Aunque pueda parecer similar a un hinchamiento de rayones de arena, ésta no pasa a las áreas reparadas. En lugar de eso, la división ocurre cuando son usadas imprimaciones de laca, y aparecerá durante o poco después de que la capa exterior sea aplicada.

feathering. A gradual tapering of the filler to a very smooth edge so it is not possible to feel or see where the filler starts and the substrate begins.

feathering. Reducción gradual del relleno en un borde muy suave, de forma que no sea posible sentir o ver cuando el relleno termina y el sustrato comienza.

fiber-optic cable. A sheathed or wrapped plastic or fiberglass stranded cable that uses a light source from one area of the vehicle and projects it through the strands to several locations throughout the vehicle.

cable de fibra óptica. Cable de filamentos de plástico o de fibra de vidrio, cubiertos o envueltos, que usa una fuente de luz de un área del vehículo y la proyecta a través de los filamentos en otros sitios a lo largo del vehículo.

fiber-reinforced plastic (FRP). Reinforced plastic that uses catalyzed resin, which is spread onto a fiberglass mat while in a mold. This process hardens the part, which is then removed from the mold, trimmed, and used.

plástico de fibras reforzadas (FRP por su acrónimo en inglés). Plástico reforzado que usa resina catalizadora, la cual es dispersada sobre una estera de fibra de vidrio mientras está en un molde. Este proceso endurece la parte, la cual es entonces removida del molde, recortada y usada.

15° Rule. Rule that states for every 15°F the temperature rises over the standard of 70°F, the time to cure is doubled; and for every 15°F the temperature drops below the standard, the cure time is cut in half.

regla de los 15 grados. Regla que establece que por cada 15 grados Fahrenheit (15°F) arriba de los estándar 70°F, el tiempo de curado se duplica; y por cada 15°F debajo de la temperatura estándar, el tiempo de curado es la mitad.

figure 8 pattern. A painting situation in which the elliptical pattern the technician is trying to achieve may become the shape of the number 8: With heavy deposits of paint in both the top and bottom, and the center light on fluid, this pattern causes streaking when sprayed.

patrón de figura 8. Situación presentada durante la aplicación de pintura en la cual el patrón elíptico que el técnico está tratando de lograr puede convertirse en la forma del número 8: con depósitos pesados de la pintura tanto en la parte de arriba como en la de abajo, y el centro se ilumina con el fluido, este patrón produce líneas cuando es rociado.

files. A tool generally used to remove burrs and sharp edges, and smooth out imperfections in metal. The three general purpose files commonly used in collision repair are flat, half round, and round.

limas. Instrumento generalmente usado para quitar abultamientos y bordes agudos, y remover imperfecciones del metal. Las tres clases de limas comúnmente usadas en la reparación de colisiones son: planas, semi-redondas, y redondas.

film build. The thickness created by multiple layers of paint on a vehicle.

acumulación de película. Grosor creado por capas múltiples de pintura en un vehículo.

film thickness. The amount of coating film that is placed on the vehicle when it is painted. It is measured in one-thousandths of an inch, normally referred to as mils.

grosor de película. Cantidad de película de recubrimiento que es colocada en el vehículo cuando es pintado. Ésta es medida en milésimas de pulgada.

film thickness gauge. A tool used to determine how thick the paint is on a vehicle's surface.

medidor de grosor de película. Instrumento usado para medir el grosor de la pintura que está en la superficie de un vehículo.

finish hammer. Hammers used to smooth low and high areas and produce a perfectly smooth surface; also called a pick hammer.

martillo de acabado. Martillos usados para suavizar áreas altas y bajas, y producir una superficie suave perfecta.

finish holdout. The quality of finishes that helps them resist marring.

durabilidad del terminado. Calidad de los acabados que ayuda a resistir abolladuras.

finish sanders. Sander designed not to leave any tell-tale swirl marks in the finish.

lijadora de terminado. Lijadoras que están diseñadas para no dejar rastros de remolinos en el acabado.

finishing hammer. Used to achieve the final panel shape and contour, the finishing hammer is lighter with a smaller, more crowned face than the bumping hammer.

martillo de acabar. Usado para conseguir la forma final y el contorno del panel, este martillo es más ligero, con una cara coronada más pequeña, que el martillo de choque.

first-degree burns. The least severe type of burn, the surface of the skin reddens and becomes tender or painful, but no skin is broken.

quemaduras de primer grado. Tipo menos severo de quemadura. La superficie de la piel se enrojece y se hace sensible o dolorosa, pero ningún tejido está roto.

fish eye. The most common contamination defect, these are caused by silicone, a very light oil found as an ingredient in many products. They look like an area on the wet paint that will not allow the paint to cover the surface, a crater that may or may not show through to the bottom of the coat. It may even take on a distinguishable shape, such as a fingerprint.

ojo de pescado. Defecto de contaminación más común que es causado por la silicona, un aceite muy ligero encontrado como ingrediente en muchos productos. Este defecto luce como un área en la pintura mojada que no permite que la pintura cubra la superficie, un cráter que puede o no mostrar por el fondo del recubrimiento. Éste puede incluso tomar una forma distinguible, tal como una huella digital.

fish plating. A process in which an extra layer of steel is welded to the inside of the frame web or fence for additional reinforcement; sometimes used when an additional piece of equipment—such as a snow plow—is attached to prevent the frame rail from cracking or bending from the additional stresses caused by the equipment.

chapa de pescado. Proceso en el cual una capa suplementaria de acero es soldada al interior de la red o cerca del marco, para lograr un refuerzo adicional; a veces esto es usado cuando una pieza adicional del equipo, como una removedora de nieve, es unida para impedir al riel del marco rajarse o doblarse por las tensiones adicionales causadas por el equipo.

fixture setup sheet. A vehicle-specific sheet utilized by the technician to attach additional fixtures to the bench at specified locations when using a dedicated fixture measuring/straightening system.

hoja de configuración de partes integrantes. Hoja específica de cada vehículo utilizada por el técnico para unir partes adicionales al banco, en sitios específicos, cuando se usa un sistema de medida o alineación de partes integrantes.

flare nut wrench. A wrench with small splits in its jaw to fit over lines and tubing. Using the flare nut wrench will reduce the chance of "rounding off" the tubing nut.

llave de tuerca acampanada. Llave con pequeñas divisiones en sus mandíbulas que se ajustan sobre bordes y tubos. Al usar esta llave el desgaste de las tuercas acampanadas será reducido.

flash point. The lowest temperature at which a solvent gives off or releases enough vapors to ignite or catch fire.

punto de llamarada. Corresponde a la temperatura más baja en la cual un solvente emite o suelta suficientes vapores como para iniciar o prender fuego.

flash time. *See* purge time.

tiempo de destello. Ver "tiempo de purga."

flexible plastic. A category of plastics, such as polyurethane, polyamide, and thermo polyurethane, that are very bendable.

plástico flexible. Categoría de plásticos, tales como poliuretano, poliamida, y termo-poliuretano, que son muy flexibles.

flexibility test. A method to classify plastic parts by their flexibility. Plastic parts can be flexible, semi-flexible, semi-rigid, or rigid.

prueba de flexibilidad. Método de clasificación de partes plásticas de acuerdo a su flexibilidad. Las partes plásticas pueden ser flexibles, semiflexibles, semirígidas, o rígidas.

float test. A method used to identify thermoplastics. A small sliver is trimmed from the plastic to be repaired and placed in a clear glass filled with water. If the plastic sliver floats, it is a thermoplastic. If the sliver sinks, the plastic may not be a thermoplastic.

prueba de flotado. Método para identificar termoplásticos. Una pequeña astilla es recortada del plástico para ser reparado y colocado en un cristal transparente lleno del agua. Si la astilla plástica flota, es un termoplástico. Si la astilla se hunde, el plástico puede no ser un termoplástico.

floating hinge plate. A piece of hardened steel plate encased in a cage larger in size than itself so it can move or float about to accommodate shifting and movement of the hinges for adjusting purposes; plate has threaded holes tapped into it to accept the bolts used to attach the door hinges to the vehicle.

placa de gozne flotante. Pieza de placa de acero endurecido revestida en una jaula más grande en tamaño que ella, de forma que se pueda mover o flotar alrededor para acomodar cambios y movimientos de los goznes. La placa tiene agujeros perforados para recibir a los pernos que unen los goznes de la puerta con el vehículo.

floor filters. Filters found in the floor panels of a spray booth that should be regularly changed.

filtros de piso. Filtros encontrados en los paneles del piso de una cabina de rociado que deben ser cambiados con regularidad.

floor-mounted system. A series of interconnecting rails installed flush to the floor that are used with special hydraulic rams that fit into the rails via special adapters or a series of steel pots encapsulated in the concrete floor used to repair structural and undercarriage repairs.

sistema montado en piso. Serie de rieles interconectados instalados alineados al piso que son usados con gatos hidráulicos especiales, que se ajustan en los rieles a través de adaptadores o una serie de postes de acero encapsulados en el piso de concreto, éste sistema es usado para hacer reparaciones estructurales y del tren de aterrizaje.

flop. Sometimes called pitch; it is the appearance of a color/finish when viewed from an angle other than straight-on. The flop or change could appear to be lightness or darkness change, or the color shade can change.

cambio de tono. A veces llamado inclinación de tono; es la apariencia de un color o acabado cuando es visto desde un ángulo diferente al directo. El cambio de tono puede ser iluminado u oscurecido, o la sombra del color puede cambiar.

flow. A liquid's ability to spread out over an area.

flujo. Capacidad de un líquido de extenderse sobre un área.

flow out. The motion of pigments and binders as they are suspended in the solvents.

flotación. Movimiento de los pigmentos y carpetas cuando ellos están suspendidos en los solventes.

flowchart. An illustration that helps chart the flow of electrons through a particular circuit in a system.

diagrama de flujo. Ilustración que ayuda a trazar el flujo de electrones en un circuito particular en un sistema.

fluid control knob. Sometimes called a trigger control knob; it is a device that adjusts the amount of paint that is allowed around the needle and through the nozzle of a spray gun when the trigger is pulled. When adjusted further out, more fluid is allowed past the needle. When adjusted further in, less fluid is allowed out.

botón de control de fluido. A veces llamado botón de control de disparo; éste es un dispositivo que ajusta la cantidad de pintura que es permitida alrededor de la aguja y por el inyector de una pistola rociadora cuando el gatillo es tirado. Cuando es reajustado posteriormente hacia afuera, más fluido puede pasar por la aguja. Cuando es reajustado hacia adentro, menos fluido es permitido.

fluid needle. With the fluid nozzle, this regulates the amount of fluid that can pass from the cup to the spray cap for atomization.

aguja de fluido. Junto con el inyector de fluido, ésta regula la cantidad de fluido que puede pasar del recipiente a la gorra de atomización.

fluid nozzle. With the fluid needle, this regulates the amount of fluid that can pass from the cup to the spray cap for atomization.

inyector de fluido. Junto con la aguja de fluido, ésta regula la cantidad de fluido que puede pasar del recipiente a la gorra de atomización.

flush-mount straight rivets. Rivets used for bonding aluminum. They may be recommended in repair work over self-piercing rivets, as they do not require a special gun to install.

remaches rectos de montaje uniforme. Remaches usados para unir aluminio. Éstos pueden ser recomendados para el trabajo de reparación sobre remaches autoperforado, ya que ellos no necesitan una pistola especial para ser instalados.

foam brush application. A method of primer application utilizing a foam brush, although it is not practical for acid etch primers.

aplicación de cepillo de espuma. Método de aplicación de imprimación utilizando un cepillo de espuma, aunque no es práctica para imprimaciones de grabado ácidas.

foam dam. A device inserted into the cavity to be filled to retain the foam within a limited area and to prevent it from flowing beyond the area where it is needed.

presa de espuma. Dispositivo insertado en la cavidad a ser ocupada para retener la espuma dentro de un área limitada y para evitar que fluya más allá del área en donde es necesitada.

frame flange. The two short "legs" that are perpendicular to the web at the top and bottom of the rail which are usually bent at a 90-degree angle facing to the inside of the engine compartment.

pestaña de marco. Son las dos "piernas" cortas que son perpendiculares a la red en la parte de arriba y abajo del riel que son usualmente dobladas a 90 grados en frente del interior del compartimiento del motor.

frame rack or bench system. A repair system that may use either a stationary bench or a portable bench onto which a vehicle is attached to accommodate the repairs on the undercarriage and structural members of the vehicle.

sistema de banco o estante de marco. Sistema de reparación que puede ser usado o como banco estacionario, o como banco portátil, sobre lo que un vehículo es puesto para hacer las reparaciones del tren de aterrizaje y miembros estructurales del vehículo.

frame replacement module. A replacement section of the frame made available by the vehicle manufacturer that includes both side rails and perhaps a cross-member between them; a module may be either the front or rear section of the frame or it may be the middle section as well.

módulo reemplazante del marco. Sección de reemplazo del marco hecha disponible por el fabricante del vehículo que incluye los rieles de ambos lados y quizá un miembro cruzado entre ellos; un módulo puede estar o en la sección del frente o atrás del marco, o puede estar en la sección de en medio de igual forma.

frame web/fence. The large vertical section of the rail on the U- or C-shaped frame rail to which most of the cross-members, mounting brackets, and mechanical parts are attached.

red o cerca del marco. Sección vertical grande del riel sobre el riel del marco en forma de U o C en la cual la mayoría de los miembros cruzados, soportes de montaje, y partes mecánicas son unidos.

franchise. A network of collision repair centers that are licensed under the same name.

franquicia. Red de centros de reparación de colisiones que funcionan bajo el mismo nombre.

front wheel setback. A condition wherein one or both front wheels may be pushed or moved back as a result of damage to the suspension parts, caused by the wheels striking an object or falling into a deep surface depression while moving.

revés de rueda frontal. Condición en la cual una o ambas ruedas delanteras pueden ser empujadas o retrocedieron a consecuencia de un daño en las partes de la suspensión; este daño pudo ser causado porque las ruedas golpearon un objeto o cayeron en una depresión superficial profunda estando en movimiento.

fulcrum. When using a lever, the center section which gives the operator additional leverage to push against.

fulcro. Cuando es usa una palanca, es la sección central que da al operador la acción de palanca adicional para empujar en contra.

full coverage. The condition of sufficient paint coverage, achieved when the basecoat no longer can be seen through.

cobertura total. Condición de cobertura suficiente de pintura conseguida cuando el capa base ya no puede ser vista.

full frame vehicle. A vehicle design that incorporates a separate and independent frame extending the entire length of the vehicle as the body's main support structure—also known as a BOF vehicle.

vehículo de marco completo. Diseño de vehículo que incorpora un marco separado e independiente que se extiende la longitud entera del vehículo como la estructura de soporte principal del cuerpo, también conocido como vehículo BOF (por su acrónimo en inglés).

full hiding. The number of coats of paint required to complete a refinish job.

ocultamiento completo. Número de capas de pintura requeridas para completar un trabajo de acabado.

full wet coat. Spraying condition in which the amount of paint applied in a single stroke is wet in appearance with full gloss. Spraying a full wet coat is one of the more difficult techniques, for the finish must be so wet that it is dangerously close to running. When applying a finish this wet, care must be taken not to rush the flash time.

recubrimiento húmedo completo. Condición de rociado en la cual la cantidad de pintura aplicada en una mano simple está húmeda en apariencia y tiene brillo total. Rociar una capa húmeda completa es una de las técnicas más difíciles, para el acabado debe estar muy húmeda por lo que es peligrosamente cercana a correrse. Cuando se aplica un terminado así de húmedo, se debe tener cuidado de no apresurar el tiempo de destello.

G

galvanic corrosion. A form of corrosion that occurs when two or more different types of metal come into contact with each other, causing the more reactive metal to break down and creating corrosion where otherwise neither would have corroded alone. This type of corrosion is often thought to be detrimental when it occurs on an automobile. *See also* **contact corrosion.**

corrosión galvánica. Forma de corrosión que ocurre cuando dos o más tipos diferentes de metales entran en contacto el uno con el otro, causando que el metal más reactivo se descomponga y cree corrosión en donde por otra parte ninguno se hubiera corroído solo. A menudo se piensa que este tipo de la corrosión es perjudicial cuando esto ocurre en un coche. Ver también "corrosión por contacto."

galvanization. A process whereby one metal is coated with a thin layer or coating of another metal, commonly by hot-dipping or electroplating the steel to create a thin layer of zinc coating.

galvanización. Proceso en el cual un metal es cubierto por una capa delgada o recubrimiento de otro metal, comúnmente por el sumergimiento en caliente o electro-laminando el acero para crear una capa delgada de recubrimiento de zinc.

gas flow meter. A device which takes high-pressure gas stored in the gas cylinder and reduces it to a usable level and flow rate to supply the GMAW welder.

medidor de flujo de gas. Dispositivo que toma gas de alta presión, almacenado en un cilindro de gas, y lo reduce a un nivel usable, y con tasa de flujo para abastecer al soldador tipo GMAW (por su acrónimo en inglés).

gas metal arc welding (GMAW). Also called metal inert gas welding; this is the most commonly used form of welding in the automotive collision repair industry. It was originally developed in 1940 as a way of fusing aluminum and other nonferrous metals together, and it was quickly adapted to steel because it was faster than other welding processes of the day.

soldadura de arco metálico de gas (GMAW por su acrónimo en inglés). También llamada soldadura de gas inerte metálica; esta es la forma de soldadura más comúnmente usada en la industria de reparación de colisiones automotrices. Fue originalmente desarrollada en 1940 como una forma de fundir el aluminio y otros metales no ferrosos juntos, y fue rápidamente adaptada al acero porque era más rápido que otros procesos de soldadura.

gas out. Evaporation of solvents into the atmosphere. That is, the solvents which have turned from liquid to gas are being released from the finish.

liberación de gas. Evaporación de solventes en la atmósfera. Es decir, los solventes que han pasado de líquido a gas son liberados al final.

gas regulator. A device that has two gauges: One shows the pressure in the gas cylinder, measured in pounds per square inch (psi), which can be very high; and the second has an adjustment valve which adjusts the gas flow.

regulador de gas. Dispositivo que tiene dos medidores: uno que muestra la presión en el cilindro de gas, medida en libras por pulgada cuadrada (psi), la cual puede ser muy alta; y el segundo medidor tiene una válvula de ajuste que regula el flujo de gas.

general purpose dolly. A dolly that contains many crown shapes and can be used for most general metal straightening.

carretilla de uso general. Carretilla que contiene muchas formas de corona y que puede ser usada para el enderezamiento metálico más general.

glaze. A coating used to protect the finish. However, glazes should not be used to correct swirl marks, as they only cover them up—not repair them.

vidriado. Recubrimiento usado para proteger el acabado. Sin embargo, el vidriado no debería ser usado para corregir marcas de remolinos porque éste sólo los cubre y no los reparan.

globular metal transfer. The least desirable method of metal transfer. In this method, as the weld is struck a ball of molten metal larger in diameter than the electrode wire forms on the electrode wire. When the ball transfers to the base metal either from gravity or short circuiting, it leaves an uneven surface on the weld and often causes spatter.

transferencia de metal globular. Este es el método menos deseable de transferencia metálica. En este método, a medida que la soldadura es golpeada, una bola de metal fundido más grande en diámetro (que el alambre del electrodo), se forma en el alambre del electrodo. Cuando la pelota se transfiere al metal base ya sea por gravedad o por un circuito corto, queda una superficie desigual en la soldadura y a menudo causa salpicaduras.

gloss. The shininess of a vehicle.

brillo. Es la brillantez de un vehículo.

glue-on dent pullers. An alternative to welding, these items are glued on a device that is used to pull up low areas on a surface.

sacador de abolladuras a base de pegante. Alternativa a la soldadura; estos artículos son pegados sobre un dispositivo que es usado para levantar áreas bajas en una superficie.

glue-on sandpaper. A self-adhesive used on some sandpapers that allows the paper to be attached and removed with little difficulty. It is generally less expensive than the hook and loop type, but can be attached only once.

papel de lija autoadhesivo. Autoadhesivo usado sobre algunos papeles de lija que permite que el papel sea adherido y quitado con poca dificultad. Éste es generalmente menos costoso que el tipo de gancho y lazada, pero sólo puede ser adherido una vez.

GMAW MIG pulse. A welding process commonly used to weld structural and nonstructural aluminum parts. The pulse feature is a relatively new concept which requires a heavier duty 220V to 230V welding machine.

pulso GMAW MIG. Proceso de soldadura comúnmente usado para soldar partes estructurales y no estructurales de aluminio. La característica del pulso es un concepto relativamente nuevo que requiere una máquina de soldar de más poder de 220V a 230V.

GMAW MIG welding. The abbreviation commonly used to indicate the American Welding Society's gas metal arc welding process using the metallic inert gas equipment.

soldadura GMAW MIG. Acrónimo comúnmente usado para indicar el proceso de soldadura de arco metálico de gas de la sociedad americana de soldadura usando el equipo de gas inerte metálico.

grating plastic. The first stage of finishing the cured plastic performed by removing the outermost layer of material from the surface, usually using a surform file or a very coarse grade of sandpaper.

plástico crispante. Primera etapa de acabado del plástico curado realizada al remover la capa más externa de material de la superficie usualmente usando una lima tipo *surform* o una lija de grano grueso.

gravity feed gun. A spray gun in which coating passes from the cup, on top of the spray gun, and down a tube to the airstream. As the trigger is pulled, the needle moves back, which allows air to pass through the gun. This movement of air—along with gravity—helps to pull paint through the gun's nozzle and out the air cap. The paint, when outside the gun, is mixed with air, atomizing the paint.

pistola de gravedad. Pistola rociadora en la cual el recubrimiento pasa de la taza, por encima de la pistola rociadora, y hacia abajo por un tubo hacia la corriente de aire. Cuando el gatillo es accionado, la aguja retrocede, lo que permite que el aire pase por la pistola. Este movimiento de aire, junto con la acción de la gravedad, ayuda a expulsar la pintura a través del inyector de la pistola y la tapa de aire. La pintura, cuando está fuera del arma, es mezclada con el aire, lo que la atomiza.

grinders. Tools used to remove items such as old finish, fillers from prior repairs, and rust from a vehicle's surface before refinishing operations are performed.

molinillos. Herramientas usadas para remover sustancias tales como viejos acabados, rellenos de anteriores reparaciones y herrumbres de la superficie de un vehículo antes de que las operaciones de refinado sean realizadas.

grit numbering. The system which assigns a number to indicate abrasive aggressiveness. The more aggressive or coarse group uses smaller numbers, such as 24-grit grinding pads, whereas the finer group uses larger numbers, like the very fine-textured and less aggressive P3000 paper.

numerado de grano. Sistema que asigna un número indicando la agresividad abrasiva. El grupo más agresivo o grueso usa números más pequeños, tales como como cojines de molinos de grano número 24; mientras que el grupo más fino usa números más grandes, como el papel de textura fina y menos agresivo P3000.

ground effects. Items on a vehicle, such as aerodynamic wings, spoilers, dams, or skirts, which not only provide cosmetic appeal but also provide down force to improve traction.

efectos de tierra. Partes de un vehículo, como alas aerodinámicas, *spoilers*, presas o faldas, que no sólo proporcionan un aspecto cosmético sino que proveen fuerza vertical para mejorar la tracción.

ground fault circuit interrupter (GFCI). Safety device which must be used with electric tools at all times when moisture is an issue in the shop.

interruptor de circuito de polo a tierra (GFCI por su acrónimo en inglés). Dispositivo de seguridad que debe ser usado con instrumentos eléctricos cuando la humedad está presente en la tienda.

gun angle. The angle at which a welding technician holds the gun in relation to the work. As the gun angle changes, the shielding gas and the quality of weld are affected. Gun angle is also relative to the direction in which the weld is traveling.

ángulo de pistola. Ángulo en el cual un técnico de soldadura sostiene la pistola con respecto al trabajo. Los cambios de ángulo de la pistola afectan el gas protector y la calidad de la soldadura. El ángulo de la pistola está también relacionado con la dirección en la cual la soldadura está siendo aplicada.

gun cleaning. Process for maintaining a spray gun, often performed with a gun washing machine.

limpieza de pistola. Proceso para mantener una pistola rociadora, a menudo realizado con una lavadora de pistolas.

gun cup. Part of the spray gun that comes in different sizes, is made from different materials, and is now available with cup liners; some cups themselves are disposable. Different cup sizes are available for the different amounts needed for refinishing.

taza de pistola. Parte de la pistola rociadora que viene en diferentes tamaños; está hecha de varios materiales, y ahora está disponible con forros de taza; algunas tazas son desechables. Tazas de varios tamaños están disponibles de acuerdo a las diferentes necesidades de barnizado.

gun fan. To avoid pitch, this area of the spray gun should not be held too closely at its top or bottom.

ventilador de pistola. Para evitar inclinaciones, esta área de la pistola rociadora no debe ser sostenida muy cerca de la parte de arriba o de abajo.

gun lubrication. Coating applied to the components of a spray gun to keep them working smoothly. This is generally done after the gun cleaning.

lubricación de pistola. Capa aplicada a los componentes de una pistola rociadora para mantenerlos trabajando suavemente. Esto es generalmente hecho después de la limpieza de la pistola.

gun rebuilding. A process where a spray gun is completely disassembled, defective parts are replaced, and the gun is reassembled. It is wise to rebuild a gun on a regular basis, before it no longer operates properly.

reconstrucción de pistola. Proceso en el cual una pistola rociadora es completamente desarmada, las partes defectuosas son reemplazadas, y la pistola es armada de nuevo. Es aconsejable reconstruir las pistolas regularmente, antes de que no funcionen correctamente.

gun setup. The setting of the proper needle and nozzle on a spray gun.

configuración de pistola. el Ajuste apropiado de la aguja e inyector en una pistola rociadora.

gusset. Any device made of plastic, metal, or a flexible rubber flap that is used to funnel airflow through the front of the radiator or core support opening over the radiator and engine compartment.

escudete. Dispositivo hecho de plástico, metal o hule flexible que es usado para canalizar las corrientes de aire a través del frente del radiador o abertura del soporte principal sobre el compartimiento del radiador y el motor.

H

hacksaw. A tool with an adjustable frame that accepts different blade lengths to cut small metal parts. When choosing a blade, select one with the tooth configuration where at least two teeth contact the material at the same time, and install it with the teeth pointed away from the handle.

sierra de arco. Herramienta con marco ajustable que acepta cuchilla de diferentes longitudes para cortar piezas pequeñas metálicas. Cuando elija una cuchilla, seleccione una con la configuración de dientes en donde al menos dos dientes entren en contacto con el material al mismo tiempo, e instálela con los dientes apuntando hacia el lado contrario del mango.

hammer-off-dolly. A technique used to raise low areas of metal. The dolly is placed beneath the metal where it is low, then is held firmly as the technician places hammer blows near where the dolly is located, using the spring method of hammering. The dolly will rebound off the metal and spring back, causing the metal to rise.

desperdicio peligroso. Sustancias químicas o materiales que han llegado al punto en el que no son más útiles y exhiben las características de ser combustibles, tóxicos, corrosivos o reactivos.

hammer-on-dolly. The most common "bumping" technique technicians use when they have access to both sides of a vehicle, this is a technique where the dolly is held on one side of the damaged panel and the hammer is swung against the opposite side. Both tools will act to either raise or lower the damaged sheet metal.

martillado en carretilla. Técnica más común de golpeado que los técnicos usan cuando ellos tienen acceso a ambos lados de un vehículo; ésta es una técnica en donde la carretilla es sostenida en un lado del panel dañado y el martillo es balanceado contra el lado opuesto. Ambas herramientas actuarán para ya sea levantar o bajar la lámina de metal dañada.

hand grinders. Tools, such as pistol grip and die grinders, which operate at a very high rpm that are used for metal removal in small confined areas where access for a larger disc is not possible.

molinillos de mano. Herramientas, tales como pistolas y molinillos, los cuales operan a un alto número de revoluciones por minuto que son usadas para remover metales en áreas confinadas en donde el acceso de discos más grandes no es posible.

hand sanding. A process where a technician manually sands a part with sandpaper, rather than using a machine.

lijado a mano. Proceso en el cual un técnico lija manualmente una parte, en lugar de usar una máquina.

hard tape line. Line seen if masking tape with paper is applied and the material is sprayed directly over the tape.

línea de cinta dura. Línea que se observa si la cinta de enmascarar con papel es aplicada y el material es rociado directamente sobre la cinta.

hard water marks. Water spots not etched into the vehicle's finish.

marcas de agua dura. Marcas de agua no grabadas en el acabado del vehículo.

hardener. A product added to polyurethane to start the cross-linking or curing process. When the hardener is added to a coating, it causes the paint to cross-link.

endurecedor. Producto añadido al poliuretano para iniciar el enlace cruzado o proceso de curado. Cuando el endurecedor es añadido a una capa, causa que la pintura reaccione.

harmful air pollutants (HAPs). Contaminants released into the atmosphere during collision repair and refinishing.

contaminadores de aire perjudiciales (HAPs por su acrónimo en inglés). Contaminantes liberados en la atmósfera durante la reparación de colisiones y barnizado.

hat channel. The result of welding a flat piece of metal over the open side of a three-sided channel; commonly used on the rear lower rails where the trunk floor is welded to the rear channels, forming a completely enclosed reinforcement.

canal de sombrero. Resultado de soldar una pieza plana de metal sobre el lado abierto de un canal trilátero; comúnmente usado sobre los rieles bajos traseros en donde el piso de la cajuela es soldada a los canales de atrás, formando un refuerzo cerrado completo.

hazard communication program. A written document created by all employers outlining the safeguards and training that must be completed to protect their employees from a potentially hazardous or dangerous work environment.

programa de comunicación de riesgos. Documento escrito creado por todos los empleadores que definen las medidas y formación que deben ser completadas para proteger a sus empleados de un ambiente de trabajo potencialmente arriesgado o peligroso.

hazardous material. Any chemicals or materials that can potentially harm the environment, the general public, or the workers who are directly handling it.

material peligroso. Cualquier producto químico o material que puede dañar potencialmente el ambiente, el público en general, o los trabajadores que lo manejan directamente.

hazardous waste. A chemical or material that has been used to the point where it is no longer viable and exhibits the characteristics of being ignitable, toxic, corrosive, or reactive.

desperdicio peligroso. Sustancias químicas o materiales que han llegado al punto en el que no son más útiles y exhiben las características de ser combustibles, tóxicos, corrosivos o reactivos.

headlight covers or bezels. A chrome or painted trim piece that surrounds the headlights to cover the otherwise unsightly framework and mounting brackets in front of the radiator support; often incorporated into—but not a part of—the grill panel.

cubierta de luces delanteras o biseles. Pieza de moldura cromada o pintada que rodea las luces delanteras para cubrir el marco que, por otra parte, es antiestético y montar los soportes en frente del apoyo del radiador; ésta es a menudo incorporada en, pero no es parte de, el panel de parrilla.

heat affect zone. When welding, the area immediately around the weld site, which may become slightly weaker from the effects of the heat.

zona de influencia del calor. Área inmediatamente cercana al sitio de soldadura, la cual puede volverse ligeramente más débil por los efectos del calor.

heat induction. Refers to tools that act like a transformer, setting up a magnetic field between the inducer tool and the metal object. A magnetic field is created and heat is produced at the metal area.

inducción de calor. Se refiere a los aparatos que actúan como un transformador, generando un campo magnético entre el aparato inductor y el objeto metálico. Un campo magnético es creado y el calor es producido en el área metálica.

heat shrinking. A process that heats stretched metal with a torch—or, more often, with a carbine tip on a welding machine—to a red-hot state. Then, before the metal has a chance to cool, it is quenched with either a wet rag or sponge, which causes the metal to shrink during the rapid cooling.

encogimiento por calor. Proceso que calienta metal estirado con una antorcha, o más a menudo, con una punta de carabina sobre una máquina de soldar, hasta que esté al rojo vivo. Luego, antes de que el metal se enfríe, éste es apagado con un trapo o esponja mojada, lo cual causa que el metal se encoja durante el enfriado rápido.

heat-treatable. Alloys that gain strength by heating; these alloys are generally used for exterior body panels such as hoods, door skins, and quarter panels.

tratamiento con calor. Aleaciones que ganan fuerza al calentarse; estas aleaciones son generalmente usadas para paneles de cuerpo exteriores, como capós, pieles de puerta, y paneles de cuarto.

heel dolly. A dolly used in sharp corners and tight places.

carretilla de talón. Carretilla usada en esquinas agudas y sitios apretados.

hepatitis B (HBV). A disease that can be transmitted through either live body fluids or by touching dried blood and then touching the eyes, mouth, nose, or an open sore.

hepatitis (HBV por su acrónimo en inglés). Enfermedad que puede ser transmitida por fluidos de cuerpos vivos o tocando la sangre secada y luego tocando los ojos, boca, nariz, o una llaga abierta.

hiding. The paint's ability to cover an existing finish or paint layer without it showing through.

cubrimiento. Capacidad de una pintura de cubrir una capa de acabado o pintura existente sin que quede visible.

high flow quick connectors. Hoses connected so they must bend downward after the connection, preventing delamination and failure.

conectores rápidos de flujo alto. Mangueras conectadas de forma que se doblen hacia abajo luego de la conexión, previniendo deshojado y falla.

high pressure guns. Sometimes referred to high pressure, low volume (HPLV) guns, these were the most common type of spray gun used from the early 1900s until the late 1970s. They provided a smooth finish when spraying low viscosity coatings, but generated a large amount of overspray and low transfer efficiency (25 to 30 percent).

pistolas de alta presión. Algunas veces llamadas pistolas de alta presión y de bajo volumen (HPLV por su acrónimo en inglés), éstas fueron el tipo más común de pistolas de rociado usadas a principios de los años 1900 hasta el final de los años 70. Ellas proveían un acabado suave cuando se rocían recubrimientos de baja viscosidad, pero generaban una larga cantidad de excesos de rociado y baja eficiencia de transferencia (de 25 a 30%).

high solids. The ingredients in finishes that do not evaporate following application and curing.

sólidos altos. Ingredientes en acabados que no se evaporan luego de la aplicación y curado.

high strength low alloy (HSLA) steel. Steels between mild and high strength steels in yield strength and heat sensitivity. They are used to manufacture numerous structural parts such as the A, B and C pillars and any area of the vehicle where strength is needed and a reduced weight is also desired.

acero de alta resistencia y baja aleación (HSLA por su acrónimo en inglés). Aceros de resistencias entre baja y alta en la resistencia a la fluencia y sensibilidad al calor. Éstos son usados para fabricar numerosas partes estructurales tales como los pilares A, B y C, y cualquier área del vehículo en donde la resistencia sea necesaria y un peso reducido sea también deseado.

high strength steel (HSS). A group of ultra strong steel forms that are very temperature sensitive; commonly used for structural parts on most models of unibody vehicles. High strength steel has a yield strength of 60,000 psi, which means that parts that are manufactured from this type of steel can be made thinner and will resist the same amount of yield stress as thicker mild steel parts.

acero de alta resistencia (HSS por su acrónimo en inglés). Grupo de aceros de alta resistencia que son muy sensibles a la temperatura; son comúnmente usados en partes estructurales en la mayoría de modelos de vehículos unicuerpo. Los aceros de alta resistencia tienen un punto de fluencia de 60.000 psi, lo cual significa que las partes que son fabricadas con este tipo de acero pueden ser fabricados más delgados y resistirán el mismo esfuerzo de fluencia que partes de acero común.

high volume, low pressure (HVLP) guns. Spray guns whose cap pressure does not exceed 10 psi.

pistolas de alto volumen y baja presión (HVLP por su acrónimo en inglés). Pistolas rociadoras cuya presión de gorra no excede 10 psi.

hinge buckle. A situation that often occurs in a crowned area of a panel where the direct damage pushes the surrounding metal.

comba de gozne. Situación que a menudo ocurre en un área coronada de un panel en donde el daño directo empuja el metal circundante.

hinge pillar. The part of the vehicle to which the door hinges are attached; on the front door the cowl side panel is considered the hinge pillar, and the B or center pillar is the hinge pillar for the rear door.

pilar de gozne. Parte del vehículo en la cual los goznes de la puerta están unidos; en la puerta principal, el panel del lado de la capucha se considera el pilar de gozne, y el pilar central ó B es el pilar de gozne de la puerta trasera.

holding turnbuckle. A holding or blocking device that neither pushes nor pulls used when repairing damaged structural panels to prevent the openings such as the doors, sunroof and windshield openings from becoming enlarged or deformed by the corrective forces.

torniquete sostenido. Dispositivo de sostenimiento o bloqueo que ni empuja, ni jala, y que es usado cuando se reparan paneles estructurales dañados para prevenir que las aberturas, tales como las puertas, el techo, y las aberturas del parabrisas, se alarguen o deformen por las fuerzas correctivas.

holdout. A new finish's long lasting color and depth of reflection (shine).

durabilidad. Color de larga duración y profundidad de reflexión (brillo) de un acabado nuevo.

hood flutter. A visible shaking and vibration of the entire hood when driving the vehicle caused by the latch and the adjusting bumpers not being properly adjusted to put the panel under pressure thus preventing the vibrating or fluttering.

sacudida del capó. Sacudida y vibración visible de todo el capó cuando se conduce el vehículo causados por no ajustar correctamente el pestillo y los parachoques, para poner el panel bajo presión y así evitar la vibración y sacudidas.

hook and loop. An attachment method for sandpaper backing performed by placing light pressure on the sandpaper when attached to the backing pad, allowing the hooks and loops to attach. Hook and loop paper can be removed and reattached later, if needed.

gancho y lazada. Método de unión para apoyo de lijas realizado al aplicar una ligera presión sobre la lija cuando es unida al cojín de apoyo, permitiendo que los ganchos y lazadas se unan. Los ganchos y lazadas pueden ser removidos y reunidos después, si es necesario.

hostile forces. Any of the forces to which a vehicle is subjected during a collision that are responsible for causing any of the collision damage the vehicle sustains.

fuerzas hostiles. Cualquiera de las fuerzas a las cuales un vehículo es sometido durante una colisión que son responsables de causar los daños por el choque.

hot air plastic welding. Usually connected to a compressed air supply such as the shop's air compressors, this welder has an air orifice that heats that base plastic in the area that will be repaired, and a tube that a plastic welding rod is passed through which preheats the rod. As the base plastic is heated, the welding tube is pulled over the repair area as the rod is pushed down into the weld area.

soldadura de plástico de aire caliente. Usualmente conectada a un abastecedor de aire comprimido tal como los compresores de aire comerciales, esta soldador tiene un orificio de aire, que calienta el plástico base en el área que será reparada, y un tubo por el que pasa una barra de soldadura plástica, el cual precalienta la barra. A medida que el plástico base es calentado, el tubo de la soldadura es colocado sobre el área de reparación, a medida que la barra es empujada en el área de soldadura.

hot rolling. A process used to form thicker frame rails in which the steel is rolled and formed to shape while red hot, allowed to cool, and then cut to the specified length for each frame application.

laminado en caliente. Proceso usado para formar rieles de marco más gruesos en los cuales el acero es laminado y formado para darle forma mientras está al rojo vivo, luego se le permite enfriarse y luego es cortado con la longitud requerida para cada aplicación de marco.

hot-rolled steel. Steel created through a process that rolls newly manufactured steel at a temperature of 1,472°F (792°C) to a thickness of 5/16 to 1/16 inch (7.9 to 1.6 mm) thick; often used to make parts such as frame cross-members and other thicker parts.

acero laminado en caliente. Acero creado por un proceso que lamina acero recién fabricado a una temperatura de 1.472°F (792°C) a un espesor de 5/16 a 1/16 de pulgada (de 7.9 a 1.6 mm); a menudo usado para hacer partes, tales como los miembros cruzados del marco y otras piezas gruesas.

hue. The quality or dimension of color, such as red, blue, or yellow.

matiz. Calidad o dimensión de color, tal como rojo, azul, o amarillo.

hue, value, chroma, and texture. The elements of a vehicle's finish that a refinish technician must match against the original vehicle.

matiz, valor, chroma y textura. Los elementos del acabado de un vehículo que un técnico de refinado debe cotejar con los del vehículo original.

human immunodeficiency virus (HIV). A disease that can be transmitted only through live body fluid, such as through blood or sexual contact.

virus de inmunodeficiencia humana (VIH). Enfermedad que sólo puede ser transmitida por fluidos de cuerpos vivos, tales como la sangre o por contacto sexual.

hydraulic lockup. Condition that may occur if a car is rolled over and left on its top for a period of time, during which the oil and other fluids may leak into the cylinders.

atrancamiento hidráulico. Condición que puede ocurrir si un coche es girado y dejado sobre su techo por un periodo de tiempo durante el cual el petróleo y otros fluidos pueden escaparse de los cilindros.

hydrocarbons. Chemicals containing hydrogen and carbon; a name often applied to products made from crude oil.

hidrocarbonos. Productos químicos que contienen hidrógeno y carbono; un nombre a menudo aplicado a productos hechos de petróleo crudo.

hydroforming. A process where water, hydraulic fluid, or another liquid is used under extremely high pressure to finish forming structural parts by placing the part into a mold, filling the rail, and subjecting it to pressure.

hidroformado. Proceso en el cual agua, fluidos hidráulicos, u otro líquidos son usados bajo extrema alta presión para terminar el formado de partes estructurales al colocar la parte en un molde, llenar el riel y someterlo a presión.

hydrophilic. A primer or material that has a tendency to absorb moisture or water, causing it to become neutralized or—in the case of any acid etch product—the moisture is absorbed into the primer, neutralizing the acid and actually causing rust or corrosion to form on the surface underneath the primer.

hidrofílico. Material de imprimación que tiene tendencia a absorber humedad o agua, lo que lo neutraliza o, en el caso de cualquier producto de grabado ácido, la humedad es absorbida por la imprimación, neutralizando el ácido y causando realmente herrumbre o corrosión para formar la superficie bajo la imprimación.

I

I-CAR (Inter-Industry Conference on Auto Collision Repair). A not-for-profit organization dedicated to researching new methods and developing programs for training, retraining, and upgrading collision repair technicians on the latest developments in collision repair.

I-CAR (Conferencia Multi-industrial de Reparación de Colisiones de Automóviles). Organización sin fines de lucro dedicada a la investigación de nuevos métodos y programas de desarrollo para formación, capacitación, y mejoramiento de los técnicos en reparación de colisiones, con los últimos avances en la reparación de colisiones.

I-CAR Education Foundation. Created in 1991, this not-for-profit organization prepares entry level candidates for careers in the collision industry, with an advanced curriculum, instructor training research, and related services.

Fundación Educativa I-CAR. Creada en 1991, esta organización sin fines de lucro prepara a candidatos novatos para carreras en la industria de la colisión, con un plan de estudios avanzado, investigación de formación de instructor, y servicios relacionados.

I-CAR Uniform Procedures for Collision Repair (UPCR). A series of guidelines developed by I-CAR through a very extensive testing process which provides recommended procedures for the collision repair technician to follow when replacing both structural and nonstructural panels and reinforcements on vehicles.

guía de procedimientos uniformes I-CAR para la reparación de colisiones. Serie de guías desarrolladas por *I-CAR* a través de un proceso de pruebas extensivas las cuales proveen procedimientos recomendados para que los técnicos en reparación de colisiones las sigan cuando reemplacen paneles estructurales y no estructurales, y refuerzos de vehículos.

impact absorber/isolator. A shock absorbing device attached to the outermost part of both the front and rear side rails for the express purpose of absorbing low speed impacts such as parking lot low bumps and isolating the effects of the impact to the bumper area only.

absorbedor de impacto o aislador. Dispositivo absorbedor de impacto unido a la parte más externa de los rieles laterales frontales y traseros con el propósito expreso de absorber impactos de baja velocidad, tales como golpes suaves por estacionamiento, y aislar los efectos de impactos en el área del parachoques únicamente.

impact object. Any solid object, ranging from a small ball to a stationary pole or oncoming vehicle, which is capable of creating damage to a vehicle.

objeto de impacto. Cualquier objeto sólido, desde una pelota pequeña hasta un poste fijo o un vehículo acercándose, que es capaz de crear daño en un vehículo.

impact socket. A socket that is thicker than a standard socket and case hardened to withstand the forces of an air-powered impact wrench.

enchufe de impacto. Enchufe que es más grueso que un enchufe estándar y de cubo reforzado que soporta la fuerza de una llave de impacto activada por aire.

improper fusion. The failure of the base metal and the filler metal to combine properly. This failure of fusion can be caused by low or inadequate heat, poor weld technique, insufficient gap, or improper edge setup.

fusión impropia. Falla del metal base y del material de relleno para combinarse. Esta falla de fusión puede ser causada por calor bajo o inadecuado, pobre técnica de soldado, holgura insuficiente o montaje de borde impropio.

included angle. A factory designed angle built into the steering knuckle and used as a diagnostic angle if damage to the suspension parts is suspected. It is the sum total of SAI and camber: When the camber is positive, the two are added together. If the camber is negative, the degrees of negative camber are subtracted from the total angle degrees.

ángulo incluido. Ángulo diseñado en fabrica incorporado al nudillo de conducción y usado como ángulo de diagnóstico si se sospecha de daño en las partes de la suspensión. Éste es la suma total del SAI y el ángulo *camber*: cuándo el *camber* es positivo, los dos se suman. Si el *camber* es negativo, los grados del *camber* negativo se restan de los grados del ángulo total.

included operations. Tasks found in the crash guides inside the corresponding *procedural explanation* ("P" Page) section that have been deemed a necessary part of an overall procedure; therefore, no additional time will need be added for that task.

operaciones incluidas. Tareas encontradas en las guías de choques dentro de la correspondiente sección *explicaciones de procedimiento* (Página "P") que ha sido juzgada como parte necesaria de un procedimiento total; por lo tanto, ningún tiempo adicional tendrá que ser añadido para dicha tarea.

independent agent. A person who, although not an employee of any specific insurance carrier, is licensed and authorized to sell insurance policies for those companies.

agente independiente. Persona que, aunque no es un empleado de alguna compañía de seguros, tiene licencia y está autorizada para vender pólizas de seguros de aquellas compañías.

indirect damage. The damage(s) caused by collision energy forces (or inertia)—that begin near the point of impact—traveling through a vehicle's structure; signs of indirect damage include ripples in sheet metal, popped spot welds, and chipping paint or undercoating; also called secondary damage.

daño indirecto. Daño causado por las fuerzas de energía de colisión (o inercia) que comienzan cerca del punto de impacto y viajan por la estructura de un vehículo; señales de daño indirecto incluyen ondulación de láminas metálicas, soldaduras salidas y desconchando de la pintura o recubrimientos bases; éste también es llamado daño secundario.

induction heater. An electromagnetic heating device used to remove apply heat for numerous operations without requiring an open flame and can be used to heat frame rails, frozen bolts adhesive bonded moldings etc.

calentador de inducción. Dispositivo de calefacción electromagnética que se usa para evitar el calor aplicado en numerosas operaciones sin requerir llama directa y puede ser usado para calentar rieles de marco, moldes unidos a adhesivos de pernos fijos, etc.

induction period. The amount of time that the reduced and hardened primer must set prior to application on the vehicle. This time period allows the paint coating and hardener to react to each other, thus providing the product's full potential for both adhesion and corrosion protection.

periodo de inducción. Cantidad de tiempo que la imprimación reducida y endurecida debe reposar antes de la aplicación sobre el vehículo. Este periodo de tiempo permite que el recubrimiento de la pintura y el endurecedor reaccionen el uno con el otro, y así lograr el potencial total del producto para la adhesión y la protección contra la corrosión.

induction removal tool. A tool with attachments for different applications of heating and removal. The tool, when turned on, sets up a magnetic field around the part being heated.

instrumento de remoción de inducción. Instrumento con accesorios para diferentes aplicaciones de calentamiento y remoción. El instrumento, cuando es encendido, genera un campo magnético alrededor de la parte calentada.

inertia cutoff switch. A fuel shutoff switch that will stop the flow of fuel when the vehicle is stopped abruptly, such as in an accident. It stops the fuel pump from continuing to pump gas out of a fuel line that may have become damaged in an accident and also prevents the engine from running after an accident.

interruptor de corte por inercia. Interruptor de corte de combustible que detiene el flujo de combustible cuando el vehículo es detenido repentinamente, como en el caso de un accidente. Éste para la bomba de combustible de seguir bombeando gasolina en la línea de combustible, que pudo haber sido dañada en un accidente, y también impide que el motor continúe funcionando luego de un accidente.

inertia energy. The force created by an object (or vehicle) hitting another stationary or moving object (or vehicle). The Law of Inertia basically states that a body in motion tends to remain in motion, while a body at rest tends to (want to) remain at rest.

energía de inercia. Fuerza creada por un objeto (o vehículo) que golpea a otro objeto inmóvil o móvil (o vehículo). La ley de la inercia básicamente establece que un cuerpo en movimiento tiende a permanecer en movimiento, mientras que un cuerpo en reposo tiende a permanecer en reposo.

inertial forces. Some of the hostile forces that act upon various parts of the vehicle during a collision. They are responsible for causing collision damage away from the part of the car that is affected by the impact object.

fuerzas de inercia. Algunas de las fuerzas hostiles que actúan sobre varias partes del vehículo durante una colisión. Éstas son responsables de causar daño por colisión lejos de la parte del vehículo que es afectada por el objeto de impacto.

infrared (IR) curing. Devices that use either short infrared or medium infrared waves to heat solid objects instead of the air around the object. Because the infrared waves pass through the finish and heat the solid object below the finish, many technicians refer to this type of curing as inside-out curing.

curado con infrarrojos (IR por su acrónimo en inglés). Dispositivos que usan ondas infrarrojas cortas o medias para calentar objetos sólidos en vez del aire alrededor del objeto. Como las ondas infrarrojas pasan a través del acabado y calientan el objeto sólido que está debajo, muchos técnicos se refieren a este tipo de curado como "curado desde el interior".

infrared light. A portion of the light spectrum outside the visible range.

luz infrarroja. Parte del espectro de luz que está fuera del rango visible.

inside diameter (ID). The measure of the width inside a hose, which must be monitored to ensure the correct air volume is allowed.

diámetro interior (ID por su acrónimo en inglés). Medida del ancho dentro de una manguera, que debe ser supervisada para verificar que el volumen de aire es el permitido.

insufficient penetration. A common defect among new welding technicians, this condition occurs when the weld fusion does not go deep enough into the base metal to produce a strong weld. The causes of poor penetration include a welding current that is too low, an arc length that is too long, an improper weld technique, and an improper joint fit-up.

penetración insuficiente. Defecto común entre nuevos técnicos de soldadura; esta condición ocurre cuando la fusión de la soldadura no es suficientemente profunda en el metal base para producir una soldadura fuerte. Las causas de la pobre penetración incluyen un flujo de soldadura que es muy bajo, una longitud de arco que es muy larga, una técnica de soldadura inapropiada y un ajuste de unión inapropiado.

International Origination for Standardization (ISO). A standard set of letter codes often molded into the back side of a part that identify either the type of plastic the part was made from, or the manufacturing process the part was made from. This is the most specific way to identify a plastic.

Origen Internacional para Estándares (ISO por su acrónimo en inglés). Grupo estándar de códigos, a menudo moldeado en la parte de atrás de una parte, que identifican ya sea el tipo de plástico de origen, o el proceso de fabricación de la parte. Este es la forma más específica de identificar un plástico.

intrusion beam. A part of the vehicle's door shell designed to prevent an impact object from penetrating into the passenger compartment.

viga de intrusión. Parte del cascarón de una puerta de vehículo diseñada para impedir que un objeto de impacto penetre en el compartimiento de pasajeros.

inverter. A mechanism used in hybrid vehicles to convert AC electrical energy to DC for purposes of storage, and again from DC to AC to operate some of the vehicle's mechanical functions.

invertidor. Mecanismo usado en vehículos híbridos para convertir energía eléctrica de corriente alterna a corriente continua para propósitos de almacenaje, y nuevamente de corriente continua a corriente alterna para operar las funciones mecánicas del vehículo.

iron oxide. A residual or by-product of steel as it deteriorates or breaks down. Over a period of time it forms a loosely anchored accumulation or coating on the steel surface, which is commonly called rust.

óxido de hierro. Residuo o subproducto del acero a medida que se deteriora o descompone. Durante un periodo de tiempo éste forma una acumulación débilmente anclada o cobertura sobre la superficie del acero, la cual es comúnmente llamada herrumbre.

isocyanates. A chemical additive used in paints to promote drying and curing; known to cause respiratory problems from long-term exposure.

isocyanates. Aditivo químico usado en pinturas para facilitar el secado y la curación; éstos son conocidos por causar problemas respiratorios por exposición de largo plazo.

J

jamming and edging. The process in which, prior to installing a replacement part onto a vehicle, the edges and underside of the panel are painted to match the color of the vehicle.

atascado y bordeado. Proceso en el cual, antes de la instalación de un repuesto en un vehículo, los bordes y la parte oculta del panel son pintados para emparejar el color del vehículo.

job costing. A method by which a repair facility breaks down the cost of a repair in order to determine if a profit will be made and, if so, how much.

costo de trabajo. Método por el cual un taller de reparación separa el costo de una reparación a fin de determinar si hay ganancias y, de ser así, cuanto es.

judgment labor time. An allowance that is written into a damage report by a damage appraiser for the purpose of determining how much money to allocate for a repair operation—usually to repair a damaged panel, part, or assembly.

estimación de tiempo de trabajo. Concesión escrita en un informe de daño por un perito, para la determinación de cuanto dinero es asignado en una operación de reparación, usualmente para reparar un panel dañado, parte, o ensamblaje.

judgment replacement time. An allowance that is written into a damage report by a damage appraiser for the purpose of determining how much money to allocate for a replacement operation that does not have a predetermined crash manual replacement time.

estimación de tiempo de reemplazo. Concesión escrita en un informe de daño por un perito, para la determinación de cuanto dinero es asignado en una operación de reemplazo que no tiene un tiempo predeterminado en un manual de choques.

K

kink. A distortion in a vehicle's frame structure, sheet metal, or other substrate that forms a tight fold whose vortex is less than a 90-degree angle.

curvado. Deformación en una estructura de marco de un vehículo, lámina metálica, u otro sustrato, que forma un pliegue apretado cuyo vórtice es menor que un ángulo de 90 grados.

kink vs. bend. *See* bend vs. kink.

curvado versus doblado. Ver "doblado versus curvado."

L

labor rates. The specific amount of money allowed for each repair or replacement time increment in every repair category (such as body, refinish, and mechanical labor, as well as other areas of collision repair).

precios de trabajo. Cantidad de dinero específica asignada para cada incremento de tiempo de reparación o reemplazo en cada categoría de reparación (tales como cuerpo, acabado y trabajo mecánico, así como otras áreas de reparación de colisiones).

lacquer. A clear or colored synthetic coating that forms a glossy film as its solvents evaporate; a finish that will refold when solvents are applied.

laca. Capa sintética transparente o coloreada que forma una película brillante a medida que sus solventes se evaporan; terminado que se redobla cuando los solventes son aplicados.

laminated steel. A process used in the trunk floor, pockets for fold- down seats, and other areas of the vehicle undercarriage that involves sandwiching a layer of either soft wood or a plastic film between two thin layers of metal to reduce road noise and vehicle weight, while at the same time considerably enhancing the metal's strength.

acero laminado. Proceso usado en el piso de la cajuela, bolsillos para asientos plegables y otras áreas del tren de aterrizaje del vehículo que involucra apilamiento de capas de madera suave, o películas plásticas entre dos capas metálicas delgadas, para reducir ruido de carretera y peso del vehículo, mientras que al mismo tiempo se mejora considerablemente la resistencia del metal.

lap weld. A weld used where two pieces of metal are overlapping themselves and the weld bead is placed to penetrate them both, placing the weld bead directly between the two base metal pieces.

soldadura doblada. Soldadura usada en donde dos piezas de metal se traslapan entre ellas, y la cuenta de la soldadura es colocada para penetrarlas a ellas dos, colocándola directamente entre las dos piezas metálicas base.

laser. A concentrated beam of electromagnetic radiation, often in the ultraviolet infrared, or visible regions of the light spectrum.

láser. Viga concentrada de radiación electromagnética, a menudo en la zona infrarroja ultravioleta o regiones visibles del espectro de luz.

laser measuring system. One of several computer-driven measuring systems that may use a solid laser beam or a series of "bouncing" laser beams projected onto or through a series of targets to diagnose damage and also monitor a vehicle's undercarriage while it is being repaired.

sistema de medición láser. Uno de varios sistemas de medición manejado por computador, que puede usar una viga láser sólida o una serie de vigas láser saltando proyectadas en o a través de una serie de objetivos para diagnosticar daños, y también monitorear el tren de aterrizaje de un vehículo mientras está siendo reparado.

last-in first-out (LIFO). A repair sequence where the last damages that affected the vehicle's structure are the first repaired, and then other damage is corrected in the reverse order of how it occurred.

último en entrar, primero en salir (LIFO por su acrónimo en inglés). Secuencia de reparación en donde los últimos daños que afectaron la estructura del vehículo son los primeros en ser reparados, y posteriormente otros daños son corregidos en el orden inverso de como ocurrieron.

latch mechanism. The mechanism responsible for latching and holding the door, trunk, or hood in a closed position once it has been actuated; it may be released manually, remotely via a connecting cable, or electronically.

mecanismo de seguro. Mecanismo responsable de asegurar y sostener la puerta, cajuela, o capó, en una posición cerrada una vez que ha sido activado; puede ser liberado a mano, o remotamente vía un cable conector, o electrónicamente.

least aggressive method first. Using the product or technique that is the least damaging to the finish first when removing a defect. Many techniques for removing defects will damage the finish slightly in the process, by leaving scratches behind after their removal. That situation requires another procedure to remove those scratches, until the finish is restored. It is always easier to move to a more aggressive method if needed than to be disappointed with the amount of work needed to "repair the repair."

el menos agresivo primero. Utilización del producto o técnica que es menos perjudicial para el acabado cuando se está removiendo un defecto. Muchas técnicas para remover defectos pueden dañar ligeramente el acabado en el proceso, al dejar rasguños después de su remoción. Esa situación requiere otro procedimiento para remover aquellos rasguños hasta que el acabado es restaurado. Es siempre más fácil cambiarse a un método más agresivo si es necesario, que estar decepcionado con la cantidad de trabajo necesario para "reparar la reparación".

letdown (test) panel. An experimental paint card or chip made by a refinish technician that layers different amounts of paint material, resulting in slightly differing shades of the same color code.

panel de matices. Tarjeta o fragmento de pintura experimental hecha por un técnico en acabados que despliega varias cantidades de material pintado, resultando en matices ligeramente diferentes del mismo código de color.

leveling. Smoothing the finish on a vehicle's surface, which can only be done when the finish is in its liquid state.

nivelación de. Suavizado del acabado de la superficie de un vehículo, lo cual puede solo ser hecho cuando el acabado está en su estado líquido.

LIFO. *See* last-in first-out.

LIFO (por su acrónimo en inglés). Ver "último en entrar, primero en salir."

liftgate. The rear closure lid between the two quarter panels on an SUV which is the equivalent of the deck lid on a sedan model car. It is usually equipped with a hinged glass window which can be opened independently of the liftgate itself.

puerta levadiza. Tapa de cierre trasera entre los dos cuartos de paneles en un vehículo deportivo (SUV por su acrónimo en inglés) la cual es equivalente a la tapa de cubierta en un carro modelo sedan. Ésta está usualmente equipada con una ventana de vidrio con bisagras que puede ser abierta independientemente de la puerta levadiza misma.

lifting. A small wrinkling of the finish below the topcoat. This defect happens when strong solvents attack the underlying finish before it has a chance to fully cross-link.

levantamiento. Pequeña arruga del acabado debajo de la capa exterior. Este defecto ocurre cuando solventes fuertes atacan el acabado subyacente antes de que éste haya podido lograr el enlace cruzado.

light meter. A tool used from time to time to test the intensity (brightness) of the spray booth lights. It is also used to test the Color Rendering Index (CRI).

medidor de la luz. Instrumento usado de vez en cuando para probar la intensidad (brillantez) de las luces de cabina de rociado. Ésta es también usada para medir el índice de interpretación de color (CRI por su acrónimo en inglés).

lightweight fillers. *See* premium fillers, lightweight, conventional/standard fillers.

rellenos ligeros. Ver "rellenos premium, ligeros y convencionales o estánder."

line of sight. A sanding technique in which the technician sands in the direction that the vehicle is normally viewed in. For example, a hood should be sanded from the front to the rear, not across the hood. Likewise, a roof is not sanded across and down the sides; instead, it is sanded from front to rear.

línea de visión. Técnica de lijado en la cual el técnico lija en la dirección en la cual el vehículo es visto. Por ejemplo, un capó debe ser lijado desde el frente hacia atrás, no en forma cruzada. De igual forma, un capó no debe ser lijado a lo largo y hacia debajo de los lados; en lugar de eso, éste debe ser lijado desde el frente hacia atrás.

linear porosity. Porosity caused by contamination within the joint, root, or interbead areas.

porosidad lineal. Porosidad causada por contaminación dentro de la unión, raíz, o áreas entre cuentas.

liquid mask. A material used to cover large areas of the vehicle, or areas that technicians feel cannot be cleaned as well as they would like. The cleaned vehicle is sprayed with liquid mask. When the mask dries, the masking can be removed. The area to be finished is then exposed using normal masking materials.

mascarilla líquida. Material usado para cubrir áreas grandes del vehículo, o áreas que los técnicos sienten que no pueden ser limpiadas tan bien como ellos quisieran. El vehículo limpio es rociado con una mascarilla líquida. Cuando la mascarilla se seca, ésta es removida. El área a ser terminada es entonces expuesta usando materiales de mascarilla normales.

LKQ (like, kind, and quality). An abbreviation that stands for Like, Kind, and Quality parts, which are commonly used to describe parts manufactured by someone other than the original manufacturer.

LKQ (por su acrónimo en inglés). Acrónimo en inglés que significa parecido, amable, y de calidad, que comúnmente son usadas para describir partes fabricadas por compañías diferentes del fabricante original.

locking pliers. A tool, like the Vise-Grip®, that has a mechanism which firmly locks the jaws closed on an object, where they will remain locked until the release lever is squeezed.

tenazas de bloqueo. Herramienta, como el *Vise-Grip*®, que tiene un mecanismo que bloquea firmemente sus mandíbulas sobre un objeto en donde permanecen firmes hasta que la palanca del seguro es apretada.

lockout/tagout. A method of controlling hazardous energy by rendering a piece of equipment not usable by either tagging it as unsafe to use or preventing the power source to be activated.

asegurado y etiquetado. Método de control de energía peligrosa al etiquetar una pieza de equipo no utilizable al ya sea colocar una etiquetarla como insegura para su uso o prevenir que esa fuente de energía sea activada.

longitudinal engine. An engine that is installed running parallel to the vehicle's centerline; mostly used on rear-wheel-drive vehicles.

motor longitudinal. Motor que es instalado funcionando en paralelo a la línea central del vehículo; éste es sobretodo usado en vehículos de tracción trasera.

low carbon or mild steel. The softest of the steels used in auto manufacturing, this steel is made from a low level of carbon and can withstand yield stresses of up to 30,000 psi (pounds per square inch). Although mild steel is the easiest to form during manufacturing (and repair), it is also thicker and thus heavier.

acero suave o de bajo o contenido de carbono. El más suave de los aceros usado en la fabricación de autos; este acero está hecho de un bajo nivel de carbono y puede soportar esfuerzos de fluencia de hasta 30.000 psi (libras por pulgada cuadrada). Aunque los aceros suaves son más fáciles de formar y trabajar en fabricación (y reparación), son también más gruesos y así más pesados.

low carbon steel. A standard form of steel used for many years in the manufacture of automobiles; on today's unibody it is used for most exterior body parts that do not have a structural function.

acero de bajo carbono. Forma estándar de acero usado durante muchos años en la fabricación de vehículos; en vehículos unicuerpo modernos, éste es usado en las partes más exteriores que no tienen una función estructural.

low volume, low pressure (LVLP) guns. An alternative to HVLP spray guns that never really became popular with technicians and are rarely used.

bajo volumen, baja presión (LVLV). Alternativa para las pistolas rociadoras tipo HVLP (por su acrónimo en inglés) que realmente nunca se hicieron populares entre técnicos y son raramente usadas.

M

machine screws. Screws that are threaded into one or more parts, which draws them together and holds them in firm contact. Machine screws can have different heads designed for different types of fastening work.

tornillos de máquina. Tornillos que son enhebrados en una o varias partes, los cuales se unen y sostienen juntos en un contacto firme. Los tornillos de máquina pueden tener diferentes cabezas para diferentes tipos de trabajo de ajuste.

MacPherson strut. A suspension system used on most unibody vehicles in which the shock absorber, spring, and upper control arm are typically incorporated into one complete unit and the steering spindle or knuckle is attached directly to this unit.

puntal tipo MacPherson. Sistema de suspensión usado en la mayoría de vehículos unicuerpo en el cual el absorbedor de impacto, resorte y brazo de control superior son típicamente incorporados en una unidad completa, y el huso de conducción o nudillo está unido directamente a esta unidad.

mallets. A hammer-like tool used for pounding out material.

mazos. Herramientas parecidas a un martillo y que son usadas para aporrear un material.

manifest. A permanent file that all hazardous waste generators must maintain as proof of properly disposing of their hazardous waste.

manifiesto. Archivo permanente que todos los generadores de desechos peligrosos deben mantener como prueba de disposición apropiada de sus residuos peligrosos.

manual cleaning. Cleaning a spray gun by hand rather than using a gun washer.

limpieza manual. Limpieza de una pistola rociadora a mano en lugar de usar una lavadora de pistolas.

manual mechanical measuring system. A system that typically includes the self-centering gauges and the strut tower gauges, which may or may not be used in conjunction with each other.

sistema de medición mecánico manual. Sistema que típicamente incluye medidores que se autocentran y medidores de torre de puntal, que pueden o no ser usados en conjunto el uno con el otro.

marine atmosphere. A corrosive environment that is laden with calcium chloride (salt) such as is found along the shorelines of the oceans.

atmósfera marina. Ambiente corrosivo que está cargado de cloruro de calcio (sal) tal como el encontrado a lo largo de las costas de los océanos.

mash. A condition which results in a shortening of either the front or rear rail length. This occurs because in a frontal or rear collision they bend, buckle, and fold under to deflect and dissipate the effects of the collision forces.

aplastado. Condición que causa un acortamiento de la longitud del riel frontal o trasero. Esto ocurre a causa de una colisión frontal o trasera que los dobla, retuerce y pliega hacia abajo, para desviar y disipar los efectos de las fuerzas de colisión.

masking for blending. A masking technique in which the panel is prepared, basecoat is applied to the repaired area of a panel, and then the basecoat is blended gradually out from the repair. The remainder of the panel—or the repaired panel and adjacent panels—are then covered with clearcoat to render the blend undetectable.

enmascarado para combinación. Técnica de enmascaramiento en la cual el panel es preparado, una capa base es aplicada al área reparada del panel, y luego la capa base es combinada gradualmente fuera de la reparación. El resto del panel, o el panel reparado y paneles adyacentes, es cubierto con un recubrimiento transparente para hacer que la mezcla sea indetectable.

masking for clear blending. A masking technique which can be accomplished by preparing and exposing the area which needs to be refinished, then finding an area suitable for this type of blending process.

enmascarado para mezcla transparente. Técnica de enmascaramiento que puede ser lograda al preparar y exponer el área que necesita ser barnizada, y luego encontrar un área apropiada para este tipo de proceso de mezclado.

masking for panel painting. A masking technique in which one or more panels—such as a hood, or hood and fender—are prepared and masked so as to only expose those panels, after which they are completely finished. Parts in the areas that need to be protected from paint are removed. The remaining parts of the vehicle are then masked for protection.

enmascarado para pintar paneles. Técnica de enmascaramiento en la cual uno o varios paneles, como un capó, o capó y guardabarros, son preparados y enmascarados para exponer sólo aquellos paneles, que luego son completamente terminados. Las partes en las áreas que tienen que ser protegidas de la pintura son quitadas. Las partes del vehículo restantes son enmascaradas entonces para su protección.

masking machines. Devices that help the technician apply masking paper more quickly. The machines apply masking tape to the paper as it is being drawn from the roll.

máquinas de enmascaramiento. Dispositivos que ayudan al técnico a colocar el papel de enmascaramiento más rápidamente. Las máquinas colocan la cinta adhesiva al papel cuando éste está siendo desenrollado.

masking materials. Items used for masking that come in many different forms, such as different sizes and types of paper for different applications. There are also liquid masks and plastic drapes for protecting larger areas. Aperture tape in different sizes is used for filling gaps to keep overspray from getting in.

materiales de enmascaramiento. Artículos usados para enmascarar que vienen en muchas formas diferentes; tales como tamaños diferentes y tipos de papel para usos diferentes. También hay enmascarados líquidos y las cortinas plásticas para proteger áreas más grandes. La cinta de abertura de varios tamaños es usada para llenar huecos e impedir que excesos de rociado de pintura entren.

masking of complete finish. An overall masking technique in which the vehicle may have parts removed, after which all the remaining finish of the vehicle will be prepared and exposed to be painted. This technique is not often done in the collision repair industry, however.

enmascaramiento de acabado final. Técnica de enmascaramiento total en la cual el vehículo puede tener partes retiradas; luego, todo el acabado restante del vehículo es preparado y expuesto para ser pintado. Esta técnica no es a menudo usada en la industria de reparación de colisiones, sin embargo.

masking paper. Paper used to cover areas of the vehicle that the paint technician does not want paint to reach. Paper comes in varied widths, including 3, 6, 9, 12, 15, 18, 24, and 36 inches.

papel de enmascarar. Papel usado para cubrir áreas del vehículo a las que el técnico no quiere que llegue la pintura. Este papel viene en diferentes anchos, incluyendo 3, 6, 9, 12, 15, 18, 24, y 36 pulgadas.

masking removal. The process of removing masking after refinishing is completed. Although it should be done soon after the vehicle has cured, some technicians—if the vehicle requires sanding and buffing—will leave some masking on until those procedures are completed. Masking paper and coverings can be removed and folded against itself in large pieces, which makes their disposal easier.

remoción del enmascarado. Proceso de quitado del enmascaramiento después de que el barnizado es completado. Aunque debe hacerse tan pronto como el vehículo ha sido curado, algunos técnicos, si el vehículo requiere lijado o pulido, dejan algunas partes enmascaradas hasta que los procedimientos sobre ellas son terminados. El papel de enmascarar y las cubiertas pueden ser quitados y doblados contra sí mismos en piezas grandes, lo que hace su disposición más fácil.

masking tapes. Adhesives used when masking a vehicle. The adhesive on automotive masking tape is "pressure sensitive." This means that once applied and pressed into place, it will adhere to a grease-free and silicone-free surface until it is released or pulled off. When it is removed, the tape will not leave an adhesive residue.

cintas de enmascarar. Adhesivos usados para enmascarar un vehículo. El pegamento en la cinta de enmascarar vehículos es "sensible a la presión"; esto significa que una vez se ha colocado y presionado en el lugar, ésta se adherirá a una superficie sin grasa y sin silicona hasta que sea removida o jalada. Cuando ésta es removida, la cinta no deja residuos adhesivos.

masking techniques. The ways a technician may mask a vehicle prior to refinishing. The four main types are masking of complete finish, masking for panel painting, masking for blending, and masking for clear blending.

técnica de enmascaramiento. Formas en las que un técnico puede enmascarar un vehículo antes de barnizado. Los cuatro tipos principales son: enmascarado de acabado completo, enmascarado para pintar paneles, enmascarado para combinación, y para combinación transparente.

mass weight movement. The property that any part or assembly on a vehicle that is not directly attached to the frame or body will continue its forward momentum in a collision, often causing damage to the attaching mounts, areas immediately adjacent to it, and so on. It may also be called mass inertial damage, which could be caused by loose and unrestrained objects that become air-borne inside the vehicle or trunk during a collision.

movimiento de masas. Propiedad de que cualquier parte o ensamblaje en un vehículo, que no está directamente unido al marco o cuerpo, seguirá su avance en una colisión, a menudo causando daño a las monturas de unión, áreas inmediatamente adyacentes, etcétera. También puede ser llamado daño por inercia de masas, que podría ser causado por objetos sueltos y desenfrenados que son aerotransportados dentro del vehículo o cajuela durante una colisión.

material safety data sheet (MSDS). A document provided by the chemical manufacturer for each chemical or product they produce indicating the material's safe use and disposal.

hojas de datos de seguridad de materiales (MSDS por su acrónimo en inglés). Documento proporcionado por el fabricante químico, para cada sustancia química o producto que ellos producen, indicando el uso seguro del material y su disposición.

mechanical adhesion. A form of adhesion that occurs when applying a primer coat over a properly prepared bare metal surface or over a previously painted surface. Mechanical adhesion is achieved by sanding and abrating the surface, allowing the finish to "bite" into or anchor itself to the walls of the sand scratches.

adherencia mecánica. Forma de adherencia que ocurre cuando se aplica una capa de imprimación sobre una superficie metálica desnuda preparada apropiadamente o sobre una superficie previamente pintada. La adhesión mecánica es lograda al lijar y limar la superficie, permitiendo que el acabado "muerda" o se ancle en las paredes de los rayones de la lija.

media blasting. Using sand blasting or other less aggressive means such as peanut shells or plastic beads to completely remove a vehicle's finish. This is sometimes done with sanding using a dual action (DA) sander, with chemical removers or blasting, or by professional media blasters.

medios de remoción por chorro. Uso de chorros de arena u otros medios menos agresivos, tales como cascarones de cacahuate o cuentas plásticas, para remover completamente el acabado de un vehículo. Esto a veces es hecho con arena usando una lijadora de acción dual (DA por su acrónimo en inglés), con removedores químicos o chorros, o con otros medios profesionales de remoción por chorro.

medium wet coat. The recommendation for applying basecoat. When applied, the surface of this finish should not be applied to a gloss finish; in fact, the finish will be dull.

recubrimiento húmedo medio. Recomendación para la aplicación del recubrimiento base. Cuando es aplicado, la superficie de este acabado no debe ser aplicada en un acabado brillante; de hecho, el acabados será opaco.

melt in. A process where additives are allowed to settle properly in a wet bed of coating.

fundido. Proceso en el que a los aditivos se les permite asentarse apropiadamente en una cama húmeda de recubrimiento.

memory presets. Programmable codes used to control electric devices, such as the buttons that are programmed for specific radio stations.

preconfiguración de memoria. Códigos programables que son usados para controlar aparatos eléctricos, como los botones que son programados para determinadas estaciones de radio.

memory steering. A condition where the vehicle wants to continue to drift or pull in the same direction as a turn that was just executed. This can be caused by a damaged or bad strut bearing or one that was installed incorrectly.

manejo de memoria. Condición en la cual el vehículo desea continuar desviándose o avanzando en la misma dirección de un giro que apenas fue ejecutado. Esto puede ser causado por soporte de puntal dañado o malo, o uno que fue incorrectamente instalado.

metal cleaner/conversion coating. A metal-treating undercoating used to help replace the factory corrosion protection, which may have been removed with repairs. It consists of two parts, a metal treatment followed by a conversion coating.

recubrimiento de conversión o limpieza metálica. Recubrimiento interno que trata metales usado para ayudar a reemplazar la protección anticorrosión de fábrica, la cual puede haber sido removida con las reparaciones. Este consiste de dos partes, un tratamiento de metales seguido por un recubrimiento de conversión.

metal cutting tools. Tools used for cutting and removing sections, as well as complete metal welded panels, such as the nibbler, air impact hammer, cutoff wheel, and reciprocating saw.

herramientas de corte de metales. Herramientas usadas para cortar y remover secciones; así como paneles soldados metálicos completos, tales como el *nibbler*, martillo de impacto de aire, rueda de corte y sierra circular.

metal inert gas (MIG). *See* gas metal arc welding.

gas inerte metálico (MIG por su acrónimo en inglés). Ver "soldadura de arco metálico de gas."

metal snips. A tool commonly used to cut straight or curved shapes in sheet metal. The jaws may be serrated to assist in cutting hard or thicker sheet metal.

tijeras metálicas. Herramienta comúnmente usada para hacer cortes rectos o curvos en láminas metálica. Las mordazas pueden ser aserradas para ayudar en el corte de láminas metálicas duras o gruesas.

metallic lie. The way metallics are oriented in a coating. When a highly metallic coating is sprayed in an arching pattern, an uneven metallic lie may cause an improper color match.

tendido metálico. Es la forma como los metales están orientados en un recubrimiento. Cuando un recubrimiento altamente metálico es rociado en un patrón de arco, un tendido metálico no uniforme puede causar disparidad en el color.

metallic toners. Coatings with flakes that are extra fine, fine, medium, medium coarse, coarse, and extra coarse. Although most flakes come in the silver color, they can also come as gold flake.

tono metálicos. Capas con escamas que son extra finas, finas, medias, medias-grueso, grueso, y extra grueso. Aunque la mayor parte de las escamas son de color plata, éstas también pueden ser doradas.

metallics. Flakes of plastics that are added to a finish to provide a glittery appearance. The light is not only reflected back, but is reflected back at slightly different angles to provide the glitter effects.

metálicos. Escamas de plásticos que son añadidas a un acabado para proporcionar un aspecto brilloso. La luz no sólo es reflejada, sino que es también reflejada con varios ángulos ligeramente diferentes para proporcionar los efectos del brillo.

metallurgical structures/lattices. A facsimile illustration of the very precise and systematic formation of the molecules in metal that shows the random spacing between each structure. This spacing—together with the natural elasticity of the lattices—allows the metal to bend and flex, yet spring back to its original configuration when the force causing it to change shape is removed. However, when a sufficient amount of force is applied, it causes the metallurgical structures to become permanently distorted.

entramados o estructuras metalúrgicas. Ilustración idéntica de la formación muy precisa y sistemática de las moléculas en un metal, que muestra el espaciado aleatorio entre cada estructura. Este espaciado, junto con la elasticidad natural de los entramados, permite al metal doblarse y flexionarse, y aún así regresar a su forma original cuando la fuerza que causa su cambio de forma es removida. Sin embargo, cuando una cantidad suficiente de fuerza es aplicada, ésta causa que las estructuras metalúrgicas sean permanentemente deformadas.

metamerism. The condition of paint colors looking different in different light sources.

metamerismo. Colores de pinturas que lucen diferentes ante fuentes de luz diferentes.

metric and inch increments. The two primary measuring scales used in America. As the future of the industry points to the continued and expanded use of the metric system, technicians would do well to use the metric system when possible for all applications.

incrementos métricos y de pulgada. Las dos escalas de medición primarias usado en Norteamérica. Aunque el futuro de la industria señala el uso continuado y ampliado del sistema métrico, los técnicos harían bien al usar el sistema métrico cuando sea posible para todas las aplicaciones.

mica (pearl). A semi-transparent crystalline material that absorbs and both reflects and refracts light in a prismatic fashion.

mica (perla). Material cristalino semitransparente que absorbe, refleja y refracta la luz en una forma prismática.

mica color additive. *See* mica (pearl).

aditivo de color de mica. Ver "mica (perla)."

micro-scratches. Very small scratches.

microrasguños. Rasguños muy pequeños.

microspheres. Small round or sphere-like material used in conjunction with talc in plastic filler to give it body and substance.

microesferas. Material pequeño redondeado o en forma de esfera usado en conjunto con talco en rellenos plásticos para darle cuerpo y sustancia.

midcoat. The paint layers applied following a ground coat in multi-stage painting.

capa intermedia. Capas de pintura aplicadas siguiendo un recubrimiento firme en una pintura multigradual.

MIG brazing. Welding method that uses much lower heat than the temperatures at which high strength steel must be melted for GMAW. Although this new type of MIG brazing has been used for some time in manufacturing, repair technicians should only use this method if the repair process is recommended by the manufacturer (and if the technician has the necessary equipment).

soldando en fuerte tipo MIG (por su acrónimo en inglés). Método de soldadura que usa mucho menos calor que la temperatura en la cual el acero de alta resistencia de ser fundido para un GMAW (por su acrónimo en inglés). Aunque este nuevo tipo de soldado en fuerte tipo MIG (por su acrónimo en inglés) ha sido usado por algún tiempo en la fabricación, los técnicos de reparación no deben usar este método si el proceso de reparación es recomendado por el fabricante (y si el técnico tiene el equipo necesario).

mil. The standard unit of paint thickness (1 mil = 1/1000 inch).

mil. Unidad estándar de medición del grosor de una pintura (1 mil = 1/1000 pulgada).

mist coat. To correct streaks, these are applied (with the correct flash time between) until the striping is no longer seen.

abrigo de niebla. Para corregir rayas, éstos son aplicados (con el correcto tiempo de destello) hasta que el desmonte no sea visible.

mixing toners. A bank used to mix the needed colors.

mezcla de tonos. Banco usado para mezclar los colores necesarios.

molding. Car parts which come in many different types and have different names as well. There are body side moldings, which can be either OEM or aftermarket, as well as belt-line moldings and even windshield-reveal moldings.

moldeado. Partes de automóvil que están disponibles en muchos tipos diferentes y tienen diferentes nombres. Hay molduras laterales del cuerpo, las cuales pueden ser o OEM (por su acrónimo en inglés), o de mercado alternativo, así como molduras de línea de correa, o incluso molduras de revelación de parabrisas.

mottling. A defect that can occur in metallic paint applications. The metallic appears inconsistent in its flow, or it has a random—not smooth—appearance to the metallic lie. The appearance sometimes will look uneven, blotchy, or cloudy.

mottling. Defecto que puede ocurrir en aplicaciones de pintura metálicas. La metálica parece inconsistente con su flujo, o tiene una apariencia aleatoria, no suave, en el tendido metálico. La apariencia algunas veces lucirá dispareja, manchada o nublada.

N

needle nose pliers. Pliers that have long, thin, tapered jaws for reaching in and grasping small parts such as pins, clips, and wires. A technician should not twist a needle nose pliers, as it will spring the jaws out of alignment.

pinzas de nariz de aguja. Pinzas que tienen mandíbulas largas, delgadas y afiladas para alcanzar y agarrar pequeñas partes como alfileres, clips y alambres. El técnico no debe torcer las pinzas, ya que esto sacara las mordazas de su alineamiento.

negative airflow. A situation in which more air is leaving the spray booth than is coming in.

corriente de aire negativa. Situación en la cual más aire deja la cabina de rociamiento que el aire que entra.

negative flushness. The opposite of positive flushness, wherein the edge of one panel is recessed slightly below that of an adjoining panel for clearance purposes and to eliminate turbulence noises.

alineado negativo. Opuesto a alineado positivo; aquí el borde de un panel es retrocedido ligeramente debajo de un panel adjunto para propósitos de despeje y para evitar ruidos turbulentos.

net building. A tool used to properly align the sheet metal parts to be welded together. These are common to both sides and can easily be used for a reference point when measuring.

construcción de redes. Herramienta usada para alinear correctamente las partes de lámina metálica para ser soldadas juntas. Éstas son comunes a ambos lados y pueden ser fácilmente usadas para un punto de referencia cuando se mide.

new plastic preparation. A process where an unprimed flexible plastic part should be inspected to check for damage, and to be sure that the part is the correct one. The part should then be cleaned to assure that all mold release agents have been removed, and scuffed to provide proper adhesion.

preparación plástica nueva. Proceso en donde una parte plástica flexible no imprimada debe ser inspeccionada para buscar daños, y asegurarse de que la parte es la correcta. La parte debe ser entonces limpiada para asegurar que los agentes de liberación del molde han sido removidos, y raspados para proveer adhesión apropiada.

NFPA placard. A four-colored placard displayed at the entrance of any business that uses hazardous chemicals.

placa NFPA (por su acrónimo en inglés). Placa de cuatro colores desplegada en la entrada de cualquier negocio que usa productos químicos peligrosos.

nickel metal hydride battery (NiMH). The type of battery used in nearly all hybrid vehicles for their high voltage system. It is used to energize the vehicle when it operates off the battery.

batería híbrida de níquel-metal (NiMH). Tipo de batería usada en casi todos los vehículos híbridos en su sistema de alta tensión. Este es usado para activar el vehículo cuando no está en funcionamiento la batería.

noise, vibration, and harshness (NVH). The sound and noise level inside the cab that is created by a vehicle when it is driven down the road.

ruido, vibración, y discordancia (NVH por su acrónimo en inglés). Nivel de sonido y ruido dentro de la cabina que es creado por un vehículo cuando es conducido en un camino.

non-heat-treatable. Alloys that gain strength by work hardening; the inner panels are generally built with this class of alloys. As the sheet aluminum is stamped into its needed forms, the non-heat-treatable alloys gain strength. This group of alloys will also continue to gain strength as it is deformed by a collision, and will also continue to become stronger and more brittle as it is work hardened during the repair process.

tratamiento sin calor. Aleaciones que ganan resistencia por trabajo de endurecimiento; los paneles interiores son generalmente construidos con esta clase de aleaciones. Cuando a la lámina de aluminio se le da la forma deseada, la aleación no tratable con calor gana resistencia. Este grupo de aleaciones también continuarán ganando resistencia a medida que son deformados por una colisión, y también continuarán volviéndose más fuertes y más frágiles a medida que son endurecidas por el trabajo duro durante el proceso de reparación.

no-sand sealers. Sealers for new e-coated parts without imperfections. After being cleaned (with both soap and water, and chemically) such a part can be coated with a sealer, then topcoated.

selladores libres de arena. Selladores para partes electro-pintadas nuevas sin imperfecciones. Después de ser limpiadas (con jabón y agua, y químicamente) tales partes pueden ser cubiertas con un sellador, y luego recubiertas con la capa final.

not included operations. Tasks found in the crash guides inside the corresponding *Procedural Explanation* ("P" page) section that are not necessarily needed to complete a specific overall procedure. If deemed necessary, additional time will need to be added for such particular task(s).

operaciones no incluidas. Tareas encontradas en la guía de colisiones dentro de la correspondiente sección *Explicación de procedimientos* (Página "P") que no son necesarias para completar un trabajo específico total. Si se considera necesario, tiempo adicional debe ser añadido para tales tareas particulares.

nozzle. A part of the welding machine that can be adjusted, ideally so it is even with the contact tip, or with the contact tip about 1/8 inch below the level of the nozzle. The nozzle should be kept clear of spatter at all times; otherwise, the shielding gas may not flow properly.

inyector. Parte de una máquina de soldar que puede ser ajustada, idealmente de forma que esté nivelada con la punta de contacto, o con la punta de contacto a aproximadamente 1/8 de pulgada debajo del nivel del inyector. El inyector debe ser mantenido libre de salpicaduras en todo momento; de otra forma, el gas protector podría no fluir apropiadamente.

nuts. The devices that thread onto a bolt and, when tightened, hold the parts being fastened together.

tuerca. Dispositivos que son atornillados en un tornillo y, cuando son apretados, sostienen las partes juntas firmemente.

NVH flexible foam. A flexible open-celled foam material commonly used inside cavities, pillars, rails, and a variety of other places throughout the vehicle to reduce the noise, vibration, and harshness inside the vehicle.

espuma flexible tipo NVH. Material de espuma con celda abierta comúnmente usada dentro de cavidades, pilares, rieles y una variedad de otras piezas a lo largo del vehículo para reducir el ruido, vibración y discordancia dentro del vehículo.

NVH rigid foam. A hard foam material used to reduce the noise, vibration, and harshness of the vehicle; often placed inside cavities, between panels, and in many of the same locations as flexible foam is used.

espuma rígida tipo NVH. Material de espuma rígida usada para reducir el ruido, vibración, y discordancia del vehículo; a menudo es colocada dentro de cavidades, entre paneles y en muchos de los mismos sitios en donde la espuma flexible es usada.

Occupational Safety and Health Administration (OSHA). The agency charged with developing and enforcing regulations to ensure a safe work environment for the American labor force.

Agencia de Seguridad y Salud Ocupacional (OSHA por su siglas en inglés). Agencia encargada de desarrollar y hacer cumplir las regulaciones para asegurar un medioambiente de trabajo seguro para la fuerza laboral de Norteamérica.

OEM. *See* original equipment manufacturer.

OEM por su acrónimo en inglés. Ver "fabricante de equipo original."

offset measuring. A measuring technique used when obstacles prevent a direct, straightline measure. When placing the pointers in an offset position becomes necessary, one should use a tape measure to measure between the two pointers to achieve an accurate measurement.

medición desplazada. Técnica de medición usada cuando obstáculos impiden una medida directa en línea recta. Cuando colocar los apuntadores en una posición desplazada se vuelve necesario, se debe usar una cinta de medición para medir entre los dos apuntadores para lograr una medida exacta.

off-triggering. Adjustment while operating a spray gun that results in less paint being applied to an area. To avoid dry spray and still have the paint applied lighter at the edges of a blend, the technician should slowly let off the trigger at the outside of a blend coat.

liberación de gatillo. Ajuste realizado mientras se opera una pistola de rociado que reduce la cantidad de pintura aplicada en un área. Para evitar rociado seco y aún así lograr la aplicación más ligera de la pintura en los bordes de una combinación, el técnico debe lentamente liberar el gatillo en la parte exterior de un recubrimiento combinado.

on-demand compressor. Compressors that, instead of running continuously, come on and stay on as long as they are needed, then shut off. Therefore, they do not generally have a holding tank. Rotary vane compressors are an example.

compresor sobre pedido. Compresores que, en lugar de funcionar continuamente, funcionan y permanecen funcionando tanto como sea necesario, y luego se apagan. Por lo tanto, ellos generalmente no tienen un tanque sostenedor. Un ejemplo son los compresores de veleta.

open coat paper. Paper that has only 50 to 70 percent of the backing covered with abrasive material. This allows the material being removed to spin off with the centrifugal force of the sanding machine without clogging the sandpaper as much. Therefore, it allows the paper to last longer.

papel de recubrimiento abierto. Papel que sólo tiene del 50 al 70% de su respaldo cubierto con material abrasivo. Esto permite que el material sea removido, al girar con la fuerza centrífuga de la lijadora, sin obstruir mucho la lija. Por lo tanto, esto permite que el papel dure más.

open time. The time before an adhesive has completely set, during which adjustment of components is possible.

tiempo abierto. Tiempo antes de que un adhesivo se haya fijado completamente durante el cual el ajuste de componentes es posible.

open-end wrench. A wrench with two flat sides that grip the flat sides of square head (four-cornered) and hex head (six-cornered) nuts. They are used to loosen bolts or nuts in confined areas when there is insufficient clearance to use a box-end wrench or socket.

llave de extremo abierto. Llave con dos lados planos que agarran los lados planos de tuercas de cabeza cuadrada (de cuatro esquinas) y hexagonal (de seis esquinas). Éstas son usadas para aflojar tornillos o tuercas en áreas confinadas cuando no hay suficiente espacio para unas una llave de cubo o enchufe cerrado.

options. The accessories and/or add-on items that are part of any motor vehicle—such as CD players, air bags, rear window defogger, transmission type, and leather seats— which can affect the estimate since they can alter the other parts that may need to be replaced due to the collision event.

opciones. Accesorios y/o partes complementarias que son parte de cualquier vehículo motorizado, tales como reproductores de CD, bolsas de aire, desempañador de ventanas traseras y asientos de cuero, los cuales pueden afectar el estimado ya que ellos pueden alterar otras partes que pueden necesitar ser reemplazadas debido a un evento de colisión.

orange peel. The most common texture that appears while spraying, this finishing texture can be caused from many different conditions, such as an improperly adjusted fluid needle. It resembles the peel of an orange, and although it looks rough, it may feel smooth.

piel de naranja. Textura más común que aparece cuando se rocía; esta textura de acabado puede ser causa por diversas condiciones, tales como ajuste inadecuado de la aguja de fluidos. Ésta se parece a la piel de una naranja, y aunque parece áspera, se puede sentir suave.

organic. A chemical that contains carbon.

orgánico. Sustancias químicas que contienen el carbono.

original equipment (OE) Parts installed on the vehicle when it was originally manufactured.

equipo original (OE por su acrónimo en inglés). Partes instaladas en el vehículo cuando fue originalmente fabricado.

original equipment manufacturer (OEM). The company that manufactured a specific vehicle, and often is used to supply replacement parts if that specific vehicle is damaged.

fabricante de equipo original (OEM por su acrónimo en Inglés). Compañía que fabrica un vehículo específico, y que a menudo suministra repuestos si que el vehículo específico es dañado.

oscillating (orbital) sander. This tool is used for plastic removal and feathering the old painted surface prior to refinishing operations.

lijadora de oscilación (u orbital). Esta herramienta es usada para la remoción de plásticos y biselar superficies viejas pintadas antes de las operaciones de acabado.

outer skin. The outside layer of a door panel; commonly referred to as a door repair panel, door skin, or outer skin.

piel externa. Capa externa de un panel de puerta; comúnmente referida como reparación de panel de puerta, piel de puerta o piel externa.

overall paint job. *See* complete paint job.

trabajo de pintura global. Ver "trabajo de pintura completo."

overlap. In terms of repair estimates, it is a labor operation that is duplicated when two or more parts are serviced, replaced, or refinished at the same time. In terms of welding, it is the protrusion beyond the toe, face, or root of the weld without fusion, sometimes called rollover.

superposición. En términos de estimaciones de reparación, es una operación de trabajo que es duplicada cuando dos o más partes son atendidas, sustituidas, o barnizadas al mismo tiempo. En términos de soldadura, es la saliente más allá del dedo del pie, cara, o raíz de la soldadura sin fusión, algunas veces llamada giro.

overspray. Generally any spray dust that settles on areas of the vehicle not protected by masking. Some technicians include as overspray any finish that has settled on molding, lights, or chrome that was not properly protected by mask.

exceso de rociado. Generalmente cualquier polvo de rociado que cae en áreas del vehículo que no están protegidas por el enmascarado. Algunos técnicos incluyen como rociado excesivo cualquier terminado que ha caído sobre molduras, luces o cromados que fueron propiamente cubiertos por el enmascarado.

oxidation. A breaking down or deterioration of steel or any other substance in the presence of oxygen and an electrolyte, which is usually water.

oxidación. Descomposición o deterioro del acero o de cualquier sustancia en la presencia de oxigeno y un electrolito, el cual es usualmente agua.

oxides. A coating that reacts when exposed to air.

óxidos. Recubrimiento que reacciona cuando es expuesto al aire.

P

paint degradation. The fading of paint as a result of time, exposure, or damage.

degradación de pintura. Decoloración de una pintura como consecuencia del paso del tiempo, exposición, o daño.

paint manufacturer's code. A code needed by the technician to retrieve the formula needed to mix a vehicle's color.

código del fabricante de pinturas. Código requerido por el técnico para recuperar la fórmula necesaria para mezclar el color de un vehículo.

paint mixing rooms. Self-contained rooms used to house mixing and cleaning equipment, and to store paint. These rooms, though controlled by different state and local regulations, are generally explosion-proof and spillproof.

cuartos de mezcla de pinturas. Cuartos independientes usados para alojar equipos de mezcla y limpieza, y para almacenar pintura. Estos cuartos, aunque controlados por diferentes Estados y regulaciones locales, son generalmente a prueba de explosiones y de fugas.

paint removal. The method of stripping paint from a finished surface prior to refinishing.

remoción de pintura. Método para remover la pintura de una superficie terminada antes de la aplicación del acabado.

paint striping. When spray painting, a condition in which the fan adjustment is set fully open, causing the spray application to be slightly heavier at the top and bottom of the spray streak. This results in a streaked finish.

retirado de pintura. Cuando se pinta por rociado, esta es una condición en la cual el ajuste del ventilador está completamente abierto causando que la aplicación por rociado sea ligeramente más pesada en la parte superior e inferior del nivel de rociado. Esto resulta en un acabado rayado.

paint thickness gauge. A device that measures the amount of paint film buildup on a vehicle.

medidor de grosor de pintura. Dispositivo que mide la cantidad de acumulación de película de pintura en un vehículo.

paint transfer. Color or pigment from another vehicle or object that is left on a damaged vehicle as a result of the two vehicles impacting against each other.

transferencia de pinturas. Color o pigmento de otro vehículo u objeto que es dejado en un vehículo dañado a consecuencia del impacto entre los dos vehículos.

painter's helper. Person who assists the refinish technician.

ayudante de pintor. Persona que ayuda al técnico en acabados.

paintless dent removal (PDR). A complete method of dent removal.

remoción de abolladuras sin pintura (PDR por su acrónimo en inglés). Método completo de reparación de abolladuras.

panel. Any large exterior body part made of sheet metal plastic or other composite material.

panel. Cualquier parte grande del cuerpo exterior hecha de hojas de plástico-metal u otro material compuesto.

parallelogram/short-arm/long-arm (SLA) system. A steering system commonly used on the body over frame vehicles that consists of separate upper and lower control arms, springs, shocks, and the spindle or knuckle mounted between these two parts.

sistema de dirección tipo paralelogramo, o brazo corto/brazo largo (SLA por su acrónimo en inglés). Sistema de dirección comúnmente usado en el cuerpo sobre vehículos de marco, el cual consiste en brazos de control superior e inferior, resortes, choques y el huso o nudillo montado entre estas dos partes.

partial film removal. A process where only part of a vehicle's finish is removed prior to refinishing, depending upon its film thickness.

remoción parcial de película. Proceso en el cual sólo una parte del acabado de un vehículo es removida antes del barnizado, dependiendo del grosor de la película.

partial frame vehicle. A vehicle that incorporates an abbreviated or shortened stub frame section under the front of the vehicle, using it in the same manner for mounting steering and suspension parts as with the full frame.

vehículo de marco parcial. Vehículo que incorpora una sección de marco acortada bajo el frente del vehículo, usándola en la misma forma para montar las partes de la dirección y suspensión como se hace con el marco completo.

particle counter. A tool that measures the amount of debris in the spray booth's air. If the booth is tested and is found to have an excessive amount of dirt, the facility can either use a different type of filter or find and eliminate the source of the dirt.

contador de partículas. Instrumento que mide la cantidad de partículas en el aire de la cabina de rociado. Si el aire de la cabina es probado y éste tiene una cantidad excesiva de residuos, el taller puede o usar un tipo diferente de filtro, o buscar y encontrar la fuente de suciedad.

parts cart. A portable cart used to carry and store parts that are removed from a vehicle. Those parts placed on the cart should be labeled as to which vehicle they belong to.

carro de partes. Carro portátil usado para llevar y almacenar partes que son removidas de un vehículo. Dichas partes colocadas en el carro deben ser etiquetadas indicando a cuál vehículo pertenecen.

parts removal. The method for taking parts off a vehicle before finishing. Technicians must consider the pros and cons of masking an area versus removing a part.

retiro de partes. Método para retirar partes de un vehículo antes del acabado. Los técnicos deben considerar los pros y contras del enmascarar un área versus remover una parte.

parts storage. Keeping removed parts clearly stored and labeled for reassembly later. Parts are often stored in a parts cart.

almacenaje de partes. Guardar partes removidas limpiamente almacenadas y etiquetadas a ser reensambladas posteriormente. Las partes son a menudo almacenadas en un carro de partes.

passivation. A corrosion process wherein the metal coats itself with its own oxides as they break down to protect the metal from further corrosion. Aluminum is a classic example of a metal on which this occurs.

pasivación. Proceso de corrosión en el cual el metal se cubre con sus propios óxidos, a medida que se descompone, para proteger el metal de corrosión futura. El aluminio es un clásico ejemplo de un metal en el cual esto ocurre.

pattern (fan) control knob. A device that adjusts the amount of air to be sent through the air cap horns. The amount of air allowed to pass through the air horns will adjust the spray pattern from round to elliptical.

botón de control del ventilador o del patrón. Dispositivo que ajusta la cantidad de aire a ser enviada por los conductos de la tapa de aire. La cantidad de aire permitida que pasa por los conductos de aire ajustará el patrón de rociado de redondo a elíptico.

peeling. Occurs when a topcoat is placed on a vehicle and the adhesion—both mechanical and chemical—fails. If the surface is not abraded or cleaned properly, the new paint will not adhere to it and will peel. The finish may separate between the old finish and the new topcoat, separate near door handles or moldings because of poor adhesion, or come off in large sheets of new finish.

pelado. Ocurre cuando un recubrimiento exterior es aplicado en un vehículo y la adhesión, mecánica y química, falla. Si la superficie no es raspada o limpiada apropiadamente, la nueva pintura no se adherirá a ella y se desconchará. El acabado puede separarse entre el acabado viejo y el nuevo recubrimiento exterior, separarse cerca de manijas de puertas o molduras por la pobre adhesión, o salirse en grandes láminas del nuevo acabado.

permanent shop equipment. Tools that may be too large or bulky to keep in a technician's personal toolbox, such as a bench grinder or chop saw.

equipo de tienda permanente. Herramientas que pueden ser demasiado grandes o abultadas para guardarse en la caja de herramientas personal de un técnico, como un molino de banco o sierra de corte.

personal protection device (PPD). Equipment the technician uses to protect them from the hazards created in work environments.

dispositivo de protección de personal (PPD por su acrónimo en inglés). Equipo que los técnicos usan para protegerse de los riesgos creados en el ambiente de trabajo.

personal protective equipment (PPE). Personal or individual protective equipment used by an individual to safeguard against health issues and personal injury, such as ventilators, gloves, steel-toed shoes, and goggles.

equipo protector personal (PPE por su acrónimo en inglés). Equipo de protección personal o individual usado por un individuo para proteger su salud y heridas personales; como ventiladores, guantes, zapatos con punta de acero y gafas de protección.

pH. The acidity of a material. pH 7 is neutral.

pH. Acidez de un material. El pH 7 es neutro.

Phillips screwdriver. A screwdriver which has two tapered crossing blades on the tip which fit the four slots in a Phillips screw. The four surfaces of the screw head make it less likely for the screwdriver to slip off of the fastener.

destornillador *Phillips*. Destornillador que tiene dos láminas afiladas que se cruzan en la punta y que encajan en las cuatro ranuras de un tornillo tipo *Phillips*. Las cuatro superficies de la cabeza del tornillo lo hacen menos vulnerable a salirse del destornillador.

phosphoric acid. An etch that will leave a hard, bright metal finish because it will etch the surface slightly, exposing new, bare metal. Often this is desirable, as it leaves an attractive and ready-to-paint surface.

ácido fosfórico. Sustancia grabadora que deja un acabado metálico duro y brillante porque reacciona con la superficie ligeramente exponiendo nuevo material desnudo. A menudo esto es deseable, ya que deja una superficie atractiva y lista para pintar.

pick hammer. Used to remove small dents, this tool has a pointed tip on one end and a round smooth face on the other end. The smooth face is used to bring down high spots on a damaged panel. The pointed (pick) end is used to gently raise the low spots from the underside of the damaged panel.

martillo de pico. Usado para quitar pequeñas abolladuras, esta herramienta tiene una extremo puntiagudo y una cara ligeramente redondeada en el otro extremo. La cara lisa es usada para rebajar puntos en un panel dañado. El extremo puntiagudo (pico) es usado para elevar suavemente los puntos bajos desde la parte inferior del panel dañado.

pick tool technique. Techniques a technician must be familiar with regarding pick use. If the pick is placed in the correct spot, a dent can be raised with less force and a lower number of impacts. If the point where the pick hits the metal is incorrect, instead of repairing the damage, more damage will be created.

técnica de la herramienta de pico. Técnicas para el uso de picos con las cuales un técnico debe estar familiarizado. Si el pico es colocado en lugar correcto, una abolladura puede ser levantada con menos fuerza y un número menor de golpes. Si el punto en donde el pico golpea el metal es incorrecto, en lugar de reparar el daño, más daño es creado.

pin punch. A punch with a straight shank which can push a pin or bolt completely out of a hole.

golpe de alfiler. Golpe con un perno recto que puede sacar un alfiler o perno completamente fuera de un agujero.

pinchweld. Sometimes referred to as the windshield fence, it is a flange or lip formed on both the A and C pillars used for bonding the windshield and rear glass to the body.

pinchweld. A veces llamada cerca de parabrisas, ésta es un reborde o labio formado en los pilares A y C usados para unir el parabrisas y el vidrio trasero al cuerpo.

pinholes and air pockets. Small holes or air pockets within the mass of plastic that are caused by excessive heat and gas given off that occurs when the plastic filler is overcatalyzed.

baches y bolsas de aire. Huecos pequeños o bolsas de aire dentro de la masa de plástico que son causado por calor excesivo y gas emitido que ocurre cuando el relleno plástico es sobrecatalizado.

pinstriping. A form of decorative exterior trim involving very thin stripes. Although they can take many forms—including vinyl application, painted on, and freehand pinstriping—most of the pinstriping that will be replaced in a collision repair shop is of the vinyl type.

pinstriping. Forma de rayas exteriores decorativas que incluyen tiras muy delgadas. Aunque ellas pueden tomar muchas formas, incluyendo aplicaciones de vinilo, pintado, y arte *pinstriping* a mano libre, la mayor parte de los trabajos de *pinstriping* que serán reemplazados durante una colisión son del tipo de vinilo.

pipe wrench. Adjustable wrench with jaws used for holding and turning round objects on damaged parts.

llave de tubo. Llave ajustable con mandíbulas usada para sostener y girar objetos redondos sobre partes dañadas.

piping porosity. Porosity caused by contamination (depending on the gas type) that escapes at the same rate as the solidity of the weld pool.

porosidad de tubería. Porosidad causada por la contaminación (según el tipo de gas) que se escapa al mismo ritmo que la solidez de la soldadura.

piston compressor. The most common type of air compressor, which works much like an automobile engine, with valves and a piston that travels up and down. When the piston travels down, the intake valve opens. Outside air is drawn into the compressor. As the piston starts to travel up, the intake valve closes and the exhaust valve opens.

compresor de pistón. Tipo más común de compresores de aire, el cual trabaja muy parecido a un motor de automóvil, con válvulas y un pistón que sube y baja. Cuando el pistón baja, la válvula de consumo se abre. El aire exterior es dirigido hacia adentro del compresor. A medida que el pistón sube, la válvula de consumo se cierra y la válvula de escape se abre.

pitch of the deck lid. The height adjustment that can be made by pivoting the hood on either the front or rear bolt on a vertical hinge, causing it to be raised or lowered at either the front or rear and causing a sympathetic reaction on the opposite end.

ajuste de la tapa de la cubierta o capó. Ajuste de la altura que puede lograrse al girar el capó sobre ya sea el perno frontal o el trasero en un gozne vertical, elevándolo o bajándolo hacia el frente o hacia atrás, y causando una reacción favorable en el extremo opuesto.

plan of attack. The way the technician decides to handle masking a vehicle prior to refinishing.

plan de ataque. La manera en que el técnico decide manejar el enmascaramiento de un vehículo antes del barnizado.

plasma arc cutting. A process that super-heats a gas (very high and very concentrated), which then can carry an electrical charge. This narrow stream of electrically charged air is forced out a small hole in the cutting head, which melts the metal and blows it away from the cut area.

cortado con arco de plasma. Proceso que supercalienta un gas (muy alto y muy concentrado), lo cual puede llevar una carga eléctrica. Esta corriente estrecha de aire eléctricamente cargado es arrancada a la fuerza por un pequeño orificio en la cabeza cortante, la cual derrite el metal y lo aleja del área de cortado.

plastic/composite substrate. A substrate that is a mixture of steel and aluminum for the structural portions, with some or all of their exterior panels manufactured from plastic.

sustrato compuesto o plástico. Sustrato que es una mezcla de acero y aluminio para las partes estructurales, con unos o todos sus paneles exteriores fabricados de plástico.

plastic drape film. Sheets that come in different sizes and qualities to cover areas where paint is to be avoided. The better drape films are made of a non-porous material that resists both paint penetration and dried paint film flaking.

película de cortina plástica. Hojas que vienen en tamaños diferentes y calidades para cubrir áreas en donde la pintura debe ser evitada. Las mejores películas de cortina son hechas de un material no poroso que resiste tanto a la penetración de pintura como a desconchado de película de pintura seca.

plastic memory. A trait of plastics that, if deformed by a collision, they will remember what their original shape was and can return to that molding. This quality helps technicians as they reshape the plastic, and heat helps plastic return to its original shape.

memoria plástica. Rasgo de los plásticos que, de ser deformado por una colisión, "recordarán" su forma original y volverán a ella. Esta cualidad ayuda a los técnicos cuando ellos dan una nueva forma al plástico, y el calor ayuda al plástico a volver a su forma original.

plastic welding. A method that uses heat—either hot air or airless welding—and often a plastic filler rod to fuse broken plastic together. The fusing of plastic takes place when the base plastic is melted by the heat source and a molten puddle is formed.

soldadura plástica. Método que usa calor, ya sea aire caliente o soldadura sin aire, y a menudo una barra de relleno plástico para fundir plástico roto. La fundición de plástico ocurre cuando el plástico base es derretido por la fuente de calor y un charco fundido es formado.

plasticity. Steel's ability to bend and form into different shapes.

plasticidad. Capacidad del acero para doblarse y tomar formas diferentes.

plenum. A case like structure attached to the fire wall that houses the AC evaporator and the heater core.

plenum. Caso como la estructura unida a la pared contrafuegos que aloja el evaporador de corriente alterna y el corazón del calentador.

pliers. A tool designed for gripping, holding, and cutting objects.

tenazas. Herramienta diseñada para agarrar, sostener y recortar de objetos.

plug weld. A weld that is placed through the drilled or punched piece or pieces. The direction of travel is push or forward due to the work angle being 90 degrees to the work at the bottom of the hole where the base metal is thinnest.

soldadura de clavija. Soldadura que es colocada a través de la pieza o piezas taladradas o perforadas. La dirección de viaje es empujada o avanzada debido al ángulo de trabajo que es 90 grados al trabajo en el fondo del agujero donde el metal base es más delgado.

pneumatic tools. Power tools driven by compressed air.

herramientas neumáticas. Herramientas eléctricas que funcionan con aire comprimido.

point of impact (POI). Usually thought to be the area of the vehicle which initially came into contact or was struck by the impact object; many times it includes any areas of the vehicle that was damaged directly by the impact object.

punto del impacto. Usualmente definida como el área del vehículo que entra en contacto o fue golpeada por el objeto de impacto; muchas veces incluye cualquier área del vehículo que fue dañado directamente por el objeto de impacto.

point-to-point measurement. A measurement derived by measuring the distances between two points on the vehicle. The dimensions may be compared to known or fixed measurements published in specification manuals or a database used for this purpose.

medida de punto a punto. Medida obtenida al medir la distancia entre dos puntos en el vehículo. Las dimensiones pueden ser comparadas con medidas conocidas o fijadas, publicadas en manuales de especificación, o una base de datos usada para este fin.

polish. A soft rubbing designed to add shine to a part or vehicle.

pulido. Frotamiento suave que añade brillo a una parte o vehículo.

polisher. Device that applies a light aggressive polish.

pulidor. Dispositivo que aplica una pulido agresivo ligero.

polyester resin. A highly refined semi-liquid material used in plastic filler to suspend the talc and all other ingredients. It also reacts with a catalyst or hardener, causing it to harden or cure.

resina de poliéster. Material semilíquido muy refinado usado en rellenos plásticos para suspender el talco y todos otros ingredientes. Este también reacciona con un catalizador o endurecedor, haciéndolo endurecer o curar.

polymers. A chemical compound formed when two or more molecules combine to form larger molecules of repeating patterns.

polímeros. Compuesto químico formado cuando dos o más moléculas se combinan para formar moléculas más grandes de patrones que se repiten.

polyolefin. Any of the polymers and copolymers of the ethylene family of hydrocarbons. This group of thermoplastic plastics is often used to make flexible bumpers called fascias.

polyolefin. Cualquiera de los polímeros y copolímeros de la familia etilena de los hidrocarbonos. Este grupo de plásticos termoplásticos a menudo es usado para hacer parachoques flexibles llamados fajas.

poor hiding. Appears when too little film thickness is applied over the repair, and in bright light the thin topcoat can be seen through.

ocultamiento pobre. Aparece cuando muy poco grosor de película es aplicado sobre la reparación, y en luz brillante se puede ver a través de la delgada capa superior.

porosity. A situation when gas becomes trapped in the weld as it is being formed. That trapped gas then develops into small holes, which can be either spherical or cylindrical in shape.

porosidad. Situación en la que un gas es atrapado en la soldadura cuando ésta está siendo formada. Aquel gas atrapado entonces se desarrolla en pequeños orificios, que pueden tener forma esférica o cilíndrica.

positive airflow. A situation where more air is coming into the spray environment than is leaving.

flujo de aire positivo. Situación en la cual más aire entra en el ambiente de rociamiento que el aire que sale.

positive flushness. Raising or moving the edge of a component panel—such as a fender—higher than the edge of a functional or movable panel—such as the door—to prevent the two edges from rubbing or brushing against each other when opening or closing the movable panel.

ajuste positivo. Levantamiento o movimiento del borde de un panel componente, tal como un guardabarros, más alto que el borde de un panel funcional o movible, tal como una puerta, para prevenir que los dos bordes se rocen o se rayen entre si cuando se abre o cierra el panel móvil.

positive/solid anchoring system. A system in which a unibody vehicle is typically anchored to the bench by clamping it at the pinchweld with special pinchweld clamps. Once the vehicle is attached to the pinch weld clamps, they are bolted to securely lock them down onto the bench. This holds the vehicle and prevents it from moving as the corrective forces are applied.

sistema de anclaje positivo o sólido. Sistema en el cual un vehículo unicuerpo es típicamente anclado al banco al sujetarlo con abrazaderas especial de pinchweld. Una vez que el vehículo es atado con las abrazaderas de pinchweld, éstas son remachadas para asegurarlas al banco. Esto sostiene el vehículo y evita su movimiento cuando las fuerzas correctivas son aplicadas.

pot-life. The time a liquid is sprayable before it hardens into a solid or semi-solid.

pot-life. Tiempo en el que un líquido se puede rociar antes de que se endurezca y quede sólido o semisólido.

pounds per square inch (psi). The measure of air pressure.

libras por pulgada cuadrada (psi). Medida de la presión atmosférica.

powder coating. A process that uses dry finish that turns to liquid when heated. Then the liquid flows over the entire part. As it cools, the finish becomes a solid, providing a very hard and durable surface.

recubrimiento en polvo. Proceso que usa un acabado en seco que se vuelve líquido cuando se calienta. Luego el líquido fluye sobre toda la parte. Cuando se enfría, el acabado se endurece, proporcionado una superficie muy dura y durable.

power supply. A device that changes alternating current to direct current for some power tool applications.

suministro de energía. Dispositivo que cambia de corriente alterna a corriente directa para el uso de algunas herramienta eléctricas.

power tools. Tools energized by electricity, batteries, air pressure, or hydraulics.

herramientas eléctricas. Herramientas activadas por electricidad, baterías, presión de aire o hidráulica.

Pozidrive screwdriver. Similar to a Phillips screwdriver, although the tip is flatter and blunter, this tool provides a more positive connection between the driver and the screw.

destornillador tipo Pozidrive. Similar a un destornillador *Phillips*, aunque con la punta más llana y embotada; ésta herramienta proporciona una conexión más positiva entre el impulsor y el tornillo.

pre-accident condition. Sometimes called pre-loss condition, it is the end goal of a repair—to have the vehicle as structurally sound and visually appealing as it was before the collision.

condición anterior al accidente. A veces llamada condición anterior a la pérdida, ésta es la meta final de una reparación, lograr que el vehículo esté estructural y visualmente bien como estaba antes de la colisión.

pre-filters. A device that screens out contaminants before the filter gets to them.

prefiltros. Dispositivo que retiene contaminantes antes de que éstos alcancen el filtro.

pre-loss condition. *See* pre-accident condition.

condición anterior a la pérdida. Ver "condición anterior al accidente."

pre-masking cleaning. A thorough cleaning performed before masking a vehicle, as adhesion works best against a clean surface.

limpieza anterior al enmascarado. Limpieza cuidadosa realizada antes de enmascarar un vehículo, ya que la adherencia trabaja mejor en una superficie limpia.

premium fillers, lightweight, conventional/standard fillers. The three classes of plastic fillers commonly used in the collision repair industry. Premium is the best quality filler, the lightweight is a mid-range quality, and the conventional filler is of lesser quality than the other two.

rellenos premium, ligeros y convencionales o estándar. Éstas son las tres clases de rellenos plásticos comúnmente usados en la industria de reparación de colisiones. El relleno *premium* es el de mejor calidad, el ligero tiene calidad media, y el convencional es de menor calidad que los otros dos.

prep decks. Areas that contain the vehicle as it is being sanded; they serve to protect other technicians from dust contamination as the vehicle is being prepared for paint.

cubiertas de preparación. Áreas que alojan al vehículo cuando es lijado; éstas sirven para proteger a otros técnicos de la contaminación de polvo cuando el vehículo está en preparación para la pintura.

pressure feed gun. A spray gun in which air is diverted from the gun to the pot (paint vessel), where pressure is applied. Pressure forces paint through a hose to the gun. This pressure delivery system allows the gun to be even more transfer efficient than either the gravity feed or the siphon feed gun.

pistola alimentada por presión. Pistola rociadora en la cual el aire es desviado de la pistola hacia el depósito (contenedor de pintura) en donde la presión es aplicada. La presión empuja a la pintura a través de un agujero en la pistola. Este sistema de entrega de presión permite que la pistola sea aún más eficiente que, ya sea el alimentador por gravedad, o la pistola alimentada por sifón.

primary colors. The colors which, when mixed in different combinations, make all other colors: red, yellow, and blue.

colores primarios. Colores que, cuando son mezclados en combinaciones diferentes, forman todos los demás colores: rojo, amarillo, y azul.

primed plastic preparation. The steps to prepare new plastic parts that are delivered in primer, which include inspection, cleaning, sanding and scuffing, sealing, final cleaning, then topcoating.

preparación plástica primaria. Pasos para preparar nuevas partes plásticas que son entregadas con imprimación; lo que incluye la inspección, limpieza, lijado y raspado, sellado, limpieza final y aplicación de recubrimiento final.

primer masking. Done mostly to prevent primer from getting on unintended parts of the vehicle, this masking process exposes the panel to be primed; all other areas are covered (especially openings like door cracks, so the primer will not get inside the door jambs).

enmascarado de imprimación. Hecho principalmente para prevenir que la imprimación llegue a partes no requeridas del vehículo; este proceso de enmascarado expone el panel que va a recibir la imprimación; todas las otras áreas son cubiertas (especialmente aberturas como grietas de puertas, de forma que la imprimación no llegue a las manijas de la puerta).

primer sealers. A coating applied following a primer, or after a previously painted surface has been prepared for paint, but before the topcoat is applied. It acts as a barrier coat, stopping the solvents in the topcoat from penetrating its surface, which could possibly cause adverse reactions.

sellador de imprimación. Recubrimiento aplicado seguido de una imprimación, o después de que una superficie pintada previamente haya sido preparada para la pintura, pero antes de que la capa sea aplicada. Ésta actúa como una barrera, evitando que los solventes en la capa superior penetren su superficie, lo cual podría causar posiblemente reacciones adversas.

primer surfacer. Undercoats that provide adhesion to the surface they are sprayed on and adhesion to that which is sprayed over them. They also provide light filling qualities, which means that when applied to an area they will provide a film build that can be sanded to remove small imperfections.

imprimación en superficie. Recubrimientos interiores que proveen adhesión a la superficie en donde han sido aplicados y adhesión a lo que sea rociado sobre ellos. Éstos también proveen calidades de relleno ligero, lo cual significa que cuando éstos son aplicados en un área proveen una película que puede ser lijada para remover pequeñas imperfecciones.

principal control points. The four corners under the passenger compartment, commonly referred to as the vehicle base, from which all measurements originate and are taken.

puntos de control principal. Las cuatro esquinas bajo el compartimiento de pasajeros, comúnmente llamado base del vehículo, desde las cuales todas las medidas se originan y son tomadas.

prismatic color additive. Manufactured from specific color-shifting products, this additive will change as much as five color shifts depending on the added material. Prismatics are not affected by orientation, and therefore do not present the same challenges as do metallics and mica.

aditivo de color prismático. Fabricado de productos específicos que cambian el color, este aditivo cambiará tanto como cinco cambios en color, según el material añadido. Los prismáticos no son afectados por la orientación, y por lo tanto no presentan los mismos desafíos que tienen los metálicos y la mica.

prismatic colors. Paints which will shift color, depending on the angle at which the sun hits the surface and on where the viewer is located. This color has layers of reflective and semi-transparent (refracting) metals suspended in it.

color prismático. Pinturas que cambiarán el color, según el ángulo en el cual el sol golpea la superficie y en donde el espectador esté localizado. Este color tiene capas de metales (de refracción) reflexivos y traslúcidos suspendidos en él.

prism-effect flakes. Semi-transparent flakes used in pigments that, depending on the angle they are viewed from, will shift color and reflect and refract different colors. Some prismatic flake pigments can shift as much as five different colors as one moves around the vehicle and view it from different angles.

escamas con efecto de prisma. Escamas semitransparentes usadas en pigmentos que, dependiendo del ángulo en que sean vistas, cambiarán el color, y reflejarán y refractarán diferentes colores. Algunos pigmentos de escamas prismáticas pueden cambiar tanto como cinco colores diferentes a medida que uno se mueve alrededor del vehículo y ve desde diferentes ángulos.

procedural/"P" pages. The section of a crash manual or database that contains explanations and information concerning the use of crash guides, such as definitions, guides to symbols, and parts information.

páginas ("P") de procedimientos. Sección de un manual de accidentes o base de datos que contiene explicaciones e información acerca del uso de guías de accidente, tales como definiciones, guías de símbolos, e información de partes.

production sandpaper. A coarse grade of sandpaper, generally from 24 to 80 grit, used for rough sanding plastic filler.

papel de lija de producción. Grado grueso de papel de lija, generalmente de grado 24 a 80, usada para relleno de plástico de lija gruesa.

prybar. Sometimes called "picks," these tools are used to reach into restricted areas. Prybars can raise low spots in enclosed components, such as doors, quarter panels, and hoods. No holes should be made to access the damaged area.

prybar. A veces llamado "elecciones", estos instrumentos son usados para meter la mano en áreas restringidas. Los *prybars* pueden levantar puntos bajos en componentes incluidos, como puertas, cuarto de paneles, y capós. Ningún agujero debe hacerse para tener acceso al área dañada.

pull rods. Similar to dent pullers, these tools have a handle on one end of the rod and a hook on the other end. Holes are drilled in the dented area. The pull rod hook is inserted into the hole and pulling force is applied while lightly tapping around the area with a hammer.

varas de tensión. Similares a las poleas dentadas, estas herramientas tienen un mango en un extremo de la vara y un gancho en el otro extremo. Los agujeros son taladrados en el área abollada. El gancho de la vara de tensión es insertado en el agujero y la fuerza de tensión es aplicada mientras se golpea ligeramente alrededor del área con un martillo.

pulsed-spray metal transfer. A welding technique that uses a pulsing current to melt the filler wire and allow one small molten droplet to fall with each pulse. The pulse provides a stable arc and no spatter, since no short-circuiting takes place.

transferencia de metal. Técnica de soldadura que usa una corriente de pulso para derretir el alambre de relleno y permite que una pequeña gotita fundida caiga con cada pulso. El pulso provee un arco estable y ninguna salpicadura, ya que ningún corto circuito ocurre.

punches. A tool used for driving out pins and bolts, as well as aligning holes during component assembly.

perforadoras. Instrumento usado para sacar alfileres y pernos, así como para alinear agujeros durante el ensamblaje de un componente.

purge cycle. The painting stage at which the booth will draw in fresh air and maintain the proper temperature, which is often slightly higher than spray cycle temperatures. This allows the booth to run in at this setting for a specific set amount of time.

ciclo de purga. Etapa de pintado en la cual la cabina dirige el aire fresco y mantendrá la temperatura apropiada, la que a menudo es ligeramente más alta que las temperaturas del ciclo de rociado. Esto permite que la cabina funcione de esta forma por un periodo de tiempo específico.

purge time. The stage during which solvents are gassed out of the finish more slowly than in the first stage. This amount of time is sometimes called the flash time.

tiempo de purga. Etapa durante la cual los solventes se evaporan del acabado más lentamente que en la primera etapa. Esta cantidad de tiempo es a veces llamada el tiempo de destello.

Q

quarter glass. Fixed windows in vehicles that do not open or close, often found in the C pillar just forward of the rear windows.

vidrio cuarto. Ventanas de vehículos fijas, que no se abren o se cierran; éstas son a menudo encontradas en el pilar C justo adelante de las ventanillas traseras.

quarter panel. The large exterior panel extending from the rear door to the end of the vehicle and upward to connect the roof.

panel cuarto. Panel exterior grande que se extiende desde la puerta trasera hasta el final del vehículo, y hacia arriba para conectar el techo.

quick connect couplers. A connecting device to join the compressor and air hose, which must be adequate for the air volume generated.

acopladores de unión rápida. Dispositivo de conexión para unir el compresor y la manguera de aire, el cual debe ser adecuado para el volumen de aire generado.

R

rack-and-pinion. A steering system that has a pinion gear on the end of the steering shaft that mates with a rack; when the steering wheel turns, the pinion turns, and the rack moves to the left or the right.

piñón y cremallera. Un sistema de dirección que tiene un engranaje piñón en la punta del eje de la columna de dirección que se acopla con una cremallera; cuando el volante gira, el piñón gira, y la cremallera se mueve a la izquierda o a la derecha.

radiator. One of the principal parts of the vehicle's cooling system, which is placed in the primary airflow area in the front of most vehicles and is used to dissipate the heat from the coolant that is circulated throughout the engine for cooling purposes.

radiador. Una de las partes principales del sistema de enfriamiento del vehículo, que es colocado en el área de corriente de aire primaria delante de la mayor parte de vehículos, y que es usado para disipar el calor desde el refrigerante que es puesto en circulación en todas partes del motor con el propósito de enfriar.

rail dust. A deposit of metal fragments into the paint on a vehicle. This occurs when the car is transported by rail or parked near railroads.

polvo ferroso. Depósito de fragmentos metálicos en la pintura en un vehículo. Esto ocurre cuando el vehículo es transportado por ferrocarril o estacionado cerca de ferrocarriles.

random orbital sander/buffer. Tools used for sanding operations ranging from feathering sand scratches to removing sand scratches where plastic filler was added. Using an orbital applicator reduces the possibility of swirl marks.

pulidora orbital o aleatoria. Herramienta usada para operaciones de lijado que van desde rayones de arena seccionados hasta la remoción de rayones de arena en donde rellenos plásticos han sido añadidos. Usar una pulidora orbital reduce la posibilidad de dejar marcas de remolino.

ratchet handle. A part of the ratchet that enables the removal or tightening of a fastener without removing the socket from the fastener. The reversing lever allows the turning direction to be changed to loosen or tighten nuts and bolts.

mango de trinquete. Parte del trinquete que permite el retiro o el apretamiento de un sujetador sin quitar el enchufe del sujetador. La palanca que pone marcha atrás permite que la dirección que da vuelta sea cambiada para soltar o apretar nueces y cerrojos.

ready to spray (RTS). A liquid that is able to be applied without any preliminary preparation.

listo para rociar (RTS por su acrónimo en inglés). Líquido que es capaz de ser aplicado sin ninguna preparación preliminar.

reciprocating. To alternately move forward and backward.

intercambiado. Moverse hacia delante y hacia atrás alternativamente.

re-collision integrity. A desired state in which a repaired vehicle, if involved in a subsequent collision, is able to withstand the same degree of collision forces. Each structural member should respond and react in the same identical manner as the manufacturer designed it to do.

integridad de nueva colisión. Estado deseado en el cual un vehículo reparado, de estar implicado en una colisión subsiguiente, es capaz de resistir el mismo grado de fuerzas de colisión. Cada miembro estructural debe responder y reaccionar en la misma manera en que fue originalmente diseñado por el fabricante.

recycled parts. Parts that have been previously utilized on—and taken off—another vehicle in order to replace parts that have been damaged (usually in lieu of utilizing new replacement parts).

partes recicladas. Partes que han sido utilizadas antes en otro vehículo, y que han sido quitadas, a fin de sustituir partes que han sido dañadas (por lo general en lugar de la utilización de repuestos nuevos).

reducer. Thinning agent used when enamel is to be thinned.

reductor. Agente reductor usado cuando un esmalte debe ser reducido.

reduction by parts. A common recommendation when two or more additives are called for. If paint requires both reducer and a hardener, the reduction requirement may be something like 2:2:1 reduction. That means two parts of paint has two parts of reducer added, then 1 part of hardener, followed by stirring or agitation to accomplish a proper reduction for application.

reducción por partes. Recomendación común cuando dos o más aditivos son requeridos. Si la pintura requiere tanto un reductor, como un endurecedor, el requerimiento de reducción puede ser algo como una reducción 2:2:1. Esto significa que dos partes de la pintura tienen dos partes del reductor añadido, y entonces 1 parte de endurecedor, luego es mezclado o agitado para lograr una reducción propia para la aplicación.

reduction by percentages. A method of reducing where the amount of paint to be reduced is measured. Then the amount of reducer that is placed in the mixture is a percentage of the paint.

reducción por porcentajes. Método de reducción en donde la cantidad de pintura a ser reducida es medida. Luego la cantidad de reductor que es colocada en la mezcla es un porcentaje de la pintura.

reduction by weight. The most accurate method of measurement, and often the method recommended by the paint manufacturer. In this method, reduction is done on the same scale by which the paint formula is measured.

reducción por peso. Método más exacto de medida, y a menudo el método recomendado por el fabricante de pinturas. En este método, la reducción es hecha sobre la misma escala en la cual la fórmula de la pintura es medida.

re-engineering. Any modifications or alterations made to a vehicle, such as reinforcing an area that has been repaired by welding another layer of material over it for strength. This must be avoided when making repairs on any structural parts as it will alter the type of reaction and the reaction timing of a part in a collision situation.

reingeniería. Cualquier alteración o modificaciones hechas a un vehículo, como el refuerzo de un área que ha sido reparada al soldar otra capa del material sobre ella para aumentar la resistencia. Esto debe ser evitado cuando haciendo reparaciones en cualquier parte estructural, la modificación cambia el tipo de reacción y el cronometraje de reacción de una parte en una situación de colisión.

reference points. Areas on the vehicle that are used when making measurements.

puntos de referencia. Áreas en el vehículo que son usadas cuando se hacen medidas.

reflection. The process in which wavelengths of sunlight are thrown back from a surface or object upon striking it, which results in the appearance of color.

reflexión. Proceso en el cual las longitudes de onda de la luz del sol son devueltas desde una superficie u objeto al golpearlo, lo que causa el aspecto de color.

refraction. The turning or bending of a wavelength of sunlight when it passes from one medium to another of different optical intensity, such as when light travels through metallic or mica flakes in a vehicle's finish.

refracción. Giro o flexión de una longitud de onda de luz solar cuando ésta pasa de un medio a otro de intensidad óptica diferente, tal como cuando la luz viaja a través de escamas metálicas o de mica en el acabado de un vehículo.

remove and install (R&I). The removal of items such as light assemblies, mirrors, and body moldings to aid in either the repair or refinish of a vehicle, and the reattachment of these items at a later date.

remueva e instale (R&I). Remoción de artículos, como ensamblajes ligeros, espejos, y molduras de cuerpo, para ayudar en la reparación o barnizado de un vehículo; y la reinstalación de estos artículos posteriormente.

remove and replace (R&R). The removal of items which will need to be replaced, rather than reinstalling the original removed piece.

remueva y reemplace (R&R). Retiro de artículos que tendrán que ser sustituidos, en lugar de instalar de nuevo la pieza original removida.

repair mapping. A subtle texture in the topcoat outlining the area where plastic filler has been applied which appears after the primers, topcoat, and clearcoat have been applied; usually caused by solvent penetration into the feathered edge of the plastic, causing it to slightly lift off the surface.

mapa de reparación. Textura sutil en el contorno del recubrimiento final en el área en donde el relleno plástico ha sido aplicado, lo cual aparece después de que la imprimación, capa superior y capa transparente han sido aplicados; ésta es usualmente causada por la penetración de un solvente en el borde seccionado del plástico, causándole un levantamiento ligero de la superficie.

repair/work order. A specific plan of action, typically created off the damage report, which the technician works off of when repairing a vehicle.

orden de trabajo o reparación. Plan específico de acción, típicamente creada aparte del informe de daño, en el cual el técnico planea cuando reparar un vehículo.

repair plan. A systematic, well-arranged procedure for repairing a collision-damaged vehicle.

plan de reparación. Procedimiento sistemático y ordenado para reparar un vehículo dañado por colisión.

repaired plastic preparation. A preparatory process that is more complicated than working with new plastic, due to the potentially different levels of the surface, which will need to be made level through using primer filler and blocking. The steps in this process are cleaning, feather edging, sanding, re-cleaning, applying adhesion promoter, priming, blocking, sanding/scuffing, and final cleaning.

preparación plástica reparada. Proceso preparatorio que es más complicado que trabajar con el nuevo plástico, debido a los niveles potencialmente diferentes de la superficie, lo cual necesitará ser hecho completamente nivelado usando un relleno de imprimación y bloqueo. Los pasos en este proceso son limpieza, biselado de bordes, lijado, nueva limpieza, aplicado del promotor de adherencia, imprimación, bloqueo, lijado/rayado, y limpieza final.

repaired to pre-accident condition. The goal of the refinish technician. Both structurally and cosmetically, the vehicle should be returned to the customer in the same condition it was prior to the collision, if at all possible.

reparado a la condición anterior al accidente. Objetivo del técnico en acabados. Tanto estructural como cosméticamente, el vehículo debe ser devuelto al cliente en la misma condición que tenía antes de la colisión, de ser posible.

reserve. A portion of finances set aside by an insurance company's claim handler exclusively to cover the financial loss suffered by an insured party as a result of a claim submitted to them.

reserva. Parte de las finanzas separadas por un agente de una compañía de seguros para exclusivamente cubrir la pérdida financiera sufrida por un asegurado como resultado de un reclamo enviado.

respirator. Personal protective equipment designed to supply clean air to the technician.

respirador. Equipo de protección personal que suministra aire limpio al técnico.

reverse blending. A process similar to standard blending, but applied in reverse. The first coat is applied to the anticipated largest area with the outer edge being off-triggered to apply a lighter amount of color. The second coat is applied to an area smaller than the first, but with full coverage over the repair area, and so on.

combinación al revés. Proceso similar a la combinación estándar, pero aplicado al revés. El primer recubrimiento es aplicado al área más grande anticipada con el borde externo fuera del objetivo de rociado para aplicar una cantidad más ligera de color. El segundo recubrimiento es aplicado a un área más pequeña que la primera, pero con completa aplicación sobre el área de reparación, etcétera.

reverse masking. Masking technique in which the masking paper is applied in the opposite direction as it would normally. Then it is brought back upon itself until a slight amount of tape is exposed. This helps eliminate a hard tape line from forming.

enmascarado al revés. Técnica de enmascaramiento en la cual el papel de enmascaramiento es aplicado en dirección contraria a la normal. Luego es regresada sobre sí misma hasta que una ligera cantidad de cinta esté expuesta. Esto ayuda a eliminar la formación de una línea de cinta dura.

revolutions per minute (rpm). A measure of vehicle rotational speed.

revoluciones por minuto (rpm). Medida de la velocidad rotatoria del vehículo.

right to know. Training designed to inform employees of the safe handling and use of all hazardous material in a work area. All employers are required to perform this training.

derecho a saber. Entrenamiento diseñado para informar a empleados el manejo seguro y el uso de todo el material peligroso en un área de trabajo. Se requiere que todos los patrones realicen esta formación.

rigid plastic. A category of plastics, such as polycarbonate, sheet moldable compound (SMC), and fiberglass, which is not very bendable.

plástico rígido. Categoría de plásticos, como el policarbonato, láminas de compuesto moldeables (SMC por su acrónimo en inglés), y fibra de vidrio, que no es muy flexible.

rivet bonding. One of the methods used to attach a replacement aluminum part using a combination of adhesives and rivets. This method may be used on exterior panels as well as select structural members.

unión con remache. Uno de los métodos usado para unir un repuesto de aluminio usando una combinación de adhesivos y remaches. Este método puede ser usado en paneles exteriores, así como en miembros estructurales selectos.

rock guard. A coating that can be applied either before topcoating or following the application of the finish. It protects the areas of potential impact with a coating which, because of its flexibility, can resist impact.

protección de roca. Capa que puede ser aplicada antes del recubrimiento superior o después de la aplicación del acabado. Esto protege las áreas de impacto potencial con una capa que, debido a su flexibilidad, puede resistir impactos.

rocker panel. The long narrow sheet metal panels immediately below the doors that extend from beneath/inside the lower front fender and meld into the lower front of the quarter panels; the primary structural and reinforcement members of the unibody design.

panel de apoyo. Paneles largos y estrechos de lámina metálica que están inmediatamente debajo de las puertas que se extienden desde debajo/adentro del guardabarros delantero inferior y se une en el frente inferior de los paneles cuartos; la estructura primaria y miembros de refuerzo del diseño unicuerpo.

rolled buckle. Secondary damage that occurs either from metal being pulled inward (collapsed rolled buckle) or by being pushed upward (simple rolled buckle) by forces from the primary damage. By relieving the pressure, the rolled buckle will be relieved.

broche enrollado. Daño secundario que ocurre ya sea cuando un metal es jalado hacia arriba (broche enrollado colapsado), o cuando está siendo empujado hacia arriba (broche enrollado simple), por fuerzas del daño primario. Al liberar la presión, el broche enrollado es liberado.

roll priming. A method proven to speed up the application process. When the material is applied with a roller, little or no masking is needed. Roller application of primer filler can be done outside of a booth.

imprimación de rollo. Método probado para acelerar el proceso de aplicación. Cuando el material es aplicado con un rodillo, poco o ningún enmascaramiento es necesario. La aplicación de rodillo del relleno de imprimación puede ser hecha fuera de una cabina.

root gap. The area between the two edges that are to be welded together securing the replacement rail to the existing one on the vehicle. This distance may be specified by the vehicle manufacturers or it may be found in the I-CAR Uniform Procedures for Collision Repair guidelines for replacing a rail section.

holgura de raíz. Área entre los dos bordes que deben ser soldados juntos asegurando el riel de reemplazo al existente en el vehículo. Esta distancia puede ser especificada por el fabricante del vehículo, o puede ser encontrado en el Guia de procedimientos uniformes *I-CAR* para reparación de colisiones para sustituir una sección de riel.

rotary screw compressor. The compressor of choice for industrial applications. They provide pulsation-free, continuous air, and are energy efficient at full load. Rotary screw compressors operate by having air enter a sealed chamber, where it is trapped by two contra-rotating rotors. With a screwing motion, the rotors compress the air as it passes through.

compresor de veleta rotatoria. Aunque no es comúnmente usado para producir los volúmenes de alta presión (arriba de 100 psi) necesarios para la reparación de colisiones en los talleres, éstos compresores a menudo son usados en la producción de aire respirable clase D para respiradores de aire suministrado.

rotary vane compressor. Although not commonly used to produce the volumes of high (above 100 psi) pressures needed for the collision repair facility's compressed air needs, these compressors are often used in producing class D breathable air for supplied air respirators.

compresor de tornillo rotatorio. Compresor de opción para aplicaciones industriales. Éstos proporcionan aire continuo y libre de pulsaciones, y son eficientes energéticamente en la carga máxima. Los compresores de tornillo rotatorios funcionan al hacer entrar el aire en una cámara sellada, en donde es atrapada por dos rotores de rotación contraria. Con un movimiento de atornillado, los rotores comprimen el aire a medida que pasa.

rotating bell. A spray application method normally used in manufacturing conditions. It is often found either (1) mounted stationary, while the vehicle is moved under it; or (2) on the arm of a robot which travels in a set pattern to apply coatings to a stationary vehicle.

campana rotativa. Método de aplicación de rociado normalmente usado en condiciones industriales. Éste es a menudo encontrado ya sea 1) montado inmóvil, mientras que el vehiculo es movido bajo él; 2) en el brazo de un robot que viaja con un patrón fijo para aplicar recubrimientos a un vehículo fijo.

rough service spoons. A tool with an offset shape which is designed to be struck with a hammer to rough out damaged sheet metal.

cucharas de uso rudo. Herramientas con forma compensada la cual está diseñada para ser golpeada con un martillo para moldear láminas metálicas dañadas.

route of entry. A path by which a hazardous chemical can get into the body, such as inhalation, ingestion, absorption, and injection.

ruta de entrada. Camino por el cual una sustancia química peligrosa puede entrar en el cuerpo; como inhalación, ingestión, absorción, e inyección.

ROY G BIV. A mnemonic device developed to remember the colors in the light spectrum and their order: red, orange, yellow, green, blue, indigo, and violet.

ROY G BIV. Dispositivo memotécnico desarrollado para recordar los colores en el espectro de luz y su orden: rojo, anaranjado, amarillo, verde, azul, añil, y violeta.

rubber mallet. A hammer-like tool that can be used to gently bump sheet metal or position trim and delicate parts without damaging the finish.

mazo de caucho. Herramienta parecida a un martillo que puede ser usada para dar golpes suavemente en láminas metálicas, o posicionar partes esbeltas y delicadas sin dañar el acabado.

runs. A paint defect caused by paint from heavy application finishes being pulled down by gravity.

carreras. Defecto de pintura causado por aplicar acabados de aplicación pesados jalados hacia abajo por la gravedad.

rust. The end result of oxidation; the term may come from the color of steel after oxidation has formed on the surface.

herrumbre. Resultado final de la oxidación; el término puede venir del color del acero después de que la oxidación se ha formado en la superficie.

S

sacrificial corrosion. A form of corrosion in which one metal—such as zinc—oxidizes, sacrificing itself to form a protective coating over another metal, thus slowing down the deterioration of the first metal.

corrosión sacrificial. Forma de corrosión en la cual un metal, como el zinc, se oxida, sacrificándose a sí mismo, para formar una capa protectora sobre el otro metal, así se hace más lenta la deterioración del primer metal.

SAE. *See* Society of Automotive Engineers.

SAE (acrónimo en inglés). Ver "Sociedad de Ingenieros Automotrices."

sag/kickup. Two very similar types of frame and undercarriage damage. The sag is characterized by the outer end of the rail on one side of the vehicle being pulled down; a kickup condition is characterized by the end of the rail being pushed up. Both give a similar visual impression that one side is raised higher than the other.

comba/kickup. Dos tipos muy similares de daño de marco y de tren de aterrizaje. La comba está caracterizada por la caída del extremo externo del riel sobre un lado del vehículo; una condición *kickup* está caracterizada por el empuje hacia arriba del extremo del riel. Ambos tipos dan una impresión visual similar que un lado es elevado más arriba que el otro.

sags. A paint defect caused by paint from heavy application finishes being pulled down by gravity.

combas. Defecto de pintura causado por aplicar acabados de aplicación pesados jalados hacia abajo por la gravedad.

sail panel. *See* C pillar.

panel de navegación. Ver "pilar C."

sand scratch swelling. A defect that appears following application of topcoat. Coarse scratches that have not been sanded smooth swell as the solvents from the basecoat or topcoat soak into them. The scratches in the paint or in the repair underneath the new paint become visible by reflecting light differently than the new topcoat.

hinchazón de rasguño de arena. Defecto que aparece después de la aplicación del recubrimiento final. Los rasguños gruesos que no han sido lijados se hinchan suavemente a medida que los solventes de la capa base o recubrimiento final los sumergen. Los rasguños en la pintura o en la reparación debajo de la nueva pintura se hacen visibles reflejando la luz diferentemente que el nuevo recubrimiento.

sand test. A method to help identify whether a plastic is a thermoplastic polyolefin. If the plastic smears, melts, and is waxy when sanded, the test indicates that the plastic is a polyolefin. If the part sands dry clean, producing dust, the plastic part may not be a polyolefin.

prueba de arena. Método que ayuda a identificar si un plástico es un termoplástico *polyolefin*. Si el plástico se corre, se derrite, y es como cera cuando se lija, la prueba indica que el plástico es un *polyolefin*. Si la parte se lija en seco, produciendo polvo, la parte plástica podría no ser un *polyolefin*.

sanding. A process of using an abrasive substance to grind off surface imperfections.

lijar. Proceso de usar una sustancia abrasiva para remover imperfecciones superficiales.

sanding paste. A substance added to a plastic abrasive pad to aid in the pad's scuffing (deglazing) ability. The scuffing can be done either wet or dry, and the added paste will help get into close areas that are otherwise difficult for the pad by itself to reach.

pasta de lijado. Sustancia añadida a una almohadilla abrasiva plástica para favorecer la capacidad de raspado (deglaseado). El raspado puede ser hecho en seco o en mojado, y la pasta añadida ayudará a llegar a áreas cerradas que de otra forma son difíciles de acceder por la almohadilla sola.

sandpaper backing. Light paper, heavier stiff paper, or even fiber used for sanding very aggressive and stiff surfaces. The different types of backings help each cutting material do its job better. On light sanding paper, the backing is rated A for the lightest with C-, D-, or E-weights being progressively heavier and stiffer.

texturas de papel de lija. Papel ligero, papel rígido más pesado, o aún fibras usadas para lijado muy agresivo, y superficies tiesas. Los diferentes tipos de texturas ayudan a cada material cortante a hacer mejor su trabajo. En el papel de lija ligera, la textura es calificada como A para la más ligera, con C, D y E la textura es más pesada y tiesa progresivamente.

scan tools. Handheld devices that link various vehicular systems via a diagnostic connector which helps the technician diagnose issues in computer-controlled electronic circuits.

instrumentos de exploración. Dispositivos de bolsillo que unen varios sistemas vehiculares a través de un conector de diagnóstico el cual ayuda al técnico a diagnosticar problemas en circuitos electrónicos controlados por computador.

scraper. A tool used to remove softened paint or adhesive.

espátula. Herramienta usada para quitar la pintura blanda o adhesiva.

screw. A fastening device with a slotted head that is twisted into a surface to connect another material to it.

tornillo. Dispositivo de ajuste con cabeza ranurada que es enroscada en una superficie para unirla con otro material.

screwdrivers. Tools that use torque to remove and reinstall items such as trim and ornamentation.

destornilladores. Herramientas que usan la torsión para quitar e instalar de nuevo artículos tales como molduras y ornamentos.

scuffing. A process designed to add roughness to a surface so the next coating that will be applied will adhere more easily. Scuffing is not intended to level any imperfections in the surface or to feather out scratches or chips. It will only take the shine off, or "deglaze," the surface of the old finish.

rayado. Proceso diseñado para añadir rugosidad a una superficie de forma que la capa siguiente a ser aplicada se adhiera más fácilmente. El rayado no se hace con el propósito de nivelar imperfecciones de la superficie o quitar grietas de rayones o astillas. Este sólo quitará el brillo u opacará la superficie del viejo acabado.

seam sealers. A variety of materials with varying viscosities used to weatherproof and seal out water, fumes, dirt, and any other undesirable material from the joints that are created by overlapping parts when the vehicle is assembled.

selladores de costuras. Variedad de materiales de diversas viscosidades usados para proteger contra la intemperie y sellar contra agua, vapores, suciedad y cualquier material no deseado, las uniones que son creadas al sobreponer las partes cuando el vehículo está siendo ensamblado.

seat belt pre-tensioners. A device, usually ignited by a gas charge in conjunction with the supplemental restraint system, that better secures a seat belt around a passenger.

pretensionador de cinturones de seguridad. Dispositivo, usualmente activado por una carga de gas en conjunto con un sistema de restricción suplementario, que asegura mejor un cinturón de seguridad alrededor de un pasajero.

secondary colors. Orange, green, and violet.

colores secundarios. Anaranjado, verde, y violeta.

secondary container label. A user-created label attached to a smaller dispensing container that identifies its contents along with other pertinent product information.

etiqueta de contenedor secundaria. Etiqueta creada por el usuario unida a un contenedor de distribución más pequeño que identifica su contenido, y provee otra información importante del producto.

second-degree burns. More serious than first degree burns, these burns cause the skin's surface to be severely damaged with blisters and possibly with breaks in the skin.

quemaduras de segundo grado. Quemaduras más serias que las de primer grado, éstas hacen que la superficie de la piel sea con dañada severamente con ampollas, y posibles rupturas en la piel.

sectional/partial replacement. A procedure used to replace a portion of a damaged panel at a location other than at the factory seams, thus minimizing the destruction of the original manufacturer's corrosion protection.

reemplazo seccional o parcial. Procedimiento usado para sustituir una parte de un panel dañado en un sitio diferente de las costuras de fábrica, y así minimizar la destrucción de la protección de corrosión del fabricante original.

sectioning. The partial replacement of a panel, in which the new part is attached by welding at an area other than the factory seam.

seccionamiento. Reemplazo parcial de un panel en el cual, la nueva parte es unida por soldadura, en un área diferente de las uniones de fábrica.

self-centering gauges. A manual gauge that indicates misalignment by allowing the technician to sight along or across a series of gauges.

medidores de autocentrado. Medidor manual que indica desalineación al permitir que el técnico observe a través de una serie de medidores.

self-etching primer. Generally used on bare metals that are smaller in size and are less likely to develop corrosion, these primers also contain an acid, which etches into the properly prepared metal surface. However, the acid that is contained in them will stop by itself if the proper airflow is available.

imprimación de autograbado. Generalmente usada en metales desnudos que son más pequeños en tamaño y que son menos vulnerables a desarrollar corrosión, estas imprimaciones tembién contienen un ácido, el cual se graba en la superficie de metal correctamente preparada. Sin embargo, el ácido que está contenido en ellas parará por sí mismo si la corriente de aire adecuada está disponible.

self-healing. A product having the ability to reflow or regenerate a self-protecting coating when it is scratched, scarred, or has a disrupted surface. Examples are galvanized steel that regenerates a protective coating and wax base anti-corrosion compounds that reflow when the surface has been scratched to recoat the affected areas.

autocurado. Producto que tiene la capacidad de refluir o regenerar una capa de autoprotección cuando es rasguñado, marcado, o cuando tiene una superficie alterada. Algunos ejemplos son los aceros galvanizados que regeneran una capa protectora, y compuestos anticorrosivos con base de cera que refluyen cuando la superficie ha sido rasguñada para cubrir de nuevo las áreas afectadas.

self-limiting. A process that occurs (for example, on aluminum exposed to oxygen) where an oxide coating develops as a result of corrosion, but as the coating is made it prevents further corrosion from taking place.

autolimitado. Proceso que ocurre (por ejemplo, en el aluminio expuesto al oxígeno) en donde una capa de óxido se desarrolla a consecuencia de la corrosión, pero porque la capa fue hecha para evitar que corrosión adicional ocurra.

self-locking nuts. Nuts that have special plastic inside them; as the bolt is tightened the plastic presses against the threads, holding the bolt in place. Self-locking nuts are a single-use fastener; once tightened and then removed, the nut should not be reused.

tuercas de autobloqueo. Tuercas que tienen un plástico especial dentro de ellas; cuando el tornillo es apretado, el plástico se presiona contra la rosca, manteniendo el tornillo en su sitio. Las tuercas de autobloqueo usan un ajustador de un solo uso; una vez apretada y luego removida, la tuerca no puede ser reusada.

self-piercing rivet (SPR). A rivet which does not require a hole for application, as the name implies; instead, a special gun is used to drive the rivet through two or more layers of aluminum. As it passes through, the rivet holds the parts together using the pressure the gun applied to the rivet as it was forced through the aluminum. Self-piercing rivets are the most common type of rivet used when a vehicle is manufactured.

remaches de autoperforado (SPR por sus siglas en inglés). Remache que no requiere un agujero para su colocación, como su nombre lo indica; en cambio, una pistola especial es usada para introducir el remache a través de dos o más capas de aluminio. Cuando esto pasa, el remache sostiene las partes juntas usando la presión que la pistola aplicó al remache cuando éste fue forzado a través del aluminio. Los remaches de autoperforado son los tipos más comunes de remaches usados cuando un vehículo es fabricado.

self-tapping/sheet metal screws. Screws generally used to attach pieces of thin sheet metal together, or to attach molding or trim to sheet metal. They come in varying types of head designs and can be "driven" or tightened with different types of tools.

tornillos autorroscantes o para metales laminados. Tornillos generalmente usados para unir piezas de láminas metálicas delgadas juntas, o unir moldes o tapicería a láminas metálicas. Éstos vienen con diversos tipos de cabeza y pueden ser "conducidos" o apretados con diferentes tipos de herramientas.

sell-back. A process where the technician tries to help the customer forget about the unpleasantness of the collision experience and replace it with confidence in their repaired vehicle; in other words, selling the vehicle back to the customer.

revender. Proceso en el cual el técnico trata de ayudar al cliente a olvidar lo desagradable de la experiencia de la colisión, y reemplazarla con confianza en su vehículo reparado; en otras palabras, revender el vehículo al cliente.

semi-downdraft spray booths. Although introduced at approximately the same time as downdraft booths, these booths are somewhat cheaper to purchase. An advantage of this style of booth is that air is brought in through the rear roof of the booth. This allows the air to be collected from outside the building, where it is cleaner and freer of debris. However, they are less efficient and less clean than downdraft booths.

cámara de rociado semidescendente. Aunque introducido en aproximadamente el mismo tiempo que las cabinas de corriente baja, estas cabinas son algo más económicas de comprar. Una ventaja de este estilo de cabina consiste en que el aire es dirigido por el techo trasero de la cabina. Esto permite que el aire sea recolectado desde fuera del edificio, en donde está más limpio y más libre de escombros. Sin embargo, éstas son menos eficientes y menos limpias que las cámaras de rociado descendente.

sensitive heat range. The temperature range at which structural parts of the automobile made with a special high strength steel that is very sensitive to heat will become very weak and lose its strength characteristic; temperature may range from 700°F to 1,200°F.

rango de sensibilidad al calor. Rango de temperaturas en el cual las partes estructurales del automóvil fabricado con un acero de alta resistencia, que es muy sensible al calor, se debilitarán y perderán sus propiedades de resistencia; la temperatura puede variar de 700°F a 1.200°F.

shear strength. Steel's ability to withstand cutting or slicing forces.

resistencia a cortante. Capacidad del acero de resistir el corte o fuerzas cortantes.

sheet metal screws *See* self-tapping/sheet metal screws.

tornillos para metales laminados. Ver "tornillos autorroscantes o para metales laminados"

sheet moldable compound (SMC). A product that combines both fiber reinforcement and resin that is placed into a die and heated. The heat activates the SMC, which hardens into the desired part.

compuestos de lámina moldeable (SMC por su acrónimo en inglés). Producto que combina refuerzos de fibra y resina que es colocada dentro de un cubo y es calentado. El calor activa el SMC (por su acrónimo en inglés), el cual se endurece en la parte deseada.

shielding gas. In GMAW, this gas is used to shield the weld and the surrounding areas, so that normal atmospheric gases such as nitrogen and oxygen do not interfere and contaminate the weld as it is formed.

gas protector. En GMAW (por su acrónimo en inglés), este gas es usado para proteger la soldadura y las áreas circundantes, de modo que los gases atmosféricos normales, como nitrógeno y oxígeno, no interfieran y contaminen la soldadura cuando es formada.

shims. A U-shaped washer-like disc placed between panels as a spacer that is used to increase the gap or height of one panel to the level of an adjacent one.

shim. Disco parecido a una arandela tipo U colocado entre paneles como un espaciador, y que es usado para aumentar la holgura o altura de un panel al nivel de otro panel adyacente.

shock towers. A reinforced area incorporated into the front inner rails used to attach the upper parts of the struts for the suspension system on the unibody.

torres de choque. Área reforzada incorporada en los rieles interiores delanteros usada para unir las partes superiores de los puntales para el sistema de suspensión en el unicuerpo.

short circuit metal transfer. In this method, as wire feed is increased, the arc is struck. When the wire electrode burns back, the gap that is developed during the globular method is bridged. This transfer produces no large irregular buildup of molten metal, since the molten filler wire is added to the base metal and thus produces a more even weld pool and a better weld with less spatter.

transferencia de metal de circuito corto. En este método, a medida que la alimentación del cable es aumentada, el arco es golpeado. Cuando el electrodo del cable se calienta de nuevo, la holgura que es lograda durante el método globular es rellenada. Esta transferencia no produce irregularidades grandes acumuladas de metal fundido, debido a que el cable de relleno fundido está unido al metal base y así produce una soldadura aún más uniforme y una mejor soldada con menos salpicaduras.

shrinking hammer. Used to help shrink stretched sheet metal, this tool is similar to a finishing hammer but it has a serrated or cross-grooved face.

martillo de acortamiento. Usado para ayudar a encoger láminas metálicas extendidas; esta herramienta es similar a un martillo de acabados pero tiene una cara aserrada o acanalada en cruz.

side curtain air bags. Protective device installed within vehicles that will deploy in the event of a side impact collision.

bolsas de aire de cortina lateral. Dispositivo de protección instalado dentro de vehículos que se desplegarán en caso de una colisión de impacto lateral.

sidesway. A damage condition that often occurs when a vehicle is struck at an angle with sufficient force for the collision energy to push one side of the front rail toward the center of the vehicle or shifting the entire front section of the vehicle—including the radiator support—off to one side.

sidesway. Condición de daño que a menudo ocurre cuando se golpea un vehículo, en un ángulo con fuerza suficiente para que la energía de colisión empuje un lado del riel delantero hacia el centro del vehículo, o cambiar toda la sección delantera del vehículo, incluyendo el apoyo del radiador, hacia un lado.

silicon bronze welding. A welding filler that melts at a relatively low temperature, resulting in a minimum amount of warpage when welding, particularly in areas such as the sail panel where a high degree of filler is required for filling a seam. *See also* MIG brazing.

soldadura de bronce de silicona. Relleno de soldadura que se derrite a una temperatura relativamente baja, causando una cantidad mínima de curvatura cuando se suelda, particularmente en áreas, tales como el panel de vela en donde se requiere un grado alto del relleno para rellenar una costura. Ver "soldadura tipo MIG."

simple buckle. A situation where pressure causes metal to bend either up or down, creating a buckle that resembles a hinge on a door.

broche simple. Situación en donde la presión hace que el metal se doble hacia arriba o hacia abajo, creando un broche que se parece a un gozne de una puerta.

single braided hose. An inner rubber hose covered with a braided mesh, then with a second covering of rubber over the braiding. The single braided type is used for general shop work.

manguera de trenzado simple. Manguera con interior de hule cubierta de una malla trenzada, que tiene un segundo recubrimiento de hule sobre el trenzado. El tipo de manguera de trenzado simple es usado para trabajo de tienda en general.

single-side repair. A repair to only one surface of a part or component.

reparación de un solo lado. Reparación de una sola superficie de una parte o componente.

single-stage compressor. Generally smaller than other compressors, these may have more than one piston, although the compressing process only occurs once.

compresor de etapa simple. Generalmente más pequeño que otros compresores, éstos pueden tener más de un pistón, aunque el proceso de compresión sólo ocurra una vez.

siphon feed gun. A spray gun that operates by creating a siphon over a tube, which extends into the paint cup. As high pressure forces air over this tube, a siphon is created. Because of the low pressure created in the tube, the paint is lifted into the airstream.

arma alimentada por sifón. Pistola rociadora que funciona al crear un sifón sobre un tubo, el cual llega al depósito de pintura. Cuando la presión alta fuerza el aire sobre este tubo, un sifón es creado. A causa de la baja presión creada en el tubo, la pintura es elevada en la corriente de aire.

slapping file. A finish tool that looks like a bent steel file. When it is used with a dolly, the slapping file is brought down directly over the top of the dolly. As it hits the dolly, it is dragged slightly before the next slap is applied. In this way, the area of high metal is both pushed down and filed off at the same time.

lima de golpeo. Herramienta de acabado que parece una lima de acero doblada. Cuando es usada con una carretilla, la lima de golpeo es bajada directamente sobre ella. A medida que la carretilla es golpeada, ésta es arrastrada ligeramente antes de que el siguiente golpe sea aplicado. De esta forma, el área metálica alta es empujada hacia abajo y limada al mismo tiempo.

slapping spoon. Sometimes called a bumping file, this tool is used for final finishing when metal finishing is performed. It can eliminate most, if not all, of the hammer marks left from the metal finishing process.

cuchara de golpeo. A veces llamada lima de aporreo, esta herramienta es usada para acabados finales, cuando el acabado metálico es realizado. Ésta puede eliminar la mayoría de, sino todas, las marcas de martillo dejadas en el proceso de acabado metálico.

sledge hammer. A large hammer used for heavy driving and reforming thicker gauge sheet metal parts such as structural components.

mazo. Martillo grande usado para trabajar y reformar partes de lámina de metal de calibre grueso tales como componentes estructurales.

sliding T-handle. An accessory that can be used for a fastener with limited access.

mango T corredizo. Accesorio que puede ser usado como un sujetador de acceso limitado.

slot welds. A type of weld bead occasionally used to secure a structural aluminum panel to the vehicle. It is very similar to the plug weld except an elongated slot is made into the top layer through which the welding is performed, instead of using the round hole commonly found when making a plug weld.

soldaduras de ranura. Tipo de cuenta de soldadura ocasionalmente usada para asegurar un panel de aluminio estructural al vehículo. Es muy similar a la soldadura de enchufe excepto que una ranura alargada es hecha en la capa superior en la cual la soldadura es aplicada, en lugar de usar el agujero redondo comúnmente encontrado cuando se coloca una soldadura de enchufe.

SMAW MIG/flux core welding. The abbreviation commonly used to indicate the American Welding Society's shielded metal arc welding process using a tubular electrode or wire. The center of the wire or electrode is filled with flux materials used in place of the inert gas used in other MIG processes, which is necessary to protect the weld from contamination.

SMAW MIG o soldadura de núcleo fluido. Abreviatura comúnmente usada para indicar el proceso de soldadura de arco metálico acorazado usando un electrodo tubular o alambre de la Sociedad Americana de Soldadores. El centro del alambre o electrodo está lleno de materiales fluidos usados en lugar del gas inerte usado en otros procesos MIG (por su acrónimo en inglés), el cual es necesario para proteger la soldadura de la contaminación.

smoke generators. A device that generates a non-toxic but visible smoke near the vehicle. The observation of the smoke will indicate if the booth has proper airflow, or where the airflow is the best and where the airflow is the least.

generadores de humo. Dispositivo que genera un humo no tóxico pero visible cerca del vehículo. La observación del humo indicará si la cabina tiene flujo de aire apropiado, o dónde la corriente de aire es mejor y dónde la corriente de aire es la menor.

snaprings. Retainers that fit into grooves on the outside of parts that the snapring is forced to open slightly. When in place, snaprings squeeze around the groove, thus not letting the parts come apart. Generally, snaprings are used for parts that do not need to be held tightly, but should not come apart.

abrazaderas. Retenedores que encajan dentro de surcos, en el exterior de partes, que la abrazadera es forzada a abrir ligeramente. Cuando están en su lugar, las abrazaderas aprietan alrededor del surco, no permitiendo que las partes se separen. Generalmente, las abrazaderas son usadas en partes que no necesitan estar muy apretadas, pero no se deben separar.

snugging the bolts. Tightening a bolt or fastener only to the sufficient point to temporarily hold a part.

ajustar pernos. Apretamiento de un perno o sujetador sólo al punto suficiente para sostener temporalmente una parte.

soap and water cleaning. A method of washing a vehicle to remove surface dirt or other typical contaminants. The technician should use warm to hot water and a pH balanced (pH 7), liquid car washing soap, rather than dish soap or powdered soap.

limpieza con jabón y agua. Método de limpieza de un vehículo para quitar suciedad superficial u otros contaminantes típicos. El técnico debe usar agua entre tibia y caliente y un pH balanceado (pH 7), jabón líquido para lavado de vehículos, en lugar de jabón de platos o en polvo.

Society of Automotive Engineers (SAE). A nonprofit educational and scientific organization dedicated to advancing mobility technology to better serve humanity.

Sociedad de Ingenieros Automotrices (SAE por su acrónimo en inglés). Organización educativa y científica sin fines de lucro dedicada a los avances tecnológicos de la movilidad, para servir mejor a la humanidad.

socket. A cylinder-shaped, box-end wrench used for removing bolts and nuts by rapid turning motion.

enchufe. Llave con extremo en forma de caja cilíndrica usada para remover tornillos y tuercas con movimientos rápidos de giro.

socket wrench. A wrench that removes fasteners more rapidly than an open-end or box-end wrench.

llave de enchufe. Llave que remueve sujetadores más rápidamente que una llave de extremo abierto o de cubo cerrada.

socket wrench accessories. Optional parts that make the socket wrench more useful, such as a breaker bar, sliding T-handle, speed handle, drive adapter, extension bars, spinner, rachet adapter, and universal joint.

accesorios de llave de enchufe. Partes opcionales que hacen la llave de enchufe más útil, tales como una barra de quiebre, mango T deslizante, mango deslizante, adaptador de manejo, barras de extensión, girador, adaptador de paso y unión universal.

socket wrench set. A tool collection that consists of a socket or drive handle and detachable sockets.

juego de llaves de enchufe. Juego de herramientas que consiste en un mango de enchufe o control y enchufes desmontables.

solder. A filler composed of a combination of lead and tin which was commonly used as a filler by both the automobile manufacturers and the repair industry prior to the introduction of plastic and other forms of present day fillers.

soldadura. Compuesto de relleno de una combinación de plomo y alquitrán la cual fue comúnmente usada como relleno por los fabricantes de autos y la industria de la reparación, antes de la introducción del plástico y otras formas de relleno modernas.

solid axle. A steering system, commonly used on trucks, SUVs, and many four-wheel-drive vehicles, in which a solid axle housing is used to attach the steering spindle, shock absorber, and spring.

eje sólido. Sistema de dirección, comúnmente usado en camiones, vehículos deportivos (SUVs por su acrónimo en inglés), y muchos vehículos de tracción en las cuatro ruedas, en el cual la cavidad de un eje sólido es usada para unir el eje de dirección, el amortiguador, y el resorte.

solid waste stream. Shop waste materials such as used masking paper that may be classified as hazardous waste, which may require the shop to comply with specific disposal methods.

generación de desechos sólidos. Materiales de desecho de talleres tales como papel de enmascarado, que pueden ser clasificados como desechos peligrosos, los cuales pueden requerir que el taller cumpla con métodos de disposición específicos.

solvent popping. A condition that occurs when trapped solvents rise to the top of a finish and burst, causing small craters in the finish's surface. The problem may be made worse from force-drying or uneven drying with infrared lights.

rotura del solvente. Condición que ocurre cuando solventes atrapados se elevan a la parte superior de un acabado y revientan, causando pequeños cráteres en la superficie del acabado. El problema puede ser peor por el forzado del secado o secado no uniforme con luces infrarrojas.

solvent test. A test performed to see if the old finish on the vehicle will reflow when a strong solvent is applied. To test a painted surface, a technician should apply thinner to a shop towel and rub it onto the surface in question. If the coating comes off on the application rag, or if the surface becomes soft following the test when scratched, the coating is a thermoplastic and will not be durable enough to accept thermoset coatings over it.

prueba del solvente. Prueba realizada para ver si el acabado viejo de un vehículo fluirá de nuevo si un solvente fuerte es aplicado. Para probar una superficie pintada, el técnico deberá aplicar *thinner* a una toalla y frotarla en la superficie en cuestión. Si el recubrimiento es removido por el trapo, o si la superficie se vuelve blanda luego de la prueba cuando es rasguñada, el recubrimiento es termoplástico y no será suficientemente durable para aceptar recubrimientos tipo *thermoset* sobre él.

sound deadening materials. Insulation materials used in strategic locations on a vehicle to reduce the road noises normally transmitted into the passenger compartment.

materiales amortiguadores de sonido. Materiales de aislamiento usados en posiciones estratégicas de un vehículo para reducir los ruidos de carretera normalmente transmitidos al compartimiento de pasajeros.

space frame. A vehicle design that utilizes a crashworthy cage-like structure made of lightweight high strength tubular steel which is covered with adhesive-bonded plastic and other composite materials.

marco espacial. Diseño de vehículo que utiliza una estructura de resistencia a colisiones, parecida a una jaula, hecha de acero tubular de alta resistencia el cual está cubierto con un plástico adhesivo y otros materiales compuestos.

specialist. A repair technician whose expertise is in a single area of repairs, such as a pinstriper or wheel alignment tech.

especialista. Técnico en reparación que es experto en una sola área de reparaciones, tales como un técnico en arte de pinstripping o en alineamiento de ruedas.

specialty coating. A category of undercoats containing items such as plastic adhesion promoters, rock guard, chip resistant undercoats, and other coatings used to perform specific tasks.

recubrimiento especial. Categoría de recubrimientos básicos que contienen sustancias, tales como promotores de adherencia plásticos, aerosoles, recubrimientos interiores resistentes a fragmentos y otras capas usadas para tareas específicas.

specialty hammers. A category of hammers, each designed to do a specific straightening job.

martillos de especialidad. Categorías de martillos, cada uno diseñado para hacer un trabajo de enderezamiento específico.

specialty shop. A facility that limits their work product to specific vehicle parts or systems, such as upholstery, wheel alignment, and supplemental restraint system facilities.

tienda de especialidades. Talleres que limitan su trabajo partes o sistemas de vehículos específicos, tales como tapicería, alineación de ruedas e instalación de sistemas de restricción suplementarios.

specialty tapes. Also called fine line tapes, these are plastic tapes designed to adhere more flatly against the surface, leaving little or no possibility of the finish bleeding under the tape. Therefore, it makes a sharp dividing line.

cintas de especialidades. También llamadas cintas de línea fina, éstas son cintas plásticas diseñadas para adherirse más firmemente contra la superficie, dejando poca o ninguna posibilidad de sangrado bajo la cinta. Por lo tanto, ésta hace una línea divisoria aguda.

speed handle. An accessory used to quickly turn off a loosened nut or bolt.

mango de velocidad. Accesorio usado para sacar rápidamente una tuerca o tornillo suelto.

speed nuts. A fastener that does not fit cleanly into either the bolt or nut category. Speed nuts come in different types such as flat, barrel, j-nut, and u-nuts, and are placed on parts before assembly for places where a technician has little or no access to use a wrench on the other side of the part.

tuercas de velocidad. Sujetador que no cae en las categorías de tornillo o tuerca. Las tuercas de velocidad vienen en diferentes tipos, tales como planas, de barril, de tuerca J, de tuerca U, y son colocadas en partes antes del ensamblaje en sitios en donde los técnicos tienen poco, o no tienen, acceso para usar llaves en el otro lado de la parte.

spoon dolly. A dolly with long handles designed to reach into deep, recessed areas.

carretilla de cuchara. Carretilla con mangos largos diseñados para alcanzar áreas ahuecadas profundas.

sport utility vehicle (SUV). A vehicle that offers many of the same features as a car but also affords the option of four-wheel-drive and other off-road features.

vehículo de uso deportivo (SUV por su acrónimo en inglés). Vehículo que ofrece muchos de los mismos rasgos que un vehículo convencional, pero que también tiene la opción de tracción en las cuatro ruedas y otras características.

spot weld. A situation in which the weld does not move. Therefore, no travel angle exists, and the work angle is 90 degrees. This type of weld is not often used in collision repair.

soldadura de punto. Situación en la cual la soldadura no se mueve. Por lo tanto, ningún ángulo de viaje existe, y el ángulo de trabajo es 90 grados. Este tipo de soldadura no es frecuentemente usado en la reparación de colisiones.

spot weld nuggets. The metal that remains on the mating flange or surface after the spot weld has been drilled out with a hole saw. This part of the panel actually holds the spot weld in place prior to the weld being drilled and removed.

trozos de soldadura de punto. Metal que permanece sobre el reborde o superficie después de que el punto de soldadura ha sido taladrado con una sierra de agujero. Esta parte del panel realmente sostiene la soldadura de punto en el lugar antes de que la soldadura sea taladrada y removida.

SPR. *See* self-piercing rivets.

SPR. Ver "remaches de autoperforado."

spray arc metal transfer. A method of GMAW welding accomplished only at relatively high voltages and amperages. It deposits small droplets of molten electrode to the weld surface at a high rate, with the size of the droplets being a smaller diameter than the electrode wire. Once the spray arc is accomplished, the weld is said to be "on" at all times.

transferencia de arco metálico por rociado. Método de soldadura tipo GMAW (por su acrónimo en inglés) logrado sólo a altos voltajes y amperajes. Éste deposita pequeñas gotitas de electrodo fundido en la superficie fundida a un ritmo alto, siendo el tamaño de las gotitas de un diámetro más pequeño que el alambre del electrodo. Una vez el arco rociado está terminado, se dice que la soldadura está "arriba" en todo momento.

spray booth balancing. A system that helps control the air movement inside the spray environment to assure that outside dirt does not come into the booth. The amount of air coming into the spray booth is controlled along with the amount of air flowing out of the booths.

balanceado de la cabina de rociado. Sistema que ayuda a controlar el movimiento del aire dentro del medioambiente de rociado para asegurar que la suciedad externa no entre en la cabina. La cantidad de aire que entra en la cabina de rociado es controlada junto con la cantidad de aire que sale de las cabinas.

spray cycle. The time needed to complete the vehicle's spray application in the booth. It is not timed, and will run continually until changed to another cycle.

ciclo de rociado. Tiempo necesario para completar la aplicación de rociado del vehículo en la cabina. Éste no es medido, y correrá continuamente hasta que sea cambiado en otro ciclo.

spray gun adjustment. One of the most important activities that a technician will perform to enable quality refinishing. It is necessary to have all controls (especially those that affect other controls) properly set.

ajuste de la pistola rociadora. Una de las actividades más importantes que un técnico realiza para lograr un acabado de calidad. Es necesario tener todos los controles (especialmente aquellos que afectan a otros controles) adecuadamente configurados.

spray gun cleaning. The regular maintenance and lubrication of a spray gun. If not cleaned properly, the gun adjustment may become impossible to adjust. Many painters not only clean their gun after each use, but also break it down, clean each part thoroughly, and reassemble and lubricate it on a regular basis.

limpieza de la pistola rociadora. Mantenimiento regular y lubricación de una pistola rociadora. Si no es limpiada correctamente, el ajuste de la pistola puede ser imposible de lograr. Muchos pintores no sólo limpian su arma después de cada uso, sino que también la desarman, limpian a fondo cada parte, y la vuelven a armar y lubricar frecuentemente.

spray pattern test. Performed after spray gun adjustments, this test is a way to verify all the settings are as the technician desires. This includes checks of the atomization, droplet size, fluid flow, and speed of flow.

prueba de patrón de rociado. Realizado después de los ajustes de la pistola rociadora, esta prueba es un modo de verificar que dichos ajustes son como los que desea el técnico. Esto incluye controles de atomización, tamaño de gotitas, flujo de fluidos y velocidad de flujo.

sprayout panel. A test performed by the technician to evaluate how many layers of paint will be needed on a vehicle, as well as to see how the final finish will appear, before actually starting the paint job.

panel de prueba de rociado. Prueba realizada por el técnico para evaluar cuántas capas de la pintura serán necesarias en un vehículo, así como para ver como quedará el acabado final, antes de comenzar realmente el trabajo de pintura.

spring hammering. A technique used when a technician needs to give a series of light blows to metal, causing it to move or relax without causing the metal to stretch.

martillado de resorte. Técnica usada cuando un técnico tiene que dar una serie de golpes ligeros al metal, causándole movimiento o relajación, sin hacer que el metal se estire.

springback. A repair situation in which, because of steel's elasticity, it is generally necessary to pull the rail or the section of the undercarriage in question beyond the normal position so that it will stay at its designated location according to the vehicle specifications.

springback. Situación de reparación en la cual, debido a la elasticidad del acero, es generalmente necesario sacar el riel o la sección del tren de aterrizaje en cuestión, más allá de la posición normal, de modo que permanezca en el lugar indicado de acuerdo a las especificaciones del vehículo.

squaring the opening (door or window). Process in which one applies corrective forces to the door and window openings and adjacent areas to restore the dimensions to those originally used by the manufacturer before removing any of the damaged panels or structures.

ajuste de abertura (puerta o ventana). Proceso en cual se aplica fuerzas correctivas a las aberturas de ventanas y puertas, y áreas adyacentes, para restaurar sus dimensiones a aquellas usadas originalmente por el fabricante, antes de quitar cualquiera de los paneles o estructuras dañadas.

squeeze-type resistance spot welding (STRSW). The robotic welding process and systems used to weld individual panels together to form assemblies and ultimately create the unibody by the manufacturer. It is similar to the process used when replacing certain panels when repairing a collision damaged vehicle.

soldadura de punto resistente a presión (STRSW por su acrónimo en inglés). Proceso y sistemas de soldadura robótica usados para soldar paneles individuales para formar ensamblajes y, por último, crear el unicuerpo del fabricante. Ésta es similar al proceso usado cuando se reemplazan ciertos paneles cuando se repara un vehículo dañado por una colisión.

stain resistant. The likelihood of the surface becoming discolored due to a chemical reaction between the catalyst of the plastic and the solvents introduced when spraying the finish coats. This trait is largely eliminated with the use of premium fillers.

resistente a manchas. Probabilidad de que la superficie se descolore debido a una reacción química entre el catalizador del plástico y los solventes involucrados cuando se aplican las capas de acabado. Este rasgo es en gran parte eliminado con el uso de rellenos de alta calidad.

staining. A discoloration in the topcoat which is a result of the unused catalyst in the plastic reacting with the solvents from the primers and topcoats.

mancha. Decoloración en la capa final la cual es resultado del catalizador, no usado en el plástico, que reacciona con los solventes de las imprimaciones y recubrimientos.

standard blending. The basic method of blending. The technician cleans, sands, and prepares the area to be blended; seals it if needed; mixes and reduces paint per manufacturer's directions; then applies the needed number of coats with adequate flash time between each application.

combinado estándar. Método básico de combinación. El técnico limpia, lija y prepara el área a ser combinada; aplica selladores si esto es necesario; mezcla y reduce la pintura siguiendo las indicaciones del fabricante; luego aplica el número necesario de capas con el tiempo de destello adecuado antes de cada aplicación.

standard tip screwdriver. A screwdriver whose tip has a single flat blade which fits a screw with a slotted head. The blade tip thickness and width must perfectly match the slot in the screw head.

destornillador de punta estándar. Destornillador cuya punta tiene una hoja plana simple la cual se ajusta en un tornillo de cabeza ranurada. El grosor de la punta de la hoja y su ancho deben hacer juego perfecto con la ranura en la cabeza del tornillo.

static mixing tube. A special mixing nozzle with a maze-like passage through which the product—such as the seam sealer or adhesive—is dispensed, which mixes the material as it moves from the tube to the end where it exits the opening.

tubo de mezcla estática. Inyector de mezcla especial con un conducto tipo laberinto a través del cual el producto, tal como un sellador de ranuras o adhesivo, es dispensado, lo cual mezcla el material a medida que se desplaza por el tubo hasta que sale por la abertura.

steel preparation. The steps needed to get steel ready for finishing. As a general rule, it is best not to strip off the factory-installed corrosion protection when performing these steps.

preparación del acero. Pasos necesarios para hacer que el acero esté listo para el acabado. Como regla general, es mejor no quitar la protección anticorrosión de fabrica cuando se realizan estos pasos.

steel sandwiched parts. A process where a layer of polypropylene plastic or balsa wood are sandwiched between two layers of steel to reduce vehicle weight and noise, and increase strength of a panel like the trunk floor.

partes intercaladas de acero. Proceso en el cual una capa de plástico de polipropileno o madera balsa son colocadas entre dos láminas de acero para reducir el peso de un vehículo y el ruido, e incrementar la resistencia de un panel como el piso de la cajuela.

steel substrate. Any material, made of steel, that makes up part of a motor vehicle's structure or outer panels.

sustrato de acero. Cualquier material, hecho de acero, que hace parte de la estructura o paneles externos de un automóvil.

steering axis inclination (SAI). The degree or amount of inward tilt of the upper pivot point from a true vertical line on the steering assembly. This inward tilt tends to keep the wheel pointed straight ahead. It is also largely responsible for returning the wheels to a straightforward position after a turn is completed.

inclinación de los ejes de conducción (SAI por su acrónimo en inglés). Grado o cantidad de inclinación hacia adentro del punto de pivote superior desde una línea vertical verdadera sobre el ensamblaje de conducción. Esta inclinación hacia adentro tiende a mantener las ruedas rectas hacia delante. Ésta también es en gran parte responsable de devolver las ruedas a su posición recta hacia delante luego de que un giro es completado.

stick-out. The amount of electrode wire that must stick out of the contact tip before the technician starts welding. The wire "stick-out" length should range from 3/8 of an inch to 1/2 inch.

sobresaliente. Cantidad de alambre de electrodo que debe sobresalir, en la punta de contacto, antes de que el técnico comience a soldar. La longitud "sobresaliente" de alambre debe extenderse de 3/8 a 1/2 pulgada.

stitch weld. A series of short overlapping MIG spot welds which, when finished, create a continuous stream. This weld is often used on thin metal where warpage and burnthrough must be controlled.

soldadura de puntada. Serie de puntos de soldadura tipo MIG (por su acrónimo en inglés) cortos sobrepuestos los cuales, cuando están terminados, crean una línea continua. Esta soldadura es a menudo usada en láminas metálicas delgadas en donde las distorsiones y quemones deben ser controlados.

straight rivets. Rivets that need to be bucked.

remaches rectos. Remaches que tienen que ser forzados (colocado al fijar una barra forzada contra el extremo del remache cuando éste es instalado con la pistola remachadora).

striker. Sometimes called the striker pin or plate, the device that engages the latch or lock to hold the door, hood, or deck lid in a closed position.

ariete. A veces llamado alfiler o placa de ariete, es un dispositivo que conecta el pestillo o candado para sostener la puerta, capó o tapa de cubierta en una posición cerrada.

striping. Uneven color or stripes of light and uneven color in the finish. It occurs when the finish is applied using poor spray technique or the finish is applied too wet. Striping usually appears in the first coat and does not correct itself with additional coats.

striping. Color disparejo o rayos de luz y color disparejo en el acabado. Este ocurre cuando el acabado es aplicado usando la técnica de rociado pobre, o cuando el acabado es aplicado muy húmedo. El striping por lo general aparece en el primer recubrimiento y no se corrige con capas adicionales.

structural foam. A foam material that has either a very closed cell structure or is totally devoid of any pores or cells in its structure. It is commonly used to reinforce the upper pillars, inner rails, and—in some instances—frame torque boxes to enhance the vehicle's crashworthiness.

espuma estructural. Material de espuma que tiene una estructura celular muy cerrada o es totalmente carente de cualquier poro o celda en su estructura. Este es comúnmente usado para reforzar los pilares superiores, rieles interiores, y, en algunos casos, cajas de torsión de marco para realzar la resistencia a colisiones del vehículo.

structural integrity. A vehicle's overall strength and ability to withstand a crash.

integridad estructural. Fuerza y capacidad total de un vehículo para resistir accidentes.

strut tower. Also called strut housing, the tower is used to attach part of the steering and suspension parts; may be an integrated part of the front rails or may be a separate piece welded onto the assembly.

torre de puntal. También llamada carcasa de puntal, la torre es usada para adjuntar parte de las partes de conducción y suspensión; puede ser una parte integrada del riel delantero o puede ser una pieza aparte soldada al ensamblaje.

strut tower gauges. In cases where damage to the strut tower is suspected, these gauges are used in conjunction with the centering gauges to check for misalignment.

calibrador de torre de puntal. En los casos en los que se sospeche daño en la torre de puntal, éstos calibradores son usados, junto con medidores de centrado, para buscar desalineaciones.

stub frame. A frame section that only extends from the front of the vehicle to an area under the passenger compartment.

marco central. Sección de un marco que sólo se extiende desde el frente del vehículo hasta el área bajo el compartimiento de pasajeros.

stud welder. A gun used to temporarily weld a stud to the outer surface of a panel, allowing a technician to attach a slide hammer and pull up the low areas.

soldador de tachuelas. Pistola usada para soldar temporalmente una tachuela a la superficie externa de un panel, permitiendo que el técnico una un martillo de moldeado y empuje hacia arriba las áreas bajas.

sublet repairs. Sometimes referred to as "farming out work," it is any repairs that are contracted to be performed by a company outside the repair facility, usually because of cost savings, lack of certain equipment or expertise, or time constraints.

reparación subcontratada. A veces referida como "encargar el trabajo"; ésta es cualquier reparación que es contratada para ser realizada por una compañía fuera del taller de reparación, por lo general debido a economía de costos, carencia de cierto equipo o experiencia, o restricciones de tiempo.

substrate. Any material that makes up part of a motor vehicle's structure or outer panels.

sustrato. Cualquier material que compone la estructura u otros paneles de un automóvil.

suction cup. A tool used to remove small dents. The cup is attached to the center of the dent and pulling force is applied. The dent may then be removed with no damage to the paint.

émbolo. Instrumento usado para quitar pequeñas abolladuras. La taza es unida al centro de la abolladura y una fuerza de jalón es aplicada. La abolladura puede ser entonces quitada sin daño en la pintura.

sulfur dioxide. A chemical compound or by-product generated as exhaust from automobiles burning fossil fuels. It is one of the principal ingredients in the formation of acid rain, a known corrosive on buildings, structures, and automobile finishes.

dióxido de azufre. Compuesto químico o subproducto generado por los gases de combustión de vehículos que queman combustibles fósiles. Es uno de los ingredientes principales en la formación de la lluvia ácida, un conocido corrosivo sobre edificios, estructuras, y acabados de coche.

supersede. When an item is more current or improves on the item that it replaces. For example, the information in a technical service bulletin will supersede the information in a repair manual.

sustitución. Cuando un artículo es más moderno o mejora el artículo que sustituye. Por ejemplo, la información de un boletín de servicio técnico reemplazará la información de un manual de reparación.

supplemental restraint system (SRS). The air bag system and all its components, designed to protect the vehicle's occupants in a collision.

sistema de restricción suplementaria (SRS por su acrónimo en inglés). Sistema bolsas de aire y todos sus componentes, diseñados para proteger a los ocupantes de un vehículo durante una colisión.

supplied air respirator. A device that is an improvement over a standard respirator as it supplies clean air for breathing, as well as temperature regulation.

respirador de aire suministrado. Dispositivo que es una mejora sobre un respirador estándar, ya que éste suministra aire limpio respirable, así como regulación de temperaturas.

surface preparation. One of the most critical operations in refinishing. If the surface is not properly prepared, all of the coating applied afterwards will fail.

preparación de superficie. Una de las operaciones más críticas para el barnizado. Si la superficie no está correctamente preparada, toda la capa aplicada después fallará.

surface temperature. The heat of the vehicle's surface, which may not be the same as the temperature of the air in the spray booth. The best way to monitor the surface temperature is with a magnetic surface monitor.

temperatura superficial. Calor de la superficie del vehículo, que puede no ser el mismo como en la temperatura del aire en la cabina de rociado. El mejor modo de supervisar la temperatura superficial es con un monitor magnético de superficies.

surfacing spoon. A nearly flat tool used to finish bumping a panel, as well as for spring hammering, prying, or slapping.

cuchara de moldeado. Herramienta casi plana usada para complementar un golpe en un panel, así como para martillar con rebote, manipular, o golpear.

surform file. A tool used to rough shape plastic body filler when the filler becomes semi-hard. This procedure reduces the amount of sanding needed to shape and level the body filler.

lima tipo surform. Herramienta usada para darle forma al relleno de cuerpo plástico cuando el relleno está semiduro. Este procedimiento reduce la cantidad de lijado necesaria para dar forma y nivel al relleno de cuerpo.

sweat/pull tab. A piece of metal, welded to a damaged panel, which is used to pull the damage out of the surface in an effort to restore the proper configuration.

etiqueta de extracción. Pieza de metal, soldada a un panel dañado, que es usada para sacar el daño fuera de la superficie en un esfuerzo por restaurar la configuración apropiada.

swirl marks. Circular marks left by aggressive buffing and, less often, during polishing. They are *defects* that should be removed.

marcas de remolino. Señales circulares dejadas por pulido agresivo y, menos frecuente, durante el brillado. Éstos son *defectos* que deben ser removidos.

swivel socket. A socket with a universal joint located between the drive end and socket body.

enchufe de eslabón giratorio. Enchufe con una unión universal localizada entre el extremo de la barra de control y el cuerpo del enchufe.

T

tack free. A characteristic of the better quality plastic fillers wherein the surface of the filler is free of the tacky uncured material which is common with the standard fillers.

libre de pegajosidad. Característica de los mejores rellenos de plástico de calidad en donde la superficie del relleno está libre de materiales no curados pegajosos, los cuales son comunes en los rellenos estándares.

tack weld. A temporary weld used to hold fit-up pieces in place so that a permanent weld can be applied. *A tack weld is only a temporary weld, and is not intended to be permanent.*

soldadura de tachuela. Soldadura temporal usada para mantener piezas de ajuste en su sitio de forma que una soldadura permanente pueda ser aplicada. *La soldadura de tachuela es sólo una soldadura temporal, y no debe ser permanente.*

tailor welded blank. A process where the manufacturer laser butt welds two or more thicknesses of steel end to end to make one flat sheet of steel, which is used to die stamp large body parts such as a uniside panel.

ensamblaje de láminas soldadas. Proceso en donde el extremo del láser del fabricante suelda dos o más espesores de acero de extremo a extremo para hacer una lámina plana de acero, la cual es usada para sellar partes de cuerpo grandes, tales como paneles unilaterales.

talc. A fine powder-like material which is one of the principal ingredients in plastic filler. It is used to provide the filler body and substance.

talco. Material fino, parecido a un polvo, que es uno de los principales ingredientes de rellenos plásticos. Éste es usado para lograr cuerpo y sustancia en el relleno.

tap and die set. A tool set that works in tandem: A tap cleans and cuts internal threads in holes, while a die cuts external threads on bolts or cleans damaged threads on bolts.

juego de roscadora interna y externa. Juego de herramientas que trabajan en parejas: una roscadora interna limpia y corta roscas internas en agujeros, mientras que una roscadora externa corta roscas externas o limpia roscas dañadas en tornillos.

tape measure. A measuring tool with a retractable blade used to measure repair areas and check frame or unibody alignment. It often includes both SAE and metric dimensions.

cinta métrica. Instrumento de medición con una lámina retractable usada para medir áreas de reparación y verificar el alineamiento de marcos o unicuerpos. Ésta a menudo incluye tanto el sistema SAE, como dimensiones métricas.

tape template. A layout plan created with masking tape on the body of the vehicle, using holes or bodylines as reference points. It is best to have the template tape on two sides of the part being removed, so that when replacing that part or putting on a new one, the technician can precisely accomplish the right/left and up-and-down placement.

plantilla de cinta. Plan de despliegue creado con cinta de enmascarar en el cuerpo del vehículo, usando agujeros o líneas de cuerpo como puntos de referencia. Es mejor tener la cinta de la plantilla en dos lados de la parte quitada, de modo que cuando se reemplaza esa parte o se coloca una nueva, el técnico puede efectivamente lograr la ubicación izquierda/derecha y arriba/abajo.

technical service bulletin (TSB). A publication released be a manufacturer that includes information updated since the manufacturer published the standard repair manual.

boletín de servicio técnico (TSB por su acrónimo en inglés). Publicación hecha por un fabricante que incluye información actualizada ya que el fabricante publica el manual estándar de reparaciones.

temperature threshold. The heat limit at which steel begins to be affected by the temperature, usually between 700°F and 1200°F.

umbral de temperatura. Límite de calor en el cual el acero comienza a ser afectado por la temperatura, por lo general entre 700°F y 1200°F.

tensile strength. The strength or ability of a piece of metal to resist ripping or tearing when tension or pressure is applied to it.

resistencia a la tensión. Fuerza o capacidad de una pieza de metal para resistir tirones o rompimiento cuando la tensión o la presión son aplicadas en ella.

tension/pressure. The two force application methods or techniques commonly used for correcting collision damage.

tensión/presión. Dos métodos o técnicas de aplicación de fuerzas comúnmente usados para corregir el daño por colisión.

test panels. *See* letdown panel; sprayout panel.

paneles de prueba. Ver "panel de matices," y "panel de prueba de rociado."

texture. A generally unwanted defect that may occur when the sealer, basecoat, or clearcoat is applied. If the texture appears under the clearcoat, the only way to repair it is to sand, smooth, and repaint the surface. If a texture was in the vehicle's original finish, however, it must be restored in the refinish.

textura. Defecto generalmente no deseado que puede ocurrir cuando el sellador, recubrimiento base o base transparente son aplicados. Si la textura aparece bajo el recubrimiento transparente, la única forma de repararla es lijando, suavizando y repintando la superficie. Si una textura estaba en el acabado original del vehículo; sin embargo, debe ser restaurada en el barnizado.

thermal expansion and contraction. A growing (expansion) or shrinking (contraction) reaction that occurs when any composite material—such as sheet molded compound (SMC)—is exposed to rising or dropping temperatures.

expansión y contracción térmica. Crecimiento (expansión) o encogimiento (contracción) que ocurre cuando cualquier material compuesto, tal como láminas de compuesto moldeado (SMC por su acrónimo en inglés), es expuesto a temperaturas crecientes y decrecientes.

thermoplastic. Plastics that will reflow or become liquid again when solvents are introduced. This category of plastics can be repeatedly softened and shaped when heated without changing their chemical makeup.

termoplástico. Plásticos que refluirán o se volverán líquidos de nuevo cuando entran en contacto con solventes. Esta categoría de plásticos puede ser repetidamente ablandada y formada cuando se calienta sin cambiar su composición química.

thermoset plastics. Plastics that will not reflow or become liquid, following their cure process, when solvents are applied to them. This category of plastics will permanently take the shape of a part by heating, using a catalyst, or applying ultraviolet light. They will not re-soften by heating or through the reapplication of a catalyst or light.

plásticos thermoset. Plásticos que no refluirán o volverán líquidos siguiendo su proceso de curado, cuando son expuestos a solventes. Esta categoría de plásticos tomará permanentemente la forma de una parte al calentarse, usando un catalizador, o aplicando luz ultravioleta. Ellos no se ablandarán de nuevo calentándose o por la nueva aplicación de un catalizador o luz.

thinner. Thinning agent used for lacquer paints.

thinner. Agente adelgazador usado para pinturas de laca.

third-degree burns. The most severe burns. In these burns, the damage extends through the top layer of the skin into the second layer, and even beyond. Immediate medical attention by a medical professional is necessary for third degree burns.

quemaduras de tercer grado. Quemaduras más severas. En estas quemaduras, el daño se extiende por la capa superior de la piel hasta la segunda capa, y hasta más allá. La asistencia médica inmediata por un profesional médico es necesaria para quemaduras de tercer grado.

third party. In terms of insurance coverage, a person involved in a collision loss with a person insured by a specific insurance company. The insured person is the first party, the insurance company is the second party, and the other claimant in a collision is the third party.

terceros. En términos de cobertura de seguros, un implicado en una pérdida de colisión con una persona asegurada por una compañía de seguros específica. La persona asegurada es el primer involucrado, la compañía de seguros es el segundo involucrado, y el otro demandante en una colisión es el tercer involucrado.

thixotropic resin. A fiberglass resin designed to be used on vertical surfaces to reduce the possibility of the repair materials sagging before they cure.

resina thixotropic. Resina de fibra de vidrio diseñada para ser usada en superficies verticales para reducir la posibilidad de pandeo en materiales de reparación antes de que ellos curen.

three-dimensional measuring system. A measuring system which monitors length, width, and height dimensions and must be used to determine the degree of misalignment that occurred.

sistema de medición tridimensional. Sistema de medición que monitorea longitud, anchura, y dimensiones de altura, y debe ser usado para determinar el grado de desalineamiento que ocurrió.

thrust angle. The direction all four wheels move in relation to the vehicle's centerline. It is determined by the position of the rear axle and the toe of the rear wheels.

ángulo de empuje. Dirección del movimiento de las cuatro ruedas con relación a la línea central del vehículo. Éste es determinado por la posición del eje trasero y el dedo del pie de las ruedas traseras.

toe dollies. Dollies used for dinging flat surfaces with low crowns.

carretillas de dedo. Carretillas usadas para golpear superficies planas con coronas bajas.

toe-in/toe-out. The difference in measurement between the front of the front wheels/tires and the rear of the front wheels. Toe is the single most critical angle that is set during an alignment. When not set correctly, it can cause extreme tire wear in a very short distance.

dedo adentro/dedo afuera. Diferencia en la medida entre el frente de las ruedas/llantas delanteras y la parte trasera de las ruedas delanteras. El dedo es el ángulo simple más crítico que es fijado durante un alineamiento. Cuando no es fijado correctamente, éste causa desgaste extremo en las llantas en una distancia muy corta.

tolerance. The amount of variance a vehicle is allowed from the manufacturer's dimensions in all three planes (length, width, and height).

tolerancia. Cantidad de varianza permitida en un vehículo a partir de las dimensiones del fabricante en los tres planos (largo, ancho y alto).

torque. A twisting motion.

torsión. Movimiento de giro.

torque wrench. A tool used to measure the amount of force required to properly tighten a nut or bolt to an exact torque specification.

llave de torque. Instrumento usado para medir la cantidad de fuerza requerida para apretar correctamente una tuerca o tornillo de acuerdo a la especificación de torsión exacta.

torsional strength. Steel's ability to withstand twisting forces.

fuerza torsional. Capacidad del acero para resistir fuerzas de torsión.

TORX. A type of fastener head.

TORX. Tipo de cabeza de sujetador.

TORX wrench or driver. A tool designed to remove a six-point socket head fastener, often available as a set.

llave o conductor TORX. Instrumento diseñado para remover un ajustador de cabeza de enchufe de seis puntos, a menudo disponible en juegos.

total loss. A loss sustained by an insured party's vehicle that encompasses the entire value of that vehicle.

pérdida total. Pérdida sufrida por el vehículo de un asegurado que equivale al valor total del vehículo.

traction control unit (TCU). A computer module that receives input from the wheel sensors to improve traction during acceleration and cornering.

unidad de control de tracción (TCU por su acrónimo en inglés). Módulo de computador que recibe datos de los sensores de las ruedas para mejorar la tracción durante la aceleración y en curvas.

tram bar. A straightline measuring device used to make various comparative measurements by comparing them against the manufacturer's printed specifications or from side to side measurements on the vehicle.

barra tram. Dispositivo de medición en línea recta usada para hacer varias medidas relativas comparándolas con especificaciones impresas del fabricante, o medidas de lado a lado en el vehículo.

transfer efficiency (TE). The measure of how much of the finish material that is placed in the gun is transferred to the target surface. Although the ideal amount is a range of about 62 to 65 percent, the average operator sprays at the 40 to 45 percent level.

eficiencia de transferencia (TE por su acrónimo en inglés). Medida de cuánto del material de acabado, que es colocado en la pistola, es transferido a la superficie objetivo. Aunque la cantidad ideal varíe entre aproximadamente del 62% al 65%, el operador promedio rocía entre el 40% y el 45%.

translucent. Clear, transparent, or able to have light shine through.

translúcido. Claro, transparente, o capaz de dejar pasar un brillo ligero.

transverse engine. An engine that is mounted sideways in the engine compartment, resulting in the crankshaft running perpendicular to the vehicle's body.

motor transversal. Motor que es montado hacia un lado en el compartimiento del motor, causando que el cigüeñal esté perpendicular al cuerpo del vehículo.

transverse viewing. Viewing a panel being sprayed at a 45-degree angle, which allows the technician to note the reflection while spraying.

vista transversal. Vista de un panel, que está siendo rociado, en un ángulo de 45 grados, que permite que el técnico note la reflexión mientras rocía.

trigger control knob. *See* fluid control knob.

botón de control del gatillo. Ver "botón de control de fluido."

trim. The coverings often found around doors, windows, and so on. They should be removed carefully when required.

moldura. Cubiertas que a menudo son encontradas alrededor de puertas, ventanas, etcétera. Ellas deben ser quitadas con cuidado cuando sea requerido.

trim tape. A masking tape with a hard band strip which is slipped under moldings such as flush mount windshields, side lights, and back light windows. The protective covering is then removed to expose the adhesive, which is rolled over, lifting the molding slightly.

cinta de recorte. Cinta de enmascarar con una tira de banda dura la cual es deslizada por debajo de molduras tales como parabrisas de montura nivelada, luces laterales y ventanas de luces traseras. El recubrimiento protector es luego removido para exponer el adhesivo, el cual es desenrollado, levantando la moldura ligeramente.

tungsten inert gas (TIG). A welding process that, unlike MIG aluminum welding, creates problems of high-frequency arc. This can be harmful to electronic equipment.

gas inerte Tungsteno (TIG por su acrónimo en inglés). Proceso de soldar que, a diferencia de la soldadura de aluminio tipo MIG (por su acrónimo en inglés), crea problemas de arco de alta frecuencia. Esto puede ser dañino para el equipo electrónico.

twin I-beam suspension. A suspension system that was once very popular on nearly all Ford Motor Company pickups, SUVs, and full size trucks, but which has not been used since approximately 1997.

suspensión de viga I gemela. Sistema de suspensión que fue alguna vez muy popular en casi todas las camionetas de la compañía Ford, vehículos de uso deportivo (SUVs por su acrónimo en inglés), y camiones de tamaño completo, pero que no ha sido usado desde aproximadamente 1997.

twist. Damage that often occurs when, after the vehicle has struck the impact object—and before it comes to a complete halt— it slides up over the impact object, resulting in the rail forward of the center section being pushed up and the rear of the rail being pushed down.

giro. Daño que a menudo ocurre cuando, después de que el vehículo ha golpeado el objeto de impacto, y antes de que se detenga completamente, éste se desliza sobre el objeto de impacto causando que el riel delantero de la sección central sea empujado hacia arriba y la parte de atrás del riel sea empujada hacia abajo.

two-bucket method of washing. A cleaning method where the technician fills two buckets with water: one with warm soapy water and one with clean water. After cleaning a section with a cloth or mitt dipped in the soapy water, the technician will rinse it out in the clear water to remove any stray debris that is clinging, which could cause scratches later.

método de lavado de los dos baldes. Método de limpieza en el cual el técnico llena dos baldes de agua: uno con agua caliente jabonosa y otro con agua pura. Después de limpiar una sección con una tela o guante sumergido en el agua jabonosa, el técnico lo enjuaga en el agua pura para quitar cualesquier residuo que se adhiera, que podría causar rayos más adelante.

two-part coating (2K). Introduced in the early 1960s, a system in which paint is reduced, then sprayed onto a vehicle.

pintura de dos partes (2K). Introducido a principios de los años 60s, es un sistema en el cual la pintura es reducida, y luego aplicada en un vehículo.

two-sided repair. A repair that is needed on a part damaged on more than one side. It often requires removal from the vehicle in order to access all damaged parts.

reparación de dos lados. Reparación que es necesaria en una parte dañada en más de un lado. Esto a menudo requiere su remoción del vehículo a fin de tener acceso a todas las partes dañadas.

two-stage compressor. Although similar to a single-stage air compressor, these are different in that there are two different-sized pistons. The first one—the larger—compresses the air the same way a single-stage compressor does. However, instead of venting air into a holding tank or to the supply lines, it sends air to another small piston to be compressed again in a second stage. This second stage compresses air at a higher pressure using less energy.

compresor de dos etapas. Aunque similar a un compresor de aire de una etapa, éstos son diferentes en que hay dos pistones de diferente tamaño. El primero, el más grande, comprime el aire de la misma forma en que un compresor de una etapa lo hace. Sin embargo, en lugar de expulsar el aire en un tanque de retención o hacia líneas de abastecimiento, éste envía el aire a otro pequeño pistón para ser comprimido otra vez en una segunda etapa. Esta segunda etapa comprime el aire a una presión más alta usando menos energía.

U

ultra high strength steel (UHSS). The strongest of the steels, with yield strength of over 100,000 psi. It is also the most sensitive to heat, and no recommendations for heating while repairing are given. If a part made of UHSS is bent and requires repair, it must be repaired cold.

acero de superalta resistencia (UHSS por su acrónimo en inglés). Éste es el más fuerte de los aceros, con una resistencia a la fluencia de más de 100.000 psi. También es el más sensible al calor, y ningunas recomendaciones de calentado son dadas durante la reparación. Si una parte hecha de UHSS (por su acrónimo en inglés) es doblada y requiere reparación, ésta debe hacerse en frío.

ultrasonic measuring system. A system that uses high-frequency sound waves in order to measure vehicle dimensions. The ultrasonic system consists of a computer to operate the system and store data, reference probes that are emitter targets, and a unit that receives the incoming sound waves.

sistema de medición ultrasónica. Sistema que usa ondas sonoras de alta frecuencia con el fin de medir las dimensiones de vehículo. El sistema ultrasónico consiste de un computador para hacer funcionar el sistema y almacenar datos, sondas de referencia que son objetivos del emisor, y una unidad que recibe las ondas sonoras que llegan.

ultraviolet (UV) curing. Using UV light with special formulations of finishes to accelerate the curing of the finish.

curado ultravioletas (UV por su acrónimo en inglés). Usar luz UV con formulaciones especiales de acabado para acelerar la curación del acabado.

ultraviolet (UV) light. Radiation from the sun that can damage a vehicle's finish.

luz ultravioleta (UV por su acrónimo en inglés). Radiación del sol que puede dañar el acabado de un vehículo.

ultraviolet (UV) rays. *See* ultraviolet (UV) light.

rayos ultravioletas (UV por su acrónimo en inglés). Ver "luz ultravioleta (UV por su acrónimo en inglés)."

undercarriage dimensions. *See* body dimensional charts.

dimensiones del tren de aterrizaje. Ver "gráfica de dimensiones vehiculares."

undercutting. Generally considered a discontinuity instead of a defect, it is located at the junction of the weld and the base metal (toe or root). As the name implies, it is a small furrowing of the weld as it meets the base metal. It can be caused by such conditions as an arc length that is too long, an improper gun angle, and a weld speed that is too fast.

corte rebajado de. Generalmente considerado una discontinuidad en vez de un defecto, está localizado en la unión de la soldadura y el metal base (pie o raíz). Como su nombre lo indica, éste es un pequeño surcado de la soldadura cuando ésta encuentra el metal base. Éste puede ser causado por condiciones tales como una longitud de arco que es demasiado larga, un ángulo de pistola inapropiado y una velocidad de soldadura que es demasiado rápida.

underhood dimensions. *See* body dimensional charts.

dimensiones interiores del capó. Ver "gráfica de dimensiones vehiculares."

unibody structure. A vehicle design wherein each individual panel, when welded into place, reinforces and supports all the others in the structure, making it one unit body or a unibody structure.

estructura unicuerpo. Diseño de vehículo en el cual cada panel individual, cuando es soldado en su lugar, refuerza y apoya todos los demás paneles en la estructura, haciéndolo un cuerpo unido o una estructura unicuerpo.

uniform color. A consistent color throughout the substrate.

color uniforme. Color consistente en todo el sustrato.

uniform holdout. Condition in which the finish will have no bleed-through, dieback (dulling of the surface overall or on repair spots), bleaching or staining (color change over areas of plastic body filler repairs), contour mapping (lifting or visible areas of rough edges around the feathered area), or other defects sometimes caused by poor substrate conditions.

durabilidad uniforme. Condición en la cual el acabado no tendrá sangrado, opacado (opacidad de toda la superficie o en puntos de reparación), manchas o decoloración (cambio de color en áreas de reparaciones con rellenos de cuerpo plásticos), mapeo de contorno (levantamiento o áreas visibles de bordes ásperos alrededor de un área seccionada), u otros defectos algunas veces causados por las condiciones pobres del sustrato.

uniform porosity. Porosity scattered throughout the weld and most often caused by poor welding techniques or dirty materials.

porosidad uniforme. Porosidad dispersa en toda la soldadura, y más a menudo causada por técnicas de soldar pobres o materiales sucios.

uniside panel. A single continuous replacement panel that extends from the rear quarter panel forward to include the rear and front door openings and the outer layer of the front door hinge pillar.

panel uniside. Panel sencillo de reemplazo continuo que se extiende desde el panel de cuarto trasero hacia delante para incluir las aberturas de las puertas traseras y delanteras, y la capa externa del pilares del gozne de la puerta delantera.

universal system. System that uses a platform to secure the vehicle at the pinchweld with special clamps attached to the framework of the bench. A variety of measuring systems may be used with the system to monitor the vehicle's undercarriage while repairs are made.

sistema universal. Sistema que usa una plataforma para asegurar el vehículo en el pinchweld con abrazaderas especiales unidas al marco del banco. Una variedad de sistemas de medir puede ser usada con el sistema para monitorear el tren de aterrizaje del vehículo mientras las reparaciones son hechas.

universal mechanical measuring system. A measuring system that uses pointers to locate the reference points.

sistema de medición mecánica universal. Sistema de medición que usa apuntadores para localizar los puntos de referencia.

upselling. The process by which a repair or service center alerts a customer to damages or services needed on their vehicle that are unrelated to the collision damages. As the vehicle is already in the shop, it may be a convenience to address all such issues at once, rather than schedule another appointment.

upselling. Proceso por el cual un centro de servicio o reparación alerta a un cliente de daños o servicios, necesarios en su vehículo, que no están relacionados con los daños por colisión. Cuando el vehículo está ya en centro de servicio, puede ser conveniente atender todos los asuntos de una vez, en lugar de programar otra cita.

urethane. A finish designed to cover a previously refinished surface.

uretano. Acabado diseñado para cubrir una superficie antes barnizada.

utility knife. A knife with a retractable blade that is used for general cutting and trimming.

cuchillo de utilidad general. Cuchillo con una lámina retractable que es usado para cortes generales y recortes.

V

value. The lightness or darkness of color.

valor. Iluminación u oscuridad del color.

valve. A mechanism used for regulation in a machine.

válvula. Mecanismo usado para regular en una máquina.

vapor barrier. The area around vehicle doors that prevents moisture from getting into the interior, where fungus and mold can grow. If removed, this causes unpleasant odors.

barrera de vapor. Área alrededor de las puertas de un vehículo que impide a la humedad entrar en el interior, donde hongos y moho pueden crecer. Si es quitada, esto causa olores desagradables.

vehicle identification number (VIN). A vehicle-specific number assigned to each motor vehicle by its manufacturer to distinguish it from all other vehicles. It also identifies characteristics like the vehicle's year, make, model, and engine size.

número de identificación de vehículo (VIN por su acrónimo en inglés). Número específico para el vehículo, asignado a cada motor de automóvil por su fabricante, para distinguirlo de todos los demás vehículos. Éste también identifica características como el año del vehículo, marca, modelo, y tamaño de motor.

vehicle nose. The area of a motor vehicle that encompasses the front bumper to the cowl panel/windshield area.

nariz del vehículo. Área de un automóvil que abarca el parachoques delantero hasta el panel/parabrisas de la capucha.

vehicle paint codes. A code found on the vehicle that is needed to get the paint manufacturer's code.

códigos de pintura de vehículo. Código encontrado en el vehículo que es necesario para conseguir el código del fabricante de la pintura.

velometer. A tool that tests the amount of air volume moving in a spray booth.

velometer. Instrumento que mide la cantidad de volumen de aire que se mueve en una cabina de rociado.

venturi. A siphon.

venturi. Un sifón.

venturi blowers. Devices that blow unheated dry air across the finish to remove moisture.

sopladores venturi. Dispositivos que expulsan aire seco no calentado a lo largo del acabado para quitar la humedad.

viscosity cup. A measuring device that will hold a precise amount of paint, and has a very specific size hole in its bottom. Viscosity is measured by how many seconds it takes for the paint to drain from the cup through the hole.

taza de viscosidad. Dispositivo de medición que mantiene una cantidad precisa de pintura, y tiene un agujero de tamaño muy específico en su fondo. La viscosidad es medida por el número de segundos que necesita la pintura para salir de la taza por el agujero.

viscous. Thickness.

viscoso. Espeso.

volatile. Unstable; the ability to vaporize at room temperature.

volátil. Inestable; la capacidad de vaporizarse en temperatura de cuarto.

volatile chemicals. Chemicals that change from a liquid to a gas at room temperature, such as the solvents used in automotive refinishing. They evaporate easily, and therefore easily pass into the atmosphere.

productos químicos volátiles. Productos químicos que cambian de líquido a gas en temperatura de cuarto, como los solventes usados en el barnizado automotor. Éstos se evaporan fácilmente, y por lo tanto pasan fácilmente a la atmósfera.

volatile organic compounds (VOCs). Solvent evaporatives that are emitted into the atmosphere as overspray during spray painting operations. They contribute to smog formation and ozone depletion.

compuestos orgánicos volátiles (VOCs por su acrónimo en inglés). Solventes que se evaporan, que son emitidos en la atmósfera como el exceso de rociado durante operaciones de aplicación de pintura. Éstos contribuyen a la formación de humo tóxico y reducción de la capa de ozono.

voltage. The force which pushes electricity through wire, measured in volts. The higher the number of volts, the greater force of the potential electricity.

voltaje. Fuerza que transporta a la electricidad a través de un cable, medida en voltios. Entre más alto el número de voltios, mayor es la fuerza de la electricidad potencial.

W

wash primer. A primer similar to the self-etching primer whose main purpose is to provide corrosion protection to exposed substrate. In contrast to self-etching primers, they must be covered with a primer or sealer before topcoating with color.

imprimación de lavado. Imprimación similar a la imprimación de autograbado cuyo principal propósito es proteger de la corrosión al sustrato expuesto. En contraste a la imprimación de autograbado, ésta debe ser cubierta con una imprimación o sellante antes de la aplicación de la capa superior con color.

wax. Protective coatings designed to clean finishes, then to shine and protect the finish. Wax will provide a thin protective coating, which with time will need to be replaced.

cera. Capas protectoras diseñadas para limpiar acabados, para luego brillar y proteger el acabado. La cera provee un recubrimiento protector delgado, el cual con el tiempo tiene que ser reemplazado.

weatherstripping. A soft rubber attached around any opening. When the opening—like a door or window—is closed, the weatherstripping fills the gaps between the vehicle and the closure, guaranteeing a tight fit that keeps wind, water, dirt, and dust from entering the vehicle.

sellos a prueba de intemperie. Caucho suave unido alrededor de cualquier abertura. Cuando la abertura, como una puerta o ventana, está cerrada, el sello a prueba de intemperie llena los huecos entre el vehículo y el cierre, garantizando un adecuado ajuste que evita que el viento, el agua, la suciedad y el polvo entre en el vehículo.

weld defects. A flaw in the weld, often caused by too many discontinuities. In structural welding, if a weld is defective it may need to be ground out and re-applied to meet the standard.

defectos de soldadura. Error en la soldadura, a menudo causado por demasiadas discontinuidades. En el soldado estructural, si una soldadura es defectuosa puede necesitar ser removida y reaplicada para cumplir las normas.

weld destructive test. A test performed to determine the integrity of a weld to see how easily it will fail. It is designed to ensure high-quality welds.

prueba destructiva de soldadura. Prueba realizada para determinar la integridad de una soldadura, para determinar qué tan fácilmente va a fallar. Está diseñada para asegurar soldaduras de alta calidad.

weld nugget. A spot where metals have been welded together.

pepita de soldadura. Punto en donde los metales han sido soldados juntos.

weld position. The approach a technician chooses to weld metal.

posición al soldar. Aproximación que el técnico elige para soldar metales.

weld-bonding. A method of attaching a replacement panel such as a roof, which incorporates the use of both adhesives and welding to ensure structural integrity has been restored to the repair area and the overall vehicle.

unión de soldadura. Método para unir un panel de reemplazo, como un techo, que incorpora el uso de pegamentos y soldadura para asegurar que la integridad estructural ha sido devuelta al área de reparación y al vehículo en general.

welders. Tools used to join two metal pieces together through the application of heat.

soldadores. Herramientas usadas para unir dos piezas metálicas juntas a través de la aplicación de calor.

welding gun. A welding device with a switch the operator activates when ready to weld. The welding electrode or wire passes through the contact tip.

pistola de soldar. Dispositivo de soldar con un interruptor que el operador activa cuando está listo para soldar. El electrodo o alambre de soldar pasa por la punta de contacto.

welding speed. The rate at which the gun travels along the face of the weld. If the welding speed is too fast, the depth of penetration and the weld bead width decrease. If the weld travel speed is too fast, weld bead undercutting may occur.

rapidez de soldado. Rapidez con la cual la pistola viaja a lo largo de la cara de la soldadura. Si la rapidez de soldado es demasiado rápida, la profundidad de penetración y la anchura de la cuenta de la soldadura disminuirán. Si la rapidez de soldado es demasiado rápida, el corte rebajado de la cuenta de la soldadura puede ocurrir.

weldment. The filling material in the form of a bead or nugget deposited onto the surface during a welding operation to join two or more pieces of metal together.

ensambladuras. Material de relleno en forma de cuenta o pepita depositada en la superficie durante una operación de soldadura para unir dos o más piezas de metal juntas.

weld-on dent pullers. Tools, often tabs, which are welded onto an area of metal that will be removed after the rough repairs are made. These tabs provide a convenient area on which to attach a hydraulic pulling tool, in order to apply force in the opposite direction from the collision direction.

sacador de abolladuras soldado en sitio. Herramientas, a menudo etiquetas, que son soldadas en un área de metal que será quitada después de que las reparaciones rudas sean hechas. Estas etiquetas proporcionan un área conveniente para sostener una herramienta de tiramiento hidráulico, a fin de aplicar la fuerza en dirección contraria a la dirección de colisión.

weld-through primer. A primer applied to maintain anticorrosion protection during MIG welding.

imprimación para aplicación de soldadura. Imprimación aplicada para mantener la protección anticorrosión durante una soldada tipo MIG (por su acrónimo en inglés).

wet bedding. Technique in which a technician sprays a coating of blending clearcoat on the panel to be blended before the color is applied. Thus, all of the color that hits the vehicle will land in a wet "bed."

cama húmeda. Técnica en la cual el técnico rocía una capa de recubrimiento transparente de mezclar sobre el panel a ser mezclado antes de que el color sea aplicado. Así, todo el color que golpea el vehículo quedará en una "cama" húmeda.

wet line. When paint spraying, the last line of paint applied. It should not be allowed to dry before the next line of paint on the panel adjoining it is applied.

línea húmeda. En el rociado de pintura, ésta es la última línea de pintura aplicada. No hay que permitir que se seque antes de que la siguiente línea de pintura en el panel aledaño sea aplicada.

wet product. The liquid items (paint, primers, sealants) used in refinishing.

producto mojado. Productos líquidos (pintura, imprimaciones, selladores) usados en acabados.

wet sanding. Sanding process in which a technician applies a bit of soap to act as a lubricant to the sandpaper and surface. Debris is rinsed away as it is created.

lija de agua. Proceso de lijado en el cual el técnico aplica un poco de jabón para que actúe como lubricante en la lija y en la superficie. Los residuos son lavados a medida que son generados.

wet-bedding blending. A technique that may be used when a color is thought to be a blendable match, but is also known to be difficult to apply. The first application of color (ready to spray) is mixed with clearcoat (ready-to-spray) at a 1:1 ratio and is applied to the vehicle as the wet bed. The clearcoat chosen to mix with the color should be a slow curing type, to allow the wet bed to remain wet as the other coats are applied over it. The second coat is applied without being mixed with clear and applied over the wet bed.

mezclado en cama húmeda. Técnica que puede ser usada cuando se cree que un color puede ser una pareja de mezclado, pero que también se sabe, es difícil de aplicar. La primera aplicación de color (lista para rociar) es mezclada con la capa transparente (listo para rociar) en una proporción 1:1 y es aplicada al vehículo como la cama mojada. La capa transparente, elegida para ser mezclada con el color, debe ser del tipo de curado lento, para permitir que la cama húmeda permanezca húmeda cuando las otras capas sean aplicadas sobre ella. El segundo recubrimiento es aplicado sin ser mezclado con transparente y aplicado sobre la cama húmeda.

wet-on-wet. Sealers that have no dry time, and just a short flash time before the topcoat can be applied. With wet-on-wet sealers, the dry time needed before topcoating is eliminated.

húmedo sobre húmedo. Selladores que no tienen tiempo de secado, y sólo un corto tiempo de destello antes de la capa superior puede ser aplicado. Con los sellantes húmedo sobre húmedo, el tiempo de secado necesario antes de la capa superior es eliminado.

wheel alignment. The procedure by which a technician aligns the geometric angles of the wheels to the original manufacturer's specifications, including items like caster, camber, steering axis inclination, and included angles.

alineación de ruedas. Procedimiento por el cual el técnico alinea los ángulos geométricos de las ruedas, de acuerdo a las especificaciones del fabricante original, incluyendo cosas como ángulo *caster*, comba, inclinación del eje de conducción y ángulos incluidos.

wheel returnability. Often influenced by caster, the ability of a vehicle's wheel to return to the straight-ahead position. For example, a vehicle with a high degree of positive caster tends to force the wheels to roll straight ahead. It also tends to help the wheel return to a straight-ahead angle.

capacidad de retorno de rueda. A menudo influenciada por el ángulo *caster*, ésta es la capacidad de la rueda de un vehículo de regresar a la posición derecha hacia delante. Por ejemplo, un vehículo con un alto grado de ángulo caster positivo tiende a forzar las ruedas a la posición recta hacia delante. Ésta capacidad también tiende a ayudar a la rueda a regresar a un ángulo directo hacia delante.

wheel well covers. A paper cover that is used to keep overspray from the wheel wells.

recubrimiento de salpicaderas. Recubrimiento de papel que es usado para proteger las salpicaduras de excesos de rociado.

wiggle wire. A tool used with a welder. After being attached with a special tool, the wiggle wire is pulled upward to help raise low areas with the force being placed over a larger area. Both the studs and the wiggle wire can be removed following the repair.

alambre de meneo. Instrumento usado con un soldador. Luego de ser unido a una herramienta especial, el alambre de meneo es jalado hacia arriba para ayudar a elevar áreas bajas con la fuerza siendo aplicada en un área más grande. Tanto las tachuelas, como el alambre de meneo, pueden ser removidos después de la reparación.

wind laces. The bead of trim that is pinched over the edges of the door opening trim. It is used to cover the edges of the B and C pillar, the dog leg trim, and other interior trim pieces.

encaje de viento. Cuenta de una moldura que es apretada sobre los bordes de la abertura de la puerta. Éste es usado para cubrir los bordes de los pilares B y C, la moldura "pata de perro" y otras piezas de molduras interiores.

wind leaks and turbulence noises. Whistling noises that commonly occur at various panel seams, joints, trim edges, and so on. They are caused by wind passing over them when a vehicle is driven down the road.

fugas de viento y ruidos por turbulencia. Ruidos silbadores que comúnmente ocurren en varias costuras de panel, uniones, bordes de molduras, etcétera. Éstos son causados por el viento que pasa sobre ellos cuando un vehículo es conducido.

window regulator. A mechanical device located inside the vehicle door that is used to raise and lower the windows. It may be operated manually or it may have an electrically operated motor.

regulador de ventana. Dispositivo mecánico localizado dentro de la puerta de un vehículo que es usada para levantar y bajar las ventanas. Puede ser operado a mano o puede tener un motor operado eléctricamente.

wire brush. A tool used to remove corrosion and dirt, as well as clean welds. A wire brush should be used sparingly on bare metal because it can leave scratches.

cepillo de alambre. Herramienta usada para quitar corrosión y suciedad, así como para limpiar soldaduras. Los cepillos de alambre deben ser usados con moderación en metales desnudos porque éste puede dejar rasguños.

wire feeder. A device which supplies a steady stream of wire to the welding gun. Some welders have the feeder inside of the machine while others have the wire feeder outside the welder case.

alimentador de alambre. Dispositivo que suministra una corriente estable de alambre a la pistola de soldar. Algunos soldadores tienen el alimentador dentro de la máquina mientras que otros tienen al alimentador de alambre fuera de la caja del soldador.

work clamp. Often incorrectly called the ground, this tool is used to complete the welding circuit. It should be fastened to the metal being welded as close as possible, so that the electrical flow of the welding circuit will not need to travel too far. In addition, when welding on a vehicle, the technician should locate the work clamp as close as possible, because the vehicle's body is used as the vehicle electricity's path back to the battery.

abrazadera de trabajo. A menudo incorrectamente llamada "tierra", este instrumento es usado para completar el circuito de soldadura. Éste debe ser sujetado al metal que está siendo soldado tan cerca como sea posible, de forma que el flujo eléctrico del circuito de soldado no tenga que viajar demasiado lejos. Además, cuando se suelda en un vehículo, el técnico debe colocar la abrazadera de trabajo tanto cerca como sea posible, porque el cuerpo del vehículo es usado como el camino de la electricidad del vehículo de regreso a la batería.

work hardening. The changing of the grain structure of steel, which causes the outer molecules to stretch and become thinner, and the inner molecules to compress and become stronger. Work hardening also is used by engineers to make relatively thin and weak metal considerably stronger. Each bend, shape, and flange work hardens the metal as it is stamped, thus making it stronger and more able to withstand the rigors of everyday operation.

trabajo de endurecimiento. Cambio de la estructura granular del acero, que hace que las moléculas externas se estiren y se hagan más delgadas, y las moléculas interiores se compriman y se hagan más fuertes. El trabajo de endurecimiento también es usado por ingenieros para hacer que metales relativamente delgados y débiles sean bastante más fuertes. Cada doblez, forma y trabajo de reborde, endurece el metal cuando es golpeado, haciéndolo más fuerte y más capaz de resistir los rigores de la operación diaria.

wrinkling. A condition on the surface of the paint film that occurs when still liquid absorbs liquids, causing the surface to swell. The surface looks leathery and prunelike.

arrugado. Condición que ocurre en la superficie de la película de pintura cuando líquidos en reposo absorben líquidos causando que la superficie se hinche. La superficie parece curtida y arrugada.

Z

zebra striping. Stripes of alternating light and dark lines in a paint job, caused by poor fan adjustment.

despintado tipo cebra. Rayas de líneas claras y oscuras alternas en un trabajo de pintura, causadas por un ajuste pobre del ventilador.

zero plane area. The four corners under the passenger compartment—more explicitly, the torque box areas—which are the primary control points from which the majority of the undercarriage measurements originate. This same area is used by the manufacturer for the fixture locations and to securely hold the vehicle while it is being manufactured.

área de plano principal. Definida por las cuatro esquinas debajo del compartimiento de pasajeros, más explícitamente el área de la caja de torsión, que son los puntos de control primarios desde los cuales la mayoría de las medidas del tren de aterrizaje son medidas. Esta misma área es usada por el fabricante para ubicar las partes integrantes y para sostener bien el vehículo mientras está siendo fabricado.

zero running toe. A situation when the wheels are moving straight ahead in perfect alignment, with neither toe-in nor toe-out.

zero running toe. Situación en la cual las ruedas se mueven derecho con alineación perfecta, ni con dedo adentro, ni con dedo afuera.

zinc. A corrosion protective coating the auto manufacturer puts on the surface of steel. It is used for both exterior and interior panels to promote a natural barrier against oxidation.

zinc. Recubrimiento anticorrosión que el fabricante de vehículos aplica en la superficie del acero. Éste es usado tanto en los paneles exteriores como interiores para crear una barrera natural contra la oxidación.

zinc phosphate coating. A corrosion resistant protective coating that forms on a steel surface after it has been chemically treated and etched with phosphoric acid and a conversion coating.

recubrimiento de fosfato de zinc. Recubrimiento protector resistente a la corrosión que se forma sobre una superficie de acero después de que ha sido químicamente tratada y grabada con ácido fosfórico y un recubrimiento de conversión.

Index

A

E

F

I

J

K

L

M

N

O

P

Q

R

S

T

W

Z